Readings in Computer Vision:
Issues, Problems, Principles, and Paradigms

Other Titles in the Morgan Kaufmann Readings Series

Readings in Artificial Intelligence
Edited by Bonnie Lynn Webber and Nils J. Nilsson (1981)

Readings in Knowledge Representation
Edited by Ronald J. Brachman and Hector J. Levesque (1985)

Readings in Artificial Intelligence and Software Engineering
Edited by Charles Rich and Richard C. Waters (1986)

Readings in Natural Language Processing
Edited by Barbara J. Grosz, Karen Sparck Jones, and Bonnie Lynn Webber (1986)

Readings in Human-Computer Interaction: A Multidisciplinary Approach
Edited by Ronald Baecker and William Buxton (Summer 1987)

Readings in Non-Monotonic Reasoning
Edited by Matthew Ginsberg (Summer 1987)

Readings in Computer Vision: Issues, Problems, Principles, and Paradigms

Edited by
Martin A. Fischler
Oscar Firschein
SRI International

MORGAN KAUFMANN PUBLISHERS, INC.
95 FIRST STREET, LOS ALTOS, CALIFORNIA 94022

Editor and President *Michael B. Morgan*
Production Manager *Jennifer Ballentine*
Cover Designer *Beverly Kennon-Kelley*
Typesetting *Aldine Press*
Production Assistant *Todd Armstrong*
Copy Editor *Jonas Weisel*

Library of Congress Cataloging-in-Publication Data

Readings in computer vision

Bibliography: p.
Includes index.
1. Computer vision. I. Fischler, Martin A.
II. Firschein, Oscar.
TA1632.R39 1987 006.3'7 86-27692
ISBN 0-934613-33-8

Morgan Kaufmann Publishers, Inc.
95 First Street, Los Altos, California 94022

Printed in the United States of America

91 90 89 88 87 5 4 3 2 1

Contents

Acknowledgments

The editors would like to thank the publishers and authors for permission to reprint copyrighted material in this volume.

Azriel Rosenfeld, "Image Analysis: Problems, Progress and Prospects," *Pattern Recognition* **17**(1):3–12, 1984. Copyright © 1984, Pergamon Press. Reprinted with permission of the publisher and the author.

Stephen T. Barnard, "A Stochastic Approach to Stereo Vision," *Proc. of the Fifth National Conf. on Artificial Intelligence*, Philadelphia, Penn., 1986, 676–680. Copyright © 1986, American Association for Artificial Intelligence. All rights reserved. Used by permission of AAAI and the author; copies of the proceedings may be obtained from Morgan Kaufmann Publishers, 95 First St., Suite 120, Los Altos, CA 94022.

Robert C. Bolles and H. Harlyn Baker, "Epipolar-Plane Image Analysis: A Technique for Analyzing Motion Sequences," *Proc. IEEE Third Workshop on Computer Vision: Representation and Control*, Bellaire, Michigan, 1985, 168–178. Copyright © 1985, IEEE. Reprinted with permission of the publisher and the authors.

Michael Brady, "Preface—The Changing Shape of Computer Vision," *Artificial Intelligence* **17**(1–3):1–15, August 1981. Copyright © 1981, North-Holland. Reprinted by permission of the editor-in-chief, Daniel Bobrow, and of the publisher, North-Holland, Amsterdam, and the author.

Berthold K. P. Horn, "Understanding Image Intensities," *Artificial Intelligence* **8**(2):201–231, 1977. Copyright © 1981, North-Holland. Reprinted by permission of the editor-in-chief, Daniel Bobrow, and of the publisher, North-Holland, Amsterdam, and of the author.

H. C. Longuet-Higgins, "A Computer Algorithm for Reconstructing a Scene from Two Projections," *Nature* **317**:314–319, Sept. 1985. Copyright © 1985, Macmillan Journals Ltd. Reprinted with permission of the publisher and the author.

H. Keith Nishihara, "Practical Real-Time Imaging Stereo Matcher," *Optical Engineering* **23**(5):536–545, 1984. Copyright © 1984, Optical Engineering. Reprinted with permission of the publisher and the author.

Kvetoslav Prazdny, "Detection of Binocular Disparities," *Biological Cybernetics* **52**:93–99, 1985. Copyright © 1985, Biological Cybernetics. Reprinted with permission of the publisher and the author.

Lynn H. Quam, "Hierarchical Warp Stereo," *Proc. DARPA Image Understanding Workshop*, New Orleans, LA, 149–155, 1984. Reprinted with permission of DARPA and the author.

Grahame B. Smith, "Stereo Integral Equation," *Proc. of the Fifth National Conf. on Artificial Intelligence*, Philadelphia, Penn., 1986, 689–694. Copyright © 1986, American Association for Artificial Intelligence. All rights reserved. Used by permission of AAAI and the author; copies of the proceedings may be obtained from Morgan Kaufmann Publishers, 95 First St., Suite 120, Los Altos, CA 94022.

Thomas M. Strat, "Recovering the Camera Parameters from a Transformation Matrix," *Proc. DARPA Image Understanding Workshop*, New Orleans, LA, 264–271, 1984. Reprinted with permission of DARPA and the author.

Thomas M. Strat and Martin Fischler, "One-Eyed Stereo: A General Approach to Modeling 3-D Scene Geometry," *IEEE PAMI* **8**(6):730–741, 1986. Copyright © 1986, IEEE. Reprinted with permission of the publisher and the authors.

Kokichi Sugihara, "An Algebraic Approach to Shape-from-Image Problems," *Artificial Intelligence* **23**(1):59–95, 1984. Copyright © 1984, North-Holland. Reprinted by permission of the editor-in-chief, Daniel Bobrow, and of the publisher, North-Holland, Amsterdam, and of the author.

Shimon Ullman, "Analysis of Visual Motion by Biological and Computer Systems," *IEEE Computer* **14**(8):57–69, 1981. Copyright © 1981, IEEE. Reprinted by permission of the publisher and the author.

Allen M. Waxman, "An Image Flow Paradigm," *Proc. IEEE Workshop on Computer Vision: Representation and Control*, Annapolis, MD, 49–55, 1984. Copyright © 1984, IEEE. Reprinted with permission of the publisher and the author.

J. Brian Burns, Allen R. Hanson, and Edward M. Riseman, "Extracting Straight Lines," *Proc. DARPA Image Understanding Workshop*, New Orleans, LA, 165–168, 1984. Reprinted with permission of DARPA and the author.

J. F. Canny, "A Computational Approach to Edge Detection," *IEEE PAMI* **8**(6):679–698, 1986. Copyright © 1986, IEEE. Reprinted with permission of the publisher and the author.

Martin A. Fischler and Helen C. Wolf, "Linear Delineation," *Proc. IEEE Computer Vision and Pattern Recognition*, Washington, D.C., 351–356, 1983. Copyright © 1983, IEEE. Reprinted with permission of the publisher and authors.

Martin A. Fischler and Robert C. Bolles, "Perceptual Organization and Curve Partitioning," *IEEE PAMI* **8**(1):100–105, 1986. Copyright © 1986, IEEE. Reprinted with permission of the publisher and the authors.

Robert M. Haralick, "Digital Stereo Edges from Zero Crossing of Second Directional Derivatives," *IEEE PAMI* **6**(1):58–68, 1984. Copyright © 1984, IEEE. Reprinted with permission of the publisher and the author.

D. D. Hoffman and W. A. Richards, "Parts of Recognition," *Cognition* **18**:65–96, 1985. Copyright © 1985, Elsevier. Reprinted with permission of the publisher and the authors.

B. Julesz and J. R. Bergen, "Textons, The Fundamental Elements in Preattentive Vision and Perception of Textures," *Bell Systems Technical Journal* **62**(6)Part II:1619–1645, 1983. Copyright © 1983, AT&T. Reprinted with permission of the publisher and the authors.

Takeo Kanade and John R. Kender, "Mapping Image Properties into Shape Constraints: Skewed Symmetry, Affine-Transformable Patterns, and the Shape-from-Texture Paradigm," Beck, J., B. Hope, and A. Rosenfeld (eds), *Human and Machine Vision*, Academic Press, New York, 1983, pp. 237–257. Copyright © 1983, Academic Press. Reprinted by permission of the publisher and the authors.

Michael Kass and Andrew Witkin, "Analyzing Oriented Patterns," *Proc. of the Ninth International Joint Conf. on Artificial Intelligence*, Los Angeles, Calif., 1985, 944–952. Copyright © 1985, International Joint Conferences on Artificial Intelligence. All rights reserved. Used by permission of IJCAI and the authors; copies of the proceedings may be obtained from Morgan Kaufmann Publishers, 95 First St., Suite 120, Los Altos, CA 94022.

Yvan Leclerc, "Capturing the Local Structure of Image Discontinuities in Two Dimensions," *Proc. IEEE Computer Vision and Pattern Recognition*, San Francisco, CA, 34–38, 1985. Copyright © 1984, IEEE. Reprinted with permission of the publisher and the author.

David G. Lowe and Thomas O. Binford, "Segmentation and Aggregation: An Approach to Figure-Ground Phenomena," *Proc. DARPA Image Understanding Workshop*, Palo Alto, CA, 168–178, 1982. Reprinted with permission of DARPA and the authors.

Laurence T. Maloney and Brian A. Wandell, "Color Constancy: A Method for Recovering Surface Spectral Reflectance," *J Optical Society of America* **3**:29–33, 1986. Copyright © 1986, Optical Society of America. Reprinted with permission of the publisher and the authors.

Shimon Ullman, "Visual Routines," *Cognition* **18**:97–106, 1984. Copyright © 1984, Elsevier. Reprinted with permission of the publisher and the author. This version reproduced by special arrangement from *Visual Cognition*, Steven Pinker, ed. Cambridge: The MIT Press, 1985, pages 97–159.

Andrew P. Witkin, "Scale-Space Filtering," *Proc. of the Ninth International Joint Conf. on Artificial Intelligence*, Karlsruhe, West Germany, 1983, 1019–1022. Copyright © 1983, International Joint Conferences on Artificial Intelligence. All rights reserved. Used by permission of IJCAI and the authors; copies of the proceedings may be obtained from Morgan Kaufmann Publishers, 95 First St., Suite 120, Los Altos, CA 94022.

Steven W. Zucker, "Early Orientation Selection: Tangent Fields and the Dimensionality of Their Support," *Computer Vision, Graphics, and Image Processing* **32**(1):74–103, 1985. Copyright © 1985, Academic Press. Reprinted with permission by the publisher and the author.

Robert C. Bolles, Patrice Howard, and Marsha Jo Hannah, "3DPO: A Three-Dimensional Part Orientation System," *Proc. of the Ninth International Joint*

Conf. on Artificial Intelligence, Karlsruhe, West Germany, 1983, 1116–1120. Copyright © 1983, International Joint Conferences on Artificial Intelligence. All rights reserved. Used by permission of IJCAI and the authors; copies of the proceedings may be obtained from Morgan Kaufmann Publishers, 95 First St., Suite 120, Los Altos, CA 94022.

Rodney A. Brooks, "Model-Based Three-Dimensional Interpretations of Two-Dimensional Images," *IEEE PAMI* **5**(2):140–149, 1983. Copyright © 1984, IEEE. Reprinted with permission of the publisher and the author.

Chris Goad, "Special Purpose Automatic Programming for 3D Model-Based Vision," *Proc. DARPA Image Understanding Workshop*, Arlington, VA, 94–104, 1983. Reprinted with permission of DARPA and the author.

W. Eric, L. Grimson, and Tomas Lozano-Perez, "Model-Based Recognition and Localization from Sparse Range or Tactile Data," *Int. Jrnl of Robotics Research* **3**(3):3–35, 1984. Copyright © 1984, MIT Press. Reprinted with permission of the publisher and the authors.

David M. McKeown, Jr., Wilson A. Harvey, Jr., and John McDermott, "Rule-Based Interpretation of Aerial Imagery," *IEEE PAMI* **7**(5):570–585, 1985. Copyright © 1985, IEEE. Reprinted with permission of the publisher and the authors.

Rodney A. Brooks, "Visual Map Making for a Mobile Robot," *1985 IEEE International Conf. on Robotics and Automation*, St. Louis, Missouri, (IEEE Cat. 85CH2152-7), 824–829, 1985. Copyright © 1985, IEEE. Reprinted with permission of the publisher and the author.

Thomas G. Evans, "A Heuristic Program to Solve Geometric-Analogy Problems," *Spring Joint Computer Conference, AFIPS* **25**:5–16, 1964. Copyright © 1964, AFIPS. Reprinted with permission of the publisher.

Brian V. Funt, "Problem-Solving with Diagrammatic Representations," *Artificial Intelligence* **13**:201–230, 1980. Copyright © 1980, North-Holland. Reprinted by permission of the editor-in-chief, Daniel Bobrow, and of the publisher, North-Holland, Amsterdam, and of the author.

Martin Herman and Takeo Kanade, "The 3D MOSAIC Scene Understanding System: Incremental Reconstruction of 3D Scenes for Complex Images," *Proc. DARPA Image Understanding Workshop*, New Orleans, LA, 137–148, 1984. Reprinted with permission of DARPA and the authors.

Daryl T. Lawton, Tod S. Levitt, Chris McConnell, and Jay Glicksman, "Terrain Models for an Autonomous Land Vehicle," *1986 IEEE International Conf. on Robotics and Automation*, San Francisco, Calif, (IEEE Cat. 86CH2282-2), 2043–2051, 1986. Copyright © 1986, IEEE. Reprinted with permission of the publisher and the authors.

Michael J. Swain and Joseph L. Mundy, "Experiments in Using a Theorem Prover to Prove and Develop Geometrical Theorems in Computer Vision," *IEEE International Conf. on Robotics and Automation*, San Francisco, Calif. (IEEE Cat. 86CH2282-2), 280–285, 1986. Copyright © 1986, IEEE. Reprinted with permission of the publisher and authors.

John K. Tsotsos, "Knowledge Organization and Its Role in Representation and Interpretation for Time-Varying Data: The ALVEN System," *Computational Intelligence* **1**:16–32, 1985. Copyright © 1985, the National Research Council of Canada. Reprinted with permission of the National Research Council of Canada and the author.

David H. Ackley, Geoffrey E. Hinton, and Terrence J. Sejnowski, "A Learning Algorithm for Boltzmann Machines," *Cognitive Science* **9**:147–169, 1985. Copyright © 1985, Ablex Publishing Corporation. Reprinted with permission of the publisher and the authors.

Dana H. Ballard, "Parameter Nets," *Artificial Intelligence* **22**:235–267, 1984. Copyright © 1984, North Holland. Reprinted by permission of the editor-in-chief, Daniel Bobrow, and of the publisher, North-Holland, Amsterdam, and of the author.

P. Carnevali, L. Coletti, and S. Patarnello, "Image Processing by Simulated Annealing," *IBM Jrnl of Research and Development* **29**(6):569–579, 1985. Copyright © 1985, International Business Machines Corporation. Reprinted with permission of the publisher and the authors.

Stuart Geman and Donald Geman, "Stochastic Relaxation, Gibbs Distributions, and the Bayesian Restoration of Images," *IEEE PAMI* **6**:721–741, 1984. Copyright © 1984, IEEE. Reprinted with permission of the publisher and the authors.

Robert A. Hummel and Steven W. Zucker, "On the Foundations of Relaxation Labeling Processes," *IEEE PAMI* **5**:267–287, 1983. Copyright © 1983, IEEE. Reprinted with permission of the publisher and the authors.

S. Kirkpatrick, C. D. Gelatt, Jr., and M. P. Vecchi, "Optimization by Simulated Annealing," *Science* **220**:671–680, 1983. Copyright © 1983, American Association for the Advancement of Science. Reprinted with permission of the AAAS and the authors.

David Marr and H. Keith Nishihara, "Visual Information Processing: Artificial Intelligence and the Sensorium of Sight," *Technology Review* **81**(1):2–23, 1978. Copyright © 1978, MIT Alumni Association. Reprinted with permission of the publisher and the authors.

Tomaso Poggio, Vincent Torre, and Christof Koch, "Computational Vision and Regularization Theory," *Nature* **317**:314–319, 1985. Copyright © 1985, Macmillan Journals Ltd. Reprinted with permission of the publisher and the authors.

D. V. Ahuja and S. A. Coons, "Geometry for Construction and Display," *IBM Systems Journal* **7**(3/4):188–205, 1968. Copyright © 1968, International Business Machines Corporation. Reprinted with permission of the publisher.

Alan H. Barr, "Global and Local Deformations of Solid Primitives," *Computer Graphics* **18**(3):21–30, 1984. Copyright © 1984, ACM. Reprinted by permission of the publisher and the author.

Peter J. Burt and Edward H. Adelson, "The Laplacian Pyramid as a Compact Image Code," *IEEE Trans. on Communications* **31**(4):532–540, 1983. Copyright © 1983, IEEE. Reprinted with permission of the publisher and the authors.

Alex P. Pentland, "Perceptual Organization and the Representation of Natural Form," *Artificial Intelligence* **28**:293–331, 1986. Copyright © 1986, North Holland. Reprinted by permission of the editor-in-chief, Daniel Bobrow, and of the publisher, North-Holland, Amsterdam, and of the author.

Whitman Richards and Donald D. Hoffman, "Codon Constraints on Closed 2D Shapes," *Computer Vision, Graphics and Image Processing* **31**(2):156–177, 1985. Copyright © 1985, Academic Press. Reprinted with permission of the publisher and the authors.

D. H. Ballard, "Generalizing the Hough Transform to Detect Arbitrary Shapes," *Pattern Recognition* **13**(2):111–122, 1981. Copyright © 1981, Pergamon Press. Reprinted with permission of the publisher and the author.

Martin A. Fischler and Robert C. Bolles, "Random Sample Consensus: A Paradigm for Model Fitting with Applications to Image Analysis and Automated Cartography," *CACM* **24**(6):381–395, 1981. Copyright © 1981, ACM. Reprinted with permission of the publisher and the authors.

M. A. Fischler, J. M. Tenenbaum, and H. C. Wolf, "Detection of Roads and Linear Structures in Low-Resolution Aerial Imagery Using a Multisource Knowledge Integration Technique," *Computer Graphics and Image Processing* **15**(3):201–223, 1981. Copyright © 1981, Academic Press. Reprinted with permission of the publisher and the authors.

Robert A. Hummel, "Representations Based on Zero-crossings in Scale-Space," *Proc. IEEE Computer Vision and Pattern Recognition*, Miami Beach, FL, 204–209, 1986. Copyright © 1986, IEEE. Reprinted with permission of the publisher and the author.

Andrew Witkin, Demetri Terzopoulis, and Michael Kass, "Signal Matching Through Scale Space," *Proc. of the Fifth National Conf. on Artificial Intelligence*, Philadelphia, Penn., 1986, 714–719. Copyright © 1986, American Association for Artificial Intelligence. All rights reserved. Used by permission of AAAI and the authors; copies of the proceedings may be obtained from Morgan Kaufmann Publishers, 95 First St., Suite 120, Los Altos, CA 94022.

Introduction

The purpose of this collection is to make explicit the key ideas and assumptions underlying recent advances in computer vision. Computer vision is the scientific discipline that addresses the problem of how we can combine sensor-derived images, previously acquired models of scene content, world knowledge, and knowledge of the imaging process to construct an explicit description or model of the surrounding environment. Computer vision incorporates techniques borrowed from many traditional disciplines, such as physics, mathematics, and psychology, but is properly considered a subfield of computer science and artificial intelligence.

Each paper in this collection satisfies one or more of the following criteria:

- It provides one of the best and most up-to-date expositions of a key idea.
- It provides the most comprehensive evaluation (theoretical or experimental) of one or more approaches to a key problem.
- It presents the best-known solution to a key problem.

Forty-eight of the selected papers were originally published in the open literature, and most appeared after 1 January 1981; these essays are supplemented by twelve additional papers that were not as easily accessible or were included for tutorial purposes. An extensive bibliography provides references to both the classic papers of the discipline and other papers that met our stated criteria but were omitted because of size, permission, or other practical constraints on inclusion. We include as our only review paper a historical perspective of computer vision written by A. Rosenfeld [**Rosenfeld 84**],[1] one of the founders of this field and still one of its central figures. An appendix on key ideas and assumptions (Appendix A), one on parallel architectures (Appendix B), a glossary, and a comprehensive index complete the volume.

This volume is structured to reflect the view that the computer vision paradigm can be portrayed as an attempt to solve seven basic problems. The first four problems are concerned with how descriptions of the environment are obtained from images:

1. How can 3-D scene geometry be recovered from one or more 2-D images?
2. How can an uninterpreted image be decomposed into independent or coherent pieces so that subsequent analysis can address a collection of simple problems, rather than the original very complex one?
3. How can the individual objects depicted in an image be recognized and assigned names or labels?

[1] Papers included in this collection are referenced in bold-faced type.

4. How can the space and time relations among the objects depicted in an image be deduced and subsequently described?

Three additional problems are concerned with the nature of the interpretative techniques and systems needed to accomplish the vision task:

1. What is the nature of a machine (or biological) *architecture* competent to support a general-purpose vision system?
2. How can we *represent* our various abstractions of the world in the memory of a computer?
3. How can we *match* our stored descriptions of things in the world to their traces in one or more images?

Each of the seven chapters of this volume corresponds to one of these problems. In each chapter we discuss the problem in some detail, describe the key ideas and techniques devised to respond to the problem, introduce the papers that embody and explicate these ideas, and finally include the papers themselves.

A number of other excellent texts and collections address the subject of computer vision, but none are so directly concerned with organizing the basic ideas into the type of coherent story presented here.

Acknowledgments

We wish to thank the following members of SRI International for their critical reviews and suggestions: Steve Barnard, Bob Bolles, Jack Goldberg, Ken Laws, Yvon Leclerc, Sandy Pentland, Grahame Smith, and Tom Strat. Thanks also to Mike Morgan and Jennifer Ballentine of Morgan Kaufmann, and to those who reviewed our early material, particularly Rod Brooks and Steve Zucker.

The Defense Advanced Research Projects Agency (DARPA) played a critical role in the development of computer vision as a scientific discipline and supported much of the research described in this volume. We especially thank the Image Understanding technical monitors with whom we have interacted: Dave Carlstrom, Larry Druffel, Ron Ohlander, and Bob Simpson.

IMAGE ANALYSIS: PROBLEMS, PROGRESS AND PROSPECTS*

AZRIEL ROSENFELD
Center for Automation Research, University of Maryland, College Park, MD 20742, U.S.A.

(*Received* 10 *February* 1983; *received for publication* 6 *May* 1983)

Abstract—Over the past 25 years, many *ad hoc* techniques for analyzing images have been developed and the subject has gradually begun to develop a scientific basis. This paper outlines the basic steps in a general image analysis process. It summarizes the state of the art with respect to each step, points out limitations of present methods and indicates potential directions for future work.

Image analysis   Scene analysis   Pictorial pattern recognition   Computer vision

1. INTRODUCTION

Over the past 25 years, much of the work on applications of pattern recognition, and a significant fraction of the work in artificial intelligence, has dealt with the analysis and interpretation of images. This subject has been variously known as pictorial pattern recognition, image analysis, scene analysis, image understanding and computer vision. Its applications include document processing (character recognition, etc.), microscopy, radiology, industrial automation (inspection, robot vision), remote sensing, navigation and reconaissance, to name only the major areas.

Many *ad hoc* techniques for analyzing images have been developed, so that a large assortment of tools is now available for solving practical problems in this field. More important, during the past few years the field has begun to develop a scientific basis. This paper outlines the basic steps in a general image analysis process. It summarizes the state-of-the-art with respect to each step, points out limitations of present methods and indicates potential directions for future work.

2. AN IMAGE ANALYSIS PARADIGM

The goal of image analysis is the construction of scene descriptions on the basis of information extracted from images or image sequences. With reference to Fig. 1, the following are some of the major steps in the image analysis process. We consider here only images obtained by optical sensors, though some of the discussion is also applicable to other types of sensors.

Many types of scenes are essentially two-dimensional (documents are an obvious example), but two-dimensional treatment is often quite adequate in applications such as remote sensing (flat terrain seen from very high altitudes), radiology (where the image is a "shadow" of the object) or microscopy (where the image is a cross-section of the object). In such situations, the image analysis process is basically two-dimensional. We extract "features", such as edges, from the image or segment the image into regions, thus obtaining a map-like representation, which Marr[1] called the "primal sketch", consisting of image features labelled with their property values. Grouping processes may then be used to obtain improved maps from the initial one. The maps may be represented by abstract relational structures in which, for example, nodes represent regions, labelled with various property values (color, texture, shape, etc.), and arcs represent relationships among regions. Finally, these structures are matched against stored models, which are generalized relational structures representing classes of maps that correspond to general types of images. Successful matches yield identifications for the image parts, and a structural description of the image, in terms of known entities.

In other situations, notably in robot vision applications, the scenes to be described are fundamentally three-dimensional, involving substantial surface relief and object occlusion. Successful analysis of images of such scenes requires a more elaborate approach in which the three-dimensional nature of the underlying scenes is taken into account. Here, the key step in the analysis is to infer the surface orientation at each image point. Clues to surface orientation can be derived directly from shading (i.e. gray level variation) in the

* The support of the Defense Advanced Research Projects Agency and the U.S. Army Night Vision Laboratory under Contract DAAG-53-76C-0138 (DARPA Order 3206) is gratefully acknowledged, as is the help of Janet Salzman in preparing this paper. A different version of this paper was presented at the Sixth International Conference on Pattern Recognition in Munich, FRG, in October 1982.

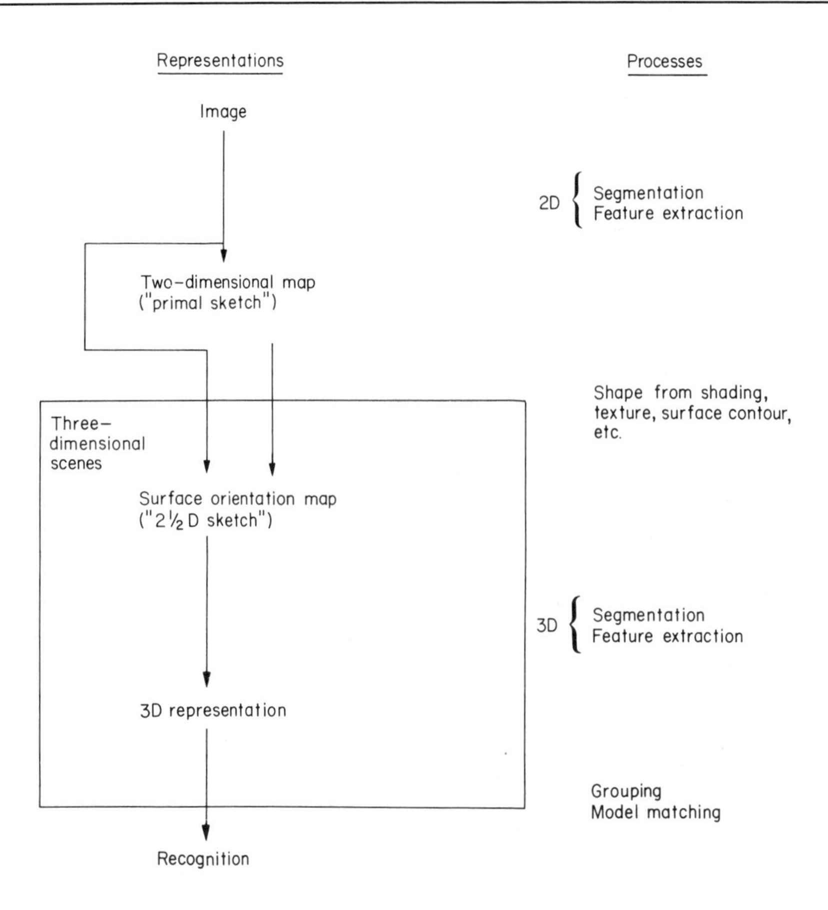

Fig. 1. Simplified diagram of an image analysis system.

image. Alternatively, two-dimensional segmentation and feature extraction techniques can first be applied to the image to extract such features as surface contours and texture primitives, and surface orientation clues can then be derived from contour shapes or from textural variations. Using the surface orientation map, which Marr called the "$2\frac{1}{2}$D sketch", feature extraction and segmentation techniques can once again be applied to yield a segmentation into (visible parts of) bodies or objects, and these can in turn be represented by a relational structure. Finally, the structure can be matched against models to yield an interpretation of the scene in terms of known objects. Note that the matching process is more difficult in the three-dimensional case, since the image only shows one side of each object and objects may partially occlude one another.

The image analysis paradigm described in the last two paragraphs, and illustrated in Fig. 1, is highly simplified in several respects. The following are some of the directions in which it needs to be extended or generalized.

(a) Ideally, the value (gray level or spectral signature) at each point of an image represents the light received from the scene along a given direction, but these values will not be perfectly accurate because of degradations arising in the process of imaging (for example blur and noise introduced by the environment or the sensor) or digitization. Image restoration techniques should be used to correct the image values before performing the steps outlined in Fig. 1. (Feature extraction may be useful as an aid in estimating the degradations in order to perform effective restoration.)

(b) We have assumed in Fig. 1 that only a single image of the scene is available as input. If two images taken from different viewpoints are available, stereomapping techniques can be used to construct the surface orientation map by matching either image values or extracted features on the two images and measuring their parallaxes. If images taken at different times are available, comparing them yields information about the motion of the sensor or of objects in the scene. In this case, the processes of segmentation, model matching, etc., should be performed on the image sequence rather than on the individual images.

(c) Figure 1 shows a "one-way" process in which we start with the image and successively construct a 2D map, a $2\frac{1}{2}$D map, etc. More realistically, the arrows in Fig. 1 should point both ways. Knowledge about the expected results of a process (segmentation, etc.) should be used to criticize the actual results and modify the process so as to improve them.

(d) The model matching process may be hierarchical, with objects composed of sub-objects, etc. Hierarchical models are extensively used in syntactic approaches to pattern recognition.

A discussion of image restoration techniques is beyond the scope of this article, but stereomapping, time-varying imagery analysis, syntactic methods and the use of feedback in image analysis will all be briefly discussed in later sections.

This paper reviews the basic stages in the image analysis process from a technique-oriented, rather than application-oriented, standpoint. Methods currently used at each stage are reviewed, their shortcomings are discussed, and approaches that show promise of yielding improved performance are described. The specific areas covered are feature extraction, texture analysis, surface orientation estimation, image matching, range estimation, segmentation, object representation and model matching.

3. FEATURE EXTRACTION

The extraction of features such as edges and curves from an image is useful for a variety of purposes. Edges and similar locally defined features play important roles in texture analysis (see Section 4). The interpretation of image edges as arising from various types of discontinuities in the scene (occluding edges are discontinuities in range; convex or concave edges are discontinuities in surface slope; shadow edges are discontinuities in illumination) plays an important role in the inference of 3D surface structure from an image (Section 5). Edges are useful in image matching for obtaining sharp matches that are insensitive to grayscale distortions (but quite sensitive to geometric distortion) (see Section 6). Edges can be used in conjunction with various segmentation techniques to improve the quality of a segmentation (Section 8). It should also be mentioned that linear features (curves) are often of importance in their own right, e.g. roads or drainage patterns on low-resolution remote sensor imagery.

The classical approach to edge detection makes use of digital (finite-difference) versions of standard isotropic derivative operators, such as the gradient or Laplacian. A closely related approach is to linearly convolve the image with a set of masks representing ideal step edges in various directions. Lines and curves can be similarly detected by linear convolutions. However, linear operators are not specific to features of a given type. They also respond in other situations involving local intensity changes. An alternative approach[2] is to use "gated" (nonlinear) operators that respond only when specific relationships hold among the local intensities, e.g. all intensities along the line higher than all the flanking intensities on both sides, and similarly for edges. In all of these approaches, the output is a quantitative edge or curve *value*. The final detection decision can be made, if desired, by thresholding this value. Similar methods can be used to detect edges defined by discontinuities in color, rather than in intensity.

Several important improvements to the edge detection process have been made over the past decade. New classes of operators have been defined based on fitting polynomial surfaces to the local image intensities and using the derivatives of these polynomials (which can in turn be expressed in terms of the local intensities) as edge value estimates. This method, which was first proposed by Prewitt,[3] allows edges to be located (at maxima of the surface gradient for example) to subpixel accuracy. Another important idea, first proposed by Hueckel,[4,5] is to find a best-fitting step edge (or edge-line) to the local intensities.

Classical edge detectors were based on small image neighborhoods, typically 3 × 3. A more powerful approach[2,6] is to use a set of first- or second-difference operators based on neighborhoods having a range of sizes (e.g. increasing by factors of 2) and combine their outputs, so that discontinuities can be detected at many different scales. Here the edges are localized at maxima of first differences or at zero-crossings of second differences. Operators based on large neighborhoods can also be used to detect texture edges, at which the statistics of varous local image properties change abruptly. Cooperation between operators in different positions can be used to enhance the feature values at points lying on smooth edges or curves. This was one of the first applications of "relaxation" methods[7,8] to image analysis at the pixel level. Cooperation among operators of different sizes is also important in characterizing edge types.

The standard approaches to edge detection are implicitly based on a very simple model in which the image is regarded as ideally composed of essentially "constant" regions separated by step edges. Recent work by Haralick[9] is based on the more general assumption of a piecewise linear, rather than piecewise constant, image, which allows simple shading effects to be taken into account. Even this approach, however, does not directly address the problem of edge detection in three-dimensional scenes. Research is needed on the development of algorithms designed to detect intensity edges resulting from specific types of scene discontinuities, including shadow edges, slope edges and range edges. Detection of texture edges (see below) should be based on models for surface texture, rather than for image texture.

4. TEXTURE ANALYSIS

Textural properties of image regions are often used for classification (e.g. of terrain types or materials) or for segmentation of the image into differently textured regions. Changes in texture "coarseness" also provide important cues about surface slant and relative range. The direction in which coarseness is changing most rapidly corresponds, for a uniformly textured surface, to the slant direction, while an abrupt change in coarseness indicates the possibility of an occluding edge.

Classically, texture properties have been derived from the autocorrelation or Fourier power spectrum, e.g. the coarser the texture (in a given direction), the slower its autocorrelation falls off in that direction from the origin (zero displacement) and the faster its power spectrum falls off in that direction from zero frequency. A related approach[10,11] characterizes textures by their second-order intensity statistics, i.e. by the frequencies with which given pairs of gray levels occur at given relative displacements. It has long been realized, however, that first order statistics of various local property values (e.g. responses to operators sensitive to local features such as edges, lines, line ends, etc.) are at least equally effective in texture discrimination.

More recent work suggests that local processes of linking between local features, giving rise to "texture elements" or "primitives", also play a significant role in the perception of texture differences. Texture discrimination based on second-order statistics of local features (e.g. occurrences of edge elements in given relative positions and orientations) has begun to be investigated.[12] Texture analysis based on explicit extraction of primitives has also been explored.[13] Here, statistics derived from properties of the primitives, or of pairs of adjacent primitives, are used as textural properties.

All of this work has dealt with texture as an image property and has been primarily concerned with uniformly textured regions, such as might arise from non-perspective views of uniformly textured surfaces. Research is needed on the development of texture analysis methods that take surface geometry into account and that perform cooperative estimation of surface slant and surface texture characteristics, leading to better estimates of both. Similarly, methods of texture-based segmentation or texture edge detection should consider both surface geometry differences and texture differences, and in stereomatching of textured regions, one should use surface slant estimates to correct for the effects of perspective on the quality of the match.

5. SURFACE ORIENTATION ESTIMATION

If only a single image of a scene is available, clues about the orientations and relative ranges of the visible surfaces in the scene can still be derived from a number of sources. One source of such information is the shapes of edges in the image, representing occluding contours at the boundaries of surfaces or contours that lie on the surfaces. The early work of Waltz[14] and his predecessors, as well as the more recent work of Kanade,[15] developed methods of inferring the nature of edges in the scene (e.g. convex, concave or occluding) from the shapes of the junctions at which the edges meet, as seen in the image. A variety of constraints on a scene can be derived from global properties of contours in the image. For example, if an edge is continuous or straight or two edges are parallel or two features coincide in the image, we assume that the same is true in the scene, and if a shape in the image could be the result of perspective distortion of a simpler shape in the scene, we assume that this is actually the case.[15,16] Other work[17,18] has dealt with the three-dimensional interpretation of occluding and surface contours. For example, given a curve in the image, we might assume that it arises from a space curve of the least possible curvature. Constraints of these types, and others still to be formulated, often yield unambiguous three-dimensional interpretations of the surfaces that appear in the images. Most of this work has as yet been applied only to idealized line drawings, but some parts of it have been successfully applied to noisy real-world images.

The inference of information about the surfaces in a scene from the shapes of edges in an image is known as "shape from shape" (i.e. 3D shape from 2D shape). A closely related problem is that of inferring surface shape from textural variations in the image. Gibson pointed out over 30 years ago that changes in texture coarseness arise from changes in range. Thus it should be possible to infer changes in range from changes in coarseness. It has recently been demonstrated[19] that the 3D orientation of a surface can be inferred from the anisotropy in its texture—note that here again, as in the case of shape from contour, we are assuming that if the anisotropy could have arisen from perspective distortion, then it actually did. To obtain good results, one should use edge-based or primitive-based, rather than pixel-based, texture descriptors. The richer the descriptors, the more likely it is that the inferences will be reliable.

In the absence of discriminable features, surface orientation in the scene can also be inferred from intensity variations ("shading") in the image. The pioneering work on the inference of "shape from shading" was done by Horn.[20,21] Given the position of the (small, distant) light source and the surface reflectance function, surface shape is still not unambiguously determined, but it is strongly constrained and can be estimated based on additional information, such as the shapes of surface contours, the restriction that the surface be a surface of revolution or the requirement that the surface curvature be as uniform as possible. Surface shape becomes unambiguous if we are given several images taken from the same position, but with light sources in different positions ("photometric stereo").[22] Much of this work has assumed diffuse reflectance and needs to be extended to reflectance functions that have strong specular components. In such cases the shapes of highlights may provide additional information.

6. IMAGE MATCHING

Image matching and registration are used for a number of different purposes. By registering two images of a scene obtained from different sensors, one can obtain the multisensor (e.g. multispectral) characteristics of each scene point, which can then be used to classify the points. By comparing images obtained from different locations, one can compute the stereoscopic parallaxes of scene points that are visible on both images and thus determine their 3D positions. By comparing images taken at different times, one can detect changes that have taken place in the scene, e.g. due to motion of the sensor or motions of objects in the scene. In all of these tasks, registration is carried out by finding pairs of subimages that match one another closely. Subimage matching is also used to detect the occurrence of specific patterns ("templates" or "control points") in an image, for the purposes of location (e.g. navigation) or object detection.

Classically, image matching has used match measures derived from cross-correlation computation, or sometimes mismatch measures based on sums of absolute differences. Both of these approaches involve point-by-point intensity comparison of the images being matched. Such processes are unsatisfactory for several reasons: they often yield unsharp matches, making it difficult to decide when a match has been detected; they are sensitive to distortions in both grayscale and geometry; they are computationally expensive. Match sharpness and grayscale insensitivity can be greatly increased by applying derivative operators, possibly followed by thresholding, to the images before matching—for example, taking first derivatives (e.g. gradient magnitudes) of both images or the second derivative (e.g. Laplacian magnitude) of one image. Geometric insensitivity can be improved by matching smaller pieces or local features (which are less affected by geometric distortion) and then searching for combinations of such matches in approximately the correct relative positions[23] or using relaxation methods to reinforce such combinations.[24–26] This hierarchical approach also serves to reduce the computational cost of the matching process. An alternative idea[27] is to segment the image into parts, represent the parts and their relationships by a graph structure and match these graph structures (see Section 10). Here too, relaxation methods are useful.

Another approach to pattern matching makes use of geometric transformations that map instances of a given pattern into peaks in a transform space. This "Hough transform" approach was originally developed to handle simple classes of shapes such as straight lines or circles, but it was recently extended[28] to arbitrary shapes in both two and three dimensions.

The matching techniques described above are all two-dimensional. They do not take into account the fact that the subimages or patterns being matched may differ not only in position and orientation, but also in size and perspective if the images are related by a three-dimensional transformation. Research is needed on matching techniques that take estimated surface orientation explicitly into account, thus making possible correction for the distorting effects of perspective, as well as for the associated intensity differences. The difficulty of matching images of objects that differ by three-dimensional rotations will be further discussed in Section 10.

7. RANGE ESTIMATION

If a high-resolution range sensor is available, the shapes of the visible surfaces in a scene can be obtained directly by constructing a range map. In this section we assume that range information is not directly available. In its absence, range can be inferred from stereopairs, by measuring stereo parallax, or relative range can be inferred from image sequences obtained from a moving sensor, by analyzing the motions of corresponding pixels from frame to frame ("optical flow").

Stereomapping is based on identifying correspnding points in the two images using image matching techniques. As indicated in Section 6, matching performance is improved if we match features such as edges, rather than intensity values. The MIT approach to stereo[29] is based on applying a set of edge operators, having a range of sizes, to the images, matching the edges produced by the coarsest operator, to yield a rough correspondence between the images, and then refining this correspondence by using successively finer edges. Edge-based approaches may still yield ambiguous results in heavily textured regions where edges are closely spaced. The ambiguity can be reduced by using intensity matching as a check or by classifying the edges into types (e.g. discontinuities in illumination, range or orientation) and requiring that corresponding edges be of the same type. In general, matching should be based on feature descriptions, rather than on raw feature response values. Work is needed on the development of matching methods based on other feature types and particularly on features derived from surface orientation maps—e.g. matching of surface patches. Matching yields a set of range values at the positions of features. Grimson[29] has developed methods of fitting smooth range surfaces to these values. For wide-angle stereo, where there is significant perspective distortion, derivation of a camera model and rectification of the images prior to matching are very desirable.

When a static scene is viewed by a moving sensor, yielding a succession of images, the relative displacements of pixels from one image to the next are known as "optical flow". If these displacements could be computed accurately, it would be possible in principle to infer the motion of the sensor relative to the scene and the relative distances of the scene points from the sensor (but note that there is an inherent speed/range ambiguity). Ideally, the displacements can be estimated by comparing the space and time rates of change of the image intensity, but in practice these estimates are quite noisy. Horn[30] has developed an iterative method of estimating a smooth displacement field, but it yields inaccurate results at object boundaries. It should be possible to obtain improved results by combining the rate of change approach with edge detection and matching. For larger displacements, a matching approach can be used to determine corresponding points in successive frames. Ullman[31] has shown that the motion of a rigid object is completely determined if we know the correspondence between a few points on the object as they appear in two or three successive images. Extensions of this work to jointed objects have also been investigated.

8. SEGMENTATION

Descriptions of an image generally refer to significant parts (regions; global features such as contours or curves) of which the image is composed. Thus image description requires segmenting it into such parts. A much more challenging task is to segment the image into parts corresponding to the surfaces or bodies of which the underlying scene is composed. This is often very hard to do, since variations in image intensity may not be good indicators of physical variations in the scene and, conversely, physical variations do not always give rise to intensity variations.

The most commonly used approach to image segmentation involves classification of the individual image points (pixels) into subpopulations. The parts obtained in this way are just the subsets of pixels belonging to each class. The classification can be done on the basis of intensity alone ("thresholding"), of color or spectral signature, or of local properties derived from the neighborhood of the given pixel. The last approach is used in feature detection (e.g. classify a pixel as on an edge if the value of some locally computed derivative operator is high in its neighborhood) and it can also be used to segment an image into differently textured regions. Pixels can be classified using a set of properties simultaneously or the properties can be used one at a time to recursively refine the segmentation.[32]

Pixels are usually classified independently, which allows fast implementation on parallel hardware. Better results can be obtained by classifying sequentially, so that regions composed of pixels belonging to a given class can be "grown" in accordance with given constraints, but such approaches are inherently slow and would not be appropriate for use in real-time systems. Another possibility is to use a relaxation approach in which pixels are classified fuzzily and the class memberships are then adjusted to favor local consistency. This approach[33, 34] requires a short sequence of iterations each of which can be implemented in parallel. In addition to local consistency, other sources of convergent evidence can be used to improve the quality of segmentation. For example, the classification criteria can be adjusted so as to maximize the edge strengths around the resulting region borders.

An alternative approach to segmentation is region-based rather than pixel-based. An example is the split-and-merge approach advocated by Pavlidis,[35] where the goal is to partition an image into homogeneous connected regions by starting with an initial partition and modifying it by splitting regions if they are not sufficiently homogeneous and merging pairs of adjacent regions if their union is still homogeneous. In this approach, "homogeneous" might mean approximately constant in intensity or, more generally, it might mean a good fit to a polynomial of some degree greater than 0, as in the facet model. Still more generally, the merging and splitting can be controlled by a "semantic" model which estimates probable interpretations of the regions and performs merges or splits so as to increase the likelihood and consistency of the resulting image interpretation.[36, 37] Note, however, that these methods still make no explicit use of surface orientation estimation. They should be based on object semantics rather than region semantics. Grouping locally detected features (edges or lines) into global contours or curves can be done on the basis of global shape (Hough transforms; see Section 6), but if this is not known in advance one can use methods analogous to split-and-merge—e.g. break a curve at branch points or sharp turns; link curves if they continue one another smoothly. Here again, it would be desirable to modify these criteria to take surrace curvature into account.

Image models should play an important role in image segmentation, but the models used in practice are usually much too simple. In segmenting an image by pixel classification, it is always assumed that the subpopulations are homogeneous, i.e. have essentially constant feature values (intensity, color, etc.). For scenes containing curved surfaces, this assumption is very unrealistic. Even if variations in illumination are ignored, changes in surface orientation will give rise to changes in feature values on the image of the surface. Similar remarks apply to region-based segmentation schemes. Haralick's facet model allows certain types of variations in feature values (e.g. linear), but the role of surface orientation needs to be made more explicit. By making local orientation estimation an integral part of the segmentation process and using these estimates to correct the feature values, it should be possible to cooperatively compute orientation estimates that optimize the clustering of feature values into subpopulations representing homogeneous surfaces (and not merely homogeneous regions). Spatial consistency constraints, as well as other types of convergent evidence, can also be incorporated into this process.

Another major drawback of segmentation based on pixel classification, particularly when it is implemented in parallel, is the difficulty of incorporating geometric knowledge about the desired regions into the segmentation process. The standard approach is to segment, measure geometric properties of the resulting regions and then attempt to improve the values of these properties by adjusting parameters of the segmentation process. However, it would be much preferable to make use of geometric constraints in the segmentation process itself. In region-based segmentation, since the units being manipulated are (pieces of) regions rather than pixels, somewhat greater control over region geometry can be achieved, by biasing the choices of splits and merges to favor the desired geometry. Another possibility is to perform segmentation using a multi-resolution ("pyramid") image representation, in which region geometry is coarsely represented by local patterns of "pixels" at the low-resolution levels. Here, segmentation is based on a cooperative process of pixel linking, which can be designed so that the linking is facilitated if it will give rise to the desired types of low-resolution local patterns. This approach too should be combined with surface orientation estimation, perhaps carried out at multiple resolutions.

9. OBJECT REPRESENTATION

The algorithms used to measure properties of image or scene parts depend on the data structures used to represent the parts. In this section we review the basic types of representations, both two- and three-dimensional, that are commonly used in image analysis systems.

Digital images are 2D arrays in which each pixel's value gives the intensity (in one or more spectral bands) of the radiation received by the sensor from the scene in a given direction. Other viewer-centered representations of the scene are also conveniently represented in array form, with the value of a "pixel" representing illumination, reflectivity, range or components of surface slant at the scene point located along a given direction. Various types of image transforms, as well as symbolic "overlay" images defining the locations of features (contours, curves, etc.) or regions, are other examples of 2D arrays that are often used in image processing.

Features and regions in any image can also be represented in other ways which are usually more compact than the overlay array representation and which also may make it easier to extract various types of information about their shapes. The following representations are all two-dimensional and are appropriate only if 3D shape information is not known. One classical approach is to represent regions by border codes, defining the sequence of moves from neighbor to neighbor that must be made in order to circumnavigate the border. Curves can also be represented by such move sequences ("chain codes").[38] Another standard way of representing regions is as unions of maximal "blocks" contained in them—for example, maximal "runs" of region points on each row of the image or maximal upright squares contained in the region; the set of run lengths on each row, or the set of centers and radii of the squares (known as the "medial axis"),[39] completely determines the region. The square centers tend to lie on a set of arcs or curves that constitute the "skeleton" of the region. If we specify each such arc by a chain code and also specify a radius function along the arc, we have a representation of the region as a union of "generalized ribbons" (see the end of this section).

There has been recent interest in the use of hierarchically structured representations that incorporate both coarse and fine information about a region or feature. One often-used hierarchical maximum-block representation is based on recursive subdivision into quadrants, where the blocks can be represented by the nodes of a degree-4 tree (a "quadtree").[40] Hierarchical border or curve representations can be defined based on recursive polygonal approximation, with the segments represented by the nodes of a "strip tree",[41] or based on quadrant subdivision.[42]

At a higher level of abstraction, a segmented image is often represented by a graph in which the nodes correspond to regions (or parts or surfaces if 3D information is available) or features, labelled with property names or values, and the arcs are labelled with relation values or names. A problem with this type of representation is that it does not preserve the details of region geometry and so can only provide simplified information about geometrical properties and relations, many of which have no simple characterizations. An ideal representation should provide information at multiple resolution, so that both gross geometry and important local features are easily available, together with the topological and locational constraints on the features' positions, where these constraints may have varying degrees of fuzziness. It should also be easy to modify the representation to reflect the effects of 3D geometrical transformations, so that representations of objects viewed from different positions can be easily compared.

Representations of surfaces and objects, i.e. "$2\frac{1}{2}$-dimensional" and "3-dimensional" scene representations, are also an important area of study. The visible surfaces in a scene can be represented by an array of slope vectors. The histogram of these vectors is known as the "gradient space" map. The range to each point in the scene is another important type of viewer-centered array representation.

In order to identify the objects in a scene, it is desirable to relate the viewer-centered representations of the visible surfaces to object-centered representations that describe the objects on a three-dimensional level. A variety of object representations can be defined, generalizing the representations of two-dimensional regions described above. An object can be represented by a series of slices and a 2D representation can be used for each slice. Alternatively, an object can be represented as a union of maximal blocks—e.g. by an "octree" (based on recursive subdivision of space into octants) or by a 3D "medial axis". If this axis is approximated by a set of space curves, each represented by a 3D chain code, and we also specify a radius function along each curve, we have a representation of the object as a union of "generalized cylinders" or "generalized cones".[43]

10. MODEL MATCHING

The image analysis processes described up to now give rise to a decomposition of the image into regions or of the scene into objects. A "literal" description of the image or scene can thus be given in the form of a relational structure in which the nodes correspond to features, regions or objects, labelled by lists of their property values (shape, texture, color, etc.). However, this type of "semantics-free" description is usually not what is wanted. Rather, one wants a description in terms of a known configuration of known objects. This requires "recognizing" the objects by comparing their descriptions to stored "models", which are generalized descriptions defining object classes.

Even in two dimensions, such models are often very difficult to formulate, since the constraints on the allowable property values and relationships are hard to define. In three dimensions, the problem is rendered even more difficult by the fact that only one side of an object can be visible in an image. The image description is two-dimensional, while the stored object models are presumably three-dimensional, object-centered representations.

The most extensive work on recognition of three-dimensional objects from images is embodied in the ACRONYM system.[44] This system incorporates methods of predicting the two-dimensional appearance (shape, shading, etc.) of a given object in an image taken from a given point of view. Conversely, it provides means of defining constraints on the three-dimensional properties of the object that could give rise to a given image and for manipulating sets of such constraints. These capabilities are incorporated in a prediction/verification process which uses the image to make predictions about the object and verifies that the image could in fact have arisen from an object that satisfies the resulting set of constraints. Thus far, ACRONYM has been implemented only in restricted domains, but it is based on very general principles and should be widely extendable.

It is often appropriate to model regions or objects hierarchically, i.e. as composed of parts arranged in particular ways, where the parts themselves are arrangements of subparts, and so on. There is an analogy between this type of hierarchical representation and the use of grammars to define languages. Here, a sentence is composed of clauses which are in turn composed of phrases, etc. Based on this observation, the process of recognizing an object as belonging to a given hierarchically defined class of objects is analogous to the process of recognizing a well-formed sentence as belonging to a given language, by *parsing* it with respect to a grammar for that language. This "syntactic" approach to object (or pattern) recognition has been extensively studied by Fu and his students.[45] It has been used successfully for recognition of two-dimensional shapes, patterns and textures, but it is less appropriate for three-dimensional object recognition, since it is not obvious how to incorporate in it mechanisms for relating 2D images to 3D objects.

Many difficult problems are associated with the model matching task. It is not trivial to define models for given classes of patterns or objects. (In the case of syntactic models, the problem of inferring them from sets of examples is known as *grammatical inference*. The pioneering work on the inference of relational structure models from examples was done by Winston.[46] Given a large set of models, it is not obvious how to determine the right one(s) with which to compare a given object. This is known as the *indexing* problem. Even if the correct model is known, comparing it with the descriptions of a given object may involve combinatorial search. (Here, however, relaxation or constraint satisfaction methods can often be used to reduce the search space.) The best approach is to use the model(s) to control the image analysis process and to design this process in such a way that most of the possible models are eliminated at early stages of the analysis. Unfortunately, there exists as yet no general theory of how to design image analysis processes based on given sets of models. The control structures used have been designed largely on heuristic grounds.

11. CONCLUDING REMARKS

We have seen that image analysis involves, in general, many different processes that incorporate many different types of information about the class of images being analyzed. There is no general theory of control in image analysis. In other words, there are no general principles that specify how these processes should interact in carrying out a given task. In particular, when a number of methods exist for performing a given task, e.g. feature detection, or inference of surface orientation, it would usually be desirable to implement several of the methods in order to obtain a consensus. However, there is no general theory of how to combine evidence from multiple sources.

Most of the successful applications of image analysis have involved relatively simple domains and have been primarily two-dimensional. For example, in robot vision, systems that recognize parts on a belt (well illuminated, non-overlapping, in specific 3D orientations) are not hard to build, but systems that recognize parts in a bin (shadowed, overlapping, arbitrarily oriented) are still a research issue. Techniques exist that will in principle handle such complex situations, but they need to be refined and extensively tested before they can be used in practice.

The discussion of the image analysis process in this paper has been quite general-purpose, without emphasis on particular domains of application. It is also possible to build "specialist" or "expert" systems tailored to a specific domain, which make use of methods especially designed for that domain. From a practical standpoint, successful applications of image analysis are likely to be of this specialized nature. It is the general approach, however, that makes image analysis at least potentially a science and that will continue to provide a theoretical background for the design of application-oriented systems.

REFERENCES

1. D. Marr, Visual information processing: the structure and creation of visual representations, *Proc. 6th Int. Joint Conf. on Artificial Intelligence*, pp. 1108–1126 (1979).
2. A. Rosenfeld and M. H. Thurston, Edge and curve detection for visual scene analysis, *IEEE Trans. Comput.* **C-20,** 562–568 (1971).
3. J. M. S. Prewitt, Object enhancement and extraction, *Picture Processing and Psychopictorics*, B. S. Lipkin and A. Rosenfeld, eds., pp. 75–149. Academic Press, New York (1970).
4. M. H. Hueckel, An operator which locates edges in digital pictures, *J. Ass. comput. Mach.* **18,** 113–125 (1971).
5. M. H. Hueckel, A local visual operator which recognizes edges and lines, *J. Ass. comput. Mach.* **20,** 634–647 (1973).
6. D. Marr and E. Hildreth, Theory of edge detection, *Proc. R. Soc.* **B207,** 187–217 (1980).
7. S. W. Zucker, R. A. Hummel and A. Rosenfeld, An application of relaxation labeling to line and curve enhancement, *IEEE Trans. Comput.* **C-26,** 394–403, 922–929 (1977).
8. B. J. Schachter, A. Lev, S. W. Zucker and A. Rosenfeld, An application of relaxation methods to edge reinforcement, *IEEE Trans. Syst. Man Cybernet.* **SMC-7,** 813–816 (1977).
9. R. M. Haralick and L. Watson, A facet model for image data, *Comput. Graphics Image Process.* **15,** 113–129 (1981).
10. B. Julesz, Visual pattern discrimination, *IRE Trans. Inf. Theory* **IT-8,** 84–92 (1962).
11. R. M. Haralick, K. Shanmugam and I. Dinstein, Textural features for image classification, *IEEE Trans. Syst. Man Cybernet.* **SMC-3,** 610–621 (1973).
12. L. S. Davis, S. A. Johns and J. K. Aggarwal, Texture analysis using generalized cooccurrence matrices, *IEEE Trans. Pattern Anal. Mach. Intell.* **PAMI-1,** 251–259 (1979).
13. J. T. Maleson, C. M. Brown and J. A. Feldman, Understanding natural texture, *Proc. DARPA Image Understanding Workshop*, pp. 19–27 (1977).
14. D. Waltz, Understanding line drawings of scenes with shadows, *The Psychology of Computer Vision*, P. H. Winston, ed., pp. 19–91. McGraw-Hill, New York (1975).
15. T. Kanade, Recovery of the three-dimensional shape of an object from a single view, *Artif. Intell.* **17,** 409–460 (1981).
16. T. O. Binford, Inferring surfaces from images, *Artif. Intell.* **17,** 205–244 (1981).
17. H. R. Barrow and J. M. Tenenbaum, Interpreting line drawings as three-dimensional surfaces, *Artif. Intell.* **17,** 75–116 (1981).
18. K. A. Stevens, The visual interpretation of surface contours, *Artif. Intell.* **17,** 47–73 (1981).
19. A. P. Witkin, Recovering surface shape and orientation from texture, *Artif. Intell.* **17,** 17–45 (1981).
20. B. Horn, Obtaining shape from shading information, *The Psychology of Computer Vision*, P. H. Winston, ed., pp. 115–155. McGraw-Hill, New York (1975).
21. K. Ikeuchi and B. K. P. Horn, Numerical shape from shading and occluding boundaries, *Artif. Intell.* **17,** 141–184 (1981).
22. R. J. Woodham, Analyzing images of curved surfaces, *Artif. Intell.* **17,** 117–140 (1981).
23. M. A. Fischler and R. A. Elschlager, The representation and matching of pictorial structures, *IEEE Trans. Comput.* **C-22,** 67–92 (1973).
24. L. S. Davis and A. Rosenfeld, An application of relaxation labelling to spring-loaded template matching, *Proc. 3rd Int. Joint Conf. on Pattern Recognition*, pp. 591–597 (1976).
25. S. Ranade and A. Rosenfeld, Point pattern matching by relaxation, *Pattern Recognition* **12,** 268–275 (1980).
26. O. D. Faugeras and K. E. Price, Semantic description of aerial images using stochastic labelling, *Proc. 5th Int. Conf. on Pattern Recognition*, pp. 352–357 (1980).
27. K. Price and R. Reddy, Matching segments of images, *IEEE Trans. Pattern Anal. Mach. Intell.* **PAMI-1,** 110–116 (1979).
28. D. H. Ballard, Generalizing the Hough transform to detect arbitrary shapes, *Pattern Recognition* **13,** 111–122 (1981).
29. W. E. L. Grimson, *From Images to Surfaces: A Computational Study of the Human Early Visual System.* MIT Press, Cambridge, MA (1981).
30. B. K. P. Horn and B. C. Schunk, Determining optical flow, *Artif. Intell.* **17,** 185–203 (1981).
31. S. Ullman, *The Interpretation of Visual Motion.* MIT Press, Cambridge, MA (1979).
32. R. Ohlander, K. Price and D. R. Reddy, Picture segmentation using a recursive region splitting method, *Comput. Graphics Image Process.* **8,** 313–333 (1978).
33. J. O. Eklundh, H. Yamamoto and A. Rosenfeld, A relaxation method for multispectral pixel classification, *IEEE Trans. Pattern Anal. Mach. Intell.* **PAMI-2,** 72–75 (1980).
34. A. Rosenfeld and R. C. Smith, Thresholding using relaxation, *IEEE Trans. Pattern Anal. Mach. Intell.* **PAMI-3,** 598–606 (1981).
35. T. Pavlidis, *Structural Pattern Recognition.* Springer, New York (1977).
36. J. A. Feldman and Y. Yakimovsky, Decision theory and artificial intelligence: I. A semantics-based region analyzer, *Artif. Intell.* **5,** 349–371 (1974).
37. J. M. Tenenbaum and H. R. Barrow, Experiments in interpretation-guided segmentation, *Artif. Intell.* **8,** 241–274 (1977).
38. H. Freeman, Computer processing of line-drawing images, *Comput. Surv.* **6,** 57–97 (1974).
39. H. Blum, A transformation for extracting new descriptors of shape, *Models for the Perception of Speech and Visual Form*, W. Wathen-Dunn, ed., pp. 362–380. MIT Press, Cambridge, MA (1967).
40. H. Samet and A. Rosenfeld, Quadtree representation of binary images, *Proc. 5th Int. Conf. on Pattern Recognition*, pp. 815–818 (1980).
41. D. Ballard, Strip trees: a hierarchical representation for curves, *Commun. Ass. comput. Mach.* **24,** 319–321 (1981).
42. M. Shneier, Two hierarchical linear feature representations: edge pyramids and edge quadtrees, *Comput. Graphics Image Process.* **17,** 211–224 (1981).

43. R. Nevatia and T. O. Binford, Description and recognition of curved objects, *Artif. Intell.* **8,** 77–98 (1977).
44. R. A. Brooks, Symbolic reasoning among 3-D models and 2-D images, *Artif. Intell.* **17,** 285–348 (1981).
45. K. S. Fu, *Syntactic Pattern Recognition and Applications.* Prentice-Hall, Englewood Cliffs, NJ (1982).
46. P. H. Winston, Learning structural descriptions from examples, *The Psychology of Computer Vision,* P. H. Winston, ed., pp. 157–209. McGraw-Hill, New York (1975).

About the Author—AZRIEL ROSENFELD received the Ph.D. in Mathematics from Columbia University in 1957. After ten years in the defense electronics industry, in 1964 he joined the University of Maryland, where he is Research Professor of Computer Science and Director of the Center for Automation Research. He is an editor of the journal *Computer Graphics and Image Processing,* an associate editor of several other journals, a past president of the International Association for Pattern Recognition and president of the consulting firm ImTech, Inc. He has published 15 books and over 300 papers, most of them dealing with the computer analysis of pictorial information.

Chapter 1

Recovering Scene Geometry

The computational theory for recovering scene geometry from one or more images is based on five distinct paradigms: stereo, motion, shading, texture, and line drawings. This overview describes these paradigms and the critical assumptions on which they depend; we also include a section on imaging geometry, which has direct relevance to all the paradigms. An outstanding collection of papers, edited by Brady [Brady 81a], summarizes the work in geometric recovery prior to 1981. Since this collection is readily available,[1] we will not reprint any of its contents beyond Brady's introduction [**Brady 81b**]. Our focus will be on later advances and new insights.

1.1 Stereo

A primary means for modeling 3-D scene geometry in both the human and the machine involves matching localized features or intensity patches in two images to determine the distance to corresponding scene points appearing in these images.[2] In correlation-based matching (see Chapter 7), image intensity patches to be matched are assumed to represent scene surface patches in which all points lie at approximately the same depth. In edge-matching techniques, matching between images is usually carried out on edges identified by the behavior of the derivatives of a smoothed version of the image—for example, zero crossings in the Laplacian of a Gaussian-smoothed image (see Chapter 2). From the correspondences one can first obtain the relative orientation and internal parameters of the two cameras that provided the images (in practical applications these quantities are often known in advance through special calibration techniques). Then, employing this knowledge and simple trigonometry, one obtains the scene depth of every pair of corresponding (matched) image points. The computational stereo paradigm is described in Barnard and Fischler [Barnard 82]. Mayhew and Frisby [Mayhew 81] describe a related computational theory of human stereopsis .

The two central problems in computational stereo are matching corresponding points, and obtaining scene depth at image locations at which matching fails (e.g., in areas not simultaneously visible in both images, or in featureless areas, or in areas where the occurrence of similar objects makes matching difficult).

Nishihara [**Nishihara 84**] describes one of the most advanced implementations of a complete system based on the computational stereo paradigm.[3] While

[1] Anyone interested in the subjects discussed in this chapter should obtain a copy of the Brady collection and see Brady's 1982 survey article on computational approaches to image understanding.

[2] While most current opinion holds this view concerning the human visual system, there are also contrary views as noted later in this section, e.g., [Jenkins 86].

[3] Two other high performance automated stereo systems are presented in Panton [Panton 84] and Hannah [Hannah 85].

strongly influenced by the Marr and Poggio "zero-crossing" theory of stereo matching [Marr 77] [Grimson 81, 85], this paper describes how the theory had to be modified to achieve goals concerned with noise tolerance, reliability, and speed.

Almost all approaches to stereo matching make use of the *continuity assumption*, which asserts that binocular disparity[4] varies smoothly almost everywhere and that only a small fraction of the area of an image is composed of boundaries discontinuous in depth. As noted earlier, continuity is required for both correlation and edge-based matching techniques. Similarly, interpolation methods based on the continuity assumptions are used to fill in the "holes" when matching fails. Prazdny [**Prazdny 85a**], in his discussion of binocular disparities, argues that continuity is too strong an assumption (e.g., it cannot explain how to deal with transparency), and is actually a special case of *coherence*. That is, objects in the world occupy well-defined 3-D volumes, but their projection onto an image plane can produce a discontinuous disparity field, which is the superposition of several interlaced continuous disparity fields. He presents a stereo algorithm that assumes coherence, rather than continuity, and uses a form of local consensus coupled with global optimization to obtain unique local matches. The algorithm was tested on random-dot stereograms and some natural images and appears to perform successfully.

In another paper, which also deviates somewhat from the conventional stereo paradigm, Quam [**Quam 84**] merges the matching and interpolation steps of the conventional stereo paradigm in the context of a coarse-to-fine hierarchical control structure. The approach uses disparity estimates from coarser levels of the hierarchy and geometrically warps one of the images to improve the performance of a cross-correlation-based matching technique. A surface interpolation algorithm by G. B. Smith ([G. B. Smith 84a], see Chapter 6) is immediately invoked to fill holes whenever the matching operator fails.

In spite of an immense body of research, and even implemented systems that operate successfully in cartographic and industrial application domains, it has yet to be demonstrated that stereo depth recovery based solely on local matching is adequate for general-purpose vision. (Criticism of correlation matching as the basis for human stereo vision can be found in Jenkins and Kolers [Jenkins 86]). Two novel approaches to stereo depth recovery, which have implications for other problems in scene analysis, have been formulated by Barnard and G. B. Smith.

Barnard [**Barnard 86**] embeds local matching at the level of individual pixels in a global optimization framework. His objective function rewards correspondences between pixels that are similar in intensity value and have disparities that are similar to those of their neighboring pixels. Simulated annealing (see Chapter 5) is used to find a complete set of correspondences that best satisfies the objective function. Because individual pixels (rather than finite areas) are matched, projective distortion is no longer a problem, nor does one have to worry about adjusting the size or shape of a correlation patch. Experiments show that this approach can successfully compile a dense model of natural 3-D (ground level) scenes.

G. B. Smith [**G. B. Smith 86**] describes a method for dense stereo compilation that entirely avoids local matching. The procedure begins with stereo images assumed to be in correspondence so that depth recovery can be accomplished for individual scan lines (i.e., the horizontal scan lines in the two images are corresponding epipolar lines). Smith shows how to set up systems of simultaneous equations whose solution is the depth profile corresponding to the intersection of the epipolar plane with the scene surfaces. Experiments with synthetically generated scenes show that the technique is theoretically sound, but the approach requires further work because in its present implementation it is overly sensitive to noise.

1.2 Motion

From a mathematical perspective the problems of recovering depth from motion or from stereo are essentially equivalent. In the depth-from-motion paradigm we analyze a sequence of closely spaced images obtained from a single camera moving through 3-D space; in the stereo paradigm we analyze individual pairs of images obtained from two cameras viewing the same scene from distinct spatial locations. Stereo can be considered to be a special case of motion where the two images to be analyzed are selected from the complete motion sequence. Similarly, motion can be considered to be a special case of stereo in which the separation between the two stereo cameras becomes vanishingly small. In this latter case, since the images are almost identical, the matching problem is greatly simplified, but the small spatial separation of the cameras makes the trigonometric method of depth estimation unreli-

[4]Disparity is the relative displacement of corresponding points in the two images of a stereo pair; disparity is inversely related to scene depth.

able. Further, since spatial disparity of corresponding image points becomes a differential quantity, it is more appropriate to divide spatial disparity by the (differential) time interval between image acquisitions and obtain a "velocity flow field" rather than a disparity array. The"velocity flow field" is a field of vectors, each expressing the instantaneous change of position in the image of a corresponding world point.

For general motion of the camera, Longuet-Higgins and Prazdny [Longuet-Higgens 80] show that scene depth information is contained solely in the translational component of the velocity flow field. Thus it is necessary to restrict camera motion to simple translation [Lawton 83] or to partition the flow field into its translational and rotational components.[5] Methods for achieving this flow field separation must deal with the aperture problem [**Ullman 81**]. This problem arises from the fact that, if we look at a moving straight-edge segment (or any small portion of a curve moving in the image plane), we can measure only the component of its velocity normal to the curve—the tangential component cannot be measured locally. The conventional optical flow paradigm thus poses two problems:

1. obtaining the velocity flow field
2. extracting depth and other scene information (e.g., detection of occlusion boundaries, object-surround separation, motion-based recognition, and deducing information about the camera motion) from the flow field. (This second problem, which cannot be solved without a solution to the first, has received relatively little attention.)

Ullman [**Ullman 81**] discusses the computational aspects of these two problems in the context of both biological and computer vision systems. In addition to reviewing relevant prior work in computer vision and describing how the primate visual system extracts motion information from its visual inputs, he suggests that sampled and continuously-sensed motion require two distinct approaches to motion detection and measurement. These two approaches give rise to different computational problems and, consequently, different kinds of processes in biological as well as in computer vision systems. He discusses the intensity-based and token-based systems that have been developed as models for motion measurement and the problems involved in interpretation of the detected motion.

The image velocity flow field can be computed in at least four different ways:

1. by matching corresponding (discrete) points or features in successive images as in Barnard and Thompson [Barnard 80].
2. by measuring an intensity gradient at a location in one image and noting the change in intensity[6] at the given location in the next image—see Limb and Murphy [Limb 75]; Fennema and Thompson [Fennema 79]; and Horn and Schunck [Horn 81].
3. by analyzing the evolution, or deformation, of continuous image curves, which, in turn, generally requires edge detection and tracking between successive image frames—see Hildreth [Hildreth 83, 84] and Waxman [**Waxman 84**].
4. by geometrically transforming an (uninterpreted) image sequence to make explicit the predictable image plane traces of scene points arising from translational or other known forms of camera motion—see Bolles, Baker, and Marimont [**Bolles 85**, 87].

Most of the research in recovering object geometry from a sequence of frames has been concerned with 2-D or 3-D rigid objects, often assuming that the image plane velocity field is locally constant and that the motion is such that the projected object contours do not deform. Nagel [Nagel 78] and Martin and Aggarwal [Martin 79] provide detailed surveys of early work; a critical view of more current work in motion is given in [Barron 84] and in a provocative paper by Aggarwal and Mitiche [Aggarwal 85].

Hildreth [Hildreth 84] derives a general constraint on the velocity field that allows the computation of the projected motion of 3-D surfaces that move freely in space, and whose projections deform over time. She relies on the physical assumption that the real world consists predominantly of solid objects, whose surfaces are generally smooth compared with their distance

[5] For uniform translation of the camera, and a forward-viewing direction, each world point flows across the image plane along a straight line emanating from a unique point called the *focus of expansion* Other simplifying assumptions, more general than simple translation, include the "fixed axis assumption" as used in Webb and Aggarwal [Webb 82], and the "parallel projection assumption" used by Ullman [Ullman 79]. For general camera motion, the optical flow field is not unique; rigidity and continuity are often assumed to obtain a unique solution.

[6] Optical flow cannot be unambiguously computed at a point in an image, independent of neighboring points, without introducing additional constraints (such as smoothness [Horn 81]). This limitation exists because the velocity field at each point has two components, while the change in image intensity at a point in the image plane due to motion yields only one constraint.

from the viewer. A smooth surface in motion usually generates a smoothly varying velocity field. Thus, she seeks a velocity field that is consistent with the motion measurements derived from the changing image, and that varies smoothly. She precisely formulates the suggestion by Horn and Schunck [Horn 81] that a single solution may be obtained by finding the velocity field exhibiting the smallest overall variation. Her formulation is motivated by an attempt to devise algorithms that are biologically feasible and lead to solutions compatible with human motion perception.

Waxman [**Waxman 84**] lays out a complete paradigm for relating the shape and motion of rigid objects to the evolution of the images of their occluding and surface contours. His approach is based on the analysis of a single surface patch in motion and its deforming neighborhood in an image sequence; the overall paradigm addresses an entire scene composed of several objects and background executing relative rigid body motion with respect to an observer.

Bolles, Baker, and Marimont [**Bolles 85**, 87] have developed geometric transformations, which, when applied to an image sequence obtained from a camera executing simple translational movement through a scene, provide a highly structured image regardless of the complexity of the given scene. These transformations are derived by "stacking up" a time sequence of frames and taking "slices" from the resulting solid "cube" of data. These slices, called "epipolar images," have a spatial dimension in one direction and time in the other. In the derived epipolar images, scene objects appear as distinctive straight-line intensity streaks; the slopes of these streaks are proportional to their distance from the camera, and their intersections mark their relative occlusions. This work is exciting because it has been successfully applied to complex natural scenes, and yet it requires no prior knowledge or assumptions about the nature of the scenes or their illumination.

1.3 Shading and Texture

Unlike the stereo and motion paradigms that employ multiple images to recover scene depth, the "shape from shading and texture" paradigms employ the intensity and texture variations in a single image to model scene geometry. A single 2-D image is an ambiguous representation of the 3-D world—many different scenes could have produced the same image—yet the human visual system is extremely successful at recovering a qualitatively correct depth model from this type of representation. Papers by Horn [**Horn 77**] and by Ikeuchi and Horn [Ikeuchi 81] present algorithms for recovering shape from shading based on the idea that the intensity at a point in an image is the result of light reflected from a small area of the surface of the imaged object. Different orientations of the surface relative to the light source and the camera will produce different image intensities. The change of image intensities, *if sufficiently smooth*, can provide information about the shape of the surface of the object. The practical utility of these algorithms is severely limited by the various assumptions and prior knowledge needed for their successful application—the continuity assumption, knowledge of location of the light source, the object-reflectivity function, and the orientation of surface normals at a suitably large set of locations (e.g., at occluding contours) to serve as boundary conditions.[7] Thus, it is important to determine the extent to which shape recovery from shading is possible using only the information directly obtainable from the image and some assumptions that will almost always be valid.

Pentland [Pentland 84b] examines the question of the limitations and uses of shading when one is constrained to using only local changes in image intensity with no prior knowledge about the viewed scene. He shows that such an analysis is incapable of determining the precise surface "slant,"[8] or whether the surface is convex or concave.[9]. However, the surface slant and tilt can be estimated if the surface is assumed to be Lambertian, the albedo and illumination constant, and the surface locally spherical.

The shape-from-texture paradigm is based on the idea that texture (the spatial distribution of surface markings) is distorted by projection in a manner that depends systematically on surface shape and orientation. If the projective distortion can be distinguished

[7] By extending the shape-from-shading paradigm to include multiple images and controlled lighting conditions, Woodham [Woodham 81] develops a concept called *photometric stereo* which eliminates the requirement for boundary conditions and continuity and which appears to be a potentially useful technique for industrial applications.

[8] Surface orientation can be represented by two angles called slant and tilt. In a formal sense, slant is the angle between the surface normal and the image plane (a number between 0 and 90 degrees); tilt is the direction of the surface normal projected onto the image plane. In an intuitive sense, a projected uniform surface texture is maximally compressed in the direction of tilt, and the amount of compression is proportional to the cosine of the slant.

[9] Other results concerning the limitations of local shading analysis can be found in Bruss [Bruss 81]. For example, she shows that patches of positive and negative Gaussian curvature cannot be distinguished locally on the basis of shading. G. B. Smith [G. B. Smith 83] provides a comprehensive mathematical description of the shape-from-shading problem.

from the properties of the texture on which the distortion acts, it is possible to recover surface shape.

The shape-from-texture paradigm has two distinct forms. The first, originally embodied in the writings of Gibson [Gibson 50], and more recently in the work of Bajcsy and Lieberman [Bajcsy 76] and K. A. Stevens [K. A. Stevens 80] among others, assumes that an observed planar surface patch is covered with uniformly dense surface markings. Based on this assumption, the gradient of texture density (change in the number of elements per unit area) can be used to deduce surface orientation. In the sense that shading can be considered to be a limiting case of texture, the shape-from-shading paradigm can be invoked [Pentland 84b, 86c] but without the need to model the illuminaton source or the reflectance function.

The second approach, due to Kender [Kender 80] and Witkin [Witkin 81], is based on the idea that natural textures do not usually conspire to mimic projective effects or to cancel those effects—that is, nonuniformity in natural texture does not mimic projective distortion. We attribute as much as possible of the observed texture variation to projection. Thus, given a portion of an image corresponding to a textured planar patch in the world, that orientation of the planar patch onto which the "back-projected"[10] image texture is most uniform is taken to be the best guess for the actual orientation of the surface. Kender requires isolation and measurement of specific attributes of the texture elements, such as diameter or orientation; Witkin measures more general properties of the image, such as local gradient direction, which does not require that the specific texture elements be analyzed. To the extent that curved surfaces can be locally approximated by planar patches, this back-projection approach is generally applicable.

In addition to shading and texture, occluding and surface contours constrain surface shape. As noted earlier, occluding contours are often used to provide boundary conditions for shape-from-shading algorithms—for example, see Ikeuchi and Horn [Ikeuchi 81] or Barrow and Tenenbaum [Barrow 81]. Also, K. A. Stevens [K. A. Stevens 81] shows how surface contours (a curve "painted" on the surface) can be used to provide information about surface shape.

[10] While not theoretically required, orthographic projection—or equivalently, perspective projection at long viewing distances—is necessary for practical implementation of this idea. The texture is also assumed to be "painted" on the surface as opposed to having a 3-D configuration.

Strat and Fischler [**Strat 86**] argue that, regardless of the specific assumptions made in the shape-from-shading, texture, and contour paradigms, if these assumptions permit recovery of a 3-D model from a single image, then they must be equivalent to providing a second (virtual) image of the scene. This paper provides a unifying framework for the various approaches to obtaining depth from a single image by showing how, in principle, they all map into the problem of stereo reconstruction.

1.4 Line Drawings

People routinely communicate with line drawings. The efficacy of such communication is based on the ability of the human visual system to recover the 3-D geometry of objects depicted in drawings, whether they are a set of precise engineering diagrams or a hand-drawn sketch.

In the case of engineering drawings restricted to planar-faced solid objects showing two or more views of such objects, algorithmic procedures exist for recovering the 3-D geometry or providing all feasible alternatives [Markowsky 81]. Although it is relatively straightforward to obtain a 3-D "wireframe" by back-projecting and intersecting the edges depicted in the line drawings, a sophisticated approach is required to avoid an exhaustive search to determine which volumes of space (defined by the wireframe) are solid and which are empty [Strat 84].

In the case of interpreting line drawings representing a single view of a 3-D scene, there is still a very large gap between the human's ability and that of the best existing computer programs. For the limited problem of perfect line drawings of polyhedral scenes, there is an elegant collection of mathematical theories concerned with deducing 3-D structure from 2-D projections. Guzman [Guzman 68] first demonstrated that a few simple heuristics could be used to decompose a single drawing of a complex collection of polyhedral objects into its separate parts. Huffman [Huffman 71] and Clowes [Clowes 71] devised formal schemes for labeling the edges and vertices of single line drawings such that (for most polyhedral objects with no more than three faces meeting at a single vertex) a valid 3-D description of the objects (in terms of these labels) could be recovered from the drawing, and drawings of impossible objects could be distinguished from drawings of real objects.

In 1973, Mackworth [Mackworth 73] introduced a new representation called "gradient space," a 2-D space in which every point represents the slope of some family of parallel planes. Mackworth discovered (or at least rediscovered [**Sugihara 84a**]) and took special advantage of the fact that if the axes of the gradient space are aligned with the x,y axis of a picture, then the line in gradient space connecting the two points corresponding to the slopes of two intersecting planes will be normal to the line of intersection of these planes in the picture (e.g., see Appendix A in Draper [Draper 81] or S. A. Shafer, Kanade, and Kender [S. A. Shafer 82]). Mackworth showed that, in addition to their importance in other areas of scene analysis, gradient space constructions could be used to generate improved interpretations of line drawings in terms of the Huffman-Clowes labels.[11]

Mackworth's program POLY recognizes the fact that a single picture (or line drawing) is an ambiguous representation of 3-D space, but that not all interpretations are possible. Every possible Huffman-Clowes labeling of a drawing is constructed and evaluated by determining if it is possible to construct a consistent gradient space diagram for it. POLY makes fewer mistakes than Huffman-Clowes, since it exploits surface-based constraints in addition to the edge-based constraints employed in the Huffman-Clowes scheme. An excellent analysis of POLY, its ultimate limits, and ways of extending its competence is given in a paper by Draper [Draper 81]. In particular he shows that gradient space is inadequate to represent all the available constraints, and he indicates that a nonintuitive algebraic solution may be required.

Waltz [Waltz 72] and Kanade [Kanade 80] also made notable extensions to the basic Huffman-Clowes line-labeling scheme.

While Waltz did not succeed in correcting the errors made by Huffman-Clowes, he extended line labeling to permit handling drawings containing shadows, cracks, missing and accidental alignments of edges, and non-trihedral vertices. By adding new labels to allow more precise descriptions, he not only increased the general competence of his program but also found that by introducing additional constraints expressible in terms of these new labels, the required computation time increased at a much slower rate than the description complexity. Perhaps his most important contribution was the introduction of a constraint propagation labeling algorithm, *Waltz filtering*, variations of which have subsequently been employed in other areas of scene analysis. In Waltz filtering, potential edge and node labels that are not locally consistent are eliminated; such local filtering, iteratively applied, can greatly simplify, or even eliminate, the need for global optimization.

Kanade extended the Huffman-Clowes label set to deal with an "origami" world of hollow shells and planar sheets, as well as with solid objects. His approach, as in the case of Waltz, does not eliminate the Huffman-Clowes errors. Kanade also introduced two additional assumptions that are good general heuristics for constraining 3-D shape: (1) lines parallel in an image are parallel in space, and (2) geometric configurations in an image that can be back-projected to symmetric figures in 3-D space should be given such an interpretation. In an important sense Kanade shifted the shape from line drawings paradigm away from qualitative description (i.e., decomposition of a polyhedral scene into primitive components, or distinguishing possible from impossible polyhedral scenes depicted as line drawings) toward quantitative reconstruction of 3-D shape. In a later paper co-authored with M. Herman [**Herman 84**], Kanade showed how solid modeling of city scenes could be effectively accomplished by a few simple assumptions about horizontal roofs and vertical faces typical of buildings.[12]

Sugihara completed the transition from the line-labeling and geometric description of Guzman, Huffman, Clowes, Mackworth, Waltz, and Kanade to algebraic analysis with the goal of quantitative 3-D reconstruction. One could also say that this research is a return to the seminal work of Roberts [Roberts 65] on the 3-D interpretation of 2-D images of polyhedral scenes. Sugihara provided a necessary and sufficient condition for a picture to represent a polyhedral scene [Sugihara 82, 84b][13] and also a general approach based on algebraic optimization to interpreting line drawings with the ability to take advantage of almost any additional type of photometric or geometric constraint [**Sugihara 84a**].

The "blocks world" of polyhedral objects dominated much of the work on line-drawing analysis because the

[11] It was known that the Huffman-Clowes scheme could produce incorrect descriptions—that is, assignment of line and vertex labels not consistent with the given drawing and the underlying assumptions—or might fail to recognize the drawing of an impossible object [Huffman 71].

[12] Kender [Kender 83] and Barnard [Barnard 83, 85] show how single-perspective views of straight lines can be used to recover scene geometry.

[13] With a correction and reduction in the scope of his original claims due to a counterexample produced by Shapira [Shapira 84].

various labeling techniques depended on the fact that a line can have only one interpretation (e.g., convex, concave, or occluding in the Huffman-Clowes scheme), which constrains the relationship between its terminating vertices. This constraint does not hold for objects with curved faces. Even though some significant work has been done with such more general objects—[Binford 81] and [Barrow 81]—extending the full competence of the blocks world techniques has not yet been achieved for general 3-D shapes.

1.5 Imaging Geometry

One generally does not know anything about the imaging geometry when looking at a picture; the focal length and location of the camera are both unknown. However, people can produce an excellent qualitative description of scene content apparently without being able to deduce these imaging parameters. How the human visual system can accomplish this feat is still a mystery. In order to recover depth, shape, albedo, and other scene properties using known computational techniques, we must first construct a quantitative model describing the geometric relationship between the scene, the camera, and the image.

Four problems in imaging geometry have received a considerable amount of attention, and all now have acceptable computational solutions:

1. Given two images of the same scene, determine the spatial relationship between the corresponding camera positions. The solution to the problem is necessary for stereo compilation as discussed in Section 1.1. Some relevant early results can be found in Kruppa [Kruppa 13]. Longuet-Higgens [**Longuet-Higgens 81**][14] and Tsai and Huang [Tsai 84] have independently derived solution techniques that involve solving a system of eight simultaneous linear equations, given that we can establish eight-point correspondences between perspective images. Tsai and Huang [Tsai 84] further show that seven-point correspondences are sufficient to obtain a unique solution under most conditions. Prazdny [Prazdny 80] and also Nagel and Neumann [Nagel 81] demonstrate that the problem for two-perspective images can be represented (but not always solved) in terms of three nonlinear equations involving five-point correspondences. Strat and Fischler [**Strat 86**] provide a linear equation solution for the case of eight correspondences between one perspective and one orthographic image. Gennery [Gennery 79] describes an iterative technique based on least-squares optimization, using as many correspondences as possible, to obtain a relative camera model.

   Since we can determine the spatial location of matched points appearing in two images for which we have a relative camera model, the above results tell us that if we have two distinct pictures of a structure with eight distinguished points (in some cases as few as five points are sufficient), we can recover the 3-D geometry of the structure.

2. Given multiple views of a rigid point configuration moving through space, determine the 3-D geometry of the configuration. If we have only two views, then this problem and the previous one are identical. Ullman [**Ullman 81**] shows that three orthographic images with established correspondences among at least four points are sufficient for recovery of both the geometry of the point configuration and its movement relative to the camera.

3. Given a single-perspective image of an object with known 3-D geometry, determine the location of the camera relative to the object. Fischler and Bolles [**Fischler 81a**] refer to a perspective image of a triangle with known dimensions, and they assume known camera parameters and correspondences between the scene and the image points. Then they show that there are at most four possible positions of the camera relative to the spatial locations of the three points defining the triangle vertices. A picture of a known configuration of four points in a plane, or six points in general position, permits unique recovery of the relative camera location.

4. Given a set of correspondences between a known 3-D configuration of points and their locations in a single 2-D image, determine the parameters of the camera that produced the image. Fischler and Bolles [**Fischler 81a**] show that if at least six correspondences between world points (with known 3-D locations) and image points can be established, then it is possible to determine the twelve coefficients of the $3x4$ "collineation" matrix that specifies the mapping (in homogeneous

[14] Longuet-Higgens [Longuet-Higgens 84] presents some examples of geometric configurations that defeat the eight-point algorithm.

coordinates)[15] from 3-space to 2-space. Strat [**Strat 84**] and Ganapathy [Ganapathy 84] provide techniques for recovering the camera parameters from the collineation matrix.

1.6 Discussion

The problem of recovering scene geometry from images has received more attention and reflects more successes than any other area of scene analysis. Because the ultimate concern here is with issues of geometry and the interaction of light with physical surfaces, not with the semantic content of the scenes being analyzed, we can phrase the questions to be answered with a reasonable degree of precision, determine when an answer has been found, and borrow solution techniques from the established disciplines of physics and mathematics. Nevertheless, it is disturbing that we should still be so far from duplicating the competence of the human visual system in performing the geometric recovery task for complex scenes.

The success, i.e., the practical utility, of such techniques as correlation-based stereo compilation may have deluded us into believing that human-level performance in geometric recovery is possible:

- by local, rather than global, analysis of the image
- without the need to solve simultaneously the other problems of scene analysis that we discuss in the following chapters: detection and recognition of scene components and their relationships, as constrained by physical and semantic considerations.

If the preceding assumptions are not valid—and there is indeed considerable doubt that they are—then geometric recovery is not just a matter of physics and mathematics. Rather it is an intelligent act that will require a radically new approach if further gains in performance are to be achieved.

Papers Included

Barnard, S.T. A stochastic approach to stereo vision. In *Proc. 5th National Conf. on AI*, pages 676–680, Philadelphia, PA, August 11–15, 1986.

Bolles, R.C. and H.H. Baker. Epipolar-plane image analysis: a technique for analyzing motion sequences. In *Proc. IEEE Third Workshop on Computer Vision: Representation and Control*, Bellaire, Michigan, pages 168–178, October 13–16, 1985.

Brady, J.M. Preface—The changing shape of computer vision. *Artificial Intelligence*, 17(1–3):1–15, August 1981.

Horn, B.K.P. Understanding image intensities. *Artificial Intelligence*, 8(2):201–231, 1977.

Longuet-Higgens, H.C. A computer algorithm for reconstructing a scene from two projections, *Nature*, 293:133–135, 1981.

Nishihara, H.K. Practical real-time imaging stereo matcher. *Optical Engineering*, 23(5):536–545, September–October 1984.

Prazdny, K. Detection of binocular disparities. *Biological Cybernetics*, 52:93–99, 1985.

Quam, L.H. Hierarchical warp stereo. In *Proc. DARPA Image Understanding Workshop*, pages 149–155, New Orleans, LA, October 1984.

Smith, G.B. Stereo integral equation. In *Proc. National Conf. on Artificial intelligence*, Philadelphia, PA, 1986

Strat, T.M. Recovering the camera parameters from a transformation matrix. In *Proc. DARPA IU Workshop*, pages 264–271, Oct. 1984.

Strat, T.M. and M.A. Fischler. One-eyed stereo: A general approach to modeling scene geometry. *IEEE PAMI*, 8(6):730–741, November 1986.

Sugihara, K. An algebraic approach to shape-from-image problems. *Artificial Intelligence*, 23(1):59–95, May 1984a.

Ullman, S. Analysis of visual motion by biological and computer systems. *IEEE Computer*, 14(8):57–69, August 1981.

Waxman, A.M. An image flow paradigm. In *Proc. IEEE Workshop on Computer Vision: Representation and Control*, Annapolis, Md., pages 49–55, April 30–May 2, 1984.

[15] This collection includes a paper by Ahuja and Coons [**Ahuja 68**], which presents an excellent tutorial discussion of homogeneous coordinates. Also, see Rogers and Adams [Rogers 76].

A STOCHASTIC APPROACH TO STEREO VISION

Stephen T. Barnard

Artificial Intelligence Center
SRI International
333 Ravenswood Avenue
Menlo Park, California 94025

Abstract

A stochastic optimization approach to stereo matching is presented. Unlike conventional correlation matching and feature matching, the approach provides a dense array of disparities, eliminating the need for interpolation. First, the stereo matching problem is defined in terms of finding a disparity map that satisfies two competing constraints: (1) matched points should have similar image intensity, and (2) the disparity map should be smooth. These constraints are expressed in an "energy" function that can be evaluated locally. A simulated annealing algorithm is used to find a disparity map that has very low energy (i.e., in which both constraints have simultaneously been approximately satisfied). Annealing allows the large-scale structure of the disparity map to emerge at higher temperatures, and avoids the problem of converging too quickly on a local minimum. Results are shown for a sparse random-dot stereogram, a vertical aerial stereogram (shown in comparison to ground truth), and an oblique ground-level scene with occlusion boundaries.

1 Introduction

To solve the stereo matching problem, one must assign correspondences between points on two lattices (the left and right images), such that corresponding points are the projections of the same point in the scene. The problem can be viewed as a complex optimization in which two criteria must be satisfied simultaneously. First, the corresponding points should have similar local features (in particular, similar intensity). Secondly, the spatial distribution of disparities, or, equivalently, the spatial distribution of depth estimates, should be plausible with respect to the depths likely to be observed in real scenes. Several authors have noted that, because surfaces are spatially coherent, the result of the stereo process should also be coherent, except at the relatively rare occlusion boundaries (for example, see Julesz [1] and Marr and Poggio [2]). The first criterion — similarity of local features — is insufficient because stereo correspondences are locally ambiguous. The second criterion, which is sometimes called the *smoothness constraint*, provides a heuristic for deciding which of the many combinations of feature-preserving correspondences are best.

The two major conventional approaches to stereo matching — feature matching and area correlation — suffer from two serious problems:

- Areas of nearly homogeneous image intensity are difficult to match because they lack local spatial structure. Edge-matching approaches never even attempt to match in such areas because no edges are found, and area correlation approaches fail because no significant peaks appear in the correlation surface. For most stereo vision applications, however, a dense matching is required. Dense estimates of depth are also more consistent with the subjective quality of human stereo experience, as revealed, for example, in random-dot stereograms. To obtain dense depth maps with the conventional approaches, one must resort to a postmatching interpolation step.
- Even where local structure is abundant, stereo correspondences may be ambiguous. Small-scale periodic structures are particularly difficult to match. To resolve these ambiguities, stereo matchers usually rely on a propagation of information, either from nearby areas, or from matching at larger scales, or both.

This paper describes an approach to stereo matching that is quite different from conventional area-based and feature-based matching. It is essentially an undirected Monte Carlo search that simulates the physical process of annealing, in which a physical system composed of a large number of coupled elements is reduced to its lowest energy configuration (or *ground state*) by slowly reducing the temperature while maintaining the system in thermal equilibrium. The system is composed of the lattice sites of the left image, and the state of each site encodes a disparity assignment. The total energy of the system is the sum of the energies of the local lattice sites. The local energy, which is a function of the states of the lattice site and its neighbors, has two terms: one term is proportional to the absolute intensity difference between the matching points, and the other term is proportional to the local variation of disparity (that is, to the *lack* of smoothness). The effect of a heat bath is simulated by considering local random state changes and accepting or rejecting them depending on the change in energy and the current temperature.

2 Simulated Annealing

Simulated annealing is a stochastic optimization technique that was inspired by concepts from statistical mechanics [3], [4]. It has been applied to a wide variety of complex problems that involve many degrees of freedom and do not have convex solution spaces. See Carnevali [5] for examples of image-processing applications. At the heart of simulated annealing is the Metropolis algorithm [6], which samples states of a system

Support for this work was provided by the Defense Advanced Research Projects Agency under contracts DACA 76-85-C-0004 and MDA903-83-C-0027. Support was also provided by FMC Corporation.

in thermal equilibrium. When a system is in thermal equilibrium, its states have a Boltzman distribution:

$$P(E) = \exp(-E/T) \quad (1)$$

where E is energy, $P(E)$ is the probability of a state having energy E, and T is the temperature of the system.[1] The Metropolis algorithm takes the system to equilibrium by considering random, local state transitions on the basis of the change in energy that they imply: if the change is negative, the transition is accepted; whereas, if the change is positive, the transition is accepted with probability $\exp(-\Delta E/T)$.

Starting at a very high temperature, simulated annealing uses the Metropolis algorithm to bring the system to equilibrium. Then the temperature is lowered slightly and the procedure is repeated until a very low temperature is achieved. If the temperature is lowered too quickly, the system may get stuck in locally optimal configurations and the ground state may not be reached. The algorithm is shown in Figure 1.

```
Select a random state S.
Select a sufficiently high starting temperature T.
while T > 0 do
    Make a random state change S' ← R(S).
    ΔE ← E(S') − E(S)
    ; Accept lower energy states.
    if ΔE ≤ 0 then S ← S'
    ; Accept higher-energy states with probability P.
    else
        P ← exp(−ΔE/T)
        x ← random number in [0,1]
        if x < P then S ← S'
    if there has been no significant decrease in E
        for many iterations
    then lower the temperature T.
```

Figure 1: The Simulated Annealing Algorithm

Simulated annealing tends to exhibit good average-case performance. It has the advantage of being a very simple algorithm that is inherently massively parallel. Furthermore, the parallelism is easily implemented because the processors need only short interconnections, may run asynchronously, and can even be unreliable. To be a good candidate for simulated annealing, a problem should follow the analogy of physical annealing. The function to be optimized should be expressed as an analog to the energy of a system composed of many local elements, and the interaction between the local elements should be short-range. A small random change in the state of the system should be possible by switching the microstate of a local element, and the resulting change in energy should be quickly computed by evaluating only the effects of the element's neighbors.

[1]The Boltzman distribution is usually written as $\exp(-kE/T)$, where k is Boltzman's constant. Because we define energy and temperature as pure numbers, no constant is necessary.

3 Stochastic Stereo Matching

If the relative positions and orientations of the two cameras are known, as well as the internal camera parameters, we can use the *epipolar constraint* to restrict the correspondences to the epipolar lines [7]. With no loss of generality, we can assume that the epipolar lines are parallel to the horizontal lines of lattice sites.[2] The correspondence problem then reduces to the assignment of a single horizontal disparity to each pixel in, say, the left image lattice.

Suppose that we have left and right image lattices, L_k and R_k, with $k = \{i,j\}, 0 \le i,j \le n-1$, that constitute a stereo pair with horizontal epipolar lines. The intensity of the left lattice point L_k is $I_L(k) = I_L(i,j)$, and similarly for right lattice points. For every k there is a (horizontal) disparity $D(k)$ such that the lattice point $L_k = L_{i,j}$ in the left image matches the point $R_{k'} = R_{i,j+D(k)}$ in the right image. The problem is to find an assignment of disparities to lattice points that satisfies the two criteria discussed in Section 1: similar intensity and smoothness. We assume that the upper and lower limits of disparity, D_{min} and D_{max}, are known. Furthermore, we consider only integer values of disparity. Even with these restrictions, the system has $N = (D_{max} - D_{min} + 1)^{n^2}$ possible states. Typical values in our examples are $D_{max} = 9$, $D_{min} = 0$, and $n = 128$, in which case $N = 10^{16384}$. Exhaustive search is obviously out of the question.

The disparity map should satisfy two criteria that are, to some extent, incompatible. The first criterion, which we call the *photometric constraint*, dictates that the disparity assignments should map points in L to points in R with comparable intensity: $I_L(k) \approx I_R(k')$. The second criterion is the *smoothness constraint*, which limits the variation in the disparity map.

Both criteria cannot be perfectly satisfied except in trivial situations. The photometric constraint can only be approximately satisfied due to sensor noise, quantization, slight lighting differences, and the presence of areas in one image that are occluded in the other. As discussed above, areas of homogeneous intensity will lead to ambiguous disparities based on photometry alone. The smoothness constraint will be perfectly satisified only with a uniform disparity map.

In an attempt to satisfy the two criteria simultaneously, we minimize a function of the form:

$$E = \sum_k \|I_L(k) - I_R(i,j + D(k))\| + \lambda\|\nabla D(k)\| \quad . \quad (2)$$

The first term inside the sum represents the photometric constraint and the second term the smoothness constraint. The constant λ determines their relative importance. We implement the $\|\nabla D(k)\|$ operator as the sum of the absolute differences between disparity $D(k)$ and the disparities of the kth lattice point's eight neighbors. Equation (2) is similar to the nonquadratic Tikhonov stabilizer proposed for stereo by Poggio, et. al. [8].

Following the simulated annealing algorithm, the system begins in a state chosen at random. Individual lattice points are considered in scan-line order, new disparities are selected at random, and the changes in energy are computed from equa-

[2]If the epipolar lines are not horizontal, the images can be mapped into a rectified stereo pair.

tion (2). Instead of monitoring the energy distribution to test for thermal equilibrium, we use a fixed annealing schedule.

4 Results

We have tested the stochastic matching algorithm on a variety of images, including random-dot stereograms, vertical aerial stereograms, and oblique ground-level stereograms. Identical parameters were used for all the examples shown in this section. In particular, the intensity ranged between 0 and 255, and we used $\lambda = 5$. We used a fixed annealing schedule: the temperature begins at $T = 100$ and is repeatedly reduced by 10% until it falls below $T = 1$. A total of ten scans through the lattice are performed for each temperature in this sequence.

Figure 2 shows a four-level "wedding cake" random-dot stereogram composed of 10% white and 90% black pixels. The background has zero disparity, and each successive layer has an additional two pixels of disparity. The figure shows the results with disparities encoded as grey values. Pixels with higher disparity are "closer" and are displayed as brighter values. Intermediate results for $T = 47$ and $T = 25$ and the final result for $T = 0$ are shown.

Figures 2c-e illustrate an important advantage in stochastic matching: the large-scale structure of the scene begins to emerge at higher temperatures, and as the temperature decreases finer structures become apparent. Temperature therefore provides a mechanism for dealing with problems of scale that is simpler than the complex search strategies employed by conventional methods. Note that the final disparity map is dense and that it corresponds very well to the three-dimensional wedding-cake shape. The errors are confined to the occlusion boundaries.

The next example, shown in Figure 3, is a vertical aerial stereogram supplied by the Engineering Topographic Laboratory (ETL). The original images have been bandpassed to remove the DC component. Again, intermediate results for $T = 47$ and $T = 25$ and the final result for $T = 0$ are shown. In addition, a disparity map supplied by ETL is shown for comparison. [3] The stochastic matching algorithm produces a result that is quite similar to the ETL data, although it is somewhat smoother. To some extent, this difference can be explained by the fact that the ETL result was produced from higher-resolution stereo images. The errors on the right border of Figure 3e are due to the fact that the stereo images do not have 100% overlap.

The final example, shown in Figure 4, is an oblique view of an outdoor scene containing a number of trees in both the foreground and background. The result in Figure 4e is certainly plausible, although we do not have a quantitative disparity model to compare it with, as in the previous examples. The matching algorithm seems to have smoothed over the foreground trees more than necessary, although we must be careful when relying on our subjective impressions of depth. When we interpret a scene like this one, we do not use stereo exclusively.

[3] The ETL disparity map was made with an interactive digital correlation device that depends on a human operator to detect and correct errors. The disparity map in Figure 3f has been sampled from a larger map compiled from much higher-resolution imagery.

(a) left image (b) right image

(c) $T = 47$ (d) $T = 25$

(e) $T = 0$

Figure 2: A 10% Random Dot Stereogram.

5 Conclusions

Stochastic stereo matching provides an attractive alternative to conventional stereo-matching techniques in several respects. The algorithm is simple, and, with suitable parallel hardware, can be very fast. Unlike conventional approaches, it produces a dense disparity map.

As noted by Geman and Geman [9], stochastic optimization by simulated annealing is in some ways similar to relaxation labeling [10]. In both approaches, objects are classified in such a way as to be consistent with a global context and to satisfy local constraints. There are, however, important differences. Relaxation labeling is a nonstochastic approach that, unlike simulated annealing, finds the local optimum closest to the initial state. Simulated annealing is intended to find the global optimum, or at least a local optimum nearly as good as the global one. Relaxation labeling has no counterparts for two important concepts in simulated annealing: temperature and thermal equilibrium.

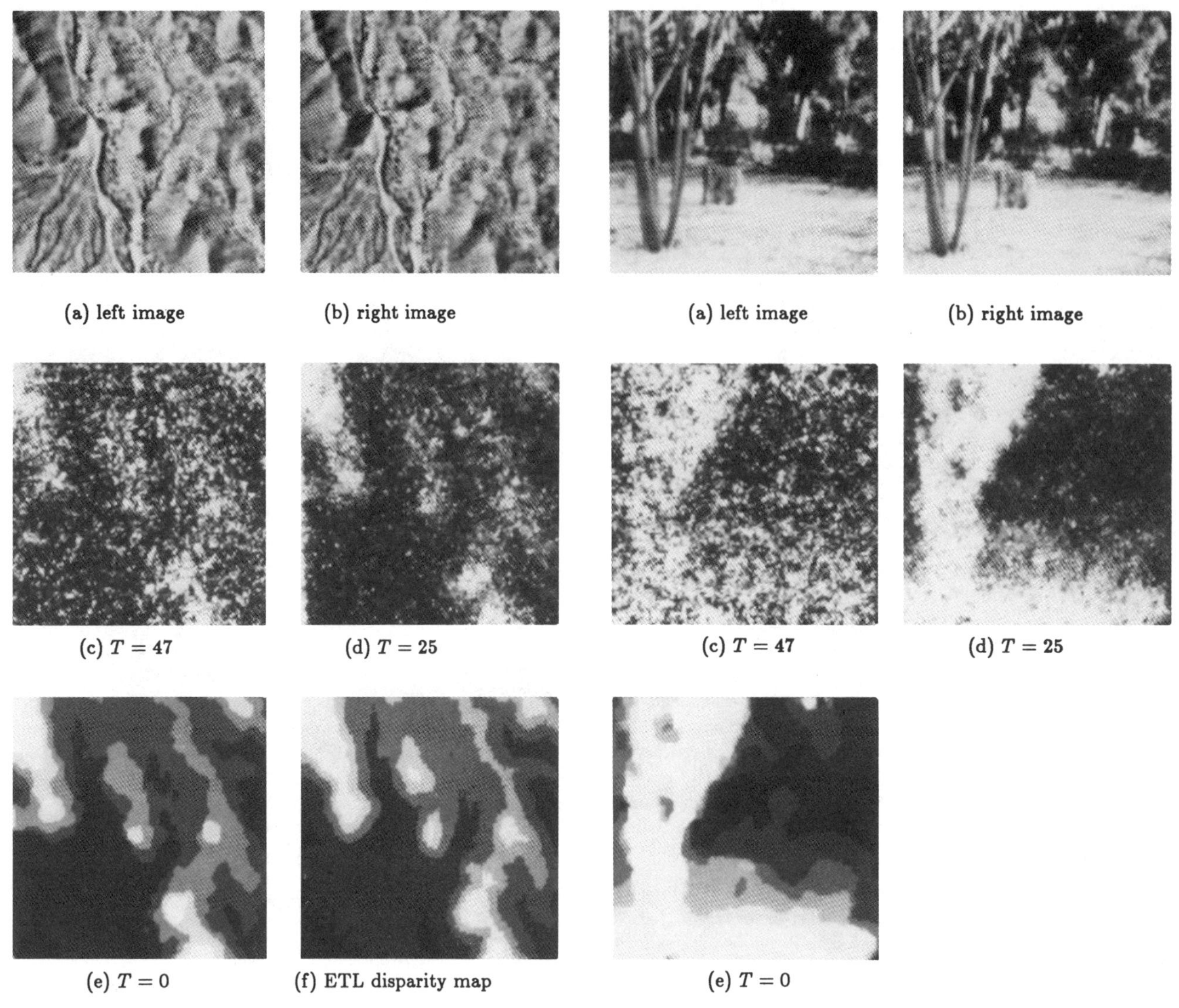

(a) left image (b) right image (a) left image (b) right image

(c) $T = 47$ (d) $T = 25$ (c) $T = 47$ (d) $T = 25$

(e) $T = 0$ (f) ETL disparity map (e) $T = 0$

Figure 3: A Vertical Aerial Stereogram.

Figure 4: An Oblique Stereogram.

The concept of temperature in simulated annealing provides a way to handle different scales in the problem instance. At higher temperatures, objects are only weakly coupled, and long-range interactions among large collections of objects can dominate the behavior of the system. At lower temperatures, local interactions take over. This effect was clearly seen in the examples of Section 4. Some physical systems exhibit a phase transition at some critical temperature. When simulating such systems, one must be careful to lower the temperature very slowly in the vicinity of the critical temperature. We have not observed phase transitions in the stereo problem and have been able to use fixed annealing schedules.

We are considering two extensions of the simple model presented here. First, the effective range of disparity could be increased by using lattices of several scales, allowing the coarser ones to bias the finer, in a manner similar to the hierarchical control structures used in many other matching techniques. Second, following Geman and Geman [9], a "line process" could be used to model depth discontinuities; although, in addition to lines, the process would also model occluded areas.

References

[1] B. Julesz, *Foundations of Cyclopean Perception*, Univ. of Chicago Press, Chicago, Ill., 1971.

[2] D. Marr and T. Poggio, Cooperative computation of stereo disparity," *Science*, **194**, (1976), pp. 283-287.

[3] S. Kirkpatrick, C.D. Gelatt, and M.P. Vecchi, Optimization by simulated annealing, *Science*, vol. 220, no. 4598, May 13, 1983, pp. 671-680.

[4] S. Kirkpatrick and R.H. Swendsen, Statistical mechanics and disordered systems, *Comm. ACM*, vol. 28, no. 4, April 1985, pp. 363-373.

[5] P. Carnevali, L. Coletti, and S. Patarnello, Image processing by simulated annealing, *IBM. J. Res. Develop.*, vol. 29, no. 6, November 1985, pp. 569-579.

[6] N. Metropolis, A.W. Rosenbluth, M.N. Rosenbluth, A.H. Teller, and E. Teller, Equations of state calculations by fast computing machines, *J. Chem. Phys.*, vol. 21, no. 6, June 1953, pp. 1087-1092.

[7] S.T. Barnard and M.A. Fischler, Computational stereo, *Computing Surveys*, vol. 14, no. 4, December 1982, pp. 553-572.

[8] T. Poggio, V. Torre, and C. Koch, Computational vision and regularization theory, *Nature*, vol. 317, September 1985, pp. 314-319.

[9] S. Geman and D. Geman, Stochastic relaxation, Gibbs distributions, and Bayesian restoration of images, *IEEE Transactions Pattern on Analysis and Machine Intelligence*, vol. PAMI-6, no. 6, November 1984, pp. 721-741.

[10] R.A. Hummel and S.W. Zucker, On the foundations of relaxation labeling processes, *IEEE Transactions Pattern Analysis and Machine Intelligence*, vol. PAMI- 5, May 1983, pp. 267-287.

EPIPOLAR-PLANE IMAGE ANALYSIS: A TECHNIQUE FOR ANALYZING MOTION SEQUENCES*

Robert C. Bolles
H. Harlyn Baker
SRI International
333 Ravenswood Avenue
Menlo Park, CA 94025.

Abstract

A technique for unifying spatial and temporal analysis of an image sequence taken by a camera moving in a straight line is presented. The technique is based on a "dense" sequence of images–images taken close enough together to form a solid block of data. Slices of this solid directly encode changes due to motion of the camera. These slices, which have one spatial dimension and one temporal dimension, are more structured than conventional images. This additional structure makes them easier to analyze. We present the theory behind this technique, describe an initial implementation, and discuss our preliminary results.

Introduction

Most motion-detection techniques analyze pairs of images, and hence are fundamentally similar to conventional stereo techniques (e.g., [1], [5], and [6]). A few researchers have considered sequences of three or more images (e.g., [8], [10], and [11]), but still the process is one of matching discrete items at discrete times. And yet, it is widely acknowledged that there is a potential benefit from unifying the analysis of spatial and temporal information. In this paper we present a technique to perform this type of unification for straight-line motions.

Motion-analysis techniques using pairs or triples of images are designed to process images that contain significant changes from one to another – features may move more than 20 pixels between views. These large changes force the techniques to tackle the difficult problem of stereo correspondence (Figure 1 shows an image triple with a typical inter-frame separation). Our idea, on the other hand, is to take a sequence of images from positions that are very close together – close enough that almost nothing changes from one image to the next. In particular, we take images close enough together that none of the image features moves more than a pixel or so (Figure 2 shows the first three images from one of our sequences containing 125 images). This sampling frequency guarantees a continuity in the temporal domain that is similar to continuity in the spatial domain. Thus, an edge of an object in one image appears temporally adjacent to (within a pixel of) its occurrence in both the preceding and following images. This temporal continuity makes it possible to construct a solid of data in which time is the third dimension and continuity is maintained over all three dimensions (see Figure 3). This solid of data is referred to as *spatio-temporal data.*

The traditional motion-analysis paradigm detects features in spatial images (i.e., the *uv* images in Figure 3), matches them from image to image, and then deduces the motion. We, however, propose an approach that is orthogonal to this. We suggest slicing the spatio-temporal data along a temporal dimension (see Figure 4), locating features in these slices, and then computing three-dimensional locations. Our reasoning is that the temporal image slices can be formed in such a way that they contain more structure than spatial images; thus, they are more predictable and, hence, easier to analyze.

To convince you of the utility of this approach, we must demonstrate that there is an interesting class of motions for which we can build structured temporal images. In the next section we show that this can be done whenever the camera moves in a straight line. We call these temporal images *epipolar-plane images,* or EPIs, from their geometric properties. In Section 3 we describe the results of our experiments in computing the depths of objects from their paths through the EPIs. And finally, in Section 4 we discuss the strengths and weaknesses of the technique and outline some current and future directions for our work.

Epipolar-Plane Images

In this section we define an epipolar-plane image (an EPI) and explain our interest in it. First, however, we review some stereo terminology. Consider Figure 5, which is a diagram of a general stereo configuration. The two cameras are modeled as pin-holes with the image planes in front of the lenses. For each point P in the scene, there is a plane, called the *epipolar plane,* which passes through the point and the line joining the two lens centers. This plane intersects the two image planes along *epipolar lines.* All the points in the epipolar plane are projected onto one epipolar line in the first image and a corresponding epipolar line in the second image. The importance of these lines for stereo processing is that they reduce the search required to find matching points from two dimensions to one. Thus, to find a match for a point along one epipolar line in an image it is only necessary to search along the corresponding epipolar line in the other image. This is termed the *epipolar constraint.*

One further definition that is essential to understanding our approach is that of an *epipole.* An *epipole* in a stereo configuration is the intersection of the line joining the lens centers and an image plane (see Figure 5). In motion analysis, an epipole is often referred to as a focus of expansion (FOE) because the epipolar lines radiate from it.

*This research was supported by DARPA Contracts MDA 903-83-C-0027 and DACA 76-85-C-0004

Consider a simple motion in which a camera moves from right to left, with its optical axis orthogonal to its direction of motion (see Figure 6). For this type of motion the epipolar plane for a point, such as P, is the same for all pairs of camera positions, and we refer to that plane as the epipolar plane for P for the whole motion.

The epipolar lines associated with one of these epipolar planes are horizontal scan lines in the images (see Figure 6). The projection of P onto these epipolar lines moves to the right as the camera moves to the left. The velocity of this movement along the epipolar line is a function of P's distance from the line joining the lens centers. The closer it is, the faster it moves.

For this motion, the epipolar lines are not only horizontal, they occur at the same vertical position in all the images. Therefore, a horizontal slice of the spatio-temporal data formed from this motion contains all the epipolar lines associated with one epipolar plane (see Figure 7).

Figure 7 shows three of the images used to form the solid of data. Typically a hundred or more images are used, making P's trajectory through the data a continuous path, as indicated in the diagram. For this type of lateral motion, if the camera moves a constant distance between images, the trajectories are straight lines (see Appendix A).

Figure 8 shows a horizontal slice through the solid of data shown in Figure 3, which was constructed from a sequence of 125 images taken by a camera moving from right to left. Figure 9 shows a frontal view of that slice. We call this type of image an epipolar-plane image (EPI) because it is composed of one-dimensional projections of the world points lying on an epipolar plane. Each horizontal line of the image is one of these projections. Thus, time progresses from bottom to top, and, as the camera moves to the left, the features move to the right.

There are several things to notice about this image. First, it contains only linear structures. In this respect it is much simpler than the spatial images used to create it (see Figure 1 for comparison). Second, the slopes of the lines determine the distances to the corresponding features in the world. The greater the slope, the farther the feature. Third, occlusion, which occurs when a closer feature moves in front of a more distant one, is immediately apparent in this representation. For example, the narrow white bar at the left center of the EPI in Figure 9 is initially occluded, then it is visible for a while until it is occluded briefly by a thin light object, then visible again before being rapidly occluded twice by two darker objects, and then is continuously visible until the end of the sequence. Thus, the same object is seen four different times.

Figure 10 shows another EPI sliced from the data in Figure 3. Its basic structure is the same as Figure 9; however, it illustrates the variety of patterns that can occur in an EPI.

The EPIs in Figures 9 and 10 were constructed from a simple right-to-left motion with the camera oriented at right angles to the motion. For what other types of motions can EPIs be constructed? The answer is that they can be constructed for any straight-line motion. As long as the lens center of the camera moves in a straight line the epipolar planes remain fixed relative to the scene. The points in each of these planes function as a unit. They are projected onto one line in the first image, another line in the second image, and so on. The camera can even change its orientation about its lens center as it moves along the line without affecting this partitioning of the scene. Orientation changes move the epipolar lines around in the image plane, significantly complicating the construction of the EPIs, but the epipolar planes remain unchanged since the line joining the lens centers remains fixed.

Figure 11 is an EPI formed from a sequence of images taken by a camera moving forward and looking straight ahead. Again the image is very structured, except that, instead of lines, it is composed of curves. For this type of motion, in fact for any straight-line motion in which the camera is at a fixed orientation relative to the direction of motion (see Figure 12), the trajectories in the EPI's are hyperbolas (see Appendix B). But not only are they hyperbolas, they are simple hyperbolas in the sense that their asymptotes are vertical and horizontal lines. A right-to-left motion, such as the one mentioned above, is just a special case in which the hyperbolas degenerate into lines.

If the lens center does not move in a line, the epipolar planes passing through a world point differ from one camera position to the next. The points in the scene are grouped one way for one pair of camera positions and a different way for another pair of positions. This makes it impossible to partition the scene into a fixed set of planes, which in turn means that it is not possible to construct EPIs for such a motion.

One last observation about EPIs: since an EPI contains all the information about the features in a slice of the world, the analysis of a scene can be partitioned into a set of analyses, one for each slice. In the case of a right-to-left motion, there is one analysis for each scanline in the image sequence. This ability to partition the analysis is one of the key properties of our motion-analysis technique. Slices of the spatio-temporal data can be analyzed independently (and possibly in parallel), and then the results can be combined into a three-dimensional rep-

Experimental Results

We have implemented a program that computes three-dimensional locations of world features by analyzing EPIs constructed from right-to-left motions. The program currently consists of the following steps:

1. 3D smoothing of the spatio-temporal data
2. Slicing the data into EPIs
3. Detecting edges, peaks, and troughs
4. Segmenting edges into linear features
5. Merging collinear features
6. Computing x-y-z coordinates
7. Building a map of free space
8. Linking x-y-z points between EPIs

In this section we illustrate the behavior of this program by applying it to the data shown in Figure 3.

The first step smooths the three-dimensional data to reduce the effects of noise and camera jitter, and to determine the temporal contours subsequently to be used as features. This is done by applying a sequence of three one-dimensional Gaussians ([3] and [4] explore other uses of spatio-temporal convolution).

The second step forms EPIs from the spatio-temporal data. For a lateral motion this is straightforward because the EPIs are horizontal slices of the data. Figure 9 shows the EPI selected to illustrate steps three through seven.

The third step detects edge-like features in the EPI. It currently locates four types of features: positive and negative zero-crossings [7] and peaks and troughs in the difference of Gaussians. The zero-crossings indicate places in the EPI where there is a sharp change in image intensity, typically at surface boundaries or surface markings, and the peaks/troughs occur between these zero-crossings. The former are generally more precisely positioned than the latter. Figure 13 shows all four types of features detected in the EPI shown in Figure 9.

The fourth step fits linear segments to the edges. It does this in two passes. The first pass partitions the edges at sharp corners by analyzing curvature estimates along the edges. The second pass applies Ramer's algorithm [9] to recursively partition the smooth segments into line segments. Figure 14 shows the line segments derived from the edges in Figure 13.

The fifth step builds a description of the line segments that links together those that are collinear. The intent is to identify sets of lines that belong to the same feature in the world. By bridging gaps caused by occlusions, the program can improve its estimates of the features' locations as well as extract clues about the nature of the surfaces in the scene. The program only links together features of the same type, except that positive and negative zero crossings may be joined, since the contrast across an edge can differ from one view to the next. Figure 15 shows the peak features from Figure 14 that are linked together by the program.

The line intersections in Figures 14 and 15 indicate temporal occlusions. For each intersection, the feature with the smaller slope is the one that occludes the other.

The sixth step computes the x-y-z locations of the world features corresponding to the EPI features. The world coordinates are uniquely determined by the location of the epipolar plane associated with the EPI and the slope and intercept of the line in the EPI. To display these three-dimensional locations, the program plots the two-dimensional coordinates of the features in the epipolar plane. Figure 16 shows the epipolar plane coordinates for the features shown in Figure 14. The shape and size of each ellipse depicts the error associated with the feature's location (this depends on the length of the line and the variance of the fit).

The seventh step builds a two-dimensional map of the world that indicates which areas are empty. The idea behind this construction is that, whenever a feature is seen by the camera, there is a clear line of sight from the camera to the feature. Therefore, if a feature is visible continuously during a portion of a motion, this line of sight sweeps out a triangle of empty space defined by the feature's location, the first position of the camera at which the feature is visible, and the last position at which the feature is visible. The program builds the map of empty space by constructing one of these triangular regions for each line segment found in an EPI, and then OR-ing them together. Figure 17 shows the map constructed for the features in Figure 16.

Figures 18 through 21 show the processing for the EPI 30 lines from the bottom of the uv images. This slice contains a plant on the left, a shirt draped over a chair, part of the top of a table, and in the right foreground, a ladder.

Figures 22 and 23 are stereo (crossed-eye) displays, showing some preliminary results in the eighth step of our analysis - combining the spatial data from the individual EPIs. Figure 22 shows the full set of x-y-z points. The display is fairly dense, since all points, including those arising from very short segments, are depicted. For spatial continuity, we link points between the various EPIs (nearest neighbors in overlapping error ellipses). Figure 23 displays those whose connected length is greater than 2.

Discussion

The following positive characteristics of this approach should be noted:

- Spatial and temporal data are treated together as a single unit;
- The acquisition and tracking steps of the conventional motion analysis paradigm are merged into one step;
- The approach is feature-based, but is not restricted to point features – linear features that are perpendicular to the direction of motion can also be used;
- There is more structure in an EPI than in a standard spatial image, which means that it is easier to analyze, and hence easier to interpret;
- Occlusion is manifested in an EPI in a way that increases the chance of detection because the edge is viewed over time against a variety of backgrounds;
- EPIs facilitate the segmentation of a scene into opaque objects occurring at different depths because they encode a *homogeneous* slice of the object over time;
- There are some obvious ways to make the analysis incremental in time, and partitionable in y (epipolar planes), for high speed performance.

With these benefits, the inherent limitations and current restrictions must be borne in mind:

- Motion must be in a straight line and (currently) the camera must be at a fixed angle relative to the direction of motion;
- Frame rate must be high enough to limit the frame-to-frame changes to a pixel or so (more specifically, such that the projective width of a surface is greater than its motion);
- Independently moving objects will either not be detected, or will be detected inaccurately.

We are currently investigating the following areas:

- Extending our analysis of connectivity between adjacent EPIs
- Identifying and interpreting spatial and temporal phenomena such as occlusions, shadows, mirrors, and highlights.
- Characterizing the appearance of curved surfaces in EPIs.
- Implementing the analysis of EPIs derived from forward motions.

We plan to write an expanded version of this paper in which we will present our results in these areas.

Appendix A: Lateral-Motion Trajectories

In this appendix we first derive an equation for the trajectory of a point in an EPI constructed from a lateral motion, and then show how to compute the (x,y,z) location of such a point. Figure 24 is a diagram of a trajectory in an EPI derived from the right-to-left motion illustrated in Figure 25. The scanline at t_1 in Figure 24 corresponds to the epipolar line l_1 in Figure 25. Similarly, the scanline at t_2 corresponds to the epipolar line l_2. (Recall that the EPI is constructed by extracting one line from each image taken by the camera as it moves along the line joining c_1 and c_2. Since the images are taken very close together in time, there would be several images taken between c_1 and c_2. However, to simplify the diagram none of these is shown.) The point (u_1, t_1) in the EPI corresponds to the point (u_1, v_1) in the image taken by the camera at time t_1 and position c_1. Thus, as the camera moves from c_1 to c_2 in the time interval t_1 to t_2, the scene point *moves* in the EPI from (u_1, t_1) to (u_2, t_2). The intent of this section is to characterize the shape of this trajectory and then compute the three-dimensional position of the corresponding scene point, given the focal length of the camera, the camera speed, and the coordinates of points along the trajectory.

For our analysis we define a left-handed coordinate system that is centered on the initial position of the camera (i.e., c_1 in Figure 25). The shape of the trajectory can be derived by analyzing the geometric relationships in the epipolar plane that passes through P. Figure 26 is a diagram of that plane.

Given the speed of the camera, s, which is assumed to be constant, the distance from c_1 to c_2, Δx, can be computed as follows:

$$\Delta x = s\,\Delta t \tag{1}$$

where Δt is $(t_2 - t_1)$. By similar triangles

$$\frac{u_1}{h} = \frac{x}{D} \tag{2}$$

$$\frac{u_2}{h} = \frac{\Delta x + x}{D} \tag{3}$$

where u_1 and u_2 have been converted from pixel values into distances on the image plane, h is the distance from the lens center to the epipolar line in the image plane, x is the x-coordinate of P in the scene coordinate system, and D is the distance from P to the line joining the lens centers. Since h is the hypotenuse of a right triangle, it can be computed as follows:

$$h = \sqrt{f^2 + v_1^2}, \tag{4}$$

where f is the focal length of the camera. From 2 and 3 we get

$$\Delta u = (u_2 - u_1) = \frac{h(\Delta x + x)}{D} - \frac{h\,x}{D} = \frac{h}{D}\Delta x \tag{5}$$

Thus, Δu is a linear function of Δx. Since Δt is also a linear function of Δx, Δt is linearly related to Δu, which means that trajectories in an EPI derived from a lateral motion are straight lines.

The (x, y, z) position of P can be computed by scaling u_1, v_1, and f appropriately. From 5 we define

$$m = \frac{D}{h} = \frac{\Delta x}{\Delta u} \tag{6}$$

which represents the slope of the trajectory computed in terms of the distance traveled by the camera (Δx as opposed to Δt) and the distance the point moved along the epipolar line (i.e., Δu). From similar triangles

$$(x, y, z) = \left(\frac{D}{h}u_1, \frac{D}{h}v_1, \frac{D}{h}f\right) \tag{7}$$

which means that

$$(x, y, z) = (mu_1, mv_1, mf) \tag{8}$$

If the first camera position, c_1, on an observed trajectory is different from the camera position, c_0, that defines a global camera coordinate system, the x coordinate can be adjusted by an amount equal to the distance traveled from c_0 to c_1. Thus,

$$(x, y, z) = ((t_1 - t_0)s + mu_1, mv_1, mf) \tag{9}$$

where t_0 is the time of the first image and s is the speed of the camera. This correction is equivalent to computing the x intercept of the line and using it as the first camera position. Therefore, for a lateral motion, the trajectories are linear and the (x, y, z) coordinates of the points can be easily computed from the slopes and intercepts of the lines.

Appendix B: Forward-Motion Trajectories

The derivation of the form of a trajectory produced by a forward motion is similar to the one used for lateral motion. Figure 27 is a diagram of a trajectory in an EPI derived from a sequence of images taken by a camera moving in a straight line at a fixed orientation relative to the principal axis of the camera (see Figure 28). Without loss of generality we have rotated the image plane coordinate systems in a uniform way so that the epipoles are on the u axes. The EPI in Figure 27 was constructed by extracting pixel intensities along epipolar lines in the images shown in Figure 28 and inserting them as scanlines in Figure 27. For example, epipolar line l_1 was placed at t_1, l_2 was placed at t_2, and so on. The point (w_1, t_1) in the EPI corresponds to the point (u_1, v_1) in the image taken at time t_1 and position c_1. Thus, as the camera moves from c_1 to c_2 over the time interval t_1 to t_2, the scene point *moves* in the EPI from (w_1, t_1) to (w_2, t_2). Our goal is to characterize the shape of this trajectory, and then compute the three-dimensional position of the corresponding scene point, given the focal length of the camera, the camera speed, the angle between the camera's axis and the direction of motion (θ), and the coordinates of points along the trajectory.

As before, we define a left-handed coordinate system that is centered on the initial position of the camera (i.e., c_1 in Figure 28). The shape of the trajectory can be derived by examining the geometric relationships in the epipolar plane that passes through P. Figure 29 is a diagram of that plane.

Given the speed of the camera, s, which is assumed to be constant, the distance from c_1 to c_2, Δe, can be computed as follows:

$$\Delta e = s\,\Delta t \tag{10}$$

where Δt is $(t_2 - t_1)$. By similar triangles

$$\frac{w_1}{h} = \frac{C}{e} \tag{11}$$

and

$$\frac{w_2}{h} = \frac{C}{(e - \Delta e)} \tag{12}$$

where w_1 and w_2 are distances on the image plane, h is the distance from the lens center to the epipole, C is the distance from P to the line joining the lens centers measured in a plane parallel to the image planes, and e is the distance along the line joining the lens centers from c_1 to the plane passing through P and parallel to the image planes. Since h is the hypotenuse of a right triangle (see Figure 28), it can be computed as follows:

$$h = \frac{f}{cos\theta} \tag{13}$$

where f is the focal length of the camera. From 11 and 12 we get

$$\Delta w = (w_2 - w_1) = \frac{hC}{(e - \Delta e)} - \frac{hC}{e} = \frac{hC\Delta e}{e(e - \Delta e)} \tag{14}$$

which can be rewritten as

$$e\Delta w\Delta e \ - \ e^2\Delta w \ + \ hC\Delta e \ = \ 0. \tag{15}$$

Using 10 to express Δe in terms of Δt, this becomes

$$se\Delta w\Delta t \ - \ e^2\Delta w \ + \ shC\Delta t \ = \ 0. \tag{16}$$

which defines a hyperbola whose asymptotes are the lines $w = 0$ and $t = e/s$ (see Figure 30). Thus, the trajectory is a hyperbola in which the point P appears arbitrarily close to the epipole when the camera is far away from it (as one would expect), and the projection of P moves away from the epipole at an increasing rate as the camera gets closer to it. This relationship agrees intuitively with the fact that a projective transformation involves a $1/z$ factor, which makes u a hyperbolic function of z.

Equation 14 can be used to compute z. First, rewrite it as follows:

$$\Delta w = \frac{hC}{(e - \Delta e)}\frac{1}{e}\Delta e \tag{17}$$

Then using Equation 12 and

$$e = \frac{z}{cos\theta} \tag{18}$$

we get

$$\Delta w = \frac{w_2 \, cos\theta \, \Delta e}{z} \tag{19}$$

or

$$z = \frac{w_2 \, cos\theta \, s\Delta t}{\Delta w} \tag{20}$$

Notice that it is NOT necessary to determine the coefficients of the hyperbola in order to compute z. Two points on the trajectory are sufficient to compute Δt and Δw, which in turn, are sufficient to compute z. Also notice, however, that it is easy to fit an hyperbola of this type because it is in the simple form

$$\Delta w\Delta t \ + \ a\Delta w \ + \ b\Delta t \ = \ 0, \tag{21}$$

which is linear with respect to the coefficients a and b. This type of fitting provides a way to increase the precision with which the scene points are located.

The expression for z in Equation 20 does not apply when $\theta = 90°$, but that is the lateral motion case covered earlier. Thus, the trajectories are always hyperbolas; they just happen to degenerate into straight lines when $\theta = 90°$, which corresponds to the case in which the epipoles are not in the image plane, but rather lie at infinity.

The x and y coordinates for P can be computed by scaling z appropriately:

$$(x, y) = (\frac{u_1}{f}z, \frac{v_1}{f}z) \tag{22}$$

Recall that u_1 and v_1 are measured in a rotated-image-plane coordinate system that was set up to place the epipole on the u axis. Therefore, in addition to converting pixel values to a standard metric, such as meters, the image coordinates of a point must be rotated about the principal axis before they can be inserted into Equation 22. To compute a world-centered position for P, the (x, y, z) position computed by Equations 20 and 22 has to be transformed for the initial position of the camera along the path.

References

[1] "Disparity Analysis of Images," S. T. Barnard and W. B. Thompson, *IEEE Trans., PAMI,* **Vol 2, No 4,** 1980.

[2] "A Discrete Spatial Representation for Lateral Motion Stereo," N. J. Bridwell and T. S. Huang, *Computer Vision, Graphics, and Image Processing,* **Vol 21,** 1983.

[3] "Monocular Depth Perception from Optical Flow by Space Time Signal Processing," B. F. and Hilary Buxton, *Proceedings of the Royal Society of London, B.,* **Vol 218,** 1983.

[4] "Computation of optic flow from the motion of edge features in image sequences," B. F. and H. Buxton, *Image and Vision Computing,* **Vol 2, No 2,** May 1985.

[5] "Detection of Moving Edges,"S. M. Haynes and R. Jain, *Computer Vision,Graphics, and Image Processing,* **Vol 21, No 3,** 1983.

[6] "Computations Underlying the Measurement of Visual Motion," E. C. Hildreth, *Artificial Intelligence,* **Vol 23,** 1984.

[7] "Theory of Edge Detection," D. C. Marr and E. Hildreth, *Proceedings of the Royal Society of London,* **B 207,** 1980.

[8] "Depth Measurement from Motion Stereo," Ramakant Nevatia, *Computer Graphics and Image Processing,* **5,** 1976.

[9] "An Iterative Procedure for the Polygonal Approximation of Plane Curves," U. Ramer, *Computer Graphics and Image Processing* **Vol 1,** 1972.

[10] *The Interpretation of Visual Motion,* S. Ullman, MIT Press, Cambridge, Mass., 1979.

[11] "Determining 3-D Motion and Structure of a Rigid Body Using the Spherical Projection," B. L. Yen and T. S. Huang, *Computer Graphics and Image Processing,* **Vol 21,** 1983.

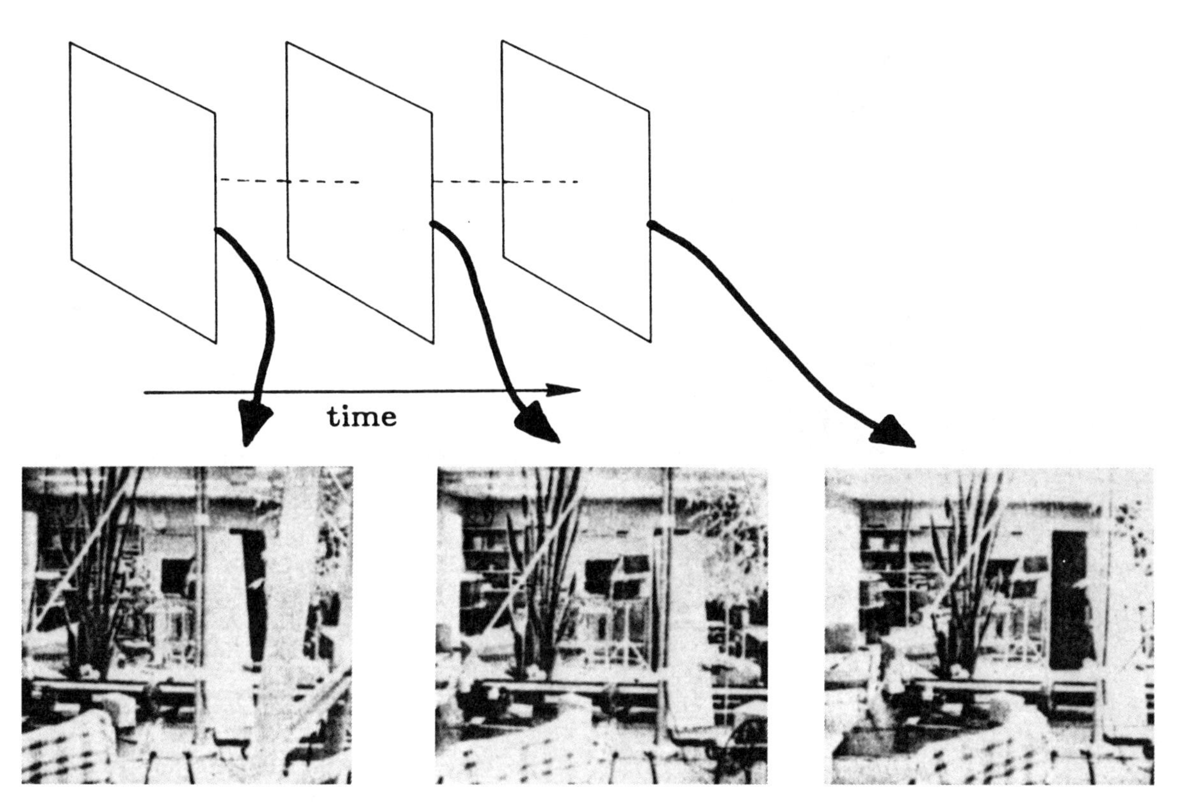

Fig. 1. Typical image separation

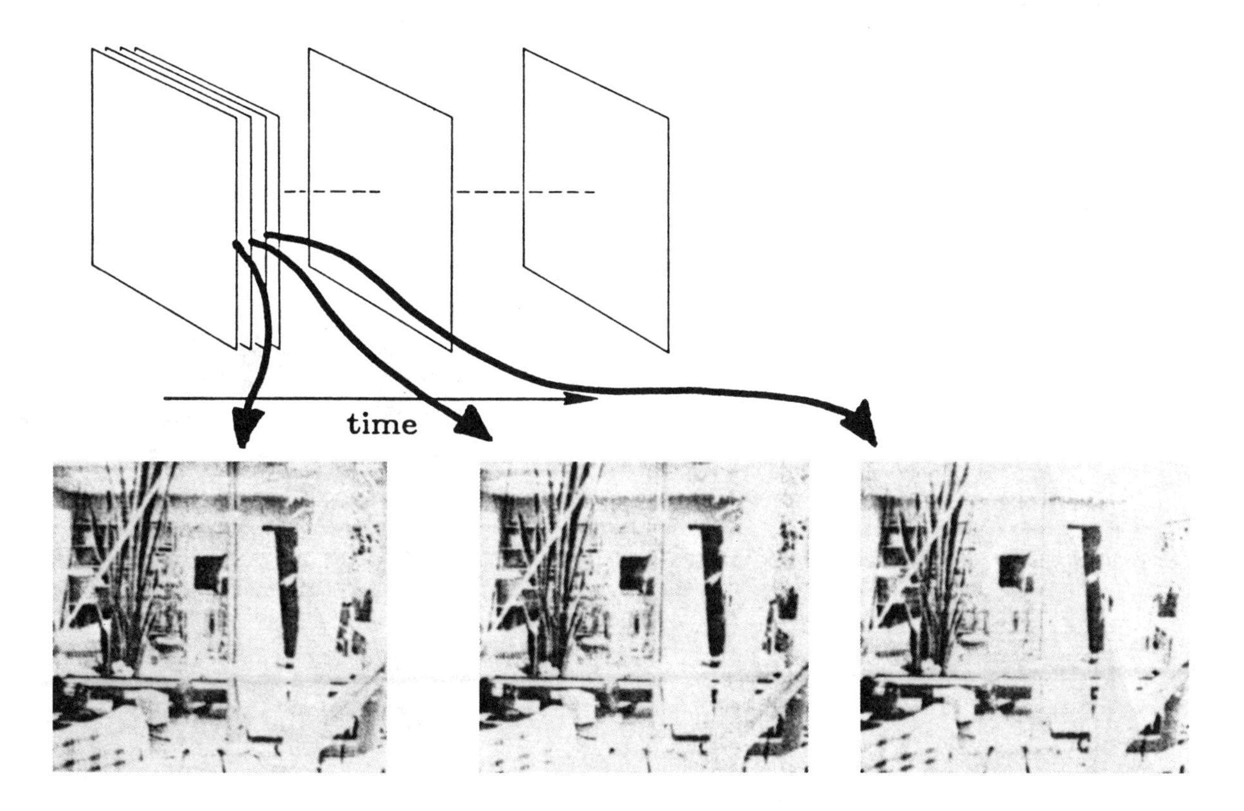

Fig. 2. Close sampling image separation

Fig. 3. Spatio-temporal solid of data

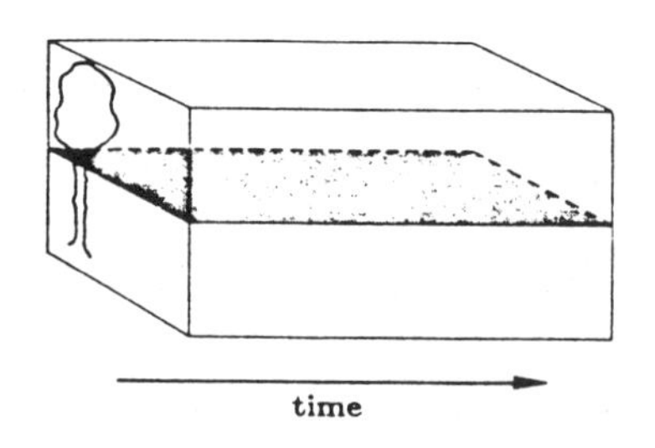

Fig. 4. Slice of the solid of data

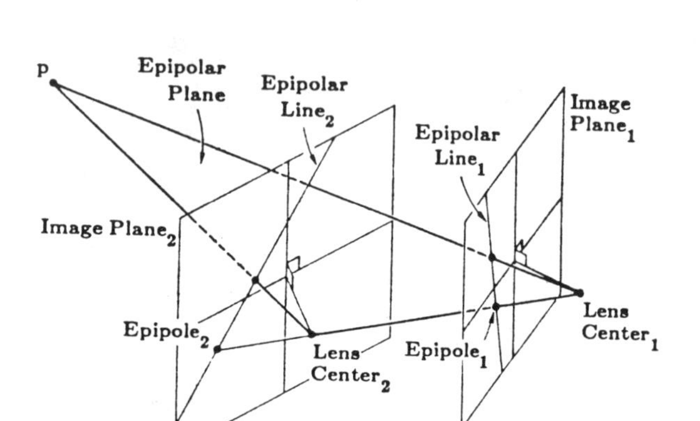

Fig. 5. General stereo configuration

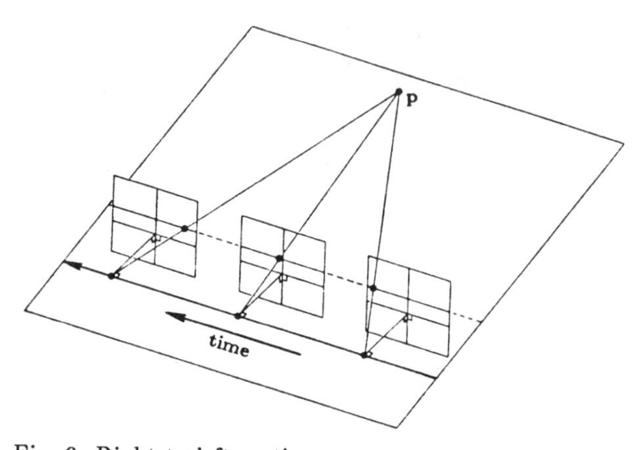

Fig. 6. Right-to-left motion

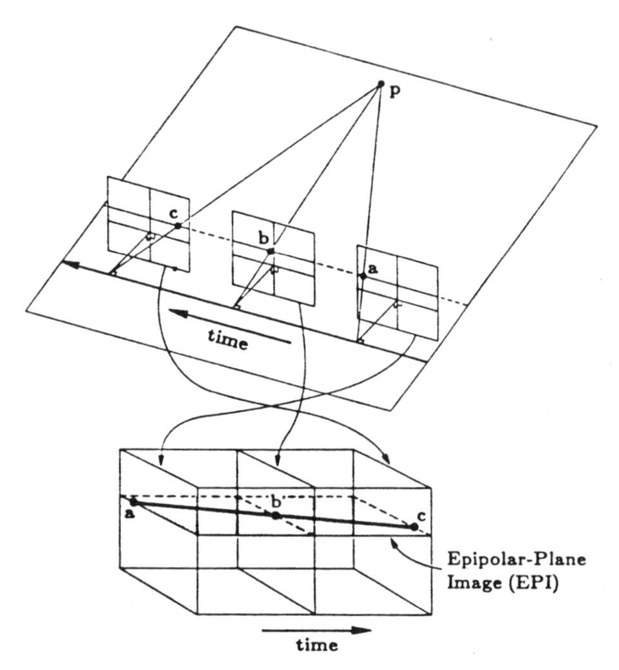

Fig. 7. Right-to-left motion with solid

Fig. 8. Sliced solid of data

Fig. 9. Frontal view of the EPI

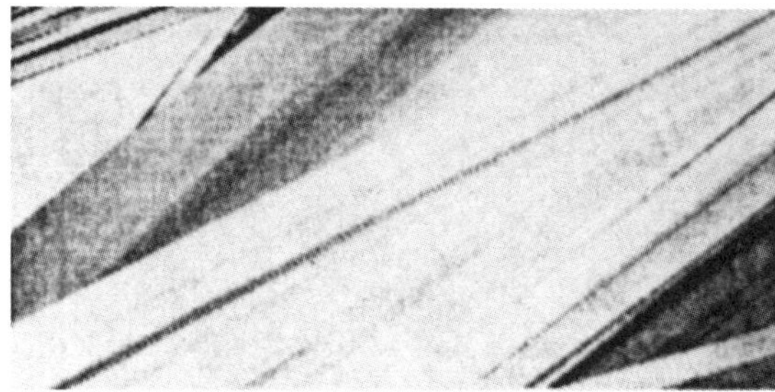

Fig. 10. A second EPI

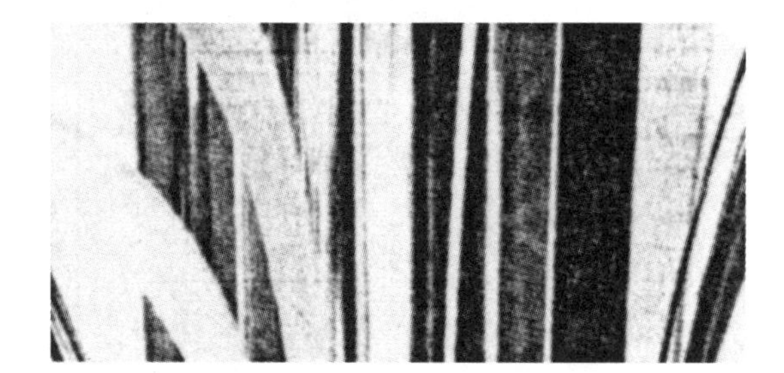

Fig. 11. EPI from forward motion

Epipolar Plane
p
θ
time
Epipolar Line
Epipole
θ
Lens Center

Fig. 12. Forward motion

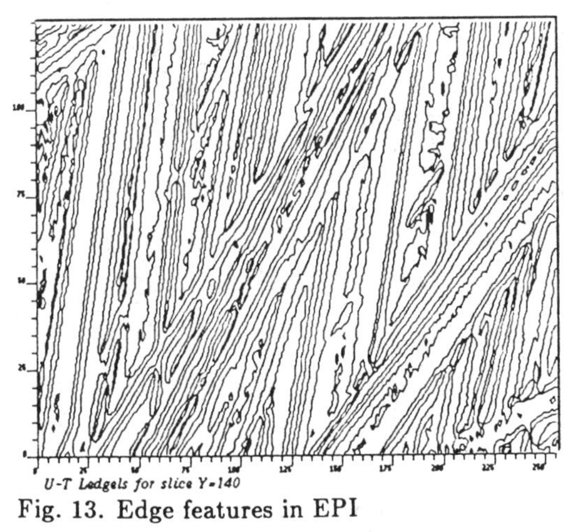

Fig. 13. Edge features in EPI

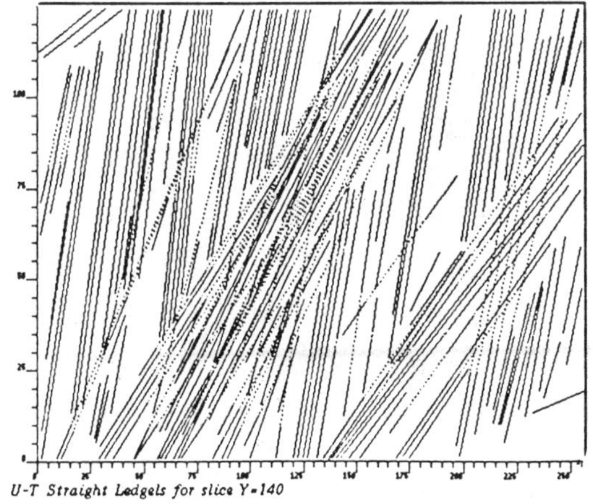

Fig. 14. Straight lines

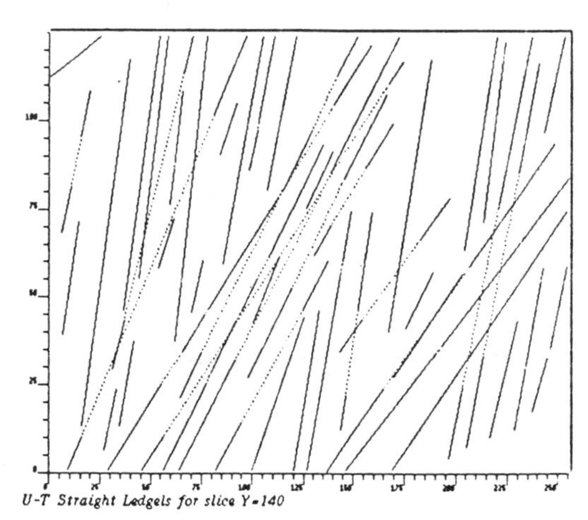

Fig. 15. Linked peak lines

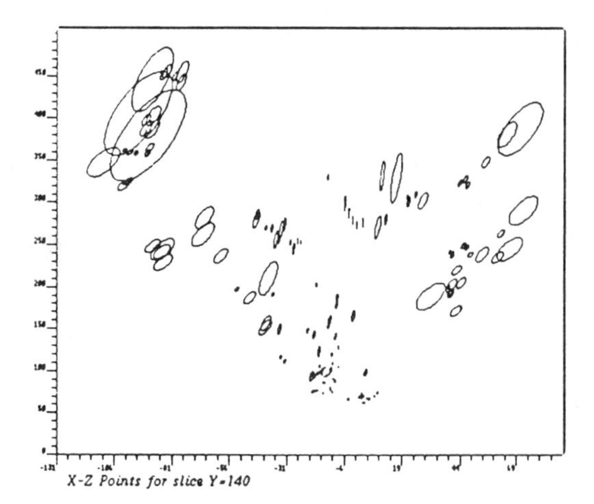

Fig. 16. xz locations

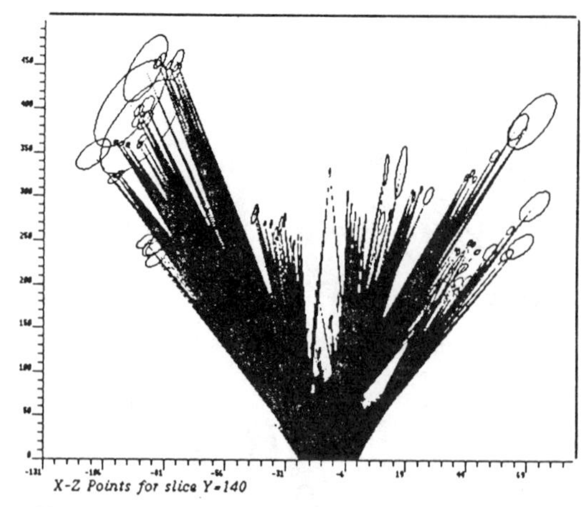

Fig. 17. Free space

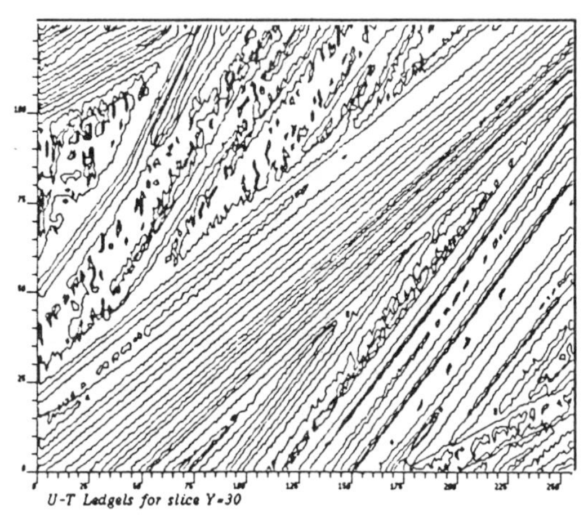

Fig. 18. Edge features in EPI

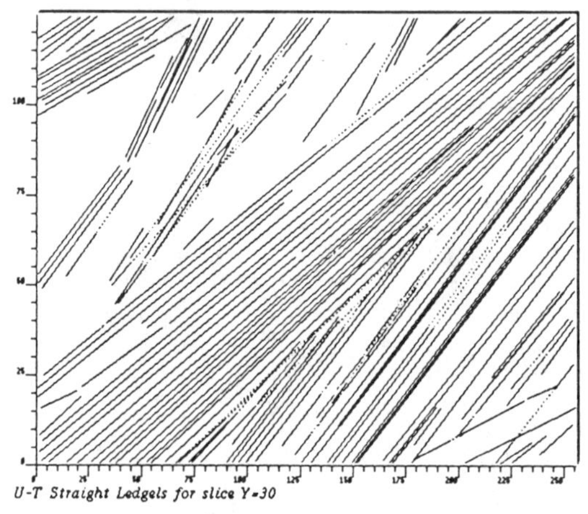

Fig. 19. Straight lines

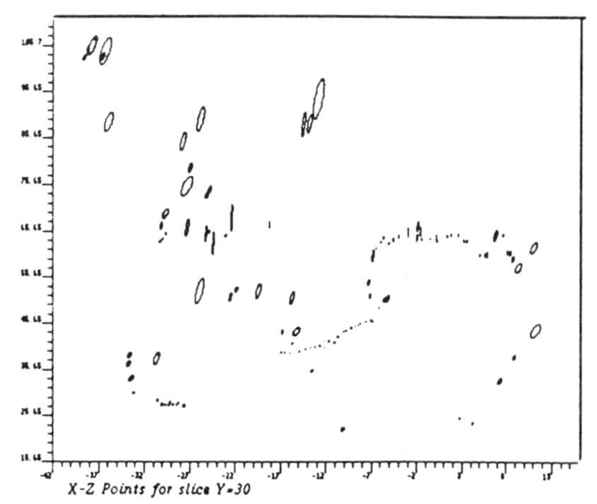

Fig. 20. xz locations

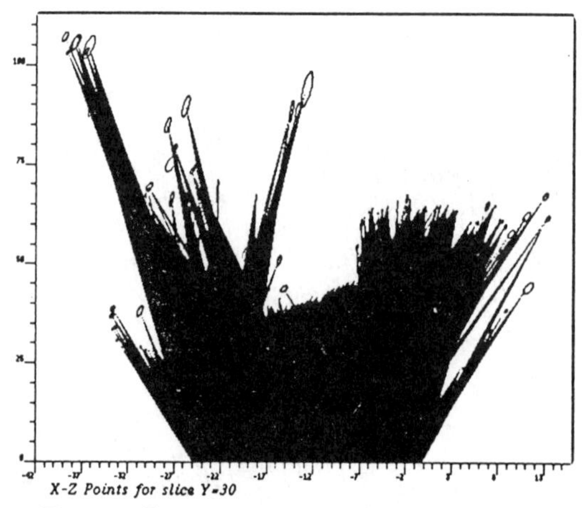

Fig. 21. Free space

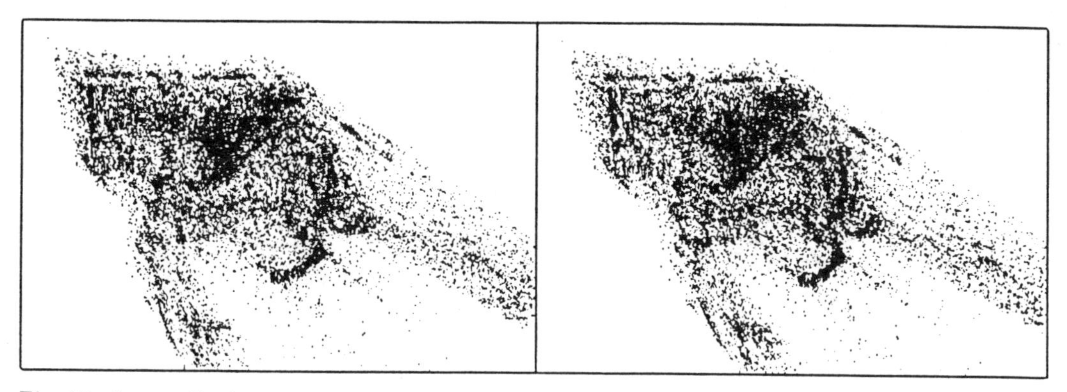

Fig. 22. Stereo display of all x-y-z points

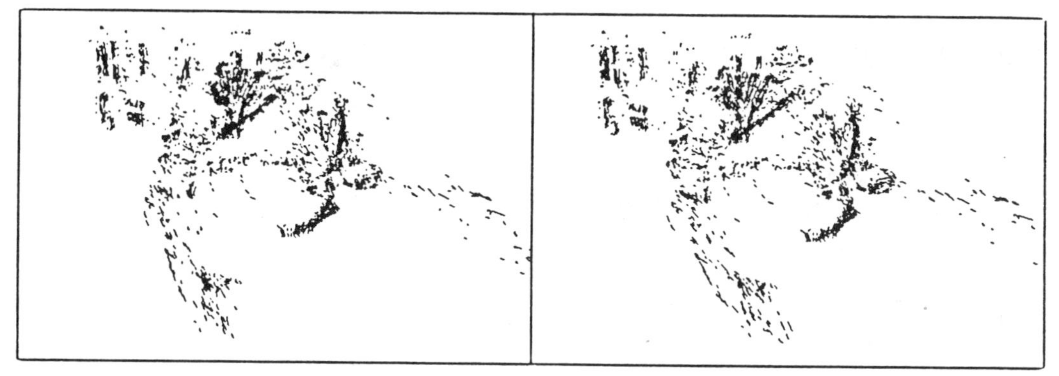

Fig. 23. Display of linked x-y-z points

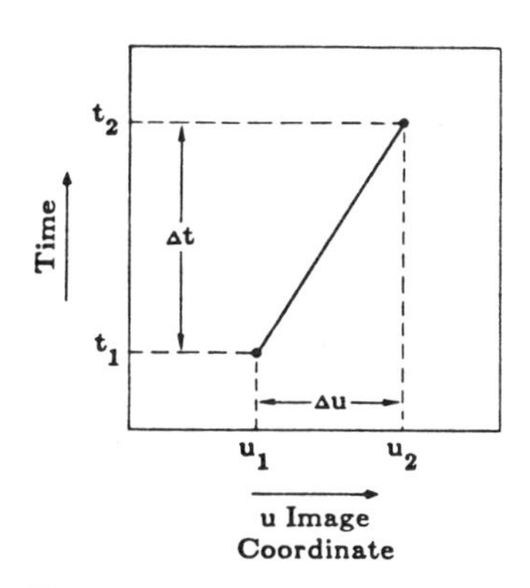

Fig. 24. Lateral motion EPI

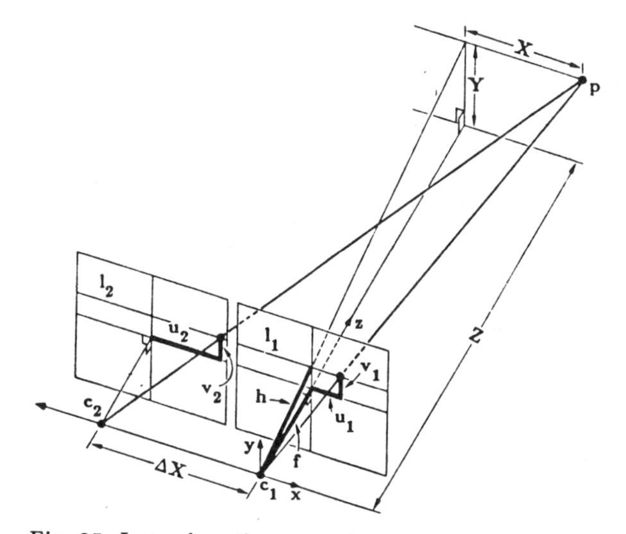

Fig. 25. Lateral motion geometry

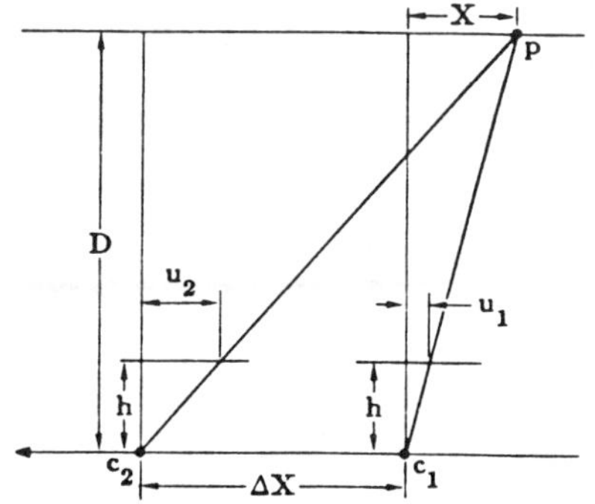

Fig. 26. Lateral motion epipolar geometry

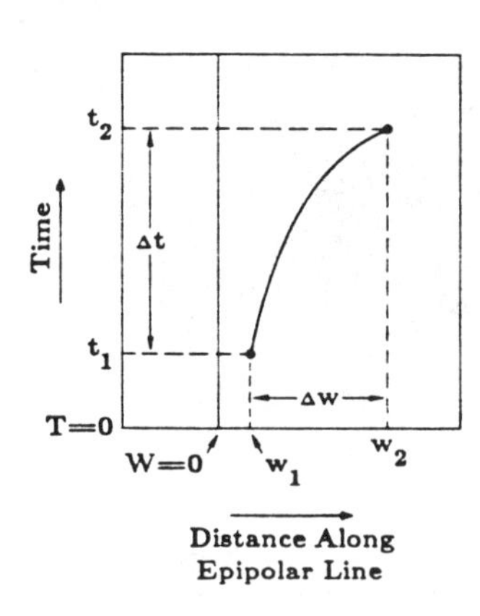

Fig. 27. Forward motion EPI

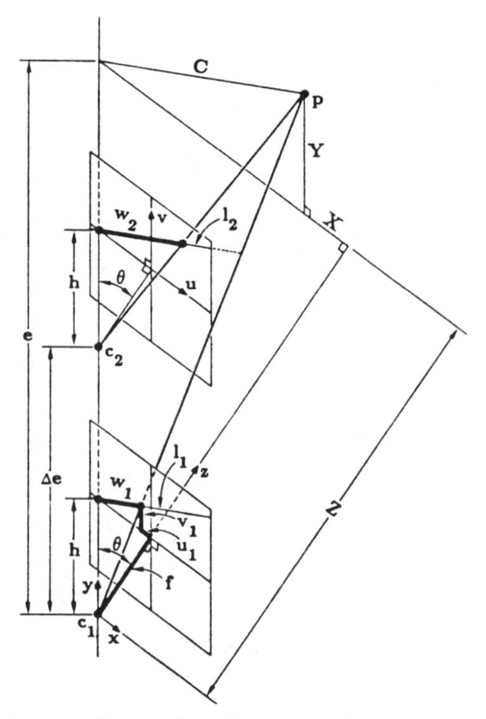

Fig. 28. Forward motion geometry

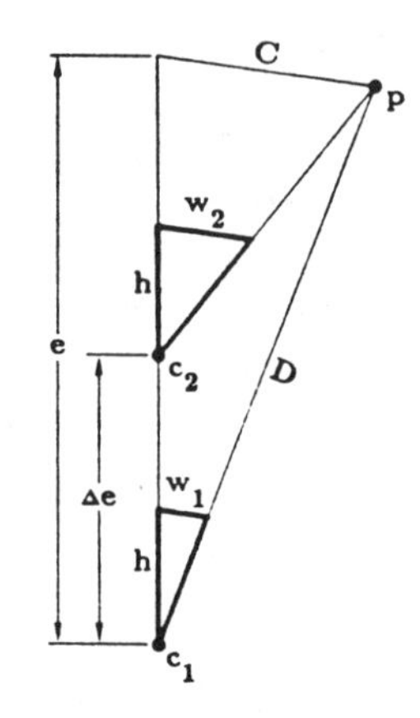

Fig. 29. Forward motion epipolar plane

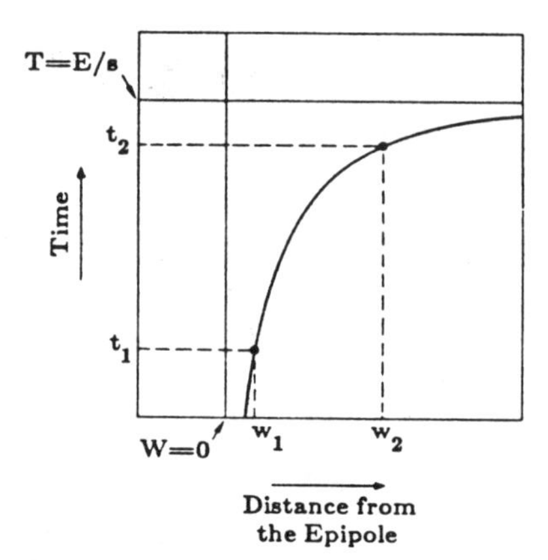

Fig. 30. Asymptotes for the hyperbola

Preface — The Changing Shape of Computer Vision

Michael Brady

Artificial Intelligence Laboratory, MIT, Cambridge, MA 02139, U.S.A.

1. Introduction

This special volume of the *Artificial Intelligence* (*AI*) Journal recognises the considerable advances that have taken place in Computer Vision over the past decade. It contains fourteen papers that are representative of the best work currently in the field. Apart from being a state-of-the-art account, the issue has been designed, as far as possible, to serve two rather different aims.

First, it is intended to give AI researchers in fields other than Vision an opportunity to become familiar with recent developments in the field. The continuing growth of AI inevitably makes it difficult to keep abreast of progress in any but a narrow area. As we shall see, the increasingly technical content of Vision, and its growing concentration on visual perception, rather than on general AI ideas, make it doubly forbidding to the casual AI reader.

Second, it is intended to enable vision researchers from fields other than AI to get a clearer picture of what an AI approach to their problem might be, or might contribute. Certainly, there is increasing interest in Computer Vision among researchers in fields as disparate as psychophysics, neurophysiology, signal and image processing, optical engineering, and photogrammetry. In addition, remote sensing, visual inspection of industrial products, and other applications in the growing field of Robotics, ensure that Vision will continue to be a topic of considerable importance for many years to come.

2. As it Was in the Beginning

Even as late as 1975, Computer Vision looked rather different than it does today. A great deal of effort had been expended on the 'blocks' microworld of scenes of polyhedra. Huffman [26] and Clowes [11] had noted the advantage of making the image forming process explicit. They observed that *picture lines* and *junctions* were the images of *scene edges* and *vertices*, and they catalogued all those interpretations of lines and junctions that were possible, given the prior assumption of planarity and the restriction that at most three surfaces were allowed to meet at a vertex. These interpretations, taking the form of 'labellings' of lines, amounted to local constraints on the volume occupied by a vertex. The local constraints propagated along picture lines since planar edges can not change their nature between two vertices. (This is not so for curved lines, as Huffman noted. Turner [59] described one possible extension to curves of the Huffman–Clowes approach. Binford, Barrow and Tenenbaum describe rather different approaches in this volume.)

Huffman [26] further pointed out that the local vertex constraints were not sufficient to capture the important restriction that picture regions were the images of planar surfaces. Mackworth's [34] development of *Gradient Space* was expressly intended to repair this deficit. Draper's article in this volume describes the greater competence of Mackworth's program, as well as its shortcomings. Despite this, most line drawings had a remarkable number of possible interpretations. Waltz's [60] work introduced the inherently global constraint afforded by shadows cast by a single distant light source, and showed that the multiple ambiguities possible without lighting were often completely resolved to a unique interpretation with lighting. Furthermore, the process by which the unique interpretation was discovered, naturally lent itself to parallel processing of a particular sort. Each vertex had an associated processor, and they all operated in strict synchrony. At each stage, the processors changed their states depending on the state of those directly connected to them. Rosenfeld, Hummel, and Zucker [54] noted the connection between this scheme and relaxation processes in numerical analysis. Actually, several authors had suggested the use of local parallel processing for Vision rather earlier, see for example the historical remarks in Subsection 5.2 of the paper by Ikeuchi and Horn in this volume.

Waltz's scheme had a number of drawbacks. For example, as Winston [63] pointed out, the program could make no use of the direction of lines in the image. On the other hand, Mackworth's program could, since gradient lines are perpendicular to image lines (see the discussions of Gradient Space in the papers of Draper, Kanade, Woodham, and Ikeuchi and Horn in this volume.) Huffman [27] later defined *Dual Space* in which this information could be made explicit (see the articles of Draper (this volume) and Spacek [58]). Again, the Waltz labellings were extremely complex and unstructured. In essence this was because they constituted interpretations that confounded many different sources of information about lighting, surface cracks, occlusion, and edge type into a single label. Clearly these different contributions to the entire percept should be made explicit and exploited separately, as they are in the human visual system. Binford's paper in this volume reconsiders the possible interpretations of edges.

A second strand in the development of Computer Vision concerns what was referred to as 'low level' processing. It was more art than science, and largely consisted of methods for the extraction of the 'important' intensity changes in an image. In the blocks world these correspond to shadow boundaries and the edges of visible surfaces, including depth discontinuities. The approach mostly consisted of convolving images with local operators (typically 3 by 3 on a 256 by 256 image) to estimate the position, contrast, and orientation of the important intensity changes. Operators were tuned to particular applications, and fared badly outside their limited domain and in the presence of noise. Little serious analysis of actual intensity changes, including the signal to noise characteristics of real images, had been carried out. A singular exception was the work of Herskovits and Binford [16]. They suggested that there were basically three qualitatively different types of intensity changes; and that particular changes often combine features of two types. This analysis formed the basis of a line finder whose performance considerably advanced the state of the art.

Other work in 'low level' vision largely consisted of the design and construction of region finders. Region finding, essentially the dual of edge finding, aimed to isolate those regions of an image that were the images of perceptual surface patches. It was thought that such regions might be isolated by defining some descriptor with respect to which they were uniform, and distinguishable from surrounding regions. It was soon clear [2, 9] that even if such descriptors existed, they were not defined simply in terms of grey level intensity values. Some researchers proposed multi-spectral descriptors [48], while others later flatly denied that it is possible to define adequate descriptors at all [36, p. 64]. Binford's paper in this volume discusses region finding in some detail.

By the early 1970's, the consensus was that 'low level' vision was inherently incapable of producing rich useful descriptions. It was observed, by analogy to the apparent need for semantics in parsing English sentences, that downward flowing knowledge of the scene could provide additional constraint. This in turn could inform local decision making. A number of program structures were proposed to effect this interaction between top down and bottom up processing of information [4, 7, 12, 43, 55, 64]. Similar ideas were advanced about natural language understanding, and speech perception. This influenced the design of, for example, Hearsay 2 [31]. To experiment with these ideas, entire systems were constructed which mobilised knowledge at all levels of the visual system as well as information specific to some domain of application. In order to complete the construction of these systems, it was inevitable that corners were cut and many over simplified assumptions were made. By and large, the performance of these systems did not give grounds for unbridled celebration. The authors of the KRL proposal (Bobrow and Winograd [6]), for example, listed several common failings (see also [7]).

3. Is Now

Perhaps the most fundamental differences between Computer Vision as it is now and as it was a decade ago, stem from the current concentration on topics corresponding to identifiable modules in the human visual system. This volume contains papers, for example, on stereopsis, the interpretation of surface contours, the determination of surface orientation from texture, and the grouping of motion primitives. To be sure, there is still a considerable amount of work oriented toward applications, but it is also increasingly based on detailed and precise analyses of specific visual abilities. The focus of research is more narrowly defined in terms of visual abilities than in terms of a domain, and the depth of analysis is correspondingly greater. This change has produced a number of far-reaching effects in the way vision is researched. This section attempts to make them explicit.

One obvious effect has been a sharp decline in the construction of entire vision systems. Most AI vision workers have thankfully abandoned the idea that visual perception can profitably be studied in the context of a priori commitment to a particular program or machine architecture. There is, for example, no more reason to believe that 'relaxation' style processing will of itself tell us more about vision than did the excursions into heterarchy. There is no obvious reason to be encouraged by Reddy's [51] claim that the Hearsay 2 model can be adapted mutatis mutandis to vision.

What identifies a particular operation as a distinguishable module in the visual system? Normal vision confronts and exploits massive redundancy. Some of the most solid evidence for the claims of individual modules is offered by psychophysical demonstrations. Care is taken, as far as possible, to isolate a particular source of information and show that the operation in question survives. One particular instance of this is the study of patients with certain disabilities resulting from brain lesions (for example [42, 57, 61]). Many psychophysical experiments, seemingly isolating particular modules of the (human) visual system, have been reported in the literature. Notable examples include Land's demonstration of the computation of lightness [19, 30] and Julesz's [28] demonstration of stereoscopic fusion without monocular cues. In some cases there is clear evidence of a human perceptual ability, although such evidence would hardly be referred to as psychophysical. Horn's work (see the papers by Woodham, and Ikeuchi and Horn in this volume) concerns the highly developed human ability to infer shape from shading. Steven's paper concerns the human three-dimensional interpretation of surface contours. On the other hand, it is equally clear that we do not have a specific module in our visual system to recognise 'yellow Volkswagens' (see for example [62]). It is less clear whether we compute depth directly, as opposed to indirectly through integrating over surface orientations, or what use we make of directional selectivity, optical flow, or texture gradients.

Not all modules work directly on the image. Indeed, it seems that few do. Instead they operate on *representations* of the information computed, or made explicit by other processes. In the case of stereopsis, Marr and Poggio [40] argue against correlating the intensity information in the left and right views.

Instead they suggest that so called zero-crossings are matched (see [15] or Nishihara's paper in this volume). The paper by Mayhew and Frisby argues that the matching actually takes place on a different representation, called the primal sketch [35]. In any case, a great deal of attention has centered on the isolation and study of individual modules, and in each case on the development of the representations on which they operate, and on those that they produce. The first of these representations, and the one whose structure is least subject to dispute, is the image itself. Not surprisingly then, most attention has centered on those modules that operate upon the image. As we shall see, the further we progress up the processing hierarchy, the less secure the story becomes, as the exact structure of the representations becomes more subject to dispute. Again, this is not surprising. The image aside any representation is one module's co-domain and another's domain. All of them shape its eventual structure.

3.1. Modules that operate on the image

A great deal of effort has been devoted to understanding how the important intensity changes in an image can be extracted, and how the information can be best represented. Marr [35] coined the term primal sketch to describe such a representation, and he described a particular algorithm by which it might be computed. A novel feature of the work was its direct reference to neurophysiological and psychophysical findings, a commitment Marr was to continue to stress in later work. His work with Poggio led to a revision of the process of construction of the Primal Sketch. Instead they advocated the use of zero-crossings of the second derivative of the filtered image. This idea was developed in turn by Marr and Hildreth [38], who propose that an image is first filtered by four Gaussians having different bandpass characteristics. Then each filtered image is convolved with a Laplacian operator (see Nishihara's paper in this volume for more detail). One of the novel features of the Marr–Hildreth account is the size of the operators involved, the smallest being roughly 35 picture elements square. This is in stark contrast to conventional operators, which are still typically on the order of 5 by 5. Such a large operator can be in much closer agreement with a Gaussian (or any other filter for that matter) than a small operator, and its effects are therefore more predictable. Unfortunately it is no longer obvious how to compute the assertions that Marr had previously advocated for inclusion in the primal sketch (see [17, p. 75]). The whole issue of constructing the primal sketch from zero-crossings is far from being resolved. Binford's paper in this volume considers this issue, as well as the choice of an optimal filter and the use of non-oriented masks, in fair detail.

Intensity changes aside, Horn and his colleagues have studied the perception of surface shape from shading. Their work is represented in the current volume by the papers of Ikeuchi and Horn, and Woodham. In brief outline, Horn has formulated a second order differential equation which he calls the image irradiance equation which relates the orientation of the local surface normal of a visible surface, the surface reflectance characteristics, and the lighting, to the intensity value recorded at the corresponding point in the image. Horn quickly realised the need for a representation which makes such surface orientations explicit. Two parameters are needed. Horn [22] observed that gradient space provides such a parameterisation, and showed how the relationship between intensity values and surface orientations could be added to gradient space to form what he called the *reflectance map*. The papers by Ikeuchi and Horn, and Woodham give details. Gradient space is by no means the only two parameter representation of surface orientations. Ikeuchi and Horn investigate *stereographic space*, which has the additional desirable property that the constraints offered by occluding boundaries can be represented and exploited. The output of shape-from-shading is a representation that makes explicit the orientation of visible surfaces, and may make other information such as depth and surface orientation discontinuities explicit also. Horn [23] suggests the name *needle map* for the representation. Other representations have been proposed which make substantially the same information explicit. Marr [36] labels this representation the *$2\frac{1}{2}D$ Sketch*, and Barrow and Tenenbaum [4] discuss *intrinsic images*. Again the exact nature of the representation (or representations) is currently far from clear. In part this is because very little work has been devoted to modules which operate upon it.

Finally, Horn and Schunck (this volume) propose a method by which the so-called *optical flow* can be determined from a sequence of images. Several authors have investigated the information that can in principle be computed from ideal optical flow fields (see the references in Horn and Schunck's paper), but no proposals have previously been made for its computation.

3.2. Modules which operate on zero-crossings and the primal sketch

We pointed out in the previous section that there remain a vast number of unresolved issues concerning the nature of the primal sketch and its computation from zero-crossings of whatever kind of filtered image. Nevertheless, the broad outlines are clear enough for work to proceed to investigate modules which are assumed to operate upon those representations. Indeed it is necessary that it does, as it will also contribute to our understanding of the information that needs to be made explicit in the primal sketch, and hence its eventual form. One area that is not represented in this volume, but that is of considerable importance, is the investigation of the processes which impose hierarchical structure on the primal sketch (what Marr [35] called the *full primal sketch*). Riley [53] has made an initial study of such processes for static scenes. Motion is an important source of information of determining structure.

The paper by Flinchbaugh and Chandrasekaran in this volume addresses grouping on the basis of motion cues. Such grouping operations play an important role on all the representations used by the visual system, and for the most part they are poorly understood. Little if any work has been done on grouping operations on what we call the *surface orientation map*.[1]

Considerable attention has been paid to *stereopsis*. Marr and Poggio's [40] theory of human stereopsis, and its implementation and refinement by Grimson [15] is discussed at length by Mayhew and Frisby, who propose a number of further refinements.

Ever since Gibson [14] stressed the importance of texture gradients for the perception of depth and surface shape, they have been the subject of detailed psychophysical and computational investigation. Pattern recognition approaches typically consist of computing crude statistics on the image intensities. This does not work at all well since, as Horn in particular has shown, an individual intensity value is a complex encoding of the lighting, the surface reflectance characteristics, and the local surface orientation. Witkin's paper in this volume once more underscores the importance of making the image forming process explicit. His approach relies upon statistical arguments but, crucially, does not require that natural textures are uniformly distributed. Rather, it requires that their non-uniformity does not mimic projection. It relies upon deriving a probability density function which relates the orientation of a scene element via projection into an image element.

The papers of Draper, Kanade, Stevens, and Tenenbaum and Barrow address various aspects of the human ability to perceive surface shape from line drawings. The first two of these assume that the scene is composed of plane-faced objects. As such, they continue the tradition of work discussed in Section 1. Kanade's paper combines the ideas of gradient space and edge labellings. It proposes the two additional assumptions of parallelism and skewed symmetry to further constrain the orientation of a planar surface. Matching the intensity profiles across two edges provides further constraint. Crucially, the program is able to make the conservative inference that two edges have the same interpretation without knowing exactly what it is. Draper's paper discusses the limitations of gradient and dual space in supporting possible processes that interpret line drawings of polyhedra. He proposes instead symbolic reasoning about 'sidedness'. Unfortunately, such inferences are inherently long range, since they rely upon the observation that the relationship between planar regions is fixed, and therefore common to all points at which they intersect. Such reasoning is likely to be of limited usefulness when applied to images of natural or curved scenes. Tenenbaum and Barrow address the subject of interpreting line drawings of curved surfaces. They use junction labellings to determine whether a bounding curve depicts an extremal boundary or a depth discontinuity. Then they propose two mechanisms: one for computing the spatial layout of the bounding curves and one for interpolating local surface orientation from the boundary values. In the remaining paper on this topic, Stevens proposes a taxonomy of interpretations of surface contours. By investigating intersecting contours in an image, a local decision can be reached about the nature of the underlying three dimensional surface.

3.3. Object representations

Considerably less is known about the modules which operate upon the surface orientation map to produce object representations, and the nature of those representations is very far from clear. Some work has been done, and it is well represented in this volume. Binford [5] proposed a volumetric primitive known as *generalized cylinders*. Nevatia and Binford [46], Hollerbach [18], and later Marr and Nishihara [39] developed representational schemes based upon such volumetric primitives. Brooks (in this volume) describes the representation of complex objects such as motors and airplanes, the incorporation of constraints such as symmetry, and the specification of affixment relations by which the local coordinate frames of two objects can be inter-related. Marr and Nishihara [39] discuss the role which such representations might play in human vision (see Nishihara's paper in this volume).

3.4. Methodological comments

The previous sections have discussed some of the modules and the important representations which have begun to emerge in Computer Vision. The broad outlines are clear, even if there are many major unresolved questions in nearly every facet of the subject. We may also note some further common themes which have crystallized over the past decade.

Most of the analyses sketched above start out with a precise description of the domain and co-domain of the visual process under scrutiny. Increasingly, 'precise' means 'mathematically precise', and so Computer Vision has become steadily more technical. This is not to say that Vision was not technical before, rather it alludes to the increasing occurrence and sophistication of mathematical analyses in Vision. Many observations about the world, as well as our assumptions about it, are naturally articulated in terms of 'smoothness' of some appropriate quantity. This intuitive idea is made mathematically precise in a number of ways in real analysis, for example in conditions for differentiability. Relationships between smoothly varying quantities give rise to differential equations, such as Horn's *image irradiance equation*. We commented several times above on the value of making the image forming process explicit. This in

[1]We sincerely hope that this name does not become established in the literature, as it only serves as a name for the intuitive notion which is rendered more or less precise in the three published versions referenced (namely the $2\frac{1}{2}$ D sketch, needle map, or intrinsic image).

turn leads to a concern with geometry, such as the properties of the gradient, stereographic, and dual spaces. Combining the considerations of geometry and smoothness leads naturally to multi-variate vector analysis and to differential geometry [13]. Mostly, a representation does not of itself contain sufficient information to guarantee that a module can uniquely arrive at the result computed so effortlessly by the human visual system. Additional assumptions, in the form of constraints, are required. This observation has led to a concern with constraint satisfaction and equation solving, using the techniques of numerical analysis such as Gauss–Seidel iteration and Lagrange multipliers (especially in the form of the calculus of variations). Examples of all of these approaches can be found in the papers in this volume.

For many authors, the changing style of research in Computer Vision has not been simply a matter of a narrowing of attention and a more highly developed technical content. Instead, greater significance is attached to the desire to make explicit the links between their work and corresponding theories in psychophysics and neurophysiology. From this perspective Computer Vision has as its goal the construction of computational theories of human visual perception. In large part, this approach stems from a series of papers written by David Marr and his colleagues at MIT. Marr's work stems from a background in neurophysiology, and is expressly addressed to psychophysicists and neurophysiologists. In particular, it is couched in terms they are accustomed to, and makes extensive reference to their literature, rather than that of Computer Vision. The work of the MIT group has excited considerable interest among psychologists and neurophysiologists, and is extensively referenced in the papers in this volume. A book summarising Marr's thoughts about human visual perception [37] and incorporating summaries of the contributions he and his colleagues have made across the entire range of the subject is currently in press.

There is considerably less diversity in emphasis, subject matter, and technical content than might be imagined between those researchers who see themselves constructing a computational theory of human visual perception and those for whom human visual perception is at most a matter of secondary concern. Compare, for example, the ACRONYM representation of objects based upon generalized cylinders (see the paper by Brooks in this volume) with that proposed by Marr and Nishihara [39], or the work on early processing of motion by Horn and Schunck (this volume) with Marr and Ullman [41]. Another common research theme is the need for local parallel processing which can discover global information through propagation. The paper by Davis and Rosenfeld (this volume) considers one such class of program structures, while others can be found in the papers by Horn and Schunck, Ikeuchi and Horn, Tenenbaum and Barrow, and Woodham. Such architectures naturally lend themselves to realization in hardware. Nishihara describes one such realization.

4. And Ever Shall Be?

As this introductory survey suggests, Computer Vision has progressed considerably on many fronts over the past decade. There has been a change in the style of research as well as in its substance. However, most issues are still poorly understood, from the exact form of representations, through the detailed understanding of the individual modules, to topics that have so far received little or no attention. A sampling of unresolved problems follows in the next few paragraphs. It is by no means exhaustive.

First, the details of what we have called the surface orientation map need to be made precise. Marr [36], Horn [23], and Barrow and Tenenbaum [4] have suggested that it records local surface orientation, as well as depth discontinuities; but it is unclear how they are recorded. Suggestions include Cartesian and polar formulations of the gradient, 'sequins' versus 'quills' [23], and the separation of various kinds of information into separate 'intrinsic' images. Nor is it obvious how accurately values are recorded. It is clear that surface information needs to be represented at different levels of resolution: a pebbled path may be considered approximately planar by a human who is walking along it. Yet an ant or person on roller skates may find the same path extremely difficult to navigate; in such cases the path is unlikely to be considered planar. As this example indicates, the level of resolution of a representation is determined largely by the process operating upon the representation, and there has been little investigation of such processes to date.

It is equally clear that grouping operations need to be defined at each level of resolution of each representation in the visual system, in order to impose hierarchical structure upon the representation. The advantages that should accrue from imposing such structure are likely to be precisely those which have inspired the development of data structures generally in computer science. Consider as an example a simple egg tray. The pattern of identical depressions to hold the eggs is immediately obvious, even though the detailed description of the individual egg cells is not.

A related set of problems concerns the determination of surface properties such as its color, manufacture, and whether or not it is wet, slippery, or prickly. Granted that we make such properties explicit, we need to determine whether they are attached as local descriptors to representations such as the surface orientation (say), or whether they are the content of separate representations. It may be that there is a separate albedo map [24, 25] or it may be that albedo information is embedded in the surface orientation map. Actually, the entire question of the computer perception of color is still very much in its infancy, despite its enormous literature.

Our current understanding of motion perception is crude. Horn and Schunck's paper is a preliminary account of the computation of optical flow from grey levels. It is less clear what information can be recovered from optical flow. Some authors are enthusiastic about the richness of the information it can provide (Clocksin [10]), while others are more sanguine (Prazdny [50]). Marr and Ullman [41], and Richter and Ullman [52] have made a start towards determining motion from the displacement of intensity changes. Ullman [65], and Flinchbaugh and Chandrasekaran (this volume) consider the grouping of primal sketch tokens in motion. Even less is known about motion computed on the surface orientation map or on object representations. It is reasonable to suppose that the description of such object motions will need to incorporate a formulation of the object's kinematics. This has proved to be quite difficult even for simple robot arms (see for example [49]).

Perhaps the most difficult problem of all concerns the perception or planning of movements through cluttered space. Space, considered as an object, typically occupies a volume and surface whose descriptions push current representational frameworks to their limits, if not far beyond them. Some progress has been made in Robotics [32]. A further important application lies in making precise the rather vague motion of cognitive map. It is usually supposed [33] that this only refers to object representations. Actually it seems that we have quite considerable navigational processes which operate on the surface orientation map.

The current rapid pace of developments in VLSI technology has further motivated research into what were referred to above as local parallel programming architectures. It is likely that our conception of computation will change as a result of such developments. Vision will be one of the first areas to benefit from such advances. It seems that it will also be a continuing source of inspiration to VLSI designers [1, 47].

Finally, we certainly need a better understanding of the extent and use of domain specific information in visual perception. Yesterday's heterarchy and today's multi-layered relaxation systems both derive from a priori commitment to a particular mechanism. The experience of the past decade should certainly have made us wary about jumping to premature conclusions regarding which phenomena appear to inevitably implicate such downward flow. This has certainly been true of our ability to compute rich useful descriptions of the information provided in an image. It also seems reasonable to suppose that the three dimensional structure of jointed rigid objects can be recovered from a time succession of images without knowing a great deal about human physiology. This would provide an explanation for, amongst other things, the demonstrations of Johannson [29] and Muybridge [44], knowing only the basic facts of dynamics.

There is every reason to believe that there will be considerable advance on these and other issues over the next few decades, probably resulting in changes in our conception of Computing and Vision at least as large as those which have occurred over the past decade. It would be a very brave person indeed who claimed to understand other than the broadest outlines of the subject now.

5. Professor David Marr

One paper which was to be written especially for this special issue of *Artificial Intelligence* will unfortunately never appear. It would have been authored by Professor David Marr, who died toward the end of 1980 after a protracted illness. The influence of the group which he founded at MIT is evident from the preceding pages.

David's background was in neurophysiology, after completing a mathematics degree at Cambridge University. His early work proposed mathematical theories of the neocortex, archicortex, and, perhaps best known of all, the cerebellum. He was to remain deeply commited to the study of human perception and memory for the rest of his life. In 1974 he was invited to spend a little time at the Artificial Intelligence Laboratory at MIT, and stayed for six years, eventually accepting a Professorship in the Department of Psychology. He quickly appreciated that computational concepts provided a further dimension for the expression of theories of human perception, and, together with a growing group of Ph.D. students, he set out to construct what he called a computational theory of human vision. The group has been enormously creative, publishing studies across the entire breadth of human vision.

David's work was notable in many ways, but in particular notable for its style. He made extensive reference to the literatures of neurophysiology and psychology, which were his background and to which he directed his contributions. In particular, he published in the journals which would be read by his intended audience, and encouraged his students to do so too. He argued for a mathematical analysis of a perceptual problem independent of, and prior to, consideration of issues concerning the choice of an algorithm. Under the heading of 'natural computation', he championed the isolation of the constraints which the world imposes upon perception, as well as the perceiver's prior beliefs about it. Though a good deal of Marr's work was mathematical in nature, its ramifications were stated in elegant prose. A book summarising his thoughts about human visual perception [37] and incorporating summaries of the contributions he and his colleagues have made across the entire range of the subject is currently in press.

David will be missed by the wide community of scholars whose work brought them in contact with his writing. He will be missed especially by those whose lives were enriched by knowing him or working with him.

ACKNOWLEDGMENT

This paper describes research done in part at the Artificial Intelligence Laboratory of the Massachusetts Institute of Technology. Support for the Laboratory's Artificial Intelligence research is provided in part by the Advanced Research Projects Agency of the Department of Defense under Office of Naval Research contract N00014-75-C-0643. I thank Ellen Hildreth, Marilyn Matz, and Demetri Terzopoulos for their comments on a draft of this paper.

REFERENCES

1. Batali, J., forthcoming S.M. dissertation, MIT, Cambridge, MA, 1981.
2. Barrow, H.G. and Popplestone, R.J., Relational descriptions in picture processing, *Machine Intelligence* **6** (1971).
3. Barrow, H.G. and Tenenbaum, J.M., Experiments in interpretation guided semantics, Tech. Note 123, SRI International (1976).
4. Barrow, H.G. and Tenenbaum, J.M., Recovering intrinsic scene characteristics from images, in: Hanson and Riseman, Eds., *Computer Vision Systems* (Academic Press, New York, 1978).
5. Binford, T.O., Visual perception by computer, *Proc. IEEE Conf. Systems and Control* (1971).
6. Bobrow, D.G. and Winograd, T., An overview of KRL, A Knowledge Representation Language, *Cognitive Science* **1** (1977).
7. Brady, J.M., The development of a computer vision system, *Recherche Psicologica* (1979).
8. Brady, J.M. and Wielinga, B.J., Reading the writing on the wall, in: Hanson and Riseman, Eds., *Computer Vision Systems* (Academic Press, New York, 1978).
9. Brice, C.R. and Fennema, C.L., Scene analysis using regions, *Artificial Intelligence* **1** (1970) 205–226.
10. Clocksin, W.F., Perception of surface slant and edge labels from optical flow: a computational approach, *Perception* **9** (1980) 253–269.
11. Clowes, M.B., On seeing things, *Artificial Intelligence* **2** (1971) 79–116.
12. Freuder, E.C., A computer vision system for visual recognition using active knowledge, Tech. Rept. 345, MIT AI Lab. (1974).
13. Faux, I.D. and Pratt, M.J., *Computational Geometry for Design and Manufacture* (Ellis Horwood, Chichester, 1979).
14. Gibson, J.J., *The Perception of the Visual World* (Houghton–Mifflin, Boston, MA, 1950).
15. Grimson, W.E.L., Computing shape using a theory of human stereo vision, Ph.D. thesis, forthcoming book published by MIT Press, MIT, 1980.
16. Herskovits, A. and Binford, T.O., On boundary detection, AI Memo 183, MIT (1970).
17. Hildreth, E.C., Implementation of a theory of edge detection, M.S. dissertation, also Tech. Rept. 579, MIT AI Lab. (1980).
18. Hollerbach, J.M., Hierarchical shape description of objects by selection and modification of prototypes, M.S. dissertation, also Tech. Rept. 346, MIT AI Lab. (1975).
19. Horn, B.K.P., Determining lightness from an image, *Comput. Graphics and Image Processing* **3** (1974) 277–299.
20. Horn, B.K.P., The Binford–Horn line-finder, AI Memo 285, MIT (1973).
21. Horn, B.K.P., Obtaining shape from shading information, in: Winston, P.H., Ed., *The Psychology of Computer Vision* (McGraw-Hill, New York, 1975).
22. Horn, B.K.P., Understanding image intensities, *Artificial Intelligence* **8** (1977) 201–231.
23. Horn, B.K.P., Sequins and quills—Representations for surface topography, in: Bajcsy, R., Ed., *Representation of 3-Dimensional Objects* (Springer, Berlin, 1982).
24. Horn, B.K.P. and Bachman, B.L., Using synthetic images to register real images with surface models, *Comm. ACM* **21** (1978) 914–924.
25. Horn, B.K.P. and Sjoberg, R.W., Atmospheric modelling for the generation of albedo images, in: Baumann, L., Ed., *Proceedings of the Image Understanding Workshop* (Science Applications, 1980).
26. Huffman, D.A., Impossible objects as nonsense sentences, in: Meltzer, B. and Michie, D., Eds., *Machine Intelligence* **6** (Edinburgh University Press, Edinburgh, 1971).
27. Huffman, D.A., A duality concept for the analysis of polyhedral scenes, in: Elcock, E.W. and Michie, D., Eds., *Machine Intelligence* **8** (Ellis Horwood, Chichester, 1977).
28. Julesz, B., *Foundations of Cyclopean Perception* (The University of Chicago Press, Chicago, 1971).
29. Johansson, G., Visual perception of biological motion and a model for its analysis, *Perception and Psychophysics* **14** (1973) 201–211.
30. Land, E.H. and McCann, J.J., Lightness and retinex theory, *J. Optical Society of America* **61** (1971) 1–11.
31. Lesser, V.R. and Erman, L.D., A retrospective view of the Hearsay-II architecture, *Proc. Int. Jt. Conf. Artificial Intelligence* **2** (1977) 790–800.
32. Lozano-Perez, T., Spatial planning: a configuration space approach, AI Memo 605, MIT (1980).
33. Lynch, K., *The Image of the City* (MIT Press, Cambridge, MA, 1960).
34. Mackworth, A.K., Interpreting pictures of polyhedral scenes, *Artificial Intelligence* **4** (1973) 121–137.
35. Marr, D., Early processing of visual information, *Philos. Trans. Roy. Soc. London B* **275** (1976) 483–524.
36. Marr, D., Representing visual information, in: Hanson and Riseman, Eds., *Computer Vision Systems* (Academic Press, New York, 1978).
37. Marr, D., *Vision* (Freeman, San Francisco, 1981).
38. Marr, D. and Hildreth, E., Theory of edge detection, *Proc. Roy. Soc. London B* **207** (1980) 187–217.
39. Marr, D. and Nishihara, H.K., Representation and recognition of the spatial organisation of three dimensional structure, *Proc. Roy. Soc. London B* **200** (1978) 269–294.
40. Marr, D. and Poggio, T., A theory of human stereo vision, *Proc. Roy. Soc. London B* **204** (1979) 301–328.
41. Marr, D. and Ullman, S., Directional selectivity and its use in early visual processing, *Proc. Roy. Soc. London B* (1981).
42. Marshall, J.C. and Newcombe, F., Patterns of Paralexia, *J. Psycholinguistic Research* **2** (1973) 175–199.
43. Minsky, M. and Papert, S., Artificial intelligence progress report, AI Memo 252, MIT (1972).
44. Muybridge, E., *Animals in Motion* (Dover, New York, 1957).
45. Nevatia, R., Computer analysis of scenes of 3-dimensional curved objects, Stanford, 1975.
46. Nevatia, R. and Binford, T.O., Description and recognition of curved objects, *Artificial Intelligence* **8** (1977) 77–98.
47. Nudd, G.R., Fouse, S.D., Nussmeier, T.A. and Nygaard, P.A., Development of custom-designed integrated circuits for image understanding, in: Baumann, L., Ed., *Proceedings of the Image Understanding Workshop* (1979).
48. Ohlander, R., Analysis of natural scenes, Carnegie–Mellon Univ., Pittsburgh, 1975.
49. Paul, R.P., Manipulator Cartesian path control, *IEEE Trans. Systems, Man Cybernet.* **9** (1979) 702–711.
50. Prazdny, K.F., Egomotion and relative depth map from optical flow, *Biological Cybernet.* **36** (1980) 87–102.
51. Reddy, R., Pragmatic aspects of machine vision, in: Hanson and Riseman, Eds., *Computer Vision Systems* (Academic Press, New York, 1978).
52. Richter, J. and Ullman, S., A model for the spatio-temporal organization of X and Y-type ganglion cells in the primate retina, AI Memo 573, MIT (1980).
53. Riley, M., Representing image structure, MIT, Stanford, 1981.
54. Rosenfeld, A., Hummel, R.A. and Zucker, S.W., Scene labelling by relaxation operations, *IEEE Trans. Systems, Man Cybernet.* **6** (1976) 420–433.
55. Shirai, Y., A context-sensitive line finder for recognition of polyhedra, *Artificial Intelligence* **4** (1973) 95–119.
56. Stevens, K.A., Surface perception by local analysis of texture and contour, Tech. Rept. 512, MIT AI Lab. (1980).
57. Stevens, K.A., Occlusion clues and subjective contours, AI Memo 363, MIT (1976).
58. Spacek, L.A., Shape from shading and more than one view, M.S. thesis, University of Essex, UK, 1979.

59. Turner, K.J., Computer perception of curved objects using a television camera, Edinburgh, 1974 (extract appeared in *Proc. 1st. AISB Conf.*, Sussex, 1974).
60. Waltz, D., Generating semantic descriptions from drawings of scenes with shadows, in: Winston, P.H., Ed., *The Psychology of Computer Vision* (McGraw–Hill, New York, 1975).
61. Weiskrantz, L., Warrington, E.K., Sanders, M.D. and Marshall, J., Visual capacity in the hemianopic field following a restricted occipital ablation, *Brain* **97** (1974) 709–728.
62. Weisstein, N., Beyond the yellow volkswagen detector and the grandmother cell: A general strategy for the exploration of operations in human pattern recognition, in: Solso, R., Ed., *Contemporary Issues in Cognitive Psychology* (Holt, Rinehart and Winston, New York, 1973).
63. Winston, P.H., *Artificial Intelligence* (Addison Wesley, New York, 1977).
64. Winston, P.H., The MIT robot, in: Meltzer, B. and Michie, D., Eds., *Machine Intelligence* 7 (Edinburgh University Press, Edinburgh, 1972).
65. Ullman, S., The Interpretation of Visual Motion (MIT Press, Cambridge, MA, 1979).

Understanding Image Intensities[1]

Berthold K. P. Horn

Artificial Intelligence Laboratory, Massachusetts Institute of Technology, Cambridge, MA 02139, U.S.A.

Recommended by Max Clowes

ABSTRACT

Traditionally, image intensities have been processed to segment an image into regions or to find edge-fragments. Image intensities carry a great deal more information about three-dimensional shape, however. To exploit this information, it is necessary to understand how images are formed and what determines the observed intensity in the image. The gradient space, popularized by Huffman and Mackworth in a slightly different context, is a helpful tool in the development of new methods.

0. Introduction and Motivation

The purpose of this paper is to explore some of the puzzling phenomena observed by researchers in computer vision. They range from the effects of mutual illumination to the characteristic appearance of metallic surfaces—subjects which at first glance may seem to take us away from the central issues of artificial intelligence. But surely if artificial intelligence research is to claim victory over the vision problem, then it has to embrace the whole domain, understanding not only the problem solving aspects, but also the physical laws than underlie image formation and the corresponding symbolic constraints that enable the problem solving.

One reason for previous neglect of the image itself was the supposition that the work must surely already have been done by researchers in image processing, pattern recognition, signal processing and allied fields. There are several reasons why this attitude was misadvised:

Image processing deals with the conversion of images into new images, usually for human viewing. Computer image understanding systems, on the other hand, must work toward symbolic descriptions, not new images.

Pattern recognition, when concerned with images, has concentrated on the classification of images of characters and other two-dimensional input, often of a binary nature. Yet the world we want to understand is three-dimensional and the images we obtain have many grey-levels.

Signal processing studies the characteristics of transformations which are amenable to mathematical analysis, not the characteristics imposed on images by nature. Yet in the end, the choice of what to do with an image must depend on it alone, not the character of an established technical discipline.

Although we can borrow some of the techniques of each of these approaches we must still understand how the world forms an image if we are to make machines see. Yet I do not mean to suggest analysis by synthesis. Nothing of the sort! I propose only that if we are to solve the problem of creating computer performance in this domain, we must first thoroughly understand that domain.

This is, of course, not without precedent. The line of research beginning with Guzman and continuing through Clowes, Huffman, Waltz, Mackworth and others, was a study of how the physical world dictates constraints on what we see—constraints that once understood can be turned around and used to analyze what is seen with great speed and accuracy relative to older techniques which stressed problem solving expertise at the expense of domain understanding.

1. Developing the Tools for Image Understanding

An understanding of the visual effects of edge imperfections and mutual illumination will be used to suggest interpretations of image intensity profiles across edges, including those that puzzled researchers working in the blocks world. We shall see that a "sharp peak" or edge-effect implies that the edge is convex, a "roof" or triangular profile suggests a concave edge, while a step-transition or discontinuity accompanied by neither a sharp peak nor a roof component is probably an obscuring edge. This last hypothesis is strengthened significantly if an "inverse peak" or negative edge-effect is also present. (See section 3.)

Next, it will be shown that the image intensities of regions meeting at a joint corresponding to an object's corner determine fairly accurately the orientation of each of the planes meeting at the corner. Thus we can establish the three-dimensional structure of a polyhedral scene without using information about the size or, support or nature of the objects being viewed. (See Section 3.4.)

Finally, we will turn to curved objects and show that their shape often can be determined from the intensities recorded in the image. The approach given here is supported by geometric arguments and does not depend on methods for solving first-order non-linear partial differential equations. It combines my previous shape-from-shading method [4, 2] with geometric arguments in gradient-space (Huffman and Mackworth [1, 2, 3, 9]). This approach to the image analysis problem enables us to establish whether or not certain features can be extracted from images. (See Sections 4 and 5.)

[1] This report describes research done at the Artificial Intelligence Laboratory of the Massachusetts Institute of Technology. Support for the laboratory's research is provided in part by the Advanced Research Projects Agency of the Department of Defence under Office of Naval Research contract N00014-75-C-0643.

1.1. Image formation

The visual world consists of opaque bodies immersed in a transparent medium. The dimensionality of the two domains match: since only the object's surfaces are important for recognition and description purposes. On one hand we have two-dimensional surfaces plus depth and on the other, a two-dimensional image plus intensity. There are two parts to the problem of exploiting this observation to understand what is being imaged: one deals with the geometry of projection, the other with the intensity of light recorded in the image.

The relation between object coordinates and image coordinates is given by the well-known perspective projection equations derived from a diagram such as Fig. 1, where f if the focal length and,

$$x' = (x/z)f \quad \text{and} \quad y' = (y/z)f$$

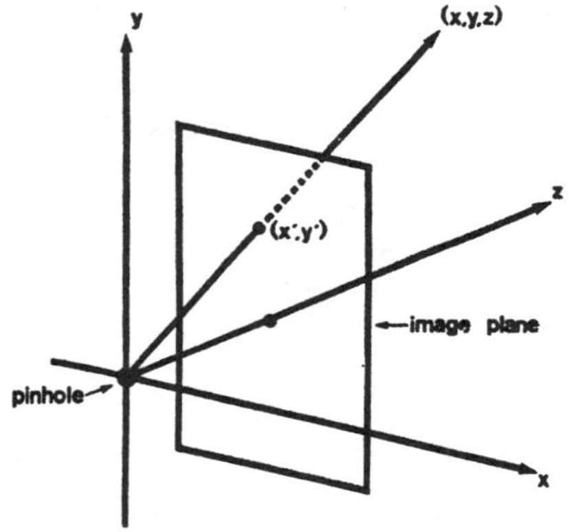

FIG. 1. Image projection geometry.

For the development presented here it will be convenient to concentrate on the case where the viewer is very far from the objects relative to their size. The resultant scene occupies a small visual angle as if viewed by a telephoto lens. This corresponds to orthographic projection, where z is considered constant in the equations above.

1.2. Surface orientation

We must understand the geometry of the rays connecting the lightsource(s), the object and the viewer in order to determine the light flux reflected to the viewer from a particular element of the object. The surface orientation in particular, plays a major role. There are, of course, various ways of specifying the surface orientation of a plane. We can give, for example, the equation defining the plane or the direction of a vector perpendicular to the surface. If the equation for the plane is $ax+by+cz = d$, then a suitable surface normal is (a, b, c).

We extend this method to curved surfaces simply by applying it to tangent planes. A local normal to a smooth surface is $(z_x, z_y, -1)$, where z_x and z_y are the first partial derivatives of z with respect to x and y. It is convenient to use the abbreviations p and q for these quantities. The local normal then becomes $(p, q, -1)$. It is clear then that the surface orientation thus defined has but two degrees to freedom. The quantity (p, q) is called the *gradient* (Fig. 2).

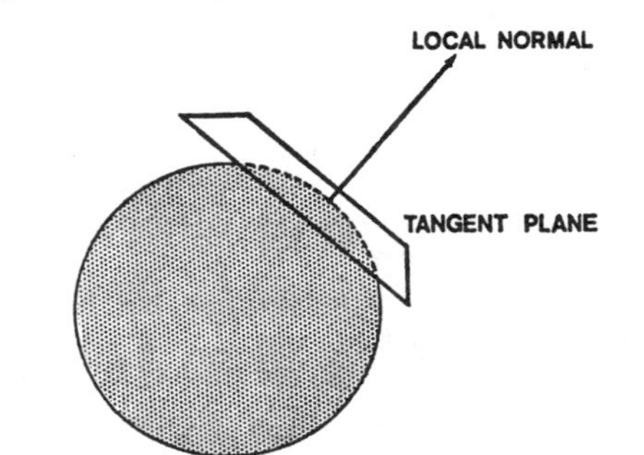

FIG. 2. Definition of surface orientation for curved objects.

1.3. Image intensity

The amount of light reflected by a surface element depends on its micro-structure and the distribution of incident light. Constructing a tangent plane to the object's surface at the point under consideration, we see that light may be arriving from directions distributed over a hemisphere. We first consider the contributions separately, from each of these directions and then superimpose the results.

For most surfaces there is a *unique value of reflectance and hence image intensity for a given surface orientation.* No matter how complex the distribution of light sources. We shall spend some time exploring this and develop the reflectance map in the process.

The simplest case is that of a single point-source where the geometry of reflection is governed by three angles: the incident, the emittance and the phase angles (Fig. 3). The incident angle, i, is the angle between the incident ray and the local normal,

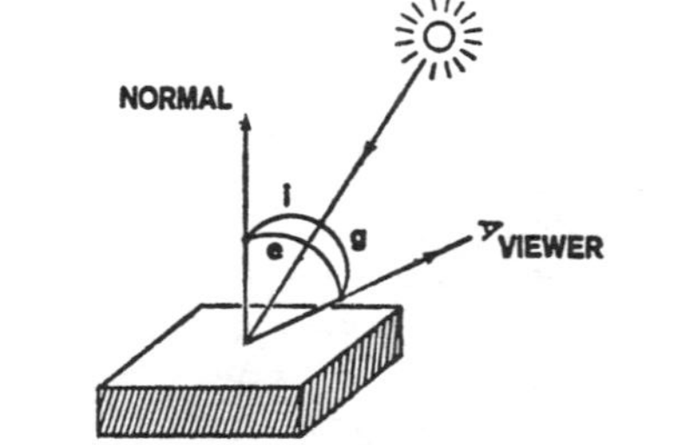

FIG. 3. The reflectivity is a function of the incident, emittance, and phase angles.

the emittance angle, e, is the angle between the ray to the viewer and the local normal, while the phase angle, g, is the angle between the incident and the emitted ray [4]. The reflectivity function is a measure of how much of the incident light is reflected in a particular direction. Superficially, it is the fraction of the incident light reflected per unit surface area, per unit solid angle in the direction of the viewer. More precisely:

Let the illumination be E (flux/area) and the resulting surface luminance in the direction of the viewer by B (flux/steradian/projected area). (The projected area is the equivalent foreshortened area as seen by the viewer.) *The reflectivity is then defined to be* B/E. It may be written $\phi(i, e, g)$, to indicate its dependence on the three angles involved.

Note that an infinitesimal surface element, dA, captures a flux $E\cos(i)\,dA$, since its surface normal is inclined i relative to the incident ray. Similarly, the intensity I(flux/steradian) equals $B\cos(e)\,dA$, since the projected area is foreshortened by the inclination of the surface normal relative to the emitted ray.

1.4 Reflectivity function

For some surfaces, mathematical models have been constructed that allow an analytical determination of the reflectivity function. Since such techniques have rarely proved successful, reflectivity functions are usually determined empirically. Often there will be more than one source illuminating the object. In this case one has to integrate the product of reflectivity and the incident light intensity per unit solid angle over the hemisphere visible from the point under consideration. This determines the total light flux reflected in the direction of the viewer.

The normal to the surface relates object geometry to image intensity because it is defined in terms of the surface geometry, yet it also appears in the equation for the reflected light intensity. Indeed two of the three angles on which the reflectivity function depends are angles between the normal and other rays. Although we could now proceed to develop a partial differential equation based on this observation, it is more fruitful to introduce first another tool—gradient space.

1.5. Gradient space

Gradient-space can be derived as a projection of dual-space or of the Gaussian sphere but it is easier here to relate it directly to surface orientation [2]. We will concern ourselves with orthographic projection only, although some of the methods can be extended to deal with perspective projection as well.

The mapping from surface orientation to gradient-space is made by constructing a normal $(p, q, -1)$ at a point on an object and mapping it into the point (p, q) in gradient-space. Equivalently, one can imagine the normal placed at the origin and find its intersection with a plane at unit distance from the origin.)

We should look at some examples in order to gain a feel for gradient-space. Parallel planes map into a single point in gradient-space. Planes perpendicular to the view-vector map into the point at the origin of gradient-space. Moving away from the origin in gradient-space, we find that the distance from the origin equals the tangent of the emittance angle e, between the surface normal and the view-vector.

If we rotate the object-space about the view-vector, we induce an equal rotation of gradient-space about the origin. This allows us to line up points with the axes and simplify analysis. Using this technique, it is easy to show that the angular position of a point in gradient-space corresponds to the direction of steepest descent on the original surface.

Let us call the orthogonal projection of the original space the image-plane. Usually this is all that is directly accessible. Now consider two planes and their intersection. Let us call the projection of the line of intersection the image-line. The two planes, of course, also correspond to two points in gradient space. Let us call the line connecting these two points the gradient-line. Thus, a line maps into a line. The perpendicular distance of the gradient-space line from the origin equals the tangent of the inclination of the original line of intersection with respect to the image plane. We show by superimposing gradient-space on the image-space [2, 11] that the gradient-space line and the image-line are *mutually perpendicular*. Mackworth's scheme for scene analysis of line drawings of polyhedra depends on this observation [2].

1.6. Trihedral corners

The points in gradient-space corresponding to the three planes meeting at a trihedral corner must satisfy certain constraints. The lines connecting these points must be perpendicular to the corresponding lines in the image-plane (Fig. 4). This provides

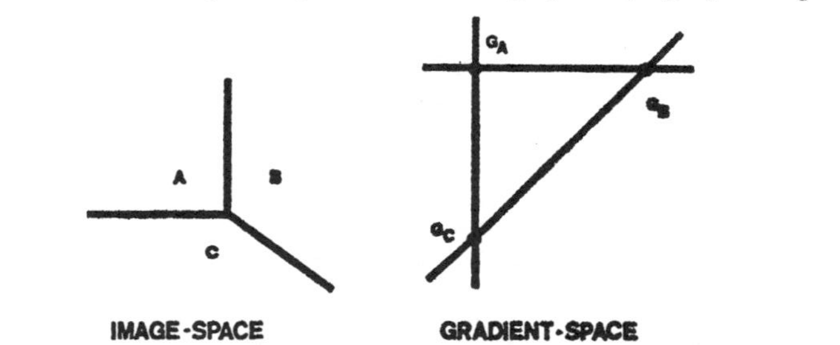

Fig. 4. Constraints on the gradient-space points corresponding to the planes meeting at a trihedral corner. The gradient-lines must be perpendicular to the image-lines.

us with three constraints but that is not enough to fix the position of three points in gradient-space. Three degrees of freedom—the *position* and *scale* of the triangle—remain undetermined. We see later that measurement of image intensities for the three planes provides enough information to specify their orientations, thus allowing a determination of the three-dimensional structure of a polyhedral scene.

2. The Reflectance Map

The amount of light reflected by a given surface element depends on the orientation of the surface and the distribution of light sources around it, as well as on the nature of the surface. For a given type of surface and distribution of light sources, there is a fixed value of reflectance for every orientation of the surface normal and hence for every point in *gradient-space*. Thus image intensity is a single-valued function of p and q. We can think of this as a map in gradient-space. This is *not* a transform of the image seen by the viewer. It is, in fact, independent of the scene and instead, only a function of the surface properties and the light source distribution (but see Section 2.6).

What can we do with this strange "image" of the world surrounding the object? If we measure a certain intensity at a given point on the object, we know that the orientation of the surface at that point is restricted to a subset of all possible orientations; we cannot, however, uniquely determine the orientation. The one constraint is that it be one of the points in gradient-space where we find this same value of intensity.

The use of the gradient-space diagram is analogous to the use of the hodogram or velocity-space diagram. The latter provides insight into the motion of particles in force field that is hard to obtain by algebraic reasoning alone. Similarly, the gradient-space allows geometric reasoning about surface orientation and image intensities.

2.1. Matte surfaces and a point-source near the viewer

A perfect lambertian surface or diffuser looks equally bright from all directions; the amount of light reflected depends only on the cosine of the incident angle. In order to postpone the calculation of incident, emittance and phase angles from p and q, we first place a single light source near the viewer. The incident angle then equals the emittance angle, the angle between the surface normal and the view-vector. The cosine of the incident angle is the dot product of the corresponding unit vectors:

$$\cos(i) = \frac{(p, q, -1)\cdot(0, 0, -1)}{|(p, q, -1)|\,|(0, 0, -1)|} = \frac{1}{\sqrt{1+p^2+q^2}}.$$

We obtain the same result by remembering that the distance from the origin in gradient-space is the tangent of the angle between the surface normal and the view-vector:

$$\sqrt{p^2+q^2} = \tan(e),$$

$$\cos(e) = \frac{1}{\sqrt{1+\tan^2(e)}},$$

and

$$e = i.$$

If we plot reflectance as a function of p and q, we obtain a central maximum of 1 at the origin and a circularly symmetric function that falls smoothly to 0 as we approach infinity in gradient-space. This is a nice, smooth reflectance map, typical of matte surfaces (Fig. 5).

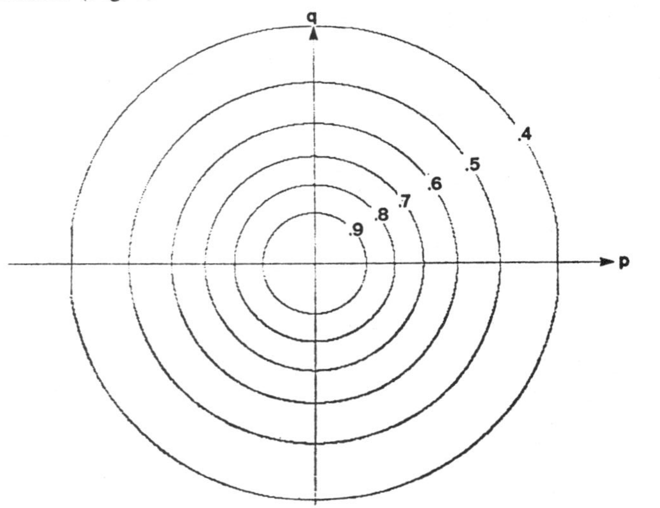

FIG. 5. Contours of $E = \cos(e)$. This is the gradient-space image for objects with lambertian surfaces when there is a single light source and the light source is near the viewer.

A given image intensity corresponds to a simple locus in gradient-space, in this case a circle centered on the origin. A measurement of image intensity tells us that the surface gradient falls on a certain circle in gradient-space.

Since the light-source is not always likely to be near the viewer we now explore the more complicated geometry of incident and emitted rays for arbitrary directions of incident light at the object.

2.2. Incident, emittance, and phase angles

For many surfaces the reflectance is a smooth function of the angles of incidence, emittance and phase. It is convenient to work with the cosines of these angles, $I = \cos(i)$, $E = \cos(e)$, and $G = \cos(g)$. (These can be obtained easily from the dot products of the unit vectors.) If we have a single distant light source whose direction is given by a vector $(p_s, q_s, -1)$, and note that the view-vector is $(0, 0, -1)$, then,

$$G = \frac{1}{\sqrt{1+p_s^2+q_s^2}}, \qquad E = \frac{1}{\sqrt{1+p^2+q^2}},$$

and

$$I = \frac{(1+p_s p+q_s q)}{\sqrt{1+p^2+q^2}\,\sqrt{1+p_s^2+q_s^2}} = (1+p_s p+q_s q)EG.$$

It is simple to calculate I, E, and G for any point in gradient-space. In fact G is *constant* given our assumption of orthogonal projection and distant light source. We saw earlier that the contours of constant E are circles in gradient-space centered on the origin. Setting I constant gives us a second-order polynomial in p and q and suggests that loci of constant I may be conic sections. The terminator—the line separating lighted from shadowed regions, for example, is a straight line obtained by setting $i = \pi/2$. That is, $I = 0$; or $1+p_sp+q_sq = 0$. Similarly, the locus of $I = 1$ is the single point $p = p_s$ and $q = q_s$.

A geometrical way of constructing the loci of constant I is to develop the cone generated by all directions that have the same incident angle. The axis of the cone is the direction to the light-source $(p_s, q_s, -1)$. We find the corresponding points in gradient-space by intersecting this cone with a plane at unit distance from the origin. Varying values of I produce cones with varying angles. The cones form a nested sheaf. The intersection of this nested sheaf with the unit plane is a nested set of conic sections (Fig. 6). Note that our previous example (Fig. 5) is merely a special case in which the axis of the sheaf of nested cones points directly at the viewer.

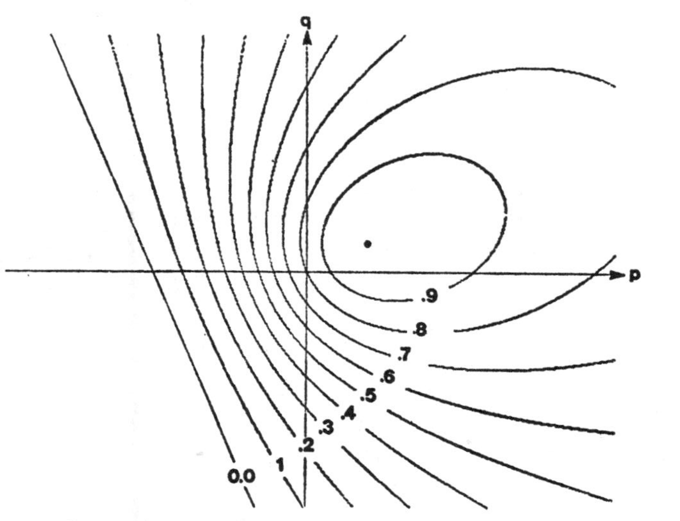

FIG. 6. Contours of $I = \cos(i)$. The direction to the single source is $(p_s, q_s) = (0.7, 0.3)$. This is the gradient-space image for objects with lambertian surfaces when the light source is not near the viewer.

If we measure a particular image intensity, we know that the gradient of the corresponding surface element must fall on a particular conic section. The possible normals are confined to a cone—in this case, merely a circular cone. In the case of more general reflectivity functions, the locus of possible normals constitutes a more general figure called the *Monge cone.*

2.3. Specularity and glossy surfaces

Many surfaces have some specular or mirror-like reflection from the outermost layers of their surface, and thus are not completely matte. This is particularly true of surfaces that are smooth on a microscopic scale. For specular reflection $i = e$, and the incident, emitted, and normal vectors are all in the same plane. Alternatively, we can say that $i+e = g$. In any case, only one surface orientation is correct for reflecting the light source towards the viewer. That is, a perfect specular reflection contributes an impulse to the gradient-space image at a particular point.

In practice, few surfaces are perfectly specular. Glossy surfaces reflect some light in directions slightly away from the geometrically correct direction [8]. It can be shown that the cosine of the angle between the direction for perfectly specular reflection and any other direction is $(2IE-G)$ [11]. This clearly equals 1 in the ideal direction and falls off towards 0 as the angle increases to a right-angle. By taking various functions of $(2IE-G)$, such as high powers, one can construct more or less compact specular contributions.

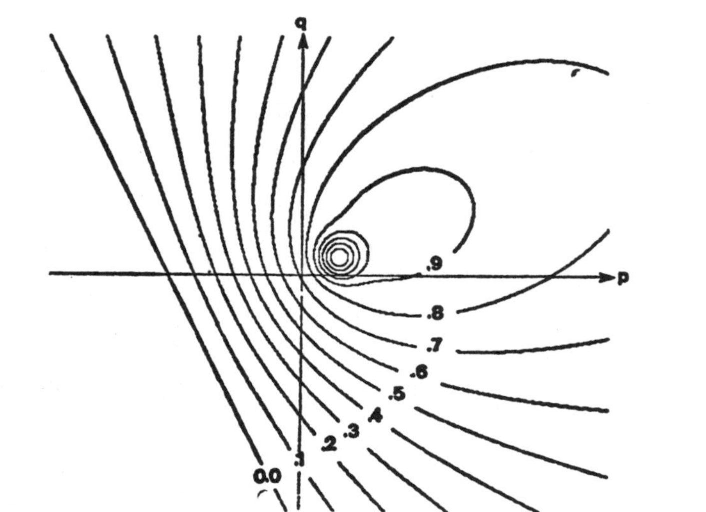

FIG. 7. Contours for $\phi(I, E, G) = \frac{1}{2}s(n+1)(2IE-G)^n+(1-s)I$. This is the gradient-space image for a surface with both a matte and a specular component of reflectivity illuminated by a single point-source.

For example, a good approximation for some glossy white paint can be obtained by combining the usual matte component with a specular component defined in this way—$\phi(I, E, G) = \frac{1}{2}s(n+1)(2IE-G)^n+(1-s)I$. Here s lies between 0 and 1 and determines the fraction of incident light reflected specularly before penetrating the surface, while n determines the sharpness of the specularity peak in the gradient-space image (Fig. 7).

2.4. Finding the gradient from the angles

In order to further explore the relation between the specification of surface orientation in gradient-space and the angles involved, we solve for p and q, given I, E, and G. We have already shown that the opposite operation is simple to perform. One approach to this problem is to solve the polynomial equations in p and q derived from the equations for I, E, and G. It can be shown that [11]:

$$p = p'\cos(\theta) - q'\sin(\theta),$$
$$q = p'\sin(\theta) + q'\cos(\theta),$$

where

$$p' = \frac{(I/E - G)}{\sqrt{1-G^2}} \quad \text{and} \quad q' = \frac{(\Delta/E)}{\sqrt{1-G^2}},$$
$$\Delta^2 = 1 + 2IEG - (I^2 + E^2 + G^2),$$
$$\cos(\theta) = \frac{p_s}{\sqrt{p_s^2+q_s^2}} \quad \text{and} \quad \sin(\theta) = \frac{q_s}{\sqrt{p_s^2+q_s^2}}.$$

It is immediately apparent that there are two solution points in gradient space for most values of I, E, and G. Notice that θ is the direction of the light source in gradient-space, that is, the line connecting (p_s, q_s) to the origin makes an angle θ with the p-axis. So p' and q' are coordinates in a new gradient-space obtained after simplifying matters by rotating the axes until $q_s = 0$. The light source is then in the direction of the x'-axis. Notice that p' is constant if I/E is constant (remembering that G is constant anyway.) Hence the loci of constant I/E are straight lines. These lines are all parallel to the terminator, for which $I = 0$. This turns out to be important since some surfaces have constant reflectance for constant I/E (Fig. 8).

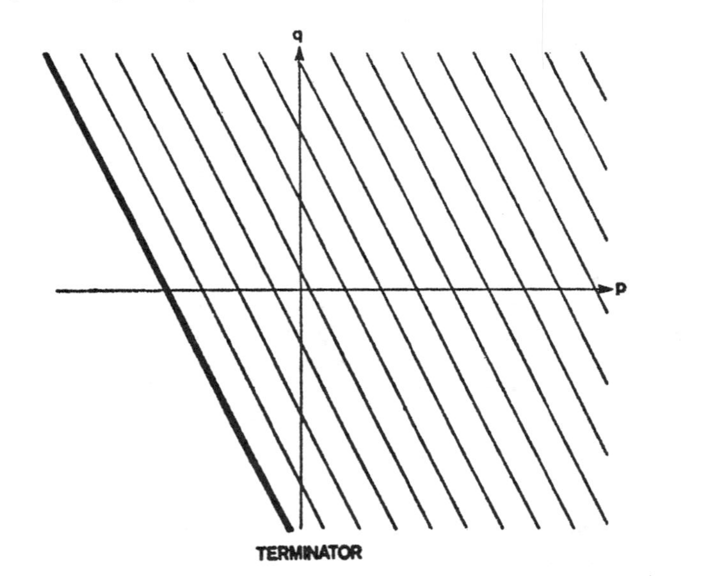

FIG. 8. Contours of $\phi(I, E, G) = I/E$. Contour intervals are 0.2 units wide. The reflectivity function for the material in the maria of the moon is constant for I/E.

2.5. Smooth metallic surfaces

Consider a smooth metallic surface: a surface with a purely specular or mirror-like reflectance. Each point in gradient-space corresponds to a particular direction of the surface normal and defines a direction from which incident light has to approach the object in order to be reflected towards the viewer. In fact, we can produce a complete map of the sphere of possible directions as seen from the object. At the origin, for example, we have the direction towards the viewer. Now for each incident direction there is a certain light-intensity depending on what objects lie that way. Consider recording these intensities at the corresponding points in gradient-space. Clearly one obtains some kind of image of the world surrounding the given metallic object. In fact, one develops a stereographic projection, a plane projection of a sphere with one of the poles as the center of projection. Another way of looking at it is that the image we construct in this fashion is like one we

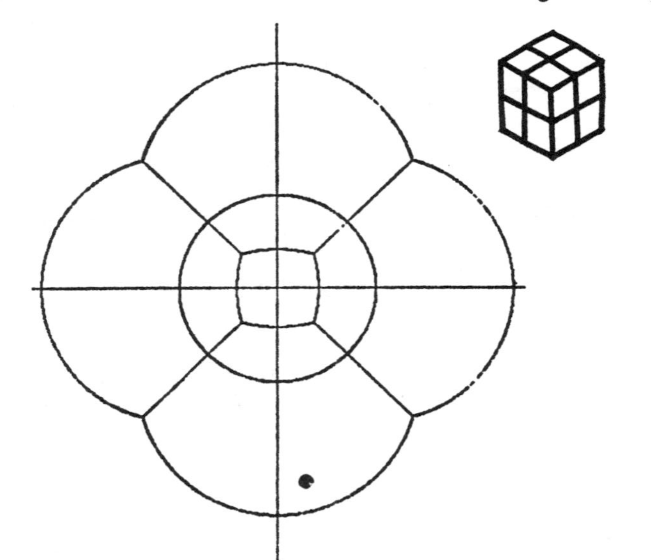

FIG. 9. Gradient-space image for a metallic object in the center of a very large wire cube. Equivalently one can think of it as the reflection of the wire cube in a paraboloid with a specularly-reflecting surface.

obtain by looking into a convex mirror—a metallic paraboloid to be precise (Fig. 9).

In order to construct reflectance maps for various surfaces and distributions of light sources, we superimpose the results in gradient-space for each light source in turn. We now examine a flaw in this approach and attempt a partial analysis of mutual illumination.

2.6. Mutual illumination

The reflectance map is based on the assumption that the viewer and all light sources are distant from the object. Only under these assumptions can we associate a unique value of image intensity with every surface orientation. If the scene consists of a single convex object these assumptions will be satisfied, but when there are several highly reflective objects placed near one another, mutual illumination may become important. That is, the distribution of incident light no longer depends only on direction, but is a function of position as well. The general case is very difficult, and we shall only study some idealized situations applicable to scenes consisting of polyhedra.

Two important effects of mutual illumination are a reduction in *contrast* between faces, and the appearance of *shading* or gradation of light on images of plane surfaces. In the absence of this effect, we would expect plane surfaces to give rise to images of uniform intensity since all points on a plane surface have the same orientation.

2.7. Two semi-infinite planes

First, let us consider a highly idealized situation of two semi-infinite planes joined at right angles and a distant light source. Let the incident rays make an angle α with respect to one of the planes (Fig. 10). Assume further that the surfaces reflect

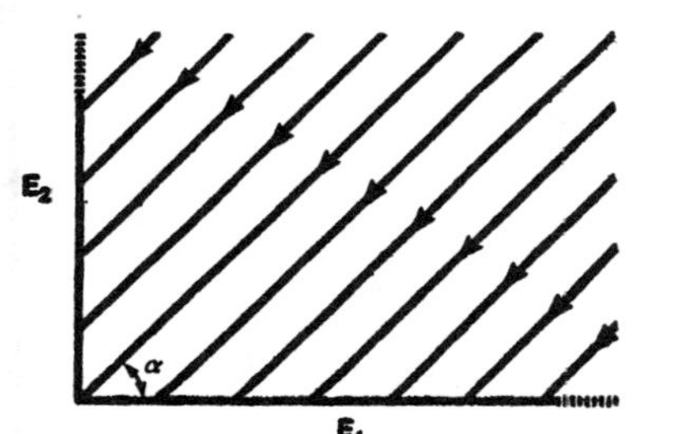

Fig. 10. Mutual illumination of two infinite lambertian planes of reflectance *r*, when illuminated by a single distant source.

a fraction r of the light falling on them, and that the illumination provided by the source is E (light flux/unit area). Picking any point on one of the half-planes, we find that one-half of its hemisphere of directions is occupied by the other plane, one-half of the light radiated from this point hits the other plane, while one-half is lost. Since both planes are semi-infinite, the geometry of this does not depend on the distance from the corner. The light incident at any point is made up of two components: that received directly from the source and that reflected from the other plane. The intensity on one plane does not vary with distance from the corner—a point receives light from one-half of its hemisphere of directions no matter what its distance from the corner. Put another way, there is no natural scale factor for a fluctuation in intensity. Let the illumination of the plane be E_1 and E_2 (light flux/unit area); then,

$$E_1 = \tfrac{1}{2}E_2 + E\cos(\alpha),$$
$$E_2 = \tfrac{1}{2}E_1 + E\sin(\alpha).$$

Solving for E_1 and E_2, we get:

$$E_1 = E[\cos(\alpha) + \tfrac{1}{2}r\sin(\alpha)]\frac{1}{[1-(\frac{1}{2}r)^2]},$$

$$E_2 = E[\sin(\alpha) + \tfrac{1}{2}r\cos(\alpha)]\frac{1}{[1-(\frac{1}{2}r)^2]}.$$

Had we ignored the effects of mutual illumination we would have found $E_1 = E\cos(\alpha)$ and $E_2 = E\sin(\alpha)$. Clearly the effect increases as reflectance r increases (it is not significant for dark surfaces). When the planes are illuminated equally, for $\alpha = \pi/4$, we find

$$E_1 = E_2 = (E/\sqrt{2})/(1-\tfrac{1}{2}r).$$

When $r = 1$, we obtain *twice* the illumination and hence twice the brightness than that obtained in the absence of mutual illumination. If the angle between the two planes varies, we find that the effect becomes larger and larger as the angle becomes more and more acute. By choosing the angle small enough, we can obtain arbitrary "amplification". Conversely, for angles larger than $\pi/2$, the effect is less pronounced.

In the derivation above, we did not make very specific assumptions about the angular distribution of reflected light, just that it is symmetrical about the normal and that it does not depend on the direction from which the incident ray comes. Hence, a lambertian surface is included, while a highly specular one is not. The effect is indeed less pronounced for surfaces with a high specular component of reflection, since most of the light is bounced back to the source after two reflections. Another important thing to note is that even if the planes are not infinite, the above calculations are approximately valid close to the corner. For finite planes we expect a variation of intensity as a function of distance from the corner; the results derived here apply asymptotically as one approaches the corner.

2.8. Two truncated planes

The geometry becomes quite complex if the planes are of finite extent, but we can develop an integral equation if we allow the planes to be infinite along their line of intersection and truncate them only in the direction perpendicular to this. Suppose two perpendicular planes extend a distance L from the corner, and that $\alpha = \pi/4$. This produces a particularly simple integral equation [11]; nevertheless I have been unable to solve it analytically. Numerical methods show that the resultant illumination falls off monotonically from the corner (Fig. 11), that the value at the corner is

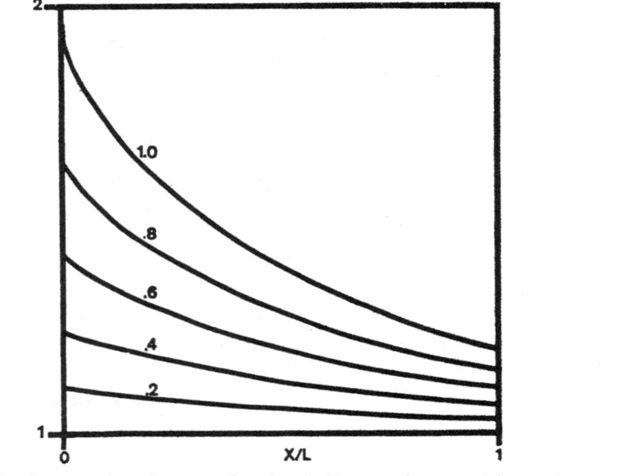

FIG. 11. Surface luminance plotted versus fractional distance from a right-angle corner. The curves are for reflectances of 0.2, 0.4, 0.6, 0.8, and 1.0.

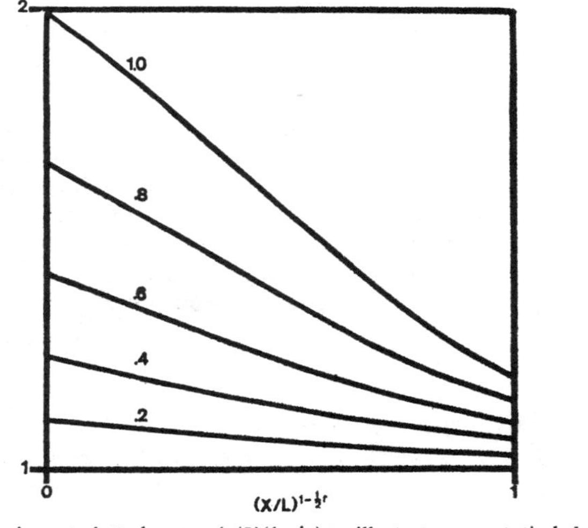

FIG. 12. Surface luminance plotted versus $(x/L)^{(1-\frac{1}{2}r)}$ to illustrate asymptotic behavior near the corner. The curves correspond to reflectances of 0.2, 0.4, 0.6, 0.8, and 1.0.

indeed what was predicted in the previous section, and the fall-off near the corner is governed by a term in $-(x/L)^{(1-\frac{1}{2}r)}$ (Fig. 12). A good approximation appears to be $1+E_{2/r}(x/L)$, where $E_n(x)$ is the elliptical integral and x is the distance along the plane measured from the edge where the planes meet.

3. The Semantics of Edge-Profiles

We are now ready to apply the tools developed so far. First let us consider the interpretation of intensity profiles taken across edges. If polyhedral objects and image sensors were perfect, if there were no mutual illumination, and if light sources were distant from the scene, images of polyhedral objects would be divided into polygonal areas, each of uniform intensity. It is well known that in real images, image intensity varies within the polygonal areas and that an intensity profile taken across an edge separating two such polygonal regions does not have a simple step-shaped intensity transition. Herskovitz and Binford determined experimentally that the most common edge transitions are step-, peak-, and roof-shaped [7] (Fig. 13). So

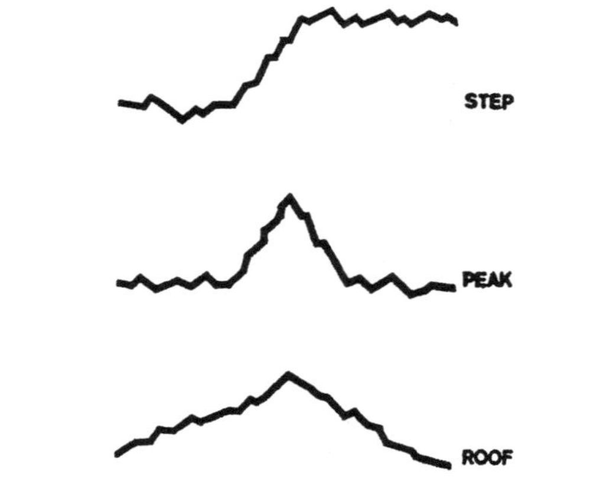

FIG. 13. Most common intensity profiles across images of polyhedral edges.

far this has been considered no more than a nuisance, because it complicates the process of finding edges. We now discuss the interpretation of these profiles in terms of the three-dimensional aspects of the scene.

3.1. Imperfections of polyhedral edges

A perfect polyhedron has a discontinuity in surface normal at an edge. In practice, edges are somewhat rounded off. A cross-section through the object's edge show that the surface normal varies smoothly from one value to the other and takes on values that are linear combinations of the surface normals of the two adjoining planes (Fig. 14). What does this mean in terms of reflected light intensity? Intensity varies smoothly at the edge, instead of jumping from one surface normal value to

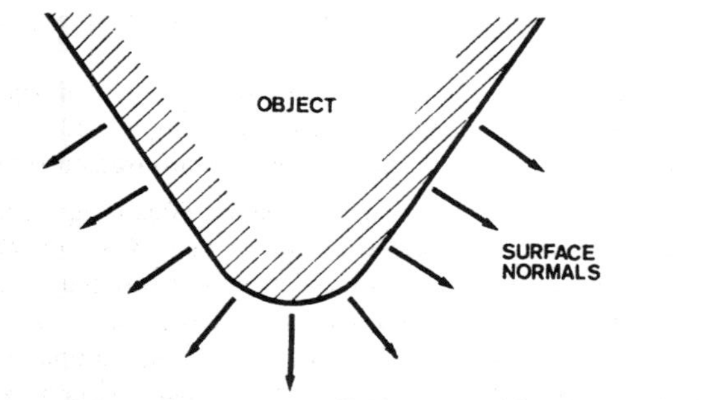

FIG. 14. The normals at an imperfect, rounded off edge are positive linear combinations of the normals of the two adjoining faces.

another. The important point is that it may take on values *outside* the range of those defined by the two planes. The best way to see this is to consider the situation in gradient-space. The two planes defined two points in gradient-space. Tangent planes on the corner correspond to points on the line connecting these two points. If the image intensity is higher for a point somewhere on this line, we will see a peak in the intensity profile across the edge. (Fig. 15.)

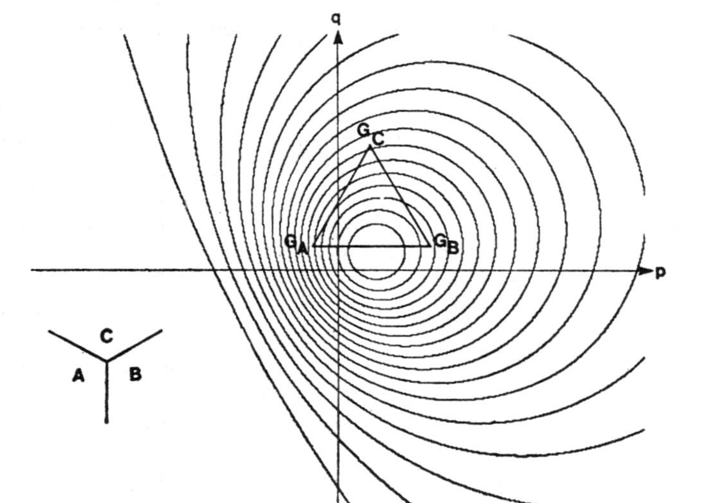

FIG. 15. Image of tri-hedral corner and corresponding gradient-space diagram. The image intensity profile across the edge between face *A* and face *B* will have a peak or highlight. The others will not.

If we find an edge profile with a peak shape or step with a peak superimposed, it is most likely that the corresponding edge is convex. The converse is not true (an edge may be convex and not give rise to a peak) if the line connecting the two points in gradient space has intensity varying monotonically along its length. The identification is also not completely certain since under peculiar lighting conditions with objects that have acute angles between adjacent faces, a peak may appear at an obscuring edge. Notice that the peak is quite compact, since it extends only as far as the rounded-off edge.

At a corner, where the planes meet, we find that surface imperfections provide surface normals that are linear combinations of the three normals corresponding to the three planes. In gradient space, this corresponds to points in the triangle connecting the three points corresponding to the planes. If this triangle contains a maximum in image intensity we expect to see a highlight right on the corner (Fig. 15).

3.2. Mutual illumination

We have seen that mutual illumination gives rise to intensity variations on planar surfaces—intensity roughly decreases linearly away from the corner. Notice that this affects the intensity profile over a large distance from the edge, quite unlike the sharp peak found due to edge imperfections. Clearly, if we find a roof-shaped profile or step with a roof-shaped superimposed we should consider labelling the edge concave.

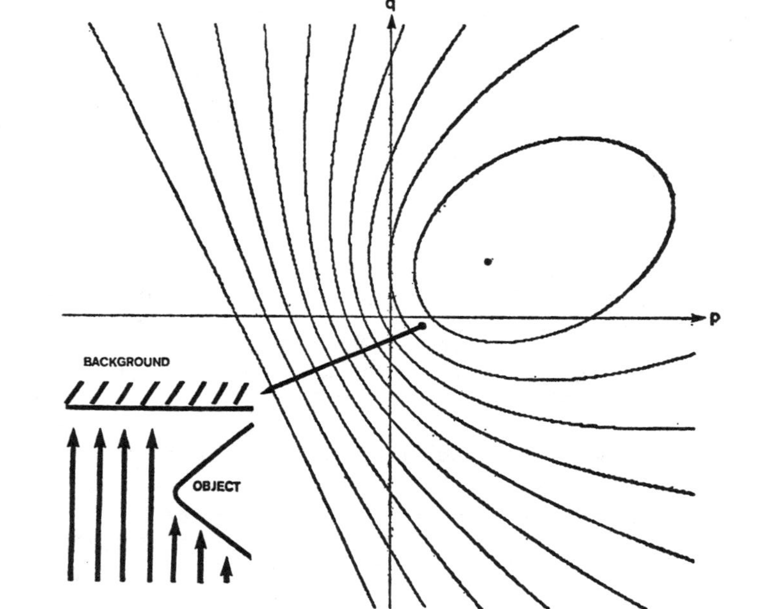

FIG. 16. Generation of a negative peak at an obscuring edge facing away from the light source.

The identification is, however, partly unreliable since some imaging device defects can produce a similar effect. Image dissectors, for example suffer from a great deal of scattering—areas further from a dark background appear brighter. So we may see a smoothed version of a roof-shape in the middle of a bright scene against a dark background. Experimentation with high quality image input devices such as a PIN-diode mirror-deflection system has confirmed that this is an artifact introduced by the image dissector. When the light source is close to the scene, significant gradients can appear on planar surfaces as pointed out by Herskovitz and Binford [7]. Lastly, the roof-shaped profiles on the two surfaces may be due to mutual illumination with other surfaces, not each other. Nevertheless, a roof-shaped profile usually suggests a concave edge.

3.3. Obscuring

Although they can be found with both convex and concave edges, step-shaped intensity profiles occur most often when objects obscure one another. If the obscuring surface adjoins a self-shadowed surface however, edge imperfections will produce a negative peak on the profile, since the line connecting the points corresponding to the two surfaces of the object passes through the terminator in gradient-space. Hence a negative peak or a step with a superimposed negative peak strongly suggests obscuration. Unfortunately it is difficult to tell on which side the obscuring plane lies (Fig. 16).

3.4. Determining the three-dimensional structure of polyhedral scenes

Mackworth's approach to understanding line drawings of polyhedra allows us to take into account some of the quantitative aspects of the three-dimensional geometry of scenes [2]. It does not, however, allow us to fully determine the orientation of all the planes. The scale and position of the gradient space diagram is undetermined by his technique. To illustrate this, consider a single trihedral corner. Here we know that the three points in gradient space representing the three planes meeting at the corner have to satisfy certain constraints: they must lie on three lines perpendicular to the image lines.

It takes six parameters to specify the position of three points on a plane, leaving three degrees of freedom after the introduction of these constraints. Measuring the three image intensities of the planes supplies another three. The constraints are due to the fact that the points in gradient-space have to lie on the correct contours of image intensity. The triangle can be stretched and moved until the points correspond to the correct image intensities as measured for the three planes.

Since this process corresponds to solving three non-linear equations for three unknowns, we can expect a finite number of solutions. Often there are but one or two—prior knowledge often eliminates some. (For a numerical example, see Appendix).

When more than three planes meet at a corner, the equations are over-determined, and the situation is even more constraint. Conversely, we cannot determine much with just two planes meeting at an edge—there are too few equations and an infinite number of solutions. The possible ambiguity at a trihedral corner is not very serious when we consider that in a typical scene there are many "connect" edges, either convex or concave as determined by Mackworth's program. In such a case the overall constraints may allow only one consistent interpretation. A practical difficulty is that it is unclear which search strategy leads most efficiently to this interpretation.

Measurements of image intensity are not very precise and surfaces have properties that vary from point to point as well as with handling. We cannot expect this method to be extremely accurate in pinning down surface orientation. In general, the equations for a typical scene are over-determined; a least-squares approach may improve matters slightly. The idea of stretching and shifting can be generalized to smooth surfaces. We know that the image of a paraboloid is the same as the reflectance map. If we can stretch and shift the reflectance map to fit the image of some object, then the same deformations turn the paraboloid into the object.

4. Determining Lunar Topography

When viewed from a great distance, the material in the maria of the moon has a particularly interesting reflectivity function. First, note that the lunar phase is the angle at the moon between the light source (sun) and the viewer (earth). This is clearly the angle we call g, and explains why we use the term phase angle. For constant phase angle, detailed measurements using surface elements whose projected area as seen from the source is a constant multiple of the projected area as seen by the viewer have shown that all such surface elements have the same reflectance. But the area appears foreshortened by $\cos(i)$ and $\cos(e)$ as seen by the source and the viewer respectively. Hence the reflectivity function is constant for constant $\cos(i)/\cos(e) = I/E$.

In this case each surface element scatters light uniformly into its hemisphere of directions, quite unlike the lambertian surface which favors directions normal to its surface. This is not an isolated incident. The surfaces of other rocky, dusty objects when viewed from great distances appear to have similar properties. For example, the surface of the planet Mercury and perhaps Mars as well as some asteroids and atmosphere-free satellites fit this pattern. Surfaces with reflectance a function of I/E thus form a third species we should add to matte surfaces where the reflectance is a function of I and glossy surfaces where the reflectance is a function of $(2IE-G)$.

4.1. Lunar reflectivity function

Returning to the lunar surface, we find an early formula due to Lommel Seelinger [6].

$$\phi(I, E, G) = \frac{\Gamma_0(I/E)}{(I/E)+\lambda(G)}.$$

Here Γ_0 is a constant and the function $\lambda(G)$ is defined by an empirically determined table. A somewhat more satisfactory fit to the data is provided by a formula of Fesenkov's [6]:

$$\phi(I, E, G) = \frac{\Gamma_0(I/E)[1+\cos^2(\alpha/2)]}{(I/E)+\lambda_0[1+\tan^2(\lambda/2)]}.$$

Where Γ_0 and λ_0 are constants and $\tan(\alpha) = -p' = -(I/E-G)/\sqrt{1-G^2}$. This formula is also supported by a theoretical model of the surface due to Hapke. Note that given I, E, and G, it is straightforward to calculate the expected reflectance. We need to go in the reverse direction and solve for I/E given G and the reflectance as measured by the image intensity. While it may be hard to invert the above equation analytically, it should be clear that by some iterative, interpolation, or hill-climbing scheme one can solve for I/E. We shall ignore for now the ambiguities that arise if there is more than one solution.

4.2. Lunar reflectance map

Next, we ask what the reflectance map looks like for the lunar surface illuminated by a single point source. The contours of constant intensity in gradient-space will be lines of constant I/E. But the contours of constant I/E are straight lines! So the gradient-space image can be generated from a single curve by shifting it along a straight-line—the shadow-line, for example (see Fig. 8). The contour lines are perpendicular to the direction defined by the position of the source (that is, the line from the origin to p_s, q_s).

Now what information does a single measurement of image intensity provide? It tells us that the gradient has to be on a particular straight line. Again, we ignore for the moment the possible existence of more than one contour for a given intensity.

What we would like to know, of course, is the orientation of the surface element. We cannot completely determine the local orientation, but we *can* determine its component in the direction perpendicular to the contour lines in gradient-space.

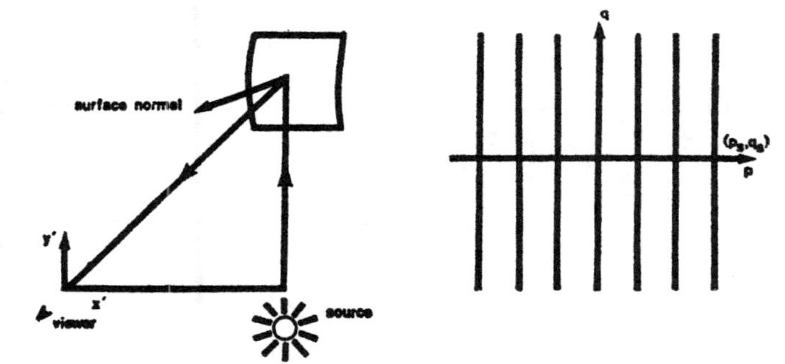

FIG. 17. Rotation of coordinate system to simplify gradient-space geometry.

We can tell nothing about it in the direction at right angles. That is, knowing I/E and G determines p', as previously defined, and tells us nothing about q'.

This favored direction lies in the plane defined by the source, the viewer, and the surface element under consideration. If we wish, we can simplify matters by rotating the viewer's coordinate system until the x axis lies in this plane as well. Then $q_s = 0$, and the contours of constant intensity in gradient-space are all vertical lines. Evidently, an image intensity measurement determines the slope of the surface in the x' direction, without telling us anything about the slope in the y' direction (Fig. 17). We are now ready to develop the surface by advancing in the direction in which we can determine locally the surface slope.

4.3. Finding a surface profile by integration

We have:

$$p' = \frac{dz}{ds} = \frac{I/E-G}{\sqrt{1-G^2}}.$$

The distance s from some starting point is measured in the object coordinate system and is related to the distance along the curve's projection in the image by $s' = s(f/z_0)$.

$$\frac{dz}{ds'} = \frac{f}{z_0}\frac{I/E-G}{\sqrt{1-G^2}}.$$

Integrating, we get:

$$z(s') = z_0 + \frac{f}{z_0}\int_0^{s'} \frac{I/E-G}{\sqrt{1-G^2}}\, ds',$$

where I/E is found from G and the image intensity $b(x', y')$. Starting anywhere in the image, we can integrate along a particular line and find the relative elevation of the corresponding points on the object.

The curves traced out on the object in this fashion are called *characteristics*, and their projections in the image-plane are called *base characteristics*. It is clear that here the base characteristics are parallel straight lines in the image, independent of the object's shape.

4.4. Finding the whole surface

We can explore the whole image by choosing sufficient starting points along a line at an angle to the favored direction. In this way we obtain the surface shape over the whole area recorded in the image (Fig. 18). Since we cannot determine the gradient at right angles to the direction of the characteristics, there is nothing to relate adjacent characteristics in the image. We have to know an initial curve, or use assumptions about reasonable smoothness. Alternatively, we can perform a second surface calculation from an image taken with a different source-surface-observer geometry. In this case, we obtain solutions along lines crossing the surface at a different angle, tying the two solutions together. This is not

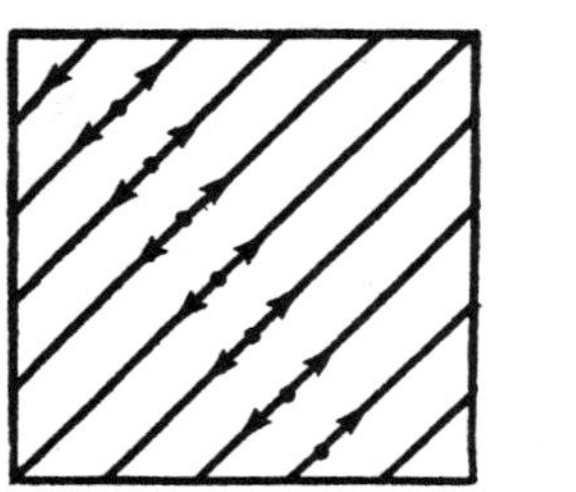

FIG. 18. Finding the shape of the surface by integrating along several base characteristics starting from different initial points.

quite as useful as one might first think, because it does not apply to pictures taken from earth. The plane of the sun, moon and earth varies little from the ecliptic plane. The lines of integration in the image vary little in inclination. This idea however *does* work for pictures taken from near the moon.

4.5. Ambiguity in local gradient

What if more than one contour in gradient-space corresponds to a given intensity? Then we cannot tell locally which gradient to apply. If we are integrating along some curve however, this is no problem, since we may assume that there is little change in gradient over small distances, and pick one close to the gradient last used. This assumption of smoothness leaves us with one remaining problem: what happens if we approach a maximum of intensity in gradient-space and then enter areas of lower intensity (Fig. 19). Which side of the local maximum do we slide down?

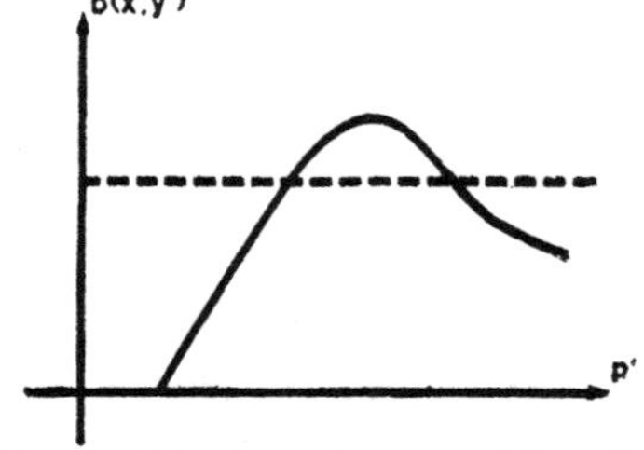

FIG. 19. Problem of ambiguity caused by non-uniqueness of slope for a particular observed image intensity.

This is an ambiguity which cannot be resolved locally, and the solution has to be terminated at this point. Under certain lighting conditions the image is divided into regions inside each of which we can find a solution. The regions are separated by ambiguity edges, which cannot be crossed without making an arbitrary choice [4].

4.6. Low sun angles

This problem can be avoided entirely if one deals only with pictures taken at low sun angles, where the gradient is a single-valued function of image intensity. This is a good idea in any case, since the accuracy of the reconstruction depends on how accurately one can determine the gradient, which in turn depends on the spacing of the contour lines in gradient-space. If they are close together, this accuracy is high (near a maximum, on the other hand, it is low). It is easy to convince oneself that pictures taken at low sun angle have "better contrast," show the "relief in more detail," and are "easier to interpret." An additional reason for interest in low sun angle images is that the contours of constant intensity near the shadow-line in gradient-space are nearly straight lines even if we are *not* dealing with the special reflectivity function for the lunar material! An early solution to the problem of determining the shape of lunar hills makes use of this fact by integrating along lines perpendicular to the terminator [5].

Working at low sun angles introduces another problem, of course, since shadows are likely to appear. Fortunately, they are easy to deal with since we simply trace the line in the image until we see a lighted area again. Knowing the direction of the source's rays, we easily determine the position of the first lighted point. The integration is then continued from there (Fig. 20). In fact, no special attention to this problem is needed since a surface element oriented for grazing incidence of light already has the correct slope. Thus, simply looking up the slope for zero intensity and integrating with this value will do. Some portion of the surface, of course, is not explored because of shadows. If we take one picture just after "sunrise" and one just before "sunset," most of this area is covered.

FIG. 20. Geometry of grazing ray needed to deal with shadow gaps in solution.

4.7. Generalization to perspective projection

All along we have assumed orthographic projection—looking at the surface from a great distance with a telephoto lens. In practice, this is an unreasonable assumption for pictures taken by artificial satellites near the surface. The first thing that changes in the more general case of perspective projection is that the sun-surface-viewer plane is no longer the same for all portions of the surface images. Since it is this plane which determines the integration lines, we expect that these lines are no longer parallel. Instead they all converge on the anti-solar point in the image which corresponds to a direction directly opposite the direction towards the source (Fig. 21).

The next change is that z is no longer constant in the projection equation. So $s' = f(s/z)$. Hence,

$$p' = \frac{dz}{ds} = \frac{f}{z}\frac{dz}{ds'} = \frac{I/E-G}{\sqrt{1-G^2}}.$$

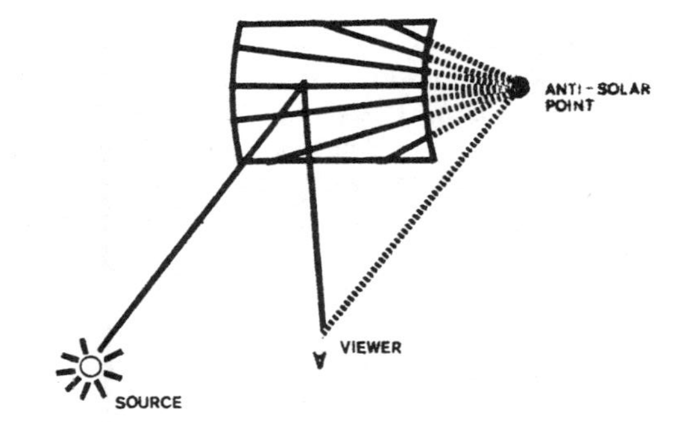

FIG. 21. The base-characteristics converge on the anti-solar point.

We can no longer simply integrate. It is easy, however, to solve the above differential equation for z by separating terms:

$$\log(z) = \frac{1}{f}\int \frac{I/E-G}{\sqrt{1-G^2}}\,ds',$$

and so

$$z(s') = z_0 \exp\left(\frac{1}{f}\int_0^{s'} \frac{I/E-G}{\sqrt{1-G^2}}\,ds'\right).$$

Finally, note that the phase angle g is no longer constant. This has to be taken into account when calculating I/E from the measured image intensity. On the whole, the process is still very simple. The paths of integration are predetermined straight lines in the image radiating from the anti-solar point. At each point we measure the image intensity, determining which value of I/E gives rise to this image intensity. Then we calculate the corresponding slope along the straight line and take a small step. Repeating the process for all lines crossing the image we obtain the surface elevation at all points in the image. The same result can be obtained by a very complex algebraic method [6].

4.8. A note on accuracy

Since image intensities are determined only with rather limited precision, we must expect the calculation of surface coordinates to suffer from errors that accumulate along characteristics. A "sharpening" method that relates adjacent characteristics reduces these errors somewhat [4].

It also appears that an object's shape is better described by the orientations of its surface normals than by the distances from the viewer to points on its surface. In part, this may be because distances to the surface undergo a more complicated transformation when the object is rotated than do surface normal directions. Note that the calculation of surface normals is *not* subject to the cumulative errors mentioned.

Finally, we should point out that the precise determination of the surface is not the main impetus for the development presented here. The understanding of how image intensities are determined by the object, the lighting, and the image forming system is of more importance and may lead to more interesting heuristic methods.

5. The Shape of Surfaces with Arbitrary Reflectance Maps

The simple method developed for lunar topography is inapplicable if the contours of constant intensity in gradient-space are not parallel straight lines. We will still be able to trace along the surface, but the direction we take at each point now depends on the image and changes along the profile. The base characteristics no longer are predetermined straight lines in the image. At each point on a characteristic curve we find that the solution can be continued only in a particular direction. It also appears that we need more information to start a solution and will have to carry along more information to proceed. Reasoning from the gradient-space diagram is augmented here by some algebraic manipulation.

Let $a(p, q)$ be the intensity corresponding to a surface element with a gradient (p, q). Let $b(x, y)$ be the intensity recorded in the image at the point (x, y). Then, for a particular surface element, we must have:

$$a(p, q) = b(x, y).$$

Now suppose we want to proceed in a manner analogous to the method developed earlier by taking a small step (dx, dy) in the image. It is clear that we can calculate the corresponding change in z as follows:

$$dz = z_x dx + z_y d_y = p\,dx + q\,dy.$$

To do this we need the values of p and q. We have to keep track of the values of the gradient as we integrate along the curve. We can calculate the increments in p and q by:

$$dp = p_x\,dx + p_y\,dy \quad \text{and} \quad dq = q_x\,dx + q_y\,dy.$$

At first, we appear to be getting into more difficulty, since now we need to know $p_x, p_y = q_x$ and q_y. In order to determine these unknowns we will differentiate the basic equation $a(p, q) = b(x, y)$ with respect to x and y:

$$a_p p_x + a_q q_x = b_x \quad \text{and} \quad a_p p_y + a_q q_y = b_y.$$

While these equations contain the right unknowns, there are only two equations, not enough to solve for three unknowns. Note, however, that we do not really need the individual values! We are only after the linear combinations $(p_x\,dx + p_y\,dy)$ and $(q_x\,dx + q_y\,dy)$.

We have to properly choose the direction of the small step (dx, dy) to allow the determination of these quantities. There is only one such direction. Let $(dx, dy) = (a_p, a_q)\,ds$, then $(dp, dq) = (b_x, dy)\,ds$. This is the solution we were after. Summarizing, we have five ordinary differential equations:

$$\dot{x} = a_p,\ \dot{y} = a_q,\ \dot{z} = a_p p + a_q q,\ \dot{p} = b_x \quad \text{and} \quad \dot{q} = b_y.$$

Here the dot denotes differentiation with respect to s, a parameter that varies along the solution curve.

5.1. Interpretation in terms of the gradient-space

As we solve along a particular characteristic curve on the object, we simultaneously trace out a base characteristic in the image and a curve in gradient-space. At each point in the solution we know to which point in the image and to which point in the gradient-space the surface element under consideration corresponds.

The intensity in the real image and in the gradient space image must, of course, be the same. The paths in the two spaces are related in a peculiar manner. The step we take in the image is perpendicular to the contour in *gradient-space* and the step we take in gradient-space is perpendicular to the intensity contour in the *image-plane*. (See Fig. 22.)

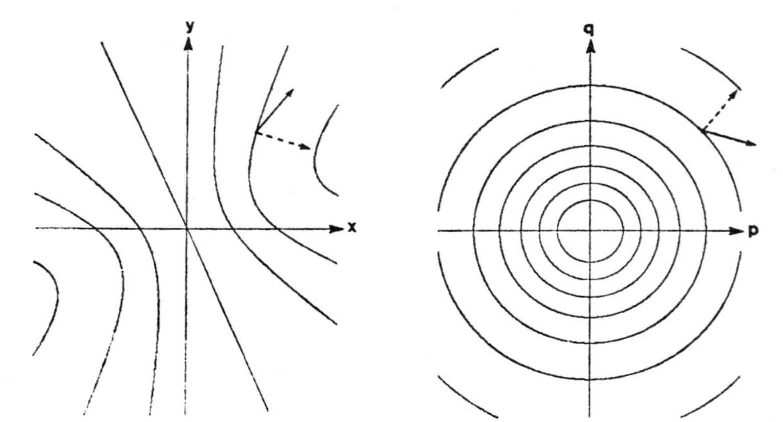

FIG. 22. The solution curve in image-space is along the line of steepest descent in gradient-spac and the solution curve in gradient-space is along the line of steepest descent in image-space.

5.2. Generalization to source and viewer near the object

The last solution method, while correct for arbitrary reflectivity functions, still assumes orthographic projection and a distant source. This is a good approximation for many practical cases. In order to take into account the effects of the nearness of the source and the viewer, we discard the gradient-space diagram, since it is based on the assumption of constant phase angle. The problem can still be tackled by algebraic manipulation, much as the last solution. It turns out that we are really trying to solve a first order non-linear partial differential equation in two independent variables. The well-known solution involves converting this equation into five ordinary differential equations, quite like the ones we obtain in the last Section [4].

Appendix

Using the reflectance map to determine three-dimensional structure of polyhedral scenes

What follows is a simple numerical example to illustrate the idea that information about surface reflectance can augment the gradient-space diagram and lead to a solution for the orientation of three planes meeting at a vertex. We will assume a lambertian reflectance for the object and assume that the light-source and viewer are far removed from the object, but close to each other. Suppose now that we are given the partial line-drawing as in Fig. 23 showing edges separating three planes

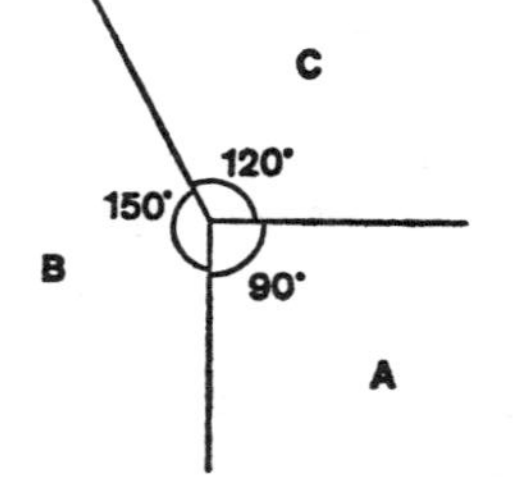

FIG. 23. Image-lines of a tri-hedral vertex. The orientations of the planes A, B and C are to be found.

A, B and C. The gradient-space diagram showing the three points G_A, G_B and G_C corresponding to these three planes will be as in Fig. 24. The scale and position of the indicated triangle are not yet determined. In fact the whole diagram could be flipped around if the scale is negative.

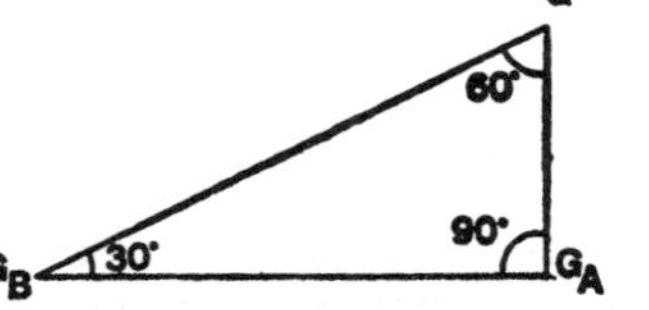

FIG. 24. The gradient-space diagram corresponding to the previous figure. The planes A, B and C map into the points G_A, G_B and G_C. The scale and position of this triangle is as yet undetermined.

Nevertheless, we now have available three linear constraints on the coordinates (p_a, q_a), (p_b, q_b) and (p_c, q_c) of the points G_A, G_B, G_C.

$$p_a = p_c, q_a = q_b, \quad \text{and} \quad (q_c - q_a) = (p_a - p_b)t.$$

Where $t = \tan(30°) = 1/\sqrt{3}$. We are now told that image measurements suggest reflectances of 0.707, 0.807 and 0.577 for the three faces respectively. From the information about the position of the source and viewer, we know that $G = 1$ and that $I = E$. Next, since we are dealing with a lambertian surface we calculate the reflectance from $\phi(I, E, G) = I$, which here equals $E = \cos(e) = 1/\sqrt{1+p^2+q^2}$. We immediately conclude that the surface normals of the three planes are inclined 45°, 36.2° and 54.8° respectively with respect to the view vector.

It also follows that the points G_A, G_B and G_C must lie on circles of radii 1.0, 0.732 and 1.415 in gradient-space, since distance from the origin in gradient space $\sqrt{p^2+q^2}$ equals tan(e). That is, the points have to lie on the appropriate contours of reflectance in the reflectance map as in Fig. 25.

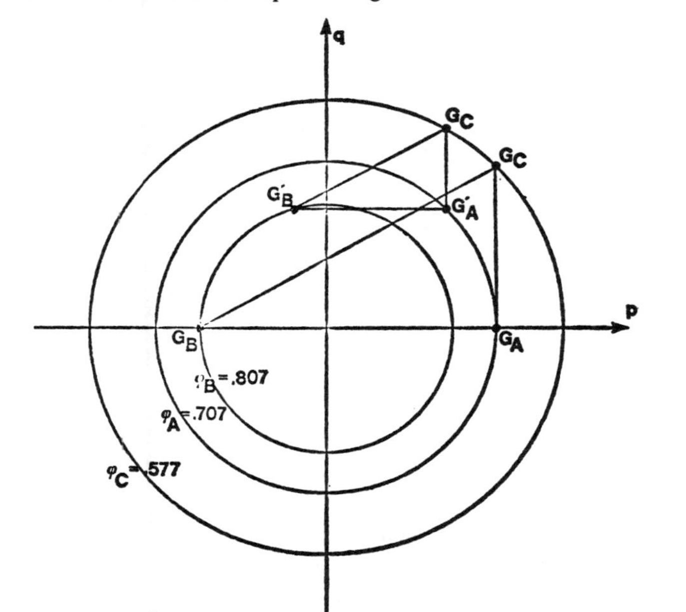

FIG. 25. Contours of constant reflectance corresponding to the reflectances of the three planes. Two solutions are superimposed on the reflectance map.

This gives us three further constraints, unfortunately non-linear ones. Let us call the radii r_a, r_b and r_c respectively, then

$$p_a^2+q_a^2 = r_a^2, \quad p_b^2+q_b^2 = r_b^2 \quad \text{and} \quad p_c^2+q_c^2 = r_c^2.$$

If the source had *not* been near the viewer, these equations would have involved linear terms in p and q as well, since then the contours of equal reflectance would have been conic sections other than circles. In the general case these three equations could be more complicated and in fact it is possible that the reflectance map is only known numerically. Then one will have to proceed iteratively from here. In our simple example however it is possible to solve the three linear equations and the three non-linear ones we have developed directly. As usual one proceeds by judiciously eliminating variables.

Let us use the three linear equations to eliminate the unknowns p_c, q_b and q_c from the three non-linear equations, then,

$$p_a^2+q_a^2 = r_a^2, \quad p_b^2+q_a^2 = r_b^2 \quad \text{and} \quad p_a^2+[q_a+t(p_a-p_b)]^2 = r_c^2.$$

We now have three non-linear equations in three unknowns. First note that

$$(p_a-p_b)(p_a+p_b) = p_a^2-p_b^2 = r_a^2-r_b^2.$$

Then expand the last of the three equations to get:

$$p_a^2+q_a^2+2tq_a(p_a-p_b)+t^2(p_a-p_b)^2 = r_c^2.$$

Now using $p_a^2+q_a^2 = r_a^2$ and the previous equation for (p_a-p_b):

$$[2tq_a+t^2(p_a-p_b)](r_a^2-r_b^2) = (p_a+p_b)(r_c^2-r_a^2).$$

This last question is linear! It is of the form $ap_a+bp_b+cq_a = 0$, where a, b and c can be evaluated and found to be -0.845, -1.155 and 0.536.

The simplest next operation is the elimination of p_a and p_b, using the equations for r_a^2 and r_b^2.

$$-a\sqrt{r_a^2-q_a^2}-b\sqrt{r_b^2-q_a^2} = cq_a.$$

A single equation in a single unknown, at last. Squaring both sides leads to:

$$2ab\sqrt{(r_a^2-q_a^2)(r_b^2-q_a^2)} = (a^2+b^2+c^2)q_a^2-(a^2r_a^2+b^2r_b^2).$$

Letting $e = (a^2+b^2+c^2)/(2ab)$ and $f = (a^2r_a^2+b^2r_b^2)/(2ab)$ and squaring again to get rid of the square-root:

$$(q_a^2)^2(1-d^2)+(q_a^2)[2de-(r_a^2+r_b^2)]+(r_a^2r_b^2-e^2) = 0.$$

This quadratic equation in q_a^2 can be further simplified by evaluating the terms to:

$$(q_a^2)^2-0.5(q_a^2) = 0.0.$$

The solutions are $q_a = 0.0$, -0.7071 and $+0.7071$. Not all of these may be solutions of the original equations, so we will have to check the results. Trying $q_a = 0$, leads to $q_b = 0$, $p_a = \pm 1$, $p_c = \pm 1$, $p_b = \pm 0.732$, and $q_c = 0.577(\pm 1 \pm 0.732)$. Two of these combinations satisfy the original equations.

$$(p_a, q_a) = (1, 0)$$
$$(p_b, q_b) = (-0.732, 0)$$
$$(p_c, q_c) = (Z, 1).$$

The other solution is the mirror image of this one, with the same numerical values, but reversed signs. One of these solutions is seen superimposed on the contours of the reflectance map in Fig. 25.

If one tries the other possible set of values for q_a, ± 0.7071 one finds, $p_a = q_b = \pm 0.7071$, $p_b = \pm 0.1895$, $p_c = \pm 0.1895$, and $q_c = \pm 0.7071 + 0.577(\pm 0.7071 \pm 0.1895)$. As before only two combinations satisfy the original equations. One of these is the following:

$$(p_a, q_a) = (0.707, 0.707),$$
$$(p_b, q_b) = (-0.189, 0.707),$$
$$(p_c, q_c) = (0.707, 1.225).$$

The other solution again simply has the signs reversed. One of these solutions is also shown in figure. The symmetry of the problem here contributes to the plethora of solutions, more usually one finds but two.

Clearly it would be desirable to avoid this tedious solution of simultaneous non-linear equations. Graphical techniques work and iterative Newton–Raphson techniques are appropriate for computer implementations of this method. For a numerical example see [11].

ACKNOWLEDGMENTS

I wish to thank Blenda Horn and Eva Kampitts for help in the preparation of the paper, Karen Prendergast for the drawings and Kathy Van Sant for an early version of the numerical solution for the truncated-plane mutual-illumination problem. David Marr and Patrick Winston provided much appreciated stimulation and discussion.

REFERENCES

1. Huffman, D. A., Impossible objects as nonsense sentences, *Machine Intelligence* **6**, Meltzer, R., and Michie, D. (Eds.), (Edinburgh University Press, 1971), 295–323.
2. Mackworth, A. K., Interpreting pictures of polyhedral scenes, *Artificial Intelligence* **4** (1973), 121–137.
3. Huffman, D. A., Curvature and creases: a primer on paper, *Proc. Conf. Computer Graphics, Pattern Recognition and Data Structures* (May 1975), 360–370.
4. Horn, B. K. P., Obtaining shape from shading information, in: Winston, P. H., (Ed.), *The Psychology of Machine Vision* (McGraw-Hill, NY, 1975), 115–155.
5. Van Diggelen, J., A photometric investigation of the slopes and heights of the ranges of hills in the maria of the moon, *Bull. Astron. Inst. Netherlands* **11** (1951), 283–289.
6. Rindfleisch, T., Photometric method for lunar topography, *Photogrammetric Eng.* **32** (1966), 262–276.
7. Herskovits, A. and Binford, T. O., On boundary detection, MIT Artificial Intelligence Memo 183 (July 1970), 19, 55, 56.
8. Phong, Bui Tuong, Illumination for Computer Generated Pictures, *CACM* **18** (1975), 311–317.
9. Hilbert, D. and Cohn-Vossen, S., *Geometry and the Imagination* (Chelsea Publishers, New York, 1952).
10. Hildebrand, F. D., *Methods of Applied Mathematics* (Prentice-Hall, New Jersey, 1952), 222–294.
11. Horn, B. K. P., Image intensity understanding, MIT Artificial Intelligence Memo 335 (August 1975).

A computer algorithm for reconstructing a scene from two projections

H. C. Longuet-Higgins

Laboratory of Experimental Psychology, University of Sussex, Brighton BN1 9QG, UK

A simple algorithm for computing the three-dimensional structure of a scene from a correlated pair of perspective projections is described here, when the spatial relationship between the two projections is unknown. This problem is relevant not only to photographic surveying[1] but also to binocular vision[2], where the non-visual information available to the observer about the orientation and focal length of each eye is much less accurate than the optical information supplied by the retinal images themselves. The problem also arises in monocular perception of motion[3], where the two projections represent views which are separated in time as well as space. As Marr and Poggio[4] have noted, the fusing of two images to produce a three-dimensional percept involves two distinct processes: the establishment of a 1:1 correspondence between image points in the two views—the 'correspondence problem'—and the use of the associated disparities for determining the distances of visible elements in the scene. I shall assume that the correspondence problem has been solved; the problem of reconstructing the scene then reduces to that of finding the relative orientation of the two viewpoints.

Photogrammetrists know that if a scene is photographed from two viewpoints, then the relationship between the camera positions is uniquely determined, in general, by the photographic coordinates of just five distinguishable points; but actually calculating the structure of the scene from five sets of image coordinates involves the iterative solution of five simultaneous third-order equations[1]. I show here that if the scene contains as many as eight points whose images can be located in each projection, then the relative orientation of the two projections, and the structure of the scene, can be computed, in general, from the eight sets of image coordinates by a direct method which calls for nothing more difficult than the solution of a set of simultaneous linear equations.

Let P be a visible point in the scene, and let (X_1, X_2, X_3) and (X'_1, X'_2, X'_3) be its three-dimensional cartesian coordinates with respect to the two viewpoints. The 'forward' coordinates X_3 and X'_3 are necessarily positive. The image coordinates of P in the two views may then be defined as

$$(x_1, x_2) = (X_1/X_3, X_2/X_3), \quad (x'_1, x'_2) = (X'_1/X'_3, X'_2/X'_3) \tag{1}$$

and it is convenient to supplement them with the dummy coordinates

$$x_3 = 1, \qquad x'_3 = 1 \tag{2}$$

so that one can then write

$$x_\mu = X_\mu/X_3, \qquad x'_\nu = X'_\nu/X'_3 \qquad (\mu, \nu = 1, 2, 3) \tag{3}$$

As the two sets of three-dimensional coordinates are connected by an arbitrary displacement, we may write

$$X'_\mu = \mathbf{R}_{\mu\nu}(X_\nu - \mathbf{T}_\nu) \tag{4}$$

where **T** is an unknown translational vector and **R** is an unknown rigid rotation matrix. (In this and subsequent equations I sum over repeated Greek subscripts.) The rotation **R** satisfies the relationships

$$\mathbf{R}\tilde{\mathbf{R}} = \mathbf{1} = \tilde{\mathbf{R}}\mathbf{R}, \qquad \det \mathbf{R} = 1 \tag{5}$$

and it is convenient to adopt the length of the vector **T** as the unit of distance:

$$\mathbf{T}^2_\nu (= \mathbf{T}^2_1 + \mathbf{T}^2_2 + \mathbf{T}^2_3) = 1 \tag{6}$$

I begin by establishing a general relationship between the two sets of image coordinates—a relationship which expresses the condition that corresponding rays through the two centres of projection must intersect in space. We define a new matrix **Q** by

$$\mathbf{Q} = \mathbf{RS} \tag{7}$$

where **S** is the skew-symmetric matrix

$$\mathbf{S} = \begin{bmatrix} 0 & \mathbf{T}_3 & -\mathbf{T}_2 \\ -\mathbf{T}_3 & 0 & \mathbf{T}_1 \\ \mathbf{T}_2 & -\mathbf{T}_1 & 0 \end{bmatrix} \tag{8}$$

Equation (8) may be written as

$$\mathbf{S}_{\lambda\nu} = \varepsilon_{\lambda\nu\sigma}\mathbf{T}_\sigma \tag{9}$$

where $\varepsilon_{\lambda\nu\sigma} = 0$ unless (λ, ν, σ) is a permutation of (1, 2, 3), in which case $\varepsilon_{\lambda\nu\sigma} = \pm 1$ depending on whether this permutation is even or odd. It follows from equations (4)–(9) that

$$\begin{aligned} X'_\mu \mathbf{Q}_{\mu\nu} X_\nu &= \mathbf{R}_{\mu\kappa}(X_\kappa - \mathbf{T}_\kappa)\mathbf{R}_{\mu\lambda}\varepsilon_{\lambda\nu\sigma}\mathbf{T}_\sigma X_\nu \\ &= (X_\lambda - \mathbf{T}_\lambda)\varepsilon_{\lambda\nu\sigma}\mathbf{T}_\sigma X_\nu \end{aligned} \tag{10}$$

but because the quantity $\varepsilon_{\lambda\nu\sigma}$ is antisymmetric in every pair of its subscripts, the right-hand side vanishes identically:

$$X'_\mu \mathbf{Q}_{\mu\nu} X_\nu = 0 \tag{11}$$

Dividing equation (11) by $X'_3 X_3$ we arrive at the desired relationship between the image coordinates:

$$x'_\mu \mathbf{Q}_{\mu\nu} x_\nu = 0 \tag{12}$$

The next step is to determine the nine elements $\mathbf{Q}_{\mu\nu}$. There will be one equation of type (12) for every point P_i, namely

$$(x'_\mu x_\nu)_i \mathbf{Q}_{\mu\nu} = 0 \tag{13}$$

and in this equation the nine quantities $(x'_\mu x_\nu)_i$ are presumed to be known. The ratios of the nine unknowns $\mathbf{Q}_{\mu\nu}$ can therefore be obtained, in general, by solving eight simultaneous linear equations of type (13), one for each of eight visible points $P_1, \ldots, P_8$. I shall not yet discuss the special circumstances under which the solution fails; for the present merely note that if the eight equations (13) are independent, their solution is entirely straightforward from a computational point of view.

The translational vector **T** must be calculated next. Multiplying **Q** on the left of equation (7) by its transpose we obtain

$$\tilde{\mathbf{Q}}\mathbf{Q} = \tilde{\mathbf{S}}\tilde{\mathbf{R}}\mathbf{R}\mathbf{S} = \tilde{\mathbf{S}}\mathbf{S} \tag{14}$$

so that, by the definition of **S**,

$$\mathbf{Q}_{\mu\nu}\mathbf{Q}_{\mu\sigma} = \mathbf{T}^2_\mu \delta_{\nu\sigma} - \mathbf{T}_\nu \mathbf{T}_\sigma \tag{15}$$

But $\mathbf{T}^2_\mu = 1$ by equation (6), and so the trace of $\tilde{\mathbf{Q}}\mathbf{Q}$ must be

$$\mathbf{Q}_{\mu\nu}\mathbf{Q}_{\mu\nu} = \delta_{\nu\nu} - \mathbf{T}^2_\nu = 2 \tag{16}$$

The nine elements of **Q** can therefore be normalized by dividing them by $\sqrt{\tfrac{1}{2}}$ trace $\tilde{\mathbf{Q}}\mathbf{Q}$; the elements of the normalized matrix $\tilde{\mathbf{Q}}\mathbf{Q}$ can then be used for computing the ratios of the components of **T**:

$$\tilde{\mathbf{Q}}\mathbf{Q} = \begin{bmatrix} 1 - \mathbf{T}^2_1 & -\mathbf{T}_1\mathbf{T}_2 & -\mathbf{T}_1\mathbf{T}_3 \\ -\mathbf{T}_2\mathbf{T}_1 & 1 - \mathbf{T}^2_2 & -\mathbf{T}_2\mathbf{T}_3 \\ -\mathbf{T}_3\mathbf{T}_1 & -\mathbf{T}_3\mathbf{T}_2 & 1 - \mathbf{T}^2_3 \end{bmatrix} \tag{17}$$

There are evidently three independent relationships between the diagonal and the off-diagonal elements of $\tilde{\mathbf{Q}}\mathbf{Q}$; these supply three independent checks on the results obtained so far. The absolute signs of the $\mathbf{T}_\mu$ and the $\mathbf{Q}_{\mu\nu}$ are still undetermined but, as we shall see, these ambiguities are easily resolved later.

We are now in a position to compute the elements of the rotation matrix **R**. First note that equation (7) has a simple interpretation in terms of vector products. If we regard each row of **Q**, and each row of **R**, as a vector, then

$$\mathbf{Q}_\alpha = \mathbf{T} \times \mathbf{R}_\alpha \qquad (\alpha = 1, 2, 3) \tag{18}$$

and the condition for $\mathbf{R}$ to represent a proper rotation can be expressed in a similar form:

$$\mathbf{R}_\alpha = \mathbf{R}_\beta \times \mathbf{R}_\gamma \tag{19}$$

for α, β, γ such that $\varepsilon_{\alpha\beta\gamma} = 1$. The problem is then to express the $\mathbf{R}_\alpha$ in terms of $\mathbf{T}$ and the $\mathbf{Q}_\alpha$.

By equation (18), $\mathbf{R}_\alpha$ is orthogonal to $\mathbf{Q}_\alpha$ and may therefore be expressed as a linear combination of $\mathbf{T}$ and $\mathbf{Q}_\alpha \times \mathbf{T}$. We therefore introduce new vectors

$$\mathbf{W}_\alpha = \mathbf{Q}_\alpha \times \mathbf{T} \qquad (\alpha = 1, 2, 3) \tag{20}$$

and write

$$\mathbf{R}_\alpha = a_\alpha \mathbf{T} + b_\alpha \mathbf{W}_\alpha \tag{21}$$

Substitution into equation (18) gives

$$\mathbf{Q}_\alpha = \mathbf{T} \times (a_\alpha \mathbf{T} + b_\alpha \mathbf{W}_\alpha) = b_\alpha (\mathbf{T} \times \mathbf{W}_\alpha) \tag{22}$$

But as $\mathbf{T}$ is a unit vector,

$$\mathbf{T} \times \mathbf{W}_\alpha = \mathbf{T} \times (\mathbf{Q}_\alpha \times \mathbf{T}) = \mathbf{Q}_\alpha \tag{23}$$

and so

$$b_\alpha = 1 \tag{24}$$

Turning to equation (19) we deduce that when $\varepsilon_{\alpha\beta\gamma} = 1$,

$$\begin{aligned} a_\alpha \mathbf{T} + \mathbf{W}_\alpha &= (a_\beta \mathbf{T} + \mathbf{W}_\beta) \times (a_\gamma \mathbf{T} + \mathbf{W}_\gamma) \\ &= a_\beta \mathbf{Q}_\gamma - a_\gamma \mathbf{Q}_\beta + \mathbf{W}_\beta \times \mathbf{W}_\gamma \end{aligned} \tag{25}$$

But in equation (25) the vectors $\mathbf{W}_\alpha$, $\mathbf{Q}_\beta$ and $\mathbf{Q}_\gamma$ are all orthogonal to $\mathbf{T}$, whereas $\mathbf{W}_\beta \times \mathbf{W}_\gamma$ is, by equation (20), a multiple of $\mathbf{T}$. It follows that in equation (25) the first term on the left equals the last term on the right,

$$a_\alpha \mathbf{T} = \mathbf{W}_\beta \times \mathbf{W}_\gamma \tag{26}$$

and equation (21) finally becomes

$$\mathbf{R}_\alpha = \mathbf{W}_\alpha + \mathbf{W}_\beta \times \mathbf{W}_\gamma \tag{27}$$

Having obtained in this way the vector $\mathbf{T}$ and the three rows of the matrix $\mathbf{R}$, we can at last find the three-dimensional coordinates X_μ, as follows:

By equation (4),

$$X'_\mu = \mathbf{R}_{\mu\nu}(X_\nu - \mathbf{T}_\nu)$$

from which it follows that

$$x'_1 = \frac{X'_1}{X'_3} = \frac{\mathbf{R}_{1\nu}(X_\nu - \mathbf{T}_\nu)}{\mathbf{R}_{3\nu}(X_\nu - \mathbf{T}_\nu)} \tag{28}$$

Introducing the vectors

$$\mathbf{X} = (X_1, X_2, X_3), \qquad \mathbf{x} = (x_1, x_2, 1) \tag{29}$$

we may write equation (28) in terms of the rows $\mathbf{R}_\alpha$ of the matrix $\mathbf{R}$:

$$x'_1 = \frac{\mathbf{R}_1 \cdot (\mathbf{X} - \mathbf{T})}{\mathbf{R}_3 \cdot (\mathbf{X} - \mathbf{T})} = \frac{\mathbf{R}_1 \cdot (\mathbf{x} - \mathbf{T}/X_3)}{\mathbf{R}_3 \cdot (\mathbf{x} - \mathbf{T}/X_3)} \tag{30}$$

from which it follows that

$$X_3 = \frac{(\mathbf{R}_1 - x'_1 \mathbf{R}_3) \cdot \mathbf{T}}{(\mathbf{R}_1 - x'_1 \mathbf{R}_3) \cdot \mathbf{x}} \tag{31}$$

The other unprimed coordinates are then given by equation (3) as

$$X_1 = x_1 X_3, \qquad X_2 = x_2 X_3 \tag{32}$$

and the primed coordinates are finally obtained from equation (4).

There are, in fact, four distinct solutions to the problem, associated with the alternative choices of sign for the components of $\mathbf{T}$ and the elements of $\mathbf{Q}$. But any doubt as to which choices to adopt is easily resolved: the condition that the forward coordinates of any point must both be positive will be satisfied if, and only if, both sets of signs are correctly chosen.

There are certain 'degenerate' eight-point configurations for which the algorithm fails because the associated equations (13) become non-independent. A configuration will be degenerate if as many as four of the points lie in a straight line, or if as many as seven of them lie in a plane. Quite unexpectedly, degeneracy also arises if the configuration includes six points at the vertices of a regular hexagon, or consists of eight points at the vertices of a cube. The 'invisibility' of such configurations to the eight-point algorithm may be demonstrated by arguments too long to be presented here; but the reasons for it are unconnected with any ambiguity in the interpretation of the resulting projections. A degenerate configuration immediately becomes 'visible', however, if one of the offending points P_i is moved slightly away from its original position.

In general, then, the three-dimensional coordinates of a set of eight or more visible points may be obtained by the following algorithm:

(1) Set up eight equations of the form (13), and solve them for the ratios of the nine unknowns $\mathbf{Q}_{\mu\nu}$.

(2) Compute the matrix $\tilde{\mathbf{Q}}\mathbf{Q}$ and normalize the elements of $\mathbf{Q}$ by dividing them by $\sqrt{\tfrac{1}{2}\,\text{trace}\,\tilde{\mathbf{Q}}\mathbf{Q}}$.

(3) Obtain the magnitudes and the relative signs of the $\mathbf{T}_\nu$ from equation (17); their absolute signs, and those of the $\mathbf{Q}_{\mu\nu}$, may have to be chosen arbitrarily at this stage.

(4) Define three new vectors by equation (20) and use equation (27) to calculate the rows of the matrix $\mathbf{R}$.

(5) Use equations (31) and (32) for computing the unprimed three-dimensional coordinates of all the visible points, and equation (4) for calculating the primed coordinates.

(6) Check that the forward coordinates X_3 and X'_3 of any point are both positive. If both signs are negative, alter the signs of the $\mathbf{T}_\nu$ and return to step (5); if X_3 and X'_3 are of opposite sign, reverse the signs of the $\mathbf{Q}_{\mu\nu}$ and return to step 4.

The algorithm yields the most accurate results when applied to situations in which the distance D between the centres of projection is not too small compared with their distances from the points P_i. If the projective coordinates are accurate to a few seconds of arc, the forward coordinates of the P_i can be estimated out to about $10D$ with great accuracy, and even as far as $100D$ if the P_i are adequately spaced in depth. This performance is comparable with that of the human visual system; but that does not, of course, imply that the eight-point algorithm is actually used in stereoscopic vision, as in binocular vision we have at least some information about the relative orientation of the two eyes. The most useful applications of the eight-point algorithm will probably be found in computer vision systems, where there is still a need for fast and accurate methods of converting two-dimensional images into three-dimensional interpretations.

I thank Drs A. L. Allan and K. B. Atkinson for the reference to Thompson[1], and the RS and SRC for support.

Received 13 April; accepted 13 July 1981.

1. Thompson, E. H. *Photogrammetric Record* **3**(14), 152–159 (1959).
2. Ogle, K. N. *Researches in Binocular Vision* (Hafner, New York, 1964).
3. Ullman, S. *The Interpretation of Visual Motion* (MIT Press, Cambridge, 1979).
4. Marr, D. & Poggio, T. *Science* **194**, 283–287 (1976).

Practical real-time imaging stereo matcher

H. K. Nishihara
Massachusetts Institute of Technology
Artificial Intelligence Laboratory
545 Technology Square
Cambridge, Massachusetts 02139

Abstract. A binocular-stereo-matching algorithm for making rapid visual range measurements in noisy images is described. This technique is developed for application to problems in robotics where noise tolerance, reliability, and speed are predominant issues. A high speed pipelined convolver for preprocessing images and an *unstructured light* technique for improving signal quality are introduced to help enhance performance to meet the demands of this task domain. These optimizations, however, are not sufficient. A closer examination of the problems encountered suggests that broader interpretations of both the objective of binocular stereo and of the zero-crossing theory of Marr and Poggio [Proc. R. Soc. Lond. B 204, 301 (1979)] are required. In this paper, we restrict ourselves to the problem of making a single primitive surface measurement —for example, to determine whether or not a specified volume of space is occupied, to measure the range to a surface at an indicated image location, or to determine the elevation gradient at that position. In this framework we make a subtle but important shift from the explicit use of zero-crossing contours (in bandpass-filtered images) as the elements matched between left and right images, to the use of the signs between zero crossings. With this change, we obtain a simpler algorithm with a reduced sensitivity to noise and a more predictable behavior. The practical real-time imaging stereo matcher (PRISM) system incorporates this algorithm with the unstructured light technique and a high speed digital convolver. It has been used successfully by others as a sensor in a path-planning system and a bin-picking system.

Keywords: robot vision; binocular stereo; stereo matching; noise tolerance; correlation; zero crossings; binarization; computer vision; obstacle avoidance; proximity sensors; structured light.

Optical Engineering 23(5), 536-545 (September/October 1984).

CONTENTS

Invited Paper RV-109 received Feb. 4, 1984; revised manuscript received March 6, 1984; accepted for publication March 10, 1984; received by Managing Editor May 29, 1984. This paper is a revision of paper 449-21 which was presented at the SPIE conference on Intelligent Robots: Third International Conference on Robot Vision and Sensory Controls RoViSeC3, Nov. 7-10, 1983, Cambridge, MA. The paper presented there appears (unrefereed) in SPIE Proceedings Vol. 449.

1. INTRODUCTION

This paper presents an approach to solving the binocular-stereo-matching problem that places special emphasis on the practical issues of noise tolerance, reliability, and speed. It is strongly influenced by Marr and Poggio's[1] zero-crossing theory, but differs from recent implementations in the way zero-crossing information is used to drive the matching and in the product the matcher is designed to produce.

1.1. Robust, high speed stereo system

Binocular stereo is a technique for measuring range, by triangulation, to selected locations in a scene imaged by two cameras. Figure 1 illustrates the imaging geometry. The vergence angle indicated in the diagram from the cameras to a selected target can be determined from the relative positions of that target in the left and right camera images in conjunction with the angle between the camera axes. The primary computational problem of binocular stereo is to identify corresponding locations in the two images. Once the position of the same physical surface point is known in both images, the vergence angle indicated in the diagram can be determined, and from that, the distance to the surface from the cameras.

1.1.1. Design goals

Four design objectives have guided this study. The first is *noise tolerance*. We want to understand how matching can be accomplished in the presence of moderate to large noise levels that occur

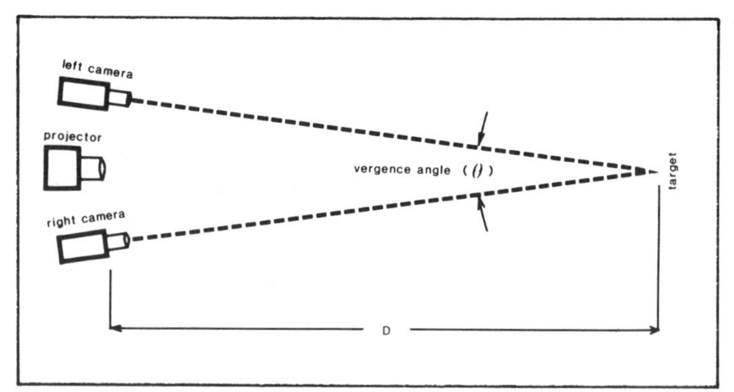

Fig. 1. The imaging geometry used in the PRISM stereo matcher. Two vidicon cameras are mounted 40 cm apart and 150 cm from the target scene. The object of binocular stereo is to identify corresponding physical surface locations in the two camera images so that the vergence angle θ can be determined. With this and the camera separation, the distance D to that surface location can be calculated. A slide projector is situated between the cameras and projects a random dot texture onto the target surfaces to provide fixed surface markings on objects lacking sufficient natural texture.

anytime surface contrast is low compared with sensor noise and other interimage distortions. The second is to achieve *competent performance* for at least one of the three stereo measurements — volume occupancy, range measurement, and detection of elevation discontinuities.[2] The third is to operate at a *practical speed* for robotics. Our emphasis here will be to streamline the computation to increase speed and use processing resources efficiently. This forces a careful consideration of the relative cost of producing a measurement in different ways vis-à-vis their contribution to the final product of the algorithm. Finally we require *simplicity*.[3-5] An algorithm is difficult to analyze extensively if it involves many serial decisions or many special cases. Of course a more complex algorithm might be necessary to obtain a desired level of performance in a specialized domain, but the strategy of this paper is to concentrate on simpler modular techniques, which have the advantage of greater generality. These can then be incorporated flexibly into the design of more complex systems for special applications over a more diverse range.

1.1.2. Overview of the practical real-time imaging stereo matcher (PRISM) system

The PRISM system was developed in light of these design considerations. It relies both on a determination of the nature of the distortions that occur in noisy images and on changing the product of the matcher to more closely agree with the requirements of specific tasks that might be presented to a robotics vision system. The algorithm we use has its roots in the zero-crossing theory of Marr and Poggio, which will be described in the next section, but does not explicitly match zero-crossing contours. Algorithms that match zero-crossing contours tend to be very sensitive to local distortions due to system noise and are prevented from operating well on signals with moderate noise levels even though a substantial amount of information may still be present. Instead, the PRISM system is based on the matching of a sign representation which is a dual of the zero-crossing representation. The initial design task of the implementation has been to rapidly detect obstacles in a robotics work space and determine their rough extents and heights.

The basic flow and organization of the computation is shown in Fig. 2. The scene (in front of a Unimation PUMA manipulator) is illuminated with an *unstructured* texture pattern by a slide projector, as indicated in Fig. 1, to provide suitable matching targets on the otherwise clean surfaces common in industrial settings. A pair of inexpensive vidicon cameras, mounted above the workspace, digitize two 576 × 454 pixel images. The digitized video signals are then fed to a high speed digital convolver that preprocesses the images to produce filtered images at three scales of resolution for both left and

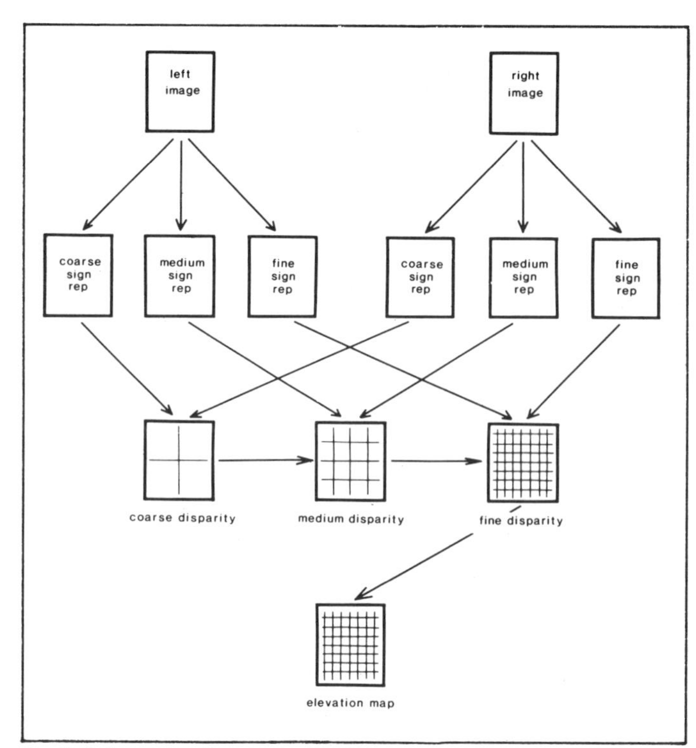

Fig. 2. Information flow in the PRISM system. Left and right images are digitized, and each is filtered to extract image information at three scales of resolution by convolution with two-dimensional $\nabla^2 G$ operators (with diameters of 32, 20, and 10 pixels). Each of these convolved images is hard-clipped to produce a binary sign representation that is stored. The left and right coarse-, medium-, and fine-scale sign representations are matched pair-wise to produce coarse-, medium-, and fine-scale disparity maps over the visual field of the cameras. The matching at the three scales is loosely coupled with coarser results guiding the search at the next finer scale. Finally, the fine-scale disparity map is translated by means of a look-up table into an elevation map.

right originals. A matching algorithm is then applied to the coarse pair to obtain a coarse 8 × 6 array of disparity measurements — disparity is defined here to be a vector in the image plane giving the translation required to bring the two images into registration at a designated location in one of the images. Then the same matching algorithm is applied to the medium resolution filtered images to produce a 17 × 13 disparity array. The algorithm in this case starts its search for each measurement using the disparity measured at the corresponding location at the coarser scale. This loose coupling speeds up the search significantly. Finally, the third pair of filtered images is matched in the same way to produce a 36 × 26 array of disparity measurements. The system then transforms the disparities in this last array into an array of absolute elevations (millimeters above the work space surface). The entire process from raw images to a 36 × 26 array of elevation measurements with a resolution of about one part in 100 takes approximately 30 s.

1.2. Background

Intensity-based area-correlation techniques have been investigated extensively for commercial applications in stereophotogrammetry (see, for instance, Ref. 6). Two of the best and most recent research efforts with correlation-based approaches are due to Moravec[7] and Gennery,[8] who developed stereo systems for vehicular autonomous navigation. Tsai[9] has recently proposed two new methods that use as many as eight perspective views with known positions and orientations of the cameras. The correlation functions obtained by his method are significantly more peaked than conventional two-frame area correlation and allow a much smaller window size for the correlation measurements.

Symbolic matching techniques abstract away from the direct

comparison of raw intensity values, permitting more explicit control over the image information used for matching.[10] Arnold and Binford[11,12] showed how constraints derived from the geometry of physical surfaces can be introduced to guide such matchers. Baker and Binford[13-15] developed a sophisticated matching system using many such constraints. Ohta and Kanade[16] have investigated dynamic programming methods for searching large spaces of possible correspondences between symbolic primitives from left and right images.

A third effort has its roots in the study of stereo matching in the human visual system.[17,18] Julesz showed conclusively with his random dot stereograms that stereo matching was an early process in the visual system that could function independently of monocular recognition. The psychophysical constraints that resulted from his work with the random dot stereogram stimulated thinking about how such computations could be accomplished. The prevailing intuition at the time had been that stereo involved some kind of parallel correlation or pattern matching process that operated on the fine detail of images.

1.2.1. Channels and zero crossings

More recently Marr and Poggio began a concerted effort to develop a computational theory of stereo matching consistent with and guided by the apparently modular biological solution. Their first computational model[19] was designed to solve Julesz's random dot stereograms and was noteworthy for its explicit formulation of the computational assumptions — *continuity* and *uniqueness* of the imaged surfaces — required for solving the otherwise underdetermined matching problem.

Later, concentrating on the differences between their cooperative model and a larger body of psychophysical and physiological data, they formulated an alternate model.[1] A key component was the observation that the matching problem was simplified if range and resolution were not both required simultaneously. Taken to either extreme, a high resolution short range matcher or a coarse resolution long range matcher could be based on the use of simple matching primitives defined at a scale appropriate to the range-resolution trade-off selected. The resolution proposed for this matching module was very low, allowing just three disparity values: crossed, near zero, and uncrossed — a target at a crossed disparity would require a further crossing of the eyes (a larger vergence angle) to cause it to appear at the same position in left and right images. The nearest compatible candidate match was to be used, and so the effective disparity range of the matcher depended directly on the spacing between neighboring matching primitives in the image.

A battery of such matching modules operating in parallel — each with a primitive size different from the others by about an octave — could produce high resolution measurements over a large range of disparities. Eye movements were proposed as a simple way to allow information obtained from matching coarse-scale primitives to bring modules operating on a finer scale into their respective ranges. The matching primitives required by their approach had to be scale-specific and sensitive to reliable surface markings at that scale. Intensity values from corresponding locations in the two stereo images can differ due to the effects of camera position, camera response characteristics, and noise. These effects make the direct comparison of intensity a relatively weak indicator of the presence of a true correspondence. Image intensities coincident with larger intensity gradients in the images will be better localized spatially than at other locations in the images. Thus the locations of maxima in the intensity gradient are good places to compare intensities. In fact, the positions of these gradient maxima can be used directly as the primitive image features used for matching. On these grounds, Marr and Poggio based the definition of their matching primitive on local maxima in the intensity gradient — or, equivalently, zero crossings in the second derivative. These directional derivatives are measured along the direction parallel to the line connecting the two cameras, or, more generally, along what are called *epipolar* lines when lens and perspective distortions are taken into account. This zero-crossing primitive provides locally optimized reference points distributed uniformly over the image.

The second derivative alone, however, is a high-pass operator, so its zeros would be correlated with the fine-scale structure of the images. To obtain the required scale specificity, a low-pass filter can be used to first attenuate the undesired high frequency detail. A convolution operator that comes close to the combined requirements of attenuating high spatial frequencies while preserving the geometric structure at coarser scales is the two-dimensional Gaussian.

The matching primitive was therefore defined as zero crossings in the second derivative of the image after Gaussian smoothing, with the primitive scale determined by the space constant of the Gaussian. Differential operators commute with convolution, so the above can also be formulated as zero crossings in the image after convolution with the function:

$$\frac{\partial^2}{\partial x^2} G_\sigma(x,y) \quad , \qquad (1)$$

where $G_\sigma(x,y)$ is a two-dimensional Gaussian with space constant σ.

Marr and Hildreth[20] later showed that zero crossings in the Laplacian of a Gaussian convolved (low-pass filtered) image avoided some technical problems associated with the use of directional linear filters. Moving the Laplacian inside the convolution, as above, yields a circularly symmetric linear operator:

$$\nabla^2 G = \left(\frac{\partial^2}{\partial x^2} + \frac{\partial^2}{\partial y^2} \right) G(x,y)$$

$$= \left(1 - \frac{4r^2}{w^2} \right) \exp \left(- \frac{4r^2}{w^2} \right) \quad , \qquad (2)$$

where G is a two-dimensional Gaussian function, w is the diameter of the positive central region of the operator and is proportional to the Gaussian's space constant, and $r^2 = x^2 + y^2$.

The circularly symmetric $\nabla^2 G$ operator models closely the impulse response of some retinal ganglion cells.[21-23] Its shape is also closely approximated by a difference of two Gaussians with different space constants but normalized volumes. This property permitted a substantial simplification in the design of a $\nabla^2 G$ convolver.[24]

1.2.2. Problems encountered

A program written by Grimson[25] to test Marr and Poggio's algorithm on random dot stereograms showed general agreement with the results obtained by Julesz for static patterns. Soon thereafter N. Larson and I designed a high speed digital convolver[24] in hardware for $\nabla^2 G$ convolutions as part of a real-time implementation of the zero-crossing theory. During simulation tests for possible hardware designs for the stereo matcher, I prepared a scene incorporating a complex but known shape — an instant coffee jar — and spattered it with black paint after painting it mat white to provide a good texture for stereo matching. This stereo pair was intended to provide a better feel for natural disparity variations and image characteristics than could be obtained with synthetic random dot patterns. Though the bottle image was carefully produced using 35 mm negatives scanned on a high resolution low noise scanning microdensitometer, several important differences with performance on noise-free random dot patterns became apparent. Early versions of Grimson's program could match most of the image, but random errors occurred and there was a marked sensitivity to the small vertical misalignments — on the order of a pixel across the image pair — introduced by the imaging and digitization process.

Research at MIT to improve performance to a level adequate for practical application took three directions from this point. Grimson continued development of his program, investigating ways to reduce the error rate in the matching and making use of repetitive runs of his algorithm at each vertical disparity to deal with the vertical alignment problem. In his present algorithm, he switches from area statistics for pruning unacceptable candidate matches to a figural continuity technique following the ideas of Baker and Binford[13-15] and Mayhew and Frisby.[26] The latter approach requires an extended correspondence along the length of a zero-crossing contour for a candidate match to be accepted. Much of the emphasis of this approach has been directed toward the question of filling in the gaps between actual disparity measurements using surface interpolation techniques.

At about the same time, Kass[27] took a different tack, attempting to explicitly handle the vertical disparity problem by using a more elaborate primitive that could be reliably located vertically as well as horizontally. He obtained encouraging results on the bottle image using zero-crossing primitives augmented by contour curvature attributes as well as the orientation information used by Grimson's early program. He later was able to extend his idea to a more general class of pixel-based primitives from a study of stochastic models of images. For example, first and second partial derivatives of the image in orthogonal directions and at scales of resolution separated by an octave or more are relatively independent measures. For three scales of resolution a 12 vector is obtained for each pixel in the image. Kass showed that pixel-to-pixel matching can yield low false positive and negative error rates over a relatively large two-dimensional search space, even in the presence of moderate noise.

Both the Grimson and Kass techniques in their present forms appear to perform reasonably well on natural images. They both, however, are computationally demanding algorithms requiring on the order of mn searches for matching each primitive, where m and n are the vertical and horizontal disparity ranges, respectively, in pixels. The processing time for both types of algorithms in optimized lisp and microcode on the MIT Lisp machine for 512^2 images with 100 pixel disparity ranges is on the order of an hour.

Marr and Poggio's idea of trading off resolution for range seems to be largely abandoned in both techniques. In addition, though not a serious issue for robotics, both approaches appear to have lost much of their biological plausibility in attempting to overcome technical problems.

2. SIGN REPRESENTATION

The explicit matching of zero-crossing points in binocular stereo may not be the best way to use the information they carry. In the presence of noise the zero-crossing positions are better modeled by probability density functions than as contours at fixed locations. The actual zero-crossing geometry may fluctuate widely within the regions where zero crossings are likely. Thus to be noise tolerant, a matching algorithm must not be sensitive to such variations. In this section an alternate representation, the convolution sign, is proposed as a more explicit representation of the stable position information present in the $\nabla^2 G$ convolution.

2.1. Stability

Noise will cause zero-crossing points to move by an amount proportional to the noise amplitude and inversely proportional to the convolution gradient at the zero crossing.[28, 29] If the spacing between zero crossings is relatively large compared with this amount of movement, the region of constant sign between zeros will be stable over a large range of signal-to-noise ratios. The precision to which the positions of these regions can be measured will vary uniformly with the signal quality.

The stability of the convolution sign degrades much more uniformly than does the coherence of zero-crossing contours. This is because the sign map representation imposes a different perspective, that of regions of likely constant sign in the convolved image. Information is still carried at locations where the sign changes, but the focus of the analysis is shifted to the regions rather than to their boundaries.

The effect of noise on sign stability for a particular $\nabla^2 G$ convolution can be calculated easily if the noise and signal are zero-mean Gaussian random processes with relative amplitudes after convolution of σ_s and σ_n, respectively. The probability of a sign change due to noise at any position in such an image is the probability that the noise value is larger than the signal and of opposite sign. The following expression gives the proportion of points in the convolution that are likely to have their sign changed by the addition of noise:

$$P = \frac{2}{\sqrt{2\pi}\sigma_s} \int_0^\infty \exp\left(\frac{-c^2}{2\sigma_s^2}\right) \frac{1}{\sqrt{2\pi}\sigma_n} \int_c^\infty \exp\left(\frac{-x^2}{2\sigma_n^2}\right) \, dx\,dc$$

$$= \frac{1}{\pi} \int_0^\infty \int_{\frac{\nu\sigma_s}{\sigma_n}}^\infty \exp\left[-\tfrac{1}{2}(u^2+v^2)\right] \, du\,dv$$

$$= \frac{\tan^{-1}\left(\frac{\sigma_n}{\sigma_s}\right)}{\pi}, \qquad (3)$$

where x and c are the noise and signal values, respectively. Note that even when the noise and signal levels are the same ($\sigma_s = \sigma_n$), P is only ¼ (P = ½ when $\sigma_n >> \sigma_s$).

2.2. Autocorrelation of the convolution sign

The $\nabla^2 G$ sign representation is characterized by the autocorrelation function. This also gives us some information about the zero-crossing representation, which is difficult to analyze directly. We show here the autocorrelation of the sign of the $\nabla^2 G$ filtered image, assuming both white noise and a pink noise model for the image.

Let the image I (x, y) be a Gaussian random process with uniform spectrum and let

$$C(x,y) = \nabla^2 G * I(x,y) \qquad (4)$$

where * denotes a two-dimensional convolution. The autocorrelation of C(x, y) when the image I(x, y) is taken to be Gaussian white noise has the form

$$R_c(\tau) = k \left(1 - \frac{4\tau^2}{w^2} + \frac{2\tau^4}{w^4}\right) \exp\left(\frac{-2\tau^2}{w^2}\right) , \qquad (5)$$

where k is a constant. The autocorrelation $R_s(\tau)$ of the sign of Eq. (4), S(x,y) = sgn[C(x, y)], obeys an arcsin law when C is a Gaussian random process[30,31]:

$$R_s(\tau) = \frac{2}{\pi} \sin^{-1}\left[\frac{R_c(\tau)}{R_c(0)}\right] . \qquad (6)$$

A more accurate model of real images[32,33] has a two-dimensional power spectrum proportional to

$$\frac{1}{(f_0^2 + f^2)^{3/2}} , \qquad (7)$$

where f_0 is a constant. We call this a *pink* noise model because it is weighted toward the longer spatial frequencies. The autocorrelation function of Eq. (7) is proportional to

$$R_p(\tau) = \exp(-\alpha|\tau|) . \qquad (8)$$

The autocorrelation of the $\nabla^2 G$ convolution of this noise model can be expressed in terms of Eq. (5) as

$$R_{c'}(\tau) = R_c(\tau) * \exp(-\alpha|\tau|) . \qquad (9)$$

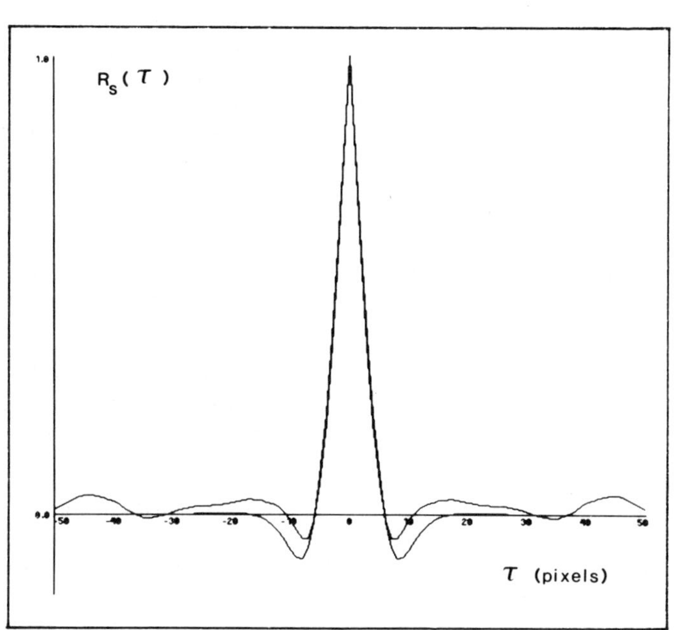

Fig. 3. $R_s(\tau)$ using the pink noise image model of Eq. (8) (with $\alpha = 0.2$ and with w = 8 pixels). The value for α was measured from the right image of Fig. 4(b). This curve is overlaid with an empirical measurement of the autocorrelation of the corresponding sign array from Fig. 5(b).

The principal effect of this convolution is that it broadens the central peak of the function. Figure 3 plots Eq. (6), the autocorrelation function of S(x,y), using $R_{c'}(\tau)$ for the pink noise image model, and compares it with an empirical measurement.

2.3. Significance of the sign representation

We can see from the above properties of the autocorrelation function that the $\nabla^2 G$ convolution sign representation is distinct from the raw intensity image and the full $\nabla^2 G$ convolution in three important aspects:

(1). First, with regard to *resolution,* $R_s(\tau)$ is sharply peaked and so is capable of a high resolution disparity measurement. This is also the case for $R_p(\tau)$, direct correlation on the intensity image, but it is not true for correlation on the full convolution C(x, y) since $R_c(\tau)$ has a Gaussian shape near the origin. The sharpness of R_s is due to the nonlinear sgn function which ties all information in the signal to the zero-crossing locations.

(2). Second, the detection *range* for finding the direction (in disparity) toward the correlation peak from single measurements is determined by the width of the autocorrelation peak. This width is controllable — a function of the convolution operator size (w) — for the full convolution and the sign representations, but not for the raw intensity image.

(3). Finally, regarding *confidence,* $R_s(0) = 1$, independent of local image properties about the point of measurement such as mean intensity or contrast. This is not the case for the autocorrelation functions of either the raw image or the raw $\nabla^2 G$ convolution. This property of R_s makes it possible to assess the significance of a correlation measurement $R_s(\tau_0)$ at an unknown τ_0 directly.

3. NEAR/FAR MODULE

The preceding results are used here to design an efficient module for determining the two-dimensional displacement τ_0 between patches out of the left and right images. This *near/far* module does not do point-by-point search. Instead, a single correlation measurement is made at a test disparity (provided as input), and a determination is made as to whether the correlation peak can be nearby (within w/2). If there is a positive result, several additional correlation measurements are made at neighboring disparities to determine the shape of the correlation function over the test disparity. From this an estimate is made for the disparity at which the correlation peak occurs. The name of the module comes from the work of G. Poggio and Fisher,[34] who described a class of neurons in primate visual cortex sensitive to either near or far disparities.

3.1. Measurement of physical surface parameters

The gradient of the two-dimensional autocorrelation function $R_s(\tau)$ points toward the proper alignment over a range of about w/2, as can be seen in Fig. 3. Thus a small set of correlation measurements is sufficient to determine whether or not a correlation peak is present within that range along with its direction and approximate distance if there is one. If, on the other hand, it is determined that a peak is not present within the detection range of the method, that fact can be indicated by the module, and we will make it the responsibility of the user's program to decide what action to take next.

The principal surface parameter is distance from the cameras, which manifests itself as a translational disparity between corresponding patches from the two images. We can also correlate against parameters other than translational disparity. For example, an elevation gradient on the physical surface viewed can be measured by correlations against compressive or shear distortions. These distortions are introduced between the left and right images by horizontal and vertical elevation gradients.

The autocorrelation surface $R_s(u,v)$ can be approximated by a cone of the form $\phi = 1 - a\sqrt{u^2 + v^2}$ for points (u,v) within w/2 of the origin. In an image coordinate frame where the correlation peak occurs at an unknown disparity (u_p,v_p) and where there is camera noise and geometric distortion, the following gives a better approximation to the shape of the correlation function near (u_p,v_p):

$$\phi = 1 - \sqrt{a(u-u_p)^2 + b(v-v_p)^2 + c} \quad . \tag{10}$$

From Eq. (10) we obtain

$$\psi^2 = (\phi - 1)^2 = a(u-u_p)^2 + b(v-v_p)^2 + c \quad . \tag{11}$$

A one-dimensional slice through this surface along the u axis has the form

$$\psi^2 = a(u-u_p)^2 + bv_p^2 + c \quad . \tag{12}$$

If we measure this function at three points along this curve, we can solve for u_p. For example, using the points (−1,0), (0,0), and (1,0), we get the equations

$$\psi_{-1,0}^2 = a(-1-u_p)^2 + bv_p^2 + c \quad , \tag{13}$$

$$\psi_{0,0}^2 = au_p^2 + bv_p^2 + c \quad , \tag{14}$$

$$\psi_{1,0}^2 = a(1-u_p)^2 + bv_p^2 + c \quad . \tag{15}$$

This yields

$$u_p = \frac{\psi_{-1,0}^2 - \psi_{1,0}^2}{2\psi_{-1,0}^2 - 4\psi_{0,0}^2 + 2\psi_{1,0}^2} \quad , \tag{16}$$

and similarly, by measuring the correlation at two additional points we get

$$v_p = \frac{\psi_{0,-1}^2 - \psi_{0,1}^2}{2\psi_{0,-1}^2 - 4\psi_{0,0}^2 + 2\psi_{0,1}^2} \quad . \tag{17}$$

3.2. Measurement of the autocorrelation

The standard deviation σ of an autocorrelation estimate is inversely proportional to $\sqrt{A}$, where A is the area over which the measurement is made. If measurements are made uniformly in a square patch of the image, σ will have a 1/d dependence, where d is the diameter of the patch.

There is also an important dependence on the size of the $\nabla^2 G$ convolution operator. The width of the central peak of the autocorrelation function is approximately w, the diameter of the $\nabla^2 G$ operator's positive center. Increasing the size of this operator increases the effective detection range of the measurement. However, it also reduces the independence of measurements at neighboring image points. Thus a larger patch diameter d is required when w is increased to maintain the same level of confidence in the correlation measurement.

If we write the estimate for the autocorrelation $R_s(\tau)$ measured on an L $\times$ M rectangle of S(x,y) as

$$V(\tau) = \frac{1}{LM} \int_0^M \int_0^L S(x,y)S(x+\tau,y)dxdy \tag{18}$$

and assume that τ is very large compared with the size of L and M, so that the patches compared are uncorrelated, the expected variance of the measurement V becomes

$$\sigma^2 = \frac{4}{LM} \int_0^M \int_0^L \left(1 - \frac{v}{M}\right)\left(1 - \frac{u}{L}\right) R_s(u,v)^2 dudv \quad . \tag{19}$$

Using the pink noise parameters from Fig. 3 and with Eq. (6) for R_s in Eq. (19), we obtain the approximation

$$\sigma \approx \frac{0.5w}{d} \tag{20}$$

for the case L = M = d. Thus a patch diameter of 8w will give a standard deviation of about 0.06 for the autocorrelation measurement with Eq. (18).

Another important consideration when estimating the correlation between left and right image patches is the constancy of the disparity over the measurement patch. Both disparity discontinuities and gradients due to surface elevation variations reduce the height of the correlation peak by an amount proportional to the size of the patch and thus limit the useful patch size.

3.3. Example: surface velocity sensor

Sequential images from a single camera can be compared using the above technique to measure uniform surface translation during the interval between exposures. The above calculation was implemented using a Hitachi CCD area camera and a memory-mapped frame grabber designed by N. Larson. The convolutions were carried out in hardware,[24] and correlations and arithmetic calculations were done on an MIT Lisp machine. The five correlation measurements required for a single disparity measurement using Eqs. (16) and (17) can be accomplished in 32 ms, including the time for convolution, when the measurements are made on a 32^2 pixel support with w = 4 for the convolution operator. This arrangement handles a maximum interframe displacement of 2 pixels reliably on textured surfaces, which corresponds to an acceleration of $\Delta x/\Delta t^2 \approx 2000$ pixels/s^2, and is capable of measuring small displacements to a resolution of about 0.1 pixel.

This example illustrates the behavior of the basic sensing module. We now apply it to binocular stereo simply by shifting to a two-camera system and designing an appropriate control algorithm for using the near/far module.

4. PRISM SYSTEM

We will now use the near/far module to produce stereo measurements tailored to the requirements of specific tasks in robotics. The first application considered has been the problem of rapidly determining an elevation map over the visual field sufficient for obstacle avoidance tasks. The prime objectives for this are speed and reliability. Spatial and depth resolution requirements, on the other hand, are less stringent than would be the case for tasks such as shape description or part position measurement. In addition to developing control strategies for operating the near/far module, this section discusses techniques for ensuring that adequate surface texture will be present on the imaged surfaces and for computing surface elevation from image disparity.

4.1. Design task

Our design goal is a system that produces a coarse surface elevation map over the camera field with a 36 $\times$ 26 tessellation. Its height range should be similar to the field diameter, and height should be resolved to about 200 levels over that range. Thus we model the actual surface using square prisms with varying heights. We seek reliable measurements and so must pay attention to the height of the correlation at each peak found by the near/far module. The nature of the correlation measurement causes this method to be blind to surface details smaller than the patch size on which measurements are made. It will also fail to find a surface — though it will not make a false measurement — when surface orientations or heights are outside the matcher's design limits. To minimize errors in such situations, we further require the algorithm to abstain from reporting a prism elevation unless its measurement is solid. A user's program can either avoid suspect regions or employ alternate methods if they are available for rechecking those surface patches. This allows greater design simplicity without jeopardizing reliability.

4.2. Unstructured light

A potential handicap of binocular stereo, as compared with structured light approaches for robotics vision, is its dependence on surface markings for making range measurements. Industrial parts and surfaces are often without dense surface textures that can be registered. Light stripe or structured light techniques have the opposite problem. They suffer ambiguity problems when dealing with objects that have high contrast surface markings. Repeated surface markings can also cause serious problems for binocular stereo matching since false matches appear as good as the correct ones locally. Both situations — clean surfaces and regular patterns — are especially common in man-made environments.

These problems can be dealt with easily in most robotics tasks by illuminating the workspace with a suitable random texture pattern with a projector situated near the cameras, as shown in Fig. 1. Unlike structured light techniques, our matching system begins with no *a priori* knowledge of the surface markings it is to use. It does not matter whether the markings are natural or artificially produced, and so the projected pattern can mix with any texture markings already present on the imaged surfaces with no adverse consequences. A binary random dot pattern was generated with a 50% dot density on a computer, and a 35mm slide was produced of the pattern from a high resolution display. A standard slide projector was then used to project this pattern onto the workspace. This produced a marked improvement in the signal-to-noise performance of the system, even for already-textured surfaces.

Figure 4 shows an example of this unstructured light technique in use. Figure 5 shows the convolution sign representations obtained from Fig. 4(b) with w = 16, 8, and 4 pixel $\nabla^2 G$ convolution operators. White and black indicate locations where the convolution was positive or negative, respectively.

4.3. Control strategies

A simple way to use the near/far module to produce a disparity map for a pair of stereo images would be to apply it iteratively in a triple-nested loop indexed over image position and disparity. The measurement patch size d must be chosen to be sufficiently small to give adequate spatial resolution. The operator size (w) must be small enough relative to d to give a good correlation estimate. These choices determine the number of steps required in the iteration. For a 32^2 pixel measurement patch, a 10 $\times$ 10 operator (w = 4) will provide a low correlation variance and has a reliable detection range

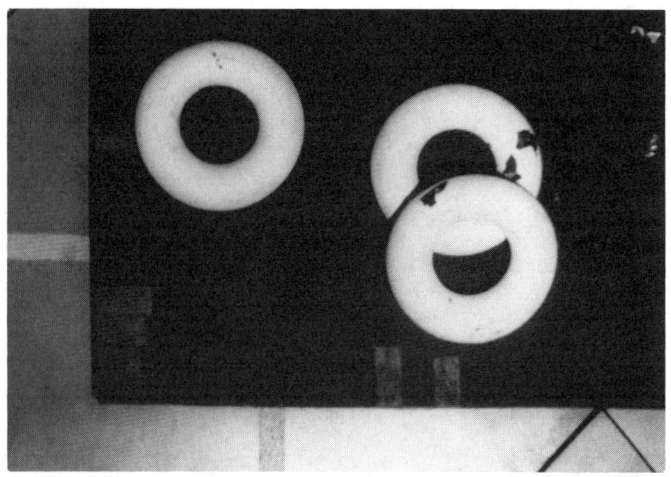
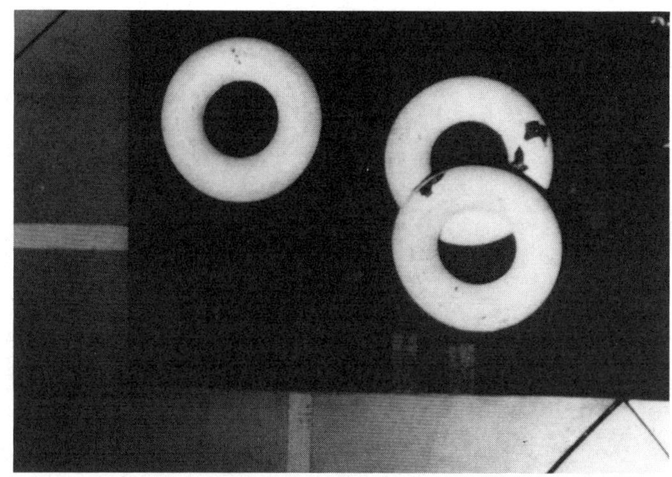

Fig. 4(a). Stereo pair of three plastic doughnuts approximately 10 cm in diameter on a dark board 2.5 cm thick. Aside from some tape marks and paint chips on the doughnuts, there is little texture present on the surfaces suitable for stereo matching. The cameras are 150 cm above the work table and separated by 40 cm. Low cost vidicons are used with 25 mm lenses. A single 1/30 s TV frame is grabbed into a 576×454 array for each image and redisplayed here.

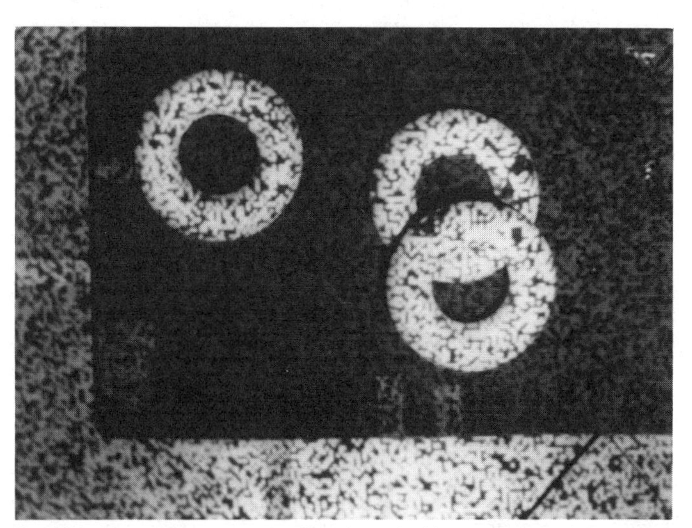
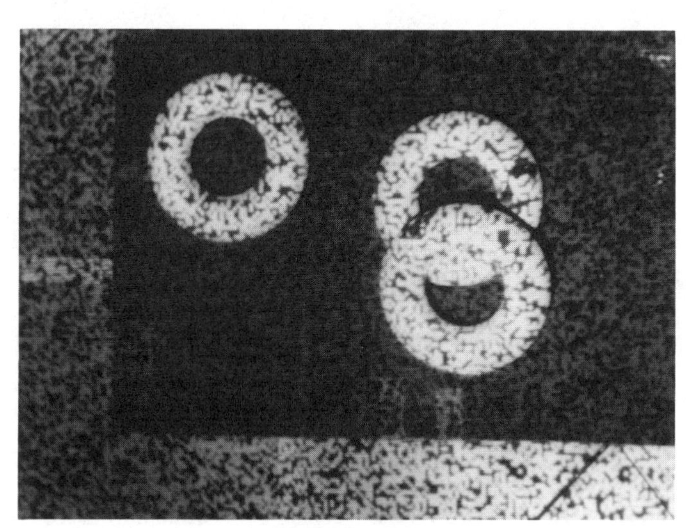

Fig. 4(b). The same stereo scene illuminated with a random dot texture pattern from a projector located near the TV cameras. This *unstructured light* technique provides a dense, high contrast surface texture to drive stereo matching. This enables the matcher to operate with little dependence on the sample material or its surface markings.

of ± 2 pixels. If the camera geometry is arranged so that the elevation range requirement corresponds to a disparity range of 200 pixels, then as many as 50 correlation checks would be required at each image patch to determine the disparity there. If each near/far check takes 32 ms, a 36×26 matrix of patches covering the image with an interpatch spacing of 16 pixels would take 1500 s to compute, in the worst case.

This search time can be reduced substantially by first doing a coarse resolution pass with a larger convolution operator and proportionally larger measurement patches, as indicated in Fig. 2. With a $w = 16$ operator on 128^2 patches, the detection range of the near/far module is ± 8 pixels, so at most only 13 checks are required at each patch location. The near/far module in its present implementation takes about the same amount of time for a check at this scale because the brunt of the computation time is in the convolution, which is fixed at 1 μs per output point, independent of the mask size. Thus in the worst case a 9×7 matrix covering the whole field can be computed in about 26 s. In practice, neighboring patches are often at similar elevations, and taking advantage of this reduces the time required considerably — down to 4 s on the average.

Once elevation measurements are obtained at the coarse scale, a second pass is made with $w = 8$ pixel convolutions and 64^2 patches, followed by a third pass with $w = 4$ pixels and 32^2 patches, as shown in Fig. 6. In most cases, a single near/far check is required at each patch location. The algorithm presently takes 30 to 40 s to produce a 36×26 matrix of disparity measurements using this three scale coarse-to-fine control.

4.4. Calibration

Disparities measured between the two cameras must be transformed to physical elevation values. This transformation follows the approximate relation

$$\frac{\Delta\theta}{\theta} \approx -\frac{\Delta D}{D} \quad , \tag{21}$$

where θ is the vergence angle of the cameras and D is the distance to the surface viewed (see Fig. 1). From this we can derive the relation

$$\frac{\Delta S}{\Delta D} \approx \frac{\ell}{D} \quad , \tag{22}$$

where ΔS is the spatial pixel resolution in mm, ΔD is the depth resolution also in mm, and ℓ is the camera-to-camera separation. The desired absolute transformation to a physical height map,

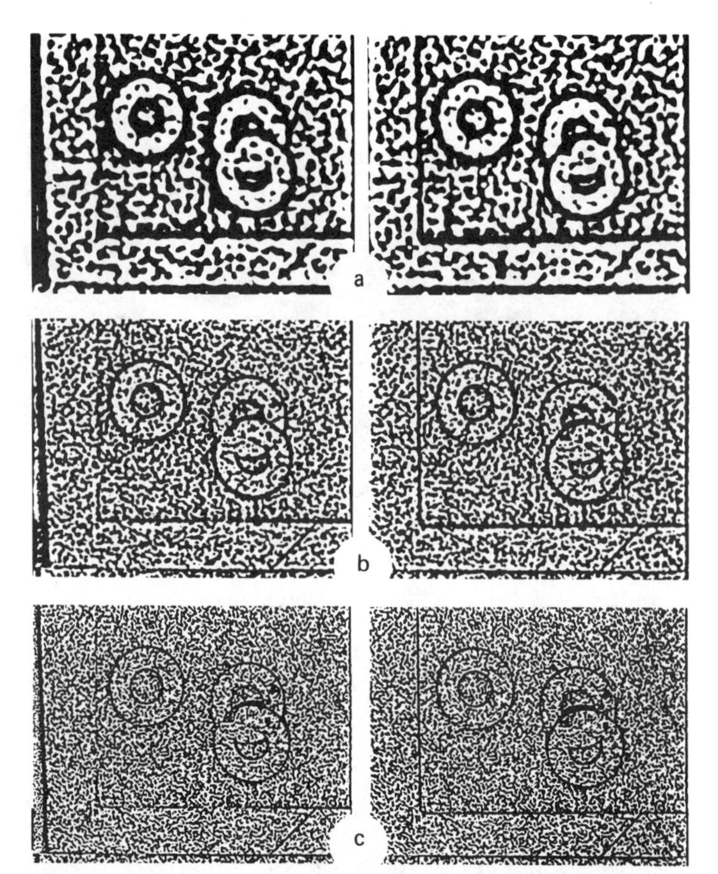

Fig. 5. The images of Fig. 4(b) convolved with a difference of Gaussian operator at three scales. The sign of the convolution is shown here using white and black to indicate positive and negative regions, respectively. The first pair (a) is with a 32^2 operator ($\omega = 16$ pixels); (b) is with a 20^2 operator ($\omega = 8$ pixels); (c) is produced with a 10^2 operator ($\omega = 4$ pixels). The convolutions are carried out digitally in a pipelined convolver designed by N. Larson and the author. This set of six convolutions is accomplished in 1.5 s.

however, is complicated by the geometric distortion introduced by the cameras and the imaging geometry. In particular, a relative magnification between the left and right images is induced by perspective effects whenever there is a difference in distance from the target to the two cameras.

The PRISM system makes a restricted set of measurements, so a simple look-up table can be used for the disparity-to-height transformation. The present calibration procedure takes two disparity maps obtained from running the system on a flat surface at 0 mm elevation and at 300 mm. The matcher produces both vertical and horizontal disparity measurements over both of these planes. A linear disparity-to-height mapping is computed from these calibration planes for each patch position of the 36×26 disparity array. A similar linear function is produced for estimating vertical disparity as a function of horizontal disparity at each patch position. The resulting system tolerates large camera and imaging distortions so long as these distortions are stable. At present, position measurements are accurate to 10 mm in all three dimensions over the entire operating volume. Figure 7 illustrates the elevation values obtained by this method from the disparity measurements displayed in Fig. 6(d). Elevation measurements are repeatable to 2 mm, and absolute calibration to that precision should be possible with a more elaborate transformation table.

4.5. Test applications

The PRISM stereo matcher has been used as a vision input to two robotics manipulation systems with different application requirements.

4.5.1. Brooks's path-planning system

In this experiment, the PRISM system was used in a two-channel configuration to provide an elevation map of a PUMA workspace to a collision avoidance system developed by Brooks.[35] The combined system was presented with the task of moving a part from one predefined workspace location to another. Large obstacles were to be placed at random in the workspace prior to the movement, and the system was responsible for measuring workspace elevation with PRISM and planning a trajectory that would safely accomplish a pick-and-place task free of collisions with any part of the manipulator or its payload. Brooks's program constructed a polyhedral model of the PRISM elevation array and mapped that into the manipulator's configuration space, where a search for a collision-free path was made.

The vision system and the path-planning system were developed separately, and they were operated in parallel on separate Lisp machines with communication over a local area network. Interfacing PRISM with Brooks's path planner required little work, and the first demonstration was successfully completed the first day the two systems were tested together.

4.5.2. Ikeuchi's bin-of-parts system

A second collaboration incorporated the PRISM system into Ikeuchi's bin-picking system.[36,37] He uses photometric stereo[38-43] to measure local surface orientation at each pixel of the image. The technique uses three images all taken with the same camera, but with the light source in a different position for each. The local surface orientation information is first used to segment the image into regions of continuous surface. A histogram of surface orientations — the *extended Gaussian image* (EGI) — can then be made for each of these regions and used to recognize parts out of a catalog of known shapes and determine their orientation in space. The photometric stereo technique, however, is not capable of measuring range, and this information was provided by the PRISM system.

There were three phases in the bin-picking study. The first was to pick up plastic doughnuts off the top of the pile and stack them on a post. Information from the PRISM system was used in this case to determine the height of the selected part so that the hand could be made to approach that part along a trajectory different from the line of sight. The highest part of the doughnut was selected for the grasp point with no other precautions taken to avoid collisions with neighboring doughnuts. Figures 4 through 7 illustrate the processing from input, convolution, matching, and final elevation for a typical scene used in this phase. In the second phase, finger clearance was measured using the PRISM data to select the best grasp point around the doughnut circumference. This was done by projecting the finger *footprints* along the direction of the approach trajectory until the first surface element in the PRISM elevation map was encountered. In the third phase, the clearance test was further elaborated to consider three different approach angles at each point around the doughnut circumference.

In both experiments, the PRISM system was used independently by Brooks and Ikeuchi as a tool in their respective systems. Except for a single change to the correlation threshold used to eliminate uncertain measurements, no significant modifications were made to the PRISM algorithm during the two month period when these demonstration systems were in operation. Over that period several hundred runs of the PRISM program were made. Failures to operate properly occurred infrequently and were due almost always to a failure of a mechanical relay to switch the frame grabber between left and right images at the proper time.

A further test was done running the PRISM program repeatedly 500 times (four hours run time) without changing the scene, though room lighting and building motion with trains passing outside could not be controlled, collecting a histogram of horizontal disparity measurements at each patch position in a scene like that shown in Fig. 4. The standard deviation about any clear surface position was less than a pixel in disparity about the central mean, and no matches occurred to disparities more than a few pixels away from a disparity

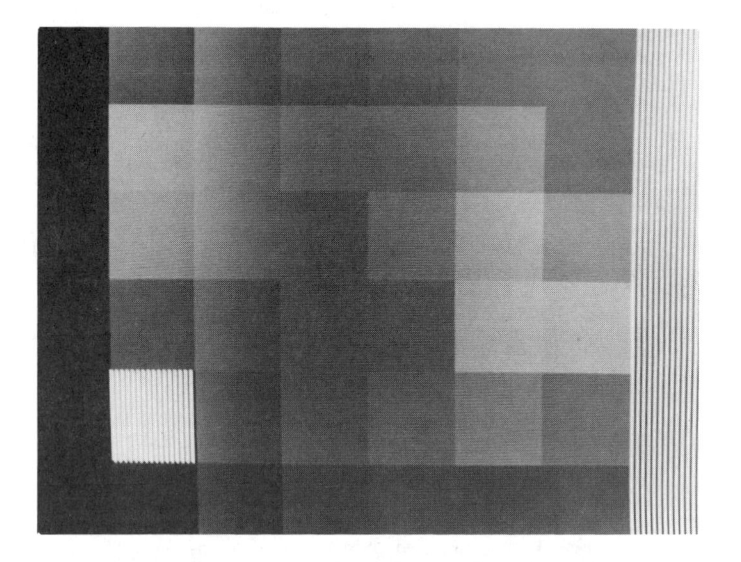
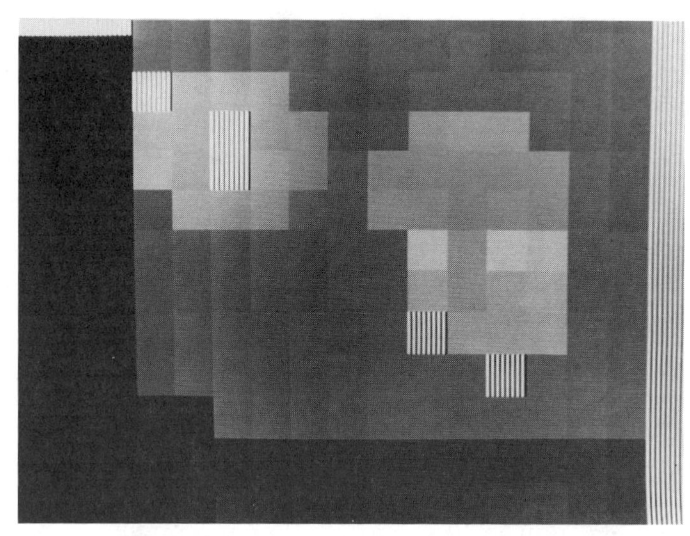
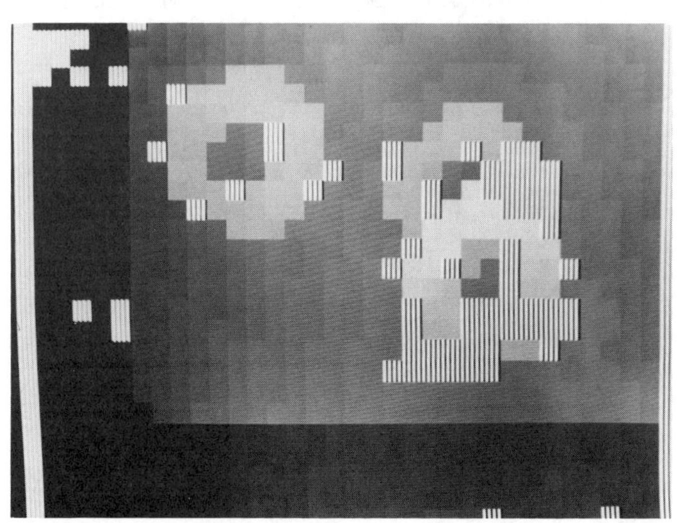
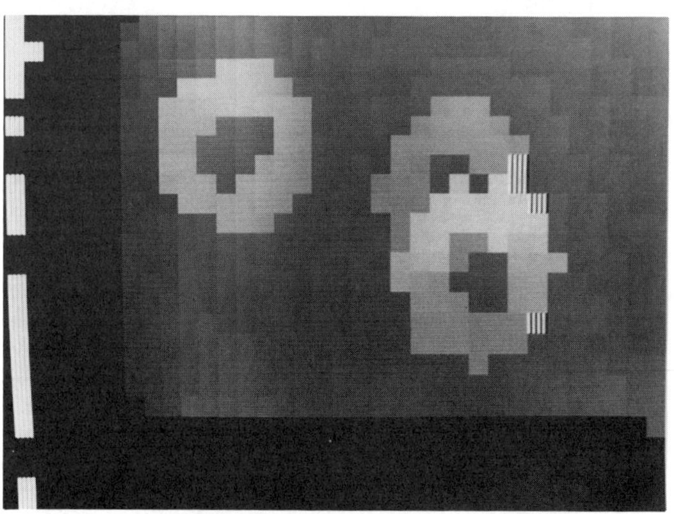

Fig. 6. Disparity measurements produced by the PRISM matcher from the data in Fig. 5. Shading indicates the disparity magnitude — lighter means near. The vertical stripe texture is used to indicate patch positions at which no satisfactory correlation was obtained. The measurement patches overlap by 50% in both dimensions. (a) is obtained from Fig. 5(a) using 128^2 patches, (b) from Fig. 5(b) using 64^2 patches, and (c) from Fig. 5(c) using 32^2 patches. In (d) the possibility that an elevation discontinuity passes through the unmatched patches in (c) is checked. In such cases, the side of the discontinuity having greater area in the patch is used.

actually present in the measurement patch. Bimodal disparity distributions occurred in patches straddling surface height discontinuities because two correlation peaks occur at such locations.

5. DISCUSSION

We have designed a simple and robust near/far stereo module following Marr and Poggio's idea of trading resolution for range. By restricting the analysis to a specific scale in the stereo images, disparity displacements comparable to that scale can be measured without search, but details much finer than that scale are lost. Operating the scale-specific module at several scales in a coarse-to-fine progression allows performance to be tailored to the range and resolution requirements of specific applications. Two results of this work are important: first, that a minimal mechanism[3] like the one proposed here may be capable of explaining much more complex aspects of stereo performance, and second, that the sign of the $\nabla^2 G$ convolution can be matched more reliably in practice than is the case for the explicit matching of zero crossings.

As illustrated in the previous section, surface topography information from a stereo vision system can be used under a broad range of operating conditions to successfully guide robotic manipulation systems in part position measurement and obstacle detection and avoidance tasks. Single range measurements can already be made in a fraction of a second, and it appears that computation of an elevation map like that shown in Fig. 7 can be brought into the same time range with relatively simple special purpose hardware for the near/far module.

We are now working on the design of a surface proximity detector based on the near/far module. It will be attached to a manipulator hand to measure hand-to-part position relations with 10 to 30 near/far measurements per second. The device will be used to measure surface range, orientation, and flatness in a small field in front of the sensor. This measurement in conjunction with an edge position measurement may be capable of obtaining very high precision position and orientation. The problem of detecting depth discontinuities to support this task is presently under study using techniques analogous to those used by the near/far module.

6. ACKNOWLEDGMENTS

This report describes research done at the Artificial Intelligence Laboratory of the Massachusetts Institute of Technology. Support for the laboratory's artificial intelligence research is provided in part by the Advanced Research Projects Agency of the Department of

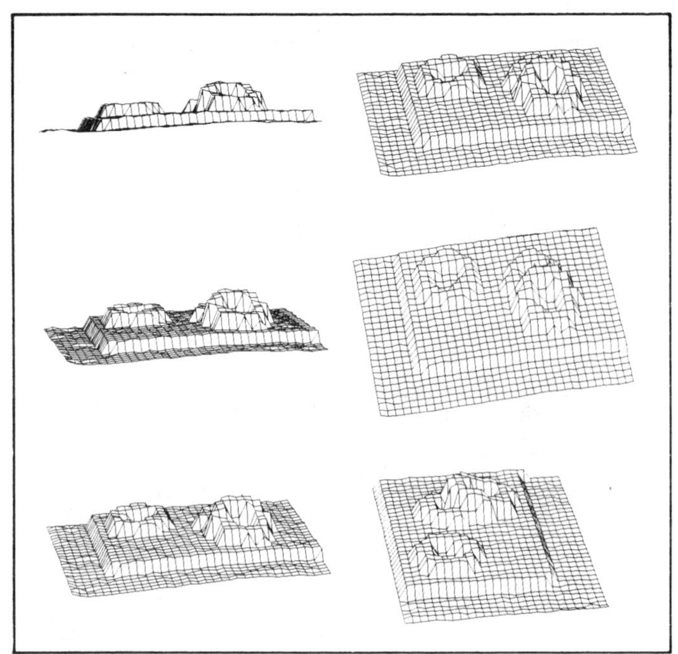

Fig. 7. Perspective displays of data from Fig. 6(d) after conversion from disparity to physical height. Height variations along the edges of flat surfaces give a rough measure of the repeatability of the measurements. The total time required for the process, from taking the pictures to having the elevation matrix shown here, was 30 s.

Defense under Office of Naval Research contract N00014-80-C-0505, in part by the System Development Foundation, and in part by National Science Foundation Grant 79-23110MCS.

7. REFERENCES

1. D. Marr and T. Poggio, Proc. R. Soc. Lond. B 204, 301 (1979).
2. H. K. Nishihara and T. Poggio, Proc. Int. Symp. Rob. Res., Bretton Woods, NH (1983), to be published by MIT Press.
3. D. Marr and H. K. Nishihara, Proc. R. Soc. Lond. B 200, 269 (1978).
4. H. K. Nishihara, Artificial Intelligence 17, 265 (1981).
5. H. K. Nishihara, in *Physical and Biological Processing of Images*, O. J. Braddick and A. C. Sleigh, eds., pp. 335-348, Springer-Verlag, New York (1983).
6. R. E. Kelly, P. R. H. McConnell, and S. J. Mildenberger, Photogramm. Eng. Rem. Sens. 43, 1407 (1977).
7. H. P. Moravec, Ph.D. Dissertation, Stanford University, Stanford, Calif., Stanford Artificial Intelligence Laboratory Memo 340 (1980).
8. D. B. Gennery, Ph.D. Dissertation, Stanford University, Stanford, Calif., Artificial Intelligence Laboratory Memo 339 (1980).
9. R. Y. Tsai, IEEE Trans. Pattern Analysis and Machine Intelligence PAMI-5(2), 159 (1983).
10. B. K. P. Horn, Photogramm. Eng. Rem. Sens. 49, 535 (1983).
11. R. D. Arnold, Proc. ARPA Image Understanding Workshop, L. Baumann, ed., Science Applications, Inc., 65 (1978).
12. R. D. Arnold and T. O. Binford, in *Image Processing for Missile Guidance*, T. F. Wiener, ed., Proc. SPIE 238, 281 (1980).
13. H. H. Baker, Proc. ARPA Image Understanding Workshop, L. Baumann, ed., Science Applications, Inc., 168 (1980).
14. H. H. Baker, Ph.D. Dissertation, University of Illinois, Urbana (1981).
15. H. H. Baker and T. O. Binford, Proc. 7th Int. Joint Conf. on AI, Vancouver, British Columbia, 631 (1981).
16. Y. Ohta and T. Kanade, Carnegie-Mellon University Dept. of Computer Science Memo CMU-CS-83-162 (1983).
17. B. Julesz, *Foundations of Cyclopean Perception*, University of Chicago Press, Chicago (1971).
18. G. F. Poggio and T. Poggio, Ann. Rev. Neurosci. 7, 379 (1984).
19. D. Marr and T. Poggio, Science 194, 283 (1976).
20. D. Marr and E. Hildreth, Proc. R. Soc. Lond. B 207, 187 (1980).
21. R. W. Rodieck and J. Stone, J. Neurophysiol. 28, 833 (1965).
22. F. Ratliff, *Mach Bands: Quantitative Studies on Neural Networks in the Retina*, Holden-Day, San Francisco (1965).
23. C. Enroth-Cugell and J. G. Robson, J. Physiol. Lond. 187, 517 (1966).
24. H. K. Nishihara and N. G. Larson, Proc. ARPA Image Understanding Workshop, L. Baumann, ed., Science Applications, Inc., 114 (1981).
25. W. E. L. Grimson, *From Images to Surfaces: A Computational Study of the Human Early Visual System*, MIT Press, Cambridge, Mass. (1981).
26. J. E. W. Mayhew and J. P. Frisby, Artif. Intell. 17, 349 (1981).
27. M. Kass, Proc. ARPA Image Understanding Workshop, L. Baumann, ed., Science Applications, Inc. (1983).
28. H. K. Nishihara, in *Robotics and Industrial Inspection*, D. P. Casasent, ed., Proc. SPIE 360, 76 (1983).
29. H. K. Nishihara and T. Poggio, Nature 300, 347 (1982).
30. J. L. Lawson and G. E. Uhlenbeck, *Threshold Signals*, McGraw-Hill, New York (1950).
31. A. Papoulis, *Probability, Random Variables, and Stochastic Processes*, McGraw-Hill, New York (1965).
32. A. Netravali and J. Limb, Proc. IEEE 68, 3 (1980).
33. M. Kass, IEEE Int. Conf. on Systems, Man, and Cybernetics, Bombay and New Delhi, India (1983).
34. G. F. Poggio and B. Fisher, J. Neurophysiol. 40, 1392 (1977).
35. R. A. Brooks, Proc. Int. Symp. Rob. Res., BrettonWoods, NH (1983), to be published by MIT Press.
36. K. Ikeuchi, B. K. P. Horn, S. Nagata, T. Callahan, and O. Feingold, Proc. Int. Symp. Rob. Res., Bretton Woods, NH, to be published by MIT Press. Also available as Massachusetts Institute of Technology Artificial Intelligence Laboratory Memo 718 (1983).
37. K. Ikeuchi, H. K. Nishihara, B. K. P. Horn, P. Sobalvarro, and S. Nagata, Massachusetts Institute of Technology Artificial Intelligence Laboratory Memo 742 (1984).
38. B. K. P. Horn, R. J. Woodham, and W. M. Silver, Massachusetts Institute of Technology Artificial Intelligence Laboratory Memo 490 (1978).
39. R. J. Woodham, in *Image Understanding Systems and Industrial Applications*, R. Nevatia, ed., Proc. SPIE 155, 136 (1978).
40. W. A. Silver, MS Thesis, Dept. Electrical Engineering and Computer Science, Massachusetts Institute of Technology, Cambridge (1980).
41. R. J. Woodham, Opt. Eng. 19, 139 (1980).
42. E. N. Coleman and R. Jain, Proc. 7th Int. Joint Conf. on AI, Vancouver, B.C. 652 (1981).
43. K. Ikeuchi, IEEE Trans. Pattern Analysis and Machine Intelligence PAMI-2(6), 661 (1981).

Detection of Binocular Disparities

K. Prazdny

Artificial Intelligence Laboratory, Schlumberger Palo Alto Research Center, 3340 Hillview Avenue, Palo Alto, CA 94304, USA

Abstract. A stereo correspondence algorithm designed to perform matching on figurally similar images (arising in normal human binocular vision) is described. It is based on the observation that the operational principles underlying biological stereo disparity detection seem to be extremely general and few in number instead of an extended set of specific "constraints". We identify one general characteristic of objects in the three dimensional world and use it to formulate a simple noniterative, parallel and local algorithm that successfully detects disparities generated by opaque as well as transparent surfaces.

Introduction

Binocular disparity is the difference between the positions of an object on the projection surfaces of the two eyes. It has been known since the invention of the stereoscope by Wheatstone (1838) that these disparities form the basis of binocular depth perception. Julesz's (1960) invention of random-dot stereogram has demonstrated that perception of stereoscopic depth is subserved by a "primitive" mechanism that does not depend on the monocular recognition and identification of shapes, objects, or other high level cues. The information about three-dimensional surfaces in these stimuli is contained only in local correlations between elements in the two images because surfaces and shapes in such stereograms appear only after stereopsis has been achieved: each half of the stereo pair exhibits only (more or less isotropic) random texture. Such displays represent probably the ultimate form of camouflage.

Random-dot stereograms exemplify more than anything else the massive ambiguity problem the visual system may face: each picture element in one eye's view can potentially correspond to many elements in the other eye's view. In fact, the correspondence problem looks even more formidable because the human visual system can interpret random-dot stereograms portraying transparent surfaces (Fig. 1). Transparent surfaces are a disparity domain analogue of co-existing multiple organizations in other visual domains (see e.g. Prazdny, 1984).

As with many other capabilities, it is now widely recognized that stereopsis is not only a research area in psychology, physiology and psychophysics but also in information processing. In this report, the problem of stereopsis is approached primarily as a complex information processing task. Although our discussion does not directly bear on the question of how a physiological mechanism may detect binocular disparities, it addresses issues which may be, we believe, common to all such mechanisms, biological or artificial.

To detect the binocular disparity the visual system has to determine which location in one image corresponds to a given location in the other image. Considered as a computational problem, this involves the answer to two questions: what to match (i.e. what are the matching primitives) and how to match (i.e. how to discover the mapping from one image to the other). Correspondence between two images can be established by matching specific features such as blobs or edges, or by matching small regions by direct correlation of image intensities without identifying features. The basic problem with direct intensity correlation has to do with the fact that things can look significantly different from different points of view. Attempts to match directly the intensity values have had limited success and are not considered to be a biologically viable hypothesis. For example, Julesz (1971) has shown that a stereo pair consisting of images with different contrast (but the same contrast polarity!) can be easily fused. Surface markings and discontinuities are more invariant with respect to the change in viewing direction. It is thus not surprizing that most computationally oriented theories (Sperling, 1970;

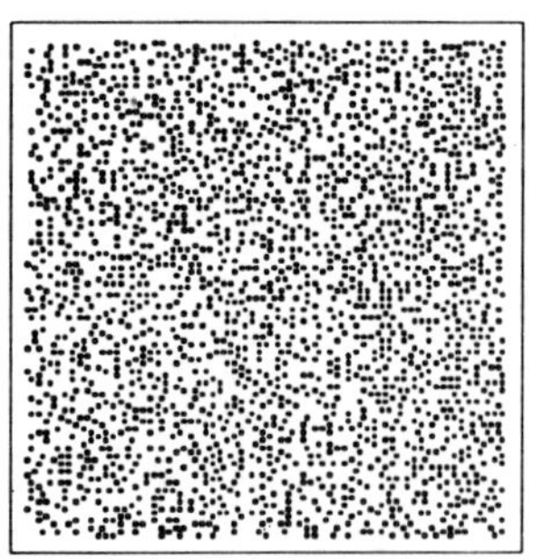

Fig. 1. A random dot stereogram portraying transparent surfaces. Stereopsis is easily obtained

Dev, 1975; Marr and Poggio, 1976, 1979; Baker and Binford, 1981; Mayhew and Frisby, 1980) use edges or some other more primitive edge precursors as the main matching primitives. Recently, some success has been reported with a correlational technique based on a set of functions of intensity values (i.e. a vector) rather than directly on the (scalar) intensity values themselves (Kass, 1984).

The question of what matching primitives are used in human stereopsis is still largely open although it is known from neurophysiological observations that binocular cortical cells respond to oriented edges. In this report we assume that the descriptors have a punctiform nature, i.e. that they can be more or less accurately localized on the projection surface. One advantage of such descriptions is that they can be thought of as carrying only positional information. This property enables one to decouple the question of matching primitives from the problem of using them to detect binocular disparties and allows us to study the matching process by itself. The analysis and the algorithm developed below does not specific the type of matching primitives to be used.

An Analysis of the Problem

A computational approach to the problem begins with an analysis of the domain in which the stereoscopic mechanism operates: the real physical world (Gibson, 1950; Marr, 1982). Physical objects exhibit one important general property relevant to our purpose: they are "cohesive", i.e. surfaces of objects are smooth relative to the viewing distance (Marr and Poggio, 1976). From the cohesivity of matter follows directly the *coherence* principle: the world is not made of points chaotically varying in depth but of (not necessarily opaque) objects each occupying a well defined 3D volume. The principle is different from the continuity constraints[1] used implicitly (Sperling, 1970; Dev, 1975; Nelson, 1975) or explicitly (Marr and Poggio, 1976, 1979; Baker and Binford, 1981; Julesz, 1971; Mayhew, 1983) in most previous theories. In fact, the cohesivity of matter cannot be used to support the Marr and Poggio (1976, 1979) type of continuity rule for processing disparities arising from transparent surfaces. The continuity rule stipulates that nearby image points are projections of nearby 3D points. This implies smoothness (in the usual sense) of the resultant disparity field. Such continuity holds, however, only for opaque surfaces and non-boundary regions. The *coherence* principle is much more general. It recognizes that for transparent surfaces where proximal points on the projection surface may arise from widely separated three-dimensional objects, image proximity does not necessarily imply disparity continuity. While the disparity field may be locally discontinuous, it must (if it is generated by an actual three-dimensional scene obeying the coherence principle) be a superposition of locally smooth disparity fields corresponding to individual three-dimensional surfaces. These smooth variations usually are apparent only when larger image regions are taken into consideration. Locally, the field may be discontinuous due to disparities originating at different depth.[2] In short, *a discontinuous disparity field may be a superposition of a number of several interlaced continuous disparity fields each corresponding to a piecewise smooth surface*. The coherence principle captures this possible state of affairs and includes continuous disparity variations associated with opaque surfaces as a special case.

The disparity detection problem can now be formulated as a decision problem. The visual system has to choose, from a set of possible disparities at a retinal position, the disparity which explains the best (in some sense) the distribution of possible correspondences generated by points (features) in a given neighbourhood. In our current work, the set of possible disparities is further restricted to matches within an area of fixed radius (the area of local stereopsis) along the epipolar line.[3] The *coherence principle* requires that neighbouring disparities of elements *corresponding to the*

1 The continuity constraint states that disparity varies smoothly almost everywhere and that only a small fraction of the area of an image is composed of boundaries that are discontinuous in depth (Marr and Poggio, 1976)

2 While truly transparent surfaces may occur infrequently in the nature, semi-transparency (fences, bushes, grass viewed by small creatures against the horizon, etc.) is a general phenomenon

3 The epipolar lines are straight lines (on a planar projection surface) that match point by point. When the observer fixates a distant object these lines are horizontal and parallel, i.e. matching can be performed raster by raster. In general, their orientation depends on the mutual direction of gaze of the two eyes. It appears possible to compute the directions of the gaze (and thus the orientation of the epipolar lines) without any extraretinal information (Prazdny, 1983). If the epipolar lines are known, the matching problem is essentially one-dimensional

same 3D object be similar, i.e. neighbouring image points corresponding to the same object should have nearly the same disparities. This suggests that the principal disambiguation mechanism should be facilitation due to disparity similarity. Dissimilar disparities should not inhibit each other because in a transparency situation, a disparity may be surrounded by a set of features corresponding to other surfaces. *Two disparities are either similar, in which case they facilitate each other because they possibly contain information about the same surface, or dissimilar in which case they are informationally orthogonal, and should not interact at all because they potentially carry information about different surfaces.*

Some Computational Considerations

In order to translate the coherence principle into a working algorithm one has to explicitly specify what is meant by disparity similarity, i.e. one has to develop a measure quantifying the similarity between neighbouring disparities. As a first approximation, we want a simple scalar function capturing the following three requirements:

1. The disparity similarity function should be inversely proportional to the difference of disparities of interacting points.

2. More distant points should exert less influence while nearby matches should have more disambiguating power.

3. The more distant the two interacting points are the less seriously should their disparity difference be considered because of the inherent uncertainty: steeply-sloped suraces will generate large disparity differences which should nevertheless contribute to disambiguation. For large separations one should probably expect a (nearly) flat support function, i.e. all disparity differences should have the same influence. For small feature distances exactly the opposite should be the case: the probability of a large disparity difference is zero in the limit, i.e. the distribution function should peak at the centre.

One similarity function capturing all of these requirements is the familiar Gaussian distribution function

$$s(i,j)=\frac{1}{c|i-j|\sqrt{2\pi}}e^{-\frac{|d_j-d_i|^2}{2c^2|i-j|^2}}. \quad (1)$$

Here, $s(i,j)$ expresses the amount of support disparity d_i at a retinal point i (a vector) receives from disparity d_j at another point j, and $|i-j|$ is the distance between the two retinal locations (scaling constant c is explained below).

There are several important points to be made about this similarity function.

a) The disparity difference in the exponent of the gaussian weighting function is scaled by the spatial separation of the two interacting points. This scaling means that the spread of the gaussian (controlled by $c|i-j|$) will be greater for widely separated points. Thus, distant point with a large disparity will be contributing some support whereas nearby points with the same large disparity will give little or no support.

b) Because of the way $c|i-j|$ controls the whole shape of the similarity function, when spatial separation is small the size of the weightning from nearby matches is greater.

c) The square root of the non-constant term of the exponent of the gaussian weighting function, $\frac{|d_j-d_i|}{|i-j|}$, is related to the disparity gradient (Burt and Julesz, 1980). In fact, this terms is the disparity gradient measured on a monocular (as opposed to cyclopean) visual manifold. Increasing the disparity difference increases the gradient and decreases the magnitude of mutual support between the two disparities. The major difference between this formulation and the notion of limiting disparity gradient proposed by Burt and Julesz (1980) is that their formulation implies the existence of inhibition between two interacting points while in our formulation two greatly different disparities simply do not interact at all.[4] Interestingly, it has been found recently that the value of the limiting disparity gradient is a function of feature similarity, i.e. it cannot be defined on purely geometrical basis (Prazdny, 1985). More dissimilar features allow larger disparity gradients.

d) Maximal support is obtained when the disparity difference is zero independently of the distance between the interacting features. This means that an algorithm based on such a weighting function is slightly biased in favour of frontoparallel surfaces. This should not, however, be regarded as a weakness. The algorithm described below can successfully process steeply-sloped surfaces simply because similar disparities provide greater support than dissimilar ones. The bias towards the frontoparallel planes can also be defended on probabilistic grounds. It has been shown (Arnold and Binford, 1980) that, because of foreshortening, surfaces with steep depth gradients occupy only a small portion of most images. Most of the area of the image is covered by surfaces with normals to the viewing direction.

In the following, we tacitly assume that the stereograms were obtained by an imaging system in which the separation between the two viewpoints is small

4 Recently, Pollard, Mayhew, and Frisby (1984, personal communication) formulated an algorithm based on the concept of the disparity gradient

relative to the viewing distance. These requirements are met by all biological visual systems because the interocular separation is small relative to distances for which the binocular stereo mechanism is useful.

The Algorithm

The coherence constraint and the reasoning pertaining to the disparity interaction can be translated into a set of explicit rules: an algorithm. A computer program implementing the algorithm is outlined below. Basically, the program proceeds in two stages. First, we find all potential disparities (i.e. allowable correspondences between a point in the left and right image) for each point in the left image. Associated with each possible disparity is an "activity cell" whose value indicates the amount of support the particular disparity receives from its neighbors. Next, the disparities are allowed to influence each other. This interaction is a simple facilitation expressed by Eq. (1). More precisely, suppose that a left image feature point at location i has a set of possible disparities D_i and that we are interested in the amount of support a particular disparity $d_i \in D_i$ receives from the feature point j (with possible disparities D_j). The algorithm locates, among the disparities in the set D_j, that disparity $\hat{d}_j$ for which the absolute value of the difference $\delta d = |d_i - \hat{d}_j|$ is the minimum, and increments the "activity cell" associated with d_i at i by $s(i, j)$ found by evaluating Eq. (1). Observe that in searching for the support for d_i among disparities D_j we consider only the best possible support that d_i can get there, i.e. only the support from $\hat{d}_j$ is considered! This is a non-linear step which not only ignores the irrelevant information but also leads to significant computational savings.

After the support for all possible disparities at a given point has been determined in this way the disparity with the largest support (the highest value in the associated "activity cell") is chosen as the most likely disparity at that point. The procedure as described above is run for the left and right image in parallel and all points for which the left and right disparities differ are marked as ambiguous (the decision about the disparity at such places is postponed and left for the later processes to decide).[5]

It is possible to implement the coherence principle in another way, using strictly local interactions. The "global" support function [Eq. (1)] can be approximated by one that results from interation of local finite difference equation. In this way, a tradeoff can be made between a large number of connections and a number of iterations necessary to propagate information (see also Szelisky and Hinton, 1984).

Algorithm's Performance

The performance of the algorithm depends on the value of c in Eq. (1). This is the only "free variable" in the algorithm. If c is small, the algorithm works very well for fronto-parallel surfaces but performs rather

5 In our algorithm, this decission takes the form of "interpolation". We simply look around the given point, make a histogram of the neighbouring disparities found by the detection process (using a rather narrow point spread function) and choose the disparity closest to the histogram peak

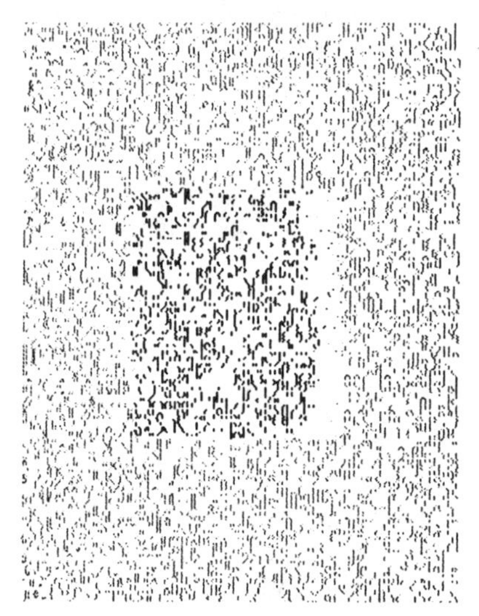

Fig. 2. A noisy stereogram obtained by digitizing a stereogram published in Julesz (1971) and a disparity field detected by the algorithm using this stereogram as input. Intensity codes disparity magnitude. The amount of incorrectly detected disparities is small (300 points out of 19,000 were assigned an incorrect disparity)

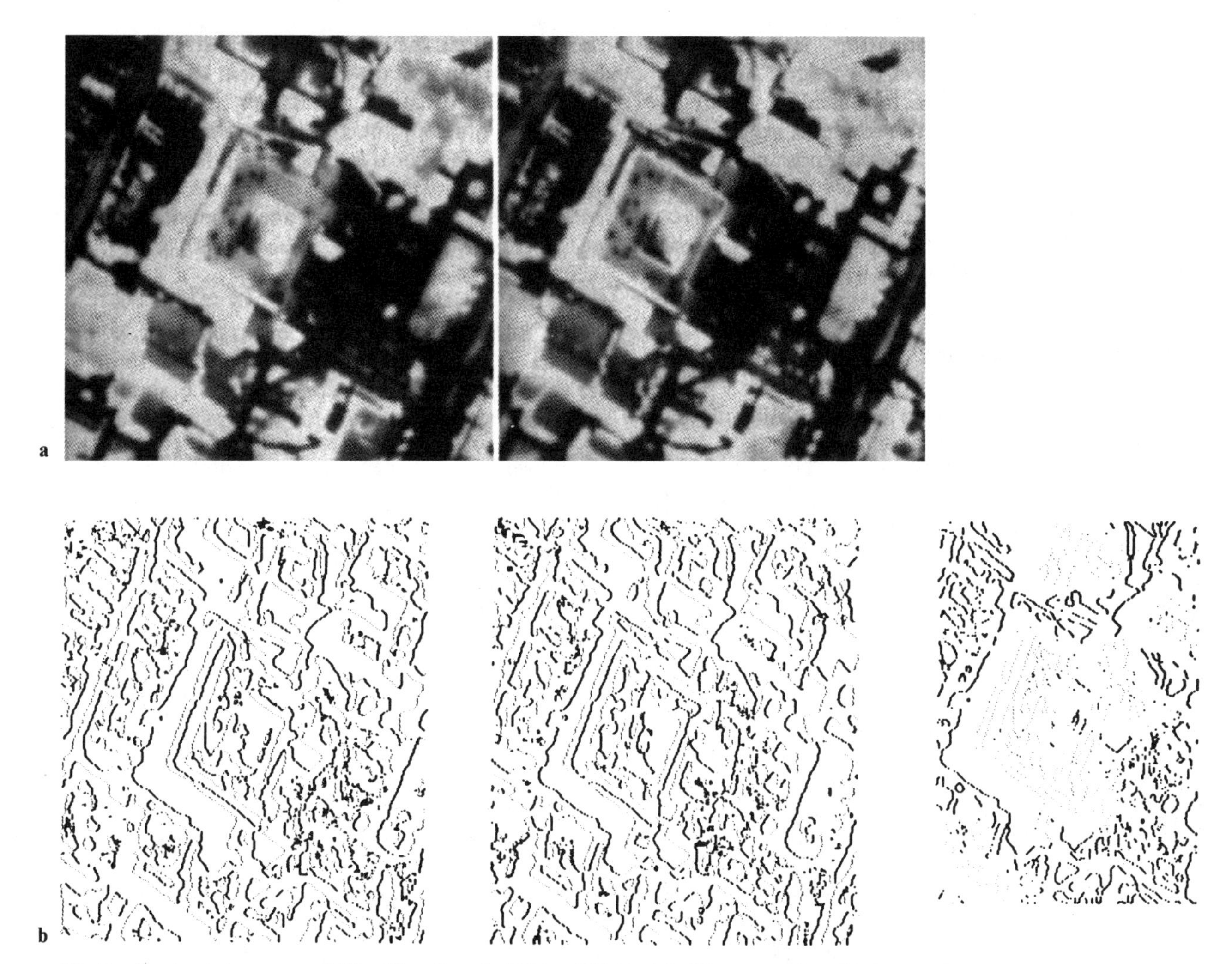

Fig. 3. a A natural stereogram digitized from Poggio (1984, p. 107). Each half-image consists of 255 × 320 picture elements. b The input to the program consisted of "edges" detected by a simple difference operator. Darker lines code "edges" with positive polarity (darker to lighter transitions) while lighter lines code "edges" with negative polarity. c Disparities detected by the program. Increasing elevation is coded by decreasing the darkness of the lines. A comparison with the photograph (viewed stereoptically e.g. by crossing the eyes) reveals a good qualitative match between obtained disparities and perceived depth. The disparity image is smaller than each stereo half-image due to border effects associated with the support neighbourhood

poorly for other surface orientations. We have found empirically that a value between 0.55 and 0.85 gives a good all round performance for a wide range of surface orientations.

The algorithm was tested on random-dot stereograms and some natural images. Random dot stereograms are a suitable vehicle for testing the performance of the algorithm because in these displays it is easy to obtain positional information needed for matching. All stereograms were pre-processed first to obtain edge information. Because the random-dot stereograms contain only two brightness levels (black and white) extracting edge information amounts to detecting the dark/light transitions and their signs. The program takes this information as input and matches only features of the same contrast polarity (edges of the same sign). The algorithm was tested on a broad range of such stimuli. It performs with accuracy close to 100% in stereograms portraying opaque surfaces. The accuracy decreases when transparent surfaces are present (about 75% of points were matched correctly). We believe that this is a rather good performance, possibly at the level achievable by human vision. Figure 2 illustrates the algorithm's performance on a noisy stereo image pair obtained by digitizing and thresholding a random-dot stereogram published by Julesz (1971). Figure 3 shows algorithm's performance on a natural image.[6] The two half-images were first convolved with an "edge" operator to obtain the

6 This stereogram is digitized directly from Poggio (1984, pp. 107)

necessary positional information for matching (Fig. 3b). Only the position of the "edges" and their polarity were used as features for matching. Figure 3c shows detected disparities. The results of experiments with the algorithm using stereograms portraying opaque surfaces compare favourably with the result obtained using other existing stereomatching algorithms, e.g. the algorithm developed by Marr and Poggio (1979) and implemented by Grimson (1981).[7]

Discussion and Conclusions

The algorithm exhibits several important and interesting properties. It is, unlike previous methods, a non-iterative, parallel and local process. The interaction mechanism itself is a form of generalized correlation which allows for (local) deformations and belongs to a class of methods based on local consensus (Julesz, 1971; Stevens, 1978; Ballard et al., 1983). In some sense, the behaviour of the algorithm can be described as a distributed (global) maximization of the total support for (unique) local matches.

Another important point is that the coherence concept used as a guiding principle in our analysis directly subsumes several matching and disambiguation rules commonly used as explicit matching predicates to solve the correspondence problem. Figural continuity and edge connectivity (Mayhew and Frisby, 1981; Baker and Binford, 1981; Mayhew, 1983) or collinearity (Mayhew and Frisby, 1980) are special instances of the coherence principle because spatially continuous surface markings directly imply disparity similarity. Human visual system apparently does not use edge connectivity as a constraint. Recently, Krol and Van Grind (1980) demonstrated that a retinally continuous bar can be seen as two bars separated in depth. In their experiments with the double nail illusion (in which two pins are presented one behind the other in the mid-sagital plane) they observed that a part of a pin appears to float above, behind and between two other pins while both pins project into two continuous "bars" on the retina, i.e. a retinally continuous edge is phenomenally split into two parts in depth (Krol and Van Grind, 1980, Fig. 4).

The form of gathering supporting evidence from the neighborhood can be extended by measuring local image intensity gradients and "deforming" the shape of the support neighbourhood (currently a horizontally extended rectangle) according to "predominant" (e.g. modal) gradient orientation in the neighbourhood. The idea is that the disparity of features along the direction perpendicular to the gradient varies potentially less than disparity of features along the local (luminance) gradient direction. This amounts to introducing a bias into local interactions in favour of local continuity. Observe that this is different from postulating a separate orientationally tuned nonlinear grouping process (Mayhew and Frisby, 1980) which explicitly uses the edge connectivity as a matching constraint.

An interesting feature of our approach is the absence of any explicit inhibitory connections (both in the spatial and disparity domain) in the algorithm, sometimes regarded as essential to disparity detection (Sperling, 1970; Dev, 1975; Nelson, 1975). The coherence principle does not penalize dissimilarities by inhibiting deviations. We believe that this is the major reason why the algorithm copes naturally and rather succesfully with transparent surfaces.

The algorithm is not intended, in its present form, to be a performance model of the fusional process although its performance is demonstrably good. It does not address, for example, the issues of multiple spatial scales and their interaction[8] the type of matching primitives to be used, or the role and control of vergence eye movements.[9] These considerations are, however, in some sense, orthogonal to the major issue addressed in this work: a fast matching mechanism that does not rely on artificial constraints and can successfully operate in a general environment. Our results demonstrate that useful disparity information can be obtained fast and reliably without a need to postulate inhibitory connections between detectors tuned to different disparities, a coarse-to-fine strategy requiring a sequence of fine vergence eye movements, or an elaborate set of matching and disambiguation rules. Instead of trying to escape the ambiguity problem the algorithm exploits regularities in the distribution of local ambiguities necessarily following from the cohesivity of matter.

Acknowledgement. I wish to thank Dr. J. Marty Tenenbaum for useful comments and discussion.

7 The result in Fig. 3 can be directly compared with the result published by Poggio (1984, figure on p. 107) who used the same stereogram

8 In one version of our algorithm, the multiple spatial scales are handled by assigning a slightly greater importance (weights) to matches at more coarse levels. Again, this introduces a bias into, rather than imposing an order on the course of the matching process

9 It is possible that vergence eye movements may represent the main method for dealing with transparent surfaces. Such accounts hold that transparent surfaces are perceived because the visual system fixates, in turn, each surface independently. The idea is that first the features belonging to one surface are fused and therafter held "locked" together while a new vergence eye movement enables the next surface, with a different disparity, to be fused

References

Arnold, R.D., Binford, T.O.: Geometric constraints in stereo vision. Proc. SPIE, Vol. 238, pp. 281–292. San Diego, CA 1980

Baker, H.H., Binford, T.O.: Depth from edge and intensity based stereo. Proc. J. Int. Conf. Art. Intell. 631–636 (1981)

Ballard, D.H., Hinton, G.E., Sejnowski, T.J.: Parallel visual computation. Nature **306**, 21–26 (1983)

Burt, P.J., Julesz, B.: A disparity gradient limit for binocular fusion. Science **208**, 615–617 (1980)

Gibson, J.J.: The perception of visual world. Boston: Houghton Mifflin 1950

Grimson, W.E.L.: From images to surfaces. Cambridge: MIT Press, 1981

Julesz, B.: Binocular depth perception of computer generated patterns. Bell Sys. Tech. J. **38**, 1001–1020 (1960)

Julesz, B.: Foundations of cyclopean perception. Chicago: Chicago University Press 1971

Kass, M.H.: Computing stereo correspondence. MS Thesis, MIT 1984

Krol, J.D., van de Grind, W.A.: The double nail illusion. Perception **9**, 651–669 (1980)

Marr, D.: Vision. San Francisco: Freeman 1982

Marr, D., Poggio, T.: A cooperative computation of stereo disparity. Science **194**, 283–287 (1976)

Marr, D., Poggio, T.: A theory of human stereopsis. Proc. R. Soc. London Ser. B**204**, 301–328 (1979)

Mayhew, J.E.W.: Stereopsis. In: Physical and biological processing of images. Sleigh, A.C. (ed.). Berlin, Heidelberg, New York: Springer 1983

Mayhew, J.E.W., Frisby, J.P.: The computation of binocular edges. Perception **9**, 69–87 (1980)

Mayhew, J.E.W., Frisby, J.P.: Psychophysical and computational studies towards a theory of human stereopsis. Artif. Intell. **17**, 349–387 (1981)

Nelson, J.I.: Globality and stereoscopic fusion in binocular vision. J. Theor. Biol. **49**, 1–88 (1975)

Poggio, T.: Vision by mean and machine. Sci. Am. **250**, 106–116 (1984)

Prazdny, K.: Stereoscopic matching, eye position, and absolute depth. Perception **12**, 151–160 (1983)

Prazdny, K.: On the perception of Glass patterns. Perception **13**, 469–478 (1984)

Prazdny, K.: On the disparity gradient limit for binocular fusion. Percept. Psychophys. (1985, in press)

Sperling, G.: Binocular vision. J. Am. Psychol. **83**, 461–534 (1970)

Stevens, K.: Computation of locally parallel structure. Biol. Cybern. **29**, 19–28 (1978)

Szelisky, R., Hinton, G.: Solving random-dot stereograms using the heat equation. Research Report, Computer Science Department, Carnegie-Mellon University, Pittsburg, PA, 1984 (December)

Wheatstone, C.: On some remarkable, and hitherto unobserved, phenomena of binocular vision. Philos. Trans. R. Soc. London **128**, 371–394 (1838)

Received: January 29, 1985

Dr. K. Prazdny
Artificial Intelligence
Laboratory
Schlumberger Palo Alto
Research Center
3340 Hillview Avenue
Palo Alto
CA 94304
USA

Hierarchical Warp Stereo

Lynn H. Quam
SRI International
333 Ravenswood Avenue
Menlo Park, California 94025

September 10, 1984

Abstract

This paper describes a new technique for use in the automatic production of digital terrain models from stereo pairs of aerial images. This technique employs a coarse-to-fine hierarchical control structure both for global constraint propagation and for efficiency. By the use of disparity estimates from coarser levels of the hierarchy, one of the images is geometrically warped to improve the performance of the cross-correlation-based matching operator. A newly developed surface interpolation algorithm is used to fill holes wherever the matching operator fails. Experimental results for the Phoenix Mountain Park data set are presented and compared with those obtained by ETL.

1 Introduction

The primary objective of this research was to explore new approaches to automated stereo compilation for producing digital terrain models from stereo pairs of aerial images. This paper presents an overview of the hierarchical warp stereo (HWS) approach , and shows experimental results when it is applied to the ETL Phoenix Mountain Park data set.

The stereo images are assumed to be typical aerial-mapping pairs, such as those used by USGS and DMA. Such pairs of images are different perspective views of a 3-D surface acquired at approximately the same time and illumination angles. Normally these views are taken with the camera looking straight downward. The major effect of non verticality is to increase the incidence of occlusion, which increases the difficulty of point correspondence.

We shall call one of these images the "reference image," and the other the "target image." We will be searching in the target image for the point that best matches a specified point in the reference image.

It is also assumed that the epipolar model for the stereo pair is known, which means that for any given point in one image we can determine a line segment in the other image that must contain the point, unless it is occluded from view by other points on the 3-D surface. This is certainly a reasonable assumption, since an approximation to the epipolar model can be derived from a relatively small number of point correspondences if the parameters of the imaging platform are not known a priori.

The primary goal is to automatically determine correspondences between points in the two images, subject to the following criteria:

- Minimize the rms difference between the disparity measurements and "ground truth." Without ground truth, we cannot measure this.
- Maximize the sensitivity of the disparity measurements to small-scale terrain features, while minimizing the effects of noise.
- Minimize the frequency of false matches.
- Minimize the frequency of match failures.

These criteria are mutually exclusive. Under ideal conditions, increasing the size of the match operator decreases the effects of noise on the disparity measurement, but it also diminishes sensitivity to small terrain features. Similarly, tightening the match acceptance criteria reduces the frequency of false matches, but results in more frequent match failures.

One of the goals of this system is to minimize the number of parameters that must be adjusted individually for each stereo pair to get optimum performance.

2 Approach

This section briefly explains the HWS approach, which consists of three major components:

- Coarse-to-fine hierarchical control structure for global constraint propagation as well as for efficiency.
- Disparity surface interpolation to fill holes wherever the matching operator fails.
- Geometric warping of the target image by using disparity estimates from coarser levels of the hierarchy to improve the performance of the cross-correlation-based matching operator.

2.1 The Use of Hierarchy and Surface Interpolation to Propagate Global Constraints

The goal of stereo correspondence is to find the point in the target image that corresponds to the same 3-D surface point as a given point in the reference image. It is often impossible to select the correct match point with only the image information that is local to the given point in the reference image in combination with the image nformation along the epipolar line segment in the target image. When the 3-D surface contains a replicated pattern, there is the likelihood of match point ambiguity. Let us consider, for example, a stereo pair that contains a parking lot

This research was supported by the Defense Advanced Research Projects Agency under Contract No. MDA 903-83-C-0027.

with repetitive markings delimiting the parking spaces. Around the edges of the lot there are image points that can be matched unambiguously. Within the parking lot, ambiguity is likely, depending on the orientation of the repetititive patterns with the epipolar line. A successful stereo correspondence system must be able to use global match information to resolve local match-point ambiguity.

HWS approaches this problem in two ways. First, global constraints on matches are propagated by the coarse-to-fine progression of the matching process. Disparities computed at lower resolution are employed to constrain the search in the target image to a small region of the epipolar line, which also greatly reduces the probability of selecting the wrong point when ambiguity is present. Second, whenever the match process fails to find a suitable match or detects a possible match ambiguity, a disparity estimate is inserted that is based on a surface interpolation algorithm, which uses information from a neighborhood around the disparity "hole," with the size of the neighborhood depending on the number of neighboring "holes."

2.2 The Use of Image Warping to Improve Correlation Operator Performance

One of the greatest problems in the use of area correlation for match point determination is the distortion that occurs because of disparity changes within the correlation window. Since area-based correlation matches areas, rather than individual points, the disparity it calculates is influenced by the disparities of all of the points in the window, not just the point at the center. When there are high disparity gradients or disparity discontinuities, the correlation calculated for the correct disparity can actually be so poor that some other disparity will have a higher correlation score.

The effect of correlation window distortion can be greatly mitigated in a hierarchical system by using the disparity estimates from the previous level of matching to warp the target image geometrically at its current resolution level into closer correspondence with the reference image.

2.3 Related Work

Norvelle [1] implemented a semi automatic stereo compilation system at the U.S. Army Engineer Topographic Laboratories (ETL) that operates in a single pass through the images. It uses disparity surface extrapolation both to predict the region of the epipolar segment for matching and to estimate the local surface orientation so as to warp the correlation window. He found that these techniques improved the performance of the system significantly, but that considerable manual intervention was needed when the surface extrapolator made bad predictions, or when the image contained areas with no information for matching, with ambiguities, or with occlusions.

3 Sequence Of Operations In Hierarchical Warp Stereo

Figure 1 illustrates the hierarchical control structure of the system.

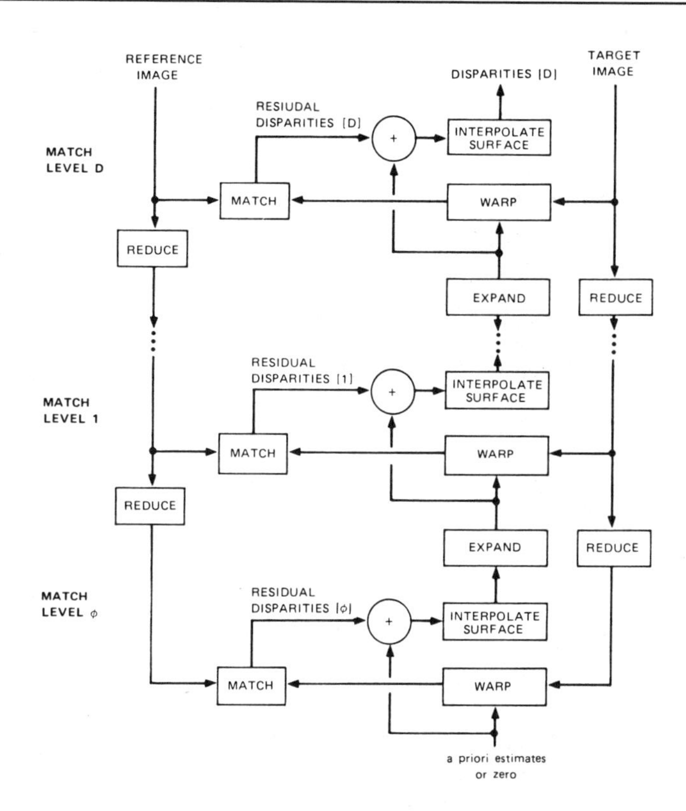

FIGURE 1

Block Diagram of Hierarchical Control Structure

1. Initialize:

   - Start with a stereo pair of images (assumed to be of the same dimensions).
   - Call one of these images the "reference image," the other the "target image."
   - Construct Gaussian pyramids (Burt [2]) $reference_i$ and $target_i$ for each image. The images at level i in these pyramids correspond to reductions of the original images by a factor of 2^i.
   - Set $disp_{-1}$ to either the a priori disparity estimates or all zeros.
   - Start the iteration at level $i = 0$.
   - Choose the pyramid depth D so that:

   $$D = ceiling(log2(uncertainty)) - 1.$$

   where $uncertainty$ is an estimate of the maximum difference between $disp_{-1}$ and the "true" disparities. This guarantees the "true" disparities will be within the range (-2 : +2) at level 0 of the matching.

2. Warp: Use the disparity estimates $2 * disp_{i-1}$ to warp $target_{D-i}$ geometrically into approximate alignment with $reference_{D-i}$. Note that the factor of two is equal to the ratio of image scales between level i and level $i-1$ of the hierarchy.

3. Match: Using the matching operator, compute the residual disparities $\Delta disp_i$ between the warped target and the reference images at level i.

4. Refine: Compute the refined disparity estimates:

$$disp_i = 2 * disp_{i-1} + \Delta disp_i.$$

5. Fill: Use the surface interpolation algorithm to fill in disparities estimates at positions where matching operator fails because of no image contrast, ambiguity, etc.

6. Increase resolution: If $i = D$, quit; otherwise let $i = i + 1$ and go to Step 2.

4 Disparity Estimation

Disparity estimation consists of three parts:

- Computing match operator scores for disparities along an epipolar segment.
- Accepting or rejecting the collection of scores according to a model for the shape of the correlation peak.
- Estimating the subpixel disparities at acceptable peaks.

4.1 Match Score Operator

The HWS approach presented here can be implemented with a variety of match operators. All results reported here were obtained with an operator that closely approximates Gaussian-weighted normalized cross correlation. The values of the Gaussian weights decrease with Euclidean distance from the center of a square correlation window. In the examples shown here, the window dimension is 13×13 pixels with a standard deviation of approximately 2 pixels in the Gaussian weights. Preliminary results indicate that the Gaussian-weighted correlation operator is better than uniformly weighted correlation operators at locating changes in disparity while maintaining a given level of disparity precision.

4.2 Evaluation of Correlation Surface Shape

The match operator reports a failure if any of the following conditions exist:

- Disparity out of range: The maximum match score is found at either extreme of the epi-polar segment.
- Multiple peaks: The best and next best match scores is found at disparities that differ by more than one pixel.

There are other models for the expected shape of the correlation surface that can be based on the autocorrelation surface shape of the windows in the reference and target images. Further investigation is needed to evaluate the utility of such models for both surface shape evaluation and disparity estimation.

4.3 Subpixel Disparity Estimation

The subpixel location of the correlation surface peak is estimated by parabolic interpolation of both the x and y directions of disparity. For each direction, three adjacent match scores – s_{i-1}, s_i, and s_{i+1}, where s_i is the maximum score – are used to compute the peak as follows:

$$.5 * \frac{s_{i+1} - s_{i-1}}{2 * s_i - s_{i+1} - s_{i-1}}$$

More complicated approaches to peak estimation, such as two-dimensional least-squares fitting of the correlation surface, might yield better estimates, but at a higher computational cost.

5 Surface Interpolation Algorithm

The goal of the surface interpolation algorithm is to estimate values for the disparity surface at points where the match operator reported failure; such points will be called "holes." The approach to filling a hole at location x, y is to model the surface by employing the disparity measurements over the set of nonholes $\overline{H}$ in the $n \times n$ pixel neighborhood centered at x, y. The set H contains the indices of all holes in the neighborhood.

This surface interpolation algorithm is based on the solution to the hyperbolic multiquadric equations described in Smith [3]. The surface is known at the set of points x_i, y_i, z_i where $i \in \overline{H}$, and can be estimated at other points $h \in H$ by the formula

$$z(x_h, y_h) = \sum_{i \in \overline{H}} c_i * g(x_h - x_i, y_h - y_i),$$

where g is the basis function for the surface respresentation, and coefficients c_i are the solutions to the set of linear equations:

$$z(x_j, y_j) = \sum_{i \in \overline{H}} c_i * g(x_i - x_j, y_i - y_j) \ \ for\ all\ j \in \overline{H}$$

Clearly, this irregular grid solution could be used to compute the surface values at the holes in the disparity data, but this involves solving for the coefficients c_j for each different configuration of holes and nonholes in the $n \times n$ neighborhoods of the disparity surface.

An alternative approach, which is used here, is to convert the quasi-regular grid problem into a regular grid problem in which each c_i at a hole is forced to be zero, and the corresponding z_i remains as an unknown. This results in the same solution that would have been obtained from the irregular grid formulation and produces the following system of linear equations:

$$\sum_{i \in H} A^{-1}_{h,i} * z_i = - \sum_{j \in \overline{H}} A^{-1}_{h,j} * z_j \ \ for\ all\ h \in H, \tag{1}$$

where A^{-1} is the inverse of the matrix $A_{i,j} = g(x_i - x_j, y_i - y_j)$ for $i, j \in H \cup \overline{H}$. This system of equations must be solved for each z_i for $i \in H$. Thus, we have reduced the size of the linear system of equations that must be solved from the number of elements in $\overline{H}$ to the number of elements in H. Of course, the matrix A must be computed and inverted once.

Areas on the disparity surface that contain large clusters of holes cause problems. The previous surface interpolation algorithm degenerates to a surface extrapolation algorithm when the nonholes in the neighborhood are not more or less isotropically distributed over the entire neighborhood. The problem can be overcome by increasing the size of the neighborhood until some spatial-distribution criterion is met, but this would require solving extremely large linear systems.

Large holes are filled by means of the following hierarchical approach:

Procedure Surface-Interpolate($surface_i$)

1. If $surface_i$ contains large holes then

(a) Compute filled-surface$_{i+1}$ = *expand*(surface-interpolate(*reduce*($surface_i$))), where *reduce* computes a Gaussian convolution reduction by a factor of two, surface-interpolate is a recursion call to this interpolation algorithm, and *expand* computes expansion by a factor of two, using bilinear interpolation.

(b) For each hole in $surface_i$ that is completely surrounded by other holes, fill the hole with the value from the filled-surface$_{i+1}$.

2. For each hole in $surface_i$ fill the hole by solving the system of linear equations (1) for the $n \times n$ pixel neighborhood centered at the hole (n = 7 in the examples).

3. Return the filled $surface_i$.

6 Examples

This section describes the experimental results achieved when the HWS technique was applied to areas of the ETL Phoenix Mountain Park data set, and compares these results to those obtained from the semiautomatic system developed by Norvelle [1].

The following components of the Phoenix Mountain Park data set were used:

- Left image: 2048 x 2048 pixels, 8 bits per pixel
- Right image: 2048 x 2048 pixels, 8 bits per pixel
- x-correspondence array: 400 x 400 points , floating point.

The left and right images had been scanned such that the epipolar lines were almostly exactly horizontal. The ETL x-correspondence array was converted to an x-disparity image to enable comparison between ETL and HWS results.

Results are shown for two different areas of the Phoenix data set. All disparity measurements are indicated in terms of pixel distances in the 2048×2048 Phoenix stereo pair, rather than the resolution of the selected windows.

- Area A is defined by two approximately aligned 150×150-pixel windows of the Phoenix pairs which were reduced by a factor of four (the windows thus corresponding to the 600×600-pixel windows of the originals). The measured disparities for area A range from -40 to +16 pixels.
- Area B is defined by two approximately aligned 125×125-pixel windows of the Phoenix pairs which were reduced by a factor of two (the windows thus corresponding to the 250×250-pixel windows of the originals). The measured disparities for area B range from -40 to -34 pixels.

Figures 2 and 3 show the inputs and outputs of three levels of the hierarchy for areas A and B, respectively. Columns 1 and 2 are the reference and target images at each level. Column 3 is a binary image that indicates the positions of match failures. Column 4 shows the resulting disparity image of each level after the match failures have been replaced by surface-interpolated disparity values.

Figures 4 and 5 contain a comparison of the HWS results with those obtained at ETL by Norvelle for areas A and B respectively. The bottom-left images of figures 4 and 5 show the pixel-by-pixel differences, after contrast enhancement, between the HWS and ETL disparities. The graphs to the right of these difference images depict the histograms of these differences.

The mean and standard-deviation values shown with the histograms provide a useful quantative comparision between the HWS and ETL results. They show that the average disparity differences were .082 and .025 pixels, and that the standard deviations of the disparity differences were .67 and .34 pixels for the A and B window pairs, respectively, in terms of pixel distances in the 2048×2048 Phoenix pairs. These standard deviations become .17 and .17 pixels when expressed relative to the scales of A and B windows, respectively.

Similar results have been achieved for other examples that include both higher resolution and larger windows.

7 Problems

HWS is still very experimental. Some of the parameters that affect the system, such as the range of disparities to compute at each level of hierarchy and the size of the correlation operator, are still specified manually.

There are problems in estimating the range of disparities to be computed at each level of the hierarchy. If the estimate is too low, there will be frequent out-of-range match failures. If, on the other hand, the estimate is too high, computation time will increase and there will be more potential for match point ambiguity.

HWS has difficulty dealing with steep terrain features that have small image projections, but large disparities. At low resolutions in the matching hierarchy, the disparities of the terrain surrounding the feature dominate those of the feature itself, resulting in a disparity estimate that is usually intermediate between that of the feature and that of the surround. At higher resolutions in matching, the disparity of the steep feature may be outside the permissible disparity range.

HWS has even greater problems with oblique stereo pairs containing many occlusions. At low matching resolution, the disparities of foreground and background in the same neighborhoods cannot be distinguished. As the matching resolution increases, foreground and background features are discernible as separate objects, but their disparities are out of range for the matcher.

Most of the difficulties caused by sudden changes in disparity might be solved by preceding the disparity surface interpolation step with an algorithm that attempts to match still unmatched regions in the reference image with regions in the target image that likewise have not yet been matched. We thus attempt to match holes with holes.

8 Conclusions

HWS produces very good results for vertical stereo pairs of rolling terrain. With the incluson of a hole-to-hole matching step, HWS should be capable of comparable performance for terrain characterized by steep slopes and frequent occlusions.

Bibliography

[1] Norvelle, F. Raye, *Interactive Digital Correlation Techniques For Automatic Compilation of Elevation Data*, U.S. Army Engineer Topographic Laboratories, Fort Belvoir, VA 22060, Report Number ETL-0272, Oct. 1981.

[2] Burt, Peter J., *Fast Filter Transforms for Image Processing*, CGIP, 16, 20-51. 1981.

[3] Smith, Grahame .B., *A Fast Surface Interpolation Technique*, Technical Note 333, Artificial Intelligence Center, SRI International, Menlo Park, California, August 1984. (Also in these proceedings).

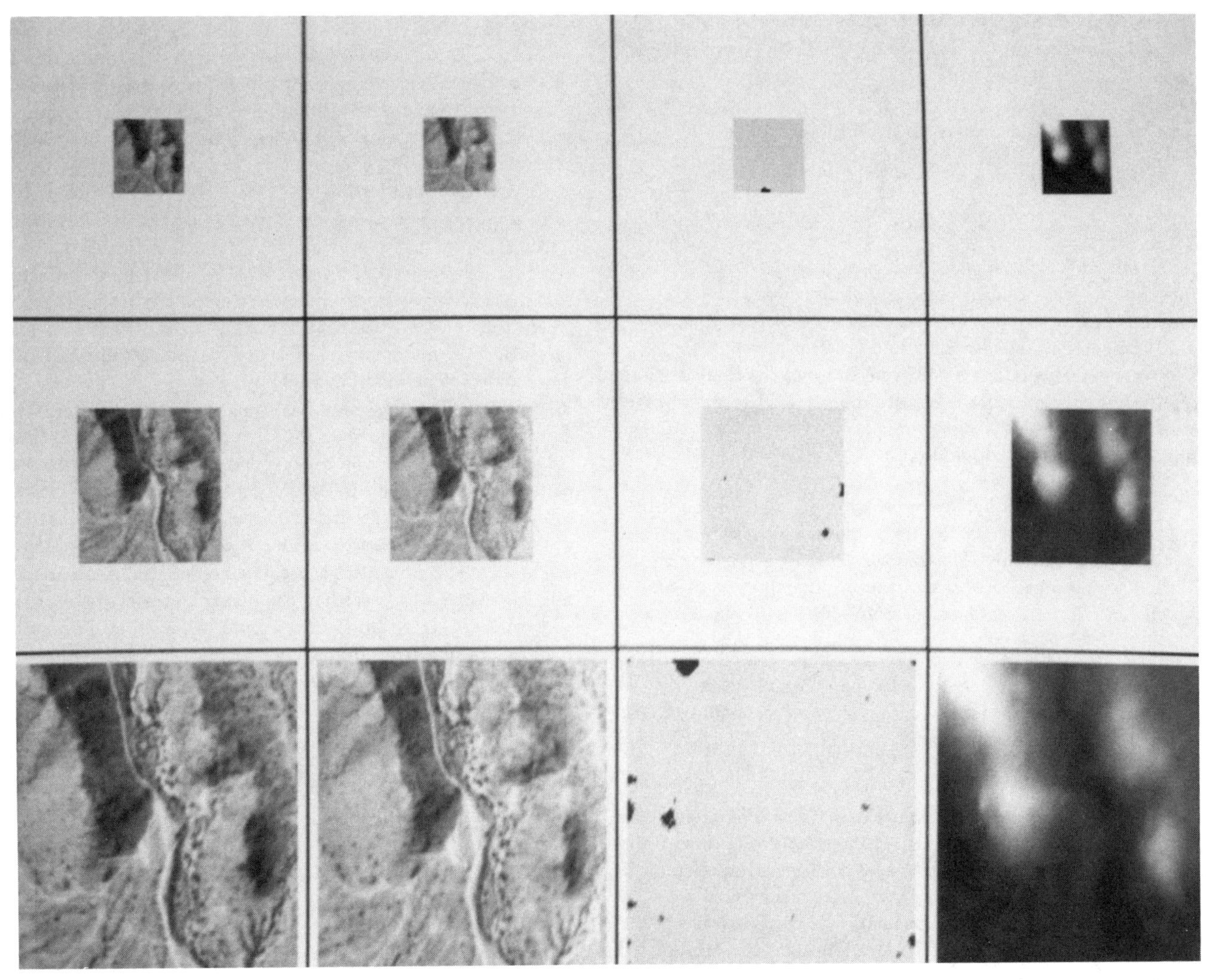

Reference images Target images Hole images Filled disparity images

FIGURE 2 HWS results for area A

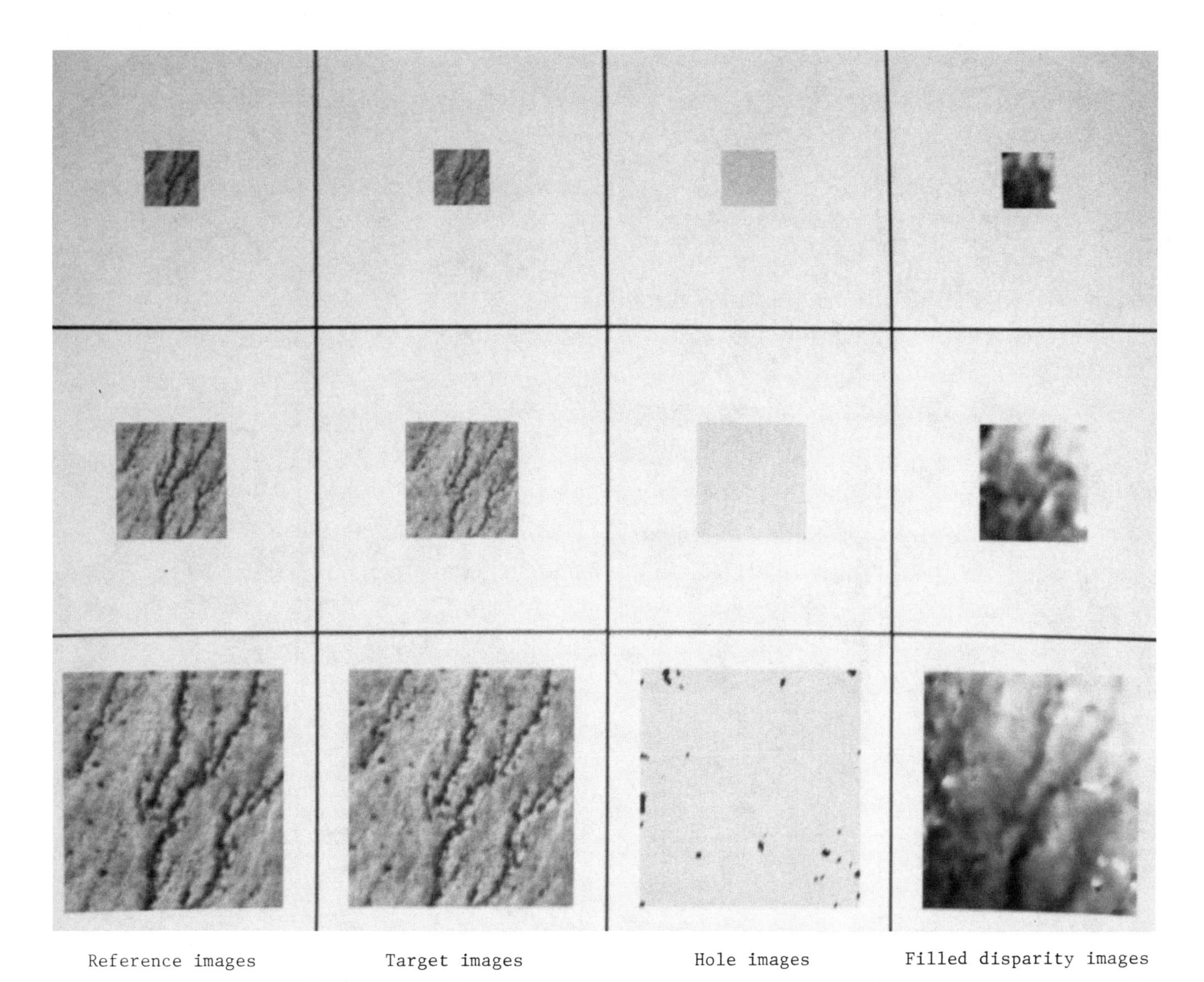

Reference images | Target images | Hole images | Filled disparity images

FIGURE 3 HWS results for area B

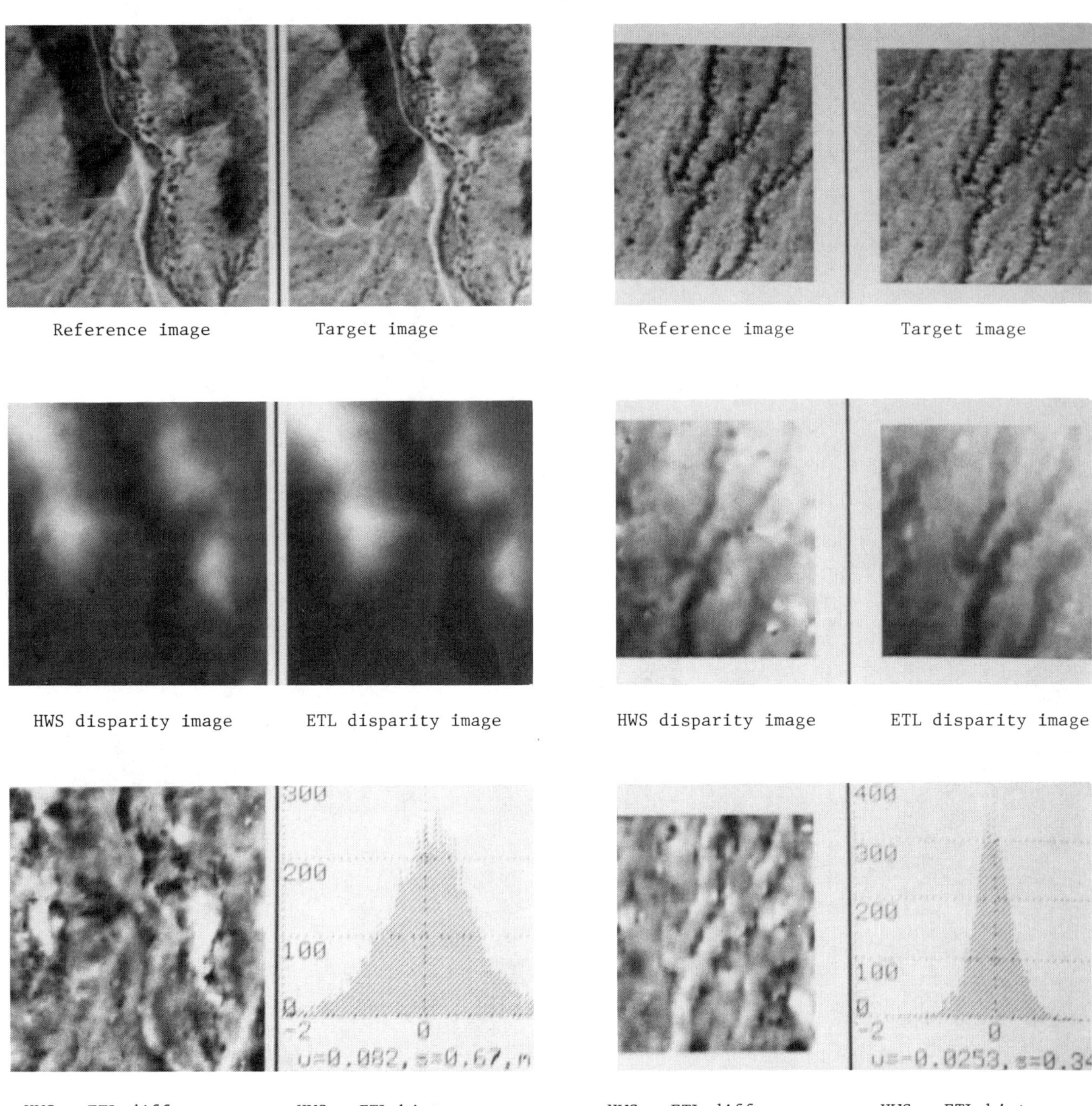

FIGURE 4 HWS vs. ETL results for area A

FIGURE 5 HWS vs. ETL results for area B

Stereo Integral Equation

Grahame B. Smith

Artificial Intelligence Center,
SRI International
Menlo Park, California 94025

Abstract

A new approach to the formulation and solution of the problem of recovering scene topography from a stereo image pair is presented. The approach circumvents the need to solve the correspondence problem, returning a solution that makes surface interpolation unnecessary. The methodology demonstrates a way of handling image analysis problems that differs from the usual linear-system approach. We exploit the use of nonlinear functions of local image measurements to constrain and infer global solutions that must be consistent with such measurements. Because the solution techniques we present entail certain computational difficulties, significant work still lies ahead before they can be routinely applied to image analysis tasks.

1 Introduction

The recovery of scene topography from a stereo pair of images has typically proceeded by three, quasi-independent steps. In the first step, the relative orientation of the two images is determined. This is generally achieved by selecting a few scene features in one image and finding their counterparts in the other image. From the position of these features, we calculate the parameters of the transformation that would map the feature points in one image into their corresponding points in the other image. Once we have the relative orientation of the two images, we have constrained the position of corresponding image points to lie along lines in their respective images. Now we commence the second phase in the recovery of scene topography, namely, determining a large number of corresponding points. The purpose of the first step is to reduce the difficulty involved in finding this large set of corresponding points.

Because we have the relative orientation of the two images, we only have to make a one-dimensional search (along the epipolar lines) to find points in the two images that correspond to the same scene feature. This step, usually called solving the "correspondence" problem, has received much attention. Finding many corresponding points in stereo pairs of images is difficult. Irrespective of whether the technique employed is area-based correlation or that of edge-based matching, the resultant set of corresponding points is usually small, compared with the number of pixels in the image. The solution to the correspondence problem, therefore, is not a dense set of points over the two images but rather a sparse set. Solution of the correspondence problem is made more difficult in areas of the scene that are relatively featureless or when there is much repeated structure, constituting local ambiguity. To generate the missing intermediate data, the third step of the process is one of surface interpolation.

Scene depth at corresponding points is calculated by simple triangulation; this gives a representation in which scene depth values are known for some set of image plane points. To fill this out and to obtain a dense set of points at which scene depth is known, an interpolation procedure is employed. Of late there has been significant interest in this problem and various techniques that use assumptions about the surface properties of the world have been demonstrated [1,3,5,8]. Such techniques, despite some difficulties, have made it possible to reconstruct credible scene topography.

Of the three steps outlined, the initial one of finding the relative orientation of the two images is really a procedure designed to simplify the second step, namely, finding a set of matched points. We can identify several aspects of these first two steps that suggest the need for an alternative view of the processes entailed in reconstructing scene topography from stereo image pairs.

The techniques employed to solve the correspondence problem are usually local processes. When a certain feature is found in one image, an attempt is made to find the corresponding point in the other image by searching for it within a limited region of that image. This limit is imposed not just to reduce computational costs, but to restrict the number of comparisons so that false matches can be avoided. Without such a limit many points may "match" the feature selected. Ambiguity cannot be resolved by a local process; some form of global postmatching process is required. The difficulties encountered in featureless areas and where repeated structure exists are those we bring upon ourselves by taking too local a view.

In part, the difficulties of matching even distinct features are self-imposed by our failure to build into the matching procedure the shape of the surface on which the feature lies. That is, when we are doing the matching we usually assume that a feature lies on a surface patch that is orthogonal to the line of sight – and it is only at some later stage that we calculate the true slope of the surface patch. Even when we try various slopes for the surface patch during the matching procedure, we rarely return after the surface shape has been estimated to determine whether that calculated shape is consistent with the best slope actually found in matching.

In the formulation presented in the following sections, the problem is deliberately couched in a form that allows us to ask the question: what is the shape of the surface in the world that can account for the two image irradiances we see when we view that surface from the two positions represented by the stereo pair? We make no assumptions about the surface shape to do the matching – in fact, we do not do any matching at all. What we are interested in is recovering the surface that explains simul-

The work reported here was supported by the Defense Advanced Research Projects Agency under Contracts MDA903-83-C-0027 and DACA76-85-C-0004.

taneously all the parts of the irradiance pattern that are depicted in the stereo pair of images. We seek the solution that is globally consistent and is not confused by local ambiguity.

In the conventional approach to stereo reconstruction, the final step involves some form of surface interpolation. This is necessary because the previous step – finding the corresponding points – could not perform well enough to obviate the need to fabricate data at intermediate points. Surface interpolation techniques employ a model of the expected surface to fill in between known values. Of course, these known data points are used to calculate the parameters of the models, but it does seem a pity that the image data encoding the variation of the surface between the known points are ignored in this process and replaced by assumptions about the expected surface.

In the following formulation we eliminate the interpolation step by recovering depth values at all the image pixels. In this sense, the image data, rather than knowledge of the expected surface shape, guide the recovery algorithm.

We previously presented a formulation of the stereo reconstruction problem in which we sought to skirt the correspondence problem and in which we recovered a dense set of depth values [6]. That approach took a pair of image irradiance profiles, one from the left image and its counterpart from the right image, and employed an integration procedure to recover the scene depth from what amounted to a differential formulation of the stereo problem. While successful in a noise-free context, it was extremely sensitive to noise. Once the procedure, which tracked the irradiance profiles, incurred an error recovery proved impossible. Errors occurred because there was no locally valid solution. It is clear that that procedure would not be successful in cases of occlusion when there are irradiance profile sections that do not correspond. The approach described in this paper attempts to overcome these problems by finding the solution at all image points simultaneously (not sequentially, as in the previous formulation) and making it the best approximation to an overconstrained system of equations. The rationale behind this methodology is based on the expectation that the best solution to the overconstrained system will be insensitive both to noise and to small discrepancies in the data, e.g., at occlusions. While the previous efforts and the work presented here aimed at similar objectives, the formulation of the problem is entirely different. However, the form of the input – image irradiance profiles – is identical.

The new formulation of the stereo reconstruction task is given in terms of one-dimensional problems. We relate the image irradiance along epipolar lines in the stereo pair of images to the depth profile of the surface in the world that produced the irradiance profiles. For each pair of epipolar lines we produce a depth profile, from which the profile for a whole scene may then be derived. The formulation could be extended directly to the two-dimensional case, but the essential information and ideas are better explained and more easily computed in the one-dimensional case.

We couch this presentation in terms of stereo reconstruction, although there is no restriction on the acquisition positions of the two images; they may equally well be frames from a motion sequence.

2 Stereo Geometry

As noted earlier, our formulation takes two image irradiance profiles – one from the left image, one from the right – and describes the relationship between these profiles and the corresponding

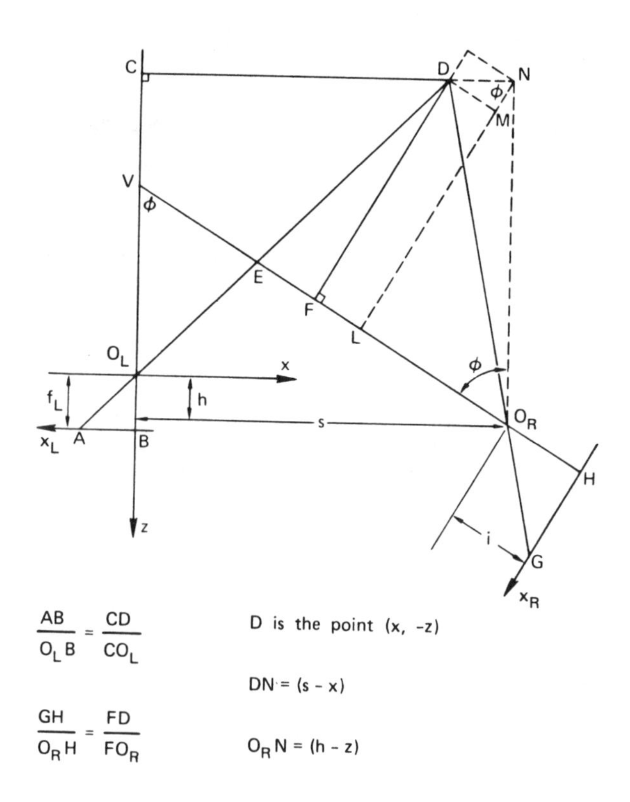

Figure 1: Stereo Geometry. The two-dimensional arrangement in the epipolar plane that contains the optical axis of the left imaging system.

depth profile of the scene. The two irradiance profiles we consider are those obtained from corresponding epipolar lines in the stereo pair of images. Let us for the moment consider a pair of cameras pointed towards some scene. Further visualize the plane containing the optical axis of the left camera and the line joining the optical centers of the two cameras, i.e., an epipolar plane. This plane intersects the image plane in each camera, and the image irradiance profiles along these intersections are the corresponding irradiance profiles that we use. Of course, there are many epipolar planes, not just the one containing the left optical axis. Consequently, each plane gives us a pair of corresponding irradiance profiles. For the purpose of this formulation we can consider just the one epipolar plane containing the left optical axis since the others can be made equivalent. A description of this equivalence is given in a previous paper [6]. Figure 1 depicts the two-dimensional arrangement. AB and GH are in the camera image planes, while O_L and O_R are the cameras' optical centers. D is a typical point in the scene and AD and GD are rays of light from the scene onto the image planes of the cameras. From this diagram we can write two equations that relate the image coordinates x_L and x_R to the scene coordinates x and z. These are standard relationships that derive from the geometry of stereo viewing. For the left image

$$\frac{x}{-z} = \frac{x_L}{f_L} \quad ,$$

while for the right image

$$\frac{x}{-z} = g_R(x_R) - \frac{(s + g_R(x_R)h)}{z} \quad ,$$

where

$$g_R(x_R) = \frac{x_R \cos\phi - i \sin\phi}{x_R \sin\phi + i \cos\phi}$$

In addition, it should be noted that the origin of the scene coordinates is at the optical center of the left camera, and therefore the z values of all world points that may be imaged are such that

$$z < 0$$

3 Irradiance Considerations

From any given point in a scene, rays of light proceed to their image projections. What is the relationship between the scene radiance of the rays that project into the left and the right images? Let us suppose that the angle between the two rays is small. The bidirectional reflectance function of the scene's surface will vary little, even when it is a complex function of the lighting and viewing geometry. Alternatively, let us suppose that the surface exhibits Lambertian reflectance. The scene radiance is independent of the viewing angle; hence, the two rays will have identical scene radiances, irrespective of the size of the angle between them. For the model presented here, we assume that the scene radiance of the two rays emanating from a single scene point is identical. This assumption is a reasonable one when the scene depth is large compared with the separation distance between the two optical systems, or when the surface exhibits approximate Lambertian reflectance. It should be noted that there are no assumptions about albedo (i.e., it is not assumed to be constant across the surface) nor, in fact, is it even necessary to know or calculate the albedo of the surface. Since image irradiance is proportional to scene radiance, we can write, for corresponding image points,

$$I_L(x_L) = I_R(x_R)$$

I_L and I_R are the image irradiance measurements for the left and right images, respectively. It should be understood that these measurements at positions x_L and x_R are made at image points that correspond to a single scene point x.

While the above assumption is used in the following formulation, we see little difficulty in being less restrictive by allowing, for example, a change in linear contrast between the image profile and the real profile.

4 Integral Equation

Let us consider a single scene point x. For this scene point, we can write $I_L(x) = I_R(x)$. This equality relation holds for any function F of the image irradiance, that is, $F(I_L(x)) = F(I_R(x))$. If we let p select the particular function we want to use from some set of functions, we shall write

$$F(p, I_L(x)) = F(p, I_R(x))$$

The set of functions we use will be the set of all nonlinear functions for which $F(p_1, I) \neq \alpha(p_1, p_2)F(p_2, I)$ for all p. A specific example of such a function is $F(p, I) = I^p$.

The foregoing functions relate to the image irradiance. We can combine them with expressions that are functions of the stereo geometry. In particular, for the as yet unspecified function T of $\frac{x}{-z}$, we can write

$$F(p, I_L(x))\frac{d}{dx}T(\frac{x}{-z(x)}) = F(p, I_R(x))\frac{d}{dx}T(\frac{x}{-z(x)}).$$

We have written z as $z(x)$ to emphasize the fact that the depth profile we wish to recover is a function of x. Should a more concrete example of our approach be required, we could select $T(\frac{x}{-z}) = \ln(\frac{x}{-z})$, which, when combined with the example for F above, gives us

$$I_L{}^p(x)\frac{d}{dx}\ln(\frac{x}{-z(x)}) = I_R{}^p(x)\frac{d}{dx}\ln(\frac{x}{-z(x)})$$

We now propose to develop the left-hand side of the above expression in terms of quantities that can be measured in the left stereo image, and develop the right-hand side in terms of quantities from the right stereo image. If we were to substitute x_L for x in the left-hand side of the above expression and x_R for x in the right-hand side, we would have to know the correspondence between x_L and x_R. This is a requirement we are trying to avoid. At first, we shall integrate both sides of the above expression with respect to x before attemping substitution for the variable x:

$$\int_a^b F(p, I_L(x))\frac{d}{dx}T(\frac{x}{-z(x)})dx = \int_a^b F(p, I_R(x))\frac{d}{dx}T(\frac{x}{-z(x)})dx \quad ,$$

where a and b are specific scene points. Now let us change the integration variable in the left-hand side of the above expression to x_L, and the integration variable in the right-hand side to x_R:

$$\int_{a_L}^{b_L} F(p, I_L(x_L))\frac{d}{dx_L}T(\frac{x_L}{f_L})dx_L = \int_{a_R}^{b_R} F(p, I_R(x_R))U(x_R)dx_R \quad , \quad (1)$$

where

$$U(x_R) = \frac{d}{dx_R}T(g_R(x_R) - \frac{(s + g_R(x_R)h)}{z(x_R)})$$

Equation (1) is our formulation of the stereo integral equation. Given that we have two image irradiance profiles that are matched at their end points – i.e., a_L and b_L in the left image correspond, respectively, to a_R and b_R in the right image – then Equation (1) expresses the relationship between the image irradiance profiles and the scene depth. It will be noted that the left-hand side of Equation (1) is composed of measurements that can be made in the left image of the stereo pair, while measurements in the right hand side are those that can be made in the right image. In addition, the right-hand side has a function of the scene depth as a variable. Our goal is to recover z as a function of the right-image coordinates x_R, not as a function of the world coordinates x. Once we have $z(x_R)$, we can transform it into any coordinate frame whose relationship to the image coordinates of the right image is known.

The recovery of $z(x_R)$ is a two-stage process. After first solving Equation (1) for $U(x_R)$, we integrate the latter to find $z(x_R)$ by using

$$T(g_R(x_R) - \frac{(s + g_R(x_R)h)}{z(x_R)}) = T(g_R(a_R) - \frac{(s + g_R(a_R)h)}{z(a_R)}) + \int_{a_R}^{x_R} U(x'_R)dx'_R \, .$$

In this expression one should note that $z(a_R)$ is known, since a_R and a_L are corresponding points.

It is instructive as regards the nature of the formulation if we look at the means of solving this equation when we have discrete data. In particular, let us take another look at an example previously introduced, namely,

$$F(p, I) = I^p$$

$$T(\frac{x}{-z}) = \ln(\frac{x}{-z})$$

and hence

$$\int_{a_L}^{b_L} \frac{I_L{}^p(x_L)}{x_L} dx_L = \int_{a_R}^{b_R} I_R{}^p(x_R) U(x_R) dx_R \quad ,$$

and then

$$z(x_R) = \frac{(s + g_R(x_R)h)}{g_R(x_R) - K \exp \int_{a_R}^{x_R} U(x'_R) dx'_R} \quad ,$$

where

$$K = (g_R(a_R) - \frac{(s + g_R(a_R)h)}{z(a_R)})$$

Suppose that we have image data at points $x_{L1}, x_{L2}, \ldots, x_{Lq}$ that lie between the left integral limits and, similarly, that we have data from the right image, between its integral limits, at points $x_{R1}, x_{R2}, \ldots, x_{Rn}$. Further, let us approximate the integrals as follows:

$$\sum_{j=1}^{q} \frac{I_L{}^p(x_{Lj})}{x_{Lj}} = \sum_{j=1}^{n} I_R{}^p(x_{Rj}) U(x_{Rj})$$

In actual calculation, we may wish to use a better integral formula than that above, (particularly at the end points), but this approximation enables us to demonstrate the essential ideas without being distracted by the details. Although the above approximation holds for all values of p, let us take a finite set of values, $p_1, p_2, \ldots, p_m$, and write the approximation out as a matrix equation, namely,

$$\begin{bmatrix} I_R{}^{p_1}(x_{R1}) & I_R{}^{p_1}(x_{R2}) & \ldots & I_R{}^{p_1}(x_{Rn}) \\ I_R{}^{p_2}(x_{R1}) & I_R{}^{p_2}(x_{R2}) & \ldots & I_R{}^{p_2}(x_{Rn}) \\ \ldots & \ldots & \ldots & \ldots \\ I_R{}^{p_n}(x_{R1}) & I_R{}^{p_n}(x_{R2}) & \ldots & I_R{}^{p_n}(x_{Rn}) \\ \ldots & \ldots & \ldots & \ldots \\ I_R{}^{p_m}(x_{R1}) & I_R{}^{p_m}(x_{R2}) & \ldots & I_R{}^{p_m}(x_{Rn}) \end{bmatrix} \begin{bmatrix} U(x_{R1}) \\ U(x_{R2}) \\ \ldots \\ U(x_{Rn}) \end{bmatrix}$$

$$= \begin{bmatrix} \sum_{j=1}^{q} \frac{I_L{}^{p_1}(x_{Lj})}{x_{Lj}} \\ \sum_{j=1}^{q} \frac{I_L{}^{p_2}(x_{Lj})}{x_{Lj}} \\ \ldots \\ \sum_{j=1}^{q} \frac{I_L{}^{p_n}(x_{Lj})}{x_{Lj}} \\ \ldots \\ \sum_{j=1}^{q} \frac{I_L{}^{p_m}(x_{Lj})}{x_{Lj}} \end{bmatrix}$$

Let us now recall what we have done. We have taken a set of image measurements, along with measurements that are just some non-linear functions of these image measurments, multiplied them by a function of the depth, and expressed the relationship between the measurements made in the right and left images. Why should one set of measurements, however purposefully manipulated, provide enough constraints to find a solution with almost the same number of variables as there are image measurements? The matrix equation helps in our understanding of this. First, we are not trying to find the solution for the scene depth at each point independently, but rather for all the points simultaneously. Second, we are exploiting the fact that, if the functions of image irradiance used by us are nonlinear, then each equation represented in the above matrix is linearly independent and constrains the solution. There is another way of saying this: even though we have only one set of measurements, requiring that the one depth profile relates the irradiance profile in the left image to the irradiance profile in the right image, and also relates the irradiance squared profile in the left image to the irradiance squared profile in the right image, and also relates the irradiance cubed profile etc., provides constraints on that depth profile.

The question arises as to whether there are sufficient constraints to enable a unique solution to the above equations to be found. This question really has three parts. Does an integral equation of the form of Equation (1) have a unique solution? This is impossible to answer when the irradiance profiles are unknown; even when they are known an exceedingly difficult problem confronts us [2,4]. Does the discrete approximation, even with an unlimited number of constraints, have the same solution as the integral equation? Again this is extremely difficult to answer even when the irradiance profiles are known. The final question relates to the finite set of constraint equations, such as those shown above. Does the matrix equation have a unique solution, and is it the same as the solution to the integral equation? Yes, it does have an unique solution – or at least we can impose solution requirements that makes a unique answer possible. But the question that asks whether the solution we find is a solution of the integral equation remains unanswered. From an empirical standpoint, we would be satisified if the solution we recover is a believable depth profile. Issues about sensitivity to noise, function type, and the form of the integral approximation will be discussed later in the section on solution methods.

Let us return to considerations of the general equation, Equation (1). We have just remarked upon the difficulty of solving this equation, so any additional constraints we can impose on the solution are likely to be beneficial. In the previous section on geometrical constraints, we noted that an acceptable solution has $z < 0$ and hence $z(x_R) < 0$. Unfortunately, solution methods for matrix equations (that have real coefficients) find solutions that are usually unrestricted over the domain of the real numbers. To impose the restriction of $z(x_R) < 0$, we follow the methods of Stockham [7]; instead of using the function itself, we formulate the problem in terms of the logarithm of the function. Consequently, in Equation (1) we usually set $T(\frac{x}{-z}) = \ln(\frac{x}{-z})$, just as we have done in our example. It should be noted that use of the logarithm also restricts $x > 0$ if $z < 0$. To construct the $x < 0$ side of the stereo reconstruction problem, we have to employ reflected coordinate systems for the world and image coordinates. Use of the logarithmic function ensures $z < 0$ and allows us to use standard matrix methods for solving the system of constraint equations. Once we have found the solution to the matrix equation, we can integrate that solution to find the depth profile.

In our previous example, we picked $F(p, I) = I^p$. In our experiments, we have used combinations of different functions to establish a particular matrix equation. For example we have used functions such as

$$\begin{aligned} F(p, I) &= |\cos pI|^p \\ &= (p + I)^{\frac{p}{2}} \\ &= p^I \\ &= \sin pI \\ &= (p + I)^{\frac{1}{2}} \end{aligned}$$

and we often use image density rather than image irradiance.

The point to be made here is that the form of the function F in the general equation is unrestricted, provided that it is nonlinear.

Equation (1) provides a framework for investigating stereo reconstruction in a manner that exploits the global nature of the solution. This framework arises from the realization that nonlinear functions provide a means of creating an arbitary number of constraints on that solution. In addition, the framework provides a means of avoiding the correspondence problem, except at the end points, for we never match points. Solutions have the same resolution as the data and this allows us to avoid the interpolation problem.

5 Solution Methods

Equation (1) is an inhomogeneous Fredholm equation of the first kind whose kernel function is the function $F(p, I_R(x_R))$. To solve this equation, we create a matrix equation in the manner previously shown in our example. We usually approximate the integral with the trapezoidal rule, where the sample spacing is that corresponding to the image resolution. Typically we use more than one functional form for the function F, each of which is parameterized by p. We have noticed that the sensitivity of the solution to image noise is affected by the choice of these functions, although we have not yet characterized this relationship. In the matrix equation, we usually pick the number of rows to be approximately twice the number of columns. However, owing to the rank-deficient nature of the matrix and hence to the selection of our solution technique, the solution we recover is only marginally different from the one obtained when we use square matrices.

Unfortunately, there are considerable numerical difficulties associated with solving this type of integral equation by matrix methods. Such systems are often ill-conditioned, particularly when the kernel function is a smooth function of the image coordinates. It is easy to see that, if the irradiance function varies smoothly with image position, each column of the matrix will be almost linearly dependent on the next. Consequently, it is advisible to assume that the matrix is rank-deficient and to utilize a procedure that can estimate the actual numerical rank. We use singular-valued decomposition to estimate the rank of the matrix; we then set the small singular values to zero and find the pseudoinverse of the matrix. Examples of results obtained with this procedure are shown in the following section.

An alternative approach to solving the integral equation is to decompose the kernel function and the dependent variable into orthogonal functions, then to solve for the coefficients of this decomposition, using the aforementioned techniques. We have used Fourier spectral decomposition for this purpose. The Fourier coefficients of the depth function were then calculated by solving a matrix equation composed of the Fourier components of image irradiance. However, the resultant solution did not vary significantly from that obtained without spectral decomposition.

While the techniques outlined can handle various cases, they are not as robust as we would like. We are actively engaged in overcoming the difficulties these solution methods encounter because of noise and irradiance discontinuities.

6 Results and Discussion

Our examples make use of synthetic image profiles that we have produced from known surface profiles. The irradiance profiles were generated under the assumptions that the surface was a Lambertian reflector and that the source of illumination was a point source directly above the surface. This choice was made so that our assumption concerning image irradiance, namely, that $I(x_L) = I(x_R)$ at matched points, would be complied with. In addition, synthetic images derived from a known depth profile allow comparison between the recovered profile and ground truth. Nonetheless, our goal is to demonstrate these techniques on real-world data. It should be noted that the examples used have smooth irradiance profiles; they therefore represent a worst case for the numerical procedures, as the matrix is most ill-conditioned under these circumstances.

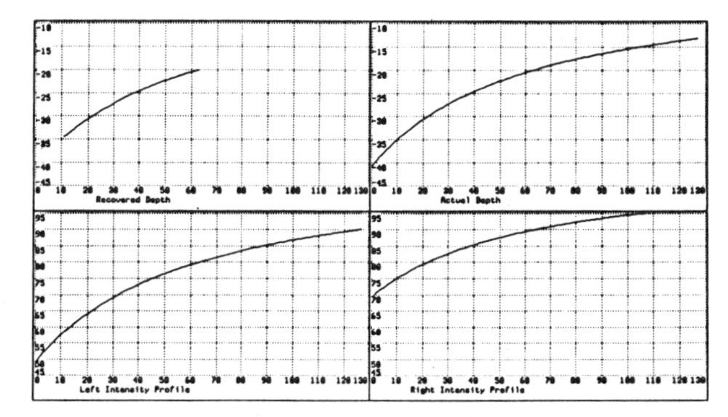

Figure 2: Planar Surface. At the upper left is depicted the recovered depth from the two irradiance profiles shown in the lower half. For comparison, the actual depth is shown in the upper right.

Our first example, illustrated in Figure 2, is of a flat surface with constant albedo. In the lower half of the figure, the left and right irradiance profiles are shown, while in the upper right, ground truth – the actual depth profile as a function of the image coordinates of the right image, x_R – is shown. The upper left of the figure contains the recovered solution. The limits of the recovered solution correspond to our selection of the integral end points. This solution was obtained from a formulation of the problem in which we used image density instead of irradiance in the kernel of the integral equation, and for which the function T was $\ln(\frac{x}{-z(x)})$.

The second example, Figure 3, shows a spherical surface with constant albedo, except for the stripe we have painted across the surface. The recovered solution was produced from the same formulation of the problem as in the previous example. The ripple effects in the recovered profile appear to have been induced by the details of the recovery procedure; the attendant difficulties are in part numerical in nature. However, any changes made in the actual functions used in the kernel of the equation do have effects that cannot be dismissed as numerical inaccuracies.

As we add noise to the irradiance profiles, the solutions tend to become more oscillatory. Although we suspect numerical problems, we have not yet ascertained the method's range of effectiveness. This aspect of our approach, however, is being actively investigated.

In the formulation presented here, we have used a particular function of the stereo geometry, $\frac{x}{-z}$, in the derivation of Equation (1) but we are not limited to this particular form. Its attractiveness is based on the fact that, if we use this particular function of the geometry, the side of the integral equation related to the left image is independent of the scene depth. We have used other functional forms but these result in more complicated integral equations. Equations of these forms have been subjected to relatively little study in the mathematical literature. Consequently, the effectiveness of solution methods on these forms remains un-

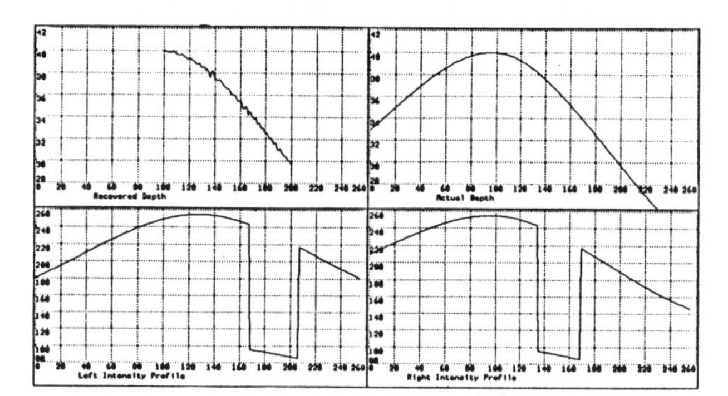

Figure 3: Spherical Surface with a Painted Stripe.

known.

In most of our study we have used $T(\frac{x}{-z})$ to be $\ln(\frac{x}{-z})$ and the properties of this particular formulation should be noted. It is necessary to process the right half of the visual field separately from the left half. The integral is more sensitive to image measurements near the optical axis than those measurements off-axis. In fact, the irradiance is weighted by the reciprocal of the distance off-axis. If we were interested in an integral approximation exhibiting uniform error across the extent of that integral, we might expect measurements that had been taken at interval spacings proportional to the off-axis distance to be appropriate. While it is obvious that two properties of a formulation that match those of the human visual system do not in themselves give cause for excitement it is worthy of note that the formulation presented is at least not at odds with the properties of the human stereo system.

On balance, we must say that significant work still lies ahead before this method can be applied to real-world images. While the details of the formulation may be varied, the overall form presented in Equation (1) seems the most promising. Nonetheless, solution methods for this class of equations are known to be difficult and, in particular, further efforts towards the goal of selecting appropriate numerical procedures are essential.

In formulating the integral equation, we took a function of the image irradiance and multiplied it by a function of the stereo geometry. To introduce image measurements, we changed variables in the integrals. If we had not used the derivative of the function of the stereo geometry, we would have had to introduce terms like $\frac{dx}{dx_R}$ and $\frac{dx}{dx_L}$ into the integrals. By introducing the derivative we avoided this. However, we did not really have to select the function of the geometry for this purpose; we could equally well have introduced the derivative through the function of image irradiance. Then we would have exchanged the calculation of irradiance gradients for the direct recovery of scene depth (thus eliminating the integration step we now use). Our selection of the formulation presented here was based on the belief that irradiance gradients are quite susceptible to noise; consequently we prefered to integrate the solution rather than differentiate the data. In a noise-free environment, however, both approaches are equivalent (as integration by parts will confirm).

7 Conclusion

The formulation presented herein for the recovery of scene depth from a stereo pair of images is based not on matching of image features, but rather on determining which surface in the world is consistent with the pair of image irradiance profiles we see. The solution method does not attempt to determine the nature of the surface locally; it looks instead for the best global solution. Although we have yet to demonstrate the procedure on real images, it does offer the potential to deal in a new way with problems associated with albedo change, occlusions, and discontinuous surfaces. It is the approach, rather than the details of a particular formulation, that distinguishes this method from conventional stereo processing.

This formulation is based on the observation that a global solution can be constrained by manufacturing additional constraints from nonlinear functions of local image measurements. Image analysis researchers have generally tried to use linear-systems theory to perform analysis; this has led them, consequently, to replace (at least locally) nonlinear functions with their linear approximation. Here we exploit the nonlinearity; "What is one man's noise is another man's signal."

While the presentation of the approach described here is focussed upon stereo problems, its essential ideas apply to other image analysis problems as well. The stereo problem is a convenient problem on which to demonstrate our approach; the formulation of the problem reduces to a linear system of equations, which allows the approach to be investigated without diversion into techniques for solving nonlinear systems. We remain actively interested in the application of this methodology to other problems, as well as in the details of the numerical solution.

References

[1] Boult, T.E., and J.R. Kender, "On Surface Reconstruction Using Sparse Depth Data," *Proceedings: Image Understanding Workshop*, Miami Beach, Florida, December 1985.

[2] Courant, R., and D. Hilbert, *Methods of Mathematical Physics*, Interscience Publishers, Inc., New York, 1953.

[3] Grimson, W.E.L., "An Implementation of a Computational Theory of Visual Surface Interpolation," *Computer Vision, Graphics, and Image Processing*, Vol. 22, pp 39-69, April 1983.

[4] Hildebrand, F.B., *Methods of Applied Mathematics*, 2nd ed., Prentice-Hall, Inc., Englewood Cliffs, New Jersey, 1965.

[5] Smith, G.B., "A Fast Surface Interpolation Technique," *Proceedings: Image Understanding Workshop*, New Orleans, Louisiana, October 1984.

[6] Smith, G.B., "Stereo Reconstruction of Scene Depth," *IEEE Proceedings on Computer Vision and Pattern Recognition*, San Francisco, California, June 1985, pp 271-276.

[7] Stockham, T.G., "Image Processing in the Context of a Visual Model," *Proceeding of IEEE*, Vol. 60, pp 828-842, July 1972.

[8] Terzopoulos, D., "Multilevel Computational Processes for Visual Surface Reconstruction," *Computer Vision, Graphics, and Image Processing*, Vol. 24, pp 52-96, October 1983.

Recovering the Camera Parameters from a Transformation Matrix

Thomas M. Strat
SRI International
333 Ravenswood Avenue
Menlo Park, California 94025

Abstract

The transformation of the three-dimensional coordinates of a point to the two-dimensional coordinates of its image can be expressed compactly as a 4 x 4 homogeneous coordinate transformation matrix in accordance with a particular imaging geometry. The matrix can either be derived analytically from knowledge about the camera and the geometry of image formation, or it can be computed empirically from the coordinates of a small number of three-dimensional points and their corresponding image points. Despite the utility of the matrix in image understanding, motion tracking, and autonomous navigation, very little is understood about the inverse problem of recovering the projection parameters from its coefficients. Previous attempts have produced solutions that require iteration or the solution of a set of simultaneous nonlinear equations. This paper shows how the location and orientation of the camera, as well as the other parameters of the image-formation process can easily be computed from the homogeneous coordinate transformation matrix. The problem is formulated as a simple exercise in constructive geometry and the solution is both noniterative and intuitively understandable.

1 Introduction

Homogeneous coordinates and the ***homogeneous coordinate transformation matrix*** are a convenient means for representing arbitrary transformations, including perspective projection in a single formalism. One such use for this matrix is as a ***camera transformation matrix*** that maps points in an object-centered coordinate system into the corresponding points in image coordinates according to a particular imaging geometry [7]. The camera transformation matrix has seen wide use in several disciplines. Rogers and Adams present numerous applications in computer graphics [8]. Other fields that have made use of the camera transformation matrix include stereo reconstruction, robot vision, photogrammetry, unmanned-vehicle guidance, and image understanding [5], [6], [12], [11]. Several techniques for computing the matrix have been derived, yet very little is understood about how to recover the projection parameters from the coefficients of the matrix.

When the location and orientation of the camera are known, the camera transformation matrix that models the image formation process can easily be derived analytically [1]. This model forms the basis for subsequent processing of images produced by that camera. On the other hand, when the location and orientation of the camera are unknown, the parameters of image formation must be derived from the correspondence between a set of image features and a set of object features. Images obtained from an unknown source or from cameras mounted on moving platforms exemplify contexts in which the imaging geometry may not be known.

One approach to computing the parameters of the image formation process directly is embodied in RANSAC, developed by Fischler and Bolles [3]. RANSAC computes the camera location directly from a set of landmarks with known three-dimensional locations when, in addition, the focal length and piercing point are known.

Alternatively, several methods exist for estimating the coefficients of the camera transformation matrix from the correspondence between image and object coordinates. Sutherland [10] describes a method to determine the matrix experimentally from the image by using a least-squares technique to obtain the coefficients from available ground truth data. A consideration of the experimental errors involved and a means for improving accuracy are described by Sobel [9].

The issue addressed in this paper is how to determine the imaging geometry from a camera transformation matrix that has been derived experimentally. For example, given a photograph taken by an unknown camera from an unknown location and which, moreover, may have been cropped and/or enlarged, how can we recover the camera's position and orientation and determine the extent to which the picture was cropped or enlarged? If some ground truth data are available, an established least-squares technique such as Sutherland's can be used to derive the camera transformation matrix, whereupon the problem reduces to that of computing the values of the desired parameters from the matrix. Ganapathy [4] recently published the first noniterative method for solving this problem by posing it as a set of eleven simultaneous nonlinear equations that can be solved to obtain the eleven independent coefficients of the camera transformation matrix. While the method is successful at solving for camera location and orientation, it is an algebraic one that provides little insight into the underlying geometry. The method to be described here is a geometric one that solves for the same parameters, but is posed as a simple problem in constructive geometry and allows an intuitively clear derivation.

This work has immediate application in several areas:

- Many algorithms in image understanding require knowledge of the camera parameters. These can be computed from an arbitrary photograph by using the method presented here when ground truth data is available.

- Autonomous navigation can be posed as a problem in deriving camera parameters. A cruise missile, for instance, could obtain the camera transformation matrix from a terrain model stored on board and then compute the camera parameters that define the vehicle's location and heading.

- A stationary camera viewing a robot arm workspace could determine the position and orientation of the arm. Conspicuous marking of several points on a part of the manipulator would allow their easy extraction from an image and provide the ground truth necessary for Sutherland's algorithm to ascertain the camera transformation matrix. The camera parameters can be derived from this matrix, and the location and orientation of the manipulator can then be obtained relative to the stationary camera.

2 The Camera Transformation Matrix

As indicated earlier, the camera transformation matrix can be used to model in a single formalism the effects of rotation, translation, perspective, scaling, and cropping—*i.e.*, all the variables associated with the normal imaging process. Here we review the fundamentals of homogeneous coordinate transformations that are essential for understanding the decomposition to be described. The presence of an ideal lens and the absence of any atmospheric distortions are assumed.

The imaging situation can be modeled as shown in Figure 1. The XYZ coordinate system represents the world or object-centered coordinates. The *center of projection* (the location of the lens) is shown as a point L in space. The *image plane* is a plane between the lens and the object onto which the object is projected to obtain the image. Each *image point* is that point in the image plane where the plane intersects the line connecting L with the corresponding *object point.* The UVW coordinate system is situated such that (u, v) are the image coordinates of an image point and $w = 0$ defines the image plane. The perpendicular distance between L and the image plane is the camera *focal length*, f.

In a homogeneous coordinate system, a three-dimensional point (x, y, z) is represented as a four-component row vector, (tx, ty, tz, t); the three-dimensional coordinates are obtained by dividing through by the fourth component. A point in the world is represented as a four-component row vector and its projection in the image is obtained by postmultiplying by the 4 x 4 camera transformation matrix:

$$\mathbf{x}M = \mathbf{u}$$

$$(x, y, z, 1)M = (su, sv, sw, s)$$

This homogeneous coordinate system is most useful for modeling the effects of perspective projection—further details can be found in Ballard and Brown [1].

The matrix M can be viewed as being composed of several simple transformations. While it is possible to decompose the matrix in a variety of ways, the particular decomposition chosen must capture all the degrees of freedom of the imaging geometry. The somewhat arbitrary choice used throughout this paper is shown below:

$$M = (translate)(rotate)(project)(scale)(crop)$$

$$M = TRPSC \tag{1}$$

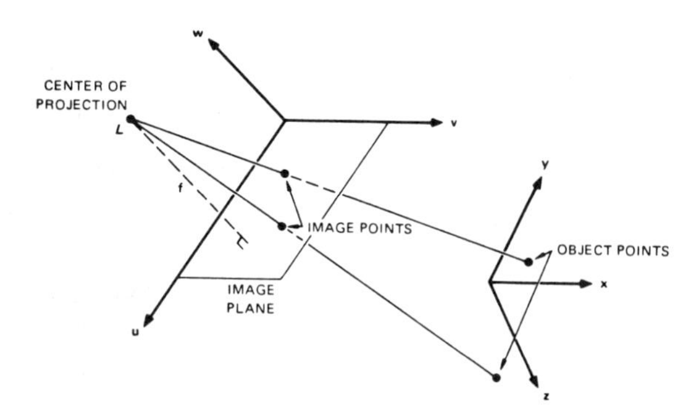

FIGURE 1 THE IMAGING GEOMETRY

Each of the component transformations can be expressed as a 4 x 4 matrix; multiplying them together produces the camera transformation matrix M. Details of the decomposition are given below.

2.1 Translation

Translation moves the image plane away from the object-centered origin. To translate the plane by (x_0, y_0, z_0) multiply by the matrix

$$T = \begin{bmatrix} 1 & 0 & 0 & 0 \\ 0 & 1 & 0 & 0 \\ 0 & 0 & 1 & 0 \\ -x_0 & -y_0 & -z_0 & 1 \end{bmatrix}.$$

2.2 Rotation

The orientation of the camera is specified by the rotation matrix R, which can be further decomposed to $R = R_x R_y R_z$, corresponding to rotation about each of the principal axes. Clockwise rotation by θ about the X axis while looking toward the origin is accomplished by

$$R_x = \begin{bmatrix} 1 & 0 & 0 & 0 \\ 0 & \cos\theta & -\sin\theta & 0 \\ 0 & \sin\theta & \cos\theta & 0 \\ 0 & 0 & 0 & 1 \end{bmatrix}.$$

Similarly, clockwise rotation by ϕ about the newly rotated Y axis is represented by

$$R_y = \begin{bmatrix} \cos\phi & 0 & \sin\phi & 0 \\ 0 & 1 & 0 & 0 \\ -\sin\phi & 0 & \cos\phi & 0 \\ 0 & 0 & 0 & 1 \end{bmatrix}.$$

and rotation by ψ about the new Z axis is given by

$$R_z = \begin{bmatrix} \cos\psi & -\sin\psi & 0 & 0 \\ \sin\psi & \cos\psi & 0 & 0 \\ 0 & 0 & 1 & 0 \\ 0 & 0 & 0 & 1 \end{bmatrix}.$$

The first two rotations, R_x and R_y, serve to align the Z axis with the line of sight defined by the W axis. The final rotation, R_z, is within the image plane about the line of sight. Together, T and $R = R_x R_y R_z$ account for the location and orientation of the camera.

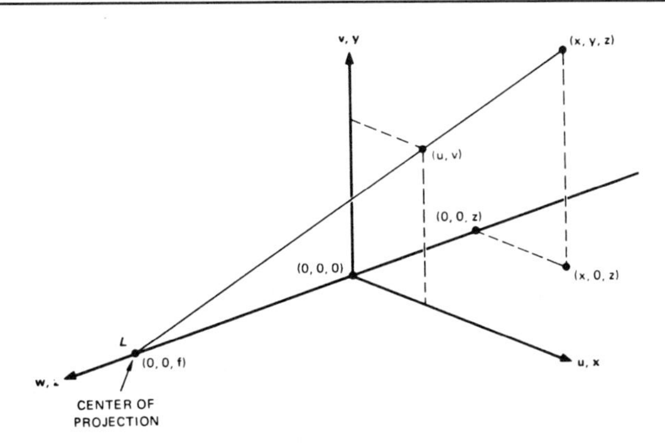

FIGURE 2 PERSPECTIVE PROJECTION AFTER ROTATION AND TRANSLATION

z = 0 is the image plane.

2.3 Projection

P provides the distortion associated with perspective projection. Figure 2 shows the simplified imaging geometry after translation and rotation have been accounted for. The camera location L is on the positive Z axis at a distance f from the origin. The image plane passes through the origin and the world to be imaged lies behind the image plane. Analysis of similar triangles shows that the image coordinates

$$(u,v) = \left(\frac{fx}{f-z}, \frac{fy}{f-z}\right).$$

Using homogeneous coordinates, this perspective projection can be obtained by multiplying the homogeneous coordinates of the world point by the matrix

$$P = \begin{bmatrix} 1 & 0 & 0 & 0 \\ 0 & 1 & 0 & 0 \\ 0 & 0 & 1 & -1/f \\ 0 & 0 & 0 & 1 \end{bmatrix}.$$

The resulting row vector is divided through by the fourth component to renormalize the homogeneous coordinates, and then projected orthographically onto the image plane $w = 0$ to yield the proper perspective projection of the world point.

2.4 Scaling

The image coordinates can be scaled to reflect an enlargement or shrinking of the image. Scaling by k_u and k_v in the U and V directions is achieved with

$$S = \begin{bmatrix} k_u & 0 & 0 & 0 \\ 0 & k_v & 0 & 0 \\ 0 & 0 & 1 & 0 \\ 0 & 0 & 0 & 1 \end{bmatrix}.$$

Scaling the W axis is meaningless because the perspective projection always requires that $w = 0$.

2.5 Cropping

The effect of cropping a photograph is obtained by translating the UV coordinates within the image plane. The following matrix is used to shift the origin by (u_0, v_0):

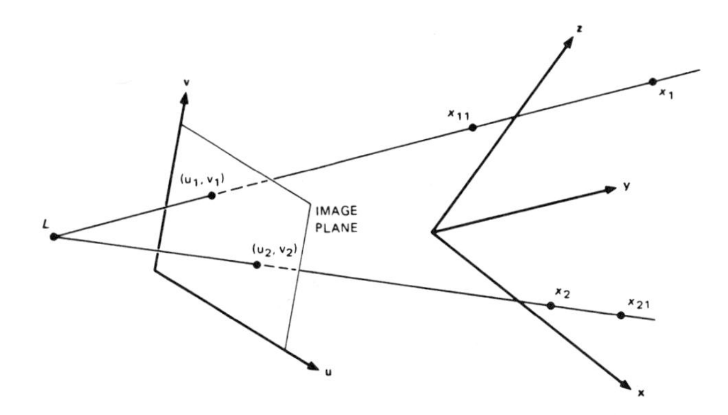

FIGURE 3 DETERMINING THE LOCATION OF THE CAMERA

$$C = \begin{bmatrix} 1 & 0 & 0 & 0 \\ 0 & 1 & 0 & 0 \\ 0 & 0 & 1 & 0 \\ -u_0 & -v_0 & 0 & 1 \end{bmatrix}.$$

Note that neither scaling nor cropping affects the w and s coordinates of the homogeneous image point, so that orthographic projection and renormalization can take place after the entire transformation has been computed.

3 Recovering the Camera Parameters

The camera transformation matrix M allows representation of all eleven degrees of freedom associated with the image formation process. These camera parameters are embedded in the matrix in a way that makes their determination difficult. This section presents a simple method that recovers the various parameters associated with image formation. Its main advantages are that it is both noniterative and geometric, enabling a clear understanding of the equations involved.

The matrix M can be viewed as a function that maps world coordinates into image coordinates according to the constraints of Figure 1. For notational simplicity, we shall assume that all matrix multiplications automatically normalize the homogeneous coordinates of the resulting row vector. For example, $\mathbf{u} = \mathbf{x}M = (su, sv, sw, s) = (u, v, w, 1)$. The image formation process can then be written as

$$(u, v, 0, 1) = orthoproject(\mathbf{x}M), \qquad (2)$$

where x is the homogeneous coordinate of a world point, and $orthoproject(\cdot)$ is a function that performs an orthographic projection along the w axis such that

$$orthoproject(u, v, w, 1) = (u, v, 0, 1).$$

3.1 Location of the Camera

Figure 3 illustrates the technique for finding the center of projection. First, compute M^{-1} for later use. Note that M will always be invertible because all its components in Equation 1 are clearly invertible. The location of the center of projection, L, can be determined as follows:

Choose an arbitrary world point $\mathbf{x}_1 = (x_1, y_1, z_1, 1)$ and compute $\mathbf{u}_1 = \mathbf{x}_1 M$. If we were to multiply $\mathbf{u}_1 = (u_1, v_1, w_1, 1)$ by M^{-1}, we would obtain the original $\mathbf{x}_1$. Instead, first project $\mathbf{u}_1$

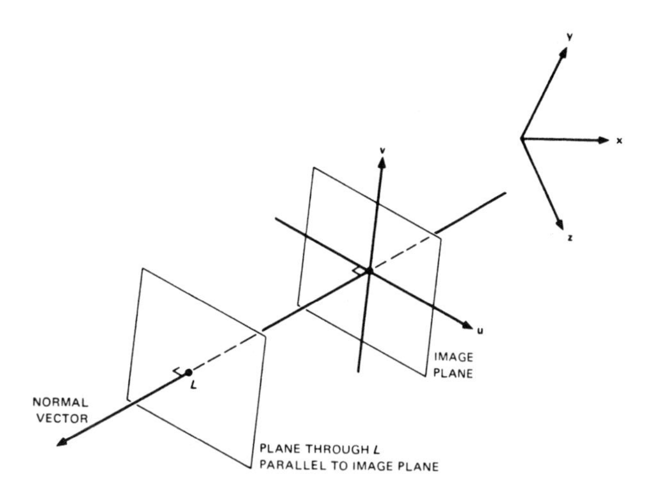

FIGURE 4 DETERMINING THE ORIENTATION OF THE CAMERA

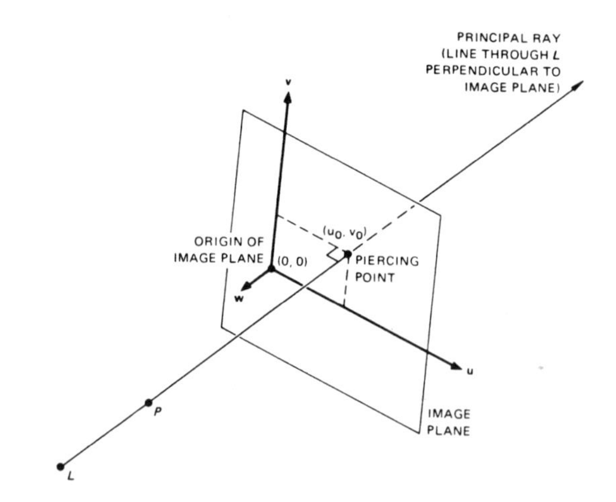

FIGURE 5 DETERMINING THE PIERCING POINT

to obtain $orthoproject(\mathbf{u}_1) = (u_1, v_1, 0, 1)$, where (u_1, v_1) are the image coordinates of $\mathbf{x}_1$. Next, backproject this image point by multiplying by M^{-1} to obtain $\mathbf{x}_{11}$. This specifies another world point, $\mathbf{x}_{11}$, which is different from the original $\mathbf{x}_1$ but constrained to lie along the line connecting $\mathbf{x}_1$ and the center of projection L. To confirm this, note that all points lying along the line connecting $\mathbf{x}_1$ and L are transformed by M to points identified by $(u_1, v_1, w, 1)$, where w varies with each point. The converse must also be true. That is, for any w, $(u_1, v_1, w, 1)M^{-1}$ specifies a point somewhere along the line connecting $\mathbf{x}_1$ and L.

Repeat the above process with another point $\mathbf{x}_2$ to obtain the point $\mathbf{x}_{21}$, which must lie on the line connecting $\mathbf{x}_2$ and L. Now $\mathbf{x}_1$ and $\mathbf{x}_{11}$, and $\mathbf{x}_2$ and $\mathbf{x}_{21}$ define two lines that pass through L; their intersection can be computed to obtain the world coordinates of L. This method will fail, of course, if either $\mathbf{x}_1$ or $\mathbf{x}_2$ lie in the image plane or if $\mathbf{x}_1$, $\mathbf{x}_2$ and L are colinear. Because their choice is arbitrary, valid points can always be found that allow the unique determination of L.

3.2 Orientation of the Camera

The orientation of the camera is defined by the orientation of the image plane (Figure 4). The latter can easily be established by observing that world points lying in the plane that is parallel to the image plane and that passes through the center of the lens will map to infinity in image coordinates. The only way this can happen for a finite world point is if the fourth component of the homogeneous image coordinate is zero.

Thus, if

$$(x, y, z, 1)M = (u, v, w, 0),$$

it follows that

$$M_{14}x + M_{24}y + M_{34}z + M_{44} = 0,$$

which is the equation of the plane through L parallel to the image plane. From this equation it is clear that the vector $\mathbf{n} = (M_{14}, M_{24}, M_{34})$ is normal to the image plane and parallel to the camera's direction of view.

The orientation in terms of rotations about the axes can be calculated by using spherical coordinates such that

$$\theta = \arctan \frac{M_{24}}{-M_{34}}$$

and

$$\phi = \arcsin \frac{-M_{14}}{\sqrt{M_{14}^2 + M_{24}^2 + M_{34}^2}}$$

where θ is the clockwise rotation about the X axis and ϕ is the clockwise rotation about the rotated Y axis. The final rotation parameter, ψ, is the rotation within the image plane about the W axis. The magnitude of ψ cannot be obtained from the normal to the image plane; instead, it requires a more complex derivation that involves determination of the piercing point and the relative scale factors. These values are derived in the following sections and the value of ψ is finally computed in Section 3.5.

3.3 Piercing Point, Principal Ray, and Cropping

Much work in image understanding requires knowledge of the *piercing point* (or *stare point*) in an image. This is the point in an image that corresponds to the world point at which the camera was aimed. It is the point at which the ***principal ray*** (the ray along which the camera was aimed) pierces the image plane (Figure 5). The principal ray is assumed to be perpendicular to the image plane.[1]

To find the piercing point, first find a point $\mathbf{p}$ along the principal ray (other than L):

$$\mathbf{p} = L + k\vec{n},$$

where k is any scalar except 0. The piercing point $\mathbf{u}_0$ is given by

$$\mathbf{u}_0 = orthoproject(\mathbf{p}M) = (u_0, v_0, 0, 1)$$

because any point along the principal ray must project to the piercing point in the image. The extent to which the image has been cropped is given by (u_0, v_0).

3.4 Focal Length and Scale

When the center of projection is held a constant distance from the scene, there is no way to tell the difference between scaling the image and varying the focal length. For example, doubling the focal length is equivalent to enlarging the picture by a factor of two. The best one can hope for is a relation between the two

[1]The image plane in some cameras used in photogrammetry is not perpendicular to the line of sight; this case, however, is not considered here.

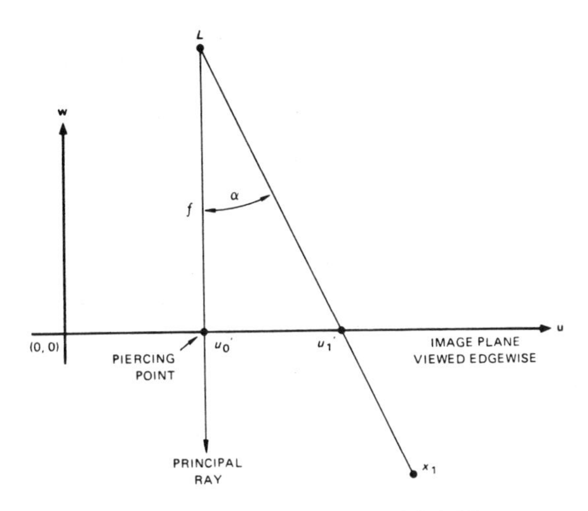

FIGURE 6 DETERMINING THE SCALE FACTOR

parameters. On the other hand, if the focal length of the camera is known, the exact scale factors can be determined.

Figure 6 shows the geometry for computing the U-component of the scale factor. First choose an arbitrary world point $\mathbf{x}_1$ (not on the principal ray) and compute its image point $\mathbf{u}_1 = orthoproject(\mathbf{x}_1 M) = (u_1, v_1, 0, 1)$. Conversion of these image units to world units requires dividing by the scale factor such that

$$u'_1 = \frac{u_1}{k_u} \quad \text{and} \quad v'_1 = \frac{v_1}{k_v},$$

where u'_1 and v'_1 are the distances of an image point from the image origin, measured in world units. Next compute α, the angle between the principal ray and the ray from L to $\mathbf{x}_1$, projected in the plane v=0. Then it is clear from the diagram that the following relation must hold

$$\tan\alpha = \frac{u'_1 - u'_0}{f} = \frac{\frac{u_1}{k_u} - \frac{u_0}{k_u}}{f},$$

where k_u is the magnification of the image in the U direction. If the focal length is known, the scale factor

$$k_u = \frac{u_1 - u_0}{f \tan\alpha}$$

or if the scale factor is known,

$$f = \frac{u_1 - u_0}{k_u \tan\alpha}.$$

The computation of k_v, the V-component of the scale factor, is identical. Neither k_u nor k_v can be determined individually without knowledge of the focal length, but their ratio can be calculated from quantities derived from the matrix:

$$\frac{k_u}{k_v} = \frac{u_1 - u_0}{f \tan\alpha_u} \Big/ \frac{v_1 - v_0}{f \tan\alpha_v} = \frac{(u_1 - u_0)\tan\alpha_v}{(v_1 - v_0)\tan\alpha_u}$$

3.5 Rotation within the Image Plane

We now return to the derivation of ψ, the rotation of the camera about the W axis. This rotation is equivalent to cropping the image at an angle to the UV coordinate system. The value of ψ is found by choosing a world point and comparing its transformation under two different situations, as illustrated in Figure 7.

First, use the coordinates of L computed earlier and an arbitrary focal length to reconstruct the translation matrix T. Use

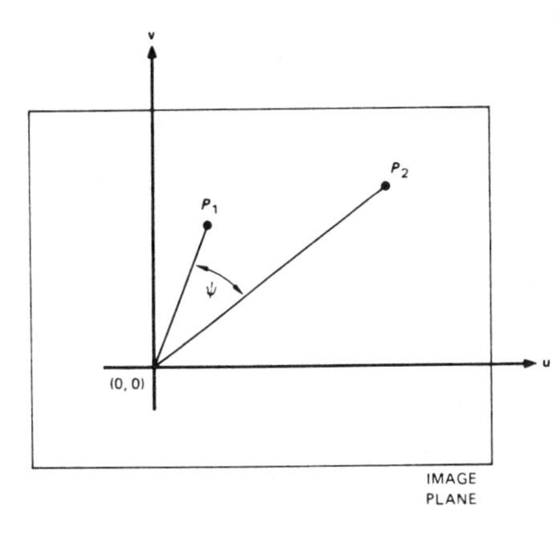

FIGURE 7 DETERMINING THE ROTATION WITHIN THE IMAGE PLANE

the values of θ and ϕ to reconstruct R_x and R_y and use the chosen focal length to construct a perspective projection matrix P'. This comprises a model $M_1 = TR_xR_yP'$, which can be employed to compute the transformation of an arbitrary world point. Call the resulting point $\mathbf{p}_1$.

Next, use the previously determined piercing point to reconstruct the matrix C. Then undo the effects of cropping from the camera model by multiplying the original camera transformation matrix by C^{-1} to obtain

$$M' = MC^{-1} = TRPSCC^{-1} = TRPS.$$

Now the effects of unbalanced scaling will be eliminated. Use the relative scale factor computed earlier to construct a scale transformation matrix:

$$S' = \begin{bmatrix} 1 & 0 & 0 & 0 \\ 0 & \frac{k_u}{k_v} & 0 & 0 \\ 0 & 0 & 1 & 0 \\ 0 & 0 & 0 & 1 \end{bmatrix}.$$

Then multiply M' by S' to obtain

$$M_2 = M'S' = TRPSS' = TRPS'',$$

where

$$S'' = SS' = \begin{bmatrix} k_u & 0 & 0 & 0 \\ 0 & k_u & 0 & 0 \\ 0 & 0 & 1 & 0 \\ 0 & 0 & 0 & 1 \end{bmatrix}.$$

Finally, use M_2 to compute the transformation of the previously chosen world point and call the result $\mathbf{p}_2$.

The angle ψ can now be determined by making use of Equation 1 and the following observation. The only differences between M_1 and M_2 are their focal lengths, a scale factor, and a rotation about the W axis. Although the scale factor is unknown, it is equal in the U and V directions because this was compensated for in computing M_2. Together the scale factor and focal length differences serve only to change the size of the image and impose no other distortions. Observe that $\mathbf{p}_1$ is the image point that would be obtained if there were no rotation about the W axis, no scaling of the image, and no cropping of the image. Similarly, $\mathbf{p}_2$ is the image point that is obtained by starting with the true image point associated with the chosen object point and undoing the effects of cropping and unbalanced scaling. Any difference between $\mathbf{p}_1$ and $\mathbf{p}_2$ must be the result of different-sized images or of rotation about the W axis. Since this rotation is centered about the origin of this coordinate space, the angle ψ can be determined by measuring the angle between $\mathbf{p}_1$ and $\mathbf{p}_2$ at the origin. The differing focal lengths and scale factors can affect only the distance of the points from the origin and cannot alter the angle between the points when measured from the origin.

4 Discussion

The method presented here provides a straightforward way of determining the parameters of the imaging process from a homogeneous coordinate transformation matrix. The geometric interpretation provides some insight into what the equations mean and when they may fail. The appendix illustrates application of the technique to several sets of real data.

In practice, we must be concerned with the robustness of such an algorithm and how it is affected by errors in the data. For example, if f or k are very large, the view angle subtended by the image is small and the projection is nearly orthogonal. In this case, the method becomes sensitive to the precision of the matrix, and only the camera's orientation can be ascertained with confidence. This property is intrinsic in the problem formulation, and any method that derives camera parameters from the correspondence between image and world coordinates is subject to this sensitivity. The parameters computed by the methods outlined in this paper can be used to reconstruct the camera transformation matrix (within a choice of focal length) when synthetic data are used. When empirical data are used, as in the appendix, instabilities in the matrix often make it impossible to reconstruct it with accuracy.

The camera model used throughout this paper is somewhat simplified. The image plane has been assumed to be perpendicular to the principal ray and the image axes are assumed to be perpendicular. Furthermore, the effects of a non-ideal lens and other nonisotropic distortions have been ignored. The accuracy of the decomposition will degrade if these assumptions are not valid. While we do not expect this technique to be any more robust than that of Ganapathy, we do feel that its geometric interpretation provides useful clues as to when it will be dependable. The method's utility has been demonstrated on actual photographs and because it is non-iterative, the computational burden is insignificant.

5 Acknowledgments

I would like to thank Robert Bolles for his comments and patience in debugging the software for least-squares determination of a camera transformation matrix, and for supplying that program in the first place. I am also grateful for discussions with Kicha Ganapathy as to the relation between this method and his own.

Bibliography

[1] Ballard, D. H., and Brown, C. M., *Computer Vision*, Prentice-Hall, New Jersey, 1982.

[2] Cameron, R., *Above San Francisco*, Cameron and Company, San Francisco, 1976.

[3] Fischler, M. A., and Bolles, R. C., "Random Sample Consensus: A Paradigm for Model Fitting with Applications to Image Analysis and Automated Cartography", Communications of the ACM, Vol. 24, No. 6, June 1981, pp. 381–395.

[4] Ganapathy, S., "Decomposition of Transformation Matrices for Robot Vision", IEEE, 1984, pp. 130–139.

[5] Gennery, D. B., "Stereo-Camera Calibration", Proceedings of the Image Understanding Workshop, November 1979, pp. 101–107.

[6] Lowe, D. G., "Solving for the Parameters of Object Models from Image Descriptions", IU Workshop, April 1980, pp. 121–127.

[7] Roberts, L. G., "Machine Perception of Three-Dimensional Solids", MIT Lincoln Lab, Technical Report No. 315, May 1963.

[8] Rogers, D. F., and Adams, J. A., *Mathematical Elements for Computer Graphics*, McGraw-Hill Book Company, New York, 1976.

[9] Sobel, I., "On Calibrating Computer Controlled Cameras for Perceiving 3-D Scenes", *Artificial Intelligence* 5, 1974, pp. 185–198.

[10] Sutherland, I. E., "Three Dimensional Data Input by Tablet", Proceedings of the IEEE, Vol. 62, No. 4, April 1974, pp. 453–461.

[11] Thompson, Morris M., *Manual of Photogrammetry*, American Society of Photogrammetry, Falls Church, Virginia, 1966.

[12] Ullman, S., *The Interpretation of Visual Motion*, The MIT Press, Cambridge, Massachusetts, 1979.

A Examples

We now present two examples to illustrate our technique.

A.1 Imagery from Robotics Applications

Ganapathy used the following experimentally determined 3 x 4 matrix to demonstrate his method [4].

$$\begin{bmatrix} -2.3819 & 0.49648 & -.039462 & 847.40 \\ -.043897 & -.062872 & -2.4071 & 882.91 \\ -.00026388 & -.00062759 & -.000071843 & 1.0 \end{bmatrix}$$

This matrix is used to obtain the image coordinates (us, vs, s) by premultiplying the world coordinates, (xt, yt, zt, t), by the matrix. To make it compatible with the notation used throughout this paper, it must be transposed and an arbitrary column vector inserted. This column is the one that determines w and does not affect the imaging process. The matrix, suitably rewritten, is

$$M_1 = \begin{bmatrix} -2.3819 & -.043897 & 0.0 & -.00026388 \\ 0.49648 & -.062872 & 0.0 & -.00062759 \\ -.039462 & -2.4071 & 1.0 & -.000071843 \\ 847.40 & 882.91 & 0.0 & 1.0 \end{bmatrix}$$

From this matrix the following values were obtained by the method presented in Section 3.

$$L = (620.51, 1295.68, 321.64)$$

This agrees closely with Ganapathy's determination of the camera's location:

$$(620.9344, 1295.476, 321.8140)$$

The orientation of the camera is computed from the normal to the image plane:

$$\vec{n} = (-.3855, -.9167, -.1049)$$

This yields (in degrees)

$$\theta = 157.1949 \quad \phi = -6.0239 \quad \psi = 359.909$$

in agreement with Ganapathy's results:

$$\theta = 157.1951 \quad \phi = -6.023912 \quad \psi = 359.6915$$

Other parameters obtained from M_1 include

piercing point: $(u_0, v_0) = (682, 478)$ in pixels

focal length: $f \cdot k_u = 3488$

$f \cdot k_v = 3485$

relative scale factor: $k_u/k_v = 1.0009$

A.2 Outdoor Imagery

Figure 8 shows a photograph taken from a book of pictures of San Francisco [2]. It was necessary to determine the imaging geometry in order to use the picture for work in image understanding. Ground truth data were obtained manually from a map of the city. A total of fifteen pairs of image and world coordinates were used to obtain the following camera transformation matrix with a least-squares program:

$$M_2 = \begin{bmatrix} .172137 & .131132 & 0.0 & .000346452 \\ -.15879 & .112747 & 0.0 & .000311253 \\ .0187902 & .291494 & 1.27976 & .0000656643 \\ 274.943 & 258.686 & 0.0 & 1.0 \end{bmatrix}$$

The results computed from this image are plotted on the map in Figure 9 and described further below. The camera location was computed to be near the intersection of California and Mason Streets at an elevation of 435 feet above sea level. The camera was oriented as shown on the map, at an angle of 8° above the horizon. Computation of the piercing point is sensitive to errors in the matrix because the projection is nearly orthographic, but the location derived is marked by the point P in the image in Figure 8. The focal length and scale factor relations were computed to be $f \cdot k_u = 495$ and $f \cdot k_v = 560$, indicating an aspect ratio of $k_u/k_v = .88$.

Figures 10 and 11 show the results for another photograph of San Francisco. The camera transformation matrix computed from 16 points of ground truth data is shown here:

$$M_3 = \begin{bmatrix} -.175451 & .0269801 & 0.0 & .000151628 \\ -.105205 & -.0963531 & 0.0 & -.00016085 \\ .0043556 & .23031 & 1.07834 & .0000159749 \\ 297.836 & 249.574 & 0.0 & 1.0 \end{bmatrix}$$

FIGURE 8 PHOTOGRAPH OF SAN FRANCISCO

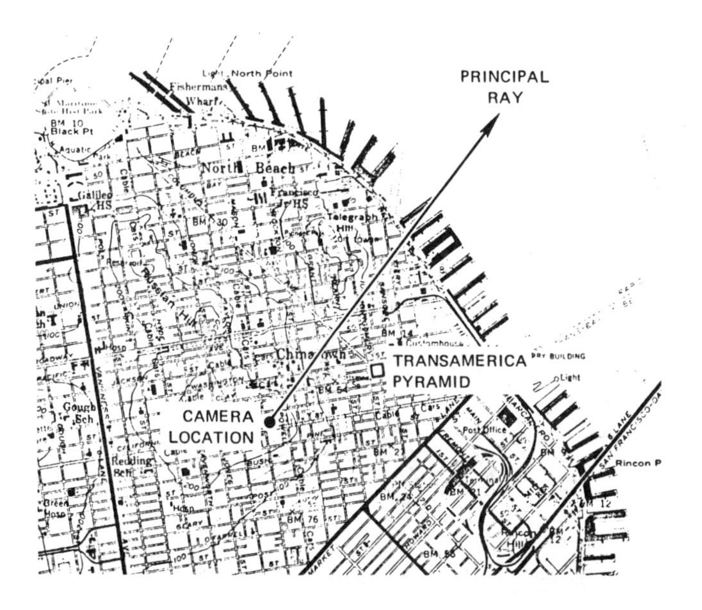

FIGURE 9 MAP OF SAN FRANCISCO

The elevation of the camera was found to be 1200 feet above sea level and the inclination was 4° above the horizon. The horizontal location and orientation are plotted in Figure 11. The piercing point was unreliably computed to be at the point P in Figure 10. The focal length and scale factor relations were $f \cdot k_u = 876$ and $f \cdot k_v = 999$, implying an aspect ratio of $k_u/k_v = .88$.

FIGURE 10 ANOTHER PHOTOGRAPH OF SAN FRANCISCO

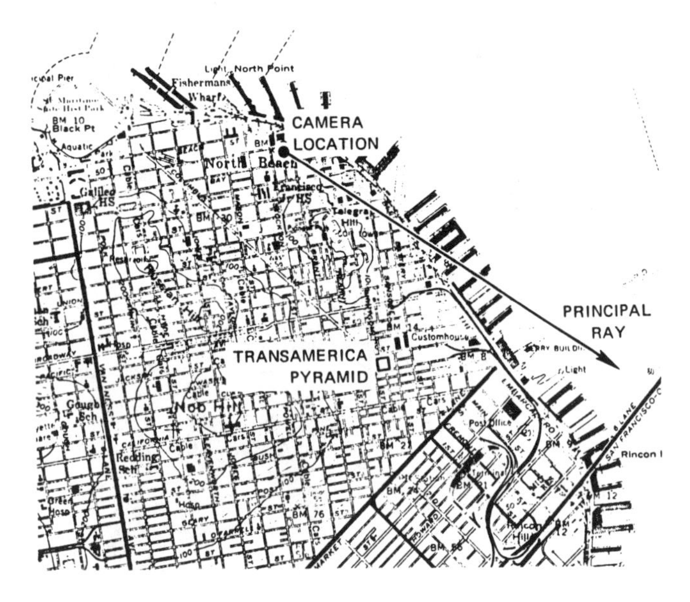

FIGURE 11 MAP OF SAN FRANCISCO

One-Eyed Stereo: A General Approach to Modeling 3-D Scene Geometry

THOMAS M. STRAT AND MARTIN A. FISCHLER

Abstract—A single two-dimensional image is an ambiguous representation of the three-dimensional world—many different scenes could have produced the same image—yet the human visual system is extremely successful at recovering a qualitatively correct depth model from this type of representation. Workers in the field of computational vision have devised a number of distinct schemes that attempt to emulate this human capability; these schemes are collectively known as "shape from . . ." methods (e.g., shape from shading, shape from texture, or shape from contour). In this paper we contend that the distinct assumptions made in each of these schemes is tantamount to providing a second (virtual) image of the original scene, and that each of these approaches can be translated into a conventional stereo formalism. In particular, we show that it is frequently possible to structure the problem as one of recovering depth from a stereo pair consisting of the supplied perspective image (the *original* image) and an hypothesized orthographic image (the *virtual* image). We present a new algorithm of the form required to accomplish this type of stereo reconstruction task.

Index Terms—Computer vision, image understanding, scene modeling, shape from shading, shape from texture, stereo.

I. Introduction

The recovery of 3-D scene geometry from one or more images, which we will call the scene-modeling problem (SMP), has solutions that appear to follow one of three distinct paradigms: stereo; optic flow; and shape from shading, texture, or contour.

In the stereo paradigm, we match corresponding world/scene points in two images and, given the relative geometry of the two cameras (eyes) that acquired the images, we can use simple trigonometry to determine the depths of the matched points [1].

In the optic-flow paradigm, we use two or more images to compute the image velocity of corresponding scene points. If the camera's motion and imaging parameters are known, we can again use simple trigonometry to convert velocity measurements in the image to depths in the scene [22].

In the shape from shading, texture, and contour (SSTC) paradigm, we must either know or make some assumptions about the nature of the scene, the illumination, and the imaging geometry. Brady's 1981 volume on computer vision [2] contains an excellent collection of papers, many of which address the problem of how to recover depth from the shading, texture, and contour information visible in a single image. Two distinct computational approaches have been employed in the SSTC paradigm: 1) integration of partial differential equations describing the relation of shading in an image to surface geometry in a scene, and 2) back-projection of planar image facets to undo the distortion in an image attribute (e.g., edge orientation) induced by the imaging process on an assumed scene property (e.g., uniform distribution of edge orientations).

Our purpose in this paper is to provide a unifying framework for the SMP and to present a new computational approach to recovering scene geometry from the shading, texture, and surface contour information in a single image. The value of a comprehensive theory is that it provides a basis for determining specific problems that remain unsolved and the inherent limitations of the existing paradigm. The ability to address a wide variety of imaging situations and scene conditions within a common framework can thus provide insight into the virtues and limitations of the currently highly specialized algorithms within the SSTC category. For example, some reflectance functions can be shown to be analogous to particular varieties of textures; hence, results appropriate to one type of modeling can be mapped onto the other. One-eyed stereo links existing approaches by providing a uniform interpretation of the diverse algorithms proposed for SSTC. It also extends the common paradigm by offering an analytic solution to the previously unsolved problem of reconstructing 3-D geometry from a pair of images when one is a true perspective projection and the other orthographic. Its value as a preferred computational technique for those SSTC problems with known solutions remains undecided.

Our main result is based on the following observation: regardless of the assumptions employed in the SSTC paradigm, if a 3-D scene model has been derived successfully, it will generally be possible to establish a large number of correspondences between image and scene (model) points. From these correspondences we can compute a collineation matrix [12] and then extract the imaging geometry from it [4], [20]. We can now construct a second image of the scene as viewed by the camera from some arbitrary location in space. It is thus obvious that any technique that is competent to solve the SMP must either be provided with at least two images or make assumptions that are equivalent to providing a second im-

Manuscript received June 20, 1985; revised February 6, 1986. Recommended for acceptance by S. W. Zucker. This work was supported by the Defense Advanced Research Projects Agency under Contract MDA903-83-C-0027.

The authors are with the Artificial Intelligence Center, SRI International, Menlo Park, CA 94025.

IEEE Log Number 8609319.

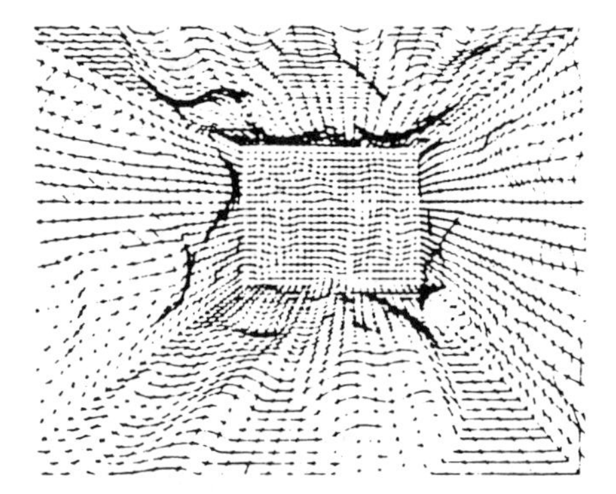

Fig. 1. Wire room.

age. We can unify the various approaches to the SMP by converting their respective assumptions and auxiliary information into the implied second image and employing the stereo paradigm to recover depth. In the case of the SSTC paradigm, our approach amounts to "one-eyed stereo."

II. Shape from One-Eyed Stereo

Most people viewing Fig. 1 get a strong impression of depth. We can recover an equivalent depth model algorithmically by assuming that we are viewing a projection of a uniform grid and employing the computational procedure to be described. In the remainder of this paper we will show how some simple modifications and variations of the uniform grid, as the implied second image, allow us to recover depth from shading, texture, and surface contours.

The one-eyed stereo paradigm can be described as a five-step process, as outlined in the paragraphs below. Some scenes with special surface markings or image-formation processes must be analyzed by variants of the algorithm described, but the general approach remains the same.

A. Partition the Image

As with all approaches to the SMP, the image must be segmented into regions prior to the application of a particular algorithm. Before the one-eyed stereo computation can be employed, the segmentation process must delineate regions that are individually in conformance with a single model of image formation. The computation can then be carried out independently in each region and the results fitted together.

B. Select a Model

For each region identified by the partitioning process, we must decide upon the underlying model of image formation that explains that portion of the image. Surface reflectance functions and texture patterns are examples of such models. Partitioning of the image and selection of the appropriate models are difficult tasks that are not addressed in this paper. Witkin and Kass [24] describe a new class of techniques that promises some answers to these questions. Generally, it will be impossible to recover depth whenever a single model cannot be associated with a region. Similarly, inaccurate or incorrect results can be expected if the partitioning or modeling is performed incorrectly.

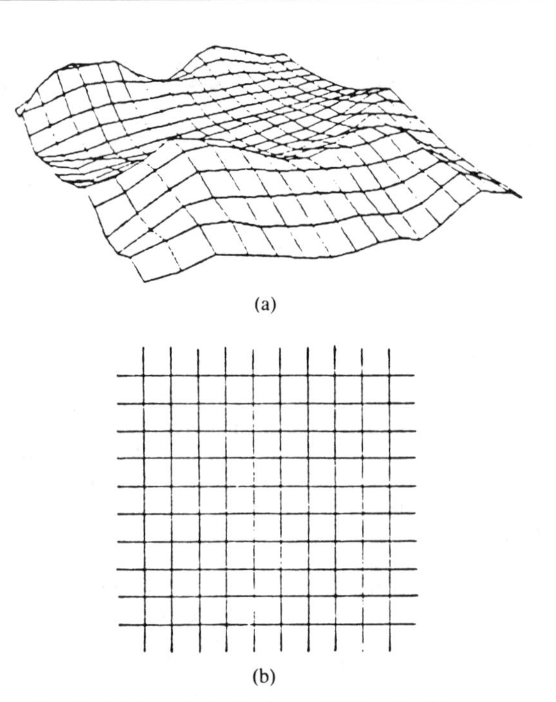

Fig. 2. (a) A projected texture. (b) Its virtual image.

C. Generate the Virtual Image

The key to one-eyed stereo is using the model of image formation to fabricate a second (virtual) image of the scene. The idea is that the model often allows one to construct an image that is independent of the actual shape of the imaged surface. This allows the virtual image to be depicted solely from knowledge of the model without making use of the original image. For example, the markings on the surface of Fig. 2(a) could have arisen from projection of a uniform grid upon the surface. For all images that fit this model, we can use a uniform grid as the virtual image. As a rule, the orientation, position, and scale of this grid will be unknown; however, we will show how this information can be recovered from the original image. Other models give rise to other forms of virtual images and are discussed in Section III.

D. Determine Correspondences

Before applying stereo techniques to calculate depths, we must first establish correspondences between points in the real image and the virtual image. When dealing with textures, the process is typified by counting texels in each image from a chosen starting point. With shaded images, the general approach is to integrate intensities. Several variants of the method for establishing correspondences are described in the next section. The difficulty of the procedure, it should be noted, will depend on the nature of the model.

E. Compute Depths Using Stereo

With two images and a number of point-to-point correspondences in hand, the techniques of binocular stereo are immediately applicable. At this point, the problem has been reduced to computing the relative camera models between the two images and using that information to compute depths by triangulation. The fact that the virtual image will normally be an orthographic projection required reformulation of existing algorithms for performing this computation. The Appendix describes a new algorithm that computes the relative camera model and reconstructs the 3-D scene from eight point correspondences between a perspective and an orthographic image, assuming knowledge of either the focal length or the principal point of the perspective image.

The problem of recovering scene and imaging geometry from two or more images has been addressed by workers not only in binocular stereo, but also in monocular perception of motion in which the two projections are separated in time as well as space. Various approaches have been employed to derive equations for the 3-D coordinates and motion parameters; these equations are generally solved by iterative techniques [5], [9], [14], [15]. Ullman [22] presents a solution for recovering 3-D shape from three orthographic projections with established correspondences among at least four points. His "polar equation" allows computation of shape when the motion of the scene is restricted to rotation about the vertical axis with arbitrary translation. Nagel and Neumann [11] have devised a compact system of three nonlinear equations for the unrestricted problem when five point correspondences between the two perspective images are known. More recently, Tsai and Huang [21] and Longuet-Higgins [10] have independently derived methods requiring only that a set of eight simultaneous linear equations be solved when eight point correspondences between two perspective images are known. In our formulation, we are faced with a stereo problem involving a perspective and an orthographic image; while the aforementioned references are indeed germane, none provides a solution to this particular problem.

The derivation described in the Appendix was inspired by the formulation of Longuet-Higgins for perspective images. When either image nears orthography, Longuet-Higgins's method becomes unstable; it is undefined if either image is truly orthographic. Moreover, his approach requires knowledge of the focal length and principal point in each image, while our method was derived specifically for one orthographic and one perspective image whose internal imaging parameters may not be fully known.

III. Variations on the Theme

In this section we illustrate how our approach is used with several models of texture, shading, and surface contour. Where these models do not match given scene characteristics, additional modifications to the algorithm may be necessary. However, a qualitatively correct answer might still be obtainable by applying one of the specific models we discuss below to a situation that appears to be inappropriate or to an image in which the validity of the assumptions cannot be established.

A. Shape from Texture

Surface shapes are often communicated to humans graphically by drawings like Fig. 2(a). Such illustrations can also be interpreted by one-eyed stereo. In this case, there is no need to partition the image; the underlying model of the entire scene consists of the intersections of lines distributed in the form of a square grid projected upon the surface. When viewed directly from above at an infinite distance, the surface would appear as shown in the virtual image of Fig. 2(b) regardless of the shape of the surface. This virtual image can be construed as an orthographic projection of the object surface from a particular, but unknown, viewing direction. Correspondences between the original and virtual images are easily established if there are no occlusions in the original image. Select any intersection in the original image to be the reference point and pair it with any intersection in the virtual image. A second corresponding pair can be found by moving to an adjacent intersection in both images. Additional pairs are found in the same manner, being careful to correlate the motions in each image consistently in both directions. When occlusions are present, it may still be possible to obtain correspondences for all visible junctions by following a nonoccluded path around the occlusion [such as the hill in the foreground of Fig. 2(a)]. If no such path can be found, the shape of each isolated region can still be computed provided eight image points are identified in each region. The reconstructed surface will remain undetermined in the gaps between the regions, but the ability to recover the same relative camera model from any region guarantees a globally consistent model across regions.

Other techniques used to represent images of 3-D shapes graphically may require other virtual images. Fig. 3(a), for example, would imply a virtual image as shown in Fig. 3(b). Methods for recognizing which model to apply are needed, but are not discussed here.

Once correspondences have been determined, we can use the algorithm given in the Appendix to recover depth. We have presumably one perspective image and one orthographic image whose scale and origin are still unknown. The depths to be recovered will be scaled according to the scale chosen for the virtual image.[1] The choice of origin for the orthographic image is arbitrary and will lead to the same solution regardless of the point chosen. The Appendix shows how to compute both the orientation and the displacement of the orthographic coordinate system, relative to the perspective imaging system. Three-dimensional coordinates of each matched point are then easily computed by means of back-projection. A unique

[1]Recall that the original image does not contain the information necessary to recover the absolute size of the scene.

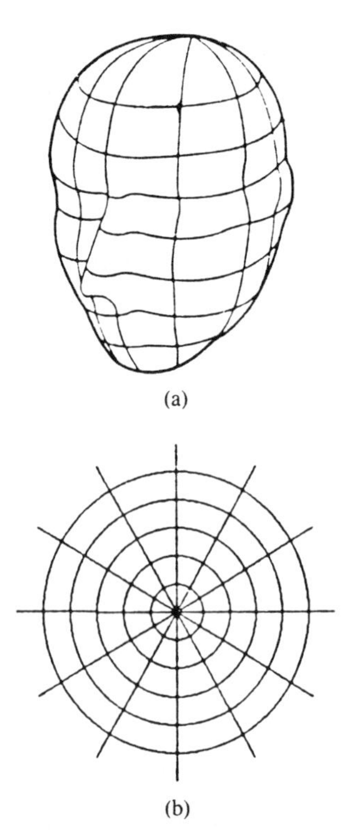

Fig. 3. (a) The original image. (b) The virtual image.

solution will be obtained whenever the piercing point or focal length of the perspective image is known.[2] A minimum of eight pairs of matched points is required to obtain a solution; depths can be computed for all matched points.

The imaging situation just described is reminiscent of that used in range finding with contrived lighting, such as light striping and grid coding. The previous discussion shows how these situations can be recast as problems to be solved by one-eyed stereo. For more information on these techniques, see Jarvis [7].

There exists a growing literature on methods to recover shape from natural textures [8], [13], [19], [23]. We will now show how the constraints imposed by one type of natural texture can be exploited to obtain similar results by using one-eyed stereo.

Consider the pattern of streets in Fig. 4. If this city were viewed from an airplane directly overhead at high altitude, the streets would form a regular grid not unlike the one used as the virtual image in Fig. 2. There are many other scene attributes that satisfy this same model. The houses in Fig. 5 would appear to be distributed in a rectangular grid if viewed from directly overhead. In an apple orchard growing on a hillside, the trees are planted in rows that are evenly spaced when measured horizontally; the vineyard in Fig. 6 exhibits this property.

Ignoring the nontrivial tasks of partitioning these images into isotextural regions, verifying that they satisfy the model, and identifying individual texels, it can be seen how these images can be interpreted with the same techniques as were described in the previous section. The vir-

[2]More precisely, the solution will be the Necker pair of interpretations of the perspective image.

Fig. 4. The streets in this scene form a natural projected texture [3].

Fig. 5. The houses can be construed as a projected texture [3].

Fig. 6. These grapevines exhibit a regular texture [3].

tual image in each case will be a rectangular grid that can be considered as an orthographic view from an unknown orientation. Correspondences can be established by counting street intersections, rooftops, or grape vines. As before, one can solve for the relative camera model and compute depths of matched points. Obviously, for the situations discussed here, we must be satisfied with a qualitatively correct interpretation—not only because of the difficulty of locating individual texels reliably and accurately, but also in view of the numerical instabilities arising from the underlying nonlinear transformation.

B. Shape from Shading

For our purposes, surface shading can be considered the limiting case of a locally uniform texture distribution (as the texels approach infinitesimal dimensions). To compute correspondences, we need to integrate image intensities appropriately in place of counting lines since the image intensities can be seen to be related to the density of lines projected on the surface. The feasibility of this procedure depends on the reflectance function of the surface.

What types of material possess the special property that allows their images to be treated like the limiting case of the projected textures of the previous section? The integral of intensity in an image region has to be proportional to the number of texels that would be projected in that region. If the angles i and e are defined as depicted in Fig. 7, it can be seen that the number of texels projected onto a surface patch will be proportional to cos i, the cosine of the incident angle. At the same time, the surface patch (as seen from the viewpoint) will be foreshortened by cos e, the cosine of the emittance angle. Thus, the integral of reflected light intensity over a region will be proportional to the flux of the light striking the surface if the intensity of the reflected light at any point is proportional to (cos i/cos e). Horn [6] has pointed out that, when viewed from great distances, the material in the maria of the moon and other rocky, dusty objects exhibit a reflectance function that allows recovery of the ratio (cos i/cos e) from the imaged intensities. That is, their reflectance functions are constant for constant (cos i/cos e). This surface property has made possible unusually simple algorithms for computing shape from shading, so it is not surprising that it submits easily to one-eyed stereo as well.

To interpret this type of shading, we can construct a virtual image whose direction of view is the lighting direction (i.e., taken from a "virtual camera" located at the light source). When the original shaded image is orthographic, we consider a family of parallel lines in which each line lies in a plane that includes both the light source and the (distant) viewpoint. When viewed from the light source, the image of the surface corresponding to these lines will also be a set of parallel lines regardless of the shape of the surface. These parallel lines constitute the virtual image. We will use the image intensities to refine these line-to-line correspondences to point-to-point correspondences. Fig. 8 shows the geometry for an individual line in the family. A little trigonometry shows that

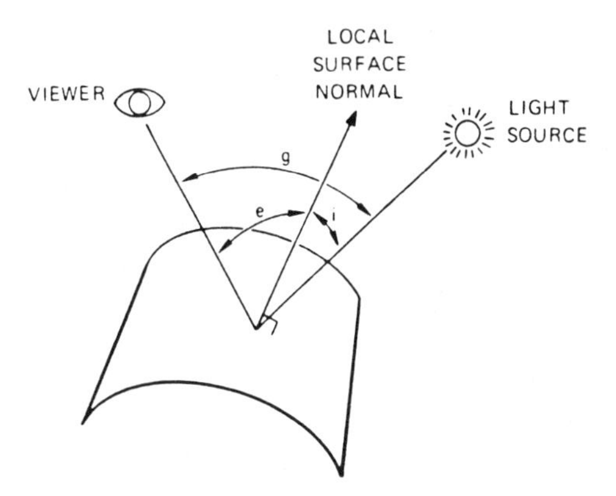

Fig. 7. The geometry of surface illumination.

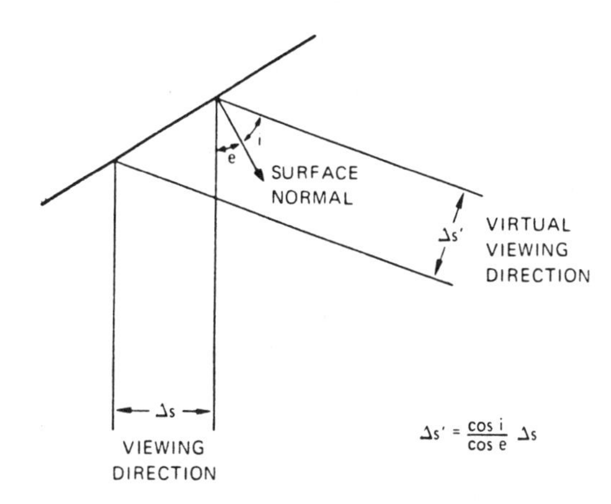

Fig. 8. The geometry along a line in the direction of the light source.

$$\Delta s' = \frac{\cos i}{\cos e} \Delta s \tag{1}$$

where Δs is a distance along the line in the real image and $\Delta s'$ is the corresponding distance along the corresponding line in the virtual image. Integrating this equation produces the following expression, which defines the point correspondences in the two images along the given line:

$$s' = s'_o + \int_0^s \frac{\cos i}{\cos e} ds. \tag{2}$$

The reader may observe that this equation is, not coincidentally, similar to that used by Horn [6] to recover shape directly. To use this equation, we must first compute (cos i/cos e) from the intensity value at each point along the line. This will, of course, be possible only when the reflectance function has the property discussed above. To construct the virtual image, we choose a starting point in the shaded image and begin integrating intensities according to (2). For any value of s, the corresponding virtual image point is found on a straight line at a distance s' from the virtual reference point. With these point-to-point correspondences in hand, all that remains to be done is to solve for their 3-D world coordinates. The eight-point algorithm given in the Appendix cannot find the solution when seven of the points are coplanar. Here, all the points lie on a single line in the image, so they are coplanar in the world. Fortunately, there is a simpler solution in this case since both images are orthographic and the relative camera model is defined by the orientation of the light source. It is a simple matter of triangulation to find the 3-D coordinates of the surface points, given that we know the direction to the light source.

We can explore the remainder of the surface by repeating the process for each of the successive parallel lines in the image. Adjacent profiles still remain unrelated to each other since their individual scale factors have not yet been ascertained. Prior knowledge of the actual depth of one point along each profile provides the necessary additional information to complete the reconstruction. It is important to note that our assumptions and initial conditions are those used by Horn; the fact that he was able to obtain a solution under these conditions predicted the existence of a suitable virtual image for the one-eyed stereo paradigm.

For shaded perspective images, we must integrate along a family of straight lines that radiate from the antisolar point (the point in the image that corresponds to the location of the light source). This ensures that the image line will be in a plane containing both the viewer and the light source and that the virtual image of each line will also be a straight line. The integration becomes a bit more complex than shown in (2) because the nonlinear effects of perspective imaging must be accommodated. Nevertheless, it remains possible to establish point-to-point correspondences between images, with the virtual image being similar in appearance to Fig. 3(b).

C. Shape from Contour

It is sometimes possible to extract a line drawing, such as the one shown in Fig. 9, from scene textures. Parallel streets like those encountered in Fig. 4 give rise to a virtual image consisting of parallel lines when the cross streets cannot be located; terraced hills also produce a virtual image of parallel lines. Correspondences between real and virtual image lines can be found by counting adjacent lines from an arbitrary starting point. This matches a virtual image line with each point in the real image, but point-to-line correspondences alone are not sufficient for reconstruction of the surface. Knowledge of the relative orientation between the two images (equivalent to knowing the orientation of the camera that produced the real image relative to the parallel lines in the scene) provides an adequate constraint; the surface can then be reconstructed uniquely through back-projection. Without knowledge of the relative orientation of the virtual image, heuristics must be employed that relate points on adjacent contours so that a regular grid can be used as the virtual image. Similar heuristics are required by all algorithms that attempt to solve the SMP for images of surface contours of arbitrarily curved surfaces [19]. The human visual system is normally able to interpret images like Fig. 9 unambiguously, although just what assumptions are

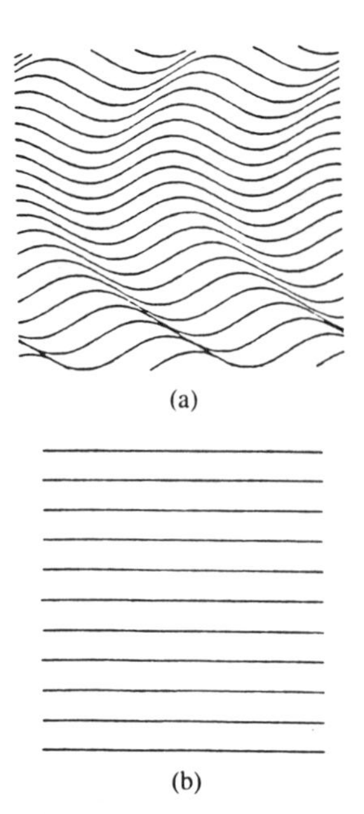

Fig. 9. (a) An image of contours. (b) Its virtual image.

being made remains unclear. Further study of this phenomenon may make it possible to extract models that are especially suited to the employment of one-eyed stereo on this type of image without requiring prior knowledge of the virtual orientation.

D. Distorted Textures and Unfriendly Shading

We have already noted that image shading can be viewed as a limiting (and, for our purposes, a degenerate) result of closely spaced texture elements. To recover depth from shading, we must use integration instead of the process of counting the texture elements that define the locations of the "grid lines" of our virtual image. The integration process depends on the existence of a "friendly" reflectance function and an imaging geometry that allows us to convert distance along a line in the actual image to a corresponding distance along a line in the virtual image.

The recovery of lunar topography from a single shaded image [6], as discussed in Section III-B is one of the few instances in which "shape from shading" is known to be possible without a significant amount of additional knowledge about the scene. Nevertheless, even here we are required to know the actual reflectance function, the location of the (point) source of illumination, and the depths along a curve on the object surface and to be dealing with a portion of the surface that has constant albedo. Furthermore, the reflectance function has to have just the property we require to replace direct counting, i.e., the reflectance function has to compensate exactly for the "foreshortening" of distance due to viewing points on the object surface from any angle. Most of the commonly encountered reflectance functions, such as Lambertian reflectance, do not possess this friendly property, and it is not clear to what extent it is possible to recover depth from shading in such cases (e.g., see Pentland [13] and Smith [16]). Additional assumptions will probably be necessary, and the qualitative nature of the recovery by

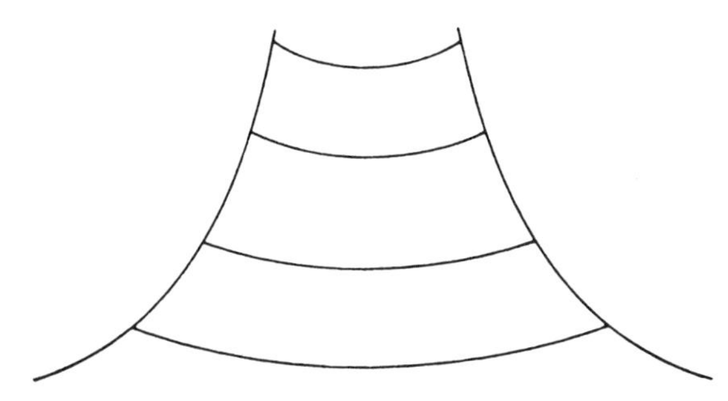

Fig. 10. This simple drawing has two reasonable interpretations. It is seen as curved roller-coaster tracks if the lines are assumed to be the projection of a rectangular grid, or as a volcano when the lines are assumed to be the projection of a circular grid.

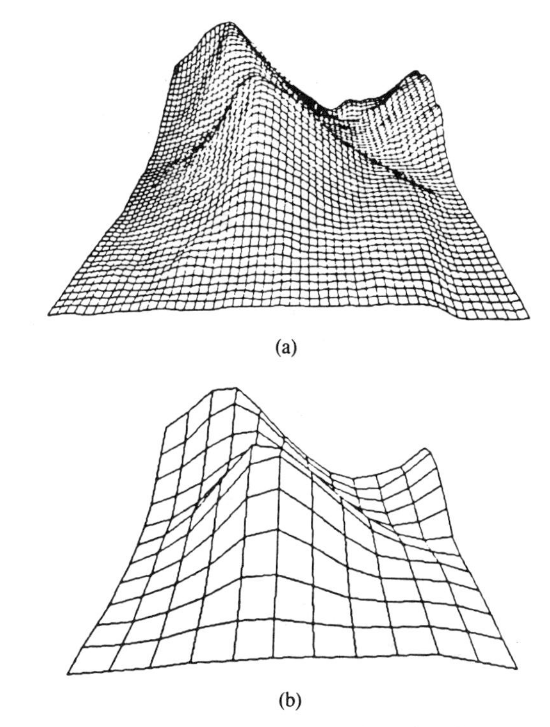

Fig. 11. (a) View of part of a DTM. (b) View of surface reconstructed from (a).

one-eyed stereo will be more pronounced. Just as in the case in which a complex function can be evaluated by making a local linear approximation and iterating the resulting solution, so it may be possible to deal with unfriendly, or even unknown, reflectance functions by assuming that they are friendly in the vicinity of some point, solving directly for local shape by using the algorithm applicable to the friendly case and then extending the solution to adjacent regions. We are currently investigating this approach.

The uniform rectangular grid and the polar grid that we used as virtual images to illustrate our approach to one-eyed stereo are effective in a large number of cases because there are processes operating in the real world that produce corresponding textures (i.e., grid-like textures that appear to be orthographically projected onto the surfaces of the scene). However, there are also textures that produce similar-appearing images, but are due to different underlying processes. For example, a uniform grid-like texture might have been created on a flat piece of terrain that is subsequently subjected to geologic deformation—in this case the virtual image (or the recovery algorithm) needed to recover depth must be different from the projective case. The iterative technique suggested in the preceding paragraph for unfriendly shading is one way to address these distorted textures within the one-eyed stereo paradigm.

We have already indicated the problem of choosing the appropriate model for the virtual image, and as noted above, image appearance alone is probably insufficient for making this determination—some semantic knowledge about the scene is undoubtedly essential. Fig. 10 shows an example in which two completely different, yet equally believable, interpretations of scene structure result, depending on whether we use the rectangular grid model or the polar grid model.

IV. Experimental Results

The stereo reconstruction algorithm described in the Appendix has been programmed and successfully tested on both real and synthetic imagery. Given a sparse set of image points and their correspondence in a virtual image, a qualitative description of the imaged surface can be obtained.

Synthetic images were created from surfaces painted with computer-generated graphic textures. Fig. 11(a) shows a synthetic image constructed from a section of a digital terrain model (DTM). The intersections of every tenth grid line constitute the set of 36 image points made available to the one-eyed stereo algorithm. Their correspondences were established by selecting an arbitrary origin and counting grid lines to obtain virtual image coordinates. When these pairs are processed by the algorithm in the Appendix, a set of 3-D coordinates is obtained in either the viewer-centered coordinate space or the virtual image coordinate space (which, if correct, is aligned with the original DTM). Fig. 11(b) was produced from the resulting 3-D coordinates expressed in the virtual image space by using Smith's surface interpolation algorithm [17] to fit a surface to these points. This yields a dense set of 3-D coordinates that can then be displayed from any viewpoint. The viewpoint used to produce Fig. 11(b) was the one computed by the one-eyed stereo algorithm. Its similarity to the original rendering [Fig. 11(a)] illustrates the successful reconstruction of the scene. The depths of the 36 chosen points were computed within a relative error of 10^{-6}. This degree of accuracy was obtained in all experiments with nondegenerate synthetic data.

The same procedure was followed when we worked with real photographs. The intersections of 31 street intersections were extracted manually from the photograph of San Francisco shown in Fig. 4. Those that were occluded or indistinct were disregarded. Virtual image coordinates were obtained by counting city blocks from the lower-left-hand intersection. The one-eyed stereo algorithm was then used to acquire 3-D coordinates of the corresponding image points in both viewer-centered and grid-

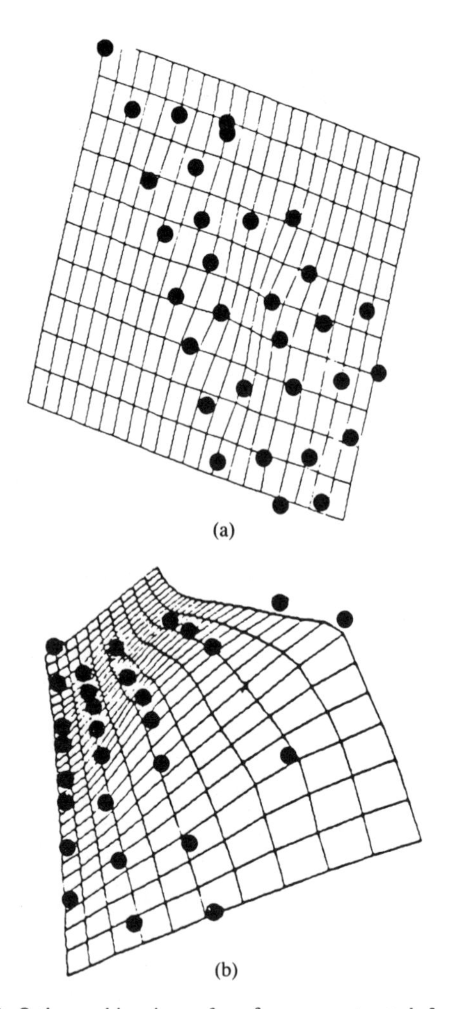

Fig. 12. (a) Orthographic view of surface reconstructed from Fig. 4. (b) Perspective view of same surface (from derived camera location).

centered coordinate systems. A continuous surface was fitted to both representations of these points. The location and orientation of the camera relative to the grid were also computed. Fig. 12(a) shows the reconstructed surface as an orthographic view from the direction computed to be true vertical. The dots superimposed are the computed locations of the original 31 points; their locations portray the rectangular grid formed by the streets. Fig. 12(b) shows the surface from the derived location of the viewpoint of the original photo. While several of the original points were badly mislocated, the general shape of the landform is apparent. The actual elevations of the street intersections range from 36 to 260 ft above sea level. The average error of the reconstructed surface was 22.3 ft.

The task remains to evaluate the effectiveness of the iterative technique, described in Section III-D, for recovering 1) shape from shading in the case of scenes possessing "unfriendly" reflectance functions, and 2) shape from nonprojective and distorted textures. Our experience with the process indicates that the key to successful reconstruction using the iterative approach lies in the similarity of the real imaging situation with the imaging model chosen to construct the virtual image. If the local approximations are good, a global solution can be obtained.

V. Conclusion

In this paper, we have shown that, in principle, it is possible to employ the stereo paradigm in place of various approaches proposed for modeling 3-D scene geometry—including the case in which only one image is available. We have further shown that, for the case of a single image, the approach could be implemented by

1) setting up correspondences between portions of the image and some variants of a uniform grid, and

2) treating each image region and its grid counterpart as a stereo pair and employing a stereo technique to recover depth. (We present a new algorithm that makes it possible to accomplish this step.)

There are several reasons the algorithm can provide only a qualitative shape description. First, the problem itself can be somewhat sensitive to slight perturbations in the estimates of the piercing point or focal length. This appears to be inherent to the problem of recovering shape from a single image. Special care must be taken when inverting a nonlinear transformation to minimize the effect of noisy data, and our algorithm ignores this issue. The second factor precluding precise, quantitative description of shape is the practical difficulty of acquiring large numbers of corresponding points. While the algorithm can proceed with as few as eight points, the location of the object will be identified at those eight points only. If a more complete model is sought, additional points will be required to constrain the subsequent surface interpolation. Our goal was not to provide a highly accurate, robust algorithm for reconstructing a scene from an arbitrary image; rather, our main contribution lies in the recognition of a common theoretical framework for relating the more specialized techniques that have been developed through the years.

Devising automatic procedures to partition the image, select the appropriate form of the virtual image, and establish the correspondences is a difficult task that was not addressed in this paper. Nevertheless, we have unified a number of apparently distinct approaches that, individually, also have to contend with these same pervasive problems (i.e., partitioning, model selection, and matching).

Appendix

The main body of this paper was devoted to showing how the problem of interpreting certain varieties of textured and shaded images can be transformed into equivalent problems in binocular stereo. Beginning with a perspective image, a second (virtual) image is hypothesized according to some presumed model of the original image. The model also specifies how to establish the correspondence between points in the two images. To compute the shape of the surfaces in the original scene, we need only compute the 3-D coordinates from the information in the two images where the actual scene is a perspective projection and the virtual image has been constructed as an orthographic projection. This Appendix shows how 3-D coordinates can be computed from point correspondences between a perspective and an orthographic projection when the relation between the imaging geometries is unknown.

We will use lower case letters to denote image coordinates and upper case letters for 3-D object coordinates. Unprimed coordinates will refer to the geometry of the perspective image, and primed coordinates will refer to the orthographic image. Let x_1 and x_2 be the image coordinates of a point in the perspective image relative to an arbitrarily selected origin. Let $-d_1$ and $-d_2$ be the (unknown) image coordinates of the principal point, and let $f\,[>0]$ be the focal length. The object coordinates associated with an image point are (X_1, X_2, X_3) where the origin coincides with the center of projection and the X_3 axis is perpendicular to the image plane. The X_3 coordinates of any object point will necessarily be positive.

The imaging geometry is as depicted in Fig. 13 and yields the following standard perspective equations:

$$x_1 + d_1 = f\frac{X_1}{X_3}; \qquad x_2 + d_2 = f\frac{X_2}{X_3}. \tag{3}$$

For the orthographic image, x'_1 and x'_2 are the image coordinates (relative to an arbitrary origin) and (X'_1, X'_2, X'_3) is the world coordinate system defined such that

$$x'_1 = X'_1; \qquad x'_2 = X'_2. \tag{4}$$

We use the unknown scale factor between orthographic image coordinates and the scene as our unit of measurement. This may cause a difference of scale between the primed and unprimed coordinate systems, but will lead to the unique reconstruction of the scene (modulo the unknown scale factor) whenever the reconstruction is possible. No distortion is introduced because the perspective image can be arbitrarily scaled without affecting its 3-D interpretation.

The two world coordinate systems can be related as follows:

$$X' = R(X - T) \tag{5}$$

where X is the column vector $[X_1, X_2, X_3]^T$, X' is the column vector $[X'_1, X'_2, X'_3]^T$, R is a 3×3 rotation matrix, and T is a translation vector from the center of perspective projection to the origin of the world coordinate system associated with the orthographic projection. For either component ($i = 1$ or 2), we can write

$$X'_i = R_i \cdot (X - T) \tag{6}$$

where R_i is the ith row of R. By substituting (3) and (4) into the above, we obtain

$$X_3 = \frac{f(x'_i + R_i \cdot T)}{R_1 \cdot [(x_1 + d_1), (x_2 + d_2), f]}. \tag{7}$$

Eliminating X_3 from the two equations in (7) yields

$$\begin{aligned} 0 = {} & x'_1x_1R_{21} + x'_1x_2R_{22} + x'_1R_2 \cdot D \\ & - x'_2x_1R_{11} - x'_2x_2R_{12} - x'_2R_1 \cdot D \\ & + x_1(R_{21}R_1 \cdot T - R_{11}R_2 \cdot T) \\ & + x_2(R_{22}R_1 \cdot T - R_{12}R_2 \cdot T) \\ & + (R_1 \cdot T)(R_2 \cdot D) - (R_2 \cdot T)(R_1 \cdot D) \end{aligned} \tag{8}$$

where D is the vector $[d_1, d_2, f]$.

The above equation relates image coordinates for corresponding points in both images. The following unknowns can be found by using eight corresponding pairs and solving the system of eight linear equations:

$$\begin{aligned} B_1 &= \frac{R_{21}}{R_{11}} \\ B_2 &= \frac{R_{22}}{R_{11}} \\ B_3 &= \frac{R_2 \cdot D}{R_{11}} \\ B_4 &= \frac{R_{12}}{R_{11}} \\ B_5 &= \frac{R_1 \cdot D}{R_{11}} \\ B_6 &= \frac{R_{21}}{R_{11}} R_1 \cdot T - R_2 \cdot T \\ B_7 &= \frac{R_{22}}{R_{11}} R_1 \cdot T - \frac{R_{12}}{R_{11}} R_2 \cdot T \\ B_8 &= \frac{1}{R_{11}}(R_1 \cdot T)(R_2 \cdot D) - \frac{1}{R_{11}}(R_2 \cdot T)(R_1 \cdot D) \end{aligned} \tag{9}$$

$$\begin{bmatrix} x'_1x_1 & x'_1x_2 & x'_1 & -x'_2x_2 & -x'_2 & x_1 & x_2 & 1 \\ \vdots & \vdots & \vdots & \vdots & \vdots & \vdots & \vdots & \vdots \end{bmatrix} \begin{bmatrix} B_1 \\ B_2 \\ B_3 \\ B_4 \\ B_5 \\ B_6 \\ B_7 \\ B_8 \end{bmatrix} = \begin{bmatrix} x'_2x_1 \\ \vdots \end{bmatrix}. \tag{10}$$

When more than eight points are available, a least-squares method can be employed to solve the system of equations. Once we have the B_i's in hand, we can solve for the components of the rotation matrix R.

We have three constraints on the coefficients of R from (9):

$$R_{21} = B_1R_{11} \qquad R_{22} = B_2R_{11} \qquad R_{12} = B_4R_{11}.$$

The requirement that R be orthonormal provides the ad-

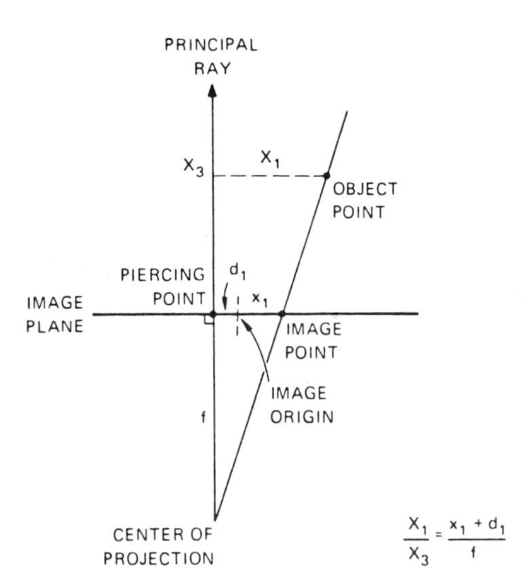

Fig. 13. Definition of coordinate system.

ditional constraints necessary:

$$\|R_1\| = 1 \quad \|R_2\| = 1 \quad R_1 \cdot R_2 = 0 \quad R_3 = R_1 \times R_2.$$

Solution of this system of nonlinear equations yields four orthonormal matrices. The physical interpretation of these solutions is that the original perspective image may be viewed as either of a pair of Necker reversals. Either of these interpretations gives rise to two solutions: one where the image plane is forward of the center of projection and the other where the image plane is behind it. Those constructions in which the image plane is behind the center of projection can be eliminated by checking the sign of the X_3's later. For now, let us carry out the following computations for each of the four choices for R and discard it later if it produces a solution whose image plane is behind the center of projection.

Now we will solve for the translation vector T. First let us note that T cannot be found uniquely because the origin of the primed world coordinate system has not been completely specified. The X_1' and X_2' coordinates of the origin were fixed by the choice of origin for the orthographic image coordinates, but the position of the origin along the X_3' axis is still unconstrained. Since we are free to choose any origin for X_3', we will choose the one for which $T_3 = 0$.

Using the expression for B_6 in (9), we find

$$B_6 = \frac{R_{21}}{R_{11}}(R_{11}T_1 + R_{12}T_2 + R_{13}T_3) - (R_{21}T_1 + R_{22}T_2 + R_{23}T_3). \quad (11)$$

Making use of the fact that $R_{33} = R_{11}R_{22} - R_{12}R_{21}$ and $T_3 = 0$, we get

$$T_2 = -B_6\frac{R_{11}}{R_{33}}. \quad (12)$$

Similarily,

$$T_1 = B_7\frac{R_{11}}{R_{33}}. \quad (13)$$

The origin of the primed coordinate system in unprimed coordinates is given by

$$T = \left[B_7\frac{R_{11}}{R_{33}}, -B_6\frac{R_{11}}{R_{33}}, 0\right]. \quad (14)$$

If the location of the principal point is known but the focal length (the scale factor of the perspective image) is not, f can easily be computed from (9):

$$f = \frac{B_5R_{11} - R_{11}d_1 - R_{12}d_2}{R_{13}}. \quad (15)$$

If the focal length is known, the principal point of the perspective image is found as follows. Use the third and fifth expressions of (9) to write two equations in the two unknowns, d_1 and d_2. Their solution yields

$$d_1 = f\frac{R_{31}}{R_{33}} + \frac{B_5R_{11}R_{22} - B_3R_{11}R_{12}}{R_{33}}$$

$$d_2 = f\frac{R_{32}}{R_{33}} + \frac{B_3R_{11}^2 - B_5R_{11}R_{21}}{R_{33}}. \quad (16)$$

The perspective image coordinates of the principal point are $[-d_1, -d_2]$.

If neither the focal length nor principal point is known beforehand, then the problem we have proposed does not have a unique solution. Equation (16) specifies the constraints between focal length and piercing point. For any choice of focal length, there exists a unique principal point and Necker pair of 3-D interpretations which distort as the focal length is varied. It is worth noting, however, that our computations of the rotation matrix R and the translation vector T did not require knowledge of either the focal length or the principal point.

We are now in a position to compute the world coordinates of all points for which we have correspondences. There may, of course, be many more than the minimum of eight points used so far. Equation (6) gives

$$x_1' = R_1 \cdot \left[\frac{X_3}{f}(x_1 + d_1), \frac{X_3}{f}(x_2 + d_2), X_3\right] - R_1 \cdot T, \quad (17)$$

which can be solved for

$$X_3 = \frac{f(x_1' + R_1 \cdot T)}{R_{11}x_1 + R_{12}x_2 + R_1 \cdot D}. \quad (18)$$

Now we must check the signs of the X_3's. It they are negative, the world points are located behind the center of projection and the result can be discarded. The correct solutions are found by choosing the pair of rotation matrices derived earlier that yield positive values of X_3. After obtaining the set of positive X_3's, we can continue.

Equation (3) gives the other unprimed world coordinates:

$$X_1 = \frac{X_3}{f}(x_1 + d_1); \qquad X_2 = \frac{X_3}{f}(x_2 + d_2). \quad (19)$$

If desired, the primed coordinates can be found by applying (5).

The above derivation makes the implicit assumption that the X_1' and X_2' axes are scaled equally. It is conceivable that the virtual image coordinates could be unequally scaled, as is the case when they are derived from a rectangular grid (e.g., Fig. 4) or with an unknown aspect ratio in an otherwise known texture pattern. If we have prior knowledge of the ratio of the sides of each rectangular grid element, then the virtual image coordinates should be normalized before applying this algorithm (i.e., by dividing X_2' by this ratio). Without knowledge of the ratio, the problem is underspecified and a unidimensional class of solutions exists. Knowledge of the piercing point, if available, can be used to constrain the problem further and to solve for the unique solution. To do this, we define the following virtual coordinate systems in place of (4):

$$x_1' = X_1'; \qquad x_2' = \frac{1}{k} X_2' \tag{20}$$

where k is the ratio of the sides of the rectangular grid elements.

The solution proceeds as before, yielding

$$\begin{aligned} 0 = {} & x_1'x_1R_{21} + x_1'x_2R_{22} + x_1'R_2 \cdot D \\ & - kx_2'x_1R_{11} - kx_2'x_2R_{12} - kx_2'R_1 \cdot D \\ & + x_1(R_{21}R_1 \cdot T - R_{11}R_2 \cdot T) \\ & + x_2(R_{22}R_1 \cdot T - R_{12}R_2 \cdot T) \\ & + R_1 \cdot TR_2 \cdot D - R_2 \cdot TR_1 \cdot D. \end{aligned} \tag{21}$$

The above equation is recast as the eight linear equations

$$\begin{bmatrix} -x_1'x_2 & -x_1' & x_2'x_1 & x_2'x_2 & x_2' & x_1 & x_2 & 1 \\ \vdots & \vdots & \vdots & \vdots & \vdots & \vdots & \vdots & \vdots \end{bmatrix} \begin{bmatrix} C_1 \\ C_2 \\ C_3 \\ C_4 \\ C_5 \\ C_6 \\ C_7 \\ C_8 \end{bmatrix} = \begin{bmatrix} x_1'x_1 \\ \vdots \end{bmatrix} \tag{22}$$

where

$$C_1 = \frac{R_{22}}{R_{21}}$$

$$C_2 = \frac{R_2 \cdot D}{R_{21}}$$

$$C_3 = \frac{kR_{11}}{R_{21}}$$

$$C_4 = \frac{kR_{12}}{R_{21}}$$

$$C_5 = \frac{kR_1 \cdot D}{R_{21}}$$

$$C_6 = \frac{R_{11}}{R_{21}} R_2 \cdot T - R_1 \cdot T$$

$$C_7 = \frac{R_{12}}{R_{21}} R_2 \cdot T - \frac{R_{22}}{R_{21}} R_1 \cdot T$$

$$C_8 = \frac{1}{R_{21}} (R_2 \cdot T)(R_1 \cdot D) - \frac{1}{R_{21}} (R_1 \cdot T)(R_2 \cdot D). \tag{23}$$

The following equalities can then be derived from (23):

$$R_{13} = \frac{R_{21}}{fk} (C_5 - C_3d_1 - C_4d_2)$$

$$R_{23} = \frac{R_{21}}{f} (C_2 - d_1 - C_1d_2) \tag{24}$$

$$f = \sqrt{\frac{-(C_5 - C_3d_1 - C_4d_2)(C_2 - d_1 - C_1d_2)}{C_3 + C_1C_4}} \tag{25}$$

$$R_{21} = \pm \frac{f}{\sqrt{f^2 + C_1^2f^2 + (C_2 - d_1 - C_1d_2)^2}} \tag{26}$$

$$k = \frac{R_{21}}{f} \sqrt{f^2C_3^2 + f^2C_4^2 + (C_5 - C_3d_1 - C_4d_2)^2}. \tag{27}$$

The solutions for the rest of R can now be computed easily from (23) and $R_1 \times R_2 = R_3$. The translation vector T is given by the expressions for C_6 and C_7 in (23):

$$T = \left[-C_7 \frac{R_{21}}{R_{33}}, C_6 \frac{R_{21}}{R_{33}}, 0 \right]. \tag{28}$$

With R and T now fully recovered, it is a simple matter to derive the object coordinates from (3), (20), and (5). Let us recall that we have four candidate matrices R hinging on the choice for R_{21}; as before, the correct pair must be selected by examining the signs of the X_3 coordinates.

To summarize, we have described an algorithm to compute the relative orientation and position between two imaging systems—perspective and orthographic—from the locations of eight (or more) corresponding image points. Either the principal point or the focal length and rectangular aspect ratio are computed along the way. With this information in hand, the world coordinates of all points in the imaged scene can be computed.

References

[1] S. T. Barnard and M. A. Fischler, "Computational stereo," *Comput. Surv.*, vol. 14, no. 4, Dec. 1982.

[2] M. Brady, Ed., *Artificial Intelligence (Special Volume on Computer Vision) Vol. 17*, nos. 1-3, Aug. 1981.

[3] R. Cameron, *Above San Francisco.* San Francisco, CA: Cameron and Company, 1976.

[4] S. Ganapathy, "Decomposition of transformation matrices for robot vision," in *Proc. IEEE Comput. Soc. Int. Conf. Robotics*, Atlanta, GA, Mar. 13-15, 1984, pp. 130-139.
[5] D. B. Gennery, "Stereo camera calibration," in *Proc. IU Workshop*, Nov. 1979, pp. 101-107.
[6] B. K. P. Horn, "Image intensity understanding," M.I.T. Artif. Intell. Memo 335, Aug. 1975.
[7] R. A. Jarvis, "A perspective on range finding techniques for computer vision," *IEEE Trans. Pattern Anal. Mach. Intell.*, vol. PAMI-5, pp. 122-139, Mar. 1983.
[8] J. R. Kender, "Shape from texture," Ph.D. dissertation, Carnegie-Mellon Univ., Pittsburgh, PA, Tech. Rep. CMU-CS-81-102, Nov. 1980.
[9] D. T. Lawton, "Constraint-based inference from image motion," in *Proc. AAAI-80*, Aug. 1980, pp. 31-34.
[10] H. C. Longuet-Higgins, "A computer algorithm for reconstructing a scene from two projections," *Nature*, vol. 293, pp. 133-135, Sept. 1981.
[11] H. Nagel and B. Neumann, "On 3-D reconstruction from two perspective views," *Proc. IEEE*, 1981.
[12] D. Nitzan, R. C. Bolles, *et al.*, "Machine intelligence research applied to industrial automation," 12th Rep. SRI Project 2996, Jan. 1983.
[13] A. P. Pentland, "Shading into texture," in *Proc. AAAI-84*, Aug. 1984, pp. 269-273.
[14] K. Prazdny, "Motion and structure from optical flow," in *Proc. IJCAI-79*, pp. 702-704.
[15] J. W. Roach and J. K. Aggarwal, "Determining the movement of objects from a sequence of images," *IEEE Trans. Pattern Anal. Mach. Intell.*, vol. PAMI-2, pp. 554-562, Nov. 1980.
[16] G. B. Smith, "The relationship between image irradiance and surface orientation," in *Proc. IEEE CVPR-83*.
[17] ——, "A fast surface interpolation technique," in *Proc. DARPA Image Understanding Workshop*, Oct. 1984, pp. 211-215.
[18] K. A. Stevens, "The line of curvature constraint and the interpretation of 3-D shape from parallel surface contours," in *Proc. AAAI-83*, Aug. 1983, pp. 1057-1061.
[19] ——, "The visual interpretation of surface contours," *Artif. Intell.*, vol. 17, nos. 1-3, pp. 47-73, Aug. 1981.
[20] T. M. Strat, "Recovering the camera parameters from a transformation matrix," in *Proc. DARPA Image Understanding Workshop*, Oct. 1984, pp. 264-271.
[21] R. Y. Tsai and T. S. Huang, "Uniqueness and estimation of three-dimensional motion parameters of rigid objects with curved surfaces," *IEEE Trans. Pattern Anal. Mach. Intell.*, vol. PAMI-6, pp. 13-27, Jan. 1984.
[22] S. Ullman, *The Interpretation of Visual Motion*. Cambridge, MA: M.I.T. Press, 1979.
[23] A. P. Witkin, "Recovering surface shape and orientation from texture," *Artif. Intell.*, vol. 17, nos. 1-3, pp. 17-45, Aug. 1981.
[24] A. Witkin and M. Kass, "Analyzing oriented patterns," in *Proc. IJCAI-85*, Aug. 1985.

Thomas M. Strat received the B.S. degree in computer science and electrical engineering in 1977, the M.S. degree in artificial intelligence in 1978, and the Electrical Engineer degree in 1979, all from the Massachusetts Institute of Technology, Cambridge.

He joined the Artificial Intelligence Center at SRI International in 1983 after spending four years with the U.S. Army. His research interests include computer vision, knowledge representation, and uncertain reasoning. He has published papers on intermediate-level image understanding and the theory of evidential reasoning. Currently, he is involved in research on the autonomous land vehicle and the application of evidential reasoning to information integration.

Mr. Strat is a member of Tau Beta Pi, Eta Kappa Nu, the American Association for Artificial Intelligence, and the U.S. Army Reserve.

Martin A. Fischler received the B.E.E. degree from the City College of New York, New York, and the M.S. and Ph.D. degrees in electrical engineering from Stanford University, Stanford, CA.

He held positions at the National Bureau of Standards and at Hughes Aircraft Corporation and was a Staff Scientist at the Lockheed Palo Alto Research Laboratory. In 1977 he joined the staff of SRI International where he is currently Program Director for Machine Perception in the Artificial Intelligence Center. He has conducted and directed research in the areas of artificial intelligence, machine vision, switching theory, computer organization, and information theory.

Dr. Fischler has published over 40 papers in these areas, has served as an officer of the IEEE, is a member of the editorial board of *Pattern Recognition Journal*, and is coauthor of a forthcoming book on the subject of intelligence (Addison-Wesley).

An Algebraic Approach to Shape-from-Image Problems*

Kokichi Sugihara
Department of Information Science, Faculty of Engineering, Nagoya University, Furo-cho, Chikusa-ku, Nagoya 464, Japan

Recommended by Michael Brady

ABSTRACT
This paper presents a new method for recovering three-dimensional shapes of polyhedral objects from their single-view images. The problem of recovery is formulated in a constrained optimization problem, in which the constraints reflect the assumption that the scene is composed of polyhedral objects, and the objective function to be minimized is a weighted sum of quadric errors of surface information such as shading and texture. For practical purpose it is decomposed into the two more tractable problems: a linear programming problem and an unconstrained optimization problem. In the present method the global constraints placed by the polyhedron assumption are represented in terms of linear algebra, whereas similar constraints have usually been represented in terms of a gradient space. Moreover, superstrictness of the constraints can be circumvented by a new concept 'position-free incidence structure'. For this reason the present method has several advantages: it can recover the polyhedral shape even if image data are incorrect due to vertex-position errors, it can deal with perspective projection as well as orthographic projection, the number of variables in the optimization problem is very small (three or a little greater than three), and any kinds of surface information can be incorporated in a unifying manner.

1. Introduction

Recovery of three-dimensional shape from two-dimensional images is one of the most fundamental and interesting problems in computer vision, and there have been found many cues that are useful to restrict the set of possible surfaces represented in images. The degree of freedom of a planar surface (or a small portion of a curved surface that can be regarded as planar) is three. This degree of freedom can be lessened by various kinds of visual information such as light intensity [1–4], apparent distortion of known patterns [5–7], statistical properties of texture [8, 9], occluding boundaries [10], vanishing points [11], and distribution of small-pattern sizes [12].

Each of these cues alone, however, is not enough to specify the local surface uniquely, and two approaches have been taken to circumvent this difficulty.

The first approach is the use of two or more images. Woodham [13] and Ikeuchi [3] used three images of different illumination to determine surface orientation at each point of the scene, and Coleman [4] used four images of different illumination to determine surface orientation and surface reflectance. In their methods, the number of images to be used is one more than the number of parameters to be determined (the surface orientation is specified by two parameters, and hence Woodham's and Ikeuchi's methods determine two parameters and Coleman's method three parameters). This is because one cue (i.e., one image) usually gives one nonlinear equation and consequently a set of cues of the same number as parameters gives a finite number of solutions, but not a unique one; one more cue is necessary to clear up this ambiguity.

The second approach is the use of global constraints. This approach was first demonstrated by Horn in his pioneering work [1], where the shape of a curved surface is extracted from light-intensity information of a single image together with the global assumption that the surface is smooth. The same constraint is also used by Ikeuchi and Horn [10].

Another important global constraint is obtained when we assume that the objects are polyhedra (that is, faces of the objects are all planar). Horn [14] pointed out that this constraint can be combined with light-intensity information in order to extract the unique shape of the objects from single images. Kanade [7] also combined this constraint with surface cues such as parallelism and skewed symmetry.

The second approach is preferable in that the surface can be recovered from only one image; from a viewpoint of artificial intelligence, this approach affords us computational models for human visual perception of photographs and paintings as well as real scenes, and from a viewpoint of engineering, it can lessen time-consuming preprocessing of image data.

In the second approach, the global constraint placed by the polyhedron assumption was represented in a gradient space. A gradient space is one of the most prevailing tools for the analysis of polyhedral scenes, and has been used also in other stages of scene analysis such as detection of inconsistency in the interpretation of pictures [15–17] and categorization of lines in pictures [18, 19].

The gradient space, however, has the following two serious difficulties. First, it does not represent the constraint completely; it can represent only a certain necessary (but not sufficient) condition for a picture to represent a polyhedral scene correctly (see [20, 21]). Hence, there is no assurance that the recovered shape forms a polyhedral scene. Secondly, the constraint represented in the gradient space is too strict; it is often disturbed even if vertices in a picture are

* This work is supported in part by the Grant in Aid for Scientific Research of the Ministry of Education, Science and Culture of Japan (Grant No. 57780035).

moved slightly from their correct positions. Since vertex-position errors are inevitable in computer processing of real images, some mechanism is required to lessen the superstrictness of the constraint.

There is another and more naive idea to represent the global constraint placed by the polyhedron assumption, that is, an algebraic representation [22, 23]. This idea, though it seems natural, has not been used so widely as the gradient space. However, in terms of linear algebra Sugihara [24, 25] gave a necessary and sufficient condition for a picture to correctly represent a polyhedral scene. While this condition itself is also superstrict to pictures with vertex-position errors, he [26, 27], furthermore, found a method for circumventing this superstrictness, that is, a method for deleting redundant equations from an overdetermined system of equations representing the condition. Using this algebraic representation, we can thus overcome the two difficulties contained in the gradient-space representation.

In the present paper, we shall combine the algebraically represented global constraint with other surface cues, and thus construct a new and powerful method for recovering polyhedral shape uniquely from single-view images. The problem of recovery is formulated in a constrained optimization problem, in which the constraints are derived from the algebraic structures of line drawings of polyhedra and the objective function to be minimized is a weighted sum of quadric differences between observed values and theoretical values of available cues. Our method has the following preferable properties.

(1) Since it is based on the necessary and sufficient condition for a correct picture of polyhedra, it is assured that the results of recovery are always polyhedra.

(2) It can recover the polyhedral shape even if images are incorrect due to vertex-position errors.

(3) It can deal with perspectively projected images as well as orthographically projected ones.

(4) The optimization problem is tractable in the sense that the number of variables is very small; it is usually three or a little greater than three.

(5) Any kind of cues that lessen the degree of freedom of surfaces can be incorporated in a unifying manner.

The fundamental idea is presented in Section 2, where the problem of recovery is formulated in a certain optimization problem. Since this optimization problem is not very tractable in its original form, in Section 3 it is decomposed into two more tractable problems (a linear programming problem and an unconstrained optimization problem) and thus a practical method is constructed. In Section 4, advantages of the present method over the gradient-space method are discussed from several points of view; this section is for readers who are familiar with the gradient space, and hence can be skipped over. In Section 5, the performance of the method is demonstrated through several kinds of image data.

2. Fundamental Scheme

We briefly review an algebraic structure of line drawings of polyhedra [25, 27], and then present our basic idea for recovering unique shape from single-view images.

2.1. Algebraic structure of line drawings of polyhedra

A *polyhedron* is a three-dimensional solid object bounded by a finite number of planar *faces*. A line-segment shared by two faces is called an *edge*, and a point shared by three or more faces a *vertex*. Suppose that a polyhedron is fixed to an (x, y, z) Cartesian coordinate system and that it is projected orthographically on the x-y plane, as is shown in Fig. 1. The projection is called a *picture* or a *line drawing*. This projection can be thought of as a picture of an object seen from a viewpoint that is infinitely far in the positive direction of the z-axis.

According to Huffman's definition [15] lines in a picture can be classified into the three categories: a *convex line* representing a convex edge whose side faces are both visible, a *concave line* representing a concave edge, and an *occluding line* representing a convex edge with exactly one visible face. These categories are usually represented by labels assigned to the lines; a convex line, a concave line, and an occluding line are labeled '+', '−', and an arrow, respectively (the direction of the arrow is chosen in such a way that the right side of the arrow corresponds to the side face and the left side to a background). See [15, 18, 19] for finding consistent assignment of the labels.

Let V and F, respectively, denote the set of visible vertices and that of visible faces (including partly visible faces) of the polyhedron, and let R denote the subset of $V \times F$ such that R contains (v, f) if and only if vertex v is on face f. We shall call an element of R an *incidence pair* and triple (V, F, R) an *incidence*

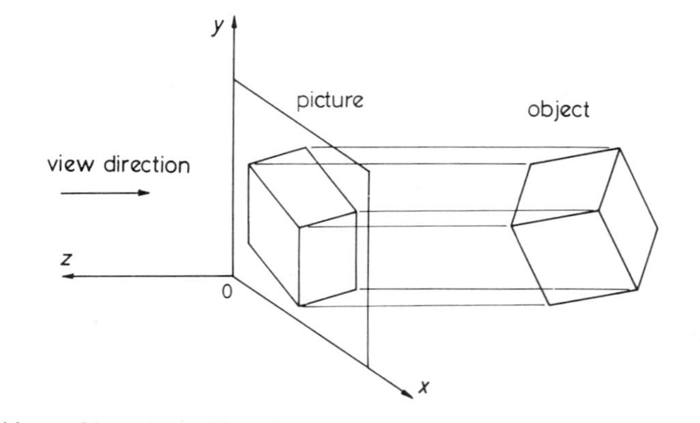

FIG. 1. Object and its orthographic projection.

structure. The incidence structure can be generated automatically when line-segments in the picture are labeled according to their categories [25].

We shall assume that a labeled picture and its incidence structure $\boldsymbol{S} = (V, F, R)$ are given but that the precise shape of the polyhedron is not known. We shall, furthermore, assume that no faces are parallel to the z-axis.

Let (x_i, y_i, z_i) be the coordinates of vertex v_i ($\in V$) and

$$a_j x + b_j y + z + c_j = 0$$

denote the equation of face f_j ($\in F$). Since the picture is given, x_i and y_i are known constants; the only variables are z_i, a_j, b_j, and c_j. Suppose that (v_i, f_j) is an element of R, that is, the vertex v_i is on the face f_j. Then, we get

$$a_j x_i + b_j y_i + z_i + c_j = 0 ,$$

which is a linear equation with respect to unknowns z_i, a_j, b_j, and c_j. A similar equation is obtained for each element of R, and, collecting them all, we get

$$A\boldsymbol{w} = \boldsymbol{0} \tag{1}$$

where $\boldsymbol{w} = {}^t(z_1 \cdots z_n a_1 b_1 c_1 \cdots a_m b_m c_m)$, $n = |V|$, $m = |F|$, t denotes a transposition, and A is a constant matrix of size $|R| \times (n + 3m)$.

A picture also contains cues for relative depth.

The first cue is bending of faces along edges. Suppose, for example, that two faces f_j and f_k share a concave edge and a vertex v_i is on f_k but not on f_j as is shown in Fig. 2(a). If the face f_j is extended, it goes behind v_i, and hence we get

$$a_j x_i + b_j y_i + z_i + c_j > 0 .$$

Similar inequalities are obtained, one for each convex or concave line.

The second cue is occlusion. In Fig. 2(b), for example, part of face f_j is occluded by face f_k along occluding line l. Let v_p, v_q, v_r be the three points on f_k that correspond, respectively, to the initial point, the terminal point, and the midpoint of l (this kind of pseudo-vertex v_r and the associated incidence pair (v_r, f_k) are added to V and R, respectively, whenever necessary; see [25] for a strict procedure for introducing pseudo-vertices). Then, the condition that f_j must be behind the line l can be represented by

$$a_j x_p + b_j y_p + z_p + c_j \geq 0 ,$$
$$a_j x_q + b_j y_q + z_q + c_j \geq 0 ,$$
$$a_j x_r + b_j y_r + z_r + c_j > 0 .$$

Note that an equality is allowed in the first two inequalities, but not in the third inequality. This is because the occluding line l also means that f_j and f_k may touch at some point on l but should not touch at all points on l (if they touch at all points on l, l must be categorized as a concave line). Similar inequalities are obtained for every occluding line.

Collecting all of such inequalities, we get

$$B\boldsymbol{w} > \boldsymbol{0} \tag{2}$$

where B is a constant matrix and the inequality is an abbreviation of componentwise inequalities, some of which allow equalities.

Then, we can prove that "a labeled picture represents a polyhedral scene correctly if and only if the system of the equations (1) together with the inequalities (2) has solutions" (see [25] for strict discussions). Therefore, there is a one-to-one correspondence between the set of all polyhedra the picture can represent and the set of all solutions to the system consisting of (1) and (2).

The equations (1) and the inequalities (2) have been derived for orthographic projections. However, the same algebraic structure can be obtained when we consider perspective projections (see [27]). Therefore, though we hereafter consider only orthographic projections, the method proposed in this paper can also be applied, with obvious modification, to perspective pictures.

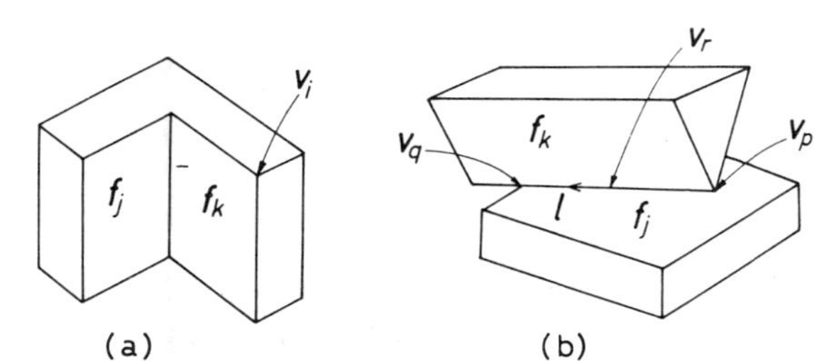

Fig. 2. Cues for relative depth.

2.2. Recovery as an optimization

The algebraic structure (i.e., the system consisting of (1) and (2)) specifies the set of all polyhedral scenes that yield the picture. That is, any $\boldsymbol{w}$ satisfying (1) and (2) corresponds to one possible scene represented by the picture, and vice versa. Hence, a solution $\boldsymbol{w}$ to (1) and (2) is also referred to as a scene $\boldsymbol{w}$. Then, our problem of recovery can be regarded as a problem of choosing from the set of solutions to (1) and (2) an element $\boldsymbol{w}$ that is most consistent with other surface information such as shading and texture. This idea leads us to an optimization problem in the following way.

We consider any cue that can lessen the degree of freedom of the faces. Let I_k denote an observed value of the kth cue, and for any $\boldsymbol{w}$ satisfying (1) and (2), let $J_k(\boldsymbol{w})$ denote a theoretical value of the kth cue that should be observed if the real scene is $\boldsymbol{w}$. For example, if a face is covered with grain texture of a known uniform density, we can adopt as I_k and $J_k(\boldsymbol{w})$, respectively, an observed value and an theoretical one of the apparent grain density in the image; if the illumination condition and surface reflectance are known, we can adopt as I_k and $J_k(\boldsymbol{w})$ an observed value and an theoretical one of light intensity on the face.

Let $g_k(\boldsymbol{w})$ be the difference of the two values:

$$g_k(\boldsymbol{w}) = I_k - J_k(\boldsymbol{w}) . \tag{3}$$

If there is no error in observation, the true scene $\boldsymbol{w}$ must satisfy $g_k(\boldsymbol{w}) = 0$; this condition lessens the degree of freedom of the face by 1. Since errors are inevitable in real image data, we cannot expect $g_k(\boldsymbol{w}) = 0$ exactly. However, if enough cues are available, we can seek a solution to the optimization problem:

Problem 1. Minimize $\sum_k s_k(g_k(\boldsymbol{w}))^2$, subject to (1) and (2), where s_k denotes a positive weight of the kth cue, and the summation is to be taken over all available cues.

The solution $\bar{\boldsymbol{w}}$ to the optimization problem can be thought of as the scene that is most consistent with the surface cues under the quadric error criterion. Thus, the fundamental scheme of our shape recovery is to adopt the solution to the optimization problem as the shape to be recovered.

The minimum number of surface cues necessary for unique recovery is usually very small. The degree of freedom of solutions to (1) and (2) is the number of unknowns minus the number of linearly independent equations, that is,

$$n + 3m - \text{rank}(A) .$$

This number is small in many cases (four or a little greater). Moreover, no matter how many cues are given, one degree of freedom usually remains, that is, the degree of freedom due to the translation along the z-axis. Hence, without loss of generality we can arbitrarily fix a z-coordinate of any one vertex. Therefore, the degree of freedom to be dissolved (i.e., the minimum number of cues necessary) equals

$$n + 3m - \text{rank}(A) - 1 ,$$

whose value is usually three or a little greater than three (see Section 5).

3. Practical Method for Shape Recovery

The fundamental scheme presented in the last section has some difficulties in practical applications. In this section we shall overcome these difficulties, and construct a practical method for shape recovery.

3.1. Difficulties to be overcome

The optimization problem, Problem 1, has the following difficulties in its original form.

(i) The set of equations (1) is not necessarily linearly independent. When it is not linearly independent, the slightest errors in numerical computation can change the rank of A and consequently change drastically the set of solutions to (1) and (2). Therefore, in order to solve Problem 1 numerically, we have to beforehand delete from (1) those equations that are linearly dependent on the rest of the equations.

(ii) The equations (1) are too strict. Fig. 3(a), for example, seems to be a top view of a truncated pyramid. In a strict sense, however, it does not represent any polyhedron; indeed, if it were a truncated pyramid, the three quadrilateral faces had a common point of intersection in a three-dimensional space (when they were extended) and hence the three edges should meet at a common point on the picture plane, but they do not as is shown in Fig. 3(b). Therefore, there is no solutions to the system of (1) and (2) associated with the picture in Fig. 3(a). We must overcome this superstrictness of (1), because vertex-position errors are inevitable in real image data. (Note that this difficulty is not equivalent to the difficulty (i); for example, the system (1) associated with the picture in Fig. 3(a) is linearly independent.)

(iii) The set of all solutions to (1) and (2) (this set is called a *constraint set*) is not a closed set. In general, a constrained optimization problem does not necessarily have a solution when its constraint set is not closed (see, for example, [28]). In this sense, Problem 1 seems unsound.

In the rest of this section we shall get over these difficulties.

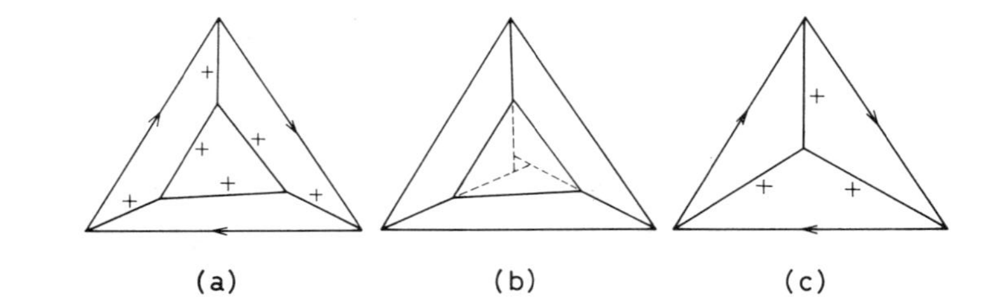

FIG. 3. Incorrectness due to vertex-position errors.

3.2. Overcoming the superstrictness of the equations

We have seen that a picture of a truncated pyramid like Fig. 3(a) becomes incorrect when a slight error occurs in its vertex positions. The picture in Fig. 3(c), on the other hand, represents a polyhedron correctly even if vertices (and consequently the lines incident to them) are moved slightly on the picture plane. We start our consideration with the distinction between these two kinds of pictures.

Let $\boldsymbol{S} = (V, F, R)$ be an incidence structure consisting of a vertex set V, a face set F, and an incidence pair set R. We assume that each vertex is at least on one face and each face has at least three vertices. The matrix A in (1) is determined by $\boldsymbol{S}$ and the coordinates of the vertices on the picture plane $x_1, y_1, \ldots, x_n, y_n$. The vertices are said to be *in general position* if $x_1, y_1, \ldots, x_n, y_n$ are algebraically independent over the rational field. Intuitively, if the vertices are in general position, there is not any special relation among the vertex positions; for example, neither three vertices are collinear nor three edges are concurrent. It follows from the definition of algebraic independence that when the vertices are in general position, a subdeterminant of A is 0 if and only if it is identically 0 when we consider $x_1, y_1, \ldots, x_n, y_n$ as variables (rather than any fixed real numbers). Therefore, if the vertices are in general position, the linear independence of (1) depends only on $\boldsymbol{S}$.

Note that the system (1) has trivial solutions in which every surface is identical, that is, $a_i = a_j$, $b_i = b_j$, and $c_i = c_j$ for any i and j $(1 \leqslant i < j \leqslant m)$. Solution $\boldsymbol{w}$ to (1) is said to be *nontrivial* if $a_i \neq a_j$ or $b_i \neq b_j$ or $c_i \neq c_j$ for any i and j $(1 \leqslant i < j \leqslant m)$. $\boldsymbol{S}$ is said to be *position-free* if the system (1) associated with $\boldsymbol{S}$ has a nontrivial solution when the vertices are in general position. In other words, $\boldsymbol{S}$ is position-free if, for any picture with incidence structure $\boldsymbol{S}$ in which the vertices are in general position, the surfaces associated with F can be arranged in a three-dimensional space in such a way that the surfaces are distinct from each other and that the projections of the vertices coincides with the points drawn on a picture plane.

If a picture represents a polyhedron correctly and its incidence structure is position-free, as in the case in Fig. 3(c), it remains correct when the vertices are moved providing that the movements are not too large; hence, even if errors occur in the vertex positions, the system of equations (1) and (2) has solutions. On the other hand, a picture whose incidence structure is not position-free, such as the one in Fig. 3(a), becomes incorrect when the vertices are moved from the correct positions. Therefore, this kind of a picture is almost always incorrect in a practical situation, because data are represented as a finite number of digits.

The concept of position-freeness admits the following theorems (see [26, 27] for the proofs).

Theorem 3.1 (Recognition of position-free structures). *Suppose that* $\boldsymbol{S} = (V, F, R)$ *is an incidence structure of a visible part of a polyhedron in which no three faces sharing a vertex have a common line of intersection (for example, a convex polyhedron or a trihedral polyhedron obviously satisfies this condition). Then,* $\boldsymbol{S}$ *is position-free if and only if*

$$|V(X)| + 3|X| \geqslant |R(X)| + 4 \tag{4}$$

is satisfied for any $X \subseteq F$ *such that* $|X| \geqslant 2$, *where* $V(X)$ *denotes the set of vertices that are on some faces in* X *and* $R(X)$ *the set of incidence pairs having elements of* X.

Theorem 3.2 (Condition for (1) to be linearly independent). *Suppose that* $\boldsymbol{S} = (V, F, R)$ *is the same as in Theorem* 3.1. *If* $\boldsymbol{S}$ *is position-free and the vertices are in general position, the system* (1) *is linearly independent.*

Theorem 3.1 enables us to judge whether an incidence structure is position-free or not by integer calculation only; we need not be annoyed by errors in numerical computation.

For example, let $\boldsymbol{S} = (V, F, R)$ be the incidence structure of the picture in Fig. 3(a). The picture has six vertices ($|V| = 6$) and four faces ($|F| = 4$). The three faces are quadrilateral and the other one is triangular, and hence $|R| = 4 \times 3 + 3 \times 1 = 15$. If we put $X = F$, the left-hand side of (4) equals $|V| + 3|F| = 18$ and the right-hand side equals $|R| + 4 = 19$, and thus (4) does not hold. Therefore, Theorem 3.1 tells us that $\boldsymbol{S}$ is not position-free. For the picture in Fig. 3(c), on the other hand, we get $|V| = 4$, $|F| = 3$, and $|R| = 3 \times 3 = 9$, and consequently $|V| + 3|F| = |R| + 4 = 13$. We can see that any subset X of F also satisfies (4) and thus the incidence structure of the picture in Fig. 3(c) is position-free.

If we simply follow Theorem 3.1, we have to check the inequality (4) 2^m times, where m is the number of faces. However, it was recently found that the condition in Theorem 3.1 can be checked in $O(m^2)$ steps [29, 30].

Theorem 3.2, on the other hand, assures us that the difficulty (i) does not occur for a position-free incidence structure.

Using these theorems, we can reduce Problem 1 to a more tractable problem and thus overcome the difficulties (i) and (ii) in the following way. The key point is that the theorems allow us to extract an essential and linearly independent subset of the equations (1), which can be solved without any fear of errors in numerical computation.

Suppose that the vertices of a picture is drawn in general position and its incidence structure $\boldsymbol{S} = (V, F, R)$ is not position-free. Then, from the definition of position-freeness $\boldsymbol{S}$ contains too many incidence pairs to represent the global constraints placed by the polyhedron assumption; if some of incidence pairs are deleted from $\boldsymbol{R}$ one by one, $\boldsymbol{S}$ eventually becomes position-free. Let $\boldsymbol{R}^*$ be a maximal subset of $\boldsymbol{R}$ such that incidence structure $\boldsymbol{S}^* = (V, F, \boldsymbol{R}^*)$ is position-free. Then, from Theorem 3.2 the system (1) associated with S^* is linearly independent, and hence the matrix A contains a nonsingular submatrix of size $|\boldsymbol{R}^*| \times |\boldsymbol{R}^*|$. Therefore, A can be transformed by some permutation of rows and that of columns into the next form:

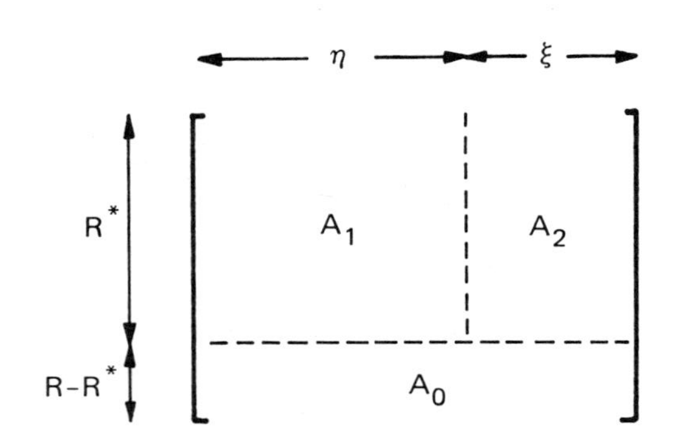

where A_0 corresponds to the incidence pairs that are deleted when we get $\boldsymbol{S}^*$ from $\boldsymbol{S}$, and A_1 is a nonsingular square matrix of size $|\boldsymbol{R}^*| \times |\boldsymbol{R}^*|$. A_0 can be obtained by Theorem 3.1, and A_2 can be obtained as a column set that corresponds to a set of vertices whose z-coordinates can be given independently when we reconstruct a polyhedral scene from the picture (see [27] for details). Let η be the vector of unknowns corresponding to the column set of A_1 and ξ be that of A_2. Then, $\boldsymbol{w}$ can be transformed by the above column permutation into (η, ξ). The system (1) associated with $\boldsymbol{S}^*$ can be represented by

$$A_1\eta + A_2\xi = 0 \tag{5}$$

and, since A_1 is nonsingular, we get

$$\eta = -(A_1)^{-1}A_2\xi .$$

This expression represents a solution to (1) as a linear combination of free parameters ξ. In other words, the set of vectors $\boldsymbol{w}$ subject to (1) can be identified with the set of vertices $(-(A_1)^{-1}A_2\xi, \xi)$ without any constraints, that is, vector $\boldsymbol{w}$ subject to (1) can be represented by

$$\boldsymbol{w} = \boldsymbol{h}(\xi) ,$$

where $\boldsymbol{h}(\xi)$ is a certain permutation of the vector $(-(A_1)^{-1}A_2\xi, \xi)$. Consequently, we can rewrite (2) into

$$B\boldsymbol{h}(\xi) > 0 . \tag{6}$$

Then, we can identify the set of vectors $\boldsymbol{w}$ subject to (1) and (2) with the set of vectors ξ subject to (6). Hence Problem 1 can be reduced to:

Problem 2. Minimize $\varphi(\xi) = \sum_k s_k(g_k(\boldsymbol{h}(\xi)))^2$, subject to (6).

Problem 2 does not have the difficulty (ii). This is because the vector $\boldsymbol{w} = \boldsymbol{h}(\xi)$ is a solution to (1) associated with $\boldsymbol{S}^*$, which is position-free and hence whose picture is correct even if vertex positions change by numerical errors. Moreover, Problem 2 does not have the difficulty (i). Indeed, since A_1 is nonsingular, $(A_1)^{-1}$ can be calculated by any method.

Note that Problem 2 is not only free from the difficulties (i) and (ii) but also much simpler than Problem 1; first, the equality constraint (1) is deleted and, secondly, the number of unknowns is smaller. The number of unknowns in Problem 2 (i.e., the dimension of the vector ξ) is $n + 3m - \text{rank}(A)$. It equals to the number of vertices whose z-coordinates can be specified independently when we reconstruct a polyhedral scene from the picture, and is usually four or a little greater than four. Thus, Problem 2 is very tractable except for the difficulty (iii).

3.3. Practical method

Transforming Problem 1 into Problem 2, we can get over the difficulties (i) and (ii), but the difficulty (iii) still remains; the constraint set of Problem 2 is not closed because some of the inequalities in (6) do not include equalities. This seems unsound when the problem is seen from a viewpoint of mathematical programming, whereas it is a natural consequence of the properties of pictures of polyhedral scenes. For example, some of inequalities in (6) represent the conditions that faces bend along edges. An edge shared by two faces implies that the two faces meet at some angle other than π. This angle may be near to π, but is not equal to π because, if so, the edge would disappear. Consequently, such inequalities do not include equalities.

The difficulty (iii) seems unsound only when we consider Problem 2 as a 'general' optimization problem. For our specific problem, the absense of equalities merely means that the shape to be recovered does not lie on the boundary of such constraints.

Let

$$B_1\boldsymbol{h}(\xi) \geqslant 0 \tag{6'}$$

be the set of all improper inequalities (i.e., inequalities including equalities) in (6), and

$$B_2\boldsymbol{h}(\xi) > 0 \qquad (6'')$$

be the set of all proper inequalities (inequalities excluding equalities). Then, the shape to be recovered may lie on the boundary of the constraints (6′), but never lies on the boundary of the constraints (6″). That is, the inequalities in (6″) are not 'active' at the optimal point. In fact, this property is not a difficulty but it makes the problem easier in the following manner.

An optimization problem is usually solved by an iterated process; starting with a certain initial point in a solution space, we repeatedly seek for a more optimal point until we reach the optimal point. One of the most difficult points in solving a constrained optimization problem is to control each step in such a way that the replaced point does not go out of the constraint set [28]. This control is especially important when the optimal point lies on the boundary of the constraint set.

In our problem the optimal point does not lie on the boundary of the constraints (6″). Hence, if an initial point is chosen to be near enough to the optimal point, we can ignore the inactive constraints (6″). Thus we can lessen the number of constraints in optimization.

Moreover, we can also ignore the other constraints (6′) in the following sense. Note that our objective function to be minimized is the sum of quadric differences between the observed values and the theoretical values of the available cues. Hence the objective function is nonnegative, and if there is no error in observation, it becomes zero at the optimal point. This means that the optimal point remains optimal even if we remove active constraints. That is, the optimal point attains the local minimum of the objective function no matter whether the constraints in (6′) are considered or ignored. Hence if an initial point is chosen to be near enough to the optimal point, we can reach the optimal point by any local optimization method (such as hill climbing) even if we skip the control for preventing the point from going out of the constraint set. Therefore, instead of solving Problem 2 directly, we search for the solution in the following manner. Starting with a certain initial point ξ_0 that satisfies (6), we search for the point $\bar{\xi}$ which yields the local minimum of the objective function of Problem 2. If $\bar{\xi}$ satisfies (6), we adopt $\bar{\xi}$ as a candidate of the shape to be recovered and analyze it in detail. If otherwise, we choose a new initial point and repeat the process. Thus, we can establish the next method for shape recovery.

Method for shape recovery.

Input: Image of a polyhedral scene and its incidence structure $\boldsymbol{S} = (V, F, R)$.

Output: Three-dimensional shape represented in the image.

Process:

Step 1. Find a maximal position-free incidence structure $\boldsymbol{S}^* = (V, F, R^*)$, where $R^* \subseteq R$.

Step 2. Construct Problem 2.

Step 3. Choose a vector $\xi = \xi_0$ that satisfies (6).

Step 4. Find, starting with ξ_0, the locally minimum point $\xi = \bar{\xi}$ of the objective function $\varphi(\xi)$.

Step 5. If $\xi = \bar{\xi}$ satisfies (6) and $\varphi(\bar{\xi})$ is smaller than a certain prespecified threshold, go to Step 6. If otherwise, replace ξ_0 with a new initial point satisfying (6) and go to Step 4.

Step 6. Construct a three-dimensional scene using vector $\bar{\boldsymbol{w}} = \boldsymbol{h}(\bar{\xi})$ (whose components represent z-coordinates of the vertices and surface equations of the faces).

Step 7. If $R^* = R$, end processing, If $R^* \neq R$ (note that, in this case, the scene constructed in Step 6 does not necessarily satisfy the incidence constraints in $R - R^*$ because these constraints have been removed), correct the positions of the vertices associated with elements in $R - R^*$ by finding intersections of the surfaces constructed in Step 6, and end processing.

Step 1 can be done by Theorem 3.1. Step 2 consists of finding a set of vertices whose z-coordinates can be given independently (a method for this is given in [27]), solving (5), and constructing an objective function that depends on what kinds of cues are available in the image. Since (6) is linear with respect to ξ, ξ_0 in Step 3 can be found, for example, by linear programming methods [31]. For a certain important class of pictures, a more efficient method is given in Appendix A. Step 4 is an unconstrained local optimization; it can be done by any method (such as a steepest descent method). Steps 5 and 6 are obvious. Step 7 is also easy; necessity of this step will be easily understood if the readers see the examples in Section 5.

4. Algebraic Approach versus Gradient-Space Approach

In the present paper the global constraints placed by the polyhedron assumption are represented in terms of linear algebra, whereas in previous works [7, 14] similar constraints have been represented by reciprocal figures in a gradient space. In this section we shall compare these two approaches and show that the algebraic approach has several advantages over the gradient-space approach.

4.1. Reciprocal figures in a gradient space

A *reciprocal figure* of a labeled picture is a two-dimensional figure consisting of points and straight line-segments where the points correspond to the faces represented in the original picture and, for every pair of faces sharing a common edge, the associated two points are connected by a straight line-segment in such a way that the line-segment is perpendicular to the original edge. A labeled picture does not always have a reciprocal figure. However, we can see that "if a picture represents a polyhedral scene, it has a reciprocal figure". This property has been known since more than a century ago, and used in structural mechanics for graphical calculus of stresses in plane truss structures [32–34]. This property was rediscovered recently, and has been used in many stages of scene analysis [7, 14–19, 35].

The most familiar interpretation of a reciprocal figure is to consider it as a figure drawn in a 'gradient space' (other interpretations are also possible; see Section 4.4). A *gradient space* is defined as a two-dimensional space with an (a, b) Cartesian coordinate system whose point $G_i = (a_i, b_i)$ represents the family of planar surfaces: $a_i x + b_i y + z + c = 0$, where c is arbitrary. Suppose that two faces $a_i x + b_i y + z + c_i = 0$ $(i = 1, 2)$ share a common edge. Eliminating z, we get the equation of the common edge projected on the picture plane:

$$(a_1 - a_2)x + (b_1 - b_2)y + c_1 - c_2 = 0 .$$

This equation implies that if the gradient space is superimposed on the picture plane so that the a-axis coincides with the x-axis and the b-axis with the y-axis, then the edge is perpendicular to the line connecting two points $G_1 = (a_1, b_1)$ and $G_2 = (a_2, b_2)$. We can thus draw a reciprocal figure of a labeled picture when we are given a real three-dimensional scene represented in the picture.

Therefore, the assumption that the scene is composed of polyhedra places the following global constraints. If our aim is to categorize lines in a picture, labels should be assigned in such a way that the resultant labeled picture has a reciprocal figure [15, 18, 19]. If our aim is to recover the shape from a picture, the faces should be arranged in a three-dimensional space in such a way that the edges in the picture are perpendicular to the associated lines in a gradient space [7, 14].

Draper [21] gives good criticism upon the gradient-space approach to the representation of the global constraints derived from the polyhedron assumption. He also criticizes the algebraic approach; the main point of his criticism seems to be based on the superstrictness of (1). However, since we have overcome the difficulties by Theorems 3.1 and 3.2, his criticism is no more valid. In the rest of this section, we discuss several advantages of our approach over the gradient-space approach.

4.2. Completeness of the constraints

The existence of a solution to the system (1) together with (2) is a necessary and sufficient condition for a picture to correctly represent a polyhedral scene. The existence of a reciprocal figure, on the other hand, is a merely necessary condition; even if a picture has a reciprocal figure, it does not always represent a polyhedral scene.

One of the simplest counterexamples is shown in Fig. 4(a). The picture differs from that in Fig. 3(a) in that the triangular top face is replaced with a vacant hole and consequently there are three visible faces, faces 1, 2, and 3. An incidence structure of this kind can be obtained when three prisms are joined together in a cyclic order. Since the three convex edges (those with '+' labels) are not concurrent, the picture does not represent any planar-surface object. However, we can draw its reciprocal figure as is shown in Fig. 4(b), where the three points correspond to the three faces and the lines connecting them are perpendicular to the associated edges.

In order to diminish this insufficiency many efforts have been done, and the condition is backed up by several new concepts such as Huffman's φ- and φ'-points [20], Kanade's spanning angles [7, 14], and Draper's sidedness reasoning [21]. These concepts can strengthen the condition to some degree, but cannot afford us a necessary and sufficient condition for correct pictures.

Whiteley [36] found that "if a polyhedron is topologically equivalent to a sphere and a picture represents its invisible part as well as visible part, then the existence of a reciprocal figure is a necessary and sufficient condition for the picture to represent a polyhedron correctly." However, this result is not available for our purpose because we cannot extract information about invisible part of objects from real image data. Note that even the incidence structure of the picture in Fig. 4(a) can be obtained from a polyhedron that is topologically equivalent to a sphere; we can do this first by cutting off the upper part of a triangular pyramid, and next by digging a hole on the top of this object so that the section of the object along the line AA' in Fig. 4(a) becomes as shown in Fig. 4(c).

Up to the present time, therefore, the existence of a solution to the system (1) and (2) is the only condition that is necessary as well as sufficient for a picture to represent a polyhedral scene and that is available for recovery of the shape from images.

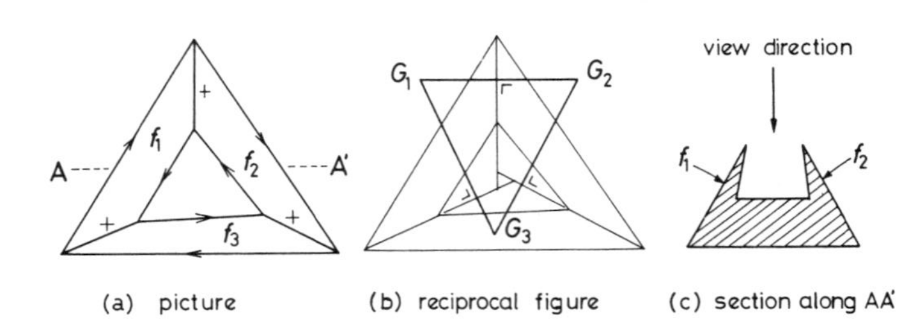

FIG. 4. Incorrect picture and its reciprocal figure.

4.3. False superstrictness phenomenon

The algebraic structure (i.e., the system consisting of (1) and (2)) is too strict to represent the polyhedron constraints, and this superstrictness has been overcome by Theorems 3.1 and 3.2. The reciprocal representation is likewise superstrict; indeed the picture in Fig. 3(a) does not have a reciprocal figure. It may seem that the superstrictness of the reciprocal representation can also be overcome by Theorems 3.1 and 3.2 or by something similar. However, it is not true.

The picture in Fig. 5(a) represents a triangular pyramid on a desk surface seen from above (face 4 is to represent the desk surface). The incidence structure of this picture is position-free and hence the picture remains correct when the positions of vertices are perturbed. Therefore, without being annoyed about numerical errors we can judge whether or not the system (1) and (2) has a solution.

Since this picture is correct, it has a reciprocal figure, shown in Fig. 5(b). However, the existence of a reciprocal figure is not obvious, because the number of constraints is greater than the number of parameters. Indeed, the four points $G_i = (a_i, b_i)$ $(i = 1, \ldots, 4)$ of the reciprocal figure are represented by 8 parameters $a_1, b_1, \ldots, a_4, b_4$. From the definition of a reciprocal figure, a reciprocal figure is invariant under translation and scaling, and hence three out of the eight parameters can be given arbitrarily; there remains only five parameters to be determined. There are, on the other hand, six constraints that lines G_iG_j $(1 \leq i < j \leq 4)$ are perpendicular to the associated original edges. The number of the constraints is one more than the degrees of freedom of the parameters. We can, for example, determine the positions of $G_1, \ldots, G_4$ in the following manner. We first fix G_1 arbitrarily, by which we dissolve the degree of freedom due to translation. Then the direction of G_1G_2 is determined uniquely from the constraint that the line G_1G_2 should be perpendicular to the associated original edge. We next choose as G_2 an arbitrary point on the line, by which we dissolve the degree of freedom due to scaling. Because of the perpendicularity constraints, the direction of G_1G_3 and G_2G_3 are determined uniquely and hence G_3 is determined as the intersection of the two lines. G_4 is determined similarly as the intersection of two lines G_1G_4 and G_2G_4. Thus we can find the positions of all the points. It should be noted that in the above plotting we use only five constraints, but not the constraint that G_3G_4 should be perpendicular to the associated original edge. In fact, Reidemeister's theorem [37] assures us that the last constraint is automatically satisfied, but it is not trivial. It is very difficult for a computer program to judge whether the last condition is satisfied automatically or by change. Thus, the existence of a reciprocal figure is a superstrict condition for the picture in Fig. 5(a), whereas the existence of a solution to the system (1) and (2) is not superstrict because the incidence structure is position-free.

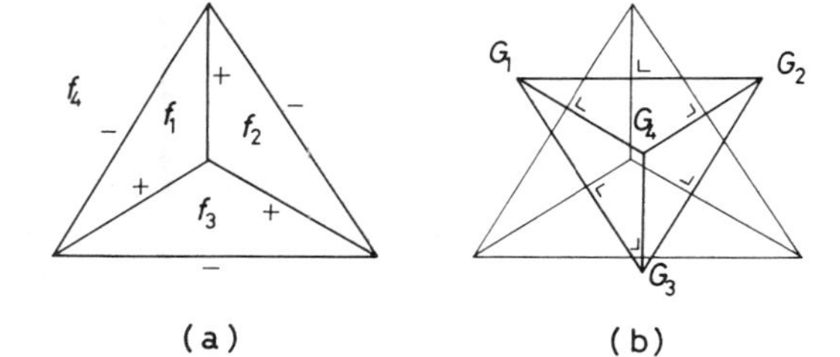

FIG. 5. False superstrictness phenomenon.

This kind of 'false' superstrictness often occurs in reciprocal figures. For this reason, the reciprocal representation seems to be a distorted and ill-natured way of representation of the polyhedron constraints.

4.4. Pseudo-gradient space for perspective projection

The algebraic approach is superior to the gradient-space approach also in the sense that the former is possible for perspectively projected images whereas the latter is not.

However, this statement is misleading. It should be noted that, although it is usually believed impossible (see, for example, [38]), reciprocal figures can also be used for perspective pictures as the representation of the global constraints placed by the polyhedron assumption. That is, we can interpret a reciprocal figure as a figure in a 'pseudo-gradient space' in the following manner.

Without loss of generality, suppose that the view point is at the origin $(0, 0, 0)$ and the picture lies on the plane $z = -1$, that is, the picture is formed as a perspective projection of an object seen from the eye at the origin, as is shown in Fig. 6. A *pseudo-gradient space* is defined as a two-dimensional space

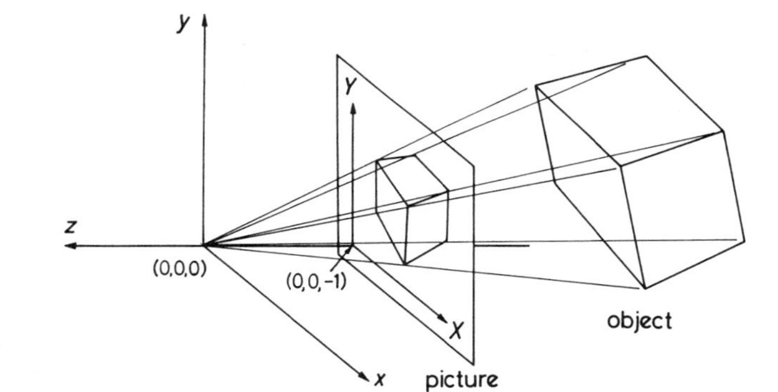

FIG. 6. Object and its perspective projection.

with an (a, b) Cartesian coordinate system whose point $H_i = (a_i, b_i)$ represents the family of planar surfaces: $a_i x + b_i y + cz + 1 = 0$, where c is arbitrary (it differs from the gradient space in the way the parameter c appears in the surface equation). Let (X, Y) be the coordinate system on the picture plane that is obtained by the translation of the (x, y) coordinate system along the z-axis. Then, by the perspective projection point (x, y, z) is transformed to point $(X, Y, -1)$ in such a way that

$$X = -x/z\,, \qquad Y = -y/z\,. \tag{7}$$

Suppose that two faces

$$a_i x + b_i y + c_i z + 1 = 0, \quad i = 1, 2 \tag{8}$$

share a common edge. Eliminating x, y and z from (7) and (8), we get the equation of the common edge projected on the picture plane:

$$(a_1 - a_2)X + (b_1 - b_2)Y = c_1 - c_2\,.$$

This equation implies that if the pseudo-gradient space is superimposed on the picture plane so that the a- and b-axes respectively coincide with the X- and Y-axes, then the edge is at right angle to the line connecting two points $H_1 = (a_1, b_1)$ and $H_2 = (a_2, b_2)$. Thus, the existence of a reciprocal figure is a necessary condition for a picture to be a correct perspective projection of a polyhedral scene.

The important point is that, although the reciprocal figure can represent global constraints for a perspective picture, it cannot be used for shape recovery. This is because surface cues are no more invariant in a pseudo-gradient space.

Suppose, for example, that an object is covered with grain texture of a known density. If an image is orthographic, an apparent grain density does not change when the surface is translated, that is, the apparent density on face $ax + bx + z + c = 0$ depends only on a and b. Therefore, each point $G = (a, b)$ in a gradient space is associated with the unique theoretical value of the apparent density of the texture, and hence we can draw contour curves by connecting points of the same density in a gradient space. When an apparent density on face f_i is observed in the image, then the face is constrained in such a way that the associated point G_i must lie on the contour curve whose density is the same as the observed one.

In the case of a pseudo-gradient space, on the other hand, each point $H = (a, b)$ represents the family of surfaces $-z = (ax + by + 1)/c$ for any c, that is, the family of surfaces that pass $(-1/a, 0, 0)$ and $(0, -1/b, 0)$. Since the apparent density of the grain texture (when seen from the eye at the origin) on the face also depends on c, point $H = (a, b)$ in a pseudo-gradient space cannot be associated with any unique value of the apparent density. Thus, it is impossible to represent texture density cues in a pseudo-gradient space. It is likewise impossible to represent other kinds of surface cues in this space.

To sum up, in the case of perspective projection, a reciprocal figure can represent the global constraints placed by the polyhedron assumption, but it cannot be incorporated with surface cues.

4.5. Unifying treatment of various kinds of surface cues

From the discussion on texture density in Section 4.4 it is obvious that surface cues can be represented in a gradient space only if the cues are invariant under translation of the surfaces; a point in a gradient space represents a family of parallel surfaces and consequently, if a value of a surface cue changes in translation, each point in a gradient space cannot be associated with the unique value of the cue.

In our method, on the other hand, the constraints conveyed by surface cues are represented in the form of the objective function to be minimized. Since the objective function contains surface parameter c_i as well as a_i and b_i, no difficulty occurs even if the value of the cue changes in translation.

4.6. Number of variables in problems

As we have often pointed out, the number of variables in our optimization problem is usually three or a little greater than three. This number is very small; other methods contain much more variables. Kanade [7] also reduced the problem of shape recovery to an optimization problem. In his formulation the optimal solution is sought by adjusting points of a reciprocal figure. Since the reciprocal figure has m points (where m is the number of faces in the original picture) and each point is represented by two coordinates, the number of variables equals to $2m$. On the other hand, the number of variables in our problem does not depend on m.

5. Experiments

We shall present results of three experiments on shape recovery in order to illustrate the performance of our method.

5.1. Computer simulation for ideal images

In the first experiment the method is applied to ideal light intensity data of a scene generated by a computer. This experiment is intended to illustrate basic behaviour of our method for a typical scene with typical surface information.

Fig. 7(a) shows a scene constructed in a computer, where a truncated

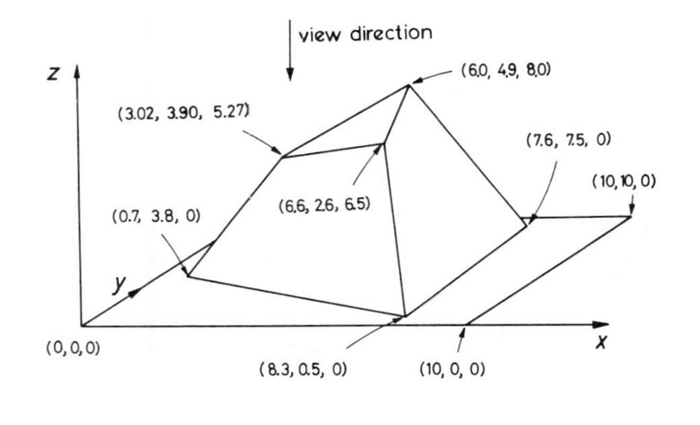

(a) scene

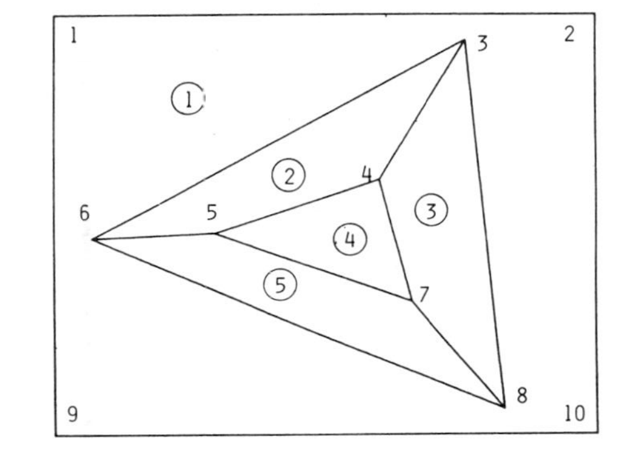

(b) line drawing of the scene

FIG. 7. Scene used in computer simulation.

pyramid lies on a desk surface. Seen from above, it is projected orthographically on the x-y plane as is shown in Fig. 7(b). Since this picture is also generated by a computer, the vertex positions are correct up to digitization errors. It has ten visible vertices and five visible faces, and they are numbered as in the figure.

Let $\boldsymbol{S} = (V, F, R)$ be the incidence structure of the picture. Then, we get

$$V = \{1, 2, \ldots, 10\},$$
$$F = \{1, 2, \ldots, 5\},$$
$$R = \{(1,1), (2,1), (3,1), (3,2), (3,3), (4,2), (4,3), (4,4), (5,2), (5,4), (5,5), (6,1), (6,2), (6,5), (7,3), (7,4), (7,5), (8,1), (8,3), (8,5), (9,1), (10,1)\},$$

where v_i and f_j are abbreviated to i and j ($i = 1, \ldots, 10$, $j = 1, \ldots, 5$) because it is obvious from the context whether a number denotes a vertex number or a face number. Since $|V| + 3|F| = 25 < |R| + 4 = 26$, $\boldsymbol{S}$ is not position-free (see Theorem 3.1). If we delete from R any one element except for (1, 1), (2, 1), (9, 1), (10, 1), we get position-free incidence structure $\boldsymbol{S}^* = (V, F, R^*)$. In this experiment we put $R^* = R - \{(5, 5)\}$. From R^* we construct the system of equations (1), consisting of 21 equations, with respect to 25 unknowns $\boldsymbol{w} = {}^t(z_1 \cdots z_{10} a_1 b_1 c_1 \cdots a_5 b_5 c_5)$. Theorem 3.2 assures us that the system (1) is linearly independent, and hence the solutions to (1) can be expressed as a linear combination of 4 free variables. According to the systematic way for finding a set of free variables [27], we can, for example, put $\xi = {}^t(z_1, z_2, z_3, z_4)$. For the present example we can easily see that when we specify ξ, the shape is determined uniquely. Thus we get the general form of solutions to (1) as

$$\boldsymbol{w} = \boldsymbol{h}(\xi), \quad \xi = {}^t(z_1, z_2, z_3, z_4). \tag{9}$$

Since the picture in Fig. 7(b) has six convex lines and three concave lines, we get the inequality system (2) consisting of nine inequalities. Substituting (9) in (2), we get (6) and thus obtain the constraint set of Problem 2 explicitly.

A light intensity image of the scene is also generated by a computer. The scene is assumed to be illuminated by a parallel light. Let $-\boldsymbol{l}$ be a direction vector along which the light is projected, that is, $\boldsymbol{l}$ is a vector from a point on the surface toward the light source. It is furthermore assumed that the scene is covered with a Lambertian surface; light intensity at a point on the surface is $L \cos \theta$ where L is a constant depending on the light source and surface reflectance, and θ is the incident angle, that is, the angle between a surface normal $\boldsymbol{n}$ and the light source direction $\boldsymbol{l}$, as is shown in Fig. 8. Then, when we have fixed the light source direction $\boldsymbol{l}$, the light intensity on the surface depends only on its normal, and hence each planar face of the object has a constant intensity.

Let $\boldsymbol{n}_k$ be a vector normal to the kth surface $a_k x + b_k y + z + c_k = 0$. Then, we get

$$\boldsymbol{n}_k = (a_k, b_k, 1)$$

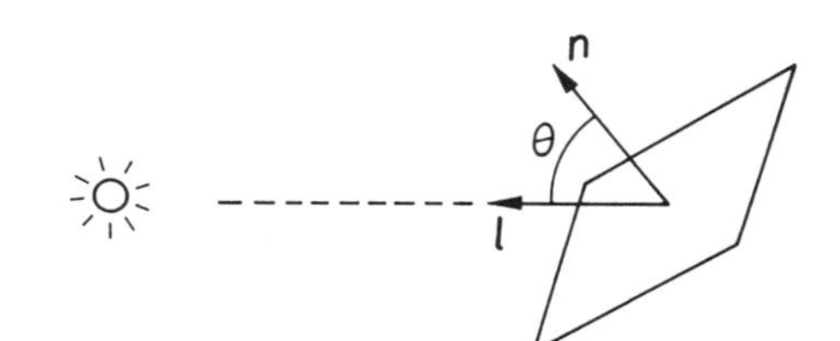

FIG. 8. Incident angle.

and consequently obtain the light intensity I_k on this surface:

$$I_k = L \cos\theta = L \cdot \boldsymbol{l} \cdot \boldsymbol{n}_k / (|\boldsymbol{l}| \cdot |\boldsymbol{n}_k|)$$
$$= L \cdot \boldsymbol{l} \cdot (a_k, b_k, 1)/(|\boldsymbol{l}|\sqrt{(a_k)^2 + (b_k)^2 + 1}) .$$

In the present experiment we put $L = 1$, $\boldsymbol{l} = (-1, 1, 3)$ and thus get the observed value I_k of the light intensity on the kth face ($k = 1, \ldots, 5$) by

$$I_k = (-a_k + b_k + 3)/\sqrt{11((a_k)^2 + (b_k)^2 + 1)} , \quad k = 1, \ldots, 5 ,$$

where a_k and b_k are fixed real numbers given in the scene in Fig. 7(a). Similarly, the theoretical value $J_k(\boldsymbol{w})$ of the light intensity on the kth face ($k' = 1, \ldots, 5$) in scene $\boldsymbol{w} = {}^t(z_1 \cdots z_{10} a_1 b_1 c_1 \cdots a_5 b_5 c_5)$ is given by

$$J_k(\boldsymbol{w}) = (-a_k + b_k + 3)/\sqrt{11((a_k)^2 + (b_k)^2 + 1)} , \quad k = 1, \ldots, 5 ,$$

in which a_k and b_k are regarded as variables. Thus we obtain the objective function in Problem 2 explicitly:

$$\varphi(\xi) = \sum_{k=1}^{5} s_k(I_k - J_k(\boldsymbol{w}))^2$$
$$= \sum_{k=1}^{5} s_k(I_k - J_k(\boldsymbol{h}(\xi)))^2 , \quad \xi = (z_1, \ldots, z_4) ,$$

where as s_k we adopt the area of the kth face on the picture plane.

Because of our assumption on the illumination condition the light intensity data is invariant under the translation of the scene along the z-axis. Hence, without loss of generality we can fix one of the free parameters; we put $z_1 = 0$. Then, our problem has only three unknowns (i.e., z_2, z_3, and z_4), whereas the number of available cues are five, one for each face. We have thus enough cues to recover the shape uniquely.

Now, we can start solving the optimization problem. In order to illustrate behaviour of our method, we have to display recovered three-dimensional shape. For this purpose we use, in what follows, the light stripe representation. Suppose that a virtual light source is set to the left of the view point and light is projected through a narrow vertical slit onto the scene, as is shown in Fig. 9. Then, an image of the slit on the scene surface forms a piecewise linear polygonal line. Changing the direction of the slit light and superimposing the resultant slit images together, we get a light stripe image which reflects the shape of objects in the scene. While the light stripe image was originally used for range finding [39–41], it is also suitable for shape display in that we can illustrate three-dimensional shape without changing the eye position.

Fig. 10(a) shows an initial shape satisfying (6), with which we start our optimization. The result of recovery is shown in Fig. 10(b). We can see that the initial shape (a) is not very near to the optimal shape (b). In our experience, we find that the correct shape can be recovered from almost any initial shape provided that the initial shape satisfies (6). Hence, we need not repeat Steps 4 and 5 of the method more than once.

On the other hand, if we start with an initial shape shown in Fig. 11(a), which does not satisfy (6), then we reach a local minimum that corresponds to the shape shown in Fig. 11(b), in which the convex–concave property is reversed at all edges. Even if we come across this shape as a local minimum of the objective function, we can easily reject it in Step 5 because it does not satisfy the constraint (6). The interesting point is that Fig. 11(b) is very similar to what is called Necker's reversal [42] or a 'negative' object [7], in which the relative depths are all reversed. It should be noted, however, that Fig. 11(b) is not the same as Necker's reversal in the following reason. A primal shape and its Necker's reversal are both correct interpretations of a picture, whereas Fig. 11(b) is incorrect because it allows the objective function to be minimum but not zero. Indeed, the value of the objective function at the local minimum point corresponding to Fig. 11(b) equals to $\varphi(\xi) = 7.21$.

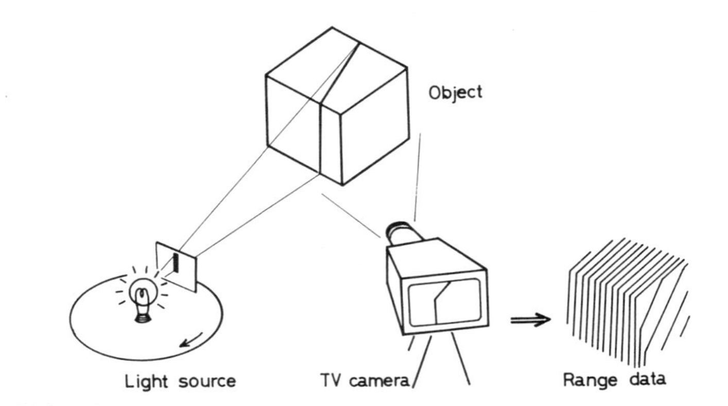

FIG. 9. Light stripe image.

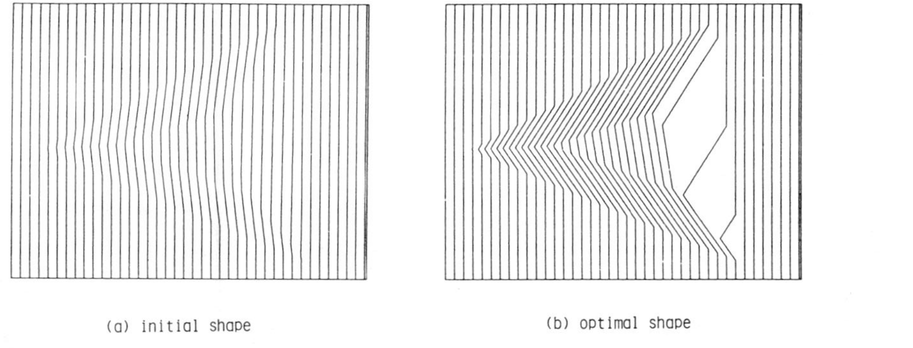

FIG. 10. Result of recovery.

The situation is revealed more clearly when we plot the values of the objective function $\varphi(\xi)$ for various $\xi = (z_1, \ldots, z_4)$. We put $z_1 = z_2 = z_3 = 0$ and move z_4 (that is, we fix the desk surface to the correct position and alter one of the top vertices of the truncated pyramid). Then, $\varphi(0, 0, 0, z_4)$ changes its value as is shown in Fig. 12. When $z_4 > 0$, $\xi = (0, 0, 0, z_4)$ satisfies (6), whereas all of the inequalities in (6) are reversed if $z_4 < 0$. As is easily seen, $\varphi(0, 0, 0, z_4)$ has two locally minimum points denoted by P and Q. P corresponds to the correct shape (Fig. 10(b)), and Q corresponds to a reversed shape like Fig. 11(b) (note that Q does not correspond to Fig. 11(b) exactly; in Fig. 11(b), z_2 and z_3 also have non-zero values). Fig. 12 seems to suggest that the correct optimal point can be attained by any local optimization method for a very large range of initial points in the constraint set.

In the above observation it is not very easy for us to understand the effect of

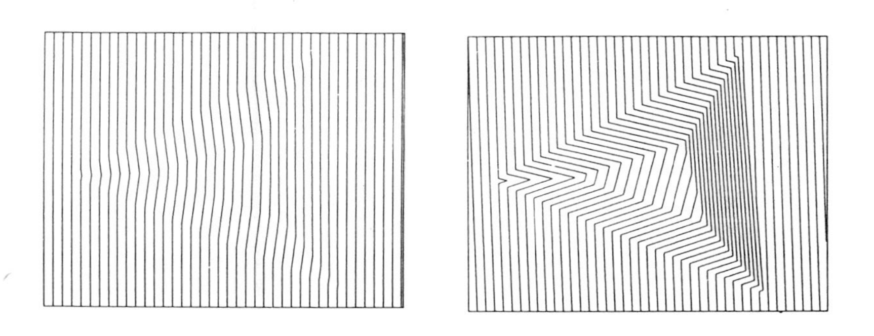

FIG. 11. Reversed shape associated with a locally optimal point.

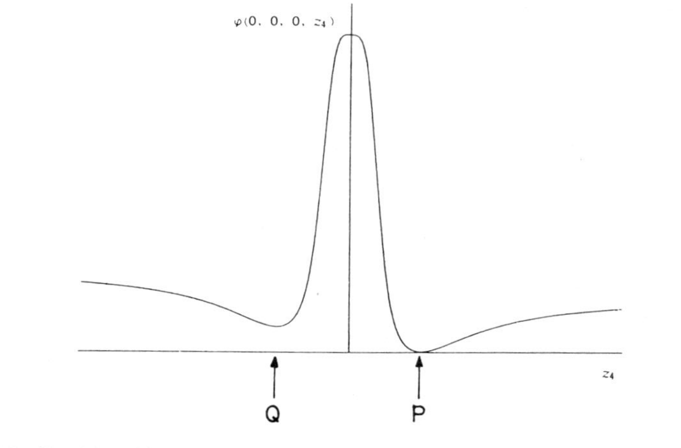

FIG. 12. Profile of the objective function.

the extraction of the position-free structure $\boldsymbol{S}^*$ from $\boldsymbol{S}$. This is because our data do not contain vertex-position errors. Now we perturb the vertex position; we change the position on the picture plane of vertex 7 from (6.6, 2.6) to (6.5, 2.5), and apply our method to the new picture. The shape corresponding to the optimal point is shown in Fig. 13(a). We can observe that the shape has gaps in z-coordinates along the edges 5–6 and 5–7 (where an edge connecting i and j is denoted by i–j). This is because we delete the incidence pair (5, 5) when we construct the position-free incidence structure $\boldsymbol{S}^*$. The incidence pair (5, 5) represents the constraint that vertex 5 should be on face 5. Since this constraint

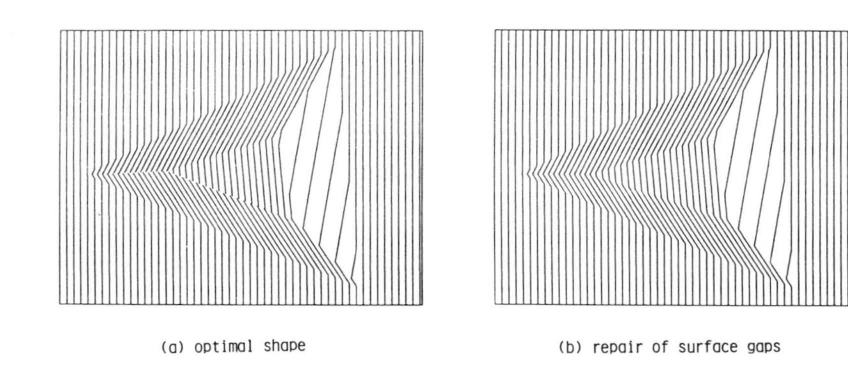

FIG. 13. Shape recovery from a picture with vertex-position errors.

has been deleted, it is not satisfied by the recovered shape in Fig. 13(a). It should be noted that, if we did not delete this constraint, we could not reconstruct any shape. Indeed, the picture is incorrect and hence Problem 1 has no solutions due to the superstrictness of the system of equations (1) associated with $\boldsymbol{S}$.

The only thing we have to do in order to repair the gaps in Fig. 13(a) is to find the exact point of intersection of three recovered faces, faces 2, 4, and 5. The result of the repair is shown in Fig. 13(b). This is the reason why Step 7 in the method is necessary.

5.2. Recovery of the shape from light intensity

The second experiment is an application of the method to real shading images of plaster-object scenes.

The scenes are set on a desk surface in an ordinary room. The desk surface is covered with light gray cloth, and objects made of plaster are put on it. Photographs are taken by a camera with a 50-mm lens, which is about 1.3 m distant from the scenes. The scenes are illuminated by six 40-watt fluorescent lights on the ceiling together with a 300-watt bulb to the left of the camera.

In order to register the theoretical values, $J_k(\boldsymbol{w})$ in (3), of the light intensity cue, we first put a sphere on the desk and take a photograph (Fig. 14(a). In a real world light intensity of a point on the surface is not determined only by the

(a) image of a sphere

b

a

(b) contours of the light intensity in a gradient space

FIG. 14. Calibration of the illumination condition.

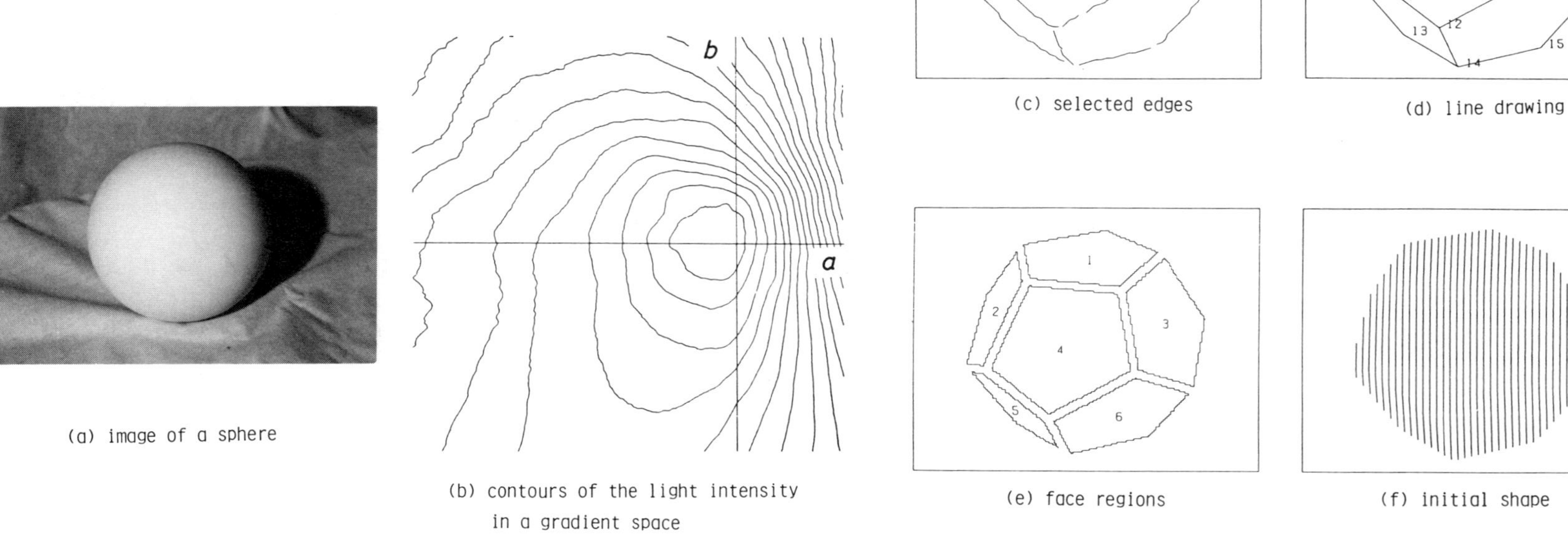

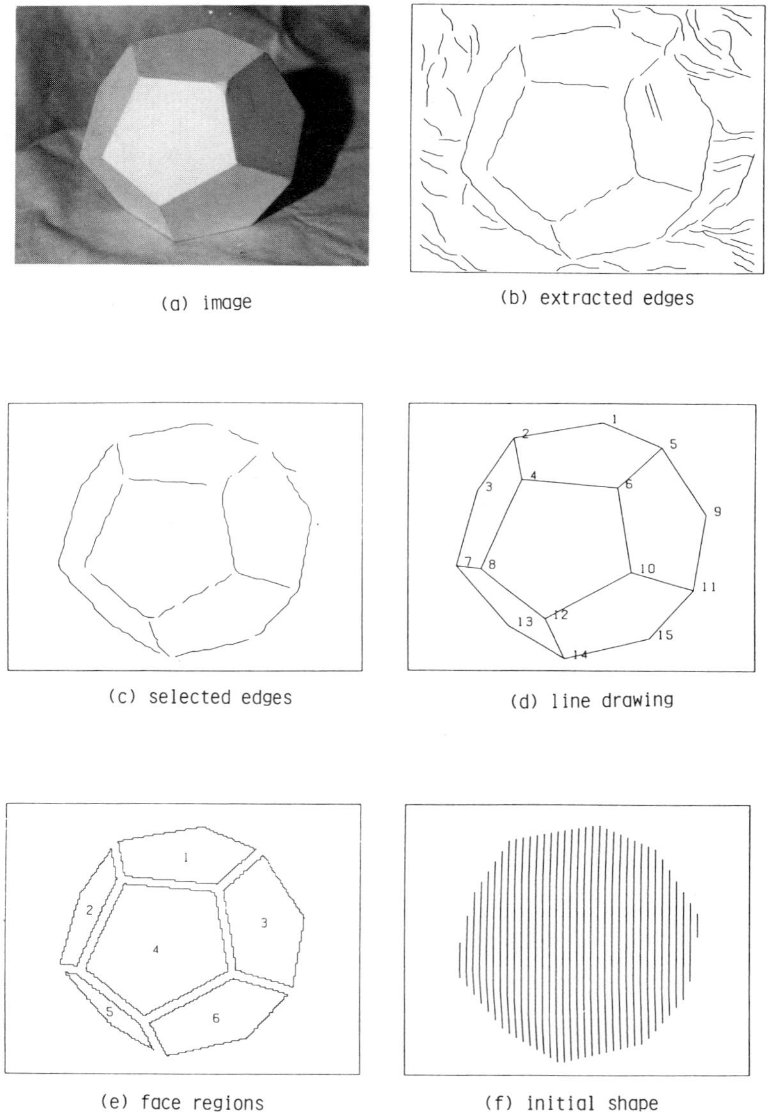

FIG. 15. Recovery from light intensity data, I.

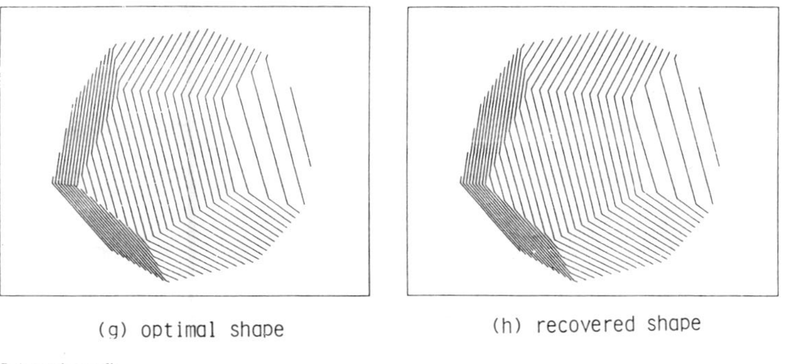

(g) optimal shape (h) recovered shape

FIG. 15 (continued).

direct light from the sources toward the point; it also depends on secondary light that reaches the point after reflecting on other surfaces, and hence depends on the shape around. As a rough approximation, however, we assume in the present experiment that the light intensity depends only on the surface normal. Since a sphere has points of any normal direction, we can read from Fig. 14(a) the light intensity value on a surface of any given normal.

Fig. 14(b) is a contour map of equi-intensity curves plotted in a gradient space. The smallest closed curve denotes a contour of the highest light intensity, and the light intensity descends as we go outward. The intensity descends rather slowly in the left lower direction in Fig. 14(b). This is mainly because secondly light reflected on the desk surface is stronger in the left lower part of the sphere.

Then, a scene to be recovered is set under the same illumination condition. Fig. 15(a) is an image of a scene in which an icosahedron lies on the desk. Using an interactive system, we process this image. The system extracts edges (Fig. 15(b)), and an operator chooses important lines (Fig. 15(c)). Then, the system organizes the lines and constructs a line drawing as shown in Fig. 15(d), where the edge 7–8 is added according to an instruction given by the operator.

The incidence structure of this line drawing is not position-free. In order to get a position-free substructure we delete the constraint that vertex 8 lies on face 5 (see Fig. 15(e) for the face numbers). We choose this constraint because the position of vertex 8 is determined as an intersection of only two lines 8–4 and 8–12 and hence it seems less reliable (recall that a vertex associated with a deleted incidence pair may be moved in Step 7 of the method).

Since the distance from the scene to the camera is not large enough, we formulate the algebraic structure of the line drawing on the basis of the perspective projection model [27], and obtain the constraint set (6).

From the line drawing in Fig. 15(d) the system also finds face regions as is shown in Fig. 15(e). For face k ($k = 1, \ldots, 6$), the system computes the average intensity I_k and the region area s_k; they are used in the objective function $\varphi(\xi)$ of Problem 2 as the observed value and the weight, respectively, of the kth cue.

The scene has four free parameters (we can, for example, choose $\xi = (z_1, z_2, z_3, z_4)$). However, similar to the last experiment, the present cues are invariant under scaling: $(x, y, z) \rightarrow (px, py, pz)$ (recall that the eye is at the origin). Hence, one of the free parameters can be fixed arbitrarily, and consequently the number of unknowns is three whereas the number of cues is six. Thus, we have enough cues to recover the shape uniquely.

Starting with an initial shape (Fig. 15(f)), we get the optimal shape as shown in Fig. 15(g), where the surface has discontinuity along edges 7–8 and 8–12. The discontinuity is repaired and the final result is obtained as is shown in Fig. 15(h).

Another example is shown in Fig. 16. The object is composed of two rectangular prisms penetrating each other. Fig. 16(a) is a light intensity image,

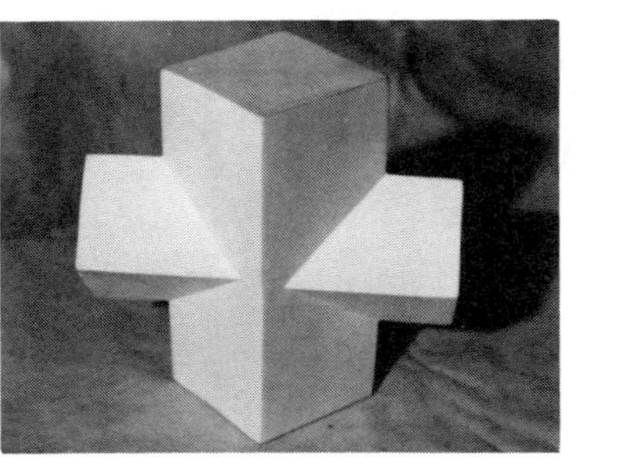

(a) image

(b) line drawing

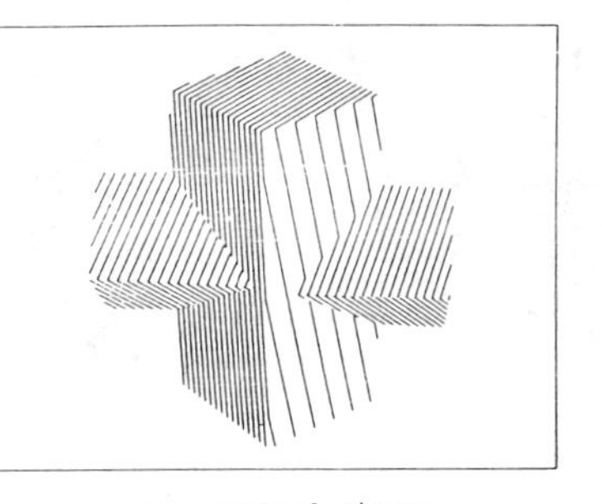

(c) optimal shape

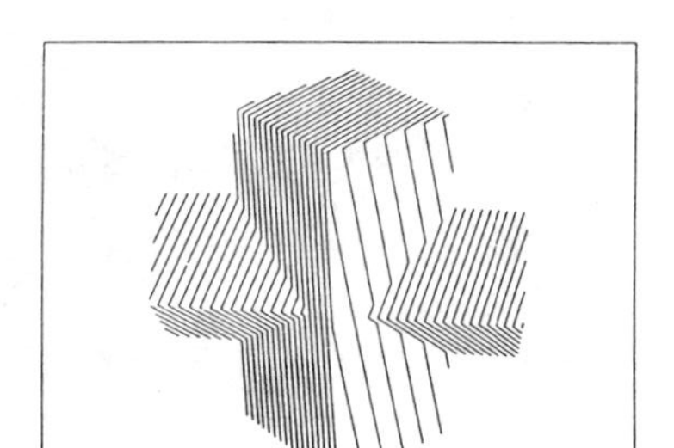

(d) recovered shape

FIG. 16. Recovery from light intensity data, II.

and (b) is a line drawing extracted from (a). The line drawing has seven faces (they are numbered as shown in (b)), but the system is told that faces 4 and 6, and faces 5 and 7 are respectively coplanar. Therefore, its incidence structure has 19 vertices, 5 distinct planar surfaces, and 34 incidence pairs, and, consequently, four incidence pairs are deleted for the construction of a position-free substructure (note that $|V| + 3|F| - |R| - 4 = 4$). Fig. 16(c) is an optimal shape. We can see surface gaps along several edges, which are due to the deletion of some incidence pairs. Repairing the gaps, we obtain the final result as is shown in Fig. 16(d).

5.3. Recovery of the shape from texture density

In the present experiment the method is applied to orthographic images of scenes in which objects are covered with grain texture of a known uniform density.

Scenes are composed of polyhedral objects covered with grain texture. Sizes of the objects are about 180 mm ~ 230 mm in their maximum diameters. Photographs are taken by a camera with a 200-mm telephoto lens that is about 5 m distant from the objects, which allows us to assume, as a rough approximation, that the photographs are orthographic projections of the scenes.

Fig. 17(a) shows an image of a textured object. A rectangular plate covered

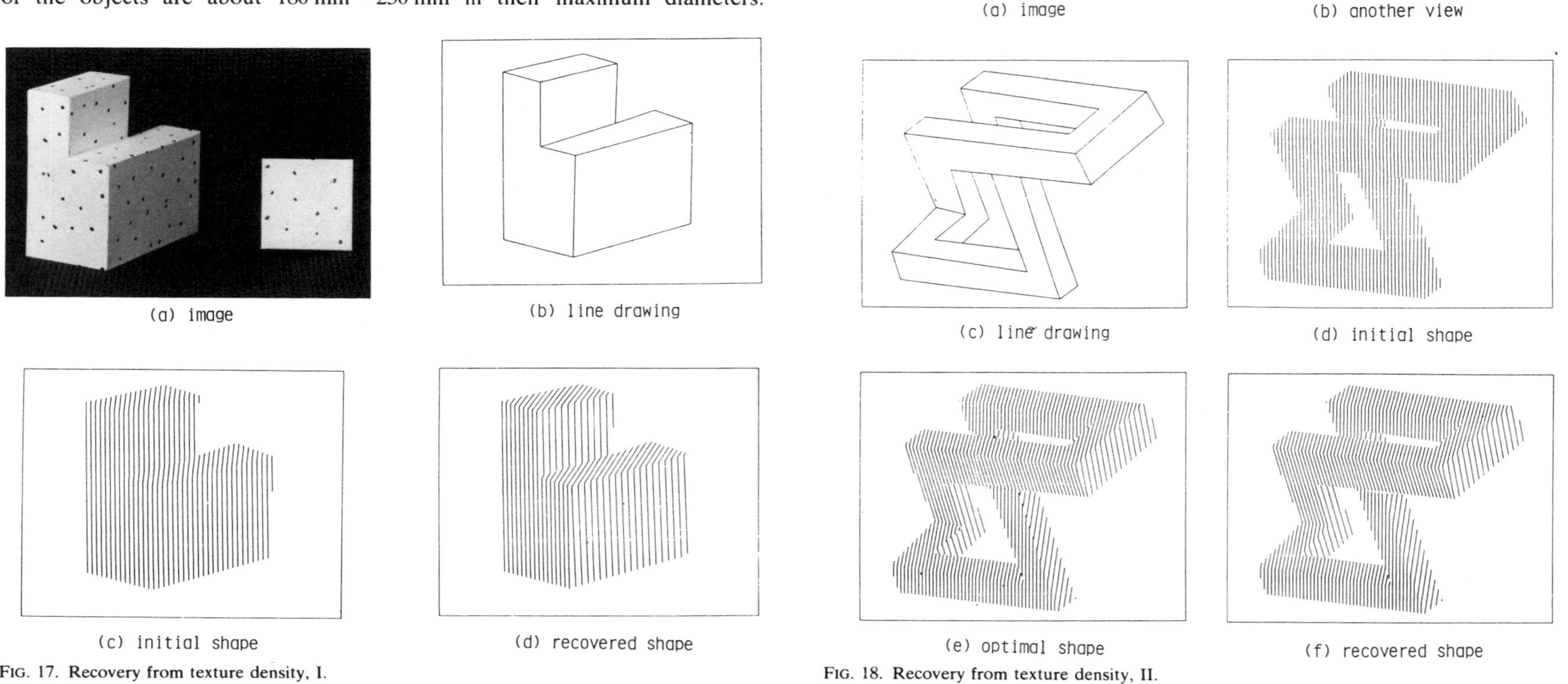

FIG. 17. Recovery from texture density, I.

FIG. 18. Recovery from texture density, II.

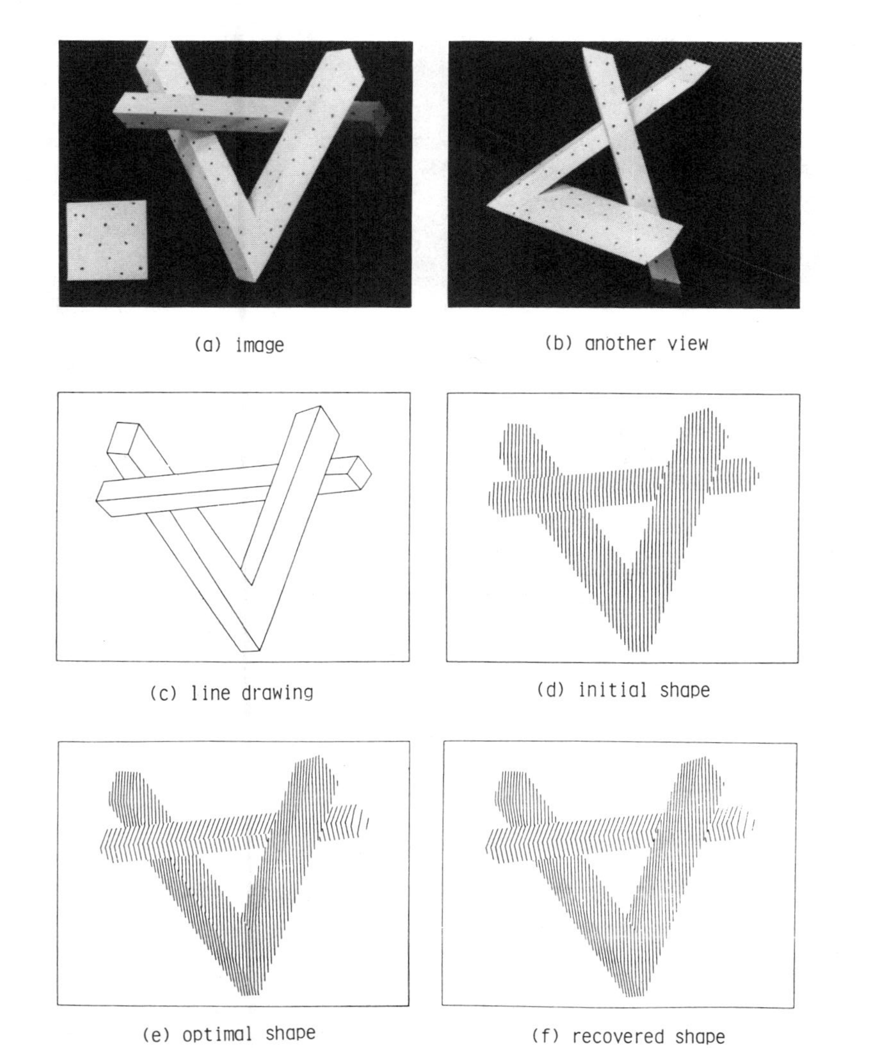

FIG. 19. Recovery from texture density, III.

with the same texture is also put in the scene. The normal to the plate faces toward the camera so that an apparent density may coincide with the real density. From the image of this plate we can easily calculate the theoretical value of the grain density of a given planar surface.

A line drawing of the object is shown in Fig. 17(b). The incidence structure of this picture is position-free, and hence we can directly construct Problem 2, in which the number of variables is four. The texture density is invariant under the translation of the scene along the z-axis, and hence one of the variables can be specified arbitrarily; the number of variables is reduced to three. The image has five cues (one on each face) and thus we have enough information to recover the shape uniquely. Starting with an initial shape (Fig. 17(c)), we get the optimal shape as is shown in Fig. 17(d).

Two more examples are shown in Figs. 18 and 19. The objects chosen in these examples are what produce 'anomalous pictures' [43, 44]. Figs. 18(a) and 19(a) are image data to which our method is applied. Figs. 18(b) and 19(b) are images of the same objects seen from another angle. They are presented only for helping readers to understand the shape of the objects; they are not used in the experiment.

In each of Figs. 18 and 19, (c) shows a line drawing, (d) an initial shape used in the optimization, (e) the optimal shape (in which the surface is discontinuous along some edges), and (f) the final result of the recovery.

Appendix A. Efficient Method for Finding Initial Shape

In Steps 3 and 5 in the method of the shape recovery, we have to find an initial point $\xi = \xi_0$ that satisfies the constraint (6). This can in general be done by linear programming methods, such as a simplex method [31]. However, if the degree of freedom of the solution to (1) is 4 (that is, $n + 3m - \text{rank}(A) = 4$), then we can find an initial point much quicker than we employ a linear programming method. The new method is based on the next theorem.

Theorem A.1. *Suppose that the system* (1) *and* (2) *associated with a labeled picture has solutions and the degree of freedom of the solutions is* 4. *Then, any nontrivial solution* $\boldsymbol{w}$ *to* (1) *satisfies either* $B\boldsymbol{w} > 0$ *or* $B\boldsymbol{w} < 0$.

Proof. Let D be the labeled picture in the theorem, and, for any nontrivial solution $\boldsymbol{w}$ to (1), let $P(\boldsymbol{w})$ be a polyhedron constructed from $\boldsymbol{w}$. Let, furthermore, $T(\alpha, \beta, \gamma, \delta)$ be a transformation in a three-dimensional space that transforms point (x, y, z) to point (x', y', z') in such a way that

$$\begin{bmatrix} x' \\ y' \\ z' \end{bmatrix} = \begin{bmatrix} 1 & 0 & 0 \\ 0 & 1 & 0 \\ \alpha & \beta & \gamma \end{bmatrix} \begin{bmatrix} x \\ y \\ z \end{bmatrix} + \begin{bmatrix} 0 \\ 0 \\ \delta \end{bmatrix}.$$

When point (x, y, z) moves in $P(\boldsymbol{w})$, point (x', y', z') defined as above sweeps a certain three-dimensional region, say P'. Since the transformation is linear, P' is also a polyhedron. Moreover, since $x' = x$ and $y' = y$, the picture D is also an orthographic projection of P'. Therefore, for any $\alpha, \beta, \gamma, \delta$ ($\gamma \neq 0$), there exists a solution $\boldsymbol{w}'$ to (1) such that $P' = P(\boldsymbol{w}')$.

Conversely, let $\boldsymbol{w}$ and $\boldsymbol{w}'$ be two solutions to (1). Since the degree of freedom of the solutions to (1) is 4, $\boldsymbol{w}$ is specified uniquely by four parameters; for example, z-coordinates of four non-coplanar vertices, say z_1, z_2, z_3, z_4. Similarly $\boldsymbol{w}'$ is specified by four parameters z'_1, z'_2, z'_3, z'_4. The transformation $T(\alpha, \beta, \gamma, \delta)$ has four parameters and they are uniquely determined if we specify four non-coplanar points and their destinations. Therefore, there exists a unique set of parameters $\alpha, \beta, \gamma, \delta$ such that $T(\alpha, \beta, \gamma, \delta)$ transforms point (x_i, y_i, z_i) to (x_i, y_i, z'_i) for $i = 1, \ldots, 4$, and hence transforms $P(\boldsymbol{w})$ to $P(\boldsymbol{w}')$. Thus, for any fixed solution $\boldsymbol{w}$ to (1) and (2), $\boldsymbol{w}'$ is a solution to (1) if and only if there exists a transformation $T(\alpha, \beta, \gamma, \delta)$ that transforms $P(\boldsymbol{w})$ to $P(\boldsymbol{w}')$.

Each inequality in (2) is of form

$$a_j x_i + b_j y_i + z_i + c_j > 0 \quad (\text{or } < 0) .$$

Let $\boldsymbol{w} = {}^t(z_1 \cdots z_n a_1 b_1 c_1 \cdots a_m b_m c_m)$ be a solution to (1) and (2), and let $v_1 = (x_1, y_1, z_1)$, $v_2 = (x_2, y_2, z_2)$, $v_3 = (x_3, y_3, z_3)$ be three vertices on face f_j. Then, the above inequality is satisfied for any solution $\boldsymbol{w}' = {}^t(z'_1 \cdots z'_n a'_1 b'_1 c'_1 \cdots a'_m b'_m c'_m)$ to (1) provided that the orientation of (v'_1, v'_2, v'_3, v'_i) is the same as that of (v_1, v_2, v_3, v_i), where $v'_k = (x_k, y_k, z'_k)$, $k = 1, 2, 3, i$. If the two orientations are converse to each other, $\boldsymbol{w}'$ satisfies the converse of the above inequality. The transformation $T(\alpha, \beta, \gamma, \delta)$ preserves the orientation if $\gamma > 0$, and reverses it if $\gamma < 0$. Therefore, any solution $\boldsymbol{w}$ to (1) satisfies either $B\boldsymbol{w} > 0$ or $B\boldsymbol{w} < 0$.

From this theorem, we can construct a simple method for finding an initial point $\xi = \xi_0$. Suppose that the degree of freedom of the solutions to (1) is 4. Without loss of generality we renumber vertices so that $\xi = (z_1, z_2, z_3, z_4)$. Then, we get an initial point $\xi = \xi_0$ (i.e., a solution ξ_0 to (6)) by the method:

Step 1. Assign any real numbers to z_1, z_2, z_3.

Step 2. Find the intersection of the plane passing $v_i = (x_i, y_i, z_i)$, $i = 1, 2, 3$, and the line that is parallel to the z-axis and that passes (x_4, y_4). Let the intersection be $(x_4, y_4, z_4^{(0)})$.

Step 3. Choose any real numbers $z_4^{(1)}$ and $z_4^{(2)}$ such that $z_4^{(1)} < z_4^{(0)} < z_4^{(2)}$. (Then, the orientation of $(v_1, v_2, v_3, v_4^{(1)})$ is converse to that of $(v_1, v_2, v_3, v_4^{(2)})$, where $v_4^{(k)} = (x_4, y_4, z_4^{(k)})$, $k = 1, 2$.)

Step 4. If $\xi = (z_1, z_2, z_3, z_4^{(1)})$ satisfies (6), return $\xi_0 = (z_1, z_2, z_3, z_4^{(1)})$. If $\xi = (z_1, z_2, z_3, z_4^{(2)})$ satisfies (6), return $\xi_0 = (z_1, z_2, z_3, z_4^{(2)})$. If otherwise (in this case there is no solution to (6)), return false.

ACKNOWLEDGMENT

The author is much indebted to Prof. N. Sugie of Nagoya University for valuable discussions. The image data used in the experiment in Section 5.2 are obtained with the help of Computer Vision Section, Electrotechnical Laboratory, Ministry of International Trade and Industry of Japan. Nagoya University Computation Center is used in the experiments.

REFERENCES

1. Horn, B.K.P., Obtaining shape from shading information, in: P.H. Winston (Ed.), *The Psychology of Computer Vision* (McGraw-Hill, New York, 1975) 115–155.
2. Woodham, R.J., A cooperative algorithm for determining surface orientation from a single view, *Proc. 5th Internat. Joint Conf. Artificial Intelligence* (1977) 635–641.
3. Ikeuchi, K., Determining surface orientation of specular surfaces by using the photometric stereo method, *IEEE Trans. Pattern Anal. Machine Intelligence* **3** (1981) 661–669.
4. Coleman, Jr., E.N. and Jain, R., Obtaining 3-dimensional shape of textured and specular surfaces using four-source photometry, *Comput. Graphics Image Process.* **18** (1982) 309–328.
5. Kender, J.R., Shape from texture: A computational paradigm, *Proc. ARPA Image Understanding Workshop* (Science Application, 1979) 134–138.
6. Ikeuchi, K., Shape from regular patterns: An example of constraint propagation in vision, AI Memo 567, Artificial Intelligence Lab., MIT, Cambridge, MA, 1980.
7. Kanade, T., Recovery of the three-dimensional shape of an object from a single view, *Artificial Intelligence* **17** (1981) 409–460.
8. Bajcsy, R. and Lieberman, L., Texture gradient as a depth cue, *Comput. Graphics Image Process.* **5** (1976) 52–67.
9. Witkin, A.P., Recovering surface shape and orientation from texture, *Artificial Intelligence* **17** (1981) 17–45.
10. Ikeuchi, K. and Horn, B.K.P., Numerical shape from shading and occluding boundaries, *Artificial Intelligence* **17** (1981) 141–184.
11. Nakatani, H. and Kitahashi, T., Inferring 3-d shape from line drawings using vanishing points (to be presented at 1st Internat. Conf. Computer Applications, Peking, 1984).
12. Ohta, Y., Maenobu, K. and Sakai, T., Obtaining surface orientation from texels under perspective projection, *Proc. 7th Internat. Joint Conf. Artificial Intelligence* (1981) 748–751.
13. Woodham, R.J., Analysing images of curved surfaces, *Artificial Intelligence* **17** (1981) 117–140.
14. Horn, B.K.P., Understanding image intensities, *Artificial Intelligence* **8** (1977) 201–231.
15. Huffman, D.A., Impossible objects as nonsense sentences, in: B. Meltzer and D. Michie (Eds.), *Machine Intelligence* **6** (Edinburgh Univ. Press, Edinburgh, 1971) 295–323.
16. Huffman, D.A., Curvature and creases: A primer on paper, *IEEE Trans. Comput.* **25** (1976) 1010–1019.
17. Huffman, D.A., A duality concept for the analysis of polyhedral scenes, in: E.W. Elcock and D. Michie (Eds.), *Machine Intelligence* **8** (Ellis Horwood, Chichester, 1977) 475–492.
18. Mackworth, A.K., Interpreting pictures of polyhedral scenes, *Artificial Intelligence* **4** (1973) 121–137.
19. Kanade, T., A theory of Origami world, *Artificial Intelligence* **13** (1980) 279–311.
20. Huffman, D.A., Realizable configurations of lines in pictures of polyhedra, in: E.W. Elcock and D. Michie (Eds.), *Machine Intelligence* **8** (Ellis Horwood, Chichester, 1977) 493–509.
21. Draper, S.W., The use of gradient and dual space in line-drawing interpretation, *Artificial Intelligence* **17** (1981) 461–508.
22. Duda, R.O. and Hart, P.E., *Pattern Classification and Scene Analysis* (Wiley, New York, 1973).
23. Falk, G., Interpretation of imperfect line data as a three-dimensional scene, *Artificial Intelligence* **3** (1972) 101–144.
24. Sugihara, K., Quantitative analysis of line drawings of polyhedral scenes, *Proc. 4th Internat. Conf. Pattern Recognition*, Kyoto (1978) 771–773.

25. Sugihara, K., A necessary and sufficient condition for a picture to represent a polyhedral scene, Tech. Research Rept. No. 8302, Dept. Inform. Sci., Nagoya Univ., Nagoya, 1983.
26. Sugihara, K., Studies on mathematical structures of line drawings of polyhedra and their applications to scene analysis, Researches Electrotech. Lab., No. 800, Electrotech. Lab., Tokyo, 1979 (in Japanese); also: Sugihara, K., An algebraic and combinatorial approach to the analysis of line drawings of polyhedra, *Discrete Appl. Math.* (1984) to appear.
27. Sugihara, K., Mathematical structures of line drawings of polyhedrons: Toward man–machine communication by means of line drawings, *IEEE Trans. Pattern Anal. Machine Intelligence* **4** (1982) 458–469.
28. Gill, P.E., Murray, W. and Wright, M.H., *Practical Optimization* (Academic Press, London, 1981).
29. Imai, H., Network-flow algorithms for lower-truncated transversal polymatroids, *J. Oper. Res. Soc. Japan* **26** (1983) 186–211.
30. Sugihara, K., Detection of structural inconsistency in systems of equations with internal degrees of freedom, *Trans. IECE Japan*, **J65-A** (1982) 911–918 (in Japanese). (The English version is also available from the author.)
31. Dantzig, G.B., *Linear Programming and Extensions* (Princeton Univ. Press, Princeton, NJ, 1963).
32. Maxwell, J.C., On reciprocal figures and diagrams of forces, *Phil. Mag., Ser.* 4, **27** (1864) 250–261.
33. Maxwell, J.C., On reciprocal figures, frames, and diagrams of forces, *Trans. Roy. Soc. Edinburgh* **26** (1870) 1–40.
34. Cremona, L. (Beare, T.H. (Trans.)), *Graphical Statics* (Clarendon Press, London, 1890).
35. Whiteley, W., Realizability of polyhedra, *Structural Topology* **1** (1979) 46–58.
36. Whiteley, W., Motions and stresses of projected polyhedra, *Structural Topology* **7** (1982) 13–38.
37. Gurevich, G.B., *Proektivnaya Geometriya* (Gosudarstvennoe Izdatel'stvo, Fiziko-Matematicheskoi Literatury, Moskva, 1960).
38. Ballard, D.H. and Brown, C.M., *Computer Vision* (Prentice-Hall, Englewood Cliffs, NJ, 1982).
39. Shirai, Y. and Suwa, M., Recognition of polyhedrons with a range finder, *Proc. 2nd Internat. Joint Conf. Artificial Intelligence* (1971) 80–87.
40. Oshima, M. and Shirai, Y., Representation of curved objects using three-dimensional information, *Proc. 2nd USA–Japan Comput. Conf.* (1975) 108–112.
41. Sugihara, K., Range-data analysis guided by a junction dictionary, *Artificial Intelligence* **12** (1979) 41–69.
42. Gregory, R.L., *The Intelligent Eye* (Weidenfeld and Nicolson, London, 3rd Ed., 1971).
43. Draper, S.W., The Penrose triangle and a family of related figures, *Perception* **7** (1978) 283–294.
44. Sugihara, K., Classification of impossible objects, *Perception* **11** (1982) 65–74.

Computer processes in the analysis of visual motion are similar to those of biological systems. An alliance of psychology and computer science is necessary to understand them.

Analysis of Visual Motion by Biological and Computer Systems

Shimon Ullman
MIT Artificial Intelligence Laboratory

Analysis of motion plays a central role in biological visual systems. Sophisticated mechanisms for observing, extracting, and utilizing motion exist even in simple animals. For example, the frog has efficient "bug detection" mechanisms that respond selectively to small, dark objects moving in its visual field.[1] The ordinary housefly can track moving objects and discover the relative motion between a target and its background, even when the two are identical in texture—and therefore indistinguishable in the absence of relative motion.[2]

In higher animals, including primates, the analysis of motion is "wired" into the visual system from the earliest processing stages. Some species, such as the pigeon[3] and the rabbit[4] (see Grusser and Grusser-Cornehls[5] for other examples) perform rudimentary motion analysis at the retinal level. In other animals, including cats and primates, the first neurons in the visual cortex to receive input from the eyes are already involved in the analysis of motion: they respond well to stimuli moving in one direction, but little, or not at all, to motion in the opposite direction.[6,7]

The central role of motion perception in biological systems is not surprising, since motion reveals valuable information about the environment. The use of motion by biological systems—in particular the human visual system—demonstrates the feasibility of carrying out certain information processing tasks and helps to establish specific goals for computer analysis of time-varying imagery. The tasks examined in this article are the recovery of structure from motion and the interpretation of Johansson-type moving light displays.

Conversely, computational studies on the interpretation of time-varying imagery can provide insight into general principles that apply to and increase our understanding of biological visual systems.

This article is about the computational problems fundamental to the analysis of time-varying imagery. These problems fall under two broad categories: motion detection and measurement and the interpretation of visual motion.

Motion detection and measurement

The motion of elements and regions in an image is not given directly, but must be computed from more elementary measurements. The initial registration of light by the eye or by electronic image digitizers can be described as producing a two-dimensional array of time-dependent light intensity values $I(x,y,t)$. Motion in an image can be described in terms of a vector field $V(x,y,t)$ that gives the direction and speed of movement of a point with image coordinates (x,y) at time t. While $I(x,y,t)$ is given directly by the initial measurements, $V(x,y,t)$ is not. The first problem in analyzing motion is therefore the computation of $V(x,y,t)$ from $I(x,y, t)$. This computation is the measurement of visual motion.

In some cases, it is sufficient to detect only certain properties of the vector field $V(x,y,t)$, rather than measure it completely and precisely. For example, it might be desirable to respond quickly to a moving object. In such cases motion must be detected, but not necessarily measured.

The category of problems discussed here, however, are those in which both detection and measurement of motion are important. As research progressed, these problems proved to be considerably more difficult than initially anticipated. The search for efficient and reliable measurement methods is therefore an important research area in the analysis of time-varying imagery.

Discrete and continuous motion. Psychological studies of motion detection and measurement by the human visual system established two types of motion, discrete and continuous. For human observers to perceive motion, the stimulus need not move continuously across the visual field. The appropriate spatial and temporal presentation parameters—such as those of motion pictures—can give the impression of smooth, uninterrupted motion to sequential stimuli. The visual system can fill-in the gaps in the discrete presentation, even when the stimuli are separated by up to several degrees of visual angle and by long (400 msec[8]) temporal intervals. The resulting motion, termed *apparent* or *beta*, is perceptually indistinguishable from continuous motion.[9] Further, the filled-in positions are available to subsequent processes such as stereopsis.[10] Apparent motion mechanisms are probably innate in both humans[11] and lower animals.[12]

The apparent motion phenomena raise the question of whether discrete and continuous motion are registered by the same or separate mechanisms. The fact that the visual system can register both types of motion does not necessarily imply separate mechanisms, since a system registering discrete motion could, in principle, register continuous motion. Recent psychophysical evidence, however, supports the existence of two different mechanisms.[13-18] Braddick[15] suggested the terms *short range* and *long range* for the two mechanisms. The short range mechanism measures continuous motion or discrete displacements of about 15 minutes of arc (in the center of the visual field) and temporal intervals of less than about 60-100 msec. The long range mechanism processes larger displacements and temporal intervals. This terminology characterizes the distinction between the two mechanisms better than the discrete/continuous dichotomy, since discrete presentation with jumps of up to 15 minutes of visual arc are processed by the short range mechanism.

There is a more fundamental distinction between the two systems than their difference in range. They seem to perform their motion measurements at different processing stages, based on different motion primitives. In measuring visual motion, it is useful to draw a distinction between two main schemes. At the lowest level, motion measurements are based directly on the local changes in light intensity values; these are called *intensity-based* schemes. Alternatively, it is possible to first identify features such as edges, lines, blobs, or regions and then measure motion by matching these features over time and detecting their changing positions. Schemes of this type are called *token-matching* schemes. In the human visual system, it appears that the short range process is an intensity-based scheme and the long range process is a token-matching scheme.

These two modes of motion detection and measurement give rise to different computational problems, and consequently to different kinds of processes in biological as well as in computer vision systems.

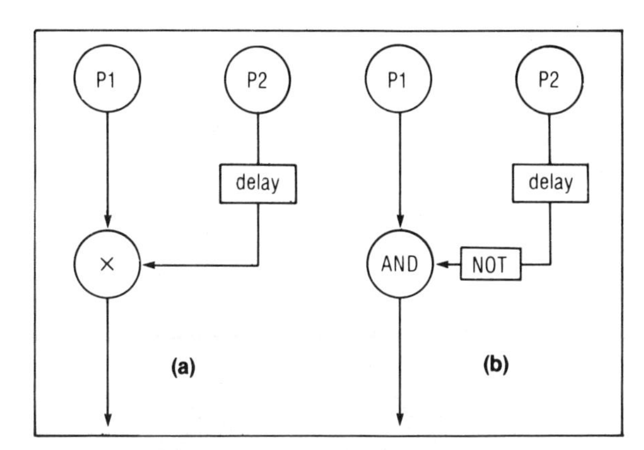

Figure 1. The delayed comparison motion detection scheme. P_1 and P_2 are detectors that respond transiently to a spot of light. (a) For a spot moving to the left at the appropriate speed, the responses of P_1 and P_2 coincide, yielding a positive output of the combined unit. (b) The veto scheme. Motion from P_2 to P_1 produces no response, since the delayed response from P_2 cancels the response from P_1.

Intensity-based schemes. A number of different intensity-based schemes have been advanced as models for motion measurement in biological systems. The various biological schemes can be divided into two main types: correlation techniques and gradient methods.

Correlation schemes. A simple motion detector can be constructed by comparing the outputs of two detectors to light increments at two adjacent positions. The output at position p_1 and time t is compared with that of position p_2 at time $t - \delta t$. Two variations of this approach—called the *delayed comparison* scheme—have been proposed as models for biological systems. The first is obtained by multiplying the two values, i.e., $D(p_1, t) \cdot D(p_2, t - \delta t)$, where D denotes the output of the subunits (Figure 1a). If a point of light moves from p_2 to p_1 in time equal to δt, it causes a light increment at p_1 and, after an interval of δt, a similar increment at p_2. Therefore, the above product is positive. In an array of such detectors, the average output is essentially equivalent to a cross-correlation of the inputs.[19] This scheme provides a successful model for the overall optomotor behavior of various insects in response to motion in their visual fields.

A similar method is the "And-Not" scheme proposed by Barlow and Levick[20] for the directionally selective units in the rabbit's retina and by Emerson and Gerstein[21] for the cat's visual cortex. These units are termed *directionally selective* because their response to stimuli moving in the so-called "preferred" direction is much stronger than their response to the same stimuli moving in the opposite, or "null," direction. Since Barlow and Levick found evidence for inhibitory interaction within the directionally selective mechanism, they proposed a model in which the motion detector computes the logical "And" of $D(p_1,t)$ and "Not" of $D(p_2,t-\delta t)$ (Figure 1b). In this scheme, a motion from p_2 to p_1 is "vetoed" by the delayed response from p_2, whereas motion from p_1 to p_2 produces a positive response.

Torre and Reichardt[22] have proposed a similar scheme for the visual system of the fly in which the delay is replaced by low-pass temporal filtering. Torre and Poggio[23] describe an elegant synaptic mechanism that implements these computations.

Some general properties of the delayed comparison schemes are worth observing. First, these detectors respond selectively not only to continuous motion, but also to discrete jumps of the stimulus between positions p_1 and p_2. Second, they have some obvious limitations. For example, the speed of motion must lie within a certain range, determined by the delay (or the low-pass filtering) and the separation between the receptors. A range of velocities can be covered by a family of several detectors with different internal delays and interreceptor separations. Finally, motion measurement cannot be determined reliably from the output of a single detector of this type. For example, in a field of many moving elements, motion detectors of this type could be activated spuriously if the detector at p_1 is activated by one moving element and the detector at p_2 by a different element. To obtain accurate and reliable motion measurements, the outputs from an array of such detectors should be combined. (This combination problem is discussed below.)

Additional correlation techniques, such as the use of cross-correlation on the raw intensity values,[17,24] were proposed as models for motion measurement in the human visual system. Anstis[25] proposed a subtraction method in which two successive images are shifted and subtracted and displacements are indicated by minima in the resulting image. In general, no precise models were described for the biological implementation of these techniques, and there seems to be no compelling evidence for their existence in the human visual system.

Gradient schemes. A gradient scheme for the detection and measurement of motion by biological systems[26] has recently been proposed as a model for motion analysis by cortical simple cells. These cells, found in the primary visual cortex of the cat[6] and the monkey,[7] respond selectively to edges and bars of light. They are also selective for orientation and, often, for direction of motion. That is, to activate such a unit, the stimulus must have the orientation preferred by the unit and must move in the preferred direction. An analysis of the structure and function of simple cells therefore suggests mechanisms for the early detection and measurement of visual motion. These could, conceivably, be utilized in computer vision systems.

To understand the operation of simple cells, we need a brief description of the input they receive. This input is provided by the fibers of the optic nerve, coming from the eyes via an intermediate station called the LGN. The operations performed on the image by the retina and the LGN are neither direction nor orientation selective. What are these operations? How they are combined to measure the motion of visual stimuli?

Retinal operations on the image. The retinal structure serves two main functions. First, it registers the incoming light on an array of light-sensitive photoreceptors. Second, it performs the initial transformations of the registered image. The transformed image is then transmitted from the last layer of retinal cells (the ganglion cells layer) along about a million nerve fibers to the LGN, and from there to the visual cortex.

Retinal operations were studied experimentally using microelectrodes that measure the electrical activity of ganglion cells in response to light stimulation. Using this technique, the pioneering studies of Hartline,[27] Barlow,[28] and Kuffler[29,30] revealed two major properties of retinal ganglion cells. First, each ganglion cell responds to light stimulation falling within a limited retinal region, or *receptive field,* of the cell. Second, that the receptive field has a center-surround organization of two complementary types, called *on-center* and *off-center* cells. In on-center cells, activity is increased by light falling in the central region of the receptive field; light falling in a surrounding annulus inhibits activity. In the off-center cells, the roles of the center and surround are reversed. The organization of an on-center cell is shown in Figure 2a. Figure 2b is a cross-section through the middle of the field, where response to light is plotted against position in the receptive field. The response is maximal in the middle of receptive field, decreases as the light stimulus is moved outward, and becomes negative in the surround.

The shape of the receptive field can be described analytically as the difference of two gaussians,[31-34] or $\nabla^2 G$, where ∇^2 is the Laplacian operator and G is a two-dimensional gaussian. The retinal operation on the image I can then be described mathematically as $(\nabla^2 G) * I$, that is, the convolution of the image I with the retinal operator $\nabla^2 G$.

What can be derived from operating on the image with $\nabla^2 G$-shaped receptive fields? It is easier to consider this question in the one-dimensional case first; the extension to two dimensions then becomes straightforward. In the one-dimensional analog, the receptive field is described as d^2G/dx^2, i.e., the second derivative of a one-dimensional gaussian G. The retinal operation is then the convolution $(d^2G/dx^2) * I$.

The order of performing differentiation and convolution can be interchanged without affecting the result, as in

$$\left(\frac{d^2G}{dx^2}\right) * I = \frac{d^2}{dx^2}(G * I)$$

We can thus view the retinal operation as a concatenation of two operations: gaussian filtering and a second derivative operation, both performed in a single stage.

The convolution $G * I$ is just a gaussian smoothing of the image. By controlling the size of the gaussian, it is possible to control the resolution at which the image is analyzed and to offset noise amplification introduced by the differentiation.

The second derivative operation that follows the gaussian smoothing is useful in locating sharp intensity

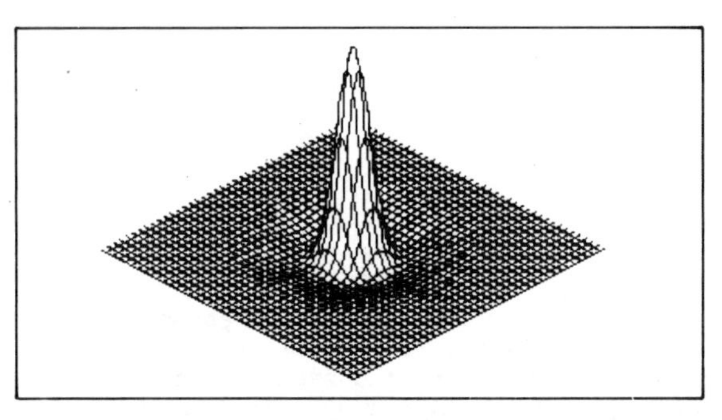

Figure 2. A plot of an on-center receptive field. This shape can be approximated by $\nabla^2 G$, where ∇^2 is the Laplacian operator and G is a two-dimensional gaussian.

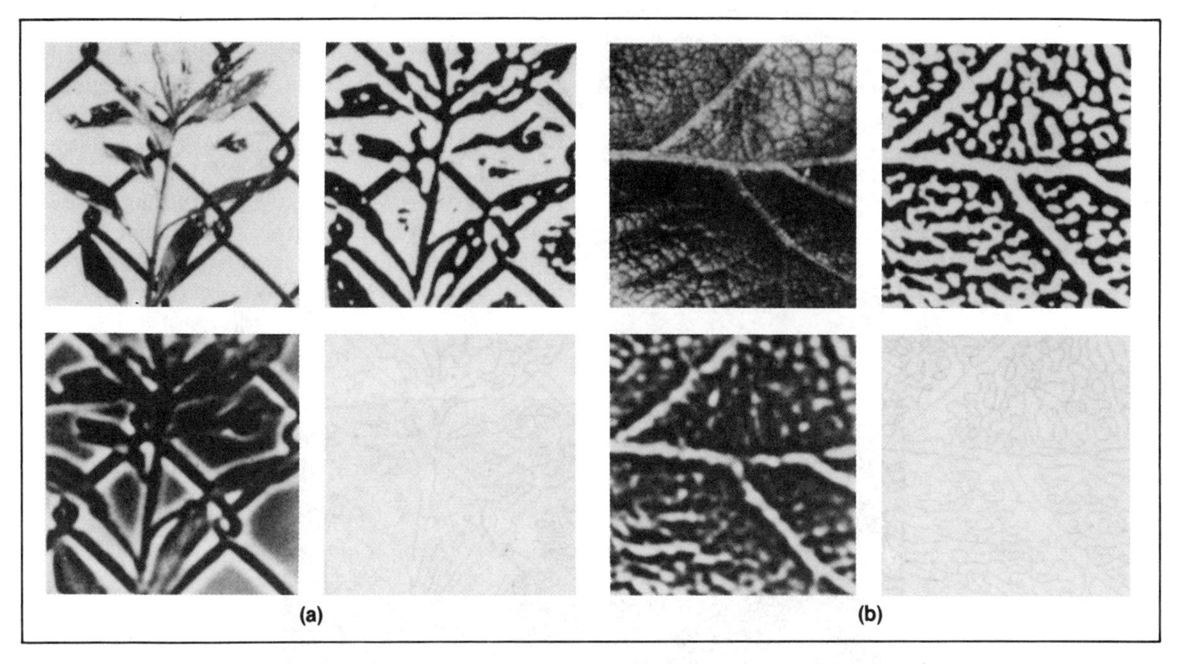

Figure 3. Results of the zero-crossing operation. Top left quadrants of both (a) and (b) show the original image. Lower left quadrants show the image convolved with $\nabla^2 G$. Top right quadrants: a binary image, where the positive convolution values are white and the negative black. Lower right quadrants: the zero-crossings in the convolved image.

changes. Steep intensity changes in the image correspond to peaks in the first derivative, or, equivalently, to zero-crossings in the second derivative. The second derivative has a further advantage: its two-dimensional analog, the Laplacian, is circularly symmetric (see Marr and Hildreth[34] for the two-dimensional case).

The conclusion is straightforward. Zero-crossings in the output of the retinal operation correspond to sharp intensity changes in the original image at the desired resolution, which is controlled in turn by the receptive field size. Figure 3 shows examples of the retinal operation and zero-crossing contours.

A zero-crossing detector can be thought of, roughly, as an edge detector. With the use of on-center and off-center units, it is simple to construct. Figure 4 shows the profile of $\nabla^2 G$ near a zero-crossing. The profile is positive on one side of the zero-crossing, and negative on the other. On-center units are activated in the positive region and off-center units in the negative region. The simultaneous activity of two adjacent units (one on-center and the other off-center) indicates a zero-crossing running between them.

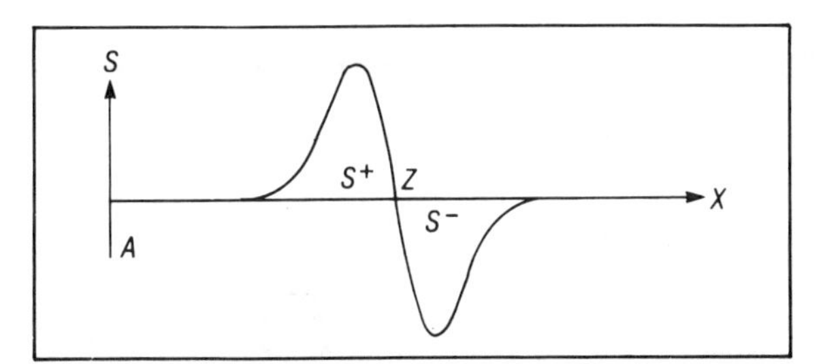

Figure 4. The convolution output (S) as a function of position (X) near a zero-crossing (Z). The zero-crossing is flanked by a positive lobe (S^+) and a negative one (S^-).

Zero-crossings analysis and their use in edge detection are detailed in earlier work.[34,35] Comment here is restricted to the analysis of motion using the zero-crossings scheme.

The motion of a zero-crossing can be determined with use of one additional subunit. Let Z denote the current retinal position of the zero-crossing. It is clear from Figure 4 that the values of the convolution at position Z increase if the zero-crossing is moving to the right and decrease if the zero-crossing is moving to the left. Hence, by inspecting the sign of the time derivative of the convolution, i.e., of $d/dt\,(\nabla^2 G) * I$ at position Z, the direction of motion can be unambiguously determined. Furthermore, the time derivative and the slope of the zero-crossing determine the velocity in the direction between the on-center and off-center units. A motion measuring unit can thus be constructed by combining on-center, off-center, and time derivative units. In Marr and Ullman's biological model,[31,36] the time derivative is measured by Y-type cells, a subpopulation of the retinal ganglion cells.

In this scheme, motion is determined from the slope of the zero-crossing and the time derivative; it is therefore an intensity-based scheme that does not require the matching of elements over time. Motion in opposite directions is detected by different units, a view of human perception supported by psychological evidence.[37]

Details of the model[34] and an outline of its computer implementation[38] have been published. The computer implementation operates on a pair of images separated by a short time interval. The $\nabla^2 G$ operator is applied to the images; the zero-crossing contours are located in the output of the first image. The time derivative is approximated by subtracting the two convolved images and then used to determine the motion of the zero-crossing contours. Figure 5 shows an example of this analysis in which

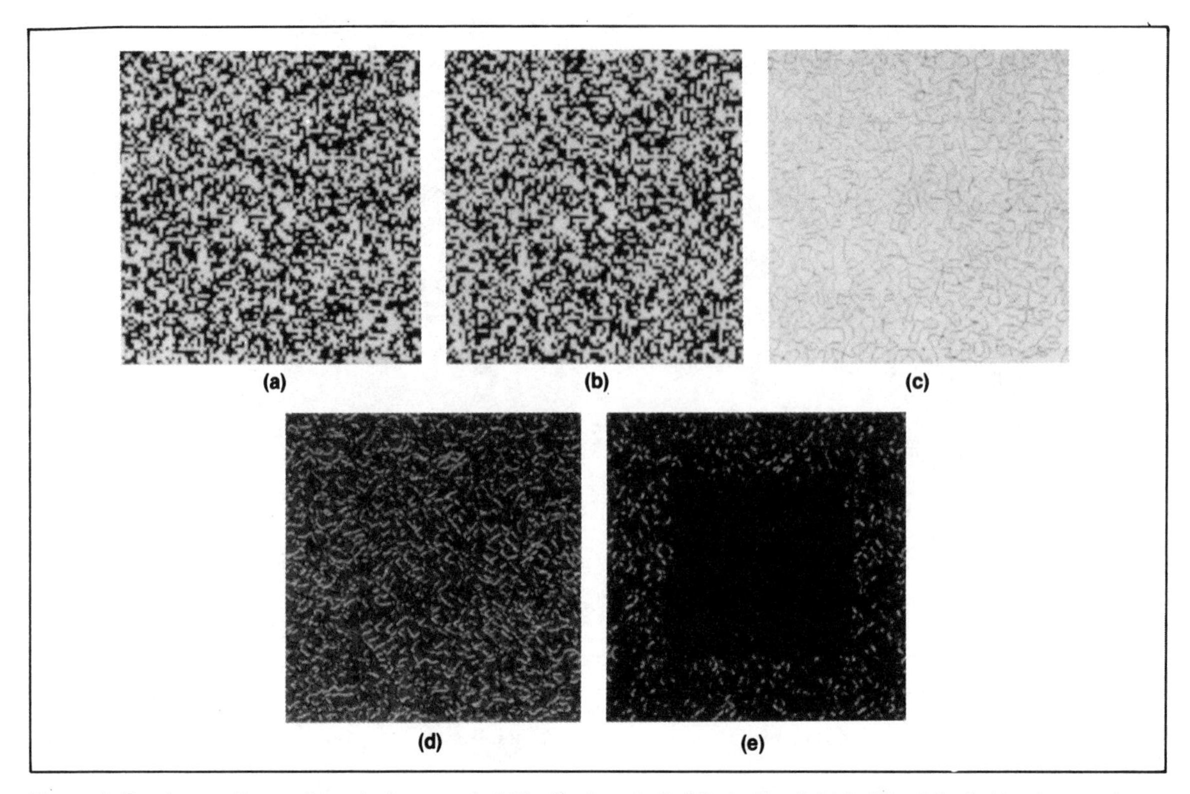

Figure 5. Random patterns. A central square in (a) is displaced slightly to the right in (b), while the backgrounds are uncorrelated. (c) The zero-crossing contours of (a) filtered through $\nabla^2 G$. The motion of the zero-crossings in (d) is in the direction of the light dots (motion assignments). (e) The central square is discovered on the basis of its motion alone. The light dots were removed from this area, except for isolated points where the directions assigned were incorrect.

an imbedded pattern is detected on the basis of its relative motion alone. A central square in Figure 5a is displaced slightly to the right in 5b. The backgrounds in the two images are uncorrelated. When the two patterns are presented in succession, human observers perceive a clearly delineated square moving against a background of uncorrelated noise. The zero-crossings of the first image filtered through $\nabla^2 G$ are shown in Figure 5c. Figure 5d represents the motion of the zero-crossings as determined by the time derivative. The small light dots attached to the contours indicate the direction of motion (the zero-crossing is moving toward the light dot). The central square was found to have a consistent common direction (to the right). Since the backgrounds are uncorrelated, no common motion exists in that region.

Using this scheme, motion measurements can be assigned to zero-crossing contours. However, neither the motion of these contours nor that of any other linear feature can be determined completely on the basis of purely local measurements due to the aperture problem illustrated in Figure 6. If motion is detected by a unit that is small compared to the overall contour, the only information that can be extracted is of the motion component perpendicular to the local orientation of the element. Motion along the element would be invisible. To determine the motion completely, a second stage must combine the local measurements, either in local neighborhoods or along the contour.[26,38]

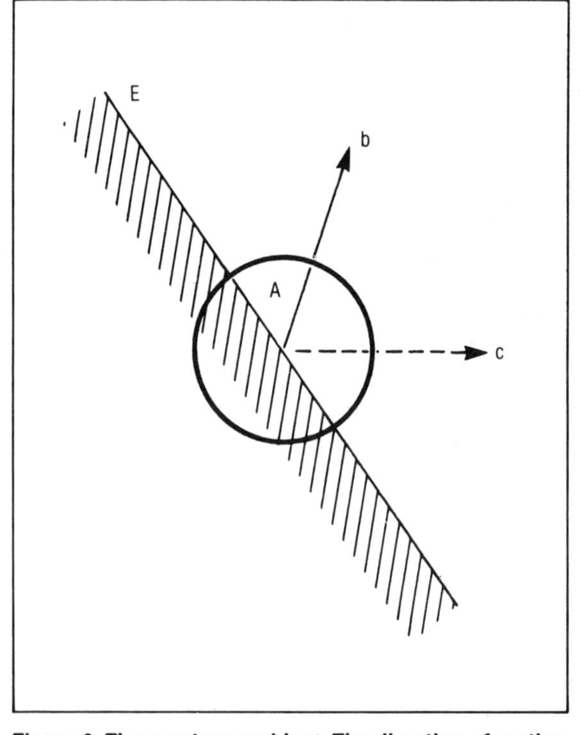

Figure 6. The aperture problem. The direction of motion of a one-dimensional profile cannot be recovered uniquely by a unit that is small in comparison to the moving element. Looking at the moving edge E through the aperture A, it is impossible to determine whether the actual motion is, e.g., toward b or toward c.

The need for a combination stage is a general one: it does not depend on the particular zero-crossings scheme outlined above, but on the fact that no localized features are detected prior to motion measurement. Due to the aperture problem, the use of linear elements as primitives for the measurement would require a subsequent combination of the local measurement. A similar problem occurs in the intensity-based delayed comparison schemes described above. The activity of a single detector of this type is insufficient for determining the motion; a combination stage is required.

Intensity-based techniques in computer vision. A number of intensity-based schemes for motion detection and measurement have been considered for use in computer vision systems. One of them is the gray-level cross-correlation technique used for measuring motion in image pairs. This technique was applied, for example, to the measurement of cloud motion from satellite image data,[39,40] to traffic control,[41] and to the comparison of SLR images.[42] A related scheme, based on the comparison of intensity distributions over small image patches, was proposed.[43] Image subtraction methods were used for motion and change detection and for motion measurement.[44,45]

A fundamental problem of most cross-correlation and subtraction schemes is that they assume the image (or a large portion of it) moves as a whole between the two frames. Images containing independently moving objects and image distortions induced by the unrestricted motion of objects in space pose difficult—perhaps insurmountable—problems for these techniques.

Gradient schemes for motion detection have been proposed by several authors.[46-49] These methods are all based on the relation between the intensity gradient at a given point and the temporal intensity change produced at that point when the intensity pattern is moving. If $I(x,y)$ denotes the light intensity in the image, then

$$\frac{dI}{dt} = I_x u + I_y v$$

where dI/dt is the temporal intensity change at position (x,y); I_x and I_y represent the intensity gradient at that image point; and u, v, are the local velocities in the x and y directions, respectively. Since dI/dt, I_x and I_y are all measurable, in principle, by the observer, a linear relation between u and v at the point in question can be determined by the above relation. However, uniquely determining the values of u and v requires more than a single measurement—it necessitates a combination stage using the local measurements. This, in turn, means that certain assumptions about the velocity field have to be made. The simplest is that u and v are constant over the image. Under such a condition, two independent measurements should suffice, but the applicability of the scheme will be restricted to uniform translations of the image.

The gradient methods outlined above (also see Thompson, pp. 20-28) and the zero-crossing scheme described previously are similar in several respects. Both use temporal change and image gradients to measure the local motion in the direction of the gradient. The method described by Fennema and Thompson[47] and that of Marr and Ullman[26] share another property—both are applied to the image after smoothing (i.e., at defocusing in the former and gaussian filtering at several resolutions in the latter).

Summary. The computation of the image velocity field using intensity-based techniques poses difficult and presently unresolved problems. Under general conditions (unrestricted motion, several objects), different parts of the image have different motions. Consequently, the preferred initial measurements are local. Such measurements are insufficient to determine the motion completely; therefore, the local measurements must be integrated in a subsequent stage. The integration stage is a major unresolved problem in both the understanding of biological systems and the construction of computer vision systems.

Token-matching schemes. In token-matching schemes for measuring motion, identifiable elements—tokens—are located and then matched over time. The apparent motion phenomena discussed above illustrate the capacity of the human visual system to establish motion by matching tokens over considerable spatial and temporal intervals. In perceiving continuous motion between successively presented elements, the visual system has to establish a *correspondence* between the elements of the two presentations. That is, a counterpart for each element in the first frame must be located in the second. A simple correspondence problem is illustrated in Figure 7. The filled circles in the figure represent the first frame, the open circles the second. There are two possible one-to-one matches between the elements of the two frames, leading to two possible patterns of perceived motion: horizontal (top) and diagonal (bottom).

In Figure 7 the match is only two-way ambiguous. In practice, each frame could contain many elements arranged in complex figures; a correspondence must then be

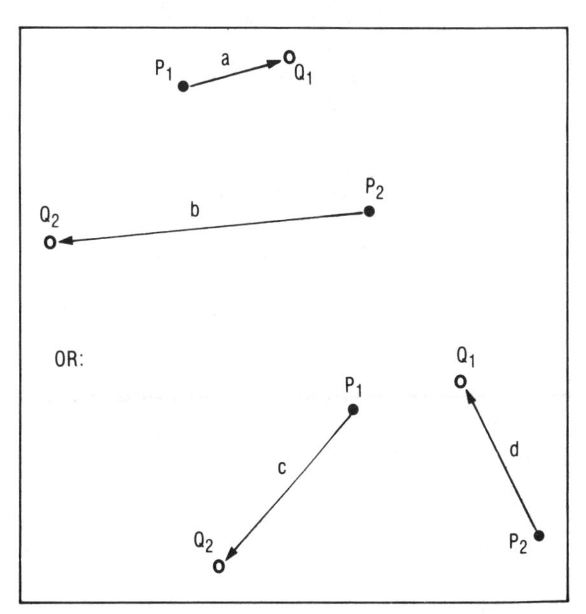

Figure 7. The correspondence problem. P_1 and P_2 are shown on the first frame, Q_1 and Q_2 on the next. Two one-to-one matches are possible, leading to two possible patterns of perceived motion.

established among them. The rules governing the correspondence process in human vision have been investigated in several studies,[17,50-57] but they are still far from being completely understood. A brief summary of some of the aspects of the correspondence process and a general discussion of problems relating to token-matching schemes is now in order.

When the elements participating in the motion are isolated dots alone (as in Figure 7), their correspondence is governed primarily by the distance between the dots. Other parameters being equal, each dot "prefers" to match its nearest neighbor in the subsequent frame.

If the elements in each frame are short line segments, additional rules apply. With line segments, correspondence depends not only upon the interelement distances, but upon their relative lengths and orientations. Other parameters being equal, a given line segment prefers to match another line of similar length and orientation. These results define an affinity metric between line segments that is a function of proximity, orientation, length, and contrast. Each element tends to match its highest-affinity neighbor. A simple and well-known example of this preference is the wagon wheel phenomenon in motion pictures, where the apparent rotation of a spoked wheel is opposite to its actual rotation. This phenomenon is a result of the visual system's preference to choose, from two competing motions, the one that involves minimal distance and angular change.

The correspondence between figures and groups of elements is not based on affinity alone. There are interactions that tend to map groups of elements on a one-to-one basis. As a result, an element might not match its highest-affinity neighbor because that neighbor already has a counterpart; two-to-one matches are ususally avoided.

Two general problems of token-matching schemes are relevant to both biological and machine motion analysis. The first concerns the level at which correspondence is established, that is, the degree of preprocessing and the complexity of the participating tokens. Matching can be established between simple tokens such as points, blobs, edge fragments, and short line segments. Alternatively, the matching process can operate on complex tokens such as structured forms, or even images of recognized objects. The use of complex tokens can simplify the correspondence process, since a complex token usually has unique counterpart in a subsequent frame. In contrast, a primitive token, such as a small edge fragment, usually has a number of competing possible matches. But the use of primitive tokens has two distinct advantages. The first is a reduced preprocessing requirement, which is of special importance in motion perception, where computation time is severely restricted. Second, primitive tokens allow the correspondence scheme to operate on arbitrary objects engaged in complex shape changes. This is because the correspondence between complex figures is established by matching the elementary blocks from which the figures are built. It seems, therefore, that the correspondence process should operate on the level of rather primitive elements, perhaps at the level of Marr's full primal sketch.[58]

The second general problem concerns the possible role of intensity-based and token-matching schemes in an integrated system for visual motion analysis. Intensity-based schemes tend to be fast and sensitive. The human visual system, for instance, can detect velocities as low as one minute of visual arc per second,[59,60] only four times the velocity of the sun across the sky. Directionally selective units in the cat's visual cortex respond reliably to displacements as small as 0.87 minutes of arc[61] (about 1mm at 4m); even smaller displacements can be detected by humans.[60] On the other hand, the ambiguity of the local measurements can make it difficult to recover the velocity field accurately with intensity-based schemes alone. A token-matching scheme can, in principle, track a sharply localized token (such as a line termination) over long distances, and thereby achieve a high degree of accuracy – at the price of more extensive processing to locate the tokens and solve the correspondence problem.

In light of the differences in their basic properties, the two motion measurement schemes could perform distinct visual tasks. The intensity-based system is useful as a peripheral, attention-attracting, "early warning" system and for separating moving objects from their background. Token-matching schemes might play an important role in the recovery of 3-D structure from motion, where accurate tracking over considerable distances is useful. Some recent psychophysical evidence[62] suggests that in the human visual system, the long range process is crucial to the recovery of structure from motion. A second possibility is that the two schemes interact to complement each other. For example, the computation of the long range correspondence could be guided by additional constraints supplied by the short range, intensity-based system.

Token-matching schemes in computer vision. Several token-matching schemes for motion detection have been investigated in the framework of computer vision research.[63,64] Motion is initially determined on the basis of matching edges or gray-level discontinuities. Velocity is assumed to be slow enough, with respect to the sampling rate, that the correspondence problem essentially disappears, and the discontinuities can be matched on the basis of proximity alone. For points not lying on the edges, motion is assigned on the basis of their distances from the edges, using specially constructed masks.

In another system,[65] region segmentation precedes motion detection. Motion is then established by matching regions, followed by cross-correlation on the objects' boundaries. Chow and Aggarwal[66] took a similar approach in analyzing the motion of planar figures, matching figures on the basis of shape descriptors. A more flexible scheme is described in other studies.[67-69] In these, matching is established not between complete figures, but between small fragments of the objects' boundaries. The use of small, primitive tokens overcame some of the difficulties in previous schemes. One such problem is object occlusion, which hinders the global matching of complete figures.

Taken together, psychophysical studies and computational considerations suggest that future research on token-matching schemes should explore, more extensively, the use of small, primitive tokens and study the integration of token-matching and intensity-based schemes.

The interpretation of visual motion

After motion has been detected and measured, it is used to recover properties of the visible environment. Three major uses, common to both biological and computer-based motion analysis systems, are listed below in order of increasing complexity:

(1) object-surround separation,
(2) recovery of three-dimensional shape, and
(3) motion-based recognition.

The following section illustrates and compares these three categories as they occur in human and computer vision systems.

Object-surround separation. The human visual system can separate a moving object from its surround on the basis of motion information alone. Figure 5 illustrates an example in which a moving subfigure is detected in a pair of random dot patterns shown in succession. The subfigure cannot be detected in either of the static frames, since it is defined solely by its frame-to-frame displacement. Experiments of this kind demonstrate that the visual system uses motion to delineate the boundaries of moving objects, even in the absence of intensity edges or texture changes at those boundaries.

In several studies,[64,65,70-72] visual motion was shown to provide useful cues for object-surround separation in computer vision systems. Reliable motion-based segmentation by computer systems proved, however, to be more difficult than anticipated. Some of these difficulties can be appreciated in light of the intensity-based schemes discussed above. In combining the results of the local measurements, continuity of the velocity field is often assumed. Therefore, considerable error can occur at object boundaries, where the continuity assumption is violated. Paradoxically, many intensity-based schemes would benefit if discontinuities in the velocity field were detected prior to the measurement of motion, rather than on the basis of the velocity measurement.

One way to cope with this difficulty is to use the fact that detection of motion discontinuities does not require precise measurements of the velocity field; rough estimates of direction and speed usually suffice. Consequently, it could prove profitable to carry out the local intensity-based motion measurements in two stages. In the first, rough estimates would be made and used to locate discontinuities in the velocity field. In the second, the velocity field would be established without propagating the smoothness constraint across discontinuity contours.

Three-dimensional structure from motion. The human visual system can recover the three-dimensional shape of moving objects even when the objects are unfamiliar and when each static view of the scene contains no 3-D information. The first systematic investigation of this remarkable capacity was carried out in the Wallach and O'Connell study of the kinetic depth effect.[73] In these experiments, an unfamiliar object was rotated behind a translucent screen, and the shadow of its orthographic projection observed from the other side of the screen. (In orthographic projection, the object is projected by parallel light rays that are perpendicular to the screen.) In most cases, the viewers were able to give a correct description of the hidden object's 3-D structure and motion in space, even when each static view was unrecognizable and contained no 3-D information. The original kinetic depth effect employed primarily wireframe objects which project as sets of connected lines. Later studies[56,74-77] established that 3-D structure can be perceived from displays consisting of unconnected elements in motion, and in continuous, as well as apparent, motion.

The computational problems underlying the recovery of 3-D structure from motion have been investigated in a number of studies. The main issues they explored were the conditions under which the structure-from-motion problem has a unique solution and the development of algorithms for the recovery of structure from motion. The main results of these studies are summarized in Table 1. As the table shows, both discrete and continuous versions of the problem have been investigated. The first formulation is discrete in both space and time. The computation is based on a number of discrete frames, or "snapshots," each of which contain a number of isolated points. It has been shown that a small number of frames and points are sufficient for a unique interpretation:[78] for a moving rigid object containing at least four noncoplanar points, the spatial motion and the 3-D coordinates of the points are determined by three frames.

The uniqueness proof is constructive, leading to a possible scheme for the recovery of structure from motion. The scene is divided into groups of about four elements each. The structure of each group is recovered independently (and therefore possibly in parallel), and, finally, the local results are combined in one additional step. This scheme is based on orthographic projection; perspective projection is treated by using the locality of the recovery scheme. That is, for four neighboring points, the two types of projections are similar. It is also possible to use perspective projection directly. There are indications[56] in the perspective case that two frames and

Table 1.
Main results of studies on the recovery of structure from motion.

DISCRETE POINTS AND VIEWS	DISCRETE POINTS, VELOCITIES	VELOCITY FIELD AND ITS DERIVATIVES
4 POINTS IN 3 ORTHOGRAPHIC VIEWS (ULLMAN, 1979)	5 POINTS AND THEIR VELOCITIES IN A SINGLE PERSPECTIVE VIEW (PRAZDNY, 1980)	UNIQUE SOLUTION FOR PURE TRANSLATION (CLOCKSIN, 1980)
5 POINTS IN 2 PERSPECTIVE VIEWS (ULLMAN, 1979)		UP TO 3 SOLUTIONS FOR GENERAL MOTION (LONGUET-HIGGINS AND PRAZDNY, 1980)

perhaps as few as five points are sufficient for a unique interpretation, but the proof of this case is not yet known.

Instead of considering the positions of the points in a number of discrete views, a single view indicating the positions, as well as the velocities of the points, can be taken as input (second column in Table 1). This might be considered a limiting case of the two-frame problem, as the time interval between the frames approaches zero. The problem then takes the following form: given the position and velocity of N points in the image, determine whether or not they belong to a single moving object and find the motion in space and their 3-D coordinates.

A preliminary theoretical problem is to determine the number N for which this recovery problem has a unique solution. Mathematically, this problem is still unresolved. A counting argument of equations and unknowns suggests that five points might suffice, an argument supported by a computer algorithm implemented by Prazdny.[79] Since the computer algorithm proved sensitive to errors in the input, especially when the viewed object was small, it seems that a robust recovery algorithm would require more than five points.

The third formulation of the problem uses velocities and spatial derivatives of the velocity field (third column of Table 1). This can be thought of as a limiting case of the previous formulation, as the distances between the points approaches zero. In a simplified version of this problem, the motion in space is restricted to pure translation.[80] Under this assumption, uniqueness of the solution is readily established. The surface orientation at each image point can be recovered from the velocity and the spatial derivatives of the velocity field at that point.

The more general case, in which the motion can contain rotation components, is considerably more complex. One analysis established that the velocity field at a point has, at most, three interpretations.[81] More precisely, it showed that for nonplanar surfaces (given the velocity and its first and second spatial derivatives at a point) there are, at most, three solutions to the surface orientation at that point. It also provided a scheme for computing the solution. The possible improvement of this result, in particular a determination of whether the solution is unique, poses an open question for future research.

Partial descriptors. The schemes described above aim at a full solution to the structure-from-motion problem. That is, they try to recover all motion parameters as well as the 3-D structure of the visible environment. In contrast to these comprehensive schemes, methods for partial descriptors seek to recover only selected parameters. For example, several mathematical and psychological studies attempted to isolate, in the image velocity field, variables that permit the separate and independent recovery of each of the rotation and translation components of the motion.[82-85] In general, such a decomposition of the problem is not possible, since the velocity field in the image is determined by the intricate interaction of all the motion parameters and the 3-D shape of the viewed objects. It therefore does not seem possible to isolate a variable in the motion field that directly determines, say, the rotation part of the motion independent of the translation components. Partial solutions do become tractable, however, under restricted motion, such as pure translation with no rotation components. Four schemes have been proposed for partial descriptors under restricted motion.

Time to collision. A motion parameter that could be particularly useful for locomoting animals (or machines) is the time-to-collision variable proposed by Lee.[86,87] He argues that the absolute speed of an approaching observer toward an object and the absolute distance between objects cannot be recovered from the image velocity field. He does believe, however, that their ratio, which yields time to collision with the object, can be recovered from the rate of expansion of the image cast by the object. The simple relation between time to collision and image expansion holds, however, only for translatory motion. Under general conditions, it probably cannot be decoupled from the other parameters of motion and shape. Nevertheless, rate of expansion might be useful as a rough estimate of approaching speed. For animals, the fast detection of an approaching object is important, even if the computation is not entirely accurate. Psychophysical and physiological evidence for the existence of mechanisms that respond selectively to expanding stimuli were reported recently.[88,89] These reports could provide the mechanism for "looming" detection suggested by Lee.

Binocular looming detection. Motion in which the right eye sees motion to the left and the left eye sees motion to the right indicates a trajectory in space that is pointing between the two eyes. This can be used for the fast detection and avoidance of moving objects.[26,90]

Focus of the optical flow. Under pure translatory motion, all the points in the image are streaming toward a common vanishing point, which can lie outside the observer's field of view. The vector to this point gives the direction of the observer's motion (assuming a stationary environment).

Gibson[91] suggested that the center of expansion could be used in controlling locomotion. A later study[92] indicated that human observers, can, to a limited extent, locate the focus of expansion, even when it lies outside their field of view.[93] The focus of expansion of a perpendicular plane can be pointed at reliably only in the last 0.5 second prior to collision. It is unclear, therefore, whether this clue is in fact used by the human visual system.

Qualitative shape descriptions. Certain qualitative shape descriptions can be derived from local properties of the image velocity field. For example, Koenderink and van Doorn have shown that the sign of the gaussian curvature at a point is derived from the velocity field around it.[94] Other studies describe a simple scheme for locating depth boundaries in a changing image[95] and a scheme for edge classification (e.g., convex, concave, occluding) under the assumption of pure translatory motion.[80]

Summary on 3-D recovery. Many of the computational problems underlying the recovery of the 3-D structure of moving objects from the changing image are now well understood. In addition to consideration of the open questions, an important next step for machine application will be the development of efficient and robust recovery algorithms. Two major obstacles to such algorithms are:

- real-time computation constraints when the number of participating elements is large and
- robustness in the face of errors and deviations from perfect rigidity. Since measurement inaccuracies and some deformations of the viewed objects might occur, algorithms for the recovery of rigid structures should be able to perform even when perfectly rigid solutions are not available.

Motion-based recognition. Human observers can recognize certain objects solely on the basis of characteristic motion patterns. Powerful demonstrations of this capacity were provided by the studies of Johansson.[76,96] These demonstrations were created by filming human actors moving in the dark with small light sources attached to their main joints. Each actor is thus represented by up to 13 moving light dots. The resulting dynamic dot patterns create a vivid impression of the moving actors. Later studies[97,98] have demonstrated that male actors can often be distinguished from females, and sometimes familiar persons can be recognized on the basis of the moving light dots alone.

These motion-based recognition phenomena have attracted considerable interest among psychologists as well as workers in the field of computer vision. A number of computer-based systems have been constructed to study and simulate motion-based recognition.[99-103]

The general strategy in these studies has been to divide the problem into two stages, one for organizing and describing the motion patterns and another for comparing the resulting representations with similar descriptions stored in memory. The first stage is usually supposed to be autonomous, that is, guided by general organization principles rather than by knowledge of specific objects. The existence of such organizing principles in human perception has been demonstrated in a number of psychological studies,[76,104-106] but their details are still far from clear. Computer vision systems have primarily used spatial proximity and similarity of motion to group moving elements into higher-order units.[63,101,107] An approach recently proposed by Hoffman and Flinchbaugh[99] showed how, based on a planarity assumption, the motion of the limbs can be used to obtain a 3-D interpretation of moving light displays. When these principles are applied to the Johansson-type configurations, perhaps the organization principles will establish a connection between the main joints, yielding a moving stick-figure representation that can be analyzed in a subsequent stage, possibly using the scheme proposed by Marr and Nishihara.[108]

The study of motion-based recognition is only in its initial stages. It is an interesting and potentially useful research topic that combines the study of motion perception with a study of the higher cognitive functions involved in visual recognition and the organization of long-term memory. Experience has shown that advances in this area depend upon the cooperation of computational studies and empirical psychological research. ■

Acknowledgments

I thank E. Hildreth for invaluable comments. This report describes research done at the Artificial Intelligence Laboratory of the Massachusetts Institute of Technology. Support for the laboratory's artificial intelligence research is provided in part by the Advanced Research Project Agency of the Department of Defense under Office of Naval Research contract N00014-75-C-0643 and in part by National Science Foundation Grant MCS77-07569.

References

1. J. Y. Lettvin, H. R. Maturana, W. S. McCulloch, and W. H. Pitts, "What the Frog's Eye Tells the Frog's Brain," *Proc. IRE,* Vol. 47, 1959, pp. 1940-1951.
2. W. Reichardt and T. Poggio, "Figure-Ground Discrimination by Relative Movement in the Visual System of the Fly. Part I: Experimental Results," *Biology and Cybernetics,* Vol. 35, 1980, pp. 81-100.
3. H. R. Maturana and S. Frenk, "Directional Movement and Horizontal Edge Detectors in Pigeon Retina," *Science,* Vol. 142, 1963, pp. 977-979.
4. H. B. Barlow and R. N. Hill, "Selective Sensitivity to Direction of Movement in Ganglion Cells of the Rabbit Retina," *Science,* Vol. 139, 1963, pp. 412-414.
5. O-J. Grusser and U. Grusser-Cornehls, "Neuronal Mechanisms of Visual Movement Perception and Some Psychophysical and Behavioral Correlation," In *Handbook of Sensory Physiology,* R. Jung, ed., Vol. VII/3A, 1973, pp. 333-429.
6. D. H. Hubel and T. N. Wiesel, "Receptive Fields, Binocular Interaction, and Functional Architecture in the Cat's Visual Cortex," *J. Physiology* (London), Vol. 160, 1962, pp. 106-154.
7. D. H. Hubel and T. N. Wiesel, "Receptive Fields and Functional Architecture of Monkey Striate Cortex," *J. Physiology* (London), Vol. 195, 1962, pp. 215-243.
8. W. Neuhaus, "Experimentelle Untersuchung der Scheinbewegung," *Arch. Ges. Psychol.,* Vol. 75, 1930, pp. 315-348.
9. M. Wertheimer, "Experimentelle Studien Über das Sehen von Bewegung," *Zeitschrift für Psychologie,* Vol. 61, 1912, p. 382.
10. M. J. Morgan, "Pulfrich Effect and the Filling-In of Apparent Motion" *Perception,* Vol. 5, 1976, pp. 187-195.
11. E. S. Tauber and S. Koffler, "Optomotor Response in Human Infants to Apparent Motion: Evidence of Innateness," *Science,* Vol. 152, 1966, p. 382.
12. I. Rock, E. S. Tauber, and D. P. Heller, "Perception of Stroboscopic Movement: Evidence for Its Innate Basis," *Science,* Vol. 147, 1964, pp. 1050-1052.
13. S. M. Anstis and B. J. Rogers, "Illusory Reversal of Visual Depth and Movement During Changes of Contrast," *Vision Research,* Vol. 15, 1975, pp. 957-961.
14. S. M. Anstis, "The Perception of Apparent Motion," *Philosophical Trans. Royal Soc. London,* Vol. B, No. 290, 1980, pp. 153-168.
15. O. J. Braddick, "A Short-Range Process in Apparent Motion," *Vision Research,* Vol. 14, 1974, pp. 519-527.
16. O. J. Braddick, "Low-Level and High-Level Processes in Apparent Motion," *Philosophical Trans. Royal Soc. London,* Vol. B, No. 290, 1980, pp. 137-151.

17. A. J. Pantle and L. Picciano, "A Multistable Display: Evidence for Two Separate Motion Systems in Human Vision," *Science,* Vol. 193, Aug. 6, 1976, pp. 500-502.

18. J. T. Petersik, K. I. Hicks, and A. J. Pantle, "Apparent Movement of Successively Generated Subjective Patterns," *Perception,* Vol. 7, 1978, pp. 371-383.

19. B. Hassenstien and W. Reichardt, "Systemtheoretische Analyse der Zeit-, Reihenfolgen-und Vorzeichenauswertung bei der Bewegungs-Perzeption der Russelkafers," *Chlorophanus. Z. Naturf.,* Vol. IIb, 1956, pp. 513-524.

20. H. B. Barlow and R. W. Levick, "The Mechanism of Directionally Selective Units in Rabbit's Retina," *J. Physiology* (London), Vol. 173, 1965, pp. 377-407.

21. R. C. Emerson and G. L. Gerstein, "Simple Striate Neurons in the Cat. II. Mechanisms Underlying Directional Asymmetry and Directional Selectivity," *J. Neurophysiology,* Vol. 40, No. 1, 1977, pp. 136-155.

22. T. Poggio and W. Reichardt, "Visual Control of Orientation Behavior in the Fly. Part II. Towards the Underlying Neural Interactions," *Quarterly Rev. Biophysics,* Vol. 9, 1976, pp. 377-438.

23. V. Torre and T. Poggio, "A Synaptic Mechanism Possibly Underlying Directional Selectivity to Motion," *Proc. Royal Soc. London,* Vol. B, No. 202, 1978, pp. 409-416.

24. H. H. Bell and J. S. Lappin, "Sufficient Conditions for the Discrimination of Motion," *Perception and Psychophysics,* Vol. 14, No. 1, 1973, pp. 45-50.

25. S. M. Anstis, "Phi Movement As a Subtraction Process," *Vision Research,* Vol. 10, 1970, pp. 1411-1430.

26. D. Marr and S. Ullman, "Directional Selectivity and Its Use in Early Visual Processing," *Proc. Royal Soc. London,* Vol. B, No. 211, 1981, pp. 151-180; available also as MIT A.I. Memo 524, MIT, Cambridge, Mass., 1979.

27. H. K. Hartline, "The Response of Single Optic Nerve Fibers of the Vertebrate Eye to Illumination of the Retina," *Am. J. Physiology,* Vol. 121, 1938, pp. 400-415.

28. H. B. Barlow, "Summation and Inhibition in the Frog's Retina," *J. Physiology* (London), Vol. 119, 1953, pp. 69-88.

29. S. W. Kuffler, "Neurons in the Retina: Organization, Inhibition and Excitation Problems," *Cold Spring Harbour Symp. Quantifiable Biology,* Vol. 17, 1952, pp. 281-292.

30. S. W. Kuffler, "Discharge Patterns and Functional Organization of Mammalian Retina," *J. Neurophysiology,* Vol. 16, 1953, pp. 37-68.

31. C. Enroth-Cugell and J. D. Robson, "The Contrast Sensitivity of Retinal Ganglion Cells of the Cat," *J. Physiology* (London), Vol. 187, 1966, pp. 517-522.

32. R. W. Rodieck, "Quantitative Analysis of Cat Retinal Ganglion Cell Responses to Visual Stimuli," *Vision Research,* Vol. 5, 1965, pp. 583-601.

33. R. W. Rodieck and J. Stone, "Analysis of Receptive Fields of Cat Retinal Ganglion Cells," *J. Neurophysiology,* Vol. 28, 1965, pp. 833-849.

34. D. Marr and E. Hildreth, "Theory of Edge Detection," *Proc. Royal Soc. London,* Vol. B, 1980, pp. 187-217.

35. E. C. Hildreth, *Implementation of a Theory of Edge Detection,* MIT A.I. Technical Report AI-TR-579, MIT, Cambridge, Mass., 1980.

36. G. Cleland, M. W. Dubin, and W. R. Levick, "Sustained and Transient Neurons in the Cat's Retina and LGN," *J. Physiology* (London), Vol. 217, 1971, pp. 473-496.

37. R. Sekuler and E. Levinson, "The Perception of Moving Targets," *Sci. Am.,* Vol. 236, No. 1, 1977, pp. 60-73.

38. J. Batali and S. Ullman, "Motion Detection and Analysis," In *ARPA Image Understanding Workshop,* L. S. Bauman, ed., Science Application Inc., Arlington, Va., pp. 69-75.

39. J. A. Leese, C. S. Novak, and V. R. Taylor, "The Determination of Cloud Pattern Motion from Geosynchronous Satellite Image Data," *Pattern Recognition,* Vol. 2, 1970, pp. 279-292.

40. E. A. Smith and D. R. Phillips, "Automated Cloud Tracking Using Precisely Aligned Digital ATS Pictures," *IEEE Trans. Computers,* Vol. 21, 1972, pp. 715-729.

41. K. Wolferts, "Special Problems in Interactive Image Processing for Traffic Analysis," *2nd Int'l Conf. on Pattern Recognition,* Vol. 1 and 2, 1974.

42. R. L. Lillestrand, "Techniques for Change Detection," *IEEE Trans. Computers,* Vol. C-21, 1972, pp. 654-659.

43. R. Jain, D. Militzer, and H.-H. Nagel, "Separating Non Stationary from Stationary Scene Components in a Sequence of Real-World TV Images," *Proc. 3rd Int'l Conf. Artificial Intelligence,* Aug. 1977, p. 612.

44. R. Jain and H.-H. Nagel, "On the Analysis of Accumulative Difference Picture from Image Sequences of Real World Scenes," *IEEE Trans. Pattern Analysis and Machine Intelligence,* Vol. 1, No. 2, 1979, pp. 206-214.

45. R. Jain, W. N. Martin, and J. K. Aggarwal, "Extraction of Moving Objects Images Through Change Detection," *Proc. 6th Int'l Conf. Artificial Intelligence,* 1979, pp. 425-428.

46. J. O. Limb and J. A. Murphy, "Estimating the Velocity of Moving Objects in Television Signals," *Computer Graphics and Image Processing,* Vol. 4, 1975, pp. 311-327.

47. C. L. Fennema and W. B. Thompson, "Velocity Determination in Scenes Containing Several Moving Objects," *Computer Graphics and Image Processing,* Vol. 9, 1979, pp. 301-315.

48. B. Hadani, G. Ishai, and M. Gur, "Visual Stability and Space Perception in Monocular Vision: Mathematical Model," *J. Optical Soc. America,* Vol. 70, No. 1, 1980, pp. 60-65.

49. B. K. P. Horn and B. G. Schunk, *Determining Optical Flow,* MIT A.I. Memo 572, MIT, Cambridge, Mass., 1980.

50. J. P. Frisby, "The Effect of Stimulus Orientation on the Phi Phenomenon," *Vision Research,* Vol. 12, 1972, pp. 1145-1166.

51. P. A. Kolers, *Aspects of Motion Perception,* Pergamon Press, New York, 1972.

52. A. J. Pantle, "Stroboscopic Movement Based Upon Global Information in Successively Presented Visual Patterns," *J. Optical Soc. America,* 1973, p. 1280.

53. J. Ternus, "Experimentelle Untersuchung Über Phanomenale Identitat," *Psychologische Forschung,* Vol. 7, 1926, pp. 81-136.

54. D. Navon, "Irrelevance of Figural Identity for Resolving Ambiguities in Apparent Motion," *J. Experimental Psychology, Human Perception and Performance,* Vol. 2, No. 1, 1976, pp. 130-138.

55. S. Ullman, "Two Dimensionality of the Correspondence Process in Apparent Motion," *Perception,* Vol. 7, 1978, pp. 683-693.

56. S. Ullman, *The Interpretation of Visual Motion,* MIT Press, Cambridge and London, 1979.

57. S. Ullman, "The Effect of Similarity Between Line Segments on the Correspondence Strength in Apparent Motion," *Perception,* Vol. 9, 1981, pp. 617-626.

58. D. Marr, "Early Processing of Visual Information," *Philosophical Trans. Royal Soc. London,* Vol. B, No. 275, 1976, pp. 483-524.

59. C. H. Graham, "Perception of Movement," In *Vision and Visual Perception,* C. H. Graham, ed., John Wiley and Sons, New York, 1965.

60. P. E. King-Smith, A. Riggs, R. K. Moore, and T. W. Butler, "Temporal Properties of the Human Visual Nervous System," *Vision Research,* Vol. 17, 1977, pp. 1101-1106.

61. A. W. Goodwin, G. H. Henry, and P.O. Bishop, "Direction Selectivity of Simple Striate Cells: Properties and Mechanism," *J. Neurophysiology,* Vol. 38, 1975, pp. 1500-1523.

62. J. T. Petersik, "The Effect of Spatial and Temporal Factors on the Perception of Stroboscopic Rotation Simulations," *Perception,* Vol. 9, 1980, pp. 271-283.

63. J. Potter, *The Extraction and Utilization of Motion in Scene Description,* PhD thesis, University of Wisconsin, Madison, Wis., 1974.

64. J. L. Potter, "Scene Segmentation Using Motion Information," *Computer Graphics and Image Processing,* Vol. 6, 1977, pp. 558-581.

65. H.-H. Nagel, "Formation of an Object Concept by Analysis of Systematic Time Variation in the Optically Perceptible Environment," *Computer Graphics and Image Processing,* Vol. 7, 1978, pp. 149-194.

66. W. K. Chow and J. K. Aggarwal, "Computer Analysis of Planar Curvilinear Moving Images," *IEEE Trans. Computers,* Vol. C-26, 1977, pp. 179-185.

67. W. N. Martin and J. K. Aggarwal, "Dynamic Scene Analysis: The Study of Moving Images," Information Systems Research Lab, Electronic Research Center, Technical Report 184, University of Texas, Austin, Tex., 1979.

68. C. J. Jacobus, R. T. Chien, and J. M. Selander, "Motion Detection and Analysis by Matching Graphs of Intermediate-Level Primitives," *IEEE Trans. Pattern Analysis and Machine Intelligence,* Vol. 2, No. 6, 1980, pp. 465-510.

69. S. Tsuuji, M. Osada, and M. Yachida, "Three-Dimensional Movement Analysis of Dynamic Line Images," *Proc. 6th Int'l Conf. Artificial Intelligence,* 1979, pp. 876-901.

70. J. L. Potter, "Velocity as a Cue for Segmentation," *IEEE Trans. Systems, Man, and Cybernetics,* Vol. SMC-5, 1975, pp. 390-394.

71. J. M. Prager, *Segmentation of Static and Dynamic Scenes,* PhD thesis, University of Massachusetts, Amherst, Mass., 1979.

72. W. B. Thompson, "Combining Motion and Contrast for Segmentation," *IEEE Trans. Pattern Analysis and Machine Intelligence,* Vol. 2, No. 26, 1980, pp. 543-549.

73. H. Wallach and D. N. O'Connell, "The Kinetic Depth Effect," *J. Experimental Psychology,* Vol. 45, 1953, pp. 205-217.

74. B. F. Green, "Figure Coherence in the Kinetic Depth Effect," *J. Experimental Psychology,* Vol. 62, No. 3, 1961, pp. 272-282.

75. G. Johansson, "Perception of Motion and Changing Form," *Scandinavian J. Psychology,* Vol. 5, 1964, pp. 181-208.

76. G. Johansson, "Visual Motion Perception," *Sci. Am.,* Vol. 232, No. 6, 1975, pp. 76-88.

77. M. L. Braunstein, "Depth Perception in Rotation Dot Patterns: Effects of Numerosity and Perspective," *J. Experimental Psychology,* Vol. 64, No. 4, 1962, pp. 415-420.

78. S. Ullman, "The Interpretation of Structure from Motion," *Proc. Royal Soc. London,* Vol. B, No. 203, pp. 405-426.

79. K. Prazdny, "Egomotion and Relative Depth Map from Optical Flow," *Biology and Cybernetics,* Vol. 36, 1980, pp. 87-102.

80. W. F. Clocksin, "Perception of Surface Slant and Edge Lables from Optical Flow: A Computational Approach," *Perception,* Vol. 9, No. 3, 1980, pp. 253-269.

81. H. C. Longuet-Higgins and K. Prazdny, "The Interpretation of a Moving Retinal Image," *Proc. Royal Soc. London,* Vol. B, No. 208, pp. 385-397.

82. Erik Borjesson and Claes von Hofsten, "Spatial Determinants of Depth Perception in Two-Dot Motion Patterns," *Perception and Psychophysics,* Vol. 11, No. 4, 1972, pp. 263-268.

83. Erik Borjesson and Claes von Hofsten, "Visual Perception of Motion in Depth: Application of a Vector Model to Three Dot Motion Patterns," *Perception and Psychophysics,* Vol. 13, No. 2, 1973, pp. 169-179.

84. J. J. Gibson, "What Gives Rise to the Perception of Motion?" *Psychology Rev.,* Vol. 75, No. 4, 1968, pp. 335-346.

85. C. J. Hay, "Optical Motions and Space Perception – An Extention of Gibson's Analysis," *Psychological Rev.,* Vol. 73, 1966, pp. 550-565.

86. D. N. Lee, "A Theory of Visual Control of Braking Based on Information About Time to Collision," *Perception,* Vol. 5, 1976, pp. 437-459.

87. D. N. Lee, "The Optic Flow Field: The Foundation of Vision" *Philosophical Trans. Royal Soc. London,* Vol. B, No. 290, 1980, 169-179.

88. D. Regan and K. I. Beverly, "Looming Detection in the Human Visual Pathway," *Vision Research,* Vol. 18, 1978, pp. 415-421.

89. D. Regan, K. I. Beverly, and M. Cynader, "The Visual Perception of Depth," *Sci. Am.,* Vol. 241, 1979, pp. 122-133.

90. D. Regan, K. I. Beverley, and M. Cynader, "Stereoscopic Subsystem for Position in Depth and for Motion in Depth," *Proc. Royal Society London,* Vol. B, No. 204, 1979, pp. 485-501.

91. J. J. Gibson, "Visually Controlled Locomotion and Visual Orientation in Animals," *British J. Psychology,* Vol. 49, 1958, pp. 182-194.

92. R. Warren, "The Perception of Motion," *J. Experimental Psychology,* Vol. 2, 1976, pp. 448-456.

93. I. R. Johnston, G. R. White, and R. W. Cumming, "The Role of Optical Expansion Patterns in Locomotor Control," *Am. J. Psychology,* Vol. 86, 1973, pp. 311-424.

94. J. J. Koenderink and A. J. van Doorn, "Invariant Properties of the Motion Parallax Field Due to the Motion of Rigid Bodies Relative to the Observer," *Optica Acta,* Vol. 22, No. 9, 1975, pp. 773-791.

95. K. Nakayama and J. M. Loomis, "Optical Velocity Patterns, Velocity Sensitive Neurons, and Space Perception: A Hypothesis," *Perception,* Vol. 3, 1974, pp. 63-80.

96. G. Johansson, "Visual Perception of Biological Motion and a Model for Its Analysis," *Perception and Psychophysics,* Vol. 14, No. 2, 1973, pp. 201-211.

97. J. E. Cutting, and L. T. Kozlowski, "Recognizing Friends by Their Walk: Gait Perception Without Familiarity Cues," *Bull. Psychonometric Soc.*, Vol. 9, No. 5, 1977, pp. 353-356.

98. L. T. Kozlowski and J. E. Cutting, "Recognizing the Sex of a Walker from Dynamic Point-Light Displays," *Perception and Psychophysiology*, Vol. 21, No. 6, 1977, pp. 575-580.

99. D. D. Hoffman, and B. E. Flinchbaugh, *The Interpretation of Biological Motion*, MIT A.I. Memo 608, MIT, Cambridge, Mass., 1981.

100. J. O'Rourke and N. I. Badler, "Model-Based Image Analysis of Human Motion Using Constraint Propagation," *IEEE Trans. Pattern Analysis and Machine Intelligence*, Vol. 3, No. 4, 1980, pp. 522-537.

101. R. Rashid, *LIGHTS: A System for the Interpretation of Moving Light Displays*, PhD dissertation, University of Rochester, Rochester, N.Y., 1980.

102. N. I. Badler, "Temporal Scene Analysis: Conceptual Descriptions of Object Movements," Technical Report 80, Dept. of Computer Science, University of Toronto, Toronto, Canada, 1975.

103. J. K. Tsotsos, *A Framework for Visual Motion Understanding*, PhD dissertation, University of Toronto, Toronto, Canada, 1980.

104. K. Koffka, *Principles of Gestalt Psychology*, Harcourt, Brace, and World, New York, 1935.

105. D. R. Proffitt, J. E. Cutting, and D. M. Stier, "Perception of Wheel-Generated Motions," *J. Experimental Psychology*, Vol. 5, No. 2, 1979, pp. 289-302.

106. H. Wallach, "The Perception of Motion," *Sci. Am.*, Vol. 201, 1959, pp. 56-60.

107. B. E. Flinchbaugh and B. Chandrasekaran, "Early Visual Processing of Spatio-Temporal Information," *Proc. IEEE Workshop on Computer Analysis of Time-Varying Imagery*, Apr. 1979, pp. 39-41.

108. D. Marr and H. K. Nishihara, "Representation and Recognition of the Spatial Organization of Three-Dimensional Shapes," *Proc. Royal Soc. London*, Vol. B, No. 200, 1978, pp. 269-294.

Shimon Ullman currently shares his time between MIT's Artificial Intelligence Laboratory and the Applied Mathematics Department at the Weizmann Institute of Science in Israel. His main research area is vision, including biological and machine processing of visual information.

Ullman received the BSc degree in mathematics, physics, and biology from the Hebrew University in Jerusalem in 1973. He completed his PhD at MIT in 1977.

AN IMAGE FLOW PARADIGM

Allen M. Waxman

Computer Vision Laboratory
Center for Automation Research
University of Maryland
College Park, MD 20742

ABSTRACT

The relationship between a three-dimensional (3-D) scene consisting of several objects in rigid body motion, and its associated two-dimensional (2-D) time-varying imagery, is discussed in the context of an *Image Flow Paradigm*. This paradigm addresses a number of theoretical issues: What is a useful representation for the 2-D flow field and how can it be obtained from the time-varying imagery? How should a flow field be segmented and how do these segmentation boundaries relate to the 3-D scene itself? How is the 3-D structure and motion of objects in the scene recovered from the 2-D flow representation and its segmentation? In attempting to answer these questions a variety of interesting concepts arise such as image neighborhood deformation, evolving contours, space-time stream tubes, virtual contours and virtual tubes, flow analyticity, boundaries of analyticity, kinematic analysis and structure-motion compatibility. The various ''elements'' of the paradigm are supported by analytic techniques, some of which have already been developed, some of which are now under investigation, and others which remain to be studied.

The support of the National Bureau of Standards under Grant NB83-NADA-4036 is gratefully acknowledged.

1. INTRODUCTION

In a single 2-D image, there are many cues which convey information about the 3-D scene such as color, shading, gray-level texture, edge texture, contour and shape. Yet, it is often the case that a scene evolves due to the motion of objects in the scene as well as of the observer. The time-varying imagery due to motion provides a rich set of cues for understanding the 3-D scene, its structure and constituent motions. Much of the psychophysical evidence for this has been discussed by Ullman[1] and Marr.[2] The importance of image motion was emphasized by Gibson[3] vis a vis its role in ecological optics. It was also considered a prime contributor to the representation known as the "2.5-D sketch" of Marr.[2]

The mathematical relationship between various representations of time-varying imagery and the recovery of 3-D structure and motion has been investigated by a number of authors. The approaches divide into two general catagories which may be termed "feature point methods"[1,4,5,6] and "flow field methods"[7,8,9]. Much of this work, however, consists of investigations into isolated aspects of the "structure from motion" problem. Such work is often enlightening, but in focusing on an isolated part of a larger problem, it is not always apparent how the work depends on, and determines, other aspects of the overall problem.

The *Image Flow Paradigm* we present here has many "elements" that may be viewed as successive stages in a computation which begins with an evolving image sequence, and ends with a variety of information relating to the structure and motion of objects in the 3-D scene under view. These elements are interconnected by a set of "methods and concepts" which are to be embodied in the algorithms which realize this computation.

In Section 2 we discuss the analytic structure of image flow fields generated by textured surfaces executing rigid body motions through space. This analyticity is manifest in the deformation of image neighborhoods which progresses as the scene evolves. Part and parcel of this analytic structure are the boundaries of analyticity, the singular contours which serve to partition the individual images into analytic regions. This segmentation reflects the physical structure of the 3-D scene itself. The central theme of the paradigm is presented in Section 3; it was the basis of the analyses developed by Waxman & Ullman[9] and Waxman & Wohn.[10] Focusing on an isolated surface patch moving through space and the associated image sequence, we explain the role of *evolving contours* in determining the *flow deformation parameters* from which the 3-D structure and motion of the surface can be recovered. The individual elements of the paradigm are then described in Section 4. The paradigm expands about the central theme of Section 3 in order to encompass the global structure of objects and the organization of the scene as a whole. Here, the segmentation of the flow field emerges as an important facet of the paradigm. This segmentation must be accomplished in parallel with the determination of 3-D structure and motion of surface patches. Global compatibility can then be imposed on the patches which comprise individual objects. Closing remarks are presented in Section 5 where we also note other avenues of research relating to image flows which are yet to be integrated into the *Image Flow Paradigm*.

2. FLOW FIELDS AND NEIGHBORHOOD DEFORMATION

As a textured, rigid object moves through space, the evolving image sequence registered by a monocular observer (e.g. a moving pin-hole camera) contains information in the form of an image flow field. This image flow is determined by the relative rigid body motion between object and observer, as well as the structure of the object's surface visible to the observer. Derivation of this flow field follows that of Waxman & Ullman.[9]

We attribute the relative rigid body motion to an observer represented by the spatial coordinate system (X, Y, Z) in Figure 1. The origin of this system is located at the vertex of perspective projection, and the Z-axis is directed along the center of the instantaneous field of view. The instantaneous rigid body motion of this coordinate system is specified in terms of the translational velocity $\mathbf{V} = (V_X, V_Y, V_Z)$ of its origin and its rotational velocity $\mathbf{\Omega} = (\Omega_X, \Omega_Y, \Omega_Z)$. The 2-D image sequence is created by the perspective projection of the object onto a planar screen oriented normal to the Z-axis. The origin of the image coordinate system (x, y) on the screen is located in space at $(X, Y, Z) = (0, 0, 1)$; that is, the image is reinverted and scaled to a focal length of unity.

Due to the observer's motion, a point P in space (located by position vector $\mathbf{R}$) moves with a relative velocity $\mathbf{U} = -(\mathbf{V} + \mathbf{\Omega} \times \mathbf{R})$. At each instant, point P projects onto the screen as point p with coordinates $(x, y) = (X/Z, Y/Z)$. The corresponding image velocities of point p are $(v_x, v_y) = (\dot{x}, \dot{y})$, obtained by differentiating the image coordinates with respect to time and utilizing the components of $\mathbf{U}$ for the time

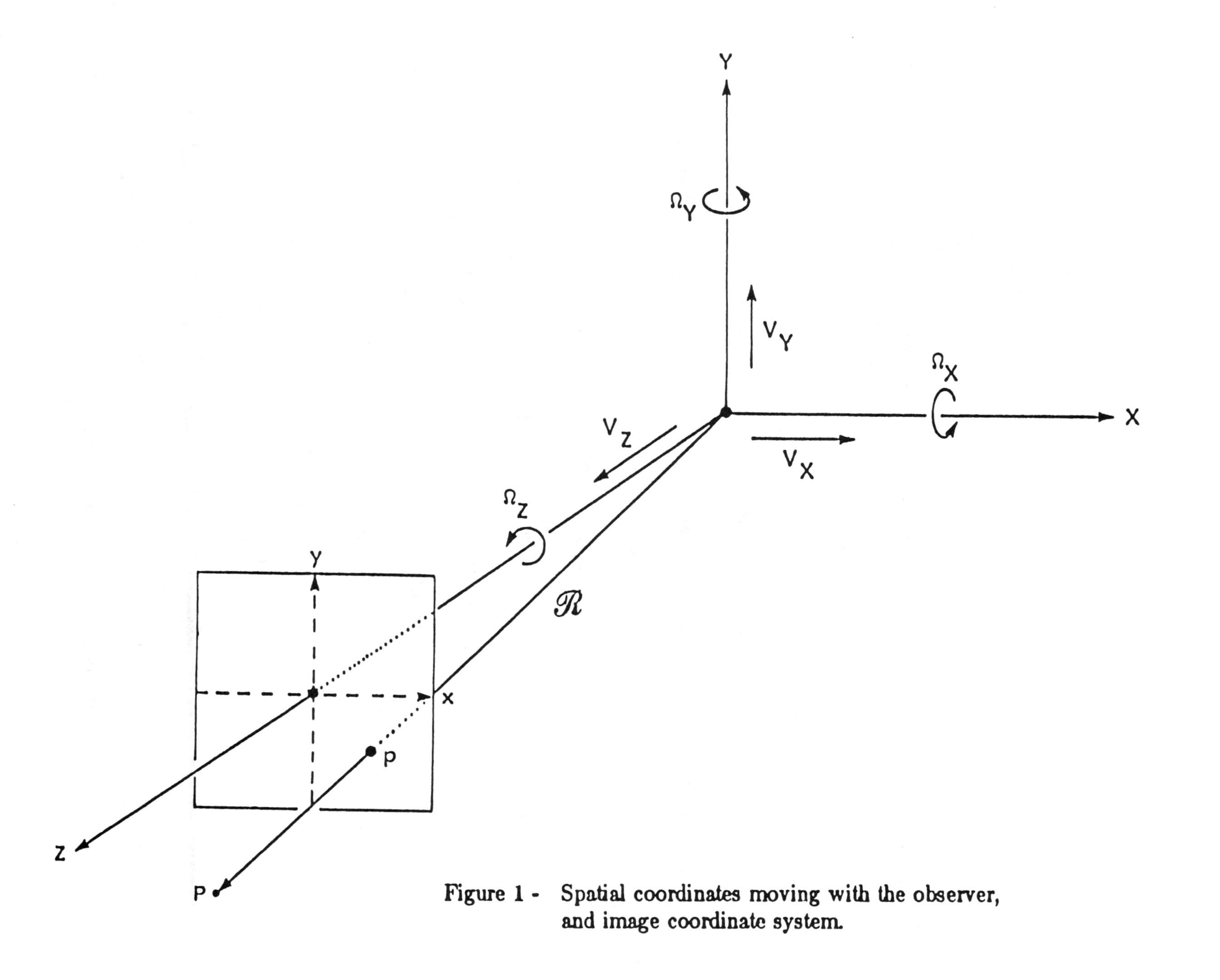

Figure 1 - Spatial coordinates moving with the observer, and image coordinate system.

derivatives of the spatial coordinates of P. The result is

$$v_x = \left\{ x\frac{V_Z}{Z} - \frac{V_X}{Z} \right\} + [xy\Omega_X - (1+x^2)\,\Omega_Y + y\Omega_Z] \,,$$

$$v_y = \left\{ y\frac{V_Z}{Z} - \frac{V_Y}{Z} \right\} + [(1+y^2)\,\Omega_X - xy\Omega_Y - x\Omega_Z] \,.$$

These equations define an instantaneous image flow field, assigning a unique 2-D image velocity $\boldsymbol{v}$ to each direction (x, y) in the observer's field of view. An example is shown in Figures 2a & b. For the moment, we shall consider only a single surface patch of some object in the field of view. A small but finite surface patch may be locally approximated by a quadric surface in space as described by six parameters: two slopes, three curvatures and an overall distance scale. If the surface patch is described in this viewer-centered spatial coordinate system by $Z = \varsigma(X, Y)$, then it is straightforward to find the corresponding local representation $Z = Z(x, y)$ as a second-order polynomial in terms of image coordinates.[9] Of these six surface parameters, only five can be recovered directly from the image flow field; the overall scale factor is lost as it always appears in ratio with the translational velocity V.[9] (In principle, this scale factor can be recovered by a method termed *Dynamic Stereo*[11] which entails the comparison of two image flow fields registered by two cameras in relative motion.)

In the recovery of surface structure and 3-D motion from image flow, it was shown by Waxman & Ullman[9] that it is sufficient to describe an image flow as a locally second-order flow field (below we shall interpret this in terms of neighborhood deformation). This has implications with regard to the surfaces which generate the flow itself. For example, consider a planar surface patch $Z = Z_0 + T_1X + T_2Y$. In

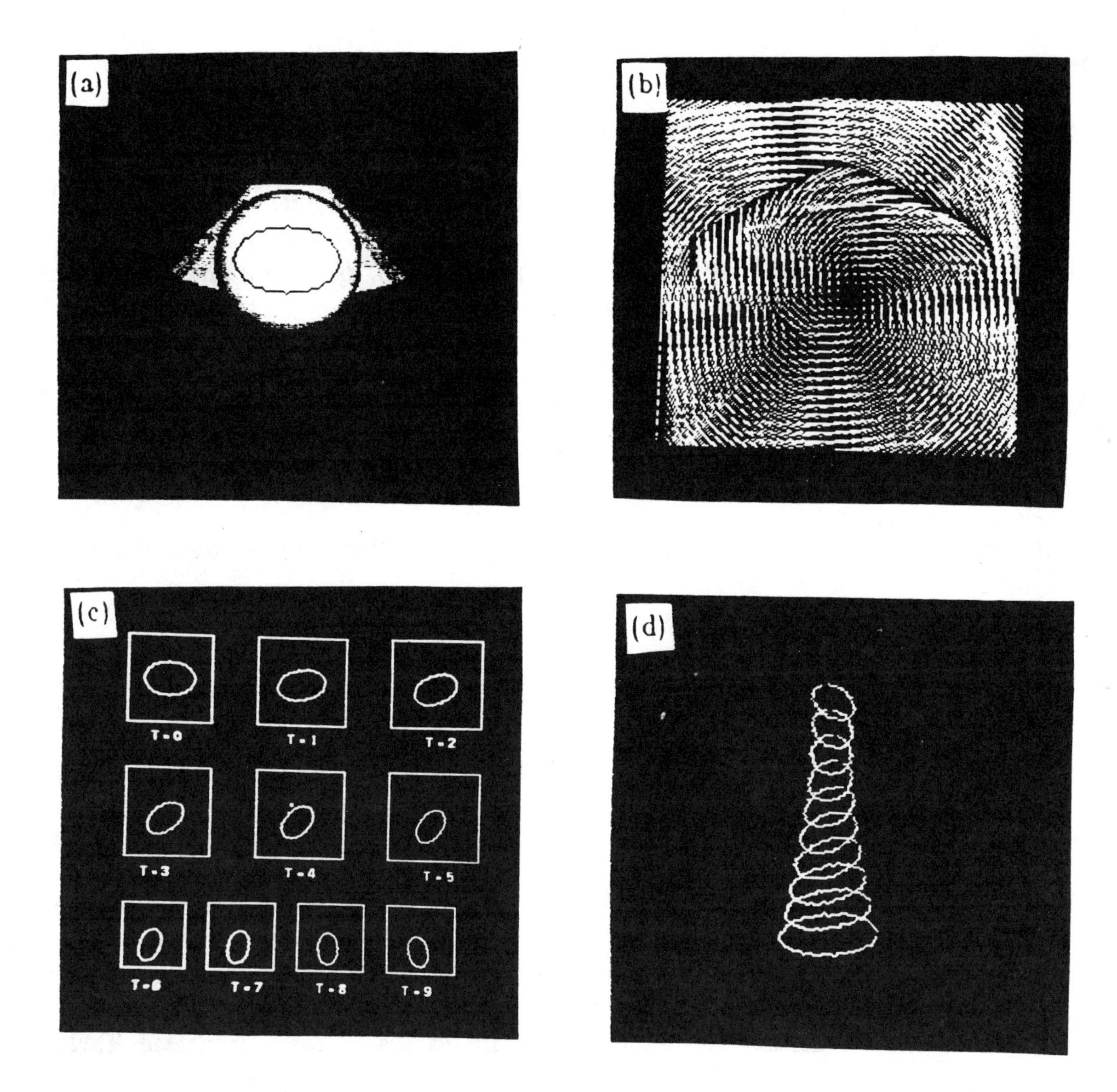

Figure 2 - (a) View of a sphere and a cone (background at infinity) as seen by an observer executing a general translation and rotation through space.
(b) The theoretical image flow field associated with the scene. Regions of flow analyticity are apparent when picture is viewed at arm's length.
(c) A sequence of evolving image contours corresponding to the contour drawn on the sphere. The motion parameters of the observer are held constant over time. (d) By "stacking" the evolving contour sequence, the associated space-time stream tube may be visualized.

terms of image coordinates this surface may be described exactly as $Z = Z_0(1 - T_1x - T_2y)^{-1}$. Substitution into the velocity equations above yields expressions in the form of a second-order flow field. For planar surfaces, such second-order flows are *globally valid.* On the other hand, quadric surfaces generate flows which are not simple polynomials in the image coordinates. However, they may be *locally approximated* as second-order flows. The coefficients of this second-order flow determine the slopes and (scaled) curvatures of the quadric surface patch as well as its (scaled) space motion. In this context, complex surfaces are viewed as a composite of planar and quadric patches.

Now approximating a local image flow as a second-order polynomial in image coordinates (x, y) amounts to studying the first six terms of the Taylor series associated with each component of the vector field. For a quadric surface patch, this truncated Taylor series has a small but finite radius of convergence in the image plane. For a planar surface patch, the radius of convergence is arbitrarily large as all higher order terms of the Taylor series vanish identically. In discussing these truncated Taylor series and their convergence properties, it becomes clear that the *analytic structure* of the image flow field is quite important in its relation to the 3-D scene.

This notion of analyticity is actually the foundation of the paradigm, as will soon become clear. It plays a role in extracting the flow representation (our *deformation parameters*) from the image sequence; it is also the basis for the recovery of surface structure and space motion (our *kinematic analysis*); it even drives the flow segmentation process. From the image velocity equations it is apparent that the flow field is "functionally analytic" (i.e. twice differentiable) wherever object surfaces $Z(x, y)$

are twice differentiable. The flow is non-analytic at points where Z or its first partials are discontinuous, and where the relative space motion parameters change. Such points occur along occluding boundaries and structural edges (e.g. the edges of a polyhedron). Thus, an image flow field is naturally partitioned into regions of analyticity separated by singular contours (our *boundaries of analyticity*). These analytic regions are, in turn, decomposed into neighborhoods in which the image flow is locally approximated as a second-order flow. It is part of a complete image flow analysis to delineate these boundaries of analyticity so that 3-D interpretations can be assigned to the regions within them. The regions of flow analyticity in Figure 2b are apparent when the figure is viewed at arm's length.

Returning to the truncated Taylor series representation of a local image flow, its interpretation in terms of "neighborhood deformation" becomes relevant. The first term of the Taylor series (for each flow component) represents the image motion of the local origin and characterizes the mean motion of the neighborhood as a whole. The second and third terms of each series (coefficients being first partials of the velocity field evaluated at the local origin) characterize the motion in an infinitesimal neighborhood of the local origin. In fact, these four terms which comprise the "velocity gradient tensor" are usually combined into symmetric and anti-symmetric parts termed the "rate-of-strain" and "spin" tensors, respectively. These tensors describe the rate of change of the geometry of infinitesimal image neighborhoods in terms of 2-D dilation, stretch and compression along orthogonal axes, and rotation.[7,9] This description is known as the "Cauchy-Stokes Decomposition Theorem" in the applied mechanics literature.[12] The remaining three terms of each Taylor series (coefficients being second partials of the velocity field) extend this notion of neighborhood deformation from infinitesimal to small but finite neighborhoods in the image plane. This interpretation motivated the ***kinematic analysis*** of Waxman & Ullman[9] for the recovery of surface structure and space motion. It should be clear that "neighborhood deformation" and "flow analyticity" go hand in hand. These concepts are the starting point for extracting the flow representation, or ***deformation parameters,*** from evolving contours in an image sequence,[10] as is explained in the next section.

3. THE CENTRAL THEME

The overall paradigm (to be described in Section 4) addresses the image flow generated by several whole objects in rigid body motion. But as the paradigm is rather involved, it is worthwhile to first consider the "central theme" which relates structure, motion and image flows for a small but finite surface patch. This theme, displayed in Figure 3, illustrates the various stages of the computation; its sequential structure suggests a pipelined computer architecture.

The theme starts with a textured surface patch (planar or quadric) moving as a rigid body through space relative to an observer (modeled as a pin-hole camera). Thus, a time-varying image sequence is generated by perspective projection. Utilizing low-level image processing techniques, one must then acquire from the image sequence (several frames at a time, say) a set of primitives that will lead to a useful representation of the image flow field.

The work of Waxman & Ullman[9] suggests that a local flow be described in terms of ***flow deformation parameters*** which consist of the following twelve quantities evaluated at the origin of the surface patch (taken to be at the image origin): the two components of image velocity, the three independent strain rates, the spin, and the six independent derivatives of strain rate and spin; essentially the first six terms of the Taylor series corresponding to the two components of image velocity. The ***kinematic analysis,*** which interprets these deformation parameters as one or more possible surface structures in rigid body motion (up to a scale factor), can tolerate significant perturbations to the input parameters.[9] This is not true of the "feature point methods" which utilize image velocity estimates directly.[4,5,6] However, these deformation parameters must then be obtained from the image sequence, and this is where neighborhood deformation and ***evolving contours*** become relevant.

Recalling the discussion in Section 2, we realize that if image flow manifests itself in neighborhood deformation, then image features which experience this deformation serve to sample the flow field. Evolving contours (which correspond to physical surface properties) seem a prime candidate for this task (feature points could be used as well, but they are usually too sparse). This was first suggested by Waxman & Ullman[9] and has since been developed in much greater detail by Waxman & Wohn.[10] These evolving contours are to be obtained from the image sequence by an edge detection scheme such as that proposed by Marr & Hildreth,[13] the zero-crossings of the Laplacian operating on a Gaussian smoothed image. Such contours generally correspond to texture edges on the surface patch. One must avoid using contours which coincide with the boundaries of analyticity as well as shadow edges, though a useful exception is the structural edge on a polyhedron where the surface patches are known to be planar.

The recovery of the deformation parameters from evolving contours is rather involved, though intuitively appealing. Figure 2c displays the contour sequence associated with the surface contour in Figure 2a. In practice, the contour sequence must be obtained by edge detection and tracking from frame to frame. As is well known, the image motion of any small edge segment yields only that component of image velocity normal to the edge; this is often called the "aperture problem,"[2] we call it the "aperture problem in the small."[10] However, entire evolving contours yield a *normal flow* around the contour which, when combined with the *requirement of analyticity,* yields the desired deformation parameters (i.e. the Taylor series coefficients) and hence, a good approximation to the local image flow field.[10] (This analytic procedure for recovering full flow fields from normal flow along contours should be compared with the heuristic smoothness criteria utilized in the past.[14,15]) In order to recover all twelve deformation parameters from a single evolving contour, it is required that the contour have sufficient geometric structure in order to fully sample the flow field. For example, biquadratic contours (the conic sections) and even bicubic contours are inadequate; they possess too much symmetry. It is necessary that the contour be at least biquartic in structure; we call this the "aperture problem in the large."[10] In practice, however, a more robust procedure is to utilize several contours simultaneously.[10]

In Figure 3, we have indicated that the path from *evolving contours* to *deformation parameters* should be based on *flow analyticity.* There are actually several methods which embody these ideas, and they employ a number of interesting con-

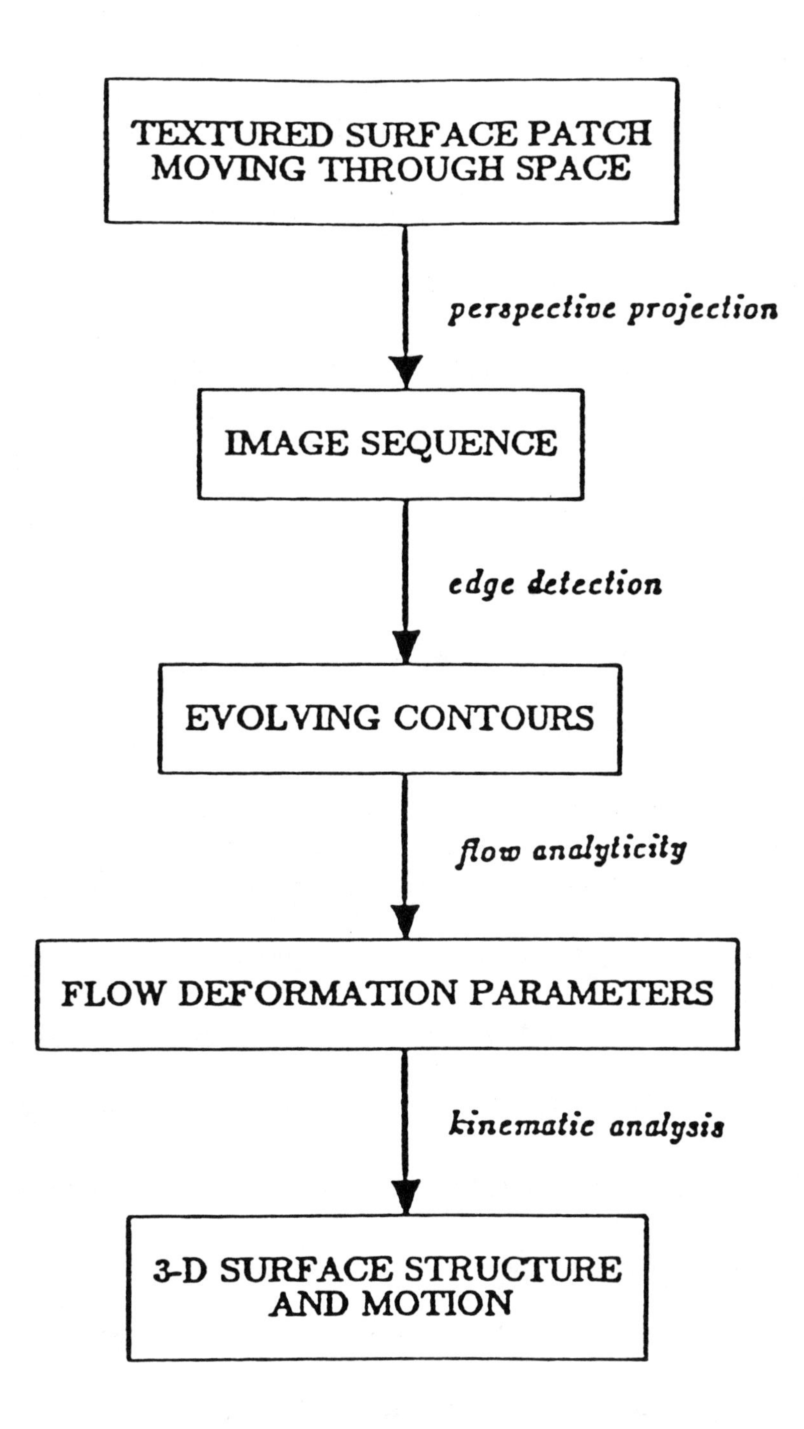

Figure 3 - The central theme of the paradigm relates the structure, motion and image flow for an isolated surface patch.

cepts. The one already described, which requires measurements of the normal flow around a contour, we call the "Velocity Functional Method."[10] The normal velocity measurements themeselves are to be obtained from the changing geometry of ***locally smoothed segments*** of the noisy, digital contours. A second approach, termed the "Contour Functional Method," utilizes functional fits to ***entire contours*** in each frame. The deformation parameters are then obained from the rates of change of the coefficients appearing in the functionals.[10] Yet a third approach emerges from the notion of a ***space-time stream tube.*** This entity is described in a three-dimensional coordinate system (x, y, t) consisting of the image space (x, y) and time t. A tube corresponds to an evolving contour sequence, such that a cross-section through the tube at $t{=}t_0$ corresponds to the image contour at $t{=}t_0$. Each evolving contour generates its own stream tube. The walls of the tubes are composed of the (initially unknown) ***pathlines*** through image space-time, of the individual points comprising a contour. The idea is then to fit a ***surface functional*** to a portion of the tube, where this functional is constrained by the fact that the various cross-sections are related by a deformation. The coefficients of the surface functional will again be related to the deformation parameters we seek. In Figure 2d we illustrate the stream tube which corresponds to the contour sequence in Figure 2c. In the absence of noise and digitization error, these various methods should yield equivalent results for the deformation parameters. Future research, in which these effects are simulated, should reveal which method is most robust.

In discussing functional fits to contour sequences and stream tubes, we must keep in mind that the general functional forms must be specified in advance (e.g. biquadratic, bicubic, biquartic, etc.) and so cannot approximate well contours of arbitrary geometry that may be extracted from an image. This is not really a problem for we may view the fuctionals as representing ***virtual contours*** or ***virtual stream tubes.*** The real contour sequence serves to sample the image flow field via the deformation it undergoes. The image motion of other features (real or virtual) in the neighborhood is then determined. Thus, the evolution of real contours guides the evolution of virtual contours (i.e., the functionals). Such concepts must be incorporated into the measure of "goodness of fit" to a contour sequence. This is to be pursued in the near future.

In summary, the central theme of the paradigm (Fig. 3) starts with the image sequence generated by the perspective projection of a textured surface patch moving through space, utilizes edge detection and tracking to obtain sets of evolving contours from the texture, imposes flow analyticity to derive the deformation parameters from the contour sequences, and then assigns one or more 3-D interpretations from a kinematic analysis of these deformation parameters.

4. ELEMENTS OF THE PARADIGM

The image flow paradigm presented here builds around the central theme described in the previous section. Whereas the central theme concerns a single surface patch in motion and its associated deforming neighborhood in an image sequence, the overall paradigm addresses an entire scene composed of several objects and background executing relative rigid body motions with respect to an observer. The task of segmenting the time-varying imagery must now be integrated into the central theme.

This segmentation aims to partition the images into *regions of analyticity* demarcated by the boundaries of flow analyticity described in Section 2, with individual regions composed of *overlapping neighborhoods* (whose boundaries are somewhat arbitrary) corresponding to quadric surface patches in space. The paradigm also suggests a process for grouping the individual surface patches into larger entities representing *3-D structure and motion models* for visible portions of rigid bodies. Various applications of the output of such a computational scheme are briefly mentioned. The organization of this paradigm, illustrated in Figure 4, lends itself to implementation on a parallel-pipelined architecture with local connections between processors in each stage of the pipeline as well as between stages.

We give here a brief explanation of each of the elements and their interconnections which comprise the *Image Flow Paradigm* of Figure 4.

Scene: The scene is composed of objects with textured surfaces and background executing rigid body motions relative to an observer (a pin-hole camera).

Image Sequence: A time-varying image sequence is generated by the perspective projection of the evolving scene.

Evolving Contours: Edge detection is performed on each image and corresponding edges must be tracked through the sequence yielding sets of evolving contours. (Individual contours with sufficient structure, or groups of several contours, are then associated into small neighborhoods.)

Normal Flow: Individual contours, observed over several frames, can be used to esti-

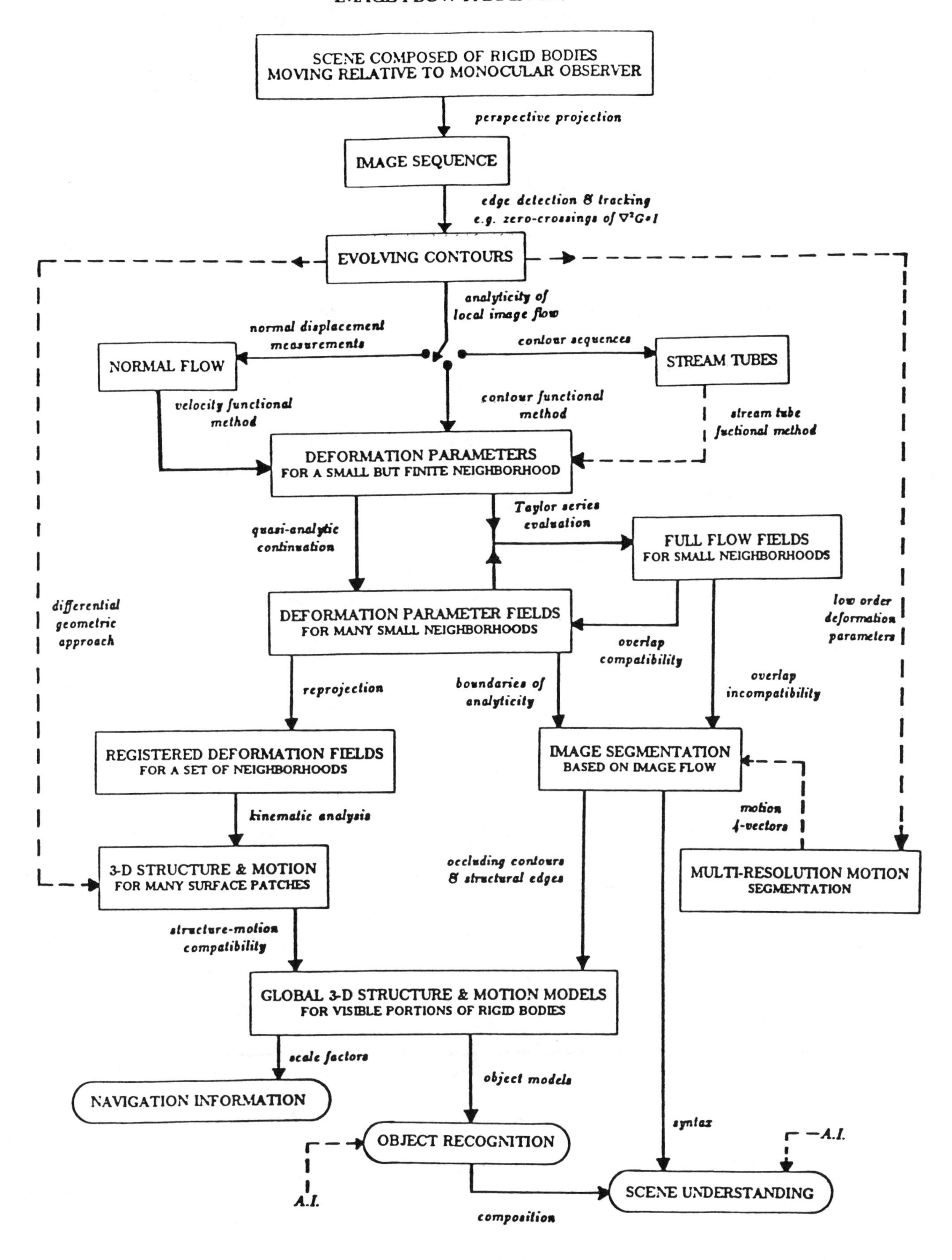

Figure 4 - Elements of the Image Flow Paradigm.

mate the normal flow component around each contour. The accuracy of the normal velocity, derived from displacements normal to a smoothed segment of contour, should improve as the number of frames considered is increased.

Stream Tubes: Each sequence of evolving contours can be "stacked" to form stream tubes through image space-time. The surface of each stream tube is composed of path-lines generated by points on a contour. The varying cross-section of the tube is a manifestation of contour deformation over time.

Deformation Parameters: By imposing analyticity of the local image flow, the set of twelve deformation parameters for the neighborhood can be derived from an evolving contour sequence. There are several alternative routes, employing the normal flow in the *velocity functional method* or entire contours in the *contour functional method.* (The *stream tube functional method* awaits further development.)

Full Flow Fields: For individual neighborhoods, the deformation parameters constitute linear combinations of the (truncated) Taylor series coefficients of the local image flow. By summing these truncated series, one obtains a second-order approximation to the flow throughout the neighborhood. As the various methods of extracting the deformation parameters employ a least-squares procedure, multiple contours that were inappropriately grouped into neighborhoods can be detected at this stage.

Deformation Parameter Fields: Each neighborhood gives rise to its own set of deformation parameters which can then be used to approximate full flow fields in each neighborhood. However, these neighborhoods in which the contours are embedded are required to overlap if they are to be parts of the same analytic region. Thus, the full flow fields derived from separate sets of deformation parameters are required to be *compatible* in the area of overlap between neighborhoods. Incompatibility is interpreted as a *boundary of analyticity* passing through the overlap area. This procedure, in which overlapping neighborhoods grow to analytic regions, is termed *quasi-analytic continuation* (in analogy with that used in the theory of analytic functions of a complex variable). The isolated parameter sets are thereby extended to parameter fields over analytic regions.

Image Segmentation: Though a variety of visual cues can drive a segmentation process,we are interested in that based on image flow. The *boundaries of analyticity* associated with *overlap incompatibility* reflect structural boundaries in the scene as descried in Section 2. We have also begun to investigate another motion segmentation scheme which is based on a multi-resolution representation of the image flow, structured in the form of a linked pyramid. At low resolutions, nodes in the pyramid represent whole neighborhoods in the image; thus the motion description should contain parameters which discribe neighborhood deformation at that resolution. We suggest employing *motion four-vectors* which comprise the two components of mean translation of the neighborhood, the neighborhood dilation and the neighborhood spin. Neighborhood stretch and compression at constant area is not retained in this low resolution description. The linking procedure embodies the analyticity of the four-vector field. The input to this stage would most likely come from the evolving contours in the form of *low order deformation parameters.*

Registered Deformation Fields: Within regions of flow analyticity, the individual overlapping neighborhoods are distributed over large portions of the field of view. However, the *kinematic analysis* for 3-D interpretation is most easily carried out for small neighborhoods located at the center of the field of view.[9] This is equivalent to reorienting the observer's space coordinates (X, Y, Z) such that the new Z-axis points towards the corresponding surface patch (a kind of *registration*). The new image sequence would be related to the original by a projective transformation (a *perspection* followed by a *section*[16]) since the vertex of projection will not have changed. This in turn, alters the values of the deformation parameters of a neighborhood. This process of *reprojection* of the deformation parameter fields is straightforward, though the projective transformations remain to be derived.

3-D Structure & Motion: Using the registered deformation parameters of each individual neighborhood, a *kinematic analysis*[9] is employed to obtain one or more 3-D interpretations of the corresponding surface patch and its relative space motion (in the new registered axes). Typically, curved surface patches give rise to a unique interpretation whereas planar patches yield dual interpretations, though both cases have their exceptions.[9] An interesting conjecture is that 3-D interpretations for neighborhoods may be derived directly from evolving image contours by employing a *differential geometric approach.* Thus, surface texture boundaries would be space curves constrained to lie on a quadric surface patch, their curvature and torsion being related to the local metric and curvature tensors of the surface.[17,18] Seen in projection on the image as plane curves, various global measures could be adopted in the spirit of "differential geometry in the large."[18] The rates of change of these measures would then be related to the space motion.

Global Models: In order to assign a global 3-D interpretation to entire regions, we must also interpret its boundaries in terms of *occluding contours* and *structural edges.* The approximate locations of these boundaries were derived in the segmentation process. The analytic regions themselves correspond to visible portions of rigid bodies; therefore, the surface patches which correspond to the neighborhoods comprising these regions (and their associated space motions) must be compatible. This *structure-motion compatibility* stems from the fact that the rigid body motion parameters of individual surface patches must agree (when converted to the same coordinate system) if these patches belong to the same rigid body. This procedure can help resolve the multiple interpretations that may have been assigned to isolated surface patches. It should also enable us to distinguish the kinds of region boundaries, i.e., which whole surfaces (joined by structural edges) share the same space motions. Object representation in terms of "curvature patches" has also been discussed by Brady.[19]

Once global 3-D structure and motion interpretations have been assigned, they can be utilized for a variety of tasks, as is indicated at the bottom of Figure 4. Such applications generally require other capabilities and knowledge as well. Nonetheless, image flow has the potential to provide useful information for purposes of autonomous navigation, object recognition and general scene understanding.

5. CLOSING REMARKS

The *Image Flow Paradigm* described here is a plan of on-going research at the Computer Vision Laboratory. As the work progresses, we shall be able to reassess the individual elements of the paradigm and this will undoubtedly lead to some revision. As the various details of the paradigm are worked out, we plan to first implement it on the *Image Flow Simulator* (used to construct Fig. 2) being developed by Sinha.[20] This will provide us with an artificial environment in which dependable contours are assured, and noise and digitization errors are controlled. If this is successful, we can then go on to experiment with simple moving objects and real cameras (with all their inherent problems).

Other areas of image flow research that we are carrying out, but that have yet to be integrated into the paradigm, include *Dynamic Stereo*[11] for recovering the scale factors, time-varying edge texture for recovery of structure and motion (here the evolving contours are treated statistically), and exploiting motion blur for the recovery of flow deformation parameters.

ACKNOWLEDGEMENTS

I am indebted to Shimon Ullman who sparked my interest in the structure-from-motion problem. This work would not be possible without the lively participation of K. Wohn, S. Sinha, and M. Subbarao, and the excellent research environment at the Center for Automation Research provided by Larry Davis and Azriel Rosenfeld. The support of the National Bureau of Standards under Grant NB83-NADA-4036 is greatfully acknowledged.

REFERENCES

[1] S. Ullman, **The Interpretation of Visual Motion** (Cambridge: MIT Press), 1979.

[2] D. Marr, **VISION: A Computational Investigation into the Human Representation and Processing of Visual Information** (San Francisco: W.H. Freeman), 1982.

[3] J.J. Gibson, **The Senses Considered as Perceptual Systems** (Boston: Houghton Mifflin), 1966.

[4] K. Prazdny, "Egomotion and Relative Depth Map from Optical Flow", *Biol. Cyber.* **36,** 87, 1980.

[5] R.Y. Tsai & T.S. Huang, "Uniqueness and Estimation of 3-D Motion Parameters of Rigid Objects with Curved Surfaces", *Univ. Illinois/Urbana-Champaign, Coordinated Science Lab. Report R-921,* October, 1981.

[6] A.M. Waxman & K. Wohn, "A Fixed-Point Iterative Algorithm for Orientation and Motion of Planar Surfaces from Image Flow", *Univ. Maryland, Center for Automation Research Tech. Report* (in preparation).

[7] J.J. Koenderink & A.J. van Doorn, "Invariant Properties of the Motion Parallax Field due to the Movement of Rigid Bodies Relative to an Observer", *Optica Acta* **22,** 773, 1975.

[8] H.C. Longuet-Higgins & K. Prazdny, "The Interpretation of a Moving Retinal Image", *Proc. Roy. Soc. London* **B 208,** 385, 1980.

[9] A.M. Waxman & S. Ullman, "Surface Structure and 3-D Motion From Image Flow: A Kinematic Analysis", *Univ. Maryland, Center for Automation Research Tech. Report 24,* October, 1983.

[10] A.M. Waxman & K. Wohn, "Contour Evolution, Neighborhood Deformation and Global Image Flow: Planar Surfaces in Motion", *Univ. Maryland, Center for Automation Research Tech. Report* (in preparation).

[11] A.M. Waxman, "Passive Ranging by Image Velocity Augmentation: Dynamic Stereo", *Univ. Maryland, Center for Automation Research Abstract,* July, 1983.

[12] R. Aris, **Vectors, Tensors and the Basic Equations of Fluid Mechanics** (Englewood Cliffs: Prentice-Hall), 1962.

[13] D. Marr & E.C. Hildreth, "Theory of Edge Detection", *Proc. Roy. Soc. London* **B 207,** 187, 1980.

[14] B.K.P. Horn & B.G. Schunck, "Determining Optical Flow", *Artificial Intelligence* **17,** 185, 1981.

[15] E.C. Hildreth, "The Measurement of Visual Motion", *M.I.T. Ph.D. Thesis,* August, 1983. Also see: "Computing the Velocity Field Along Contours" in Proceed. of Workshop on *MOTION: Representation and Perception,* pp. 26-32, (Toronto) April, 1983.

[16] R.M. Winger, **Projective Geometry** (New York: Dover), 1962.

[17] A.J. McConnell, **Applications of Tensor Analysis** (New York: Dover), 1957.

[18] J.J. Stoker, **Differential Geometry** (New York: Wiley-Interscience), 1969.

[19] M. Brady, "Criteria for Representations of Shape," in **Human and Machine Vision,** (eds.) J. Beck, B. Hope & A. Rosenfeld (New York: Academic), pp. 39-84, 1983.

[20] S. Sinha, "Graphics Algorithms for Image Flow Simulation and Experimentation", *Univ. Maryland M.S. thesis* (research in progress).

Chapter 2

Image Partitioning and Perceptual Organization

Obtaining a semantic description of scene content, from one or more images, implies a process in which the sensor-derived signals are translated into the vocabulary of a symbolic language and interpreted in the context of prior knowledge and experience. While it is conceivable that the translation from image pixels to the names and the relationships between depicted objects can be accomplished in a single step, the complexity of this problem suggests that a hierarchy of successively more abstract and goal-oriented descriptions are required as intermediaries between the signals and the final description. If this is indeed the case, what is the nature of the most primitive description—the one that first organizes the array of intensity values representing the image into something more manageable and useful? Some important issues are:

1. What are the possible purposes of this first primitive description (FPD)?
2. What should it be able to represent? For example, what are the visual primitives in the language of early vision; should the FPD be primarily concerned with describing the intensity variations in the image, or with describing properties of the 3-D world implied by the image content?
3. Should the FPD be influenced by contextual knowledge and the intended purpose of the desired scene description? For example, should it be able to take advantage of a 3-D model of scene geometry, if available?
4. How can the FPD actually be produced? For example, should it be based on local or global processes; what is the explicit nature of these processes, what information do they require, and in what order do they operate? Should there be a single FPD, or do different purposes imply that multiple FPDs are generated for different visual channels?
5. How does the human visual system address the preceding problems of initial description (often called the problem of perceptual organization).

Because of the potential for essentially unbounded scene complexity, a primary purpose of the first few stages of visual processing must be to simplify the interpretation problem by decomposing or partitioning the image into coherent components that can be considered separate things that can be independently characterized. Perceptual organization also involves extracting a description for each image component that is suitable for indexing into a knowledge base, providing a basis for higher level description and understanding. While there is probably little disagreement about the *purpose* of the initial description, each of the other issues listed above leads to contending viewpoints and approaches.

2.1 Perceptual Organization in the Human Visual System

Theories of human perceptual organization can be divided into two categories: those that assume a fixed vocabulary for grouping and describing image content, and those that primarily emphasize some set of organizing principles. The Gestalt psychologists such as Wertheimer [Wertheimer 38] provided one of the first and most convincing sets of organizing principles; image components are grouped on the basis of closeness, similar form, continuity, and simplicity. Since the Gestaltists did not offer explicit mechanisms for expressing their principles as operational procedures, their work has had only indirect influence on computational vision. We will return to perceptual organization techniques based on Gestalt-type criteria in a later section of this chapter.

Some of the most interesting recent psychological theories of human perceptual organization have been based on the detection of specific sets of perceptual primitives in images.

Julesz and Bergen [**Julesz 83**] suggest that the first stages of human visual processing are carried out by a separate "preattentive" system, which can, without effort or scrutiny, detect differences between textured regions distinguished by the presence of various configurations of elongated blobs called "textons."[1] Textons include rectangles, ellipses, and line segments, with specific properties such as color, angular orientation, length, and width; the ends-of-lines (terminators) and crossings of line segments are also textons. Only differences in types of textons and their density can be preattentively detected; the positional relationship between neighboring textons is ignored. Positional information, required for form perception, is extracted in the human visual system by a serial, time-consuming, and spatially restricted process called "focal attention." While Julesz and Bergen present impressive demonstrations to support their theory, their demonstrations are based on artificially generated binary images. It is not clear how important textons are for natural continuous-tone images. There is also the open problem of devising computational procedures for reliably detecting the textons.

The idea of a functionally distinct, preattentive visual system that is isolated from the observer's knowledge base, and that operates automatically and in parallel across the entire visual field, is also suggested in the work of Treisman [Treisman 85]. She attributes to Neisser [Neisser 67] the concept of an initial preattentive stage of visual processing which accomplishes texture segregation and figure-ground grouping, and a second stage in which focused attention is used to conjoin separate features into complex objects.[2] She also points to Beck's observation [Beck 67] that texture segregation is preattentive when based on color, brightness, and line orientation. In her own work she confirms that segregation is easy when areas differ in simple properties like shape of the texturing primitive or color. On the other hand, a boundary defined solely by a conjunction of properties (e.g., green triangles and red circles on the left, and red triangles and green circles on the right) is much harder to find. She also provides evidence that texture boundaries are not computed on the basis of strictly local information. She offers the hypothesis that the preattentive visual system does not produce a single representation such as a single partitioned image. Rather, it provides different initial partitions to support distinct channels in the human visual system, which analyze imagery along a number of separate dimensions to extract such information as depth, movement, color, orientation, and so on.

Texture-based partitioning, the focus of the work just described, is only one of many distinct forms of perceptual grouping within the competence of the human visual system; grouping also occurs on the basis of proximity, shape, and motion. The descriptive information in the FPD, which provides the basis for these additional grouping processes, undoubtedly extends beyond the vocabulary of textons.

Zucker [**Zucker 85**] proposes a level of description between the raw image and the perceptual grouping processes. This description makes explicit the oriented intensity patterns underlying the perception of 1-D contours (e.g., occlusion edges, shadow boundaries) and 2-D flows (e.g., hair, fur, wheat fields, water, and so on.). He describes two separate computational processes for extracting these descriptions.[3] In both cases local flow orientation is evaluated by convolving

[1] It is known that the primate visual system has special machinery for detecting such elongated blobs [Hubel 68].

[2] The idea that the human visual system is functionally partitioned—that it first extracts features from the raw imagery, and then employs symbolic processes to deduce scene content—can be traced back to the work of Helmholtz (1821-94).

[3] For 1-D contours he employs a relaxation-based optimization scheme that finds a sparse set of 1-D flow lines that minimize change of curvature; for 2-D flow, he employs relaxation to find a vector field that maximizes local uniformity of direction.

an elongated Gaussian-shaped mask with the image. This mask, whose parameters are justified on physiological grounds, leads to flow descriptions of dot patterns, which provide an explanation of known psychological grouping phenomena. For example, it explains how close together a chain of dots must be to be perceived as a coherent line rather than as a collection of isolated dots.

An important assumption implicit in the investigations discussed thus far is that perceptual organization is independent of context and immediate purpose; also, to the extent that perceptual organization involves separate modalities, these modalities operate independently of each other. While adhering to these assumptions, Marr [**Marr 78b**, 82] further assumes that perceptual organization is responsive to a specific set of goals. It provides a basis for disambiguating (separating) the effects of scene illumination, surface reflectance, and viewing location, and for extracting a geometric surface description (the $2\frac{1}{2}$-D sketch). Marr asserts that the first composite description (the "primal sketch") makes explicit the intensity variations in the image (e.g., the location of edges, oriented lines, line endings, and homogeneous intensity patches) and provides a primitive representation of local geometric relationships. The main distinguishing characteristics of his theory are: (1) a specific set of computational mechanisms for constructing the primal sketch, and (2) a complete architecture for the human visual system (see Chapter 5), which provides a strong context for the description embedded in the primal sketch.

A number of additional research efforts have focused on determining the architectural structure of the human early vision system: what information is employed, the nature of the procedures used to transform sensory data into assertions about scene content, and the order in which the procedures are invoked.

Hoffman and Richards [**Hoffman 85**] offer a partial solution to the problem of how complex objects can be decomposed into their component parts without knowing the explicit shape of those parts or the identity of the composite object. They note that interpenetrating convex shapes have negative curvature extrema at their joins, and they illustrate the fact that the human visual system uses such curvature cues for partitioning. Furthermore, since extrema of principal curvature are invariant under changes in imaging geometry, the authors suggest that coherent parts of the world can be isolated by breaking scene surfaces into pieces along loci of negative minima of principal curvature. Their theory provides a unified explanation for a number of well-known visual illusions. It differs from much of the previously discussed work on perceptual organization because their partitioning decisions depend on global context—that is, knowing which side of a boundary is figure and which side is ground. In related work (see Chapter 6), Richards and Hoffman [**Richards 85**] define a vocabulary of six simple primitives, called *codons*, for partitioning and describing planar curves. Codons lie between minima of curvature and are distinguished by the number of curvature zeros (inflections) that fall on the intervening segment. An important property of any description based on the singular points of curvature—minima, maxima, and zeros—is that their ordinal relations remain invariant under projective transformations, including translations, rotations, and dilations. Thus, regardless of the 3-D orientation and size of a part of some object, the codon description of its projection onto a 2-D image retains a high degree of invariance.

With his invention of the "random-dot stereogram," Julesz [Julesz 71, 74] was able to demonstrate that stereoscopic depth perception does not depend on first recognizing known features in the individual images.[4] This work, together with demonstrations by others (e.g., subjective contours,[5] the Ponzo illusion) suggests that depth is computed very early in the perceptual system, possibly in a "special" channel, and conditions the decisions made by other processes involved in perceptual organization. Such conjecture would be in contrast to Marr's view that depth (i.e., the $2\frac{1}{2}$-D sketch) is determined after much of the process of perceptual organization has been completed, i.e., after the formation of the primal sketch.

Cavanagh [Cavanagh 87] presents the results of psychological studies that address the question of what information is used by the human visual system to extract depth information from images. He observes that the perception of subjective contours and shadows, cues to depth determination, depend only on luminance differences and are not produced by color,

[4] In a random-dot stereogram, portions of a random-dot field depicted in one image are coherently displaced to form a second image of a stereo pair.

[5] Subjective contours are formed when the visual system is led to believe that an occluding surface, not physically different in intensity from some portions of the occluded background, is being viewed. Under these conditions an intensity transition is perceived to define the boundary of the occluding surface, even though there is no local physical evidence to support this perception.

texture, stereo,[6] or motion. Furthermore, natural constraints that could have ruled against the perception of shadows or subjective contours, such as impossible color differences across a shadow boundary, do not appear to be taken into consideration by the visual system. "The perceptual hypotheses of the observer appeared to play a far more important role in determining the interpretation of the figure than did natural constraints ... there does not seem to be a constraint that a strong perceptual hypothesis can't override." A further confirmation of this viewpoint can be found in the experiments by Gilchrist [Gilchrist 77] in which perceived lightness is conditioned by the observer's assumption of spatial relationships, even when these assumptions are incorrect. See also the section on unconscious inference in Rock [Rock 83, Chapter 9].

Ullman [**Ullman 84**] argues that the computation of spatial relations in the human visual system consists of two main stages. The first stage is the bottom-up creation of certain representations of the visible environment. The second stage involves the application of processes called "visual routines" to the representations constructed in the first stage. The second stage of processing is necessary because there is an open-ended variety of visual relations and properties that may be needed for different purposes. The number of such relations and the difficulty of computing even some of the seemingly simple ones (such as the inside-outside relationship) makes it infeasible for the visual system to attempt to compute and store all such relations on all occasions. Ullman believes that the visual system can assemble sequences of basic operations such as "boundary tracing," "shift of focus," "indexing," and "marking" to extract objects and parts and their spatial relationships from the base representation to satisfy specific interpretation goals.

Tsotsos [Tsotsos 82] discusses the kinds of knowledge the visual system requires in order to interpret its sensory input so as to correspond to objects in the world. He first presents a short overview of the current knowledge of the human visual process to indicate the concepts that must be considered in the design of computer vision systems. These concepts are then related to representational formalisms and control structures currently used in artificial intelligence applications. In particular, he discusses the role of the "is-a" relationship (generalization-specialization), the "part-of" relationship (part-whole), and the "instance-of" relationship (the relation between type and token). He finds that current representational research can provide well-defined tools for only some of the required concepts, and notes that our greatest weakness lies in dealing with dynamic structures.

[6] Prazdny [Prazdny 85b] notes that subjective contours are not perceived when they conflict with a binocularly fused interpretation of the scene.

2.2 Computational Approaches to Perceptual Organization

In this and the following section we describe computational approaches and specific techniques for perceptual organization. Although the primary purpose of this computationally oriented work is to build effective computer vision systems rather than to model the human visual system, we cannot avoid direct reference to, and comparison with, human visual system capabilities and preferences. This is because perceptual organization is an intermediate step in the interpretation process; its effectiveness cannot be assessed on the basis of immediate performance, but only on how this performance contributes to later stages in the interpretation hierarchy. The human visual system is the only existing complete general vision system in which we can assess the effectiveness of perceptual organization techniques.

It is possible to identify two general computational vision approaches to perceptual organization: those approaches that attempt to translate image data into a primitive description of *scene* content (e.g., geometric properties of the visible surfaces or information about the sources of illumination), and those approaches that describe *image* content or simply partition the image into coherent structures and regions.

2.2.1 Scene Content Approaches

Approaches in which the FPD describes scene content as opposed to image content must be explicit about specific scene attributes. Three alternatives have been suggested:

1. Describe the *local* geometric and photometric *properties* of the visible scene surface.
2. Describe scene geometry in terms of a vocabulary of simple 3-D *geometric models* (e.g., polyhedra, spheres, cylinders, and so on), which can be associated with relatively primitive image content.
3. Identify basic *physical processes* that underly the scene structures and produce coherent patterns of intensity variation in the image (e.g., fractal textures, flow patterns).

Barrow and Tenenbaum [Barrow 78], in work paralleling that of Marr, proposed an explicit computational framework for recovering point properties of the visible scene surfaces.[7] Bypassing an intermediate "primal sketch," they show how a relaxation-based approach could be used to compute cooperatively (as opposed to Marr's separate "channels") point arrays of such local surface properties as depth, orientation, and albedo.

Perhaps inspired by the realistic rendering techniques developed by the graphics community (see Chapter 6), which employ detailed models of physical processes, Witkin and Tenenbaum [Witkin 83b], Pentland [**Pentland** 84a, **86**a,**b**], and Kass and Witkin [**Kass 85**] have proposed that the FPD be constructed in terms of a relatively small set of processes such as flows, accretion and deletion (fractals), branching, and so on. The problem for computational vision is to detect the presence of these processes in the intensity variations of a new image and then to extract the essential model parameters.

The use of 3-D geometric primitives to model scene structure is fairly common in much of the work in computational vision, but the introduction of such models generally occurs late in the interpretative process. For example, Marr suggests the use of generalized cylinders after the primal and $2\frac{1}{2}$-D sketches have been computed. Noting that speech is based on a vocabulary of approximately fifty phonemes, and that artists often start their sketching process employing only triangles, circles, and quadralaterals, Biederman [Biederman 85] suggests that a vocabulary of about fifty 3-D primitives could be used to describe scene content; the presence of these primitives could be determined after a preliminary processing step has located edges in the raw image.

2.2.2 Image Content Approaches

The second general computational vision approach to perceptual organization has the goal of describing the intensity variations in the image, or of partitioning the image into coherent regions. The image, rather than the corresponding scene, defines the domain of discourse. Specific techniques in this category, divorced from concepts of scene geometry and illumination, are typically based on Gestalt-type principles. Coherent regions are identified either as being homogeneous in some set of local attributes (e.g., intensity, color, and texture), or as being bounded by discontinuities in such attributes. Alternative interpretations are adjudicated in favor of the simpler and more stable of the competing descriptions.

Thus techniques in this second category generally require a preliminary descriptive step in which pixels are labeled as to their color, local texture, edge strength, and so on. The pixels are then grouped into contiguous regions, or coherent contours; in some cases, these coherent structures are then themselves partitioned into simpler units. For example, contours are partitioned into segments at locations where their shape or curvature undergoes some type of discontinuity.

Some of the earliest work in grouping pixels into regions or in finding the boundaries of regions involved binary or discretely labeled images, where the organizing criterion was simply defined—adjacent pixels with identical labels are clustered together. Elegant computational techniques were devised to accomplish the grouping efficiently, e.g. the modified union-find algorithm described in Firschein and Fischler [Firschein 78], or the contour-finding algorithm of Pavlidis [Pavlidis 78]. As soon as the grouping criteria extended beyond strict label identity to more general forms of homogeneity, such as feature-vector similarity, the problem shifted from one of computational efficiency to one of providing adequate definitions for the intuitive grouping criteria.

The classic approach to clustering employing a feature-vector similarity criterion was based on statistical decision theory [Duda 73], but when the required statistical information is not available, the method described in Ohlander [Ohlander 75] and Ohlander, Price, and Reddy [Ohlander 78] is still typical of the current state-of-the-art. However, in examining images of real scenes, it is apparent that most semantically meaningful regions, or even perceptually salient regions, cannot be adequately defined by homogeneity of such local attributes as intensity, color, texture, and so on. We must invoke some of the more abstract Gestalt criteria such as simplicity or stability. Much of the current research in this area has focused on the problem of partitioning 1-D curves.

Witkin [**Witkin 83a**] provides a representation that makes explicit the stability of structures appearing in 1-D curves when viewed at different scales of resolution. His scale-space diagrams allow hierarchical decomposition of arbitrary curves in a perceptually satisfying manner. The 1-D signal is first "ex-

[7] Also see Chapter 5.

panded" by convolution with Gaussian masks over a continuum of sizes. The zero crossings of the second derivative of the Gaussian-filtered signal are then plotted on a "scale-space image" that uses as coordinate axes the coarseness σ of the Gaussian filter and the time (or distance) axis of the original curve. (The Gaussian convolution has the property that, as the standard deviation σ decreases, additional zeroes may appear, but existing ones cannot in general disappear.) The scale-space image is then reduced to a simple tree using the connectivity of extremal points (zero crossings) tracked through the scale space representation; the singular points at which new extrema appear are the nodes of the tree. The resulting tree describes the qualitative structure of the signal over all scales of observation. This representation can be used to match signals (Chapter 7) and to explain perceptual grouping phenomena.

In Chapter 5, we discuss methods for image partitioning based on the simplest (most compact) description in some given descriptive formalism.[8] This work, while very interesting, is still preliminary since efficient minimization techniques have not yet been defined, and the success of this general approach is strongly dependent on the efficacy of the descriptive formalism that is assumed to be given. In a somewhat related vein, Lowe and Binford [**Lowe 82**] suggest that the main task of early vision is to find meaningful groupings in the imagery to be processed, where meaningfulness is defined to be domain independent and independent of world knowledge. Examples of such groupings readily detected by the human visual system in an otherwise random field of dots include colinearity, predominant orientation, bilateral and rotational symmetry, and repetition. The central theme of the paper is that it is possible to calculate a statistical measure of the likelihood that a given grouping truly reflects an interdependence of its subparts and is not the result of a random alignment of independent elements. Lowe and Binford propose a nonparametric rank test of statistical significance to determine if a selected grouping in the image could have arisen from a random perturbation of the surrounding image features. (An important issue not adequately treated here is how to select the neighborhood over which the statistics are to be computed.) Since it is not computationally feasible to examine all possible groupings in an image, the authors propose that a search be carried out for only certain classes of patterns, particularly those patterns known to be easily handled by the human visual system. These ideas have been implemented in a computer program and the authors present results showing the detection of meaningful linear groupings embedded in a random field of dots.

Fischler and Bolles [**Fischler 86**] argue that the underlying justification for Gestalt partitioning criteria is that such criteria provide the elements of a believable explanation of how the image was assembled from coherent parts. To be believable, explanations must be concise, complete, and stable; schemes satisfying just one of these criteria will generally be inadequate. Fischler and Bolles provide computational techniques for 1-D curve partitioning and these techniques embody multiple Gestalt criteria.

2.3 Computational Techniques for Detecting and Describing Perceptual Primitives

Most of the partitioning formalisms described in the preceding section require a description of the raw image in terms of color, texture, edge structure, flow, depth, and motion. In the following subsections we describe some of the computational techniques for detecting and describing these perceptual primitives. (Depth and motion were extensively covered in Chapter 1.) In each case a huge literature exists, and we limit our discussion to a few innovative efforts.

The two questions of primary concern in this section are:

1. How can the various perceptual primitives be extracted from the pixel array of image intensities?
2. How can associated *scene* properties be deduced from image-derived attributes? (For example, how can surface albedo be deduced from pixel intensity or color?) How can we distinguish between shadow, material, and occlusion edges based on discontinuities in image intensity?

2.3.1 Intensity and Color

Intensity and color[9] typically provide the basic information specifying the "raw" image. While a large body of techniques exists for altering the given image

[8] Barnard [Barnard 84], G. B. Smith and Wolf [G. B. Smith 84], and Georgeff and Wallace [Georgeff 84].

[9] For practical and psychological purposes, color can be represented by three intensity arrays corresponding to the red, green, and blue spectral components of reflected light. A conventional black-and-white image can be reconstructed from the three "color" arrays.

intensities for such purposes as noise removal and enhancement required for human viewing, or as heuristic procedures to simplify subsequent automated analysis, our discussion here will be limited to the question of how to recover surface albedo (spectral reflectance) from the raw image information. The problem of assigning stable colors to objects (color constancy) is difficult because image intensity is a function of the illumination, viewing and scene geometry, as well as surface reflectance [**Horn 77**]. In most practical problems we have no prior knowledge of the ambient illumination, surface albedo, or scene geometry, but yet the human visual system is able to achieve a high degree of color constancy. What is the computational basis for this ability?[10]

Land [Land 77a] developed a theory of color vision to explain how the effects of surface reflectance changes can be separated from the vagaries of the prevailing illumination. Land suggested that changes in illumination are usually gradual, whereas changes in reflectance of a surface are often quite sharp. By filtering out slow changes, one could then isolate the changes due to surface reflectance alone. Because it was uncertain as to whether the processing occurred in the retina or the cortex (or partially in both), Land called his theory the "retinex theory." He claimed that the retinex system created a lightness image for each of the three sensitive systems and compared them in order to generate color. Land and McCann [Land 77b] implemented this scheme as follows: Multiple paths between (nominally all) pairs of points in an image are investigated. For each path, in each of the three color systems, we compute the successive ratios of intensity of adjacent points along the path. If a ratio is less than some given fixed threshold value, we set it equal to 1. The product of the ratios along each path approximates the relative color-specific reflectances of the area containing the initial and final points on the path. Absolute reflectances are obtained by assuming that the brightest region in all three color systems is white; a region which is brightest in exactly one or two systems is used to identify other colors.

Maloney and Wandell [**Maloney 86**], like Land, present an approach that does not require prior knowledge of ambient illumination, surface albedo, or scene geometry to calibrate a reflectance model. They express surface reflectance and ambient illumination (assumed locally constant) as weighted sums of a small number of basis functions. The basis functions are fixed; they do not vary with location in the scene and are predetermined. Evaluation of three to four weights each for both the reflectance and illumination basis functions were experimentally found to provide very good approximations to most surface spectral reflectances and illumination conditions. Thus, a linear model can be constructed and solved by making enough spectrally distinguished measurements over the scene. This analysis has some interesting implications for human color vision. If there are only three spectrally distinguished classes of photoreceptors in the human eye, their analysis shows that the weights for only two spectral reflectance basis functions can be recovered, which is probably not enough for good color constancy. Furthermore, to solve the linear system without prior knowledge of the ambient illumination, there must be a variety of different surface reflectances visible in the scene. Thus, compared to a very simple scene, a more "complicated" scene permits better recovery of surface albedo.

2.3.2 Texture and Symmetry

While texture can refer to both statistical variation in intensity (micro-structure texture) as well as patterns of lines or shaped tokens (macro-structure texture), most work in computational vision has focused on microstructure texture [Laws 85] [Haralick 79]. Early work in texture description simply measured statistical parameters or described image textures in terms of their similarity to natural textures such as those catalogued by Brodatz [Brodatz 66]. More recent efforts, especially those involving fractal textures, are discussed in other parts of this book (see Chapter 6).

As noted in Chapter 1, regular texture patterns in 3-D space are distorted by the perspective imaging process in such a way that they encode information about scene geometry. The human visual system simultaneously recovers the original spatial texture pattern and the underlying scene geometry. Computational techniques that ignore the effects of scene geometry on texture appearing in an image have little hope of using texture homogeneity for partitioning the scene into semantically meaningful components. Thus, the "back-projection" techniques of Witkin [Witkin 81] and Kender [Kender 80], or the fractal model approach of Pentland [Pentland 84a], which interpret image textures in terms of their associated 3-D scene geometry, hold the greatest promise for extracting useful information from image microtexture.

The human visual system has a preattentive ability to perceive "regular" structure in an image. When the

[10] See the review of human color vision by Gershon [Gershon 85].

elements of an image pattern have resolvable shape, the human visual system is especially sensitive to the presence of symmetric and skew-symmetric[11] configurations. We have no general methods for detecting symmetric image structures—except, possibly, for the case of repeated patterns which can be detected by Fourier techniques. However, considerable progress has been made in determining whether a given closed contour is symmetric or skew-symmetric. Both Friedberg and Brown [Friedberg 84, 86] and Kanade and Kender [**Kanade 83b**] note the following: A bilaterally symmetric figure in an arbitrarily oriented plane *P*, viewed under orthography, yields a skew symmetric figure whose axes of symmetry and skew constrain the orientation of *P* to a one-parameter subspace of the spherical space of orientation vectors. One can first back-project the given image shape to all possible planes permitted by the skew-symmetric constraint and then evaluate the resulting figure with respect to a criterion for the presence of symmetric form.

2.3.3 Flow

In Chapter 1 we described how a sequence of images can give rise to a "velocity flow field" that encodes 3-D scene geometry. Earlier in this chapter we discussed Zucker's work on describing two basic types of image intensity flow patterns as the basis for perceiving contours and surfaces. A third concept, discussed in Kass and Witkin [**Kass 85**] is that flow is a basic natural process that can be detected and described as a perceptual primitive. They believe that, just as we need to understand the image-forming process to recover scene geometry, we need to understand the physical processes that generate image patterns in order to decompose them into their intrinsic parts. They investigate the class of oriented flow patterns produced by propagation (the streaked train left by a paint brush dipped in multicolored paint), accretion (the grain in wood that is built up over time; the laminar patterns of geological strata), and deformation (a bent rod).

The authors model a flow pattern in terms of a deformation applied to a uniform grid (i.e., parallel flow lines), and a superimposed residual pattern (e.g., a knot in a wood grain pattern) which has an intrinsic description in the straight flow pattern. The analysis process thus involves determining the image flow field, computing the deformation, and extracting residual patterns. The technique underlying the Kass and Witkin approach is related to work by Marr and Hildreth [Marr 80] concerned with determining the local density of edge elements as a function of orientation, and to Zucker's work on flow [**Zucker** 83, **85**]. However, the specific Kass and Witkin flow analysis technique is based on analysis of the local power spectrum. The high frequency energy in the Fourier domain will tend to cluster along "lines" in the image perpendicular to the flow orientation. A simple way to detect this clustering is to sum the energy in an appropriate region of the power spectrum and examine how the sum is affected by rotations of an orientation-selective filter. Examples of the Kass and Witkin decomposition include using the flow coordinates to provide preferred directions for edge detection, detecting anomalies, and fitting simple models to the straightened patterns.

2.3.4 Edges, Lines, and Ribbons

Edges and contours delimit image regions and encode shape information. Because of the importance of edges, few problems in computer vision have been more intensively studied than that of edge detection. Even so, there are probably only modest differences in performance between the most sophisticated modern techniques[12] and edge operators originally described fifteen to twenty years ago.[13] Furthermore, none of the available techniques can produce an edge map that can reasonably approximate a human-produced sketch for a complex scene.

Some of the problems are:

1. We do not really know how the human visual system identifies an edge. For example, under what conditions do we perceive texture edges, color edges, or subjective contours? While some simple types of intensity discontinuities such as an intensity step, are obviously seen as edges, we do not know what percentage of visible edges are of this type.

---

[11] Skew symmetry was introduced by Kanade [Kanade 81] as characterizing 2-D shapes that are 2-D affine transformations of truly symmetric patterns. Truly symmetric 2-D shapes are left unaltered by operations such as rotation or reflection about a point or axis.

[12] For example, [Burns 84], [**Canny** 83, **86**], [Haralick 84], [**Leclerc 85**], [Nalwa 86].

[13] For example, [L. S. Davis 75], [Herskovitz 70], [Hueckel 71], [Kirsch 71], [Prewitt 70], [Roberts 65], [Rosenfeld 71], and Sobel in [Ballard 82].

2. Most edge operators look for local discontinuities in intensity. However, if the operator examines too small an area, its output is very unstable. Large-sized operators tend to overlap more than one edge event in the image, and their output is often confused because the image data do not correspond to the designer's implicit model of how intensity should vary in the presence of an edge. Furthermore, subjective edges [Gregory 78] [Kanisza 76] as perceived by the human visual system are illusory; they have no distinguishing image content over much of their extent.
3. There are few, if any, reliable methods for interpreting image intensity discontinuities as specific scene events such as occlusion, shadow, or material boundaries.
4. It is probable that the operations of edge linking and filtering (the elimination of "accidental" edge structures), rather than local edge detection, are the critical steps in producing a good edge map for an image. Yet, there is no formal basis for these more global operations.

The early edge operators generally embodied a specific edge model; they were designed to locate step edges, or measured local intensity gradients. These operators are generally of a fixed size and do not address the issues of edge width and resolution. The "zero-crossing operator," introduced by Marr and his co-workers [Marr 80, 82] has one important new feature: it explicitly deals with scale of resolution by smoothing, as well as mathematically differentiating, the image-intensity surface; it smooths out the intensity variations below the spatial frequency of the edges it finds.

We will describe some recent edge detection methods here; for a review of work in finding edges, see Blicher [Blicher 84]. Most of the existing approaches are based on the design of a detector that identifies likely edge pixels. Nalwa and Binford [Nalwa 86] seek to identify *edgels*—short linear edge elements—rather than individual edge pixels. The papers by Haralick and Leclerc deal with obtaining *descriptions* of intensity surfaces, leading to the determination of likely edges.

Canny [**Canny** 83, **86**] describes what is generally purported to be one of the best-performing variations of the many proposed zero-crossing implementations. He describes a procedure for the design of edge detectors for arbitrary edge profiles based on the specification of detection and localization criteria augmented by a criterion to ensure that the detector has only one response to a single edge. He specializes the analysis to step edges in Gaussian noise and derives a single operator shape that is optimal at any scale. The main attributes of Canny's approach that distinguish it from the more traditional zero-crossing implementations[14] are (1) non-maximum suppression to insure that the detector has only one response to a single edge, (2) a detection scheme that uses several directionally sensitive elongated (rather than circular) operators at each point, and (3) a linking scheme that requires a connected edge segment have all of its points above some minimal detector response, and at least one point above some specified higher threshold.[15] Thus, Canny's scheme finds edge segments rather than closed zero-crossing contours.

Torre and Poggio [Torre 86] argue that the numerical differentiation of images required for edge detection is an ill-posed problem that must be *regularized* by a filtering operation before differentiation; they discuss the behavior and information content of zero crossings obtained with filters of different sizes. The Burns operator [**Burns 84**] is interesting in that it chooses intensity gradient direction, rather than gradient magnitude, as the essential local evidence for the presence of an edge.

Haralick [**Haralick 84**] analytically models a patch of the intensity surface and then deduces the presence of edges from the coefficients of his model.[16] The explicit nature of the model should, in principle, allow rejection of its conclusions if it does not adequately describe the image data. Leclerc [**Leclerc 85**], like Haralick, obtains an analytic model of the local intensity surface, but his approach uses local statistics rather than fixed analytic criteria for deducing the presence of a discontinuity.[17] Nalwa and Binford [Nalwa 86] detect *edgels*—short, linear edge segments, each characterized by a direction and position. They approach the problem by fitting a series of 1-D surfaces[18] to each patch of the image and accepting the surface description that is adequate in a least-squares

[14] For example, see Ullman [**Ullman 81**]

[15] The ratio of the high to low threshold is between one and three.

[16] Haralick compares the performance of his scheme with that of the Prewitt gradient operator [Prewitt 70] and the Marr and Hildreth zero-crossing operator [Marr 80]. An excellent discussion of this evaluation is contained in Grimson, Hildreth, and Haralick [Grimson 85].

[17] Other differences include Leclerc's use of differently-sized neighborhoods and a scheme for combining the results obtained from analyzing these neighborhoods.

[18] A 1-D surface is a surface constrained to be constant along some dimension.

sense and has the fewest parameters. The authors present edge results for a variety of images and compare their results with those of the Marr-Hildreth operator [Marr 80].

Witkin [Witkin 82] describes an *edge classification* technique that relies on structural, rather than quantitative, photometric properties of the image. The method is capable of making judgments about the scene structures that produced the edge in the image. For example, it can distinguish between edges corresponding to occluding contours and shadows. The solution is based on (1) the continuity of surfaces, surface markings, and illumination and (2) the assumption that widely separated scene constituents may be regarded as causally independent. Witkin uses as his basic measure the correlation between image intensities along lines a small distance on either side of an image curve. The procedure is to construct a family of curves parallel to a hypothesized edge and to perform a sequence of linear regressions of the intensity values along each curve onto those along its neighbor. A precipitous drop in correlation at the nominal edge location signifies an occluding contour. High correlation with an abrupt shift in the regression parameters signifies a shadow. Sustained high correlation with additive and multiplicative regression parameters near zero and unity, respectively, implies no edge. Low correlation throughout prevents drawing any conclusions.

Rubin and Richards [Rubin 82, 84] note that when the relative ordering of the intensities of two spectral components is interchanged in crossing an edge in a color image, or when one spectral component becomes more intense while another spectral component decreases in intensity, the edge will usually denote a boundary between different material types in the scene. Gershon, Jepson, and Tsotsos [Gershon 86] discuss the efficacy of these tests when the ambient and incident illumination have different spectral characteristics.

Pairs of edges that are a fixed distance apart, such as the pair of edges that constitute a road in an aerial photograph, are known as "ribbons." Nevatia and Babu [Nevatia 80] describe a technique for locating such ribbons. Their approach consists of determining edge magnitude and direction by convolving an image with a number of edge masks, thinning and thresholding these edge magnitudes, linking the edge elements based on proximity and orientation, and finally approximating the linked elements by piecewise linear sigments. Ribbons are detected by finding "antiparallel" pairs of linear segments—that is, parallel edge segments with opposite intensity gradients. In the context of tracking roads in aerial imagery, Quam [Quam 78] used correlation of intensities as a ribbon-finding technique. Given a starting point in an image identified as being on the centerline of the ribbon, the technique obtains a cross section normal to the centerline, and then uses correlation to align new cross sections with a cumulative cross-section model to find the extensions of the ribbon.

With very few exceptions, computational vision techniques treat lines and edges as similar entities, yet we know that the human visual system makes some important distinctions between these image events. For example, subjective edges arise from occlusion cues, while subjective lines almost never result from 3-D interpretation. Fischler and Wolf [**Fischler 83**] describe a procedure, specifically tailored for line detection, which is very effective in complex scenes where edge detection methods generally fail. Their technique involves two steps: (1) a nonlocal, but computationally effective, method for finding "line points"; and (2) a clustering procedure, based on the use of a minimum spanning tree, which first links the line points into coherent structures and then extracts the perceptually salient lines.

2.4 Discussion

We have discussed image partitioning and perceptual organization and have looked to the human visual system for clues about what to measure and how to combine these measurements into statements about scene content. The most interesting recent theories are based on (1) postulating a vocabulary of image or scene primitives and providing computational techniques for detecting these primitives, or (2) employing Gestalt-type criteria for decomposing an image into coherent, independently analyzable components. We reviewed perceptual organization approaches with special emphasis on techniques that translate image data into a description of *scene* content and techniques that describe *image* content. Finally, we reviewed computational techniques for detecting and describing perceptual primitives, such as texture, flows, edges, and lines, which are used by the partitioning formalisms. Understanding the nature of perceptual organization is probably the most important open issue currently being addressed by researchers in the computer vision field.

Papers Included

Burns, J. B., A. Hanson, and E. Riseman. Extracting straight lines. In *Proc. DARPA IU Workshop*, New Orleans, La., pages 165–168, October 1984.

Canny, J.F. A Computational Approach to Edge Detection. *IEEE PAMI* 8(6)679–698, November 1986.

Fischler, M.A. and H.C. Wolf. Linear delineation. In *Proc. IEEE Conf. on Computer Vision and Image Processing*, pages 351–356, Washington, D.C., June 19–23, 1983.

Fischler, M.S. and R.C. Bolles. Perceptual organization and curve partitioning. *IEEE PAMI* 8(1)100–105, January 1986.

Haralick, R.M. The digital step edge from zero crossings of second directional derivatives. *IEEE PAMI* 6(1)58–68, January 1984.

Hoffman, D. and W. Richards. Parts of recognition. *Cognition* 18:65–96, 1985. (Also In *From Pixels to Predicates*, Pentland, A. (ed.), Ablex, Norwood, N.J., 1985).

Julesz, B. and J.R. Bergen. Textons, the fundamental elements in preattentive vision and the perception of textures. *Bell System Technical Journal*, 62(6):1619–1644, July–August 1983.

Kanade, T. and J.P. Kender. Mapping image properties into shape constraints: skewed symmetry, affine-transformable patterns, and the shape-from-texture paradigm. In *Human and Machine Vision*, Beck, Hope, and Rosenfeld (eds.), Academic Press, New York, 1983.

Kass, M. and A. Witkin. Analyzing oriented patterns. In *Proc. Ninth Int Joint Conf. on Artificial Intelligence*, pages 944–952, Los Angeles, California, August 18–23, 1985.

Leclerc, Y. Capturing the local structure of image discontinuities in two dimensions. In *Proc. IEEE CVPR*, pages 34–38, San Francisco, California, June 19–23, 1985.

Lowe, D.G. and T.O. Binford. Segregation and aggregation: an approach to figure-ground phenomena. In *Proc. DARPA IU Workshop*, Palo Alto, California, pages 168–178, September 1982.

Maloney, L.T. and B.A. Wandell. Color constancy: a method for recovering surface spectral reflectance. *J. Optical Society Am.*, 3:29–33, 1986.

Ullman, S. Visual routines. *Cognition* 18:97–160, 1984.

Witkin, A.P. Scale-space filtering. In *Proc. 8th Int. Joint Conf. on Artificial Intelligence*, pages 1019–1022, Karlsruhe, West Germany, Aug. 8–12, 1983.

Zucker, S. Early orientation selection: tangent fields and the dimensionality of their support. *CVGIP*, 32(1):74–103, Oct. 1985.

EXTRACTING STRAIGHT LINES

J. Brian Burns, Allen R. Hanson, Edward M. Riseman

Computer and Information Science Department
University of Massachusetts
Amherst, Massachusetts 01003

ABSTRACT

This paper presents a new approach to the extraction of straight lines in intensity images. Pixels are grouped into edge support regions of similar gradient orientation, and then the structure of the associated intensity surface is used to determine the location and properties of the edge. The resulting regions and extracted parameters form two separate representations of a straight line segment, pixel-based and symbolic, that can be used together for a variety of purposes. The algorithm appears to be far more effective than previous techniques for two key reasons: 1) the gradient orientation is used as the initial organizing criterion in the extraction of straight lines, instead of the gradient magnitude; and 2) data directed organization of the complete context of a straight line is determined prior to any local decisions about participating edge points.

1.0 INTRODUCTION

The organization of significant local intensity changes into the more global abstractions called "lines" or "global intensity boundaries" is an early, but important, step in the transformation of the visual signal into useful intermediate constructs for interpretation processes. Despite the large amount of research appearing in the literature, effective extraction of linear boundaries has remained a difficult problem in many image domains. There are two goals of this paper: a) the development of mechanisms for extracting straight lines from complex images, including intensity discontinuities of arbitrarily low contrast; and b) the construction of an intermediate representation of edge/line information which allows high-level mechanisms efficient access to relevant lines. A more detailed presentation can be found in [BUR84].

We contend that the major failings of line extraction algorithms are twofold: the relegation of information about edge orientation to a secondary role in the processing, and the lack of a comprehensive global view of the underlying image structure prior to making local decisions about edge features. In most edge and line extraction algorithms, the magnitude of the intensity change has been used in some manner as the dominant measure of importance of the local edge. It is our view that edge orientation carries important information about the spatial extent of the straight line.

The technique presented here was motivated by a need for a straight line extraction method which can find straight lines in reasonably complex images, particularly those lines that are long but not necessarily of high contrast. A key characteristic of the approach presented here that distinguishes it from most previous work is the global organization of the supporting edge context prior to any decisions about the relevance of local intensity changes. An estimate of the local gradient orientation at each pixel is the basis of these first organizing processes. Grouping pixels into edge support regions avoids the plethora of magnitude responses from masks at varying sizes and orientations, as well as unnecessary complexity in the subsequent organizing mechanisms. The approach presented here has its roots in the "gradient collection" process of Hanson et al [HAN80], as well as [EHR78].

Our approach allows the extraction of straight lines despite weaknesses in line clarity due to local edge inconsistencies or deficiencies in width and contrast. It directly addresses the problems associated with the size of the edge operators and determines the extent of support to be given to edges and lines directly from the underlying data.

2.0 A REPRESENTATION AND PROCESS FOR EXTRACTING STRAIGHT LINES

2.1 Overview

There are four basic steps in extracting straight lines:

1. Group pixels into edge-support regions based on similarity of gradient orientation. This allows data directed organization of edge contexts without commitment to masks of a particular size.
2. Approximate the intensity surface by a weighted planar fit. The fit is weighted by the gradient magnitude associated with the pixels so that intensities in the steepest part of the edge will dominate.

3. Extract attributes from the edge-support region and the plane fit. The attributes extracted include the representative line and its length, contrast, width, location, orientation, and straightness.

4. Filter on the attributes to isolate various image events such as long straight lines; high contrast short lines (heavy texture); low contrast short lines (light texture); and lines at particular orientations and postitions.

2.2 Grouping Pixels into Edge Support Regions Via Gradient Orientation

Figure 1 shows two representative images used to illustrate the process. Figure 2(a) is a surface plot of a 32x32 subimage from another house image; results for the full images are shown in subsequent sections. The vector field drawn in figure 2(b) shows the corresponding gradient image where the length of the vector encodes gradient magnitude. The gradient estimates were formed by convolving the image with two-by-two edge masks (figure 2(b) inset). Note that the sign of the gradient is relevant.

An extremely simple process was employed to group the local gradients into regions on the basis of the orientation estimates. The 360 degree range of gradient directions is arbitrarily partitioned into a small set of regular intervals, say eight 45 degree partitions or sixteen 22.5 degree partitions. If our conjectures about edge orientation are correct, then pixels participating in the edge-support context of a straight line will tend to be in the same edge orientation partitions and adjacent pixels that are not part of a straight line will tend to have different orientations. A simple connected components algorithm can be used to form distinct region labels for groups of adjacent pixels with the same orientation label (Figure 2(c)). Note that in Figure 2(c) the great degree of fragmentation into many small regions of very low gradient magnitude could be grouped into a homogeneous region later, rather than interpreting them as edge elements.

To make the fixed partition technique more sensitive to edges of any orientation, the current algorithm uses two overlapping sets of partitions, with one set rotated a half-partition interval. Thus, if a 45 degree partition is used starting at 0 degrees, then a second set of 45 degree partitions starting at 22.5 is also used. The critical problem of this approach is merging the two representations in such a way that a single edge is principly associated with a single gradient region. The following scheme is used to select such regions for each edge: first the lengths are determined for the regions; then, since each pixel is a member of exactly two regions (one in each segmentation), the pixel decides which one provides the longest interpretation; finally each region counts up the number of pixels within its boundaries that voted for it as opposed to regions of the other segmentation. The 'support' a region is given is the number of votes for it over the total number of pixels in the region. The regions selected are those that have a majority, i.e., the support is greater than 50%. For further discussions on grouping see [BUR84].

2.3 Interpreting the Edge-Support Region as a Straight Line

The underlying intensity surface of each gradient region is a candidate for a straight line structure; the key problem is to use this information to find the line. The positional parameters extracted will serve as the core of the structure's symbolic description as well as a coordinate system about which other attributes will be measured.

In this section, we will examine a simple process for computing the parameters of a planar fit to the intensity surface of the pixels in each edge-support region. The region depicted in figure 3(a) and as dots in the surface plot of figure 2(a) will serve as our example. Note that it includes pixels outside the group of gradients depicted in figure 2(c), since the two-by-two masks incorporated them in the gradient estimation. Haralick [HAR81] modelled the local intensity surface in the neighborhood of a pixel as a planar surface patch called a 'sloped facet'. This planar fit served as a model of the region structure and was used to determine if the pixel was at a region boundary or not. In our application the planar fit will be applied to all pixels in a support region instead of an a priori fixed geometric configuration. If a direct least-square planar fit to all pixels in the support region is computed, then many pixels which might be at the tail of the intensity change could dominate the fit. Therefore, the pixels were weighted by local gradient magnitude to enhance the effect of points near the edge center.

An obvious constraint on the orientation of the line is that it be perpendicular to the gradient of the fitted plane. Thus, this leaves the problem of locating the line along the projection of the gradient. A simple approach is to intersect the fitted plane with a horizontal plane representing the average intensity of the region weighted by local gradient magnitude as shown in Figure 3(b); the straight line resulting from the intersection of the two planes is shown in Figure 3(a).

2.4 Extracting Attributes of the Support Context

The gradient region and the planar fit of the associated intensity surface provides the basic information necessary to quantify a variety of attributes beyond the basic orientation and position parameters. Length is simply the distance between the two endpoints. Other attributes of the line include properties of the intensity profile perpendicular to the line and its behavior along the length of the line. Analysis of the profile of the line can provide a measure of the edge's contrast and width (fuzziness), while behavior along the length determines it's straightness; see [BUR84].

3.0 EXPERIMENTAL RESULTS

The algorithm described in the preceding sections was applied to the full images shown in Figure 1. The algorithm utilized overlapping partitions as described in Section 2.2; the partition size was 45 degrees, staggered by 22.5 degrees. Figures 4-5 demonstrate the performance of the algorithm.

Figure 4(a) shows the unfiltered output of the algorithm applied to the first house image. Note that all of the small and low contrast edges are still present. Figures 4(b-c) show the result of filtering 4(a) on the basis of gradient steepness (change in gray-levels per pixel) followed by a filtering on length that separates the edges into two disjoint sets, one corresponding to short texture edges (Figure 4b) and the other to longer lines related to the surface structure of objects in the image (Figure 4c). We are also examining ways in which texture descriptors may be constructed from the edge set remaining when a filter similar to that which produced 4(c) is applied to the initial edge data. In Figure 4c, the structural edges representing the telephone wires were extracted from a thin, one pixel wide diagonally oriented region, a difficult problem for many line extraction processes.

Figure 5 shows results on another typical house image, this time filtered on length alone (all edges greater than 5 pixels).

4.0 CONCLUSIONS

This paper has presented a low-level representation for straight lines. The technique for extracting straight lines is effective because it globally organizes the spatial extent of a straight line without local decisions about the meaningfulness of an edge feature. It does this by utilizing gradient orientation to provide a gradient segmentation of the pixels in the formation of edge-support regions. Analysis of the intensity surface of the pixels in these regions yields the information required to extract lines and characterize the intensity variations in a variety of ways. The algorithm is very robust and accurately extracts many low contrast long lines.

While the extracted lines, such as long straight lines, might be directly useful, the underlying edge-support region has a wealth of information useful to intermediate processing strategies. These include additional grouping mechanisms for linking co-linear straight line segments and linking piecewise-linear approximations to curved lines bounding areas of similar properties. It also contains information useful for grouping lines with common properties into textured regions. The representation can serve as a simple yet rich edge-line "primal sketch" [MAR77]. The edge-support regions might be useful in separating the straight lines into intrinsic images [BAR78] representing boundaries of different types such as illumination, texture, reflectance, orientation, etc.

5.0 REFERENCES

[BAR78] Barrow, H.G. and Tenenbaum, J.M., "Recovering Intrinsic Scene Characteristics from Images," in Computer Vision Systems (A. Hanson and E. Riseman, eds.), Academic Press, 1978, pp. 3-26.

[BUR84] Burns, J., Hanson, A. and Riseman, E., "Extracting Straight Lines," COINS Technical Report, University of Massachusetts, January 1984.

[EHR78] Ehrich, R.W. and Foith, J.P., "Topology and Semantics of Intensity Arrays," in Computer Vision Systems (A. Hanson and E. Riseman, eds.), Academic Press, 1978, pp. 111-127.

[HAN80] Hanson, A., Riseman, E., and Glazer, F., "Edge Relaxation and Boundary Continuity," COINS Technical Report 80-11, University of Massachusetts, May 1980.

[HAR81] Haralick, R.M. and Watson, L., "A Facet Model for Image Data," Computer Graphics and Image Processing, Volume 15, 1981, pp 113-129.

[MAR77] Marr, D. and Nishihara, H.K., "Representation and Recognition of the Spatial Organization of Three-Dimensional Shapes," Proc. Royal Society B. 200, 1977, pp. 269-294.

Figure 1. Two images used to demonstrate the straight line finder.

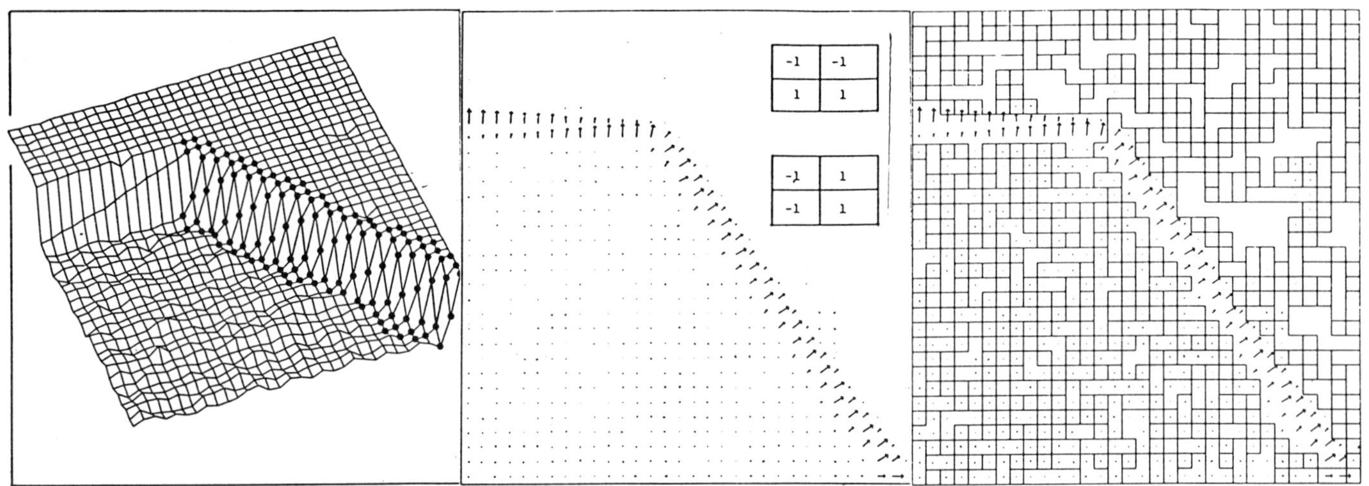

Figure 2. Forming gradient regions. (a) A 32x32 subarea of a house image which will be used to illustrate the process. (b) The 2x2 operators used to estimate dI/dy and dI/dx, from which the local gradient orientation is obtained and the resulting gradient vectors. (c) Gradient regions formed by a regular partitioning of the data into eight orientation classes.

Figure 3. Obtaining straight lines. (a) A group of pixels associated with a gradient region from figure 2 and its interpreted straight line. (b) The straight line obtained by intersecting the L.S.E. plane fit to the intensities, weighted by the gradient magnitudes, with the horizontal plane representing the average intensity, weighted again by magnitude.

Figure 5. Results of the line extraction algorithm on the house image of Figure 1(b), after filtering on the length alone ($\geq$ 5).

Figure 4. Results of the line extraction algorithm on the house image of Figure 1(a). (a) Initial edge set. (b) Filtering (a) on gradient steepness ($\geq$ 10 gray levels per pixel) and for shortness (length $\leq$ 5). (c) Steep, long lines (length $\geq$ 15 and gradient steepness $\geq$ 10 levels per pixel).

A Computational Approach to Edge Detection

JOHN CANNY, MEMBER, IEEE

Abstract—**This paper describes a computational approach to edge detection. The success of the approach depends on the definition of a *comprehensive* set of goals for the computation of edge points. These goals must be precise enough to delimit the desired behavior of the detector while making minimal assumptions about the form of the solution. We define detection and localization criteria for a class of edges, and present mathematical forms for these criteria as functionals on the operator impulse response. A third criterion is then added to ensure that the detector has only one response to a single edge. We use the criteria in numerical optimization to derive detectors for several common image features, including step edges. On specializing the analysis to step edges, we find that there is a natural uncertainty principle between detection and localization performance, which are the two main goals. With this principle we derive a single operator shape which is optimal at any scale. The optimal detector has a simple approximate implementation in which edges are marked at maxima in gradient magnitude of a Gaussian-smoothed image. We extend this simple detector using operators of several widths to cope with different signal-to-noise ratios in the image. We present a general method, called feature synthesis, for the fine-to-coarse integration of information from operators at different scales. Finally we show that step edge detector performance improves considerably as the operator point spread function is extended along the edge. This detection scheme uses several elongated operators at each point, and the directional operator outputs are integrated with the gradient maximum detector.**

Index Terms—**Edge detection, feature extraction, image processing, machine vision, multiscale image analysis.**

I. INTRODUCTION

EDGE detectors of some kind, particularly step edge detectors, have been an essential part of many computer vision systems. The edge detection process serves to simplify the analysis of images by drastically reducing the amount of data to be processed, while at the same time preserving useful structural information about object boundaries. There is certainly a great deal of diversity in the applications of edge detection, but it is felt that many applications share a common set of requirements. These requirements yield an abstract edge detection problem, the solution of which can be applied in any of the original problem domains.

We should mention some specific applications here. The Binford–Horn line finder [14] used the output of an edge detector as input to a program which could isolate simple geometric solids. More recently the model-based vision system ACRONYM [3] used an edge detector as the front end to a sophisticated recognition program. Shape from motion [29], [13] can be used to infer the structure of three-dimensional objects from the motion of edge contours or edge points in the image plane. Several modern theories of stereopsis assume that images are preprocessed by an edge detector before matching is done [19], [20]. Beattie [1] describes an edge-based labeling scheme for low-level image understanding. Finally, some novel methods have been suggested for the extraction of three-dimensional information from image contours, namely shape from contour [27] and shape from texture [31].

In all of these examples there are common criteria relevant to edge detector performance. The first and most obvious is low error rate. It is important that edges that occur in the image should not be missed and that there be no spurious responses. In all the above cases, system performance will be hampered by edge detector errors. The second criterion is that the edge points be well localized. That is, the distance between the points marked by the detector and the "center" of the true edge should be minimized. This is particularly true of stereo and shape from motion, where small disparities are measured between left and right images or between images produced at slightly different times.

In this paper we will develop a mathematical form for these two criteria which can be used to design detectors for arbitrary edges. We will also discover that the first two criteria are not "tight" enough, and that it is necessary to add a third criterion to circumvent the possibility of multiple responses to a single edge. Using numerical optimization, we derive optimal operators for ridge and roof edges. We will then specialize the criteria for step edges and give a parametric closed form for the solution. In the process we will discover that there is an uncertainty principle relating detection and localization of noisy step edges, and that there is a direct tradeoff between the two. One consequence of this relationship is that there is a single unique "shape" of impulse response for an optimal step edge detector, and that the tradeoff between detection and localization can be varied by changing the spatial width of the detector. Several examples of the detector performance on real images will be given.

II. ONE-DIMENSIONAL FORMULATION

To facilitate the analysis we first consider one-dimensional edge profiles. That is, we will assume that two-

Manuscript received December 10, 1984; revised November 27, 1985. Recommended for acceptance by S. L. Tanimoto. This work was supported in part by the System Development Foundation, in part by the Office of Naval Research under Contract N00014-81-K-0494, and in part by the Advanced Research Projects Agency under Office of Naval Research Contracts N00014-80-C-0505 and N00014-82-K-0334.

The author is with the Artificial Intelligence Laboratory, Massachusetts Institute of Technology, Cambridge, MA 02139.

IEEE Log Number 8610412.

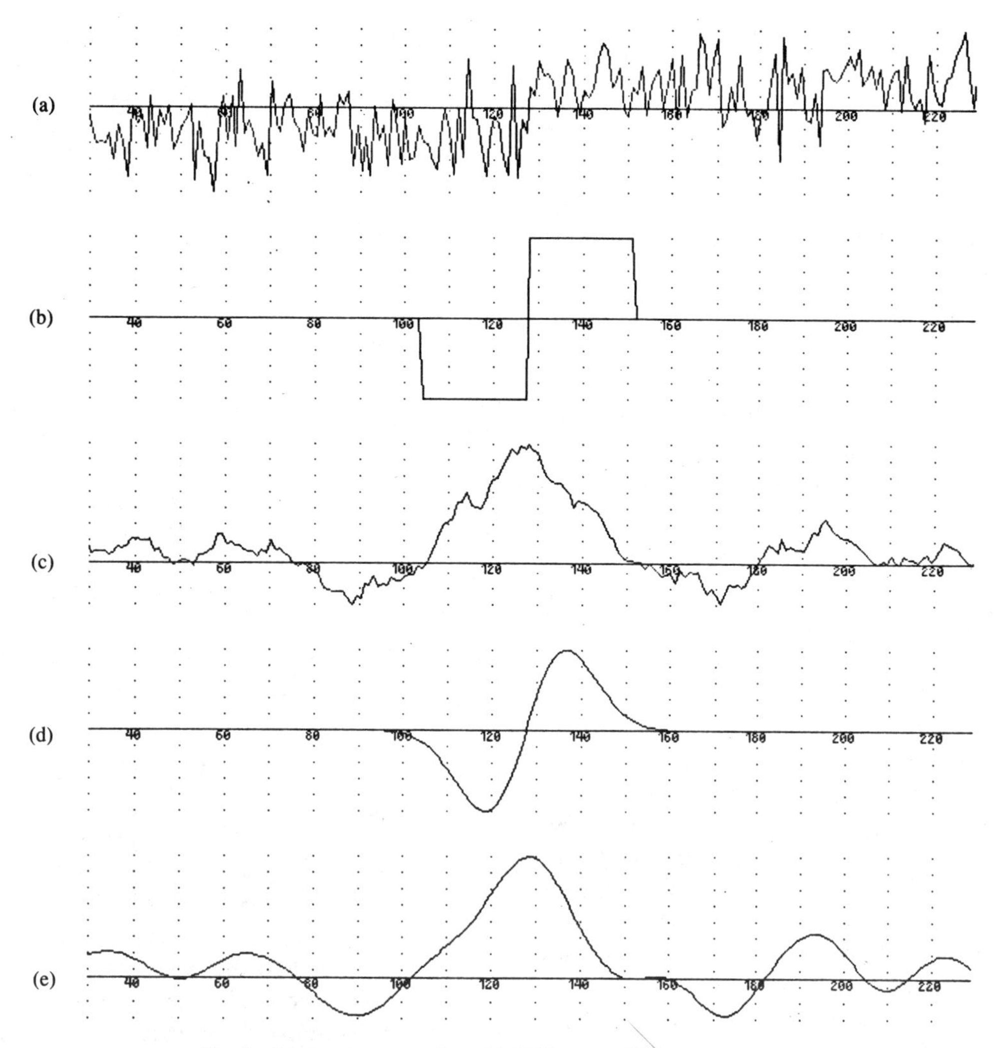

Fig. 1. (a) A noisy step edge. (b) Difference of boxes operator. (c) Difference of boxes operator applied to the edge. (d) First derivative of Gaussian operator. (e) First derivative of Gaussian applied to the edge.

dimensional edges locally have a constant cross-section in some direction. This would be true for example, of smooth edge contours or of ridges, but not true of corners. We will assume that the image consists of the edge and additive white Gaussian noise.

The detection problem is formulated as follows: We begin with an edge of known cross-section bathed in white Gaussian noise as in Fig. 1(a), which shows a step edge. We convolve this with a filter whose impulse response could be illustrated by either Fig. 1(b) or (d). The outputs of the convolutions are shown, respectively, in Fig. 1(c) and (e). We will mark the center of an edge at a local maximum in the output of the convolution. The design problem then becomes one of finding the filter which gives the best performance with respect to the criteria given below. For example, the filter in Fig. 1(d) performs much better than Fig. 1(b) on this example, because the response of the latter exhibits several local maxima in the region of the edge.

In summary, the three performance criteria are as follows:

1) Good detection. There should be a low probability of failing to mark real edge points, and low probability of falsely marking nonedge points. Since both these probabilities are monotonically decreasing functions of the output signal-to-noise ratio, this criterion corresponds to maximizing signal-to-noise ratio.

2) Good localization. The points marked as edge points by the operator should be as close as possible to the center of the true edge.

3) Only one response to a single edge. This is implicitly captured in the first criterion since when there are two responses to the same edge, one of them must be considered false. However, the mathematical form of the first criterion did not capture the multiple response requirement and it had to be made explicit.

A. Detection and Localization Criteria

A crucial step in our method is to capture the intuitive criteria given above in a mathematical form which is readily solvable. We deal first with signal-to-noise ratio and localization. Let the impulse response of the filter be $f(x)$, and denote the edge itself by $G(x)$. We will assume that the edge is centered at $x = 0$. Then the response of the filter to this edge at its center H_G is given by a convolution integral:

$$H_G = \int_{-W}^{+W} G(-x)\, f(x)\, dx \tag{1}$$

assuming the filter has a finite impulse response bounded by $[-W, W]$. The root-mean-squared response to the noise $n(x)$ only, will be

$$H_n = n_0 \left[\int_{-W}^{+W} f^2(x)\, dx \right]^{1/2} \tag{2}$$

where n_0^2 is the mean-squared noise amplitude per unit length. We define our first criterion, the output signal-to-noise ratio, as the quotient of these two responses.

$$\mathrm{SNR} = \frac{\left| \int_{-W}^{+W} G(-x)\, f(x)\, dx \right|}{n_0 \sqrt{\int_{-W}^{+W} f^2(x)\, dx}} \tag{3}$$

For the localization criterion, we want some measure which increases as localization improves, and we will use the reciprocal of the root-mean-squared distance of the marked edge from the center of the true edge. Since we have decided to mark edges at local maxima in the response of the operator $f(x)$, the first derivative of the response will be zero at these points. Note also that since edges are centered at $x = 0$, *in the absence of noise* there should be a local maximum in the response at $x = 0$.

Let $H_n(x)$ be the response of the filter to noise only, and $H_G(x)$ be its response to the edge, and suppose there is a local maximum in the total response at the point $x = x_0$. Then we have

$$H_n'(x_0) + H_G'(x_0) = 0. \tag{4}$$

The Taylor expansion of $H_G'(x_0)$ about the origin gives

$$H_G'(x_0) = H_G'(0) + H_G''(0)x_0 + O(x_0^2). \tag{5}$$

By assumption $H_G'(0) = 0$, i.e., the response of the filter in the absence of noise has a local maximum at the origin, so the first term in the expansion can be ignored. The displacement x_0 of the actual maximum is assumed to be small so we will ignore quadratic and higher terms. In fact by a simple argument we can show that if the edge $G(x)$ is either symmetric or antisymmetric, all even terms in x_0 vanish. Suppose $G(x)$ is antisymmetric, and express $f(x)$ as a sum of a symmetric component and an antisymmetric component. The convolution of the symmetric component with $G(x)$ contributes nothing to the numerator of the SNR, but it does contribute to the noise component in the denominator. Therefore, if $f(x)$ has any symmetric component, its SNR will be worse than a purely antisymmetric filter. A dual argument holds for symmetric edges, so that if the edge $G(x)$ is symmetric or antisymmetric, the filter $f(x)$ will follow suit. The net result of this is that the response $H_G(x)$ is always symmetric, and that its derivatives of odd orders [which appear in the coefficients of even order in (5)] are zero at the origin. Equations (4) and (5) give

$$H_G''(0)x_0 \approx -H_n'(x_0). \tag{6}$$

Now $H_n'(x_0)$ is a Gaussian random quantity whose variance is the mean-squared value of $H_n'(x_0)$, and is given by

$$E[H_n'(x_0)^2] = n_0^2 \int_{-W}^{+W} f'^2(x)\, dx \tag{7}$$

where $E[y]$ is the expectation value of y. Combining this result with (6) and substituting for $H_G''(0)$ gives

$$E[x_0^2] \approx \frac{n_0^2 \int_{-W}^{+W} f'^2(x)\, dx}{\left[\int_{-W}^{+W} G'(-x)\, f'(x)\, dx \right]^2} = \delta x_0^2 \tag{8}$$

where δx_0 is an approximation to the standard deviation of x_0. The localization is defined as the reciprocal of δx_0.

$$\text{Localization} = \frac{\left| \int_{-W}^{+W} G'(-x)\, f'(x)\, dx \right|}{n_0 \sqrt{\int_{-W}^{+W} f'^2(x)\, dx}}. \tag{9}$$

Equations (3) and (9) are mathematical forms for the first two criteria, and the design problem reduces to the maximization of both of these simultaneously. In order to do this, we maximize the product of (3) and (9). We could conceivably have combined (3) and (9) using any function that is monotonic in two arguments, but the use of the product simplifies the analysis for step edges, as should become clear in Section III. For the present we will make use of the product of the criteria for arbitrary edges, i.e., we seek to maximize

$$\frac{\left| \int_{-W}^{+W} G(-x)\, f(x)\, dx \right|}{n_0 \sqrt{\int_{-W}^{+W} f^2(x)\, dx}} \; \frac{\left| \int_{-W}^{+W} G'(-x)\, f'(x)\, dx \right|}{n_0 \sqrt{\int_{-W}^{+W} f'^2(x)\, dx}}. \tag{10}$$

There may be some additional constraints on the solution, such as the multiple response constraint (12) described next.

B. Eliminating Multiple Responses

In our specification of the edge detection problem, we decided that edges would be marked at local maxima in the response of a linear filter applied to the image. The detection criterion given in the last section measures the effectiveness of the filter in discriminating between signal and noise at the center of an edge. It does not take into account the behavior of the filter *nearby* the edge center. The first two criteria can be trivially maximized as follows. From the Schwarz inequality for integrals we can show that SNR (3) is bounded above by

$$n_0^{-1}\sqrt{\int_{-W}^{+W} G^2(x)\,dx}$$

and localization (9) by

$$n_0^{-1}\sqrt{\int_{-W}^{+W} G'^2(x)\,dx}.$$

Both bounds are attained, and the product of SNR and localization is maximized when $f(x) = G(-x)$ in $[-W, W]$.

Thus, according to the first two criteria, the optimal detector for step edges is a truncated step, or difference of boxes operator. The difference of boxes was used by Rosenfeld and Thurston [25], and in conjunction with lateral inhibition by Herskovits and Binford [11]. However it has a very high bandwidth and tends to exhibit many maxima in its response to noisy step edges, which is a serious problem when the imaging system adds noise or when the image itself contains textured regions. These extra edges should be considered erroneous according to the first of our criteria. However, the analytic form of this criterion was derived from the response at a single point (the center of the edge) and did not consider the interaction of the responses at several nearby points. If we examine the output of a difference of boxes edge detector we find that the response to a noisy step is a roughly triangular peak with numerous sharp maxima in the vicinity of the edge (see Fig. 1).

These maxima are so close together that it is not possible to select one as the response to the step while identifying the others as noise. We need to add to our criteria the requirement that the function f will not have "too many" responses to a single step edge in the vicinity of the step. We need to limit the number of peaks in the response so that there will be a low probability of declaring more than one edge. Ideally, we would like to make the distance between peaks in the noise response approximate the width of the response of the operator to a single step. This width will be some fraction of the operator width W.

In order to express this as a functional constraint on f, we need to obtain an expression for the distance between adjacent noise peaks. We first note that the mean distance between adjacent maxima in the output is twice the distance between adjacent zero-crossings in the derivative of the operator output. Then we make use of a result due to Rice [24] that the average distance between zero-crossings of the response of a function g to Gaussian noise is

$$x_{\text{ave}} = \pi\left(\frac{-R(0)}{R''(0)}\right)^{1/2} \tag{11}$$

where $R(\tau)$ is the autocorrelation function of g. In our case we are looking for the mean zero-crossing spacing for the function f'. Now since

$$R(0) = \int_{-\infty}^{+\infty} g^2(x)\,dx \quad\text{and}\quad R''(0) = -\int_{-\infty}^{+\infty} g'^2(x)\,dx$$

the mean distance between zero-crossings of f' will be

$$x_{zc}(f) = \pi\left(\frac{\int_{-\infty}^{+\infty} f'^2(x)\,dx}{\int_{-\infty}^{+\infty} f''^2(x)\,dx}\right)^{1/2} \tag{12}$$

The distance between adjacent maxima in the noise response of f, denoted $x_{\max}$, will be twice x_{zc}. We set this distance to be some fraction k of the operator width.

$$x_{\max}(f) = 2x_{zc}(f) = kW. \tag{13}$$

This is a natural form for the constraint because the response of the filter will be concentrated in a region of width $2W$, and the expected number of noise maxima in this region is N_n where

$$N_n = \frac{2W}{x_{\max}} = \frac{2}{k}. \tag{14}$$

Fixing k fixes the number of noise maxima that could lead to a false response.

We remark here that the intermaximum spacing (12) scales with the operator width. That is, we first define an operator f_w which is the result of stretching f by a factor of w, $f_w(x) = f(x/w)$. Then after substituting into (12) we find that the intermaximum spacing for f_w is $x_{zc}(f_w) = wx_{zc}(f)$. Therefore, if a function f satisfies the multiple response constraint (13) for fixed k, then the function f_w will also satisfy it, assuming W scales with w. For any fixed k, the multiple response criterion is invariant with respect to spatial scaling of f.

III. Finding Optimal Detectors by Numerical Optimization

In general it will be difficult (or impossible) to find a closed form for the function f which maximizes (10) subject to the multiple response constraint. Even when G has a particularly simple form (e.g., it is a step edge), the form of f may be complicated. However, if we are given a candidate function f, evaluation of (10) and (12) is straightforward. In particular, if the function f is represented by a discrete time sequence, evaluation of (10) requires only the computation of four inner products between sequences. This suggests that numerical optimization can be done directly on the sampled operator impulse response.

The output will not be an analytic form for the operator, but an implementation of a detector for the edge of interest will require discrete point-spread functions anyway. It is also possible to include additional constraints by using a *penalty method* [15]. In this scheme, the constrained optimization is reduced to one, or possibly several, unconstrained optimizations. For each constraint we define a penalty function which has a nonzero value when one

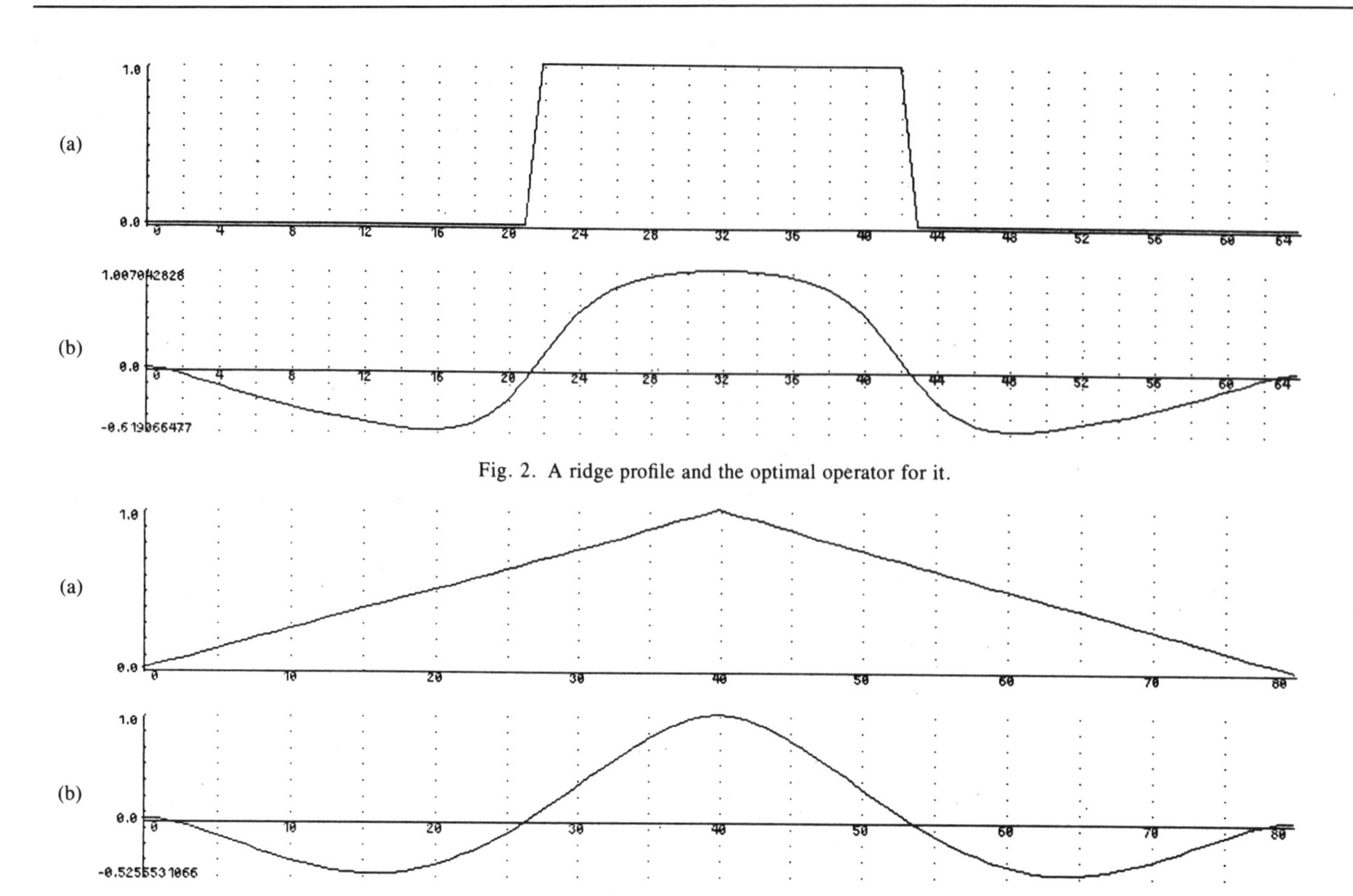

Fig. 2. A ridge profile and the optimal operator for it.

Fig. 3. A roof profile and an optimal operator for roofs.

of the constraints is violated. We then find the f which maximizes

$$\text{SNR}(f) * \text{Localization}\ (f) - \sum \mu_i P_i(f) \qquad (15)$$

where P_i is a function which has a positive value only when a constraint is violated. The larger the value of μ_i the more nearly the constraints will be satisfied, but at the same time the greater the likelihood that the problem will be ill-conditioned. A sequence of values of μ_i may need to be used, with the final form of f from each optimization used as the starting form for the next. The μ_i are increased at each iteration so that the value of $P_i(f)$ will be reduced, until the constraints are "almost" satisfied.

An example of the method applied to the problem of detecting "ridge" profiles is shown in Fig. 2. For a ridge, the function G is defined to be a flat plateau of width w, with step transitions to zero at the ends. The auxiliary constraints are

• The multiple response constraint. This constraint is taken directly from (12), and does not depend on the form of the edge.

• The operator should have zero dc component. That is it should have zero output to constant input.

Since the width of the operator is dependent on the width of the ridge, there is a suggestion that several widths of operators should be used. This has not been done in the present implementation however. A wide ridge can be considered to be two closely spaced edges, and the implementation already includes detectors for these. The only reason for using a ridge detector is that there are ridges in images that are too small to be dealt with effectively by the narrowest edge operator. These occur frequently because there are many edges (e.g., scratches and cracks or printed matter) which lie at or beyond the resolution of the camera and result in contours only one or two pixels wide.

A similar procedure was used to find an optimal operator for roof edges. These edges typically occur at the concave junctions of two planar faces in polyhedral objects. The results are shown in Fig. 3. Again there are two subsidiary constraints, one for multiple responses and one for zero response to constant input.

A roof edge detector has not been incorporated into the implementation of the edge detector because it was found that ideal roof edges were relatively rare. In any case the ridge detector is an approximation to the ideal roof detector, and is adequate to cope with roofs. The situation may be different in the case of an edge detector designed explicitly to deal with images of polyhedra, like the Binford–Horn line-finder [14].

The method just described has been used to find optimal operators for both ridge and roof profiles and in addition it successfully finds the optimal step edge operator derived in Section IV. It should be possible to use it to find operators for arbitrary one-dimensional edges, and it should be possible to apply the method in two dimensions to find optimal detectors for various types of corner.

IV. A Detector for Step Edges

We now specialize the results of the last section to the case where the input $G(x)$ is step edge. Specifically we set $G(x) = Au_{-1}(x)$ where $u_n(x)$ is the nth derivative of a delta function, and A is the amplitude of the step. That is,

$$u_{-1}(x) = \begin{cases} 0, & \text{for } x < 0; \\ 1, & \text{for } x \geq 0; \end{cases} \tag{16}$$

and substituting for $G(x)$ in (3) and (9) gives

$$\text{SNR} = \frac{A\left|\int_{-W}^{0} f(x)\, dx\right|}{n_0\sqrt{\int_{-W}^{+W} f^2(x)\, dx}} \tag{17}$$

$$\text{Localization} = \frac{A|f'(0)|}{n_0\sqrt{\int_{-W}^{+W} f'^2(x)\, dx}}. \tag{18}$$

Both of these criteria improve directly with the ratio A/n_0 which might be termed the signal-to-noise ratio of the image. We now remove this dependence on the image and define two performance measures Σ and Λ which depend on the filter only:

$$\text{SNR} = \frac{A}{n_0}\Sigma(f) \qquad \Sigma(f) = \frac{\left|\int_{-W}^{0} f(x)\, dx\right|}{\sqrt{\int_{-W}^{+W} f^2(x)\, dx}} \tag{19}$$

$$\text{Localization} = \frac{A}{n_0}\Lambda(f') \qquad \Lambda(f') = \frac{|f'(0)|}{\sqrt{\int_{-W}^{+W} f'^2(x)\, dx}}. \tag{20}$$

Suppose now that we form a spatially scaled filter f_w from f, where $f_w(x) = f(x/w)$. Recall from the end of Section II that the multiple response criterion is unaffected by spatial scaling. When we substitute f_w into (19) and (20) we obtain for the performance of the scaled filter:

$$\Sigma(f_w) = \sqrt{w}\,\Sigma(f) \quad \text{and} \quad \Lambda(f'_w) = \frac{1}{\sqrt{w}}\Lambda(f'). \tag{21}$$

The first of these equations is quite intuitive, and implies that a filter with a broad impulse response will have better signal-to-noise ratio than a narrow filter when applied to a step edge. The second is less obvious, and it implies that a narrow filter will give better localization than a broad one. What is surprising is that the changes are inversely related, that is, both criteria either increase or decrease by $\sqrt{w}$. There is an uncertainty principle relating the detection and localization performance of the step edge detector. Through spatial scaling of f we can trade off detection performance against localization, but we cannot improve both simultaneously. This suggests that a natural choice for the composite criterion would be the product of (19) and (20), since this product would be invariant under changes in scale.

$$\Sigma(f)\,\Lambda(f') = \frac{\left|\int_{-W}^{0} f(x)\, dx\right|}{\sqrt{\int_{-W}^{+W} f^2(x)\, dx}} \frac{|f'(0)|}{\sqrt{\int_{-W}^{+W} f'^2(x)\, dx}}. \tag{22}$$

The solutions to the maximization of this expression will be a *class of functions* all related by spatial scaling. In fact this result is independent of the method of combination of the criteria. To see this we assume that there is a function f which gives the best localization Λ for a particular Σ. That is, we find f such that

$$\Sigma(f) = c_1 \quad \text{and} \quad \Lambda(f') \text{ is maximized.} \tag{23}$$

Now suppose we seek a second function f_w which gives the best possible localization while its signal-to-noise ratio is fixed to a different value, i.e.,

$$\Sigma(f_w) = c_2 \quad \text{while} \quad \Lambda(f'_w) \text{ is maximized.} \tag{24}$$

If we now define $f_1(x)$ in terms of $f_w(x)$ as $f_1(x) = f_w(xw)$ where

$$w = c_2^2/c_1^2$$

then the constraint on f_w in (24) translates to a constraint on f_1 which is identical to (23), and (24) can be rewritten as

$$\Sigma(f_1) = c_1 \quad \text{and} \quad \frac{1}{\sqrt{w}}\Lambda(f'_1) \text{ is maximized} \tag{25}$$

which has the solution $f_1 = f$. So if we find a single such function f, we can obtain maximal localization for any fixed signal-to-noise ratio by scaling f. The design problem for step edge detection has a *single unique* (up to spatial scaling) solution regardless of the absolute values of signal to noise ratio or localization.

The optimal filter is implicitly defined by (22), but we must transform the problem slightly before we can apply the calculus of variations. Specifically, we transform the maximization of (22) into a constrained minimization that involves only integral functionals. All but one of the integrals in (22) are set to undetermined constant values. We then find the extreme value of the remaining integral (since it will correspond to an extreme in the total expression) as a function of the undetermined constants. The values of the constants are then chosen so as to maximize the original expression, which is now a function only of these constants. Given the constants, we can uniquely specify the function $f(x)$ which gives a maximum of the composite criterion.

A second modification involves the limits of the integrals. The two integrals in the denominator of (22) have limits at $+W$ and $-W$, while the integral in the numerator has one limit at 0 and the other at $-W$. Since the function f should be antisymmetric, we can use the latter limits for all integrals. The denominator integrals will have half the value over this subrange that they would have over the full range. Also, this enables the value of $f'(0)$ to be set as a boundary condition, rather than expressed as an integral of f''. If the integral to be minimized shares the same limits as the constraint integrals, it is possible to exploit the *isoperimetric constraint* condition (see [6, p. 216]). When this condition is fulfilled, the constrained optimization can be reduced to an unconstrained optimization using Lagrange multipliers for the constraint functionals. The problem of finding the maximum of (22) reduces to the minimization of the integral in the denominator of the SNR term, subject to the constraint that the other integrals remain constant. By the principle of reciprocity, we could have chosen to extremize any of the integrals while keeping the others constant, and the solution should be the same.

We seek some function f chosen from a space of *admissible* functions that minimizes the integral

$$\int_{-W}^{0} f^2(x)\,dx \tag{26}$$

subject to

$$\int_{-W}^{0} f(x)\,dx = c_1 \qquad \int_{-W}^{0} f'^2(x)\,dx = c_2$$

$$\int_{-W}^{0} f''^2(x)\,dx = c_3 \qquad f'(0) = c_4. \tag{27}$$

The space of admissible functions in this case will be the space of all continuous functions that satisfy certain boundary conditions, namely that $f(0) = 0$ and $f(-W) = 0$. These boundary conditions are necessary to ensure that the integrals evaluated over finite limits accurately represent the infinite convolution integrals. That is, if the nth derivative of f appears in some integral, the function must be continuous in its $(n-1)$st derivative over the range $(-\infty, +\infty)$. This implies that the values of f and its first $(n-1)$ derivatives must be zero at the limits of integration, since they are zero outside this range.

The functional to be minimized is of the form $\int_a^b F(x, f, f', f'')$ and we have a series of constraints that can be written in the form $\int_a^b G_i(x, f, f', f'') = c_i$. Since the constraints are isoperimetric, i.e., they share the same limits of integration as the integral being minimized, we can form a composite functional using Lagrange multipliers [6]. The functional is a linear combination of the functionals that appear in the expression to be minimized and in the constraints. Finding a solution to the unconstrained maximization of $\Psi(x, f, f', f'')$ is equivalent to finding the solution to the constrained problem. The composite functional is

$$\Psi(x, f, f', f'') = F(x, f, f', f'') + \lambda_1 G_1(x, f, f', f'') + \lambda_2 G_2(x, f, f', f'') + \cdots$$

Substituting,

$$\Psi(x, f, f', f'') = f^2 + \lambda_1 f'^2 + \lambda_2 f''^2 + \lambda_3 f. \tag{28}$$

It may be seen from the form of this equation that the choice of which integral is extremized and which are constraints is arbitrary, the solution will be the same. This is an example of the *reciprocity* that was mentioned earlier. The choice of an integral from the denominator is simply convenient since the standard form of variational problem is a minimization problem. The Euler equation that corresponds to the functional Ψ is

$$\Psi_f - \frac{d}{dx}\Psi_{f'} + \frac{d^2}{dx^2}\Psi_{f''} = 0 \tag{29}$$

where Ψ_f denotes the partial derivative of Ψ with respect to f, etc. We substitute for Ψ from (28) in the Euler equation giving:

$$2f(x) - 2\lambda_1 f''(x) + 2\lambda_2 f''''(x) + \lambda_3 = 0. \tag{30}$$

The solution of this differential equation is the sum of a constant and a set of four exponentials of the form $e^{\gamma x}$ where γ derives from the solution of the corresponding homogeneous differential equation. Now γ must satisfy

$$2 - 2\lambda_1\gamma^2 + 2\lambda_2\gamma^4 = 0$$

so

$$\gamma^2 = \frac{\lambda_1}{2\lambda_2} \pm \frac{\sqrt{\lambda_1^2 - 4\lambda_2}}{2\lambda_2}. \tag{31}$$

This equation may have roots that are purely imaginary, purely real, or complex depending on the values of λ_1 and λ_2. From the composite functional Ψ we can infer that λ_2 is positive (since the integral of f''^2 is to be minimized) but it is not clear what the sign or magnitude of λ_1 should be. The Euler equation supplies a necessary condition for the existence of a minimum, but it is not a sufficient condition. By formulating such a condition we can resolve the ambiguity in the value of λ_1. To do this we must consider the second variation of the functional. Let

$$J[f] = \int_{x0}^{x1} \Psi(x, f, f', f'')\,dx.$$

Then by Taylor's theorem (see also [6, p. 214]),

$$J[f + \epsilon g] = J[f] + \epsilon J_1[f, g] + \tfrac{1}{2}\epsilon^2 J_2[f + \rho g, g]$$

where ρ is some number between 0 and ϵ, and g is chosen from the space of admissible functions, and where

$$J_1[f, g] = \int_{x0}^{x1} \Psi_f g + \Psi_{f'} g' + \Psi_{f''} g''\,dx$$

$$J_2[f, g] = \int_{x0}^{x1} \Psi_{ff} g^2 + \Psi_{f'f'} g'^2 + \Psi_{f''f''} g''^2 + 2\Psi_{ff'} g g' + 2\Psi_{f'f''} g' g'' + 2\Psi_{ff''} g''\,dx. \tag{32}$$

Note that J_1 is nothing more than the integral of g times the Euler equation for f (transformed using integration by parts) and will be zero if f satisfies the Euler equation. We can now define the second variation $\delta^2 J$ as

$$\delta^2 J = \frac{\epsilon^2}{2} J_2[f, g].$$

The necessary condition for a minimum is $\delta^2 J \geq 0$. We compute the second partial derivatives of Ψ from (28) and we get

$$J_1[f + g] = \int_{x0}^{x1} 2g^2 + 2\lambda_1 g'^2 + 2\lambda_2 g''^2 \, dx \geq 0. \tag{33}$$

Using the fact that g is an admissible function and therefore vanishes at the integration limits, we transform the above using integration by parts to

$$2 \int_{x0}^{x1} g^2 - \lambda_1 g g'' + \lambda_2 g''^2 \, dx \geq 0$$

which can be written as

$$2 \int_{x0}^{x1} \left(g^2 - \frac{\lambda_1}{2} g''\right)^2 + \left(\lambda_2 - \frac{\lambda_1^2}{4}\right) g''^2 \, dx \geq 0.$$

The integral is guaranteed to be positive if the expression being integrated is positive for all x, so if

$$4\lambda_2 > \lambda_1^2$$

then the integral will be positive for all x and for arbitrary g, and the extremum will certainly be a minimum. If we refer back to (31) we find that this condition is precisely that which gives complex roots for γ, so we have both guaranteed the existence of a minimum and resolved a possible ambiguity in the form of the solution. We can now proceed with the derivation and assume four complex roots of the form $\gamma = \pm\alpha \pm i\omega$ with α, ω real. Now $\gamma^2 = \alpha^2 - \omega^2 \pm 2i\alpha\omega$ and equating real and imaginary parts with (31) we obtain

$$\alpha^2 - \omega^2 = \frac{\lambda_1}{2\lambda_2} \quad \text{and} \quad 4\alpha^2\omega^2 = \frac{4\lambda_2 - \lambda_1^2}{4\lambda_2^2}. \tag{34}$$

The general solution in the range $[-W, 0]$ may now be written

$$f(x) = a_1 e^{\alpha x} \sin \omega x + a_2 e^{\alpha x} \cos \omega x + a_3 e^{-\alpha x} \cdot \sin \omega x + a_4 e^{-\alpha x} \cos \omega x + c. \tag{35}$$

This function is subject to the boundary conditions

$$f(0) = 0 \qquad f(-W) = 0 \qquad f'(0) = s \qquad f'(-W) = 0$$

where s is an unknown constant equal to the slope of the function f at the origin. Since $f(x)$ is asymmetric, we can extend the above definition to the range $[-W, W]$ using $f(-x) = -f(x)$. The four boundary conditions enable us to solve for the quantities a_1 through a_4 in terms of the unknown constants α, ω, c, and s. The boundary conditions may be rewritten

$$\begin{aligned}
&a_2 + a_4 + c = 0 \\
&a_1 e^{\alpha} \sin \omega + a_2 e^{\alpha} \cos \omega + a_3 e^{-\alpha} \sin \omega + a_4 e^{-\alpha} \cos \omega + c = 0 \\
&a_1 \omega + a_2 \alpha + a_3 \omega - a_4 \alpha = s \\
&a_1 e^{\alpha}(\alpha \sin \omega + \omega \cos \omega) + a_2 e^{\alpha}(\alpha \cos \omega - \omega \sin \omega) + a_3 e^{-\alpha}(-\alpha \sin \omega + \omega \cos \omega) \\
&\quad + a_4 e^{-\alpha}(-\alpha \cos \omega - \omega \sin \omega) = 0.
\end{aligned} \tag{36}$$

These equations are linear in the four unknowns a_1, a_2, a_3, a_4 and when solved they yield

$$\begin{aligned}
a_1 &= c(\alpha(\beta - \alpha) \sin 2\omega - \alpha\omega \cos 2\omega + (-2\omega^2 \sinh \alpha + 2\alpha^2 e^{-\alpha}) \sin \omega + 2\alpha\omega \sinh \alpha \cos \omega \\
&\quad + \omega e^{-2\alpha}(\alpha + \beta) - \beta\omega)/4(\omega^2 \sinh^2 \alpha - \alpha^2 \sin^2 \omega) \\
a_2 &= c(\alpha(\beta - \alpha) \cos 2\omega + \alpha\omega \sin 2\omega - 2\alpha\omega \cosh \alpha \cdot \sin \omega - 2\omega^2 \sinh \alpha \cos \omega + 2\omega^2 e^{-\alpha} \sinh \alpha \\
&\quad + \alpha(\alpha - \beta))/4(\omega^2 \sinh^2 \alpha - \alpha^2 \sin^2 \omega) \\
a_3 &= c(-\alpha(\beta + \alpha) \sin 2\omega + \alpha\omega \cos 2\omega + (2\omega^2 \sinh \alpha + 2\alpha^2 e^{\alpha}) \sin \omega + 2\alpha\omega \sinh \alpha \cos \omega \\
&\quad + \omega e^{2\alpha}(\beta - \alpha) - \beta\omega)/4(\omega^2 \sinh^2 \alpha - \alpha^2 \sin^2 \omega) \\
a_4 &= c(-\alpha(\beta + \alpha) \cos 2\omega - \alpha\omega \sin 2\omega + 2\alpha\omega \cosh \alpha \cdot \sin \omega + 2\omega^2 \sinh \alpha \cos \omega - 2\omega^2 e^{\alpha} \sinh \alpha \\
&\quad + \alpha(\alpha - \beta))/4(\omega^2 \sinh^2 \alpha - \alpha^2 \sin^2 \omega)
\end{aligned} \tag{37}$$

where β is the slope s at the origin divided by the constant c. On inspection of these expressions we can see that a_3 can be obtained from a_1 by replacing α by $-\alpha$, and similarly for a_4 from a_2.

The function f is now parametrized in terms of the constants α, ω, β, and c. We have still to find the values of these parameters which maximize the quotient of integrals that forms our composite criterion. To do this we first express each of the integrals in terms of the constants. Since these integrals are very long and uninteresting, they are not given here but may be found in [4]. We have reduced the problem of optimizing over an infinite-dimensional space of functions to a nonlinear optimization in three variables α, ω, and β (not surprisingly, the combined criterion does not depend on c). Unfortunately the resulting criterion, which must still satisfy the multiple response constraint, is probably too complex to be solved analytically, and numerical methods must be used to provide the final solution.

The shape of f will depend on the multiple response constraint, i.e., it will depend on how far apart we force the adjacent responses. Fig. 5 shows the operators that result from particular choices of this distance. Recall that there was no single best function for arbitrary ω, but a class of functions which were obtained by scaling a prototype function by ω. We will want to force the responses further apart as the signal-to-noise ratio in the image is lowered, but it is not clear what the value of signal-to-noise ratio will be for a single operator. In the context in which this operator is used, several operator widths are available, and a decision procedure is applied to select the smallest operator that has an output signal-to-noise ratio above a fixed threshold. With this arrangement the operators will spend much of the time operating close to their output Σ thresholds. We try to choose a spacing for which the probability of a multiple response error is comparable to the probability of an error due to thresholding.

A rough estimate for the probability of a spurious maximum in the neighborhood of the true maximum can be formed as follows. If we look at the response of f to an ideal step we find that its second derivative has magnitude $|Af'(0)|$ at $x = 0$. There will be only one maximum near the center of the edge if $|Af'(0)|$ is greater than the second derivative of the response to noise only. This latter quantity, denoted s_n, is a Gaussian random variable with standard deviation

$$n_0 \sigma_s = n_0 \left(\int_{-W}^{+W} f''^2(x)\, dx \right)^{1/2}.$$

The probability p_m that the noise slope s_n exceeds $Af'(0)$ is given in terms of the normal distribution function Φ

$$p_m = 1 - \Phi\left(\frac{A|f'(0)|}{n_0 \sigma_s} \right). \tag{38}$$

We can choose a value for this probability as an acceptable error rate and this will determine the ratio of $f'(0)$ to σ_s. We can relate the probability of a multiple response p_m to the probability of falsely marking an edge p_f which is

$$p_f = 1 - \Phi\left(\frac{A}{n_0} \Sigma \right) \tag{39}$$

by setting $p_m = p_f$. This is a natural choice since it makes a detection error or a multiple response error equally likely. Then from (38) and (39) we have

$$\frac{|f'(0)|}{\sigma_s} = \Sigma. \tag{40}$$

In practice it was impossible to find filters which satisfied this constraint, so instead we search for a filter satisfying

$$\frac{|f'(0)|}{\sigma_s} = r\Sigma \tag{41}$$

where r is as close as possible to 1. The performance indexes and parameter values for several filters are given in Fig. 4. The a_i coefficients for all these filters can be found from (37), by fixing c to, say, $c = 1$. Unfortunately, the largest value of r that could be obtained using the constrained numerical optimization was about 0.576 for filter number 6 in the table. In our implementation, we have

Filter Parameters						
n	x_{max}	$\Sigma\Lambda$	r	α	ω	β
1	0.15	4.21	0.215	24.59550	0.12250	63.97566
2	0.3	2.87	0.313	12.47120	0.38284	31.26860
3	0.5	2.13	0.417	7.85869	2.62856	18.28800
4	0.8	1.57	0.515	5.06500	2.56770	11.06100
5	1.0	1.33	0.561	3.45580	0.07161	4.80684
6	1.2	1.12	0.576	2.05220	1.56939	2.91540
7	1.4	0.75	0.484	0.00297	3.50350	7.47700

Fig. 4. Filter parameters and performance measures for the filters illustrated in Fig. 5.

approximated this filter using the first derivative of a Gaussian as described in the next section.

The first derivative of Gaussian operator, or even filter 6 itself, should not be taken as the final word in edge detection filters, even with respect to the criteria we have used. If we are willing to tolerate a slight reduction in multiple response performance r, we can obtain significant improvements in the other two criteria. For example, filters 4 and 5 both have significantly better $\Sigma\Lambda$ product than filter 6, and only slightly lower r. From Fig. 5 we can see that these filters have steeper slope at the origin, suggesting that the performance gain is mostly in localization, although this has not been verified experimentally. A thorough empirical comparison of these other operators remains to be done, and the theory in this case is unclear on how best to make the tradeoff.

V. An Efficient Approximation

The operator derived in the last section as filter number 6, and illustrated in Fig. 6, can be approximated by the first derivative of a Gaussian $G'(x)$, where

$$G(x) = \exp\left(-\frac{x^2}{2\sigma^2} \right).$$

The reason for doing this is that there are very efficient ways to compute the two-dimensional extension of the filter if it can be represented as some derivative of a Gaussian. This is described in detail elsewhere [4], but for the present we will compare the theoretical performance of a first derivative of a Gaussian filter to the optimal operator. The impulse response of the first derivative filter is

$$f(x) = -\frac{x}{\chi^2} \exp\left(-\frac{x^2}{2\sigma^2} \right) \tag{42}$$

and the terms in the performance criteria have the values

$$|f'(0)| = \frac{1}{\sigma_s}$$

$$\int_{-\infty}^{0} f(x)\, dx = 1 \qquad \int_{-\infty}^{+\infty} f^2(x)\, dx = \frac{\sqrt{\pi}}{2\sigma}$$

$$\int_{-\infty}^{+\infty} f'^2(x)\, dx = \frac{3\sqrt{\pi}}{4\sigma^3} \qquad \int_{-\infty}^{+\infty} f''^2(x)\, dx = \frac{15\sqrt{\pi}}{8\sigma^5}. \tag{43}$$

The overall performance index for this operator is

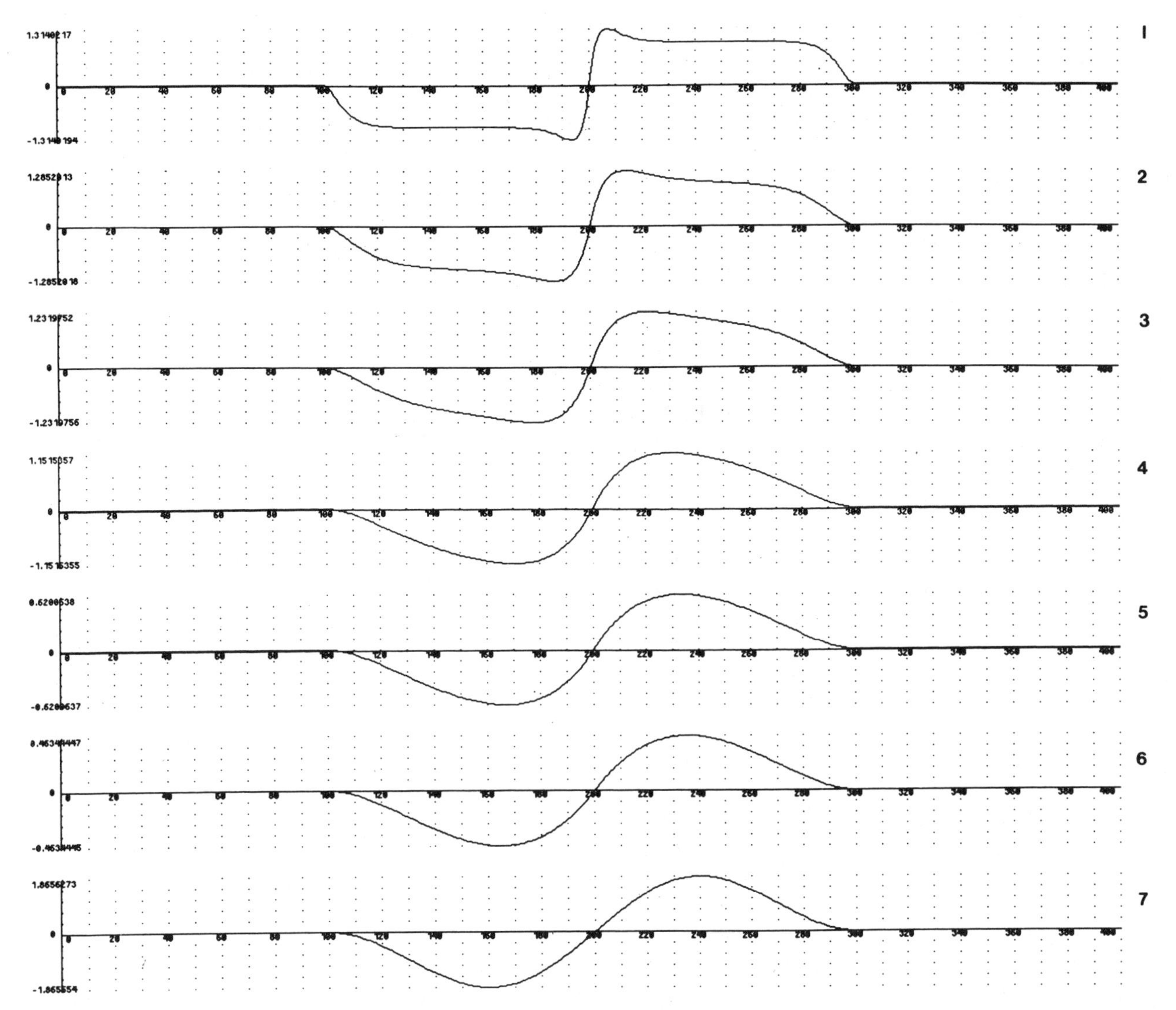

Fig. 5. Optimal step edge operators for various values of x_{max}. From top to bottom, they are x_{max} = 0.15, 0.3, 0.5, 0.8, 1.0, 1.2, 1.4.

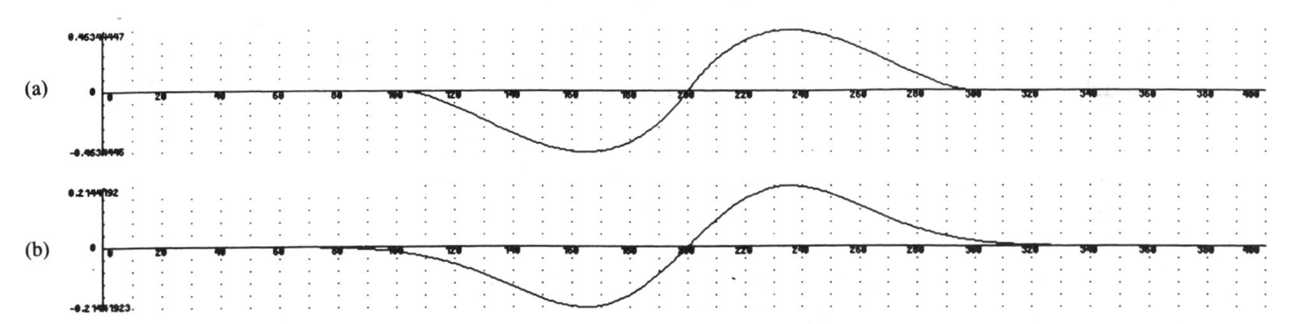

Fig. 6. (a) The optimal step edge operator. (b) The first derivative of a Gaussian.

$$\Sigma\Lambda = \sqrt{\frac{8}{3\pi}} \approx 0.92 \qquad (44)$$

while the r value is, from (41),

$$r = \sqrt{\frac{4}{15}} \approx 0.51.$$

The performance of the first derivative of Gaussian operator above is worse than the optimal operator by about 20 percent and its multiple response measure r, is worse by about 10 percent. It would probably be difficult to detect a difference of this magnitude by looking at the performance of the two operators on real images, and because the first derivative of Gaussian operator can be computed with much less effort in two dimensions, it has been used exclusively in experiments. The impulse responses of the two operators can be compared in Fig. 6.

A close approximation of the first derivative of Gaussian operator was suggested by Macleod [16] for step edge detection. Macleod's operator is a difference of two displaced two-dimensional Gaussians. It was evaluated in Fram and Deutsch [7] and compared very favorably with several other schemes considered in that paper. There are also strong links with the Laplacian of Gaussian operator suggested by Marr and Hildreth [18]. In fact, a one-dimensional Marr–Hildreth edge detector is almost identical with the operator we have derived because maxima in the output of a first derivative operator will correspond to zero-crossings in the Laplacian operator as used by Marr and Hildreth. In two dimensions however, the directional properties of our detector enhance its detection and localization performance compared to the Laplacian. Another important difference is that the amplitude of the response at a maximum provides a good estimate of edge strength, because the SNR criterion is the ratio of this response to the noise response. The Marr–Hildreth operator does not use any form of thresholding, but an adaptive thresholding scheme can be used to advantage with our first derivative operator. In the next section we describe such a scheme, which includes noise estimation and a novel method for thresholding edge points along contours.

We have derived our optimal operator to deal with known image features in Gaussian noise. Edge detection between textured regions is another important problem. This is straightforward if the texture can be modelled as the response of some filter $t(x)$ to Gaussian noise. We can then treat the texture as a noise signal, and the response of the filter $f(x)$ to the texture is the same as the response of the filter $(f * t)(x)$ to Gaussian noise. Making this replacement in each integral in the performance criteria that computes a noise response gives us the texture edge design problem. The generalization to other types of texture is not as easy, and for good discrimination between known texture types, a better approach would involve a Markov image model as in [5].

VI. Noise Estimation and Thresholding

To estimate noise from an operator output, we need to be able to separate its response to noise from the response due to step edges. Since the performance of the system will be critically dependent on the accuracy of this estimate, it should also be formulated as an optimization. Wiener filtering is a method for optimally estimating one component of a two-component signal, and can be used to advantage in this application. It requires knowledge of the autocorrelation functions of the two components and of the combined signal. Once the noise component has been optimally separated, we form a global histogram of noise amplitude, and estimate the noise strength from some fixed percentile of the noise signal.

Let $g_1(x)$ be the signal we are trying to detect (in this case the noise output), and $g_2(x)$ be some disturbance (paradoxically this will be the edge response of our filter), then denote the autocorrelation function of g_1 as $R_{11}(\tau)$ and that of g_2 as $R_{22}(\tau)$, and their cross-correlation as $R_{12}(\tau)$, where the correlation of two real functions is defined as follows:

$$R_{ij}(\tau) = \int_{-\infty}^{+\infty} g_i(x)\, g_j(x+\tau)\, dx.$$

We assume in this case that the signal and disturbance are uncorrelated, so $R_{12}(\tau) = 0$. The optimal filter is $K(x)$, which is implicitly defined as follows [30]:

$$R_{11}(\tau) = \int_{-\infty}^{+\infty} (R_{11}(\tau - x) + R_{22}(\tau - x))\, K(x)\, dx.$$

Since the autocorrelation of the output of a filter in response to white noise is equal to the autocorrelation of its impulse response, we have

$$R_{11}(x) = k_3\left(\frac{x^2}{2\sigma^2} - 1\right) \exp\left(-\frac{x^2}{4\sigma^2}\right)$$

If g_2 is the response of the operator derived in (42) to a step edge then we will have $g_2(x) = k \exp(-x/2\sigma^2)$ and

$$R_{22}(x) = k_2 \exp\left(-\frac{x^2}{4\sigma^2}\right).$$

In the case where the amplitude of the edge is large compared to the noise, $R_{22} + R_{11}$ is approximately a Gaussian and R_{11} is the second derivative of a Gaussian of the same σ. Then the optimal form of K is the second derivative of an impulse function.

The filter K above is convolved with the output of the edge detection operator and the result is squared. The next step is the estimation of the mean-squared noise from the local values. Here there are several possibilities. The simplest is to average the squared values over some neighborhood, either using a moving average filter or by taking an average over the entire image. Unfortunately, experience has shown that the filter K is very sensitive to step edges, and that as a consequence the noise estimate from any form of averaging is heavily colored by the density and strength of edges.

In order to gain better separation between signal and noise we can make use of the fact that the amplitude distribution of the filter response tends to be different for edges and noise. By our model, the noise response should have a Gaussian distribution, while the step edge response will be composed of large values occurring very infrequently. If we take a histogram of the filter values, we should find that the positions of the low percentiles (say less than 80 percent) will be determined mainly the noise energy, and that they are therefore useful estimators for noise. A global histogram estimate is actually used in the current implementation of the algorithm.

Even with noise estimation, the edge detector will be susceptible to streaking if it uses only a single threshold. Streaking is the breaking up of an edge contour caused by the operator output fluctuating above and below the

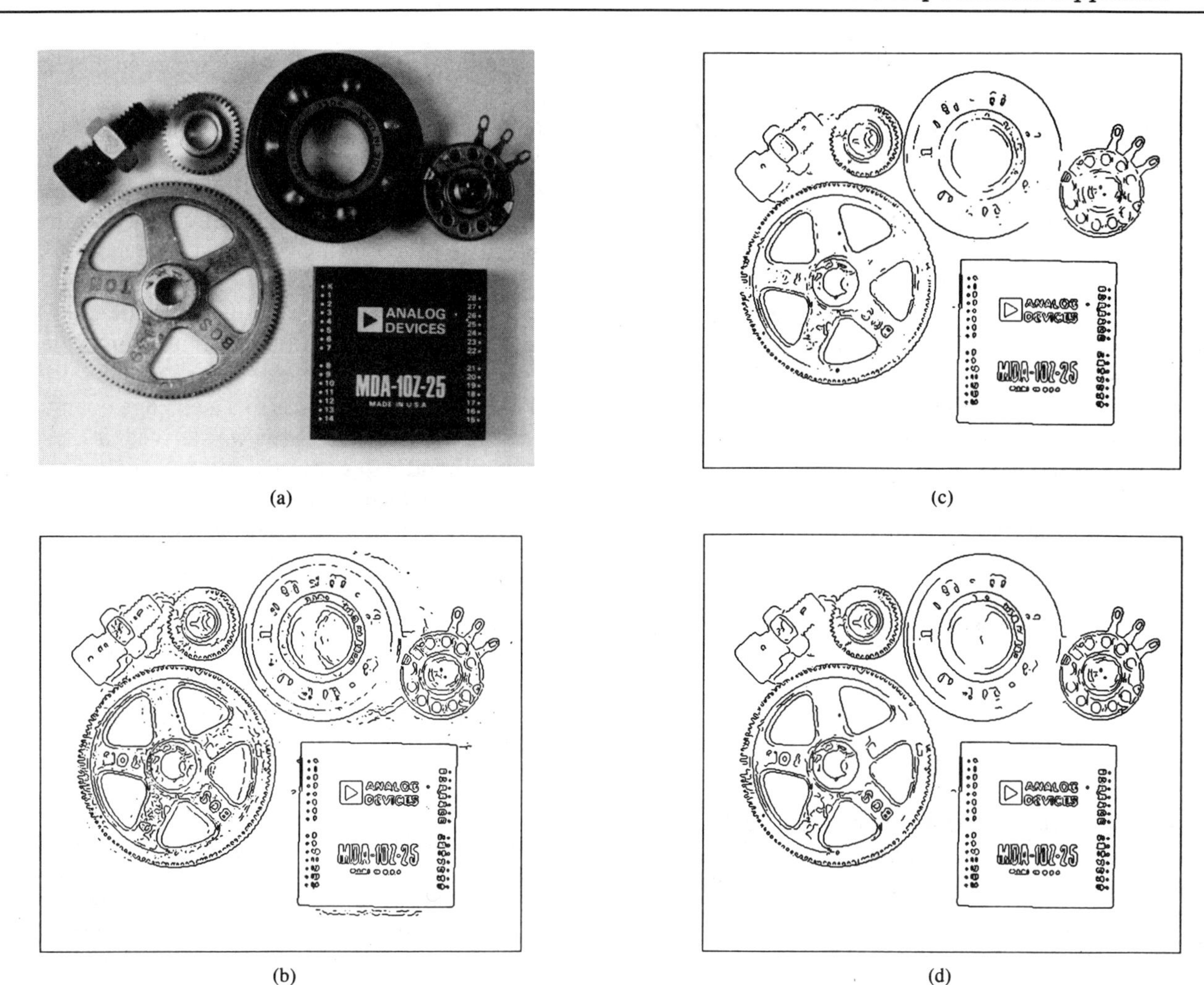

Fig. 7. (a) Parts image, 576 by 454 pixels. (b) Image thesholded at T_1. (c) Image thresholded at 2 T_1. (d) Image thresholded with hysteresis using both the thresholds in (a) and (b).

threshold along the length of the contour. Suppose we have a single threshold set at T_1, and that there is an edge in the image such that the response of the operator has mean value T_1. There will be some fluctuation of the output amplitude due to noise, even if the noise is very slight. We expect the contour to be above threshold only about half the time. This leads to a broken edge contour. While this is a pathological case, streaking is a very common problem with edge detectors that employ thresholding. It is very difficult to set a threshold so that there is small probability of marking noise edges while retaining high sensitivity. An example of the effect of streaking is given in Fig. 7.

One possible solution to this problem, used by Pentland [22] with Marr–Hildreth zero-crossings, is to average the edge strength of a contour over part of its length. If the average is above the threshold, the entire segment is marked. If the average is below threshold, no part of the contour appears in the output. The contour is segmented by breaking it at maxima in curvature. This segmentation is necessary in the case of zero-crossings since the zero-crossings always form closed contours, which obviously do not always correspond to contours in the image.

In the current algorithm, no attempt is made to presegment contours. Instead the thresholding is done with hysteresis. If any part of a contour is above a high threshold, those points are immediately output, as is the entire connected segment of contour which contains the points and which lies above a low threshold. The probability of streaking is greatly reduced because for a contour to be broken it must now fluctuate above the high threshold and below the low threshold. Also the probability of isolated false edge points is reduced because the strength of such points must be above a higher threshold. The ratio of the high to low threshold in the implementation is in the range two or three to one.

VII. Two or More Dimensions

In one dimension we can characterize the position of a step edge in space with one position coordinate. In two dimensions an edge also has an orientation. In this section we will use the term "edge direction" to mean the direction of the tangent to the contour that the edge defines in two dimensions. Suppose we wish to detect edges of a particular orientation. We create a two-dimensional mask for this orientation by convolving a linear edge detection function aligned normal to the edge direction with a projection function parallel to the edge direction. A substantial savings in computational effort is possible if the projection function is a Gaussian with the same σ as the (first derivative of the) Gaussian used as the detection function. It is possible to create such masks by convolving the image with a symmetric two-dimensional Gaussian and then differentiating normal to the edge direction. In fact we do not have to do this in every direction because the slope of a smooth surface in any direction can be determined exactly from its slope in two directions. This form of directional operator, while simple and inexpensive to compute, forms the heart of the more elaborate detector which will be described in the next few sections.

Suppose we wish to convolve the image with an operator G_n which is the first derivative of a two-dimensional Gaussian G in some direction $\boldsymbol{n}$, i.e.,

$$G = \exp\left(-\frac{x^2 + y^2}{2\sigma^2}\right)$$

and

$$G_n = \frac{\partial G}{\partial \boldsymbol{n}} = \boldsymbol{n} \cdot \nabla G. \tag{45}$$

Ideally, $\boldsymbol{n}$ should be oriented normal to the direction of an edge to be detected, and although this direction is not known *a priori*, we can form a good estimate of it from the smoothed gradient direction

$$\boldsymbol{n} = \frac{\nabla(G * I)}{|\nabla(G * I)|} \tag{46}$$

where $*$ denotes convolution. This turns out to be a very good estimator for edge normal direction for steps, since a smoothed step has strong gradient normal to the edge. It is exact for straight line edges in the absence of noise, and the Gaussian smoothing keeps it relatively insensitive to noise.

An edge point is defined to be a local maximum (in the direction $\boldsymbol{n}$) of the operator G_n applied to the image I. At a local maximum, we have

$$\frac{\partial}{\partial \boldsymbol{n}} G_n * I = 0$$

and substituting for G_n from (45) and associating Gaussian convolution, the above becomes

$$\frac{\partial^2}{\partial \boldsymbol{n}^2} G * I = 0. \tag{47}$$

At such an edge point, the edge strength will be the magnitude of

$$|G_n * I| = |\nabla(G * I)|. \tag{48}$$

Because of the associativity of convolution, we can first convolve with a symmetric Gaussian G and then compute directional second derivative zeros to locate edges (47), and use the magnitude of (48) to estimate edge strength. This is equivalent to detecting and locating the edge using the directional operator G_n, but we need not know the direction $\boldsymbol{n}$ before convolution.

The form of nonlinear second derivative operator in (47) has also been used by Torre and Poggio [28] and by Haralick [10]. It also appears in Prewitt [23] in the context of edge enhancement. A rather different two-dimensional extension is proposed by Spacek [26] who uses one-dimensional filters aligned normal to the edge direction but without extending them along the edge direction. Spacek starts with a one-dimensional formulation which maximizes the product of the three performance criteria defined in Section II, and leads to a step edge operator which differs slightly from the one we derived in Section IV. Gennert [8] addresses the two-dimensional edge detector problem directly, and applies a set of directional first derivative operators at each point in the image. The operators have limited extent along the edge direction and produce good results at sharp changes in edge orientation and corners.

The operator (47) actually locates either maxima or minima by locating the zero-crossings in the second derivative in the edge direction. In principle it could be used to implement an edge detector in an arbitrary number of dimensions, by first convolving the image with a symmetric n-dimensional Gaussian. The convolution with an n-dimensional Gaussian is highly efficient because the Gaussian is separable into n one-dimensional filters.

But there are other more pressing reasons for using a smooth projection function such as a Gaussian. When we apply a linear operator to a two-dimensional image, we form at every point in the output a weighted sum of some of the input values. For the edge detector described here, this sum will be a difference between local averages on different sides of the edge. This output, before nonmaximum suppression, represents a kind of moving average of the image. Ideally we would like to use an infinite projection function, but real edges are of limited extent. It is therefore necessary to window the projection function [9]. If the window function is abruptly truncated, e.g., if it is rectangular, the filtered image will not be smooth because of the very high bandwidth of this window. This effect is related to the Gibbs phenomenon in Fourier theory which occurs when a signal is transformed over a finite window. When nonmaximum suppression is applied to this rough signal we find that edge contours tend to "wander" or that in severe cases they are not even continuous.

The solution is to use a smooth window function. In statistics, the Hamming and Hanning windows are typically used for moving averages. The Gaussian is a reasonable approximation to both of these, and it certainly has very low bandwidth for a given spatial width. (The Gaussian is the unique function with minimal product of bandwidth and frequency.) The effect of the window function becomes very marked for large operator sizes and it is probably the biggest single reason why operators with large support were not practical until the work of Marr and Hildreth on the Laplacian of Gaussian.

It is worthwhile here to compare the performance of

this kind of directional second derivative operator with the Laplacian. First we note that the two-dimensional Laplacian can be decomposed into components of second derivative in two arbitrary orthogonal directions. If we choose to take one of the derivatives in the direction of principal gradient, we find that the operator output will contain one contribution that is essentially the same as the operator described above, and also a contribution that is aligned along the edge direction. This second component contributes nothing to localization or detection (the surface is roughly constant in this direction), but increases the output noise.

In later sections we will describe an edge detector which incorporates operators of varying orientation and aspect ratio, but these are a superset of the operators used in the simple detector described above. In typical images, most of the edges are marked by the operators of the smallest width, and most of these by nonelongated operators. The simple detector performs well enough in these cases, and as detector complexity increases, performance gains tend to diminish. However, as we shall see in the following sections, there are cases when larger or more directional operators should be used, and that they do improve performance when they are applicable. The key to making such a complicated detector produce a coherent output is to design effective decision procedures for choosing between operator outputs at each point in the image.

VIII. The Need for Multiple Widths

Having determined the optimal shape for the operator, we now face the problem of choosing the width of the operator so as to give the best detection/localization tradeoff in a particular application. In general the signal-to-noise ratio will be different for each edge within an image, and so it will be necessary to incorporate several widths of operator in the scheme. The decision as to which operator to use must be made dynamically by the algorithm and this requires a local estimate of the noise energy in the region surrounding the candidate edge. Once the noise energy is known, the signal-to-noise ratios of each of the operators will be known. If we then use a model of the probability distribution of the noise, we can effectively calculate the probability of a candidate edge being a false edge (for a given edge, this probability will be different for different operator widths).

If we assume that the *a priori* penalty associated with a falsely detected edge is independent of the edge strength, it is appropriate to threshold the detector outputs on probability of error rather than on magnitude of response. Once the probability threshold is set, the minimum acceptable signal-to-noise ratio is determined. However, there may be several operators with signal-to-noise ratios above the threshold, and in this case the smallest operator should be chosen, since it gives the best localization. We can afford to be conservative in the setting of the threshold since edges missed by the smallest operators may be picked up by the larger ones. Effectively the global tradeoff between error rate and localization remains, since choosing a high signal-to-noise ratio threshold leads to a lower error rate, but will tend to give poorer localization since fewer edges will be recorded from the smaller operators.

In summary then, the first heuristic for choosing between operator outputs is that *small operator widths should be used whenever they have sufficient* Σ. This is similar to the selection criterion proposed by Marr and Hildreth [18] for choosing between different Laplacian of Gaussian channels. In their case the argument was based on the observation that the smaller channels have higher resolution, i.e., there is less possibility of interference from neighboring edges. That argument is also very relevant in the present context, as to date there has been no consideration of the possibility of more than one edge in a given operator support. Interestingly, Rosenfeld and Thurston [25] proposed exactly the opposite criterion in the choice of operator for edge detection in texture. The argument given was that the larger operators give better averaging and therefore (presumably) better signal-to-noise ratios.

Taking the fine-to-coarse heuristic as a starting point, we need to form a local decision procedure that will enable us to decide whether to mark one or more edges when several operators in a neighborhood are responding. If the operator with the smallest width responds to an edge and if it has a signal-to-noise ratio above the threshold, we should immediately mark an edge at that point. We now face the problem that there will almost certainly be edges marked by the larger operators, but that these edges will probably not be exactly coincident with the first edge. A possible answer to this would be to suppress the outputs of all nearby operators. This has the undesirable effect of preventing the large channels for responding to "fuzzy" edges that are superimposed on the sharp edge.

Instead we use a "feature synthesis" approach. We begin by marking all the edges from the smallest operators. From these edges, we synthesize the large operator outputs that would have been produced if these were the only edges in the image. We then compare the actual operator outputs to the synthetic outputs. We mark additional edges only if the large operator has significantly greater response than what we would predict from the synthetic output. The simplest way to produce the synthetic outputs is to take the edges marked by a small operator in a particular direction, and convolve with a Gaussian normal to the edge direction for this operator. The σ of this Gaussian should be the same as the σ of the large channel detection filter.

This procedure can be applied repeatedly to first mark the edges from the second smallest scale that were not marked by at the first, and then to find the edges from the third scale that were not marked by either of the first two, etc. Thus we build up a cumulative edge map by adding those edges at each scale that were not marked by smaller scales. It turns out that in many cases the majority of edges are picked up by the smallest channel, and the later channels mark mostly shadow and shading edges, or edges between textured regions.

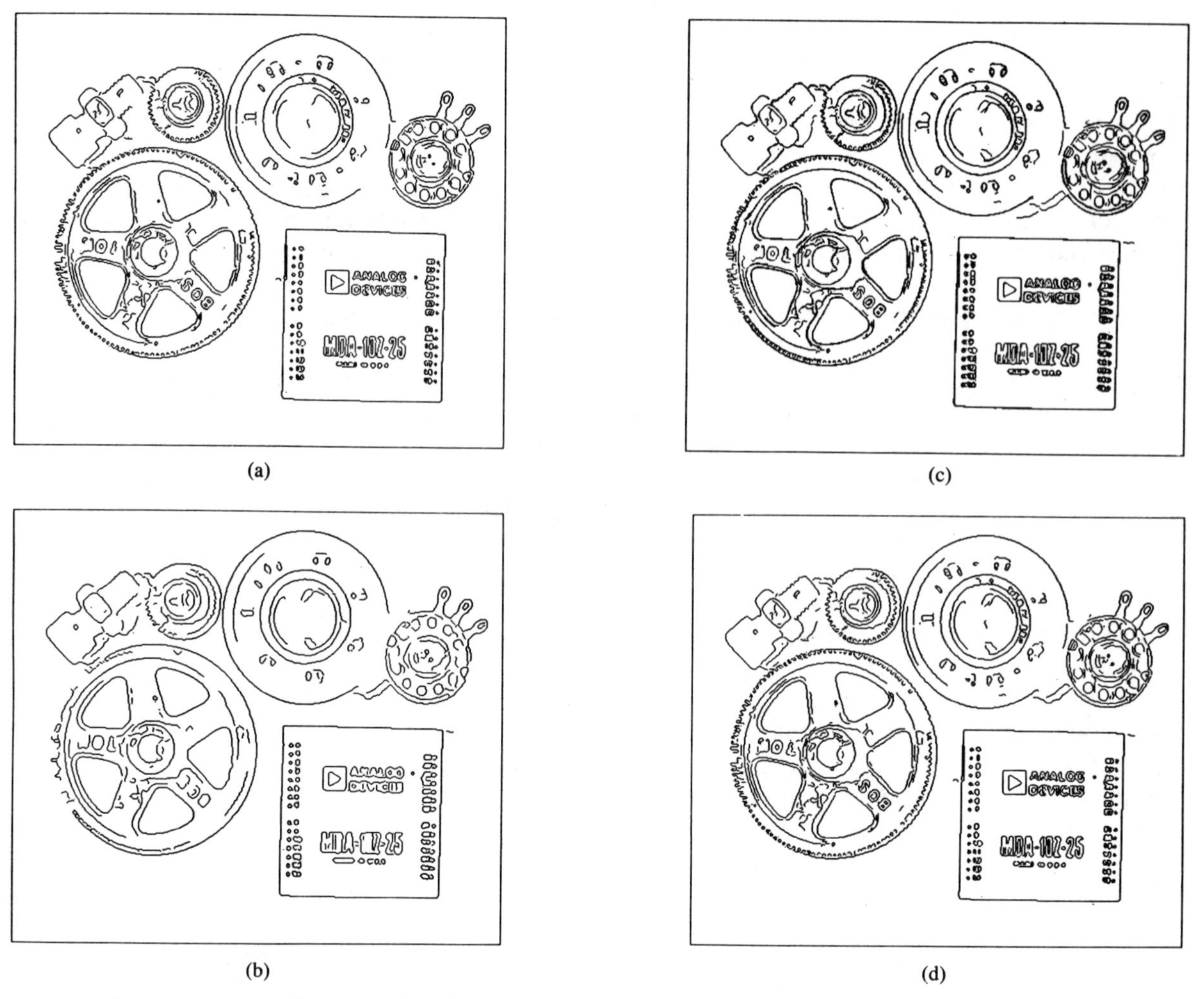

Fig. 8. (a) Edges from parts image at $\sigma = 1.0$. (b) Edges at $\sigma = 2.0$. (c) Superposition of the edges. (d) Edges combined using feature synthesis.

Some examples of feature synthesis applied to some sample images are shown in Figs. 8 and 9. Notice that most of the edges in Fig. 8 are marked by the smaller scale operator, and only a few additional edges, mostly shadows, are picked up by the coarser scale. However when the two sets of edges are superimposed, we notice that in many cases the responses of the two operators to the same edge are not spatially coincident. When feature synthesis is applied we find that redundant responses of the larger operator are eliminated leading to a sharp edge map.

By contrast, in Fig. 9 the edges marked by the two operators are essentially independent, and direct superposition of the edges gives a useful edge map. When we apply feature synthesis to these sets of edges we find that most of the edges at the coarser scale remain. Both Figs. 8 and 9 were produced by the edge detector with exactly the same set of parameters (other than operator size), and they were chosen to represent opposing extremes of image content across scale.

IX. The Need for Directional Operators

So far we have assumed that the projection function is a Gaussian with the same σ as the Gaussian used for the detection function. In fact both the detection and localization of the operator improve as the length of the projection function increases. We now prove this for the operator signal-to-noise ratio. The proof for localization is similar. We will consider a step edge in the x direction which passes through the origin. This edge can be represented by the equation

$$I(x, y) = Au_{-1}(y)$$

where u_{-1} is the unit step function, and A is the amplitude of the edge as before. Suppose that there is additive Gaussian noise of mean squared value n_{00}^2 per unit area. If we convolve this signal with a filter whose impluse response is $f(x, y)$, then the response to the edge (at the origin) is

$$\int_{-\infty}^{0} \int_{-\infty}^{+\infty} f(x, y)\, dx\, dy.$$

The root mean squared response to the noise only is

$$n_{00} \left(\int_{-\infty}^{+\infty} \int_{-\infty}^{+\infty} f^2(x, y)\, dx\, dy \right)^{1/2}$$

The signal-to-noise ratio is the quotient of these two

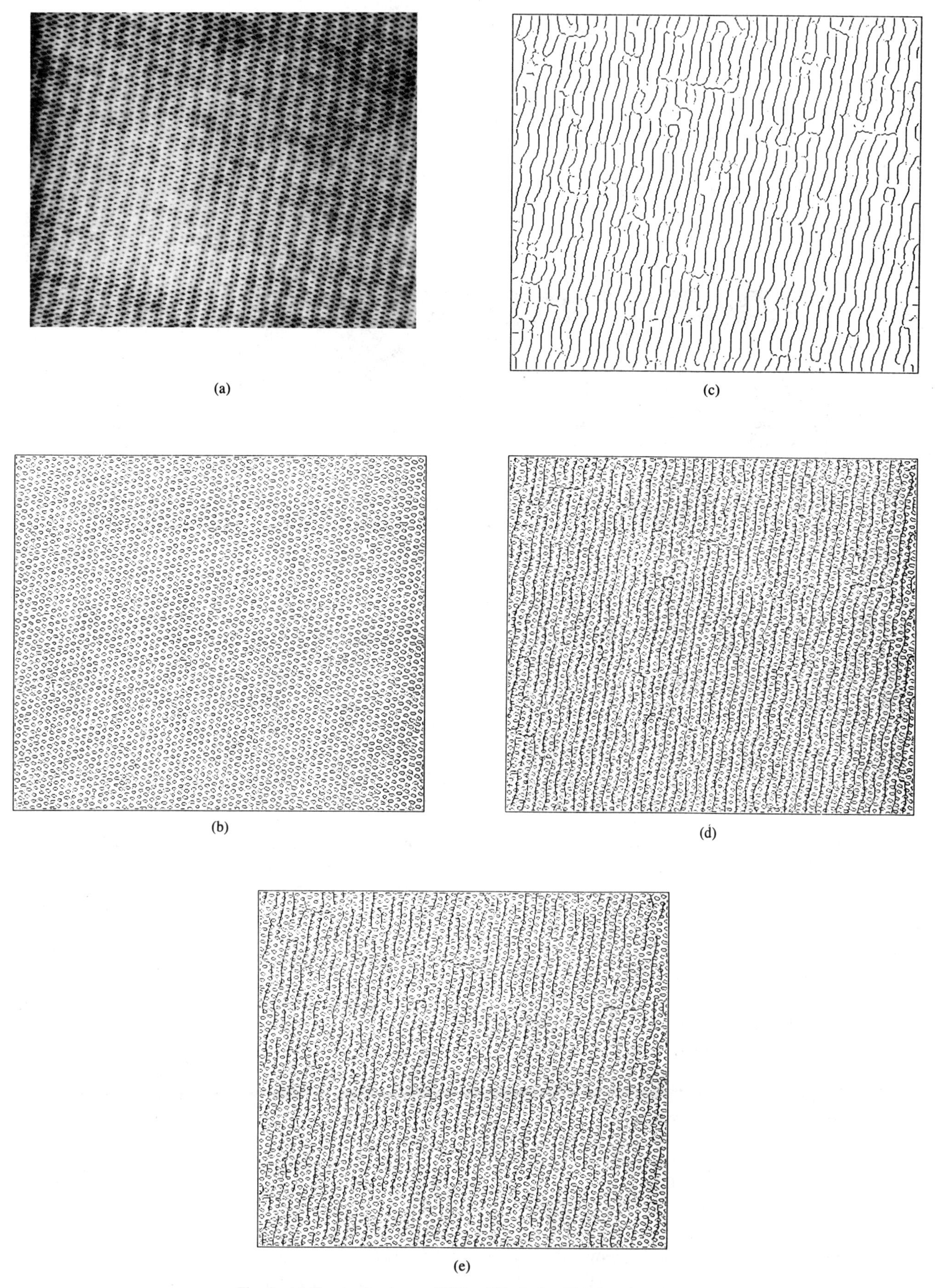

Fig. 9. (a) Handywipe image 576 by 454 pixels. (b) Edges from handywipe image at $\sigma = 1.0$. (c) $\sigma = 5.0$. (d) Superposition of the edges. (e) Edges combined using feature synthesis.

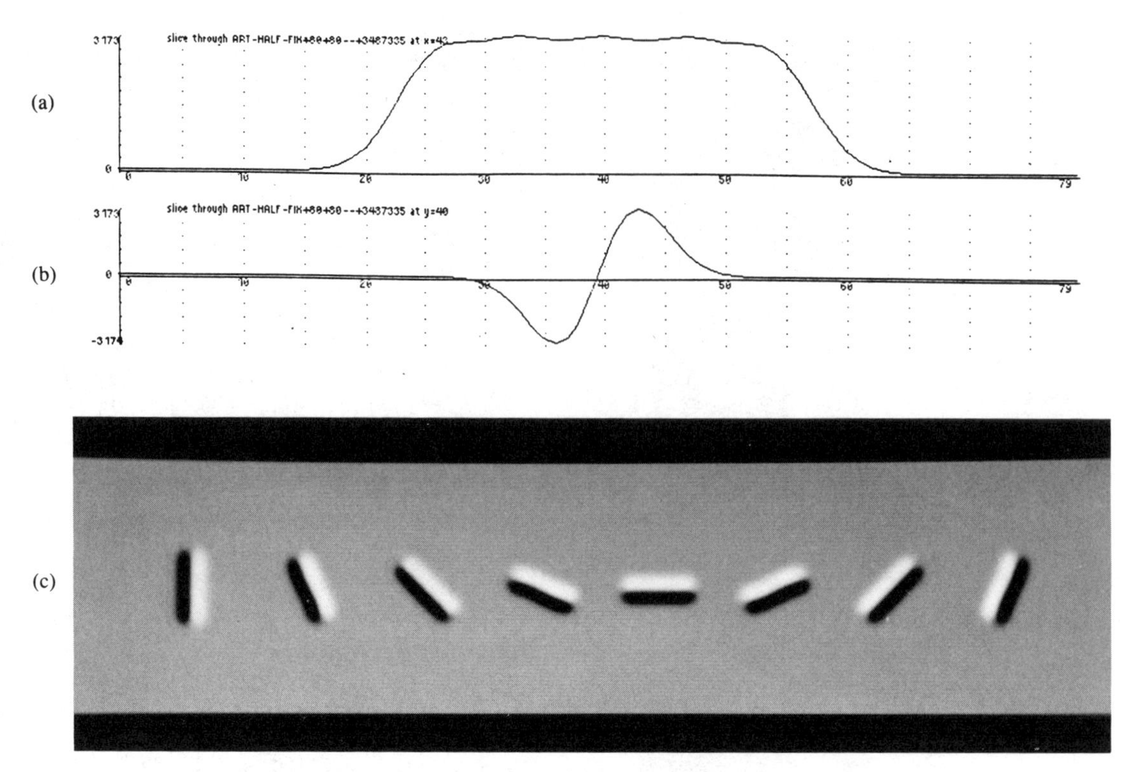

Fig. 10. Directional step edge mask. (a) Cross section parallel to the edge direction. (b) Cross section normal to edge direction. (c) Two-dimensional impulse responses of several masks.

integrals, and will be denoted by Σ. We have already seen what happens if we scale the function normal to the edge (21). We now do the same to the projection function by replacing $f(x, y)$ by $f_l(x, y) = f(x, (y/l))$. The integrals become

$$\int_{-\infty}^{0}\int_{-\infty}^{+\infty} f\left(x, \frac{y}{l}\right) dx\, dy = \int_{-\infty}^{0}\int_{-\infty}^{+\infty} f(x, y_1) l\, dx\, dy_1$$

$$n_{00}\left(\int_{-\infty}^{+\infty}\int_{-\infty}^{+\infty} f^2\left(x, \frac{y}{l}\right) dx\, dy\right)^{1/2} = n_{00}\left(\int_{-\infty}^{+\infty}\int_{-\infty}^{+\infty} f^2(x, y_1) l\, dx\, dy_1\right)^{1/2} \quad (49)$$

And the ratio of the two is now $\sqrt{l}\Sigma$. The localization Λ also improves as $\sqrt{l}$. It is clearly desirable that we use as large a projection function as possible. There are practical limitations on this however, in particular edges in an image are of limited extent, and few are perfectly linear. However, most edges continue for some distance, in fact much further than the 3 or 4 pixel supports of most edge operators. Even curved edges can be approximated by linear segments at a small enough scale. Considering the advantages, it is obviously preferable to use directional operators whenever they are applicable. The only proviso is that the detection scheme must ensure that they are used only when the image fits a linear edge model.

The present algorithm tests for applicability of each directional mask by forming a goodness-of-fit estimate. It does this at the same time as the mask itself is computed. An efficient way of forming long directional masks is to sample the output of nonelongated masks with the same direction. This output is sampled at regular intervals in a line parallel to the edge direction. If the samples are close together (less than 2σ apart), the resulting mask is essentially flat over most of its range in the edge direction and falls smoothly off to zero at its ends. Two cross sections of such a mask are shown in Fig. 10. In this diagram (as in the present implementation) there are five samples over the operator support.

Simultaneously with the computation of the mask, it is possible to establish goodness of fit by a simple squared-error measure. The mask is computed by summing some number of circular mask outputs (say 5) in a line. If the mask lies over a step edge in its preferred direction, these 5 values will be roughly the same. If the edge is curved or not aligned with the mask direction, the values will vary. We use the variance of these values as an estimate of the goodness of fit of the actual edge to an ideal step model. We then suppress the output of a directional mask if its variance is greater than some fraction of the squared output. Where no directional operator has sufficient goodness of fit at a point, the algorithm will use the output of the nonelongated operator described in Section VII. This simple goodness-of-fit measure is sufficient to eliminate the problems that traditionally plague directional operators, such as false responses to highly curved edges and extension of edges beyond corners; see Hildreth [12].

This particular form of projection function, that is a

function with constant value over some range which decays to zero at each end with two roughly half-Gaussians, is very similar to a commonly used extension of the Hanning window. This latter function is flat for some distance and decays to zero at each end with two half-cosine bells [2]. We can therefore expect our function to have good properties as a moving average estimator, which as we saw in Section VII, is an important role fulfilled by the projection function.

All that remains to be done in the design of directional operators is the specification of the number of directions, or equivalently the angle between two adjacent directions. To determine the latter, we need to determine the angular selectivity of a directional operator as a function of the angle θ between the edge direction and the preferred direction of the operator. Assume that we form the operator by taking an odd number $2N + 1$ of samples. Let the number of a sample be n where n is in the range $-N \cdots +N$. Recall that the directional operator is formed by convolving with a symmetric Gaussian, differentiating normal to the preferred edge direction of the operator, and then sampling along the preferred direction. The differentiated surface will be a ridge which makes an angle θ to the preferred edge direction. Its height will vary as $\cos\theta$, and the distance of the nth sample from the center of the ridge will be $nd \sin\theta$ where d is the distance between samples. The normalized output will be

$$O_n(\theta) = \frac{\cos\theta}{2N+1}\left[\sum_{n=-N}^{N} \exp\left(-\frac{(nd\sin\theta)^2}{2\sigma^2}\right)\right]. \quad (50)$$

If there are m operator directions, then the angle between the preferred directions of two adjacent operators will be $180/m$. The worst case angle between the edge and the nearest preferred operator direction is therefore $90/m$. In the current implementation the value of d/σ is about 1.4 and there are 6 operator directions. The worst case for θ is 15 degrees, and for this case the operator output will fall to about 85 percent of its maximum value. Directional operators very much like the ones we have derived were suggested by Marr [17], but were discarded in favor of the Laplacian of Gaussian [18]. In part this was because the computation of several directional operators at each point in the image was thought to require an excessive amount of computation. In fact the sampling scheme described above requires only five multiplications per operator. An example of edge detection using five-point directional operators is given in Fig. 11.

(a)

(b)

(c)

Fig. 11. (a) Dalek image 576 by 454 pixels. (b) Edges found using circular operator. (c) Directional edges (6 mask orientations).

X. Conclusions

We have described a procedure for the design of edge detectors for arbitrary edge profiles. The design was based on the specification of detection and localization criteria in a mathematical form. It was necessary to augment the original two criteria with a multiple response measure in order to fully capture the intuition of good detection. A mathematical form for the criteria was presented, and numerical optimization was used to find optimal operators for roof and ridge edges. The analysis was then restricted to consideration of optimal operators for step edges. The result was a class of operators related by spatial scaling. There was a direct tradeoff in detection performance versus localization, and this was determined by the spatial width. The impulse response of the optimal step edge operator was shown to approximate the first derivative of a Gaussian.

A detector was proposed which used adaptive thresholding with hysteresis to eliminate streaking of edge contours. The thresholds were set according to the amount of noise in the image, as determined by a noise estimation scheme. This detector made use of several operator widths to cope with varying image signal-to-noise ratios, and operator outputs were combined using a method called feature synthesis, where the responses of the smaller operators were used to predict the large operator responses. If the actual large operator outputs differ significantly from the predicted values, new edge points are marked. It is therefore possible to describe edges that occur at different scales, even if they are spatially coincident.

In two dimensions it was shown that marking edge points at maxima of gradient magnitude in the gradient direction is equivalent to finding zero-crossings of a certain nonlinear differential operator. It was shown that when edge contours are locally straight, highly directional operators will give better results than operators with a circular support. A method was proposed for the efficient generation of highly directional masks at several orientations, and their integration into a single description.

Among the possible extensions of the work, the most interesting unsolved problem is the integration of different edge detector outputs into a single description. A scheme which combined the edge and ridge detector outputs using feature synthesis was implemented, but the results were inconclusive. The problem is much more complicated here than for edge operators at different scales because there is no clear reason to prefer one edge type over another. Each edge set must be synthesized from the other, without a bias caused by overestimation in one direction.

The criteria we have presented can be used with slight modification for the design of other kinds of operator. For example, we may wish to design detectors for nonlinear two-dimensional features (such as corners). In this case the detection criterion would be a two-dimensional integral similar to (3), while a plausible localization criterion would need to take into account the variation of the edge position in both the x and y directions, and would not directly generalize from (9). There is a natural generalization to the detection of higher-dimensional edges, such as occur at material boundaries in tomographic scans. As was pointed out in Section VII, (47) can be used to find edges in images of arbitrary dimension, and the algorithm remains efficient in higher dimensions because n-dimensional Gaussian convolution can be broken down into n linear convolutions.

Acknowledgment

The author would like to thank Dr. J. M. Brady for his influence on the course of this work and for comments on early drafts of this paper. Thanks to the referees for their suggestions which have greatly improved the presentation of the paper. In particular thanks to the referee who suggested the simple derivation based on the Schwarz inequality that appears on p. 682.

References

[1] R. J. Beattie, "Edge detection for semantically based early visual processing," Ph.D. dissertation, Univ. Edinburgh, 1984.
[2] C. Bingham, M. D. Godfrey, and J. W. Tukey, "Modern techniques of power spectrum estimation," *IEEE Trans. Audio Electroacoust.*, vol. AU-15, no. 2, pp. 56–66, 1967.
[3] R. A. Brooks, "Symbolic reasoning among 3-D models and 2-D images," Dep. Comput. Sci., Stanford Univ., Stanford, CA, Rep. AIM-343, 1981.
[4] J. F. Canny, "Finding edges and lines in images," M.I.T. Artificial Intell. Lab., Cambridge, MA, Rep. AI-TR-720, 1983.
[5] F. S. Cohen, D. B. Cooper, J. F. Silverman, and E. B. Hinkle, "Simple parallel hierarchical and relaxation algorithms for segmenting textured images based on noncasual Markovian random field models," in *Proc. 7th Int. Conf. Pattern Recognition and Image Processing*, Canada, 1984.
[6] R. Courant and D. Hilbert, *Methods of Mathematical Physics*, vol. 1. New York: Wiley-Interscience, 1953.
[7] J. R. Fram and E. S. Deutsch, "On the quantitative evaluation of edge detection schemes and their comparison with human performance," *IEEE Trans. Comput.*, vol. C-24, no. 6, pp. 616–628, 1975.
[8] M. Gennert, "Detecting half-edges and vertices in images," in *IEEE Conf. Comput. Vision and Pattern Recognition*, Miami Beach, FL, June 24–26, 1986.
[9] R. W. Hamming, *Digital Filters*. Englewood Cliffs, NJ: Prentice-Hall, 1983.
[10] R. M. Haralick, "Zero-crossings of second directional derivative edge operator," in *SPIE Proc. Robot Vision*, Arlington, VA, 1982.
[11] A. Herskovits and T. O. Binford, "On boundary detection," M.I.T. Artificial Intell. Lab., Cambridge, MA, AI Memo 183, 1970.
[12] E. C. Hildreth, "Implementation of a theory of edge detection," M.I.T. Artificial Intell. Lab., Cambridge, MA, Rep. AI-TR-579, 1980.
[13] —, *The Measurement of Visual Motion*. Cambridge, MA: M.I.T. Press, 1983.
[14] B. K. P. Horn, "The Binford–Horn line-finder," M.I.T. Artificial Intell. Lab., Cambridge, MA, AI Memo 285, 1971.
[15] D. G. Luenberger, *Introduction to Linear and Non-Linear Programming*. Reading, MA: Addison-Wesley, 1973.
[16] I. D. G. Macleod, "On finding structure in pictures," in *Picture Language Machines*, S. Kaneff, Ed. New York: Academic, 1970, p. 231.
[17] D. C. Marr, "Early processing of visual information," *Phil. Trans. Roy. Soc. London*, vol. B 275, pp. 483–524, 1976.
[18] D. C. Marr and E. Hildreth, "Theory of edge detection," *Proc. Roy. Soc. London.*, vol. B 207, pp. 187–217, 1980.
[19] D. C. Marr and T. Poggio, "A theory of human stereo vision," *Proc. Roy. Soc. London.*, vol. B 204, pp. 301–328, 1979.
[20] J. E. W. Mayhew and J. P. Frisby, "Psychophysical and computational studies toward a theory of human stereopsis," *Artificial Intell. (Special Issue on Computer Vision)*, vol. 17, 1981.
[21] T. Poggio, H. Voorhees, and A. Yuille, "A regularized solution to edge detection," M.I.T. Artificial Intell. Lab., Cambridge, MA, Rep. AIM-833, 1985.
[22] A. P. Pentland, "Visual inference of shape: Computation from local features," Ph.D. dissertation, Dep. Psychol., Massachusetts Inst. Technol., Cambridge, MA, 1982.
[23] J. M. S. Prewitt, "Object enhancement and extraction," in *Picture Processing and Psychopictorics*, B. Lipkin and A. Rosenfeld, Eds. New York: Academic, 1970, pp. 75–149.
[24] S. O. Rice, "Mathematical analysis of random Noise," *Bell Syst. Tech. J.*, vol. 24, pp. 46–156, 1945.
[25] A. Rosenfeld and M. Thurston, "Edge and curve detection for visual scene analysis," *IEEE Trans. Comput.*, vol. C-20, no. 5, pp. 562–569, 1971.
[26] L. Spacek, "The computation of visual motion," Ph.D. dissertation, Univ. Essex at Colchester, 1984.
[27] K. A. Stevens, "Surface perception from local analysis of texture and contour," M.I.T. Artificial Intell. Lab., Cambridge, MA, Rep. AI-TR-512, 1980.

[28] V. Torre and T. Poggio, "On edge detection," M.I.T. Artificial Intell. Lab., Cambridge, MA, Rep. AIM-768, 1984.
[29] S. Ullman, *The Interpretation of Visual Motion.* Cambridge, MA: M.I.T. Press, 1979.
[30] N. Wiener, *Extrapolation, Interpolation and Smoothing of Stationary Time Series.* Cambridge, MA: M.I.T. Press, 1949.
[31] A. P. Witkin, "Shape from contour," M.I.T. Artificial Intell. Lab., Cambridge, MA, Rep. AI-TR-589, 1980.

John Canny (S'81–M'82) was born in Adelaide, Australia, in 1958. He received the B.Sc. degree in computer science and the B.E. degree from Adelaide University in 1980 and 1981, respectively, and the S.M. degree from the Massachusetts Institute of Technology, Cambridge, in 1983.

He is with the Artificial Intelligence Laboratory, M.I.T. His research interests include low-level vision, model-based vision, motion planning for robots, and computer algebra.

Mr. Canny is a student member of the Association for Computing Machinery.

LINEAR DELINEATION

Martin A. Fischler and Helen C. Wolf
Artificial Intelligence Center

SRI International
333 Ravenswood Avenue
Menlo Park, California 94025

ABSTRACT

In this paper we address a basic problem in machine perception: the tracing of perceptually obvious "line-like" structure appearing in an image; we present some new ways of looking at this problem and provide techniques that are significantly more general and effective than previously reported methods for this task.

Our approach is a departure from the procedures usually employed in the following important respects: (1) We recognize the typically overlooked distinction between lines and edges; finding lines offers significant simplifications not available when searching for edges. (2) The perceptual primitive we extract from the image to indicate the local presence of linear structure is a function of total intensity variation over an extended portion of the image rather than intensity gradient information which can be extracted with a mask-like "local image operator." While easy to compute, this new primitive appears to be more effective and stable than the almost universally employed gradient based primitives. (3) The overall procedure can be applied to an arbitrary image with essentially no human intervention (parameter tuning or attention focusing) and produces excellent results. Examples of the delineation procedure are shown for aerial, industrial, and radiographic imagery.

I INTRODUCTION

For many tasks in scene analysis, there may not exist general solutions independent of purpose or intended application. However, for the task of linear delineation, one can easily find image subsets for which a panel of human observers would be almost unanimous in their interpretation without having to agree on the explicit criteria underlying their decisions; our goal is to produce a computer system that can perform the delineation task at close to human levels for at least these more obvious cases, especially where semantic knowledge is not required. In this paper we present some new ways of looking at the problem of linear delineation and provide techniques that are significantly more general and effective than previously reported methods for this task.

II PROBLEM DEFINITION

For our purposes here, we define linear delineation as the task of generating a set of lists of points, for a given 2-D image, such that the points in each list fall sequentially along what any reasonable human observer would describe as a clearly visible "line-like" structure in the image. Practical examples of this task might be to delineate the roads, rivers, and rail lines in an aerial photograph, or to trace the paths taken by blood vessels in a radiographic angiogram, or to locate the wiring paths on a printed circuit board; however, our goal in this paper is not to look for specific real-world objects or to assign semantic labels to the detected linear structures, but rather to find what a human observer might choose as the most perceptually obvious occurrences of such structures. We further distinguish between the problems of (1) detecting the edges or contours of extended objects, and (2) delineating those objects whose appearance is adequately represented by a central skeleton -- only the second problem is addressed.

III LINES AND EDGES

While most approaches to linear delineation do not distinguish between lines and edges, and even use edge detection as a necessary first step in the delineation task, a critical concept advanced in this paper is the distinction between line and edge detection.

Edge detection is intuitively based on the concept of finding a discontinuity (in intensity or some other locally measurable attribute such as color or texture) between two adjacent but distinct regions in an image. However, in a digital representation of an image, a smooth surface can always be fit to the sample values of the integer raster. Thus, edge detection must be based on parameters or thresholds set by assumptions about the nature of the image. Even if edge points are marked only at those locations at which there are first derivative maxima, or zeros of the second derivative, ultimately, an arbitrary decision is made in deciding when the corresponding gradient is large enough to be called a discontinuity.

Intuitively, a "line-like" or linear structure is a (connected) region that is very long relative to its width, and has a ridge or skeleton along which the intensities change slowly and are distinguished from intensities outside the region; the width need not be constant, but any changes in width should occur in a smooth manner. To simplify the discussion, we will assume the linear structures are distinguished by their ridgepoints being brighter than the surrounding background, but any other specified attribute, which is locally detectable, would be an acceptable substitute. It is important to recognize the fact that a clearly visible line in an image may not have locally detectable edges and thus no locally measurable width, or possibly only one detectable edge, or even two edges which are significantly separated and nonparallel as might occur in a local widening of a river. It is also generally the case that linear structures have no visible internal detail that is essential to their delineation.

As a point of interest, it might be noted that the human mechanisms which lead to the perception of subjective edge and line illusions are quite different; subjective edges appear to require a 3-D interpretation, while subjective lines appear to be produced by adaptation phenomena.

IV SMOOTHING, DISTANCE TRANSFORMS, AND THE GRAY SCALE SKELETON

If we can find the edges of a linear structure, we can generate a distance transform and extract a skeleton as the desired delineation (e.g., Rosenfeld [1], Fischler [2]). However, as noted in the preceding section, linear structures do not always have locally detectable edges, and, since all of the generally known techniques for deriving a skeleton require a complete contour, some other approach is required. The classical skeletonizing techniques intimately link the contour/edges of a region and its skeleton, and it is just this linkage that we wish to break.

Surprisingly, something equivalent to a distance transform that works on gray scale images, and on binary images as well, is already available. To achieve our purpose, we need only observe that the intensities in a properly smoothed image can be considered to be the values of an approximate distance transform. What is the best smoothing function for general use? Actually, it doesn't seem to make much difference in many cases. Most digital images have been processed by low-pass optical and electronic systems that have inserted the required minimum level of smoothing. The viewpoint that the smoothed image can be considered to be a distance transform is the essential element. However, if we start with a binary image, or a very noisy image, then additional smoothing is often desirable. Since we are not concerned with blurring edges, and we would like to eliminate or blur any structure or texture internal to the linear regions, we want the smoothing function to have a width of approximately that of the region to be delineated. If the width of the smoothing function is increased further, the thinner linear structures are eventually eliminated. Thus, if we wish to find all possible linear structures without prior knowledge of the content of the image, the processing should be repeated with a set of smoothing filters having a spectrum of widths. Actually, no more than two or three filtering steps should ever be required. For example, to trace all the linear structures (diameters up to 20 pixels) in a noisy radiographic angiogram, a single filter of width 20 was used (see Figure 1d). Smoothing introduced by the acquisition process was sufficient to produce excellent results in tracing the linear structures in aerial imagery (see Figures 1a and 1b).

V RIDGES (OR VALLEYS), OPERATORS, AND NEIGHBORHOODS

Having produced an approximate distance transform via smoothing, we now must deal with the problem of locating the ridgepoints that denote the spines (skeletons) of the linear structures. When an exact distance transform is derived from a complete contour, noise is not a problem and the skeleton has assured geometric properties that make it easy to detect; finding the ridgepoints of an approximate distance transform is considerably more complex.

We traditionally distinguish between locally and globally detectable features: local features are detectable by an intensity pattern which can be observed through a small peep-hole centered on the feature, while global features are ambiguous in a small area. The model or description of the local feature is generally compiled into an intensity patch (matched filter or operator) which can be convolved with the image to detect the corresponding feature. In the case of an exact distance transform, a 3X3 pixel operator is sufficient to detect ridgepoints (a 2X2 operator is sufficient for the Labeled Distance Transform (Fischler [2])); for the approximate distance transform, a small fixed-size operator is ineffective.

The principal utility of a local operator is that the number of data patterns the operator might encounter is small enough to allow one to enumerate a decision for each such pattern. If we further agree to use a small square window of the image as our local domain, and to use either table look-up or convolution as the basis for decision making, then a uniform mechanization can be employed to implement a large number of distinct (and generally unrelated) local operators. The attractiveness of this second implementation aspect has led to the situation that almost all low level (local) scene analysis is done using such peep-hole type operators. The disadvantage of this approach is that the concept of local is relative to the size of the entity of interest, and either one must know this size in advance, or use a whole family of operators of increasing size, where the larger operators lose the advantages that led to their use in the first place. In the case of line detection, where the line width can vary over a wide range of values, the conventional operator concept is inappropriate.

Based on these general issues (even more than on the immediate problem at hand), we have considered other realizations of general "local" decision-making processes that satisfy the previously stated conditions, but do not necessarily lend themselves to a convolution type mechanization. In particular, restricting our attention to finding the maxima and minima of functions of the displacement along a space curve defined over the image, satisfies our requirements for computational and decision-making simplicity even when the curve traverses the entire image. While the space curve might assume any shape (e.g., follow the contour of an object), the analysis itself is independent of the shape; for the linear delineation problem, we used image intensity as a function of displacement along horizontal and vertical scan lines. Since maxima and minima are symmetrical attributes, we will only discuss the problem of labeling maximal points along the curve.

The problem of finding the ridgepoints of an approximate distance transform can be viewed as the problem of finding the ridgepoints (local maxima) of an exact distance transform to which some amount of noise has been added. We are not concerned about the possibility of making isolated (incoherent) incorrect decisions, because we have developed effective linking and pruning methods, described in the next section, that are capable of eliminating such errors. Our main problem is that we can't count on finding either large local gradients or using known line width to determine some minimum significant gradient threshold to identify valid ridgepoints; additionally, noise will introduce many false local maxima. Thus, we must use total intensity change, rather than rate of change, to detect valid ridgepoints, and we must have an effective way of determining such total change even in the presence of local variation introduced by noise. (While it is not immediately obvious that total intensity change, rather than rate of change, will recover the perceptually obvious linear features, our experiments indicate that this is indeed the case.)

Our approach is to evaluate two attributes of each of the detected intensity maxima along the space curves (in this case, horizontal and vertical scan lines), which we call the "local" and "global" maxima values. The local maxima value of a point is the total intensity difference from the point to the highest of its immediate left and right intensity mimima along the curve. The left (right) global maxima value of a point is the total intensity difference from the point to the lowest intensity value found moving to the left (right) prior to encountering a point with an intensity value greater than that of the given point; the global maxima value of the point is the smaller of its left and right global values. In the case of a plateau, only the center point is treated as a maximal point and evaluated as previously described. If a point (or points) on a plateau has an immediate neighbor with a higher intensity value, it is not a maximal point and it is not assigned either a local or global value (actually, for implementation purposes, non-maximal points are assigned zero values); on the other hand, every maximal point will have both a local and global value where the global value equals or exceeds the local value. Figure 2 provides some examples illustrating the operations just described.

What is a large-enough value, of either the local intensity maxima (LIM) or global intensity maxima (GIM), to indicate significance. In our unsmoothed image data, about 1/3 of the points were maximal points, and, in a smoothed image, this percentage is much smaller. Given the linking and pruning techniques we describe in the next section, it might be possible to return all the maximal points in a binary mask and still extract the desired line structure from the background noise contained in such a mask. However, it makes much more sense to first eliminate those maximal points that do not have enough intensity variation to be perceptually distinguishable from a flat background.

Rather than attempting to find some optimal threshold setting (in the sense of maximum noise elimination without any loss of linear structure), which would be difficult or impossible to automatically determine at this level of information organization, we iteratively adjust our threshold settings to satisfy a constraint based on a complexity measure. These program-determined thresholds typically allow at least two to 10 times the number of ridgepoints (maxima) to be retained above that which would result by manually setting the thresholds to achieve visually acceptable results. The final elimination of "non-significant" maxima is achieved later in the processing at a higher level of organization.

VI CLUSTERING, LINKING, PRUNING, NODE ANALYSIS, RANKING, AND FINAL DELINEATION

Based on the availability of a binary overlay depicting the locations of the major linear structures contained in the given gray scale image, obtained as described in the preceding section, we have been able to demonstrate that.

(1) The linking step in the delineation process can effectively be based on the single attribute of geometric proximity, and that a clustering or association step, followed by the construction of a Minimum Spanning Tree (MST) through the points of each cluster (Fischler [4]), will correctly link the ridgepoints along the skeletons of the linear structures.

(2) The desired delineations will be embedded in trees containing additional branches that are either minor linear structures or noise, and that simple pruning techniques can eliminate most of this unwanted detail (see Figure 3; note that tree pruning can effectively achieve simplifications that would be difficult, if not impossible, at lower levels of organization of the information).

(3) Having properly linked the ridgepoints and pruned some of the smaller branches of the

resulting trees, we can extract long coherent paths by a decision procedure applied at each node of each tree. This decision procedure, based on the local branch attributes of intensity, connectivity, and directionality at each node, assigns path connectivity through a node by splitting off incompatible branches; any remaining ambiguities (more than two branches entering a node) are resolved by choosing those pairings that result in the longest paths.

(4) The paths obtained in the tree partitioning step can be rank ordered with respect to perceptual quality by a metric based on the path attributes of total length, contrast, and continuity.

Details of the procedures mentioned above are described in a journal length paper now in preparation [3].

VII EXPERIMENTAL RESULTS

Figure 1 presents the results of applying the delineation algorithm to aerial, industrial, and medical images. Our objective was to be able to take imagery from arbitrary domains, and without any parameter adjustment or attention focusing, produce high quality delineations of the obvious linear structures. The results show that we have accomplished much of what we originally intended. The delineations achieved by the uniform parameter settings are quite good in all the images with the exception of the angiogram (which is extremely noisy, and does not really satisfy the criteria of having "clearly visible" linear structure); and even here, by making an appropriate selection of two parameters, the smoothing diameter and the association distance (required in the clustering step), we obtain very good results. By using the values produced by the ranking step of the delineation algorithm, we believe it should be possible to automatically search the parameter space (of approximately 10 to 100 parameter combinations) to optimize the processing for any given image. We are currently investigating this possibility.

VIII CONCLUDING COMMENTS

We have presented the viewpoint that the problem of delineating the obvious linear structures in an image is distinct from that of finding edges or contours, and is best viewed as the process of finding skeletons in gray scale images (i.e., that line detection does not necessarily depend on gradient information, but rather is approachable from the standpoint of detecting total intensity variation); as a necessary step in this process, we have suggested that an approximate gray scale distance transform can be attained by smoothing the original image. We have described an effective technique for finding ridgepoints (points on the delineating skeleton), and, in the process, raised some important questions about the conventional approach to designing "local operators."

The competence of the ridgepoint based algorithm to abstract the linear structure of an image is apparent by inspection of Figure 1. That ridgepoints should be so effective is, in a sense, a "psychological discovery." It would appear that ridgepoints are important perceptual primitives which may play a significant role in a variety of other tasks (e.g., perception of surface shape). A significant feature of "ridepoints" is that their locations are independent of any monotonic intensity transformation of the image, and their geometric configuration is independent of a change of scale. These are precisely the properties we would expect of a perceptual primitive, but are lacking in the commonly employed gradient dependent primitives, such as "edgepoints."

Starting with both the binary overlay (produced as discussed in the main body of this paper) and the original gray scale image, we have demonstrated via examples that the remaining steps in the delineation process can be effectively achieved.

Our goal in this work has been to approach human levels of performance in finding perceptually obvious delineations in images selected at random from a reasonably broad class of scene domains, and without any human intervention or prior knowledge about the image content. We believe that this goal can be achieved through extension and refinement of the techniques described in this paper.

ACKNOWLEDGEMENT

The research reported herein was supported by the Defense Advanced Research Projects Agency under Contract No. MDA903-83-C-0027, and by the National Science Foundation under contract No. ECS-7917028.

REFERENCES

1. A. Rosenfeld and J.L. Pfalz, "Sequential Operations in Digital Picture Processing," J. ACM, Vol. 13(4), October 1966, pp. 471-494.

2. M.A. Fischler and P. Barrett, "An Iconic Transform for Sketch Completion and Shape Abstraction," Computer Graphics and Image Processing, Vol. 13(4), August 1980, pp. 334-360.

3. M.A. Fischler and H.C. Wolf, "A General Approach to Machine Perception of Linear Structure in Imaged Data," (in preparation).

4. M.A. Fischler, J.M. Tenenbaum, and H.C. Wolf, "Detection of Roads and Linear Structures in Low-Resolution Aerial Imagery Using a Multisource Knowledge Integration Technique," Computer Graphics and Image Processing, Vol. 15(3), March 1981, pp. 201-223.

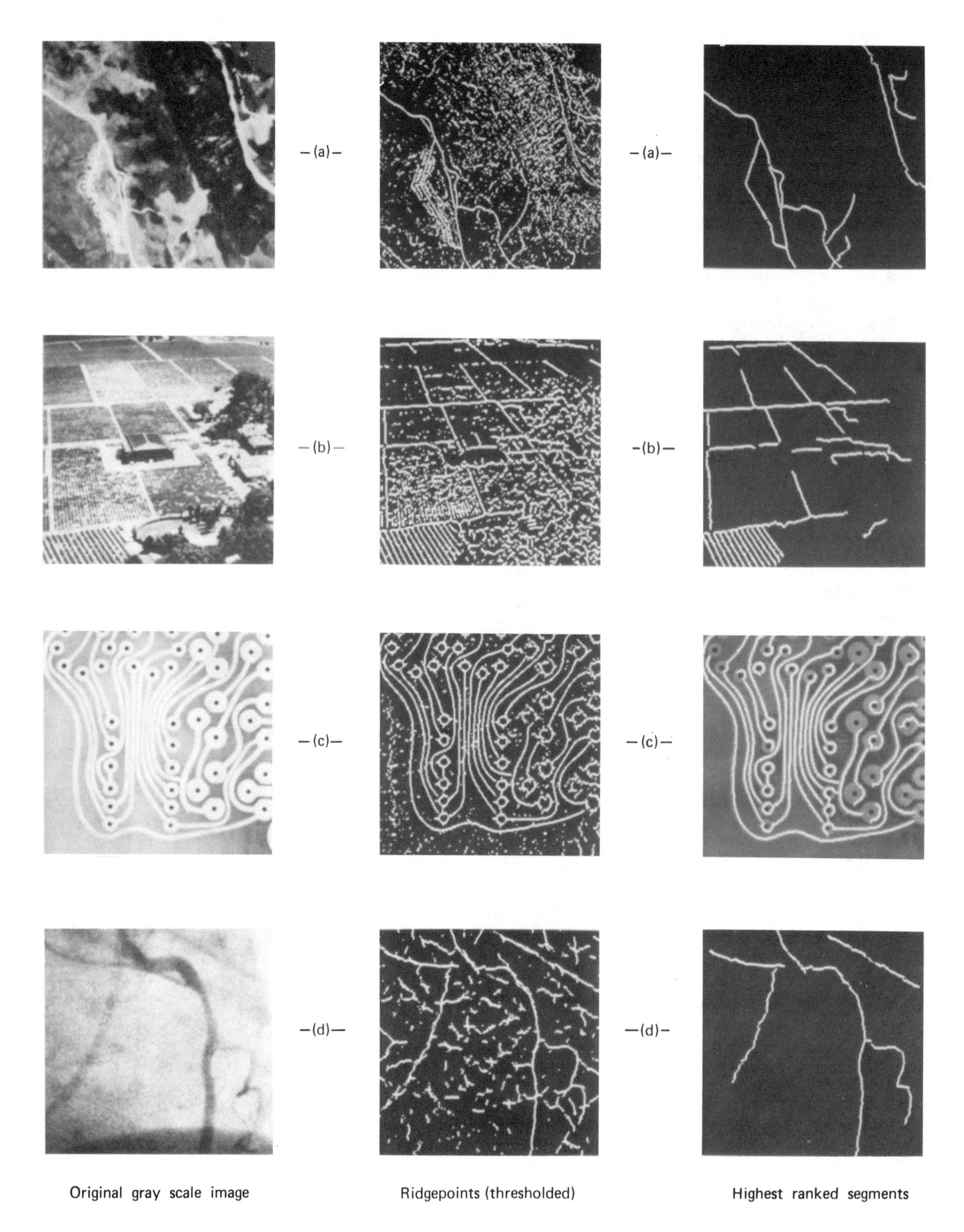

FIGURE 1 EXAMPLES OF THE LINEAR DELINEATION PROCESS

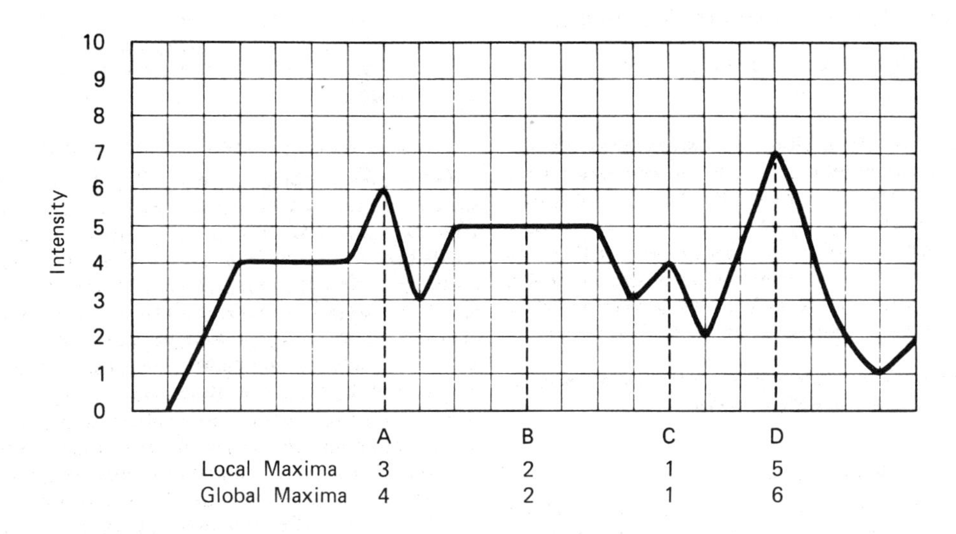

FIGURE 2 EXAMPLES OF LOCAL AND GLOBAL MAXIMA COMPUTATION

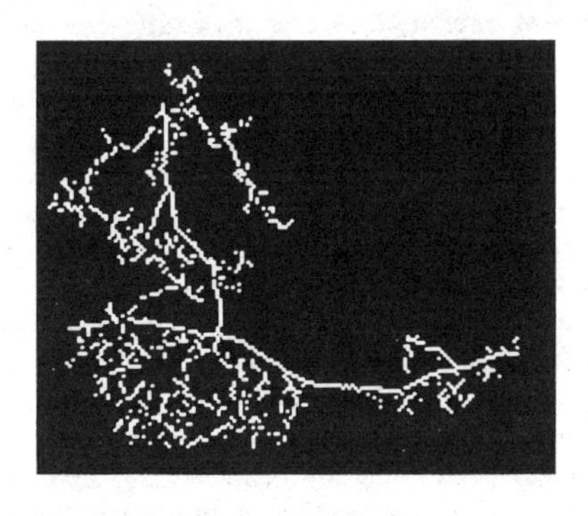

(a) A single cluster of linear feature points

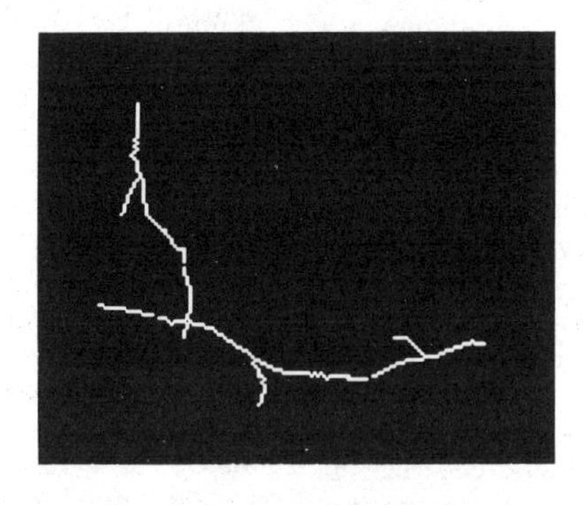

(b) Cluster after pruning

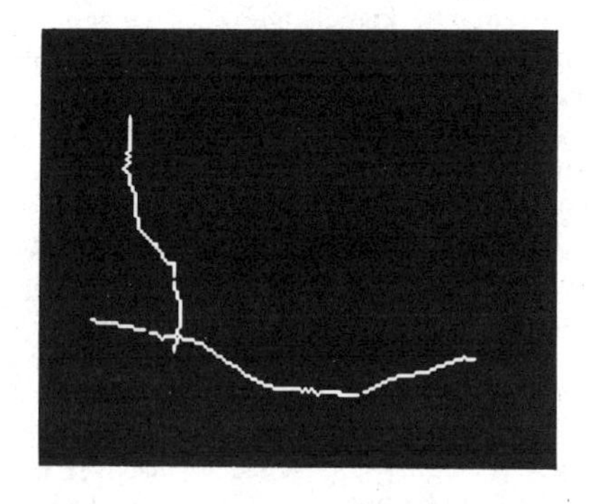

(c) The two linear segments extracted (One vertical line and one horizontal line)

FIGURE 3 PRUNING AND LINEAR SEGMENT EXTRACTION FROM A SINGLE CLUSTER

Perceptual Organization and Curve Partitioning

MARTIN A. FISCHLER AND ROBERT C. BOLLES

***Abstract*—In this paper we offer a critical evaluation of the partitioning (perceptual organization) problem, noting the extent to which it has distinct formulations and parameterizations. We show that most partitioning techniques can be characterized as variations of four distinct paradigms, and argue that any effective technique must satisfy two general principles. We give concrete substance to our general discussion by introducing new partitioning techniques for planar geometric curves, and present experimental results demonstrating their effectiveness.**

***Index Terms*—Computer vision, critical points, partitioning, perceptual grouping, two-dimensional curve segmentation.**

I. Introduction

A basic attribute of the human visual system is its ability to group elements of a perceived scene or visual field into meaningful or coherent clusters; in addition to clustering or partitioning, the visual system generally imparts structure and often a semantic interpretation to the data. In spite of the apparent existence proof provided by human vision, the general problem of scene partitioning remains unsolved for computer vision. Furthermore, there is even some question as to whether this problem is meaningful (or a solution verifiable) in its most general form.

Part of the difficulty resides in the fact that it is not clear to what extent semantic knowledge (e.g., recognizing the appearance of a straight line or some letter of the English alphabet), as opposed to generic criteria (e.g., grouping scene elements on the basis of geometic proximity), is employed in examples of human performance. It would not be unreasonable to assume that a typical human has on the order of tens of thousands of iconic primitives in his visual vocabulary; a normal adult's linguistic vocabulary might consist of from 10 000 to 40 000 root words, and iconic memory is believed to be at least as effective as its linguistic counterpart. Since, at present, we cannot hope to duplicate human competence in semantic interpretation, it would be desirable to find a task domain in which the influence of semantic knowledge is limited. In such a domain it might be possible to discover the generic criteria employed by the human visual system and to duplicate human performance. One of the main goals of the research effort described in this correspondence is to find a set of generic rules and models that will permit a machine to duplicate human performance in partitioning planar curves.

II. The Partitioning Problem: Issues and Considerations

Even if we are given a problem domain in which explicit semantic cues are missing, to what extent is partitioning dependent on the purpose, vocabulary, data representation, and past experience of the "partitioning instrument," as opposed to being a search for context independent "intrinsic structure" in the data? We argue that rather than having a unique formulation, the partitioning problem must be paramaterized along a number of basic dimensions. In the remainder of this section we enumerate some of these dimensions and discuss their relevance.

A. Intent (Purpose) of the Partitioning Task

In the experiment described in Fig. 1, human subjects were presented with the task of partitioning a set of two-dimensional curves with respect to three different objectives: 1) choose a set of contour points that best mark those locations at which curve segments produced by different processes were "glued" together; 2) choose a set of contour points that best allow one to reconstruct the complete curve; 3) choose a set of contour points that would best allow one to distinguish the given curve from others. Each person was given only one of the three task statements. Even though the point selections within a task varied from subject to subject, there was significant overlap and the variations were easily explained in terms of recognized strategies invoked to satisfy the given constraints; however, the points selected in the three tasks were significantly different. Thus, even in the case of data with almost no semantic content, the partitioning problem is NOT a generic task independent of purpose.

B. Partitioning Viewed as an Explanation of Curve Construction

With respect to "process partitioning" (partitioning the curve into segments produced by different processes), a partition can be viewed as an explanation of how the curve was constructed. Explanations have the following attributes which, when assigned different "values," lead to different explanations and thus different partitions.

- *Vocabulary (primitives and relations):* What properties of our data should be represented, and how should these properties be computed? That is, we must select those aspects of the problem domain we consider relevant to our partition decisions (e.g., geometric shape, gray scale, line width, semantic content), and enable their computation by providing models for the corresponding structures (e.g., straight-line segment, circular arc, wiggly segment). We must also allow for the appropriate "viewing" conditions; e.g., symmetry, repeated structure, parallel lines, are global concepts that imply that the curve has finite extent and can be viewed as a "whole," as opposed to only permitting computations that are based on some limited interval or neighborhood of (or along) the curve.
- *Definition of Noise:* In a generic sense, any data set that does not have a "simple (concise)" description is noise. Thus, noise is relative to both the selected descriptive language and an arbitrary level of complexity. The particular choices for vocabulary and the acceptable complexity level determine whether a point is selected as a partition point or considered to be a noise element.
- *Believability:* Depending on the competence (completeness) of our vocabulary to describe any curve that may be encountered, the selected metric for judging similarity, and the arbitrary threshold we have chosen for believing that a vocabulary term corresponds to some segment of a given curve, partition points will appear, disappear, or shift.

C. Representation

The form in which the data are presented (i.e., the input representation), as well as the type of data, are critical aspects of the problem definition, and will have a major impact on the decisions made by different approaches to the partitioning task. Some of the key variables are as follows:

Manuscript received January 24, 1984; revised July 9, 1985. This work was supported by the Defense Advanced Research Projects Agency under Contract MDA 903-83-C-0027 and by the National Science Foundation under Contract ECS-7917028.

The authors are with SRI International, 333 Ravenswood Avenue, Menlo Park, CA 94025.

IEEE Log Number 8406511.

Task 1 Select AT MOST five points to describe this line drawing so that you will be able to reconstruct it as well as possible ten years from now, given just the sequence of selected points.

Since five points were sufficient to form an approximate convex hull of the figure, virtually everyone did so, selecting the five points indicated below.

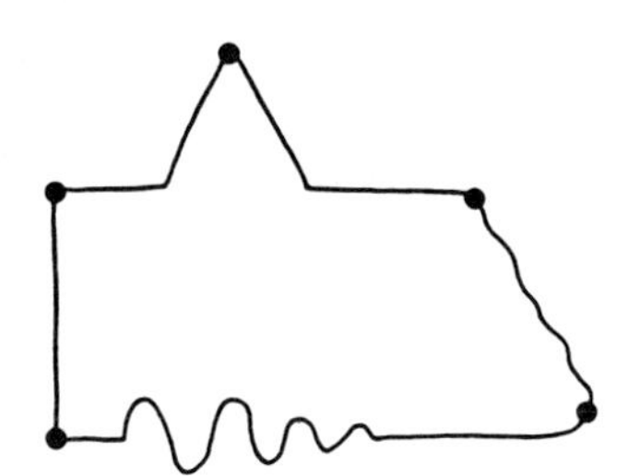

Task 2 Assume that a friend of yours is going to be asked to recognize this line drawing on the basis of the information you supply him about it. He will be presented with a set of drawings, one of which will be a rotated and scaled version of this curve. You are only allowed to provide him with A SEQUENCE OF AT MOST FIVE POINTS. Mark the points you will select.

Since five points were not enough to outline all the key features of the figure, the subjects had to decide what to leave out. They seemed to adopt one of two general strategies: 1) use the limited number of points to describe one distinct feature well (illustrated by the selection on the left), or 2) use the points to outline the basic shape of the figure (illustrated by the selection on the right).

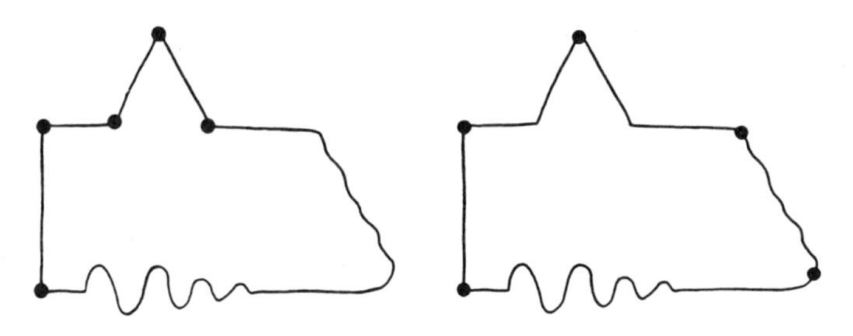

Task 3 This line drawing was constructed by piecing together segments produced by different processes. Please indicate where you think the junctions between segments occur AND VERY BRIEFLY DESCRIBE EACH SEGMENT. Use as few points as possible, but no more than five.

The constraint of being limited to five points forced the subjects to consider the whole curve and develop a consistent, global explanation. The basic strategy seemed to be recursive one in which they first partitioned the curve into two segements by placing a breakpoint at position 1 and another one at either position 2 or position 3 to separate the smooth curves from the sharp corners. Then they used the remaining points to subdivide these segments according to a vocabulary they selected that included such things as triangles, rectangles, and sinusoids. For example, almost everyone placed breakpoints at positions 3 and 4 and described the enclosed segment as part of a triangle. Similarly, the segment between positions 1 and 5 was generally described as a decaying sinusoid. It is interesting to note that in task 1 the subjects consistently placed a point close to position 5 but always farther to the right, because they were trying to approximate a convex hull. The different purpose led to different placements.

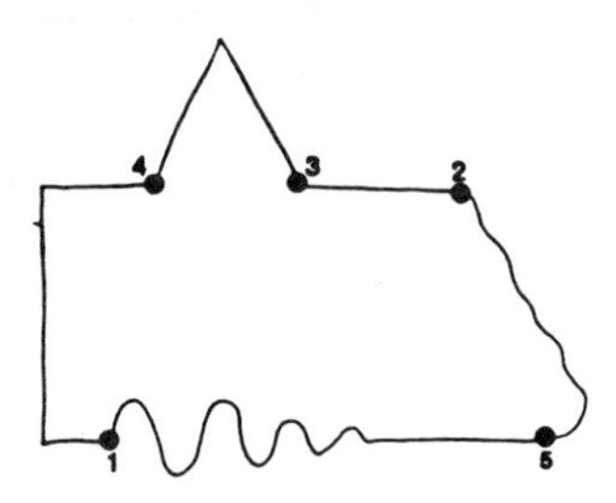

Fig. 1. Experiments in which human subjects were asked to segment a curve.

• analog (pictorial) versus digital (quantized) versus analytic description of the curves.

• single versus multiple "views" (e.g., single versus multiple quantizations of a given segment).

• input resolution versus length of smallest segment of interest.

• simply connected (continuous) curves versus self-intersecting curves or curves with "gaps."

• for complex situations, is connectivity provided, or must it be established?

• if a curve possesses attributes (e.g., gray scale, width) other than "shape" that are to serve as partitioning criteria, how are they obtained—by measurement on an actual "image,"or as symbolic tags provided as part of the given data set?

D. Evaluation

How do we determine if a given technique or approach to the partitioning problem is successful? How can we compare different techniques? We have already observed that, to the extent that partitioning is a "well-defined" problem at all, it has a large number of alternative formulations and parameterizations. Thus, a technique that is dominant under one set of conditions may be inferior under a different parameterization. Nevertheless, any evaluation procedure must be based on the following considerations.

• Is there a known "correct" answer (e.g., because of the way the curves were constructed)?

• Is the problem formulated in such a way that there is a "provably" correct answer?

• How good is the agreement of the partitioned data with the descriptive vocabulary (models) in which the "explanation" is posed?

• How good is the agreement with (generic or "expert") subjective human judgement?

• What is the tradeoff between "false alarms" and "misses" in the placement of partition points. To the extent that it is not possible to ensure a perfect answer (in the placement of the partition points), there is no way to avoid such a tradeoff. Even if the relative weighting between these two types of errors is not made explicit, it is inherent in any decision procedure—including the use of subjective human judgement.

In spite of all of the previous discussion in this section, it might still be argued that if we take the union of all partition points obtained for all reasonable definitions and parameterizations of the partition problem, we would still end up with a "small" set of partition points for any given curve, and further, there may be a generic procedure for obtaining this covering set. While a full discussion of this possibility is not feasible here, we can construct a counterexample to the unqualified conjecture based on selecting a very high ratio of the cost of a miss to a false alarm in selecting the partition points. A (weak) refutation can also be based on the observation that if a generic covering set of partition points exists, then there should be a relatively consistent way of ordering all the points on a given curve as to their being acceptable partition points; the experiment presented in Fig. 1 indicates that, in general, such a consistent ordering does not exist.

III. Paradigms for Curve Partitioning

Almost all algorithms employed for curve partitioning appear to be special cases (instantiations) of one or more of the following paradigms.

• *Local Detection of Distinguished Points:* A partition point is inserted at locations along the curve at which one or more of the descriptive attributes (e.g., curvature, distance from a coordinate axis or centroid) is determined to have a discontinuity, an extreme value (maxima or minima) or a zero value separating intervals of positive and negative values.

• *Best Global Description:* A set of partition points is inserted at those locations along a curve that allow the "best" description of the associated segements in terms of some *a priori* set of models (e.g., the set of models might consist of all first and second degree polynomials, with only one model permitted to explain the data between two adjacent partition points; the quality of the description might be measured by the mean square deviation of the data points from the fitting polynomials).

• *Confirming Evidence:* Given a number of "independent procedures (or possibly different parameterizations of a given procedure) for locating potential partition points, we retain only those partition points that are common to some subset of the different procedures or their parameterizations.

• *Recursive Simplification:* The input data are subjected to repeated applications of some transformation that monotonically reduces some measurable aspect of the data to one of a finite number of terminal states (e.g.,differentiation, smoothing, projection, thresholding). The hierarchy of data sets thus produced is then processed with an algorithm derived from the previous three paradigms.

IV. Principles of Effective (Robust) Model-Based Interpretation

What underlies our choice of partitioning criteria? We assert that any competent partitioning technique, regardless of which of the above paradigms is employed, will incorporate the following principles.

A. Stability

The "principle of stability" is the assertion that any valid perceptual desision should be stable under at least small perturbations cof both the imaging conditions and the decision algorithm parameters. This generalization of the assumption of "general position" also subsumes the assertion (often presented as an assumption) that most of a scene must be described in terms of continuous variables if meaningful interpretation is to be possible.

It is interesting to observe that many of the constructs in mathematics (e.g., the derivative) are based on the concepts of convergence and limit, also subsumed under the stability principle. Attempts to measure the digital counterparts of the mathematical concepts have traditionally employed window type "operators" that are not based on a limiting process; it should come as no suprise that such attempts have not been very effective.

In practice, if we perturb the various imaging and decision parameters, we observe relatively stable decision regions separated by obviously unstable intervals (e.g., the two distinct percepts produced by a Necker cube). The stable regions represent alternative hypotheses that generally cannot be resolved without recourse to either additional and more restrictive assumptions, or semantic (domain-specific) knowledge.

B. Complete, Concise, and Complexity Limited Explanation

The decision-making process in image interpretation, i.e., matching image derived data to *a priori* models, not only must be stable, but must also explain the structure observable in the data. Equally important, the explanation must satisfy specific criteria for believability and complexity. Believability is largely a matter of offering the simplest possible description of the data and, in addition, explaining any deviation of the data from the models (vocabulary) used in the description. Even the simplest description, however, must also be of limited complexity; otherwise it will not be understandable and thus not be believable.

By making the foregoing principles explicit, we can directly invoke them (as demonstrated in the following section) to formulate effective algorithms for perceptual organization.

V. Instantiation of the Theory: Specific Techniques for Curve Partitioning

In this section we offer two effective new algorithms for curve partitioning (program listings available from the authors). In each case, we first describe the algorithm, and later indicate how it was motivated and constrained by the principles just presented. In both algorithms, the key ideas are: 1) to view each point, or segment of a curve, from as many perspectives as possible, retaining only those partition points receiving the highest level of multiple confirmation; and 2) inhibiting the further selection of partition points when the density of points already selected exceeds a preselected or computed limit.

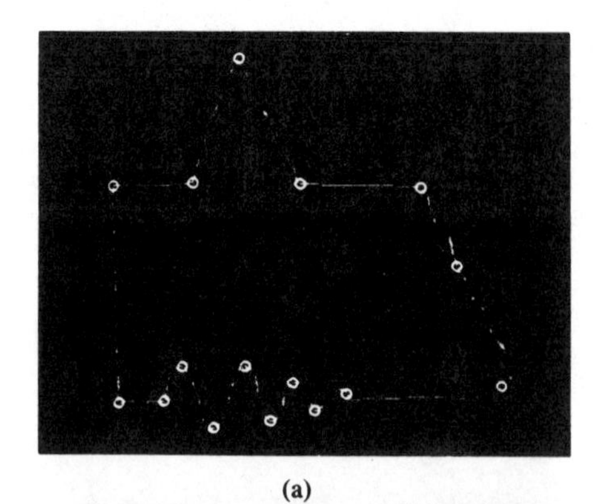

(a)

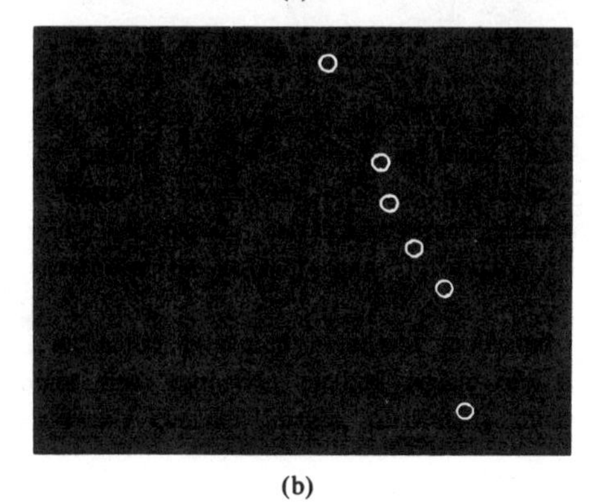

(b)

Fig. 2. (a) Estimation of curvature from discrete approximations. (a) The results of applying the "improved angle detection" procedure described in [13] to a digitized version of the curve in Fig. 1. The procedure works quite well, except for the introduction of a breakpoint in the middle of the right side and the merging of two small bumps at the right of the sinusoidal segment. (b) If we extract a portion of the curve and apply the algorithm, it introduces several additional breakpoints because the change in curve length causes some of the algorithm parameters to change.

A. Curve Partitioning Based on Detecting Local Discontinuity

In this subsection we present a new approach to the problem of finding points of discontinuity ("critical points") on a curve. Our criterion for success is whether we can match the performance of human subjects given the same task (e.g., see Fig. 1). The importance of this problem from the standpoint of the psychology of human vision dates back to the work of [1]. However, it has long been recognized as a very difficult problem, and no satisfactory computer algorithm currently exists for this purpose. An excellent discussion of the problem may be found in [3]; other pertinent references included [4], [6], [10], and [13]. Results and observations akin and complementary to those presented here can be found in [5] and [14].

Most approaches equate the search for critical points with looking for points of high curvature. Although this intuition seems to be correct, it is incomplete as stated (i.e., it does not explicitly take into account "explanation" complexity); further, the methods proposed for measuring curvature are often inadequate in their selection of stability criteria. In Fig. 2 we show some results of measuring curvature using discrete approximation to the mathematical definition.

We have developed an algorithm for locating critical points that invokes a model related to, but distinct from, the mathematical concept of curvature. The algorithm labels each point on a curve as belonging to one of the three categories: 1) a point in a smooth

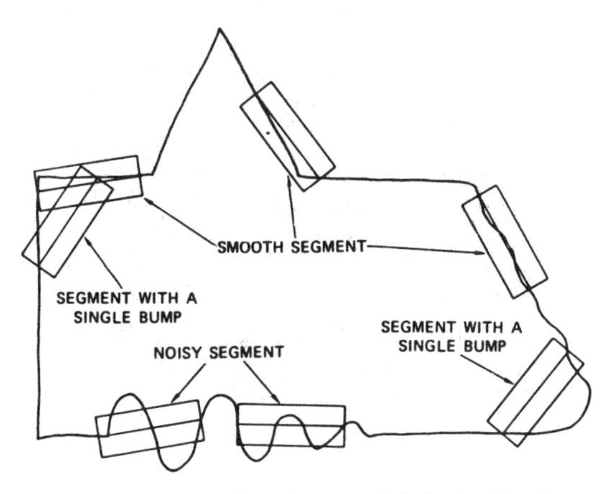

Fig. 3. Example curve segments and their classifications.

interval, 2) a critical point, or 3) a point in a noisy interval. To make this choice, the algorithm analyzes the deviations of the curve from a chord or "stick" that is iteratively advanced along the curve (this will be done for a variety of lengths, which is analogous to analyzing the curve at different resolutions). If the curve stays close to the chord, points in the interval spanned by the chord will be labeled as belonging to a smooth section. If the curve makes a single excursion away from the chord, the point in the interval that is farthest from the chord will be labeled a critical point (actually, for each placement of the chord, an accumulator associated with the farthest point will be incremented by the distance between the point and the chord). If the curve makes two or more excursions, points in the interval will be labeled as noise points.

We should note here that "noisy" intervals at low resolution (large chord length) will have many critical points at higher resolution (small chord length). Fig. 3 shows examples of curve segments and their classifications. The distance from a chord that defines a significant excursion (i.e., the width of the boxes in Fig. 3) is a function of the expected noise along the curve and the length of the chord.

At each resolution (i.e., stick size), the algorithm orders the critical points according to the values in their accumulators and selects the best ones first. To avoid setting an arbitrary "goodness" threshold for distinguishing critical from ordinary points, we use a complexity criterion. To halt the selection process, we stop when the points being suggested are too close to those selected previously at the given resolution. In our experiments we define "too close" as being within a quarter of the stick length used to suggest the point.

After the critical points have been selected at the coarsest resolution, the algorithm is applied at higher resolutions to locate additional critical points that are outside the regions dominated by previously selected points. Fig. 4(a) shows the critical points determined at the coarsest level (stick length of 100 pixels; approximately 1/10 of the length of the curve). Fig. 4(b) shows all the critical points labeled with the stick lengths used to determine them. (We note that this critical point detection procedure does not locate inflection points or smooth transitions betweem segments, such as the transition from an arc of a circle to a line tangent to the circle.)

The above algorithm appears to be very effective, especially for finding obvious partition points and in not making "ugly" mistakes (i.e., choosing partition points at locations that none of our human subjects would pick). Its ability to find good partition points is based on evaluating each point on the curve from multiple viewpoints (placement of the stick)—a direct application of the principle of stability. Requiring that the partition points remain stable under changes in resolution (i.e., small changes in the stick length) did not appear to be effective and was not employed; in fact, stick length was altered by a significant amount in each iteration, and partition points found at these different scales of resolution were not expected to support each other, but were assumed to be due to distinct phenomena.

The avoidance of ugly mistakes was due to our method of limiting the number of partition points that could be selected at any level of resolution, or in any neighborhood of a selected point (i.e., limiting the explanation complexity). One concept we invoked here, related to that of complete explanation, was that the detection procedure could not be trusted to provide an adequate explanation when more than a single critical point was in its field of view, and in such a situation, any decision was deferred to later iteration at higher levels of resolution (i.e., shorter stick lengths).

Finally, in accord with our previous discussion, the algorithm has two free parameters that provide control over its definition of noise (i.e., variations too small or too close together to be of interest), and its willingness to miss a good partition point so as to be sure it does not select a bad one.

B. Curve Partitioning Based on Detecting Process Homogenity

To match human performance in partitioning a curve, by recognizing those locations at which one generating process terminates and another begins, is orders of magnitude more difficult than partitioning based on local discontinuity analysis. As noted earlier, a critical aspect of such performance is the size and effectiveness of the vocabulary (of *a priori* models) employed. Explicitly providing a general purpose vocabulary to the machine would entail an unreasonably large amount of work—we hypothesize that the only effective way of allowing a machine to acquire such knowledge is to provide it with a learning capability.

For our purposes in this investigation, we chose a problem in which the relevant vocabulary was extremely limited: the curves to be partitioned are composed exclusively of straight lines and arcs of circles. (Two specific applications we were interested in here were the decomposition of silhouettes of industrial parts, and the decomposition of the line scans returned by a "structured light" ranging device viewing scenes containing various diameter cylinders and planar faced objects lying on a flat surface.) Our goal here was to develop a procedure for locating critical points along a curve in such a way that the segments between the critical points would be satisfactorily modeled by either a straight-line segment or a circular arc. Relevant work addressing this problem has been done by [7]-[9], [11], and [12].

Our approach is to analyze several "views" of a curve, construct a list of possible critical points, and then select the optimum points between which models from our vocabulary can be fitted. For our experiments we quantized an analytic curve at several positions and orientations (with respect to a pixel grid), then attempted to recover the original model.

For each view (quantization) of the curve we locate occurrences of lines and arcs, marking their ends as prospective partition points. This is accomplished by randomly selecting small seed segments from the curve, fitting to them a line or arc, examining the fit, and then extending as far as possible those models that exhibit a good fit. After a large number of seeds have been explored in the different views of the curve, the histogram (frequency count as a function of path length) of beginnings and endings is used to suggest critical points (in order of their frequency of occurrence). Each new critical point, considered for inclusion in the explanation of how the curve is constructed, introduces two new segments which are compared to both our line and circle models. If one or both of the segments have acceptable fits, the corresponding curve segments are marked as explained. Otherwise, the segments are left to be explained by additional critical points and the partitions they imply. The addition of critical points continues until the complete curve is explained. Fig. 5 shows an example of the operation of this algorithm.

While admittedly operating in a relatively simple environment, the above algorithm exhibits excellent performance. This is true even in the difficult case of finding partition points along the smooth interface between a straight line and a circle to which the line is tangent.

Both basic principles, stability and complete explanation, are deeply embedded in this algorithm. Retaining only those partition

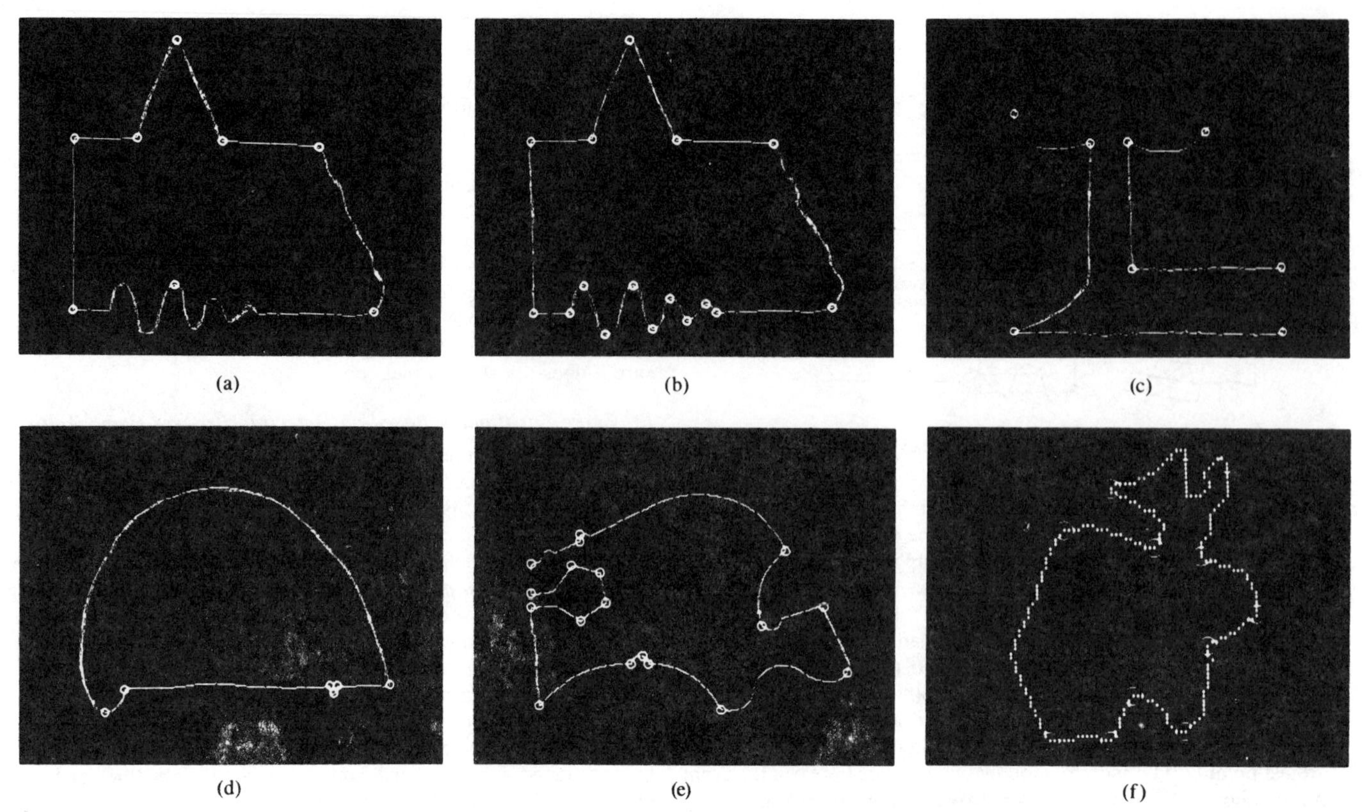

Fig. 4. Local discontinuity analysis. (a) Results of the analysis at the coarsest resolution (i.e., with a stick length of 100 pixels, which is approximately a tenth of the length of the curve), (b) Results from all resolutions (100, 80, 60, 40, 20, 15, 10). (c)–(f) Additional examples of the local discontinuity analysis.

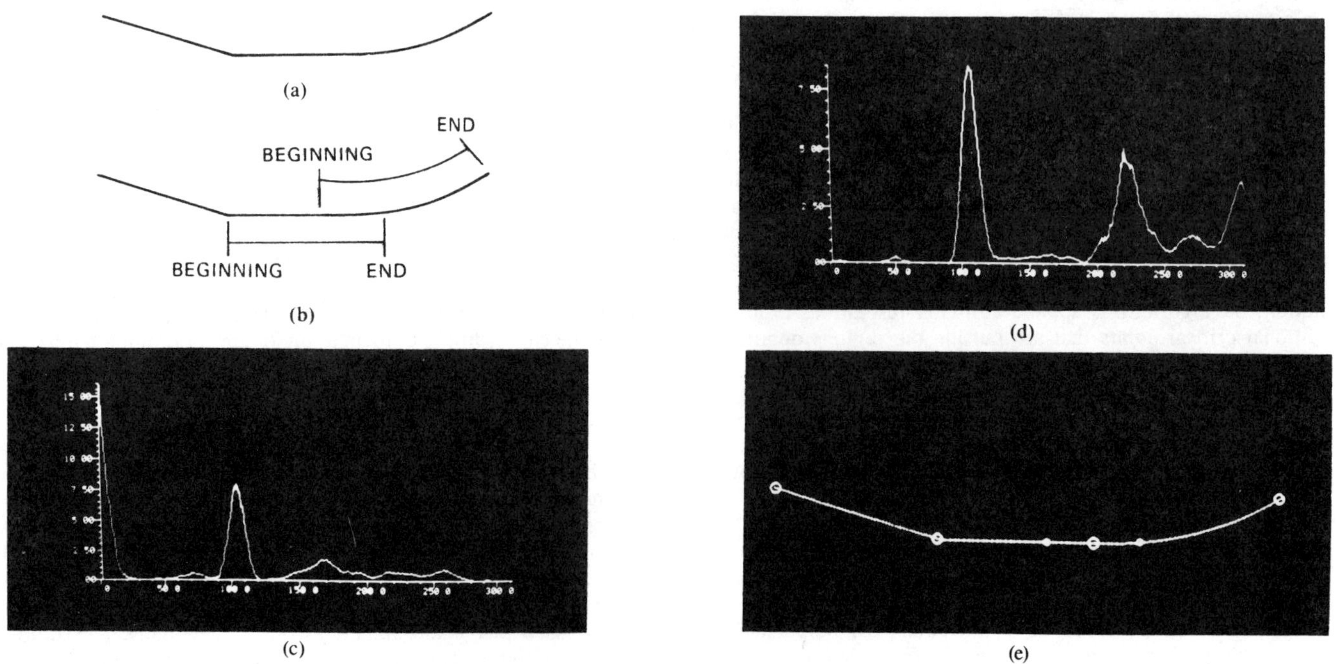

Fig. 5. Homogeneity analysis. (a) An analytic curve consisting of two straight segments and a circular arc. (b) A multiply explained segment of the curve formed by the extension of the arc and one of the line segments to include as many compatible points as possible. (c) The smoothed histogram of the starting points of the segments detected along the curve. (d) The smoothed histogram of the ending points of the segments detected along the curve. (e) The breakpoints suggested by the histograms. The breakpoint between the arc and the line was placed at the center of the multiply explained region, which is bounded by asterisks.

points which persist under different "viewpoints" was motivated by the principle of stability. Our technique for evaluating the fit of the segment of a curve between two partition points, to both the line and circle models, requires that the deviations from an acceptable model have the characteristics of "white" (random) noise; this is an instantiation of the principle of complete explanation, and is based on our previous work presented in [2].

VI. Discussion

We can summarize our key points as follows:

- The partition problem does not have a unique definition, but is parameterized with the respect to such items as purpose, data representation, tradeoff between different error types (false-alarms versus misses), etc.
- Psychologically acceptable partitions are associated with an implied explanation that must satisfy criteria for accuracy, complexity, and believability. These criteria can be formulated in terms of a set of principles, which, in turn, can guide the construction of effective partitioning algoithms (i.e., they provide necessary conditions).

One implication contained in these observations is that a purely mathematical definition of "intinsic structure" (i.e., a definition justified solely by appeal to mathematical criteria or principles) cannot, by itself, be sufficiently selective to serve as a basis for duplicating human performance in the partitioning task; generic partitioning (i.e., partitioning in the absence of semantic content) is based on psychological "laws" and physiological mechanisms, as well as on correlations embedded in the data.

In this correspondence we have looked at a very limited subset of the class of all scene partitioning problems; nevertheless, it is interesting to speculate on how the human performs so effectively in the broader domain of interpreting single images of natural scenes. The speed of response in a human's ability to interpret a sequence of images of dissimilar scenes makes it highly questionable that there is some mechanism by which he simultaneously matches all his semantic primitives against the imaged data, even if we assume that some independent process has already presented him with a "camera model" that resolves some of the uncertainties in image scale, orientation, and projective distortion. How does the human index into the large semantic data base to find the appropriate models for the scene at hand?

Consider the following paradigm: first a set of coherent components is recovered from the image on the basis of very general (but parameterized) clustering criteria of the type described earlier; next, a relatively small set of semantic models, which are components of many of the objects in the complete semantic vocabulary, are matched against the extracted clusters; successful matches are then used to index into the full database and the corresponding entries are matched against both the extracted clusters and adjacent scene components; these additional successful matches will now trigger both iconic and symbolic associations that result in further matching possibilities as well as perceptual hypotheses that organize large portions of the image into coherent structures (gestalt phenomena).

If this paradigm is valid, then, even though much of the perceptual process would depend on an individual's personal experience and immediate goals, we might still expect "hard wired" algorithms (genetically programmed, but with adjustable parameters) to be employed in the initial partitioning steps.

In this correspondence, we have attempted to give computational definitions to some of the organizing criteria needed to approach human level performance in the partitioning task. However, we believe that our more important contribution has been the explcit formulation of a set of principles that we assert must be satisfied by any effective procedure for perceptual grouping.

References

[1] F. Attneave, "Some aspects of visual perception," *Psychol. Rev.*, vol. 61, pp. 183–193, 1954.

[2] R. C. Bolles, and M. A. Fischler, "A RANSAC-based approach to model fitting and its application to finding cylinders in range data," in *Proc. 7th Int. Joint Conf. Artificial Intell.*, Vancouver, B.C., Canada, Aug. 1981, pp. 637–643.

[3] L. S. Davis, "Understanding shape: Angles and sides," *IEEE Trans. Comput.*, vol. C-26, pp. 236–242, Mar. 1977.

[4] H. Freeman and L. S. Davis, "A corner-finding algorithm for chain-coded curves," *IEEE Trans. Comput.*, vol. C-26, pp. 297–303, Mar. 1977.

[5] D. D. Hoffman and W. A. Richards, "Representing smooth plane curves for recognition: Implications for figure-ground reversal," in *Proc. 2nd Nat. Conf. Artificial Intell.*, Pittsburgh, PA, Aug. 1982, pp. 5–8.

[6] B. Kruse and C. V. K. Rao, "A matched filtering technique for corner detection," in *Proc. 4th Int. Joint Conf. Pattern Recognition*, Kyoto, Japan, Nov. 1978, pp. 642–644.

[7] Y. Liao, "A two-stage method of fitting conic arcs and straight-line segments to digitized contours," in *Proc. Pattern Recognition and Image Processing Conf.*, Dallas, TX, Aug. 1981, pp. 244–229.

[8] D. G. Lowe and T. G. Binford, "Segmentation and aggregation; an approach to figure-ground phenomena," in *Proc. Image Understanding Workshop*, Stanford Univ., Standford, CA, Sept. 1982.

[9] U. Montanari, "A note on minimal length polygonal approximation to a digitized contour," *Commun. ACM*, vol. 13, pp. 41–47, Jan. 1970.

[10] T. Pavlidis, "Algorithms for shape analysis of contours and waveforms," *IEEE Trans. Pattern Anal. Mach. Intell.* vol. PAMI-2, pp. 301–312, July 1980.

[11] T. Pavlidis and S. L. Horowitz, "Segmentation of plane curves," *IEEE Trans. Comput.*, vol. C-23, pp. 860–870, Aug. 1974.

[12] U. Ramer, "An iterative procedure for the polygonal approximation of plane curves," *Comput. Graphics Image Processing*, vol. 1, pp. 244–256, 1972.

[13] A. Rosenfeld and J. S. Weszka, "An improved method of angle detection on digital curves," *IEEE Trans. Comput.* vol. C-24, pp. 940–941, Sept. 1975.

[14] A. Witkin, "Scale-dependent qualitative signal description," in *Proc. 8th Int. Joint Conf. Artificial Intell.*, Karlsruhe, West Germany, Aug. 1983, pp. 1019–1022.

Digital Step Edges from Zero Crossing of Second Directional Derivatives

ROBERT M. HARALICK, FELLOW, IEEE

Abstract—We use the facet model to accomplish step edge detection. The essence of the facet model is that any analysis made on the basis of the pixel values in some neighborhood has its final authoritative interpretation relative to the underlying gray tone intensity surface of which the neighborhood pixel values are observed noisy samples. With regard to edge detection, we define an edge to occur in a pixel if and only if there is some point in the pixel's area having a negatively sloped zero crossing of the second directional derivative taken in the direction of a nonzero gradient at the pixel's center. Thus, to determine whether or not a pixel should be marked as a step edge pixel, its underlying gray tone intensity surface must be estimated on the basis of the pixels in its neighborhood. For this, we use a functional form consisting of a linear combination of the tensor products of discrete orthogonal polynomials of up to degree three. The appropriate directional derivatives are easily computed from this kind of a function.

Upon comparing the performance of this zero crossing of second directional derivative operator with the Prewitt gradient operator and the Marr-Hildreth zero crossing of the Laplacian operator, we find that it is the best performer; next is the Prewitt gradient operator. The Marr-Hildreth zero crossing of the Laplacian operator performs the worst.

Index Terms—Edge operator, facet model, image processing, image segmentation, zero crossings of second directional derivative.

I. Introduction

WHAT IS an edge in a digital image? The first intuitive notion is that a digital edge occurs on the boundary between two pixels when the respective brightness values of the two pixels are significantly different. Here "significantly different" may depend upon the distribution of brightness values around each of the pixels.

We often point to a region on an image and say this region is brighter than its surrounding area. Having noticed this we would then say that an edge exists between each pair of neighboring pixels where one pixel is inside the brighter region and the other is outside the region. *Such edges are referred to as step edges.*

Step edges are not the only kind of edge. If we scan through a region in a left-right manner observing the brightness values steadily increasing and then after a certain point observe that the brightness values are steadily decreasing, we are likely to say that there is an edge at the point of change from increasing to decreasing brightness values. *Such edges are called roof edges.*

Therefore, it is clear from our use of the word "edge" that edge refers to places in the image where there appears to be a jump in brightness value or a local extremum in brightness value derivative. Jumps in brightness values are the kinds of edges originally detected by Roberts [18]. Relative extrema of first derivative in a one-dimensional form is used by Ehrich and Schroeder [6] and in an isotropic two-dimensional suboptimal form by Marr and Hildreth [13].

In some sense this summary statement about edges is quite revealing since in a discrete array of brightness values there are jumps, between neighboring brightness values if the brightness values are different, even if only slightly different. Perhaps more to the heart of the matter, there exists no calculus definition of derivative for a discrete array of brightness values.

One clear way to interpret jumps in value or local extrema of derivatives when referring to a discrete array of values is to assume that the discrete array of values comes about by sampling a real-valued function f defined on the domain of the image which is a bounded and connected subset of the real plane R^2. The finite difference typically used in the numerical approximation of first-order derivatives are usually based on assuming the function f to be linear. From this point of view, the jumps in value or extrema in derivative really must refer to points of high first derivative of f or to points of relative extrema in the second derivatives of f. Edge detection must then involve fitting a function to the sample values. Prewitt [17] was the first to suggest the fitting idea. Hueckel [11], [12], Brooks [3], Haralick [8], Haralick and Watson [10], Morgenthaler and Rosenfeld [15], Zucker and Hummel [20], and Morgenthaler [14] all use the surface fit concept in determining edges.

Edge finders should then regard the digital picture function as a sampling of the underlying function f, where some kind of random noise has been added to the true function values. To do this, the edge finder must assume some parametric form for the underlying function f, use the sampled brightness values of the digital picture function to estimate the parameters, and finally make decisions regarding the locations of discontinuities and the locations of relative extrema of partial derivatives based on the estimated values of the parameters.

Of course, it is impossible to determine the true locations of discontinuities in value or relative extrema in derivatives directly from a sampling of the function values. The locations are estimated analytically after doing function approximation. Sharp discontinuities can reveal themselves in high values for estimates of first partial derivatives. Relative extrema in first directional derivative can reveal themselves as zero crossings of the second directional derivative. Thus, if it were known that the first and second partial derivatives of any possible underly-

Manuscript received February 9, 1982; revised December 20, 1982.

The author is with the Departments of Electrical Engineering and Computer Science, Virginia Polytechnic Institute and State University, Blacksburg, VA 24061.

ing image function had known bounds, then any estimated first- or second-order partials which exceed these known bounds must be due to discontinuities in value or in derivative of the underlying function. This is the basis for the gradient magnitude and Laplacian magnitude edge detectors which have appeared in the literature: detect an edge if the gradient is high enough.

However, edges can be weak but well localized. Such edges, as well as the strong edges just discussed, manifest themselves as local extrema of the directional derivative of the estimated gray tone intensity function taken across the edge. This idea for edges is the basis of the edge detector discussed here. In this paper, we assume that in each neighborhood of the image the underlying gray tone intensity function f takes the parametric form of a polynomial in the row and column coordinates and that the sampling producing the digital picture function is a regular equal interval grid sampling of the square plane which is the domain of f. Thus, in each neighborhood f takes the form

$$f(r,c) = k_1 + k_2 r + k_3 c + k_4 r^2 + k_5 rc + k_6 c^2 + k_7 r^3 + k_8 r^2 c + k_9 rc^2 + k_{10} c^3.$$

As just mentioned, we place edges not at locations of high gradient, but at locations of spatial gradient maxima. More precisely, a pixel is marked as an edge pixel if in the pixel's immediate area there is a zero crossing of the second directional derivative taken in the direction of the gradient [9] and the slope of the zero crossing is negative. Thus, this kind of edge detector will respond to weak but spatially peaked gradients.

The underlying functions from which the directional derivatives are computed are easy to represent as linear combinations of the polynomials in any polynomial basis set. A polynomial basis set which permits the independent estimation of each coefficient would be the easiest to use. Such a basis is discussed in the appendix.

Section II discusses how the discretely sampled data values are used to estimate the coefficients of the linear combinations: coefficient estimates for exactly fitting or estimates for least square fitting are calculated as linear combinations of the sampled data values.

Having used the pixel values in a neighborhood to estimate the underlying polynomial function we can now determine the value of the partial derivatives at any location in the neighborhood and use those values in edge finding. Having to deal with partials in both the row and column directions makes using these derivatives a little more complicated than using the simple derivatives of one-dimensional functions. Section III discusses the directional derivative, how it is related to the row and column partial derivatives, and how the coefficients of the fitted polynomial get used in the edge detector. In Section IV we discuss the statistical confidence of the estimate of edge existence and the edge angle. In Section V we show results comparing the directional derivative zero crossing edge operator with the generalized Prewitt gradient operator and the related Marr-Hildreth zero crossing of the Laplacian operator.

II. Fitting Data with Discrete Orthogonal Polynomials

Let an index set R with the symmetry property $r \in R$ implies $-r \in R$ be given. Let the number of elements in R be N. Using the construction technique in the Appendix, we may construct the set $\{P_0(r), \cdots, P_{N-1}(r)\}$ of discrete orthogonal polynomials over R. Using the tensor product technique also discussed in the Appendix we can construct discrete orthogonal polynomials over a two-dimensional neighborhood. Some one- and two-dimensional discrete orthogonal polynomials are shown as follows:

Index Set	Discrete Orthogonal Polynomial Set
$\{-1/2, 1/2\}$	$\{1, r\}$
$\{-1, 0, 1\}$	$\{1, r, r^2 - 2/3\}$
$\{-3/2, -1/2, 1/2, 3/2\}$	$\{1, r, r^2 - 5/4, r^3 - 41/20r\}$
$\{-2, -1, 0, 1, 2\}$	$\{1, r, r^2 - 2, r^3 - 17/5, r^4 + 3r^2 + 72/35\}$
$\{-1, 0, 1\} \times \{-1, 0, 1\}$	$\{1, r, c, r^2 - 2/3, rc, c^2 - 2/3, r(c^2 - 2/3), c(r^2 - 2/3), (r^2 - 2/3)(c^2 - 2/3)\}$.

For each $r \in R$, let a data value $d(r)$ be observed. The exact fitting problem is to determine coefficients $a_0, \cdots, a_{N-1}$ such that

$$d(r) = \sum_{n=0}^{N-1} a_n P_n(r). \tag{1}$$

The approximate fitting problem is to determine coefficients $a_0, \cdots, a_K, K \leqslant N-1$ such that

$$e^2 = \sum_{r \in R} \left[d(r) - \sum_{n=0}^{K} a_n P_n(r) \right]^2$$

is minimized. In either case the result is

$$a_m = \sum_{r \in R} P_m(r) d(r) / \sum_{s \in R} P_m^2(s). \tag{2}$$

The exact fitting coefficients and the least squares coefficients are identical for $m = 0, \cdots, K$. Similar equations hold for the two-dimensional case.

Equation (2) means that each fitting coefficient a_m can be computed as linear combination of the data values. For each index r in the index set, the data value $d(r)$ is multiplied by the weight

$$P_m(r) / \sum_{s \in R} P_m^2(s)$$

which is just an appropriate normalization of an evaluation of the polynomial P_m at the index r. Figs. 1 and 2 show these weights for the 3×3 and 4×4 neighborhoods.

Once the fitting coefficients a_k, $k = 1, \cdots, K$, have been computed, the estimated polynomial $Q(r)$ is given by

$$Q(r) = \sum_{n=0}^{K} a_n P_n(r).$$

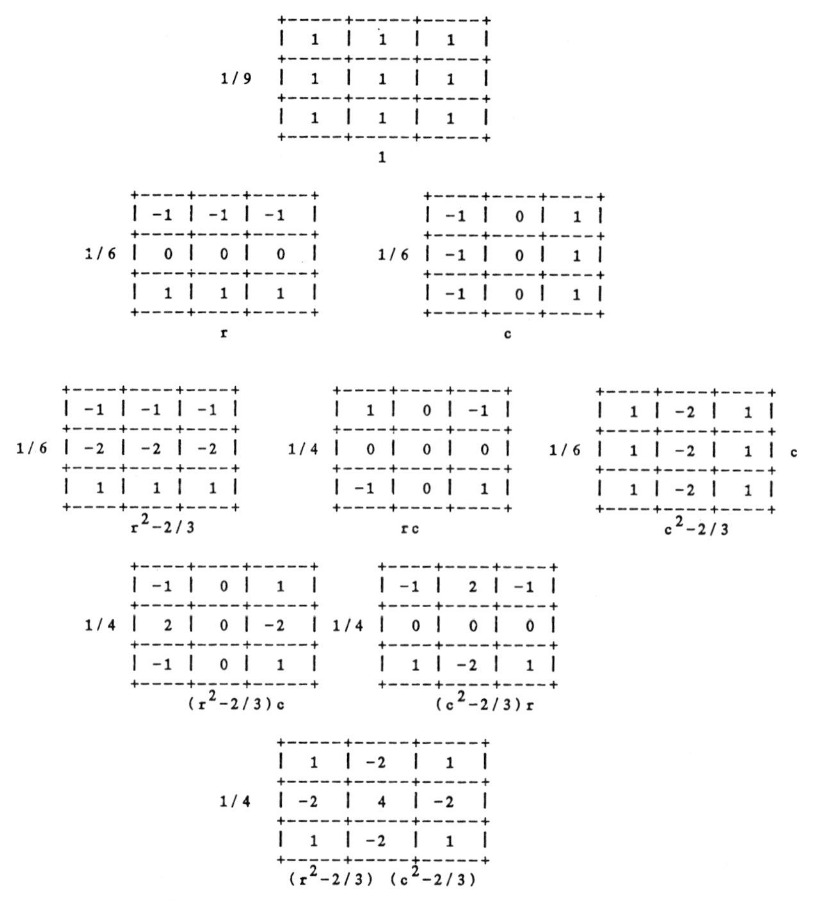

Fig. 1. Illustrates the 9 masks for the 3 X 3 window.

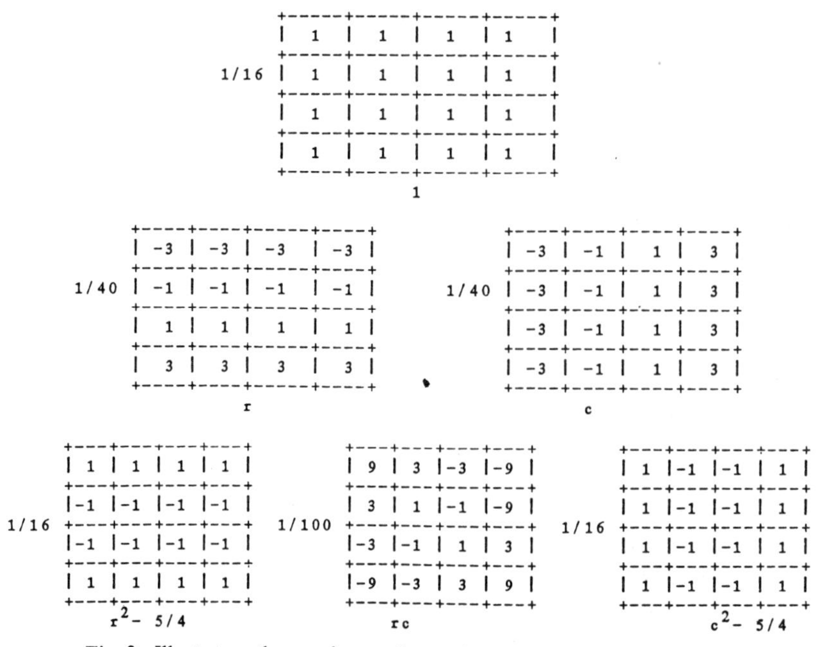

Fig. 2. Illustrates the masks used to obtain the coefficients of all polynomials up to the quadratic ones for a 4 X 4 window.

This equation permits us to interpret $Q(r)$ as a well-behaved real-valued function defined on the real line. To determine

$$\frac{dQ}{dr}(r_0)$$

we need only to evaluate

$$\sum_{n=0}^{N} a_n \frac{dP_n}{dr}(r_0). \tag{3}$$

In this manner, the estimate for any derivative at any point may be obtained. Similarly for any definite integrals. Section IV gives a statistical analysis of the errors of the estimate.

$g(r,c) = a_{00} + a_{10}r + a_{01}c$

	-1	-1	-1
1/6	0	0	0
	1	1	1

$g(r,c) = a_{00} + a_{10}r + a_{01}c$
$a + a_{20}(r^2-2/3) + a_{11}rc + a_{02}(c^2-2/3)$
$+ a_{21}(r^2-2/3)c + a_{12}(c^2-2/3)r$

	0	-1	0
1/2	0	0	0
	0	1	0

Fig. 3. Illustrates that the kernel mask used to estimate a quantity such as row derivative can depend on the order of the assumed model.

Beaudet [2] uses this technique for estimating derivatives employed in rotationally invariant image operators.

It should be noted that the kernel used to estimate a derivative depends on the neighborhood size, the order of the fit, and the basis functions used for the fit. These parameters constitute the assumed model. The importance of the model is illustrated by the example of Fig. 3. This difference means that the model used must be justified, the justification being that it is a good fit to the data. From this we learn that the use of first-order models to estimate first-order partials as is often done may not produce correct results if the first-order fit is not good enough. In particular, it may require a third-order model to get good estimates of first-order partial derivatives.

III. The Directional Derivative Edge Finder

We stated in the introduction that a digital step edge occurs at pixels having a negatively sloped zero crossing of the second directional derivative taken in the direction of the gradient. In this section we discuss the relationship between the directional derivatives and the coefficients from the polynomial fit.

We denote the directional derivative of f at the point (r, c) in the direction α by $f'_\alpha(r, c)$. It is defined as

$$f'_\alpha(r,c) = \lim_{h \to 0} \frac{f(r + h \sin\alpha, c + h\cos\alpha) - f(r,c)}{h}. \tag{4}$$

The direction angle α is the clockwise angle from the column axis. It follows directly from this definition that

$$f'_\alpha(r,c) = \frac{\partial f}{\partial r}(r,c)\sin\alpha + \frac{\partial f}{\partial c}(r,c)\cos\alpha. \tag{5}$$

We denote the second directional derivative of f at the point (r, c) in the direction α by $f''_\alpha(r, c)$ and it quickly follows by substituting f'_α for f in (5) that

$$f''_\alpha = \frac{\partial^2 f}{\partial r^2}\sin^2\alpha + \frac{2\partial^2 f}{\partial rc}\sin\alpha\cos\alpha + \frac{\partial^2 f}{\partial c^2}\cos^2\alpha. \tag{6}$$

Taking f to be a cubic polynomial in r and c which can be estimated by the discrete orthogonal polynomial fitting procedure, we can compute the gradient of f and the gradient direction angle α at the center of the neighborhood used to estimate f. In order for our notation to be invariant to the different discrete orthogonal polynomials which result from different neighborhood sizes, we rewrite this cubic in canonical form as

$$\begin{aligned} f(r,c) = {} & k_1 + k_2 r + k_3 c \\ & + k_4 r^2 + k_5 rc + k_6 c^2 \\ & + k_7 r^3 + k_8 r^2 c + k_9 rc^2 + k_{10} c^3. \end{aligned} \tag{7}$$

We obtain the angle by α by

$$\sin\alpha = k_2 / \sqrt{(k_2^2 + k_3^2)}$$

$$\cos\alpha = k_3 / \sqrt{(k_2^2 + k_3^2)}. \tag{8}$$

At any point (r, c), the second directional derivative in the direction α is given by

$$\begin{aligned} f''_\alpha(r,c) = {} & (6k_7 \sin^2\alpha + 4k_8 \sin\alpha\cos\alpha + 2k_9\cos^2\alpha)r \\ & + (6k_{10}\cos^2\alpha + 4k_9\sin\alpha\cos\alpha + 2k_8\sin^2\alpha)c \\ & + (2k_4\sin^2\alpha + 2k_5\sin\alpha\cos\alpha + 2k_6\cos^2\alpha). \end{aligned} \tag{9}$$

We wish to only consider points (r, c) on the line in direction α. Hence, $r = \rho\sin\alpha$ and $c = \rho\cos\alpha$. Then

$$\begin{aligned} f''_\alpha(\rho) = {} & 6[k_7\sin^3\alpha + k_8\sin^2\alpha\cos\alpha \\ & + k_9\sin\alpha\cos^2\alpha + k_{10}\cos^3\alpha]\,\rho \\ & + 2[k_4\sin^2\alpha + k_5\sin\alpha\cos\alpha + k_6\cos^2\alpha] \\ = {} & A\rho + B. \end{aligned} \tag{10}$$

If for some ρ, $|\rho| < \rho_0$, where ρ_0 is slightly smaller than the length of the side of a pixel, $f'''_\alpha(\rho) < 0$, $f''_\alpha(\rho) = 0$ and $f'_\alpha(\rho) \neq 0$ we have discovered a negatively sloped zero crossing of the estimated second directional derivative taken in the estimated direction of the gradient and we mark the center pixel of the neighborhood as an edge pixel.

IV. Statistical Analysis

Noise induces a randomness in the least squares coefficients which then induces a randomness in the estimated gradient value, the estimated angle of the gradient, and the estimated location of the zero crossing. In this section we show how to handle this randomness by appropriate hypothesis tests or confidence interval estimation.

A. General Model

We let P_n, $n = 1, \cdots, N$, denote the discrete orthonormal basis functions, η denote the independent and identically distributed noise, and g denote the gray tone intensity function. Under this model, each neighborhood of the observed image can be written as

$$g(r,c) = \sum_{n=1}^{N} a_n P_n(r,c) + \eta(r,c) \tag{11}$$

where

$$\sum_{r,c} P_n(r,c)\,P_m(r,c) = \begin{cases} 0, & n \neq m \\ 1 & n = m \end{cases}$$

and the least squares estimates $a'_1, \cdots, a'_N$ for the unknown coefficients $a_1, \cdots, a_n$ are given by

$$a'_n = \sum_{r,c} g(r,c)\,P_n(r,c). \tag{12}$$

Substituting the formula for $g(r, c)$ into the equation for a'_n and simplifying results in

$$a'_n = a_n + \sum_{r,c} P_n(r, c)\, \eta(r, c) \tag{13}$$

clearly showing that a'_n has a deterministic part and a random part, the randomness due to the noise. We assume that the noise is independent normal having mean 0 and variance σ^2. Therefore, the estimated coefficient a'_n has mean a_n, variance σ^2 and is uncorrelated with every other coefficient:

$$E[a'_n] = a_n$$
$$E[a'_m a'_n] = a_m a_n,\ m \neq n$$
$$E[a'^2_n] = a^2_n + \sigma^2$$
$$V[a'_n] = \sigma^2.$$

The residual error e is defined as the difference between the observed values and fitted values. It too is a random variable:

$$e(r, c) = g(r, c) - \sum_{n=1}^{N} a'_n P_n(r, c)$$

$$= \sum_{n=1}^{N} (a_n - a'_n) P_n(r, c) + \eta(r, c). \tag{14}$$

It is not difficult to see that at each (r, c), the residual error has mean zero and is uncorrelated with each estimated coefficient a'_n since

$$E[a'_n\, e(r, c)] = 0.$$

After some algebraic substitutions and manipulation, the total residual error S^2 can be written as

$$S^2 = \sum_{r,c} e^2(r, c) = \sum_{r,c} \eta^2(r, c) - \sum_{n=1}^{N} (a_n - a'_n)^2. \tag{15}$$

Thus, if the noise is assumed normal and there are K pixels in a window

$$\sum_{r,c} \eta^2(r, c)/\sigma^2 \text{ has } X^2_K,$$

a chi-squared variate with K degree of freedom,

$$\sum_{n=1}^{N} (a_n - a'_n)^2/\sigma^2 \text{ has } X^2_N$$

which makes

$$\sum_{r,c} e^2(r, c)/\sigma^2 \text{ have } X^2_{K-N}.$$

B. Estimating the First Partials

If the discrete orthogonal basis functions are polynomials then each first partial derivative at (0, 0) in the row and column directions is given as some linear combination of the estimated coefficients. Furthermore, the linear combination for the row partial will be orthogonal to the linear combination in the column partial. It is not difficult to derive the expected value and variance of the first row partial μ'_r and the first column partial μ'_c. They are

$$E[\mu'_r] = \mu_r$$
$$E[\mu'_c] = \mu_c$$
$$V[\mu'_r] = \sigma^2 k$$
$$V[\mu'_c] = \sigma^2 k$$
$$E[\mu'_r \mu'_c] = \mu_r \mu_c.$$

Hence, the estimates for the row and column partial derivatives are uncorrelated.

C. Hypothesis Testing for Zero Gradient

To see the effect of the randomness on the estimate of the gradient magnitude, consider testing the hypothesis that $\mu'_r = \mu'_c = 0$. This hypothesis must be rejected if there is to be a zero crossing of second directional derivative. Under this hypothesis,

$$\frac{\mu'^2_r + \mu'^2_c}{k\sigma^2}$$

has a X^2_2 distribution.

The total residual error normalized by the noise variance, S^2/σ^2, has a X^2_{K-N} distribution. Hence

$$\frac{(\mu'^2_r + \mu'^2_c)/2}{k S^2/(K - N)}$$

has an $F_{2,K-N}$ distribution and the hypothesis of $\mu'_r = \mu'_c = 0$ would be rejected for suitably large values.

D. Confidence Interval for Gradient Direction

To see the effect of the randomness on the estimate of the direction of the gradient, consider the relationships portrayed in Fig. 4. The axes are the row and column partials μ'_r and μ'_c. The direction angle Θ of the gradient is given by

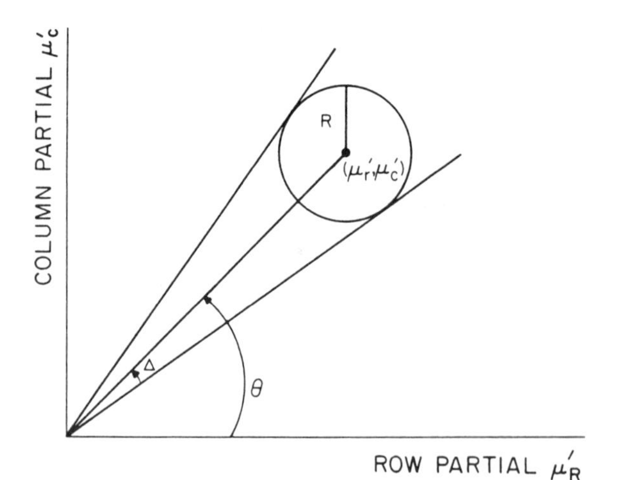

Fig. 4. Illustrates the geometry of the confidence interval estimation for the edge angle.

$$\cos\Theta = \mu_r/(\mu^2_r + \mu^2_c)^{1/2}$$

$$\sin\Theta = \mu_c/(\mu^2_r + \mu^2_c)^{1/2} \tag{16}$$

The center of the circle is at the estimate (μ_r', μ_c'). Upon substituting the estimates μ_r' and μ_c' for μ_r and μ_c, we obtain the estimated direction angle Θ' by

$$\cos\Theta' = \mu_r'/(\mu_r'^2 + \mu_c'^2)^{1/2}$$

$$\sin\Theta' = \mu_c'/(\mu_r'^2 + \mu_c'^2)^{1/2}. \tag{17}$$

From a Bayesian point of view, the area of the circle represents the conditional probability that the unknown (μ_r, μ_c) lies within a distance R from the observed (μ_r', μ_c') given that the variance of μ_r' and μ_c' is known and equal to $k\sigma^2$. Assuming a normal distribution for the noise, this conditional probability is $q = 1 - e^{-R^2/2k\sigma^2}$. Hence, if probability q is given, the corresponding radius R is

$$R = k\sigma[-2\log(1-q)]^{1/2}. \tag{18}$$

To determine a confidence interval for Θ of the form $\Theta' - \Delta \leqslant \Theta \leqslant \Theta' + \Delta$, we have from Fig. 4 that

$$\sin^2\Delta = \frac{k\sigma^2(-2\log(1-q))}{\mu_r'^2 + \mu_c'^2}. \tag{19}$$

Note that the 2Δ confidence interval length depends on the probability q of the circle confidence region for (μ_r, μ_c) and the unknown noise variance σ^2. Although σ^2 is not known, we do know S^2 which has a $\sigma^2\, X^2_{K-N}$ distribution. We can handle the problem of the unknown σ^2 by determining a joint confidence region for (μ_r, μ_c) and σ^2 [7]. Taking p to be the probability that a chi-squared random variable with $K - N$ degrees of freedom has an observed value greater than $X^2_{K-N,p}$ we have the confidence interval $[0, S^2/X^2_{K-N,p}]$ for σ^2 having at least probability p. Replacing σ^2 in (19) by $S^2/X^2_{N-K,p}$ we obtain

$$\sin^2\Delta = \frac{kS^2(-2\log(1-q))}{X^2_{K-N,p}(\mu_r'^2 + \mu_c'^2)}. \tag{20}$$

A confidence interval for Θ having at least probability pq is then $(\Theta' - \Delta, \Theta' + \Delta)$.

E. Edge Hypothesis Testing

In this section we first take the edge direction α to be a fixed constant. We let μ_A and μ_B be the expected values of the random variables A and B appearing in (10):

$$f_\alpha''(\rho) = A\rho + B.$$

The null hypothesis is that an edge exists and it is satisfied if for some ρ, $0 \leqslant \rho \leqslant \rho_0$, where ρ_0 is just smaller than the length of a side of a pixel, we have $\mu_A\rho + \mu_B = 0$ and $\mu_A < 0$.

The observed random variables are A, B, and the residual fitting error S^2. The bivariate random variable

$$\begin{pmatrix} A \\ B \end{pmatrix}$$

is normal having mean

$$\begin{pmatrix} \mu_A \\ \mu_B \end{pmatrix}$$

and covariance

$$\sigma^2 \begin{bmatrix} k_A & 0 \\ 0 & k_B \end{bmatrix}$$

where k_A and k_B are known constants. For a window of K pixels and a cubic fit, S^2/σ^2 has a X^2_{K-10}.

From this it follows that

$$Z(\mu_A, \mu_B) = \frac{[(A-\mu_A)/k_A]^2 + [(B-\mu_B/k_B^2)]/2}{S^2/(K-10)}$$

has an F_{2K-10} distribution.

We define $R = \{(x, y) |$ for some ρ, $0 \leqslant \rho \leqslant d$, $x\rho + y = 0\}$ Then the null hypothesis is rejected at the p significance level if

$$\min_{(\mu_A, \mu_B) \in R} Z(\mu_A, \mu_B)$$

is larger than $F_{2K-10\ 1-p}$.

An edge strength probability can be defined by q where q satisfies

$$\min_{(\mu_A, \mu_B) \in R} Z(\mu_A, \mu_B) = F_{2k-10\ q}.$$

Of course, the edge direction α is not fixed. But as derived at the end of Section IV-C, we do have a confidence interval for it. And for each value of α in the confidence interval, the random variable $A(\alpha)$ and $B(\alpha)$ can be computed and the null hypothesis tested. If for all α in the confidence interval the null hypothesis is rejected, then the existence of an edge is also rejected.

In practice, we can perform a nonexact hypothesis test selecting only the left end, middle, and right end values of α from its confidence interval. If for each of these three values of α the null hypothesis is rejected, then the existence of an edge is also rejected.

V. Experimental Results

To understand the performance of the second directional derivative zero crossing digital step edge operator we examine its behavior on a well structured simulated data set and on a real aerial image. For the simulated data set, we use a 100 × 100 pixel image of a checkerboard, the checks being 20 × 20 pixels. The dark checks have gray tone intensity 75 and the light checks have gray tone intensity 175. To this perfect checkerboard we add independent Gaussian noise having mean zero and standard deviation 50. Defining the signal to noise ratio as 10 times the logarithm of the range of signal divided by rms of the noise, the simulated image has a 3 dB signal to noise ratio. The perfect and noisy checkerboards are shown in Fig. 5.

Section V-A illustrates the performance of the classic 3 × 3 edge operators with and without preaveraging compared against the generalized Prewitt operator. Section V-B illustrates the performance of the 11 × 11 Marr-Hildreth zero crossing of the Laplacian operator, the 11 × 11 Prewitt operator, and the 11 × 11 zero crossing of second directional derivative operator. The zero crossing of second directional derivative surpasses the performance of the other two on the twofold basis of probability of correct assignment and error distance which is defined as the average distance to closest true edge pixel of pixels which are assigned nonedge but which are true edge pixels.

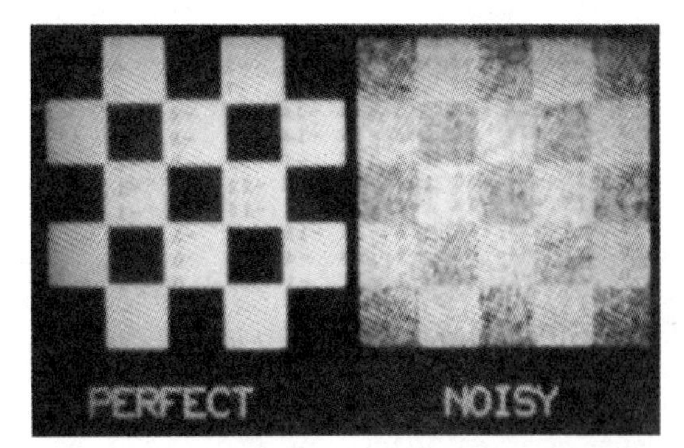

Fig. 5. Illustrates the controlled perfect and noisy checkerboard images.

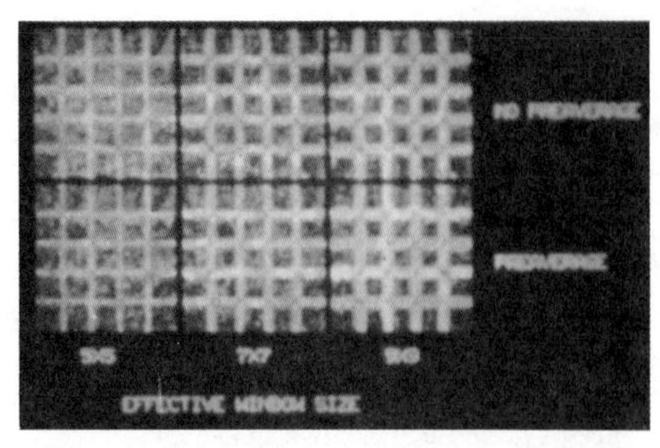

Fig. 7. Illustrates the Prewitt operator done by using a least squares quadratic fit in the neighborhood versus doing preaveraging and using a smaller fitting neighborhood size. The no preaveraging results show slightly higher contrast.

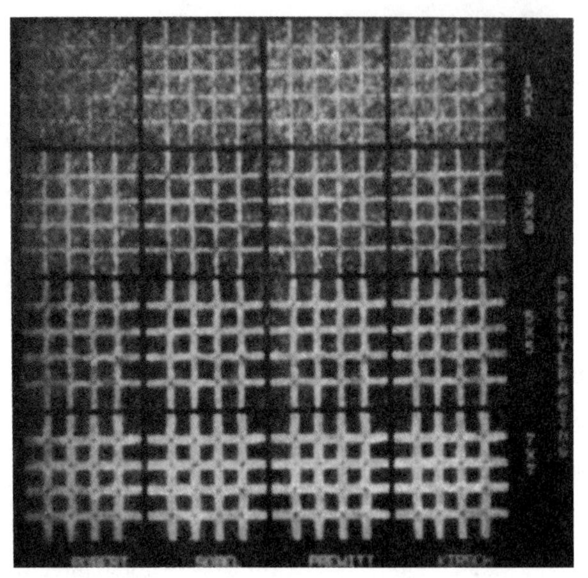

Fig. 6. Illustrates the 3 X 3 Roberts, Sobel, Prewitt, and Kirsch edge operators with a box filter preaveraging of 1 X 1, 3 X 3, 5 X 5, and 7 X 7.

Fig. 8. Compares the Nevatia and Babu compass operator with the Prewitt operator in a 5 X 5 neighborhood.

A. The Classic Edge Operators

The classic 3 X 3 gradient operators all perform badly as shown in Fig. 6. Note that the usual definition of the Roberts operator has been modified in the natural way so that it uses a 3 X 3 mask.

Averaging before the application of the gradient operator is considered [19] to be one cure for such bad performance on noisy images. Fig. 6 also shows the same operators applied after a box filtering with a 3 X 3, 5 X 5, and 7 X 7 neighborhood sizes.

An alternative to the preaveraging is to define the gradient operator with a larger window. This is easily done with the Prewitt operator [17] which fits a quadratic surface in every window and uses the square root of the sum of the squares of the linear term coefficients to the estimate the gradient. (A linear fit would actually yield the same result. A cubic fit is the first higher order fit which would yield a different result.) This is illustrated in Fig. 7. A 3 X 3 preaverage followed by a 3 X 3 gradient operator yields a resulting neighborhood size of 5 X 5. Thus in Fig. 8 we also show the 3 X 3 preaverage followed by a 3 X 3 gradient under the 5 X 5 Prewitt and we show the 5 X 5 preaverage followed by the 3 X 3 gradient under the 7 X 7 Prewitt. The noise is higher in the preaverage edge-detector. For comparison purposes the 5 X 5 [16] compass operator is shown alongside the 5 X 5 Prewitt in Fig. 8. They give virtually the same result. The Prewitt operator has the advantage of requiring half the computation.

It is obvious from these results that good gradient operators must have larger neighborhood sizes than 3 X 3. Unfortunately, the larger neighborhood sizes also yield thicker edges.

To detect edges, the gradient value must be thresholded. In each case, we chose a threshold value which makes the conditional probability of assigning an edge given that there is an edge equal to the conditional probability of there being a true edge given that an edge is assigned. True edges are defined to be the two pixel wide region in which each pixel neighbors some pixel having a value different from it on the perfect checkerboard. Fig. 9 shows the thresholded Prewitt operator (quadratic fit) for a variety of neighborhood sizes. Notice that because the gradient is zero at the saddle points (the corner where four checks meet) any operator depending on the gradient to detect an edge will have trouble there.

B. The Second Derivative Zero Crossing Edge Operators

Marr and Hildreth [13] suggest an edge operator based on the zero crossing of a generalized Laplacian. In effect, this is a nondirectional or isotropic second derivative zero crossing

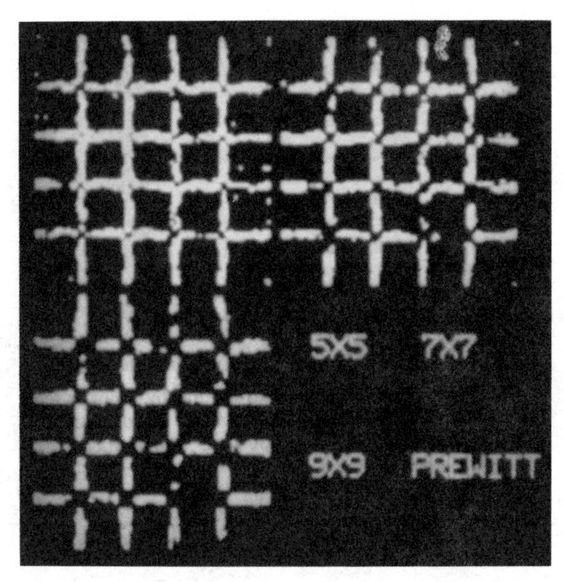

Fig. 9. Illustrates the edges obtained by thresholding the results of the Prewitt operator.

operator. The mask for this generalized Laplacian operator is given by sampling the kernel

$$A\left(1-k\frac{r^2+c^2}{\sigma^2}\right)e^{-1/2\,\frac{r^2+c^2}{\sigma^2}}$$

at the integer row column coordinates (r, c) designating the center of each pixel position in the neighborhood and then setting the value k so that the sum of the resulting weights is zero. Fig. 10 shows the resulting kernels for $\sigma = 1.4$, 5, and 10. The value $\sigma = 1.4$ is near the value usually used.

Edges are detected at all pixels whose generalized Laplacian value is of one sign and one of whose neighbor's generalized Laplacian value is of the opposite sign. A zero crossing threshold strength can be introduced here by insisting that the difference between the positive value and the negative value must exceed the threshold value before the pixel is declared to be an edge pixel. Fig. 11 illustrates the edge images produced by this technique for a variety of threshold values and a variety of values for σ for an 11×11 window. It is apparent that if all edge pixels are to be detected, there will be many pixels declared to be edge pixels which are really not edge pixels. And if there are to be no pixels which are to be declared edge pixels which are not edge pixels, then there will be many edge pixels which are not detected. Its performance appears to be poorer than the Prewitt operator.

The directional second derivative zero crossing edge operator introduced in this paper is shown in Fig. 12 for a variety of gradient threshold values. If the gradient exceeds the threshold value and a zero crossing occurs in a direction of $\pm 14.9^\circ$ of the gradient direction within a circle of one pixel length centered in the pixel, then the pixel is declared to be an edge pixel. This technique performs worst at the saddle points, the corners where four checks meet, because the gradient in zero there. The second derivative zero crossing edge operator compares favorably with the more computationally expensive maximum likelihood boundary estimation technique of Cooper *et al.* [5]

σ=1.4

0	0	0	-1	-1	-2	-1	-1	0	0	0
0	0	-2	-4	-8	-9	-8	-4	-2	0	0
0	-2	-7	-15	-22	-23	-22	-15	-7	-2	0
-1	-4	-15	-24	-14	-1	-14	-24	-15	-4	-1
-1	-8	-22	-14	52	103	52	-14	-22	-8	-1
-2	-9	-23	-1	103	178	103	-1	-23	-9	-2
-1	-8	-22	-14	52	103	52	-14	-22	-8	-1
-1	-4	-15	-24	-14	-1	-14	-24	-15	-4	-1
0	-2	-7	-15	-22	-23	-22	-15	-7	-2	0
0	0	-2	-4	-8	-9	-8	-4	-2	0	0
0	0	0	-1	-1	-2	-1	-1	0	0	0

σ=5.0

-24	-21	-17	-13	-10	-10	-10	-13	-17	-21	-24
-21	-16	-10	-4	0	2	0	-4	-10	-16	-21
-17	-10	-1	6	11	13	11	6	-1	-10	-17
-13	-4	6	15	22	24	22	15	6	-4	-13
-10	0	11	22	29	31	29	22	11	0	-10
-10	2	13	24	31	34	31	24	13	2	-10
-10	0	11	22	29	31	29	22	11	0	-10
-13	-4	6	15	22	24	22	15	6	-4	-13
-17	-10	-1	6	11	13	11	6	-1	-10	-17
-21	-16	-10	-4	0	2	0	-4	-10	-16	-21
-24	-21	-17	-13	-10	-10	-10	-13	-17	-21	-24

σ=10.00

-34	-25	-18	-12	-8	-7	-8	-12	-18	-25	-34
-25	-15	-7	-1	3	4	3	-1	-7	-15	-25
-18	-7	2	8	12	14	12	8	2	-7	-18
-12	-1	8	15	20	21	20	15	8	-1	-12
-8	3	12	20	24	26	24	20	12	3	-8
-7	4	14	21	26	27	26	21	14	4	-7
-8	3	12	20	24	26	24	20	12	3	-8
-12	-1	8	15	20	21	20	15	8	-1	-12
-18	-7	2	8	12	14	12	8	2	-7	-18
-25	-15	-7	-1	3	4	3	-1	-7	-15	-25
-34	-25	-18	-12	-8	-7	-8	-12	-18	-25	-34

Fig. 10. Shows that 11×11 kernels used for Mexican hat or generlized Laplacian operator defined by sampling

$$A\left(1-k\frac{r^2+c^2}{\sigma^2}\right)e^{-1/2\,\frac{r^2+c^2}{\sigma^2}}$$

at integer coordinates (r, c) $r, c - -5, \cdots, +5$. The constant k is chosen to make the sum of the values in the kernel to be zero, within the quantization error. The constant A just scales the values so that integer arithmetic can be used.

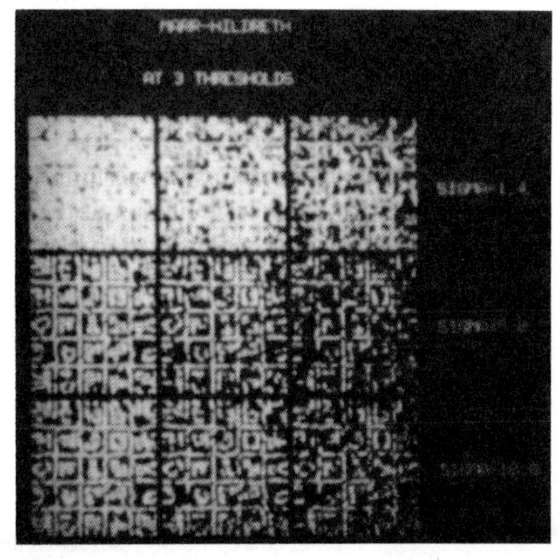

Fig. 11. Illustrates the edges obtained by the 11×11 Marr-Hildreth zero crossing of Laplacian operator set for three different zero crossing thresholds and three different standard deviations for the associated Mexican hat filter.

who show results using a higher signal to noise ratio (5.05 dB) synthetic image.

Table I shows the comparison among the Prewitt operator and the directional and the Marr-Hildreth nondirectional second derivative zero crossing edge operators. As before, the

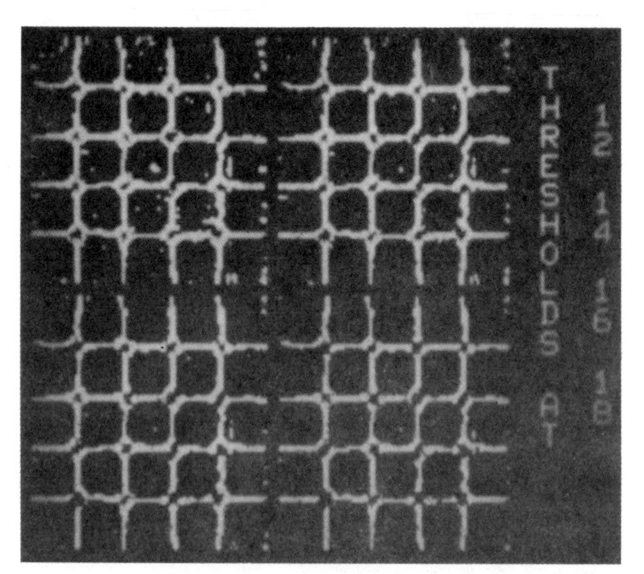

Fig. 12. Illustrates the directional derivative edge operator for a window size of 11 X 11 and deciding that the true gradient is nonzero when the estimated gradient is higher than the thresholds of 12, 14, 16, or 18.

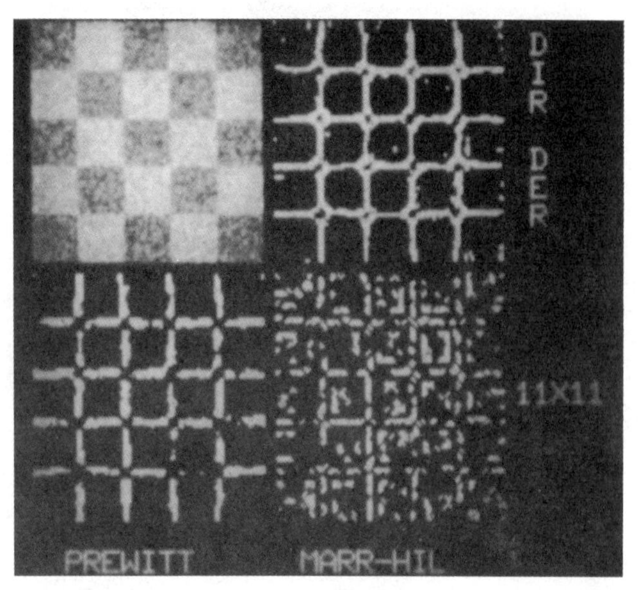

Fig. 13. Compares the directional derivative edge operator with the Marr-Hildreth edge operator and the Prewitt edge operator. The thresholds chosen were the best possible ones.

TABLE I

COMPARES THE PERFORMANCE OF THREE EDGE OPERATORS USING AN 11 X 11 WINDOW ON THE NOISY CHECKERBOARD IMAGE. THRESHOLDS ARE CHOSEN TO EQUALIZE, AS BEST AS POSSIBLE, $P(AE/TE)$, THE CONDITIONAL PROBABILITY OF ASSIGNED EDGE GIVEN TRUE EDGE AND THE CONDITIONAL PROBABILITY, $P(TE/AE)$ OF TRUE EDGE GIVEN ASSIGNED EDGE. THE ERROR DISTANCE IS THE AVERAGE DISTANCE TO CLOSEST TRUE EDGE PIXELS OF PIXELS WHICH ARE ASSIGNED NONEDGE BUT WHICH ARE TRUE EDGE. A VISUAL EVALUATION ALSO LEAVES THE IMPRESSION THAT THE DIRECTIONAL DERIVATIVE OPERATOR PRODUCES BETTER EDGE CONTINUITY AND HAS LESS NOISE THAN THE OTHER TWO.

	Prewitt	Marr-Hildreth	Directional Derivative
Parameters	Gradient Threshold = 18.5	Zero Crossing Strength = 4.0 $\sigma = 5.0$	Gradient Threshold = 14.0 $\rho = 0.5$
$P(AE\|TE)$	0.6738	0.3977	0.7207
$P(TE\|AE)$	0.6872	0.4159	0.7197
Error Distance	1.79	1.76	1.16

threshold used is the one equalizing the conditional probability of assigned edge given true edge and the conditional probability of true edge given assigned edge. It appears that the performance of the directional derivative operator is better than the Prewitt operator and the Marr-Hildreth operator, both on the basis of the correct assignment probability and the error distance which is the average distance to closest true edge pixels of pixels which are assigned nonedge labels but which are true edge pixels.

Fig. 13 shows the corresponding edge images of the 11 X 11 Prewitt operator using a cubic fit rather than a quadratic fit, the 11 X 11 Marr-Hildreth operator, and the 11 X 11 directional derivative zero crossing operator. The thresholds used are the ones to equalize the conditional probabilities as given in Table I.

For the case of constant variance additive noise, thresholding on the basis of the hypothesis test of Section IV-C yields essentially the same results as simply thresholding the gradient value.

Fig. 14 illustrates the second directional derivative zero crossing operator on an aerial image which has been median filtered and then enhanced by replacing each pixel with the closer of its 3 X 3 neighborhood minimum or maximun. The technique is so good that it is possible to determine region boundaries essentially by doing a connected components on nonedge pixels. Fig. 14(b) shows the cleaned edge image which is obtained by doing a connected components on the non edge pixels, then removing all pixels whose region has fewer than 20 pixels. The resulting boundaries are given as pixels which have a neighbor with a different label than its own.

Initial raw edges which leave gaps in a region boundary will in effect make the regions merge in the connected components step. Thus the small number of missing boundaries is surprising. To be sure, we are not advocating connected components as an image segmentation technique. The fact that it works as well as it does is an indication of the strength of the edge detector.

VI. CONCLUSIONS

We have argued that numeric digital image operations should be explained in terms of their actions on the underlying gray tone intensity surface of which the digital image is an observed noisey sample. We called this model the local facet model for digital image processing and showed how the facet model can be used to estimate in each neighborhood the underlying gray tone intensity surface.

We described a digital step edge operator which detects edges at all pixels whose estimated second directional derivative taken in the direction of the gradient has a zero crossing within the pixel's area. We discussed the statistical analysis of this technique, illustrating how to determine confidence intervals for the direction of the gradient and how this interval de-

(a)

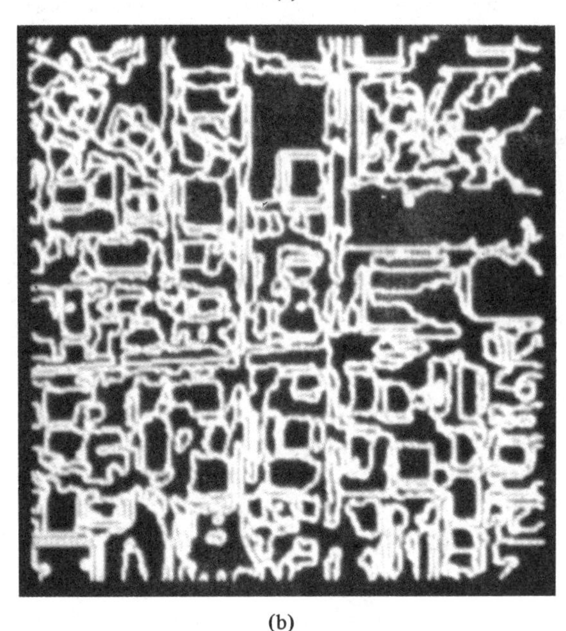

(b)

Fig. 14. (a) Illustrates an aerial photograph. (b) Illustrates the directional derivative edges obtained from the aerial photograph by first 3 × 3 median filtering, then replacing each pixel by the closer of its 3 × 3 neighborhood minimum or maximum, then taking the directional derivative edges using a 7 × 7 window, then doing a connected components on the nonedge pixels, and removing all regions having fewer than 20 pixels, and then displaying any pixel neighboring an pixel different than it as an edge pixel.

termines a confidence interval for the placement of the zero crossing.

We have compared the performance of the directional derivative zero crossing edge operator with that of some classic edge operators, the generalized Prewitt gradient operator, and the Marr-Hildreth zero crossing edge operator. We found that in both the simulated and real image data sets the directional derivative zero crossing edge operator had superior performance.

We have illustrated that for good performance it is important to use larger neighborhood sizes than 3 × 3 and have shown that better results are achieved by defining the edge operator naturally in the large neighborhood rather than preaveraging and then using a smaller neighborhood edge operator on the averaged image.

There is much work yet to be done. A comparison should be done between the image segmentation discussed in Chen and Pavlidis [4] and the segmentation achievable by the edge technique presented in this paper. We need to explore the relationship of basis function kind (polynomial, trigonometric polynomial, etc.), order of fit, and neighborhood size to the goodness of fit. Evaluation must be made of the confidence intervals produced by the technique. The technique needs to be generalized so that it works on saddle points created by two edges crossing. A suitable edge linking method needs to be developed which uses these confidence intervals. Ways of incorporating semantic information and ways of using variable resolution need to be developed. An analogous technique for roof edges needs to be developed. We hope to explore these issues in future papers.

Appendix

The Discrete Orthogonal Polynomials

The discretely orthogonal polynomial basis set of size N which we use has polynomials from degree zero through degree $N - 1$. These polynomials are unique and sometimes called the discrete Chebyshev polynomials [1]. In this Appendix we show how to construct them for one or two variables.

A. Discrete Orthogonal Polynomial Construction Technique

Let the discrete integer index set R be symmetric in the sense that $r \in R$ implies $-r \in R$. Let $P_n(r)$ be the nth order polynomial. We define the construction technique for discrete orthogonal polynomials iteratively.

Define $P_0(r) = 1$. Suppose $P_0(r), \cdots, P_{n-1}(r)$ have been defined. In general, $P_n(r) = r^n + a_{n-1}r^{n-1} + \cdots + a_1 r + a_0$. $P_n(r)$ must be orthogonal to each polynomial $P_0(r), \cdots, P_{n-1}(r)$. Hence, we must have the n equations

$$\sum_{r \in R} P_k(r)(r^n + a_{n-1}r^{n-1} + \cdots + a_1 r + a_0) = 0,$$

$$k = 0, \cdots, n-1. \tag{A1}$$

These equations are linear equations in the unknown $a_0, \cdots, a_{n-1}$ and are easily solved by standard techniques.

The first five polynomial functions formulas are

$$P_0(r) = 1$$

$$P_1(r) = r$$

$$P_2(r) = r^2 - \mu_2/\mu_0$$

$$P_3(r) = r^3 - (\mu_4/\mu_2)r$$

$$P_4(r) = \frac{r^4 + (\mu_2\mu_4 - \mu_6)r^2 + (\mu_2\mu_6 - \mu_4^2)}{\mu_0\mu_4 - \mu_2}$$

where

$$\mu_k = \sum_{s \in R} s^k.$$

B. Two-Dimensional Discrete Orthogonal Polynomials

Two-dimensional discrete orthogonal polynomials can be created from two sets of one-dimensional discrete orthogonal polynomials by taking tensor products. Let R and C be index sets satisfying the symmetry condition $r \in R$ implies $-r \in R$ and $c \in C$ implies $-c \in C$. Let $\{P_0(r), \cdots, P_N(r)\}$ be a set of discrete polynomials on R. Let $\{Q_0(c), \cdots, Q_M(c)\}$ be a set of discrete polynomials on C. Then the set $\{P_0(r)Q_0(c), \cdots, P_n(r)Q_m(c), \cdots, P_N(r)Q_M(c)\}$ is a set of discrete polynomials on $R \times C$.

The proof of this fact is easy. Consider whether $P_i(r)Q_j(c)$ is orthogonal to $P_n(r)Q_m(c)$. When $n \neq i$ or $m \neq j$. Then

$$\sum_{r \in R} \sum_{c \in C} P_i(r)Q_j(c)P_n(r)Q_m(c)$$

$$= \sum_{r \in R} P_i(r)P_n(r) \sum_{c \in C} Q_j(c)Q_m(c).$$

Since by assumption $n \neq i$ or $m \neq j$ the first sum or second sum must be zero, thereby proving the orthogonality.

REFERENCES

[1] P. Beckmann, *Orthogonal Polynomials for Engineers and Physicists.* Boulder, CO: Golem, 1973.
[2] P. Beaudet, "Rotationally invariant image operators," in *Proc. 4th Int. Joint Conf. Pattern Recognition*, Tokyo, Japan, Nov. 1978, pp. 579-583.
[3] M. J. Brooks, "Rationalizing edge detectors," *Comput. Graphics Image Processing*, vol. 8, pp. 277-285, 1978.
[4] P. C. Chen and T. Pavlidis, "Image segmentation as an estimation problem," in *Image Modeling*, A. Rosenfeld, Ed. New York: Academic, 1981.
[5] D. B. Cooper, H. Elliott, F. Cohen, L. Reiss, and P. Symosek, "Stochastic boundary estimation and object recognition," in *Image Modeling*, A. Rosenfeld, Ed. New York: Academic, 1981.
[6] R. Ehrich and F. Schroeder, "Contextual boundary formation by one-dimensional edge detection and scan line matching," *Comput. Graphics Image Processing*, vol. 16, pp. 116-149, 1981.
[7] R. Foutz, personal communication, 1981.
[8] R. Haralick, "Edge and region analysis for digital image data," *Comput. Graphics Image Processing*, vol. 12, pp. 60-73, 1980.
[9] —, "Zero-crossing of second directional derivative edge operator," presented at the SPIE Symp. on Robot Vision, Washington, DC, May 1982.
[10] R. Haralick and L. Watson, "A facet model for image data," *Comput. Graphics Image Processing*, vol. 15, pp. 113-129, 1981.
[11] M. Hueckel, "A local visual operator which recognizes edges and lines," *J. Ass. Comput. Mach.*, vol. 20, pp. 634-647, 1973.
[12] —, "An operator which locates edges in digitized pictures, *J. Ass. Comput. Mach.*, vol. 18, pp. 113-125, 1971.
[13] D. Marr and E. Hildreth, "Theory of edge detection," *Proc. Royal Soc. London*, ser. B, vol. 207, pp. 187-217, 1980.
[14] D. Morgenthaler, "A new hybrid edge detector," *Comput. Graphics Image Processing*, vol. 16, pp. 166-176, 1981.
[15] D. Morgenthaler and A. Rosenfeld, "Multidimensional edge section by hypersurface fitting," *IEEE Trans. Pattern Anal. Machine Intell.*, vol. PAMI-3, pp. 482-486, July 1981.
[16] R. Nevatia and R. Babu, "Linear feature extraction and description," *Comput. Graphics Image Processing*, vol. 13, pp. 257-269, 1980.
[17] J. Prewitt, "Object enhancement and extraction," in *Picture Processing and Psychopictorics*, B. Lipkin and A. Rosenfeld, Eds. New York: Academic, 1970, pp. 75-149.
[18] L. G. Roberts "Machine perception of three-dimensional solids," in *Optical and Electrooptical Imformation Processing*, J. T. Trippett *et al.*, Eds. Cambridge, MA: MIT Press, 1965, pp. 159-197.
[19] A. Rosenfeld and A. Kak, *Digital Picture Processing.* New York: Academic, 1976.
[20] S. Zucker and R. Hummel, "An optimal three-dimensional edge operator," in *Proc. Pattern Recognition and Image Processing Conf.*, Chicago, IL, Aug. 1979, pp. 162-168.

Robert M. Haralick (M'69-SM'76-F'84) was born in Brooklyn, NY, on September 30, 1943. He received the B.A., B.S., M.S., and Ph.D. degrees from the University of Kansas, Lawrence, in 1964, 1966, 1967, and 1969, respectively.

He has worked at Autonetics and IBM. In 1965 he worked for the Center for Research, University of Kansas, as a Research Engineer and in 1969 he joined the faculty of the Department of Electrical Engineering, where he served as a Professor from 1975 to 1978. In 1979 he joined the faculty of the Departments of Electrical Engineering and Computer Science, Virginia Polytechnic Institute and State University, Blacksburg, where he is now a Professor and Director of the Spatial Data Analysis Laboratory. He has done research in pattern recognition, multiimage processing, remote sensing, texture analysis, data compression, clustering, artificial intelligence, and general systems theory. He is responsible for the development of GIPSY (general image processing system), a multiimage processing package which runs on a minicomputer system.

Dr. Haralick is a member of the Association for Computing Machinery, Sigma Xi, the Pattern Recognition Society, and the Society for General Systems Research.

Parts of recognition*

D.D. HOFFMAN
University of California, Irvine

W.A. RICHARDS
Massachusetts Institute of Technology

Abstract

We propose that, for the task of object recognition, the visual system decomposes shapes into parts, that it does so using a rule defining part boundaries rather than part shapes, that the rule exploits a uniformity of nature—transversality, and that parts with their descriptions and spatial relations provide a first index into a memory of shapes. This rule allows an explanation of several visual illusions. We stress the role inductive inference in our theory and conclude with a précis of unsolved problems.

1. Introduction

Any time you view a statue, or a simple line drawing, you effortlessly perform a visual feat far beyond the capability of the most sophisticated computers today, through well within the capacity of a kindergartener. That feat is shape recognition, the visual identification of an object using only its shape. Figure 1 offers an opportunity to exercise this ability and to make several observations. Note first that, indeed, shape alone is sufficient to recognize the objects; visual cues such as shading, motion, color, and texture are not present in the figure. Note also that you could not reasonably predict the contents of the figure before looking at it, yet you recognized the objects.

Figure 1. *Some objects identifiable entirely from their profiles.*

Clearly your visual system is equipped to describe the shape of an object and to guess what the object is from its description. This guess may just be a first guess, perhaps best thought of as a first index into a memory of shapes, and might not be exactly correct; it may simply narrow the potential matches and trigger visual computations designed to narrow them further.

This first guess is more precisely described as an inference, one the truth of whose premises—the descriptions of shape—does not logically guarantee the truth of its conclusion—the identity of the object. Because the truth of the conclusion does not follow logically from the truth of the premises, the strength of the inference must derive from some other source. That source, we claim, is the regularity of nature, its uniformities and general laws. The design of the visual system exploits regularities of nature in two ways: they underlie the mental categories used to represent the world and they permit inferences from impoverished visual data to descriptions of the world.

Regularities of nature play both roles in the visual task of shape recognition, and both roles will be examined. We will argue that, just as syntactic analysis decomposes a sentence into its constituent structure, so the visual system decomposes a shape into a hierarchy of parts. Parts are not chosen arbitrarily; the mental category 'part' of shapes is based upon a regularity of nature discovered by differential topologists—transversality. This is an example of a regularity in the first role. The need arises for a regularity in the second role because although parts are three-dimensional, the eye delivers only a two-dimensional projection. In consequence the three-dimensional parts must be inferred from their two-dimensional projections. We propose

*We are grateful to Thomas Banchoff, Aaron Bobick, Mike Brady, Carmen Egido, Jerry Fodor, Jim Hodgson, Jan Koenderink, Jay Lebed, Alex Pentland, John Rubin, Joseph Scheuhammer, and Andrew Witkin for their helpful discussions and, in some cases, for reading earlier drafts. We are also grateful to Alan Yuille for comments and corrections on the mathematics in the appendices. Preparation of this paper was supported in part by NSF and AFOSR under a combined grant for studies in Natural Computation, grant 79-23110-MCS, and by the AFOSR under an Image Understanding contract F49620-83-C-0135. Technical support was kindly provided by William Gilson; artwork was the creation of Julie Sandell and K. van Buskirk. Reprints may be obtained from D. Hoffman, School of Social Sciences, University of California, Irvine, CA 92717, U.S.A.

that this inference is licensed by another regularity, this time from the field of singularity theory.

2. Why parts?

Before examining a part definition and its underlying regularity, we should ask: Given that one wants to recognize an object from its shape, why partition the shape at all? Could template matching or Fourier descriptors rise to the occasion? Possibly. What follows is not so much intended to deny this as to indicate the usefulness of parts.

To begin, then, an articulation of shapes into parts is useful because one never sees an entire shape in one glance. Clearly the back side is never visible (barring transparent objects), but even the front side is often partially occluded by objects interposed between the shape and the observer. A Fourier approach suffers because all components of a Fourier description can change radically as different aspects of a shape come into view. A part theory, on the other hand, can plausibly assume that the parts delivered by early vision correspond to the parts stored in the shape memory (after all, the contents of the shape memory were once just the products of early visual processing), and that the shape memory is organized such that a shape can be addressed by an inexhaustive list of its parts. Then recognition can proceed using the visible parts.

Parts are also advantageous for representing objects which are not entirely rigid, such as the human hand. A template of an outstretched hand would correlate poorly with a clenched fist, or a hand giving a victory sign, etc. The proliferation of templates to handle the many possible configurations of the hand, or of any articulated object, is unparsimonious and a waste of memory. If part theorists, on the other hand, pick their parts prudently (criteria for prudence will soon be forthcoming), and if they introduce the notion of spatial relations among parts, they can decouple configural properties from the shape of an object, thereby avoiding the proliferation of redundant mental models.

The final argument for parts to be considered here is phenomenological: we see them when we look at shapes. Figure 2, for instance, presents a cosine surface, which observers almost uniformly see organized into ring-like parts. One part stops and another begins roughly where the dotted circular contours are drawn. But if the figure is turned upside down the organization changes such that each dotted circular contour, which before lay between parts, now lies in the middle of a part. Why the parts change will be explained by the partitioning rule to be proposed shortly; the point of interest here is simply that our visual systems do in fact cut surfaces into parts.

Figure 2. *The cosine surface at first appears to be organized into concentric rings, one ring terminating and the next beginning approximately where the dashed circular contours are drawn. But this organization changes when the figure is turned upside down.*

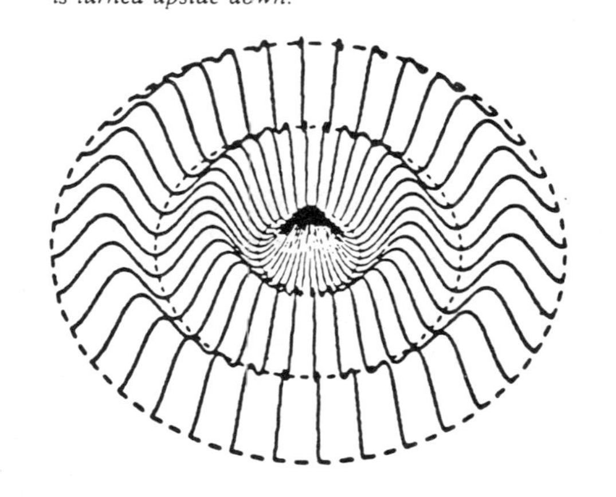

3. Parts and uniformities of nature

Certainly any proper subset of a surface is a part of that surface. This definition of part, however, is of little use for the task of shape recognition. And although the task of shape recognition constrains the class of suitable part definitions (see Section 5), it by no means forces a unique choice. To avoid an *ad hoc* choice, and to allow a useful correspondence between the world and mental representations of shape, the definition of part should be motivated by a uniformity of nature.[1]

One place not to look for a defining regularity is in the shapes of a part. One could say that all parts are cylinders, or cones, or spheres, or polyhedra, or some combination of these; but this is legislating a definition, not discovering a relevant regularity. And such a definition would have but limited applicability, for certainly not all shapes can be decomposed into just cylinders, cones, spheres, and polyhedra.

If a defining regularity is not to be found in part shapes, then another place

[1]Unearthing an appropriate uniformity is the most creative, and often most difficult, step in devising an explanatory theory for a visual task. Other things being equal, one wants the most general uniformity of nature possible, as this grants the theory and the visual task the broadest possible scope.

Figure 3. *An illustration of the transversality regularity. When any two surfaces interpenetrate at random they always meet in concave discontinuities, as indicated by the dashed contours.*

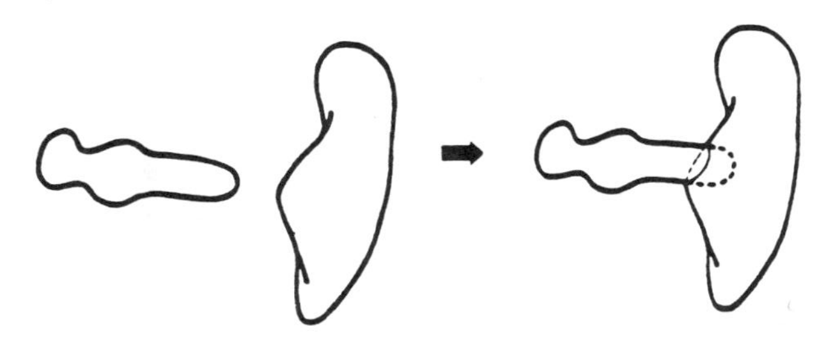

to look is part intersections. Consider the two three-dimensional blobs depicted in the left of Fig. 3. Certainly these two spatially separated shapes are different parts of this figure. Indeed, each spatially distinct object in a visual scene is a part of that scene. Now if two such separate objects are interpenetrated to form one new composite object, as shown in the right of Fig. 3, then the two objects, which were before separate parts of the visual scene, are surely now prime candidates to be parts of the new composite shape. But can we tell, simply by examining the new composite shape, what the original parts are? That is, is there a way to tell where one part stops and the next part begins on the new composite shape? Fortunately there is a way, one which depends on a regularity in the way two shapes generically intersect. This regularity is called transversality (for a detailed discussion of transversality see Guillemin and Pollack (1974)).

- *Transversality regularity*. When two arbitrarily shaped surfaces are made to interpenetrate they always[2] meet in a contour of concave discontinuity of their tangent planes.

To see this more clearly, observe the silhouette of the composite shape shown in the right of Fig. 3. Notice that this composite silhouette is not smooth at the two points where the silhouette of one of its parts intersects the silhouette of the other part. At these two points the direction of the silhouette's outline (i.e., its tangent direction) changes abruptly, creating a concave cusp (i.e., a cusp which points into the object, not into the background) at each of the two points. In fact, such concave discontinuities arise at every point on the surface of the composite shape where the two parts meet. These contours of concave discontinuity of the tangent plane of the composite shape will be the basis for a partitioning rule in the next section. But three observations are in order.

First, though it may sound esoteric, transversality is a familiar part of our everyday experience. A straw in a soft drink forms a circular concave discontinuity where it meets the surface of the drink. So too does a candle in a birthday cake. The tines of a fork in a piece of steak, a cigarette in a mouth, all are examples of this ubiquitous regularity.

Second, transversality does not double as a theory of part growth or part formation (D'Arcy Thompson, 1968). We are not claiming, for example, that a nose was once physically separated from the face and then got attached by interpenetration. We simply note that when two spatially separated shapes are interpenetrated, their intersection is transversal. Later we will see how this regularity underlies the visual definition of separate parts of any composite shape, such as the nose of a face or a limb of a tree, regardless of how the composite shape was created.

Finally, transversality does encompass movable parts. As mentioned earlier, one attraction of parts is that, properly chosen, they make possible a decoupling of configuration and shape in descriptions of articulated objects. But to do this the parts must cut an object at its articulations; a thumb–wrist part on the hand, for instance, would be powerless to capture the various spatial relations that can exist between the thumb and the wrist. Now the parts motivated by transversality will be the movable units, fundamentally because a transversal intersection of two surfaces remains transversal for small pertubations of their positions. This can be appreciated by reviewing Fig. 3. Clearly the intersection of the two surfaces remains a contour of concave discontinuity even as the two surfaces undergo small independent rotations and translations.

4. Partitioning: The minima rule

On the basis of the transversality regularity we can propose a first rule for dividing a surface into parts: divide a surface into parts along all contours of concave discontinuity of the tangent plane. Now this rule cannot help us with the cosine surface because this surface is entirely smooth. The rule must be generalized somewhat, as will be done shortly. But in its present form the rule can provide insight into several well-known perceptual demonstrations.

[2]The word *always* is best interpreted "with probability one assuming the surfaces interpenetrate at random"

4.1. Blocks world

We begin by considering shapes constructed from polygons. Examine the staircase of Fig. 4. The rule predicts that the natural parts are the steps, and not the faces on the steps. Each step becomes a 'part' because it is bounded by two lines of concave discontinuity in the staircase. (A face is bounded by a concave and a convex discontinuity.) But the rule also makes a less obvious prediction. If the staircase undergoes a perceptual reversal, such that the 'figure' side of the staircase becomes 'ground' and *vice versa*, then the step boundaries must change. This follows because only *concave* discontinuities define step boundaries. And what looks like a concavity from one side of a surface must look like a convexity from the other. Thus, when the staircase reverses, convex and concave discontinuities must reverse roles, leading to new step boundaries. You can test this prediction yourself by looking at the step having a dot on each of its two faces. When the staircase appears to reverse note that the two dots no longer lie on a single step, but lie on two adjacent steps (that is, on two different 'parts').

Similar predictions from the rule can also be confirmed with more complicated demonstrations such as the stacked cubes demonstration shown in Fig. 5. The three dots which at first appear to lie on one cube, lie on three different cubes when the figure reverses.

Still another quite different prediction follows from our simple partitioning rule. If the rule does not define a unique partition of some surface, then the division of that surface into parts should be perceptually ambiguous (unless, of course, there are additional rules which can eliminate the ambiguity). An elbow-shaped block provides clear confirmation of this prediction (see Fig. 6). The only concave discontinuity is the vertical line in the crook of the elbow; in consequence, the rule does not define a unique partition of the block. Perceptually, there are three plausible ways to cut the block into parts (also shown in Fig. 6). All three use the contour defined by the partitioning rule, but complete it along different paths.

Figure 4. *The Schroder staircase, published by H. Schroder in 1858, shows that part boundaries change when figure and ground reverse. The two dots which at first appear to lie on one step suddenly seem to lie on two adjacent steps when the staircase reverses.*

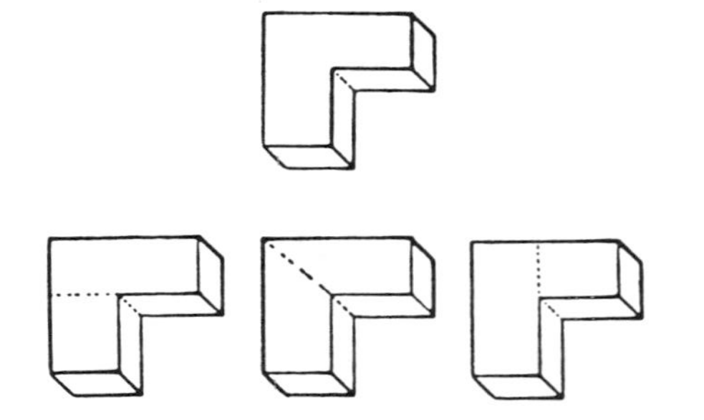

Figure 5. *Stacked cubes also show that parts change when figure and ground reverse. Three dots which sometimes lie on one cube will lie on three different cubes when the figure reverses.*

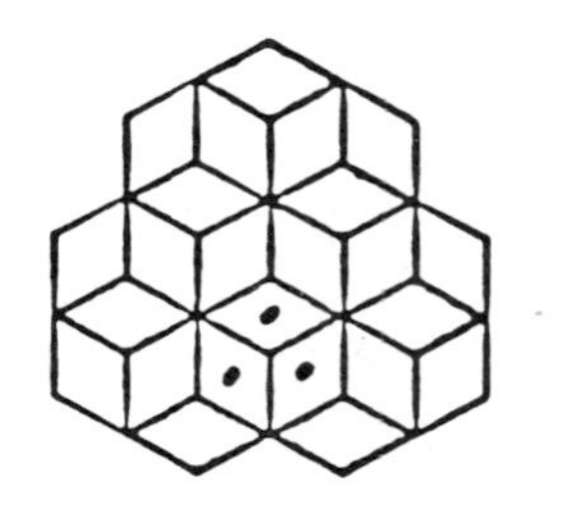

Figure 6. *Elbow-shaped blocks show that a rule partitioning shapes at concave discontinuities is appropriately conservative. The rule does not give a closed contour on the top block, and for good reason. Perceptually, three different partitions seem reasonable, as illustrated by the bottom three blocks.*

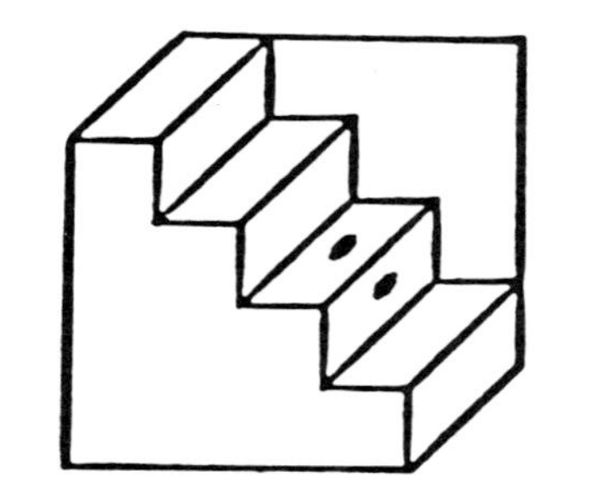

4.2. *Generalization to smooth surfaces*

The simple partitioning rule directly motivated by transversality leads to interesting insights into our perception of the parts of polygonal objects. But how can the rule be generalized to handle smooth surfaces, such as the cosine surface? To grasp the generalization, we must briefly digress into the differential geometry of surfaces in order to understand three important concepts: surface normal, principal curvature, and line of curvature. Fortunately, although these concepts are quite technical, they can be understood intuitively.

The surface normal at a point on a surface can be thought of as a unit length needle sticking straight out of (orthogonal to) the surface at that point, much like the spines on a sea urchin. All the surface normals at all points on a surface are together called a field of surface normals. Usually there are two possible fields of surface normals on a surface—either outward pointing or inward pointing. A sphere, for instance, can either have the surface normals all pointing out like spines, or all pointing to its center. Let us adopt the convention that the field of surface normals is always chosen to point into the figure (i.e., into the object). Thus a baseball has inward normals whereas a bubble under water, if the water is considered figure, has outward normals. Reversing the choice of figure and ground on a surface implies a concomitant change in the choice of the field of surface normals. And, as will be discussed shortly, a reversal of the field of surface normals induces a change in sign of each principal curvature at every point on the surface.

It is often important to know not just the surface normal at a point but also how the surface is curving at the point. The Swiss mathematician Leonhard Euler discovered around 1760 that at any point on any surface there is always a direction in which the surface curves least and a second direction, always orthogonal to the first, in which the surface curves most. (Spheres and planes are trivial cases since the surface curvature is identical in all directions at every point.) These two directions at a point are called the principal directions at that point and the corresponding surface curvatures are called the principal curvatures. Now by starting at some point and always moving in the direction of the greatest principal curvature one traces out a line of greatest curvature. By moving instead in the direction of the least principal curvature one traces out a line of least curvature. On a drinking glass the family of lines of greatest curvature is a set of circles around the glass. The lines of least curvature are straight lines running the length of the glass (see Fig. 7).

With these concepts in hand we can extend the partitioning rule to smooth surfaces. Suppose that wherever a surface has a concave discontinuity we smooth the discontinuity somewhat, perhaps by stretching a taut skin over it.

Figure 7. *Lines of curvature are easily depicted on a drinking glass. Lines of greatest curvature are circles. Lines of least curvature are straight lines.*

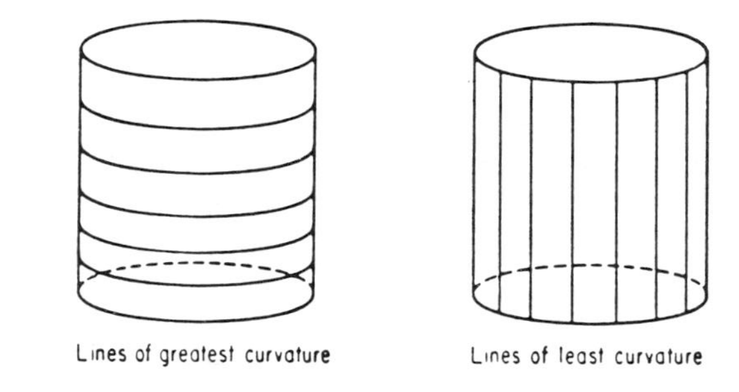

Then a concave discontinuity becomes a contour where, locally, the surface has greatest negative curvature. In consequence we obtain the following generalized partitioning rule for surfaces.

- *Minima rule.* Divide a surface into parts at loci of negative minima of each principal curvature along its associated family of lines of curvature.

The minima rule is applied to two surfaces in Fig. 8. The solid contours indicate members of one family of lines of curvature, and the dotted contours are the part boundaries defined by the minima rule. The bent sheet of paper on the right of Fig. 8 is particularly informative. The lines of curvature shown for this surface are sinusoidal, whereas the family of lines not shown are perfectly straight and thus have zero principal curvature (and no associated minima). In consequence, the product of the two principal curvatures at each point, called the *Gaussian curvature*, is always zero for this surface. Now if the Gaussian curvature is always zero on this surface, then the Gaussian curvature cannot be used to divide the surface into parts. But we see parts on this surface. Therefore whatever rule our visual systems use to partition surfaces cannot be stated entirely in terms of Gaussian curvature. In particular, the visual system cannot be dividing surfaces into parts at loci of zero Gaussian curvature (parabolic points) as has been proposed by Koenderink and van Doorn (1982b).

The minima rule partitions the cosine surface along the circular dotted contours shown in Fig. 2. It also explains why the parts differ when figure and ground are reserved. For when the page is turned upside down the visual system reverses its assignment of figure and ground on the surface (perhaps

Figure 8. *Part boundaries, as defined by the smooth surface partitioning rule, are indicated by dashed lines on several different surfaces. The families of solid lines are the lines of curvature whose minima give rise to the dashed partitioning contour.*

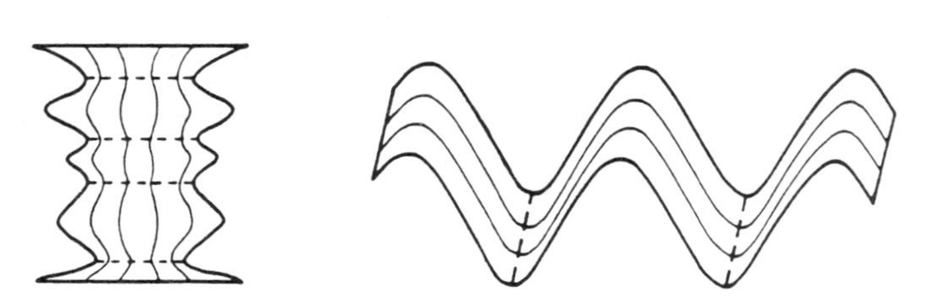

due to a preference for an interpretation which places the object below rather than overhead). When figure and ground reverse so does the field of surface normals, in accordance with the convention mentioned earlier. But simple calculations show that when the normals reverse so too does the sign of the principal curvatures. Consequently minima of the principal curvatures must become maxima and *vice versa*. Since minima of the principal curvatures are used for part boundaries, it follows that these part boundaries must also move. In sum, parts appear to change because the partitioning rule, motivated by the transversality regularity, uses minima of the principal curvatures, and because these minima relocate on the surface when figure and ground reverse. A more rigorous treatment of the partitioning rule is provided in Appendix 1.

5. Parts: Constraints from recognition

The task of visual recognition constrains one's choice of parts and part descriptions. We evaluate the part scheme proposed here against three such constraints—*reliability, versatility*, and *computability*—and then note a non-constraint, *information preservation*.

Reliability. Recognition is fundamentally a process of matching descriptions of what one sees with descriptions already in memory. Imagine the demands on memory and on the matching process if every time one looked at an object one saw different parts. A face, for example, which at one instant appeared to be composed of eyes, ears, a nose, and a mouth, might at a later instant metamorphose into a potpourri of eye–cheek, nose–chin, and mouth–ear parts—a gruesome and unprofitable transmutation. Since no advantage accrues for allowing such repartitions, in fact since they are uniformly deleterious to the task of recognition, it is reasonable to disallow them and to require that the articulation of a shape into parts be invariant over time and over change in viewing geometry. This is the constraint of reliability (see Marr, 1982; Marr and Nishihara, 1978; Nishihara, 1981; Sutherland, 1968): the parts of a shape should be related reliably to the shape. A similar constraint governs the identification of linguistic units in a speech stream (Liberman *et al*, 1967; Fodor, 1983). Apparently the shortest identifiable unit is the syllable: shorter units like phones are not related reliably to acoustic parameters.

The minima rule satisfies this reliability constraint because it uses only surface properties, such as extrema of the principal curvatures, which are independent (up to a change in sign) of the coordinate system chosen to parametrize the surface (Do Carmo, 1976). Therefore the part boundaries do not change when the viewing geometry changes. (The part boundaries do change when figure and ground reverse, however.)

Versatility. Not all possible schemes for defining parts of surfaces are sufficiently versatile to handle the infinite variety in shape that objects can exhibit. Other things being equal, if one of two partitioning schemes is more versatile than another, in the sense that the class of objects in its scope properly contains the class of objects in the scope of the other scheme, the more versatile scheme is to be preferred. A partitioning scheme which can be applied to any shape whatsoever is most preferable, again other things being equal. This versatility constraint can help choose between two major classes of partitioning schemes: boundary-based and primitive-based. A *boundary-based* approach defines parts by their contours of intersection, not by their shapes. A *primitive-based* approach defines parts by their shapes, not by their contours of intersection (or other geometric invariants, such as singular points).

Shape primitives currently being discussed in the shape representation literature include spheres (Badler and Bajcsy, 1978; O'Rourke and Badler, 1979), generalized cylinders (Binford, 1971; Brooks *et al.*, 1979; Marr and Nishihara, 1978; Soroka, 1979), and polyhedra (Baumgart, 1972; Clowes, 1971; Guzman, 1969; Huffman, 1971; Mackworth, 1973; Waltz, 1975), to name a few (see Ballard and Brown, 1982). The point of interest here is that, for all the interesting work and conceptual advances it has fostered, the primitive-based approach has quite limited versatility. Generalized cylinders, for instance, do justice to animal limbs, but are clearly inappropriate for faces, cars, shoes, ... the list continues. A similar criticism can be levelled against each proposed shape primitive, or any conjunction of shape primitives. Perhaps a large enough conjunction of primitives could handle most shapes we do in fact encounter, but the resulting proposal would more resemble a restaurant menu than a theory of shape representation.

A boundary-based scheme on the other hand, if its rules use only the geometry (differential or global) of surfaces, can apply to any object whose bounding surface is amenable to the tools of differential geometry—a not too severe restriction.[3] Boundary rules simply tell one where to draw contours on a surface, as if with a felt marker. A boundary-based scheme, then, is to be preferred over a primitive-based scheme because of its greater versatility.

The advantage of a boundary-based scheme over a primitive-based scheme can also be put this way: using a boundary-based scheme one can locate the parts of an object without having any idea of what the parts look like. This is not possible with the primitive-based scheme. Of course one will want descriptions of the parts one finds using a boundary-based scheme, and one may (or may not) be forced to a menu of shapes at this point. Regardless, a menu of part shapes is not necessary for the task of locating parts. In fact a menu-driven approach restricts the class of shapes for which parts can be located. The minima rule, because it is boundary-based and uses only the differential geometry of surfaces, satisfies the versatility constraint—all geometric surfaces are within its scope.[4]

Computability. The partitioning scheme should in principle be computable using only information available in retinal images. Otherwise it is surely worthless. This is the constraint of *computability*. Computability is not to be confused with efficiency. Efficiency measures how quickly and inexpensively something can be computed, and is a dubious criterion because it depends not only on the task, but also on the available hardware and algorithms. Computability, on the other hand, states simply that the scheme must in principle be realizable, that it use only information available from images.

We have not yet discussed whether our parts are computable from retinal images (but see Appendix 2). And indeed, since minima of curvature are third derivative entities, and since taking derivatives exaggerates noise, one might legitimately question whether our part boundaries are computable. This concern for computability brings up an important distinction noted by Marr and Poggio (1977), the distinction between theory and algorithm. A theory in vision states what is being computed and why; an algorithm tells how. Our partitioning rule is a theoretical statement of what the part boundaries should be, and the preliminary discussion is intended to say why. The rule is not intended to double as an algorithm so the question of computability is still open. Some recent results by Yuille (1983) are encouraging though. He has found that directional zero-crossings in the shading of a surface are often located on or very near extrema of one of the principal curvatures along its associated lines of curvature. So it might be possible to read the part boundaries directly from the pattern of shading in an image, avoiding the noise problems associated with taking derivatives (see also Koenderink and van Doorn, 1980, 1982a). It is also possible to determine the presence of part boundaries directly from occluding contours in an image (see Appendix 2).

Information preservation: A non-constraint. Not just any constraints will do. The constraints must follow from the visual task; otherwise the constraints may be irrelevant and the resulting part definitions and part descriptions inappropriate. Because the task of recognition involves classification, namely the assignment of an individual to a class or a token to a type, not all the information available about the object is required. Indeed, in contrast to some possible needs for machine vision (Brady, 1982b, 1982c), we stress that a description of a shape for recognition need not be information preserving, for the goal is not to reconstruct the image. Rather it is to make explicit just what is key to the recognition process. Thus, what is critical is the form of the representation, what it makes explicit, how well it is tailored to the needs of recognition. Raw depth maps preserve all shape information of the visible surfaces, but no one proposes them as representations for recognition because they are simply not tailored for the task.

6. Projection and parts

We have now discussed how 'parts' of shapes may be defined in the three-dimensional world. However the eye sees only a two-dimensional projection. How then can parts be inferred from images? Again, we proceed by seeking a regularity of nature. As was noted earlier, the design of the visual system exploits regularities of nature in two ways: they underlie the mental categories

[3]Shapes outside the purview of traditional geometric tools might well be represented by fractal-based schemes (Mandelbrot, 1982; Pentland 1983). Candidate shapes are trees, shrubs, clouds—in short, objects with highly crenulate or ill-defined surfaces.

[4]One must, however, discover the appropriate scales for a natural surface (Hoffman, 1983a, b; Witkin, 1983). The locations of the part boundaries depend, in general, on the scale of resolution at which the surface is examined. In consequence an object will not receive a single partitioning based on the minima rule, but will instead receive a nested hierarchy of partitions, with parts lower in the hierarchy being much smaller than parts higher in the hierarchy. For instance, at one level in the hierarchy for a face one part might be a nose. At the next lower level one might find a wart on the nose. The issue of scale is quite difficult and beyond the scope of this paper.

used to represent the world and they license inferences from impoverished visual data to descriptions of the world. The role of transversality in the design of the mental category 'part' of shape is an example of the first case. In this section we study an example of the second case. We find that lawful properties of the singularities of the retinal projection permit an inference from retinal images to three-dimensional part boundaries. For simplicity we restrict attention to the problem of inferring part boundaries from silhouettes.

Consider first a discontinuous part boundary (i.e., having infinite negative curvature) on a surface embedded in three dimensions (Fig. 3). Such a contour, when imaged on the retina, induces a concave discontinuity in the resulting silhouette (notice the concave cusps in the silhouette of Fig. 3). Smooth part boundaries defined by the minima partitioning rule can also provide image cusps, as shown in the profiles of Fig. 1. It would be convenient to infer the presence of smooth and discontinuous part boundaries in three dimensions from concave discontinuities in the two-dimensional silhouette, but unfortunately other surface events can give rise to these discontinuities as well. A torus (doughnut), for instance, can have two concave discontinuities in its silhouette which do not fall at part boundaries defined by the minima rule (see Fig. 9).

Fortunately, it is rare that a concave discontinuity in the silhouette of an object does not indicate a part boundary, and when it does not this can be detected from the image data. So one can, in general, correctly infer the presence or absence of part boundaries from these concave discontinuities. The proof of this useful result (which is banished to Appendix 2) exploits regularities of the singularities of smooth maps between two-dimensional manifolds. We have seen how a regularity of nature underlies a mental category, *viz.*, 'part' of shape; here we see that another regularity (e.g., a singularity regularity) licenses an inference from the retinal image to an instance of this category.

Figure 9. *A torus can have concave discontinuities (inducated by the arrows) which do not correspond to part boundaries.*

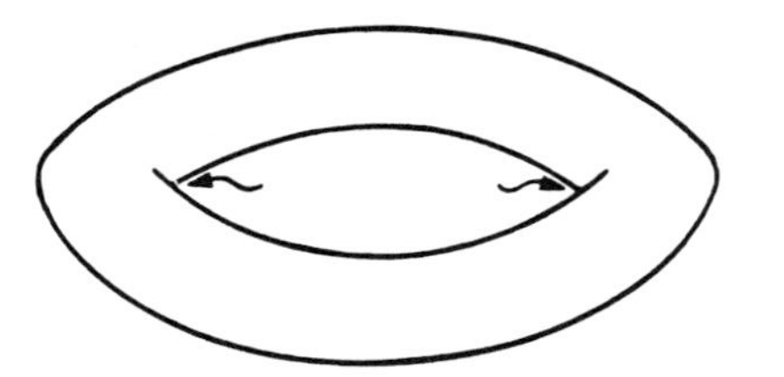

Figure 10. *A reversing figure, similar to Attneave (1974), appears either as an alternating chain of tall and short mountains or as a chain of tall mountains with twin peaks.*

The singularity regularity, together with transversality, motivates a first partitioning rule for plane curves: *Divide a plane curve into parts at concave cusps.* Here the word *concave* means concave with respect to the silhouette (figure) side of the plane curve. A concavity in the figure is, of course, a convexity in the ground.

This simple partitioning rule can explain some interesting perceptual effects. In Fig. 10, for instance, the same wiggly contour can look either like valleys in a mountain range or, for the reversed figure–ground assignment, like large, twin-peaked mountains. The contour is carved into parts differently when figure and ground reverse because the partitioning rule uses only concave cusps for part boundaries. And what is a concave cusp if one side of the contour is figure must become a convex cusp when the other side is figure, and *vice versa*. There is an obvious parallel between this example and the reversible staircase discussed earlier.

6.1. Geometry of plane curves

Before generalizing the rule to smooth contours we must briefly review two concepts from the differential geometry of place curves: principal normal and curvature. The principal normal at a point on a curve can be thought of as a unit length needle sticking straight out of (orthogonal to) the curve at that point, much like a tooth on a comb. All the principal normals at all points on a curve together form a field of principal normals. Usually there are two possible fields of principal normals—either leftward pointing or rightward pointing. Let us adopt the convention that the field of principal normals is always chosen to point into the figure side of the curve. Reversing the choice of figure and ground on a curve implies a concomitant change in the choice of the field of principal normals.

Curvature is a well-known concept. Straight lines have no curvature, circles have constant curvature, and smaller circles have higher curvature than larger circles. What is important to note is that, because of the convention forcing the principal normals to point into the figure, concave portions of a smooth curve have negative curvature and convex portions have positive curvature.

6.2. Parts of smooth curves

It is an easy matter now to generalize the partitioning rule. Suppose that wherever a curve has a concave cusp we smooth the curve a bit. Then a concave cusp becomes a point of negative curvature having, locally, the greatest absolute value of curvature. This leads to the following generalized partitioning rule: *Divide a plane curve into parts at negative minima of curvature.*[5]

Several more perceptual effects can be explained using this generalized partitioning rule. A good example is the reversing figure devised by Attneave (see Fig. 11). He found that by simply scribbling a line through a circle and separating the two halves one can create two very different looking contours. As Attneave (1971) points out, the appearance of the contour depends upon which side is taken to be part of the figure, and does not depend upon any prior familiarity with the contour.

Figure 11. *Attneave's reversing figure, constructed by scribbling a line down a circle. The apparent shape of a contour depends on which side is perceived as figure.*

Now we can explain why the two halves of Attneave's circle look so different. For when figure and ground reverse, the field of principal normals also reverses in accordance with the convention. And when the principal normals reverse, the curvature at every point on the curve must change sign. In particular, minima of curvature must become maxima and *vice versa*. This repositioning of the minima of curvature leads to a new partitioning of the curve by the partitioning rule. In short, the curve looks different because it is organized into fundamentally different units or chunks. Note that if we chose to define part boundaries by inflections (see Hollerbach, 1975; Marr, 1977), or by both maxima and minima of curvature (see Duda and Hart, 1973), or by all tangent and curvature discontinuities (Binford, 1981), then the chunks would not change when figure and ground reverse.

A clear example of two very different chunkings for one curve can be seen in the famous face–goblet illusion published by Turton in 1819. If a face is taken to be figure, then the minima of curvature divide the curve into chunks corresponding to a forehead, nose, upper lip, lower lip, and chin. If instead the goblet is taken to be figure then the minima reposition, dividing the curve into new chunks corresponding to a base, a couple of parts of the stem, a bowl, and a lip on the bowl. It is probably no accident that the parts defined by minima are often easily assigned verbal labels.

Demonstrations have been devised which, like the face–goblet illusion, allow more than one interpretation of a single contour but which, unlike the face–goblet illusion, do not involve a figure–ground reversal. Two popular examples are the rabbit–duck and hawk–goose illusions (see Fig. 13). Because these illusions do not involve a figure–ground reversal, and because in consequence the minima of curvature never change position, the partitioning rule

Figure 12. ***The reversing goblet can be seen as a goblet or a pair of facial profiles (adapted from Turton, 1819). Defining part boundaries by minima of curvature divides the face into a forehead, nose, upper lip, lower lip, and chin. Minima divide the goblet into a base, a couple parts of the stem, a bowl, and a lip on the bowl.***

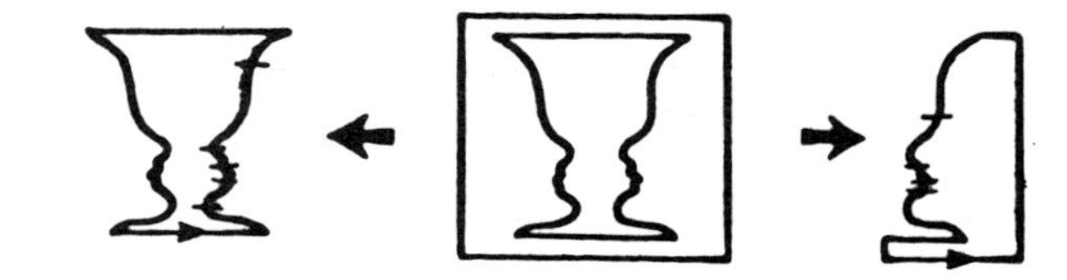

[5]Transversality directly motivates using concave cusps as part boundaries. Only by smoothing do we include minima as well (both in the case of silhouette curves and in the case of part boundaries in three dimensions). Since the magnitude of the curvature at minima decreases with increased smoothing, it is useful to introduce the notion of the strength or goodness of a part boundary. The strength of a part boundary is higher the more negative the curvature of the minimum. Positive minima have the least strength, and deserve to be considered separately from the negative minima, a possibility suggested to us by Shimon Ullman.

Figure 13. *Some ambiguous shapes do not involve a reversal of figure and ground. Consequently, the part boundaries defined by minima of curvature do not move when these figures change interpretations. In this illustration, for instance, a rabbit's ear turns into a duck's bill without moving, and a hawk's head turns into a goose's tail, again without moving.*

must predict that the part boundaries are identical for both interpretations of each of these contours. This prediction is easily confirmed. What is an ear on the rabbit, for instance, becomes an upper bill on the duck.

If the minima rule for partitioning is really used by our visual systems, one should expect it to predict some judgments of shape similarity. One case in which its prediction is counterintuitive can be seen in Fig. 14. Look briefly at the single half-moon on the right of the figure. Then look quickly at the two half-moons on the left and decide which seems more similar to the first (go ahead). In an experiment performed on several similar figures, we found that nearly all subjects chose the bottom half-moon as more similar. Yet if you look again you will find that the bounding contour for the top half-moon is identical to that of the right half-moon, only figure–ground reversed. The bounding contour of the bottom half-moon, however., has been mirror reversed, and two parts defined by minima of curvature have been swapped. Why does the bottom one still look more similar? The minima rule gives a simple answer. The bottom contour, which is not figure–ground reversed from the original contour, has the same part boundaries. The top contour, which is figure–ground reversed from the original, has entirely different part boundaries.

7. Holes: A second type of part

The minima rule for partitioning surfaces is motivated by a fact about generic intersections of surfaces: surfaces intersect transversally. As Fig. 3 illustrates, this implies that if two surfaces are interpenetrated and left together to form a composite object then the contour of their intersection is a contour of concave discontinuity on the composite surface. Now suppose instead that after the two surfaces are interpenetrated one surface is pulled out of the other, leaving behind a depression, and then discarded. The depression created in this manner has just as much motivation for being a 'part' on the basis of transversality as the parts we have discussed up to this point.

Figure 14. *A demonstration that some judgments of shape similarity can be predicted by the minima partitioning rule. In a quick look, the bottom left half-moon appears more similar to the right half-moon than does the top left one. However the bounding contour of the top left half-moon is identical to that of the right half-moon, whereas the bounding contour of the bottom left half-moon has been mirror reversed and has had two parts interchanged.*

As can be seen by examining the right side of Fig. 3, the contour that divides one part from the other on the composite object is precisely the same contour that will delimit the depression created by pulling out the penetrating part. But whereas in the case of the composite object this contour is a contour of *concave* discontinuity, in the case of the depression this contour is a contour of *convex* discontinuity. And smoothing this contour leads to positive extrema of a principal curvature for the case of a depression. We are led to conclude that a shape can have at least two kinds of parts—'positive parts' which are bounded by negative extrema of a principal curvature, and 'negative parts' (holes) bounded by positive extrema of a principal curvature.

This result presents us with the task of finding a set of rules that determine when to use positive extrema or negative extrema as part boundaries. We do not have these rules yet, but here is an example of what such rules might look like. If a contour of negative extrema of a principal curvature is not a closed contour, and if it is immediately surrounded (i.e., no intervening extrema) by a closed contour of positive extrema of a principal curvature, then take the contour of positive extrema as the boundary of a (negative) part.

Note in any case that what we will not have are single parts bounded by both negative and positive extrema of a principal curvature.

8. Perception and induction

Inferences and regularities of nature have cropped up many times in the theory and discussions presented here. It is useful to explore their significance more fully.

Perceptual systems inform the perceiver about properties of the world she needs to know. The need might be to avoid being eaten, to find what is edible, to avoid unceremonious collisions, or whatever. The relevant knowledge might be the three-dimensional layout of the immediate surrounds, or that ahead lies a tree loaded with fruit, or that crouched in the tree is an unfriendly feline whose perceptual systems are also at work reporting the edible properties of the world. Regardless of the details, what makes the perceptual task tricky is that the data available to a sensorium invariably underdetermine the properties of the world that need to be known. That is, in general there are infinitely many states of the world which are consistent with the available sense data. Perhaps the best known example is that although the world is three-dimensional, and we perceive it as such, each retina is only two-dimensional. Since the mapping from the world to the retina is many-to-one, the possible states of the world consistent with a retinal image, or any series of retinal images, are many. The upshot of all this is that knowledge of the world is inferred. Inference lies at the heart of perception (Fodor and Pylyshyn, 1981; Gregory, 1970; Helmholtz, 1962; Hoffman, 1983b; Marr, 1982).

An inference, reduced to essentials, is simply a list of premises and a conclusion. An inference is said to be *deductively valid* if and only if the conclusion is logically guaranteed to be true given that the premises are true. So, for example, the following inference, which has three premises and one conclusion, is deductively valid: "A mapping from 3-D to 2-D is many-to-one. The world is 3-D. A retinal image is 2-D. Therefore a mapping from the world to a retinal image is many-to-one." An inference is said to be *inductively strong* if and only if it is unlikely that the conclusion is false while its premises are true, and it is not deductively valid (see Skyrms, 1975).[6] So the following inference is inductively strong: "The retinal disparities across my visual field are highly irregular. Therefore whatever I am looking at is not flat." Though this inference is inductively strong, it can prove false, as is in fact the case whenever one views a random dot stereogram.

In perceptual inferences the sensory data play the role of the premises, and the assertions about the state of the world are the conclusions. Since the state of the world is not logically entailed by the sensory data, perceptual inferences are not of the deductive variety—therefore they are inductive.

This is not good news. Whereas deductive inference is well understood, inductive inference is almost not understood at all. Induction involves a morass of unresolved issues, such as projectibility (Goodman, 1955), abduction (Levi, 1980; Peirce, 1931), and simplicity metrics (Fodor, 1975). These problems, though beyond the scope of this paper, apply with unmitigated force to perceptual inferences and are thus of interest to students of perception (Nicod, 1968).

But, despite these difficulties, consider the following question: If the premises of perceptual inferences are the sensory data and the conclusion is an assertion about the state of the world, what is the evidential relation between perceptual premises and conclusions? Or to put it differently, how is it possible that perceptual interpretations of sensory data bear a nonarbitrary (and

[6] The distinction between deductively valid and inductively strong inferences is not mere pedantry; the distinction has important consequences for perception, but is often misunderstood. Gregory (1970, p. 160), for instance, realizes the distinction is important for theories of perception, but then claims that "Inductions are generalizations of instances." This is but partly true. Inductive inferences may proceed from general premises to general conclusions, from general premises to particular conclusions, as well as from particular premises to general conclusions (Skyrms, 1975). The distinction between inductive and deductive inferences lies in the evidential relation between premises and conclusions.

even useful) relation to the state of the world? Or to put it still differently, why are perceptual inferences inductively strong?

Surely the answer must be, at least in part, that since the conclusion of a perceptual inference is a statement about the world, such an inference can be inductively strong only if it is motivated by laws, regularities, or uniformities of nature. To see this in a more familiar context, consider the following inductively strong inference about the world: "If I release this egg, it will fall". The inference here is inductively strong because it is motivated by a law of nature—gravity. Skeptics, if there are any, will end up with egg on their feet.

Laws, regularities, and uniformities in the world, then, are crucial for the construction of perceptual inferences which have respectable inductive strength. Only by exploiting the uniformities of nature can a perceptual system overcome the paucity of its sensory data and come to useful conclusions about the state of the world.

If this is the case, it has an obvious implication for perceptual research: identifying the regularities in nature which motivate a particular perceptual inference is not only a good thing to do, but a *sine qua non* for explanatory theories of perception.[7] An explanatory theory must state not only the premises and conclusion of a particular perceptual inference, but also the lawful properties of the world which license the move from the former to the latter. Without all three of these ingredients a proposed theory is incomplete.

[7] At least two conditions need to be true of a regularity, such as rigidity, for it to be useful: (1) It should in fact be a regularity. If there were not rigid objects in the world, rigidity would be useless. (2) It should allow inductively strong inferences from images to the world, by making the 'deception probability', to be defined shortly, very close to zero. For instance, let w (world) stand for the following assertion about four points in the world: "are in rigid motion in 3-D". Let i (image) stand for the following assertion about the retinal images of the same four points: "have 2-D positions and motions consistent with being the projections of rigid motion in 3-D". Then what is the probability of w given i? The existence of rigid objects does not in itself make this conditional probability high. Using Bayes' theorem we find that $P(w|i) = P(w) \cdot P(i|w)/[P(w) \cdot P(i|w) + P(-w) \cdot P(i|-w)]$. Since the numerator and the first term of the denominator are identical, this conditional probability is near one only if $P(w) \cdot P(i|w) \gg P(-w) \cdot P(i|-w)$. And since $P(-w)$, though unknown is certainly much greater than zero, $P(w|i)$ is near one only if $P(i|-w)$—let's call this the 'deception probability'—is near zero. Only if the deception probability is near zero can the inference from the image to the world be inductively strong. A major goal of 'structure from motion' proofs (Bobick, 1983; Hoffman and Flinchbaugh, 1982; Longuet-Higgins and Prazony, 1981; Richards *et al.*, 1983; Ullman, 1979) is to determine under what conditions this deception probability is near zero. Using an assumption of rigidity, for instance, Ullman has found that with three views of three points the deception probability is one, but with three views of four points it is near zero.

9. Conclusion

The design of the visual system exploits regularities of nature in two ways: they underlie the mental categories used to represent the world and they license inferences from incomplete visual data to useful descriptions of the world. Both uses of regularities underlie the solution to a problem in shape recognition. Transversality underlies the mental category 'part' of shape; singularities of projection underlie the inference from images to parts in the world.

The partitioning rules presented in this paper are attractive because (1) they satisfy several constraints imposed by the task of shape recognition, (2) they are motivated by a regularity of nature, (3) the resulting partitions look plausible, and (4) the rules explain and unify several well-known visual illusions.

Remaining, however, is a long list of questions to be answered before a comprehensive, explanatory theory of shape recognition is forthcoming. A partial list includes the following. How are the partitioning contours on surfaces to be recovered from two-dimensional images? How should the surface parts be described? All we have so far is a rule for cutting out parts. But what qualitative and metrical descriptions should be applied to the resulting parts? Can the answer to this question be motivated by appeal to uniformities and regularities in the world? What spatial relations need to be computed between parts? Although the part definitions don't depend upon the viewing geometry, is it possible or even necessary that the predicates of spatial relations do (Rock, 1974; Yin, 1970)? How is the shape memory organized? What is the first index into this memory?

The task of vision is to infer useful descriptions of the world from changing patterns of light falling on the eye. The descriptions can be reliable only to the extent that the inferential processes which build them exploit regularities in the visual world, regularities such as rigidity and transversality. The discovery of such regularities, and the mathematical investigation of their power in guiding particular visual inferences, are promising directions for the researcher seeking to understand human vision.

Appendix 1

Surface partitioning in detail

This appendix applies the surface partitioning rule to a particular class of surfaces: surfaces of revolution. The intent is to convey a more rigorous understanding of the rule and the partitions it yields. Since this section is quite mathematical, some readers might prefer to look at the results in Fig. 16 and skip the rest.

Notation. Tensor notation is adopted in this section because it allows concise expression of surface concepts, (see Dodson and Poston, 1979; Hoffman, 1983a; Lipschutz, 1969). A vector in $\Re^3$ is $\mathbf{x} = (x^1, x^2, x^3)$. A point in the parameter plane is (u^1, u^2). A surface patch is $\mathbf{x} = \mathbf{x}(u^1, u^2) = (x^1(u^1, u^2), x^2(u^1, u^2), x^3(u^1, u^2))$. Partial deriviatives are denoted by subscripts:

$$\mathbf{x}_1 = \frac{\partial \mathbf{x}}{\partial u^1}, \mathbf{x}_2 = \frac{\partial \mathbf{x}}{\partial u^2}, \mathbf{x}_{12} = \frac{\partial^2 \mathbf{x}}{\partial u^1 \partial u_2}, \text{ etc.}$$

A tangent vector is $d\mathbf{x} = \mathbf{x}_1 du^1 + \mathbf{x}_2 du^2 = \mathbf{x}_i du^i$ where the Einstein summation convention is used. The first fundamental form is

$$\text{I} = d\mathbf{x} \cdot d\mathbf{x} = \mathbf{x}_i \cdot \mathbf{x}_j du^i du^j = g_{ij} du^i du^j$$

where the g_{ij} are the first fundamental coefficients and $i, j = 1, 2$.

The differential of the normal vector is the vector $d\mathbf{N} = \mathbf{N}_i du^i$ and the second fundamental form is

$$\text{II} = d^2\mathbf{x} \cdot \mathbf{N} = \mathbf{x}_{ij} \cdot \mathbf{N} du^i du^j = b_{ij} du^i du^j$$

where the b_{ij} are the second fundamental coefficients and $i, j = 1, 2$.

A plane passing through a surface S orthogonal to the tangent plane of S at some point P and in a direction $du^i{:}du^j$ with respect to the tangent plane intersects the surface in a curve whose curvature at P is the *normal curvature* of S at P in the direction $du^i{:}du^j$. The normal curvature in a direction $du^i{:}du^j$ is $k_n = \text{II}/\text{I}$. The two perpendicular directions for which the values of k_n take on maximum and minimum values are called the *principal directions*, and the corresponding curvatures, k_1 and k_2, are called the *principal curvatures*. The *Gaussian curvature* at P is $K = k_1 k_2$. A *line of curvature* is a curve on a surface whose tangent at each point is along a principal direction.

Partitions of a surface of revolution. A surface of revolution is a set $S \subset \Re^3$ obtained by rotating a regular plane curve α about an axis in the plane which does not meet the curve. Let the x^1x^3 plane be the plane of α and the x^3 axis the rotation axis. Let

$$\alpha(u^1) = (x(u^1), z(u^1)), \ a < u^1 < b, \ z(u^1) > 0.$$

Let u^2 be the rotation angle about the x^3 axis. Then we obtain a map

$$\mathbf{x}(u^1, u^2) = (x(u^1)\cos(u^2), x(u^1)\sin(u^2), z(u^1))$$

Figure 15. *Surface of revolution.*

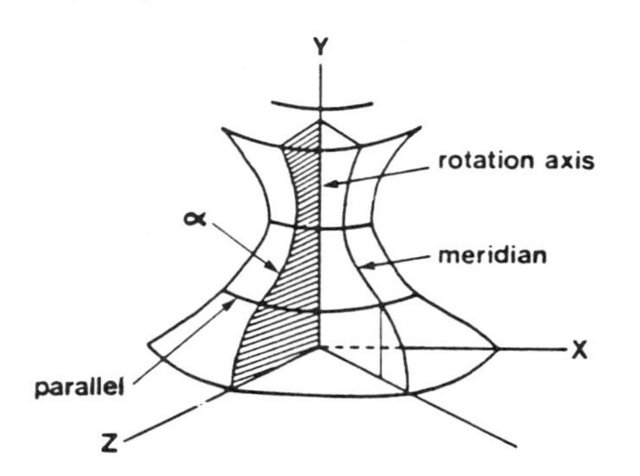

from the open set $U = \{(u^1, u^2) \in \Re^2; 0 < u^2 < 2\pi, a < u^1 < b\}$ into S (Fig. 15). The curve α is called the *generating curve* of S, and the x^3 axis is the *rotation axis* of S. The circles swept out by the points of α are called the *parallels* of S, and the various placements of α on S are called the *meridians* of S.

Let $\cos(u^2)$ be abbreviated as c and $\sin(u^2)$ as s. Then $\mathbf{x}_1 = (x_1 c, x_1 s, z_1)$ and $\mathbf{x}_2 = (-xs, xc, 0)$. The first fundamental coefficients are then

$$g_{ij} = \mathbf{x}_i \cdot \mathbf{x}_j = \begin{pmatrix} x_1^2 + z_1^2 & 0 \\ 0 & x^2 \end{pmatrix}.$$

The surface normal is

$$\mathbf{N} = \frac{\mathbf{x}_1 \times \mathbf{x}_2}{|\mathbf{x}_1 \times \mathbf{x}_2|} = \frac{(z_1 c, z_1 s, -x_1)}{\sqrt{z_1^2 + x_1^2}}.$$

If we let u be arc length along α then $\sqrt{z_1^2 + x_1^2} = 1 = g_{11}$ and

$$\mathbf{N} = (z_1 c, z_1 s, -x_1).$$

The second fundamental coefficients are

$$b_{ij} = \mathbf{x}_{ij} \cdot \mathbf{N} = \begin{pmatrix} x_{11} z_1 - x_1 z_{11} & 0 \\ 0 & -x z_1 \end{pmatrix}.$$

Since $g_{12} = b_{12} = 0$ the principal curvatures of a surface of revolution are

$$k_1 = b_{11}/g_{11} = x_{11}z_1 - x_1z_{11}$$
$$k_2 = b_{22}/g_{22} = -z_1/x.$$

The expression for k_1 is identical to the expression for the curvature along α. In fact the meridians (the various positions of α on S) are lines of curvature, as are the parallels. The curvature along the meridians is given by the expression for k_1 and the curvature along the parallel is given by the expression for k_2. The expression for k_2 is simply the curvature of a circle of radius x multiplied by the cosine of the angle that the tangent to α makes with the axis of rotation.

Observe that the expressions for k_1 and k_2 depend only upon the parameter u^1, not u^2. In particular, since k_2 is independent of u^2 there are no extrema or inflections of the normal curvature along the parallels. The parallels are circles. Consequently no segmentation contours arise from the lines of curvature associated with k_2. Only the minima of k_1 along the meridians are used for segmentation. Fig. 16 shows several surfaces of revolution with the minima of curvature along the meridians marked. The resulting segmentation contours appear quite natural to human observers.

As a surface of revolution is flattened along one axis, the partitioning contours which are at first circles become, in general, more elliptical and bow slightly up or down.

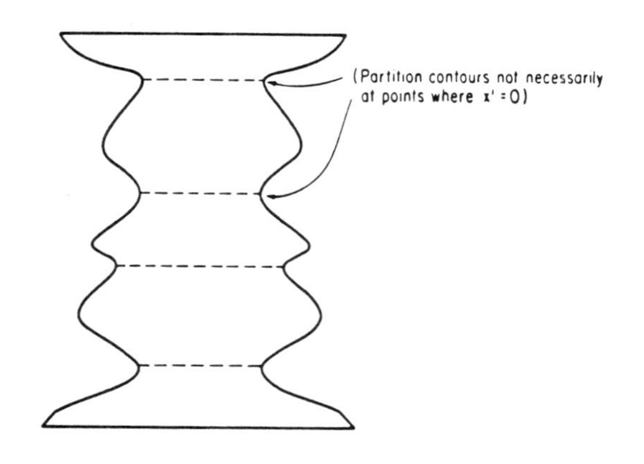

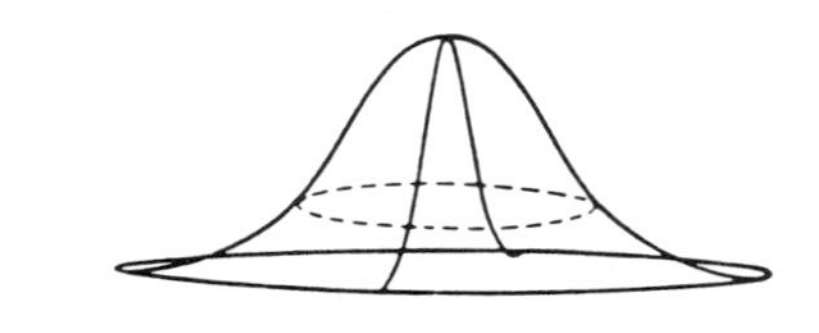

Figure 16. *Partitions on surfaces of revolution.*

Appendix 2

Inferring part boundaries from image singularities

In general, a concave discontinuity in a silhouette indicates a part boundary (as defined by the minima rule) on the imaged surface. This appendix makes this statement more precise and then examines a special case.

Only two types of singularity can arise in the projection from the world to the retina (Whitney, 1955). These two types are *folds* and *spines* (see Fig. 17). Intuitively, folds are the contours on a surface where the viewer's line of sight would just graze the surface, and a spine separates the visible portion of a fold from the invisible. A contour on the retina corresponding to a fold on a surface is called an *outline* (Koenderink and van Doorn, 1976, 1982b). A *termination* is a point on the retina corresponding to a spine on a surface. A *T-junction* (see Fig. 17) occurs where two outlines cut each other.

We wish to determine the conditions in which a T-junction indicates the presence of a part boundary. Two results are useful here. First, the sign of curvature of a point on an outline (projection of a fold) is the sign of the Gaussian curvature at the corresponding surface point (Koenderink and van Doorn, 1976, 1982b). Convex portions of the outline indicate positive Gaussian curvature, concave portions indicate negative Gaussian curvature, and inflections indicate zero Gaussian curvature. Second, the spine always occurs

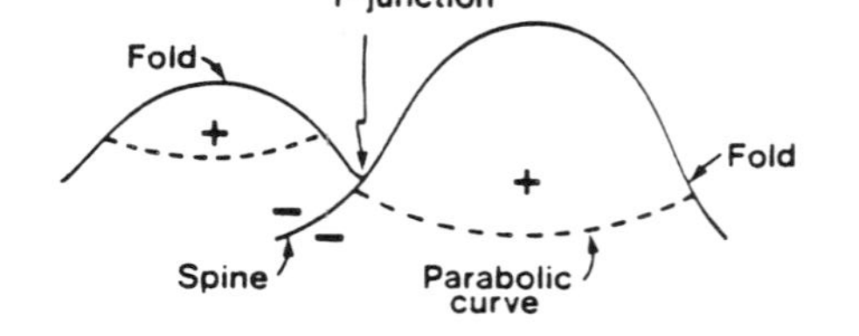

Figure 17. *Singularities of the retinal projection.*

at a point of negative Gaussian curvature. That is, the visible portion of a fold always ends in a segment whose projected image is concave (Koenderink and van Doorn, 1982b).

The scheme of the proof is the following. Suppose that the folds on both sides of a T-junction have convex regions, as shown in Fig. 17. Then the sign of the Gaussian curvature is positive, and in fact both principal curvatures are positive, in these two regions. Now the presence of a spine indicates that these regions of positive Gaussian curvature are separated by a region of negative Gaussian curvature. This implies that the principal curvature associated with one family of lines of curvature is negative in this region. But then the principal curvature along this family of lines of curvature must go from positive to negative and back to positive as the lines of curvature go from one hill into the valley and back up the other hill. If this is true, then in the generic case the principal curvature will go through a negative minimum somewhere in the valley—and we have a part boundary.

There are two cases to consider. In the first the loci where one principal curvature goes from positive to negative (parabolic curves) surround each hill. In the second case the parabolic curve surrounds the valley between the two hills. We consider only the first case.

In the first case there are two ways that the lines of curvature entering the valley from one parabolic curve might fail to connect smoothly with lines of curvature entering the valley from the other parabolic curve: they might intersect orthogonally or not at all. If they intersect orthogonally then the two principal curvatures must both be negative, and the Gaussian curvature, which is the product of the two principal curvatures, must be positive. But the valley between the parabolic contours has negative Gaussian curvature, a contradiction.

If the lines of curvature fail to intersect then there must be a singularity in the lines of curvature somewhere in the region having negative Gaussian curvature. However, "The net of lines of curvature may have singular properties at umbilical points, and at them only." (Hilbert and Cohn-Vossen, 1952, p. 187). Umbilical points, points where the two principal curvatures are equal, can only occur in regions of positive Gaussian curvature—again a contradiction. (Here we assume the surface is smooth. A singularity could also occur if the surface were not smooth at one point in the valley. But in the generic case part boundaries would still occur.)

This proof requires that the two folds of a T-junction each have a convex region. The two folds of T-junctions on a torus do not satisfy this condition—they are always concave. Thus it is a simple matter to determine from an image when a T-junction warrants the inference of a part boundary.

The proof outlined here is a special case. A general proof is needed which specifies when a concave cusp in a silhouette indicates the presence of a part boundary or two different objects. The more general proof would not use the relation between spine points and Gaussian curvature. The proof might run roughly as follows: a concave cusp is a double point in the projection. A line connecting the two points on the surface which project to the cusp necessarily lies outside the surface between the two points. But then the surface is not convex everywhere between these two points. Consequently there is a concave discontinuity (part boundary) between the points or the Gaussian curvature must go negative. If the Gaussian curvature goes from positive (convex) to negative and then back to positive (convex), one of the principal curvatures must also. But this implies it has a negative minimum, in the general case, and so we have a smooth part boundary.

References

Attneave, F. (1974) Multistability in perception. *Scient. Am.*, *225*, 63–71.

Badler, N. and Bajcsy, R. (1978) Three-dimensional representations for computer graphics and computer vision. *Comp. Graph.*, *12*, 153–160.

Ballard, D. and Brown, C. (1982) *Computer Vision*. Englewood Cliffs, N.J., Prentice-Hall.

Baumgart, B. (1972) *Winged edge polyhedron representation*. STAN-CS-320, AIM-179, Stanford AI Lab.

Binford, T. (1971) Visual perception by computer. IEEE Conf. Syst. Cont., Miami.

Binford, T. (1981) Inferring surfaces from images. *Art. Intell.*, *17*, 205–244.

Bobick, A. (1983) A hybrid approach to structure-from-motion. Association for Computing Machinery Workshop on Motion: Representation and Perception.

Brady, J.M. (1982a) Parts description and acquisition using vision. *Proc. Soc. Photo-opt. Instrument. Eng.*

Brady, J.M. (1982b) Criteria for representations of shape. In A. Rosenfeld and J. Beck (eds.), *Human and Machine Vision*.

Brady, J.M. (1982c) Describing visible surfaces. In A. Hanson and E. Riseman (eds.), *Computer Vision Systems*.

Brooks, R., Greiner Russell, and Binford, T. (1979) The ACRONYM model based vision system. *Proc. Int. Joint Conf. Art. Intell.*, *6*, 105–113.

Clowes M. (1971) On seeing things. *Art. Intell.*, *2*, 79–116.

Dennett, D. (1978) Intentional systems. In *Brainstorms*. Montgomery, VT, Bradford.

Do Carmo, M. (1976) *Differential Geometry of Curves and Surfaces*. Englewood Cliffs, NJ, Prentice-Hall.

Dodson, D. and Poston, T. (1977) *Tensor Geometry*. London, Pitman.

Duda, R. and Hart, P. (1973) *Pattern Classification and Scene Analysis*. New York, Wiley.

Fodor, J. (1975) *The Language of Thought*. Cambridge, MA, Harvard University Press.

Fodor, J. (1983) *The Modularity of Mind*. Cambridge, MA, MIT Press.

Fodor, J. and Pylyshyn, Z. (1981) How direct is visual perception?: Some reflections on Gibson's "Ecological Approach". *Cog.*, *9*, 139–196.

Goodman, N. (1955) *Fact, Fiction and Forecast*. Cambridge, MA, Harvard University Press.

Gregory, R. (1970) *The Intelligent Eye*. New York, McGraw-Hill.

Guillemin, V. and Pollack, A. (1974) *Differential Topology*. Englewood Cliffs, NJ, Prentice-Hall.

Guzman, A. (1969) Decomposition of a visual scene into three-dimensional bodies. In A. Grasseli (ed.), *Automatic Interpretation and Classification of Images*, New York, Academic Press.

Helmholtz, H. (1962) *Treatise on Physiological Optics, Volume 3*. Dover reprint.
Hilbert, D. and Cohn-Vossen, S. (1952) *Geometry and the Imagination*. New York, Chelsea.
Hoffman, D. (1983a) Representing Shapes for Visual Recognition. MIT Ph.D. thesis.
Hoffman, D. (1983b) The interpretation of visual illusions. *Scient. Am.*, *249*, 154–162.
Hoffman D. and Flinchbaugh, B. (1982) The interpretation of biological motivation. *Biol. Cybernet.*, *42*, 195–204.
Hoffman, D. and Richards, W. (1982) Representing smooth plane curves for visual recognition: Implications for figure–ground reversal. *Proc. Am. Ass. Art. Intell.*, 5–8.
Hollerbach, J. (1975) *Hierarchical Shape Description of Objects by Selection and Modification of Prototypes*. MIT AI-TR-346.
Huffman, D. (1971) Impossible objects as nonsense sentences. *Mach. Intell.*, *6*.
Koenderink, J. and van Doorn, A. (1976) The singularities of the visual mapping. *Biol. Cybernet.*, *24*, 51–59.
Koenderink, J. and van Doorn, A. (1979) The internal representation of solid shape with respect to vision. *Biol. Cybernet.*, *32*, 211–216.
Koenderink, J. and van Doorn, A. (1980) Photometric invariants related to solid shape. *Optica Acta*, *7*, 981–996.
Koenderink, J. and van Doorn, A. (1982a) Perception of solid shape and spatial lay-out through photometric invariants. In R. Trappl (ed.), *Cybernetics and Systems Research*, Amsterdam, North-Holland.
Koenderink, J. and van Doorn, A. (1982b) The shape of smooth objects and the way contours end. *Perception*, *11*, 129–137.
Levi, I. (1980) *The Enterprise of Knowledge*. Cambridge, MA, MIT Press.
Liberman, A., Cooper, F., Shankweiler, D., and Studdert-Kennedy, M. (1967) The perception of the speech code. *Psychol. Rev.*, *74*, 431–461.
Lipschutz, M. (1969) *Differential Geometry*. (Schaum's Outline), New York, McGraw-Hill.
Longuet-Higgins, H.C. and Prazdny, K. (1981) The interpretation of a moving retinal image. *Proc. R. Soc. Lond.*, *B208*, 385–397.
Mackworth, A. (1973) Interpreting pictures of polyhedral scenes. *Art. Intell.*, *4*, 121–137.
Mandelbrot, B. (1982) *The Fractal Geometry of Nature*. San Francisco, Freeman.
Marr, D. (1977) Analysis of occluding countour. *Proc. R. Soc. Lond.*, *B197*, 441–475.
Marr, D. (1982) *Vision*. San Francisco, Freeman.
Marr, D. and Nishihara, H.K. (1978) Representation and recognition of the spatial organization of three-dimensional shapes. *Proc. R. Soc. Lond.*, *B200*, 269–294.
Marr, D. and Poggio, T. (1977) From understanding computation to understanding neural circuitry. *Neurosci. Res. Prog. Bull.*, *15*, 470–488.
Nicod, J. (1968) *Geometry and Induction*, Berkeley, University of California Press.
Nishihara, H.K. (1981) Intensity, visible-surface, and volumetric representations. *Art. Intell.*, *17*, 265–284.
O'Rourke, J. and Badler, N. (1979) Decomposition of three-dimensional objects into spheres. *IEEE Trans. Pattern Anal. Mach. Intell.*, *1*.
Pentland, A. (1983) Fractal-based description. *Proc. Int. Joint Conf. Art. Intell.*
Peirce, C. (1931) *Collected Papers*, Cambridge, MA, Harvard University Press.
Richards, W., Rubin, J.M. and Hoffman, D.D. (1983) Equation counting and the interpreation of sensory data. *Perception*, *11*, 557–576, and MIT AI Memo 618 (1981).
Rock, I. (1974) The perception of disoriented figures. *Scient. Am.* *230*, 78–85.
Skyrms, B. (1975) *Choice and Chance*. Belmont, Wadsworth Publishing Co.
Soroka, B. (1979) Generalized cylinders from parallel slices. *Proc. Pattern Recognition and Image processing*, 421–426.
Spivak, M. (1970) *Differential Geometry, Volume 2*, Berkeley, Publish or Perish.
Sutherland, N.S. (1968) Outlines of a theory of visual pattern recognition in animals and man. *Proc. R. Soc. Lond.*, *B171*, 297-317.
Thompson, D'Arcy (1968) *On Growth and Form*, Cambridge, University of Cambridge Press.
Turton, W. (1819) *A Conchological Dictionary of the British Islands*, (frontispiece), printed for John Booth, London. [This early reference was kindly pointed out to us by J.F.W. McOmie.]
Ullman, S. (1979) *The Interpretation of Visual Motion*. Cambridge, MA, MIT Press.
Waltz, D. (1975) Understanding line drawings of scenes with shadows. In P. Winston (ed.), *The Psychology of Computer Vision*, New York, McGraw-Hill.
Whitney, H. (1955) On singularities of mappings of Euclidean spaces. I. Mappings of the plane into the plane. *Ann. Math.*, *62*, 374–410.
Witkin, A.P. (1983) Scale-space filtering. *Proc. Int. Joint Conf. Artificial Intelligence.*
Yin, R. (1970) Face recognition by brain-injured patients: A dissociable ability? *Neuropsychologia*, *8*, 395–402.
Yuille, A. (1983) *Scaling theorems for zero-crossings*. MIT AI Memo 722.

Résumé

Les auteurs suggèrent que le système visuel pour la reconnaissance des objets, décompose les formes en éléments et qu'il utilise pour cela une règle définissant les frontères de ces éléments plutôt que leurs formes. Cette règle exploite une régularité de la nature: la transversalité. Les éléments, leurs descriptions et leurs relations spatiales fournissent un premier index dans la mémoire des formes. On peut avec cette règle rendre compte de plusieurs illusions visuelles. Les auteurs insistent sur le rôle de l'inférence inductive et concluent en indiquant les problèmes non résolus.

Human Factors and Behavioral Science:

Textons, The Fundamental Elements in Preattentive Vision and Perception of Textures

By B. JULESZ* and J. R. BERGEN*

(Manuscript received September 23, 1981)

Recent research in texture discrimination has revealed the existence of a separate "preattentive visual system" that cannot process complex forms, yet can, almost instantaneously, without effort or scrutiny, detect differences in a few local conspicuous features, regardless of where they occur. These features, called "textons", are elongated blobs (e.g., rectangles, ellipses, or line segments) with specific properties, including color, angular orientation, width, length, binocular and movement disparity, and flicker rate. The ends-of-lines (terminators) and crossings of line segments are also textons. Only differences in the textons or in their density (or number) can be preattentively detected while the positional relationship between neighboring textons passes unnoticed. This kind of positional information is the essence of form perception, and can be extracted only by a time-consuming and spatially restricted process that we call "focal attention". The aperture of focal attention can be very narrow, even restricted to a minute portion of the fovea, and shifting its locus requires about 50 ms. Thus preattentive vision serves as an "early warning system" by pointing out those loci of texton differences that should be attended to. According to this theory, at any given instant the visual information intake is relatively modest.

* Bell Laboratories.

I. INTRODUCTION

In this article we give an overview of some insights into the workings of the human visual system gained during two decades of research at Bell Laboratories, and culminating in the discovery of a few local conspicuous features that we call textons. Textons appear to be the basic units of preattentive texture perception,[1] when textures are viewed in a quick glance with no further effort or analysis. Although this article goes beyond texture perception into preattentive vision in general, studies of texture discrimination led to the basic insights presented here and provide excellent demonstrations of the main findings. Based on our findings we propose a novel theory of vision in which the preattentive visual system inspects a large portion of the visual field in parallel and detects only density differences in textons. It then directs focal attention to these loci of texton differences for detailed scrutiny.

Now, after 20 years of research, when we know what textons are and their role in vision is clarified, we can save the reader from following the rather difficult steps that led to their discovery. [The reader interested in the history of these developments, and in the sophisticated mathematics necessary to generate textures with certain stochastic constraints, should turn to the original articles referred to in a recent review by one of us[1] and to the Appendix.] Here we follow an axiomatic treatment. The main findings are presented in Section II as heuristics (similar to axioms, but not necessarily totally independent), immediately followed by many demonstrations. The reader can test the power of these newly acquired heuristics by being able to predict and then verify which texture pairs will be perceived to be different, and which will appear as a single texture. The reader can thus understand the new theory of vision without mathematical knowledge.

Section III emphasizes the essentially local nature of texture perception. Section IV relates the psychologically identified textons to some neurophysiological results concerning local feature analyzers in primate cortex. Section V extends the texton theory from texture perception to the discrimination of briefly presented patterns. In Section VI a model of human vision is proposed that postulates two different modes of visual system function. Section VII discusses some implications of this model.

II. HEURISTICS: DEFINITION OF TEXTONS AND THEIR INTERACTIONS IN PREATTENTIVE VISION

Visual textures are defined as aggregates of many small elements. The elements can be either dots of certain colors (e.g., black, white, grey, red) or simple patterns. For purposes of this article, we consider only elements that do not overlap, and are placed at either regular or random positions, in identical or in random angular orientations.

Usually in our demonstrations two textures (composed of two different elements) are placed side-by-side, or one is embedded in the other, as shown in Fig. 1. When the reader cursorily inspects Fig. 1, an area made up of +'s will appear to stand out from the surrounding texture composed of L's. Indeed, without scrutiny, that is without detailed element-by-element inspection, the reader might not notice that a third area composed of T-shaped elements is also embedded in the texture of L's. We call this effortless perceptual segregation of the texture composed of +'s from the surrounding texture of L's *preattentive texture perception.* On the other hand, if texture discrimination requires element-by-element scrutiny, as is the case of finding the T's in the L's, we call this way of looking with scrutiny *focal attention.* We will show many other preattentively indiscriminable texture pairs (e.g., Figs. 3c and 6b), which, because they do not segregate, often are not even perceived as containing different elements until this is pointed out.

Although in all texture perception the preattentive system is dominant, the role of focal attention can be even further reduced by brief presentation. The reader who is not convinced by the qualitative difference between preattentive and attentive texture discrimination might inspect Fig. 1 through a camera shutter set at 1/10 second exposure time.

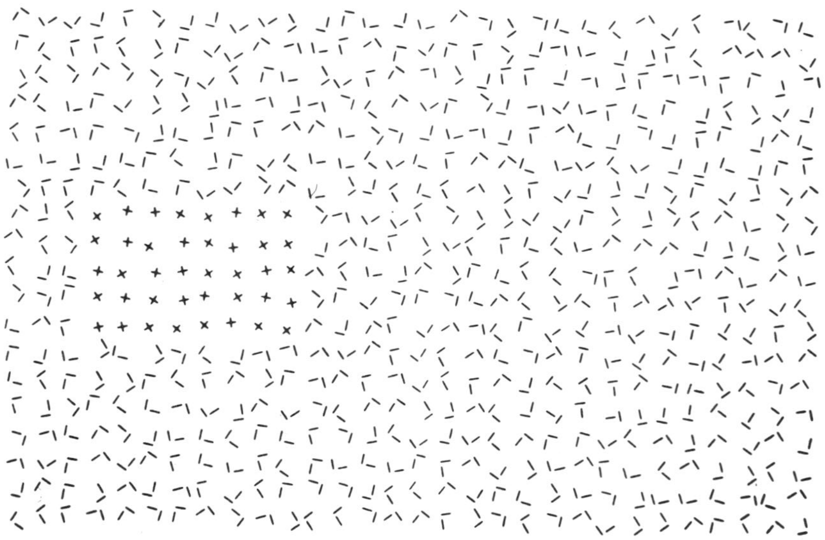

Fig. 1—"Preattentive texture discrimination" is shown between areas composed of +'s and L's, while element-by-element scrutiny, called "focal attention" is required to find the T's embedded in the L's.

Heuristic 1: Human vision operates in two distinct modes

1. Preattentive vision—parallel, instantaneous, without scrutiny, independent of the number of patterns, covering a large visual field, as in texture discrimination.
2. Attentive vision—serial search by focal attention in 50-ms steps limited to a small aperture, as in form recognition.

Heuristic 2: Textons

1. Elongated blobs—e.g., rectangles, ellipses, line segments with specific colors, angular orientations, widths, and lengths.
2. Terminators--ends-of-line segments
3. Crossings of line segments

Heuristic 3: Preattentive vision directs attentive vision to the locations where differences in the density (number) of textons occur, but ignores the positional relationships between textons.

Before we discuss the implications of these heuristics, let us apply them to a few pairs of elements and predict whether the texture pairs formed from these elements will yield preattentive texture discrimination or not. This application of the rules also helps to clarify them. For instance, elongated blobs of different widths or lengths are different textons, as Fig. 2a demonstrates. The larger sized R's containing longer and wider line segments form a texture that segregates (i.e., is preattentively discriminable) from its surround, which is composed of smaller R's with shorter and narrower line segments.

Similarly, elongated blobs of different orientations are different

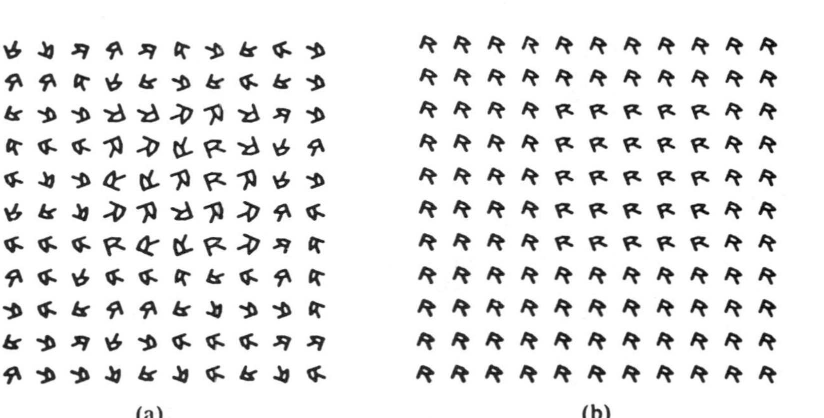

Fig. 2—Preattentive texture discrimination based on texton differences between line segments of (a) length and width and (b) angular orientation. (*Nature*, March 12, 1981[1])

textons. Indeed, in Fig. 2b the texture pair composed of the same sized R's having two different orientations in the two textures, yields preattentive discrimination. Obviously, the same elongated blob shape with the same orientation yields different textons if the colors (e.g., black, gray, white, red, green, etc.) are different.

Now, let us predict what would happen if we took an R and a mirror-image R, as shown in Fig. 3a, and formed a texture pair by throwing them in random orientations. Obviously, without randomizing the orientations, the two textures would yield texture discrimination, since even though their widths and lengths agree, some of the line segment textons have different orientations in the R and in its mirror image, though the widths and lengths agree, as shown in Fig. 3b. However, if the two elements are thrown at random orientations, then the two textures formed have the same average density of textons (i.e., in some area of integration the number of line segments with the same color, width, length, and orientation is identical). Therefore, the preattentive visual system should not be able to direct focal attention to loci of texton differences that form the boundary between the two regions. Indeed, an inspection of Fig. 3c yields a single, uniform texture. It requires laborious element-by-element inspection for several seconds to find the boundary between the array of R's and mirror-image R's. Obviously, in a 100-ms presentation discrimination of these textures is impossible.

Let us note that if one were to select a pair of elements without knowing the rules given above, most probably the resulting texture pair would be discriminable. Only through the joint effort of our colleagues (D. Slepian, M. Rosenblatt, E. Gilbert, L. Shepp, H. Frisch, T. Caelli, and J. Victor) from 1962 to 1978 were some elegant methods found that yielded indistinguishable textures, even though their elements appeared very different.

In the next examples we stress the importance of terminator textons. For instance, in Fig. 4a the two elements are composed of three identical line segments (i.e., same orientation, width, and length). The only difference is in the number of their ends-of-lines (terminators). The triangle-shaped element has no open ends, while the "dual" element has three ends-of-lines. One should expect texture segregation, given such a large difference in terminator number, and as Fig. 4b demonstrates, this is the case.

As a matter of fact, discrimination is so strong that a single element can be preattentively detected among 35 dual elements, as shown in Fig. 4c. This arrangement is now routinely used by us in studying

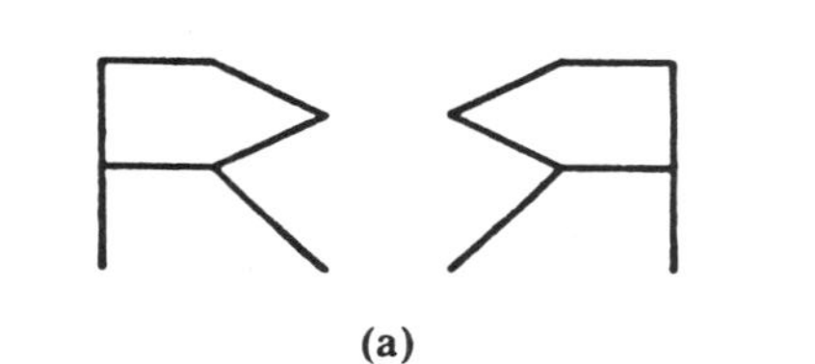

(a)

(b) (c)

Fig. 3—Demonstration of how the heuristics given in text predict why (a) R and its mirror image in aggregates yield texture discrimination (b), or are indistinguishable (c). (*Perception*, 1973[3])

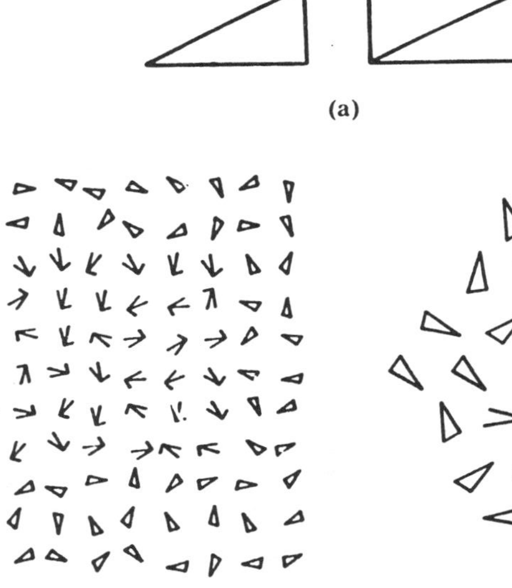

(a)

(b)

(c)

Fig. 4—Demonstration of how the heuristics given in text predict preattentive texture discrimination (b) and even discrimination of a single element among many (c), based on terminator number difference (zero versus three) between elements (a). (*Nature*, March 12, 1981[1])

pattern discrimination in preattentive vision, as discussed in Section V. Here we note only that when there is a texton difference (as in Fig. 4c) detecting one element in the midst of 35 other elements is almost as easy as detecting the difference between two elements (shown in Fig. 4a) for presentation times as brief as 100 ms.

In the next example, both members of the element pair of Fig. 5a are again composed of the same five line segments (each corresponding line segment in the two elements has identical width, length, and orientation, respectively) but one element contains only two ends-of-lines, whereas the other contains five. This large difference in terminator numbers should yield texture segregation, and inspection of Fig. 5b demonstrates that it does. Figure 5c consists of the same texture pair as Fig. 5b, except that the texture containing the five terminators is now the surround. Although, as predicted, the large difference in terminator numbers again yields texture segregation, the appearance of the boundary between the two regions is different for Fig. 5b and 5c.

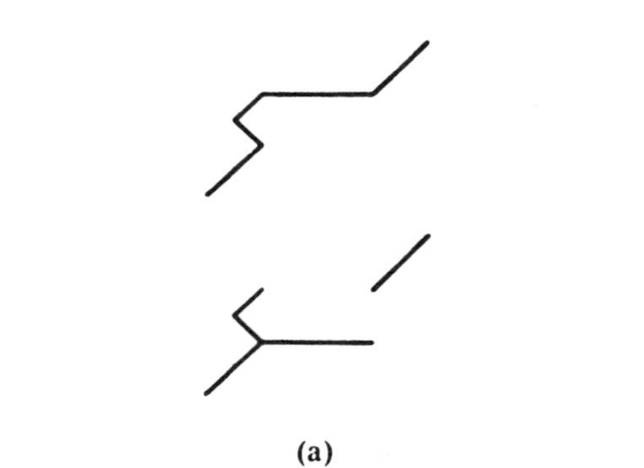

(a)

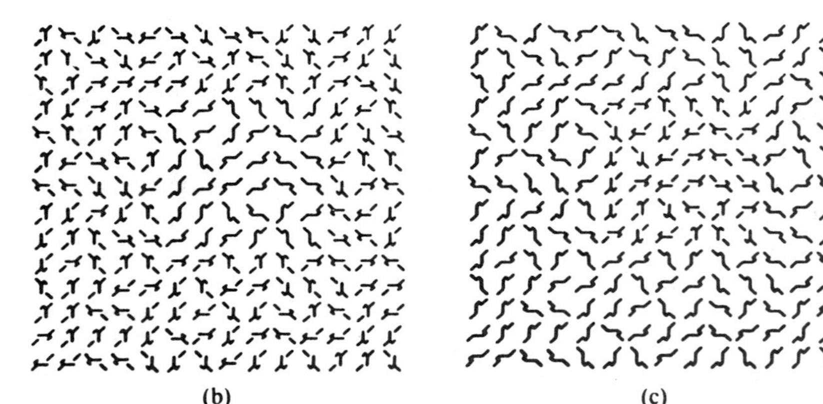

(b) (c)

Fig. 5—Similar to Fig. 4 except the terminator number between elements is two versus five.

The next example, shown in Fig. 6a, consists of the "S"- and "10"-shaped elements, that in isolation appear quite different. However, the two contain the same number of line segment textons (three identical horizontal and two identical vertical line segments) and both contain two ends-of-lines. The fact that the positional relationship between these textons is different (as it is in Fig. 3b) can be perceived only by the attentive visual system (yielding the percept of an S versus a 10). However, according to Heuristic 3 the preattentive system can count only the density (number) of textons and ignores their relative positions. So, according to our rules, a texture pair composed of these elements contains the same average density (number) of textons, and thus should be indistinguishable. Surprising as it may seem, the texture pair is indeed preattentively indistinguishable as demonstrated by Fig. 6b. [Readers who find this demonstration of the distinction between preattentive and focal vision not adequately convincing without brief presentation should note the contrast between the attentively different percepts of Fig. 6a, and the texture pair in Fig. 6b, which remains difficult to distinguish even with element-by-element scrutiny.]

Finally, let us demonstrate the third texton, the crossing of elongated blobs (line segments). Figure 7a shows the conspicuous difference between a texture pair that segregates based on the presence or absence of elements having crossing versus not-crossing line segments.

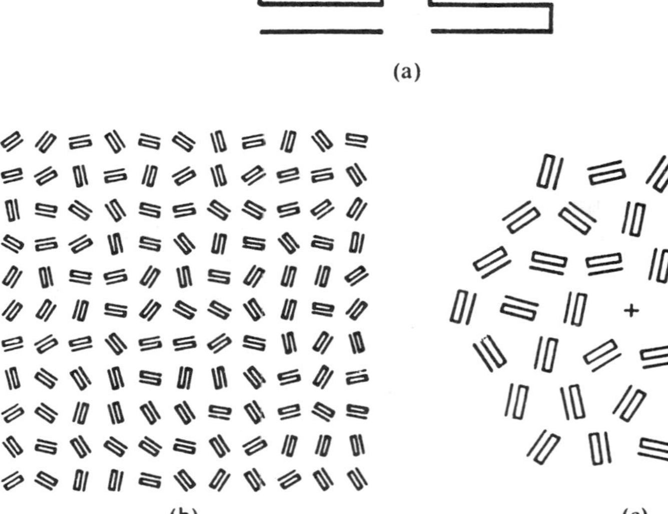

(a)

(b) (c)

Fig. 6—Demonstration of how the heuristics given in text predict why (a) the differently appearing S- and 10- shaped elements in aggregates (b) and one S in 10's (c) are indistinguishable. (*Nature*, March 12, 1981[1])

If the elements have identical textons, including crossing (or not-crossing line segments) the texture pairs become preattentively indistinguishable. The positional relationship between the line segment textons is unnoticed by the preattentive system. The difference in gap size between the L-shaped elements in Fig. 7b yields a preattentively indistinguishable texture pair. Particularly interesting is the demonstration in Fig. 7c where T- versus L-shaped elements yield an indistinguishable texture pair. Although we have kept a small gap between the perpendicular line segments that make up the L's and T's, preattentive discrimination of texture pairs composed of these elements is impossible even when the gaps are not resolvable. Apparently, the difference of a single end-of-line terminator is not adequate to yield texture segregation. Finally, Fig. 7d depicts a preattentively indistinguishable texture pair, where, with scrutiny, it is obvious that the elements contain line segments that either cross at midpoint or cross far from the midpoint.

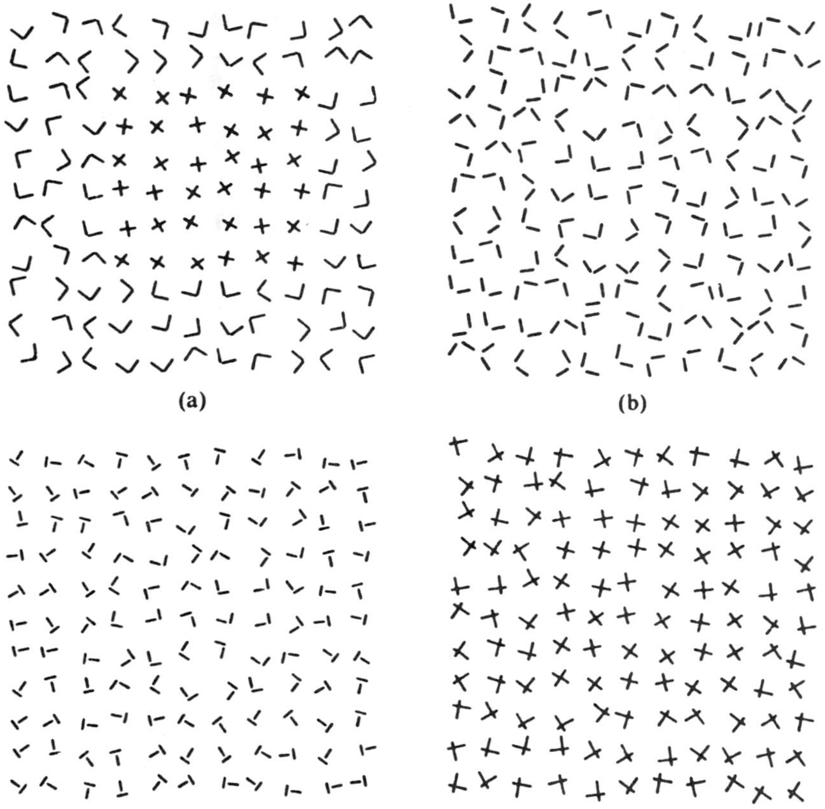

(a) (b)

(c) (d)

Fig. 7—Demonstration that crossing of line segments is a texton.

The last two examples are given in Figs. 8a and b and Figs. 9a, b, and c. From the element pairs containing the same textons, the reader can predict that although their elements in isolation appear very different, the resulting texture pairs will be indistinguishable.

In all these demonstrations the texture elements consisted of line segments. For line segments the definition of terminators (ends-of-lines) and their crossings are straightforward. For elongated bars with substantial width these definitions are less direct. Particularly difficult is the notion of terminators, because instead of terminators some combination of white elongated bars in a black surround with black elongated bars in white surround might suffice. So, we are not certain whether terminators are independent textons. Nevertheless, as a first approximation these three heuristics work remarkably well.

III. PREATTENTIVE TEXTURE PERCEPTION IS ESSENTIALLY A LOCAL PROCESS

The essence of all the findings reported in the previous section can be summed up as follows: In texture perception the preattentive visual system utilizes only local conspicuous features, textons, and these textons are not coupled to each other (i.e., a vertical and horizontal line segment do not cohere to form an L or T). The preattentive system utilizes globally only the textons in the simplest possible way by counting their numbers (densities). This might surprise many of our readers who assume that texture perception utilizes complex global statistical interactions between textural elements.

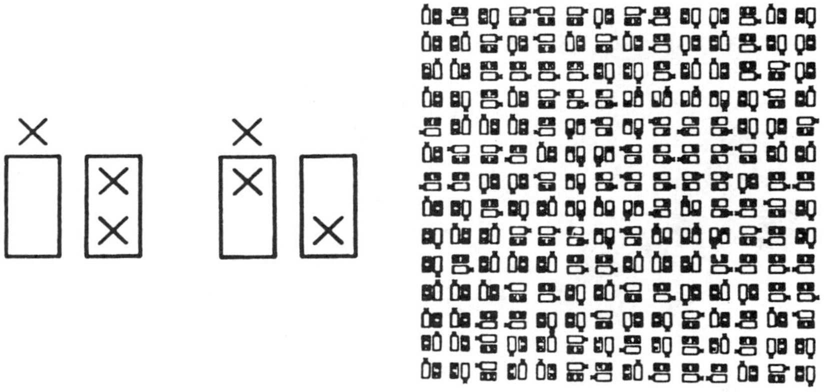

(a) (b)

Fig. 8—Since the element pair in (a) is composed of the same textons, the texture pair (b) composed of these elements is preattentively indistinguishable. (*Philosophical Transactions*, 1980[25])

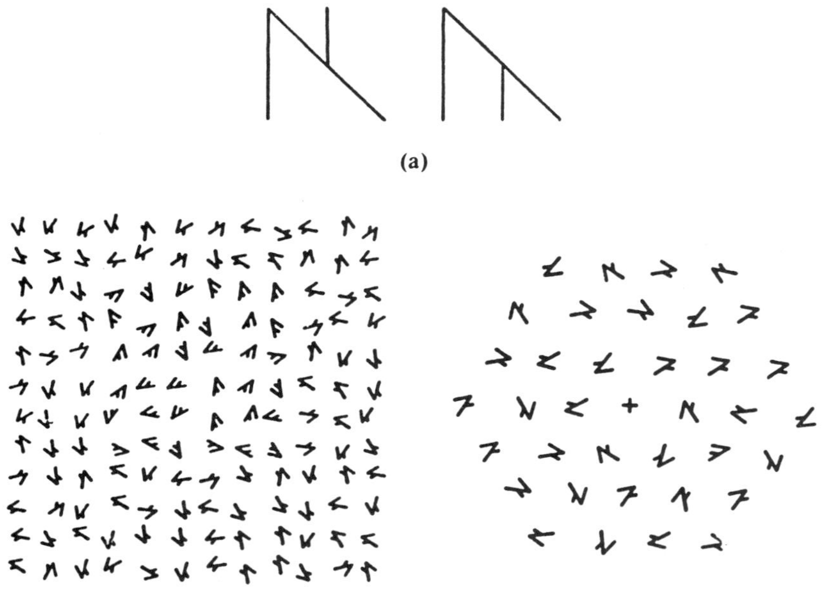

(a)

(b) (c)

Fig. 9—Similar to Fig. 8, showing that aggregates of elements composed of the same textons cannot be preattentively discriminated. (*Philosophical Transactions*, 1[illegible][25])

One of the simplest *global* computations routinely performed on images by engineers, and recently by psychologists in vision research, is to determine the images' Fourier power spectra. This process involves the decomposition of the images into one-dimensional sinusoidal luminance gratings whose specific amplitudes, spatial frequencies, phases, and angular orientations depend on the spatial characteristics of luminance distributions across the entire image. The amplitude of the spectral components ignoring phase determine the power spectra. When Fourier power spectra of textures are taken, it is a common misconception that differences in these will reveal differences in texture granularity. That the preattentive visual system does not perform Fourier analysis is demonstrated next.

Figure 10a consists of three areas that have identical Fourier power spectra (invented by Julesz, Gilbert and Victor[2]) and yet appear as very distinct textures. [The mathematically sophisticated reader might appreciate that the three areas have identical third-order statistics, and differ only in their fourth-order statistics. Those interested in the definition of *n*th-order statistics should consult Refs. 3 and 4 and the Appendix.] Figure 10b also consists of three areas with identical Fourier power spectra, and again these areas appear conspicuously different. Conversely, in Fig. 11a the lower left quadrant of the bottom

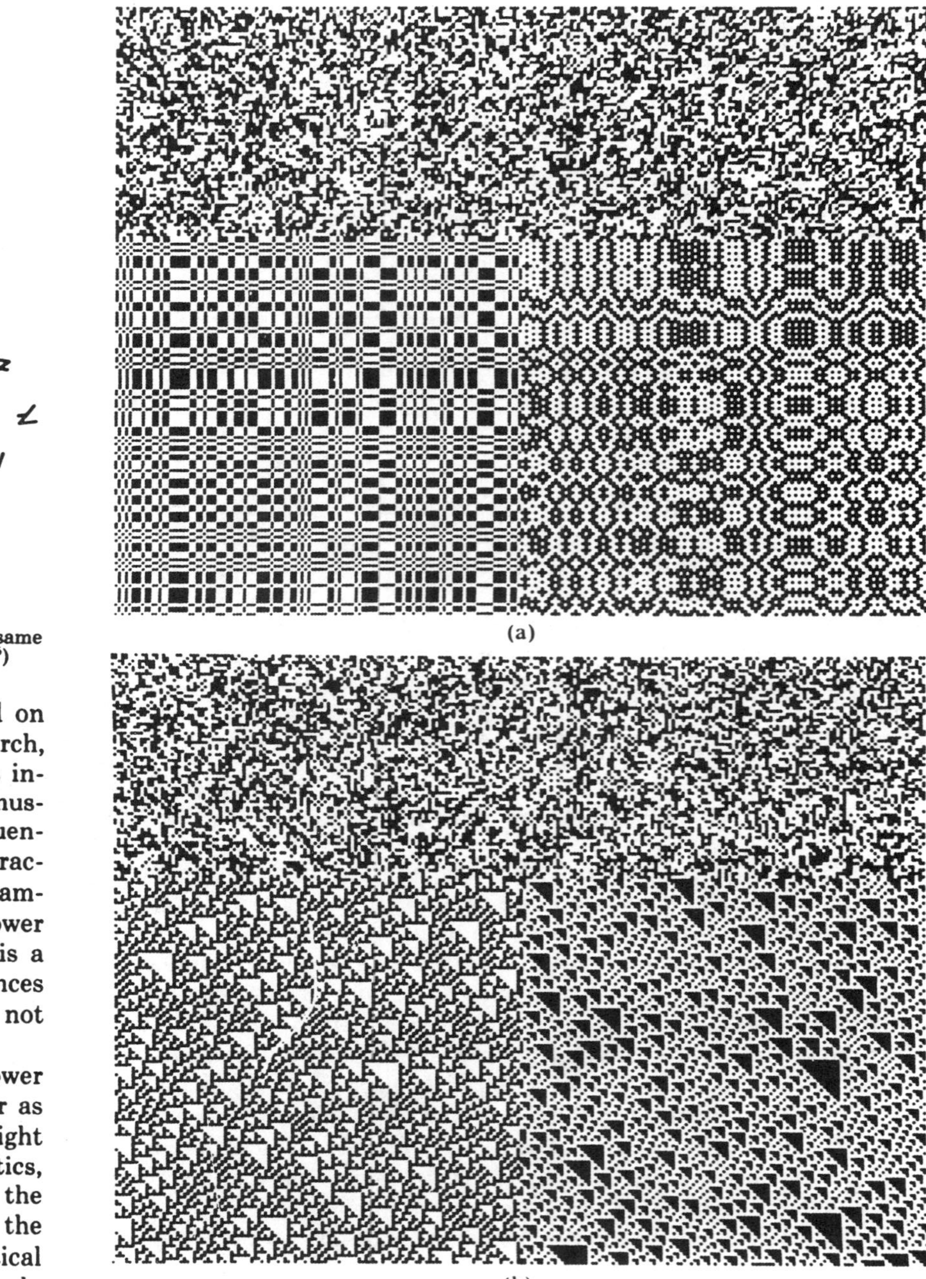

Fig. 10—Discriminable texture pairs with identical Fourier power spectra (*a* has even identical third-order statistics) based on local granularity (texton) differences. (*Biological Cybernetics*, 1981[2])

right array has a very different power spectrum from the remainder of the array, yet no preattentive texture discrimination results.[5] The derivation of this texture pair is presented in three steps. The top left array, in Fig. 11a, consists of 4x4 dot elements (8 black and 8 white) with 6-dot periodicity. The bottom left array contains this periodic array in one quadrant, but the 2-dot-wide gaps are filled by a checkerboard screen, while the rest is covered with uniformly random black and white dots. The bottom right array is similar to the bottom left array, except the 2-dot-wide gaps between the periodic patterns are now randomly speckled with dots. Obviously, the periodic patterns in the lower quadrant of the bottom right array in Fig. 11a yield a very different Fourier power spectrum from the rest, which has a flat (white noise) spectrum. The reason that this texture pair is indistinguishable can be easily understood in the light of the texton theory. The periodic patterns are not different from the surrounding random-dot array in the density of elongated blob textons, and therefore are indistinguishable. Indeed, if the 4x4 dot micropattern consists of vertical stripes, which contain textons different from the surrounding random-dot array, as shown in Fig. 11b, the periodic quadrant embedded in randomness is easily perceived.

In all these densely packed dot textures, discrimination is based on local granularity differences that correspond to differences in the density (number) of elongated blobs of certain sizes and orientations. Global statistical descriptors of textures, including the Fourier power spectrum, apparently are ignored in preattentive vision.

IV. TEXTURES AND NEUROPHYSIOLOGICAL FEATURE ANALYZERS

We have seen how elongated blob textons are crucial in preattentive texture perception. These human psychological findings have a parallel in primate neurophysiology. Neural units have been found by Hubel and Wiesel[6] in the visual cortex of monkeys that fire optimally for elongated blobs of specific width, length, and orientation. These neural units in the cortex have *retinal receptive fields* consisting of elongated, blob-shaped, excitatory regions, which are surrounded by inhibitory regions. Some of these elongated blob detecting units—which fire optimally for black elongated blobs surrounded by white flanking areas—are called simple "off" detectors. Other neural units are excited optimally by white elongated blobs surrounded by black. These are called simple "on" detectors. The exact shape of the receptive fields of these simple neural units varies a great deal, and is of secondary importance. The important property of these cortical units is that the weighting of the excitatory and inhibitory areas of their receptive fields is about equal, so that for homogeneous stimuli they do not fire.

It should be stressed that the textons reported here were found by

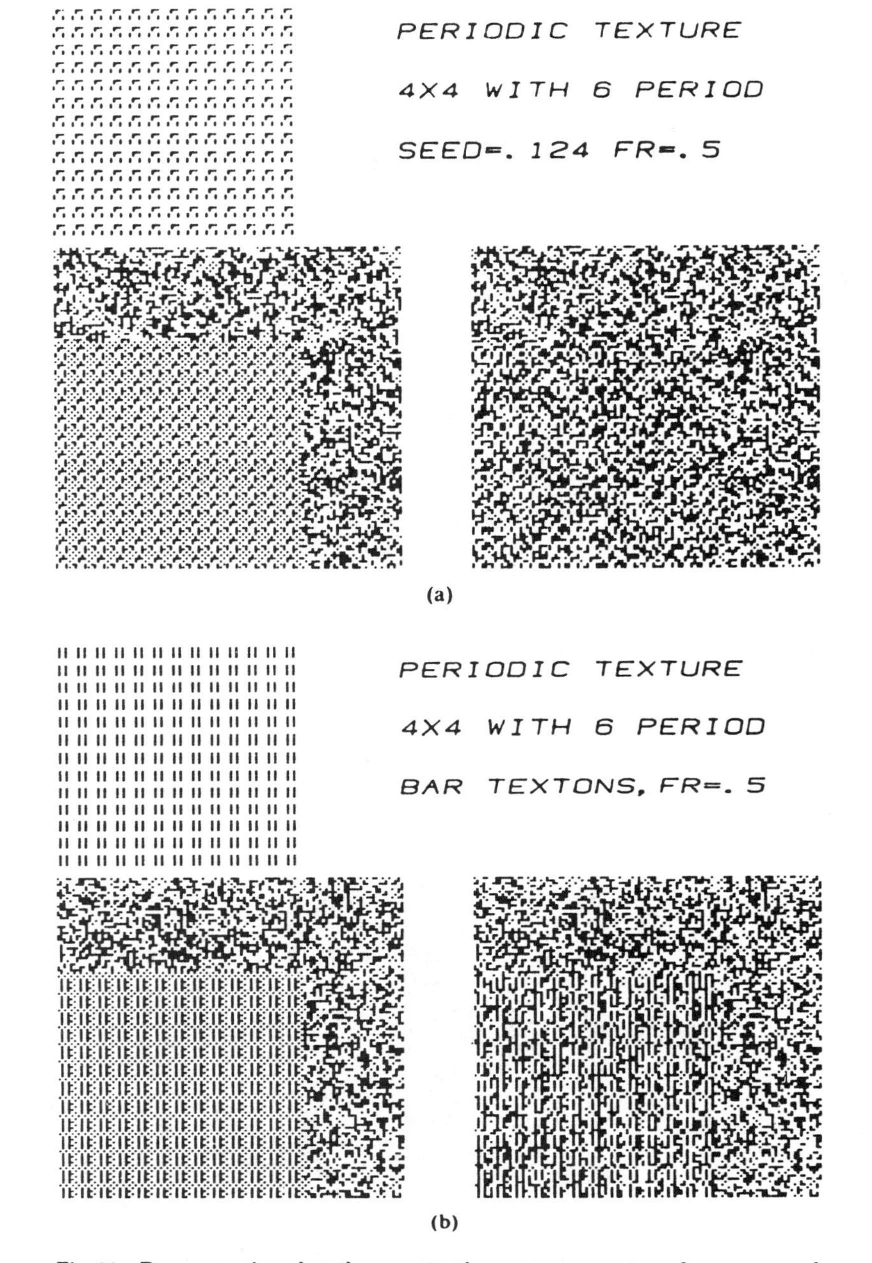

Fig. 11—Demonstration that the preattentive system cannot perform even such a simple global computation as Fourier power spectra, as described in text. (*Biological Cybernetics*, 1978[5])

psychological methods, and imply that simple neural units found as early as the striate cortex of the monkey might be used in texture perception. However, the relationship between a texton—for example, a perceived line segment—and a Hubel and Wiesel type of neural feature analyzer with a receptive field whose excitatory center matches the shape of the line segment is not a simple isomorphism. As we pointed out years ago (Ref. 7, p. 3), a single simple neural unit might respond equally for a broad line of high contrast or a narrow line of low contrast, while perceptually one can preattentively perceive both the width and contrast of a line segment. Thus, obviously a perceived line segment is encoded by many neural units of similar orientations but tuned to different widths, and having different firing thresholds. It is some combination of these units that would correspond to a perceived line segment. Until more is known about the relationship between perception and neurophysiology, the textons must be defined as perceptual entities, that is conspicuous local features as we actually perceive them. Nevertheless, even though textons and neural units are not simply related, one can easily conceptualize how a "perceptual analyzer" could be built from known neural analyzers that could extract, say, a line segment texton. The question of whether terminators and crossings of line segments—which have been regarded as textons—could be related to the complex and hypercomplex neural analyzers found by the neurophysiologists remains to be seen.[6]

David Marr, in his primal-sketch model of machine vision, also incorporated such elongated blob detectors, by assuming that the neurophysiological findings had direct relevance to vision.[8] The work reported here followed an opposite trend. It took almost two decades to find evidence for the utilization of simple cortical units in texture perception. Caelli and Julesz found the first elongated blob textons that could account for texture discrimination locally, when all global statistical properties of the texture pairs were kept identical.[9] Later demonstrations such as Figs. 10a and b illustrate even more strikingly the importance of local blob textons.

To demonstrate the possible role of the Hubel and Wiesel type of neural units in preattentive texture perception, we developed a computer program called TEXTONS that filters any image with a pool of elongated bar-shaped receptive fields. Each pool of filters consists of "on" and "off" types having the same width, length, orientation, and firing threshold and placed at each point of the array. Figure 12 (bottom) shows the three largest response levels of a pool of 3x3 dot square-shaped receptive fields as this pool processes the texture pair of Fig. 10a, shown also in Fig. 12 (top). These filters have 2x2 dot excitatory centers flanked by one-dot-wide inhibitory margins, as shown in Fig. 12 (right). Of course, there are several pools consisting

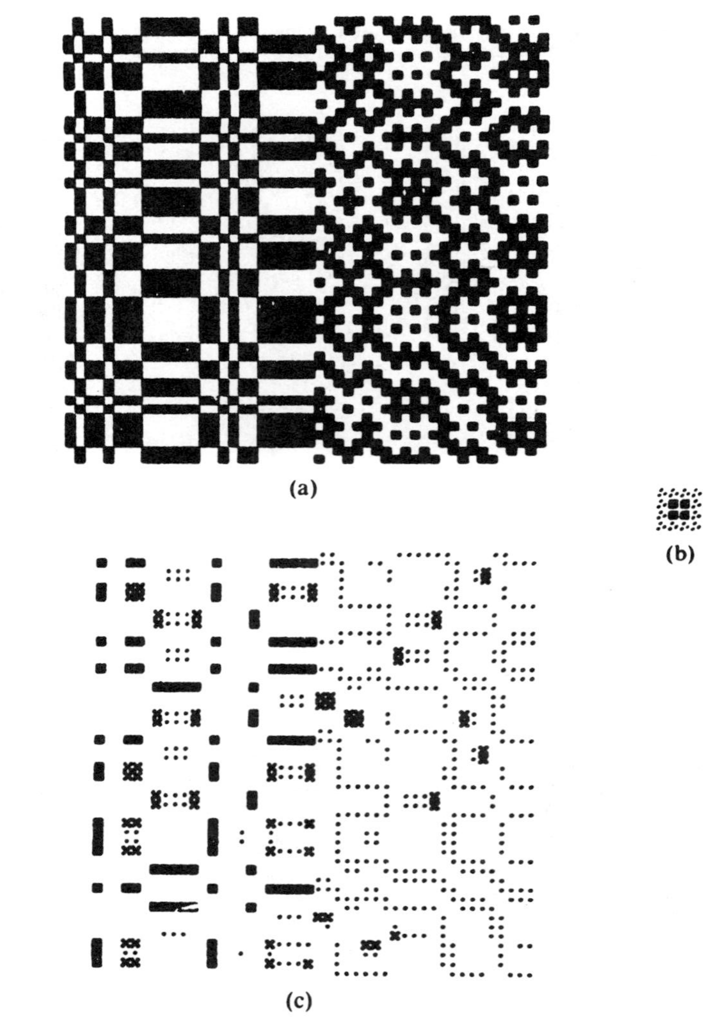

Fig. 12—Automatic texture segregation, shown in (c), by applying a texton filter (b) to texture pair (a) (also shown in Fig. 10a).

of filters having receptive fields with some other dimensions and orientations that would be even more effective in segregating the two textures of Fig. 12 (top). Here we stress again that the combination of several filters would be required to yield the best texture segregation, corresponding to human texture discrimination. This combination of filters would correspond to a texton detector.

What our psychological findings show, however, could not have been guessed by physiologists and theoreticians of artificial intelligence. In preattentive texture perception the various textons are not coupled, that is their relative positions are ignored. T- and L-shaped pairs of line segments cannot be discriminated preattentively in textures. Marr thought that elongated blobs and terminators would form some higher molar unit, which he called place tokens.[8] However, in preattentive vision no such higher interactions are found; the textons appear to be independent of each other.

V. EXTENSION OF THE TEXTON THEORY TO RAPID PATTERN DISCRIMINATION

The success of the texton theory in predicting phenomena of texture perception is the result of the spatial complexity of the patterns. This complexity over a large area exceeds the capacity of focal attention and thus allows the preattentive system to dominate. This same deemphasis of focal attention can be achieved in simpler patterns by very brief presentation. We will show that under these conditions the same texton theory can be applied.[10]

Because brief temporal presentation is required, the stimuli used in these experiments can be produced only in the laboratory. Consequently, they cannot be demonstrated as the texture discrimination results have been. Thus, in this section we present the main findings as curves describing observers' performance.

The stimuli used in these experiments are shown in Fig. 13. In Fig. 13a there are 35 T's and one L arranged on a hexagonal grid with slight random positional jitter added. In Fig. 13b the T's have been replaced by + elements, and in Fig. 13c only two of the 36 possible positions actually contain an element. In all cases, a disk surrounding the central fixation marker is kept empty. Stimuli of this type are presented for 40 ms, followed by a blank interval of variable duration and a 40-ms erasing field. This erasing field consists of elements, which are the union of the two being discriminated, arranged in the same way as the test field. Use of this erasing technique allows restriction of the inspection interval to times shorter than the duration of the retinal afterimage. The times used are all too short to allow eye movements to be initiated during the presentation. In half of the presentations the test field consists of all identical elements, while in the other half one element is different, as in the examples shown. The task of the observer is to discriminate between these two conditions.

Results obtained using the three stimuli of Fig. 13 are shown in Fig. 14. On the abscissa is the time in milliseconds between the onsets of the test and erasing fields, or the stimulus onset asynchrony (SOA). On the ordinate is the percentage of correct discrimination. The results

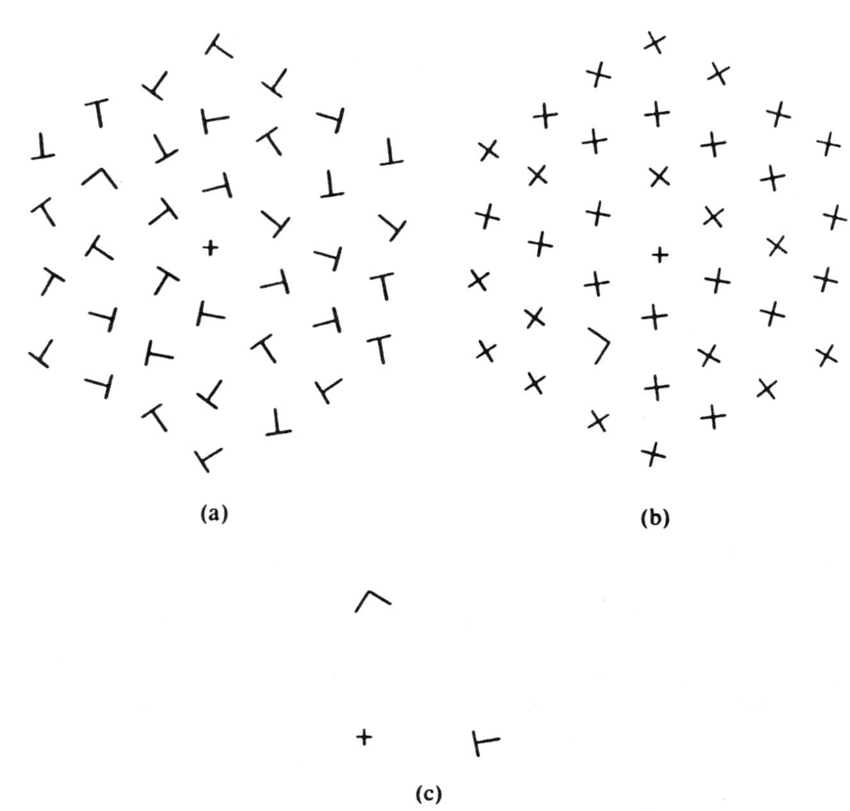

Fig. 13—The three types of stimuli used to obtain the results of Fig. 14. In all cases the task is to discriminate these from a control stimulus in which all elements are identical.

are very different for the case in which the elements share the same textons (T vs L. solid circles) from that in which they contain different textons (+ vs L, open circles). Note that in the T vs L case, not only does performance never exceed 65 percent correct, but it takes over 300 ms to reach this asymptote, while in the + vs L case the asymptote is reached within 200 ms. In fact, by the time the asymptote in the T vs L case is reached, the afterimage resulting from the test flash has largely disappeared. In the case in which only one T and one L are presented (filled squares), the results closely resemble those for 35 +'s and one L. Perceptually, the difference between the same-textons and different-textons cases is simply that the L in the field of T's stands out almost as if presented alone on a blank field, while the same L in a field of T's must be sought out. Attention is rapidly shifted to the L in the former case, while in the latter the search process is apparently still going on after 300 ms have passed. When there are only two elements to choose from, this search time is very brief.

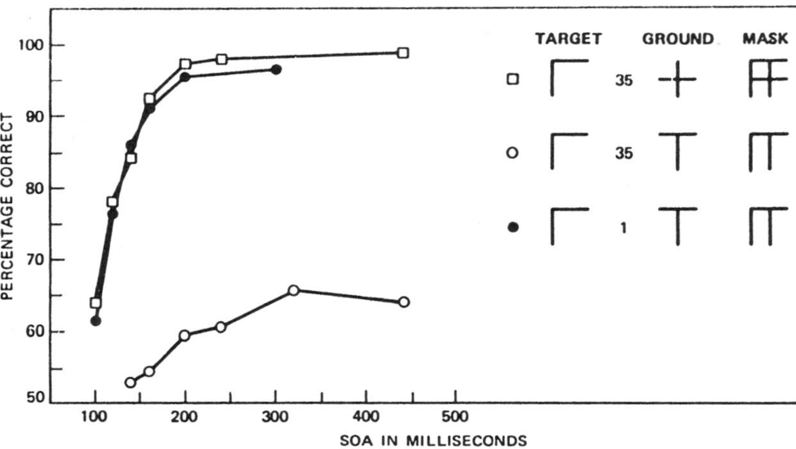

Fig. 14—Results of discrimination experiments using Fig. 13. The open circles, filled circles, and squares correspond to stimuli of Fig. 13a,c,b, respectively. SOA is the abscissa, and percent correct discrimination is the ordinate.

It is interesting to note that the observed asymptotic level of about 65 percent correct is what would be expected if seven or eight of the possible positions could be searched in the time available. Combining this number with the afterimage persistence time of 300–400 ms[11,12] gives a figure of about 50 ms per position inspected.[10]

This process of sequential inspection seems to be essentially independent of the overall angular subtense of the stimulus. Figure 15 shows results from an experiment in which the observer is required to distinguish a stimulus consisting of six T's and one L, or vice versa, from one in which all elements are identical. The stimulus was uniformly contracted so as to fall entirely within the fovea (<3 degrees across), or dilated to extend almost 14 degrees across, with no systematic variation in performance.

Another way of describing this is to say that the measurements are independent of the distance from which the stimulus is viewed, assuming that all of the elements remain resolvable: This independence suggests two important points. First, the fovea is not better than the near periphery in the extraction of this type of visual information. Second, the aperture of attention changes its spatial scale according to the size of the feature being sought. Thus, the same number of sequential fixations of attention are needed when the stimulus is reduced in size uniformly, because the sizes of the features upon which the discrimination is based are proportionally reduced. This extension of the scope of the texton theory from texture perception to rapid pattern discrimination suggests a model of vision in general as described in the following section.

VI. A MODEL OF THE "TWO VISUAL SYSTEMS"

When a visual scene changes suddenly in time or space, and our attention encompasses the entire visual scene, only those areas in which density differences in textons occur are conspicuous. These textons are elongated blobs with specific colors, widths, lengths, orientations, terminators, and crossings between them. Furthermore, because binocular disparity, movement disparity, and flicker are locally conspicuous features that can be detected in a brief presentation,[7,11,13] they, like color, are also properties of elongated blob textons.

Focal attention is directed to areas of spatial or temporal texton changes. The preattentive process appears to work in parallel and extends over a wide area of the visual field, while scrutiny by local or foveal attention is a serial process, which at any given time is restricted to a small patch. Focal attention can be shifted in 50-ms steps, four times faster than the fastest scanning eye movements. Furthermore, the aperture of focal attention can vary in size and can be a minute portion of the fovea, that is, extending to only a few minutes of arc (as shown in Fig. 15). Therefore, if the visual environment is rich in detail even when slowly changing in time, or is rather lacking in spatial detail but changes rapidly, we perform the major portion of our spatio-temporal processing in the preattentive state.

The focus of visual attention seems to be characterized by a texton class as well as a spatial locus. In particular, just as it apparently is impossible to attend simultaneously to two different places, it also seems impossible simultaneously to attend to very different sizes of features. This fact has been noted previously by other psychologists.[14] Stimuli widely separated in space produce cortical responses which are far apart. Similarly, stimuli of differing sizes often generate re-

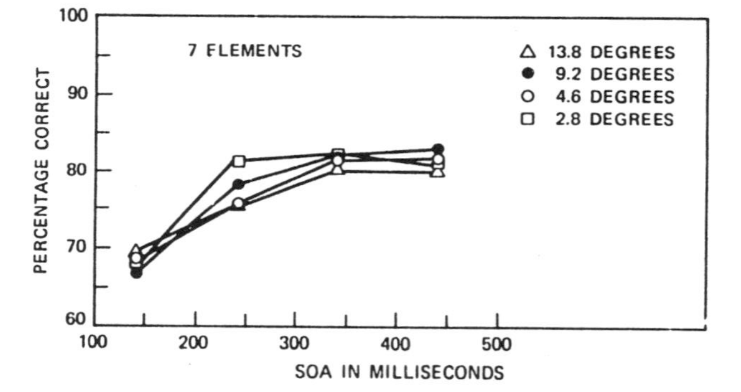

Fig. 15—Size invariance while the angular subtense of seven elements is varied from 2.8 degree of arc diameter to 13.8 degree of arc. Findings imply that the aperture of focal attention can be as small as a few minutes of arc.

sponses in different cortical areas.[6,15] These results seem to imply that the focus of attention is restricted to a very small region of visual cortex, and that stimuli producing responses far apart in the cortex cannot be attended simultaneously.

The essence of our findings is illustrated in Fig. 16. The left array contains a texture composed of "L"-shaped elements (formed by two perpendicular line segments with a gap), except for one "+" shaped and one "T" shaped element (formed by two perpendicular line segments which cross, or have a gap, respectively). The "+" shaped element (target) differs from the many surrounding L's in one texton, namely the "crossing", and perceptually stands out immediately. On the other hand, the T-shaped target can be detected only after some search, by directing the aperture of attention to the target itself.

The right array of Fig. 16 is identical to the left but illustrates our model of vision. The parallel preattentive system instantly detects the location of texton differences and directs the aperture of focal attention to this location, as indicated by the dotted disk around the "+". Since the T contains the same textons as its surround, its detection requires the aperture of attention (symbolized as a "cone" of a searchlight) to scrutinize the texture elements in sequence. Therefore, this

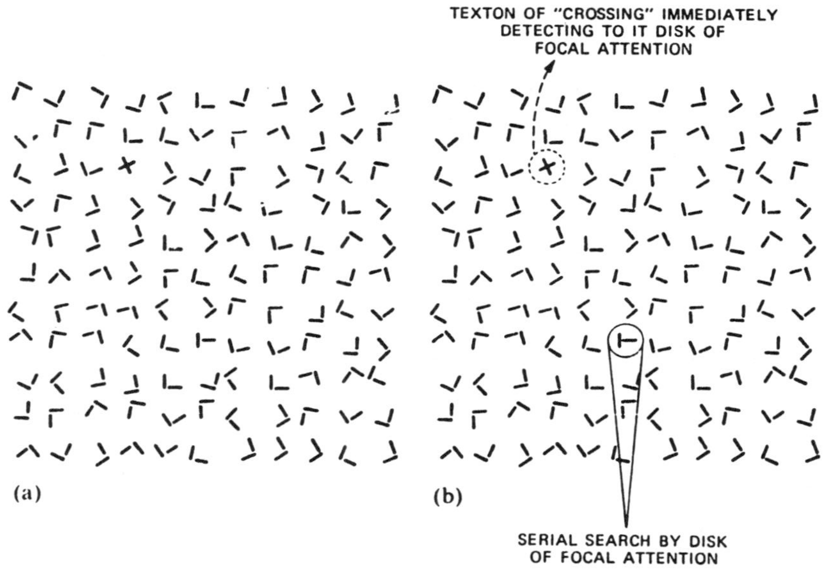

Fig. 16—Model of the two visual systems (b), showing how the preattentive system directs the aperture of focal attention to the loci of texton differences [the + in the L's in (a)], while without such texton differences [the T in the L's in (a)] focal attention requires time-consuming search.

serial search for the T-shaped target depends on the number of texture elements and may take considerable effort and time. However, after the T has been found, and the aperture of focal attention surrounds it, both the + and the T targets are seen with the same clarity. Obviously, form recognition, restricted to the aperture of focal attention, does not depend on the way attention has been directed to the targets. Whether a local difference in textons quickly directed focal attention to the target, or in the absence of texton differences it required time-consuming search to find the target, is immaterial for processing of the target by the attentive visual system.

This mode of behavior of the preattentive and attentive visual systems can also be observed in texture perception, when the reader inspects Fig. 1. The preattentive system immediately detects the texton differences at the boundary of the + and L aggregates, and a quick inspection by focal attention of a few elements on the two sides of the boundary lets the observer conclude that the two areas must contain +'s and L's. Only detailed scrutiny will reveal that the area believed to contain L's only has a region of T's as well.

In summary, the reason that texture discrimination is such a revealing process for showing the workings of the two visual systems is that textures usually cover wide areas of the visual field, while the texture elements are a small portion of the textural area. When the observer is inspecting an extended field, there is an "uncertainty region" in which the relative spatial position of local features is ignored. This is very different from a resolution limit due to visual acuity. In all of the indistinguishable texture pairs, the line segments which make up the texture elements are clearly resolved; nevertheless, if these textons fall within this uncertainty region, it is impossible to tell a T from an L. Many physiologists and psychologists have proposed two visual systems, one ambient and the other focal.[11-20] Yet, without the notion of textons, whose spatial and temporal changes are detected by the preattentive system, which in turn directs focal attention to these loci, the model of the two visual systems is not complete. We hope that the model outlined here gives some useful insights into human vision.

VII. IMPLICATIONS AND CONCLUSIONS

Some conspicuous local features called textons have been identified by psychological means. These textons, particularly the elongated blobs, are quite similar to features found to stimulate the simple neural units in the striate cortex of the monkey, which are selectively tuned to elongated blobs of certain colors, orientations, width, and length.

Our findings, that in preattentive vision objects are distinguished only through their texton decompositions, might be of considerable importance. Since in preattentive vision these textons are not coupled, and furthermore the resolution of texton properties—i.e., the perceptual threshold for color, width, length, and orientation differences—is rather limited, the number of distinguishable textons is within practically useful bounds. (For example, the width of periodic bars can be judged with an error of 4 to 6 percent,[21] while accuracy of bar orientation is measured to be only 6 degrees of arc.[22]) This limitation makes practical the devices that simulate preattentive vision. This contrasts with attentive vision for which virtually an infinite number of recognizable patterns exist whose biological, social, or intellectual interest to the observer is unknown. Whether additional textons will be discovered remains to be seen. But as long as they remain independent of the previously isolated textons, the model outlined here will not be importantly affected.

The main implication of our findings is as follows: A considerable amount of vision is carried out by the preattentive system whose workings appear to be much simpler than that of the attentive system. This is important in judging the information requirements of the human visual system realistically. Furthermore, it is important to realize that even in the attentive mental state, with all its prodigious processing powers, complex feats of form recognition are restricted to a small spatial aperture, often as small as a few minutes of arc. Also, changing the position or extent of the aperture of focal attention requires considerable time. The shortest time is about 50 ms when eye movements are prevented, and as long as about 200 ms if saccadic eye movements are necessary.

This dichotomy between preattentive and attentive mental states, the first limited in its power of information processing, the latter limited in its spatial extent, gives a model of human vision that could be exploited in visual communication. Here we do not want to invent specific methods, but only indicate some obvious possibilities. With the advent of fast, perhaps parallel computers, the textons that direct the human observer's attention could be simultaneously extracted by hardware. Detailed images need only be presented in such areas.

Also, one could program computers to extract local features other than textons. For instance, a parallel computer might rapidly detect the difference between an L and a T, rather than between a + and an L. If an observer's attention were directed by such a machine, whose capabilities are very different from human preattentive vision, perhaps a new way of inspecting the visual environment could be made available and possibly learned.

The textons reported here help to discriminate textures, mainly surfaces of objects, without the need of complex familiarity cues. Such an early separation of the visual environment into figure and ground, or objects and their backgrounds, is a fundamental operation of visual perception. Lack of understanding of this process is, as of now, the greatest bottleneck in machine vision, which in turn is necessary in extending the capabilities of robots.

Regardless of the feasibility of such ambitious schemes, the finding that texton differences can be almost instantaneously perceived over large areas of the visual field can be practically exploited in traffic signs and in directing attention to select areas of visual displays. Traditionally, flickering or static colored lights have been used as traffic signs, or in instrument panels. Now we can add other texton classes—for instance, gaps to increase the terminator number—to enhance visibility. For example, in Fig. 17 we show how a single gap introduced in the conventional alphabet draws attention to the word STOP, which otherwise would require a long time to be segmented and detected. Such slight modification of the alphanumeric characters (amounting to a new "font") might be beneficial in improving legibility. For instance, dyslexic children—children who cannot distinguish well between similar characters with different symmetric transformations such as b, d, or p—might greatly benefit if a gap or stroke were added to one of the characters, so that all characters would differ in at least one texton.

It should be stressed that the textons of preattentive vision only draw attention to certain areas, and we do not claim that these same textons are also the building blocks of form vision. If they were, our findings would prove preattentive vision to be the basis of attentive vision. Even if textons are restricted to vision in the preattentive state, we feel that to know those conspicuous features that grab our attention, wherever they appear, is of interest to everyone who wants to communicate through visual means.

VIII. ACKNOWLEDGMENTS

We thank our many colleagues who contributed to this research effort throughout the years, and who are mentioned in Section II and in the references. We also thank Max V. Mathews for his helpful comments while reading the manuscript. We are indebted to Walter

SPTOTOPSSOPTTPSO
OSPTTSPOSTOPTSOP
PSTOSTOPTSPOPOTS
OTSPSPTOPSOTTOSP

Fig. 17—Demonstration that the introduction of textons into the alphabet (here through increasing the terminator numbers by adding a gap) can help to segment and detect certain areas in a dense letter array.

Kropfl who developed the display hardware, and to Peter Burt who wrote the GENTEX program permitting the rapid generation, display, and manipulation of texture arrays. We thank our summer student, Franklin Schmidt, for developing the TEXTONS program.

REFERENCES

1. B. Julesz, "Textons, the Elements of Texture Perception, and Their Interactions," Nature, *290* (March 12, 1981), pp. 91–7.
2. B. Julesz, E. N. Gilbert, and J. D. Victor, "Visual Discrimination of Textures with Identical Third-Order Statistics," Biol. Cybernetics, *31* (1978), pp. 137–40.
3. B. Julesz et al., "Inability of Humans to Discriminate Between Visual Textures that Agree in Second-Order Statistics—Revisited," Perception, *2* (1973), pp. 391–405.
4. B. Julesz, "Experiments in the Visual Perception of Texture," Sci. Am., *232* (April 1975), pp. 34–43.
5. B. Julesz, "A Theory of Preattentive Texture Discrimination Based on First-Order Statistics of Textons," Biol. Cybernetics, *41* (1981), pp. 131–8.
6. D. H. Hubel and T. N. Wiesel, "Receptive Fields and Functional Architecture of Monkey Striate Cortex," J. Physiol., *195* (1968), pp. 215–43.
7. B. Julesz, *Foundations of Cyclopean Perception*, Chicago: University of Chicago Press, 1971.
8. D. Marr, "Early Processing of Visual Information," Philos. Trans. R. Soc. London Ser. B, *275* (1976) pp. 483, 524.
9. T. Caelli and B. Julesz, "On Perceptual Analyzers Underlying Visual Texture Discrimination: Part I," Biol. Cybernetics, *28* (1978), pp. 167–75.
10. J. R. Bergen and B. Julesz, "Discrimination with Brief Inspection Times," J. Opt. Soc. Am., *71*, No. 12 (December 1981), p. 1570. Also see "Parallel Versus Serial Processing in Rapid Pattern Discrimination," Nature, *303* (June 23–29, 1983), pp. 696–8.
11. B. Julesz, "Binocular Depth Perception of Computer-Generated Patterns," B.S.T.J., *39* No. 5 (September 1960), pp. 1125–62.
12. E. Averbach and G. Sperling, "Short-term Storage of Information in Vision," in C. Cherry (ed.) *Information Theory, Fourth London Symposium*, London: *Butterworth, 1961*, pp. 196–211.
13. B. Julesz and J. J. Chang, "Interaction Between Pools of Binocular Disparity Detectors Tuned to Different Disparities," Biol. Cybernetics, *22* (1976), pp. 107–19.
14. G. Sperling and M. J. Melchner, "Visual Search, Visual Attention and the Attention Operating Characteristics," in J. Requin (ed.) *Attention and Performance VII*, Hillsdale, NJ: Erlbaum, 1978, pp. 675–86.
15. S. M. Zeki, "The Functional Organization of Projections From Striate to Prestriate Cortex in the Rhesus Monkey," Cold Spring Harbor Symposia on Quantitative Biology, *15* (1976), pp. 591–600.
16. R. Held et al., "Locating and Identifying: Two Modes of Visual Processing, *31*, Psychol. Forsch., pp. 44–62; 299–348 (1967–1968).
17. C. B. Trevarthen, "Two Mechanisms of Vision in Primates," Psychol. Forsch. *31* (1968), pp. 299–337.
18. J. E. Hoffmann, "Hierarchical Stages in the Processing of Visual Information," Perception and Psychophysics, *18* (1975), pp. 348–54.
19. A. Treisman and G. Gelade, "A Feature-Integration Theory of Attention," Cognitive Psychol., *12* (1980), pp. 97–136.
20. B. Julesz, "Visual Pattern Discrimination," IRE Trans. Inform. Theory, *IT-8* (February 1962), pp. 84–92.
21. F. W. Campbell, J. Nachmias, and J. Jukes, "Spatial-Frequency Discrimination in Human Vision," J. Opt. Soc. Am., *60* No. 4 (April 1970), pp. 555–9.
22. J. P. Thomas and J. Gille, "Bandwidths of Orientation Channels in Human Vision," J. Opt. Soc. Am., *69*, No. 5 (May 1973), pp. 652–60.
23. M. Rosenblatt and D. Slepian, "Nth Order Markov Chains With Any Set of N Variable Independent," J. Soc. Indust. Appl. Math., *10* (1962), pp. 537–49.
24. T. Caelli, B. Julesz, and E. N. Gilbert, "On Perceptual Analyzers Underlying Visual Texture Discrimination: Part II," Biol. Cybernetics, *29* (1978), pp. 201–14.
25. B. Julesz, "Spatial Nonlinearities in the Instantaneous Perception of Textures with Identical Power Spectra," Phil. Trans. R. Soc. Lond. B, *290* (1980), pp. 83–94.
26. B. Julesz, "Perceptual Limits of Texture Discrimination and Their Implications to Figure-Ground Separation," in E.L.J. Leeuwenberg and H.F.J.M. Buffart (eds.), *Formal Theories of Visual Perception*, New York: Wiley, 1978, pp. 205–16.
27. H. L. Frisch and F. H. Stillinger, "Contribution to the Statistical Geometric Basis of Radiation Scattering," J. Chem. Phys., *38* (1963), pp. 2200–7.
28. N. Wiener, "Extrapolation, Interpolation and Smoothing of Stationary Time Series, With Engineering Applications," New York: Cambridge University Press.

APPENDIX

It required two decades of research efforts to discover that preattentive texture perception depends on local features alone and that global higher-order statistical parameters can be ignored. In 1962, Julesz asked mathematicians to generate stochastic texture pairs that would be identical in their first $(n\text{-}1)$th order statistics, but different in the nth- and higher than nth-order statistics.[20] The nth-order statistics are similar to the well-known nth-order joint probability distribution of n samples. The n samples are n points of a texture selected at random. However, in random geometry the shape of the n samples is of importance.

These n points can be regarded as the vertices of an n-gon. The n-gon (or nth-order) statistics are obtained when these n points (having the same n-gon shape) are selected at random, and statistics indicate that these n points have certain color values. For instance, the second-order statistics can be obtained if a 2-gon (dipole, or needle) is randomly thrown at the texture and the probability is determined that the two end-points of the dipole—of given lengths and orientations—fall on certain color combinations: e.g., black and black; or black and white; or black and gray, etc.

In the intervening years many such stochastic textures were discovered, particularly with identical first- and second-order statistics, but different third- and higher-order statistics.[3,23–25] As a matter of fact, the texture pairs in Figs. 3–6, and 8–10 have this property. The finding that many of these *iso-second-order* texture pairs differing only in third- and higher-orders are indistinguishable suggests that the preattentive visual system cannot compute statistical difference beyond the second order. The recent finding by Julesz, demonstrated in Fig. 11a, suggests that the preattentive visual system cannot even process second-order statistical parameters.[4] From the second-order statistics the autocorrelation function can be uniquely determined—as a matter of fact, for two-tone textures composed of black and white dots, the second-order (dipole) statistic *is* the autocorrelation function[26,27]—and the Fourier transform of the autocorrelation is the Fourier power spectrum.[28] Therefore, all the texture pairs with identical second-order statistics also have identical power spectra. The finding that texture segregation can be obtained in iso-second-order textures, after it was established that the preattentive system cannot process third-order statistics (and, as Fig. 11a demonstrates, not even second-order statistics), implies that this segregation must be based on *local* density differences. Finally, it was proposed that the density changes of certain local conspicuous features, the textons, explain preattentive texture discrimination.[1,25]

AUTHORS

James R. Bergen, A.B. (Mathematics and Psychology), 1975, University of California, Berkeley; Ph.D. (biophysics and theoretical biology), 1981, University of Chicago; Postdoctoral Fellow, Bell Laboratories, 1981–82; RCA, 1983—. Mr. Bergen's work concerns the quantitative analysis of information processing in the human visual system. At the University of Chicago he was involved in the development of a model of the spatial and temporal processing which occurs in the early stages of the system. At Bell Laboratories, his work has concentrated on the effect of visual system structure on the extraction of information from a visual image. Mr. Bergen recently joined RCA Laboratories in Princeton, NJ.

Bela Julesz, Diploma, 1950 (Electrical Engineering), Technical University, Budapest; Ph.D., 1956, Hungarian Academy of Sciences; Bell Laboratories, 1956—. Mr. Julesz taught and did research in communications systems for several years prior to 1956. Since joining Bell Laboratories, he has devoted himself to visual research, particularly depth perception and pattern recognition. He is the originator of the random-dot stereoimage technique and of the method of studying texture discrimination by constraining second-order statistics. He has written extensively in the area of visual and auditory perception and is the author of *Foundations of Cyclopean Perception.* Mr. Julesz was Head of the Sensory and Perceptual Processes Department from 1964 to 1982, and in 1983 was made Head of the Visual Perception Research Department. He has been visiting professor of experimental psychology at M.I.T. and other universities. In February 1983 he received the MacArthur Prize Fellow Award in Experimental Psychology and Artificial Intelligence. He was a Fairchild Distinguished Scholar at the California Institute of Technology from 1977 to 1979. Fellow, AAAS, OSA, and the American Academy of Arts and Sciences; Corresponding Member of the Goettingen Academy of Sciences.

Mapping Image Properties into Shape Constraints: Skewed Symmetry, Affine-Transformable Patterns, and the Shape-from-Texture Paradigm

Takeo Kanade
John R. Kender

Carnegie-Mellon University
Pittsburgh, Pennsylvania

Abstract

In this paper we demonstrate two new approaches to deriving three-dimensional surface orientation information ("shape") from two-dimensional image cues. The two approaches are the method of affine-transformable patterns and the shape-from-texture paradigm. They are introduced by a specific application common to both: the concept of skewed symmetry. Skewed symmetry is shown to constrain the relationship of observed distortions in a known object regularity to a small subset of possible underlying surface orientations. Besides this constraint, valuable in its own right, the two methods are shown to generate other surface constraints as well. Some applications are presented of skewed symmetry to line drawing analysis, to the use of gravity in shape understanding, and to global shape recovery.

1. Introduction

Certain image properties, such as parallelisms, symmetries, and repeated patterns, provide cues for perceiving 3-D shape from a 2-D picture. This paper demonstrates how we can map these image properties into 3-D shape constraints by associating appropriate assumptions with them and by using appropriate computational and representational tools.

We begin with the exploration of how one specific image property, "skewed symmetry," can be defined and formulated to serve as a cue to the determination of surface orientations. Then we will discuss the issue from two new, broader viewpoints. One is the class of affine-transformable patterns. It has various interesting properties, and includes skewed symmetry as a special case. The other is the computational paradigm of shape-from-texture. Skewed symmetry is derived in a second, independent way, as an instance of the application of the paradigm. Also, it is proven that the same skewed-symmetry constraint can arise from greatly different image conditions.

This paper further claims that the ideas and techniques presented here are applicable to many other properties under a general framework of the shape-from-texture paradigm with the underlying meta-heuristic of non-accidental image properties.

2. Skewed Symmetry

In this section we assume the standard orthographic projection from scene to image, and a knowledge of the gradient space (see Mackworth, 1973).

2.1. Definition, Assumption and Constraints

Symmetry in a 2-D picture has an axis for which the opposite sides are reflective; in other words, the symmetrical properties are found along the transverse lines perpendicular to the symmetry axis. The concept *skewed symmetry* was introduced by Kanade (1979) by relaxing this condition a little. It means a class of 2-D shapes in which the symmetry is found along lines not necessarily perpendicular to the axis, but at a fixed angle to it. Formally, such shapes can be defined as 2-D affine transforms of real symmetries. Figures 1a-c show a few examples.[1]

Stevens (1980) presents a number of psychological experiments which suggest that human observers can perceive surface orientations from figures with this property. This is probably because such qualitative symmetry in the image is often due to real symmetry in the scene. Thus let us associate the following assumption with this image property:

> A skewed symmetry depicts a real symmetry viewed from some unknown view angle.

Note that the converse of this assumption is always true under orthographic projection.

We can transform this assumption into constraints in the gradient space. As shown in Figure 1, a skewed symmetry defines two directions; let us call them the skewed-symmetry axis and the skewed-transverse axis, and denote

[1] The mouse hole example of Figure 1c is due to K. Stevens (1980).

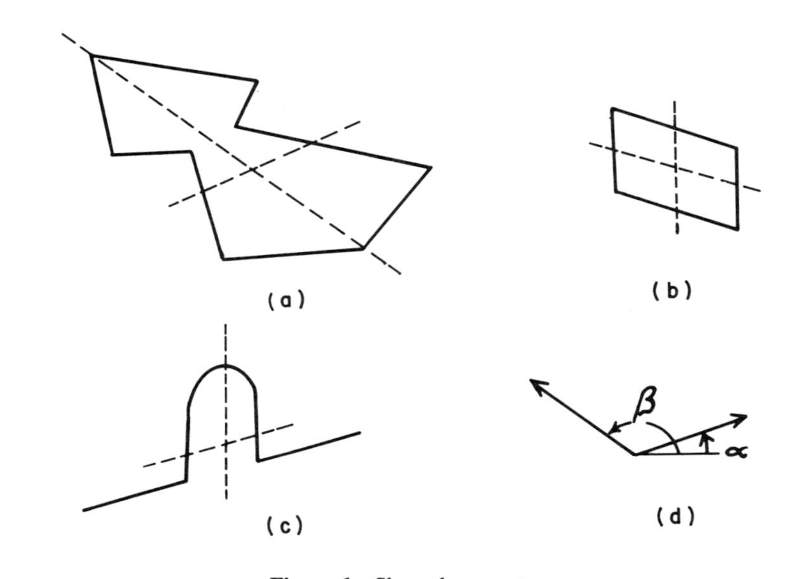

Figure 1. Skewed symmetry.

their directional angles in the picture by α and β, respectively (Figure 1d). Let $G = (p, q)$ be the gradient of the plane which includes the skewed symmetry. The 3-D vectors on the plane corresponding to the directions α and β are

$$(\cos\alpha, \sin\alpha, -p\cos\alpha - q\sin\alpha) \text{ and } (\cos\beta, \sin\beta, -p\cos\beta - q\sin\beta).$$

The assumption demands that these two vectors be perpendicular; their inner product vanishes:

$$\cos(\alpha-\beta) + (p\cos\alpha + q\sin\alpha)(p\cos\beta + q\sin\beta) = 0. \tag{1}$$

By rotating the p-q coordinates into the p'-q' coordinates so that the new p'-q' axes are the bisectors of the angle made by the skewed symmetry and skewed-transverse axes, it is easy to show that

$$p'^2\cos^2\left(\frac{\alpha-\beta}{2}\right) - q'^2\sin^2\left(\frac{\alpha-\beta}{2}\right) = -\cos(\alpha-\beta) \tag{2}$$

where

$$p' = p\cos\lambda + q\sin\lambda$$
$$q' = -p\sin\lambda + q\cos\lambda$$
$$\lambda = (\alpha+\beta)/2.$$

Thus, the (p, q)'s are on the hyperbola shown in Figure 2. That is, the skewed symmetry defined by α and β in the picture can be a projection of a real symmetry if and only if the gradient is on this hyperbola. The skewed symmetry thus imposes a one-dimensional family of constraints on the underlying surface orientation (p, q). As we will see in Section 5, other constraints can be exploited for the unique determination of surface orientation.

The tips or vertices G_T and G_T' of the hyperbola represent special orientations with interesting properties. First, since they are closest to the origin of the gradient space, and since the distance from the origin to a gradient represents the magnitude of the surface slant, G_T and G_T' correspond to the least slanted orientations that can produce the skewed symmetry in the picture from a real symmetry in the scene.

Second, since they are on the line (the axis of the hyperbola) which bisects the obtuse angle made by α and β, they correspond to the orientations for which the rates of depth change along the directions of α and β in the picture are the same. In other words, the apparent ratio of length to width of the object in the picture represents the real ratio in the scene (see Kanade [1979] for the proof.)

2.2. Rationale and Justification

Skewed symmetry has straightforward applications to scenes containing objects that have been manufactured, whether naturally or artificially. Many constructed items exhibit symmetry, occasionally about many axes.

Some symmetries are introduced due to economies of the manufacturing process: an object is often composed of identically formed component parts (fibers, cells, bricks, etc.). The symmetries result from the three-dimensional tessellation of the components into the whole. Often the tessellation is effectively two-dimensional, in laminae (cloth, honeycombs, walls, etc.), and the application of the skewed symmetry method is straightforward. Further, the requirement for a close symmetric packing of the components occasionally imposes a local symmetry on the individual components, too. The method can then be applied to individual parts (such as the bricks themselves). Notice the method does *not* assume 3-D symmetry of the whole object; what is assumed is *local* 2-D symmetry.

A further source of symmetry is the bilateral symmetry that results from biological manufacture (growth). It not only contributes symmetric objects to the environment; it may also be responsible for an imitative esthetic bias in human manufacture. If the extent of a bilaterally symmetric pattern into the third dimension is not too great (a face, a leaf, an airplane), the skewed symmetry method can be approximately applied also.

3. Affine-Transformable Patterns

In texture analysis we often consider small patterns (texels=texture elements) whose repetition constitutes "texture." Suppose we have a pair of texel

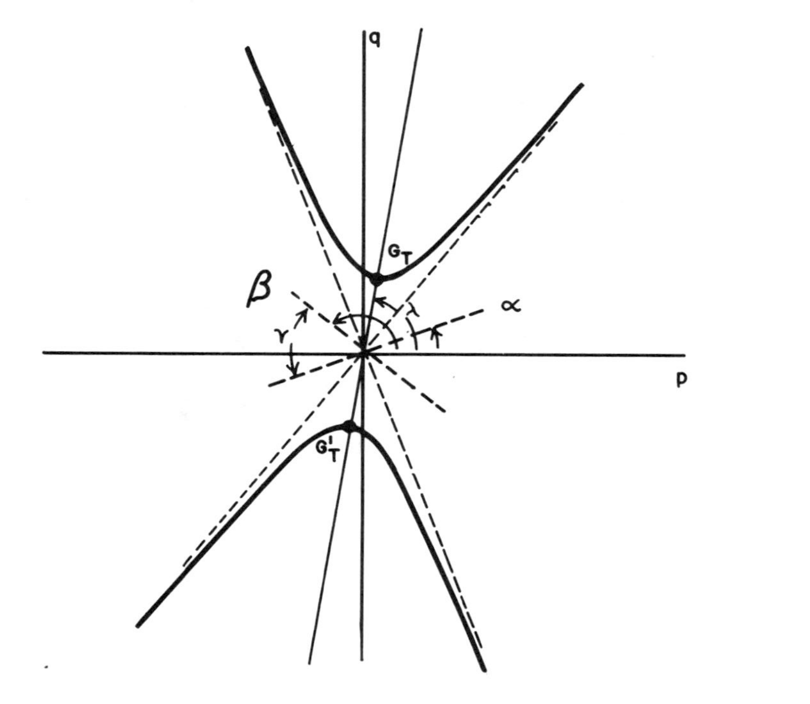

Figure 2. The hyperbola determined by a skewed symmetry defined by α and β.

patterns in which one is a 2-D affine transform of the other; we call them a pair of *affine-transformable* patterns. Let us assume:

> A pair of affine-transformable patterns in the picture are projections of similar patterns in the 3-D space (i.e., they can be overlapped by scale change, rotation, and translation).

Note that, as in the case of skewed symmetry, the converse of this assumption is always true under orthographic projection. The above assumption can be schematized by Figure 3. Consider two texel patterns P_1 and P_2 in the picture, and place the origins of the x-y coordinates at their centers, respectively. The transform from P_2 to P_1 can be now expressed by a regular 2×2 matrix $A = (a_{ij})$. P_1 and P_2 are projections of patterns P_1' and P_2' which are drawn on the 3-D surfaces. We assume that P_1' and P_2' are small enough so that we can regard them as being drawn on small planes. Let us denote the gradients of those small planes by $G_1 = (p_1,q_1)$ and $G_2 = (p_2,q_2)$, respectively; i.e., P_1' is drawn on a plane $-z = p_1x + q_1y$ and P_2' on $-z = p_2x + q_2y$.

Now, our assumption amounts to saying that P_1' is transformable from

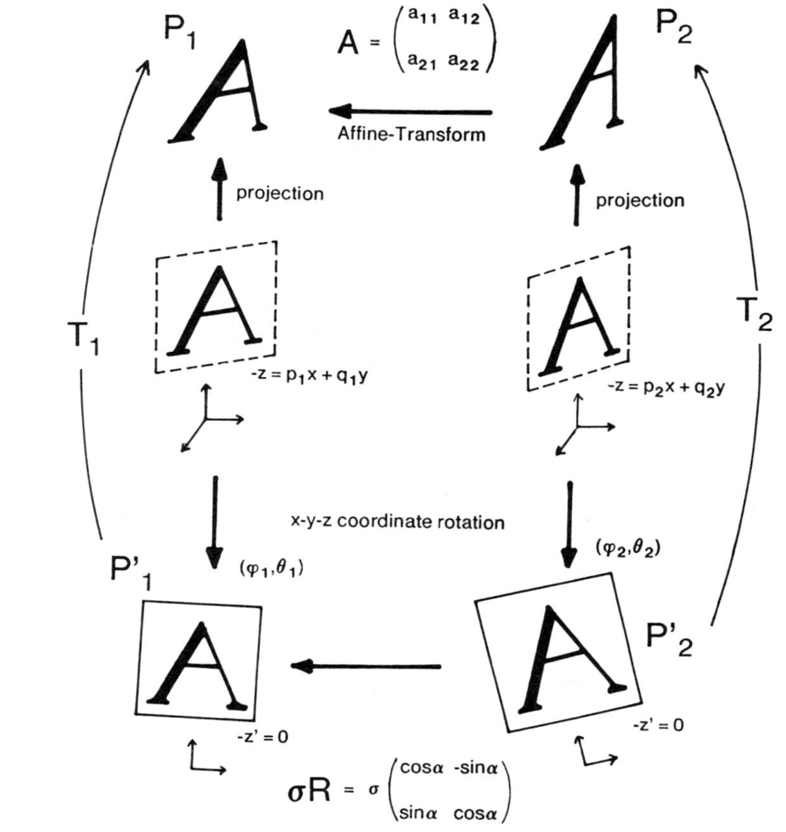

Figure 3. A schematic diagram showing the assumption about the affine-transformable patterns.

P_2' by a scalar scale factor σ and a rotation matrix:

$$R = \begin{pmatrix} \cos\alpha & -\sin\alpha \\ \sin\alpha & \cos\alpha \end{pmatrix}.$$

(We can omit the translation from our consideration, since for each pattern the origin of the coordinates is placed at its gravity center, which is preserved under the affine transform.) Thinking about a pattern drawn on a small plane, $-z = px + qy$, is equivalent to viewing the pattern from directly overhead; that is, rotating the x-y-z coordinates so that the normal vector of the plane is along the new z-axis (line of sight). For this purpose we rotate the coordinates first by ϕ around the y-axis and then by θ around the x'-axis (Figure 4). We have

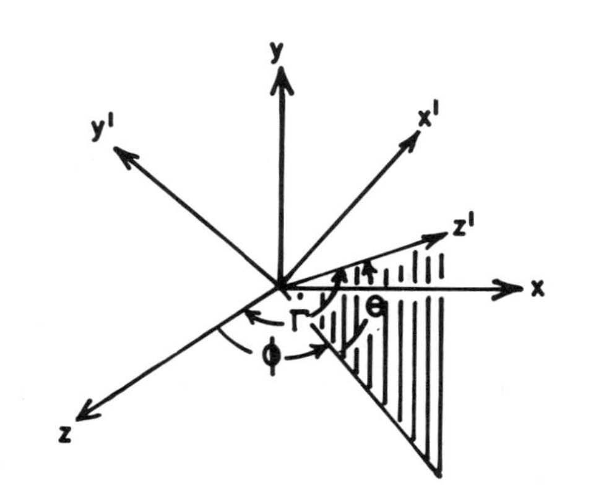

Figure 4. Rotation of the x-y-z coordinates.

the following relations among ϕ, θ, p, and q:

$$\sin\phi = p/\sqrt{p^2+1}, \qquad \cos\phi = 1/\sqrt{p^2+1},$$
$$\sin\theta = q/\sqrt{p^2+q^2+1}, \qquad \cos\theta = \sqrt{p^2+1}/\sqrt{p^2+q^2+1}. \tag{3}$$

Further, let Γ denote the angle of slant of the pattern, i.e., the angle between the old and the new z axes. Then

$$\cos\Gamma = 1/\sqrt{p^2+q^2+1}. \tag{4}$$

The plane which was represented as $-z = px+qy$ in the old coordinates is, of course, now represented as $-z' = 0$ in the new coordinates.

Let us denote the angles of the coordinate rotations to obtain P_1' and P_2' in Figure 3 by (ϕ_1,θ_1) and (ϕ_2,θ_2), respectively. The 2-D mapping from P_i' (x'-y' plane) to P_i (x-y plane) can be conveniently represented by the following 2×2 matrix T_i which is actually a submatrix of the usual 3-D rotation matrix:

$$T_i = \begin{pmatrix} \cos\phi & -\sin\phi\sin\theta \\ 0 & \cos\theta \end{pmatrix}.$$

Now, in order for the schematic diagram of Figure 3 to hold, what relationships have to be satisfied among the matrix $A = (a_{ij})$, the gradients $G_i = (p_i, q_i)$ for $i = 1,2$, the angles (ϕ_i, θ_i) for $i = 1,2$, the scale factor σ, and the matrix R? We equate the two transforms that start from P_2' to reach P_1: one following the diagram counter-clockwise, $P_2' \rightarrow P_2 \rightarrow P_1$, and the other clockwise, $P_2' \rightarrow P_1' \rightarrow P_1$. We obtain

$$AT_2 = T_1\sigma R.$$

That is,

$$\begin{aligned}
a_{11}\cos\phi_2 &= \sigma(\cos\alpha\cos\phi_1 - \sin\alpha\sin\phi_1\sin\theta_1) \\
a_{12}\cos\theta_2 - a_{11}\sin\phi_2\sin\theta_2 &= -\sigma(\sin\alpha\cos\phi_1 + \cos\alpha\sin\phi_1\sin\theta_1) \\
a_{21}\cos\phi_2 &= \sigma\sin\alpha\cos\theta_1 \\
a_{22}\cos\theta_2 - a_{21}\sin\phi_2\sin\theta_2 &= \sigma\cos\alpha\cos\theta_1.
\end{aligned} \tag{5}$$

By eliminating σ and α and substituting for $\sin\phi_i$, $\cos\phi_i$, $\sin\theta_i$, and $\cos\theta_i$ from (3), we have the following equations in p_1, q_1, p_2, and q_2:

$$\begin{aligned}
&\sqrt{p_2^2+q_2^2+1}\,(a_{11}(p_1^2+1)+a_{21}p_1q_1) \\
&\qquad = \sqrt{p_1^2+q_1^2+1}\,(a_{22}(p_2^2+1)-a_{21}p_2q_2) \\
&(-a_{12}(p_2^2+1)+a_{11}p_2q_2)(p_1^2+1)-(a_{22}(p_2^2+1)-a_{21}p_2q_2)p_1q_1 \\
&\qquad = a_{21}\sqrt{p_1^2+q_1^2+1}\sqrt{p_2^2+q_2^2+1}.
\end{aligned} \tag{6}$$

We thus find that the assumption of affine-transformable patterns yields the constraint represented by (6) on surface orientations. The constraint is determined solely by the matrix $A = (a_{ij})$, which is determined by the relation between P_2 and P_1 *observable* in the picture without knowing either the original patterns (P_1' and P_2') or their relationships (σ and R) in the 3-D space.

In order to have an idea about the degree of the constraint represented by (6), if we assume that the orientation of P_2' is known (i.e., $G_2 = (p_2, q_2)$ is known), then (6) gives two simultaneous equations for $G_1 = (p_1, q_1)$. The system appears to be of degree 4, but it can be shown that there are only two solutions; they are of the form (p_0, q_0) and $(-p_0, -q_0)$, which are symmetrical around the origin of the gradient space (see the Appendix).

From (5) we can also derive the following relationship:[2]

$$\frac{\det(A)}{\sigma^2} = \frac{\sqrt{p_2^2+q_2^2+1}}{\sqrt{p_1^2+q_1^2+1}} = \frac{\cos\Gamma_1}{\cos\Gamma_2}. \tag{7}$$

This means that the ratio of cosines of the slant angles of the patterns is equal to the ratio $\det(A)/\sigma^2$. If we assume $\sigma = 1$ (the original patterns are of the same size) or that σ is known, (7) shows that we can order the texel patterns according to the magnitude of slant, Γ_i or $\sqrt{p_i^2+q_i^2}$, using the values of $\det(A)$.

[2] This indicates that $\det(A)$ should be positive. But if it is negative, then we can assume that P_1' and P_2' are mirrored patterns, and put $R = \begin{pmatrix}\cos\alpha & \sin\alpha \\ \sin\alpha & -\cos\alpha\end{pmatrix}$.

3.1. Skewed Symmetry from Affine-transformable Patterns

The affine transform from P_2 to P_1 is more intuitively understood by how a pair of perpendicular unit-length vectors (typically along the x and y coordinate axes) are mapped into their transformed vectors. As shown in Figure 5, two angles (α and β) and two lengths (τ and ρ) can characterize the transform. The components of the transformation matrix $A = (a_{ij})$ are represented by

$$a_{11} = \tau\cos\alpha \qquad a_{12} = \rho\cos\beta$$
$$a_{21} = \tau\sin\alpha \qquad a_{22} = \rho\sin\beta. \tag{8}$$

Suppose, for simplicity, the orientation of P_2 in Figure 3 is known to be $(p_2,q_2) = (0,0)$. This simplifies equation (6) to

$$a_{11}(p_1^2+1)+a_{21}p_1q_1 = a_{22}\sqrt{p_1^2+q_1^2+1} \tag{9}$$
$$-a_{12}(p_1^2+1)-a_{22}p_1q_1 = a_{21}\sqrt{p_1^2+q_1^2+1}.$$

If we assume that α, β, τ, and ρ are known, then (p_1,q_1) has two possible solutions. This is essentially the case which Ikeuchi (1980b) investigated in his shape recovery method by assuming a known standard pattern, even though he used the constraint only partially.

Let us consider the case where α and β are known, but τ and ρ are not. One can substitute a_{ij} in (9) by (8), and eliminate τ and ρ. Then we obtain

$$(p_1\cos\alpha+q_1\sin\alpha)(p_1\cos\beta+q_1\sin\beta)+\cos(\alpha-\beta) = 0$$

which reduces to the same as the hyperbola (1). This can be interpreted as follows.

As was noted in the previous subsection, a pair of affine-transformable patterns impose the constraints (6) between their surface orientations, in which, if one is fixed, the other has only two possible orientations. However, if we loosen the transform in such a way that the angular (rotational) correspondence (α and β) is known while the length relationship is not known (or arbitrary), then the one-dimensional constraint of the skewed-symmetry hyperbola is obtained.

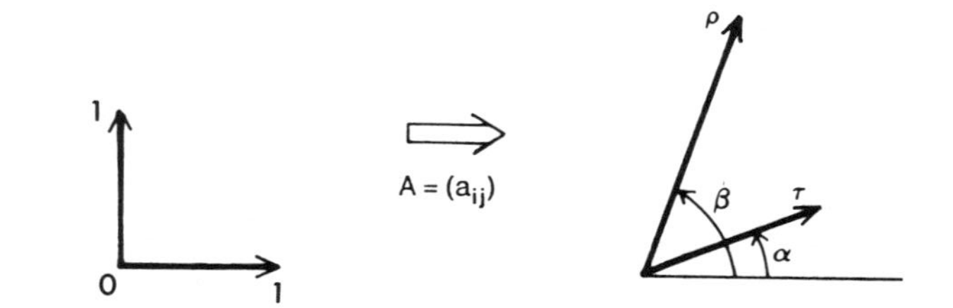

Figure 5. An affine transform (without translation) as characterized by two angles and two lengths.

4. The Shape-from-texture Paradigm

This section derives the same skewed-symmetry constraints from a second theory, different from that of the affine-transformable patterns. The shape-from-texture paradigm is a method of relating image texture properties to scene object properties, by explicitly incorporating assumptions about the imaging phenomenon into a computational framework. The paradigm is briefly presented here, but a fuller discussion can be found in (Kender, 1980).

The paradigm has two major portions. In the first, a given image textural property is "normalized" to give a general class of surface orientation constraints. In the second, the normalized values are used in conjunction with assumed scene relations to refine the constraints. If there are sufficiently many textural elements ("texels") in the image to be normalized, and if enough assumptions are made about their scene counterparts, then the underlying surface's orientation can be specified uniquely. Somewhat more weakly, only two texels are required, and only one assumption (equality of scene textural objects, or some other simple relation), to generate a well-behaved one-dimensional family of possible surface orientations. The method of skewed symmetry – the use of qualitative symmetries in the image to create a perspectively distorted right angle – is an example of such a weak method.

The first step in the paradigm is the normalization of a given texel property. The goal is to create a normalized texture property map (NTPM), which is a representational and computational tool relating image properties to scene properties. The NTPM summarizes the many different conditions that may have occurred in the scene leading to the formation of the given textural element. In general, the NTPM of a certain property is a scalar-valued function of two variables. The two input variables describe the postulated surface orientation in the scene (top-bottom and left-right slants: (p,q) when we use the gradient space). The NTPM for a horizontal unit line length in the image summarizes the lengths of lines that would have been necessary in 3-D space under various orientations: at surface orientation (p,q), it would have to be $\sqrt{p^2+1}$.

More specifically, the NTPM is formed by selecting a texel and a texel property, back-projecting the texel through the known imaging geometry onto all conceivable surface orientations, and measuring the texel property there. The representation chosen for the two-dimensional space of orientations is important; we will, however, only use the gradient space here.

In the second phase of the paradigm, the NTPM is refined in the following way. Texels usually have various orientations in the image, and there are many different texel types. Each texel generates its own image-scene relationships, summarized in its NTPM. If, however, assumptions can be made to relate one texel to another, then their NTPMs can also be related; in most cases only a few scene surface orientations can satisfy *both* texels' requirements. Some examples of the assumptions that relate texels are: both lie in the same plane, both are equal in textural measure (length, area, etc.), one is k times the other in measure, etc. Relating texels in the manner forces more stringent demands on the scene. If enough relations are invoked, the orientation of the local surface supporting two or more related texels can be very precisely determined.

4.1. Skewed Symmetry from the Paradigm Applied to Slope

What we now show is that the skewed symmetry method is a special case of the shape-from-texture paradigm; it can be derived from considerations of texel slope.

To normalize the slope of a texel, it is back-projected onto a plane with the postulated orientation. The back-projected texel now has a new shape on this new surface. Its exact value, however, depends upon the coordinate system on this surface plane. Many coordinate systems are possible; we chose here a coordinate system whose x-axis lies along the gradient direction. The normalized slope is then the angle that the back-projected texel makes with respect to the surface coordinate system x-axis. The calculation is a bit involved, especially under perspective, which requires a knowledge of both the location of the center of focus and the length of the focal distance.

Using the construction in Figure 6, together with several lemmas relating surfaces in perspective to their local vanishing lines, slope is normalized as follows. Assume a slope is parallel to the p-axis; the image and gradient space can always be rotated into such a position. (If rotation is necessary, the resulting NTPM can be de-rotated into the original position using the standard two-by-two orthonormal matrix.) Also assume that the slope is somewhere along the line $y = y_s$, where the unit of measurement in the image is equal to one focal length. The normalized value of the slope is equal to the tangent of the 3-D space angle η, whose base (of length R) is parallel to the surface plane, and is in the direction of the gradient. R is determined from the focal distance, and from the point of the nearest approach of the vanishing line of the plane. This line has equation $px + qy = 1$ (or $G \cdot P = 1$) and its nearest approach is $G/\|G\|^2$. The distance d is given by the intersection of the line $y = y_s$ with the vanishing line. Then, the normalized slope value – the Normalized Texture Property Map – is given by

$$\frac{q - y_s(p^2+q^2)}{p\sqrt{1+p^2+q^2}}. \tag{10}$$

This normalized value can be exploited in several ways. Most important is the result that is obtained when one has two slopes in the image that are assumed to arise from equal slopes in the scene. Under this assumption, their normalized property maps can be equated. The resulting constraint, surprisingly, is a simple straight line in the gradient space. It is intimately related to the vanishing

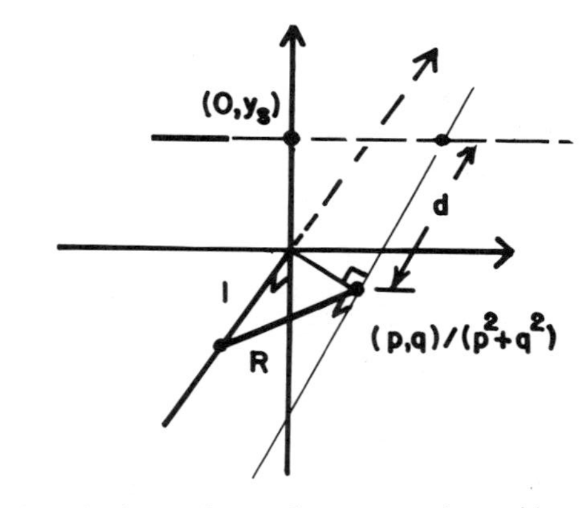

Figure 6. Back-projecting an image slope onto a plane with gradient (p, q).

point formed by the intersection of the extensions of the two image slopes (Kender, 1980).

The constraint equations resulting from assuming that the two slopes arose from perpendicular lines in the scene is, however, enormously complex. It unfortunately does not appear to have many tractable forms or special cases.

Under orthography, nearly everything simplifies. The normalized slope of a texel becomes

$$\frac{q}{p\sqrt{1+p^2+q^2}}. \tag{11}$$

It is independent of y_s; in effect, all slopes are at the focal point.

Considering two image slopes to have arisen from parallel lines in the scene has a trivial solution. If the image slopes are parallel, the entire gradient space is a solution. If they are not, there is no solution at all. This corresponds to the projective geometry theorem that under orthography, parallels are taken into parallels regardless of surface orientation.

In the case where the scene slopes are assumed to be perpendicular, we again get a simplification, but this time a useful one. Not only is the solution tractable, it is the skewed symmetry method of Section 2. We derive it as follows.

Consider Figure 7. Note that under orthography, texels can be translated arbitrarily, since the focal length is infinite and the focal point is effectively everywhere; there is no information in image position. Given the angle that the two texels form, rotate the gradient space so that the positive p-axis bisects the angle. Call this adjustment angle λ; we will use it to de-adjust our results into the original position after they have been computed.

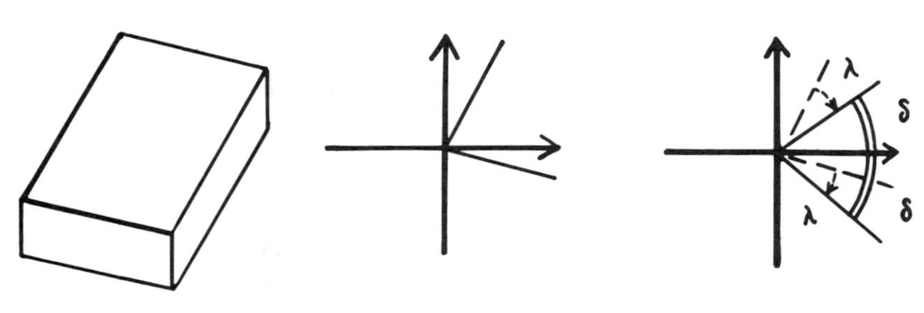

Figure 7. Two image texels assumed to be perpendicular in the scene.

Let the angle that is bisected be 2δ. The normalized value of either slope is obtained directly from the standard normalized slope formula, corrected for the displacement of $+\delta$ and $-\delta$ respectively. That is, for the slope at the positive δ orientation, instead of formula (11), we use the formula under the substitution $p\cos\delta + q\sin\delta$ for p, $-p\sin\delta + q\cos\delta$ for q. We proceed similarly for the slope at $-\delta$. Note that the factor $\sqrt{1+p^2+q^2}$ is invariant under this transformation (it is the length of the normal vector of the surface).

The fact that the normalized slopes are assumed to be perpendicular in the scene allows us to set one of the normalized values equal to the negative reciprocal of the other. The resultant equation becomes

$$p^2\cos^2\delta - q^2\sin^2\delta = \sin^2\delta - \cos^2\delta = -\cos 2\delta. \tag{12}$$

This is exactly the hyperbola in Section 2 with $2\delta = |\alpha - \beta|$.

4.2. Skewed Symmetry from the Paradigm Applied to Length and Angle

The paradigm is similarly applicable to other texture measures. Using texel length as the property to be normalized, we find that under perspective, lengths must lie on the same line in order for the resultant equations to be simpler than the fourth order. If they are collinear, again the resultant gradient space constraint is a simple straight line.

Under orthography and the assumption that image lengths have arisen from equal scene lengths, the constraint equation is again a hyperbola – the skewed-symmetry hyperbola, somewhat offset. In fact, the geometric construction in Figure 8 shows that the assumption of equal length can be made equivalent to skewed symmetry.

First, a triangle is formed by translating one or the other of the lengths so that they meet at a common endpoint. Under orthography, such translations do not affect the resulting constraints. Connecting the remaining endpoints creates a triangle which must be isosceles in the scene. Further, under orthography,

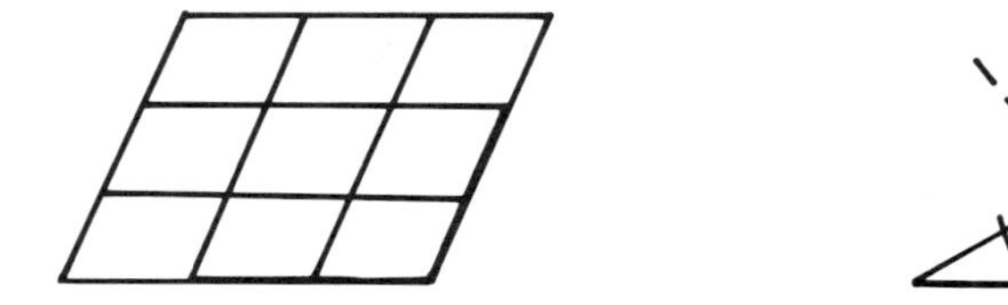

Figure 8. Assuming lengths are equal generates the skewed symmetry constraint.

midpoints of lines are preserved (the midpoint of the base of the scene triangle is imaged as the midpoint of the base of the image triangle). The line connecting the vertex and this midpoint has the property that, in the scene, it must form a right angle with the base. Its distortion to something other than a right angle in the image – the induced angle 2δ – is precisely the distortion which characterizes skewed symmetry. Therefore, the same methods apply.

One other case is worth mentioning. Suppose the image has two angles such that one leg of the first is parallel to one leg of the second. See Figure 9. In this case, again the constraint is equivalent to skewed symmetry, as the construction shows. Choosing one of the angles, extend its non-parallel leg until it intersects both legs of the other angle. (If it cannot do so, then first translate the angle before extending.) The resulting triangle must be isosceles in the scene, since the angles are assumed equal in the scene. However, this is the same situation encountered above with the construction involving lengths. Therefore, the altitude from the midpoint of the base (here, the midpoint of the parallel side) to the vertex must form a right angle. Again, the distortion observed in the image is the skewed symmetry distortion.

5. Applications of Skewed Symmetry and Affine-transformable Patterns

5.1. Quantitative Shape Recovery from Line Drawings

Given the line drawing of Figure 10a, we usually perceive a right-angled parallelepiped. The Huffman-Clowes-Waltz labeling scheme for the trihedral

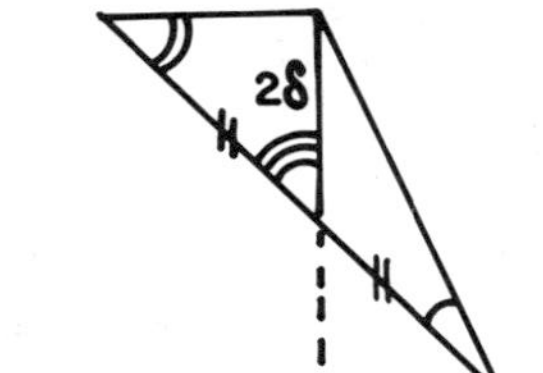

Figure 9. Assuming angles are equal generates the skewed symmetry constraint.

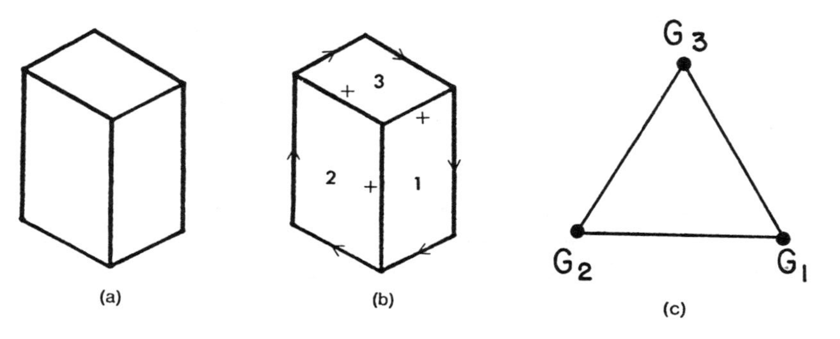

Figure 10. (a) A line drawing of a block; (b) Huffman-Clowes-Waltz labeling; (c) constraints in the gradient space.

world gives the labeling shown in Figure 10b, which signifies that the three edges meeting at the central FORK vertex are all convex, i.e., the object is a convex corner of a block. However, it does *not* specify a particular quantitative shape. In fact, the labeling indicates only that the gradients of the three surfaces should be placed in the gradient space so as to form the triangle shown in Figure 10c. The edges of the triangle should be perpendicular to the picture edges separating the corresponding regions, but the location and size of the triangle are arbitrary in the gradient space. Therefore, the object is not necessarily right-angled.

We can use skewed symmetry here to provide additional constraints. The three regions are skewed-symmetrical with the axes shown in Figure 11a. The hyperbolas corresponding to these regions are shown in Figure 11b. Thus the problem is now how to place the triangle of Figure 10c in Figure 11b so that each vertex is on the corresponding hyperbola. Kanade (1979) proves that the combination of locations shown in Figure 11b is the only possibility, and that the resultant shape is a right-angled block.

It is interesting to note that if we apply the same procedure to the line drawing of Figure 12, we find that there is no way for all the three regions to satisfy the skewed symmetry assumptions. That is, at least one of them has to be non-symmetrical (skewed) in the 3-D space; in other words, the object *cannot* be right angled, but should be rhomboid (a prism). Remember that Figure 10a *can* be either right-angled or rhomboid, but it is usually perceived as right-angled.

Figure 13 demonstrates how the above procedure results in the interpretation of the drawing as a trapezoidal block in this case.

5.2. Skewed Symmetry under Gravity

One principal influence toward symmetry seems to be an object's structural necessity to oppose the gravitational field. Objects that must support

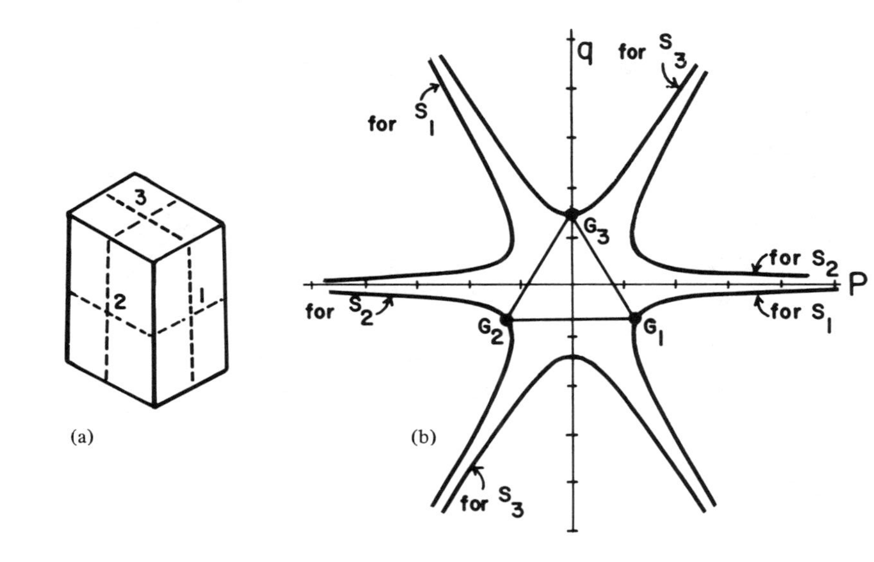

Figure 11. (a) Axes of the skewed symmetry of the regions of Figure 10a; (b) corresponding hyperbolas and allocations of the gradients.

themselves tend to have structural members aligned parallel to the direction of force, that is, vertically. Such members are mutually parallel – a type of symmetry. The base of such an object is often perpendicular to gravity to distribute weight and provide balance. Together, then, the base and structural members provide a local symmetry frame that can also be exploited by the skewed-symmetry method. One can show that in this last case it is usually possible to specify surface orientation uniquely.

We will assume that the direction of the gravity field is known, say the top-to-bottom lines in the image frame are assumed to be true projections of a line of gravity force. The gradient space is also considered to be aligned in the direction of the gravity field; $-q$ is also "down."

Under such conditions, suppose we do find a portion of the image that is assumed to be a vertical, symmetric surface: say, a building face as in Figure

Figure 12. A line drawing of a rhomboid: this *cannot* be a right-angled block. Notice that Figure 10a can be a rhomboid.

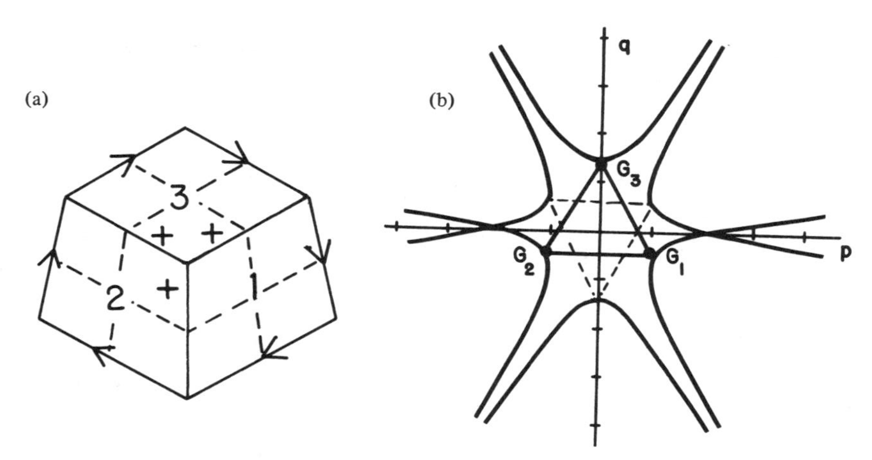

Figure 13. Shape recovery of a trapezoid block: (a) axes; (b) gradient allocations.

14. Using skewed symmetry (or even direct observation), it is not hard to obtain an angle in the image that corresponds to a right angle in the scene. Suppose one of the legs of the angle is parallel to the known gravity field as in Figure 14. The skewed-symmetry method generates the following constraint hyperbola:

$$p = -(q + 1/q)\cot\gamma. \qquad (13)$$

This constraint is somewhat interesting: it expresses p (left-right slant) as a *function* of q (top-bottom slant). The value of q itself is easily obtained.

If gravity points in the $-q$ direction, the ground plane must have as its orientation $(0,q_g)$, for a value of q_g determinable through sensing. Since all vertical planes are perpendicular to the ground plane, all vertical planes must have the orientation $(p_v, -1/q_g)$, for variable p_v. (A quick check shows that the dot product of the corresponding normals is zero: $(0,q_g,1)\cdot(p_v,-1/q_g,1) = 0$.) Note that the value of q for *any* vertical plane is fixed at $-1/q_g$. Thus, in our example, p is also determined: it is $-(q_g + 1/q_g)\cot\gamma$. Since q_g is a constant, p varies simply with γ. Figure 14 shows the constraints graphically.

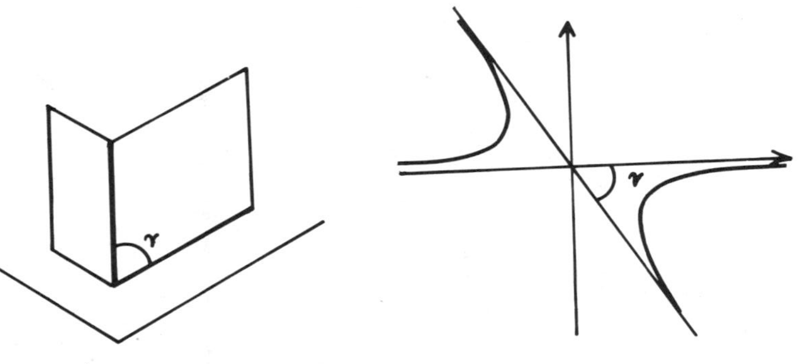

Figure 14. Assumptions about gravity can uniquely specify surface orientations.

5.3. Shape Recovery of an Object with Many Patterns Stamped

Consider the problem of recovering the shape from a picture of a ball with a number of patterns stamped on it (see Figure 15). For each pair of texel patterns, if they are affine-transformable, we compute a transformation matrix A. Thus we obtain many constraints on the gradients of texels. From these, however, we cannot uniquely determine the surface orientation of each texel.

We need more assumptions or data. We will suppose we know the gradients of some particular texels, and assume that the surface is smooth (together, maybe, with an assumption of global concavity or convexity). Then a relaxation or cooperative technique similar to the one for shape-from shading (Woodham, 1977; Ikeuchi, 1980a) will allow us to determine consistent assignments of gradients to the texels which satisfy those many constraints. Notice that we need not assume that the original pattern is known, nor that the patterns are stamped in a particular manner. Even other patterns can be mixed together with them.

One of the plausible methods of determining the gradient of one particular texel is to use equation (7). Assuming $\sigma = 1$, we order the texels by the magnitude of $\sqrt{p_i^2 + q_i^2}$, and assign $(p, q) = (0,0)$ (the orientation that is directly facing the viewer) to the least slanted texel. This is analogous to a similar hypothesis in shape from shading. That is, we tend to assign to the brightest point the orientation directly facing the light source, even though under the assumptions of parallel lights and a matte surface, one can only say that the brightest pixels have the minimum incident angle of light, not necessarily 0°.

6. Conclusion

The assumptions we used for skewed symmetry, affine-transformable patterns, and texture analysis can be generalized as:

> Properties observable in the picture are not accidental, but are projections of some preferred corresponding 3-D properties.

This provides a useful meta-heuristic for exploiting image properties: we can call it the meta-heuristic of *non-accidental image properties*. It can be regarded as a generalization of general view directions, often used in the blocks world, to exclude the cases of accidental line alignments.

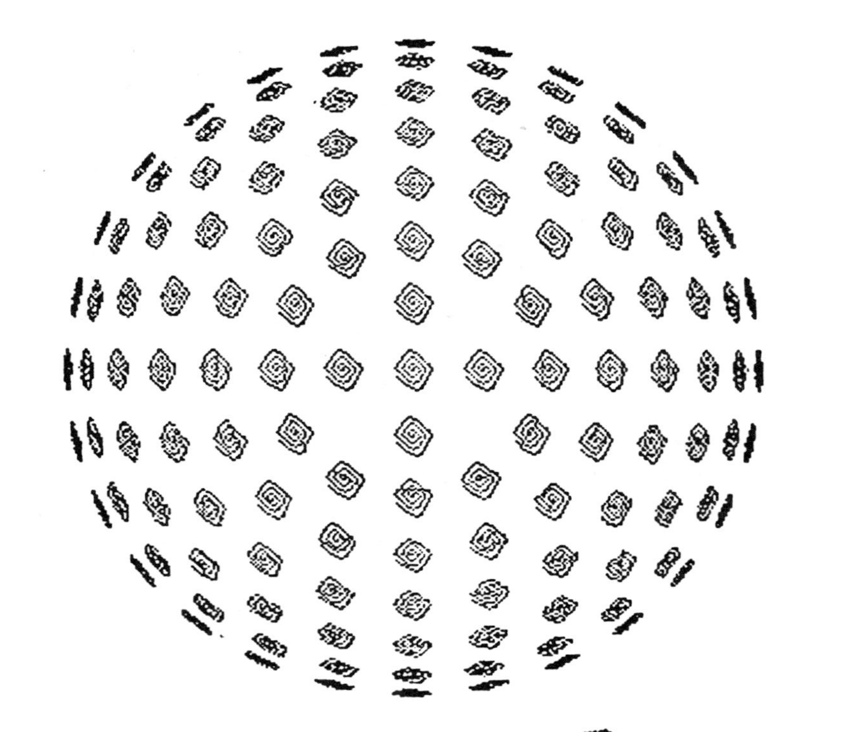

Figure 15. A picture of a ball with a number of s stamped.

Instances that can fall within this meta-heuristic include: parallel lines in the picture vs. parallel lines in the scene, texture gradients due to distance, and sets of lines convergent to a vanishing point.

The most essential point of our technique is that we relate certain image properties to certain 3-D space properties, and that we map the relationships into convenient representations of shape constraints. We explicitly incorporate assumptions based either on the meta-heuristic or on *a priori* knowledge of the world. The shape-from-texture paradigm provides a computational framework for our technique. In most of our discussion we assumed orthography. Similar (though more involved and less intuitive) results can be obtained under perspective projection.

Acknowledgment

This research was sponsored by the Defense Advanced Research Projects Agency (DOD) under ARPA Order No. 3597, and monitored by the Air Force Avionics Laboratory under Contract F33615-78-C-1551. The views and conclusions in this document are those of the author and should not be interpreted as representing the official policies, either expressed or implied, of the Defense Advanced Research Projects Agency or the U.S. Government.

Appendix

Proof that (6) has two symmetrical solutions:

We will try to solve (6) for p_1 and q_1, assuming that p_2, q_2, and $A = (a_{ij})$ are known. We assume $\det(A) > 0$. Let us put $\gamma = \sqrt{p_1^2 + q_1^2 + 1}$. Then (6) can be rewritten as

$$\begin{aligned} Ca_{11}(p_1^2+1)+Ca_{21}p_1q_1 &= B\gamma \\ A(p_1^2+1)-Bp_1q_1 &= Ca_{21}\gamma \end{aligned} \qquad (14)$$

where

$$A = a_{11}p_2q_2 - a_{12}(p_2^2+1)$$

$$B = a_{22}(p_2^2+1) - a_{21}p_2q_2$$

$$C = \sqrt{p_2^2+q_2^2+1}.$$

We can derive a quadratic equation on γ from (14):

$$f(\gamma) = DE\gamma^2 - (D^2+E^2+F^2)\gamma + DE = 0, \qquad (15)$$

where

$$D = C(Ba_{11}+Aa_{21}) = C\ \det(A)(p_2^2+1) > 0$$

$$E = B^2 + (Ca_{21})^2 > 0$$

$$F = -C^2a_{11}a_{21} + AB.$$

The discriminant of (15) is

$$\begin{aligned} \text{disc} &= (D^2+E^2+F^2)^2 - 4(DE)^2 \\ &= F^4 + 2F^2(D^2+E^2) + (D-E)^2 \\ &\geqslant 0. \end{aligned}$$

Thus, $f(\gamma)$ has real roots. Now, notice that $\gamma \geqslant 1$ and thus we are interested in the root greater than or equal to 1. Let us check the sign of $f(1)$ multiplied by the coefficient of γ^2:

$$
\begin{aligned}
f(1)DE &= (2DE-(D^2+E^2+F^2))DE \\
&= -(F^2+(D-E)^2)DE \\
&\leqslant 0.
\end{aligned}
$$

This means that one and only root of $f(\gamma)$ is greater than or equal to 1. Let us denote this root by γ_0. By substituting γ_0 into (14), we can solve it as a simultaneous quadratic equation on p_1 and q_1, and know that (p_1,q_1) has two solutions in the form of (p_0,q_0) and $(-p_0,-q_0)$, which are symmetrical to the gradient space origin.

References

Ikeuchi, K. Numerical shape from shading and occluding contours in a single view. MIT AI Memo 566, 1980. (a)

Ikeuchi, K. Shape from regular patterns (an example of constraint propagation in vision). MIT AI Memo 567, 1980. (b)

Kanade, T. Recovery of the 3-dimensional shape of an object from a single view. *Artificial Intelligence,* 1981, **17**, 409-460.

Kender, J. R. Shape from texture. Doctoral dissertation, Carnegie-Mellon University, 1980.

Mackworth, A. K. Interpreting pictures of polyhedral scenes. *Artificial Intelligence,* 1973, **4**, 121-137.

Stevens, K. A. Surface perception from local analysis of texture and contour. MIT AI Memo 512, 1980.

Woodham, R.J. A cooperative algorithm for determining surface orientation from a single view. Proceedings, 5th International Joint Conference on Artificial Intelligence, 1977, 635-641.

ANALYZING ORIENTED PATTERNS

Michael Kass
Andrew Witkin

Schlumberger Palo Alto Research
3340 Hillview Ave.
Palo Alto, CA 94304

ABSTRACT

Oriented patterns, such as those produced by propagation, accretion, or deformation, are common in nature and therefore an important class for visual analysis. Our approach to understanding such patterns is to decompose them into two parts: a flow field, describing the direction of anisotropy, and the residual pattern obtained by describing the image in a coordinate system built from the flow field. We develop a method for the local estimation of anisotropy and a method for combining the estimates to construct a flow coordinate system. Several examples of the use of these methods are presented. These included the use of the flow coordinates to provide preferred directions for edge detection, detection of anomalies, fitting simple models to the straightened pattern, and detecting singularities in the flow field.

I Introduction

A central focus in recent computational vision has been the decomposition of the original intensity image into intrinsic images (Horn 1977; Barrow & Tenenbaum, 1978; Marr, 1982), representing such properties as depth, reflectance, and illuminance. These intrinsic properties are believed to be more meaningful than image intensity because they describe basic independent constituents of the image formation process. Thus, for example, in separating shape from illumination, we can recognize an invariance of shape regardless of changing illumination.

The advantages of decomposing what we see into its more-or-less independent parts extends beyond the image formation process to the shapes and patterns on which that process operates. For instance, decomposing a bent rod into a straight rod and a bending transformation reveals the similarity between a bent rod and one that hasn't been bent, or some other solid that's been bent the same way (Barr, 1984).

Just as we need to understand the image-forming process to decompose an image into intrinsic images, we need to understand the processes that generate patterns to decompose them into their intrinsic parts. But, while there is only one image-forming process, a staggering variety of processes shape and color the world around us. Our only hope of dealing with this complexity is to begin with some basic pattern classes that recur in nature, and understand how to decompose and describe them.

One such class are oriented patterns, notably those produced by propagation, accretion, or deformation. To understand an oriented pattern we must be able to say (1) what is propagating, accreting, or deforming, and (2) which way and how much. More precisely, we must estimate everywhere the direction and magnitude of anistropy (which we will call the flow field,) and describe the residual pattern, independent of that field. Why this decomposition leads to simpler, more regular descriptions is best illustrated by example:

- A typical oriented pattern created by propagation is the streaked trail left by a paint brush dipped in variegated paint. The flow field describes the trajectory of the brush, the residual pattern depending only on the distribution of paint on the brush.
- Accretion typically results in laminar structures, such as wood grain. Here, the flow field gives isochrones (the moving accretion boundary,) and the residual pattern describes the change in color or brightness of the accreting material over time.
- If an isotropic body is deformed, the flow field principally describes the bending and stretching it has undergone, while the residual pattern describes the undeformed body.

In all these cases, separate descriptions of the flow field and the residual pattern are appropriate because they describe different processes. The path of propagation for many physical processes is controlled by very different mechanisms than control the coloration of the trail left behind. Similarly, the mechanisms which control the shape of an accretion boundary are frequently unrelated to the processes controlling the color of the accreted material. Finally, the forces which deform a piece of material are often completely unrelated to the process which created the piece of material in the first place. By separately describing these processes, we can create descriptions of the whole which are often simpler than is possible without the separation because each of the pieces may have different regularities.

Orientation selective mechanisms have been extensively studied by physiologists since Hubel and Wiesel's (1962) discovery of orientation selective cells in mammalian visual cortex (see Schiller et. al. (1976) for a comprehensive example). There has also been considerable interest among psychologists in the perception of oriented patterns, particularly dot patterns (Glass, 1969). Only recently have the computational issues involved received attention. Stevens (1978) examined the grouping of tokens in Glass patterns based on orientation. While successful with Glass patterns, his methods were never extended to natural imagery. Zucker (1983) investigated the estimation of orientation by combining the outputs of linear operators. Zucker's estimation method for what he calls "Type II" patterns, while differing in many respects, is quite close in spirit to our own.

Little progress has been made in using local orientation estimates to interpret patterns, perhaps because reliable estimates have proved difficult to obtain. The key difference between our work and earlier efforts lies in our use of the flow field to build a natural coordinate system for analyzing the pattern.

The remainder of the paper covers the computation of the flow field by local estimation of orientation, the construction of a coordinate system using the flow field, and some examples of analysis and description using flow coordinates.

In Section 2 we develop an estimator for the local flow direction, that direction in which intensity tends to vary most slowly due to an underlying anisotropic process. The estimator, based on the direction of least spatial variance in the output of an oriented filter, is computed as follows: After initial filtering, the intensity gradient is measured at each point in the image. The gradient angle, θ, is then doubled (by treating the gradient vectors as complex numbers and squaring them) to map directions differing by π into a single direction. The transformed vectors are then summed over a weighted neighborhood around the point of interest. The angle of the summed vector is halved, undoing the previous transformation. This gives an estimate for the direction of *greatest* variance, which is then rotated by $\pi/2$ to yield the flow direction.

In section 3, we describe the construction and use of coordinate systems based on the result of local estimation. Integral curves in the flow field are computed numerically, by following the estimated vectors from point to point. A coordinate system is constructed in which the integral curves are parameter lines. Transforming the image into these "flow coordinates" straightens the pattern, removing the effects of changing orientation. We present several examples of analysis and description of the flow field and the straightened pattern.

II Flow Computation

For intensity patterns created by anisotropic processes such as propagation, accretion, or deformation, variation in the flow direction is much slower than variation in the perpendicular direction. Anisotropy in such patterns will be evident in the local power spectrum. The high frequency energy will tend to cluster along the line in the Fourier domain perpendicular to the flow orientation.

A simple way to detect this clustering is to sum the energy in an appropriate region of the power spectrum and examine how the sum is affected by rotations. This can be done by examining the energy in the output of an appropriate orientation-selective linear filter. The orientation at which the energy is maximal can be expected to be perpendicular to the flow orientation.

Selection of the filter involves a number of tradeoffs. Very low spatial frequencies are affected more strongly by illumination effects than surface coloration, so they are inappropriate for measuring textural anisotropy. Very high spatial frequencies are sensitive to noise and aliasing effects so they too are inappropriate. Hence some type of roughly bandpass filtering is required. The orientation specificity of the filter is also quite important. If the filter is too orientation-specific then a large spatial neighborhood will be required in order to make a reliable measurement of the energy. Conversely, if the filter responds over a wide range of orientations then it will be difficult to localize the orientation very accurately. Thus there is a trade-off between angular- and spatial- resolution.

One reasonable choice for the frequency response of the filter is

$$F(r,\theta) = [e^{r^2\sigma_1^2} - e^{r^2\sigma_2^2}]2\pi i r\cos(\theta). \tag{1}$$

The filter is bandpass with passband determined by σ_1 and σ_2. In our experience, ratios of the sigmas in the range of 2.0 to 10.0 work well. The orientation specificity or *tuning curve* is provided by the cosine dependence of the filter on θ. This appears to strike a reasonable balance between angular- and spatial- resolutions for the range of patterns we have examined. The filter's power spectrum is shown in figure 1

The cosine orientation tuning-curve of the filter has some unusually good properties for computing the filter output at different orientations. The impulse response $S(x,y)$ of the filter

Figure 1: The power spectrum of the filter in equation 1 for $\sigma_2 = 2\sigma_1$.

is

$$S(x,y) = \frac{\partial}{\partial x}H(x,y)$$

where

$$H(x,y) = [\sigma_1^{-2}e^{r^2/\sigma_1^2} - \sigma_2^{-2}e^{r^2/2\sigma_2^2}]$$

is an isotropic filter. Let $C = H * I$ and let $R_\theta[S]$ denote a counter-clockwise rotation of S by an angle θ. Then the convolution $R_\theta[S] * I$ is just the directional derivative of $H * I$ in the θ direction. The directional derivative can easily be written in terms of the gradient so we have

$$R_\theta[S] * I = (\cos\theta, \sin\theta) \cdot \nabla H * I. \tag{2}$$

Thus a single convolution suffices for all orientations.

Since the filter S severely attenuates very low frequencies $R_\theta[S] * I$ can be safely regarded as zero-mean. Thus the variance in its output can be estimated by the expression

$$V(\theta) = W * (R_\theta[S] * I)^2$$

where $W(x,y)$ is a local weighting function with unit integral. We use Gaussian weighting functions $W(x,y)$ because approximate Gaussian convolutions can be computed efficiently (Burt 1979).

Using the gradient formulation of the filter output in equation 2, we can write the variance $V(\theta)$ as

$$V(\theta) = W * [\cos(\theta)C_x + \sin(\theta)C_y]^2. \tag{3}$$

A. Interpretation of Filter Output

There remains the issue of interpreting $V(\theta)$. Assume that there is only one axis of anisotropy. Then $V(\theta)$ will have two extrema π apart corresponding to that axis. Let $V_2(\theta) = V(\theta/2)$. Then $V_2(\theta)$ will have a single extremum in the interval $0 < \theta < 2\pi$. A computationally inexpensive way of estimating the position of this extremum is to consider V_2 as a distribution and compute its mean. Since θ is periodic, V_2 should be considered as a distribution on the unit circle. Hence its mean is the vector integral $(\alpha,\beta) = \int_0^{2\pi} V_2(\theta)(\cos\theta,\sin\theta)d\theta$. The angle $\tan^{-1}(\beta/\alpha)$ is an estimate of the angle of the peak in V_2 and hence twice the angle of the peak in V. Thus the angle ϕ of *greatest* variance can be written

$$\begin{aligned}\phi &= \tan^{-1}\frac{\beta}{\alpha} \\ &= \tan^{-1}\left(\frac{\int_0^{2\pi} V_2(\theta)\sin(\theta)d\theta}{\int_0^{2\pi} V_2(\theta)\cos(\theta)d\theta}\right)/2 \\ &= \tan^{-1}\left(\frac{\int_0^{\pi} V(\theta)\sin(2\theta)d\theta}{\int_0^{\pi} V(\theta)\cos(2\theta)d\theta}\right)/2\end{aligned} \tag{4}$$

These integrals are evaluated in Appendix A to show that the angle of anisotropy ϕ can be written

$$\phi = \tan^{-1}\left(\frac{W * 2C_xC_y}{W * (C_x^2 - C_y^2)}\right)/2 \tag{5}$$

which directly yields a simple algorithm for computing ϕ.

B. Combining Gradient Orientations

Notice that the right hand side of equation 5 can be regarded as the orientation of a locally weighted sum of the vectors of the form $J(x,y) = (C_x^2 - C_y^2, 2C_xC_y)$. These vectors are related in a simple way to the gradient vectors $G(x,y) = (C_x, C_y)$. The magnitude of $J(x,y)$ is just the square of the magnitude of $G(x,y)$ and the angle between $J(x,y)$ and the x-axis is twice the angle between $G(x,y)$ and the x-axis. This follows easily from the observation that $(C_x + C_y i)^2 = C_x^2 - C_y^2 + 2C_xC_y i$.

One might be tempted to believe that smoothing the gradient vectors $G(x,y)$ would be nearly as good a measure of anisotropy as smoothing the rotated squared gradient vectors $J(x,y)$. This is emphatically not the case. Consider an intensity ridge such as $I(x,y) = \exp(-x^2)$. The gradient vectors on the left half-plane all point to the right and the gradient vectors on the right half-plane all point to the left. Adding them together results in cancellation. By contrast, if they are first rotated to form the J vectors, they reinforce. The types of oriented patterns we are concerned with often have nearly symmetric distributions of gradient directions around the axis of anisotropy. In such patterns, if the gradients are added together directly, the cancellation is so severe that the result often has little relation to the direction of anisotropy. Thus the difference between rotating the gradient vectors or leaving them be is often the difference between being able or unable to detect the anisotropy. Note also that smoothing the image first and then computing the gradients is exactly the same as computing the gradients and then smoothing. It will not avoid the difficulties of cancellation.

C. Coherence

In addition to finding the direction of anisotropy, it is important to determine how strong an anisotropy there is. If the orientation of the local J vectors are nearly uniformly distributed between 0 and 2π, then the orientation ϕ of slight anisotropy is not very meaningful. Conversely, if all the J vectors are pointing the same way then the indication of anisotropy is quite strong and ϕ is very meaningful. A simple way of measuring the strength of the peak in the distribution of J vectors is to look at the ratio $\chi(x,y) = |W * J|/W * |J|$ which we will call the *coherence* of the flow pattern. If the J vectors are close to uniformly distributed, then the ratio will be nearly zero. If the J vectors all point the same way, the ratio will be one. In between, the ratio will increase as the peak gets narrower.

D. Summary

The computation of the flow direction and local coherence can be summarized as follows. First the image $I(x,y)$ is convolved with the isotropic portion $H(x,y)$ of the filter response. The result $C(x,y)$ is then differentiated (by finite differences) to form $C_x(x,y)$ and $C_y(x,y)$. The resulting vectors $(C_x(x,y), C_y(x,y))$ are rotated by computing $J_1(x,y) = 2C_x(x,y)C_y(x,y)$ and $J_2(x,y) = C_x^2(x,y) - C_y^2(x,y)$. The gradient magnitude $J_3(x,y) = [C_x^2(x,y) + C_y^2(x,y)]^{1/2}$ also has to be computed in order to measure the coherence. The next step is to convolve $J_1(x,y), J_2(x,y)$, and $J_3(x,y)$ with the weighting function $W(x,y)$ to obtain $J_1^*(x,y), J_2^*(x,y)$, and $J_3^*(x,y)$. The angle $\phi(x,y)$ of anisotropy and the coherence $\chi(x,y)$ can then be computed from the formulas

$$\phi(x,y) \approx \tan^{-1}(J_1^*(x,y)/J_2^*(x,y))/2$$

and

$$\chi(x,y) = (J_1^*(x,y)^2 + J_2^*(x,y)^2)^{1/2}/J_3^*(x,y).$$

An example of this computation applied to a picture of a piece of wood is shown in figure 2. The flow direction $\phi(x,y) + \pi/2$ is displayed by the orientation of small needles superimposed on the image. The lengths of the needles is proportional to the coherence $\chi(x,y)$. Note that the pattern is strongly oriented except near the knot in the middle.

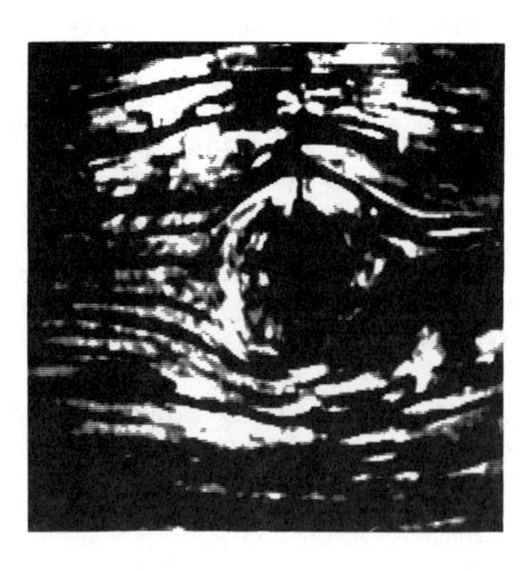

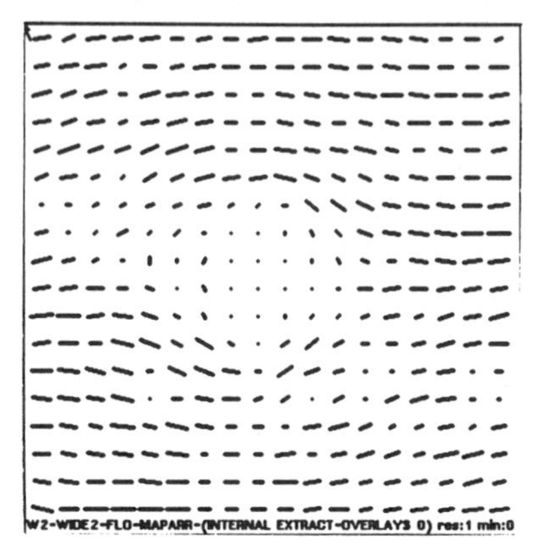

Figure 2: An image of wood grain with its flow field. Estimated flow directions are given by the black needles. The length of the needle encodes coherence. Notice that coherence is low within the knot at the center.

E. Relation To Prior Work

The flow computation just described bears an interesting relation to an early proposal of David Marr that information about local distributions of oriented edge elements be included in the *primal sketch* (Marr 76). If this proposal is combined with his later work with Hildreth on edge detection (Marr & Hildreth, 1980) it results in a special case of the above computation. Marr and Hildreth define edges as zero-crossings in the Laplacian of the Gaussian smoothed image. The natural combination of Marr's proposal with this definition of edge elements calls for examining the local density of zero crossings as a function of orientation. For stationary zero-mean Gaussian processes the square of the oriented zero-crossing density is approximately $V(\theta)$ (see appendix B). Thus in the special case where the point spread function of the filter is $S = (\partial/\partial x)\nabla^2 \exp(-(x^2+y^2)/2\sigma^2)$ our computation can be viewed as computing the direction of minimal edge density in the Marr-Hildreth theory.

Zucker's work on flow (Zucker 1983) is also related to a special case of the above computation. For biological reasons, he prefers to use oriented second derivatives of Gaussians as the initial filters. These have $F(r,\theta) = r^2 \exp(-r^2/2)\cos^2(\theta)$. Instead of looking at the variance of the filter outputs as the orientation is changed, he combines the outputs in a biologically motivated relaxation process. Although quite different in detail, the computation described here has much in common with his technique.

III Flow coordinates.

The orientation field is an abstraction from the anisotropic pattern that defines it. We can, for example, get the same spiral field from a pattern composed of bands, irregular streaks, dot pairs, etc. In addition to measuring the orientation field, it is useful to be able to produce a description of the underlying pattern independent of the changing direction of anisotropy. Such a description would make it possible to recognize, for example, that two very different orientation fields are defined by the same kind of bands or streaks.

A powerful way to remove the effects of changing orientation is to literally "straighten" the image, subjecting it to a deformation that maps the flow lines into straight, parallel lines in a canonical (e.g. horizontal) orientation. Performing this deformation is equivalent to viewing the image in a coordinate system (u, v), with $u = u(x, y)$ and $v = v(x, y)$ that everywhere satisfies

$$\nabla u \cdot (\sin\phi, -cos\phi) = 0. \quad (6)$$

Equation 6 does not determine a unique coordinate system. An additional constraint may be imposed by choosing lines of constant v orthogonal to those of constant u, i.e,

$$\nabla v \cdot (\cos\phi, \sin\phi) = 0 \quad (7)$$

which has the desirable effect of avoiding the introduction of spurious shear in the deformation.

Even with equation 7, an additional constraint is needed, because we are free to specify arbitrary scaling functions for the u and v axes. In the spirit of equation 7, we want to choose these functions to avoid the introduction of spurious stretch or dilation. Although difficult to do globally (one might minize total stretch,) we will usually want to construct a fairly local coordinate frame around some point of interest. For this purpose, it suffices to take that point as the origin, scaling the axes $u = 0$ and $v = 0$ to preserve arc-length along them.

Intuitively, the flow field describes the way the pattern is bent, and viewing the image in these *flow coordinates* straightens the pattern out. Figure 3 shows the flow coordinate grid for the wood-grain image from figure 1. The grid lines were computed by taking steps of fixed length in the direction $(cos\phi, sin\phi)$ or $(-sin\phi, cos\phi)$ for lines across and along the direction of flow respectively, using bilinear interpolation on the orientation field. Since ϕ is always computed between 0 and π, we must assume that there are no spurious discontinuities in direction to track smoothly.

Figure 3: A flow coordinate grid obtained for the image of figure 2.

For many purposes it is unnecessary to compute the deformed image explicitly, but doing so vividly illustrates the flow coordinates' ability to simplify the pattern. Figure 4 shows the deformation from image coordinates to flow coordinates in several stages. As the grain lines straighten the knot shrinks and finally vanishes. The deformed images were anti-aliased using texture-map techniques (Williams, 1983) The deformed image shows, to a reasonable approximation, what the grain would have looked like had it not been subjected to the deforming influence of the knot.

Thus far, we have separated the image into a flow field, and a pattern derived by viewing the image in flow coordinates. We argued earlier that the advantage of this decomposition, like the decomposition of an image into intrinsic images, is that the components are liable to be simpler and more closely tied to independent parts of the pattern-generating process than is the original image. To exploit the decomposition, we need ways of analyzing, describing, and comparing both the flow field and the straightened pattern. These are difficult problems. In the remainder of this section, we present several examples illustrating the utility of the decomposition.

A. A coordinate frame for edge detection.

Oriented measurements have been widely used in edge detection. For example Marr & Poggio (1979) employed directional second derivative operators, whose zero-crossings were taken to denote rapid intensity changes. Due to the difficulty in selecting an orientation, Marr & Hildreth (1980) later abandoned this scheme, in favor of zero-crossings of the Laplacian, a non-directional operator.

The flow field provides two meaningful directions—along and across the direction of flow—in which to look for edges within an oriented pattern. Zero-crossings in the second directional derivative in the direction of ϕ (against the grain) should highlight edges that contribute to defining the flow field, while zero-

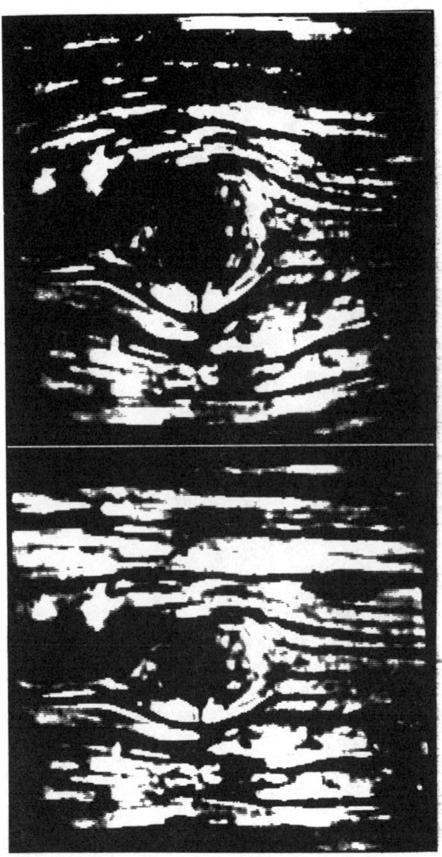

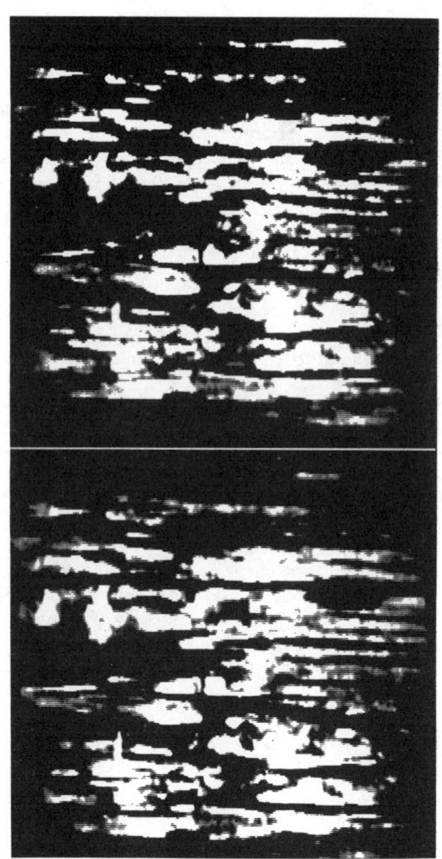

Figure 4: Deformation, in stages, from image coordinates to flow coordinates. Upper left: the original image; Lower left, upper right: two intermediate stages, in which the grain's curvature has diminshed, and the knot compressed; Lower right: the image as seen in flow coordinates: the grain lines are straight and the knot has vanished, showing approximately what the grain would have looked like had it not been deformed by the intrusion of the knot.

crossings in the second derivative perpendicular to ϕ (with the grain) should highlight anomalous elements or terminations. The sum of these two derivatives is the Laplacian.

The two directional derivatives of the wood grain image are shown, with the Laplacian, in figure 5. Indeed, the derivative against the grain captures all the elements comprising the grain pattern, while the derivative with the grain does not appear meaningful. The Laplacian confuses these very different signals by adding them together.

The derivative along the grain can also be meaningful, where anomalous elements are present. In addition to being perceptually salient, such anomalies are often physically significant, with origins such as cracks, intrusions, or occlusions, that are distinct from those of the main pattern. In man-made structures, anamolies are often important because they indicate some variety of flaw.

Figure 6 shows a pattern of aligned elements (straw) with some anomalous elements. The directional second derivatives along and across the flow direction are shown, together with their sum (which is just the Laplacian.) Differentiating along the grain highlights anomalous elements, attenuating the rest (thus finding the "needles" in the haystack.) Differentiating across the grain supresses the anomalies. The Laplacian shows both.

A related demonstration is shown in figure 7, in which the anomalous elements have actually been removed by directional median filtering in the flow direction.

B. Singularities.

We have shown several ways in which viewing an oriented pattern in flow coordinates facilitates analysis and description of the pattern. Describing and anlyzing the flow field itself is the other side of the coin. The topology of a flow field, as of any vector field, is determined by the structure of its singularities, those points at which the field vanishes. Identifying and describing singularities is therefore basic to describing the flow field. The singularities provide the framework around which metric properties, such as curvature, may be described. Singularities are also perceptually salient (see figure 8.)

A robust basis for identifying singularities is the index or winding number (Spivak, 1979.) Suppose we follow a closed curve on a vector field. As we traverse the circuit, the vector rotates continuously, returning to its original orientation when the circuit is completed. The index or winding number of the curve is the number of revolutions made by the vector in traversing the curve. The index of a point is the index of a small circle as we shrink it around the point:

$$\begin{aligned}\text{ind}(x,y) &= \lim_{\epsilon\to 0} \tfrac{1}{2\pi} \int_0^{2\pi} \tfrac{\partial}{\partial\theta}\phi(x+\epsilon\cos\theta, y+\epsilon\sin\theta)\, d\theta \\ &= \lim_{\epsilon\to 0} \tfrac{1}{2\pi} \int_0^{2\pi} (-\sin\theta, \cos\theta) \\ &\quad \cdot \nabla\phi(x+\epsilon\cos\theta, y+\epsilon\sin\theta)\, d\theta.\end{aligned}$$

To compute the winding number numerically, we divide the flow field into suitably small rectangles, summing the rotation of ϕ around each rectangle. As in computing the flow lines, we assume

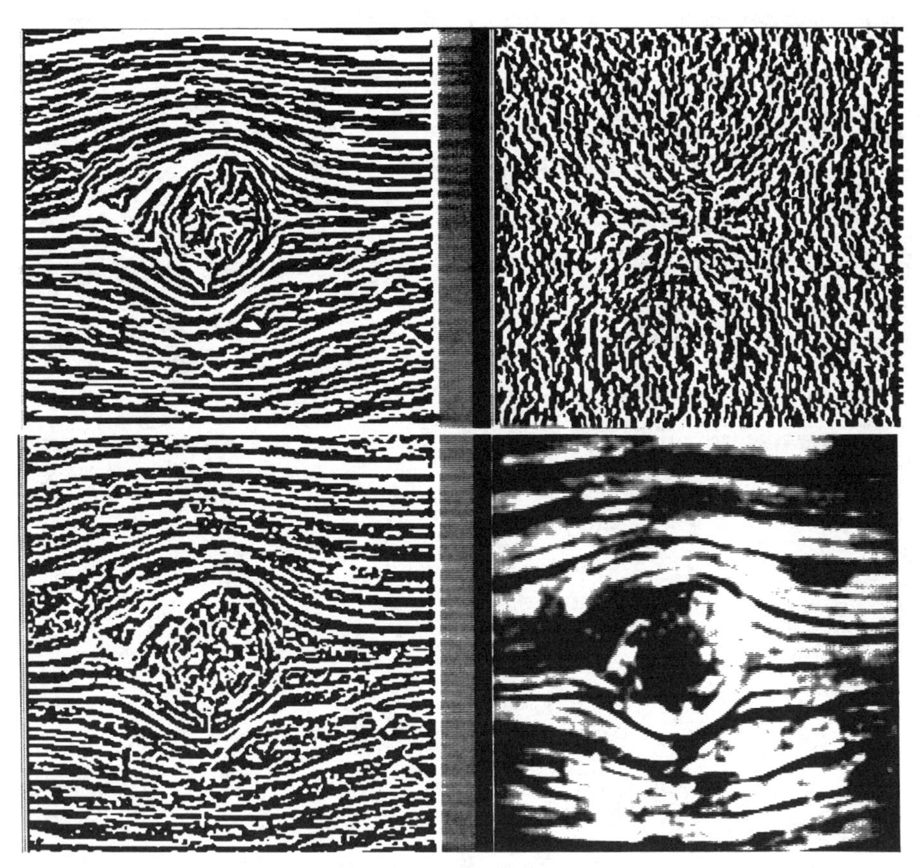

Figure 5: Using flow coordinates for edge detection. Upper left: 2nd directional derivative across the flow direction. Upper right: 2nd directional derivative along the flow direction. The first of these highlights the oriented structure, the second supresses it. Lower left: the sum of the directional derivatives is the Laplacian.

that ϕ has no spurious discontinuities. Where the result is non-zero, the rectangle surrounds a singularity. Figure 9 shows an example of the detection of singularities using winding number, for a fingerprint. We are currently working on classifying the singularities, and using them to describe the topology of the flow field.

IV Conclusion

We addressed the problem of analyzing oriented patterns by decomposing them into a flow field, describing the direction of anisotropy, and describing the pattern independent of changing flow direction.

A specific computation for estimating the flow direction was proposed. The computation can be viewed as a) finding the direction of maximal variance in the output of a linear filter, b) combining gradient directions locally, or c) finding the direction of maximal edge density. The computation has been applied to a number of natural and man-made patterns with consistent success.

The flow field was then used to form a coordinate system in which to view the pattern. Two orthogonal familes of curves—along and across the direction of flow—form the coordinate system's parameter lines. Viewing the pattern in these flow coordinates amounts to deforming the pattern so that the flow lines become parallel straight lines. This deformation produces a pattern that is simpler, more regular, and therefore more amenable to analysis and description than the original one.

Several examples of the use of this decomposition were presented. These included the use of the flow coordinates to provide preferred directions for edge detection, detection of anomalies, fitting simple models to the straightened pattern, and detecting singularities in the flow field.

Our ongoing work focuses on the analysis of patterns with multiple axes of anisotropy, statistical modeling and resynthesis of straightened patterns, and richer description of the structure of the flow field.

Appendix A: Derivation of Equation 5

The integrals in equation 4 can be evaluated fairly easily by expansion using equation 3 for V. The numerator of $\tan(2\phi)$ can be written

$$\int_0^\pi V(\theta)\sin(2\theta)d\theta = W * [2C_x^2\int_0^\pi \sin(\theta)\cos^3(\theta)d\theta$$
$$+4C_xC_y\int_0^\pi \sin^2(\theta)\cos^2(\theta)d\theta$$
$$+2C_y^2\int_0^\pi \sin^3(\theta)\cos(\theta)d\theta].$$

The first and third integrals are zero and the middle integral is $\pi/8$. Hence

$$\int_0^\pi V(\theta)\sin(2\theta)d\theta = (\pi/2)W * C_xC_y$$

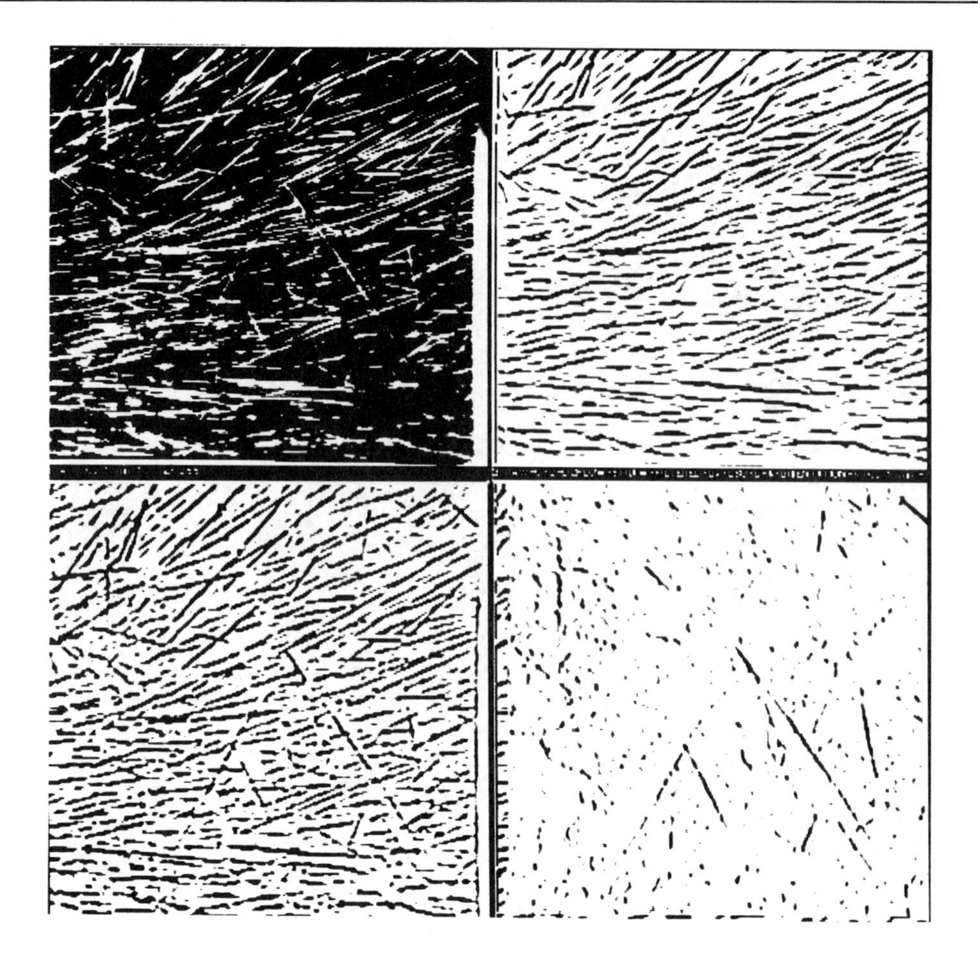

Figure 6: "Finding the needle in the haystack." In this straw pattern, directional derivatives across the flow direction show elements aligned with the pattern (upper right.) Those along the flow direction show anomalous elements (lower right.) The Laplacian (lower left) shows both.

Figure 7: Left: the straw picture from figure 6. Right: the anomalous elements have been removed by directional median filtering in the flow direction. (Following a suggestion by Richard Szeliski.)

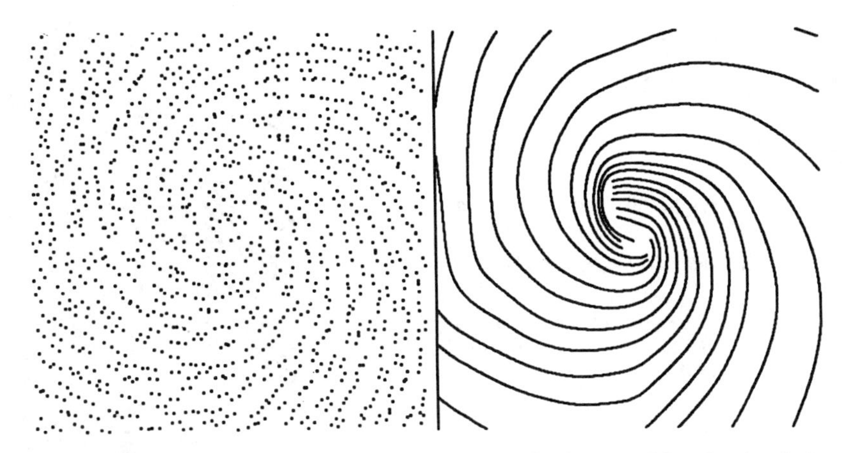

Figure 8: A spiral Glass pattern and its flow lines. The pattern is perceptually dominated by the singularity at the center. Since the flow field vanishes at a singularity, the flow lines obtained by integrating the flow field tend to become ill behaved as they approach one.

The denomenator of $\tan(2\phi)$ similarly is

$$\int_0^\pi V(\theta)\cos(2\theta)d\theta = W * 2C_x^2 \int_0^\pi 2\cos^4(\theta) - \cos^2(\theta)d\theta$$

$$+4C_xC_y \int_0^\pi \sin(4\theta)d\theta$$

$$+2C_y^2 \int_0^\pi 2\cos^2(\theta)\sin^2(\theta) - sin^2(\theta)d\theta].$$

Here the first integral is $\pi/4$, the second integral is zero and the third is $-\pi/4$. Hence the above expression can be simplified to

$$\int_0^\pi V(\theta)\cos(2\theta)d\theta = (\pi/4)W * (C_x^2 - C_y^2).$$

Substituting for both the numerator and denomenator of $\tan(2\phi)$ in equation 4 yields

$$\phi = \tan^{-1}\left(\frac{W * 2C_xC_y}{W * (C_x^2 - C_y^2)}\right)/2$$

There is a close connection between the flow computation described in section 2 and the density of edges in the Marr-Hildreth theory of edge detection. Suppose the image $I(x,y)$ is a stationary zero-mean Gaussian process and let $D(x,y)$ be the Laplacian of a Gaussian. The density of zero-crossings in $B_\theta(x,y) = D * R_\theta[I]$ along the x-axis can be approximated by (Papoulis, 1965; Rice, 1944-45)

$$\lambda^2(\theta) \approx \frac{1}{\pi^2}\frac{\int \omega^2 K(\omega)d\omega}{\int K(\omega)d\omega}$$

where $K(\omega)$ is the power spectrum of a slice through B_θ along the x-axis. Using Papoulis' Fourier transform conventions, the numerator can be converted to a spatial integral through the identity

$$\int B_x^2 dx = \frac{1}{2\pi}\int \omega^2 K(\omega)d\omega.$$

Similarly, the denomenator can be converted with the identity

$$\int B^2 dx = \frac{1}{2\pi}\int K(\omega)d\omega.$$

If the integrals are computed locally with the windowing function W, then we have the following estimate for zero crossing density:

$$\lambda^2(\theta) \approx \frac{1}{\pi^2}\frac{W * B_x^2}{W * B^2}$$

If W is radially symmetric, $W * B^2$ will not depend on θ so the maximum zero-crossing density will occur at the maximum of $W * B_x^2$. By assumption, the mean of the process is zero, so $W * B_x^2$ is the variance of $D_x * R_\theta[I]$. Thus, for stationary zero-mean Gaussian processes, if $S = D_x$, $\lambda^2(\theta) \approx V(\theta)$.

References

Horn, B.K.P "Understanding image intensities," *Aritificial Intelligence*, **8**, 1977, 201–231.

Barrow, H., & Tenenbaum, J. M. "Recovering intrinsic scene charactristics from images. In Hanson & Riseman (Eds.), *Computer Vision Systems*. New York: Academic Press, 1978.

Marr, D. *Vision*, 1982, Freeman, San Fransisco CA.

Barr, A. "Global and local deformation of solid primitives." *Computer Graphics*, **18**, pp. 21–30, July 1984.

Brodatz, P. Textures. New York: Dover, 1966.

Glass, L. "Moire effect from random dots." *Nature*, 1969, **243**, 578-580.

Hubel, D. H. & Wiesel, T. N. "Receptive fields, binocular interaction and functional architecture in the cat's visual cortex." *J. Physiol., Lond.*, 1962, **166**, 106–154

Papoulis, *Probability, Random Variables and Stochastic Processes*. New York: McGraw-Hill, 1965.

Marr, D. "Early processing of visual information." *Proc. Royal Soc.*, 1976, **B 275**, 484-519.

Marr, D., & Hildreth, E. "Theory of edge detection." *Proc. Royal Soc.*, 1980, **B 207**, 187-217.

Marr, D., & Poggio, T. "A computational theory of human stereo vision." Proc. Royal Soc., 1979, **204**, 301-328.

Rice, S. O. "Mathematical Analysis of Random Noise." *Bell Sys. Tech. J.*, **23-24**, 1944-1945.

Schiller, P. H., Finlay, B. L. & Volman, S. F. "Quantitative studies of single-cell properties in monkey striate cortex. II. Orientation specificity and ocular dominance." *J. Neurophysiology* **39**, 1976, 1320-1333.

Spivak, *Differential Geometry*. Berkely, California: Publish or Perish, 1979.

Stevens, K. "Computation of locally parallel structure." *Biological Cybernetics*, 1978, **29** 29-26.

Williams, L. "Pyramidal Parametrics" *Computer Graphics,* **17** No. 3, 1983

Zucker, S. "Computational and psychophysical experiments in grouping." In Beck, Hope, Rosenfeld (Eds.),*Human and Machine Vision*, New York: Academic Press, 1983.

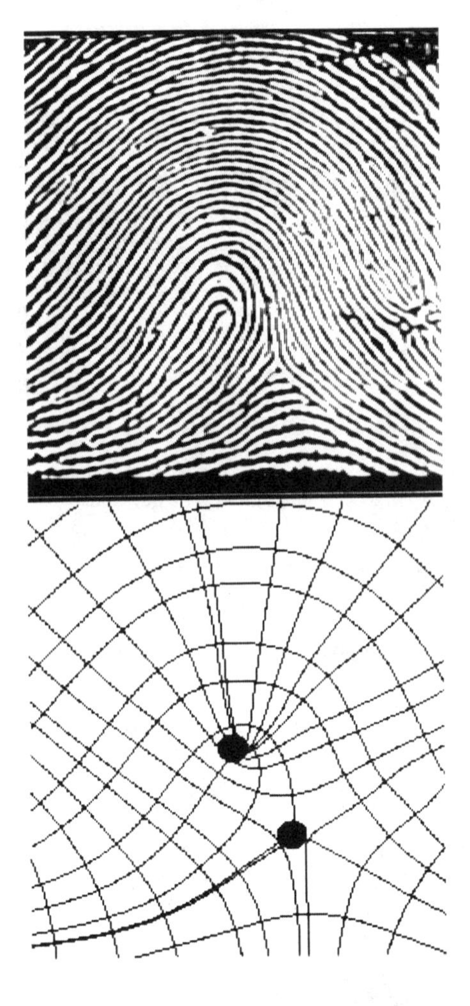

Figure 9: A fingerprint and its flow coordinate grid. The two white circles represent the major singularities, which were detected by measuring the flow field's winding number.

Capturing the Local Structure of Image Discontinuities in Two Dimensions

Yvan Leclerc

Computer Vision and Robotics Laboratory
Department of Electrical Engineering
McGill University
Montreal, Quebec, Canada

Abstract

We generalize our previous work on the local structure of discontinuities from one- to two-dimensional piecewise smooth functions. By local structure, we mean the structure of a function in a local neighbourhood about a discontinuity. For the case of one-dimensional functions, the local neighbourhood can always be partitioned into a left and right half, with each half necessarily containing a smooth function. Thus, the limiting values of these two functions and their derivatives comprise the local structure of a discontinuity in one-dimension. The generalization to two dimensions brings forward a level of complexity not inherent in the one-dimensional case. Namely, the local neighbourhood can now be comprised of an indeterminate number of partitions, each of which contains a smooth function (imagine, for example, the vertex of a checkerboard). Thus, in order to adequately capture the local structure, we must now determine the partitioning of the neighbourhood prior to estimating each of the smooth functions.

Introduction

Our basic premise is that the structure of the intensities within the local neighbourhood of a discontinuity is important. This *local structure* is oftentimes characteristic of the underlying physical event, and can thus be used in guiding the interpretation of the discontinuity [Horn 1977, Witkin 1982]. Further, local structure can be used as an additional constraint in stereo and motion correspondence (e.g., requiring that discontinuities be put in correspondence only if their local structures match to within some predefined tolerance). These and other motivations for capturing local structure are described in greater detail in our preceding paper, [Leclerc and Zucker 1984].

Our starting point is to model idealized images as piecewise smooth functions corrupted by additive noise. Such a function would arise, for example, by imaging an idealized world in which the surfaces and their reflectance properties are piecewise smooth, with only point sources of light and no diffraction, using an idealized camera in which everything is perfectly in focus. In the preceding paper we showed that discontinuities in one-dimensional cuts through such piecewise smooth functions cannot be correctly located without simultaneously determining their local structure. The same holds true, of course, for discontinuities in the full two-dimensional functions.

For example, various edge detection schemes, such as those proposed in [Haralick 1984, Horn 1971, Hueckel 1971, Marr and Hildreth 1979] all implicitly assume that a discontinuity has the local structure of a step function. This assumption can be violated in many ways, some of which have been explored in our preceding paper and by Berzin [1984] in relation to the Marr-Hildreth scheme. As Berzin notes, such violations can cause the zero-crossings to wander a considerable distance from the location of the discontinuities, and can also lead to spurious zero-crossings. By way of summary, Figure 1a illustrates the results of this scheme when applied to an infinitely long edge with a linear variation of intensity in the right half-plane, and Figure 1b illustrates the results when applied to a checkerboard with an assymetric intensity distribution. The latter figure illustrates well the danger of using symmetric patterns to test symmetric operators.

Thus, it is clearly imperative that the local structure of discontinuities be determined at the same time that the discontinuities are localized, and not assumed to be fixed *a priori*.

Our ultimate goal is to generalize the results presented in [Leclerc and Zucker 1984] to the case of discontinuities in two-dimensional functions. That is, we wish to locate discontinuites and simultaneously determine their local structure in two dimensions. In this paper, we present a robust procedure for doing the latter and some proposals for accomplishing the former. In the next section, we quickly review the work done for the one dimensional case, since the results are required for solving the two-dimensional problem. Following this we present our definition of local structure in two dimensions, and a robust procedure for capturing this local structure. Finally, we present a proposal for locating discontinuities based on the above procedure.

A Review of the One-Dimensional Case

For the case of one-dimensional functions, the local neighbourhood of a discontinuity can always be partitioned in two—the left and right halves. For a sufficiently small neighbourhood, each half necessarily contains a smooth function. Thus, the limiting values of these two functions and their derivatives comprise the local structure of the discontinuity.

To capture this local structure, and simultaneously determine the location of the discontinuities, we apply the following procedure at every sample for some smallest neighbourhood size (typically 6-8 samples):

1. **perform a minimum error fit to the data in each half of the neighbourhood (we used a least squares fit for**

convenience, with the order of the fit increasing with the neighbourhood size);

2) determine whether the fits are valid estimates of the underlying function by verifying whether certain necessary conditions for estimation are satisfied; and

3) if so, determine whether the difference between the estimated limits of the two functions is statistically significant.

To improve sensitivity, we then apply the procedure to increasingly larger neighbourhoods, with the constraint that the larger half-neighbourhoods must not contain discontinuities found at the smaller neighbourhood size.[†] By starting with a small neighbourhood, and then going on to increasingly larger neighbourhoods with the above constraint, we can locate both closely packed high-contrast discontinuities and very low contrast, noisy, discontinuities.

The key to this approach is the second step above—verifying that the fits are valid estimates of the underlying function. Without going into all the details, the idea is that the neighbourhoods close to, but not exactly centered on, the discontinuity will have one or the other half-neighbourhood straddling the discontinuity. Thus, the average error of fit for such a neighbourhood (the sum of residuals squared in the case of least-squares fitting) will be larger than for the neighbourhood exactly centered at the discontinuity, since then both half-neighbourhoods will contain only underlying smooth functions. Thus, the necessary condition for the fits to be valid estimates is that their average error of fit be a spatial local minimum. We called this verification step *intra-scale inhibition*, since it involves inhibiting estimates derived from the same neighbourhood size, or scale. See Figure 2 for an illustration of this principle.

Local Structure in Two Dimensions

The generalization to two dimensions brings forward a level of complexity not inherent in the one-dimensional case. Namely, the local neighbourhood can now be comprised of an indeterminate number of partitions, each of which contains a smooth function (imagine, for example, the vertex of a checkerboard). Thus, in order to adequately capture the local structure, we must now determine the partitioning of the neighbourhood prior to estimating each of the smooth functions. Note that one-dimensional cuts simply cannot capture the full extent of the local structure in two dimensions.

In order to determine this partitioning, we further restrict the class of idealized images (piecewise smooth functions) to those in which the boundaries between the smooth parts are themselves piecewise smooth curves. This would rule out, for example, images in which the boundaries are fractals, such as the coastline of Britain [Mandlebrot 1977].

Given the above restriction, every discontinuous point now falls into one of three categores; either the point is at:

1) an endpoint of a boundary;
2) a midpoint along a boundary; or
3) a vertex where two or more boundaries meet.

Now, consider a sufficiently small neighbourhood centered at a point of discontinuity. For such a neighbourhood, each boundary can be approximated by a straight line going to (case 1 above) or through (cases 2 and 3) the center of the neighbourhood. Thus, a sufficiently small neighbourhood can always be partitioned into "pie slices" containing only smooth parts of the function, with the endpoint of a boundary being a degenerate 360° slice (see Figure 3). This leads to the following definition:

Definition The local structure of a point of discontinuity of a piecewise smooth function with piecewise smooth boundaries comprises:

1) the orientation of the boundaries within the local neighbourhood of the point, and
2) the limiting values and appropriate directional derivatives of each of the smooth functions delimited by these boundaries.

That is, we require the angular extent and orientation of each "pie slice," and the differential structure of the smooth function within each of the slices.

Determining Local Structure in Two Dimensions

The above definition tells us that, for a sufficiently small neighbourhood, we need only consider partitioning the neighbourhood into "pie slices" in order to guarantee that each partition contains only a smooth part of the underlying function. However, we still need to determine the orientation of each boundary of the center vertex in order to estimate the underlying smooth functions, and thus determine the local structure.

This is where the results of the one-dimensional case come into play. The analogy is that the problem of determining the position of discontinuities in a one-dimensional function is now the problem of determining the orientation of discontinuities (the boundaries above) over a range of angles. By further restricting our class of piecewise smooth functions to those in which the boundaries meet at angles greater than some minimum angle, θ_{min}, we can determine the orientation of the boundaries by using an angular neighbourhood divided into two angular half-neighbourhoods (see Figure 4) as follows.

Choose an angular neighbourhood of width $2\theta_{min}$. This guarantees that each angular half-neighbourhood contains only a smooth part of the function when the full neighbourhood has the same orientation as any of the boundaries. Then use use the same steps for determining the correct orientations as we did for the one-dimensional case, specifically:

1) perform some kind of minimum error surface fit to the data in each half of the angular neighbourhood;

[†] This is almost, but not quite true. Noisy, low contrast discontinuities found at the smaller neighbourhood size do not always inhibit the larger neighbourhood. We developed an appropriate statistical test which we called *inter-scale inhibition* to determine whether the larger neighbourhood should be inhibited. For the whole truth, see the preceding paper.

2) keep only those surface fits that are valid estimates of the underlying function, i.e., those that are at an angular minimum in average error; and

3) determine whether the difference between the estimated limits of the two functions (and the appropriate directional derivatives) is statistically significant.

As for the one-dimensional case, we can then apply the above steps using larger angular neighbourhoods, with the proviso that they not contain boundaries found at the smaller neighbourhoods. The particular statistical tests are identical to the ones used for the one-dimensional case.

Figure 5 illustrates the result of this procedure on several synthetic images with varying amounts of noise.

Locating Discontinuities in Two Dimensions

The preceding section described a procedure for determining the local structure of a discontinuity in two dimensions given that the center point of the neighbourhood is located at a discontinuity. The remaining problem is to determine whether or not the center point actually is at a discontinuity. Preliminary empirical tests indicate that one possibility is to apply the above procedure at every point in the image, compute the average of the error within each of the pie slices, and then inhibit all points whose computed average is not a spatial local minimum. This technique works quite well in low noise situations. We are currently examining other mechanisms for greater robustness.

Summary and Conclusions

The local structure of image intensity discontinuities can be used to great advantage in the interpretation of images. Yet, the techniques that have been advocated by other researchers for finding discontinuities cannot recover such structure. Indeed, they assume a predetermined structure, and this produces inadequate results when applied to more general domains. A new approach in which the local structure is determined concurrently with the localization of the discontinuity is therefore required. We have developed such an approach in this paper.

We have shown that the local structure of discontinuities in two dimensions comprises the orientation and differential structure of the smooth functions within "pie slices" of the local neighbourhood of a point. We have developed a procedure for capturing this local structure in two dimensions that is robust in the face of a considerable amount of noise. Finally, we have proposed a mechanism for localizing discontinuities based on this procedure.

References

V. Berzins, "Accuracy of Laplacian Edge Detectors," *CVGIP*, (**27**, 1984), pp. 195-210.

R.M. Haralick, "Digital Step Edges from Zero Crossings of Second Directional Derivatives," *PAMI*, (**1**, 1984), pp. 58-68.

B.K.P. Horn, "The Binford-Horn LINE-FINDER," *AI Memo 285*, December, 1973.

B.K.P. Horn, "Understanding Image Intensities," *Artificial Intelligence*, (**8**, 1977), pp. 201-231.

M.H. Hueckel, "An Operator Which Locates Edges in Digitized Pictures," *JACM*, (**18**, 1971), pp. 113-225.

Y. Leclerc and S.W. Zucker, "The Local Structure of Image Discontinuities in One Dimension," TR-83-19R, Dept. Electrical Eng., McGill U., May, 1984.

B. Mandelbrot, "The Fractal Geometry of Nature," Freeman Press, N.Y., 1977.

D. Marr and E. Hildreth, "Theory of Edge Detection," *MIT AI Memo 518*, April, 1979.

A.W. Witkin, "Intensity-based Edge Classification," *Proceedings: AAAI-82*, 1982.

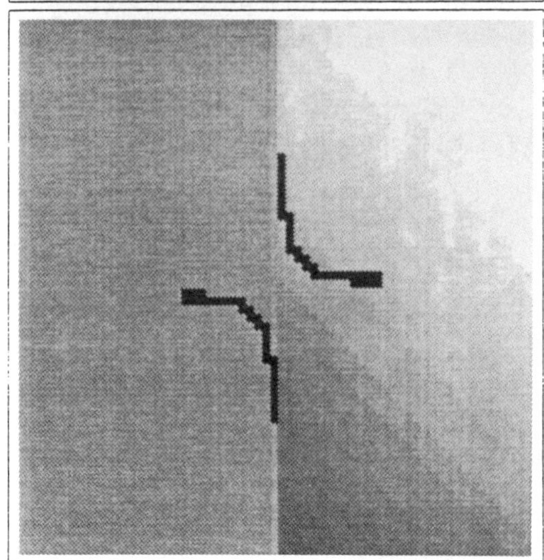

1(a) The Marr-Hildreth zero-crossings algorithm applied to an infinitely long vertical discontinuity with a linear variation of intensity in the right half-plane.

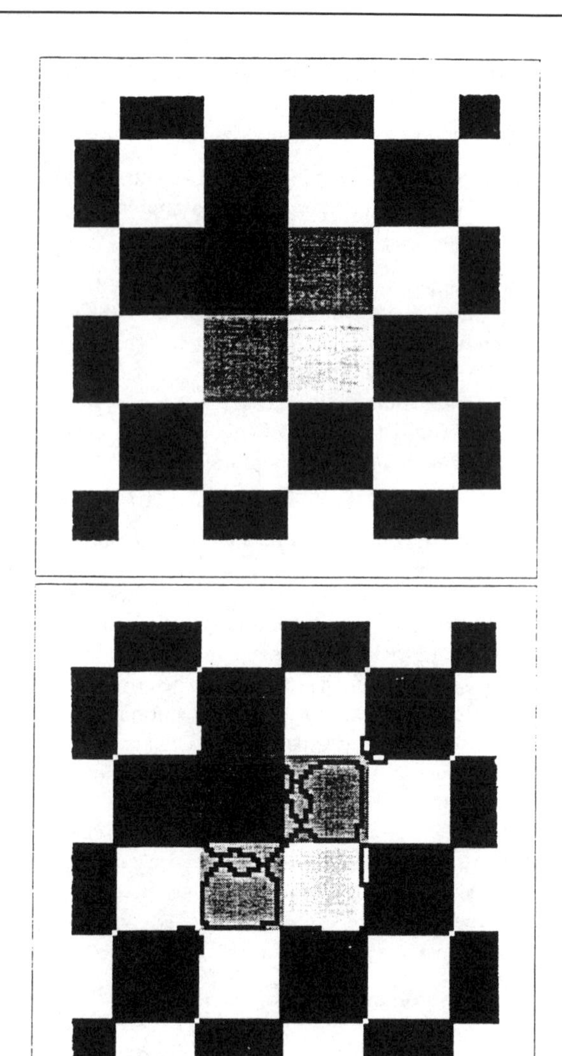

1(b) The Marr-Hildreth zero-crossings algorithm applied to a checkerboard with an assymetric intensity distribution in the center 4 squares.

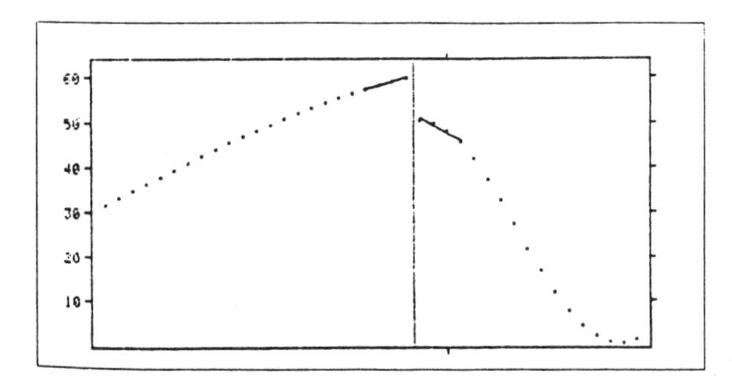

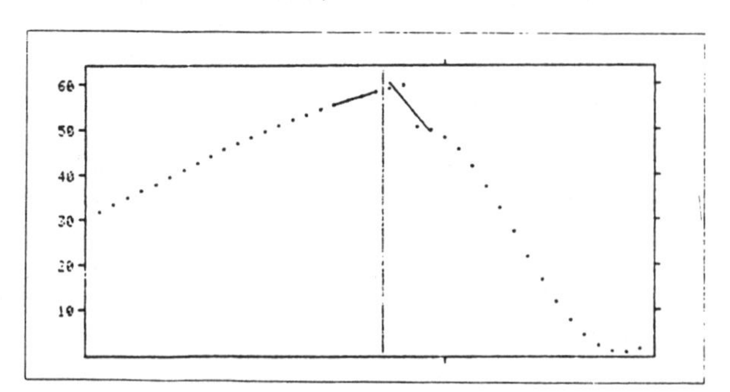

2 An illustration of intra-scale inhibition — the neighbourhood on the right is inhibited because it's right half-neighbourhood straddles a discontinuity, and hence it's average error of fit is larger than for the neighbourhood on the left.

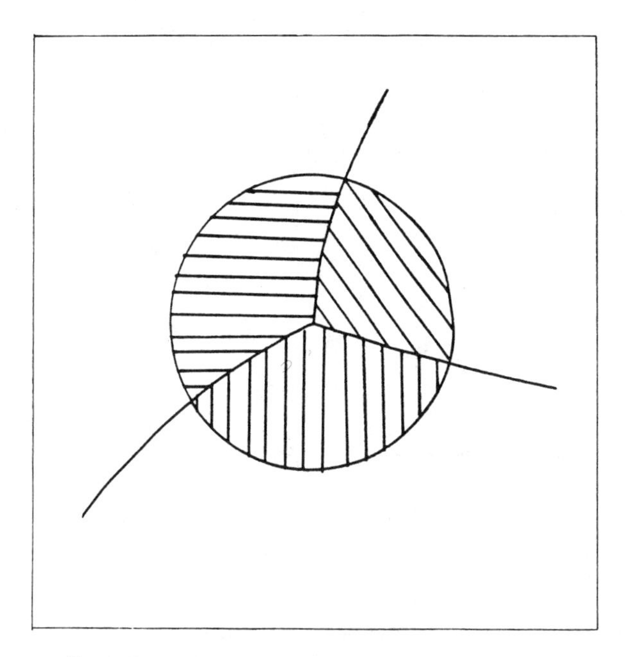

3 An illustration of the local neighbourhood of a discontinuity divided into 3 "pie slices."

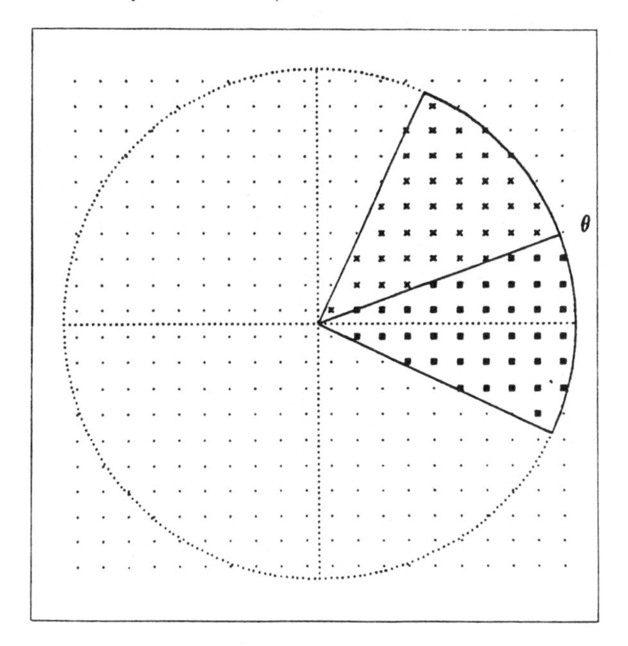

4 An angular neighbourhood of orientation θ divided into 2 angular half-neighbourhoods (x's and filled squares, respectively).

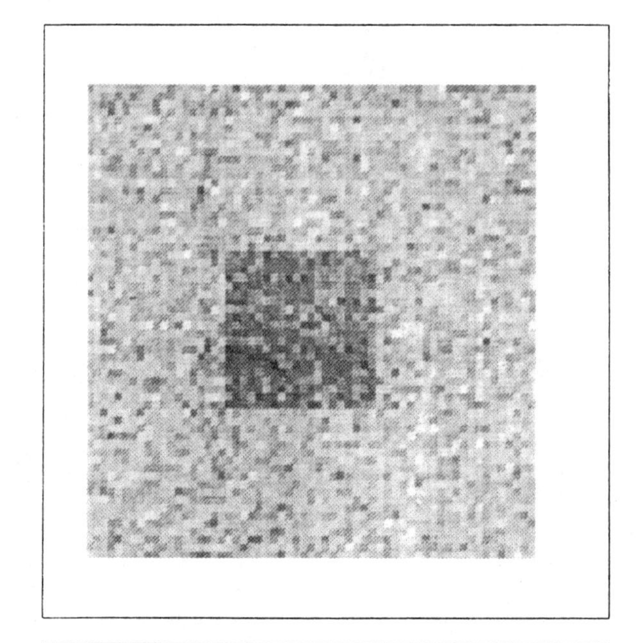

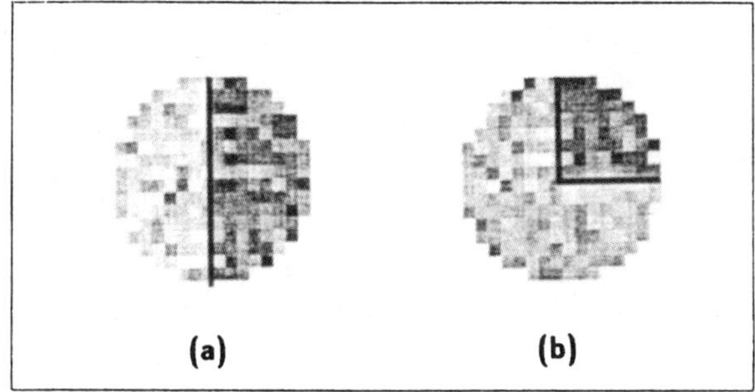

5 Results of the procedure for capturing the local structure. **(a)** Point mid-way along the left boundary. **(b)** Lower left hand corner.

SEGMENTATION AND AGGREGATION: AN APPROACH TO FIGURE-GROUND PHENOMENA

David G. Lowe and Thomas O. Binford

Computer Science Department
Stanford University, Stanford, California 94305

Abstract

We describe a new approach to low-level vision in which the task of image segmentation is to distinguish meaningful relationships between image elements from a background distribution of random alignments. Unlike most previous approaches, which start from idealized models of what we wish to detect in the world, this approach is not based on prior world knowledge and uses measurements which can be computed directly from the input signal. Groupings of image elements are formed over a wide range of sizes and classes while attempting to make use of all available statistical information at each level of the grouping hierarchy, resulting in far more sensitive discrimination than is possible from just local measurements. This paper explores the range of grouping capabilities and discriminations exhibited by the human visual system and discusses the application of the meaningfulness measure to each of them.

Introduction

The human visual system has the capability of spontaneously detecting many very general classes of patterns in an image, even when there is no high-level or semantic knowledge available to guide the interpretation. Figure 1 gives some examples of the range of this capability, including the detection of colinearity, predominant orientation, bilateral and rotational symmetry, and repetition in an otherwise random field of dots. Computer vision systems currently lack almost all of these early vision capabilities, with the exception of edge detection. Even the edge detection capabilities of current computer programs are far below the level of human performance.

The importance of early vision and spontaneous image organization has long been recognized, and has gone under names such as image segmentation, figure/ground phenomena, perceptual grouping, and gestalt organization, all of which emphasize the selection of subsets of image elements which somehow naturally belong together. A grouping is successful to the extent that it brings together elements in the image that have arisen from the same process or belong to the same object in the three-dimensional world being viewed. These groupings can then greatly reduce the combinatorics of the search space when forming higher levels of grouping, matching against world knowledge, making use of texture properties, or searching for correspondences as in stereo vision.

Previous methods for image segmentation have usually been derived from an idealized model of the world (such as the step edge model often used in edge detection or regular texture models used in texture description). The approach taken here is very different and is largely free of prior expectations about the structure of the world. The central concept of this paper is that it is possible to calculate—in a domain-independent way—a statistical measure of the likelihood that some grouping truly reflects an interdependence of its subparts in the world (i.e., that the grouping is not the result of a random alignment of independent elements). This measure can be applied to groupings at all resolutions and positions in the image to determine which ones are the most meaningful and are most likely to lead to further correct interpretations. There is no need to assume a certain level of "noise" in the image, since groupings at all resolutions (allowing all ranges of variations) can be examined, and those which are most meaningful (carry the strongest statistical implications) can be selected as the most useful description of the structure. To give a practical example, it may be important for further interpretation to recognize that the edge of a tree trunk is essentially straight, even though our eye is able to resolve many small perturbations along its length. In this case it would be important to generate at least two different meaningful descriptions for the same curve in the image, corresponding to different resolutions of grouping.

The importance of making precise measurements of the meaningfulness of each grouping is most apparent when attempting to derive strong global information from locally weak information. For example, an edge in a digitized image may be indistinguishable over each small neighborhood along the edge from background sensor noise, as many researchers attempting to derive local edge detectors have discovered. However, if we make many local measurements of meaningfulness at different orientations, and then measure the statistical likelihood that different groupings of local measures would happen to align themselves into a longer smooth edge, we are able to derive much higher measures of meaningfulness for the overall edge than for the local measurements. This implies that a long edge is more detectable than a short edge and that a straight edge

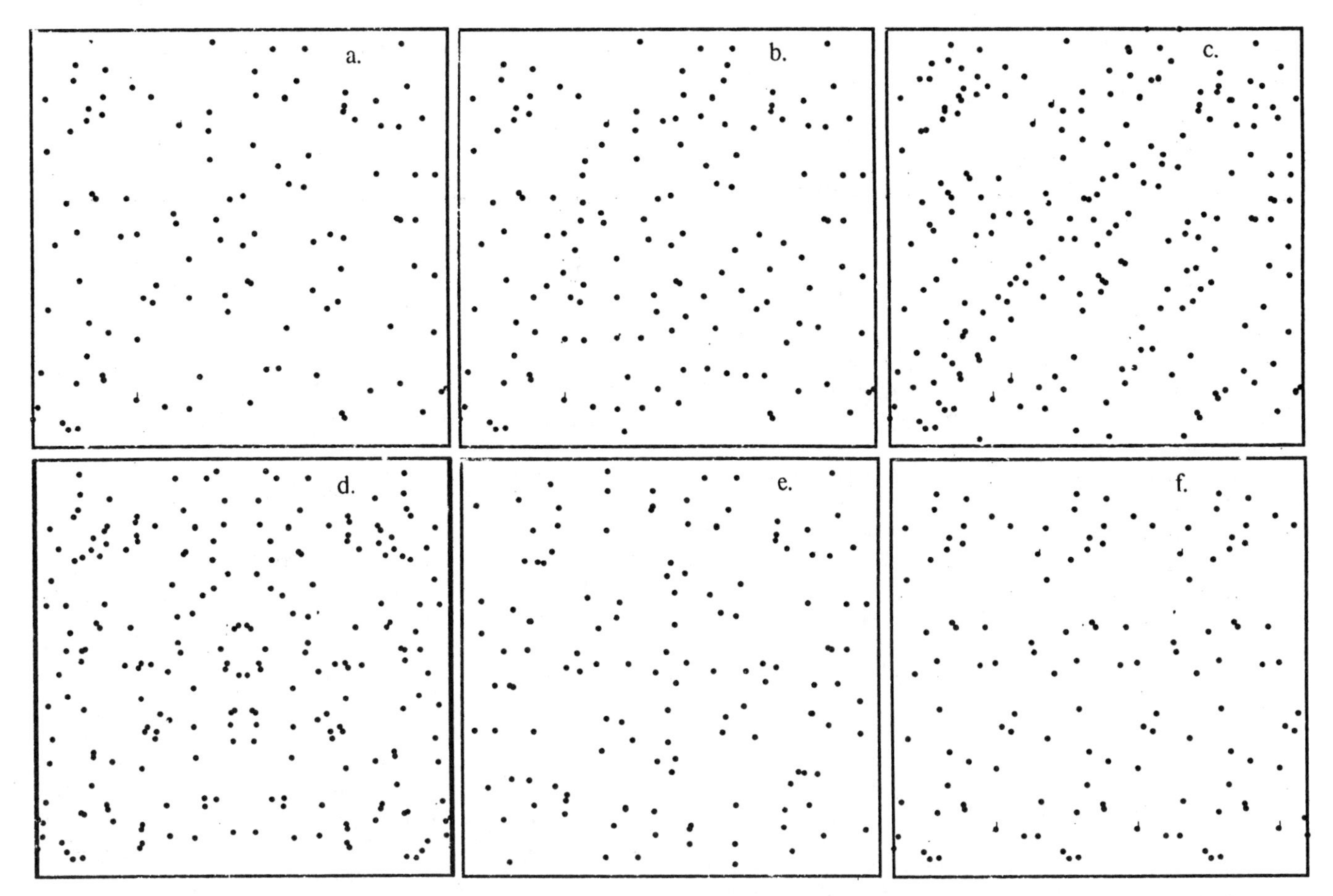

Figure 1: Some examples of the human ability to detect various classes of patterns in an otherwise random field of dots: (a) the basic pattern of 110 dots positioned at random; (b) a string of dots is added which has nearest neighbor distances approximately the same as the background average; (c) the field in (a) is shifted diagonally and overlayed on itself, resulting in a statistically predominant peak in orientation; (d) the field of dots is reflected about a vertical axis, producing bilateral symmetry; (e) the upper right quadrant is rotated into the other quadrants to produce four-fold circular symmetry; (f) the left third of the image is shifted and reproduced twice to form a repetitive pattern.

is more detectable than one with many sharp twists and turns, something apparently also true of human vision. A similar need for global combination of weak local measures applies to all of the examples in Figure 1.

Although the meaningfulness measures applied to each grouping are domain-independent and based on purely geometrical measures of non-randomness, there is still considerable leeway in choosing the groupings to subject to these tests. To test all possible combinations of image elements would be an exponentially expensive process. The human visual system certainly does not recognize all meaningful groupings: as is shown in Figure 2, a pattern such as five equally-spaced dots aligned in a row (which is very unlikely to have arisen by random) will not be spontaneously detected if it is surrounded by enough similar elements. Our approach to defining the sets of groupings to be considered is to use several general principles (such as scale invariance and linear time computational complexity) and to otherwise rely on data regarding human performance.

The Measurement of Meaningfulness

By *meaningfulness* we mean a measure of how likely some grouping is to have arisen from an underlying physical relationship between the constituent features rather than through some accident of viewpoint or location. It is important to have quantitative measures of meaningfulness so that more global combinations of local measures will have precise information to work with in evaluating the significance of each combination. Some commonly used operations, such as thresholding and linear convolutions, destroy a great deal of this information.

We calculate the probability that each selected grouping in the image could have arisen from a random perturbation of the surrounding distribution of similar features. If this is unlikely, then as is commonly done in inferential statistics, we infer that nonchance factors are probably responsible for the grouping. It is common in inferential statistics to specify a threshold at which something is considered to be meaningful (e.g., a 0.05 significance level, or alpha level, corresponding to one chance in 20 that such an unusual result would have arisen from random data). We use a threshold such as this when deciding which results to display during output, but otherwise there is no need for any type of thresholding. The significance values can be combined into higher levels with their own significance based on the fact that two events which are unlikely to have occured at random are even more unlikely to have occured together.

The most common way to measure the likelihood of some value is to look at the shape of the distribution in which it is embedded and measure what proportion of the distribution has values at least as extreme as the one under consideration (for example, we may assume a Gaussian distribution and compare the value to the standard deviation). However, in the types of vision problems we are considering there is no particular knowledge of the shape of the distribution and we do not have enough data to reliably derive the shape from the data for each region. Therefore, we have chosen to use a distribution-free way of calculating likelihoods based on the methods of nonparametric statistics. A nonparametric test of significance makes no assumptions concerning the shape of the parent distribution or population. Its measure of likelihood is based on a *rank test* which measures what proportion of all combinations of rankings have values ranked higher in the surrounding distribution than the values under consideration.

We use discrete statistics above because that is the nature of the vision problem. The image data is not a continuous function but a set of discrete values, and further groupings consist of discrete sets of these initial values. That is one reason why we emphasize the grouping of isolated points in our examples rather than working with gray-scale images that superficially appear continuous—the discrete versions are probably more accurate reflections of what the individual stages of human vision have to work with.

Which Groupings Should be Tested?

As mentioned previously, it would be combinatorially expensive to examine all possible groupings in an image. It would be prohibitive to even form all pairs of features, let alone the larger sets. The human visual system clearly detects only certain classes of meaningful patterns, and experimental work has been carried out to explore this range of performance. Within the computer vision community, Marr [8, 9] in conjunction with Riley [11] and Stevens [13] has discussed the importance of grouping operations and has demonstrated many informal psychophysical experiments testing human performance. Marr emphasized the need for grouping on the basis of length, orientation, size, contrast, and spatial density. However, very little of this work was fully specified or implemented in computer programs. Within the psychology community there have been many investigations of human performance on specific grouping problems. Julesz [6] has carried out many experiments on human performance in distinguishing different textures. Glass [4, 5] has examined the perception of Moiré

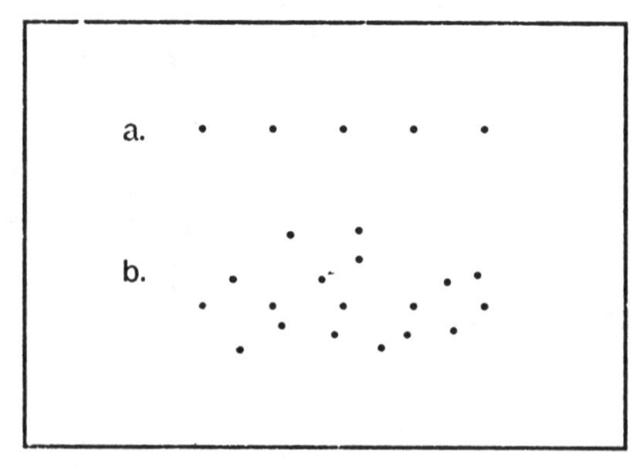

Figure 2: Some patterns, such as the five equally spaced dots in (a) are not spontaneously detected by human vision if they are surrounded by enough similar elements, as in (b), even though the five dots remain highly meaningful in the statistical sense.

patterns of the sort shown in Figure 1c. Many researchers have been intrigued by the human ability to detect bilateral symmetry in otherwise random images (as in Figure 1d), and quantitative experiments on human performance in the face of perturbations in the symmetry have been carried out by Barlow and Reeves [1] and Bruce and Morgan [3]. With the exception of some of Marr's work, all of this research has focussed on performance rather than mechanism.

In addition to data on human performance there are several other constraints on the classes of groupings which should be formed. One is the principle of scale invariance, which means that the same groupings should be formed over a wide range of different sizes in the image. This is just another case of the principle of viewpoint invariance, which also implies the obvious position and rotation invariance. The practical implication of scale invariance is that the same grouping operators must be applied at a range of different sizes in the image (usually chosen to increase by powers of two). Another constraint is that the classes of groupings attempted should be of linear time complexity in terms of the number of items being grouped. Since we are dealing with such large numbers of elements, any attempt at higher order complexity would seem to be too computationally intensive. Therefore, we can only examine groupings between each feature and a fixed number of neighboring features, relying on lower resolution operators to connect features which are more physically distant in the image. When features cannot be grouped at a lower resolution and are too distant to be grouped at a high resolution (as in Figure 2), then the grouping will not be detected.

Only the very lowest levels of grouping operate directly on the image intensity data. Other levels combine the results of previous groupings, looking for meaningful groupings of meaningful values, and in this way build up a layered description of the image. Since there must is some overlap between selected groupings along each dimension in order to minimize the effects of discretization, it is possible to interpolate between neighboring values to precisely locate the best description for each feature.

Linear groupings. One level of image segmentation that has received a great deal of attention is the detection of "edges," usually defined as the detection of extended intensity discontinuities in the original scene. In keeping with our philosophy of looking for statistically meaningful groupings in the image rather than for the image of some idealized feature in the world, we prefer to think of edge detection as the detection of meaningful linear or curvilinear groups of points, where the values of the points have already been detected by some earlier stage. In this case the earlier stage should probably be an isotropic (circularly symmetric) operator which calculates the ratio of center to surround intensity (the strongest evidence for this stage of processing comes from neurophysiological experiments measuring the output of center-surround neurons in the retina). This isotropic operator would be applied at a range of resolutions over the entire image, and at each resolution we would look at all orientations and positions for linear sets of isotropic values that were significant with respect to the surrounding isotropic values.

When combining meaningfulness values from independent regions of the image (such as when combining the values of isotropic operators that lie in a linear arrangement), the independent likelihood values are multiplied together to produce the meaningfulness measure for the new combination. For example, if there is only one chance in 10 that some feature would arise at random from its surrounding distribution, and there is a likelihood of 5 for some other independent feature aligned with it, then there is only one chance in 50 that a specific combination with such unusual values would have arisen from the distribution. However, if we attempt to make many groupings of some feature with its neighbors, then we must divide the likelihood of the combination by the number of attempted groupings of each feature to compensate for the increased number of groupings being tested. These considerations all derive from basic probability theory when calculating the likelihood that some grouping would have arisen randomly from a background distribution.

In addition to looking for linear groupings at a full range of sizes (resolutions) in the image, we also need to look at a range of elongations (ratios of length to width). We do not know in advance what the length of a linear grouping will be, and attempting to form groupings at all elongations allows us to combine the statistical information over the entire length of an edge before deciding upon its significance. As we increase the elongation it is necessary to increase the number of orientations being examined by the same ratio in order to cover the full space of possibilities. However, longer elongations need to be sampled less frequently in the direction of elongation, so the overall computational requirements remain constant as elongation is increased. This example of detecting linear features is worked out in full in the following section of this paper, and a computer implementation of the algorithm is described.

Some important features of this method for detecting meaningful linear structures are that it does not require edges to be continuous (it works well with dotted lines or lines with gaps) and it makes no prior assumptions about the amount of "noise" (variation) in the linear structure. Since it tests for statistical meaningfulness over all possible lengths of an edge, using a much more sensitive test than the typical linear mask, it is possible to accomodate any amount of noise in the linear description so long as the length of the edge is sufficient to make the overall linearity statistically meaningful. After testing all resolutions and elongations, it is possible to select the resolution and elongation with the highest meaningfulness as the most useful description of the grouping. Many groupings may have more than one peak in meaningfulness at different resolutions or elongations, as shown in Figure 3.

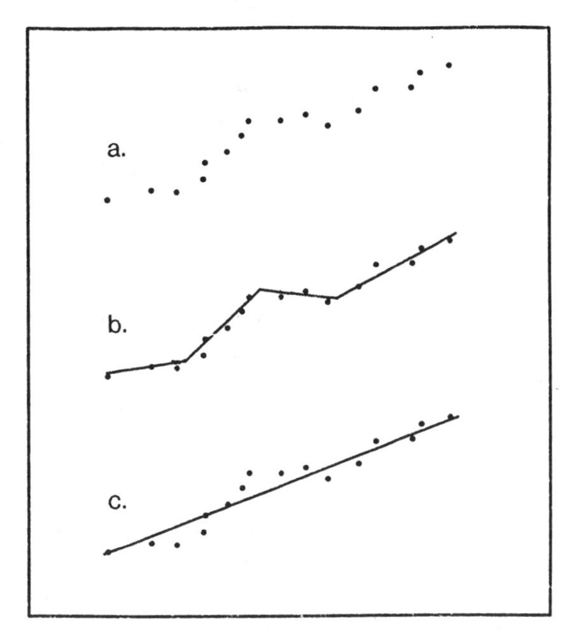

Figure 3: There may be more than one peak in meaningfulness at different resolutions for the same data. The dots shown in (a) could be described as a series of straight line segments at one resolution, as in (b), but also be represented as a single linear grouping at a lower resolution, as in (c).

Curvature and corners. We have dealt so far with only straight edges. One possible extension would be to apply the same grouping techniques to regions of the image corresponding to all possible arcs of constant curvature. However, the increase in computation this requires is rather large, and human performance does not seem to make full use of the statistical information over the length of a curve of constant curvature (however, careful experiments have not yet been carried out to test this point). It seems more likely that human vision only detects smoothness between locally linear groupings. In other words, a smooth curve contains a sequence of meaningful linear segments—the length of each segment depending on the resolution being used—and neighboring linear segments at slightly different orientations are grouped into a description of curvature at that point. The allowable range of orientations for smooth curvature depends on the elongation of the linear segments,

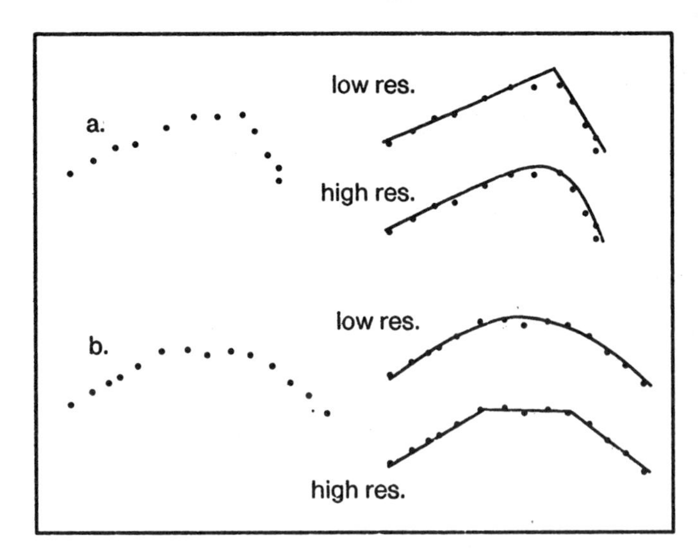

Figure 4: The decision as to whether there is a corner (tangent discontinuity) along a curve can change at different resolutions depending on the length of the support for the curves on each side of the potential corner. In (a) there is a corner when grouping at a low resolution but a smooth curve at a higher resolution. In (b) the roles are reversed.

with longer elongations allowing less change in orientation between segments.

The dual of a smooth-curve continuation from a segment is the detection of a termination. In other words, if there is no smooth continuation from a meaningful linear grouping, then this implies the detection of a meaningful termination of the curve. A corner occurs where two terminations coincide. The same set of points may have a termination at one resolution and may be grouped as a smooth curve at another, as shown in Figure 4.

Marr [8], Schatz [12] and others have shown the importance of detecting terminations and grouping them in further stages. Many of their examples of human texture discrimination are best explained by assuming that "virtual lines" are formed between curve terminations in the same way that we have formed linear groupings of other points in the image. Figure 5a shows an example of an edge formed by a linear sequence of terminations, and Schatz gives many other examples where grouping of terminations is necessary for texture discrimination. Each termination can be treated as a point and fed back into the colinearity detection stage. This is one place where the multiple descriptions of Figure 4 become important, since human vision can detect alignments of low-resolution corners or terminations, even when a curve is smooth at a high resolution.

Orientation and size. Almost all work on edge detection within the computer vision community has dealt with the detection of edges between regions of different intensity. However, as Figure 5 demonstrates, there are many cases in which human vision detects edges between regions with the same average intensity but with properties differing in other ways. One of the strongest effects is produced by changes in orientation, as shown in Figure 5b. The discrimination of the more subtle differences in Figure 5c can be explained similarly by assuming that virtual lines are constructed between the line terminations and the differing orientations of these virtual lines are discriminated. Another significant dimension of variation is size, including both length and width of elongated elements. In Figure 5d each dot in one region is smaller than the other, although their number has been increased to produce equal image intensity in both regions. Figure 5e has the same number and size of dots in both regions, but with different spatial distributions. Discrimination of this example can be explained by the same mechanism as for Figure 5d, since at some lower resolution the dots in the central region will clump into clusters with greater "size" but lower density than the other region. Figure 5f shows the effect of differing lengths, which is much less pronounced.

Size and orientation parameters need to be calculated for regions at all resolutions in the image just as average intensity was calculated, and these results can be fed into the edge grouping process in the same way as the intensity information was. However, there are a number of important issues to be resolved in this process. First of all, how detailed are the characterizations of the distributions of element parameters in each region? It seems that they are not very detailed at all, as is shown in an example by Riley (reproduced in Marr [9]) in which people are unable to distinguish a region with equal numbers of edges at two opposite orientations from a region with randomly distributed orientations. This is another example of a highly meaningful grouping which human vision fails to detect. It appears that a single parameter specifying predominant orientation is all that is required to explain human performance. We intend to carry out a series of similar experiments to precisely determine the characterization that human vision gives to orientation and size statistics. A second important issue is how to determine meaningfulness of these orientation or size measures. For example, if a region only contains one element there is no way to determine whether a particular orientation or size value is random or meaningful. In general, as long as a region contains at least two elements it is possible to calculate some meaningfulness value, and the more elements it contains the higher is the potential meaningfulness.

Symmetry and repetition. Assuming that the above calculations are carried out in a fairly complete way, we conjecture that the detection of symmetry and repetition will require no further mechanisms. Note that the linear grouping of objects as described above will group even a single nearby pair of objects into a linear grouping, although the meaningfulness assigned to a single pair will be low. However, if we look for peaks in the orientation distribution of each region in the image, then a number of low-meaningfulness linear groupings can have a high meaningfulness if they form a large peak in orientation. Repetition of a pattern (including the case where the repeated pattern is reflected about some axis as in bilateral or other symmetries) will result in many parallel matches between similar features at various resolutions. These parallel matches will not be significant unless

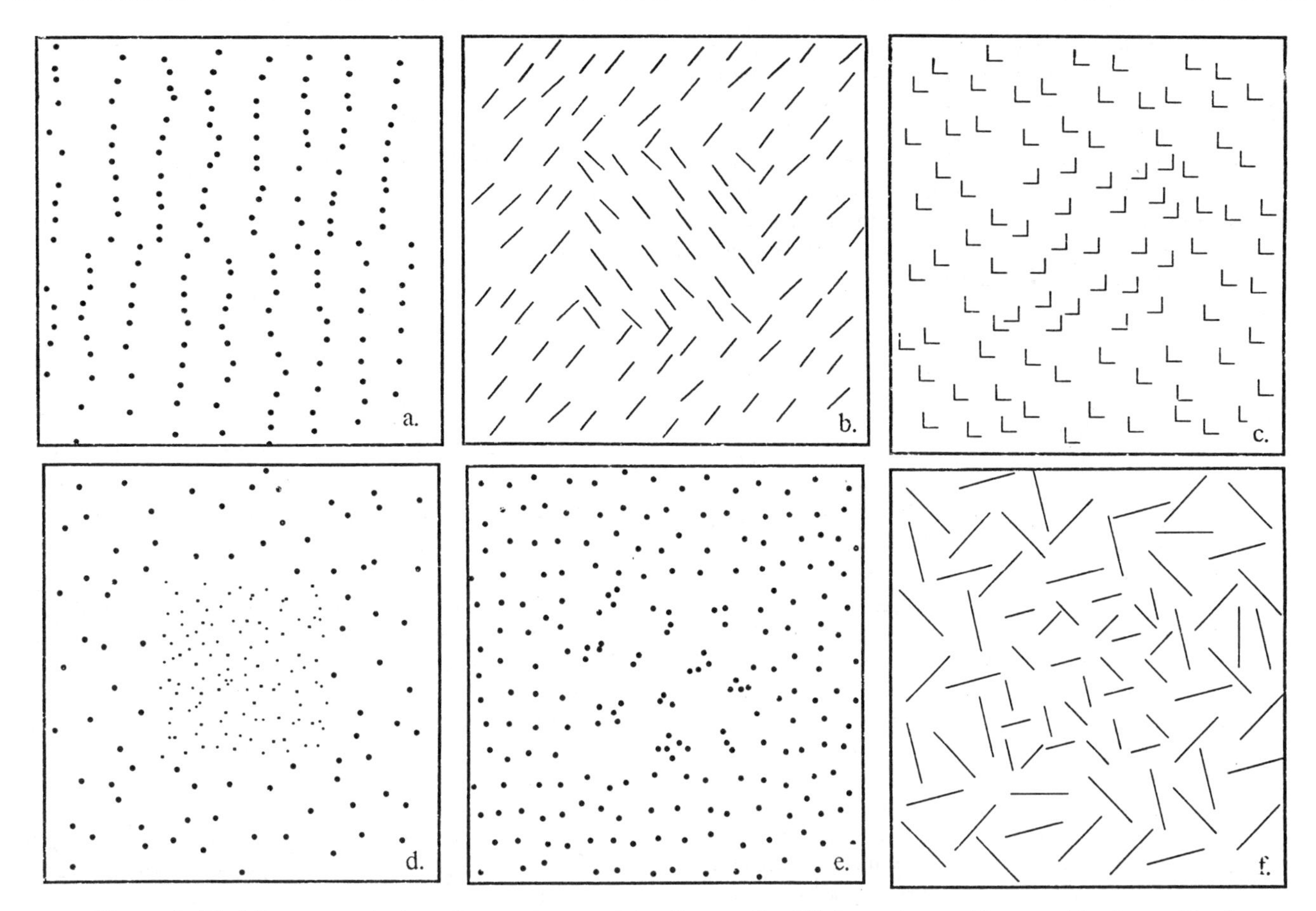

Figure 5: All of these examples have the same average intensity over the differing regions, yet human vision is capable of detecting edges based on changes in other properties. In (a) there is a linear alignment of edge terminations horizontally through the center of the image. Example (b) demonstrates the detection of a change in orientation, and (c) demonstrates a change in orientation of virtual lines. In (d) there is a change in element size, in (e) there is a change in clustering statistics, and in (f) there is a change in line length.

the components are located close to one another at their particular resolution; however, as many psychological experiments have shown (see Julesz [6]) human performance deteriorates very rapidly as the distance to be spanned in making these correspondences increases. Given this conjecture, we intend to carry out other experiments to test its implications for human performance.

The case of symmetry points up the role of high-level knowledge in detecting meaningful groupings. Symmetry is easier to detect when the orientation of the symmetry is parallel to some strong local reference (e.g., the edges of the figure or the page). For operations of the type we have described, the only role that high-level knowledge can play is to bias the meaningfulness results calculated by lower level operations. This can lead to improved performance in such tasks as symmetry detection where the low level results are very weak to begin with. However, this is still very different than the heterarchical approach often adopted in artificial intelligence, which is to calculate only the easiest results bottom-up and to use this knowledge in a top-down fashion to guide computation of the rest. The reason this heterarchical approach fails in many cases for low-level vision is that *all* the low-level results may be weak so that none of them can be used to guide computation of the rest. The only solution in this case is the computationally intensive one we have adopted: form all potential groupings bottom-up and test each one for meaningfulness of the entire combination.

Three-dimensional groupings. It has long been recognized that an important function of early vision is the derivation of the three-dimensional structure of the scene. In previous papers by Binford [2] and Lowe and Binford [7], we have described how various classes of meaningful alignments in a monocular image carry implications for the three-dimensional structure of the scene. For example, if elements of the image are colinear then they are also colinear in three-space, barring an accident in viewpoint. If two edges terminate at a point in the image, or three or more edges converge to a common point, then they must also terminate at a common point in three-space unless the viewpoint is restrictively aligned to produce the coincidence. If one curve terminates at another continuous curve, the terminating curve cannot be closer to the viewer than the continuous curve, or the termination would be unlikely to occur at that location. Curves which are parallel in the image are probably parallel in three-space. There are other similar inferences for interpreting cast shadows or the boundaries of a region. In all these cases, the inferences are based on measures of meaningfulness, where a meaningful grouping in the image leads to statistical inferences for the three-dimensional structure. By combining the meaningfulness of image groupings as described above with the assumptions of general camera position and general light-source position, it is possible to precisely quantify the strength of each inference.

One much studied mechanism for deriving three dimensional information is stereopsis, which depends on a general-purpose mechanism for matching between two images. The groupings we have described can serve as preliminary descriptions for forming matches between images, and the resulting matches can be treated as new groupings with meaningfulness measures of their own. For example, in performing stereo interpretation of random-dot stereograms, at each resolution about one element in 10 should form a grouping with 0.1 significance, one element in 100 should form a grouping with 0.01 significance, etc., and if two images contain significant features from the same class within small corresponding fusional areas, it would be possible to calculate the meaningfulness of this correspondence. Mayhew and Frisby [10] describe a stereo interpretation system that detects edges by grouping isotropic point measures in linear three-space groupings, similar to the techniques outlined here but without an explicit measure of meaningfulness. Stereo processing presents yet another opportunity to derive highly meaningful larger groupings from weak local groupings, in addition to its role in providing explicit depth information.

Implementation of the Linear Meaningfulness Algorithm

In order to illustrate these methods in more detail, we have written a program for detecting meaningful linear groupings of points in an image at all resolutions, orientations, and elongations.

The program takes as input a set of dots like those in Figure 9a (although they need not all be of the same size). The first stage of the program accumulates the density of points falling into square regions at all resolutions from 1/256 the width of the image to 1/8 the width of the image, each resolution twice as course the previous one (a total of 6 resolutions). Each region overlaps with its neighbors by 50% in the vertical and horizontal directions, so that each point falls into four of these regions at each resolution. We then compute the center-surround values at each resolution, which is done by subtracting the average intensity of the surrounding 8 square regions from the intensity of each central region. This first stage of processing in our implementation is very crude—there should at least be smooth transitions between neighboring regions rather than abrupt boundaries. However, the novel part of the algorithm is not in this stage but in the way these initial isotropic values are combined to produce meaningful linear groupings.

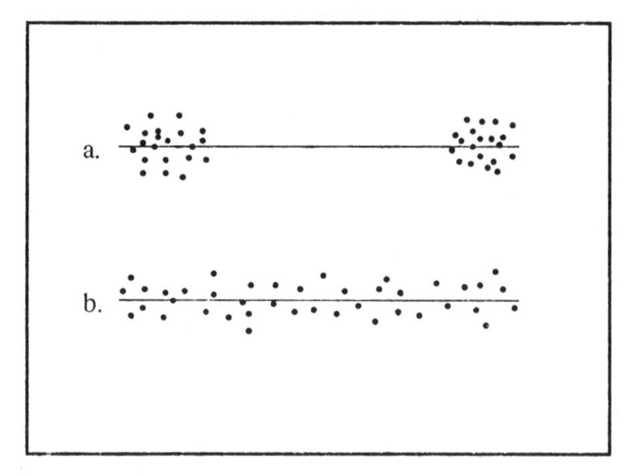

Figure 6: These two sets of points have the same standard deviations from best-fit lines of the same length, yet (b) is much more meaningful as a linear feature than (a).

One commonly-used method for measuring the degree of linearity among data points is to measure the standard deviation of the data points from the line with the best least-squares fit, with a lower standard deviation indicating a better fit. However, as Figure 6 demonstrates, two groups of points with the same standard deviation to a line of the same length can exhibit very different degrees of meaningful linearity. That is the importance of the two-stage process we have used here, where isotropic meaningfulness values are calculated first and their effects separated from the linear meaningfulness values.

As discussed in the previous section, we examine sets of isotropic values in linear arrangements in the image and calculate the probability that each set of such meaningful values would arise at random. This is done by first comparing each isotropic value to a surrounding set of similar values, and calculating a likelihood for it based upon its rank in the surround. Then, given that the isotropic values are computed over independent regions of the image, we multiply their likelihood values together to compute the likelihood that values with a meaningfulness at least that high would happen to occur together. Finally, we divide this value by the number of groupings of that class attempted from each point, since the more groupings which are attempted the more coincidences we expect to find in a random distribution.

We start by computing linear meaningfulness of just pairs of isotropic values and then combine these in stages into longer elongations. We compute linear meaningfulness for pairs at eight orientations as shown in Figure 7, following our rule that adjacent orientations should overlap by 50% to minimize the effects of discretization. We compare each value in these pairs to those of eight surrounding values in sidebars alongside the pair as shown in Figure 8. Meaningfulness is calculated by taking the total number of surrounding values plus one and dividing by the number which are ranked higher than the center value plus one. Therefore, if one of the center values is higher than all 8 of

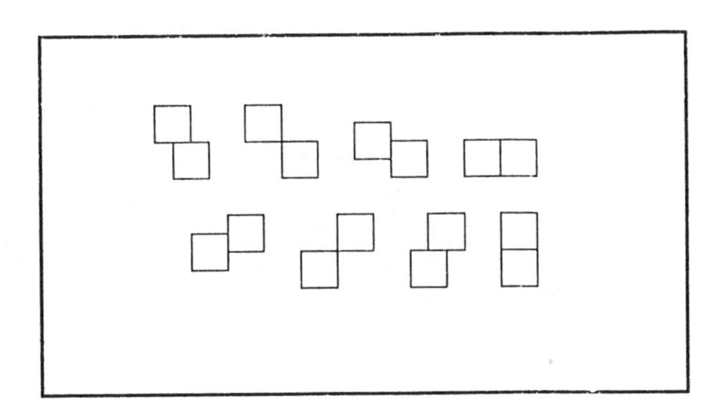

Figure 7: Meaningfulness values for pairs of isotropic values are computed at eight orientations, making use of the independent overlaps in two directions.

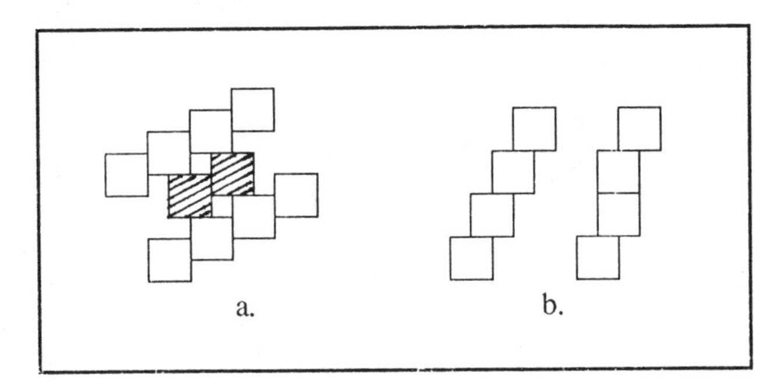

Figure 8: The meaningfulness of each pair of isotropic values is computed with respect to eight surrounding values as shown in (a). In (b), a pair is combined with two colinear pairs of the same orientation but slightly different positions.

the surrounding values, it is assigned a meaningfulness of $(8+1)/(0+1) = 9$, whereas if it is only higher than seven of them it is assigned a value of $(8+1)/(1+1) = 4.5$. Note that one of these linear operators can be assigned a certain degree of meaningfulness just from a single meaningful isotropic value—this is necessary so that combinations into longer elongations will have some measure by which to compute the meaningfulness of the combination even if there is incomplete evidence for linearity in the shorter operator.

Each of these linear operators is combined with its neighboring colinear operators to produce operators with longer elongations. Each stage of combination produces elongations which are twice as long and therefore need to be tested at twice as many orientations to cover the space of possibilities with the same degree of overlap. Therefore, each linear operator is combined with each of two operators with the same orientation but slightly different positions perpendicular to the direction of linearity—to approximate linear operators of different orientations—as is shown in Figure 8b.

The results of carrying out these computations on an image of dots is shown in Figure 10. Figure 9a is the original input to the program, which contains a linear feature immediately apparent to human vision but which nonetheless is based on weak local evidence. Figures 10a though 10c show the results of computing linear meaningfulness at three different resolutions, each one twice as course as the previous ones. Each line shows position and length of a linear operator and the circles at the end of each line are proportional to the log of the likelihood computed for that operator. The large amount of output in these figures includes many linear groupings which are only of marginal meaningfulness, although data of this sort would be necessary in many situations for grouping on the basis of orien-

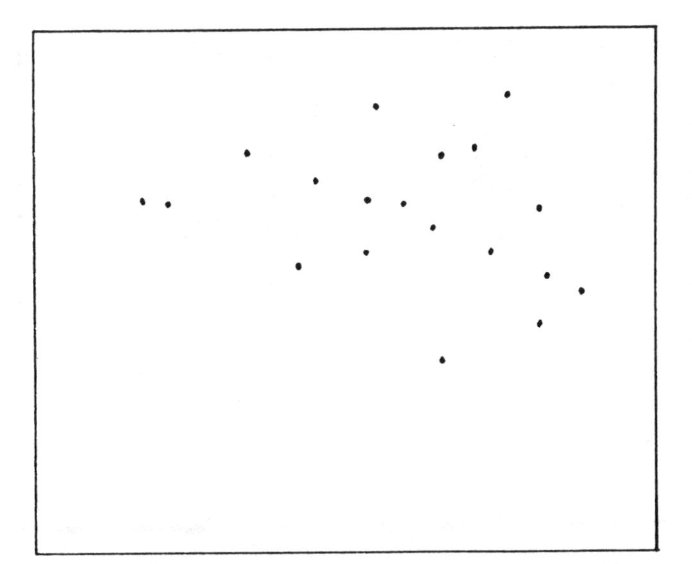

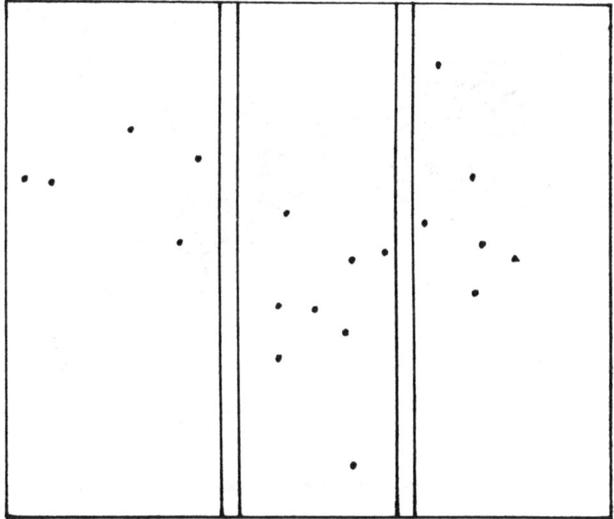

Figure 9: The dots shown in (a) are the input for the computation shown in Figure 10. Although there is a prominent linear feature in (a), this feature is not easily detectable on the basis of local evidence if we separate the image into thirds and displace the center third downwards, as shown in (b).

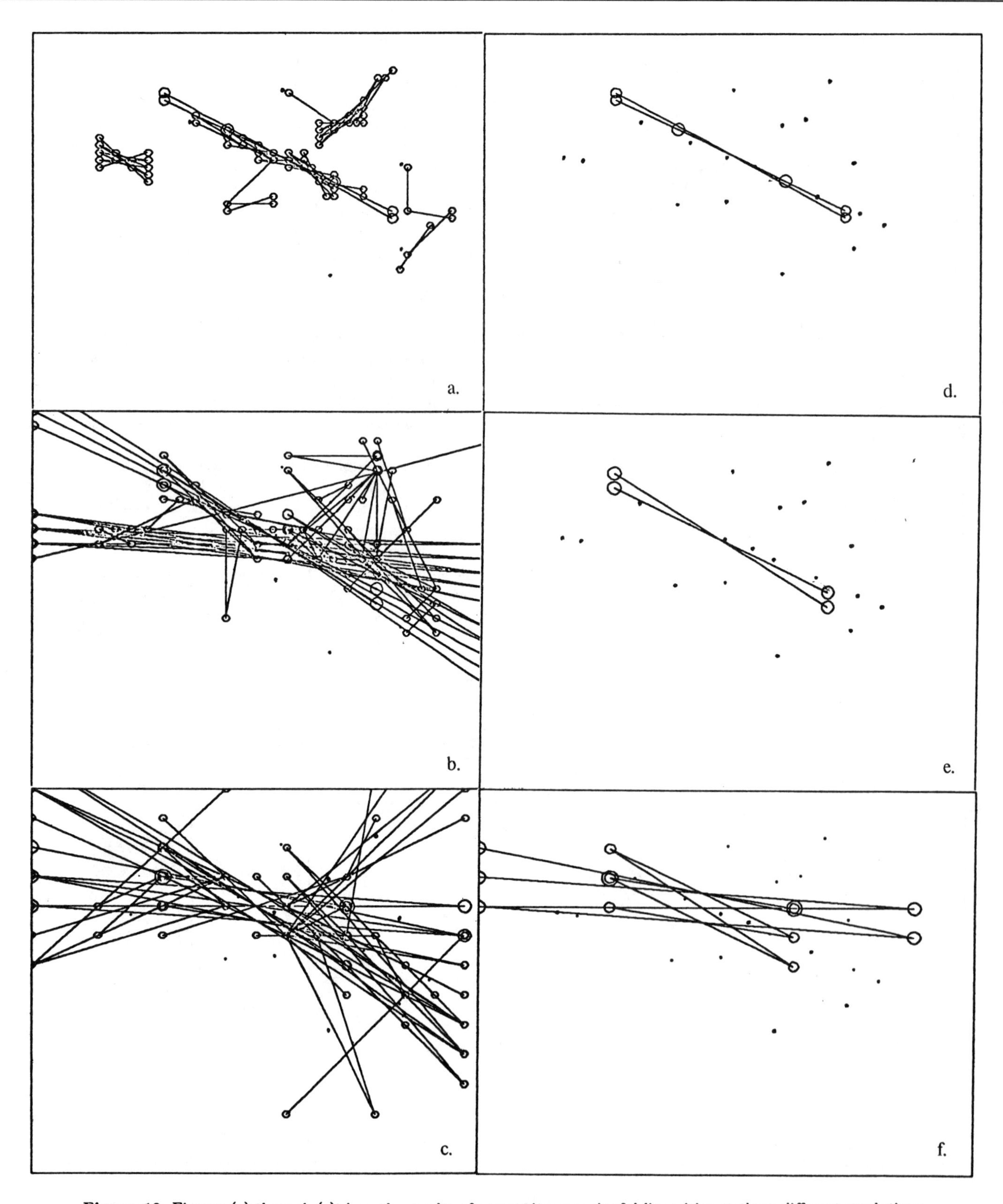

Figure 10: Figures (a) through (c) show the results of computing meaningful linearities at three different resolutions (of increasing powers of 2) on the data of Figure 9. The lines represent linear groupings at various elongations, and the circles at the end of each line are proportional to the log of the likelihood value. If we reduce the significance threshold to the 0.01 level, we are left with the results displayed in (d) through (f).

tation or other higher level grouping. However, when we reduce the meaningfulness threshold for display to the 0.01 significance level, we get the results shown in Figures 10d through 10f, which give only linear groupings of elements corresponding to the single prominent diagonal line.

This initial demonstration is quite crude in many respects, and there are many issues remaining to be resolved. Nevertheless, there are some ways in which this example has impressive performance. As shown in Figure 9b, if we examine small regions of the input data there is insufficient information to reliably detect a linear arrangement of dots. Therefore, the program has in some sense combined the statistical information from along the full length of the feature before arriving at its conclusion of meaningfulness. This is a simple example of the detection of globally significant features from locally weak information that was discussed earlier. Also, unlike most other edge detection programs, it has made no assumptions about the amount of "noise" (deviation of the dots from a straight line), and would have detected the grouping over a very wide range of sizes in the image. Unlike linear convolutions, it is relatively insensitive to a few large dots placed near the linear grouping in the image, since it uses a non-parametric rank test rather than making assumptions about the surrounding distribution. This approach is also very different than the simple detection of zero crossings, since zero crossings are a local measure that may not be supported by any meaningful information (for example, zero crossings may wander randomly under the influence of sensor noise in regions of the image which do not have some sufficiently strong changes in intensity).

Computation Time Considerations

The methods discussed in this paper—testing groupings for meaningfulness at all possible positions, resolutions, and parameter values—are more computationally expensive than those used in most current computer vision programs. However, the difference is only a moderate linear increase over other methods and is not as large as might be feared at first. For example, examining the image at 6 resolutions can be less than a factor of 2 more expensive than examining it at the finest resolution, since halving the resolution means that only one quarter as many groupings need to be examined over the area of the image. When using a digital computer there are techniques that allow us to not even consider groupings that do not have any potentially meaningful constituents. The program described in the previous section used hash coding to access each grouping (each dot was hashed into all groupings which contained it), so that groupings for positions in the image which did not contain any dots were not even considered.

However, regardless of these implementation considerations, it seems likely that there is no alternative to examining these large numbers of groupings if machine vision is to rival human visual capabilities. As emphasized earlier, there may be no information available at an early stage to indicate which of a large number of possible groupings will turn out to be meaningful. From what we know of the human visual system, it seems that our brains have opted for a brute-force, parallel approach to carrying out these computations. A surprisingly large proportion of the human brain seems to be devoted to simply computing local results over all positions in visual images.

Summary

Our derivation of the function of early vision does not start from a specific model of what we expect to find in the world. Given that the world is very general and variable, any prior knowledge about its structure at this level would probably be rather weak. Instead we see the task of early vision to be the formation of meaningful groupings in the image, where meaningfulness can be tested in a domain-independent self-verifying way. It is possible to approach this task as a signal detection problem which makes maximum possible use of the statistical information in the image to distinguish significant relationships from the background of accidentals.

Therefore, meaningfulness does not depend on prior world knowledge. For example, we emphasize the detection of linearity as one useful basis not because it is a common structure in the world so much as because it is a particularly simple basis, where simplicity is required to make maximum use of the available information. Therefore, linearity is a useful way to segment and describe random fields (for the purpose, say, of comparing them to similar random fields) even though the fields were generated without respect to any linearities. Although we cannot claim to have fully done so in this paper, it seems possible to base this method on a sound mathematical derivation.

One result of the domain-independence assumption is that control issues become less important. We compute all possibilities in a bottom-up exhaustive manner, and have given reasons why there may be no computationally less-expensive way to achieve these results. For example, there is probably nothing to gain from the often mentioned technique of detecting the strong parts of an edge and "extending" these segments to look for weaker segments, since we also want to be able to detect the edge when all parts are locally weak. Further progress in low-level perception is more likely to come from attempts to precisely specify what we want to measure than from improvements in control of the computation.

Of course, the above is not meant to imply that all aspects of early vision can be derived from a few abstract principles. Fortunately, we are working in an area in which it is comparatively easy to perform experiments to test the capabilities and parameters of the human visual system. The measurement of meaningfulness we have described is a theory that can be subjected to empirical tests and can be used to guide experimentation. One encouraging result is that a few fairly simple computations seem to cover what initially appear to be a very diverse set of capabilities. There are numerous avenues along which to pursue further research.

Acknowledgements

We would like to thank David Marimont, Brian Wandell, and our other colleagues for their help. This work was supported under ARPA contract N00039-82-C-0250. The first author was also supported by a postgraduate scholarship from the National Research Council of Canada.

References

[1] Barlow, H.B. and R.C. Reeves, "The versatility and absolute efficiency of detecting mirror symmetry in random dot displays," *Vision Research,* **19** (1979), 783-793.

[2] Binford, Thomas O., "Inferring surfaces from images," *Artificial Intelligence,* **17** (1981), 205-244.

[3] Bruce, Vicky G. and Michael J. Morgan, "Violations of symmetry and repetition in visual patterns," *Perception,* **4** (1975), 239-249.

[4] Glass, Leon, "Perception of random dot interference patterns," *Nature,* **246** (1973), 360-362.

[5] Glass, Leon and Eugene Switkes, "Pattern recognition in humans: Correlations which cannot be perceived," *Perception,* **5** (1976), 67-72.

[6] Julesz, B., "Experiments in the visual perception of texture," *Scientific American,* April, 1975.

[7] Lowe, David G. and Thomas O. Binford, "The interpretation of three-dimensional structure from image curves," *Proceedings IJCAI-7,* (Vancouver, Canada, August 1979), 613-618.

[8] Marr, David, "Early processing of visual information," *Philosophical Transactions of the Royal Society of London, Series B,* **275** (1976), 483-524.

[9] Marr, David, *Vision,* (San Francisco: W.H. Freeman, 1982).

[10] Mayhew, John E.W., and John P. Frisby, "The computation of binocular edges," *Perception,* **9** (1980), 69-86.

[11] Riley, M., "The representation of image texture," Master's Thesis, MIT, September 1981.

[12] Schatz, Bruce R., "The computation of immediate texture discrimination," MIT AI Memo 426, August, 1977.

[13] Stevens, K.A., "Computation of locally parallel structure," *Biological Cybernetics,* **29** (1978), 19-28.

Color constancy: a method for recovering surface spectral reflectance

Laurence T. Maloney* and Brian A. Wandell

Department of Psychology, Stanford University, Stanford, California 94305

Received July 18, 1985; accepted August 9, 1985

Human and machine visual sensing is enhanced when surface properties of objects in scenes, including color, can be reliably estimated despite changes in the ambient lighting conditions. We describe a computational method for estimating surface spectral reflectance when the spectral power distribution of the ambient light is not known.

INTRODUCTION

When evening approaches, and daylight gives way to artificial light, we notice little change in the colors of objects around us. The perceptual ability that permits us to discount spectral variation in the ambient light and assign stable colors to objects is called color constancy. Much of human color-vision research focuses on the adaptational mechanisms underlying color constancy. Yet, given the kinds of information available in the initial stages of biological vision, it is not known how color constancy is even possible. Indeed, without restrictions on the range of lights and surfaces that the visual system will encounter, color constancy is not, in general, possible.[1]

In this paper we describe an algorithm for estimating the surface reflectance functions of objects in a scene with incomplete knowledge of the spectral power distribution of the ambient light. We assume that lights and surfaces present in the environment are constrained in a way that we make explicit below. An image-processing system using this algorithm can assign colors that are constant despite changes in the lighting on the scene. This capability is essential to correct color rendering in photography, in television, and in the construction of artificial visual systems for robotics. We describe how constraints on lights and surfaces in the environment make color constancy possible for a visual system and discuss the implications of the algorithm and these constraints for human color vision.

PRELIMINARY DEFINITIONS

Consider a visual sensing device consisting of a lens that focuses light from a scene onto a planar array of sensors, analogous to a retina. We introduce the following definitions and assumptions. We begin by restricting attention to a region of the scene where the spectral power distribution of the light is constant.[2] At any location in the scene, the ambient light is specified by its spectral power distribution, $E(\lambda)$, which describes the energy per second at each wavelength, λ. The ambient light is reflected from surfaces and focused onto the sensor array. The proportion of light of wavelength λ reflected from an object toward location x on the sensor array is determined by the surface spectral reflectance, $S^x(\lambda)$. The superscript x denotes the spatial position on the two-dimensional sensor array at which the object is imaged.[3] The light arriving at each location x on the sensor array is described by the function $E(\lambda)S^x(\lambda)$.

We assume that there are p distinct classes of sensors at each location x. In human vision, there are four photoreceptor classes (rods and cones), of which three (cones) are known to be active in daylight vision. We denote the relative wavelength sensitivity of the visual color sensors of the kth class as $R_k(\lambda)$

Each of the p sensors at location x records a sensor quantum catch

$$\rho_k^{\,x} = \int E(\lambda)S^x(\lambda)R_k(\lambda)\mathrm{d}\lambda, \qquad k = 1, 2, \ldots, p, \qquad (1)$$

where the integral is taken over the entire spectrum. The information about the scene available to the visual system is contained in the p sensor quantum catches at each location x. The spectral reflectance at each location $S^x(\lambda)$ is assumed to be unknown.

Given only the sensor responses $\rho_k^{\,x}$, we show how to recover the surface spectral reflectances $S^x(\lambda)$ over a range of possible ambient lights $E(\lambda)$. Knowledge of $S^x(\lambda)$ permits us to compute color descriptors that are independent of the ambient light $E(\lambda)$.

PREVIOUS WORK

In their early and important work on color constancy, Helson and Judd[4] studied and formally modeled the ability of human observers to achieve this goal. Land and McCann[5] proposed a theory of color vision (the retinex theory) and a method for computing color-constant color descriptors given only the kinds of information available in the sensor responses. The retinex algorithm performs this task only under a limited set of physical conditions.[6] Buchsbaum[7] demonstrated that it is possible to compute color descriptors that are completely independent of the ambient light in an image if the average spectral reflectance of the objects in the image is known. Buchsbaum's result is the strongest that has been obtained until now. His result is useful for many applications, but it is of limited utility in visual sensing applications such as photography and satellite remote sensing in which it is not possible to know in advance the average spectral reflectance function of the objects in the image.

Our purpose here is to improve on Buchsbaum's result. We describe how to recover surface spectral reflectance from an image without knowledge of the average spectral reflectance function.[8]

MODELS OF LIGHTS AND SURFACES REFLECTANCES

We express surface reflectance as a weighted sum of basis spectral reflectance functions $S_j(\lambda)$ as suggested by Sällström,[1] Buchsbaum,[7] and Brill,[9]

$$S^x(\lambda) = \sum_{j=1}^{n} \sigma_j^x S_j(\lambda) \tag{2}$$

and term this representation a linear model of surface reflectance. The basis reflectances are fixed. They do not vary with location in the scene and are assumed known. The number of basis elements, n, is referred to as the number of degrees of freedom in the model. Knowledge of the weights σ_j^x corresponding to a surface reflectance $S^x(\lambda)$ described by the finite linear model amounts to complete knowledge of $S^x(\lambda)$.

Any finite set of surface spectral reflectances can be reproduced by a linear model of this kind if n is large enough. What is surprising is that models with only a few basis reflectances provide excellent approximations to many naturally occurring spectral reflectances. Stiles *et al.*[10] suggest that spectral reflectances may be treated as band-limited functions. A collection of band-limited functions is perfectly captured by a linear model. The number of basis elements is proportional to the band limit.[11] The range of limiting frequencies that they suggest corresponds to three to five basis reflectances in Eq. (2). Buchsbaum and Gottschalk[12] demonstrate that band-limited reflectances generated using as few as three basis reflectances provide metamers to most naturally occurring spectral reflectances.

The Munsell collection includes color chips spanning a wide range of colors. Cohen[13] used a characteristic vector decomposition of the spectral reflectances of 150 Munsell color chips selected at random from a full set of 433 chips to compute the linear models with from 1 to 5 basis reflectances that best approximated the surface reflectances of the Munsell chips.[14] He found that a linear model using as few as three properly chosen basis reflectances captured 99.2% of the overall variance. He states that model reflectances generated by these three basis reflectances provided a good approximation to the spectral reflectances of the Munsell chips. Reflectances generated by these same three basis reflectances also provide good approximations to 337 surface spectral reflectances of naturally occurring objects measured by Krinov.[15]

We also represent the ambient light by a linear model,

$$E(\lambda) = \sum_{i=1}^{m} \epsilon_i E_i(\lambda), \tag{3}$$

with fixed, known basis lights $E_i(\lambda)$. It is natural to inquire how well such a model captures the range of spectral variation of natural lights such as daylight. Judd *et al.*[16] performed a characteristic vector analysis of 622 functions describing the spectral distribution of natural daylight measured over a range of weather conditions and times of day. They determined that, for practical purposes, three to four basis lights provide essentially perfect matches to the spectral distributions measured. The measurements by Judd *et al.* suggest that the number of parameters required to have an adequate linear model of the ambient light may often be small. Dixon and others have independently measured and analyzed spectral power distributions of daylight and drawn similar conclusions.[17]

REFORMULATION OF THE PROBLEM

Next we describe the consequences of assuming linear models of the ambient light and surface reflectances for the problem of color constancy. The m values of ϵ_i in Eq. (3) form a (column) vector ϵ that specifies the light $E(\lambda)$. The n values of σ_j in Eq. (2) form a (column) vector σ that specifies the surface reflectance $S^x(\lambda)$. Substituting Eqs. (2) and (3) into Eq. (1) permits us to express the relationship between the daylight surface reflectances and sensor responses by the matrix equation

$$\rho^x = \Lambda_\epsilon \sigma^x, \tag{4}$$

where ρ^x is a (column) vector formed from the quantum catches of the p sensors at location x. The matrix Λ_ϵ is p by n, and its kjth entry is of the form $\int E(\lambda)S_j(\lambda)R_k(\lambda)d\lambda$. The matrix Λ_ϵ captures the role of the light in transforming surface reflectances at each location x into sensor quantum catches.

Various limits on surface reflectance recovery are dictated by Eq. (4). We consider the limits on recovery (1) when the light on the scene is assumed to be known and (2) when the light on the scene is unknown.

In the simple case in which the ambient light and (therefore) the lighting matrix Λ_ϵ is known, we see that to recover the n weights that determine the surface reflectance we need merely solve a set of simultaneous linear equations. The recovery procedure reduces to matrix inversion when $p = n$. If p is less that n Eq. (4) is underdetermined and there is no unique solution.

If the ambient light is unknown then it easy to show that we cannot do so well: we cannot in general recover the ambient light vector ϵ or the spectral reflectances even when $p = n$. The matrix Λ_ϵ is square. For any ϵ such that Λ_ϵ is nonsingular there is a set of surface reflectances that satisfy Eq. (4). Any such choice of a light vector ϵ and corresponding surface reflectances σ^x could have produced the observed surface reflectances. No unique solution is possible without additional information concerning lights and surfaces in the scene.

Any solution method must therefore resolve the unfavorable ratio of unknown parameters to observed data points. We do so by assuming that there are more classes of sensors than degrees of freedom in surface reflectances: $p > n$. Suppose that there are $p = n + 1$ linearly independent sensors to sample the image at each location spectrally. In this case from s different spatial locations $s(n + 1)$ data values are obtained on the left-hand side of the equation. The number of unknown parameters is only sn unknowns from the different surface vectors and m unknowns from the light vector. After sampling at $s > m$ locations we finally obtain more data values than unknowns.

In principle, then, the response of the $n + 1$ sensors can

contain enough information to permit exact recovery of both the lighting parameter ϵ and the surface reflectance σ^x at each location. Next we outline a method for computing the light ϵ and the n-dimensional surface reflectance vector σ^x given the $n + 1$ dimensional sensor response vector ρ^x at each location.

COMPUTATIONAL METHOD

The ambient light vector ϵ and the surface vectors σ^x contribute to the value of the sensor vectors in different ways. The ambient light vector specifies the value of the light transformation matrix, Λ_ϵ. The matrix Λ_ϵ is a linear transformation from the n-dimensional space of surface reflectances σ^x into the $n + 1$-dimensional space of sensor quantum catches ρ^x. The sensor response to any particular surface, σ^x, is the weighted sum of the n column vectors of Λ_ϵ. Consequently, the sensor responses must fall in a proper subspace of the sensor space determined by Λ_ϵ and therefore by the lighting parameter ϵ.

Figure 1 illustrates the situation when there are three classes of sensors ($p = 3$) and two degrees of freedom in the surface reflectances ($n = 2$). In the particular example shown in Fig. 1, the two-dimensional surface vectors span a plane (passing through the origin) in the three-dimensional sensor space. The light ϵ determines the plane.

We propose a two-step procedure to estimate light and surface reflectances. First, we determine the plane spanning the sensor quantum catches: knowledge of the plane in Fig. 1 permits us to recover the ambient light vector ϵ. Second, once we know the light vector ϵ, we determine the lighting matrix Λ_ϵ and achieve our goal of recovering the surface vectors simply by inverting this transformation. The exact mathematical conditions that must obtain in order for our procedure to yield the unique correct result are analyzed by Maloney.[18]

We have shown that the formal problem of estimating ambient light and surface spectral reflectance from image data may be reduced to a simple computational procedure. Our procedure is also applicable to the construction of automatic sensor systems capable of discounting fluctuations in the ambient light in unusual working environments.

In some natural scenes, the spectral composition of the ambient light varies with spatial location. The computation above can be extended in a straightforward manner to the problem of estimating and discounting a slowly varying (spatial-low-pass) ambient light. We will describe the details of how to do so elsewhere.

The recovery procedure that we developed is exact when the actual physical conditions fall within the bounds defined by the finite-dimensional models of lights and of surface reflectances. The method is readily extended to the case in which the finite-dimensional models only approximate actual lights and surface reflectances. We can no longer guarantee, for example, that the sensor quantum catches in Fig. 1 will lie exactly in the plane determined by the light. Under these circumstances, we estimate the plane that best fits the sensor quantum catches in the least-squares sense and derive an estimate of the light vector $\hat{\epsilon}$. Given $\Lambda_{\hat{\epsilon}}$, we continue by computing the best estimates of surface reflectance σ^x in the least-squares sense.[19] Small deviations from the assumptions of the models produce small errors in estimation.

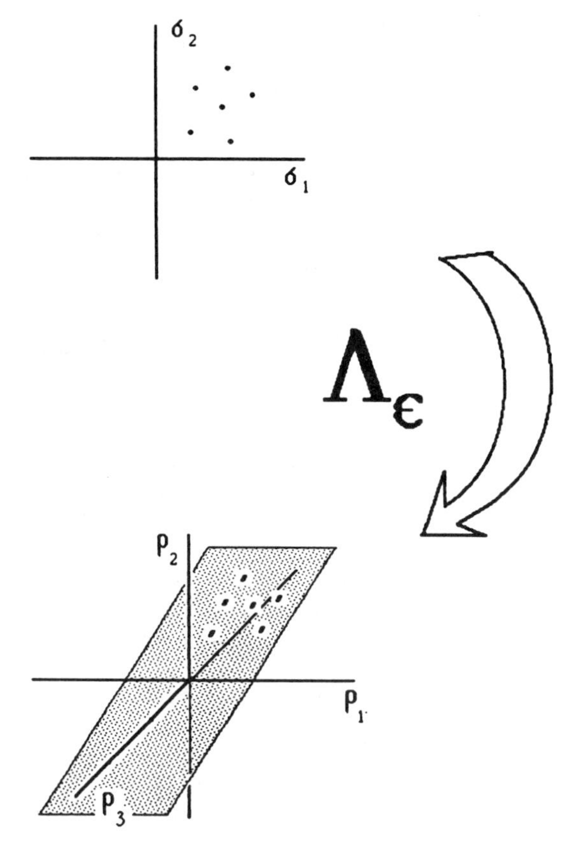

Fig. 1. Outline of the solution method in the case when there are three classes of sensors ($p = 3$) and two degrees of freedom in the model of surface reflectances ($n = 2$). The sensor quantum catches lie on a plane through the origin in the three-dimensional space of sensor quantum catches below. Knowledge of the plane determines the light ϵ and the matrix Λ_ϵ. The matrix Λ_ϵ is inverted to recover the surface reflectances σ^x above.

By increasing the dimensionality of the finite-dimensional approximation, exact solutions may be approached to within any desired degree of precision. A trichromatic visual system can approximate color constancy when reflectances in the visual environment are predominantly captured by a linear model with two degrees of freedom.

IMPLICATIONS

Our calculation has several implications for human color vision. First, it suggests that perfect color constancy is possible only for ranges of lights and reflectances that can be described by a small number of parameters. The human visual system is known to have better color constancy over some ranges of lights than others. Our formulation provides a framework that may be used to determine, experimentally, the range of ambient lights and surfaces over which human color constancy succeeds and over which it fails.

We can test whether a linear model characterizes the range of lights permitting essentially perfect color constancy in human vision as follows. Suppose that the color appearance of a set of surfaces is preserved when measured with two ambient lights, say $E(\lambda)$ and $E'(\lambda)$. If the lights for which human color constancy succeeds form a linear model, then color appearance should also be preserved when the surfaces are viewed under weighted mixtures of the ambient lights $E(\lambda)$ and $E'(\lambda)$.

Second, our results show how the number of classes of photoreceptors active in color vision limits the number of degrees of freedom in the surface reflectances that can be recovered. With three classes of photoreceptors, we can exactly recover surface reflectances drawn from a fixed model of surface reflectance with at most two degrees of freedom. Additional degrees of freedom in the surface reflectance will, in general, introduce error into the estimates of surface reflectance obtained, precluding perfect color constancy.

Third, the parameter counting argument indicates that a minimum number of distinct surface reflectances must be present in the scene to permit recovery. In the illustration of the solution above, at least two distinct surface reflectances are needed to specify the plane (which must pass through the origin) that determines the light. In general, at least $p - 1$ distinct surfaces are needed in order to determine uniquely the light vector ϵ. In the presence of small deviations from the linear models of light and surface reflectances, an increase in the number of distinct surface reflectances will, in general, improve the estimate of the light and the corresponding surface reflectance estimates. Our analysis suggests that color constancy should improve with the number of distinct surfaces in a scene.[20]

Fourth, our algorithm may explain the reduced spatial sampling of the short-wavelength receptors in human vision.[21] Note that incorporating an additional sensor class reduces the spatial sampling density within any single type of sensor class. This creates a conflict between two goals of the visual system. Better color correction for the ambient light can be obtained by including more sensor classes. But including additional sensor classes reduces the spatial sampling density within individual sensor classes. A reasonable trade-off between these two goals may be obtained by the following observation.

Once the light vector ϵ has been estimated, we may calculate the inverse of the lighting matrix Λ_ϵ. Once this matrix is known, the $n + 1$ quantum catches at each location are redundant: only n sensor values are needed to compute the value of σ^x. It follows that if the ambient light varies slowly across the scene, the $n + 1$th sensor class need not be present at as high a sampling density as the other sensor classes.

In a three-sensor system there is no need to place the third sensor with as high a spatial sampling density as the first two sensors. It follows that reducing the retinal space occupied by the short-wavelength sensor class permits the system to obtain a higher spatial resolution of the estimated surface spectral reflectance function with virtually no deterioration in its ability to correct for variation in the spectral power distribution of the ambient light.

ACKNOWLEDGMENTS

This research was supported by grant no. 2 RO1 EY03164 from the National Eye Institute and NASA grant NCC-2-44. We thank M. Pavel, R. N. Shepard, and D. Varner for their advice and suggestions. Stanford University has applied for a patent directed to the matter described herein.

* Lawrence T. Maloney is now at the Human Performance Center, University of Michigan, 330 Packard Road, Ann Arbor, Michigan 48104.

REFERENCES AND NOTES

1. See P. Sällström, "Colour and physics: some remarks concerning the physical aspects of human colour vision," Institute of Physics Rep. 73-09 (University of Stockholm, Stockholm, 1973).
2. The computational method that we develop requires only that the ambient light be approximately constant over small local patches of the image. The method is easier to explain if we restrict attention to a region of the image across which the ambient light does not change.
3. In general the surface reflectance function may depend on the geometry of the scene, the angle of incidence of the light on the surface, and the angle between the surface and the line of sight. We are concerned here with the analysis of a single image drawn from a scene with fixed geometric relations among objects, light sources, and the visual sensor array. $S^x(\lambda)$ refers to the proportion of light returned from the object toward the sensor array within that fixed geometrical framework.
4. D. B. Judd, "Hue saturation and lightness of surface colors with chromatic illumination," J. Opt. Soc. Am. **30,** 2 (1940); H. Helson, "Fundamental problems in color vision. I. The principle governing changes in hue saturation and lightness of non-selective samples in chromatic illumination," J. Exp. Psychol. **23,** 439 (1938).
5. E. H. Land and J. J. McCann, "Lightness and retinex theory," J. Opt. Soc. Am. **61,** 1 (1971); E. H. Land, "Recent advances in retinex theory and some implications for cortical computations: color vision and the natural image," Proc. Nat. Acad. Sci. U.S. **80,** 5163 (1983); E. H. Land, D. H. Hubel, M. Livingston, S. Perry, and M. Burns, "Colour-generating interactions across the corpus callosum," Nature **303,** 616 (1983).
6. D. Brainard and B. Wandell, "An analysis of the retinex theory of color vision," Stanford Applied Psychology Lab. Tech. Rep. 1985-04 (Stanford University, Stanford, Calif., 1985).
7. G. Buchsbaum, "A spatial processor model for object colour perception," J. Franklin Inst. **310,** 1 (1980).
8. It is not possible to recover $E(\lambda)$ better than to within a multiplicative constant given only the sensor quantum catches. If, for example, the intensity of the light is doubled to $2E$ but all reflectances are halved to $\frac{1}{2}S^x(\lambda)$, it is easy to verify that the sensor quantum catches in Eq. (1) are unchanged. When we speak of recovering the ambient light and surface reflectances, we mean recovery up to this unknown mutiplicative constant.
9. M. H. Brill, "A device performing illuminant-invariant assessment of chromatic relations," J. Theor. Biol. **71,** 473 (1978).
10. W. S. Stiles, G. Wyszecki, and N. Ohta, "Counting metameric object-color stimuli using frequency-limited spectral reflectance functions," J. Opt. Soc. Am. **67,** 779 (1977).
11. For a discussion of band-limited functions, see R. Bracewell, *The Fourier Transform and Its Application*, 2nd ed. (McGraw-Hill, New York, 1978), Chap. 10.
12. G. Buchsbaum and A. Gottschalk, "Chromaticity coordinates of frequency-limited functions," J. Opt. Soc. Am. A **1,** 885 (1984).
13. J. Cohen, "Dependency of the spectral reflectance curves of the Munsell color chips," Psychonomic Sci. **1,** 369 (1964).
14. See K. V. Mardia, J. T. Kent, and J. M. Bibby, *Multivariate Analysis* (Academic, London, 1979), Chap. 8, for a discussion of characteristic vector analysis (also known as principal-components analysis or the Karhunen–Loève decomposition).
15. E. Krinov, *Spectral Reflectance Properties of Natural Formations*, Technical translation TT-439 (National Research Council of Canada, Ottawa, 1947); details of the fit of Cohen's characteristic vectors to the Munsell surface reflectances are given in L. Maloney, "Computational approaches to color constancy," Stanford Applied Psychology Lab. Tech. Rep. 1985-01 (Stanford University, Stanford, Calif., 1985).
16. D. B. Judd, D. L. MacAdam, and G. Wyszecki, "Spectral distribution of typical daylight as a function of correlated color temperature," J. Opt. Soc. Am. **54,** 1031 (1964).
17. E. R. Dixon, "Spectral distribution of Australian daylight," J. Opt. Soc. Am. **68,** 437 (1978); G. T. Winch, M. C. Boshoff, C. J. Kok, and A. G. du Toit, "Spectroradiometric and colorimetric characteristics of daylight in the southern hemisphere: Pre-

toria, South Africa," J. Opt. Soc. Am. **56,** 456 (1966); S. R. Das and V. D. P. Sastri, "Spectral distribution and color of tropical daylight," J. Opt. Soc. Am. **55,** 319 (1965); V. D. P. Sastri and S. R. Das, "Spectral distribution and color of north sky at Delhi," J. Opt. Soc. Am. **56,** 829 (1966); "Typical spectral distributions and color for tropical daylight," J. Opt. Soc. Am. **58,** 391 (1968).

18. L. Maloney, "Computational approaches to color constancy," Stanford Applied Psychology Lab. Tech. Rep. 1985-01 (Stanford University, Stanford, Calif., 1985).
19. See Ref. 18, Chap. 4, for details.
20. R. B. MacLeod, "An experimental investigation of brightness constancy," Arch. Psychol. **135,** 1 (1932), reports that human brightness constancy does.
21. E. N. Willmer and W. D. Wright, "Colour sensitivity of the fovea centralis," Nature **156,** 119 (1945); D. R. Williams, D. I. A. MacLeod, and M. Hayhoe, "Punctate sensitivity of the blue-sensitive mechanism," Vision Res. **21,** 1357 (1981); F. M. de Monasterio, S. J. Schein, and E. P. McCrane, "Staining of blue-sensitive cones of the macaque retina by a fluorescent dye," Science **213,** 1278 (1981).

Visual routines*

SHIMON ULLMAN
Massachusetts Institute of Technology

Abstract

This paper examines the processing of visual information beyond the creation of the early representations. A fundamental requirement at this level is the capacity to establish visually abstract shape properties and spatial relations. This capacity plays a major role in object recognition, visually guided manipulation, and more abstract visual thinking.

For the human visual system, the perception of spatial properties and relations that are complex from a computational standpoint nevertheless often appears deceivingly immediate and effortless. The proficiency of the human system in analyzing spatial information far surpasses the capacities of current artificial systems. The study of the computations that underlie this competence may therefore lead to the development of new more efficient methods for the spatial analysis of visual information.

The perception of abstract shape properties and spatial relations raises fundamental difficulties with major implications for the overall processing of visual information. It will be argued that the computation of spatial relations divides the analysis of visual information into two main stages. The first is the bottom-up creation of certain representations of the visible environment. The second stage involves the application of processes called 'visual routines' to the representations constructed in the first stage. These routines can establish properties and relations that cannot be represented explicitly in the initial representations.

Visual routines are composed of sequences of elemental operations. Routines for different properties and relations share elemental operations. Using a fixed set of basic operations, the visual system can assemble different routines to extract an unbounded variety of shape properties and spatial relations.

At a more detailed level, a number of plausible basic operations are suggested, based primarily on their potential usefulness, and supported in part by empirical evidence. The operations discussed include shifting of the processing focus, indexing to an odd-man-out location, bounded activation, boundary tracing, and marking. The problem of assembling such elemental operations into meaningful visual routines is discussed briefly.

*This report describes research done at the Artificial Intelligence Laboratory of the Massachusetts Institute of Technology. Support for the laboratory's artificial intelligence research is provided in part by the Advanced Research Projects Agency of the Department of Defense under Office of Naval Research contract N00014-80-C-0505 and in part by National Science Foundation Grant 79-23110MCS. Reprint requests should be sent to Shimon Ullman Department of Psychology and Artificial Intelligence Laboratory, M.I.T., Cambridge, MA 02139, U.S.A.

1. The perception of spatial relations

1.1. Introduction

Visual perception requires the capacity to extract shape properties and spatial relations among objects and objects' parts. This capacity is fundamental to visual recognition, since objects are often defined visually by abstract shape properties and spatial relations among their components.

A simple example is illustrated in Fig. 1a, which is readily perceived as representing a face. The shapes of the individual constituents, the eyes, nose, and mouth, in this drawing are highly schematized; it is primarily the spatial arrangement of the constituents that defines the face. In Fig. 1b, the same components are rearranged, and the figure is no longer interpreted as a face. Clearly, the recognition of objects depends not only on the presence of certain features, but also on their spatial arrangement.

The role of establishing properties and relations visually is not confined to the task of visual recognition. In the course of manipulating objects we often rely on our visual perception to obtain answers to such questions as "is *A* longer than *B*", "does *A* fit inside *B*", etc. Problems of this type can be solved without necessarily implicating object recognition. They do require, however,

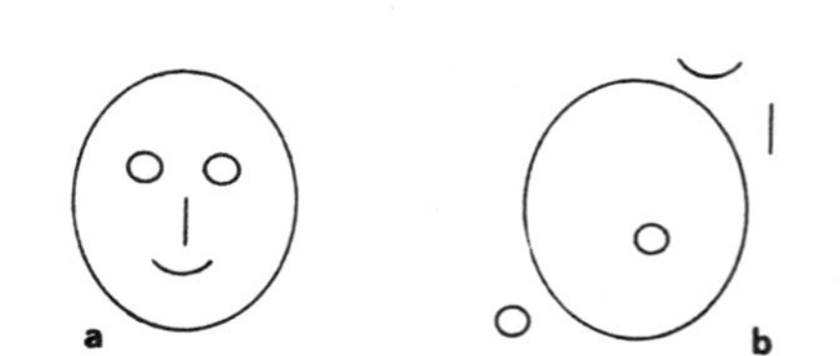

Figure 1. *Schematic drawings of normally-arranged* (a) *and scrambled* (b) *faces. Figure 1a is readily recognized as representing a face although the individual features are meaningless. In 1b, the same constituents are rearranged, and the figure is no longer perceived as a face.*

the visual analysis of shape and spatial relations among parts.[1] Spatial relations in three-dimensional space therefore play an important role in visual perception.

In view of the fundamental importance of the task, it is not surprising that our visual system is indeed remarkably adept at establishing a variety of spatial relations among items in the visual input. This proficiency is evidenced by the fact that the perception of spatial properties and relations that are complex from a computational standpoint, nevertheless often appears immediate and effortless. It also appears that some of the capacity to establish spatial relations is manifested by the visual system from a very early age. For example, infants of 1–15 weeks of age are reported to respond preferentially to schematic face-like figures, and to prefer normally arranged face figures over 'scrambled' face patterns (Fantz, 1961).

The apparent immediateness and ease of perceiving spatial relations is deceiving. As we shall see, it conceals in fact a complex array of processes that have evolved to establish certain spatial relations with considerable efficiency. The processes underlying the perception of spatial relations are still unknown even in the case of simple elementary relations. Consider, for instance, the task of comparing the lengths of two line segments. Faced with this simple task, a draftsman may measure the length of the first line, record the result, measure the second line, and compare the resulting measurements. When the two lines are present simultaneously in the field of view, it is often possible to compare their lengths by 'merely looking'. This capacity raises the problem of how the 'draftsman in our head' operates, without the benefit of a ruler and a scratchpad. More generally, a theory of the perception of spatial relations should aim at unraveling the processes that take place within our visual system when we establish shape properties of objects and their spatial relations by 'merely looking' at them.

The perception of abstract shape properties and spatial relations raises fundamental difficulties with major implications for the overall processing of visual information. The purpose of this paper is to examine these problems and implications. Briefly, it will be argued that the computation of spatial relations divides the analysis of visual information into two main stages. The first is the bottom up creation of certain representations of the visible environment. Examples of such representations are the primal sketch (Marr, 1976) and the 2½-D sketch (Marr and Nishihara, 1978). The second stage involves the top–down application of visual routines to the representations constructed in the first stage. These routines can establish properties and relations that cannot be represented explicitly in the initial base representations. Underlying the visual routines there exists a fixed set of elemental operations that constitute the basic 'instruction set' for more complicated processes. The perception of a large variety of properties and relations is obtained by assembling appropriate routines based on this set of elemental operations.

The paper is divided into three parts. The first introduces the notion of visual routines. The second examines the role of visual routines within the overall scheme of processing visual information. The third (Sections 3 and 4) examines the elemental operations out of which visual routines are constructed.

1.2. An example: The perception of inside/outside relations

The perception of inside/outside relationships is performed by the human perceptual system with intriguing efficiency. To take a concrete example, suppose that the visual input consists of a single closed curve, and a small 'X' figure (see Fig. 2), and one is required to determine visually whether the X lies inside or outside the closed curve. The correct answers in Fig. 2a and 2b appear to be immediate and effortless, and the response would be fast and accurate.[2]

Figure 2. *Perceiving inside and outside. In 2a and 2b, the perception is immediate and effortless; in 2c, it is not.*

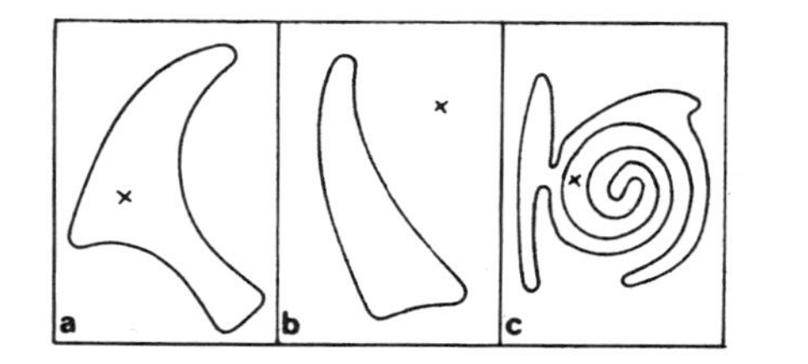

[1]Shape properties (such as overall orientation, area, etc.) refer to a single item, while spatial relations (such as above, inside, longer–than, etc.) involve two or more items. For brevity, the term spatial relations used in the discussion would refer to both shape properties and spatial relations.

[2]For simple figures such as 2a, viewing time of less than 50 msec with moderate intensity, followed by effective masking is sufficient. This is well within the limit of what is considered immediate, effortless perception (e.g., Julesz, 1975). Reaction time of about 500 msec can be obtained in two–choice experiments with simple figures (Varanese, 1981). The response time may vary with the presentation conditions, but the main point is that in/out judgments are fast and reliable and require only a brief presentation.

One possible reason for our proficiency in establishing inside/outside relations is their potential value in visual recognition based on their stability with respect to the viewing position. That is, inside/outside relations tend to remain invariant over considerable variations in viewing position. When viewing a face, for instance, the eyes remain within the head boundary as long as they are visible, regardless of the viewing position (see also Sutherland (1968) on inside/outside relations in perception).

The immediate perception of the inside/outside relation is subject to some limitations (Fig. 2c). These limitations are not very restrictive, however, and the computations performed by the visual system in distinguishing 'inside' from 'outside' exhibit considerable flexibility: the curve can have a variety of shapes, and the positions of the X and the curve do not have to be known in advance.

The processes underlying the perception of insde/outside relations are entirely unknown. In the following section I shall examine two methods for computing 'insideness' and compare them with human perception. The comparison will then serve to introduce the general discussion concerning the notion of visual routines and their role in visual perception.

1.2.1. Computing inside and outside

The ray-intersection method. Shape perception and recognition is often described in terms of a hierarchy of 'feature detectors' (Barlow, 1972; Milner, 1974). According to these hierarchical models, simple feature detecting units such as edge detectors are combined to produce higher order units such as, say, triangle detectors, leading eventually to the detection and recognition of objects. It does not seem possible, however, to construct an 'inside/outside detector' from a combination of elementary feature detectors. Approaches that are more procedural in nature have therefore been suggested instead. A simple procedure that can establish whether a given point lies inside or outside a closed curve is the method of ray-intersections. To use this method, a ray is drawn, emanating from the point in question, and extending to 'infinity'. For practical purposes, 'infinity' is a region that is guaranteed somehow to lie outside the curve. The number of intersections made by the ray with the curve is recorded. (The ray may also happen to be tangential to the curve without crossing it at one or more points. In this case, each tangent point is counted as two intersection points.) If the resulting intersection number is odd, the origin point of the ray lies inside the closed curve. If it is even (including zero), then it must be outside (see Fig. 3a, b).

This procedure has been implemented in computer programs (Evans, 1968; Winston, 1977, Ch. 2), and it may appear rather simple and straightforward. The success of the ray-intersection method is guaranteed, however, only if rather restrictive constraints are met. First, it must be assumed that the curve is closed, otherwise an odd number of intersections would not be indicative of an 'inside' relation (see Fig. 4a). Second, it must be assumed that the curve is isolated: in Figs. 4b and 4c, point p lies within the region bounded by the closed curve c, but the number of intersections is even.[3]

Figure 3. *The ray intersection method for establishing inside/outside relations. When the point lies inside the closed curve, the number of intersections is odd (a); when it lies outside, the number of intersections is even (b).*

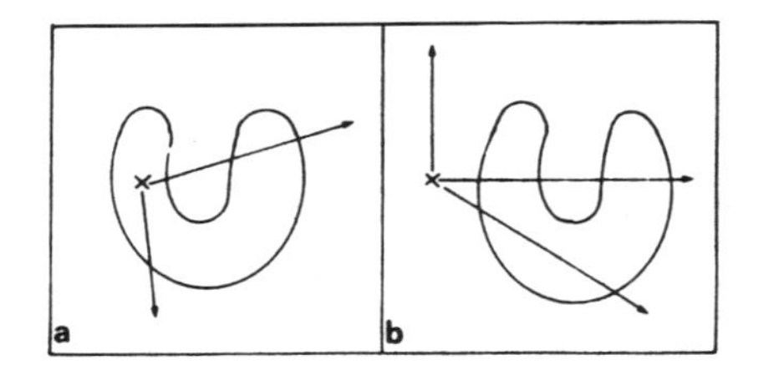

These limitations on the ray-intersection method are not shared by the human visual system: in all of the above examples the correct relation is easily established. In addition, some variations of the inside/outside problem pose almost insurmountable difficulties to the ray-intersection procedure, but not to human vision. Suppose that in Fig. 4d the problem is to determine whether any of the points lies inside the curve C. Using the ray-intersection procedure, rays must be constructed from all the points, adding significantly to the complexity of the solution. In Figs. 4e and 4f the problem is to determine whether the two points marked by dots lie inside the same curve. The number of intersections of the connecting line is not helpful in this case in establishing the desired relation. In Fig. 4g the task is to find an innermost point—a point that lies inside all of the three curves. The task is again straightforward, but it poses serious difficulties to the ray-intersection method.

It can be concluded from such considerations that the computations employed by our perceptual system are different from, and often superior to the ray-intersection method.

[3]In Fig. 4c region p can also be interpreted as lying inside a hole cut in a planar figure. Under this interpretation the result of the ray-interaction method can be accepted as correct. For the original task, however, which is to determine whether p lies within the region bounded by c, the answer provided by the ray-intersection method is incorrect.

Figure 4. *Limitations of the ray-intersection method. a, An open curve. The number of intersections is odd, but* p *does not lie inside* C. *b—c, Additional curves may change the number of intersections, leading to errors. d—g, Variations of the inside/outside problem that render the ray-intersection method in ineffective. In* d *the task is to determine visually whether any of the dots lie inside* C, *in (—f), whether the two dots lie inside the same curve; in* g *the task is to find a point that lies inside all three curves.*

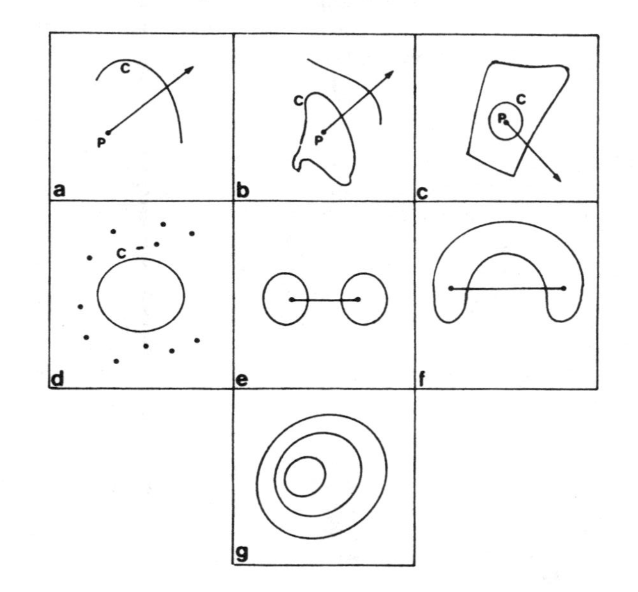

The 'coloring' method. An alternative procedure that avoids some of the limitations inherent in the ray-intersection method uses the operation of activating, or 'coloring' an area. Starting from a given point, the area around it in the internal representation is somehow activated. This activation spreads outward until a boundary is reached, but it is not allowed to cross the boundary. Depending on the starting point, either the inside or the outside of the curve, but not both, will be activated. This can provide a basis for separating inside from outside. An additional stage is still required, however, to complete the procedure, and this additional stage will depend on the specific problem at hand. One can test, for example, whether the region surrounding a 'point at infinity' has been activated. Since this point lies outside the curve in question, it will thereby be established whether the activated area constitutes the curve's inside or the outside. In this manner a point can sometimes be determined to lie outside the curve without requiring a detailed analysis of the curve itself. In Fig. 5, most of the curve can be ignored, since activation that starts at the X will soon 'leak out' of the enclosing corridor and spread to 'infinity'. It will thus be determined that the X cannot lie inside the curve, without analyzing the curve and without attempting to separate its inside from the outside.[4]

Alternatively, one may start at an infinity point, using for instance the following procedure: (1) move towards the curve until a boundary is met; (2) mark this meeting point; (3) start to track the boundary, in a clockwise direction, activating the area on the right; (4) stop when the marked position is reached. If a termination of the curve is encountered before the marked position is reached, the curve is open and has no inside or outside. Otherwise, when the marked position is reached again and the activation spread stops, the inside of the curve will be activated. Both routines are possible, but, depending on the shape of the curve and the location of the X, one or the other may become more efficient.

The coloring method avoids some of the main difficulties with the ray-intersection method, but it also falls short of accounting for the performance of human perception in similar tasks. It seems, for example, that for human perception the computation time is to a large extent scale independent. That

Figure 5. *That the* x *does not lie inside the curve* C *can be established without a detailed analysis of the curve.*

[4]In practical applications 'infinity points' can be located if the curve is known in advance not to extend beyond a limited region. In human vision it is not clear what may constitute an 'infinity point', but it seems that we have little difficulty in finding such points. Even for a complex shape, that may not have a well-defined inside and outside, it is easy to determine visually a location that clearly lies outside the region occupied by the shape.

is, the size of the figures can be increased considerably with only a small effect on the computation time.[5] In contrast, in the activation scheme outlined above computation time should increase with the size of the figures.

The basic coloring scheme can be modified to increase its efficiency and endow it with scale independence, for example by performing the computation simultaneously at a number of resolution scales. Even the modified scheme will have difficulties, however, competing with the performance of the human perceptual system. Evidently, elaborate computations will be required to match the efficiency and flexibility exhibited by the human perceptual system in establishing inside/outside relationships.

The goal of the above discussion was not to examine the perception of inside/outside relations in detail, but to introduce the problems associated with the seemingly effortless and immediate perception of spatial relations. I next turn to a more general discussion of the difficulties associated with the perception of spatial relations and shape properties, and the implications of these difficulties to the processing of visual information.

1.3. Spatial analysis by visual routines

In this section, we shall examine the general requirements imposed by the visual analysis of shape properties and spatial relations. The difficulties involved in the analysis of spatial properties and relations are summarized below in terms of three requirements that must be faced by the 'visual processor' that performs such analysis. The three requirements are (i) the capacity to establish abstract properties and relations (abstractness), (ii) the capacity to establish a large variety of relations and properties, including newly defined ones (open-endedness), and (iii) the requirement to cope efficiently with the complexity involved in the computation of spatial relations (complexity).

1.3.1. Abstractness

The perception of inside/outside relations provides an example of the visual system's capacity to analyze abstract spatial relations. In this section the notion of abstract properties and relations and the difficulties raised by their perception will be briefly discussed.

Formally, a shape property P defines a set S of shapes that share this property. The property of closure, for example, divides the set of all curves into the set of closed curves that share this property, and the complementary set of open curves. (Similarly, a relation such as 'inside' defines a set of configurations that satisfy this relation.)

Clearly, in many cases the set of shapes S that satisfy a property P can be large and unwieldy. It therefore becomes impossible to test a shape for property P by comparing it against all the members of S stored in memory. The problem lies in fact not simply in the size of the set S, but in what may be called the size of the *support* of S. To illustrate this distinction, suppose that given a plane with one special point X marked on it we wish to identify the black figures containing X. This set of figures is large, but, given an isolated figure, it is simple to test whether it is a member of the set: only a single point, X, need be inspected. In this case the relevant part of the figure, or its support, consists of a single point. In contrast, the set of supports for the property of closure, or the inside/outside relation, is unmanageably large.

When the set of supports is small, the recognition of even a large set of objects can be accomplished by simple template matching. This means that a small number of patterns is stored, and matched against the figure in question.[6] When the set of supports is prohibitively large, a template matching decision scheme will become impossible. The classification task may nevertheless be feasible if the set contains certain regularities. This roughly means that the recognition of a property P can be broken down into a set of operations in such a manner that the overall computation required for establishing P is substantially less demanding than the storing of all the shapes in S. The set of all closed curves, for example, is not just a random collection of shapes, and there are obviously more efficient methods for establishing closure than simple template matching. For a completely random set of shapes containing no regularities, simplified recognition procedures will not be possible. The minimal program required for the recognition of the set would be in this case essentially as large as the set itself (cf. Kolmogorov, 1968).

The above discussion can now serve to define what is meant here by 'abstract' shape properties and spatial relations. This notion refers to properties and relations with a prohibitively large set of supports that can nevertheless be established efficiently by a computation that captures the regularities in the set. Our visual system can clearly establish abstract properties and

[5] The dependency of inside/outside judgments on the size of the figure is currently under empirical investigation. There seems to be a slight increase in reaction time as a function of the figure size.

[6] For the present discussion, template-matching between plane figures can be defined as their cross-correlation. The definition can be extended to symbolic descriptions in the plane. In this case at each location in a plane a number of symbols can be activated, and a pattern is then a subset of activated symbols. Given a pattern P and a template T, their degree of match m is a function that is increasing in $P \cap T$ and decreasing in $P \cup T - P \cap T$ (when P is 'positioned over' T so as to maximize m).

relations. The implication is that it should employ sets of processes for establishing shape properties and spatial relations. The perception of abstract properties such as insideness or closure would then be explained in terms of the computations employed by the visual system to capture the regularities underlying different properties and relations. These computations would be described in terms of their constituent operations and how they are combined to establish different properties and relations.

We have seen in Section 1.2 examples of possible computations for the analysis of inside/outside relations. It is suggested that processes of this general type are performed by the human visual system in perceiving inside/outside relations. The operations employed by the visual system may prove, however, to be different from those considered in Section 1.2. To explain the perception of inside/outside relations it would be necessary, therefore, to unravel the constituent operations employed by the visual system, and how they are used in different judgments.

1.3.2. Open-endedness

As we have seen, the perception of an abstract relation is quite a remarkable feat even for a single relation, such as insideness. Additional complications arise from the requirement to recognize not only one, but a large number of different properties and relations. A reasonable approach to the problem would be to assume that the computations that establish different properties and relations share their underlying elemental operations. In this manner a large variety of abstract shape properties and spatial relations can be established by different processes assembled from a fixed set of elemental operations. The term 'visual routines' will be used to refer to the processes composed out of the set of elemental operations to establish shape properties and spatial relations.

A further implication of the open-endedness requirement is that a mechanism is required by which new combinations of basic operations can be assembled to meet new computational goals. One can impose goals for visual analysis, such as "determine whether the green and red elements lie on the same side of the vertical line". That the visual system can cope effectively with such goals suggests that it has the capacity to create new processes out of the basic set of elemental operations.

1.3.3. Complexity

The open-endedness requirement implied that different processes should share elemental operations. The same conclusion is also suggested by complexity considerations. The complexity of basic operations such as the bounded activation (discussed in more detal in Section 3.4) implies that different routines that establish different properties and relations and use the bounded activation operation would have to share the same mechanism rather than have their own separate mechanisms.

A special case of the complexity consideration arises from the need to apply the same computation at different spatial locations. The ability to perform a given computation at different spatial positions can be obtained by having an independent processing module at each location. For example, the orientation of a line segment at a given location seems to be performed in the primary visual cortex largely independent of other locations. In contrast, the computations of more complex relations such as inside/outside independent of location cannot be explained by assuming a large number of independent 'inside/outside modules', one for each location. Routines that establish a given property or relation at different positions are likely to share some of their machinery, similar to the sharing of elemental operations by different routines.

Certain constraints will be imposed upon the computation of spatial relations by the sharing of elemental operations. For example, the sharing of operations by different routines will restrict the simultaneous perception of different spatial relations. The application of a given routine to different spatial locations will be similarly restricted. In applying visual routines the need will consequently arise for the sequencing of elemental operations, and for selecting the location at which a given operation is applied.

In summary, the three requirements discussed above suggest the following implications.

(1) Spatial properties and relations are established by the application of visual routines to a set of early visual representations.
(2) Visual routines are assembled from a fixed set of elemental operations.
(3) New routines can be assembled to meet newly specified processing goals.
(4) Different routines share elemental operations.
(5) A routine can be applied to different spatial locations. The processes that perform the same routine at different locations are not independent.
(6) In applying visual routines mechanisms are required for sequencing elemental operations and for selecting the locations at which they are applied.

1.4. Conclusions and open problems

The discussion so far suggests that the immediate perception of seemingly simple spatial relations requires in fact complex computations that are difficult to unravel, and difficult to imitate. These computations were termed above 'visual routines'. The general proposal is that using a fixed set of basic operations, the visual system can assemble routines that are applied to the visual representations to extract abstract shape properties and spatial relations.

The use of visual routines to establish shape properties and spatial relations raises fundamental problems at the levels of computational theory, algorithms, and the underlying mechanisms. A general problem on the computational level is which spatial properties and relations are important for object recognition and manipulation. On the algorithmic level, the problem is how these relations are computed. This is a challenging problem, since the processing of spatial relations and properties by the visual system is remarkably flexible and efficient. On the mechanism level, the problem is how visual routines are implemented in neural networks within the visual system.

In concluding this section, major problems raised by the notion of visual routines are listed below under four main categories.

(1) *The elemental operations*. In the examples discussed above the computation of inside/outside relations employed operations such as drawing a ray, counting intersections, boundary tracking, and area activation. The same basic operations can also be used in establishing other properties and relations. In this manner a variety of spatial relations can be computed using a fixed and powerful set of basic operations, together with means for combining them into different routines that are then applied to the base representation. The first problem that arises therefore is the identification of the elemental operations that constitute the basic 'instruction set' in the composition of visual routines.

(2) *Integration*. The second problem that arises is how the elemental operations are integrated into meaningful routines. This problem has two aspects. First, the general principles of the integration process, for example, whether different elemental operations can be applied simultaneously. Second, there is the question of how specific routines are composed in terms of the elemental operations. An account of our perception of a given shape property or relation such as elongation, above, next-to, inside/outside, taller-than etc. should include a description of the routines that are employed in the task in question, and the composition of each of these routines in terms of the elemental operations.

(3) *Control*. The questions in this category are how visual routines are selected and controlled, for example, what triggers the execution of different routines during visual recognition and other visual tasks, and how the order of their execution is determined.

(4) *Compilation*. How new routines are generated to meet specific needs, and how they are stored and modified with practice.

The remainder of this paper is organized as follows. In Section 2 I shall discuss the role of visual routines within the overall processing of visual information. Section 3 will then examine the first of the problems listed above, the elemental operations problem. Section 4 will conclude with a few brief comments pertaining to the other problems.

2. Visual routines and their role in the processing of visual information

The purpose of this section is to examine how the application of visual routines fits within the overall processing of visual information. The main goal is to elaborate the relations between the initial creation of the early visual representations and the subsequent application of visual routines. The discussion is structured along the following lines.

The first half of this section examines the relation between visual routines and the creation of two types of visual representations: the bare representation (Section 2.1) that precedes the application of visual routines, and the incremental representations that are produced by them (Section 2.2). The second half examines two general problems raised by the nature of visual routines as described in the first half. These problems are: the initial selection of routines (Section 2.3) and the parallel processing of visual information (Section 2.4).

2.1. Visual routines and the base representations

In the scheme suggested above, the processing of visual information can be divided into two main stages. The first is the 'bottom-up' creation of some base representations by the early visual processes (Marr, 1980). The second stage is the application of visual routines. At this stage, procedures are applied to the base representations to define distinct entities within these representations, establish their shape properties, and extract spatial relations among them. In this section we shall examine more closely the distinction between these two stages.

2.1.1. The base representations

The first stage in the analysis of visual information can usefully be described as the creation of certain representations to be used by subsequent visual processes. Marr (1976) and Marr and Nishihara (1978) have suggested a division of these early representations into two types: the primal sketch, which is a representation of the incoming image, and the 2½-D sketch, which is a representation of the visible surfaces in three-dimensional space. The early visual representations share a number of fundamental characteristics: they are unarticulated, viewer-centered, uniform, and bottom-up driven. By 'unarticulated' I mean that they are essentially local descriptions that represent properties such as depth, orientation, color, and direction of motion at a point. The definition of larger more complicated units, and the extraction and description of spatial relationships among their parts, is not achieved at this level.

The base representations are spatially uniform in the sense that, with the exception of a scaling factor, the same properties are extracted and represented across the visual field (or throughout large parts of it). The descriptions of different points (e.g., the depth at a point) in the early representations are all with respect to the viewer, not with respect to one another. Finally, the construction of the base representations proceeds in a bottom-up fashion. This means that the base representations depend on the visual input alone.[7] If the same image is viewed twice, at two different times, the base representations associated with it will be identical.

2.1.2. Applying visual routines to the base representations

Beyond the construction of the base representations, the processing of visual information requires the definition of objects and parts in the scene, and the analysis of spatial properties and relations. The discussion in Section 1.3 concluded that for these tasks the uniform bottom-up computation is no longer possible, and suggested instead the application of visual routines. In contrast with the construction of the base representations, the properties and relations to be extracted are not determined by the input alone: for the same visual input different aspects will be made explicit at different times, depending on the goals of the computation. Unlike the base representations, the computations by visual routines are not applied uniformly over the visual field (e.g., not all of the possible inside/outside relations in the scene are computed), but only to selected objects. The objects and parts to which these computations apply are also not determined uniquely by the input alone; that is, there does not seem to be a universal set of primitive elements and relations that can be used for all possible perceptual tasks. The definition of objects and distinct parts in the input, and the relations to be computed among them may change with the situation. I may recognize a particular cat, for instance, using the shape of the white patch on its forehead. This does not imply, however, that the shapes of all the white patches in every possible scene and all the spatial relations in which such patches participate are universally made explicit in some internal representation. More generally, the definition of what constitutes a distinct part, and the relations to be established often depends on the particular object to be recognized. It is therefore unlikely that a fixed set of operations applied uniformly over the base representations would be sufficient to capture all of the properties and relations that may be relevant for subsequent visual analysis.[8] A final distinction between the two stages is that the construction of the base representations is fixed and unchanging, while visual routines are open-ended and permit the extraction of newly defined properties and relations.

In conclusion, it is suggested that the analysis of visual information divides naturally into two distinct successive stages: the creation of the base representations, followed by the application of visual routines to these representations. The application of visual routines can define objects within the base representations and establish properties and spatial relations that cannot be established within the base representations.

It should be noted that many of the relations that are established at this stage are defined not in the image but in three-dimensional space. Since the base representations already contain three-dimensional information, the visual routines applied to them can also establish properties and relations in three-dimensional space.[9]

---

[7] Although 'bottom–up' and 'top–down' processing are useful and frequently used terms, they lack a precise, well–accepted definition. As mentioned in the text, the definition I adopt is that bottom–up processing is determined entirely by the input. Top–down processing depends on additional factors, such as the goal of the computation (but not necessarily on object–specific knowledge).

Physiologically, various mechanisms that are likely to be involved in the creation of the base representation appear to be bottom–up: their responses can be predicted from the parameters of the stimulus alone. They also show strong similarity in their responses in the awake, anesthetized, and naturally sleeping animal (e.g., Livingstone and Hubel, 1981).

[8] The argument does not preclude the possibility that some grouping processes that help to define distinct parts and some local shape descriptions take place within the basic representations.

[9] Many spatial judgments we make depend primarily on three dimensional relations rather than on projected, two–dimensional ones (see e.g., Joynson and Kirk, 1960; Kappin and Fuqua, 1983). The implication is that various visual routines such as those used in comparing distances, operate upon a three–dimensional representation, rather than a representation that resembles the two–dimensional image.

2.2. *The incremental representations*

The creation of visual representations does not stop at the base representations. It is reasonable to expect that results established by visual routines are retained temporarily for further use. This means that in addition to the base representations to which routines are applied initially representations are also being created and modified in the course of executing visual routines. I shall refer to these additional structures as 'incremental representations', since their content is modified incrementally in the course of applying visual routines. Unlike the base representations, the incremental representations are not created in a uniform and unguided manner: the same input can give rise to different incremental representations, depending on the routines that have been applied.

The role of the incremental representations can be illustrated using the inside/outside judgments considered in Section 1. Suppose that following the response to an inside/outside display using a fairly complex figure, an additional point is lit up. The task is now to determine whether this second point lies inside or outside the closed figure. If the results of previous computations are already summarized in the incremental representation of the figure in question, the judgment in the second task would be expected to be considerably faster than the first, and the effects of the figure's complexity might be reduced.[10] Such facilitation effects would provide evidence for the creation of some internal structure in the course of reaching a decision in the first task that is subsequently used to reach a faster decision in the second task. For example, if area activation or 'coloring' is used to separate inside from outside, then following the first task the inside of the figure may be already 'colored'. If, in addition, this coloring is preserved in the incremental representation, then subsequent inside/outside judgments with respect to the same figure would require considerably less processing, and may depend less on the complexity of the figure.

This example also serves to illustrate the distinction between the base representations and the incremental representations. The 'coloring' of the curve in question will depend on the particular routines that happened to be employed. Given the same visual input but a different visual task, or the same task but applied to a different part of the input, the same curve will not be 'colored' and a similar saving in computation time will not be obtained. The general point illustrated by this example is that for a given visual stimulus but different computational goals the base representations remain the same, while the incremental representations would vary.

Various other perceptual phenomena can be interpreted in a similar manner in light of the distinction between the base and the incremental representations. I shall mention here only one recent example from a study by Rock and Gutman (1981). Their subjects were presented with pairs of overlapping red and green figures. When they were instructed to attend selectively to the green or red member of the pair, they were later able to recognize the 'attended' but not the 'unattended' figure. This result can be interpreted in terms of the distinction between the base and the incremental representations. The creation of the base representations is assumed to be a bottom-up process, unaffected by the goal of the computation. Consequently, the two figures would not be treated differently within these representations. Attempts to attend selectively to one sub-figure resulted in visual routines being applied preferentially to it. A detailed description of this sub-figure is consequently created in the incremental representations. This detailed description can then be used by subsequent routines subserving comparison and recognition tasks.

The creation and use of incremental representations imply that visual routines should not be thought of merely as predicates, or decision processes that supply 'yes' or 'no' answers. For example, an inside/outside routine does not merely signal 'yes' if an inside relation is established, and 'no' otherwise. In addition to the decision process, certain structures are being created during the execution of the routine. These structures are maintained in the incremental representation, and can be used in subsequent visual tasks. The study of a given routine is therefore not confined to the problem of how a certain decision is reached, but also includes the structures constructed by the routine in question in the incremental representations.

In summary, the use of visual routines introduces a distinction between two different types of visual representations: the base representations and incremental representations. The base representations provide the initial data structures on which the routines operate, and the incremental representations maintain results obtained by the application of visual routines.

The second half of Section 2 examines two general issues raised by the nature of visual routines as introduced so far. Visual routines were described above as sequences of elementary operations that are assembled to meet specific computational goals. A major problem that arises is the initial selection of routines to be applied. This problem is examined briefly in Section 2.3. Finally, sequential application of elementary operations seems to stand in contrast with the notion of parallel processing in visual perception. (Biederman *et al.*, 1973; Donderi and Zelnicker, 1969; Egeth *et al.*, 1972; Jonides and Gleitman, 1972; Neisser *et al.*, 1963). Section 2.4 examines the distinction

[10] This example is due to Steve Kosslyn. It is currently under empirical investigations.

between sequential and parallel processing, its significance to the processing of visual information, and its relation to visual routines.

2.3. Universal routines and the initial access problem

The act of perception requires more than the passive existence of a set of representations. Beyond the creation of the base representations, the perceptual process depends upon the current computational goal. At the level of applying visual routines, the perceptual activity is required to provide answers to queries, generated either externally or internally, such as: "is this my cat?" or, at a lower level, "is *A* longer than *B*?" Such queries arise naturally in the course of using visual information in recognition, manipulation, and more abstract visual thinking. In response to these queries routines are executed to provide the answers. The process of applying the appropriate routines is apparently efficient and smooth, thereby contributing to the impression that we perceive the entire image at a glance, when in fact we process only limited aspects of it at any given time. We may not be aware of the restricted processing since whenever we wish to establish new facts about the scene, that is, whenever an internal query is posed, an answer is provided by the execution of an appropriate routine.

Such application of visual routines raises the problem of guiding the perceptual activity and selecting the appropriate routines at any given instant. In dealing with this problem, several theories of perception have used the notion of schemata (Bartlett, 1932; Biederman *et al.*, 1973; Neisser, 1967) or frames (Minsky, 1975) to emphasize the role of expectations in guiding perceptual activity. According to these theories, at any given instant we maintain detailed expectations regarding the objects in view. Our perceptual activity can be viewed according to such theories as hypothesizing a specific object and then using detailed prior knowledge about this object in an attempt to confirm or refute the current hypothesis.

The emphasis on detailed expectations does not seem to me to provide a satisfactory answer to the problem of guiding perceptual activity and selecting the appropriate routines. Consider for example the 'slide show' situation in which an observer is presented with a sequence of unrelated pictures flashed briefly on a screen. The sequence may contain arbitrary ordinary objects, say, a horse, a beachball, a printed letter, etc. Although the observer can have no expectations regarding the next picture in the sequence, he will experience little difficulty identifying the viewed objects. Furthermore, suppose that an observer does have some clear expectations, e.g., he opens a door expecting to find his familiar office, but finds an ocean beach instead. The contradiction to the expected scene will surely cause a surprise, but no major perceptual difficulties. Although expectations can under some conditions facilitate perceptual processes significantly (e.g. Potter, 1975), their role is not indispensable. Perception can usually proceed in the absence of prior specific expectations and even when expectations are contradicted.

The selection of appropriate routines therefore raises a difficult problem. On the one hand, routines that establish properties and relations are situation-dependent. For example, the white patch on the cat's forehead is analyzed in the course of recognizing the cat, but white patches are not analyzed invariably in every scene. On the other hand, the recognition process should not depend entirely on prior knowledge or detailed expectations about the scene being viewed. How then are the appropriate routines selected?

It seems to me that this problem can be best approached by dividing the process of routine selection into two stages. The first stage is the application of what may be called *universal routines*. These are routines that can be usefully applied to any scene to provide some initial analysis. They may be able, for instance, to isolate some prominent parts in the scene and describe, perhaps crudely, some general aspects of their shape, motion, color, the spatial relations among them etc. These universal routines will provide sufficient information to allow initial indexing to a recognition memory, which then serves to guide the application of more specialized routines.

To make the motion of universal routines more concrete, I shall cite one example in which universal routines probably play a role. Studying the comparison of shapes presented sequentially, Rock *et al.* (1972) found that some parts of the presented shapes can be compared reliably while others cannot. When a shape was composed, for example, of a bounding contour and internal lines, in the absence of any specific instructions only the bounding contour was used reliably in the successive comparison task, even if the first figure was viewed for a long period (5 sec). This result would be surprising if only the base representations were used in the comparison task, since there is no reason to assume that in these representations the bounding contours of such line drawings enjoy a special status. It seems reasonable, however, that the bounding contour is special from the point of view of the universal routines, and is therefore analyzed first. If successive comparisons use the incremental representation as suggested above, then performance would be superior on those parts that have been already analyzed by visual routines. It is suggested, therefore, that in the absence of specific instructions, universal routines were applied first to the bounding contour. Furthermore, it appears that in the absence of specific goals, no detailed descriptions of the entire figure are generated even under long viewing periods. Only those aspects analyzed by the universal routines are summarized in the incremental representation. As a result, a description of the outside boundary alone has been created in the incremental representation. This description could then be compared against the second figure. It is of interest to note that the description generated in this task appears to be not just a coarse structural description of the figure, but has template-like quality that enable fine judgments of shape similarity.

These results can be contrasted with the study mentioned earlier by Rock and Gutman (1981) using pairs of overlapping and green figures. When subjects were instructed to "attend" selectively to one of the subfigures, they were subsequently able to make reliable shape comparisons to this, but not the other, subfigure. Specific requirements can therefore bias the selection and application of visual routines. Universal routines are meant to fill the void when no specific requirements are set. They are intended to acquire sufficient information to then determine the application of more specific routines.

For such a scheme to be of value in visual recognition, two interrelated requirements must be met. The first is that with universal routines alone it should be possible to gather sufficiently useful information to allow initial classification. The second requirement has to do with the organization of the memory used in visual recognition. It should contain intermediate constructs of categories that are accessible using the information gathered by the universal routines, and the access to such a category should provide the means for selecting specialized routines for refining the recognition process. The first requirement raises the question of whether universal routines, unaided by specific knowledge regarding the viewed objects, can reasonably be expected to supply sufficiently useful information about any viewed scene. The question is difficult to address in detail, since it is intimately related to problems regarding the structure of the memory used in visual recognition. It nonetheless seems plausible that universal routines may be sufficient to analyze the scene in enough detail to allow the application of specialized routines.

The potential usefulness of universal routines in the initial phases of the recognition process is supported in part by Marr and Nishihara's (1978) study of shape recognition. This work has demonstrated that at least for certain classes of shapes crude overall shape descriptions, which can be obtained by universal routines without prior knowledge regarding the viewed objects, can provide a powerful initial categorization. Similarly, the "perceptual 20 question game" of W. Richards (1982) suggests that a small fixed set of visual attributes (such as direction and type of motion, color, etc.) is often sufficient to form a good idea of what the object is (e.g., a walking person) although identifying a specific object (e.g., who the person is) may be considerably more difficult [cf. Milner, 1974]. These examples serve to illustrate the distinction in visual recognition between universal and specific stages. In the first, universal routines can supply sufficient information for accessing a useful general category. In the second, specific routines associated with this category can be applied.

Figure 6. *The routine processor acts as an intermediary between the visual representations and higher level components of the system.*

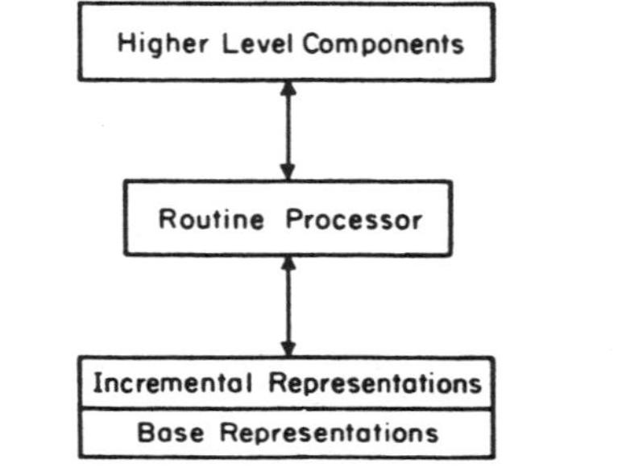

The relations between the different representations and routines can now be summarized as follows. The first stage in the analysis of the incoming visual input is the creation of the base representations. Next, visual routines are applied to the base representations. In the absence of specific expectations or prior knowledge universal routines are applied first, followed by the selective application of specific routines. Intermediate results obtained by visual routines are summarized in the incremental representation and can be used by subsequent routines.

2.3.1. Routines as intermediary between the base representations and higher-level components

The general role of visual routines in the overall processing of visual information as discussed so far is illustrated schematically in Fig. 6. The processes that assemble and execute visual routines (the 'routines processor' module in the figure) serve as an intermediary between the visual representations and higher level components of the system, such as recognition memory. Communication required between the higher level components and the visual representations for the analysis of shape and spatial relations are channeled via the routine processor.[11]

[11] Responses to certain visual stimuli that do not require abstract spatial analysis could bypass the routine processor. For example, a looming object may initiate an immediate avoidance response (Regan and Beverly, 1978). Such 'visual reflexes' do not require the application of visual routines. The visual system of lower animals such as insects or the frog, although remarkably sophisticated, probably lack routine mechanisms, and can perhaps be described as collections of 'visual reflexes'.

Visual routines operate in the middle ground that, unlike the bottom-up creation of the base representations, is a part of the top-down processing and yet is independent of object-specific knowledge. Their study therefore has the advantage of going beyond the base representations while avoiding many of the additional complications associated with higher level components of the system. The recognition of familiar objects, for example, often requires the use of knowledge specific to these objects. What we know about telephones or elephants can enter into the recognition process of these objects. In contrast, the extraction of spatial relations, while important for object recognition, is independent of object-specific knowledge. Such knowledge can determine the routine to be applied: the recognition of a particular object may require, for instance, the application of inside/outside routines. When a routine is applied, however, the processing is no longer dependent on object-specific knowledge.

It is suggested, therefore, that in studying the processing of visual information beyond the creation of the early representations, a useful distinction can be drawn between two problem areas. One can approach first the study of visual routines almost independently of the higher level components of the system. A full understanding of problems such as visually guided manipulation and object recognition would require, in addition, the study of higher level components, how they determine the application of visual routines, and how they are affected by the results of applying visual routines.

2.4. *Routines and the parallel processing of visual information*

A popular controversy in theories of visual perception is whether the processing of visual information proceeds in parallel or sequentially. Since visual routines are composed of sequences of elementary operations, they may seem to side strongly with the point of view of sequential processing in perception. In this section I shall examine two related questions that bear on this issue. First, whether the application of visual routines implies sequential processing. Second, what is the significance of the distinction between the parallel and sequential processing of visual information.

2.4.1. *Three types of parallelism*

The notion of processing visual information 'in parallel' does not have a unique, well-defined meaning. At least three types of parallelism can be distinguished in this processing: spatial, functional, and temporal. Spatial parallelism means that the same or similar operations are applied simultaneously to different spatial locations. The operations performed by the retina and the primary visual cortex, for example, fall under this category. Functional parallelism means that different computations are applied simultaneously to the same location. Current views of the visual cortex (e.g., Zeki, 1978a, b) suggest that different visual areas in the extra-striate cortex process different aspects of the input (such as color, motion, and stereoscopic disparity) at the same location simultaneously, thereby achieving functional parallelism.[12] Temporal parallelism is the simultaneous application of different processing stages to different inputs (this type of parallelism is also called 'pipelining'.[13]

Visual routines can in principle employ all three types of parallelism. Suppose that a given routine is composed of a sequence of operations O_1, O_2, ... O_n. Spatial parallelism can be obtained if a given operation O_i is applied simultaneously to various locations. Temporal parallelism can be obtained by applying different operations O_i simultaneously to successive inputs. Finally, functional parallelism can be obtained by the concurrent application of different routines.

The application of visual routines is thus compatible in principle with all three notions of parallelism. It seems, however, that in visual routines the use of spatial parallelism is more restricted than in the construction of the base representations.[14] At least some of the basic operations do not employ extensive spatial parallelism. The internal tracking of a discontinuity boundary in the base representation, for instance, is sequential in nature and does not apply to all locations simultaneously. Possible reasons for the limited spatial parallelism in visual routines are discussed in the next section.

2.4.2. *Essential and non-essential sequential processing*

When considering sequential *versus* spatially parallel processing, it is useful to distinguish between essential and non-essential sequentially. Suppose, for example, that O_1 and O_2 are two independent operations that can, in principle, be applied simultaneously. It is nevertheless still possible to apply them in sequence, but such sequentiality would be non-essential. The total computation required in this case will be the same regardless of whether the operations are performed in parallel or sequentially. Essential sequentiality, on the other hand, arises when the nature of the task makes parallel processing impossible or highly wasteful in terms of the overall computation required.

---

[12]Disagreements exist regarding this view, in particular, the role of area V4 in the rhesus monkey i processing color (Schein *et al.*, 1982). Although the notion o "one cortical area for each function" is too simplistic, the physiological data support in general the notion of functional parallelism.

[13]Suppose that a sequence of operations O_1, O_2 ... O_k is applied to each input in a temporal sequence I_1, I_2, I_3 .... First, O_1 is applied to I_1. Next, as O_2 is applied to I_1, O_1 can be applied to I_2. In general O_i, $1 < i < k$ can be applied simultaneously to I_{n-i}. Such a simultaneous application constitute temporal parallelism.

[14]The general notion of an extensively parallel stage followed by a more sequential one is in agreement with various findings and theories of visual perception (e.g., Estes, 1972; Neisser, 1967; Shiffrin *et al.*, 1976).

Problems pertaining to the use of spatial parallelism in the computation of spatial properties and relations were studied extensively by Minsky and Papert (1969) within the perceptrons model.[15] Minsky and Paper have established that certain relations, including the inside/outside relation, cannot be computed at all in parallel by any diamater-limited or order-limited perceptrons. This limitation does not seem to depend critically upon the perceptron-like decision scheme. It may be conjectured, therefore, that certain relations are inherently sequential in the sense that it is impossible or highly wasteful to employ extensive spatial parallelism in their computation. In this case sequentiality is essential, as it is imposed by the nature of the task, not by particular properties of the underlying mechanisms. Essential sequentiality is theoretically more interesting, and has more significant ramifications, than non-essential sequential ordering. In non-essential sequential processing the ordering has no particular importance, and no fundamentally new problems are introduced. Essential sequentiality, on the other hand, requires mechanisms for controlling the appropriate sequencing of the computation.

It has been suggested by various theories of visual attention that sequential ordering in perception is non-essential, arising primarily from a capacity limitation of the system (see, e.g., Holtzman and Gazzaniga, 1982; Kahneman, 1973; Rumelhart, 1970). In this view only a limited region of the visual scene (1 degree, Eriksen and Hoffman, 1972; see also Humphreys, 1981; Mackworth, 1965) is processed at any given time because the system is capacity-limited and would be overloaded by excessive information unless a spatial restruction is employed. The discussion above suggests, in contrast, that sequential ordering may in fact be essential, imposed by the inherently sequential nature of various visual tasks. This sequential ordering has substantial implications since it requires perceptual mechanisms for directing the processing and for concatenating and controlling sequences of basic operations.

Although the elemental operations are sequenced, some of them, such as the bounded activation, employ spatial parallelism and are not confined to a limited region. This spatial parallelism plays an important role in the inside/outside routines. To appreciate the difficulties in computing inside/outside relations without the benefit of spatial parallelism, consider solving a tactile version of the same problem by moving a cane or a fingertip over a relief surface. Clearly, when the processing is always limited to a small region of space, the task becomes considerably more difficult. Spatial parallelism must therefore play an important role in visual routines.

In summary, visual routines are compatible in principle with spatial, temporal, and functional parallelism. The degree of spatial parallelism employed by the basic operations seems nevertheless limited. It is conjectured that this reflects primarily essential sequentiality, imposed by the nature of the computations.

3. The elemental operations

3.1. Methodological considerations

In this section, we examine the set of basic operations that may be used in the construction of visual routines. In trying to explore this set of internal operations, at least two types of approaches can be followed. The first is the use of empirical psychological and physiological evidence. The second is computational: one can examine, for instance, the types of basic operations that would be useful in principle for establishing a large variety of relevant properties and relations. In particular, it would be useful to examine complex tasks in which we exhibit a high degree of proficiency. For such tasks, processes that match in performance the human system are difficult to devise. Consequently, their examination is likely to provide useful constraints on the nature of the underlying computations.

In exploring such tasks, the examples I shall use will employ schematic drawings rather than natural scenes. The reason is that simplified artificial stimuli allow more flexibility in adapting the stimulus to the operation under investigation. It seems to me that insofar as we examine visual tasks for which our proficiency is difficult to account for, we are likely to be exploring useful basic operations even if the stimuli employed are artificially constructed. In fact, this ability to cope efficiently with artificially imposed visual tasks underscores two essential capacities in the computation of spatial relations. First, that the computation of spatial relations is flexible and open-ended: new relations can be defined and computed efficiently. Second, it demonstrates our capacity to accept non-visual specification of a task and immediately produce a visual routine to meet these specifications.

The empirical and computational studies can then be combined. For example, the complexity of various visual tasks can be compared. That is, the theoretical studies can be used to predict how different tasks should vary in

---

[15] In the perceptron scheme the computation is performed in parallel by a large number of units ϕ_i. Each unit examine a restricted part of the 'retina' R. In a diamater-limited perceptron, for instance, the region examined by each unit is restricted to lie within a circle whose diameter is small compared to the size of R. The computation performed by each unit is a predicate of its inputs (i.e., $\phi_i = 0$ or $\phi_i = 1$). For example, a unit may be a 'corner detector' at a particular location, signalling 1 in the presence of a corner and 0 otherwise. All the local units then feed a final decision stage, assumed to be a linear threshold device. That is, it tests whether the weighted sum of the inputs $\Sigma_i\ \omega_i\ \phi_i$ exceeds a predetermined threshold θ.

complexity, and the predicted complexity measure can be gauged against human performance. We have seen in Section 1.2 an example along this line, in the discussion of the inside/outside computation. Predictions regarding relative complexity, success, and failure, based upon the ray-intersection method prove largely incompatible with human performance, and consequently the employment of this method by the human perceptual system can be ruled out. In this case, the refutation is also supported by theoretical considerations exposing the inherent limitations of the ray-intersection method.

In this section, only some initial steps towards examining the basic operations problem will be taken. I shall examine a number of plausible candidates for basic operations, discuss the available evidence, and raise problems for further study. Only a few operations will be examined; they are not intended to form a comprehensive list. Since the available empirical evidence is scant, the emphasis will be on computational considerations of usefulness. Finally, some of the problems associated with the assembly of basic operations into visual routines will be briefly discussed.

3.2. Shifting the processing focus

A fundamental requirement for the execution of visual routines is the capacity to control the location at which certain operations take place. For example, the operation of area activation suggested in Section 1.2 will be of little use if the activation starts simultaneously everywhere. To be of use, it must start at a selected location, or along a selected contour. More generally, in applying visual routines it would be useful to have a 'directing mechanism' that will allow the application of the same operation at different spatial locations. It is natural, therefore, to start the discussion of the elemental operations by examining the processes that control the locations at which these operations are applied.

Directing the processing focus (that is, the location to which an operation is applied) may be achieved in part by moving the eyes (Noton and Stark, 1971). But this is clearly insufficient: many relations, including, for instance, the inside/outside relation examined in Section 1.2, can be established without eye movements. A capacity to shift the processing focus internally is therefore required.

Problems related to the possible shift of internal operations have been studied empirically, both psychophysically and physiologically. These diverse studies still do not provide a complete picture of the shift operations and their use in the analysis of visual information. They do provide, however, strong support for the notion that shifts of the processing focus play an important role in visual information processing, starting from early processing stages. The main directions of studies that have been pursued are reviewed briefly in the next two sections.

3.2.1. Psychological evidence

A number of psychological studies have suggested that the focus of visual processing can be directed, either voluntarily or by manipulating the visual stimulus, to different spatial location in the visual input. They are listed below under three main categories.

The first line of evidence comes from reaction time studies suggesting that it takes some measurable time to shift the processing focus from one location to another. In a study by Eriksen and Schultz (1977), for instance, it was found that the time required to identify a letter increased linearly with the eccentricity of the target letter, the difference being on the order of 100 msec at 3° from the fovea center. Such a result may reflect the effect of shift time, but, as pointed out by Eriksen and Schultz, alternative explanations are possible.

More direct evidence comes from a study by Posner *et al.* (1978). In this study a target was presented seven degrees to the left or right of fixation. It was shown that if the subjects correctly anticipated the location at which the target will appear using prior cueing (an arrow at fixation), then their reaction time to the target in both detection and identification tasks were consistently lower (without eye movements). For simple detection tasks, the gain in detection time for a target at 70 eccentricity was on the order of 30 msec.

A related study by Tsal (1983) employed peripheral rather than central cueing. In his study a target letter could appear at different eccentricities, preceded by a brief presentation of a dot at the same location. The results were consistent with the assumption that the dot initiated a shift towards the cued location. If a shift to the location of the letter is required for its identification, the cue should reduce the time between the letter presentation and its identification. If the cue precedes the target letter by k msec, then by the time the letter appears the shift operation is already k msec under way, and the response time should decrease by this amount. The facilitation should therefore increase linearly with the temporal delay between the cue and target until the delay equals the total shift time. Further increase of the delay should have no additional effect. This is exactly what the experimental results indicated. It was further found that the delay at which facilitation saturates (presumably the total shift time) increases with eccentricity, by about 8 msec on the average per 1° of visual angle.

A second line of evidence comes from experiments suggesting that visual sensitivity at different locations can be somewhat modified with a fixed eye position. Experiments by Shulman *et al.* (1979) can be interpreted as indicating that a region of somewhat increased sensitivity can be shifted across the visual field. A related experiment by Remington (1978, described in Posner, 1980), showed an increase in sensitivity at a distance of 8° from the fixation point 50–100 msec after the location has been cued.

A third line of evidence that may bear on the internal shift operations comes from experiments exploring the selective readout from some form of short term visual memory (e.g., Shiffrin *et al.*, 1976; Sperling, 1960). These experiments suggest that some internal scanning can be directed to different locations a short time after the presentation of a visual stimulus.

The shift operation and selective visual attention. Many of the experiments mentioned above were aimed at exploring the concept of 'selective attention'. This concept has a variety of meanings and connotations (cf. Estes, 1972), many of which are not related directly to the proposed shift of processing focus in visual routines. The notion of selective visual attention often implies that the processing of visual information is restricted to small region of space, to avoid 'overloading' the system with excessive information. Certain processing stages have, according to this description, a limited total 'capacity' to invest in the processing, and this capacity can be concentrated in a spatially restricted region. Attempts to process additional information would detract from this capacity, causing interference effects and deterioration of performance. Processes that do not draw upon this general capacity are, by definition, pre-attentive. In contrast, the notion of processing shift discussed above stems from the need for spatially-structured processes, and it does not necessarily imply such notions as general capacity or protection from overload. For example, the 'coloring' operation used in Section 1.2 for separating inside from outside started from a selected point or contour. Even with no capacity limitations such coloring would not start simultaneously everywhere, since a simultaneous activation will defy the purpose of the coloring operation. The main problem in this case is in coordinating the process, rather than excessive capacity demands. As a result, the process is spatially structured, but not in a simple manner as in the 'spotlight model' of selective attention. In the course of applying a visual routine, both the locations and the operations performed at the selected locations are controlled and coordinated according to the requirement of the routine in question.

Many of the results mentioned above are nevertheless in agreement with the possible existence of a directable processing focus. They suggest that the redirection of the processing focus to a new location may be achieved in two ways. The experiments of Posner and Shulman *et al.* suggest that it can be 'programmed' to move along a straight path using central cueing. In other experiments, such as Remmington's and Tsal's, the processing focus is shifted by being attracted to a peripheral cue.

3.2.2. Physiological evidence

Shift-related mechanisms have been explored physiologically in the monkey in a number of different visual areas: the superior colliculus, the posterior parietal lobe (area 7) the frontal eye fields, areas V1, V2, V4, MT, MST, and the inferior temporal lobe.

In the superficial layers of the superior colliculus of the monkey, many cells have been found to have an enhanced response to a stimulus when the monkey uses the stimulus as a target for a saccadic eye movement (Goldberg and Wurtz, 1972). This enhancement is not strictly sensory in the sense that it is not produced if the stimulus is not followed by a saccade. It also does not seem strictly associated with a motor response, since the temporal delay between the enhanced response and the saccade can vary considerably (Wurtz and Mohler, 1976a). The enhancement phenomenon was suggested as a neural correlate of "directing visual attention", since it modifies the visual input and enhances it at selective locations when the sensory input remains constant (Goldberg and Wurtz, *op. cit.*). The intimate relation of the enhancement to eye movements, and its absence when the saccade is replaced by other responses (Wurtz and Mohler, *op. cit.*, Wurtz *et al.*, 1982) suggest, however, that this mechanism is specifically related to saccadic eye movements rather than to operations associated with the shifting of an internal processing focus. Similar enhancement that depends on saccade initiation to a visual target has also been described in the frontal eye fields (Wurtz and Mohler, 1976b) and in prestriate cortex, probably area V4 (Fischer and Boch, 1981).

Another area that exhibits similar enhancement phenomena, but not exclusively to saccades, is area 7 of the posterior parietal lobe of the monkey. Using recordings from behaving monkeys, Mountcastle and his collaborators (Mountcastle, 1976, Mountcastle *et al.*, 1975) found three populations of cells in area 7 that respond selectively (i) when the monkey fixates an object of interest within its immediate surrounding (fixation neurons), (ii) when it tracks an object of interest (tracking neurons), and (iii) when it saccades to an object of interest (saccade neurons). (Tracking neurons were also described in area MST (Newsome and Wurtz, 1982).) Studies by Robinson *et al.* (1978) indicated that all of these neurons can also be driven by passive sensory stimulation, but their response is considerably enhanced when the stimulation is 'selected' by the monkey to initiate a response. On the basis of such findings it was suggested by Mountcastle (as well as by Posner, 1980; Robinson *et al.*, 1978; Wurtz *et al.*, 1982) that mechanisms in area 7 are responsible for "directing visual attention" to selected stimuli. These mechanisms may be primarily related, however, to tasks requiring hand-eye coordination for manipulation in the reachable space (Mountcastle, 1976), and there is at present no direct evidence to link them with visual routines and the shift of processing focus discussed above.[16]

In area TE of the inferotemporal cortex units were found whose responses depend strongly upon the visual task performed by the animal. Fuster and Jervey (1981) described units that responded strongly to the stimulus' color, but only when color was the relevant parameter in a matching task. Richmond and Sato (1982) found units whose responses to a given stimulus were enhanced when the stimulus was used in a pattern discrimination task, but not in other tasks (e.g., when the stimulus was monitored to detect its dimming).

In a number of visual areas, including V1, V2, and MT, enhanced responses associated with performing specific visual tasks were not found (Newsome and Wurtz, 1982; Wurtz *et al.*, 1982). It remains possible, however, that task-specific modulation would be observed when employing different visual tasks. Finally, responses in the pulvinar (Gattas *et al.*, 1979) were shown to be strongly modulated by attentional and situational variables. It remains unclear, however, whether these modulations are localized (i.e., if they are restricted to a particular location in the visual field) and whether they are task-specific.

Physiological evidence of a different kind comes from visual evoked potential (VEP) studies. With fixed visual input and in the absence of eye movements, changes in VEP can be induced, for example, by instructing the subject to "attend" to different spatial locations (e.g., van Voorhis and Hillyard, 1977). This evidence may not be of direct relevance to visual routines, since it is not clear whether there is a relation between the voluntary 'direction of visual attention' used in these experiments and the shift of processing focus in visual routines. VEP studies may nonetheless provide at least some evidence regarding the possibility of internal shift operations.

In assessing the relevance of these physiological findings to the shifting of the processing focus it would be useful to distinguish three types of interactions between the physiological responses and the visual task performed by the experimental animal. The three types are task-dependent, task-location dependent, and location-dependent responses.

A response is task-dependent if, for a given visual stimulus, it depends upon the visual task being performed. Some of the units described in area TE, for instance, are clearly task-dependent in this sense. In contrast, units in area V1 for example, appear to be task-independent. Task-dependent responses suggest that the units do not belong to the bottom-up generation of the early visual representations, and that they may participate in the application of visual routines. Task-dependence by itself does not necessarily imply, however, the existence of shift operations. Of more direct relevance to shift operations are responses that are both task- and location-dependent. A task-location dependent unit would respond preferentially to a stimulus when a given task is performed at a given location. Unlike task-dependent units, it would show a different response to the same stimulus when an identical task is applied to a different location. Unlike the spotlight metaphor of visual attention, it would show different responses when different tasks are performed at the same locations.

There is at least some evidence for the existence of such task-location dependent responses. The response of a saccade neuron in the superior colliculus, for example, is enhanced only when a saccade is initiated in the general direction of the unit's receptive field. A saccade towards a different location would not produce the same enhancement. The response is thus enhanced only when a specific location is selected for a specific task.

Unfortunately, many of the other task-dependent responses have not been tested for location specificity. It would be of interest to examine similar task-location dependence in tasks other than eye movement, and in the visual cortex rather than the superior colliculus. For example, the units described by Fuster and Jervey (1981) showed task-dependent response (responded strongly during a color matching task, but not during a form matching task). It would be interesting to know whether the enhanced response is also location specific. For example, if during a color matching task, when several stimuli are presented simultaneously, the response would be enhanced only at the location used for the matching task.

Finally, of particular interest would be units referred to above as location-dependent (but task-independent). Such a unit would respond preferentially to a stimulus when it is used not in a single task but in a variety of different visual tasks. Such units may be a part of a general 'shift controller' that selects a location for processing independent of the specific operation to be applied. Of the areas discussed above, the responses in area 7, the superior colliculus, and TE, do not seem appropriate for such a 'shift controller'. The pulvinar remains a possibility worthy of further exploration in view of its rich pattern of reciprocal and orderly connections with a variety of visual areas (Beneveneto and Davis, 1977; Rezak and Beneveneto, 1979).

[16]A possible exception is some preliminary evidence by Robinson *et al.* (1978) suggesting that, unlike the superior colliculus, enhancement effects in the parietal cortex may be dissociated from movement. That is, a response of a cell may be facilitated when the animal is required to attend to a stimulus even when the stimulus is not used as a target for hand or eye movement.

3.3. Indexing

Computational considerations strongly suggest the use of internal shifts of the processing focus. This notion is supported by psychological evidence, and to some degree by physiological data.

The next issue to be considered is the selection problem: how specific locations are selected for further processing. There are various manners in which such a selection process could be realized. On a digital computer, for instance, the selection can take place by providing the coordinates of the next location to be processed. The content of the specified address can then be inspected and processed. This is probably not how locations are being selected for processing in the human visual system. What determines, then, the next location to be processed, and how is the processing focus moved from one location to the next?

In this section we shall consider one operation which seems to be used by the visual system in shifting the processing focus. This operation is called 'indexing'. It can be described as a shift of the processing focus to special 'odd-man-out' locations. These locations are detected in parallel across the base representations, and can serve as 'anchor points' for the application of visual routines.

As an example of indexing, suppose that a page of printed text is to be inspected for the occurrence of the letter 'A'. In a background of similar letters, the 'A' will not stand out, and considerable scanning will be required for its detection (Nickerson, 1966). If, however, all the letters remain stationary with the exception of one which is jiggled, or if all the letters are red with the exception of one green letter, the odd-man-out will be immediately identified.

The identification of the odd-man-out items proceeds in this case in several stages.[17] First the odd-man-out location is detected on the basis of its unique motion or color properties. Next, the processing focus is shifted to this odd-man-out location. This is the indexing stage. As a result of this stage, visual routines can be applied to the figure. By applying the appropriate routines, the figure is identified.

Indexing also played a role in the inside/outside example examined in Section 1.2. It was noted that one plausible strategy is to start the processing at the location marked by the X figure. This raises a problem, since the location of the X and of the closed curve were not known in advance. If the X can define an indexable location, that is, if it can serve to attract the processing focus, then the execution of the routine can start at that location. More generally, indexable locations can serve as starting points or 'anchors' for visual routines. In a novel scene, it would be possible to direct the processing focus immediately to a salient indexable item, and start the processing at that location. This will be particularly valuable in the execution of universal routines that are to be applied prior to any analysis of the viewed objects.

The indexing operation can be further subdivided into three successive stages. First, properties used for indexing, such as motion, orientation, and color, must be computed across the base representations. Second, an 'odd-man-out operation' is required to define locations that are sufficiently different from their surroundings. The third and final stage is the shift of the processing focus to the indexed location. These three stages are examined in turn in the next three subsections.

3.3.1. Indexable properties

Certain odd-man-out items can serve for immediate indexing, while others cannot. For example, orientation and direction of motion are indexable, while a single occurrence of the letter 'A' among similar letters does not define an indexable location. This is to be expected, since the recognition of letters requires the application of visual routines while indexing must precede their application. The first question that arises, therefore, is what the set of elemental properties is that can be computed everywhere across the base representations prior to the application of visual routines.

One method of exploring indexable properties empirically is by employing an odd-man-out test. If an item is singled out in the visual field by an indexable property, then its detection is expected to be immediate. The ability to index an item by its color, for instance, implies that a red item in a field of green items should be detected in roughly constant time, independent of the number of green distractors.

Using this and other techniques, A. Treisman and her collaborators (Treisman, 1977; Treisman and Gelade, 1980; see also Beck and Ambler, 1972, 1973; Pomerantz *et al.*, 1977) have shown that color and simple shape parameters can serve for immediate indexing. For example, the time to detect a target blue X in a field of brown T's and green X's does not change significantly as the number of distractors is increased (up to 30 in these experiments). The target is immediately indexable by its unique color. Similarly, a target green S letter is detectable in a field of brown T's and green X's in constant time. In this case it is probably indexable by certain shape parameters, although it cannot be determined from the experiments what the relevant parameters are. Possible candidates include (i) curvature, (ii) orientation, since the S contains some orientations that are missing in the X and

[17]The reasons for assuming several stages are both theoretical and empirical. On the empirical side, the experiments by Posner, Treisman, and Tsal provide support for this view.

T, and (iii) the number of terminators, which is two for the S, but higher for the X and T. It would be of interest to explore the indexability of these and other properties in an attempt to discover the complete set of indexable properties.

The notion of a severely limited set of properties that can be processed 'pre-attentively' agrees well with Julesz' studies of texture perception (see Julesz (1981) for a review). In detailed studies, Julesz and his collaborators have found that only a limited set of features, which he termed 'textons', can mediate immediate texture discrimination. These textons include color, elongated blobs of specific sizes, orientations, and aspect ratios, and the terminations of these elongated blobs.

These psychological studies are also in general agreement with physiological evidence. Properties such as motion, orientation, and color, were found to be extracted in parallel by units that cover the visual field. On physiological grounds these properties are suitable, therefore, for immediate indexing.

The emerging picture is, in conclusion, that a small number of properties are computed in parallel over the base representations prior to the application of visual routines, and represented in ordered retinotopic maps. Several of these properties are known, but a complete list is yet to be established. The results are then used in a number of visual tasks including, probably, texture discrimination, motion correspondence, stereo, and indexing.

3.3.2. Defining an indexable location

Following the initial computation of the elementary properties, the next stage in the indexing operation requires comparisons among properties computed at different locations to define the odd-man-out indexable locations.

Psychological evidence suggests that only simple comparisons are used at this stage. Several studies by Treisman and her collaborators examined the problem of whether different properties measured at a given location can be combined prior to the indexing operation.[18] They have tested, for instance, whether a green T could be detected in a field of brown T's and Green X's. The target in this case matches half the distractors in color, and the other half in shape. It is the combination of shape and color that makes it distinct. Earlier experiments have established that such a target is indexable if it has a unique color or shape. The question now was whether the conjunction of two indexable properties is also immediately indexable. The empirical evidence indicates that items cannot be indexed by a conjunction of properties: the time to detect the target increases linearly in the conjunction task with the number of distractors. The results obtained by Treisman *et al.* were consistent with a serial self-terminating search in which the items are examined sequentially until the target is reached.

The difference between single and double indexing supports the view that the computations performed in parallel by the distributed local units are severely limited. In particular, these units cannot combine two indexable properties to define a new indexable property. In a scheme where most of the computation is performed by a directable central processor, these results also place constraints on the communication between the local units and the central processor. The central processor is assumed to be computationally powerful, and consequently it can also be assumed that if the signals relayed to it from the local units contained sufficient information for double indexing, this information could have been put to use by the central processor. Since it is not, the information relayed to the central processor must be limited.

The results regarding single and double indexing can be explained by assuming that the local computation that precedes indexing is limited to simple local comparisons. For example, the color in a small neighborhood may be compared with the color in a surrounding area, employing, perhaps, lateral inhibition between similar detectors (Estes, 1972; Andriessen and Bouma, 1976; Pomerantz *et al.*, 1977). If the item differs significantly from its surround, the difference signal can be used in shifting the processing focus to that location. If an item is distinguishable from its surround by the conjunction of two properties such as color and orientation, then no difference signal will be generated by either the color or the orientation comparisons, and direct indexing will not be possible. Such a local comparison will also allow the indexing of a local, rather than a global, odd-man-out. Suppose, for example, that the visual field contains green and red elements in equal numbers, but one and only one of the green elements is completely surrounded by a large region of red elements. If the local elements signaled not their colors but the results of local color comparisons, then the odd-man-out alone would produce a difference signal and would therefore be indexable. To explore the computations performed at the distributed stage it would be of interest, therefore, to examine the indexability of local odd-men-out. Various properties can be tested, while manipulating the size and shape of the surrounding region.

3.3.3. Shifting the processing focus to an indexable location

The discussion so far suggests the following indexing scheme. A number of elementary properties are computed in parallel across the visual field. For each property, local comparisons are performed everywhere. The resulting difference signals are combined somehow to produce a final odd-man-out signal at each location. The processing focus then shifts to the location of the strongest signal. This final shift operation will be examined next.

[18]Triesman's own approach to the problem was somewhat different from the one discussed here.

Several studies of selective visual attention likened the internal shift operation to the directing of a spotlight. A directable spotlight is used to 'illuminate' a restricted region of the visual field, and only the information within the region can be inspected. This is, of course, only a metaphor that still requires an agent to direct the spotlight and observe the illuminated region. The goal of this section is to give a more concrete notion of the shift in processing focus, and, using a simple example, to show what it means and how it may be implemented.

The example we shall examine is a version of the property-conjunction problem mentioned in the previous section. Suppose that small colored bars are scattered over the visual field. One of them is red, all the others are green. The task is to report the orientation of the red bar. We would like therefore to 'shift' the processing focus to the red bar and 'read out' its orientation.

A simplified scheme for handling this task is illustrated schematically in Fig. 7. This scheme incorporates the first two stages in the indexing operation discussed above. In the first stage (*S1* in the figure) a number of different properties (denoted by P_1, P_2, P_3 in the figure) are being detected at each location. The existence of a horizontal green bar, for example, at a given location, will be reflected by the activity of the color- and orientation-detecting units at that location. In addition to these local units there is also a central common representation of the various properties, denoted by CP_1, CP_2, CP_3, in the figure. For simplicity, we shall assume that all of the local detectors are connected to the corresponding unit in the central representation. There is, for instance, a common central unit to which all of the local units that signal vertical orientation are connected.

Figure 7. *A simplified scheme that can serve as a basis for the indexing operation. In the first stage (S_1), a number of properties (P_1, P_2, P_3 in figure) are detected everywhere. In the subsequent stage (S_2), local comparisons generate difference signals. The element generating the strongest signal is mapped onto the central common representations (CP_1, CP_2, CP_3).*

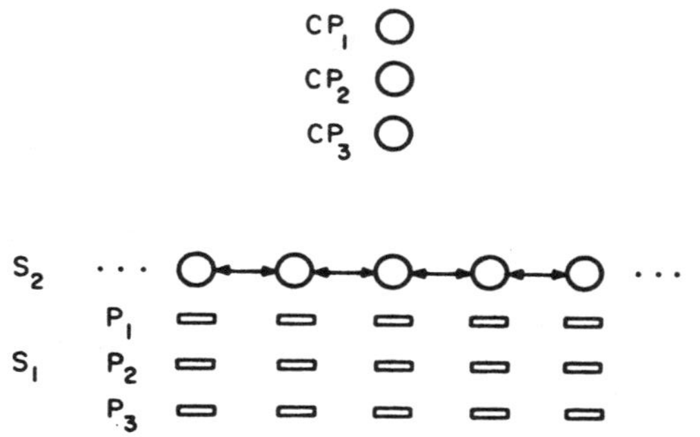

It is suggested that to perform the task defined above and determine the orientation of the red bar, this orientation must be represented in the central common representation. Subsequent processing stages have access to this common representation, but not to all of the local detectors. To answer the question, "what is the orientation of the red element", this orientation alone must therefore be mapped somehow into the common representation.

In section 3.3.2, it was suggested that the initial detection of the various local properties is followed by local comparisons that generate difference signals. These comparisons take place in stage *S2* in Fig. 7, where the odd-man-out item will end up with the strongest signal. Following these two initial stages, it is not too difficult to conceive of mechanisms by which the most active unit in *S2* would inhibit all the others, and as a result the properties of all but the odd-man-out location would be inhibited from reaching the central representation.[19] The central representations would then represent faithfully the properties of the odd-man-out item, the red bar in our example. At this stage the processing is focused on the red element and its properties are consequently represented explicitly in the central representation, accessible to subsequent processing stages. The initial question is thereby answered, without the use of a specialized vertical red line detector.

In this scheme, only the properties of the odd-man-out item can be detected immediately. Other items will have to await additional processing stages. The above scheme can be easily extended to generate successive 'shifts of the processing focus' from one element to another, in an order that depends on the strength of their signals in *S2*. These successive shifts mean that the properties of different elements will be mapped successively onto the common representations.

Possible mechanisms for performing indexing and processing focus shifts would not be considered here beyond the simple scheme discussed so far. But even this simplified scheme illustrates a number of points regarding shift and indexing. First, it provides an example for what it means to shift the processing focus to a given location. In this case, the shift entailed a selective

[19]Models for this stage are being tested by C. Koch at the M.I.T. A.I. Lab. One interesting result from this modeling is that a realization of the inhibition among units leads naturally to the processing focus being shifted continuously from item to item rather than 'leaping', disappearing at one location and reappearing at another. The models also account for the phenomenon that being an odd-man-out is not a simple all or none property (Engel, 1974). With increased dissimilarity, a target item can be detected immediately over a larger area.

readout to the central common representations. Second, it illustrates that shift of the processing focus can be achieved in a simple manner without physical shifts or an internal 'spotlight'. Third, it raises the point that the shift of the processing focus is not a single elementary operation but a family of operations, only some of which were discussed above. There is, for example, some evidence for the use of 'similarity enhancement': when the processing focus is centered on a given item, similar items nearby become more likely to be processed next. There is also some degree of 'central control' over the processing focus. Although the shift appears to be determined primarily by the visual input, there is also a possibility of directing the processing focus voluntarily, for example to the right or to the left of fixation (van Voorhis and Hillyard, 1977).

Finally, it suggests that psychophysical experiments of the type used by Julesz, Treisman and others, combined with physiological studies of the kind described in Section 3.2, can provide guidance for developing detailed testable models for the shift operations and their implementation in the visual system.

In summary, the execution of visual routines requires a capacity to control the locations at which elemental operations are applied. Psychological evidence, and to some degree physiological evidence, are in agreement with the general notion of an internal shift of the processing focus. This shift is obtained by a family of related processes. One of them is the indexing operation, which directs the processing focus towards certain odd-man-out locations. Indexing requires three successive stages. First, a set of properties that can be used for indexing, such as orientation, motion, and color, are computed in parallel across the base representation. Second, a location that differs significantly from its surroundings in one of these properties (but not their combinations) can be singled out as an indexed location. Finally, the processing focus is redirected towards the indexed location. This redirection can be achieved by simple schemes of interactions among the initial detecting units and central common representations that lead to a selective mapping from the initial detectors to the common representations.

3.4. Bounded activation (coloring)

The bounded activation, or 'coloring' operation, was suggested in Section 1.2. in examining the inside/outside relation. It consisted of the spread of activation over a surface in the base representation emanating from a given location or contour, and stopping at discontinuity boundaries.

The results of the coloring operation may be retained in the incremental representation for further use by additional routines. Coloring provides in this manner one method for defining larger units in the unarticulated base representations: the 'colored' region becomes a unit to which routines can be applied selectively. A simple example of this possible role of the coloring operation was mentioned in Section 2.2: the initial 'coloring' could facilitate subsequent inside/outside judgments.

A more complicated example along the same line is illustrated in Fig. 8. The visual task here is to identify the sub-figure marked by the black dot. One may have the subjective feeling of being able to concentrate on this sub-figure, and 'pull it out' from its complicated background. This capacity to 'pull out' the figure of interest can also be tested objectively, for example, by testing how well the sub-figure can be identified. It is easily seen in Fig. 8 that the marked sub-figure has the shape of the letter G. The area surrounding the sub-figure in close proximity contains a myriad of irrelevant features, and therefore identification would be difficult, unless processing can be directed to this sub-figure.

The sub-figure of interest in Fig. 8 is the region inside which the black dot resides. This region could be defined and separated from its surroundings by using the area activation operation. Recognition routines could then concentrate on the activated region, ignoring the irrelevant contours. This examples uses an artificial stimulus, but the ability to identify a region and process it selectively seems equally useful for the recognition of objects in natural scenes.

3.4.1. Discontinuity boundaries for the coloring operation

The activation operation is supposed to spread until a discontinuity bound-

Figure 8. *The visual task here is to identify the subfigure containing the black dot. This figure (the letter 'G') can be recognized despite the presence of confounding features in close proximity to its contours, the capacity to 'pull out' the figure from the irrelevant background may involve the bounded activation operation.*

ary is reached. This raises the question of what constitutes a discontinuity boundary for the activation operation. In Fig. 8, lines in the two-dimensional drawing served for this task. If activation is applied to the base representations discussed in Section 2, it is expected that discontinuities in depth, surface orientation, and texture, will all serve a similar role. The use of boundaries to check the activation spread is not straightforward. It appears that in certain situations the boundaries do not have to be entirely continuous in order to block the coloring spread. In Fig. 9, a curve is defined by a fragmented line, but it is still immediately clear that the X lies inside and the black dot outside this curve.[20] If activation is to be used in this situation as well, then incomplete boundaries should have the capacity to block the activation spread. Finally, the activation is sometimes required to spread across certain boundaries. For example, in Fig. 10, which is similar to Fig. 8, the letter G is still recognizable, in spite of the internal bounding contours. To allow the coloring of the entire sub-figure in this case, the activation must spread across internal boundaries.

In conclusion, the bounded activation, and in particular, its interactions with different contours, is a complicated process. It is possible that as far as the activation operation is concerned, boundaries are not defined universally, but may be defined somewhat differently in different routines.

3.4.2. A mechanism for bounded activation and its implications

The 'coloring' spread can be realized by using only simple, local operations. The activation can spread in a network in which each element excites all of its neighbors.

Figure 9. *Fragmented boundaries. The curve is defined by a dashed line, but inside/outside judgments are still immediate.*

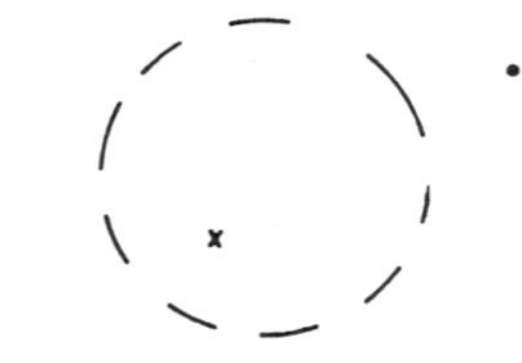

[20]Empirical results show that inside/outside judgments using dashed boundaries require somewhat longer times compared with continuous curves, suggesting that fragmented boundaries may require additional processing. The extra cost associated with fragmental boundaries is small. In a series of experiments performed by J. Varanese at Harvard University this cost averaged about 20 msec. The mean response time was about 540 msec (Varanese, 1983).

Figure 10. *Additional internal lines are introduced into the G-shaped subfigure. If bounded activation is used to 'color' this figure, it must spread across the internal contours.*

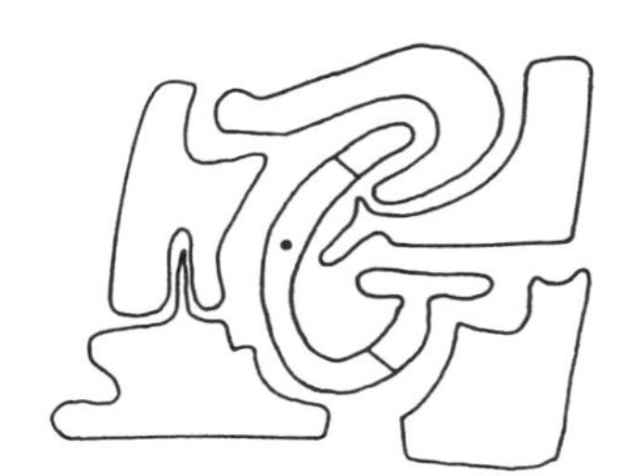

A second network containing a map of the discontinuity boundaries will be used to check the activation spread. An element in the activation network will be activated if any of its neighbors is turned on, provided that the corresponding location in the second, control network, does not contain a boundary. The turning on of a single element in the activation network will thus initiate an activation spread from the selected point outwards, that will fill the area bounded by the surrounding contours. (Each element may also have neighborhoods of different sizes, to allow a more efficient, multi-resolution implementation.)

In this scheme, an 'activity layer' serves for the execution of the basic operation, subject to the constraints in a second 'control layer'. The control layer may receive its content (the discontinuity boundaries) from a variety of sources, which thereby affect the execution of the operation.

An interesting question to consider is whether the visual system incorporates mechanisms of this general sort. If this were the case, the interconnected network of cells in cortical visual areas may contain distinct subnetworks for carrying out the different elementary operations. Some layers of cells within the retinotopically organized visual areas would then be best understood as serving for the execution of basic operations. Other layers receiving their inputs from different visual areas may serve in this scheme for the control of these operations.

If such networks for executing and controlling basic operations are incorporated in the visual system, they will have important implications for the interpretation of physiological data. In exploring such networks, physiological studies that attempt to characterize units in terms of their optimal stimuli would run into difficulties. The activity of units in such networks would be better understood not in terms of high-order features extracted by the units, but in terms of the basic operations performed by the networks. Elucidating the basic operations would therefore provide clues for understanding the activity in such networks and their patterns of interconnections.

3.5. *Boundary tracing and activation*

Since contours and boundaries of different types are fundamental entities in visual perception, a basic operation that could serve a useful role in visual routines is the tracking of contours in the base representation. This section examines the tracing operation in two parts. The first shows examples of boundary tracing and activation and their use in visual routines. The second examines the requirements imposed by the goal of having a useful, flexible, tracing operation.

3.5.1. *Examples of tracing and activation*

A simple example that will benefit from the operation of contour tracing is the problem of determining whether a contour is open or closed. If the contour is isolated in the visual field, an answer can be obtained by detecting the presence or absence of contour terminators. This strategy would not apply, however, in the presence of additional contours. This is an example of the 'figure in a context' problem (Minsky and Papert, 1969): figural properties are often substantially more difficult to establish in the presence of additional context. In the case of open and closed curves, it becomes necessary to relate the terminations to the contour in question. The problem can be solved by tracing the contour and testing for the presence of termination points on that contour.

Another simple example which illustrates the role of boundary tracing is shown in Fig. 11. The question here is whether there are two X's lying on a common curve. The answer seems immediate and effortless, but how is it achieved? Unlike the detection of single indexable items, it cannot be mediated by a fixed array of two-X's-on-a-curve detectors. Instead, I suggest that this simple perception conceals, in fact, an elaborate chain of events. In response to the question, a routine has been compiled and executed. An appropriate routine can be constructed if the repertoire of basic operations included the indexing of the X's and the tracking of curves. The tracking provides in this task a useful identity, or 'sameness' operator: it serves to verify that the two X figures are marked on the same curve, and not on two disconnected curves.

This task has been investigated recently by Jolicoeur *et al.* (1984, Reference note 1) and the results strongly supported the use of an internal contour tracing operation. Each display in this study contained two separate curves. In all trials there was an X at the fixation point, intersecting one of the curves. A second X could lie either on the same or on the second curve, and the observer's task was to decide as quickly as possible whether the two X's lay on the same or different curves. The physical distance separating the two X's was always 1.8° of visual angle. When the two X's lay on the same curve, their distance along the curve could be changed, however, in increments of 2.2° of visual angle (measured along the curve).

The main result from a number of related experiments was that the time to detect that the two X's lay on the same curve increased monotonically, and roughly linearly, with their separation along the curve. This result suggests the use of a tracing operation, at an average speed of about 24 msec per degree of visual angle. The short presentation time (250 msec) precluded the tracing of the curve using eye movements, hence the tracing operation must be performed internally.

Although the task in this experiment apparently employed a rather elaborate visual routine, it nevertheless appeared immediate and effortless. Response times were relatively short, about 750 msec for the fastest condition. When subjects were asked to describe how they performed the task, the main response was that the two X's were "simply seen" to lie on either the same curve or on different curves. No subject reported any scanning along a curve before making a decision.

The example above employed the tracking of a single contour. In other cases, it would be advantageous to activate a number of contours simultaneously. In Fig. 12a, for instance, the task is to establish visually whether there is a path connecting the center of the figure to the surrounding contour. The solution can be obtained effortlessly by looking at the figure, but again, it must involve in fact a complicated chain of processing. To cope with this

Figure 11. *The task here is to determine visually whether the two X's lie on the same curve. This simple task requires in fact complex processing that probably includes the use of a contour tracing operation.*

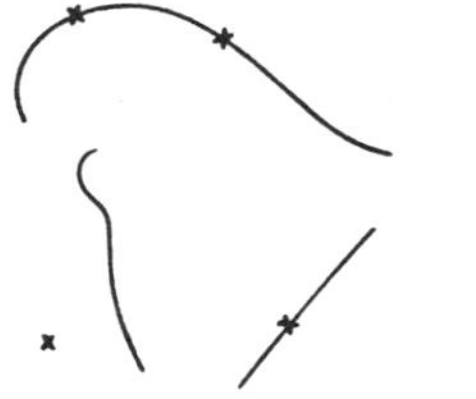

Figure 12. *The task in* a *is to determine visually whether there is a path connecting the center of the figure to the surrounding circle. In* b *the solution is labeled. The interpretation of such labels relys upon a set of common natural visual routines.*

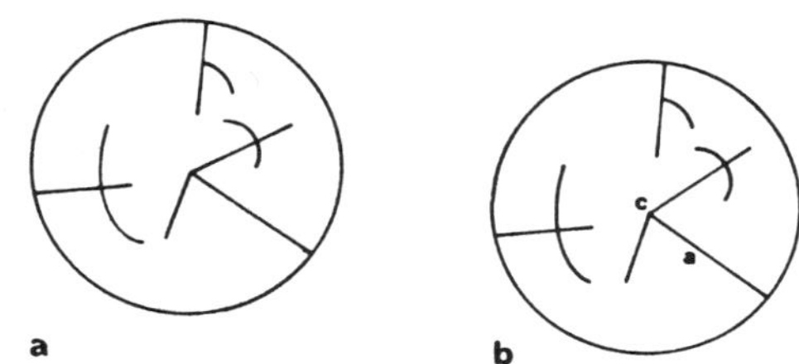
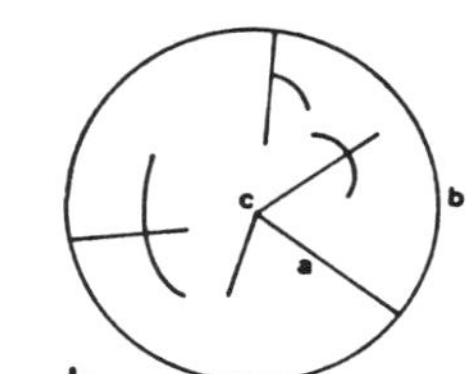

seemingly simple problem, visual routine must (i) identify the location referred to as "the center of the figure", (ii) identify the outside contour, and (iii) determine whether there is a path connecting the two. (It is also possible to proceed from the outside inwards.) By analogy with the area activation, the solution can be found by activating contours at the center point and examining the activation spread to the periphery. In Fig. 12b, the solution is labeled: the center is marked by the letter *c*, the surrounding boundary by *b*, and the connecting path by *a*. Labeling of this kind is common in describing graphs and figures. A point worth noting is that to be unambiguous, such notations must reply upon the use of common, natural visual routines. The label *b*, for example, is detached from the figure and does not identify explicitly a complete contour. The labeling notation implicitly assumes that there is a common procedure for identifying a distinct contour associated with the label.[21]

In searching for a connecting contour in Fig. 12, the contours could be activated in parallel, in a manner analogous to area coloring. It seems likely that at least in certain situations, the search for a connecting path is not just an unguided sequential tracking and exploration of all possible paths. A definite answer would require, however, an empirical investigation, for example, by manipulating the number of distracting culy-de-sac paths connected to the center and to the surrounding contour. In a sequential search, detection of the connecting path should be strongly affected by the addition of distracting paths. If, on the other hand, activation can spread along many paths simultaneously, detection will be little affected by the additional paths.

[21]It is also of interest to consider how we locate the center of figures. In Noton and Stark's (1971) study of eye movements, there are some indications of an ability to start the scanning of a figure approximately at its center.

Tracking boundaries in the base representations. The examples mentioned above used contours in schematic line drawings. If boundary tracking is indeed a basic operation in establishing properties and spatial relations, it is expected to be applicable not only to such lines, but also to the different types of contours and discontinuity boundaries in the base representations. Exper-

Figure 13. *Certain texture boundaries can delineate effectively shape for recognition (a), while others cannot (b). Micropatterns that are ineffective for delineating shape boundaries can nevertheless give rise to discriminable textures (c). (From Riley, 1981).*

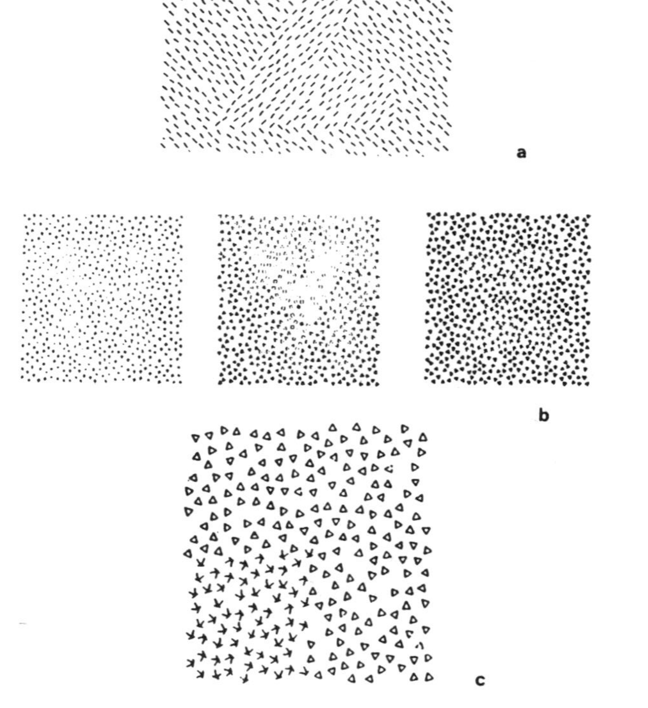

iments with textures, for instance, have demonstrated that texture boundaries can be effective for defining shapes in visual recognition. Figure 13a (reproduced from Riley (1981)) illustrates an easily recognizable Z shape defined by texture boundaries. Not all types of discontinuity can be used for rapid recognition. In Fig. 13b, for example, recognition is difficult. The boundaries defined for instance by a transition between small k-like figures and triangles cannot be used in immediate recognition, although the texture generated by these micropatterns is easily discriminable (Fig. 13c)).

What makes some discontinuities considerably more efficient than others in facilitating recognition? Recognition requires the establishment of spatial properties and relations. It can therefore be expected that recognition is facilitated if the defining boundaries are already represented in the base representations, so that operations such as activation and tracking may be applied to them. Other discontinuities that are not represented in the base representations can be detected by applying appropriate visual routines, but recognition based on these contours will be considerably slower.[22]

3.5.2. Requirements on boundary tracing

The tracing of a contour is a simple operation when the contour is continuous, isolated, and well defined. When these conditions are not met, the tracing operation must cope with a number of challenging requirements. These requirements, and their implications for the tracing operation, are examined in this section.

(a) Tracing incomplete boundaries. The incompleteness of boundaries and contours is a well-known difficulty in image processing systems. Edges and contours produced by such systems often suffer from gaps due to such problems as noise and insufficient contrast. This difficulty is probably not confined to man–made systems alone; boundaries detected by the early processes in the human visual system are also unlikely to be perfect. The boundary tracing operation should not be limited, therefore, to continuous boundaries only. As noted above with respect to inside/outside routines for human perception, fragmented contours can indeed often replace continuous ones.

(b) Tracking across intersections and branches. In tracing a boundary crossings and branching points can be encountered. It will then become necessary to decide which branch is the natural continuation of the curve. Similarity of color, contrast, motion, etc. may affect this decision. For similar contours, collinearity, or minimal change in direction (and perhaps curvature) seem to be the main criteria for preferring one branch over another.

Tracking a contour through an intersection can often be useful in obtaining a stable description of the contour for recognition purposes. Consider, for example, the two different instances of the numeral '2' in Fig. 14a. There are considerable differences between these two shapes. For example, one contains a hole, while the other does not. Suppose, however, that the contours are traced, and decomposed at places of maxima in curvature. This will lead to the decomposition shown in Fig. 14b. In the resulting descriptions, the

Figure 14. *The tracking of a contour through an intersection is used here in generating a stable description of the contour. a, Two instances of the numeral '2'. b, In spite of the marked difference in their shape, their eventual decomposition and description are highly similar.*

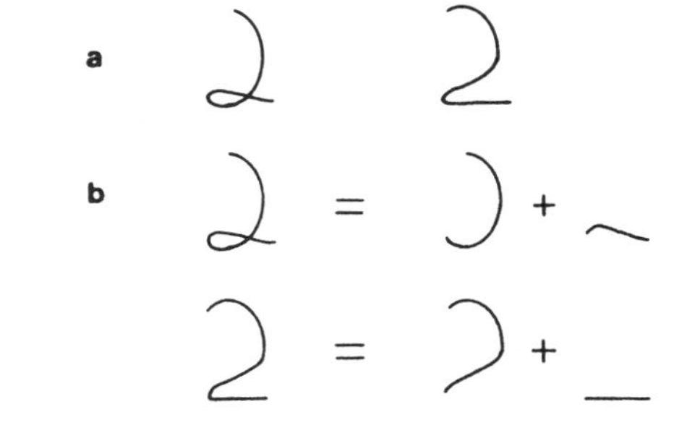

Figure 15. *Tracing a skeleton. The overall figure can be traced and recognized without recognizing first all of the individual components.*

[22]M. Riley (1981) has found a close agreement between texture boundaries that can be used in immediate recognition and boundaries that can be used in long–range apparent motion (cf. Ullman, 1979). Boundaries participating in motion correspondence must be made explicit within the base representations, so that they can be matched over discrete frames. The implication is that the boundaries involved in immediate recognition also preexist in the base representations.

decomposition into strokes, and the shapes of the underlying strokes, are highly similar.

(c) Tracking at different resolutions. Tracking can proceed along the main skeleton of a contour without tracing its individual components. An example is illustrated in Fig. 15, where a figure is constructed from a collection of individual tokens. The overall figure can be traced and recognized without tracing and identifying its components.

Examples similar to Fig. 15 have been used to argue that 'global' or 'holistic' perception precedes the extraction of local features. According to the visual routines scheme, the constituent line elements are in fact extracted by the earliest visual process and represented in the base representations. The constituents are not recognized, since their recognition requires the application of visual routines. The 'forest before the trees' phenomenon (Johnston and McLelland, 1973; Navon, 1977; Pomerantz *et al.*, 1977) is the result of applying appropriate routines that can trace and analyze aggregates without analyzing their individual components, thereby leading to the recognition of the overall figure prior to the recognition of its constituents.

The ability to trace collections of tokens and extract properties of their arrangement raises a question regarding the role of grouping processes in early vision. Our ability to perceive the collinear arrangement of different tokens, as illustrated in Fig. 16, has been used to argue for the existence of sophisticated grouping processes within the early visual representations that detect such arrangements and make them explicit (Marr, 1976). In this view, these grouping processes participate in the construction of the base representations, and consequently collinear arrangements of tokens are detected and represented throughout the base representation prior to the application of visual routines. An alternative possibility is that such arrangements are identified in fact as a result of applying the appropriate routine. This is not to deny the existence of certain grouping processes within the base representations. There is, in fact, strong evidence in support of the existence of such processes.[23] The more complicated and abstract grouping phenomena such as in Fig. 16 may, nevertheless, be the result of applying the appropriate routines, rather than being explicitly represented in the base representations.

Finally, from the point of view of the underlying mechanism, one obvious possibility is that the operation of tracing an overall skeleton is the result of applying tracing routines to a low resolution copy of the image, mediated by low frequency channels within the visual system. This is not the only possibility, however, and in attempting to investigate this operation further, alternative methods for tracing the overall skeleton of figures should also be considered.

Figure 16. *The collinearity of tokens (items and endpoints) can easily be perceived. This perception may be related to a routine that traces collinear arrangements, rather than to sophisticated grouping processes within the base representations.*

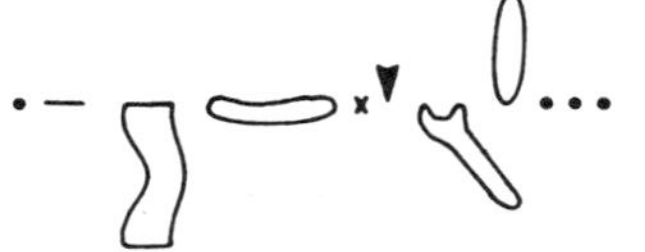

In summary, the tracing and activation of boundaries are useful operations in the analysis of shape and the establishment of spatial relations. This is a complicated operation since flexible, reliable, tracing should be able to cope with breaks, crossings, and branching, and with different resolution requirements.

3.6. *Marking*

In the course of applying a visual routine, the processing shifts across the base representations from one location to another. To control and coordinate the routine, it would be useful to have the capability to keep at least a partial track of the locations already processed.

A simple operation of this type is the marking of a single location for future reference. This operation can be used, for instance, in establishing the closure of a contour. As noted in the preceding section, closure cannot be tested in general by the presence or absence of terminators, but can be established using a combination of tracing and marking. The starting point of the tracing operation is marked, and if the marked location is reached again the tracing is completed, and the contour is known to be closed.

Figure 17 shows a similar problem, which is a version of a problem examined in the previous section. The task here is to determine visually whether there are two X's on the same curve. Once again, the correct answer is perceived immediately. To establish that only a single X lies on the closed curve c, one can use the above strategy of marking the X and tracking the

[23] For evidence supporting the existence of grouping processes within the early creation of the base representations using dot–interference patterns see Glass (1969), Glass and Perez (1973), Marroquin (1976), Stevens (1978). See also a discussion of grouping in early visual processing in Barlow (1981).

Figure 17. *The task here is to determine visually whether there are two X's on a common curve. The task could be accomplished by employing marking and tracing operations.*

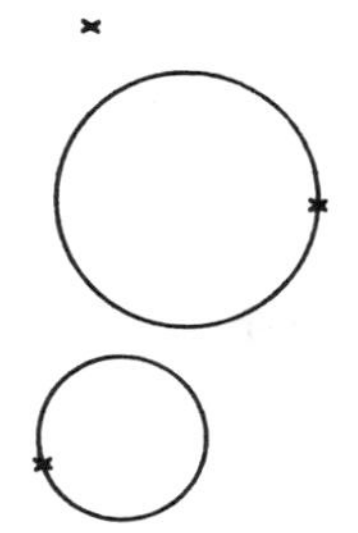

curve. It is suggested that the perceptual system has marking and tracing in its repertoire of basic operations, and that the simple perception of the X on the curve involved the application of visual routines that employ such operations.

Other tasks may benefit from the marking of more than a single location. A simple example is visual counting, that is, the problem of determining as fast as possible the number of distinct items in view (Atkinson *et al.*, 1969; Kowler and Steinman, 1979).

For a small number of items visual counting is fast and reliable. When the number of items is four or less, the perception of their number is so immediate, that it gave rise to conjecture regarding special *Gestalt* mechanisms that can somehow respond directly to the number of items in view, provided that this number does not exceed four (Atkinson *et al.*, 1969).

In the following section, we shall see that although such mechanisms are possible in principle, they are unlikely to be incorporated in the human visual system. It will be suggested instead that even the perception of a small number of items involves in fact the execution of visual routines in which marking plays an important role.

3.6.1. Comparing schemes for visual counting

Perception-like counting networks. In their book *Perceptrons*, Minsky and Papert (1969, Ch. 1) describe parallel networks that can count the number of elements in their input (see also Milner, 1974). Counting is based on computing the predicates "the input has exactly M points" and "the input has between M and N points" for different values of M and N. For any given value of M, it is thereby possible to construct a special network that will respond only when the number of items in view is exactly M. Unlike visual routines which are composed of elementary operations, such a network can adequately be described as an elementary mechanism responding directly to the presence of M items in view. Unlike the shifting and marking operations, the computation is performed by these networks uniformly and in parallel over the entire field.

Counting by visual routines. Counting can also be performed by simple visual routines that employ elementary operations such as shifting and marking. For example, the indexing operation described in Section 3.3 can be used to perform the counting task provided that it is extended somewhat to include marking operations. Section 3.3 illustrated how a simple shifting scheme can be used to move the processing focus to an indexable item. In the counting problem, there is more than a single indexable item to be considered. To use the same scheme for counting, the processing focus is required to travel among all of the indexable items, without visiting an item more than once.

A straightforward extension that will allow the shifting scheme in Section 3.3 to travel among different items is to allow it to mark the elements already visited. Simple marking can be obtained in this case by 'switching off' the element at the current location of the processing focus. The shifting scheme described above is always attracted to the location producing the strongest signal. If this signal is turned off, the shift would automatically continue to the new strongest signal. The processing focus can now continue its tour, until all the items have been visited, and their number counted.

A simple example of this counting routine is the 'single point detection' task. In this problem, it is assumed that one or more points can be lit up in the visual field. The task is to say 'yes' if a single point is lit up, and 'no' otherwise. Following the counting procedure outlined above, the first point will soon be reached and masked. If there are no remaining signals, the point was unique and the correct answer is 'yes'; otherwise, it is 'no'.

In the above scheme, counting is achieved by shifting the processing focus among the items of interest without scanning the entire image systematically. Alternatively, shifting and marking can also be used for visual counting by scanning the entire scene in a fixed predetermined pattern. As the number of items increases, programmed scanning may become the more efficient strategy. The two alternative schemes will behave differently for different numbers of items. The fixed scanning scheme is largely independent of the number of items, whereas in the traveling scheme, the computation time will depend on the number of items, as well as on their spatial configuration.

There are two main differences between counting by visual routines of one type or another on the one hand, and by specialized counting networks on the other. First, unlike the perception-like networks, the process of determining the number of items by visual routines can be decomposed into a sequence of elementary operations. This decomposition holds true for the perception of a small number of items and even for the single item detection. Second, in contrast with a counting network that is specially constructed for the task of detecting a prescribed number of items, the same elementary operations employed in the counting routine also participate in other visual routines.

This difference makes counting by visual routines more attractive than the counting networks. It does not seem plausible to assume that visual counting is essential enough to justify specialized networks dedicated to this task alone. In other words, visual counting is simply unlikely to be an elementary operation. It is more plausible in my view that visual counting can be performed efficiently as a result of our general capacity to generate and execute visual routines, and the availability of the appropriate elementary operations that can be harnessed for the task.

3.6.2. Reference frames in marking

The marking of a location for later reference requires a coordinate system, or a frame of reference, with respect to which the location is defined. One general question regarding marking is, therefore, what is the referencing scheme in which locations are defined and remembered for subsequent use by visual routines. One possibility is to maintain an internal 'egocentric' spatial map that can then be used in directing the processing focus. The use of marking would then be analogous to reaching in the dark: the location of one or more objects can be remembered, so that they can be reached (approximately) in the dark without external reference cues. It is also possible to use an internal map in combination with external referencing. For example, the position of point q in Fig. 18 can be defined and remembered using the prominent X figure nearby. In such a scheme it becomes possible to maintain a crude map with which prominent features can be located, and a more detailed local map in which the position of the marked item is defined with respect to the prominent feature.

Figure 18. *The use of an external reference. The position of point* q *can be defined and retained relative to the predominant X nearby.*

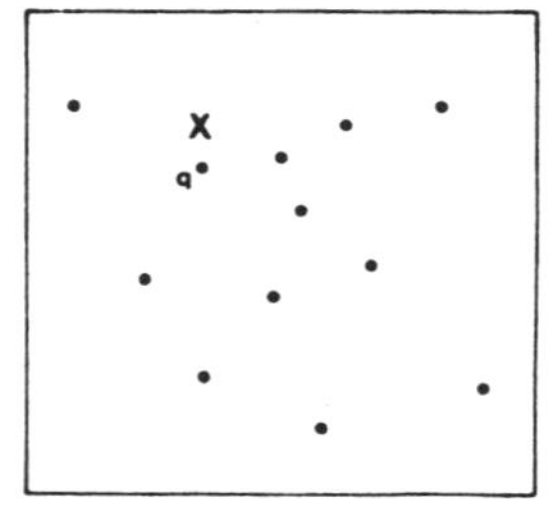

The referencing problem can be approached empirically, for example by making a point in figures such as Fig. 18 disappear, then reappear (possibly in a slightly displaced location), and testing the accuracy at which the two locations can be compared. (Care must be taken to avoid apparent motion.) One can test the effect of potential reference markers on the accuracy, and test marking accuracy across eye movements.

3.6.3. Marking and the integration of information in a scene

To be useful in the natural analysis of visual scenes, the marking map should be preserved across eye motions. This means that if a certain location in space is marked prior to an eye movement, the marking should point to the same spatial location following the eye movement. Such a marking operation, combined with the incremental representation, can play a valuable role in integrating the information across eye movements and from different regions in the course of viewing a complete scene.[24]

Suppose, for example, that a scene contains several objects, such as a man at one location, and a dog at another, and that following the visual analysis of the man figure we shift our gaze and processing focus to the dog. The visual analysis of the man figure has been summarized in the incremental representation, and this information is still available at least in part as the gaze is shifted to the dog. In addition to this information we keep a spatial map, a set of spatial pointers, which tell us that the dog is at one direction, and the man at another. Although we no longer see the man clearly, we have a clear notion of what exists where. The 'what' is supplied by the incremental representations, and the 'where' by the marking map.

In such a scheme, we do not maintain a full panoramic representation of the scene. After looking at various parts of the scene, our representation of it will have the following structure. There would be a retinotopic representation of the scene in the current viewing direction. To this representation we can apply visual routines to analyze the properties of, and relations among, the items in view. In addition, we would have markers to the spatial locations of items in the scene already analyzed. These markers can point to peripheral

[24]The problem considered here is not limited to the integration of views across saccadic eye motions, for which an 'integrative visual buffer' has been proposed by Rayner (1978).

objects, and perhaps even to locations outside the field of view (Attneave and Pierce, 1978). If we are currently looking at the dog, we would see it in detail, and will be able to apply visual routines and extract information regarding the dog's shape. At the same time we know the locations of the other objects in the scene (from the marking map) and what they are (from the incremental representation). We know, for example, the location of the man in the scene. We also know various aspects of his shape, although it may now appear only as a blurred blob, since they are summarized in the incremental representation. To obtain new information, however, we would have to shift our gaze back to the man figure, and apply additional visual routines.

3.6.4. On the spatial resolution of marking and other basic operations

In the visual routines scheme, accuracy in visual counting will depend on the accuracy and spatial resolution of the marking operation. This conclusion is consistent with empirical results obtained in the study of visual counting.[25] Additional perceptual limitations may arise from limitations on the spatial resolution of other basic operations. For example, it is known that spatial relations are difficult to establish in peripheral vision in the presence of distracting figures. An example, due to J. Lettvin (see also Andriessen and Bouma, 1976; Townsend *et al.*, 1971), is shown in Fig. 19. When fixating on the central point from a normal reading distance, the N on the left is recognizable, while the N within the string TNT on the right is not. The flanking letters exert some 'lateral masking' even when their distance from the central letter is well above the two-point resolution at this eccentricity (Riggs, 1965).

Interaction effects of this type may be related to limitations on the spatial resolution of various basic operations, such as indexing, marking, and boundary tracking. The tracking of a line contour, for example, may be distracted by the presence of another contour nearby. As a result, contours may interfere with the application of visual routines to other contours, and consequently with the establishment of spatial relations. Experiments involving the establishment of spatial relations in the presence of distractors would be useful in investigating the spatial resolution of the basic operations, and its dependence on eccentricity.

Figure 19. *Spatial limitations of the elemental operations. When the central mark is fixated, the N on the left is recognizable, while the one on the right is not. This effect may reflect limitations on the spatial resolution of basic operations such as indexing, marking, and boundary tracing.*

N TNT

•

The hidden complexities in perceiving spatial relationships. We have examined above a number of plausible elemental operations including shift, indexing, bounded activation, boundary tracing and activation, and marking. These operations would be valuable in establishing abstract shape properties and spatial relations, and some of them are partially supported by empirical data. (They certainly do not constitute, however, a comprehensive set.)

The examination of the basic operations and their use reveals that in perceiving spatial relations the visual system accomplishes with intriguing efficiency highly complicated tasks. There are two main sources for these complexities. First, as was illustrated above, from a computational standpoint, the efficient and reliable implementation of each of the elemental operations poses challenging problems. It is evident, for instance, that a sophisticated specialized processor would be required for an efficient and flexible bounded activation operation, or for the tracing of contours and collinear arrangements of tokens.

In addition to the complications involved in the realization of the different elemental operations, new complications are introduced when the elemental operations are assembled into meaningful visual routines. As illustrated by the inside/outside example, in perceiving a given spatial relation different strategies may be employed, depending on various parameters of the stimuli (such as the complexity of the boundary, or the distance of the X from the bounding contour). The immediate perception of seemingly simple relations often requires, therefore, selection among possible routines, followed by the coordinated application of the elemental operations comprising the visual routines. Some of the problems involved in the assembly of the elemental operations into visual routines are discussed briefly in the next section.

4. The assembly, compilation, and storage of visual routines

The use of visual routines allows a variety of properties and relations to be established using a fixed set of basic operations. According to this view, the establishment of relations requires the application of a coordinated sequence of basic operations. We have discussed above a number of plausible basic operations. In this section I shall raise some of the general problems as-

[25] For example, Kowler and Steinman (1979) report a puzzling result regarding counting accuracy. It was found that eye movements increase counting accuracy for large (2°) displays, but were not helpful, and sometimes detrimental, with small displays. This result could be explained under the plausible assumptions that marking accuracy is better near fixation, and that it deteriorates across eye movements. As a result, eye movements will improve marking accuracy for large, but not for small, displays.

sociated with the construction of useful routines from combinations of basic operations.

The appropriate routine to be applied in a given situation depends on the goal of the computation, and on various parameters of the configuration to be analyzed. We have seen, for example, that the routine for establishing inside/outside relations may depend on various properties of the configuration: in some cases it would be efficient to start at the location of the X figure, in other situations it may be more efficient to start at some distant locations.

Similarly, in Treisman's (1977, 1980) experiments on indexing by two properties (e.g., a vertical red item in a field of vertical green and horizontal red distractors) there are at least two alternative strategies for detecting the target. Since direct indexing by two properties is impossible, one may either scan the red items, testing for orientation, or scan the vertical items, testing for color.[26] The distribution of distractors in the field determines the relative efficiency of these alternative strategies. In such cases it may prove useful, therefore, to precede the application of a particular routine with a stage where certain relevant properties of the configuration to be analyzed are sampled and inspected. It would be of interest to examine whether in the double indexing task, for example, the human visual system tends to employ the more efficient search strategy.

The above discussion introduces what may be called the 'assembly problem'; that is, the problem of how routines are constructed in response to specific goals, and how this generation is controlled by aspects of the configuration to be analyzed. In the above examples, a goal for the computation is set up externally, and an appropriate routine is applied in response. In the course of recognizing and manipulating objects, routines are usually invoked in response to internally generated queries. Some of these routines may be stored in memory rather than assembled anew each time they are needed.

The recognition of a specific object may then use pre-assembled routines for inspecting relevant features and relations among them. Since routines can also be generated efficiently by the assembly mechanism in response to specific goals, it would probably be sufficient to store routines in memory in a skeletonized form only. The assembly mechanism will then fill in details and generate intermediate routines when necessary. In such a scheme, the perceptual activity during recognition will be guided by setting pre-stored goals that the assembly process will then expand into detailed visual routines.

[26]There is also a possibility that all the items must be scanned one by one without any selection by color or orientation. This question is relevent for the shift operation discussed in Section 3.2. Recent results by J. Rubin and N. Kanwisher at M.I.T. suggest that it is possible to scan only the items of relevant color and ignore the others.

The application of pre-stored routines rather then assembling them again each time they are required can lead to improvements in performance and the speed-up of performing familiar perceptual tasks. These improvements can come from two different sources. First, assembly time will be saved if the routine is already 'compiled' in memory. The time saving can increase if stored routines for familiar tasks, which may be skeletonized at first, become more detailed, thereby requiring less assembly time. Second, stored routines may be improved with practice, for example, as a result of either external instruction, or by modifying routines when they fail to accomplish their tasks efficiently.

Summary

1. Visual perception requires the capacity to extract abstract shape properties and spatial relations. This requirement divides the overall processing of visual information into two distinct stages. The first is the creation of the base representations (such as the primal sketch and the 2½-D sketch). The second is the application of visual routines to the base representations.

2. The creation of the base representations is a bottom-up and spatially uniform process. The representations it produces are unarticulated and viewer-centered.

3. The application of visual routines is no longer bottom-up, spatially uniform, and viewer-centered. It is at this stage that objects and parts are defined, and their shape properties and spatial relations are established.

4. The perception of abstract shape properties and spatial relations raises two major difficulties. First, the perception of even seemingly simple, immediate properties and relations requires in fact complex computation. Second, visual perception requires the capacity to establish a large variety of different properties and relations.

5. It is suggested that the perception of spatial relation is achieved by the application to the base representations of visual routines that are composed of sequences of elemental operations. Routines for different properties and relations share elemental operations. Using a fixed set of basic operations, the visual system can assemble different routines to extract an unbounded variety of shape properties and spatial relations.

6. Unlike the construction of the base representation, the application of visual routines is not determined by the visual input alone. They are selected or created to meet specific computational goals.

7. Results obtained by the application of visual routines are retained in the incremental representation and can be used by subsequent processes.

8. Some of the elemental operations employed by visual routines are applied to restricted locations in the visual field, rather than to the entire field in parallel. It is suggested that this apparent limitation on spatial parallelism reflects in part essential limitations, inherent to the nature of the computation, rather than non-essential capacity limitations.

9. At a more detailed level, a number of plausible basic operations were suggested, based primarily on their potential usefulness, and supported in part by empirical evidence. These operations include:

9.1. *Shift of the processing focus*. This is a family of operations that allow the application of the same basic operation to different locations across the base representations.

9.2. *Indexing*. This is a shift operation towards special odd-man-out locations. A location can be indexed if it is sufficiently different from its surroundings in an indexable property. Indexable properties, which are computed in parallel by the early visual processes, include contrast, orientation, color, motion, and perhaps also size, binocular disparity, curvature, and the existence of terminators, corners, and intersections.

9.3. *Bounded activation*. This operation consists of the spread of activation over a surface in the base representation, emanating from a given location or contour, and stopping at discontinuity boundaries. This is not a simple operation, since it must cope with difficult problems that arise from the existence of internal contours and fragmented boundaries. A discussion of the mechanisms that may be implicated in this operation suggests that specialized networks may exist within the visual system, for executing and controlling the application of visual routines.

9.4. *Boundary tracing*. This operation consists of either the tracing of a single contour, or the simultaneous activation of a number of contours. This operation must be able to cope with the difficulties raised by the tracing of incomplete boundaries, tracing across intersections and branching points, and tracing contours defined at different resolution scales.

9.5. *Marking*. The operation of marking a location means that this location is remembered, and processing can return to it whenever necessary. Such an operation would be useful in the integration of information in the processing of different parts of a complete scene.

10. It is suggested that the seemingly simple and immediate perception of spatial relations conceals in fact a complex array of processes involved in the selection, assembly, and execution of visual routines.

References

Andriessen, J.J. and Bouma, H. (1976) Eccentric vision: adverse interactions between line segments. *Vis. Res., 16*, 71–78.

Atkinson, J., Campbell, F.W. and Francis, M.R. (1969) The magic number 4 ± 0: A new look at visual numerosity judgments. *Perception, 5*, 327–334.

Attneave, F. and Pierce, C.R. (1978) The accuracy of extrapolating a pointer into perceived and imagined space. *Am. J. Psychol., 91(3)*, 371–387.

Barlow, H.H. (1972) Single units and sensation: A neuron doctrine for perceptual psychology? *Perception, 1*, 371–394.

Barlow, H.B. (1981) Critical limiting factors in the design of the eye and the visual cortex. The Ferrier Lecture 1980. *Proc. Roy. Soc. Lond. B, 212*, 1–34.

Bartlett, F.C. (1932) *Remembering*. Cambridge, Cambridge University Press.

Beck, J. and Ambler, B. (1972) Discriminability of differences in line slope and in line arrangement as a function of mask delay. *Percep. Psychophys. 12(1A)*, 33–38.

Beck, J. and Ambler, B. (1973) The effects of concentrated and distributed attention on peripheral acuity. *Percept. Psychophys., 14(2)*, 225–230.

Beneveneto, L.A. and Davis, B. (1977) Topographical projections of the prestriate cortex to the pulvinar nuclei in the macaque monkey: an autoradiographic study. *Exp. Brain Res., 30*, 405–424.

Biederman, I., Glass, A.L. and Stacy, E.W. (1973) Searching for objects in real-world scenes. *J. exp. Psychol., 97(1)*, 22–27.

Donderi, D.C. and Zelnicker, D. (1969) Parallel processing in visual same–different decisions. *Percep. Psychophys., 5(4)*, 197–200.

Egeth, H., Jonides, J. and Wall, S. (1972) Parallel processing of multi-element displays. *Cog. Psychol., 3*, 674–698.

Engel, F.L. (1971) Visual conspecuity, directed attention and retinal locus. *Vis. Res., 11*, 563–576.

Eriksen, C.W. and Hoffman, J.E. (1972) Temporal and spatial characteristics of selective encoding from visual displays. *Percep. Psychophys., 12(2B)*, 201–204.

Eriksen, C.W. and Schultz, D.W. (1977) Retinal locus and acuity in visual information processing. *Bull. Psychon. Soc., 9(2)*, 81–84.

Estes, W.K. (1972) Interactions of signal and background variables in visual processing. *Percep. Psychophys., 12(3)*, 278–286.

Evans, T.G. (1968) A heuristic program to solve geometric analogy problems. In M. Minsky (ed.), *Semantic Information Processing*. Cambridge, MA, M.I.T. Press.

Fantz, R.L. (1961) The origin of form perception. *Scient. Am., 204(5)*, 66–72.

Fischer, B. and Boch, R. (1981) Enhanced activation of neurons in prelunate cortex before visually guided saccades of trained rhesus monkey. *Exp. Brain Res., 44*, 129–137.

Fuster, J.M. and Jervey, J.P. (1981) Inferotemporal neurons distinguish and retain behaviorally relevant features of visual stimuli. *Science, 212*, 952–955.

Gattas, R., Osealdo Cruz, E. and Sousa, A.P.B. (1979) Visual receptive fields of units in the pulvinar of cebus monkey. *Brain Res., 160*, 413–430.

Glass, L. (1969) Moire effect from random dots. *Nature, 243*, 578–580.

Glass, L. and Perez, R. (1973) Perception of random dot interference patterns. *Nature, 246*, 360–362.

Goldberg, M.E. and Wurtz, R.H. (1972) Activity of superior colliculus in behaving monkey. II. Effect of attention of neural responses. *J. Neurophysiol, 35*, 560–574.

Holtzman, J.D. and Gazzaniga, M.S. (1982) Dual task interactions due exclusively to limits in processing resources. *Science, 218*, 1325–1327.

Humphreys, G.W. (1981) On varying the span of visual attention: evidence for two modes of spatial attention. *Q. J. exp. Psychol., 33A*, 17–31.

Johnston, J.C. and McClelland, J.L. (1973) Visual factors in word perception. *Percep. Psychophys., 14(2)*, 365–370.
Jonides, J. and Gleitman, H. (1972) A conceptual category effect in visual search: O as a letter or as digit. *Percep. Psychophys., 12(6)*, 457–460.
Johnson, R.B. and Kirk, N.S. (1960) The perception of size: An experimental synthesis of the associationist and gestalt accounts of the perception of size. Part III. *Q. J. exp. Psychol., 12*, 221–230.
Julesz, B. (1975) Experiments in the visual perception of texture. *Scient. Am., 232(4)*, April 1975, 34–43.
Julesz, B. (1981) Textons, the elements of texture perception, and their interactions. *Nature, 290*, 91–97.
Kahneman, D. (1973) *Attention and Effort*. Englewood Cliffs, NJ, Prentice-Hall.
Kolmogorov, A.N. (1968) Logical basis for information theory and probability theory. *IEEE Trans. Info. Theory, IT-14(5)*, 662–664.
Kowler, E. and Steinman, R.M. (1979) Miniature saccades: eye movements that do not count. *Vis. Res., 19*, 105–108.
Lappin, J.S. and Fuqua, M.A. (1983) Accurate visual measurement of three-dimensional moving patterns. *Science, 221*, 480–482.
Livingstone, M.L. and Hubel, D.J. (1981) Effects of sleep and arousal on the processing of visual information in the cat. *Nature, 291*, 554–561.
Mackworth, N.H. (1965) Visual noise causes tunnel vision. *Psychon. Sci., 3*, 67–68.
Marr, D. (1976) Early processing of visual information. *Phil. Trans. Roy. Soc. and B, 275*, 483–524.
Marr, D. (1980) Visual information processing: the structure and creation of visual representations. *Phil. Trans. Roy. Soc. Lond. B, 290*, 199–218.
Marr, D. and Nishihara, H.K. (1978) Representation and recognition of the spatial organization of three–dimensional shapes. *Proc. Roy. Soc. B, 200*, 269–291.
Marroquin, J.L. (1976) Human visual perception of structure. MSc. Thesis, Department of Electrical Engineering and Computer Science, Massachusetts Institute of Technology.
Milner, P.M. (1974) A model for visual shape recognitio. *Psychol. Rev. 81(6)*, 521–535.
Minsky, M. and Papert, S. (1969) *Perceptrons*. Cambridge, MA and London: The M.I.T. Press.
Minsky, M. (1975) A framework for representing knowledge. In P.H. Winston (ed.), *The Psychology of Computer Vision*. New York, Prentice Hall.
Mountcastle, V.B. (1976) The world around us: neural command functions for selective attention. The F.O. Schmitt Lecture in Neuroscience 1975. *Neurosci. Res. Prog. Bull., 14*, Supplement 1–37.
Mountcastle, V.B., Lynch, J.C., Georgopoulos, A., Sakata, H. and Acuna, A. (1975) Posterior parietal association cortex of the monkey: command functions for operations within extrapersonal space. *J. Neurophys., 38*, 871–908.
Navon, D. (1977) Forest before trees: the precedence of global features in visual perception. *Cog. Psychol., 9*, 353–383.
Neisser, U., Novick, R. and Lazar, R. (1963) Searching for ten targets simultaneously. *Percep. Mot. Skills, 17*, 955–961.
Neisser, U. (1967) *Cognitive Psychology*. New York, Prentice-Hall.
Newsome, W.T. and Wurtz, R.H. (1982) Identification of architectonic zones containing visual tracking cells in the superior temporal sulcus of macaque monkeys. *Invest. Ophthal. Vis. Sci., Suppl. 3*, 22, 238.
Nickerson, R.S. (1966) Response times with memory–dependent decision task. *J. exp. Psychol., 72(5)*, 761–769.
Noton, D. and Stark, L. (1971) Eye movements and visual perception. *Scient. Am., 224(6)*, 34–43.
Pomerantz, J.R., Sager, L.C. and Stoever, R.J. (1977) Perception of wholes and of their component parts: some configural superiority effects. *J. exp. Psychol., Hum. Percep. Perf., 3(3)*, 422–435.
Posner, M.I. (1980) Orienting of attention. *Q. J. exp. Psychol., 32*, 3–25.
Posner, M.I., Nissen, M.J. and Ogden, W.C. (1978) Attended and unattended processing modes: the role of set for spatial location. In Saltzman, I.J. and H.L. Pick (eds.), *Modes of Perceiving and Processing Information*. Hillsdale, NJ, Lawrence Erlbaum.
Potter, M.C. (1975) Meaning in visual search. *Science, 187*, 965–966.
Rayner, K. (1948) Eye movements in reading and information processing. *Psychol. Bull., 85(3)*, 618–660.
Regan, D. and Beverley, K.I. (1978) Looming detectors in the human visual pathway. *Vis. Res., 18*, 209–212.
Rezak, M. and Beneveneto, A. (1979) A comparison of the organization of the projections of the dorsal lateral geniculate nucleus, the inferior pulvinar and adjacent lateral pulvinar to primary visual area (area 17) in the macaque monkey. *Brain Res., 167*, 19–40.
Richards, W. (1982) How to play twenty questions with nature and win. *M.I.T.A.I. Laboratory Memo 660*.
Richmond, B.J. and Sato, T. (1982) Visual responses of inferior temporal neurons are modified by attention to different stimuli dimensions. *Soc. Neurosci. Abst., 8*, 812.
Riggs, L.A. (1965) Visual acuity. In C.H. Grahan (ed.), *Vision and Visual Perception*. New York, John Wiley.
Riley, M.D. (1981) The representation of image texture. M.Sc. Thesis, Department of Electrical Engineering and Computer Science, Massachusetts Institute of Technology.
Robinson, D.L., Goldberg, M.G. and Staton, G.B. (1978) Parietal association cortex in the primate: sensory mechanisms and behavioral modulations. *J. Neurophysiol., 41(4)*, 910–932.
Rock, I., Halper, F. and Clayton, T. (1972) The perception and recognition of complex figures. *Cog. Psychol., 3*, 655–673.
Rock, I. and Gutman, D. (1981) The effect of inattention of form perception. *J. exp. Psychol.: Hum. Percep. Perf., 7(2)*, 275–285.
Rumelhart, D.E. (1970) A multicomponent theory of the perception of briefly exposed visual displays. *J. Math. Psychol., 7*, 191–218.
Schein, S.J., Marrocco, R.T. and De Monasterio, F.M. (1982) Is there a high concentration of color–selective cells in area V4 of monkey visual cortex? *J. Neurophysiol., 47(2)*, 193–213.
Shiffrin, R.M., McKay, D.P. and Shaffer, W.O. (1976) Attending to forty-nine spatial positions at once. *J. exp. Psychol.: Human Percep. Perf., 2(1)*, 14–22.
Shulman, G.L., Remington, R.W. and McLean, I.P. (1979) Moving attention through visual space. *J. exp. Psychol.: Huma. Percep. Perf., 5*, 522–526.
Sperling, G. (1960) The information available in brief visual presentations. *Psychol. Mono., 74*, (11, Whole No. 498).
Stevens, K.A. (1978) Computation of locally parallel structure. *Biol. Cybernet., 29*, 19–28.
Sutherland, N.S. (1968) Outline of a theory of the visual pattern recognition in animal and man. *Proc. Roy. Soc. Lond. B, 171*, 297–317.
Townsend, J.T., Taylor, S.G. and Brown, D.R. (1971) Latest masking for letters with unlimited viewing time. *Percep. Psycholphys., 10(5)*, 375–378.
Treisman, A. (1977) Focused attention in the perception and retrieval of multidimensional stimuli. *Percep. Psychophys., 22*, 1–11.
Treisman, A. and Celade, G. (1980) A feature integration theory of attention. *Cog. Psychol., 12*, 97–136.
Tsal, Y. (1983) Movements of attention across the visual field. *J. exp. Psychol.: Hum. Percep. Perf.* (In Press).
Ullman, S. (1979) *The Interpretation of Visual Motion*. Cambridge, MA, and London: The M.I.T. Press.
Varanese, J. (1983) Abstracting spatial relations from the visual world B.Sc. thesis in Neurobiology and Psychology. Harvard University.
van Voorhis, S. and Hillyard, S.A. (1977) Visual evoked potentials and selective attention to points in space. *Percep. Psychophys., 22(1)*, 54–62.
Winston, P.H. (1977) *Artificial Intelligence*. Reading, MA., Addison-Wesley.
Wurtz, R.H. and Mohler, C.W. (1976a) Organization of monkey superior colliculus: enhanced visual response of superficial layer cells. *J. Neurophysiol., 39(4)*, 745–765.
Wurtz, R.H. and Mohler, C.W. (1976b) Enhancement of visual response in monkey striate cortex and frontal eye fields. *J. Neurophysiol., 39*, 766–772.
Wurtz, R.H., Goldberg, M.E. and Robinson D.L. (1982) Brain mechanisms of visual attention. *Scient. Am., 246(6)*, 124–135.
Zeki, S.M. (1978a) Functional specialization in the visual cortex of the rhesus monkey. *Nature, 274*, 423–428.
Zeki, S.M. (1978b) Uniformity and diversity of structure and function in rhesus monkey prestriate visual cortex. *J. Physiol., 277*, 273–290.

Reference Note

1. Joliceur, P., Ullman, S. and Mackay, M. (1984) Boundary Tracing: a possible elementary operation in the perception of spatial relations. Submitted for publication.

SCALE-SPACE FILTERING

Andrew P. Witkin

Fairchild Laboratory for Artificial Intelligence Research

ABSTRACT—The extrema in a signal and its first few derivatives provide a useful general-purpose qualitative description for many kinds of signals. A fundamental problem in computing such descriptions is scale: a derivative must be taken over some neighborhood, but there is seldom a principled basis for choosing its size. Scale-space filtering is a method that describes signals qualitatively, managing the ambiguity of scale in an organized and natural way. The signal is first expanded by convolution with gaussian masks over a continuum of sizes. This "scale-space" image is then collapsed, using its qualitative structure, into a tree providing a concise but complete qualitative description covering all scales of observation. The description is further refined by applying a stability criterion, to identify events that persist of large changes in scale.

1. Introduction

Hardly any sophisticated signal understanding task can be performed using the raw numerical signal values directly; some description of the signal must first be obtained. An initial description ought to be as compact as possible, and its elements should correspond as closely as possible to meaningful objects or events in the signal-forming process. Frequently, local extrema in the signal and its derivatives—and intervals bounded by extrema—are particularly appropriate descriptive primitives: although local and closely tied to the signal data, these events often have direct semantic interpretations, e.g. as edges in images. A description that characterizes a signal by its extrema and those of its first few derivatives is a *qualitative* description of exactly the kind we were taught to use in elementary calculus to "sketch" a function.

A great deal of effort has been expended to obtain this kind of primitive qualitative description (for overviews of this literature, see [1,2,3].) and the problem has proved extremely difficult. The problem of *scale* has emerged consistently as a fundamental source of difficulty, because the events we perceive and find meaningful vary enormously in size and extent. The problem is not so much to eliminate fine-scale noise, as to separate events at different scales arising from distinct physical processes.[4] It is possible to introduce a *parameter of scale* by smoothing the signal with a mask of variable size, but with the introduction of scale-dependence comes ambiguity: every setting of the scale parameter yields a different description; new extremal points may appear, and existing ones may move or disappear. How can we decide which if any of this continuum of descriptions is "right"?

There is rarely a sound basis for setting the scale parameter. In fact, it has become apparent that for many tasks no one scale of description is categorically correct: the physical processes that generate signals such as images act at a variety of scales, none intrinsically more interesting or important than another. Thus the ambiguity introduced by scale is inherent and inescapable, so the goal of scale-dependent description cannot be to eliminate this ambiguity, but rather to manage it effectively, and reduce it where possible.

This line of thinking has led to considerable interest in multi-scale descriptions [5,2,6,7]. However, merely computing descriptions at multiple scales does not solve the problem; if anything, it exacerbates it by increasing the volume of data. Some means must be found to organize or simplify the description, by relating one scale to another. Some work has been done in this area aimed at obtaining "edge pyramids" (e.g. [8]), but no clear-cut criteria for constructing them have been put forward. Marr [4] suggested that zero-crossings that coincide over several scales are "physically significant," but this idea was neither justified nor tested.

How, then, can descriptions at different scales be related to each other in an organized, natural, and compact way? Our solution, which we call *scale-space filtering,* begins by continuously varying the scale parameter, sweeping out a surface that we call the *scale-space image.* In this representation, it is possible to track extrema as they move continuously with scale changes, and to identify the singular points at which new extrema appear. The scale-space image is then collapsed into a tree, providing a concise but complete qualitative description of the signal over all scales of observation.[1]

2. The Scale-Space Image

Descriptions that depend on scale can be computed in many ways. As a primitive scale-parameterization, the gaussian convolution is attractive for a number of its properties, amounting to "well-behavedness": the gaussian is symmetric and strictly decreasing about the mean, and therefore the weighting assigned to signal values decreases smoothly with distance. The gaussian convolution behaves well near the limits of the scale parameter, σ, approaching the un-smoothed signal for small σ, and approaching the signal's mean for large σ. The gaussian is also readily differentiated and integrated.

The gaussian is not the only convolution kernel that meets these criteria. However, a more specific motivation for our choice is a property of the gaussian convolution's

[1]A complementary approach to the "natural" scale problem has been developed by Hoffman [9].

Figure 1. A sequence of gaussian smoothings of a waveform, with σ decreasing from top to bottom. Each graph is a constant-σ profile from the scale-space image.

zero-crossings (and those of its derivatives): as σ decreases, additional zeroes may appear, but existing ones cannot in general disappear; moreover, of convolution kernels satisfying "well behavedness" criteria (roughly those enumerated above,) the gaussian is the ***only*** one guaranteed to satisfy this condition [12]. The usefulness of this property will be explained in the following sections.

The gaussian convolution of a signal $f(x)$ depends both on x, the signal's independent variable, and on σ, the gaussian's standard deviation. The convolution is given by

$$F(x,\sigma) = f(x) * g(x,\sigma) = \int_{-\infty}^{\infty} f(u) \frac{1}{\sigma\sqrt{2\pi}} e^{-\frac{(x-u)^2}{2\sigma^2}} \, du, \tag{1}$$

where "$*$" denotes convolution with respect to x. This function defines a surface on the (x,σ)-plane, where each profile of constant σ is a gaussian-smoothed version of $f(x)$, the amount of smoothing increasing with σ. We will call the (x,σ)-plane ***scale space***, and the function, F, defined in (1), the ***scale-space image*** of f.[2] Fig. 1 graphs a sequence of gaussian smoothings with increasing σ. These are constant-σ profiles from the scale-space image.

At any value of σ, the extrema in the nth derivative of the smoothed signal are given by the zero-crossings in the $(n+1)$th derivative, computed using the relation

$$\frac{\partial^n F}{\partial x^n} = f * \frac{\partial^n g}{\partial x^n},$$

where the derivatives of the gaussian are readily obtained. Although the methods presented here apply to zeros in any derivative, we will restrict our attention to those in the second. These are extrema of slope, i.e. inflection points. In terms of the scale-space image, the inflections at ***all*** values of σ are the points that satisfy

$$F_{xx} = 0, F_{xxx} \neq 0, \tag{2}$$

[2]It is actually convenient to treat $\log\sigma$ as the scale parameter, uniform expansion or contraction of the signal in the x-direction will cause a translation of the scale-space image along the $\log\sigma$ axis.

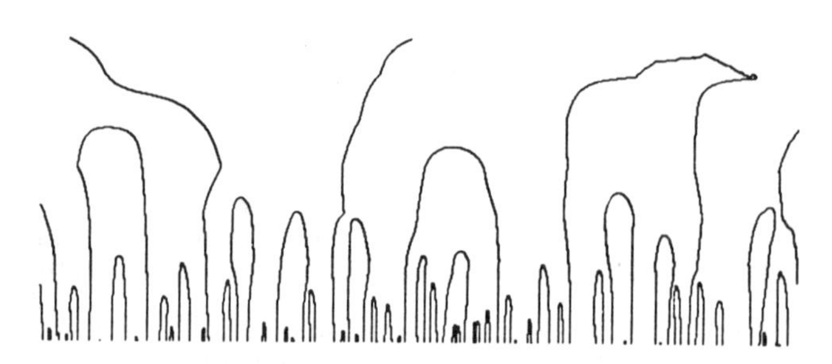

Figure 2. Contours of $F_{xx} = 0$ in a scale-space image. The x-axis is horizontal; the coarsest scale is on top. To simulate the effect of a continuous scale-change on the qualitative description, hold a straight-edge (or better still, a slit) horizontally. The intersections of the edge with the zero-contours are the extremal points at some single value of σ. Moving the edge up or down increases or decreases σ.

using subscript notation to indicate partial differentiation.[3]

3. Coarse-to-fine Tracking

The contours of $F_{xx} = 0$ mark the appearance and motion of inflection points in the smoothed signal, and provide the raw material for a qualitative description over all scales, in terms of inflection points. Next, we will apply two simplifying assumptions to these contours: (1) the ***identity*** assumption, that extrema observed at different scales, but lying on a common zero-contour in scale space, arise from a single underlying event, and (2) the ***localization*** assumption, that the true location of an event giving rise to a zero-contour is the contour's x location as $\sigma \to 0$.

Referring to fig. 2, notice that the zero contours form arches, closed above, but open below. The restriction that zero-crossings may never disappear with with decreasing σ (see section 2) means that the contours may ***never*** be closed below. Note that at the apexes of the arches, $F_{xxx} = 0$, so by eq. (2), these points do not belong to the contour. Each arch consists of a pair of contours, crossing zero with opposite sign.

The ***localization assumption*** is motivated by the observation that linear smoothing has two effects: qualitative simplification—the removal of fine-scale features—and spatial distortion—dislocation, broadening and flattening of the features that survive. The latter undesirable effect may be overcome, by tracking coarse extrema to their fine-scale locations. Thus, a coarse scale may be used to ***identify*** extrema, and a fine scale, to ***localize*** them. Each zero-contour therefore reduces to an (x,σ) pair, specifying its fine-scale location on the x-axis, and the coarsest scale at which the contour appears.

A coarse-to-fine tracking description is compared to the

[3]Note that the second condition in (2) excludes zero-crossings that are parallel to the x-axis, because these are not zero-crossings in the convolved signal.

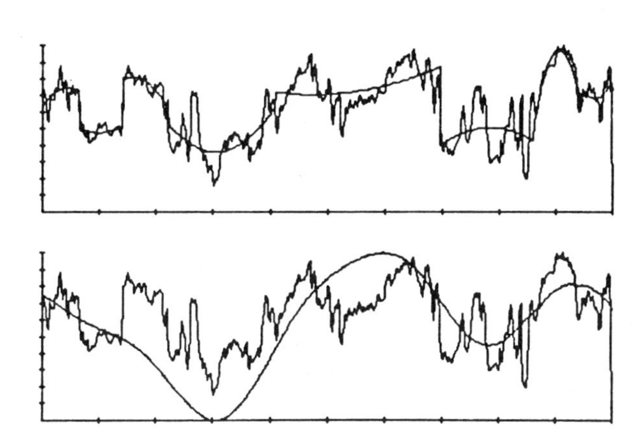

Figure 3. Above is shown a signal with a coarse-to-fine tracking approximation superimposed. The approximation was produced by independent parabolic fits between the localized inflections. Below is shown the corresponding (qualitatively isomorphic) gaussian smoothing.

corresponding linear smoothing in Fig. 3.[4]

4. The Interval Tree

While coarse-to-fine tracking solves the problem of localizing large-scale events, it does not solve the multi-scale integration problem, because the description still depends on the choice of the continuous global scale parameter, σ, just as simple linear filtering does. In this section, we reduce the scale-space image to a simple tree, concisely but completely describing the qualitative structure of the signal over all scales of observation.

This simplification rests on a basic property of the scale-space image: as σ is varied, extremal points in the smoothed signal appear and disappear at singular points (the tops of the arches in fig. 2.) Passing through such a point with decreasing σ, a pair of extrema of opposite sign appear in the smoothed signal. At these points, and only these points, the *un*distinguished interval (i.e. an interval bounded by extremal points but containing none) in which the singularity occurs splits into three subintervals. In general, each undistinguished interval, observed in scale space, is bounded on each side by the zero contours that define it, bounded above by the singular point at which it merges into an enclosing interval, and bounded below by the singular point at which it divides into sub-intervals.

Consequently, to each interval, I, corresponds a node in a (generally ternary-branching) tree, whose parent node denotes the larger interval from which I emerged, and whose offspring represent the smaller intervals into which I subdivides. Each interval also defines a rectangle in scale-space, denoting its location and extent on the signal (as defined by coarse-to-fine tracking) and its location and extent on the scale dimension. Collectively, these rectangles

[4]In this and all illustrations, approximations were drawn by fitting parabolic arcs independently to the signal data on each interval marked by the description. This procedure is crude, particularly because continuity is not enforced across inflections. Bear in mind that this procedure has been used only to display the qualitative description.

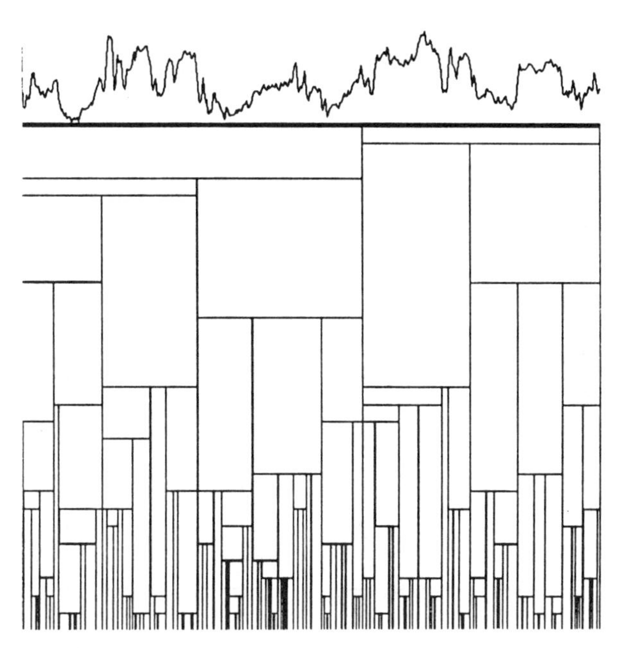

Figure 4. A signal with its interval tree, represented as a rectangular tesselation of scale-space. Each rectangle is a node, indicating an interval on the signal, and the scale interval over which the signal interval exists.

tesselate the (x, σ)-plane. See fig. 4 for an illustration of the tree.

This ***interval tree*** may be viewed in two ways: as describing the signal simultaneously at all scales, or as generating a family of single-scale descriptions, each defined by a subset of nodes in the tree that cover the x-axis. On the second interpretation, one may move through the family of descriptions in orderly, local, discrete steps, either by choosing to subdivide an interval into its offspring, or to merge a triple of intervals into their parent.[5]

We found that it is in general possible, by moving interactively through the tree and observing the resulting "sketch" of the signal, to closely match observers' spontaneously perceived descriptions. Thus the interval tree, though tightly constrained, seems flexible enough to capture human perceptual intuitions. Somewhat surprisingly, we found that the tree, rather than being too constraining, is not constrained enough. That is, the perceptually salient descriptions can in general be duplicated within the tree's constraints, but the tree also generates many descriptions that plainly have no perceptual counterpart. This observation led us to develop a ***stability*** criterion for further pruning or ordering the states of the tree, which is described in the next section.

5. Stability

Recall that to each interval in the tree corresponds a rectangle in scale space. The x boundaries locate the interval on the signal. The σ boundaries define the scale range over which the interval exists, its ***stability*** over scale changes. We have observed empirically a marked correspondence between the stability of an interval and its perceptual salience: those intervals that survive over a broad range of scales

[5]For previous uses of hierarchic signal descriptions see e.g. [10,11,2].

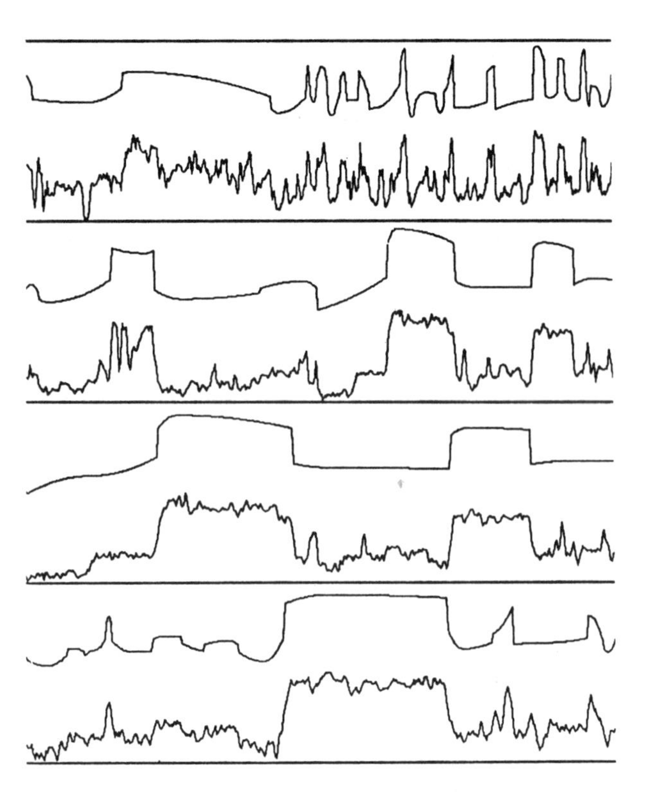

Figure 5. Several signals, with their maximum-stability descriptions. These are "top-level" descriptions, generated automatically and without thresholds. You should compare the descriptions to your own first-glance "top-level" percepts. (the noisy sine and square waves are synthetic signals.)

tend to leap out at the eye, while the most ephemeral are not perceived at all. To capture this relation, we have devised several versions of a stability criterion, one of which picks a "top-level" description by descending the tree until a local maximum in stability is found. Another iteratively removes nodes from the tree, splicing out nodes that are less stable than any of their parents and offspring. Both of these radically improve correspondence between the interval tree's descriptions and perceptual features (see fig. 5.)

6. Summary

Scale-space filtering is a method that describes signals qualitatively, in terms of extrema in the signal or its derivatives, in a manner that deals effectively with the problem of scale—precisely localizing large-scale events, and effectively managing the ambiguity of descriptions at multiple scales, without introducing arbitrary thresholds or free parameters. The one-dimensional signal is first expanded into a two-dimensional ***scale-space image,*** by convolution with gaussians over a continuum of sizes. This continuous surface is then collapsed into a discrete structure, using the connectivity of extremal points tracked through scale-space, and the singular points at which new extrema appear. The resulting tree representation is a a concise but complete qualitative description of the signal over all scales of observation. The tree is further constrained using a maximum-stability criterion to favor events that persist over large changes in scale.

We are currently developing applications of scale-space filtering to several signal matching and interpretation problems, and investigating its ability to explain perceptual grouping phenomena. The method is also being extended to apply to two-dimensional images: the scale-space image of a 2-D signal occupies a volume, containing zero-crossing surfaces.[6]

REFERENCES

[1] A. Rosenfeld and A. C. Kak Digital Picture Processing. *Academic Press, New York, New York,* 1976.

[2] D. H. Ballard and C. M. Brown Computer Vision. *Prentice Hall, Englewood Cliffs, New Jersey,* 1982.

[3] Pavlidis, T. ***Structural Pattern Recognition.*** Springer, 1977

[4] D. Marr Vision, *W. H. Freeman, San Fransisco,* 1982.

[5] Rosenfeld, A. and Thurston, M. "Edge and curve detection for visual scene analysis." *IEEE Transactions on computers, Vol C-20 pp. 562-569.* (May 1971).

[6] Marr, D and Poggio, T. "A computational theory of human stereo vision." *Proc. R. Soc. Lond.,* B. 204 (1979) pp. 301-328

[7] D. Marr and E. C. Hildreth Theory of Edge Detection. *M.I.T. Artificial Intelligence Memo Number 518, Cambridge, Massachusetts,* April 1979.

[8] Hong, T. H., Shneier, M. and Rosenfeld, A. Border Extraction using linked edge pyramids. *TR-1080, Computer Vision Laboratory, U. Maryland,* July 1981

[9] Hoffman, D. Representing Shapes For Visual Recognition *Ph.D. Thesis, MIT, forthcoming.*

[10] Erich, R. and Foith, J., "Representation of random waveforms by relational trees," *IEE Trans. Computers, Vol C-26, pp. 725-736,* (July 1976).

[11] Blumenthal A., Davis, L., and Rosenfeld, R., "Detecting natural 'plateaus' in one-dimensional patterns." *IEEE Transactions on computers,* (Feb. 1977)

[12] J. Babaud, R. Duda, and A. Witkin, ***in preparation.***

[6] **Acknowledgments**—I thank my colleagues at FLAIR, particularly Richard Duda and Peter Hart, as well as J. Babaud of Schlumberger, ltd., for their help and encouragement.

Early Orientation Selection: Tangent Fields and the Dimensionality of Their Support

STEVEN W. ZUCKER

Computer Vision and Robotics Laboratory, Department of Electrical Engineering, McGill University, Montréal, Québec, Canada

Received May 7, 1985; revised May 15, 1985

Orientation selection is the inference of orientation information out of images. It is one of the foundations on which other visual structures are built, since it must precede the formation of contours out of pointillist data and surfaces out of surface markings. We take a differential geometric view in defining orientation selection, and develop algorithms for actually doing it. The goal of these algorithms is formulated in mathematical terms as the inference of a vector field of tangents (to the contours), and the algorithms are studied in both abstract and computational forms. They are formulated as matching problems, and algorithms for solving them are reduced to biologically plausible terms. We show that two different matching problems are necessary, the first for 1-dimensional contours (which we refer to as Type I processes) and the second for 2-dimensional flows (or Type II processes). We conjecture that this difference is reflected in the response properties of "simple" and "complex" cells, respectively, and predict several other psychophysical phenomena.

1. INTRODUCTION

Function within our visual systems has evolved to reflect, at least in part, ecologically relevant aspects of the structure of the physical world. Perhaps the most salient of these structures is orientation, whose importance is underlined by the fact that a major fraction of the visual systems in primates are somehow involved in analyzing it. Orientation is clearly essential for inferring contours, as they arise from creases, occlusions, and any number of other distinct physical situations. It also plays an intimate role in inferring (descriptions of) other kinds of physical structures. We therefore begin this paper with an examination of physical structures rich in orientation information—including flow patterns such as wheat fields and hair coverings as well as contours—and note how their structure is reflected in images. Our immediate goal is to discover whether these physical structures map into well-defined classes of image structure. If so, then they provide viable candidates to be inferred out of images and, once available, interpreted [36].

Given this background, the majority of the paper is concerned with developing and establishing the plausibility of a model for early orientation selection. The development begins along mathematical lines, with a consideration of how contours can be built up from more local elements—arrangements of tangents—and how abstract ideas from functional minimization and variational theory can be used to develop an algorithm for actually accomplishing the inference of tangents. Thus the model differs from most previous ones, in the sense that the concentration is not on so-called *line detectors*, but rather on a mathematical abstraction of what it means to actually detect lines and contours.

To establish the model's plausibility, we show how it leads to new psychophysical predictions, all of which have been confirmed, and to a novel conjecture about how the computations that it requires might be reduced to physiological terms.

1.1 *Occluding Contours and Surface Markings*

Contours and flow patterns are abstractions of some of the most prevalent structures in the physical world. Contours are produced every time an occlusion takes place between coherent objects. They are also produced by well-defined highlights cast from surface creases and other such physical events. Flow patterns, such as the hair on one's head or the surface markings on the skin of an apple, are also ubiquitous and can offer powerful clues about 3-dimensional shapes. But flow patterns arise from physical situations of a fundamentally different structure than distinct contours, as the following example shows.

1.2. *The Fox and the Forest*

Imagine a fox chasing a rabbit through the forest. The fox will focus its eyes on the rabbit, trying to keep most of it in full view. The fox must be sensitive to small changes in the path of the rabbit, or the rabbit could affect evasive manoeuvres. This is analogous to contour inferencing, in the sense that the fox must infer the rabbit's path through space/time.

Now consider the forest, which we shall suppose to be a dense collection of grasses and trees. As the fox runs the image cast by the forest will change, with the grasses and trees going into and out of occlusion relationships. This variability and unpredictability will be further complicated by changes in lighting and cast shadows. No matter how important the structure of the forest is, there is no way for the fox to accurately infer it in detail. Particular objects project into the fox's image for only a short time before another object obscures them. Unlike the rabbit, the forest is more like a waterfall flowing past him. The fox must make do by observing only gross changes. These two events, the rabbit and the forest, define two conceptually different optical flow problems for the fox engaged in a chase [30].

1.3. *Two Structural Classes*

We are now in a position to relate the metaphor of the fox and the forest to the opening examples of contours and surface markings. From now on we shall consider only spatial patterns.[1] Like the path swept out by stable highlights on the rabbit,

[1] The full extension of this to space/time, or optical flow, will be dealt with in a subsequent paper.

occluding contours are usually well defined *almost everywhere*. However, like the forest or a waterfall, the pattern of hair on one's head is well defined *almost nowhere* in relative terms. Each hair passes into and out of occlusion relationships with neighboring hairs so rapidly that no clear hair contours are seen: rather the impression is one of a *flow* pattern supported primarily by occasional highlights.

Both types of pattern are rich in orientation structure, but they differ in spatial support. We take this difference to be fundamental for both computational and mathematical reasons. There is a precise sense in which contours are 1-dimensional entities, while hair patterns and (the static projection of) waterfalls are 2-dimensional. Consider, in particular, an infinitesimal neighborhood in the plane. The contours with which we shall be concerned will cover only a small portion of this neighborhood as they pass through it; in fact, a 1-dimensional subneighborhood.[2] The flow of a waterfall, however, will cover the entire neighborhood with "locally parallel" entities; i.e., with entities that are pointing in roughly the same orientation.

Precisely what these entities are will become clear shortly, as will their basic computational differences. But the following example may illustrate the differences more intuitively. Consider a family of 1-dimensional surface contours, such as the pin-stripes on a shirt. Pinstripes are well defined almost everywhere, except possibly in a neighborhood around shirt creases (or self-occlusions). Now, let the pinstripes become more dense, say by smoothly decreasing the distance between them. As they become dense enough to completely cover the shirt, with no intervening space, they will completely cover the shirt: in any image of them the indication of orientation will be gone.

The difference with a hair pattern can now be stated: the image support of the hairs is given by occasional cast shadows and highlights. While the hairs cover the surface completely, the image of its orientation structure does not. Somehow the orientation structure in areas where the hairs are occluded or where there are no highlights will have to be filled in by our visual systems.

In summary, while both contours and flows provide examples of orientation structure, there is a basic difference between them: their dimensionality. But this is just the beginning. As we now show, such differences are indicative of different functional roles in recovering structure from images. For example contours (of discontinuities) delimit image areas, while flows are defined over image areas. Such distinctions are reminiscent of "edge" and "region" differences in classical computer vision, and can be expressed within a framework of partial differential equations. This framework can be specialized for the recovery of orientation-based structure and generalized to allow the integration of other kinds of structure, as follows.

2. THE FRAMEWORK

The way in which occluding contours and hair patterns relate to the physical world leads to another conceptual difference between them: Occluding contours arise when surfaces intersect projectively; they give rise to *inter-surface constraints*. Hair patterns, on the other hand, provide information about the particular surface on which they lie: they give rise to *intra-surface constraints*. This difference is fundamental to the processes that must put various sources of information together.

[2] We shall not be considering either fractal curves or space filling curves in this paper.

The difference can be captured metaphorically by differential equations whose solution is governed by two distinct classes of constraints:[3] the differential operator, which constrains how the solution varies over its domain, and the boundary conditions, which constrain the solution value at certain points in the domain. Note that, for 2-dimensional differential equations, such as Laplace's equation, the differential operator is an *infinitesimal* function of two variables, while the boundary condition is given by, say, the values along a 1-dimensional contour. To be more specific, recall that Laplace's equation is

$$\Delta u(x, y) = 0,$$

where Δ denotes the differential operator $(\partial^2/\partial x^2 + \partial^2/\partial y^2)$. If Ω is an open neighborhood, then a well-defined problem would be to: Find u in Ω from prescribed values of Δu in Ω and of u on $\delta\Omega$, the boundary of Ω. Such problems are known as *Dirichlet problems*. Within Ω, u is completely determined by the constraints provided by Laplace's equation and by the value of u along the boundary $\delta\Omega$.

Infinitesimally such differential operators represent the kind of constraint available from flow patterns; while the boundary conditions resemble 1-dimensional contours. In terms of our previous examples, the Laplacian corresponds to the orientation information provided by an infinitesimal "piece" of a waterfall, while the boundary condition corresponds to its bounding contour. Boundary conditions can also be specified within the domain of support Ω, as is the case with surface contours and certain highlights.

The above example is, of course, metaphorical. We do not intend to say that waterfalls are Laplacians. Rather, it is the abstract mathematical form that concerns us, and it is not limited to waterfalls. Other sources of static intra-surface constraint come from monocular shape cues, such as shape-from-shading [6, 18]: from binocular stereo disparities [16, 9, 24], and so on. Intra-surface cues only hold for particular surfaces, when coordinates are imposed on them, they are 2-dimensional. However, the values of these intra-surface cues undergo an abrupt transition as the projected image passes through a jump from one surface to another. Similar arguments hold for transitions in lighting, say from an illuminated area to one in a cast shadow [13]. Note that all of these transitions lie along 1-dimensional contours; they are topologically different from the intra-surface cues. The 1-dimensional contours can be viewed as the boundary conditions that constrain the area over which the other, 2-dimensional constraints can be integrated.

2.1. *Generalization of the Framework*

The framework provided by inter- and intra-surface information, or, in different terms, by differential equations, holds not only for the features described here, but for abstractions over them. Contours arising from abrupt changes in, say, a flow or hair pattern could provide the boundary constraint to a higher-level process. This could correspond to a physical situation in which the underlying surface changed orientation abruptly, but the surface markings *smoothed* it over somewhat. Thus

[3] A third class of constraint is also needed, which delimits the domain of the operator, but which we shall not consider in this paper. Its effects are often linked to the other boundary conditions.

issues of how to differentiate flows become as important as the flows themselves. As we shall show, the inference of a flow field may even smooth over a discontinuity (Sect. 15)! In motion, or optical flow, even more such cases arise. For it is here that the fox had to detect abrupt changes in the flow of the rabbit from smooth ones. These changes, of course, relate to the rabbit's acceleration.

It is important to stress that the *difference between the constraints is in their dimensionality*, not in the fact that they correspond to intra- and inter-surface information. While this is often the case, it is not necessarily always so, as in the case of well-defined linear surface markings or highlights. Although these are *intra-surface* events, they do arise from *inter-light-source* arrangements. But they still serve nicely as boundary conditions.

3. STATIC AND DYNAMIC FLOWS

We are about ready to launch into the body of the paper, although one caveat is necessary. For the remainder of the paper we shall concentrate on static images, understanding completely that this is an idealization for biological vision systems. Natural systems respond to images both as functions of space and of time [2]. However, these systems do have a response in normal spatial frequency and low temporal frequency ranges; this is what we are studying. The extension to more dynamic image flows is in progress [30].

4. TANGENTS, VECTOR FIELDS, AND THE RECOVERY OF CONTOURS

Since orientation selection plays such an intimate role in contour recovery, we begin with the fundamental question: *what is a contour*? It is from the answer to this question that a definition of orientation selection can be obtained.

According to differential geometry, a contour is a locus of points that satisfies a given (but perhaps unknown) functional relationship. Smooth contours arise when the function is differentiable. In such cases we locally require that the tangent, or the direction in which the contour is going as it passes through each point, must be known as well. Therefore we propose that: *Orientation selection is the process of inferring (a representation of) these tangents*. The particular representation that we adopt is one of vector fields, or a collection of unit length "needles" arranged over a 2-dimensional, planar region. That is, a vector field is a mapping that assigns a vector to each point in the region. Each of these unit vectors points in a direction that is tangent to the contour at that point.

A physical example often helps to motivate the idea of a vector field. Consider a particle of dust moving in a dust storm. Clearly it sweeps out a curve through space such that, at every point, the velocity is given by a vector at that point. The length of this vector is proportional to the speed of the particle, and its orientation indicates the direction in which it is going just as it passes through that point. The velocity vector is always tangent to the path of the particle; the tangents at every point along a curve are an arrangement of vectors in $\Re^2$. Such an assignment of vectors to points in the plane is a vector field. The naturalness of this choice as a representation for contour inferencing will become clear shortly.

5. OUTLINE OF A MODEL FOR ORIENTATION SELECTION

We propose a specific computational model for orientation selection that is motivated by the differential geometry of contours. It consists of two distinct stages, the first of which amounts to orientation selection, or the inference of a vector field of tangents, and the second of which is one of curve synthesis, or the formation of integrals through this vector field. Since this second stage is a global one, it is not unrelated to computations of shear and compression in the vector field as well.

We concentrate on the first stage of the model; for a discussion of how to find integral curves through such vector fields, as well as how they could relate to the processes by which objects are physically formed, see Kass and Witkin [11].

While the overall goal of inferring a tangent field is not sufficiently constrained by the differential geometry to lead to an algorithm, the consideration of physiological evidence suggests that there are actually two conceptual steps to this first stage. The first of these steps is one of measurement, or the acquisition of information carrying signals about orientation information. We formulate this stage as a series of convolutions that model known receptive field structures. While this first step is quite standard, the next one is not. The second step is one of interpretation of these convolution values, and we show that it cannot be done by the most commonly applied algorithm: namely, select the maximal response. Rather, more complex interactions are required, and these are formulated into what we refer to as a *response matching problem*. While this leads to an abstract formulation in optimization terms, we next sketch how this formulation can be reduced to a cooperative network, and demonstrate that the network works. A much more detailed presentation of the network and its properties is in Parent and Zucker [17].

The real scientific test of any model is in its predictions, however, and several of these are discussed. The first two relate to the convolutions that provide the initial measurements, and predictions about curvature sensitivity and density are shown to be consistent with human psychophysics. The third class of predictions leads to functional speculations about how the structure of our visual systems reflects the structure of the world. It leads to what we refer to as the *Type I* and *Type II* distinctions. The numerals here refer to the dimensionality of support of the tangent fields.

Throughout the discussion, however, it is important to remember that orientation selection in particular, and grouping in general, are complex processes. There is a wonderful diversity in what appears, on the surface, to require only one process [35]. Throughout the entire paper we shall only be talking about early orientation selection. In order to delimit when it might be active, we shall require some control over sample image patterns.

6. DOT PATTERNS AND THE GEOMETRY OF ORIENTATION SELECTION

The problem of orientation selection is inextricably connected to many aspects of early vision. There are first-order applications in which it may be thought of being applied directly to image information, such as the recovery of linear surface markings, and second-order applications in which it can be applied to the result of local processes of intensity discontinuity detection. How, then, can orientation selection be studied in isolation? Our answer to this question consists in finding a class of patterns that captures the essential geometry of orientation selection without the added complication of the interconnections with other processes. Arrangements of dots provide just such a class of patterns, as we shall now explain.

Suppose a process of discontinuity detection has operated upon an image. Three qualifiers are necessary to characterize the output: (i) the spatial positions (in, say, retinal coordinates); (ii) the value at that position; and (iii) the process responsible for producing it. Note that point (i) is common to every orientation selection process; the arrangement of positions that define the contour. Since we are concentrating on early orientation selection, as was alluded to above, it is further convenient to assume that the positions are represented in a pointillist fashion—that is, as an array of dots. Note that the intensity values associated with these dots will not be considered here; they correspond to the information produced by the process (point (ii) above), nor will the particulars of the process. Rather, we shall assume that one composite process is responsible for conveying the (high contrast) dot patterns back to the orientation selective machinery (see Zucker and Hummel, [32], for one physiologically motivated model of such a process). Details of this preliminary process, such as the fact that positive and negative contrast data are separated, change only the details of our model, not its substance. In summary, then, it is dot patterns that capture the geometry underlying orientation selection, and which permit its study (to a large extent) independently of contrast, size, and other complications. These other feature dimensions must still be investigated, however. For some preliminary research along these lines, see [34, 20].

While dot patterns may seem an unnecessary abstraction of more elementary notions such as a dark contour drawn on a light background, this is not the case. Since the contour will be sensed by an array of photoreceptors, its first representation will be pointillist; it will be an arrangement of dark points within a field of light ones. Orientation selection is the process of recovering this contour from the pointillist array. As the dots become slightly less dense, the task of orientation selection begins to resemble the task of perceptual grouping. One of the consequences of the model presented in the next section is that similar (in fact, identical) mechanisms are sufficient for a range of orientation selection and grouping tasks, but not for all. It is this difference, which is a function of both the size and the density of the dots, that in principle separates early processing from later processing.

7. A MODEL FOR EARLY ORIENTATION SELECTION AND GROUPING

Since our ultimate goal is the recovery of curves, any theory of orientation selection must begin with the definition of a curve. This, of course, raises issues of differential geometry, which we begin in the next section. Our goal is to provide the formal motivation for our model which consists of two distinct stages, the first one aimed at producing a local representation based on quantized measurements, and the second one actually recovering (a representation of) the contour. The stages are as follows:

Stage 1. Estimate a vector field of tangents. This is a spatial arrangement of unit *arrows* touching the contour at exactly one position and pointing in the direction that the curve is going as it passes through that point. The tangent can thus be interpreted as the best linear approximation to the contour at a point, an interpretation that motivates our decomposition of this first stage of the model into two steps:

Step 1. Perform measurements on a representation of the dot patterns. These measurements will turn out to be related to physiologically observed receptive field structures. But their interpretation is not unique, so we must:

Step 2. Interpret the results of the measurements. This step is formulated both abstractly as a functional minimization problem (the *response matching problem*), and concretely as a cooperative network that computes solutions to the problem. Physiologically this amounts to interactions between the various receptive fields.

Stage 2. Find integral curves through the vector field. The second stage of the model is based on the fact that the tangent is the first derivative of the contour. This suggests that contours can be recovered by a process of integration, which for our model is primarily a matter of numerical analysis and will not be treated in this paper.

These ideas will now be developed in more detail, both intuitively and formally. We concentrate, in this paper, on the two steps comprising the first stage.

7.1. *What is a Curve?*

The mathematical definition of a (parameterized, differentiable) curve is a differentiable map $\alpha: I \mapsto \Re^2$, where $I \subset \Re$ is an open interval, $I = (a, b)$. Here t is a parameter that runs along the curve, and for each $t \in I$ there corresponds a point $\mathbf{x}(t) = ((x(t), y(t))$ on the curve. The image set $\alpha(I) \in \Re^2$ is called the *trace* of α; it consists of just the points through which the curve passes. In the dot examples, all we are given is an aproximation to this trace; our task is to infer the curve from it.

The first derivative of the curve $\alpha'(t) = (x'(t), y'(t))$ at t is called the tangent vector to α at t. It indicates the direction in which the curve is going as it passes through t. *By the definition of a first derivative, the tangent gives the best linear approximation to the curve in the neighborhood of a point.* This observation will be important shortly when we try to estimate the tangents. Since we shall require such tangents to exist, we shall assume that our curves are (at least) piecewise C^1.[4] If the tangent vector is of unit length, $|\alpha'(t)| = 1$, then the parameter t measures *arc length* along the curve. Unit tangents are convenient for us, because our approach to orientation selection will estimate the direction of the tangent, not its length (which we shall take to be unity). It is further convenient because the derivative of the tangent $|\alpha''(t)|$, the second derivative of the curve, is the *curvature*. Viewing the tangent as the best linear approximation to the curve at t, the curvature measures how rapidly the curve is deviating from this linear approximation.

It is important to differentiate between the curve $\alpha(t)$, which is a map, and the *trace*$(\alpha(t))$, which is a set. *The reason that orientation selection is not altogether trivial is that, in the vision context, the given information consists entirely of discrete samples of the trace of the curve, not of the curve itself.* Our plan will be to use estimates of the tangent made from observations on the trace as a basis for inferring the curve. The next question, then, is what sort of measurement can provide information about the tangents, and how should it be represented.

7.2. *Estimating the Tangent*

Suppose that two points along the curve are given, $\alpha(t)$ and $\alpha(t + \delta)$. For δ close to zero, the two points will be *neighbors* along the curve, and the line joining them

[4] The issue of how to detect the breaks, or discontinuities, between the smooth curves relates to the tangent field and, hence, to how it is formed. We report some experiments on estimating human sensitivity to such breaks later in the paper.

will be an approximation to the tangent, $\alpha'(t)$. Since this is what we are after, the problem of estimating the tangent thus reduces to estimating the line joining nearby points. When the parameter t is discretized so that points lie a unit arc distance apart, the line connecting them will be a unit tangent. In actual vision applications, however, noise will be introduced by the sensors, by quantization, and so on. Therefore the formulation that we seek should allow some further degree of flexibility, as in the following re-statement of the problem: given the trace of the curve, match a template for a unit line segment (i.e., a unit tangent) with the trace of the curve; or find the unit line segment that agrees most closely with the trace samples in a neighborhood around each point. This is a mathematical problem, and it can be solved in many ways; what, for example, does "most closely" mean? These and other points must be settled, so additional motivation and constraint are required. We shall seek them in an abstraction of early visual physiology, in order to introduce a sense of biological plausibility into the model. As we show, this leads not only to a successful algorithm, but to new psychophysical predictions as well. It is these latter predictions that provide a kind of check on the biological plausibility; should any of them turn out to be false, the plausibility would be questionable.

7.2.1. *Orientationally Selective Receptive Fields*

Our principle biological constraint comes from orientationally selective receptive fields. Recall that receptive fields of cells in the visual system are methodologically defined by the class of stimuli that influence them. *Orientation selectivity* is a property that requires non-isotropic stimulus patterns, such as bars and lines, and indicates that the response is a function of the orientation of the stimulus.

The striate cortex, area 17, is the earliest location in the primate visual system at which any orientation selectivity can be evoked. When receptive fields here are mapped, some exhibit an elongated structure with well defined excitatory and inhibitory regions. We shall be particularly interested in the subclass of cells that Hubel and Wiesel [7] referred to as *simple cells*, or (some of) which Schiller *et al.* [21] referred to as S-cells. These cells exhibit receptive fields with a single, usually elongated center and antagonistic surrounds on either side. We shall model these as an elongated difference of Gaussians with the form shown in Fig. 1. Electrophysiologically it has been reported that these cells respond maximally to straight line stimuli with a given orientation and location; if either of these are varied, then the response drops. (We shall exhibit a precise model for this shortly.) More recently, Wilson [26] has estimated the properties of these cells psychophysically, and they agree with the model as well.

The first step of our model is a convolution of operators whose structure resembles the receptive field of certain simple cells against a dot pattern image. Although there are certainly non-linearities in the responses of cells such as these, within the first step of our model and for the class of stimuli that we are using we shall assume that they are linear. There is at least some physiological evidence for this [22], and non-linearities will be introduced by the second step of the algorithm. There is, moreover, recent psychophysical evidence that the task requires a contrast difference, which lends further support to the convolutions [20].

Another property that has been widely observed both electrophysiologically and psychophysically is that these receptive fields span a range of sizes. While this is usually explained by introducing a notion of "scale" according to which small

FIG. 1. A display of the operators used to model the first matching step. It is composed of a difference of two 2-dimensional Gaussians with 3 : 1 aspect ratio, and approximates the Gaussian envelope measured psychophysically.

receptive fields are sensitive to thin contours and large ones to wide contours, it will turn out that multiple size receptive fields are necessary for a more fundamental, though subtle, reason.

Although there is more to the behavior of orientationally selective cells than that listed above,[5] the above properties are sufficient for extracting the essentials: (i) the receptive field models a local line template and (ii) these templates are applied to (a representation of) the image data by a process of convolution. The templates are discrete, so orientation is quantized into explicit directions, and each of these can be interpreted as a model for a tangent to a putative contour passing through its spatial support. *Thus the receptive field templates both measure the presence of tangents and provide a medium within which they can be represented.* And the computational model is specified: define a collection of line-like templates of different orientations and sizes. Mathematically these can be specified by a collection of operators $S_x(\theta, \sigma)$, where θ indicates the orientation of the operator and σ its size. It is instructive to specify the operator at the vertical orientation, i.e., pointing along the y axis:

$$S_x(0, \sigma) = \left[c_1 G(x, \sigma) - c_2 G(x, 1.75\sigma)\right] \cdot G(y, 3\sigma) \qquad (OP)$$

where

$$G(x, \sigma) = \frac{1}{\sqrt{2\pi\sigma}} e^{-|x|^2/4\sigma}.$$

The constants were estimated by Wilson and Gelb [26].

Now, at each point in the image, perform a convolution of each of these templates against the image. In mathematical terms, if we interpret the $S_x(\theta, \sigma)$ as (non-orthoginal) basis functions, then the above procedure is tantamount to performing an L_2 match of the image against them. Although the amount of computation required for this match is enormous, it can be carried out completely in parallel. Presumably primate visual systems take full advantage of this parallelism.

[5] Such as contrast non-linearities, etc,; see Wilson and Gelb [26].

7.3. *Orientation Estimation for Straight Lines*

From a computational point of view, the matching scheme outlined in the previous section—convolutions against $S_{\mathbf{x}}$ basis functions—amounts to an initial local measurement of orientation information. But this measurement scheme is *not* sufficently powerful to *detect* contours in any but the simplest of circumstances. Detection requires processing subsequent to the measurement. Detection requires not only measurements, but a decision to determine what the measurements mean. Such decision processes are inherently non-linear. As long as the contours are perfectly straight and far apart (with respect to the width of the operator's spatial support), and as long as the operator is centered over the contour, this detection can be straightforward: since the convolution $S_{\mathbf{x}}(\theta, \sigma)$ decreases monotonically with the difference in orientation between the operator and the contour, simply select the orientation at a point as that of the operator with the maximal response at that point. We shall call this scheme *straight maximum selection*, or *SMS*, since the selection of the maximal response at each position works only for straight lines. In symbols, for each $\mathbf{x}$, let the tangent $\lambda(\mathbf{x})$ be given by

$$\lambda(\mathbf{x}) = \max_{\theta} \{ S_{\mathbf{x}}(\theta, \sigma) * I(\mathbf{x}) \}. \qquad \text{(SMS)}$$

7.4. *The Curvature Constraint*

Now, assume the operator is still lying directly on the contour but suppose the contour is not straight within the spatial support of $S(\sigma)$. That is, let the contour curve. This introduces an inherent variation in response with the size of the operator that will totally disrupt the performance of (SMS). Somewhat surprisingly, this point is almost always missed in the literature, and slight variations on (SMS) are by far the most commonly expressed mechanisms for orientation selection.

There is one circumstance in which (SMS) is likely to work when the smallest operator is evaluated over thin contours. It will then cover only a discrete approximation to a "differential" element of the curve, which differential geometry tells us must be straight. The smallest operator thus defines the spatial precision of the vector field, and localizes the curve as precisely as possible within it. But larger operators will cover more of the curve. To illustrate how the response structure now differs from the straight case, consider an operator twice the smallest, unit size. This one covers about two unit tangent lengths. If the contour is straight, they will both be oriented identically, and the response profile will be the same as for the smallest one. However, if the contour curves, adjacent tangents will differ and the response will loose its monotonic structure.[6] Straight maximum selection cannot work.

But these response profiles—the variation in response as a function of orientation change—do contain the information necessary to generalize (SMS) into a scheme so that it can work. The key observation is that they contain information about curvature, or how the tangent changes direction within the spatial support of an operator. This information cannot be represented by the orientation at which the maximal response occurred—as it was for straight lines—but now requires information from all of the responses as a function of orientation at each point. We shall refer to this function as the **response profile**:

$$R^S(\theta, \sigma), \qquad \theta = \theta_1, \theta_2, \ldots, \theta_N;\ \sigma = \sigma_1, \sigma_2, \ldots, \sigma_M,$$

denoting the response of the operator S, size σ, at orientation θ, where

$$R^S(\theta_i, \sigma_j) = S_{\mathbf{x}}(\theta_i, \sigma_j) * I.$$

The response profiles are necessary for the generalization of straight maximum selection (SMS) into a full response matching strategy.

8. RESPONSE MATCHING PROBLEMS AND FUNCTIONAL MINIMIZATION

To generalize straight maximum selection, we shall shift our perspective to re-interpret what it was doing. Our immediate goal is to show that it worked because it was attempting to select a pattern which, if it were actually present, would give a response profile as close as possible (within the relevant universe of straight line segments) to the one actually measured. It is this matching of response profiles to known patterns with measured response profiles from unknown patterns that we refer to as a *response matching problem*.

The subsequent analysis is differential, so think in terms of small open neighborhoods around a point. We shall first indicate the structure of a response matching problem for estimating the tangent, then one for estimating curvature.

Suppose we are given an image I_{λ_1} of a straight line passing through the point $\mathbf{x}$ at orientation λ_1, say vertical. After convolutions against S, we would *expect* to get the response profile:

$$R^S_{\text{expected}}(\lambda_1 : \theta, \sigma) = \{ S(\theta, \sigma) * I_{\lambda_1}, \forall\theta, \forall\sigma \}.$$

Now, suppose we are given another image I_{λ_2} of a line passing through the same point but rotated one quantum in orientation. The convolution would now yield the *expected response profile*:

$$R^S_{\text{expected}}(\lambda_2; \theta, \sigma) = \{ S(\theta, \sigma) * I_{\lambda_2}, \forall\theta, \forall\theta \}.$$

which is just a shifted version of the previous one. Such *expected response profiles* could be obtained for all possible orientations of the line λ in images $I_{\lambda_1}, I_{\lambda_2}, \ldots, I_{\lambda_N}$. They signal what the perfect response profile would look like for all possible tangent patterns; i.e., for all distinct, quantized orientations of a straight line.

The above calculations can now be used to reformulate the selection process from one of selecting a maximal response to one of finding a "best" match, according to the following observation: Should any one of the line patterns λ_i, $i = 1, 2, \ldots, N$, be present in an image, then the result of the convolution should resemble one of the calculated response profiles $R^S_{\lambda_i}(\theta, \sigma)$. To be precise, suppose that we are given an image $I_{\hat{\lambda}}$ containing a line at an *unknown* orientation $\hat{\lambda}$. If we convolve the masks $S(\theta, \sigma)$ against $I_{\hat{\lambda}}$ the result will be an *observed response profile* $R^S_{\text{observed}}(\hat{\lambda}; \theta, \sigma)$. This is of the same form as the expected response profiles except that the unknown variable $\hat{\lambda}$ appears in it. Clearly $R^S_{\text{observed}}(\hat{\lambda}; \theta, \sigma)$ will match one of the

[6] For graphs of how these responses change as a function of orientation change, see [29, 35].

$R^S_{\text{expected}}(\lambda_i; \lambda, \sigma)$ most closely, and it is this value of λ_i that we should choose as the best instantiation of $\hat{\lambda}$. This is the line image which, if present, would give an expected response profile as close as possible to the one actually observed from the image with a line at an unknown orientation.

In symbols, then, we can formulate an abstract *response matching problem for inferring tangents to straight lines*:

> Let a collection of expected response profiles $R^S_{\text{expected}}(\lambda_i; \mathbf{x}; \theta, \sigma)$ be given for a universe of primitive tangent patterns λ_i, $i = 1, 2, \ldots, P$ centered at position $\mathbf{x}$. Further, let an image $I_{\hat{\lambda}}(\mathbf{x})$ which contains unknown tangent patterns $\hat{\lambda}(\mathbf{x})$ be given. The best estimate λ^* of $\hat{\lambda}(\mathbf{x})$ is the one that minimizes the norm difference between the expected and observed response profiles; i.e., the one that

$$\min_{\lambda_i} \left\| R^S_{\text{observed}}(\hat{\lambda}; \mathbf{x}; \theta, \sigma) - R^S_{\text{expected}}(\lambda_i; \mathbf{x}; \theta, \sigma) \right\| \qquad (1)$$

at each position $\mathbf{x}$,

The estimation of (un-normalized) curvature at a point $\mathbf{x}$ can also be formulated as a response matching problem, but this time the expected responses must be built up from larger operators whose receptive field or spatial support covers (at least) two unit tangents. If we think of these tangents as λ_i and λ_j as the tangents located at neighboring positions $\mathbf{x}_i$ and $\mathbf{x}_j$, then we arrive at an analagous formulation to the one above: select λ_i^* and λ_j^* such that

$$\min_{\lambda_i, \lambda_j} \left\| R^{S_{\text{large}}}_{\text{observed}}(\hat{\lambda}, \hat{\lambda}'; \mathbf{x}; \theta, \sigma) - R^{S_{\text{large}}}_{\text{expected}}(\lambda_i, \lambda_j; \mathbf{x}; \theta, \sigma) \right\| \qquad (2)$$

at each position $\mathbf{x}$.

8.1. *From Functional Minimization To Cooperative Network*

While the abstract formulation above is motivating, there are serious problems in trying to implement solutions to it directly. First, two tangents spanning three points are too local for accurately estimating curvature given discretization problems. Second, with larger neighborhoods the number of combinations of primitive tangents becomes overwhelming. Third, large operators will smooth over differences between local events, such as two nearby "c"-shaped curves or one "s"-shaped curve, so that inappropriate estimates will result. Finally, with large numbers of primitive patterns specified, the differences in the response profiles will become insignificant.

In general, the issue is how can we be guaranteed that a solution to minimization problem (1) above will be consistent with one from minimization problem (2) unless they are both solved simultaneously. Clearly the structure of the formulation must be used more directly in obtaining a solution. For example, recall that all of the convolutions and expected responses considered previously assume that the operator is centered on a piece of contour; responses from operators off the contour, but which overlap it to some extent at some orientation, will give responses also. These, too, must be dealt with.

A solution to these problems comes from realizing that variational and functional minimization problems can be solved using relaxation labeling techniques [8], and that the key to formulating a relaxation network consists in specifying the compatibility network. While the details here become significant, and are provided in [17], it is worthwhile to sketch a rough outline of the solution.

Associated with each possigle tangent λ at each positon $\mathbf{x}_i$ is a continuous variable[7] $p_i(\lambda)$. Neighboring tangents (in arbitrarily large neighborhoods) interact through compatibility functions $r_{i,j}(\lambda_i, \lambda_j)$ to maximize the support $S_i(\lambda_i)$ for tangent λ_i according to the formula:

$$S_i(\lambda_i) = \sum_{j, \lambda_j} r_{i,j}(\lambda_i, \lambda_j) p_j(\lambda_j)$$

where the above sum is taken over all tangent orientations λ_j at all positions $\mathbf{x}_j$ in the neighborhood of $\mathbf{x}_i$.

More precisely, the nature of the interaction is to modify the values of $p_i(\lambda)$ depending on how consistent λ at $\mathbf{x}_i$ is with its neighbors. A labeling is said to be consistent—in particular, solves a minimization problem like the ones sketched—when it maximizes all

$$\sum_{\lambda} S_i(\lambda) p_i(\lambda)$$

for all positions $\mathbf{x}_i$ and tangents λ simultaneously. The result is represented in the usual case by $p_i(\lambda *) = 1$, and $p_i(\lambda)$, $\lambda \neq \lambda^* = 0$. Appropriate formulae for accomplishing this iteratively are in [8].

What remains, then, is to specify the compatibility coefficients $r_{i,j}(\lambda_i, \lambda_j)$. As we have shown previously, tangents interact through curvature, or the change in the tangent's orientation as it is approached from either direction along a curve. Another way to characterize curvature is by the *osculating circle*, or the circle that just "kisses" the curve at a given point. We use the definition of an osculating circle to motivate the compatibility between two tangents: if a circle exists to which they are both tangent, then the tangents are said to be co-circular. Note that co-circularity is a function of both the orientation of the tangents and of their positions. It is a generalization of co-linearity in which the line is replaced by a circle. Compatibility is a simple function of co-circularity; it is one for co-circular tangents, and drops off smoothly with departure from co-circularity. The precise formulae and their derivations are in [17].

Co-circularity is a relation that can hold between putative tangents anywhere in an image. Their local neighborhood of interaction still remains to be specified. Here we use a notion of *curvature classes*, or a partitioning of an open neighborhood around each point into the discrete positions and orientations in which appropriate osculating circles can lie. Each curvature class amounts to an allowable estimate of curvature at each point. The resultant compatibilities are shown in Fig. 2 for the case of 7 curvature classes. The tangent being updated is the vertical one at the center of each display. These compatibilities generalize the earlier ones suggested on heuristic

[7] In biological terms, a firing rate.

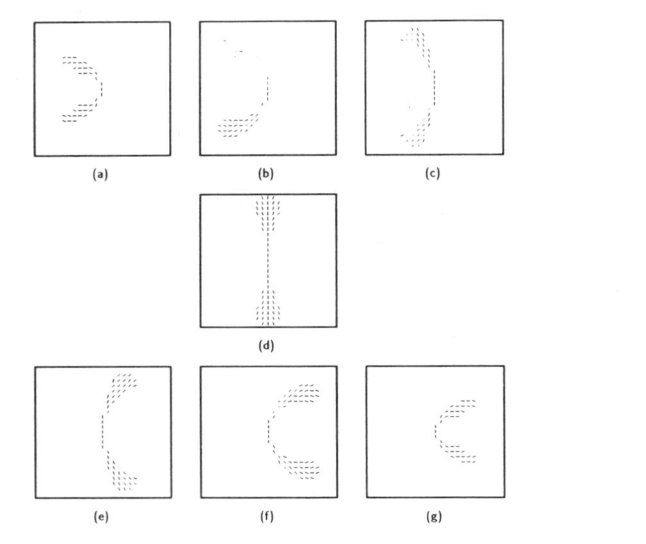

FIG. 2. The curvature classes and compatible tangent labels for the Type I process. Note how each of them contains a discrete approximation to a section of an osculating circle to the vertical tangent in the center of the displays (after Parent and Zucker [17]).

grounds by Zucker, Hummel, and Rosenfeld [33] which agree only for long straight lines that do not intersect or end. The current compatibilities extend them to curves.

9. RESULTS OF THE ORIENTATION SELECTION ALGORITHM

Now that we have formulated the orientation selection problem as two abstract steps, we must reduce those steps to realizable computations. The first step, the operator convolutions, only requires selecting discrete values for all of the parameters. We used a difference of two Gaussians, with a 3 : 1 aspect ratio and 1 : 1.75 ratio between central σ's, as suggested by Wilson and Gelb's [26] psychophysical data. Orientation (i.e., θ) was quantized into eight-directions, which is certainly too coarse for direct comparison with biological systems. The compatibilities for the relaxation process comprising the second step were the ones discussed above.

The result of applying this relaxation process to an image of a fingerprint is shown in Fig. 3. Note that the result is an arrangement of tangents that align exactly with the fingerprints, and whose orientation is consistent with theirs. This fingerprint pattern was chosen because it is more complex than the pure dot patterns that were used as motivation in the beginning of the paper. There are real variations in thickness, intensity, and contrast that have to be dealt with. The fact that the algorithm functions so robustly over this class of patterns suggests that it will work perfectly on pure dot patterns.

10. PREDICTIONS

The algorithm for orientation selection is necessarily abstract. It is derived primarily from the differential geometry of curves within the framework that visual

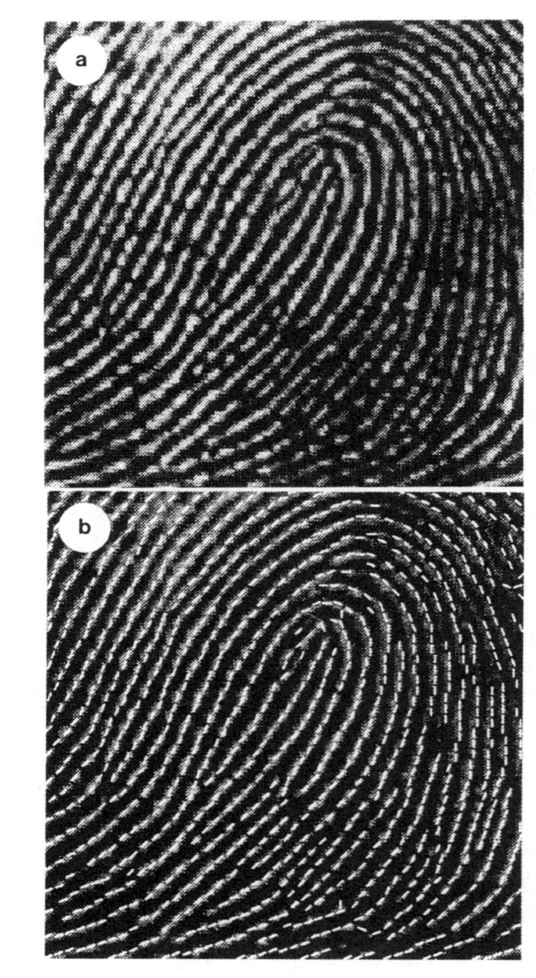

FIG. 3. An image of a fingerprint (a) and the computed tangent field (b). Each entry in the tangent field is a unit-length vector (with no difference between the head and the tail) pointing in one of 8 possible directions. The small amount of thickening at some orientations is due entirely to the coarse orientation quantization.

tasks this early are nicely formulated as matching problems. The algorithm works in a certain engineering sense. But does it have any psychophysical plausibility? Our goal in this section is to argue that it does. We shall simply sketch results, since they are all being expanded in separate papers.

The orientation selection algorithm has two steps; (i) convolutions with the oriented operators $S_x(\theta, \sigma)$, and (ii) the non-linear interpretation of these convolutions through relaxation matching. We shall concentrate on the first of these steps, since this is where the data for the response matching comes from. Should it be distorted, then we would expect the output of the response matching algorithm to be distorted. This is exactly what happens in the following two situations, both of which are designed to cause variations in the response solely with variations in the pattern. But it should be emphasized that the second processing step is also required, so that, e.g., the operator responses lying just off the contours are treated appropriately.

We shall now return to considering dot patterns as stimuli, because these will permit detailed control over the geometry of the patterns. Our particular goal is to vary the number and/or arrangement of dots that fall within the spatial support of the operators without varying the geometric structure of the patterns.

10.1. *Prediction 1: The Size/Spacing Constraint*

It takes two dots to define a line segment. It therefore follows that, for one of the operators $S_x(\theta, \sigma)$ to signal an oriented segment, there must be (at least) two properly oriented dots within the excitatory part of its spatial support (or receptive field).[8] Should this not be the case, then the operator will not signal the correct orientation. Perhaps the easiest way to check this is by varying the density of dots within a given pattern, leaving other structural aspects of the pattern identical. As long as two (or more) dots fall within the excitatory region, we would expect the percept to remain invariant.[9] However, as soon as the density falls below this level, so that there are less than two dots, we would expect the percept to change. For a given operator, moreover, this change should be abrupt. The prediction, to give it a first formulation, is that the percept should vary as a function of dot spacing.

But specifying spacing is not enough. Unless we also specify the size of the operators, one can be chosen to make the density take on any value. For dot patterns there is a natural manner in which the size of the operator can be chosen: by the size of the dots. For a given operator, dots that are much smaller than the excitatory center will have little effect, and those that are much larger will be canceled out by the inhibitory side lobes. There is a natural operator scale established by the size of the dots. The above prediction must be refined, then, to state that the percept should vary as a function of dot size and spacing.

We can use the Gaussian structure of the operators to quantitatively predict the function. By the above argument, the natural unit of distance is dot diameter. In the Gaussian context, take the excitatory center to be given by a Gaussian with $\sigma_{\text{excitatory}} = 1$ dot diameter. Since the other dimension is also given by a Gaussian weighting function, how far apart can two dots (with diameter $\sigma_{\text{excitatory}}$) be pushed before they no longer lie under significant support. That is, how does the integrated value of the Gaussian vary with the spacing between them? When the dots are 4.5 diameters apart, then the integrated Gaussian is about 96%. When they are 5 diameters apart, it is 99%, and more than 5 it is negligable. This means that, in the first case, when the dots are 4.5 diameters apart, there is something less than 4% of the Gaussian support left; for more than 5 diameters, there is essentially nothing. We thus have *a quantative prediction of the size/spacing constraint: the percept will change at a spacing of* 1 *dot*: 5 *spaces*. Below this point it should appear one way, and above it another.

[8] More precisely, there must be at least two more dots in the excitatory part than in the inhibitory part.
[9] Modulo perceived variations in density.

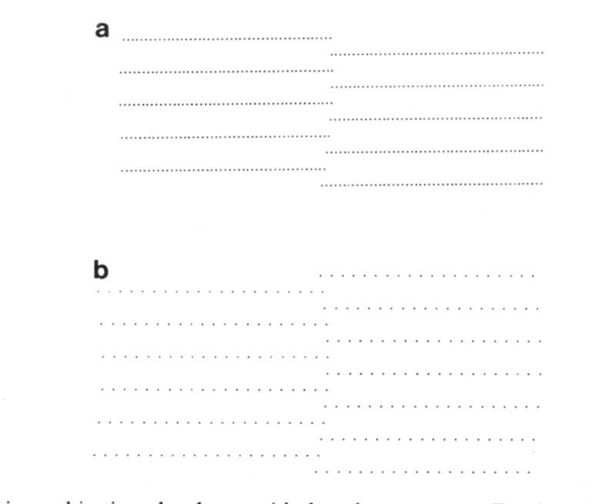

FIG. 4. A Kanizsa subjective edge drawn with dotted contours (a). For densely spaced dots, note the apparent difference in depth of the sides, and the brightness change along the edge. For sparse dots (b), however, note how these subjective effects disappear.

This is precisely what occurs, although a specific class of patterns are necessary to reveal it. The requirement on these patterns is that they signal something different when the dots are grouped according to the above, early mechanism as compared with other, higher-level processes. We have discovered that patterns which contain *endpoints*, such as the Kanisza [10] subjective edge or the "sun illusion" work perfectly. When the dots are denser than the size/density constraint, they give rise to apparent "endpoint"; otherwise, they do not. See Fig. 4. These results are discussed in more detail in Zucker [35].

10.2. *Prediction 2: Dot Positioning Effects*

In our examination of the size/density constraint we varied the number of dots falling within the spatial support of an operator; we did not vary the geometry of the dots. In this next series of experiments we shall vary the positioning of the dots to determine the effects on operator responses. Now the geometry will be varying in a smooth way, and we would expect the percept to vary as well. However, as we now show, this happens only insofar as the structure of the receptive fields allows it.

Consider, to be specific, a dot pattern in which the dots are constrained to fall on the arms of a "$\wedge$". How necessary is the *top dot* for defining the abrupt change in orientation? As we shift the dots along this pattern, so that the spacing[10] between them remains constant but their positions change, will there be a point at which the sharp wedge appears smoothly rounded at the top; i.e., will there be a point at which the "$\wedge$" will appear like a "$\cap$" in the neighborhood of the top?[11]

[10] Assume that the density of dots is on the dense side of the size/spacing constraint, so that the above mechanism is plausible.
[11] When the experiments were actually performed, the angles of the "arms" remained constant, unlike these graphical symbols.

Many distinct lines of reasoning are possible to answer this question. The first might argue that the top corner will always be apparent, since the two arms of the "$\wedge$" are always well supported by other dots. This is not the case in practice, however; there is a point at which the "$\wedge$" appears like a "$\cap$." How much shift is required, then? A second line of reasoning might suggest that any amount of shift is sufficent (provided it is beyond the limits of hyperacuity), because any of these would anchor the corner incorrectly. Or, one might argue, the dot must be shifted half of the inter-dot spacing, so that the pattern becomes symmetric and evenly curved on each side. Experimentally, neither of these holds either. Finally, our last line of reasoning is based on the model of orientation selection. It predicts that the dot must be moved far enough to properly inhibit the operators that signal the orientations on either side of the discontinuity; by the above geometric arguments demonstrating how dot size selects operators, one would have to predict a movement of one dot diameter. This is precisely what occurs in practice [Link and Zucker, in preparation; Link, in preparation]. Movements of $\frac{1}{4}$ or $\frac{1}{2}$ of a dot diameter cause no change.

11. FROM CONTOURS TO HAIR PATTERNS

The final test of the model comes from a class of patterns rich in orientation structure but possibly of a different sort than we have just considered. We shall now be concerned with the "flows" of waterfalls and surface markings discussed in the introductory sections. Recall that there we argued for basic structural and dimensionality differences. Our present goal is to use the model to determine whether in fact these differences exist.

In Fig. 5 we show an image of a fur pattern; in Fig. 6 we show the result of applying the algorithm to a window of it. Note that, although the predominant

FIG. 5. An image of a fur pattern. Note that the impression is one of a flowing hair pattern.

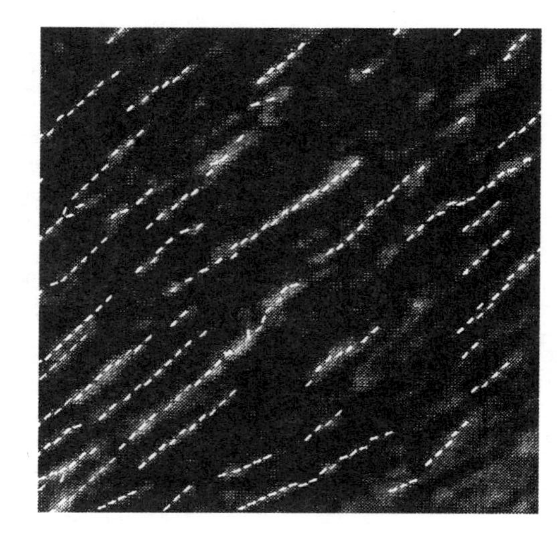

FIG. 6. The tangent field resulting from the contour finding algorithm running on a sub-window of the fur image (Fig. 5). Note that the tangents, while indicating the correct orientation are sparse and incomplete.

orientations agree with the image pattern, they are sparse and short; it is not the dense arrangement of tangents that we would expect to underlie such patterns. Something else must be done.

12. TYPE I AND TYPE II PROCESSES

To analyze the differences between contours, such as those in the fingerprint pattern, and flows, such as the fur pattern above, we must return to basics and ask: what kinds of physical structure can give rise to what kinds of orientation structure. The most common examples cited are connected with (so-called) edge detection, or the detection of discontinuities in physical surface orientation, depth, or lighting. The salience of these physical events is manifest in the fact that they often result in intensity changes in images. It is important to realize, however, that while objects are coherent in the physical world, the ray properties of light and the arrangement of photoreceptors in our retinas destroy this coherence; rather, they result in pointillist arrays of values such as intensity discontinuities. The role of early orientation selection is to reconstruct contours through these points, contours which, when properly interpreted, will be the precursors to those which bound objects or separate light from shadows. We shall refer to the processes designed to recover patterns in this class as *Type I processes* and to the patterns themselves as *Type I patterns*.

The defining characteristics of Type I contours are that they are 1-dimensional and that they have extended and well-defined spatial support. While this notion will be made more precise shortly, to understand it intuitively just look around at any of the bounding contours in your immediate surround. Most will have clear definition almost everywhere (but not necessarily everywhere), and are quite extensive. Some may even form closed curves, although this is a point that in our view is utilized, if at all, by processes more abstract than those that we have considered. Other examples in this category come from extended highlights. This is the class of patterns that our orientation selection algorithm functioned on, so we can conclude that it is a Type I algorithm.

The physical genesis of Type I contours is that they arise from the topological intersection of 2-dimensional events; *inter-surface* occlusions; *inter-lightsource* shadows, or *inter-surface-orientation* highlights. Except for unlikely circumstances,[12] Type I contours will be 1-dimensional.

But there is another class of physical patterns rich in orientation structure that have a completely different sort of spatial support; these are more intuitively generated by *intra-surface* events such as surface markings. Consider, for example, a pattern of hair or fur [23] or a field of wheat. While these are clearly rich—even dense—in orientation information, their physical structure suffers a complexity totally unlike bounding contours. Instead of the photometry giving rise to well organized intensity events, now the photometry almost defies categorization. The image results from highlights and reflectance changes of objects (hairs) that are going into and out of occlusion relationships so often that almost no long contours are visible; rather, the impression is more one of a *flow* than of a contour. Because the subjective appearance of these *flow* patterns is so different from the Type I patterns above, we shall refer to them as *Type II patterns*; the processes that recognize and recover them are *Type II processes*. The fur image is archetypal Type II, and we have already seen that our Type I model of orientation selection does not work on it, which lends further support to the distinction. Our principle remaining goal is to formalize the difference between Type I and Type II patterns by developing a model for inferring Type II orientation structure.

13. ESTIMATING TANGENTS IN TYPE II PATTERNS

If we take a fingerprint contour as a Type I pattern, and a fur pattern as a Type II pattern, their structural difference is immediate. The first consists of single contours, densely supported and distinct. The other consists of "fragments" of contours (imaged hairs) distributed homogeneously over a surface. This difference must be incorporated into the orientation selection algorithm. Two possibilities are open. First, we could alter the initial convolutions. This is unsatisfactory since we would then have to develop another set of operators sufficent for initial orientation measurements.

Thus we are left with a second alternative, to modify the response matching and relaxation. Since the random nature of hair patterns can be shown to be equivalent to randomly skewing the positions of short hair perpendicularly to their axis of orientation, within the matching process *each response should be interpolated across several spatial positions perpendicular to the orientation of the opreator*. This will have the effect of spreading the orientation formerly associated with just one position to a small region of positions, so that "implicit contours" could then be said to exist. That is, *Type II patterns consist of dense families of implicit contours*. It is these implicit contours that give rise to the flows.

[12] I.e., those of measure 0.

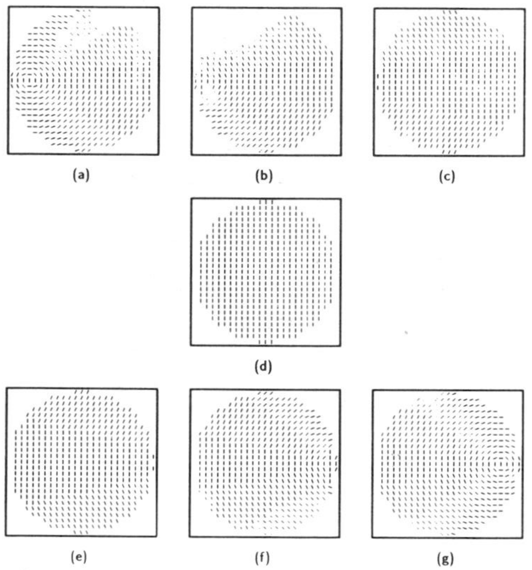

(a) (b) (c) (d) (e) (f) (g)

FIG. 7. The compatibility relationships for fur, and, more generally, for flow patterns. Note that they consist not only of distinct osculating circles, but of laterally interpolated families of them.

The basic difference between the response matching problems for Type I and Type II patterns, then, is the alignment and density of the tangent fields orthogonal to the direction in which the tangents are aligned. Intuitively, some notion of "spreading" of tangents orthogonal to their orientation is required, and this can be accomplished by supporting putative tangents not only by others in the direction they are "pointing," but also skewed slightly to either side. The lateral interpolation can be seen within the new compatibilities shown in Fig. 7 for Type II patterns. The result of applying it to the fur image is shown in Fig. 8.

14. THE DIFFERENCE BETWEEN TYPE I AND TYPE II PROCESSES

The difference in the relaxation networks for Type I and Type II patterns gives additional insight into where the names come from. Since the first, Type I matching process takes place only along the curve, it is essentially a 1-dimensional process. The Type II process, however, which incorporates interpolation i the other direction, is essentially a 2-dimensional process. Within the discretely quantized grid, this difference is apparent when the compatibilities are compared. However, the difference can be expressed more elegantly within the language of dynamical systems and differential flows.

14.1. Vector Fields and Flows

Recalling that the tangent is the first derivative of a contour, it follows that vector fields are intimately associated with differential equations. For

$$x' = f(x)$$

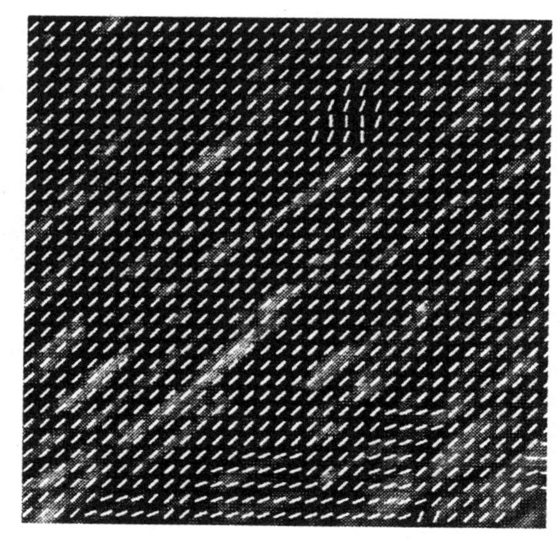

FIG. 8. The result of running the relaxation process with the new compatibilities incorporating lateral interpolation on the sparse tangent field in Fig. 6. Note that now the tangent field is dense.

$f = \alpha'$ is a mapping from $S \subset \Re^2$ into $\Re^2$. The space S is an open subset of Eucledian space that we shall refer to as the *spatial support for the vector field*; it is the collection of points over which it is defined. Another way to think of this trajectory is as a map from $\Re \times S \mapsto S$ defined by $(t, x) \mapsto x_t$. The map $\phi_t : S \mapsto S$ that takes x into x_t is what is normally referred to as a *dynamical system*. These, too, are intimately connected with differential equations by

$$f(x) = \frac{d}{dt}\phi_t(x)\bigg|_{t=0}.$$

Thus, for $x \in S$, $f(x)$ is a vector in $\Re^2$ which one can think of as the tangent vector to the curve $t \mapsto \phi_t(x)$ at $t = 0$. $\phi_t(x)$ is the flow of all points over which the vector field is defined, starting with x and parameterized by t.

14.2. The Dimensionality Difference Between Type I and Type II Processes

With this background on differential flows, we are now in a position to precisely specify the dimensionality difference between Type I and Type II patterns. Mathematically, a Type I pattern is a flow along the spatial support S which is a 1-dimensional sub-manifold of $\Re^2$; in a Type II pattern, S is fully 2-dimenstional. Conceptually, in Type I patterns, the points flow along a contour; in Type II patterns, entire regions flow along "locally parallel" families of contours.

This is not to say that families of Type I contours cannot be arranged along the second dimension; they can. This is exactly what happened in the case of the fingerprint. But the kinds of patterns that they create are structurally different, as can be seen by running the Type II interpretation process in the fingerprint pattern; see Fig. 9. Note that now, although the tangents are oriented appropriately, they no longer align with the individual contours; rather, they cover the surface of the fingerprint uniformly.

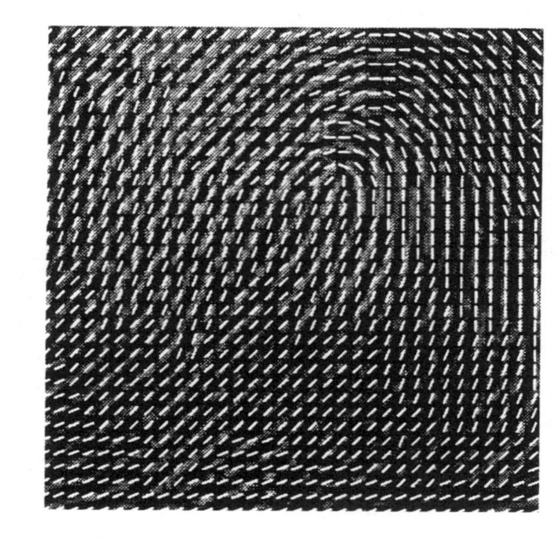

FIG. 9. The tangent field computed with the Type II process for the fingerprint pattern. Note that now the tangent field is dense (only every sixth label vertically and horizontally is shown) and that they do not align with the specific print contours.

15. PSYCHOPHYSICAL PREDICTIONS FOR TYPE II PATTERNS

To study human sensitivity to Type II patterns, it is convenient to have a dotted version of them. There is precisely such a class, known as random dot Moiré patterns (RDMPs), or Glass patterns [5]. Unlike those that we have been considering up till now, these have an orientation structure that is supported "randomly." There is no trace of a well-defined (albeit unknown) contour. Rather, there is a uniform distribution of short traces from a family of contours.

The structure of RDMPs becomes more clear when the mechanism for generating them is specified. It starts with a uniform distribution of random dots. A copy of this random dot pattern is then made. This copy is subjected to a transformation (such as a rotation, dilation, or vertical shift), and finally superimposed on the original. The result is striking for small transformations; see Fig. 10. The orientation structure imposed by the transformation is clear—in this case circular—although the pattern has a drastically different quality to it. Instead of consisting of clear, well-defined, 1-dimensional contours, now it consists of a 2-dimensional flow. *No contours are apparent running through it; just the impression of a flow.*

15.1. Prediction 3: Orientation Change Sensitivity in Type II Patterns

The additional lateral interpolation in Type II processing should effect human sensitivity to them and we can use RDMPs to study it. These differences are

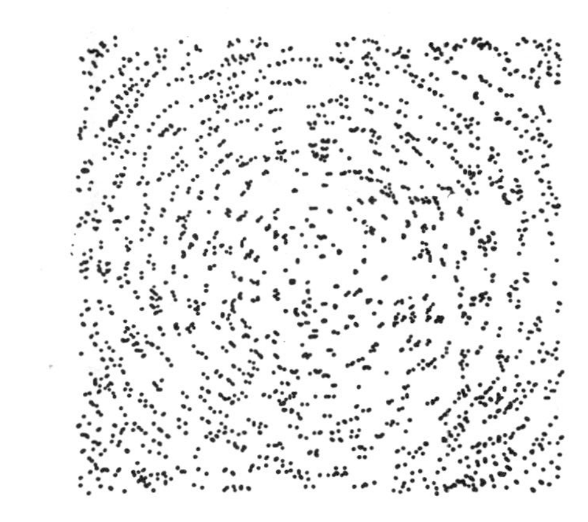

FIG. 10. A random dot Moiré (or Glass) pattern made with a circular transformation. Note how the impression is one of flow, and that no clear circular contours are apparent.

especially clear around discontinuities, as we show in the next figures. In Fig. 11, we show a parallel arrangement of Type I contours with a small orientation discontinuity horizontally through the center. These contours consist of pairs of dots at the correct orientation arranged randomly along the contour. Note that the horizontal discontinuity is immediately apparent. In Fig. 10b, however, we show the same dot pairs at the same orientation, but no longer constrained to lie along the straight contours; they can fall uniformly across the plane. The horizontal discontinuity is no longer apparent, and the display seems to change orientation smoothly from the top to the bottom. Larger differences in orientation are apparent both in arrangements of Type I patterns and in Type II patterns; see Fig. 12.

The reason why orientation discontinuities are smoothed over by the lateral interpolation is illustrated in Fig. 13. Further analysis and the effects of segment length (or number of overlays in RDMPs) are discussed in Link and Zucker [14].

The difference between the patterns in Fig. 11 can be used to illustrate the computational difference between them. Imagine applying the orientated operators in Fig. 1 to the top pattern in Fig. 11. It would give correct responses provided the lines did not get too close; at the point where the lines were really dense (less than 1 dot diameter apart) they would null the operators' response. The field would be essentially homogeneous. There is *no* way in which the operators could signal 2-dimensional versions of Type I patterns. In Type II patterns, however, the random skewing of dots allows the operators to function. The extra interpolation step, or the smearing introduced by the matching process, then allows the second dimension to be "filled in." This filling in accounts for a number of sensitivity differences in the perception of Type I and Type II patterns.

16. SIMPLE AND COMPLEX CELLS

The motivation for the first step of the orientation selection algorithm was based loosely on the physiologically observed performance of simple cells. These are cells that respond to, say, line stimuli as a function of both position and orientation. If

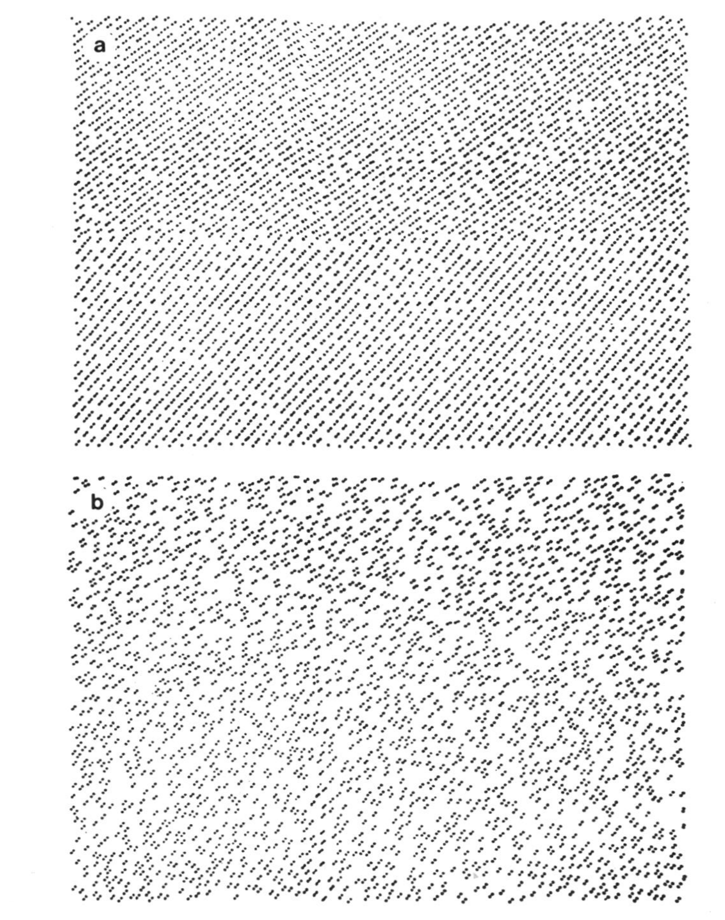

FIG. 11. An illustration of the difference between Type I and Type II patterns. (a) A pattern composed of parallel arrangements of Type I contours. The dots are in properly oriented pairs dropped randomly along the contours. There is a small orientation difference between the top half and the bottom half that is readily apparent. But the lines could not be moved much closer before the dots would interfere with one another along perpendicular directions. (b) A random dot Moiré pattern with the same orientation structure as the top pattern. Note that the line of orientation discontinuity is not visible.

either of these are varied from the maximal value, the response of the cell decreases. We are now in a position to re-examine this connection.

If we just consider the initial operator convolutions, then the above performance criteria hold. But it is significant to observe that when we add on the interactions between receptive fields necessary to implement the interpretation, they still hold. This is not, however, the case for Type II patterns and processes. In this latter situation, since the match takes place over a 2-dimensional neighborhood rather than just point-to-point, different performance criteria emerge. The response will still drop off with changes in orientation, since this would alter the initial convolutions. However, it will *not* drop off with positon until this variation takes it out of the effective matching neighborhood! That is, if one were to perform what might be thought of as the equivalent to the electrophysiology on the Type II algorithm, orientation change sensitivity would be high but positonal sensitivity (within a range) low. This is precisely what Hubel and Wiesel [7] have observed for complex cells, and leads us to conjecture that, for some simple and complex cells, *simple cells are responsible for Type I processing and complex cells are responsible for Type II*. The fact that some Type I/simple cell processing preceeds Type II/complex cell processing is also consistent with the physiology, since a large number of complex cells receive input through what appear to be simple cells [4]. Also, none of the tangent field computations shown here required more than 4 iterations; i.e., more than 4 layers of interconnected simple and simple/complex cells. Hence there is more than ample "machinery" available in V1 and V2.

While the precise mapping onto physiology remains to be investigated, there is one class of predictions that would greatly accelerate it. According to the model, we predict that simple cells should only respond sparsely and occasionally to RDMPs and flow patterns. Simple cells near to the layers that receive the first innervation from the lateral geniculate nucleus should show the most responses, since these are providing the earliest orientation measurements, while those further along in the processing hierarchy should show less. It is these higher order cells that are presumably implementing something like the relaxation matching. Complex cells, on the other hand, should show strong responses to RDMPs and flow patterns.

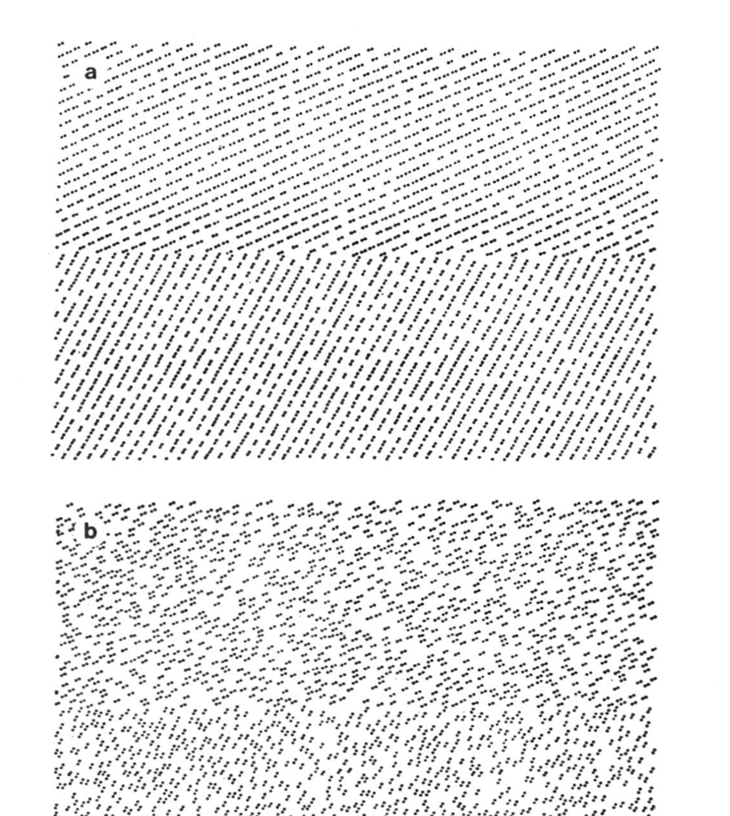

FIG. 12. A pair of patterns identical to the previous ones, except that now the orientation difference is larger.

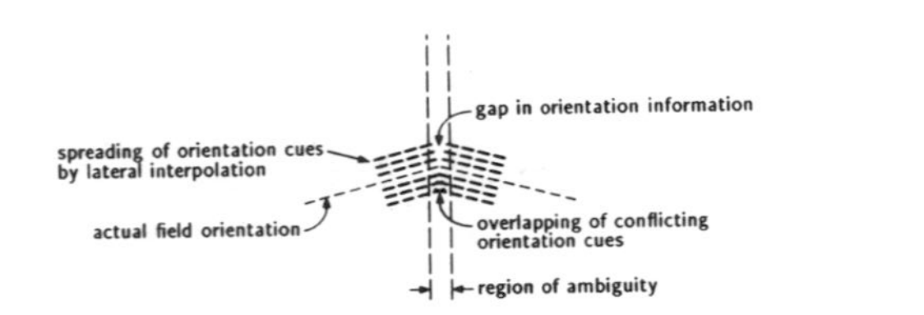

FIG. 13. An illustration of how the lateral spreading of orientation information causes problems in the neighborhood of a discontinuity [14].

17. CONCLUSIONS

How can contours be inferred from a collection of dots, or, more generally, from the intensity data available in biological vision systems? More generally, what is the role of contours in vision systems, and, more specifically, how is orientation connected with contour inference? The answers to these questions are all connected with the study of orientation selection, and they illustrate the scope of the investigation that we have begun.

Orientation selection was defined in abstract terms as the inference of a vector field of tangents. Since the tangent is the first derivative of a contour (at a point), this led to a 2-step algorithm for orientation selection: (1) try to estimate the tangent with a local measurement; and (2) since these measurements are ambiguous, interpret them. Both of these steps were given precise formulations as matching problems. The first match was further motivated by the rough structure of receptive fields early in vision. The second, called a response matching problem, amounted to a relaxation labeling network that can be thought of as excitatory and inhibitory interactions between the receptive field convolutions.

The algorithm worked, and, even more importantly, has led to a number of new psychophysical predictions about curvature sensitivity. But the emphasis in this paper was on orientation selection as an early process in vision. As such it should be syntactic, or data directed, and can provide only rough "first guesses" about significant orientation information. Another prediction, a size/spacing constraint, was confirmed psychophysically that seems to delimit when this syntactic processing could be reasonable. Heuristically it can be interpreted as defining conditions under which dotted lines are dense enough to function as solid ones. This heuristic was further supported by showing that densely dotted lines can trigger subjective figures and smooth curves, but sparse ones cannot. The size/density constraint is further connected to the model, because it constrains the operators that provide the initial orientation signals.

But not all patterns rich in orientation information can be handled by the contour detection process. Another class of patterns was uncovered that differ in a fundamental way. Waterfalls are an archetypal example of this class, in that they appear as dense flows rather than well-delimited families of contours. These flows arise from the fleeting highlights cast by the water, persist only briefly, and contain a random component. Random dot Moiré patterns model this class wonderfully, and we referred to them as Type II patterns. Type I patterns are the contours. The names, Type I and Type II, reflect the basic difference in dimensionality of the spatial support of the resultant tangent fields: Type I patterns are 1-dimensional, and Type II patterns are 2-dimensional. This dimensionality difference was reflected in the second matching step. It also leads to various sensitivity differences which have physiological counterparts. Thus we conjectured that Type I processes are supported by simple cells, and Type II processes by complex ones.

For orientation selection, one of the earliest and most general of the inverse problems comprising vision, the constraints necessary for a solution were derived from differential geometry. An algorithm for inferring oriented entities was developed that led, in examining its biological plausibility, to a number of predictions that have since been confirmed. Thus the paper began the cycle of theory, prediction, and confirmation in the context of orientation selection. Our next steps are to extend the theory into motion, or early optical flow. All indications are that analagous results will hold, and we remain excited about the shape of the predictions.

ACKNOWLEDGMENTS

An earlier version of this paper was presented at the Workshop on Human and Machine Vision held in Montreâl, August, 1984. Research was sponsored by NSERC Grant A4470. The simulation was formulated with and implemented by Pierre Parent and the psychophysics were done with Norah Link. I thank them and Yvan Leclerc for comments on the manuscript.

REFERENCES

1. T. Binford, Inferrring surfaces from images, *Artif. Intell.* 1981.
2. D. Fleet, A. Jepson, and P. Hallett, *A Spatio-Temporal Model for Early Visual Processing*. RCBV-TR-84-1, Computer Science Dept., University of Toronto, 1984.
3. J. Gibson, *Perception of the Visual World*, Houghton Mifflin, New York, 1950.
4. C. Gilbert, Microcircuitry of the visual cortex, *Ann. Rev. Physiol.* 1983.
5. L. Glass, Moire effect from random dots, *Nature (London)* **243**, 1969, 578–580.
6. B. Horn, Obtaining shape from shading information, in *The Psychology of Computer Vision*, (P. Winston, Ed.), McGraw-Hill, New York, 1975.
7. D. Hubel and T. Wiesel, Functional architecture of macaque monkey visual cortex. *Proc. R. Soc. (London), Ser. B*, 1977, **198**, 1–59.
8. R. A. Hummell and S. W. Zucker, On the foundations of relaxation labeling processes, *IEEE Trans. Pattern Anal. Mach. Intell.* **PAMI-5**, 1983, 267–287.
9. B. Julesz, *Foundations of Cyclopean Perception*, Univ. of Chicago Press, Chicago, 1971.
10. G. Kanizsa, *Organization in Vision*, Praeger, New York, 1979.
11. M. Kass and A. Witkin, Analyzing oriented textures, in *Proc. Int. Joint Conference on Artificial Intelligence*, Los Angeles, 1985.
12. K. Koffka, *Principles of Gestalt Psychology*, Harcourt, Brace & World, New York, 1935.
13. Y. Leclerc and S. W. Zucker, The local structure of intensity discontinuities in one dimension, *Proc. 7ICPR*, Montreal, 1984.
14. N. Link and S. Zucker, *Sensitivity to Corners in Flow Patterns*, Technical Report 85-3, Computer Vision and Robotics Lab., McGill University, Montreal, 1985.
15. D. Marr, *Vision*, Freeman, San Francisco, 1982.
16. J. Mayhew and J. Frisby, Psychophysical and computational studies towards a theory of human stereopsis, *Artif. Intell.* **17**, 1981, 349–385.
17. P. Parent and S. Zucker, *Curvature Consistency and Curve Detection*, Technical Report 85-12R, Computer Vision and Robotics Lab., McGill University, Montreal, May, 1985.
18. A. Pentland, *Local Shading Analysis*, Tech. Note 272, SRI International, Palo Alto, Calif., 1982.
19. A. Pentland, Fractal-based description of natural scenes, in *Proc. Image Understanding Workshop*, 1983, pp. 184–192.
20. S. Pradzny, Some new phenomena in the perception of Glass patterns, draft, Schlumberger Palo Alto Research Center, 1985.
21. P. Schiller, B. Finlay, and S. Volman, Quantative studies of single-cell properties of monkey striate cortex. I. Spatiotemporal organization of receptive fields, *J. Neurophysiol.* **6**, 1976, 1288–1319.
22. R. Schumer and A. Movshon, Length summation in simple cells of cat striate cortex, *Vision Res.* **24**, 1984, 565–571.
23. K. Stevens, Computation of locally parallel structure. *Biol. Cybern.* **29**, 1978, 19–26.
24. D. Terzopoulos, Multilevel computational processes for visual surface reconstruction. *Computer Vision Graphics Image Process.* **24**, 1983, 52–96.
25. M. Wertheimer, Laws of organization in perceptual forms, *Psychol. Forsch.* **4**, 1923, 301–350; transl. in *A Source Book of Gestalt Psychology*, pp. 71–88, W. Ellis, Routledge & Kegan Paul, London, 1938.
26. H. Wilson and D. Gelb, Modified line-element theory for spatial-frequency and width discrimination, *J. Opt. Soc. Amer.* **A1**, 1984, 124–131.
27. A. Witkin and J. M. Tenenbaum, The role of structure in vision, in *Human and Machine Vision*, (J. Beck, B. Hope, and A. Rosenfeld, Eds.), Academic Press, New York, 1983.
28. S. W. Zucker, On the structure of texture, *Perception*, **5**, 1976, 419–436.
29. S. Zucker, *Early Orientation Selection and Grouping: Evidence for Type I and Type II Processes*, Technical Report 82-8, Department of Electrical Engineering, McGill University, August, 1982.
30. S. W. Zucker, *The Fox and the Forest: A Type I, Type II Constraint for Early Optical Flow*. Technical Report 83-11, McGill University, Montreal; presented at the *ACM Workshop on Motion: Representation and Control*, Toronto, 1983.
31. S. W. Zucker and R. A. Hummel, Toward a low-level description of dot clusters: labelling edge, interior, and noise points, *Comput. Graphics Image Process.* **9**, 1979, 213–233.
32. S. W. Zucker and R. A. Hummel, Receptive fields and the reconstruction of visual information, *Seventh International Conference on Pattern Recognition*, Montreal, July 1984.
33. S. W. Zucker, R. A. Hummel, and A. Rosenfeld, An application of relaxation labelling to line and curve enhancement, *IEEE Trans. Comput.* **C-26**, 1977, 393–403, 922–929.
34. S. Zucker, K. Stevens, and P. Sander, *Similarity, Proximity, and the Perceptual Grouping of Dots*, A.I. Memo 670, Artificial Intelligence Lab., MIT, April, 1982; *Percept. and Psychophys.* **34**, 1983, 513–522.
35. S. W. Zucker, Cooperative grouping and early orientation selection, in *Physical and Biological Processing of Images*, (O. Braddick and A. Sleigh, Eds.), Springer, New York, 1983.
36. S. W. Zucker, A. Rosenfeld, and L. S. Davis, General purpose models: Expectations about the unexpected, in *Adv. Papers Fourth International Joint Conference on Artificial Intelligence*, Vol. 2, Tblisi, USSR, September, 1975.

Chapter 3

Recognition and Labeling of Scene Objects

This chapter deals with the problems of recognizing and assigning names or labels to the individual objects depicted in an image. Difficulties arise in: (1) describing the way objects can appear in an image as a result of occlusions, shadows, projective distortions, and so on; (2) describing the appearance of formless objects, such as a crumpled sweater or a cloud, or nonrigid objects such as a person; (3) deducing purpose and function from image appearance when these semantic attributes are the primary criteria for classification—for example, a bridge is recognized primarily by function rather than by geometric shape—and (4) selecting the relevant set of reference models (from a potentially infinite set) to label imaged objects.

There is no known general computer vision solution to this collection of problems. Existing approaches are dependent on controlling the viewing context and/or the variability of the scene content. Four main paradigms are currently used for accomplishing the recognition and labeling task:

1. Attribute-based classification in which a set of measurements of an object is used to assign it to one of a small number of predefined classes. For example, an orange can be distinguished from a watermelon on the basis of size, color, and weight.

2. Matching of an explicit geometric or photometric description of a reference object to image data in order to find instances of the reference object. For example, one may have stored in the computer 3-D models of several manufactured objects and use these stored models to determine which of the objects is currently being viewed.

3. Location or map-guided interpretation (or monitoring) in which physical location of the object is a primary attribute in identification. The critical knowledge is not a description of the object itself, as in the second paradigm, but rather a method of finding the object relative to other scene components; recognition of the object is then often quite simple. For example, when trying to determine whether a label has been properly placed on a bottle in a manufacturing process, one might first locate the bottle quite precisely and then determine the label placement using the bottle as a reference.

4. Domain-constrained interpretation in which specialized knowledge of a limited domain is used to determine the identity of a few selected objects central to that domain. The contextual constraints peculiar to the application domain are the critical knowledge items explicitly provided to the recognition system. For example, in a medical Xray of the chest, one can use general knowledge of the rib structure to identify the complete outlines of organs, such as the heart, that may be partially occluded by the ribs.

These paradigms are discussed in the following four sections.

3.1 Attribute-based Classification

Suppose we let each coordinate axis of a multidimensional space represent a physically measurable quantity, such as size, color, or orientation. A set of such measurements for a particular object can then be represented as a point in this "attribute" or "feature" space. If the measurements capture the essential similarity between members of the same class and essential distinctions between classes, then points representing objects in the same class will be clustered together—that is, will have small spatial separation in the feature space—and clusters representing distinct classes will be widely separated. Typical examples of the use of this approach can be found in the literature on classification of (multispectral) remotely sensed data [Bernstein 78] and the literature on character recognition [Nagy 82]. The important theoretical issues are:

- how to select an effective set of measurements.
- how to partition the attribute space so that appropriate classification decisions can be made for previously unseen objects.

No formal procedure exists for deciding on the features to be used in the initial description of a recognition problem. However, there are formal procedures for reducing the dimensionality of a given feature space (selecting a smaller number of features). Such procedures are motivated by the need to reveal hidden dependencies in the measurements and by the desire to reduce the computational requirements. Procedures have been developed to determine those subspaces of the original measurement space that will preserve or nearly preserve the classification accuracy of the original space. An extensive bibliography of feature selection procedures prior to 1978 is given by Decell and Guseman [Decell 79], and recent developments are given in [Tubbs 82] and [Young 84].

The partitioning of feature space can be based on either "supervised" or "unsupervised" techniques. In the supervised approach a set of classes is defined, and the experimenter provides a training set of labeled objects from each of these classes; the statistical parameters extracted from the training set are used to formulate a classification rule. The classification accuracy depends on the selected measurements and the experimenter's skill in choosing the training set. In the unsupervised approach a set of classes is not predefined. The basic premise is that the degree of similarity will be high among members of a class and low between members belonging to different classes. "Natural" clusters will therefore form and provide a basis for partitioning the feature space.

Most of the work in *statistical pattern recognition*[1] is based on the attribute space paradigm. The voluminous literature includes texts such as Duda and Hart [Duda 73], Tou and Gonzalez [Tou 74], Castleman [Castleman 79], and Watanabe [Watanabe 85]. The journal *Pattern Recognition* publishes many papers in this field. A good collection of readings is [Agrawala 76].

In addition to the problem of obtaining the required multivariate statistical distributions, the basic problem with attribute-based classification is that the descriptive power of a set of measurements is limited—there is no explicit mechanism for describing relationships among the attributes that do not correspond to other (individual) predefined attributes. This approach is, therefore, not suitable for recognizing complex objects for which such interrelationships are critical in the classification process. As a result, statistical pattern recognition plays a very limited role in current computer vision research and its applications.

3.2 Interpretation Based on Explicit-shape Models

Explicit-shape models can be used for recognition and labeling if a fairly simple geometric description of an object can be based on attributes that are readily measured or computed. Such models are most often employed in situations where the exact geometry of expected objects can be prespecified, as in industrial vision. This approach is generally unable to deal with flexible objects such as the human body, formless objects such as clouds, or complex natural objects such as vegetation.

The basic problem in the explicit-shape recognition paradigm is how to determine efficiently when one of a number of stored models appears in an image—given a set of image measurements that may be noisy, may be derived from a partially occluded object, may

[1] The term statistical pattern recognition denotes a methodology for solving classification (naming) problems as decision problems posed in probabilistic terms; it is assumed that all of the relevant probability values are known.

come from some combination of objects, or may come from objects having unknown locations and orientations. The papers described in this section differ along two important dimensions: (1) the complexity of the features used, and (2) the choice of using discovery or search to locate these features. These dimensions are not unrelated. For example, simple features such as surface normals, which are pervasive and individually not too informative, are often used in a discovery-based approach in which models are defined in terms of constraints and consistency measures on these features. An interpretation is found that is consistent with the many image measurements. Methods at the other end of the spectrum depend on prespecification of significant and distinguishing model features such as holes and unique object shapes and clusters of such features, followed by search for these features in the image. An interpretation is obtained by verifying the existence of a small number of these more complex features. In general, a technique that precomputes model characteristics for use in a predictive mode will be much faster computationally than a discovery approach. Sometimes a combination of search and discovery is used, with discovery first employed to find relevant features. Then constraints are used to identify feature combinations that satisfy an object model, and search is used to verify the presence of predicted additional features of the model describing the hypothesized object. Another distinguishing characteristic of the methods discussed is whether the recognition process is based on the analysis of 3-D data (obtained from stereo or from an active ranging device), or the analysis of 2-D photographic images. In the latter case there must be some effective method for projecting the stored 3-D model shapes and relational information into a 2-D form compatible with the 2-D imagery.

It should be noted that all the papers here assume that a limited set of object types are to be recognized (a dozen or less, rather than thousands), and that most existing techniques would be impractical (and perhaps ineffective) if large numbers of object types were required to be recognized.

Taking a constraint-based approach, Grimson and Lozano-Perez [**Grimson 84**] show how features consisting of 3-D surface point locations and surface normals derived from range sensor data may be used to identify and locate objects from among a set of known objects. The objects are modeled as polyhedra having up to six degrees of freedom relative to the sensor. The authors show that inconsistent hypotheses about pairings between sensed points and hypothesized object surfaces can be discarded by using local constraints on distances between faces and angles between face normals. The possible assignments of sensed points to model faces for an object are represented in the form of an "interpretation tree," an exhaustive enumeration of all the faces on which each sensed point can lie. The heart of the technique is the efficient pruning of the interpretation tree using local constraints. The approach of Faugeras and Hebert [Faugeras 83] is based on the use of more complex features. They approximate each object by a set of planar and quadratic faces, resulting in a relational graph model in which nodes correspond to faces and arcs connect adjacent faces. Dense range data obtained using a laser range finder are analyzed to segment the surface of the object into planar and quadratic patches. Matching is performed using a best-first search for a consistent pairing between observed and model faces. The use of more complex features simplifies the graph search problem.

Mulgaonkar, Shapiro, and Haralick [Mulgaonkar 84] describe a recognition scheme using relational and rough geometric shape information about 3-D man-made objects such as chairs and tables to recognize instances of the objects in single images of scenes. The features used are the boundaries of 2-D image regions that should correspond to the projections of the boundaries of 3-D model parts. The regions are decomposed into constituent parts using a clustering procedure, and the decomposed 2-D views are then compared with the rough 3-D model descriptions. The technique is based on the fact that the camera position constrains the 2-D projection of various parts of the object. The propagation of these constraints from one planar object surface to another through the projection equations is derived, and this constraint propagation guides the matching scheme in interpreting the scene.

While conceptually similar, the preceding papers used distinct approaches to deal with constraints. Grimson and Lozano-Perez used pruning of the interpretation tree describing relations between point properties of planar surfaces. Faugeras and Hebert analyzed a graph of relations between object faces. Mulgaonkar, Shapiro, and Haralick used projection equations to propagate constraints about visibility of projected planar face contours.

Early versions of ACRONYM [Brooks 81] were discovery-based, but the scope of ACRONYM was later extended to include prediction of image features and their relations based on 3-D geometric models [**Brooks 83**]. The 3-D models are represented by hierarchies of generalized cones that project as ribbons[2] or as ellipses. The ability to project the 3-D generalized cone to 2-D as a simple ribbon and to retain the relationships between component parts of the object is an important feature of the generalized cone representation. Given an image, the interpretation module uses predictions of what shapes will be visible and predicted ranges of shape parameters to identify object components in the image (ribbons and ellipses derived from an edge image) and to verify and make more precise the 3-D information given by the model concerning the shape, location, and orientation of the object. To identify L-1011 and B-747 aircraft, ACRONYM matches observed ribbons to predicted ribbons for general descriptions of wings and fuselage and then finds clusters of ribbons consistent with the combined wings and fuselage of a general 3-D aircraft model [Brooks 83]. When carrying out the initial prediction and interpretation, ACRONYM first uses the most general set of constraints—those associated with wide-bodied jets. Once a consistent match or partial match to a geometric model has been found, ACRONYM determines the particular aircraft subclass (L-1011 or B-747). The ACRONYM system has no special knowledge of aerial scenes, airports, or aircraft; all its rules are about geometric relationships.

The main contribution of ACRONYM is the nonlinear constraint manipulation system that allows constraint implications to be propagated downward to predict the shape and relationships of the generalized cylinder-based image features. During interpretation, measurements are used to put constraints on the parameters of the 3-D models; the local matches are retained only if these constraints are consistent with the model. In their review of 3-D object recognition, Besl and Jain [Besl 85] consider ACRONYM to be an "open loop" system because there are no feedback connections between the final decision-making mechanisms and the original data: "ACRONYM's problems provide a reminder for us that any open-loop system is only as robust as its most limited component. Even the best possible geometric reasoning system cannot be successful if its input is consistently unreliable and no feedback paths exist."

A combination of search and discovery is used in the Three Dimensional Part Orientation (3DPO) system described in [**Bolles 83**]. The basic premise here is that hypothesis generation and matching should be based on two or three key features or feature clusters rather than on many features. The discovery process should concentrate on finding these few features. If a few features are sufficient to distinguish between possible interpretations of an observed object, then almost any matching approach, even exhaustive search, can effectively determine the best match between image data and object models. In 3DPO the model of an object consists of two components—an extended computer-aided design (CAD) model and a set of feature classification networks. The extended CAD model contains a standard volume-surface-edge-vertex description as well as pointers linking topologically connected features. For example, each edge points to the two surfaces that form it, and each surface contains an ordered list of its boundary edges and a list of its holes. The feature networks described in the paper organize the features according to their types and sizes; a future version will include other groupings such as surface elements that share a common normal, or cylinders that share a common axis.

The first step in designing a 3DPO classification system is to enumerate potential feature clusters and to evaluate their utility. This process is accomplished semiautomatically. Local features to be matched are selected, and tables of relative positions are precomputed. Object recognition by 3DPO consists of a low-level feature detection phase followed by the growth of clusters of features. Then hypotheses about part types and locations are generated and verified by further image analysis.

A similar approach is employed in Bolles and Cain [Bolles 82] for recognizing 2-D object configurations (with possible occlusions). Here the design step required to select and evaluate feature clusters is accomplished automatically. A sophisticated "maximal clique" matching scheme is used to combine and evaluate inconsistent evidence when recognizing objects in imaged data.

The computationally efficient approach for explicit-shape matching by Goad [**Goad 83**] also concentrates on the design or planning stage in which useful information about the model is compiled. In the execution, or runtime stage, the precomputed information is exploited for the rapid recognition of the objects in the image. An object feature is a line along the object surface at which either a surface normal or a reflectivity discontinuity occurs. Image features are straight-line segments along which an intensity discontinuity occurs. The recognition algorithm performs a simple

[2] A ribbon is a 3-D object part that projects to a thin, elongated structure in the 2-D image.

depth-first search for a match between object and image features. A current hypothesis about the position and orientation of the object relative to the camera is used to predict the location of object edges, look for these edges in the image, and if successful, back-project to obtain a better estimate of camera location. Of particular interest is a scheme that precomputes for each edge (and efficiently represents) camera locations from which the edge can be seen. Goad's principal contribution is the development of algorithms that exploit this kind of precomputation to yield the best possible runtime performance. The paper also describes many nuances that make the approach effective.

An excellent survey article by Besl and Jain [Besl 85] describes most of the existing literature in 3-D object recognition. They note that range-image understanding has become an important and recognized branch of computer vision, since high-quality range images contain a wealth of explicit information obscured in intensity images. They feel that range-image vision systems will soon surpass the capabilities of intensity-images systems in many environments, perhaps leading to new insights about the recognition process as well as improved performance.

3.3 Location- and Context-Constrained Interpretation

If we can always locate an object of interest relative to a map or some other known object, and if we know the relationship of the map to the image, or the position in the image of the other known object, then we can find the object of interest, when visible, and deduce its state. In the case of a map-referenced object, some examples of object states are water height in a reservoir, the presence of a ship at a dock, or the level of pollution at an industrial site. The key idea is that the map constrains where to look in an image and what to look for. In an industrial context a part may be defined in terms of its position on a major assembly; most of the effort goes into finding the major assembly. For example, one may locate the valve spring assembly on an engine head precisely to determine whether or not the stem lock has been attached [Perkins 81]. Note that location-constrained interpretation is more useful for monitoring or inspection than for recognition and that such a system can easily be fooled since its expectations are so strong.

The map-guided interpretation system of Tenenbaum [Tenenbaum 80] can be used to monitor predefined targets by locating the monitoring sites in an image. The critical step in exploiting the given map database is to establish a geometric correspondence between image and map coordinates, which then allows known ground sites to be located in the image. A symbolic map containing explicit ground coordinates and elevations for all monitoring sites, as well as landmarks, is used as the reference in conjunction with an analytic camera model. The camera model is first calibrated in terms of known landmarks and then used to transform between map coordinates of designated sites and their corresponding image coordinates. An interesting aspect of this work is the relatively small amount of symbolic information needed to accomplish a fairly sophisticated set of interpretation tasks.

It is not necessary to use a precise map to locate structures in an aerial photograph. For example, Nevatia and Price [Nevatia 82a] locate structures in aerial images by using a rough sketch of a geographic region. The sketch is converted into a graph representation; objects and feature values are at the nodes, and relations between objects are the arcs. The system analyzes the image to find objects and generates a symbolic description using the same graph-like representation as the model. The model and the derived symbolic descriptions are then matched by first finding corresponding model and image objects and then matching the relationships between these objects. For example, given a sketch of the San Francisco area showing the land-water boundaries, the bridges, and the airports, this system is able to locate these entities in a corresponding aerial photograph.

3.4 Domain-Constrained Interpretation

In domain-constrained interpretation attention is restricted to a small subset of objects that can appear in the world. Interpretation is greatly simplified by limiting attention to objects associated with a particular application or purpose, such as airports or medical Xrays. The critical information elements in this recognition paradigm are the contextual constraints and assumptions peculiar to the application domain, which are explicitly provided to the recognition system. The success of this approach depends on the extent to which one can ignore objects not predefined to be in the domain, or the extent to which one can control the appearance of nondomain objects in the image.

The System for Photointerpretation of Airports using Maps (SPAM) [**McKeown 85**] integrates map knowledge, image-processing tools, and rule-based control and recognition strategies to label objects and regions appearing in aerial images of airports. A rule-based system is used to control the image processing and interpretation of results. Some of the rules help to create hypotheses about the identity of each region based on its size, shape, texture, and elevation; other rules control region growing and image resegmentation. Consistency rules define groups of mutually consistent fragments and encode simple observations that are generally true for airport scenes. Finally, some rules specify situations inconsistent with general knowledge of airport layout. Currently, the system can extract and identify runways, taxiways, grassy areas, and buildings from region-based segmentations for several images of National Airport in Washington, D.C. This experimental OPS5-based system requires many CPU hours on a VAX 11/780 to carry out its analysis.

A medical apllication [Ballard 78, 79] involves the use of model-directed detection of ribs in chest radiographs. The system contains an executive program that can extract information from a declarative structure (a relational graph) and can invoke components of a procedural structure. The declarative structure is primarily used to represent gross geometrical relations between the ribs in the ribcage. The procedural structure contains low-level programs that can find the detailed representations of the ribs in the image. Key aspects of the approach are the use of the relational graph to reduce the area to be searched and the use of planning to select efficient rib-finding procedures for a given local context. Other interesting domain-specialized labeling systems include Nagao and Matsuyama [Nagao 80]; Belknap, Riseman, and Hansen [Belknap 86]; and Nazif and Levine [Nazif 84].

3.5 Discussion

In this chapter we described the most advanced systems and approaches for the specific task of naming the discrete objects appearing in single images of real-world scenes. These systems must solve the problem of perceptual organization where, as noted in Chapter 2, there is some exciting progress but still no indication of the shape of a comprehensive solution. They must obtain and use information about scene geometry—often by employing active sensors rather than by using the still immature methods[3] discussed in Chapter 1. Such systems must also select some formalism for describing the classes and objects of interest (Chapter 6) and then compare sensed data to these stored descriptions (Chapter 7). They are generally limited to basing their final decisions strictly on appearance, since methods are not yet available for describing and reasoning about context and purpose (although some progress has been made, see Chapter 4).

In the face of all the preceding challenges, the object recognition problem can be solved with today's technology only by limiting the scope of the world in which these systems operate. In the 2-D world, commercial systems are in common use for character recognition and other aspects of document interpretation [Nagy 82]. The controlled world of the factory offers the best environment for existing 3-D object recognition technology, and indeed, commercial robotic systems [Beni 85] are already playing an important role in modern manufacturing plants.[4]

In the next chapter we discuss what appears to be a more difficult problem—describing complete environments rather than just single objects. However, it is often the case that the additional context more than pays for the complexity it introduces.

Papers Included

Bolles, R. C., R. Horaud, and M.J. Hannah. 3DPO: A 3-D part orientation system. In *Proc. 8th Int. Joint Conf. on Artificial Intelligence*. Karlsruhe, West Germany, pages 1116–1120, Aug. 1983.

Brooks, R. A. Model-based 3-D interpretations of 2-D images. *IEEE PAMI* 5(2):140–150, March 1983.

Goad, C. Special purpose automatic programming for 3D model-based vision. In *Proc. Image Understanding Workshop*, pages 94–104, June 1983. (Also see the version in S. Pentland (editor). *Pixels to Predicates*. Ablex, Norwood, New Jersey, 1986.)

Grimson, W. E. L. and L. Lozano-Perez. Model-based recognition from sparse range or tactile data. *International Journal of Robotics Research*, 3(3):3–35, 1984.

McKeown, Jr., D.M., W.A. Harvey, Jr. and J. McDermott. Rule-Based Interpretation of Aerial Imagery. *IEEE PAMI*, 7(5):570–585, September 1985.

[3]Even automated stereo techniques that are routinely used in cartographic (terrain modeling) applications involving aerial imagery are neither fast enough nor fully competent to deal with the complexities of the high resolution, 3-D world at ground level.

[4]Thus, our selection of papers is heavily weighted toward recognition based on specific object models; computer-aided manufacturing already provides such models.

3DPO: A Three-Dimensional Part Orientation System*

Robert C. Bolles, Patrice Horaud, and Marsha Jo Hannah

SRI, International
333 Ravenswood Ave.
Menlo Park, CA 94025

ABSTRACT

A system for recognizing and locating three-dimensional objects in range data is presented. Design considerations are discussed, followed by a description of a new object representation scheme, designed to facilitate object recognition. Experimental results are given for a preliminary implementation of the object modeling scheme and for an initial version of the object recognition system.

INTRODUCTION

One factor inhibiting the application of industrial automation is the lack of general methods for acquiring a part from unstructured storage so that it can be presented to the work station in a known position and orientation. A classic example of this limitation is the "bin-picking" problem--automating the job of a worker who reaches into a bin of jumbled parts and retrieves one part at a time, as needed. Solving this problem is difficult because a part in a bin may lie in infinitely many locations and may be partially occluded or buried.

Our research goal is to develop general-purpose techniques for recognizing and locating partially visible parts with three-dimensional (3-D) uncertainties in their locations. To achieve this goal, our approach is to develop 3-D part models that facilitate interpretation of range and intensity information on the basis of reasoning. Our rationale is that, first of all, range data simplify the locational analysis because the 3-D geometric information is encoded directly in such data. Second, familiarity with the model of a part will add enough new constraints to make it practical to recognize and locate different parts jumbled in a bin.

Most previous work in the area of range data processing has either concentrated on the recognition of simple, generic objects, such as polyhedra, cylinders, and spheres (e.g., see [1,2,3]) or on the description of a scene in terms of generic primitives, such as planar patches (e.g., see [4,5]) or in terms of generalized cylinders (e.g., see [6,7]). Only a few groups have investigated techniques for recognizing specific objects (e.g., see [8,9,10]). Previous work in bin picking has often pursued a two-step recognition and location strategy: (1) physically acquire a part from the bin, and (2) determine its part's location in the hand or on some intermediate holding place, such as a table (e.g., see [11,12]).

Our ultimate goal is to develop a program that can understand a configuration of parts well enough to determine the location of a part and direct an arm to grasp it, then predict which parts will move when the selected one is lifted out of the pile. This approach will result in a one-step process in which the arm system knows the approximate location of each part when it picks it up. To reach this goal, we are simultaneously exploring ways to represent specific features of parts in order to facilitate the recognition process, while developing a recognition procedure that capitalizes on the information available in these extended parts models.

* The work reported herein was supported by the National Science Foundation under grant DAR-7922248.

FIGURE 1 CASTING TO BE LOCATED

We are considering moderately complex parts, such as the casting in Figure 1, for two reasons. First, they are typical of a large class of industrial parts. Second, we expect (and have found) the abundance of features in a complex part to be helpful rather than confusing because distinctive clusters of local features can be used to recognize the part from a partial view of it. For example, two concentric circles connected by a planar surface strongly suggest the end of a thick-walled pipe, such as the portion of the casting in Figure 1.

One of our basic assumptions is that hypothesis generation and matching should be based on a few (e.g., 2 or 3) features or feature clusters rather than on many (e.g., 30) features, each with several (e.g., 5 to 10) possible interpretations. Using this approach, almost any matching algorithm, even an exhaustive search, can efficiently determine the best match. The problem is to be able to find those few features that are sufficient to distinguish the part. The solution to this problem is based on feature representation and preliminary planning.

The use of a small number of feature clusters is also appropriate because each feature provides a substantial amount of information. For example, the position and orientation of a dihedral edge determines all but one (position along the edge) of the six degrees of freedom associated with an object; the position and orientation of the end of a cylinder also determines all but one degree of freedom (rotation around its axis).

The function of the preliminary planning is to enumerate the occurrences of feature clusters and evaluate their utility so that the execution system can efficiently use each detected feature. This goal-oriented planning system may even provide additional information, such as a list of nearby features that may help identify a feature. A significant amount of computer time performing this type of analysis is justified because it is only done once and because it increases the efficiency of the run-time system, which is executed many times.

One goal of our research is to develop a program that automatically converts computer-aided-design (CAD) models of 3-D parts into a recognition-based representation that will help the system recognize and locate the parts. To achieve this goal we expect to extend the techniques we developed for the analysis of two-dimensional (2-D) parts [13].

We have so far concentrated on the representation of 3D objects and on the detection of feature clusters in range data. In this paper we first outline the recognition process and then briefly describe our progress in the development of a representation and the detection of feature clusters.

THE 3D PART ORIENTATION (3DPO) SYSTEM

We partitioned the recognition process into five steps:

(1) Primitive feature detection
(2) Feature cluster formation
(3) Hypothesis generation
(4) Hypothesis verification
(5) Parameter refinement.

For the first step we envision a combination of special-purpose hardware that performs the computationally intensive low-level detection of primitive features, such as discontinuities, and software that locates object-specific features, such as a circular arc with a specific radius. The hardware analysis will be applied to all images in a uniform way except for possibly changing a few parameters for different types of parts. The software processing involves the application of general-purpose techniques that use lists of part features to obtain specific feature parameters.

The purpose of the cluster formation step is to "grow" clusters of features that can be used to hypothesize the type, position, and orientation of a part. In this step a feature cluster is grown by locating features that are directly related to ones in the cluster. Each feature in a part model has a list of related features. The types of relationships include "topologically connected" and "in the same plane."

In the third step, one or more clusters of features are used to generate hypotheses about the part types and locations. If individual clusters are sufficient, they are used; otherwise, a strategy similar to the one used in the local-feature-focus method [13] is used to recognize clusters of clusters. Note that this process involves more searching than the cluster formation step because there are no relationships between the clusters except their relative positions that can be used to suggest good candidates for extension. It is this step in the recognition process that must be minimized to avoid the problems of combinatorial explosion.

The purpose of the fourth step is to check a hypothesis by looking for additional features that are consistent with it. There are two ways to do this: one is to use the hypothesis to predict the locations of features to be found; the second is to see if the features already detected agree with the hypothesis. The first way is sounder, but it requires a prediction system that includes computationally expensive operations, such as hidden-line elimination. Finally, given a verified hypothesis, the fifth step applies an averaging technique, such as a least-squares, to compute a more precise location of the part.

In the following sections we briefly describe the part-modeling system and techniques for locating feature clusters. Throughout this discussion, we use examples from the recognition of the casting shown in Figure 1 in range data obtained by a projector-camera range sensor [14]. Figure 2 shows a perspective image of the 3-D intersections of the sensor's light plane with two castings of the type shown in Figure 1, one up-side-down relative to the other.

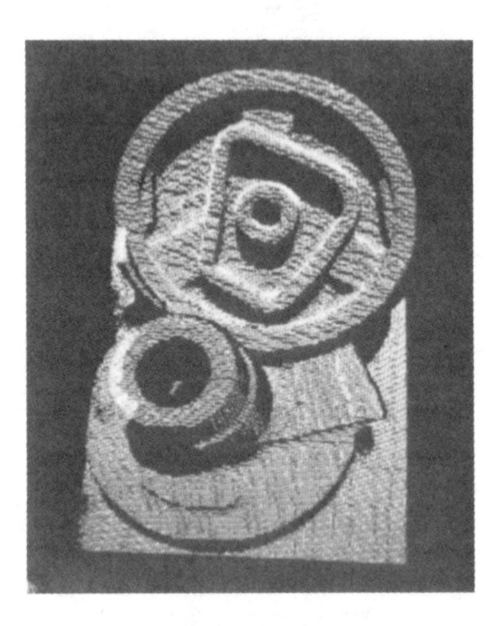

FIGURE 2 PERSPECTIVE IMAGE OF PLANAR LIGHT INTERSECTIONS WITH TWO CASTINGS (137 SLICES WITH 240 POINTS PER SLICE)

OBJECT MODELS

CAD systems and their representations are designed for constructing and displaying parts, not for recognizing them. Even though a CAD model may be complete in the sense that it contains all the 3-D information about a part, there are more convenient representations for a recognition system than a CAD model. For example, a CAD model might state the size and position of a hole, but a recognition procedure needs a list of all holes that size. In general, a recognition system must be able to provide answers quickly to such questions as:

How many features are there of a given type and size?
Which surfaces intersect to form this edge?
What other features lie in this plane?
What neighboring features could be used to distinguish this feature from others like it?

To answer these questions efficiently, each part feature should be listed under several different classifications. For example, a dihedral edge should be in the lists of edges bounding the two surfaces that meet and form the edge; it should be in the list of all dihedral edges (classified according to their included angles); and it may also be in other lists, such as the list of features in a specific plane. This redundancy is the key to efficient recognition.

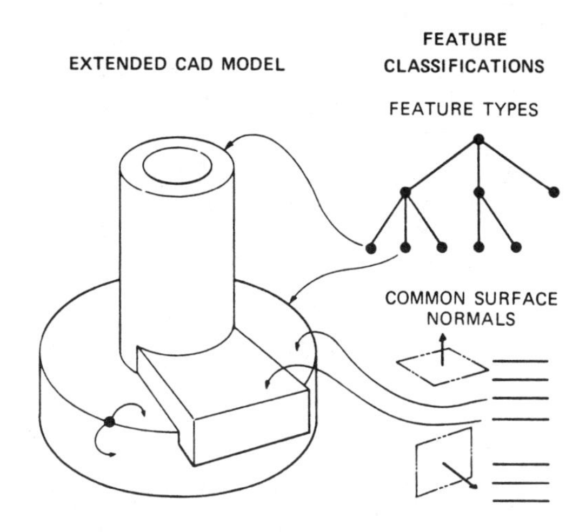

FIGURE 3 MODELING A 3-D PART BY EXTENDING ITS CAD MODEL AND CLASSIFYING ITS FEATURES

In the 3DPO system the model of a part consists of two components--an extended CAD model and a set of feature classification networks (see Figure 3). The extended CAD model contains a standard volume-surface-edge-vertex description as well as pointers linking topologically connected features. For example, each edge points to the two surfaces that form it, and each surface contains an ordered list of its boundary edges and a list of its holes. We have implemented a preliminary version of this type of model, using a pointer structure similar to Baumgart's winged-edge representation [15].

The feature networks in the current implementation classify the features according to their types and sizes. We plan to include other groupings, such as lists of surface elements that share a common normal or lists of cylinders that share a common axis. Each representation will be designed for a set of special-purpose procedures that analyze the data in terms of one property, and yet all the representations will contain pointers back to the extended CAD model, which will serve as the core representation.

We implemented a set of routines that can answer some of the questions mentioned earlier by analyzing the topology of a part and extracting entries from the classification lists. For example, the current system can quickly produce a list of all circular edges that are concave dihedral angles and have radii between .8 and 1.0 inch.

In this modeling system we differentiate between view-independent relationships, such as topological connectivity, and view-dependent relationships, such as image proximity (in pixels). The view-independent relationships are functions of the inherent geometric characteristics of the part, such as its size and topology, and can easily be enumerated by an analysis of the part model. The view-dependent relationships, on the other hand, are functions of the sensor, its type, and its location; they are significantly more difficult to enumerate. Even a simple part can have fifty or more structurally different views [16]. We plan to investigate techniques similar to those used in ACRONYM [17] for representing classes of similar views.

FEATURE DETECTION AND HYPOTHESIS GENERATION

The current, partial implementation of the 3DPO system can detect features, form simple clusters, and hypothesize objects. The system has two methods for detecting edges in range data. One method analyzes the discontinuities occurring in slices through the scene, and the other applies a zero-crossing edge detection method to a "height" image (e.g., see [18]). In the first method each discontinuity is labelled according to its geometric and photometric significance and then the labelled discontinuities are linked together. Figure 4 illustrates the types of discontinuities used in this method. The position of the light source is used to distinguish an object edge that casts a shadow from the edge of that shadow. Figure 5 shows the classified discontinuities in one slice of the data shown in Figure 2. Figure 6 shows the chains of discontinuities found in Figure 2 by the line-by-line linking process. Figure 7 shows only the chains that may correspond to physical edges in the scene. These are the chains that the program later tries to identify as specific object features.

The second method of detecting edges treats the range data in Figure 2 as a gray-scale image, locates zero crossings, and then filters out the insignificant ones (i.e., those that are too weak or too short). Figure 8 shows a grey scale version of the data in Figure 2, and Figure 9 shows the zero crossings detected at one resolution that passed the significance tests.

After the edges have been found, the system tries to identify object-specific features, such as a cylinder with a radius of .9 inches, by performing the following operations: (1) partitioning each edge into sequences of points that lie in a single plane; (2) partitioning each planar sequence into linear segments and circular arcs whose radii match radii in the part models; (3) characterizing the surfaces that intersect to form each segment in terms of their types, approximate sizes, and the relative angle between them; and (4) constructing a list of model features that match the characteristics of each observed segment. Figure 10 shows a plane fitted to a sequence of points and the position of the matching cylindrical body from the model. Figure 11 shows all the circular features found in Figure 2. Each one matches only one or two model features.

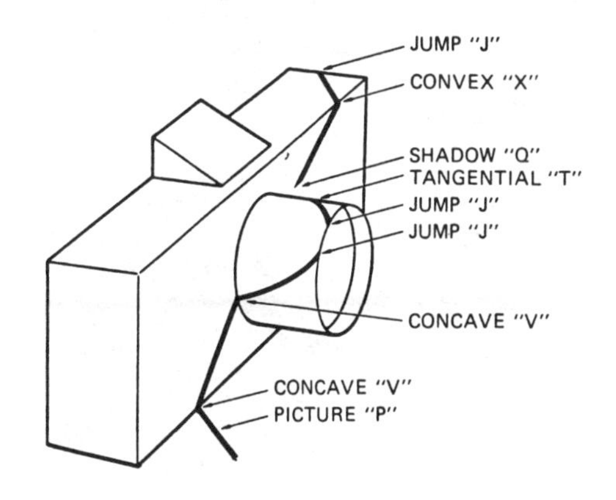

FIGURE 4 DISCONTINUITY CLASSIFICATIONS ALONG A SLICE

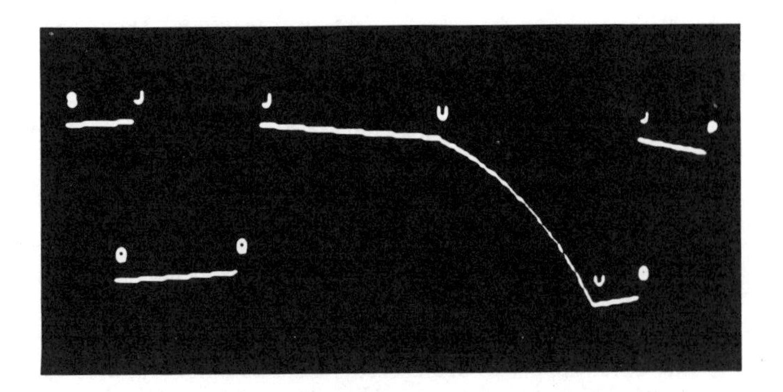

FIGURE 5 CLASSIFIED DISCONTINUITIES IN ONE SLICE

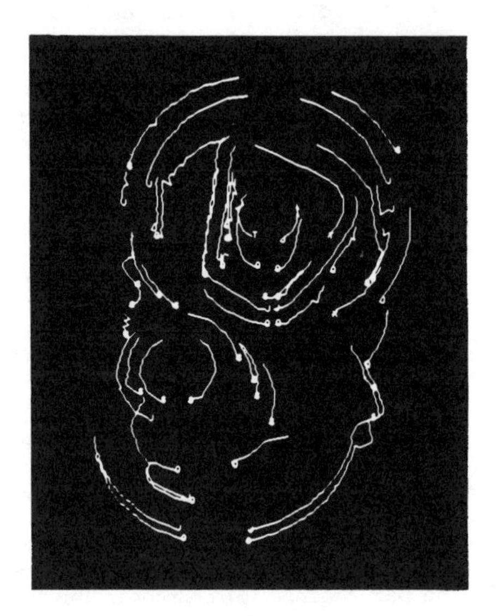

FIGURE 6 CHAINS OF DISCONTINUITIES FOUND IN FIGURE 2

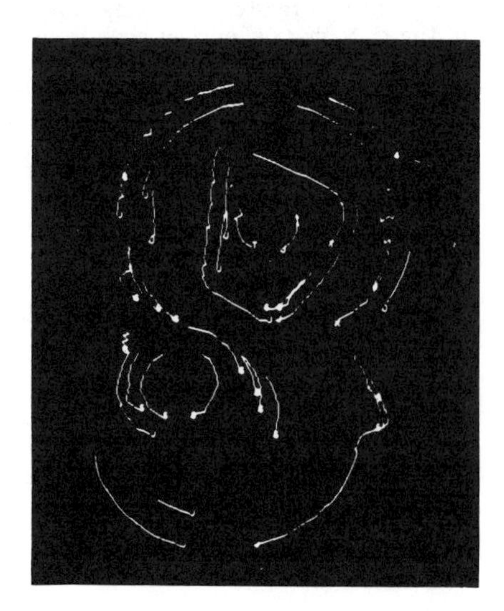

FIGURE 7 CHAINS THAT MAY CORRESPOND TO PHYSICAL EDGES

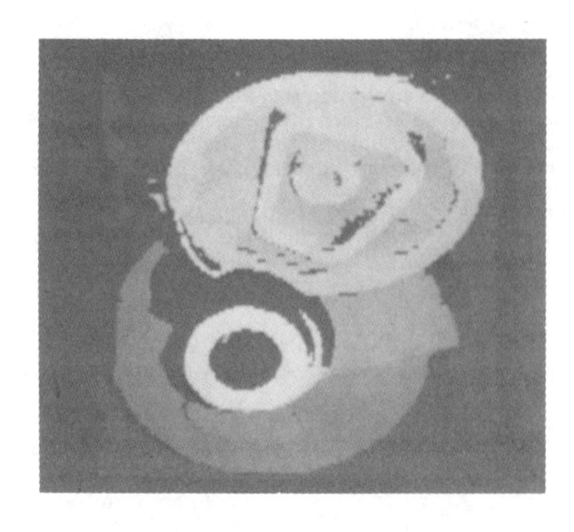

FIGURE 8 AN INTENSITY IMAGE IN WHICH THE BRIGHTNESS CORRESPONDS TO THE HEIGHT ABOVE THE TABLE

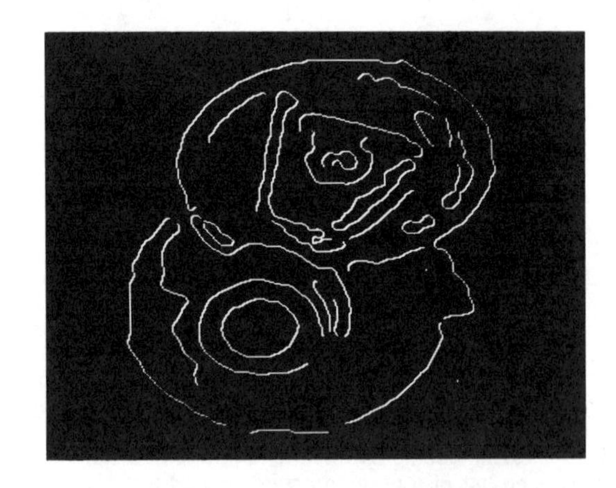

FIGURE 9 FILTERED ZERO CROSSINGS

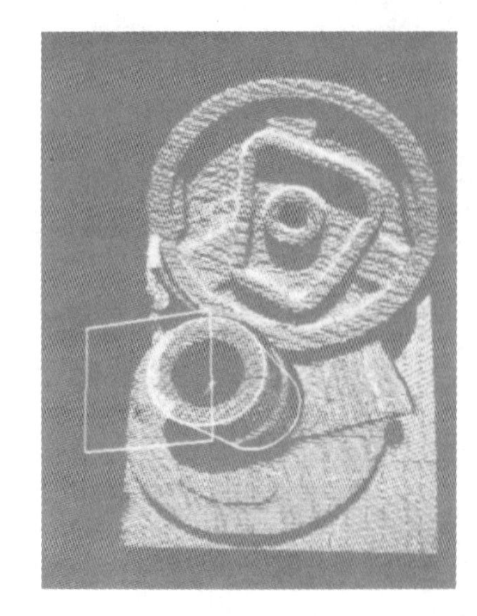

FIGURE 10 CYLINDER IMPLIED BY ONE EDGE

FIGURE 11 POSSIBLE ENDS OF CYLINDRICAL FEATURES

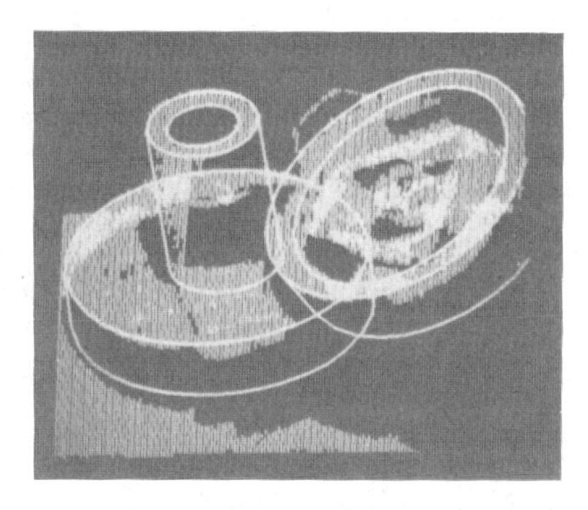

FIGURE 12 TWO HYPOTHESIZED CASTINGS

To hypothesize objects, the program searches for sets of mutually consistent features by examining all pairs of features. Figure 12 shows two hypothesized objects, one based on three cylinders and one based on two cylinders.

DISCUSSION

We have described a system for locating complex parts in range data and have shown the results of some of its components. We have argued for a recognition strategy that concentrates on a few feature clusters and a planning process that selects the best features to use. We have only partially demonstrated the effectiveness of this approach. We plan to improve the 3DPO system and continue the comparison of matching strategies. We also plan to explore the use of an existing CAD system, such as PADL-2 [19], as the basis for a more complete implementation of our representation scheme.

REFERENCES

[1] Shirai, Y., and M. Suwa, "Recognition of Polyhedrons with a Range Finder," *Proc. of the Second IJCAI*, pp. 80-87, London, England (August 1971).

[2] Popplestone, R.J., C.M. Brown, A.P. Ambler, and G.F. Crawford, "Forming Models of Plane-and-cylinder Faceted Bodies from Light Stripes," *Proc. of the Fourth IJCAI*, Tbilisi, Georgia, USSR, pp. 664-668 (September 1975).

[3] Nitzan, D., A.E. Brain, and R.O. Duda, "The Measurement and Use of Registered Reflectance and Range Data in Scene Analysis," *Proc. of the IEEE 65*, pp. 206-220 (1977).

[4] Oshima, M., and Y. Shirai, "A Scene Description Method Using Three-Dimensional Information," *Pattern Recognition*, Vol. 11, pp. 9-17 (1978).

[5] Faugeras, O.D., et al, "Toward a Flexible Vision System," *Robot Vision*, Alan Pugh, editor (IFS Publications, Ltd., United Kingdom, 1983).

[6] Agin, G.J. and T.O. Binford, "Computer Description of Curved Objects," *Proc. of the Third IJCAI*, Stanford University, Stanford, California, pp. 629-640 (August 1973).

[7] Nevatia, R., and T.O. Binford, "Description and Recognition of Curved Objects," *Artificial Intelligence*, Vol. 8, pp. 77-98 (1977).

[8] Shneier, M., "A Compact Relational Structure Representation," *Proc. of the Sixth IJCAI*, pp. 818-826, Tokyo, Japan (August 1979).

[9] Sugihara, K., "Range-data Analysis Guided by a Junction Dictionary," *Artificial Intelligence 12*, pp. 41-69 (1979).

[10] Oshima, M., and Y. Shirai, "Object Recognition using Three-Dimensional Information," *Proc. of the Seventh IJCAI*, University of British Columbia, Vancouver, B.C., Canada, pp. 601-606 (August 1981).

[11] Rosen, C.A., and D. Nitzan, et al., "Exploratory Research in Advanced Automation," Fourth Report, NSF Grant GI38100X1, SRI Project 2591, Stanford Research Institute, Menlo Park, California (June 1975).

[12] Birk, J.R., R.B. Kelley, and J. Dessimoz, "Visual Control for Robot Handling of Unoriented Parts," *Proc. of the SPIE Technical Symposium East*, Vol. 281, Washington, D.C. (April 1981).

[13] Bolles, R.C., and R.A. Cain, "Recognizing and Locating Partially Visible Objects, The Local-feature-focus Method," *International Journal of Robotics Research*, Vol. 1, No. 3, pp. 57-82 (1982).

[14] Bolles, R.C., J.H. Kremers, and R.A. Cain, "A Simple Sensor to Gather Three-Dimensional Data," A.I. Center Technical Note 249, SRI International, Menlo Park, California (July 17, 1981).

[15] Baumgart, B.G., "Winged Edge Polyhedron Representation," Stanford A.I. Memo AIM-179, Stanford University, Stanford, California (1972).

[16] Chakravarty, I., "The Use of Characteristic Views as a Basis for Recognition of Three-dimensional Objects," Ph.D. thesis, Rensselaer Polytechnic Institute (October 1982).

[17] Brooks, R.A., "Symbolic Reasoning Amound 3-D Models and 2-D Images," *Artificial Intelligence Journal*, Vol. 17, pp. 285-348 (1981).

[18] Marr, D., and E.C. Hildreth, "Theory of Edge Detection," *Proc. R. Soc. Lond. B. 207*, pp. 187-217 (1980).

[19] Brown, C.M., "Some Mathematical and Representational Aspects of Solid Modeling," *IEEE Transactions on Pattern Analysis and Machine Intelligence*, Vol. PAMI-3, No. 4, pp. 444-453 (July 1981).

Model-Based Three-Dimensional Interpretations of Two-Dimensional Images

RODNEY A. BROOKS

Abstract—ACRONYM is a comprehensive domain independent model-based system for vision and manipulation related tasks. Many of its submodules and representations have been described elsewhere. Here the derivation and use of invariants for image feature prediction is described. Predictions of image features and their relations are made from three-dimensional geometric models. Instructions are generated which tell the interpretation algorithms how to make use of image feature measurements to derive three-dimensional size, structural, and spatial constraints on the original three-dimensional models. Some preliminary examples of ACRONYM's interpretations of aerial images are shown.

Index Terms—Algebraic vision, computer vision, constraint systems, geometric models, model-based vision, spatial reasoning.

I. INTRODUCTION

THE ACRONYM system has always relied on detailed three-dimensional geometric models to direct image understanding. Originally, it used models of specific objects to direct a qualitative geometric reasoning system in labeling images [8].

The scope of ACRONYM has now been increased to include object class recognition and extraction of three-dimensional information from images (including monocular images), reasoning about how to grasp objects [3], and real-time simulation of multiple manipulator work stations for purposes of off-line programming and the design and analysis of new manipulators [11]. ACRONYM has moved from using a purely geometric representation and qualitative geometric reasoning system to a system with a combined algebraic and geometric representation and a geometric reasoning system which can make precise deductions about partially specified situations. The geometric and algebraic aspects of the representation complement each other during image interpretation.

To support these extended capabilities a large number of new techniques had to be developed. They include the addition of a class and subclass relation representation scheme to the geometric modeling system. This is based on the use of symbolic algebraic constraints. In support of this a constraint manipulation system which includes a partial decision procedure on consistency of sets of nonlinear inequalities was formulated and implemented. A geometric reasoning system which can deal with underconstrained spatial relations was developed. These are all detailed in [7].

This paper deals with techniques developed for image feature and feature-relation prediction, and a matching process which is directed by both geometric relations and algebraic constraints during image interpretation. Some first examples of the performance of the re-built ACRONYM system on some real images are presented. The low-level bottom-up processes used in these experiments (for reasons of expedience) provide either little or noisy data. Nevertheless, ACRONYM makes strong and accurate deductions about the objects appearing in the images. Even better performance is expected when more accurate low-level descriptive processes become available.

A. Model Domain

The reader is referred to Brooks [7] for a complete description of the ACRONYM modeling system. A short overview is given here to make the paper reasonably self-contained.

An *a priori* model of the world is given to ACRONYM. The world is modeled as a coordinate system. Objects are modeled as subpart hierarchies of generalized cones, each with a local coordinate system related to an object coordinate system. Cameras are modeled as coordinate systems (with viewing direction along the $-z$ axis) and a focal ratio. Objects and cameras are placed in the world by constraining the transforms between their local coordinate systems and the world coordinate system.

The particular domain considered in this paper is aerial views of airfields. However, ACRONYM has no particular knowledge of airfields and is applicable to other domains, and camera geometries other than aerial cameras—an example of such is given in Section II.

Generalized cones, introduced by Binford [2], describe a three-dimensional volume. A generalized cone is represented by a curve through space, called the spine, along which a two-dimensional shape, called the cross section, is swept. The cross section is kept at a constant angle to the tangent of the spine, and is deformed according to a deformation function called the sweeping rule.

In ACRONYM generalized cones are restricted to having straight line segments or circular arcs as spines, cross sections bounded by straight lines and circular arcs, and sweeping rules which are linear magnification functions about the spine, or

Manuscript received December 12, 1981; revised September 8, 1982. This work was supported in part by the Defense Advanced Research Projects Agency under Contract MDA903-80-C-0102, in part by the National Science Foundation under Contract DAR78-15914, and in part by a grant from the ALCOA Corporation. An earlier version of this paper was presented at the International Joint Conference on Artificial Intelligence, Vancouver, Canada, August 1981.

The author was with the Stanford Artificial Intelligence Laboratory, Stanford University, Stanford, CA 94305. He is now with the Artificial Intelligence Laboratory, Massachusetts Institute of Technology, Cambridge, MA 02139.

linear in two orthogonal directions about the spine. By convention in ACRONYM the coordinate systems of generalized cones have their origins at one end of the spine, with the cross section laying in the y-z plane at that point, and the spine extending into the positive x half-space. For cross sections which are to be kept normal to the spine this implies that the tangent of the spine at the origin lies along the x-axis.

In ACRONYM, generalized cones, the subpart hierarchy of objects, and the affixment tree representing spatial relationships of objects are all represented as *units* with *slots* and *fillers* (e.g., Bobrow and Winograd [5]). Thus, a generalized cone is represented by a unit with slots called *SPINE*, *CROSS-SECTION*, and *SWEEPING-RULE*, each of which is filled with a unit of the appropriate type. Eventually, the units hierarchy bottoms out in units with slots filled with numbers. For instance, a spine unit for a straight spine has a slot called *LENGTH* which can be filled with a number representing the length of the spine.

The above describes a representation scheme sufficiently rich to model a large class of specific objects where all parameters are completely specified. However, for image interpretation tasks it is often the case that not all details of objects in the world are known *a priori*. Thus, it is desirable to be able to represent classes of objects. More particularly, it is often desirable to be able to talk about classes of objects classified according to their function. For instance, an image interpretation task might involve identification of automobiles, buildings, or oil refineries.

Fortunately, the functionality of many man-made objects is reflected in the geometric structure. For instance, automobiles have four cylindrical sections (wheels) in contact with the ground and with their axes parallel to the ground plane. The cylindrical sections are paired with colinear axes, and the centroids of the cylinders form a rectangle (or perhaps an isosceles trapezium). An essentially box-like structure is above the ground with base parallel to the ground and below the rectangle formed by the wheelbase. In addition there may be a section of the box-like structure which extends higher than the rest which is as wide as the main body but not as long. Such general geometric descriptions of functional classes will be referred to as *generic* descriptions throughout the rest of the paper. ACRONYM includes mechanisms for representing such generic object classes.

Representation of geometric classes is achieved by representing commonality of class members by shared structure, and variations across class members by generalizing the fillers of slots which ordinarily might contain a number. Generalizing the slots of the generalized cones allows representation of variations in size and shape, the slots in the subpart hierarchy allow representation of variations in structure, and those in the affixment tree allow variations in spatial relationships. The generalization is to allow numbers to be replaced by formal variables (referred to below as *quantifiers*) and algebraic expressions over formal variables. Additionally, constraints are placed on the formal variables (generally as inequalities, not necessarily linear). The constraints and the geometric unit slot representations together represent the class of all objects which have the given geometric structure and whose parameters satisfy the constraints. The reader is referred to Brooks [7] for details of this modeling scheme.

As an example of a use of quantifiers consider the problem of representing the fact that airplanes on the ground are usually upright with their wings and fuselage parallel to the ground, but be headed in arbitrary directions. Also they can be found at many places at ground level within the confines of an airfield. Suppose the world coordinate system has $z = 0$ as the ground plane, with positive z above the ground. Then the coordinate transform relating the coordinate system of the airplane to the ground would consist of a translation and a rotation. The translation would have its x and y component slots filled by quantifiers bounded only to lie within the confines of the airfield (for a polygonal shaped airfield this would simply be a conjunction of linear inequalities over the two quantifiers), and its z slot filled by some constant dependent on the airplane (the height of the airplane origin above the ground). The rotation would be about the z axis with its magnitude slot filled by a quantifier bound only to lie in the range 0 to 2π.

B. Image Domain

ACRONYM is given images which have been preprocessed by a line finder due to Nevatia and Babu [9]. Figs. 3(b), 4(b), and 5(b) show examples of this processing.

From the edges, a rather poor performance module extracts descriptions of the images as ribbons and ellipses. A ribbon is a two-dimensional analog of a generalized cone. In the ACRONYM implementation they are restricted to having straight spines and linear sweeping rules. Ellipses are described by the lengths of their major and minor axes.

In addition to the ribbons and ellipses ACRONYM is told the resolution in pixels of the original digitization. It uses this to estimate pixel error magnitudes.

In the examples of this paper ACRONYM is given preprocessed images of airfields with a number of airplanes in view on the ground.

C. The Image Interpretation Task

This paper is concerned with the image interpretation tasks performed by ACRONYM. It is given some classes of generic geometric models. The world is described in terms of the typical coordinate transforms of those objects to the world coordinate system (as in the example of the airplane on the ground, above). A camera is also modeled by the relationship of its coordinate system to the world coordinate system.

ACRONYM is given a preprocessed image as above. Its task is to identify instances of the object model classes in the image, along with their location and orientation in world coordinates, to make any subclass identification which is possible, to determine constraints implied by the image on the quantifiers in the models (i.e., determine three-dimensional parameters of the objects from the image) and finally to determine the location and orientation of the camera if that was not completely determined *a priori*.

In the examples of Section III the task which ACRONYM must carry out is to locate airplanes and identify their type. In addition it determines parameters of the camera.

II. Prediction

In the ACRONYM system generic object classes and specific objects are represented by volumetric models based on generalized cones along with a partial order on sets of nonlinear algebraic inequalities relating model parameters. Image features and relations between them which are invariant over variations in the models and camera parameters are identified by a geometric reasoning system. Such predictions are combined first to give guidance to low-level image description processes, then to provide coarse filters on image features which are to be matched to local predictions. Predictions also contain instructions on how to use noisy measurements from identified image features to construct algebraic constraints on the original three-dimensional models. Local matches are combined subject both to consistently meeting predicted image feature relations, and the formation of consistent sets of algebraic constraints derived from the image. The result is a three-dimensional interpretation of the image.

This section describes some of the invariants that are identified by the reasoning system, and gives examples of how the back constraints are set up giving three-dimensional information about the instances of the models which appear in images.

A. Constraints

To illuminate the discussion in succeeding subsections the uses and capabilities of ACRONYM's constraint mechanism are briefly discribed along with the allowed structure of constraints themselves.

ACRONYM's three-dimensional models are represented by *units* and *slots* (e.g., Bobrow and Winograd [5]). Any slot which admits numeric *fillers* also admits *quantifiers* (predeclared variable names) and expressions over quantifiers using the operators +, −, ×, /, and $\sqrt{}$.

Constraints can be put on quantifiers. They take the form of inequalities between expressions as defined above, along with the possibility of including max and min (on the left and right of $\leqslant$, respectively). Equality can be encoded as two inequalities. For instance suppose a cylinder is represented as a generalized cone whose straight spine has its length defined by the quantifier CYL-LENGTH and whose cross section is a circle with radius CYL-RADIUS. Then the class of all cylinders of volume 5 (in some units) can be represented by the two constraints:

$$5 \geqslant \text{CYL-LENGTH} \times \text{CYL-RADIUS} \times \text{CYL-RADIUS} \times \pi$$

$$5 \leqslant \text{CYL-LENGTH} \times \text{CYL-RADIUS} \times \text{CYL-RADIUS} \times \pi$$

The ACRONYM constraint manipulation system (CMS), described in detail in [7], operates on sets of constraints. A set of constraints (implicitly conjunctive) defines a subset of n-dimensional space for which all constraints are true (where n is the number of quantifiers mentioned in the constraint set). This is called the satisfying set, and is empty if the constraints are inconsistent. The CMS is used for three tasks related to this constraint set.

1) Given a set of constraints partially decide whether their satisfying set is empty. The outcomes are "*empty*" or "*I don't know.*"

2) Find numeric (or $\pm\infty$) upper and lower bounds on an expression in quantifiers over the satisfying set of a constraint set. This uses procedures called SUP and INF.

3) (A generalization of task 2.) For an expression E and a set of quantifiers V find expressions L and H in V such that $L \leqslant E \leqslant H$ identically over the satisfying set of the constraint set.

In tasks 2 and 3 the expressions being bounded can include trigonometric functions such as sin, cos, and arcsin.

The CMS implemented in ACRONYM is a nonlinear generalization [7] of the linear SUP-INF method described by Bledsoe [4] and Shostak [10]. In [7] it was shown that ACRONYM's CMS behaves identically to that described by Shostak for purely linear sets of constraints and linear expressions, and determines least upper and greatest lower bounds.

For nonlinear expressions and constraints there is not such a well delineated characterization of the behavior of the CMS. The proofs of [7] can be extended to cases of sums of independent terms, each of which is a product or quotient of terms whose sign can be determined purely from the subset of constraints which are linear, i.e., in those cases the CMS produces the best bounds possible.

Informal empirical evidence suggests that there are other cases where the CMS produces good bounds. No good characterization of these cases exists. It is also possible to demonstrate cases where the bounds found by the CMS are not good. If all terms have signs determinable from the subset of constraints which are linear then the bounds are no worse than considering all terms to be independent. If some terms have indeterminate signs then the bounds may be quite poor.

Much of the expertise in the ACRONYM system lies in being able to reduce sets of constraints by projection into subspaces where better bounds can be found on expressions. Sometimes, of course, it is not possible to find such a projection, and then poor bounds must suffice. The rules used for prediction and interpretation were written with these limitations in mind.

B. The Prediction Process

The prediction module of ACRONYM is implemented as a set of approximately 280 product-type rules.

The major control paradigm is backward chaining, i.e., rules set up subgoals and recursively invoke the rule mechanism to satisfy those subgoals. Both subgoals and rule capabilities are represented as Lisp *s*-expressions. Rules are invoked if they unify with a stated subgoal. The unification process allows passing of multiple parameters to rules, and receiving multiple results from them. In addition, the pattern matching aspect of unification provides rule selection criteria.

A global assertional database, and additional local databases provide the means for recording the state of the prediction computation. The data structures embodying the predictions are built as side effects of the rules firing.

During a typical prediction phase, e.g., in the example of Section III of this paper, there are on the order of 6000 rule firings.

The order of rule firings, and flow of control can not be characterized at a local level as it is completely dependent on the models given to the system. At a global level the order of computations can be roughly described as follows.

A breadth first walk down the subpart hierarchy of the

models is made. At each level prediction is carried out, followed by partial interpretations of the images. Then refined predictions are made at the next level of the hierarchy. (In the examples of Section III only the first level of interpretations are carried out).

At each level the coordinate transforms of the generalized cones, relative to the camera are computed and examined. Visibility conditions and implications of visibility are computed. Individual shape predictions are made for the generalized cones. Relationships between the generalized cones are examined for invariant characterizations. These four steps are summarized in more detail in the following paragraphs.

The affixment tree is transversed to get a symbolic expression for the coordinate transform relating a generalized cone's local coordinate system to that of the camera. The transform is represented as a product of rotations and translations. Typically, these products are long (i.e., over ten terms) and contain many quantifiers. The reader is referred to [7] for the details of some rules which make such expressions manageable. During later phases of prediction these rules are applied in a goal directed manner to find a simplication of the products in the form most suitable for the task at hand. (Again examples are given in [7].)

One can associate with a camera a volume of space (an infinite pyramid with rectangular cross section and apex at the focal point of the camera) in which objects can be visible. Outside of that volume an object is definitely invisible to a camera. Thus, for the purposes of predicting the appearance of a generalized cone it can be assumed that it lies within the sight of the camera, since otherwise it will not be seen anyway. Therefore, constraints are added to the object model which confine its coordinates so that it lies within the visible volume. If these constraints are inconsistent with those that already exist then the object is definitely invisible and no further prediction need be made. Otherwise it provides possibly tighter constraints on the possible range of positions and orientations possible for the object. Once the visibility conditions for individual generalized cones are established, pairwise comparisons are made to see if it can definitely be established that one cone always occludes another. This process is carried out by tentatively adding constraints that imply the converse (i.e., one cone never occludes the other). If these constraints are inconsistent then the obscuration must always occur. If they cannot be shown to be inconsistent (recall that the decision procedure is only partial) then it can be concluded that perhaps the obscuration will not always occur, so both cones may be visible.

Actual predictions of shapes proceeds in five phases. These are described in detail in Section II-C.

An exhaustive pairwise examination of shapes produced by different generalized cones is carried out. The geometric reasoning system tries to find invariant characterizations of the relationships between the shapes in terms of the relationships detailed in Section II-D.

C. Shape Prediction

Shapes are predicted as *ribbons* (the two-dimensional analog of three-dimensional generalized cones) and *ellipses*. These are also the features which are found by the low-level descriptive process which are temporarily being used in ACRONYM.

Ribbons are a good way of describing the images generated by generalized cones. Consider a ribbon which corresponds to the image of the swept surface of a generalized cone. For straight spines, the projection of the cone spine into the image would closely correspond to the spine of the ribbon. Thus, a good approximation to the observed angle between the spines of two generalized cones is the angle between the spines of the two ribbons in the image corresponding to their swept surfaces. A quantitative theory of these correspondences is not used. Ellipses are a good way of describing the shapes generated by the ends of generalized cones. The perspective projections of ends of cones with circular cross-sections are exactly ellipses.

Shape prediction involves deciding what shapes will be visible, predicting ranges for shape parameters (to be used as a coarse filter during interpretation and also to guide the low level descriptive processes) and deriving instructions about how to locally invert the perspective transform and hence use image measurements to generate constraints on the original three-dimensional models.

To predict the shapes generated by a single generalized cone, ACRONYM does not explicitly predict all possible quantitatively different viewpoints. Rather, it predicts what shapes may appear in the image, and associates with them methods to compute constraints on the model that are implied by their individual appearance in an image. For example, identification of the image of the swept surface of a right circular cone constrains the relative orientation of the cylinder to the camera (these are called back constraints). Identification of an end face of the cylinder provides a different set of constraints. If both the swept surface and an end face are identified then both sets of constraints apply. Also predicted are specific relations between shapes that will be true if they are both observed correctly. For more complex cones, the payoff is even greater for predicting individual shapes rather than exhaustive analysis of which shapes can appear together.

At other times during prediction invariant cases of obscuration are noticed. For instance, it may be noticed that one cone abuts another so that its end face will never be visible. The consequences of such realizations are propagated through the predictions.

Prediction of shapes proceeds in five phases. First, all the contours of a generalized cone which could give rise to image shapes are indentified by a set of special purpose rules. These include occluding contours and contours due purely to internal cone faces. Thus, for instance, a right square cylinder will generate contours for the end faces, the swept faces, and contours generated by the swept edges at diagonally vertices of the square cross section. The contours are generated independently of camera orientation, and in terms of object dimensions rather than image quantities.

Second, the orientation of the generalized cone relative to the camera (this is done by the geometric reasoning system; see [7]) is then examined to decide which contours will be visible and how their image shapes will be distorted over the range of variations in the model parameters which appear in the orientation expressions.

The third phase predicts relations between contours of a single generalized cone (see Section II-D).

Fourth, the actual shapes are then predicted. The expected

values for shape parameters in the image are estimated as closed intervals (see below).

Finally, the back constraints which will be instantiated during interpretation are constructed.

1) *Back Constraints:* Consider the following simple camera geometry. Suppose that it is desired to predict the length of an observable feature which is generated by something of length l lying in a plane parallel to the camera image plane, at distance d from the camera. Furthermore, suppose the camera has a focal ratio of f. Then the measured length of the observed feature is given by $p = (l \times f)/d$. Any or all of l, f, and d may be expressions in quantifiers, rather than numbers. Using the CMS bounds can be obtained on the above expression for image feature length, giving that it will lie in some range $P = [p_l, p_h]$ where p_l and p_h are either numbers or $\pm\infty$. For more complex geometries the expression for p will be more complex, but the method is the same (trigonometric functions are usually involved).

Now, given an image feature, which is hypothesized to correspond to the prediction it must be decided whether it is acceptable on the basis of its parameters. The low-level descriptive processes are noisy and provide an error interval, rather than an exact measurement for image parameters. Suppose the interval is $M = [m_l, m_h]$ for a feature parameter predicted with expression p. Then the parameter is acceptable if $P \cap M$ is nonempty. This is the coarse filtering used during initial hypothesis of image feature to feature prediction matches.

But note also that it must be true that the true value of p for the particular instance of the model which is being imaged must lie in the range M. Thus the constraints

$$m_l \leqslant (l \times f)/d$$

$$m_h \geqslant (l \times f)/d$$

can be added to the instance of the model being hypothesized, where l, f, and d are numbers or expressions in quantifiers.

The above is the analysis for the simplest possible camera geometry. In general the predicted geometry is much more complex and requires stronger symbolic analysis methods. ACRONYM has individual rules which are capable of handling all cases of up to three orthogonal degrees of freedom in orientation of objects relative to the camera.

2) *Trigonometric Back Constraints:* When the expression p involves trigonometric functions the above method of generating back constraints will not work. It would generate constraints involving trogonmetric functions, which ACRONYM's CMS cannot handle.

One approach to this problem is to bound expression p above and below by expressions involving no quantifiers contained in arguments to trigonometric functions, and then use these expressions in setting up the back constraints. This has the unfortunate side effect of losing all information implied by the image feature about the quantifiers eliminated from the bounds.

A second approach is sometimes applicable. If a trigonometric function has as its argument e, an expression, and if the CMS determines that e is bounded to lie within a region of the function's domain where it is strictly monotonic and hence invertible, then specific back constraints on e can be computed at interpretation time (as distinct from during prediction). An example illustrates this. A cylinder with length CYL-LENGTH is sitting upright on a table. A camera with unknown but constrained pan and tilt (the latter is constrained to lie in the interval $[\pi/12, \pi/6]$) is looking across from the side of the table, and it is elevated above table top height. The geometric details and numeric constants are not important here. Suffice it to say that the geometric reasoning system deduces that the pan of the camera is irrelevant to the prediction of the length of the ribbon corresponding to the swept surface of the cylinder. It predicts that the length of the ribbon in the image will in fact be

$$\frac{-2.42 \times \text{CYL-LENGTH} \times \cos(-\text{TILT})}{\text{CYLINDER.CAMZ}}$$

where 2.42 is the focal ratio of the camera and CYLINDER.CAMZ is an internal quantifier generated by the prediction module. Since cosine is invertible over the range $[\pi/12, \pi/6]$ the expression can be solved for TILT.

Both of the above approaches are used to generate back constraints to ensure coverage of all the relevant quantifiers. They are

$$m_h \geqslant -2.096 \times \text{CYL-LENGTH} \times (1/\text{CYLINDER.CAMZ})$$

$$m_l \leqslant -2.338 \times \text{CYL-LENGTH} \times (1/\text{CYLINDER.CAMZ})$$

$$-\text{TILT} \leqslant -\arccos(\sup(-0.413 \times m_h \times \text{CYLINDER.CAMZ} \times (1/\text{CYL-LENGTH})))$$

$$-\text{TILT} \geqslant -\arccos(\inf(-0.413 \times m_l \times \text{CYLINDER.CAMZ} \times (1/\text{CYL-LENGTH})))$$

The first two are nontrigonometric back constraints and at interpretation time a simple substitution of the measured numeric quantities for m_l and m_h is done. The latter two require further computation at interpretation time. After the substitution, expressions must be bounded over the satisfying set of all the known constraints, and the function arccos applied to give numeric upper and lower bounds on the quantifier TILT.

The technique described here work for a more general class of functions than trigonometric functions (in the current implementation of ACRONYM it is used for functions sin, cos, and arcsin). The requirement is that the domain of the function (e.g., the interval $[-\pi, \pi]$ for sin and cos), can be subdivided into a finite number of intervals over which the function is strictly monotonic, and hence locally invertible.

D. Feature Relation Prediction

Image feature (shape) predictions are organized as the nodes of the *prediction graph*. The arcs of the graph predict image-domain relations between the features. During interpretation correspondences are constructed which match image features and prediction nodes. More global interpretations are derived by taking pairs of such correspondences and trying to instantiate any prediction arcs linking the two prediction nodes. The semantics of the arc types used are now described in detail.

As with shape predictions (Sections II-C1 and 2) many relation predictions involve measurable parameters. For each parameter associated with a relation prediction, both a range of acceptable values and a set of back constraints are computed. When instantiating the prediction of measured parameters can be quickly checked against the value range prediction as a coarse filter to eliminate grossly inconsistent instantiations. The back constraints are then used both to check for more global consistency and to compute what the particular instantiation of the prediction arc implies about the model.

Prediction arcs are generated between pairs of shapes arising in two ways.

1) Prediction arcs are generated to relate multiple shapes predicted for a single cone. For instance a right circular cylinder prediction includes shapes for the swept surface and perhaps each of the end faces (depending on whether the camera geometry is known well enough to determine *a priori* exactly which faces will be visible). It can be predicted that a visible end face will be coincident at least one point in the image with a visible swept surface. (In fact, a stronger prediction can be made: the straight spine of the swept surface image ribbon can be extended through the center of mass of the elliptical image of the end face.)

2) Prediction arcs are also generated between shapes associated with predictions for different generalized cones. These are actually of more importance in arriving at a consistent global interpretation of collections of image features as complex objects.

The semantics of the arc types currently used are as follows.

1) *Exclusive:* If a generalized cone has a straight spine, and during sweeping, the cross section is kept at a constant angle to the spine, then at most one of the cone's end faces can be visible in a single image. *Exclusive* arcs relate image features which are mutually exclusive for this or other reasons. (Note that in this case, instantiations of the two end faces would probably result in inconsistent back constraints being applied to the spatial orientation of the original model, so that eventually the CMS would detect an inconsistency. However, checking for the existence of a simple arc at an early stage is computationally much cheaper than waiting to invoke the decision procedure.)

2) *Collinear:* If two straight line segments in three-space are *collinear* then any two-space image of them will either be a single degenerate point or two collinear line segments. As was pointed out earlier, the spine of the image shape corresponding to the swept surface of a cone is usually a good approximation to the projection of the spine of the cone into the image. Thus, if two cones are known to have collinear spines in three dimensions, a collinear spine arc between the prediction of their swept surfaces can be included.

3) *Coincident:* If two cones are physically *coincident* at some point(s) in three-space, then for any camera geometry, if they are both visible then their projections will be coincident at some point(s) (except for some cases of obscuration). Failure to match predicted coincident arcs turns out to be the strongest pruning process during image interpretation.

4) *Angle:* If the *angle* between the spines of two generalized cones as viewed from the modeled camera is invariant over all the rotational variations in the model, or if an expression for the observed angle can be symbolically computed and is sufficiently simple, then a prediction of the observed angle can be made. For example, wing-wing and wing-fuselage angles are invariant when an aircraft is viewed from above—that is because the only rotational freedom of an aircraft on the ground is about an axis parallel to the direction of view of an overhead camera. Again the fact that the projections of model spines correspond to image spines is used here. This arc type includes (trigonometric) back constraints which make use of the observed angle. Some such constraints constrain relative spatial orientations of generalized cones. Others provide constraints on the orientation of the plane of rotation, which generated the angle, relative to the camera, and hence constraints on an object's orientation relative to the camera.

5) *Approach-Ratio:* Suppose a cone B is affixed at one end of its spine to another cone A, with a straight spine, somewhere along its length. The spines need not be coincident, but the cones must be. Suppose the spine of cone A has endpoints a_1 and a_2, and let a_3 be the point on the spine of A closest to the end of the spine of B. Then the *approach-ratio* is the ratio of the length of the spine segment from a_1 to a_3 and the length of the complete spine from a_1 to a_2. If the spines of A and B are both observable, then the approach-ratio is invariant under a normal projection for all camera geometries. Thus it is a quasi-invariant for a perspective projection for a camera sufficiently far from the object. For example, the ratio of the distance from the rear of the fuselage to the point of wing attachment, to the length of the fuselage, is almost invariant over all viewing angles for objects sufficiently far from the camera. Again this relies on the correspondences between the projection of a cone spine and the spine of the ribbon generated by the image of its swept surface. Approach-ratios arcs are only generated for pairs of image features which have a coincident arc. They provide back constraints on the model via the symbolic expression which describes the modeled spine approach ratio.

6) *Distance:* Sometimes symbolic expressions for the image distance between two image features can be computed. *Distance* arcs are only generated for pairs of image features which also have an angle arc, but no coincident arc. Distance arcs generate back constraints on the original model.

7) *Ribbon-Contains:* This is a directed arc type which two dimensionally relates two predicted ribbons, one of which will contain the other in the image. For instance, *ribbon-contains* arcs are built between the ribbon predicted from the occluding contour of a generalized cone with rectangular cross section, and each of the ribbons generated by the two visible swept faces.

III. Some Image Interpretations

The image interpretations reported here are of a rather preliminary nature. They are based on a low-level descriptive module [6] chosen for its availability rather than its performance and an environment where experimentation has been hampered by address space limitations—the current system occupies two 256K address spaces on a DEC-10. The system

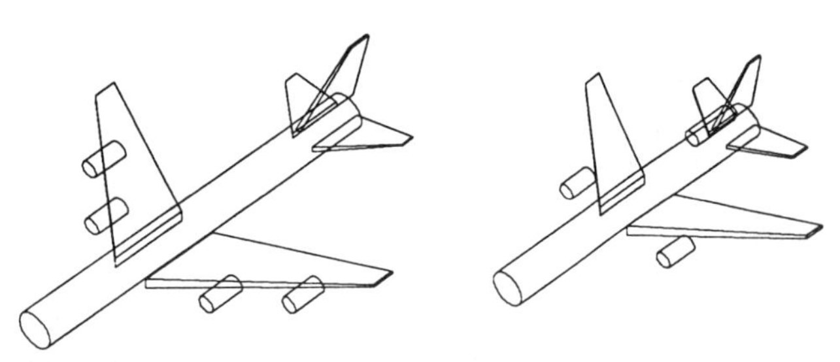

Fig. 1. Instances drawn by ACRONYM of 747's and L-1011's, which are themselves both subclasses of the generic class of wide-bodied passenger jet aircraft.

has been transported to a VAX. Further experimentation is planned on that version by a number of people.

In the examples to be described here ACRONYM was given a generic model of wide-bodied passenger jet aircraft, along with class specializations to L-1011's and Boeing-747's. The Boeing-747 class had further subclass specializations to Boeing-747B and Boeing-747SP. The subclasses do not completely partition their parent classes. The classes are described by sets of constraints on some 30 quantifiers. Fig. 1 shows instances of the two major modeled classes of jet aircraft. These diagrams were draw by ACRONYM from the models given it to carry out the image interpretations. The constraints for the generic class of wide bodied jets are given in Fig. 2. Units are in meters. The diagrams of Fig. 1 demonstrate the range of variations represented in the generic model. The geometric structure consists of a cylindrical fuselage, two symmetrically placed wings perhaps with rudder, and perhaps a centrally mounted cylindrical rear engine.

The camera was modeled as being between 1000 and 12 000 m above the ground. Thus there is little *a priori* knowledge of the scale of the images. A specific focal ratio was given: 20. (Similar interpretations have been carried out with a variable focal ratio, but then the final constraints on camera height and focal ratio are coupled, and not as clear for illustrative purposes–no accuracy is lost due to the nonlinearities that are introduced into the constraints, although both computation time and garbage collection time are increased.)

The aircraft models, the camera model and the number of pixels in each dimension of the image (512 X 512 in these examples) were the only pieces of world knowledge input to ACRONYM. It has no special knowledge of aerial scenes: all its rules are about geometry and algebraic manipulation. These were applied to the particular generic models it was given, to make predictions and then to carry out interpretations.

Figs. 3-5 show three examples of interpretations carried out by ACRONYM. In each case part (a) is a half-tone of the original gray level image. The (b) version is the result of applying the line finder of Nevatia and Babu [9]. That line finder was designed to find linear features such as roads and rivers in aerial photos. Close examination of results on these images indicate many errors, and undue enlargement in width of narrow linear features. It also produces many noise edges in smooth brightness gradients (not visible at the resolution of the reproductions of these figures). These edges are the lowest level input to ACRONYM.

An edge linker [6] is directed by the predictions to look

```
ENG-DISP-GAP ∈ [6, 10]
ENG-DISP ∈ [0, 4]
ENG-GAP ∈ [7, 10]
STAB-ATTACH ∈ [3, 5]
R-ENG-ATTACHMENT ∈ [3, 5]
ENG-OUT ∈ [5, 12]
WING-ATTACHMENT ∈ [20, 40]
        WING-ATTACHMENT ≥ 0.4*FUSELAGE-LENGTH
        WING-ATTACHMENT ≤ 0.6*FUSELAGE-LENGTH
STAB-RATIO ∈ [0.2, 0.55]
STAB-SWEEP-BACK ∈ [3, 7]
STAB-LENGTH ∈ [7.6, 13]
STAB-THICK ∈ [0.7, 1.1]
STAB-WIDTH ∈ [5, 11]
RUDDER-RATIO ∈ [0.3, 0.4]
RUDDER-SWEEP-BACK ∈ [3, 9]
RUDDER-LENGTH ∈ [8.5, 14.2]
RUDDER-X-HEIGHT ∈ [7, 13]
RUDDER-X-WIDTH ∈ [0.7, 1.1]
WING-RATIO ∈ [0.35, 0.45]
WING-THICK ∈ [1.5, 2.5]
WING-WIDTH ∈ [7, 12]
        WING-WIDTH ≤ 0.5*WING-LENGTH
WING-LIFT ∈ [1, 2]
WING-SWEEP-BACK ∈ [13, 18]
WING-LENGTH ∈ [22, 33.5]
        WING-LENGTH ≥ 2*WING-WIDTH
        WING-LENGTH ≥ 0.43*FUSELAGE-LENGTH
        WING-LENGTH ≤ 0.65*FUSELAGE-LENGTH
REAR-ENGINE-LENGTH ∈ [6, 10]
ENGINE-LENGTH ∈ [4, 7]
ENGINE-RADIUS ∈ [1, 1.8]
FUSELAGE-RADIUS ∈ [2.5, 4]
FUSELAGE-LENGTH ∈ [40, 70]
        FUSELAGE-LENGTH ≥ 1.66666666*WING-ATTACHMENT
        FUSELAGE-LENGTH ≥ 1.53846154*WING-LENGTH
        FUSELAGE-LENGTH ≤ 2.5*WING-ATTACHMENT
        FUSELAGE-LENGTH ≤ 2.3255814*WING-LENGTH
R-ENG-QUANT ∈ [0, 1]
        R-ENG-QUANT ≤ 2 + −1*F-ENG-QUANT
F-ENG-QUANT ∈ [1, 2]
        F-ENG-QUANT ≤ 2 + −1*R-ENG-QUANT
```

Fig. 2. The constraints implied by the model given to ACRONYM for the generic class of wide-bodied passenger aircraft.

for ribbons and ellipses. In this case there is very little *a priori* information about the scale of the images. The (c) versions of each figure show the ribbons fitted to the linked edges when it is searching for candidate matches for the fuselage and wings of aircraft. There is even further degradation of image information at this stage. These are the only data which the ACRONYM reasoning system is given to interpret. Notice that in the Fig. 5 almost all the shapes corresponding to aircraft are lost. Quite a few aircraft in Fig. 4 are lost also. Besides losing many shapes, the combination of the edge finder and edge linker conspire to give very inaccurate image measurements. It is assumed that all image measurements have a ±30 percent error, except that for very small measurements, it is assumed that pixel noise swamps even those error estimates. Then the error is estimated to be inversely proportional to the measurement with a 2 pixel measurement admitting a 100 percent error. Thus the data which ACRONYM really gets to work with are considerably more fuzzy than indicated by the (c) series of Figs. 3-5.

It is intended to make use of new and better low-level de-

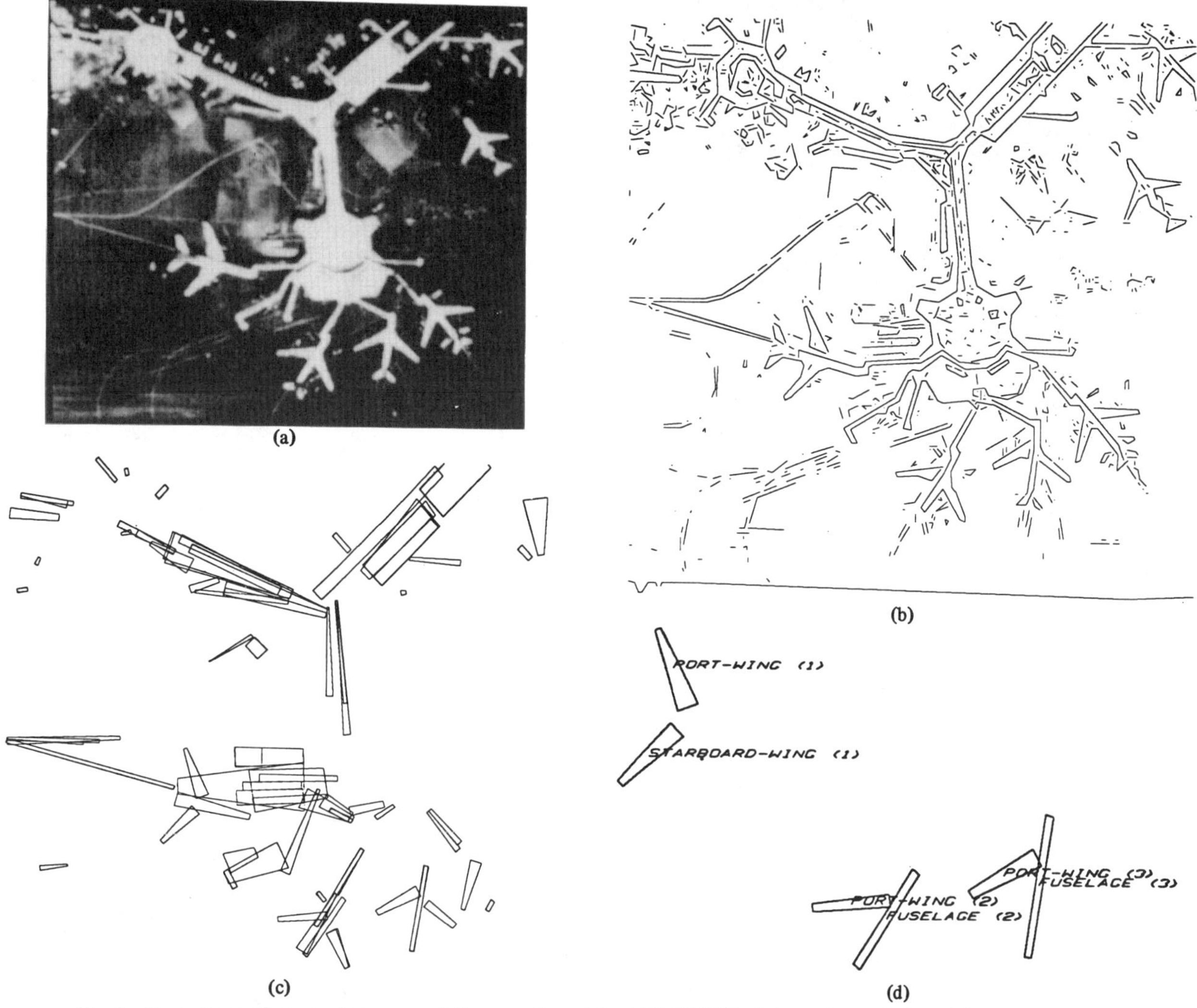

Fig. 3. Illustrations of some of the computations performed by ACRONYM in interpreting an image. The text contains the details.

scriptive processes being developed by other researchers as soon as they become robust enough for every day use (e.g., Baker [1] whose descriptions from stereo will also include surface and depth information).

Despite this very noisy descriptive data ACRONYM makes good interpretations of the images. Figs. 3(d), 4(d), and 5(d) show their interpretations with the ribbons labeled by what part of the model they were matched to. (The numbers which may be unreadable in Fig. 3(d) show the groupings into individual aircraft.)

ACRONYM first uses the most general set of constraints, those associated with the generic class of wide-bodied jets, when carrying out initial prediction and interpretation. Interpretation adds additional constraints for each hypothesized aircraft instance. For example, in finding the correspondences in Fig. 4(d) constraints were added which eventually constrained the WING-WIDTH (the width of the wings where they attach to fuselage) to lie in the range [7, 10.5677531] compared to the modeled bounds of [7, 12]. The height of the camera, modeled to lie in the range [1000, 12 000] is constrained by the interpretation to the range [2199, 3322].

Once a consistent match or partial match to a geometric model has been found in the context of some set of constraints (model class), it is easy to check whether it might also be an instance of a subclass. Only the extra constraints associated with the subclass need be added and checked for consistency with those already implied by the interpretation using the CMS as described in Section II-A. The aircraft located in Fig. 4(d) is consistent with the constraints for an L-1011, but not for a Boeing-747. Examination of the images by the author had previously indicated that the aircraft was an L-1011. The additional symbolic constraints implied by accepting that the aircraft is in fact an L-1011 propagate through the entire constraint set. Although the constraints describing an L-1011 do not include constraints on camera height, the back constraints deduced during interpretation relate quantifiers representing such quantities as length of the wings to the height (and focal ratio in the more general case). Thus the height of the camera is further constrained in Fig. 4(d) to lie in the range [2356, 2489]. Recall that all image measurements were subject to ±30 percent errors, and that this estimate has taken all such errors into account.

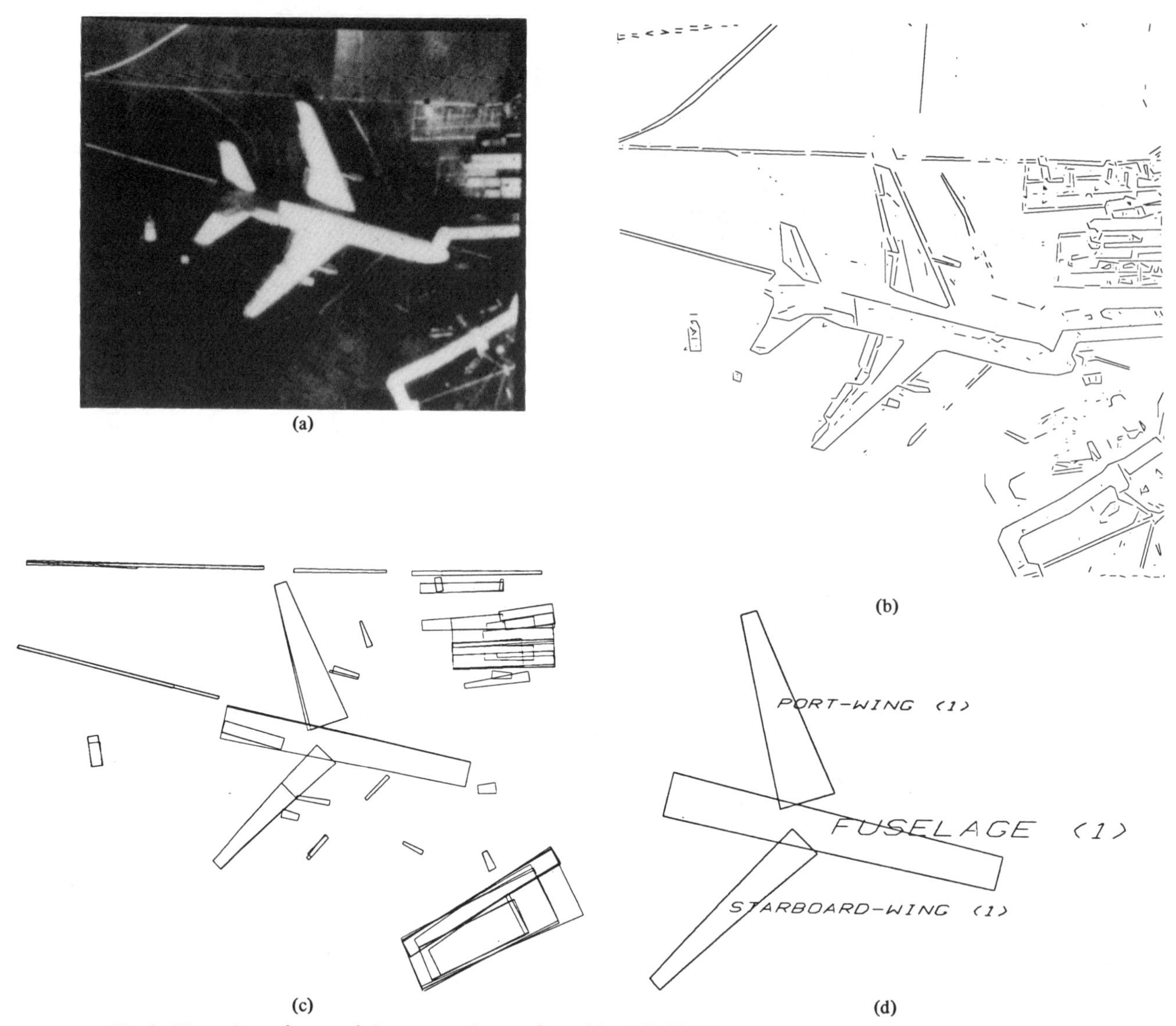

Fig. 4. Illustrations of some of the computations performed by ACRONYM in interpreting an image. The text contains the details.

Fig. 3(d) indicates matches were found for three airplanes. Examination of the data in Fig. 3(c) indicates that this is the best that could be expected. Note, however, that only partial matches were found in all three cases. For such small ribbons errors were apparently larger than the generous estimate used. The fuselage ribbon in the leftmost aircraft (number 1), for instance, fails to pass the coarse filtering stage. Despite the partial match, this particular aircraft is found to be consistent with the constraints for an L-1011, but not consistent with those of a Boeing-747. Again this is correct.

The other two aircraft identified are even more interesting. The author had thought from casual inspection of the grey level image that they were instances of Boeing-747's. They both gave matches consistent with the class of wide-bodied jets. As expected neither was consistent with the extra constraints of an L-1011. However, although each individual parameter range from the interpretation constraint sets was consistent with the individual parameter value or range for the class of Boeing-747's, neither set of constraints was consistent with that subclass (the constraints contain much finer information than just the parameter ranges–in the same manner as in the example above where constraints on wing length propagate to constrain the camera height). On close examination of the gray level image it was determined that the aircraft were not in fact Boeing-747's. The author used the fact that they were much smaller than the L-1011 to make that deduction, but ACRONYM made the deduction at the local level before considering comparisons between aircraft.

It also found the inconsistency at a more global level. The aircraft (probably Boeing-707's) are in fact too small to be wide-bodied jets of any type. Since the scale of the image is unknown *a priori* this cannot be deduced locally. However, it

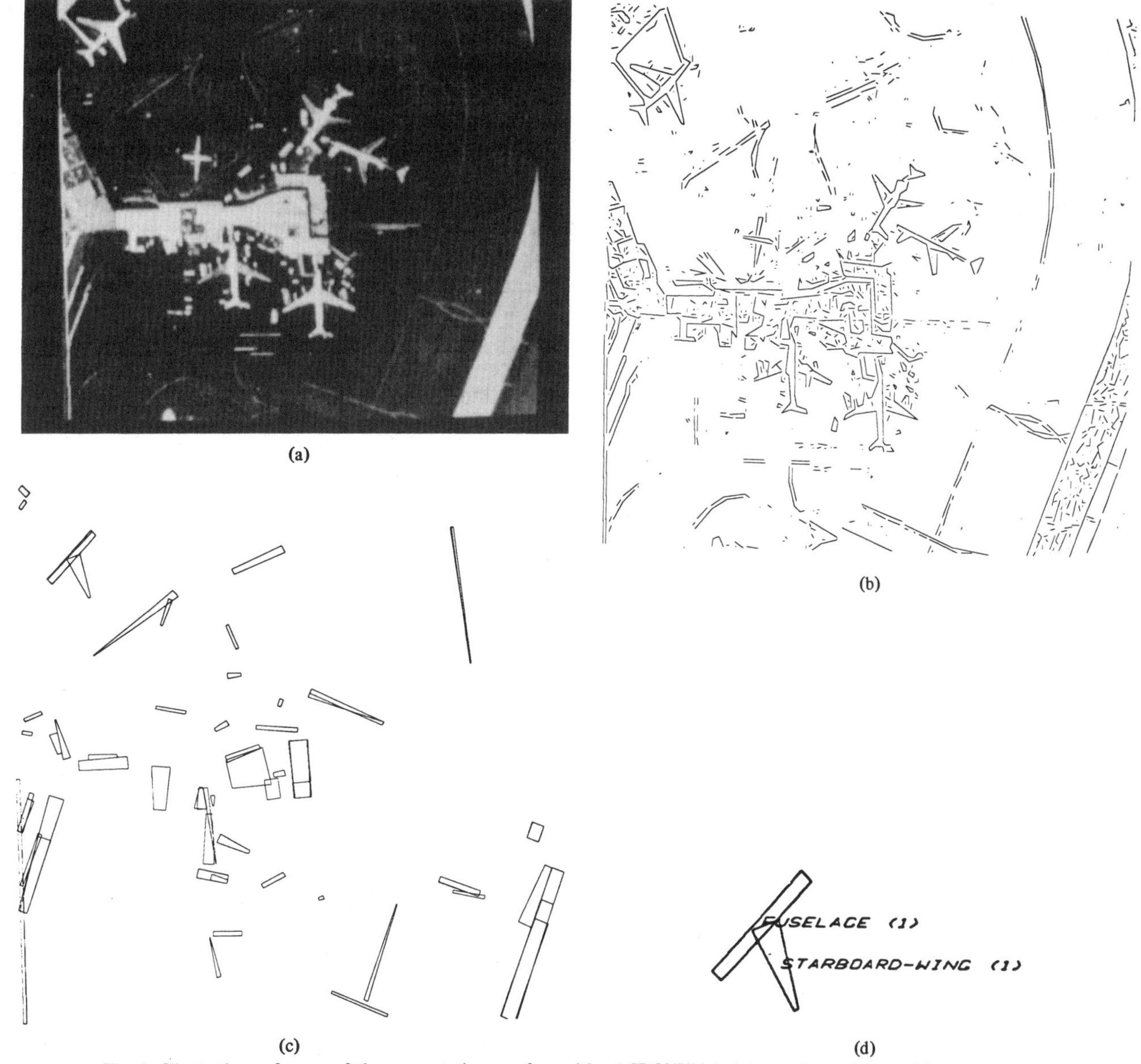

(a) (b) (c) (d)

Fig. 5. Illustrations of some of the computations performed by ACRONYM in interpreting an image. The text contains the details.

is reflected in the height estimates derived at the local level—[5400, 8226] interpreting the L-1011 just as a generic wide-body, ([5786, 6170] as an L-1011), and [9007, 11846] for the rightmost aircraft. Thus ACRONYM deduces that either the left aircraft is a wide-body and the others are not, or the right two are wide-bodies and the left one is not (it is too big).

Finally, note that geometrically there were other candidates for aircraft in the ribbons of Fig. 3(c). For instance, the wing of the aircraft just to the right of those identified and a ribbon found for its passenger ramp could be the two wings of an aircraft with a fuselage missing between them. In fact, these two ribbons were instantiated as an aircraft on the basis of the coarse filters on the nodes and arcs. However, the set of back constraints they generated were mutually inconsistent.

Thus, it can be seen from the examples that even with very poor and noisy data the combined use of geometry and symbolic algebraic constraints can lead to accurate image interpretations. They system should be tested on more accurate low-level data to fully evaluate the power of this approach.

References

[1] H. H. Baker and T. O. Binford, "Edge based stereo correlation," in *Proc. 7th Joint Int. Conf. Artificial Intell.*, Vancouver, Canada, Aug. 1981, pp. 631–636.

[2] T. O. Binford, "Visual perception by computer," presented at the IEEE Syst., Sci., Cybern. Conf., Miami, FL, invited paper, Dec. 1971.

[3] —, "Computer integrated assembly systems," in *Proc. NSF Grantees Conf. Industrial Automation*, Cornell Univ., Sept. 1979.

[4] W. W. Bledsoe, "The sup-inf method in Presburger arithmetic," Dep. Math. and Comput. Sci., Univ. Texas, Austin, Memo. ATP-18, Dec. 1974.

[5] D. G. Bobrow and T. Winograd, "An overview of KRL: A knowledge representation language," *Cognitive Sci.*, vol. 1, pp. 3-46, 1977.

[6] R. A. Brooks, "Goal-directed edge linking and ribbon finding," in *Proc. ARPA Image Understanding Workshop*, Menlo Park, Apr. 1979, pp. 72-76.

[7] —, "Symbolic reasoning among 3-D models and 2-D images," *Artificial Intell. J.*, vol. 17, pp. 285-348, 1981; a longer version is available as Stanford AIM-343, Dep. Comput. Sci., Stanford, CA.

[8] R. A. Brooks, R. Greiner, and T. O. Binford, "The ACRONYM model-based vision system," in *Proc. IJCAI-6*, Tokyo, Japan, Aug. 1979, pp. 105-113.

[9] R. Nevatia and K. R. Babu, "Linear feature extraction and description," *Comput. Graphics and Image Processing*, vol. 13, pp. 257-269, 1980.

[10] R. E. Shostak, "On the sup-inf method for proving Presburger formulas," *J. Ass. Comput. Mach.*, vol. 24, pp. 529-543, 1977.

[11] B. I. Soroka, "Debugging manipulator programs with a simulator," presented at the Autofact West Conf., Soc. Manuf. Eng., Anaheim, CA, Nov. 1980.

Rodney A. Brooks was born in Adelaide, South Australia, on December 30, 1954. He received the B.Sc. and M.Sc. degrees in mathematics from Flinders University, Adelaide, Australia, in 1974 and 1977, respectively, and the Ph.D. degree in computer science from Stanford University, Stanford, CA, in 1981.

He was a Visiting Scientist in the Department of Computer Science, Carnegie-Mellon University, Pittsburgh, PA, from June through September 1981. Since then he has been a Research Associate in the Artificial Intelligence Laboratory, Massachusetts Institute of Technology, Cambridge. His research interests span computer vision, robotics, spatial reasoning and planning, and issues in Lisp language and system development.

SPECIAL PURPOSE AUTOMATIC PROGRAMMING FOR 3D MODEL-BASED VISION

Chris Goad
Department of Computer Science
Stanford University,Stanford Ca 94305

Abstract

A method for the automatic construction of fast special purpose vision programs is described. The starting point for the automatic construction process is a description of a particular 3D object. The result is a fast special purpose program for recognizing and locating that object in images, without restriction on the orientation of the object in space. The method has been implemented and tested on a variety of images with good results. Some of the tests involved images in which the target objects appear in a jumbled pile. The current implementation is not fully optimized for speed. However, evidence is given that image analysis times on the order of a second or less can be obtained for typical industrial recognition tasks. (This time estimate excludes edge finding).

1. Introduction

In many practical applications of automated vision, the vision task takes the form of recognizing and locating a particular three dimensional object in a digitized image. The exact shape of the object to be perceived is known in advance; the purpose of the act of perception is only to determine its position and orientation relative to the viewer. This is model based vision in its strict form.

Most industrial applications of vision have this property, and also the property that the same object (or, more precisely, objects of the same shape), must be located in many images. In this kind of situation, it is desirable to split the computation into two stages: an analysis or precomputation stage, in which useful information about the (unchanging) object is compiled, and an execution or runtime stage, in which this information is exploited for the rapid recognition of the object in an image. The reason for breaking up the computation in this way is of course that the analysis only needs to be done once, whereas its results can be exploited repeatedly. Bolles[Bolles 1982] among others, has taken this general approach to the model based vision problem.

The advance analysis stage may take a variety of forms. In our work, this stage involves a kind of automatic programming. A description of the object to be recognized is "compiled" into a special purpose program whose only function is to recognize that one object in digitized images. In the second, runtime stage, individual images are processed by the special purpose program produced at the first stage.

This formulation of the work accomplished by the advance analysis is very unrestrictive. It makes no commitment as to the algorithm which is used to process images; rather the algorithm may be chosen according to the object which is to be recognized. The problem of finding the best algorithm among all algorithms for a given object is intractable. However, we may attempt to construct special purpose algorithms for object recognition within a restricted class of algorithms, and hope for good, though not optimal, results.

In this paper, we describe a method for automatically constructing special purpose programs for 3D object recognition. To be precise, we mean by 3D object recognition the recognition of three dimensional objects in ordinary light intensity images (not, eg, range images), where no restriction is made on the orientation of the object with respect to the camera. Although we speak of recognition, the process of recognition delivers information not only about the presence or absence of the object in the image, but also the position and orientation of the object if it is present. The method relies on matching object features to image edges. We will not concern ourselves here with how image edges are extracted from pixel data. The method does not rely on perfect results from the edge finder (if it did, it would be of no practical interest). The special purpose programs generated are quite fast. To give a very rough idea of how fast, our method should allow the recognition of ordinary industrial objects in moderately complex background in a second or less on a 1 MIP computer; this is the time required for the matching process, and excludes the time required for edge finding. The data on which this kind of general speed estimate is based will be given later in the paper.

2. A general strategy for special purpose automatic programming

The general features of the model based vision problem which make it a candidate for special purpose automatic programming are shared by a wide variety of computational problems. The features in question are that the inputs to the computation are delivered in two stages, and that for each first stage input (here, the object to be recognized), many second stage inputs (here, images) must be treated. The special purpose automatic programming problem in its general form can be stated as follows.

Let us suppose that a function f of two inputs x and y must be computed repeatedly under conditions where many values of y must be treated for each value of x. Then we attempt to devise, for each x, a special purpose program P_x with $P_x(y) = f(x,y)$. More precisely, what is wanted is an automatic process for constructing the programs P_x - that is, synthesis method M with $M(x) = P_x$.

The informal strategy which we use for the model based vision problem can also be described at this level of generality. The strategy is just that of starting with a general purpose program for doing the computation in one stage, and using specialized variants of this program as the templates from which special purpose programs are developed. Suppose, again, that $f(x,y)$ must be computed repeatedly with slowly changing x and rapidly changing y. Suppose also that we have in hand a program $G(x,y)$ which computes $f(x,y)$ in one stage. We begin by "unwinding" G as it applies to the value x for which a special purpose program P_x is wanted; this unwound program $G'_x(y)$ is then optimized to get the desired result. The unwinding is done by a kind of symbolic execution, in which loops are unwound when possible, and recursions are unfolded. If the procedures for unwinding and optimization are mechanical in nature, then together they constitute what we have called a synthesis method for f. The unwinding and optimization procedures need not be applicable to arbitrary programs; they can be custom designed for the particular program G at hand.

In general, the program G does not have to be completely specified - G may itself be a template for an algorithm, with many details left out. The details may be filled in after - rather than before - G has been unwound.

In our work on vision, we begin with a one stage algorithm template $G(x,y)$ which takes an object description x, and an image y, and identifies instances of x in y. G is a template for an algorithm in the sense of the last paragraph; many of the details of its operation will be specified only after it has been unwound for particular objects.

Model-based vision is the second problem to which we have applied this style of special purpose automatic programming. [Goad 82] describes the earlier application to hidden surface elimination in 3D computer graphics, and also contains a general discussion of special purpose automatic programming as it relates to other work on the automatic construction and manipulation of programs.

3. A one stage algorithm for model based vision

The one stage algorithm $G(x,y)$ which we start with is a simple sequential matching procedure. The kind of object description expected by G is a list of object features, together with conditions on their visibility. For the current purposes, an object feature is taken to be a curve along the object surface at which either a surface normal or a reflectivity discontinuity occurs. We will restrict ourselves to straight line segments rather than considering arbitrary curves. So, informally, an object feature is just a straight edge on the object. For each object feature, G also needs to know the range positions in space from which that feature is visible (means for representing this information will be described later). There is no need for the list of features making up an object description to be exhaustive; it is sufficient that enough features be included to make reliable recognition possible. As a result, the kind of description of an object which is needed for its recognition is much less extensive than that needed for displaying it.

The image description expected by G is of the same general kind as the object description: it is a list of features. In particular, the image features employed by G are of exactly the kind which the object features give rise to in the imaging process: they are straight segments in the image along which an intensity discontinuity occurs. The process by which this kind of image description is generated from raw pixel data will not be discussed in this paper. In our experiments, we used an edge detection program written by David Marimont[Marimont 1982] The program convolves the image with a lateral inhibition operator, detects zero-crossings in the laterally inhibited image, and then performs linking. Straight edges are arrived at by applying a simple segmentation scheme (we added this last step to Marimont's algorithm).

Although we will restrict ourselves in this paper to treating edge features, the same methods would apply to any kind of object feature which gives rise in a predictable way to an image feature.

The operation of G may be described in general terms as follows. G performs a simple depth-first search for a match between object and image edges. At any given time in the search, G's state includes a currently hypothesized match M, and a current hypothesis about the position and orientation of the object relative to the camera. The hypothesis concerning the location of the object gives bounds on the location parameters, and not exact values. In the main loop of the algorithm, G attempts to extend and refine its current hypothesis by means of the following three steps (which lie at the heart of many algorithms for perception):

(1) Predict: An object edge is selected which has not yet been matched by any image feature. Based on the current hypothesis, the position and orientation of its projection in the image is predicted.

(2) Observe: The list of image edges is checked to see whether any has the predicted qualities.

(3) Back-project: If an edge with predicted qualities was found in step (2), then extend the match to include this edge, and use the measured position and orientation of the edge to refine the current hypothesis as to the location of the camera.

The algorithm repeats this loop until either a satisfactory match is found, or until the algorithm fails to observe a predicted edge. In the latter case, the algorithm backtracks to the last choice point. Choice points arise when more than one image edge appears in a predicted position and orientation.

This is a sketch of the algorithm. Before supplying further details, some definitions are needed.

It will be convenient to work in object centered coordinates: The object will be thought of as fixed, and the position and orientation of the viewer as the unknown to be determined. An object edge is always regarded as an oriented segment (given by the *ordered*) rather than unordered pair of its end-points), while an image edge may or may not be oriented. The imaging process is given by the ordinary perspective transformation. The parameters of this transformation which derive from the camera model are the distance from the point of projection to the image plane, and the field of view, given by a rectangle on the image plane. Assume that these parameters are fixed. Let p be a 3D position, q a 3D orientation, and x an object edge. Then $\Downarrow([p,q],x)$ denotes the oriented image edge which results from viewing x from camera position and orientation $[p,q]$. The edge x may not be visible from $[p,q]$, either because x is occluded, lies on the "wrong side" of the object, or because the projection of x onto the image plane lies outside of the field of view. In these cases, $\Downarrow([p,q],x)$ is undefined. We will write $\Downarrow(x)$ to denote the projection of x when the parameters $[p,q]$ of the projection are clear from context. At this point we make several assumptions about the imaging geometry.

First we assume that, in the images which are to be analyzed, the object sought is either not visible at all, or lies entirely within the field of view. Second, we assume that the field of view is sufficiently narrow that changes of the orientation parameter q at a fixed position p have only negligible effects on the lengths of projected edges, and the angles between them. Thus, while a change in q will in general cause some of the image to move out of the field of view, that part of the image which remains visible will have undergone only a 2D rotation and translation - to within a small tolerance. This criterion is met in typical industrial imaging situations. Finally, we make the more restrictive assumption that the distance from the camera to the object - or more precisely from the perspective projection point to the origin of the object centered coordinate system - is known in advance. (This restriction is made to simplify the exposition, and does not apply to the implementation described later). Thus the position parameter p is restricted to lie on a sphere about the origin. Without loss of generality, we may assume that this is the unit sphere.

We will refer to a set of positions on the unit sphere as a ***locus***. A locus is to be thought of as a set of possible camera positions. Let X be an object description and Y an image description. Recall that X is a list of edges together with visibility conditions. The visibility condition for each edge e in X is given by the locus of points from which that edge is wholly visible. This is called the *visibility locus* of e. (Methods for representing such loci will be given later). Now, a *match* M between object edges X and image edges Y is an assignment of image edges to object edges. A match also assigns orientations to the otherwise unoriented image edges. For any object edge, $M(e)$ denotes the oriented image edge (if any) assigned to it by M. The assignment may be partial; that is, for some e, $M(e)$ may be undefined.

A match M is consistent with a camera position and orientation $[p,q]$ if for each object edge e, the projection $\Downarrow([p,q],e) \approx M(e)$ to within errors in measurement. A match M is consistent with a camera position p if there is some orientation q such that M is consistent with $[p,q]$. A match is consistent with a locus L if it is consistent with every position in the locus.

As indicated earlier, the algorithm G conducts its search for a match by attempting at each point to extend and refine its *current hypothesis* about the imaging situation. This hypothesis has two parts: the match M found so far, and the locus L of possible positions of the camera. In the course of the matching process, the consistency of L with M is maintained (modulo errors in measurement).

Now we can be more explicit about how the basic predict-observe-back-project loop is carried out. We may restrict ourselves to considering the case where at least one edge has already been matched, since prediction and back-projection do not apply to the matching of the first edge; all acceptable candidates for matches to the first edge must be considered, wherever they appear in the image. Let M be the current match, and L the current locus. G selects an object edge e which has not as yet been matched. For the sake of brevity, when we refer to the "position" of an edge, we will henceforth mean its position *and* orientation. Bounds on the position of the image $\Downarrow(e)$ of the new edge e can be predicted simply by selecting an already matched edge e_0, and computing the bounds on the possible position of $\Downarrow([p,q],e)$ relative to $\Downarrow([p,q],e_0)$ as p ranges over the current locus L (recall that the value of q does not affect relative measurements). This prediction, together with the known position of $M(e_0)$, give predicted bounds on the position of the image $\Downarrow(e)$ of e.

A similar method can be used for back-projection. Suppose that an image edge $M(e)$ has been matched to the object edge edge e. Back-projection consists of restricting the current locus L to the smaller locus L'

which is consistent with the measured position of $M(e)$. Let e_0 be some already matched edge, and $M(e_0)$ its match in the image. L' is then just the set of camera positions p in L from which the predicted position of $\Downarrow([p,q],e)$ relative to $\Downarrow([p,q],e_0)$ is the same as the measured position of $M(e)$ relative to $M(e_0)$, to within measurement error.

This scheme preserves consistency of matches if measurement errors are negligible.

The algorithm G as described so far does not take into account the fact that any given object edge may or may not be visible depending on the position of the camera. The following extension to G deals with this aspect of the matching problem.

In the prediction step, rather than selecting an arbitrary unmatched edge as was done before, G selects an edge e whose visibility is consistent with the current hypothesis (formally, and edge whose locus of visibility intersects the currently hypothesized locus). Then, a case analysis according to whether the edge is actually visible is performed.

On one side of the case analysis, G assumes that e is visible, and restricts the currently hypothesized locus accordingly: the new restricted locus is the intersection of the current locus with the locus of visibility of the edge. Then G proceeds as before: it predicts the position of e, looks for it in the predicted position, and back-projects if found.

On the other side of the case analysis, G assumes that the edge is invisible, and again restricts the currently hypothesized locus accordingly: this time the restricted locus is the intersection of the current locus with the complement of the locus of visibility of e. If the restricted locus is empty, then the current attempt at a match has failed, and G backtracks to the last choice point. If the restricted locus is non-empty, G proceeds by selecting another object edge to match.

This case analysis step constitutes a choice point for back-tracking. Thus, for each edge selected for matching, G assumes first that the edge is visible and looks for it. If this course of action leads to a good match, then all is well. Otherwise G backtracks and looks for a match under the assumption that the edge is invisible.

In the above, an edge should be considered visible only if - in addition to meeting the usual criteria - its projection is long enough to allow detection by the edge finding program. An edge which presents itself end-on to a given camera position should not be considered visible from that camera position.

Among other details about G which have been suppressed so far is the method by which loci of camera positions are represented. A very simple representation is adequate for our purposes. Suppose that we have a scheme for partitioning the unit sphere into an arbitrary number patches such that the diameters of the patches go to zero as their number increases. Then a locus can be represented to any desired resolution by a set of patches from a partition of adequate size. More precisely, a locus L is to be represented by the set of patches from the partition which contain some point of L. The resolution of this representation is bounded by the maximum diameter of a patch. Thus loci are represented by subsets of a finite set. These in turn may be represented by bit maps: one bit is allocated to each patch on the sphere. Bit maps are a particularly good representation for the current application, since the operation most frequently performed on loci is intersection, and intersection of bit maps is very fast on any computer. The particular scheme which we have chosen for partitioning the sphere is not the best but the simplest. The partition is generated by first imposing a regular grid on the faces of a cube. The cube is then projected radially onto the sphere. The patches on the sphere which we end up with are simply the projections of grid elements from the faces of the cube. In recent experiments, we have used 6 by 6 grids on the faces of the cube, yielding a total of 218 patches. This representation is depicted in figure 1. One 36 bit PDP-10 machine word is allocated to each face. So, a locus is represented by 6 machine words.

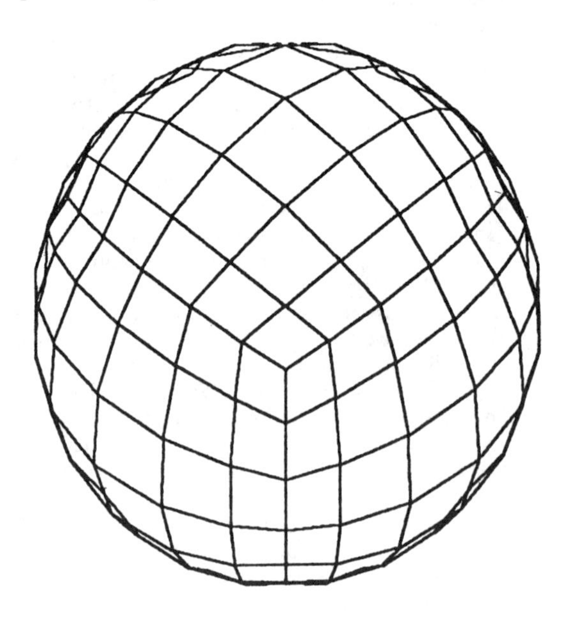

figure 1

We are still not done with the development of G. A major shortcoming of G as described so far is that it relies on perfect performance by the edge finder - each edge on the object which is in view must be detected by edge finder if the method to function properly. This kind of perfect performance is not obtained by existing edge detection programs, nor can it be obtained by any edge detector which relies on local image intensity discontinuities to detect edges, since object edges do not always give rise to such intensity discontinuities.

The matching algorithm may be modified in order to take into account the imperfections of the edge finder by accepting matches in which only a fraction of the expected edges are present. If such a modification is made, the criteria of match success and match failure become more complicated. It is necessary to determine conditions under which a partial match should be dropped because of inadequate success in finding predicted edges, and also conditions under which a match should be accepted as reliable evidence that the object being looked for has actually been found.

In order to do this, we would like to be able to assess the probability that a given partial match arose from the object in the manner claimed. If this probability is very low, then the match should be dropped, and if very high, it should be accepted. For intermediate values, more edges should be matched, if possible, in order to accumulate further evidence.

Direct estimates of this probability are difficult to make in the usual cases. Nonetheless, we can proceed by estimating conditional probabilities, and use qualitative considerations to get from these conditional probabilities to the needed conclusions concerning a given match.

On the one hand, it is possible to estimate the probability that the configuration of edges making up a match arose by chance given any particular assumed "background" distribution of edges giving rise to chance matches, since the specificity of each prediction involved in the match is known, as is the number of such predictions which have been met. We will refer to the inverse of this probability as the "reliability" of the match. The background distribution of edges usually cannot be derived from first principles, but is best determined by gathering statistics on sample images of the kind on which the algorithm is to be used.

On the other hand, suppose that we have a partial match in which some fraction of the predicted edges are missing. Then we can estimate the probability that the given set of edges would be missed by the edge detector under the assumption that the partial match did arise as claimed. We will refer to this probability as the "plausibility" of the match. Estimating plausibility requires information about the performance of the edge detector. As in the case of edge distributions, deriving this kind of information from first principles is difficult; again, compiling statistics from sample images is a better idea. Note that underestimating the performance of the edge detector will lead to robust performance by the matching algorithm.

So the reliability of a match measures how unlikely it is to have arisen assuming it is in fact incorrect, and its plausibility measures how likely it is to have arisen assuming it is in fact correct. Assuming that the presence of the object in the field of view is moderately likely and there are unlikely to be impostors of the object in view, it follows from Bayes' Rule that high reliability provides strong evidence that the match is correct, while very low plausibility provides strong evidence that the match is incorrect. (Notes: (a) Low reliability does not provide evidence that the match is incorrect, nor does high plausibility provide evidence that the match is correct (b) By an "impostor" in the above, we mean an object which is regarded as distinct from the target object, but looks nearly the same.)

The following modifications to the algorithm G are needed to deal with imperfect edge finding. First of all, G must maintain estimates of the reliability R and the plausibility P of the current match in the course of its search. When R exceeds a predetermined threshold, the match should be accepted, and when P falls below another predetermined threshold, backtracking should occur. Second, G needs to perform a case analysis according to whether each expected edge is detected by the edge finder, in addition to the case analysis which it already performs concerning whether the edge is in view. This case analysis will also constitute a choice point for the purpose of back-tracking. Thus, G will proceed as follows. For each new edge e which it selects for matching, it will (1) assume that e is in view, (2) assume that e is detected, (3) look for e, (4) continue the match. If the match failed, then it will backtrack to (2) and assume that e, though in view, was not detected, and will proceed to the matching of other edges. Finally, if this last match fails, it will backtrack to (1) in the manner described earlier.

The algorithm as it now stands is no more than an elaboration on the simplest of matching algorithms: sequential matching with backtracking. Nonetheless, if we judge the efficiency of a matching method by the number of matching steps which it goes through in searching for a correct match, the algorithm does not come out badly. The principal reason for this is that only a few edges on an object of known shape need to be identified in order to determine the position and orientation of the object. In fact, identification of the image projections of three non-colinear points on the object is sufficient to narrow the set of possible positions and orientations of the object to at most two distinct possibilities. (This last statement holds exactly for orthographic projection, and applies to perspective projection as well unless the camera is very close to the object, or the precision of measurement is very high, in which case one of the two possibilities may be eliminated). A fourth non-coplanar point suffices to remove the remaining ambiguity. The identification of three pairwise non-parallel lines (without specified end points) will accomplish the same task. Thus, a match does not need to proceed very far before the locus of possible positions of the camera will have been narrowed to only a few grid points by back-projection. Thereafter, the matching of additional edges serves to check the correctness of the match, but not to further refine the estimate of camera position. The positions of these additional edges will be predicted accurately, and as a consequence the probability that many such edges will be found in the case of an incorrect match will be exceedingly low. Thus, bad matches are likely to fail very early. Conversely, the reliability of a good match will rise quickly as more expected edges are found, so that the cost of achieving a very reliable match is low. So, the effectiveness of the current algorithm relies on the fact that it requires exact quantitative matching of object edges to image edges at all stages during a match.

4. Specialization and instantiation of the one stage algorithm

The result of unwinding the main loop of the schematic algorithm G described in the last section may be diagramed informally in the following way.

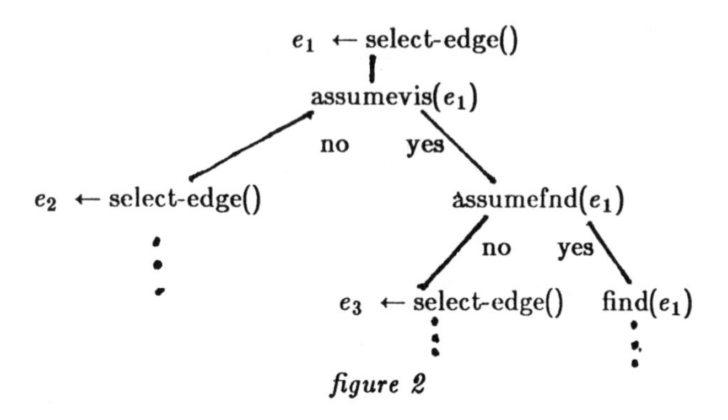

figure 2

Here, "find(e)" represents the predict-observe-back-project operation by which new edges are matched, "$e \Leftarrow$select-edge()" represents the selection of an unmatched edge to look for, "assumevis(e)" represents the case analysis according to visibility of edges, and "assumefnd(e)" represents the case analysis according to whether the edge has been detected. Our job now is to fill in the details in this unwound schematic algorithm, making use as appropriate of the fact that the object description is available in advance. In the following discussion, we will employ the "compile-time/run-time" terminology familiar from compiler design. Operations which are carried out in the course of constructing specialized variants of G will be refered to as "compile-time" operations, while operations carried out by those specialized variants in the analysis of images will be refered to an "run-time" operations.

The compile-time process by which the above schematic search tree is filled in may be thought of as moving from the root of the tree down, fully instantiating nodes as it goes. Imagine for the moment that the first k levels of the tree have been filled in, and that the task at hand is to fill in a particular node at next level. As will be seen in a moment, selection of the object edge to be matched at each point in the tree is done at compile-time. That is, the select-edge() operation is "executed" at compile-time, so that each node in the instantiated search tree will refer to a particular object edge to be matched. The tree developed to level k will look something like this:

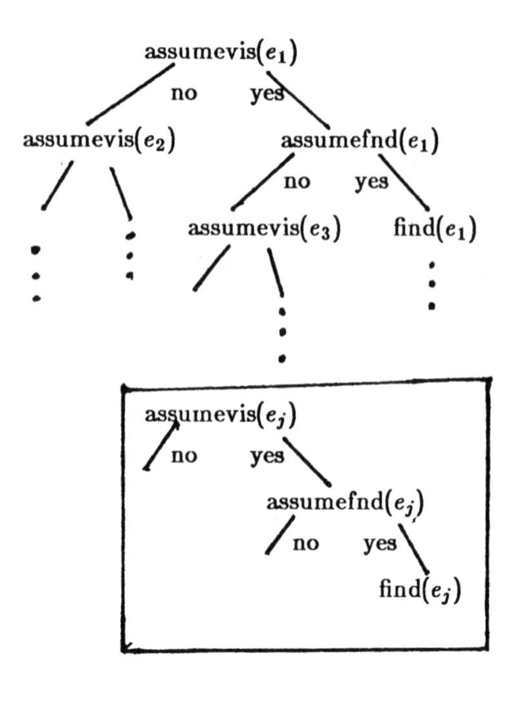

figure 3

The box indicates nodes to be filled in at the current stage. We will deal with all the nodes involved in matching a particular object edge at once, rather than following a strict level by level order. In what follows we will specify how each of the operations in the boxed nodes are instantiated.

(1) The find operation. This involves prediction of the position of e_j, a check to see if any image edges lie in the predicted position, and back projection. Recall that prediction is carried out by computing bounds on the location of e_j relative to an already matched edge e_0, and then using the known position of e_0 to get bounds on the position of e_j in the image. The position of one edge e_j relative to another e_0 may be specified in various ways. The only requirement here is that the relative position be given in a way that is invariant under translations and rotations. In any case, a vector of four numbers $[a_1, a_2, a_3, a_4]$ suffices. For example, a_1,a_2 might give the coordinates of the center of e_j relative to the image coordinate system with origin at the center of e_0 and x-axis directed along e_0, a_3 the length of e_j, and a_4 the orientation of e_j relative to e_0. Let $relpos(e_j, e_0, p)$ denote the position of e_j relative to e_0 from camera position p in whatever representation is chosen. More generally, let $relpos(e_j, e_0, K)$ denote bounds on the components of $relpos(e_j, e_0, p)$ as p ranges over locus K. Let L be the currently hypothesized locus at the time that the find operation is executed. We

want to compute $relpos(e_j, e_0, L)$. The following very simple scheme is adequate. Namely, at compile time, $relpos(e_j, e_0, g)$ is computed for each grid element g, and stored in a table. Then, $relpos(e_j, e_0, L)$ is computed at run-time simply by taking the union of the bounds $relpos(e_j, e_0, g)$ for $g \epsilon L$. These unions are taken component-wise, so that the end result of the process is a set of numerical upper and lower bounds on each of the components of the relative position. Note that this manner of computing bounds loses some information since constraints relating different components of the relative position are not expressed by simple bounds on the components. As a result, edges with positions predicted in this way may not be consistent with any camera position in the locus. But this is comparatively unlikely, and, in any case, bad edges of this kind will be thrown out in the back-projection stage.

The above method is a very crude, but it can be made quite fast. For example, less than 50 machine instructions per grid element are required to carry out the prediction operation in an aggressively coded implementation on the PDP-10, VAX, or Motorola 68000. The number of grid elements which need to be considered in a given prediction step of course depends on the details of the match in progress. In most matches, the size of position loci to be considered decreases rapidly as the match proceeds. For the experiments described later, the average prediction step involved less than 10 grid elements.

Efficient implementation of the observation stage is a standard exercise in computational geometry. The problem is to design a data structure for storing image edges such that the set of edges satisfying a given prediction can be retrieved rapidly. Any of a variety of methods involving binary search on the parameters of the prediction will do.

The tables of relative positions constructed at compile time for the prediction step can be used for back-projection as well. In order to determine which grid elements of the current locus are consistent with a given set of measurements, it is only necessary to compare the measured values to the bounds $relpos(e_j, e_0, g)$ which have been pre-computed for each grid element g.

(2) Selection of the next edge to look for.

In the course of a match, the currently hypothesized locus of camera positions is refined in two ways: by back-projection, and by making assumptions about the visibility or invisibility of particular edges. The data necessary for the latter kind of refinement is available at compile time. As a result, each point in the instantiated search tree has an associated compile-time locus of possible camera positions - namely, the set of camera positions which are consistent with the visibility assumptions made on the path leading from the root to the current node.

Our task is to select at compile-time an appropriate object edge to look for next at the current stage of the match. There are three considerations which are relevant to this selection. First, the likelihood that the selected edge is visible should be as high as possible. We don't wish to select an edge which will be visible from only a small fraction of the current locus or which, even if visible, the edge finder is unlikely to detect, since the computation time spent looking for it will then be unlikely to pay off. Second, the prediction of the position of the edge should be as specific as possible, since this will lead to a lower likelihood of false matches for the edge. Third, it is desirable that measurements on the image position of the observed edge should provide as much information as possible about the camera position. Each of these factors can be evaluated in a quantitative manner at compile time. Assuming a uniform probability distribution on the position of the camera (or more generally, assuming any particular prior distribution on camera positions) and statistics about the performance of the edge detector, the probability of the visibility of any edge over any locus can be computed. Similarly, the specificity of a prediction is naturally measured by the inverse of the probability that a randomly chosen edge will meet the prediction. This in turn can be computed in a straight-forward way from the bounds involved in the prediction. The numerical values of the bounds for the prediction can be computed for each possible camera position in the compile-time locus at compile time. By averaging these bounds (using the weighting of the prior distribution on camera positions), an expected specificity for a prediction of a given edge can be determined at compile time. Finally, in a similar manner, the information obtained from measuring the position of any given edge can be evaluated for each camera position at compile time.

By this method, the best edge to match next can be determined at compile time, assuming that a way of combining the factors listed above has been chosen. The question of exactly what weight should be given to each factor is a complex one. In qualitative terms, the order in which the considerations have been stated here reflects their order of importance for the efficiency of the matching process. Any weighting function which respects this order of importance is likely to be acceptable. We regard the detailed analysis of this question as an open research topic.

Note that the determination of the best edge to match next at any stage is computationally expensive, since it involves calculations concerning each edge at each camera position. But, as remarked above, large amounts of computation time at compile time are often justified for smaller gains at runtime.

(3) The visibility case analysis. The method by which visibility case analyses are performed has already been fully specified, so no compile time instantiation of the node is needed. However, certain visibility case analyses can be dispensed with entirely based on information available at compile time. It will often happen that the particular edge e_j whose visibility is in question is in fact visible throughout the compile-time locus, so that no case analysis according to its visibility need be performed. That is, it will often happen that along the

path taken from the root of the search tree to the current node, visibility assumptions have been made which together guarantee the visibility of the current edge. In this case, the case analysis node is simply left out.

Note that this optimization is a very important one in that it greatly reduces the size of the search tree (though usually has only a minor effect on its depth). If n edges are involved in a match, then in principle, there are 2^n distinct combinations of visibility and invisibility for the edges to be considered. Of course, only a small fraction of these cases can actually occur, since since the visibility of edges are not determined independently. For example, there are 26 distinct visibility/invisibility combinations for the edges of a cube (one for each face, edge, and vertex which the viewer might be "facing"), rather than $2^{12} = 4096$.

(4) The detection case analysis. We treat detection case analyses in a similar manner to visibility case analyses: we drop a detection case analysis if the assumption that the current edge is not detected causes the plausibility of the current match to drop below threshold. The information about the current match which is needed to compute its plausibility - namely, the information as to which edges have been detected and matched and which have not - is available at compile-time. It can be read off by following the path from the root of the search tree to the current node.

We can sum up the speed gains achieved by specialization in a very rough way by noting that the time required for an average matching step in the specialized program will be on the order of a few miliseconds on a 1 MIP machine. As a consequence, several hundred matching steps can be executed per second in the search for a match. This speed is much greater than that obtained by existing methods, and is adequate for a variety of practical applications.

5. Experiments

The scheme for generating special purpose vision programs described above has been implemented in MacLisp running on the PDP-10 at the Stanford Artificial Intelligence Laboratory. A few refinements not described earlier are included in the implementation. For example, in this account, we have not considered the effect of measurement error, nor have we described any method for matching partially visible (or partially detected) segments. The implementation includes machinery for dealing with both of these matters. Also, until now we have required that the distance to the object be known exactly in advance. This requirement is weakened in the implementation; it is generally sufficient if the distance is known to within a factor of 2.

On the other hand, the implementation does not yet fully automate the selection of the order in which edges are treated; in the experiments we chose the order by hand. Nor does it come close to realizing the potential for speed of the underlying algorithm. For example, efficient data structures and accessing methods for the set of image edges have not been implemented. The speed figures given earlier are estimates of what could be obtained in an aggressive implementation, not measurements of current performance.

So far, tests involving three different objects have been run. The objects treated were a connecting rod casting, a universal joint casting, and a key-cap (key-caps are the plastic keys which make up typewriter and terminal keyboards). In each test, a special purpose program was generated automatically from a description of the object; this program was then applied to images of the object digitized from a television camera. In the case of the connecting rod and universal joint, the pictures contained only one instance of the object against a relatively uncluttered background. These images were successfully analyzed with relatively little effort by the vision programs; correct matches were obtained in each case after fewer than 50 matching steps.

The special purpose program for recognizing key-caps was subjected to a more arduous test. We digitized an image of a jumbled pile of key-caps (see figure 4). The edges found in this image by David Marimont's edge finder [Marimont 1982] are displayed in figure 5. The task of the key-cap recognition program was to find instances of key-caps which were - roughly speaking - within 45 degrees of right-side-up. More precisely, the locus of allowable orientations of the camera relative to the key-cap was the locus making up the top face of the cube in the scheme for representing loci described earlier. Key-caps of a variety of shapes appear in the image; only key-caps with square upper faces were sought by the program. This is a severe test for the matching method for several reasons: (1) Objects of the desired kind must be recognized in a complex background - a background in which many objects similar to the target object appear. (2) The target object has only a limited number of features on which the match can be based. (3) Resolution is quite low. Although the entire image has a resolution of 240 by 240 pixels, each object to be recognized occupies only a 40 by 40 region. Also, lighting and contrast were not particularly good.

The program was run in a mode in which not just one, but every match meeting the reliability criteria was returned. Further, the reliability threshold was set at a very low level, so that every plausible match was found. The matches found were then ranked by reliability. The top three matches in this ranking were in fact correct. They are displayed in figure 6. Most of the remaining matches - which had been assigned lower reliability - were incorrect. The total number of matching steps required in this experiment was 960; so, in the hypothetical "aggressive implementation" mentioned earlier, the whole process would take a couple of seconds.

This experiment indicates that the matching algorithm can find matches under difficult circumstances.

In this case, due to the small number of matching features, it is not possible to achieve very high reliability of matches. Typical industrial objects, such as the castings mentioned above, have many more features on which a match can be based, and hence allow very good reliability of detected matches.

figure 4

figure 5

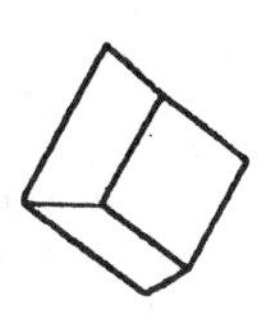
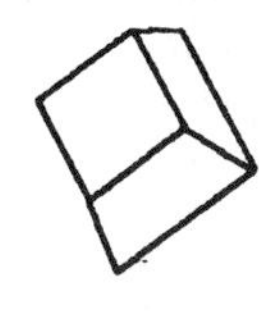
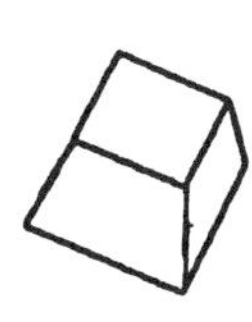

figure 6

6. Extensions

We have described in some detail the construction of specialized variants of a comparatively simple matching algorithm. It should be evident that the same general scheme can be applied to more complex matching algorithms which handle a wider class of problems, or which exploit additional structure in the matching situation in order to enhance performance.

An elaboration of the current algorithm which is particularly useful and to which our scheme for specialization extends easily is as follows. If the target object has symmetries, or if more generally there are recurring patterns of edges on the object, then the matching process may proceed by first seeking an instance of the recurring pattern, and then performing a case analysis according to which of several instances of the pattern has been encountered. This modification will speed up the matching process in the cases to which it is relevant, since a good match is likely to be found sooner, and since extensive bad matches to the "wrong" instance of the pattern will be avoided. The same technique can be used for matching of multiple objects which share common patterns of features. Again, the common pattern is matched first, and then a case analysis as to which object the pattern arose from is performed. The technique may be applied recursively to very large sets of target objects which have been classified in a hiearchical manner according to a taxonomy of common features. The taxonomy is exploited by a matching method which performs a kind of binary search down the hierarchy until a complete match is found.

More generally, the matching algorithm which we have considered is an instance of an extremely common kind perception algorithm. Such algorithms are built up from interleaved observation steps, in which some detectable quality of the world is predicted, observed, and used to refine the current world-model, and case analysis steps, in which assumptions are made about the world - assumptions which are not justified by any data or argument, but which are necessary to decide on what observation to make next, and which can be withdrawn later if necessary. This kind of algorithm appears - usually in elaborated form - in many areas of computing. Any such algorithm can be specialized according to the general plan which we used here. To perform the specialization, we proceed by first unwinding the case analysis steps into a full tree of possibilities. Then, we use the context of assumptions available at any point in this tree to optimize the work performed at that point.

7. Related Work

The particular vision problem which we have chosen to attack is 3D model-based vision in its strict form: the exact shape of the object to be recognized is assumed to be known in advance, and no restriction is placed on the 3D orientation of the object relative to the camera. Comparatively little work has been devoted to this form of the vision problem. There appear to be two traditions of work in model-based vision, one of which might be referred to as 2D vision from exact models, and the other as 3D vision from inexact models. The first tradition includes work focused directly on industrial problems such as that of Perkins [Perkins 1982], where the exact shape of the object is specified in advance, and where the orientation of the object is restricted in such a way as to reduce the problem to a "nearly" 2D form. The second tradition treats problems in which orientation is (comparatively) unrestricted, but where the previously available information about the model is less complete. Examples of this kind of work include [Garvey 1976] and [Shirai 1978]; here the matching processes used tend to employ qualitative rather than quantitative restrictions on matches. Acronym[Brooks 1981] is an exception to the above, in that it does exploit quantitative restrictions on the parameters involved in a match. Acronym uses a considerably more ornate matching scheme than ours. Also it uses a method for generating numerical constraints which is much more general and consequently much slower than ours - by a factor of at least 100. (Acronym does not "compile" the object model into a fast program as we do). Still, there are strong similarities in approach between our work and the work on Acronym.

Thus our work is less ambitious than some previous work in 3D model-based vision, in that we restrict ourselves to exactly specified models, and in that we are investigating comparatively simple algorithms. Nonetheless, it seems to us that the problems and processes involved in simple sequential matching of exactly specified models are not yet well understood, and that, as a research strategy, it makes some sense to concentrate on this limited domain before attacking matching problems of a more general kind.

The general strategy by which we have obtained an efficient implementation of matching - namely special purpose automatic programming - has been followed in technically different form by [Bolles 1982]. Bolles has attacked the problems of matching 2D models to images, and more recently, 3D models to range images, by what he calls the local feature focus method. The method involves selecting a class of "focus" features of similar shape on the object from which the match is to begin. Then maximal sets of mutually consistent interpretations for features near a given candidate match to a focus feature are sought. Such a "maximal clique" of consistent interpretations forms the seed for a more complete match, which is done sequentially. (See [Bolles 1982] for a description of the method). This method is compiled into a very fast matching program by the same kinds of methods we have used. Local features to be matched are selected in advance, and tables of relative positions are compiled. Experiments have shown that the maximal clique method is robust and fast for 2D matching.

The maximal clique method does not extend easily to the problem of 3D matching from intensity (rather than range) images. There are two reasons for this. First, the maximal clique method depends on the transitivity of the consistent-interpretation relation: if interpretation A for point a is consistent with interpretation B for point b, and if interpretation B for point b is consistent with interpretation C for point c, then interpretation A for point a is consistent with interpretation C for point c. This transitivity holds for 2D, and for 3D points from range data, but not for 2D projections from a 3D model with arbitrary orientation. (Still, the clique method might be used for more complicated structures for which this transitivity does hold). The second and more decisive reason for the difficulty of extending the maximal clique method to 3D is this. The method depends on the possibility of selecting a reasonably small set of image features which are candidates for matching features near the already matched focus feature, and which together uniquely identify the match. In 2D or 3D from range data, the identification of such a small set of features is aided by the fact that the positions and orientations of the local features relative to the focus feature are known in advance. In 3D intensity images this kind of advance information is not available, or is much weaker, so that the set of nearby features in the image which require consideration may be quite large. Equally importantly, the number of possible interpretations for each such feature will be large as well (the size of the graph of possible interpretations in which cliques must be found is of order $k \times n$, where k is the number of local features considered, and n is the average number of interpretations of each feature). In the keycap example, the interpretation graph would be so large that clique finding would be impractical. Sequential matching is less vulnerable to this kind of problem because at each stage in the match all of the information derived from the match so far is used to restrict the number of candidates for match at the next stage.

Bolles' work is very closely related to ours in general aim and stategy; his results support the idea that the feature matching approach to vision is feasible and robust.

Acknowledgements

We would like to thank David Marimont and David Lowe for many useful discussions. This research was supported in part by the National Science Foundation under Grant MCS81-04873 and in part by the Advanced Research Projects Agency of the Department of Defense under Contract MDA903-80-C-0102

References

[Bolles 1981]
Bolles, R.C., and Cain, R.A., *Recognizing and locating partially visible objects: the local-feature-focus method*, International Journal of Robotics Research, Vol 1,No. 3,Fall 1982

[Brooks 1981]
Brooks, R.A., *Symbolic reasoning among 3-D models and 2-D images*, Artificial Intelligence Journal, August, 1981.

[Garvey 1976]
Garvey, T.D., *Perceptual strategies for purposive vision*, Technical Note 117, AI Center, SRI International, 1976

[Goad 1982]
Goad, C. A., *Automatic construction of special purpose programs for hidden surface elimination*, Computer Graphics, vol. 16 No. 3, July 1982, pp. 167-178

[Marimont 1982]
Marimont, D.H., *Segmentation in Acronym*, Proc. Image Understanding Workshop, September 1982

[Perkins 1978]
Perkins, W.A., *A model-based vision system for industrial parts*, IEEE Trans. Comput. C-27:126-143.

[Shirai 1978]
Shirai, Y., *Recognition of man-made objects using edge cues* in **Computer Vision Systems**, A. Hanson, E. Riseman, eds, Academic Press, New York, 1978.

Model-Based Recognition and Localization from Sparse Range or Tactile Data

W. Eric L. Grimson
Tomás Lozano-Pérez
Artificial Intelligence Laboratory
Massachusetts Institute of Technology
545 Technology Square
Cambridge, Massachusetts 02139

Abstract

This paper discusses how local measurements of three-dimensional positions and surface normals (recorded by a set of tactile sensors, or by three-dimensional range sensors), may be used to identify and locate objects from among a set of known objects. The objects are modeled as polyhedra having up to six degrees of freedom relative to the sensors. We show that inconsistent hypotheses about pairings between sensed points and object surfaces can be discarded efficiently by using local constraints on distances between faces, angles between face normals, and angles (relative to the surface normals) of vectors between sensed points. We show by simulation and by mathematical bounds that the number of hypotheses consistent with these constraints is small. We also show how to recover the position and orientation of the object from the sensory data. The algorithm's performance on data obtained from a triangulation range sensor is illustrated.

1. The Problem and the Approach

A central characteristic of advanced applications in robotics is the presence of significant uncertainty about the identities and positions of objects in the workspace of the robot. It is this characteristic that makes sensing of the external environment an essential component of robot systems. The process of sensing can be loosely divided into two stages: first, the measurements of properties of the objects in the environment, and second, the interpretation of those measurements. In the present paper, we will concentrate on the interpretation of sensory data. In investigating this problem, we make only a few simple assumptions about available sensory measurements, rather than considering specific details of a particular sensor. As a consequence, the interpretation technique that is developed here should be applicable to a wide range of sensing modalities. The technique may have implications for the design of three-dimensional sensors as well.

This report describes research done at the Artificial Intelligence Laboratory of the Massachusetts Institute of Technology. Support for the Laboratory's Artificial Intelligence research is provided in part by a grant from the System Development Foundation, and in part by the Advanced Research Projects Agency under Office of Naval Research contracts N00014-80-C-0505 and N00014-82-K-0334.

The International Journal of Robotics Research,
Vol. 3, No. 3, Fall 1984,
0278-3649/84/030003-33 $05.00/00,

1.1. Problem Definition

The specific problem we consider in this paper is to identify an object from among a set of known objects and to locate it relative to the sensor. The object sensed is assumed to be a single, possibly nonconvex, polyhedral object (for which we have an accurate geometric model). The object may have up to six degrees of freedom relative to the sensor (three translational and three rotational). The sensor, which could be tactile or range, is assumed to be capable of providing three-dimensional information about the position and local surface orientation of a small set of points on the object. Each sensor point is processed to obtain

1. Surface points. On the basis of sensor readings, the positions of some points on the sensed object can be determined to lie within some small volume relative to the sensor.
2. Surface normals. At the sensed points, the surface normal of the object's surface can be recovered to within some cone of uncertainty.

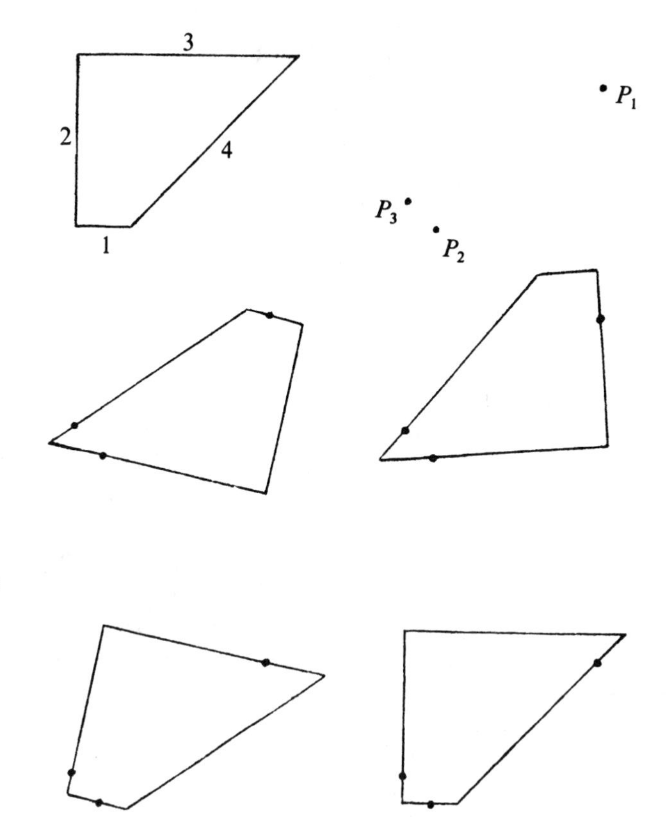

Fig. 1. An example of the approach.

Our goal is to use local information about sensed points to determine the set of positions and orientations of an object that are consistent with the sensed data. If there are no consistent positions and orientations, the object is excluded from the set of possible objects.

In this paper, we do not discuss how surface points and normals may be obtained from actual sensor data, since this process is highly sensor-dependent (for references to existing measurement methods, see Section 1.3). Our aim is to show, instead, how such data may be used in conjunction with object models to recognize and localize objects. The method, in turn, suggests criteria for the design of sensors and sensor-processing strategies.

Our only assumption about the input data is that fairly accurate positions of surface points are obtainable from the sensor, but that significant errors may be associated with the surface normals. This assumption reflects the type of data obtainable from tactile sensors. Range sensors based on triangulation can be used to obtain high-quality measurements of normals from patches of depth data. The availability of good normal data merely increases the efficiency of the method.

1.2. Approach

A recent paper (Gaston and Lozano-Pérez 1983) introduced a new approach to tactile recognition and localization for polyhedra with three degrees of positional freedom (two translational and one rotational). The present paper generalizes that approach to polyhedra with six degrees of positional freedom. The inputs to the recognition process are (1) a set of sensed points and normals and (2) a set of geometric object models for the known objects. The recognition process, as outlined in the earlier paper, proceeds in the following two steps:

1. Generate feasible interpretations. A set of feasible interpretations of the sensory data is constructed. Interpretations consist of pairings of each sensed point with some object surface of one of the known objects. Interpretations inconsistent with local constraints (derived from the model) on the sensory data are discarded.
2. Model test. The feasible interpretations are tested for consistency with surface equations obtained from the object models. An interpretation is legal if it is possible to solve for a rotation and translation that would place each sensed point on an object surface. The sensed point must lie *inside* the object face, not just on the surface.

The first step is the key to this process. The number of possible interpretations given s sensed points and n surfaces is n^s. Therefore, it is not feasible to carry out a model test on all possible interpretations. The goal of the recognition algorithm is to exploit the local constraints on the sensed data so as to minimize the number of interpretations that need testing. This approach is an instance of a classic paradigm of artificial intelligence: generate and test (e.g., Buchanan, Sutherland, and Feigenbaum 1969).

Consider a simple example of the approach, illustrated in Fig. 1. The model is a four-sided figure, with

Table 1. Distance Ranges Between Edges

	1	2	3	4
1	[0, 1]	[0, $\sqrt{10}$]	[3, 5]	[0.5]
2	[0, $\sqrt{10}$]	[0, 3]	[0, 5]	[1, 5]
3	[3, 5]	[0, 5]	[0, 4]	[0, $3\sqrt{2}$]
4	[0, 5]	[1, 5]	[0, $3\sqrt{2}$]	[0, $3\sqrt{2}$]

edge sizes of 1, 3, 4, and $3\sqrt{2}$, respectively. From this model, we can construct a table of ranges of distances between pairs of points on the edges (Table 1).

Now suppose we know the positions of the three sensed points, P_1 through P_3, shown in Fig. 1. The measured distances between those points are $dist(P_1, P_2) = 4.03$; $dist(P_1, P_3) = 3.91$; and $dist(P_2, P_3) = 0.71$. From this we see that any interpretation of the sensed points that assigns both P_1 and P_2 to edge 1 is inconsistent with the model. Similarly, assigning P_1 and P_2 to edges 1 and 2 is not consistent. Many other pairwise assignments of points to edges can be discarded simply by comparing the measured distances to the ranges in the table. Note that the sensed positions are subject to error, so that a range of actual distances is consistent with the measured positions. It is these distance ranges that must be compared against the ranges in the table. For this example, only 26 of the 64 possible assignments of the three points to the three model edges are legal.

Of the 26 interpretations consistent with the distance ranges, the four shown in Fig. 1 are completely consistent once the line equations of the edges are taken into account. Each of these interpretations leads to a solution for the position and orientation of the triangle relative to the sensor. Furthermore, these positions and orientations of the triangle place the measured points inside the finite edges, not just on the infinite line.

This paper discusses both steps of the recognition process, focusing first on the generate step and then considering the model-testing stage. We show, by mathematical analysis and by simulation, that the number of feasible interpretations can be reduced to manageable numbers by the use of local geometric constraints. In particular, we investigate the effectiveness of the different local constraints and the impact of measurement errors on their effectiveness. We further show that the few remaining feasible interpretations can efficiently be subjected to an explicit model test, generally resulting in a single interpretation of the sensory data (up to symmetries). We also illustrate the performance of the algorithm on range data obtained by triangulation.

1.3. Three-Dimensional Sensing

Sensors can be roughly divided into two categories: *noncontact* and *contact*. Noncontact sensing, especially visual sensing, has received extensive attention in the robotics and artificial intelligence literature. Contact sensing, such as tactile or haptic sensing, plays an equally important role in robotics, but has received much less attention. In this paper, our aim is to develop a sensory interpretation method that is applicable to data from both contact and noncontact sensors.

While two-dimensional sensing, for example, silhouette or binary vision, may be adequate for restricted situations, such as problems with three degrees of freedom in positioning, the general localization and recognition problem requires three-dimensional sensing. Throughout this paper, we will concentrate on the six-degree-of-freedom recognition and localization problem and the use of three-dimensional sensing. Restrictions of the method to the simpler case of three degrees of freedom are straightforward.

1.3.1. Previous Work in Visual Range Sensing

The measurement stage of visual sensing has received extensive attention in the literature. Of particular interest here are methods for obtaining three-dimensional position and surface-normal information; see the paper by Jarvis (1983) for a detailed survey. Possible methods include edge-based stereo systems (Grimson 1981*a*; Baker and Binford 1981; Mayhew and Frisby 1981), which provide three-dimensional positions of sparse sets of points in the image. This sparse data can be used to reconstruct a dense surface representation, from which surface normals can be estimated (Grimson 1982; 1983; Terzopoulos 1983). Other methods for obtaining three-dimensional positions are laser range-finding (Nitzan, Brain, and Duda

1977; Lewis and Johnston 1977) and structured-light systems (Shirai and Suwa 1971; Popplestone et al. 1975). Many other visual processes can be used to obtain surface-normal information directly, for example, photometric stereo (Woodham 1978; 1980; 1981; Ikeuchi and Horn 1979) and texture gradients (Bajcsy 1973; Bajcsy and Liebermann 1976; Kender 1980; Stevens 1981). In fact, there is no constraint that the sensory data for one problem must come from one sensory modality. Data from visual sensors and tactile sensors may be combined in one run of the algorithm.

The interpretation stage of visual recognition has received less attention, especially with respect to three-dimensional objects with six degrees of freedom. Much of the previous work in the area of interpretation of three-dimensional data has focused on the recognition of such simple generic objects as planar patches, regular polyhedra, generalized cylinders, and spheres (Shirai and Suwa 1971; Agin and Binford 1973; Popplestone et al. 1975; Nitzan, Brain, and Duda 1977; Nevatia and Binford 1977; Oshima and Shirai 1978; Faugeras et al. 1983). Some authors have examined the problem we deal with here of recognizing specific objects from three-dimensional data (Shneier 1979; Sugihara 1979; Oshima and Shirai 1983; Bolles, Horaud, and Hannah 1983; Brou 1983; Ikeuchi et al. 1983). The principal difference between previous work on recognition and the approach described here is our reliance on *sparse* data acquired at points. This makes our approach adaptable to contact sensing as well as visual sensing. The sparseness of the data does make the problem of *segmentation,* determining which data are drawn from which objects in a scene, more difficult. Further research on this topic is currently under way.

In the final stages of preparing this paper, we became aware of the work of Faugeras and Hebert (1983), which adopts an approach that is similar in many respects to the one described here. Their work, however, focuses on deriving an accurate model test. Their method does not emphasize the problem of enumerating all the legal interpretations of the data. Instead, a measure of the accuracy of the model test (and a simple angle-pruning heuristic) is used to drive a best-first search for a good interpretation. This method does not ensure that the interpretation found is the only one consistent with the data, however. Their method and ours are complementary in this respect. Their approach also does not assume sparse data, but it is in fact applicable to that problem.

1.3.2. Previous Work in Tactile Sensing

Contact sensors measure the locus of contact and the forces generated during contact with an object. We make the distinction between *tactile sensors,* which measure forces over small areas, such as a fingertip, and *force sensors,* which measure the resultant forces and torques on some larger structure, such as a complete gripper. A microswitch, for example, can serve as a simple tactile sensor capable of detecting when the force over a small area (e.g., an elevator button) exceeds some threshold. The most important type of tactile sensors are the *matrix tactile sensors,* composed of an array of sensitive points. The simplest example of a matrix tactile sensor is an array of microswitches. Much more sophisticated tactile sensors, with much higher spatial and force resolution, have been designed; see the paper by Harmon (1982) for a review and works by Hillis (1982), Overton and Williams (1981), Purbrick (1981), Raibert and Tanner (1982), and Schneiter (1982) for some recent designs.

For descriptions of previous work in tactile sensing, we refer the reader to two very thorough surveys by Harmon (1980; 1982). A more detailed discussion of previous work on tactile recognition can be found elsewhere (Gaston and Lozano-Pérez 1983). In this section, we briefly survey the two major alternative approaches to tactile recognition: statistical pattern recognition and description building and matching.

Much of the existing work on tactile recognition has been based on statistical pattern recognition or classification. Some researchers have used pressure patterns on matrix sensors primarily (Briot 1979; Okada and Tsuchiya 1977). Others have used the joint angles of fingers grasping the object as their data (Stojilkovic and Saletic 1975; Okada and Tsuchiya 1977; Briot, Renaud, and Stojilkovic 1978; Marik 1981). A related approach uses the pattern of activation of on-off contacts placed on the finger links (Kinoshita, Aida, and Mori 1975).

The range of possible contact patterns between multiple sensors and complex objects is highly variable and seems to require detailed geometric analysis. Tactile recognition methods based on statistical pattern recognition are limited to dealing with simple objects because they do not exploit the rich geometric data available from object models.

Several proposed recognition methods build a partial description of the object from the sensory data and match this description to the model. One approach emulates the feature-based descriptions in vision systems, for example, identification of holes, edges, vertices, pits, and burrs (Binford 1972; Snyder and St. Clair 1978; Hillis 1982). Another approach is to build surface models, either from pressure distributions on matrix sensors (Overton and Williams 1981) or from the displacements of an array of needlelike sensors (Takeda 1974; Page, Pugh, and Heginbotham 1976). A related approach builds a representation of an object's cross section (Kinoshita, Aida, and Mori 1975; Ozaki et al. 1982).

Description-based methods are more general than statistical methods but must solve two formidable problems: building accurate object descriptions from tactile data and matching the descriptions to the models. One major difficulty is that existing sensors do not have the spatial or force resolution needed to build nearly complete object descriptions. Furthermore, there are few methods for matching the partial descriptions obtainable from tactile sensors to object models. In our opinion, part of the problem in tactile data interpretation has been the tendency to adapt the techniques developed for two-dimensional vision, where dense data is readily obtainable, to tactile data, which is naturally sparse.

One lesson from the simulations described later is that some estimate of surface normal is an extremely powerful constraint on recognition and localizaton. The estimate need not be very tight for performance to improve drastically. There has been little previous emphasis on measuring surface normals with tactile sensors. Accuracy in measuring normals requires some attention to engineering trade-offs in sensor design, especially the sensor stiffness. In a stiff sensor (one that deforms very little under contact), the normal to the sensor surface at the point of contact directly gives an estimate of the object's surface normal, so a stiff sensor with high spatial resolution can be used to measure normals. In a soft sensor, the pattern of forces can be analyzed to determine the shape of the object surface, so a soft sensor with good force-measurement accuracy can also be used. Today, it is probably easier to build stiff sensors with poor force resolution than soft sensors with good force resolution (Snyder and St. Clair 1978). This argues that a stiff, very-large-scale-integration (VLSI) sensor (Raibert and Tanner 1982) may be acceptable. Another factor is that the method used here, since it is based on local information, does not require large sensor areas; it can function better with many small sensors.

The approach used in this paper is an instance of a description-based recognition method. The basic departure from previous methods is the reliance on sparse three-dimensional positions and surface normals obtained at *points*.[1] This contrasts with the dense *area* data needed in global feature-based or surface-based description methods. The point-based data we use are more readily obtainable from simple tactile sensors, and the process of matching such data to models is relatively straightforward. Therefore, the method described here could be a powerful addition to approaches based on more complete descriptions.

2. Generating Feasible Interpretations

After sensing an object, we have the positions of up to s points, P_i, known to be on the surface of one of the m known objects, O_j, having n_j faces. The range of possible pairings of sensed points and model faces for one object can be cast in the form of an *interpretation tree* (IT) (Gaston and Lozano-Pérez 1983). The root node of the IT_j, for object O_j, has n_j descendants, each representing an interpretation in which P_1 is on a different face of O_j. There are a total of s levels in the tree, level i indicating the possible pairings of P_i with the faces of object O_j (see Fig. 2). Note that there may be multiple points on a single face, so that the number of branches remains constant at all levels.

A k-interpretation is any path from the root node to a node at level k in the IT; it is a list of k pairings of points and faces. The set of ITs contains a very large

1. Very different approaches to tactile recognition based on this type of data are outlined in the papers by Ivancevic (1974) and Dixon, Salazar, and Slagle (1979).

Fig. 2. Interpretation tree.

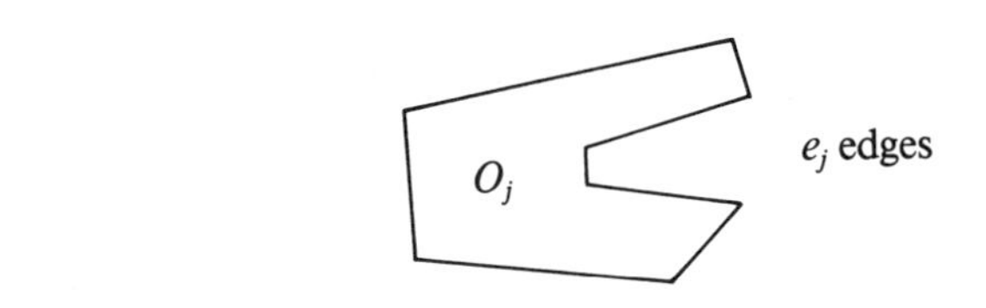

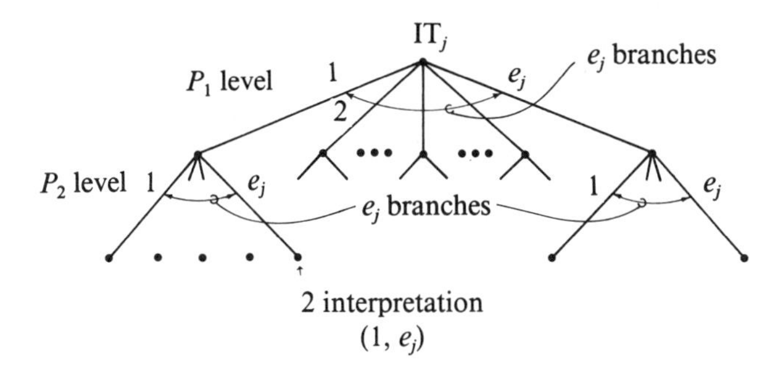

number of possible s-interpretations

$$\sum_{j=1}^{m} (n_j)^s.$$

In an object with symmetries, of course, the IT is highly redundant (Gaston and Lozano-Pérez 1983). The m ITs, one for each known object, represent the search space for the recognition problem discussed here.

2.1. Pruning the IT by Local Constraints

Only a very few interpretations in an IT are consistent with the input data. We can exploit the following local constraints to prune inconsistent interpretations:

1. Distance constraint. The distance between each pair of P_is must be a possible distance between the faces paired with them in an interpretation.
2. Angle constraint. The range of possible angles between measured normals at each pair of P_is must include the known angle between surface normals of the faces paired with them in an interpretation.
3. Direction constraint. The range of values for the component of a vector between sensed points $(P_i \mapsto P_j)$ in the direction of the sensed normal at P_i and at P_j must intersect the range of components of possible vectors between points on the faces assigned to P_i and P_j by the interpretation.
4. Triple-product constraint. The sign of the triple product of the measured normals at three points must agree with the sign of the triple product of the corresponding face normals.

These constraints typically serve to prune most of the nonsymmetric s-interpretations of the data. Other constraints are possible; for example, the area of the triangle defined by three sensed points must be contained within the range of areas defined by the faces paired with them, and the pairing of sensed points with faces must not be such as to require that the path of the sensor (beam) pass through some portion of the object before sensing that face (Gaston and Lozano-Pérez 1983). We will focus on the four constraints listed above, primarily because they are simple to implement, while being quite effective.

Note that the distance, angle, and direction constraints can be used to prune k-interpretations, for $k \geq 2$, thereby collapsing whole subtrees of the IT. This is a crucial point, worth dwelling on for a moment.

Recall that the overall problem we are considering is to determine the position and orientation of an object, using sparse sensory data. In principle, one could consider all possible interpretations of the data, and for each one determine whether there is a transformation from model coordinates to sensor coordinates that would account for the sensory data. Unfortunately, this is computationally extremely expensive. In order to compute such a model test, we need three points whose corresponding face normals are linearly independent, as well as the measured normals at those points. Clearly, we would in general need k sensory points to ensure this, where $k \geq 3$. Thus, if n is the number of faces in the object, we would need to consider on the order of n^k model tests, each of which requires considerable computational effort.

On the other hand, using the simple geometric constraints outlined above requires only a straightforward table lookup, and, as we shall see, can drastically reduce the number of interpretations to which a model test must be applied. Since the constraints can be applied near the root of the tree, it is possible to prune whole subtrees from the IT, at virtually no computational expense. We consider each of the constraints in more detail below.

2.1.1. Distance Pruning

If an interpretation calls for pairing two of the sensed points with two object faces, the distance between the sensed points must be within the range of distances between the faces (Bolles and Cain 1982). Note that the distances between *all* pairs of sensed points must be consistent, that is, there are three distances between three sensed points, and in general $\binom{k}{2}$ distances between k sensed points. Because of this, the distance constraint typically becomes more effective as more sensed points are considered.

Given two faces on a three-dimensional object, we can compute the range of distances between points on the faces. The minimum distance may be determined as the minimum of the shortest distance between all pairs of edges and the perpendicular distances between vertices of one face and the plane of the other face (when the vertex projects inside the face polygon). The maximum requires examining distances between pairs of vertices. Note that we can also compute the range of distances between points on *one* face (zero up to the diameter of the face). Sophisticated algorithms may be used to reduce the complexity of these computations, but since they are to be performed off-line, once for each model, their efficiency is not critical to the approach.

The distance constraint can be implemented in the following manner. For object O_j, with f_j faces, we construct an f_j by f_j table whose entries determine the range of possible distances between pairs of faces. In particular, for a pair of faces (i, k), $i \neq k$, the maximum distance between the faces is stored in table location $\mathbf{dtable}_j[\max(i, k), \min(i, k)]$ and the minimum distance between the faces is stored in table location $\mathbf{dtable}_j[\min(i, k), \max(i, k)]$. If $i = k$, we simply store the maximum distance in the diagonal entry $\mathbf{dtable}_j[i, i]$, since the minimum distance defaults to 0. This representation makes checking a distance constraint straightforward, since the set of all pairs of faces (i, k) on object O_j consistent with some measured distance d is given by

$$\{(i, k) \mid \mathbf{dtable}_j[\min(i, k), \max(i, k)] \leq d \leq \mathbf{dtable}_j[\max(i, k), \min(i, k)]\}$$

plus the pair (i, i) if $d \leq \mathbf{dtable}_j[i, i]$.

Given any $k-1$-interpretation, represented by the set of faces $(i_1, \ldots, i_{k-1})$, and a new kth sensed point, the generation of the next level of the IT below this interpretation can be easily computed by checking the appropriate portions of the distance tables. In particular, if the measured distance between one of the previous sensed points, i_l, and the new one is given by d_{i_l}, the set of possible faces that can be assigned to sensed point P_k is given by

$$\bigcap_{l=1}^{k-1} \{i \mid \mathbf{dtable}_j[\min(i, i_l), \max(i, i_l)] \leq d_{i_l} \leq \mathbf{dtable}_j[\max(i, i_l), \min(i, i_l)]\},$$

unioned with the set

$$\bigcap_{l=1}^{k-1} \{i_l \mid 0 \leq d_{i_l} \leq \mathbf{dtable}_j[i_l, i_l]\}.$$

For very complex objects, more time-efficient ways of representing and searching for faces that satisfy a distance constraint are possible. A full discussion of these methods is beyond the scope of this paper, however.

It may frequently be the case, for example, with a flat tactile sensor, that the sensor makes contact along an edge or at a vertex, rather than in the interior of a face. The method described above would still work unchanged under these circumstances. But if the sensor is capable of detecting that contact is at a vertex or edge, then tighter constraints can be applied. This is accomplished by constructing tables of distance ranges between vertices and between edges and applying the pruning algorithm based on those tables when appropriate.

Similarly, in the case of visual sensing, if the edges and vertices of an object can be reliably determined from the sensory data, the recognition process is greatly simplified. (Note the relationship to the recognition method described by Bolles and Cain [1982].)

2.1.2. Angle Pruning

Sensed points are associated with a range of legal surface normals consistent with the sensory data. If an interpretation calls for pairing two of the sensed points (and normals) with two object faces, the range of angles between the sensed normals must include the angle between the normals of the corresponding object faces. (Note that Bolles and Cain [1982] used a similar orientation constraint for point-to-point matching, as opposed to the point-to-surface matching used here.)

To see how this information can be used to prune the IT, we first consider the case in which the object has three degrees of freedom (two translational and one rotational). Under this restriction on degrees of freedom, the range of surface normals can be represented as a range of angles relative to the hand frame.

At a sensed point P, we can measure the local surface normal as lying in the range of angles

$$\phi \in [\omega - \epsilon, \omega + \epsilon],$$

where ω is the actual measurement, and ϵ defines the range of possible angles about this measurement. We are given a sensor point P_1, with measured normal ω_1, which has been assigned to face i, with associated model coordinate surface normal given by ψ_i. Next, we record a second point P_2, with measured normal ω_2, which has been assigned to face k, with associated model coordinate surface normal given by ψ_k. For these assignments to be consistent, it must be the case that the angle between the model faces must be included in the range of angles between the ranges of normals determined from the measured normals and the error bounds

$$(\omega_2 - \omega_1) - (\epsilon_1 + \epsilon_2) \leq \psi_k - \psi_i \leq (\omega_2 - \omega_1) + (\epsilon_1 + \epsilon_2).$$

It is clear that an implementation similar to that used for distance pruning will also suffice here. For object O_j, with e_j edges, we can set up an e_j by e_j, lower diagonal table, **atable**$_j$, such that **atable**$_j$[max (i, k), min (i, k)] $= \psi_k - \psi_i$. This representation makes checking a surface-normal constraint straightforward, since the set of all pairs of faces (i, k) on object O_j consistent with some measured ranges of surface normals is given by

$$\{(i, k) | (\omega_2 - \omega_1) - (\epsilon_1 + \epsilon_2) \leq \mathbf{atable}_j[\max(i, k), \min(i, k)] \leq (\omega_2 - \omega_1) + (\epsilon_1 + \epsilon_2)\}.$$

Given any $k - 1$-interpretation, and a new kth sensed point, the generation of the next level of the IT below this interpretation can easily be computed by checking the appropriate portions of the angle tables. Note that the kth edge must be consistent with the angles between all previous faces.

In the two-dimensional (three-degree-of-freedom) case, the range of possible surface normals at a sensed point was represented by the pair (ω_1, ϵ_1), where ω_1 denoted the sensed normal, and ϵ_1 denoted the range of error about that sensed point. In three dimensions, the obvious generalization is to use angle cones, so that if $\mathbf{u}_1$ denotes the unit sensed surface normal, the range of possible values for the actual surface normal will be denoted by the right circular cone

$$\{\mathbf{n}_1 | \mathbf{n}_1 \cdot \mathbf{u}_1 \geq \epsilon_1\}.$$

We could proceed identically to the two-dimensional case by noting that the cone of sensed normals constrains the set of possible three-dimensional rotations between the hand and model coordinate systems. Then, given a second sensed point P_2 with some sensed normal, the set of feasible faces would be restricted by the range of possible rotations. This method is quite difficult to implement, however. There is a much simpler alternative method.

Suppose that at the second sensed point, the set of possible surface normals in hand coordinates is given by

$$\{\mathbf{n}_2 | \mathbf{n}_2 \cdot \mathbf{u}_2 \geq \epsilon_2\}.$$

Then, in order for faces i and k, with associated surface normals $\mathbf{v}_i$ and $\mathbf{v}_k$ to be consistent, it must be the case that

$$\mathbf{v}_i \cdot \mathbf{v}_k \in \{\mathbf{n}_1 \cdot \mathbf{n}_2 | \mathbf{n}_1 \cdot \mathbf{u}_1 \geq \epsilon_1, \mathbf{n}_2 \cdot \mathbf{u}_2 \geq \epsilon_2\}.$$

We can rephrase this in the following manner. Let $\cos \alpha_1 = \epsilon_1$, $\cos \alpha_2 = \epsilon_2$, $\alpha_{12} = \alpha_1 + \alpha_2$ and $\cos \gamma_{12} = \mathbf{u}_1 \cdot \mathbf{u}_2$. Then, we claim that the set

$$\{\mathbf{n}_1 \cdot \mathbf{n}_2 | \mathbf{n}_1 \cdot \mathbf{u}_1 \geq \epsilon_1, \mathbf{n}_2 \cdot \mathbf{u}_2 \geq \epsilon_2\}$$

is contained in the set

$$\{\mathbf{n}_1 \cdot \mathbf{n}_2 | \cos[\min(\pi, \gamma_{12} + \alpha_{12})] \leq \mathbf{n}_1 \cdot \mathbf{n}_2 \leq \cos[\max(0, \gamma_{12} - \alpha_{12})]\}. \quad (1)$$

Fig. 3. Angle ranges.

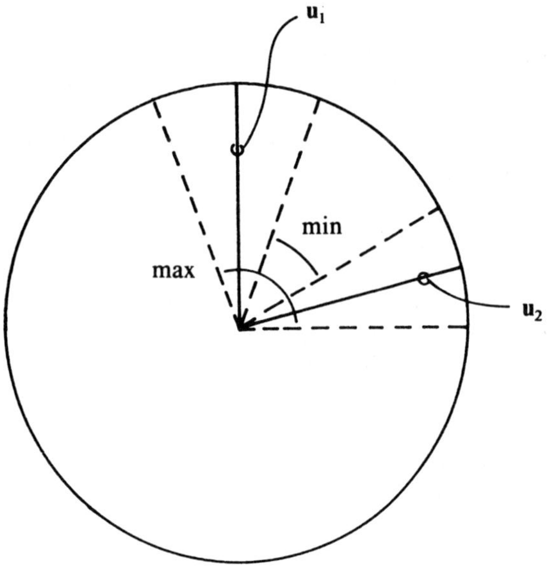

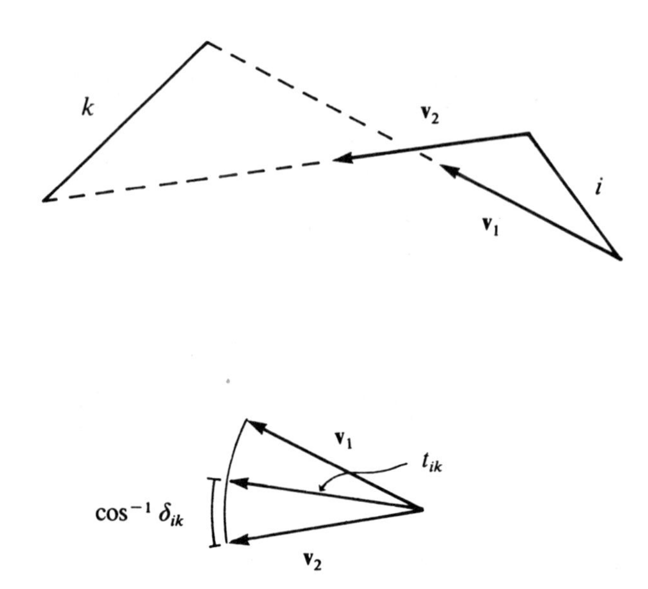

Fig. 4. Range of directions between sensed points.

A proof of this is found in Appendix 1. Figure 3 illustrates this result in two dimensions.

An implementation of angle pruning similar to that used for distance pruning is now also possible. For object O_j, with f_j faces, we can set up an f_j by f_j, lower diagonal table, $\mathbf{atable}_j[\max(i, k), \min(i, k)] = \mathbf{v}_i \cdot \mathbf{v}_k$, where $\mathbf{v}_i$ denotes the unit normal to face i in the model.

2.1.3. Direction Pruning

Consider a pair of sensed points P_1 and P_2, and let $\mathbf{u}_{12}$ be the unit direction vector between them. Suppose that we know the measured surface normal at point P_1 to within some cone of error, for example, the measured value is $\mathbf{w}_1$, and the range of possible values for the surface normal is

$$\{\mathbf{v}_1 | \mathbf{v}_1 \cdot \mathbf{w}_1 \geqslant \epsilon_1\}.$$

Then the set of possible "angles" between the direction vector and the surface normal of the face is given by

$$\{\mathbf{v}_1 \cdot \mathbf{u}_{12} | \mathbf{v}_1 \cdot \mathbf{w}_1 \geqslant \epsilon_1\}.$$

In an interpretation, suppose that point P_1 has been assigned to face i, with normal $\mathbf{n}_i$ in the model, and we now consider possible faces k to assign to point P_2. Let the range of possible unit vectors (directions) from face i to face k be denoted by the cone

$$\{\mathbf{s}_{ik} | \mathbf{s}_{ik} \cdot \mathbf{t}_{ik} \geqslant \delta_{ik}\}$$

for some pair $\mathbf{t}_{ik}$ and δ_{ik}. Figure 4 illustrates this cone in a two-dimensional example. Appendix 2 shows how this cone may be computed from models of the object faces. In the model, the set of possible angles between legal directions and the surface normal is

$$\{\mathbf{n}_i \cdot \mathbf{s}_{ik} | \mathbf{s}_{ik} \cdot \mathbf{t}_{ik} \geqslant \delta_{ik}\}. \quad (2)$$

Thus, assume that point P_1 is on face i, with normal $\mathbf{n}_i$; that we have measured $\mathbf{w}_1$; that we know ϵ_1; and that we have also measured P_2. A face k, whose direction range from face i is given by the pair $(\mathbf{t}_{ik}, \delta_{ik})$, is a feasible face for point P_2 if the set in (Eq. 2) intersects the cone

$$\{\mathbf{v}_1 \cdot \mathbf{u}_{12} | \mathbf{v}_1 \cdot \mathbf{w}_1 \geqslant \epsilon_1\}. \quad (3)$$

If $\cos \gamma_{ik} = \delta_{ik}$, and $\cos \phi_{ik} = \mathbf{n}_i \cdot \mathbf{t}_{ik}$, then we know from the derivation in Appendix 1 that the set of (Eq. 2) is contained in the set

$$\{\mathbf{n}_i \cdot \mathbf{s}_{ik} | \cos(\gamma_{ik} + \phi_{ik}) \leqslant \mathbf{n}_i \cdot \mathbf{s}_{ik} \leqslant \cos(\gamma_{ik} - \phi_{ik})\}.$$

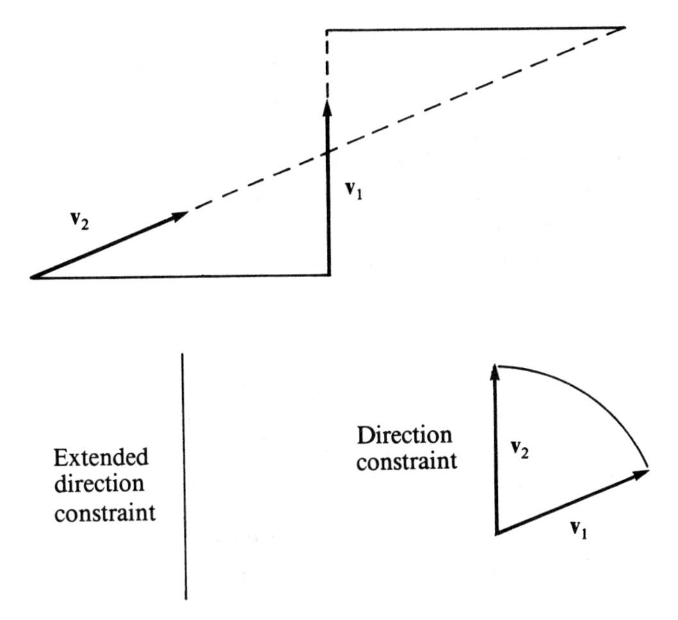

Fig. 5. Extended direction constraint.

Similarly, if $\cos \alpha_1 = \epsilon_1$ and $\cos \omega_{12} = \mathbf{v}_1 \cdot \mathbf{u}_{12}$, then the set of (Eq. 3) is contained in the set

$$\{\mathbf{v}_1 \cdot \mathbf{u}_{12} | \cos(\alpha_1 + \omega_{12}) \leq \mathbf{v}_1 \cdot \mathbf{u}_{12} \leq \cos(\alpha_1 - \omega_{12})\}.$$

Therefore, for the pairings of P_1 with face i and P_2 with face k to be consistent with the direction constraint, it must be the case that the intersection of the numerical ranges of dot products is not null, that is,

$$[\cos(\alpha_1 - \omega_{12}), \cos(\alpha_1 + \omega_{12})] \cap [\cos(\gamma_{ik} - \phi_{ik}), \cos(\gamma_{ik} + \phi_{ik})] \neq \emptyset.$$

The direction constraint can also be implemented in a form similar to that used for distance and angle pruning. For object O_j, with f_j faces, we can set up an f_j by f_j table $\mathbf{ctable}_j$ such that $\mathbf{ctable}_j[i, k] = [\cos(\gamma_{ik} - \phi_{ik}), \cos(\gamma_{ik} + \phi_{ik})]$. Again, the set of all pairs of faces (i, k) on object O_j consistent with some measured ranges of surface normals is given by

$$\{(i, k) | [\cos(\alpha_1 - \omega_{12}), \cos(\alpha_1 + \omega_{12})] \cap \mathbf{ctable}_j[i, k] \neq \emptyset\}.$$

Note that the direction constraint is not symmetric, as are the distance and angle constraints, so before pairing P_2 to face k, we must repeat the test above, interchanging the roles of i and k. Similarly, the test must be applied to each pairing of sensed points and faces in an interpretation.

The constraint described above places constraints on the angle between a surface normal and unit vectors from one face to another. In addition to constraining the angles of unit vectors, we may constrain the magnitude of the component along the surface normal of the vector between the sensed points. The statement and implementation of the constraint is essentially unchanged, except that $\mathbf{u}_{12}$ and $\mathbf{t}_{ik}$ are no longer unit vectors, but the actual vector between the sensed points. The effectiveness of the constraint is in general improved, however, since it now captures some distance and some angular constraint. The difference between this extended direction constraint and the simple direction constraint is illustrated in Fig. 5. Two parallel faces (faces 1 and 2 in Fig. 5) displaced relative to each other give rise to a cone of directions, but a single value for the normal component of vectors connecting the faces. Note that an interpretation that assigns P_1 to face 1 and P_2 to face 3 is consistent with all the previously mentioned constraints except for the extended direction constraint. The figure also illustrates that the extended direction constraint does not subsume the distance constraint, since direction only constrains the normal component of distance.

An alternate form of the direction constraint is useful when no bound on the surface normal is available. It can briefly be described as follows. Given two faces h and i on an object, we can compute the range of directions between points on the faces, forming a cone of possible directions. Similarly, for faces i and j, we can compute the cone of possible directions. The combination of these two cones defines a range of possible angles for the triplet of faces h, i, j.

If an interpretation calls for pairing three of the sensed points with three object faces, the angle formed by this triplet of sensed points must be within the range of possible angles between the triplets of faces. Note that the angles formed by *all* triplets of sensed points must be consistent, that is, for three sensed points, there are three angles; for k sensed points, there are $3\binom{k}{3}$ angles. Hence, this constraint also becomes more effective as more sensed points are considered.

This form of the direction constraint can be used when only vertices and edges are touched, as it does not require sensing surface normals. Note that this form of the constraint can also be extended to use the magnitude of the vectors between sensed points as well as their direction. This form of the direction constraint allows pruning of the IT for $k \geqslant 3$. The previous formulation of the constraint allows pruning of the IT for $k \geqslant 2$. This form of the constraint would also require an n^3 table, as opposed to an n^2 one for the previous formulation. Given the size of n to be expected for typical objects, this is a critical difference.

2.1.4. *Triple-Product Pruning*

In the case of six degrees of freedom, measuring angles between pairs of normals still leaves a potential ambiguity. In particular, it does not distinguish between interpretations corresponding to reflections of the coordinate system and rotations of the coordinate system. To do this, we require that the sign of the triple product be preserved between sensed normals and associated face normals.

This is a three-valued function, taking on the value 1 if the sign of the triple product is positive, -1 if the sign is negative, and 0 if the three vectors are not independent.

When computing the sign of the triple product for three sensed unit normals, we must account for possible errors in the measurements. If the angular error in the measurements is given by γ, then the following criteria are applied. If the angle between any two of the three unit vectors is less than 2γ, or if the magnitude of the triple product is less than $\sin 2\gamma$, then the sign of that triple product is set to zero. Otherwise, it takes on the value ± 1. Finally, when matching the triple products between sensed normals and corresponding model normals, we use the following rules. If the sensed triple product is 0, it agrees with any value of model triple product. Otherwise, the sign of the sensed triple product and the model triple product must agree. In this way, possible changes in the sign of the sensed triple product, due to error in the measured normals, are avoided.

3. Model Testing

Once the interpretation tree has been pruned by the local constraints, there will be some set of possible interpretations of the sensed data, each one consisting of a set of triples $(\mathbf{p}_i, \mathbf{n}_i, f_i)$, where $\mathbf{p}_i$ is the vector representing the sensed position, $\mathbf{n}_i$ is the vector representing the sensed normal, and f_i is the face assigned to this sensed data for that particular interpretation. In the model-test stage of the processing, we want to

1. Determine the actual transformation from model coordinates to sensor coordinates, corresponding to the interpretation
2. Check that under this transformation, not only are the sensed points transformed to lie on the appropriate planes, but that the sensed points actually lie within the bounds of the assigned faces

We will assume that a vector in the model coordinate system is transformed into a vector in the sensor coordinate system by the following transformation:

$$\mathbf{v}_s = \mathbf{R}\mathbf{v}_m + \mathbf{v}_0,$$

where $\mathbf{R}$ is a rotation matrix, and $\mathbf{v}_0$ is some translation vector. We need to solve for $\mathbf{R}$ and $\mathbf{v}_0$. We note that a solution could be obtained using a least-squares method, such as is used by Faugeras and Hebert (1983). This type of solution can be computationally expensive, however, and in the following sections, we develop an alternative method.

3.1. Rotation Component

We consider first the rotation component of the transformation. Consider the first triple of a particular interpretation, $(\mathbf{p}_i, \mathbf{n}_i, f_i)$. The sensed normal is given by $\mathbf{n}_i$, and corresponding to face f_i is a face normal $\mathbf{m}_i$. For $\mathbf{R}$ to be a legitimate rotation, it should take the normal $\mathbf{m}_i$ into $\mathbf{n}_i$ (ignoring issues of error in the measurements for now).

Now, any rotation can be represented by a direction about which the rotation takes place, and an angle of rotation about that direction. What is the set of possible directions of rotation $\mathbf{r}$ consistent with $\mathbf{n}_i$ and $\mathbf{m}_i$? Any rotation will preserve the angle between the transformed vector and the direction of rotation. Hence, any legitimate rotation direction must be equiangular with $\mathbf{n}_i$ and $\mathbf{m}_i$. Thus, the set of potential directions is given by

$$\{\mathbf{r}_{ij} | \mathbf{r}_{ij} \cdot \mathbf{m}_i = \mathbf{r}_{ij} \cdot \mathbf{n}_i\}$$

or, equivalently,

$$\{\mathbf{r}_{ij} | \mathbf{r}_{ij} \cdot (\mathbf{m}_i - \mathbf{n}_i) = 0\}.$$

That is, $\mathbf{r}_{ij}$ is perpendicular to $(\mathbf{m}_i - \mathbf{n}_i)$.

Now, consider a second triple in the interpretation, $(\mathbf{p}_j, \mathbf{n}_j, f_j)$ and let $\mathbf{m}_j$ be the normal to face f_j. Provided $\mathbf{m}_j \neq \pm \mathbf{m}_i$, and $\mathbf{n}_i - \mathbf{m}_i$ is not (anti-)parallel to $\mathbf{n}_j - \mathbf{m}_j$, we can constrain $\mathbf{r}_{ij}$ to a second set:

$$\{\mathbf{r}_{ij} | \mathbf{r}_{ij} \cdot (\mathbf{m}_j - \mathbf{n}_j) = 0\}.$$

Since the rotation is the same, $\mathbf{r}_{ij}$ must lie in both sets, that is, it must be perpendicular to both vectors. Hence, $\mathbf{r}_{ij}$ is given by the unit vector in the direction

$$(\mathbf{m}_i - \mathbf{n}_i) \times (\mathbf{m}_j - \mathbf{n}_j)$$

to within an ambiguity of 180°.

This derivation can be recast in geometric terms in the following manner. Any unit rotation vector $\mathbf{r}$ taking $\mathbf{m}_i$ into $\mathbf{n}_i$ must lie on the perpendicular bisector of the line connecting $\mathbf{n}_i$ to $\mathbf{m}_i$. Similarly, it must also lie on the perpendicular bisector of the line connecting $\mathbf{n}_j$ to $\mathbf{m}_j$. Since the rotation is the same, it must lie in the intersection of the two perpendicular bisector planes, as above, and hence is given by the specified unit vector

$$(\mathbf{m}_i - \mathbf{n}_i) \times (\mathbf{m}_j - \mathbf{n}_j).$$

If there were no error in the sensed normals, we would be done. With error included in the measurements, however, the computed rotation direction $\mathbf{r}$ could be slightly wrong. One way to reduce the effect of this error is to compute all possible $\mathbf{r}_{ij}$ as i and j vary over the faces of the interpretation, and then cluster these computed directions to determine a value for the direction of rotation $\mathbf{r}$.

Once we have computed a direction of rotation $\mathbf{r}$, we need to determine the angle θ of rotation about it. It is straightforward to show that

$$\mathbf{m}_i = \cos\theta \mathbf{n}_i + (1 - \cos\theta)(\mathbf{r} \cdot \mathbf{n}_i)\mathbf{r} + \sin\theta(\mathbf{r} \times \mathbf{n}_i).$$

(See, for example, the book by Korn and Korn [1968], p. 473.) Simple algebraic manipulation, using the fact that $\mathbf{r} \cdot \mathbf{m}_i = \mathbf{r} \cdot \mathbf{n}_i$, yields

$$\cos\theta = 1 - \frac{1 - (\mathbf{n}_i \cdot \mathbf{m}_i)}{1 - (\mathbf{r} \cdot \mathbf{n}_i)(\mathbf{r} \cdot \mathbf{m}_i)},$$

$$\sin\theta = \frac{(\mathbf{r} \times \mathbf{n}_i) \cdot \mathbf{m}_i}{1 - (\mathbf{r} \cdot \mathbf{n}_i)(\mathbf{r} \cdot \mathbf{m}_i)}.$$

Hence, given $\mathbf{r}$, we can solve for θ. Note that if $\sin\theta$ is zero, there is a singularity in determining θ, which could be either 0 or π. In this case, however, $\mathbf{r}$ lies in the plane spanned by $\mathbf{n}_i$ and $\mathbf{m}_i$; hence only the $\theta = \pi$ solution is valid. As before, in the presence of error, we may want to cluster the $\mathbf{r}$ vectors, and then take the average of the computed values of θ over this cluster.

Finally, given values for $\mathbf{r}$ and θ, we can determine the rotation matrix $\mathbf{R}$. Let r_x, r_y, r_z denote the components $\mathbf{r}$. Then

$$\mathbf{R} = \cos\theta \begin{bmatrix} 1 & 0 & 0 \\ 0 & 1 & 0 \\ 0 & 0 & 1 \end{bmatrix} + (1 - \cos\theta) \begin{bmatrix} r_x^2 & r_x r_y & r_x r_z \\ r_y r_x & r_y^2 & r_y r_z \\ r_z r_x & r_z r_y & r_z^2 \end{bmatrix} + \sin\theta \begin{bmatrix} 0 & -r_z & r_y \\ r_z & 0 & -r_x \\ -r_y & r_x & 0 \end{bmatrix}.$$

Note that in computing the rotation component of the transformation, we have ignored the ambiguity inherent in the computation. That is, there are two solutions to the problem, $(\mathbf{r}, \theta)$ and $(-\mathbf{r}, -\theta)$. We assume that a simple convention concerning the sign of the rotation is used to choose one of the two solutions.

3.2. Translation Component

Next, we need to solve for the translation component of the transformation. We know that $\mathbf{v}_s = \mathbf{R}\mathbf{v}_m + \mathbf{v}_0$, where $\mathbf{v}_m$ is a vector in model coordinates, $\mathbf{v}_s$ is the corresponding vector in sensor coordinates, and $\mathbf{R}$ has been computed as above. Given a triple $(\mathbf{p}_i, \mathbf{n}_i, f_i)$ from the interpretation, let $\mathbf{m}_i$ be the normal of face f_i, with offset d_i; that is, the face is defined by the set of vectors

$$\{\mathbf{v} | \mathbf{v} \cdot \mathbf{m}_i = d_i\}.$$

Then the point in model coordinates corresponding to $\mathbf{p}_i$ is

$$\mathbf{R}^{-1}(\mathbf{p}_i - \mathbf{v}_0),$$

and the following equation holds:

$$\mathbf{m}_i \cdot (\mathbf{R}^{-1}(\mathbf{p}_i - \mathbf{v}_0)) = d_i.$$

Or, equivalently,

$$(\mathbf{Rm}_i) \cdot (\mathbf{p}_i - \mathbf{v}_0) = d_i.$$

This equation essentially constrains the component of the translation vector in the direction of $\mathbf{Rm}_i$.

Suppose we consider three triplets from the interpretation, $(\mathbf{p}_i, \mathbf{n}_i, f_i)$, $(\mathbf{p}_j, \mathbf{n}_j, f_j)$, and $(\mathbf{p}_k, \mathbf{n}_k, f_k)$ such that the triple product $\mathbf{m}_i \cdot (\mathbf{m}_j \times \mathbf{m}_k)$ is nonzero, (i.e., the three face normals are independent). Then, we can construct three independent equations

$$(\mathbf{Rm}_i) \cdot \mathbf{v}_0 = (\mathbf{Rm}_i) \cdot \mathbf{p}_i - di,$$

$$(\mathbf{Rm}_j) \cdot \mathbf{v}_0 = (\mathbf{Rm}_j) \cdot \mathbf{p}_j - dj,$$

$$(\mathbf{Rm}_k) \cdot \mathbf{v}_0 = (\mathbf{Rm}_k) \cdot \mathbf{p}_k - d_k.$$

Each of these equations constrains a different, independent component of the translation vector $\mathbf{v}_0$, and hence the three equations together determine the actual vector. Straightforward algebraic manipulation then yields the following solution for the translation component $\mathbf{v}_0$:

$$\begin{aligned}[\mathbf{m}_i \cdot (\mathbf{m}_j \times \mathbf{m}_k)]\mathbf{v}_0 = {} & ((\mathbf{Rm}_i) \cdot \mathbf{p}_i - d_i)((\mathbf{Rm}_j) \times (\mathbf{Rm}_k)) \\ & + ((\mathbf{Rm}_j) \cdot \mathbf{p}_j - d_j)((\mathbf{Rm}_k) \\ & \times (\mathbf{Rm}_i)) + ((\mathbf{Rm}_k) \cdot \mathbf{p}_k \\ & - dk)((\mathbf{Rm}_i) \times (\mathbf{Rm}_j)).\end{aligned}$$

As in the case of rotation, if there is no error in the measurements, then we are done. The simplest means of attempting to reduce the effects of error on the computation is to average $\mathbf{v}_0$ over all possible trios of triplets from the interpretation. Note that for numerical stability, one may want to restrict the computation to triplets such that $\mathbf{m}_i \cdot (\mathbf{m}_j \times \mathbf{m}_k)$ is greater than some threshold.

Finally, we have computed the transform $(\mathbf{R}, \mathbf{v}_0)$ from model coordinates to sensor coordinates. To check a possible interpretation, we consider all triples $(\mathbf{p}_i, \mathbf{n}_i, f_i)$ in the interpretation and compute

$$\mathbf{R}^{-1}(\mathbf{p}_i - \mathbf{v}_0).$$

We then check that this point lies within the bounds of face f_i (to within some error range). If it does not, then the interpretation is invalid, and may be pruned. If all such triples satisfy this check, the interpretation is still valid.

We have assumed above that three independent face normals have been measured. When only one normal is available, neither rotation nor translation can be determined. When only two independent normals are available, the rotation can be determined as before, but only a direction of translation can be determined, not the actual magnitude of the translation. A range of possible translations can be determined, however, by intersecting the line, determined by the position of a sensed point and the translation direction, with the face assigned to the point by the interpretation. Of course, further sensing along this line to discover the position of the edge would determine the actual translation.

After the model test has been applied to all leaves of the interpretation tree, there may still be several interpretations remaining. Upon examination, one usually finds that these interpretations differ only in the assignment of one or two faces, all other faces being identical. In our models, some of the faces are coplanar and abutting; these faces arise because the models were originally built as a union of solids. When a point is sensed near the boundary of one of these abutting face pairs, there is very little difference in the computed transformation from assigning the sensed point to one face or the other. This inability to distinguish between such nearly identical interpretations is a result of the error bounds on the sensing. Thus, as a final stage, we cluster the remaining interpretations in terms of their computed transformations; that is, we cluster the interpretations in terms of the computed orientation of the object in space. Here, we generally find very few such clusters. Indeed, in general there is only one computed orientation for the object (the correct one), although occasionally two or more clusters survive, usually corresponding to symmetric interpretations of the sensed data.

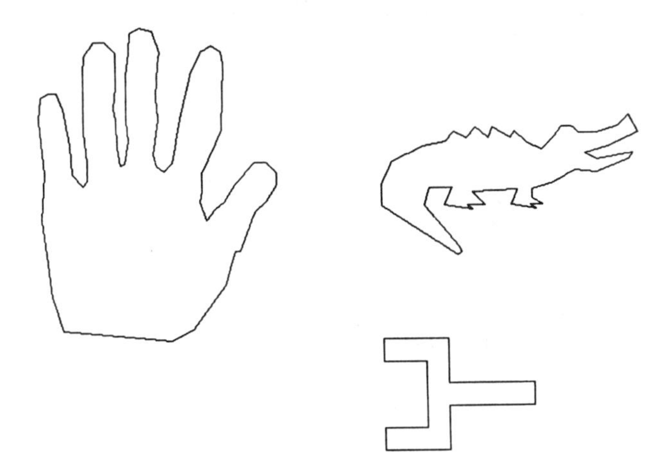

4. Simulation Data

In order to test the efficacy of the algorithm in pruning the interpretation tree, we ran a large number of simulations. Some simulations for objects with three degrees of freedom (two translational and one rotational) have been described elsewhere (Gaston and Lozano-Pérez 1983). We include additional simulation data for objects with three positional freedoms, including the direction constraint. We also provide data for the more general case of three-dimensional objects with six degrees of freedom. Our goals are (1) to demonstrate that effective pruning of the interpretation tree is possible, at low computational expense, and (2) to explore the sensitivity of the algorithm to error in measuring the surface normal and the position of the sensed points.

4.1. Three Positional Freedoms

We begin by considering objects with two degrees of translational freedom and one degree of rotational freedom, using sample objects first considered by Gaston and Lozano-Pérez (1983) and illustrated in Fig. 6. The addition of the direction constraint greatly reduces the extent of the set of possible interpretations. To demonstrate this, a series of 250 runs of the algorithm was executed for each of the objects. Each run determined the number of interpretations consistent with a set of five sensed points. The points were determined by first randomly rotating the object about its centroid and then intersecting the object with five lines from its centroid along five evenly spaced directions. The points of intersection farthest from the centroid along each line were used as the sensed point. The (simulated) error in measuring the sensed position was bounded by 0.1 (i.e., a randomly oriented offset vector of random magnitude bounded by 0.1 was added to the point on the object), and the (simulated) error in measuring the angle of the surface normal was $\pi/8$ (i.e., a random vector was chosen whose dot product with the actual normal was bounded by $\cos^{-1} \pi/8$). To place these error ranges in perspective, the diameters of the models in Fig. 6 were 9 units for the wrench, 14 for the alligator, and 12 for the hand.

Table 2 describes the results of this set of simulations through histograms of the number of interpretations found. Thus, for $i \leq 10$, the number in the ith column is the number of trial runs that resulted in i possible interpretations. Beyond this point, the histogram is compressed into units of tens. For example, the column labeled 20 lists the number of trial runs resulting in k interpretations, where $10 < k \leq 20$. In order to examine the effectiveness of adding the direction constraint to the algorithm described by Gaston and Lozano-Pérez (1983), the simulations were run both with and without this constraint. For each object in the table, the first histogram corresponds to the case of using the direction constraint, and the second histogram to the case of not using it. Note that the number of edges for the wrench (W), alligator (G) and hand (H) is 12, 50, and 67, respectively.

The results shown in Table 2 are striking in a number of different ways. First, note that the maximum number of possible interpretations observed for any of the objects was 20 (in the case of using the direction

Table 2. Histograms of Two-Dimensional Objects

	1	2	3	4	5	6	7	8	9	10	20	30	40	50	60	70	80	90	100	>100
W		242		7				1												
W		31		25		2		2			3	19	29	20	34	27	13	20	23	2
G	22	62	31	38	20	23	10	6	8	10	20									
G	1	12	9	11	8	23	10	17	15	14	63	34	7	7	7	5	5		1	1
H	15	61	36	29	21	20	22	14	11	9	12									
H	4	13	17	21	17	17	12	16	18	11	86	18								

NOTE: W = wrench; G = alligator; H = hand; errors = $\pi/8$, 0.1.

constraint), which is exceptionally low when we consider that the total number of possible interpretations for the alligator was 50^5, or 312,500,000. Second, the median number of possible interpretations was only 2 for the wrench, and 4 for the alligator and hand, when using the direction constraint. Without this constraint, the median number of interpretations rose to 48, 12, and 9 for the wrench, alligator, and hand, respectively. Of course, the results of the simulations depend to a certain extent on the error ranges, a point that will be explored in some detail in the next section. We note that a 0.1-in sensitivity in distance over a 10–20-in range is within the range of current tactile sensors. The positioning accuracy of many current manipulators is within 0.01 in. The Purbrick tactile sensor (1981) has a matrix element separation of 0.06 in, and the Hillis sensor (1982) has an element separation of 0.025 in.

4.2. Six Positional Freedoms

In considering the full three-dimensional problem of objects with six degrees of freedom, we have run extensive simulations on the models illustrated in Fig. 7. The diameters of these objects (i.e., the maximum separation of two points on the object) were roughly 4, 7, and 8 in for the housing, cylinders, and hands, respectively. In running simulations of the recognition algorithm on these objects, we used two different sensing strategies, reflecting in part the difference between range and tactile sensing capabilities.

It should be noted that in all the following simulations, the efficiency of the tree-pruning mechanism was improved by sorting the sensed points. In particular, rather than using the sensory data in arbitrary order, the points were sorted on the basis of pairwise separation, with the more distant points being ordered first. This sorting on distance tends to place the most effective constraints at the beginning of the process, a point that will be illustrated in Section 4.5.

4.3. Grid Sensing

In the first sensing method, the sensory data were generated by projecting a regular grid of points along three orthogonal directions and noting where contact was made with an arbitrarily oriented model of the object. This arbitrary orientation was obtained by randomly choosing values for the three Euler angles, computing a rotation transformation based on this, and applying the rotation to the model. Note that this method does not produce a uniform sampling of the space of rotations, but for our purposes it is a sufficiently random sampling. No translation offset was added, since this would not affect the process. The three-dimensional positions of the sensed points and the associated surface normals were then corrupted by noise within some specified bounds. For the simulations discussed below, the number of sensed points on each trial was 12.

The results of the first set of simulations are shown in Tables 3 and 4. Table 3 lists statistics of the number

Fig. 7. Three-dimensional test models.

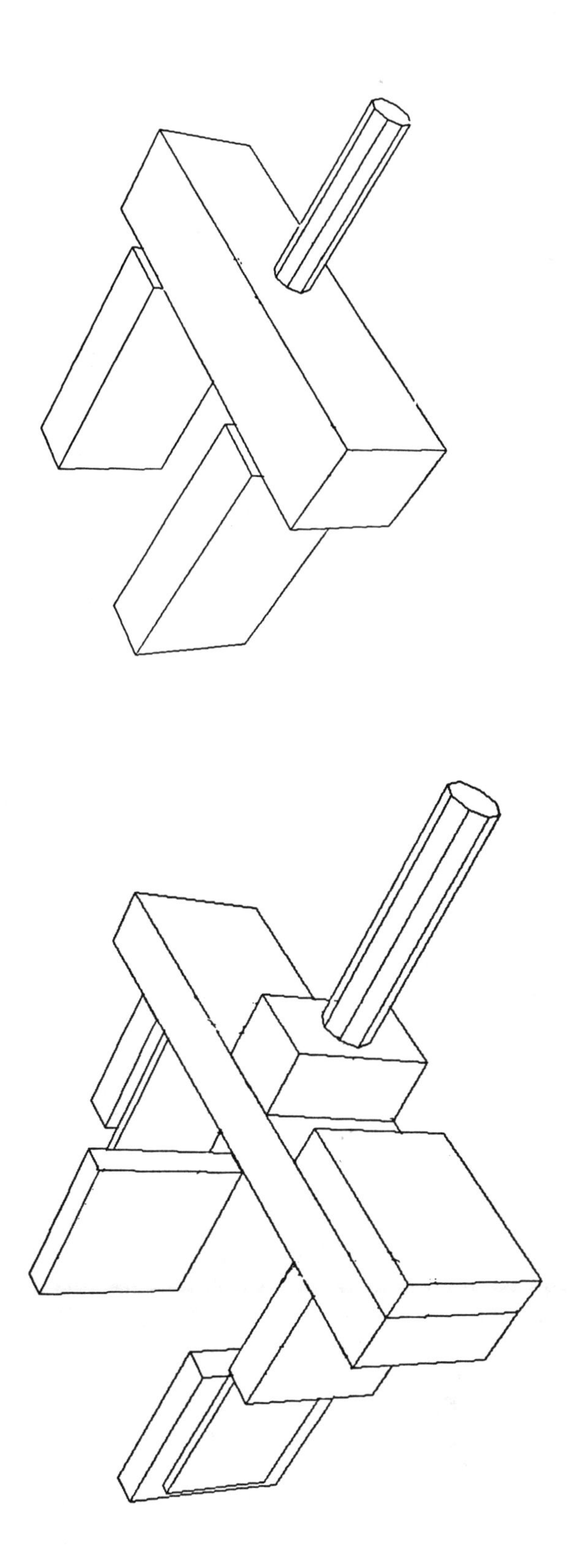

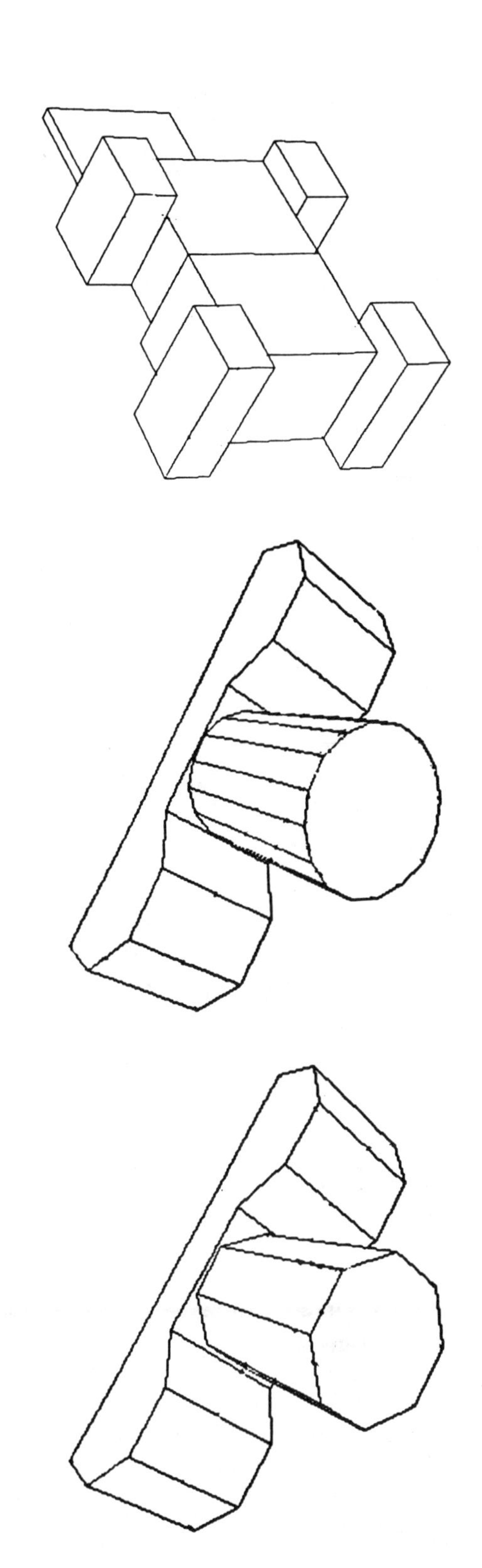

Table 3. Number of Interpretations after Local Pruning

Object	Normal	Dist	Min	50th	95th	Faces
Housing	$\pi/15$	0.01	1	1	2	40
		0.05	1	1	4	40
		0.10	1	4	17	40
	$\pi/10$	0.01	1	1	3	40
		0.05	1	2	9	40
		0.10	1	8	43	40
	$\pi/8$	0.01	1	1	4	40
		0.05	1	3	16	40
		0.10	1	12	96	40
Simple hand	$\pi/12$	0.01	2	2	8	28
		0.05	2	4	12	28
	$\pi/10$	0.01	2	2	16	28
		0.05	2	4	16	28
	$\pi/8$	0.01	2	4	32	28
		0.05	2	4	32	28
Complex hand	$\pi/12$	0.01	1	4	24	64
		0.05	1	10	64	64
	$\pi/10$	0.01	1	4	48	64
		0.05	1	15	94	64
	$\pi/8$	0.01	2	12	58	64
		0.05	2	18	136	64
Eight cylinders	$\pi/10$	0.08	2	4	12	21
		0.16	2	4	20	21
	$\pi/8$	0.08	2	8	38	21
		0.16	2	16	106	21
	$\pi/7$	0.08	2	12	68	21
		0.16	4	40	336	21
Sixteen cylinders	$\pi/10$	0.08	4	36	216	29
		0.16	4	60	504	29
	$\pi/8$	0.08	4	84	642	29

NOTE: The *normal* column lists the radius of the error cone about the measured surface normal; the *dist* column lists the error range of the distance sensing; the *min* column lists the minimum number of interpretations observed; the *50th* column lists the median point of the set of simulations; the *95th* column lists the 95th percentile of the set of simulations; and the *faces* column lists the number of faces in the model.

of interpretations in the tree following local pruning, for a variety of sensing accuracies. Each simulation consisted of 100 trials, and the minimum number of interpretations is recorded over this set of trials, as well as the 50th and 95th percentile of the distribution of number of interpretations. When model transformations were computed for each interpretation, it was observed that while the number of interpretations was not reduced to 1, as might be expected, the surviving interpretations generally tended to differ only in one or two faces. Moreover, the computed transformation parameters were nearly identical, indicating that the multiple interpretations surviving a model test actually corresponded to a single interpretation, to within the error ranges of the algorithm. Thus, Table 4 lists statistics of the number of separate transformations com-

Table 4. Number of Interpretations after Clustering

Object	Normal	Dist	Min	50th	95th	Faces
Housing	π15	0.01	1	1	1	40
		0.05	1	1	1	40
		0.10	1	1	2	40
	π/10	0.01	1	1	2	40
		0.05	1	1	2	40
		0.10	1	2	4	40
	π/8	0.01	1	1	2	40
		0.05	1	1	3	40
		0.10	1	2	5	40
Simple hand	π/12	0.01	2	2	2	28
		0.05	2	2	2	28
	π/10	0.01	2	2	2	28
		0.05	2	2	4	28
	π/8	0.01	2	2	4	28
		0.05	2	2	4	28
Complex hand	π/12	0.01	1	1	2	64
		0.05	1	1	3	64
	π/10	0.01	1	1	2	64
		0.05	1	1	3	64
	π/8	0.01	1	2	3	64
		0.05	1	2	5	64
Eight cylinders	π/10	0.08	2	2	4	21
		0.16	2	2	6	21
	π/8	0.08	2	4	12	21
		0.16	2	7	22	21
	π/7	0.08	2	8	18	21
		0.16	2	14	48	21
Sixteen cylinders	π/10	0.08	2	7	26	29
		0.16	2	11	42	29
	π/8	0.08	3	20	110	29

NOTE: The *normal* column lists the radius of the error cone about the measured surface normal; the *dist* column lists the error range of the distance sensing; the *min* column lists the minimum number of interpretations observed; the *50th* column lists the median point of the set of simulations; the *95th* column lists the 95th percentile of the set of simulations; and the *faces* column lists the number of faces in the model.

puted for each trial. In particular, transformations whose direction of rotation differed by more than 1.5° were judged to be different, yielding a very tight clustering of the computed transformations. This clustering ignores differences in the translation component, a point that is addressed later, in Table 6.

The first point to stress is that all of these numbers are remarkably low, given that the total number of possible interpretations of 12 sensed points on an object with 40 faces is roughly 1.678×10^{19}. Thus, the local geometric constraints are very effective in reducing the combinatorics of feasible interpretations.

As might be expected, the number of interpretations in both tables (Tables 3 and 4) tends to rise with increasing error in the measured parameters. The distributions also tend to be strongly clustered near the low

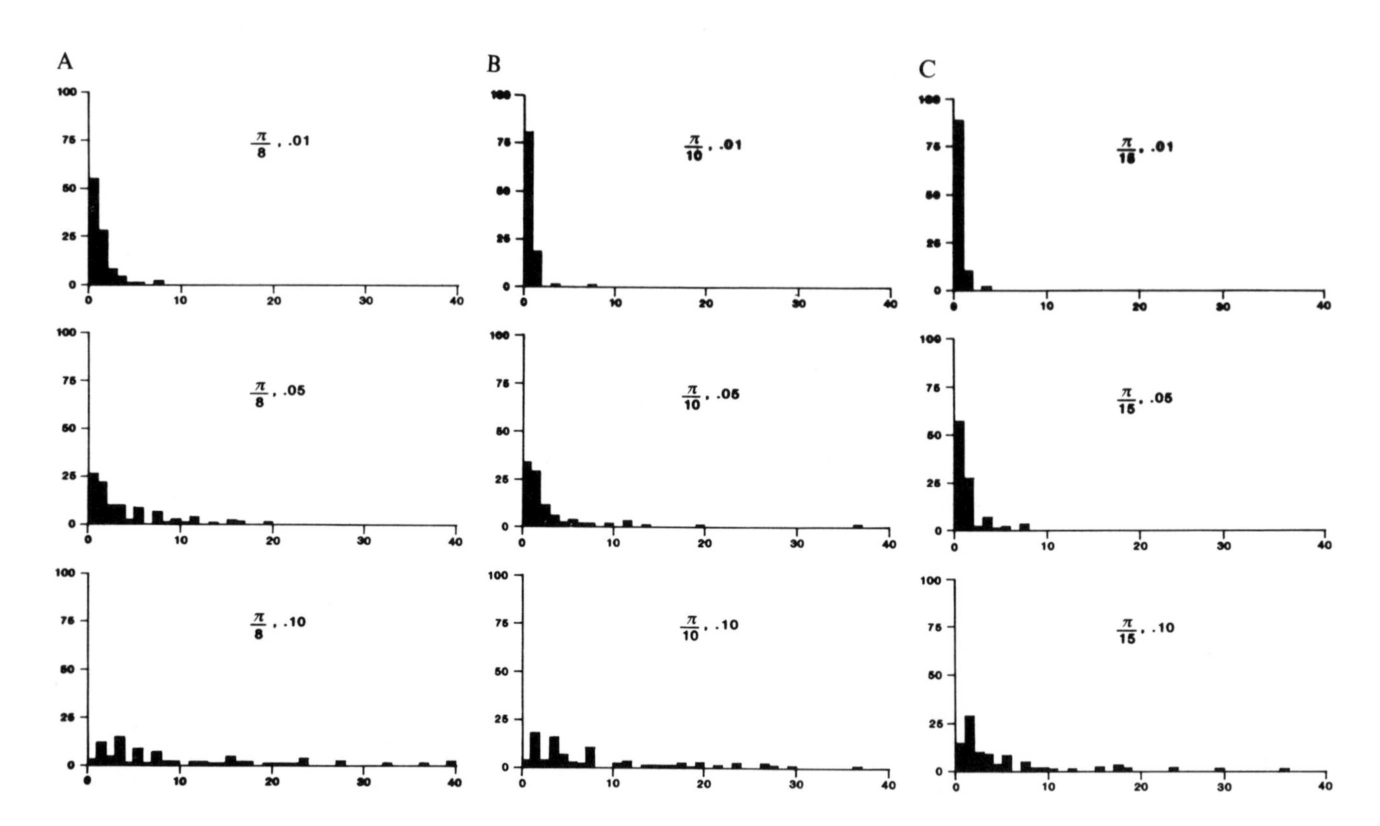

Fig. 8. Histograms of number of interpretations for different errors: π/8 (A); *π/10* (B); *π/15* (C).

end of the scale, with a very shallow tail on the high end of the distribution. Thus, while the maximum number of interpretations can be high (e.g., 1,454 for surface normal error cone of $\pi/8$ and distance error of 0.10), the median point and even the 95th percentile of the distribution are generally much smaller. Sample distributions for the number of interpretations surviving tree pruning are shown in Fig. 8. The abscissa of each graph records the number of interpretations for a given trial, while the ordinate records the number of trials for which a specific number of interpretations was observed. Thus each histogram displays a distribution of the frequency with which a particular number of interpretations was observed. Each histogram is labeled with the error associated with the measurements for that set of trials, specifically, the radius of the error cone about the sensed normal, in radians, and a bound on the magnitude of the position error, in inches.

One reason that the maximum number of feasible interpretations can be significantly larger than the median of the distribution is the occasional occurrence of dependent sensor information. For example, if most of the sensed points happen to lie on a single face, the amount of independent information about the object's position is much smaller than when the same number of sensed points lie on different faces. While the sensing strategy used here will reduce the probability of this occurring, there is still a nonzero chance of such redundant sensing taking place, resulting in an occasional case of a large number of feasible interpretations.

The number of distinct transformations is almost always 1 in these simulations. It was also observed that the computed transformation was generally very close to the actual one. For example, each row of Table 5 illustrates the average error in the computed transformations over 100 runs of the algorithm. The *direction* column lists the average angle between the correct and the computed direction of rotation, the *angle* column lists the average angle between the correct and the computed magnitude of rotation about the rotation direction, and the *translation* column lists the average magnitude of the difference between the correct and the computed translation component of the

Table 5. Average Errors in Computed Transformation

Object	Normal	Dist	Direction (degrees)	Angle (degrees)	Translation (in)
Simple hand	$\pi/12$	0.01	2.17	2.33	0.08
		0.05	2.08	2.62	0.09
	$\pi/10$	0.01	3.13	2.93	0.11
		0.05	3.58	3.15	0.12
	$\pi/8$	0.01	4.43	3.64	0.16
		0.05	5.26	3.03	0.17
Housing	$\pi/15$	0.05	2.18	2.17	0.11
	$\pi/10$	0.01	3.42	3.70	0.12
		0.05	3.64	3.22	0.14
		0.10	3.77	3.62	0.20
	$\pi/8$	0.05	4.28	5.07	0.19

NOTE: See text for definition of column headings.

transformation. It can be seen from the table that the average error is remarkably low, generally on the order of 2–3°, even for different objects and different amounts of sensor error. As might be expected, the average error does tend to rise with increases in the sensor error. In no case did the algorithm discard the correct interpretation. Note that the errors illustrated in Table 5 were recorded from the difference between the correct transformation and the computed transformation for the correct interpretation. There will be other, erroneous interpretations, with much larger differences between the computed and correct transformations.

In the few cases in which more than one transformation was found, two factors generally are observed. The first is that the noise in the measured data can result in transformations differing by only a few degrees, although these transformations are counted as being distinct. The second, more interesting, factor is the possibility of symmetric interpretations of the data due, for example, to a rotation of the object relative to the sensor. Consider first the case of a completely symmetric object, such as the simple hand, which has a rotational symmetry of 180°. Here, the algorithm always found at least two distinct transformations of the model that were consistent with the sensed data. For such objects as the motor housing, portions of the object are symmetric, for example, the base of the housing, ignoring the projecting lip. If all the sensed points happen to fall only on such a portion of the object, then symmetric interpretations of the data are possible. In general these symmetric interpretations account for most of the cases of multiple transformations, especially when the sensor error is small. The few remaining cases arise when the error in the measurements yields two nearly identical (i.e., differing by only a few degrees of rotation) transformations that account for the data. As the error in the measured data decreases, these multiple interpretations tend to disappear.

The simulation data listed in Table 4 are derived from a clustering of the interpretations based strictly on the rotation component of the transformation; that is, two transformations whose direction of rotation differed by less than 1.5° were considered to be part of the same cluster. This clustering technique, while very tight in the rotation component, ignores possible differences in the translation component of the transformation. To examine such differences, a number of the simulations were run, using a clustering of the interpretations with a rotation sensitivity of 1.5° and a translation sensitivity of either 0.05 or 0.01. The number of distinct transformations under this clustering scheme is indicated in Table 6. Note that while the number of distinct transformations does increase relative to the corresponding entries in Table 4, the change is not significant.

Table 6. Number of Transforms after Clustering

Object	Normal	Dist	Cluster	Min	50th	90th	Max
Housing	$\pi/10$	0.05	0.01	1	2	4	6
			0.05	1	1	2	4
Simple hand	$\pi/10$	0.05	0.01	2	4	6	16
			0.05	2	2	4	12
Complex hand	$\pi/10$	0.01	0.01	1	5	15	20
			0.05	1	3	6	16

4.4. Random Sensing

All of the previous simulations have generated the sensed data by projecting a regular grid of points along three orthogonal directions, generally resulting in between 12 and 20 contact points. Such a sensing strategy would be consistent with visual sensing modalities. A second set of simulations has been run using a sensing strategy more consistent with tactile sensors. Consider a set of three mutually orthogonal, directed rays, which intersect at a point. Suppose this point is taken to be some arbitrary point $(x, y, 0)$, chosen on the $x - y$ plane (note that by the definition of the object models, this plane will intersect the object). Each ray is traced along its preferred direction, (with decreasing z component), until either the object or the support plane is contacted. We repeated this operation for several different approaches, using randomly generated values of x and y, until between 7 and 9 different contact points were made on the object. Tables 7 and 8 summarize the results of running sets of simulations, using sensory data generated in this fashion.

As in the case of the earlier simulations, the effectiveness of the local constraints in reducing the number of feasible interpretations is clearly demonstrated. Interestingly, the number of distinct transformations tends to be somewhat higher than the earlier cases, especially for the motor housing. This results in part from the following situation. With the exception of one projecting portion (see Fig. 6), the housing is essentially a symmetric object with respect to two different axes. As a consequence, if the sampled data points do not lie on this distinguishing projection, there could be several consistent, symmetric interpretations of the data. In the case of sensory sampling on a regular grid of points, it is likely that at least one point will lie on this projection, and the symmetric ambiguity will not arise. In the case of fewer sample points, generated by random approaches to the object, it is much more likely that the feasible transformations will reflect this symmetry and thus be higher in number.

In cases of ambiguity in interpretation (e.g., when several orientations of the motor housing are consistent with the sensed data, due to a partial symmetry of the object), it would be useful to have effective means for distinguishing between the possible solutions. A straightforward method would be to add sensory points generated at random until only one interpretation is consistent. This, of course, could be very inefficient, since it could take the addition of several points before a solution is found. In the case of the motor housing, for example, one would need to consider additional sensory points until one lying on the projecting lip of the housing is recorded. A more effective solution is to use the difference in feasible interpretations to find directions along which the points of contact of the different interpretations are widely separated. Such directions then constitute good candidates for generating the next sensed point (Gaston and Lozano-Pérez 1983). Extensions of the method to the six-degree-of-freedom problem are currently under investigation.

4.5. Tree Pruning

Tables 9 and 10 contain a final set of statistics that demonstrates the effectiveness of the local constraints in reducing the number of feasible interpretations in

Table 7. Number of Interpretations after Local Pruning

Object	Normal	Dist	Min	50th	95th	Faces
Simple hand	$\pi/12$	0.01	2	2	16	28
		0.05	2	4	60	28
		0.10	2	4	32	28
	$\pi/10$	0.01	2	2	32	28
		0.05	2	4	60	28
		0.10	2	4	160	28
	$\pi/8$	0.01	2	4	74	28
		0.05	2	4	64	28
		0.10	2	8	128	28
Housing	$\pi/12$	0.01	1	2	4	40
		0.05	1	2	22	40
		0.10	1	11	154	40
	$\pi/10$	0.01	1	2	6	40
		0.05	1	3	26	40
		0.10	1	12	144	40
	$\pi/8$	0.01	1	2	12	40
		0.05	1	4	52	40
		0.10	2	21	329	40
Eight cylinders	$\pi/12$	0.08	2	2	18	21
		0.16	2	4	16	21
	$\pi/10$	0.08	2	4	62	21
		0.16	2	4	36	21
	$\pi/8$	0.08	2	10	152	21
		0.16	2	10	162	21

NOTE: The *normal* column lists the radius of the error cone about the measured surface normal; the *dist* column lists the error range of the distance sensing; the *min* column lists the minimum number of interpretations observed; the *50th* column lists the median point of the set of simulations; the *95th* column lists the 95th percentile of the set of simulations; and the *faces* column lists the number of faces in the model.

the IT. The regular grid approach is used to generate the sensory data. For the data in Table 9, the points are sampled in random order as the IT is generated and pruned. For the data in Table 10, the sensed points are sorted on the basis of pairwise separation, with the more distant points being ordered first. This sorting on distance tends to place the most effective constraints at the beginning of the process. Since the point of the local constraints is to prune the IT as efficiently as possible, applying the most effective constraints first should result in pruning out entire subtrees at as early a stage in the tree-generation process as possible. Using the sorted sense data, the interpretation tree was generated and pruned. Tables 9 and 10 list statistics for the number of interpretations at each level of the tree, (i.e., the number of k-interpretations for different values of k), based on trials of 100 simulations each.

It can be seen that the median number of feasible interpretations is quite small at all levels of the tree, even as the number of contact points is increased. These data imply that one of the strengths of the approach is the ability to prune out whole subtrees of the IT at a very early stage, thereby ensuring that the total number of tests to be applied is significantly smaller than the size of the entire tree. This leads to very efficient processing of the feasible interpretations.

Sorting the points on distance is extremely effective,

Table 8. Number of Transforms after Clustering on Rotation

Object	Normal	Dist	Min	50th	95th	Faces
Simple hand	$\pi/12$	0.01	2	2	4	28
		0.05	2	2	4	28
		0.10	2	2	6	28
	$\pi/10$	0.01	2	2	4	28
		0.05	2	2	4	28
		0.10	2	2	6	28
	$\pi/8$	0.01	2	2	6	28
		0.05	2	2	6	28
		0.10	2	2	8	28
Housing	$\pi/12$	0.01	1	1	3	40
		0.05	1	1	5	40
		0.10	1	2	12	40
	$\pi/10$	0.01	1	1	4	40
		0.05	1	1	6	40
		0.10	1	2	14	40
	$\pi/8$	0.01	1	1	4	40
		0.05	1	2	8	40
		0.10	1	4	15	40
Eight cylinders	$\pi/12$	0.08	2	2	6	21
		0.16	2	2	6	21
	$\pi/10$	0.08	2	2	18	21
		0.16	2	2	14	21
	$\pi/8$	0.08	2	6	50	21
		0.16	2	6	50	21

NOTE: The *normal* column lists the radius of the error cone about the measured surface normal; the *dist* column lists the error range of the distance sensing; the *min* column lists the minimum number of interpretations observed; the *50th* column lists the median point of the set of simulations; the *95th* column lists the 95th percentile of the set of simulations; and the *faces* column lists the number of faces in the model.

as can be seen from the results reported in Table 10 of the same set of runs as those in Table 9, with the exception that the points were sorted prior to pruning. The effect on running times of the pruning program is also quite drastic.

5. Performance on Range Data

We have performed limited testing of the algorithms described above, using high-quality range data obtained from a laser-based triangulation system developed by Philippe Brou at our laboratory. Two samples of the data we used are shown in Fig. 9. The data are obtained at high resolution, approximately 0.04-cm grid spacing along x and 0.08-cm spacing along y. A small number of points were obtained from the dense data by choosing points where a least-squares fit to a plane over a 5×5 patch produced very low normalized residue errors. Points were chosen that included at least three independent normals. Note that the actual object includes a protrusion that was not present in the model; no data were taken from that region. In the data from Fig. 9A, 11 points were used; in the data from figure 9B, 9 points were used. The accuracy bounds we employed were ± 0.02-in position accuracy and $\pm\pi/15$ accuracy in measuring the normal.

Figure 9 shows the results obtained from running

Table 9. Feasible Interpretations — Unsorted Points

Points	Min	50th	95th	Max
2	13	90	278	340
3	10	115	400	689
4	2	82	490	750
5	2	47	362	736
6	2	27	208	907
7	1	20	209	1077
8	1	18	170	1314
9	1	14	131	540
10	1	14	90	436
11	1	14	81	566
12	1	12	68	380
13	1	12	78	372
14	1	12	88	144
15	1	12	76	141

Table 10. Feasible Interpretations — Sorted Points

Points	Min	50th	95th	Max
2	4	22	80	122
3	2	24	75	159
4	2	14	54	94
5	1	11	48	134
6	1	10	37	64
7	1	9	36	148
8	1	8	44	63
9	1	8	48	71
10	1	8	64	122
11	1	10	43	112
12	1	9	56	108
13	1	10	75	139
14	1	12	64	170
15	1	12	72	324

the algorithm on the data described above. There were only 9 and 11 interpretations, respectively, left in the tree after pruning with the local constraints. From these, three valid transformations were found in one case (Fig. 9A) and two in the other (Fig. 9B). The correct transformation was found each time. The other transformations correspond to rotations that place the sensed points on parallel faces. Note, however, that disambiguations between the valid transformations would be straightforward once the transformations were known.

The quality of the data used in the experiments illustrated in Fig. 9 corresponds to nearly the best error conditions used in the simulations. Results with larger error bounds, using data from sections where the data are less accurate, showed results similar to those in the simulations, that is, more legal interpretations in the tree and more valid transformations, but always including the correct one. It tends to reinforce the validity of the conclusions found in the simulations.

6. The Combinatorics of Pruning the IT

In the previous sections, we outlined the basic interpretation algorithm. The crucial issue that determines the viability of this algorithm is the effectiveness of pruning the interpretation tree. Our goal has been to demonstrate that one can use simple local constraints to prune the interpretation tree, so that only a few of the relatively expensive model tests need to be made. The simulation results, under a variety of conditions, and the results on range data provide support for this claim.

It is also possible to provide a combinatorial analysis of the pruning of interpretation trees provided by local constraints. Here, we demonstrate the scope of the combinatorial analysis by presenting a detailed discussion of the use of the distance constraint in pruning interpretation trees. Similar results hold for the other constraints. We stress that the results given below are actually weak bounds on the number of interpretations to be expected after pruning. In practice, numbers close to these bounds are observed only when the sensors are arranged so as to obtain a minimum of information about the object. This is in part due to the proof technique used, which only considers the constraints between the first point and the kth point in constraining the number of possible faces consistent with the kth point. Much tighter bounds can be obtained by using the constraints between all previous points and the kth point in considering the number of faces consistent with the kth point. A detailed presentation of such an analysis is contained in a companion paper (Grimson and Lozano-Pérez, in press).

Fig. 9. Sample range data and computed interpretations.

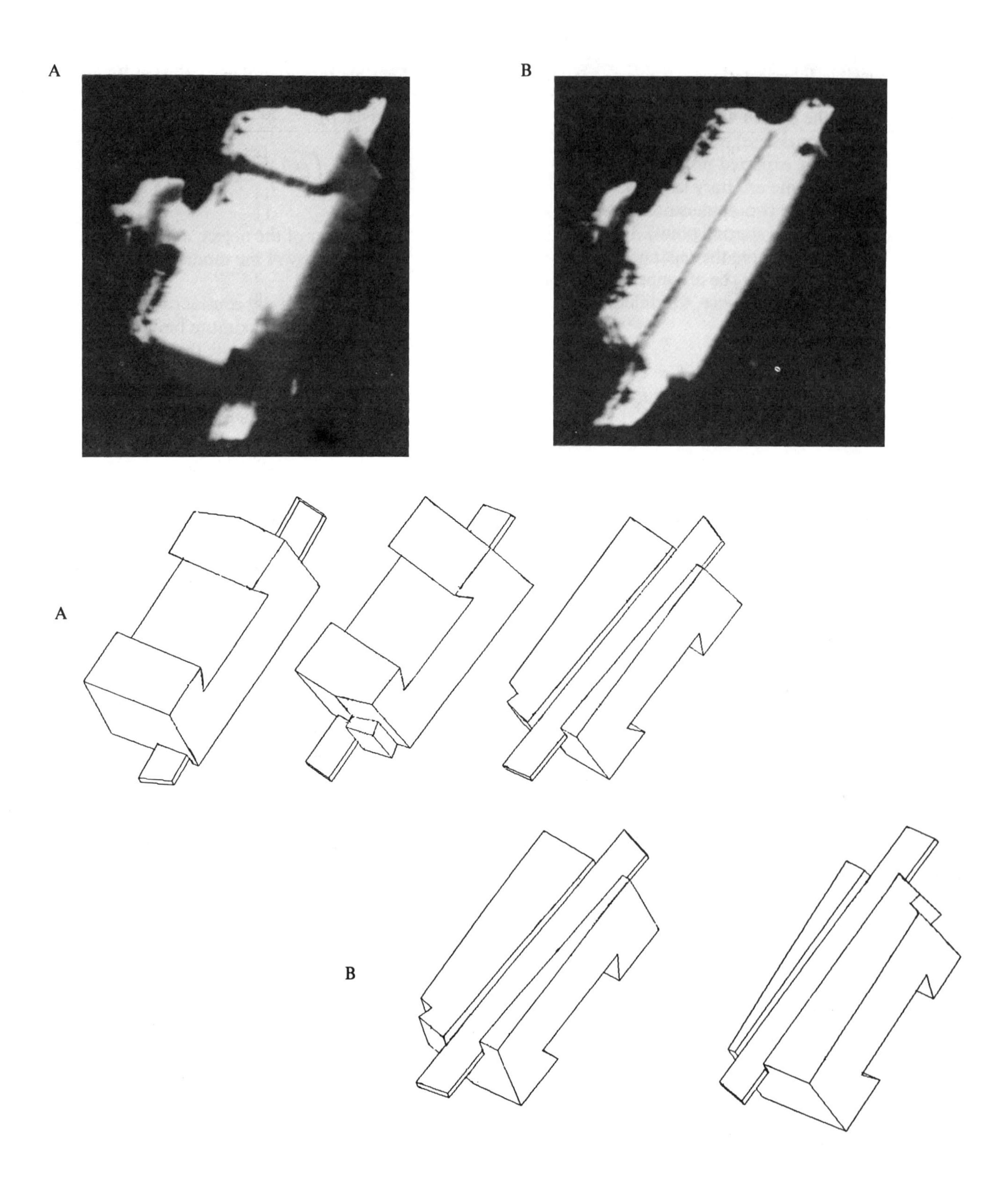

6.1. Combinatorics of Distance Pruning

We will consider the case in which all faces (or edges in the two-dimensional case) have the same size and derive bounds on the expected pruning of the IT. Assume that we have some arbitrary labeling of the faces from 1 to n (e.g., in the two-dimensional case, based on arc length from some starting point). For each pair of faces, i and j, let d_{ij} denote the separation of the midpoints of the faces. let ϵ_{ij} be an upper bound on the range of variation in distance, for different sensed points on the two faces, that is

$$\epsilon_{ij} = \lim\sup\{\epsilon\colon d_{ij} - \epsilon \leq |\mathbf{x} - \mathbf{y}| \leq d_{ij} + \epsilon, \quad \forall \mathbf{x} \text{ on face } i, \forall \mathbf{y} \text{ on face } j\},$$

where $|\mathbf{x} - \mathbf{y}|$ is the distance between point $\mathbf{x}$ on face i and point $\mathbf{y}$ on face j. Let ϵ be defined as the maximum over all i, j of ϵ_{ij}, plus some estimate of the maximum error of the sensed distance.

Now assume that we have recorded the position of two sensor points, P_1 and P_2, and let s_{12} be the measured distance between them. Assume that the first point has been arbitrarily assigned to some face i of the object. We want to determine how many faces j of the object can consistently be assigned to the second point, given the separation s_{12} and the known distribution of distances. Moreover, we want to be able to continue this for k sensor points, determining an upper bound on the number of assignments of faces to sensor points that are consistent with the sensed separation between the faces.

Let the distribution of faces with respect to face i as a function of distance be denoted by $\rho_i(s)$. In other words, $\rho_i(s)$ records the number of faces whose midpoint separation from face i is given by the distance s. As a consequence,

$$\int_{s=0}^{d} d\rho_i(s) = n,$$

where n is the total number of faces, and d is the diameter, or maximum separation of the object. Note that because $d\rho_i$ is a distribution, this is a Lebesgue-Stieltjes integral. The following bound on the number of nodes at the kth level of the IT holds for both two-dimensional and three-dimensional objects.

Proposition 1: An upper bound on the expected number of nodes at the kth level of the interpretation tree, $k \geq 2$, is given by

$$\left(\frac{2\epsilon n}{d}\right)^{k-1} n,$$

where d is the diameter of the object, and ϵ is a bound on the distance sensitivity of the model.

Proof: The proof proceeds by considering an iterative application of the expected maximum branching factor at each level of the tree. We assume that b_{k-1} denotes a bound on the number of consistent nodes at the $k-1$st level of the interpretation tree, and consider the branching factor obtained when adding a kth sensed point. Assume that sensor point P_{k-1} has been assigned to face i, and that the measured separation of sensor point P_{k-1} and P_k is s_k. This implies that the midpoint separation of the corresponding faces is within ϵ of s_k. Hence, an upper bound on the number of possible faces consistent with s_k, given face i assigned to point P_{k-1}, is

$$\int_{x=-\epsilon}^{\epsilon} d\rho_i(s_k + x).$$

Since the number of nodes at the $k-1$st level of the tree is bounded by b_{k-1}, an upper bound on the total number of nodes at the kth level of the tree is

$$b_{k-1} \max_i \int_{x=-\epsilon}^{\epsilon} d\rho_i(s_k + x).$$

We now wish to determine a bound on the expected number of nodes, evaluated over the range of possible values for s_k. If $\Psi(s)$ denotes the distribution of sensed distances, then an upper bound on the expected number of nodes is

$$\frac{\int_{s=0}^{d} \left[b_{k-1} \max_i \int_{x=-\epsilon}^{\epsilon} d\rho_i(s + x) \right] d\Psi(s)}{\int_{s=0}^{d} d\Psi(s)}.$$

If we know which object is being sensed, we can derive

an explicit form for $d\Psi(s)$. Since we are considering the case of sensing from a set of possible objects, the best we can do is to consider the distribution of sensed distances over all possible orientations of all objects, and this is best given by a uniform distribution. Thus,

$$d\Psi(s) = \frac{1}{d}\,ds,$$

and an upper bound on the expected number of nodes becomes

$$\frac{b_{k-1}}{d} \max_i \int_{s=0}^{d} \int_{x=-\epsilon}^{\epsilon} dp_i(s+x)ds.$$

Note that this double integration can essentially be considered as a counting problem. That is, we want to count the number of faces whose separation from the sampled face lies in an ϵ range about some point s, with this number being accumulated over all possible ϵ ranges (i.e., vary the midpoint s). Reversing the order of integration basically reverses the order of counting. Thus, rather than counting the number of faces lying within a range and summing over the set of ranges, we count the number of ranges in which each face is included and sum over the number of faces. Clearly, each face can be counted in at most 2ϵ ranges (as the midpoint of the range moves from $s-\epsilon$ to $s+\epsilon$), and the total number of faces is n. Thus, the branching factor at this level yields the iterative expression

$$b_k = b_{k-1}\frac{2\epsilon n}{d}.$$

The base case of $k=1$ yields the bound of $b_1 = n$, since the initial assignment of the point P_1 is arbitrary.

Evaluation of the iterative expression yields

$$b_k = \left(\frac{2\epsilon n}{d}\right)^{k-1} n,$$

thereby concluding the proof by induction.

While this proposition gives us an upper bound on the expected number of nodes, in order to evaluate it we need some estimate on ϵ. The following two propositions provide this for the two- and three-dimensional cases.

Fig. 10. Illustration for proof of proposition 2.

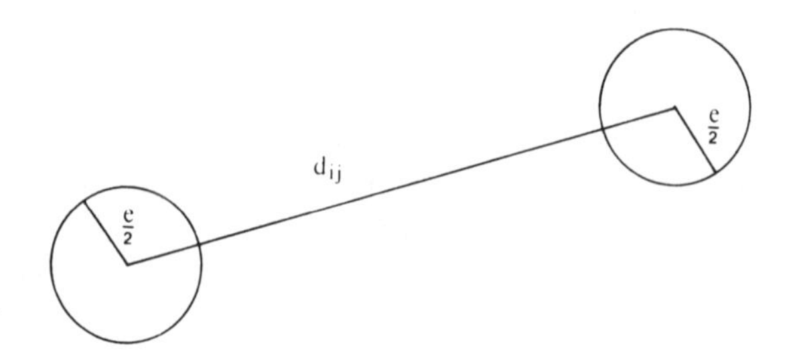

Proposition 2: If all the edges of a two-dimensional object have the same length e, then $\forall i, j\ \epsilon_{ij} \leq e$.

Proof: Connect the midpoints of two arbitrary faces, i and j, with a line of length d_{ij}. Consider first the case of $d_{ij} > e$. The set of all possible orientations of each of the edges about its midpoint describes a circle of radius $e/2$ about that end point. We are interested in the extrema in separation of points in these disks (see Fig. 10). We claim that the maximum and minimum separation of points in the disks occur for the case of the edges parallel to the midpoint connector, giving a minimum of $d_{ij} - e$ and a maximum of $d_{ij} + e$.

While this can be shown algebraically, there is also a simple geometric proof. Construct a coordinate system with origin at the midpoint of edge i and with x axis along the midpoint connector. Now construct a circle of radius $d_{ij} - e$ about the point $(d_{ij} - e/2, 0)$. Clearly, this circle grazes the first disk at the point $(e/2, 0)$. Now, in order for any other point in the second disk to have a shorter distance, we must be able to position a circle of the same radius about that point and still intersect the first disk. This is not possible, by the following argument. The envelope of possible points can be formed by sweeping a circle of radius $d_{ij} - e$ through a series of positions such that the center of the circle lies at the limit of the second disk. This envelope only intersects the first disk at the above-mentioned point, and hence the minimum possible separation between the two edges is given by $d_{ij} - e$. Similarly, the maximum separation can be shown to be $d_{ij} + e$ by constructing a circle of radius $d_{ij} + e$ about the point $(d_{ij} + e/2, 0)$ and using the same argument.

If $d_{ij} \leq e$, then the minimum distance is clearly bounded below by 0. The construction for the bound on the maximum distance is identical to that above. Hence, we see that $\epsilon_{ij} \leq e, \forall i, j$.

Corollary: If all the edges of a two-dimensional object have the same length, and the sensor error in measuring distances is much smaller than the length of an edge, then the expected number of nodes at the kth level of the interpretation tree, $k \geq 2$, that survive distance pruning is bounded above by

$$\left(\frac{2p}{d}\right)^{k-1} n,$$

where p is the perimeter of the object, and d is its diameter.

Proof: Since the sensor error is much less than the edge length, we see that ϵ is essentially given by the maximum over all i, j of ϵ_{ij}. From the previous proposition, this is bounded by the edge size e, and since all the edges are of the same length, $e = p/n$. The corollary follows naturally.

Note that for convex objects, $p \leq \pi d$, so that the bound becomes linear in n:

$$(2\pi)^{k-1} n.$$

In general, the perimeter for nonconvex objects can be much larger. We note, however, that for highly convoluted objects, if sensing is along straight lines, then much of the perimeter of the object is "invisible" to the sensor. This follows from the observation that sensing at such a face would require sensing through some other portion of the object. Thus, in practice, the perimeter term in the above expression for nonconvex objects should be replaced by an "effective perimeter," which will generally correspond to the perimeter of a nearly convex object.

Proposition 3: If all the faces of a three-dimensional object have the same diameter e, and the same area A_f, then $\forall i, j\ \epsilon_{ij} \leq e$.

Proof: The proof is almost identical to that of the two-dimensional case. Here the geometric construction consists of two spheres of radius $e/2$ centered about the endpoints of a line of length d_{ij}, and we seek the minimum and maximum separations of points on the two spheres. As in the previous case, a geometric construction shows that the extremal cases occur when the diameters of the faces are parallel to the midpoint connectors, and hence $\epsilon_{ij} \leq e$.

Corollary: If all the faces of a three-dimensional object have the same diameter and the same surface area, and the sensor error in measuring distances is much smaller than the diameter of a face, then the expected number of nodes at the kth level of the interpretation tree, $k \geq 2$, that survive distance pruning is bounded above by

$$\left(4\sqrt{\frac{A}{\pi d^2}}\right)^{k-1} n^{\frac{k+1}{2}},$$

where A is the surface area of the object and d is its diameter.

Proof: Since the sensor error in measuring distance is much less than the diameter of a face, we see that ϵ is essentially given by the maximum over all i, j of ϵ_{ij}. From the previous proposition, this is bounded by the face diameter e. If A_f is the surface area of the face, than $A_f \geq \pi(e/2)^2$. Moreover, $A_f = A/n$, so that $\epsilon \leq e \leq 2\sqrt{A/n\pi}$ and the corollary follows.

If the object is convex, then the area A is bounded above by πd^2, and the upper bound reduces to

$$4^{k-1} n^{\frac{k+1}{2}}.$$

As in the two-dimensional case, nonconvex objects can essentially be treated as convex ones, where the surface area of a convoluted object is replaced by the "effective surface area" of a nearly convex one, and a similar bound will hold.

6.2. The Relevance of the Combinatorics

The key point to be stressed here is that the use of distance pruning can be shown to reduce the interpretation problem significantly. In principle, the problem of k sensor points against a model of n faces would result in n^k possible interpretations that must be tested. We have shown that for two-dimensional objects, distance pruning reduces this to a number linear in n, and for three-dimensional objects, the number is reduced to at most one proportional to $n^{(k+1)/2}$.

We also stress that this is a weak upper bound, in particular because the analysis does not consider the full constraint of distance pruning. The analysis given considers the sequential pruning obtained by iteratively applying the constraint imposed by the sensed distance between the $(k+1)$st sensed point and the kth one. Clearly, given k sensed points, there are $\binom{k}{2}$ different distance constraints, and taking all of these into account should provide a tighter bound. Moreover, the bounds derived refer to the pruning due to a single type of constraint. Clearly, when all three constraints are used, we would expect the number of possible interpretations to be further reduced. Indeed, in a companion paper (Grimson and Lozano-Pérez, in press), we show that the number of interpretations can be bounded by an expression of the form

$$c_1 n \left[\frac{c_2 n}{k^5} \right]^k.$$

It was surprising to realize that weak upper bounds on the number of interpretations would be less than exponential in the number of sensed points, k (e.g., in the three-degree-of-freedom case, where the number of interpretations is linear in the number of sensed points). Many people find it surprising that any of the bounds should grow with k. Most people expect them to *decrease* with k, thinking that as more points are acquired, the constraint should be tighter. Recall, however, that the simple bounds derived above do not take into account the fact that there are $\binom{k}{2}$ distance constraints at the kth level of the tree; they only apply a single constraint at each level of the tree. There is another important effect that (partially) accounts for the growth in the number of interpretations with k. Namely, that for $k < 6$ each interpretation corresponds to a continuous range of positions and orientations. For example, for $k = 1$, each interpretation corresponds to the whole space of positions and orientations. As more points are added, the "volume" in the space of positions and orientations consistent with each interpretation decreases, but the number of these interpretations may increase (as they do between $k = 1$ and $k = 2$).[2]

7. Discussion

It is important to note that the algorithm described in this paper has a quite low computational cost. The pruning algorithm is particularly efficient. The range tables store all the model information needed, and pruning is done by simply comparing the ranges of values measured (plus or minus error estimates) with those in the tables. Therefore, no arithmetic is done during pruning (except for indexing into tables). It is only the model test that requires any significant computation; therefore, it is desirable to minimize the number of times it must be performed.

To illustrate this point, we have recorded actual run times for a number of simulations. While the times clearly depend on a number of factors, such as the type of machine, the specific algorithm, the object sensed, and so on, the order of magnitude of the run times helps illustrate the computational efficiency of the method. For example, using an implementation in Lisp running on a Symbolics 3600 Lisp Machine, simulations on the motor housing with angular error range of $\pi/10$ and positional error range of 0.05 took an average of 1.27 s to generate and prune the interpretation tree and an average of 3.17 s to perform the model check. The time required to generate and prune the tree is clearly dependent on the number of plausible interpretations and grows nonlinearly with an increase in this number. The time required to perform model checking grows linearly with the number of interpretations to which such a check must be applied. The average time expended on each model check was 0.24 s. In general, the average time to complete the computation was under 5 s for this particular implementation, although this number would occasionally be exceeded in sensing situations in which a large number of interpretations were possible.

The local constraint method developed here requires that all the sensory data be drawn from one object. This is difficult to guarantee, in the tactile or visual

2. We are indebted to John Canny for this observation.

domain, when the object is in a bin among other objects. Of course, if a hypothesis is made that all the points belong to one object and no feasible interpretations are found, then one can tell that the hypothesis is wrong. Much more research is needed in this area, however.

Throughout the paper, we have limited our attention to the number of interpretations, relative to one model, of data obtained from that object. To carry out recognition between several objects, one determines the number of legal interpretations of one set of data relative to multiple object models. This process can simply be performed sequentially on each model. One simple improvement is clearly possible. If one stores with each model the maximum distance between any of the faces, then if one of the measured distances is greater than this upper bound, the model can be discarded at once. This technique quickly separates large objects from small ones. Unfortunately, very small measured distances do not rule out large objects. A second method would be to use direction histograms to rule out certain models. For example, if the angle between two sensed normals were 30°, then a model of a cube would not be consistent with this data and could quickly be excluded.

Having generated and pruned the interpretation tree and performed the model test on each of the known objects, we have a listing of all the positions and orientations of all objects consistent with the measured data. At this point, further discrimination can be carried out by additional unguided sensing, as before, or by considering the alternatives and choosing a good place to sense next. The recognition problem that remains is now amenable to other techniques as well, since it has been reduced to the much more tractable problem of differentiating among a class of objects in known positions and orientations.

8. Acknowledgments

Philippe Brou contributed freely of his time and effort to provide us with the range data used in our experiments; we are very grateful. We also appreciated his comments on the presentation of this paper. We are very thankful to Bob Bolles and Berthold Horn for their detailed comments and suggestions on an earlier draft. We have also benefited from discussions with Olivier Faugeras.

Appendix 1

Here, we establish the claim of Section 2.1.2 that the set

$$\{\mathbf{n}_1 \cdot \mathbf{n}_2 | \mathbf{n}_1 \cdot \mathbf{u}_1 \geqslant \epsilon_1, \mathbf{n}_2 \cdot \mathbf{u}_2 \geqslant \epsilon_2\}$$

is contained in the set

$$\{\mathbf{n}_1 \cdot \mathbf{n}_2 | \cos[\min(\pi, \theta_{12} + \phi_1 + \phi_2)] \leqslant \mathbf{n}_1 \cdot \mathbf{n}_2 \leqslant \cos[\max(0, \theta_{12} - \phi_1 + \phi_2)]\},$$

where

$$\cos\phi_1 = \epsilon_1,$$
$$\cos\phi_2 = \epsilon_2,$$
$$\cos\theta_{12} = \mathbf{u}_1 \cdot \mathbf{u}_2 = \gamma.$$

While it is possible to prove this algebraically, it is simpler to see it by the following geometric construction (see Fig. 11). We wish to determine the extremal values of the dot product between unit vectors in the two cones, or equivalently, extremal values in the angle between any two such vectors. If the cones about $\mathbf{u}_1$ and $\mathbf{u}_2$ intersect, clearly the maximum value of the dot product is 1. If the cones are antipodal, clearly the minimum value is -1.

We now consider cases in which the cones do not overlap. We claim that the extremal values for the dot product occur when the two vectors lie in the plane spanned by $\mathbf{u}_1$ and $\mathbf{u}_2$, with the vectors lying at the limits of the cone within this plane. That is, if we let

$$\rho_i = \sqrt{\frac{1 - \epsilon_i^2}{1 - \gamma^2}},$$

then the extrema occur at

$$\mathbf{n}_1 = (\epsilon_1 - \gamma\rho_1)\mathbf{u}_1 + \rho_1\mathbf{u}_2,$$
$$\mathbf{n}_2 = \rho_2\mathbf{u}_1 + (\epsilon_2 - \gamma\rho_2)\mathbf{u}_2,$$

and

$$\mathbf{n}_1 = (\epsilon_1 + \gamma\rho_1)\mathbf{u}_1 - \rho_1\mathbf{u}_2,$$
$$\mathbf{n}_2 = -\rho_2\mathbf{u}_1 + (\epsilon_2 + \gamma\rho_2)\mathbf{u}_2.$$

Fig. 11. Extremal values of dot products between two cones.

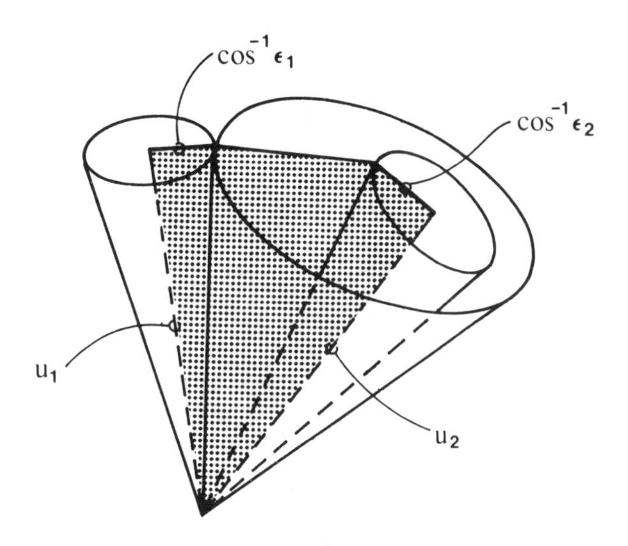

The first case can be shown to correspond to the minimal angle between vectors in the two cones, by the following construction. Construct a cone centered about $\mathbf{n}_1$ with radius such that $\mathbf{n}_2$ lies on the boundary of the cone; that is, the new cone grazes the $\mathbf{u}_2$ cone at $\mathbf{n}_2$. If there is a smaller angle, it must be possible to reposition this cone so that it is centered at some other point in the $\mathbf{u}_1$ cone and yet still intersects the $\mathbf{u}_2$ cone. This is clearly not possible, and hence the minimum value of the dot product is given by the stated choice of $\mathbf{n}_1$ and $\mathbf{n}_2$. Expanding the dot product for this case, and making the appropriate trigonometric substitutions, yields the required expression. A similar construction holds for the maximum angle (or minimum dot product).

Appendix 2

Here, we show how to compute the range of possible direction vectors between $face_i$ and $face_j$ in the object model. Let us erect a coordinate system on $face_i$ at the centroid of the face and whose z axis points along the normal of the face. Then, it is clear that the set of possible direction vectors is the set

$$\{\mathbf{v}_j - \mathbf{v}_i | \mathbf{v}_j \in face_j \ \& \ \mathbf{v}_i \in face_i\},$$

where both $\mathbf{v}_i$ and $\mathbf{v}_j$ are expressed relative to the frame on $face_i$. Assume, for now, that both faces are convex. It can be shown (Lozano-Pérez 1983) that this set is equivalent to

$$ch(\{\mathbf{v}_j - \mathbf{v}_i | \mathbf{v}_j \in vert(face_j) \ \& \ \mathbf{v}_i \in vert(face_i)\}),$$

where $ch()$ is the convex hull of a set of points, and $vert()$ is the set of vertices of a face. Because of convexity, the extrema of the component of the direction vectors along the normal of $face_i$ occur at the vertices of this convex hull. Clearly, the vertices of the convex hull of a set of points are drawn from the set of points itself. Therefore, we need only find the extrema of the finite set

$$\{\mathbf{n}_i \cdot (\mathbf{v}_j - \mathbf{v}_i) | \mathbf{v}_j \in vert(face_j) \ \& \ \mathbf{v}_i \in vert(face_i)\},$$

where $\mathbf{n}_i$ is the normal to $face_i$.

When the faces are nonconvex, the procedure above will generate a conservative bound.

References

Agin, G. J., and Binford, T. O. 1973 (Aug. 1973, Stanford). Computer description of curved objects. *Proc. Third Int. Joint Conf. Artificial Intell.* Los Altos, Calif.: Kaufmann, pp. 629–640.

Bajcsy, R. 1973. Computer identification of visual surface. *Comput. Graphics Image Processing* 2(2):118–130.

Bajcsy, R., and Liebermann, L. 1976. Texture gradient as a depth cue. *Comput. Graphics Image Processing* 5(1):52–67.

Baker, H. H., and Binford, T. O. 1981 (Aug. 1981, Vancouver). Depth from edge and intensity based stereo. *Proc. Seventh Int. Joint Conf. Artificial Intell.* Los Altos, Calif.: Kaufmann, pp. 631–636.

Binford, T. O. 1972. Sensor systems for manipulation. Paper delivered at Remotely Manned Systems Conference, Miami, Florida.

Bolles, R. C., and Cain, R. A. 1982. Recognizing and locating partially visible objects: The local-feature-focus method. *Int J. Robotics Res.* 1(3):57–82.

Bolles, R. C., Horaud, P., and Hannah, M. J. 1983 (Aug.). 3DPO: A three-dimensional part orientation system. Paper delivered at First Int. Symp. Robotics Res., Bretton Woods, N.H.

Briot, M. 1979 (Mar., Washingron, D.C.). The utilization of an "artificial skin" sensor for the identification of solid objects. *Proc. Ninth Int. Symp. Industrial Robots* Dearborn, Mich.: Society of Manufacturing Engineers, pp. 529–548.

Briot, M., Renaud, M., and Stojilkovic, Z. 1978 (Sept.). An approach to spatial pattern recognition of solid objects. *IEEE Trans. Syst. Man Cybern.* SMC-8(9):690–694.

Brou, P. 1983 (Aug.). Finding the orientation of the objects in vector maps. Ph.D. thesis, Massachusetts Institute of Technology Department of Electrical Engineering and Computer Science.

Buchanan, B., Sutherland, G., and Feigenbaum, E. A. 1969. Heuristic DENDRAL: A program for generating explanatory hypotheses in organic chemistry. *Machine Intelligence 4,* ed. B. Melzer and D. Michie. New York: American Elsevier.

Dixon, J. K., Salazar, S., and Slagle, J. R. 1979 (Mar., Washington, D.C.). Research on tactile sensors for an intelligent robot. *Proc. Ninth Int. Symp. Industrial Robots.* Dearborn, Mich.: Society of Manufacturing Engineers, pp. 507–518.

Faugeras, O. D., et al. 1983. Toward a flexible vision system. *Robot Vision,* ed. A. Pugh. Berlin: Springer-Verlag, pp. 129–142.

Faugeras, O. D., and Hebert, M. 1983 (Aug., Karlsruhe, W. Germany). A 3-D recognition and positioning algorithm using geometrical matching between primitive surfaces. *Proc. Eighth Int. Joint Conf. Artificial Intell.* Los Altos, Calif.: Kaufmann, pp. 996–1002.

Gaston, P. C., and Lozano-Pérez, T. 1984. Tactile recognition and localization using object models: The case of polyhedra on a plane. *IEEE Trans. Pattern Anal. Mach. Intell.* PAMI-6 (3):257–265. AIM-705. Cambridge, Mass.: Massachusetts Institute of Technology Artificial Intelligence Laboratory.

Grimson, W. E. L. 1981*a*. A computer implementation of a theory of human stereo vision. *Philosophical Trans. Royal Soc. London, B*292:217–253.

Grimson, W. E. L. 1981*b*. *From images to surfaces: A computational study of the human early vision system.* Cambridge, Mass.: MIT Press.

Grimson, W. E. L. 1982. A computational theory of visual surface interpolation. *Philosophical Trans. Royal Soc. London B*298:395–427.

Grimson, W. E. L. 1983. An implementation of a computational theory of visual surface interpolation. *Comput. Vision Graphics Image Processing* 22:39–69.

Grimson, W. E. L., and Lozano-Pérez, T. In press. A combinatorial analysis of recognition and localization using object models. Cambridge, Mass.: Massachusetts Institute of Technology Artificial Intelligence Laboratory.

Harmon, L. D. 1980. Touch-sensing technology: A review. MSR80-03. Dearborn, Mich.: Society of Manufacturing Engineers.

Harmon, L. D. 1982. Automated tactile sensing. *Int. J. Robotics Res.* 1(2):3–32.

Hillis, W. D. 1982. A high-resolution image touch sensor. *Int. J. Robotics Res.* 1(2):33–44.

Ikeuchi, K., et al. 1983 (Aug.). Picking up an object from a pile of objects. Paper delivered at First Int. Symp. Robotics Res., Bretton Woods, N.H.

Ikeuchi, K., and Horn, B. K. P. 1979 (Aug., Tokyo). An application of photometric stereo. *Proc. Sixth Int. Joint Conf. Artificial Intell.* Los Altos, Calif.: Kaufmann, pp. 413–415.

Ivancevic, N. S. 1974. Stereometric pattern recognition by artificial touch. *Pattern Recognition* 6:77–83.

Jarvis, R. A. 1983. A perspective on range finding techniques for computer vision. *IEEE Trans. Pattern Anal. Machine Intell.* PAMI-5(2):122–193.

Kender, J. R. 1980. Shape from texture. CMU-CS-81-102, Pittsburgh: Carnegie-Mellon University Computer Science Report.

Kinoshita, G., Aida, S., and Mori, M. 1975. A pattern classification by dynamic tactile sense information processing. *Pattern Recognition* 7:243.

Korn, G. A., and Korn, T. M. 1968. *Mathematical handbook for scientists and engineers.* New York: McGraw-Hill.

Lewis, R. A., and Johnston, A. R. 1977 (Aug., Cambridge, Mass.). A scanning laser range finder for a robotic vehicle. *Proc. Fifth Int. Joint Conf. Artificial Intell.* Los Altos, Calif.: Kaufmann, pp. 762–768.

Lozano-Pérez, T. 1983. Spatial planning: A configuration space approach. *IEEE Trans. Comput.* C-32(2):108–120.

Marik, V. 1981 (Aug., Vancouver). Algorithms of the complex tactile information processing. *Proc. Seventh Int. Joint Conf. Artificial Intell.* Los Altos, Calif.: Kaufmann, pp. 773–774.

Mayhew, J. E. W., and Frisby, J. P. 1981. Psychophysical and computational studies towards a theory of human stereopsis. *Artificial Intell.* 17:349–385.

Nevatia, R., and Binford, T. O. 1977. Description and recognition of curved objects. *Artificial Intell.* 8:77–98.

Nitzan, D., Brain, A. E., and Duda, R. O. 1977 (Feb.). The measurement and use of registered reflectance and range data in scene analysis. *Proc. IEEE* 65:206–220.

Okada, T., and Tsuchiya, S. 1977. Object recognition by grasping. *pattern Recognition* 9(3):111–119.

Oshima, M., and Shirai, Y. 1978. A scene description method using three-dimensional information. *Pattern Recognition* 11:9–17.

Oshima, M., and Shirai, Y. 1983 (July). Object recognition

using three-dimensional information. *IEEE Trans. Pattern Anal. Machine Intell.* PAMI-5(4):353–361.

Overton, K. J., and Williams, T. 1981 (Aug., Vancouver). Tactile sensation for robots. *Proc. Seventh Int. Joint Conf. Artificial Intell.* Los Altos, Calif.: Kaufmann, pp. 791–795.

Ozaki, H., et al. 1982 (May/June). Pattern recognition of a grasped object by unit-vector distribution. *IEEE Trans. Syst. Man Cybern.* SMC-12(3):315–324.

Page, C. J., Pugh, A., and Heginbotham, W. B. 1976 (Mar., University of Nottingham). Novel techniques for tactile sensing in a three-dimensional environment. *Proc. Sixth Int. Symp. Industrial Robots.* Dearborn, Mich.: Society of Manufacturing Engineers.

Popplestone, R. J., et al. 1975 (Sept., Tbilisi, Georgia, USSR). Forming models of plane and cylinder faceted bodies from light stripes. *Proc. Fourth Int. Joint Conf. Artificial Intell.* Los Altos, Calif.: Kaufmann, pp. 664–668.

Purbrick, J. A. 1981 (Apr.). A force transducer employing conductive silicone rubber. Paper delivered at the First Int. Conf. Robot Vision Sensory Contr., Stratford-upon-Avon, United Kingdom.

Raibert, M. H., and Tanner, J. E. 1982. Design and implementation of a VLSI tactile sensing computer. *Int. J. Robotics Res.* 1(3):3–18.

Shirai, Y., and Suwa, M. 1971 (Sept., London, England). Recognition of pohyhedrons with a range finder. *Proc. Second Int. Joint Conf. Artificial Intell.* Los Altos, Calif.: Kaufmann.

Schneiter, J. L. 1982. An optical tactile sensor for robots. S. M. thesis, Massachusetts Institute of Technology Department of Mechanical Engineering.

Shneier, M. 1979 (Aug., Tokyo). A compact relational structure representation. *Proc. Sixth Int. Joint Conf. Artificial Intell.* Los Altos, Calif.: Kaufmann, pp. 818–826.

Snyder, W. E., and St. Clair, J. 1978 (Mar.). Conductive elastomers as sensor for industrial parts handling equipment. *IEEE Trans. Instrumentat. Meas.* IM-27(1):94–99.

Stevens, K. A. 1981. The information content of texture gradients. *Biol. Cybern.* 42:95–105.

Stojilkovic, Z., and Saletic, D. 1975 (IIT Research Institute, Chicago). Learning to recognize patterns by Belgrade hand prosthesis. *Proc. Fifth Int. Symp. Indust. Robots.* Dearborn, Mich.: Society of Manufacturing Engineers, pp. 407–413.

Sugihara, K. 1979. Range-data analysis guided by a junction dictionary. *Artificial Intell.* 12:41–69.

Takeda, S. 1974. Study of artificial tactile sensors for shape recognition: Algorithm for tactile data input. *Proc. Fourth Int. Symp. Industrial Robots.* Dearborn, Mich.: Society of Manufacturing Engineers, pp. 199–208.

Terzopoulos, D. 1983. Multilevel computational processes for visual surface reconstruction. *Comput. Vision Graphics Image Processing* 24(1):52–96.

Woodham, R. J. 1978. Photometric stereo: A reflectance map technique for determining surface orientation from image intensity. *Image Understanding Syst. Industrial Applications, Proc. SPIE 155.* Bellingham, Wash.: SPIE.

Woodham, R. J. 1980. Photometric method for determining surface orientation from multiple images. *Optical Engineering* 19(1):139–144.

Woodham, R. J. 1981. Analysing images of curved objects. *Artificial Intell.* 17:117–140.

Rule-Based Interpretation of Aerial Imagery

DAVID M. McKEOWN, JR., MEMBER, IEEE, WILSON A. HARVEY, JR., AND JOHN McDERMOTT

***Abstract*—In this paper, we describe the organization of a rule-based system, SPAM, that uses map and domain-specific knowledge to interpret airport scenes. This research investigates the use of a rule-based system for the control of image processing and interpretation of results with respect to a world model, as well as the representation of the world model within an image/map database. We present results on the interpretation of a high-resolution airport scene where the image segmentation has been performed by a human, and by a region-based image segmentation program. The results of the system's analysis is characterized by the labeling of individual regions in the image and the collection of these regions into consistent interpretations of the major components of an airport model. These interpretations are ranked on the basis of their overall spatial and structural consistency. Some evaluations based on the results from three evolutionary versions of SPAM are presented.**

***Index Terms*—Artificial intelligence, computer vision systems, knowledge utilization, production systems, rule-based aerial photo interpretation.**

I. INTRODUCTION

SPAM, System for Photo interpretation of Airports using MAPS, is an image-interpretation system. It coordinates and controls image segmentation, segmentation analysis, and the construction of a scene model. It provides several unique capabilities to bring map knowledge and collateral information to bear during all phases of the interpretation. These capabilities include the following.

- The use of domain-dependent spatial constraints to restrict and refine hypothesis formation during analysis.
- The use of explicit camera models that allow for the projection of map information onto the image.
- The use of image-independent metric models for shape, size, distance, absolute and relative position computation.
- The use of multiple image cues to verify ambiguous segmentations. Stereo pairs or overlapping image sequences can be used to extract information or to detect missing components of the model.

A. *The Nature of the Task*

The task of airport image analysis has many interesting properties. First, airports are a complex organization of man-made structures placed over a large ground area. While the actual spatial arrangement of typical structures such as runways, terminal buildings, parking lots, etc., varies greatly between airports, the types of structures normally found in an airport scene are well understood. The airport task provides a "knowledge rich" environment, where functional relationships between structures preclude arbitrary spatial arrangements and provide spatial constraints.

Second, a body of literature [1], [2] on airport planning is readily accessible and provides general design constraints. Knowledge acquisition for spatial constraints, therefore, does not involve examination of large numbers of sample airports. It can be observed that there are two major classes of airports, commercial and military, whose organization varies widely from large-scale international airports to small county and private airstrips. Both general knowledge (class-specific) and site-specific knowledge can be expected to help in the interpretation process.

Image-processing techniques are inherently errorful. For complex, uncontrived natural scenes, image segmentation results are highly ambiguous. The correspondence between regions in the segmented image and physical objects in the scene is generally many to one. Boundaries between objects may not be distinguishable due to occlusions, objects with similar spectral properties, and the intrinsic resolution of the image. Thus, the assumption that regions in the segmented image *directly* correspond to objects in the scene is not useful unless we are able to reason about the segmentation process and reconstruct a meaningful portion of the original object.

Finally, for aerial photo-interpretation tasks, it is crucial that the metrics used by the analysis system be defined in cartographic coordinates, such as ⟨*latitude/longitude/elevation*⟩, rather than in an image-based coordinate system. Systems that rely on descriptions such as "the runway has area 12 000 pixels" or "hangars are between 212 and 345 pixels" are useless except for (perhaps) the analysis of one image. Further, spatial analysis based on the semantics of *above*, *below*, *left-of*, *right-of*, etc., are also inappropriate for general interpretation systems. To operationalize metric knowledge one must relate the world model to the image under analysis. This should be done through image-to-map correspondence using camera models which is the method used in SPAM. We can directly measure ground distances, areas, absolute compass direction, and recover crude estimates of height using a camera model computed for each image under analysis. Direct projection of known map information is also possible in SPAM. For example, the position and orientations

Manuscript received June 12, 1985. This work was supported in part by the Defense Advanced Research Projects Agency under ARPA Order 3597, and monitored by the Air Force Avionics Laboratory under Contract F33615-78-C-1551. The views and conclusions in this paper are those of the authors and should not be interpreted as representing the official policies, either expressed or implied, of the Defense Advanced Projects Agency or the U.S. Government.

The authors are with the Department of Computer Science, Carnegie-Mellon University, Pittsburgh, PA 15213.

of known airport features such as runways and terminals could be directly factored into the scene interpretation as *a priori* knowledge. We are working on a system that can bring both class-specific and site-specific knowledge to bear; however, here we report on the use of class-specific knowledge.

B. Related Work

There is a long history of research in model-based vision. Binford [3] provides a good summary of a variety of work including Brooks [4], the VISIONS project [5], SRI [6], [7], and work by Matsuyama [8] and Ohta [9].

Work by Bullock [10], based on the ACRONYM system developed by Brooks and Binford [11], uses image registration to a geographic model, and identifies preselected regions of interest and attempts to locate and identify predefined object instances within these areas. ACRONYM is an example of a model-based system which incorporates viewpoint-insensitive mechanisms in terms of its model description. Its recognition process is to map edge-based image properties to instances of object models. So far, results have been reported for the recognition of a small number of models (3) for wide-bodied jets in aerial photographs. It is not clear how spatial knowledge would be directly integrated into the ACRONYM framework.

Matsuyama [12], [13] has demonstrated a system for segmentation and interpretation of color-infrared aerial photographs containing roads, rivers, forests, and residential and agricultural areas. It uses rules to make assignments based on region adjacency and multispectral properties. It uses two-dimensional (2-D) shape descriptions and performs region merging to generate object descriptions. It generates good descriptions of a variety of fairly complex aerial scenes getting a great deal of constraint from directly mapping the multispectral properties of regions into scene hypotheses.

In his dissertation, Selfridge [14] proposes using adaptive threshold selection for region extraction by histogramming and region growing using an image-based "appearance model." Recently, Hwang [15] has explored this method, coupled with the use of domain knowledge to guide interpretation of suburban house scenes in monochromatic aerial imagery. Hwang uses a test-hypothesize-act cycle to generate large numbers of potential hypotheses which are then grouped into consistent interpretations.

II. SPAM Components

SPAM is composed of three major system components: a Washington, DC, image/map database (MAPS), image-processing tools, and the rule-based system. In this paper, we describe the rule-based component, but it is important to emphasize the necessity of a complete system design in order to realistically explore aerial photo-interpretation tasks. Fig. 1 is a block diagram of the current system. SPAM is organized to view information extracted from image(s) uniformly, that is, without knowledge of what method was used to extract the image features. Further, all *image-based* descriptions are converted into *map-*

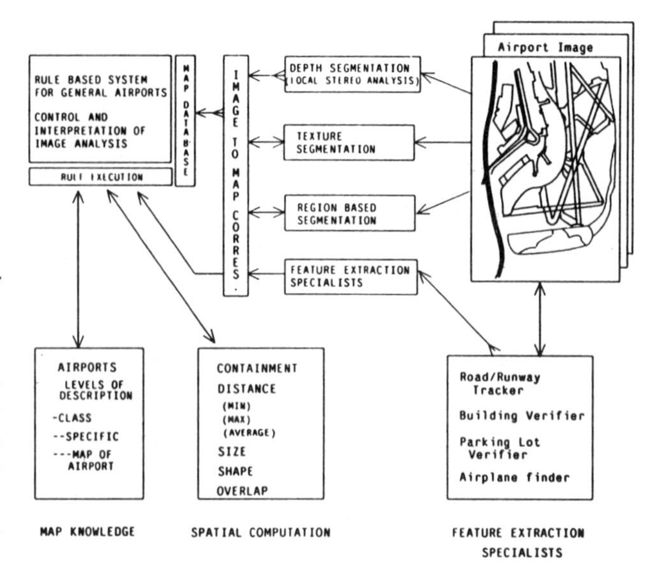

Fig. 1. SPAM system organization.

based descriptions using an image-to-map transformation based on the camera model stored for the image(s) in the MAPS database. Thus, it is possible to integrate image feature information from different segmentation or analysis methods, as well as from multiple scenes of the same airport. This architecture allows us to add and delete various sources of knowledge and to measure their effect on the quality of the resulting interpretation.

A. Image/Map Database

The MAPS [16], [17] database stores facts about man-made or natural feature existence and location; this allows SPAM to perform geometric computation in *map space* rather than *image space*. Differences in scale, orientation, and viewpoint can be handled in a consistent manner using a simple camera model. The function of the image/map database is to tie database feature descriptions to a geodetic coordinate system ⟨*latitude*, *longitude*, *elevation*⟩, and to use camera models (image-to-map correspondence) to predict their location and appearance in the aerial photography. MAPS also provides facilities to compute geometric properties and relationships of map database features such as *containment*, *adjacency*, *subsumed by*, *intersection*, and *closest point*.

B. Image-Processing Tools

There are currently four image-processing tools that can be invoked with actions specified by the rule-based system. They are a region-growing segmentation program [18], a road/road-like feature follower [19], a stereo analysis program [20], and an interactive human segmentation system. These tools perform low-level and intermediate-level feature extraction. Processing primitives are based on linear feature extraction and region extraction using edge-based and region-growing techniques. The goal of the image processing component is to identify feature-based islands of interest and extend those islands constrained by the geometric model provided by MAPS and model-based goals established by the rule-based component.

C. Rule-Based System

The rule-based component[1] provides the image-processing system with the best next task based on the strength/promise of expectations and with constraints from the image/map database system. It also guides the scene interpretation by generating successively more specific expectations based on image-processing results.

Airport scene interpretation is a classic case of the knowledge-based system approach to signal (image) analysis. This class of problems is characterized by uncertainty in the underlying data, and multiple, ambiguous interpretations for any particular hypothesis. It is our contention that hypotheses generated from raw region data should be able to be reliably verified by looking at other interpretations in its vicinity. Those interpretations emerging with the highest confidence should be those regions that can be collected into consistent cliques that will model the majority of the airport. These cliques can then be manipulated as individual pieces of a puzzle, putting appropriate ones together to form a complete interpretation (or model) of the airport in question. Multiple consistent models may be present. In SPAM, hypotheses are grouped into cliques according to their function and spatial proximity.

Region data can be derived from a number of sources. Typical sources are image-based segmentation, texture-based segmentation, multispectral analysis, and segmentation of depth map images compiled from stereo image pairs. SPAM has been tested with both hand-generated and machine-generated image-based segmentations. Obviously, perfect data are not generated by image analysis methods, so SPAM must have shape constraints flexible enough to compensate for this. The system, then, must perform the following two basic tasks:

1) generate a complete and consistent model of the airport given "perfect" hand-generated data; and

2) extract reasonable hypotheses for actual features from nonideal data.

In Sections III, V, and VI, we describe the organization of the rule-based system. In Section IV, we present several examples of SPAM processing both hand- and machine-generated data. Some performance statistics are presented in Section VIII.

III. Interpretation Primitives

The task of the rule-based system is to generate plausible, complete analyses of the airport image by building successively more specific interpretations based on the initial image segmentation. Rules to generate and evaluate such interpretations manipulate four simple working memory elements: *regions*, *fragments*, *functional areas*, and *models*.

[1] We use the OPS5 production system language [21] as the implementation language for our rule-based system.

A. Regions

The region working memory element contains low-level image properties of each segmentation region, such as shape, texture, spectral properties, etc. Regions represent the low-level data extracted from the image or specified by a human interpreter. The region properties are used to determine which classes of airport features a particular region best represents.

Since the criteria for initial class determination are weak discriminators, multiple hypotheses may be generated for each region as a different plausible interpretation. Current classes and subclass specializations are given in Fig. 2. These interpretations are represented as fragments, which are described in Section III-B. Fig. 3 illustrates a region working memory element and gives a general description of each of the attribute names. The *symbolic-name* is a unique identifier that allows functions external to the interpreter access to the segmentation region, the image, and the map database.

B. Fragments

A fragment is an interpretation for an image region. Fig. 4 shows a sample fragment working memory element. Each fragment contains class and subclass interpretation information, as well as interpretation status information used to control rule execution. Status information includes whether the fragment has been extended, aligned with other fragments, and whether consistency has been performed. Fragments which represent alternative interpretations of regions can be identified by their region-token attribute name.

C. Functional Areas

A functional area (FA) is a distinct spatial subdivision of the airport scene. It comprises a collection of man-made and natural features, normally found in close physical proximity and often related in function. For example, runways, taxiways, tarmac, and grassy areas form one functional area. Some of the characteristics of this FA are: the landing and takeoff of planes; routes of planes from terminal to takeoff or landing points; a buffer area with minimal 3-D structures (obstructions); and is centrally located within the airport area. SPAM currently supports four functional area types:

- *FA1*: *terminal, parking apron, parking lots, roads
- *FA2*: *roads and grassy areas
- *FA3*: *hangars, roads, tarmac, and parking apron
- *FA4*: *runways, taxiways, grassy areas, tarmac.

Interpretations marked with * are seed fragments for the creation of each functional area. For example, if there are no hangar fragment interpretations then no *FA3* functional areas will be generated. Fig. 5 shows a sample functional area working memory element. Included in the functional area is its confidence, its coverage in the image, and a list of fragments which support the interpretation.

D. Models

The airport-model working memory element represents a mutually consistent collection of functional areas. At

```
Class:           Sub-class:
linear:          runway, taxiway, access road
compact:         terminal building, hangar
small-blob:      parking lots, parking aprons
large-blob:      tarmac, grassy areas
```

Fig. 2. Class and subclass interpretations.

```
(region
 ^token g00005                          ;Machine generated unique id.
 ^seg-number 6471                       ;Machineseg region number.
 ^statefile ALL                         ;Machineseg state file.
 ^symbolic-name ALL-L.6471_0            ;Database access id.
 ^spectral-characteristics nil          ;Currently unknown.
 ^texture nil                           ;Currently unknown.
 ^shadow-type nil                       ;Currently unknown.
 ^shadow-class nil                      ;Currently unknown.
 ^location-lat 139844.728507            ;latitude of feature center.
 ^location-lon 277380.289332            ;longitude of feature center.
 ^orientation 1.132534                  ;Orientation of fourier ellipse.
 ^ellipse-width 10.046023               ;Width, fourier ellipse (meters).
 ^ellipse-length 344.298377             ;length, fourier ellipse (meters).
 ^mbr-length 370.092492                 ;Length, MBR (meters).
 ^height nil                            ;Height, feature (meters).
 ^eclipsed t                            ;Partially covered by another region.
 ^curvature curved                      ;Curvature wrt curvature criterion.
 ^ellipse-linearity 34.272108           ;Linearity using fourier ellipse
 ^mbr-linearity 44.59714                ;Linearity using MBR
 ^compactness 0.00413                   ;Compactness (ratio).
 ^fractional-fill 0.047454              ;Fractional-fill (ratio).
 ^area 3441.397917                      ;Area (square meters).
 ^perimeter 912.836763                  ;Perimeter (meters).
 ^linear t                              ;fits criterion for linear regions?
 ^compact nil                           ;     compact regions?
 ^large-blob nil                        ;     large-blob regions?
 ^small-blob nil                        ;     small-blob regions?
 ^road t                                ;     roads?
 ^runway nil                            ;     runways?
 ^taxiway nil                           ;     taxiways?
 ^terminal nil                          ;     terminal-buildings?
 ^hangar nil                            ;     hangars?
 ^parking-lot nil                       ;     parking-lots?
 ^parking-apron nil                     ;     parking-aprons?
 ^grassy-area nil                       ;     grassy-areas?
 ^tarmac nil                            ;     tarmac?
 ^special-region nil                    ;Special region (shadow or given)
 ^region-origin bottom-up               ;Generation: bottom-up, top-down.
 ^region-status interpreted)            ;One of: active, to-unknown,
                                        ;uninterpretable, to-be-interpreted,
 )                                      ;deleted, or interpreted.
```

Fig. 3. Region working memory element.

least one member of each functional area type must be present in order to produce a complete model. Additional functional areas that are consistent with the four primary functional areas are also represented. A confidence is assigned to each model based on the confidence of each functional area, and how well the collection of functional areas "explain" or cover the airport scene.

The conflict working memory element is used to keep track of conflict recognized in the model generation process. This structure represents conflicts of interpretation between two functional areas that may share common fragment interpretations. It is generated as SPAM attempts to group functional areas together as constituents of a model. The presence of conflict working memory elements invoke rules to resolve conflicts and are specific to particular types of conflicts (see Fig. 6). This is discussed in more detail in Section VI-G.

IV. Some Examples

Fig. 7 shows one of the high resolution (1:12000) images of the National Airport in Washington, DC, stored in the MAPS database. This is one of several test images that we are currently using to develop the SPAM rule-based component. The image is 2280 rows by 2280 columns digitized to 8 bits of intensity per pixel. Approximately 12 views of the airport at image scales from 1:12000 to 1:60000 taken over a six-year period are available in the database.

```
(vector-attribute frag-list)
(literalize fragment
 fragment-token     ;Unique ID for this fragment.
 name               ;A human readable name for the fragment.
 object-type        ;LINEAR, SMALL BLOB, LARGE-BLOB, or COMPACT.
 hypothesis         ;Sub-class specialization for object-type.
 confidence         ;goodness of this interpretation 0.0< x <1.0
 region-token       ;ID of the region this fragment represents.
 symbolic-name      ;The region symbolic name.
 extension-flag     ;Current status during extension trials.
 frag-status        ;The 'age' of the fragment.
 evaluation-flag    ;The current state of evaluation.
 eval-history       ;Number of times fragment has been evaluated.
 frag-origin        ;Origin, machineseg, handseg.
 consistency-flag   ;Consistency strategy: checked or unchecked.
 aligned            ;Has fragment been used in an alignment.
 assoc-shadow       ;sym-name of associate shadow region for frag
 fa-seed            ;Is this a functional area seed.
 in-fa              ;Whether this fragment is in a function area.
 eclipse            ;Is this fragment eclipsed by another region.
 generic-flag       ;A general-purpose flag.
 frag-cnt           ;A count of the # of consistent fragments.
 frag-list          ;Fragment ids consistent with this fragment.
)
```

Fig. 4. Fragment working memory element.

```
(vector-attribute elements)
(literalize functional-area
 fa-id          ;unique id for this element.
 fa-type        ;functional area type.
 fa-gen         ;generation number.
 confidence     ;the goodness of this area.
 coverage       ;the percentage that this FA is explained.
 count          ;the number of fragments represented.
 elements       ;the list of fragments ids contained in FA.
)
```

Fig. 5. Functional area working memory element.

```
(vector-attribute other-fas)
(literalize airport-model
    mo-id         ;the model id.
    confidence    ;the goodness of this model.
    count         ;the number of functional areas in model.
    terminal-fa   ;terminal functional area.
    runway-fa     ;runway functional area.
    hangar-fa     ;hangar functional area.
    road-fa       ;road functional area.
    other-fas     ;other functional areas included.
)

(literalize conflict
    cid             ;so we can identify this conflict.
    status          ;whether or not this conflict has been resolved.
    in-favor-of     ;which interpretation was the decision in favor of.
    symbolic-name   ;the region id.
    fid1            ;the id of the 1st interpretation.
    type1           ;and its interpretation type.
    fid2            ;the id of the 2nd interpretation.
    type2           ;and its interpretation type.
    confidence-difference ;the difference in confidence between the
                          ;alternative interpretations.
    fa1             ;the id of the FA where the 1st fragment occurs.
    fa2             ;the id of the FA where the 2nd fragment occurs.
    model           ;the id of the model in which this conflict occurs.
)
```

Fig. 6. Model and conflict working memory elements.

Fig. 8 shows the results of applying our region-growing segmentation system to the image in Fig. 7. The segmentation is run over multiple windows of the image, each window is approximately 512 × 512 pixels. Artifacts of these windows can be seen as the straight boundaries in Fig. 8. The segmenter is supplied with image scale-dependent criteria from the MAPS database and performs region growing while searching for linear, compact, and blob regions [18].

It is evident that the image segmentation is quite errorful. Few features are completely segmented, many are merged together or broken into arbitrary fragments, and some are missing entirely. For example, in the runway beginning in the lower left and running to the upper right,

Fig. 7. SPAM: National Airport, Washington, DC.

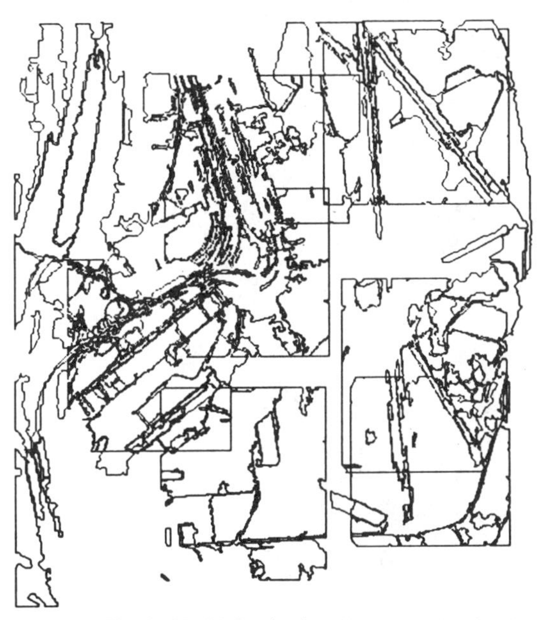

Fig. 8. SPAM: Region-based segmentation for Fig. 7.

many of the small taxiways are merged with the adjacent tarmac. In spite of this, the results are not atypical of the current state of the art in region-based segmentations using monocular views and monochrome photography with complex scenes.

The segmentation in Fig. 9 illustrates the problem of broken and missing regions for the access roads directly behind the hangars in the center-right of Fig. 7. Fig. 10 shows the area of the image that corresponds to Fig. 9. Note that even the nearly homogeneous bright parking apron is broken due to aircraft and shadows cast by aircraft on the apron. Interpretation systems simply must be able to accommodate errorful segmentations in order to perform scene analysis tasks.

Fig. 11 shows a set of 280 machine segmentation regions selected from the original 477 regions shown in Fig. 8. The 280 regions were selected to cover a representative number of airport features in the scene.

Fig. 12 is a human segmentation of the airport scene containing 95 regions. All of the road features have been broken into several segments to crudely simulate the problems noted in Fig. 9. SPAM has been tested using machine-generated and human segmentations using the identical rule base in both cases. We have also run SPAM on collections of machine-generated segmentations ranging from 20 to 280 regions per collection for the purpose of timings and rule base validation. The segmentations in Figs. 11 and 12 are the most complex tested to date.

Results are presented for two versions of the SPAM system, version 2 (V2), and version 3 (V3). The versions differ in that V3 has texture and height information supplied by associating a vector of probabilities with each of the 95 hand-segmented regions. These probabilities give estimates for *highly-textured*, *moderately-textured*, and *lightly-textured* as well as *height* > 15 *meters*, *height* > 5 *meters*, and *height* < 5 *meters*. Machine-segmented regions acquire their texture and height probabilities through calculation of their proportional overlap with the hand-

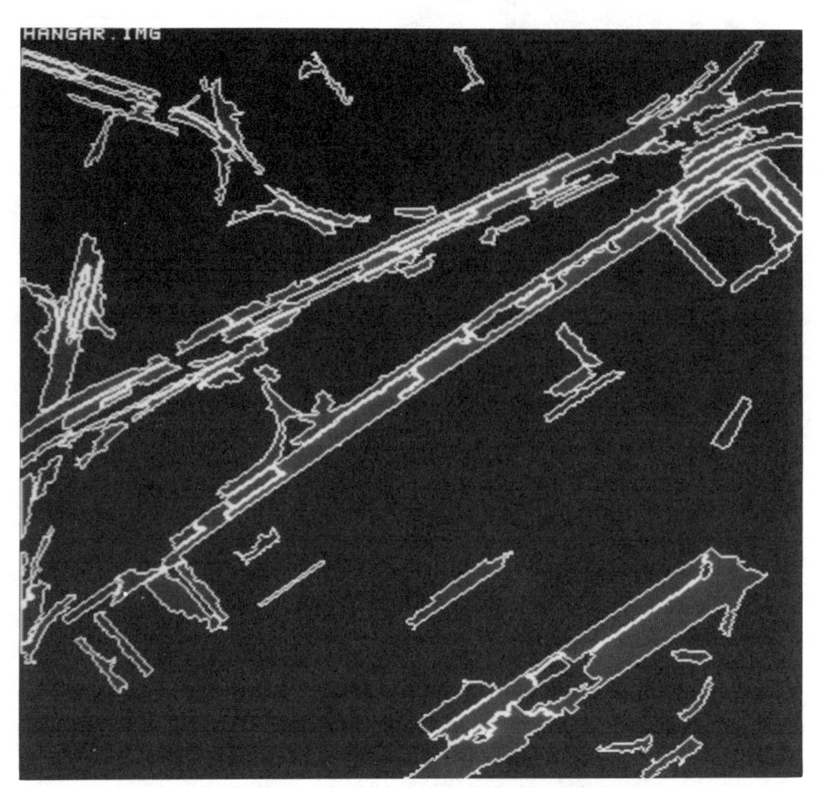

Fig. 9. SPAM: Detail of segmentation near hangar area.

Fig. 10. SPAM: Full resolution image for segmentation in Fig. 9.

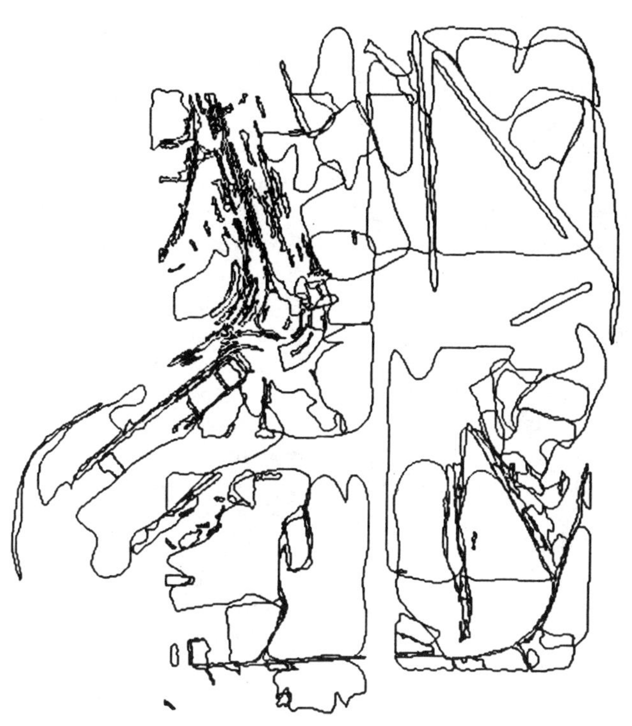

Fig. 11. SPAM: 280 region test machine segmentation.

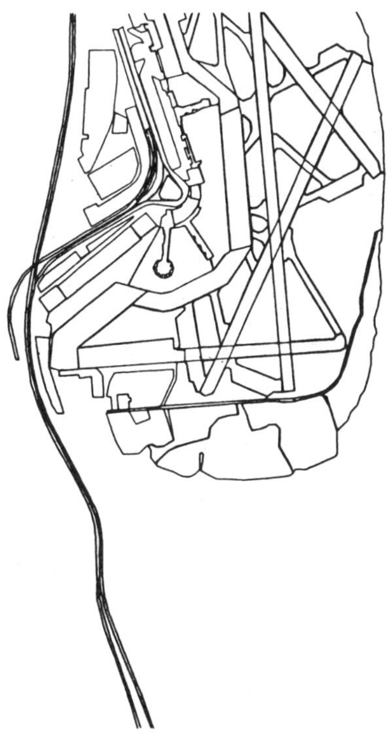

Fig. 12. SPAM: 95 region human segmentation.

segmented regions. V3 contains no additional rules to utilize the simulated texture and height information, the existing region-to-interpretation rules simply make use of the information when it is available.

Figs. 13–15 show the results of SPAM V2 on the machine segmentation data in Fig. 11. Fig. 13 shows the five highest rated fragments for each of the nine subclass interpretations. Fig. 14 shows the fragments that comprise the model in Fig. 16. Fig. 15 shows the fragments that were compatible with the model in Fig. 16, but were not included as part of the model. Fig. 14 is the collection of functional areas that comprise the "best" model generated by SPAM in terms of model confidence.

Our method for generating functional areas relies on the identification of seed fragments. These seed fragments are ordered by relative confidence and evaluated in turn. Therefore, it is important that 1) interpretations of terminal, access roads, hangars, and runway be generated and that 2) the "correct" regions be highly rated. The emergence of correctly identified runways and hangars in each of Figs. 13, 17, 21, and 25 is an important factor in the generation of runway and hangar FA's for the models in Figs. 16, 20, 24, and 28.

Figs. 17–20 show the same results for SPAM V2 run on the hand-segmentation data in Fig. 12. Figs. 21–24 show the results for SPAM V3 run on the machine-segmentation data. Figs. 25–28 show the same results for SPAM V3 run on the hand-segmentation data.

V. Interpretation Phases

SPAM is loosely organized into five processing phases. These phases are *build*, *local evaluation*, *consistency*, *functional area*, and *model evaluation*. The processing strategy is to move sequentially through the five phases. There is no prescribed relationship with the rule classes described in Section VI and the phases described in this section. For example, rules to determine the consistency of a runway with a taxiway might be activated during the functional area phase as well as during the consistency phase. This is achieved by generating the appropriate evaluation context for the fragment hypotheses.

Fig. 29 outlines the paradigm used in SPAM at each interpretation phase. At each phase, knowledge is used to check for consistency among hypotheses, to predict missing components using context, and to create contexts based on collections of consistent hypotheses. Prediction is restrained in SPAM in that a hypotheses cannot predict missing components at their own representation level. A collection of hypotheses must combine to create a context from which a prediction can be made. These contexts are refinements or spatial aggregations in the scene. For example, a collection of mutually consistent runways and taxiways might combine to generate a runway functional area. Rules that encode knowledge that runway functional areas often contain grassy areas or tarmac may predict that certain subareas within that functional area are good candidates for finding such regions. However, an isolated runway or taxiway hypothesis cannot directly make these predictions. In SPAM the context determines the prediction. This serves to decrease the combinatorics of hypothesis generation and to allow the system to focus on those areas with strong support at each level of the interpretation. In the following sections, we describe each of the interpretation phases.

A. Build

During the build phase, region data, instantiated by an initialization rule, invoke the region-to-interpretation rules. A region is "chosen" by the interpreter and tested against each class (one of small blob, large blob, linear, or compact). If that region fits a particular class, a generic fragment interpretation for that class is created.

For each of the generic class interpretations, the region is tested against more specific criteria for each subclass. For example, a region may fit the criterion for the linear class, so it is tested against the criteria for access road, taxiway, and runway. Some of the criteria are listed in Fig. 33. A region may fit the criterion for a class, but not fit the criteria for any of the subclasses. Similarly, a region may also fit several classes (and therefore several subclasses), so several class (and subclass) fragment interpretations are created. The existence of generic class interpretations, in addition to more specific subclass interpretations for the same region, allows SPAM to reevaluate regions by changing the criteria for subclass due to changed expectations.

An *initial confidence value* is calculated based on how well a region fits the criterion for a class or subclass. A confidence is assigned to each interpretation. For example, if a region is very long, wide, and straight, then it is a good candidate for a runway and, because of its length, not such a good candidate for a taxiway. These facts are reflected in the confidence values for each of the two fragment interpretations for that region. This phase terminates when all instantiated regions have been tested for interpretation.

B. Local Evaluation

The local evaluation phase allows some processing to take place on the fragment interpretations generated by the build phase. For example, straight and curved alignment of linear segments is performed and the resulting regions are added to working memory. They are recognized as new regions and are evaluated by region-to-interpretation rules. This process continues until no new alignments occur.

Currently, regions are assumed to be geometrically distinct; that is, no two regions occupy the same two-dimensional space. This assumption may be violated when alignments or further image processing is performed. Regions resulting from these operations can occupy nearly the same portion of the image. SPAM determines possible candidates for overlap using knowledge stored in the region descriptors. Rules test those candidates to determine whether an overlap is present, and what percentage of each feature has been "eclipsed." For simple cases, where one region subsumes another and both have the same class interpretation, we inactivate the subsumed fragment's subclass interpretation. Inactivation (instead of deletion) allows the system to go back to the original hypotheses in the event that an aligned region is not justifiable in the image. Thus, if two runway hypotheses are generated through multiple pairwise alignments, the "smaller" hypothesis will be inactivated. Its linear class interpretation will remain, but is already marked to preclude further alignments. Other strategies that prune interpretations under these conditions need to be explored, especially within the context of choosing the "best" representative for a spatial area using more sophisticated analysis.

C. Consistency

The consistency phase is invoked once the build/evaluation phase terminates. Only fragments with subclass interpretations participate in this phase. A fragment interpretation is chosen, and a consistency strategy is created. Consistency strategies comprise of a list of applicable tests specific to the subclass. As each test is applied, an incremental confidence score is computed based on the consistency test.

Each test is a collection of rules that execute the test, as well as routines outside of the rule-based system that actually perform spatial evaluation. At the end of the consistency phase, each fragment contains a list of the fragments with which it is consistent. Its confidence reflects how well it satisfied the tests that were applied to it. Fig. 30 shows a road fragment interpretation and the interpretations with which it was found consistent. The subclass name, region identification, fragment identification, and fragment confidence are displayed.

D. Functional Area

The functional area (FA) phase evaluates fragments that have specific subclass interpretations. These *seed fragments* are terminals, access roads, runways, and hangars. For each seed fragment, the list of fragments with which it is consistent is traced to determine whether there is sufficient evidence for a functional area to be formed.

The current criterion for sufficient evidence is if the consistency trace finds one or more fragments of compat-

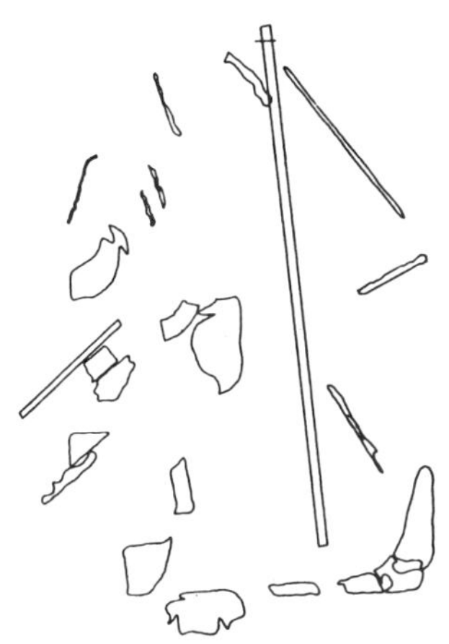

Fig. 13. V2MACH: Best 5 fragment interpretations in each subclass.

Fig. 14. V2MACH: Fragments comprising the model in Fig. 16.

Fig. 15. V2MACH: Fragments compatible with the model in Fig. 16.

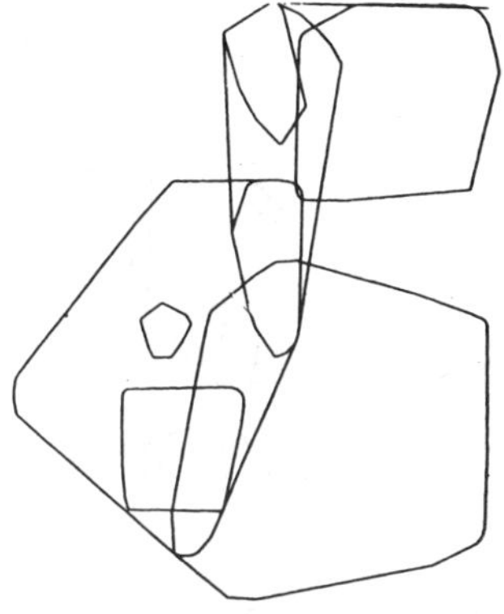

Fig. 16. V2MACH: Functional areas comprising the best model.

Fig. 21. V3MACH: Best 5 fragment interpretations in each subclass.

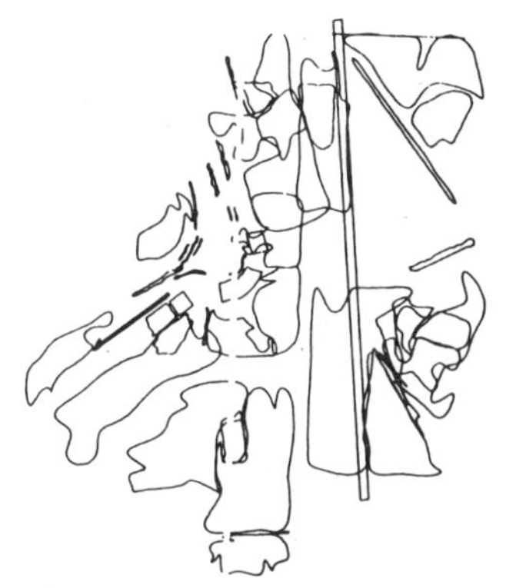

Fig. 22. V3MACH: Fragments comprising the model in Fig. 24.

Fig. 23. V3MACH: Fragments compatible with the model in Fig. 24.

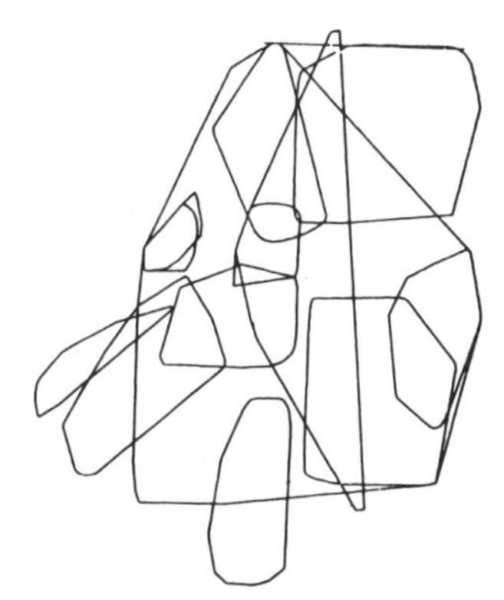

Fig. 24. V3MACH: Functional areas comprising the best model.

ible type. If satisfied, a region is created whose boundary is the convex hull of all the regions represented by the fragment interpretations. An initial confidence value is assigned to the FA based on the average of the confidences of the member fragments. The computation of the FA boundary allows us to compute two factors. The first is the area "explained" by the fragments that compose the FA; the second is the area which is "uninterpreted," or not covered by fragments in the functional area. Further, a list of all fragments lying on or within the boundary of the FA that have a *compatible* interpretation with the FA type is calculated. For example, for a runway functional area, tarmac and hangars lying inside the boundary would be in the compatible list. These fragments will not have been found to be consistent with the functional area, but this does not preclude a weaker compatibility. A list of *incompatibles* consisting of those fragments lying on or within the boundary of the FA that have an incompatible interpretation with the FA type is also calculated. A final confidence value for the functional area is based on the relative strength of these compatible and incompatible fragments. Fig. 31 is an example of the best runway functional area calculated for the hand segmentation data in version V3.

E. Model Evaluation

The model evaluation phase begins when there are sufficient functional areas generated by the previous phase. At a minimum, one of each of the four functional area types must be present in order to begin model generation. Much of model evaluation is organized to resolve inconsistencies between overlapping functional areas. In general, when an overlap is recognized, one of three following situations may occur:

- *case 1:* the area of overlap does not contain any fragment interpretations;
- *case 2:* the fragment interpretations within the area of overlap are the same, or compatible; and
- *case 3:* the fragment interpretations within the area of overlap are incompatible.

In the third case, reasoning about which fragment interpretations are in error is possible by recognizing chains of inconsistency or preferred interpretations. For example, if a hangar interpretation falls inside of a highly rated runway functional area, and the region to which the hangar is attached has an alternative taxiway interpretation as well, it may be determined that the hangar interpretation should be removed. One side effect of the deletion is to improve the confidence of the runway functional area. Another method for resolving the conflict is to simply select the fragment with higher confidence.

If a functional area is lacking support in a certain subarea, SPAM could use this information to verify the functional area by specifically looking for missing fragment interpretations. For example, if a terminal functional

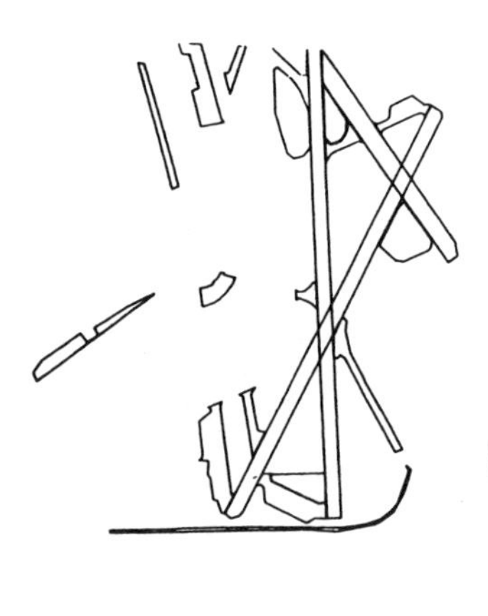
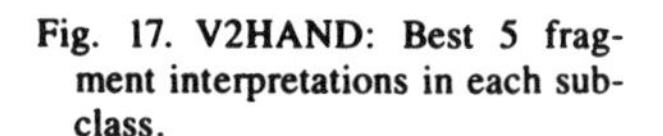

Fig. 17. V2HAND: Best 5 fragment interpretations in each subclass.

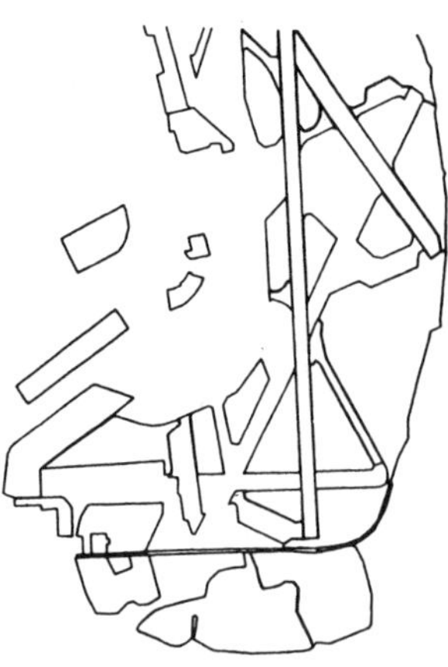

Fig. 18. V2HAND: Fragments comprising the model in Fig. 20.

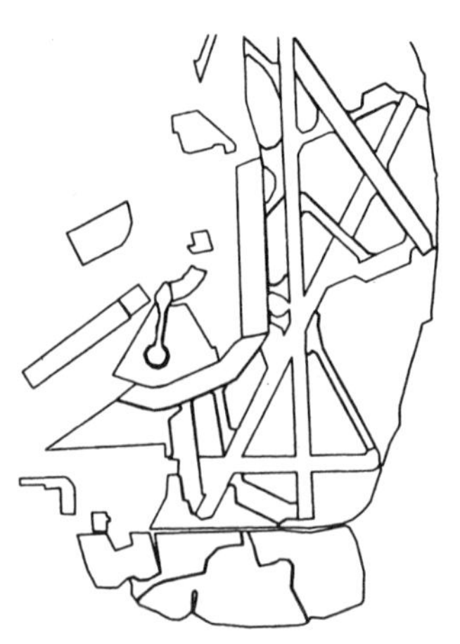

Fig. 19. V2HAND: Fragments compatible with the model in Fig. 20.

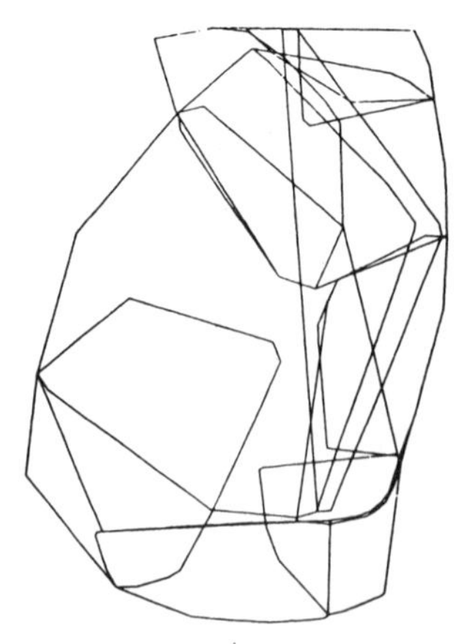

Fig. 20. V2HAND: Functional areas comprising the best model.

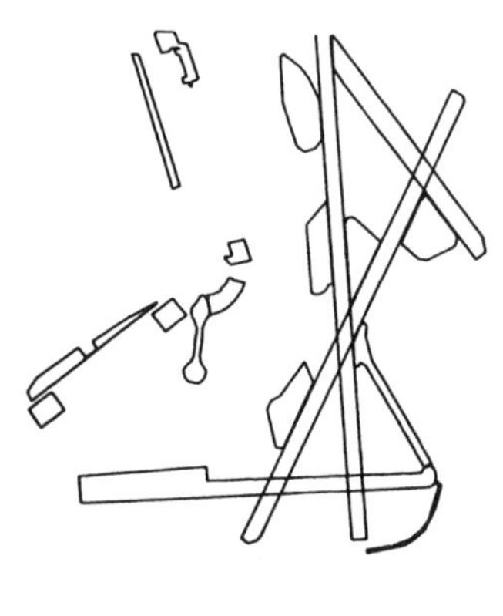

Fig. 25. V3HAND: Best 5 fragment interpretations in each subclass.

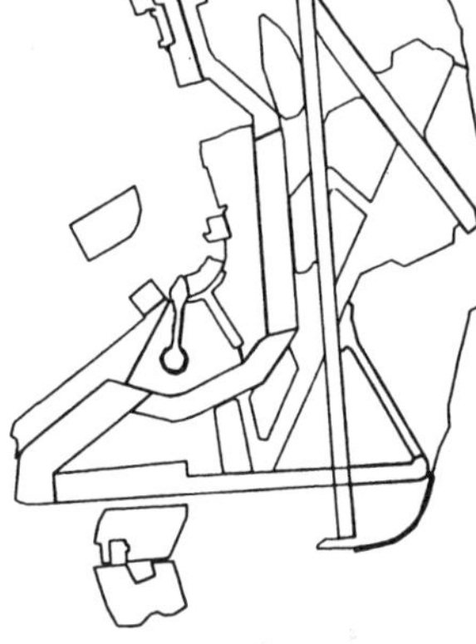

Fig. 26. V3HAND: Fragments comprising the model in Fig. 28.

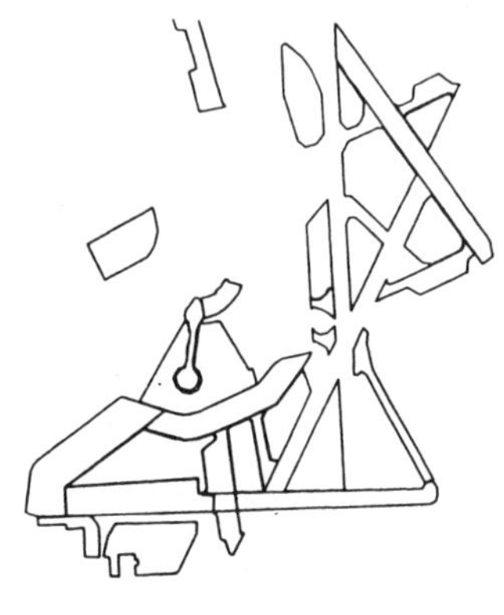

Fig. 27. V3HAND: Fragments compatible with the model in Fig. 28.

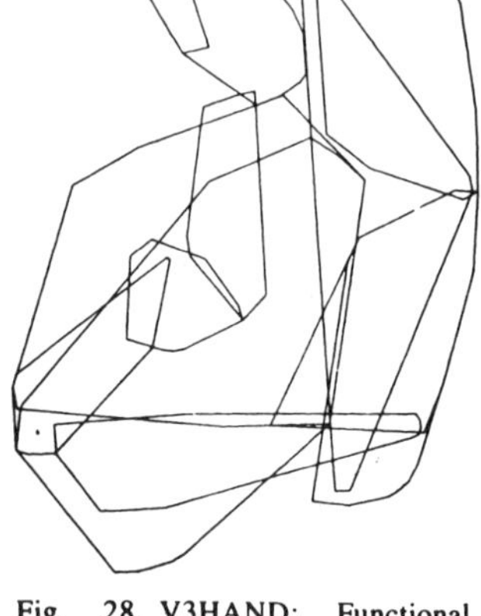

Fig. 28. V3HAND: Functional areas comprising the best model.

area contains roads and parking lots, but no parking apron, rules invoke image analysis tools to go look for regions whose shape and texture properties match with SPAM's *build* and *local evaluation* model for parking aprons. This form of prediction is consistent with our strategy outlined in Fig. 29.

When a large number of functional areas is generated, due to multiple consistent interpretations of fragments that have not been rejected by consistency tests, SPAM will generate multiple alternative models. We can make two observations about this. First, the difference between models appears to be small, the inclusion or exclusion of one or two functional areas can generate two or four different models. Second, the models all share a core set of functional areas indicating that the general organization of the airport appears to be relatively stable. In the following section, we will describe classes of rules that are used by SPAM during the interpretation phases discussed in this section.

VI. Rule Classes

Rules, as currently implemented in SPAM, are loosely organized into six classes. Within each class, sets of rules operate on various specific airport hypotheses. The classes generally correspond to the interpretation phases discussed in Section V, although the effects of interpretation in one phase can cause rules in another phase to become active. There are currently about 450 OPS5 rules in the system. Fig. 32 is a schematic that contrasts the interpretation phases in SPAM with the approximate number of rules in each class. An estimate of the potential for growth in the number of rules is given and discussed in Section IX.

A. Initialization Rules

The initialization rules (2 rules) initialize the interpretation goal states, *a priori* knowledge from map database, airport class expectations, and load working memory with the low-level image region segmentation. *A priori* knowledge such as the number of runways or the size or type of airport can be instantiated at the beginning of a run. They supply SPAM with a method to handle expectations input by the user. Alternatively, this allows one to experiment with SPAM by setting external goals and allowing the system to attempt to satisfy them. Fig. 33 shows a subset of the interpretation constants and gives a sample specification that can be used to bias interpretations for a large urban area around the airport.

B. Region-to-Interpretation Rules

The region-to-interpretation rules (35 rules) create initial fragment hypotheses for each region based on its intrinsic properties of size, shape, texture, and elevation

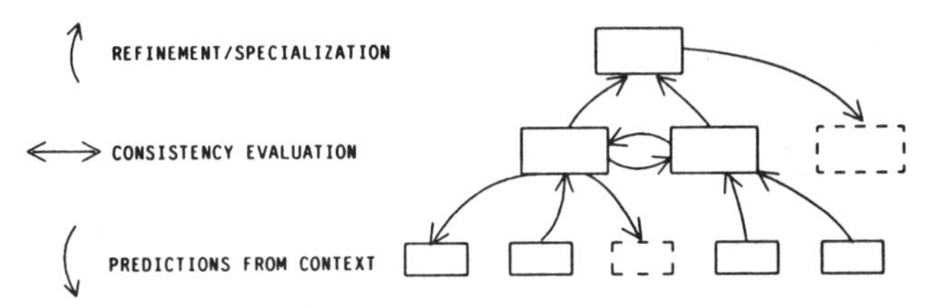

Fig. 29. Refinement, consistency, and prediction in SPAM.

```
ALL-L.6471_14
  road  g00354  0.87101
          perimeter    GIVEN-N.129_0      N129    1.0
          terminal     Hand36809-N.75_0  g00134  0.92802
          terminal     Hand36809-N.73_0  g00136  0.89112
          hangar       TERM2-C.7205_0    g00245  0.91008
          parking-lot  TERM2-B.1134_0    g00262  0.79519
          hangar       HANGAR-C.3832_0   g00295  0.71099
          hangar       HANGAR-C.2983_0   g00297  0.84008
```

Fig. 30. Consistency relationships.

```
FUNCAREA-N.7_0
  runway 1 0.7395 0.64524 4
          runway       Hand36809-N.3_0   g00260 0.82994
          grassy-area  Hand36809-N.26_0  g00158 0.63281
          grassy-area  Hand36809-N.5_0   g00162 0.65813
          taxiway      Hand36809-N.16_0  g00241 0.83732
```

Fig. 31. Functional area description.

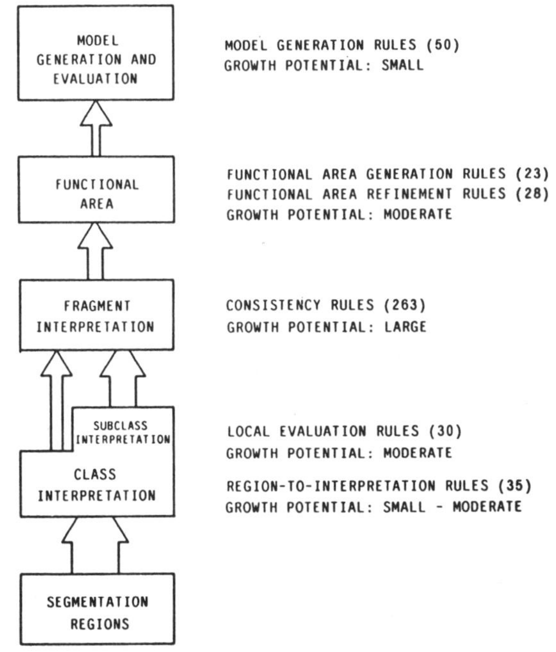

Fig. 32. Interpretation phases in SPAM.

```
(literalize constants    ;Stores constants used in productions.
 GENERAL-area-filter     ;Used to filter out small regions.
 COMPACT-min-compact-filter  ;Compactness filter.
 COMPACT-max-area-filter
 COMPACT-min-area-filter
 COMPACT-perimeter-filter  ;Perimeter filter, compact regions.
 LINEAR-ell-lin-filter     ;Ellipse-linearity filter.
 LINEAR-mbr-lin-filter     ;MBR-linearity filter, linear regions.
 confidence-filter         ;min confidence, eliminate all sub-class
                           ;interpretations lower than this.
 runway-minlength-filter   ;min length, linear region as runway.
 runway-maxlength-filter   ;max length, linear region as runway.
 runway-minwidth-filter    ;min width, linear region as runway.
 runway-maxwidth-filter    ;max width, linear region as runway.
 adjacency-criterion       ;max distance, two features are adjacent.
 curvature-criterion       ;threshold, linear regions are curved.
 curvature-minimum         ;length threshold, to compute curvature.
 texture-criterion         ;threshold, region considered untextured.
 orientation-criterion     ;distance threshold, extension along
 )                         ;orientation direction.

(AP-specs
 ^spec-type                ;Type of specification.
   urban-area-around-airport
 ^operation nil            ;Hypothesis only, no oper performed.
 ^prediction-value nil     ;Expected value for specification.
 ^probability 90           ;Probability expectation is correct.
 )

(AP-specs
 ^spec-type number-of-runways
 ^operation equal          ;Exact number of runways to be found.
 ^prediction-value 5       ;Expected value for runways.
 ^probability 100          ;Probability expectation is correct.
 )
```

Fig. 33. Evaluation criteria and expectations.

```
region-to-fragment::initialize-region
  Interpretation of each region begins by creating a context where matching
  against specific shape attributes for each class and subclass occurs.
region-to-fragment::generate-linear-class-match
  Test the shape attributes of this region against linear class prototype.
region-to-fragment::make-linear-fragments
  If region matched the linear characteristics successfully make an instance
specific::get-region-curvature::uncalculated-1
  If a linear fragment, we need the curvature to do any further matching.
region-to-fragment::generate-compact-class-match
  Match this region against attributes of the "ideal" compact fragment.
region-to-fragment::failing-compact-match
  The match against the compact attributes failed. No further interpretation
  in the compact class is performed.
region-to-fragment::generate-small-blob-class-match
  Match this region against attributes of a small-blob region.
region-to-fragment::failing-small-blob-match
  The small-blob match has failed.
region-to-fragment::generate-large-blob-class-match
  Match this region against attributes of a large-blob region.
region-to-fragment::failing-large-blob-match
  The large-blob match has also failed.
region-to-fragment::generate-road-subclass-match
  We match this region against all the subclasses of those classes that
  matched successfully.
region-to-fragment::failing-road-match
  This region should not be interpreted as a road.
region-to-fragment::generate-taxiway-subclass-match
  Generate the match against the second linear subclass, taxiway.
region-to-fragment::interpret-as-taxiway
  The match against the "ideal" taxiway succeeds. Create a fragment
  hypothesis with this region as a taxiway.
region-to-fragment::generate-runway-subclass-match
  Generate the match against the last linear subclass, runway.
region-to-fragment::interpret-as-runway
  The match against the "ideal" runway succeeds. Create a separate fragment
  hypothesis with this region as a runway.
region-to-fragment::exit-context
  No more matching to be done on this region. The system chooses another
  region to interpret.
```

Fig. 34. Region-to-interpretation rules.

(depth). Linear regions are evaluated with respect to curvature. These rules hypothesize a class interpretation for the image region as described in Section III-A. Multiple fragment interpretations can be created for each image region. Once a class interpretation has been created, a specialization of the class interpretation is attempted. New fragment interpretations are created for those fragments that satisfy the subclass specialization. Fig. 34 gives a sequence of region to interpretation rules operating on a linear region.

C. Local Evaluation Rules

Local fragment evaluation rules (30 rules) are used to invoke image processing tools for region enlargement, extension, and join/merge. These rules are specialized within feature classes. For example, given several linear image regions, hypothesize a new linear region that encompasses each original region. These rules attempt to verify this hypothesis by invoking a linear feature extraction module using the new (hypothesized) linear region as a guide. In lieu of image analysis, other linear alignment rules are used to search for possible linear fragment interpretations that have not been specialized into runway or road interpretations, and to hypothesize (hallucinate) a connected runway fragment. Thus, fragment interpretations can be generated in cases where complete low-level support is lacking. Such fragment interpretations are marked as to whether they have support in the image, as well as maintaining a list of supporting fragments. Alignments are specialized for nearly straight and curved regions. The actual evaluations are performed outside of the rule interpreter using a procedure that attempts to align linear fragments and bridge gaps in the segmentation, using splines for local interpolation [22].

From a practical standpoint, external calls by the production system need only determine the operation to be performed and the region elements participating in that operation. A process outside of the interpreter can determine the actual image data using the symbolic name and extract any geometric information it might need. Since all of the spatial computation and image analysis is organized outside of the OPS5 rule interpreter, it is important to keep this interface simple.

D. Consistency Rules

Consistency rules (263 rules) recognize groups of potentially mutually consistent fragment interpretations. For each subclass specialization, a consistency strategy is defined. This strategy contains a list of rules that apply spatial and contextual constraints to modify the confidence of the fragment hypothesis. These rules are relatively simple tests which look for supporting evidence for the fragment hypothesis. Each strategy is implemented as a rule set that may invoke geometric or image-processing routines. The following is a portion of the consistency strategy for fragments having the runway interpretation:

- runways-are-oriented-parallel-to-terminal-building
- check-runway-distance-from-terminal
- taxiways-are-oriented-orthogonal-to-runways
- grassy-areas-border-runways-or-taxiways
- taxiways-are-close-to-runways
- parking-lots-are-far-from-runways
- runways-do-not-have-curved-segments
- taxiways-intersect-runways.

Consistency rules are the major source of knowledge in SPAM. They encode simple observations that are generally true about airport scenes. For example, hangars are usually adjacent to other hangars, access roads lead to parking lots and the terminal building, runways are (usually) not oriented into the terminal building, and are further away than some tarmac and most parking aprons. In SPAM, the effect of a large number of relatively simple rules is that correct fragment interpretations are supported much more frequently by correctly interpreted partners than the random support acquired by incorrect interpretations due to fortuitous spatial alignments. However, since these rules only test local consistency, it is necessary to select the "best" fragment interpretations and aggregate them together to form more global pieces of the airport model. This is the role of the functional area rules discussed in the following section. Other consistency rules include the following sample:

- access-roads-generally-have-curved-segments
- access-roads-are-close-to-perimeter
- access-roads-lead-to-terminal-building
- taxiways-orient-toward-terminal
- terminal-building-is-centrally-located
- terminal-building-is-near-parking-lots
- hangars-adjacent-to-tarmac
- grassy-areas-separate-runways-and-taxiways
- tarmac-between-taxiways-terminal.

E. Functional Area Rules

Functional area rules (23 rules) recognize situations where fragment interpretations can be grouped together based on propagation of compatibility into more global collections. These collections of fragment interpretations are called functional areas. Rules identify those fragments that are consistent with at least one other fragment and are potentially members of the same functional area. The convex hull of these fragments forms the initial functional area. Fragments that fall within the functional area and are compatible with the functional area type, but not necessarily consistent with the initial set of fragments, can be added to the functional area. These fragments are marked as compatible. Similarly, fragments that fall within the convex hull of the functional area that are inconsistent, i.e., hangar within runway functional area, are noted. Rules evaluate the confidence of the functional area in light of the compatibles and incompatibles. If there is not sufficient support for the new functional area, or if an existing functional area of the same type covers nearly the same spatial area, it is not activated. Otherwise, fragments are marked as being a member of a functional area, or having failed to generate one. A typical sequence of functional area rules is given in Fig. 35.

F. Goal-Generation Rules

We have incorporated goal-generation rules (28 rules) which recognize small numbers of situations that are inconsistent with general knowledge of airport layout. These rules recognize situations that may cause large numbers of weakly consistent fragment interpretations to build functional area interpretations. The combinatorics dictate that pruning these weakly consistent fragments can greatly reduce the number of resulting functional area interpretations. These rules monitor the global state of working memory and use goal specification as the focus of attention. For example, since few airports have more than 10 runways, if there are more than 10 valid runway fragments in working memory, we invoke routines to coalesce or possibly eliminate weak fragment interpretations. Methods rely on the invocation of existing rules from the local evaluation and consistency classes with specific fragments as contexts, and with more stringent spatial constraints. For example,

- search for and attempt to join aligned fragments with compatible subclass interpretations;
- remove weak runway fragments that have significantly stronger support as taxiways or roads; and
- reevaluate runway fragments with respect to their position in the scene using more stringent proximity constraints.

Similar situations arise with large numbers of competing terminal and hangar hypotheses. Another type of goal-directed pruning occurs upon recognition that an abnormal percentage of the overall scene is explained by fragment interpretations of a particular subclass. For example, we do not expect that the entire airport is covered with grassy areas, tarmac, or parking lots. Rules to recognize

```
FA::fragment-initialization    Choosing fragment hypothesis
FA::trace-consistents          All the hypotheses that are consistent.
FA::spawn-FA-creation          If at least one valid start creation.
FA::create-functional-area     Instantiate into working-memory
FA::compute-convex-hull        Physical representation of FA
FA::reduce-fa-size             Spatial outlying hypotheses are eliminated.
FA::compute-convex-hull        With the outlyers removed.
FA::find-compatibles           Inside and compatible with FA definition.
FA::find-incompatibles         Inside and incompatible with FA definition
FA::re-evaluate-confidence     Re-evaluate the confidence of the FA
FA::get-fa-superset            Another FA of the same type?
FA::insufficient-overlap       Yes, but not enough overlay to deactivate.
FA::activate-functional-area   Add to working memory as active.
```

Fig. 35. Functional area rule execution.

these situations can further prune weak hypotheses making use of spatial constraints, and looking for overlaps with incompatible fragment interpretations having highly rated hypotheses.

G. Model-Generation Rules

Model-generation rules (50 rules) gather mutually consistent functional areas and build an *airport model*. Initially, the functional area with the greatest confidence is selected. These rules then recognize situations where they can enumerate conflicts between the current model and a newly chosen functional area. The methods for resolution of those conflicts, the decision to augment a model or spawn a new model, are also encoded in this rule set. Model generation rules continue to be active until all the active functional areas have been included in some model. Fig. 36 gives an example of a sequence of model-generation rule execution.

H. Rule Distribution By Fragment Class

In the previous sections, we have given the number of rules in each interpretation class. It is interesting to note the actual distribution of these rules in terms of the types of fragments that they recognize. Fig. 37 gives a breakdown of the frequency of rules by class and subclass.

Overall, linear fragments provide more constraints in the airport scene and can be more reliably determined than the other classes of interpretations. For example, runways can be distinguished fairly reliably based on shape. Roads are harder to distinguish based solely on shape since they are combinations of straight and curving linear regions, which must often be aligned before they can be hypothesized. Roads give many key constraints to the overall scene, such as determining plausible locations of the terminal and hangar areas. We believe that the relative importance of each class is reflected in the number of rules in that class as is our ability to recognize and employ appropriate tests.

Within some classes, it is more difficult to distinguish members of the subclass, therefore, more knowledge is needed to eliminate faulty hypotheses. This fact is clearly represented in the cases of roads and hangars. Since roads can be curved or straight, some small runways and most taxiways have alternative road interpretations that are faulty. Similarly, many small compact regions which are found toward the center of the runway/taxiway area are mistaken for hangars. However, as we will see in the following sections, these alternative interpretations rarely have global support. Consequently, their confidence is significantly weaker than the correct interpretations.

```
MG::initialize
  Create context where only model-generation rules are active.
MG::choose-best-FA
  Choose highest-rated active functional-area.
MG::end-conflict-test
  There were no other functional-areas being considered.
MG::choose-best-FA
  Choose the next highest-rated active functional-area.
MG::test-best-for-conflicts
  Check if there are conflicts between hypotheses.
MG::resolve-conflicts::resolve-by-confidence-1
  Choose the interpretation with the highest confidence.
MG::resolve-conflicts::remove-duplicates-1
  This conflict has been resolved before, therefore remove it.
MG::resolve-conflicts::parking-apron/parking-lot-delay
  For this conflict, there is specific knowledge for evaluating the two
  hypotheses in light of the current model.  Therefore, because we know
  that we need a complete model to resolve this conflict, delay the
  resolution of the conflict until that model is created.
MG::unset-delayed-conflicts
  Activate those conflicts that were delayed.
MG::resolve-conflicts::parking-aprons-and-parking-lots
MG::resolve-conflicts::choose-parking-lot-over-parking-apron
  Case-specific knowledge for this conflict is applied and, for this
  instance, the parking-lot hypothesis was determined to be more likely
  in the context of this model.  For this conflict, the knowledge applied
  was that a parking-apron will occur between the terminal functional-area
  and the runway functional-area, whereas a parking-lot will occur as
  far as possible from the runway functional-area.
MG::enumerate-conflicts
  A new, active functional-area of any type is chosen and is evaluated with
  respect to the functional-areas in the current model(s).
MG::resolve-conflicts::unknown-conflict-type-2
  Unknown conflicts are those interpretations that cannot safely be
  discriminated by confidence alone because the difference in the
  confidence values is very small (less than 0.1).  These conflicts are
  marked as irreconcilable.
MG::fork-new-model2
  Because we have fouhd at least one irreconcilable conflict, the current
  model is split into two models: one containing the first conflicting
  interpretation and one containing second.  The models are tested for
  completeness with respect to the definition of the airport-model.
MG::fork-new-model:failed-to-find-replacement-terminal
  The current model could not be split because the conflict occurred
  with the model's terminal functional-area, and there was no alternate
  terminal functional-area in the model to replace the first one.
MG::resolve-conflicts::grassy-areas-and-tarmac
MG::resolve-conflicts::choose-grassy-area-over-tarmac
MG::augment-airport-model
  This functional-area can be added to this particular model because all
  conflicts were reconcilable.
```

Fig. 36. Model rule execution.

Class / Subclass	Number of Rules	
Linear	80	
Runway		24
Taxiway		29
Road		38
Compact	64	
Hangar		58
Terminal		22
Small-blob	40	
Parking-apron		16
Parking-lot		32
Large-blob	32	
Grassy-area		24
Tarmac		16

Fig. 37. Rule distribution by interpretation of class.

VII. Evaluation Functions for Hypotheses

As briefly discussed in Sections V-A and C, SPAM assigns confidence values to fragment interpretations. This evaluation occurs in two stages. During the initialization and local evaluation phases, the fragment is assigned a confidence value based on how well the region it represents fits the criteria for class and subclass specialization. During the consistency phase, confidence values are assigned based on the results of compatibility with other fragment interpretations.

The initial confidence function is as follows:

$$\text{confidence} = (io) * (t/v)\ \hat{}\ 2 + mcv$$

where:

io initial offset:
$io + mcv =$ minimum confidence value (-0.3)

t the threshold (criterion) for this class
v the actual region value being tested ($v\ != 0$)
mcv maximum allowable confidence value (currently 0.5).

This produces a confidence value between 0.2 and mcv, with the function asymptotically approaching mcv.[2] There are some problems with this, however. Currently, this function allows us to rate an interpretation based only on one criterion, although some class tests require the regions to fit several criteria at once. A better confidence function would accommodate from 1 to n different threshold tests where n is not known *a priori*.

The incremental confidence function used during consistency checking is as follows:

$$\text{increment} = Cr * Cs * [Fr * (1.0 - Co)]$$

where

Cr is the confidence of the test itself $0 < Cr < 1$
Cs is the normalized score received from the test $0 < Cs < 1$
Fr is the fraction of the maximum allowable increment $0 < Fr < 1$; that is, given a perfect score ($Cs = 1$) and a perfect test ($Cr = 1$), this determines the fraction of the maximum increment
Co is the old confidence value ($0 < Co < 1$).

The increment function factors a relative goodness or selectivity of the test, the current goodness of the fragment being evaluated, and how well the fragment scored on the test. The use of a relative goodness factor allows us to weight tests according to our subjective belief as to how well it disambiguates competing hypotheses. Under this model, a test has weight 1.0 if it uniquely distinguishes one subclass from all of the other members of the class. In practice, test weights range from 0.5 to 0.8. This reflects a lack of particularly strong compatibility tests in a domain where the underlying certainty of the signal data is weak and the belief that real action is a result of the application of many simple tests. Another property of this evaluation function, which asymptotically approaches 1, is that as the score of the fragment increases, it becomes increasingly difficult to increase fragment goodness.

VIII. Performance Analysis

In this section, we give some performance statistics for the examples described in Section IV. Statistics for fragment, functional area, and model generation compare the relative effects of the texture/depth experiments, the effect of hand versus machine segmentations. An additional set of statistics for MACH200 using both the V2 and V3 system is given. MACH200 is a subset of MACH280 reported in Section IV in that it has 80 fewer regions. Most of these regions were small linear features, comprising roads around the terminal building in the left-central area

[2] It is assumed in this equation that $t >= v$, therefore, if a test requires that a value be above a threshold ($t <= v$), then one must replace the quotient (t/v) with (v/t).

Fragments Generated

Version	Linear	Compact	S-Blob	L-Blob	Total	# regions
MACH200v2	30	27	26	30	113	197
MACH280v2	40	27	26	30	123	277
HANDv2	15	30	50	44	139	95
MACH200v3	37	9	25	32	103	197
MACH280v3	59	9	25	32	125	277
HANDv3	15	7	48	43	113	95

Linear Class Interpretations

Version	Road	Runway	Taxiway	Uninterpreted
MACH200v2	6	14	19	11
MACH280v2	6	21	24	16
HANDv2	7	4	1	8
MACH200v3	10	10	29	8
MACH280v3	10	16	45	14
HANDv3	7	5	1	8

Compact Class Interpretations

Version	Hangar	Term	Uninterpreted
MACH200v2	20	25	2
MACH280v2	20	25	2
HANDv2	25	30	0
MACH200v3	2	5	4
MACH280v3	2	5	4
HANDv3	4	4	3

Small-Blob Class Interpretations

Version	PApron	Plot	Uninterpreted
MACH200v2	24	26	0
MACH280v2	24	26	0
HANDv2	40	50	0
MACH200v3	25	21	0
MACH280v3	25	21	0
HANDv3	47	33	1

Large-Blob Class Interpretations

Version	GrassA	Tarmac	Uninterpreted
MACH200v2	29	28	1
MACH280v2	29	28	1
HANDv2	35	41	3
MACH200v3	30	20	2
MACH280v3	30	20	2
HANDv3	39	19	4

Fig. 38. Fragment generation statistics.

Functional Areas And Models Generated

Version	Road	Runway	Hangar	Term	Total	Models
MACH200v2	4	5	24	19	52	9
MACH280v2	6	3	24	19	52	12
HANDv2	1	4	4	11	20	5
MACH200v3	7	8	5	2	22	1
MACH280v3	12	8	5	2	27	1
HANDv3	1	5	4	4	14	3

Fig. 39. Total functional area interpretations.

Rule Executions by Phase

Version	Build+LE	Consist	Functional	Model
MACH200v2	1816	36528	3941	1174
MACH280v2	2355	39637	3981	1619
HANDv2	1469	8079	1354	357
MACH200v3	1876	11606	1140	507
MACH280v3	2515	14879	1476	969
HANDv3	1405	4318	590	684

Fig. 40. Rule execution statistics by interpretation phase.

of Fig. 7. These statistics show the effect on the execution time and number of productions fired as the number of interpretation regions is varied.

A. Fragment/Functional Area/Model Distributions

Fig. 38 gives a breakdown of the total number of fragments generated by class and subclass. Fig. 39 gives a breakdown of the types of functional areas generated and the total number of models that were generated from these functional areas. Fig. 40 gives rule execution frequencies by interpretation phase.

As described in Section IV, SPAM versions V2 and V3 differ only in that V3 has elevation and texture knowledge supplied. This information is applied by rules which recognize, for example, man-made objects with elevation above the ground plane, as well as those that distinguish textured grassy areas from nearly smooth tarmac and parking aprons. The effects of such knowledge is more evident in the reduction of subclass interpretations than in the reduction of generic class interpretations.

Version V3 generated a much reduced set of functional areas. This is encouraging since a proliferation of functional areas can pose a major problem for global model recognition. As seen in the earlier examples, both versions generated good sets of functional areas, so that using our relative "best" model we may also be able to accommodate large numbers of errorful functional areas.

B. Timings

The timings in Figs. 41 and 42 illustrate the effects of increasing the number of initial regions and, more interestingly, the effect of adding elevation (depth) and texture knowledge to the system. Timings are averaged over several runs of the system for each machine segmentation data set on a VAX 11/780, with 8 Mbytes of memory running the UNIX operating system.

# regions	Avg User time	Avg System time
Build and Local Evaluation Phases:		
95	27:09.2	2:48.7
200	27:21.0	4:24.5
280	47:54.2	5:41.2
Local Consistency Phase:		
95	2:23:41.8	50:47.6
200	14:18:31.6	4:03:59.7
280	17:32:58.3	5:19:18.3
Functional Area Phase:		
95	2:02:56.1	16:31.3
200	4:58:52.6	37:55.3
280	5:10:51.3	42:25.2
Model Generation Phase:		
95	10:09.6	1:21.7
200	18:03.4	1:00.5
280	59:23.0	5:14.7

Fig. 41. V2 timings (CPU h:min:s.tenths).

# regions	Avg User time	Avg System time
Build and Local Evaluation Phases:		
95	20:24.2	2:31.9
200	32:48.7	3:38.9
280	1:02:22.9	5:47.3
Consistency Phase:		
95	1:04:16.6	21:13.4
200	3:56:10.9	1:10:09.6
280	5:32:58.3	1:26:36.9
Functional Area Phase:		
95	37:05.2	5;31.2
200	1:01:44.8	10:31.4
280	1:19:18.3	14:06.3
Model Generation Phase:		
95	8:16.5	35.4
200	14:25.2	43.9
280	32:23.0	1:19.3

Fig. 42. V3 timings (CPU h:min:s.tenths).

A prototype version, V1, spent much of its time (70 CPU hours) in the OPS5 match and conflict resolution phases. This very inefficient implementation structured rules such that the matcher matched *all* fragments and *all* regions against each production rule. Examples of pragmatic issues in building OPS5-based systems can be found in Brownston *et al.* [23]. We performed some optimizations on rule evaluation order and restructured all rules to make them context dependent, matching only when an appropriate context was generated. We also allowed the region and fragment working memory elements to overlap by putting redundant interpretation information in the region structure and low-level database information in the fragment structure. With these changes, the OPS5 matcher had significantly fewer match candidates since it is matching (effectively) on one working memory element.

The significant decrease in evaluation times in V3 over V2 is primarily due to the decrease in the total number of subclass fragment interpretations that participate in consistency and functional area phases. Further, certain particular interpretations such as terminal, hangar, and taxiway, that have large number of consistency rules, were greatly reduced. This is easily noted in Fig. 40.

IX. Conclusions

SPAM represents research in progress. Currently, the system can extract and identify some runways, taxiways, grassy areas, and buildings from region-based segmentations for several images of National Airport in Washington, DC. It uses these primitives to build multiple plausible functional areas descriptions which support the runway and terminal building, hangar, and access road components of the scene model. These components generate multiple models that represent the spatial layout of the airport. The models generated by SPAM using both machine and hand segmentations compare favorably and appear to give a good description of the airport scene. Many problems remain, some in reliable low-level feature extraction from the imagery, others in the design and implementation of effective recognition strategies using the rule-based approach.

As mentioned in Section VI, Fig. 32 gives an indication as to the potential for growth within each rule class as SPAM is engineered to accommodate a variety of airport organizations. Based on preliminary experimentation with hand segmentation data for the interpretation of Los Angeles and Dulles airports, we believe that the largest growth of rules will be due to the expansion of subclass interpretations to include features such as maintenance, air cargo, and service buildings, and a greater variety of runway/taxiway spatial organizations. The implication is that spatial relationships between all existing and newly created subclass interpretations must be expressed in terms of consistency rules, hence the potential for large growth. Rules that aggregate fragment interpretations into functional areas will also be effected by the increase of new subclasses or more general spatial organizations. However, we do not expect a large increase in the number of model generation and region-to-fragment rules.

A near-term agenda for continuing our work in rule-based aerial image interpretation is as follows.

- Continue to measure the effect of the number of machine-generated regions and various image domain cues such as texture and depth with respect to the quality and completeness of the models generated.
- SPAM needs to be able to refine a more complete and consistent airport interpretation through analysis of the initial models that it generates. In some sense, the model generation phase could be viewed as the beginning of a cycle of analysis phases that would focus image analysis tools with very specific goals in promising portions of the scene. For example, within a hangar functional area, look for planes on the parking apron.

• Explore techniques to determine what information is currently lacking in the airport model(s) and methods for acquiring that knowledge from the image or by further evaluation of the fragment interpretations. Develop more sophisticated goal generation rules to detect and explain missing components of the scene model.

• Apply the SPAM system to other airport scenes.

Given the results to date, we believe that the integration of map knowledge, image processing tools, and rule-based control and recognition strategies will be shown to be a powerful computational organization for automated scene interpretation in aerial imagery.

REFERENCES

[1] C. Froesch and W. Prokosch, *Airport Planning*. New York: Wiley, 1946.

[2] R. Horonjeff and F. X. McKelvey, *Planning and Design of Airports*, 3rd ed. New York: McGraw-Hill, 1983.

[3] T. O. Binford, "Survey of model-based image analysis systems," *Int. J. Robot. Res.*, vol. 1, no. 1, pp. 18-64, Spring 1982.

[4] R. A. Brooks, "Symbolic reasoning among 3-D models and 2-D images," *Artif. Intell.*, vol. 17, pp. 285-348, 1981.

[5] A. R. Hanson and E. M. Riseman, *VISIONS: A Computer System for Interpreting Scenes*. New York: Academic, 1978, pp. 303-333.

[6] J. M. Tennenbaum and H. R. Barrow, "Experiments in interpretation-guided segmentation," *Artif. Intell.*, vol. 8, pp. 241-274, 1977.

[7] M. A. Fischler, J. M. Tenenbaum, and H. C. Wolf, "Detection of roads and linear structures in low resolution aerial imagery using a multisource knowledge integration techniques," *Comp. Graph. Image Processing*, vol. 14, pp. 201-223, Mar. 1981.

[8] M. Nagao and T. Matsuyama, *A Structural Analysis of Complex Aerial Photographs*. New York: Plenum, 1980.

[9] Y. Ohta, "A region-oriented image-analysis system by computer," Ph.D. dissertation, Kyoto Univ., Kyoto, Japan, Mar. 1980.

[10] B. L. Bullock *et al.*, "Image understanding application project: Status report," in *Proc. DARPA Image Understanding Workshop*, Sept. 1982, pp. 29-41.

[11] R. A. Brooks, "Symbolic reasoning among 3-D models and 2-D images," *Artif. Intell.*, vol. 17, pp. 285-349, 1981.

[12] T. Matsuyama, "A structural analysis of complex aerial photographs," Ph.D. dissertation, Dep. Elec. Eng., Tech. Rep., Apr. 1980.

[13] M. Nagao, T. Matsuyama, and H. Mori, "Structural analysis of complex aerial photographs," in *Proc. 6th Int. Joint Conf. Artif. Intell., IJCAI*, Tokyo, Japan, Aug. 1979, pp. 610-616.

[14] P. G. Selfridge, "Reasoning about success and failure in aerial image understanding," Ph.D. dissertation, Univ. Rochester, Rochester, NY, Tech. Rep. 103, May 1982.

[15] S.-S. V. Hwang, "Evidence accumulation for spatial reasoning in aerial image understanding," Ph.D. dissertation, Univ. Maryland, College Park, MD, 1984.

[16] D. M. McKeown, "MAPS: The organization of a spatial database system using imagery, terrain, and map data," in *Proc. DARPA Image Understanding Workshop*, June 1983, pp. 105-127; Dep. Comput. Sci., Carnegie-Mellon Univ., Pittsburgh, PA, Tech. Rep. CMU-CS-83-136.

[17] —, "Digital cartography and photo interpretation from a database viewpoint," in *New Applications of Databases*, G. Gargarin and E. Golembe, Eds. New York: Academic, 1984, pp. 19-42.

[18] D. M. McKeown and J. L. Denlinger, "Map-guided feature extraction from aerial imagery," in Proc. Second IEEE Comput. Soc. Workshop Comput. Vis. Represent. Contr., May 1984; Dep. Comput. Sci., Carnegie-Mellon Univ., Pittsburgh, PA, Tech. Rep. CMU-CS-84-117.

[19] —, "Image analysis using cooperating knowledge sources: Road tracking," Dep. Comput. Sci., Carnegie-Mellon Univ., Pittsburgh, PA, Tech. Rep. in preparation, 1985.

[20] C. A. McVay, B. D. Lucas, and D. M. McKeown, "Stereo verification in aerial image analysis," Dep. Comput. Sci., Carnegie-Mellon Univ., Pittsburgh, PA, Tech. Rep. in preparation, 1985.

[21] C. L. Forgy, "The OPS5 user's manual," Dep. Comput. Sci., Carnegie-Mellon Univ., Pittsburgh, PA Tech. Rep., 1981.

[22] D. M. McKeown and J. F. Pane, "Alignment and connection of fragmented linear features in aerial imagery," in *Proc. IEEE Comp. Vis. Pattern Recogn. Conf.*, June 1985; Dep. Comput. Sci., Carnegie-Mellon Univ., Tech. Rep. CMU-CS-85-122.

[23] L. Brownston, R. Farrell, E. Kant, and N. Martin, *Programming Expert Systems in OPS5*. Reading, MA: Addison-Wesley, 1985.

David M. McKeown, Jr. (S'71-M'72) received the B.S. degree in physics and the M.S. degree in computer science from Union College, Schenectady, NY.

From 1972 to 1974 he was an Instructor in Computer Science and electrical engineering at Union College. During 1974-1975 he was a Research Associate at George Washington University, Washington, DC, and a member of the Technical Staff at Goddard Space Flight Center, Greenbelt, MD. He is currently a Senior Project Scientist with the Department of Computer Science, Carnegie-Mellon University, Pittsburgh, PA. His research interests include image understanding for aerial photointerpretation, digital mapping and image/map database systems, computer graphics, and artificial intelligence. He is the author of over 25 papers and technical reports, and is an active consultant for government and industry in these areas.

Mr. McKeown is a member of the Association for Computing Machinery, American Association of Artificial Intelligence, American Society for Photogrammetry and Remote Sensing, and Sigma Xi.

Wilson A. Harvey, Jr., was born in Baltimore, MD, in 1962. He is currently completing the B.S. degree in both physics and mathematics at Carnegie-Mellon University, Pittsburgh, PA.

Since 1983 he has been a Research Programmer for the Department of Computer Science, Carnegie-Mellon University. His interests include artificial intelligence, especially knowledge representation and organization, distributed processing, and quantum mechanics.

Chapter 4

Relational Description

The ultimate purpose of a vision system is to provide the information that allows an organism to interact with its surrounding environment in order to achieve some set of goals. The low-level information compiled by the techniques discussed in the first three chapters falls far short of an environmental model suitable for planning or similar purposes. Therefore, the perceptual system is still faced with the problem of integrating such lower-level information into a more comprehensive environmental model.

This chapter deals with techniques for describing and reasoning about scenes, images, and diagrams in which the objects of interest are at a level of organization above that of image intensities or feature arrays. We are primarily concerned with problems that cannot be solved by local analysis and thus we emphasize techniques that generate or use descriptions for expressing relationships between objects. Relevant tasks in applications such as robot navigation and object manipulation involve the need to:

- employ a compact high-level description of a scene or scene domain to guide low-level scene analysis techniques in the task of sensor interpretation
- describe an ongoing scenario from the information obtained from a single image or from a sequence of images
- answer questions about scene content that are implied, but not necessarily explicit, in an image or other representation of the scene (e.g., a sketch map or diagram)

A critical distinction between the object recognition problems of Chapter 3 and the relational description problems of this chapter is the variety of things that can be described. In the case of recognition, description is limited to the individual terms of a fixed vocabulary, while relational description is potentially unbounded in its descriptive competence. The key problems that must be solved in relational descriptions are: (1) how to represent the space and time relationships among objects; (2) how to use these representations to carry out reasoning processes; and (3) how to deal with ambiguous, vague, or contradictory information.

4.1 Navigation, Object Manipulation, and Planning in Autonomous Robotic Systems

The key to designing a robotic system that can successfully navigate through and manipulate its environment is the ability to create, maintain, and use effectively an adequate model of the external world.

To what extent are the techniques discussed in the first three chapters suitable for these purposes, and what new issues arise? Integration poses one obvious set of problems. Different knowledge sources, such as different labeling algorithms, may provide contradictory interpretations. How do we deal with such conflicts? What is a suitable architecture for communication and control. For example, should all interpretation be "orchestrated" by a single "executive" process, or should interpretation be carried out by a community of interacting "expert" processes?

A second important set of problems is associated with using stored knowledge and acquiring new knowledge. The techniques discussed in the first three chapters were almost exclusively stand-alone algorithmic procedures. All the external knowledge they require—other than the data they interpret—is precompiled into fixed instruction sequences. Even when some parameter adjustment is permitted such techniques have no effective way of improving their performance based on past experience, or even of accepting advice—they will repeatedly make the same mistakes. The need to learn, accept advice, and reason from stored knowledge might be abilities that can be superimposed on the base-level techniques of the first three chapters. Because there has been little progress in this direction, it may eventually be necessary to examine again many low-level vision problems previously thought to be solved.

4.1.1 Autonomous Vehicles

As we just discussed, several problems in perception, learning, and reasoning must be solved by a robotic device that can autonomously move about and interact with its environment. In a typical scenario for investigating such problems, an autonomous vehicle navigates in unexplored terrain, incrementally building up a map of the obstacles and terrain features that it senses as it undertakes a number of goal-directed traversals. During the early years of such research, a laboratory environment rather than the outdoor world was the usual setting for these experiments. In the still impressive work carried out at SRI in the late 1960s and early 1970s, a simple vehicle, Shakey [Nilsson 84], was able to travel from one room to another, visually sensing objects and avoiding them. Shakey expected its vision system to identify things in its simplified indoor world. The STRIPS problem-solving system [Fikes 71] represented the world as a collection of first-order predicate calculus formulas. Given a goal (e.g., move object X from location A to location B) and Shakey's initial location, STRIPS constructed sequences of intermediate-level actions needed to carry out the required task. It could also suggest appropriate alternatives when unexpected obstacles were encountered. Alternatives were found by using a mechanism that expressed the preconditions required for invoking an operator, and the effects of each of Shakey's operators on the world; each action resulted in the addition or deletion of logical assertions in the "world" database. An interesting innovation in the planner for Shakey was a representation for storing a generalized plan such that subsequences of operators relevant to a new plan could be identified.[1] Thus, if there was a generalized plan to accomplish goal A and goal B, and only goal B was desired in some new task, then it was possible to extract the operations specified for attaining goal B.

Recently there has been much interest in autonomous vehicles able to navigate in outdoor terrain and to acquire new information about the environment. Early work by Moravec [Moravec 79] involved a mobile cart that used stereo analysis to move along a road. The cart moved at an extremely slow rate; sometimes five minutes would be required for the processing of the image data acquired at a single cart position. Given the computer abilities of that time, the images has to be transmitted to a large central processor. Some of the contributions of the Moravec approach were (1) the coarse-to-fine matching technique that was initially designed as a speedup mechanism, but actually led to a more reliable stereo analysis, and (2) the development of an "interest operator," which was used to find reliable points in the image for stereo matching.

Advances in computer technology have permitted research vehicles to carry the computers on board and have allowed processing rates adequate to deal with a vehicle speed of 10+ miles per hour [Lowrie 86]. Such vehicles generally use a database that describes the topography and features of the environment. Lawton [**Lawton 86**] describes the architecture for a system employing significant amounts of prior knowledge about its environment and explains how this knowledge can be used in perceptual processing. Various inference mechanisms are provided to deal with image analysis, to resolve ambiguities in the location of the vehicle, to define tasks for the perceptual processes, and to relate the image and knowledge base items. Other work typical of recent approaches are Kuan [Kuan 84], who describes a hierarchical terrain map knowledge representation suitable for spatial planning; and S.A. Shafer, Stenz, and Thorpe [S.A. Shafer 86],

[1] The generalized plan is stored in a "triangle table," as described in [Nilsson 84].

who describe an architecture for sensor fusion in a mobile robot being constructed at Carnegie-Mellon University. Rosenfeld and L. Davis [Rosenfeld 85] summarize research at the University of Maryland on image understanding techniques for an autonomous vehicle, including analysis of time-varying imagery.

A difficult problem arises when only the direction and approximate distance to the goal are known, the map of the terrain is sketchy or unknown, the goal is not visible, and obstacles emerge unexpectedly. For such problems we must be able to represent inexact information about the environment in a way that allows the system to reason about its situation. E. Davis [Davis 83] developed a theory of how a 2-D cognitive map can be learned from a sequence of scene descriptions and then used. This theory has been implemented in MERCATOR, a program that employs an inexact representation of 2-D geography to carry out geographic reasoning for a simulated robot wandering in a simulated world. Some of the interesting aspects of MERCATOR descriptions are that objects that are only partially known can be described, description can be provided on various levels of granularity, and distances are given in terms of upper and lower bounds on uncertainty. Building on this work, McDermott and E. Davis [McDermott 84] present a design for a spatial database, assimilation and retrieval algorithms for the database, and a framework for planning routes using these algorithms. They employ representations in which objects are inexactly located with respect to various frames of reference. Spatial data are represented and stored in several forms (1) as tables of coordinates, for relating groups of objects; (2) as "discrimination trees," for retrieving objects for various uses; and (3) as propositions, for performing "topological" inferences. They use these data structures, combined with planning and inference algorithms, to build a flexible route planner. Because inexactness is tolerated in their maps, the plan produced can be more or less detailed. Of particular interest is their discussion of the "assimilation problem,"—that is, the problem of modifying a "fuzzy" map based on new facts. McDermott and E. Davis consider the assimilation problem to be the most difficult problem encountered: "It has proved more difficult than we expected to automate the process of building up such a map from symbolic facts as they come in .... The problem of what to do when a contradiction is detected is still unsolved."

More recently, Brooks [**Brooks 85**] examined the navigation problem for a mobile robot subject to control inaccuracies in the process of exploring an unknown environment. The robot has sonar sensors and stereo TV cameras on board and is required to construct a consistent and accurate world model from shaft encoder readings, visibility analysis, and landmark recognition. This paper has an interesting discussion of the design approaches taken in other autonomous robot projects, and it considers how these decisions affect the representations used. In particular, the author argues against the use of an absolute coordinate system to represent the world. "We do not believe that a robot can perform many useful tasks if it models its world as a projection in a two-dimensional plane. Nor does it suffice to decorate those projections with 'height-of-feature' information." Brooks describes the graph structures to be used in his robot system and the methods he would use for representing the uncertainties of vehicle location and orientation.

Given the complexity of the task facing autonomous vehicles capable of operating in outdoor environments, we can expect to see representations of the environment incorporating uncertainty, levels of granularity, and both spatial and semantic knowledge. Two common themes in all of the preceding papers are how to represent inexact information about the environment and how to update a database incrementally in the presence of conflicts between new and existing data. An important question omitted in all these discussions is how to represent and reason about environmental features that cannot easily be described in propositional form—for example, the configuration of natural terrain. Also needed is work that assimilates actual sensor data into such complex databases.

4.1.2 Manipulation of Objects

The work in vision-assisted manipulation of objects has generally assumed an industrial environment in which we must deal with a single object or objects whose relationships are known. In addition, engineering solutions can be used to avoid difficult problems. For instance, the lighting and positioning of objects can be controlled. Thus, techniques tend to be model-based and do not require more general forms of relational representation—for example, see [**Bolles 83**] and [**Brooks 83**] in Chapter 3. In future applications, such as autonomous repair of space vehicles where a controlled environment is not always possible, relational representations will be required to support more difficult types of deductive analysis.

4.2 Analysis of Image Sequences

People see a continuously changing world due to their eye and body movements and the movement of objects around them. Yet most vision research has dealt with static images. In Chapter 1 we discussed the problem of recovering object and surface geometry from image sequences. In this section we discuss some of the problems involved in semantic "understanding" of a sequence of images. Tsotsos et al. [Tsotsos 80] use the term "motion understanding" to denote a motion concept, such as a man walking, that describes the higher-level changes exhibited in several consecutive images. The authors present a framework for visual motion understanding that includes (1) the representation of knowledge for motion concepts based on semantic networks, and (2) associated algorithms for recognizing instances of the motion concepts in image sequences.

4.2.1 Analysis of Rigid and Jointed Objects

We gain some insight into the human ability to interpret motion sequences by attaching reflective spots to key features of an object, such as the body joints and the ends of limbs of a person, and then having the person move in a dark environment. By filming the motion, we create a "moving light display" (MLD), a sequence of binary images where a few discrete points on a moving object are represented as binary ones and everything else as zeros. Motion interpretation from a MLD is interesting because an MLD isolates and presents geometric evidence of motion divorced from such factors as edges, texture, color, and lighting. Psychological tests have shown that although individual frames of an MLD cannot be recognized by human subjects, a person can "see" a walking man when viewing a sequence of frames with only twelve points given per frame [Johansson 68, 73].

Most computer approaches to MLD interpretation are not model-based. An exception is the approach used by O'Rourke and Badler [O'Rourke 80] in which a known model of a human figure, including the connectedness structure, was assumed. Their system employs a feedback cycle between high-level predictions and low-level verifications and analysis. More typical approaches are based on grouping of points through an analysis of the relative velocities of point pairs. Rashid's LIGHTS program [Rashid 80] uses the relative velocities of all pairs of points to group points. The connectedness structure is determined by representing every point in an MLD frame by the four-vector consisting of the x and y coordinates and velocities. A graph is constructed with each point as a node and the edges representing a cost equal to the euclidean distance between the nodes. A minimum spanning tree of the graph then determines the object clusters. Webb and Aggarwal [Webb 82] propose a solution based on a "fixed axis" assumption that states that every rigid object movement consists of a translation plus a rotation about an axis that is fixed in direction for short periods of time. Lee and Chen [Lee 84] use two assumptions in their analysis: (1) motion continuity that assumes the motion trajectory of each joint is piecewise smooth, and (2) motion consistency that assumes the movements of every pair of joints of rigid segments must have compatible motion vectors. All these approaches appear competent to deal with objects that satisfy the assumptions of rigidity and connectedness of parts; however, they would have difficulty with flexible objects, such as a hose, that violate both of these assumptions.

4.2.2 Medical Image Analysis

All the motion research discussed so far has used the image sequences to identify objects; no attempt was made to analyze the actual motion of the objects to obtain higher-level concepts. For example, in the MLD work there was no attempt to identify whether the person was a man or a woman by analyzing the motion. However, in such applications as the automatic detection of defects in the human heart using a sequence of Xray images, higher-level analysis of motion must be addressed. Here two major problems have to be solved: (1) the problem of understanding visual motion given a sequence of primitive image tokens, and (2) the problem of reasoning about spatio-temporal relationships in the context of interpreting the dynamics of the heart. The heart diagnosis problem requires the use of "real-world" sensor data (radiographs) combined with very sophisticated knowledge structures and reasoning processes.

Niemann et al. [Niemann 84] use model-guided analysis of the heart motion image sequences. The declarative part of the model is captured in a semantic net structure that describes structural properties of the heart, the phases of a heart cycle, and medical evidence that can be inferred from observations concerning heart structure and motion. The procedural part consists of algorithms for computing attributes, structures, and certainty factors. The declarative knowledge is used to guide analysis of an image sequence. Initially the user postulates a specific heart defect, and the system attempts to use the information obtained from the image sequence to verify this hypothesis.

Tsotsos [Tsotsos 81][**Tsotsos 85**] is interested in expert systems that can deal with continuous time-varying data. A particular incarnation of his approach, the ALVEN system, is concerned with the dynamics of the ventricular system of the heart. In this work he develops several interesting ideas: (1) a "temporal cooperative process," a modification of relaxation labeling, that can be used to accumulate and integrate the evidence [Tsotsos 84]; and (2) his basic theme that goal-directed, data-directed, and model-directed inference mechanisms can compensate for the deficiencies of each other if used in concert.

4.2.3 Geometric Modeling from Image Sequences

Herman and Kanade [**Herman 84**] describe 3D MOSAIC, a vision system that uses both monocular and stereo analysis to reconstruct 3-D scenes incrementally. Using aerial photographs of urban scenes, they demonstrate how domain-specific knowledge can be used for interpreting complex images. Knowledge of block-shaped objects in an urban scene is integrated into the processes of monocular analysis, stereo analysis, and the reconstruction of shapes from the wire frames. By using images obtained from multiple viewpoints, the system deals with real-world scenes containing objects with a variety of shapes, textures, and reflectance characteristics. Multiple views are required to provide the information needed for interpretation since surfaces occluded in one image may become visible in another, and similarly features of surfaces that are difficult to analyze may become more apparent in another image because of a different viewpoint and lighting conditions. To combine new views with a current model, a wire-frame model is constructed from the new data; this wire frame is then matched to the current model to obtain the transformation between the new data and the existing model. The new data are then merged with the current model.

An important issue is how to deal with discrepancies between the new wire-frame data and the current model. Slight registration discrepancies are handled by a weighted averaging procedure; elements in the model inconsistent with newly obtained elements are deleted. An interesting aspect of the approach is that when previously confirmed elements in the model are modified or deleted, the elements that gave rise to the hypotheses are considered again to determine whether they must be modified or deleted. Tracing back to the original elements is done by means of dependency pointers that are explicitly recorded at the time of creation of the hypothesized elements.

4.3 Diagram and Picture Understanding

Graphical representation is an important aid in problem solving. For example, diagrams are used extensively in geometry to elaborate the problem description, filter hypotheses, and gain insight about the space of the solutions. Here we are interested in the situation in which we have a diagram or picture and want to understand what these are "about" and then reason about the relationships between the depicted objects. The following steps are necessary:

- Recognize the components or objects of the diagram or picture.
- Describe them in some suitable representation that captures their relationships with other objects relevant to some intended purpose.
- Draw conclusions or answer questions about the diagram or picture (possibly invoking auxiliary knowledge, such as naive physics).

This is a very difficult problem that corresponds to converting natural language expressions into a representation that captures meaning, and then extracting information from the sentences relative to some given context.

Evans's ANALOGY paper [**Evans 64**] is included in this collection even though it is more than twenty years old because it is one of the few papers to deal successfully with a significant problem in which it is necessary to reason about pictures. (This effort was one of the first to use the language LISP and ran on a punched card machine!) Evans addresses the following problem: Given a set of images consisting of lines and geometric shapes, say images A, B, C, D, E, and F—and given that image A is related to image B in an unspecified manner, then the problem is to determine which of images D, E, and F is similarly related to image C. Of particular interest are: (1) the mechanisms for decomposing overlapping patterns based both on Gestalt criteria of "good figure" and on context; (2) the property and relationship calculations (e.g., A is above B, X is to the left of Y); (3) the technique for finding similarity of patterns based on topological matching; and (4) techniques for generating a rule (or rules) that specifies how objects of Figure A are removed, added to, or altered in their properties and relations to other objects to generate Figure B. It should be noted that the images were not derived from a scanned scene; rather the line drawings were entered manually. The more complete paper [Evans 68] is well worth reading.

Mackworth's MAPSEE [Mackworth 77] is a program for interpreting maps sketched freehand on a graphics tablet. The program emphasizes discovering cues that invoke descriptive models encoding the requisite cartographic and geographic knowledge, and it shows how a network consistency algorithm can be used to resolve ambiguities in model instantiation.

Funt's WHISPER program [**Funt 80**] uses diagrams to evaluate the stability of a "blocks world" structure and to predict the event sequences that will occur if the structure collapses. The initial problem data are manually supplied in the form of silhouettes of geometric objects, and all operations are simulated. In a typical experiment a triangular-shaped block might be shown in an unstable position—balanced on a rectangular block. The fall of the triangular block and its effect on other objects in the scene is then determined by the program. Funt has included several innovative ideas in his approach. First, he simulates what will happen in the real world by "carrying out experiments" using a diagram that represents a situation in the world. Second, he uses a special "retina," a simulated collection of processors arranged in concentric circles, which mirrors some of the properties of a human eye. The resolution of the retina is higher at the center and is intended to be directed to the "focus of attention" of a region to carry out a detailed analysis. Information is transferred between retinal processors along both a circular and a radial path. This arrangement allows certain specialized tests and operations to be performed very efficiently. For example, when the retina's focus of attention is at a pivot point of an object, the rotation of the object around the pivot point and collisions with other objects can be rapidly analyzed. The diagram and the retina are driven by a high-level reasoner that has a qualitative understanding of mechanics. The diagram portion of WHISPER represents the sequence of "states" of the world during an experiment, and the retina answers questions posed by the reasoner by analyzing the diagram.

Swain and Mundy [**Swain 86**] are concerned with the problem of geometric reasoning to prove theorems in computer vision. The proof algorithm they exploit, developed by Wu in 1979 and described in [Wu 84], is based on expressing a target theorem, as well as hypotheses, as algebraic expressions; the proof is then constructed by expressing the theorem in terms of a polynomial basis developed from the hypothesis. Their goal is to construct a proof system, Algeprover, that can support vision research and can be used to reason about scenes. The system user will be able to ask questions about geometrical concepts and receive a reply in terms of geometrical concepts as well. One application for this system would be in the area of matching models to objects found in an image. Knowledge of theorems about 3-D scene geometry and the constraints imposed by perspective viewing can drastically reduce the space of possible matches. Because Algeprover is still mathematically incomplete and its efficiency needs improvement, it is not currently a practical tool suitable for use in computer vision.

4.4 Discussion

As autonomous vehicles begin to explore the world outside the laboratory, we can expect that new and more sophisticated descriptive schemes will be required to deal with the additional complexities. Researchers are faced with the problem of how to update and enhance previous descriptions, reason about them, and yet remain within the limits imposed by available computational systems.

Some of the key problems are:

- How can the system know what aspects of the sensed scene are worthy of analysis and what level of detail should be used?
- In the analysis of image sequences, how can one express "high-level" understanding such as "the viewed object is not moving normally," or "we can tell that the viewed object is a man because of his stride."
- What is the nature of suitable mechanisms for representing and reasoning about complex physical environments, given that sensory data are imprecise and often erroneous?
- In the presence of conflicts between existing and newly acquired knowledge, how can an incrementally updated model of the environment be constructed and maintained? Given limited storage capacity, on what basis should information be retained or discarded?

Papers Included

Brooks, R.A. Visual map making for a mobile robot. In *Proc. 1985 IEEE International Conference on Robotics and Automation*, pages 824-829, St. Louis, Missouri, March, 25–28, 1985, (IEEE Cat. 85CH2152-7).

Evans, T.G. A heuristic program to solve geometric-analogy problems. In *Spring Joint Computer Conference*, Vol. 25, 1964. (Reprinted in *Artificial Intelligence*. O. Firschein (ed.), Vol. VI, Information Technology Series, pages 5–16, American Federation of Information Processing Societies, Inc., Reston, Virginia 22091, 1984.)

Funt, B. V. Problem solving with diagrammatic representations. *Artificial Intelligence* 13:201–230, 1980.

Herman, M. and T. Kanade. The 3D MOSAIC scene understanding system: incremental reconstruction of 3D scenes from complex images. In *Proc. DARPA IU Workshop*, New Orleans, Louisiana, pages 137–148, October 3–4, 1984. Also see *Artificial Intelligence*, 30(3):289–341, December 1986.

Lawton, D. T., T. S. Levitt, C. McConnell, and J. Glicksman. Terrain models for an autonomous land vehicle. In *Proc. 1986 IEEE International Conf. on Robotics and Automation.* San Francisco, California, pages 2043–2051 (IEEE Cat. 86CH2282-2), April 7–10, 1986.

Swain, M.J. and J.L. Mundy. Experiments in using a theorem prover to prove and develop geometrical theorems in computer vision. In *Proc. 1986 IEEE Int. Conf. on Robotics and Automation.* pages 280–285, San Francisco, California, April, 1986.

Tsotsos, J.K. Knowledge organization and its role in representation and interpretation for time-varying data: the ALVEN system, *Computational Intelligence*, 1:16–32, 1985.

Visual Map Making for a Mobile Robot

Rodney A. Brooks
MIT Artificial Intelligence Lab
545 Technology Square, Cambridge, Mass 02173.

Abstract. Mobile robots sense their environment and receive error laden readings. They try to move a certain distance and direction, and do so only approximately. Rather than try to engineer these problems away it may be possible, and may be necessary, to develop map making and navigation algorithms which explicitly represent these uncertainties, but still provide robust performance. The key idea is to use a relational map, which is rubbery and stretchy, rather than try to place observations in a 2-d coordinate system.

1. Introduction

We are interested in building mobile robot control systems useful for cheap robots (i.e., on the order of the price of an automobile) working in unstructured domains such as the home, material handling in factories, street cleaning, office and hotel cleaning, mining and agriculture. The same capabilities can be useful for robots, which do not have to be so cheap to be economically feasible, and which do tasks like: planetary exploration, space station maintenance and construction, asteroid mining, nuclear reactor operations, military reconaissance and general military operations.

For all these applications a robot must wander around an unstructured environment. If it is able to build and maintain a map by itself its functionality within the environment will be much improved. In this paper we examine some problems faced by such a robot.

We first state some starting points for the research we are engaged in. In some cases they are quite different from the working assumptions made by other mobile robot researchers. As we go we will compare our approach to those of [Moravec 1983]'s rover project and [Giralt et al 1983]'s Hilaire. These are the most advanced examples of mobile robot projects.

2. Dogma

1. We do not believe that a robot can perform many useful tasks if it models its world as a projection in a two dimensional plane. Nor does it suffice to decorate those projections with "height-of-feature" information. The world a robot must operate in is inherently three dimensional even if it can locally move in only a two dimensional plane. Many objects in the world are small enough that the robot can see over the top of them, but not so small that it can roll or step over them. On the other hand some objects hang over space that is navigable by the robot. Such objects can not always be ignored. They may on occasion block the robot's view of landmarks for which it is searching. They may have surfaces that the robot wishes to interact with; e.g., a table top that it can reach up to and dust, or a deck that the robot can get to via some stairs or a ramp.

2. Almost all mobile robot projects have had as one of their underlying assumptions that it is desirable to produce a world model in an absolute coordinate system. However all sensors and control systems have both systematic and random errors. The former can be dealt with by calibration techniques (although these are often time consuming and are confounded on mobile robots by the fact that the robot itself is not fixed to any coordinate system). The latter are always present. It is usual to model some worse case bounds on such errors but this will not always suffice (e.g. mismatches in stereo vision can produce depth measurements with error magnitude the full range of depths which can be measured). In any case the bounded errors at least must be dealt with in building models of the world and using them. A number of approaches have been taken to this problem:

a. Ignore it. This has only been successful in the most toylike of worlds.

b. Use fixed reference beacons. This implies that the environment is either structured for the robot's benefit in the case that beacons are explicitly installed, or that the environment has been pre-surveyed for the robot's benefit in the case that known positions of existing beacons (e.g. power outlets) are used.

c. Some sort of inertial navigation device is used to determine accurately where the robot is in the global coordinate system. These sensors have a large number of disadvantages, including high cost (although proponents say they will be down from $100K to about $5K in only a few years), drift requiring recalibration to some global reference every few hours, long startup time for gyroscope based sensors, errors too large for use on a small scale such as within a factory or house, and uncertain behavior when subject to high jerk components (such as a tank crossing ditches, or your child playing tag with the vacuum cleaner robot).

d. Almost all wheeled mobile robots come equipped with shaft encoders on their steering and drive mechanisms. In a perfect world with no slip or slide between wheels and ground surface readings from these sensors would provide an extremely accurate estimate of robot position. Unfortunately wheels slip and slide with magnitudes that are functions of, at least, wheel velocity and acceleration, exact ground surface composition and shape, wheel load, and tire wear. These aspects can all be modelled but the surface the robot runs on must be restricted or there will still be errors with essentially the same magnitude. Of course such techniques will be less accurate on legged vehicles.

e. Landmarks are chosen by the robot and used as reference points when next seen, to update the robot position estimate. The landmarks must be chosen for recognizability within some expected uncertainty area.

Approaches a, b, and c above are ruled out for us because of costs or our unstructured domain. We are left with inaccurate motion estimates and the need for visual landmark recognition. Compare this to the *Hilare* project at Toulouse [Giralt et al 1983].

The Hilare project is probably the most complete and advanced of any mobile robot projects. They have produced by far the best non-vision based results. It uses methods b (infrared beacons), d (shaft encoders), and e (a directable laser range finder) from above to try to produce as accurate as possible a model of the environment in an absolute coordinate system. Even with an elaborate error tracking system and the use of fixed beacons errors still accumulate to the point where they are forced to break up observed objects into pieces and let the pieces change relative coordinates (i.e. the observed objects get deformed in the model relative to what was observed), to maintain a consistent position model [Chatila and Laumond 1985] The underlying problem is that worse case error needs to be assumed in placing things in an absolute coordinate system, and cumulative worse cases soon lead to useless models globally.

We use no global or absolute coordinate system. We do not ignore errors nor do we use beacons or inertial navigation systems. Instead we will use only local coordinate systems with relative transforms and error estimates. We use shaft encoder readings and visibility analysis and landmark recognition to build a consistent and accurate world model. In the future we might use a compass as these are cheap and from theoretical considerations it appears that a compass may provide a large amount of useful information.

3. Some mobile robot projects make no real attempt to model the environment on the basis of their perceptions. Such robots will necessarily have limited capabilities. Other projects, especially those based on indoor robots, which have made serious attempts to model the environment from perceptions have often assumed that the world is made up of polyhedra. This assumption manifests itself in two ways:

a.The perception system produces models of the world whose primitives are polyhedra.

b.The environment in which the robot is allowed to roam is constructed out of large planar surfaces with typically less than 100 planes in the complete environment of the robot.

In fact since most projects have used 2 dimensional representations of the world they often actually model the world as polygons, and the artificial worlds are constructed from vertical planes reaching from the ground to above the range of the sensors.

When algorithms developed under such assumptions and test in such domains are applied to more realistic worlds, they break down in two ways:

a.Time performance degrades drastically if there are any algorithms with even moderate time complexity as a function of the number of perceived surfaces. A quadratic algorithm can be fatal and even a linear algorithm can have unpleasant consequences.

b.More seriously, the mapping between the world and its perceived representation usually becomes very unstable over slight changes in position, orientation or, in the case of vision, illumination.

The second of these points can be disastrous for two reasons:

i.It becomes much more difficult to match a new perception with an existing partial model of the world. Straightforward matching of edges to edges and vertices to vertices no longer suffices.

ii.If the representation of free space is based on object faces (e.g., [Chatila 1982] has the most complete example of such a scheme) providing defining edges of convex polygons, then such representations get almost hopelessly fragmented into many, many convex polygons which extend distances many times the size of the object, and which have no semantic relevance, besides being completely unstable.

For the reasons we will build no artificial environments. The robot must exist in a "real" environment that humans inhabit and have constructed for their own use. We need to tackle the representation problems with multiple scales of representation to filter out the small high frequency perturbations of the environment which have no real effect on tasks to be performed within that environment.

It is important to remember the following:

A representation of the world is not something from which the world need be reconstructable. Rather a representation of the world is a statement of facts deducable from observations, and ideally includes enough facts that anything deducable from past observations is also deducable from the representation. A representation is not an analgous structure to the world; it is a collection of facts about the world.

4.Many groups building mobile robots are relying on sonar sensors (usually based on the Polaroid sensor used for autofocusing their cameras) as their primary source of thhree dimensional information about the world. The reasons for choosing such sensors seem to be:

a.they give direct digital readings of depth, or distance to first echo, and as such

b.they seem to require very little processing to produce a two dimensional description of the world, which means

c.they produce almost real-time data with very cheap processors, and

d.the sensors themselves are cheap so they can be hung all over the robot giving 360° sensing without the need for mechanical scanning.

Unfortunately sonar sensors also have many drawbacks which force many such groups to spend a great deal of effort overcoming them. The major drawbacks are:

a.the beam is either wide giving ambiguous and sometimes weak returns or when the beams are more focused it is necessary to have either very many sensors or to mechanically scna witha smaller number of them.

b.sonar returns come from specular reflections (i.e. mirror-like reflections int he visible spectrum) and thus the emitted sound pulse often skims off a surface leading to either no return, or a secondary return giving the depth of an object seen (heard) ina mirror, rather than in the line of sight (hearing), and

c.because of these problems large amounts of processing turn out to be necessary to get even moderately low noise levels in the data, defeating one of the original reasons for choosing sonar sensing, and lastly

d.sonar is not useable over large distances without much higher energy outputs than that of the standard cheap sensors making it much more hardware expensive and also subject to errors due to atmospheric effects.

A final problem with sonar is that it is not useable on the surface of the Moon or Mars nor in space outside of a pressurized vehicle.

For these reasons we prefer to use the natural alternative of passively sensing electromagnetic radiation in the near visible range. It suffers from none of the drawbacks we have listed above, although it does have its own; such as the need for large amounts of computation for even the simplest of data extractions.

We will however consider using sonar for tasks for which it is suited. For example it is an excellent sensor for monitoring local obstacles, missed by the visual sensor, as the robot moves along. Thus it acts as a backup safety sensor only. The Hilare project [Giralt et al 1983] is another projects which has come to similar conclusions concerning the proper role of sonar.

5. We are interested in building *artificial beings* in the sense described by [Nilsson 1983]. A robot purchased from Sears must work reliably for at least many months, and hopefully many years. It must work under the "control" of a completely unskilled, unsophisticated and untrained operator; i.e., the purchaser (you!). It must operate sensibly and safely under a large class of unforseen circumstances (some quite bizarre). To achieve such performance we may make only a minimal set of assumptions about the environment. The control software must be reliable in a sense orders of magnitude more complex and sophisticated than that which is normally considered under the banner of "software reliability". We suspect new hardware architectures will be necessary, besides software developments to meet these challenges.

2.1 *The task*

Our mobile robot is a circular platform with three wheels which always steer together and can point int any direction. It has sonar sensors and stereo TV cameras on board. Its environment is restricted to indoors but we plan to let it wander anywhere within our Lab. We want to use vision to build reliable maps of the world which are suitable for navigating about the world. We plan on using a system of [Khatib 1983] to provide a robust lower level control system avoiding local obstacles through acoustic sensing.

The only assumptions made about the robot in the rest of this paper are that it can move in any direction and it uses stereo vision to provide a sparse depth map (in particular we consider the use of a [Moravec 1983] style feature point depth map). No consideration is given to the use of acoustic sensors.

The algorithms described in this paper have not been implemented and therefore have not been tested on real data. In some cases only a computational problem has been formulated and no algorithm is given here.

3. Major Issues

A visually guided mobile robot inhabits a mental world somewhat different from the real world. Its observations do not exactly match the real world. Its physical actions do not occur exactly as intended. The task of navigating around the world can be eased by having the world map based on primitives suitable for navigation. The map should also be constructable from visual observations.

3.1 *Uncertainty*

Observations of the world are uncertain in two senses, providing two sources of uncertainty. Action in the world is also uncertain leading to a third source of uncertainty.

(a) There is inherent error due to measuring physical quantities. In particular, a pixel raster as used in computer vision enforces a minimal spatial resolution at the sensor beyond which there is no direct information. Physical constraints, such as continuity of surfaces in the world or averaging an identified event over a number of pixels, can be used to obtain sub-pixel accuracy of certain measurements, but the finite amount of information present ultimately can not be escaped. For the purposes of this paper we will only consider single pixel accuracy at best. For stereo vision this discretizes the possible depth measurements into a small number of possibilities, based on the maximum disparity considered by the algorithm. The error in depth measurements, increases with the distance from the cameras and the possible nominal values become more sparse.

(b) There may be errors in stereo matching. Thus a depth point may be completely wrong.

(c) When a robot is commanded to turn or go forward it does not carry out the action completely accurately. If drive shaft encoders are used then the best possible accuracy is to within on encoder count. If this were the only source of error it could almost be ignored as shaft encoders can be built with many thousands of marks resulting in very small physical errors. The real source of error is wheel slippage on the ground surface. This can be reduced somewhat with very accurate dynamic models, and accurate knowledge of the ground surface and wheel characteristics.

Our approach is to take explicit account of the first source of observational uncertainty in the design of all algorithms and to use the principle of least commitment to handle it in a precise and correct manner. The second source is handled more heuristically, by making the algorithms somewhat fault tolerant.

3.2 *Map primitives*

[Brooks 1983a] introduced a new representation for free space, useful for solving the find-path problem for a convex polygon. (See [Brooks 1983b] for an extension to the case of a manipulator with revolute joints.) The key idea is to represent free space as *freeways*, elongated regions of free space which naturally describe a large class of collision-free straight line motions of the object to be moved. Additionally there is a simple computation for determining the legal orientations (i.e. those where the moving object stays completely within the freeway) of the object while it is moved along a freeway.

Using freeways as a map representation for a mobile robot has a number of positive aspects. These are:

1. Visual observations provide natural freeway descriptions. If the robot can see some point in the distance, then it must be the case that there are no obstacles along the line of sight from the robot to that observed point. We will call this the *Visibility Constraint.*

2. For a simple circular robot, such as ours, navigability of a freeway depends only on its minimum width.

3. Freeway descriptions of free space do not rely on having each point of free space represented uniquely. Freeway representations are naturally overlapping. It is not necessary to know that two descriptions of pieces of free space are refering to the same place; the map can still be useful for navigation tasks.

It should be noted however that not all of free space is best described by elongated primitives. Some places are best described as convex regions. We will include such regions in our map, and refer to them as *meadows.*

3.3 *Combining these ideas*

These ideas can be combined in a map representation by avoiding the use of a 2-d coordinate system. Instead, only relationships between parts of the map are stored, in a graph representation. The relationships include estimates on their associated uncertainties.

Consider figure 1. It is one aspect of a map built by a mobile robot as it has moved from some place A, to place B, and so on to place E. It tried to travel in straight lines between places. The representation uses a 2-d coordinate system, so that straight lines in the map correspond to straight line paths in the world. It did not know the places beforehand but has been labelling them as it goes. Suppose it has been using nominal distances travelled and nominal angles turned, to give nominal 2-d coordinates for A, B, etc. Given that these values include errors there may be three physical positions for E, that give rise to the same measurements mad by the robot in moving from D. Any coordinates chosen for E implicitly add unsupported conclusions to the map. It can not be known whether the path from D to E crossed the path from A to B, went via place B, or crossed the path from B to C.

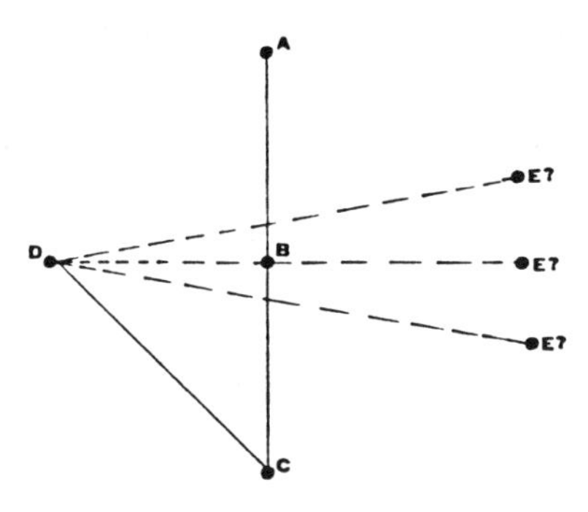

Figure 1. Metric information leads to incorrect represenatations when the information contains uncertainty.

A better approach then is to use an abstract graph as in figure 2 where the arcs, which represent straight line motions in the real world, are not represented as straight lines in the model, but are simply arcs with labels. For instance, arc BC is labelled with the robot's estimate of the distance it travelled. Intersections of arcs in this representation are purely an artifact of our attempt to draw the graph on paper. It will be possible however, to determine from the arc labels that the path represented by arc DE somewhere crossed the path followed by the robot from A to C.

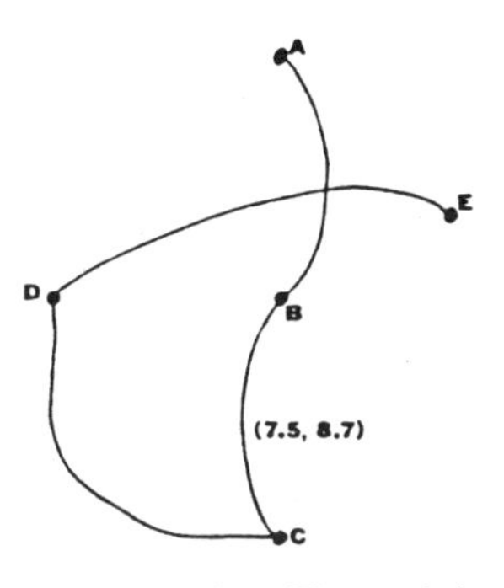

Figure 2. A better representation. The graph is not embedded in the 2-d plane, except for the purpose of drawing this diagram.

4. Dealing with Uncertainty

If a mobile robot is moving in a flat two dimensional world, and if it has a labelled direction as forward then its space of possible locations and orientations is a three dimensional *configuration space* [Lozano-Pérez 1983]. We can label its axes x, y, and θ. When the robot is at two dimensional coordinates (x_0, y_0) with orientation θ_0, its configuration corresponds to point (x_0, y_0, θ_0) in configuration space. For now lets refer to such a configuration as P_0.

Suppose the robot has configuration P_0, and it re-orients by angle η then travels distance d. Its new configuration would be:

$$P_1 = (x_0 + d\cos(\theta_0 + \eta), y_0 + d\sin(\theta_0 + \eta), \theta_0 + \eta).$$

However there is always error associated with the robot's motion. Typical errors might be $\pm 5°$ angular error and $\pm(5 + 0.05d)$ centimeters, in distance error, as a function of the distance travelled.

4.1 *Uncertainty manifolds*

We have shown that P_1 can not be uniquely identified. Instead P_1 can range over an *uncertainty manifold* (see figure 3) in configuration space. If the range of possible values for d is $[d_l, d_h]$ and for η is $[\eta_n - \alpha, \eta_n + \alpha]$ then the uncertainty manifold is:

$$\begin{aligned} M_1(x_0, y_0, \theta_0) = & \qquad (1) \\ \{ (x_0 + d\cos(\theta_0 + \eta), y_0 + d\sin(\theta_0 + \eta), \theta_0 + \eta) \\ \mid d \in [d_l, d_h], \eta \in [\eta_n - \alpha, \eta_n + \alpha] \}. \end{aligned}$$

Notice that this is a two dimensional manifold in three dimensional space.

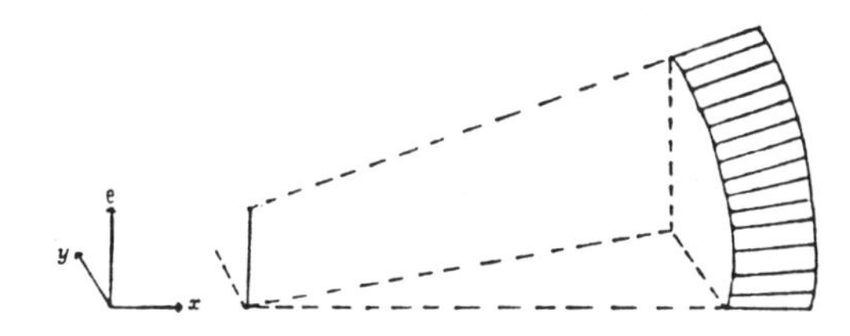

Figure 3. An uncertainty manifold arising from an uncertain motion in an uncertain direction.

A simple question, but nevertheless a useful one to ask while exploring the world, is "Am I back some place I've already been?", which, without loss of generality, can be simplified to "Am I back where I started?". Given that each individual motion is uncertain, it will be necessary to take into account the cumulative uncertainty, and in fact, without further sensing, the toughest question that can be answered is "Is it plausible that I am back where I started?". We will call this the *am-I-there-yet* question.

The uncertainty manifold resulting from two motions can be written as

$$C_{12}(x_0, y_0, \theta_0) = \bigcup_{(x,y,\theta) \in M_1(x_0,y_0,\theta_0)} M_2(x, y, \theta) \qquad (2)$$

and from two motions:

$$C_{123}(x_0, y_0, \theta_0) = \bigcup_{(x,y,\theta) \in C_{12}(x_0,y_0,\theta_0)} M_3(x, y, \theta)$$

(Note that this is different from an incorrect forumlation given in [Brooks 1984].) We call the manifolds C_{12}, C_{123}, etc., "cascaded manifolds". These manifolds are three dimensional and "solid"; i.e., they have a non-empty interior. The surfaces of these manifolds become progressively harder to express as the number of motions increase.

One approach to answering the am-I-there-yet question is to introduce three more variables for each motion (besides nominal angles and distances and bounds on errors in each), and write explicit conjunctions of symbolic inequalities of the form

$$P_1 \in M_1(x_0, y_0, \theta_0),$$

$$P_2 \in C_{12}(x_0, y_0, \theta_0),$$

etc. Note that each P_i introduces three more variables. Then we can add the constraints

$$(x_0, y_0, \theta_0) \in C_{12\ldots n}(x_0, y_0, \theta_0), \quad (3)$$

and, using the methods of [Brooks 1983] ask whether all the constraints are together satisfiable. This is *forward reasoning* with constraints. Additionally one would like to be able to use auxiliary information, such as from landmarks, to be able to assert inequalities such as (3). Then one would use the symbolic bounding algorithms from the above paper to determine the implications in terms of constraints on the actual physical values of angles of re-orientation and distances travelled. This is *backward reasoning* with constraints. Unfortunately this approach doesn't work well because of the presence of so many trigonometric terms.

4.2 *Approximating uncertainty manifolds*

A second approach, used more successfully in reasoning about uncertainties in assembly processes, is to use bounds on trigonometric functions, such as

$$\sin\alpha \le \sin(\eta - \eta_n) \le \eta - \eta_n$$

for $\eta \ge \eta_n$ and

$$1 - \frac{1}{2}(\eta - \eta_n)^2 \le \cos(\eta - \eta_n) \le 1$$

This has the effect of making an individual 2-d manifold M_i a little fuzzy, giving it some three dimensional volume. This makes the resulting manifolds (e.g. equation (2)) a little nicer in form. Unfortunately, again the constraint propagation methods fail because of the large number of cascading variables.

Broader bounding volumes for the uncertainty manifolds, with fewer parameters, and with simpler interactions, are needed if we are to make use of them in either forward or backward reasoning.

4.3 *Cylinders in configuration space*

The projection of the uncertainty manifold (1) into the x–y plane is can be bounded in the x–y plane by a circle with radius

$$\frac{(d_l + d_h)^2}{4\cos^2\alpha} - d_l d_h$$

centered distance

$$\frac{d_l + d_h}{2\cos\alpha}$$

from the start of the motion. We can then bound the complete manifold in configuration space by a cylinder sitting above the bounding circle with have constructed, and ranging from $\eta_n - \alpha$ to $\eta_n + \alpha$ in height.

A bounding cylinder B has four parameters and is a function of three variables (just as a manifold M_i itself is a function of position and orientation of its root point in configuration space). The four parameters are distance d, radius r, central orientation η, and orientation radius α.

Suppose that $B_{d_1,r_1,\eta_1,\alpha_1}(x, y, \theta)$ bounds $M_1(x, y, \theta)$, and that $B_{d_2,r_2,\eta_2,\alpha_2}(x, y, \theta)$ bounds $M_2(x, y, \theta)$. Then it turns out we can bound the cascaded manifold

$$C_{12}(x_0, y_0, \theta_0)$$

by the bounding cylinder:

$$B_{d_{12},r_{12},\eta_{12},\alpha_{12}}(x_0, y_0, \alpha_0)$$

where

$$d_{12} = \sqrt{{d_1}^2 + {d_2}^2\cos^2\alpha_2 + 2d_1 d_2 \cos\eta_2 \cos\alpha_2}$$

$$r_{12} = d_2 \sin\alpha_2 + r_1 + r_2$$

$$\eta_{12} = \eta_1 + \eta_2$$

$$\alpha_{12} = \alpha_1 + \alpha_2$$

Now we are able to cascade many motions together without increasing the complexity of our representation for the amount of uncertainty which results. The penalty we pay is that our bounds are sometimes rather generous.

Figure 4 shows the projection of bounds on uncertainty manifolds produced by four successive motions into the x–y plane. After the last, the am-I-there-yet question has an affirmative answer, i.e. "maybe", relative to the point of origin.

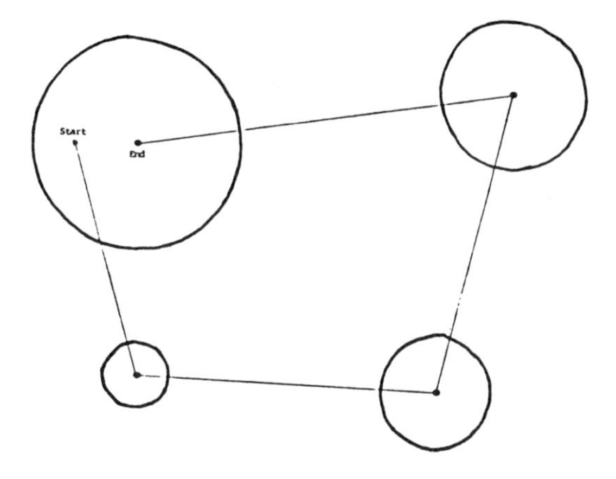

Figure 4. An example of forward reasoning, deciding that it is plausible that the robot is back where it started.

In figure 5 we see an example of backward reasoning. Suppose we have used forward reasoning to determine that it is plausible that the robot is in the same circular meadow from which the journey commenced. It might be plausible that the robot is in other meadows too. In any case the plausibility condition cuts down the number of possible meadows the robot might be in. If there are some visual landmarks associated with the meadow then perhaps one of the hypotheses can be verified (if a similar landmark is visible in another meadow then the forward reasoning will sometimes provide the disambiguation). So suppose such reasoning does indeed determine that the robot is in the original meadow. The intersection of the uncertainty cylinder cross section and the meadow determines a smaller region where the robot really might be. We bound that with a circle to produce a new uncertainty cylinder. If orientation information was gleaned from the landmark observation then that can be used to cut down the cylinder's range in the θ direction. This projection of this cylinder is illustrate in figure 5.

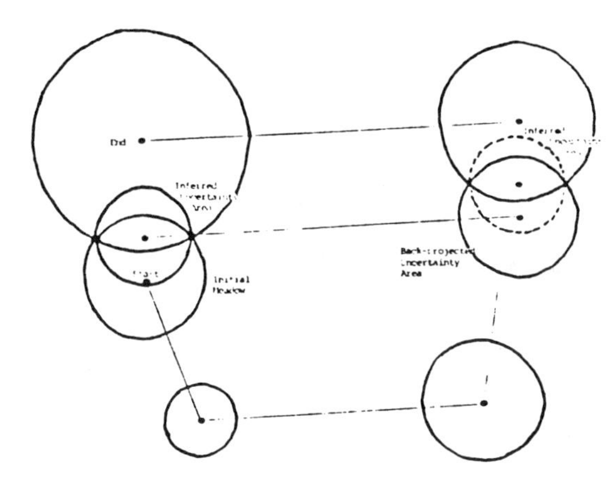

Figure 5. Backward reasoning. Additional information indicated the robot was in the meadow. The dashed circle shows bounds where the robot really must have been before.

Now we can use backward reasoning, and ask "What are all the points I could have been at, before the last motion and gotten into this new uncertainty cylinder?". This does not involve shrinking the cylinder by subtracting the uncertainly cylinder of the last motion, rather it involves cascading its inversion through the origin of configuration space. The resulting uncertainty cylinder, is shown in figure 5. Since the robot got to the third observation point by travelling forward around the path, it must have been in the intersection of the two cylinders before the fourth motion. Now a new uncertainty cylinder can be constructed to bound that intersection. It specifies an uncertainty relationship between the start of the journey, and any observations which were made after the third stop. Thus it ties down some of the rubbery map just a little bit better, and the effects of this additional constraint will be able to be used in later references to the map. Notice that the backward reasoning can continue from the last uncertainty cylinder we constructed. Soon however the uncertainties will become so large that no new information will be gained.

This technique was independently developed under another name by [Chatila and Laumond 1985] They call it *fading*.

5. Conclusion

In this paper we examined some problems which must be solved by a mobile robot which explores an unknown environment, building a map from visual observations. In particular we introduced a symbolic map representation whose primitives are suited to the task of navigation, and which is explicitly grounded on the assumption that observations of the world are inaccurate and control of the robot is inaccurate.

Acknowledgements

This research was partially supported by DARPA under contract N00014-80C-0505 and partially supported by an IBM Junior Faculty Development Award.

References

[Brooks 1982] Rodney A. Brooks, *Symbolic Error Analysis and Robot Planning* in **International Journal of Robotics Research**, vol 1, no. 4, 29-68.

[Brooks 1983a] Rodney A. Brooks, *Solving the Find-Path Problem by Good Representation of Free Space* in **IEEE Systems, Man and Cybernetics**, SMC-13, 190-197.

[Brooks 1983b] Rodney A. Brooks, *Planning Collision Free Motions for Pick and Place Operations* in **International Journal of Robotics Research**, vol 2, no. 4, 19-44.

[Brooks 1984] Rodney A. Brooks, *Aspects of Mobile Robot Visual Map Making* in **Preprints of the Second International Symposium of Robotics Research**, Kyoto, 287-293.

[Chatila and Laumond 1985] Raja Chatila and Jean-Paul Laumond, *Position References and Consistent World Modeling for Mobile Robots* in these proceedings.

[Giralt et al 1983] Georges Giralt, Raja Chatila and Marc Vaisset, *An Integrated Navigation and Motion Control System for Autonomous Multisensory Mobile Robots* in **Proceedings of the First International Symposium on Robotics Research**, Bretton Woods, New Hampshire, to be published by MIT press.

[Khatib 1983] Oussama Khatib, *Dynamic Control of Manipulators in Operational Space* in **Sixth IFTOMM Congress on the Theory of Machines and Mechanisms**, New Delhi.

[Laumond 1983] Jean-Paul Laumond, *Model Structuring and Concept Recognition: Two Aspects of Learning for a Mobile Robot* in **Proceedings IJCAI-83**, Karlsruhe, West Germany, 839–841.

[Lozano-Pérez 1983] Tomás Lozano-Pérez, *Spatial Planning: A Configuration Space Approach* in **IEEE Transactions on Computers**, (C-32):108–120.

[Moravec 1983] Hans P. Moravec, *The Stanford Cart and the CMU Rover* in **Proceedings of the IEEE**, (71)872–884.

[Nilsson 1983] Nils J. Nilsson, *Artificial Intelligence Prepares for 2001* in **The AI Magazine**, 7-14

A HEURISTIC PROGRAM TO SOLVE GEOMETRIC-ANALOGY PROBLEMS

Thomas G. Evans
Air Force Cambridge Research Laboratories (OAR)
Bedford, Massachusetts

INTRODUCTION

The purpose of this paper is to describe a program now in existence which is capable of solving a wide class of the so-called 'geometric-analogy' problems frequently encountered on intelligence tests. Each member of this class of problems consists of a set of labeled line drawings. The task to be performed can be concisely described by the question: 'figure A is to figure B as figure C is to which of the given answer figures?' For example, given the problem illus-

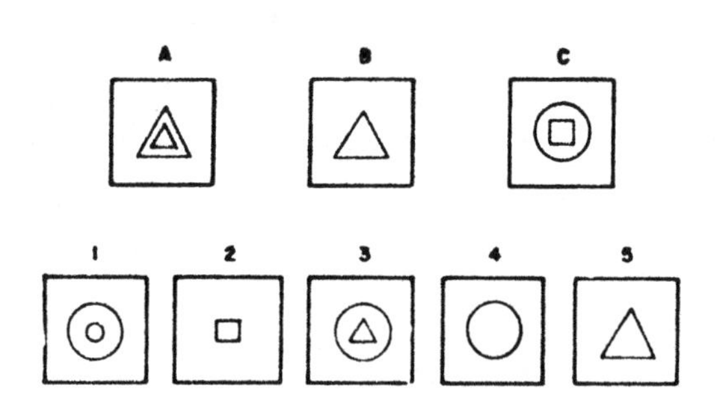

Figure 1.

trated as Fig. 1, the geometric-analogy program (which we shall subsequently call ANALOGY, for brevity) selected the problem figure labeled 4 as its answer. It seems safe to say that most people would agree with ANALOGY's answer to this problem (which, incidentally, is taken from the 1942 edition of the *Psychological Test for College Freshmen* of the American Council on Education). Furthermore, if one were required to make explicit the reasoning by which he arrived at his answer, prospects are good that the results would correspond closely to the description of its 'reasoning' produced by ANALOGY.

At this point, a large number of questions might reasonably be asked by the reader. Four, in particular, are:

(i) Why were problems of this type chosen as subject matter?

(ii) How does ANALOGY go about solving these problems?

(iii) How competent is ANALOGY at its subject matter, especially in comparison to human performance?

(iv) What has been learned in the construction of ANALOGY and what implications might this study have for the further development of problem-solving programs in general?

The remainder of this paper constitutes an attempt to answer these questions in some detail. We first deal with a variety of motivations for this investigation and attempt to place it in the context of other work in related areas. Next we turn to detailed consideration of the problem type and of the mechanism of the ANALOGY program. Finally, we present some answers to the remaining two questions raised above. (A more detailed discussion of all these issues can be found in Ref. 1).

Motivations and Background

In our opinion ample general justification for the development and study of large heuristic problem-solving programs has been provided (both through argument and through example) by previous workers in this area. We shall not attempt to add to it. Given that one is interested in the construction of such programs, a number of reasons can be advanced for the choice of geometric-analogy problems as a suitable subject matter. Some of these are:

(i) Problems of this type require elaborate processing of complex line drawings: in particular, they require an analysis of each picture into parts and the determination and use of various relationships among these parts. This is an interesting problem *per se* and one which can reasonably be expected to be of great practical importance in the near future.

(ii) The form of the problems requires one to find a transformation that takes figure A into figure B and takes figure C into exactly one of the answer figures. This situation provides a natural opportunity for trying out certain ideas about the use of explicit internal 'descriptions' (here, of both figures and transformations) in a problem-solving program. Furthermore, more speculatively, it presents an interesting paradigm of 'reasoning by analogy,' a capacity which may play a large role in far more sophisticated problem-solving programs in the future. (In Section 5 we discuss the possible relevance of ANALOGY to the introduction into problem-solving programs of more powerful learning mechanisms than have yet been achieved.)

(iii) Problems of this type are widely regarded as requiring a considerable degree of intelligence for their solution and in fact are used as a touchstone of intelligence in various general intelligence tests used for college admission and other purposes. This suggests a non-trivial aspect of any attempt to mechanize their solution.

We shall now attempt very briefly to place ANALOGY in the context of earlier work in related areas. Two aspects of ANALOGY must be considered:

(i) ANALOGY contains a substantial amount of machinery for the processing of representations of line drawings, including decomposition into subfigures, calculation of relations between figures, and 'pattern-matching' computations. Thus we must relate it to other work in picture processing and pattern recognition.

(ii) ANALOGY is a complex heuristic problem-solving program, containing an elaborate mechanism for finding and 'generalizing' transformation rules. Thus we must relate it to other work on the development of problem-solving programs.

We turn first to the picture-processing aspect. The essential feature of the treatment of line drawings by ANALOGY is the construction, from relatively primitive input descriptions, of more 'abstract' descriptions of the problem figures in a form suitable for input to the rule-finding program. The fundamental programming technique underlying this method is the use of a list-processing language, in this case LISP,[2,3] to represent and process the figures in question. Work in picture processing, for pattern-recognition purposes, involving some elements of description, is found in Grimsdale *et al.*,[4] Marill *et al.*,[5] and Sherman,[6] among others. Sutherland[7] and Roberts[8] have used, for quite different purposes, internal representations of line drawings similar in some respects to those used in ANALOGY. Kirsch[9] has worked with complex line drawings primarily as a vehicle for programs involving the analysis of English-language sentences pertaining to such pictures. Hodes[10] and Canaday[11] have used LISP expressions for figure description in much the same way that we have, though the development of machinery for manipulating such descriptions was, of necessity, carried much further in ANALOGY. Evidently the first advocacy of 'scene description' ideas (for use in pattern recognition) occurs in Minsky.[12]

To place ANALOGY with respect to other work with problem-solving programs, we shall simply list a number of developments in the construction of problem-solving programs which have influenced, in a general way, our approach to the design of ANALOGY. These include LT (the Logic Theorist)[13] and, more recently, GPS (the General Problem Solver)[14] of Newell, Simon, and Shaw, the plane-geometry theorem-prover[15] of Gelernter and Rochester, and SAINT, the formal integration program of Slagle.[16]

Summary of the Solution Process, with Example

To exhibit as clearly as possible the entire process carried out by ANALOGY, we now sketch this process, then examine its operation on an example. The sample problem we shall be considering is shown as Fig. 2 (where the Pi's are not part of the problem figures but labels keying the corresponding parts of the figures to

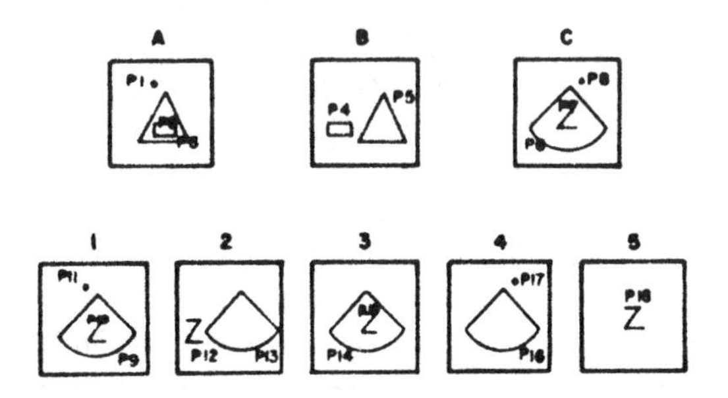

Figure 2.

expressions we shall give below). Before treating the example, we shall summarize the entire solution process. Given a problem such as that above, ANALOGY proceeds as follows: First, the input descriptions of the figures are read. Currently these descriptions, given as LISP expressions in a format to be illustrated below, are hand-made; however, they could well be mechanically generated from scanner or light-pen input by a relatively straightforward, quite 'unintelligent' program embodying line-tracing techniques already described in the literature. The descriptions represent the figures in terms of straight line segments and arcs of circles (to any desired accuracy, at the cost of longer and longer expressions). Examples of the descriptions are given below.

The first step taken by ANALOGY is to decompose each problem figure into 'objects' (subfigures). The decomposition program originally written, which was sufficient to handle many

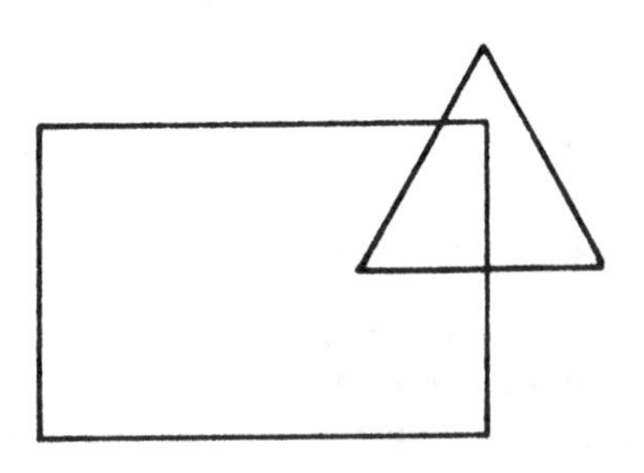

Figure 3a.

cases, including the example to be discussed below, was quite simple. It merely separated a problem figure into its connected subfigures; e.g., figure A of the above example consists of the three objects labeled P1, P2, and P3. It later became desirable to have a more sophisti-

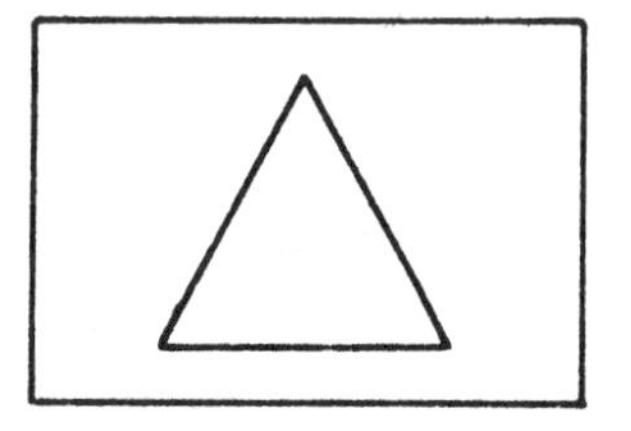

Figure 3b.

cated decomposition program with, in particular, the capability of separating overlapped objects on appropriate cues. For example, suppose problem figure A is as in Fig. 3a and figure B is as in Fig. 3b. The decomposition program should be able to separate the single object of figure A into the triangle and rectangle on the basis that they appear in figure B, from which point the remaining mechanism of parts I and II could proceed with the problem. While a decomposition program of the full generality desirable has not yet been constructed, the most recent version of the program is capable, in particular, of finding all occurrences of an arbitrary simple closed figure x in an arbitrary connected figure y; for each such occurrence the program can, if required, separate y into two objects: that occurrence of x and the rest of y (described in the standard figure format—note that this 'editing' can be rather complex: connected figures can be split into non-connected parts, etc.).

The type of decomposition illustrated above might be called 'environmental,' in that, e.g., figure A is separated into subfigures on the information that these subfigures are present, already separated, in figure B. An interesting extension to the present part I of ANALOGY might be to incorporate some form of 'intrinsic'

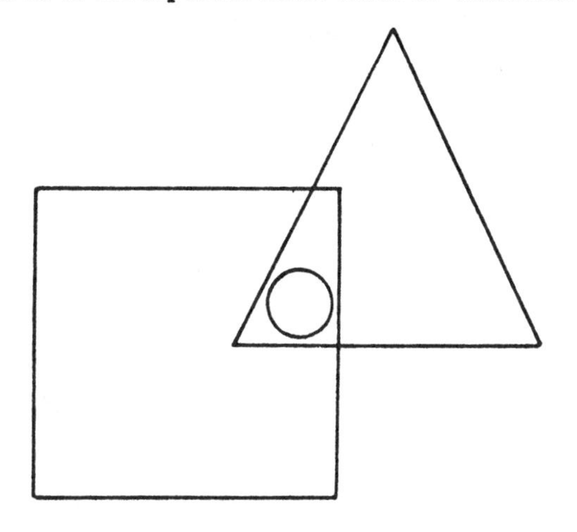

Figure 4a.

decomposition in which 'most plausible' decompositions are generated according to Gestalt-like criteria of 'good figure.' Such an extension could widen the problem-solving scope of ANALOGY considerably to include many cases where the appropriate subfigures do not appear already 'decomposed' among the problem figures. For example, suppose problem figures A and B are as shown in Figs. 4a and 4b, respectively. A decomposition into the square, triangle, and circle seems necessary to state a reasonable transformation rule. This example, incidentally, illustrates one potentially useful 'intrinsic' decomposition heuristic: roughly, choose decompositions into subfigures which have as much internal symmetry (in some precise sense) as possible.

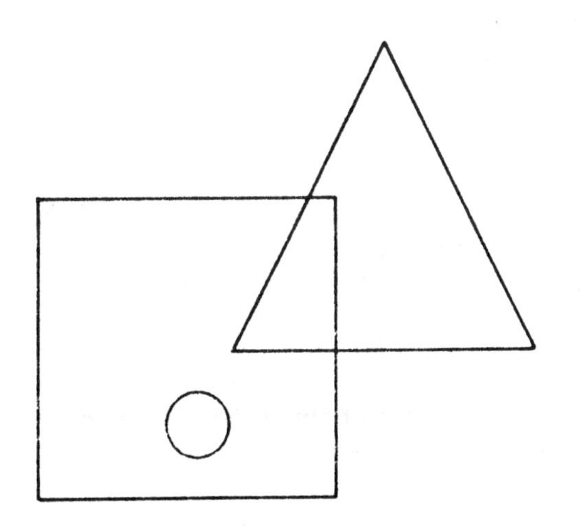

Figure 4b.

Next, the 'objects' generated from the decomposition process are given to a routine which calculates a specified set of properties of these objects and relations among them. The program is designed so that this set can be changed easily. As a sample of a relation-calculating subroutine, we cite one that calculates, for figure A of our example, that the object labeled P2 lies inside that labeled P3 and generates a corresponding expression (INSIDE P2 P3) to be added to the part I output description of figure A. The method used involves calculating all intersections with P3 of a line segment drawn from a point on P2 to the edge of the field (all figures are considered as drawn on a unit square). In this case P2 lies inside P3 since the number of such intersections is odd, namely one (and P3 is known to be a simple closed curve—if it were not, the calculation just described would be performed for each closed curve contained in P3). To do this, a substantial repertoire of 'analytic geometry' routines is required for part I, to determine, for example, intersections of straight line segments and arcs of circles in all cases and combinations. Other relation routines available in part I calculate, for example, that in figure A of our example P1 is above P2 and P3 and in figure B that P4 is to the left of P5.

The principal business of part I, aside from decomposition and the property and relation calculations, is a set of 'similarity' calculations. Here, part I determines, for each appropriate pair of objects, all members from a certain class T of transformations which carry one object of the pair into the other. The elements of T are compositions of Euclidean similarity transformations (rotation and uniform scale change) with horizontal and vertical reflections. (Given descriptions of virtually any pair of arbitrary line-drawings x and y, the routines of part I will calculate the parameters of all instances of transformations from T that 'map' x into y.

More precisely, an acceptable 'map' is a member of T for which T(x) is congruent to y up to certain metric tolerances which are parameters in the corresponding programs.

This routine is, in effect, a pattern-recognition program with built-in invariance under scale changes, rotations, and certain types of reflections. It consists essentially of a topological matching process, with metric comparisons being made between pairs of lines selected by the topological process. In Ref. 6 Sherman introduced some topological classification into a sequential decision tree program for the recognition of hand-printed letters, but the notion of systematically using the topological information to determine which metric comparisons are to be made seems to be new. This type of organization for pattern recognition has its own advantages (e.g., flexibility—the metric parts can be changed easily with no effect on the overall structure) and difficulties (e.g., sensitivity to metrically small changes in a figure which affect the connectivity—but this sensitivity can be largely removed by suitable pre-processing). Incidentally, it may be worth noting that if we suppress the metric comparisons entirely we get a general, and reasonably efficient, topological equivalence algorithm for graphs (networks).

The set of techniques we have just been describing, based on the use of a list-processing language to perform processing of line drawings by manipulating their list-structured descriptions, is by no means limited in applicability to the uses to which we have put it in part I of ANALOGY. To the contrary, it is our view that the representation of line drawings used here and the corresponding processing routines form a suitable basis for the development of a quite powerful 'line-drawing-manipulation language' with potential usefulness in a wide variety of applications. Regardless of whether the present investigation turns out to have a measurable influence on the art of designing problem-solving programs, it seems probable that the principal short-range contribution of ANALOGY is in the picture-processing by-products just described. (Incidentally, these techniques were discussed briefly from an ANALOGY-independent point of view in Ref. 17.)

After the similarity information is computed for every required pair of objects, both within a problem figure and between figures, this information, together with the decomposition and property and relation information, is punched out on cards in a standard format for input to part II. (For a typical set of figures, the total output of part I, punched at up to 72 columns/card, might come to 15 to 20 cards.)

Part II is given these cards as input. Its final output is either the number of the solution figure or a statement that it failed to find an answer. Its first step is to generate a rule (or, more frequently, a number of alternate rules) transforming figure A into figure B. Such a rule specifies how the objects of figure A are removed, added to, or altered in their properties and their relations to other objects to generate figure B. Once this set of rule possibilities has been generated, the next task is to 'generalize' each rule just enough so that the resulting rules still take figure A into figure B and now take figure C into exactly one of the answer figures. More precisely, for each 'figure A → figure B' rule and for each answer figure, part II attempts to construct a 'common generalization' rule which both takes figure A into figure B and figure C into the answer figure in question. This process may produce a number of rules, some very weak in that virtually all the distinguishing detail has been 'washed out' by 'generalization.' Hence it is necessary at this point to pick the 'strongest' rule by some means. This entire process requires a complex mechanism for manipulating and testing the rules and deciding which of the several rule candidates, the results of different initial rules or of different 'generalizations,' is to be chosen.

The principal method embodied in part II at present is able to deal quite generally with problems in which the numbers of parts added, removed, and matched in taking figure A into figure B are the same as the numbers of parts added, removed, and matched, respectively, in taking figure C into the answer figure. A substantial majority of the questions on the tests we have used are of this type, as is our present example; virtually all would be under a sufficiently elaborate decomposition process in part I; this restriction still permits a wide variety of transformation rules. It should be mentioned that the methods of part II have been kept subject-matter free; no use is made of any geometric meaning of the properties and relations appearing in the input to part II.

The more detailed workings of both parts I and II are best introduced through examining the process sketched above at work on our example. To convey some further feeling for the nature of the input to part I, we exhibit part of it, namely, the input description of figure A. The LISP expressions look like:

```
(
(DOT (0.4 . 0.8))
(SCC ((0.3 . 0.2) 0.0 (0.7 . 0.2) 0.0 (0.5 .
    0.7) 0.0 (0.3 . 0.2)))
(SCC ((0.4 . 0.3) 0.0 (0.6 . 0.3) 0.0 (0.6 .
    0.4) 0.0 (0.4 . 0.4) 0.0 (0.4 . 0.3)))
)
```

The first line above corresponds to the dot (at coordinates x = 0.4 and y = 0.8 on the unit square; the coordinate pairs in the other expressions are interpreted analogously). The next two lines correspond to the triangle (SCC stands for simple closed curve. All connected figures are divided into three classes—dots (DOT), simple closed curves (SCC), and all the rest (REG). This is solely for reasons of programming convenience; no other use is made of this three-way classification). Each non-connected figure is represented simply by a list of descriptions of its connected parts.

A curve (which may consist of an arbitrary sequence of elements chosen from straight line segments and arcs of circles) is represented by a list in which coordinate pairs alternate with the curvatures of the line elements between (all curvatures are zero here since the lines in question are all straight). Similarly, the next two lines above correspond to the rectangle; the entire description of figure A is then a list of the descriptions of these three parts. The format corresponding to the non-SCC figures like the Z-shaped subfigure of figure C is similar though somewhat more complex; it looks like:

```
(REG ((VI V2 (0.0 (0.55 . 0.5) 0.0 (0.45 .
    0.3) 0.0))
  (V2 V1 (0.0 (0.45 . 0.3) 0.0 (0.55 . 0.5)
    0.0))))
```

where V1 and V2 are the two vertices (here, endpoints) of the figure. The coordinates of V1 and V2 are given to part I in a separate list. They are V1 = (0.45 . 0.5), V2 = (0.55 . 0.3). Here, the top-level list describes the connectivity by stating which vertices are connected to which and how often—sublists describe in detail the curves making these connections. (By vertex we mean either an endpoint of a curve or a point at which three or more curves come together.) The complete details of the input format are given in Ref. 1, along with many examples.

When the input shown above corresponding to problem figure A and the corresponding inputs for the other seven figures are processed, the output from part I is, in its entirety, the ten LISP expressions shown below. For brevity, all similarity information concerning non-null reflections has been deleted. Also, we have replaced the actual arbitrary symbols generated internally by ANALOGY as names for the parts found by the decomposition program by the names P1, P2, etc., which appear as labels on our example figures above. The ten output expressions are:

```
(1) ((P1 P2 P3) . ((INSIDE P2 P3)
      (ABOVE P1 P3) (ABOVE P1 P2)))
(2) ((P4 P5) . ((LEFT P4 P5)))
(3) ((P6 P7 P8) . ((INSIDE P7 P6)
      (ABOVE P8 P6) (ABOVE P8
      P7)))
(4) ((P2 P4 (((1.0 . 0.0) . (N.N)) ((1.0 .
      3.14) . (N.N)))) (P3 P5 (((1.0 .
      0.0) . (N.N)))))
(5) ((P1 P8 (((1.0 . 0.0) . (N.N)))))
(6) NIL
(7) ( (P9 P10 P11) (P12 P13) (P14 P15)
      (P16 P17) (P18) )
(8) ( ((INSIDE P10 P9) ABOVE P11 P9)
      (ABOVE P11 P10)) ((LEFT P12
      P13)) ((INSIDE P15 P14))
      ((ABOVE P17 P16)) NIL)
(9) ( ((P6 P9 (((1.0 . 0.0) . (N.N)))) (P7
      P10 (((1.0 . 0.0) . (N.N)) ((1.0 .
      —3.14) . (N.N)))) (P8 P11 (((1.0
      . 0.0) . (N.N)))))
```

```
      ((P6 P13 (((1.0 . 0.0) . (N.N))))
        (P7 P12 (((1.0 . 0.0) . (N.N))
        ((1.0 . —3.14) . (N.N)))))
      ((P6 P14 (((1.0 . 0.0) . (N.N))))
        (P7 P15 (((1.0 . 0.0) . (N.N))
        ((1.0 . —3.14) . (N.N)))))
      ((P6 P16 (((1.0 . 0.0) . (N.N))))
        (P8 P17 (((1.0 . 0.0) . (N.N)))))
      ((P7 P18 (((1.0 . 0.0) . (N.N)) ((1.0 .
        —3.14) . (N.N.))))) )
(10) ( ( ((P1 P11 (((1.0 . 0.0) . (N.N)))))
        NIL NIL
      ((P1 P17 (((1.0 . 0.0) . (N.N)))))
        NIL ) . (NIL NIL NIL NIL NIL) )
```

To explain some of this: The first expression corresponds to figure A. It says figure A has been decomposed into three parts, which have been given the names P1, P2, and P3. Then we have a list of properties and relations and similarity information internal to figure A, namely, here, that P2 is inside P3, P1 is above P2, and P1 is above P3. The next two expressions give the corresponding information for figures B and C. The fourth expression gives information about Euclidean similarities between figure A and figure B. For example, P3 goes into P5 under a 'scale factor = 1, rotation angle = 0, and both reflections null' transformation. The next two expressions contain the corresponding information between figure A and figure C and between figure B and figure C, respectively. The seventh list is a five-element list of lists of the parts of the five answer figures; the eighth a five-element list of lists, one for each answer figure, giving their property, relation, and similarity information. The ninth is again a five-element list, each a 'similarity' list from figure C to one of the answer figures. The tenth, and last, expression is a dotted pair of expressions, the first again a five-element list, a 'similarity' list from figure A to each of the answer figures, the second the same from figure B to each of the answer figures. This brief description leaves certain loose ends, but it should provide a reasonably adequate notion of what is done by part I in processing our sample problem.

The ten expressions above are given as arguments to the top-level function of part II (optimistically called *solve*). The basic method employed by *solve*, which suffices to do this problem, begins by matching the parts of figure A and those of figure B in all possible ways compatible with the similarity information. From this process, it concludes, in the case in question, that P2 → P4, P3 → P5, and P1 is removed in going from A to B. (The machinery provided can also handle far more complicated cases, in which alternate matchings are possible and parts are both added and removed.) On the basis of this matching, a statement of a rule taking figure A into figure B is generated. It looks like:

```
(
(REMOVE A1 ((ABOVE A1 A3) (ABOVE
    A1 A2) (SIM OB3 A1 (((1.0 . 0.0) .
    (N.N))))))
(MATCH A2 (((INSIDE A2 A3) (ABOVE
    A1 A2) (SIM OB2 A2 (((1.0 . 0.0) .
    (N.N))))) . ((LEFT A2 A3) (SIM
    OB2 A2 (((1.0 . 0.0) . (N.N)) ((1.0 .
    3.14) . (N.N)))) (SIMTRAN (((1.0 .
    0.0) . (N.N)) ((1.0 . 3.14) . (N.N)
    ))))))
(MATCH A3 (((INSIDE A2 A3) (ABOVE
    A1 A3) (SIM OB1 A3 (((1.0 . 0.0) .
    (N.N))))) . ((LEFT A2 A3) (SIM
    OB1 A3 (((1.0 . 0.0) . (N.N))))
    (SIMTRAN (((1.0 . 0.0) . (N.N)
    ))))))
)
```

The A's are used as 'variables' representing objects. The format is rather simple. For each object added, removed, or matched, there is a list of the properties, relations and similarity information pertaining to it. (In the case of a matched object, there are two such lists, one pertaining to its occurrence in figure A and the other to its occurrence in figure B.) There are two special devices; the (SIM OB1 . . .) — form expressions give a means of comparing types of objects between, say, figure A and figure C; the other device is the use of the SIMTRAN expressions in the figure-B list for each matched object. This enables us to handle conveniently some additional situations which we shall omit from consideration, for brevity. They are treated in detail in Ref. 1.

The above rule expresses everything about figures A and B and their relationship that is used in the rest of the process. (The reader may verify that the rule does, in some sense, describe the transformation of figure A into figure B of our example.)

Next, a similarity matching is carried out between figure C and each of the five answer figures. Matchings which do not correspond to the ones between figure A and figure B in numbers of parts added, removed, and matched, are discarded. If all are rejected this method has failed and *solve* goes on to try a further method. In the present case, figures 1 and 5 are rejected on this basis. However, figures 2, 3, and 4 pass this test and are examined further, as follows. Choose an answer figure. For a given matching of figure C to the answer figure in question (and *solve* goes through all possible matchings compatible with similarity) we take each 'figure A → figure B' rule and attempt to fit it to the new case, making all matchings between the A's of the rule statement and the objects of figure C and the answer figures which are compatible with preserving add, remove, and match categories, then testing to see which information is preserved, thus getting a new, 'generalized' rule which fits both 'figure A → figure B' and 'figure C → the answer figure in question.' In our case, for each of the three possible answer figures we get two reduced rules in this way (since there are two possible pairings between A and C, namely, P1 ⟷ P8, P2⟷P6, and P3 ⟷ P7, or P1 ⟷ P8, P2 ⟷ P7, and P3 ⟷ P6).

In some sense, each of these rules provides an answer. However, as pointed out earlier, we want a 'best' or 'strongest' rule, that is, the one that says the most or is the least alteration in the original 'figure A → figure B' rule and that still maps C onto exactly one answer figure. A simple device seems to approximate human opinion on this question rather well; we define a rather crude 'strength' function on the rules and sort them by this. If a rule is a clear winner in this test, the corresponding answer figure is chosen; if the test results in a tie, the entire method has failed and *solve* goes on to try something else. In our case, when the values for the six rules are computed, the winner is one of the rules corresponding to figure 2, so the program, like all humans so far consulted, chooses it as the answer. The rule chosen looks like this:

```
(
(REMOVE A1 ((ABOVE A1 A3) (ABOVE
    A1 A2) (SIM OB3 A1 (((1.0 . 0.0) .
    (N.N))))))
(MATCH A2 (((INSIDE A2 A3) (ABOVE
    A1 A2)) . ((LEFT A2 A3) (SIMTRAN
    (((1.0 . 0.0) . (N.N)) ((1.0 . 3.14) .
    (N.N)))))))
(MATCH A3 (((INSIDE A2 A3) (ABOVE
    A1 A3)) . ((LEFT A2 A3) (SIMTRAN
    (((1.0 . 0.0) . (N.N)))))))
)
```

Again, it is easy to check that this rule both takes figure A into figure B and figure C into figure 2, but not into any of the other answer figures.

Further Examples and Comments

(a) *Examples*

We first exhibit several additional examples of problems given to ANALOGY:

(i) (See Fig. 5)

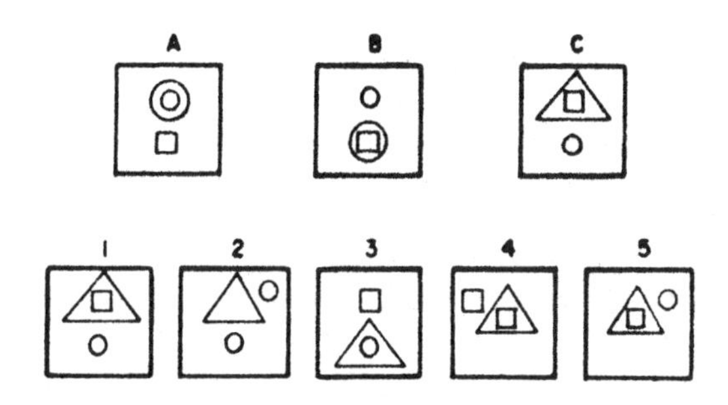

Figure 5.

Here the rule involves changes in the relations of the three parts. ANALOGY chose answer figure 3.

(ii) (See Fig. 6)

This case involves both addition and removal of objects. ANALOGY chose answer figure 2.

(iii) (See Fig. 7)

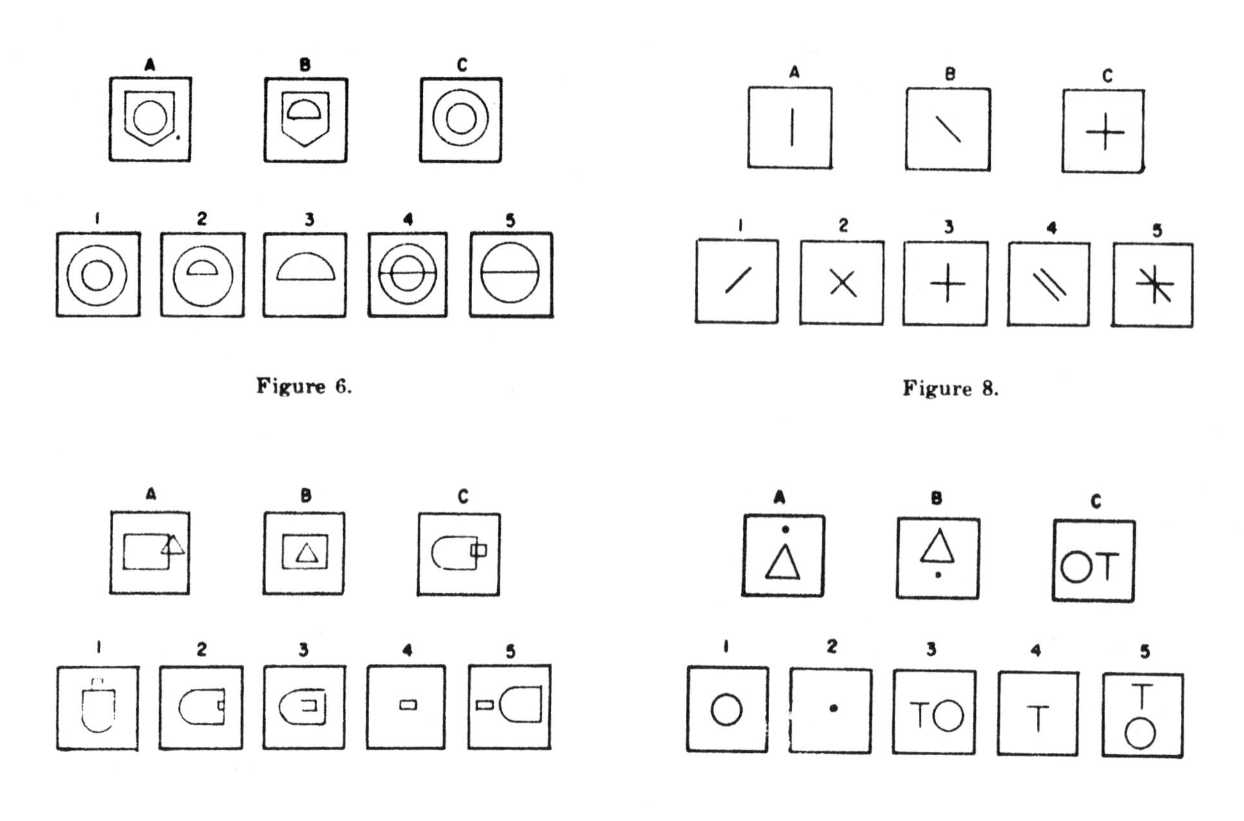

Figure 6.

Figure 8.

Figure 7.

Figure 9.

Note that this case required the more powerful decomposition program. Here ANALOGY chose figure 3.

(iv) (See Fig. 8)

The rule here simply involved a rotation. ANALOGY chose figure 2.

(v) (See Fig. 9)

Here ANALOGY chose figure 3, using an extension of the part II techniques discussed above. This extension, employed after failure of the basic process, involves systematic substitution of certain specified relations (e.g., LEFT for ABOVE) for others in the part II input descriptions, thus making it possible for ANALOGY to relate the 'vertical' transformation taking A into B to the 'horizontal' transformation of C into 3.

(vi) In the problem of Fig. 1, the large circle of answer figure 4 was replaced by a large square and the problem rerun. Again figure 4 was chosen but by a different rule. Now, instead of the inner object being removed, as before, the outer object is removed and the inner one enlarged. This illustrates some of the flexibility of the procedure and the dependence of the answer choice on the range of allowed answers as well as on A, B, and C.

(vii) (See Fig. 10)

Here is an example of a failure by ANALOGY to agree with the human consensus which favors figure 5. ANALOGY chose figure 3.

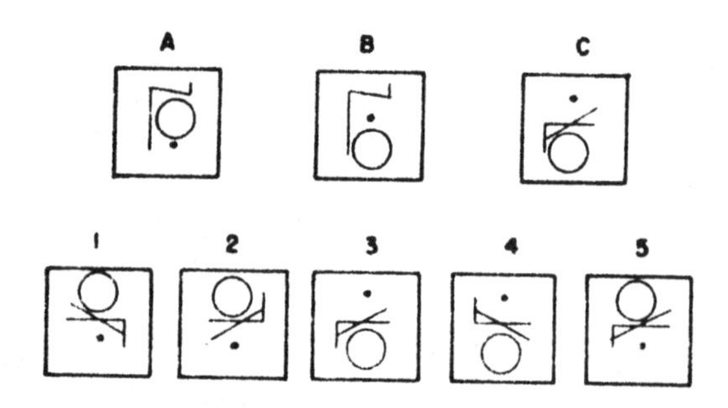

Figure 10.

(b) *Comparison with Human Performance*

We can only roughly compare the performance of ANALOGY with that of humans on geometric-analogy problems, since ANALOGY has not yet been given the complete set of such problems from any test for which scores are available. However, as some indication, we cite scores on the ACE tests based on a period of years including those editions of the test from which most of the problems on which ANALOGY was tested were selected. These scores are for a large population of college-preparatory students; the median score, on a test consisting of 30 such questions, ranged from 17 for 9th grade to 20 for 12th grade. Our estimate is that, on the same tests, ANALOGY, as it currently exists, could solve between 15 and 20 problems. Given, in addition, certain changes (mostly in part I, e.g., a more powerful decomposition program and additional properties and relations) for which we have reasonably well-worked-out implementations in mind, ANALOGY should be capable of perhaps 25 successful solutions.

(c) *The Use of LISP*

The use of a list-processing language to construct the ANALOGY program appears to have been a suitable choice; most notably, its capability at handling intermediate expressions of unpredictable size and 'shape' (such as our figure descriptions and transformation rules) is of great value. We especially wish to praise LISP as a convenient and elegant language in which to write and debug complex programs. The ease of composition of routines, the highly mnemonic nature of the language, and the good tracing facilities all contribute greatly to effective program construction. In return for the use of such a language one pays a certain price in speed and storage space, which, in the case of ANALOGY, at least, was a very acceptable bargain, since the necessity of machine-language coding would have made the entire project unfeasible. Incidentally, the ANALOGY program (apparently the largest program written in LISP to date) is so large that parts I and II must occupy core separately. The consequent limited (and one-way) communication between the parts was a serious design constraint but proved to have some compensating advantages in simplicity.

ANALOGY and Pattern-Recognition in Problem-Solving Programs

In this section we shall consider certain aspects of the design of problem-solving machines. To aid this discussion we shall specify (rather loosely) a subclass of problem-solving machines and carry out our discussion in terms of these though the ideas involved are by no means limited in applicability to this class. The machines we have in mind are typified by GPS[14] in that the problem to be solved by the machine is to transform one specified 'object' or 'situation' (whatever this may mean in a particular subject-matter context) into another by applying an appropriate sequence of transformations chosen from a class available to the machine. A wide variety of problems may be cast in this form (again see Ref. 14 or other discussions of GPS by the same authors). As in GPS, subgoals may be generated and attacked by such a machine and elaborate schemes of resource allocation may be required. However, these aspects do not concern us here. Our interest lies in the basic task of the machine; given a pair of 'objects,' it must choose an 'appropriate' transformation, i.e., one contributing to the goal of transforming one of the given 'objects' into the other.

It is a widely-held view, with which we agree completely, that for a machine to be capable of highly intelligent behavior on a task of this kind, in a rich environment of objects and transformations (and, in particular, to be capable of learning at a level more advanced than that of present machines), the critical factor is that it have a good internal representation of both its subject matter ('objects') and its methods ('transformations'), as well as an elaborate set of 'pattern-recognition' techniques for matching transformations to object pairs. Probably this means a quite 'verbal' representation of both objects and transformations as expressions in suitable 'description languages.' Furthermore, these matching techniques must be represented in a form in which they themselves are capable of being improved as the machine gains experience. The central role which 'pattern-recognition' techniques must play in sophisticated problem-solving programs and the corresponding importance for effective learning of autonomous improvement in the performance of these techniques are well expressed in Minsky.[12] There we find:

In order not to try all possibilities a resourceful program must classify problem situations into categories associated with the domains of effectiveness of the machine's different methods. These pattern-recognition methods must extract the heuristically significant features of the objects in question. Again from Ref. 12 we have:

Again from [12] we have:

In order to solve a new problem one uses what might be called the basic learning heuristic—first try using methods similar to those which have worked, in the past, on similar problems.

Here, the problem is, of course, to have pattern-recognition techniques possessing, or able themselves to learn, criteria of 'similarity' appropriate to the subject matter in question.

The 'fixed-length property-list' schemes (see Ref. 12) which characteristically have been used to perform this pattern-recognition task in current problem-solving programs have two principal defects which limit their extension to harder problems:

(i) While, in principle, given enough sufficiently elaborate properties, one can make arbitrarily fine discriminations, in practice a given set of properties will begin to fail rapidly as situations become more complex. In particular, for 'situations' which must be treated as consisting of interrelated parts, the 'global' nature of the scheme in question leaves it helpless.

(ii) Such a scheme is very limited in its learning capabilities, since it has access to very little information about its component properties; in particular, it is incapable of "knowledgeably' modifying its tests or adding new ones—it can only modify the weightings given to the results of these tests in its 'decisions.'

In view of the limitations of the 'property-list' pattern-recognition scheme just mentioned, we can formulate some requirements for a pattern-recognition scheme suitable tc replace it as a 'transformation-selecting' mechanism. First, the scheme must have access to a representation of each 'object' in terms of a 'descriptive framework' for the subject matter in question which is suitable in that useful relationships between 'objects' can be extracted relatively simply from the corresponding representations. Furthermore, the transformation-selecting rules of the pattern-recognition apparatus should themselves be expressed in a representation suitable for a 'learning mechanism' to revise the set of rules (i) by adding new rules and deleting those old ones which prove least useful as experience associates certain object pairs with certain transformations and (ii) by replacing a set of particular rules by a 'common generalization' rule again represented in the same language. Such facilities could go far toward removing the limitations of which we have spoken and providing both a powerful rule language (the rules can be stated in terms of the 'descriptive framework' we have postulated for the 'objects') and a learning mode more sophisticated than any yet incorporated in such a general problem-solving program.

So far we have been enumerating desirable features in a 'pattern-recognition' mechanism to be used as a transformation-selection device within a large problem-solver. What has all this to do with ANALOGY, which is not even a problem-solving program of the class we have been considering? We suggest that ANALOGY can, under a suitable (rather drastic) reinterpretation, be to some extent viewed as a pattern-recognition program having, to the limited degree appropriate for its particular environment, all the features we have listed. First, the 'objects' are the problem figures of ANALOGY and the suitable 'descriptive framework' appropriate to these objects is the 'subfigure and relation' representation used as the input part I generates for part II of ANALOGY. (Thus part I of ANALOGY corresponds to the apparatus that generates this representation for each object; that is, it goes from a representation of the 'problem objects' which is convenient for input to the problem-solver to one which is in a form suitable for internal use.) The generation in ANALOGY of a transformation rule taking one answer figure into another can be thought of as corresponding to the first kind of learning we listed above, namely, the adding of rules as, with experience, the machine associates certain object pairs with certain simple or composite transformations. Finally, the common generalization of two rules in ANALOGY corresponds to the second kind of learning we mentioned, namely, the generation of a common generalization of several rules associating 'objects' and 'transformations.' Furthermore, ANALOGY's process of choosing between 'common generalizations' of different rule pairs mirrors a process of selectively incorporating only those generalizations with the greatest discriminatory power. Under this interpretation, ANALOGY appears as a model for a pattern-recognition process with all the characteristics mentioned. The potential value of ANALOGY, viewed in this way, as a suggestive model for the construction of such pattern-recognition mechanisms for use within problem-solving programs may prove to be the chief product of our work with ANALOGY and the best justification for having carried it out.

References

1. T. G. EVANS, PH.D. Thesis, Department of Mathematics, MIT, June, 1963 (soon to be available as an AFCRL Technical Report).
2. J. MCCARTHY, "Recursive functions of symbolic expressions," Comm. ACM, Vol. 3, April, 1960.
3. J. MCCARTHY *et al.*, LISP 1.5 Programmer's Manual, MIT, revised edition, August, 1962.
4. R. L. GRIMSDALE, F. H. SUMNER, C. J. TUNIS, and T. KILBURN *et al.*, "A system for the automatic recognition of patterns," Proc. IEE, March, 1959, Vol. 106, pt. B, pp. 210-221.
5. T. MARILL, A. K. HARTLEY, T. G. EVANS, B. H. BLOOM, D. M. R. PARK, T. P. HART, and D. L. DARLEY, "CYCLOPS-1: a second-generation recognition system," FJCC, Las Vegas, Nevada, November, 1963.
6. H. SHERMAN, "A quasi-topological method for the recognition of line patterns," Proc. ICIP, Paris, France, June, 1959, pp. 232-238.
7. I. SUTHERLAND, "Sketchpad: a man-machine graphical communication system," SJCC, Detroit, Michigan, May, 1963.
8. L. ROBERTS, PH.D. Thesis, Department of Electrical Engineering, MIT, June, 1963.
9. R. KIRSCH, personal communication.
10. L. HODES, "Machine processing of line drawings," Lincoln Laboratory Technical Memorandum, March, 1961.
11. R. CANADAY, M.S. Thesis, Department of Electrical Engineering, MIT, February, 1962.
12. M. L. MINSKY, "Steps toward artificial intelligence," Proc. IRE, January, 1961, pp. 8-30.
13. A. NEWELL and H. A. SIMON, "The logic theory machine," IRE Trans. on Information Theory, Vol. IT-2, #3, September, 1956, pp. 61-79.
14. A. NEWELL, J. C. SHAW, and H. A. SIMON, "Report on a general problem-solving program," Proc. ICIP, Paris, France, June, 1959, pp. 256-264.
15. H. GELERNTER and N. ROCHESTER, "Intelligent behavior in problem-solving machines," IBM J. Res. and Dev., Vol. 2, #4, October, 1958, pp. 336-345.
16. J. SLAGLE, PH.D. Thesis, Department of Mathematics, MIT, June, 1961.
17. T. G. EVANS, "The use of list-structured descriptions for programming manipulations on line drawings," ACM National Conference, Denver, Colorado, August, 1963.

Acknowledgements

The assistance of the Cooperative Test Division of the Educational Testing Service, Princeton, New Jersey, in providing a large set of geometric-analogy questions from its files is gratefully acknowledged.

Thanks are also due to the Educational Records Bureau, New York, N.Y., for the statistics on human performance on geometrical-analogy questions cited in Sec. 4b.

Most of the computation associated with the development and testing of ANALOGY was performed at the MIT Computation Center.

Problem-Solving with Diagrammatic Representations*

Brian V. Funt

*Computer Science Department, State University of New York at Buffalo, Buffalo, NY 14226, U.S.A.***

Recommended by Daniel G. Bobrow and Aaron Sloman

ABSTRACT*

Diagrams are of substantial benefit to WHISPER, a computer problem-solving system, in testing the stability of a "blocks world" structure and predicting the event sequences which occur as that structure collapses. WHISPER's components include a high level reasoner which knows some qualitative aspects of Physics, a simulated parallel processing "retina" to "look at" its diagrams, and a set of re-drawing procedures for modifying these diagrams. Roughly modelled after the human eye, WHISPER's retina can fixate at any diagram location, and its resolution decreases away from its center. Diagrams enable WHISPER to work with objects of arbitrary shape, detect collisions and other motion discontinuities, discover coincidental alignments, and easily update its world model after a state change. A theoretical analysis is made of the role of diagrams interacting with a general deductive mechanism such as WHISPER's high level reasoner.

1. Introduction

Diagrams are very important tools which we use daily in communication, information storage, planning and problem-solving. Their utility is, however, dependent upon the existence of the human eye and its perceptual abilities. Since human perception involves a very sophisticated information processing system, it can be argued that a diagram's usefulness results from its suitability as an input to this powerful visual system. Alternatively, diagrams can be viewed as containing information similar to that contained in the real visual world, the canonical entity the human visual system was presumably designed through evolution to interpret. From this latter perspective, diagrams are a natural representation of certain types of primarily visual information, and the perceptual system simply provides an appropriate set of database accessing functions. Both these viewpoints underly the work described in this paper.

The role of diagrams is explored in a computer problem-solving program, named WHISPER, which refers to diagrams during its processing. WHISPER's high-level reasoning component (HLR), built along the lines of traditional procedural AI problem-solving programs, has the additional option of requesting observations in a diagram. It does this by asking its "perceptual system" to "look at" the diagram with its parallel processing "retina". The questions that the perceptual system can answer are called *perceptual primitives.* If necessary, the HLR can also make changes to the current diagram. Fig. 1 shows WHISPER's overall structure.

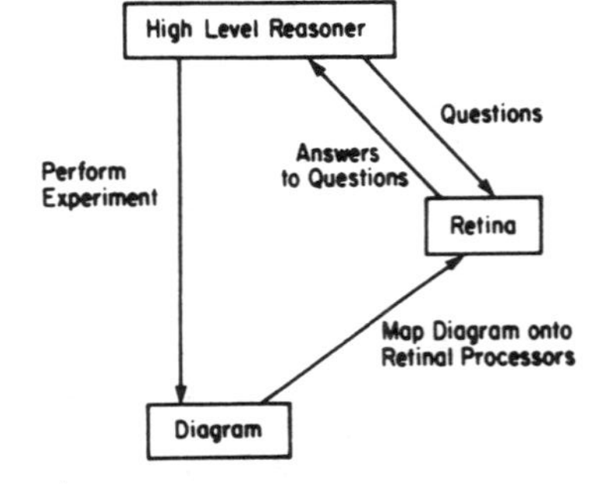

FIG. 1. The WHISPER proposal.

Upon receiving a diagram of a blocks world structure, WHISPER outputs a set of diagrams representing the sequence of events which occur as the structure collapses. The HLR contains knowledge about stability and the motion of falling objects. Using the retina to locate objects and their supports, it checks the stability of each object shown in the diagram. Unstable objects may either rotate or slide. In cases where one is rotationally unstable, the HLR asks the retina to "visualize" it rotating and thereby determine at what point it will hit some other object. Using this information WHISPER outputs an updated diagram showing the object rotated into its new position. Then with this new diagram, it restarts the problem-solving process from the beginning—rechecking the stability of each object, moving one of them, outputting another diagram, and restarting again. The process terminates when either all the objects are stable or the problem becomes too complex for the stability tester. A detailed discussion of the HLR will be postponed until Section 3.

1.1. Motivation

A strong case for computer use of diagrams as models for Geometry has been made by Gelernter (1963), and as general analogical representations by Sloman (1971). Networks with nodes representing "ideal integers" and arcs representing relationships between them were used as models for statements in arithmetic by Bundy (1973). Hayes (1974) and Bobrow (1975) comment on the theoretical nature of analogical representations; Hesse (1969) and Nagel (1961) discuss analogical reasoning.

* This paper is a substantially lengthened version of a similar paper appearing in IJCAI-5, [6].

** Now with: Dept. of Computing Science, Simon Fraser University, Vancouver, B.C., Canada V5A 1S6.

There is a variety of reasons for using diagrams in computer problem-solving. Diagrams such as maps, architectural plans, and circuit diagrams routinely facilitate human problem-solving. Perhaps diagrams function not merely to extend memory capacity, but rather present the important information in a particularly useable form. If they do, then the human visual system provides a paradigmatic example of a system for accessing these representations. Since it exploits a high degree of parallelism, it leads us into the realm of a different type of hardware. This is an exciting step, however, because we can see how much hardware characteristics influence our thinking about the difficulty of various problems and the feasibility of their solution. For example, we know we could compute with Turing Machines—but would we? Because WHISPER's retina harnesses parallelism, it in effect extends the available machine instruction set with special ones for diagram feature recognition. WHISPER is primarily an exploration of the question: to what extent can problem-solving be simplified through experiment and observation with diagrams? This is in contrast (but not in opposition) to the usual method of deduction within a formal theory as explained in the next section.

1.2. Theoretical framework

Any problem-solving system needs a representation of the problem situation. The standard approach in AI is to formalize the domain. We choose a language and write down a set of statements (axioms, productions, assertions, or a semantic network) describing the world. So that the problem-solver can generate new statements from this initial set, we provide a general deductive mechanism (theorem prover, programming language control structure, network algorithm). In terms of the predicate calculus the axioms define a theory, T, and so long as it is not self-contradictory there will be at least one model M (an assignment of predicates to the predicate symbols, functions to the function symbols, and individuals to the constant symbols) which satisfies it. Since our intention in axiomatizing the world was to accurately describe it, we expect it to be one of the models satisfying T.

We may find a second model M' satisfying T (in general there will be many such models). Now—and this is the main thrust of WHISPER—in some cases we can use M' to provide information about M without deriving it from T. What is required is that some of the predicates, functions and individuals of M' correspond to some of the predicates, functions and individuals of M in such a way that it is possible to translate the results obtained when these predicates and functions are applied to individuals in M' into the results that would be obtained if the corresponding predicates and functions were to be applied to the corresponding individuals in M. The similarity between M and M' means that experiments and observations made in M' yield results similar to those that would be obtained in M. As shown in Fig. 2, for WHISPER M' is the combination of its diagram and diagram re-drawing procedures. WHISPER obtains information about the blocks world by using its retina to observe the results of experimental changes made to its diagrams by the re-drawing procedures.

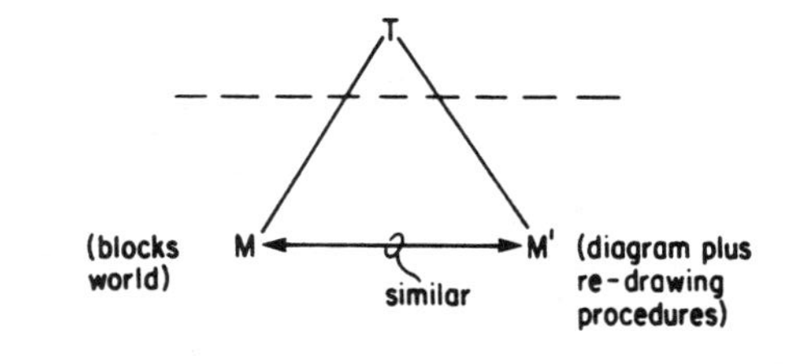

FIG. 2.

WHISPER is a prototype system designed to explore the extent to which problem-solving can be carried out below the dashed line of Fig. 2; however, it does do some reasoning above the line. WHISPER's success argues for working below the line, but not against working above the line. WHISPER's HLR is an above-the-line component.

A natural question is why use M' instead of M? If M is readily accessible then there is no reason not to use it; but, frequently it will not be. For example if we want to determine the stability of a pile of blocks on the surface of the moon, then we could construct a similar pile of blocks on earth and determine the result by experiment. In this case M, the pile of blocks on the moon is inaccessible. We can see that a lot can be learned from the blocks on earth, but some above-the-line inference must be done to handle the discrepancies arising as a result of the difference in gravity.[1]

2. Mechanisms for Diagram Interaction

The retina and perceptual primitives are designed to provide WHISPER with a new set of operations whose execution times are of the same order of magnitude as conventional machine instructions. To achieve this a high degree of parallelism has been incorporated into the system. The retina is a parallel processor, and the perceptual primitives are the algorithms it executes. (Do not be misled by the term "retina"; it refers to a general system of receptors and processors for the early stages of perceptual processing, rather than implying any close resemblance to the human retina.) Each perceptual primitive, when executed by the retina, determines whether some particular feature is present in the diagram. WHISPER's retina mixes parallel and sequential computation, so the features it can recognize are not subject to the same theoretical limitations as perceptrons (Minsky and Papert (1969)).

2.1. The retina

WHISPER's retina is a software simulation of hardware which, given the rapidly advancing state of LSI technology, should soon be possible to build. It consists of a collection of processors, each processor having its own input device called a *receptor*. There is a fixed number of processors, and they are all identical. As with the human eye, WHISPER's retina can be shifted to fixate at a new diagram location (also a feature

[1] I am grateful to Raymond Reiter for many of the ideas in Section 1.2.

of a program by Dunlavey (1975)), so that each processor's receptor receives a different input from the diagram. This fixation facility is important because the resolution of the retina decreases from its center to its periphery. Without being able to fixate, it would be impossible for WHISPER to examine the whole diagram in detail. Economy of receptors and processors dictates the use of decreasing resolution. (A declining resolution is also a characteristic of the human eye.) Each receptor covers a separate segment of the diagram and transmits a single value denoting the color of that region. The geometrical arrangement of the receptors and the area each covers is shown in Fig. 3.[2] The "circles" in the figure are called *bubbles*, and they are

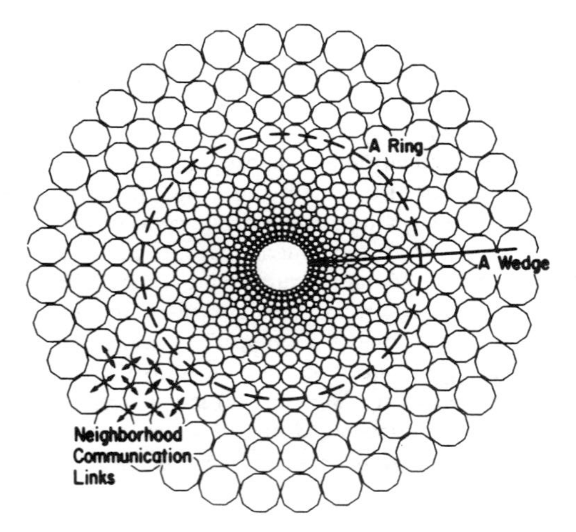

FIG. 3. WHISPER's retina.

arranged in *wedges* (rays emanating from the center) and *rings* (concentric circles of bubbles). The resolution varies across the retina because a larger portion of the underlying diagram is mapped onto a bubble depicted by a larger circle. Since the complete group of receptors is assumed to sense and transmit all signals in parallel, fixations are fast.

Each retinal processor has direct communication links to its nearest neighbors plus one additional link via a common databus connecting all the processors to a supervisory processor called the *retinal supervisor*. The communication topology has been restricted in this simple way to ensure a feasible future hardware implementation.

[2] There are more receptors filling the central blank area of Fig. 3; however, there is still one special case receptor in the very center which must be handled separately. In order to speed up the retinal simulation the bubbles lying in the blank central area can be fixated separately so they are mapped onto only when they are needed.

The bubble processors are each small computers with independent memory. They all simultaneously execute the same procedure; however, each bubble does not necessarily execute the same instruction at the same time. In the current implementation, a call to the LISP evaluator simulates a processor; and LISP MAPping functions simulate the parallel control structure.

Although the bulk of the processing of the perceptual primitives is done in parallel, there is also a small amount of sequential processing which is performed by the retinal supervisor. The retinal supervisor also directs the parallel processing by choosing which procedure the bubbles should execute next and broadcasting this common procedure to them.

2.2. The perceptual primitives

Each perceptual primitive detects a *problem domain independent* diagram feature. The HLR assigns these features interpretations pertinent to the problem it is solving. The current set of implemented perceptual primitives include ones to: find the center of area of a shape; find the points of contact between a shape of one color and a shape of another; examine curves for abrupt slope changes; test a shape for symmetry; test the similarity of shapes; and visualize the rotation of a shape while watching for a collision with another shape.

The CENTER-OF-AREA perceptual primitive is an illustrative example of the general operation of the perceptual primitives. It computes the center of area of a shape relative to the origin defined by the center of the retina. For each piece, ΔA, of the total area we need to compute the x and y components of its contribution to the total area. Dividing the vector sum of these contributions by the total area yields the coordinates of the center of area. Since each retinal bubble receives its input from a fixed sized area of the diagram and is at a fixed location relative to the retina's center, each bubble can independently compute the components of its contribution to the total area. The bubbles whose receptors do not lie over any part of the shape simply do not contribute. The retinal supervisor performs the summation and the division by the total area. A separate primitive computes the total area. It simply totals the area of all the contributing bubbles. If the computed center of area is far from the retina's center its accuracy can be improved by fixating the retina on the estimated center of area and then recomputing. The decision to iterate is made by the retinal supervisor. The accuracy improves because more of the central, high-resolution portion of the retina is used.

It is possible that systematic errors might lead to a discrepancy between the center of area as seen by the retina and the actual center of area of the object in the diagram. This is the case because the diagram-to-retina mapping does not take into account what fraction of a bubble's picture region is covered by an object. The bubble is simply marked whenever any portion of its region is covered. In practice, the accuracy of the center of area test was more than adequate for WHISPER; if necessary the accuracy could always be improved by adding more bubbles to the retina, increasing its resolution.

The center of area is used for more than simply providing the center of gravity of the objects in WHISPER's problem domain. Other primitives (symmetry, similarity, and contact finding) fixate on a shape's center of area before beginning their calculations. For example, if a shape is symmetrical its center of area will be on its axis of symmetry.

Another important primitive is RETINAL-VISUALIZATION. What is "visualized" is the rigid rotation of a shape about the retinal center. While the shape is rotating the collision detection primitive can be called as a demon to watch whether the rotation causes the shape to overlap with another stationary shape. This is useful both in "blocks world" environments involving moving objects and in testing whether two shapes are equivalent under rotation. The process is termed *visualization* because it does not involve modifying the diagram, but instead is totally internal to the retina itself. It simply entails an organized and uniform exchange of information amongst neighboring bubbles.

The geometrical arrangement of the bubble receptors facilitates the visualization of rotations. From Fig. 3 it can be seen that aligning the bubble centers along wedges results in a constant angular separation between bubbles of the same ring when they are from neighboring wedges, and that this constant is independent of the ring chosen. Thus, to rotate a shape clockwise each bubble marked by the shape simply sends a message to its clockwise ring neighbor asking it to mark itself. The sender then erases its own mark. A collision is detected if a bubble receives a message to mark when it is already marked by a shape other than the rotating one. Although the shape is rotated in sequential steps, the time required is still short because

(i) there are, as a maximum, only as many steps to be made as there are wedges on the retina (currently 36); and

(ii) all the message passing and collision checking occurs in parallel during each step.

The coarse retinal resolution means that the visualization process is much faster than the alternative of rotating the object by small increments directly in the diagram. However, the coarse resolution also means that the collision test may falsely predict a collision. Although the collision test may occasionally generate such "false alarms", it will never fail to correctly predict a true collision. The reason for this is that during the diagram-to-retina mapping a point in the diagram is blurred to fill a whole bubble on the retina with the result that the objects in the diagram appear slightly enlarged on the retina. To check out a possible false alarm the HLR

(i) calls the re-drawing procedures to rotate the object in the diagram to the point where the collision is expected,

(ii) fixates the retina at the predicted collision point,

(iii) asks the retina (now with its high resolution center) to see if the colliding objects are touching.

The CONTACT-FINDER primitive establishes the points at which an object touches other objects. The retina is first fixated on the center of area of the object and then the retinal supervisor directs each retinal bubble to execute the following steps:

Step 1. If the bubble value is not the color of the object then stop.

Step 2. For each of its neighboring bubbles do Step (3).

Step 3. If neighbor's value is the color of a different object send a "contact-found" message to the retinal supervisor.

Step 4. Stop.

The retinal supervisor may receive quite a number of messages from bubbles in the contact regions. It must sort these into groups—one for each distinct area of contact. To do this the retinal supervisor sequentially follows the chain of neighborhood links from one contact bubble to another. Each bubble in the chain is put in the same contact group. If no neighboring bubble is a contact bubble, then the chain is broken. Long chains indicate that the objects touch along a surface while short ones indicate that they touch only at a point. The bubble coordinates of the end-points of the chain represent the extremities of a contact surface, and the average of the coordinates of all the bubbles in the group is a good place at which to fixate the retina for a more detailed analysis of the contact.

When two objects touch there is a good chance that one supports the other unless they are just sitting side by side. To determine which object is the supporter and which the supportee, the coordinates of the touching bubbles are compared to find which is "above" the other in the diagram. The assignment of "up" is problem domain dependent and so is made by the HLR.

Another perceptual primitive, FIND-NEAREST, finds the bubble closest to the retinal center satisfying a given condition. For example, to find the object nearest to point P in the diagram the retina is fixated at P and then asked for the nearest marked bubble. The organization of the retina into rings, each an increasing distance from the center, facilitates the search for the required nearest bubble. To find the nearest bubble to the center of the retina satisfying condition C, the retinal supervisor executes the following algorithm:

Step 1. Direct each bubble to test C and save the result (either 'true' or 'false').

Step 2. For $n = 1$ to the number of rings on the retina do Steps 3 and 4.

Step 3. Direct each bubble to report its wedge and ring coordinates as a message to the retinal supervisor if the following hold: (a) it belongs to ring n, (b) its saved value is 'true'.

Step 4. If there is a message pending for the retinal supervisor from step (3), return the coordinates specified in that message (if there is more than one message pick any one of them—all bubbles in a ring are equidistant from the retinal center) to the calling procedure.

This algorithm is a good example of the difference between efficiency in sequential and parallel computation. Since testing C could be a lengthy computation, it is more efficient in terms of elapsed time to simultaneously test C on all bubbles as in Step 1, than to test it for only those bubbles in the scanned rings of Step 3. On a sequential processor it would be best to test C as few times as possible; whereas, on a parallel processor the total number of times C is tested is irrelevant (assuming the time to compute $C(x)$ is independent of x). It is the number of times C is tested sequentially which is important.

The SYMMETRY primitive tests for symmetry about a designated vertical axis by comparing the values of symmetrically positioned bubbles. An object is symmetrical (WHISPER tests for vertical and horizontal reflective symmetry), if each bubble having its "color" value has a symmetrically located bubble with the same value. If when testing the vertical reflective symmetry of a blue object, say, the bubble in the third wedge clockwise from the vertical axis and in the fourth ring from the center has the value 'blue', then the value of the bubble in the third wedge counterclockwise from the vertical axis and in the fourth ring must be checked to see if it is also 'blue'. If it is not, then possibly the discrepancy can be ruled out as insignificant; otherwise, the object is asymmetrical. Neighborhood message passing is used to bring together the values from bubbles on opposite sides of the proposed axis. The technique is to cause whole wedges to shift in a manner perhaps best described as analogous to the closing of an Oriental hand fan. All the bubbles to the left of the axis send their values clockwise, while all those to the right send theirs counterclockwise. Messages which meet at the axis are compared and will be equal if the object is symmetrical.

The symmetry test must be supplied a proposed axis of symmetry. The center of area offers partial information on determining this axis since it must lie on it if the object is symmetrical. This does not, however, provide the orientation of the axis. Although the simplest solution may be to test the object in all of the wedge orientations by using the rotational visualization, if one more point on the axis of symmetry could be found the axis would be uniquely determined. Such a point is the center of the circumscribing circle of the object. The only problem is that thus far I have not managed to devise a quick parallel algorithm for finding this center. Although in some cases they may coincide, in general I expect the center of area and the center of the circumscribing circle to be distinct for objects with only a single axis of symmetry.

An unexpected and interesting property of WHISPER's retinal geometry leads to a simple method, employing neighborhood communication, for scaling the retinal 'image' of an object. The primitive is RETINAL-SCALING. An object is scaled correctly (i.e. without distorting its shape) if each bubble having its value, sends this value to a bubble in the same wedge, but a fixed number of rings away. As long as each value is moved the same number of rings either inwards or outwards from the bubble which originally holds it, the size of the 'image' of the object is changed but its shape is preserved (Fig. 4). This is the case because the constraint of aligning the bubbles into wedges such that each bubble touches all of its immediate neighbors is satisfied by increasing the bubble diameters by a constant factor from ring to ring. For a proof of this see Funt (1976). Scaling an object by neighborhood communication is implemented by having each bubble simultaneously send its value as a message to its neighbor in the same wedge in either the appropriate inwards or outwards direction, and repeating this message passing process sequentially as many times as necessary to bring about the required scaling.

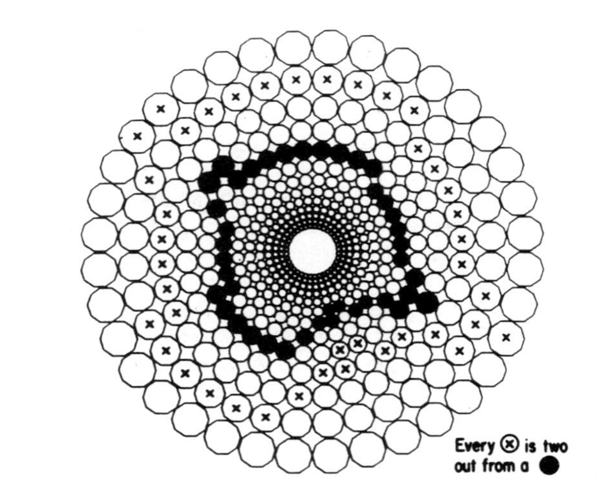

FIG. 4.

The SIMILARITY PRIMITIVE determines whether two objects, A and B, are similar under some combination of translation, rotation and scaling, and if so returns the angle of rotation, direction and distance of translation, and the scale factor. It works by taking one object, say A, and translating, scaling and rotating it so it can be matched with the other. Since the center of area of an object is unique the centers of area of A and B must be aligned if they are to match. Thus the first step is to find the centers of area, and then to translate A. Rather than call the re-drawing transformations to move A in the diagram, its translation can be accomplished entirely on the retina by:

(i) fixating on the center of area of A,
(ii) asking all bubbles not containing A to mark themselves as empty space,
(iii) fixating on the center of area of B while superimposing this new image on the old one.

After translation A must be scaled. If A and B are to match, then their areas will need to be the same; therefore, we must scale A by a factor equal to the square root of the ratio of the areas of the two objects (i.e. scalefactor = squareroot(area(B)/area(A)). The areas of A and B are available as a by-product of the center of area calculation. Now that the objects are aligned and the same size, A is rotated about its center of area using retinal visualization to see if there is any orientation at which it matches B.

CURVE-FEATURES analyses curves. In order to begin, it must first find the

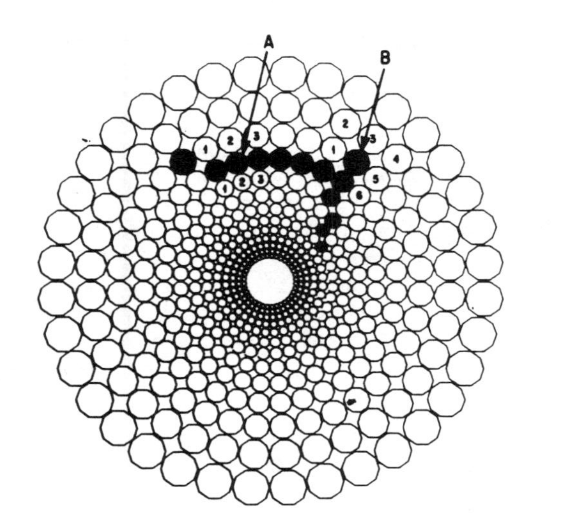

FIG. 5.

retinal bubbles on the curve. Given one bubble on the curve, the others can be found by following the chain of bubbles each having the same value. In WHISPER's diagrams the contours of objects are "colored" a different shade from their interiors, and this helps prevent the curve following process from getting lost tracing chains of bubbles which are part of an object's interior. It is not strictly necessary to color code the object contours, since a contour bubble can be determined by the type of neighbors surrounding it, but coding is cheaper and easier.

Once the set of bubbles on the curve is found, each bubble in the set can individually test for the occurrence of a particular feature; therefore, the whole curve is tested in parallel. A bubble detects a sharp bend in the curve if there is an imbalance in the number of its neighbors on opposite sides of the curve which are themselves not members of the curve. This is illustrated by Fig. 5 in which bubble A has three neighbors on each side of the curve, whereas bubble B has six neighbors on one side and none on the other. Thus, a bubble tests for bends by:

(i) asking its neighbors whether or not they are on the curve, and

(ii) comparing the number of responses originating from opposite sides of the curve.

For a simple closed curve, if the bubble knows which responding neighbors are interior and which are exterior, then it can additionally classify the bend as convex or concave.

The slope of a curve at any curve bubble is determined as the perpendicular to the bisector of the angle between the centers of its neighboring bubbles on the curve. This yields a rough approximation to the actual slope, but it is sufficient for quickly testing whether any drastic slope change occurs over the length of the curve. To more accurately determine the slope at a particular point, the retina is fixated on it for higher resolution. The curve tangent is then the perpendicular to the bisector of the angle between wedges with the most bubbles on the curve. The angle between wedges can be used because they emanate directly from the center of the retina, just as the curve must when the retina is centered on it. This method is more accurate than measuring the angle between neighboring bubbles because there are more wedges than neighbors. The **HLR** mainly uses this test to measure the slope of surfaces at contact points to decide whether or not an object will slide.

2.3. The underlying diagram

We began with the view that the retina is a special purpose parallel processor designed to detect diagrammatic features without saying anything about the precise nature of the diagrams themselves. With the retinal processor in hand, we can now see that the representation of the diagrams is unimportant as long as each bubble receives its correct input. This is analogous to a program which issues a READ command without caring whether the input is coming from a card reader, a file, or a terminal. The method of mapping from the diagram to the retinal bubbles' input must be fast, however, because the retina is re-filled everytime it is fixated at a new diagram location.

There are at least two different types of representing media for the underlying diagram. The first is the conventional medium of visible marks on a two-dimensional surface, usually paper. The map from diagram to human retina is accomplished by the lens of the eye focusing the incoming light. Since there is simultaneous stimulation of the receptors, it is a very fast process.

The second possible type of diagram representation is similar to that used in generating computer graphics. The diagram is specified as a list of primitive elements (in graphics applications, usually line segment equations). In a similar vein, Kosslyn (1975) proposes that human visual imagery is in some ways analogous to the storage and display of graphics images. The parallel processing capacity of WHISPER's retina can be used to quickly map each primitive element into the proper bubble inputs. To mark all bubbles lying on line segment, S, the retinal supervisor directs every bubble to determine independently if it is on S, and if so, to mark itself. Since this simple test—do a circle and a line segment intersect?—is performed by all bubbles simultaneously, the time required is independent of the length of S. The same method can mark all bubbles within any simple shape such as a circle, square or triangle in time independent of its area. Regardless of the type of primitive element, the time taken to "draw" the diagram on the retina is, however, proportional to the number of primitives in its description. They must be processed sequentially.

Due to the lack of true parallel processing, neither of the above two types of diagram representations is used in WHISPER. Instead, the diagram is implemented as a square array. Each array cell denotes a point on a real world, pencil and paper diagram.

2.4. The re-drawing transformations

The re-drawing transformations are the procedures the HLR can call to change the underlying diagram. In WHISPER there are transformations for adding and removing lines, and for rigidly translating and rotating shapes. Other non-linear transformations could be added if required. These re-drawing procedures are of course dependent upon the representation of the diagram they modify, and the ease and efficiency with which they can be implemented could affect the choice of diagram representation.

3. WHISPER in Operation

With the basic mechanisms for interaction with the diagram now understood, it is appropriate to see how they are used in the course of solving a problem. We will consider problems of the type: predict the sequence of events occurring during the collapse of a "blocks world" structure. The structure will be a piled group of *arbitrarily* shaped objects of uniform density and thickness. If the structure is stable, there are no events to describe; if it is unstable, then the events involve rotations, slides, falls, and collisions. WHISPER accepts a diagram of the initial problem state, and produces a sequence of diagrams, called *snapshots*, as its qualitative solution. A quantitative solution specifying precise locations, velocitites, and times is not found; however, deriving one from a qualitative solution should not be too difficult (deKleer (1975)).

Fig. 6 is a typical example of WHISPER's input diagrams. They all depict a side view of the structure. Each object is shaded a different "color" (alphanumeric value) so it can be easily distinguished and identified. Objects' boundaries are also distinctly colored. The diagram depicts a problem, called the "chain-reaction problem", which is particularly interesting because the causal connection between objects B and D must be discovered.

3.1. The Qualitative HLR

The HLR is the top level of the WHISPER system. It is solely responsible for solving each problem; the diagram and retina are simply tools at its disposal. It consists of procedural specialists which know about stability, about the outcome of different varieties of instability, how to interpret each perceptual primitive, and how to call the transformation procedures to produce the solution snapshots. There are two types of instabilites—rotational and sliding. For clarity, sliding instabilities will not be discussed for the present. Operation of the system follows the steps:

Step 1. Determine all instabilities.
Step 2. Pick the dominant instability.
Step 3. Find pivot point for rotation of unstable object.
Step 4. Find termination condition of rotation using retinal visualization.
Step 5. Call transformation procedure to modify diagram as determined in Step 4.

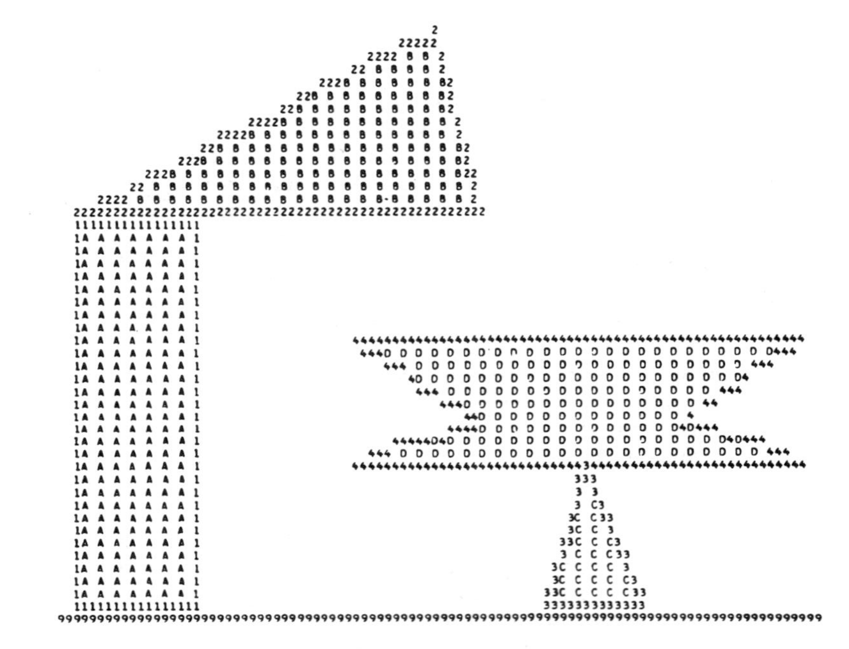

FIG. 6. Chain reaction problem.

Step 6. Output modified diagram as a solution snapshot.
Step 7. Use snapshot from Step 6 as input and restart from Step 1.

In what follows we elaborate on each of these steps.

The diagram and retina are an invaluable aid to the HLR in discovering what stops an object's motion, and in accomplishing the necessary state change. The chain reaction problem demonstrates this. The stability specialist directs the retina to fixate at numerous locations while perusing the diagram, and from an analysis (discussed below) of the visible support relationships determines that B is the only unstable object. B will pivot about the support point closest to its center of gravity. The retina is fixated there (the top right corner of A), so B's rotation can be visualized. As the object rotates, two events are possible. It may collide with another object, or it may begin to fall freely. The conditions under which either of these occur are monitored during the visualization. From this simulation of B's rotation, its collision with D is discovered, and its angle of rotation and location of first contact with D are found. Because of the coarseness of the retinal resolution, this angle of rotation is only approximate. This approximate value is used in conjunction with feedback from the diagram to refine the angle of rotation as follows. First the re-drawing transformations are called to produce a new diagram (Fig. 7) in which B

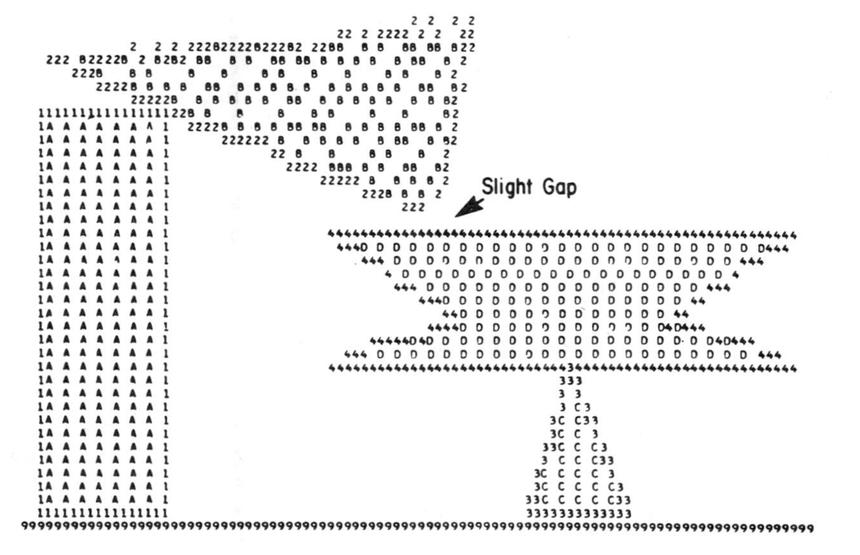

FIG. 7.

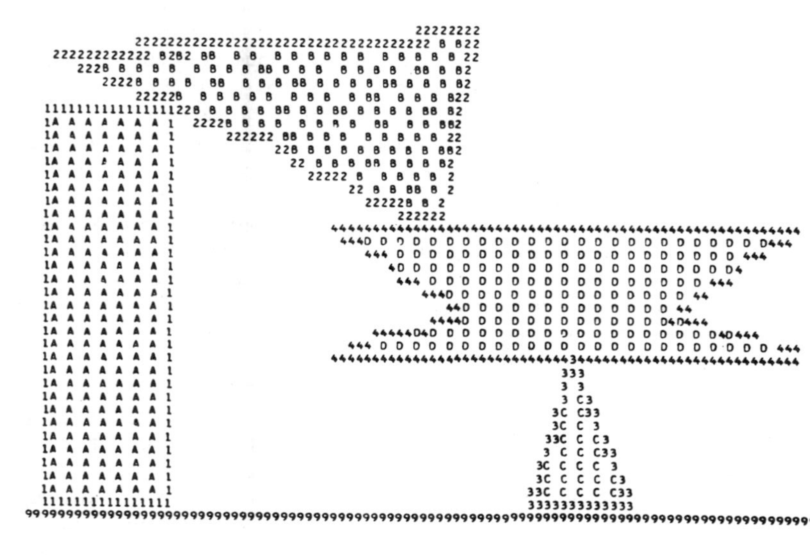

FIG. 8. First snapshot.

is rotated by slightly less than the estimated value. The rotation is made on the short side so that *B* will not overshoot. The retina is then fixated on the anticipated point of collision so that the gap between the two objects can be examined. If there is none, then the update is complete; however, if there is, then *B* is rotated again until the gap is closed. The resulting diagram (Fig. 8) is output as WHISPER's first snapshot of the solution sequence.

3.2. Motion discontinuities and experimental feedback

There are several important observations to be made at this point. One is that discovering the reason for the interruption of an object's motion, accomplished so simply here for *B* through visualization, is generally found to be a very difficult problem. Physics provides equations for object motions, but these equations describe a condition which theoretically lasts indefinitely. They do not indicate when new boundary conditions should take effect. Certainly it is possible to design a set of special heuristics specifying when and where collisions are most likely to occur (e.g. below the rotating object). However, it is quite probable that the collision occurring in Fig. 9 would be overlooked, whereas WHISPER's visualization process would detect it as a matter of course.

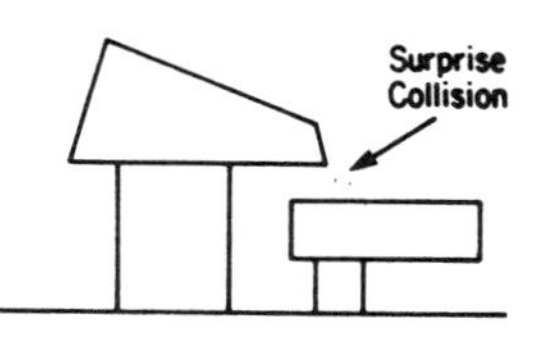

FIG. 9.

WHISPER relies on *experimental feedback* to successfully update its diagram in its method of visualization followed by gap closure. This method is basically a pragmatic equivalent to the unfeasible experiment of rotating the object in the diagram by very small increments until a collision occurs. Usually feedback is thought of in terms of a robot immersed in a real world environment. In WHISPER's case, however, the feedback is from a situation analogous to that in the real world—the diagram and diagram transformations—rather than from observation of actual falling objects. Alternatively, we can say that WHISPER is using M' to derive results about M. Using this feedback WHISPER is able to find when and where discontinuous changes in an object's motion occur without being forced to use sophisticated, "number-crunching" algorithms for touch tests (see Fahlman (1973)) for arbitrary shapes.

3.3. The frame problem

Once WHISPER has produced the first snapshot, it is ready to compute the next one. All the information the HLR needs for this is contained in the first snapshot diagram. Thus to produce the next snapshot, the HLR takes its last output snapshot as input, and begins processing exactly as if it were working on a fresh problem. Although some results derived while working on the previous snapshot remain valid (e.g. some contact relationships will still hold), many will be inapplicable to the new problem. It is easier to disregard this old information than to sort it out and update it, since the retina provides a fast and efficient method of fetching it from the new diagram.

The problem of updating a system's representation of the state of the world to reflect the effects of actions performed in the world is the *frame problem*. Raphael (1971) and Hayes (1971, 1976) discuss it in detail. The transition between WHISPER's snapshots is exactly the type of situation in which we expect the frame problem to arise. It involves the representation of action, the effects of action, and the discovery of chains of causal connection. However, because WHISPER relies on a diagram as a representation of the state of the world, it remains under control. For WHISPER the state of the world is represented by the state of the diagram, and action in the world is represented by corresponding action in the diagram. The corresponding action is the application of the appropriate transformation, and the effects of the action are correctly represented by the resulting state of the diagram.

In WHISPER's current problem, the HLR knows that the action of B's rotation is represented by calling the rotation transformation procedure to re-draw B at its new location in the diagram. Almost all of the information that it needs to continue its problem solving is correctly represented by the updated diagram. It can proceed just as if the new snapshot (the updated diagram) were its original input and it were starting a brand new problem. The most important information which has changed in the transition between the states as a result of the rotation is: the position and orientation of object B; the position of its center of area; the contacts it makes with other objects; and the shape of the areas of empty space. There are also a multitude of things that will not have changed in the world and are correctly left unchanged by the rotational transformation, such as the position of all the other objects, the shape of all objects, the area of all objects, and the contact relationships of other objects not involving B. All of these things work out correctly without the need of any deduction or inference on WHISPER's part. All that it need do is to use its retina to look at the diagram and extract whatever information it needs from the updated diagram.

An expanded WHISPER system could not completely avoid the pitfalls of the frame problem because not all of the information about the current state of the world (e.g. velocities) can be represented by the state of the diagram.

3.4. Subsequent snapshots of the chain reaction problem

The analysis producing the second and third snapshots is very similar to that for the first. In Fig. 8, B's weight on D causes D to be unstable. Its rotation is visualized with the retina fixated at the peak of C leading to the discovery of its collision with the table. The diagram is updated to produce the second snapshot, Fig. 10, which is again input for further analysis. B now lacks sufficient support, and topples to hit D again as shown in Fig. 11. The complexity of the problem rises sharply at this point, and WHISPER's analysis ends, as, I believe, would most peoples'.

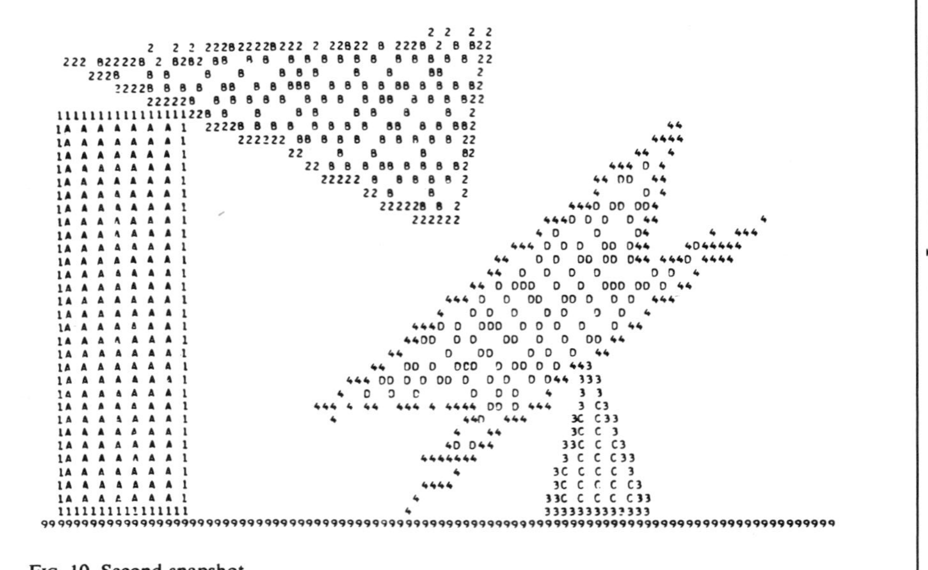

FIG. 10. Second snapshot.

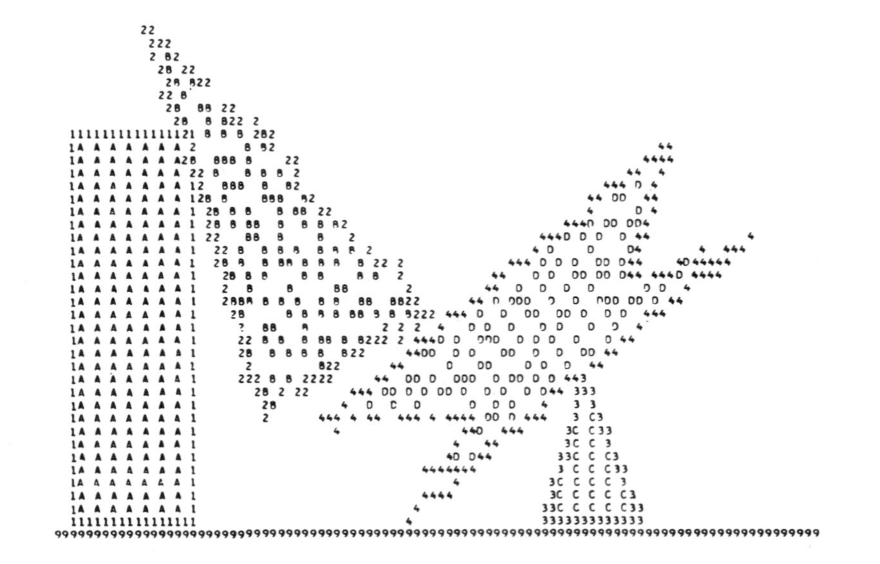

FIG. 11. Final snapshot.

B and D could be shown to fall simultaneously (WHISPER currently does not) by rotating D only part of the way to the table before allowing B to catch up, and then iterating this process a few times until D reaches the table.

3.5. Some limitations of WHISPER's qualitative knowledge

WHISPER's knowledge of Physics is far from comprehensive. As mentioned above, one obvious limitation is that a snapshot by its very nature portrays all objects as stationary, whereas some may be moving. To take velocities into account requires the addition of a quantitative reasoning component to the HLR's qualitative knowledge. Knowledge of velocity, acceleration, momentum and moments of inertia would have to be represented in terms of equations. The HLR's current qualitative predictions can be used to guide the application of these equations in the search for a quantitative solution.

Another limitation is that WHISPER approximates simultaneity by moving objects one after another. This process works for problems like the one discussed above; but this approximation is insufficient in some cases where two or more objects move at a time. In Fig. 12, for example, if B is moved after A is moved, then they will not

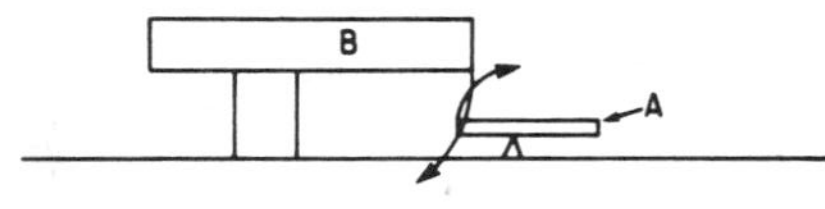

FIG. 12.

collide; however, if they are moved simultaneously they will collide. Again we can make use of the diagram by shading the areas each object will sweep through. If no two shaded areas overlap then there will not be a collision; wherever they do overlap a collision might occur and further quantitative analysis of the angular velocities of the objects is required.

3.6. Slide problems

Unstable objects may also slide. When Fig. 13 is given to WHISPER its reasoning up to the point where it generates the first snapshot, Fig. 14, is the same as for the chain reaction problem. At this point it is faced with a problem involving a sliding object. Although the basic outline of the solution process for slide problems—test stability, find termination point of motion, update diagram, output snapshot, restart with the output as input—is the same, there are some essential differences in handling sliding objects. The most important arises because it is not possible to visualize the slide of an object down an arbitrary curve. What WHISPER does instead is examine the curve itself with its retina.

A variety of conditions can terminate an object's slide. For example, there may be a sharp rise (a bump), a sharp fall (a cliff), or a hill which is higher than the starting point. Also the object may slide into another object resting on or near the surface.

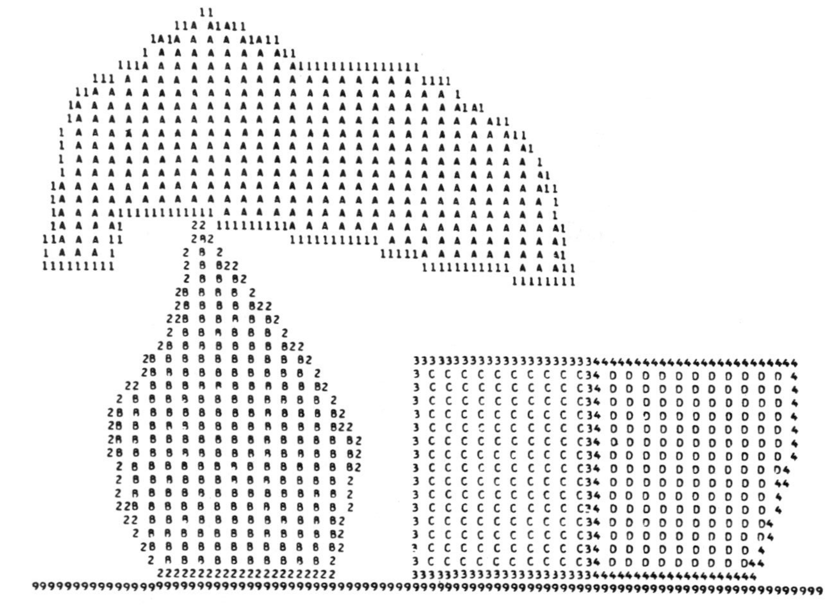

FIG. 13. A problem with a sliding object.

These conditions are illustrated in Fig. 15. In the current implementation WHISPER fixates only once (at the starting point of the slide) to test for these conditions. Multiple fixations at regular intervals along the curve would improve the accuracy of the tests.

The case shown in Fig. 15(e) of a "surprise collision" is one which requires multiple fixations. Although WHISPER does not yet handle this situation, it is clear how it easily could as illustrated in Fig. 16. In the figure an x indicates a fixation point, a semi-circle indicates the area of the diagram to be checked by the retina at each fixation (checking a circular region is easy because of the retina's ring structure), and the space between the dashed line and the surface indicates a clear "corridor" for the object. The radius of the semi-circle is a function of the object's size and the fixation interval. The same sized corridor can be examined with fewer fixations by using a larger radius. The only disadvantage is that the probability of false alarms is increased because the distance between the dashed line and the circumference of the semi-circles is greater. A false alarm can be investigated by making more fixations in the region where it occurs. This method of detecting collisions is good for two reasons:

(i) because the retina can check large segments of space in a single glance, the number of fixations required to examine the space near the surface is relatively small;

(ii) a collision will never be missed.

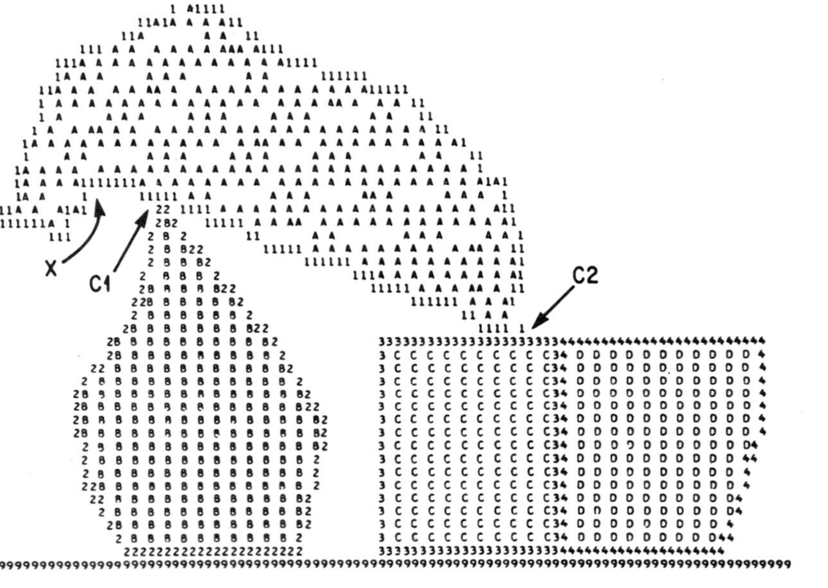

FIG. 14.

The HLR, in addition to specifying which conditions the retina should look for, must specify which curve segments it should look at. There are two kinds of curves to test:

FIG. 15.

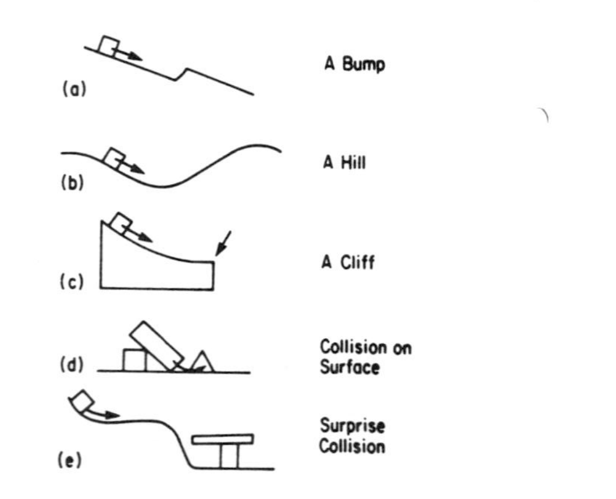

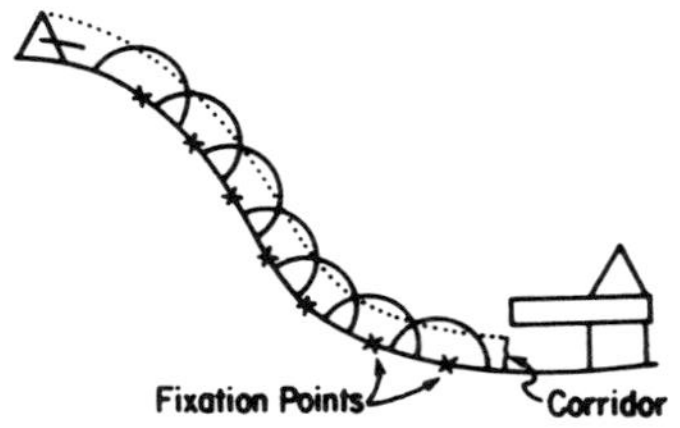

FIG. 16.

(i) those representing surfaces on the moving object (its underside) which will slide past a point on a stationary object, and

(ii) those representing surfaces of stationary objects (their topsides) which will have a point on the moving object ride over them.

Thus in the current example (Fig. 14) the HLR directs the retina to examine the surface of A from C1 to the left, and the surface of C, possibly continuing over to the surface of D, from C2 to the right. If there is more than one reason why the object's slide will end, then only the condition which occurs first (i.e. after the object has slid the shortest distance) is relevant. One more finepoint is that tests for some conditions, for example collisions, need only be made when the surface is a topside and not an underside.

3.7. Updating the diagram to reflect a slide

After the curve examination is complete and the spot where the object's slide will end is known, the next step is to update the diagram so that it will show the object at its new location. First the HLR calls the re-drawing procedures to translate the object. This is shown in the change from Fig. 14 to Fig. 17 in which point X is aligned with C1. This does not complete the diagram update however, since the object's orientation will most likely change during its slide. The contacts between the object and the surface it slides along should be the same when the slide ends as when it began. Knowing this the HLR can determine the object's correct orientation using retinal visualization. It directs the retina to fixate at the object's new location and then visualize its rotation while watching for the original contact relationships to be re-established. The angle of rotation is returned to the HLR which then calls the re-drawing procedures to rotate the object by that amount in the diagram. As before, the angle returned by the retina is only approximate so the HLR directs the retina to fixate on the expected point of contact and check for any remaining gap. If there is, a second corrective rotation is made. The resulting snapshot is Fig. 18. This two-step method—translation followed by a corrective rotation—works for curved as well as straight surfaces.

What we can see from all this is how experimental feedback combined with a first order theory of sliding motions results in a very natural form of qualitative reasoning.

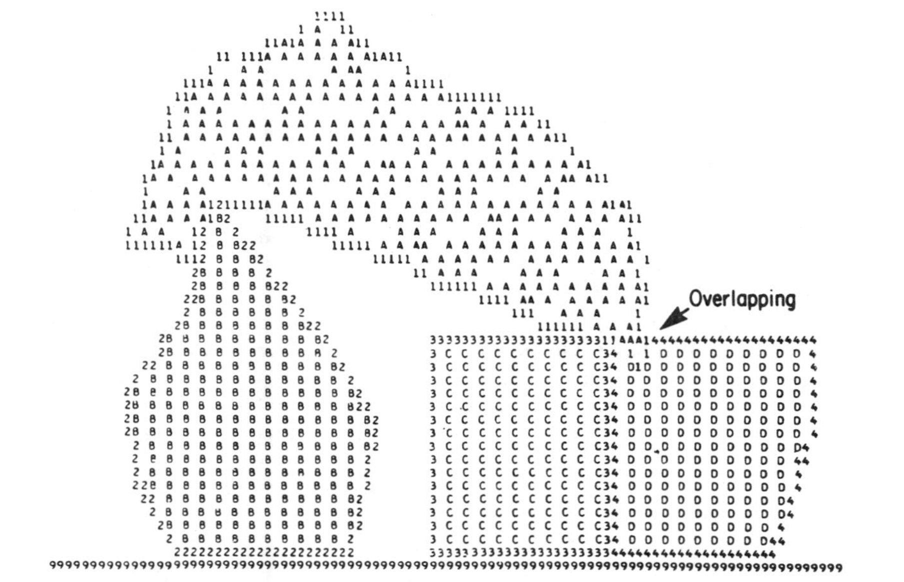

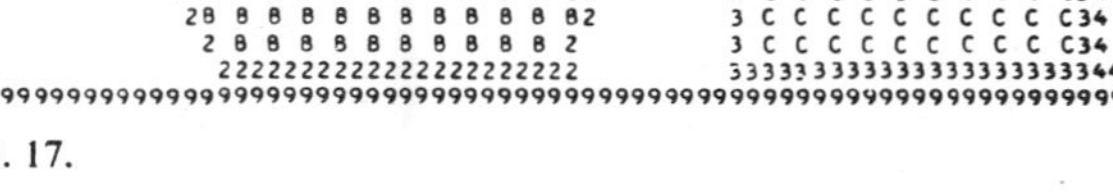

FIG. 17.

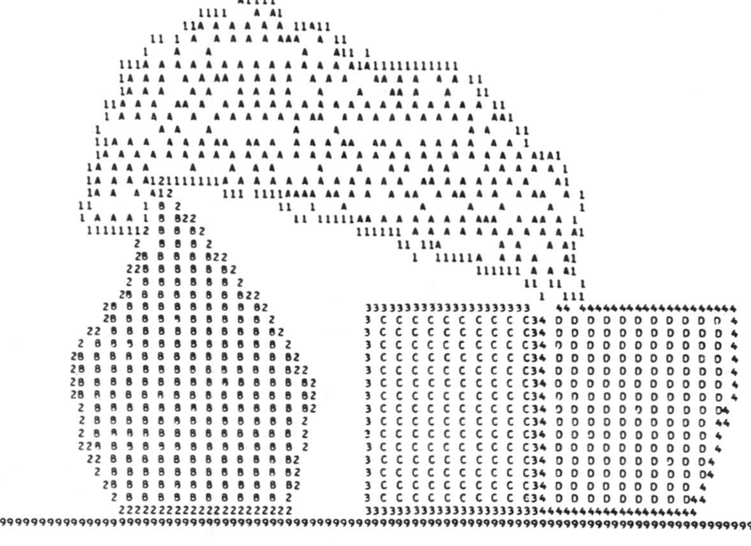

FIG. 18. Final snapshot of the sliding-object problem.

3.8. Benefit of the diagram during slide analysis

In the curve examination and diagram update process, the diagram is very useful to the HLR in the course of curve following, and it also provides feedback as it did in the case of rotations. The main pitfall in curve following is the possibility that two objects will coincidentally align so that a smooth curve is formed across them both. An object could then begin its slide on one object and continue sliding along the other as A did when it slid across C and onto D. This *emergent* property of the curve must be noticed, and the two curve segments appended. In a system relying on an independent description of each object, this would pose a significant problem because one would require:

(i) that it have a built in expectation that the situation might arise;

(ii) that it continually check for the situation;

(iii) that its check involve testing whether the first object touches any other object in the universe; and

(iv) that it know how to amalgamate the two separate curve segment descriptions into a new curve description.

For WHISPER it does not create a problem because two aligned surfaces of neighboring objects form a continuous curve in the diagram; WHISPER only has to look at this curve, rather than, in a sense, discover or construct it.

4. The Stability Test

Rather than solve the stability problem with a sophisticated general method, as Fahlman (1973) did in his BUILD system, WHISPER seeks qualitative solutions using rules corresponding to those a person untrained in Physics might apply. The HLR has specialists which express rules like: "If an object hangs over too far, it will topple"; and "If an object and one of its supporters make contact along a surface (rather than at a single point) and if this surface is not horizontal, then the object will slide." A frictionless environment is assumed.

4.1. Sub-structuring

Overall organisation of the stability test is based on the observation that a complete structure is stable if each of its independent subparts is stable whenever their supporters are stable. Thus the initial structure is broken down into smaller substructures whose stability as individual units is easier to test than the stability of the structure as a whole. To perform the stability test the HLR first asks the retina for a list of the names of all objects in the scene. Each object, O, is then handled in turn. The retina is used to find whether or not O supports any other object(s). If it does not, then its stability is tested by SINGLE-STABLE, a specialist in individual object stability. At this point the assumption is that O's supporters are themselves stable. If O supports other objects then its stability is not tested, but rather it is amalgamated with its supportees as if it and they were all glued together to form a single conglomerate object, C. If C does not support anything, then its stability

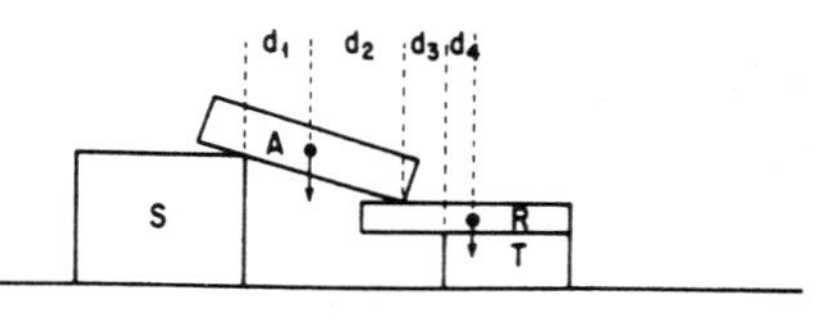

FIG. 19.

is tested by SINGLE-STABLE; if it does support something, then recursively it is combined with its supportees to form a new conglomerate, C', which is then also checked for supportees.

There is an important exception to the above description. When an object is a cosupporter, as one pillar in an arch for example, then it is not amalgamated with its supportee. In this case the object is sent to SINGLE-STABLE for testing with an addendum specifying the point of contact between it and the supportee.

The dotted curves in Fig. 19 encircle the sub-structures which are passed to SINGLE-STABLE. In (b), Q and RS are cosupporters of X, so they are not combined with it.

Incidentally, treating two objects such as A and B as a single object AB is another example of a situation in which two descriptions must be amalgamated. It is a trivial task for WHISPER to amalgamate two object descriptions, since all it need do is interpret their two color codings as the same color.

4.2. Single object stability

As we have seen, the problem of determining the stability of a complete structure is reduced at each stage to the determination of the stability of individual objects. For a single object there are only three basic types of instability. It can either rotate about some support point (rotational instability), slide along a surface (translational instability), or simply fall freely (free fall instability).

SINGLE-STABLE considers the relative positions of an object's center of gravity and its supporting contacts to decide on rotational stability. Consider first the case of an object with nothing on top of it. One with a single support must have its center of gravity positioned directly above the contact region. One with multiple supports must be positioned so that a vertical dropped from its center of gravity passes through either a contact region or the space between two contact regions. The restrictions of uniform density and thickness of objects mean that an object's center of area can be substituted for its center of gravity. SINGLE-STABLE thus sees that an object "hangs over too far" when its center of area falls outside its supports.

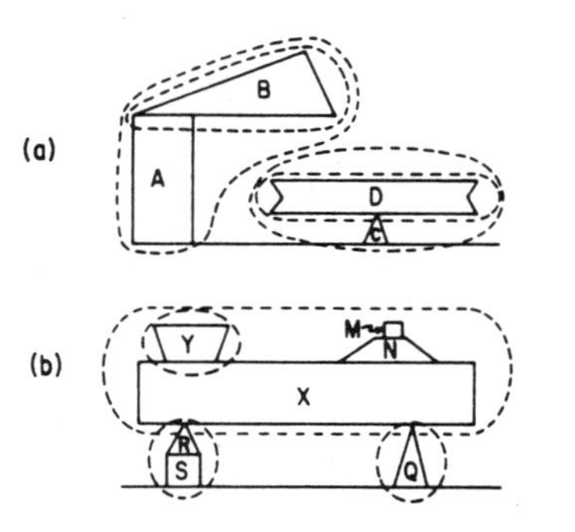

R is stable if $(d_4/d_3)w_R \geqslant (d_1/(d_1+d_2))w_A$
where: w_A is the weight of A
w_R is the weight of R
↓ is an object's center of gravity

FIG. 20.

The stability of an object with something on top of it will be affected by the extra weight. Because of the way in which objects are formed into conglomerates before they are passed to SINGLE-STABLE, if one supports something then it must in fact be one of two or more cosupporters. Let us say SINGLE-STABLE is testing the stability of an object R which, along with cosupporter S, supports A. First it checks the easy cases:

(i) if ignoring A, R is already rotationally unstable and A's weight will only add to this instability, then R rotates;

(ii) if ignoring A, R is already rotationally unstable, but A's weight might counteract its rotation, then this is a counterbalancing type problem which is too difficult to handle without further quantitative investigation;

(iii) if A, no matter how heavy it is, will not topple R (i.e. test R's stability under the assumption that its center of gravity is located at the contact point between it and A) and R ignoring A is stable, then R remains stable.

The most difficult case is when ignoring A, R by itself is stable, but A may or may not be heavy enough to cause it to rotate. In this situation the location of A's center of gravity relative to its support must be considered. These distances are shown in Fig. 20. If $w_R(d_4/d_3) \geqslant w_A d_1/(d_1+d_2)$ then R is stable; otherwise, it will rotate. SINGLE-STABLE cannot handle objects which participate in two or more cosupport relationships. Fig. 21 shows two problems the stability test does not handle.

Objects, such as D in Fig. 6, which are balancing in an unstable equilibrium provide a special problem. Since the slightest deviation in the location of D's center

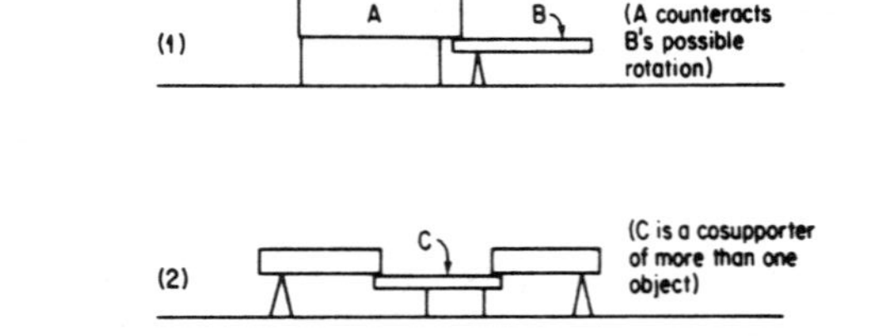

FIG. 21. Stability problems WHISPER cannot handle.

of gravity would upset the balance, it must be known precisely. The CENTER-OF-AREA's estimation is not sufficiently accurate. We expect D to balance because it is symmetrical. Using the retina's symmetry test, WHISPER draws the same conclusion. For it to balance, D must be symmetrical about a vertical axis through its support point. Since D is, the stability test reports it as stable; if it were not, then the stability test would have to report that it was unable to decide.

4.3. The eye movement protocol for the chain reaction problem

During the problem-solving process the retina is constantly moving from place to place in the diagram. A trace of the eye movements is given by the circled numbers

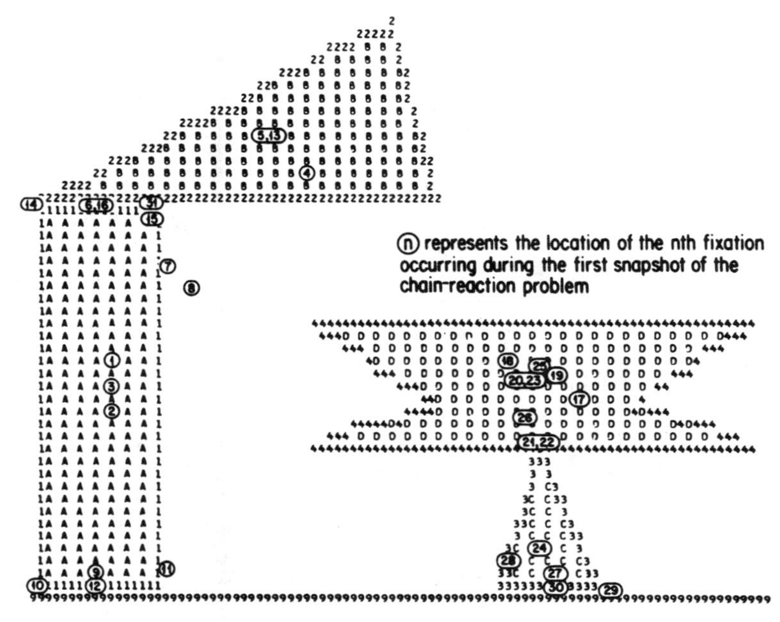

FIG. 22.

in Fig. 22 and Fig. 7. Each circle represents a fixation of the retina. The numbers give the order in which the fixations occurred. The retina was split so that its central and peripheral portions could be fixated separately. For some fixations in Fig. 22 only the center of the retina was used, while for others only the periphery. Although moving the two parts of the retina separately would be unnecessary if there actually were many processors operating in parallel, it saves a considerable amount of computation in the pseudo-parallel simulation. What follows is a list of the plotted fixations accompanied by the HLR's reasons for requesting them.

(1) Move to center of diagram; return names of all the objects in the scene.

(2–3) Find the center of gravity of A; find supportees of A.

(4–5) Find the center of gravity of B; find supportees and supporters of B.

(6) Move central section of retina; find exact contact point of A and B.

(7–8) Find center of gravity of AB; find supporters of AB.

(9) Move central section; find exact contact point of AB with table.

(10–11) Move central section; find extremities of contact surface.

(12) Find the slope of the contact surface.

(13) Move to center of gravity of B and look at contact between A and B.

(14–15) Move central section; find extremities of contact surface between A and B; (5, 72) and (19, 72) are returned.

(16) Determine the slope of the contact surface.

(17–20) Find center of gravity of D; look for supporters and supportees.

(21–22) Move both the central section and the periphery; find the exact point of contact with C. Discovers that support is a point not a surface indicating possible equilibrium situation.

(23) Move back to center of gravity of D to check for symmetry of D; equilibrium is found to be o.k.

(24) Finding center of gravity of C; look for supportees of C.

(25–26) Finding center of gravity of CD; find supporters of CD; finds the table.

(27) Move central section; find exact point of contact of CD with table.

(28–29) Move ceptral section; find extremities of contact surface; returns (64, 22) and (76, 21).

(30) Determine the type of contact and its slope.

(31) Move to the pivot point of the rotation of B to visualize the rotation. **** The rotation is then carried out in the diagram, see Fig. 7.****

(32) Move central section to estimated point of collision between B and A to see if they touch; the gap is seen; the amount of the next rotation is estimated. **** Another rotation is carried out in the diagram, see Fig. 8.****

(33) Move central section to estimated point of collision between B and A (the same as (32)); now they are seen to touch.

Although it would be rash to claim that WHISPER accurately models human problem-solving, the eye-movement protocol provides an unusual possibility for testing such a conjecture. An eye-tracking system could be used to record the eye movements of a human subject while he solves one of WHISPER's problems. This record could then be compared with WHISPER's protocol.

4.4 Translational stability

WHISPER decides translational stability by examining the object's contacts. There are three types of contact that are considered: surface-to-surface, surface-to-point, and point-to-surface. The stability criterion for a particular contact is whether or not the tangent to the surface involved in the contact is horizontal at the point of contact. (Tangents are found by the CURVE-FEATURES perceptual primitive.) If the tangent is not horizontal, then the direction of downward tilt is taken as the resultant direction of motion of the object. If a conflict in the direction arises—one contact indicating leftward motion and another indicating rightward motion—then WHISPER reports that it is unable to decide what the motion will be. In these situations a quantitative investigation is needed in order to resolve the qualitative ambiguity. (Resolving qualitative ambiguities by quantitative reasoning is discussed by deKleer (1975).) There is, of course, no conflict between a horizontal contact slope and a non-horizontal contact slope, the former simply does not contribute to the motion. After A rotates to hit C (Fig. 14), the HLR asks the retina to find and classify all the contacts. The A-to-B contact is classified as surface-to-point, with the rightward tilt of the surface of A at the contact noted as contributing to a rightward motion for A. Similarly, the A-to-C contact is classified as point-to-surface with no contribution to the motion of A because the slope of C is horizontal at the contact point. Thus WHISPER concludes that A will slide to the right along the surface of B.

5. Conclusion

WHISPER demonstrates the advantages and feasibility of using diagrams as an aid in problem-solving. We see from a theoretical standpoint that their role is one of a model M' which is similar to the model M—a blocks world structure in the problem domain. More simply stated, WHISPER's diagrams and diagram re-drawing procedures, M', are analogous to blocksworld situations, M. A fundamental component of the system is the retina which blends sequential and parallel processing while limiting the quantity of processors and processor interconnections to a fixed, not too large number. By asking questions of the retina, the HLR is able to obtain experimental feedback from M', and hence results about M—information it would otherwise have to deductively infer from general principles and assertions describing M.

ACKNOWLEDGMENT

I have benefited from the inspiration and assistance of Raymond Reiter, Alan K. Macworth, Richard S. Rosenberg, Gordon McCalla, Peter Rowat, Jim Davidson and Stuart C. Shapiro. I have also received valuable criticism from Zenon Pylyshyn and E. W. Elcock.

REFERENCES

1. Bobrow, D., Dimensions of Representation, in: D. G. Bobrow and A. Collins (Eds.), *Representation and Understanding* (Academic Press, New York, 1975), pp. 1–35.
2. deKleer, J., *Qualitative and Quantitative Knowledge in Classical Mechanics*, M.Sc. Thesis (1975), MIT, Cambridge, MA.
3. Dunlavey, M., An Hypothesis-Driven Vision System, *Advance Papers of the Fourth International Conference on Artificial Intelligence*, Tibilisi, Georgia, USSR, September 3–8, 1975, pp. 616–619.
4. Fahlman, S., *A Planning System for Robot Construction Tasks*, AI TR 283, MIT (May 1973).
5. Funt, B., WHISPER: *A Computer Implementation Using Analogues in Reasoning*, Tech. Report 76-9, Department of Computer Science, Univ. of British Columbia (1976).
6. Funt, B., WHISPER: A Problem-Solving System Utilizing Diagrams and a Parallel Processing Retina, *Advance Papers of the Fifth International Joint Conference on Artificial Intelligence*, MIT, August 1977.
7. Gelernter, H., Realization of a Geometry-Theorem Proving Machine, in: E. A. Feigenbaum and J. Feldman (Eds.), *Computers and Thought* (McGraw-Hill, New York, 1963), pp. 134–152.
8. Hayes, P., A Logic of Actions, in: B. Meltzer and D. Michie (Eds.), *Machine Intelligence 6* (American Elsevier, New York, 1971), pp. 495–520.
9. Hayes, P., Some Problems and Non-Problems in Representation Theory. *AISB Summer Conference Proceedings* (July 1974), pp. 63–79.
10. Hesse, Mary, *Models and Analogies in Science* (University of Notre Dame Press, 1966).
11. Kosslyn, S. M., Information Representation in Visual Images, *Cognitive Psychology* **7** (1975), pp. 341–370.
12. Minsky, M. and Papert, S., *Perceptrons: An Introduction to Computational Geometry* (MIT Press, Cambridge, MA, 1969).
13. Nagel, E., *The Structure of Science* (Harcourt, Brace and World, 1961).
14. Sloman, A., Interactions Between Philosophy and Artificial Intelligence: The Role of Intuition and Non-Logical Reasoning in Intelligence, *Artificial Intelligence* **2** (1971), 209–225.
15. Raphael, B., The Frame Problem in Problem-Solving Systems, in: N. V. Findler and B. Meltzer (Eds.), *Artificial Intelligence and Heuristic Programming* (Edinburgh University Press, Edinburgh, 1971), pp. 159–169.

The 3D MOSAIC Scene Understanding System: Incremental Reconstruction of 3D Scenes from Complex Images

Martin Herman
Takeo Kanade

Computer Science Department
Carnegie-Mellon University
Pittsburgh, PA 15213

Abstract

The 3D Mosaic *system is a vision system that incrementally reconstructs complex 3D scenes from multiple images. The system encompasses several levels of the vision process, starting with images and ending with symbolic scene descriptions. This paper describes the various components of the system, including stereo analysis, monocular analysis, and constructing and modifying the scene model. In addition, the representation of the scene model is described. This model is intended for tasks such as matching, display generation, planning paths through the scene, and making other decisions about the scene environment. Examples showing how the system is used to interpret complex aerial photographs of urban scenes are presented.*

Each view of the scene, which may be either a single image or a stereo pair, undergoes analysis which results in a 3D wire-frame description that represents portions of edges and vertices of objects. The model is a surface-based description constructed from the wire frames. With each successive view, the model is incrementally updated and gradually becomes more accurate and complete. Task-specific knowledge, involving block-shaped objects in an urban scene, is used to extract the wire frames and construct and update the model.

1. Introduction

It is important for a general vision system to derive three-dimensional (3D) information about a given scene from images and store the information in a coherent manner so that it can be used for various matching, planning, and display tasks. Our goal in developing the **3D Mosaic** system has been to build a full vision system, that is, one that goes all the way from images to symbolic 3D descriptions. Further, we wanted to investigate this process in the context of complex scenes. The result is really a first pass at such a system, and provides us with a better understanding of the components required. This paper describes the system and presents examples of how it is used to interpret complex aerial photographs of urban scenes.

2. The 3D MOSAIC System

The goal of the **3D Mosaic** system is to obtain an understanding of the 3D configuration of surfaces and objects in a scene. The significance of this goal may be demonstrated by the following tasks.

1. **Model-based image interpretation.** A known 3D scene model can provide significant aid in interpreting arbitrary images of the scene [7, 19, 24]. The **3D Mosaic** system performs the task of acquiring such a model of the scene.

2. **3D change detection.** Change detection is a task that determines how the geometry and structure of a scene changes over time. The conventional approach to this task involves comparing and detecting changes in images. However, because of different viewpoints and lighting conditions, changes in the images do not necessarily correspond to changes in the geometry and structure of the scene. If 3D scene descriptions were obtained from the images first, such descriptions could be compared in 3D to determine changes in the scene.

3. **Simulating the appearance of the scene.** If a 3D description of the scene were to be obtained, displays as seen from arbitrary viewpoints could be generated from it. This is useful for tasks such as familiarizing personnel with a given area, and flight planning by generating the scene appearance along hypothetical flight paths.

4. **Robot navigation.** Three-dimensional descriptions of complex environments may be used to make decisions dealing with path planning or determining which parts of the environment to analyze in more detail.

The **3D Mosaic** system deals with complex, real-world scenes (e.g., Fig. 4). That is, the scenes contain many objects with a variety of shapes, the object surfaces have a variety of textures and reflectance characterisics, and the scenes are imaged under outdoor lighting conditions. Because of the complexity, there are many difficulties in interpreting the images, including:

1. Any particular image contains only partial information about the scene because many surfaces are occluded.

2. Even portions of the scene that are visible are often difficult to recover. For example, surfaces with dark shadows cast across them, or with highlights, may be difficult to interpret. Highly oblique surfaces may be difficult to analyze if their resolution in the image is poor.

Our approach to the problems of complexity is to use multiple images obtained from multiple viewpoints. This approach aids interpretation in two ways. First, surfaces occluded in one image may become visible in another. Second, features of surfaces that are difficult to analyze and interpret in one image (such as scene edges and texture) may become more apparent in another image because of different viewpoint and/or lighting conditions.

2.1. Incremental Approach

A large number of views will, in general, be required to obtain a fully accurate and complete description of a complex scene. Typically, all these views will not be simultaneously available, while some may never become available. Many of them will only be obtained gradually through interaction with the scene environment. Our system must therefore have the ability to utilize partial descriptions and incrementally update them with new information whenever a new view happens to become available. As a practical example, consider a robot (perhaps a mobile ground robot or an automatically guided airplane) which is attempting to navigate through an unknown environment. The robot would sequentially acquire images of the environment as it moves about. Information derived from each new image would serve to update its internal model, and this partial model would be used to decide where to go next, or where to analyze in more detail.

We have adopted an approach in which the 3D scene model is incrementally acquired over the multiple views. The views of the scene are sequentially acquired and processed. Partial 3D information is derived from each view. The initial model is constructed from 3D information obtained from the first view, and represents an initial approximation of the scene. As each successive view is processed, the model is incrementally updated and gradually becomes more accurate and complete.

Most previous research efforts at acquiring 3D scene descriptions from multiple views have dealt with relatively simple scenes in controlled environments [2, 8, 9, 18, 22, 25]. This has led, in some cases, to only utilizing occluding contours in the image to form the 3D description [2, 8, 9, 18]. The work of Moravec [20] deals with complex indoor and outdoor scenes, but the 3D descriptions generated by his system consist of sparse sets of feature points. Our system, on the other hand, generates full, surface-based descriptions.

2.2. Overview

A flowchart for the **3D Mosaic** system, showing the major modules and data structures, is displayed in Fig. 1. The input is a new view of the scene, which may be either a stereo image pair or a single image. The stereo pair undergoes stereo analysis, while the single image undergoes monocular analysis. The purpose of these analyses is to obtain 3D scene features such as portions of surfaces, edges, and corners.

The central scene model is a surface-based description which is constructed and modified from these features. Before modifications to the scene model can occur, the 3D features from the new view must be matched to the current model. The scene model may, at any point along its development, be used for tasks such as image interpretation, planning, or display generation. A new view may then be acquired which may further modify the model.

For example, when the stereo analysis component is applied to the images in Fig. 4, the result is the set of wire frames in Fig. 9. The scene model constructed from these wire frames is shown in Fig. 20. When the monocular analysis component is applied to the image in Fig. 10, the result is the set of wire frames in Fig. 17. These, in turn, are converted into the scene model in Fig. 21. Finally, the result of modifying the model in Fig. 20 with a new view is shown in Fig. 27.

3. Stereo Analysis

Most stereo matching methods involve matching low-level image features, such as image intensities [3, 13, 17, 21] or image edge points [3, 12, 21]. Points to be matched may also be chosen as "interesting points", e.g., those with high variance in all directions [6, 20]. Our method involves matching structural features -- i.e., junctions -- extracted from the images. There are several reasons for this.

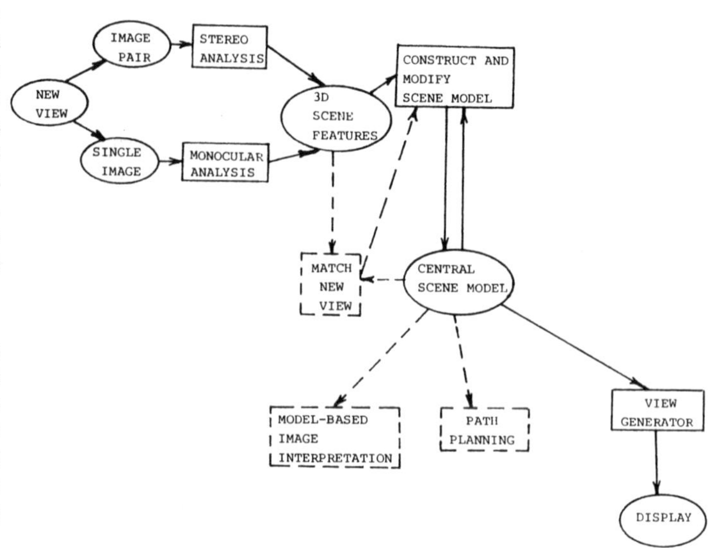

Figure 1: 3D Mosaic flowchart. The dashed lines represent components that have not yet been implemented; the solid lines represent components already implemented.

First, feature-based matching results in more accurate 3D positions for occlusion boundaries than gray scale area matching. Second, by extracting 3D information dealing with scene vertices and edges emanating from them, we obtain portions of boundaries of scene buildings, particularly building corners. These boundaries are then used to construct 3D approximations of the buildings.

Finally, because of our wide-angle stereo images, there are large disparity jumps and large portions of the scene are visible in one image but not the other. Because most stereo systems do not distinguish these from other regions of the image, they try to find matches for them and therefore have trouble [3, 5, 6, 12, 13, 17].

In our approach, rather than attempting to find matches for scene faces occluded in one of the images, we match face boundaries visible in both images. We do this by explicitly taking into account the way junction appearances change from one image to the other, using the knowledge that in urban scenes, roofs of buildings tend to be parallel to the ground plane, while walls tend to be perpendicular to this plane. Edges in the scene perpendicular to the ground will appear in each image to be directed towards the vertical vanishing point [16].

If a feature in an image lies on a roof, its appearance in the other image as a function of position along the epipolar line can be predicted if the normal to the ground plane is known. To see why, consider Fig. 2. Suppose the junction $P_3P_1P_2$ in image1 is given, and our goal is to predict the junction $Q_3Q_1Q_2$ in image2, where the point Q_1 lies anywhere (inside the infinity point) on the epipolar line corresponding to P_1. For the position Q_1, the 3-space position of V_1 can be computed as the intersection of the rays through P_1 and Q_1. This uniquely determines the position of the plane parallel to the ground that contains V_1. The 3-space positions of the points V_2 and V_3 can now be computed as the intersections of this plane with the rays corresponding to the points P_2 and P_3, respectively. Finally, the points Q_2 and Q_3 are uniquely determined as central projections of the points V_2 and V_3, respectively.

Therefore, when an L junction is found in one image, it is initially assumed to arise from a corner of a roof, and its appearance in the other image can be predicted. When an ARROW or FORK junction is found,

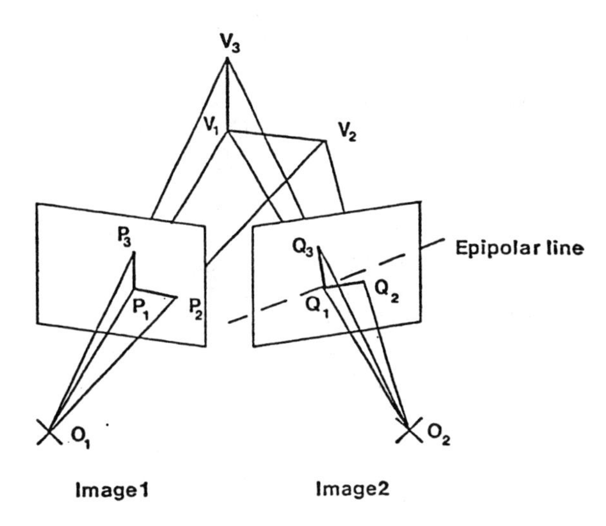

Figure 2: For junction $P_3P_1P_2$, its appearance in image2 can be predicted as a function of position Q_1 along the epipolar line. The normal to plane $V_3V_1V_2$ must be known.

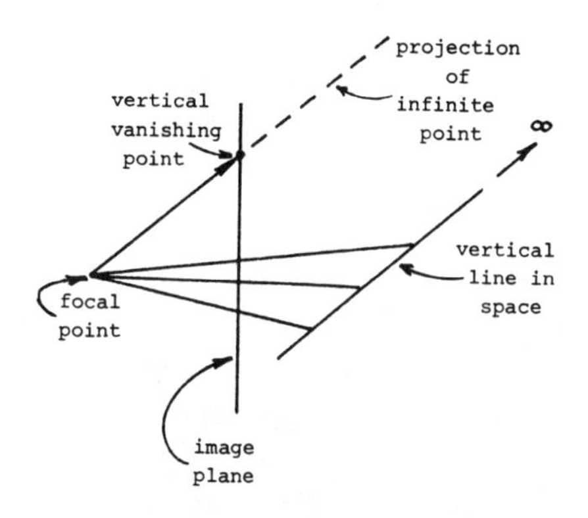

Figure 3: The vector from the focal point to the vertical vanishing point is a 3-space vector in the vertical direction.

the leg of the junction directed towards the vertical vanishing point is initially assumed to arise from a scene edge perpendicular to the ground, while the other two legs are initially assumed to arise from scene edges lying on a roof or on the ground. Again its appearance can be predicted.

Structural relationships between scene vertices are also used to aid in the matching. If two junctions in an image arise from scene vertices at the same height above the ground, the positions of the corresponding junctions in the other image, as a function of position along the epipolar line, can be predicted if the normal to the ground plane is known. This can be shown using similar arguments as before. In Fig. 2, pretend that the points P_i, Q_i, and V_i correspond to positions of separate junctions and vertices. For example, if P_1 and P_3 are two separate junctions in image1, then for some point Q_1 on the epipolar line corresponding to P_1, the position of the junction Q_3, corresponding to P_3, can be predicted if V_1 and V_3 are assumed to lie at the same height. We make the assumption that junctions close to one another in the image often correspond to vertices lying on top of the same building and therefore have approximately the same height.

These matching techniques assume that the vector normal to the ground plane is known. To obtain this vector, we form a vector from the focal point to the vertical vanishing point. As shown in Fig. 3, this results in a 3-space vector in the vertical direction [4], since a line containing the focal point and vertical vanishing point intersects any vertical line at infinity. The focal length and vertical vanisning point are currently manually obtained.

3.1. Steps in Stereo Analysis

We now provide an example showing how the stereo analysis is performed on the stereo pair of images in Fig. 4. First, linear features are extracted by finding edge points, thinning and linking them, and fitting piecewise linear segments. The resulting line images are shown in Fig. 5. Next, junctions are extracted by placing a 5x5 window around each end point of each line and searching for ends of other lines. Junctions that have been found are labeled in Fig. 5. (See [15] for more details.) Notice that many of the junctions correspond to building corners.

We now want to find potential junction matches between the two images. Let us consider how L junctions are matched. Each L junction is initially assumed to lie on a horizontal scene plane. The shape and orientation of its corresponding junction in the other image, as a function of position along the epipolar line, can therefore be predicted. Each L junction in the first image may therefore usually be matched with several junctions in the second image that have, within tolerance, the predicted shape and orientation. However, we do not try to match only with junctions in the second image that have been previously found. Rather, for every point on the epipolar line (on the appropriate side of the infinity point), a search is made within a pre-specified window for lines that might correspond to the predicted junction. The requirements, however, for two lines to form a junction is more relaxed than the requirements during initial junction search. The matching is performed in two directions, from the first image to the second, and vice versa.

At this point, each junction in one image is associated with a set of potentially matching junctions in the other image. The next step is to find the best of the potential matches, resulting in a single match for each junction. Two criteria are used in determining the best matches:

1. If the image intensities inside two potentially matching junctions are similar, the likelihood that they really match is increased. This is because the two junctions will often have similar intensities if they arise from the same face corner. To measure the degree of similarity, we compute the average intensities of regions along the two legs of the L junction in each image. As depicted in Fig. 6, let A and B be the average intensities of these regions in one image, and let A' and B' be the average intensities of corresponding regions in the other image. Then the degree of similarity, called the *local cost*, is defined as

$$C_{local} = |A - A'| + |B - B'|.$$

2. As described previously, if two junctions in an image arise from scene vertices that are at the same height, the relative positions of the corresponding junctions in the other image, as a function of position along the epipolar line, can be predicted. We use this to determine whether two sets of junction matches are consistent with one another. Suppose, in Fig. 7, that the junctions J_1 and J_2 in image1 arise from scene vertices that are at the same height. Suppose also that the junction matches (J_1, J'_1) and (J_2, J'_2) have been hypothesized. To measure the degree of consistency between these two sets of matches, we predict the position of the junction in image2 that corresponds to (say) J_2. Let us refer to the predicted position as J''_2. If the vector from J'_1 to J''_2 is (a_1, b_1) and the vector from J'_1 to J'_2 is (a_2, b_2), then the degree of consistency between the two sets of matches, called the *global cost*, is defined as

$$C_{global} = |a_1 - a_2| + |b_1 - b_2|.$$

To arrive at a unique set of junction matches, the space of potential matches is searched using a beam search [24], which is guided by the above two criteria. The results of this search are displayed in Fig. 8, which shows junctions in one image that have matches in the other image.

Finally, 3D coordinates of vertices and equations of edges are derived using triangulation. Fig. 9 shows a perspective view of the 3D vertices and edges that result. We call this a wire-frame description of the scene.

4. Monocular Analysis

Although stereo is a major source of 3D information, some views of the scene will be only single images. We can also extract 3D information from these images by exploiting task-specific knowledge. We assume that the objects in the scene are trihedral polyhedra containing only vertical and horizontal faces, i.e., faces perpendicular and parallel, respectively, to the ground plane. Our monocular analysis extracts linear structures in the image that represent boundaries of buildings, and then converts these structures into 3D wire frames.

4.1. Steps in Monocular Analysis

This section provides an example showing how the monocular analysis is performed on the image in Fig. 10.

Exracting lines and junctions. The first step is to extract linear segments and junctions from the image. The method used here is the same as that used during stereo analysis (as previously described). First thinned edge points are found, and then lines and junctions are extracted, as shown in Fig. 11.

Locating 2D structures. Next we form linear connected structures in the image by hypothesizing new lines to connect the previously extracted junctions. These connected structures are meant to represent building boundaries and the hypothesized lines are meant to correspond to building edges. The process of hypothesizing connecting lines consists of two steps. First, two junctions may be connected only if a leg of one points at the other, that is, the extended leg meets the other junction. For each pair of junctions that passes this test, a line showing the connection between the two junctions is drawn in Fig. 12.

The second step involves determining which connections shown in Fig. 12 appear as connections in the line image (Fig. 11). For each pair of connected junctions J_i and J_k, we find all segments in the line image that are contained within a thin rectangular window connecting J_i and J_k, and project these segments onto the line connecting the two junctions. Then we consider how much of this line is covered by projected segments. The connection between J_i and J_k is retained only if the percentage of coverage exceeds a threshold. After this pruning step, the junction legs originally extracted in the junction finding step are added, and extraneous legs are deleted. The final connected structures are displayed in Fig. 13.

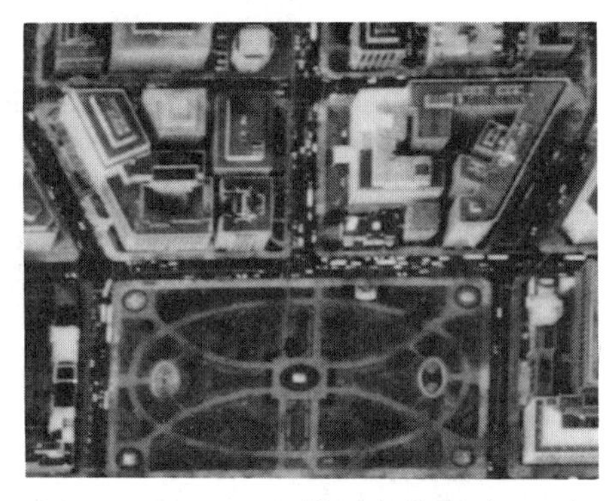
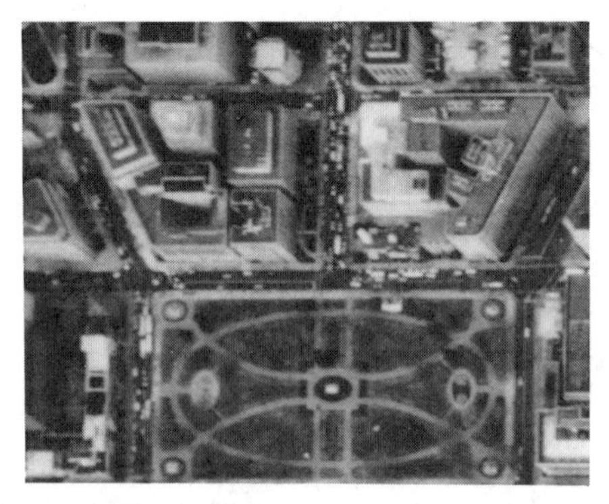

Figure 4: Gray scale stereo images of a region of Washington, D. C.

Figure 5: Fitting linear segments after extracting, thinning, and linking edge points. Junctions are classified as L, A (arrow), F (fork), or T.

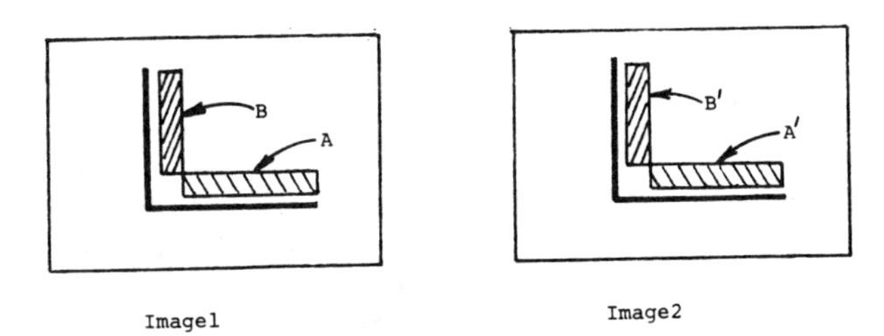

Figure 6: Intensities of corresponding regions of L junctions in the two images are used to compute the local matching cost.

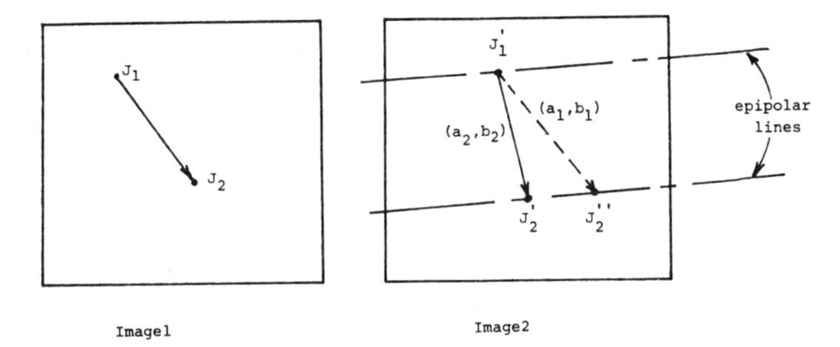

Figure 7: Positional vectors of predicted and actual positions of two junction matches are used to compute the global matching cost.

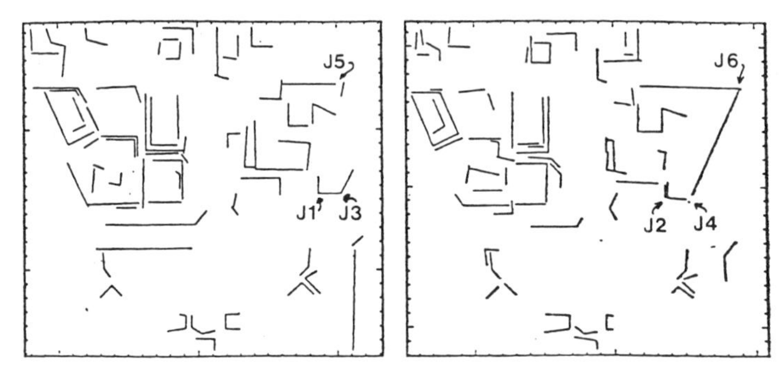

Figure 8: Matches that have been found for the junctions in Fig. 5. Actually, not all matches are correct. For example, although the junction matches (J1,J2) and (J3,J4) are correct, the match (J5,J6) is incorrect.

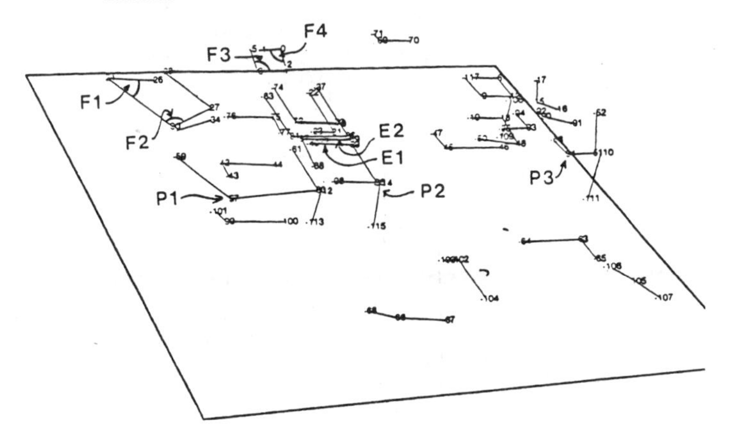

Figure 9: Perspective view of 3D wire-frame description derived from a set of matches similar to those in Fig. 8.

Obtaining 3D wire frames. The next step is to convert the 2D structures into 3D wire frames. First, the lines that form the 2D structures are labeled as either "vertical" or "horizontal" depending on whether or not they are directed toward the vertical vanishing point [16]. Next, we use the position of the vertical vanishing point to calculate the vector in the vertical direction, as described in an earlier section. Let us now consider how to recover the 3D configuration of the junction $p_1p_2p_3p_4$ in Fig. 14.

Suppose that line p_2p_4 has been labeled "vertical" and lines p_1p_2 and p_2p_3 have been labeled "horizontal". Let $\boldsymbol{u}$ be the unit vector in the vertical direction. This vector is normal to all horizontal planes. First we would like to determine the 3-space position of v_2, corresponding to the junction point p_2. Since it is impossible to determine the actual position of this point from a single image without special information, the position is determined as some arbitrary point lying on the ray through p_2, i.e., the depth a of v_2 is arbitrarily chosen. The horizontal plane $v_1v_2v_3$ can now be established, since it contains v_2 and its normal vector is $\boldsymbol{u}$. The 3-space positions of the points v_1 and v_3 can then be computed as the intersections of this plane with the rays through p_1 and p_3, respectively. Finally, the 3-space position of the point v_4 is computed as the intersection of the ray through p_4 with the line through v_2 along the vector $\boldsymbol{u}$.

Although this technique permits us to recover the 3D configuration of any junction relative to some arbitrary depth, it is not useful to apply it directly to the junctions in the original line image (Fig. 11) because the relative heights above the ground plane of the corresponding vertices cannot be determined; the height of each vertex is arbitrarily chosen without relation to the heights of other vertices. It is more useful, however, to apply the technique to the 2D structures in Fig. 13, since the heights of the vertices within each structure can be related. To see how this is done, consider the example in Fig. 15, which shows a 2D structure. Suppose lines p_1p_6 and p_3p_4 have been labeled "vertical", while the other solid lines have been labeled "horizontal". Applying our technique to (say) point p_1, the 3-space positions of the vertices corresponding to points p_1, p_2, and p_6 can be determined relative to some arbitrary depth a for p_1. If the technique is applied next to point p_2, the 3-space position of point p_3 can be determined as a function of the depth a. This procedure continues with points p_6, p_4, and so on, until the 3D configuration of the whole structure has been determined, relative to some arbitrary depth.

In order to obtain a coherent scene description, the depths of the different structures in the scene must be related. We use two methods to do this. The first method involves finding structures that lie on the ground plane. Suppose a junction point p of such a structure is hypothesized to arise from a vertex lying on the ground. Then the 3-space position of the vertex may be obtained as the intersection of the ground plane with the ray through p. The normal vector $\boldsymbol{u}$ to the ground plane is known, but the distance d from the focal point to the ground plane is arbitrarily chosen. Since the 3-space position of all junctions arising from ground points can be calculated in this manner, the depths of all structures containing such points can be related to one another through the parameter d.

To hypothesize junctions that arise from vertices lying on the ground plane, we use the observation that if a line labeled "vertical" connects two junctions (e.g., line p_1p_6 in Fig. 15), the line is directed toward the vertical vanishing point with respect to one junction, but away from this vanishing point with respect to the other junction. The latter junction is assumed to represent a vertex lying on the ground plane. Points p_1 and p_3 in Fig. 15 are examples of such junctions. The 3-space positions of these junctions are then calculated, and their values are propagated throughout their structures as described previously.

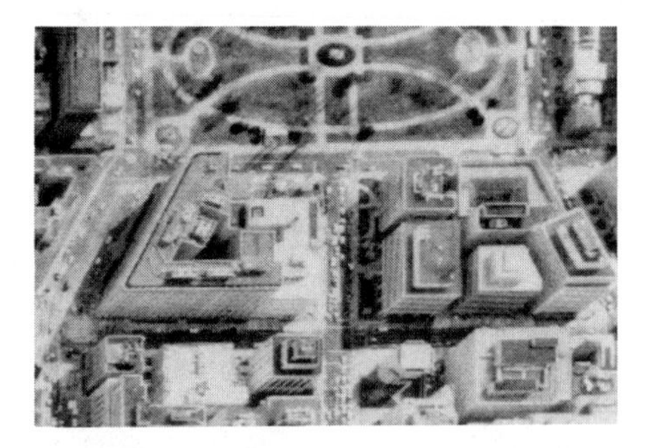

Figure 10: Aerial photograph showing part of Washington, D.C. This is a different view of the same scene as in Fig. 4.

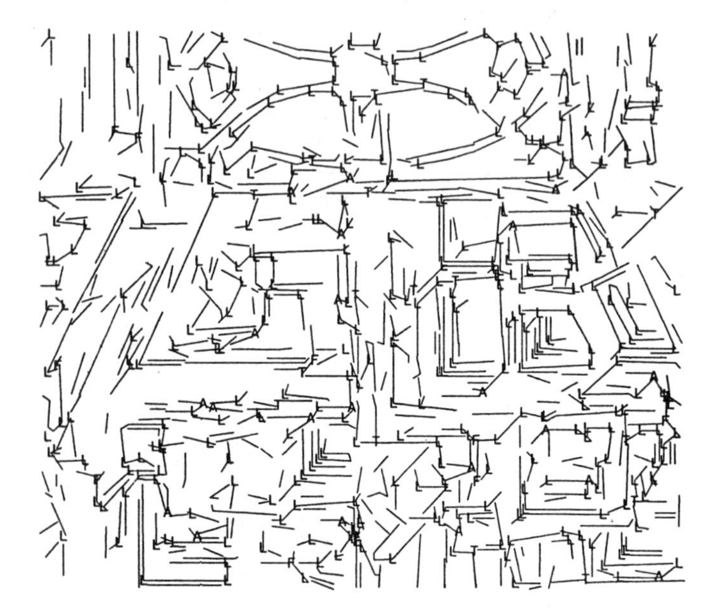

Figure 11: Lines fitted to edge points extracted from Fig. 10. Junctions in the image are classified as L, A (arrow), F (fork), or T.

Figure 12: Each line represents a possible connection between the junctions at its two end points. Each end point corresponds to a junction in Fig. 11.

Figure 13: Result after pruning junction connections in Fig. 12, and adding junction legs originally extracted in the junction finding step.

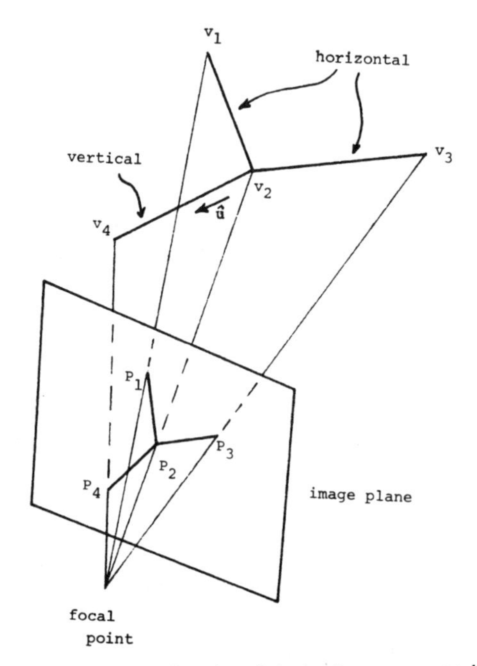

Figure 14: The 3D configuration of the junction $p_1p_2p_3p_4$ can be recovered under assumptions explained in the text.

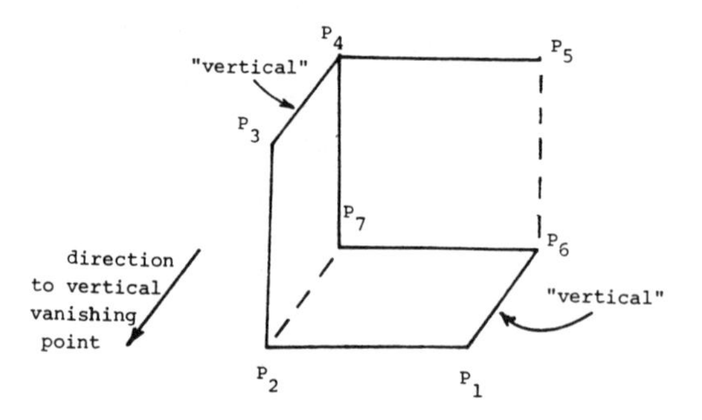

Figure 15: The solid lines represent a connected 2D structure. The dashed lines are for the reader's convenience to make the 3D shape more apparent.

There are many structures in Fig. 13 that do not contain points lying on the ground plane. Nevertheless, the heights of some of these structures can be determined using the rule that if two lines are aligned in the image, they are often aligned in 3-space. Suppose, in Fig. 16, that points p_1 through p_7 have already been assigned 3D coordinates, and we want to obtain the 3-space position of the 2D structure $p_8p_9p_{10}p_{11}$. Since the lines p_6p_7 and p_8p_{11} are aligned in the image and both are labeled "horizontal", they are assumed to be aligned in the scene and to lie in the same horizontal plane. The 3-space position of (say) point p_8 is therefore determined as the intersection of this plane with the ray through p_8. The 3D coordinates of this point may then be propagated to points p_9, p_{10}, and p_{11} as described previously. Note that all 3D positions are functions of the parameter d, which is arbitrarily chosen for the equation of the ground plane.

Fig. 17 depicts a perspective view of the 3D wire frames obtained using these methods.

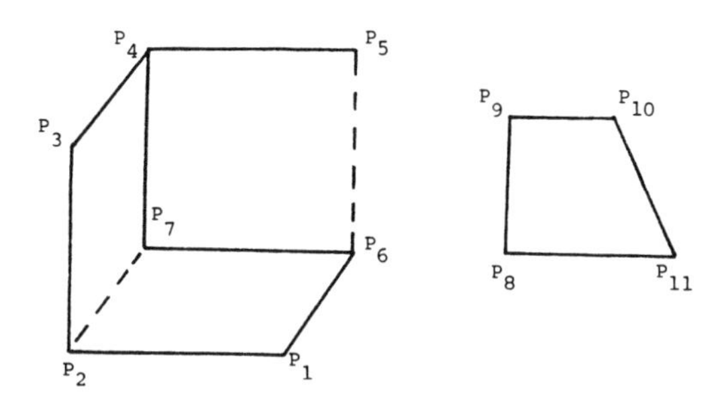

Figure 16: If the 3D configuration of the structure on the left has been determined, the relative 3D position of the structure on the right may also be determined because lines p_6p_7 and p_8p_{11} are aligned.

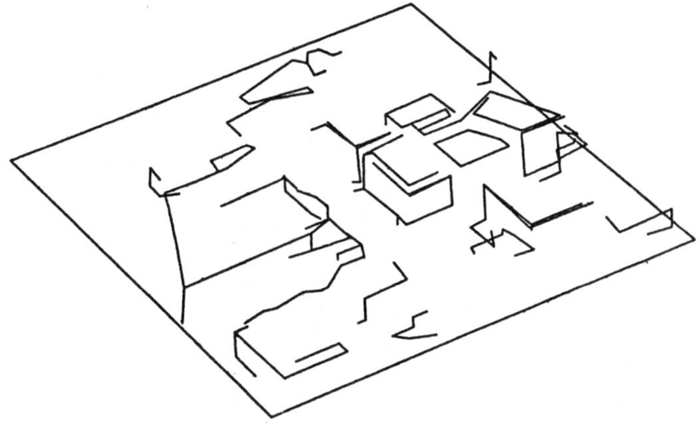

Figure 17: Perspective view of 3D wire frames generated from Fig. 13.

5. Representing and Manipulating the 3D Scene Model

The representation we have developed for the 3D scene model draws on ideas from geometric modelling used in computer-aided design systems [1, 23]. In these systems, however, the 3D models are usually derived through interaction with a user. Our case is different in that (1) the 3D models are derived automatically from 2D images, and (2) many portions of the scene are unknown or recovered with errors because of occlusions or unreliable analysis.

The following factors have determined how the scene model is represented and manipulated.

1. Partially complete, planar-faced objects must be efficiently described by the model. It is therefore represented as a graph in terms of symbolic primitives such as faces, edges, vertices, and their topology and geometry. Information is added and deleted by means of these primitives.

2. The model must be easy to use in matching.

3. Because scene approximations are often more useful if they contain reasonable hypotheses for parts of the scene for which there are partial data, we introduce mechanisms that permit hypotheses to be generated, added, and deleted.

4. Because incremental modifications to the model must be easy to perform, we introduce mechanisms to (a) add primitives to the model in a manner such that constraints on geometry imposed by these additions are propagated throughout the model, and (b) modify and delete primitives if discrepancies arise between newly derived and current information.

The 3D structure in the scene is represented in the form of a graph, called the *structure graph.* The nodes and links represent primitive topological and geometric constraints. The structure graph is incrementally constructed through the addition of these constraints. As constraints are accumulated, their effects are propagated to other parts of the graph so as to obtain globally consistent interpretations.

Nodes in the structure graph represent either primitive topological elements (i.e., faces, edges, vertices, objects, and edge-groups (which are rings of edges on faces)) or primitive geometric elements (i.e., planes, lines, and points). Face, edge, and vertex nodes are tagged as either *confirmed* or *unconfirmed.* Confirmed means that the element represented by the node has been derived directly from images. Unconfirmed means that the element has only been hypothesized.

The primitive geometric elements serve to constrain the 3-space locations of faces, edges, and vertices. Plane and line nodes contain plane and line equations, respectively. Point nodes contain coordinate values. The structure graph contains two types of links: the *part-of* link, representing the part/whole relation between two topological nodes, and the *geometric constraint* link, representing the constraint relation between a geometric and topological node.

Fig. 18 shows a simple example of a structure graph consisting of two objects, *ob1* and *ob2.* Arrows with single lines represent part-of links, and arrows with double lines represent geometric constraint links. The faces are represented as f_i, the edge-groups as g_i, the edges as e_i, and the vertices as v_i. The graph shows one point node *pt* and one plane node *pl.* Further details on representing and manipulating the 3D scene model may be found in [15, 14].

6. Generating the 3D Scene Model

The result of image analysis is a 3D wire-frame description that represents 3D vertices and edges which correspond to portions of boundaries of objects in the scene. We construct a surface-based description -- the 3D scene model -- from these boundaries by hypothesizing new vertices, edges, and faces using task-specific knowledge. Some of the rules used here will be described next, and will be illustrated on the wire frames in Fig. 9.

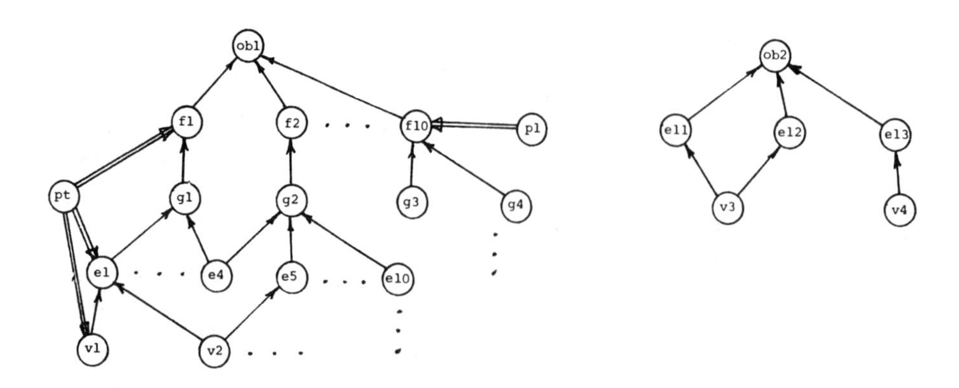

Figure 18: Simple example of a structure graph consisting of two objects, *ob1* and *ob2*.

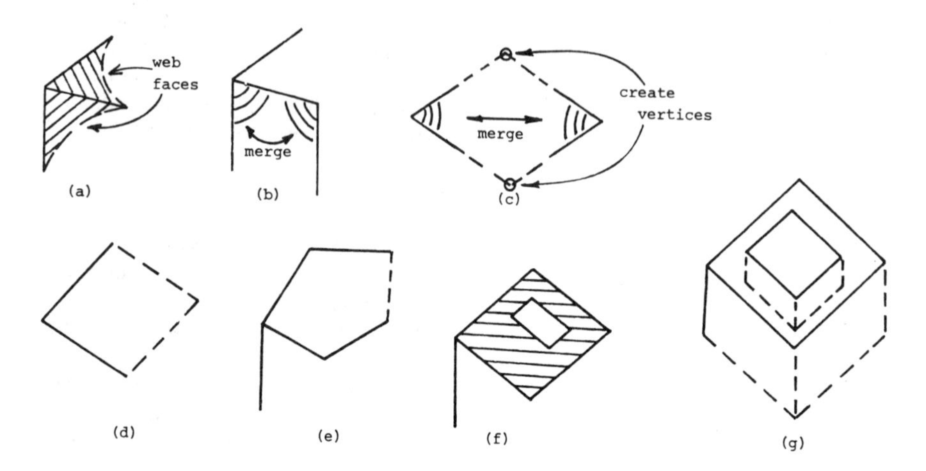

Figure 19: Obtaining a surface-based description from wire frames.

Each adjacent pair of legs ordered around a wire-frame vertex is assumed to correspond to the corner of a planar face. A partial face, called a *web face*, is generated for each such pair (Fig. 19a). Next, web faces that represent corners of a single face are merged. Web faces may either be touching (e.g., Fig. 19b, and F1 and F2 in Fig. 9) or non-touching (e.g., Fig. 19c, and F3 and F4 in Fig. 9). When merging two non-touching faces, the two edges on which each matching pair of end points lie are extended in space and intersected. The intersection points form two new vertices on the resulting face.

Incomplete faces are then completed either as parallelograms (for faces consisting of a single corner (Fig. 19d)) or as polygons (for faces containing three or more connected edges (Fig. 19e)). Next, one face is assumed to represent a hole in another face if (1) the planes of the faces are nearly parallel and close to each other, and (2) the boundary of the first face, when projected onto the plane of the second face, falls inside the boundary of that face (Fig. 19f).

At this point, many objects will be only partially complete because they are not closed. Since we are dealing with urban scenes, faces that lie high enough above the ground are assumed to represent roofs of buildings. A hypothesized vertical wall is dropped towards the ground from each edge of such faces, unless the edge is already part of another face (Fig. 19g). Each wall is dropped either to the ground or to the first face it intersects on the way down.

Fig. 20 shows perspective views of the resulting scene model. Notice that one of the buildings has a hole in it, through the roof. The planar patches at the "front" of the scene are part of the ground. Fig. 21 shows the scene model generated when these techniques are applied to the wire-frame description obtained using monocular analysis (Fig. 17).

In order to render more realistic displays, gray scale is added to them [10]. This is useful for realistically simulating the appearance of the scene from arbitrary viewpoints. We associate with each face in the model an intensity patch obtained from the image. For faces that are partially occluded in the image, the intensity patch is associated with the visible portions. Figs. 22 and 23 show the results of adding gray scale to the faces of the models in Figs. 20 and 21, respectively.

7. Combining New Views with Current Model

The process of incorporating a 3D wire-frame description extracted from a new view into the current scene model can be divided into three main steps:

1. The wire-frame data must first be matched to the current model. This process provides (a) the scale transformation and coordinate transformation from the wire-frame data to the model, and (b) corresponding elements (i.e., vertices and edges) in the two.

Figure 20: Perspective views of buildings reconstructed from wire-frame data in Fig. 9.

Figure 21: Perspective views of buildings reconstructed from wire-frame data in Fig. 17.

Figure 22: Reconstructed buildings of Fig. 20 with gray scale, derived from the left image in Fig. 4, mapped onto faces.

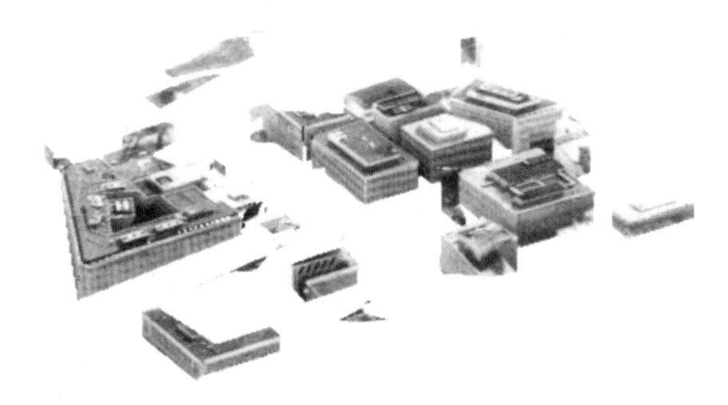

Figure 23: Reconstructed buildings of Fig. 21 with gray scale, derived from Fig. 10, mapped onto faces.

2. The new wire-frame data is then merged with the current model. This process includes (a) merging pairs of corresponding elements, and (b) adding to the model wire-frame elements for which no correspondences were found. During the merging process, hypothesized parts of the model that are inconsistent with the new wire-frame data are deleted.

3. At this point, many objects in the model may be incomplete because (a) new wire-frame data has been added, and/or (b) some hypothesized elements have been deleted. These objects are completed using the techniques described in the previous section.

To see how these steps are carried out, consider the example of incorporating the information from a second view into the scene model of Fig. 20. This scene model was constructed from the set of wire frames (Fig. 9) automatically extracted from a "front" view of the scene (Fig. 4). The second set of wire frames, shown in Fig. 24, was manually generated to simulate information available from an opposing point of view (viewing the scene from the "back"). Notice that the information in Fig. 9 emphasizes edges and vertices facing the front of the scene, while those facing the back of the scene are emphasized in Fig. 24.

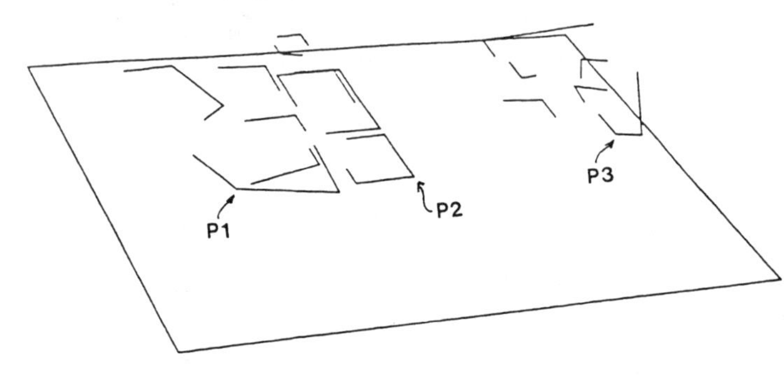

Figure 24: Perspective view of manually generated vertices and edges. The viewpoint for this drawing is chosen to be similar to Fig. 9. Points P1, P2, and P3, for example, correspond to points P1, P2, and P3 in Fig. 9.

We assume in this example that the scale and coordinate transformations from the new wire-frame data to the current model is known. Next, corresponding edges and vertices in the data and model are obtained, as described elsewhere [15, 14].

7.1. Discrepancies

We must now merge the new wire-frame data into the model. An important issue here is how to handle discrepancies between the two. We consider the following two types of discrepancies:

1. After the coordinate system of the wire-frame data has been transformed to that of the model and scale adjustments have been made, corresponding pairs of confirmed vertices and edges may not register perfectly in 3-space. In order to merge them into single elements, we perform a "weighted averaging" of their positions.

2. Hypothesized elements in the model may be inconsistent with newly obtained elements. We handle this by deleting such hypothesized elements.

To determine whether or not hypotheses are still valid when confirmed elements in the model are modified or deleted, we consider the elements which gave rise to the hypotheses. A hypothesis is dependent on all elements whose existence directly resulted in the creation of the hypothesis. If one of these elements is modified or deleted, the hypothesis must also be modified or deleted since the conditions under which it was created are no longer valid. The dependency relationships for hypothesized elements are explicitly recorded at the time of their creation using dependency pointers [11].

The following examples show how some of these relationships are recorded:

1. When two non-touching partial faces are merged, (Fig. 25a) each face has two edges which are intersected with their counterparts in the other face. The intersection points form two new hypothesized vertices, each of which is dependent on the two edges whose intersection gave rise to it. In Fig. 25a, vertex *v1* is dependent on edges *e1* and *e3*, and vertex *v2* is dependent on edges *e2* and *e4*. If one of the edges were to be modified (e.g., if its position were to be displaced), the vertex that depends on that edge would no longer be a valid hypothesis, and would therefore be deleted. A new vertex might then be hypothesized.

2. When a face is completed by connecting its two end points (Fig. 25b), two new vertices and one new edge are hypothesized. The new edge *e4* is dependent on both *e1* and *e3*, while the new vertices *v1* and *v2* are dependent on the edges on which they lie.

When a confirmed edge or vertex in the model is modified or deleted, the set of all hypothesized elements that depend on it are deleted. Recursively, elements depending on deleted ones are also deleted.

7.2. Merging

The procedure that merges corresponding wire-frame and model objects takes into account the fact that the 3-space positions of end points of edges that are confirmed vertices are generally much more accurate than the positions of non-vertex end points. Therefore, confirmed vertices are given more weight during merging. As an example, consider Fig. 26. Suppose the wire-frame object in (b) is to be merged with the model object in (a), and the corresponding edges and vertices are as follows: *(v2, v100), (v3, v101), (e2, e100), (e3, e101), (e4,e102), (e12, e104).* We assume the wire-frame object has been transformed to register with the model object.

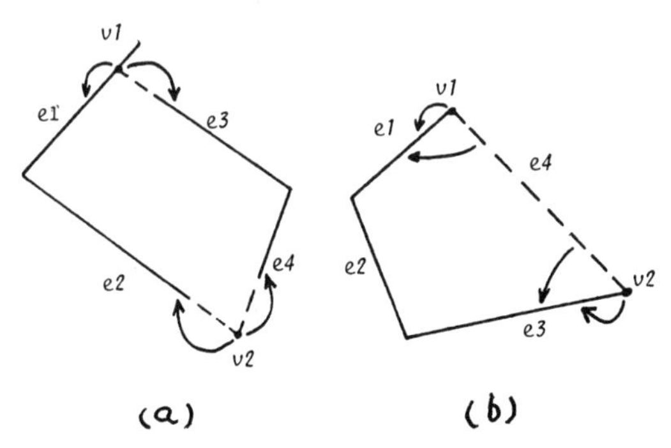

Figure 25: Generating dependencies for hypothesized edges and vertices. The dependence of an element on another is depicted as an arrow from the former to the latter. (a) Two non-touching partial faces are merged. (b) A face is completed.

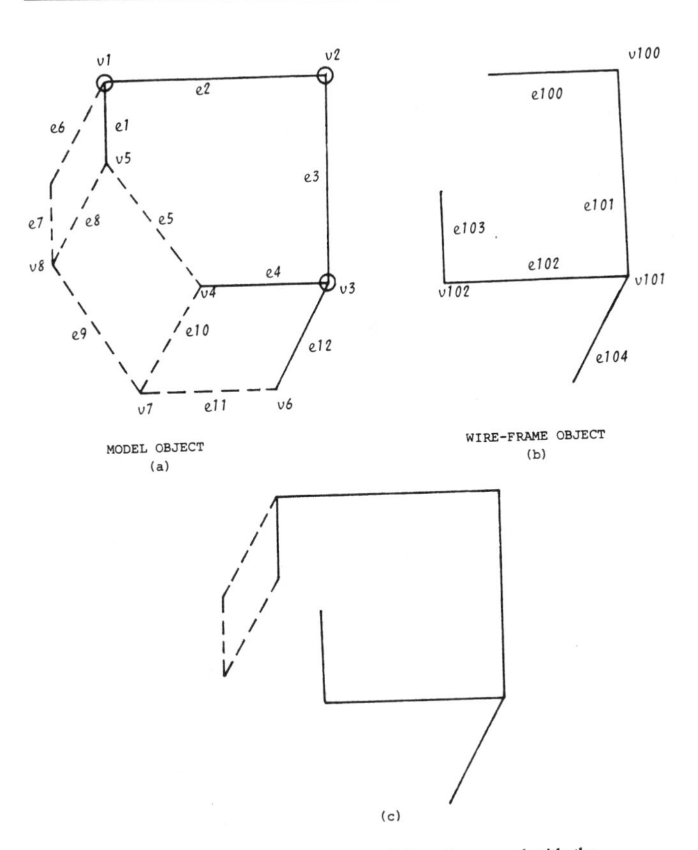

Figure 26: The wire-frame object in (b) is to be merged with the model object in (a). The confirmed edges of the model object (indicated by solid lines) are *e1*, *e2*, *e3*, *e4*, and *e12*; the confirmed vertices (indicated by circles) are *v1*, *v2*, and *v3*. Dashed lines represent hypothesized edges. (c) The result after merging.

The merging procedure starts by merging corresponding vertices. Pairs of vertices (*(v2, v100)* and *(v3, v101)* in Fig. 26) are combined into single vertices with coordinates of the midpoint between them. At this point, all corresponding pairs of edges will share at least one vertex. The corresponding edges are merged next as follows:

1. If the two edges share both their vertices (*(e3, e101)* in Fig. 26), the new edge connects the two new vertices already generated.

2. If one edge has two confirmed vertices but the other does not (*(e2, e100)* and *(e4, e102)* in Fig. 26), the new edge is the same as the former. Notice that the non-vertex end point in this case is given zero weight.

3. If the two edges share one vertex and the other end points are not confirmed (*(e12, e104)* in Fig. 26), the new edge is the "average" of the two edges.

If a model edge to be merged contains only one confirmed vertex (e.g., *e4* and *e12* in Fig. 26), then all hypothesized elements that recursively depend on this edge are deleted. For example, the hypothesized elements that recursively depend on *e4* in Fig. 26 are the vertices *v4* and *v7*, and the edges *e5*, *e10*, *e9*, and *e11*.

Figure 27: Perspective views of buildings derived by incorporating the wire-frame data in Fig. 24 into the model in Fig. 20.

After all corresponding elements of the two objects have been merged, the edges and vertices remaining in the wire-frame object that were not merged (*e103* in Fig. 26) are added to the model object. The final configuration after merging is shown in Fig. 26c. This object is incomplete and must be completed using the techniques described in an earlier section.

7.3. Results of Merging

When these procedures are applied to the wire-frame data in Fig. 24 and the scene model in Fig. 20, we obtain the updated scene model shown in Fig. 27. The updated version has two important improvements over the initial version. First, the updated model contains more buildings since new wire-frame data, some of which represent new buildings, have been incorporated into the initial model. Second, for many buildings described in both versions of the model, the positions of vertices and edges are more accurate in the updated version. This is because many hypothesized vertices and edges are replaced by accurate ones obtained from the new data, and many confirmed vertices and edges are merged with corresponding ones in the data by "averaging" their positions, generally decreasing the amount of error.

The shape of the large hole in the roof of one of the buildings has changed from a rectangle in the initial model to an almost triangular quadrilateral in the updated version. When compared with the source images in Fig. 4, the rectangular shape would seem more accurate. However, the positions of the edges and vertices that form the hole are more accurate in the updated model in the sense that they are more faithful to the wire-frame descriptions derived from the images.

This experiment demonstrates how information provided by each additional view allows the model to be incrementally made more complete and accurate.

8. Conclusions

We set out to develop an entire vision system to interpret complex images, one that goes all the way from images to symbolic 3D descriptions. The following are some conclusions we can draw from this project.

1. Complex images usually cannot be fully interpreted. Difficulties in interpretation arise not only from occlusions, but also from variations in surface texture and reflectance, variations in shape, and complex lighting conditions. Our vision systems must therefore have the capability to deal with approximate, imperfect scene descriptions when performing tasks such as matching, path planning, or model-based image interpretation.

2. Incremental reconstruction of complex scenes will often be necessary. Multiple views are required to effectively reconstruct complex scenes. A system that moves about and interacts with its environment in order to obtain the multiple views will be able to gradually add more information to its scene model at the same time that it carries out its other tasks.

3. Scene descriptions are often more useful if they contain reasonable hypotheses for parts of the scene for which there are only partial or no data. For example, path planning cannot be done for occluded regions of the scene without a good guess about what lies in these regions. If the hypotheses turn out to be incorrect, they should eventually be modified. Our vision systems must therefore have mechanisms for intelligently generating hypotheses, verifying them, and modifying them.

4. Task-specific knowledge is very useful at all levels of complex image interpretation, from low-level image analysis to high-level formation of symbolic descriptions. Knowledge of block-shaped objects in an urban scene is used in the **3D Mosaic** system for stereo analysis, monocular analysis, and reconstructing shapes from the wire frames.

5. Stereo matching of 2D structural features (such as junctions) may be important for complex images and should be further investigated.

References

1. Baer, A., Eastman, C., and Henrion, M. "Geometric Modelling: a Survey." *Computer-Aided Design 11* (September 1979), 253-272.

2. Baker, H. H. "Three-Dimensional Modelling." *Proc. IJCAI-77* (August 1977), 649-655.

3. Baker, H. H., and Binford, T. O. "Depth from Edge and Intensity Based Stereo." *Proc. IJCAI-81* (1981), 631-636.

4. Barnard, S. T. "Methods for Interpreting Perspective Images." *Proc. ARPA Image Understanding Workshop* (September 1982).

5. Barnard, S. T. and Fischler, M. A. "Computational Stereo." *Computing Surveys 14*, 4 (December 1982).

6. Barnard, S. T. and Thompson, W. B. "Disparity Analysis of Images." *IEEE Trans. on Pattern Analysis and Machine Intelligence PAMI-2*, 4 (July 1980), 333-340.

7. Barrow, H. G., Bolles, R. C., Garvey, T. D., Kremers, J. H., Tenenbaum, J.M., and Wolf, H. C. "Experiments in Map-guided Photo Interpretation." *Proc. IJCAI-77* (August 1977), 696.

8. Baumgart, B. G. Geometric Modeling for Computer Vision. Tech. Rept. STAN-CS-74-463, Department of Computer Science, Stanford University, Stanford, CA, October, 1974.

9. Bourne, D. A., Milligan, R., Wright, P. K. "Fault Detection in Manufacturing Cells Based on Three Dimensional Visual Information." *Proc. SPIE* (May 1982).

10. Devich, R. N., and Weinhaus, F. M. "Image Perspective Transformations." *Proc. SPIE 238* (July 1980), 322-332.

11. Doyle, J. Three Short Essays on Decisions, Reasons, and Logics. Tech. Rept. STAN-CS-81-864, Department of Computer Science, Stanford University, Stanford, CA, May, 1981.

12. Grimson, W.E.L. "A Computer Implementation of a Theory of Human Stereo Vision." *Phil. Trans. Roy. Soc. Lond. B292* (1980), 217-253.

13. Hannah, M. J. Computer Matching of Areas in Stereo Images. Tech. Rept. AIM-239, Stanford University, July, 1974.

14. Herman, M. Representation and Incremental Construction of a Three-Dimensional Scene Model. In *Sensors and Algorithms for 3-D Machine Perception*, A. Rosenfeld, Ed.,, 1984, to appear.

15. Herman, M. and Kanade, T. The 3D MOSAIC Scene Understanding System: Incremental Reconstruction of 3D Scenes from Complex Images. In *From Pixels to Predicates: Recent Advances in Computational and Robotic Vision*, A.P. Pentland, Ed.,Ablex Publishing Company, 1984, to appear.

16. Kender, J. R. "Environmental Labelings in Low-Level Image Understanding." *Proc. IJCAI-83* (August 1983), 1104-1107.

17. Lucas,B. D., and Kanade, T. "An Iterative Image Registration Technique With an Application to Stereo Vision." *Proc. IJCAI-81* (August 1981), 674-679.

18. Martin, W. N. and Aggarwal, J. K. "Volumetric Descriptions of Objects from Multiple Views." *IEEE Trans. on Pattern Analysis and Machine Intelligence PAMI-5*, 2 (March 1983), 150-158.

19. McKeown, D. M. MAPS: The Organization of a Spatial Database System Using Imagery, Terrain, and Map Data. Tech. Rept. CMU-CS-83-136, Department of Computer Science, Carnegie-Mellon University, Pittsburgh, PA, July, 1983.

20. Moravec, H.P. Obstacle Avoidance and Navigation in the Real World by a Seeing Robot Rover. Tech. Rept. CMU-RI-TR-3, The Robotics Institute, Carnegie-Mellon University, Pittsburgh, PA, September, 1980.

21. Ohta, Y., and Kanade, T. Stereo by Intra- and Inter-scanline Search Using Dynamic Programming. Tech. Rept. CMU-CS-83-162, Department of Computer Science, Carnegie-Mellon University, Pittsburgh, PA, October, 1983.

22. Potmesil, M. "Generating Models of Solid Objects by Matching 3D Surface Segments." *Proc. IJCAI-83* (August 1983), 1089-1093.

23. Requicha, A. A. G. "Representations for Rigid Solids: Theory, Methods, and Systems." *Computing Surveys 12*, 4 (December 1980), 437-464.

24. Rubin, S. "Natural Scene Recognition Using Locus Search." *Computer Graphics and Image Processing 13* (1980), 298-333.

25. Underwood, S. A. and Coates, C. L. "Visual Learning from Multiple Views." *IEEE Transactions on Computers*, 6 (June 1975), 651-661.

TERRAIN MODELS FOR AN AUTONOMOUS LAND VEHICLE

Daryl T. Lawton, Tod S. Levitt, Chris McConnell, Jay Glicksman

Advanced Decision Systems
201 San Antonio Circle, Suite 286
Mountain View, California 94040

Abstract

We present an architecture for terrain recognition for an autonomous land vehicle. Basic components of this are a set of data bases for generic object models, perceptual structures, temporary memory for the instantiation of object and relational hypothesis, and a long term memory for storing stable hypothesis which are affixed to the terrain representation. Different inference processes operate over these data bases. We describe components of this architecture: the perceptual structure data base, the grouping processes that operate over this, and schemas. We conclude with a processing example for matching predictions from the long term terrain model to imagery and extracting significant perceptual structures for consideration as potential landmarks.

1. INTRODUCTION

Terrain models for autonomous land vehicles (ALVs) are required for a wide range of applications such as route and tactical planning, location verification through the recognition of terrain features and objects, and acquiring new information about the environment as it is explored. Important criteria for terrain modeling techniques are:

Descriptive Adequacy: The modeling technique should be capable of describing all the objects and situations in the environment necessary for the vehicle to function. This includes representing natural as well as man-made objects. It should be a consistent representation that supports modular system development and uniform inference procedures that can operate over different types of objects at different levels of detail. Uniform shape, object subpart and surface attribute affixments are necessary to do this.

Recognition Adequacy: Much of the activity of an ALV is concerned with determining where it is and what is around it. Terrain models should be manipulable for determining the sensor-based appearances of world objects and for controlling recognition processing. This involves the formation of general predictions of sensor derived features from the terrain model. Such predictions will often be inexact and qualitative due to incomplete prior knowledge of the terrain.

Handling Uncertainty: The existence and exact environmental location of objects will often not be known with complete certainty. Locations will generally be determined relative to other known locations and not with respect to a globally consistent terrain map. This is especially the case when the sensor displacement parameters are not well determined. It is necessary to represent this uncertainty explicitly in the terrain model so incrementally acquired information can be used for disambiguation.

Learning: A vehicle will learn about the environment as it moves through it. Associating new information with the terrain representation should be straightforward. This is difficult to do, for example, by changing values in a raw elevation array. Among the types of information to be affixed to the terrain representation include newly discovered objects, details of expected objects, and the processing used in the recognition of of an object.

Fusion of Information: The ALV has to build a consistent environmental model over time from different sensors. As an object is approached, its image appearance and scale will change considerably, yet it has to be recognized as the same object, with newly acquired information associated with the unique instance of the general object type. In a typical situation, a distant dark terrain patch will be partially recognized based upon distinctive visual characteristics, but may be either a building or a road segment. As it is approached, its image appearance changes considerably, making disambiguation possible. This requires the representation of multiple hypotheses, each formated with respect to the properties of the potential world objects. The structure of the object description should direct the accumulation of information.

A further consideration in evaluating terrain models is that there is not a generic ALV. Instead, there are a wide range of autonomous vehicles, indexed by a diverse range of active and passive sensors and assumptions about a priori data. There is a continuum from systems having a complete initial model of the terrain and perfect sensors to those with no a priori model, and highly imperfect sensors. For example, a robot with no a priori data and only an unstabilized optical sensor will probably model the environment in terms of a sequence of views related by landmarks and distinct visual events embedded in a representation which is more topological than metric. An ALV solely dependent on optical imagery will have to deal with the huge variability in the appearance of objects. Experience has shown that even road surfaces have highly variable visual characteristics. Alternatively, a few pieces of highly pre-selected visual information can serve to verify predictions from a reliable and detailed terrain model and precise position and range sensors.

There has been much relevant work in artificial intelligence, computer vision, and graphics which satisfy these requirements individually, but little has been done in terms of integrating them. To date, there is no vision system which can interpret general natural scenes, although some can deal with restricted environments [Hanson et al., 1978] while other systems are restricted to artificial objects and environments. Brooks' [Brooks, 1984] representation based on generalized cylinders meets or could be extended to deal with many of these functions. It has well defined shape attribute inheritance along a set of progressively more complex object models and affixment relations which probably could be generalized to handle uncertainty. It can also be used to generate constraints on image features from object models. Nonetheless, the system built around this representation has had limited success, beyond dealing with essentially orthographic views of geometrically well defined man-made objects. This appears to be partially because the constraints on image structures generated from the abstract instances of object models are too general to generate initial correspondences between models and image structures. Brook's system also used an impoverished set of image descriptions, and the object models could not direct the segmentation process directly during their instantiation. The majority of work in terrain modeling deals with how well a representation can realistically model three dimensional terrain, but not how it is used for recognition. The simplicity of a model which is described by a few parameters is not useful for recognition unless it can direct constrained searches against image data. For example, Pentland's [Pentland, 1983] use of fractals satisfies aspects of descriptive adequacy for natural terrain, but has been less effective for recognition. Kuipers [Kuipers, 1982] has produced an interesting terrain model for learning and handling uncertainty, but it is non visual. Related to this is Kuan's [Kuan, 1984] object based terrain representation for planning that is organized in terms of distinct, modifiable objects, but is also not associated with sensor derived processing results.

We are developing an architecture for terrain and object recognition to be compatible with the wide range of potential sensor configurations on an ALV and the different qualities of a prior data. A major organizational component of the architecture is a general object model called a schema. It is related to similar concepts found in [Hanson et al., 1978 and Ohta, 1980]. Schemas correspond to objects in the world and are used as basic units in the long term terrain representation. The short term terrain representation consists of schema instantiations which represent accumulated recognition evidence for objects as attributes and relations which are hypothesized with varying levels of certainty. Schemas are also used to organize perceptual processing by integrating descriptive representations with recognition and segmentation control. One aspect of this is the use of different types of attributes and inheritance relations between generic schemas for such things as IS-A and PART-OF. A particular schema attribute relates three dimensional world properties of an object and sensor dependent view information, either by a set of generic views or viewing procedures. These viewing attributes are also inherited and modified corresponding to different types of schemas. In many systems, schemas are treated as lists of attributes which are matched against extracted image features. We treat them as specifying an active control process which directs image segmentation by specifying grouping procedures to extract and organize image structures. Our use of schemas is meant to address the criteria of descriptive, recognition, and information fusion adequacy. Another critical aspect of the architecture is the various types of affixment relations to deal with uncertainty and learning by associating different types of perceptually derived information with the terrain model. For example, local multi-sensor frames affix sets of schemas and un-recognized perceptual structures into local views of an ALV's environment to model perception of a view based environment for a vehicle with no range sensing capability. Path-affixments between local multi-sensor frames support fusion of information in time without any correspondence to an a prior grid. Location relations between schema can be specified with different levels of certainty to each other or on elevation grids (or both).

2. ARCHITECTURE

The system architecture consists of several databases and inference processes. The inference processes transform the databases, creating additional data structures, and modifying the existing ones. The task interface focuses attention in system processing and monitors progress toward system task goals. This high level architecture is depicted in Figure 2-1. The boxes with square corners in this figure represent databases, the ellipses represent inference processes, and arrows indicate dataflow.

2.1 SYSTEM DATABASES

At the highest level there are three databases. These are the short term memory (STM), long term memory (LTM), and generic models.

The STM acts as a dynamic scratchpad for the vision system. It has two sub-areas, a perceptual structures database (PSDB) and a hypothesis space. The PSDB includes incoming imagery from sensors, immediate results of extracting image structures such as curves, regions and surfaces, spatial/temporal groupings of these structures, and results of inferring 3D information.

The hypotheses space contains statements about objects and terrain in the world. A basic type of hypothesis is an instantiated schema. The schema points to the various perceptual structures in the PSDB which provide evidence that the object represented by the schema (such as a terrain patch, road, tree, etc.) actually exists in the world. Other types of hypotheses include grids, affixments, and local multisensor frames. Grids are a special type of terrain representation which contain elevation information and are derived from range data or successive depth maps from motion stereo. Affixments give space/time relationships between hypotheses. Local multisensor frames are sets of hypotheses which correspond to what can be seen from a localized position. A hypothesis with no associated perceptual structures is a prediction. As structures and affixments are incrementally added to a hypothesis, it progresses on the continuum from predicted to recognized. Hypotheses that have enough evidence associated with them to be considered recognized and stable, are moved to the LTM.

The LTM stores a priori terrain representations and hypotheses with enough associated evidence to be worth remembering. A priori data concerning elevation and terrain type information, as well as knowledge of specific landmarks are stored in the LTM. Another standard example is a hypothesis of a local multisensor frame at a certain location in the world which is stored in the LTM if the schemas and other evidence associated with the local multisensor frame are such that they could be re-used to recognize the local environment if it was re-encountered. Consistency of one hypothesis with another is not required for storage in the LTM.

The model space stores generic schemas, the inheritance relations of the (model) schema network, and a set of image structure grouping processes and rules for evaluating image structure interestingness. Generic schemas are used dynamically to instantiate and guide search processes to associate evidence to a schema instance. Inheritance relations are used by various schema inference procedures to propagate structures, attributes and relations between schema instantiations. For instance, the generic two-lane-road schema has an "IS-A" relationship to the generic road schema. It follows, based on the inheritance models, that an instantiation of the two-lane-road schema will inherit the

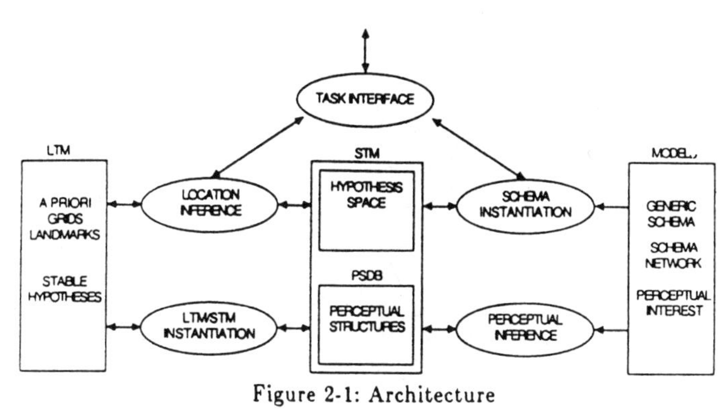

Figure 2-1: Architecture

more general characteristics of the generic road schema which in turn inherits the more general characteristics of a terrain patch. Unlike the STM and LTM, the model space is not modified by inference processes.

2.2 INFERENCE PROCESSES

At the highest level, there are five different sorts of inference processes in the vision system. These are perceptual inference, location inference, schema instantiation, LTM/STM instantiation, and the task interface.

The PSDB is initialized with the output of standard multi-resolution image processing operations for smoothing, edge extraction, flow field determination, etc. Much subtler inference is required for grouping processes which produce such things as connected curves, textures, surfaces, and temporal matches between image structures. These grouping operations are typically model guided. There are generic models (which may be task dependent) of what constitutes "interestingness" of an image structure.

The schema inference processes produce tasks for the perceptual processes. These may be satisfied by simple queries over the perceptual structure database such as "find all long lines in this region of image", where "long", "line" and "region" are suitably interpreted. Queries can be more complex also, requiring, for instance, temporal stability, such as "find all homogeneous green texture regions which are matched (i.e., remain in the field of view) over at least two seconds of imagery", where, again, qualitative descriptors are rigorously defined. Alternatively, the requested perceptual structures may be dynamically extracted. In this case, a history of the processing attempts and results are maintained. If similar requests are made later, such as if we were to view the same environment from a different perspective, these processing histories could be used to recall a processing sequence which successfully produced results.

Location processes include a number of different modes of spatial location representation and inference. While exact location information is used when it is available, a key concept is qualitative representation of relative location. This is fundamental, because the problem of acquiring terrain knowledge from moving sensors involves handling perceptual information which arises from multiple coordinate systems which are transforming in time. The basic approach to location inference is to represent the location of world objects in a qualitative manner which does not require the full knowledge of continuous transformations of sensor coordinates relative to the vehicle the sensors are mounted on, or of transformations of vehicle coordinates relative to the terrain.

Figure 2-2: IS-A Hierarchy-1

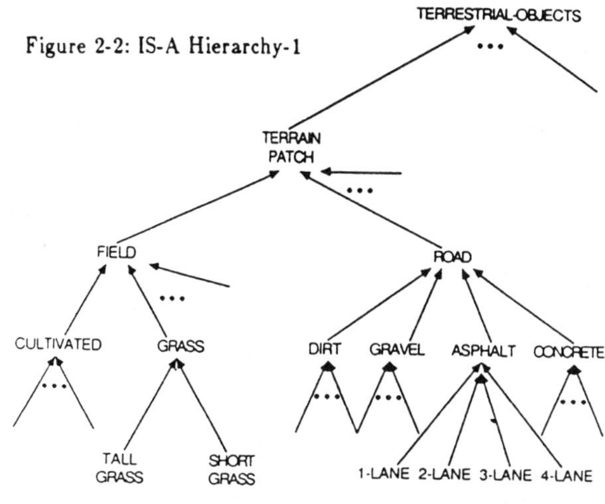

Figure 2-3: Part of Hierarchy-1

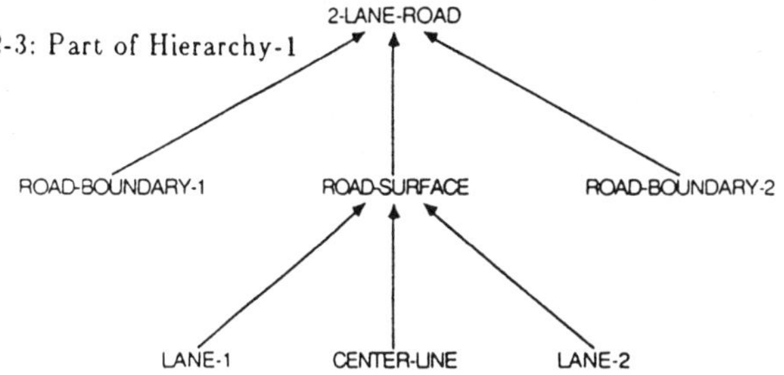

2-LANE-ROAD
ROAD-BOUNDARY-1
ROAD-SURFACE
ROAD-BOUNDARY-2
LANE-1
CENTER-LINE
LANE-2

The main structures involved in location inference are local multi-sensor frames, affixments, and grids. Local multi-sensor frames represent both metric location information about world objects derived from range sensors or view-based location information about the directions in which objects are found derived from passive sensor data.

Schema instantiation is the system workhorse for terrain understanding. Generic schemas are models of world objects that include information and procedures on how to predict and match the schema in the available sensor data. Besides representing 3D geometric constraints, 2D-3D sensor view appearance including effects of change in resolution and environmental effects such as season, weather, etc., schemas also indicate contextual relationships with other schemas, type and spatial constraints, similarity and conflict relations, affixments, and appearance in local multi-sensor frames.

Schema instantiation may occur by model-driven prediction from a priori knowledge, or directly from another schema instantiation and a PART-OF relation. The other instantiation process may also occur by matching a distinctive perceptual structure to a schema appearance instance. This sort of schema "triggering" is more common in situations where there is little a priori information to guide prediction. Schema instantiations generate queries to the perceptual structures database/grouping process in order to complete matching.

A key idea in schema instantiation processing is inference over the model schema network hierarchies. Direct representation and inference over a large enough body of world objects to accomplish outdoor terrain understanding would require very large memory and proportionately lengthy inference procedures over that memory space. Hierarchical representation makes a significant reduction in storage requirements; furthermore, it lends itself naturally to matching schema to world objects at multiple levels of abstraction, thus speeding the inference process. Two basic hierarchies are the IS-A and PART-OF trees.

IS-A hierarchies represent the refinement of object classification. Figure 2-2 shows part of an IS-A hierarchy for terrain representation. The level of terrestrial-object tells us that we will not see evidence of any schema instance below this node as perceptual structures surrounded by sky. At the level of terrain-patch we pick up the geometric knowledge of adherence to the ground plane, while information stored at the level of a road schema constrains the boundaries of a terrain patch to be locally linear (with other constraints). Types beneath road add critical appearance constraints in color and texture, while the final refinement level in the IS-A hierarchy, the number of lanes, further constrains size parameters inherited from the road schema.

PART-OF hierarchies represent the decomposition of world objects into components, each of which is, itself, another world object. Figure 2-3 shows a PART-OF hierarchy decomposition for a generic 2-lane-road. PART-OF hierarchies contain relative geometric information which is useful in prediction and search.

As schema instantiation inference reasons up and down schema network hierarchies, incrementally matching perceptual structures and other data to instances of schema appearance in the world, a history mechanism records the inference processing steps, parameters and results. This dynamic data structure is called the schema instantiation structure. One important aspect of this structure is that it can used to extract the inference and processing sequence(s) which worked earlier to see the same object, or ones that are similar. This accounts for the fact that distinctiveness in image appearance is an idiosyncratic process which depends upon many factors which are difficult to model and control, such as current motion, wind, varying outdoor illumination, etc. The schema instantiation structure provides a methodology for the vision system to dynamically adapt to the current viewing conditions.

3. PERCEPTUAL PROCESSING AND THE PSDB

The Perceptual Structure Data Base (PSDB), Figure 3-1, contains several different types of information. These are classified as images, perceptual objects, and groupings. Images are the arrays of numbers obtained from the different sensors and the results of low level image processing (such as contour extraction and region growing routines) which produce such arrays. It is difficult for the symbolic/relational representations used for object models such as schemas and the processing rules in computer vision systems to work directly with an array of numbers. Therefore, there are many spatially-tagged, symbolic representations used in image understanding systems which describe extracted image structures such as the primal sketch [Marr, 1982], the RSV structure of the VISIONS system [Hanson et al., 1978], and the patchery data structure of Ohta [Ohta, 1980]. We have found it useful to build such a representation around a set of basic perceptual objects corresponding to points, curves, regions, surfaces, and volumes. Groupings are recursively defined to be a set of perceptual objects or sub-groupings which are related. The relation may be exactly determined as in representing which edges are directly adjacent to a region, or they may require an active grouping process implemented as a search procedure to determine the set of objects which satisfy the relationship. A common example is joining curves based upon some global shape or contrast requirement. Groupings can occur over space, as in linking texture elements under some shape criteria such as compactness and density, or over time

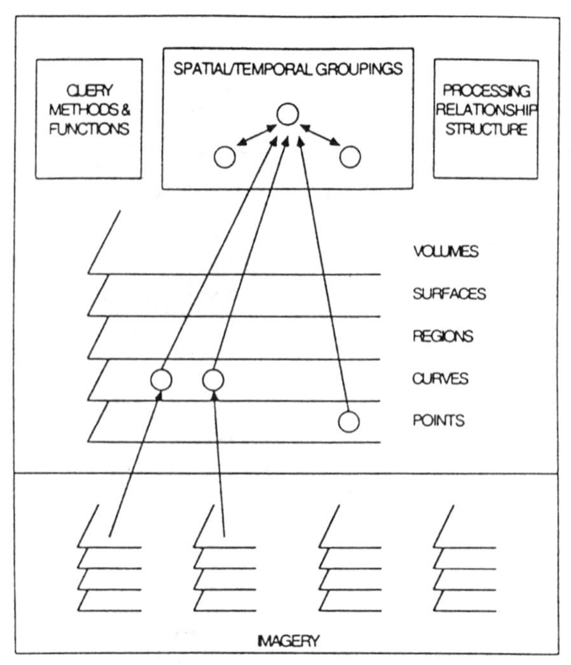

Figure 3-1: PSDB

as in associating instances of perceptual structures in successive images. We stretch the concept a bit so that groupings also refer to general non-image registered perceptual information, such as histograms.

3.1 IMAGES

Images are the data arrays derived from the optical and laser range sensors and the results of image processing routines for operations including histogram-based segmentation, different edge operators, optic flow field computations, and so forth. Associated with images are several attributes for time of acquisition, relevant sensor parameters, etc. Processing history is maintained in the Processing Relationship Structure (PRS) which keeps track of the processing history of all objects in the PSDB.

3.2 PERCEPTUAL OBJECTS

Points, curves, regions, surfaces, and volumes are basic types of perceptual structures which are accessible to schema instantiations and grouping processes. An example instance of a curve structure is shown in Figure 3-2. This figure shows many common representational characteristics of perceptual objects. There are default attributes associated with particular objects, such as endpoints, length and positions for a curve. There is also an associated attribute-list mechanism for incorporating more general properties with an object. This list is accessible by keywords and a general query mechanism using methods specific to the particular associated attribute. The associated attributes in the example are shown in capital letters. There are many types of attributes which can be consistently associated with a curve using this mechanism.

Scalar attributes can be computed along the curve with respect to some image, such the as average or variance of contrast, optical flow magnitude, or the chamfer value computed with respect to some set of selected image objects. A list of values can be extracted from some image at

```
Curve: #<CURVE 175505263>
Point1: (205 291)
Point2: (274 285)
Length: 81
Grid: 6
Points: (205 291) (206 292) (207 293) ...... (276 287) (275 286) (274 285)
AVG-INTENSITY: 159.27315
SUCCESSIVE-GRADIENT-DIFFERENCE: 1.7660371 1.4782858 1.0026073 ......
        0.5625856 0.6442404 0.7240415
GRADIENT: (-2.8555908 3.1515503) (-4.0343933 1.0365173) (-5.415619 1.3096924) ......
        (-0.03842163 -2.917389) (0.56552124 -2.6931152) (1.0051135 -2.1008733)
CONTRAST: 5.7423205
AVG-GRADIENT: -5.7336416 0.31559697
LDC: (#<CURVE 175505520>
      (#<CURVE 175327156>)
      (#<CURVE 175505525>
         (#<CURVE 175327163>)
         (#<CURVE 175505532>
          (#<CURVE 175327170>)
          (#<CURVE 175327175>)))))
```

Figure 3-2: Curve Example

positions along the curve, (the GRADIENT: values along the curve). This is one way in which three dimensional information can be associated with a curve. Different types of shape descriptions can also be associated. This example shows the linear decomposition hierarchy associated with the particular curve. Relationships to other objects can be represented, such as the set of regions to which a given curve is adjacent.

Another useful representation we use for performing geometric operations and queries over objects is the OBJECT LABEL-GRID (or GRID: in the example. The number 6 indicates this structure). This is an image where each pixel contains a vector of pointers back to the set of perceptual objects and groups which occupy that position. This allows geometric operations to be performed directly on the grid. Filtering operations can be applied to the OBJECT LABEL GRID to restrict processing based upon attributes associated with objects. Various types of masks can be associated with objects to reflect a directional or uniform neighborhood to intersect with objects in the OBJECT LABEL GRID. The generalization of the label grid to three and higher dimensions is straightforward, but is very costly and n-dimensional, multi-resolution representation techniques such as oct-trees are required.

3.3 GROUPINGS

A grouping is a set of related objects. The relation can be determined directly by a query over an object and those surrounding it as in finding the set of curves within some distance of a given region. Alternatively, it may require a search process to find the set of objects meeting some, potentially complex, criteria. For example, an ordered set of curves can be grouped together using thresholds on allowable changes in the average contrast and orientation of successive elements. By expressing the grouping process as a search over a state space of potential groups, each such group becomes a potential hypothesis in the PSDB. Groups can also reflect temporal relationships as in matching structures in successive images. A basic grouping procedure is shown in Figure 3-3 for the determination of nearby parallel lines with opposite contrast directions. This is done for a linear segment by first extracting nearby neighbors using a narrow mask oriented perpendicular from the segment at its mid-point. The intersection of this mask with points in the label grid are determined and then each candidate is evaluated by checking if is within allowable thresholds for length, contrast, and orientation. It is then ordered with respect to the smallest magnitude of the difference vector computed from the average gradients. The grouping processes can either produce the best candidate as a potential grouping or some set of them.

We have organized grouping processes so they can be modularly described with respect to different types of parameterized functions, evaluation criteria, and structural descriptions. This is so a predicted image relation or type of image relation can be mapped onto a corresponding grouping operation for extracting that structure. The grouping

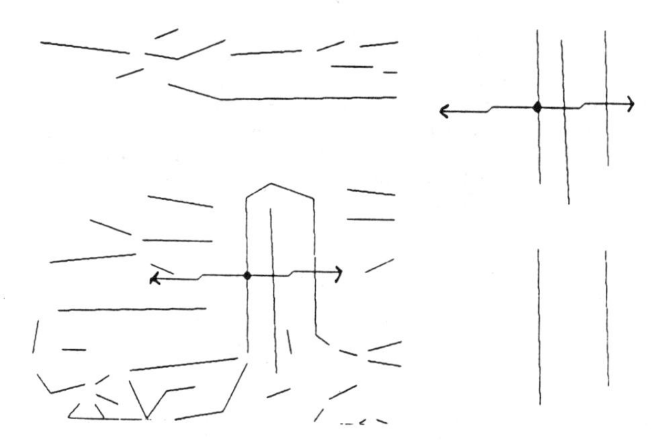

Figure 3-3: Parallel Grouping

operations are similar to edge linkers [Martelli, 1976], but generalized with respect to what is grouped. These can be pixels, perceptual objects, or groups themselves.

Organizing segmentation in terms of grouping processes has many advantages for a model based vision system. The grouping processes can be run automatically from extracted significant structures based upon perceptually significant, though non-semantic criteria. Thus, connected curves of slowly changing orientation or compact, homogeneous regions can be extracted purely on perceptual criteria. Such image structures tend to correspond to world structure and events, and they are useful for initializing schema instantiations. They correspond well to the qualitative image predictions associated with more general schemas. An inference process for the general compilation from an object model into grouping processes allows model based vision to have a very active character quite different than simple attribute matching.

3.4 INITIALIZATION OF THE PSDB

Whenever new sensor data is obtained, a default set of operations are performed to initialize the PSDB. Edges are extracted at multiple spatial frequencies and decomposed into linear subsegments. The edges are extracted into distinct connected curves and general attributes such as average intensity, contrast, and variance are associated with them. Similar processing is performed for regions extractions. Histograms are computed with respect to a wide range of object based and image based characteristics in a pyramid like structure. These default operations are used to initialize bottom-up grouping processes and schema instantiations by the determination of significant structures using heuristic interestingness rules [Lawton et al., in preparation] to prioritize the structures for the application of grouping processes or keying schema instantiations.

4. SCHEMAS

Schemas represent hypotheses about objects in the world. The process of schema instantiation creates an instance of a schema together with evidence for that schema. Evidence consists of structures in the PSDB, a priori knowledge stored in the LTM, predictions derived from location inference and relations to already instantiated schema.

Table 4-1 shows the various slots and relationships in a generic schema. Although this data structure has a frame-like appearance, it is useful to view the schema as a semantic net structure, with slots representing nodes in the net and relationships representing arcs. Schema instantiation inference reasons from a (partially) instantiated node, follows arcs, and infers procedures to execute from the sum of its acquired information in order to obtain more evidence to further instantiate the schema.

The schema network is a generic set of data structures which indicate the a priori relationships between schemas. A key part of this network are the inheritance hierarchies which indicate which descriptions and relationships can be inherited from schema to schema. Inheritance hierarchies allow efficient matching of objects in the world against sensor evidence from progressively coarser to finer levels. As we move from coarser to finer levels of description in model-based schema instantiations, the schemas inherit descriptive bounds and add new descriptions and also add constraints to inherited ones. For example, we may first recognize an object as a terrain patch (because it lies on the ground plane). A road is a type of terrain patch, see Figure 2-2, which adds linear boundary description, and constrains the visual image appearance of the terrain patch schema in the

Table 4-1: Generic Schema Data Structure

- SCHEMA TYPE
- SCHEMA NAME
- SCHEMA INSTANTIATION STRUCTURE
- 3D DESCRIPTION
    - o SHAPE
    - o SIZE
    - o COLOR
    - o TEXTURE
    - o INDEX TO SENSOR VIEWS
- SENSOR VIEWS
  (FOR EACH SENSOR:)
  (FOR EACH VIEW:)
    - o PROJECTION RELATIONS
        - PROJECTION FUNCTION
        - 3D BACK CONSTRAINTS
    - o DISTINCTIVE IMAGE BASED EVIDENCE
    - o PERCEPTUAL STRUCTURE
- COMPONENTS
    - o MUST HAVE
    - o MAY HAVE
    - o 3D SPATIAL RELATIONSHIPS
    - o VIEW DEPENDENT RELATIONSHIPS
- PART OFS
- CLASSIFICATIONS
  (POINTS UP THE IS-A HIERARCHY ONE LEVEL)
- CONTEXTUAL RELATIONSHIPS
    - o ALWAYS OCCURS WITH
    - o SOMETIMES OCCURS WITH
    - o NEVER OCCURS WITH
    - o CONFUSED WITH
    - o SIMILAR TO
- LOCATIONAL INFORMATION
    - o LOCAL MULTI-SENSOR FRAME AFFIXMENTS
    - o GRID AFFIXMENTS
    - o 3D SPATIAL RELATIONSHIPS WITH
- RECOGNITION STRATEGIES

color and texture descriptors. The two basic types of schema network inheritance hierarchies are IS-A and PART-OF.

We briefly explain each of the slots and relationships in the generic schema data structure. Schema type refers to the generic name of the schema in the IS-A hierarchy. Schema name is the identification of the schema instance, e.g., if the schema type is "road" then the schema name might be "highway 101".

The schema instantiation structure maintains the control history of the schema recognition inference processes for this schema.

The 3D description is an object-centered view of the world object represented by the schema. It includes its 3D geometry and shape description, actual size, and inherent color and texture (as opposed to how its color and texture might appear to a particular sensor). Note that this is the description which matches the schema-object before looking at its structure refined into components. For example, the 3D geometric description of a tree schema does not separate the canopy from the trunk, but gives a single enclosing volume as its representation. The volumetric descriptions of the trunk and canopy appear as the 3D descriptors on their schema further down the PART-OF hierarchy. Thus, inferring down the PART-OF hierarchy corresponds to increasing the resolution of the view of the object represented by the schemas.

The sensor views are descriptions of the stable or frequently occurring appearances of the schema object in imagery. This description is intended to be used for image appearance prediction, evidence accrual for instance recognition, 3D shape inference, and location inference. The reason for storing explicit (parametrized) image views is that the perceptual evidence matches to these descriptions, not to the three dimensional ones.

The distinctive image appearance slot holds descriptions of perceptual structures which are likely to occur bottom-up in the PSDB. They provide coarse triggers for instantiating the schema object hypothesis without prediction.

The perceptual structure is the dynamically created PSDB query history generated by the schema instantiation as it attempts to fill in evidence matching the various schema slots and relations. The instantiator can re-use successful branches of such structures to improve its recognition speed as it continues to view other instances of the same generic schema type.

Components are pointers to other schema which represent sub-parts of the schema object. They are finer resolution description of the schema, one level down on the PART-OF hierarchy. The MUST-HAVE components are assumed to be parts which the represented object must have to exist, although the schema may be instantiated without observing them all. Occasionally occurring components, such as center-lines on roads, can be stored in the MAY-HAVE slot. Spatial relationships between components as they make up the schema object are listed at this level also. Relationships can also be stored on a view dependent basis. These relationships access the sensor-view dependent data in that slot. PART-OF's point upward one level on the PART-OF hierarchy, indicating that this schema is a component of another schema.

Classification points upward and downward one level on the IS-A hierarchy. There may be more than one such pointer, which is to say that the IS-A hierarchy may be partially ordered.

Contextual relationships indicate spatial/temporal consonance or disonance between groups of schema types, omitting those which are already indicated in the PART-OF and IS-A hierarchies. Schema which ALWAYS or never-occur with the given one can be used strongly for belief or dis-belief in the schema instance and as focus of attention mechanisms within the instantiation process. SOMETIMES occurs with relationships are used to store the spatio-temporal aspects of schemas relative appearance in the viewed environment.

Confused-with and similar to relationships indicate schema which may be mistaken for the given one, but for different reasons. One schema may be confused with another because they share common evidence pieces, but which sufficient descriptors will disambiguate. Two schema are similar if there is sufficient ambiguity in their appearances, and therefore the available perceptual evidence, that they may be indistinguishable without contextual reasoning. For example, tall grass may be confused with wheat from coarse shape and texture evidence, but can often be disambiguated by color descriptors or finer resolution examination of structure (because of wheat berries, for example). However, roads are similar to runways because they cannot necessarily be distinguished by their intrinsic appearance, no matter how detailed or accurate the descriptors and evidence. Contextual reasoning, e.g., the presence of aircraft on the runway, global curvature of the road, etc. is required.

Locational information points at the various affixments of this schema instance to other structures, such as local multi-sensor frame appearances, grid affixments and inferred 3D relationships with other world objects.

Recognition strategies are prioritization cues for the schema instantiation processes which suggest inference chains likely to pay off to match this schema instance against sensor evidence.

The recognition strategies slot in the schema data structure prioritizes inference approaches relevant to this schema. These approaches include search for components, search for part of schema instance, search on weaker classification, relations with other schema instances, and PSDB matching.

Search for components and search for part of are both inference along the PART-OF hierarchy in different directions. The instantiator searches the relevant slot to see if there are components to search for or another object of which this schema is a component. If the component or part of schemas exist, they can be accessed to continue the inference. Otherwise, each causes an instantiation of the missing schema to be generated as a prediction. Instantiation control can be transferred at this point to the component or PART-OF schema. The schema inference process maintains its thread of reasoning relevant to the schema in the schema instantiation structure slot.

5. PROCESSING EXAMPLE

The following processing example demonstrates the behavior of some implemented system components. These include the format of predictions from the long term terrain model, the extraction of perceptually significant groupings from the PSDB, how an instantiated schema uses grouping processes and queries over the PSDB, and extracting relevant cues for making view-based affixments to the long term terrain representation.

Figure 5-1 shows the elevation contours and road network in the a priori terrain data. This data comes from the Martin Marietta ALV test site in Denver and was supplied by the U.S. Army Engineering Topographic Laboratory (ETL). The vehicle position on the road is indicated by the arrow in the figure. From this, we are able to roughly determine the correspondence between an image taken from the road and the terrain data, (n.b., elevation data and the sensor parameters are not available). The initial instantiation of terrain schemas into LTM is done directly from the a priori data by image processing techniques. The road network is stored as a set of curve objects which is decomposed into linear segments with supplied attributes such as road material and width. Terrain patches are extracted as regions from terrain type information and parametric surface fits to the a priori elevation data.

Figure 5-2 shows the predicted segmentation derived from the terrain model. This is a qualitative description of predicted image features, which assumes that the vehicle is on a flat plane and that its field of view consists of road and grassy field terrain patches with some mountains in the distance. Predictions of the dirt road off to the right and the intersection are made from the road-network and the elevation information stored along with it. The mountain base corresponds to a terrain feature discontinuity. These predictions come from line-of-sight calculations given an estimate of the current vehicle position relative to the extracted objects in the long term terrain model. The predictions are in terms of rough constraints on region adjacencies across boundaries and the shape and attributes, such as color contrasts, of the boundaries themselves. The horizon line constraints are that it will tend to have smoothly changing orientation and be adjacent to a large homogeneous region (the sky). Detailed image predictions are not required and would be difficult without precise position estimates. The line-of-sight based predictions of image features inferred from intersection with schemas in the LTM, can associate multiple predictions of image features with different schemas at the same image location. More precise predictions can be obtained from an elevation grid using surface display techniques or from the processing results from previous images. From such predictions, objects can be matched directly against the image. Here the predicted features are described with constrained attributes determined from the visibility components of schemas. These specify allowable feature contrasts for objects like roads.

Figures 5-3a-b show some of the contour related structures in the initialized PSDB. Figure 5-3a shows the edges extracted at one spatial resolution using the Canny edge operator [Canny, 1983]. We have found it useful not to apply noise suppression to extracted segments in order to base filtering on structural properties of the contours such as linear deviation and relationships to other image structures. Different linear segment fits for this extracted edge images are shown in Figure 5-3b.

Figure 5-4 shows the results of grouping processes applied to a set of selected curves in Figure 5-2b with multiple associated attributes for orientation and color contrasts. The grouping processes were constrained by the predicted segmentation in Figure 5-2 using constraints on allowable color contrasts, changes in linear segment orientation, and rough image position and extent.

Figure 5-5 shows the results of a road schema instantiation based upon matches to extracted road boundaries in accounting for road surface properties through PART-OF relations. Texture elements adjacent to the road boundary which are consistent with a road surface, such as low contrast, parallel edges corresponding to tread marks, are used to direct queries to instantiate potential road area. Queries are also used to determine the presence of anomalous structures in the road such as anything which is high contrast or oriented perpendicular to the road direction. Such structures require disambiguation through instantiation of another schema (it could be a road marking) cued by the anomaly or elevation estimates derived from motion displacements or range sensing.

Significant image structures near the horizon line are particularly important for landmark extraction. Figure 5-6 shows extracted interesting image structures near and above the horizon line.

Acknowledgements

Work for this paper has been sponsored by the Defense Advanced Research and Projects Agency and the U.S. Army Engineer Topographic Laboratories under Government Contract #DACA76-85-C-0005.

Figure 5-1: Terrain Data

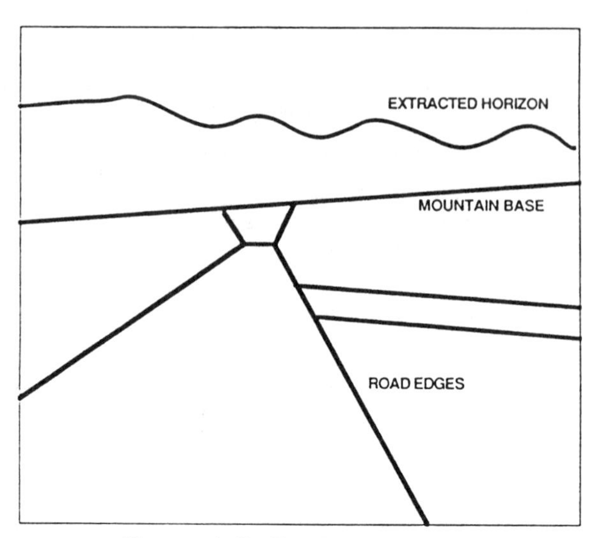

Figure 5-2: Predicted Segmentation

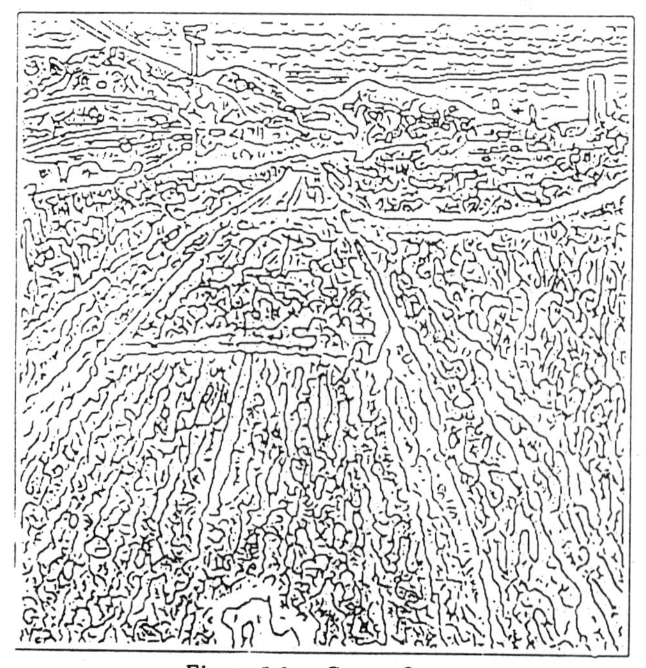

Figure 5-3a: Canny Operator

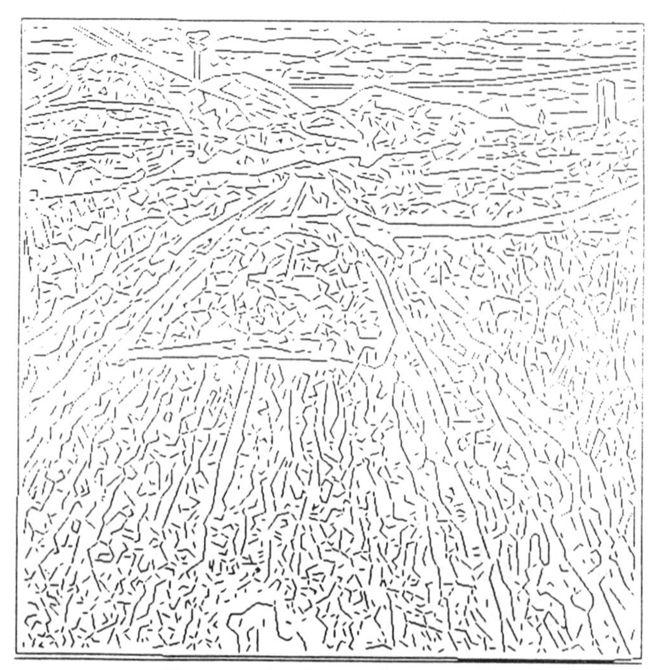

Figure 5-3b: Linear Segment Fits

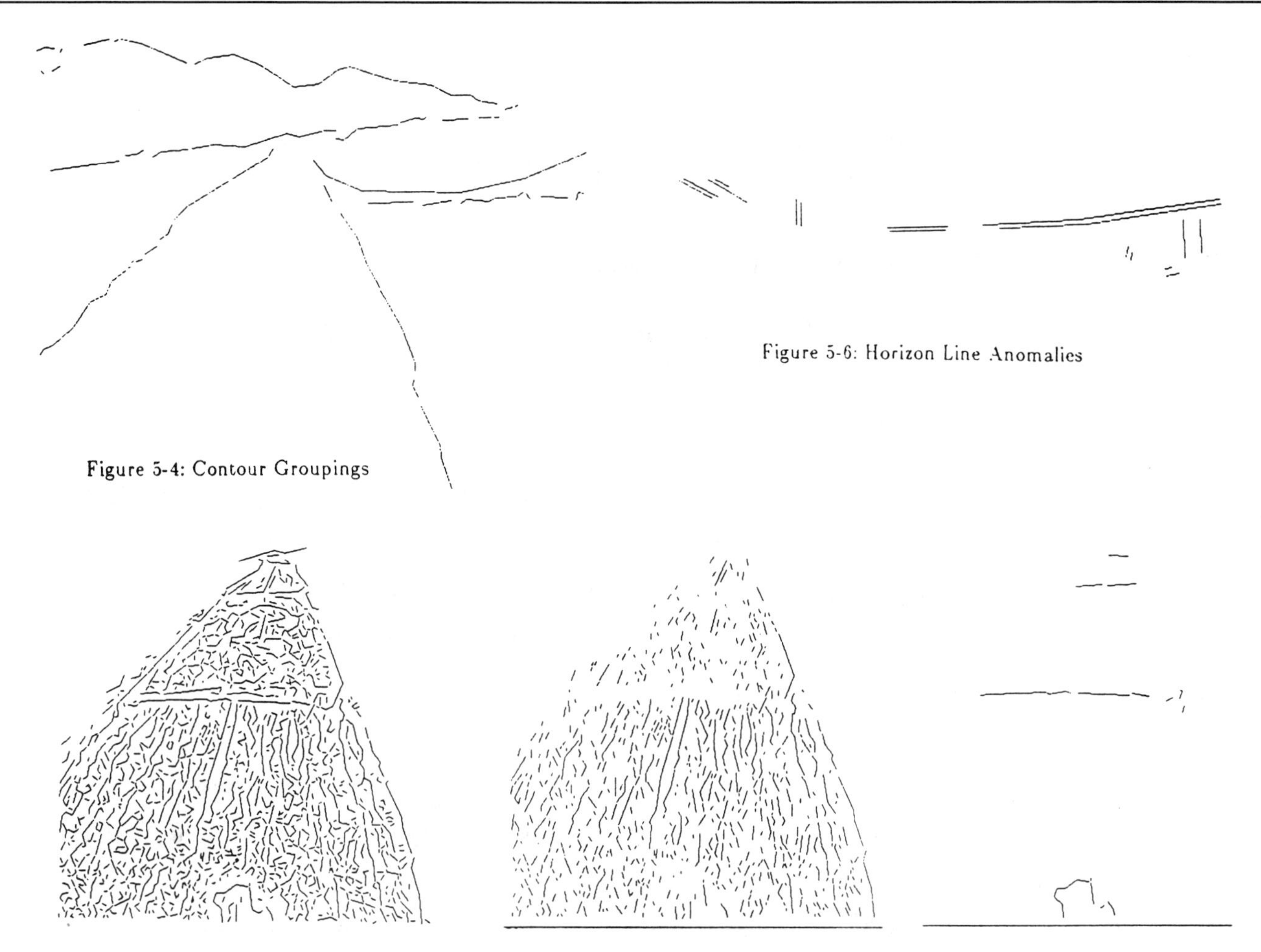

Figure 5-4: Contour Groupings

Figure 5-6: Horizon Line Anomalies

Figure 5-5: Road Schema Instantiation

6. REFERENCES

[Brooks, 1984] - R.A. Brooks, "Model-Based Computer Vision", Computer Science: Artificial Intelligence, No. 14, UMI Research Press, 1984.

[Canny, 1983] - J. Canny, "A Variational Approach to Edge Detection", In Proceedings of the National Conference on Artificial Intelligence (AAAI-83), pp. 54-58, August 1983.

[Hanson et al., 1978] - A.R. Hanson and E.M. Riseman, "VISIONS: A Computer System for Interpreting Scenes", In Computer Vision Systems, Academic Press, 1978.

[Kuan, 1984] - D.T. Kuan, "Terrain Map Knowledge Representation for Spatial Planning", 1st National Conference on AI Applications, pp. 578-584, 1984.

[Kuipers, 1982] - B.J. Kuipers, "Getting the envisionment right", In Proceedings of the National Conference on Artificial Intelligence (AAAI-82), Pittsburgh, Pennsylvania, August 1982.

[Lawton et al., in preparation] - D.T. Lawton and C.C. McConnell, "Segmentation by Heuristic Interestingness", in preparation.

[Marr, 1982] - D. Marr, "Vision", W.H. Freeman, San Francisco, 1982.

[Martelli, 1976] - A. Martelli, "An application of heuristic search methods to edge and contour detection", Commun. ACM, Vol. 19, No. 2, February 1976, pp. 73-83.

[Ohta, 1980] - Y. Ohta, "A Region-Oriented Image-Analysis System by Computer", Ph.D. Thesis, Kyoto University, Department of Information Science, Kyoto, Japan, 1980.

[Pentland, 1983] - A. Pentland, "Fractal-Based Description of Natural Scenes", In Proceedings Image Understanding Workshop, Arlington, Virginia, June, 1983, pp. 184-192.

Experiments in Using a Theorem Prover to Prove and Develop Geometrical Theorems in Computer Vision

Michael J. Swain and Joseph L. Mundy

General Electric Research and Development
Schenectady, New York, 12345

1. Abstract

A geometrical theorem prover based on algebraic techniques has been used to prove and derive theorems in computer vision. The results show that theorems relevant to persepective viewing and the properties of shadows can be proven, including some which contain inequalities. Deriving theorems by stating incomplete hypotheses and letting the prover suggest the missing constraints can provide valuable new insights into vision problems. However, there is no guarantee that the suggestions made by the proof algorithm are interpretable in terms of primitive geometric concepts.

2. Introduction

In 1977, a Chinese mathematician Wu Wen-Tsun discovered an algorithm which can prove significant, non-trivial theorems in geometry. The proof algorithm is based on expressing a theorem as well as hypotheses as algebraic expressions. The proof is then carried by expressing the theorem in terms of a polynomial basis developed from the hypotheses. In 1984 both Wu [Wu84] and Shang-Ching Chou [Cho84] published articles describing implementations of the prover. Chou simplified the algorithm in his implementation so as to further increase the speed of the prover in many cases, although it was less *complete*, i.e. there is a larger class of theorems it is unable to solve given unlimited computational requirements. More recently Hai-ping Ko [Ko85a] has implemented a version of the prover called Alge-Prover This version has been implemented using the MACSYMA algebraic utilities and command language. This Macsyma implementation allows the user to examine and process the results of the proof after the proof has finished, and also to easily define functions that allow translation from geometric concepts into polynomials -- the objects that must be given as input to the prover.

This paper describes experiments performed in the summer of 1985 whose intent were to examine the limits of the prover in proving geometrical theorems that arise in computer vision work, especially those related to the perspective transformation.

3. Motivation

Algeprover is being studied as an initial prototype of a proof system that can be used to support vision research [KMM85] and perhaps act directly in reasoning about scenes. The emaphasis is on the geometric properties of vision since that aspect is the best developed in axiomatic form. When using this system the user will be able to ask questions about geometrical concepts and receive a reply in terms of geometrical concepts as well. The language for discussing the concepts, called GRAIL, is currently under development.

One application for this system would be in the area of matching models to objects found in an image. Knowledge of theorems about the three-dimensional geometry present in the scene and the constraints of perspective viewing can drastically reduce the space of possible matches necessary to search.

The possibility of human error makes a strong argument for the use of a prover when designing programs. A prover reasons flawlessly and can also point out degenerate conditions that make a 'true' theorem false (see below). These degenerate cases are often overlooked in developing vision programs and are manifested as "bugs" or "crashes" of the resulting program.

4. A Desciption of the Prover

For a complete description of the prover see [Cho84], [KMM85], [Ko85a], [Ko85b], [Ko85d], [Wu78]. These papers have been collected into one volume entitled "Studies of Algebraic Theorem Proving in Geometry," [GGG85,GGG85]. Here only a brief outline of the prover will be given.

Input to the prover consists of a set of hypotheses, a set of conclusions and optionally, a list of triangulating variables. The triangulating variables guide the operation of the prover.

Suppose a polynomial (the conclusion) can be expressed as a linear sum of the hypothesis polynomials multiplied by other polynomials. Then given that the hypothesis polynomials are all zero, the conclusion polynomial is as well. That is, suppose we call the hypotheses $h_1..h_n$ and the conclusion c. Then if $c=p_1h_1+p_2h_2+\cdots+p_nh_n$ for some polynomials $p_1..p_n$, $h_1=h_2=..=h_n=0$ implies c=0. Chou's method determines something close to this. By a series of pseudo divisions it determines whether

$$s_1s_2..s_nc=q_1g_1+q_2g_2+\cdots+q_ng_n.$$

The multiplicative factors $s_1..s_n$ are introduced at each pseudo-division to prevent fractional coefficients from forming in the remainders. If after pseudo-dividing by all the triangulated polynomials the remainder is zero, the theorem is true providing $s_1..s_n$ are not zero. If one or more of them are, there can be no conclusion about the truth of the theorem, since the left-hand side of the above equation is zero independent of the value of c. The factors $s_1..s_n$ are called degenerate conditions.

The g polynomials are derived from the h's by triangulation. The h's are transformed in such a way that if $h_1=h_2=..=h_n=0$ then $g_1=g_2=..=g_n=0$ and there is a set of variables $x_1..x_n$ such that the polynomials have the following triangular form:

$g_1(x_1)$

$g_2(x_1,x_2)$

$gn(x_1,x_2,....x_n)$

When pseudo division is performed, it is first performed using the polynomial g_n in variable x_n, then with g_{n-1} in variable x_{n-1}, etc. This ordering of divisions tends to produce a zero remainder whenever possible.

The theorem prover is not guaranteed to produce a zero remainder when presented with a conclusion that follows from the hypotheses. However the number of theorems that cannot be proved because of finite computation resources far outweighs the number that remain unproved due to the incompleteness of the prover. For this reason provers based on Wu's work are interesting and useful, since they are significantly more efficient for geometric problems than provers based on resolution or other methods.

When asked to prove theorems a prover can be of aid mainly as a device for checking previous work. It cannot be of aid in suggesting new ideas but only in eliminating some of the labour and errors involved in proving a theorem by hand. But if the theorem prover is given a conclusion and incomplete hypotheses it can be of aid in generating new hypotheses that make the theorem true. The remainder, when added to the hypotheses, makes the theorem true. Therefore, the remainder is the missing hypothesis. In this way Algeprover can help discover new theorems, although there is no guarantee that the remainder has an easily detected geometric interpretation.

5. Experiments

The prover was applied to a selection of problems from computer vision. A first simple example of the capabilities of the prover can be seen in an example concerning perpendicular lines seen in perspective and their vanishing points. Suppose two different lines intersect in the world or *scene* at right angles. In the image each of these lines will have a vanishing point, the point onto which the line converges as the distance from the viewer tends towards infinity. Define the first line using the two points p and q and the second using q and r. The vanishing point of the line through any two arbitrary points a and b is defined to be

$$\left[\frac{b_1-a_1}{b_3-a_3}, \frac{b_2-a_2}{b_3-a_3}, 1\right]$$

where the focal length is taken to be 1 and the perspective transformation defined as P(X,Y,Z) = (X/Z, Y/Z, 1). Now if we state that s and t are the vanishing points of the lines qp and qr respectively as hypotheses 1 and 2 and that the lines pq and qr are perpendicular as the conclusion, the remainder is:

$$(s_2t_2+s_1t_1+1)(q_3-p_3)(r_3-q_3)$$

so if $s_2t_2+s_1t_1+1$ is added to the hypotheses the theorem is true. This equation places a simple constraint on the locations of the vanishing points in the image. It states that the vectors from the origin to the vanishing points must have a dot product equal to -1 when considering the viewplane to be at $z=1$ (see Appendix A).

Under orthography a well-known heuristic is Kanade's skew symmetry constraint. [Kan81] When using the skew symmetry constraint one assumes that the skew-symmetrical objects in the image are symmetrical in the scene. Kanade shows that the symmetries present in a cube considerably constrain the relationships that can occur in a orthographic view of the cube. (see Figure 1)

The prover can prove the same result using the same assumptions. The subsidiary conditions were found to be complex in this case. The simple ones could be interpreted and were found to be degenerate cases. One stated that an edge and an adjacent face were

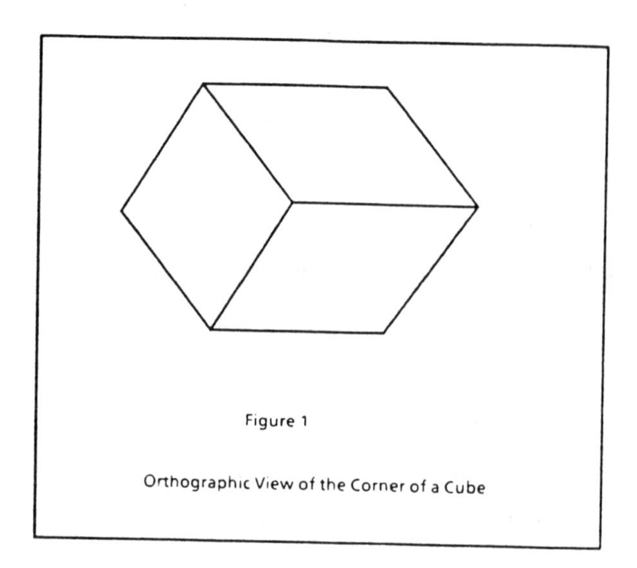

Figure 1

Orthographic View of the Corner of a Cube

such that the gradient of the face was perpendicular to the edge. In this case the face would not be visible to the viewer (see Appendix B).

6. Experiments With Problems Involving Inequalities

Inequalities can be handled by the prover using a method due to Seidenberg. To state that a quantity q is nonzero one introduces a new variable k and writes the equation $1-kq=0$. Similarly to state that q is positive (negative) one writes $q-k^2=0$ ($q+k^2=0$).

A problem that has been solved using these methods is the following, set in a two-dimensional world we may call 'Flatland'.

Consider a floor with a barrier above and parallel to it. Without loss of generality say the barrier is one unit of distance above the floor in the y-direction. Also without loss of generality place the left-hand end of the barrier at x=0. Let the barrier extend to infinity on the right. At some point above the floor there is a light source shining down on the floor and barrier. Place an observer somewhere in this two-dimensional world above the floor. When can the observer see the shadow? (see Figure 2)

Let S be the position of the light source, **O** the position of the observer, **I** the tip of the barrier, **B** some point in the barrier and **R** some point in the shadow. We propose the solution to be that if **O** lies on the side away from the barrier of the line joining S and **I** the shadow can be seen.

The hypotheses are:

h1: $O_2 - k_O$ {O is above the floor}

h2: $S_2 - 1 - k_S$ {S is above the barrier}

h3: $I_2 - 1$ {I is in the barrier}

h4: I_1 {I is at the tip of the barrier -- x = 0}

h5: $(O_1 - S_1)(O_2 - S_2) - (O_2 - S_2)(I_1 - S_1) - k_{OSI}^2$ {(O−S)×(I−S)>0 i.e. O is to the left of the line defined by I and S}

h6: $(R_1 - I_1)(S_2 - I_2) - (I_2 - R_2)(S_1 - I_1) - k_{RSI}^2$ {(I−S)×(R−S)>0 i.e. R is to the right of the line defined by I and S; in other words R lies in the shadow}

h7: R_2 {R is a point on the floor}

h8: $B_1 - k_B^2$ {B is a point in the barrier}

h9: $B_2 - 1$

The conclusion is:

concl: $1 - k_{BOR}\,collinear\_2(O,B,R)$ {B is not collinear with O and R, i.e. R can be seen} where collinear_2(a,b,c) := $(b_2 - a_2)(c_1 - b_1) - (c_2 - b_2)(b_1 - a_1)$

For the theorem to be true the conclusion must follow from the hypotheses no matter how S, O or B are place and for some placement of R. Therefore there must be some k_{OBR} and some k_{RSI} such that for all values of the rest of the k's the remainder is zero. The remainder returned by Algeprover was such that when solved for k_{BOR} and k_{RSI} the following expressions were obtained:

$$k_{RSI} = t$$

$$k_{BOR} = \frac{k_S^2}{k_B^2 k_O^2 k_S^2 + k_{OSI}^2 + t^2(k_O^2 - 1)}$$

For any value of $(k_O^2 - 1)$, k_{RSI} may be chosen such that the term containing these two expressions is smaller than the sum of squares that makes up the rest of the denominator. Therefore there exist real values of k_{RSI} and k_{BOR} that satisfy these equations and the theorem is true.

7. Conclusions

Algeprover has shown a capability to prove and aid in developing geometrical theorems encountered in computer vision. The proving capability is certainly much easier to use since the usual result is the unmistakable zero remainder returned by the prover. When incomplete hypotheses are provided there is the added problem of interpreting the remainder in a geometrical way. This problem can be difficult, as suggested by the remainders which remain uninterpreted in some of the problems above. There is no guarantee of any simple geometric interpretation at all to the remainders that are obtained. Perhaps a computer program could make use of the remainders themselves, so it is not necessarily the case that all remainders must be interpretable to be useful.

The success of a prover is generally measured by two measures: its completeness and its efficiency. A complete prover is desirable not only because more theorems can be proven but because the results of the prover can be used to show that a theorem is untrue as well as confirm its veracity. Algeprover is currently incomplete but developments are being made in the direction of finding algorithms which are complete [Wu84].

Progress is being made in the direction of improving the efficiency of Algeprover as well. H.P. Ko [Ko85c] has discovered a heuristic method of directing the order of triangulation in Chou's triangulation procedure. The method has reduced the amount of

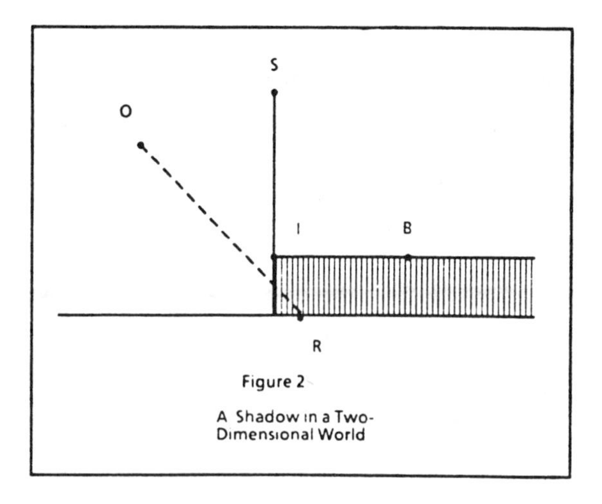

Figure 2

A Shadow in a Two-Dimensional World

computation by up to 8 times that necessary even when the selection of triangulating variables was done by Chou using geometric intuition.

Not all geometric problems can be easily represented in a form suitable for the prover. A major problem is found in representing summations of lengths, for example. Here one would like to use the square root of a polynomial quantity rather than the polynomial quantity itself. The prover only accepts polynomials.

Algeprover and its sister programs are proving many theorems in geometry that have never been proved mechanically before. Two theorems have even been proven that have never been proven by a human [Wu84] although one would not call the theorems mathematically significant. Current capabilities do not produce a reliable tool to use in applications such as real-time computer vision but work is being done in the area that will move us closer to that goal. Advances will have to be made in interpreting the remainders produced by Algeprover before it can be integrated into a geometrical query system.

8. Acknowledgements We are indebted to H.P. Ko for implementing, explaining and improving Algeprover.

References

[Cho84] S. Chou, Proving Elementary Geometry Theorems Using Wu's Algorithm, *Contemporary Mathematics 29*, (1984), 243-286.

[GGG85] Studies of Algebraic Theorem Proving in Geometry, Internal Report, General Electric Corp., Schenectady, N.Y., May 29, 1985.

[Kan81] T. Kanade, Reconstruction of the Three-Dimensional Shape of an Object from a Single View, *Artificial Intelligence 17*, (1981), 409-460, North-Holland.

[KMM85] D. Kapur, J. Mundy, D. Musser and P. Narendran, *Reasoning About Three Dimensional Space*, General Electric Corporate Research and Development, Schenectady, 1985.

[Ko85a] H. Ko and M. Hussain, *ALGE-Prover: an Algebraic Geometry Theorem Proving Software*, General Electric Corporate Research and Development, Schenectady, N.Y., 1985.

[Ko85b] H. Ko and M. A. Hussain, *A Study of Wu's Method -- A Method to Prove Certain Theorems in Elementary Geometry*, General Electric Corporate Research and Development, Schenectady, N.Y., 1985.

[Ko85c] H. Ko, , Unpublished Manuscript, July, 1985.

[Ko85d] H. Ko, *On Algebraic formulation of Wu's Method and Chou's Method -- Methods to Prove Certain Theorems in Elementary Geometry*, General Electric Corporate Research and Development, Schenectady, N.Y., 1985.

[Wu78] W. Wu, On the Decision Problem and the Mechanization of Theorem-Proving in Elementary Geometry, *Scientia Sinica 21*, (1978), 159-172.

[Wu84] W. Wu, Basic Principles of Mechanical Theorem Proving in Elementary Geometries, *J. Sys. Sci. & Math. Scis. 4 3*, (1984), 207-235.

Appendices

The appendices contain the inputs given to ALGE-Prover and the remainders and subsidiary conditions that were returned by the program for the examples studied in the text of the article.

Conventional mathematical notation is used in the examples. The hypotheses are abbreviated hyp1, hyp2, etc. The conclusion is abbreviated concl.

Appendix A -- Vanishing Points of Perpendicular Lines

Derive the conditions that are satisfied by vanishing points in an image when their respective lines meet at right angles.

Input:

$$\text{vanishing(q,r)} = \left(\frac{b_1-a_1}{b_3-a_3}, \frac{b_2-a_2}{b_3-a_3}, 1\right)$$

hyp1 = s - vanishing(q,p)

hyp2 = t - vanishing(q,r)

concl = perpendicular(q-p,q-r)

Output:

remainder = $(s_2t_2+s_1t_1+1)(q_3-p_3)(r_3-q_3)$

subsidiary_condition = 1

Appendix B -- Skew Symmetry Applied to a Corner View of a Cube

Notation:

(0,0) = the coordinates of the image point corresponding to the corner of the cube.

aa = $(cos\theta_a, \sin\theta_a)$

ab = $(cos\theta_b, \sin\theta_b)$

ac = $(cos\theta_c, \sin\theta_c)$

where θ_i measures the counter-clockwise angle in the image from the x-axis to the i'th edge.

ga = the gradients (p,q) of the plane defined by aa and ab.

gb = the gradients of the plane defined by ab and ac.

gc = the gradients of the plane defined by ac and aa.

The gradients (p,q) are defined:

$$p=\frac{\partial z}{\partial x} \quad q=\frac{\partial z}{\partial y}$$

Proof That Figure 1 is the Corner of a Cube, Assuming Skew Symmetry

Input:

hyp1 = $ga \cdot ab - gb \cdot ab$

hyp2 = $gb \cdot ac - gc \cdot ac$

hyp3 = $gc \cdot aa - ga \cdot aa$

hyp4 = $(ga \cdot aa)(ga \cdot ab)+(aa \cdot ab)$

hyp5 = $(gb \cdot ab)(gb \cdot ac)+(ab \cdot ac)$

hyp6 = $(gc \cdot ac)(gc \cdot ac)+(ac \cdot aa)$

Hypothesis 1 results from Mackworth's constraint applied to planes a and b and line a, their line of intersection. Hypotheses 2 and 3 are generated similarly. Hypotheses 4-6 come from the skew symmetry assumptions.

concl: $ga \cdot gb+1$

Output:

remainder = 0

The subsidiary condition had 60 terms. With aa_2 set to zero in the subsidiary condition (one edge lies on the x-axis) the subsidiary condition simplified to some extent and was:

$$-aa_a{}^3 ab_1{}^2 ab_2 (ab_2 gb_2 + ab_1 gb_1)(gb_1 ac_2 gb_2 + ac_1 gb_1^2 + ac_1)^5$$
$$(ac_2 gb_2^2 + ac_1 gb_1 gb_2 + ac_2)^2 (ac_2 gb_2{}^2 - ac_1 gc_1 gb_2 + ac_1 gb_1 gb_2 + gb_1 gc_1 ac_2 + ac_2)$$

only the first four factors of which are geometrically interpretable. The statement that the tourtn factor is zero may be rewritten:

$$\left[\frac{ab_2}{ab_1}\right]\left[\frac{gb_2}{gb_1}\right]=-1$$

i.e. ab and gb are perpendicular, which implies that plane b cannot be seen and the view of the cube is a degenerate one.

Knowledge organization and its role in representation and interpretation for time-varying data: the ALVEN system

JOHN K. TSOTSOS[1]

Department of Computer Science, 10 King's College Road, University of Toronto, Toronto, Ontario, Canada M5S 1A4

Received August 28, 1984
Revision accepted December 5, 1984

The so-called "first generation" expert systems were rule-based and offered a successful framework for building applications systems for certain kinds of tasks. Spatial, temporal, and causal reasoning, knowledge abstractions, and structuring are among topics of research for "second generation" expert systems. It is proposed that one of the keys for such research is *knowledge organization*. Knowledge organization determines control structure design, explanation and evaluation capabilities for the resultant knowledge base, and has strong influence on system performance. We are exploring a framework for expert system design that focuses on knowledge organization, for a specific class of input data, namely, continuous, time-varying data (image sequences or other signal forms). Such data are rich in temporal relationships as well as temporal changes of spatial relations, and are thus a very appropriate testbed for studies involving spatio-temporal reasoning. In particular, the representation formalism specifies the semantics of the organization of knowledge classes along the relationships of generalization/specialization, decomposition/aggregation, temporal precedence, instantiation, and expectation-activated similarity. A hypothesize-and-test control structure is driven by the class organizational principles, and includes several interacting dimensions of search (data-driven, model-driven, goal-driven temporal, and failure-driven search). The hypothesis ranking scheme is based on temporal cooperative computation, with hypothesis "fields of influence" being defined by the hypothesis' organizational relationships. This control structure has proven to be robust enough to handle a variety of interpretation tasks for continuous temporal data. A particular incarnation, the ALVEN system, for left ventricular performance assessment from X-ray image sequences, will be summarized in this paper.

Key words: knowledge representation, expert systems, medical consultation systems, time-varying interpretation, knowledge-based vision

Les systèmes experts dits de "première génération" étaient basés sur des règles et offraient un cadre intéressant pour la construction de systèmes d'application effectuant certaines tâches particulières. Le raisonnement spatial, temporel et causal, l'extraction et la structuration de la connaissance sont parmi les axes de recherche considérés pour les systèmes experts de "seconde génération". On suggère que l'une des clés de ce type de recherche soit l'*organisation de la connaissance*. L'organisation de la connaissance conditionne l'élaboration de la structure de contrôle, les capacités d'évaluation et d'explicitation de la base de données qui en résulte, et a une très forte incidence sur les performances du système. Nous explorons un cadre d'élaboration des systèmes experts qui s'intéresse particulièrement à l'organisation de la connaissance pour une classe spécifique de données analysées: les données continues temporalisées (séquences d'images ou autres formes de signaux). De telles données sont riches en relations temporelles de même qu'en modifications temporelles des relations spatiales et offrent ainsi un cadre d'étude approprié pour les recherches impliquant le raisonnement spatio-temporel. En particulier, la représentation facilite et renforce la sémantique dans l'organisation des catégories de savoir en fonction des rapports entre généralisation/spécification, décomposition/agrégation, ordre temporel, instantiation, et prévision des similarités. Une structure de contrôle par hypothèses et tests est guidée par les principes d'organisation catégorielle et comprend plusieurs dimensions complémentaires de recherche (recherche guidée par les données, par modèle, par but, par échec, et temporelle). Le schéma principal d'hypothèse est basé sur une évaluation prenant en compte la temporalité, où les "champs d'influence" d'une hypothèse sont définis par ses liens organisationnels. Cette structure s'est avérée suffisamment solide pour effectuer une variété de tâches d'interprétation de données temporelles continues. Une réalisation particulière, le système ALVEN, qui évalue le fonctionnement du ventricule gauche à partir de séquences d'images radiographiées, sera présenté dans cet article.

[Traduit par la revue]

Mots clés: représentation de la connaissance, systèmes experts, systèmes de diagnostic médical, interprétation temporalisée, vision raisonnée.

Comput. Intell. 1, 16–32 (1985)

1.0 Introduction

A brief overview of the ALVEN application domain and the solution strategy is in order before detailed discussions are presented.

The domain of application of ALVEN is that of the evaluation of the dynamics of left ventricular tantalum marker implants from X-ray image sequences. ALVEN is thus both a visual motion understanding system as well as an example of artificial intelligence applications in medicine. The goal is to analyse both pre-operative (without markers, using contrast media) and post-operative marker films (following coronary bypass surgery), to evaluate the efficacy of surgery, locally and globally, quantitatively and qualitatively, over the recovery period (several months), and to evaluate the effects of drug interventions. It is crucial for such comparisons of perhaps subtle changes that a rich representation involving both qualitative and quantitative be obtained for each film. Other examples of computer analysis of marker implants are presented in Gerbrands *et al.* (1979), and Alderman *et al.* (1979), which addresses the problem of point of reference.

The evaluation of left ventricular (LV) performance by computer from cine representations of LV dynamics is a difficult and long-studied problem. A large number of heuristics have been proposed for measuring shape changes (Brower and Meester 1981), following anatomical landmarks (Slager *et al.*

[1]Currently a Fellow of the Canadian Institute for Advanced Research.

1981), computing segmental volume contributions (for a comparison, see Gelberg *et al.* (1979)), etc., all performing with varying degrees of success, but being applied independently of each other. Although such heuristics are indeed valuable quantitative measures, we propose that their limited performance is due to two key considerations: (1) it is unlikely, given the complexity of the domain of LV dynamics and the amount of training that a clinical specialist in this area receives, that any single heuristic can capture all the important facets of the evaluation and be successful in all applications; (2) the heuristics are purely quantitative in nature, contrasting with the fact that clinicians, and for that matter humans in general, deal in qualitative or descriptive terms combined with numerical quantities. That is, relational quantities are necessary components of the interpretation process, while numerical ones are secondary. The key here is that a computer system that is to solve the difficult problems present in the domain of LV dynamics interpretation must integrate the above-mentioned numerical heuristics as well as consider the symbolic processing aspects of the interpretation.

We propose that the current limited success of computer assisted analysis of left ventricular dynamics is due to three main reasons: (1) there is a strong tendency to remain within the realm of mathematical modelling for LV dynamics, and it is not at all clear that this is an adequate approach; (2) in places where mathematical models alone may be insufficient, current research into more sophisticated schemes is not yet complete, and thus, more basic research is required, particularly into representations of knowledge and interpretation control structures, before applications such as LV performance can be solved, a view also stated in Boehm and Hoehne (1981); (3) there is a distinct lack of knowledge about LV dynamics, in conjunction with disagreements about what is important to model and what terminology is to be used. We distinguish our approach from those whose goal is to provide some intermediate visual representation that must still be subjectively interpreted by a clinician (the work described in Hoehne *et al.* (1980) is a particularly good example of such a representation). Our goal is to perform this interpretation, in much the same way as the clinician does, and to do it in an objective and consistent manner.

The strategy we adopted for the solution of this problem follows. In reality, there are two major problems to be solved: (1) the problem of understanding visual motion given a set of primitive image tokens over time, and (2) the problem of reasoning about spatio-temporal relationships in the context of human left ventricular dynamics.

There are three major points to be made regarding the ALVEN methodology. The first deals with the construction of the knowledge base. ALVEN's knowledge base is made up of frame-like objects called *classes*, which are organized using the relationships *IS-A*, *PART-OF*, *SIMILARITY*, and *Temporal Precedence*. These are described later in the paper. The definition of the knowledge classes requires two stages. The first is to define the general knowledge that pertains to the task of motion understanding. This must be done in such a way as to satisfy the following: (a) a set of motion classes must be found that is sufficient for use in defining the motion classes of the problem domain—problem specific motion classes are defined in terms of the general ones; and (b) the interface between image-specific concepts and general knowledge concepts is defined to be the leaves of the PART-OF hierarchy of motion concepts, and the image-specific procedures must be known. That is, the only concepts that can be leaves of the PART-OF hierarchy are those that are directly extractable from the signal input. The second stage of knowledge class definition is to use the general purpose motion knowledge (which is a knowledge base, fully usable, in its own right), and define concepts specific to left ventricular dynamics in terms of the general ones. The problem specific knowledge base of ALVEN therefore simply 'hooks' onto the general purpose one through the knowledge organization relationships. Other motion problem domains could be handled in a similar manner.

The second aspect of ALVEN to be introduced is the control strategy. A cyclic process was defined that integrates the different search schemes into a coherent whole (see Fig. 1). Definitions of these search schemes appear later in the paper. Perhaps the most interesting aspect is how the different search schemes interact and are coordinated, and this is clear in Fig. 1. The basic cycle involves the extraction of tokens from the input signal, instantiating those tokens as leaves of the concept PART-OF hierarchy, following that hierarchy in a data-directed fashion to activate new hypotheses that are aggregates of the input tokens, thus obtaining an initial set of hypotheses. That initial set is then specialized by downward traversal of the IS-A hierarchy (one level), selecting either the most likely of the specialization, through *a priori* knowledge, or a random one. This intermediate set is elaborated by the activation of all component hypotheses (elements of the model specified by each class). Each hypothesis of this refined set is then matched with data or other instances. Matching successes lead to further specializations and elaborations, while failures lead to selection of alternate hypotheses via the failure-directed mechanism, which are to be considered in parallel with the failing hypotheses. Once refinements of this sort are complete (there are no more model elaborations, that is, each PART-OF subtree for each hypothesis has been fully activated) and there are no specializations (no successful matches are found for any specializations of any hypothesis), the hypotheses are ranked within competing sets by their updated goodness of fit measure, or certainty. For some hypotheses, that goodness of fit displays sufficient confidence so that the hypothesis is instantiated. Other hypotheses are deleted from further consideration. The best hypotheses are used to produce a set of predictions for the next data sample. Those predictions may be of events that are expected to occur next, or may be components of hypotheses that should have been observed but have not (differentiating between temporally-directed predictions in the former case and model-directed ones in the latter). The predictions are projected from hypothesis space to the signal domain and are used by the token extraction process as guidance. If tokens cannot be found from some prediction, a relaxation of constraints occurs, thus generalizing the prediction by moving upwards along the IS-A hierarchy from the hypothesis responsible for the faulty prediction. A 'blind' token-finding procedure is available for cases where no predictions allow for successful token finding.

The process by which time-varying matching evidence is accumulated and integrated over time is the final aspect to be introduced. This is based on relaxation labelling processes (Zucker 1978); the relaxation process described by Zucker, however, is modified in several important ways. Details of the *temporal cooperative process* are beyond the scope of this paper but are presented in Tsotsos (1984). The important points follow. Firstly, it should be clear that it is meaningless to talk about the certainty of a hypothesis using just its spatial evidence—space and time are considered together and cannot

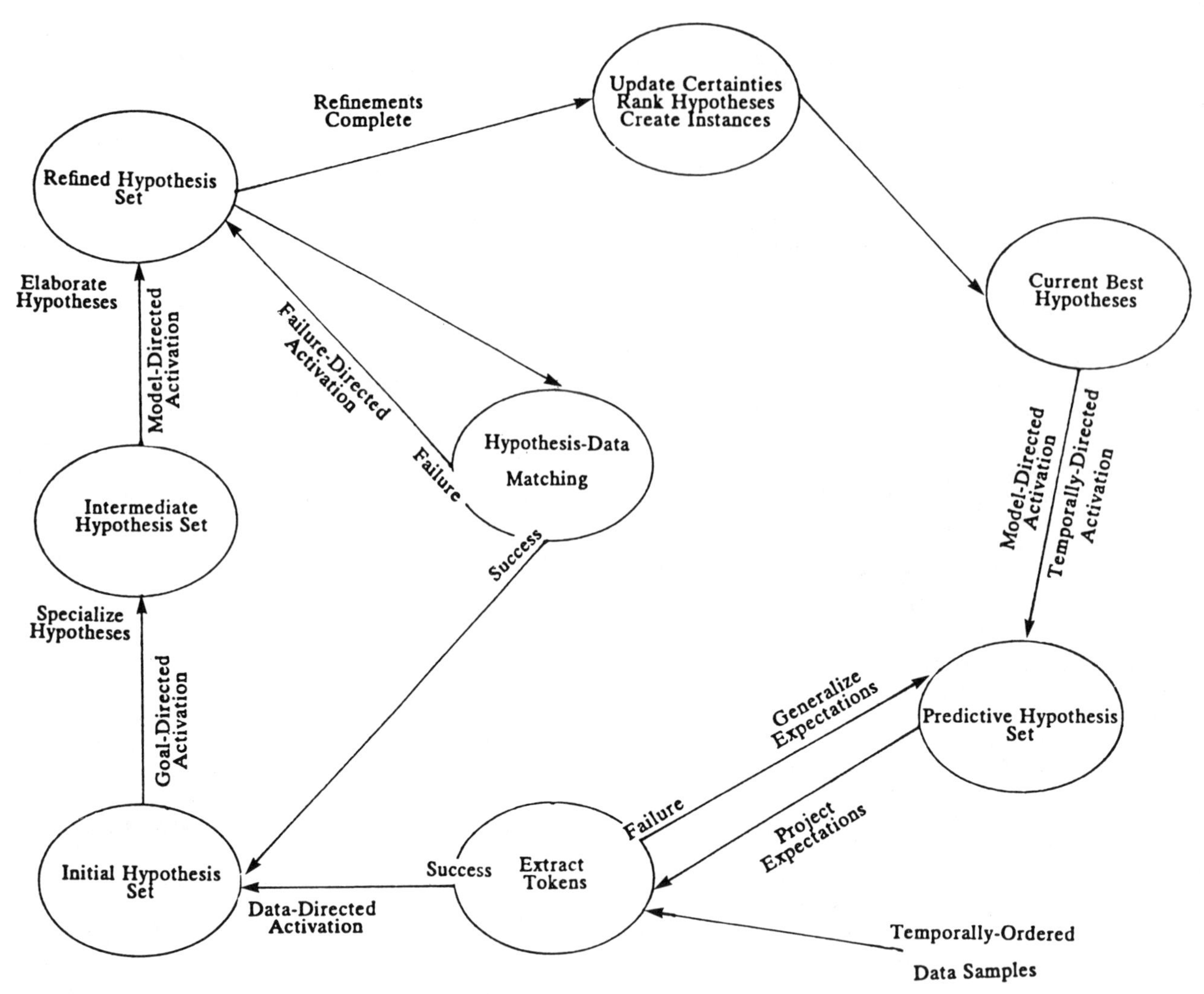

FIG. 1. The control strategy

be separated. Hypothesis certainty depends on the accumulation of spatial evidence exhibiting temporal consistency and continuity over a time interval. It also depends not only on individual hypothesis consistency but also on comparison with the evidence for other hypotheses—in other words, both local and global evaluations play a role. Relaxation methods operate over networks of local processes where processes are connected to other processes via weighted links or communication pathways. The determination of the weights is a difficult problem. In addition, the one that we use is non-linear. Relaxation schemes typically require many iterations before they converge to stable solutions (they are hill-climbing schemes). In a time-varying interpretation situation, this is not a desirable characteristic. We require that the process converge to solutions within a small fixed number of iterations and thus must discover the conditions under which relaxation schemes satisfy this requirement. This must be so because new data are being added to the interpretation as the interpretation proceeds. Our temporal cooperative process is unified with the remainder of the control scheme because the local processes are hypotheses, and the communication pathways are the knowledge organization relationships. Each relationship has an associated weight (whose value is time-varying) that is related to the semantics of that relationship. Thus, each local process (hypothesis) has a *field of influence* (or neighbourhood, using relaxation terminology) that is defined by its semantic relationships to other processes. A simple numerical inequality that was determined empirically must be satisfied by the weights in order for the required convergence property to be satisfied.

The remainder of the paper describes details of the representation, the control and reasoning strategies, the domain of left ventricular dynamics, and briefly presents an example of a complete analysis. Implications of the methodology conclude the discussion.

2.0 Overview of the representational scheme

2.1 Knowledge packages: classes

Packaging up knowledge leads to a modular representation, with all the advantages of modularity, particularly the enhancement of clarity and flexibility. Most knowledge-package representation schemes borrow strongly from Minsky (1975). Our frames are called *classes*. A class provides a generalized definition of the components, attributes, and relationships that must be confirmed of a particular concept under consideration in order to be able to make the deduction that the particular concept is an instance of the prototypical concept. Classes also have embedded, declarative control information, namely exceptions and similarity links. These features will be described shortly. Note that there is a distinction between the 'prerequisites' of the class, those components that must be observed in order to instantiate the class, and the 'dependents' of a class, those components that must be derived on instantiation. Dependent slots carry their own computation information. Classes exhibit large grain size, and translating their contents to rules would require many rules. An obvious advantage over the rule scheme is that elements that conceptually belong together are packaged together into a class, with some control information included. Other frame-based schemes for medical consultation systems include the MDX system (Chandrasekaran *et al.* 1979) and CADUCEUS (Pople 1982).

2.2 Multi-dimensional levels of detail

The term 'level of detail' seems to denote different things to different people. In most schemes, it is used to express problem decomposition only (Nilsson 1971). We present two separate views of abstraction 'level'. These views are related to the fact that all concepts have both IS-A and PART-OF relationships with other concepts. Thus the level of specificity of detail can be controlled by, or examined by traversing, the IS-A hierarchy, while the level of resolution of detail (decomposition in other schemes) is reflected in the PART-OF hierarchy. In Patil *et al.* (1982) only the decomposition view of level is present, while in CADUCEUS (Pople 1982), it seems that the level of specificity is employed and level of resolution is restricted to causal connections. In Wallis and Shortliffe (1982) rule complexity is used, which may be likened to our view of level of resolution; however, its use is restricted to explanation.

2.3 Time and its representation

Several interacting mechanisms are available for the representation of temporal information. This multi-pronged approach differs from other schemes that embody a single type of construct for handling temporal information. The complexity of time necessitates several special mechanisms. Our approach differs from others (Allen 1981; Mittal and Chandrasekaran 1980) in that we were motivated by problems in signal analysis rather than in representing natural language temporal descriptions and their inherent ambiguity and vagueness. It is not clear, for example, what kind of control strategy can be employed along with Allen's scheme of temporal representation. Fagan (1980) is concerned with a temporal interpretation situation. However, there are a number of issues, primarily in control, that are not considered by his system, VM:

- using the rule-based approach, only a data-driven recognition scheme is incorporated, and thus, VM cannot initiate a search for temporally expected events;
- the handling of noise is not formalized, but is rather ad hoc;
- the complexity of temporal relationships among rules seems limited, and arbitrary groupings of temporal events and their recognition are not addressed;
- expectations in time are table-driven, and no distinction is made between them and default values or expected ranges. Expectations in ALVEN are computed from such information, but current context is taken into account as well so that expectations are tailored for the task at hand;
- partial satisfiability of temporal event groupings cannot be handled.

In addition, Long and Russ (1983) also address the problem of time-dependent reasoning. Their scheme is closer to Fagan's than to ours. The control is data-driven exclusively. Their representation of time, however, shares some similarities with ours in that both points and intervals are used, and special meaning is assigned to the variable 'now'.

A brief description of the representation of time used by ALVEN follows. A TIME_INTERVAL class is defined that contains three slots, namely, start time, end time, and duration. This class can then be included in the structure of any other class and would define its temporal boundaries and uncertainty in those values. Using those slots, the relations before, after, during, etc., (similar to Allen (1981)) are provided. In constraint or default definition, sequences of values (or ranges of values) may be specified using an 'at' operator, so that in effect a piecewise linear approximation to a time-varying function can be included. In this case of course, constraint evaluation must occur at the proper point in time. Tokens of values such as volume or velocity for which use of this operator is appropriate, have two slots, one for the actual value and the other for the time instant at which that value is true. The time instant slot is a dependent slot whose value is set to the value of the special variable 'now' (current time slice). Note that this kind of mechanism could easily be expanded if required to multi-dimensional functions.

Finally, arbitrary groupings of events can be represented. The set construct (which may be used for any type of class grouping, not only for events) specifies elements of a group, names the group as a slot, and has element selection criteria represented as constraints on the slot. Patil *et al.* (1982) described a version of temporal aggregation similar to ours, but do not seem to have a time-line along which selection of values can occur, nor do they distinguish between aggregations of events and sequences of measurements.

Since knowledge classes are organized using the IS-A and PART-OF relations, their temporality is as well. By constructing a PART-OF hierarchy of events, one implicitly changes the temporal resolution of knowledge classes (as long as not only simultaneous events are considered). For example, suppose that the most primitive events occur with durations on the order of seconds. Then groupings of those may define events that occur with durations in the minute range, and then groupings of those again on the order of hours, and so on. Events whose durations are measured using months can be so built up. Yet, many kinds of events cannot be so decomposed, and there is no requirement that all events have such a complete decomposition. Those events, however, are not left hanging, since they will also be related to others in the knowledge base via the IS-A relationship. The control scheme makes use of the temporal resolution with respect to sampling rates and convergence of certainties.

In the following examples, first the TIME_INTERVAL class is shown, followed by the class for the concept of SEQUENCE, followed by a constraint on volume of the left ventricle from the normal left ventricle class, showing the use of the 'at' mechanism for both default and constraint definition, as well as an example of a piecewise linear approximation to the volume vs time function.

Example 1

```
class TIME_INTERVAL with
prerequisites
   st : TIME_V such that [st > = 0];
   et : TIME_V such that [et > = st];
dependents
   dur : TIME_V with dur ← et - st;
end $
```

Example 2

```
class SEQUENCE is-a MOTION with
prerequisites
   motion_set : set of MOTION such that [
```

```
    for all m : (MOTION such that
      [m element-of motion_set])
      verify [
        m.subj = self.subj,
        ~find m1 : MOTION where [
        m1 element-of motion_set,
        (m1.time_int.st during m.time_int or
        m.time_int.st during m1.time_int)],
        find m2 : MOTION where [
          m2 element-of motion_set,
          (m.time_int.st = m2.time_int.et or
          m2.time_int.st = m.time_int.et)]],
        card(motion_set) > 1,
        strict_order_set(motion_set,time_int.st)];

dependents
  first_mot : MOTION with
    first_mot <- earliest_st(motion_set);
  last_mot : MOTION with
    last_mot <- latest_st(motion_set;
  time_int : with time_int <-
    (st of TIME_INTERVAL with st <- first_mot.time_int.st,
    et of TIME_INTERVAL with et <- last_mot.time_et);
end $
```

Example 3

```
volume : VOLUME_V with
  volume <- (vol of VOLUME_V with
    vol <- (minaxis.length @ now) ** 3
      default (117 @ m.systole.time_int.st,
          22 @ m.systole.time_int.et,
          83 @ m.diastole.rapid_fill.time_int.et,
         100 @ m.diastole.diastasis.time_int.et,
         117 @ m.diastole.atrial_fill.time_int.et)
      such that [
    volume @ m.diastole.time_int.et > = 97
      exception [TOO_LOW_EDV with volume <- volume ],
    volume @ m.diastole.time_int.et < = 140
      exception [TOO_HIGH_EDV with volume <- volume ],
    volume @ m.systole.time_int.et > = 20
      exception [TOO_LOW_ESV with volume <- volume],
    volume @ m.systole.time_int.et < = 27
      exception[TOO_HIGH_ESV with volume <- volume]],
  time_inst of VOLUME_V with time_inst <- now);
```

A few words of explanation are in order. The key words "verify", "find", and "strict_order_set" appear. Their meanings are straightforward: "verify" means match constraints, "find" is the equivalent of "does there exist", and "strict_order_set" is a function that checks to see if a potential motion_set's elements are strictly ordered in time.

2.4 Exceptions and similarity relations

The recording of exceptions to slot filling and constraint matching has proven to be valuable. Exceptions are classes in their own right, with slots to be filled on instantiation, i.e., when raised. Each slot constraint (or group of constraints) of a class may have an associated exception clause. This clause names the type of exception that would be raised on matching failure, and provides a definition for filling the exception's slots, since these slot fillers identify the context within which the exception occurred and play an important role in the determination of the action to take on the exception. Each slot has an implicit exception associated with it for cases where a slot filler cannot be found. Exceptions are used in two ways: (1) to record the matching failures of current hypotheses, recording the failures of the reasoning process; and (2) to assist in directing system attention to other, perhaps more viable hypotheses. The prototypical exception class is shown below along with one of its specializations, followed by an example from a stroke volume slot. Other examples have already appeared in example 3.

Example 4

```
class EXCEPTION with
dependents
  subj : PHYS_OBJ ;
  time_int : TIME_INTERVAL ;
  source_type : CLASS ;
  source_id : INTEGER ;
end $
```

Example 5

```
class TOO_MUCH_MOTION is-a EXCEPTION with
dependents
  seg : STRING ;
  disp : LENGTH_VAL with disp <-
      (len of LENGTH_VAL with
    len <- dist(subj.centroid @ source_id.time_int.st,
         subj.centroid @ source_id.time_int.et),
    time_inst of LENGTH_VAL with time_inst <- now);
end $
```

Example 6

```
stroke_vol : VOLUME_V with
  stroke_vol <- (vol of VOLUME_V with
    vol <- self.volume @ m.diastole.time_int.et -
             self.volume @ m.systole.time_int.et
      default(95) such that [
      vol > = 70
        exception [LOW_STROKE_VOLUME with
          volume <- vol ],
      vol < = 120
        exception [HIGH_STROKE_VOLUME with
          volume <- vol ] ],
    time_inst of VOLUME_V with time_inst <- now);
```

Similarity measures that can be used to assist in the selection of other relevant hypotheses on hypothesis matching failure are useful in the control of growth of the hypothesis space. These measures usually relate classes that together comprise a discriminatory set, i.e., only one of them can be instantiated at any one time. As such, they relate classes that are at the same level of specificity of the IS-A hierarchy, and that have the same IS-A parent classes. Similarity links are components of the frame scheme of Minsky (1975), and a realization of SIMILARITY links as an exception-handling mechanism is presented in Tsotsos *et al.* (1980) based on a representation of the common and differing portions between two classes. This view is contrasted with the sets of competitors described for the ABEL system (Patil *et al.* 1982). In that formulation, the level of specificity of the competing set is not represented. Similarity links enable explicit discussion of class comparisons, not only between the connected classes, but also by traversals of several links (Gershon 1982). Thus they are an element of embedded declarative control, and add a different view of class representation, thereby enhancing redundancy of the representation.

The three major components of a SIMILARITY link are (1) the list of target classes (given first), (2) the 'similarities' expression; and (3) the "differences" expression, the time-course of exceptions that would be raised through inter-slot constraints of the source class or in parts of the source class. The similarities represent the important common portions between the source and target classed—during interpretation, the target classes are not active when the SIMILARITY link is being evaluated; thus, in time-dependent reasoning situations, the components of the target class that are the same as in the source class before activation of the SIMILARITY link, or that the source class may not care about that have already 'passed in time', can be verified using the similarities expression. There is an implicit conjunction of the differences in the exception record, while the similarities form a disjunction. Many SIMILARITY links will be shown in subsequent examples.

2.5 Partial results and levels of description

Partial instances are permitted with an accompanying exception record. More importantly, since instance tokens are produced for each verified hypothesis, and since hypotheses maintain the organization exhibited by the classes that they are formed from, interpretation results also exhibit the same structure. That is, there are levels of description that may be examined as appropriate by a user.

It is important to realize that the instantiation of a hypothesis is achieved only when its certainty has reached a threshold value. (The thresholds are not set in an *ad hoc* fashion, but rather depend on a number of factors relating to the context of interpretation and knowledge structure; see Tsotsos (1984) for details). Thus, even though not all components of a hypothesis have been verified, instantiation may still take place if that hypothesis has significantly more successes than its competitors over the same time period. This would then create a partial instance, including the verified components, the final certainty, and a set of exception records specifying what was not observed.

3.0 The interpretation control structure

3.1 Hypothesize and test: Parallelism and levels of attention

The ALVEN system employs hypothesize and test as the basic recognition paradigm. The activation of a hypothesis sets up an internal goal, that is, that the class from which the hypothesis was formed, try to verify itself. However, activation of hypotheses proceeds along each of five dimensions concurrently, and hypotheses are considered in parallel rather than sequentially. These dimensions are the same class organization axes that are described above. Specifically, we define: *goal-directed* search to be movement from general to specialized classes along the IS-A dimension, the goal being to find the appropriate sub-class definition for the data in question; *model-directed* search to be movement from aggregate to component classes along the PART-OF dimension; *temporally directed* search to be a specific form of model-directed search in that a temporal ordering among components controls the time of activation; *failure-directed* search to be movement along the SIMILARITY dimension; and *data-directed* search to be movement from components to aggregates of components upwards along the PART-OF dimension. For a given set of input data, in a single time slice, activation is terminated when none of the activation mechanisms can identify an un-activated viable hypothesis. Termination is guaranteed by virtue of the finite size of the knowledge and the explicit prevention of re-activation of already active hypotheses. The activation of one hypothesis has implications for other hypotheses as well, as will be described below. Because of the multi-dimensional nature of hypothesis activation, the 'focus' of the system also exhibits levels of attention. That is, in its examination, the focus can be stated according to desired level of specificity or resolution (the two are related), discrimination set, or temporal slice.

Each newly activated hypothesis is recorded in a structure that is similar to the class whose instance it has hypothesized. This structure includes the class slots awaiting fillers, the relationships that the hypothesis has with other hypotheses, and an initial certainty value determined by sharing the certainty with the hypothesis that activates the new hypothesis.

3.2 Goal-directed and model-directed search

Top-down traversal of an IS-A hierarchy, moving downward when concepts are verified implies a constrained form of hypothesize-and-test for more specialized concepts. Similarily, top-down traversal of the PART-OF hierarchy implies a constrained form of hypothesize-and-test for components of classes that reflect greater resolution of detail. These search dimensions are success-driven, as shown in Fig. 2.

Verification of an IS-A parent concept implies that perhaps one of its IS-A siblings applies, while the confirmation of an IS-A sibling implies that its parents must also be true. Multiple IS-A siblings can be activated, but a more efficient scheme would be to activate one of the siblings if all siblings form a mutually exclusive set, or one from several such sets, and then allow failure-directed search to take over. This mechanism will then determine how many siblings in a discriminatory set are viable possibilities. Note that hypotheses are activated for each class in a particular IS-A branch as the hierarchy is being traversed, and thus tokens will be created for each on instantiation. The activation of a hypothesis implies activation of all of its PART-OF components as hypotheses as well. Cycles are avoided since at most one hypothesis for a particular class can exist for each time interval and set of structural components.

In the case of top-down PART-OF hierarchy traversal, the activation of a hypothesis forces activation of hypotheses corresponding to each of its components, i.e., slots. Note that slots may have a temporal ordering, a feature handled by the temporal search mechanism interacting with this one. The search is therefore for all components of a class, increasing the resolution of the class definition.

The MYCIN system (Shortliffe 1976) has only a single search dimension, namely that of depth-first search of the AND/OR tree of rules, while the INTERNIST system (Pople 1977) employs both of these mechanisms in addition to the data-directed search about to be described.

3.3 Data-directed search

The PART-OF hierarchy can also be traversed bottom-up in aggregation mode as shown in Fig. 3. Bottom-up traversal implies a form of hypothesize-and-test, where hypotheses activate other hypotheses that may have them as components, i.e., data-directed search. This form of search is success-driven as well. Activation of hypotheses in this direction implies activation of all IS-A ancestors of new hypotheses as well. Arbitrary hypothesis groupings can be accomplished, but specific groupings can only be recognized if defined as a class.

3.4 Failure-directed search

Failure-directed search is along the SIMILARITY dimension as in Fig. 4, and depends on the exceptions of a particular hypothesis. Typically, several SIMILARITY links will be activated for a given hypothesis, and the resultant set of hypotheses is considered as a discriminatory set, i.e., at most, one of them may be the correct one. Similarity interacts with the PART-OF relationship in that exceptions raised that specify missing slot tokens are handled by the hypothesis' PART-OF parent, the hypothesis that contains the context within which the exception occurred.

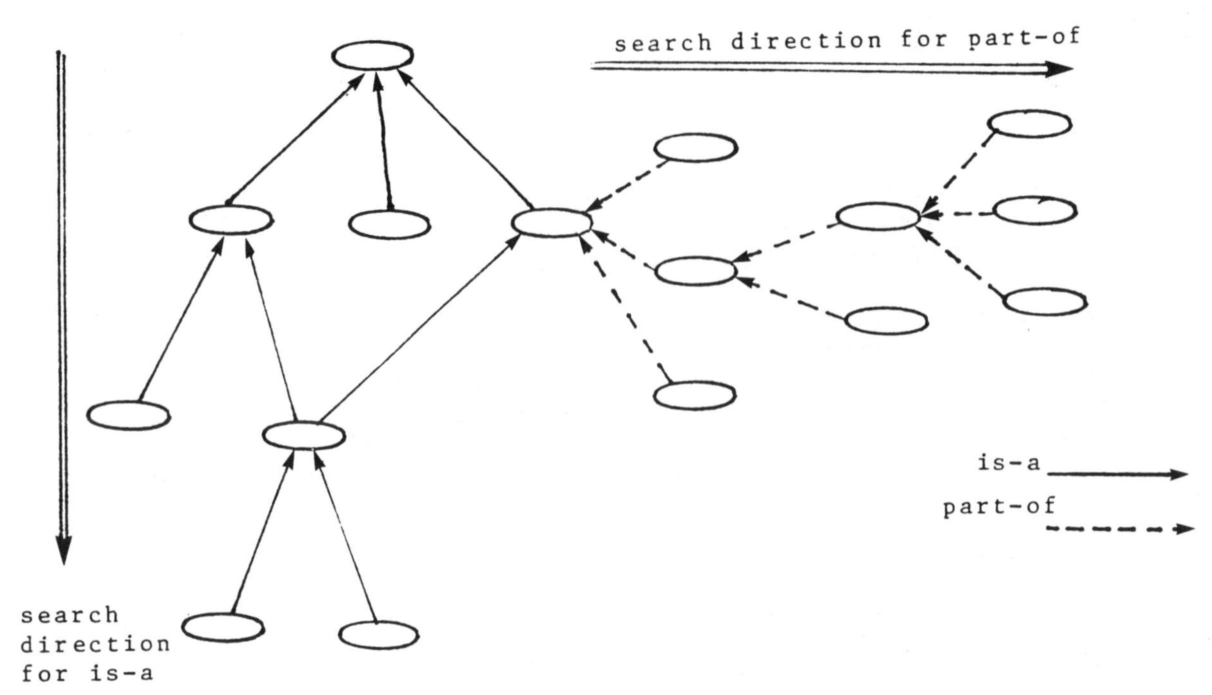

FIG. 2. Goal-directed and model-directed search.

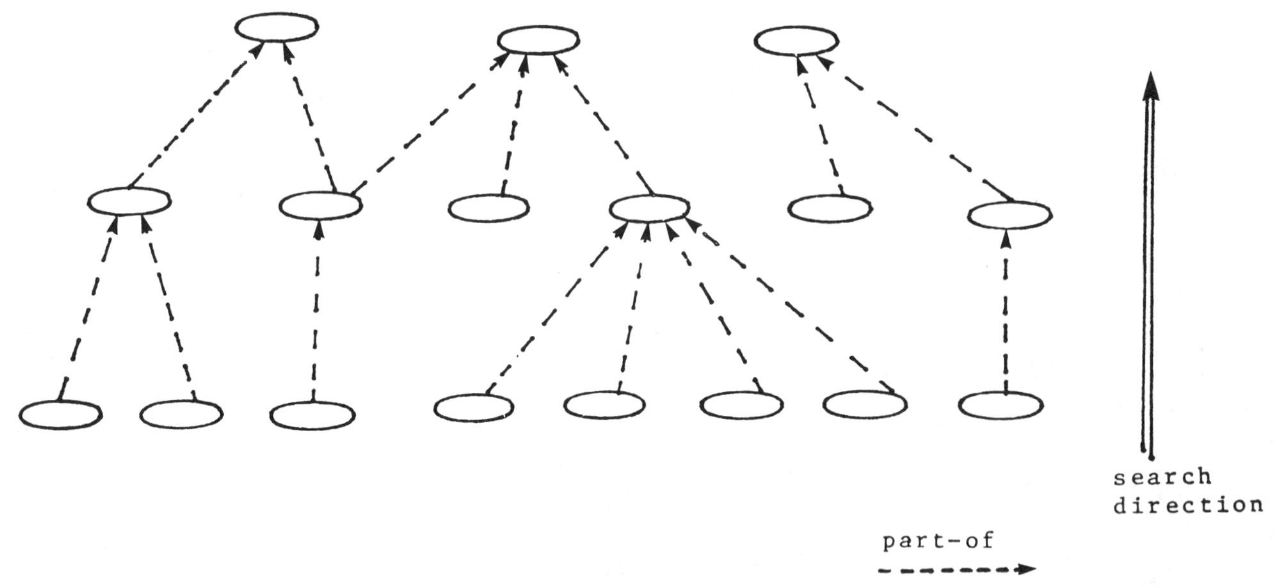

FIG. 3. Data-directed search.

3.5 Temporally directed search

Temporally directed search is automatically activated whenever a class has an IS-A relationship with the SEQUENCE class, and this includes sequences of measurement values. It is, in other words, a special case of model-directed search along the PART-OF dimension (see Fig. 5). (Causal search is a special case of temporal search, since causality implies an existensional dependency as well as a temporal relation. This is present in the CAA system (Shibahara *et al.* 1983)). Note that elements of a sequence may be compound events, such as other sequences, simultaneous events, or overlapping events. In a sequence, each element of the sequence has a PART-OF relationship with the event class. Thus, on activation of the class, it is meaningless to activate all parts of a sequence at the same time or to expect all measurements of a sequence at the same time. This dimension of search is crucial for the ordering of expectations in time.

3.6 Hypothesis conceptual adjacency

Active hypotheses are related to one another by their 'conceptual adjacencies'. If a knowledge organization relation (IS-A, PART-OF, SIMILARITY, Temporal Precedence) exists between two classes, and hypotheses are active for those two classes such that the hypotheses involve the same set of structural components and time interval (they are attempting to explain the same phenomenon), then the hypotheses also have that same relation. The conceptual adjacency is one of the major components of hypothesis ranking, in that it specifies what kinds of global and local consistencies play a role for a given hypothesis. In fact, the certainty updating scheme only uses information about conceptual adjacency and hypothesis matching.

An interesting result of the use of conceptual adjacencies in hypothesis ranking is that performance of the ranking scheme is

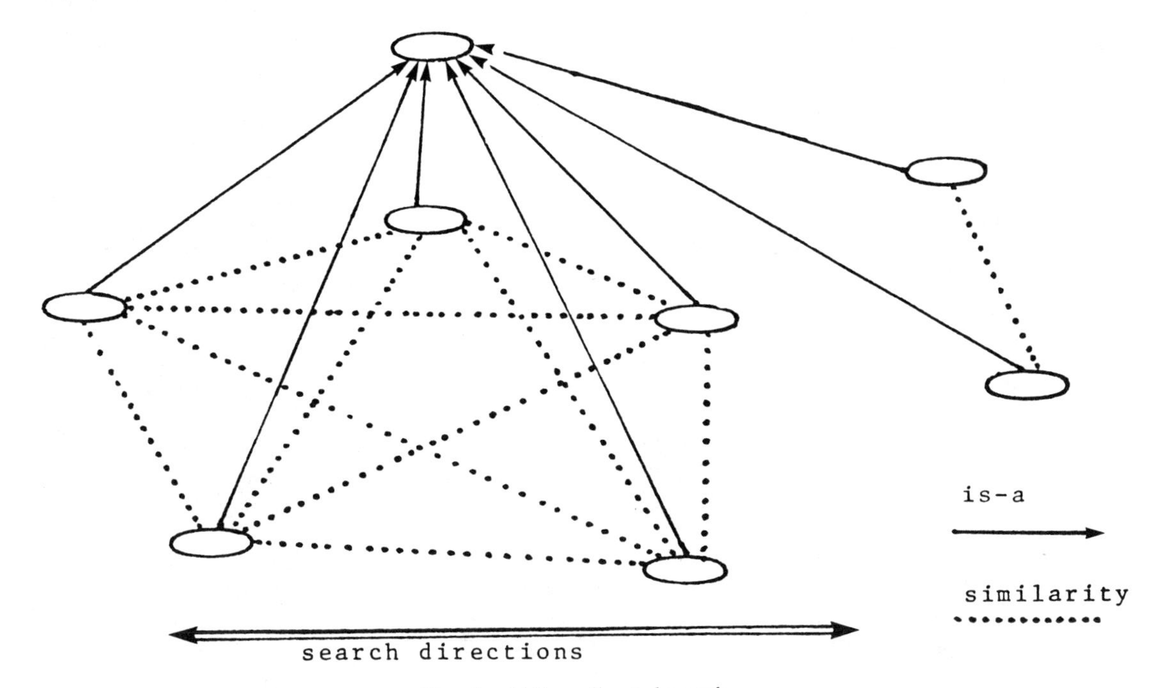

FIG. 4. Failure-directed search.

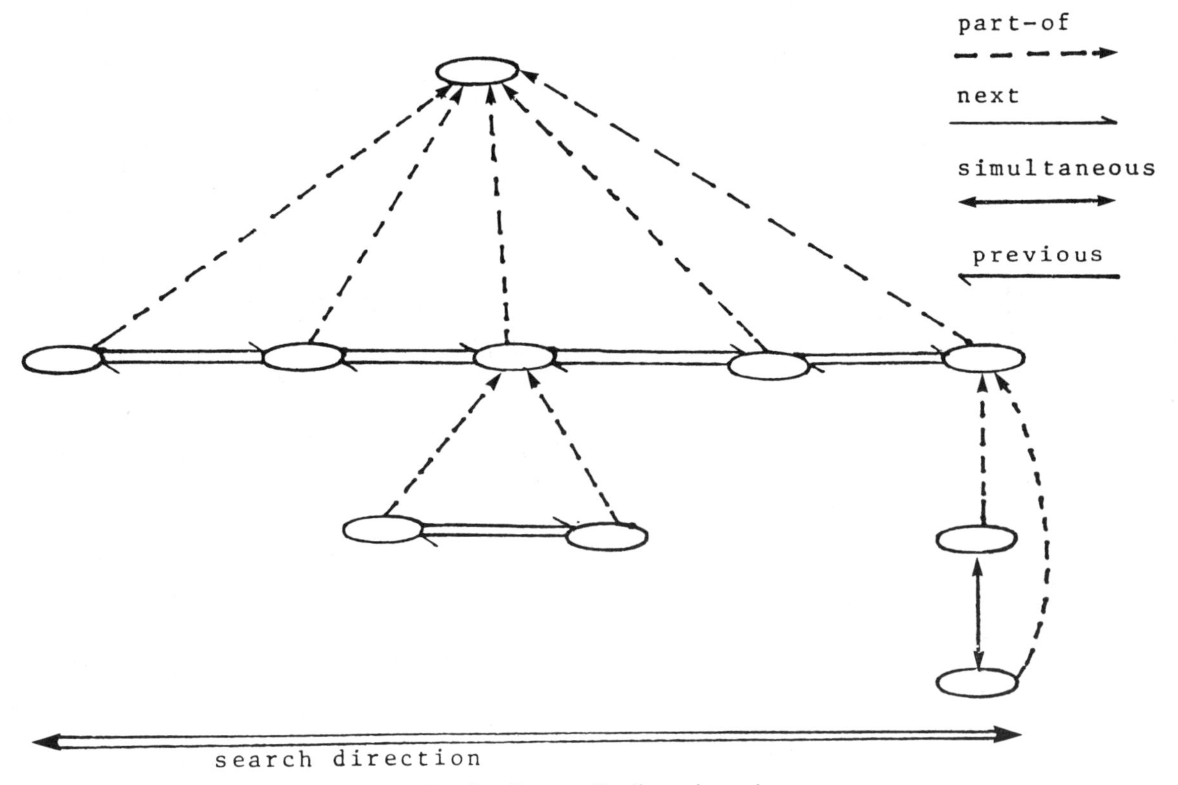

FIG. 5. Temporally directed search.

dramatically improved by their use. In Tsotsos (1984), experiments on the control structure are reported. Those experiments show that for this particular hypothesis certainty updating scheme (and indeed for relaxation labelling schemes in general), the addition of the global constraints to competition exhibited in a discriminatory set of hypotheses, via the IS-A relation, speeds up convergence to correct instantiation.

3.7 Hypothesis matching and hypothesis ranking

The matching result of a hypothesis for the purpose of hypothesis ranking is summarized as either success or failure. Matching is defined as successful if all slots that should be considered for filling are filled and no matching exceptions are raised. Otherwise, the match is unsuccessful. Using this binary categorization of matching, and the conceptual adjacencies amongst hypotheses, a certainty updating scheme based on relaxation processes (Zucker 1978) is used. Details of this scheme appear in Tsotsos (1984). Basically, hypotheses that are connected by conceptual adjacencies that imply consistency support one another, and those linked by adjacencies that imply inconsistency compete with one another by removing support. The IS-A relationship is in the former group, while the

SIMILARITY relationship is in the latter group. The focus of the system is defined as the set of best hypotheses, at each level of specificity, for each set of structural components being considered in the given time slice. The focus, because of the slow change of certainties inherent in relaxation schemes, exhibits inertia or procrastination, i.e., it does not alter dramatically between certainty updates. Both global and local consistency is enforced through the contributions of hypotheses to one another via their conceptual adjacencies.

3.8 Hypothesis instantiation and deletion thresholds

The use of thresholds is necessitated by the numerical nature of certainties for the instantiation and deletion of hypotheses. These thresholds are not fixed for the lifetime of the system, but rather are dynamic, in that they depend on the number of competing hypotheses in a discriminatory set, and on whether or not the same hypothesis is present in more than one discriminatory set. Sampling rate and noise considerations are also included (see Tsotsos 1984). In particular, the sampling rate depends on the temporal resolution of the discriminatory set, and can thus be variable. Events whose durations are described in months must be samples at that resolution, and those of durations in seconds must be sampled accordingly.

4.0 LV Dynamics knowledge and its representation

Although there is still much work to be done in the determination of the knowledge of LV dynamics, much can be found in current literature that can be incorporated into our formalism. Two examples will be given. This knowledge is used as a starting point for knowledge-base construction only. Moreover, although the exact numerical quantities may differ between imaging techniques, the *qualitative* descriptions do not.

In the series of papers by Gibson and his colleagues (i.e., Gibson *et al.* (1976); Doran *et al.* (1978)), several investigations were carried out that determined quantitative aspects of specific LV motions. In the Gibson *et al.* paper, the segmental motions of the LV during isovolumic relaxation were examined in normal and ischemic LVs using echocardiography in order to determine dynamic differences between these two cases. Without describing technical details of their method, we will briefly summarize their findings. They discovered that in normal LV's an outward wall motion of 1.5–3.0 mm could be present in any region during isovolumic relaxation. In abnormal cases, i.e., patients with coronary artery disease, affected areas show inward motion, 2 mm or more for posterior or apical segments, and any at all for anterior regions, and non-affected areas, because of a compensatory mechanism, may exhibit an increased outward motion of up to 6 mm over normal. The key feature to note here is that the description given does not have a mathematical form at all—it is a combination of quantitative and qualitative measures. The term 'outwards' does not specify any precise direction as long as the motion of the segment is away from the inside of the LV. It is not impossible to set up a mathematical model of this; however, the model will be both cumbersome and will bury the pertinent facts in its equations, so that inspection by a non-sophisticated user becomes impossible. The knowledge class for this information (and more) follows:

```
class N_ISORELAX is-a NO_VOLUME_CHANGE with
prerequisites
  subj : N_LV such that [

    (find ant_mot : NO_TRANSLATION where [
       and_mot.subj = self.subj.anterior,
       ant_mot.time_int = self.time_int
       ]
    or
    find ant_mot : OUTWARD where [
       ant_mot.subj = self.subj ,
       ant_mot.time_int = self.time_int,
       dist(ant_mot.subj.centroid @ ant_mot.time_int.st,
          ant_mot.subj.centroid @ ant_mot.time_int.et) < 3
       exception [TOO_MUCH_MOTION with seg ← "anterior",
            direction ← "outward",
          disp ← dist(ant_mot.subj.centroid @ ant_mot.time_int.st,
            ant_mot.subj.centroid @ ant_mot.time_int.et)]
       ]
    ) exception [TOO_MUCH_MOTION with seg ← "anterior",
            direction ← "inward"],

    (find post_mot : NO_TRANSLATION where [
       post_mot.subj = self.subj.posterior,
       post_mot.time_int = self.time_int
       ]
    or
    find post_mot : INWARD where [
       post_mot.subj = self.subj,
       post_mot.time_int = self.time_int,
       dist(post_mot.subj.centroid @ post_mot.time_int.st,
          post_mot.subj.centroid @ post_mot.time_int.et) < 2
          exception [TOO_MUCH_MOTION with seg ←
          "posterior",
            direction ← "inward",
       disp ← dist(post_mot.subj.centroid @ post_mot.time_int.st,
          post_mot.subj.centroid @ post_mot.time_int.et)]
    ]
    or
    find post_mot : OUTWARD where [
       post_mot.subj = self.subj,
       post_mot.time_int = self.time_int,
       dist(post_mot.subj.centroid @ post_mot.time_int.st,
          post_mot.subj.centroid @ post_mot.time_int.et) < 3
       exception [TOO_MUCH_MOTION with seg ← "posterior",
            direction ← "outward",
       dist(post_mot.subj.centroid @ post_mot.time_int.et,
          post_mot.subj.centroid @ post_mot.time_int.et)] ]
    ),

     (find ap_mot : NO_TRANSLATION where [
        ap_mot.subj = self.subj.apical,
        ap_mot.time_int = self.time_int
        ]
     or
     find ap_mot : INWARD where [
        ap_mot.subj = self.subj,
        ap_mot.time_int = self.time_int,
        dist(ap_mot.subj.centroid @ ap_mot.time_int.st,
           ap_mot.subj.centroid @ ap_mot.time_int.et) < 2
           exception [TOO_MUCH_MOTION with seg ← "apical",
             direction ← "inward",
          ·disp ← dist(ap_mot.subj.centroid @ ap_mot.time_int.st,
             ap_mot.subj.centroid @ ap_mot.time_int.et)]
        ]
     or
     find ap_mot : OUTWARD where [
        ap_mot.subj = self.subj,
        ap_mot.time_int = self.time_int,
        dist(ap_mot.subj.centroid @ ap_mot.time_int.st,
        ap_mot.subj.centroid @ ap_mot.time_int.et) < 3
     exception [TOO_MUCH_MOTION with seg ← "apical"
             direction ← "outward",
        disp ← dist(ap_mot.subj.centroid @ ap_mot.time_int.st,
             ap_mot.subj.centroid @ ap_mot.time_int.et) ] ]
        )
        ];
```

```
dependents
  time_int : with time_int ← (dur of TIME_INTERVAL with
  dur ← default(0.093*(30/(0.8*HR))))
    such that [
      time_int.st ≥ 0.24*(30/(0.8*HR)),
      tim_int.et ≤ 0.43*(30/(0.8*HR)),
      time_int.dur ≥ 0.08*(30/(0.8*HR)),
      time_int.dur ≤ 0.12*(30/(0.8*HR))
        exception [TOO_LONG_ISORELAX]
    ];

similarity links
  sim_link1 : ISCH_AP_ISOVOL_RELAX
    for differences :
      d1 : TOO_MUCH_MOTION where [
        seg = "apical",
        direction = "inwards",
        time_int = ap_mot.time_int];
      d2 : TOO_MUCH_MOTION where [
        seg = "anterior"
        direction = "outwards",
        disp < 9,
        time_int = ant_mot.time_int];
      d3 : TOO_MUCH_MOTION where [
        seg = "posterior",
        direction = "outwards",
        disp < 9,
        time_int = post_mot.time_int];;

  sim_link2 : ISCH_ANT_ISOVOL_RELAX
    for differences :
      d1 : TOO_MUCH_MOTION where [
        seg = "anterior",
        direction = "inwards",
        time_int = ant_mot.time_int];
      d2 : TOO_MUCH_MOTION where [
        seg = "apical",
        direction = "outwards",
        disp < 9,
        time_int = ap_mot.time_int];
      d3 : TOO_MUCH_MOTION where [
        seg = "posterior",
        direction = "outwards",
       disp < 9,
       time_int = post_mot.time_int];;

 sim_link3 : ISCH_POST_ISOVOL_RELAX
   for differences :
     d1 : TOO_MUCH_MOTION where [
       seg = "posterior",
       direction = "inwards",
       time_int = post_mot.time_int];
     d2 : TOO_MUCH_MOTION where [
       seg = "anterior"
       direction = "outwards",
       disp < 9,
       time_int = ant_mot.time_int];
     d3 : TOO_MUCH_MOTION where [
       seg = "apical",
       direction = "outwards",
       disp < 9,
       time_int = ap_mot.time_int];;
end
```

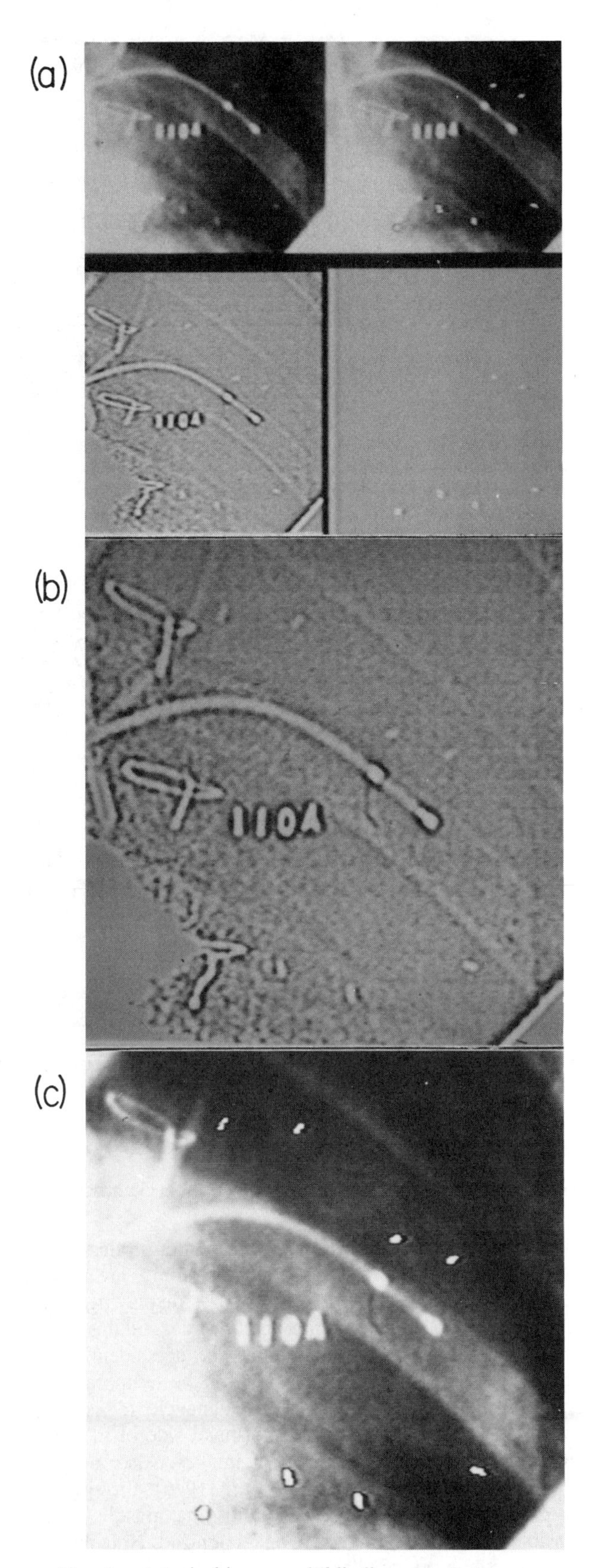

FIG. 6. A typical image and 'blind' marker finding.

The definition states that for a normal isovolumic relaxation phase to be recognized, normal motions for each segment must be present. There are three main clauses in the definition. The first defines the expected normal motion of the anterior segment, the second for the posterior segment, and the third for the remaining segment, the apical one. So for example, in the first clause, the definition reflects Gibson's characterization: the anterior segment during this phase must either not display any translational movement, or could display an outward motion of displacement of less than 3 mm. A larger displacement than this in the outwards direction would be recorded as the exception TOO_MUCH_MOTION, with specific additional contextual

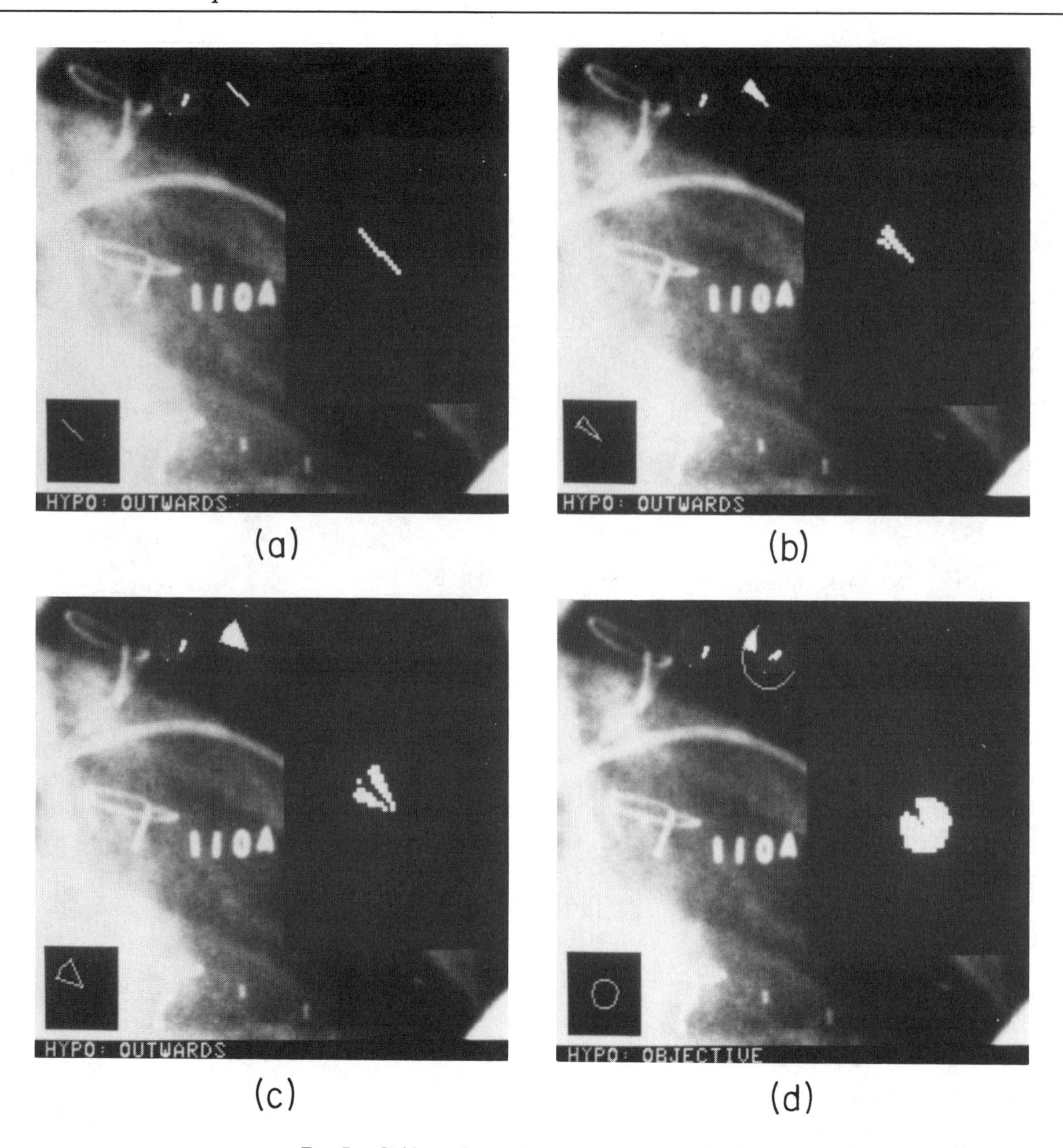

FIG. 7. Guidance for marker finding from hypotheses.

information recorded as well. In the matching of class definitions to actual observed motions, matching failures are recorded as exceptions. If the anterior segment was displaying motion and was not outwards, then it must be inwards, and this fact too would be recorded as an exception. The dependent portion specifies relevant timing information for the temporal placement of the phase within the left ventricular cycle. *HR* is in units of beats per second, so that the right-hand side of the timing expressions is in units of number of images. Also, using the information derived from Gibson *et al.* (1976), the similarity links provide definitions of the constraints that must be found if a possible ischemic segment is to be recognized. Note that only the connections to possible ischemic states detectable by considering only the characteristics of the isovolumic relaxation phase are included above; a set of similarly formed constraints would have to be present for other disease states as well, for those cases where the isovolumic relaxation phase plays a role in their definition. 'sim_link2' relates the normal phase to the motion of an abnormal apical segment exhibiting the effects of ischemia. This, according to Gibson's definition, is shown by either the apical region itself having too much inward motion during this phase, and (or) one of the other regions (posterior or anterior) exhibiting too much outward motion during the phase. Note that the set of differences does not define a necessary set; any one of the conditions is sufficient.

If should be clear that the above is not complete; it requires the remainder of the definitions for the other phases and motions, since the entire definition of each class of LV motion is defined as a hierarchy of abstraction, each level adding more detail to the previous one. Some of the types of information that are represented are: volume changes where known for normal phases, ejection fractions, for example; measures of degrees of abnormalities, derived heuristically; and others.

A second body of knowledge of the form necessary for interpretation can be found in Fujii *et al.* (1979). These researchers investigated eight different clinical cardiac disease

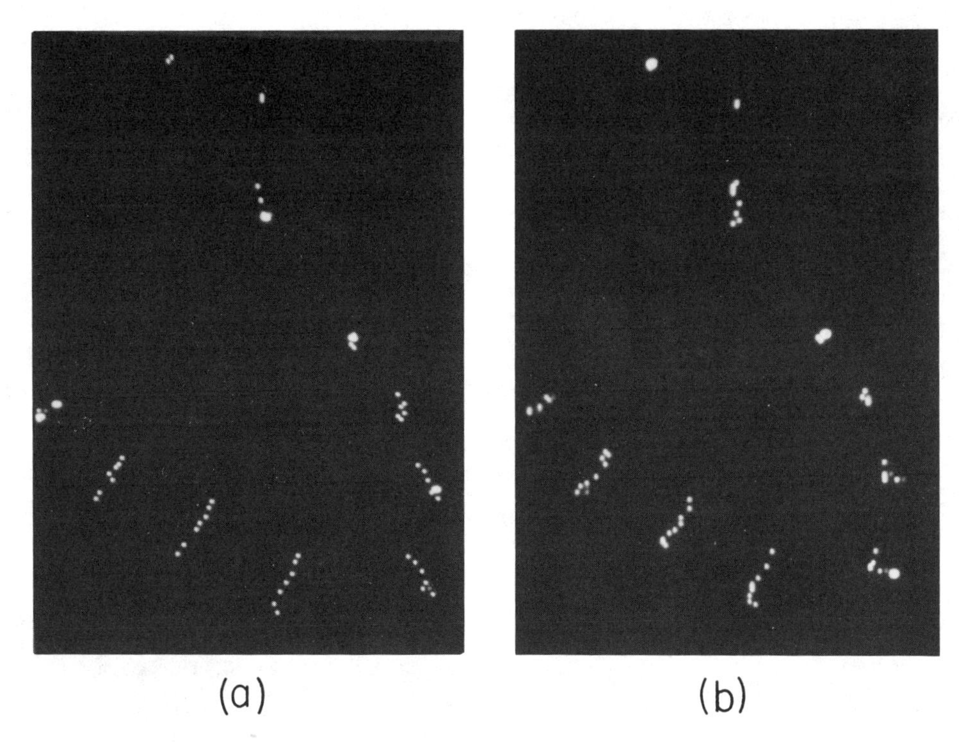

FIG. 8. Inward and outward motions of an LV cycle.

states with the intent of discovering posterior-wall motion differences and similarities among the diseases, as well as global LV characteristics. The diseases were: pericarditis, congestive cardiomyopathy, hypertrophic cardiomyopathy, valvular aortic stenosis, aortic insufficiency, mitral stenosis, mitral insufficiency, and systemic hypertension. Normal LV's were also studied. The measurements made for each of the above LV states were: stroke volume, rapid filling volume, slow filling volume, atrial filling volume, the percent filling for each of the previous three phases with respect to the stroke volume, posterior wall excursion in total, and for each of the three phases of diastole, as well as the percentage excursion in each phase diastolic posterior wall velocity, rapid filling rate, LV end diastolic dimension, and ejection fraction. It is, of course, difficult to verify their results. However, they are important for they provide at least a starting point for the further elaboration and verification of such detailed dynamic information. In addition to the large amount of numerical information that they derived, they attached to the significant findings qualitative descriptors, such as whether or not this quantity should be higher or lower than in the normal case. This was rather fortunate from our point of view: the representational formalism that we had designed can handle description via common components and differences very well, and uses such information to advantage during the decision phases of the interpretation. It should be clear from the previous example how such information would be included into the representation, and this fact alone raises another important advantage of this scheme. The addition of information into a mathematical model may require a complete re-definition of the model. In our case, information is easily inserted, as long as one understands the semantics of the representation.

5.0 An example

The initial evaluation is done on a cine contrast representation; each patient has a permanent volume correction factor for both diastole and systole that accounts for the shell of muscle enclosed by the contour created by connecting the markers (Alderman *et al.* 1976). Interpolation is used for variations in this correction over time. Nine markers on the LV wall, and two on the aortic valve edges constitute the LV outline from which volume calculations are done, using an area–length formula that was devised for this purpose. Figure 6 displays an actual image with the stages of image analysis that lead to 'blind' marker finding, that is, without any sort of guidance as to expected marker location. The first stage involves filtering the image with a Marr—Hildreth like operator (Hildreth 1980). Zero-crossings with their standard definition, however, do not lead to useful image tokens because of the nature of the X-ray images and their low contrast. A specially tuned version of the Marr–Hildreth operator was then used to extract the markers. This operator was tuned such that the size and shape of the marker was reflected in the center of the operator with the surround enveloping this center. The results of this are then superimposed on the original image, in order to highlight the markers. These two steps are expanded in Fig. 6*b* and *c*.

Guidance, however, is an integral feature of the framework, namely, during the hypothesization of motion classes, the hypotheses themselves can be used to predict expected motion characteristics for the markers, segment, and entire left ventricle. Figure 7 then shows the kind of predictions that an 'outwards' motion hypothesis creates and the guidance it provides. Note that for this example 'outwards' refers to outwards motion of the marker with respect to the segment, not to the ventricle. Clearly, for this case the marker is not found on that path. The hypothesis structure is then modified to enclose a larger space, corresponding to a relaxation of the constraints of the hypotheses, until it is found. The same marker-finding process described earlier is used, but only in the prediction window. Four images are shown corresponding to the four predictions generated until this marker is found. In addition to the examination of a very small image subset for each marker,

marker 5 exhibits:
TRANSLATING—time interval (0, 5)
rate (mm/s) → 60, 21, 33, 45, 51
trajectory (radians) → 4.71, 2.36, 0.46, 1.24, 2.18
specializations:
OUTWARDS wrt ANTERIOR during (0, 1)
INWARDS wrt ANTERIOR during (1, 2)
OUTWARDS wrt ANTERIOR during (2, 3)

TRANSLATING—time interval (6, 10)
rate (mm/s) → 15, 33, 42, 15
trajectory (radians) → 1.24, 4.19, 5.50, 1.24
specializations:
INWARDS wrt ANTERIOR during (6, 7)
OUTWARDS wrt ANTERIOR during (7, 9)

TRANSLATING—time interval (14, 15)
rate (mm/s) → 15
trajectory (radians) → 1.24
specializations:
INWARDS wrt ANTERIOR during (14, 15)

others
NO MOTION during (5, 6)
NO MOTION during (10, 14)
NO MOTION during (15, 16)
exceptions to normal detected :
MODERATELY HYPOKINETIC—CONTRACTION
wrt ANTERIOR during (1, 2)

ANTERIOR segment exhibits:
TRANSLATING—time interval (0, 1)
rate (mm/s) → 45
trajectory (radians) → 4.71
specializations:
INWARDS wrt VENTRICLE during (0, 1)

TRANSLATING—time interval (3, 8)
rate (mm/s) → 30, 15, 21, 15, 15
trajectory (radians) → 1.24, 1.24, 2.36, 2.36, 4.71
specializations:
INWARDS wrt VENTRICLE during (3, 8)

TRANSLATING—time interval (9, 11)
rate (mm/s) → 15, 15
trajectory (radians) → 1.24, 1.24,
specializations:
OUTWARDS wrt VENTRICLE during (9, 11)

TRANSLATING—time interval (13, 16)
rate (mm/s) → 15, 15, 15, 15
trajectory (radians) → 0.00, 3.14, 0.00, 0.00
specializations:
OUTWARDS wrt VENTRICLE during (14, 16)

VOLUME CHANGE—time interval (0, 16)
rate (ml/s) → −1.2, −66, 33, −48, −30, −12, −3, 27, 12, −12, 21, 33, −6, −6, 2.1, 39, 39
specializations:
CONTRACTING during (0, 1)
UNIFORMLY CONTRACTING during (0, 2)
SYSTOLE during (3, 6)
EXPANDING during (2, 3)
UNIFORMLY CONTRACTING during (3, 5)
CONTRACTING during (6, 7)
DIASTOLE during (7, 9)
CONTRACTING during (9, 10)
DIASTOLE during (10, 12)
CONTRACTING during (12, 14)
DIASTOLE during (14, 16)

PERIMETER CHANGE—time interval (0, 8)
rate (mm/s) → 45, −60, 30, −45, −45, −30, −30, 90
specializations:
LENGTHENING during (0, 1)
SHORTENING during (1, 2)
LENGTHENING during (2, 3)
SHORTENING during (3, 7)
LENGTHENING during (7, 8)

PERIMETER CHANGE—time interval (9, 10)
rate (mm/s) → −30
specializations:
SHORTENING during (9, 10)

PERIMETER CHANGE—time interval (13, 16)
rate (mm/s) → 30, −30, 45, 45
specializations:
LENGTHENING during (13, 14)
SHORTENING during (14, 15)
LENGTHENING during (15, 16)

others
NO TRANSLATION during (1, 3)
NO TRANSLATION during (8, 9)
NO PERIMETER CHANGE during (8, 9)
NO PERIMETER CHANGE during (10, 13)
NO TRANSLATION during (11, 13)
exceptions to normal detected:
SEVERELY HYPOKINETIC—CONTRACTION
wrt VENTRICLE during (2, 6)
TOO SHORT SYSTOLE during (7, 7)
MILDLY POOR SYSTOLE during (7, 7)
SEVERELY HYPOKINETIC—EXPANSION
wrt VENTRICLE during (8, 14)

APICAL segment exhibits:
TRANSLATING—time interval (1, 6)
rate (mm/s) → 33, 33, 60, 48, 33
trajectory (radians) → 2.08, 1.05, 1.24, 1.99, 2.08
specializations:
INWARDS wrt VENTRICLE during (1, 6)

TRANSLATING—time interval (7, 10)
rate (mm/s) → 60, 51, 15
trajectory (radians) → 4.71, 4.09, 3.14
specializations:
OUTWARDS wrt VENTRICLE during (7, 9)
INWARDS wrt VENTRICLE during (9, 10)

TRANSLATING—time interval (11, 14)
rate (mm/s) → 33, 15, 21
trajectory (radians) → 5.81, 4.71, 5.50
specializations:
OUTWARDS wrt VENTRICLE during (11, 14)

TRANSLATING—time interval (15, 16)
rate (mm/s) → 33, 33
trajectory (radians) → 5.81, 5.81
specializations:
OUTWARDS wrt VENTRICLE during (15, 16)

VOLUME CHANGE—time interval (0, 6)
rate (ml/s) → −12, −72, −24, −60, −42, −36
specializations:
CONTRACTING during (0, 1)
UNIFORMLY CONTRACTING during (0, 2)
SYSTOLE during (1, 6)
UNIFORMLY CONTRACTING during (3, 6)

FIG. 9. ALVEN'S descriptive output for the motions in Fig. 8.

VOLUME CHANGE—time interval (7, 16)
rate (ml/s) → 54, 24, 15, 15, 48, 36, 9, −15, 45, 45
specializations:
DIASTOLE during (7, 14)
UNIFORMLY EXPANDING during (7, 8)
UNIFORMLY EXPANDING during (9, 13)
CONTRACTING during (14, 15)
DIASTOLE during (15, 16)

PERIMETER CHANGE—time interval (0, 6)
rate (mm/s) → 15, −75, −30, −60, −45, −45
specializations:
LENGTHENING during (0, 1)
SHORTENING during (1, 6)

PERIMETER CHANGE—time interval (7, 16)
rate (mm/s) → 15, 30, 30, 30, 60, 45, 15, 15, 15, 15
specializations:
LENGTHENING during (7, 13)
SHORTENING during (13, 14)
LENGTHENING during (14, 16)

others
NO TRANSLATION during (0, 1)
NO MOTION during (6, 7)
NO TRANSLATION during (10, 11)
NO TRANSLATION during (14, 15)
exceptions to normal detected:
SEVERELY HYPOKINETIC—EXPANSION wrt VENTRICLE during (10, 11)
SEVERELY HYPOKINETIC—EXPANSION wrt VENTRICLE during (14, 15)

POSTERIOR segment exhibits:
TRANSLATING—time interval (0, 6)
rate (mm/s) → 15, 48, 33, 21, 33, 33
trajectory (radians) → 1.24, 0.95, 1.05, 0.77, 1.05, 1.05
specializations:
INWARDS wrt VENTRICLE during (0, 3)
OUTWARDS wrt VENTRICLE during (3,4)
INWARDS wrt VENTRICLE during (4, 6)

TRANSLATING—time interval (7, 16)
rate (mm/s) → 15, 30, 21, 48, 21, 21, 15, 15, 15, 15
trajectory (radians) → 4.71, 4.71, 3.92, 3.92, 3.92, 3.92, 4.71, 0.00, 3.14, 3.14
specializations:
OUTWARDS wrt VENTRICLE during (8, 14)
INWARDS wrt VENTRICLE during (14, 15)

VOLUME CHANGE—time interval (0, 6)
rate (ml/s) → −33, −90, −15, −75, −96, −78
specializations:
SYSTOLE during (1, 6)

VOLUME CHANGE—time interval (7, 16)
rate (ml/s) → 5, 6, 27, 75, 111, 21, 15, 60, 21, 21
specializations:
DIASTOLE during (7, 16)
UNIFORMLY EXPANDING during (9, 12)

PERIMETER CHANGE—time interval (0, 2)
rate (mm/s) → −45, −15
specializations:
SHORTENING during (0, 2)

PERIMETER CHANGE—time interval (3, 6)
rate (mm/s) → −30, −90, −15
specializations:
SHORTENING during (3, 6)

PERIMETER CHANGE—time interval (7, 12)
rate (mm/s) → −45, −30, 30, 75, 75
specializations:
SHORTENING during (7, 9)
LENGTHENING during (9, 12)

PERIMETER CHANGE—time interval (13, 16)
rate (mm/s) → 15, 60, 15, 15
specializations:
LENGTHENING during (13, 16)

others
NO PERIMETER CHANGE during (2, 3)
NO MOTION during (6, 7)
NO PERIMETER CHANGE during (12, 13)

LEFT VENTRICLE exhibits:
TRANSLATING—time interval (0, 6)
rate (mm/s) → 15, 33, 15, 33, 1, 21
trajectory (radians) → 4.71, 1.05, 1.24, 1.05, 1.24, 2.36

TRANSLATING—time interval (7, 15)
rate (mm/s) → 15, 15, 15, 15, 15, 15, 15, 15
trajectory (radians) → 4.71, 4.71, 4.71, 3.14, 4.71, 3.14, 0.00, 4.71

VOLUME CHANGE—time interval (0, 16)
rate (ml/s) → −57, −216, −75, −168, −186, −138, 2, 120, 57, 54, 120, 162, 90, 27, 45, 90, 90
specializations:
UNIFORMLY CONTRACTING during (0, 1)
SYSTOLE during (1, 6)
UNIFORMLY CONTRACTING during (2, 6)
UNIFORMLY EXPANDING during (7, 11)
DIASTOLE during (7, 16)
UNIFORMLY EXPANDING during (12, 14)
UNIFORMLY EXPANDING during (15, 16)

PERIMETER CHANGE—time interval (0, 6)
rate (mm/s) → 15, −150, 15, −165, −165, −105
specializations:
LENGTHENING during (0, 1)
SHORTENING during (1, 2)
LENGTHENING during (2, 3)
SHORTENING during (3, 6)

PERIMETER CHANGE—time interval (7, 8)
rate (mm/s) → 90
specializations:
LENGTHENING during (7, 8)

PERIMETER CHANGE—time interval (9, 16)
rate (mm/s) → 30, 75, 150, 60, 15, 60, 60, 60
specializations:
LENGTHENING during (9, 16)

WIDTH CHANGE—time interval (0, 16)
rate (mm/s) → −15, −15, −60, −15, −60, −60, −60, −60, −60, 60, 75, 45, 45, 45, 45, −15, −15

LENGTH CHANGE—time interval (0, 16)
rate (mm/s) → 30, −45, −15, −60, −60, −30, −30, 45, 15, 15, 15, 45, 45, 45, 45, 45, 45

others
ISOMETRIC CONTRACTION during (0, 1)
NO TRANSLATION during (6, 7)
NO PERIMETER CHANGE during (6, 7)
NO PERIMETER CHANGE during (8, 9)

NO TRANSLATION during (15, 16)
exceptions to normal detected :
MILDLY DYSKINETIC—CONTRACTION during (3, 4)
ISCHEMIC ANTERIOR ISOMETRIC RELAXATION during (6, 7)
SEVERELY POOR SYSTOLE during (7, 7)
MODERATELY DYSKINETIC—EXPANSION during (9, 15)

FIG. 9. *(Concluded).*

this process of prediction-verification also provides important feedback for other levels of the system. This marker-finding process is guaranteed to always find a marker because the default process is the 'blind' one referred to above. Figure 8 shows the sequence of marker motions for a complete cycle (a different cycle than the one from which the preceding images were taken.)

ALVEN is capable of reporting on LV performance at marker, segment, and global LV levels of detail. Relative directions, motion extents, rates of change, and temporal relationships are described both numerically and symbolically. Anomalies are detected by using the appropriate heuristic or by comparisons to accepted normal performance. Anomalies such as asynchrony, hypokinesis, dyskinesis, too slow or too fast rate of change of volume with respect to the LV phase, too long or too short phase duration, or degree of anomaly are considered.

An example of marker motions is shown in Fig. 8 *a* and *b*, for a patient from the Cardiovascular Unit at Toronto General Hospital. Figure 8*a* shows the contraction phase, while Fig. 8*b* shows the expansion phase. This particular example was assessed by the radiologists with respect to motion anomalies: the radiologist reported that the anterior segment was hypokinetic, and the remaining segments exhibited normal motion. A portion of the output of the ALVEN system for this particular film (taken at 30 images per second, 17 images in all) is shown in Fig. 9. Let us highlight some of the important points of this analysis. Firstly, a short summary of how to read the example is necessary. For each physical entity that the system knows about, which is in this case the markers, the segments, and the LV as a whole, a short summary of the motions observed is produced. This has been abbreviated because of space limitations in the following way: descriptions for the aortic clips were deleted, as were the descriptions for all of the markers save for marker 5. The remaining motions would have a form similar to those for the other markers. Each motion has a descriptive term, a possible referent where necessary (for example, 'INWARD' motion is not semantically complete without saying inwards with respect to some other object that has an inside, usually defined by the geometric centroid), quantitative values where appropriate (clearly a calibration phase is necessary), and a time interval or instant at which it was recognized. Time is noted in image units. The range of descriptive terms that ALVEN can understand is apparent from the example. Descriptions are shown for only one marker (5), and for each segment, and for the left ventricle. The motion of marker 5 is of particular interest for this example.

Secondly, the example of the knowledge for isovolumic relaxation given earlier is relevant here. The motions exhibited by the anterior segment (that is, there is a small inward motion during that phase, as shown by the description at time interval (6, 7), because of an inward motion of marker 5 during that interval, and further evidenced by the volume contraction noted in the segment description), cause that chunk of knowledge to be activated and verified. The result is the descriptive term 'ISCHEMIC ANTERIOR ISOVOLUMIC RELAXATION', which can be found in the description of the motions of the left ventricle. In addition, it will usually be true that if one segment is not performing up to par (notice the number of HYPOKINESIS instances detected—the great majority are present for the anterior segment thus confirming the radiologist's report), then the overall performance of the ventricle must be impaired as well. This can be seen by the instances of 'POOR SYSTOLE' that appear. These are confirmed independently using volume change information.

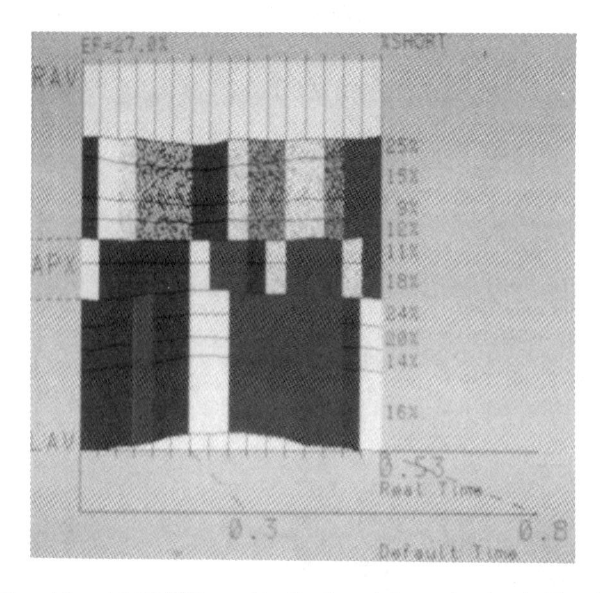

FIG. 10. ALVEN'S graphic display of the evaluation in Fig. 9.

Some other interesting descriptive terms are briefly described. UNIFORM CONTRACT/EXPAND—for this to be detected, the object considered must have a decreasing volume, and all of its markers/segments (depending on the level of description) must be moving in the proper direction. So for a uniform contraction at the LV level, the three segments must all be moving inwards and the overall volume of the LV must be decreasing. HYPOKINESIS—can only be noted if all markers/segments are moving in the same direction, and a comparison of their relative motions reveals one that is lagging behind. Note that the use of the term hypokinesis does not make sense if all markers are not moving in the same direction, since this is a term describing anomalies of motion extent. If they are all moving too slowly, then no anomaly is detected at the marker level, but it is detected at the segment level. If in turn, all segments are exhibiting small motion extents, no hypokinesis is noted at all, however, serious performance problems will be noted, because the volume changes will be lower than normal. The detection of hypokinesis is purely relational. Note, however, that it is not necessarily so. The data in Fujii *et al.* (1979) do provide some quantitative information on normal and abnormal extents for the posterior segment; these will be incorporated into the representation. However, the relational approach is a valid one when one is lacking information.

No constraints are currently in place for length changes, i.e., normal or abnormal circumferential shortening, although examples are shown of how such changes are detected.

It should be apparent that the amount of information reported is large, and that this is not a desirable characteristic for a medical consultation system. Therefore, a simple, graphic display has been devised that captures much of the important information required for appropriate analysis. This display is presented in Fig. 10. A brief explanation is in order. Imagine that the ventricle is opened up along the circumference and laid flat along the vertical axis with the right side of the aorta on the bottom, the apex in the middle, and the left side of the aorta at the top. Time is the horizontal axis. For each time interval (image pair of the film), and for each segment, a summary is displayed in terms of whether or not the segment was moving inwards (blue), outwards (red), or was not moving (white). Remember that these are motions relative to the ventricular centroid. The yellow dotting represents hypokinesis, with the more densely concentrated dots representing increasing levels of severity. The black lines traversing the plot horizontally are the marker paths in time, useful for viewing circumferential shortening effects. Finally, percent shortening for each marker with respect to the ventricular centroid are provided on the right side, along with ejection fraction. This display is particularly clear in revealing temporal relationships of a variety of types.

If these evaluations are compared with those of the radiologist, it can be seen that there is infinitely more detail present in ALVEN's evaluation, yet it is completely consistent with the radiologist's opinion. This has also been borne out in several other examples. Moreover, this analysis is repeatable and objective. Although there is much knowledge refinement required before ALVEN's knowledge base is as competent in general as a good radiologist/cardiologist, the value of the enhanced evaluation is clear.

6.0 Discussion

The knowledge organization dimensions of generalization/specialization (IS-A), aggregation/decomposition (PART-OF), mutual exclusion (SIMILARITY), and Temporal Precedence have all appeared previously in the representational literature, with the first two receiving the lion's share of attention (see Brachman 1979, 1982; Levesque and Mylopoulos 1979). The arguments for their use have been mostly qualitative, that is, these dimensions seem to have desirable formal properties and lend themselves naturally to the construction of large knowledge bases. In this work, we have shown that not only are these aspects important, but also that each representational dimension has a distinct role to play in an interpretation scheme. In fact, they each have two important roles. One role is that of enabling multiple, interacting search mechanisms. This function should not be underestimated. Rule-based recognition paradigms, for example, only offer a single dimension of search. As pointed out in Aiello (1983), such systems suffer from serious problems owing to the one-dimensionality of the inference procedure. The conclusion that we draw from this report is that goal-directed, data-directed, and model-directed inference mechanisms most effectively can compensate for the deficiencies of each other if used in concert. For example, a data-directed scheme considers all the data and tries to follow through on every event generated. It can be non-convergent, can only produce conclusions that are derivable directly or indirectly from the input data, and cannot focus or direct the search toward a desired solution. The goal-directed strategy is easy to understand and implement, and at each step of the execution the next step is pre-determined. Rules are evaluated in the same order regardless of input data. It is thus inefficient and cannot exhibit a focus with respect to the problem being solved since there is no mechanism that determines what is important and what is not. Finally, the model-directed approach, although the most efficient and the one that exhibits correct foci of problem-solving activity, has the disadvantage that its conclusions depend heavily on the availability of the correct model and initial focus. An incorrect initial focus will lead it to the examination of useless and incorrect analyses and will cause some perhaps relevant data to be ignored.

In our scheme, several dimensions of inference are included, each driven by the semantics of one of the organizational dimensions. They are integrated with one another so that each dimension of search compensates for the failings of another, and thus as a whole offers a rich and robust framework.

Further, each organization dimension offers distinct and necessary contributions to the updating of hypothesis certainty within a relaxation framework, and to the maintenance of consistency within an interpretation. Not only have the knowledge organization dimensions been integrated within a relaxation labelling process, driving the definitions of neighbourhood, compatibilities, weights, and consistency in an intuitive yet concrete fashion, but also the following results on the relaxation process have emerged:

- IS-A, besides offering a definition of global consistency of hypothesis' certainty, plays the role of speeding up the convergence of results. This is an important role, since it allows smaller temporal sampling rates. Because of inheritance, the problem posed by the propagation of results through the network disappears. IS-A also has an important part in the graceful recovery from poor predictions. Finally, feedback imposed by the IS-A hierarchy increases the stability of the cooperative process and partly compensates for the effects of noise disturbances and parameter variations, important considerations for the non-linear relaxation scheme.
- SIMILARITY plays the discrimination role, and is the only mechanism that allows for competition between hypotheses, enabling 'best choice' selection. In conjunction with the exceptions that drive SIMILARITY activations, this is a strong feedback mechanism, enhancing the stability of the cooperative process. Moreover, it is central to the definition of temporal sampling rate and of compatibility values.
- PART-OF is the mechanism that permits the selection of the stronger of two equally consistent hypotheses based on the strength of their components.
- Temporal Precedence assists in the discrimination of proper temporal order, important for temporal predictions, and temporal 'gluing' of events into higher order ones.

Such strong evidence for the use of knowledge organization axes, besides for knowledge access, during interpretation has been lacking in past works, and supports our claim that knowledge organization has far more to offer than has previously been evident.

7.0 Conclusions

A methodology has been presented for the knowledge-based interpretation of continuous time-varying signals. The key idea is that knowledge organization dimensions play several major roles beyond their access and structuring properties. We have shown that significant advantages result in both sophistication of the reasoning process for spatio-temporal data as well as in the formalization of a certainty-updating process rooted in relaxation processes. An example of the practical implementation of this methodology was shown with the examples of ALVEN's knowledge base and analysis. ALVEN is currently implemented on a VAX 11/780 running Berkeley UNIX 4.2 in the C language. The example shown above requires about 30 min. of CPU time, including the generation of the displays.

8.0 Acknowledgments

The application to the problem of left ventricular performance assessment would not have been possible without the constant support and encouragement of E. Douglas Wigle, Chief of Cardiology, Toronto General Hospital. Dominic Covvey, Peter McLaughlin, Robert Burns, Peter Liu, and Maurice Druck, all of the Division of Cardiology, at Toronto General Hospital at the time, provided much useful guidance and data. Programming assistance for the ALVEN system was provided by Brian Down, Andrew Gullen, Michael Jenkin, Niels Lobo, Ron Gershon, and Yawar Ali. Financial support was provided by the Natural Science and Engineering Research Council of Canada, the Connaught Foundation of the University of Toronto, Defense Research Establishment Atlantic, and the Ontario Heart Foundation. During the course of this work, the author was a recipient of a Canadian Heart Foundation Research Scholarship.

AIELLO, N. 1983. A comparative study of control strategies for expert systems: AGE implementation of three variations of PUFF. Proceedings of the American Association for Artificial Intelligence, 1983, Washington, D.C., pp. 1–4.

ALDERMAN, E., SIMPSON, C., STINSON, E., DAUGHTERS, G., and INGELS, N. 1976. Computer processing of intramyocardial marker dynamics for the measurement of ventricular function. Proceedings Computers in Cardiology, St. Louis, Missouri, pp. 75–78.

ALDERMAN, E., SCHWARZKOPF, A., INGELS, N., DAUGHTERS, G., STINSON, C., and SANDERS, W. 1979. Application of an externally referenced, polar coordinate system for left ventricular wall motion analysis. Proceedings Computers in Cardiology, Geneva, Switzerland, pp. 207–210.

ALLEN, J. 1981. Maintaining knowledge about temporal intervals. Department of Computer Science, University of Rochester, Rochester, N.Y., Report TR-86.

BOEHM, M., and HOEHNE, K. 1981. The processing and analysis of radiographic image sequences. *In* Digital image processing in medicine. *Edited by* K. Hoehne. Springer–Verlag, New York, N.Y., pp. 15–41.

BRACHMAN, R. 1979. On the epistemological status of semantic networks. *In* Associative networks. *Edited by* Findler. Academic Press, New York, N.Y., pp. 3–50.

——— 1982. What IS-A is and isn't. Proceedings of the Canadian Society for Computational Studies of Intelligence 1982, Saskatoon, Saskatchewan, pp. 212–221.

BROWER, R., and MEESTER, G. 1981. The shape of the human left ventricle: quantification of symmetry. Proceedings Computers in Cardiology, Florence, Italy, pp. 211–214.

CHANDRASEKARAN, B., GOMEZ, F., MITTAL, S., and SMITH, J. 1979. An approach to medical diagnosis based on conceptual structures. Proceedings of the International Joint Conference on Artificial Intelligence, Tokyo, Japan, pp. 134–142.

DAUGHTERS, G., ALDERMAN, E., STINSON, E., and INGELS, N. 1979. Methods for ventricular wall motion assessment: towards a uniform terminology. Proceedings Computers in Cardiology, Geneva, Switzerland, pp. 145–148.

DORAN, J., TRAILL, T., BROWN, D., and GIBSON, D. 1978. Detection of abnormal left ventricular wall movement during isovolumic contraction and early relaxation. British Heart Journal, **40**, pp. 367–371.

DOWN, B. 1983. Using feedback in understanding motion. M.Sc. thesis, Department of Computer Science, University of Toronto, Toronto, Canada.

FAGAN, L. 1980. VM: Representing time-dependent relations in a clinical setting. Ph.D. dissertation, Heuristic Programming Project, Stanford University, Stanford, California.

FUJII, J., WATANABE, H., KOYAMA, S., and KATO, K. 1979. Echocardiographic study on diastolic posterior wall movement and left ventricular filling by disease category. American Heart Journal, **98**, pp. 144–152.

GERBRANDS, J., BOOMAN, F., and REIBER, J. 1979. Computer analysis of moving radiopaque markers from X-ray films. Computer graphics and image processing, **11**, pp. 143–152.

GERSHON, R. 1982. Explanation methods for visual motion understanding systems. M.Sc. thesis, Department of Computer Science, University of Toronto, Toronto, Canada.

GIBSON, D., PREWITT, T., and BROWN, D. 1976. Analysis of left ventricular wall movement during isovolumic relaxation and its relation to coronary artery disease. British Heart Journal, **38**, pp. 1010–1019.

HILDRETH, E. 1980. Implementation of a theory of edge detection. Massachusetts Institute of Technology, Artificial Intelligence Laboratory, Report TR-579, April 1980.

HOEHNE, K., BOEHM, M., and NICOLAE, G. 1980. The processing of X-ray image sequences. *In* Advances in digital image processing. *Edited by* Stucki. Plenum Publishing Corporation, New York, N.Y., pp. 147–163.

LEVESQUE, H., and MYLOPOULOS, J. 1979. Procedural semantic networks. *In* Associative networks. *Edited by* Findler. Academic Press, New York, N.Y., pp. 92–120.

LONG, W., and RUSS, T. 1983. A control structure for time-dependent reasoning. Proceedings of the International Joint Conference on Artificial Intelligence 1983, Karlsruhe, Germany, pp. 230–232.

MINSKY, M. 1975. A framework for representing knowledge. *In* Psychology of computer vision. *Edited by* Winston. McGraw-Hill Book Company, New York, N.Y., pp. 211–277.

MITTAL, S., and CHANDRASEKARAN, B. 1980. Organizing data bases involving temporal information. Proceedings of the Institute of Electrical and Electronics SMC Society, pp. 1–5.

NILSSON, H. 1971. Problem solving methods in artificial intelligence. McGraw-Hill Book Company, New York, N.Y.

PATIL, R., SZOLOVITS, P., and SCHWARTZ, W. 1982. Modeling knowledge of the patient in acid-base and electrolyte disorders. *In* Artificial intelligence in medicine. Edited by P. Szolovits. Westview Press, pp. 191–226.

POPLE, H. 1977. The formation of composity hypotheses in diagnostic problem solving: an exercise in synthetic reasoning. Proceedings of the International Joint Conference on Artificial Intelligence 1977, Cambridge, Massachusetts, pp. 1030–1037.

——— 1982. Heuristic methods for imposing structure on ill-structured problems: the structuring of medical diagnostics. *In* Artificial intelligence in medicine. Edited by P. Szolovits. Westview Press, pp. 119–190.

SLAGER, C., HOOGHOUDT, T., REIBER, J., SCHUURBIERS, J., BOOMAN, F., and MEESTER, G. 1979. Left ventricular contour segmentation from anatomical landmark trajectories and its application to wall motion analysis. Proceedings Computers in Cardiology, Geneva, Switzerland, pp. 347–350.

SHIBAHARA, T., TSOTSOS, J., MYLOPOULOS, J., and COVVEY, H. 1983. CAA: a knowledge-based system using causal knowledge to diagnose cardiac rhythm disorders. Proceedings of the International Joint Conference on Artificial Intelligence 1983, Karlsruhe, Germany, pp. 242–245.

SHORTLIFFE, E. 1976. MYCIN: computer-based clinical consultations. Elsevier Press, Amsterdam, Netherlands.

TSOTSOS, J.K., MYLOPOULOS, J., COVVEY, H.D., and ZUCKER, S.W. 1980. A framework for visual motion understanding. Institute of Electrical and Electronics Engineering Transactions on pattern analysis and machine intelligence, November, pp. 563–573.

TSOTSOS, J. 1981. Temporal event recognition: an application to left ventricular performance assessment. Proceedings of the International Joint Conference on Artificial Intelligence 1981, Vancouver, Canada, pp. 900–907.

——— 1984. Representational axes and temporal cooperative processes. Department of Computer Science, University of Toronto, Toronto, Canada, Report RBCV-TR-2.

WALLIS, J., and SHORTLIFFE, E. 1982. Explanatory power for medical expert systems: studies in the representation of causal relationships for clinical consultations. Stanford University, Stanford, California, Report STAN-CS-82-923.

ZUCKER, S. 1978. Production systems with feedback. *In* Pattern-directed inference systems. Edited by Waterman and Hayes-Roth. Academic Press, New York, N.Y., pp. 539–556.

Chapter 5

Vision System Architectures and Computational Paradigms

This chapter addresses the question of how a general-purpose vision system should be structured—how the solution to the individual problems of visual interpretation can be assembled into a composite architecture. We also consider the alternative of monolithic procedures sufficiently powerful to solve broad classes of visual processing and interpretation tasks.

5.1 Vision System Architecture

A widely-held belief in computational vision is that one cannot proceed in a single step from the intensity information of an image directly to spatial or semantic understanding. Almost all currently proposed general vision system architectures are based on the "signals-to-symbols" paradigm—that is, they successively transform the scene information through a series of representations, from the initial grey-level image to the final identification of 3-D objects and their relationships. The differences in architectural concepts are typically due to differences in the intermediate representations.

5.1.1 Marr Architecture

Much of the vision research at MIT is consistent with the approach defined by David Marr [**Marr 78b**, 82] for the low and intermediate stages of computational vision. This work is strongly motivated by biological considerations. Starting with a raw image, represented as an intensity array, the first set of transformations produces the "primal sketch." This representation records point properties of the image, such as the direction and magnitude of local intensity gradients, as well as the stereo disparity or image velocity as computed from multiple images. The primal sketch computation is assumed to be carried out mechanically—independent of context or immediate purpose. It is also assumed that the primal sketch is a complete recoding of the intensity information present in the original image, and that complete recovery of this intensity information is possible from the primal sketch representation.

Using the primal sketch as input, a second set of transformations constructs the "$2\frac{1}{2}$-D sketch" that makes explicit the visible surfaces of the objects—that is, describes their shape and orientation. Finally, 3-D shapes are derived from the $2\frac{1}{2}$-D sketch. Volumetric primitives are used to express the shape of complex objects.

Some of the work that has been motivated by the Marr model includes the zero-crossings research of Marr and Hildreth [Marr 80] and Torre [Torre 86]; the Marr and Poggio stereo analysis [Marr 79]; interpolation of surfaces from visual information [Grimson 81]; and the theory of structure from motion [Ullman 79] [Hildreth 84].

5.1.2 Intrinsic Images

Barrow and Tenenbaum [Barrow 78] propose that early vision involves describing a scene in terms of "intrinsic" characteristics of the surface element visible at each point in the image.[1] Typical intrinsic properties of scene surfaces are range, orientation, reflectance, and incident illumination. The intrinsic characteristics are represented as "images" that are in registration with the intensity image. Thus, the Barrow and Tenenbaum parallel computational model for recovering intrinsic images consists of a stack of registered arrays, representing the original intensity image and the desired intrinsic image "overlays."

Processing is initialized by detecting intensity edges in the original image, interpreting them according to a "catalog," and then inserting the appropriate edges in the intrinsic images. Parallel local operations modify the values in each intrinsic image to make them consistent with assumed constraints such as reflectance constancy and depth continuity. Simultaneously, a second set of processes operates to make the values consistent with constraints relating the different intrinsic properties. A third set of processes operates to insert and delete edge elements that locally inhibit continuity constraints. The constraint and edge modification processes operate "continuously" and interact in an attempt to recover accurate intrinsic scene characteristics and to perfect the initial edge interpretation.

5.1.3 The VISIONS System

The Visual Integration by Semantic Interpretation of Natural Scenes (VISIONS) system of Hansen and Riseman [Hansen 78], University of Massachusetts, employs an empirical approach for interpreting static, monocular, color images of natural scenes to obtain a semantic description. The system is composed of low-level filtering and segmentation processes that operate on numeric arrays in registration with the intensity image, and high-level interpretation processes for constructing a description of the portion of the world portrayed in the image. Interpretation of an image involves the construction of a model that includes a description of the major conceptual entities present and a description of the volumes and surfaces in the 3-D space of the scene. VISIONS is one of the best examples of a complete implementation of the signals-to-symbols paradigm.

Some of the work motivated by the VISIONS model includes inferencing mechanisms based on the Shafer-Dempster-Lowrance idea of evidential reasoning [Wesley 86] [Belknap 86] [Weymouth 83] and schema networks as a representation of knowledge [Weymouth 86]. Also see the paper by Lawton [Lawton 86] in Chapter 4.

5.1.4 Connectionist Approach

In the *connectionist* approach the conventional computer mechanism of communicating information by passing complex symbolic structures is not used; rather, the burden of the computation is embedded in the interconnections of a network of computational elements. The main characteristic of the connectionist approach is the use of massive parallelism, involving simple computational units having a high degree of connectivity. The approach is motivated by analogy to the human brain: comparatively slow (millisecond) neural computing elements with complex, parallel connections form a structure that is dramatically different from a high-speed, predominantly serial, digital computer.

Various viewpoints have been expressed concerning the nature of the basic functional element used in the network. In one view, concepts are captured by a pattern of activity in a large group of unspecialized units. Here research is generally concerned with general properties of connectionist networks, particularly the construction of "training" algorithms for obtaining desired behavior. At the other end of the spectrum a basic computing unit specifically represents each item of interest such as a concept, a line segment, and so on, and this is called a *localist* model.[2] Feldman and Ballard [Feldman 82] introduce the general connectionist model, and Feldman [Feldman 85] lays out a computational framework in which all levels of the vision task can be carried out in a highly parallel fashion. A key localist idea is the representation of all visual information needed for high-level processing as discrete parameter values that can be cooperatively extracted from the image by simple computational units. Thus, one would have a separate unit dedicated to each possible value of each parameter of interest—for example, a separate unit for each value of depth, size, and color at each image location.

[1] Their more recent papers have shifted to a "process-oriented" view of early vision (see Chapter 2).

[2] The extreme of the localist point of view would be the "grandmother cell" approach that used cells so specialized that there would be a specific cell whose only function was to recognize one's grandmother.

Work inspired by the localist connectionist model includes Ballard on parameter networks [**Ballard 84**], an extension of the Hough transform by Ballard and Sabah [Ballard 83], and the representation of the Origami world in connectionist terms by Sabah [Sabah 82]. Research in the other connectionist school of thought, sometimes called "parallel distributed processing (PDP)," is described in a two-volume set prepared by Rumelhart, McClelland, and the PDP Research Group at the University of California, San Diego [Rumelhart 86] and in an excellent tutorial overview by Fahlman and Hinton [Fahlman 87]; a learning algorithm for massively parallel networks is given in [**Ackley 85**].

5.1.5 Discussion

Each of the preceding architectures has led to a different research focus:

- The Marr-based research tends to concentrate on understanding the nature of individual low-level cues, such as intensity discontinuities and motion. The idea is to perfect the primal sketch before undertaking any higher-level analysis.
- Research based on the Barrow and Tenenbaum model stresses the integration of multiple cues in which the 3-D nature of the scene comes into play in the earliest set of transformations. Three-dimensional interpretations are assigned to intensity edges in order to initialize processing of the intrinsic images, and the relationship between original intensities and interpretation of scene properties is maintained as tightly as possible.
- The VISIONS approach is strongly knowledge-based; effort is concentrated on finding additional evidence for improving the partitioning of a scene, and on methods for integrating this evidence into higher-level interpretations.
- The localist school of connectionism attempts to devise parameterizations of vision problems that are suitable for network mechanizations. Global optimization and learning are major concerns of the PDP approach to connectionism.

When comparing the preceding architectures, we note the following significant distinctions.

1. **The extent to which they are extendable to the entire set of problems faced by a general-purpose vision system**: The intrinsic images paradigm has little to say about high-level vision, and while the Marr paradigm includes a high-level component, it is weakly coupled to the lower levels and very limited in its competence. Connectionism has not yet presented a convincing argument as to how it can deal with the problem of relational description. VISIONS does provide a framework for a complete general-purpose vision system, but it has no strong integrating theme—the various levels are largely independent of each other.
2. **The level at which scene constraints are introduced**: Both Marr and VISIONS start with an analysis of the image independent of its interpretation as the projection of a 3-D scene; intrinsic images and connectionism invoke 3-D constraints at every level of processing.
3. **The level at which context and purpose are introduced:** Both Marr and intrinsic images suggest that the lowest levels of analysis are independent of any higher level knowledge. Both VISIONS and connectionism are compatible with having high-level knowledge influence all stages of processing, but neither of these approaches has taken a strong stand on this issue.

5.2 Computational Paradigms

The computational paradigms described in this section represent general approaches for performing the visual interpretation tasks required in the signals-to-symbols paradigm. They can also serve as stand-alone alternatives to signals-to-symbols. Many of the algorithms use variational or relaxation techniques and are based on the idea of local activities influencing some overall behavior; a common theme is global optimization.

5.2.1 "Regularization"

When using a 2-D image to draw conclusions about the 3-D world, we may find that a consistent solution does not exist, is not unique, or is overly sensitive to the initial data. An obvious approach for solving ill-posed problems such as these is to restrict the class of admissible solutions by introducing suitable a priori knowledge. While much of the work in machine vision supplies additional knowledge based on physical principles or specific knowledge of the scene, an approach called "variational regularization" focuses on methods that reformulate an ill-posed problem in terms of a variational principle—that is, as an optimization problem. In preference to employing domain-dependent knowledge to constrain the space of possible solutions, the underlying physical consideration typically invoked by the variational regularization approach is that the real world consists of solid objects with smooth surfaces.

The regularization approach seeks to unify the computational paradigms addressing early vision within a single framework. This perspective provides a link between the computational nature of early vision problems, the structure of the algorithms for solving them, and the parallel hardware that can be used for efficient visual information processing. Poggio, Torre, and Koch [**Poggio 85**] review the work in variational regularization and show the relationship between the problems, the algorithms, and the possible mechanizations.

5.2.2 Relaxation

Relaxation processes are iterative techniques, either deterministic or stochastic, that use adjustments on a local level to achieve an overall goal. Examples of goals are determining a set of labels for the objects depicted in an image that is "best" in some sense or finding the values of a set of variables that minimize some cost function.

Deterministic Relaxation: Relaxation Labeling Processes

Suppose we are analyzing a picture or scene and have located a number of objects but have not identified them unambiguously. The relationships existing among the objects can often be used to reduce or eliminate the ambiguity. The problem has two forms: (1) only a single label is allowed for an object, or (2) each object can be assigned a set of labels, each with a weight such that the sum of label weights for an object equals 1.

Relaxation labeling processes (RLP), used in constraint satisfaction problems, iteratively change the weightings given to object labels based on some criterion of consistency with their neighbors. Groups of labeled objects that are mutually consistent should survive the iteration process, whereas inconsistent labelings are given less and less weight. RLP are generally intended to operate in a situation in which there are a large number of objects to be labeled with a relatively small set of labels. For example, an image might be described as consisting of many regions, each of which is treated as an object to be labeled. Consistency constraints act between neighboring objects or regions.

A classic paper on relaxation-based scene labeling by Rosenfeld, Hummel, and Zucker [Rosenfeld 1976a] introduced several different RLP models and discussed their properties.[3]

A characteristic of many of the relaxation labeling techniques is that they tend to be ad hoc. Addressing this problem, Hummel and Zucker [**Hummel 83**] show that finding consistent labelings is equivalent to solving a variational inequality. Their theory answers some important questions about convergence.

Stochastic Relaxation: "Simulated Annealing"

"Simulated annealing," a statistical mechanics method for solving complex optimization problems, can be used in problems arising in image processing and scene analysis.[4] Simulated annealing is especially useful in those cases where lack of continuity prevents the application of the differential calculus. This approach avoids being trapped in local minima by two techniques: (1) the adjustments are suggested by a random process that will occasionally propose very large changes, and (2) an adjustable "temperature" parameter controls the acceptance of any change in a variable—the corresponding solution will not always be better than the previous one.

Specifically, to minimize a function $E(x_i)$ of the many variables, x_i, the system is treated as a physical system, with the "energy" given by $E = E(x_i)$. One then looks for the minimum energy state of the physical system by starting with an arbitrary configuration, x_i, and introducing random perturbations, Δx_i. If a perturbation decreases E, it is accepted. If a perturbation increases E, it is accepted with a probability $e^{-\Delta E/T}$, where T is the temperature parameter. The random acceptance of "bad" perturbations as a function of ΔE and T is the mechanism that drives the system out of local minima.

In the simulated annealing approach a solution is reached by simulating the slow cooling of a physical system, starting with a very high "temperature" and monotonically reducing the temperature to zero. The "cooling" must be slow enough to ensure that the system does not get stuck in local minima of $E(x_i)$. Kirkpatrick, Gelatt, and Vecchi [**Kirkpatrick 83**] review the relevent constructs in combinatorial optimization and in statistical mechanics and then establish the relationships between the two fields. They

[3]For a recent review of relaxation labeling algorithms, see Kittler and Illingworth [Kittler 85].

[4]See [**Barnard 86**] in Chapter 1.

show that four ingredients are needed to apply simulated annealing to optimization problems: (1) a concise description of any configuration of the system; (2) a mechanism to generate random rearrangements of the elements in a configuration; (3) a quantitative objective function that permits selection of the "better" of two proposed system configurations; and (4) an annealing schedule of temperatures and length of times for which the system is to be evolved.

Geman and Geman [**Geman 84**] show how annealing techniques can be used for image restoration at low signal-to-noise ratios. They form an analogy between images and statistical mechanical systems: pixel gray levels and the presence and orientation of edges are viewed as states of atoms or molecules in a latticelike physical system. The essence of their approach[5] is the use of a stochastic relaxation algorithm that generates a sequence of images that converges to a restored image. This sequence evolves by local changes in the pixel gray levels and in locations and orientations of boundary elements. From our perspective the importance of this work is their formulation of the restoration problem as a prototype for other image-related applications of annealing, rather than the specific application of annealing to restoration.

Carnevali, Coletti, and Patarnello [**Carnevali 85**] also use simulated annealing for image-processing problems. They deal with estimating the parameters necessary to describe a geometrical pattern corrupted by noise, the smoothing of binary images, and the process of half-toning a gray-level image. They show how to formulate these problems in terms of a minimization problem that can be solved using simulated annealing. In addition, they show that for many problems simulated annealing is more than an arbitrary global optimization technique; when it is analogous to a well understood physical process, one can benefit from knowledge of the original process.[6] This paper presents an especially clear exposition of how annealing can be applied to image analysis problems.

5.2.3 Connectionist Computational Paradigms

Ballard [**Ballard 84**] describes a connectionist theory of low-level and intermediate-level vision. Given a set of intrinsic images of edges, optic flow, surface orientation, and so on, the basic problem is to segment (determine groupings in) each intrinsic image. The groupings are determined by using massively parallel cooperative computations among the images and a set of "feature networks" at different levels of abstraction.

The theory is based on extensive use of the Hough transform idea that measurements at an individual pixel location in an image can be used to increment one or more compatible "cells" in a feature space histogram. Histogram peaks are then used to detect groupings in the image (see Chapter 7). When the Hough transform is used for detecting prominent lines in an image, each edge-labeled pixel in the image increases the confidence of corresponding line units (cells) in the "feature space network." The line units in the feature space that receive the most support (the cells with a large number of "votes") identify the prominent lines in the intrinsic images.

In general, if parts of an intrinsic image have a meaningful organization or partitioning, this organization can be detected by a general Hough transform that uses a multidimensional feature space. Ballard describes how this type of "constraint transform" can be used to organize the parts of various types of intrinsic images. As one goes to higher-dimensional feature spaces, the number of units required increases exponentially. Ballard describes two approaches that have been used to deal with this increase: (1) "coarse coding" of feature measurements—that is, rather than using a single unit per pixel of the intrinsic image, one uses several overlapping units that each cover a larger number of pixels—and (2) partitioning a very large feature space into subspaces to decrease the combinatorial explosion. In [Ballard 86], Ballard explores the hypothesis that an important part of the cortex of the human brain can be modeled as a connectionist computer.

In the localist approach to connectionism, as exemplified by the work of Feldman and Ballard, semantic concepts relevant to a given problem domain are embodied in specific computational units as part of the network design. In the parallel-distributed-processing (PDP) approach, weighted connections intermix the stored knowledge; semantic concepts—to the extent that they exist at all—are embodied in the global patterns of activity of the network. Because of such complex mixing of information, it is generally impractical to explicitly design PDP networks. Such networks are adapted to specific problem domains by "learning" techniques. Learning in PDP networks in-

[5] A special tutorial overview, written by G. Smith, is included as a preface to the Geman paper.

[6] They do this by showing that various problems involving binary images are formally equivalent to a physics problem known as the "ground state problem for two-dimensional Ising spin systems embedded in an external field."

volves adjusting network parameters to produce a desired behavior specified by a table of input/output pairs. For example, Ackley, Hinton, and Sejnowski [**Ackley 85**] describe a domain-independent learning rule, based on simulated annealing, for modifying the connection "strength" between the computing units in a multilayered network. Training the network involves entering strings of characters and modifying the connection strengths until the desired codes are produced as output. The authors employ a special type of parallel-constraint-satisfaction network, a "Boltzman machine," in which primitive computing elements called *units* are connected to each other by bidirectional links. Link weights are symmetric, having the same strength in both directions.[7] A unit is always of two states, *on* or *off*, and it adopts these states as a probabilistic function of the states of its neighboring units and the interconnecting weights. The Bolzman machine uses simulated annealing to minimize a cost function introduced by Hopfield [Hopfield 82] called "energy," which is defined for each state of the machine. Hopfield's energy function assigns a cost equal to the connection weight between two units if: (1) they are in opposite states when the connection weight is positive,[8] or (2) they are in the same state when the connection weight is negative. Thus, for any input/output pair, there will be a minimal energy configuration of the network compatible with the corresponding configuration of the input/output units.

A critical issue for a learning machine is how to appropriately generalize its decision rule to input/output patterns that have not been prespecified. In spite of some successful experiments, there is no reason to believe that Hopfield's energy function will force a desired generalization when the number of specified input/output pairs is—as is typically the case in a practical problem—a vanishingly small fraction of all such possible pairs.

5.2.4 Simplest Description

The ability to form a good theory or explanation for a given body of data is an essential attribute of a competent image-understanding system. Recent work in perceptual organization (Chapter 2) and in AI has suggested a mechanism based on the concept of *simplest* description for obtaining structural theories that better match human conceptualizations.

A given set of data can be described by providing both the model parameters and the relationship of each data point to the model. For example, in the case of a model consisting of a straight line, one would provide the line parameters, the coordinates of the intersection of a normal from each data point to the line, and the length of each normal. A given body of data can be described in many possible ways by such a system, and choosing the simplest description corresponds to minimizing an objective function on the model components and the deviations of the data points from the instantiated model.

Georgeff and Wallace [Georgeff 84] consider a *theory* to be an abstraction of a class of data. When there are several possible theories for representing some particular element of data, they propose the following measure: one theory is better than another if it yields a *more compact* explanation. What constitutes a good theory will always be dependent on our expectations about the world. Thus, if we know a priori that most of the objects under consideration are of shape X rather than Y, our language for describing theories should make it harder to describe Y than X; that is, for equally accurate descriptions of a body of data, the explanation X should be more compact than Y.

G. B. Smith and Wolf [G. B. Smith 84a] used the Georgeff and Wallace approach in a problem concerned with generating descriptions of line structures appearing in an image. They show that the usual technique of first extracting a set of line segments and then describing the individual segments can be replaced by a procedure that finds the "best" description of the data on the basis of a global view of that data. This technique simultaneously extracts and describes the lines. "Best" is defined as the cheapest encoding of the data when we consider the trade-off between the quality of explanation[9] of the data and the complexity of that explanation.

Barnard [Barnard 84] deals with the problem of how to compose good descriptions of images of abstract figures in terms of a given vocabulary of geometric primitives. He argues that neither rule-based deductive inference nor model-based matching are satisfactory computational paradigms since these approaches can only deal with only a small set of anticipated situations. The author introduces an inductive approach for describing shapes of curves and surfaces that relies

[7]Hummel and Zucker [**Hummel 83**] show that it is possible to assure a monotonic decrease in a cost function defined over a constraint satisfaction network with symmetric connections.

[8]A positive connection weight can be interpreted as a constraint that says that if one unit is on, the other should also be on.

[9]The degree to which the data fits the proposed model

on a simplicity criterion for deciding which description is preferred. Preferrence is based on entropy, the average amount of information per symbol in a description; descriptions having minimum entropy are selected. One can gain some appreciation of the basic idea by considering the histogram of curvature and torsion values along a curve. The "randomness" of the histogram is a measure of the simplicity of the curve. For example, a circle would have a very peaked histogram, with a small measure of randomness or entropy, while a more complex curve would have more scatter in parameter space, with correspondingly increased entropy. Thus, to the extent that a circle was a reasonably good fit to the data set, this description would be preferred to a description consisting of fitting many small, straight-line segments to the circular contour.

5.3 Discussion

The vision system "architectures" and computational paradigms discussed in this chapter have traditionally served the purpose of providing a philosophical framework for guiding research in computational vision, rather than being serious candidates for the design of an actual vision system. Historically, two contending themes can be discerned:

1. Those approaches that emphasize intermediate levels of description as characterized by the signals-to-symbols paradigm[10] and the VISIONS and Marr architectures of Section 5.1.

2. Those approaches in which interpretation is based on satisfying a set of constraints, or in which interpretation corresponds to an extreme value of some global objective function. These techniques, the monolithic computational paradigms discussed in Section 5.2, compute their interpretation (output) directly from the input data. They do not produce a hierarchy of semantically meaningful intermediate representations.

Since we have not yet solved the vision problem, we cannot assert that one of these approaches is superior to the other. Monolithic computational approaches dominated vision research until the early 1970s. Signals-to-symbols held sway from the mid-1970s to the mid-1980s. The monolithic computational approaches now appear to be returning to favor.

Papers Included

Ackley, D. H., G. E. Hinton, and T. J. Sejnowski. A learning algorithm for Bolzmann machines. *Cognitive Science*, 9:147–169, 1985.

Ballard, D. H. Parameter Nets. *Artificial Intelligence*, 22(3):235–267, April 1984.

Carnevali, P., L. Coletti, and S. Patarnello. Image processing by simulated annealing *IBM J of Research and Development*, 29(6):569–579, November 1985.

Smith, G.B. *Preface to the Geman and Geman paper.*

S. Geman and D. Geman. Stochastic relaxation, Gibbs distribution, and the Bayesian restoration of images. *IEEE PAMI*, 6:721–741, 1984.

Hummel, R.A., and Zucker, S. W. On the foundation of relaxation labeling processes. *IEEE PAMI*, 5:267–287, 1983.

Kirkpatrick, S., C. D. Gelatt, Jr., and M. P. Vecchi. Optimization by simulated annealing. *Science*, 220(4598):671–680, May 13, 1983.

Marr, D. and Nishihara, H.K. Visual information processing: artificial intelligence and the sensorium of sight. *Technology Review*, 81(1):2–23, Oct. 1978.

Poggio, T. V. Torre, and C. Koch. Computational vision and regularization theory. *Nature*, 317(26):314–319, Sept. 1985.

[10] An extended description of this paradigm is presented in M.A. Fischler and O. Firschein, *Intelligence: The Eye, the Brain, and the Computer* (Reading: Addison-Wesley, 1987).

A Learning Algorithm for Boltzmann Machines*

David H. Ackley
Geoffrey E. Hinton

Computer Science Department
Carnegie-Mellon University

Terrence J. Sejnowski

Biophysics Department
The Johns Hopkins University

The computational power of massively parallel networks of simple processing elements resides in the communication bandwidth provided by the hardware connections between elements. These connections can allow a significant fraction of the knowledge of the system to be applied to an instance of a problem in a very short time. One kind of computation for which massively parallel networks appear to be well suited is large constraint satisfaction searches, but to use the connections efficiently two conditions must be met: First, a search technique that is suitable for parallel networks must be found. Second, there must be some way of choosing internal representations which allow the preexisting hardware connections to be used efficiently for encoding the constraints in the domain being searched. We describe a general parallel search method, based on statistical mechanics, and we show how it leads to a general learning rule for modifying the connection strengths so as to incorporate knowledge about a task domain in an efficient way. We describe some simple examples in which the learning algorithm creates internal representations that are demonstrably the most efficient way of using the preexisting connectivity structure.

1. INTRODUCTION

Evidence about the architecture of the brain and the potential of the new VLSI technology have led to a resurgence of interest in "connectionist" systems (Feldman & Ballard, 1982; Hinton & Anderson, 1981) that store their long-term knowledge as the strengths of the connections between simple neuron-like processing elements. These networks are clearly suited to tasks like vision that can be performed efficiently in parallel networks which have physical connections in just the places where processes need to communicate. For problems like surface interpolation from sparse depth data (Grimson, 1981; Terzopoulos, 1984) where the necessary decision units and communication paths can be determined in advance, it is relatively easy to see how to make good use of massive parallelism. The more difficult problem is to discover parallel organizations that do not require so much problem-dependent information to be built into the architecture of the network. Ideally, such a system would adapt a given structure of processors and communication paths to whatever problem it was faced with.

This paper presents a type of parallel constraint satisfaction network which we call a "Boltzmann Machine" that is capable of learning the underlying constraints that characterize a domain simply by being shown examples from the domain. The network modifies the strengths of its connections so as to construct an internal *generative* model that produces examples with the same probability distribution as the examples it is shown. Then, when shown any particular example, the network can "interpret" it by finding values of the variables in the internal model that would generate the example. When shown a partial example, the network can complete it by finding internal variable values that generate the partial example and using them to generate the remainder. At present, we have an interesting mathematical result that guarantees that a certain learning procedure will build internal representations which allow the connection strengths to capture the underlying constraints that are implicit in a large ensemble of examples taken from a domain. We also have simulations which show that the theory works for some simple cases, but the current version of the learning algorithm is very slow.

The search for general principles that allow parallel networks to learn the structure of their environment has often begun with the assumption that networks are randomly wired. This seems to us to be just as wrong as the view that *all* knowledge is innate. If there are connectivity structures that are good for particular tasks that the network will have to perform, it is much more efficient to build these in at the start. However, not all tasks can be foreseen, and even for ones that can, fine-tuning may still be helpful.

Another common belief is that a general connectionist learning rule would make sequential "rule-based" models unnecessary. We believe that this view stems from a misunderstanding of the need for multiple levels of description of large systems, which can be usefully viewed as either parallel or serial depending on the grain of the analysis. Most of the key issues and questions that have been studied in the context of sequential models do not magically disappear in connectionist models. It is still necessary to perform

* The research reported here was supported by grants from the System Development Foundation. We thank Peter Brown, Francis Crick, Mark Derthick, Scott Fahlman, Jerry Feldman, Stuart Geman, Gail Gong, John Hopfield, Jay McClelland, Barak Pearlmutter, Harry Printz, Dave Rumelhart, Tim Shallice, Paul Smolensky, Rick Szeliski, and Venkataraman Venkatasubramanian for helpful discussions.

Reprint requests should be addressed to David Ackley, Computer Science Department, Carnegie-Mellon University, Pittsburgh, PA 15213.

searches for good solutions to problems or good interpretations of perceptual input, and to create complex internal representations. Ultimately it will be necessary to bridge the gap between hardware-oriented connectionist descriptions and the more abstract symbol manipulation models that have proved to be an extremely powerful and pervasive way of describing human information processing (Newell & Simon, 1972).

2. THE BOLTZMANN MACHINE

The Boltzmann Machine is a parallel computational organization that is well suited to constraint satisfaction tasks involving large numbers of "weak" constraints. Constraint-satisfaction searches (e.g., Waltz, 1975; Winston, 1984) normally use "strong" constraints that *must* be satisfied by any solution. In problem domains such as games and puzzles, for example, the goal criteria often have this character, so strong constraints are the rule.[1] In some problem domains, such as finding the most plausible interpretation of an image, many of the criteria are not all-or-none, and frequently even the best possible solution violates some constraints (Hinton, 1977). A variation that is more appropriate for such domains uses weak constraints that incur a cost when violated. The quality of a solution is then determined by the total cost of all the constraints that it violates. In a perceptual interpretation task, for example, this total cost should reflect the implausibility of the interpretation.

The machine is composed of primitive computing elements called *units* that are connected to each other by bidirectional *links.* A unit is always in one of two states, *on* or *off,* and it adopts these states as a probabilistic function of the states of its neighboring units and the *weights* on its links to them. The weights can take on real values of either sign. A unit being on or off is taken to mean that the system currently accepts or rejects some elemental hypothesis about the domain. The weight on a link represents a weak pairwise constraint between two hypotheses. A positive weight indicates that the two hypotheses tend to support one another; if one is currently accepted, accepting the other should be more likely. Conversely, a negative weight suggests, other things being equal, that the two hypotheses should not both be accepted. Link weights are *symmetric,* having the same strength in both directions (Hinton & Sejnowski, 1983).[2]

[1] But, see (Berliner & Ackley, 1982) for argument that, even in such domains, strong constraints must be used only where absolutely necessary for legal play, and in particular must not propagate into the determination of *good* play.

[2] Requiring the weights to be symmetric may seem to restrict the constraints that can be represented. Although a constraint on boolean variables A and B such as "$A \equiv B$ with a penalty of 2 points for violation" is obviously symmetric in A and B, "$A \Rightarrow B$ with a penalty of 2 points for violation" appears to be fundamentally asymmetric. Nevertheless, this constraint can be represented by the combination of a constraint on A alone and a symmetric pairwise constraint as follows: "Lose 2 points if A is true" and "Win 2 points if both A and B are true."

The resulting structure is related to a system described by Hopfield (1982), and as in his system, each global state of the network can be assigned a single number called the "energy" of that state. With the right assumptions, the individual units can be made to act so as to *minimize the global energy.* If *some* of the units are externally forced or "clamped" into particular states to represent a particular input, the system will then find the minimum energy configuration that is compatible with that input. The energy of a configuration can be interpreted as the extent to which that combination of hypotheses violates the constraints implicit in the problem domain, so in minimizing energy the system evolves towards "interpretations" of that input that increasingly satisfy the constraints of the problem domain.

The energy of a global configuration is defined as

$$E = -\sum_{i<j} w_{ij} s_i s_j + \sum_i \theta_i s_i \qquad (1)$$

where w_{ij} is the strength of connection between units i and j, s_i is 1 if unit i is on and 0 otherwise, and θ_i is a threshold.

2.1 Minimizing Energy

A simple algorithm for finding a combination of truth values that is a *local* minimum is to switch each hypothesis into whichever of its two states yields the lower total energy given the current states of the other hypotheses. If hardware units make their decisions asynchronously, and if transmission times are negligible, then the system always settles into a local energy minimum (Hopfield, 1982). Because the connections are symmetric, the difference between the energy of the whole system with the kth hypothesis rejected and its energy with the kth hypothesis accepted can be determined locally by the kth unit, and this "energy gap" is just

$$\Delta E_k = \sum_i w_{ki} s_i - \theta_k \qquad (2)$$

Therefore, the rule for minimizing the energy contributed by a unit is to adopt the *on* state if its total input from the other units and from outside the system exceeds its threshold. This is the familiar rule for binary threshold units.

The threshold terms can be eliminated from Eqs. (1) and (2) by making the following observation: the effect of θ_i on the global energy or on the energy gap of an individual unit is identical to the effect of a link with strength $-\theta_i$ between unit i and a special unit that is by definition always held in the *on* state. This "true unit" need have no physical reality, but it simplifies the computations by allowing the threshold of a unit to be treated in the same manner as the links. The value $-\theta_i$ is called the *bias* of unit i. If a permanently active "true unit" is assumed to be part of every network, then Eqs. (1) and (2) can be written as:

$$E = -\sum_{i<j} w_{ij} s_i s_j \quad (3)$$

$$\Delta E_k = \sum_i w_{ki} s_i \quad (4)$$

2.2 Using Noise to Escape from Local Minima

The simple, deterministic algorithm suffers from the standard weakness of gradient descent methods: It gets stuck in *local* minima that are not globally optimal. This is not a problem in Hopfield's system because the local energy minima of his network are used to store "items": If the system is started near some local minimum, the desired behavior is to fall into that minimum, not to find the global minimum. For constraint satisfaction tasks, however, the system must try to escape from local minima in order to find the configuration that is the global minimum given the current input.

A simple way to get out of local minima is to occasionally allow jumps to configurations of higher energy. An algorithm with this property was introduced by Metropolis, Rosenbluth, Rosenbluth, Teller, & Teller (1953) to study average properties of thermodynamic systems (Binder, 1978) and has recently been applied to problems of constraint satisfaction (Kirkpatrick, Gelatt, & Vecchi, 1983). We adopt a form of the Metroplis algorithm that is suitable for parallel computation: If the energy gap between the *on* and *off* states of the kth unit is ΔE_k then regardless of the previous state set $s_k = 1$ with probability

$$p_k = \frac{1}{(1 + e^{-\Delta E_k / T})} \quad (5)$$

where T is a parameter that acts like temperature (see Figure 1).

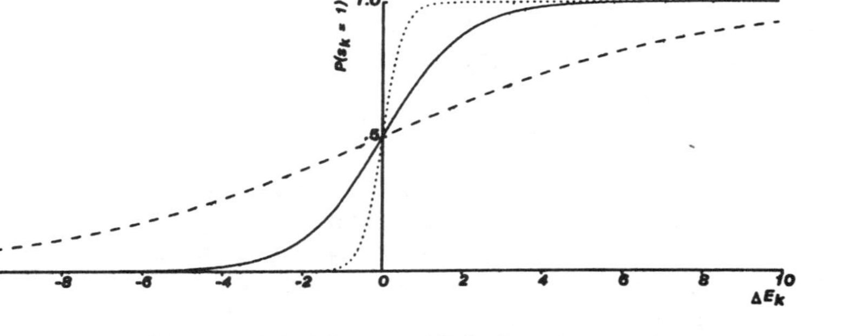

Figure 1. Eq. (5) at T=1.0 (*solid*), T=4.0 (*dashed*), and T=0.25 (*dotted*).

The decision rule in Eq. (5) is the same as that for a particle which has two energy states. A system of such particles in contact with a heat bath at a given temperature will eventually reach thermal equilibrium and the probability of finding the system in any global state will then obey a Boltzmann distribution. Similarly, a network of units obeying this decision rule will eventually reach "thermal equilibrium" and the relative probability of two global states will follow the Boltzman distribution:

$$\frac{P_\alpha}{P_\beta} = e^{-(E_\alpha - E_\beta)/T} \quad (6)$$

where P_α is the probability of being in the αth global state, and E_α is the energy of that state.

The Boltzmann distribution has some beautiful mathematical properties and it is intimately related to information theory. In particular, the difference in the log probabilities of two global states is just their energy difference (at a temperature of 1). The simplicity of this relationship and the fact that the equilibrium distribution is independent of the path followed in reaching equilibrium are what make Boltzmann machines interesting.

At low temperatures there is a strong bias in favor of states with low energy, but the time required to reach equilibrium may be long. At higher temperatures the bias is not so favorable but equilibrium is reached faster. A good way to beat this trade-off is to start at a high temperature and gradually reduce it. This corresponds to annealing a physical system (Kirkpatrick, Gelatt, & Vecchi, 1983). At high temperatures, the network will ignore small energy differences and will rapidly approach equilibrium. In doing so, it will perform a search of the coarse overall structure of the space of global states, and will find a good minimum at that coarse level. As the temperature is lowered, it will begin to respond to smaller energy differences and will find one of the better minima within the coarse-scale minimum it discovered at high temperature. Kirkpatrick et al. have shown that this way of searching the coarse structure before the fine is very effective for combinatorial problems like graph partitioning, and we believe it will also prove useful when trying to satisfy multiple weak constraints, even though it will clearly fail in cases where the best solution corresponds to a minimum that is deep, narrow, and isolated.

3. A LEARNING ALGORITHM

Perhaps the most interesting aspect of the Boltzmann Machine formulation is that it leads to a domain-independent learning algorithm that modifies the connection strengths between units in such a way that the whole network develops an internal model which captures the underlying structure of its environment. There has been a long history of failure in the search for such algorithms (Newell, 1982), and many people (particularly in Artificial Intelligence) now believe that no such algorithms exist. The major technical stumbling block which prevented the generalization of simple learning algorithms to more complex networks was this: To be capable of interesting computations, a network must contain nonlinear elements that are not directly constrained by the input, and when such a network does the wrong thing it appears to be impossible to decide which of the many connection strengths is at fault. This "credit-assignment" problem was what led to the demise of perceptrons (Minsky & Papert, 1968; Rosenblatt, 1961). The perceptron convergence theorem guarantees that the weights of a single layer of decision units can be trained, but it could not be generalized to networks of such units when the task did not directly specify how to use all the units in the network.

This version of the credit-assignment problem can be solved within the Boltzmann Machine formulation. By using the right stochastic decision rule, and by running the network until it reaches "thermal equilibrium" at some finite temperature, we achieve a mathematically simple relationship between the probability of a global state and its energy. For a network that is running freely without any input from the environment, this relationship is given by Eq. (6). Because the energy is a *linear* function of the weights (Eq. 1) this leads to a remarkably simple relationship between the log probabilities of global states and the individual connection strengths:

$$\frac{\partial \ln P_\alpha}{\partial w_{ij}} = \frac{1}{T}[s_i^\alpha s_j^\alpha - p'_{ij}] \quad (7)$$

where s_i^α is the state of the ith unit in the αth global state (so $s_i^\alpha s_j^\alpha$ is 1 only if units i and j are both on in state α), and p'_{ij} is just the probability of finding the two units i and j on at the same time when the system is at equilibrium.

Given Eq. (7), it is possible to manipulate the log probabilities of global states. If the environment directly specifies the required probabilities P_α for each global state α, there is a straightforward way of converging on a set of weights that achieve those probabilities, provided any such set exists (for details, see Hinton & Sejnowski, 1983a). However, this is not a particularly interesting kind of learning because the system has to be given the required probabilities of *complete* global states. This means that the central question of what internal representation should be used has already been decided by the environment. The interesting problem arises when the environment implicitly contains high-order constraints and the network must choose internal representations that allow these constraints to be expressed efficiently.

3.1 Modeling the Underlying Structure of an Environment

The units of a Boltzmann Machine partition into two functional groups, a nonempty set of *visible* units and a possibly empty set of *hidden* units. The visible units are the interface between the network and the environment; during training all the visible units are clamped into specific states by the environment; when testing for completion ability, any subset of the visible units may be clamped. The hidden units, if any, are never clamped by the environment and can be used to "explain" underlying constraints in the ensemble of input vectors that cannot be represented by pairwise constraints among the visible units. A hidden unit would be needed, for example, if the environment demanded that the states of three visible units should have even parity—a regularity that cannot be enforced by pairwise interactions alone. Using hidden units to represent more complex hypotheses about the states of the visible units, such higher-order constraints among the visible units can be reduced to first and second-order constraints among the whole set of units.

We assume that each of the environmental input vectors persists for long enough to allow the network to approach thermal equilibrium, and we ignore any structure that may exist in the *sequence* of environmental vectors. The structure of an environment can then be specified by giving the probability distribution over all 2^v states of the v visible units. The network will be said to have a perfect model of the environment if it achieves exactly the same probability distribution over these 2^v states when it is running freely at thermal equilibrium with all units unclamped so there is no environmental input.

Unless the number of hidden units is exponentially large compared to the number of visible units, it will be impossible to achieve a *perfect* model because even if the network is totally connected the $(v+h-1)(v+h)/2$ weights and $(v+h)$ biases among the v visible and h hidden units will be insufficient to model the 2^v probabilities of the states of the visible units specified by the environment. However, if there are regularities in the environment, and if the network uses its hidden units to capture these regularities, it may achieve a good match to the environmental probabilities.

An information-theoretic measure of the discrepancy between the network's internal model and the environment is

$$G = \sum_\alpha P(V_\alpha) \ln \frac{P(V_\alpha)}{P'(V_\alpha)} \quad (8)$$

where $P(V_\alpha)$ is the probability of the αth state of the visible units when their states are determined by the environment, and $P'(V_\alpha)$ is the corresponding probability when the network is running freely with no environmental input. The G metric, sometimes called the asymmetric divergence or information gain (Kullback, 1959; Renyi, 1962), is a measure of the distance from the distribution given by the $P'(V_\alpha)$ to the distribution given by the $P(V_\alpha)$. G is zero if and only if the distributions are identical; otherwise it is positive.

The term $P'(V_\alpha)$ depends on the weights, and so G can be altered by changing them. To perform gradient descent in G, it is necessary to know the partial derivative of G with respect to each individual weight. In most cross-coupled nonlinear networks it is very hard to derive this quantity, but because of the simple relationships that hold at thermal equilibrium, the partial derivative of G is straightforward to derive for our networks. The probabilities of global states are determined by their energies (Eq. 6) and the energies are determined by the weights (Eq. 1). Using these equations the partial derivative of G (see the appendix) is:

$$\frac{\partial G}{\partial w_{ij}} = -\frac{1}{T}(p_{ij} - p'_{ij}) \tag{9}$$

where p_{ij} is the average probability of two units both being in the *on* state when the environment is clamping the states of the visible units, and p'_{ij}, as in Eq. (7), is the corresponding probability when the environmental input is not present and the network is running freely. (Both these probabilities must be measured at equilibrium.) Note the similarity between this equation and Eq. (7), which shows how changing a weight affects the log probability of a single state.

To minimize G, it is therefore sufficient to observe p_{ij} and p'_{ij} when the network is at thermal equilibrium, and to change each weight by an amount proportional to the difference between these two probabilities:

$$\Delta w_{ij} = \epsilon(p_{ij} - p'_{ij}) \tag{10}$$

where ϵ scales the size of each weight change.

A surprising feature of this rule is that it uses only *locally available* information. The change in a weight depends only on the behavior of the two units it connects, even though the change optimizes a global measure, and the best value for each weight depends on the values of all the other weights. If there are no hidden units, it can be shown that G-space is concave (when viewed from above) so that simple gradient descent will not get trapped at poor local minima. With hidden units, however, there can be local minima that correspond to different ways of using the hidden units to represent the higher-order constraints that are implicit in the probability distribution of environmental vectors. Some techniques for handling these more complex G-spaces are discussed in the next section.

Once G has been minimized the network will have captured as well as possible the regularities in the environment, and these regularities will be enforced when performing completion. An alternative view is that the network, in minimizing G, is finding the set of weights that is most likely to have generated the set of environmental vectors. It can be shown that maximizing this likelihood is mathematically equivalent to minimizing G (Peter Brown, personal communication, 1983).

3.2 Controlling the Learning

There are a number of free parameters and possible variations in the learning algorithm presented above. As well as the size of ϵ, which determines the size of each step taken for gradient descent, the lengths of time over which p_{ij} and p'_{ij} are estimated have a significant impact on the learning process. The values employed for the simulations presented here were selected primarily on the basis of empirical observations.

A practical system which estimates p_{ij} and p'_{ij} will necessarily have some noise in the estimates, leading to occasional "uphill steps" in the value of G. Since hidden units in a network can create local minima in G, this is not necessarily a liability. The effect of the noise in the estimates can be reduced, if desired, by using a small value for ϵ or by collecting statistics for a longer time, and so it is relatively easy to implement an annealing search for the minimum of G.

The objective function G is a metric that specifies how well two probability distributions match. Problems arise if an environment specifies that only a small subset of the possible patterns over the visible units ever occur. By default, the unmentioned patterns must occur with probability zero, and the only way a Boltzmann Machine running at a non-zero temperature can guarantee that certain configurations *never* occur is to give those configurations infinitely high energy, which requires infinitely large weights.

One way to avoid this implicit demand for infinite weights is to occasionally provide "noisy" input vectors. This can be done by filtering the "correct" input vectors through a process that has a small probability of reversing each of the bits. These noisy vectors are then clamped on the visible units. If the noise is small, the correct vectors will dominate the statistics, but every vector will have some chance of occurring and so infinite energies will not be needed. This "noisy clamping" technique was used for all the examples presented here. It works quite well, but we are not entirely satisfied with it and have been investigating other methods of preventing the weights from growing too large when only a few of the possible input vectors ever occur.

The simulations presented in the next section employed a modification of the obvious steepest descent method implied by Eq. (10). Instead of changing w_{ij} by an amount proportional to $p_{ij} - p'_{ij}$, it is simply incremented by a fixed "weight-step" if $p_{ij} > p'_{ij}$ and decremented by the same amount if $p_{ij} < p'_{ij}$. The advantage of this method over steepest descent is that it can cope with wide variations in the first and second derivatives of G. It can make significant progress on dimensions where G changes gently without taking very large divergent steps on dimensions where G falls rapidly and then rises rapidly again. There is no suitable value for the ϵ in Eq. (10) in such cases. Any value large enough to allow progress along the gently sloping floor of a ravine will cause divergent oscillations up and down the steep sides of the ravine.[3]

4. THE ENCODER PROBLEM

The "encoder problem" (suggested to us by Sanjaya Addanki) is a simple abstraction of the recurring task of communicating information among various components of a parallel network. We have used this problem to test out the learning algorithm because it is clear what the optimal solution is like and it is nontrivial to discover it. Two groups of visible units, designated V_1 and V_2, represent two systems that wish to communicate their states. Each group has v units. In the simple formulation we consider here, each group has only one unit on at a time, so there are only v different states of each group. V_1 and V_2 are not connected directly but both are connected to a group of h hidden units H, with $h < v$ so H may act as a limited capacity bottleneck through which information about the states of V_1 and V_2 must be squeezed. Since all simulations began with all weights set to zero, finding a solution to such a problem requires that the two visible groups come to agree upon the meanings of a set of codes without any *a priori* conventions for communication through H.

To permit perfect communication between the visible groups, it must be the case that $h \geq log_2 v$. We investigated minimal cases in which $h = log_2 v$, and cases when h was somewhat larger than $log_2 v$. In all cases, the environment for the network consisted of v equiprobable vectors of length $2v$ which specified that one unit in V_1 and the corresponding unit in V_2 should be on together with all other units off. Each visible group is completely connected internally and each is completely connected to H, but the units in H are not connected to each other.

Because of the severe speed limitation of simulation on a sequential machine, and because the learning requires many annealings, we have primarily experimented with small versions of the encoder problem. For example, Figure 2 shows a good solution to a "*4-2-4*" encoder problem in which $v = 4$ and $h = 2$. The interconnections between the visible groups and H have developed a binary coding—each visible unit causes a different pattern of *on* and *off* states in the units of H, and corresponding units in V_1 and V_2 support identical patterns in H. Note how the bias of the second unit of V_1 and V_2 is positive to compensate for the fact that the code which represents that unit has all the H units turned off.

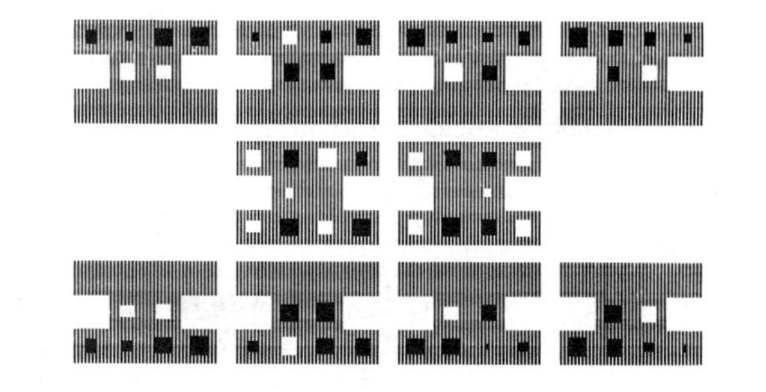

Figure 2. A solution to an encoder problem. The link weights are displayed using a recursive notation. Each unit is represented by a shaded 1-shaped box; from top to bottom the rows of boxes represent groups V_1, H, and V_2. Each shaded box is a map of the entire network, showing the strengths of that unit's connections to other units. At each position in a box, the size of the white (positive) or black (negative) rectangle indicates the magnitude of the weight. In the position that would correspond to a unit connecting to itself (the second position in the top row of the second unit in the top row, for example), the bias is displayed. All connections between units appear twice in the diagram, once in the box for each of the two units being connected. For example, the black square in the top right corner of the leftmost unit of V_1 represents the same connection as the black square in the top left corner of the rightmost unit of V_1. This connection has a weight of -30.

4.1. The 4-2-4 Encoder

The experiments on networks with $v = 4$ and $h = 2$ were performed using the following learning cycle:

1. *Estimation of* p_{ij}: Each environmental vector in turn was clamped over the visible units. For each environmental vector, the network was allowed to reach equilibrium twice. Statistics about how often pairs of units were both on together were gathered at equilibrium. To prevent the weights from growing too large we used the "noisy" clamping technique described in Section 3.2. Each *on* bit of a clamped vector was set to *off* with a probability of 0.15 and each *off* bit was set to *on* with a probability of 0.05.
2. *Estimation of* p'_{ij}: The network was completely unclamped and allowed to reach equilibrium at a temperature of 10. Statistics about

[3] The problem of finding a suitable value for ϵ disappears if one performs a line search for the lowest value of G along the current direction of steepest descent, but line searches are inapplicable in this case. *Only* the local gradient is available. There are bounds on the second derivative that can be used to pick conservative values of ϵ (Mark Derthick, personal communication, 1984), and methods of this kind are currently under investigation.

co-occurrences were then gathered for as many annealings as were used to estimate p_{ij}.

3. *Updating the weights:* All weights in the network were incremented or decremented by a fixed weight-step of 2, with the sign of the increment being determined by the sign of $p_{ij} - p'_{ij}$.

When a settling to equilibrium was required, all the unclamped units were randomized with equal probability on or off (corresponding to raising the temperature to infinity), and then the network was allowed to run for the following times at the following temperatures: [2@20, 2@15, 2@12, 4@10].[4] After this annealing schedule it was assumed that the network had reached equilibrium, and statistics were collected at a temperature of 10 for 10 units of time.

We observed three main phases in the search for the global minimum of *G*, and found that the occurrence of these phases was relatively insensitive to the precise parameters used. The first phase begins with all the weights set to zero, and is characterized by the development of negative weights throughout most of the network, implementing two winner-take-all networks that model the simplest aspect of the environmental structure—only one unit in each visible group is normally active at a time. In a *4-2-4* encoder, for example, the number of possible patterns over the visible units is 2^8. By implementing a winner-take-all network among each group of four this can be reduced to 4×4 low energy patterns. Only the final reduction from 2^4 to 2^2 low energy patterns requires the hidden units to be used for communicating between the two visible groups. Figure 3a shows a *4-2-4* encoder network after four learning cycles.

Although the hidden units are exploited for inhibition in the first phase, the lateral inhibition task can be handled by the connections within the visible groups alone. In the second phase, the hidden units begin to develop positive weights to some of the units in the visible groups, and they tend to maintain symmetry between the sign and approximate magnitude of a connection to a unit in V_1 and the corresponding unit in V_2. The second phase finishes when every hidden unit has significant connection weights to each unit in V_1 and analogous weights to each unit in V_2, and most of the different codes are being used, but there are some codes that are used more than once and some not at all. Figure 3b shows the same network after 60 learning cycles.

Occasionally, all the codes are being used at the end of the second phase in which case the problem is solved. Usually, however, there is a third and longest phase during which the learning algorithm sorts out the remaining conflicts and finds a global minimum. There are two basic mechanisms involved in the sorting out process. Consider the conflict between the first and fourth units in Figure 3b, which are both employing the code $< -, + >$. When the system is running without environmental input, the two units will be on together quite frequently. Consequently, $p'_{1,4}$ will be higher than $p_{1,4}$ because the environmental input tends to prevent the two units from being on together. Hence, the learning algorithm keeps decreasing the weight of the connection between the first and fourth units in each group, and they come to inhibit each other strongly. (This effect explains the variations in inhibitory weights in Figure 2. Visible units with similar codes are the ones that inhibit each other strongly.) Visible units thus compete for "territory" in the space of possible codes, and this repulsion effect causes codes to migrate away from similar neighbors. In addition to the repulsion effect, we observed another process that tends to eventually bring the unused codes adjacent (in terms of hamming distance) to codes that are involved in a conflict. The mechanics of this process are somewhat subtle and we do not take the time to expand on them here.

The third phase finishes when all the codes are being used, and the weights then tend to increase so that the solution locks in and remains stable against the fluctuations caused by random variations in the co-occurrence statistics. (Figure 2 is the same network shown in Figure 3, after 120 learning cycles.)

In 250 different tests of the *4-2-4* encoder, it always found one of the global minima, and once there it remained there. The median time required to discover four different codes was 110 learning cycles. The longest time was 1810 learning cycles.

4.2. The 4-3-4 Encoder

A variation on the binary encoder problem is to give *H* more units than are absolutely necessary for encoding the patterns in V_1 and V_2. A simple example is the *4-3-4* encoder which was run with the same parameters as the *4-2-4* encoder. In this case the learning algorithm quickly finds four different codes. Then it always goes on to modify the codes so that they are optimally spaced out and no pair differ by only a single bit, as shown in Figure 4. The median time to find four well-spaced codes was 270 learning cycles and the maximum time in 200 trials was 1090.

4.3. The 8-3-8 Encoder

With $v = 8$ and $h = 3$ it took many more learning cycles to find all 8 three-bit codes. We did 20 simulations, running each for 4000 learning cycles using the same parameters as for the *4-2-4* case (but with a probability of 0.02 of reversing each *off* unit during noisy clamping). The algorithm found all 8

[4] One unit of time is defined as the time required for each unit to be given, on average, one chance to change its state. This means that if there are *n* unclamped units, a time period of 1 involves *n* random probes in which some unit is given a chance to change its state.

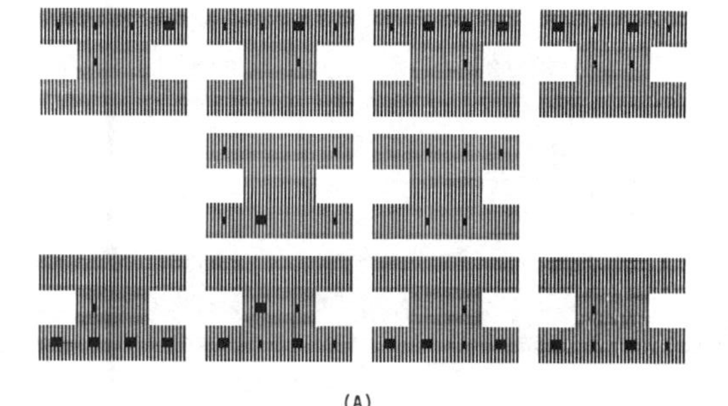

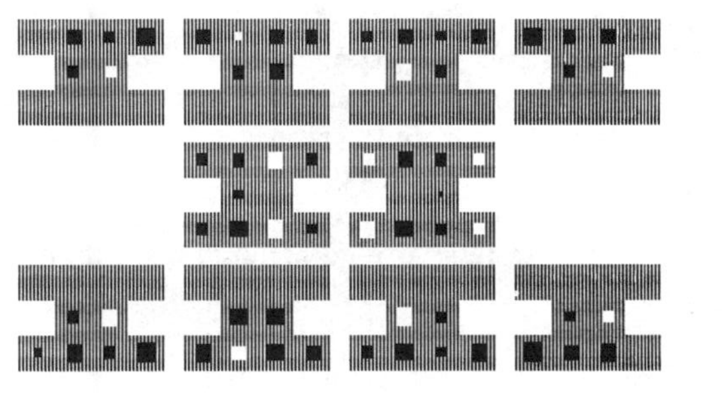

Figure 3. Two phases in the development of the perfect binary encoding shown in Figure 2. The weights are shown (A) after 4 learning trials and (B) after 60 learning trials.

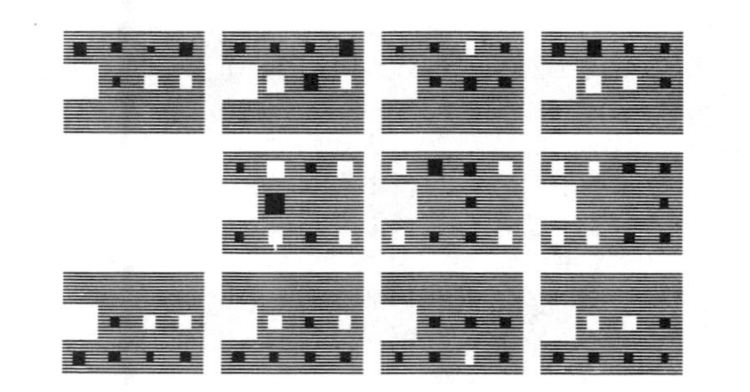

Figure 4. A 4-3-4 encoder that has developed optimally spaced codes.

codes in 16 out of 20 simulations and found 7 codes in the rest. The median time to find 7 codes was 210 learning cycles and the median time to find all 8 was 1570 cycles.

The difficulty of finding all 8 codes is not surprising since the fraction of the weight space that counts as a solution is much smaller than in the *4-2-4* case. Sets of weights that use 7 of the 8 different codes are found fairly rapidly and they constitute local minima which are far more numerous than the global minima and have almost as good a value of G. In this type of G-space, the learning algorithm must be carefully tuned to achieve a global minimum, and even then it is very slow. We believe that the G-spaces for which the algorithm is well-suited are ones where there are a great many possible solutions and it is not essential to get the very best one. For large networks to learn in a reasonable time, it may be necessary to have enough units and weights and a liberal enough specification of the task so that no single unit or weight is essential. The next example illustrates the advantages of having some spare capacity.

4.4. The 40-10-40 Encoder

A somewhat larger example is the *40-10-40* encoder. The 10 units in H are almost twice the theoretical minimum, but H still acts as a limited bandwidth bottleneck. The learning algorithm works well on this problem. Figure 5 shows its performance when given a pattern in V_1 and required to settle to the corresponding pattern in V_2. Each learning cycle involved annealing once with each of the 40 environmental vectors clamped, and the same number of times without clamping. The final performance asymptotes at 98.6% correct.

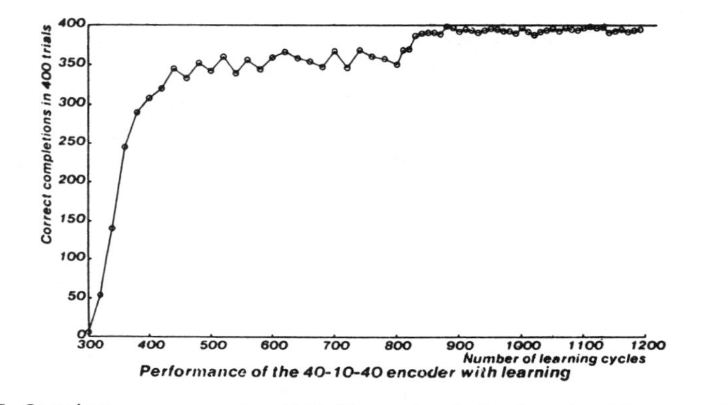

Figure 5. Completion accuracy of a 40-10-40 encoder during learning. The network was tested by clamping the states of the units in V_1 and letting the remainder of the network reach equilibrium. If just the correct unit was on in V_2, the test was successful. This was repeated 10 times for each of the 40 units in V_1. For the first 300 learning cycles the network was run without connecting up the hidden units. This ensured that each group of 40 visible units developed enough lateral inhibition to implement an effective winner-take-all network. The hidden units were then connected up and for the next 500 learning cycles we used "noisy" clamping, switching *on* bits to *off* with a probability of 0.1 and *off* bits to *on* with a probability of 0.0025. After this we removed the noise and this explains the sharp rise in performance after 800 cycles. The final performance asymptotes at 98.6% correct.

The codes that the network selected to represent the patterns in V_1 and V_2 were all separated by a hamming distance of at least 2, which is very unlikely to happen by chance. As a test, we compared the weights of the connections between visible and hidden units. Each visible unit has 10 weights connecting it to the hidden units, and to avoid errors, the 10 dimensional weight vectors for two different visible units should not be too similar. The cosine of the angle between two vectors was used as a measure of similarity, and no two codes had a similarity greater than 0.73, whereas many pairs had similarities of 0.8 or higher when the same weights were randomly rearranged to provide a control group for comparison.

To achieve good performance on the completion tests, it was necessary to use a very gentle annealing schedule during testing. The schedule spent twice as long at each temperature and went down to half the final temperature of the schedule used during learning. As the annealing was made faster, the error rate increased, thus giving a very natural speed/accuracy trade-off. We have not pursued this issue any further, but it may prove fruitful because some of the better current models of the speed/accuracy trade-off in human reaction time experiments involve the idea of a biased random walk (Ratcliff, 1978), and the annealing search gives rise to similar underlying mathematics.

5. REPRESENTATION IN PARALLEL NETWORKS

So far, we have avoided the issue of how complex concepts would be represented in a Boltzmann machine. The individual units stand for "hypotheses," but what is the relationship between these hypotheses and the kinds of concepts for which we have words? Some workers suggest that a concept should be represented in an essentially "local" fashion: The activation of one or a few computing units is the representation for a concept (Feldman & Ballard, 1982); while others view concepts as "distributed" entities: A particular pattern of activity over a large group of units represents a concept, and different concepts corresponds to *alternative* patterns of activity over the same group of units (Hinton, 1981).

One of the better arguments in favor of local representations is their inherent modularity. Knowledge about relationships between concepts is localized in specific connections and is therefore easy to add, remove, and modify, if some reasonable scheme for forming hardware connections can be found (Fahlman, 1980; Feldman, 1982). With distributed representations, however, the knowledge is diffuse. This is good for tolerance to local hardware damage, but it appears to make the design of modules to perform specific functions much harder. It is particularly difficult to see how new distributed representations of concepts could originate spontaneously.

In a Boltzmann machine, a distributed representation corresponds to an energy minimum, and so the problem of creating a good collection of distributed representations is equivalent to the problem of creating a good "energy landscape." The learning algorithm we have presented is capable of solving this problem, and it therefore makes distributed representations considerably more plausible. The diffuseness of any one piece of knowledge is no longer a serious objection, because the mathematical simplicity of the Boltzmann distribution makes it possible to manipulate all the diffuse local weights in a coherent way on the basis of purely local information. The formation of a simple set of distributed representations is illustrated by the encoder problems.

5.1. Communicating Information between Modules

The encoder problem examples also suggest a method for communicating symbols between various components of a parallel computational network. Feldman and Ballard (1982) present sketches of two implementations for this task; using the example of the transmission of the concept "wormy apple" from where it is recognized in the perceptual system to where the phrase "wormy apple" can be generated by the speech system. They argue that there appears to be only two ways that this could be accomplished. In the first method, the perceptual information is encoded into a set of symbols that are then transmitted as messages to the speech system, where they are decoded into a form suitable for utterance. In this case, there would be a set of general-purpose communciation lines, analogous to a bus in a conventional computer, that would be used as the medium for all such messages from the visual system to the speech system. Feldman and Ballard describe the problems with such a system as:

- Complex messages would presumably have to be transmitted sequentially over the communication lines.
- Both sender and receiver would have to learn the common code for each new concept.
- The method seems biologically implausible as a mechanism for the brain.

The alternative implementation they suggest requires an individual, dedicated hardware pathway for each concept that is communicated from the perceptual system to the speech system. The idea is that the simultaneous activation of "apple" and "worm" in the perceptual system can be transmitted over private links to their counterparts in the speech system. The critical issues for such an implementation are having the necessary connections available between concepts, and being able to establish new connection pathways as new concepts are learned in the two systems. The main point of this approach is that the links between the computing units carry simple, nonsymbolic information such as a single activation level.

The behavior of the Boltzmann machine when presented with an encoder problem demonstrates a way of communicating concepts that largely combines the best of the two implementations mentioned. Like the second approach, the computing units are small, the links carry a simple numeric value, and the computational and connection requirements are within the range of biological plausibility. Like the first approach, the architecture is such that many different concepts can be transmitted over the same communication lines, allowing for effective use of limited connections. The learning of new codes to represent new concepts emerges automatically as a cooperative process from the *G*-minimization learning algorithm.

6. CONCLUSION

The application of statistical mechanics to constraint satisfaction searches in parallel networks is a promising new area that has been discovered independently by several other groups (Geman & Geman, 1983; Smolensky, 1983). There are many interesting issues that we have only mentioned in passing. Some of these issues are discussed in greater detail elsewhere: Hinton and Sejnowski (1983b) and Geman and Geman (1983) describe the relation to Bayesian inference and to more conventional relaxation techniques; Fahlman, Hinton, and Sejnowski (1983) compare Boltzmann machines with some alternative parallel schemes, and discuss some knowledge representation issues. An expanded version of this paper (Hinton, Sejnowski, & Ackley, 1984) presents this material in greater depth and discusses a number of related issues such as the relationship to the brain and the problem of sequential behavior. It also shows how the probabilistic decision function could be realized using gaussian noise, how the assumptions of symmetry in the physical connections and of no time delay in transmission can be relaxed, and describes results of simulations on some other tasks.

Systems with symmetric weights form an interesting class of computational device because their dynamics is governed by an energy function.[5] This is what makes it possible to analyze their behavior and to use them for iterative constraint satisfaction. In their influential exploration of perceptrons, Minsky and Papert (1968, p. 231) concluded that: "Multilayer machines with loops clearly open up all the questions of the general theory of automata." Although this statement is very plausible, recent developments suggest that it may be misleading because it ignores the symmetric case, and it seems to have led to the general belief that it would be impossible to find powerful learning algorithms for networks of perceptron-like elements.

[5] One can easily write down a similar energy function for asymmetric networks, but this energy function does not govern the behavior of the network when the links are given their normal causal interpretation.

We believe that the Boltzmann Machine is a simple example of a class of interesting stochastic models that exploit the close relationship between Boltzmann distributions and information theory.

> All of this will lead to theories [of computation] which are much less rigidly of an all-or-none nature than past and present formal logic. They will be of a much less combinatorial, and much more analytical, character. In fact, there are numerous indications to make us believe that this new system of formal logic will move closer to another discipline which has been little linked in the past with logic. This is thermodynamics, primarily in the form it was received from Boltzmann, and is that part of theoretical physics which comes nearest in some of its aspects to manipulating and measuring information.
>
> (John Von Neumann, *Collected Works* Vol. 5, p. 304)

APPENDIX: DERIVATION OF THE LEARNING ALGORITHM

When a network is free-running at equilibrium the probability distribution over the visible units is given by

$$P'(V_\alpha) = \sum_\beta P'(V_\alpha \wedge H_\beta) = \frac{\sum_\beta e^{-E_{\alpha\beta}/T}}{\sum_{\lambda\mu} e^{-E_{\lambda\mu}/T}} \tag{11}$$

where V_α is a vector of states of the visible units, H_β is a vector of states of the hidden units, and $E_{\alpha\beta}$ is the energy of the system in state $V_\alpha \wedge H_\beta$

$$E_{\alpha\beta} = -\sum_{i<j} w_{ij} s_i^{\alpha\beta} s_j^{\alpha\beta}.$$

Hence,

$$\frac{\partial e^{-E_{\alpha\beta}/T}}{\partial w_{ij}} = \frac{1}{T} s_i^{\alpha\beta} s_j^{\alpha\beta} e^{-E_{\alpha\beta}/T}.$$

Differentiating (11) then yields

$$\frac{\partial P'(V_\alpha)}{\partial w_{ij}} = \frac{\frac{1}{T}\sum_\beta e^{-E_{\alpha\beta}/T} s_i^{\alpha\beta} s_j^{\alpha\beta}}{\sum_{\alpha\beta} e^{-E_{\alpha\beta}/T}} - \frac{\sum_\beta e^{-E_{\alpha\beta}/T} \frac{1}{T} \sum_{\lambda\mu} e^{-E_{\lambda\mu}/T} s_i^{\lambda\mu} s_j^{\lambda\mu}}{\left(\sum_{\lambda\mu} e^{-E_{\lambda\mu}/T}\right)^2}$$

$$= \frac{1}{T}\left[\sum_{\beta} P'(V_\alpha \wedge H_\beta) s_i^{\alpha\beta} s_j^{\alpha\beta} - P'(V_\alpha) \sum_{\lambda\mu} P'(V_\lambda \wedge H_\mu) s_i^{\lambda\mu} s_j^{\lambda\mu}\right].$$

This derivative is used to compute the gradient of the G-measure

$$G = \sum_{\alpha} P(V_\alpha) \ln \frac{P(V_\alpha)}{P'(V_\alpha)}$$

where $P(V_\alpha)$ is the clamped probability distribution over the visible units and is independent of w_{ij}. So

$$\frac{\partial G}{\partial w_{ij}} = -\sum_{\alpha} \frac{P(V_\alpha)}{P'(V_\alpha)} \frac{\partial P'(V_\alpha)}{\partial w_{ij}}$$

$$= -\frac{1}{T}\sum_{\alpha} \frac{P(V_\alpha)}{P'(V_\alpha)}\left[\sum_{\beta} P'(V_\alpha \wedge H_\beta) s_i^{\alpha\beta} s_j^{\alpha\beta} - P'(V_\alpha) \sum_{\lambda\mu} P'(V_\lambda \wedge H_\mu) s_i^{\lambda\mu} s_j^{\lambda\mu}\right].$$

Now,

$$P(V_\alpha \wedge H_\beta) = P(H_\beta | V_\alpha) P(V_\alpha),$$

$$P'(V_\alpha \wedge H_\beta) = P'(H_\beta | V_\alpha) P'(V_\alpha),$$

and

$$P'(H_\beta | V_\alpha) = P(H_\beta | V_\alpha). \qquad (12)$$

Equation (12) holds because the probability of a hidden state given some visible state must be the same in equilibrium whether the visible units were clamped in that state or arrived there by free-running. Hence,

$$P'(V_\alpha \wedge H_\beta) \frac{P(V_\alpha)}{P'(V_\alpha)} = P(V_\alpha \wedge H_\beta).$$

Also,

$$\sum_{\alpha} P(V_\alpha) = 1.$$

Therefore,

$$\frac{\partial G}{\partial w_{ij}} = -\frac{1}{T}[p_{ij} - p'_{ij}]$$

where

$$p_{ij} \stackrel{\text{def}}{=} \sum_{\alpha\beta} P(V_\alpha \wedge H_\beta) s_i^{\alpha\beta} s_j^{\alpha\beta}$$

and

$$p'_{ij} \stackrel{\text{def}}{=} \sum_{\lambda\mu} P'(V_\lambda \wedge H_\mu) s_i^{\lambda\mu} s_j^{\lambda\mu}$$

as given in (9).

The Boltzmann Machine learning algorithm can also be formulated as an input-output model. The visible units are divided into an input set I and an output set O, and an environment specifies a set of conditional probabilities of the form $P(O_\beta | I_\alpha)$. During the "training" phase the environment clamps both the input and output units, and p_{ij}s are estimated. During the "testing" phase the input units are clamped and the output units and hidden units free-run, and p'_{ij}s are estimated. The appropriate G measure in this case is

$$G = \sum_{\alpha\beta} P(I_\alpha \wedge O_\beta) \ln \frac{P(O_\beta | I_\alpha)}{P'(O_\beta | I_\alpha)}$$

Similar mathematics apply in this formulation and $\partial G / \partial w_{ij}$ is the same as before.

REFERENCES

Berliner, H. J., & Ackley, D. H. (1982, August). The QBKG system: Generating explanations from a non-discrete knowledge representation. *Proceedings of the National Conference on Artificial Intelligence AAAI-82,* Pittsburgh, PA, 213–216.

Binder, K. (Ed.) (1978). *The Monte-Carlo method in statistical physics.* New York: Springer-Verlag.

Fahlman, S. E. (1980, June). The Hashnet Interconnection Scheme. (Tech. Rep. No. CMU-CS-80-125), Carnegie-Mellon University, Pittsburgh, PA.

Fahlman, S. E., Hinton, G. E., & Sejnowski, T. J. (1983, August). Massively parallel architectures for AI: NETL, Thistle, and Boltzmann Machines. *Proceedings of the National Conference on Artificial Intelligence AAAI-83,* Washington, DC, 109–113.

Feldman, J. A. (1982). Dynamic connections in neural networks. *Biological Cybernetics, 46,* 27–39.

Feldman, J. A., & Ballard, D. H. (1982). Connectionist models and their properties. *Cognitive Science, 6,* 205–254.

Geman, S., & Geman, D. (1983). Stochastic relaxation, Gibbs distributions, and the Bayesian restoration of images. Unpublished manuscript.

Grimson, W. E. L. (1981). *From images to surfaces.* Cambridge, MA: MIT Press.

Hinton, G. E. (1977). *Relaxation and its role in vision.* Unpublished doctoral dissertation, University of Edinburgh. Described in D. H. Ballard & C. M. Brown (Eds.), *Computer Vision.* Englewood Cliffs, NJ: Prentice-Hall, 408–430.

Hinton, G. E. (1981). Implementing semantic networks in parallel hardware. In G. E. Hinton & J. A. Anderson (Eds.), *Parallel Models of Associative Memory*. Hillsdale, NJ: Erlbaum.

Hinton, G. E., & Anderson, J. A. (1981). *Parallel models of associative memory*. Hillsdale, NJ: Erlbaum.

Hinton, G. E., & Sejnowski, T. J. (1983a, May). Analyzing cooperative computation. *Proceedings of the Fifth Annual Conference of the Cognitive Science Society*. Rochester, NY.

Hinton, G. E., & Sejnowski, T. J. (1983b, June). Optimal perceptual inference. *Proceedings of the IEEE Computer Society Conference on Computer Vision and Pattern Recognition*. Washington, DC, pp. 448–453.

Hinton, G. E., Sejnowski, T. J., & Ackley, D. H. (1984, May). *Boltzmann Machines: Constraint satisfaction networks that learn*. (Tech. Rep. No. CMU-CS-84-119). Pittsburgh, PA: Carnegie-Mellon University.

Hopfield, J. J. (1982). Neural networks and physical systems with emergent collective computational abilities. *Proceedings of the National Academy of Sciences USA*, *79*, 2554–2558.

Kirkpatrick, S., Gelatt, C. D., & Vecchi, M. P. (1983). Optimization by simulated annealing. *Science*, *220*, 671–680.

Kullback, S. (1959). *Information theory and statistics*. New York: Wiley.

Metropolis, N., Rosenbluth, A., Rosenbluth, M., Teller, A., & Teller, E. (1953). Equation of state calculations for fast computing machines. *Journal of Chemical Physics*, *6*, 1087.

Minsky, M., & Papert, S. (1968). *Perceptrons*. Cambridge, MA: MIT Press.

Newell, A. (1982). *Intellectual issues in the history of artificial intelligence*. (Tech. Rep. No. CMU-CS-82-142). Pittsburgh, PA: Carnegie-Mellon University.

Newell, A., & Simon, H. A. (1972). *Human problem solving*. Englewood Cliffs, NJ: Prentice-Hall, 1972.

Ratcliff, R. (1978). A theory of memory retrieval. *Psychological Review*, *85*, 59–108.

Renyi, A. (1962). *Probability theory*. Amsterdam: North-Holland.

Rosenblatt, F. (1961). *Principles of neurodynamics: Perceptrons and the theory of brain mechanisms*. Washington, DC: Spartan.

Smolensky, P. (1983, August). Schema selection and stochastic inference in modular environments. *Proceedings of the National Conference on Artificial Intelligence AAAI-83*, Washington, DC. 109–113.

Terzopoulos, D. (1984). *Multiresolution computation of visible-surface representations*. Unpublished doctoral dissertation, MIT, Cambridge, MA.

Waltz, D. L. (1975). Understanding line drawings of scenes with shadows. In P. Winston (Ed.), *The Psychology of Computer Vision*. New York: McGraw-Hill.

Winston, P. H. (1984). *Artificial Intelligence*. (2nd ed.) Reading, MA: Addison-Wesley.

Parameter Nets*

Dana H. Ballard
Computer Science Department, University of Rochester, Rochester, NY 14627, U.S.A.

Recommended by Michael Brady

ABSTRACT

This paper describes the nucleus of a connectionist theory of low-level and intermediate-level vision. The theory explains segmentation in terms of massively parallel cooperative computation among intrinsic images and a set of feature networks at different levels of abstraction.

Explaining how parts of an image are perceived as a meaningful whole or gestalt is a problem central to vision. A stepping stone towards a solution is recent work showing how to calculate images of physical parameters from intensity data. Such images are known as intrinsic images, and examples are images of velocity (optical flow), surface orientation, occluding contour, and disparity. Intrinsic images show great promise; they are distinctly easier to work with than the original intensity image, but they are not grouped into objects. A general way in which such groupings can be detected is to represent possible groupings as networks whose nodes signify explicit parameter values. In this case the relation between parts of an intrinsic image and the gestalt parameters can be specified by active, two-way connections between an intrinsic image network and a gestalt parameter network. The active connections will be many-to-one onto parameter value nodes that represent object features. The virtues of the methodology are that it can handle occlusion and noise and can specify intrinsic image boundaries. Furthermore, it can be made practical for high-dimensional feature spaces.

1. Introduction

Vision is the ability to build useful descriptions of objects from images. Realizing this ability is technically difficult since the physical transformations which represent objects as images are degenerate. Information about shape, surface reflectance, motion, texture, and color is confounded in a time-varying array of intensities. Thus a major first step in description-building is to undo these degeneracies and compute image-like representations which make important physical features explicit. These representations have been termed intrinsic images [10]. While they are closer in abstraction to useful object descriptions which also mention features, they are still spatio-temporally indexed. What must also happen is a grouping of the parts of the domains of these functions that represent features that in turn can be related to physical objects.

The process of grouping spatially indexed representations is known as *segmentation*. The result of this process is an explicit representation of a gestalt [40]. A portion of an image may give rise to a segment (gestalt) if its corresponding physical or geometric properties have some appropriate organization. For example, if a connected component of the image has a single color, say red, then it may be seen as a segment. The patch of red normally arises from a physical object's surface reflectance. Usually there are not one but several features which have the same spatial registration. In the case of a moving, red cube, sensors at registered spatial locations have appropriate values for velocity, color, and form. Previously segmentation has been viewed as grouping parts of intensity arrays. We generalize this notion of segmentation to include parts of intrinsic images, and, more importantly, parts of explicit representations of possible features and objects.

In any given image, two factors combine to make segmentation difficult. First, an object may be occluded; one wants to be able to explain how an object is seen as a segment when the features are only partially registered or incomplete. Second, image data is also noisy and many segments are only perceived owing to the combination of weak evidence of several features. The evidence may be so weak that each feature, if viewed in isolation, would be uninterpretable.

Recently much progress has been made on segmentation by adopting a *computational view*. The two key components of this view are: (1) *environmental*—the specification of the physical and geometrical constraints of the world [12, 44]; and (ii) *implementational*—the specification of an abstract computational machine. The environmental component studies parametrizations of the world and how they are related through *constraints*. Since these constraints are usually in the form of large sets of relations, one must also specify mathematical methods for solving them. But even after the mathematical methods have been discovered, there is more work to do. One must also design architectures that implement the mathematical solutions in real time. A parallel method for handling constraints that addresses all these issues, but especially the architecture issue, is the Hough transform. This transform is particularly useful in relating intrinsic images to global (non-spatially indexed) features, i.e., capturing an explicit notion of a gestalt. The Hough transform has been regarded as limited to analyzing algebraic functions of few variables; however, recent work has overcome these restrictions with the result that it now has major implications in understanding segmentation [2, 3, 5, 15, 54]. The Hough transform is also naturally insensitive to the two major environmental difficulties: occlusion and noise. The Hough transform has typically been

*This research was supported in part by NIH under Public Health Service Grant 2R01 HL21253-04 and in part by NSF under Grant MCS 8203290.

implemented on von Neumann machines; we show that it has special virtues in specifying connections in a parallel, non-von Neumann machine architecture.

Interest in the underlying abstract machine architecture that can perform vision computations has waxed and waned. At one time it was thought that one could accomplish segmentation on a general-purpose computer. More recently it was believed that segmentation could be analyzed as an abstract physics/geometrical problem (the environmental tack) without considering the issues of implementing the solution in hardware. In a sense this belief was that competence issues arising from well-defined solutions in mathematical physics could be separated from performance issues involved in a specific implementation. This paper suggests that special-purpose parallel architectures must be developed to accomplish general-purpose vision and that innovative implementations of mathematical solutions will be required.

The computational viewpoint has blurred the sharp distinctions that used to exist between machine and human models. The result is that designers of vision machine architectures are looking increasingly at the evolving knowledge of the brain for insights. Evidence from psychology suggests that the minimum time for human beings to respond to a visual stimulus is approximately 100 ms. If the time constant for a neural unit is approximately 2 ms, this means that about 50 cycles are available for the complete perceptual processing and motor response in this case. The human brain has about 10^{11} neurons, any of which may be involved in the processing. If connections must go through intermediate units, the best strategies involve on the order of the logarithm of the number of neurons. Thus there is little time to do more than link up the right neurons. Arguments like these support *connectionist* theories of brain processing: "The connectionist view of brain and behavior is that all encodings of importance in the brain are in terms of the relative strengths of synaptic connections. The fundamental premise of connectionism is that individual neurons do not transmit large amounts of symbolic information. Instead they compute by being appropriately connected to large numbers of similar units." [21]. Like Paul Revere's signaller, individual units can stand for complex concepts, but the way they communicate information about these concepts is simple. Much effort is being spent on parallel machine architectures that are inspired by connectionist views of human brain architectures [18, 20, 21, 27, 28].

2. The Theory

This paper develops the nucleus of a connectionist theory of low-level and intermediate-level vision that is based on an abstract model of "neural-like" units [20, 21]. To summarize:

> The theory explains segmentation in terms of *massively-parallel* cooperative computation [47, 53, 65] between two groups of interconnected networks. One group, *intrinsic images* [10], can be computed primarily in terms of local constraints. The other, termed *feature spaces*, can be computed primarily in terms of global constraints between itself and the intrinsic images. A feature space is also distinguished from intrinsic image space in that it is not spatio-temporally indexed. Each of these two groups of networks may have many different levels of abstraction. Intrinsic images and feature spaces are collectively called *parameter networks* because they both have a common organization. That is, the network is an organization of basic units, each representing a value of a particular parameter. The basic element of a parameter network is a *parameter unit.* A parameter unit will represent a small range of parameter *values* and has an associated *confidence* between zero and unity. The values are numerical measurements; confidence can be (loosely) thought of as a measure of whether or not the value describes the image.

2.1. A familiar example

Consider the case of detecting lines in edge images. The importance of edges as a preliminary description was recognized by Attneave [1]. Marr [45] called the explicit representation of edges a primal sketch. In our formalism, primal sketch edges are represented as parameter units. If there is an edge at (10, 10) with orientation 30° and length 5, the vector value of the parameter unit representing the edge is $(x, y, \alpha, s) = (10, 10, 30°, 5)$. The associated confidence is a measure of how well the unit describes the input. For edges, confidence may be initialized to normalized edge magnitude. One way a confidence may be increased is if there are nearby edges of the same orientation which align. Thus in Fig. 1 the dark-circled edges in (a) and (b) have the same orientation value but we can be more confident in case (b). Prager [51] and Zucker [64] have shown ways of allowing neighboring aligned edges to increase each other's confidence. Their methods are compatible with the parameter unit formalism.

The primal sketch is an intrinsic image since it represents spatially indexed parameters. Now consider the representation of a feature such as a line. With respect to the edge intrinsic image lines are a global feature. Possible lines can be explicitly represented with units for different parameter values in a line feature space. One nice parametrization represents lines as a distance from the origin ρ and an orientation θ where, for a point (x, y) on the line, $x \cos \theta + y \sin \theta = \rho$. In our formalism each high confidence edge unit (x, y, α, s) raises the confidence of a corresponding line unit $(\rho, \theta) = (x \cos \alpha + y \sin \alpha, \alpha)$ by virtue of a direct connection to that unit. Fig. 1 (middle) illustrates some representative connections. Fig. 1 (bottom) shows a computer implementation of these two networks. On the left is shown edge units for an image of Rubik's cube; on the right, line units. Edge units are shown as short line segments of

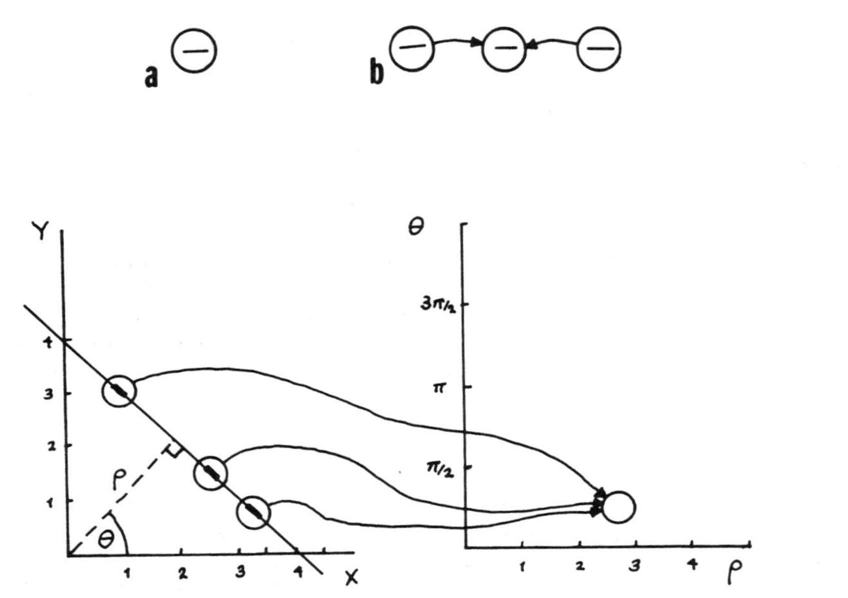

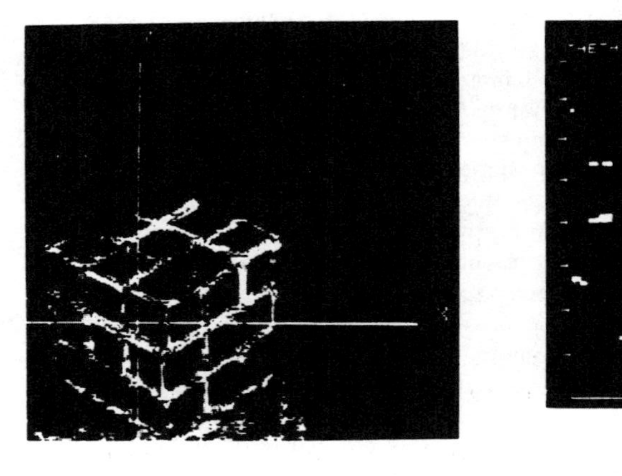

FIG. 1. (Top) (a) An isolated edge has an initial confidence due to direct measurements of intensity, but (b) its confidence can be increased when it is between edges of similar orientations. (Middle) Lines may be represented by a set of line units of different (ρ, θ) values. Connections from edge units to line units allow colinear edge units to raise the confidence of the appropriate line unit. (Bottom) Results of a simulation on an image of Rubik's cube. Edge units are on the left; (ρ, θ) units are on the right.

the appropriate orientation; line units are shown as cells in a (ρ, θ) array. In both cases confidence is proportional to intensity. If many edge units are aligned, high-confidence line units will result. A high-confidence line unit is a simple example of a segment. The line detection example shows two parameter networks: the primal sketch network of edge units and the line parameter network of (ρ, θ) units. The idea of transforming edge input to a line parameter space will be recognized as the Hough transform [17]. However, the parameter network implementation has several attributes that distinguish it from the von Neumann implementation:

- computations are performed in parallel;
- it is implemented in units that represent the values of parameters;
- the meaning of the units is determined by the connection patterns.

In general:

- each value unit is connected to a *subset* of other value units, and can alter only the confidence of those units;
- underlying *physical principles* determine the appropriate connection subsets and abstraction levels;
- the confidence updating is done by non-linear multi-layer relaxation involving many different parameter nets;
- a segment *is a set of high-confidence units which are connected*;
- at any instant, only a *subset* of the value units have high confidence values.

Although the line detection example is the nub of the theory, many other elements are necessary to extend this idea to low-level vision. Several of these are not manifested in the line detection example owing to its simplicity. The principal elements of the theory are the following.

(1) *A lowest level of abstraction consisting of registered intrinsic images which are computed simultaneously.* Recent work has shown how to calculate intrinsic images from intensity data. Examples are images of velocity (optical flow) [30, 61, 62], surface orientation [31, 35], occluding contour [51, 53], and disparity [9, 25, 48]. Intrinsic images are in concert with the hypothesis that the visual system builds many intermediate descriptions from image data. These descriptions represent important parameters such as velocity, depth, surface reflectance explicitly, since in the explicit form they are easier to map into object descriptions. Intrinsic images can be computed independently under special conditions, but in general they are interdependent. Parts of one intrinsic image specify boundary conditions for another, and vice versa.

(2) *Intermediate levels of abstraction consisting of non-spatially indexed feature spaces.* If parts of the intrinsic image are organized in some way, this organization can be detected by a general Hough transform technique [2, 17, 38, 50] termed a constraint transform. This is done by describing the organization in terms of more abstract parameters and then mapping the intrinsic image points into parameter space. The transformation will be many-to-one onto parameter values which represent meaningful segments. Its major advantages

are that it is relatively insensitive to occlusion and noise. The transform provides parsimonious explanations of many gestalt grouping phenomena.

(3) *Hierarchies involving several levels of abstraction.* The constraint transform is a way of seeing spatial information as a unit. However, if the unit has a complex structure the mapping from space to unit can be unmanageably complex. A way around this is to introduce units at several levels of abstraction [38, 54]. This reduces a complex transform to several simpler transforms between units at successively higher levels of abstraction.

(4) *Sparse coding techniques.* The rich visual input of any given image is in fact an instance of the many possible image inputs that could have occurred. Our experiments have revealed that the active inputs in abstract representation spaces are normally *extremely sparsely distributed.* This allows the development of many encoding methods to represent high-dimensional feature spaces. Two important techniques are: coarse coding [28] and subspaces or projections [6].

(5) *Coupling between intrinsic images and feature spaces; segmentation as a convergence property of networks.* In general, intrinsic images *cannot be computed* without global parameters. At the same time, these global parameters are what we mean by seeing parts of the intrinsic image as a segment. In these cases the intrinsic image and parameters are said to be *tightly coupled*; although each cannot be computed independently, they can be computed simultaneously [5, 15]. Tight coupling refines the notion of segment to: *that part of the network which converges.* An example is shape from shading: the direction of the light source, a global parameter, must be computed at the same time as the surface normals.

(6) *Focus-of-attention via spatial coupling.* Visual focus-of-attention can be partly explained as the conjunction of two mechanisms: (i) the use of constraint transforms to modify sensor input; and (ii) the *sequential* application of constraint transforms. We reemphasize that our interest here is low-level vision. Focus of attention is interpreted in a narrow sense: visual features which are clear can help the recognition of other features (or perhaps direct eye movements). For example, the texture of a blue textured object may be identified by limiting processing to only blue parts of the image. We do not attempt to explain general plans and goals.

2.2. Parameter networks and constraint satisfaction

Before pursuing the details of the theory further it will help to show how parameter networks can be related to the familiar constraint satisfaction or labeling problem. To develop this relationship, we work through a familiar toy example of map coloring. Consider the map in Fig. 2, with four regions. Each region may be colored with one of the colors shown. This problem is characteristic of a labeling problem in that it may be given a graphical interpretation. Regions are nodes and constraints may be encoded as arcs. The restrictions on colors at certain nodes are unary constraints. A binary constraint is that

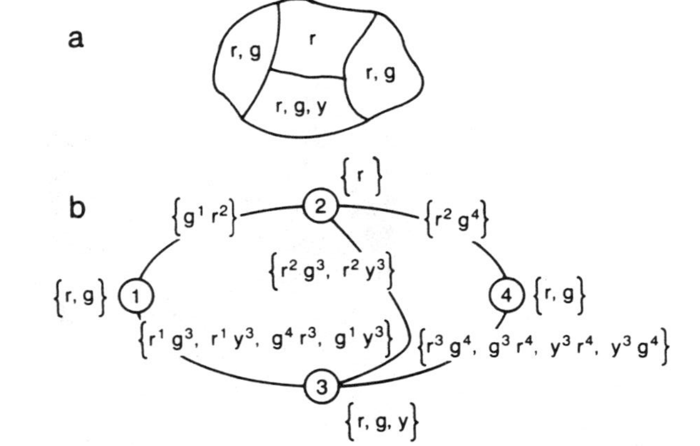

FIG. 2. (a) Four-region map with possible colorings red (r), green (g), and yellow (y) for different regions. (b) The map coloring problem represented as a graph.

adjacent regions may not have the same color. These constraints may *imply* constraints of order greater than two, and Freuder [23] has shown a systematic way of building and propagating such constraints. In the example we will not be systematic but will illustrate the general idea.

From Fig. 2 it is seen that region one cannot be red, as region two can only be red. Similarly, red is not a possibility for regions three and four. This means that corresponding binary pairings involving red labels for regions one, three, and four are ruled out. Since region four can only be green, then region three can only be yellow. This analysis of the constraints results in a stable unique labeling for the regions. With tighter constraints there could have been no possible labeling and looser constraints may have implied several labelings. The lesson of constraint satisfaction is that *local* constraints can imply a *global* labeling.

Now consider the same problem translated to parameter network notation. Each color of each region is a separate value unit. If the confidence of a color unit is high, then that unit represents its corresponding region's color. The constraints are represented as links between units which raise or lower confidences. There are many different ways to do this. We choose to let unary constraints be inhibitory (lower confidences) and binary constraints be excitatory (raise confidences). These links are shown in Fig. 3. For brevity two symmetric links are drawn as a single double-ended link. Links lower or raise confidences proportionate to their own current confidence. While we will not offer a formal proof that the above network converges, informally one can see that if the confidences are initially equal and inhibitory links are weighted at least twice as great as excitatory links, the preponderance of excitatory links at

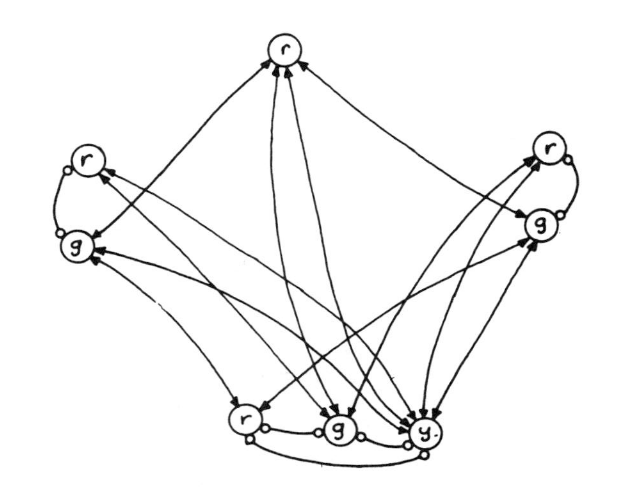

FIG. 3. Map coloring problem cast in parameter networks formalism.

the appropriate units will favor the solution described above. The example has been simulated and has proven surprisingly stable to variations in initial confidence. However, guaranteeing convergence in large networks is in general a tricky problem, and in the near future will probably be tested by simulations of specific examples. If the updating function can be linearized then the cooperative computations can be shown to be equivalent to linear programming [29]. Convergence in the linear case has been extensively analyzed by [33]. Their formulation shows that relaxation labeling is a form of constrained optimization.

In the parameter nets version of constraint satisfaction, labels are explicit units. As we show later, the requirements of vision lead to groups of many thousands of units (labels) and the units are modified through heterogeneous connections. The large number of units postulated make simulations particularly expensive. In our studies we distinguish three levels of abstraction in simulations. At the most basic level, level one, all units and connections are explicitly represented. Such a level is implied by Fig. 3. At level two, units are explicitly represented but connections are simulated with programs. At level three, arrays may be used to represent intrinsic image values. Each of the increasing levels of abstraction leads to more efficient von Neumann simulations at the price of losing detailed behavior of the networks. At level two it is difficult to represent details in the behavior of connections. At level three multiple-valued intrinsic images (analogous to reversible figures) are not possible. Since the focus of this paper is vision, we frequently describe our simulations at levels two and three. The network details are described in [21].

3. Intrinsic Images

Intrinsic images have several important properties. First, an intrinsic parameter is *in registration* with the original viewer-centered intensity image [10, 47], that is, the parameter is indexed by the same spatial coordinates. For example, in the velocity (optical flow) image, one is able to compute at each point in time and for each spatial position a local velocity vector $\boldsymbol{u}(\boldsymbol{x}, t)$. The positional and temporal arguments correspond to those in the image intensity function $f(\boldsymbol{x}, t)$. Fig. 4 shows Horn's example for a rotating sphere [30]. Second, intrinsic images may only be computable over certain *parts* of the image. For example, the image is usually only smoothly shaded in parts. Lastly, over those parts the

FIG. 4. Horn and Schunck's rotating sphere. An optic flow image calculated from primal sketch data. (a)–(c). Four frames of successive intensities. (d) Optic flow image. (Courtesy of B.K.P. Horn)

parameters are *continuously varying*; except for boundary points, there are no abrupt discontinuities with respect to space or time. While intrinsic images are not segmented into parts of objects, they are distinctly easier to segment than the original intensity image.

Somewhat surprisingly, intrinsic images can all be computed in a similar manner. Two constraints, one derived from physical principles and the other from a constraint that the resultant images should be locally smooth, suffice to specify a parallel-iterative algorithm. Table 1 shows this commonality but is not an exhaustive list of approaches. In each algorithm, estimates for the parameters are updated based on local measures until a convergence criterion is satisfied.

The demonstration that intrinsic images can be computed represents a tremendous conceptual advance for our understanding of vision. At the same time, it is important to realize that the algorithms require that the test images be constrained to satisfy most or all of the following conditions:

(i) boundary conditions are pre-specified;

(ii) certain free parameters (such as light source direction in shape-from-shading calculations) must also be pre-specified;

(iii) the variation in the image intensity can be entirely explained by the intrinsic parameter being computed;

(iv) the image resolution is appropriately chosen.

In this paper we show cases in which assumptions (i) and (ii) can be relaxed; the additional information is computed in different parameter networks. Our hope is that parameter networks may also provide a formalism for dealing with (iii) and (iv).

TABLE 1. Intrinsic images

Parameter	Physical constraint	Smoothness constraint
Edge orientation θ	Boundaries are locally linear	Nearby edges should align [51]
Disparity d	If x corresponds to x', then $f(x+\Delta)=f(x'+\Delta)$	Neighboring points should have similar disparities [48]
Surface orientation, θ, φ	$f(x)=R(\theta,\varphi,\theta_s,\varphi_s)$ θ_s, φ_s is the light source direction	$\theta_x^2+\theta_y^2+\varphi_x^2+\varphi_y^2=0$[a] [35]
Optical flow u, v	$df/dt=0$	$u_x^2+u_y^2+v_x^2+v_y^2=0$ [30]

[a]Subscripts x and y denote partial derivatives, e.g., $u_x=\partial u/\partial x$.

3.1. Intrinsic images and parameter networks

Two models have been used to compute intrinsic images: (i) the value unit defined in Section 2 [48, 51]; and (ii) a variable unit [30, 35]. In the first model there is a unit for every value of every variable; in effect the representation has only constants. Value units may have outputs which are confidences between zero and one. In the second model, each unit represents a variable which can take on values (the standard implementation method is to use an array for these units). The output is the value; there is no explicit notion of confidence.

In general the unit/value representation should be sufficient since problems formulated to use variables can be transformed into unit/value problems in the following manner. Suppose x, y, and z satisfy a relation $R(x,y,z)=0$. Let us use a set of values A for x, B for y, and C for z. Where $a\in A$, we would like (in the limit) $\mathrm{Conf}(a)$ to be 1 if there exist $b\in B$ and $c\in C$ such that $\mathrm{Conf}(b)=1$, $\mathrm{Conf}(c)=1$, and $R(a,b,c)=0$. To implement this in a parameter network connect all pairs of $(b,c)\in B\times C$ to a value (a) if $R(a,b,c)=0$. Then starting with initial confidences, increment $\mathrm{Conf}(a)$ if there exist (b,c) such that $R(a,b,c)=0$ and $\mathrm{Conf}(b)+\mathrm{Conf}(c)>$ some threshold. The individual values b and c may be treated similarly. Note that this updating scheme implies a special relationship between connections from b and c to the unit a. This relationship is termed a *conjunctive connection* [21].

4. Feature Spaces

What does it mean to perceive parts of an image as a segment? In our theory, this perception takes place if each of the image parts can have the same set of *parameter values* in a *feature space*. This general idea is illustrated by the following examples.

(1) Parts of a *color* image may be seen as a segment if they have the same hue. In this case the feature space is a space of colors and the parts connect to a common unit representing the common hue.

(2) Parts of an *optical flow* image may be seen as a segment if they are part of a rigid body that is moving. In this case the feature space represents the rigid body motion parameters of translational and rotational velocity and parts of the image connect to a common unit in that space.

(3) Parts of *edge* and *surface orientation* images may be seen as a segment if they are part of the same shape. This is more complicated as there must exist some internal representation of the shape. Given this, part of the feature space represents the transformation (scale, rotation, translation) from the internal representation to the (viewer-centered) image representation. Parts of the image which are seen as the shape have common values for these parameters [2, 6, 28, 58].

A general way of describing this relationship between parts of an intrinsic image and the associated parameters is a connectionist interpretation of the

Hough transform [17, 32, 41, 56]. A nice metaphor for understanding the Hough transform is that of *voting*. Intrinsic image data votes for consistent values of global parameters. The parameter value with the most votes is selected as the global feature. For example, in the case of lines, one can think of edges voting for compatible line parameters. Selecting the maximum vote feature is equivalent to computing the *mode* of a distribution in feature space, and this helps to explain why the Hough transformation is so effective: in vision, it is generally true that most of the input does not fit a particular model at all, but a small portion of it fits the model very well. Contrast this to least squares techniques which have trouble distinguishing model and non-model points. The voting metaphor also makes it easy to understand why the Hough transform is so resilient to occlusion and noise: only a relatively significant number of model points need be present.

In the parameter net architecture the Hough transform has additional importance as it defines connections between units. To highlight this particular point, we term the connection patterns *constraint maps*. The example of line detection used a constraint map between edge units and line units. Individual edge units were connected to appropriate line units. Line units receiving a large number of active inputs from edge units describe lines in the image. As another example of a constraint map, we describe how a patch of red in an image may be seen as a unit. For this to happen, an association is made between red points in the image and the particular value 'red' in a parameter space of colors. There are essentially three dimensions to color space: Fig. 5(a)–(c) shows a red–green–blue encoding for a color image. (Although r–g–b is widely used in computer applications, humans seem to use an opponents-process basis (r–g, y–b, white–black) [34].) The constraint map has three elements: a less abstract network of units $\boldsymbol{a}$, where

$$\boldsymbol{a} = (x, y, \boldsymbol{r}(x, y), b(x, y), g(x, y)),$$

a more abstract network of units given by

$$\boldsymbol{b} = (r, g, b),$$

and the relationship between the two networks of units,

$$f = \boldsymbol{a}' - \boldsymbol{b} \quad \text{where } \boldsymbol{a}' = \text{the last three components of } \boldsymbol{a}.$$

In other words, each spatial color unit is connected to its non-spatial counterpart. The confidence of a red non-spatial unit will be high if there are several high-confidence spatially indexed red units. In our study we used a color cube where individual (*rgb*) color components are represented with four bits each. This means that 16^3 color units are used. Fig. 5 shows a color image as relative intensities of red (a), green (b) and blue (c). In Fig. 5(d), high confidences are indicated by high intensities. This idea was first used with color subspaces by [26, 50].

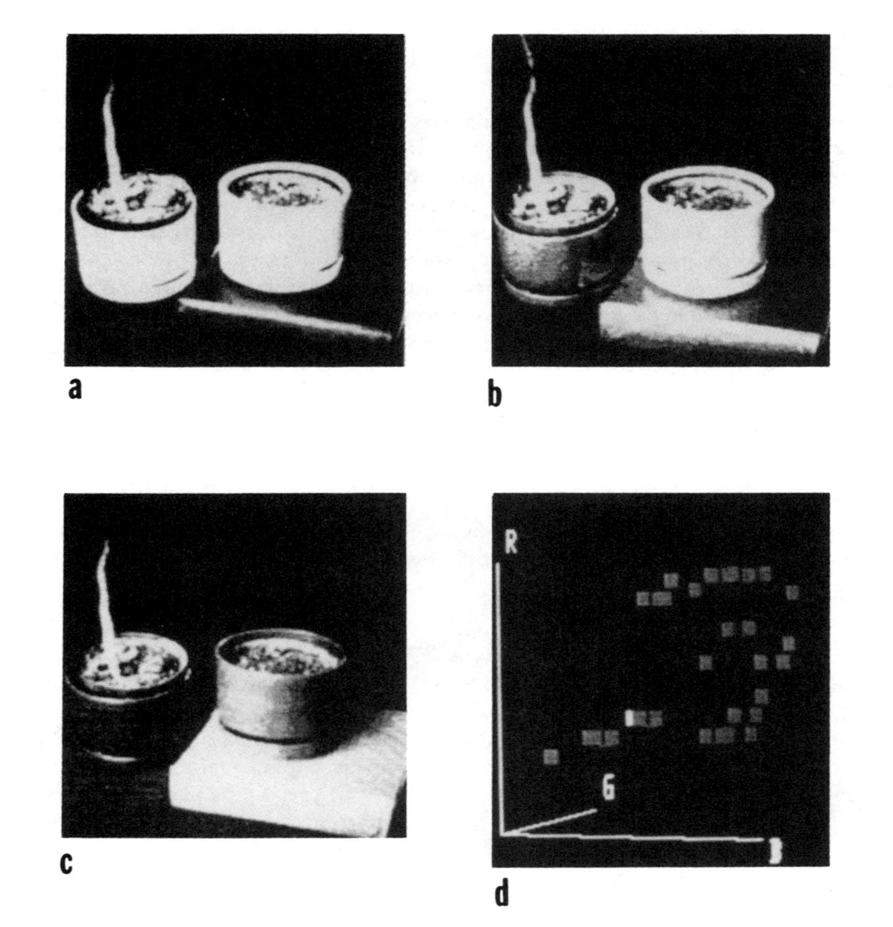

FIG. 5. Color image represented as relative intensities of red (a), green (b), and blue (c). (d) Color units in a 16^3 high-confidence (local maxima) red–green–blue color space.

As shown in Fig. 5(d), the transformation results in a set of *very sparsely distributed* high-confidence feature space units. Only approximately 1% of the units have maximum confidence values. This figure is also typical of other modalities.

Do these high-confidence units in fact represent reasonable color segments? This can be answered affirmatively by labeling image space using the high-confidence non-spatial color units. Each local maximum in the color cube is given an intensity label. Spatial units are assigned the label that the corresponding color spatial unit is closest to. This shows that the non-spatial color units have in some sense captured the notion of color segments. However, note that a red and green checkerboard image is indistinguishable from any other half-red and half-green image. Color space can only do so much; these other differences must be resolved by other *form* units that explicitly represent spatial information [2, 6, 28, 29].

4.1. Formalism for constraint maps

Constraint maps relate intrinsic image units to feature units. If an intrinsic image parameter is a vector unit $\boldsymbol{a}$ in a discrete intrinsic image space A and an element of feature space is a vector $\boldsymbol{b}$ in a discrete feature space B, then there is usually a *physical constraint* that relates $\boldsymbol{a}$ and $\boldsymbol{b}$, i.e., some relation $f(\boldsymbol{a}, \boldsymbol{b})$ such that $f(\boldsymbol{a}, \boldsymbol{b}) = 0$. The general form of $f(\boldsymbol{a}, \boldsymbol{b})$ is a table since f may not have an algebraic form. This physical constraint can be exploited in the following manner. For each value of $\boldsymbol{a}$ compute the set

$$\{\boldsymbol{b} \mid f(\boldsymbol{a}, \boldsymbol{b}) \leq \delta_b\}$$

where the constant δ_b above is related to quantization in the space B. This is the set of units in the feature space network B to which an $\boldsymbol{a}$ unit must connect to partially represent the constraint relation f.

The space A represents all possible intrinsic image values. A *particular* intrinsic image is described by a set of high confidence value units $\{\boldsymbol{a}_k\}$. Note that each $\boldsymbol{a}_k$ represents a *specific* set of high-confidence values, e.g., for a depth image, a single high-confidence unit might be: range = 15, $x = 120$, $y = 31$. Other units at the same position, e.g., {(20, 120, 31), (30, 120, 31)} should have low confidences (if the object is opaque). Each element from the set $\{\boldsymbol{a}_k\}$ is only consistent with certain elements in the space B, owing to the constraint imposed by the relation f.

Given that only a subset of units $\{\boldsymbol{a}_k\}$ are in a high-confident state, then the set of units in B that will receive input from a unit $\boldsymbol{a}_k$ in the network A is given by:

$$B_k = \{\boldsymbol{b} \mid \text{high-confident}(\boldsymbol{a}_k) \text{ and } f(\boldsymbol{a}_k, \boldsymbol{b}) \leq \delta_B\}$$

Define $H(\boldsymbol{b})$ as the number of times the value $\boldsymbol{b}$ occurs in $U_k B_k$ (the union of all sets B_k). $H(\boldsymbol{b})$ is the number of units in intrinsic image space which are consistent with the parameter value $\boldsymbol{b}$. Define $\text{Conf}(\boldsymbol{b}) := H(\boldsymbol{b})/\sum_b H(\boldsymbol{b})$. In that case, the value Conf($\boldsymbol{b}$) can stand for the initial confidence that the segment with feature value $\boldsymbol{b}$ is present in the image. There are several intuitive interpretations of Conf($\boldsymbol{b}$). Conf($\boldsymbol{b}$) can be thought of as a normalized histogram. Also, Conf($\boldsymbol{b}$) is analogous to an impulse response of the network. It simply represents an instantaneous normalized sum of active network links without taking into account any more complicated dynamic interactions such as lateral inhibition that occur when the network relaxes.

A constraint map need not originate from intrinsic image space but can be defined between any two spaces A and B as long as there is some relation $f(\boldsymbol{a}, \boldsymbol{b}) = 0$ for $\boldsymbol{a} \in A$ and $\boldsymbol{b} \in B$. To avoid describing the above computations in detail each time we need to use them, we use a shorthand notation for constraint maps. Each map can be described as the triple $\langle \boldsymbol{a}, \boldsymbol{b}, f\rangle$ where the necessary connections between units are implicit. Note that the order of $\boldsymbol{a}$ and $\boldsymbol{b}$ is important in the notation; in general, $\langle \boldsymbol{a}, \boldsymbol{b}, f\rangle$ is not equivalent to $\langle \boldsymbol{b}, \boldsymbol{a}, f\rangle$. In the case of a Hough transform, one can think of the triple as representing a transformation from less abstract, more spatially organized units to more abstract, less spatially organized units. Thus the triple generally represents

⟨spatial parameters,
non-spatial parameters,
the relation between elements in the two parameter spaces⟩

but many variants are possible. As we shall see, (i) some of the parameters in $\boldsymbol{a}$ need not be spatial, (ii) different triples may have overlapping subspaces in $\boldsymbol{a}$ or $\boldsymbol{b}$, and (iii) the relationship f may be expressed as a set of relations. Where the space $\boldsymbol{a}$ is split between two (or more) subspaces $\boldsymbol{a}_1$, $\boldsymbol{a}_2$, then conjunctive connections or auxilliary units are required to specify the constraint.

The color example was a special case in that each spatial color unit is connected to a single non-spatial color unit. In general, each $\boldsymbol{a}_k$ and the relationship f will not determine a single unit in B_k but there still will be isolated high-confidence units. Fig. 6 shows why this is the case; different $\boldsymbol{a}_k$ units connect to common units in the feature space B.

TABLE 2. Constraint maps

Intrinsic image	Feature space
Color	Segments of constant color [26, 50]
Disparity	Segments of constant disparity [22]
Optical flow	Heading [42, 51]
	Rotation of 3-d rigid body [5]
Surface orientation	Illumination angle [15]
	Shape [6]
Occluding contour	Shape [2, 6, 17, 41, 58]
Texels	Surface orientation [39]

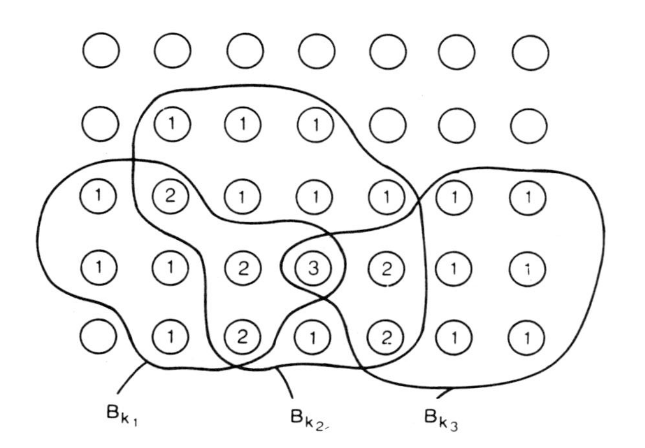

FIG. 6. How the Hough transform works even when the sets B_k are not singletons. High-confidence spatial units a_{k1}, a_{k2}, a_{k3} connect to overlapping sets of units in the non-spatially indexed network. The result is a single maximum confidence unit in B.

Of course segmentation must involve ways of associating peaks in several different feature spaces and methods for doing this are described in Section 8, but the cornerstone of the techniques are high-confidence units (histogram maxima) in the individual-modality feature spaces.

One might think that this is a clustering technique (e.g., [24]) presented with a different formalism. The constraint map technique is similar to clustering but differs in an important way. In general, feature space units are not independent but may interact through mutual connections. One can appreciate the need for lateral inhibition but there may also be the need for mutual excitation. Aligned edges is one example. A similar idea occurs in L-joint units in an origami world feature space [54]. Opposing L-joint units increase the confidence of each other. These kinds of effects are not expressible in clustering models.

Table 2 shows some other constraint maps.

4.2. Hierarchies of abstraction levels

The advantages of using several hierarchical levels of abstraction in parameter networks are: (i) the interaction between levels is simplified; (ii) the same levels can be used by different feature spaces; and (iii) more possibilities are allowed with the same number of units. Abstraction levels do not mean that high-level descriptions cannot influence low-level descriptions, or that the entire computations are not carried out in parallel. Rather, each descriptive level can only influence nearby levels. In [54], the limitation is to levels directly above and below. Other levels are influenced indirectly. The implication for the constraint maps, which specify the constraints between levels, is that the constraint relationships between levels involve only a few parameters. This is an especially important feature, since the space required by the constraint transform is exponential in the number of parameters, as are the sets $\{B_k\}$. Different levels of abstraction have been used by [26]. Examples using the constraint map idea may be found in [38, 54]. Sabbah uses four levels to recognize origami world figures. Fig. 7 shows a partial organization which attempts to integrate the results of previous work.

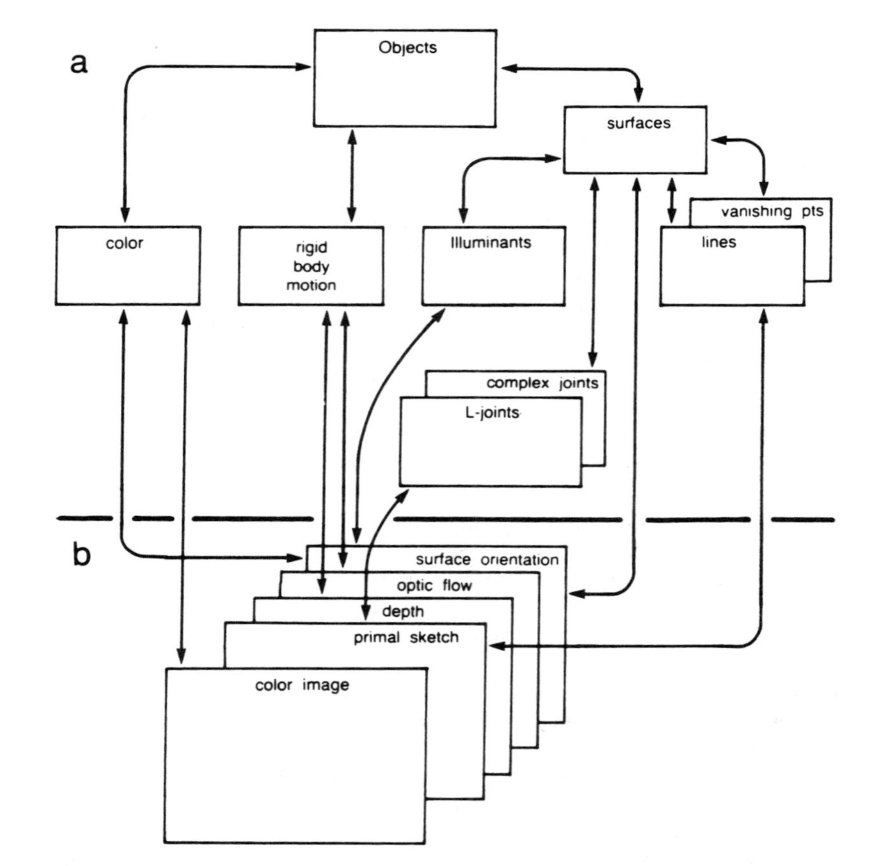

FIG. 7. Representation hierarchies showing a progression from spatially indexed, viewer-centered intrinsic images (b) and less-spatially indexed topological features to non-spatially indexed features (a). (Form perception as denoted by the relation of objects and surfaces is more complicated than shown and involves relations between object-centered and viewer-centered representations.)

4.3. Intrinsic images at different levels of abstraction

The survey of intrinsic images (Table 1) excluded the fact that intrinsic images may have fine structure involving several levels of abstraction. In fact, it seems likely that multiple abstraction levels are necessary in many cases. For example, Zucker [64] uses two levels of abstraction in computing orientation intrinsic images, one for points of high gradients and the other for edge segments. The computation of a velocity image in 3-d could involve three levels of abstraction:

- a *change detection* level where units are used for variations in intensity over space and time $\partial f/\partial x'$, $\partial f/\partial y'$, $\partial f/\partial t$ (primes denote retinal coordinates);
- an *optical flow* level where units correspond to retinal velocities $(u(x', y'), v(x', y'))$;
- a *3-d flow* level where units correspond to 3-d velocities $(v_x(x, y, z), v_y(x, y, z), v_z(x, y, z))$.

The feasibility of computing the optical flow from change measures has been studied by [9, 30, 51]. The feasibility of computing 3-d flow is explored in [5].

5. Towards Generality in Intrinsic Images

As previously discussed, boundary conditions are an important, open issue in intrinsic image computations. The parameter network, as a representation, does not contribute to the discovery of such constraints. It does, however, provide an easy way of implementing such constraints when they are discovered.

5.1. Multi-resolution relaxation methods

One notion of 'boundary condition' is image resolution. Previous methods for computing intrinsic images have used a single image resolution, but in most situations this is unrealistic. What is the correct resolution? At high resolution: (i) noise is a factor; (ii) convergence is slow; and (iii) basic assumptions may not hold. To see the last point, imagine trying to compute shape from shading using a surface with a micro-texture. At low resolution the surface structure is blurred and simple reflectance models hold, but at high resolution the microstructure can render such models useless. Low resolution may not always be appropriate, however. Even though noise is less of a factor and convergence is fast, other basic assumptions may not hold. The last point arises from the fact that most intrinsic images are computed from constraints which assume local variations are smooth. With increasing grid size, these assumptions are less likely to be valid.

Hence there is a range of resolutions for which the computations will be valid. Furthermore, this range is expected to be spatially variant. A tool for exploring this conjecture is multigrid relaxation techniques [13], which have proven very useful for solving differential equations. This model, together with reasoning from physical first principles, should allow the determination of image-dependent grid resolutions for which intrinsic image computations are valid. Multigrid techniques are of course related to pyramids [26, 57, 59].

In parameter networks, multiple resolution images are represented as explicit units. A complete pyramid only requires twice the number of units of the highest resolution layer.

5.2. Cooperative computation of multiple intrinsic images

Intrinsic images are logically computed simultaneously. In fact, they have to be; otherwise each intrinsic image is underdetermined in the general case. (Only on certain synthetic images is the computation well-defined.) Furthermore, they are highly interdependent, particularly at points of discontinuity [10]. For example:

- intensity edges can be indicative of depth discontinuities; thus the edge image is coupled to the disparity image;
- an abrupt change in surface orientation is also indicative of depth discontinuity;
- different objects which are moving relative to each other produce discontinuities in the flow field.

By incorporating these couplings in the intrinsic image computations, one can find general cases where the computations will converge. For example, we have found that an edge in a primal sketch is necessary to partially specify a boundary curve which in turn constrains a surface normal [15]. Fig. 8 shows a simplified case where the boundary is in the image plane. Barrow and Tenenbaum [11] have shown psychophysical evidence consistent with this result: if the boundary information is inconsistent with the shading information, shape

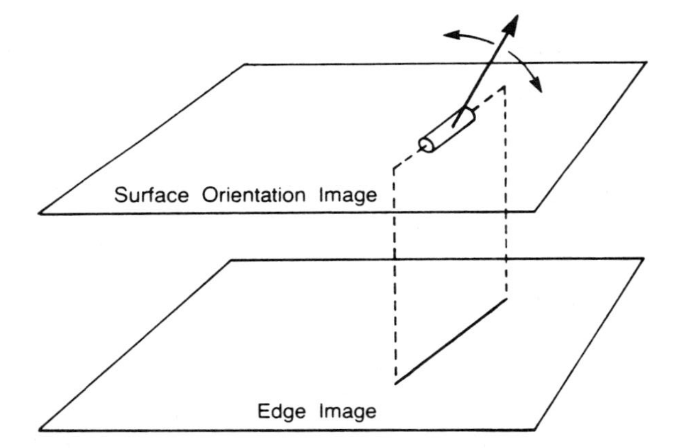

FIG. 8. Boundary condition in shape-from-shading computations.

from shading is not perceived. Bruss [16] has shown that a boundary contour (a more rigorous constant than Fig. 8) is a sufficient condition. A separate open issue in combined intrinsic image computations is the behavior of the coupled computations in the face of conflicting information. While the parameter networks will not resolve this issue, which depends on the specific constraint interactions, it does provide for the explicit representation of boundaries in terms of units. A competition of different constraints for a single interpretation occurs naturally in these networks.

6. Coarse Coding and Decompositions

A completely separate issue from the representation of environmental constraints in networks is the issue of different structuring of the networks themselves. One huge problem with value units is the exponential growth of units required to represent high-dimensional spaces. Here we describe two ideas to structure units that drastically reduce this requirement: coarse coding and decompositions.

6.1. Coarse coding

Coarse coding [28] is a way of encoding sparse measurements with overlapping ranges of feature measurements. To see how coarse coding works, consider the two-dimensional measurement space in Fig. 9(a). The presence of a measurement value is indicated by turning on the appropriate unit and there are N^2 total units. Now consider the coarse coding solution of Fig. 9(b). Three overlapping arrays of size $(\frac{1}{3}N)^2$ of coarse units are used, of which only one unit of each array is drawn. Now the presence of the same measurement vector can be signaled by turning on three coarse units simultaneously. Analysis shows that where the diameter of the coarse unit is D, and the dimension of the

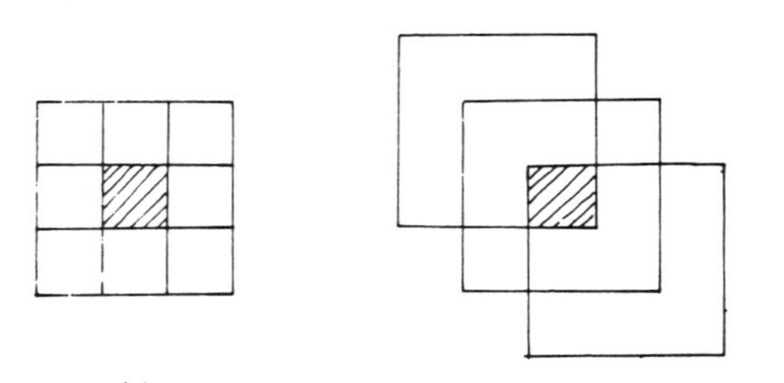

FIG. 9. Coarse coding example. (a) In a two-dimensional measurement space, the presence of a measurement can be encoded by making a single unit in the fine resolution space have a high-confidence value, or (b) the same measurement can be encoded by making overlapping coarse units in three distinct coarse arrays have high-confidence values.

measurement is k, the savings in units is proportional to $1/D^{k-1}$. In the example $k = 2$ and $D = 3$, so approximately $\frac{1}{3}$ as many units are required in the coarse coding case. This reduction in units is particularly attractive since many feature spaces have high dimensionalities.

Coarse coding does have some disadvantages. For example, it depends on sparse measurements, since if the coarse fields overlap false signalings can result. Also, in feature data many measurements initially exist with differing confidences. Getting sparse data depends on an appropriate threshold or convergence to high confidence local maxima. Thus the viability of coarse coding depends on k, D and threshold settings, and also the sparseness of the parameter space.

6.2. Feature space decompositions

A general feature of constraint transforms is that if the algorithms are completely parallel, the space required is exponential in the number of parameters. This can lead to immense space requirements. For example, consider an eight-parameter space of 100 discrete values for each parameter. The number of parameter nodes required to represent the space is 100^8. Fortunately this problem can generally be alleviated by *decomposing a high-dimensional space into lower dimensional spaces*. The advantage of this extremely powerful decomposition technique is that the dimensionality of the computation at each stage is much less than the single computation involving all of the parameters simultaneously [6].

In terms of the notation, the feature space B can be partitioned into two subspaces (B_1, B_2). Then the corresponding computation, denoted by $\langle \boldsymbol{a}, \boldsymbol{b}, f\rangle$, can be decomposed into two successive computations. First, compute $\langle \boldsymbol{a}_1, \boldsymbol{b}_1, f_1\rangle$ which has a set of maxima b^*, followed by $\langle(\boldsymbol{a}, \boldsymbol{b}_1^*), \boldsymbol{b}_2, f_2\rangle$. Naturally it follows that $H_1(\boldsymbol{b}_1) = \sum \boldsymbol{b}_2 H(\boldsymbol{b})$ and $H_2 = H(\boldsymbol{b}_1^*, \boldsymbol{b}_2)$.

To illustrate the idea of subspaces, we use the example of detecting rigid body motion in a three-dimensional flow field. In [5], it is shown that the three-dimensional velocity field can be recovered from depth and optic flow images. If the 3-d flow field is derived from a moving rigid body, then the motion can be described by nine parameters: three for a body-centered origin $\boldsymbol{x}_b$, three for a translational velocity $\boldsymbol{v}_T$, and three for a rotation vector $\boldsymbol{\Omega}$. This is easily seen since for every 3-d flow vector $\boldsymbol{v}$ at $\boldsymbol{x}$,

$$\boldsymbol{v}(x) = \boldsymbol{v}_T + \boldsymbol{\Omega} \times (\boldsymbol{x} - \boldsymbol{x}_b) .$$

Rather than solve for nine parameters for the object, one can find six by assuming that at a reference time $t = 0$ the origins of the reference coordinate frame and that of the rigid body coincide. For times greater than zero, the origin position can be determined by knowing the motion parameters at the previous time step, i.e., $\boldsymbol{x}_b(t + \Delta t) = \boldsymbol{x}_b(t) + \boldsymbol{v}_b \Delta t$. (The drawback of this assumption is that the origin will not, in general, be in the best place for the most intuitive description of the motion. In fact there are many equivalent descriptions of the motion which arise from different body coordinate frame locations. The most intuitive are those on the axis of rotation.)

The point of this example is that the remaining six parameters can be represented in four networks, each representing a subspace of the original nine-parameter space. The networks are: (i) units of (ω_x, ω_y) where $\omega = (\omega_x, \omega_y, \omega_z)$ is a unit vector in the direction of Ω; and (ii) units of $(|\Omega|, v_{Tx})$, $(|\Omega|, v_{Ty})$, and $(|\Omega|, v_{Tz})$. Thus, rather than having a single six-dimensional parameter network, there are four two-dimensional networks. Using 40 units for each linear dimension results in $4 \times (40)^2$ units versus $(40)^6$ in the six-dimensional case. Constraint maps from the velocity image $\boldsymbol{v}(\boldsymbol{x})$ to these parameter spaces[1] determine the appropriate parameter values as local maxima.

In an experiment, 3-d flow vectors were generated from a rigid body model whose parameters were determined a priori (Fig. 10(a)). These were used as input data. In a more complete experiment these vectors would be determined from spatially registered depth and optical flow fields. Fig. 10(b) shows the results of detecting ω, the unit vector in the direction of the rotation vector. Since ω is a unit vector, only two components need be determined, so the figure shows only Conf(ω_y, ω_z). The single maximum shows that different pairs of acceleration vectors give rise to a common rotation vector, as expected. Fig. 10(c) shows the results of using a constraint transform to detect $|\Omega|$ and v_T. The figure shows that each flow vector $\boldsymbol{v}(\boldsymbol{x})$ is connected to a linear set of $(v_{Tx}, |\Omega|)$ units but that different flow vectors are connected to different linear sets. Hence a high-confidence $(v_{Tx}, |\Omega|)$ unit is seen which is at the intersection of the lines of active units. A similar result is seen for $(v_{Ty}, |\Omega|)$ and $(v_{Tz}, |\Omega|)$. Furthermore, the units all have the same value of $|\Omega|$. In this transform, ω is a known constant vector. Also the decomposition technique is clearly shown. The point of this example is not the details of the constraints but rather the idea of using subspaces.

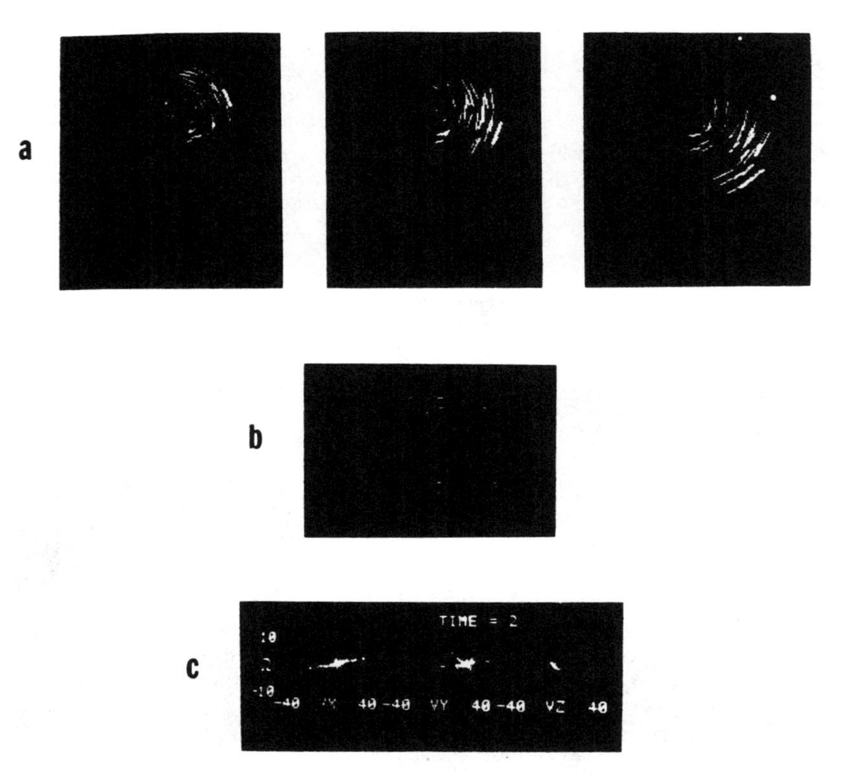

Fig. 10. Representation of rigid body motion with subspaces. (a) Three frames of computer-generated motion vectors from a rigid body model. (b) Single high-confidence unit denoting rotation about y-axis in a two-dimensional space of rotation direction units. (c) Three two-dimensional spaces of (Ω, v_x), (Ω, v_y), and (Ω, v_z) (progressing from left to right).

One problem that the subspaces introduce is that if two bodies are present, pairs of high confidence units appear in the subspaces. How can one determine the correct associations? This question is taken up in Section 8.

7. Coupled Computations

Most of the previous examples imply that the various constraint transforms are relatively independent. That is, once the intrinsic images are computed, the

[1] If one assumes that no large external forces are acting on the body, then pairs of successive accelerations derived from the time-varying 3-d flow field allow the determination of a value for ω. If these measurements all come from the same rigid body, then the values for ω should all be the same within the accuracy of the measurement

$$\langle (\boldsymbol{a}_1, \boldsymbol{a}_2), \omega, (\omega = \boldsymbol{a}_1 \times \boldsymbol{a}_2 / (|\boldsymbol{a}_1 \times \boldsymbol{a}_2|)) \rangle .$$

(Recall the notation for constraint maps introduced in Section 4.) Once ω has been determined, the remaining parameters can be found via a second constraint map. In the equation

$$\boldsymbol{v} = \boldsymbol{v}_T + |\Omega| \omega \times \rho \qquad (*)$$

only $\boldsymbol{v}_T$ and $|\Omega|$ are unknown. These can be determined by three constraint maps

$$\langle (\rho, \omega), (|\Omega| v_{Tk}), k\text{th component of } (*) \rangle$$

where $k = x, y, z$. The derivation of these constraints may be found in [5].

transforms can be computed. The general case is that this is not true; the intrinsic image contains global parameters which must be computed using constraint maps. Since the feature network required an intrinsic image it might seem that neither could be computed. In fact, both the feature network and the intrinsic images can be computed by incorporating the constraint maps into the parallel-iterative scheme used to compute the intrinsic images. If the combined problem is well-conditioned: (1) the partial result for the intrinsic image will be sufficient to produce a partial result for the feature network, and vice versa; and (ii) this process of using partial results in a parallel-iterative manner will converge. We term this interdependence *tight coupling* and illustrate it with an example.

The example shows how a surface orientation intrinsic image can be computed from intensity information. Given the orientation of a surface with respect to a viewer, its reflectance properties and the location of a single light source, that the brightness at a point of the viewer's retina can be determined. That is, the reflectance function $R(\theta, \varphi, \theta_s, \varphi_s)$, where θ, φ and θ_s, φ_s are orientations of the surface and source respectively, allows us to determine $I(x, y)$, the normalized intensity in terms of retinal coordinates [31]. The form of R is assumed to be known. However, the perceptual problem is the reverse: given $I(x, y)$ and $R(\cdot, \cdot)$, determine $\theta(x, y)$, $\varphi(x, y)$ and θ_s, φ_s.

In general, the perceptual problem of deriving $\theta(x, y)$, $\varphi(x, y)$ and θ_s, φ_s is underdetermined. However, Ikeuchi [35] showed that the surface could be determined locally once $\theta_s \varphi_s$ was specified. This method has been extended [15] to the case where θ_s, φ_s is initially unknown. Fig. 11 shows the result of the combined surface normal and illuminant direction calculations.

This example seems paradoxical at first since to compute surface orientation one must know the location of the source of illumination and vice versa. However, both these computations can be conducted simultaneously with the partial result for the surface orientation helping the illumination angle determination, and the partial result for the illumination angle helping the surface orientation determination. The illumination angle is determined by a constraint map $\langle(\theta(x, y), \varphi(x, y), I(x, y)), (\theta_s, \varphi_s), R\rangle$. In this simulation, the surface normals are simulated at level three (with arrays) and the illuminant direction at level two (with value units). Initially the illuminant is assumed to be in the direction of the viewer. This allows for an estimate of surface normals near boundaries which, in turn, allows for a new estimate of illuminant direction. This process is repeated and settles to good estimates of both surface normals and illuminant direction. The level three representation of surface normals does not allow for the simultaneous determination of the two possibilities for surface normals and illuminant direction. In a value unit simulation both solutions could be represented.

Rather than being an isolated example, tight coupling is believed to be the general case. Extending the scope of the parallel-iterative computation is the general solution.

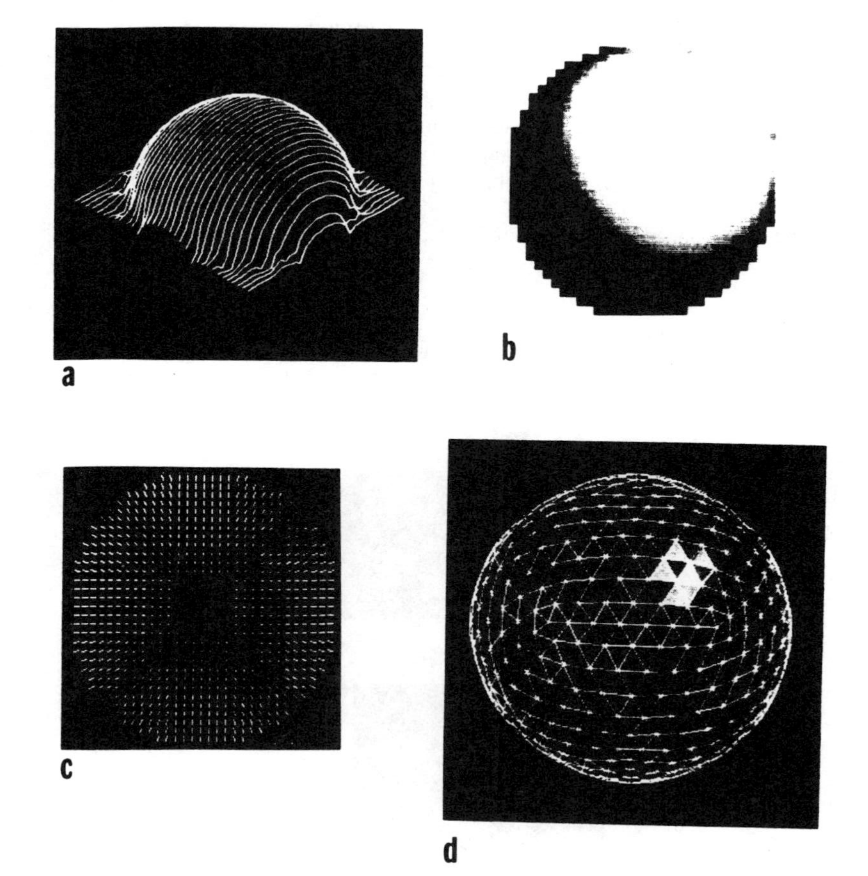

FIG. 11. (a) Representation of a simple physical surface. (b) Matte shading. Tightly coupled connections between (c) surface normal units and (d) illuminant direction value units.

8. Focus of Attention

The earlier examples of intrinsic image to feature space transforms used constraint maps between just two spaces. Two main issues arise when multiple feature spaces are involved. First, when multiple constraint maps are invoked in parallel to detect a segment with multiple features, some modalities may not have a high-confidence unit. This problem is solved via the mechanism of a *context* constraint map which allows an ambiguity in one feature space to be resolved by another.

8.1. Spatial context

If a unit has multiple spatially registered features, these can be detected by applying two different sets of constraint transforms. The constraint transform defined in Section 4 is bottom-up: points in the intrinsic image space determine plausible sets of points in feature space. The complementary transform is top-down: points in feature space determine plausible sets of points in intrinsic image space. Formally, given a set $\{\boldsymbol{b}_k\} \in B$, compute

$$A_k = \{\boldsymbol{a} \mid f(\boldsymbol{a}, \boldsymbol{b}_k) \leq \delta_A\} .$$

$H(\boldsymbol{a})$ is the number of times the value $\boldsymbol{a}(\boldsymbol{x})$ occurs in $U_k A_k$. The mapping which defines $H(\boldsymbol{a})$ is likely to be one to many and furthermore, for a given feature, different b_ks should give rise to disjoint subsets of A. Owing to this last point, it is intuitively appealing to deal with $\text{Conf}_{\boldsymbol{a}}(\boldsymbol{x})$ which is simply the normalized sum of the confidences of different values of the parameters a_1, $a_2, \ldots$ which are at the same spatial location x, y, i.e.,

$$\text{Conf}_{\boldsymbol{a}}(\boldsymbol{x}) = \sum \boldsymbol{a} \, \text{Conf}(\boldsymbol{a}, \boldsymbol{x})$$

Example 8.1. Consider an image of a red spot on a green background, where the spot takes up one-third of the image pixels. Then the transform $\text{Conf}(\boldsymbol{b})$ where $\boldsymbol{b} = (r, g, b)$ has two peaks and is zero everywhere else, i.e., for four-bit color scale accuracy

$$\text{Conf}(\boldsymbol{b}) = \begin{cases} 1, & \text{if } \boldsymbol{b} = (0, 15, 0), \\ \frac{1}{2}, & \text{if } \boldsymbol{b} = (15, 0, 0), \\ 0, & \text{otherwise} . \end{cases}$$

Now consider $\boldsymbol{b}_1 = (15, 0, 0)$ and compute $\text{Conf}(\boldsymbol{a}, \boldsymbol{x})$. This is given by

$$\text{Conf}(\boldsymbol{a}, \boldsymbol{x}) = \begin{cases} 1, & \text{if } \boldsymbol{x} \text{ part of the spot and } \boldsymbol{a} = (15, 0, 0), \\ 0, & \text{otherwise} . \end{cases}$$

A point in $\boldsymbol{A}$ represents the single color red and so $\text{Conf}(\boldsymbol{x})$ in this case is

$$\text{Conf}(\boldsymbol{x}) = \begin{cases} 1, & \text{if } x \text{ is part of the spot}, \\ 0, & \text{otherwise} . \end{cases}$$

The transform $\text{Conf}(\boldsymbol{x})$ is called the *spatial context* transform for reasons that will become more apparent when we discuss focus of attention. The *effect* of this transform is to place an imaginary filter or *mask* in front of the sensors. In the above case, only sensors that are spatially registered with RED sensors would receive input.

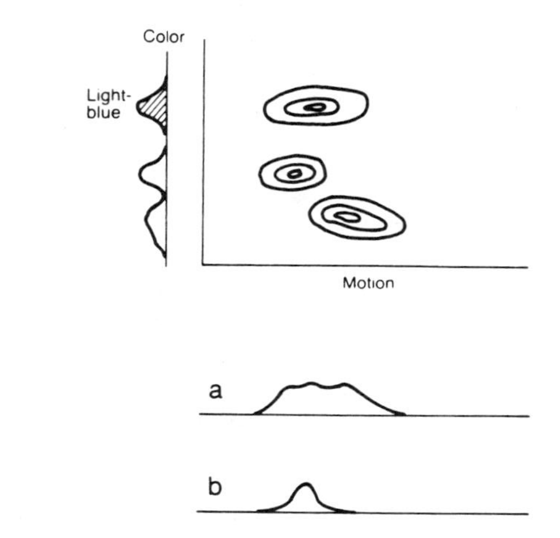

FIG. 12. Iso-confidence contours in a hypothetical color–velocity space. This space is too large to represent and must instead be represented as subspaces in the different modalities of velocity and color. The figure shows how spatial coupling between the subspaces can allow a clear color maximum to focus on a particular part of space and thus produce a clear velocity maximum. (a) Velocity space before color focusing. (b) Velocity space after color focusing. The units are not drawn, but they represent discrete tesselations of the spaces.

8.2. Spatial coherence

A segment in an image is ideally represented as a conjunction of constraint transform maxima. Each set of maxima corresponds to an organization with respect to a given modality: color, motion, etc. The previous section showed how the parallel generation of these maxima could be used to discover regions in the image corresponding to multi-modal units. One problem that can come up is that a gestalt is not manifested as a clear maximum in all the modalities. As an example, consider a light-blue, moving frisbee, against a background of other frisbees, none of which are light-blue, but which are moving. In the color space, the frisbee is clearly revealed; light-blue units corresponding to the frisbee's color have high confidence values. In velocity space, however, there may be no clear maximum owing to the motion of the other frisbees which map to the same motion unit space. This situation is schematized in Fig. 12 using one-dimensional axes for color and motion (even though these are higher-dimensional spaces). The figure shows iso-confidence curves in a hypothetical color-motion space. In the example, we assume that the dimension of this combined space is too large to be represented with explicit units.

The fundamental problem is that each modality consists of a *projection* of multi-modal feature space. In the high-dimensional space consisting of the concatenation of all the individual dimensions of each modality, each unit would appear as a distinct maximum. The visual system model is structured to examine only the subspaces of the individual modalities. The principal reason for this is economy; the space requirement increases exponentially with the number of modalities.

This problem can be surmounted if the different parameter spaces are examined sequentially. First the parameter spaces are examined for maxima. The most distinct maxima is picked and the response to its inverse constraint transform, $\mathrm{Conf}(\boldsymbol{x})$, is generated. This transform can be used to block input from sensors positioned at its low confidence values. To see how this might work, let us reconsider the previous example of the light-blue, moving unit. In color space there is a clear maximum corresponding to light-blue. This value is used to generate $\mathrm{Conf}_{\text{light-blue}}(\boldsymbol{x})$ and block input from all sensors that are not spatially registered with light-blue color input. The net effect is that in velocity space there is now a clear maxima as input from other units has been blocked.

The frisbee example may be interpreted in a different way. In Section 6 we encountered another variation of the subspaces problem: that of associating maximally confident units in different spaces. For example, suppose there are two maximally confident units in color space and two in motion space. Which color unit corresponds to which motion unit? The above example shows that one can sequentially link multi-modal units by focusing on a distinct part of visual space. The problem of *maintaining* such correspondences is still not settled, but some elegant possibilities are described in [19]. The subspace peak-linking problem is interesting as it has appeared in tests with human subjects. Given brief tachistoscopic presentations of different kinds of features, subjects often report illusory conjunctions [60]. For example, subjects viewing a green triangle and a red circle will often report having seen a red triangle. Although there is no evidence linking these two results, the interesting possibility is that the perceptual phenomena are related to a connectionist architecture.

9. Conclusion

This paper has advanced the beginnings of a theory of low-level vision with six main elements:

(1) spatial intrinsic images;
(2) non-spatial feature spaces;
(3) hierarchies;
(4) sparse encodings;
(5) coupled computations between intrinsic images and feature spaces;
(6) spatial coherence as a way to associate multi-modal features.

Most of the ideas in the paper are not original. The importance of intrinsic images and boundary conditions was emphasized by [10] and similar ideas comprise the $2\frac{1}{2}$-d sketch [44, 46]. The unit value idea is due to [8], and used as part of a computational scheme in [21], which in turn evolved from [20]. The crucial notion of coarse coding is due to Hinton [28]. Some of the newer ideas are: the significance of the Hough transform in explaining the layer of visual processing beyond intrinsic images; the use of subspaces in Hough transform computations; the idea of coupled computations between feature spaces and intrinsic images; and the unification of all the above ideas into the beginings of a coherent theory of low-level vision that addresses the problem of implementation. Marr [47], in his theory of visual processing, would agree with many of the tenets advanced here but with one class of exceptions. As emphasized in [44], he favored a postponement of the study of the architectural structure of vision. The example used to emphasize his point is "to study how birds fly one should study aerodynamics." The contention of this paper and also of [21] is that mathematics and physics alone are insufficient to understand vision, and that the design of the vision machine itself involves a new set of illuminating constraints. In this paper the main ones are value units, and the methods available to handle sparse inputs such as coarse coding and explicit subspaces.

It is also important to understand the emphasis on the particular *physical constraints* used in this paper. They are by no means unique, and other constraints may in fact be more likely in the human visual system. For example, (i) humans use an opponents-color basis; (ii) many other shape-from-shading parametrizations are possible [35]; and (iii) many other motion constraints are possible [49, 52]. However, having the right constraints is still not sufficient to explain certain perceptual behaviors. We propose that these can be explained by considering the architecture of a connectionist machine that uses the constraints, and illustrate architectural points via different examples: colored objects can be represented by a non-spatially indexed parameter space, shape-from-shading computations are inherently coupled, and rigid body computations require subspaces of parameters.

The paper proposes a direct approach to the perception of gestalts: represent possible perceptions in terms of feature spaces of explicit units. This does not require one feature space per gestalt since: (a) different gestalts can appear as different parameters of a given feature space; and (b) units from different feature spaces may be linked. The second ability allows for extremely large varieties of inputs to be perceived as different. The restriction to visual gestalts is in fact helpful as it limits the computations to a single, massively parallel cooperative computation among different parameter networks with only a limited amount of spatial focusing. More sequential processes that use high-level plans and goals will require sophisticated mechanisms for memory management and program control that go beyond the material presented herein. Future work may show that connectionist networks are a helpful formalism for modeling these concepts also.

Throughout the paper we have used confidence of the unit to mean that the unit can be thought of as being part of a perception. A more refined view has recently been advanced by Hinton and Sejnowski [55]. They show that regarding the output of a unit as a probability can lead to interesting network convergence properties. Also, this view is attractive as a better biological model of neurons. These ideas are still being developed, but are very promising.

To what extent can all these ideas be tested? Evidence can come from three principal sources: psychological tests, neurobiological data, and mathematical models. The arguments for intrinsic images can be made on the third ground, that such information can be computed from the image by exploiting physical constraints, but also on the second ground; evidence exists for neurons that have spatial mappings and are differentially sensitive to disparity, edges, color, and intensity change. These neurons are usually coarse coded; that is, they respond to multidimensional inputs. A neuron that is sensitive to edges may also respond to intensity changes. However, this is not surprising, given the representational efficiencies of coarse coding.

The arguments for non-spatial parameter spaces are based mainly on mathematical necessity: without such parameters, certain intrinsic images cannot be computed [5]. However, there is evidence for neurons that respond to surface reflectance (perceived color) independent of illumination and spatial position [63], and the idea is an important component of current neurobiological models [7, 43].

ACKNOWLEDGMENT

This paper benefited greatly from feedback from the vision research groups at Carnegie-Mellon University, Universities of Massachusetts and Rochester. Thanks go to Peggy Meeker for typing the manuscript and its revisions.

REFERENCES

1. Attneave, F., Some informational aspects of visual perception, *Psych. Rev.* **61** (1954).
2. Ballard, D.H., Generalizing the Hough transform to detect arbitrary shapes, *Pattern Recognition* **13**(2) (1981) 111–122.
3. Ballard, D.H., Parameter networks: Towards a theory of low-level vision, *Proc. 7th IJCAI*, Vancouver, B.C., 1981, *Artificial Intelligence* (submitted).
4. Ballard, D.H. and Brown, C.M., *Computer Vision* (Prentice-Hall, Englewood Cliffs, NJ, 1982).
5. Ballard, D.H. and Kimball, O.A., Rigid body motion from depth and optical flow, TR70, Computer Science Dept., Univ. Rochester, 1981.
6. Ballard, D.H. and Sabbah, D., On shapes, *Proc. 7th IJCAI*, Vancouver, B.C., 1981.
7. Barlow, H.B., Critical limiting factors in the design of the eye and visual cortex, *Proc. Roy. Soc. London B* **212** (1981) 1–34.
8. Barlow, H.B., Single units and sensation: A neuron doctrine for perceptual psychology? *Perception* **1** (1972) 371–394.
9. Barnard, S.T. and Thompson, W.B., Disparity analysis of images, TR79-1, Computer Science Dept., Univ. Minnesota, 1979.
10. Barrow, H.G. and Tenenbaum, J.M., Recovering intrinsic scene characteristics from images, in: A.R. Hanson and E.M. Riseman (Eds.), *Computer Vision Systems*, Academic Press (New York, 1978).
11. Barrow, H.G. and Tenenbaum, J.M., Interpreting line drawings as three-dimensional surfaces, *Artificial Intelligence* **17** (1981) 75–116.
12. Brady, M., Computational approaches to image understanding, AI Lab., MIT, Cambridge, MA, 1982.
13. Brandt, A., Multi-level adaptive solutions to boundary-value problems, *Math. Comp.* **31** (1977) 333–390.
14. Bribiesca, E. and Guzman, A., How to describe pure form and how to measure differences in shapes using shape numbers, *Proc. IEEE Computer Society Conf. on Pattern Recognition and Image Processing*, Chicago, IL, (1979) 427–436.
15. Brown, C.M., Ballard, D.H. and Kimball, O.A., Constraint interaction in shape-form-shading algorithms, *Proc. DARPA Image Understanding Workshop*, Palo Alto, CA, 1982; *1982-83 Research Review*, Computer Science Dept., Univ. Rochester, 1982.
16. Bruss, A.R., The image irradiance equation: its solution and application, Ph.D. Thesis, TR623, AI Lab., MIT, Cambridge, MA, 1981.
17. Duda, R.O. and Hart P.E., Use of the Hough transform to detect lines and curves in pictures, *Comm. ACM* **15**(1) (1972) 11–15.
18. Fahlman, S.E., The Hashnet interconnection scheme, Computer Science Dept. Carnegie-Mellon Univ., 1980.
19. Feldman, J.A., Memory and change in connection networks, TR96, Computer Science Dept., Univ. Rochester, 1981.
20. Feldman, J.A., A connectionist model of visual memory, in: G.E. Hinton and J.A. Anderson (Eds.), *Parallel Models of Associative Memory* (Lawrence Erlbaum Associates, Hillsdale, NJ, 1981).
21. Feldman, J.A. and Ballard, D.H., Connectionist models and their properties, *Cognitive Sci.* **6** (1982) 205–254.
22. Fischler, M.A. and Barrett, P., An iconic transform for sketch completion and shape abstraction, *Comput. Graphics Image Processing* **13** (1980) 334–360.
23. Freuder, E.C., Synthesizing constraint expressions, (*Comm. ACM* **21** (1978) 958–965.
24. Fukunaga, K., *Introduction to Statistical Pattern Recognition* (Academic Press, New York, 1972).
25. Grimson, W.E.L., *From Images to Surfaces: A Computational Study of the Human Early Visual System* (MIT, Cambridge, MA, 1981).
26. Hanson, A.R. and Riseman, E.M., Segmentation of natural scenes, in: A.R. Hanson and E.M. Riseman (Eds.), *Computer Vision Systems* (Academic Press, New York, 1978).
27. Hillis, W.D., The connection machine (Computer architecture for the new wave), AI Memo 646, MIT, Cambridge, MA, 1981.
28. Hinton, G.E., Shape representation in parallel systems, *Proc. 7th IJCAI*, Vancouver, B.C. (1981) 1088–1096.
29. Hinton, G.E., Relaxation and its role in vision, Ph.D. Dissertation, Univ. Edinburgh, 1979.
30. Horn, B.K.P. and Schunck, B.G., Determining optical flow, AI Memo 572, AI Lab., MIT, Cambridge, MA, 1980.
31. Horn, B.K.P. and Sjoberg, R.W., Calculating the reflectance map, *Proc. DARPA IU Workshop*, Pittsburgh, PA (1978) 115–126.
32. Hough, P.V.C., Method and means for recognizing complex patterns, U.S. Patent 3 069 654, 1962.
33. Hummel, R. and Zucker, S., On the foundations of relaxation labeling processes, TR, Dept. Electrical Engineering, McGill Univ. 1980.
34. Hurvich, L.M. and Jameson, D., An opponent-process theory of color vision, *Psych. Rev.* **64** (1957) 384–390.
35. Ikeuchi, K., Numerical shape from shading and occluding contours in a single view, AI Memo 566, AI Lab., MIT, Cambridge, MA, 1980.
36. Kanade, T., Recovery of the three-dimensional shape of an object from a single view, CMU-CS-79-153, Computer Science Dept., Carnegie-Mellon Univ. 1979.
37. Kanade, T., A theory of Origami world, CMU-CS-78-144, Computer Science Dept., Carnegie-Mellon Univ., 1978.

38. Kender, J.R., Shape from texture: A brief overview and a new aggregation transform, *Proc. DARPA IU Workshop*, Pittsburgh, PA (1978) 79–84.
39. Kender, J.R. and Kanade, T., Mapping image properties into shape constraints: Skewed symmetry, affine-transformable patterns, and the shape-from-texture paradigm, *Proc. 1st Annual Nat. Conf. on Artificial Intelligence*, Stanford Univ., Stanford, CA (1980) 4–6.
40. Koffka, K., *Principles of Gestalt Psychology* (Harcourt, Brace, and World, New York, 1935).
41. Kimme, C., Ballard, D.H. and Sklansky, J., Finding circles by an array of accumulators, *Comm. ACM* **18**(1) February (1975) 120–122.
42. Lawton, D.T., Constraint-based inference from image motion, *Proc. 1st Nat. Conf. on Artificial Intelligence*, Stanford Univ., Stanford, CA (1980) 31–34.
43. Lee, D.N. and Reddish, P.E., Plummeting gannets: A paradigm of ecological optics, *Nature* **293** (1981) 293–294.
44. Marr, D., *Vision* (Freeman, San Francisco, CA, 1982).
45. Marr, D., Early processing of visual information, *Phil. Trans. Roy. Soc. B* **275** (1976) 483–524.
46. Marr, D., Representing visual information, in: A.R. Hanson and E.M. Riseman (Eds.), *Computer Vision Systems* (Academic Press, New York, 1978).
47. Marr, D., Representing and computing visual information, in: P.H. Winston and R.H. Brown (Eds.), *Artificial Intelligence: An MIT Perspective* (MIT, Cambridge, MA, 1979).
48. Marr, D. and Poggio, T., Cooperative computation of stereo disparity, *Science* **194** (1976) 283–287.
49. Nagel, H.-H. and Neumann, B., On 3d reconstruction from two perspective views, *Proc. 7th IJCAI*, Vancouver, B.C. (1981) 661–663.
50. Ohlander, R., Price, K. and Reddy, D.R., Picture segmentation using a recursive region splitting method, *Comput. Graphics Image Processing* **8** (1979).
51. Prager, J.M., Extracting and labeling boundary segments in natural scenes, *IEEE Trans. Pattern Anal. Machine Intelligence* **2**(1) (1980) 16–27.
52. Prazdny, K., Determining the instantaneous direction of motion from optical flow generated by a curvilinearly moving observer, *Comput. Graphics Image Processing* **17** (1981) 238–248.
53. Rosenfeld, A., Hummel, R.A. and Zucker, S.W., Scene labelling by relaxation operations, *IEEE Trans. Systems Man Cybernet.* **6** (1976).
54. Sabbah, D., Design of a highly parallel visual recognition system, *Proc. 7th IJCAI*, Vancouver, B.C., 1981.
55. Hinton, G.E., and Sejnowski, T.J. Analyzing cooperative computation, *Proc. 5th Annual Conference of the Cognitive Science Society*, Rochester, NY, 1983.
56. Shapiro, S.D., Generalization of the Hough transform for curve detection in noisy digital images, *Proc. 4th IJCPR*, Kyoto, Japan, (1978) 710–714.
57. Sloan, K.R., Jr., Dynamically quantized pyramids, *Proc. 7th IJCAI*, Vancouver, B.C., 1981.
58. Sloan, K.R., Jr. and Ballard, D.H., Experience with the Generalized Hough Transform, *Proc. 5th Internat. Conf. on Pattern Recognition and Image Processing*, Miami Beach, FL, 1980.
59. Tanimoto, S. and Pavlidis, T., A hierarchical data structure for picture processing, *Comput. Graphics Image Processing* **4**(2) (1975) 104–119.
60. Triesman, A.M. and Gelade, G., A feature-integration theory of attention, *Cognitive Psychology* **12** (1980) 97–136.
61. Ullman, S., *Interpretation of Visual Motion* (MIT, Cambridge, MA, 1979).
62. Ullman, S., Relaxation and constrained optimization by local processes, *Comput. Graphics Image Processing* **10** (1979) 115–125.
63. Zeki, S., The representation of colours in the cerebral cortex, *Nature* **284** (1980).
64. Zucker, S.W., Labeling lines and links: An experiment in cooperative computation, TR80-5, Computer Vision and Graphics Lab., McGill Univ., 1980.
65. Zucker, S.W., Relaxation labelling and the reduction of local ambiguities, TR451, Computer Science Dept, Univ. Maryland, College Park, MD, 1976.

Image processing by simulated annealing

by P. Carnevali
L. Coletti
S. Patarnello

It is shown that simulated annealing, a statistical mechanics method recently proposed as a tool in solving complex optimization problems, can be used in problems arising in image processing. The problems examined are the estimation of the parameters necessary to describe a geometrical pattern corrupted by noise, the smoothing of bi-level images, and the process of halftoning a continuous-level image. The analogy between the system to be optimized and an equivalent physical system, whose ground state is sought, is put forward by showing that some of these problems are formally equivalent to ground state problems for two-dimensional Ising spin systems. In the case of low signal-to-noise ratios (particularly in image smoothing), the methods proposed here give better results than those obtained with standard techniques.

Introduction

Simulated annealing is a technique recently introduced [1, 2] to solve very complex optimization problems. A quick description of this method is given here to make this paper self-contained, but the reader is referred to [1] for a complete discussion.

Consider the problem of minimizing the function $E(x_i)$ of the many variables x_i, i.e., of looking for the values of x_i that yield the absolute minimum of the function $E(x_i)$. The basic idea of simulating annealing consists of treating the system to be optimized as a physical system described by the degrees of freedom x_i, with the energy given by $E = E(x_i)$. One then looks for the state of minimum energy of the physical system, i.e., what physicists call the *ground state*.

With simulated annealing, the ground state is reached by simulating a slow cooling of the physical system, starting from a very high temperature T down to $T = 0$. The cooling must be slow enough that the system does not get stuck into thermodynamically metastable states that are *local* minima of $E(x_i)$. This slow cooling process (called annealing from the analogy with metallurgic processes) is simulated using a standard method proposed by Metropolis et al. [3].

For a given temperature, the Metropolis method is a way to sample states of the physical system with the Boltzmann distribution

$$f = e^{-\frac{E}{T}}, \tag{1}$$

which is the distribution that properly describes the state of thermodynamical equilibrium for a given temperature T. One starts with a random configuration x_i. One then chooses (again, randomly) a small perturbation Δx_i in the system and calculates the energy change ΔE caused by the perturbation

$$\Delta E = E(x_i + \Delta x_i) - E(x_i). \tag{2}$$

If $\Delta E < 0$, then the perturbation is "accepted," for it means that it is energetically favorable for the system; otherwise, it is accepted with probability $e^{-\Delta E/T}$. When the perturbation is accepted, one continues the process with the perturbed state $x_i + \Delta x_i$ replacing the old one; otherwise a new perturbation Δx_i is attempted. It can be shown that the sequence of states obtained in this way is distributed according to (1). The Metropolis method is widely used in physics to study numerically the thermodynamical properties of large systems that cannot be treated with analytical methods.

In simulated annealing, one starts with a high value of T, so that the probability of the system being in a given state is independent of the energy of that state. One then slowly reduces T, by making sure that at each new value of T enough steps of the Metropolis procedure are made to guarantee that thermodynamical equilibrium has been reached. One continues the procedure until $T = 0$. If the cooling has been slow enough, the final state reached is the ground state of the physical system being considered; i.e., the values of x_i so obtained realize the absolute minimum of the function E. In practice, in many cases one is not really interested in finding the absolute minimum. Rather, in many interesting situations the minimum configuration is highly degenerate. In other words, there are many minima with values of E very close to its absolute minimum value, and one looks for one of the very many of them.

For problems of constrained minimum, the method can still be applied. One simply has to make sure that all the perturbations Δx_i that are generated during the Metropolis procedure continue to satisfy the constraints of the problem. In particular, the constraints could consist of prescribing discrete values for the x_i. Thus, simulated annealing applies as well in problems of discrete optimization (as shown by Kirkpatrick [1] in his various examples).

In this paper, we study the application of simulated annealing to various optimization problems arising in image processing. The method seems to be well suited for problems involving very low signal-to-noise ratios. In addition, we show that the analogy between the system to be optimized and a physical system can be even stronger than that implied by the process of simulated annealing. In particular, we show that various problems involving bi-level images (bitmaps) are formally equivalent to ground state problems for two-dimensional Ising spin systems imbedded in an external field.

In the following section we analyze the problem of the recognition of a regular pattern of rectangles out of an initial configuration corrupted by noise. The next section is devoted to the process of smoothing a bi-level image by means of the annealing procedure. In this context we find that the "cost function" for this problem can be easily related to the energy of a statistical model defined on a lattice (the given bitmap), known as the Ising model. As we have already mentioned, this procedure works remarkably well in the presence of high noise levels. Finally, in the fourth section we generalize this analogy by showing that the problem of halftoning a gray-level image is again equivalent to a spin system in which the internal interaction is of a different nature than the previous one. Correspondingly, the annealing procedure has to be a bit more accurate, since the physical system turns out to have a more complex ground state. We are also able to provide quantitative estimates of the gray and spatial resolutions in terms of the physical quantities involved in the spin system.

Estimation of parameters

The first problem of image processing we want to solve using simulated annealing is that of parameter estimation.

Suppose we have an image represented by the rectangular array of real numbers α_{ij}, with $1 \le i \le N_x$, $1 \le j \le N_y$. In addition, assume that a parametric model for the image is available. In other words, we know that the given image represents a scene containing objects of shapes and sizes that are partially known and that can be precisely described in terms of the parameters $x_l (1 \le l \le N_p)$. This model is assumed to be well known, so that, given any set of values x_l, it is possible to calculate the corresponding image $\alpha_{ij} = \alpha'_{ij}(x_l)$ that would be observed, in the absence of noise, for a scene described by the values x_l for the parameters.

The problem then consists of estimating "good" values x_l that describe the given image. Since noise or other sources of error can be present, it cannot be expected that a precise fit of the observed image can be obtained; i.e., in general one will not be able to find a set of values x_l^* such that

$$\alpha'_{ij}(x_l^*) = \alpha_{ij}. \tag{3}$$

Rather, one will introduce a measure E of the difference between the left and the right side of (3) and minimize it.

For example, choosing the L^2 norm to measure the distance between α and α', one has

$$E(x_l) = \sum_{i=1}^{N_x} \sum_{j=1}^{N_y} [\alpha'_{ij}(x_l) - \alpha_{ij}]^2, \tag{4}$$

whereas the expression for E in the case that an L^1 norm is preferred is

$$E(x_l) = \sum_{i=1}^{N_x} \sum_{j=1}^{N_y} |\alpha'_{ij}(x_l) - \alpha_{ij}|. \tag{5}$$

Thus the problem of parameter estimation is a minimization problem that can be solved using simulated annealing.

As an example, we chose to apply simulated annealing to the following parameter estimation problem. The given image α_{ij} is assumed to be a bi-level image (bitmap); i.e., it can only have the values 0 and 1, with 0 conventionally representing white (or background) and 1 black (or foreground). The image is a square one, with $N_x = N_y = N$. The scene observed is assumed to be a white sheet of paper, with an unknown number of black rectangles on it. The rectangles are at unknown positions on the sheet and their sizes are also unknown. The rectangles are allowed to overlap each other. To simplify the programming, it is assumed that the rectangles have their sides parallel to the sides of the image. In this way, the lth rectangle is a black area composed of the pixels with $x_{\min}^{(l)} \le i \le x_{\max}^{(l)}$ and $y_{\min}^{(l)} \le j \le y_{\max}^{(l)}$ and each rectangle is described by the four parameters $x_{\min}^{(l)}$, $x_{\max}^{(l)}$, $y_{\min}^{(l)}$, and $y_{\max}^{(l)}$. If N_R is the (unknown) number of rectangles, the image α'_{ij} corresponding to a given set of rectangles is given by

$$\alpha'_{ij} = \sum_{l=1}^{N_R} \theta(i - x_{\min}^{(l)})\theta(x_{\max}^{(l)} - i)\theta(j - y_{\min}^{(l)})\theta(y_{\max}^{(l)} - j), \tag{6}$$

where the sum is understood to be a boolean sum (OR) and θ is a step function of integer argument defined to be 0 if the argument is negative and 1 otherwise.

The unknowns of the problem are N_R and the $4N_R$ parameters describing the rectangles. Using (4) or (5) to define E is equivalent, since for these bi-level images we have $(\alpha' - \alpha)^2 = |\alpha' - \alpha|$. Thus one obtains

$$E = \sum_{i=1}^{N}\sum_{j=1}^{N}\left|\alpha_{ij} - \sum_{l=1}^{N_R}\theta(i - x_{\min}^{(l)})\theta(x_{\max}^{(l)} - i) \times \theta(j - y_{\min}^{(l)})\theta(y_{\max}^{(l)} - j)\right|. \quad (7)$$

Note that this expression represents nothing other than the number of pixels on which α and α' are in disagreement.

However, this expression for E is useless because of the variable number of degrees of freedom (which is $4N_R + 1$). By taking a large enough number of rectangles, one can obtain a solution with $E = 0$ for any given α (in fact, one such solution consists of having a 1×1 rectangle for each black pixel of the given image). To overcome this problem, one needs to penalize solutions involving too many rectangles, in such a way that if two solutions give the same value of E using (7), the solution with the smaller N_R is to be preferred. This is easily achieved by adding to the expression for E a term proportional to N_R. The new expression for E then becomes

$$E = \kappa N_R + \sum_{i=1}^{N}\sum_{j=1}^{N}\left|\alpha_{ij} - \sum_{l=1}^{N_R}\theta(i - x_{\min}^{(l)})\theta(x_{\max}^{(l)} - i) \times \theta(j - y_{\min}^{(l)})\theta(y_{\max}^{(l)} - j)\right|. \quad (8)$$

The parameter κ is interpreted as follows: For small values of κ, small rectangles present in the input image α are considered to be "real." When κ is made larger, small rectangles start to be considered as noise because their inclusion in the solution would increase E rather than decrease it. In fact, suppose the inclusion of a rectangle causes an improvement of n pixels in the agreement between α and α'. Then the second term of (8) decreases by n. However, the corresponding change in E is $\kappa - n$, and this is energetically favorable (negative) only if $n > \kappa$. Thus, only rectangles implying an improvement of at least κ pixels in the agreement between α and α' are included in the solution; κ then acts as a control parameter describing, roughly, how much noise one is willing to keep in the solution.

The solution of the problem of the rectangles using simulated annealing has been implemented for a 128×128 bitmap. The input image α_{ij} has been constructed starting from a noiseless configuration including 50 rectangles (shown in **Figure 1**) and adding random noise to it: Each pixel has been changed from 0 to 1 (or vice versa) with probability r, r taking the values 0.2, 0.3, and 0.4. The images obtained in this way are shown in **Figure 2**.

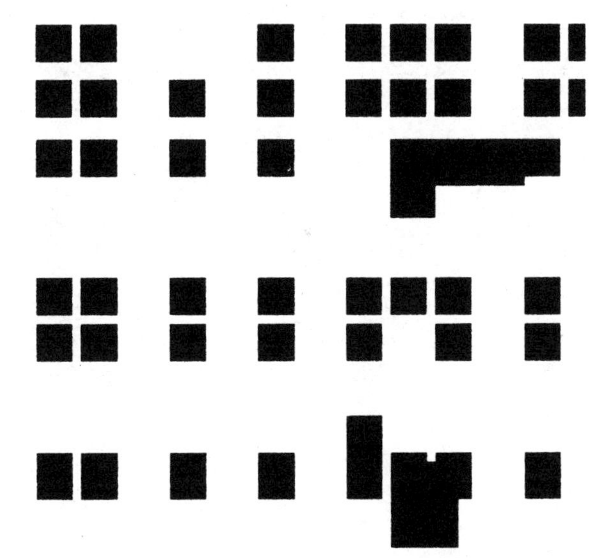

Figure 1

Parameter estimation: noiseless configuration with 50 rectangles.

In each of these images, the average value of α is r in originally white areas, $1 - r$ in originally black areas. The difference between the two average values is $1 - 2r$. The standard deviation of α in both the originally white and originally black areas is given by $\sqrt{r(1 - r)}$. Thus we can define a signal-to-noise ratio as the ratio of the difference between the two average values and the standard deviation, given by $(1 - 2r)/\sqrt{r(1 - r)}$. This gives signal-to-noise ratios of 1.5, 0.87, and 0.41 for the three values of r considered.

The annealing process has been started using a random configuration of rectangles, like the one shown in **Figure 3**. At each step, a change has been picked up randomly among the following classes of possible changes:

- Creation of a new rectangle at a random position and with a random size (smaller than a fixed maximum size).
- Removal of an existing rectangle.
- Stretching of one side of an existing rectangle in one direction by a small, random number of pixels.
- Splitting of an existing rectangle into two rectangles, leaving a one-pixel-wide strip between the two. The splitting can occur at a random point and can be either in the horizontal or in the vertical direction.

Results of the simulated annealing process for the three values of r considered are shown in **Figure 4**. Despite the low signal-to-noise ratios considered, reconstruction of the original image is quite good: Even for the almost hopeless case $r = 0.4$ (corresponding to a signal-to-noise ratio of only

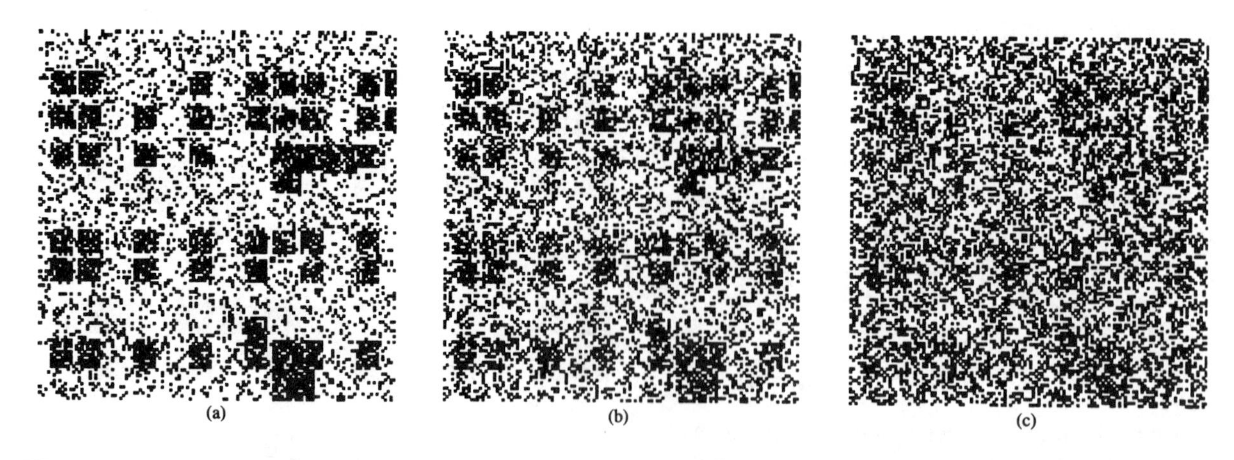

Figure 2

Parameter estimation: noise-corrupted configurations. The image in Fig. 1 has been corrupted by adding random noise as explained in the text. Signal-to-noise ratios are 1.5, 0.87, and 0.41 in (a), (b), and (c), respectively.

0.41), many features of the original image are reconstructed correctly.

Image smoothing and Ising spin systems

The second problem we want to study is that of smoothing images in a sense to be made precise below. We show that a large class of smoothing methods similar to those proposed in [4] are equivalent to ground state problems for two-dimensional Ising spin systems.

For an introduction on Ising spin systems, we refer the reader to textbooks (such as, for example, [5] or [6]). Here we only mention some of the points we need for the following.

Consider a two-dimensional network of points arranged in a square lattice with each point, labeled by its integer coordinates i, j, connected to its four nearest neighbors. Suppose we place in each point a particle with a magnetic moment (spin) and that each particle can be in one of two states, conventionally labeled $\mu = -1$ and $\mu = +1$ or called "spin down" and "spin up," respectively. Suppose that each particle interacts with its four neighbors, and assume that the interaction is translation-invariant and isotropic. Then the energy of the system can be written as

$$E = -\frac{J}{2} \sum_{i_1 j_1 i_2 j_2} C_{i_1 j_1 i_2 j_2} \mu_{i_1 j_1} \mu_{i_2 j_2}, \tag{9}$$

where $C_{i_1 j_1 i_2 j_2}$ is a connection matrix which is 1 if the points i_1, j_1 and i_2, j_2 are nearest neighbors to each other and 0 otherwise. This is a very simplified but physically reasonable model for a two-dimensional substance exhibiting ferromagnetic behavior which was introduced by Ising [7] in 1925. If the system is infinite (or if border effects are neglected), the expression for E can also be written as

$$E = -J \sum_{ij} (\mu_{ij}\mu_{i+1,j} + \mu_{ij}\mu_{i,j+1}), \tag{10}$$

from which it is clear that J (a positive constant) is the contribution to the total energy given by a pair of adjacent spins, the sign of the contribution being negative if the two adjacent spins are aligned (both up or both down) and positive otherwise. The system, in order to minimize its energy, tends to align all its spins in the same direction. Therefore, the ground state configuration is very simple, with all the spins of the lattice pointing in one direction (up or down). Antiferromagnetic models are also possible, in which J is negative, and they are considered in the following section.

An additional feature one can include in Ising models is the presence of an external magnetic field γ_{ij} which tends to align the spins in the direction prescribed by its sign. This introduces an additional term $-\frac{1}{2}\gamma_{ij}\mu_{ij}$ for each spin in the expression for the energy, so that we have

$$E = -\frac{1}{2} \sum_{ij} \gamma_{ij}\mu_{ij} - \frac{J}{2} \sum_{i_1 j_1 i_2 j_2} C_{i_1 j_1 i_2 j_2} \mu_{i_1 j_1} \mu_{i_2 j_2} \tag{11}$$

or

$$E = -\frac{1}{2} \sum_{ij} \gamma_{ij}\mu_{ij} - J \sum_{ij} (\mu_{ij}\mu_{i+1,j} + \mu_{ij}\mu_{i,j+1}). \tag{12}$$

The factor $\frac{1}{2}$ has been introduced simply for convenience and its only effect is to change the units in which the field γ is measured. It is clear that each spin is under the influence of two (possibly competing) forces: one due to the interaction with its neighbors and that tends to align the spin with its neighbors; the second, due to the external field, which tends to align the spin with the external field (in fact, the external field contribution to E is negative, i.e., energetically favorable, for spins which have μ with the same sign as γ). The introduction of such an external field strongly increases the complexity of the model under study; in particular, it may be very difficult to find the ground state if the value of γ_{ij} changes significantly with the position on the lattice.

An additional generalization could consist of introducing an interaction that does not involve only pairs of nearest neighbor points in the lattice. In this case the expression for E would contain a contribution for every pair of points that are close enough to each other. This can be obtained simply by keeping all the above expressions as they are, but allowing the connection matrix $C_{i_1 j_1 i_2 j_2}$ to have values different from 0 (and not constrained to be 1) if the points (i_1, j_1) and (i_2, j_2) are close enough to each other. If one wants to keep an interaction term which is translation invariant, C must be of the form

$$C_{i_1 j_1 i_2 j_2} = C(i_1 - i_2, j_1 - j_2), \tag{13}$$

with the function C being nonzero only for sufficiently low values of its arguments.

Since their introduction in the literature [7], a large amount of work has been done on these models. Onsager [8] found an exact analytical solution for the model described by (10) for all values of the coupling constant J; he also showed the existence of a critical temperature (for which an expression, depending only on J, can be written) at which a phase transition occurs. This has important physical interpretations, and is consistent with the behavior of ferromagnetic systems as a function of the temperature, in particular the existence of the Curie temperature at which ferromagnetism is known to disappear. Unfortunately, efforts to find analytical solutions for the model with an external field have been so far unsuccessful; even for the uniform case (γ_{ij} position independent) only approximate solutions are available.

Figure 3

Parameter estimation: initial random configuration of rectangles.

We now turn to the problem of smoothing an image. We consider a bi-level (0, 1) square image α_{ij} with $1 \le i \le N$ and $1 \le j \le N$. Given α_{ij}, we want to find a smoothed image β_{ij}. Following the approach suggested in [4], β is chosen to be the image that minimizes a two-part cost function. The first part of the cost function, R, measures the "roughness" of the smoothed image β: Ideally the smoothed image should not be "rough" at all. The second part of the cost function, D, is a measure of the discrepancy between the smoothed image β

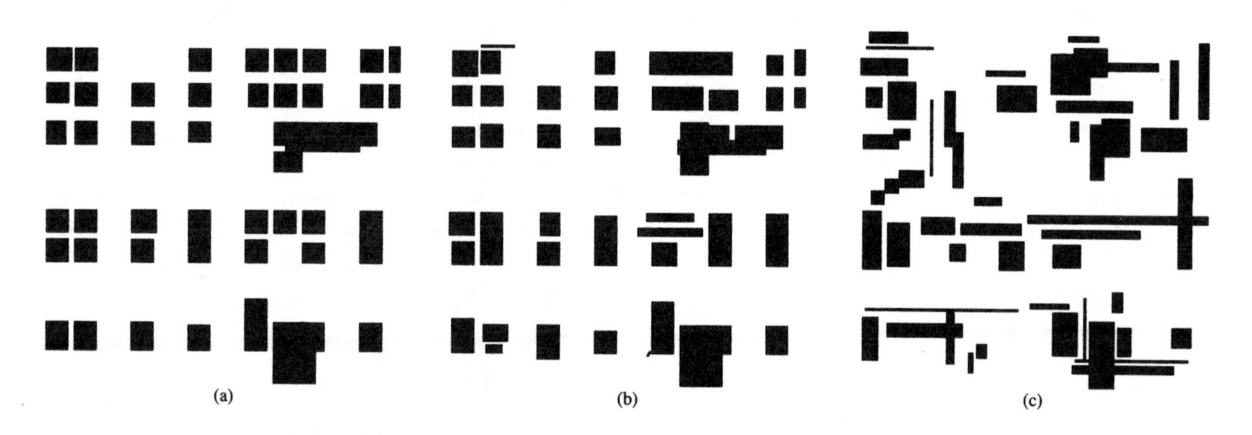

Figure 4

Parameter estimation: results of simulated annealing on images of Fig. 2.

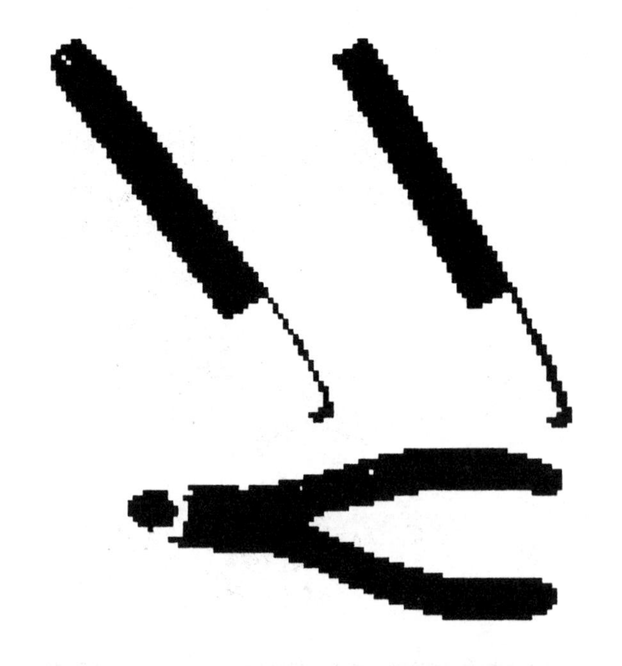

Figure 5

Image smoothing: noiseless bitmap.

and the original image α, and has the effect of preventing the solution β from differing too much from the given image to be smoothed. The cost function to be minimized is then taken to have the form

$$E = R + \lambda D, \tag{14}$$

where λ is a number that parametrizes the desired trade-off between roughness and discrepancy for the smoothed image β.

For D we can take either the L^1 or the L^2 distance between α and β: As we have already mentioned, they are equivalent for these bi-level images since, for any values for α and β in (0, 1), one has trivially $|\alpha - \beta| = (\alpha - \beta)^2$. Thus we can choose

$$D = \sum_{ij} (\alpha_{ij} - \beta_{ij})^2. \tag{15}$$

For R, the two simplest choices proposed in [4] (the digital Laplacian and the digital gradient magnitude) happen to be equivalent. Both choices result in

$$R = \sum_{i_1=1}^{N} \sum_{j_1=1}^{N} \sum_{i_2=1}^{N} \sum_{j_2=1}^{N} C_{i_1j_1i_2j_2}(\beta_{i_1j_1} - \beta_{i_2j_2})^2, \tag{16}$$

where, as in (9), $C_{i_1j_1i_2j_2}$ is a connection matrix. In this way the problem of finding β is well defined, and consists of minimizing E given by (14) for given α_{ij} and λ, with D and R defined as in (15) and (16).

This problem is formally equivalent to finding the ground state of an Ising system imbedded in an external field. To show this, we have to reformulate the problem in terms of a bi-level image with values $(-1, 1)$ instead of (0, 1). Thus we define

$$\gamma_{ij} = 2\alpha_{ij} - 1 \tag{17}$$

and

$$\mu_{ij} = 2\beta_{ij} - 1. \tag{18}$$

Since α and β can only take the values 0 and 1, γ and μ correspondingly take the values -1 and $+1$. We need to express D and R in terms of these new variables.

We first write the following identities, which can be easily verified:

$$\gamma_{ij}^2 = 1; \tag{19}$$

$$\mu_{ij}^2 = 1; \tag{20}$$

$$(\alpha_{ij} - \beta_{ij})^2 = \frac{1 - \gamma_{ij}\mu_{ij}}{2}; \tag{21}$$

$$(\beta_{i_1j_1} - \beta_{i_2j_2})^2 = \frac{1 - \mu_{i_1j_1}\mu_{i_2j_2}}{2}. \tag{22}$$

Using these identities, we get the following new expressions for D and R:

$$D = \frac{1}{2}N^2 - \frac{1}{2}\sum_{ij} \gamma_{ij}\mu_{ij}; \tag{23}$$

$$R = \frac{1}{2}\sum_{i_1j_1i_2j_2} C_{i_1j_1i_2j_2} - \frac{1}{2}\sum_{i_1j_1i_2j_2} C_{i_1j_1i_2j_2}\beta_{i_1j_1}\beta_{i_2j_2}. \tag{24}$$

The first term in R is simply half the number of neighbor pairs and equals $N^2(N^2 - 1)$. Thus,

$$R = N^2(N^2 - 1) - \frac{1}{2}\sum_{i_1j_1i_2j_2} C_{i_1j_1i_2j_2}\mu_{i_1j_1}\mu_{i_2j_2}. \tag{25}$$

Expression (14) now becomes

$$E = -\frac{1}{2}\sum_{i_1j_1i_2j_2} C_{i_1j_1i_2j_2}\mu_{i_1j_1}\mu_{i_2j_2} - \frac{\lambda}{2}\sum_{ij} \gamma_{ij}\mu_{ij}, \tag{26}$$

where we have dropped the constant terms which do not depend on μ. We can do that because addition of a term independent of μ does not change the point where E reaches its absolute minimum. This is also consistent with the well-known fact that the energy of a physical system can be redefined by a constant (or by a quantity independent from the degrees of freedom that describe the system) without changing the physics. Moreover, we can also multiply the energy by a positive constant without altering its physical features; in our case, if we choose the factor $1/\lambda$, we obtain the expression

$$E = -\frac{1}{2}\sum_{ij} \gamma_{ij}\mu_{ij} - \frac{1}{2\lambda}\sum_{i_1j_1i_2j_2} C_{i_1j_1i_2j_2}\mu_{i_1j_1}\mu_{i_2j_2}, \tag{27}$$

which is identical to (11) provided that $J = 1/\lambda$. Thus we have shown the complete equivalence of the smoothing

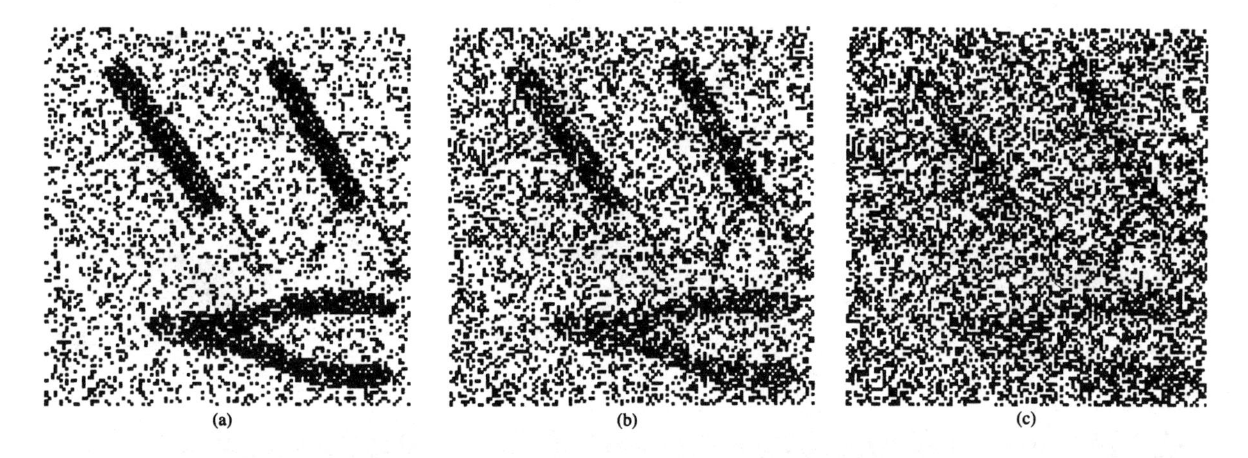

Figure 6

Image smoothing: noise-corrupted bitmaps. The image in Fig. 5 has been corrupted by adding random noise. Signal-to-noise ratios are 1.5, 0.87, and 0.41 in (a), (b), and (c), respectively.

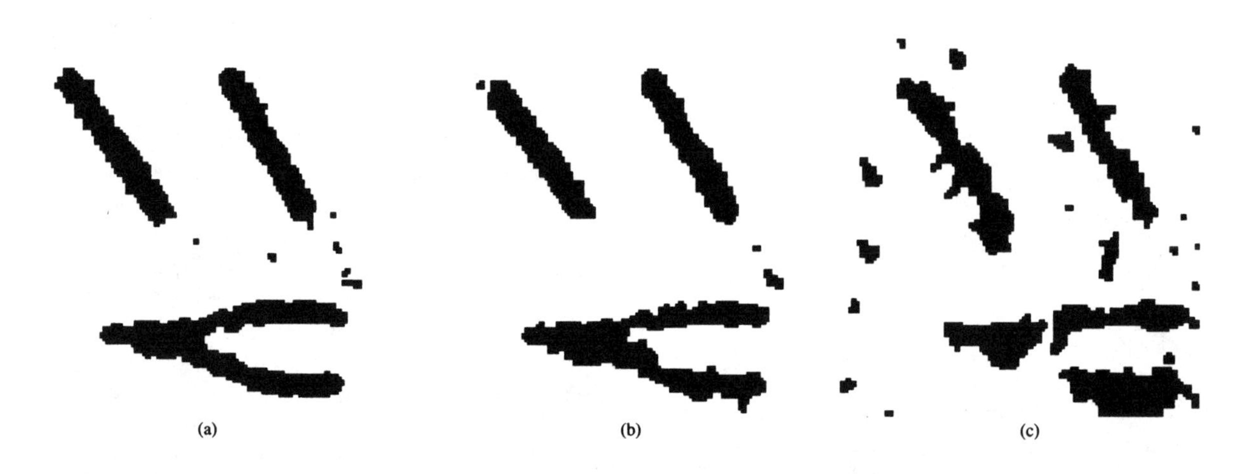

Figure 7

Results of simulated annealing on images of Fig. 6 obtained with $\lambda = 1$.

problem and the ground state problem for an Ising system imbedded in an external field. The image to be smoothed determines γ and thus plays the role of the external field.

The method has been tested using the original bitmap shown in **Figure 5** and corrupting it as in the problem of parameter estimation previously described. **Figure 6** shows the corrupted versions corresponding to $r = 0.2$, $r = 0.3$, and $r = 0.4$ (signal-to-noise ratios 1.5, 0.87, and 0.41). **Figures 7** and **8** show the results of simulated annealing, with different values of λ.

For comparison, **Figure 9** shows the results of applying the usual image processing algorithms to the same bitmap.

Image halftoning and antiferromagnetic systems

In the previous section we have shown how the process of smoothing a bitmap is formally equivalent to a ground state problem for a ferromagnetic Ising model imbedded in an external field. In this section we show that the same problem, but with a reversed sign for the spin-spin interaction (antiferromagnetic coupling), is equivalent to the problem of *halftoning*, i.e., of calculating a bi-level image (bitmap) whose average density mimics the one of a given continuous image.

Let γ_{ij} be the given continuous image with values in the interval $(-1, 1)$ and μ_{ij} the resulting bitmap with values -1,

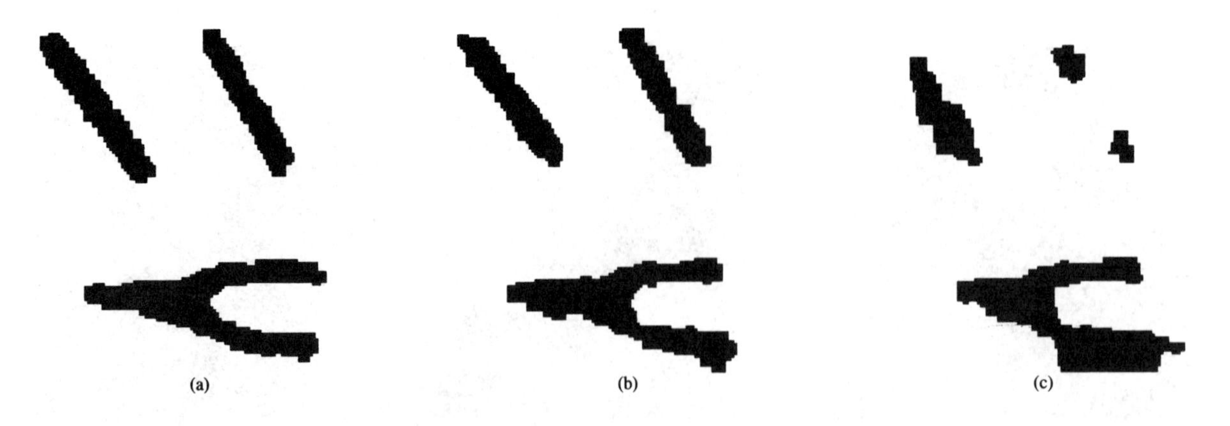

Figure 8

Results of simulated annealing on images of Fig. 6 obtained with $\lambda = 2$.

1. If one wishes to define the cost function for such a problem, one may try to minimize the difference between the continuous-level image and an average density of the bitmap, which can be suitably defined through the introduction of a filter V:

$$\rho_{ij} = \sum_k \sum_l V_{ijkl}\mu_{kl} \,. \tag{28}$$

The general properties of V_{ijkl}, in order to deal with a proper average, are the following:

$$\sum_k \sum_l V_{ijkl} = 1 \qquad V_{ijkl} \geq 0. \tag{29}$$

Moreover, if we ask for an average operation which is the same for the whole lattice, we have $V_{ijkl} = V_{i-k,j-l}$. In principle one can choose to extend the average over a large spatial extent, but it is more practical to restrict it to a small region around the spin μ_{ij}. With this choice V is a square matrix whose dimension is related to the spatial size of the average. The form of the cost function, choosing again the L^2 norm to measure the distance between the original image and the average introduced, is the following:

$$E = \sum_i \sum_j (\gamma_{ij} - \rho_{ij})^2. \tag{30}$$

Developing this expression, neglecting terms which do not depend on μ_{ij} and rescaling the energy by a factor $\frac{1}{2}$ (as in the previous paragraph), we obtain

$$E = -\frac{1}{2}\sum_{ij} \tilde{\gamma}_{ij}\mu_{ij} + \frac{1}{2}\sum_{i_1j_1i_2j_2} L_{i_1j_1i_2j_2}\mu_{i_1j_1}\mu_{i_2j_2}, \tag{31}$$

where we defined

$$\tilde{\gamma}_{ij} = \sum_k \sum_l V_{ijkl}\gamma_{kl} \tag{32}$$

and

$$L_{ijkl} = \sum_n \sum_m V_{ijnm}V_{nmkl} \,, \tag{33}$$

which is the convolution of the filter V_{ijkl} with itself. We recognize that the expression in (31) is of the general type (11); in fact, $\tilde{\gamma}_{ij}$ plays the role of the external field, L_{ijkl} is a connection matrix, and the constant J has the value -1. This shows the equivalence between image halftoning and the ground state problem for an Ising system. Due to the sign of J and to the properties of V_{ijkl}, this term in the energy is to be regarded as an antiferromagnetic interaction, which tends to align two neighboring spins in opposite directions. In the context of image processing the two terms have a straightforward interpretation: While the magnetic field term has the function of keeping the bitmap as similar as possible to the original image, the antiferromagnetic part of the energy is intended to produce the diffusion effect proper of halftoning. We performed the annealing on a 256 × 256 image with 256 gray levels; the resulting bitmaps for different filters V are shown in **Figure 10**. We discuss later the details of the filters V used and the influence of the choice of the filter on the result.

From the statistical mechanics point of view, we are in the presence of a complex general interaction: an average magnetic field, strongly depending on the position, which would tend to align the spins according to its direction, and the internal field which might be in conflict with it. In other words, it is possible to observe for this system the frustration effects already mentioned in [2]: A system subjected to different physical constraints may show very peculiar behavior in the annealing procedure; in fact, due to this competition, low-energy states may be highly degenerate and a great many local minima very near to the ground state energy may appear.

This statistical model is interesting *per se*, so we devote part of this section to the study of its thermodynamical features. A discussion of the issues concerning image processing follows.

By applying the annealing procedure in the case $\gamma = 0$ (no magnetic field), we observed that no phase transition occurs going from high to low temperature regardless of the filter V used. This is quite a common feature in two-dimensional antiferromagnetic spin systems.

To investigate the critical behavior of the system, we first analyzed the simplified case of uniform magnetic fields (i.e., $\gamma_{ij} \equiv \gamma$, constant over the whole lattice) and then observed the implications of such analysis on a real image with varying gray level.

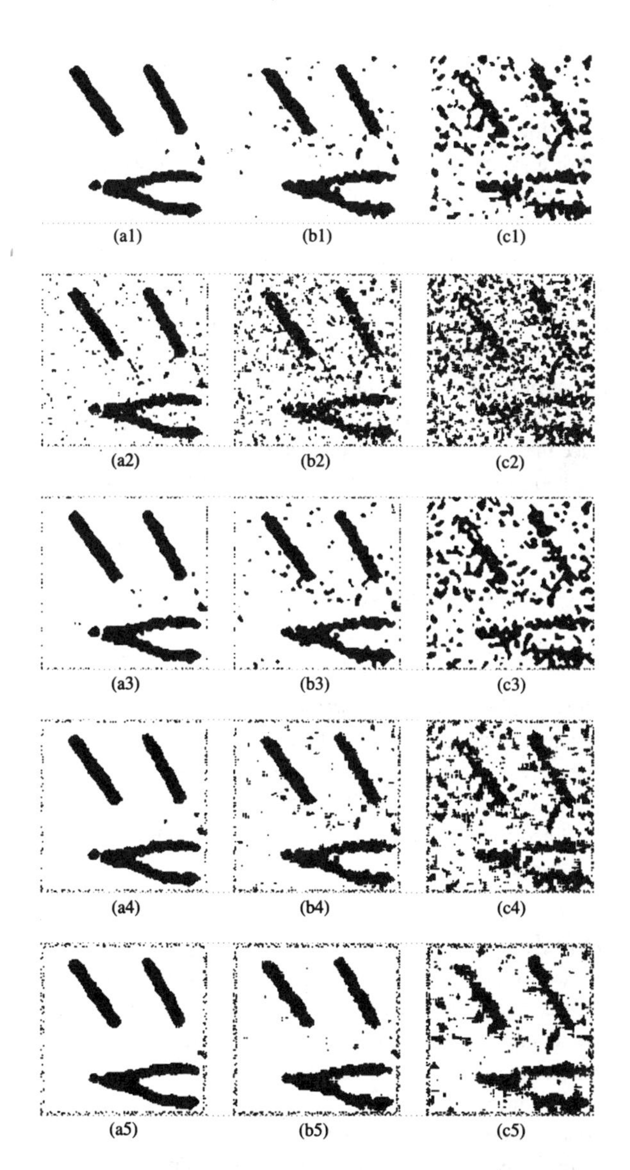

Figure 9

Image smoothing: application of standard algorithms. The bitmaps in Fig. 6 have been smoothed using the following standard algorithms: 3×3 low-pass filter iterated three times (a1, b1, c1); 3×3 median filter (a2, b2, c2); 3×3 median filter iterated three times (a3, b3, c3); 5×5 (a4, b4, c4), and 7×7 (a5, b5, c5) median filters. Comparison of these results with those of Figs. 7 and 8 shows that in the presence of low signal-to-noise ratios, simulated annealing yields comparatively better smoothing.

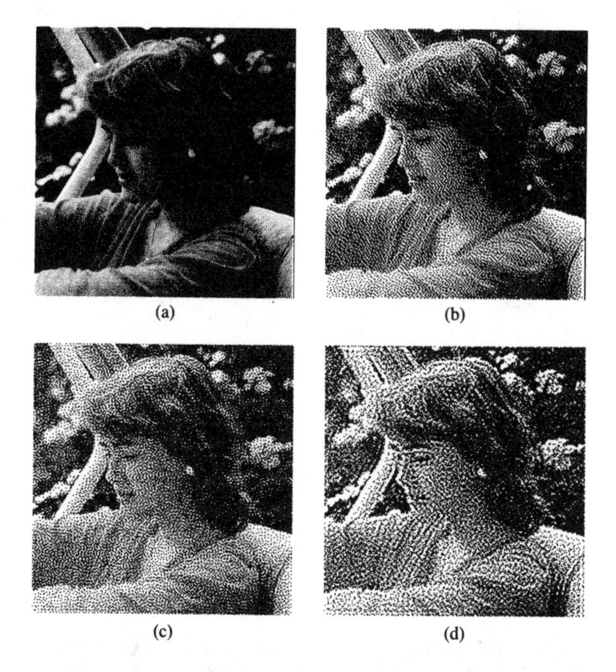

Figure 10

Image halftoning by simulated annealing. (a) Original 256×256, 8-bit/pel image; (b) result of halftoning by the standard MECCA algorithm [9]; (c) result of simulated annealing with a uniform 3×3 filter V; (d) result of simulated annealing with a uniform 5×5 filter V. Comparison of (c) and (d) shows the trade-off between spatial and tonal resolution.

A very useful parameter for understanding the response of the system to an applied magnetic field is the so-called order parameter (also called net magnetization), which is defined as

$$M \equiv \left| \sum_i \sum_j \frac{\mu_{ij}}{N^2} \right|. \tag{34}$$

If M equals 1, then the system is in a completely ordered state, i.e., all spins point to the same direction; conversely, in the absence of order, M would equal 0. We fixed γ and performed the annealing, bringing the system toward zero temperature; thus we reached approximately the ground state for the given γ and we measured its magnetization. Repeating this procedure for increasing values of γ, we found the occurrence of a phase transition driven by the external field: There is a critical value γ_c under which the magnetization grows linearly with the strength of the field; for $\gamma > \gamma_c$ there is an abrupt change in this response, and the magnetization jumps to the maximum value, i.e., $M = 1$ (**Figure 11**). To determine the location of this transition, we

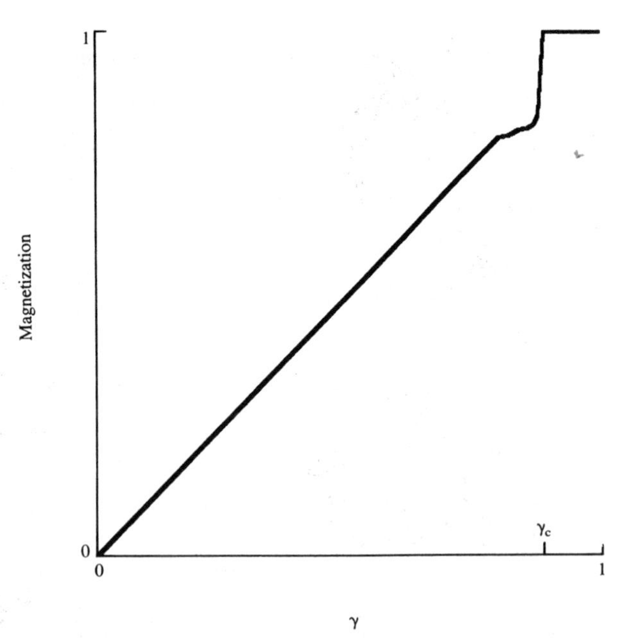

Figure 11

Net magnetization versus external field at $T=0$. This plot has been obtained by simulated annealing using a position-independent γ and a uniform 3×3 filter V.

present the following argument. Let us consider a magnetic field γ very near to 1, i.e., $\gamma = 1 - 2\epsilon$, with ϵ very small and positive (due to the symmetry of the problem, we could have chosen $\gamma = -1 + 2\epsilon$ with identical results). We can argue on intuitive grounds that if ϵ is chosen very near to 0, the state with all $\mu_{ij} = 1$ is the most favorable energetically; i.e., it is the ground state of the system. To confirm this picture, we compare the energy E_0 of this configuration with that of the state with a single spin reversed, E_1 (this is often indicated as an elementary excitation of the system); to be definite, we say that $\mu_{ij} = -1$ if $i = j = 0$ and $\mu_{ij} = 1$ otherwise, but translational symmetry guarantees the generality of the argument. Using Eq. (30) to compute the two energies, we find

$$\Delta E \equiv E_0 - E_1 = 2\epsilon - \sum_i \sum_j V_{ij}^2 . \tag{35}$$

Thus $\Delta E \leq 0$ if $\epsilon \leq \frac{1}{2} \sum_i \sum_j V_{ij}^2$; defining $\epsilon_c = \frac{1}{2} \sum_i \sum_j V_{ij}^2$, we have $\gamma_c = 1 - 2\epsilon_c$. In fact, when $\gamma > \gamma_c$ the state with all $\mu_{ij} = 1$ is the stable phase of the system, and $M = 1$; when γ is lowered under γ_c, this configuration becomes energetically unstable with respect to an elementary excitation, so we must conclude that the ground state will have a more complex structure, due to the increasing relevance of antiferromagnetic interaction. We notice that the critical field is near to 1 as $\sum_i \sum_j V_{ij}^2$ gets smaller. For the case in Fig. 11, where V is a uniform 3×3 filter, one obtains $\gamma_c = 0.89$, which agrees with the result in the figure, obtained by simulating annealing.

Let us see what this analysis implies for image processing. A uniform region of the original image with a gray level very near to black (white) corresponds to a subset of the spin lattice under the influence of a field very near to 1 (-1). The location of the phase transition (and its dependence on the specific filter used) tells us to which extent intermediate levels may be controlled: Thus a gray region very near to black ($\gamma > \gamma_c$) will be converted by the annealing procedure into a region of the bitmap completely black (with all $\mu_{ij} = 1$); in other words, the expression $\sum_i \sum_j V_{ij}^2$ is a measure of the tonal resolution of the filter.

For a given spatial extent of the filter (that is to say, for a given dimension of the matrix V), the best filter in terms of tonal resolution is the one in which the average is equally weighted for all the neighboring pixels, i.e., a matrix V with all the elements equal. In this case we have $\epsilon_c = 1/2m^2$ (where m is the size of V), from which we can deduce that the tonal resolution increases quadratically with the spatial extent of the average (or linearly with the number of pixels involved in the average). Of course, one is forced to pay a price in terms of spatial resolution of the image. The loss of spatial resolution due to the averaging can be measured by

$$\sigma^2 \equiv \sum_i \sum_j V_{ij}(i^2 + j^2); \tag{36}$$

we obtain for this class of filters

$$\sigma^2 = 2m \frac{(m+1)}{3} . \tag{37}$$

Thus, spatial resolution decreases when tonal resolution increases and vice versa. For large m, spatial resolution is inversely proportional to tonal level resolution. Difficulties deriving from this trade-off could be overcome by using a modified filter (with larger m) only in regions for which $|\gamma| > \gamma_c$.

Conclusions

It has been shown that seemingly unrelated problems in image processing, such as smoothing or halftoning a bitmap, are analogous to ground state problems for spin systems with different kinds of interaction: ferromagnetic and antiferromagnetic, respectively. Simulated annealing is a good tool to solve these problems, as well as other problems in image processing, such as the one of parameter estimation described in this paper. In the case of highly noise-corrupted bitmaps (both in parameter estimation and image smoothing problems), this technique gives better results than existing methods. As far as halftoning is concerned, results are comparable with those of standard algorithms. The main drawback of the annealing procedure is that computer requirements are larger and, in general, not easily predictable. On the other hand, the connection between image processing and Ising spin systems is important because the extensive results on Ising spin systems might provide useful hints when approaching the corresponding problems in image processing.

References

1. S. Kirkpatrick, C. D. Gelatt, and M. P. Vecchi, "Optimization by Simulated Annealing," *Science* **220,** 671 (1983).
2. S. Kirkpatrick, "Optimization by Simulated Annealing: Quantitative Studies," *J. Statist. Phys.* **34,** 975 (1984).
3. N. Metropolis, A. Rosenbluth, M. Rosenbluth, A. Teller, and E. Teller, "Equation of State Calculations by Fast Computing Machines," *J. Chem. Phys.* **21,** 1087 (1953).
4. K. A. Narayanan, D. O'Learly, and A. Rosenfeld, "Image Smoothing and Segmentation by Cost Minimization," *IEEE Trans. Syst., Man, Cyber.* **SMC-12,** 91 (1982).
5. K. Huang, *Statistical Physics*, John Wiley & Sons, Inc., New York, 1963, pp. 329–373.
6. H. E. Stanley, *Introduction to Phase Transition and Critical Phenomena*, Oxford University Press, Oxford, England, 1971.
7. E. Ising, "A Contribution to the Theory of Ferromagnetism," *Z. Phys.* **31,** 253 (1925).
8. L. Onsager, "A Two-Dimensional Model with an Order-Disorder Transition," *Phys. Rev.* **65,** 117 (1944).
9. P. Stucki, "MECCA—A Multiple-Error Correction Computation Algorithm for Bi-Level Image Hardcopy Reproduction," *Research Report RZ-1060*, IBM Research Laboratory, Zurich, Switzerland, 1981.

Received February 11, 1985; revised June 10, 1985

Paolo Carnevali *IBM Italy, Scientific Center, Via Giorgione, 129, 00147 Rome, Italy.* Dr. Carnevali is currently working at the Rome Scientific Center in the parallel processing group. He obtained his degree in physics from the University of Rome in 1979 and subsequently spent two years at the IBM Thomas J. Watson Research Center in Yorktown Heights, New York, as part of a postdoctoral fellowship program, working on the dynamics of galaxies and clusters of galaxies. Since joining IBM Italy in 1982, he has worked in the fields of solid-state physics, parallel computing, and image processing. In 1983 Dr. Carnevali received an IBM Outstanding Technical Achievement Award for his work on the IBM 3838 Array Processor.

Lattanzio Coletti *IBM Italy, Scientific Center, Via Giorgione, 129, 00147 Rome, Italy.* Dr. Coletti joined the Rome Scientific Center in 1982 as a systems engineer. His current research is in image acquisition and processing, specifically in image acquisition devices connected to IBM computers and in restoration of blurred images. He was involved in the development of the IBM 7350 Image Processing System (previously known as HACIENDA). Dr. Coletti received his degree in electronics from Rome University in 1980.

Stefano Patarnello *IBM Italy, Scientific Center, Via Giorgione, 129, 00147 Rome, Italy.* Dr. Patarnello obtained his degree in physics from the University of Pisa in 1983. His main interest has been in statistical physics, with emphasis on the application of renormalization group techniques to the critical behavior of systems involving multivalued spins. Dr. Patarnello joined the IBM Rome Scientific Center in 1985, and at present is involved in large-scale numerical computations in fluid dynamics.

Stuart Geman and Donald Geman, "Stochastic Relaxation, Gibbs Distributions, and the Bayesian Restoration of Images"

IEEE PAMI, **6(6):721-41, November 1984.**

Grahame B. Smith
SRI International, Menlo Park, California 94025

An original image has been blurred by some point spread function, and this blurred image has been corrupted by noise. The goal of processing is to recover the original image. The processing is based on a user-provided model that specifies the likelihood of a pixel having an intensity value similar to those of its neighbors (i.e., the spatial coherence in images), and on the data provided by the observed degraded image. The Geman and Geman paper is more general than this, showing how to reconstruct an original data set from the degraded observation of that data set using a model of the interaction between neighboring elements of the data set.

The Geman and Geman paper can be divided into three parts: (1) The first part draws together ideas and theorems from the literature on Markov Random Fields (MRF) that allow one to specify the maximum *a posteriori* estimate of the state of the MRF given degraded observations of that MRF. The MRF is used as the formalism for describing images, and established theorems provide a means for specifying the probability of a particular original image given the observed degraded image. (2) The second part introduces the technique of *simulated annealing* as a mechanism for finding the image that maximizes the probability of it being a replica of the original image given the observations, and establishes convergence properties for these annealing procedures. (3) The third part introduces a model of spatial coherence in images. This model explicitly permits the placement of image boundaries that terminate this coherence.

1. Ideas and Theorems

A MRF is a lattice of pixels – for example, an image. Each pixel can be assigned any of its allowed values. Consequently, any configuration of the system is possible – that is, all "images" can be generated. Also, in a MRF the conditional probability that a pixel has a certain value, given the values of all the other pixels in the image, is only a function of the pixels in a finite neighborhood of that pixel, not a function of all the pixels in the entire image.

Known theorems now state that these conditional probabilities are sufficient to specify uniquely the probability that the system is in a particular state, and that this probability is a Gibbs distribution whose form is particularly simple to calculate. The form is simple since it involves a sum, one term for each pixel where each term is only a function of neighboring pixels. Hence the probability that the system is in a particular state can be calculated by summing the effects of local interactions between a pixel and the pixels in its neighborhood.

Probability (system is in a particular state) = $(\exp(-U/T))/\text{normalizer}$,
where T is a parameter and U is a sum of functions, one for each pixel in the image, that specify how neighboring values influence the probability that the given pixel has certain values. This form of probability distribution is known as the Gibbs distribution. U is a sum of *potentials* that specify how each neighbor, pair of neighbors, triple of neighbors, and so on contributes to the probability that the pixel in question has a certain value. A physical analogy may be helpful. In the case of magnetism these potentials specify how the spin of a particle is affected by the 2-body interaction between the particle and each of its neighbors, how the spin is affected by the 3-body interaction with its neighbors (i.e., the particle under question and neighbors taken two at a time), and so on for higher order interactions. In the case of image restoration these potentials amount to our understanding of spatial coherence, and the potentials reflect how neighboring pixels taken one at a time and two at a time and so on change the probability that the central pixel has certain values.

The probability discussed here is the *a priori* probability – the probability before we have observed the data. If we were to find the state that maximized this probability, we would find the state that is most likely given the interaction model we have between neighboring pixels, and given the restriction that the probability that a pixel has a certain state is only a function of a limited neighborhood of that pixel, not a function of the state of the whole image. This limited neighborhood is just the neighborhood that we used to specify the potentials that affected each pixel. So if we use just nearest neighbors to express the potentials, we are saying that only nearest neighbors influence the state of the central pixel.

Of course, we would like to find the state that maximizes the *a posteriori* probabilities – that is, the maximum probability state given the observed data. Fortunately, there is a theorem that helps in this regard; it says the the probability that a MRF system is in a particular state given the observed data is Gibbs distributed where the U just calculated has added to it an extra term involving the difference between the actual observed data and what would be observed if the current state was the original image. This extra term accounts for the difference between observed and predicted data in terms of noise in the observed image. The neighborhood over which the interaction is calculated is changed a little from the *a priori* case supplementing that neighborhood with one related to the neighborhood over which the point spread function blurs the image.

Hence the ideas and theorems culled from the literature of MRF allows one to say: If you can describe an image as a MRF in which the probability that a pixel has a certain value is only a function of a finite neighborhood of pixels and you can express this function as an interaction potential between neighboring pixels, then the probability that the system is in a particular state given the observed degraded image is a Gibbs distribution in which the U function, the *energy* function, consists of two terms – one associated with the interaction potential and one associated with the difference between the predicted image and the actual observed data. Reconstructing the original image then means finding the MRF state that maximizes the probability of that state given the observed data. The paper uses an annealing technique to find this maximum.

2. Simulated Annealing and its Convergence Properties

In the expression

Probability (system is in a particular state) = $(\exp(-U/T))/\text{normalizer}$

T is a parameter that alters the peakiness of the probability distribution. The state that has the maximum probability is independent of T. If the number of possible states is large, it may take a long time to find the state that maximizes the probability simply by examining random states. The state that has maximum probability is the one for which U is a minimum. As we decrease T, the relative probability of the optimum state to the probability of suboptimal states is greater than at higher T. Consequently, if we want to generate a sequence of states that will converge on the optimum state somewhat faster than a random sequence, we might start with a high value of T when the distribution is broad and the relative probability of many states are similar, pick a sequence of states from the $\exp(-U/T)$ distribution until we are "in the ball park" of the distribution peak, lower T to sharpen the peak and pick a further sequence of states, and so on. Occasionally we need to sequence through a state that is less likely than a previous one to avoid being caught in a local minimum of the $\exp(-U/T)$ surface.

Simulated annealing is a procedure for deciding what states to sequence through and which ones to reject from being in the sequence. Annealing is an algorithm that generates a sequence of states that are supposed to be a representative sample drawn from the $\exp(-U/T)$ distribution. As this sequence progresses, T is decreased – the notion being that one will find a state for which U is minimum and hence the probability of that state will be a maximum. The Geman and Geman paper proves two theorems. One shows that the algorithm they use (the usual algorithm in the literature) does indeed draw samples representative of the $\exp(-U/T)$ distribution irrespective of the starting state. The second shows that if T is decreased logarithmically with respect to the number of states that have been sequenced through, then the procedure will converge on the state that minimizes U and hence maximizes the probability of that state, again, irrespective of the starting state. These theorems formalize previous intuitions.

(A third theorem says that if you want to calculate an measurable quantity of the system, such as a function of image intensity like its mean, then you can calculate that function by averaging the values of the function calculated for each of the last n states of the annealing sequence. n should be large.)

While the convergence theorem formalizes previous intuitions, it does so only for an annealing procedure that is "too slow" – that is, T is decreased too slowly to make the algorithm practical. However, it is a lower bound that justifies in part the notion of finding a optimum state by simulated annealing.

3. Image Models

When selecting the form of the potentials used to calculate U (and hence to determine the probability that the central pixel has a certain value given the state of the neighborhood) we can include in these potentials features of the neighborhood that are not just functions of the intensity values of neighboring pixels. If a neighborhood consists of both neighboring pixels and intervening edges, then a potential can be formed that is a function of both these aspects. The paper uses two properties of a neighborhood in the model it uses for spatial coherence. First, it uses the expected similarity between intensity values of neighboring pixels, and, second, it assumes that edges may be present between pixels; if they are present, they prevent pixels across that edge from influencing one another.

The problem solved in the paper is one of image restoration in which the original image consists of pixel values and lines. The aim is to find the state – a set of pixel values and a set of line values – that maximizes the *a posteriori* probability of that state given the observed data.

The model introduced is a general one: Any aspects of a neighborhood that are representable as interaction potentials can be modelled. The original system state – that is, the source of the observed data – is the maximum *a posteriori* estimate from the Gibbs distribution based on those interaction potentials.

Stochastic Relaxation, Gibbs Distributions, and the Bayesian Restoration of Images

STUART GEMAN AND DONALD GEMAN

Abstract—We make an analogy between images and statistical mechanics systems. Pixel gray levels and the presence and orientation of edges are viewed as states of atoms or molecules in a lattice-like physical system. The assignment of an energy function in the physical system determines its Gibbs distribution. Because of the Gibbs distribution, Markov random field (MRF) equivalence, this assignment also determines an MRF image model. The energy function is a more convenient and natural mechanism for embodying picture attributes than are the local characteristics of the MRF. For a range of degradation mechanisms, including blurring, nonlinear deformations, and multiplicative or additive noise, the posterior distribution is an MRF with a structure akin to the image model. By the analogy, the posterior distribution defines another (imaginary) physical system. Gradual temperature reduction in the physical system isolates low energy states ("annealing"), or what is the same thing, the most probable states under the Gibbs distribution. The analogous operation under the posterior distribution yields the maximum *a posteriori* (MAP) estimate of the image given the degraded observations. The result is a highly parallel "relaxation" algorithm for MAP estimation. We establish convergence properties of the algorithm and we experiment with some simple pictures, for which good restorations are obtained at low signal-to-noise ratios.

Index Terms—Annealing, Gibbs distribution, image restoration, line process, MAP estimate, Markov random field, relaxation, scene modeling, spatial degradation.

I. Introduction

THE restoration of degraded images is a branch of digital picture processing, closely related to image segmentation and boundary finding, and extensively studied for its evident practical importance as well as theoretical interest. An analysis of the major applications and procedures (model-based and otherwise) through approximately 1980 may be found in [47]. There are numerous existing models (see [34]) and algorithms and the field is currently very active. Here we adopt a Bayesian approach, and introduce a "hierarchical," stochastic model for the original image, based on the *Gibbs distribution*, and a new restoration algorithm, based on stochastic relaxation and *annealing*, for computing the maximum *a posteriori* (MAP) estimate of the original image given the degraded image. This algorithm is highly parallel and exploits the equivalence between Gibbs distributions and *Markov random fields* (MRF).

Manuscript received October 7, 1983; revised June 11, 1984. This work was supported in part by ARO Contract DAAG-29-80-K-0006 and in part by the National Science Foundation under Grants MCS-83-06507 and MCS-80-02940.

S. Geman is with the Division of Applied Mathematics, Brown University, Providence, RI 02912.

D. Geman is with the Department of Mathematics and Statistics, University of Massachusetts, Amherst, MA 01003.

The essence of our approach to restoration is a stochastic relaxation algorithm which generates a sequence of images that converges in an appropriate sense to the MAP estimate. This sequence evolves by *local* (and potentially *parallel*) changes in pixel gray levels and in locations and orientations of boundary elements. Deterministic, iterative-improvement methods generate a sequence of images that monotonically increase the posterior distribution (our "objective function"). In contrast, stochastic relaxation permits changes that *decrease* the posterior distribution as well. These are made on a *random* basis, the effect of which is to avoid convergence to *local maxima*. This should not be confused with "probabilistic relaxation" ("relaxation labeling"), which is deterministic; see Section X.

The stochastic relaxation algorithm can be informally described as follows.

1) A local change is made in the image based upon the current values of pixels and boundary elements in the immediate "neighborhood." This change is *random*, and is generated by sampling from a local conditional probability distribution.

2) The local conditional distributions are dependent on a global control parameter T called "temperature." At *low* temperatures the local conditional distributions concentrate on states that *increase* the objective function, whereas at high temperatures the distribution is essentially uniform. The limiting cases, $T = 0$ and $T = \infty$, correspond respectively to greedy algorithms (such as gradient ascent) and undirected (i.e., "purely random") changes. (High temperatures induce a loose coupling between neighboring pixels and a chaotic appearance to the image. At low temperatures the coupling is tighter and the images appear more regular.)

3) Our image restorations avoid local maxima by beginning at high temperatures where many of the stochastic changes will actually decrease the objective function. As the relaxation proceeds, temperature is gradually lowered and the process behaves increasingly like iterative improvement. (This gradual reduction of temperature simulates "annealing," a procedure by which certain chemical systems can be driven to their low energy, highly regular, states.)

Our "annealing theorem" prescribes a schedule for lowering temperature which guarantees convergence to the global maxima of the posterior distribution. In practice, this schedule may be too slow for application, and we use it only as a guide in choosing the functional form of the temperature-time dependence. Readers familiar with Monte Carlo methods in statistical physics will recognize our stochastic relaxation algorithm as a "heat bath" version of the *Metropolis algorithm* [42]. The idea of introducing temperature and simulating annealing is due to Černý [8] and Kirkpatrick *et al.* [40], both of whom used it for combinatorial optimization, including the traveling salesman problem. Kirkpatrick also applied it to computer design.

Since our approach is Bayesian it is model-based, with the "model" captured by the prior distribution. Our models are "hierarchical," by which we mean layered processes reflecting the type and degree of *a priori* knowledge about the class of images under study. In this paper, we regard the original image as a pair $X = (F, L)$ where F is the matrix of observable pixel intensities and L denotes a (dual) matrix of unobservable edge elements. Thus the usual gray levels are considered a marginal process. We refer to F as the *intensity process* and L as the *line process*. In future work we shall expand this model by adjoining other, mainly geometric, attribute processes.

The degradation model allows for noise, blurring, and some nonlinearities, and hence is characteristic of most photochemical and photoelectric systems. More specifically, the degraded image G is of the form $\phi(H(F)) \odot N$, where H is the blurring matrix, ϕ is a possibly nonlinear (memoryless) transformation, N is an independent noise field, and $\odot$ denotes any suitably invertible operation, such as addition or multiplication. Surprisingly, these nonlinearities do not affect the computational burden.

To pin things down, let us briefly discuss the Markovian nature of the intensity process; similar remarks apply to the line process, the pair (F, L), and the distribution of (F, L) conditional on the "data" G. Of course, all of this will be discussed in detail in the main body of the paper.

Let $Z_m = \{(i, j): 1 \leq i, j \leq m\}$ denote the $m \times m$ integer lattice; then $F = \{F_{i,j}\}$, $(i, j) \in Z_m$, denotes the gray levels of the original, digitized image. Lowercase letters will denote the values assumed by these (random) variables; thus, for example, $\{F = f\}$ stands for $\{F_{i,j} = f_{i,j}, (i, j) \in Z_m\}$. We regard F as a sample realization of a random field, usually isotropic and homogeneous, and with significant correlations well beyond nearest neighbors. Specifically, we model F as an MRF, or, what is the same (see Section IV), we assume that the probability law of F is a Gibbs *distribution*. Given a *neighborhood system* $\mathcal{F} = \{\mathcal{F}_{i,j}, (i, j) \in Z_m\}$, where $\mathcal{F}_{i,j} \subseteq Z_m$ denotes the neighbors of (i, j), an MRF over $(Z_m, \mathcal{F})$ is a stochastic process indexed by Z_m for which, for every (i, j) and every f,

$$P(F_{i,j} = f_{i,j} | F_{k,l} = f_{k,l}, (k, l) \neq (i, j))$$
$$= P(F_{i,j} = f_{i,j} | F_{k,l} = f_{k,l}, (k, l) \in \mathcal{F}_{i,j}). \quad (1.1)$$

The MRF-Gibbs equivalence provides an explicit formula for the *joint* probability distribution $P(F = f)$ in terms of an *energy function*, the choice of which, together with $\mathcal{F}$, supplies a powerful mechanism for modeling spatial continuity and other scene features.

The relaxation algorithm is designed to maximize the conditional probability distribution of (F, L) given the data $G = g$, i.e., find the mode of the *posterior distribution* $P(X = x | G = g)$. This form of Bayesian estimation is known as *maximum a posteriori* or MAP estimation, or sometimes as *penalized maximum likelihood* because one seeks to maximize $\log P(G = g | X = x) + \log P(X = x)$ as a function of x; the second term is the "penalty term." MAP estimation has been successfully employed in special settings (see, e.g., Hunt [31] and Hansen and Elliott [25]) and we share the opinion of many that the MAP formulation (and a Bayesian approach in general; see also [24], [43], [45]) is well-suited to restoration, particularly for handling general forms of spatial degradation. Moreover, the distribution of G itself need not be known, which is fortunate due to its usual complexity. On the other hand, MAP estimation clearly presents a formidable computational problem. The number of possible intensity images is L^{m^2}, where L denotes the number of allowable gray levels, which rules out any direct search, even for small ($m = 64$), binary ($L = 2$) scenes. Consequently, one is usually obliged to make simplifying assumptions about the image and degradation models as well as compromises at the computational stage. Here, the computational problem is overcome by exploiting the pivotal observation that the posterior distribution is again Gibbsian with approximately the same neighborhood system as the original image, together with a sampling method which we call the *Gibbs Sampler*. Indeed, our principal theoretical contribution is a general, practical, and mathematically coherent approach for investigating MRF's by sampling (Theorem A), and by computing modes (Theorem B) and expectations (Theorem C).

The Gibbs Sampler generates realizations from a given MRF by a "relaxation" technique akin to site-replacement algorithms in statistical physics, such as "spin-flip" and "exchange" systems. The prototype is due to Metropolis *et al.* [42]; see also [7], [18], and Section X. Cross and Jain [12] use one of these algorithms invented for studying binary alloys. ("Relaxation labeling" in the sense of [13], [30], [46], [47] is different; see Section X.) The Markov property (1.1) permits parallel updating of the line and pixel sites, each of which is "refreshed" according to a simple recipe determined by the governing distribution. Thus, both parts of the MRF-Gibbs equivalence are exploited, for computing and modeling, respectively. Moreover, minimum mean-square error (MMSE) estimation is also feasible by using the (temporal) ergodicity of the relaxation chain to compute *means* w.r.t. the posterior distribution. However, we shall not pursue this approach.

We have used a comparatively slow, raster scan-serial version of the Gibbs Sampler to generate images and restorations (see Section XIII). But the algorithm is parallel; it could be executed in essentially one-half the time with two processors running simultaneously, or in one-third the time with three, and so on. The full parallel potential is realized by assigning one (simple) processor to each site of the intensity process and to each site of the line process. Whatever the number of processors, parallel implementation is made feasible by a small communications requirement among processors. The communications burden is related to the neighborhood size of the graph associated with the image model, and herein lies much of the power of the hierarchical structure: although the field model $X = (F, L)$ has a local graph structure, the *marginal* distribution on the observable intensity process F has a *completely connected graph*. The introduction of a hierarchy dramatically expands the richness of the model of the observed process while only moderately adding to the computational burden. We shall return to these points in Sections IV and XI.

The MAP algorithm depends on an *annealing schedule*, which refers to the (sufficiently) slow decrease of a ("control") parameter T that corresponds to *temperature* in a physical system. As T decreases, samples from the posterior distribution are forced towards the minimal energy configurations; these correspond to the mode(s) of the distribution. Theorem B makes this precise, and is, to our knowledge, the first theoretical result of this nature. Roughly speaking, it says that if the temperature $T(k)$ employed in executing the kth site replacement (i.e., the kth image in the iteration scheme) satisfies the bound

$$T(k) \geqslant \frac{c}{\log(1+k)}$$

for every k, where c is a constant independent of k, then with probability converging to one (as $k \to \infty$), the configurations generated by the algorithm will be those of minimal energy. Put another way, the algorithm generates a Markov chain which converges *in distribution* to the uniform measure over the minimal energy configurations. (It should be emphasized that *pointwise* convergence, i.e., convergence *with probability one*, is in general not possible.) These issues are discussed in Section XII, and the algorithm is demonstrated in Section XIII on a variety of degraded images. We also discuss the nature of the constant c in regard to practical convergence rates. Basically, we believe that the logarithmic rate is best possible. However, the best (i.e., smallest) value of c that we have obtained to date (see the Appendix) is far too large for computational value and our restorations are actually performed with small values of c. As yet, we do not know how to bring the theory in line with experimental results in this regard.

The role of the Gibbs (or Boltzmann) distribution, and other notions from statistical physics, in the construction of "expert systems" is expanding. To begin with, we refer the reader to [21] for the original formulation of our computational method and of a general approach to expert systems based on maximum entropy extensions. As previously mentioned, Černý [8] and Kirkpatrick *et al.* [40] introduced annealing into combinatorial optimization. Other examples include the work of Cheeseman [9] on maximum entropy and diagnosis and of Hinton and Sejnowski [29] on neural modeling of inference and learning.

This paper is organized as follows. The degradation model is described in the next section, and the undegraded image models are presented in Section IV after preliminary material on graphs and neighborhood systems in Section III. In particular, Section IV contains the definitions of MRF's, Gibbs distributions, and the equivalence theorem. Due to the plethora of Markovian models in the literature, we pause in Section V to compare ours to others, and in Section VI to explain some connections with maximum entropy methods. In Section VII we raise the issues of parameter estimation and model selection, and indicate why we are avoiding the former for the time being. The posterior distribution is computed in Section VIII and the corresponding optimization problem is addressed in Section IX. The concept of stochastic relaxation is reviewed in Section X, including its origins in physics. Sections XI and XII are devoted to the Gibbs Sampler, dealing, respectively, with its mechanical and mathematical workings. Our experimental results appear in Section XIII, followed by concluding remarks.

II. Degraded Image Model

We follow the standard modeling of the (intensity) image formation and recording processes, and refer the reader to [31] or [47] for better accounts of the physical mechanisms.

Let H denote the "blurring matrix" corresponding to a shift-invariant point-spread function. The formation of F gives rise to a blurred image $H(\mathrm{F})$ which is recorded by a sensor. The latter often involves a nonlinear transformation of $H(\mathrm{F})$, denoted here by ϕ, in addition to random sensor noise $\mathrm{N} = \{\eta_{i,j}\}$, which we assume to consist of independent, and for definiteness, Gaussian variables with mean μ and standard deviation σ.

Our methods apply to essentially arbitrary noise processes $\mathrm{N} = \{\eta_{i,j}\}$, discrete or continuous. However, computational feasibility requires that the description of N as an MRF (this can always be done; see Section IV) has an associated graph structure that is approximately "local"; the same requirement is applied to the image process $\mathrm{X} = (\mathrm{F}, \mathrm{L})$. For clarity, we forgo full generality and focus on the traditional Gaussian white noise case. Extension to a general noise process is mostly a matter of notation.

The degraded image is then a function of $\phi(H(\mathrm{F}))$ and N, say $\psi(\phi(H(\mathrm{F})), \mathrm{N})$, for example, addition or multiplication. (To compute the posterior distribution, we only need to assume that $b \to \psi(a, b)$ is invertible for each a.) For notational ease, we will write

$$\mathrm{G} = \phi(H(\mathrm{F})) \odot \mathrm{N}. \tag{2.1}$$

At the pixel level, for each $(i,j) \in Z_m$,

$$G_{i,j} = \phi\left(\sum_{(k,l)} H(i-k, j-l) F_{k,l}\right) \odot \eta_{i,j}. \tag{2.2}$$

The mathematical results require an additional assumption, namely, that F and N be independent as stochastic processes (and likewise for L and N) and we assume this henceforth. This is customary, although we recognize the limitation in certain contexts, e.g., for nuclear scan pictures.

For computational purposes, the degree of locality of F should be approximately preserved by (2.1), so that the neighborhood systems for the prior and posterior distributions on (F, L) are comparable. This is achieved when H is a simple convolution over a small window. For instance, take

$$H(k,l) = \begin{cases} \frac{1}{2}, & k=0, l=0 \\ \frac{1}{16}, & |k|, |l| \leqslant 1, (k,l) \neq (0,0) \end{cases} \tag{2.3}$$

so that the intensity at (i,j) is weighted equally with the average of the eight nearest neighbors. The function ϕ is unrestricted, bearing in mind that the true noise level depends on ϕ, $\odot$, and σ. Typically, ϕ is logarithmic (film) or algebraic (TV).

An important special case, which occurs in two-dimensional (2-D) signal theory, is the segmentation of noisy images into coherent regions. The usual model is

$$\mathrm{G} = \mathrm{F} + \mathrm{N} \tag{2.4}$$

where N is white noise and the number of intensity levels is small. This is the model entertained by Hansen and Elliott [25] for simple, binary MRF's F, and by many other workers with varying assumptions about F; see [14], [16], [17]. In this case, namely (2.4), we can extract simple images under extremely low signal-to-noise ratios.

The full degraded image is (G, L); that is, the "line process" is not transformed.

III. Graphs and Neighborhoods

Here and in Section IV we present the general theory of MRF's on graphs, focusing on the aspects and examples which figure in the experimental restorations. The level of abstraction is warranted by the variety of MRF's, graphs, and probability distributions simultaneously under discussion.

Let $S = \{s_1, s_2, \cdots, s_N\}$ be a set of *sites* and let $\mathcal{G} = \{\mathcal{G}_s, s \in S\}$ be a *neighborhood system* for S, meaning any collection of subsets of S for which 1) $s \notin \mathcal{G}_s$ and 2) $s \in \mathcal{G}_r \Leftrightarrow r \in \mathcal{G}_s$. Obviously, $\mathcal{G}_s$ is the set of *neighbors* of s and the pair $\{S, \mathcal{G}\}$ is a graph in the usual way. A subset $C \subseteq S$ is a *clique* if every pair of distinct sites in C are neighbors; $\mathcal{C}$ denotes the set of cliques.

The special cases below are especially relevant.

Case 1: $S = Z_m$. This is the set of pixel sites for the intensity process F; $\{s_1, s_2, \cdots, s_N\}$, $N = m^2$, is any ordering of the lattice points. We are interested in homogeneous neighborhood systems of the form

$$\mathcal{G} = \mathcal{F}_c = \{\mathcal{F}_{i,j}, (i,j) \in Z_m\};\ \mathcal{F}_{i,j} = \{(k,l) \in Z_m : 0 < (k-i)^2 + (l-j)^2 \leq c\}.$$

Notice that sites at or near the boundary have fewer neighbors than interior ones; this is the so-called "free boundary" and is more natural for picture processing than torodial lattices and other periodic boundaries. Fig. 1(a), (b), (c) shows the (interior) neighborhood configurations for $c = 1, 2, 8$; $c = 1$ is the first-order or nearest-neighbor system common in physics, in which $\mathcal{F}_{i,j} = \{(i, j-1), (i, j+1), (i-1, j), (i+1, j)\}$, with adjustments at the boundaries. In each case, (i, j) is at the center, and the symbol $\circ$ stands for a neighboring pixel. The cliques for $c = 1$ are all subsets of Z_m of the form $\{(i, j)\}$, $\{(i, j), (i, j+1)\}$ or $\{(i, j), (i+1, j)\}$, shown in Fig. 1(d). For $c = 2$, we have the cliques in Fig. 1(d) as well as those in Fig. 1(e). Obviously, the number of clique types grows rapidly with c. However, only small cliques appear in the model for F actually employed in this paper; indeed, the degree of progress with only *pair* interactions is somewhat surprising. Nonetheless, more complex images will likely necessitate more complex energies. Our experiments (see Section XIII) suggest that much of this additional complexity can be accommodated while maintaining modest neighborhood sizes by further developing the hierarchy.

Case 2: $S = D_m$, the "dual" $m \times m$ lattice. Think of these sites as placed midway between each vertical or horizontal pair of pixels, and as representing the possible locations of "edge elements." Shown in Fig. 1(f) are six pixel sites together with seven line sites denoted by an X. The six surrounding X's are the neighbors of the middle X for the neighborhood system we denote by $\mathcal{L} = \{\mathcal{L}_d, d \in D_m\}$. Fig. 1(g) is a segment of a realization of a binary line process for which, at each line site, there may or may not be an edge element. We also consider line processes with more than two levels, corresponding to edge elements with varying orientations.

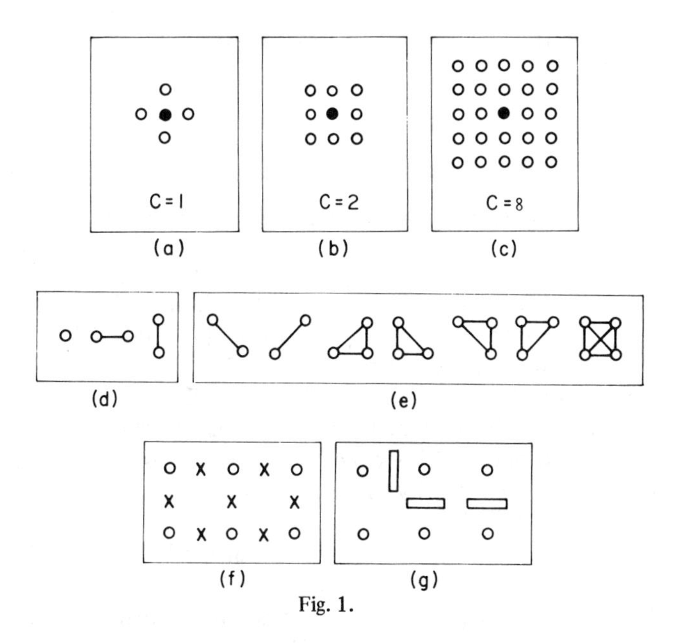

Fig. 1.

Case 3: $S = Z_m \cup D_m$. This is the setup for the field (F, L). Z_m has neighborhood system $\mathcal{F}_1$ (nearest-neighbor lattice) and D_m has the above-described system. The pixel neighbors of sites in D_m are the two pixels on each side, and hence each (interior) pixel has four line site neighbors.

IV. Markov Random Fields and Gibbs Distributions

We now describe a class of stochastic processes that includes both the prior and posterior distribution on the original image. In general, this class of processes (namely, MRF's) is neither homogeneous nor isotropic, assuming the index set S has enough geometric structure to even *define* a suitable family of translations and rotations. However, the *particular* models we choose for prior distributions on the original image are in fact both homogeneous and isotropic in an appropriate sense. (This is not the case for the *posterior* distribution.) We refer the reader to Section XIII for a precise description of the prior models employed in our experiments, and in particular for specific examples of the role of the line elements.

As in Section III, $\{S, G\}$ denotes an arbitrary graph. Let $\mathrm{X} = \{X_s, s \in S\}$ denote *any* family of random variables indexed by S. For simplicity, we can assume a common state space, say $\Lambda \doteq \{0, 1, 2, \cdots, L-1\}$, so that $X_s \in \Lambda$ for all s; the extension to site-dependent state spaces, appropriate when S consists of both line and pixel sites, is entirely straightforward (although not merely a notational matter due to the "positivity condition" below). Let Ω be the set of all possible *configurations:*

$$\Omega = \{\omega = (x_{s_1}, \cdots, x_{s_N}) : x_{s_i} \in \Lambda, 1 \leq i \leq N\}.$$

As usual, the event $\{X_{s_1} = x_{s_1}, \cdots, X_{s_N} = x_{s_N}\}$ is abbreviated $\{X = \omega\}$.

X is an MRF with respect to $\mathcal{G}$ if

$$P(X = \omega) > 0 \quad \text{for all} \quad \omega \in \Omega; \tag{4.1}$$

$$P(X_s = x_s | X_r = x_r, r \neq s) = P(X_s = x_s | X_r = x_r, r \in \mathcal{G}_s) \tag{4.2}$$

for every $s \in S$ and $(x_{s_1}, \cdots, x_{s_N}) \in \Omega$. Technically, what is meant here is that the pair $\{X, P\}$ satisfies (4.1) and (4.2) relative to some probability measure on Ω. The collection of functions on the left-hand side of (4.2) is called the *local characteristics* of the MRF and it turns out that the (joint) probability distribution $P(X = \omega)$ of *any* process satisfying (4.1) is *uniquely* determined by these conditional probabilities; see, e.g., [6, p. 195].

The concept of an MRF is essentially due to Dobrushin [15] and is one way of extending Markovian dependence from 1-D to a general setting; there are, of course, many others, some of which will be reviewed in Section V.

Notice that *any* X satisfying (4.1) is an MRF if the neighborhoods are large enough to encompass the dependencies. The utility of the concept, at least in regard to image modeling, is that priors are available with neighborhoods that are small enough to ensure feasible computational loads and yet still rich enough to model and restore interesting classes of images (and textures: [12]).

Ordinary 1-D Markov chains are MRF's relative to the nearest-neighbor system on $S = \{1, 2, \cdots, N\}$ (i.e., $\mathcal{G}_1 = \{2\}$, $\mathcal{G}_i = \{i-1, i+1\}$ $2 \leq i \leq N-1$, $\mathcal{G}_N = \{N-1\}$) if we assume all positive transitions and the chain is started in equilibrium. In other words, the "one-sided" Markov property

$$P(X_k = x_k | X_j = x_j, j \leq k-1) = P(X_k = x_k | X_{k-1} = x_{k-1})$$

and the "two-sided" Markov property

$$P(X_k = x_k | X_j = x_j, j \neq k) = P(X_k = x_k | X_j = x_j, j \in \mathcal{G}_k)$$

are equivalent. Similarly for an rth order Markov process on the line with respect to the r nearest neighbors on one side and on both sides. (This appears to be doubted in [1] but follows, eventually, from straightforward calculations or immediately from the Gibbs connection.)

Gibbs models were introduced into image modeling by Hassner and Sklansky [28], although the treatment there is mostly expository and limited to the binary case.

A *Gibbs distribution* relative to $\{S, \mathcal{G}\}$ is a probability measure π on Ω with the following representation:

$$\pi(\omega) = \frac{1}{Z} e^{-U(\omega)/T} \tag{4.3}$$

where Z and T are constants and U, called the *energy function*, is of the form

$$U(\omega) = \sum_{C \in \mathcal{C}} V_C(\omega). \tag{4.4}$$

Recall that $\mathcal{C}$ denotes the set of cliques for $\mathcal{G}$. Each V_C is a function on Ω with the property that $V_C(\omega)$ depends only on those coordinates x_s of ω for which $s \in C$. Such a family $\{V_C, C \in \mathcal{C}\}$ is called a *potential.* Z is the normalizing constant:

$$Z \doteq \sum_{\omega} e^{-U(\omega)/T} \tag{4.5}$$

and is called the *partition function.* Finally, T stands for "temperature"; for our purposes, T controls the degree of "peaking" in the "density" π. Choosing T "small" exaggerates the mode(s), making them easier to find by sampling; this is the principle of annealing, and will be applied to the posterior distribution $\pi(\boldsymbol{f}, \boldsymbol{l}) = P(F = \boldsymbol{f}, L = \boldsymbol{l} | G = \boldsymbol{g})$ in order to find the MAP estimate. Of course, we will show that $\pi(\boldsymbol{f}, \boldsymbol{l})$ is Gibbsian and identify the energy and neighborhood system in terms of those for the priors. The *choice* of the prior distributions, i.e., of the particular functions V_C for the image model $\pi(\omega) = P(X = \omega)$, will be discussed later on; see Section VII for some general remarks and Section XIII for the particular models employed in our experiments.

The terminology obviously comes from statistical physics, wherein such measures are "equilibrium states" for physical systems, such as ferromagnets, ideal gases, and binary alloys. The V_C functions represent contributions to the total energy from external fields (singleton cliques), pair interactions (doubletons), and so forth. Most of the interest there, and in the mathematical literature, centers on the case in which S is an *infinite*, 2-D or 3-D lattice; singularities in Z may then occur at certain ("critical") temperatures and are associated with "phase transitions."

Typically, several free parameters are involved in the specification of U, and Z is then a function of those parameters—notoriously intractable. For more information see [3], [5], [6], [23], [32], and [39].

The best-known of these lattice systems is the Ising model, invented in 1925 by E. Ising [33] to help explain ferromagnetism. Here, $S = Z_m$ and $\mathcal{G} = \mathcal{F}_1$, the nearest-neighbor system. The most general form of U is then

$$U(\omega) = \sum V_{\{i,j\}}(x_{i,j}) + \sum V_{\{(i,j),(i+1,j)\}}(x_{i,j}, x_{i+1,j}) + \sum V_{\{(i,j),(i,j+1)\}}(x_{i,j}, x_{i,j+1}) \tag{4.6}$$

where the sums extend over all $(i, j) \in Z_m$ for which the indicated cliques make sense. The Ising model is the special case of (4.6) in which X is binary ($L = 2$), homogeneous (= strictly stationary), and isotropic (= rotationally invarient):

$$U(\omega) = \alpha \sum x_{i,j} + \beta \left(\sum x_{i,j} x_{i+1,j} + \sum x_{i,j} x_{i,j+1} \right) \tag{4.7}$$

for some parameters α and β, which measure, respectively, the external field and bonding strengths.

Returning to the general formulation, recall that the local characteristics

$$\pi(x_s | x_r, r \neq s) = \frac{\pi(\omega)}{\sum_{x_s \in \Lambda} \pi(\omega)} \qquad s \in S, \omega \in \Omega$$

uniquely determine π for any probability measure π on Ω, $\pi(\omega) > 0$ for all ω. The difficulty with the MRF formulation *by itself* is that

i) the *joint* distribution of the X_s is not apparent;

ii) it is extremely difficult to spot local characteristics, i.e., to determine when a given set of functions $\psi(x_s|x_r, r \neq s)$, $s \in S$, $(x_{s_1}, \cdots, x_{s_N}) \in \Omega$, are conditional probabilities for some (necessarily unique) distribution on Ω.

For example, Chellappa and Kashyap [10] allude to i) as a disadvantage of the "conditional Markov" models. See also the discussion in [6]. In fact, these apparent limitations to the MRF formulation have been noted by a number of authors, many of whom were obviously not aware of the following theorem.

Theorem: Let $\mathcal{G}$ *be a neighborhood system. Then* X *is an MRF with respect to* $\mathcal{G}$ *if and only if* $\pi(\omega) = P(X = \omega)$ *is a Gibbs distribution with respect to* $\mathcal{G}$.

Among other benefits, this equivalence provides us with a simple, practical way of specifying MRF's, namely by specifying potentials, which is easy, instead of local characteristics, which is nearly impossible. In fact, with some experience, one can choose U's in accordance with the desired *local* behavior, at least at the intensity level. In short, the modeling and consistency problems of i) and ii) are eliminated.

Proofs may be found in many places now; see, e.g., [39] and the references therein, or the approach via the Hammersley-Clifford expansion in [6]. An influential discussion of this correspondence appears in Spitzer's work, e.g., [48]. Explicit formulas exist for obtaining U from the local characteristics. Conversely, the local characteristics of π are obtained in a straightforward way from the potentials: use the defining ratios and make the allowable cancellations. Fix $s \in S$, $\omega = (x_{s_1}, \cdots, x_{s_N}) \in \Omega$, and let ω^x denote the configuration which is x at site s and agrees with ω everywhere else. Then if $\pi(\omega) = P(X = \omega)$ is Gibbsian,

$$P(X_s = x_s | X_r = x_r, r \neq s) = Z_s^{-1} \exp - \frac{1}{T} \sum_{C: s \in C} V_C(\omega) \tag{4.8}$$

$$Z_s \doteq \sum_{x \in \Lambda} \exp - \frac{1}{T} \sum_{C: s \in C} V_C(\omega^x). \tag{4.9}$$

Notice that the right-hand side of (4.8) only depends on x_s and on x_r, $r \in \mathcal{G}_s$, since any site in a clique containing s must be a neighbor of s. These formulas will be used repeatedly to program the Gibbs Sampler for local site replacements.

For the Ising model, the conditional probability that $X_{i,j} = x_{i,j}$, given the states at $S \backslash \{i, j\}$, or equivalently, just the four nearest neighbors, reduces to

$$\frac{e^{-x_{ij}(\alpha + \beta v_{i,j})}}{1 + e^{-(\alpha + \beta v_{i,j})}}$$

where $v_{i,j} = x_{i,j-1} + x_{i-1,j} + x_{i,j+1} + x_{i+1,j}$. This is also known as the autologistic model and has been used for texture modeling in [12]. More generally, if the local characteristics are given by an exponential family and if $V_C(\omega) \equiv 0$ for $|C| > 2$, then the pair potentials always "factor" into a product of two like terms; see [6].

We conclude with some further discussion of a remark made in Section I: that the hierarchical structure introduced with the line process L expands the graph structure of the *marginal* distribution of the intensity process F. Consider first an arbitrary MRF X with respect to a graph $\{S, \mathcal{G}\}$. Fix $r \in S$ and let $\hat{X} = \{X_s, s \in S, s \neq r\}$. The marginal distribution $\hat{P}$ of $\hat{X}$ is derived from the distribution P of X by summing over the range of X_r. Use the Gibbs representation for P and perform this summation: the resulting expression for $\hat{P}$ can be put in the Gibbs form, and from this the neighborhood system on $\hat{S} \doteq S \backslash \{r\}$ can be inferred. The conclusion of this exercise is that $s_1, s_2 \in \hat{S}$ are, in general, neighbors if either i) they were neighbors in S under $\mathcal{G}$ or ii) each is a neighbor of $r \in S$ under $\mathcal{G}$. Now let $X = (F, L)$, with neighborhood system defined at the end of Section III. Successive summations of the distribution of X over the ranges of the elements of L yields the marginal distribution of the observable intensity process F. Each summation leaves a graph structure associated with the marginal distribution of the remaining variables, and this can be related to the original neighborhood system by following the preceding discussion of the general case. It is easily seen that when all of the summations are performed, the remaining graph is completely connected; under the marginal distribution of F, all sites are neighbors. This calculation suggests that significant long-range interactions can be introduced through the development of hierarchical structures without sacrificing the computational advantages of local neighborhood systems.

V. Related Markov Image Models

The use of neighborhoods is, of course, pervasive in the literature: they offer a geometric framework for the clustering of pixel intensities and for many types of statistical models. In particular, the Markov property is a natural way to formalize these notions. The result is a somewhat bewildering array of Markov-type image models and it seems worthwhile to puase to relate these to MRF's. The process under consideration is $F = \{F_{i,j}, (i, j) \in Z_m\}$, the gray levels, or really any pixel attribute.

An early work in this direction is Abend, Harley and Kanal [1] about pattern classification. Among many novel ideas, there is the notion of a *Markov mesh* (MM) process, in which the Markovian dependence is *causal:* generally, one assumes that, for all (i, j) and f,

$$\begin{aligned} &P(F_{i,j} = f_{i,j} | F_{k,l} = f_{k,l}, (k,l) \in A_{i,j}) \\ &\quad = P(F_{i,j} = f_{i,j} | F_{k,l} = f_{k,l}, (k,l) \in B_{i,j}) \end{aligned} \tag{5.1}$$

where $B_{i,j} \subseteq A_{i,j} \subseteq \{(k, l): k < i \text{ or } l < j\}$. A common example is $B_{i,j} = \{(i-1, j), (i-1, j-1), (i, j-1)\}$. Besag [6], Kanal [37], and Pickard [44] also discuss such "unilateral" processes, which are usually a subclass of MRF's, although the resulting (bilateral) neighborhoods can be irregular. Anyway, for MM models the emphasis is on the causal, iterative aspects, including a recursive representation for the joint probabilities. Incidentally, a Gibbs type description of rth order Markov chains is given in [1]; of course, the full Gibbs-MRF equivalence is not perceived and was not for about five years. Derin *et al.* [14] model F as an MM process and use recursive Bayes smoothing to recover F from a noisy version $F + N$; the algorithms exploit the causality to maximize the univariate posterior distribution at each pixel based on the data over a strip containing it, and are very effective at low S/N ratios for some simple images.

Motivated by a paper of Lévy [41], Woods [51] defined "P-Markov" processes for the resolution of wavenumber spectra. The definition involves two spatial regions separated by a "boundary" of width P, and correspond to the past, future, and present in 1-D. Woods also considers a family of "wide-sense" Markov fields of the form

$$F_{i,j} = \sum_{(k,l) \in W_p} \theta_{k,l} F_{i-k,j-l} + U_{i,j} \tag{5.2}$$

where $W_p = \{(k, l): 0 < k^2 + l^2 \leq P\}$, $\theta_{k,l}$ are the MMSE coefficients for projecting $F_{i,j}$ on $\{F_{k,l}, (k, l) \in (i, j) + W_p\}$, and $\{U_{i,j}\}$ is the error, generally nonwhite. The main theoretical result is that if $\{U_{i,j}\}$ is homogeneous, Gaussian, and satisfies a few other assumptions, then $\mathbb{F}$ is Gaussian, P-Markov and vice-versa. In general, there are consistency problems and the P-Markov property is hard to verify. In the nearest-neighbor case, one gets a Gaussian MRF.

Other "wide-sense" Markov processes appear in Jain and Angel [35] and Stuller and Kurz [49]. The assumptions in [35] are a nearest-neighbor system, white noise, and no blur; restoration is achieved by recursively filtering the rows $\{F_{i,j}\}_{j=1}^{m}$, which form a vector-valued, second-order Markov chain, to find the optimal interpolator of each row. In [49], causality is introduced and earlier work is generalized by considering an arbitrary "scanning pattern."

The "spatial interaction models" in Chellappa and Kashyap [10], [38] satisfy (5.2) for general coefficients and W's. The model is causal if W lies in the third quadrant. The authors consider "simultaneous autoregressive" (SAR) models, wherein the noise is white, and "conditional Markov" (CM) models, wherein the "bilateral" Markov property holds (i.e., (1.1) with $\mathcal{F}_{i,j} = (i, j) + W$) in addition to (5.2), and the noise is nonwhite. Thus, the CM models are MRF's, although in [10], [38] the boundary of Z_m is periodic, and hence boundary conditions must be adjoined to (5.2). Given any (homogeneous) SAR process there exists a unique CM process with the same spectral density, although different neighborhood structure. The converse holds in the Gaussian case but is generally false (see the discussion in Besag [6]). MMSE restoration of blurred images with additive Gaussian noise is discussed in [10]; the original image is SAR or CM, usually Gaussian.

Finally, Hansen and Elliott [25] and Elliott *et al.* [17] design MAP algorithms for the segmentation of remotely sensed data with high levels of additive noise. The image model is a nearest-neighbor, binary MRF. However, the autologistic form of the joint distribution is not recognized due to the lack of the Gibbs formulation. The conditional probabilities are approximated by the product of four 1-D transitions, and segmentation is performed by dynamic programming, first for each row and then for the entire images. More recent work in Elliott *et al.* [16] is along the same lines, namely MAP estimation, via dynamic programming, of very noisy but simple images; the major differences are the use of the Gibbs formulation and improvements in the algorithms. Similar work, applied to boundary finding, can be found in Cooper and Sung [11], who use a Markov boundary model and a deterministic relaxation scheme.

VI. Maximum Entropy Restoration

There are several contact points. The Gibbs distribution can be derived (directly from physical principles in statistical mechanics) by maximizing entropy: basically, it has maximal entropy among all probability measures (equilibrium states) on Ω with the same *average* energy. Thus it is no accident that, like maximum entropy (ME) methods, ours are well-suited to nonlinear problems; see [50]. Moreover, based on the success of ME restoration (along the lines suggested by Jaynes [36]) for recovering randomly pulsed objects (cf. Frieden [19]), we intend in the future to analyze such data (e.g., starfield photographs) by our methods.

We should also like to mention the interesting observation of Trussell [50] that conventional ME restoration is a special case of MAP estimation in which the prior distribution on $\mathbb{F}$ is

$$P(\mathbb{F} = f) = \exp\left(-\beta \sum f_{i,j} \log f_{i,j}\right) \Big/ (\text{normalizing constant}).$$

By "conventional ME," we refer to maximizing the entropy $\Sigma f_{i,j} \log f_{i,j}$ subject to $\Sigma \eta_{i,j}^2 = \text{constant}$ ($\eta_{i,j}$ is here again the noise process); see [2]. Other ME methods (e.g., [19]) do not appear to be MAP-related.

VII. Model Selection and Parameter Estimation

The quality of the restoration will clearly depend on choices made at the modeling stage, in our case about specific energy types, attribute processes, and parameters. Cross and Jain [12] use maximum likelihood estimation in the context of Besag's [6] "coding scheme," as well as standard goodness-of-fit tests, for matching realizations of autobinomial MRF's to real textures. Kashyap and Chellappa [38] introduce some new methods for parameter estimation and the choice of neighborhoods for the SAR and CM models, mostly in the Gaussian case. These are but two examples.

For uncorrupted, simple MRF's, the coding methods do finesse the problem of the partition function. However, for more complex models and for corrupted data, we feel that the coding methods are ultimately inadequate due to the complexity of the distribution of $\mathbb{G}$. This view seems to be shared by other authors, although in different contexts. Of course, for MRF's, the obstacles facing conventional statistical inference due to Z have often been noted. Even for the Ising model, analytical results are rare; a famous exception is Onsager's work on the correlational structure.

At any rate, we have developed a new method [20] for estimating clique parameters from the "noisy" data, and this will be implemented in a forthcoming paper. For now, we are obliged to choose the parameters on an ad hoc basis (which is common), but hasten to add that the quality of restoration does not seem to have been adversely affected, probably due to the relative simplicity of the MRF's we actually use for the line and intensity processes; see Section XIII.

One should also address the *general* choice of π and $\mathcal{G}$. This is really quite different than parameter estimation and somewhat related to "image understanding": how does one incorporate "real-world knowledge" into the modeling process? In image interpretation systems, various semantical and hierarchical models have been proposed (see, e.g., [26]). We have begun our study of hierarchical Gibbs models in this paper. A *general theory* of interactive, self-adjusting models that is practical and mathematically coherent may lie far ahead.

VIII. Posterior Distribution

We now turn to the posterior distribution $P(\mathrm{F} = f, \mathrm{L} = l \mid \mathrm{G} = g)$ of the original image given the "data" g. In this section we take $S = Z_m \cup D_m$, the collection of pixel and line sites, with some neighborhood system $\mathcal{G} = \{\mathcal{G}_s, s \in S\}$; an example of such a "mixed" graph was given in Section III. The configuration space is the set of all pairs $\omega = (f, l)$ where the components of f assume values among the allowable gray levels and those of l among the (coded) line states.

We assume that X is an MRF relative to $\{S, \mathcal{G}\}$ with corresponding energy function U and potentials $\{V_C\}$:

$$P(\mathrm{F} = f, \mathrm{L} = l) = e^{-U(f,l)/T}/Z$$

$$U(f,l) = \sum_C V_C(f,l).$$

For convenience, take $T = 1$

Recall that $\mathrm{G} = \phi(H(\mathrm{F})) \odot \mathrm{N}$, where N is white Gaussian noise with mean μ and variance σ^2 and is independent of X.

We emphasize that what follows is easily extended to processes N that are more general MRF's, although we still require that N be independent of X. The operation $\odot$ is assumed invertible and we will write $\mathrm{N} = \Phi(\mathrm{G}, \phi(H(\mathrm{F}))) = \{\Phi_s, s \in Z_m\}$ to indicate this inverse.

Let $\mathcal{H}_s$, $s \in Z_m$, denote the pixels which affect the blurred image $H(\mathrm{F})$ at s. For instance, for the H in (2.3), $\mathcal{H}_s$ is the 3×3 square centered at s. Observe that Φ_s, $s \in Z_m$, depends only on g_s and $\{f_t, t \in \mathcal{H}_s\}$. By the shift-invariance of H, $\mathcal{H}_{r+s} = s + \mathcal{H}_r$ where $\mathcal{H}_r \subseteq Z_m$, $s + r \in Z_m$, and $s + \mathcal{H}_r$ is understood to be intersected with Z_m, if necessary. In addition, we will assume that $\{\mathcal{H}_s\}$ is "symmetric" in that $r \in \mathcal{H}_0 \Rightarrow -r \in \mathcal{H}_0$. Then the collection $\{\mathcal{H}_s \backslash \{s\}, s \in Z_m\}$ is a neighborhood system over Z_m. Let $\mathcal{H}^2$ denote the second-order system, i.e.,

$$\mathcal{H}_s^2 = \bigcup_{r \in \mathcal{H}_s} \mathcal{H}_r, \quad s \in Z_m.$$

Then it is not hard to see that $\{\mathcal{H}_s^2 \backslash \{s\}, s \in Z_m\}$ is also a neighborhood system. Finally, set $\mathcal{G}^P = \{\mathcal{G}_s^P, s \in S\}$ where

$$\mathcal{G}_s^P = \begin{cases} \mathcal{G}_s, & s \in D_m \\ \mathcal{G}_s \cup \mathcal{H}_s^2 \backslash \{s\}, & s \in Z_m. \end{cases} \tag{8.1}$$

The "P" stands for "posterior"; some thought shows that $\mathcal{G}^P$ is a neighborhood system on S.

Let $\boldsymbol{\mu} \in \mathbb{R}^M (M = N^2)$ have all components $= \mu$ and let $\|\cdot\|$ denote the usual norm in $\mathbb{R}^M$: $\|V\|^2 = \Sigma_1^M V_i^2$.

Theorem: For each g fixed, $P(\mathrm{X} = \omega \mid \mathrm{G} = g)$ is a Gibbs distribution over $\{S, \mathcal{G}^P\}$ with energy function

$$U^P(f,l) = U(f,l) + \|\boldsymbol{\mu} - \Phi(g, \phi(H(f)))\|^2/2\sigma^2. \tag{8.2}$$

Proof: Using standard results about "regular conditional expectations," we can and do assume that

$$P(\mathrm{X} = \omega \mid \mathrm{G} = g) = \frac{P(\mathrm{G} = g \mid \mathrm{X} = \omega) P(\mathrm{X} = \omega)}{P(\mathrm{G} = g)} \tag{8.3}$$

for all $\omega = (f, l)$, for each g.

Since $P(\mathrm{G} = g)$ is a constant and $P(\mathrm{X} = \omega) = e^{-U(\omega)}/Z$, the key term is

$$\begin{aligned} P(\mathrm{G} = g \mid \mathrm{X} = \omega) &= P(\phi(H(\mathrm{F})) \odot \mathrm{N} = g \mid \mathrm{F} = f, \mathrm{L} = l) \\ &= P(\mathrm{N} = \Phi(g, \phi(H(f))) \mid \mathrm{F} = f, \mathrm{L} = l) \\ &= P(\mathrm{N} = \Phi(g, \phi(H(f)))) \end{aligned}$$

(since N is independent of F and L)

$$= (2\pi\sigma^2)^{-M/2} \exp - \left(\frac{1}{2\sigma^2}\right) \|\boldsymbol{\mu} - \Phi\|^2.$$

We will write Φ for $\Phi(g, \phi(H(f)))$. Collecting constants we have, from (8.3),

$$P(\mathrm{X} = \omega \mid \mathrm{G} = g) = e^{-U^P(\omega)}/Z^P$$

for U^P as in (8.2); Z^P is the usual normalizing constant (which will depend on g). It remains to determine the neighborhood structure.

Intuitively, the line sites should have the *same* neighbors whereas the neighbors $\mathcal{G}_s$ of a pixel site $s \in Z_m$ should be augmented in accordance with the blurring mechanism.

Take $s \in D_m$. The local characteristics at s for the posterior distribution are, by (8.2),

$$\begin{aligned} &P(L_s = l_s \mid L_r = l_r, r \neq s, r \in D_m, \mathrm{F} = f, \mathrm{G} = g) \\ &\quad = \frac{e^{-U^P(f,l)}}{\sum_{l_s} e^{-U^P(f,l)}} = \frac{e^{-U(f,l)}}{\sum_{l_s} e^{-U(f,l)}} \end{aligned}$$

where the sum extends over all possible values of L_s. Hence $\mathcal{G}_s^P = \mathcal{G}_s$.

For $s \in Z_m$, the term in (8.2) involving Φ does not cancel out. Now $\Phi(g, \phi(H(f))) = \{\Phi_s, s \in Z_m\}$ and let us denote the dependencies in Φ_s by writing $\Phi_s = \Phi_s(g_s; f_t, t \in \mathcal{H}_s)$. Then

$$\begin{aligned} &P(F_s = f_s \mid F_r = f_r, r \neq s, r \in Z_m, \mathrm{L} = l, \mathrm{G} = g) \\ &\quad = \frac{e^{-U^P(f,l)}}{\sum_{f_s} e^{-U^P(f,l)}}; \; U^P(f,l) \\ &\quad = U(f,l) + \sum_{r \in Z_m} (\Phi_r - \mu)^2/2\sigma^2. \end{aligned} \tag{8.4}$$

Decompose U^P as follows:

$$\begin{aligned} U^P(f,l) = &\sum_{C: s \in C} V_C(f,l) \\ &+ (2\sigma^2)^{-1} \sum_{r: s \in \mathcal{H}_r} (\Phi_r(g_r; f_t, t \in \mathcal{H}_r) - \mu)^2 \\ &+ \sum_{C: s \notin C} V_C(f,l) \\ &+ (2\sigma^2)^{-1} \sum_{r: s \notin \mathcal{H}_r} (\Phi_r(g_r; f_t, t \in \mathcal{H}_r) - \mu)^2. \end{aligned}$$

Since the last two terms do not involve f_s (remember that V_C only depends on the sites in C), the ratio in (8.4) depends only on the first two terms above. The first term depends only on coordinates of (f, l) for sites in $\mathcal{G}_s (s \in C \Rightarrow C \subseteq \mathcal{G}_s)$ and the second term only on sites in

$$\bigcup_{r: s \in \mathcal{H}_r} \mathcal{H}_r = \bigcup_{r \in \mathcal{H}_s} \mathcal{H}_r \doteq \mathcal{H}_s^2.$$

Hence, $\mathcal{G}_s^P = \mathcal{G}_s \cup \mathcal{H}_s^2 \backslash \{s\}$, as asserted in the theorem. □

IX. The Computational Problem

The posterior distribution $P(X = \omega | g)$ is a powerful tool for image analysis; in principle, we can construct the optimal (Bayesian) estimator for the original image, examine images sampled from $P(X = \omega | g)$, estimate parameters, design near-optimal statistical tests for the presence or absence of special objects, and so forth. But a conventional approach to any of these involves prohibitive computations. Specifically, our job here is to find the value(s) of ω which maximize the posterior distribution for a fixed g, i.e., *minimize*

$$U(f, l) + ||\mu - \Phi(g, \phi(H(f)))||^2/2\sigma^2, (f, l) \in \Omega \tag{9.1}$$

where (see Section VIII) Φ is defined by $\phi(H(f)) \odot \Phi = g$. Even without L, the size of Ω is at least 2^{4000}, corresponding to a binary image on a small (64 × 64) lattice. Hence, the identification of even near-optimal solutions is extremely difficult for such a relatively complex function.

In Sections XI and XII we will describe our stochastic relaxation method for this kind of optimization. The same method works for sampling and for computing expectations (and hence forming likelihood ratios), as will be explained in Section XI. The algorithm is highly parallel, but our current implementation is serial: it uses a single processor. The restoration of more complex images than those in Section XIII, probably involving more levels in the hierarchy, may necessitate *some* parallel processing.

X. Stochastic Relaxation

There are many types of "relaxation," two of them being the type used in statistical physics and the type developed in image processing called "relaxation labeling" (RL), or sometimes "probabilistic relaxation." Basically, ours is of the former class, referred to here as SR, although there are some common features with RL.

The "Metropolis algorithm" (Metropolis *et al.* [42]) and others like it [7], [18] were invented to study the equilibrium properties, especially ensemble averages, time-evolution, and low-temperature behavior, of very large systems of essentially identical, interacting components, such as molecules in a gas or atoms in binary alloys.

Let Ω denote the possible configurations of the system; for example, $\omega \in \Omega$ might be the molecular positions or site configuration. If the system is in thermal equilibrium with its surroundings, then the probability (or "Boltzmann factor") of ω is given by

$$\pi(\omega) = e^{-\beta \mathcal{E}(\omega)} \Big/ \sum_{\omega} e^{-\beta \mathcal{E}(\omega)}, \quad \omega \in \Omega$$

where $\mathcal{E}(\omega)$ is the potential energy of ω and $\beta = 1/KT$ where K is Boltzmann's constant and T is absolute temperature. We have already seen an example in the Ising model (4.7). Usually, one needs to compute ensemble averages of the form

$$\langle Y \rangle = \int_{\Omega} Y(\omega)\, d\pi(\omega) = \frac{\sum_{\omega} Y(\omega)\, e^{-\beta \mathcal{E}(\omega)}}{\sum_{\omega} e^{-\beta \mathcal{E}(\omega)}}$$

where Y is some variable of interest. This cannot be done analytically. In the usual Monte Carlo method, one restricts the sums above to a *sample* of ω's drawn uniformly from Ω. This, however, breaks down in the situation above: the exponential factor puts most of the mass of π over a very small part of Ω, and hence one tends to choose samples of very low probability. The idea in [42] is to choose the samples from π instead of uniformly and then weight the samples evenly instead of by $d\pi$. In other words, one obtains $\omega_1, \omega_2, \cdots, \omega_R$ from π and $\langle Y \rangle$ is approximated by the usual ergodic averages:

$$\langle Y \rangle \approx \frac{1}{R} \sum_{r=1}^{R} Y(\omega_r). \tag{10.1}$$

Briefly, the sampling algorithm in [42] is as follows. Given the state of the system at "time" t, say $X(t)$, one randomly chooses another configuration η and computes the energy change $\Delta\mathcal{E} = \mathcal{E}(\eta) - \mathcal{E}(X(t))$ and the quantity

$$q = \frac{\pi(\eta)}{\pi(X(t))} = e^{-\beta \Delta \mathcal{E}}. \tag{10.2}$$

If $q > 1$, the move to η is allowed and $X(t+1) = \eta$, whereas if $q \leqslant 1$, the transition is made *with probability* q. Thus we choose $0 \leqslant \xi \leqslant 1$ uniformly and set $X(t+1) = \eta$ if $\xi \leqslant q$ and $X(t+1) = X(t)$ if $\xi > q$. (A "parallel processing variant" of this for simulating certain binary MRF's is given by Berger and Bonomi [4].)

In binary, "single-flip" studies, $\eta = X(t)$ except at one site, whereas in "spin-exchange" [18] systems, a pair of neighboring sites is selected. In either case, the "flip" or "exchange" is made with probability $q/(1+q)$, where q is given in (10.2). In special cases, the single-flip system is equivalent to our Gibbs Sampler. The exchange algorithm in Cross and Jain [12] is motivated by work on the evolution of binary alloys. The samples generated are used for visual inspection and statistical testing, comparing the real and simulated textures. The model is an autobinomial MRF; see [6] or [12]. The algorithm is not suitable (nor intended) for restoration: for one thing, the intensity histogram is constant throughout the iteration process. This is necessarily the case with exchange systems which depend heavily on the initial configuration.

The algorithm in Hassner and Sklansky [28] is apparently a modification of one in Bortz *et al.* [7]. Another application of these ideas outside statistical mechanics appears in Hinton and Sejnowski [29], a paper about neural modeling but a spiritual cousin of ours. In particular, the parallel nature of these algorithms is emphasized.

The essence of every SR scheme is that changes ($\omega \to \eta$) which *increase* energy, i.e., *lower* probability, are permitted.

By contrast, deterministic algorithms only allow jumps to states of lower energy and invariably get "stuck" in *local* minima. To get to samples from π, we must occasionally "backtrack."

All of these algorithms can be cast in a general theory involving Markov chains with state space Ω. See Hammersley and Handscomb [27] for a readable treatment. The goal is an irreducible, aperiodic chain with equilibrium measure π. If $\omega_1, \omega_2, \cdots, \omega_R$ is a realization of such a chain, then standard results yield (10.1), in fact at a rate $O(R^{-1/2})$ as $R \to \infty$. In this setup an auxiliary transition matrix is used to go from ω to η, and the general replacement recipe involves the same ratio $\pi(\eta)/\pi(\omega)$. The Markovian properties of the Gibbs Sampler will be described in the following sections.

Chemical annealing is a method for determining the low energy states of a material by a gradual lowering of temperature. The process is delicate: if T is lowered too rapidly and insufficient time is spent at temperatures near the freezing point, then the process may bog down in nonequilibrium states, corresponding to flaws in the material, etc. In *simulated* annealing, Kirkpatrick *et al.* [40] identify the solution of an optimal (computer) design problem with the ground state of an imaginary physical system, and then employ the Metropolis algorithm to reach "steady-state" at each of a decreasing sequence of temperatures $\{T_n\}$. This sequence, and the time spent at each temperature, is called an "annealing schedule." In [40], this is done on an ad hoc basis using guidelines developed for chemical annealing. Here, we prove the existence of annealing schedules which guarantee convergence to minimum energy states (see Section XII for formal definitions), and we identify the *rate* of decrease relative to the number of full sweeps.

Turning to RL, there are many similarities with SR, both in purpose and, at least abstractly, in method. RL was designed for the assignment of numeric or symbolic labels to objects in a visual system, such as intensity levels to pixels or geometric labels to cube edges, in order to achieve a "global interpretation" that is consistent with the context and certain "local constraints." Ideally, the process evolves by a series of *local* changes, which are intended to be simple, homogeneous, and performed in parallel. The local constraints are usually so-called "compatibility functions," which are much like statistical correlations, and often defined in reference to a graph. We refer the reader to Davis and Rosenfeld [13] for an expository treatment, to Rosenfeld *et al.* [46] for the origins, to Hummel and Zucker [30] for recent work on the logical and mathematical foundations, and to Rosenfeld and Kak [47] for applications to iterative segmentation.

But there are also fundamental differences. First, most variants of RL are rather ad hoc and heuristic. Second, and more importantly, RL is essentially a *nonstochastic* process, both in the interaction model and in the updating algorithms. (Indeed, various probabilistic analogies are often avoided as misleading; see [30], for example.) There is nothing in RL corresponding to an equilibrium measure or even a joint probability law over configurations, whereas there is no analogue in SR of the all-important, iterative updating *formulas* and corresponding sequence of "probability estimates" for various hypotheses involving pixel or object classification.

In summary, there are shared goals and shared features (locality, parallelism, etc.) but SR and RL are quite distinct, at least as practiced in the references made here.

XI. Gibbs Sampler: General Description

We return to the general notation of Section IV: $\setminus = \{X_s, s \in S\}$ is an MRF over a graph $\{\mathcal{G}_s, s \in S\}$ with state spaces Λ_s, configuration space $\Omega = \prod_s \Lambda_s$, and Gibbs distribution $\pi(\omega) = e^{-U(\omega)/T}/Z$, $\omega \in \Omega$.

The general computational problems are

A) sample from the distribution π;
B) minimize U over Ω;
C) compute expected values.

Of course, we are most concerned with B), which corresponds to MAP estimation when π is the posterior distribution. The most basic problem is A), however, because A) together with annealing yields B) and A) together with the ergodic theorem yields C). We will state three theorems corresponding to A), B), and C) above. Theorem C is not used here and will be proven elsewhere; we state it because of its potential importance to other methods of restoration and to hypothesis testing.

Let us imagine a simple processor placed at each site s of the graph. The connectivity relation among the processors is determined by the bonds: the processor at s is connected to each processor for the sites in $\mathcal{G}_s$. In the cases of interest here (and elsewhere) the number of sites N is very large. However, the size of the neighborhoods, and thus the number of connections to a given processor, is modest, only eight in our experiments, including line, pixel and mixed bonds.

The state of the machine evolves by discrete changes and it is therefore convenient to discretize time, say $t = 1, 2, 3, \cdots$. At time t, the state of the processor at site s is a random variable $X_s(t)$ with values in Λ_s. The total configuration is $X(t) \doteq (X_{s_1}(t), X_{s_2}(t), \cdots, X_{s_N}(t))$, which evolves due to state changes of the individual processors. The starting configuration, $X(0)$, is arbitrary. At each epoch, only *one* site undergoes a (possible) change, so that $X(t-1)$ and $X(t)$ can differ in at most one coordinate. Let $n_1, n_2, \cdots$ be the sequence in which the sites are "visited" for replacement; thus, $n_t \in S$ and $X_{s_i}(t) = X_{s_i}(t-1)$, $i \neq n_t$. Each processor is programmed to follow the same algorithm: at time t, a sample is drawn *from the local characteristics* of π for $s = n_t$ and $\omega = X(t-1)$. In other words, we choose a state $x \in \Lambda_{n_t}$ from the conditional distribution of X_{n_t} given the observed states of the neighboring sites $X_r(t-1)$, $r \in \mathcal{G}_{n_t}$. The new configuration $X(t)$ has $X_{n_t}(t) = x$ and $X_s(t) = X_s(t-1)$, $s \neq n_t$.

These are *local* computations, and *identical* in nature when π is homogeneous. Moreover, the actual calculation is *trivial* since the local characteristics are generally very simple. These conditional probabilities were discussed in Section IV and we refer the reader again to formulas (4.8) and (4.9). Notice that Z does not appear.

Given an initial configuration $X(0)$, we thus obtain a sequence $X(1), X(2), X(3), \cdots$ of configurations whose convergence properties will be described in Section XII. The limits obtained do not depend on $X(0)$. The sequence (n_t) we actually use is simply the one corresponding to a raster scan, i.e., repeatedly visiting all the sites in some "natural" fixed order. Of course, in this case one does not actually need a processor at each site. But the theorems are valid for very general (not necessarily periodic) sequences (n_t) allowing for *asychronous* schemes in which each processor could be driven *by its own clock*. Let us briefly discuss such a parallel implementation of the Gibbs Sampler and its advantage over the serial version.

Computation is parallel in the sense that it is realized by simple and alike units operating largely independently. Units are dependent only to the extent that each must transmit its current state to its neighbors. Most importantly, the amount of time required for one complete update of the entire system is *independent of the number of sites*. In the raster version, we simply "move" a processor from site to site. Upon arriving at a site, this processor must first load the local neighborhood relations and state values, perform the replacement, and move on. The time required to refresh S grows linearly with $N = |S|$. Thus, for example, for the purposes at hand, the parallel procedure is potentially at least 10^4 times faster than the raster version we used, and which required considerable CPU time on a VAX 780. Of course, we recognize that the fully parallel version will require extremely sophisticated new hardware, although we understand that small prototypes of similar machines are underway at several places.

A more modest degree of parallelism can be simply implemented. Since the convergence theorems are independent of the details of the site replacement scheme $n_1, n_2, \cdots$ the graph associated with the MRF X can be divided into collections of sites with each collection assigned to an independently running (asynchronous) processor. Each such processor would execute a raster scan updating of its assigned sites. Communication requirements will be small if the division of the graph respects the natural topology of the scene, provided, of course, that the neighborhood systems are reasonably local. Such an implementation, with five or ten micro- or minicomputers, represents a straightforward application of available technology.

XII. Gibbs Sampler: Mathematical Foundations

As in Section XI, (n_t), $t = 1, 2, \cdots$, is the sequence in which the sites are visited for updating, and $X_s(t)$ denotes the state of site s after t replacement opportunities, of which only those for which $n_\tau = s$, $1 \leq \tau \leq t$, involve site s. For simplicity, we will assume a common state space $\Lambda_s \equiv \Lambda = \{0, 1, \cdots, L-1\}$, and as usual that $0 < \pi(\omega) < 1$ for all $\omega \in \Omega$ or, what is the same, that $\sup_\omega |U(\omega)| < \infty$. The initial configuration is $X(0)$.

We now investigate the statistical properties of the random process $\{X(t), t = 0, 1, 2, \cdots\}$. The evolution $X(t-1) \to X(t)$ of the system was explained in Section XI. In mathematical terms,

$$\begin{aligned} P(X_s(t) = x_s, s \in S) &= \pi(X_{n_t} = x_{n_t} | X_s = x_s, s \neq n_t) P(X_s(t-1) \\ &= x_s, s \neq n_t) \end{aligned} \tag{12.1}$$

where, of course, $\pi = e^{-U/T}/Z$ is the Gibbs measure which drives the process. Our first result states that the distribution of $X(t)$ converges to π as $t \to \infty$ regardless of $X(0)$. The only assumption is that we continue to visit every site, obviously a necessary condition for convergence.

Theorem A (Relaxation): Assume that for each $s \in S$, the sequence $\{n_t, t \geq 1\}$ contains s infinitely often. Then for every starting configuration $\eta \in \Omega$ and every $\omega \in \Omega$,

$$\lim_{t \to \infty} P(X(t) = \omega | X(0) = \eta) = \pi(\omega). \tag{12.2}$$

The proof appears in the Appendix, along with that of Theorem B. Like the Metropolis algorithm, the Gibbs Sampler produces a Markov chain $\{X(t), t = 0, 1, 2, \cdots\}$ with π as equilibrium distribution. The only complication is that the transition probabilities associated with the Gibbs Sampler are nonstationary, and their matrix representations do not commute. This precludes the usual algebraic treatment. These issues are discussed in more detail at the beginning of the Appendix.

We now turn to annealing. Hitherto the temperature has been fixed. Theorem B is an "annealing schedule" or rate of temperature decrease which forces the system into the lowest energy states. The necessary programming modification in the relaxation process is trivial, and the *local* nature of the calculations is preserved.

Let us indicate the dependence of π on T by writing π_T, and let $T(t)$ denote the temperature at stage t. The annealing procedure generates a different process $\{X(t), t = 1, 2, \cdots\}$ such that

$$\begin{aligned} P(X_s(t) = x_s, s \in S) &= \pi_{T(t)}(X_{n_t} = x_{n_t} | X_s = x_s, s \neq n_t) \\ &\quad \cdot P(X_s(t-1) = x_s, s \neq n_t). \end{aligned} \tag{12.3}$$

Let

$$\Omega_0 = \{\omega \in \Omega : U(\omega) = \min_\eta U(\eta)\}, \tag{12.4}$$

and let π_0 be the uniform distribution on Ω_0. Finally, define

$$\begin{aligned} U^* &= \max_\omega U(\omega), \\ U_* &= \min_\omega U(\omega), \\ \Delta &= U^* - U_*. \end{aligned} \tag{12.5}$$

Theorem B (Annealing): Assume that there exists an integer $\tau \geq N$ such that for every $t = 0, 1, 2, \cdots$ we have

$$S \subseteq \{n_{t+1}, n_{t+2}, \cdots, n_{t+\tau}\}.$$

Let $T(t)$ be any decreasing sequence of temperatures for which

a) $T(t) \to 0$ as $t \to \infty$;
b) $T(t) \geq N\Delta/\log t$ for all $t \geq t_0$ for some integer $t_0 \geq 2$.

Then for any starting configuration $\eta \in \Omega$ and for every $\omega \in \Omega$,

$$\lim_{t \to \infty} P(X(t) = \omega | X(0) = \eta) = \pi_0(\omega). \tag{12.6}$$

The first condition is that the individual "clocks" do not slow to an arbitrarily low frequency as the system evolves, and imposes no limitations in practice. For raster replacement,

$\tau = N$. The major practical weakness is b); we cannot truly follow the "schedule" $N\Delta/\log t$. For example, with $N = 20{,}000$ and $\Delta = 1$, it would take $e^{40{,}000}$ site visits to reach $T = 0.5$. We single out this temperature because we have obtained good results by making T decrease from approximately $T = 4$ to $T = 0.5$ over 300–1000 sweeps ($= 300N - 1000N$ replacements), using a schedule of the form $C/\log(1 + k)$, where k is the number of full sweeps. (Notice that the condition in b) is then satisfied provided C is sufficiently large.) Apparently, the bound in b) is far from optimal, at least as concerns the constant $N\Delta$. (In fact, the proof of Theorem B does establish something stronger, namely that Δ can be taken as the largest absolute difference in energies associated with pairs ω and ω^* which differ at only one coordinate. But this improvement still leaves $N\Delta$ too large for actual practice.) On the other hand, the logarithmic rate is not too surprising in view of the widespread experience of chemists that T must be lowered very slowly, particularly near the freezing point. Otherwise one encounters undesirable physical embodiments of *local* energy minima.

Concerning ergodicity, in statistical physics one attempts to predict the observable quantities of a system in equilibrium; these are the "time averages" of functions on Ω. Under the "ergodic hypothesis," one assuumes that (10.1) is in force, so that time averages approach the corresponding "phase averages" or expected values. The analog for our system is the assertion that, in some suitable sense,

$$\lim_{n \to \infty} \frac{1}{n} \sum_{t=1}^{n} Y(X(t)) = \int_{\Omega} Y(\omega)\, d\pi(\omega). \tag{12.7}$$

(Here again T is fixed.) As we have already stated, a direct calculation of the righthand side of (12.7), namely,

$$\sum_{\omega} Y(\omega)\, e^{-U(\omega)/T} \Big/ \sum_{\omega} e^{-U(\omega)/T}$$

is impossible in general. The left-hand side of (12.7) suggests that we use the Gibbs Sampler and compute a time average of the function Y. For most physical systems, the ergodic hypothesis is just that–a *hypothesis*–which can rarely be verified in practice. Fortunately, for our system it is not too difficult to directly establish ergodicity.

Theorem C (Ergodicity): Assume that there exists a τ *such that* $S \subseteq \{n_{t+1}, \cdots, n_{t+\tau}\}$ *for all t. Then for every function Y on* Ω *and for every starting configuration* $\eta \in \Omega$, (12.7) *holds with probability one.*

XIII. Experimental Results

There are three groups of pictures. Each contains an original image, several degraded versions, and the corresponding restorations, usually at two stages of the annealing process to illustrate its evolution. The degradations are formed from combinations of

i) ϕ absent or $\phi(x) = \sqrt{x}$;
ii) multiplicative or additive noise;
iii) signal-to-noise levels.

The signal-to-noise ratios are all very low. For blurring, we always took the convolution H in (2.3). The restorations are all MAP estimates generated by the serial Gibbs Sampler with annealing schedule

$$T(k) = \frac{C}{\log(1 + k)}, \qquad 1 \leqslant k \leqslant K$$

where $T(k)$ is the temperature during the kth *iteration* (= full sweep of S), so that K is the total number of iterations. In each case, $C = 3.0$ or $C = 4.0$. No pre- or postfiltering, nor anything else was done. The models for the intensity and line processes were kept as simple as possible; indeed, only cliques of size two appear in the intensity model.

Group 1: The original image [Fig. 2(a)] is a sample of an MRF on Z_{128} with $L = 5$ intensities and the eight-neighbor system (Fig. 1, $c = 2$). The potentials $V_C = 0$ unless $C = \{r, s\}$, in which case

$$V_C(f) = \begin{cases} \frac{1}{3}, & f_s = f_r \\ -\frac{1}{3}, & f_s \neq f_r. \end{cases}$$

Two hundred iterations (at $T \equiv 1$) were made to generate Fig. 2(a).

The first degraded version is Fig. 2(b), which is simply Fig. 2(a) plus Gaussian noise with $\sigma = 1.5$ *relative to* gray levels f, $1 \leqslant f \leqslant 5$. Fig. 2(c) is the restoration of Fig. 2(b) with $K = 25$ iterations only, i.e., early in the annealing process. In Fig. 2(d), $K = 300$.

The second degraded image [Fig. 3(b)] uses the model

$$G = H(F)^{1/2} \cdot N \tag{13.1}$$

where $\mu = 1$ and $\sigma = 0.1$, again relative to intensities $1 \leqslant f \leqslant 5$. Fig. 3(c) and 3(d) shows the restorations of Fig. 3(b) with $K = 25$ and $K = 300$, respectively.

Group 2: Fig. 4(a) is "hand-drawn." The lattice size is 64×64 and there are three gray levels. Gaussian noise ($\mu = 0$, $\sigma = 0.7$) was added to produce Fig. 4(b). We tried two types of restoration on Fig. 4(b). First, we used the "blob process" which generated Fig. 2(a) for the F-model. There was no line process and $K = 1000$. Obviously these are flaws; see Fig. 4(c).

A line process L was then adjoined to F for the original image model, and the corresponding restoration after 1000 iterations is shown in Fig. 4(d). L itself was described in Case 2 of Section III and the neighborhood system for (F, L) on $Z_{64} \cup D_{64}$ was discussed in Case 3 of Section III. The (prior) distribution on X = (F, L) was as follows. The range of F is $\{0, 1, 2\}$ ($L = 3$ intensities). The energy $U(f, l)$ consists of two terms, say $U(f|l) + U(l)$. To understand the interaction term $U(f|l)$, let d denote a line site, say between pixels r and s. If $L_d = 1$, i.e., an edge element is "present" at d, then the bond between s and r is "broken" and we set $V_{\{r,s\}}(f_r, f_s) = 0$ regardless of f_r, f_s; otherwise ($L_d = 0$) $V_{\{r,s\}}$ is as before except that $\pm\frac{1}{3}$ are replaced by ± 1. As for $U(l)$, only cliques of size four are nonzero, of which there are six distinct types up to rotations. These are shown in Fig. 5(a) with their associated energy values.

Then we corrupted the hand-drawn figure using (13.1) with the same noise parameters as Fig. 3(b), obtaining Fig. 6(b), which is restored in Fig. 6(c) using the same prior on (F, L) as above and with $K = 1000$ iterations.

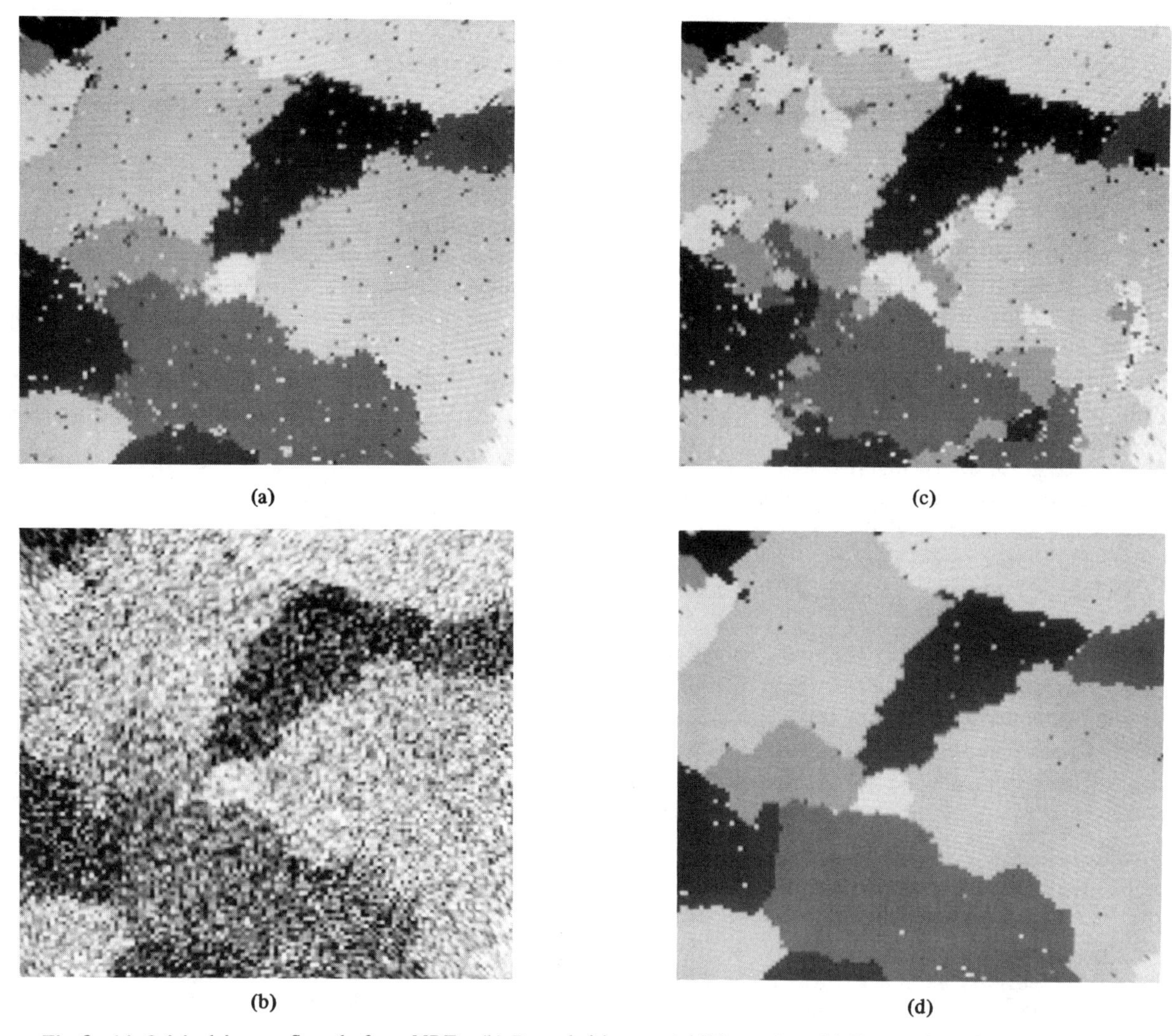

Fig. 2. (a) Original image: Sample from MRF. (b) Degraded image: Additive noise. (c) Restoration: 25 iterations. (d) Restoration: 300 iterations.

Group 3: The results in Group 2 suggest a boundary-finding algorithm for general shapes: allow the line process more directional freedom. Group 3 is an exercise in boundary finding at essentially 0 dB. Fig. 7(a) is a 64×64 segment of a roadside photograph that we obtained from the Visions Research Group at the University of Massachusetts. The levels are scaled so that the (existing) two peaks in the histogram occur at $f = 0$ and $f = 1$. We regard Fig. 7(a) as the *blurred image* $H(\mathbb{F})$. Noise is added in Fig. 7(b); the standard error is $\sigma = 0.5$ *relative to the two main gray levels* $f = 0, 1$.

Figs. 7(c) and 7(d) are "restorations" of Fig. 7(b) for $K = 100$ and $K = 1000$ iterations, respectively. The outcome of the line process is indicated by painting black any pixels to the left of or above a "broken bond." The two main regions, comprising the sign and the arrow, are perfectly circumscribed by a continuous sequence of line elements.

The model for $\mathbb{X}$ is more complex than the one in Group 2. There are now four possible states for each line site corresponding to "off" ($l = 0$) and three directions, shown in Fig. 5(b). The $U(f|l)$ term is the same as before in that the pixel bond between r and s is broken whenever $l_d \neq 0$. The range of $\mathbb{F}$ is $\{0, 1\}$ ($L = 2$).

Only cliques of size four are nonzero in $U(l)$, as before. However, there are now many combinations for $(l_{d_1}, l_{d_2}, l_{d_3}, l_{d_4})$ given such a clique $C = \{d_1, d_2, d_3, d_4\}$ of line sites, although the number is substantially reduced by assuming rotational invariance, which we do. Fig. 5(c) shows the convention we will use for the ordering and an example of the notation. The energies for the possible configurations $(l_{d_i}, 1 \leq i \leq 4)$ range from 0 to 2.70. (Remember that high energies correspond to low probability, and that the exponential exaggerates differences.) We took $V(0, 0, 0, 0) = 0$ and $V(l_{d_i}, 1 \leq i \leq 4) = 2.70$ otherwise, except when exactly two of the l_{d_i} are nonzero. Parallel segments [e.g., (1, 0, 1, 0)] receive energy 2.70; sharp turns [e.g., (0, 2, 1, 0)] and other "corner" types get 1.80; mild turns [e.g., (0, 2, 3, 0)] are 1.35; and continuations [e.g., (2, 0, 2, 0) or (0, 1, 3, 0)] are 0.90.

XIV. Concluding Remarks

We have introduced some new theoretical and processing methods for image restoration. The models and estimates are noncausal and nonlinear, and do not represent extensions into two dimensions of one-dimensional filtering and smoothing

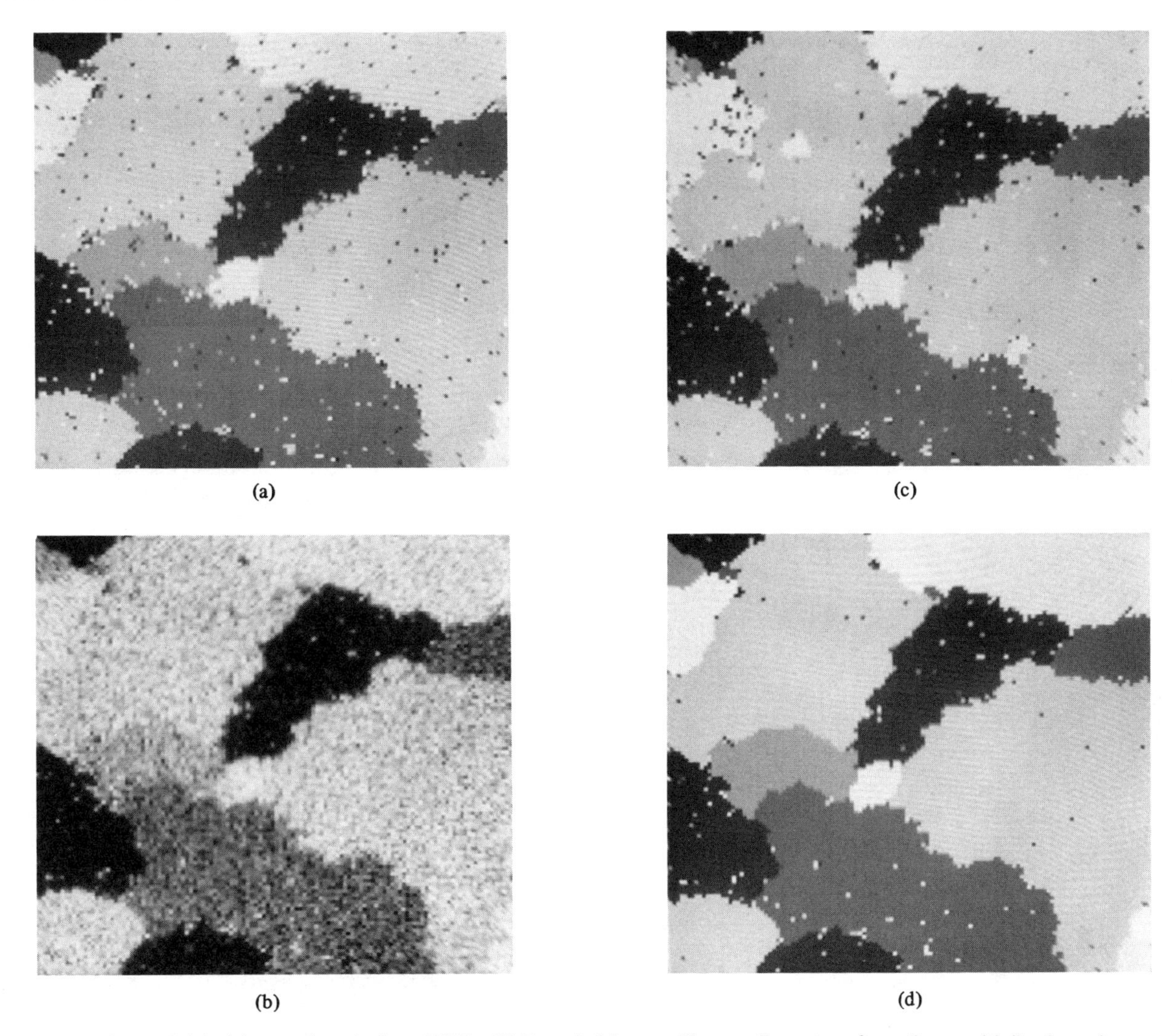

Fig. 3. (a) Original image: Sample from MRF. (b) Degraded image: Blur, nonlinear transformation, multiplicative noise. (c) Restoration: 25 iterations. (d) Restoration: 300 iterations.

algorithms. Rather, our work is largely inspired by the methods of statistical physics for investigating the time-evolution and equilibrium behavior of large, lattice-based systems.

There are, of course, *many* well-known and remarkable features of these massive, homogeneous physical systems. Among these is the evolution to minimal energy states, regardless of initial conditions. In our work posterior (Gibbs) distribution represents an *imaginary* physical system whose lowest energy states are exactly the MAP estimates of the original image given the degraded "data."

The approach is very flexible. The MRF-Gibbs class of models is tailor-made for representing the dependencies among the intensity levels of nearby pixels as well as for augmenting the usual, pixel-based process by other, unobservable attribute processes, such as our "line process," in order to bring exogenous information into the model. Moreover, the degradation model is almost unrestricted; in particular, we allow for deformations due to the image formation and recording processes. All that is required is that the posterior distribution have a "reasonable" neighborhood structure as a MRF, for in that case the computational load can be accommodated by appropriate variants (such as the Gibbs Sampler) of relaxation algorithms for dynamical systems.

Appendix

Proofs of Theorems

Background and Notation

Recall that $\Lambda = \{0, 1, 2, \cdots, L-1\}$ is the common state space, that η, η', ω, etc. denote elements of the configuration space $\Omega = \Lambda^N$, and that the sites $S = \{s_1, s_2, \cdots, s_N\}$ are visited for updating in the order $\{n_1, n_2, \cdots\} \subset S$. The resulting stochastic process is $\{X(t), t = 0, 1, 2, \cdots\}$, where $X(0)$ is the initial configuration.

For Theorem A, the transitions are governed by the Gibbs distribution $\pi(\omega) = e^{-U(\omega)/T}/Z$ in accordance with (12.1), whereas, for Theorem B (annealing), we use $\pi_{T(t)}$ (see Section XII) for the transition $X(t-1) \rightarrow X(t)$ [see (12.3)].

Let us briefly discuss the process $\{X(t), t \geqslant 0\}$, restricting attention to constant temperature; the annealing case is essentially the same. To begin with, $\{X(t), t \geqslant 0\}$ is indeed a Markov chain; this is apparent from its construction. Fix t and

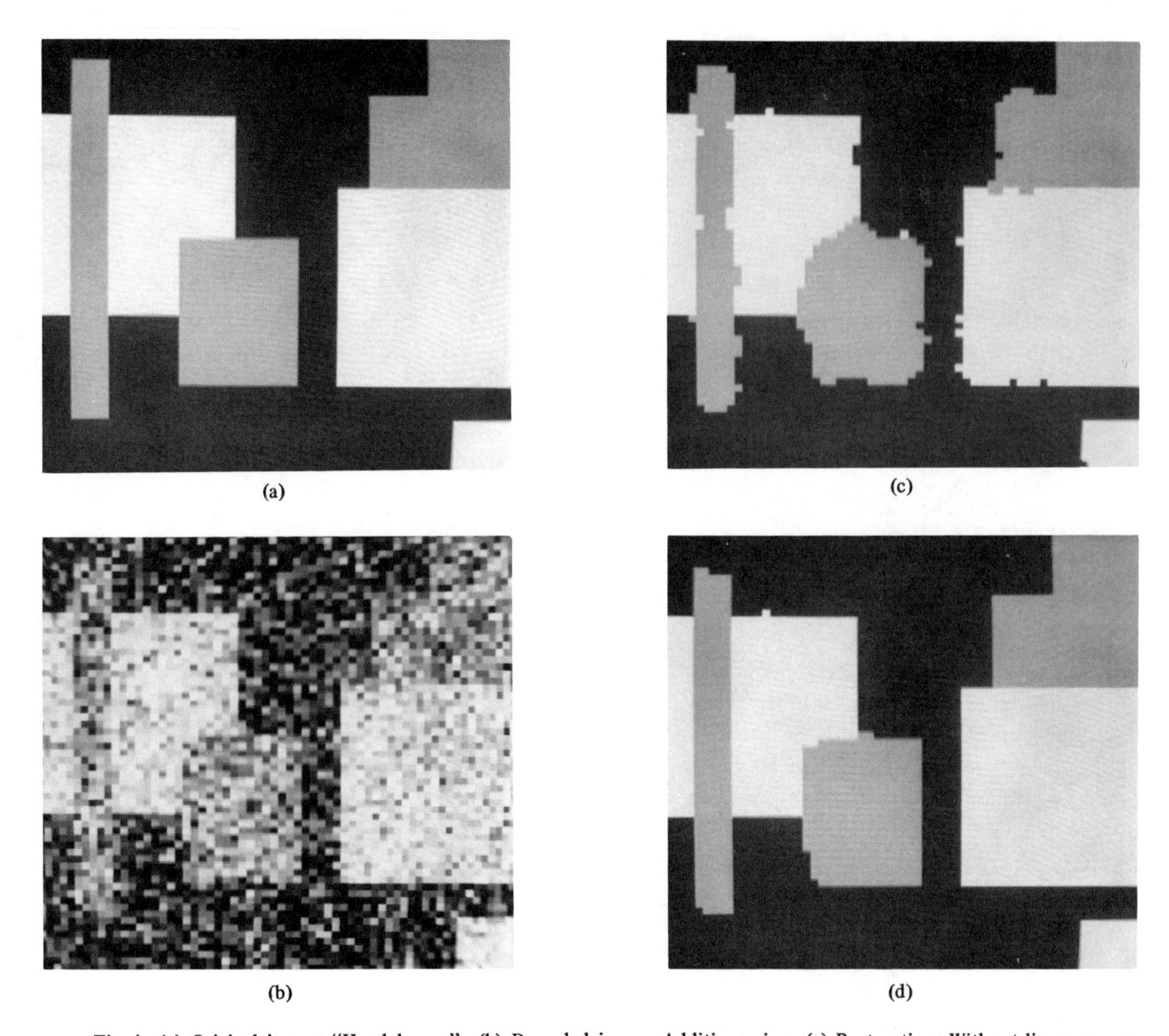

Fig. 4. (a) Original image: "Hand-drawn." (b) Degraded image: Additive noise. (c) Restoration: Without line process; 1000 iterations. (d) Restoration: Including line process; 1000 iterations.

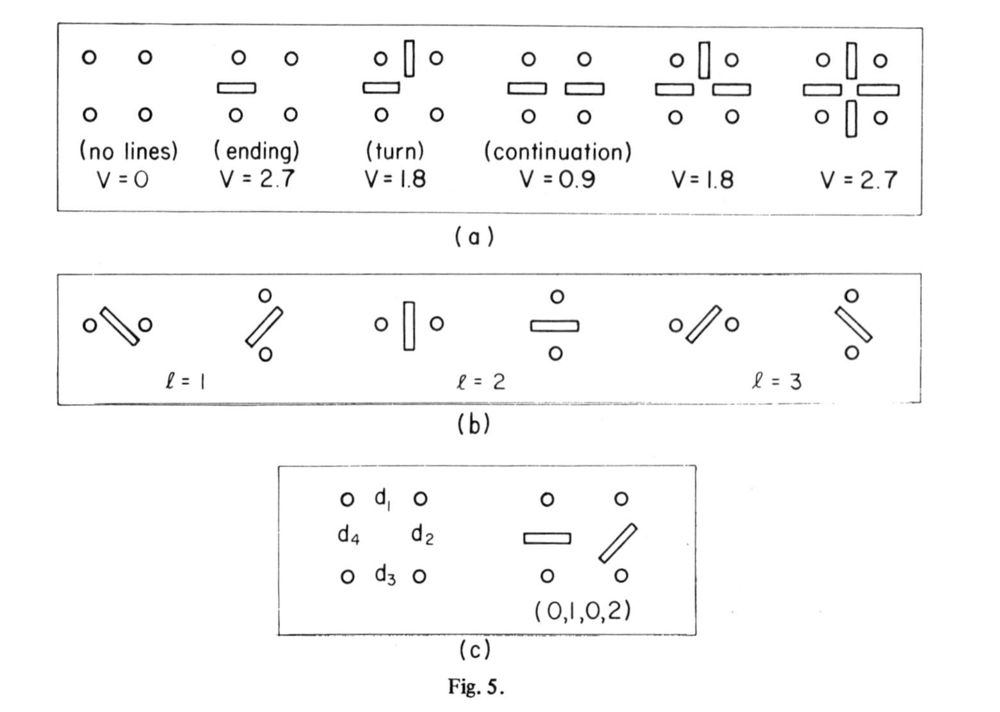

Fig. 5.

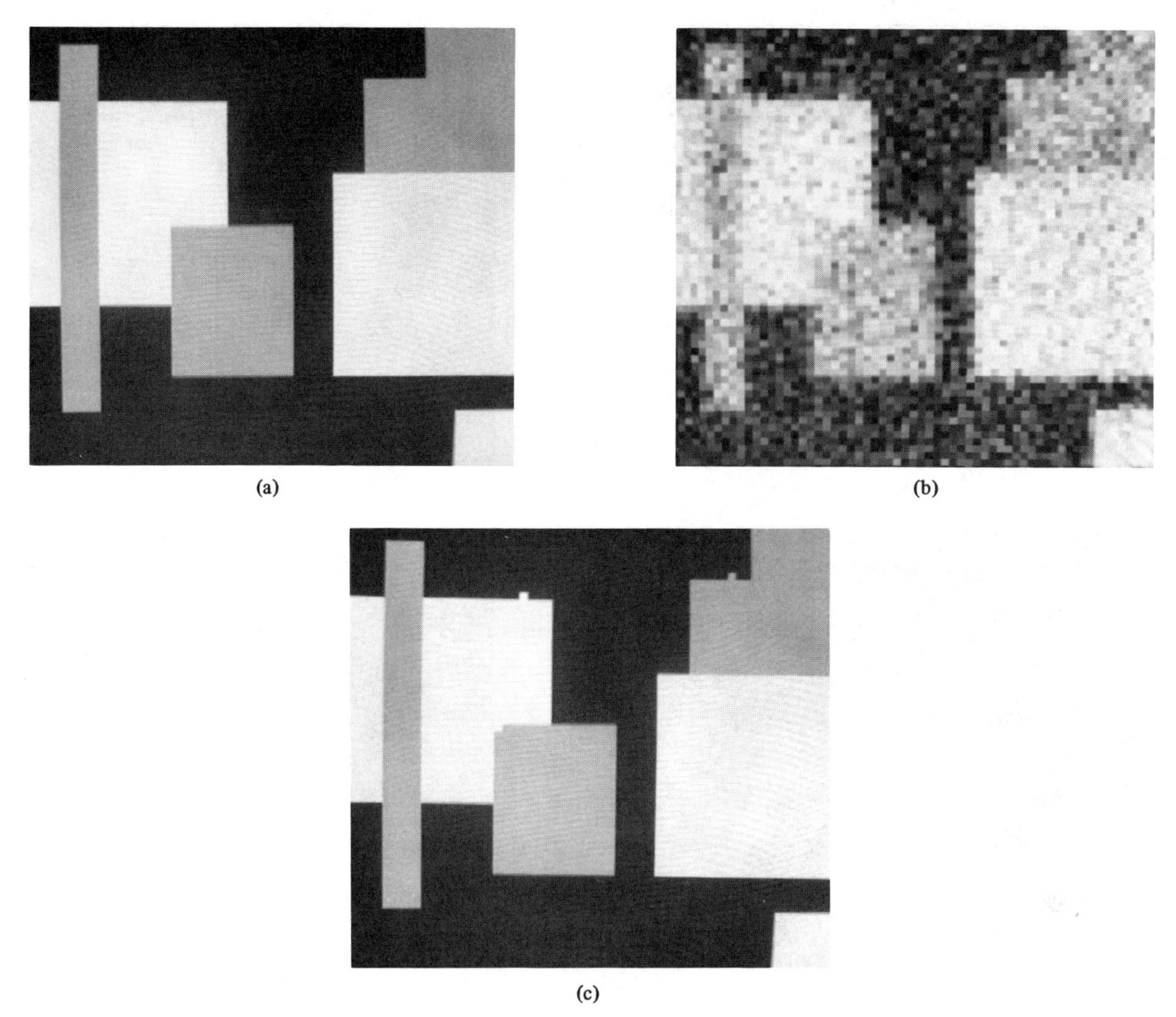

Fig. 6. (a) Original image: "Hand-drawn." (b) Degraded image: Blur, nonlinear transformation, multiplicative noise. (c) Restoration: including line process; 1000 iterations.

$\omega \in \Omega$. For any $x \in \Lambda$, let ω^x denote the configuration which is x at site n_t and agrees with ω elsewhere. The transition matrix *at time t* is

$$(M_t)_{\eta,\omega} = \begin{cases} \pi(X_{n_t} = x_{n_t} \mid X_s = x_s, s \neq n_t), \\ \qquad \text{if} \quad \eta = \omega^x \quad \text{for some} \quad x \in \Lambda \\ 0, \quad \text{otherwise} \end{cases}$$

where $(M_t)_{\eta,\omega}$ denotes the row η, column ω entry of M_t, and $\omega = (x_{s_1}, x_{s_2}, \cdots, x_{s_N})$. In particular, the chain is *nonstationary*, although clearly *aperiodic* and *irreducible* (since $\pi(\omega) > 0 \ \forall \omega$). Moreover, given any starting vector (distribution) μ_0, the distribution of $X(t)$ is given by the vector $\mu_0 \prod_{j=1}^t M_j$, i.e.,

$$P_{\mu_0}(X(t) = \omega) = \left(\mu_0 \times \prod_{j=1}^{t} M_j\right)_\omega$$

$$= \sum_\eta P(X(t) = \omega \mid X(0) = \eta)\, \mu_0(\eta).$$

Notice that π is the (necessarily) unique invariant vector, i.e., for every $t = 1, 2, \cdots,$

$$\pi(\omega) = (\pi M_t)_\omega = \sum_\eta P(X(t) = \omega \mid X(0) = \eta)\, \pi(\eta). \tag{A.1}$$

To see this, fix t and $\omega = \{x_s\}$, and write

$$(\pi M_t)_\omega = \sum_\eta \pi(\eta)(M_t)_{\eta,\omega}$$

$$= \sum_{x \in \Lambda} \pi(\omega^x)(M_t)_{\omega^x,\omega}$$

$$= (M_t)_{\omega^{x'},\omega} \sum_{x \in \Lambda} \pi(\omega^x) \qquad (\text{for } any \quad x' \in \Lambda)$$

$$= \pi(X_{n_t} = x_{n_t} \mid X_s = x_s, s \neq n_t)\, \pi(X_s = x_s, s \neq n_t)$$

$$= \pi(\omega).$$

It will be convenient to use the following, semistandard notation for transitions. For nonnegative integers $r < t$ and $\omega, \eta \in \Omega$, set

$$P(t, \omega \mid r, \eta) = P(X(t) = \omega \mid X(r) = \eta)$$

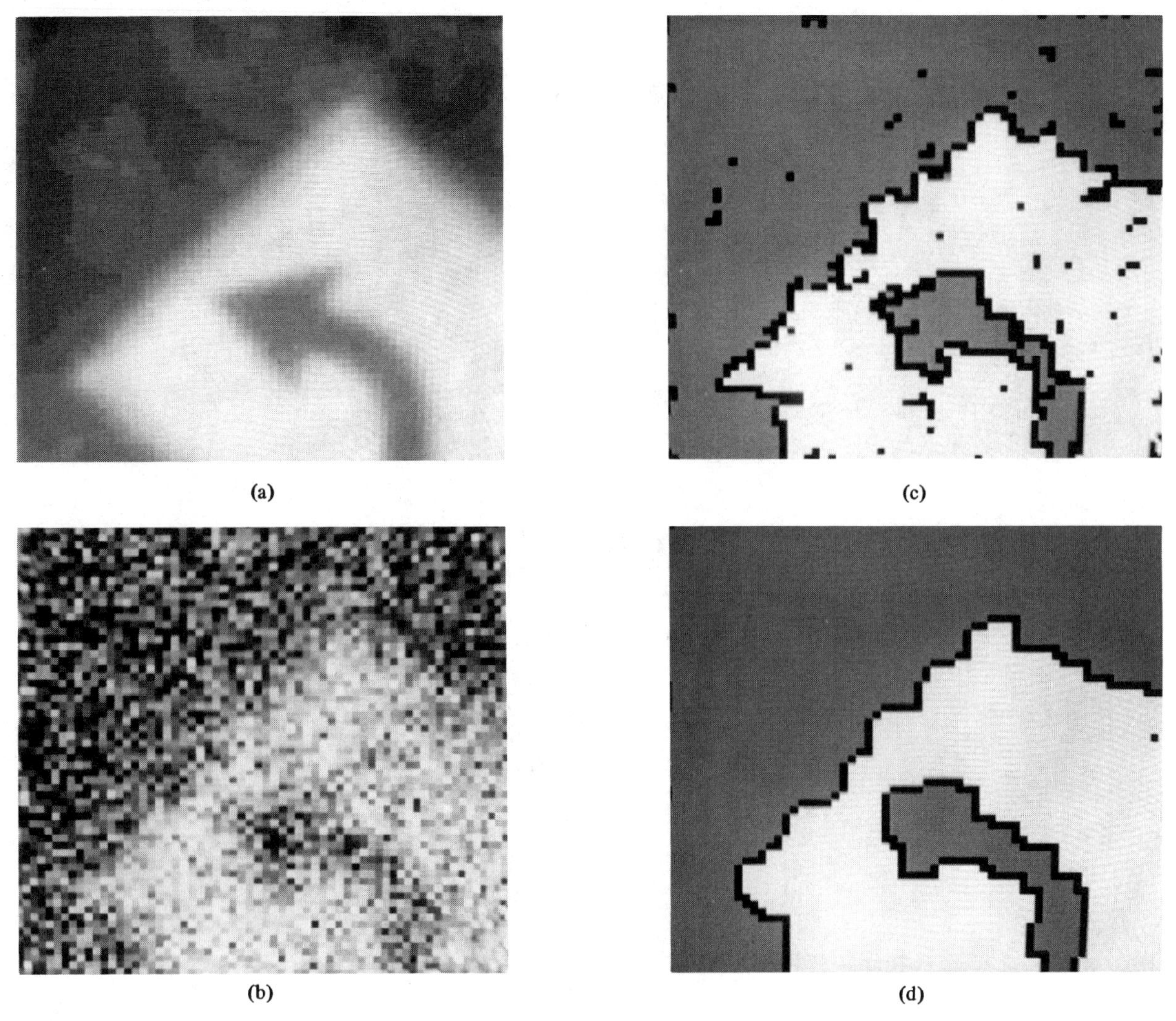

Fig. 7. (a) Blurred image (roadside scene). (b) Degraded image: Additive noise. (c) Restoration including line process; 100 iterations. (d) Restoration including line process; 1000 iterations.

and, for any *distribution* μ on Ω, set

$$P(t, \omega | r, \mu) = \sum_{\eta} P(t, \omega | r, \eta) \, \mu(\eta).$$

Finally, $\|\mu - \nu\|$ denotes the L^1 distance between two distributions on Ω:

$$\|\mu - \nu\| = \sum_{\omega} |\mu(\omega) - \nu(\omega)|.$$

Obviously, $\mu_n \to \mu (n \to \infty)$ in distribution (i.e., $\mu_n(\omega) \to \mu(\omega)$ $\forall \omega$) if and only if $\|\mu_n - \mu\| \to 0$, $n \to \infty$. (Remember that Ω is finite.)

Proof of Theorem A: Set $T_0 = 0$ and define $T_1 < T_2 < \cdots$ such that

$$S \subseteq \{n_{T_{k-1}+1}, n_{T_{k-1}+2}, \cdots, n_{T_k}\}, \qquad k = 1, 2, \cdots.$$

This is possible since every site is visited infinitely often. Clearly (at least) k iterations or full sweeps have been completed by "time" T_k. In particular, $kN \leqslant T_k < \infty \; \forall k$. Let

$$K(t) = \sup \{k : T_k < t\}.$$

Obviously $K(t) \to \infty$ at $t \to \infty$. The proof of Theorem A is based on the following lemma, which also figures in the proof of the annealing theorem.

Lemma 1: There exists a constant r, $0 \leqslant r < 1$, *such that for every* $t = 1, 2, \cdots$,

$$\sup_{\omega, \eta', \eta''} |P(X(t) = \omega | X(0) = \eta') - P(X(t) = \omega | X(0) = \eta'')| \leqslant r^{K(t)}.$$

Assume for now that the lemma is true. Since π is an invariant vector for the chain:

$$\overline{\lim_{t \to \infty}} \sup_{\omega, \eta} |P(X(t) = \omega | X(0) = \eta) - \pi(\omega)|$$

$$= \overline{\lim_{t \to \infty}} \sup_{\omega, \eta} \left| \sum_{\eta'} \pi(\eta') \{P(X(t) = \omega | X(0) = \eta) - P(X(t) = \omega | X(0) = \eta')\} \right|$$

[by (A.1)]

$$\leqslant \overline{\lim_{t\to\infty}} \sup_{\omega,\eta',\eta''} |P(X(t)=\omega|X(0)=\eta') - P(X(t)=\omega|X(0)=\eta'')|$$

= 0, by Lemma 1.

So it suffices to prove Lemma 1.

Proof of Lemma 1: For each $k = 1, 2, \cdots$ and $1 \leqslant i \leqslant N$, let m_i be the time of the last replacement of site s_i before $T_k + 1$, i.e.,

$$m_i = \sup \{t : t \leqslant T_k, n_t = s_i\}.$$

We can assume, without loss of generality, that $m_1 > m_2 > \cdots > m_N$; otherwise, relabel the sites. For any $\omega = (x_{s_1}, \cdots, x_{s_N})$ and ω',

$$\begin{aligned} P(X(T_k) = \omega | X(T_{k-1}) = \omega') &= P(X_{s_1}(m_1) = x_{s_1}, \cdots, X_{s_N}(m_N) = x_{s_N} | X(T_{k-1}) = \omega') \\ &= \prod_{j=1}^{N} P(X_{s_j}(m_j) = x_{s_j} | X_{s_{j+1}}(m_{j+1}) = x_{s_{j+1}}, \cdots, X_{s_N}(m_N) = x_{s_N}, X(T_{k-1}) = \omega'). \end{aligned}$$

Let δ be the smallest probability among the local characteristics:

$$\delta = \inf_{\substack{(x_{s_1}, \cdots, x_{s_N}) \in \Omega \\ 1 \leqslant i \leqslant N}} \pi(X_{s_i} = x_{s_i} | X_{s_j} = x_{s_j}, j \neq i).$$

Then $0 < \delta < 1$ and a little reflection shows that every term in the product above is at least δ. Hence,

$$\inf_{\substack{k=1,2,\cdots \\ \omega, \omega'}} P(X(T_k) = \omega | X(T_{k-1}) = \omega') \geqslant \delta^N. \qquad \text{(A.2)}$$

Consider now the inequality asserted in Lemma 1. It is trivial for $t \leqslant T_1$ since in this case $K(t) = 0$. For $t > T_1$,

$$\begin{aligned} &\sup_{\omega,\eta',\eta''} |P(X(t)=\omega|X(0)=\eta') - P(X(t)=\omega|X(0)=\eta'')| \\ &= \sup_{\omega} \{\sup_{\eta} P(X(t)=\omega|X(0)=\eta) - \inf_{\eta} P(X(t)=\omega|X(0)=\eta)\} \\ &= \sup_{\omega} \Big\{\sup_{\eta} \sum_{\omega'} P(X(t)=\omega|X(T_1)=\omega') P(X(T_1)=\omega'|X(0)=\eta) \\ &\quad - \inf_{\eta} \sum_{\omega'} P(X(t)=\omega|X(T_1)=\omega') P(X(T_1)=\omega'|X(0)=\eta)\Big\} \\ &\doteq \sup_{\omega} Q(t,\omega). \end{aligned}$$

Certainly, for each $\omega \in \Omega$,

$$\begin{aligned} &\sup_{\eta} \sum_{\omega'} P(X(t)=\omega|X(T_1)=\omega') P(X(T_1)=\omega'|X(0)=\eta) \\ &\leqslant \sup_{\mu} \sum_{\omega'} P(X(t)=\omega|X(T_1)=\omega') \mu(\omega') \end{aligned}$$

where the supremum is over all probability measures μ on Ω which, by (A.2), are subject to $\mu(\omega') \geqslant \delta^N \; \forall \omega'$. Suppose $\omega' \to P(X(t)=\omega|X(T_1)=\omega')$ is maximized at $\omega' = \omega^*$ (which depends on ω). Then the last supremum is attained by placing mass δ^N on each ω' and the remaining mass, namely, $1 - |\Omega|\delta^N = 1 - L^N\delta^N$, on ω^*. The value so obtained is

$$(1 - (L^N - 1)\delta^N) P(X(t) = \omega | X(T_1) = \omega^*) + \delta^N \sum_{\omega' \neq \omega^*} P(X(t)=\omega|X(T_1)=\omega').$$

Similarly,

$$\begin{aligned} &\inf_{\eta} \sum_{\omega'} P(X(t)=\omega|X(T_1)=\omega') P(X(T_1)=\omega'|X(0)=\eta) \\ &\geqslant (1 - (L^N - 1)\delta^N) P(X(t) = \omega | X(T_1) = \omega_*) + \delta^N \sum_{\omega' \neq \omega_*} P(X(t)=\omega|X(T_1)=\omega_*) \end{aligned}$$

where $\omega' \to P(X(t)=\omega|X(T_1)=\omega')$ is *minimized* at ω_*. It follows immediately that

$$Q(t,\omega) \leqslant (1 - L^N\delta^N)\{P(X(t) = \omega|X(T_1)=\omega^*) - P(X(t)=\omega|X(T_1)=\omega_*)\},$$

and hence,

$$\begin{aligned} &\sup_{\omega,\eta',\eta''} |P(X(t)=\omega|X(0)=\eta') - P(X(t)=\omega|X(0)=\eta'')| \\ &\leqslant (1 - L^N\delta^N) \sup_{\omega,\eta',\eta''} |P(X(t) = \omega|X(T_1)=\eta') - P(X(t) = \omega|X(T_1)=\eta'')|. \end{aligned}$$

Proceeding in this way, we obtain the bound

$$(1 - L^N\delta^N)^{K(t)} \sup_{\omega,\eta',\eta''} |P(X(t) = \omega|X(T_{K(t)})=\eta') - P(X(t)=\omega|X(T_{K(t)})=\eta'')|$$

and the lemma now follows with $r = 1 - L^N\delta^N$. Notice that $r = 0$ corresponds to the (degenerate) case in which $\delta = L^{-1}$, i.e., all the local characteristics are uniform on Λ. Q.E.D.

Proof of Theorem B: We first state two lemmas.

Lemma 2: For every $t_0 = 0, 1, 2, \cdots$,

$$\lim_{t\to\infty} \sup_{\omega,\eta',\eta''} |P(X(t) = \omega|X(t_0)=\eta') - P(X(t)=\omega|X(t_0)=\eta'')| = 0.$$

Lemma 3:

$$\lim_{t_0 \to \infty} \sup_{t \geq t_0} \|P(t, \cdot \,|t_0, \pi_0) - \pi_0\| = 0.$$

Recall that π_0 is the uniform probability measure over the minimal energy states $\Omega_0 = \{\omega : U(\omega) = \min_\eta U(\eta)\}$.

First we show how these lemmas imply Theorem B, which states that $P(X(t) = \cdot \,|X(0) = \eta)$ converges to π_0 as $t \to \infty$. For any $\eta \in \Omega$,

$$\overline{\lim_{t \to \infty}} \|P(X(t) = \cdot \,|X(0) = \eta) - \pi_0\|$$

$$= \overline{\lim_{t_0 \to \infty}} \ \overline{\lim_{\substack{t \to \infty \\ t \geq t_0}}} \Big\|\sum_{\eta'} P(t, \cdot \,|t_0, \eta') \cdot P(t_0, \eta'|0, \eta) - \pi_0\Big\|$$

$$\leq \overline{\lim_{t_0 \to \infty}} \ \overline{\lim_{\substack{t \to \infty \\ t \geq t_0}}} \Big\|\sum_{\eta'} P(t, \cdot \,|t_0, \eta') \cdot P(t_0, \eta'|0, \eta) - P(t, \cdot \,|t_0, \pi_0)\Big\|$$

$$+ \overline{\lim_{t_0 \to \infty}} \ \overline{\lim_{\substack{t \to \infty \\ t \geq t_0}}} \|P(t, \cdot \,|t_0, \pi_0) - \pi_0\|.$$

The last term is zero by Lemma 3. Furthermore, since $P(t_0, \cdot \,|0, \eta)$ and π_0 have total mass 1, we have

$$\Big\|\sum_{\eta'} P(t, \cdot \,|t_0, \eta') P(t_0, \eta'|0, \eta) - P(t, \cdot \,|t_0, \pi_0)\Big\|$$

$$= \sum_\omega \sup_{\eta''} \Big|\sum_{\eta'} (P(t, \omega|t_0, \eta') - P(t, \omega|t_0, \eta'')) \times (P(t_0, \eta'|0, \eta) - \pi_0(\eta'))\Big|$$

$$\leq 2 \sum_\omega \sup_{\eta', \eta''} |P(t, \omega|t_0, \eta') - P(t, \omega|t_0, \eta'')|.$$

Finally, then,

$$\overline{\lim_{t \to \infty}} \|P(X(t) = \cdot \,|X(0) = \eta) - \pi_0\|$$

$$\leq 2 \sum_\omega \overline{\lim_{t_0 \to \infty}} \ \overline{\lim_{\substack{t \to \infty \\ t \geq t_0}}} \sup_{\eta', \eta''} |P(t, \omega|t_0, \eta') - P(t, \omega|t_0, \eta'')|$$

$$= 0 \quad \text{by Lemma 2.}$$

Q.E.D.

Proof of Lemma 2: We follow the proof of Lemma 1. Fix $t_0 = 0, 1, \cdots$ and define $T_k = t_0 + k\tau$, $k = 0, 1, 2, \cdots$. Recall that $S \subseteq \{n_{t+1}, \cdots, n_{t+\tau}\}$ for all t by hypothesis, that $\pi_{T(t)}(\omega) = e^{-U(\omega)/T(t)}/Z$ and that U^*, U_* are the maximum and minimum of $U(\omega)$, respectively, the range being $\Delta = U^* - U_*$. Let

$$\delta(t) = \inf_{\substack{1 \leq i \leq N \\ (x_{s_1}, \cdots, x_{s_N}) \in \Omega}} \pi_{T(t)}(X_{s_i} = x_{s_i}|X_{s_j} = x_{s_j}, j \neq i).$$

Observe that

$$\delta(t) \geq \frac{e^{-U^*/T(t)}}{Le^{-U_*/T(t)}} = \frac{1}{L} e^{-\Delta/T(t)}.$$

Now fix k for the moment and define the m_i as before:

$$m_i = \sup \{t : t \leq T_k, n_t = s_i\}, \qquad 1 \leq i \leq N.$$

We again assume that $m_1 > m_2 > \cdots > m_N$. Then

$$P(X(T_k) = \omega|X(T_{k-1}) = \omega')$$
$$= P(X_{s_1}(m_1) = x_{s_1}, \cdots, X_{s_N}(m_N) = x_{s_N}|X(T_{k-1}) = \omega')$$
$$= \prod_{j=1}^N P(X_{s_j}(m_j) = x_{s_j}|X_{s_{j+1}}(m_{j+1}) = x_{s_{j+1}}, \cdots, X_{s_N}(m_N) = x_{s_N}, X(T_{k-1}) = \omega')$$
$$\geq \prod_{j=1}^N \delta(m_j) \quad \text{(using (12.3) and the definition of } \delta)$$
$$\geq L^{-N} \prod_{j=1}^N e^{-\Delta/T(m_j)}$$
$$\geq L^{-N} \exp - \left\{\frac{\Delta N}{T(t_0 + k\tau)}\right\} \quad (\text{since } m_j \leq T_k = t_0 + k\tau, j = 1, 2, \cdots, N, \text{ and } T(\cdot) \text{ is decreasing})$$
$$\geq L^{-N}(t_0 + k\tau)^{-1}$$

wherever $t_0 + k\tau$ is sufficiently large. In fact, for a sufficiently small constant C, we can and do assume that

$$\inf_{\omega, \omega'} P(X(T_k) = \omega|X(T_{k-1}) = \omega') \geq \frac{CL^{-N}}{t_0 + k\tau} \tag{A.3}$$

for every $t_0 = 0, 1, 2, \cdots$ and $k = 1, 2, \cdots$, bearing in mind that T_k depends on t_0.

For each $t > t_0$, define $K(t) = \sup \{k : T_k < t\}$ so that $K(t) \to \infty$ as $t \to \infty$. Fix $t > T_1$ and continue to follow the argument in Lemma 1, but using (A.3) in place of (A.2), obtaining

$$\sup_{\omega, \eta', \eta''} |P(X(t) = \omega|X(t_0) = \eta') - P(X(t) = \omega|X(t_0) = \eta'')|$$

$$\leq \prod_{k=1}^{K(t)} \left(1 - \frac{C}{t_0 + k\tau}\right).$$

Hence it will be sufficient to show that

$$\lim_{m \to \infty} \prod_{k=1}^m \left(1 - \frac{C}{t_0 + k\tau}\right) = 0 \tag{A.4}$$

for every t_0. However, (A.4) is a well-known consequence of the divergence of the series $\sum_k (t_0 + k\tau)^{-1}$ for all t_0, τ. This completes the proof of Lemma 2.

Proof of Lemma 3: The probability measures $P(t, \cdot \,|t_0, \pi_0)$ figure prominently in the proof, and for notational ease we prefer to write $P_{t_0, t}(\cdot)$, so that for any $t \geq t_0 > 0$ we have

$$P_{t_0, t}(\omega) = \sum_\eta P(X(t) = \omega|X(t_0) = \eta) \, \pi_0(\eta).$$

To begin with, we claim that for any $t > t_0 \geq 0$,

$$\|P_{t_0, t} - \pi_{T(t)}\| \leq \|P_{t_0, t-1} - \pi_{T(t)}\|. \tag{A.5}$$

Assume for convenience that $n_t = s_1$. Then

$$\begin{aligned}
&\|P_{t_0,t} - \pi_{T(t)}\| \\
&= \sum_{(x_{s_1},\cdots,x_{s_N})} \left|\pi_{T(t)}(X_{s_1} = x_{s_1} | X_s = x_s, s \neq s_1) \cdot P_{t_0,t-1}(X_s = x_s, s \neq s_1) - \pi_{T(t)}(X_s = x_s, s \in S)\right| \\
&= \sum_{x_{s_2},\cdots,x_{s_N}} \left\{ \sum_{x_{s_1}\in\Lambda} \pi_{T(t)}(X_{s_1} = x_{s_1} | X_s = x_s, s \neq s_1) \times \left|P_{t_0,t-1}(X_s = x_s, s \neq s_1) - \pi_{T(t)}(X_s = x_s, s \neq s_1)\right| \right\} \\
&= \sum_{x_{s_2},\cdots,x_{s_N}} \left|P_{t_0,t-1}(X_s = x_s, s \neq s_1) - \pi_{T(t)}(X_s = x_s, s \neq s_1)\right| \\
&= \sum_{x_{s_2},\cdots,x_{s_N}} \left|\sum_{x_{s_1}} \{P_{t_0,t-1}(X_s = x_s, s \in S) - \pi_{T(t)}(X_s = x_s, s \in S)\}\right| \\
&\leq \sum_{(x_{s_1},\cdots,x_{s_N})\in\Omega} \left|P_{t_0,t-1}(X_s = x_s, s \in S) - \pi_{T(t)}(X_s = x_s, s \in S)\right| \\
&= \|P_{t_0,t-1} - \pi_{T(t)}\|.
\end{aligned}$$

Observe that $\|\pi_0 - \pi_{T(t)}\| \to 0$ as $t \to \infty$. To see this, let $|\Omega_0|$ be the size of Ω_0. Then

$$\begin{aligned}
\pi_{T(t)}(\omega) &= \frac{e^{-U(\omega)/T(t)}}{\sum_{\omega'\in\Omega_0} e^{-U(\omega')/T(t)} + \sum_{\omega'\in\Omega\backslash\Omega_0} e^{-U(\omega')/T(t)}} \\
&= \frac{e^{-(U(\omega)-U_*)/T(t)}}{|\Omega_0| + \sum_{\omega'\in\Omega\backslash\Omega_0} e^{-(U(\omega')-U_*)/T(t)}} \\
&\xrightarrow{t\to\infty} \begin{cases} 0, & \omega \notin \Omega_0 \\ \dfrac{1}{|\Omega_0|}, & \omega \in \Omega_0. \end{cases}
\end{aligned} \tag{A.6}$$

Next, we claim that

$$\sum_{t=1}^{\infty} \|\pi_{T(t)} - \pi_{T(t+1)}\| < \infty. \tag{A.7}$$

Since

$$\sum_{t=1}^{\infty} \|\pi_{T(t)} - \pi_{T(t+1)}\| = \sum_{\omega} \sum_{t=1}^{\infty} |\pi_{T(t)}(\omega) - \pi_{T(t+1)}(\omega)|$$

and since $\pi_{T(t)}(\omega) \to \pi_0(\omega)$ for every ω, it will be enough to show that, for every ω, $\pi_T(\omega)$ is monotone (increasing or decreasing) in T for all T sufficiently small. But this is clear from (A.6): if $\omega \notin \Omega_0$, then a little calculus shows that $\pi_T(\omega)$ is strictly increasing for $T \in (0, \epsilon)$ for some ϵ, whereas if $\omega \in \Omega_0$, then $\pi_T(\omega)$ is strictly decreasing for all $T > 0$.

Lemma 3 can now be obtained from (A.5) and (A.7) in the following way. Fix $t > t_0 \geq 0$:

$$\begin{aligned}
&\|P_{t_0,t} - \pi_0\| \\
&\leq \|P_{t_0,t} - \pi_{T(t)}\| + \|\pi_{T(t)} - \pi_0\| \\
&\leq \|P_{t_0,t-1} - \pi_{T(t)}\| + \|\pi_{T(t)} - \pi_0\|, \qquad \text{by (A.5)} \\
&\leq \|P_{t_0,t-1} - \pi_{T(t-1)}\| + \|\pi_{T(t-1)} - \pi_{T(t)}\| + \|\pi_{T(t)} - \pi_0\| \\
&\leq \|P_{t_0,t-2} - \pi_{T(t-1)}\| + \|\pi_{T(t-1)} - \pi_{T(t)}\| + \|\pi_{T(t)} - \pi_0\| \\
&\leq \|P_{t_0,t-2} - \pi_{T(t-2)}\| + \|\pi_{T(t-2)} - \pi_{T(t-1)}\| + \|\pi_{T(t-1)} - \pi_{T(t)}\| + \|\pi_{T(t)} - \pi_0\|.
\end{aligned}$$

Proceeding in this way,

$$\|P_{t_0,t} - \pi_0\| \leq \|P_{t_0,t_0} - \pi_{T(t_0)}\| + \sum_{k=t_0}^{t-1} \|\pi_{T(k)} - \pi_{T(k+1)}\| + \|\pi_{T(t)} - \pi_0\|.$$

Since $P_{t_0,t_0} = \pi_0$ and $\|\pi_{T(t)} - \pi_0\| \to 0$ as $t \to \infty$, we have,

$$\begin{aligned}
&\overline{\lim_{t_0\to\infty}} \sup_{t\geq t_0} \|P_{t_0,t} - \pi_0\| \\
&\leq \overline{\lim_{t_0\to\infty}} \sup_{t>t_0} \sum_{k=t_0}^{t-1} \|\pi_{T(k)} - \pi_{T(k+1)}\| \\
&= \overline{\lim_{t_0\to\infty}} \sum_{k=t_0}^{\infty} \|\pi_{T(k)} - \pi_{T(k+1)}\| \\
&= 0 \qquad \text{due to (A.7).}
\end{aligned}$$

Q.E.D.

ACKNOWLEDGMENT

The authors would like to acknowledge their debt to U. Grenander for a flow of ideas; his work on pattern theory [23] prefigures much of what is here. They also thank D. E. McClure and S. Epstein for their sound advice and technical assistance, and V. Mirelli for introducing them to the practical side of image processing as well as arguing for MRF scene models.

REFERENCES

[1] K. Abend, T. J. Harley, and L. N. Kanal, "Classification of binary random patterns," *IEEE Trans. Inform. Theory*, vol. IT-11, pp. 538-544, 1965.

[2] H. C. Andrews and B. R. Hunt, *Digital Image Restoration*. Englewood Cliffs, NJ, Prentice-Hall, 1977.

[3] M. S. Bartlett, *The Statistical Analysis of Spatial Pattern*. London: Chapman and Hall, 1976.

[4] T. Berger and F. Bonomi, "Parallel updating of certain Markov random fields," preprint.

[5] J. Besag, "Nearest-neighbor systems and the auto-logistic model for binary data," *J. Royal Statist. Soc.*, series B, vol. 34, pp. 75-83, 1972.

[6] —, "Spatial interaction and the statistical analysis of lattice systems (with discussion)," *J. Royal Statist. Soc.*, series B, vol. 36, pp. 192-326, 1974.

[7] A. B. Bortz, M. H. Kalos, and J. L. Lebowitz, "A new algorithm

for Monte Carlo simulation of Ising spin systems," *J. Comp. Phys.*, vol. 17, pp. 10–18, 1975.
[8] V. Cerný, "A thermodynamical approach to the travelling salesman problem: an efficient simulation algorithm," preprint, Inst. Phys. & Biophys., Comenius Univ., Bratislava, 1982.
[9] P. Cheeseman, "A method of computing maximum entropy probability values for expert systems," preprint.
[10] R. Chellappa and R. L. Kashyap, "Digital image restoration using spatial interaction models," *IEEE Trans. Acoust., Speech, Signal Processing*, vol. ASSP-30, pp. 461–472, 1982.
[11] D. B. Cooper and F. P. Sung, "Multiple-window parallel adaptive boundary finding in computer vision," *IEEE Trans. Pattern Anal. Machine Intell.*, vol. PAMI-5, pp. 299–316, 1983.
[12] G. C. Cross and A. K. Jain, "Markov random field texture models," *IEEE Trans. Pattern Anal. Machine Intell.*, vol. PAMI-5, pp. 25–39, 1983.
[13] L. S. Davis and A. Rosenfeld, "Cooperating processes for low-level vision: A survey," 1980.
[14] H. Derin, H. Elliott, R. Christi, and D. Geman, "Bayes smoothing algorithms for segmentation of images modelled by Markov random fields," Univ. Massachusetts Tech. Rep., Aug. 1983.
[15] R. L. Dobrushin, "The description of a random field by means of conditional probabilities and conditions of its regularity," *Theory Prob. Appl.*, vol. 13, pp. 197–224, 1968.
[16] H. Elliott, H. Derin, R. Christi, and D. Geman, "Application of the Gibbs distribution to image segmentation," Univ. Massachusetts Tech. Rep., Aug. 1983.
[17] H. Elliott, F. R. Hansen, L. Srinivasan, and M. F. Tenorio, "Application of MAP estimation techniques to image segmentation," Univ. Massachusetts Tech. Rep., 1982.
[18] P. A. Flinn, "Monte Carlo calculation of phase separation in a 2-dimensional Ising system," *J. Statist. Phys.*, vol. 10, pp. 89–97, 1974.
[19] B. R. Frieden, "Restoring with maximum likelihood and maximum entropy," *J. Opt. Soc. Amer.*, vol. 62, pp. 511–518, 1972.
[20] D. Geman and S. Geman, "Parameter estimation for some Markov random fields," Brown Univ. Tech. Rep., Aug. 1983.
[21] S. Geman, "Stochastic relaxation methods for image restoration and expert systems," in *Proc. ARO Workshop: Unsupervised Image Analysis*, Brown Univ., 1983; to appear in *Automated Image Analysis: Theory and Experiments*, D. B. Cooper, R. L. Launer, and D. E. McClure, Eds. New York: Academic, 1984.
[22] U. Grenander, *Lectures in Pattern Theory*, Vols. I-III. New York: Springer-Verlag, 1981.
[23] D. Griffeath, "Introduction to random fields," in *Denumerable Markov Chains*, Kemeny, Knapp and Snell, Eds. New York: Springer-Verlag, 1976.
[24] A. Habibi, "Two-dimensional Bayesian estimate of images," *Proc. IEEE*, vol. 60, pp. 878–883, 1972.
[25] F. R. Hansen and H. Elliott, "Image segmentation using simple Markov field models," *Comput. Graphics Image Processing*, vol. 20, pp. 101–132, 1982.
[26] A. R. Hanson and E. M. Riseman, "Segmentation of natural scenes," in *Computer Visions Systems*. New York: Academic, 1978.
[27] J. M. Hammersley and D. C. Handscomb, *Monte Carlo Methods*. London: Methuen, 1964.
[28] M. Hassner and J. Sklansky, "The use of Markov random fields as models of texture," *Comput. Graphics Image Processing*, vol. 12, pp. 357–370, 1980.
[29] G. E. Hinton and T. J. Sejnowski, "Optimal perceptual inference," in *Proc. IEEE Conf. Comput. Vision Pattern Recognition*, 1983.
[30] R. A. Hummel and S. W. Zucker, "On the foundations of relaxation labeling processes," *IEEE Trans. Pattern Anal. Machine Intell.*, vol. PAMI-5, pp. 267–287, 1983.
[31] B. R. Hunt, "Bayesian methods in nonlinear digital image restoration," *IEEE Trans. Comput.*, vol. C-23, pp. 219–229, 1977.
[32] V. Isham, "An introduction to spatial point processes and Markov random fields," *Int. Statist. Rev.*, vol. 49, pp. 21–43, 1981.
[33] E. Ising, *Zeitschrift Physik*, vol. 31, p. 253, 1925.
[34] A. K. Jain, "Advances in mathematical models for image processing," *Proc. IEEE*, vol. 69, pp. 502–528, 1981.
[35] A. K. Jain and E. Angel, "Image restoration, modeling and reduction of dimensionality," *IEEE Trans. Comput.*, vol. C-23, pp. 470–476, 1974.
[36] E. T. Jaynes, "Prior probabilities," *IEEE Trans. Syst. Sci. Cybern.*, vol. SSC-4, pp. 227–241, 1968.
[37] L. N. Kanal, "Markov mesh models," in *Image Modeling*. New York: Academic, 1980.
[38] R. L. Kashyap and R. Chellappa, "Estimation and choice of neighbors in spatial interaction models of images," *IEEE Trans. Inform. Theory*, vol. IT-29, pp. 60–72, 1983.
[39] R. Kinderman and J. L. Snell, *Markov Random Fields and Their Applications*. Providence, RI: Amer. Math. Soc., 1980.
[40] S. Kirkpatrick, C. D. Gellatt, Jr., and M. P. Vecchi, "Optimization by simulated annealing," IBM Thomas J. Watson Research Center, Yorktown Heights, NY, 1982.
[41] P. A. Levy, "A special problem of Brownian motion and a general theory of Gaussian random functions," in *Proc. 3rd Berkeley Symp. Math. Statist. and Prob.*, vol. 2, 1956.
[42] N. Metropolis, A. W. Rosenbluth, M. N. Rosenbluth, A. H. Teller, and E. Teller, "Equations of state calculations by fast computing machines," *J. Chem. Phys.*, vol. 21, pp. 1087–1091, 1953.
[43] N. E. Nahi and T. Assefi, "Bayesian recursive image estimation," *IEEE Trans. Comput.*, vol. C-21, pp. 734–738, 1972.
[44] D. K. Pickard, "A curious binary lattice process," *J. Appl. Prob.*, vol. 14, pp. 717–731, 1977.
[45] W. H. Richardson, "Bayesian-based iterative method of image restoration," *J. Opt. Soc. Amer.*, vol. 62, pp. 55–59, 1972.
[46] A. Rosenfeld, R. A. Hummel, and S. W. Zucker, "Scene labeling by relaxation operations," *IEEE Trans. Syst., Man, Cybern.*, vol. SMC-6, pp. 420–433, 197.
[47] A. Rosenfeld and A. C. Kak, *Digital Picture Processing*, vols. 1, 2, 2nd ed. New York: Academic, 1982.
[48] F. Spitzer, "Markov random fields and Gibbs ensembles," *Amer. Math. Mon.*, vol. 78, pp. 142–154, 1971.
[49] J. A. Stuller and B. Kruz, "Two-dimensional Markov representations of sampled images," *IEEE Trans. Commun.*, vol. COM-24, pp. 1148–1152, 1976.
[50] H. J. Trussell, "The relationship between image restoration by the maximum a posteriori method and a maximum entropy method," *IEEE Trans. Acoust., Speech, Signal Processing*, vol. ASSP-28, pp. 114–117, 1980.
[51] J. W. Woods, "Two-dimensional discrete Markovian fields," *IEEE Trans. Inform. Theory*, vol. IT-18, pp. 232–240, 1972.

Stuart Geman received the B.A. degree in physics from the University of Michigan in 1971, the M.S. degree in physiology from Dartmouth College in 1973, and the Ph.D. degree in applied mathematics from the Massachusetts Institute of Technology in 1977.

Since 1977 he has been a member of the Division of Applied Mathematics at Brown University, Providence, RI, where he is currently an Associate Professor. His research interests include statistical inference, parallel computing, image processing, and stochastic processes.

Dr. Geman is an Associate Editor of *The Annals of Statistics* and is a recipient of the Presidential Young Investigator Award.

On the Foundations of Relaxation Labeling Processes

ROBERT A. HUMMEL, MEMBER, IEEE, AND STEVEN W. ZUCKER, MEMBER, IEEE

Abstract—A large class of problems can be formulated in terms of the assignment of labels to objects. Frequently, processes are needed which reduce ambiguity and noise, and select the best label among several possible choices. Relaxation labeling processes are just such a class of algorithms. They are based on the parallel use of local constraints between labels. This paper develops a theory to characterize the goal of relaxation labeling. The theory is founded on a definition of consistency in labelings, extending the notion of constraint satisfaction. In certain restricted circumstances, an explicit functional exists that can be maximized to guide the search for consistent labelings. This functional is used to derive a new relaxation labeling operator. When the restrictions are not satisfied, the theory relies on variational calculus. It is shown that the problem of finding consistent labelings is equivalent to solving a variational inequality. A procedure nearly identical to the relaxation operator derived under restricted circumstances serves in the more general setting. Further, a local convergence result is established for this operator. The standard relaxation labeling formulas are shown to approximate our new operator, which leads us to conjecture that successful applications of the standard methods are explainable by the theory developed here. Observations about convergence and generalizations to higher order compatibility relations are described.

Index Terms—Consistency, constraint satisfaction, cooperative processes, labeling, probabilistic relaxation, relaxation labeling.

Manuscript received June 1, 1981; revised July 19, 1982. This work was supported by ARO Grant DAAG29-81-K-0043 and NSERC Grant A4470.

R. A. Hummel is with the Courant Institute of Mathematical Sciences, New York University, New York, NY 10012.

S. W. Zucker is with the Computer Vision and Graphics Laboratory, Department of Electrical Engineering, McGill University, Montreal, P.Q., Canada H3A 2A7.

MOTIVATION

RELAXATION labeling processes are a class of mechanisms that were originally developed to deal with ambiguity and noise in vision systems. The general framework, however, has far broader potential applications and implications. The structure of relaxation labeling is motivated by two basic concerns: 1) the decomposition of a complex computation into a network of simple "myopic," or local, computations; and 2) the requisite use of context in resolving ambiguities. Given impetus by successes in constraint propagation, particularly in "blocks-world" vision [16], a broad class of algorithms was introduced and subsequently explored under the generic heading of relaxation labeling operations [12].

Although many technical details in the design and implementation of relaxation labeling algorithms have been subject to ad hoc, or heuristic choices, the general structure has attracted substantial interest. Our goal in this paper is to provide a formal foundation. The algorithms are conceptually parallel, allowing each process to make use of the context to assist in a labeling decision. Further, there is an appealing qualitative agreement between the parallel structure of relaxation algorithms and the distributed appearance of the neural machinery in the early stages of the human visual system. There has been a lot of hope and some evidence that relaxation labeling operations are robust—that the resulting processes are relatively insensitive to large changes in design parameters. The lure of relaxation labeling is largely based on a desire to achieve a globally consistent interpretation by using a fixed simple control structure together with some common sense appraisals of local constraints. Since it can be applied to any problem which can be posed as a labeling problem, interest in the algorithms has inspired applications in domains very different from computer vision [4].

When relaxation operations are used to solve systems of linear equations, or equivalently, discretized partial differential equations, many of the parameters and choices about the graph structure are dictated by the problem domain and the underlying equations. However, sophisticated relaxation applications left a number of choices to the "computors," or project leaders [13]. For example, the use of overrelaxation and block relaxation methods are "labor-saving devices" whose utility is demonstrated by practical experience.

Relaxation labeling is a natural extension of relaxation operations to the class of problems whose solutions involve symbols rather than functions. Constraints between neighboring labels replace finite difference equations used to represent the local behavior imposed by differential equations. The relaxation of labels occurs by manipulating assignment weights attached to separate labels, as opposed to adjusting the estimated function value up or down. Indeed, the main difference between relaxation labeling and relaxation is that the labels do not necessarily have a natural ordering. One might be literally comparing apples and oranges, and relaxation labeling might point to an adjustment from the label of apples to the label of oranges. When using classical relaxation to find solution functions, the adjustments of the function values are always either up or down.

Most applications of relaxation labeling, however, have not been guided by an underlying differential equation or system of linear equations. As a result, most of the parameters and much of the control structure had to be justified by empirical support. As long as applications are successful, researchers generally feel little motivation to further justify the approach with an analysis of underlying equations. When applications are less than perfectly successful, it is impossible to ascribe failure. Is the problem with the application of the algorithm, or is the problem with the algorithm? Usually, one documents the pattern of failure, and understands the behavior in light of the particular design parameters. Without a reasonable, abstract characterization of what the algorithm is doing, it is impossible to attribute the cause of failure to an inadequate theory. In order to assist in the development of relaxation labeling applications with a predictable domain of success, a complete characterization of the computation underlying the algorithm, independent of the application, is necessary. This paper is intended as a contribution toward such a characterization. Using this foundation, the design of relaxation processes need no longer be based entirely on ad hoc principles and heuristic choices.

In order to accomplish this characterization, the treatment is necessarily abstract. We also need to cover a lot of material: to relate discrete relaxation to a description of the usual relaxation labeling schemes, to develop a theory of consistency, and to formalize its relationship to optimization. Several mathematical results also follow from the formulation.

Other frameworks for the foundations of relaxation labeling have been attempted. We briefly compare our approach to some of these alternative viewpoints in Section II. Our framework is derived from variational calculus, i.e., from a generalization of standard optimization techniques, and central to this framework is an explicit notion of consistency. It is this notion of consistency that provides the link between discrete and continuous relaxation labeling, and serves as a foundation from which relaxation updating formulas can be derived. This central notion is developed in Section III. We begin by introducing the domain of problems to which relaxation labeling is applicable.

I. Introduction to Labeling Problems

In a labeling problem, one is given:

1) a set of objects;

2) a set of labels for each object;

3) a neighbor relation over the objects; and

4) a constraint relation over labels at pairs (or n-tuples) of neighboring objects.

Generally speaking, a solution to a labeling problem is an assignment of labels to each object in a manner which is consistent with respect to the constraint relation 4) above. Noticing that 1) and 3) above define a graph, the problem can also be described as one of assigning labels to nodes in a graph. With the graph structure in mind, we will sometimes refer to objects as nodes.

To make these terms more precise, it is useful to consider the historically important example of labeling edges in an ideal image of polyhedral solids [16]. The objects in this example represent the line segments of a line drawing of the polyhedra. Each object is a single line segment, and arises from an edge or portion of an edge of a polyhedron. Two objects, or nodes, are considered neighbors, and therefore joined by an arc in the graph structure, if the two line segments represented by the objects meet at a junction or vertex. The labels indicate the interpretation of the physical configuration giving rise to the edge: edges can be formed by the convex joining of two surfaces, concave joining, or as an occlusion boundary with the object to the right or left of the directed edge. Constraints are provided by lists of physically realizable edge vertices. For example, three lines representing convex edges can meet, but three occlusion edges cannot meet at a single junction. The goal of the labeling process is to associate with each node a label that describes the actual correct interpretation of each corresponding edge. In cases when the line drawing is ambiguous and could conceivably be used to represent two or more different polyhedral solids, each node should be labeled with the set of all labels which can arise from a correct interpretation.

The determination of which physically realizable scenes are described by a given line drawing can be viewed as a combinatorial search. However, by requiring that labels be consistent locally, i.e., by demanding that the edges form legitimate vertices, it is usually possible to prune the space that must be searched. Discrete relaxation labeling is a process which performs this pruning in a parallel, local, fashion [6].

To abstract and formalize the situation, consider a graph with a set of labels attached to each node. We shall denote the nodes by the variable i, which can take on integer values between 1 and n (the number of nodes), the set of labels attached to node i by Λ_i, and the individual label (elements of Λ_i) by the variable λ.[1] For simplicity, we will assume that the number of labels at each node is m, independent of i, so that the variable λ takes on integer values from 1 to m. The constraint Λ_{ij} is the set of all pairs (λ, λ') such that label λ at object i is compatible with label λ' at object j. Label pairs in $\Lambda_i \times \Lambda_j$ which are not in Λ_{ij} represent pairs of incompatible labels at the corresponding objects i and j. Constraints are only defined over neighboring nodes.

Discrete relaxation is accomplished by means of the *label discarding rule:* discard a label λ at a node i if there exists a neighbor j of i such that every label λ' currently assigned to j is incompatible with λ at i, i.e., $(\lambda, \lambda') \notin \Lambda_{ij}$ for all λ' assigned to j. The discrete relaxation labeling process is defined by the iterative application of the label discarding rule, applied in parallel at each node, until limiting label sets are obtained. Note that the label discarding rule prescribes that a label is retained if at every neighboring node there exists at least one compatible label, and that this property will hold for all labels in the limit sets.

A numerical formulation can be used to give an alternative characterization of labels in the limit sets. Define variables $p_i(\lambda)$ that indicate whether label λ is associated with node i, according to

$$p_i(\lambda) = \begin{cases} 1 & \text{if } \lambda \text{ is associated with object } i \\ 0 & \text{if } \lambda \text{ is not associated with object } i. \end{cases}$$

Let the variables $R_{ij}(\lambda, \lambda')$ represent the constraints by

$$R_{ij}(\lambda, \lambda') = \begin{cases} 1 & \text{if } (\lambda, \lambda') \in \Lambda_{ij} \\ 0 & \text{if } (\lambda, \lambda') \notin \Lambda_{ij}. \end{cases}$$

[1] A glossary of symbols is included in Appendix B.

We can count the number of neighbors of an object i which has labels compatible to a given label λ at i by the support function

$$S_i(\lambda) = \sum_{j \text{ neighboring } i} \max_{\lambda' \in \Lambda_j} \{R_{ij}(\lambda, \lambda') p_j(\lambda')\}.$$

In the limit sets, a label λ which is associated with a node i has support from all neighbors, and thus

$$S_i(\lambda) \geqslant S_i(\lambda'), \quad \text{for all } \lambda' \in \Lambda_i.$$

That is, labels that have been discarded have support which is strictly less than $S_i(\lambda)$, because at least one of the terms in the sum obtained from a max over $\lambda' \in \Lambda_j$ is zero.

We can define the map which assigns sets of labels to each object in a manner which is invarient with respect to the label discarding rule (i.e., a limit set) to be a *consistent labeling* with respect to the constraints. We see that if support is suitably defined, labels in a consistent labeling have maximal support at each object. This is the notion which serves as the basis for our subsequent studies of continuous relaxation labeling processes.

II. Continuous Relaxation Labeling Processes

The constraints used in the labeling problem described in the previous section do not allow for labels to express a preference or relative dislike for other labels at neighboring nodes. Instead, pairs of labels are either compatible or completely incompatible. Continuous relaxation labeling attempts to allow greater flexibility in the constraints by replacing these logical assertions about compatibilities with weighted values representing relative preferences. That is, the constraints are generalized to real-valued compatibility functions $r_{ij}(\lambda, \lambda')$ signifying the relative support for label λ at object i that arises from label λ' at object j. This support can be either positive or negative. (Some formulations require that the compatibility values satisfy $-1 \leqslant r_{ij}(\lambda, \lambda') \leqslant 1$, but we will make no such restriction.) Generally, positive values indicate that labels form a locally consistent pair, whereas a negative value indicates an implied inconsistency. The magnitude of $r_{ij}(\lambda, \lambda')$ is proportional to the strength of the constraint. When there is no interaction between labels, or when i and j are not neighbors, the compatibility $r_{ij}(\lambda, \lambda')$ is zero.

Having given the compatibilities weights, continuous relaxation also uses weights for label assignments. We denote the weight with which label λ is assigned to node i by $p_i(\lambda)$, and will require that

$$0 \leqslant p_i(\lambda) \leqslant 1, \quad \text{all } i, \lambda$$

and

$$\sum_{\lambda=1}^{m} p_i(\lambda) = 1, \quad \text{all } i = 1, \cdots, n.$$

As an example of these ideas, consider the problem of detecting and labeling lines in digital imagery [17], [18]. Initial assertions about the presence of lines can be established at every pixel location on the basis of the response of a local line detector. However, the responses are usually ambiguous to some degree: there may be gaps, weak responses, and multiple responses in different orientations at some pixels. In a relaxation labeling approach, each pixel would represent a distinct object, and the label sets are formed from assertions about the existence of a line at that pixel in a given orientation, or the nonexistence of a line at that pixel. Initial assignments of the values $p_i(\lambda)$ are accomplished by inspecting the local line detector responses. Constraints can be obtained from a local model for good continuation of line elements. Thus, a horizontal line label supports a horizontal line label positioned (horizontally) next to it. The relaxation process iteratively updates the weighted label assignments to be more consistent with neighboring labels, ideally so that the weights designate a unique label at each node.

A conceptual difficulty remains in attempting to abstract this situation to the general problem of formulating a consistent labeling. We have not defined the precise meaning of consistency to replace the notion used for discrete relaxation labeling. Generally speaking, a consistent labeling is one in which the constraints are satisfied. But since we have replaced logical constraints by weighted assertions, a new foundation is required to describe the structural framework and the precise meaning of the goal of consistency.

Many such structural frameworks have been attempted. Some have defined consistency as the stopping points of a standard relaxation labeling algorithm. This approach is circular, however, and gives no clue as to the weaknesses in the standard algorithm.

Many researchers have regarded the label weights as probabilities, and attempted to describe the relaxation labeling process in terms of a Bayesian analysis [11]. The constraints are generally interpreted as statistical quantities, related to conditional probabilities or correlations between, e.g., pairs of labels. From our perspective, analysis of relaxation labeling operations within the probabilistic framework has been unsuccessful. Various independence assumptions are required, and the analysis at best leads to an approximate understanding of one and only one iteration of the process. Our approach is very different, and at no time do we interpret the $p_i(\lambda)$'s as probabilities.

An alternate development, based on optimization theory, has used as a definition of consistency the norm of the difference between a vector composed of the current label weights and an evidence vector obtained from a computation involving each label's neighborhood weights [2], [5]. This measure can be minimized by standard methods (see also [1], [14]). We believe that this approach is fundamentally the more appropriate one, although the goal of finding the best functional to minimize is limiting. In this paper we extend the optimization viewpoint to treat relaxation from foundations that are more basic than the presupposition of such functionals.

A related approach develops relaxation labeling entirely within the constructs of linear programming [7]. In this theory, the constraints are obtained from arithmetical equivalents to logical constraints, and preferences can be incorporated only by adding new labels. Consistent labelings form a convex subset of assignment space. This viewpoint is different from our theory, but interesting and not incompatible with our development.

Each of these different structural frameworks for explaining the goal of relaxation labeling leads to a variant algorithm. In the next section, we begin with a formulation of continuous labelings and compatibilities, and then define consistency. These notions eventually yield (Section VIII) yet another relaxation labeling algorithm. However, all of these variants share similarities with the original algorithm defined in [12] (see Section XI), which was justified purely by heuristic arguments. This prototype algorithm is an iterative, parallel procedure analogous to the label discarding rule used in discrete relaxation. Specifically, for each object and each label, one computes

$$q_i(\lambda) = \sum_{j=1}^{n} \sum_{\lambda'=1}^{m} r_{ij}(\lambda, \lambda') p_j(\lambda')$$

using the current assignment values $p_i(\lambda)$. Then new assignment values are defined to replace the current values according to the ad hoc formula

$$p_i(\lambda) := \frac{p_i(\lambda) [1 + q_i(\lambda)]}{\sum_{l=1}^{m} p_i(l) [1 + q_i(l)]}.$$

In light of the foundations developed here, new formulas will be offered that have important advantages over these original ones.

III. Consistency

The principal contribution of this paper is a new definition of a consistent labeling. Rather than defining consistency in terms of a set of logical constraints, we will require a system of inequalities. In a sense, then, we are permitting the logical constraints to be ordered, or weighted, with respect to their relative importance. This leads to much more flexibility in defining and analyzing consistent labelings than earlier treatments. Furthermore, our approach allows an analytic, rather than logical or symbolic, study. For example, in Section IX, we use techniques from the theory of ordinary differential equations to prove a local convergence theorem.

Our definition of consistency is considerably different from more familiar treatments of consistency in labeling problems [6]. However, we will attempt to show that our formulation is a natural one, embodying an intuitive notion of consistency, and leading to a rich mathematical theory. The term will be defined for both unambiguous labelings and for weighted labeling assignments. We begin by giving a formal definition of these two spaces of labelings.

An *unambiguous labeling assignment* is a mapping from the set of objects into the set of all labels, so that each object is associated with exactly one label. The mapping can be represented (inefficiently) by a collection of binary digits, $p_i(\lambda)$, indicating whether label λ is assigned to object i. We set

$$p_i(\lambda) = \begin{cases} 1 & \text{if object } i \text{ maps to label } \lambda \\ 0 & \text{if object } i \text{ does not map to } \lambda. \end{cases}$$

Note that for each object i,

$$\sum_{\lambda=1}^{m} p_i(\lambda) = 1.$$

The variables $p_i(1), \cdots, p_i(m)$ can be viewed as composing an m-vector $\bar{p}_i$, and the concatenation of the vectors $\bar{p}_1, \bar{p}_2, \cdots, \bar{p}_n$ can be viewed as forming an assignment vector $\bar{p} \in \mathbb{R}^{nm}$. The space of unambiguous labelings is defined by

$$\mathbb{K}^* = \Big\{\bar{p} \in \mathbb{R}^{nm}: \ \bar{p} = (\bar{p}_1, \cdots, \bar{p}_n); \quad \bar{p}_i = (p_i(1), \cdots, p_i(m)) \in \mathbb{R}^m; \quad p_i(\lambda) = 0 \text{ or } 1, \quad \text{all } i, \lambda; \quad \sum_{\lambda=1}^{m} p_i(\lambda) = 1 \quad \text{for } i = 1, \cdots, n\Big\}.$$

Given a vector $\bar{p}$ in $\mathbb{K}^*$, the corresponding unambiguous assignment is determined exactly. That is, the set of vectors in $\mathbb{K}^*$ is in one-to-one correspondence with the set of mappings from objects to labels.

The extension to *weighted labeling assignments* can be done by replacing the condition $p_i(\lambda) = 0$ or 1 by the condition $0 \leqslant p_i(\lambda) \leqslant 1$ for all i and λ. This extension yields the space of weighted labeling assignments

$$\mathbb{K} = \Big\{\bar{p} \in \mathbb{R}^{nm}: \ \bar{p} = (\bar{p}_1, \cdots, \bar{p}_n); \quad \bar{p}_i = (p_i(1), \cdots, p_i(m)) \in \mathbb{R}^m; \quad 0 \leqslant p_i(\lambda) \leqslant 1 \quad \text{for all } i, \lambda; \quad \sum_{\lambda=1}^{m} p_i(\lambda) = 1 \quad \text{for } i = 1, \cdots, n\Big\}.$$

We claim that $\mathbb{K}$ is simply the convex hull of $\mathbb{K}^*$. To see this, let $\bar{e}_k$ denote the standard unit m-vector with a 1 in the kth component. Then any labeling assignment $\bar{p}$ in $\mathbb{K}$ can be expressed by

$$\bar{p} = \sum_{l_1=1}^{m} \sum_{l_2=1}^{m} \cdots \sum_{l_n=1}^{m} p_1(l_1) p_2(l_2) \cdots p_n(l_n) \cdot (\bar{e}_{l_1}, \bar{e}_{l_2}, \cdots, \bar{e}_{l_n}).$$

Since each nm-vector $(\bar{e}_{l_1}, \cdots, \bar{e}_{l_n})$ is in $\mathbb{K}^*$, the above sum can be interpreted as a convex combination of the elements of $\mathbb{K}^*$. Note that the sum over all of the coefficients is 1.

The notion of consistency depends upon constraints between label assignments. These will be represented by a matrix of real numbers—*the compatibility matrix*—the elements of which can indicate both positive and negative constraints. Compatibility between pairs, for example, is denoted by $r_{ij}(\lambda, \lambda')$, representing how label λ' at object j influences label λ at object i. If object j having label λ' lends high support to object i having label λ, then $r_{ij}(\lambda, \lambda')$ should be large and positive. If the constraint is such that object j having label λ' means that label λ at object i is highly unlikely, then $r_{ij}(\lambda, \lambda')$ should be negative. No restrictions are placed on the magnitudes of the constraints. If there is no interaction between (i, λ) and (j, λ'), then $r_{ij}(\lambda, \lambda') = 0$.

In Section I, we defined the support $S_i(\lambda)$ of label at node i by $\Sigma \max_\lambda \{r_{ij}(\lambda, \lambda') p_j(\lambda')\}$. A maximum operation was used because more than one label with assignment value $p_j(\lambda') = 1$ could occur at each node j. Since our labeling spaces now require $\Sigma_\lambda p_i(\lambda) = 1$, we replace the max operator with a sum.

Definition 3.1: Let a matrix of compatibilities and a labeling assignment (unambiguous or not) be given. We define the *support* for label λ at object i by the assignment $\bar{p}$ by

$$s_i(\lambda) = s_i(\lambda; \bar{p}) = \sum_{j=1}^{n} \sum_{\lambda'=1}^{m} r_{ij}(\lambda, \lambda') p_j(\lambda'). \qquad \square$$

Note that the support value is a linear function of the components of the labeling $\bar{p}$. In this case, the support values depend on the coefficients $\{r_{ij}(\lambda, \lambda')\}$ comprising the matrix of compatibilities, which we assume given along with the objects and labels as part of the problem definition. In fact, the support for label λ at object i is a weighted average of the compatibilities of currently assigned labels at neighboring objects with the label λ at node i.

The exact functional dependence of the support values on the assignment $\bar{p}$ is in fact unimportant to the theory which follows. It only matters that support values $s_i(\lambda; \bar{p})$ depending on $\bar{p}$ are given. For example, it is possible to define the support function in terms of higher order combinations of object labels. Suppose that $r_{ijk}(\lambda, \lambda', \lambda'')$ represents the compatibility of object i having label λ with the configuration of label λ' at object j and label λ'' at object k. Then, given the multidimensional matrix of compatibilities $\{r_{ijk}(\lambda, \lambda', \lambda'')\}$, we can define the support at object i for label λ by the assignment $\bar{p} \in \mathbb{K}$ as

$$s_i(\lambda) = \sum_{j,\lambda'} \sum_{k,\lambda''} r_{ijk}(\lambda, \lambda', \lambda'') p_j(\lambda') p_k(\lambda'').$$

More generally, the support values $s_i(\lambda)$ combine to give a support vector $\bar{s}$ that is a function of the total $\bar{s} = \bar{s}(\bar{p})$. The form of the vector valued function serves as a representation for the (*a priori*) knowledge of the relative compatibility of all configurations of labelings. In the following, $s_i(\lambda)$ are the components of any one of these support functions, calculated for the specified assignment $\bar{p}$. Whenever a formula for $s_i(\lambda)$ is needed, we find it convenient to use the linear dependence defined in Definition 3.1. Historically, and because of its simplicity, the case when $\bar{s}$ is linear is of special interest, but the entire theory presented in this paper can be extended to general nonlinear support functions.

We can now define consistency for unambiguous labelings.

Definition 3.2: Let $\bar{p} \in \mathbb{K}^*$ be an unambiguous labeling. Suppose that $\lambda_1, \cdots, \lambda_n$ are the labels which are assigned to objects $1, \cdots, n$, respectively, by the labeling $\bar{p}$. That is, $\bar{p} = (\bar{e}_{\lambda_1}, \bar{e}_{\lambda_2}, \cdots, \bar{e}_{\lambda_n})$. The unambiguous labeling $\bar{p}$ is *consistent* (in $\mathbb{K}^*$) providing

$$s_i(\lambda_i) \geqslant s_i(\lambda), \quad \text{all } \lambda, \quad \text{for } i = 1, \cdots, n. \qquad \square$$

It is important to realize that consistency in $\mathbb{K}^*$ corresponds to satisfying a system of inequalities:

$$s_1(\lambda_1; \bar{p}) \geqslant s_1(\lambda; \bar{p}), \qquad 1 \leqslant \lambda \leqslant m$$

$$s_2(\lambda_2;\bar{p}) \geqslant s_2(\lambda;\bar{p}), \qquad 1 \leqslant \lambda \leqslant m$$

$$s_n(\lambda_n;\bar{p}) \geqslant s_n(\lambda;\bar{p}), \qquad 1 \leqslant \lambda \leqslant m$$

The consistency condition is a reasonable one. At a consistent unambiguous labeling, the support, at each object, for the assigned label is the maximum support at that object. However, also note that consistency is *not* a pure maximization problem (although see Section V for a partial retraction). If we change the assignment of labels, it may happen that some or all of the maximum supports at individual objects may increase. Such a labeling is not judged to be better; indeed, if the maximum support is no longer attained by the instantiated label at one or more of the nodes, then consistency is spoiled.

Given a set of objects, labels, and support functions, there may be many consistent labelings. Our notion of consistency does not place an objective ranking on labelings. The system of inequalities is simply a criterion for determining whether a given unambiguous labeling (one of the n^m elements in $\mathbb{K}^*$) is consistent. There is an analogy with mathematical programing, in that a consistent labeling is a feasible point in $\mathbb{K}^*$. However, there is no objective function, $\mathbb{K}^*$ is discrete, and the constraints are nonlinear. We can view consistency as a "locking-in" property. That is, since the support for a given label at a given node depends upon the assigned labels at all other nodes, these assigned labels dictate, through the compatibilities and support functions, values for the support which are consistent with the current assignment. Further, this agreement between the assigned label and the ordering of the support values among the label set must hold at every object.

The condition for consistency in $\mathbb{K}^*$ can be restated as follows:

$$\sum_{\lambda=1}^{m} p_i(\lambda)\, s_i(\lambda;\bar{p}) \geqslant \sum_{\lambda=1}^{m} v_i(\lambda)\, s_i(\lambda;\bar{p}), \qquad i = 1, \cdots, n,$$

for all unambiguous labelings $\bar{v} \in \mathbb{K}^*$.

Note that the condition consists of n inequalities which must hold simultaneously for every competing labeling $\bar{v} \in \mathbb{K}^*$. We emphasize that in this formulation, and in the next definition, the supports are calculated for the given assignment $\bar{p}$, and do not change as the vector $\bar{v}$ varies. Accordingly, if $\bar{p}$ assigns label λ_i to object i, then the left side of this inequality evaluates to $s_i(\lambda_i)$, whereas the right-hand side is simply $s_i(\lambda)$ for some λ, depending on which label $\bar{v}$ designates to label object i.

Consistency for weighted labeling assignments is defined analogously.

Definition 3.3: Let $\bar{p} \in \mathbb{K}$ be a weighted labeling assignment. Then $\bar{p}$ is *consistent* (in $\mathbb{K}$) providing

$$\sum_{\lambda=1}^{m} p_i(\lambda)\, s_i(\lambda;\bar{p}) \geqslant \sum_{\lambda=1}^{m} v_i(\lambda)\, s_i(\lambda;p), \qquad i = 1, \cdots, n,$$

for all labelings $\bar{v} \in \mathbb{K}$.

Once again, consistency in $\mathbb{K}$ is not a maximization problem, since n quantities, each depending on $\bar{p}$, simultaneously satisfy maxima problems: $\bar{p}_i \cdot \bar{s}_i = \max \bar{v}_i \cdot \bar{s}_i$, for $i = 1, \cdots, n$. Changing $\bar{p}$ might increase some of the n quantities on the left-hand side, but it might decrease some of the others. One could maximize the sum or product of the n quantities, but this does not (generally) guarantee that each quantity is individually maximized.

In Section IX, we will find it useful to consider labelings which are strictly consistent.

Definition 3.4: Let $\bar{p} \in \mathbb{K}$. Then $\bar{p}$ is *strictly consistent* providing

$$\sum_{\lambda=1}^{m} p_i(\lambda)\, s_i(\lambda;\bar{p}) > \sum_{\lambda=1}^{m} v_i(\lambda)\, s_i(\lambda;\bar{p}), \qquad i = 1, \cdots, n,$$

for all labeling assignments $\bar{v} \in \mathbb{K}$, $\bar{v} \neq \bar{p}$. □

In Section IX we will see that a strictly consistent labeling $\bar{p}$ is necessarily unambiguous, i.e., $\bar{p} \in \mathbb{K}^*$.

The intuitive meaning of consistency is easier to consider at a strictly consistent labeling. Suppose one were to discard knowledge of the label assigned to a given object, but retain such knowledge at all neighboring objects. Then by computing the support values, the original labeling at the object could be deduced by maxima selection among the support values. We see that the support values and the labeling are in complete agreement. This is a very special property, since the instantiated label contributes to the calculation of the support values of neighboring object-labels, which must independently satisfy the same agreement principle.

Note that an unambiguous assignment that is consistent in $\mathbb{K}$ will also be consistent in $\mathbb{K}^*$, since $\mathbb{K}^* \subseteq \mathbb{K}$. The converse is also true.

Proposition 3.5: An unambiguous labeling which is consistent in $\mathbb{K}^*$ is also consistent in $\mathbb{K}$.

Proof: This follows because any weighted assignment $\bar{v} \in \mathbb{K}$ can be written as

$$\bar{v}_i = \sum_{l=1}^{m} v_i(l)\, \bar{e}_l$$

where $\bar{e}_k$ is the standard unit m-vector with a 1 in the kth coordinate. Since

$$\sum_{\lambda=1}^{m} p_i(\lambda)\, s_i(\lambda) \geqslant \sum_{\lambda=1}^{m} e_l(\lambda)\, s_i(\lambda), \qquad l = 1, \cdots, m, \quad i = 1, \cdots, n,$$

the same inequality holds true for any convex combination:

$$\sum_{\lambda=1}^{m} p_i(\lambda)\, s_i(\lambda) \geqslant \sum_{l=1}^{m} v_i(l) \sum_{\lambda=1}^{m} e_l(\lambda)\, s_i(\lambda) = \sum_{\lambda=1}^{m} v_i(\lambda)\, s_i(\lambda), \qquad i = 1, \cdots, n$$

which proves that $\bar{p}$ is consistent in $\mathbb{K}$. □

However, there may be consistent assignments in $\mathbb{K}$ which are ambiguous, and thus not consistent in $\mathbb{K}^*$.

IV. Overview of Results

Having established the notion of consistency, we wish to develop algorithms for converting a given labeling into a consistent one. We will pursue two approaches, one based on

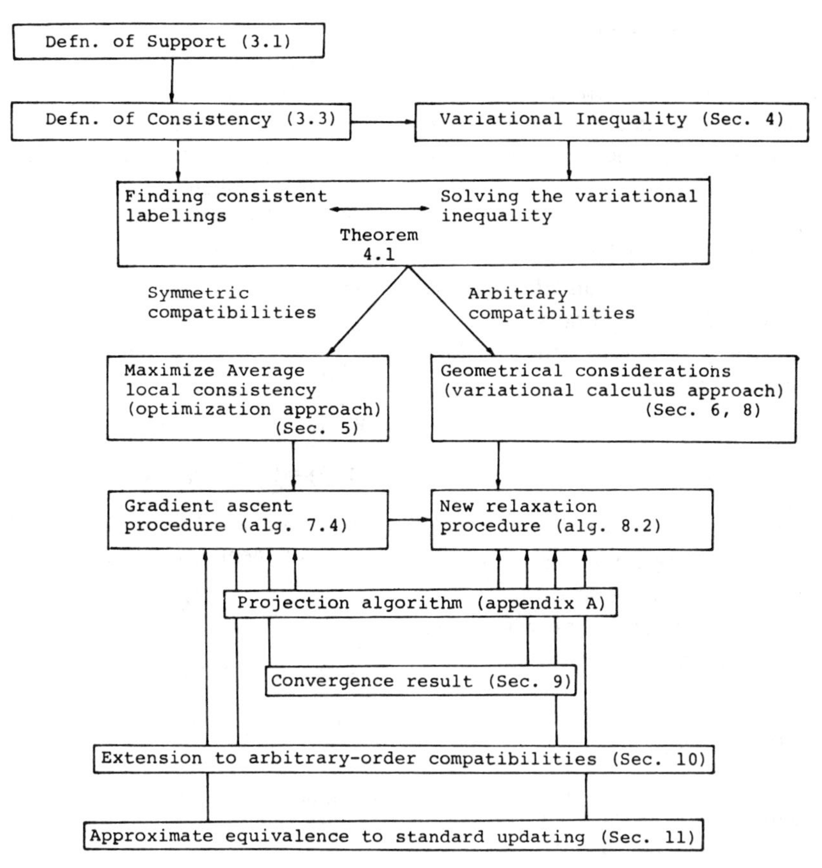

Fig. 1. Organization of the paper.

optimization theory, and the other founded in finite variational calculus. Surprisingly, the two methods will lead to the same algorithm, although the optimization approach applies only to a restricted class of compatibility matrices.

Fundamental to the analytic study of consistency is Theorem 4.1, which shows that achieving consistency is equivalent to solving a variational inequality. This inequality is important because it suggests an algorithm for finding solutions. Moreover, it will be shown that, in some cases, all local maxima of a certain functional on $\mathbb{K}$ solve the variational inequality. Thus one can find solutions, and hence consistent labelings, by finding local maxima.

In the following, we assume that consistency is defined (Definition 3.3) in terms of the linear support function $\bar{s}$ (Definition 3.1).

Theorem 4.1: A labeling $\bar{p} \in \mathbb{K}$ is consistent if and only if

$$\bar{p} \in \mathbb{K}: \sum_{i,\lambda,j,\lambda'} r_{ij}(\lambda, \lambda') p_j(\lambda') [v_i(\lambda) - p_i(\lambda)] \leqslant 0$$

for all $\bar{v} \in \mathbb{K}$.

Proof: If $\bar{p}$ is consistent, then

$$\sum_{\lambda} p_i(\lambda) s_i(\lambda) \geqslant \sum_{\lambda} v_i(\lambda) s_i(\lambda)$$

for any $\bar{v} \in \mathbb{K}$. But $s_i(\lambda) = \Sigma_{j,\lambda'} r_{ij}(\lambda, \lambda') p_j(\lambda')$. Substituting, and summing over i, one obtains the above variational inequality.

Conversely, if the variational inequality is solved by $\bar{p}$, then, for any $\bar{v} \in \mathbb{K}$, fix k, $1 \leqslant k \leqslant n$, and set

$$\bar{v}' = (\bar{p}_1, \cdots, \bar{v}_k, \cdots, \bar{p}_n).$$

Then $\bar{v}' \in \mathbb{K}$, and so

$$\sum_{i,\lambda,j,\lambda'} r_{ij}(\lambda, \lambda') p_j(\lambda') [v_i'(\lambda) - p_i(\lambda)]$$

$$= \sum_{i,\lambda} s_i(\lambda) [v_i'(\lambda) - p_i(\lambda)]$$

$$= \sum_{\lambda} s_k(\lambda) [v_k(\lambda) - p_k(\lambda)] \leqslant 0.$$

Since k is arbitrary, the above expression implies that $\bar{p}$ is consistent. □

Later, in Section VIII, we will derive another equivalent formulation of consistency. This final formulation is in terms of the stopping criteria for algorithms that solve the variational inequality in Theorem 4.1.

The study of consistency, and the derivation of algorithms for achieving it, will be pursued along two paths. These paths, as well as the organization of the rest of the paper, are illustrated in Fig. 1. One path involves showing how the variational inequality can be solved by an iterative procedure, as indicated by the right path in the diagram. Since this is a rather formal procedure, we shall postpone it to consider the case of symmetric compatibility matrices first (the left path through Fig. 1). This assumption allows us to structure the derivation more intuitively, in that it allows us to prove that maximizing a natural measure of the consistency of a labeling also yields solutions to the variational inequality. This maximization, which is posed as a gradient ascent algorithm, leads to a form of relaxation labeling with a modified updating formula (see Section XI for a comparison of this new iterative procedure with the more standard ones). Finally, in Section X, we show how higher (and lower) order support functions modify the theory.

V. Average Local Consistency

The notion of consistency defined in Section III suggests a measure useful for guiding the updating of a nearly consistent labeling into a consistent one. Specifically, the n values $\{\sum_\lambda p_i(\lambda) s_i(\lambda)\}$ should each be large. Thus,

$$A(\bar{p}) = \sum_{i=1}^{n} \sum_{\lambda} p_i(\lambda) s_i(\lambda)$$

should also be large. We will refer to $A(\bar{p})$ as the *average local consistency*, because each of the terms in the sum represents the local consistency of $\bar{p}$ from the viewpoint of an object/label, weighted by the labeling weight. In the case when $s_i(\lambda)$ is given by a linear sum of assignment values as in Definition 3.1, we note that

$$A(\bar{p}) = \sum_{i,\lambda} \sum_{j,\lambda'} r_{ij}(\lambda, \lambda') p_i(\lambda) p_j(\lambda').$$

In this case, $A(\bar{p})$ is proportional to the average of the local consistency of the labeling $\bar{p}$ based on the pair of object labels (i, λ) and (j, λ'), averaged over all such pairs.

Since the average local consistency $A(\bar{p})$ should be large, it would seem natural to attempt to maximize it. Two problems deserve immediate comment. First, maximizing a sum does not necessarily maximize each of the individual terms. Secondly, the individual components $s_i(\lambda)$ depend on $\bar{p}$ (which varies during the maximization process), whereas consistency occurs when $\sum v_i(\lambda) s_i(\lambda; \bar{p})$ is maximized by $\bar{v} = \bar{p}$. That is, the $s_i(\lambda)$ should be fixed during the maximization. In summary, maximizing $A(\bar{p})$ is the same as maximizing

$$\sum_i \sum_\lambda p_i(\lambda) s_i(\lambda; \bar{p})$$

which is not the same as maximizing the n quantities

$$\sum_\lambda p_i(\lambda) s_i(\lambda; \bar{p}), \qquad i = 1, \cdots, n.$$

In fact, since the n quantities are not independent, there is no such thing as the problem of finding a single $\bar{p}$ which maximizes these n values simultaneously. However, finding consistent labelings $\bar{p}$ is a real problem, corresponding to the statement that each of the n values

$$\sum_\lambda v_i(\lambda) s_i(\lambda; \bar{p}), \qquad i = 1, \cdots, n.$$

are independently maximized among $\bar{v} \in \mathbb{K}$ when $\bar{v} = \bar{p}$.

Despite these caveats, we still find it interesting to study average local consistency. This is because of the independent interest of average local consistency, and because of the following theorem.

Theorem 5.1: Suppose that the matrix of compatibilities $\{r_{ij}(\lambda, \lambda')\}$ is symmetric, i.e.,

$$r_{ij}(\lambda, \lambda') = r_{ji}(\lambda', \lambda) \qquad \text{for all } i, j, \lambda, \lambda'.$$

If $A(\bar{p})$ attains a local (relative) maximum at $\bar{p} \in \mathbb{K}$, then $\bar{p}$ is a consistent labeling.

Proof: Let $\bar{v} \in \mathbb{K}$. Since $\mathbb{K}$ is convex, $t\bar{v} + (1-t)\bar{p} \in \mathbb{K}$ for $0 \leq t \leq 1$. Since $A(\bar{p})$ does not increase as one moves away from $\bar{p}$

$$\left. \frac{d}{dt} \right|_{t=0} A(t\bar{v} + (1-t)\bar{p}) \leq 0.$$

Thus grad $A(\bar{p}) \cdot (\bar{v} - \bar{p}) \leq 0$, by the chain rule. Using

$$A(\bar{p}) = \sum \sum r_{ij}(\lambda, \lambda') p_i(\lambda) p_j(\lambda'),$$

a calculation which is given later in this section shows that the (i, λ) component, $q_i(l)$, of grad $A(\bar{p})$ is given by

$$q_i(\lambda) = \sum_{j,\lambda'} [r_{ij}(\lambda, \lambda') + r_{ji}(\lambda', \lambda)]\, p_j(\lambda').$$

Since the $r_{ij}(\lambda, \lambda')$'s are symmetric,

$$q_i(\lambda) = 2 \sum_{j,\lambda'} r_{ij}(\lambda, \lambda') p_j(\lambda').$$

Thus grad $A(\bar{p}) \cdot (\bar{v} - \bar{p}) \leq 0$ implies that

$$2 \sum_{i,\lambda} \sum_{j,\lambda'} r_{ij}(\lambda, \lambda') p_j(\lambda') (v_i(\lambda) - p_i(\lambda)) \leq 0.$$

This holds true for all $\bar{v} \in \mathbb{K}$, so $\bar{p}$ satisfies the variational inequality of Theorem 4.1. According to Theorem 4.1, $\bar{p}$ is therefore a consistent labeling. □

Theorem 5.1 is surprising in light of all of the caveats about maximizing average local consistency, and its apparent dissimilarity with finding consistent labelings. Even so, in the special case when the compatibility matrix is symmetric, maximizing $A(\bar{p})$ leads to consistent labeling assignments!

In the general case when the compatibility matrix is not symmetric, local maxima of $A(\bar{p})$ still exist, and may be of limited interest, but will not be consistent labeling assignments in the sense of Definition 3.3. In Section VIII, we present an algorithm which generally leads to consistent labelings regardless of whether the compatibilities are symmetric or not. However, if the compatibilities happen to be symmetric, the algorithm in Section VIII is equivalent to finding local maxima of average local consistency, which, according to our previous theorem, are consistent labelings.

The proof of Theorem 5.1 makes clear the need for the symmetry condition. In general, local maxima of $A(\bar{p})$ will satisfy a variational inequality (as in Theorem 4.1) in which the $r_{ij}(\lambda, \lambda')$ terms are replaced by symmetrical terms $(r_{ij}(\lambda, \lambda') + r_{ji}(\lambda', \lambda))$:

$$\sum_{i,\lambda} \sum_{j,\lambda'} (r_{ij}(\lambda, \lambda') + r_{ji}(\lambda', \lambda)) p_j(\lambda') \cdot (v_i(\lambda) - p_i(\lambda)) \leq 0$$

for all $\bar{v} \in \mathbb{K}$. That is, a local maximum of $A(\bar{p})$ will be a consistent labeling with respect to a symmetrized matrix of compatibilities. If one begins with a nonsymmetric matrix of compatibilities, and then locally maximizes $A(\bar{p})$, the result is the same as if the original matrix had been symmetrized. Indeed,

$$A(\bar{p}) = \sum r_{ij}(\lambda, \lambda') p_i(\lambda) p_j(\lambda')$$
$$= \sum \tfrac{1}{2} (r_{ij}(\lambda, \lambda') + r_{ji}(\lambda', \lambda)) p_i(\lambda) p_j(\lambda').$$

Clearly, the calculation of $A(\bar{p})$ discards any information contained in the nonsymmetry of the compatibility matrix. We emphasize, however, that in Section VIII we present a technique for finding consistent labelings that preserves the information in the nonsymmetric (skew) part of the compatibility matrix.

A classical approach to finding local maxima of a smooth functional $A(\bar{p})$ is by means of gradient ascent [9]. The method prescribes that we should successively move the current position $\bar{p}$ by a small step to a new position $\bar{p}'$, so that the functional $A(\bar{p})$ is increased as much as possible. The amount of increase in $A(\bar{p})$ is related to the directional derivative of A in the direction of the step, which in turn is related to the gradient of A at $\bar{p}$. As promised in the proof of Theorem 5.1, we now compute grad $A(\bar{p})$. Since $\bar{p} \in \mathbb{K} \subseteq \mathbb{R}^{nm}$, the gradient, which we will denote by $\bar{q}$, contains nm components. The (i, λ) component is given by

$$q_i(\lambda) = \frac{\partial}{\partial p_i(\lambda)} \left[\sum_{\alpha, l} \sum_{\beta, l'} r_{\alpha\beta}(l, l') p_\alpha(l) p_\beta(l') \right]$$
$$= \sum_{\alpha, l} \sum_{\beta, l'} [r_{\alpha\beta}(l, l') \delta_{i\alpha} \delta_{\lambda l} p_\beta(l') + r_{\alpha\beta}(l, l') p_\alpha(l) \delta_{i\beta} \delta_{\lambda l'}]$$
$$= \sum_{\beta, l'} r_{i\beta}(\lambda, l') p_\beta(l') + \sum_{\alpha, l} r_{\alpha i}(l, \lambda) p_\alpha(l)$$
$$= \sum_{j, \lambda'} r_{ij}(\lambda, \lambda') p_j(\lambda') + \sum_{j, \lambda'} r_{ji}(\lambda', \lambda) p_j(\lambda')$$
$$= \sum_{j, \lambda'} (r_{ij}(\lambda, \lambda') + r_{ji}(\lambda', \lambda)) p_j(\lambda').$$

As noted before, when the compatibilities are symmetric, this expression simplifies to

$$q_i(\lambda) = 2 \sum r_{ij}(\lambda, \lambda') p_j(\lambda').$$

The right-hand side of this formula is a familiar expression. Comparing with Definition 3.1, we see that $q_i(\lambda) = 2\, s_i(\lambda)$. Also, the $q_i(\lambda)$ components are essentially the same values that were used as an intermediate updating "direction" in the earlier treatments of relaxation labeling [12].

Our goal is now to find a consistent labeling nearby a given initial weighted labeling assignment. By Theorem 5.1, when the compatibilities are symmetric, this can be accomplished by locally maximizing $A(\bar{p})$. Ordinarily, gradient ascent proceeds by moving a small step in the direction of the gradient. In the case of the problem at hand, the labeling assignment $\bar{p}$ must be constrained to lie in $\mathbb{K}$, whereas grad $A(\bar{p})$ may "point out of the surface." Instead, the assignment $\bar{p}$ should be updated by moving a small step in the direction which maximizes the directional derivative. In Section VII, we consider the problem of maximizing the directional derivative, and describe the complete gradient ascent algorithm. Despite its restricted applicability to the case of symmetric compatibilities, the algorithm leads naturally to the more general case considered in Section VIII. Before presenting the gradient ascent algorithm, however, we address in Section VI the geometric configuration and terminology to be used when describing updating assignment labelings in $\mathbb{K}$.

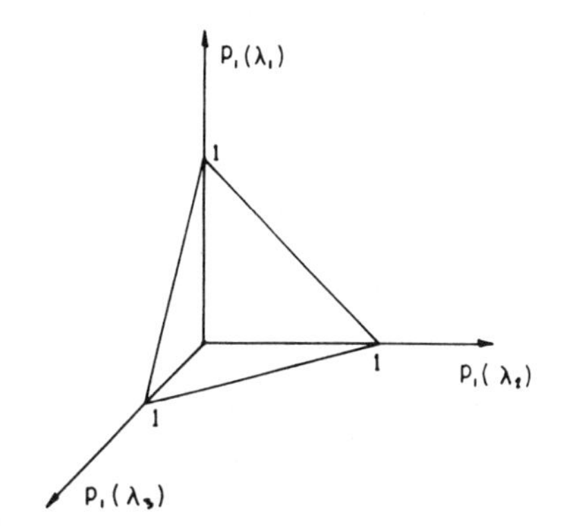

Fig. 2. The space of possible weighted labeling assignments for a single object with three labels.

VI. Geometric Structure of Assignment Space

To discuss gradient ascent on $\mathbb{K}$, and to visualize the more general updating algorithms, it is useful to consider the geometric structure of the weighted labeling space. The labeling space $\mathbb{K}$ was introduced in Section III as the convex hull of of the unambiguous labeling assignment space $\mathbb{K}^*$. In order to visualize the picture of the space $\mathbb{K}$, let us consider a simple example. Suppose that there are two objects, with three possible labels for each object. A labeling assignment then consists of six nonnegative numbers:

$$\bar{p} = (\bar{p}_1, \bar{p}_2) = (p_1(1), p_1(2), p_1(3), p_2(1), p_2(2), p_2(3))$$

satisfying

$$\sum_{\lambda=1}^{3} p_i(\lambda) = 1, \quad i = 1, 2.$$

If we plot the locus of possible subvectors $\bar{p}_1$ in $\mathbb{R}^3$, the result is shown in Fig. 2. It consists of the portion of an affine subspace in the positive quadrant, has the shape of a simplex (a 2-simplex), and might be called "a probability space." We hasten to add, however, that the latter association with probabilities is particularly unhelpful. In any case, the vector $\bar{p} = (\bar{p}_1, \bar{p}_2)$ can be regarded as two points, each lying in a copy of the space shown in Fig. 2. Thus $\mathbb{K}$, for the particular example under consideration, can be identified with the set of all pairs of points in two copies of the triangular space in Fig. 2.

In more general situations, say with n objects each with m labels, $\mathbb{K}$ is more complicated. Now, the space consists of n copies of an $(m-1)$-simplex each formed from the positive quadrant portion of a flat $(m-1)$-dimensional affine subspace lying in $\mathbb{R}^m$. Then $\mathbb{K}$ can be identified as the set of all n-tuples of points, each point lying in a copy of the $(m-1)$-dimensional surface. A weighted labeling assignment is a point in the assignment space $\mathbb{K}$, and $\mathbb{K}$ is in turn the convex hull of the set of unambiguous labeling assignments $\mathbb{K}^*$. An unambiguous labeling also lies in $\mathbb{K}$, and can be thought of as

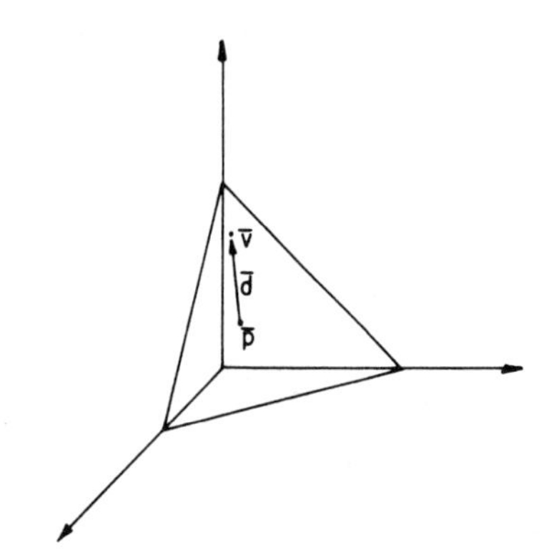

Fig. 3. A tangent vector $\bar{d}$ at a point $\bar{p} \in \mathbb{K}$.

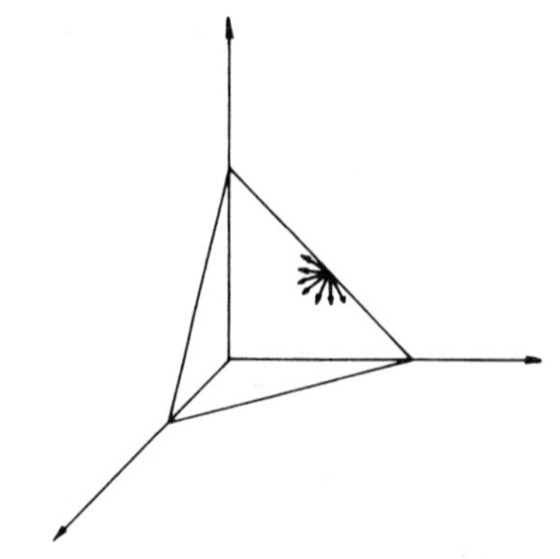

Fig. 4. The set of tangent directions at a point $\bar{p} \in \mathbb{K}$ that lies on a boundary.

one of the "corners," or extreme points, of the set. As an n-tuple of points, each lying in an $(m - 1)$-simplex, an unambiguous assignment is composed of points which lie at vertices of their respective surfaces. Of course, any combination of n vertices gives rise to an unambiguous labeling—it is not necessary that each point represent the same vertex. In fact, each simplex has m corners, corresponding to the m possible labels for that particular object.

In differential geometry, the tangent space to a point of a multidimensional surface has a well defined meaning in terms of the set of all directions, in a limiting sense, along which a curve can move away from the given point. The tangent space is a surface, which when placed at the given point, lies "tangent" to the entire surface. For the assignment space $\mathbb{K}$, the initial surface $\mathbb{K}$ and tangent space at any interior point are flat, and so coincide when the tangent space is placed at its base point. The tangent space to a point in the interior of a surface is in fact a vector space. However, at a point of the boundary of a surface, the set of possible directions is restricted by the boundary, and one is forced to speak of the tangent set, which is simply a convex subset of a vector space.

More precisely, suppose that $\bar{p}$ is a labeling assignment in $\mathbb{K}$, and that $\bar{v}$ is any other assignment in $\mathbb{K}$. The difference vector $\bar{d} = \bar{v} - \bar{p}$, when placed at $\bar{p}$, points toward $\bar{v}$ (see Fig. 3). Thus $\bar{d}$ indicates a direction at $\bar{p}$ which points toward $\bar{v}$. Moreover, $\bar{d}$ and all positive scalar multiples of $\bar{d}$ are tangent vectors to $\mathbb{K}$ at $\bar{p}$. As $\bar{v}$ roams around $\mathbb{K}$, the set of all possible tangent directions at $\bar{p}$ is swept out. The set of all tangent vectors at $\bar{p}$ is therefore given by

$$T_{\bar{p}} = \{\bar{d}: \bar{d} = \alpha(\bar{v} - \bar{p}),\ \bar{v} \in \mathbb{K},\ \alpha \geq 0\}.$$

Note that any tangent vector is composed of n subvectors $\bar{d}_i$, so that $\bar{d} = (\bar{d}_1, \cdots, \bar{d}_n)$, and

$$\sum_{\lambda=1}^{m} d_i(\lambda) = \sum_{\lambda=1}^{m} \alpha(v_i(\lambda) - p_i(\lambda))$$

$$= \alpha \cdot (1 - 1) = 0.$$

When $\bar{p}$ is a point in the interior of $\mathbb{K}$ (i.e., no components are zero), the vector $\bar{v}$ may be chosen from a neighborhood that completely surrounds $\bar{p}$ in the n copies of the affine subspace. The result is that the set of tangent vectors at the interior point $\bar{p}$ consists of an entire subspace, which is given by

$$T_{\bar{p}} = \left\{\bar{d} = (\bar{d}_1, \cdots, \bar{d}_n):\ \bar{d}_i \in \mathbb{R}^m,\ \sum_{\lambda=1}^{m} d_i(\lambda) = 0\right\}$$

$$(\bar{p} \text{ interior to } \mathbb{K}).$$

Observe that $T_{\bar{p}}$ and $\mathbb{K}$ are parallel flat surfaces.

When $\bar{p}$ lies on a boundary of $\mathbb{K}$, the tangent set is a proper subset of the above space $T_{\bar{p}_o}$, for any interior point $\bar{p}_o$. That is, when the assignment $\bar{p}$ has some zero components, the set of vectors of the form $\alpha \cdot (\bar{v} - \bar{p})$ is restricted to

$$T_{\bar{p}} = \left\{\bar{d} = (\bar{d}_1, \cdots, \bar{d}_n):\ \bar{d}_i \in \mathbb{R}^m,\ \sum_{\lambda=1}^{m} d_i(\lambda) = 0, \right.$$

$$\left. \text{and} \quad d_i(\lambda) \geq 0 \quad \text{if} \quad p_i(\lambda) = 0 \right\}.$$

Fig. 4 shows a set of tangent directions that can arise from a boundary point in a single object, three label, assignment space. The tangent set $T_{\bar{p}}$ changes as $\bar{p}$ moves from one boundary edge to another. However, the tangent space $T_{\bar{p}}$ is always the same when $\bar{p}$ is an interior point.

VII. Maximizing Average Local Consistency

Now we can return to the problem of finding relative maxima of average local consistency using gradient ascent. We reiterate that maximizing $A(\bar{p})$ corresponds to finding a consistent labeling when the constraints are symmetric (see Section V). A different analysis must be applied when the constraints are not symmetric, but leads to essentially the same algorithm (Section VIII).

The increase in $A(\bar{p})$ due to a small step of length α in the direction $\bar{u}$ is approximately the directional derivative:

$$A(\bar{p} + \alpha\bar{u}) - A(\bar{p}) \approx \left.\frac{d}{dt}\right|_{t=0} A(\bar{p} + t\alpha\bar{u}) = \operatorname{grad} A(\bar{p}) \cdot \alpha\bar{u}$$

where $\|u\| = 1$. In general, the greatest increase in $A(\bar{p})$ can be expected if a step is taken in the tangent direction $\bar{u}$ which maximizes the directional derivative. However, if the directional derivative is negative or zero for all nonzero tangent directions, then $A(\bar{p})$ is a local maximum and no step should be taken.

Accordingly, to find a direction of steepest ascent, grad $A(\bar{p}) \cdot \bar{u}$ should be maximized among the set of tangent

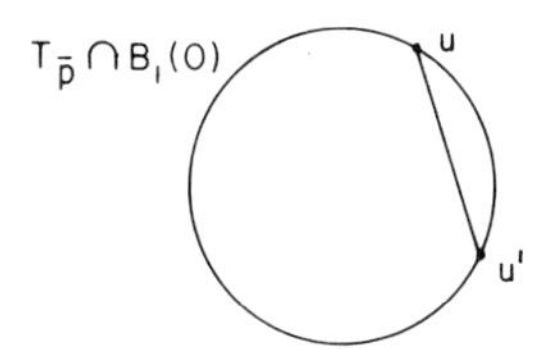

Fig. 5. Convex combinations of two distinct solutions $\bar{u}$ and $\bar{u}'$ lie on the straight line joining them.

vectors. However, it suffices to consider only those tangent vectors with Euclidean norm $\|\bar{u}\| = 1$ together with $\bar{u} = 0$. If unrestricted vectors are included in the maximization, then the problem has no solution since the vector dot product grad $A(\bar{p}) \cdot \bar{u}$ is scaled by $\|\bar{u}\|$. The problem is equivalent to maximizing $\bar{q} \cdot \bar{u}$ among tangent vectors with $\|\bar{u}\| \leq 1$, where $\bar{q} = \frac{1}{2}$ grad $A(\bar{p})$. Thus the direction of steepest ascent can be found by solving the following.

Problem 7.1: Find $\bar{u} \in T_{\bar{p}} \cap B_1(0)$ such that

$$\bar{u} \cdot \bar{q} \geq \bar{v} \cdot \bar{q} \qquad \text{for all } \bar{v} \in T_{\bar{p}} \cap B_1(0).$$

Here $B_1(0) = \{\bar{v} \in \mathbb{R}^{nm}: \|\bar{v}\| \leq 1\}$. □

A solution to Problem 7.1 always exists because the problem requires one to maximize a continuous functional over a compact set. In many cases, the solution will be unique.

Observation 7.2: If $\bar{u}$ solves Problem 7.1, and if $\bar{u} \cdot \bar{q} > 0$, then $\bar{u}$ is the unique solution.

Proof: First, note that $\bar{u} \neq 0$, since $\bar{u} \cdot \bar{q} \neq 0$. Thus $\|u\| = 1$, since otherwise $\bar{w} = \bar{u}/\|\bar{u}\|$ is a vector in $T_{\bar{p}} \cap B_1(0)$ satisfying $\bar{u} \cdot \bar{q} < \bar{w} \cdot \bar{q}$, in contradiction to $\bar{u}$ being a solution. Suppose $\bar{u}'$ is any other solution. Then $\bar{u}' \cdot \bar{q} = \bar{u} \cdot \bar{q} \neq 0$, and so $\|\bar{u}'\| = 1$ for the same reason that $\|u\| = 1$. We will show that $\bar{u}' = \bar{u}$.

Suppose that $\bar{u}$ and $\bar{u}'$ are two *distinct* solutions. Then $\bar{u} \cdot \bar{q} = \bar{u}' \cdot \bar{q} = (t\bar{u} + (1-t)\bar{u}') \cdot \bar{q}$ for all t, $0 \leq t \leq 1$ (see Fig. 5). Since $T_{\bar{p}} \cap B_1(0)$ is convex, all the vectors on the straight line from $\bar{u}$ to $\bar{u}'$ are solutions. But by the same argument given in the previous paragraph, all such solutions must lie on the surface of the unit ball $B_1(0)$. Since $\|\bar{u}\| = \|\bar{u}'\| = 1$, and $B_1(0)$ is strictly convex, the line between $\bar{u}$ and $\bar{u}'$ must lie in the interior of the ball, and so cannot contain any solutions. This contradicts the existence of two distinct solutions. □

The zero vector is always in $T_{\bar{p}} \cap B_1(0)$, so the maximum of $\bar{v} \cdot \bar{q}$ is nonnegative. When the maximum $\bar{u} \cdot \bar{q}$ is positive, then $\bar{u}$ is the unique solution. When the maximum $\bar{u} \cdot \bar{q} = 0$, the zero vector is among possibly many solutions. In this case, we will agree that $\bar{u} = 0$ is the best solution for Problem 7.1. In the solution method presented in Appendix A, the zero vector is always the solution returned by the algorithm when more than one solution is possible.

Conceptually, our method for finding relative maxima of average local consistency is very simple. Starting at an initial labeling $\bar{p}$, we compute $\bar{q} = \frac{1}{2}$ grad $A(\bar{p})$, and solve Problem 7.1. If the resulting $\bar{u}$ is nonzero, we take a small step in the direction $\bar{u}$, and repeat the process. The algorithm terminates when $\bar{u} = 0$. This is the algorithm given below (Algorithm 7.4). Clearly, however, we need a way to solve Problem 7.1 given a $\bar{p} \in \mathbb{K}$ and a $\bar{q} \subset \mathbb{R}^{nm}$.

A simple, finite algorithm for solving Problem 7.1 is presented in Appendix A. When $\bar{p}$ is in the interior of the assignment space $\mathbb{K}$, solving Problem 7.1 is a triviality, corresponding to projecting $\bar{q}$ onto the tangent space $T_{\bar{p}}$, and then normalizing. Lemma 7.3, which follows, proves that this procedure works. When $\bar{p}$ is on a boundary of $\mathbb{K}$, the situation is considerably more complicated. Fortunately, the algorithm given in Appendix A handles all cases.

Lemma 7.3: If $\bar{p}$ lies in the interior of $\mathbb{K}$, then the following algorithm solves Problem 7.1:

1) set $c_i = \frac{1}{m} \sum_{l=1}^{m} q_i(l)$, $\qquad i = 1, \cdots, n$.

2) set $w_i(\lambda) = q_i(\lambda) - c_i$, $\qquad$ all i, λ,

3) set $u_i(\lambda) = w_i(\lambda)/\|\bar{w}\|$, $\qquad$ all i, λ.

Here

$$\|\bar{w}\| = \left[\sum_{i,\lambda} w_i(\lambda)^2\right]^{1/2}.$$

Proof: First observe that $\bar{u} \in T_{\bar{p}} \cap B_1(0)$, since $\|\bar{u}\| = 1$ and

$$\sum_{\lambda=1}^{m} u_i(\lambda) = \sum_{\lambda=1}^{m} (q_i(\lambda) - c_i)/\|\bar{w}\|$$

$$= \left[\sum_{\lambda=1}^{m} q_i(\lambda) - mc_i\right] \Big/ \|\bar{w}\| = 0, \qquad \text{for all } i.$$

Since $\bar{p}$ is in the interior of $\mathbb{K}$, membership in $T_{\bar{p}}$ requires no further conditions.

Next observe that $\bar{w}$ is the projection of $\bar{q}$ onto $T_{\bar{p}}$, i.e., $(\bar{q} - \bar{w}) \cdot \bar{v} = 0$ for all $\bar{v} \in T_{\bar{p}}$. To see this, we simply calculate

$$\sum_i \sum_\lambda (q_i(\lambda) - w_i(\lambda)) \cdot v_i(\lambda) = \sum_i c_i \sum_\lambda v_i(\lambda) = 0$$

(since $\bar{v} \in T_{\bar{p}}$). Thus $\bar{v} \cdot \bar{q} = \bar{v} \cdot \bar{w}$ for all $\bar{v} \in T_{\bar{p}}$. Since $\bar{u} \in T_{\bar{p}}$,

$$\bar{u} \cdot \bar{q} = \bar{u} \cdot \bar{w} = \frac{\bar{w}}{\|\bar{w}\|} \cdot \bar{w} = \|\bar{w}\| \geq \|\bar{w}\| \cdot \|\bar{v}\|$$

for any $\bar{v} \in T_{\bar{p}} \cap B_1(0)$ (note that $\|\bar{v}\| \leq 1$). By the Cauchy-Schwarz inequality, $\bar{v} \cdot \bar{w} \leq \|\bar{v}\| \, \|\bar{w}\|$, so we have

$$\bar{u} \cdot \bar{q} \geq \bar{v} \cdot \bar{w} = \bar{v} \cdot \bar{q} \qquad \text{for all } \bar{v} \in T_{\bar{p}} \cap B_1(0).$$

That is, $\bar{u}$ solves Problem 7.1. □

The algorithm in Lemma 7.3 may fail when $\bar{p}$ is a boundary point of $\mathbb{K}$, since there is no guarantee that $w_i(\lambda) \geq 0$ when $p_i(\lambda) = 0$. It works out that Problem 7.1 is still solved by performing a projection of $\bar{q}$ followed by length normalization, but when $\bar{p}$ is a boundary point the projection onto $T_{\bar{p}}$ is a projection onto a convex set, and not onto a subspace (see the discussion of tangent sets in Section VI). The algorithm in Appendix A gives a method for computing this projection. However, the theory which shows that this method solves Problem 7.1, even when $\bar{p}$ is a boundary point, is rather involved, and is the topic of a companion paper [10].

Combining these results, we obtain the following algorithm for finding local maxima of $A(\bar{p})$.

Algorithm 7.4:

Initialize:

1) Start with an initial labeling assignment $\bar{p}^o \in \mathbb{K}$.
Set $k = 0$.

Loop until a stop is executed:

2) Compute $\bar{q}^k = \frac{1}{2}$ grad $A(\bar{p}^k)$.
3) Use the algorithm in Appendix A, with $\bar{p} = \bar{p}^k, \bar{q} = \bar{q}^k$, to find the solution $\bar{u}^k$ to Problem 7.1.
4) If $\bar{u}^k = 0$, stop.
5) Set $\bar{p}^{k+1} = \bar{p}^k + h\bar{u}^k$, where $0 < h \leqslant \alpha_k$ is determined so that $\bar{p}^{k+1} \in \mathbb{K}$. The maximum step size α_k is some predetermined small value, and may decrease as k increases to facilitate convergence.
6) Replace k by $k + 1$.

End loop. □

In summary, successive iterates are obtained by moving a small step in the direction of the projection of the gradient $\bar{q}$ onto the convex set of tangent directions $T_{\bar{p}}$. The algorithm stops when this projection is zero. Note also that the gradient, calculated in Step 2, yields the formula

$$q_i^k(\lambda) = \sum_j \sum_{\lambda'} r_{ij}(\lambda, \lambda') p_j^k(\lambda')$$

according to the calculation in Section V (assuming symmetric compatibilities).

Proposition 7.5: Suppose $\bar{p}$ is a stopping point of Algorithm 7.4. Then if the matrix of compatibilities is symmetric, $\bar{p}$ is consistent.

Proof: At a stopping point, $\bar{u} = 0$ solves Problem 7.1. Thus $\bar{v} \cdot \bar{q} \leqslant 0$ for all tangent vectors $\bar{v} \in T_{\bar{p}}$. If we choose any $\bar{v} \in \mathbb{K}$, then $\bar{v} - \bar{p}$ is a tangent vector, and so $(\bar{v} - \bar{p}) \cdot \bar{q} \leqslant 0$. Using the formula for $\bar{q}$,

$$\sum_{i,\lambda} \sum_{j,\lambda'} r_{ij}(\lambda, \lambda') p_j(\lambda') \cdot (v_i(\lambda) - p_i(\lambda)) \leqslant 0 \qquad \text{for all} \quad \bar{v} \in \mathbb{K}.$$

Thus, $\bar{p}$ satisfies the variational inequality of Section IV, and so by Theorem 4.1, $\bar{p}$ is consistent in $\mathbb{K}$. □

We now have a method for finding consistent labelings, given an initial labeling assignment. Whether the resulting consistent labeling is an improvement over the initial assignment depends upon the extent to which it makes sense to increase average local consistency.

The property that $\bar{p}$ yields a local maximum of the average local consistency is actually stronger than is required by the definition of consistency. In particular, while all local maxima of $A(\bar{p})$ are at consistent labelings (Theorem 5.1), the gradient ascent algorithm 7.4 may stop at one of a number of pathological points. This behavior can occur because, for symmetric compatibilities, consistency is a property that depends on the first derivatives of $A(\bar{p})$, whereas a local maximum must include second derivatives in its characterization. The pathologies can include local minima, saddle points, and boundary points with local minima or saddles [see Fig. 6(b) and (c)]. Note that if a relative maximum occurs at a boundary point, the gradient does not necessarily vanish [Fig. 6(d)].

In practice, the gradient ascent algorithm will generally find a local maximum of $A(\bar{p})$. The algorithm will stop at one of the pathological points only if one of the iterates happens to land exactly on such a point. However, all of these points, according to Proposition 7.5, are consistent.

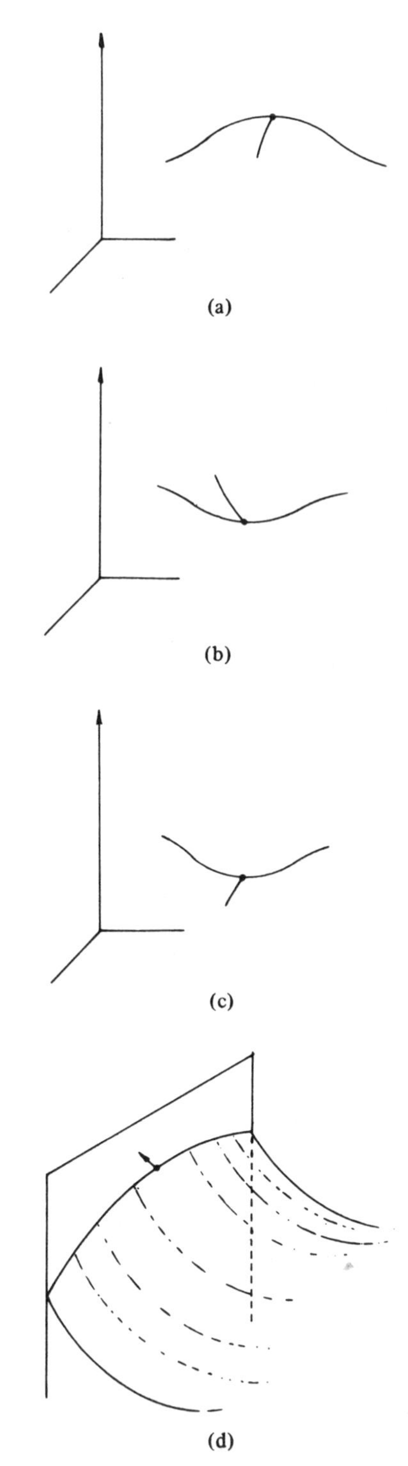

Fig. 6. Examples of stopping points: (a) local max, (b) local min, (c) saddle point, (d) local max at the boundary. Note that the gradient is nonzero at the stopping point in (d).

VIII. The Relaxation Labeling Algorithm

Algorithm 7.4 gives us a method of finding consistent labelings when the matrix of compatibilities is symmetric. Our entire analysis of average local consistency relies on the assumption of symmetric compatibilities. The assumption is unreasonable. In general, we believe that compatibility coefficients need not be symmetric.

Consider, for example, the constraints between letters comprising words in the English language. The occurrence of a "q" in a word lends strong support to the labeling of the following letter as a "u," whereas the occurrence of a "u" only weakly supports the proposition that the previous letter is a "q."

Fortunately, our theory of consistency does not rely on symmetric compatibilities. The characterization of consistent labelings in Theorem 4.1 is completely general, valid whether or not the compatibilities are symmetric. The purpose of this section is to present an algorithm that finds consistent labelings based on the variational inequality in Theorem 4.1. Although the point of view is now much more general, the resulting algorithm is nearly identical to Algorithm 7.4! Further, both algorithms share some similarities with the ad hoc methods [12], as we discuss in Section XI. The algorithm presented in this section is based on a theory of consistency, and allows one to interpret what the relaxation labeling process accomplishes.

We begin by recalling the variational inequality for consistency:

$$\sum_{i,\lambda}\sum_{j,\lambda'} r_{ij}(\lambda,\lambda')p_j(\lambda')(v_i(\lambda)-p_i(\lambda))\leqslant 0 \qquad \text{for all } \bar{v}\in \mathbb{K}$$

or, more generally,

$$\sum_{i,\lambda} s_i(\lambda;\bar{p})\cdot(v_i(\lambda)-p_i(\lambda))\leqslant 0 \qquad \text{for all } \bar{v}\in \mathbb{K}.$$

In Section VII, we defined $\bar{q}=\frac{1}{2}\operatorname{grad} A(\bar{p})$, so that for symmetric compatibilities, we had

$$q_i(\lambda)=\sum_{j,\lambda'} r_{ij}(\lambda,\lambda')p_j(\lambda')=s_i(\lambda;\bar{p}).$$

Hereafter, we define $\bar{q}$ by

$$q_i(\lambda)=\sum_{j,\lambda'} r_{ij}(\lambda,\lambda')p_j(\lambda')$$

whether or not the compatibilities are symmetric. That is, we have set $\bar{q}=\bar{s}(\bar{p})$. It is important to realize, however, that $\bar{q}$ is not in general the gradient of any functional on $\mathbb{K}$.

Observation 8.1: With $\bar{q}$ defined as above, the variational inequality is equivalent to the statement

$$\bar{q}\cdot\bar{t}\leqslant 0 \qquad \text{for all } \bar{t}\in T_{\bar{p}}.$$

That is, a labeling $\bar{p}$ is consistent if and only if $\bar{q}$ points away from all tangent directions.

Proof: We have $\bar{q}=\bar{s}$, and any tangent vector $\bar{t}$ at $\bar{p}$ can be written as a positive scalar multiple of $\bar{v}-\bar{p}$, where $\bar{v}\in\mathbb{K}$. The observation follows immediately. □

Observation 8.1 suggests a way of finding a consistent labeling. If, at a labeling $\bar{p}$, the associated vector $\bar{q}$ points in the same direction as some tangent vector, then $\bar{p}$ is not consistent. So $\bar{p}$ should be moved in the direction of that tangent vector. This process may be repeated until $\bar{q}$ evaluated at the current assignment points away from all tangent directions. Then $\bar{p}$ will be a consistent labeling.

Note that $\bar{q}$ varies as $\bar{p}$ moves, but that generally $\bar{q}$ will change smoothly and gradually. Thus if $\bar{q}$ points away from the surface $\mathbb{K}$ at a vertex (and is therefore an unambiguous consistent labeling, then $\bar{q}$ will point generally toward the vertex at nearby assignments in $\mathbb{K}$. Accordingly, if $\bar{p}$ is near the unambiguous consistent labeling, moving $\bar{p}$ in a tangent direction $\bar{u}$ that points in the same direction as $\bar{q}$, should cause $\bar{p}$ to converge to the vertex (see Theorem 9.1). To present these ideas more formally, we begin by defining the algorithm.

If $\bar{q}\cdot\bar{t}>0$ for some tangent direction t, then the current assignment $\bar{p}$ is not consistent, and should be updated. In which direction should we move $\bar{p}$? In analogy with the gradient ascent algorithm it makes sense to move $\bar{p}$ in the direction $\bar{u}$ that maximizes $\bar{q}\cdot\bar{u}$. This is exactly the vector $\bar{u}$ returned by the algorithm of Appendix A as the solution to Problem 7.1. This method of updating $\bar{p}$ is identical to the gradient ascent method (Algorithm 7.4), applicable for symmetric compatibilities, except that $\bar{q}$ has a new conceptual meaning. Instead of setting $\bar{q}=\frac{1}{2}\operatorname{grad} A(\bar{p})$ in Step 2, we set $\bar{q}=\bar{s}(\bar{p})$. If the gradient ascent method is applied only when the compatibilities are symmetric, and providing $\bar{s}(\bar{p})$ is linear, the formulas are identical.

Accordingly, the relaxation labeling algorithm is given by the following.

Algorithm 8.2: Replace Step 2 in Algorithm 7.4 with:

2') Compute $\bar{q}=\bar{s}(\bar{p})$. That is,

$$q_i(\lambda)=\sum_{j,\lambda'} r_{ij}(\lambda,\lambda')p_j(\lambda').$$

All other steps remain the same. □

Compare the following result with Proposition 7.5.

Proposition 8.3: Suppose $\bar{p}$ is a stopping point of Algorithm 8.2. Then $\bar{p}$ is consistent.

Proof: A point $\bar{p}$ is a stopping point of Algorithm 8.2 if and only if $\bar{u}=0$ solves Problem 7.1. If $\bar{u}=0$, then $\bar{t}\cdot\bar{q}\leqslant \bar{0}\cdot\bar{q}=0$ for all tangent vectors $\bar{v}\in T_{\bar{p}}$. On the other hand, if $\bar{t}\cdot\bar{q}\leqslant 0$ for all $\bar{t}\in T_{\bar{p}}$, then $\bar{u}=0$ maximizes $\bar{u}\cdot\bar{q}$ for $\bar{u}\in T_{\bar{p}}\cap B_1(0)$. According to Observation 8.1, $\bar{t}\cdot\bar{q}\leqslant 0$ for all $\bar{t}\in T_{\bar{p}}$ is equivalent to the variational inequality, which is in turn equivalent to $\bar{p}$ being consistent (Theorem 4.1). □

At this point, we have introduced a definition of consistency and presented the relaxation labeling algorithm in such a way that the stopping points of the algorithm are consistent labelings. A number of questions remain to be answered. First, are there any consistent labelings for the relaxation labeling algorithm to find? Second, assuming that such points exist, will the algorithm find them? And finally, even if a relaxation labeling process converges to a consistent labeling, is the final labeling better than the initial assignment?

The first question is answered affirmatively by Proposition 8.4 below. The second question is far more subtle, and is substantially answered by a local convergence result in Section IX. The third question, concerning the significance of the final result, is not really well defined. If the compatibilities are symmetric, then the functional $A(\bar{p})$ provides a quantitative measure of how much the consistent labeling improves the initial, inconsistent one. In the very special case that $A(\bar{p})$ is convex, then the final labeling will be a global maximum. But if the compatibilities are not symmetric, then no such functional exists. The only statement that can be made is that the labeling is now consistent with respect to the compatibility relations; it may be substantially different from the initial labeling.

In general, the space of consistent labelings will be quite rich. Consider, once again, the labeling of letters comprising words in the English language. If we are given five letters, then every five letter word constitutes a consistent labeling. Thus, if an initial estimate is derived from realistic measurements, there will normally be a consistent labeling nearby to which the process will converge.

Proposition 8.4: The variational inequality of Theorem 4.1 always has at least one solution. Thus consistent labelings always exist, for arbitrary compatibility matrices.

Proof: We invoke a version of the Brouwer fixed point theorem, due to Stampacchia [8]. The result states that if $\mathbb{K} \subseteq \mathbb{R}^N$ is a compact convex set, and if $\bar{F}(\bar{x})$ is a continuous function from $\mathbb{K}$ into $\mathbb{R}^N$, then there exists an $\bar{x} \in \mathbb{K}$ such that $\bar{F}(\bar{x}) \cdot (\bar{y} - \bar{x}) \leqslant 0$ for all $\bar{y} \in \mathbb{K}$. For the variational inequality of Theorem 4.1, we set

$$(F(\bar{p}))_i(\lambda) = -\sum_{j,\lambda'} r_{ij}(\lambda, \lambda') p_j(\lambda').$$

Note that F is a continuous (linear, in fact), and that $\mathbb{K}$ is a compact and convex subset of $\mathbb{R}^{nm}$. Thus Stampacchia's result applies, and the proof is complete. □

Usually, more than one solution will exist, however, which is desirable for most relaxation labeling applications. It is only in highly specialized situations, such as when the matrix of compatibilities is negative definite, that the solution is unique.

IX. A Local Convergence Result

As the step size of the relaxation labeling algorithm 7.4 or 8.2 becomes infinitesimal, these discrete algorithms approximate dynamical systems [3]. The iterates $\bar{p}^k$ become a parameterized curve $\bar{p}(t)$, $t \in \mathbb{R}$, lying in $\mathbb{K}$. The tangent to the curve at every point is the updating direction $\bar{u}$, where $\bar{u}$ is the solution to Problem 7.1 at that point. Note that $\bar{u}$ is the normalized projection of $\bar{q}$, which in turn is computed from the matrix of compatibilities and the current labeling assignment $\bar{p}$. Thus the dynamical system obeys a differential equation

$$\frac{d}{dt}\bar{p}(t) = \bar{u}(\bar{p}(t)).$$

Inside the interior of $\mathbb{K}$, $\bar{u}(\bar{p})$ is a linear function of $\bar{p}$. As $\bar{p}$ moves from the interior to a boundary point of $\mathbb{K}$, $\bar{u}(\bar{p})$ may be discontinuous. For this reason, the dynamical system may be very complex.

The main mathematical result about relaxation labeling in this paper concerns the convergence of the above dynamical system. We already know that the dynamical system has a stopping point at $\bar{p}$ if and only if $\bar{p}$ is a consistent labeling (i.e., $\bar{u}(\bar{p}) = 0$). Thus, if the dynamical system, which is approximated by the relaxation labeling algorithm, converges to a stopping point, then we have found a consistent labeling. However, in general, a dynamical system need not converge. Even though $\bar{p}(t)$ always lies in the compact assignment space $\mathbb{K}$, the process might approach an infinite cycle (an orbit), or, it might approach an Ω-limit set that forms a toroidal shape or higher dimensional set. We cannot guarantee that orbits or other examples of nonconvergence will never happen. However, we claim that by using the proper updating rule (Appendix A) and reasonable compatibility values, such behavior is extremely unlikely. Theorem 9.1 argues in justification of this claim.

Recall that a labeling is strictly consistent if

$$\sum_{\lambda} p_i(\lambda)\, s_i(\lambda) > \sum_{\lambda} v_i(\lambda)\, s_i(\lambda), \qquad i = 1, \cdots, n$$

whenever $\bar{v} \neq \bar{p}$, $\bar{v} \in \mathbb{K}$. As a result, the variational inequality can be replaced by the statement

$$\sum_{i,\lambda,j,\lambda'} r_{ij}(\lambda, \lambda')\, p_j(\lambda')\, (v_i(\lambda) - p_i(\lambda)) < 0$$

$$\text{for all } \bar{v} \in \mathbb{K}, \bar{v} \neq \bar{p}$$

for a strictly consistent labeling. In particular, $\bar{q} \cdot \bar{u} < 0$ for all nonzero tangent directions $\bar{u}$ at a strictly consistent labeling $\bar{p}$. We claim that $\bar{p} \in \mathbb{K}^*$ (i.e., that $\bar{p}$ is an unambiguous labeling). Suppose, for contradiction, that $0 < p_{i_o}(\lambda_o) < 1$ for some (i_o, λ_o). Then for some other λ'_o, $0 < p_{i_o}(\lambda'_o) < 1$. We consider two tangent directions,

$$u_1(i, \lambda) = \begin{cases} 0 & i \neq i_o \\ (0, \cdots, 0, 1, \cdots, -1, \cdots, 0) & i = i_o, \end{cases}$$

and $\bar{u}_2 = -\bar{u}_1$.

That is, $\bar{u}_1$ has a 1 in the (i_o, λ_o) position and a -1 in the (i_o, λ'_o) position, and $\bar{u}_2$ is the other way around. These are valid tangent directions according to the formulation of $T_{\bar{p}}$ in Section VI. However, $\bar{q} \cdot \bar{u}_1 = -\bar{q} \cdot \bar{u}_2$, so they cannot both be negative. Hence, we have shown that a strictly consistent labeling $\bar{p}$ must be unambiguous.

Our main result is the following.

Theorem 9.1: Suppose $\bar{e} \in \mathbb{K}^*$ is strictly consistent. Then there exists a neighborhood of $\bar{e}$ such that if $\bar{p}(t)$ enters the neighborhood, then $\lim_{t \to \infty} \bar{p}(t) = \bar{e}$.

In fact, once $\bar{p}(t)$ enters the neighborhood, $\bar{p}(t) \equiv \bar{e}$ after a finite length of time.

Proof: We will make use of the Euclidean norm

$$\|\bar{v}\| = \left[\sum_{i,\lambda} (v_i(\lambda))^2\right]^{1/2}$$

for vectors in $\mathbb{R}^{nm}$. Our first task is to show that if $\bar{p} \in \mathbb{K}$, $\bar{p} \neq \bar{e}$, then there is an assignment $\bar{p}' \in \mathbb{K}$ such that

$$\bar{p}' - \bar{e} = \frac{\bar{p} - \bar{e}}{\|\bar{p} - \bar{e}\|}.$$

We begin by setting $\alpha = \|\bar{p} - \bar{e}\|$, and define

$$\bar{p}' = \left(1 - \frac{1}{\alpha}\right)\bar{e} + \frac{1}{\alpha}\bar{p}.$$

If $\alpha \geqslant 1$, then $\bar{p}' \in \mathbb{K}$ since $\mathbb{K}$ is convex. (Here we use the convexity of $\mathbb{K}$ in a crucial way!) If $\alpha \leqslant 1$, a separate argument is needed. Clearly,

$$\sum_{\lambda} p_i'(\lambda) = 1, \qquad i = 1, \cdots, n.$$

It remains to be shown that $0 \leqslant p_i'(\lambda) \leqslant 1$ when $\alpha \leqslant 1$, in order to conclude that $\bar{p}' \in \mathbb{K}$. Suppose that $e_i(\lambda) = 0$. Then since $\alpha = \|\bar{p} - \bar{e}\| \geqslant |p_i(\lambda) - e_i(\lambda)| = p_i(\lambda)$, we have $p_i(\lambda)/\alpha \leqslant 1$. But $p_i'(\lambda) = p_i(\lambda)/\alpha$, so $0 \leqslant p_i'(\lambda) \leqslant 1$. Next suppose that $e_i(\lambda) = 1$. Then $\alpha \geqslant 1 - p_i(\lambda)$, and so

$$p_i'(\lambda) = (1 - 1/\alpha) + p_i(\lambda)/\alpha = \frac{\alpha - 1 + p_i(\lambda)}{\alpha}$$

satisfies $p_i'(\lambda) \geqslant 0$. Further, since $p_i(\lambda) \leqslant 1$, therefore $\alpha - 1 + p_i(\lambda) \leqslant \alpha$, and so $p_i'(\lambda) \leqslant 1$. The conclusion is that $\bar{p}' \in \mathbb{K}$, and satisfies the required equation.

Next, according to the variational inequality,

$$\sum_{i,\lambda} \sum_{j,\lambda'} r_{ij}(\lambda, \lambda') e_j(\lambda') (p_i(\lambda) - e_i(\lambda)) < 0$$

for any $\bar{p} \in \mathbb{K}$, $\bar{p} \neq \bar{e}$. In particular, the left-hand side has a negative maximum on any compact subset of $\mathbb{K}$ not including $\bar{e}$. Thus there exists a $\beta < 0$ such that

$$\bar{q}(\bar{e}) \cdot (\bar{p}' - \bar{e}) \leqslant \beta < 0,$$

for all $\bar{p}' \in \mathbb{K}$ such that $\|\bar{p}' - \bar{e}\| = 1$.

Here $\bar{q}(\bar{p}) = \bar{q}$, where

$$q_i(\lambda) = \sum_{j,\lambda'} r_{ij}(\lambda, \lambda') p_j(\lambda').$$

Now suppose that $\bar{p} \in \mathbb{K}$, $\bar{p} \neq \bar{e}$, and $\bar{p}'$ is the corresponding assignment in $\mathbb{K}$ satisfying

$$\bar{p}' - \bar{e} = \frac{\bar{p} - \bar{e}}{\|\bar{p} - \bar{e}\|}.$$

Note that $\|\bar{p}' - \bar{e}\| = 1$, so that

$$\bar{q}(\bar{e}) \cdot \frac{(\bar{p} - \bar{e})}{\|\bar{p} - \bar{e}\|} \leqslant \beta < 0.$$

Since $\bar{q}(\bar{p})$ is a continuous function, we have $\|\bar{q}(\bar{p}) - \bar{q}(\bar{e})\| \leqslant |\beta|/2$ for $\bar{p}$ in a neighborhood of $\bar{e}$, and thus

$$\bar{q}(\bar{p}) \cdot \frac{(\bar{p} - \bar{e})}{\|\bar{p} - \bar{e}\|} \leqslant \beta/2 < 0$$

for $\bar{p} \in \mathbb{K}$ in the neighborhood of $\bar{e}$.

If $\bar{p}$ is in the interior of $\mathbb{K}$, then $\bar{u}(\bar{p}) = \bar{w}/\|\bar{w}\|$ where $\bar{w}$ is the projection of $\bar{q}$ onto $T_{\bar{p}}$. By the projection theorem, $(\bar{q} - \bar{w}) \cdot \bar{t} = 0$ for any tangent vector $\bar{t} \in T_{\bar{p}}$. Thus $\bar{q} \cdot (\bar{p} - \bar{e}) = \bar{w} \cdot (\bar{p} - \bar{e})$, and so for $\bar{p}$ in the interior of $\mathbb{K}$ sufficiently near $\bar{e}$, $\bar{p} \neq \bar{e}$,

$$\bar{u}(\bar{p}) \cdot \frac{(\bar{p} - \bar{e})}{\|\bar{p} - \bar{e}\|} = \frac{1}{\|w\|} \cdot \bar{q}(\bar{p}) \cdot \frac{(\bar{p} - \bar{e})}{\|\bar{p} - \bar{e}\|} \leqslant \frac{\beta}{2\|w\|}.$$

Further, since $\bar{w}$ is the projection of $\bar{q}$, $\|\bar{w}\| \leqslant \|\bar{q}(\bar{p})\|$. Set

$$\gamma = \max_{\bar{p} \in \mathbb{K}} \|\bar{q}(\bar{p})\|$$

whence $\|w\| \leqslant \gamma$. Since β is negative,

$$\bar{u}(\bar{p}) \cdot \frac{(\bar{p} - \bar{e})}{\|\bar{p} - \bar{e}\|} \leqslant \frac{\beta}{2\gamma} < 0.$$

We will next show that the same inequality holds if $\bar{p}$ is on a boundary of $\mathbb{K}$, but sufficiently near $\bar{e}$.

According to Appendix A, the vector $\bar{u}$ at $\bar{p}$ is obtained by normalizing a vector $\bar{w}$, which in turn is obtained from $\bar{q} = \bar{q}(\bar{p})$ by setting some of the components to zero, and subtracting constants from the other components: $w_i(\lambda) = q_i(\lambda) - c_i$. The indexes (i, λ) of the components which are set to zero are a subset of the set for which $p_i(\lambda) = 0$. Denote this subset of indexes by S. Since $\bar{p}$ is near $\bar{e}$, the components of $\bar{e}$ are close to the components of $\bar{p}$, so we must have $e_i(\lambda) = 0$ whenever $p_i(\lambda) = 0$. Hence $p_i(\lambda) - e_i(\lambda) = 0$ for $(i, \lambda) \in S$. Thus,

$$\begin{aligned}
\bar{w} \cdot (\bar{p} - \bar{e}) &= \sum_{(i,\lambda) \notin S} w_i(\lambda) \cdot (p_i(\lambda) - e_i(\lambda)) \\
&= \sum_{(i,\lambda) \notin S} (q_i(\lambda) - c_i)(p_i(\lambda) - e_i(\lambda)) \\
&= \sum_{i,\lambda} q_i(\lambda) \cdot (p_i(\lambda) - e_i(\lambda)) \\
&\quad - \sum_{i=1}^{n} c_i \sum_{\lambda=1}^{m} (p_i(\lambda) - e_i(\lambda)) \\
&= \bar{q} \cdot (\bar{p} - \bar{e}) + 0.
\end{aligned}$$

Having shown that $\bar{q} \cdot (\bar{p} - \bar{e}) = \bar{w} \cdot (\bar{p} - \bar{e})$ even when $\bar{p}$ is a boundary point of $\mathbb{K}$ (provided $\bar{p}$ is near $\bar{e}$), the proof of the inequality

$$\bar{u}(\bar{p}) \cdot \frac{(\bar{p} - \bar{e})}{\|\bar{p} - \bar{e}\|} \leqslant \frac{\beta}{2\gamma}, \qquad \bar{p} \text{ near } \bar{e}$$

proceeds exactly as in the above case when $\bar{p}$ is interior to $\mathbb{K}$.

We recognize the left hand side of this inequality as the derivative of $\|\bar{p}(t) - \bar{e}\|$ with respect to t. So

$$\frac{d}{dt} \|\bar{p}(t) - \bar{e}\| \leqslant \frac{\beta}{2\gamma} < 0$$

providing $\bar{p}(t)$ is sufficiently near $\bar{e}$, but not equal to $\bar{e}$. Once $\bar{p}(t)$ enters this neighborhood of $\bar{e}$, the distance $\|\bar{p}(t) - \bar{e}\|$ must decrease to zero within a period of time equal to $2\gamma/|\beta|$. When the distance drops to zero, we have $\bar{p}(t) = \bar{e}$, which is a stopping point of the dynamical system, since $\bar{e}$ is consistent. This completes the proof. □

In Theorem 9.1, we used the assumption that $\bar{e}$ is strictly consistent in order to prove that it is a local attractor of the relaxation labeling dynamical system. As a bonus, convergence to $\bar{e}$ occurs in finite time. One might object that the rapid convergence is an artifact of the normalization process in the projection operator. That is, since either $\bar{u} = 0$ or $\|\bar{u}(\bar{p})\| = 1$, the dynamical system must always move with unit speed before convergence. In fact, however, with the assumption of strict consistency, Theorem 9.1 would still hold true if the length normalization step were omitted from the projection algorithm.

If $\bar{p}$ is consistent, but not strictly consistent, then $\bar{p}$ may be a local attractor of the dynamical system, or it may be a saddle point, or even an unstable stopping point. We will not pursue these topics here.

We consider a local convergence result like Theorem 9.1 to be preferable to a universal convergence result, because the resulting consistent labeling is related to the initial labeling assignment. In Theorem 9.1, the hypothesis that a labeling is close to the consistent assignment requires that the labeling assignment at every object is close to the designated assignment. It would be desirable to extend Theorem 9.1 to the case when the initial labeling assignment is completely incorrect in a few "noncritical" object labelings.

X. Generalizations to Higher Order Compatibilities

In Section III, we stated that consistency could be defined in terms of support functions which depend on arbitrary orders of compatibilities. For example, for third-order compatibilities, we need a matrix of values $\{r_{ijk}(\lambda, \lambda', \lambda'')\}$, from which the support values $s_i(\lambda)$ are calculated:

$$s_i(\lambda) = \sum_{j,\lambda'} \sum_{k,\lambda''} r_{ijk}(\lambda, \lambda', \lambda'') p_j(\lambda') p_k(\lambda'').$$

In general, compatibilities of order k can be used to define the support components

$$s_i(\lambda) = \sum_{i_2,\lambda_2} \sum_{i_3,\lambda_3} \cdots \sum_{i_k,\lambda_k} r_{i,i_2,i_3,\cdots,i_k}(\lambda, \lambda_2, \cdots, \lambda_k) \cdot p_{i_2}(\lambda_2) \cdots p_{i_k}(\lambda_k).$$

Of special note is the case of first order compatibilities. In this case, the compatibilities are given by a vector $\{r_i(\lambda)\}$ and the support $s_i(\lambda)$ is independent of $\bar{p}$, and given by

$$s_i(\lambda) = r_i(\lambda).$$

The analog to the variational inequality, which serves as a characterization of consistent labelings for higher order compatibilities, is given by $\bar{p} \in \mathbb{K}$ such that $\Sigma_{i,\lambda}\, s_i(\lambda)\,(v_i(\lambda) - p_i(\lambda)) \leqslant 0$ for all $\bar{v} \in \mathbb{K}$. Of course, $s_i(\lambda)$ depends on $\bar{p}$ according to the appropriate formula.

For second order compatibilities, we showed that a symmetry condition leads to the existence of a potential $A(\bar{p})$, satisfying $\bar{s} = \bar{q} = 1/2 \operatorname{grad} A(\bar{p})$. For compatibilities of first order, such a function always exists, namely,

$$A(\bar{p}) = \sum_{i,\lambda} r_i(\lambda) p_i(\lambda).$$

In the general case of kth order, the appropriate function $A(\bar{p})$ will satisfy the condition $\bar{s} = 1/k \operatorname{grad} A(\bar{p})$ when a certain symmetry condition is satisfied. In this case the average local consistency is given by

$$A(\bar{p}) = \sum_{i_1,\lambda_1} \sum_{i_2,\lambda_2} \cdots \sum_{i_k,\lambda_k} r_{i_1 \cdots i_k}(\lambda_1, \cdots, \lambda_k) \cdot p_{i_1}(\lambda_1) \cdots p_{i_k}(\lambda_k).$$

The symmetry condition, in its most general form, states that

$$\sum_{\sigma \in C_k} r_{i_{\sigma(1)}, i_{\sigma(2)}, \cdots, i_{\sigma(k)}}(\lambda_{\sigma(1)}, \cdots, \lambda_{\sigma(k)}) = k \cdot r_{i_1, \cdots, i_k}(\lambda_1, \cdots, \lambda_k)$$

for all sets of indexes $\{(i_1, \lambda_1), \cdots, (i_k, \lambda_k)\}$, where C_k is the group of cyclic permutations on k objects. For example, for order 3 the symmetry condition is

$$r_{ijk}(\lambda, \lambda', \lambda'') + r_{kij}(\lambda'', \lambda, \lambda') + r_{jki}(\lambda', \lambda'', \lambda) = 3 r_{ijk}(\lambda, \lambda', \lambda'')$$

for all i, λ, j, λ', k, and λ''.

The same algorithms serve to find consistent labelings. For the general nonsymmetric case, Algorithm 8.2 yields consistent labelings by finding solutions to the analog of the variational inequality, except that in Step 2′, we set $q_i(\lambda) = s_i(\lambda)$. Here $s_i(\lambda)$ is calculated using $\bar{p}$ and the appropriate formula for the kth order compatibilities.

As mentioned in Section III, more general support functions are possible by combining supports of different orders. In full generality, the support vector $\bar{s}$ is a function $\bar{s}(\bar{p})$, where $s_i(\lambda)$ is computed from a nonlinear function of current assignment values in $\bar{p}$. Presumably, $s_i(\lambda)$ depends on the components of $\bar{p}_j$ for objects j near object i, and is relatively independent of the values of $\bar{p}_j$ for objects distant from i. The extent to which a particular problem yields local dependence of the support value is outside the scope of the abstract analysis given here. Once functions have been formulated from problem-specific considerations, the formal theory could allow one to determine whether a lower order approximation, perhaps in terms of a Taylor series expansion, is approximately sufficient.

When the compatibilities are first-order and the supports $s_i(\lambda)$ are constant and equal to the given coefficients $r_i(\lambda)$, a consistent labeling can be found immediately, i.e., without an iterative procedure. A simple argument shows that the unambiguous labeling $\bar{e} \in \mathbb{K}^*$ defined by

$$e_i(\lambda) = \begin{cases} 1 & \text{if } r_i(\lambda) > r_i(\lambda'), \text{ for all } \lambda', \\ 0 & \text{otherwise} \end{cases}$$

is always consistent in $\mathbb{K}$. If at some object i, there is no single label λ that maximizes $r_i(\lambda)$, then $e_i(\lambda)$ may be set to 1 for exactly one of them, and the result is a consistent labeling. In short, finding consistent labelings for first order compatibilities amounts to local maxima selection (cf. [19]).

It is interesting to note that Ullman's scheme for motion correspondence [15] is a problem of just this (unary) form, but for which the compatibilities are variable as a function of the underlying data. This more general situation requires an algorithm, such as gradient ascent, to obtain (local) maxima.

For compatibilities higher than second order, or nonpolynomial compatibilities, the difficulty becomes one of a combinatorial growth in the number of required computations. At this date, most implementations of conceptually similar relaxation labeling processes have limited the computations to second-order compatibilities.

XI. Comparisons with Standard Relaxation Labeling Updating Schemes

Algorithm 8.2 updates weighted labeling assignments by computing an intermediate vector $\bar{q}$, where

$$q_i(\lambda) = \sum_j \sum_{\lambda'} r_{ij}(\lambda, \lambda') p_j(\lambda')$$

and then updating $\bar{p}$ in the direction defined by the projection of $\bar{q}$ onto $T_{\bar{p}}$. As we shall show, the original updating formula introduced in [12] is an approximation to this new one. The intermediate vector $\bar{q}$ is identical in the two algorithms. The principal difference lies in the manner in which $\bar{q}$ is projected onto a tangent vector. In Algorithm 8.2, the tangent direction was obtained by maximizing $\bar{u} \cdot \bar{q}$ among $\bar{u} \in T_{\bar{p}} \cap B_1(0)$.

Other formulas have been proposed and used for relaxation labeling. We will concentrate on two standard formulas. The first was suggested by Rosenfeld *et al.* [12], and is given by

$$p_i(\lambda) := \frac{p_i(\lambda)\,[1 + q_i(\lambda)]}{\sum_{l=1}^{m} p_i(l)\,[1 + q_i(l)]}.$$

(It is assumed, when using this formula, that the $r_{ij}(\lambda, \lambda')$ values are sufficiently small that one can be sure that $|q_i(\lambda)| < 1$.) To consider the behavior of this standard formula, first assume that $\bar{p}$ is near the center of the assignment space, so that very approximately $p_i(\lambda) \approx 1/m$ for all i, λ. The updating can then be regarded as consisting of two steps. First, the vector $\bar{p}$ is changed into an intermediate $\hat{p}$, where

$$\hat{p}_i(\lambda) = p_i(\lambda)\,[1 + q_i(\lambda)]$$
$$\approx p_i(\lambda) + q_i(\lambda)/m.$$

Next, $\hat{p}$ is normalized using a scalar constant for each object $\hat{p}_i$. When $\bar{p}$ is near the center of $\mathbb{K}$, this rescaling process shifts $\hat{p}$ in a direction essentially perpendicular to $\mathbb{K}$. That is, $\bar{p}$ is reset to approximately the projection of $\hat{p}$ onto $\mathbb{K}$. Denoting the orthogonal projection operator by $\mathcal{O}_K$, we have

$$\bar{p} := \bar{p}' \approx \mathcal{O}_K(\hat{p}) \approx \mathcal{O}_K(\bar{p} + \bar{q}/m)$$

by virtue of the continuity of $\mathcal{O}_K$. Further, assuming that $\bar{p}$ is in the interior of $\mathbb{K}$, and $\bar{q}$ is sufficiently small, then

$$\mathcal{O}_K(\bar{p} + \bar{q}/m) = \bar{p} + \frac{1}{m}\mathcal{O}_T(\bar{q})$$

where $\mathcal{O}_T$ is the orthogonal projection onto the linear subspace $T_{\bar{p}}$. However, by Lemma 7.3, the solution $\bar{u}$ to Problem 7.1 is obtained by normalizing $\mathcal{O}_T(\bar{q})$. Combining, we have that

$$\bar{p} := \bar{p}' \approx \bar{p} + \alpha\bar{u}$$

for some scalar α. Thus, $\bar{p}$ is reset to a vector which is approximately the updated vector that one would obtain by Algorithm 8.2. However, we have assumed that $\bar{p}$ is near the center of the assignment space.

When $\bar{p}$ is close to an edge or corner, the situation is somewhat more complicated. The first step in standard updating can be viewed as an initial operation changing $\bar{q}$, since the components of $\bar{q}$ corresponding to small components of $\bar{p}$ have minimal effect. That is, the operation $p_i(\lambda) + p_i(\lambda)\,q_i(\lambda)$ differs from adding $\bar{q}$ to $\bar{p}$ (or a small multiple of $\bar{q}$ to $\bar{p}$) in that the motions in directions perpendicular to the nearby edges are scaled down. Furthermore, the normalization step is no longer equivalent to a simple projection. Rather, it can be seen that the rescaling is equivalent to modifying the updating direction back toward the center of the assignment space. This further constricts motions perpendicular to nearby edges, when that motion is outward. Qualitatively, this behavior of the standard updating formula near the edges is reminiscent of the projection operator on the same edges. However, because the formula results in attenuation of motion perpendicular to an edge, even when that motion is directed back toward the surface, the standard updating formula will tend to round corners more smoothly and more slowly. Further, a zero component can never become nonzero, even if the evidence supports increasing the value. In summary, there is strong agreement between the projection operator and standard updating in the interior of $\mathbb{K}$, but some differences are possible near the edges. One of the probable effects of these differences is to show convergence when the older updating formula is used.

An alternative updating formula is now common, based on its relationship to a Bayesian analysis of a single iteration of relaxation labeling [11]. In this formulation, the compatibility coefficients are nonnegative, and the updating formula is given by

$$p_i(\lambda) := \frac{p_i(\lambda) \cdot \left(\prod_{j=1}^{n} \sum_{\lambda'=1}^{m} r_{ij}(\lambda, \lambda')\, p_j(\lambda') \right)}{\sum_l p_i(l) \prod_j \left(\sum_{\lambda'} r_{ij}(\lambda, \lambda')\, p_j(\lambda') \right)}.$$

Once again, the denominator is a normalization term. The numerator can be rewritten as

$$p_i(\lambda) \prod_{j=1}^{n} \left[1 + \sum_{\lambda'=1}^{m} (r_{ij}(\lambda, \lambda') - 1)\, p_j(\lambda') \right]$$
$$= p_i(\lambda) \left[1 + \sum_j \sum_{\lambda'} (r_{ij}(\lambda, \lambda') - 1)\, p_j(\lambda') \right.$$
$$+ \text{quadratic terms of } \bar{p}$$
$$\left. + \text{higher order terms} \right].$$

We can view this terms as $p_i(\lambda)\,[1 + q_i(\lambda)]$, where $q_i(\lambda)$ is a complicated nonlinear function of the assignment vector $\bar{p}$. When viewed in this fashion, we see that the "product of sums" updating formula is identical to the earlier updating formula, except that the formulation of the updating direction $\bar{q}$ is more complicated, and disguised. In this sense, the same comments as before apply to the relationship between Algorithm 8.2, which uses the projection operator, and more classical relaxation labeling procedures using other updating formulas.

XII. Summary and Conclusions

Relaxation labeling processes were introduced and studied as mechanisms for employing context and constraints in labeling problems. Over the past few years, a number of researchers have applied relaxation labeling processes to a variety of different tasks. Some of these applications were more successful than others. In all cases, there was an intuition that the compatibility coefficients ought to be doing something useful to the labeling assignment (at least for the first few iterations), but there has never been a succinct and reasonable statement of what useful properties were being enhanced. Lacking a proper model characterizing the process and its stopping points, the choice of the coefficient values and the updating formula are subject only to empirical justification.

In this paper, we have attempted to develop the foundations of a theory explaining what relaxation labeling accomplishes. This theory is based on an explicit new definition of consistency, and leads to a relaxation algorithm with an updating formula which uses a projection operator. It is our hope that these results will help explain why some applications of relaxation labeling have been successful and others have been less impressive, and will assist in a more proper and systematic design of relaxation labeling processes in the future.

In discrete relaxation, a label is discarded if it is not supported by the local context of assigned labels. In the limiting assignment of labels, every retained label is unsupported, whereas every label that has been discarded is unsupported. Of course, in discrete relaxation, support is an all-or-nothing proposition, and is based on a system of logical conditions on neighboring assignments.

In extending this idea to weighted label assignments, support values become ordered real numbers, and are expressed by potentially complicated functions of local weighted assignment values. An unambiguous labeling is consistent, according to our definition, if the support for the instantiated label at each object is greater than or equal to the support for all other labels at that object. A generalization of this idea leads to a definition of consistency for weighted labeling assignments.

We showed that the relaxation labeling process defined by Algorithm 8.2 together with the projection operator specified in Appendix A stops at consistent labelings. We further showed that if one begins sufficiently near a consistent labeling, the dynamic process will then converge to that labeling. The sense in which a consistent labeling constitutes an improvement over an initial labeling assignment relates to the proximity of the two labelings and the precise meaning of the terms in the definition of consistency. We also showed that our relaxation algorithm is approximated by some of the more standard formulas used for labeling, but offer our algorithm as an alternative formulation that is based on a much more mathematically well-founded theory.

When the support values are given by specific polynomial formulas, and provided certain symmetry properties hold among the coefficients, the relaxation labeling algorithm given here is equivalent to gradient ascent using a functional which we have called average local consistency. We used this optimization viewpoint to introduce the algorithm, and were in part motivated to study the more general case by earlier work of others using optimization to assist in labeling problems. However, the symmetry assumption is restrictive, and, as we showed by giving up the optimization functional, unnecessary.

Relaxation labeling processes were originally conceived in terms of cooperation, as a system of local, simple, and sparsely interacting processes. Its current form is still structured in this fashion. Now, however, we can interpret the meaning of the compatibility matrices in terms of the permissible local configurations of consistent labelings.

Much work remains to be done to analyze, extend, and apply the theory presented in this paper. Practical considerations, such as efficient implementations of the projection operator, choice of the step size, and normalization methods need to be addressed. More consideration should be placed on design criteria for particular relaxation applications, as well as on the formal relationships between relaxation labeling and other families of algorithms for solving similar tasks. The incorporation of logical constraints, and the relative merits of using more complicated support functions are intriguing topics for further study. We hope that the answers to some of these questions are facilitated by the foundations presented here.

Appendix A
Updating Direction Algorithm

In Algorithm 7.4 and the relaxation labeling Algorithm 8.2, a solution to Problem 7.1 is required. Problem 7.1, as stated in the text, amounts to maximizing the vector dot product $\bar{q} \cdot \bar{u}$ among all tangent vectors $\bar{u} \in T_{\bar{p}}$ of length less than or equal to one. Both $\bar{p}$ and $\bar{q}$ are given. Thus, we have the following.

Problem: Find $\bar{u} \in T_{\bar{p}} \cap B_1(0)$ such that

$$\bar{u} \cdot \bar{q} \geqslant \bar{v} \cdot \bar{q} \qquad \text{for all } \bar{v} \in T_{\bar{p}} \cap B_1(0).$$

If $\bar{u} = 0$ is a possible solution, then $\bar{u} = 0$ is the solution that should be returned.

We recall that

$$T_{\bar{p}} = \left\{ \bar{v} \in \mathbb{R}^{nm}: \sum_{\lambda=1}^{m} v_i(\lambda) = 0, \quad i = 1, \cdots, n, \quad \text{and} \right.$$

$$\left. v_i(\lambda) \geqslant 0 \quad \text{whenever } p_i(\lambda) = 0 \right\}$$

and $B_1(0) = \{\bar{v} \in \mathbb{R}^{nm}: \|\bar{v}\| \leqslant 1\}$.

The algorithm given below is intended to replace the updating formulas in common use in relaxation labeling processes. The projection operator, as given below, has the advantage of being based on a theory of consistency, and permits the analytic proof of convergence results.

A complete discussion of the algorithm, as well as a proof of its correctness, is given in an accompanying paper, "A Feasible Direction Operator for Relaxation Methods," which appears in this issue [10]. Here, we merely give a formal specification of the algorithm. We state the procedure

```
PROCEDURE Projection_Operator(p,q,n,m)
  $
  $  This procedure returns the vector u used to update the weighted
  $  labeling assignment p in a relaxation labeling process.
  $
  $  p  is an element in the assignment space IK, formed as a vector
  $    of vectors.  That is, p = [p1,p2,..,pn], each pi = [pi1,..,pim],
  $    each pik is a nonnegative real, the sum over k of pik is 1.
  $  q  is the direction vector, obtained from the support values at the
  $    current labeling assignment p, represented as an n-vector of
  $    m-vectors (in the same form as p).
  $  n  is the number of objects.
  $  m  is the number of labels.
  $
  $  The updating direction u is returned through the procedure name,
  $  and is an n-vector of m-vectors.  A typical call will appear as
  $                    u := Projection_Operator(p,q,n,m)
  $
  $  In SETL, n and m are not needed in the parameter list.  In that
  $  case, the procedure would begin with the following statements:
  $                    n := #p ;  m := #p(1) ;
  $

  $  Begin algorithm:

   u := [ ];                                  $ Defines u as a vector

   LOOP for i in [1..n] DO

      D := { k in [1..m] | p(i)(k) = 0. };
      S := {};

      LOOP DO          $ Indefinite loop terminates when QUIT is executed.

        ns := #S
        t  := +/[ q(i)(k) : k in [1..m] | k NOTIN S] / (m-ns);

  $        i.e., t is the sum of q(i)(k) over k not in S, divided by m-#S

        S  := { k in D |  q(i)(k) < t  }
        IF #S = ns  THEN  QUIT;  END IF;

      END LOOP DO;

      u(i) := [ (IF k in S then 0. ELSE q(i)(k) - t ) : k in [1..m] ];

  $                  i.e., u(i)(k) = 0 if k IN S, = q(i)(k)-t otherwise.

   END LOOP for i;

  $   Normalize the vector u :

   Norm := SQRT( +/[uik**2 : ui = u(i), uik = ui(k)] );
   u :=   (IF   Norm = 0. then    u
           ELSE            [ [uik/Norm : uik = ui(k)] : ui = u(i)] );

   RETURN u;

  END PROCEDURE Projection_Operator;
```

Fig. 7.

in the computer language SETL, a very high level programming language which has dynamic allocation and includes set and tuples in its set of primitives [20]. The operators in the SETL language closely resemble standard mathematical sentences (see Fig. 7). A more typical mathematical description of the algorithm is given in the accompanying paper.

The returned vector u will be a solution to the projection problem. It can also be shown that the potentially infinite loop (LOOP DO in the SETL code) will execute at most $m+1$ times for each value of i in $[1, \cdots, n]$. That is, the projection operator is a finite iterative algorithm. Although the SETL specification given above is executable code, in practical applications the projection algorithm will be implemented in Fortran, assembler, or perhaps even special purpose hardware.

We have also included the normalization $\|\bar{u}\| = 1$ (or $\bar{u} = 0$) as part of the algorithm, as required by the statement of the problem. When n is large, however, the calculation of the norm of the unnormalized vector $\bar{u}$ may be costly. As mentioned after Theorem 9.1, the local convergence result does not actually require that $\bar{u}$ be normalized. In fact, the theorem holds true if any norm measure is used, as long as small steps are taken (with respect to that norm) in the direction $\bar{u}$. However, as specified by Step 6 of Algorithms 7.4 and 8.2, the step size must be dynamically adjusted to make sure that step never forces any component of the current assignment labeling to become negative. Both the normalization process and the dynamic step size adjustment are greatly simplified if applied to individual subvectors $\bar{u}_i$ and $\bar{p}_i$ separately. In this case, $\|\bar{u}_i\| = 1$ for $i = 1, \cdots, n$, and the step length α_i is adjusted for each i so that $\alpha_i \leq h$, α_i maximized, and $\bar{p}_i + \alpha_i \bar{u}_i$ has nonnegative components. However, this changes the relaxation labeling process, since the direction of updating may not be parallel to $\bar{u}$. Nonetheless, one can still prove that stopping points are consistent labelings and strictly consistent labelings are local attractors.

APPENDIX B
GLOSSARY OF SYMBOLS

$A(\bar{p})$ Average local consistency of a labeling (introduced in Section V).

$B_1(0)$ The ball of vectors of radius 1 centered at 0 (Section VII).

$\delta_{i\alpha}$ Kronecker delta, $\delta_{i\alpha} = 0$ if $i \neq \alpha$, and $= 1$ if $i = \alpha$ (Section V).

$\bar{e}$ An unambiguous labeling vector (Section IX).

$\bar{e}_k$ An m-vector of the form $\bar{e}_k = [0, \cdots, 1, \cdots, 0]$, where the 1 appears in the kth component (Section III).

grad $A(\bar{p})$ Gradient of A evaluated at $\bar{p}$ (Section V).

i, j Variables to indicate nodes in the labeling graph, or indexes through sets of nodes. j typically indicates a neighbor of i (Section I).

$\mathbb{K}^*$ The space of unambiguous labelings (Section III).

$\mathbb{K}$ Convex space of weighted labeling assignments (Section III).

λ Variable to either denote a label or to serve as an index through a set of labels (Section I).

Λ_i Set of labels attached to node i (Section I).

Λ_{ij} Constraint relation listing all pairs (λ, λ') such that λ at i is consistent with λ' at j.

m Number of labels in Λ_i (Section I).

n Number of nodes in G (Section I).

$\mathcal{O}_W \bar{v}$ Projection operator, indicating the projection of a vector $\bar{v}$ onto a space W (Section XI).

$p_i(\lambda)$ Weight indicating the strength with which label λ is associated with node i (Section III).

$\bar{p}_i$ The labeling vector associated with node i: $\bar{p}_i = [p_i(1), p_i(2), \cdots, p_i(m)]$ (Section III).

$\bar{p}$ The complete labeling assignment vector (or, for short, labeling) $\bar{p} = [\bar{p}_1, \bar{p}_2, \cdots, \bar{p}_n]$ (Section III).

$r_{ij}(\lambda, \lambda')$ Compatibility matrix over pairs of labels on pairs of nodes (Section III).

$r_{ijk}(\lambda, \lambda', \lambda'')$ Compatibility matrix over triples of labels on triples of nodes (Section III).

$R_{ij}(\lambda, \lambda')$ Indicator function for the numerical representation of Λ_{ij} (Section I).

$\mathbb{R}^m$ m-dimensional Euclidean space (Section III).

$s_i(\lambda)$ Support given by an (unambiguous or ambiguous) labeling to label λ on node i (Section III).

$S_i(\lambda)$ Support for label λ on i from a discrete labeling (Section I).

$T_{\bar{p}}$ Tangent space at $\bar{p} \in \mathbb{K}$ (Section VI).

$\bar{u}$ The tangent direction in which updating takes place (the projection of $\bar{q}$) (Section VII).

$\bar{v}$ An arbitrary weighted labeling assignment in $\mathbb{K}$ (Section III).

$\#S$ The number of elements in the set S (Appendix A).

$a := b$ The value of b replaces the current value of a (Section II).

$\bar{u} \cdot \bar{v}$ Vector dot product of $\bar{u}$ and $\bar{v}$ (Section V).

$\|\bar{u}\|$ Euclidean norm of $\bar{u}$ (Section VII).

REFERENCES

[1] K. Arrow, L. Hurwicz, and K. Uzawa, *Studies in Linear and Nonlinear Programming*. Stanford, CA: Stanford Univ. Press, 1960.

[2] M. Berthod and O. Faugeras, "Using context in the global recognition of a set of objects: An optimization approach," in *Proc. IFIP*, 1980.

[3] N. P. Bhatia and G. P. Szego, *Dynamical Systems: Stability Theory and Applications* (*Lecture Notes in Mathematics*), vol. 35. New York: Springer-Verlag, 1967.

[4] L. Davis and A. Rosenfeld, "Cooperating processes for low-level vision: A survey," *Artificial Intell.*, vol. 17, p. 412, 1981.

[5] O. Faugeras and M. Berthod, "Improving consistency and reducing ambiguity in stochastic labeling: An optimization approach," *IEEE Trans. Pattern Anal. Machine Intell.*, vol. PAMI-3, p. 245, 1981.

[6] R. M. Haralick and L. Shapiro, "The consistent labeling problem: Part 1," *IEEE Trans. Pattern Anal. Machine Intell.*, vol. PAMI-1, p. 173, 1979.

[7] G. E. Hinton, "Relaxation and its role in vision," Ph.D. dissertation, Univ. Edinburgh, Dec. 1979.

[8] D. Kinderlehrer and G. Stampacchia, *An Introduction to Variational Inequalities and Their Applications*. New York: Academic, 1980.

[9] D. Luenberger, *Optimization by Vector Space Methods*. New York: Wiley, 1969.

[10] J. Mohammed, R. Hummel, and S. Zucker, "A feasible direction operator for relaxation methods," this issue, pp. 330–332.

[11] S. Peleg, "A new probabilistic relaxation scheme," *IEEE Trans. Pattern Anal. Machine Intell.*, vol. PAMI-2, p. 362, 1980.

[12] A. Rosenfeld, R. Hummel, and S. Zucker, "Scene labeling by relaxation operations," *IEEE Trans. Syst., Man, Cybern.*, vol. SMC-6, p. 420, 1976.

[13] R. Southwell, *Relaxation Methods in Engineering Science*. Clarendon, 1940.

[14] S. Ullman, "Relaxation and constrained optimization by local processes," *Comput. Graphics Image Processing*, vol. 10, p. 115, 1979.

[15] —, *The Interpretation of Structure from Motion*. Cambridge, MA: MIT Press, 1979.

[16] D. Waltz, "Understanding line drawings of scenes with shadows," in *The Psychology of Computer Vision*, P. Winston, Ed. New York: McGraw-Hill, 1975.

[17] S. Zucker, "Labeling lines and links: An experiment in cooperating computation," in *Consistent Labeling Problems in Pattern Recognition*, R. Haralick, Ed. New York: Plenum, 1980.

[18] S. Zucker, R. Hummel, and A. Rosenfeld, "An application of relaxation labeling to line and curve enhancement," *IEEE Trans. Comput.*, vol. C-26, pp. 394–403, 922–929, 1977.

[19] S. Zucker, Y. Leclerc, and J. Mohammed, "Relaxation labeling and local maxima selection: Conditions for equivalence," *IEEE Trans. Pattern Anal. Machine Intell.*, to be published.

[20] R. B. K. Dewar *et al.*, "Programming by refinement," *TOPLAS*, vol. 1, p. 27, 1979.

Robert A. Hummel (M'82) received the B.A. degree in mathematics from the University of Chicago, Chicago, IL, and the Ph.D. degree in mathematics from the University of Minnesota, Minneapolis.

While in school, he spent summers at the Picture Processing Lab of the Computer Science Center at the University of Maryland. He has also been employed at Stanford Research Institute and the Signal and Image Processing section of Honeywell's Systems and Research

Center in Minneapolis, MN. From 1980 to 1982 he was a Courant Institute Instructor in the Department of Mathematics, New York University's Courant Institute of Mathematical Sciences. He is currently an Assistant Professor of Computer Science in the Courant Institute, and a member of the Robotics and Vision Research Group there. He also serves as a consultant to Martin Marietta Aerospace in Orlando, and is an adjunct faculty member of IBM's System Research Institute in New York. In mathematics, his research interests include variational methods for the study of partial differential equations, fluid mechanics, and foundations of thermodynamics. Other research interests include computer vision and image processing.

Dr. Hummel is a member of Phi Beta Kappa and the American Mathematical Society.

Steven W. Zucker (S'71-M'75) received the B.S. degree in electrical engineering from Carnegie-Mellon University, Pittsburgh, PA, in 1969, and the M.S. and Ph.D. degrees in biomedical engineering from Drexel University, Philadelphia, PA, in 1972 and 1975, respectively.

From 1974 to 1976 he was a Research Associate at the Picture Processing Laboratory, Computer Science Center, University of Maryland, College Park. He is currently an Assistant Professor in the Department of Electrical Engineering, McGill University, Montreal, P.Q., Canada. His research interests include image processing, computer vision, and artificial intelligence.

Dr. Zucker is a member of Sigma Xi and the Association for Computing Machinery.

Optimization by Simulated Annealing

S. Kirkpatrick, C. D. Gelatt, Jr., M. P. Vecchi

Summary. There is a deep and useful connection between statistical mechanics (the behavior of systems with many degrees of freedom in thermal equilibrium at a finite temperature) and multivariate or combinatorial optimization (finding the minimum of a given function depending on many parameters). A detailed analogy with annealing in solids provides a framework for optimization of the properties of very large and complex systems. This connection to statistical mechanics exposes new information and provides an unfamiliar perspective on traditional optimization problems and methods.

In this article we briefly review the central constructs in combinatorial optimization and in statistical mechanics and then develop the similarities between the two fields. We show how the Metropolis algorithm for approximate numerical simulation of the behavior of a many-body system at a finite temperature provides a natural tool for bringing the techniques of statistical mechanics to bear on optimization.

We have applied this point of view to a number of problems arising in optimal design of computers. Applications to partitioning, component placement, and wiring of electronic systems are described in this article. In each context, we introduce the problem and discuss the improvements available from optimization.

Of classic optimization problems, the traveling salesman problem has received the most intensive study. To test the power of simulated annealing, we used the algorithm on traveling salesman problems with as many as several thousand cities. This work is described in a final section, followed by our conclusions.

Combinatorial Optimization

The subject of combinatorial optimization (*1*) consists of a set of problems that are central to the disciplines of computer science and engineering. Research in this area aims at developing efficient techniques for finding minimum or maximum values of a function of very many independent variables (*2*). This function, usually called the cost function or objective function, represents a quantitative measure of the "goodness" of some complex system. The cost function depends on the detailed configuration of the many parts of that system. We are most familiar with optimization problems occurring in the physical design of computers, so examples used below are drawn from that context. The number of variables involved may range up into the tens of thousands.

The classic example, because it is so simply stated, of a combinatorial optimization problem is the traveling salesman problem. Given a list of N cities and a means of calculating the cost of traveling between any two cities, one must plan the salesman's route, which will pass through each city once and return finally to the starting point, minimizing the total cost. Problems with this flavor arise in all areas of scheduling and design. Two subsidiary problems are of general interest: predicting the expected cost of the salesman's optimal route, averaged over some class of typical arrangements of cities, and estimating or obtaining bounds for the computing effort necessary to determine that route.

All exact methods known for determining an optimal route require a computing effort that increases exponentially with N, so that in practice exact solutions can be attempted only on problems involving a few hundred cities or less. The traveling salesman belongs to the large class of NP-complete (nondeterministic polynomial time complete) problems, which has received extensive study in the past 10 years (*3*). No method for exact solution with a computing effort bounded by a power of N has been found for any of these problems, but if such a solution were found, it could be mapped into a procedure for solving all members of the class. It is not known what features of the individual problems in the NP-complete class are the cause of their difficulty.

Since the NP-complete class of problems contains many situations of practical interest, heuristic methods have been developed with computational requirements proportional to small powers of N. Heuristics are rather problem-specific: there is no guarantee that a heuristic procedure for finding near-optimal solutions for one NP-complete problem will be effective for another.

There are two basic strategies for heuristics: "divide-and-conquer" and iterative improvement. In the first, one divides the problem into subproblems of manageable size, then solves the subproblems. The solutions to the subproblems must then be patched back together. For this method to produce very good solutions, the subproblems must be naturally disjoint, and the division made must be an appropriate one, so that errors made in patching do not offset the gains

S. Kirkpatrick and C. D. Gelatt, Jr., are research staff members and M. P. Vecchi was a visiting scientist at IBM Thomas J. Watson Research Center, Yorktown Heights, New York 10598. M. P. Vecchi's present address is Instituto Venezolano de Investigaciones Cientificas, Caracas 1010A, Venezuela.

obtained in applying more powerful methods to the subproblems (4).

In iterative improvement (5, 6), one starts with the system in a known configuration. A standard rearrangement operation is applied to all parts of the system in turn, until a rearranged configuration that improves the cost function is discovered. The rearranged configuration then becomes the new configuration of the system, and the process is continued until no further improvements can be found. Iterative improvement consists of a search in this coordinate space for rearrangement steps which lead downhill. Since this search usually gets stuck in a local but not a global optimum, it is customary to carry out the process several times, starting from different randomly generated configurations, and save the best result.

There is a body of literature analyzing the results to be expected and the computing requirements of common heuristic methods when applied to the most popular problems (1–3). This analysis usually focuses on the worst-case situation—for instance, attempts to bound from above the ratio between the cost obtained by a heuristic method and the exact minimum cost for any member of a family of similarly structured problems. There are relatively few discussions of the average performance of heuristic algorithms, because the analysis is usually more difficult and the nature of the appropriate average to study is not always clear. We will argue that as the size of optimization problems increases, the worst-case analysis of a problem will become increasingly irrelevant, and the average performance of algorithms will dominate the analysis of practical applications. This large number limit is the domain of statistical mechanics.

Statistical Mechanics

Statistical mechanics is the central discipline of condensed matter physics, a body of methods for analyzing aggregate properties of the large numbers of atoms to be found in samples of liquid or solid matter (7). Because the number of atoms is of order 10^{23} per cubic centimeter, only the most probable behavior of the system in thermal equilibrium at a given temperature is observed in experiments. This can be characterized by the average and small fluctuations about the average behavior of the system, when the average is taken over the ensemble of identical systems introduced by Gibbs. In this ensemble, each configuration, defined by the set of atomic positions, $\{r_i\}$, of the system is weighted by its Boltzmann probability factor, $\exp(-E(\{r_i\})/k_BT)$, where $E(\{r_i\})$ is the energy of the configuration, k_B is Boltzmann's constant, and T is temperature.

A fundamental question in statistical mechanics concerns what happens to the system in the limit of low temperature—for example, whether the atoms remain fluid or solidify, and if they solidify, whether they form a crystalline solid or a glass. Ground states and configurations close to them in energy are extremely rare among all the configurations of a macroscopic body, yet they dominate its properties at low temperatures because as T is lowered the Boltzmann distribution collapses into the lowest energy state or states.

As a simplified example, consider the magnetic properties of a chain of atoms whose magnetic moments, μ_i, are allowed to point only "up" or "down," states denoted by $\mu_i = \pm 1$. The interaction energy between two such adjacent spins can be written $J\mu_i\mu_{i+1}$. Interaction between each adjacent pair of spins contributes $\pm J$ to the total energy of the chain. For an N-spin chain, if all configurations are equally likely the interaction energy has a binomial distribution, with the maximum and minimum energies given by $\pm NJ$ and the most probable state having zero energy. In this view, the ground state configurations have statistical weight $\exp(-N/2)$ smaller than the zero-energy configurations. A Boltzmann factor, $\exp(-E/k_BT)$, can offset this if k_BT is smaller than J. If we focus on the problem of finding empirically the system's ground state, this factor is seen to drastically increase the efficiency of such a search.

In practical contexts, low temperature is not a sufficient condition for finding ground states of matter. Experiments that determine the low-temperature state of a material—for example, by growing a single crystal from a melt—are done by careful annealing, first melting the substance, then lowering the temperature slowly, and spending a long time at temperatures in the vicinity of the freezing point. If this is not done, and the substance is allowed to get out of equilibrium, the resulting crystal will have many defects, or the substance may form a glass, with no crystalline order and only metastable, locally optimal structures.

Finding the low-temperature state of a system when a prescription for calculating its energy is given is an optimization problem not unlike those encountered in combinatorial optimization. However, the concept of the temperature of a physical system has no obvious equivalent in the systems being optimized. We will introduce an effective temperature for optimization, and show how one can carry out a simulated annealing process in order to obtain better heuristic solutions to combinatorial optimization problems.

Iterative improvement, commonly applied to such problems, is much like the microscopic rearrangement processes modeled by statistical mechanics, with the cost function playing the role of energy. However, accepting only rearrangements that lower the cost function of the system is like extremely rapid quenching from high temperatures to $T = 0$, so it should not be surprising that resulting solutions are usually metastable. The Metropolis procedure from statistical mechanics provides a generalization of iterative improvement in which controlled uphill steps can also be incorporated in the search for a better solution.

Metropolis *et al.* (8), in the earliest days of scientific computing, introduced a simple algorithm that can be used to provide an efficient simulation of a collection of atoms in equilibrium at a given temperature. In each step of this algorithm, an atom is given a small random displacement and the resulting change, ΔE, in the energy of the system is computed. If $\Delta E \leq 0$, the displacement is accepted, and the configuration with the displaced atom is used as the starting point of the next step. The case $\Delta E > 0$ is treated probabilistically: the probability that the configuration is accepted is $P(\Delta E) = \exp(-\Delta E/k_BT)$. Random numbers uniformly distributed in the interval (0,1) are a convenient means of implementing the random part of the algorithm. One such number is selected and compared with $P(\Delta E)$. If it is less than $P(\Delta E)$, the new configuration is retained; if not, the original configuration is used to start the next step. By repeating the basic step many times, one simulates the thermal motion of atoms in thermal contact with a heat bath at temperature T. This choice of $P(\Delta E)$ has the consequence that the system evolves into a Boltzmann distribution.

Using the cost function in place of the energy and defining configurations by a set of parameters $\{x_i\}$, it is straightforward with the Metropolis procedure to generate a population of configurations of a given optimization problem at some effective temperature. This temperature is simply a control parameter in the same units as the cost function. The simulated annealing process consists of first "melting" the system being optimized at a high effective temperature, then lowering the temperature by slow stages until the system "freezes" and no further changes occur. At each temperature, the simulation must proceed long enough for the system to reach a steady state. The sequence of temperatures and the number of rearrangements of the $\{x_i\}$ attempted to reach equilibrium at each temperature can be considered an annealing schedule.

Annealing, as implemented by the Metropolis procedure, differs from iterative improvement in that the procedure need not get stuck since transitions out of a local optimum are always possible at nonzero temperature. A second and more important feature is that a sort of adaptive divide-and-conquer occurs. Gross features of the eventual state of the system appear at higher temperatures; fine details develop at lower temperatures. This will be discussed with specific examples.

Statistical mechanics contains many useful tricks for extracting properties of a macroscopic system from microscopic averages. Ensemble averages can be obtained from a single generating function, the partition function, Z,

$$Z = \mathrm{Tr}\exp\left(\frac{-E}{k_B T}\right) \quad (1)$$

in which the trace symbol, Tr, denotes a sum over all possible configurations of the atoms in the sample system. The logarithm of Z, called the free energy, $F(T)$, contains information about the average energy, $<E(T)>$, and also the entropy, $S(T)$, which is the logarithm of the number of configurations contributing to the ensemble at T:

$$-k_B T \ln Z = F(T) = <E(T)> - TS \quad (2)$$

Boltzmann-weighted ensemble averages are easily expressed in terms of derivatives of F. Thus the average energy is given by

$$<E(T)> = \frac{-d\ln Z}{d(1/k_B T)} \quad (3)$$

and the rate of change of the energy with respect to the control parameter, T, is related to the size of typical variations in the energy by

$$C(T) = \frac{d <E(T)>}{dT} = \frac{[<E(T)^2> - <E(T)>^2]}{k_B T^2} \quad (4)$$

In statistical mechanics $C(T)$ is called the specific heat. A large value of C signals a change in the state of order of a system, and can be used in the optimization context to indicate that freezing has begun and hence that very slow cooling is required. It can also be used to determine the entropy by the thermodynamic relation

$$\frac{dS(T)}{dT} = \frac{C(T)}{T} \quad (5)$$

Integrating Eq. 5 gives

$$S(T) = S(T_1) - \int_T^{T_1} \frac{C(T')\,dT}{T} \quad (6)$$

where T_1 is a temperature at which S is known, usually by an approximation valid at high temperatures.

The analogy between cooling a fluid and optimization may fail in one important respect. In ideal fluids all the atoms are alike and the ground state is a regular crystal. A typical optimization problem will contain many distinct, noninterchangeable elements, so a regular solution is unlikely. However, much research in condensed matter physics is directed at systems with quenched-in randomness, in which the atoms are not all alike. An important feature of such systems, termed "frustration," is that interactions favoring different and incompatible kinds of ordering may be simultaneously present (*9*). The magnetic alloys known as "spin glasses," which exhibit competition between ferromagnetic and antiferromagnetic spin ordering, are the best understood example of frustration (*10*). It is now believed that highly frustrated systems like spin glasses have many nearly degenerate random ground states rather than a single ground state with a high degree of symmetry. These systems stand in the same relation to conventional magnets as glasses do to crystals, hence the name.

The physical properties of spin glasses at low temperatures provide a possible guide for understanding the possibilities of optimizing complex systems subject to conflicting (frustrating) constraints.

Physical Design of Computers

The physical design of electronic systems and the methods and simplifications employed to automate this process have been reviewed (*11, 12*). We first provide some background and definitions related to applications of the simulated annealing framework to specific problems that arise in optimal design of computer systems and subsystems. Physical design follows logical design. After the detailed specification of the logic of a system is complete, it is necessary to specify the precise physical realization of the system in a particular technology.

This process is usually divided into several stages. First, the design must be partitioned into groups small enough to fit the available packages, for example, into groups of circuits small enough to fit into a single chip, or into groups of chips and associated discrete components that can fit onto a card or other higher level package. Second, the circuits are assigned specific locations on the chip. This stage is usually called placement. Finally, the circuits are connected by wires formed photolithographically out of a thin metal film, often in several layers. Assigning paths, or routes, to the wires is usually done in two stages. In rough or global wiring, the wires are assigned to regions that represent schematically the capacity of the intended package. In detailed wiring (also called exact embedding), each wire is given a unique complete path. From the detailed wiring results, masks can be generated and chips made.

At each stage of design one wants to optimize the eventual performance of the system without compromising the feasibility of the subsequent design stages. Thus partitioning must be done in such a way that the number of circuits in each partition is small enough to fit easily into the available package, yet the number of signals that must cross partition boundaries (each requiring slow, power-consuming driver circuitry) is minimized. The major focus in placement is on minimizing the length of connections, since this translates into the time required for propagation of signals, and thus into the speed of the finished system. However, the placements with the shortest implied wire lengths may not be wirable, because of the presence of regions in which the wiring is too congested for the packaging technology. Congestion, therefore, should also be anticipated and minimized during the placement process. In wiring, it is desirable to maintain the minimum possible wire lengths while minimizing sources of noise, such as cross talk between adjacent wires. We show in this and the next two sections how these conflicting goals can be combined and made the basis of an automatic optimization procedure.

The tight schedules involved present major obstacles to automation and optimization of large system design, even when computers are employed to speed up the mechanical tasks and reduce the chance of error. Possibilities of feedback, in which early stages of a design are redone to solve problems that became apparent only at later stages, are greatly reduced as the scale of the overall system being designed increases. Op-

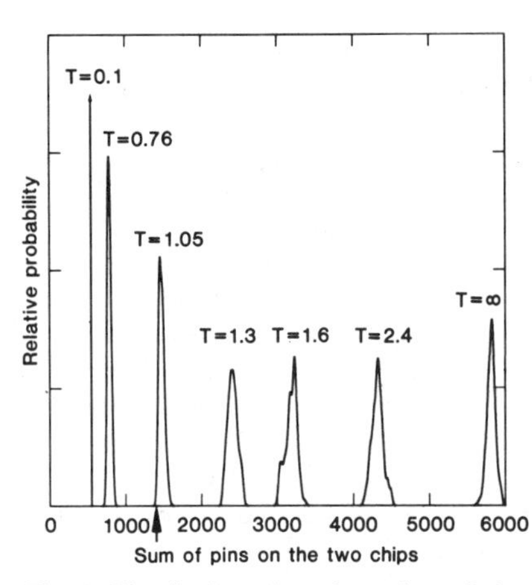

Fig. 1. Distribution of total number of pins required in two-way partition of a microprocessor at various temperatures. Arrow indicates best solution obtained by rapid quenching as opposed to annealing.

timization procedures that can incorporate, even approximately, information about the chance of success of later stages of such complex designs will be increasingly valuable in the limit of very large scale.

System performance is almost always achieved at the expense of design convenience. The partitioning problem provides a clean example of this. Consider N circuits that are to be partitioned between two chips. Propagating a signal across a chip boundary is always slow, so the number of signals required to cross between the two must be minimized. Putting all the circuits on one chip eliminates signal crossings, but usually there is no room. Instead, for later convenience, it is desirable to divide the circuits about equally.

If we have connectivity information in a matrix whose elements $\{a_{ij}\}$ are the number of signals passing between circuits i and j, and we indicate which chip circuit i is placed on by a two-valued variable $\mu_i = \pm 1$, then N_c, the number of signals that must cross a chip boundary is given by $\Sigma_{i>j}(a_{ij}/4)(\mu_i - \mu_j)^2$. Calculating $\Sigma_i \mu_i$ gives the difference between the numbers of circuits on the two chips. Squaring this imbalance and introducing a coefficient, λ, to express the relative costs of imbalance and boundary crossings, we obtain an objective function, f, for the partition problem:

$$f = \sum_{i>j} \left(\lambda - \frac{a_{ij}}{2}\right)\mu_i\mu_j \qquad (7)$$

Reasonable values of λ should satisfy $\lambda \lesssim z/2$, where z is the average number of circuits connected to a typical circuit (fan-in plus fan-out). Choosing $\lambda \simeq z/2$ implies giving equal weight to changes in the balance and crossing scores.

The objective function f has precisely the form of a Hamiltonian, or energy function, studied in the theory of random magnets, when the common simplifying assumption is made that the spins, μ_i, have only two allowed orientations (up or down), as in the linear chain example of the previous section. It combines local, random, attractive ("ferromagnetic") interactions, resulting from the a_{ij}'s, with a long-range repulsive ("antiferromagnetic") interaction due to λ. No configuration of the $\{\mu_i\}$ can simultaneously satisfy all the interactions, so the system is "frustrated," in the sense formalized by Toulouse (*9*).

If the a_{ij} are completely uncorrelated, it can be shown (*13*) that this Hamiltonian has a spin glass phase at low temperatures. This implies for the associated magnetic problem that there are many degenerate "ground states" of nearly equal energy and no obvious symmetry. The magnetic state of a spin glass is very stable at low temperatures (*14*), so the ground states have energies well below the energies of the random high-temperature states, and transforming one ground state into another will usually require considerable rearrangement. Thus this analogy has several implications for optimization of partition:

1) Even in the presence of frustration, significant improvements over a random starting partition are possible.

2) There will be many good near-optimal solutions, so a stochastic search procedure such as simulated annealing should find some.

3) No one of the ground states is significantly better than the others, so it is not very fruitful to search for the absolute optimum.

In developing Eq. 7 we made several severe simplifications, considering only two-way partitioning and ignoring the fact that most signals connect more than two circuits. Objective functions analogous to f that include both complications are easily constructed. They no longer have the simple quadratic form of Eq. 7, but the qualitative feature, frustration, remains dominant. The form of the Hamiltonian makes no difference in the Metropolis Monte Carlo algorithm. Evaluation of the change in function when a circuit is shifted to a new chip remains rapid as the definition of f becomes more complicated.

It is likely that the a_{ij} are somewhat correlated, since any design has considerable logical structure. Efforts to understand the nature of this structure by analyzing the surface-to-volume ratio of components of electronic systems [as in "Rent's rule" (*15*)] conclude that the

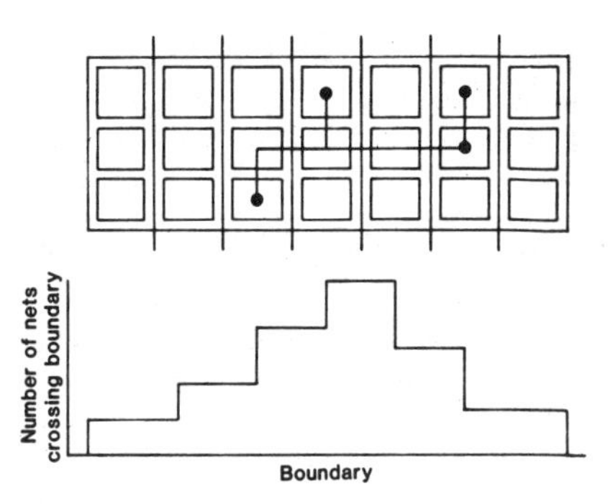

Fig. 2. Construction of a horizontal net-crossing histogram.

circuits in a typical system could be connected with short-range interactions if they were embedded in a space with dimension between two and three. Uncorrelated connections, by contrast, can be thought of as infinite-dimensional, since they are never short-range.

The identification of Eq. 7 as a spin glass Hamiltonian is not affected by the reduction to a two- or three-dimensional problem, as long as $\lambda N \simeq z/2$. The degree of ground state degeneracy increases with decreasing dimensionality. For the uncorrelated model, there are typically of order $N^{1/2}$ nearly degenerate ground states (*14*), while in two and three dimensions, $2^{\alpha N}$, for some small value, α, are expected (*16*). This implies that finding a near-optimum solution should become easier, the lower the effective dimensionality of the problem. The entropy, measurable as shown in Eq. 6, provides a measure of the degeneracy of solutions. $S(T)$ is the logarithm of the number of solutions equal to or better than the average result encountered at temperature T.

As an example of the partitioning problem, we have taken the logic design for a single-chip IBM "370 microprocessor" (*17*) and considered partitioning it into two chips. The original design has approximately 5000 primitive logic gates and 200 external signals (the chip has 200 logic pins). The results of this study are plotted in Fig. 1. If one randomly assigns gates to the two chips, one finds the distribution marked $T = \infty$ for the number of pins required. Each of the two chips (with about 2500 circuits) would need 3000 pins. The other distributions in Fig. 1 show the results of simulated annealing.

Monte Carlo annealing is simple to implement in this case. Each proposed configuration change simply flips a randomly chosen circuit from one chip to the other. The new number of external connections, C, to the two chips is calculated (an external connection is a net with circuits on both chips, or a circuit connected to one of the pins of the original single-chip design), as is the new balance score, B, calculated as in deriving Eq. 7. The objective function analogous to Eq. 7 is

$$f = C + \lambda B \quad (8)$$

where C is the sum of the number of external connections on the two chips and B is the balance score. For this example, $\lambda = 0.01$.

For the annealing schedule we chose to start at a high "temperature," $T_0 = 10$, where essentially all proposed circuit flips are accepted, then cool exponentially, $T_n = (T_1/T_0)^n T_0$, with the ratio $T_1/T_0 = 0.9$. At each temperature enough flips are attempted that either there are ten accepted flips per circuit on the average (for this case, 50,000 accepted flips at each temperature), or the number of attempts exceeds 100 times the number of circuits before ten flips per circuit have been accepted. If the desired number of acceptances is not achieved at three successive temperatures, the system is considered "frozen" and annealing stops.

The finite temperature curves in Fig. 1 show the distribution of pins per chip for the configurations sampled at $T = 2.5$, 1.0, and 0.1. As one would expect from the statistical mechanical analog, the distribution shifts to fewer pins and sharpens as the temperature is decreased. The sharpening is one consequence of the decrease in the number of configurations that contribute to the equilibrium ensemble at the lower temperature. In the language of statistical mechanics, the entropy of the system decreases. For this sample run in the low-temperature limit, the two chips required 353 and 321 pins, respectively. There are 237 nets connecting the two chips (requiring a pin on each chip) in addition to the 200 inputs and outputs of the original chip. The final partition in this example has the circuits exactly evenly distributed between the two partitions. Using a more complicated balance score, which did not penalize imbalance of less than 100 circuits, we found partitions resulting in chips with 271 and 183 pins.

If, instead of slowly cooling, one were to start from a random partition and accept only flips that reduce the objective function (equivalent to setting $T = 0$ in the Metropolis rule), the result is chips with approximately 700 pins (several such runs led to results with 677 to 730 pins). Rapid cooling results in a system frozen into a metastable state far from the optimal configuration. The best result obtained after several rapid quenches is indicated by the arrow in Fig. 1.

Placement

Placement is a further refinement of the logic partitioning process, in which the circuits are given physical positions (*11, 12, 18, 19*). In principle, the two stages could be combined, although this is not often possible in practice. The objectives in placement are to minimize

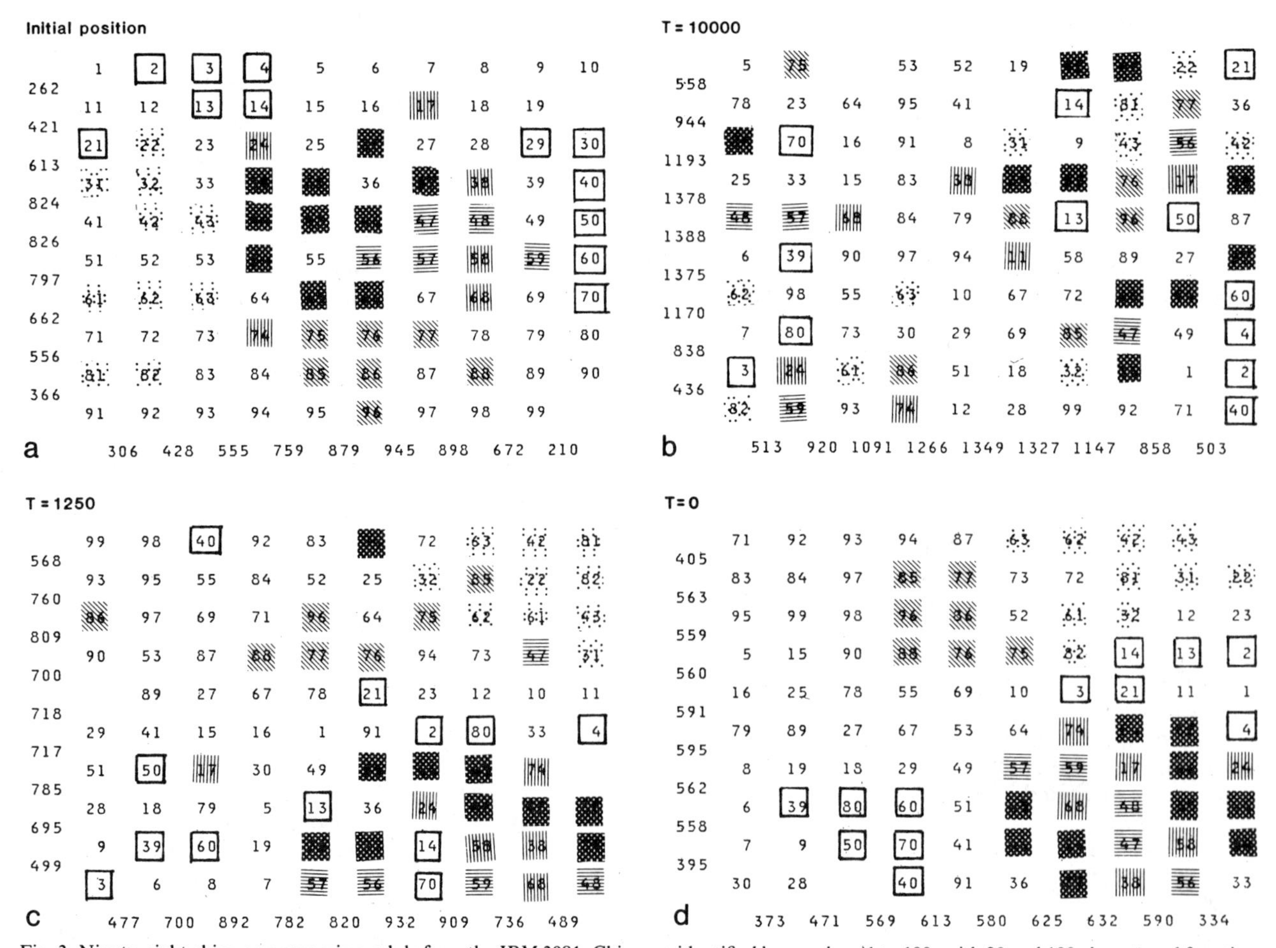

Fig. 3. Ninety-eight chips on a ceramic module from the IBM 3081. Chips are identified by number (1 to 100, with 20 and 100 absent) and function. The dark squares comprise an adder, the three types of squares with ruled lines are chips that control and supply data to the adder, the lightly dotted chips perform logical arithmetic (bitwise AND, OR, and so on), and the open squares denote general-purpose registers, which serve both arithmetic units. The numbers at the left and lower edges of the module image are the vertical and horizontal net-crossing histograms, respectively. (a) Original chip placement; (b) a configuration at $T = 10{,}000$; (c) $T = 1250$; (d) a zero-temperature result.

signal propagation times or distances while satisfying prescribed electrical constraints, without creating regions so congested that there will not be room later to connect the circuits with actual wire.

Physical design of computers includes several distinct categories of placement problems, depending on the packages involved (*20*). The larger objects to be placed include chips that must reside in a higher level package, such as a printed circuit card or fired ceramic "module" (*21*). These chip carriers must in turn be placed on a backplane or "board," which is simply a very large printed circuit card. The chips seen today contain from tens to tens of thousands of logic circuits, and each chip carrier or board will provide from one to ten thousand interconnections. The partition and placement problems decouple poorly in this situation, since the choice of which chip should carry a given piece of logic will be influenced by the position of that chip.

The simplest placement problems arise in designing chips with structured layout rules. These are called "gate array" or "master slice" chips. In these chips, standard logic circuits, such as three- or four-input NOR's, are preplaced in a regular grid arrangement, and the designer specifies only the signal wiring, which occupies the final, highest, layers of the chip. The circuits may all be identical, or they may be described in terms of a few standard groupings of two or more adjacent cells.

As an example of a placement problem with realistic complexity without too many complications arising from package idiosyncrasies, we consider 98 chips packaged on one multilayer ceramic module of the IBM 3081 processor (*21*). Each chip can be placed on any of 100 sites, in a 10 × 10 grid on the top surface of the module. Information about the connections to be made through the signal-carrying planes of the module is contained in a "netlist," which groups sets of pins that see the same signal.

The state of the system can be briefly represented by a list of the 98 chips with their x and y coordinates, or a list of the contents of each of the 100 legal locations. A sufficient set of moves to use for annealing is interchanges of the contents of two locations. This results in either the interchange of two chips or the interchange of a chip and a vacancy. For more efficient search at low temperatures, it is helpful to allow restrictions on the distance across which an interchange may occur.

To measure congestion at the same time as wire length, we use a convenient intermediate analysis of the layout, a net-crossing histogram. Its construction is summarized in Fig. 2. We divide the package surface by a set of natural boundaries. In this example, we use the boundaries between adjacent rows or columns of chip sites. The histogram then contains the number of nets crossing each boundary. Since at least one wire must be routed across each boundary crossed, the sum of the entries in the histogram of Fig. 2 is the sum of the horizontal extents of the rectangles bounding each net, and is a lower bound to the horizontal wire length required. Constructing a vertical net-crossing histogram and summing its entries gives a similar estimate of the vertical wire length.

The peak of the histogram provides a lower bound to the amount of wire that must be provided in the worst case, since each net requires at least one wiring channel somewhere on the boundary. To combine this information into a single objective function, we introduce a threshold level for each histogram—an amount of wire that will nearly exhaust the available wire capacity—and then sum for all histogram elements that exceed the threshold the square of the excess over threshold. Adding this quantity to the estimated length gives the objective function that was used.

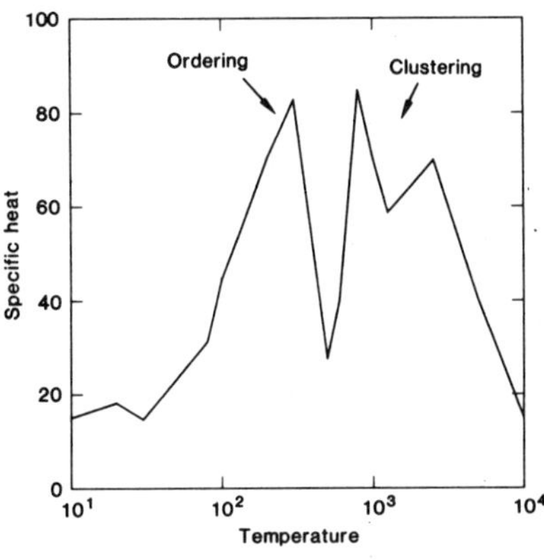

Fig. 4. Specific heat as a function of temperature for the design of Fig. 3, a to d.

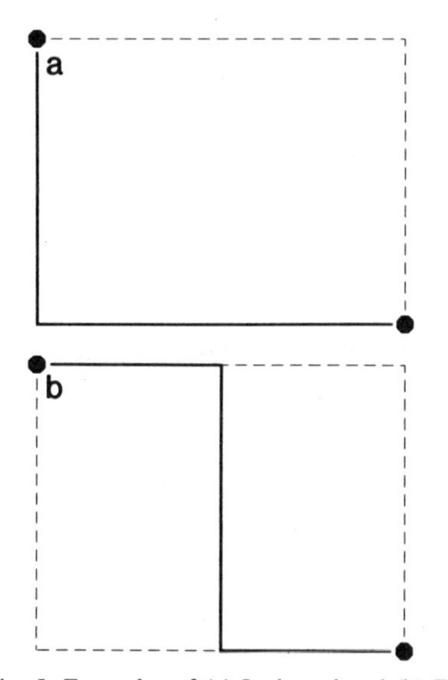

Fig. 5. Examples of (a) L-shaped and (b) Z-shaped wire rearrangements.

Figure 3 shows the stages of a simulated annealing run on the 98-chip module. Figure 3a shows the chip locations from the original design, with vertical and horizontal net-crossing histograms indicated. The different shading patterns distinguish the groups of chips that carry out different functions. Each such group was designed and placed together, usually by a single designer. The net-crossing histograms show that the center of the layout is much more congested than the edges, most likely because the chips known to have the most critical timing constraints were placed in the center of the module to allow the greatest number of other chips to be close to them.

Heating the original design until the chips diffuse about freely quickly produces a random-looking arrangement, Fig. 3b. Cooling very slowly until the chips move sluggishly and the objective function ceases to decrease rapidly with change of temperature produced the result in Fig. 3c. The net-crossing histograms have peaks comparable to the peak heights in the original placement, but are much flatter. At this "freezing point," we find that the functionally related groups of chips have reorganized from the melt, but now are spatially separated in an overall arrangement quite different from the original placement. In the final result, Fig. 3d, the histogram peaks are about 30 percent less than in the original placement. Integrating them, we find that total wire length, estimated in this way, is decreased by about 10 percent. The computing requirements for this example were modest: 250,000 interchanges were attempted, requiring 12 minutes of computation on an IBM 3033.

Between the temperature at which clusters form and freezing starts (Fig. 3c) and the final result (Fig. 3d) there are many further local rearrangements. The functional groups have remained in the same regions, but their shapes and relative alignments continue to change throughout the low-temperature part of the annealing process. This illustrates that the introduction of temperature to the optimization process permits a controlled, adaptive division of the problem through the evolution of natural clusters at the freezing temperature. Early prescription of natural clusters is also a central feature of several sophisticated placement programs used in master slice chip placement (*22, 23*).

A quantity corresponding to the thermodynamic specific heat is defined for this problem by taking the derivative with respect to temperature of the average value of the objective function observed at a given temperature. This is plotted in Fig. 4. Just as a maximum in the specific heat of a fluid indicates the onset of freezing or the formation of clusters, we find specific heat maxima at two temperatures, each indicating a different type of ordering in the problem. The higher temperature peak corresponds to the aggregation of clusters of functionally related objects, driven apart by the congestion term in the scoring. The lower temperature peak indicates the further decrease in wire length obtained by local rearrangements. This sort of measurement can be useful in practice as a means of determining the temperature ranges in which the important rearrangements in the design are occurring, where slower cooling with be helpful.

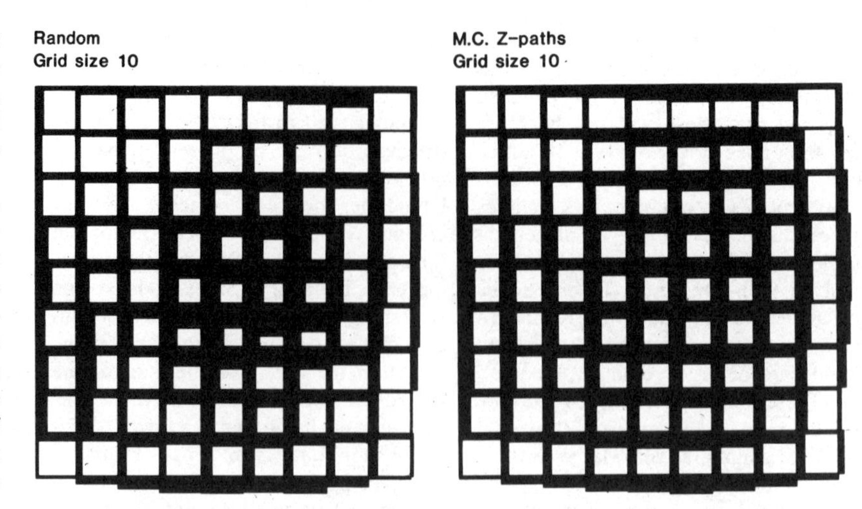

Fig. 6 (left). Wire density in the 98-chip module with the connections randomly assigned to perimeter routes. Chips are in the original placement. Fig. 7 (right). Wire density after simulated annealing of the wire routing, using Z-shaped moves.

Wiring

After placement, specific legal routings must be found for the wires needed to connect the circuits. The techniques typically applied to generate such routings are sequential in nature, treating one wire at a time with incomplete information about the positions and effects of the other wires (*11, 24*). Annealing is inherently free of this sequence dependence. In this section we describe a simulated annealing approach to wiring, using the ceramic module of the last section as an example.

Nets with many pins must first be broken into connections—pairs of pins joined by a single continuous wire. This "ordering" of each net is highly dependent on the nature of the circuits being connected and the package technology. Orderings permitting more than two pins to be connected are sometimes allowed, but will not be discussed here.

The usual procedure, given an ordering, is first to construct a coarse-scale routing for each connection from which the ultimate detailed wiring can be completed. Package technologies and structured image chips have prearranged areas of fixed capacity for the wires. For the rough routing to be successful, it must not call for wire densities that exceed this capacity.

We can model the rough routing problem (and even simple cases of detailed embedding) by lumping all actual pin positions into a regular grid of points, which are treated as the sources and sinks of all connections. The wires are then to be routed along the links that connect adjacent grid points.

The objectives in global routing are to minimize wire length and, often, the number of bends in wires, while spreading the wire as evenly as possible to simplify exact embedding and later revision. Wires are to be routed around regions in which wire demand exceeds capacity if possible, so that they will not "overflow," requiring drastic rearrangements of the other wires during exact embedding. Wire bends are costly in packages that confine the north-south and east-west wires to different layers, since each bend requires a connection between two layers. Two classes of moves that maintain the minimum wire length are shown in Fig. 5. In the L-shaped move of Fig. 5a, only the essential bends are permitted, while the Z-shaped move of Fig. 5b introduces one extra bend. We will explore the optimization possible with these two moves.

For a simple objective function that will reward the most balanced arrangement of wire, we calculate the square of the number of wires on each link of the network, sum the squares for all links, and term the result F. If there are N_L links and N_W wires, a global routing program that deals with a high density of wires will attempt to route precisely the average number of wires, N_W/N_L, along each link. In this limit F is bounded below by N_W^2/N_L. One can use the same objective function for a low-density (or high-resolution) limit appropriate for detailed wiring. In that case, all the links have either one or no wires, and links with two or more wires are illegal. For this limit the best possible value of F will be N_W/N_L.

For the L-shaped moves, F has a relatively simple form. Let $\epsilon_{i\nu} = +1$ along the links that connection i has for one orientation, -1 for the other orientation, and 0 otherwise. Let $a_{i\nu}$ be 1 if the ith connection can run through the νth link in either of its two positions, and 0 otherwise. Note that $a_{i\nu}$ is just $\epsilon_{i\nu}^2$. Then if $\mu_i = \pm 1$ indicates which route the ith connection has taken, we obtain for the number of wires along the νth link,

$$n_\nu = \sum_i \frac{a_{i\nu}(\epsilon_{i\nu}\mu_i + 1)}{2} + n_\nu(0) \qquad (9)$$

where $n_\nu(0)$ is the contribution from straight wires, which cannot move without increasing their length, or blockages.

Summing the n_ν^2 gives

$$F = \sum_{i,j} J_{ij}\mu_i\mu_j + \sum_i h_i\mu_i + \text{constants} \qquad (10)$$

which has the form of the Hamiltonian for a random magnetic alloy or spin glass, like that discussed earlier. The "random field," h_i, felt by each movable connection reflects the difference, on the average, between the congestion associated with the two possible paths:

$$h_i = \sum_\nu \epsilon_{i\nu}\,[2n_\nu(0) + \sum_j a_{j\nu}] \qquad (11)$$

The interaction between two wires is proportional to the number of links on which the two nets can overlap, its sign depending on their orientation conventions:

$$J_{ij} = \sum_\nu \frac{\epsilon_{i\nu}\epsilon_{j\nu}}{4} \qquad (12)$$

Both J_{ij} and h_i vanish, on average, so it is the fluctuations in the terms that make up F which will control the nature of the low-energy states. This is also true in spin glasses. We have not tried to exhibit a functional form for the objective function with Z-moves allowed, but simply calculate it by first constructing the actual amounts of wire found along each link.

To assess the value of annealing in wiring this model, we studied an ensemble of randomly situated connections, under various statistical assumptions. Here we consider routing wires for the 98 chips on a module considered earlier. First, we show in Fig. 6 the arrangement of wire that results from assigning each wire to an L-shaped path, choosing orientations at random. The thickness of the links is proportional to the number of wires on each link. The congested area that gave rise to the peaks in the histograms discussed above is seen in the wiring just below and to the right of the center of the module. The maximum numbers of wires along a single link in Fig. 6 are 173 (x direction) and 143 (y direction), so the design is also anisotropic. Various ways of rearranging the wiring paths were studied. Monte Carlo annealing with Z-moves gave the best solution, shown in Fig. 7. In this example, the largest numbers of wires on a single link are 105 (x) and 96 (y).

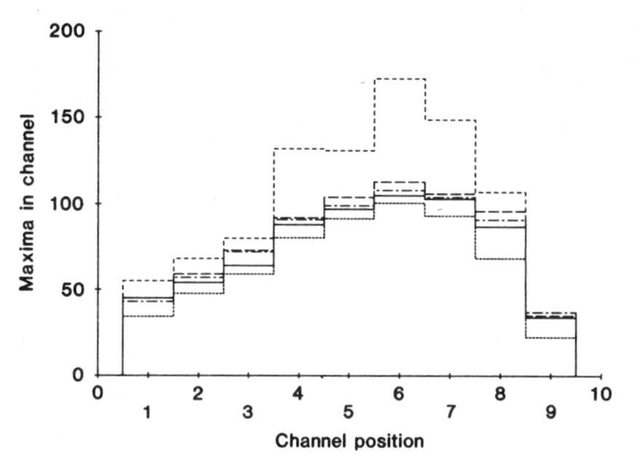

Fig. 8. Histogram of the maximum wire densities within a given column of x-links, for the various methods of routing.

We compare the various methods of improving the wire arrangement by plotting (Fig. 8) the highest wire density found in each column of x-links for each of the methods. The unevenness of the density profiles was already seen when we considered net-crossing histograms as input information to direct placement. The lines shown represent random assignment of wires with L-moves; aligning wires in the direction of least average congestion—that is, along h_i—followed by cooling for one pass at zero T; simulated annealing with L-moves only; and annealing with Z-moves. Finally, the light dashed line shows the optimum result, in which the wires are distributed with all links carrying as close to the average weight as possible. The optimum cannot be attained in this example without stretching wires beyond their minimum length, because the connections are too unevenly arranged. Any method of optimization gives a significant improvement over the estimate obtained by assigning wire routings at random. All reduce the peak wire density on a link by more than 45 percent. Simulated annealing with Z-moves improved the random routing by 57 percent, averaging results for both x and y links.

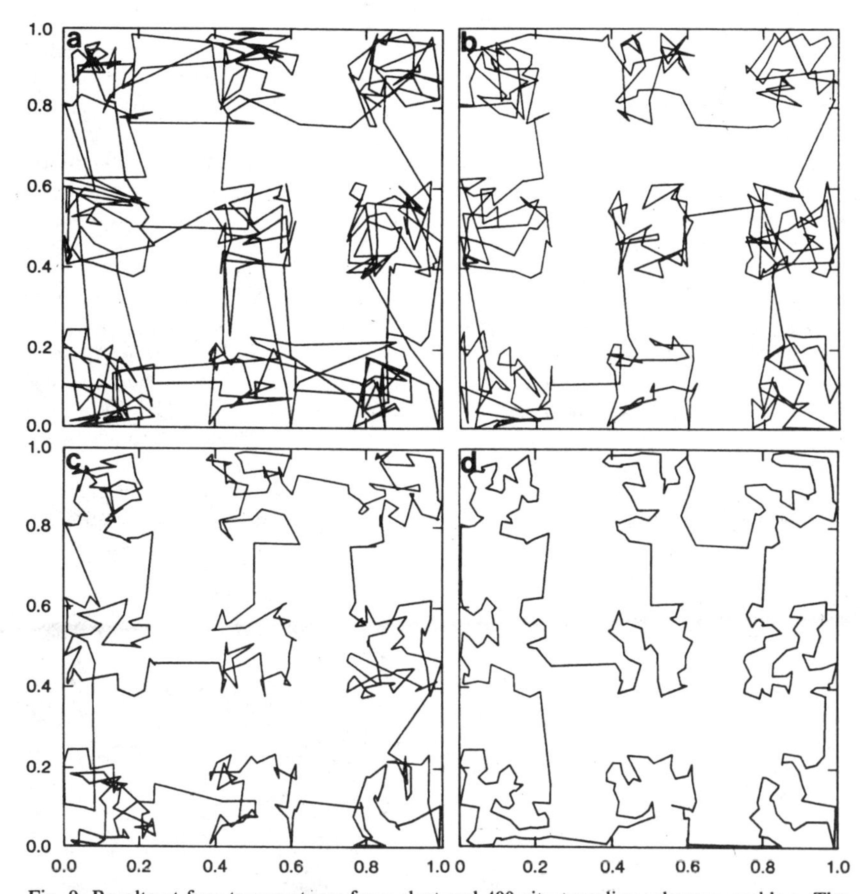

Fig. 9. Results at four temperatures for a clustered 400-city traveling salesman problem. The points are uniformly distributed in nine regions. (a) $T = 1.2$, $\alpha = 2.0567$; (b) $T = 0.8$, $\alpha = 1.515$; (c) $T = 0.4$, $\alpha = 1.055$; (d) $T = 0.0$, $\alpha = 0.7839$.

Traveling Salesmen

Quantitative analysis of the simulated annealing algorithm or comparison between it and other heuristics requires problems simpler than physical design of computers. There is an extensive literature on algorithms for the traveling salesman problem (*3, 4*), so it provides a natural context for this discussion.

If the cost of travel between two cities is proportional to the distance between them, then each instance of a traveling salesman problem is simply a list of the positions of N cities. For example, an arrangement of N points positioned at random in a square generates one instance. The distance can be calculated in either the Euclidean metric or a "Manhattan" metric, in which the distance between two points is the sum of their separations along the two coordinate axes. The latter is appropriate for physical design applications, and easier to compute, so we will adopt it.

We let the side of the square have length $N^{1/2}$, so that the average distance between each city and its nearest neighbor is independent of N. It can be shown that this choice of length units leaves the optimal tour length per step independent of N, when one averages over many instances, keeping N fixed (*25*). Call this average optimal step length α. To bound α from above, a numerical experiment was performed with the following "greedy" heuristic algorithm. From each city, go to the nearest city not already on the tour. From the Nth city, return directly to the first. In the worst case, the ratio of the length of such a greedy tour to the optimal tour is proportional to $\ln(N)$ (*26*), but on average, we find that its step length is about 1.12. The variance of the greedy step length decreases as $N^{-1/2}$, so the situation envisioned in the worst case analysis is unobservably rare for large N.

To construct a simulated annealing algorithm, we need a means of representing the tour and a means of generating random rearrangements of the tour. Each tour can be described by a permuted list of the numbers 1 to N, which represents the cities. A powerful and general set of moves was introduced by Lin and Kernighan (*27, 28*). Each move consists of reversing the direction in which a section of the tour is traversed. More complicated moves have been used to enhance the searching effectiveness of iterative improvement. We find with the adaptive divide-and-conquer effect of annealing at intermediate temperatures that the subsequence reversal moves are sufficient (*29*).

An annealing schedule was determined empirically. The temperature at which segments flow about freely will be of order $N^{1/2}$, since that is the average bond length when the tour is highly random. Temperatures less than 1 should be cold. We were able to anneal into locally optimal solutions with $\alpha \leq 0.95$ for N up to 6000 sites. The largest traveling salesman problem in the plane for which a proved exact solution has been obtained and published (to our knowledge) has 318 points (*30*).

Real cities are not uniformly distributed, but are clumped, with dense and sparse regions. To introduce this feature into an ensemble of traveling salesman problems, albeit in an exaggerated form, we confine the randomly distributed cities to nine distinct regions with empty gaps between them. The temperature gives the simulated annealing method a means of separating out the problem of the coarse structure of the tour from the local details. At temperatures, such as $T = 1.2$ (Fig. 9a), where the small-scale structure of the paths is completely disordered, the longer steps across the gaps are already becoming infrequent and steps joining regions more than one gap are eliminated. The configurations studied below $T = 0.8$ (for instance, Fig. 9b) had the minimal number of long steps, but the detailed arrangement of the long steps continued to change down to $T = 0.4$ (Fig. 9c). Below $T = 0.4$, no further changes in the arrangement of the long steps were seen, but the small-scale structure within each region continued to evolve, with the result shown in Fig. 9d.

Summary and Conclusions

Implementing the appropriate Metropolis algorithm to simulate annealing of a combinatorial optimization problem is straightforward, and easily extended to new problems. Four ingredients are needed: a concise description of a configuration of the system; a random generator of "moves" or rearrangements of the elements in a configuration; a quantitative objective function containing the trade-offs that have to be made; and an annealing schedule of the temperatures and length of times for which the system is to be evolved. The annealing schedule may be developed by trial and error for a given problem, or may consist of just warming the system until it is obviously melted, then cooling in slow stages until diffusion of the components ceases. Inventing the most effective sets of moves and deciding which factors to incorporate into the objective function require insight into the problem being solved and may not be obvious. However, existing methods of iterative improvement can provide natural elements on which to base a simulated annealing algorithm.

The connection with statistical mechanics offers some novel perspectives on familiar optimization problems. Mean field theory for the ordered state at low temperatures may be of use in estimating the average results to be obtained by optimization. The comparison with models of disordered interacting systems gives insight into the ease or difficulty of finding heuristic solutions of the associated optimization problems, and provides a classification more discriminating than the blanket "worst-case" assignment of many optimization problems to the NP-complete category. It appears that for the large optimization problems that arise in current engineering practice a "most probable" or average behavior analysis will be more useful in assessing the value of a heuristic than the traditional worst-case arguments. For such analysis to be useful and accurate, better knowledge of the appropriate ensembles is required.

Freezing, at the temperatures where large clusters form, sets a limit on the energies reachable by a rapidly cooled spin glass. Further energy lowering is possible only by slow annealing. We expect similar freezing effects to limit the effectiveness of the common device of employing iterative improvement repeatedly from different random starting configurations.

Simulated annealing extends two of the most widely used heuristic techniques. The temperature distinguishes classes of rearrangements, so that rearrangements causing large changes in the objective function occur at high temperatures, while the small changes are deferred until low temperatures. This is an adaptive form of the divide-and-conquer approach. Like most iterative improvement schemes, the Metropolis algorithm proceeds in small steps from one configuration to the next, but the temperature keeps the algorithm from getting stuck by permitting uphill moves. Our numerical studies suggest that results of good quality are obtained with annealing schedules in which the amount of computational effort scales as N or as a small power of N. The slow increase of effort with increasing N and the generality of the method give promise that simulated annealing will be a very widely applicable heuristic optimization technique.

Dunham (*5*) has described iterative improvement as the natural framework for heuristic design, calling it "design by natural selection." [See Lin (*6*) for a fuller discussion.] In simulated annealing, we appear to have found a richer framework for the construction of heuristic algorithms, since the extra control provided by introducing a temperature allows us to separate out problems on different scales.

Simulation of the process of arriving at an optimal design by annealing under control of a schedule is an example of an evolutionary process modeled accurately by purely stochastic means. In fact, it may be a better model of selection processes in nature than is iterative improvement. Also, it provides an intriguing instance of "artificial intelligence," in which the computer has arrived almost uninstructed at a solution that might have been thought to require the intervention of human intelligence.

References and Notes

1. E. L. Lawlor, *Combinatorial Optimization* (Holt, Rinehart & Winston, New York, 1976).
2. A. V. Aho, J. E. Hopcroft, J. D. Ullman, *The Design and Analysis of Computer Algorithms* (Addison-Wesley, Reading, Mass., 1974).
3. M. R. Garey and D. S. Johnson, *Computers and Intractability: A Guide to the Theory of NP-Completeness* (Freeman, San Francisco, 1979).
4. R. Karp, *Math. Oper. Res.* **2**, 209 (1977).
5. B. Dunham, *Synthese* **15**, 254 (1963).
6. S. Lin, *Networks* **5**, 33 (1975).
7. For a concise and elegant presentation of the basic ideas of statistical mechanics, see E. Shrödinger, *Statistical Thermodynamics* (Cambridge Univ. Press, London, 1946).
8. N. Metropolis, A. Rosenbluth, M. Rosenbluth, A. Teller, E. Teller, *J. Chem. Phys.* **21**, 1087 (1953).
9. G. Toulouse, *Commun. Phys.* **2**, 115 (1977).

10. For review articles, see C. Castellani, C. DiCastro, L. Peliti, Eds., *Disordered Systems and Localization* (Springer, New York, 1981).
11. J. Soukup, *Proc. IEEE* **69**, 1281 (1981).
12. M. A. Breuer, Ed., *Design Automation of Digital Systems* (Prentice-Hall, Engelwood Cliffs, N.J., 1972).
13. D. Sherrington and S. Kirkpatrick, *Phys. Rev. Lett.* **35**, 1792 (1975); S. Kirkpatrick and D. Sherrington, *Phys. Rev. B* **17**, 4384 (1978).
14. A. P. Young and S. Kirkpatrick, *Phys. Rev. B* **25**, 440 (1982).
15. B. Mandelbrot, *Fractals: Form, Chance, and Dimension* (Freeman, San Francisco, 1979), pp. 237–239.
16. S. Kirkpatrick, *Phys. Rev. B* **16**, 4630 (1977).
17. C. Davis, G. Maley, R. Simmons, H. Stoller, R. Warren, T. Wohr, in *Proceedings of the IEEE International Conference on Circuits and Computers*, N. B. Guy Rabbat, Ed. (IEEE, New York, 1980), pp. 669–673.
18. M. A. Hanan, P. K. Wolff, B. J. Agule, *J. Des. Autom. Fault-Tolerant Comput.* **2**, 145 (1978).
19. M. Breuer, *ibid.* **1**, 343 (1977).
20. P. W. Case, M. Correia, W. Gianopulos, W. R. Heller, H. Ofek, T. C. Raymond, R. L. Simek, C. B. Steiglitz, *IBM J. Res. Dev.* **25**, 631 (1981).
21. A. J. Blodgett and D. R. Barbout, *ibid.* **26**, 30 (1982); A. J. Blodgett, in *Proceedings of the Electronics and Computers Conference* (IEEE, New York, 1980), pp. 283–285.
22. K. A. Chen, M. Feuer, K. H. Khokhani, N. Nan, S. Schmidt, in *Proceedings of the 14th IEEE Design Automation Conference* (New Orleans, La., 1977), pp. 298–302.
23. K. W. Lallier, J. B. Hickson, Jr., R. K. Jackson, paper presented at the European Conference on Design Automation, September 1981.
24. D. Hightower, in *Proceedings of the 6th IEEE Design Automation Workshop* (Miami Beach, Fla., June 1969), pp. 1–24.
25. J. Beardwood, J. H. Halton, J. M. Hammersley, *Proc. Cambridge Philos. Soc.* **55**, 299 (1959).
26. D. J. Resenkrantz, R. E. Stearns, P. M. Lewis, *SIAM (Soc. Ind. Appl. Math.) J. Comput.* **6**, 563 (1977).
27. S. Lin, *Bell Syst. Tech. J.* **44**, 2245 (1965).
28. ——— and B. W. Kernighan, *Oper. Res.* **21**, 498 (1973).
29. V. Černy has described an approach to the traveling salesman problem similar to ours in a manuscript received after this article was submitted for publication.
30. H. Crowder and M. W. Padberg, *Manage. Sci.* **26**, 495 (1980).
31. The experience and collaborative efforts of many of our colleagues have been essential to this work. In particular, we thank J. Cooper, W. Donath, B. Dunham, T. Enger, W. Heller, J. Hickson, G. Hsi, D. Jepsen, H. Koch, R. Linsker, C. Mehanian, S. Rothman, and U. Schultz.

Visual Information Processing: Artificial Intelligence and the Sensorium of Sight

David Marr
H. Keith Nishihara

For human vision to be explained by a computational theory, the first question is plain: What are the problems the brain solves when we see?

Modern neurophysiology has learned much about the operation of the individual nerve cell, but unpleasantly little about the meaning of the circuits they compose in the brain. The reason for this can be attributed, at least in part, to a failure to recognize what it means to understand a complex information-processing system; for a complex system cannot be understood as a simple extrapolation from the properties of its elementary components. One does not formulate, for example, a description of thermodynamical effects using a large set of equations, one for each of the particles involved. One describes such effects at their own level, that of an enormous collection of particles, and tries to show that in principle, the microscopic and macroscopic descriptions are consistent with one another.

The core of the problem is that a system as complex as a nervous system or a developing embryo must be analyzed and understood at several different levels. Indeed, in a system that solves an information processing problem, we may distinguish four important levels of description. (We here are following a formulation published in 1977 by Marr and Tomaso Poggio.) At the lowest, there is basic component and circuit analysis — how do transistors (or neurons), diodes (or synapses) work? The second level is the study of particular mechanisms: adders, multipliers, and memories, these being assemblies made from basic components. The third level is that of the algorithm, the scheme for a computation; and the top level contains the *theory* of the computation. A theory of addition, for example, would encompass the meaning of that operation, quite independent of the representation of the numbers to be added — say Arabic versus Roman. But it would also include the realization that the first of these representations is the more suitable of the two. An algorithm, on the other hand, is a particular method by which to add numbers. It therefore applies to a particular representation, since plainly an algorithm that adds Arabic numerals would be useless for Roman. At still a further level down, one comes upon a mechanism for addition — say a pocket calculator — which simply implements a particular algorithm. As a second example, take the case of Fourier analysis. Here the computational theory of the Fourier transform — the decomposition of an arbitrary mathematical curve into a sum of sine waves of differing frequencies — is well understood, and is ex-

The beginning of vision: a gray-level intensity array which will serve to approximate an input to the retina. The processing of such an image by the brain proceeds so naturally — so unconsciously, in a sense — that we are seldom aware that it begins with only this: a two-dimensional play of light upon the receptors of either eye. Our facility suggests the existence of a well-defined computational method, and makes vision a promising field of investigation in artificial intelligence. The image shown here was originally an array of 128 by 128, with each of its elements — numbers, actually — signifying one of 256 possible brightnesses. But the actual image as seen in this magazine is affected by the process by which it first was displayed on the screen of an imaging system resembling the technology of television, then by the processes that copied and printed it, and finally by limitations in human discrimination of brightness. The largest dots in the image are noise in the imaging system: each serves to suggest the actual size of each picture element, or "pixel." Smaller dots are patterns of stippling used in this figure to denote the shades of gray.

8/15/78 15:56:57 44 NIS

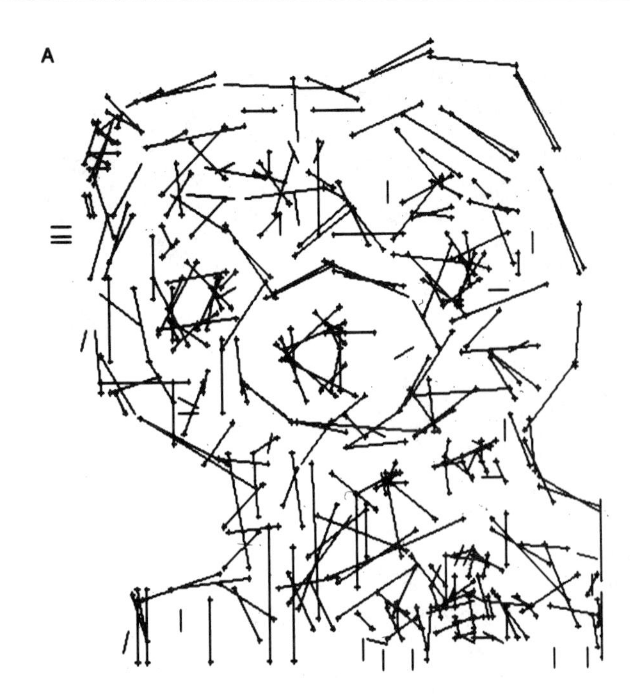

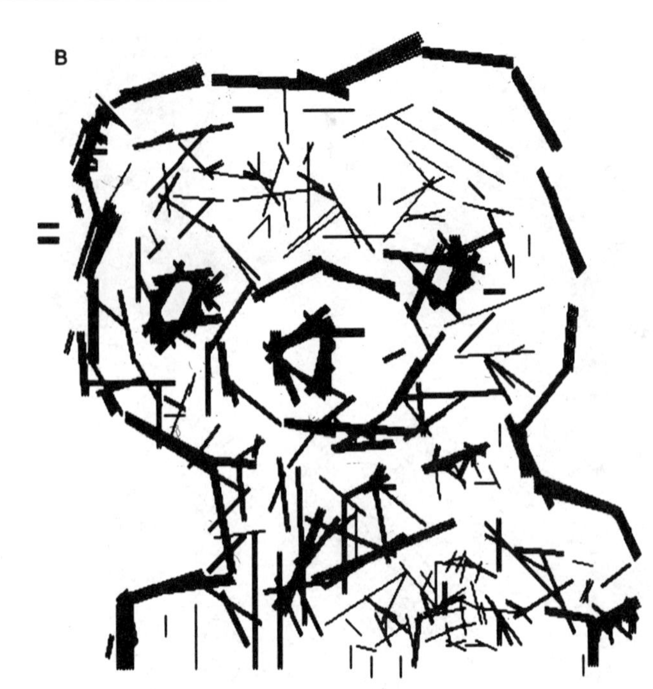

pressed independently of the particular way in which it might be computed. One level down, there are several algorithms for computing a Fourier transform, among them the so-called Fast Fourier Transform (FFT), which comprises a sequence of mathematical operations, and the so-called spatial algorithm, a single, global operation that is based on the mechanisms of laser optics. All such algorithms produce the same result, so the choice of which one to use depends upon the particular mechanisms that are available. If one has fast digital memory, adders, and multipliers, one will use the FFT, and if one has a laser and photographic plates, one will use an "optical" method.

Now each of the four levels of description will have its place in the eventual understanding of perceptual information processing, and of course there are logical and causal relations among them. But the important point is that the four levels of description are only loosely related. Too often in attempts to relate psychophysical problems to physiology there is confusion about the level at which a problem arises — is it related, for instance, mainly to the physical mechanisms of vision (like the after-images such as the one you see after staring at a lightbulb) or mainly to the computational theory of vision (like the ambiguity of the Necker cube as it appears on page 6)? More disturbingly, although the top level is the most neglected, it is also the most important. This is because the nature of the computations that underlie perception depend more upon the computational *problems* that have to be solved than upon the particular hardware in which their solutions are implemented. To phrase the matter another way, an algorithm is likely to be understood more readily by understanding the nature of the problem that it deals with than by examining the mechanism (and the hardware) by which it is embodied. There is, after all, an analog to all of this in physics, where a thermodynamical approach represented, at least historically, the first stage in the study of matter: it succeeded in producing a theory of gross properties such as temperature. A description in terms of mechanisms or elementary components — in this case atoms and molecules — appeared some decades afterwards.

Our main point, therefore, is that the topmost of our four levels, that at which the necessary structure of computation is defined, is a crucial but neglected one. Its study is separate from the study of particular algorithms, mechanisms, or hardware, and the techniques needed to pursue it are new. In the rest of this article, we summarize some examples of vision theories at the uppermost level. We will conclude with some remarks on the development of the field of which these theories are part: the field called artificial intelligence.

Conventional Approaches

The problems of visual perception have attracted the curiosity of scientists for many centuries. Important early contributions were made by Newton, who laid the foundations for modern work on color vision, and Helmholtz, whose treatise on physiological optics maintains its interest even today. Early in this century, Wertheimer noticed the apparent motion not of individual dots but instead of wholes, or "fields," in images presented sequentially, as if in a movie. In much the same way we perceive the migration across the sky of a flock of geese, the flock somehow constituting a single entity, and not individual birds. This observation started the Gestalt school of psychology, which was concerned with describing the qualities of wholes, including solidarity and distinctness, and trying to formulate the laws that governed their creation. The attempt failed for various reasons, and the Gestalt school dissolved into the fog of subjectivism. With the death of the school, many of its early and genuine insights were unfortunately lost to the mainstream of experimental psychology.

The next developments of importance were recent and technical. The advent of electrophysiology in the 1940s

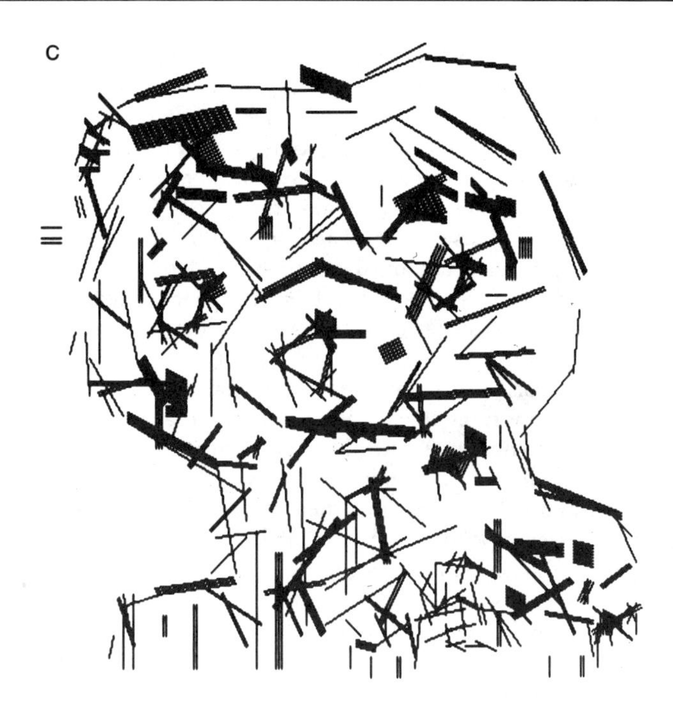

The so-called primal sketch is shown in three of its aspects; each is a representation of intensity changes in a gray-level image such as appears on page 3. Its creation constitutes the earliest stage in the authors' theory of visual information processing. Sketch A shows only "edge-assertions": each line represents the position and orientation at which a change in intensity is found. The cross-bars at the ends of each line show the terminations of the change. This is not to say that each line necessarily denotes a sharp edge in the gray-level array, but rather that each denotes the existence of a gradient in intensity. Sketch *B* adds information about differing contrasts: each line in *A* is transformed into a set of parallel lines in proportion to the logarithm of the intensity change. In other words, a large magnitude of transition from light to dark or dark to light leads to a bolder edge-assertion. Notice, accordingly, that the margin of the bear tends to have more prominent assertions than those to be found within. In the displays shown here, the spacing between parallel lines in a set is simply proportional to their length. For short assertions, therefore, the multiple lines have tended to overlap, so as to create the impression of a single, thick bar. Sketch *C* shows the fuzziness of each of the edge assertions — that is to say, the widths of the variations in intensity, as opposed to their magnitudes. The thicker sets of lines now tend to lie in the interior of the image, which reflects the circumstance that broad gradients in shading tend not to appear at its edges. As held in computer storage, the primal sketch includes the information shown in *A*, *B*, and *C* alike: that is to say, the positions, directions, magnitudes, and spatial extents of intensity gradients in the gray-level array.

and '50s made single cell recording possible, and with Stephen W. Kuffler's study of retinal ganglion cells — the neurons of the eye that give rise to the optic nerve — a new approach to the problem was born. Its most renowned practitioners are David H. Hubel and Torsten N. Wiesel, who since 1959 have conducted an influential series of investigations on single cell responses at various points along the visual pathway in the cat and the monkey.

Hubel and Wiesel used the notion of a cell's "receptive field" to classify cells in the so-called primary and secondary visual areas of the cerebral cortex into simple, complex, and hypercomplex types. Simple cells are orientation-sensitive and roughly linear. That is to say, the simple cell monitors a particular district of visual space, a so-called receptive field, in this case divided into parallel elongated excitatory and inhibitory parts; events in the first of these promote the cell's electrical activity, events in the second tend to inhibit it; the two opposing phenomena act simultaneously on the cell — in a word, they summate; and finally, a simple cell's response to a stimulating pattern is roughly predictable from its receptive field's geometry. Complex cells, on the other hand, apparently respond to edges and bars over a wider range than a simple cell's field. Hypercomplex cells seem to respond best to points where an edge or bar terminates. How the different types of cell are connected and why they behave as they do is controversial.

Students of the psychology of perception were also affected by a technological advance, the advent of the digital computer. Most notably, it allowed Bela Julesz in 1959 to devise random-dot stereograms, which are image pairs constructed of dot patterns that appear random when viewed monocularly, but which fuse when viewed one through each eye to give a percept of shapes and surfaces with a clear three-dimensional structure. An example is shown on page 9. Here the image for the left eye is a matrix of black and white squares generated at random by a computer program. The image for the right is made by copying the left image and then shifting a square-shaped region at its center slightly to the left, providing a new random pattern to fill in the gap that the shift must create. If each of the eyes sees only one matrix, as if they were both in the same physical place, the result is the sensation of a square floating in space. Plainly such percepts are caused solely by the stereo disparity between matching elements in the images presented to each eye.

Very recently, considerable interest has been attracted by a rather different approach. In 1971, Roger N. Shepard and Jacqueline Metzler made line drawings of simple objects that differed from one another either by a three-dimensional rotation, or by a rotation plus a reflection (see the illustration on page 14). They asked how long it took to decide whether two depicted objects differed by a rotation and a reflection, or merely a rotation. They found that the time taken depended on the 3-D angle of rotation necessary to bring the two objects into correspondence. Indeed, it varied linearly with this angle. One is led thereby to the notion that a mental rotation of sorts is actually being performed: that a mental description of the first shape in a pair is being adjusted incrementally in orientation until it matches the second, such adjustment requiring greater time when greater angles are involved.

Interesting and important though these findings are, one must sometimes be allowed the luxury of pausing to reflect upon the overall trends that they represent, in order to take stock of the kind of knowledge that is accessible through these techniques. For we repeat: perhaps the most striking feature of neurophysiology and psychophysics at present is that they *describe* the behavior of cells or of subjects, but do not *explain* it. What are the visual areas of the cerebral cortex actually doing? What are the problems in doing it that need explaining, and at what level of description should such explanations be sought?

A Computational Approach to Vision

In trying to come to grips with these problems, our group at the M.I.T. Artificial Intelligence Laboratory has adopted a point of view that regards visual perception as a problem primarily in information processing. The problem commences with a large, gray-level intensity array, which suffices to approximate an image such as the world might cast upon the retinas of the eyes (an example appears on page 3), and it culminates in a *description* that depends on that array, and on the purpose that the viewer brings to it. Our particular concern in this article will be with a description well suited for the recognition of three-dimensional shapes.

The Primal Sketch. It is a commonplace that a scene and a drawing of the scene appear very similar, despite the completely different gray-level images to which they give rise. This suggests that the artist's symbols correspond in some way to natural symbols that are computed out of the image during the normal course of its interpretation. Our theory therefore asserts that the first operation on an image is to transform it into a primitive but rich description of the way its intensities change over the visual field, as opposed to a description of its particular intensity values in and of themselves. This yields a description of markedly reduced size that still captures the important aspects required for image analysis. We call it a *primal sketch.* Consider, for example, an intensity array of 1,000 by 1,000, or a million points in all. Even if the possible intensity at any one point were merely black or white — two different brightnesses — the number of all possible arrays would still be $2^{1,000,000}$. In a real image, however, there tend to be continuities of intensity — areas where brightness varies uniformly — and this tends to eliminate possibilities in which the black and white oscillate wildly. It also tends to simplify the array. Typically, therefore, a primal sketch need not include a set of values for every point in an image. As stored in a computer, it will instead constitute an array with numbers representing the directions, magnitudes, and spatial extents of intensity changes assigned to certain specific points in an image — points that tend to be places of locally high or low intensity. The positions of these points, particularly their arrangement amongst their immediate neighbors — that is to say, the local geometry of the image — must also be made explicit in the primal sketch, as it would otherwise be lost. (It was implicit, of course, in the 1,000-by-1,000 array, but we are no longer retaining data for each of those million places.) One way to do this is to specify "virtual lines" — directions and distances — between neighboring points in the sketch.

The process of computing the primal sketch involves several important steps, but let it suffice that the first of them is comparable with the measurements that are apparently made by simple cells in the visual cortex. A well-defined interaction then takes place between simple-cell-type measurements made at the same orientation and position in the visual field but with different receptive field sizes, so that any given intensity gradient can be succinctly described by measurements made with receptive fields of commensurate size. This is in direct contrast to theories which assert that every simple cell acts as a "feature detector" whose output is freely available to subsequent processes.

Modules of Early Visual Processing. The primal sketch of an image is typically a large and unwieldy collection of data, even despite its simplification relative to a gray-level array; for this is the unavoidable consequence of the irregularity and complexity of natural images. The next computational problem is thus its decoding. Now the traditional approach to machine vision assumes that the essence of such a decoding is a process called *segmentation,* whose purpose is to divide a primal sketch, or more generally an image, into regions that are meaningful, perhaps as physical objects. Tenenbaum and Barrow, for example, applied knowledge about several different types of scene to the segmentation of images of landscapes, an office, a room, and a compressor. Freuder used a similar approach to identify a hammer in a simple scene. Upon

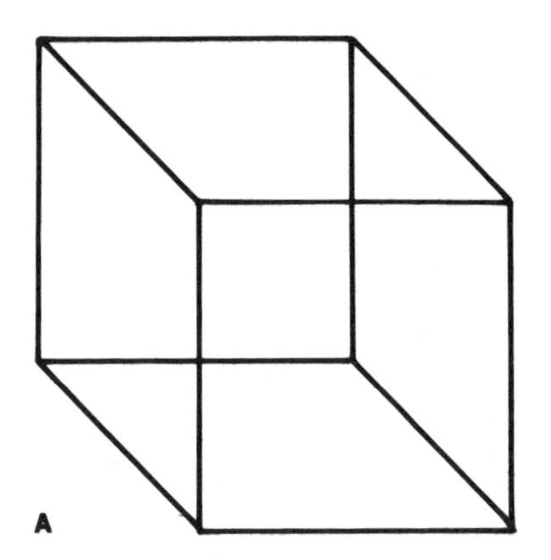

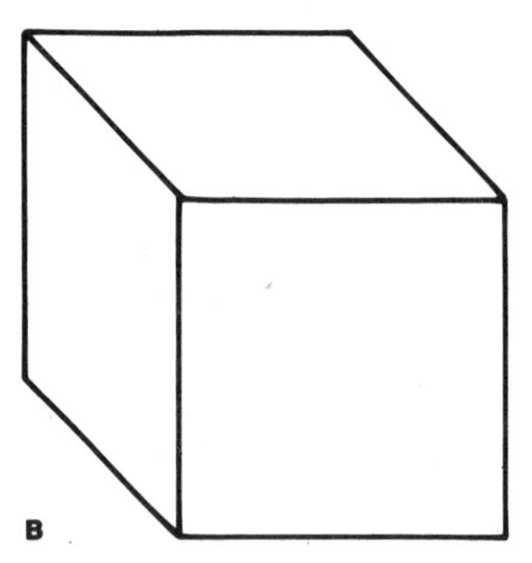

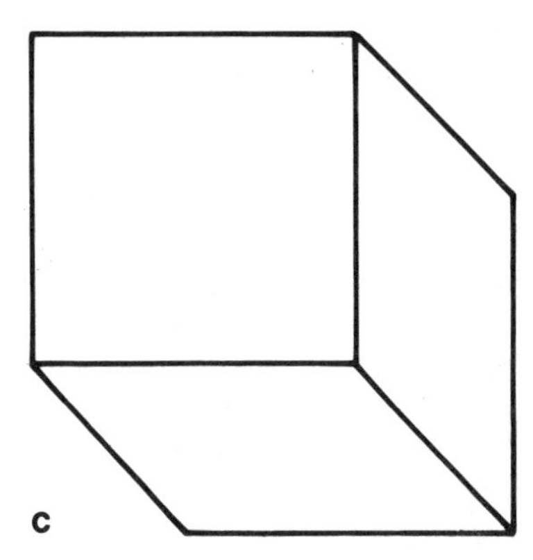

The so-called Necker illusion, named after L. A. Necker, the Swiss naturalist who developed it in 1832. The essence of the matter is that the two-dimensional representation which appears as part (a) of the figure has collapsed the depth out of a cube, and that a certain aspect of human vision is thus to recover this missing third dimension. It develops that the depth of the cube (or rather in its image) can indeed be perceived, but only to the extent that two interpretations are possible: the two shown as (b) and (c). Your perception of (a) characteristically "flips" from one to the other. Understanding why this should be so is a part of devising a computational theory of vision. By contrast, the understanding of the afterimage that you see when you stare at a lightbulb seems simply to be a matter of understanding the characteristics of the visual "hardware" — in this instance, that a sustained stimulus will fatigue the light-receptive cells of the retina.

finding a blob, his computer program would tentatively label it as the head of a hammer, and begin a search for confirmation in the form of an appended shaft. If this approach were correct, it would mean that a central problem for vision is arranging for the right piece of specialized knowledge to be made available at the appropriate time in the segmentation of an image. Freuder's work, for example, was almost entirely devoted to the design of a system that made this possible. But despite considerable efforts over a long period, the theory and practice of segmentation remain rather primitive, and here again we believe that the main reason lies in the failure to formulate precisely the goals of this stage of the processing — a failure, in other words, to work at the topmost level of visual theory. What, for example, is an object? Is a head an object? Is it still an object if it is attached to a body? What about a man on horseback?

We shall argue that the early stages of visual information processing ought instead to squeeze the last possible ounce of information from an image before taking recourse to the descending influence of "high-level" knowledge about objects in the world. Let us turn, then, to a brief examination of the physics of the situation. As we noted earlier, the visual process begins with arrays of intensities projected upon the retinas of the eyes. The principal factors that determine these intensities are (1) the illuminant, (2) the surface reflectance properties of the objects viewed, (3) the shapes of the visible surfaces of these objects, and (4) the vantage point of the viewer. Thus if the analysis of the input intensity arrays is to operate autonomously, at least in its early stages, it can only be expected to extract information about these four factors. In short, early visual processing must be limited to the recovery of localized physical properties of the visible *surfaces* of a viewed object — particularly local surface dispositions (orientation and depth) and surface material properties (color, texture, shininess, and so on). More abstract matters such as a description of overall three-dimensional shape must come after this more basic analysis is complete.

Four examples of "receptive fields" for so-called simple cells of the primary visual cortex. Each field circumscribes the part of the world that is monitored, so to speak, by the cell. But within that locus are bands in which the appearance of light will excite the neuron's ongoing electrical activity (plus signs) and parallel bands that inhibit it (minus signs). The best possible stimulus for the fourth of these examples is a sharp edge in an image with brightness at the left and darkness at the right; for the shining of light on the right-hand side of the receptive field would inhibit, not excite, the associated neuron.

An example of early processing is stereopsis. Imagine that images of a scene are available from two nearby points at the same horizontal level — the analog of the images that play upon the retinas of your left and right eyes. The images are somewhat different, of course, in consequence of the slight difference in vantage. Imagine further that a particular location on a surface in the scene is chosen from one image; that the corresponding location is identified in the other image; and that the relative positions of the two versions of that location are measured. This information will suffice for the calculation of depth — the distance of that location from the viewer. Notice that methods based on gray-level correlation between the pair of images fail to be suitable because a mere gray-level measurement does not reliably define a point on a physical surface. To put the matter plainly, numerous points in a surface might fortuitously be the same shade of gray, and differences in the vantage points of the observer's eyes could change the shade as well. The matching must evidently be based instead on objective markings that lie upon the surface, and so one has to use changes in reflectance. One way of doing this is to obtain a primitive description of the intensity changes that exist in each image (such as a primal sketch), and then to match these descriptions. After all, the line segments, edge segments, blobs, and edge termination points included in such a description correspond quite closely to boundaries and reflectance changes on physical surfaces. The stereo problem — the determination of depth given a stereo pair of images — may thus be reduced to that of matching two primitive descriptions, one from each eye; and to help in this task there are physical constraints that translate into two rules for how the left and right descriptions are combined:

Uniqueness. Each item from each image may be assigned at most one disparity value — that is to say, a unique position relative to its counterpart in the stereo pair. This condition rests on the premise that the items to be matched have a physical existence, and can be in only one place at a time.

Continuity. Disparity varies smoothly almost everywhere. This condition is a consequence of the cohesiveness of matter, and it states that only a relatively small fraction of the area of an image is composed of discontinuities in depth.

In the case of random-dot stereograms, the computational problem is rather well-defined, essentially because of Julesz's demonstration that random-dot stereograms, containing no monocular information, still yield stereopsis. In 1976 Marr and Poggio developed a method for computing local disparities in a pair of random-dot stereograms by an iterative, parallel procedure known technically as a cooperative algorithm. This sort of algorithm has the property that it can be defined completely in terms of simple local interactions because at each of its iterations, each point is affected only by a calculation performed on its immediate neighborhood. Yet all points are so affected during each successive iteration, so the transformations take on a complex global nature. Subsequent comparison of the algorithm's performance with psychophysical data showed that it did not hold up well as a model for human stereopsis. To be sure, it performed better than people do on the standard stereograms like that shown at the right; but it did not explain people's ability to see stereograms in which one of the two images is defocused slightly or enlarged slightly relative to the other. These observations led Marr and Poggio in 1977 to devise another algorithm, this one based on the human use of so-called vergence eye movements, in which the two eyes cross to a greater or lesser extent without changing their average direction of view. This algorithm is consistent with all of the currently known psychophysical data.

A second example of early visual processing concerns the derivation of structure from motion. It has long been known that as an object moves relative to the viewer, the way its appearance changes provides information that we use to determine its shape. The problem decomposes into two parts: matching the elements that occur in consecutive images; and deriving shape information from measurements of their changes in position. Shimon Ullman has shown that these problems can be solved mathematically. His idea is that in general, nothing can be inferred about the shape of an object given only a set of sequential views of it; for some extra assumptions have to be made. Accordingly, he formulates an assumption of rigidity, which states that if a set of moving points has a *unique* interpretation as a rigid body in motion, that interpretation is correct. (The assumption is based on a theorem which he proves, stating that three distinct views of four non-coplanar points on a rigid body are sufficient to determine uniquely their three-dimensional arrangement in space.) From this he derives a method for computing structure from motion. The method gives results that are quantitatively superior to the ability of humans to determine shape from motion, and which fail in qualitatively similar circumstances. Ullman has also devised a set of simple algorithms by which the method may be implemented.

Recovering the Depth of the Three Dimensional World

The following pages provide two pairs of so-called random-dot stereograms as developed at Bell Telephone Laboratories by Bela Julesz. The reader is urged to experiment with them; all that you need, in addition to the images themselves, is a hand mirror that you can hold against either side of your nose. First an explanation: Hold your thumb at various distances from your eyes against a more distant background. Closing first one eye and then the other will then convince you that objects in the world have somewhat different positions in the images that the world casts upon each of your retinas. (The magnitude of the difference is inversely proportional to the distance of the object.) The point of the stereo pair is to show that such disparities are sufficient for your brain to recover the lost third dimension from such two-dimensional images (except for a small percentage of people who lack stereo vision). After all, each of the patterns printed on subsequent pages shows nothing recognizable. Each is a computer-generated assembly of black and white picture-elements ("pixels"). The pair labelled A, however, have a square-shaped region shifted in one of the images several pixels relative to its placement in the other. In short, then, the stereo pair contains *no information whatever about visible surfaces* — except for left-right disparities of the aforementioned sort.

Will this be sufficient for 3-D perception? First of all, cut out the images. Place the pair labelled A on a table-top with a few inches between them. Adhere to "left" and "right" as printed beneath the arrays. Position your head a foot or more above them. Hold the mirror with its back against the left side of your nose, in such a way that your right eye looks directly at the right member of the pair, but your left eye sees the mirror reflection of the left. By suitable maneuvering of the images, your head, and the mirror, get the left and right images to appear to be in the same place — in a word, to coincide. (They would coincide, too, if they were placed in a stereo viewer, but these images have been designed to be used with a mirror, and thus take account of a mirror reflection. They therefore won't work in a stereo viewer.) A few minutes of trying are likely to be required before the image-fusion occurs. Try to relax, and not to strain your eyes. If you achieve stereopsis, there will be no mistaking the effect: it is a striking one. With the mirror reflecting the left image into your left eye, your perception is of a square floating in space above the plane of the background. With the mirror reflecting the right image into the right eye (or alternatively by turning both images upside-down and keeping the mirror as before), the square floats behind. (Many people find this harder to see.) For the effect involving the left-right pair labelled B, consult the illustration and caption on page 12. We wish to express our appreciation to Bela Julesz for the suggestion that a mirror might be used in place of more complicated apparatus to achieve stereopsis.

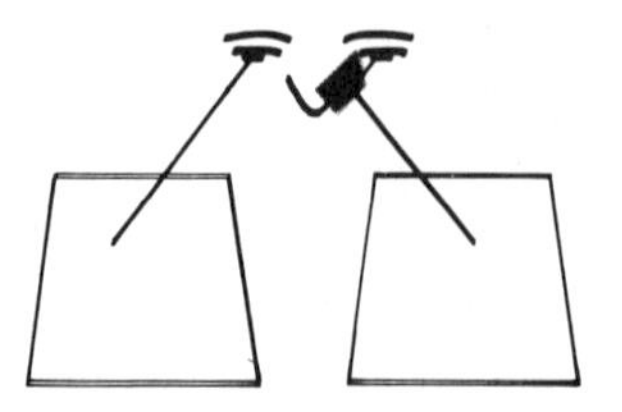

Pair A: Right
Pair A: Left

Pair B: Left
Pair B: Right

The 2½-Dimensional Sketch. Both of the techniques of image analysis discussed in the preceding paragraphs provide information about the relative distances to various places in an image. In the case of stereopsis, it is the matching of points in a stereo pair that leads to such information. In the case of structure from motion, it is the matching of points in successive images. More generally, however, we know that vision provides several sources of information about shapes in the visual world. The most direct, perhaps, are the aforementioned stereo and motion, but texture gradients in a single image are nearly as effective. Furthermore, the theatrical techniques of facial make-up reveal the sensitivity of perceived shapes to shading, and color sometimes suggests the manner in which a surface reflects light. It often happens that some parts of a scene are open to inspection by some of these techniques, and other parts to inspection by others. Yet different as the techniques are, they all have two important characteristics in common: they rely on information from the image rather than *a priori* knowledge about the shapes of the viewed objects; and the information they specify concerns the depth or surface orientation at arbitrary points in an image, rather than the depth or orientation associated with particular objects.

In order to make the most efficient use of different and often complementary channels of information deriving from stereopsis, from motion, from texture, from color, from shading, they need to be combined in some way. The computational question that now arises is thus how best to do this, and the natural answer is to seek some representation of the visual scene that makes explicit just the information these processes can deliver. We seek, in other words, a representation of surfaces in an image that makes explicit their shapes and orientations, much as the Arabic representation of a number makes explicit its composition by powers of ten. It might be contrasted with the representation of a surface as a mathematical expression, in which the orientation is only implicit, and not at all apparent. We call such a representation the 2½-dimensional sketch, and in the particular candidate for it shown on page 15, surface orientation is represented by covering an image with needles. The length of each needle defines the dip of the surface at that point, so that zero length corresponds to a surface that is perpendicular to the vector from the viewer to the point, and increasing lengths denote surfaces that tilt increasingly away from the viewer. The orientation of each needle defines the local direction of dip.

Our argument is that the 2½-D sketch is useful because it makes explicit information about the image in a form that is closely matched to what image analysis can deliver. To put it another way, we can formulate the goals of this stage of visual processing as being primarily the construction of this representation, discovering, for example, what are the surface orientations in a scene, which of the contours in the primal sketch correspond to surface discontinuities and should therefore be represented in the 2½-D sketch, and which contours are missing in the primal sketch and need to be inserted into the 2½-D sketch in order to bring it into a state that is consistent with the nature of three-dimensional space. This formulation avoids the difficulties associated with the terms "region" and "object" — the difficulties inherent in the image segmentation approach; for the gray level intensity array, the primal sketch, the various modules of early visual processing, and finally the 2½-dimensional sketch itself deal only with discovering the properties of *surfaces* in an image. One is pleased about that, for we know of ourselves as perceivers that surface orientation can be associated with unfamiliar shapes, so its representation probably precedes the decomposition of the scene into objects. One is thus free to ask precise questions about the computational structure of the 2½-D sketch and of processes to create and maintain it. We are currently much occupied with these matters.

Later Processing Problems

The final components of our visual processing theory concern the application of visually derived surface information for the representation of three-dimensional shapes in a way that is suitable specifically for recognition. By this we mean the ability to recognize a shape as being the same as a shape seen earlier, and this in essence depends on being able to describe shapes consistently each time they are seen, whatever the circumstances of their positions relative to the viewer. The problem with local surface representations such as the 2½-D sketch is that the description depends as much on the viewpoint of the observer as it does on the structure of the shape. In order to factor out a description of a shape that depends on its structure alone, the representation must be based on readily identifiable geometric features of the overall shape, and the dispositions of these features must be specified relative to the shape in itself. In brief, the coordinate system must be "object-centered," not "viewer-centered." One aspect of this deals with the nature of the representation scheme that is to be used, and another with how to obtain it from the 2½-D sketch. We begin by discussing the first, and will then move on to the second.

The 3-D Model Representation. The most basic geometric properties of the volume occupied by a shape are (1) its average location (or center of mass); (2) its overall size, as exemplified, for example, by its mean diameter or volume; and (3) its principal axis of elongation or symmetry, if one exists. A description based on these qualities would certainly be inadequate for an application such as shape recognition; after all, one can tell little about the three-dimensional structure of a shape given only its position, size, and orientation. But if a shape itself has a natural decomposition into components that can be so described, this volumetric scheme is an effective means for describing the relative spatial arrangement of those components. The illustration on page 17 shows a familiar version of this type of description, the stick figure. The recognizability of the animal shapes depicted in the illustration is surprising considering the simplicity of representation used to describe them.

The reason such a description works so well lies, we think, in (1) the volumetric (as opposed to surface-based)

definition of the primitive elements — the sticks — used by the representation; (2) the relatively small number of elements used; and (3) the relation of elements to each other rather than to the viewer. In short, this type of shape representation is volumetric, modular, and can be based on object-centered coordinates. The figure on page 18 illustrates the scheme of representation that was developed from these ideas. Here the description of a shape is composed of a hierarchy of stick-figure specifications we call 3-D models. In the simplest, a single axis element is used to specify the location, size, and orientation of the entire shape; the human body displayed in the illustration will serve as an instance. This element is also used to define a coordinate system that will specify the dispositions of subsidiary axes, each of these specifying in turn a coordinate system for 3-D models of "arm," "hand," and so on. This hierarchical structure makes it possible to treat any component of a shape as a shape in itself. It also provides flexibility in the detail of a description.

Shapes Admitting 3-D Model Descriptions. If the scheme for a given shape is to be uniquely defined and stable over unimportant variations such as viewpoint — if, in a word it is to be canonical — its definition must take advantage of any salient geometrical characteristics that the shape inherently possesses. If a shape has natural axes, then those should be used. The coordinate system for a sausage should take advantage of its major axis, and for a face, of its axis of symmetry.

The decoding of a random-dot stereogram pair, as performed by an algorithm devised by David Marr and Tomaso Poggio in 1976. The nature of the problem is to determine which point in one image is a match to any given point in the other. (Remember that the same surface markings have differing placements in the images cast upon either retina.) After that, a depth can be assigned for any given stereo disparity. The algorithm begins by creating a series of parallel planes to represent possible depths — that is to say, possible distances of various surfaces from the viewer. It then marks this three-dimensional matrix to indicate each location where a local patch of surface could conceivably lie, based on varying construals of the patterns of picture elements in the stereo pair. To phrase it another way, any given pixel in one stereogram is temporarily assumed to match with any of a number of candidate pixels in the other, within a limit of wide angular differences. At this stage, a pair of conditions are applied to each mark in the array thus created: first, that only a single match for any one point in either stereogram will ultimately be accepted; and second, that real 3-D surfaces tend to be continuous in depth. More particularly, at each iteration of the processing that now takes place, the number of marks at a given position but various depths is compared with the number of marks that lie nearby at similar depth. The more of the former, the more likely it is that the mark is incorrectly placed. The more of the latter, the more likely it is that its placement is correct. Thus a weighting of these two factors determines if the mark will be preserved to the next iteration. The illustration shows the original stereo pair, the original matrix computed therefrom, and also the results after one, two, three, four, five, six, eight, and fourteen iterations. Shades of gray are employed to signify marks at greater or lesser depths in the 3-D matrix. The algorithm therefore progressively reveals a nested set of tiers — the pattern, in essence, of a rectangular wedding cake. The reader who succeeds in achieving stereopsis with stereo pair B (printed on page 10 of this article) will see the actual effect that the algorithm here uncovers. It turns out that the algorithm fails under conditions in which human vision is known to succeed — for example, a slight defocusing of the left or right stereogram. In 1977, Marr and Poggio devised a second algorithm, based on different principles, that closely matches human abilities.

Highly symmetrical objects, like a sphere, a square, or a circular disc, will inevitably lead to ambiguities in the choice of coordinate systems. For a shape as regular as a sphere this poses no great problem, because its description in all reasonable systems is the same. One can even allow other factors, like the direction of motion or spin, to influence the choice of coordinate frame. For other shapes, the existence of more than one possible choice probably means that one has to represent the object in several ways, but this is acceptable provided that their number is small. For example, there are four possible axes on which one might wish to base the coordinate system for representing a door, namely the midlines along its length, its width, and its thickness, and also the axis of its hinges. (This last would be especially useful to represent how the door opens.) For a typewriter, there are two reasonable choices, an axis parallel to its width, because that is usually its largest dimension, and the axis about which a typewriter is roughly symmetrical.

In general, if an axis can be distinguished in a shape, it can be used as the basis for a local coordinate system. One approach to the problem of defining object-centered coordinates is therefore to examine the class of shapes having an axis as an integral part of their structure. Consider, accordingly, the class of so-called *generalized cones,* each of these being the surface swept out by moving a cross-section of constant shape but smoothly varying size along an axis, as shown on page 20. Thomas O. Binford has drawn attention to this class of constructions, suggesting that it might provide a convenient way of describing three-dimensional surfaces for the purposes of computer vision. We regard it as an important class not because the shapes themselves are easily describable, but because the presence of an axis allows one to define a canonical local coordinate system. Fortunately, many objects, especially those whose shape was achieved by growth, are described quite naturally in terms of one or more generalized cones. The animal shapes on page 43 provide some examples; the individual sticks are simply the axes of generalized cones that approximate the shapes of parts of these creatures. Many artifacts can also be described in this way — say a car (a small box sitting atop a longer one) or a building (a box with a vertical axis).

It is perhaps worth mentioning the following curious point that has emerged from this way of representing three-dimensional shapes. In 1973 Warrington and Taylor described patients with right parietal lobe lesions (that is to say, damage to a particular part of the cerebral cortex) who had difficulty in recognizing objects seen in "unconventional" views, such as the view of a water pail seen from above in the figure on page 19. The researchers did not attempt to define what makes a view "unconventional." But according to our theory, the most troublesome views of an object will likely be those in which its intrinsic coordinate axes cannot easily be recovered from the image. Our theory therefore predicts that unconventional views in the Warrington and Taylor sense will correspond to those views in which an important axis in the object's 3-D model representation is foreshortened. Such views are by no means uncommon. If a 35mm camera is

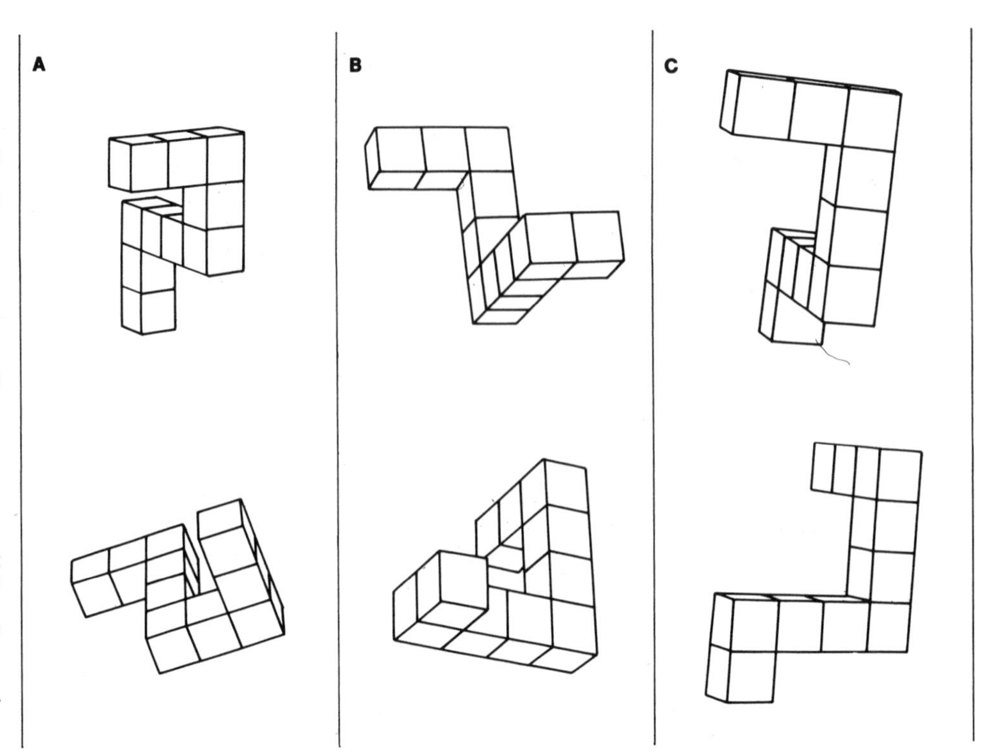

Some drawings similar to those used in Shepard and Metzler's 1971 experiments on mental rotation. The ones shown in (A) are identical, as a clockwise turning of this magazine by 80 degrees will readily prove. Those in (B) are also identical, and again the relative angle between the two is 80 degrees. Here, however, it is a rotation in depth that will make the first coincide with the second. Finally, those in (C) are not at all identical, for no rotation will bring them into congruence. The time taken to decide whether a pair is the same was found to vary linearly with the angle through which one figure must be rotated to be brought into correspondence with the other. This suggested to the investigators that a stepwise mental rotation was in fact being performed by the subjects of their experiments.

directed towards you, you are seeing an unconventional view of it, since the axis of its lens is foreshortened.

It is important to remember, however, that there exist surfaces that cannot conveniently be approximated by generalized cones, for example a cake that has been transected at some arbitrary plane, or the surface formed by a crumpled newspaper. Cases like the cake could be dealt with by introducing a suitable surface primitive for describing the plane of the cut, in much the same way as an axis in the 3-D model representation is a primitive that describes a volumetric element. But the crumpled newspaper poses apparently intractable problems.

Finding the Natural Coordinate System. Even if a shape possesses a canonical coordinate frame, one still is faced with the problem of finding it from an image. Our own interest in this problem grew from the question of how to interpret the *outlines* of objects as seen in a two-dimensional image, and our starting point was the observation that when one looks at the silhouettes in Picasso's "Rites of Spring" (reproduced here on page 22), one perceives them in terms of very particular three-dimensional shapes, some familiar, some less so. This is quite remarkable, because the silhouettes could in theory have been generated by an infinite variety of three-dimensional shapes which, from other viewpoints, would have no discernible similarities to the shapes we perceive. One can perhaps attribute part of the phenomenon to a familiarity with the depicted shapes, but not all of it, because one can use the medium of a silhouette to convey a new shape, and because even with considerable effort it is difficult to imagine the more bizarre three-dimensional surfaces that could have given rise to the same silhouettes. The paradox, then, is that the bounding contours in Picasso's "Rites" apparently tell us more than they should about the shape of the figures. For example, neighboring points on such a contour could in general arise from widely separated points on the original surface, but our perceptual interpretation usually ignores this possibility.

The first observation to be made is that the contours that bound these silhouettes are contours of surface discontinuity, which are precisely the contours with which the 2½-D sketch is concerned. Secondly, because we can interpret the silhouettes as three-dimensional shapes, then implicit in the way we interpret them must lie some *a priori* assumptions that allow us to infer a shape from an outline. If a surface violates these assumptions, our analysis will be wrong, in the sense that the shape we assign to the contours will differ from the shape that actually caused them. An everyday example is the shadowgraph, where the appropriate arrangement of one's hands can, to the surprise and delight of a child, produce the shadow of a duck or a rabbit.

What assumptions is it reasonable to suppose that we make? In order to explain them, we need to define the four constructions that appear in the figure on page 21. These are (1) a three-dimensional surface Σ; (2) its image or silhouette S_V as seen from a viewpoint V; (3) the bounding contour C_V of S_V; and (4) the set of points on the surface Σ that project onto the contour C_V. We shall call this last the *contour generator* of C_V, and we shall denote it by Γ_V.

Observe that the contour C_V, like the contours in the work of Picasso, imparts very little information about the three-dimensional surface that caused it. Indeed, the only obvious feature available in the contour is the distinction between convex and concave places — that is to say, the presence of inflection points. In order that these inflections be "reliable," one needs to make some assumptions about the way the contour was generated, and we choose the following restrictions:

1. Each point on the contour generator Γ_V projects to a different point on the contour C_V.
2. Nearby points on the contour C_V arise from nearby points on the contour generator Γ_V.
3. The contour generator Γ_V lies wholly in a single plane.

The first and second restrictions say that each point on the contour of the image comes from one point on the surface (which is an assumption that facilitates the analysis but is not of fundamental importance), and that where the surface looks continuous in the image, it really is continuous in three dimensions. The third restriction is simply the demand that the difference between convex and concave contour segments reflects properties of the surface, rather than of the imaging process.

It turns out that the following theorem is true, and it is a result that we found very surprising.

Theorem. If the surface is smooth (for our purposes, if it is twice differentiable with continuous second derivitive) and if restrictions 1 through 3 hold for all distant viewing positions in any one plane, as illustrated on page 21, then the viewed surface is a generalized cone. The converse is also true: if the surface is a generalized cone, then conditions 1 through 3 will be found to be true.

This means that if the convexities and concavities of a bounding contour in an image are actual properties of a surface, then that surface is a generalized cone or is composed of several such cones. In brief, the theorem says that a natural link exists between generalized cones and the imaging process itself. The combination of these two must mean, we think, that generalized cones will play an intimate role in the development of vision theory.

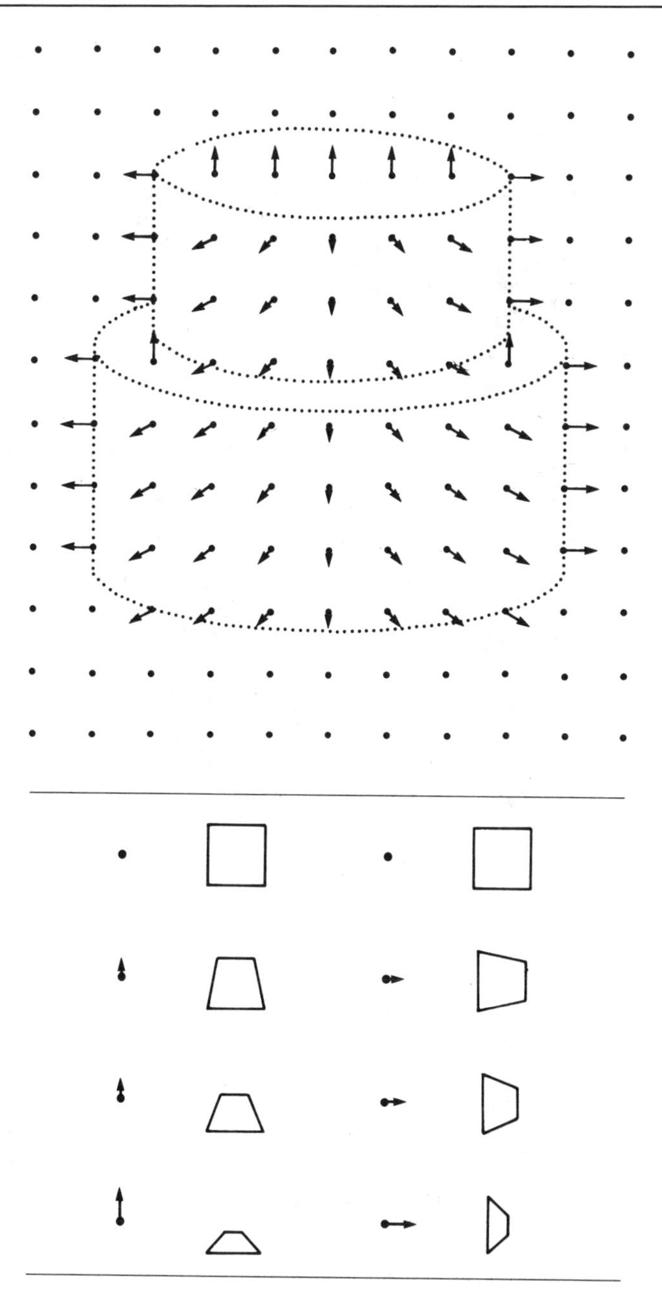

A candidate for the so-called 2½-dimensional sketch, which encompasses local determinations of the depth and orientation of surfaces in an image, as derived from processes that operate upon the primal sketch or some other representation of changes in gray-level intensity. The lengths of the needles represent the degree of tilt at various points in the surface; the orientations of the needles represent the directions of tilt — some examples are shown in the insert. Dotted lines show contours of surface discontinuity. No explicit representation of depth appears in this figure.

The Search for a Theory

We have tried in this survey of visual information processing to make two principal points. The first is methodological: namely that it is important to be very clear about the nature of the understanding we seek. The results we try to achieve should be precise ones, at the level of what we call a computational theory. The critical act in formulating computational theories turns out to be the discovery of valid constraints on the way the world is structured — constraints that provide sufficient information to allow the processing to succeed. Consider stereopsis, which presupposes continuity and uniqueness in the world, or structure from visual motion, which presupposes rigidity, or shape from contour, which presupposes the three restrictions just discussed. The discovery of constraints that are valid and universal leads to results about vision that have the same quality of permanence as results in other branches of science.

The second point is that the critical issues for vision seem to us to revolve around the nature of the representations and the nature of the processes that create, maintain, and eventually interpret them. We have suggested an overall framework for visual information processing that includes three categories of representation upon which the processing is to operate. The first encompasses representations of intensity variations and their local geometry in the input to the visual system. One among these, the primal sketch, is expressly intended to be an efficient description of these variations which captures just that information required by the image analysis to follow. The second category encompasses the representations of visible surfaces — the descriptions, in other words, of the physical properties of the surfaces that

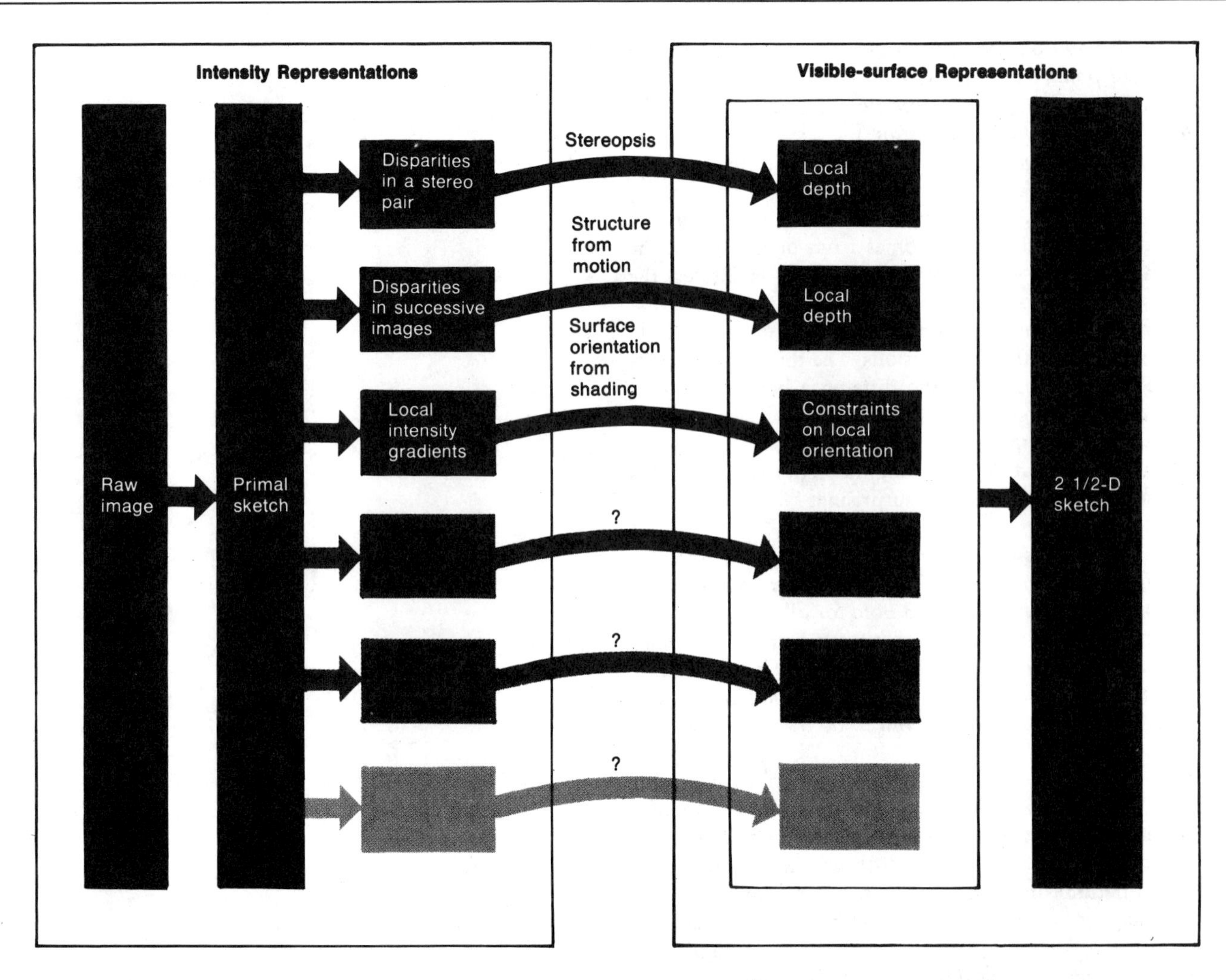

A framework for early and intermediate stages in a theory of visual information processing as proposed by the authors. The computations begin with representations of the intensities in an image — first the image itself, such as the gray-level intensity array shown on page 3, and then the primal sketch, a representation of spatial variations in intensity. Next comes the operation of a set of modules, each employing certain aspects of the information contained in the image to derive information about local orientation, local depth, and the boundaries of surfaces (Further details on the two uppermost modules are supplied in the text.) From this is constructed the so-called 2½-dimensional sketch, as shown on page 15. Note that no "higher-level" information is yet brought to bear: the computations proceed by utilizing only what is available in the image itself.

caused the images in the first place. The nature of these representations — the 2½-dimensional sketch in particular — is determined primarily by what information can be extracted by modules of image analysis such as stereopsis and structure from motion. Like the primal sketch of the previous category, the 2½-dimensional sketch is intended to be a final or output representation: this is where the separate contributions from the various image-analysis modules can be combined into a unified description. The third category encompasses all representations which are subsequently constructed from information contained in the 2½-D sketch. The designs of these tertiary representations are determined largely by the use to which they are to be put, as was the case for the 3-D model representation, to be used for shape recognition. If one had wanted instead, for example, to represent a shape simply for later *reproduction,* say by the milling of a block of metal, then the 2½-D sketch would itself have been sufficient, as the milling process depends explicitly on information about local depth and orientation, such as that sketch can provide.

We conclude with some observations on artificial intelligence in general. First a definition: "Artificial Intelligence" is (or ought to be) the study of information processing problems that characteristically have their roots in some aspect of biological information processing. The goal of the subject is to identify useful information processing problems, and give an abstract account of how to solve them. Such an account is essentially what we have been calling a computational theory — the uppermost of the four levels of understanding described at the outset of this article — and it corresponds to a theorem in mathematics. Once a computational theory has been discovered for solving a problem, the final stage is to develop algorithms that suit it. The choice of an algorithm usually depends upon the hardware available, and there may be many algorithms that implement the same computation. This is not to say that devising suitable algorithms will typically be easy once the computational theory is known, but it is to insist that before one can devise them, one has

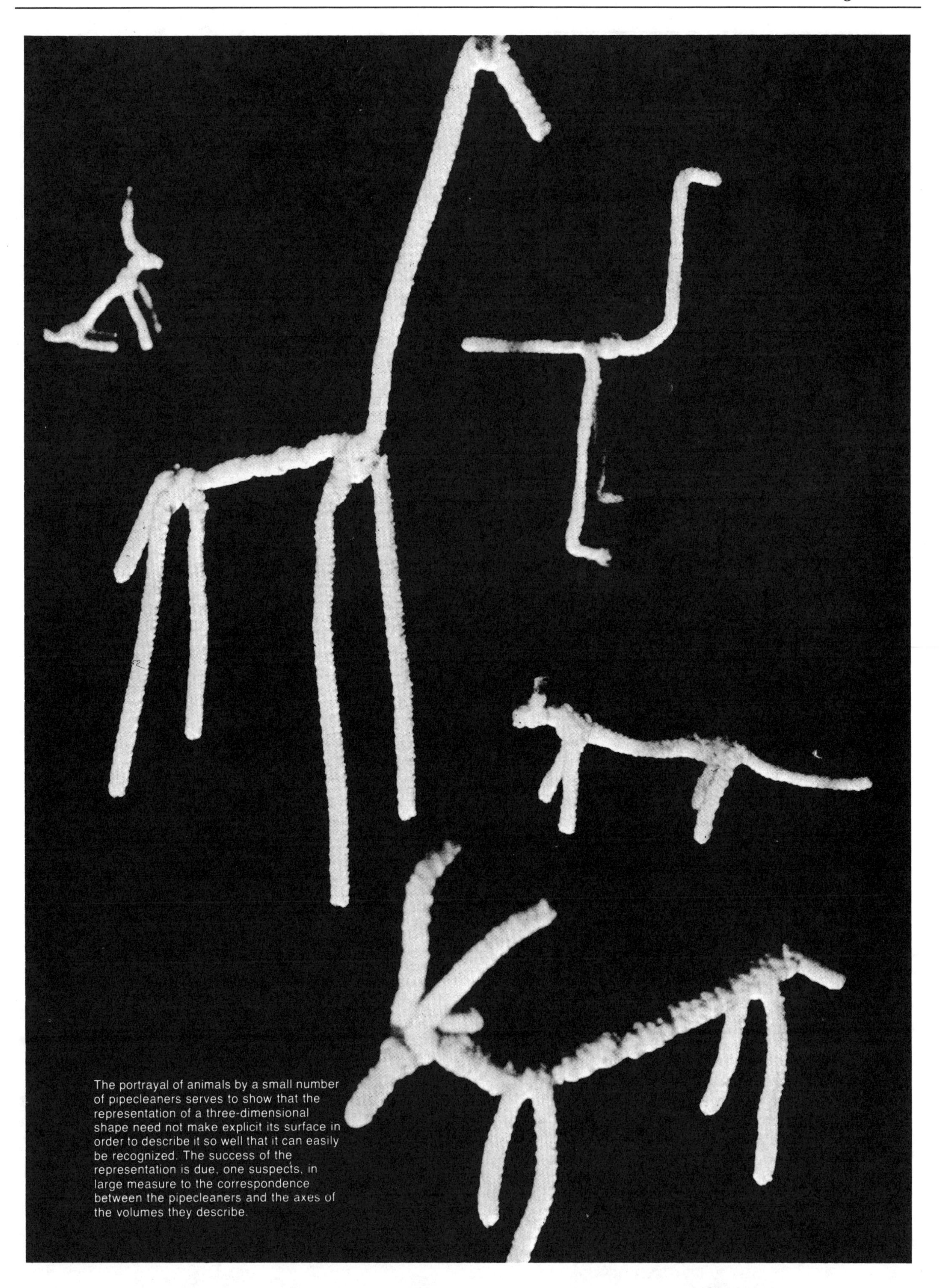

The portrayal of animals by a small number of pipecleaners serves to show that the representation of a three-dimensional shape need not make explicit its surface in order to describe it so well that it can easily be recognized. The success of the representation is due, one suspects, in large measure to the correspondence between the pipecleaners and the axes of the volumes they describe.

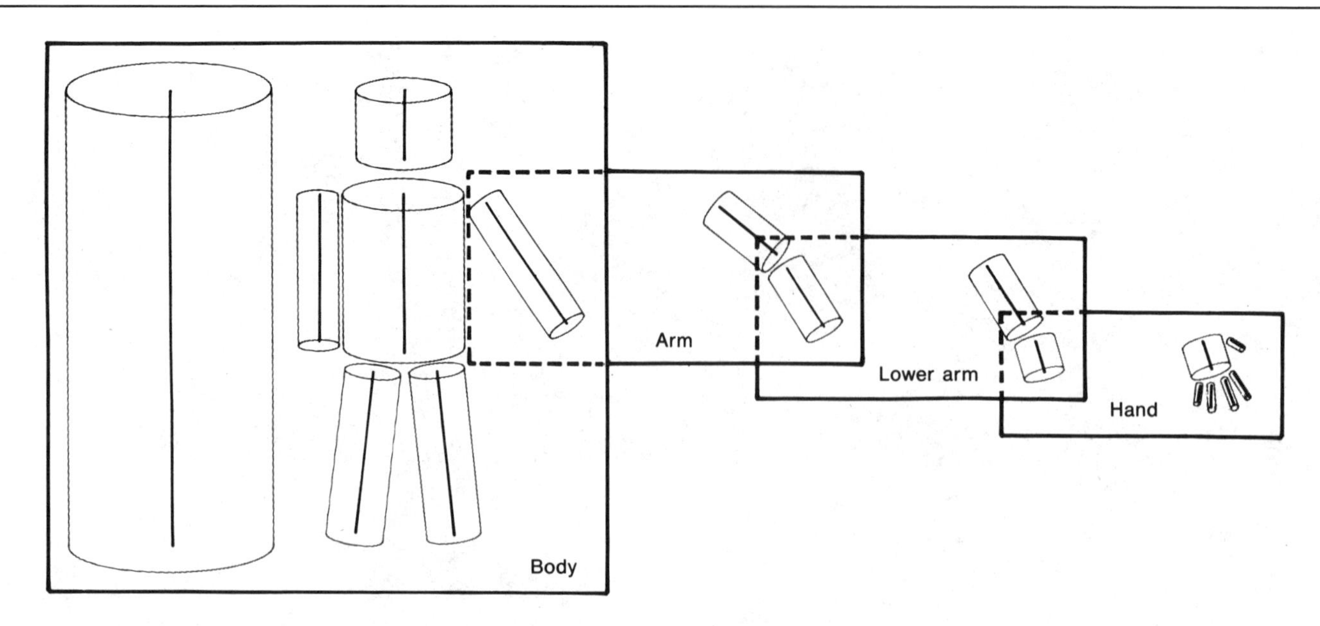

to know what exactly it is that they are supposed to be doing. When a problem in biological information processing decomposes in this way, we shall refer to it as having a *Type I* theory.

The fly in the ointment is that while many problems of biological information processing may turn out to have a Type I theory, there is no reason why all of them should. Consider in particular a problem that is solved by the simultaneous activity of a considerable number of processes *whose interaction is their own simplest description.* One possible example is the problem of predicting how a protein will fold, since it appears that a large number of influences act concurrently upon a large polypeptide chain as it flaps and flails in a medium. To be sure, only a few of the possible interactions will be important at any one moment, and any attempt to construct a simplified theory must ignore some of the conceivable interactions; but if most interactions are crucial at some stage during the folding, then the simplified theory will prove to be inadequate. As it happens, the most promising studies of protein folding are currently those that take a brute-force approach, setting up a rather detailed model of the amino acids, the geometry associated with their sequence, interactions with the circumambient fluid, random thermal perturbations, *etc.,* and letting the whole set of processes run until a stable configuration is achieved. We shall refer to such a situation as a *Type II* theory.

Now the principal difficulty in artificial intelligence is that one can never be quite sure whether a problem has a Type I solution. If one is found, well and good; but failure to find one does not mean that it does not exist. In particular, if one produces a large and clumsy set of processes that solves a problem, one cannot always be sure that there isn't a simple underlying computational theory whose formulation has somehow been lost in the fog. This danger is most acute in premature assaults on a high-level problem, for which few or none of the concepts that underlie its eventual decomposition into Type I

		Origin location			**Part orientation**		
Shape	Part	ρ	r	θ	i	ϕ	s
Human	head	DE	AB	NN	NN	NN	AB
	arm	DE	CC	EE	SE	EE	BC
	arm	DE	CC	WW	SE	WW	BC
	torso	CC	AB	NN	NN	NN	BC
	leg	CC	CC	EE	SS	NN	CC
	leg	CC	CC	WW	SS	NN	CC
Arm	upper arm	AA	AA	NN	NN	NN	CC
	lower arm	CC	AA	AA	NE	NN	CC
Lower Arm	forearm	AA	AA	NN	NN	NN	DD
	hand	DD	AA	NN	NN	NN	BB
Hand	palm	AA	AA	NN	NN	NN	CC
	thumb	AA	BB	NN	NE	NN	BC
	finger	CC	BB	NN	NN	NN	CC
	finger	CC	AB	NN	NN	NN	CC
	finger	CC	AB	SS	NN	NN	CC
	finger	CC	BB	SS	NN	NN	CC

The arrangement of 3-D models into the representation of a human shape. First the overall form — the "body" — is given an axis. This yields an object-centered coordinate system which can then be used to specify the arrangement of the "arms," "legs," "torso," and "head." The position of each of these is specified by an axis of its own, which in turn serves to define a coordinate system for specifying the arrangement of further subsidiary parts. This gives us a hierarchy of 3-D models: we show it extending downward as far as the fingers. The shapes in the figure are drawn as if they were cylindrical, but that is purely for illustrative convenience: it is the axes alone that stand for the volumetric qualities of the shape, much as the pipecleaners on page 17 serve in themselves to describe the various animals. The illustration also includes a printout of the 3-D model representation as it is stored for use in a computer. The essence of the coding is to express how the various subsidiary axes relate to the shape as whole: where are they, which way are they pointing, and how long are they? For each of the modules, the first three quantities shown in the computer code specify the location of the proximal end of the axis: ρ gives its position along the length of the axis of the overall shape, r gives its distance outward therefrom, and θ gives the angle at which it is found. The last three quantities specify the orientation of the subsidiary axis. Two angles, i and ϕ, serve to give its direction, and a number, s, gives its length. In all cases, angles are specified by a set of compass directions, and lengths by a system of line-segment names; the details need not concern us. Note, however, that there is no *a priori* reason why this scheme ought to be favored; it is simply a possible way to describe a shape in a form that is volumetric, modular, and independent of vantage point.

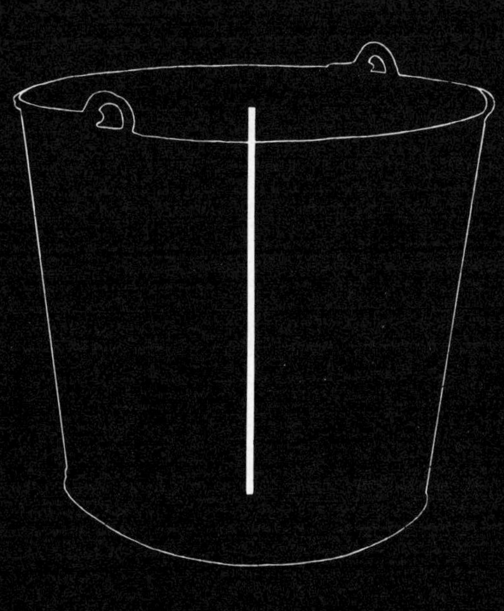

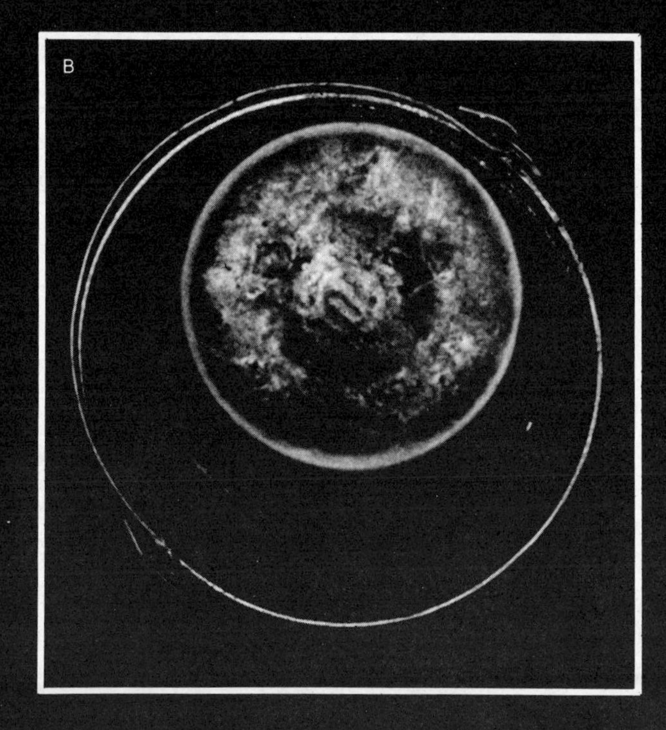

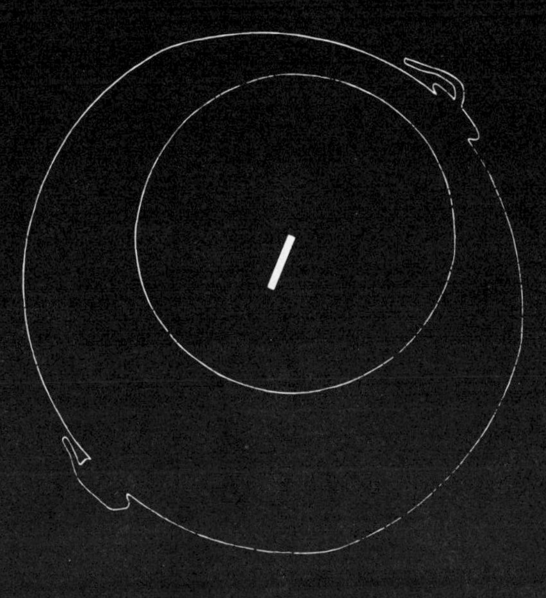

Two views of a water-pail. We display them because Warrington and Taylor reported in 1973 that patients with certain lesions in the right parietal lobe have difficulty in recognizing objects in views such as the one shown in (B). Consider, therefore, that the axis of the water-pail is directly recoverable from an image such as (A), but not from (B), where it is severely foreshortened, as shown by the line drawings that compose the right half of the figure. Consider also that in the 3-D model representation the recognition of a three-dimensional shape relies on the explicit representation of just such an axis. One thus is led by the theory itself to conclude that the recognition of views such as (B) will require considerably more computation than that required for (A).

theories have yet been developed, and one runs the risk of failing to formulate correctly the problems that in fact are involved. In the work of our own group, it first appeared, for example, that image analysis would require a Type II theory. But as more information came to light, we began to see how the analysis might decompose into separate modules for computing certain aspects of visual information — motion, stereoscopy, fluorescence, color — each one of *these* with a theory of Type I. After all, there is no reason why a single theory should encompass the whole. Indeed, one would *a priori* expect the opposite; that as evolution progresses, new modules come into existence that can cope with yet more aspects of the data, and as a result keep the animal alive in ever more widely ranging circumstances. The only important constraint is that the system as a whole should be roughly modular, so that new facilities can be added easily.

Yet even if there turns out to be a Type I theory, or a set of Type I theories, for the extraction of information from sensory data, there would still be no reason why that theory or theories should bear much relation to the theory of more central phenomena. In vision, for example, the theory that says 3-D representations are based on stick-figure coordinate systems and shows how to manipulate them is independent of the theory of the primal sketch, or for that matter of most other stages *en route* from the image to that representation. In short, it is dangerous to suppose that a theory of a peripheral process has any significance for higher level operations.

What, then, shall we say of intelligence? Many people in the field expect that, deep in the heart of our understanding, there will eventually lie at least one and probably several important principles about how to organize and represent knowledge that in some sense captures what is important about the *general* nature of our intellectual abilities. While still somewhat cloudy, the ideas that seem to be emerging are the following:

1. *That the "chunks" of related knowledge for reasoning, language, memory, or perception ought to be larger and have more flexibility in their structure than most recent theories in psychology have allowed.*
2. *That the perception of an object or of an event must include the simultaneous computation of several different descriptions — descriptions that capture diverse aspects of the use, purpose, or circumstances of the object or event.*
3. *That the various descriptions include coarse versions as well as fine ones; for the coarse descriptions are a vital link in establishing correctly the roles played by objects and events.*

An example will help to make these points clear. If one reads

A. *The fly buzzed irritatingly on the window-pane.*
B. *John picked up the newspaper.*

the immediate inference is that John's intentions towards the fly are fundamentally malicious. If he had picked up

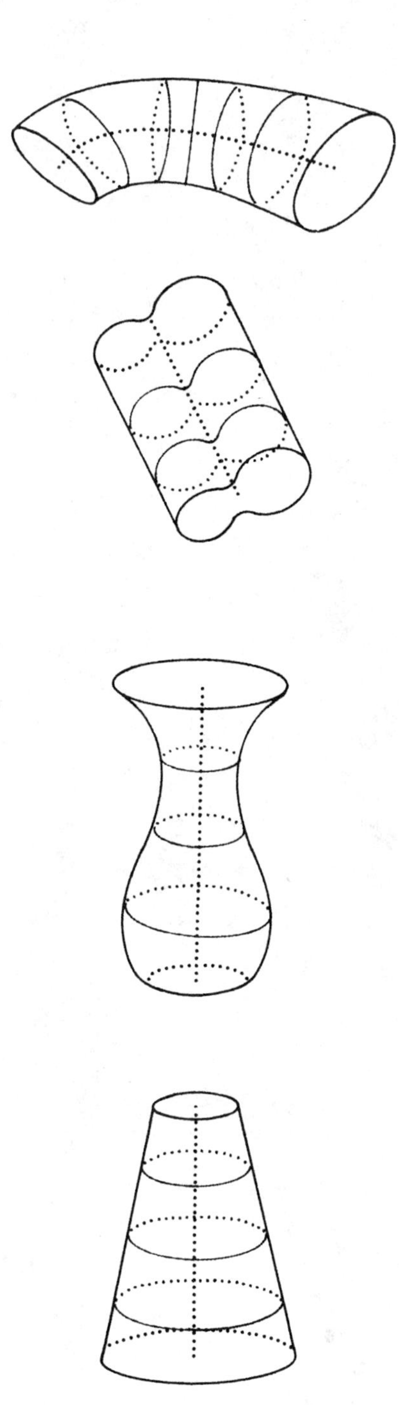

The definition of a generalized cone. In this article, it is the surface created by moving a cross-section along a given straight axis. The cross-section may vary smoothly in size, but its shape remains constant. We here show several examples. In each, the cross-section is shown at several positions along the trajectory that spins out the construction.

the telephone, the inference would be less secure. It is generally agreed that an "insect-damaging" scenario is somehow deployed during the reading of these sentences, being suggested in its coarsest form by the fly buzzing irritatingly. Such a scenario will contain a reference to something that can squash an insect on a window's brittle surface — a description which fits a newspaper but not a hammer. We might therefore conclude that when the newspaper is mentioned (or in the case of vision, seen) not only is it described internally as a newspaper, and some rough 3-D description of its shape and axes set up; it also is described as a light, flexible object with area. Indeed, because sentence (B) might have continued with the words "and sat down to read," the newspaper may also be being described as reading-matter; similarly, as a combustible article, and so forth. It follows that most of the time, a given object or event will give rise to several different coarse internal descriptions. After all, one seldom knows in advance what aspect of an object or event is important. Notice that the description of fly-swatting or reading or fire-lighting does not have to be attached to the newspaper; merely that a description of the newspaper is available that will match its role in each scenario.

The importance of a primitive, coarse catalogue of objects and events lies in the role such coarse descriptions play in the ultimate construction of exquisitely tailored specific scenarios, rather in the way that a general 3-D model description of a shape — one in which only the first level in a hierarchy of stick axes has been specified — can be elaborated as further visual information becomes available to produce eventually a very specific interpretation. What after sentence (A) existed as little more than malicious intent towards the innocent fly becomes, with the additional information about the newspaper, a very specific case of fly-squashing.

Exactly what descriptions should accompany different words or perceived objects is not yet known. In fact, the problems to which we now are led have yet to be precisely formulated, let alone satisfactorily solved. But it seems certain that some problems of this kind do exist and are important; and it seems likely that a fairly respectable theory of them will eventually emerge.

One last observation. It sometimes happens that researchers postulate a particular mechanism or programming style as a central element of the human processor. They then use this mechanism to mimic some small aspect of human performance, for example by writing a language-understanding program, a problem-solving program, or an associative-memory program — each of these applicable only in a highly specialized domain. We believe that such studies are misguided, and the reason is this. If one believes that the aim of information-processing studies is to formulate and understand particular information-processing problems, then it is the structure of those problems that is central, not the mechanisms through which their solutions are implemented. Therefore, the first thing to do is to find operations that we as human beings perform well, fluently, reliably, and hence unconsciously, since it is difficult to see how reliability could be achieved if there were no sound underlying

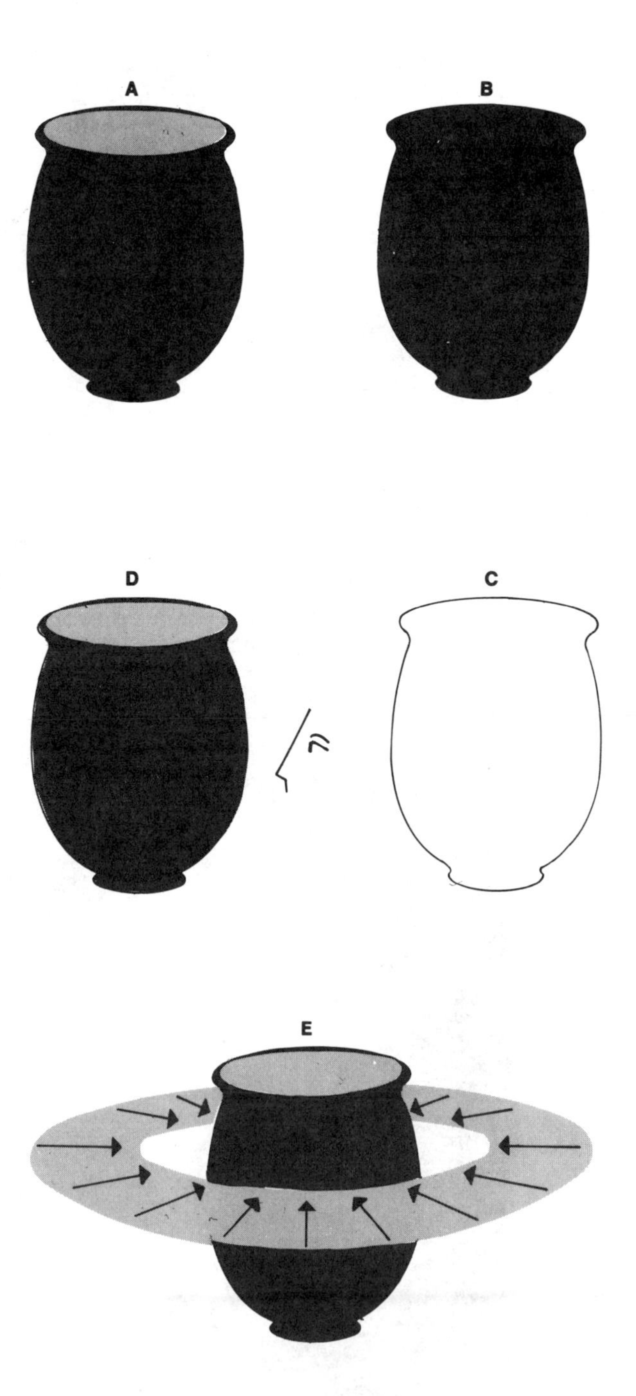

Four structures of importance in studying the *a priori* conditions that we bring to bear on the analysis of a contour. Part (A) shows a three-dimensional surface, Σ. Part (B) shows its silhouette S_V as seen from viewpoint V. Part (C) shows the contour C_V of S_V. Part (D) shows the set of points Γ that project onto the contour. A further part of the illustration shows a condition for a theorem discussed in the text. Here, in particular, the meaning of "all distant viewing directions that lie in a plane" is schematically shown.

computational theory. The next thing is to find out how to do them, and the next after that is to examine our performance in the light of our new understanding. In contrast to all this, current problem-solving research has tended to concentrate on problems that we understand well intellectually, but in fact *perform* poorly, such as mental arithmetic, in which one tries to add, multiply, *etc.* without aids such as pencil and paper; or crypt-arithmetic (for instance, DONALD plus GERALD equals ROBERT, where each letter stands for a digit whose identity is to be found). In other intances the research centers on theorem-proving in geometry or on games such as chess, in which human skills seem to rest on a huge base of knowledge and expertise. We argue that these are exceptionally good grounds for *not* studying how we carry out such tasks — at least not yet. There can be no doubt that when we do mental arithmetic we are doing *something* well, but it is not arithmetic, and we seem far from understanding even one component of what that something is. Let us therefore concentrate on the simpler problems first, for there we have some genuine hope of advancing.

For Further Reading

Levels of Understanding
Marr, D. & Poggio, T. (1977) "From Understanding Computation to Understanding Neural Circuitry." *Neurosciences Res. Prog. Bull. 15,* 470-488. Also available as M.I.T. A.I. Lab. Memo 357.

Conventional Approaches
Newton, I. (1704) *Optics.* London.
Helmholtz, H. L. F. von (1910) *Treatise on Physiological Optics.* Translated by J. P. Southall, 1925. New York: Dover Publications.
Werthheimer, M. (1923) "Principles of Perceptual Organization." In W. H. Ellis, *Source Book of Gestalt Psychology.* London and New York, 1938.
Kuffler, S. W. (1953) "Discharge Patterns and Functional Organization of Mammalian Retina." *J. Neurophysiol. 16,* 37-68.
Kuffler, S. W. & Nicholls, J. G. (1976) *From Neuron to Brain.* Sunderland, Massachusetts: Sinauer Associates.
Hubel, D. H. & Wiesel, T. N. (1962) "Receptive Fields, Binocular Interaction and Functional Architecture in the Cat's Visual Cortex." *J. Physiol., Lond. 160,* 106-154.
Julesz, B. (1971) *Foundations of Cyclopean Perception.* Chicago: The University of Chicago Press.
Shepard, R. N. & Metzler, J. (1971). "Mental Rotation of Three-Dimensional Objects." *Science 171,* 701-703.

A Computational Approach to Vision
Marr, D. (1976) "Early Processing of Visual Information." *Phil. Trans. Roy. Soc. B. 275,* 483-524. Also available as M.I.T. A.I. Lab. Memo 340.

Image Segmentation
Tenenbaum, J. M. & Barrow, H. G. (1976) "Experiments in Interpretation-Guided Segmentation." *Stanford Research Institute Technical Note 123.*
Freuder, E. C. (1975). "A Computer Vision System for Visual Recognition Using Active Knowledge." M.I.T. A.I. Lab. Technical Report 345.

"Rites of Spring," by Pablo Picasso. We immediately interpret such silhouettes in terms of particular three-dimensional surfaces — this despite the paucity of information in the image itself. In order to do this, we plainly must unconsciously invoke certain a priori assumptions and constraints about the nature of the shapes. Further details are discussed in the text.

Image-Analysis Modules
Marr, D. & Poggio, T. (1976) "Cooperative Computation of Stereo Disparity." *Science 194,* 283-287. Also available as M.I.T. A.I. Lab. Memo 364.
Marr, D. & Poggio, T. (1977) "A Theory of Human Stereo Vision." To appear in *Proc. Roy. Soc. Lond.* Also available as M.I.T. A.I. Lab. Memo 451.
Marr, D., Poggio, T. & Palm, G. (1977) "Analysis of a Cooperative Stereo Algorithm." *Biol. Cybernetics 28,* 223-239. Also available as M.I.T. A.I. Lab. Memo 446.
Marr, D. (1977) "Representing Visual Information." *AAAS 143rd Annual Meeting, Symposium on Some Mathematical Questions in Biology,* February (in press). Also available as M.I.T. A.I. Lab. Memo 415.
Ullman, S. (1976) "On Visual Detection of Light Sources." *Biol. Cybernetics 21,* 205-212.
Ullman, S. (1977) "The Interpretation of Visual Motion." M.I.T. Ph. D. Thesis, June.
Ullman, S. (1978) "The Interpretation of Structure from Motion." *Proc. Roy. Soc.,* forthcoming.
Ullman, S. (1978) "Two-Dimensionality of the Correspondence Process in Apparent Motion." *Perception 5,* forthcoming.
Ullman, S. (1978) *The Interpretation of Visual Motion.* M.I.T. Press, forthcoming.
Horn, B. K. P. (1975) "Obtaining Shape from Shading Information." In *The Psychology of Computer Vision,* ed. P. H. Winston. McGraw-Hill, New York, pp. 115-155.
Land, E. H. & McCann, J. J. (1971) "Lightness and Retinex Theory." *J. Opt. Soc. Am. 61,* 1-11.

Later Processing Problems
Blum, H. (1973) "Biological Shape and Visual Science (part 1)." *J. Theor. Biol. 38,* 205-287.
Binford, T. O. (1971) "Visual Perception by Computer." Presented to the I.E.E.E. Conference on Systems and Control, Miami, December.
Agin, G. J. (1972) "Representation and Description of Curved Objects." Stanford Artificial Intelligence Project, Memo AIM-173, Stanford University.
Nevatia, R. (1974) "Structured Descriptions of Complex Curved Objects for Recognition and Visual Memory." Stanford Artificial Intelligence Project, Memo AIM-250, Stanford University.

3-D Model Representation
Marr, D. & Nishihara, H. K. (1978) "Representation and Recognition of the Spatial Organization of Three-Dimensional Shapes." *Proc. Roy. Soc. B. 200,* 269-294. Also available as M.I.T. A.I. Lab. Memo 416.
Warrington, E. K. & Taylor, A. M. (1973). "The Contribution of the Right Parietal Lobe to Object Recognition." *Cortex 9,* 152-164.
Marr, D. (1977) "Analysis of Occluding Contour." *Proc. Roy. Soc. B. 197,* 441-475. Also available as M.I.T. A.I. Lab. Memo 372.

The Search for a Theory
Marr, D. (1977) "Artificial Intelligence — A Personal View." *Artificial Intelligence 9,* 37-48.
Levitt, M. & Warshel, A. (1975) "Computer Simulation of Protein Folding." *Nature 253,* 694-698.
Minsky, M. (1975) "A Framework for Representing Knowledge." In *The Psychology of Computer Vision,* ed. P. H. Winston. New York: McGraw-Hill, pp. 211-277.
Newell, A. & Simon, H. A. (1972) *Human Problem Solving.* New Jersey: Prentice Hall.

David Marr is Associate Professor of Psychology at the Massachusetts Institute of Technology. He was educated at Trinity College in the University of Cambridge, receiving his B.A. and M.A. in mathematics and his Ph.D. in neurophysiology, but his education also included training in neuroanatomy, psychology, and biochemistry. After research positions in Britain, he came to the U.S. as an invited visitor at M.I.T. and at the California Institute of Technology. He is author of numerous publications, some of them on theories of neocortex, cerebellar cortex, hippocampus, and the retina. **H. Keith Nishihara** is a Research Associate at the M.I.T. Artificial Intelligence Laboratory. He received his B.A. and M.A., both in mathematics, from the University of Hawaii, and his Ph.D., again in mathematics, from M.I.T., where his thesis, under Marr, was on the "Representation of the Spatial Organization of Three-Dimensional Shapes for Recognition." From 1972 to 1975 he was a National Science Foundation Graduate Fellow.

Computational vision and regularization theory

Tomaso Poggio, Vincent Torre* & Christof Koch

Artificial Intelligence Laboratory and Center for Biological Information Processing, Massachusetts Institute of Technology, 545 Technology Square, Cambridge, Massachusetts 02193, USA
* Istituto di Fisica, Universita di Genova, Genova, Italy

Descriptions of physical properties of visible surfaces, such as their distance and the presence of edges, must be recovered from the primary image data. Computational vision aims to understand how such descriptions can be obtained from inherently ambiguous and noisy data. A recent development in this field sees early vision as a set of ill-posed problems, which can be solved by the use of regularization methods. These lead to algorithms and parallel analog circuits that can solve 'ill-posed problems' and which are suggestive of neural equivalents in the brain.

COMPUTATIONAL vision denotes a new field in artificial intelligence, centred on theoretical studies of visual information processing. Its two main goals are to develop image understanding systems, which automatically construct scene descriptions from image input data, and to understand human vision.

Early vision is the set of visual modules that aim to extract the physical properties of the surfaces around the viewer, that is, distance, surface orientation and material properties (reflectance, colour, texture). Much current research has analysed processes in early vision because the inputs and the goals of the computation can be well characterized at this stage (see refs 1-4 for reviews). Several problems have been solved and several specific algorithms have been successfully developed. Examples are stereomatching, the computation of the optical flow, structure from motion, shape from shading and surface reconstruction.

A new theoretical development has now emerged that unifies much of these results within a single framework. The approach has its roots in the recognition of a common structure of early vision problems. Problems in early vision are 'ill-posed', requiring specific algorithms and parallel hardware. Here we introduce a specific regularization approach, and discuss its implications for computer vision and parallel computer architectures, including parallel hardware that could be used by biological visual systems.

Early vision processes

Early vision consists of a set of processes that recover physical properties of the visible three-dimensional surfaces from the two-dimensional intensity arrays. Their combined output roughly corresponds to Marr's 2-1/2D sketch[1], and to Barrow and Tennenbaum's intrinsic images[5]. Recently, it has been customary to assume that these early vision processes are general and do not require domain-dependent knowledge, but only generic constraints about the physical word and the imaging stage (see box). They represent conceptually independent modules that can be studied, to a first approximation, in isolation. Information from the different processes, however, has to be combined. Furthermore, different modules may interact early on. Finally, the processing cannot be purely 'bottom-up': specific knowledge may trickle down to the point of influencing some of the very first steps in visual information processing.

Examples of early vision processes

- **Edge detection**
- **Spatio–temporal interpolation and approximation**
- **Computation of optical flow**
- **Computation of lightness and albedo**
- **Shape from contours**
- **Shape from texture**
- **Shape from shading**
- **Binocular stereo matching**
- **Structure from motion**
- **Structure from stereo**
- **Surface reconstruction**
- **Computation of surface colour**

Computational theories of early vision modules typically deal with the dual issues of representation and process. They must specify the form of the input and the desired output (the representation) and provide the algorithms that transform one into the other (the process). Here we focus on the issue of processes and algorithms for which we describe the unifying theoretical framework of regularization theories. We do not consider the equally important problem of the primitive tokens that represent the input of each specific process.

A good definition of early vision is that it is inverse optics. In classical optics or in computer graphics the basic problem is to determine the images of three-dimensional objects, whereas vision is confronted with the inverse problem of recovering surfaces from images. As so much information is lost during the imaging process that projects the three-dimensional world into the two-dimensional images, vision must often rely on natural constraints, that is, assumptions about the physical world, to derive unambiguous output. The identification and use of such constraints is a recurring theme in the analysis of specific vision problems.

Two important problems in early vision are the computation of motion and the detection of sharp changes in image intensity (for detecting physical edges). They illustrate well the difficulty of the problems of early vision. The computation of the two-dimensional field of velocities in the image is a critical step in several schemes for recovering the motion and the three-dimensional structure of objects. Consider the problem of determining the velocity vector $\boldsymbol{V}$ at each point along a smooth contour in the image. Following Marr and Ullman[6], one can assume that the contour corresponds to locations of significant intensity change. Figure 1 shows how the local velocity vector is decomposed into a normal and a tangential component to the curve. Local motion measurements provide only the normal component of velocity. The tangential component remains 'invisible' to purely local measurements (unless they refer to some discontinuous features of the contour such as a corner). The problem of estimating the full velocity field is thus, in general, underdetermined by the measurements that are directly available from the image. The measurement of the optical flow is inherently ambiguous. It can be made unique only by adding information or assumptions.

The difficulties of the problem of edge detection are somewhat different. Edge detection denotes the process of identifying the

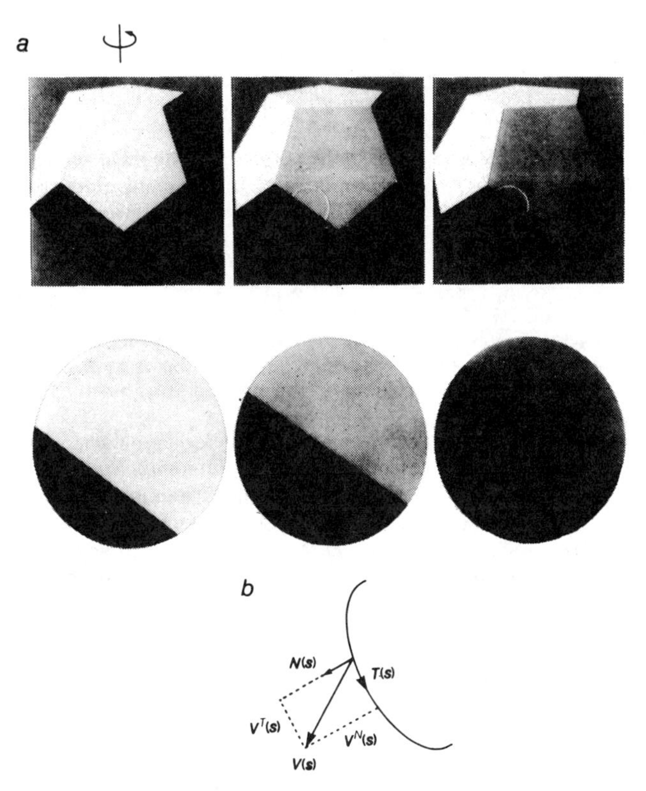

Fig. 1 Ambiguity of the velocity field. *a*, Local measurements cannot measure the full velocity field in the image plane, originated here by three-dimensional rotation of a solid object (three frames are shown). Any process operating within the aperture (shown as a white circle) can compute only the component of motion perpendicular to the contour. *b*, Decomposition of the velocity vector along the contour, parametrized by the arc length s into components normal ($V^N(s)$) and tangential ($V^T(s)$) to the curve. The computer drawing was kindly provided by Karl Sims.

physical boundaries of three-dimensional surfaces from intensity changes in their image. What is usually intended with edge detection is a first step towards this goal, that is, detecting and localizing sharp changes in image intensity. This is a problem of numerical differentiation of image data, which is plagued by the noise unavoidable during the imaging and the sampling processes. Differentiation amplifies noise and this process is thus inherently unstable. Figure 3 shows an example of an edge profile and its second derivative, where noise is significantly amplified. Most problems in early vision present similar difficulties. They are mostly underconstrained, as in the computation of the optical flow, or not robust against noise, as in edge detection.

Ill-posed problems

The common characteristics of most early vision problems (in a sense, their deep structure) can be formalized: most early vision problems are ill-posed problems in the precise sense defined by Hadamard[7,8]. This claim captures the importance of constraints and reflects the definition of vision as inverse optics.

Hadamard first introduced the definition of ill-posedness in the field of partial differential equations[9]. Although ill-posed problems have been considered for many years as almost exclusively mathematical curiosities, it is now clear that many ill-posed problems, typically inverse problems, are of great practical interest (for instance, computer tomography). A problem is well-posed when its solution exists, is unique and depends continuously on the initial data. Ill-posed problems fail to satisfy one or more of these criteria. Note that the third condition does not imply that the solution is robust against noise in practice. For this, the problem must not only be well-posed but also be well conditioned to ensure numerical stability[10].

It is easy to show formally that several problems in early vision are ill-posed in the sense of Hadamard[8]: stereo matching, structure from motion, computation of the optical flow, edge detection, shape from shading, the computation of lightness and surface reconstruction. Computation of the optical flow is ill-posed because the 'inverse' problem of recovering the full velocity field from its normal component along a contour fails to satisfy the uniqueness condition. Edge detection, intended as numerical differentiation, is ill-posed because the solution does not depend continuously on the data.

The main idea for 'solving' ill-posed problems, that is for restoring 'well-posedness', is to restrict the class of admissible solutions by introducing suitable *a priori* knowledge. *A priori* knowledge can be exploited, for example, under the form of either variational principles that impose constraints on the possible solutions or as statistical properties of the solution space. We will use the general term regularization for any method used to make an ill-posed problem well-posed. Variational regularization will indicate the regularization methods that reformulate an ill-posed problem in terms of a variational principle. We will next outline specific variational methods that we will denote as the standard regularization methods, attributable mainly to Tikhonov[11,12] (see also refs 13, 14). We will also outline future extensions of the standard theory from the perspective of early vision.

The regularization of the ill-posed problem of finding z from the 'data' y

$$Az = y \tag{1}$$

requires the choice of norms $\|\cdot\|$ and of a stabilizing functional $\|Pz\|$. In standard regularization theory, A is a linear operator, the norms are quadratic and P is linear. Two methods that can be applied are[8,13]: (1) among z that satisfy $\|Az - y\| \leq \varepsilon$ find z that minimizes (ε depends on the estimated measurement errors and is zero if the data are noiseless)

$$\|Pz\|^2 \tag{2}$$

(2) find z that minimizes

$$\|Az - y\|^2 + \lambda \|Pz\|^2 \tag{3}$$

where λ is a so-called regularization parameter.

The first method computes the function z that is sufficiently close to the data and is most 'regular', that is minimizes the 'criterion' $\|Pz\|^2$. In the second method, λ controls the compromise between the degree of regularization of the solution and its closeness to the data. Standard regularization theory provides techniques for determining the best λ[12,15]. Thus, standard regularization methods impose the constraints on the problem by a variational principle, such as the cost functional of equation (3). The cost that is minimized reflects physical constraints about what represents a good solution: it has to be both close to the data and regular by making the quantity $\|Pz\|^2$ small. P embodies the physical constraints of the problem. It can be shown for quadratic variational principles that under mild conditions the solution space is convex and a unique solution exists. It must be pointed out that standard regularization methods have to be applied after a careful analysis of the ill-posed nature of the problem. The choice of the norm $\|\cdot\|$, of the stabilizing functional $\|Pz\|$ and of the functional spaces involved is dictated both by mathematical properties and by physical plausibility. They determine whether the precise conditions for a correct regularization hold for any specific case.

Variational principles are used widely in physics, economics and engineering. In physics, for instance, most of the basic laws have a compact formulation in terms of variational principles that require minimization of a suitable functional, such as the energy or the lagrangian.

Examples

Variational principles of the form of equation (3) have been used in the past in early vision[16-25]. Other problems have now been approached in terms of standard regularization methods

Table 1 Regularization in early vision

Problem	Regularization principle
Edge detection	$\int [(Sf-i)^2+\lambda(f_{xx})^2]\,dx$
Optical flow (area based)	$\int [i_x u+i_y v+i_t)^2+\lambda(u_x^2+u_y^2+v_x^2+v_y^2)]\,dx\,dy$
Optical flow (contour based)	$\int [(\boldsymbol{V}\cdot\boldsymbol{N}-V^N)^2+\lambda((\partial/\partial_s)\boldsymbol{V})^2]\,ds$
Surface reconstruction	$\int [S\cdot f-d)^2+\lambda(f_{xx}^2+2f_{xy}^2+f_{yy}^2)^2]\,dx\,dy$
Spatiotemporal approximation	$\int [(S\cdot f-i)^2+\lambda(\nabla f\cdot \boldsymbol{V}+ft)^2]\,dx\,dy\,dt$
Colour	$\Vert I^\nu-Az\Vert^2+\lambda\Vert Pz\Vert^2$
Shape from shading	$\int [(E-R(f,g))^2+\lambda(f_x^2+f_y^2+g_x^2+g_y^2)]\,dx\,dy$
Stereo	$\int \{[\nabla^2 G*(L(x,y)-R(x+d(x,y),y))]^2 + \lambda(\nabla d)^2\}\,dx\,dy$

Some of the early vision problems that have been solved in terms of variational principles. The first five are standard quadratic regularization principles. In edge detection[26,27] the data on image intensity ($i=i(x)$) (for simplicity in one dimension) are given on a discrete lattice: the operator S is the sampling operator on the continuous distribution f to be recovered. A similar functional may be used to approximate time-varying imagery. The spatio-temporal intensity to be recovered from the data $i(x,y,t)$ is $f(x,y,t)$; the stabilizer imposes the constraint of constant velocity $\boldsymbol{V}$ in the image plane (ref. 61). In area-based optical flow[18], i is the image intensity, u and v are the two components of the velocity field. In surface reconstruction[21,22] the surface $f(x,y)$ is computed from sparse depth data $d(x,y)$. In the case of colour[32] the brightness is measured on each of three appropriate colour coordinates $I^\nu(\nu=1,2,3)$. The solution vector z contains the illumination and the albedo components separately; it is mapped by A into the ideal data. Minimization of an appropriate stabilizer enforces the constraint of spatially smooth illumination and either constant or sharply varying albedo. For shape from shading[19] and stereo (T.P. and A. Yuille, unpublished), we show two non-quadratic regularization functionals. R is the reflectance map, f and g are related to the components of the surface gradient, E is the brightness distribution[19]. The regularization of the disparity field d involves convolution with the laplacian of a gaussian of the left (L) and the right (R) images and a Tikhonov stabilizer corresponding to the disparity gradient.

(see Table 1). Most stabilizing functionals used so far in early vision are of the Tikhonov type, being linear combinations of the first p derivatives of the desired solution z (ref. 12). The solutions arising from these stabilizers correspond to either interpolating or approximating splines. We return now to our examples of motion and edge detection, and show how standard regularization techniques can be applied.

Intuitively, the set of measurements of the normal component of velocity over an extended contour should provide considerable constraint on the global motion of the contour. Some additional assumptions about the nature of the real world are needed, however, in order to combine local measurements at different locations. For instance, the assumption of rigid motion on the image plane is sufficient to determine V uniquely[23,24]. In this case, local measurements of the normal component at different locations can be used directly to find the optical flow, which is the same everywhere. The assumption, however, is overly restrictive, because it does not cover the case of motion of a rigid object in three-dimensional space (see Fig. 1). Hildreth suggested[23,24], following Horn and Schunck[18], a more general smoothness constraint on the velocity field. The underlying physical consideration is that the real world consists of solid objects with smooth surfaces, whose projected velocity field is usually smooth. The specific form of the stabilizer (a Tikhonov stabilizer) was dictated by mathematical considerations, especially uniqueness of the solution. The two regularizing methods correspond to the two algorithms proposed and implemented by Hildreth[23]. The first one, which assumes that the measurements of the normal velocity components $V^N(s)$ are exact, minimizes

$$\Vert P\boldsymbol{V}\Vert^2=\int\left(\frac{\partial \boldsymbol{V}}{\partial s}\right)^2 ds \tag{4}$$

subject to the measurements of the normal component of velocity (where s is arc length). The integral is evaluated along the contour. For non-exact data the second method provides the solution by minimizing

$$\Vert \boldsymbol{V}\cdot\boldsymbol{N}-V^N\Vert^2+\lambda\int\left(\frac{\partial \boldsymbol{V}}{\partial s}\right)^2 ds \tag{5}$$

where $\boldsymbol{N}$ is the normal unit vector to the contour and λ^{-1} expresses the reliability of the data. Figure 2*a* shows an example of a successful computation of the optical flow by the first algorithm.

Recently, regularization techniques have been applied to edge detection[26,27]. The problem of numerical differentiation can be regularized by the second method with a Tikhonov stabilizer that reflects a constraint of smoothness on the image (see Table 1). The physical justification is that the image is an analytical function with bounded derivatives, because of the band-limiting properties of the optics that cuts off high spatial frequencies. This regularized solution is equivalent, under mild conditions, to convolving the intensity data with the derivative of a filter similar to the gaussian[26] (see Fig. 3), proposed earlier[28,29].

Other early vision problems can be solved by standard regularization techniques. Surface reconstruction, for example, can be performed from a sparse set of depth values by imposing smoothness of the surface[20-22]. Optical flow can be computed at each point in the image, rather than along a contour, using a constraint of smooth variation, in the form of a Tikhonov stabilizer[17]. Variational principles that are not exactly quadratic but have the form of equation (3) can be used for other problems in early vision. The main results of Tikhonov can in fact be extended to the case in which the operators A and P are nonlinear, provided they satisfy certain conditions[30]. The variation of an object's brightness gives clues to its shape: the surface orientation can be computed from an intensity image in terms of the variational principle shown in Table 1, which penalizes orientations violating the smoothness constraint and the irradiance constraint[18]. Stereo matching is the problem of inferring the correct binocular disparity (and therefore depth) from a pair of binocular images, by finding which feature in one image corresponds to which feature in the other image. This is an ill-posed problem which, under some restrictive conditions corresponding to the absence of occlusions, can be regularized by a variational principle that contains a term measuring the discrepancy between the feature maps extracted from the two images and a stabilizer that penalizes large disparity gradients (see Table 1) and effectively imposes a disparity gradient limit. The algorithm can reduce to an area-based correlation algorithm of the Nishihara type[31] if the disparity gradient is small. A standard regularization principle has been proposed for solving the problem of separating a material reflectance from a spatially varying illumination in colour images[32]. The algorithm addresses the problem known in visual psychophysics as colour constancy[33].

Physical plausibility and illusions

Physical plausibility of the solution, rather than its uniqueness, is the most important concern in regularization analysis. A physical analysis of the problem, and of its significant constraints, plays the main role[8]. The *a priori* assumptions required to solve ill-posed problems may be violated in specific instances where the regularized solution does not correspond to the physical solution. The algorithm suffers an optical illusion. A good example is provided by the computation of motion. The smoothness assumption of equation (5) gives correct results under some general conditions (for example, when objects have images consisting of connected straight lines[34]). For some classes of motion and contours, the smoothness principle will not yield

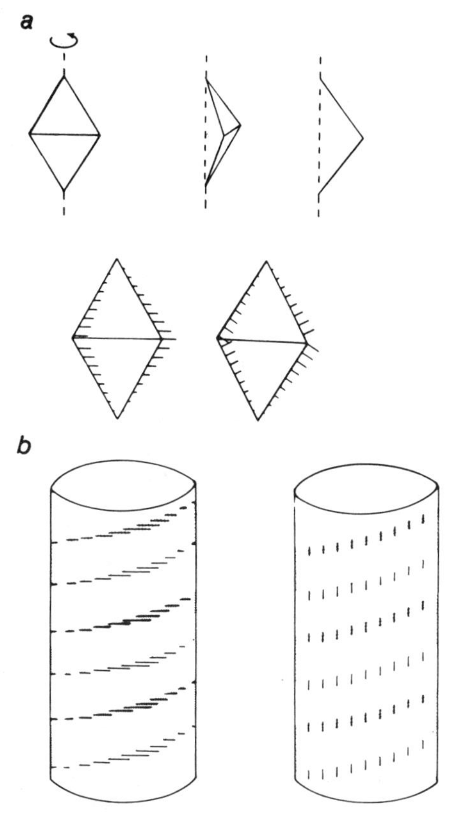

Fig. 2 Computing the smoothest velocity field along contours. *a*, Three-dimensional stimulus first used by Wallach[62] to demonstrate the ability of the human visual system to derive three-dimensional structure from the projected two-dimensional motion of an object (kinetic depth effect). The top part shows three views of a figure as it is rotated around the vertical axis. The initial measurements of the normal velocity components V_i^N are shown on the lower right. The velocity field computed using equation (4) is shown on the lower left. The final solution corresponds to the physical correct velocity distribution. Recent electrophysiological evidence implicates the middle temporal area of the monkey as a site where a similar motion integration may occur[63]. *b*, Circular helix on an imaginary three-dimensional cylinder, rotating about its vertical axis (barber pole). The projection of the curve onto the image plane, together with the resulting two-dimensional velocity vectors are drawn on the left. Although the true velocity field V is strictly horizontal (left), the smoothest velocity field (right) is vertical. This example illustrates a case where both the algorithm and the human visual system suffer the same optical illusion. Adapted from ref. 23.

the correct velocity field. In several of these cases, however, the human visual system also seems to derive a similar, incorrect velocity field, thereby possibly revealing *a priori* assumptions the brain is making about the world. A striking instance is the barber-pole illusion[23] (illustrated in Fig. 2*b*).

Analog networks

One of the mysteries of biological vision is its speed. Parallel processing has often been advocated as the answer to this problem. The model of computation provided by digital processes is, however, unsatisfactory, especially given the increasing evidence that neurones are complex devices, very different from simple digital switches. It is, therefore, interesting to consider whether the regularization approach to early vision may lead to a different type of parallel computation. We have recently suggested that linear, analog networks (either electrical or chemical) are, in fact, a natural way of solving the variational principles dictated by standard regularization theory[7] (see also refs 22, 35).

The fundamental reason for such a mapping between variational principles and electrical or chemical networks is Hamilton's least action principle. The class of variational principles that can be computed by analog networks is given by Kirchhoff's current and voltage laws, which represent conservation and continuity restrictions satisfied by each network component (appropriate variables are usually voltage and current for electrical networks and affinity and turnover rate for chemical systems[36]; see also ref. 37). There is in general no unique network but possibly many networks implementing the same variational principle. For example, graded networks of the type proposed by Hopfield in the context of associative memory[38] can solve standard regularization principles[39].

From Kirchhoff's law, it can be proved[7] that for every quadratic variational problem with a unique solution (which is usually the case[8]), there exists a corresponding electrical network consisting of resistances and voltage or current sources having the same solution. In other words, the steady-state current (or voltage) distribution in the network corresponds to the solution, for example to the tangential velocity distribution $V^T(s)$, of the standard regularization problem (Fig. 4). Furthermore, when capacitances are added to the system, thereby introducing dynamics, the system is stable. The data are supplied by injecting currents or by introducing batteries, that is by constant current or voltage sources[7].

This analog parallel model of computation is especially interesting from the point of view of the present understanding of the biophysics of neurones, membranes and synapses. Increasing evidence shows that electrotonic potentials play a primary role in many neurones[40]. Mechanisms as diverse as dendrodendritic synapses[41,42], gap junctions[43], neurotransmitters acting over different times and distances[44], voltage-dependent channels that can be modulated by neuropeptides[45] and interactions between synaptic conductance changes[46] provide neurones with various different circuit elements. Patches of neural membrane are equivalent to resistances, capacitances and phenomenological inductances[47]. Synapses on dendritic spines mimic voltage sources, whereas synapses on thick dendrites or the soma act as current sources[48,49]. Thus, single neurones or small networks of neurones could implement analog solutions of regularization principles. Hypothetical neuronal implementations of the analog circuits of Fig. 4 have been devised, involving only one or two separate dendrites[7].

Beyond standard regularization theory

The new theoretical framework for early vision clearly shows the attractions and the limitations that are intrinsic to the standard Tikhonov form of regularization theory. The main problem is the degree of smoothness required for the unknown function that has to be recovered. For instance, in surface interpolation, the degree of smoothness corresponding to the so-called thin-plate splines smoothes depth discontinuities too much, and often leads to unrealistic results[20] (discontinuities may, however, be detected and then used in a second regularization step[66]).

Standard regularization theory deals with linear problems and is based on quadratic stabilizers. It leads therefore to the minimization of quadratic functionals and to linear Euler-Lagrange equations. Non-quadratic functionals may be needed to enforce the correct physical constraints (Table 1 shows the non-quadratic case of shape-from-shading). Even in this case, methods of standard regularization theory can be used[30], but the solution space is no longer convex and many local minima can be found in the process of minimization.

A non-quadratic stabilizer has been proposed for the problem of preserving discontinuities in the reconstruction of surfaces from depth data[50]. The stabilizer, in its basic form attributable to Geman and Geman[51] (a similar principle but without a rigorous justification was proposed by Blake[52]; see also the variational continuity control of Terzopoulos[67]), embeds prior knowledge about the geometry of the discontinuities (the line process) and, in particular, that they are continuous and often straight contours. In standard regularization principles, the search space has only one local minimum to which suitable algorithms always converge. For non-quadratic functionals, the search space may be similar to a mountain range with many

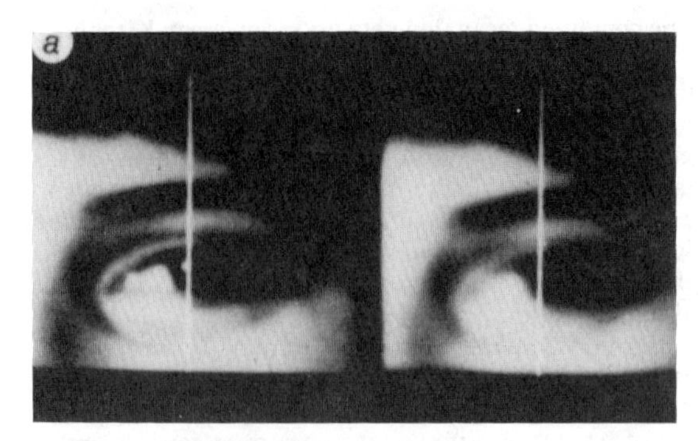

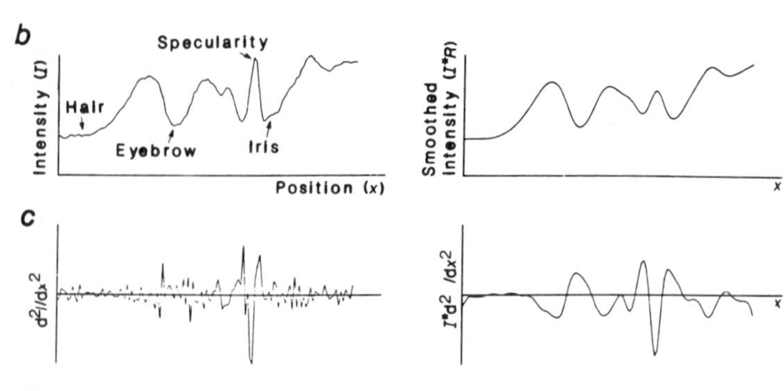

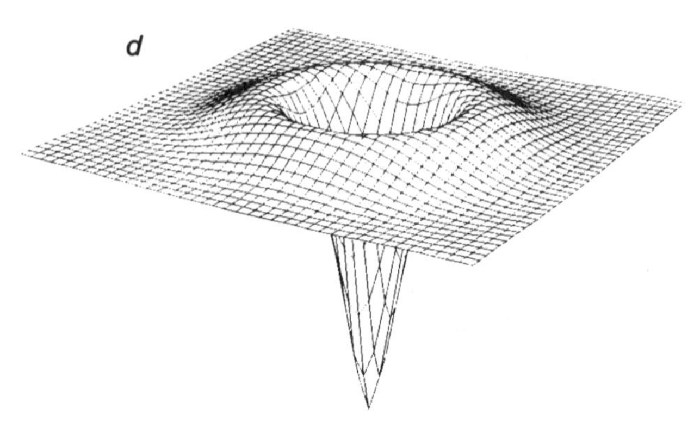

Fig. 3 A regularized solution to edge detection. *a*, Digital image (256×256 pixels) without any filtering (left) and filtered (right) with the two-dimensional operator provided by standard regularization theory[26] shown in *d*. *b*, Intensity profile along the scan line indicated in *a* without (left) and with (right) the regularizing operation provided by filtering with the one-dimensional regularized filter[26]. *c*, Second derivative of the profile shown in *b* without (left) and with (right) the one-dimensional regularizing filtering. *d*, Two-dimensional filter obtained by regularizing the ill-posed problem of edge detection[26]. It is a quintic spline, very similar to a gaussian distribution. The parameter λ controls the scale of the filter. Its value depends on the signal-to-noise ratio in the image. The spatial receptive field of most ganglion cells in the vertebrate retina have a very similar structure, with a central excitatory region and an inhibitory surround[64,65]. Drawings kindly provided by Harry Vorhees.

local minima. Stochastic algorithms for solving minimization problems of this type have been proposed recently, to escape from local minima at which simple hill-climbing algorithms would be trapped[53–55]. The basic idea is somewhat similar to adding a forcing noise term to the search algorithm. If the non-quadratic variational principle can be represented in a nonlinear analog network (as in ref. 39), an appropriate source of gaussian noise could drive the analog network. The dynamics of the system would then be described by a nonlinear stochastic differential equation, representing a diffusion process.

The challenge now for the regularization theory of vision is to extend it beyond standard regularization methods. The universe of computations that can be performed in terms of quadratic functionals is rather restricted. To see this, it is sufficient to realize that minimization of quadratic cost functionals leads to a linear regularization operator, that is, to a linear mapping of the input data into the solution space. In the special case when the data are on a regular grid and obey suitable conditions, the linear operator may become a convolution, that is, a simple filtering operation on the data. Similar to linear models in physics, standard regularization theory is an extremely useful approximation in many cases, but cannot deal with the full complexity of vision.

Stochastic route to regularization

A different rigorous approach to regularization is based on Bayes estimation and Markov random fields models. In this approach the *a priori* knowledge is represented in terms of appropriate probability distributions, whereas in standard regularization *a priori* knowledge leads to restrictions on the solution space. Consider as an example the case of surface reconstruction. *A priori* knowledge can be formulated in terms of a Markov random field (MRF) model of the surface. In a MRF the value at one discrete location depends only on the values within a given neighbourhood. In this approach the best surface maximizes some likelihood criterion such as the maximum *a posteriori* estimate or the *a posteriori* mean of the MRF. It has been pointed out[50] that the maximum *a posteriori* estimate of a MRF is equivalent to a variational principle of the general form of equation (3); the first term measures the discrepancy between the data and the solution, the second term is now an arbitrary potential function of the solution (defined on a discrete lattice). The overall variational principle, in general not quadratic, reduces to a quadratic functional of the standard regularization type when the noise is additive and gaussian and first-order differences of the field are zero-mean, independent, gaussian random variables. In this case the maximum *a posteriori* estimate (MAP) coincides with all estimates and, in particular, with the *a posteriori* mean. But Marroquin[56] has shown recently that this is not true in general: in most cases the MAP estimate is not optimal with respect to natural error measures and better estimates such as the *a posteriori* mean can be found. In these cases the problem is not equivalent to finding the global minimum of an energy functional: simulated annealing is not needed, and a Metropolis-type algorithm[55] can be used instead.

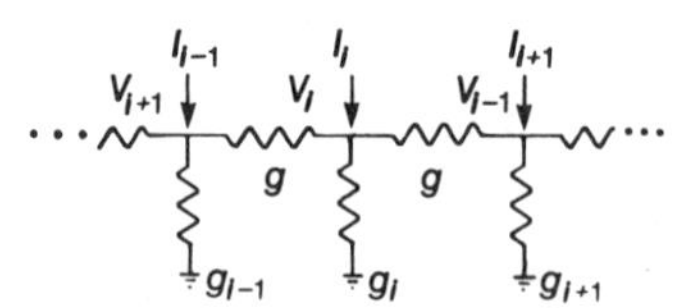

Fig. 4 Analog networks. A resistive network computing the smoothest velocity field[23]. The network corresponds to the situation where the measurements of the normal velocity component V^N are assumed to be exact. Discretizing the associated variational equation (4) along the contour yields the Euler–Lagrange equations $(2+\kappa_i^2)V_i^T - V_{i+1}^T - V_{i-1}^T = d_i$, where κ_i is the curvature of the contour at location i, d_i is a function of the data V_i^N and the contour and V_i^T is the unknown tangential component of the velocity V at location i along the contour. The equation describing the ith node in the electrical circuit is $(2g+g_i)V_i - gV_{i+1} - g_{i-1} = I_i$, where V_i is the voltage corresponding to the unknown V_i^T, and I_i is the injected current at node i depending on the measurement V_i^N. A slightly more complicated circuit can be designed for the case when the measurements of V_i^N are not exact[7] (equation (5)). Uniqueness of the regularized solution always ensures stability of the corresponding network, even if capacities are introduced. Equivalent analog networks can be implemented by diffusion-reaction systems, where the interaction between neighbouring locations are mimicked using diffusion or chemical reactions with first-order kinetics. Hypothetical neuronal implementations may be envisaged. The conductance g may correspond to a small segment of a dendrite, the variable conductance g_i to a synaptic input with a reversal potential close or equal to the resting potential of the dendrite (that is, silent or shunting inhibition) and the current source to a conventional chemical synapse injecting current I_i into the dendrite. The output is sampled at location i by a chemical synapse. Adapted from ref. 7.

In the case of Hildreth's motion computation[23] the smoothness assumption corresponds to the hypothesis that the changes in velocity between neighbouring points along the contour are zero-mean, independent, gaussian random variables. This connection between the stochastic approach and standard regularization methods gives an interesting perspective on the nature of the constraints and the choice of the stabilizer. The variational principles used to solve the inverse problems of vision correspond to the Markov structure that generates plausible solutions.

A related area of future investigation concerns the problem of learning a regularizing operator. In the case of standard regularization, the corresponding linear operator mapping the data into the solution may be learned by an associative learning scheme[57], of the type proposed in connection with biological memory[58].

Towards symbolic descriptions

So far, we have restricted our discussion to the early stages of vision that create image-like representations of the physical three-dimensional surfaces around the viewer. The step beyond these representations, also called intrinsic images[5], or 2-1/2D sketches[1], is a large one. Intrinsic images are still image-like numerical representations, not yet described in terms of objects. They are already sufficient for some of the high-level tasks of a vision system such as manipulation and navigation. They cannot be used directly for the tasks of recognition and description that require the generation and use of more symbolic representations. It seems at first difficult to see how the computation of symbolic representations may fit at all in the perspective of regularizing ill-posed problems.

The basic idea of all regularization methods is to restrict the space of possible solutions. If this space is constrained to have finite dimensions, there is a good chance that an inverse problem will be well-posed. Thus, a representation based on a finite set of discrete symbols regularizes a possibly ill-posed problem. From this point of view, the problem of perception (regularizing an otherwise underconstrained problem using generic constraints of the physical world) becomes practically equivalent to the classical artificial intelligence problem of solving and inference, that is, of finding ways of solving intractable problems (such as chess) by limiting the search for solutions.

Conclusions

We suggest a classification of vision algorithms that maps naturally into parallel digital computer architectures now under development. Standard regularization, when sufficient, leads to two classes of parallel algorithms. Algorithms for finding minima of a convex functional such as steepest descent or the more efficient multigrid algorithms developed for vision[59] can always been used. They can be replaced by convolution algorithms if the data are given on a regular grid and A in equation (1) is space invariant. In the latter case, the regularized solution is obtained by convolving the data through a precomputed filter.

All these algorithms may be implemented by parallel architectures of many processors with only local connections. Problems that cannot be approached in terms of regularization and that require symbolic representations and operations on them, may need parallel architectures with a global communication facility, such as the Connection Machine currently under development[60].

The concept of ill-posed problems and the associated old and new regularization theories seem to provide a satisfactory theoretical framework for much of early vision. This new perspective also provides a link between the computational (ill-posed) nature of early vision problems, the structure of the algorithms for solving them and the parallel hardware that can be used for efficient visual information processing. It also shows the intrinsic limitations of the variational principles used so far in early vision, indicating at the same time how to extend regularization analysis beyond the standard theory.

We thank E. Hildreth, A. Hurlbert, J. Marroquin, G. Mitchison, D. Terzopoulos, H. Voorhees and A. Yuille for discussions and suggestions. Mario Bertero first pointed out to us that numerical differentiation is an ill-posed problem. E. Hildreth, L. Ardrey and especially H. Voorhees, K. Sims and M. Drumheller helped with some of the figures. Support for the Artificial Intelligence Laboratory's research in artificial intelligence is provided in part by the Advanced Research Projects Agency of the Department of Defense under Office of Naval Research contract N00014-80-C-0505. The Center for Biological Information Processing is supported in part by the Sloan Foundation and in part by Whitaker College. C.K. is supported by a grant from the Office of Naval Research, Engineering Psychology Division.

1. Marr, D. *Vision* (Freeman, San Francisco, 1982).
2. Brady, J. M. *Computing Surv.* **14**, 3-71 (1982).
3. Ballard, D. H., Hinton, G. E. & Sejnowski, T. J. *Nature* **306**, 21-26 (1983).
4. Brown, C. M. *Science* **224**, 1299-1305 (1984).
5. Barrow, H. G. & Tennenbaum, J. M. *Artif. Intell.* **17**, 75-117 (1981).
6. Marr, D. & Ullman, S. *Proc. R. Soc.* **B211**, 151-180 (1981).
7. Poggio, T. & Koch, C. *Proc. R. Soc.* B (in the press).
8. Poggio, T. & Torre, V. *Artif. Intell. Lab. Memo* No. 773 (MIT, Cambridge, 1984).
9. Hadamard, J. *Lectures on the Cauchy Problem in Linear Partial Differential Equations* (Yale University Press, 1923).
10. Bertero, M., Del Mol, C. & Pike, E. R. *J. inverse Prob.* (in the press).
11. Tikhonov, A. N. *Sov. Math. Dokl.* **4**, 1035-1038 (1963).
12. Tikhonov, A. N. & Arsenin, V. Y. *Solutions of Ill-posed Problems* (Winston, Washington, DC, 1977).
13. Bertero, M. in *Problem non ben posti ed inversi* (Istituto di Analisi Globale, Firenze, 1982).
14. Nashed, M. Z. (ed.) *Generalized Inverses and Applications* (Academic, New York, 1976).
15. Wahba, G. *Tech. Rep.* No. 595 (University of Wisconsin, 1980).
16. Horn, B. K. P. *Computer Graphics Image Processing* **3**, 111-299 (1974).
17. Horn, B. K. P. *Robot Vision* (MIT Press & McGraw-Hill, Cambridge & New York, 1985).
18. Horn, B. K. P. & Schunck, B. G. *Artif. Intell.* **17**, 185-203 (1981).
19. Ikeuchi, K. & Horn, B. K. P. *Artif. Intell.* **17**, 141-184 (1981).
20. Grimson, W. E. L. *From Images to Surfaces: A Computational Study of the Human Early Visual System* (MIT, Cambridge, 1981).
21. Grimson, W. E. L. *Phil. Trans. R. Soc.* **B298**, 395-427 (1982).
22. Terzopoulos, D. Computer Graphics Image Processing **24**, 52-96 (1983).
23. Hildreth, E. C. *The Measurement of Visual Motion* (MIT Press, Cambridge, 1984).
24. Hildreth, E. C. *Proc. R. Soc.* **B221**, 189-220 (1984).
25. Horn, B. K. P. & Brooks, M. J. *Artif. Intell. Lab. Memo* No. 813 (MIT, Cambridge, 1985).
26. Poggio, T., Voorhees, H. & Yuille, A. *Artif. Intell. Lab. Memo* No. 833 (MIT, Cambridge, 1985).
27. Torre, V. & Poggio, T. *IEEE Trans. Pattern Analysis Machine Intelligence* (in the press).
28. Marr, D. & Poggio, T. *Proc. R. Soc.* **B204**, 301-328 (1979).
29. Marr, D. & Hildreth, E. C. *Proc. R. Soc.* **B207**, 187-217 (1980).
30. Morozov, V. A. *Methods for Solving Incorrectly Posed Problems* (Springer, New York, 1984).
31. Nishihara, H. K. *Artif. Intell. Lab. Memo* No. 780 (MIT, Cambridge, 1984).
32. Hurlbert, A. *Artif. Intell. Lab. Memo* No. 814 (MIT, Cambridge, 1985).
33. Land, E. H. *Proc. natn. Acad. Sci. U.S.A.* **80**, 5163-5169 (1984).
34. Yuille, A. *Artif. Intell. Lab. Memo* No. 724 (MIT, Cambridge, 1983); *Advances in Artificial Intelligence* (ed. O'Shea, T. M. M.) (Elsevier, Amsterdam, in the press).
35. Ullman, S. *Computer Graphics Image Processing* **9**, 115-125 (1979).
36. Eigen, M. in *The Neurosciences: 3rd Study Program* (eds Schmitt, F. O. & Worden, F. G.) xix-xxvii (MIT Press, Cambridge, 1974).
37. Oster, G. F., Perelson, A. & Katchalsky, A. *Nature* **234**, 393-399 (1971).
38. Hopfield, J. J. *Proc. natn. Acad. Sci. U.S.A.* **81**, 3088-3092 (1984).
39. Koch, C., Marroquin, J. & Yuille, A. *Artif. Intell. Lab. Memo* No. 751 (MIT, Cambridge, 1985).
40. Schmitt, F. O., Dev, P. & Smith, B. H. *Science* **193**, 114-120 (1976).
41. Graubard, K. & Calvin, W. H. in *The Neurosciences: 4th Study Program* (eds Schmitt, F. O. & Worden, F. G.) 317-332 (MIT Press, Cambridge, 1979).
42. Shepherd, G. M. & Brayton, R. K. *Brain Res.* **175**, 377-382 (1979).
43. Bennett, M. V. L. in *Handbook of Physiology*, 221-250 (American Physiological Society, Bethesda, 1977).
44. Marder, E. *Trends Neurosci.* **7**, 48-53 (1984).
45. Schmitt, F. D. *Neuroscience* **13**, 991-1002 (1984).
46. Koch, C., Poggio, T. & Torre, V. *Phil. Trans. R. Soc.* **B298**, 227-268 (1982).
47. Cole, K. S. *Membranes, Ions and Impulses* (University of California Press, Berkeley, 1968).
48. Jack, J. J., Noble, D. & Tsien, R. W. *Electric Current Flow in Excitable Cells* (Clarendon, Oxford, 1975).
49. Koch, C. & Poggio, T. *Proc. R. Soc.* **B218**, 455-477 (1983).
50. Marroquin, J. *Artif. Intell. Lab. Memo* No. 792 (MIT, Cambridge, 1984).
51. Geman, S. & Geman, D. *IEEE Trans. Pattern Analysis Machine Intelligence* **6**, 721-741 (1984).
52. Blake, A. *Pattern Recognition Lett.* **1**, 393-399 (1983).
53. Hinton, G. E. & Sejnowski, T. J. *Proc. IEEE 1983 Conf. Computer Vision and Pattern Recognition* (Washington, DC, 1983).
54. Kirkpatrick, S., Gelatt, C. D. Jr & Vecchi, M. P. *Science* **220**, 671-680 (1983).
55. Metropolis, N., Rosenbluth, A., Rosenbluth, M., Teller, A. & Teller, E. *J. chem. Phys.* **21(6)**, 1087-1092 (1953).
56. Marroquin, J. *Artif. Intell. Lab. Memo* No. 839 (MIT, Cambridge, 1985).
57. Poggio, T. & Hurlbert, A. *Artif. Intell. Lab. Working Pap.* No. 264 (MIT, Cambridge, 1984).
58. Kohonen, T. *Self-Organization and Associative Memory* (Springer, Berlin, 1984).
59. Terzopoulos, D. *IEEE Trans. Pattern Analysis Machine Intelligence* (in the press).
60. Hillis, W. D. *The Connection Machine* (MIT Press, Cambridge, 1985).
61. Fahle, M. & Poggio, T. *Proc. R. Soc.* **B213**, 451-477 (1981).
62. Wallach, H. & O'Connell, D. N. *J. exp. Psychol.* **45**, 205-217 (1953).
63. Movshon, J. A., Adelson, E. H., Gizzi, M. S. & Newsome, W. T. in *Pattern Recognition Mechanisms* (eds Chagas, C., Gattar, R. & Gross, C. G.) 95-107 (Vatican, Rome, 1984); *Expl Brain Res.* (in the press).
64. Barlow, H. B. *J. Phsiol., Lond.* **119**, 69-88 (1953).
65. Kuffler, S. W. *J. Neurophysiol.* **16**, 37-68 (1953).
66. Terzopoulos, D. thesis, Massachusetts Inst. Technol. (1948).
67. Terzopoulos, D. *Artif. Intell. Lab. Memo.* No. 800 (MIT, Cambridge, 1985).

Chapter 6

Representations and Transformations

In an important sense a computational domain is distinguished by the representations it employs and a specific set of transformations defined over these representations. Computational vision is characterized by its use of an unusually large number of distinct representations, some of which are unique to this field of study, such as iconic or picturelike numerical arrays. Our competence for dealing with objects in the world depends on the adequacy of these representations; they are critical to our ability to describe and recognize objects and events, to predict how a scene changes when observed from a different viewpoint in space or time, and to combine different views of the world. The requirement for multiple representations in a general-purpose vision system arises from the needs of the signals-to-symbols paradigm (Chapter 5) in which we start with pixel-level signals and must describe successively more organized and goal-dependent attributes of the data. At each level in this hierarchy we require the definition of a natural vocabulary that allows us to make desired information explicit, and we need corresponding data structures that permit efficient coding and storage.

Although we have selected specific papers for this chapter (most of which are relevant to low- and intermediate-level vision), the question of representation pervades all the other chapters, and many papers described in other chapters are important because they introduce new or novel representations or transformations.

6.1 Classes of Representations used in Scene Analysis

The following material summarizes the key representations and transformations used in computational vision; see the Glossary for additional definitions.[1]

6.1.1 Geometric and Attribute "Spaces"

"Spaces" are multidimensional, generally continuous representations in which image or scene attributes are made explicit by the geometry or topology of their transformed configurations. Some examples are:

- *Gradient space* (Chapter 1) [Mackworth 73] [Huffman 71]. A 2-D space explicitly depicting the orientation of scene surfaces.
- *Hough space* (Chapter 7) [Duda 72] [Duda 73]. A multi-dimensional histogram used to estimate model parameters; each histogram cell corresponds to a complete specification of the model parameters. For example, for each edge point detected in an image, the "Hough transform" will increment the counters in each cell in Hough space that corresponds to the parameters of a line that could pass through the given point. All the points in the image on the same line contribute a vote to

[1] The references provided for the items in this section are intended to "point" the reader to an obvious source but not necessarily the most complete or most recent source.

the same histogram cell, allowing this line to be detected.

- *Scale space* (Chapter 2) [**Witkin 83a**]. A 1-D signal is convolved with a Gaussian filter and the zeros of the second derivative (actually, the "zero crossings") are found and followed as the size of the filter increases. Scale space is the plot of the zero-crossing contours on the (x, σ) plane, where σ measures the size of the Gaussian filter.

- *Projective space* [**Ahuja 68**]. The result of a perspective transform—for example, a 2-D image representing a 3-D scene. In general, a projective transform is the result of one or more "perspective" transforms; each perspective transform can intuitively be thought of as an imaging operation in which points in a scene or image are, in turn, imaged by rays passing through a single point in space (e.g., a simple lens). The resulting image is formed by the intersection of the rays and an "image plane."

- *Frequency space* [Castleman 79]. An explicit representation of the spatial variation of some image attribute, such as intensity, in terms of a set of basis functions that use frequency as a parameter—for example, as produced by the Fourier transform.

- *Dot product space.* [Firschein 79]. Given a vector A and a set of unit vectors that span the unit sphere (unit circle in two dimensions), dot product space is the plot of the dot products of A with the unit vectors versus the unit vector directions. For example, to determine the "diameter" of a region in an image, one can take the dot product of the span of unit vectors and the vector from some arbitrary origin to each pixel in the given region. The diameter is found in dot product space as the largest difference between the maximum dot product and the minimum dot product at a particular unit vector direction.

6.1.2 Arrays

Arrays play an important role in scene analysis because they implicitly represent scene geometry in a numerical form suitable for manipulation in a digital computer.

- A *pixel array* is a numerical array in geometric correspondence with the picture (image) elements depicting a scene. The numerical values are typically light intensity, range, or other sensor-derived "point" measurements.

- A *feature array* is a pixel array whose elements are the values of an attribute of some image or scene feature—for example, intensity gradient, surface orientation, material type, and so on.

- A *cost array* is a feature array whose elements represent the probability of the presence of a feature at the given pixel location, or the cost of assuming that the feature is present at the given pixel location.

6.1.3 Trees and Pyramids

Trees and pyramids are symbolic structures that can be used to represent both iconic and relational information at multiple levels of resolution. Some examples are:

- A *pyramid* is a data structure consisting of image information at successively increasing levels of resolution [Uhr 72] [Samet 80] [Rosenfeld 80]. Quad trees and strip trees, described next, are examples of pyramids.

- A *quad tree* [Samet 80] is constructed by repeatedly subdividing a $2^n \times 2^n$ array of pixels into quadrants, subquadrants, and so on until we obtain single pixel blocks. This process is represented as a tree in which the root node corresponds to the entire array, the four sons of the root node correspond to the quadrants, and the terminal nodes correspond to those blocks of the array for which no further subdivision is desired.

- A *strip tree* [Ballard 81b] is a binary tree used to represent a curve. The datum at each node is an eight-tuple, of which six entries define a rectangle and two denote the addresses of higher resolution sons, if any exist. (While only five parameters are necessary to define an arbitrary rectangle, the redundant representation is useful in union and intersection algorithms.)

- A *minimum spanning tree* (*MST*) [Zahn 71] [Pavlidis 77] is the most compact (with respect to sum of edge lengths) tree that spans a graph and is often used in edge connection algorithms.

6.1.4 Shape Representations

Shape representations encode the form of a curve or surface in a way that captures some intrinsic attribute for computational, storage, or display purposes. For example:

- A *chain code* [Freeman 74] is a representation of a curve using line segments that must lie on a fixed grid with a fixed set of orientations. A curve can be compactly expressed as the sequence of numbers representing the orientation of each line segment.
- A *spline* [Pavlidis 77] is a piecewise polynomial curve with smoothness constraints on how the polynomial segments are joined.
- *Coons patch* [Coons 74] is a four-sided surface patch for approximating a piece of a surface. The four sides of the patch are specified by polynomials, and these polynomials are used to interpolate interior points.
- *Fourier descriptors* [Persoon 74]. The boundary of a region can be represented as an analytic function, such as tangent angle versus arc length, and expanded in a Fourier series. The Fourier descriptors of the boundary are the coefficients of the Fourier series; these descriptors can be used to describe and compare boundary shapes.

6.1.5 Representation of Semantic Knowledge

Representations of semantic knowledge attempt to capture aspects of the world that may be important in interpreting, understanding, and coming to conclusions about what has, or will happen, in a scene. For example, such representations may deal with descriptions of the physical characteristics of objects and their environments, and the purposes and goals of scene entities.

- *Expert (production rule) systems* [**McKeown 85**]. Knowledge about the world is expressed in rules of the form, "If condition *A* is satisfied, then carry out operation *B*." A control system examines all the rules to determine those whose "if" portion is satisfied, and decides which rule is to be activated when more than one is satisfied. Rule-based systems have the attraction of allowing addition and deletion of rules without requiring extensive reprogramming.
- *Propositional and predicate calculus* [Nilsson 84] [E. Davis 83]. Knowledge is represented in a formalism that permits explicit information to be derived from implicit information. The advantage of this formalism is that only consistent results can be deduced—that is, it is not possible to deduce both a fact and its contradiction. Disadvantages are that there is no guarantee that reasonable descriptions of real-world problems can be constructed, and for the more expressive predicate calculus, that the desired explicit information can be deduced or that practical algorithms can be found to perform the deduction. An interesting discussion on these themes is given in [Nilsson 83] (an advocate of logic) and [Pentland 83] (Pentland and Fischler discuss the limitations of logic).
- *Relational or semantic nets* [Ballard 82]. A common representation used in high-level vision, these graph structures represent objects as nodes and relationships between objects as labeled arcs.

6.1.6 Miscellaneous

One of the most important representations used in both computer graphics and computational vision is that of homogeneous coordinates in which a point in Cartesian n-space is represented as a line in homogeneous $(n+1)$ space. Homogeneous coordinates allow many important geometrical transformations to be represented uniformly and elegantly—for example, as matrix multiplications rather than as nonlinear functional mappings. They allow large integer numbers to be represented in a computer of limited word size, and they can be used to represent points at infinity. Because of the importance of this representation, we have included the excellent tutorial paper by Ahuja and Coons [**Ahuja 68**]. Some other important representations are the following.

- The *Gaussian sphere* is a representation in which some geometric property of surfaces in a Cartesian space is mapped onto a unit sphere. The finite extent of the Gaussian or unit sphere is an advantage when searching for some property, as opposed to searching the infinite extent of Cartesian space. See [Ikeuchi 81] for an example of the use of the Gaussian sphere.
- *Epipolar images* (Chapter 1) [**Bolles 85**] are obtained from a "dense" set of images—images taken close enough together in time to form a solid block of data. An epipolar image is a 2-D plane with one spatial and one time dimension, extracted from the solid block of image data.
- *Triangular irregular networks* [Peucker 78] [Fowler 79] [Watson 84] [Boissonnat 84] are triangular patches joined together to approximate a surface.

- A *contour map* [Merrill 73] [Morse 69] [Ward 78] is a 2-D representation of the topography of a (terrain) surface using a set of closed curves, each of which represents a constant surface elevation.
- A *logarithmic spiral grid* [Weiman 79] is obtained by a transform from rectangular to polar coordinates; the representation of an object is invariant under object rotation, and a change of scale for the object simply causes a translation of its representation.

6.2 Representation of the Contours and Surfaces of Objects

Computer-based techniques for shape description have traditionally employed analytic representations based on points,[2] lines, and surface patches. For example, terrain information is often represented by a point array of elevation values, contour maps, profile arrays, and networks of surface patches. Compact 3-D objects are represented by wire frames and surface patches—see [Faux 79]. While such representations are adequate for purposes such as manufacturing and visual rendering, they generally do not adequately represent functional or conceptual structures in a manner suitable for manipulation by machine-reasoning techniques. Further, these representations are inappropriate for representing such natural objects as trees, bushes, clouds, and so on, or for automatically recognizing such objects in images. This section describes some of the formalisms used to represent the contours and surfaces of man-made objects; the next section describes representations for "natural" objects such as vegetation and terrain features. In both cases we are particularly interested in representations that can be used as the basis for matching, recognition, or reasoning.

6.2.1 Representation of Contours

An obvious way to approximate a curve is with a set of linear or polynomial segments. However, finding these approximations involves segmentation issues—that is, determining the most suitable breakpoints along the curve. One of the strategies is that of merging and splitting; points are merged into a segment as long as the fit is within predefined limits, and they are split into a new segment when the limits are exceeded [Pavlidis 77]. Duda and Hart [Duda 73] describe a simple but effective recursive procedure. However, often what is important is not that the resulting fit falls within error bounds, or that the procedure finds the most economical set of segments, but rather that the resulting set of segments captures the shape of the curve in an intuitively appealing and semantically relevant manner.

Representations of 2-D curves are often based on the changes in orientation that occur as one travels along the curve. A chain code [Freeman 74] describes a curve in terms of a sequence of line segments that can have a limited number of orientations. Various "derivative" and code compaction operations can be carried out on a chain code. Two curves can be compared by comparing their chain codes.

Richards and Hoffman [**Richards 85**] use a representation based on the maxima, minima, and zeros of curvature along the curve, since ordinal relations among these are invariant under translation, rotation, and orientation. Six primitive shapes or "codons" are used to describe plane curves. These codons capture important information about the 3-D world such as making part-boundaries explicit. The constraints between pairs and triples of codons are so strong for smooth closed boundaries that sequences of six codons yield only 33 legal generic shapes.

The medial axis transform [Blum 73] [Rosenfeld 76b] represents the boundary of a thin region by a stick figure or "skeleton." The skeleton can also be derived by thinning algorithms that preserve connectivity of an enclosed region [Pavlidis 77][Arcelli 85]. While skeleton-generating techniques can be quite sensitive to slight perturbations in the boundary shape, the labeled distance transform of Fischler and Barrett [Fischler 80] provides a mechanism for avoiding this problem.

6.2.2 Representation of Surfaces

The surfaces of an object can be represented by a set of planar (e.g., triangles) or curved surface patches. An example of the use of both planar and quadratic surface patches to model and then recognize objects is given in Faugeras [Faugeras 83], as discussed in Chapter 3. Besl and Jain [Besl 85] review the many surface representations based on this idea. Binford [Binford 71], Binford and Agin [Binford 73], Nevatia and Binford [Nevatia 77], and later Marr and Nishihara [Marr 78b] proposed schemes using hierarchies of cylinderlike modeling primitives called "generalized cones" to describe volumetric forms. Generalized cones represent object shape by a 3-D space curve that acts as the

[2] Cellular or voxel representations [Srihari 81], volume elements analogous to pixels in two dimensions, are employed in such specialized areas of scene analysis as computational tomography, biological modeling, and space planning. These representations are just beginning to play a more important role in general-purpose machine vision.

spine or axis of the cone, a 2-D cross-sectional figure, and a sweeping rule that defines how the cross section is to be swept and possibly modified along the space curve. This representation captures many of our intuitions about axes of symmetry and hierarchical description and can be quite useful for industrial-style machine vision systems. However, it is often inadequate to accurately and succinctly capture animate or natural forms.

6.3 Representation of Natural Form

To extract cognitive building blocks from an unstructured array of image intensities, we must have a representation of objects in the world that not only models their 3-D structure but also allows us to partition and describe these objects in a psychologically satisfying, compact, and semantically relevant way.

Many forms in nature can be described mathematically or statistically. P.S. Stevens [P.S. Stevens 74] has noted that, "In matters of visual form we sense that nature plays favorites. Among her darlings are spirals, meanders, branching patterns, and 120-degree joints." As discussed below, physical models of natural processes have provided some of the most effective existing representations for describing natural forms.

Many natural-occurring surfaces are produced by physical processes that randomly modify shape through local action—for example, accretion or deletion of material independent of scale. Mandelbrot [Mandelbrot 77, 82] developed a mathematics of such surfaces known as "fractal forms." The fractal dimension of a surface corresponds to our intuitive notion of surface roughness. Mandelbrot's contribution was to propose that a geometric description could be based on a statistical process, and that one could create a 2-D rendering of a 3-D object by a similar process. Pentland [Pentland 84a] used a fractal model for describing natural surface shapes. He showed that the fractal parameters of physical surfaces can be estimated from their projective images and thus the geometry of real-world objects could be modeled by making fractal measurements based on the intensities in their 2-D images. Pentland also demonstrated that fractal textures can be effectively used for image segmentation.

Barr [**Barr 84**] introduces globally and locally defined deformations of geometric objects as new hierarchical operations for use in solid modeling. These operations simulate twisting, bending, tapering, or similar transformations of geometric objects. Using a similar philosophy, Pentland [**Pentland 86b**] provides a "part and process" model[3] in which the primitives may be thought of as "lumps of clay," which may be deformed and shaped, and which are intended to correspond roughly to our naive perceptual notion of "part." The process description tells how to distort and combine these primitives to form the desired composite object. The basic modeling element is a parameterized family of shapes known as superquadrics [Barr 81].[4] The corresponding equations define a 3-D surface parameterized in latitude and longitude; the shape of the surface is controlled by two additional parameters. This family of functions includes cubes, cylinders, spheres, diamonds, and pyramidal shapes, as well as the round-edge forms intermediate between these standard shapes. These basic "lumps of clay" are used as prototypes that can be deformed by stretching, bending, twisting, or tapering. Then they can be combined using Boolean operations to form new, composite prototypes that may, recursively, again be subjected to deformation and Boolean combination.

Thus, shapes are described in a manner that could correspond to a formative history—for example, how one would create a given shape by combining lumps of clay. Using such a process-oriented representation results in descriptions that group points that have similar causal histories, thus obtaining "parts" that interact with the surrounding world in a coherent manner. This type of representation simplifies many reasoning tasks, because the parameters and components that affect interactions tend to be explicitly represented rather that being some complex or difficult-to-calculate function of the variables of the description.

Natural forms, such as trees, can be represented by using a generative procedure analogous to the generation of sentences from a grammar [A.R. Smith 84], particle systems [W.T. Reeves 85], and polygonal description or simple volume primitives plus texture mapping [Bloomenthal 85] [Gardner 84]. Wyvill [Wyvill 86] discusses the representation of water and presents results for ray tracing and "displacement mapping" techniques. To model formless objects such as fire, clouds, and water, W.T. Reeves [W.T. Reeves 83] uses clouds of primitive particles that define its volume. Over a period of time, stochastic processes cause the particle clouds to move, change form, die, and finally disappear from the system. The resulting model is able to

[3] Also, see [Leyton 86] and the papers by Richards and Hoffman [**Richards 85**] and by Kass and Witkin [**Kass 86**] in Chapter 2.

[4] In some very recent work Hanson [Hanson 86] describes an important generalization of the superquadric family.

represent motion, changes of form, and dynamics not possible with classical surface-based representations.

Of all synthetic images, those rendered by ray tracing are unsurpassed for realism. For each pixel in an image the path of the ray from surfaces that could contribute to the intensity of the pixel is traced, first to the surface and then to the light source. Thus, there is a large computational burden. Kajiya [Kajiya 83] presents algorithms for efficient ray tracing of fractal surfaces, prisms, and surfaces of revolution. Basically, he presents new ways of computing the intersection of a ray and certain objects by ignoring all the regions that will not contribute to the final result.

6.4 Spatial Representation at Multiple Levels of Resolution

An effective strategy for scene analysis, the "coarse-fine" stategy, first views the world globally in low resolution and then uses the resulting interpretation to constrain the analysis of higher resolution but spatially more restricted views. A "pyramid" of images, where each successive image is lower in resolution than the one below, has been used in many image analysis applications. The pyramid approach is described in [Rosenfeld 80], [Uhr 72], [Samet 80], [Moravec 77], and [Tanimoto 75].[5] Burt and Adelson [**Burt 83a**] provide a very effective technique for obtaining a pyramidtype image representation; they use local operators of many scales but identical shape to generate the hierarchy of images. With this technique, the authors obtain a sequence of band-pass filtered images by repeatedly convolving a small mask with an image. The computations are local and may be performed in parallel, and the same computations are iterated to build each lower resolution pyramid level from its predecessor. Image sample density is decreased with each iteration so that the bandwidth is reduced in uniform one-octave steps; sample reduction also reduces the cost of computation.

6.5 Merging Different Views of the World

We never see the entire world in a single image; instead we must montage pieces of the world to form a whole. We are then faced with the problem of making images fit at the edges. Burt and Adelson [**Burt 83a**] define a multiresolution technique (using the Laplacian pyramid discussed in Section 6.4) for combining two or more images into a larger image mosaic. According to this procedure, the images to be montaged are first decomposed into a set of band-pass-filtered component images. Next, the component images in each frequency band are assembled into a corresponding band-pass mosaic. In this step component images are joined using a weighted average within a transition zone, which is proportional in size to the wavelengths represented in the band. Finally, these band-pass mosiac images are summed to obtain the final composite image. In this way the merging function is matched to the scale of features within the images themselves. When coarse features occur near borders, these are blended gradually over a relatively large distance without blurring or otherwise degrading finer image details in the neighborhood of the border.

6.6 Transformations that Fill "Holes" in a Spatial Representation

Replacing sampled spatial data with a fitted surface can be considered as deriving a new representation for the data. Techniques whose resultant surfaces do not conform exactly to the data are known as approximation methods; methods that produce surfaces passing through the input data points are called interpolation models.

To obtain a suitable model we need a specification of the properties required of the fitted surface, such as degree of differentiability, and the characteristics of the data, such as accuracy. When fitting the surfaces, we balance the influence exerted by the data values against that exerted by the implicit surface model embedded in the fitting procedure. If our data values are inaccurate and we know the specific class of surfaces that should fit the data, we can usually let the surface model dominate the construction process. Least-square methods are typical of approximation procedures that prefer a model to data.

Although we are often satisfied with a smooth interpolating function, as in the case of obtaining a terrain model from sparse elevation values, we should also have mechanisms for dealing with discontinuities, such as buildings and cliffs. An interpolation algorithm must be able to detect and deal with such discontinuities.

Terzopoulis [Terzopoulis 83] addresses the surface interpolation problem in the context of providing a computational framework for integrating depth and orientation information, dealing with surface discontinuities, and ensuring computational efficiency and speed. His approach extends the work of Grimson

[5] Witkin's scale space concept [**Witkin 83a**] is another example of the general approach.

[Grimson 83], who used minimization of "quadradic variation" of surface shape as a smoothness constraint, and also the work of Duchon [Duchon 76], who suggested the physical analogy between quadratic variation and the potential energy of thin plates. The Terzopoulis algorithm (actually an approximation rather than an interpolation technique) can be intuitively viewed as representing the given sparse depth information by vertical pins with lengths proportional to depth.[6] A nominally horizontal, thin flexible plate representing the desired surface is shaped by being connected to the vertical pins by ideal springs with zero natural lengths. The springs pull the plate toward the pin tips; the individually controllable stiffness of the springs represents confidences in the input information.

Based on psychological justification, Terzopoulis assumes that discontinuities in the first partial derivatives of surface curvature should be explicitly detected and treated as interpolation boundaries.[7] The human visual system perceives such curvature discontinuities, and thin plates are too rigid to represent them—that is, a thin plate, in principle, cannot form a surface with a crease. Higher-order curvature discontinuities do not form perceptual boundaries and are adequately handled by the thin plate formulation.

Terzopoulis suggests that the difficult problem of discontinuity detection (see Chapter 2) be accomplished before surface reconstruction begins. However, he presents a method for automatically detecting orientation discontinuities during reconstruction by looking for local extrema in the bending moments of the thin plate surface, and for detecting depth discontinuities by looking for significant inflections (locations where the bending moment undergoes a change in sign). Terzopoulis has recently proposed a general class of "controlled-continuity constraints" that provide the necessary control over smoothness in order to deal with discontinuities [Terzopoulis 86].

The multiresolution surface reconstruction algorithm devised by Terzopoulis is claimed to be orders of magnitude faster than results obtained using algorithms operating at a single level of resolution. For example, one problem with earlier work by Grimson [Grimson 83] was the extremely slow convergences of his iterative algorithm on large (256 × 256) single-resolution grids.

G.B. Smith [G.B. Smith 84a] addresses the problem of fitting a surface composed of arbitrary basis functions to a large data set that contains mostly regularly spaced data points but that also includes grid points at which there is no data, and nongrid points where data values are known. His contribution is to provide a computationally efficient solution to the problem of solving the large system of equations necessary for global interpolation. (Each data point, a constraint on the solution, introduces one equation.) The Smith algorithm produces a smooth solution from which one can calculate both first and second derivatives, but it does not explicitly address the surface discontinuity problem.

6.7 Discussion

Machine vision can be viewed as a science primarily concerned with inventing representations for expressing spatial knowledge, and for finding effective transformations between these representations. In this chapter we described more than thirty different representations and a few of the more general transformations in current use; this list is far from exhaustive.

In the first section of this chapter we categorized representations with respect to their associated data structures. An alternative categorization could have been in terms of their competence to describe analytical, statistical, or procedural (process-oriented) knowledge. Most of the more traditional representations were originally developed to express the analytic relationships required for mathematical abstraction of the world—unfortunately, this is a very small part of the vision problem. The statistical and process-oriented representations are relatively new, and while their potential is still not fully understood, they offer an exciting frontier for new research.

Papers Included

Ahuja, D. V. and S.A. Coons. Geometry for construction and display, *IBM Systems Journal*, 7(3-4):188–205, 1968.

Barr, Alan H. Global and local deformations of solid primitives, *ACM Computer Graphics, SIGGRAPH '84*, 18(3):21–30, July 1984.

Burt, P.J. and E.H. Adelson. The Lapacian pyramid as a compact image code, *IEEE Trans. on Communications*, 31(4):532–540, April 1983a.

Pentland, A.P. Perceptual organization and the representation of natural form, *Artificial Intelligence*, 28:293–331, May 1986.

Richards, W. and D.D. Hoffman. Codon constraints on closed 2-D shapes, *CVGIP*, 31(3):265–281, September 1985.

[6] Surface orientation constraints are handled by pins positioned at appropriate angles.

[7] Terzopoulis distinguishes between occlusion boundaries at which the thin plates are "broken," and orientation discontinuities, where the thin plate model is replaced by a more flexible membrane model to allow some propagation of constraints.

Matric notation is used to develop geometric concepts for computer-controlled graphics.

This natural form of geometric expression leads to homogeneous coordinates which form the basis of an algorithm used for geometric construction. In this way, three-dimensional objects and other pictures can be displayed on a graphics console.

The paper also briefly discusses the notation and development of functions for the construction of surfaces.

INTERACTIVE GRAPHICS IN DATA PROCESSING

Geometry for construction and display

by D. V. Ahuja and S. A. Coons

The design and delineation of arbitrary shapes, and the display of these shapes from arbitrarily chosen viewpoints are more easily accomplished if done on the graphic display console of a computer rather than by pencil and paper methods. In this paper, we show modern methods for geometry, so constructed and organized that geometric manipulations can be performed in a way natural to the computer, and can yield results that are natural to man.

We discuss three major ideas: matric methods for algebra, homogeneous coordinates, and curves and surfaces from a parametric standpoint. First, we introduce matric methods for algebra, because matrices exhibit in a very transparent way the nature of geometric entities. They are also natural forms for computer implementation, since matrix manipulations are already programmed, and routines exist that are easily exploited.

Matrices lead naturally to the notion of coordinate transformations, and this notion leads into homogeneous coordinates. On this basis, we are able to show that all possible pictures, or displays, of three-dimensional objects, including the projections employed by draftsmen, so-called "axonometric pictorials," and pictures in full perspective can be constructed on the display console of the computer by a single, simple algorithm.

Finally, we briefly discuss curves and surfaces from a parametric standpoint. This form of curve and surface equations is used in "differential" geometry, and it lends itself to computer implementation in a very natural way.

Matric notation

We begin with a very elementary discussion of matric notation. In the plane, a line can be represented by the equation:

$$Ax + By = C$$

In matric notation, this equation becomes

$$[x \;\; y]\begin{bmatrix} A \\ B \end{bmatrix} = C$$

The brackets enclose two matrices, and their entries are the quantities that appear in the ordinary algebraic equation. The bracketed matrices are juxtaposed, which suggests that they are to be multiplied together; the "product" of the matrices is just $Ax + By$.

Now assume there are *two* lines in the plane, given by the linear equations:

$$Ax + By = C$$
$$Dx + Ey = F$$

In matric notation, these equations become the single matric equation

$$[x \;\; y]\begin{bmatrix} A & D \\ B & E \end{bmatrix} = [C \;\; F]$$

Ordinarily, there exists a point of intersection of these lines. This is to say that provided the lines are not parallel (we dispose of this case later, and show that it is not exceptional), it is possible, for $A\ B\ D\ E\ C\ F$ given numbers, to choose a pair of numbers x and y that "satisfy" both equations simultaneously. This algebraic process is very elementary and well-known, but it is interesting to see how it is done with matrices.

We can illustrate this process by an example. We wish to find the intersection of the lines:

$$x + y = 1$$
$$y = \tfrac{2}{3}x$$

The last equation can be written as

$$2x - 3y = 0$$

In matric form, we have

$$[x \;\; y]\begin{bmatrix} 1 & 2 \\ 1 & -3 \end{bmatrix} = [1 \;\; 0]$$

matrix operators

If we multiply both sides of this equation by the *inverse* of matrix $\begin{bmatrix} 1 & 2 \\ 1 & -3 \end{bmatrix}$ we get without destroying its validity

$$[x \;\; y]\begin{bmatrix} 1 & 2 \\ 1 & -3 \end{bmatrix}\begin{bmatrix} \frac{3}{5} & \frac{2}{5} \\ \frac{1}{5} & -\frac{1}{5} \end{bmatrix} = [1 \;\; 0]\begin{bmatrix} \frac{3}{5} & \frac{2}{5} \\ \frac{1}{5} & -\frac{1}{5} \end{bmatrix}$$

or

$$[x \quad y]\begin{bmatrix}1 & 0\\0 & 1\end{bmatrix} = [\tfrac{3}{5} \quad \tfrac{2}{5}]$$

The matrix $\begin{bmatrix}1 & 0\\0 & 1\end{bmatrix}$ is called the *identity matrix*. Any matrix multiplied by the identity matrix is unchanged. Hence, $[x \quad y] = [\tfrac{3}{5} \quad \tfrac{2}{5}]$, which is the point of intersection of these lines. However, not all matrices have inverses. For instance, the matrix $\begin{bmatrix}1 & 3\\2 & 6\end{bmatrix}$ has no inverse. It is called a *singular matrix*. Its determinant is equal to zero, and indeed, the vanishing of the determinant of a matrix is a necessary and sufficient condition to test singularity. But even when matrices do not have inverses, we can obtain solutions to equations.

The equation for a line,

$$[x \quad y]\begin{bmatrix}A\\B\end{bmatrix} = C$$

can be written

$$[x \quad y \quad 1]\begin{bmatrix}A\\B\\-C\end{bmatrix} = 0$$

without destroying the equality. When we multiply the matrices, we obtain

$$Ax + By - C = 0$$

Now forgetting the sign of C, consider the expression

$$[x \quad y \quad 1]\begin{bmatrix}A\\B\\C\end{bmatrix} = 0$$

coordinates

The matrix $[x \quad y \quad 1]$ consists of the coordinates of points that lie on a line; any pair of numbers x and y that satisfy this equation lie on this line. For this reason, we call $[x \quad y \quad 1]$ point coordinates. But suppose we were to fix x and y; then any set of numbers $A\ B\ C$, properly chosen, would represent lines passing through the fixed x-y point. For this reason, $[A \quad B \quad C]$ are called *line coordinates*. There are only two coordinates for the point, x and y, but there are apparently three coordinates for the line. We are already aware of the two statements, "two lines determine a point" and "two points determine a line." These statements are in a sense symmetric (in geometry, they are known as "dual" statements), and it would be more satisfying if we could observe a similar symmetry in the equation. We might write:

$$[x \quad y \quad 1]\begin{bmatrix}A/C\\B/C\\1\end{bmatrix} = 0 \qquad \text{or} \qquad [x \quad y \quad 1]\begin{bmatrix}P\\Q\\1\end{bmatrix} = 0$$

This representation would be acceptable, except that it would not permit us to write an equation for a line passing through the origin. Such a line is:

$$Ax + By = 0 \qquad \text{or} \qquad [x \quad y \quad 1]\begin{bmatrix}A\\B\\0\end{bmatrix} = 0$$

Since we need $A\ B\ C$, three coordinates for the line, it appears that perhaps we really need three coordinates for the point as well:

$$[x \quad y \quad w]\begin{bmatrix}A\\B\\C\end{bmatrix} = 0$$

For ordinary points we can set $w = 1$, and for lines not through the origin, we can set $C = 1$ also. When $C = 0$, we have an equation for lines through the origin. When $w = 1$, x and y are the *ordinary* coordinates of points in the plane. When $w \neq 1$, we can obtain the ordinary coordinates by putting $[x \quad y \quad w]$ into the form $[x_o \quad y_o \quad 1]$ where x_o, y_o are the ordinary coordinates. Obviously we can do this without destroying the validity of the equation by multiplying by $1/w$ provided w is not zero, i.e., if

$$[x \quad y \quad w]\begin{bmatrix}A\\B\\C\end{bmatrix} = 0$$

then

$$\frac{1}{w}[x \quad y \quad w]\begin{bmatrix}A\\B\\C\end{bmatrix} = \begin{bmatrix}\frac{x}{w} & \frac{y}{w} & 1\end{bmatrix}\begin{bmatrix}A\\B\\C\end{bmatrix} = 0$$

Then $x/w = x_o$ and $y/w = y_o$, the sought-for ordinary coordinates. Since the ordinary coordinates are obtained by $x_o = x/w$ and $y_o = y/w$, we modify the notation slightly. We can thus write $[wx \quad wy \quad w]$ instead of $[x \quad y \quad w]$.

In this way, we are able to keep track of the ordinary coordinates of a point; we consider wx and wy as biliteral symbols throughout all calculations and perform the division wx/w and wy/w only at the end. With this notation, we can retain x, y as ordinary coordinates instead of x_o, y_o. But if $w = 0$, we cannot perform the division. What does the set of numbers $[wx \quad wy \quad w]$ then mean?

To experiment with this question, let us try a specific example. Let wx and wy be 2 and 3, but let w vary, and calculate the ordinary x and y coordinates. The results appear in Table 1.

From this table, it is clear that as w decreases to 0, x and y increase without bound, but always in the ratio 2 : 3. Evidently

Table 1 x and y coordinates

w	x	y
1	2	3
$\frac{1}{2}$	4	6
$\frac{1}{3}$	6	9
.	.	
.	.	
.	.	
$\frac{1}{10}$	20	30
.		
.		
.		
$\frac{1}{100}$	200	300

$[2 \quad 3 \quad 0]$ represents a point at infinity. The ordinary coordinates of such a point are of no use to us, because they are $[\infty \quad \infty \quad 1]$, and the infinity signs cannot be manipulated algebraically.

Homogeneous coordinates

The set of numbers $[2 \quad 3 \quad 0]$ is a matrix of coordinates for a point at infinity that is not algebraically exceptional, as is shown later. The set of coordinates

$$\begin{bmatrix} A \\ B \\ C \end{bmatrix}$$

and the set of coordinates $[wx \quad wy \quad w]$ are called the *homogeneous* coordinates of line and point, respectively. Homogeneous refers to the algebraic forms that they constitute; all entries are of the same dimension. Indeed, it is appropriate to think of $[wx \quad wy \quad w]$ as a three-dimensional matrix.

We return to the two lines in the plane, which give the homogeneous equations:

$$[wx \quad wy \quad w] \begin{bmatrix} 1 & 2 \\ 1 & -3 \\ -1 & 0 \end{bmatrix} = [0 \quad 0]$$

We cannot solve this equation for wx and wy by finding the inverse of the 3×2 matrix because it is not a square matrix and has no ordinary inverse. However, we can arrange it so that the matrix does have an inverse by adjoining a column to it:

$$[wx \quad wy \quad w] \begin{bmatrix} 1 & 2 & 0 \\ 1 & -3 & 0 \\ -1 & 0 & 1 \end{bmatrix} = [0 \quad 0 \quad w]$$

This step is harmless and does not violate the equality. The inverse of this new square matrix can be found and is the matrix

$$\tfrac{1}{5}\begin{bmatrix} 3 & 2 & 0 \\ 1 & -1 & 0 \\ 3 & 2 & 5 \end{bmatrix}$$

Then

$$[wx \quad wy \quad w] = \tfrac{1}{5}[0 \quad 0 \quad w] \begin{bmatrix} 3 & 2 & 0 \\ 1 & -1 & 0 \\ 3 & 2 & 5 \end{bmatrix} = \tfrac{1}{5}[3w \quad 2w \quad 5w]$$

$$= \begin{bmatrix} \frac{3w}{5} & \frac{2w}{5} & w \end{bmatrix}$$

Dividing both sides by w to obtain the ordinary coordinates of the point of intersection, we have $[x \quad y \quad 1] = [\frac{3}{5} \quad \frac{2}{5} \quad 1]$ as before.

Consider the following two parallel lines. (In ordinary algebraic notation, they are: $x + y = 1$ and $y = -x$.)

parallel lines

$$[wx \quad wy \quad w] \begin{bmatrix} 1 & 1 \\ 1 & 1 \\ -1 & 0 \end{bmatrix} = [0 \quad 0]$$

We can now proceed to determine their point of intersection. We prepare the matrix to read

$$[wx \quad wy \quad w] \begin{bmatrix} 1 & 1 & 0 \\ 1 & 1 & 0 \\ -1 & 0 & 1 \end{bmatrix} = [0 \quad 0 \quad w]$$

The matrix on the left is still singular, and we cannot invert it. (It is singular because two rows are identical, and its determinant still vanishes.) We change the modification of the matrix, so that it reads

$$[wx \quad wy \quad w] \begin{bmatrix} 1 & 1 & 1 \\ 1 & 1 & 0 \\ -1 & 0 & 0 \end{bmatrix} = [0 \quad 0 \quad wx]$$

Now one of the unknowns, wx, appears on the right side of the equation.

The matrix has an inverse, and we obtain the solution:

$$[wx \quad wy \quad w] = [0 \ 0 \ wx] \begin{bmatrix} 0 & 0 & -1 \\ 0 & 1 & 1 \\ 1 & -1 & 0 \end{bmatrix} = [wx \quad -wx \quad 0]$$

Consider the point represented by the matrix $[wx \quad -wx \quad 0]$. This matrix is the same as $wx \; [1 \quad -1 \quad 0]$, where wx is any number whatever. It is, as we have seen, a point at infinity.

The preceding elementary discussion indicates that homogeneous coordinates are useful as well as aesthetically satisfying. They make equations more symmetric in form, which appeals to our sense of structure and completeness; they permit us to deal with infinitely distant points in the plane as easily as we deal with local points; and they permit us to remove the exceptional cases. In the following discussion of projective transformations, we see that they have even greater value.

perspective transformation

Let us first consider the perspective projections of a plane on to a plane by taking an example. From the theory of conic sections, we know that the circle, ellipse, hyperbola, and parabola are perspective projections of one another as shown in Figure 1.[1] We accomplish this transformation mathematically using homogeneous coordinates and the matric notation.

Figure 1 Perspective projections of a circle*

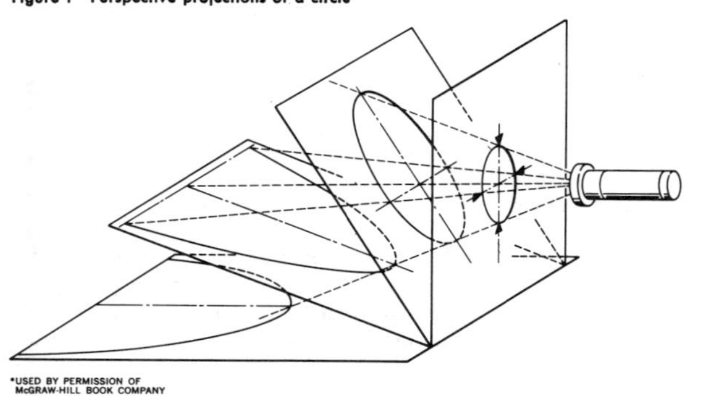

*USED BY PERMISSION OF McGRAW-HILL BOOK COMPANY

Figure 2 Perspective transformation

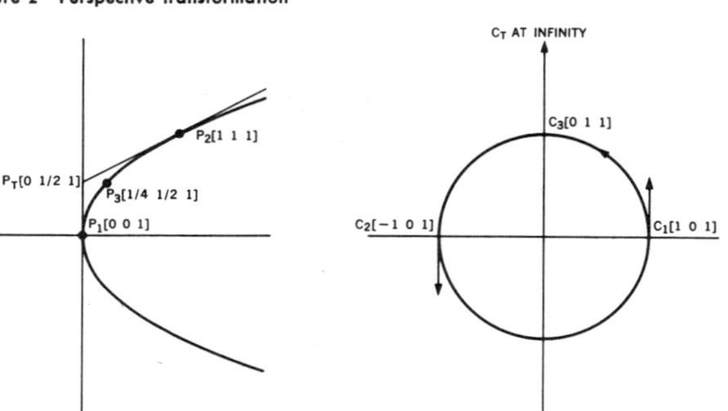

For simplicity, let us consider a parabola $x = y^2$, shown in Figure 2, as our starting curve, and transform this parabola into any other conic section. The parabola $x = y^2$ may be expressed parametrically as $x = u^2$ where $y = u$.

Any point $[x \quad y \quad 1]$ on this parabola is given in matric notation by:

$$[x \quad y \quad 1] = [u^2 \quad u \quad 1]\begin{bmatrix} 1 & 0 & 0 \\ 0 & 1 & 0 \\ 0 & 0 & 1 \end{bmatrix} = [u^2 \quad u \quad 1]$$

We transform this base parabola into any other conic section (including another parabola) by transforming four arbitrarily chosen points from the plane "R" of the parabola into four points in another plane "C" of any conic section. Let us choose points P_1 $(u = 0)$, P_2 $(u = 1)$, P_3 $(u = \frac{1}{2})$, and P_T (the point of intersection of tangents at P_1 and P_2) which transform into corresponding points C_1, C_2, C_3, and C_T in the plane C. We use the previously established homogeneous coordinate notation for the general point in plane C — for example, $C_1 = [w_1x_1 \quad w_1y_1 \quad w_1]$ which corresponds to the ordinary point $[x_1 \quad y_1 \quad 1]$.

We state that there exists a matrix A that accomplishes the previously described transformation as follows:

$$PA = C$$

or

$$[u^2 \quad u \quad 1]A = [wx \quad wy \quad w]$$

Evidently A is a 3×3 matrix. Let us proceed to evaluate A. For the transformation of P_1 into C_1, the above equation is

$$[0 \quad 0 \quad 1]A = [w_1x_1 \quad w_1y_1 \quad w_1] \tag{1}$$

Similarly for points P_2 and P_T

$$[1 \quad 1 \quad 1]A = [w_2x_2 \quad w_2y_2 \quad w_2] \tag{2}$$

$$[0 \quad \tfrac{1}{2} \quad 1]A = [w_Tx_T \quad w_Ty_T \quad w_T] \tag{3}$$

We can combine Equations 1, 2, and 3 and write

$$\underbrace{\begin{bmatrix} 0 & 0 & 1 \\ 0 & \frac{1}{2} & 1 \\ 1 & 1 & 1 \end{bmatrix}}_{B} A = \begin{bmatrix} w_1x_1 & w_1y_1 & w_1 \\ w_Tx_T & w_Ty_T & w_T \\ w_2x_2 & w_2y_2 & w_2 \end{bmatrix}$$

$$= \begin{bmatrix} w_1 & 0 & 0 \\ 0 & w_T & 0 \\ 0 & 0 & w_2 \end{bmatrix}\begin{bmatrix} x_1 & y_1 & 1 \\ x_T & y_T & 1 \\ x_2 & y_2 & 1 \end{bmatrix}$$

or

$$A = \underbrace{\begin{bmatrix} 1 & -2 & 1 \\ -2 & 2 & 0 \\ 1 & 0 & 0 \end{bmatrix}}_{D}\begin{bmatrix} w_1 & 0 & 0 \\ 0 & w_T & 0 \\ 0 & 0 & w_2 \end{bmatrix}\begin{bmatrix} x_1 & y_1 & 1 \\ x_T & y_T & 1 \\ x_2 & y_2 & 1 \end{bmatrix} \tag{4}$$

where matrix D is the inverse of B.

The homogeneous coordinates w_1, w_T, and w_2 are unknown. We utilize the fourth point transformation $P_3 \rightarrow C_3$ to determine these quantities for a particular transformation. Let us illustrate this unique determination of w_1, w_T, w_2 by transforming the base parabola into a unit circle, shown in Figure 2, such that there is the correspondence

$$\begin{bmatrix} x_1 & y_1 & 1 \\ x_T & y_T & 1 \\ x_2 & y_2 & 1 \end{bmatrix} \Rightarrow \begin{bmatrix} 1 & 0 & 1 \\ 0 & 1 & 0 \\ -1 & 0 & 1 \end{bmatrix} \quad \text{and} \quad C_3 = w_3[0 \quad 1 \quad 1]$$

(Note the correspondence $[x_T \quad y_T \quad 1] \Rightarrow [0 \quad 1 \quad 0]$, the point of intersection of tangents at C_1 and C_2, which is at infinity in the y direction.)
Now

$C_3 = P_3A$ or

$$w_3[0 \quad 1 \quad 1] = [\tfrac{1}{4} \quad \tfrac{1}{2} \quad 1]\begin{bmatrix} 1 & -2 & 1 \\ -2 & 2 & 0 \\ 1 & 0 & 0 \end{bmatrix}\begin{bmatrix} w_1 & 0 & 0 \\ 0 & w_T & 0 \\ 0 & 0 & w_2 \end{bmatrix}\begin{bmatrix} 1 & 0 & 1 \\ 0 & 1 & 0 \\ -1 & 0 & 1 \end{bmatrix}$$

which on solution yields

$$w_T = 2w_3 = w_1 = w_2$$

We arbitrarily set $w_3 = \frac{1}{2}$. (Reader may verify that this is harmless in homogeneous coordinates because w_3 is a common factor and could have any value.) Substitution of these values in Equation 4 yields

$$A = \begin{bmatrix} 0 & -2 & 2 \\ -2 & 2 & -2 \\ 1 & 0 & 1 \end{bmatrix}$$

which transforms the base parabola into the unit circle (center at origin).

By similar methods we can evaluate A for any conic section. Hence, any conic section could be generated by the equation

$$[wx \quad wy \quad w] = [u^2 \quad u \quad 1]\begin{bmatrix} 3 \times 3 \\ \text{matrix} \end{bmatrix} \qquad (5)$$

Specifically, we have evaluated the matrix that transforms four points from the plane of the parabola into the plane of the circle.

According to the fundamental theorem of plane perspectivity, four points in one coordinate system and four corresponding points in a transformed coordinate system completely define a projective transformation. Let us see if our matrix A, just obtained from the correspondence of four points, does indeed completely define the transformation of curves.

According to Equation 5, a general point on the circle should be

$$[wx \quad wy \quad w] = [u^2 \quad u \quad 1]\begin{bmatrix} 0 & -2 & 2 \\ -2 & 2 & -2 \\ 1 & 0 & 1 \end{bmatrix}$$

or the ordinary coordinates x, y are given by

$$x = \frac{-2u + 1}{2u^2 - 2u + 1} \qquad y = \frac{-2u^2 + 2u}{2u^2 - 2u + 1}$$

(Reader may verify that $x^2 + y^2 = 1$.) Thus our Equation 5 is in conformance with the fundamental theorem of perspectivity.

affine transformation

A general affine transformation can be defined as one in which a space coordinate frame $0 \; x \; y \; z$ is transformed into some other frame $0' \; x' \; y' \; z'$, generally speaking, with a different "metric," i.e., with unit segments of different lengths and with different angles between them, and in which a point M is sent into point M' having the same coordinates relative to the new frame as those of the point M relative to the old frame[2] as illustrated in Figure 3. It can be shown that under an affine transformation every straight line is sent into a straight line, parallel lines are mapped into parallel lines, and if a point divides a segment in a given ratio, its image divides the image of this segment in the same ratio.

It follows from this definition that all circles and ellipses are affinely related to one another, i.e., one can be obtained from another by an affine transformation. In the last section, we obtained the parametric equation for a unit circle with its center at the origin. Now let us write the affine transformation to generate any circle of radius r and center (h, k) from the unit circle.

$$[wx \quad wy \quad w] = [u^2 \quad u \quad 1]\begin{bmatrix} 0 & -2 & 2 \\ -2 & 2 & -2 \\ 1 & 0 & 1 \end{bmatrix}\begin{bmatrix} r & 0 & 0 \\ 0 & r & 0 \\ h & k & 1 \end{bmatrix}$$

The last matrix describes the scale change to radius r followed by translation of the center to (h, k). The reader may verify the result by performing the multiplication so that $(x - h)^2 - (y - k)^2 = r^2$, (x, y) being the ordinary coordinates of a point on the circle. Furthermore, we can write the equation of any ellipse whose major axis is $2a$, whose minor axis is $2b$, whose center is at (h, k), and whose major axis makes an angle Θ with the x axis:

$$[wx \quad wy \quad w] = [u^2 \quad u \quad 1]\underbrace{\begin{bmatrix} 0 & -2 & 2 \\ -2 & 2 & -2 \\ 1 & 0 & 1 \end{bmatrix}}_{\text{unit circle}}\underbrace{\begin{bmatrix} a & 0 & 0 \\ 0 & b & 0 \\ 0 & 0 & 1 \end{bmatrix}}_{\text{scale}}\underbrace{\begin{bmatrix} \cos\Theta & \sin\Theta & 0 \\ -\sin\Theta & \cos\Theta & 0 \\ h & k & 1 \end{bmatrix}}_{\text{rotate-translate}}$$

We can show by similar methods that all hyperbolas are affine to one another, and so are all parabolas.

In passing, we remark that using the aforementioned methods for the generation of vectors for curves (conics in this case) eliminates the use of trigonometric functions and, hence, improves the "response time" for display.

Intersections of conic sections using these methods are discussed elsewhere.[3]

Figure 3 Affine transformatiom

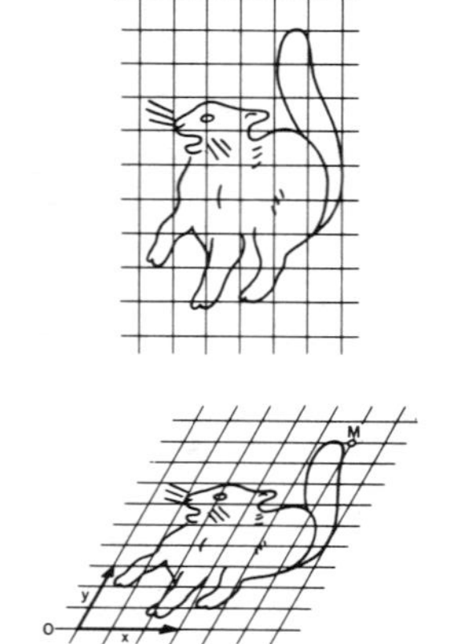

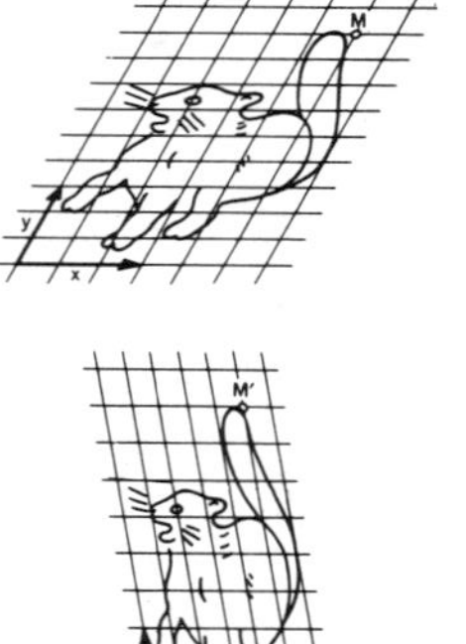

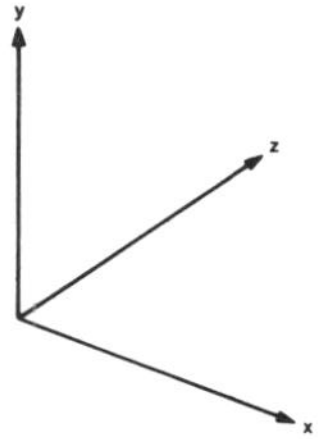

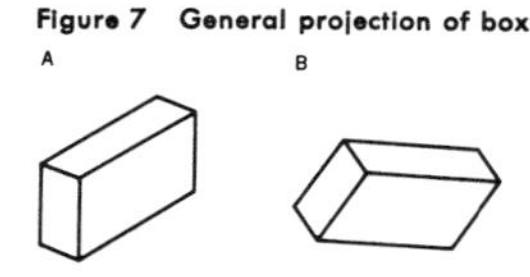

Figure 4 Coordinate system

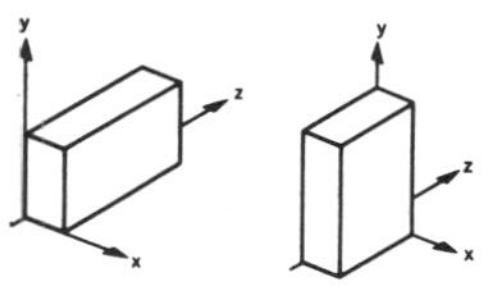

Figure 5 Top view by rotation

BEFORE ROTATION AFTER ROTATION

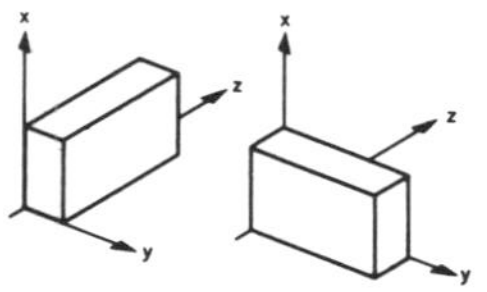

Figure 6 Right-side view by rotation

Projection of three-dimensional objects into two-dimensional pictures

An important aspect of computer graphics is the projection of objects onto image planes.[4] All drawings and pictures are examples of special cases of projection. The orthographic, isometric, cavalier, and cabinet projections employed by draftsmen fall into this category, as do the perspective pictures used by architects. We can show that *all* of these two-dimensional images can be produced by a single 4 × 4 matrix, whose 16 elements are easy to determine.

Before we examine the general case, let us look briefly at the orthographic projections of an object as used by draftsmen—the ordinary top, front, and side views of an object. We establish a coordinate system such as in Figure 4 where x is horizontal and increases to the right, y is vertical and increases upward, and z is horizontal and increases as indicated.

We shall think of the plane of xy, or the $z = 0$ plane, as the picture plane. Now imagine some object related to this coordinate system, and imagine that points on the object are represented by matrices of the form $[x \quad y \quad z]$. We might consider in particular a rectangular box with edges parallel to the three coordinate axes and with one corner at the origin of coordinates. There will be one corner of this box that lies in none of the three coordinate planes; let it be at $[1 \quad 2 \quad 3]$. Evidently the projection of this point on the picture plane is given by the point whose coordinates are $[1 \quad 2 \quad 0]$; that is, the z coordinate after projection has become zero. The projection is given by the matric product

$$[1 \quad 2 \cdot 3]\begin{bmatrix}1 & 0 & 0\\0 & 1 & 0\\0 & 0 & 0\end{bmatrix} = [1 \quad 2 \quad 0]$$

Now imagine that we wish to obtain a "top view" of this box. If xy remains the picture plane, we must rotate the object with respect to this picture plane, and *after* the rotation we must project it into the picture plane.

rotation

Let us assume that rotation takes place as shown in Figure 5. The rotation is given by the matric product

$$[1 \quad 2 \quad 3]\begin{bmatrix}1 & 0 & 0\\0 & 0 & -1\\0 & 1 & 0\end{bmatrix} = [1 \quad 3 \quad -2]$$

It consists of an interchange of the y and z coordinates, together with a sign change. We can also obtain a right-side view of the object by the rotation shown in Figure 6. Here the matric transformation is

$$[1 \quad 2 \quad 3]\begin{bmatrix}0 & 0 & -1\\0 & 1 & 0\\1 & 0 & 0\end{bmatrix} = [3 \quad 2 \quad -1]$$

Again note the interchange of coordinates and the change of sign.

It works out that the determinants of both these rotation matrices are equal to +1. Without the change of sign, the determinants would be −1, which would be equivalent to producing a "reflection" of the object as well as a rotation. Right-hand objects would turn into left-hand objects. Indeed, it can be shown in general that the determinant of any rigid-rotation matrix has a value of +1, and this is a necessary (although not sufficient) condition on the matrix.

Finally, after either of the rotations, we obtain the projection on the x-y plane by multiplying by the projection matrix:

$$\begin{bmatrix}1 & 0 & 0\\0 & 1 & 0\\0 & 0 & 0\end{bmatrix}$$

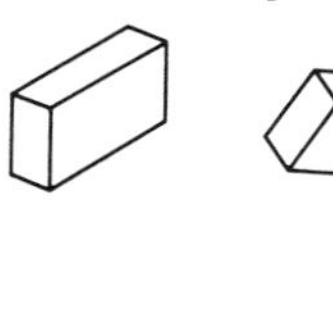

Figure 7 General projection of box

axonometric pictorials

Now a *general* projection of an object can, as is well known, be obtained by drawing two auxiliary views of the object. This is essentially equivalent to making two arbitrary rotations in sequence, and then projecting the figure into the picture plane after the two rotations have been performed. The result will be an "axonometric pictorial" of the object.

Suppose we wish to produce a picture of an object in a general projection, but in addition, we wish to make vertical edges *appear* vertical after the transformation. We wish to achieve a picture of the box that looks like the one in Figure 7A and not like the one in Figure 7B.

We achieve the desired result by first rotating about the vertical y axis. The rotation matrix is, in part,

$$\begin{bmatrix} & 0 & \\0 & 1 & 0\\ & 0 & \end{bmatrix}$$

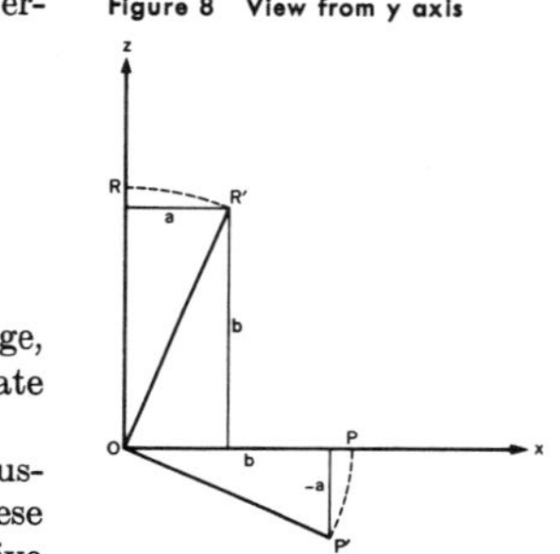

Figure 8 View from y axis

since y dimensions do not change. But x and z locations change, and we need to examine the rotation to determine the appropriate matric entries.

Looking down the y axis, we see the x and z axes as illustrated in Figure 8. Consider points P and R on each of these axes. If these points rotate rigidly about the origin O, they arrive at points P' and R'. If the triangle OPR is rigid, it is congruent to triangle $OP'R'$, and the right angle at O is preserved, as well as the lengths. Let us say that the coordinates (x and z) of point R' after the rotation are $[a \quad b]$. If the coordinates of R before rotation are $[0 \quad 1]$, and the length OR = the length OR', then $a^2 + b^2 = 1$; (b can be the cosine of the angle of rotation, and a can be the sine of this angle).

Again, the coordinates of P before rotation can be $[1 \quad 0]$, and after rotation, the coordinates of P' are necessarily $[b \quad -a]$ in order to preserve the right angle. When we examine this *plane* rotation, we have:

$$\begin{bmatrix} P \\ R \end{bmatrix} [T] = \begin{bmatrix} b & -a \\ a & b \end{bmatrix}$$

But

$$\begin{bmatrix} P \\ R \end{bmatrix} = \begin{bmatrix} 1 & 0 \\ 0 & 1 \end{bmatrix}$$

and the rotation transformation matrix T is then simply the matrix $\begin{bmatrix} b & -a \\ a & b \end{bmatrix}$. Note that the determinant of this matrix is $+1$, since $a^2 + b^2 = 1$.

We introduce this result into the three-dimensional transformation to obtain

$$\begin{bmatrix} b & 0 & -a \\ 0 & 1 & 1 \\ a & 0 & b \end{bmatrix}$$

We now rotate the resulting figure about the x axis; the x coordinates do not change this time, and so the rotation matrix, in part, is

$$\begin{bmatrix} 1 & 0 & 0 \\ 0 & & \\ 0 & & \end{bmatrix}$$

The missing partition is obtained as before and is equivalent to the plane rotation represented by the matrix $\begin{bmatrix} -d & c \\ c & d \end{bmatrix}$ where again $c^2 + d^2 = 1$, and c and d can be cosine and sine of the rotation angle. The complete three-dimensional rotation matrix is

$$\begin{bmatrix} 1 & 0 & 0 \\ 0 & -d & c \\ 0 & c & d \end{bmatrix}$$

The combination of rotations can be represented by the matric product of the separate primitive rotations taken in their proper order, and we evaluate it:

$$\begin{bmatrix} b & 0 & -a \\ 0 & 1 & 0 \\ a & 0 & b \end{bmatrix} \begin{bmatrix} 1 & 0 & 0 \\ 0 & -d & c \\ 0 & c & d \end{bmatrix} = \begin{bmatrix} b & -ac & -ad \\ 0 & -d & c \\ a & bc & bd \end{bmatrix}$$

Observe the occurrence of the zero in the first (or new x-generating) column of the matrix. This zero occurs in the position that *will* be multiplied by the y coordinate of the original point; it tells us that x coordinates are independent of the heights of points on the object. A moment's reflection confirms that this statement is equivalent to saying that vertical lines appear vertical after the rotation, although they are certainly foreshortened. This is exactly what we want.

We now can achieve the projection, as before, by post-multiplication with the matrix:

$$\begin{bmatrix} 1 & 0 & 0 \\ 0 & 1 & 0 \\ 0 & 0 & 0 \end{bmatrix}$$

translation

The foregoing procedure has left the corner of the object still attached to the origin. We wish to examine next the translation of the object to some new position in space. We now need homogeneous coordinates. If we wish to slide the object e units to the right (in the x direction), f units upward (in the y direction), and g units back from the picture plane (in the z direction), we can accomplish this by the transformation:

$$[x \quad y \quad z \quad 1] \begin{bmatrix} 1 & 0 & 0 & 0 \\ 0 & 1 & 0 & 0 \\ 0 & 0 & 1 & 0 \\ e & f & g & 1 \end{bmatrix} = [(x+e) \quad (y+f) \quad (z+g) \quad 1]$$

The new point coordinates exhibit the translation. We can again compound the pure rotation transformation with this translation transformation, and we obtain the matric product

$$\left[\begin{array}{ccc|c} b & -ac & -ad & 0 \\ 0 & -d & c & 0 \\ a & bc & bd & 0 \\ \hline e & f & g & 1 \end{array}\right]$$

Notice that the rotation matrix and the translation matrix essentially appear in the compound matrix. The projection matrix is now

$$\begin{bmatrix} 1 & 0 & 0 & 0 \\ 0 & 1 & 0 & 0 \\ 0 & 0 & 0 & 0 \\ 0 & 0 & 0 & 1 \end{bmatrix}$$

Incidentally, the projection matrix is trivial; it represents a mathematical way of saying that we simply ignore the z coordinates of the rotated and translated object when we construct the picture. But we see next that this matrix is nontrivial when we consider perspective pictorials, of which the preceding axonometric pictorials are a subclass.

perspective pictorials

We next consider the projection of an object on the plane $z = 0$, but from a local point, say the point $[0 \quad 0 \quad -h]$ in ordinary coordinates. The situation can be pictured as in Figure 9. In this figure, the y axis rises vertically out of the page.

We can write, by similar triangles, that

$$\frac{x'}{h} = \frac{x}{z + h},$$

Figure 9 Projection on plane z = 0

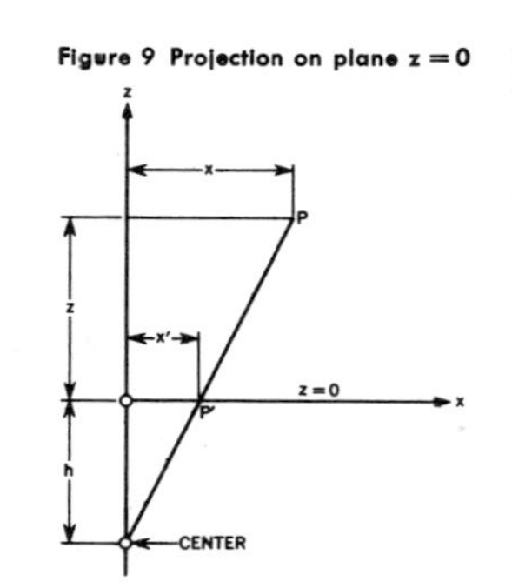

where x' is the picture plane coordinate of P', the image of P. This equation leads to

$$x' = \frac{xh}{z + h} = \frac{x}{(z/h) + 1}$$

A similar expression can be written for the

$$y' = \frac{y}{(z/h) + 1}\,.$$

Now consider the matric product:

$$[x \quad y \quad z \quad 1]\begin{bmatrix} 1 & 0 & 0 & 0 \\ 0 & 1 & 0 & 0 \\ 0 & 0 & 0 & 1/h \\ 0 & 0 & 0 & 1 \end{bmatrix} = [x \quad y \quad 0 \quad ((z/h) + 1)]$$

Obviously this matrix can be interpreted as equal to the homogeneous coordinate matrix

$$[wx \quad wy \quad 0 \quad w]$$

where $w = (z/h) + 1$, and

$$[x \quad y \quad 0 \quad 1] = (1/w)[wx \quad wy \quad 0 \quad w]$$

This relationship shows that the matrix

$$\begin{bmatrix} 1 & 0 & 0 & 0 \\ 0 & 1 & 0 & 0 \\ 0 & 0 & 0 & 1/h \\ 0 & 0 & 0 & 1 \end{bmatrix}$$

serves to project the object by rays from the center of projection, and the sectioning of this bundle of rays by the plane $z = 0$ produces the picture, which is accomplished by dividing the matrix $[wx \quad wy \quad 0 \quad w]$ by the quantity w.

We now see the nontrivial nature of the projection matrix and the need for homogeneous coordinates. Incidentally, if the distance h from the center of projection to the picture plane is increased in the limit as h approaches infinity, then $1/h$ approaches zero, and we obtain axonometric projection as a special case since the matrix becomes the trivial one already described.

distortion

The point $[0 \quad 0 \quad -h]$, the center of projection, is also the point from which the picture should be viewed. Any other viewing position yields more or less "distortion of perspective." This fact is very imperfectly understood, particularly by nontechnical artists and architects; this is obvious in much that has been written about distortion of perspectives and what empirical measures to take to avoid it. However, *any* perspective picture looks distorted unless viewed from this single point in space—but conversely any picture appears undistorted if this point is known and the picture is viewed from there.

If, in the construction of the picture, this point is, say, three inches from the picture, but the picture is viewed from a normal distance of, say, 15 inches, the perspective picture of necessity appears distorted, and violently so. It is difficult for most people to accommodate (or focus) the eye on a picture held three inches from the eye; however, if a person looks at the picture through a three-inch focal length magnifying glass, the picture will appear undistorted.

rational functions

We have seen that all conics can be generated by a transformation of a simple base conic by the formula, in homogeneous coordinates:

$$[wx \quad wy \quad w] = [u^2 \quad u \quad 1]\,A\,,$$

where A is a 3×3 matrix. Then

$$x = \frac{wx}{w} \qquad \text{and} \qquad y = \frac{wy}{w}$$

Since wx, wy, and w are each quadratic in u, we could call the coordinates x and y "rational quadratic functions" of u. Similarly, we can write[5]

$$[wx \quad wy \quad wz \quad w] = [u^3 \quad u^2 \quad u \quad 1]\,A$$

where A is a 4×4 matrix.

In this case, the coordinates x, y, and z are rational cubic functions of u. If the top row of A consists of zeros, we have rational quadratics, or ordinary conics, a special case.

If the A matrix is chosen so that its last column is

$$\begin{bmatrix} 0 \\ 0 \\ 0 \\ 1 \end{bmatrix}$$

the curve is $[x \quad y \quad z \quad 1]$, and the denominator is always 1. This matrix represents an ordinary parametric cubic curve. Thus, conics and cubics are special cases of rational cubic functions, and a computer can generate circles, ellipses, hyperbolas, parabolas, and cubics, as well as more general curves, simply by proper choice of the A matrix, without the necessity for having special and distinct routines for these curve forms. A specific application for generating spline-like curves utilizing this formula is discussed in this issue in a paper by Ahuja.

Surfaces

A surface is the locus of a point that moves in space with two degrees of freedom. A point V on a surface may be written in matric notation as:

$$[x \quad y \quad z] = [f(u, s) \quad g(u, s) \quad h(u, s)]$$

where u and s are independent parameters. Before proceeding further we shall compact the notation. We write

$$us \text{ for } [f(u, s) \quad g(u, s) \quad h(u, s)], \quad us_u \text{ for } \frac{\partial(us)}{\partial u},$$

$$us_s \text{ for } \frac{\partial(us)}{\partial s}, \quad us_{us} \text{ for } \frac{\partial^2(us)}{\partial u \partial s}, \quad us_{uu} \text{ for } \frac{\partial^2(us)}{\partial u^2},$$

and likewise for other derivatives.

patches

We build complicated surfaces by adjoining small surface "patches." Accordingly, we focus our attention on one such surface patch. For computational simplicity, we restrict the variation of parameters in the range 0 to 1 for each patch, i.e., $0 \leq u, s \leq 1$. With this notation in mind, a surface patch can be considered to be a surface segment bounded by four space curves, $0s$, $1s$, $u0$, $u1$ as shown in Figure 10. (Note that symbol $u0$ stands for the vector describing the $(x \quad y \quad z)$ coordinates of points along the curve generated by holding $s = 0$ constant and varying u.) We wish to blend such patches (for example, $A1$ and $A2$ in Figure 11) into one surface with any desired characteristics at common boundaries. The surface equation for a slope-matching, slope-continuous surface patch with entirely arbitrary boundaries and entirely arbitrary slopes across these boundaries may be written in matric notation:[5]

Figure 10 Surface patch

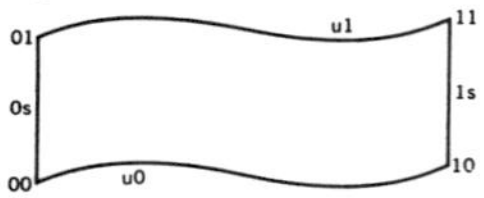

$$us = -[-1 \quad F_0u \quad F_1u \quad G_0u \quad G_1u] \cdot$$

$$\times \begin{bmatrix} 0 & u0 & u1 & u0_s & u1_s \\ 0s & 00 & 01 & 00_s & 01_s \\ 1s & 10 & 11 & 10_s & 11_s \\ 0s_u & 00_u & 01_u & 00_{us} & 01_{us} \\ 1s_u & 10_u & 11_u & 10_{us} & 11_{us} \end{bmatrix} \begin{bmatrix} -1 \\ F_0s \\ F_1s \\ G_0s \\ G_1s \end{bmatrix} \quad (6)$$

where F_0, F_1, G_0, G_1, are scalar functions of a single variable with the following end conditions:

$$F_0(0) = F_1(1) = 1,$$

$$F_0(1) = F_1(0) = G_0(0) = G_1(0) = G_1(1) = G_0(1) = 0,$$

$$F_0'(0) = F_1'(0) = F_1'(1) = F_0'(1) = G_0'(1) = G_1'(0) = 0,$$

$$G_0'(0) = G_1'(1) = 1$$

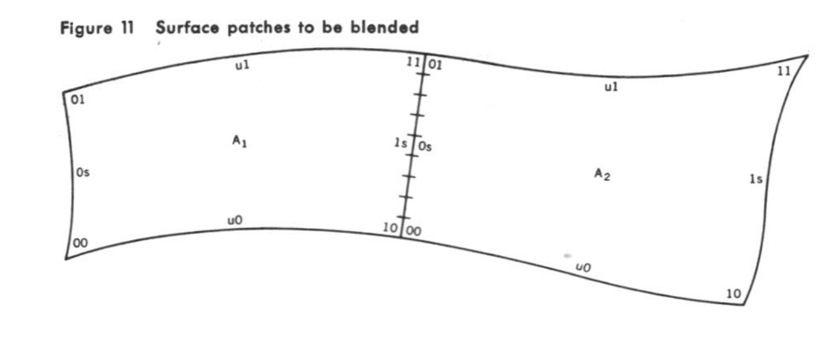

Figure 11 Surface patches to be blended

These functions serve to blend the aforementioned characteristics in the surface patch and are hence called *blending* functions. Equation 6 can be easily expanded for higher derivatives continuity.[5] Blending functions with the previously described stipulations can be used to define curves in terms of their end points and end-point tangent vectors, e.g.,

$$u0 = [F_0u \quad F_1u \quad G_0u \quad G_1u] \begin{bmatrix} 00 \\ 10 \\ 00_u \\ 10_u \end{bmatrix}$$

Furthermore, we can relate the blending function vector to a so-called *basis vector* $[u_1 \quad u_2 \quad u_3 \quad u_4]$ in the following way:

$$[F_0u \quad F_1u \quad G_0u \quad G_1u] = [u_1 \quad u_2 \quad u_3 \quad u_4]\, M$$

With an appropriate choice of the basis vector, Equation 6 can be used to generate a very wide class of surfaces. Specifically, if the basis vector is chosen to be $[u^3 \quad u^2 \quad u \quad 1]$, if $u0$, $u1$, $u0_s$, $u1_s$, 0_s, $1s$, $0s_u$, $1s_u$ are linear combinations of the elements of the basis vector, and if the expression is written in homogeneous form, then we obtain a $4 \times 4 \times 4$ tensor as descriptive of the boundary conditions. As a special case, this tensor leads to a parametric description of quadric surfaces.

CITED REFERENCES

1. J. T. Rule and R. F. Watts, *Engineering Graphics*, McGraw-Hill Book Company, New York, New York, 49 (1951).
2. A. D. Alexendrov, A. N. Kolmgrove and M. A. Lavrent'ev, *Mathematics: Its Content, Methods and Meaning*, American Mathematical Society, New York, New York (1963).
3. L. G. Roberts, *Homogeneous Matrix Representation and Manipulation of N-Dimensional Constructs*, Document MS 1405, Lincoln Laboratory, Massachusetts Institute of Technology, Cambridge, Massachusetts (May 1965).
4. L. G. Roberts, *Machine Perception of Three-Dimensional Solids*, Technical Report No. 315, Massachusetts Institute of Technology, Cambridge, Massachusetts (May 1963).
5. S. A. Coons, *Surfaces for Computer-Aided Design of Space Forms*, MAC-TR-41, Clearinghouse for Federal, Scientific, and Technical Information, Springfield, Virginia (June 1967).

GLOBAL AND LOCAL DEFORMATIONS OF SOLID PRIMITIVES

Alan H. Barr
Computer Science Department †
California Institute of Technology
Pasadena, California

Abstract

New hierarchical solid modeling operations are developed, which simulate twisting, bending, tapering, or similar transformations of geometric objects. The chief result is that the normal vector of an arbitrarily deformed smooth surface can be calculated directly from the surface normal vector of the undeformed surface and a transformation matrix. Deformations are easily combined in a hierarchical structure, creating complex objects from simpler ones. The position vectors and normal vectors in the simpler objects are used to calculate the position and normal vectors in the more complex forms; each level in the deformation hierarchy requires an additional matrix multiply for the normal vector calculation. Deformations are important and highly intuitive operations which ease the control and rendering of large families of three-dimensional geometric shapes.

KEYWORDS: Computational Geometry, Solid Modeling, Deformation

Introduction

Modeling hierarchies are a convenient and efficient way to represent geometric objects, allowing users to combine simpler graphical primitives and operators into more complex forms. The leaf-nodes in the hierarchy are the hardware/firmware commands on the equipment which draws the vectors, changes the colors of individual pixels, and operates on lists of line segments or polygons. With the appropriate algorithms and interfaces, users can develop a strong intuitive feeling for the results of a manipulation, can think in terms of each operation, and are able to create the objects and scenes which they desire.

In this paper, we introduce globally and locally defined deformations as new hierarchical operations for use in solid modeling. These operations extend the conventional operations of rotation, translation, Boolean union, intersection and difference. In section one, the transformation rules for tangent vectors and for normal vectors are shown. In section two, several examples of deformation functions are listed. A method is shown in section three to convert arbitrary local representations of deformations to global representations, for space curves and surfaces. Finally, in section four, applications of the methods to the rendering process are described, opening future research directions in ray-tracing algorithms. Appendix A contains a derivation of the normal vector transformation rule.

Deformations allow the user to treat a solid as if it were constructed from a special type of topological putty or clay, which may be bent, twisted, tapered, compressed, expanded, and otherwise transformed repeatedly into a final shape. They are highly intuitive and easily visualized operations which simulate some important manufacturing processes for fabricating objects, such as the bending of bar stock and sheet metal. Deformations can be incorporated into traditional CAD/CAM solid modeling and surface patch methods, reducing the data storage requirements for simulating flexible geometric objects, such as objects made of metal, fabric or rubber.

† Previous address, Raster Technologies Inc., N. Billerica, Mass.

Although it is possible to use these techniques to accurately model the physical properties of different elastic materials with the partial differential equations of elasticity and plasticity theory, simpler mathematical deformation methods exist. These simpler methods have reduced computational needs, are widely applicable in modeling, and are described in the examples section. It is beyond the scope of this paper to formulate the mathematical details of exact mechanical descriptions of physical deformation properties of materials.

1.0 Background and Derivations.

A **globally specified deformation** of a three dimensional solid is a mathematical function $\underline{F}$ which explicitly modifies the global coordinates of points in space. Points in the undeformed solid are called (small) $\underline{x}$, while points in the deformed solid are called (capital) $\underline{X}$. Mathematically, this is represented by the equation

$$\underline{X} = \underline{F}(\underline{x}). \qquad [\textit{Equation } 1.1a]$$

The x, y, and z components of the three dimensional vector $\underline{x}$ are designated x_1, x_2, and x_3. (For notational convenience, x_1, x_2, and x_3 and x, y, and z are used interchangably. A similar convention holds for the upper case forms.)

A **locally specified deformation** modifies the tangent space of the solid. Differential vectors in the substance of the solid are rotated and/or skewed; these vectors are integrated to obtain the global position. The differential vectors can be thought of as separate chain-links which can rotate and stretch; the local specification of the deformation is the rotation and skewing matrix function. The position of the end-link in the chain is the vector sum of the previous links, as shown in section three.

Tangent vectors and normal vectors are the two most important vectors used in modeling — the former for delineating and constructing the local geometry, and the latter for obtaining surface orientation and lighting information. Tangent and normal vectors on the undeformed surface may be transformed into the tangent and normal vectors on the deformed surface; the algebraic manipulations for the transformation rules involve a single multiplication by the Jacobian matrix $\underline{\underline{J}}$ of the transformation function $\underline{F}$. In this paper, the term "tangent transformation" substitutes for "contravariant transformation" and is the transformation rule for the tangent vectors. The term "normal transformation" substitutes for "covariant transformation" and is the transformation rule for the normal vectors.

The Jacobian matrix $\underline{\underline{J}}$ for the transformation function $\underline{X} = \underline{F}(\underline{x})$ is a function of $\underline{x}$, and is calculated by taking partial derivatives of $\underline{F}$ with respect to the coordinates x_1, x_2, and x_3:

$$\underline{J}_i(\underline{x}) = \frac{\partial \underline{F}(\underline{x})}{\partial x_i} \qquad [\textit{Equation } 1.1b]$$

In other words, the ith column of $\underline{\underline{J}}$ is obtained by the partial derivative of $\underline{F}(\underline{x})$ with respect to x_i.

When the surface of an object is given by a parametric function of two variables u and v,

$$\underline{x} = \underline{x}(u, v), \qquad [\textit{Equation } 1.1c]$$

any tangent vector to the surface may be obtained from linear combinations of partial derivatives of $\underline{x}$ with respect to u and v. The normal vector direction may be obtained from the cross product of two linearly independent surface tangent vectors.

The **tangent vector transformation rule** is a restatement of the chain rule in multidimensional calculus. The new vector derivative is equal to the Jacobian matrix times the old derivative.

In matrix form, this is expressed as:

$$\frac{\partial \underline{X}}{\partial u} = \underline{\underline{J}} \frac{\partial \underline{x}}{\partial u} \qquad [\textit{Equation } 1.2a]$$

This is equivalent in component form to:

$$X_{i,u} = \sum_{j=1}^{3} J_{ij} x_{j,u} \qquad [\textit{Equations } 1.2b]$$

In other words, the new tangent vector $\partial \underline{X}/\partial u$ is equal to the Jacobian matrix $\underline{\underline{J}}$ times the old tangent vector $\partial \underline{x}/\partial u$

The **normal vector transformation rule** involves the inverse transpose of the Jacobian matrix. A derivation of this result is found in Appendix A.

[*Equation* 1.3]

$$\underline{n}^{(X)} = \det J \underline{\underline{J}}^{-1T} \underline{n}^{(x)}$$

Of course, since only the direction of the normal vector is important, it is not necessary to compute the value of the determinant in practice, although it sometimes is implicitly calculated as shown in Appendix A. As is well known from calculus, the determinant of the Jacobian is the local volume ratio at each point in the transformation, between the deformed region and the undeformed region.

2.0 Examples of Deformations.

Example 2.1: Scaling. One of the simplest deformations is a change in the length of the three global components parallel to the coordinate axes. This produces an orthogonal scaling operation :

$$\begin{aligned} X &= a_1 x \\ Y &= a_2 y \\ Z &= a_3 z \end{aligned} \qquad [\textit{Equation } 2.1a]$$

The components of the Jacobian matrix are given by

$$J_{ij} = \frac{\partial X_i}{\partial x_j},$$

so

$$\underline{\underline{J}} = \begin{pmatrix} a_1 & 0 & 0 \\ 0 & a_2 & 0 \\ 0 & 0 & a_3 \end{pmatrix} \qquad [\textit{Equation } 2.1b]$$

The volume change of a region scaled by this transformation is obtained from the Jacobian determinant, which is $a_1 a_2 a_3$. The normal transformation matrix is the inverse transpose of the Jacobian matrix (optionally times the determinant of the Jacobian matrix), and is given by:

$$\det J \ \underline{\underline{J}}^{-1T} = \begin{pmatrix} a_2 a_3 & 0 & 0 \\ 0 & a_1 a_3 & 0 \\ 0 & 0 & a_1 a_2 \end{pmatrix}$$

Without the factor of the determinant, the normal transformation matrix is:

$$\underline{\underline{J}}^{-1T} = \begin{pmatrix} 1/a_1 & 0 & 0 \\ 0 & 1/a_2 & 0 \\ 0 & 0 & 1/a_3 \end{pmatrix}$$

To obtain the new normal vector at any point on the surface of an object subjected to this deformation, we multiply the original normal vector by either of the above normal transformation matrices. The new **unit** normal vector is easily obtained by dividing the output components by the magnitude of the vector.

For instance, consider converting a point $[x_1, x_2, x_3]^T$ lying on a roughly spherical surface centered at the origin, with normal vector $[n_1, n_2, n_3]^T$. The transformed surface point on the resulting ellipsoidal shape is $[a_1 x_1, a_2 x_2, a_3 x_3]^T$ and the transformed normal vector is parallel to $[n_1/a_1, n_2/a_2, n_3/a_3]^T$. The volume ratio between the shapes is $a_1 a_2 a_3$.

The scaling transformation is a special case of general affine transformations, in which the Jacobian matrix is a constant matrix. Affine transformations include skewing, rotation, and scaling transformations. When the transformation consists of pure rotation, it is interesting to note that the inverse of the matrix is equal to its transpose. For pure rotation, this means that the tangent vector and the normal vector are transformed by a single matrix. For more general affine transformations, pairs of constant matrices are required.

Example 2.2: Global Tapering along the Z Axis. Tapering is similar to scaling, by differentially changing the length of two global components without changing the length of the third. In figure 2.2, the function $f(z)$ is a piecewise linear function which decreases as z increases (from page bottom to the top). The magnitude of the tapering rate progressively increases from figure 2.2 a through figure 2.2 d. When the tapering function $f(z) = 1$, the portion of the deformed object is unchanged; the object increases in size as a function of z when $f'(z) > 0$, and decreases in size when $f'(z) < 0$. The object passes through a singularity at $f(z) = 0$ and becomes everted when $f(z) < 0$.

$$\begin{aligned} r &= f(z), \\ X &= rx, \\ Y &= ry, \\ Z &= z \end{aligned} \qquad [\textit{Equation } 2.2a]$$

The tangent transformation matrix is given by:

$$\underline{\underline{J}} = \begin{pmatrix} r & 0 & f'(z)x \\ 0 & r & f'(z)y \\ 0 & 0 & 1 \end{pmatrix} \qquad [\textit{Equation } 2.2b]$$

The local volumetric rate of expansion, from the determinant, is r^2.

The normal transformation matrix is given by:

$$r^2 \underline{\underline{J}}^{-1T} = \begin{pmatrix} r & 0 & 0 \\ 0 & r & 0 \\ -rf'(z)x & -rf'(z)y & r^2 \end{pmatrix}$$

The inverse transformation is given by:

$$\begin{aligned} r(Z) &= f(Z), \\ x &= X/r, \\ y &= Y/r, \\ z &= Z \end{aligned} \qquad [\textit{Equation } 2.2c]$$

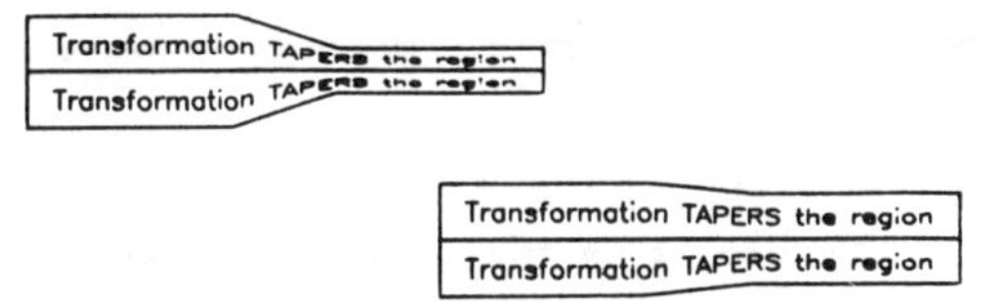

Figure 2.2 Progressive Tapering of a Ribbon

Example 2.3: Global Axial Twists. For some applications, it is useful to simulate global twisting of an object. A twist can be approximated as differential rotation, just as tapering is a differential scaling of the global basis vectors. We rotate one pair of global basis vectors as a function of height, without altering the third global basis vector. The deformation can be demonstrated by twisting a deck of cards, in which each card is rotated somewhat more than the card beneath it.

The global twist around the z axis is produced by the following equations:

$$\theta = f(z)$$
$$C_\theta = cos(\theta)$$
$$S_\theta = sin(\theta)$$

$$X = xC_\theta - yS_\theta,$$
$$Y = xS_\theta + yC_\theta, \qquad [Equation\ 2.3a]$$
$$Z = z.$$

The twist proceeds along the z axis at a rate of $f'(z)$ radians per unit length in the z direction.

The tangent transformation matrix is given by

$$\underline{\underline{J}} = \begin{pmatrix} C_\theta & -S_\theta & -xS_\theta f'(z) - yC_\theta f'(z) \\ S_\theta & C_\theta & xC_\theta f'(z) - yS_\theta f'(z) \\ 0 & 0 & 1 \end{pmatrix}$$

Note that the determinant of the Jacobian matrix is unity, so that the twisting transformation preserves the volume of the original solid. This is consistent with our "card-deck" model of twisting, since each individual card retains its original volume.

The normal transformation matrix is given by:

$$\underline{\underline{J}}^{-1T} = \begin{pmatrix} C_\theta & -S_\theta & 0 \\ S_\theta & C_\theta & 0 \\ yf'(z) & -xf'(z) & 1 \end{pmatrix}$$

Our original deck of cards is a rectangular solid, with orthogonal normal vectors. We can see from the above transformation matrix that the normal vectors to the twisted deck will generally tilt out of the x-y plane.

Figures 2.3.1 a–d show the effect of a progressively increasing twist. In these line drawings of solids, vectors are hidden by the normal vector criterion—if the normal vector (as calculated by the above transformation matrix) faces the viewer, the line is drawn, otherwise, the line segment is not drawn. Figure 2.3.3 shows an object which has been twisted and tapered, while figures 2.3.4 and 2.3.2 show the results from twisting an object around an axis not within the object itself.

The inverse transformation is given by:

$$[Equation\ 2.3b]$$

$$\theta = f(Z),$$
$$x = XC_\theta + YS_\theta,$$
$$y = -XS_\theta + YC_\theta,$$
$$z = Z$$

which is basically a twist in the opposite direction.

Figure 2.3.1 Progressive Twisting of a Ribbon

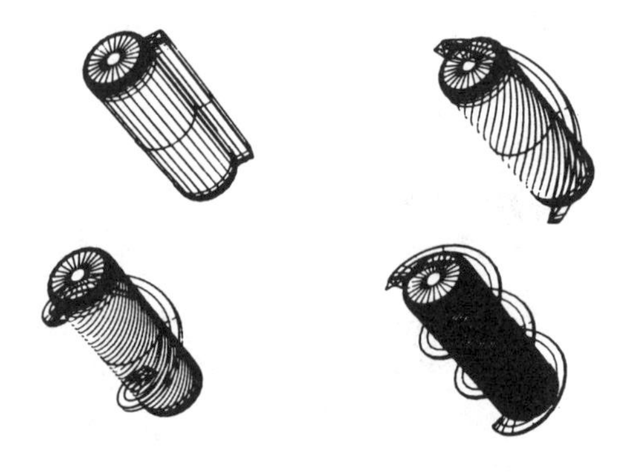

Figure 2.3.2 Progressive Twisting of Two Primitives

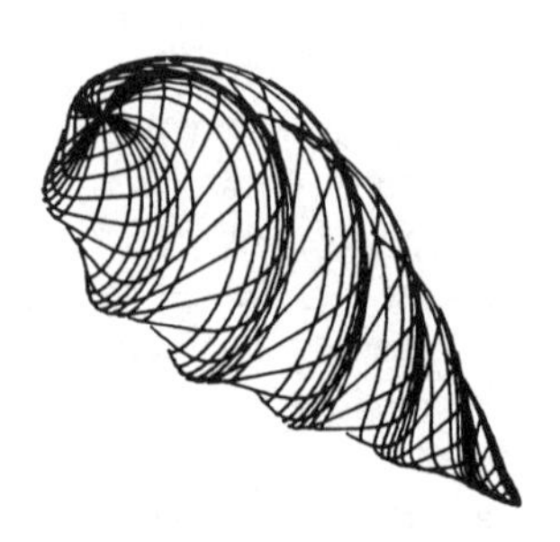

Figure 2.3.3 Twisting of a Tapered Primitive

Figure 2.3.4 Tapering of a Twisted offset Primitive

Example 2.4: Global Linear Bends along the Y-Axis. For other applications, it is useful to have a simple simulation of bending.

The following equations represent an isotropic bend along a centerline parallel to the y-axis: the length of the centerline does not change during the bending process. The bending angle θ, is constant at the extremities, but changes linearly in the central region. In the bent region, the bending rate k, measured in radians per unit length, is constant, and the differential basis vectors are simultaneously rotated and translated around the third local basis vector. Outside the bent region, the deformation consists of a rigid body rotation and translation. The range of the bending deformation is controlled by y_{min}, and y_{max}, with the bent region corresponding to values of y such that $y_{min} \leq y \leq y_{max}$. The axis of the bend is located along $[s, y_0, 1/k]^T$, where s is the parameter of the line. The center of the bend occurs at $y = y_0$—i.e., where one would "put one's thumbs" to create the bend. The radius of curvature of the bend is $1/k$.

The bending angle θ is given by:

$$\theta = k(\hat{y} - y_0),$$
$$C_\theta = cos(\theta),$$
$$S_\theta = sin(\theta),$$

where

$$\hat{y} = \begin{cases} y_{min}, & \text{if } y \leq y_{min} \\ y, & \text{if } y_{min} < y < y_{max} \\ y_{max}, & \text{if } y \geq y_{max} \end{cases}$$

The formula for this type of bending along the y axis centerline is given by the following relations:

[*Equation* 2.4a]

$$X = x$$

$$Y = \begin{cases} -S_\theta(z - \frac{1}{k}) + y_0, & y_{min} \leq y \leq y_{max}, \\ -S_\theta(z - \frac{1}{k}) + y_0 + C_\theta(y - y_{min}), & y < y_{min} \\ -S_\theta(z - \frac{1}{k}) + y_0 + C_\theta(y - y_{max}), & y > y_{max} \end{cases}$$

$$Z = \begin{cases} C_\theta(z - \frac{1}{k}) + \frac{1}{k}, & y_{min} \leq y \leq y_{max}, \\ C_\theta(z - \frac{1}{k}) + \frac{1}{k} + S_\theta(y - y_{min}), & y < y_{min} \\ C_\theta(z - \frac{1}{k}) + \frac{1}{k} + S_\theta(y - y_{max}), & y > y_{max} \end{cases}$$

These functions have continuous values at the boundaries of each of the three regions for y, and in the limit, for $k = 0$. However, there is a jump in the derivative of the bending angle θ at the $y = y_{min}$ and $y = y_{max}$ boundaries. The discontinuities may be eliminated by using a smooth function for θ as a function of y, but the transformation matrices would need to be re-derived.

The tangent transformation matrix is given by:

$$\underline{\underline{J}} = \begin{pmatrix} 1 & 0 & 0 \\ 0 & C_\theta(1 - \hat{k}z) & -S_\theta \\ 0 & S_\theta(1 - \hat{k}z) & C_\theta \end{pmatrix}$$

where

$$\hat{k} = \begin{cases} k, & \text{if } \hat{y} = y \\ 0, & \text{if } \hat{y} \neq y. \end{cases}$$

The local rate of expansion, as obtained from the determinant, is $1 - \hat{k}z$.

The normal transformation matrix is given by:

$$(1 - \hat{k}z)\underline{\underline{J}}^{-1T} = \begin{pmatrix} 1 - \hat{k}z & 0 & 0 \\ 0 & C_\theta & -S_\theta(1 - \hat{k}z) \\ 0 & S_\theta & C_\theta(1 - \hat{k}z) \end{pmatrix}$$

The inverse transformation is given by:

[*Equation* 2.4b]

$$\theta_{min} = k(y_{min} - y_0)$$

$$\theta_{max} = k(y_{max} - y_0)$$

$$\hat{\theta} = -tan^{-1}\left(\frac{Y - y_0}{Z - \frac{1}{k}}\right)$$

$$\theta = \begin{cases} \theta_{min}, & \text{if } \theta < \hat{\theta}_{min} \\ \hat{\theta}, & \text{if } \theta_{min} \leq \hat{\theta} \leq \theta_{max} \\ \theta_{max}, & \text{if } \hat{\theta} > \theta_{max} \end{cases}$$

$$x = X$$

$$\hat{y} = \frac{\theta}{k} + y_0$$

$$y = \begin{cases} \hat{y}, & y_{min} < \hat{y} < y_{max} \\ (Y - y_0)C_\theta + (z - \frac{1}{k})S_\theta + \hat{y}, & \hat{y} = y_{min} \text{ or } y_{max} \end{cases}$$

$$z = \begin{cases} \frac{1}{k} + ((Y - y_0)^2 + (Z - \frac{1}{k})^2)^{1/2}, & y_{min} < \hat{y} < y_{max} \\ -(Y - y_0)S_\theta + (z - \frac{1}{k})C_\theta + \hat{y}, & \hat{y} = y_{min} \text{ or } y_{max} \end{cases}$$

In figure 2.4.2, a constant 90° bend is produced by varying the range and the bend rate. In other words, $k(y_{max} - y_{min}) = \pi/2$ in each of the examples. In figure 2.4.3, a twisted object is subjected to a progressive bend to produce a Moebius band. Figures 2.4.4 a and b show a hierarchy of tapering, twisting, and bending, by superimposing a bend on the objects in figures 2.3.2 and 2.3.3. In figure 2.4.5, a chair is made from six primitives using seven bends. The details of the crimp in the coordinate systems is shown in figures 2.4.6 a - b.

However, the type of bending shown in the figures does not retain all of the generality that true bending requires. Some materials are anisotropic and have an intrinsic "grain" or directionality in them. Although this is beyond the scope of this paper, it is interesting to note that the tangent and normal transformation rules may still be utilized.

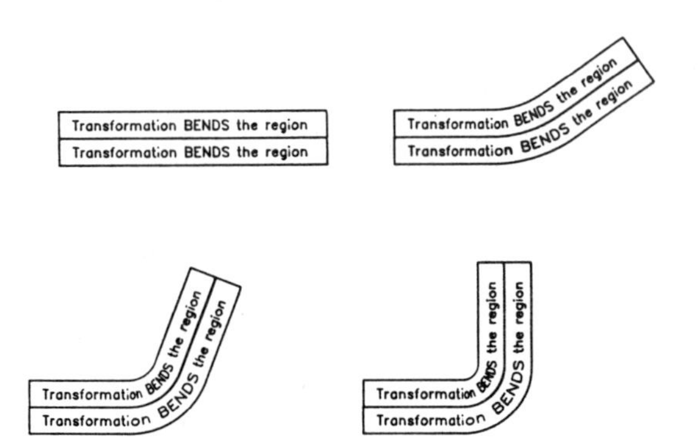

Figure 2.4.1 Progressive Bending of a Region

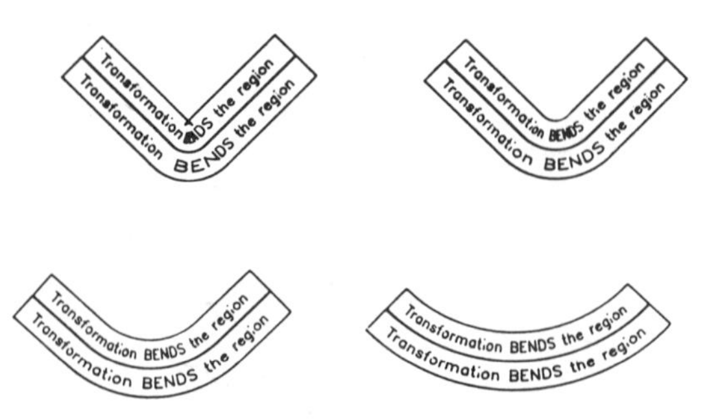

Figure 2.4.2 Progressive Change in Bending Range of a Region

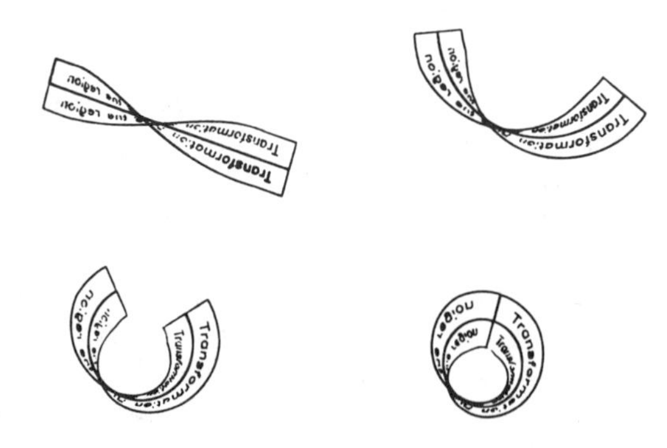

Figure 2.4.3 Moebius band is produced with a twist and a bend

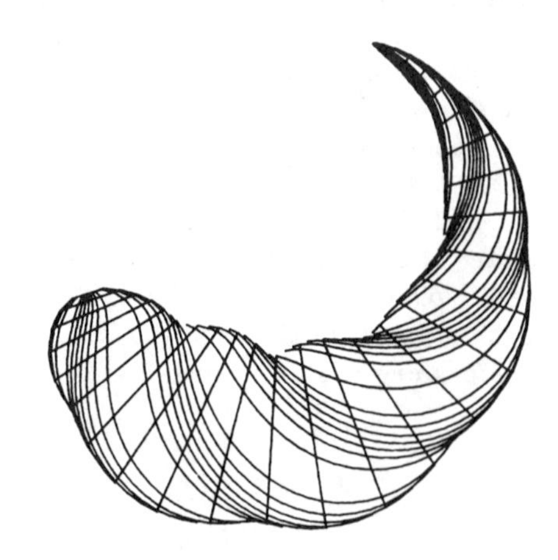

Figure 2.4.4 a Bent, Twisted, Tapered Primitive

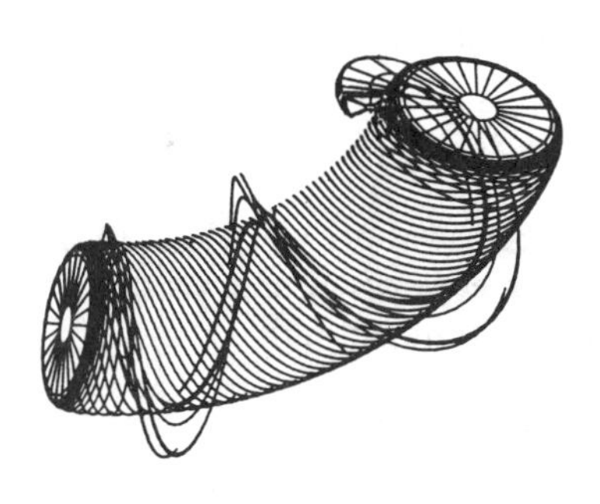

Figure 2.4.4 b Bent, Twisted Primitive

Figure 2.4.5 Chair Model, with six primitives and seven bends.

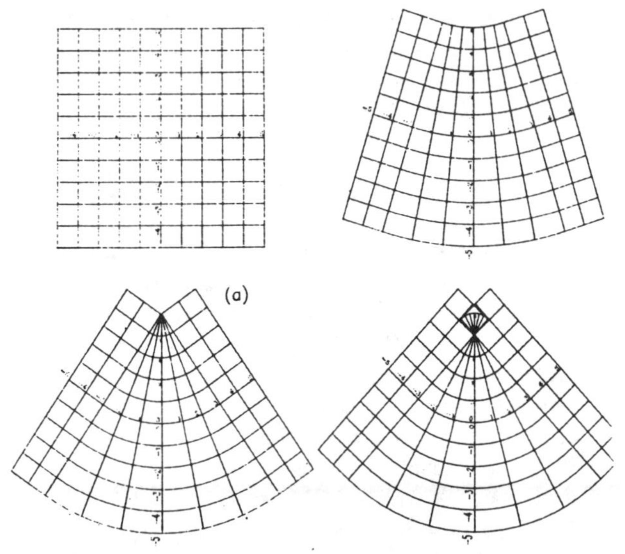

Figure 2.4.6 Details of the Bend near the Crimp

3.0 Converting Local Representations to Global Representations.

In this section, a method for generating more general shapes is addressed. The Jacobian matrix $\underline{\underline{J}}(\underline{x})$ is assumed to be known as a function of x_1, x_2, and x_3, but a closed form expression for the corresponding coordinate deformation function $\underline{X} = \underline{F}(\underline{x})$ is not known (i.e., in terms of standard mathematical functions). The basic method involves

(1) the conversion of the undeformed input shape into its tangent vectors by differentiation,

(2) transforming the tangent vectors via the tangent transformation rule into the tangent vectors of the deformed object, and then

(3) integrating the new tangent vectors to obtain the new position vectors of the deformed space curve, surface, or solid.

This "local-to-global" operation converts the local tangent vectors and Jacobian matrix into the global position vectors. The absolute position in space of the deformed object is defined within an arbitrary integration constant vector .

The above method provides a completely general description of deformation, and may be directly coupled to the output from the elasticity equations, finite element analysis, or other advanced mathematical models of deformable entities describing a profoundly general collection of shapes. The integrations outlined above need not be calculated explicitly in a ray-tracing environment: a multidimensional Newton's method can use the Jacobian matrix directly.

3.1 Transformations of Space Curves. Given a space curve, parameterized by a single variable s,

$$\underline{x} = \underline{x}(s), \quad s_0 \leq s \leq s_1$$

a new curve $\underline{X}(s)$ is desired which is the deformed version of $\underline{x}(s)$. The Jacobian matrix $\underline{\underline{J}}(s)$ or $\underline{\underline{J}}(\underline{x}(s))$ is assumed to be known, but the coordinate transformation function $\underline{X} = \underline{F}(\underline{x})$ is assumed to be unavailable. As stated above, the equation for $\underline{X}(s)$ may bé derived from the fact that,

(1) by definition, the position $\underline{X}(s)$ is a constant vector plus the integral of the derivative of the position, i.e.,

[*Equation* 3.1a]

$$\underline{X}(s) = \int_0^s \underline{X}'(\tilde{s})d\tilde{s} + \underline{x}_0,$$

(2) the derivative of the position is obtained via the tangent transformation rule, Equation 1.2 a, so

[*Equation* 3.1*b*]

$$\underline{X}(s) = \int_0^s \underline{\underline{J}}(\underline{x}(\tilde{s}))\underline{x}'(\tilde{s})d\tilde{s} + \underline{x}_0$$

where $\underline{\underline{J}}(\underline{x}(s))$ is the Jacobian matrix which depends upon the value of s, and $\underline{x}'(s)$ is the arclength derivative (a tangent vector) of the input curve $\underline{x}(s)$. At each point in the untransformed curve, $\underline{x}(s)$, the tangent vectors $\underline{x}'(s)$ are rotated and skewed to a new orientation in the transformed curve: the curve can be bent and twisted with or without being being stretched. For this case, any matrix function which allows the integral to be evaluated may serve as a Jacobian, since there is only one path along which to integrate.

For inextensible bending and twisting transformations of the space curve, with no stretching at any point of the curve, the Jacobian matrix $\underline{\underline{J}}(s)$ must be a varying rotation matrix function. (Even though this is not a constant affine rotation, the matrix function for the tangent vector transformation rule is identical to that used for the normal vector transformation rule.)

3.2 Transformations of 3-D surfaces and solids. The representation of a transformed surface or solid can be obtained much in the same manner as a space curve. First, an origin $\underline{O}$ is chosen in the object to be deformed. For each point $\underline{x}$ in the surface of the object, a piecewise smooth space curve is chosen, which connects the origin $\underline{O}$ to the input point $\underline{x}$. The space curve is then subjected to the deformation as in section 3.1. If $\underline{\underline{J}}(\underline{x})$ is in fact the Jacobian of some (unspecified) deformation function $\underline{X} = \underline{F}(\underline{x})$, the transformation from $\underline{x}$ to $\underline{X}$ is unique: all smooth paths connecting $\underline{O}$ and $\underline{x}$ will be equivalent. Since the equation of the surface is given by $\underline{x} = \underline{x}(u, v)$, the space curve in the surface may be obtained by selecting two functions of a single variable, say s , for u and for v. i.e.,

$$u = u(s)$$

$$v = v(s)$$

so that the space curve in the surface $\hat{\underline{x}}(s)$ is obtained by substituting the values of u and v into the equation for $\underline{x}$.

$$\hat{\underline{x}}(s) = \underline{x}(u(s), v(s))$$

This space curve is then transformed as shown above, in Equation 3.1 b. The space curve should be piecewise differentiable, so that the derivatives can be evaluated and integrated. The equation for the deformed curve is

[*Equation* 3.2.1]

$$\underline{X}(u(s), v(s)) =$$

$$\int_0^s \underline{\underline{J}}(\underline{x}(u(\hat{s}), v(\hat{s})))\underline{x}'(u(\hat{s}), v(\hat{s}))d\hat{s} + \underline{x}_0$$

Expanding the above equation, using the fact that the symbol $'$ means d/ds, and using the multidimensional chain rule, we obtain

$$\underline{X}(u(s), v(s)) =$$

$$\int_0^s \underline{\underline{J}}(\underline{x}(u(\hat{s}), v(\hat{s})))(\frac{\partial \underline{x}}{\partial u}u'(\hat{s}) + \frac{\partial \underline{x}}{\partial v}v'(\hat{s})))d\hat{s} + \underline{x}_0$$

As stated before, for consistency, $\underline{\underline{J}}$ must be the Jacobian matrix of some global function $\underline{F}(\underline{x})$, so that the results are independent of the path connecting $\underline{O}$ and $\underline{x}$, and so that the tangent and normal vector transformation rules apply. The test for the "Jacobian-ness" of the matrix, (in the absence of a prespecified deformation function $\underline{F}(\underline{x})$) depends on the partial derivatives of the columns of $\underline{\underline{J}}(\underline{x})$

The columns must satisfy

$$\underline{J}_{i,j} = \underline{J}_{j,i} \qquad \text{[Equation 3.2.2]}$$

In other words, the partial derivative of the ith column of $\underline{\underline{J}}$ with respect to x_j must be equal to the partial derivative of the jth column of $\underline{\underline{J}}$ with respect to x_i. (The underlying principle to prove this result is a multiple-integration path consistency requirement. The integrand must be an exact differential.) The values of the Jacobian may be directly related to the material properties of the substance to be modeled, and may utilize the plasticity and elasticity equations.

4.0 Applications to Rendering

To obtain a set of control points and normal vectors with which to create surface patches like polygons or spline patches, we sample the deformed surface parametrically, With the appropriate sampling, the patches can faithfully tesselate the desired object, with more detail where the surface is highly curved, and less detail where the surface is flat.

First, the object is sampled with a raw grid of parametric u-v values. This raw parametric sampling of the surface is then refined using normal vector criteria, as calculated by the transformation rule: the surface is recursively subdivided when the adjacent normal vectors diverge too greatly. Dot products which are far enough from unity indicate that more recursive detail is necessary in that region.

In this way, patch-oriented methods like depth-buffer and scan-line encoding schemes are effective. These algorithms are linear in terms of the total surface area and total number of patches. The direct subdivison approach is not as well-suited to ray tracing, since the total number of operations is quadratic in the number of ray comparisons and objects.

The incident ray can be intersected with the deformed primitive analytically, to reduce the number of objects. In addition, it is possible to use the inverse deformation to undeform the primitives and trace along the deformed rays. (See figures 4.1 and 4.2). This reduces the dimensionality of the parameter search from three to one, indicating a tremendous saving in numerical complexity.

The Jacobian techniques in this paper aid the traditional solution methods to find roots of non-linear ray equations (in the context of ray-tracing deformed objects), including the multidimensional Newton-Raphson method, the method of regula falsi, and the one-dimensional Newton's methods in N-space. (See [ACTON].) The analysis of rendering deformed primitives using these techniques is left to a future study.

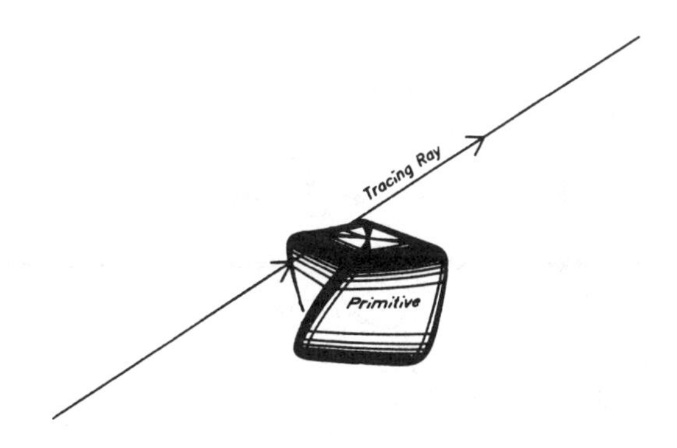

Figure 4.1 Deformed primitive, in undeformed space.

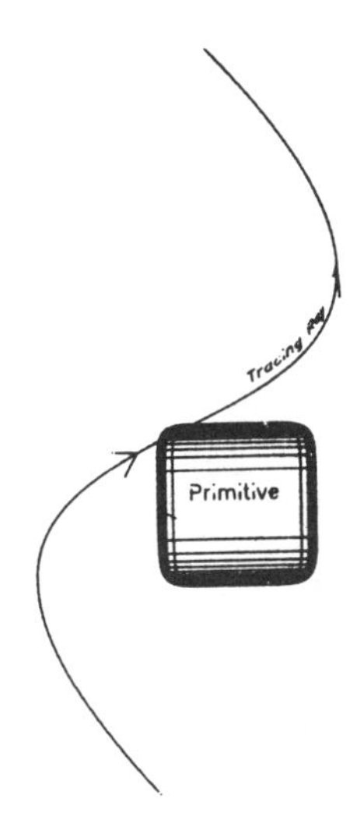

Figure 4.2 Undeformed primitive, in its undeformed coordinate system, showing path of ray

Appendix A: Proof of the normal vector transformation rule.

A short derivation in cross product and dot product style demonstrates the normal vector transformation rule.

The surface of an undeformed object is given by a parametric function of two variables u and v, $\underline{x} = \underline{x}(u,v)$. The goal is to discover an expression for the normal vector to the surface after it has been subjected to the deformation $\underline{X} = \underline{F}(\underline{x})$.

We note that the inverse of an arbitrary three by three matrix $\underline{\underline{M}}$ may be obtained from the cross-products of pairs of its columns via:

$$[\underline{M}_1, \underline{M}_2, \underline{M}_3]^{-1} = \frac{[\underline{M}_2 \wedge \underline{M}_3, \underline{M}_3 \wedge \underline{M}_1, \underline{M}_1 \wedge \underline{M}_2]^T}{\underline{M}_1 \cdot (\underline{M}_2 \wedge \underline{M}_3)}.$$

We start the derivation using the fact that the normal vector is the cross product of independent surface tangent vectors:

$$\underline{n}^{(X)} = \frac{\partial \underline{X}}{\partial u} \wedge \frac{\partial \underline{X}}{\partial v} \qquad [\textit{Equation B.1d}]$$

The tangent vectors for $\underline{X}(u,v)$ are expanded in terms of $\underline{x}(s,t)$.

$$\underline{n}^{(X)} = \left(\underline{\underline{J}} \frac{\partial \underline{x}}{\partial u}\right) \wedge \left(\underline{\underline{J}} \frac{\partial \underline{x}}{\partial v}\right)$$

Matrix multiplication is expanded, yielding

$$\underline{n}^{(X)} = \left(\sum_{i=1}^{3} \underline{J}_i x_{i,u}\right) \wedge \left(\sum_{j=1}^{3} \underline{J}_j x_{j,v}\right)$$

The summations are combined together:

$$= \sum_{i=1}^{3} \sum_{j=1}^{3} \left(\underline{J}_i \wedge \underline{J}_j\right) x_{i,s} x_{j,t}$$

Since the cross product of a vector with itself is the zero vector, and since for any vectors $\underline{b}$ and $\underline{c}$, $\underline{b} \wedge \underline{c} = -\underline{c} \wedge \underline{b}$, this expands to:

$$\underline{n}^{(X)} = \left(\underline{J}_2 \wedge \underline{J}_3, \underline{J}_3 \wedge \underline{J}_1, \underline{J}_1 \wedge \underline{J}_2\right) \begin{pmatrix} x_{2,u}x_{3,v} - x_{3,u}x_{2,v} \\ x_{3,u}x_{1,v} - x_{1,u}x_{3,v} \\ x_{1,s}x_{2,v} - x_{2,u}x_{1,v} \end{pmatrix}$$

Thus,

$$\underline{n}^{(X)} = [\underline{J}_2 \wedge \underline{J}_3, \underline{J}_3 \wedge \underline{J}_1, \underline{J}_1 \wedge \underline{J}_2] \underline{n}^{(x)}$$

Since $\det M = \underline{M}_1 \cdot (\underline{M}_2 \wedge \underline{M}_3)$ for an arbitrary matrix $\underline{\underline{M}}$,

$$\underline{n}^{(X)} = \det J \underline{\underline{J}}^{-1T} \underline{n}^{(x)}$$

In other words, the new normal vector $\underline{n}^{(X)}$ is expressed as a multiplication of matrix $\underline{\underline{J}}^{-1T}$ and the old normal vector $\underline{n}^{(x)}$.

Since only the direction of the normal vector is important, it is not necessary to compute the value of the determinant in practice, unless one needs the local volume ratio between corresponding points in the deformed and undeformed objects.

The fact that the normal vector follows this type of transformation rule makes it less expensive to calculate, increasing its applicability in a variety of modeling circumstances.

Acknowledgements

I would like to thank Dan Whelan, of the California Institute of Technology, and Olin Lathrop, of Raster Technologies Inc., for technical help with the typography and the illustrations.

Bibliography

1. Acton, F.S., Numerical Methods that Work, Harper and Row, 1970.
2. Barr, A.H., "Superquadrics and Angle-Preserving Transformations," IEEE Computer Graphics and Applications, Volume 1 number 1 1981.
3. Buck, R. C., Advanced Calculus, McGraw-Hill, 2nd edition, 1965
4. Faux, I.D., and M.J. Pratt, Computational Geometry for Design and Manufacture, Ellis Horwood Ltd., Wiley and Sons, 1979.
5. Franklin, W.R., and A.H. Barr, "Faster Calculation of Superquadric Shapes," IEEE Computer Graphics and Applications, Volume 1 number 3, 1981.
6. Kajiya, J.T., "Ray Tracing Parametric Patches," SigGraph 82 Conference Proceedings, Computer Graphics, Volume 16, Number 3, 1982.
7. Segel, L.A., Mathematics Applied to Continuum Mechanics, Macmillan Publishing Co., 1977.
8. Solkolnikoff, I.S., Mathematical Theory of Elasticity, McGraw Hill, 1956.

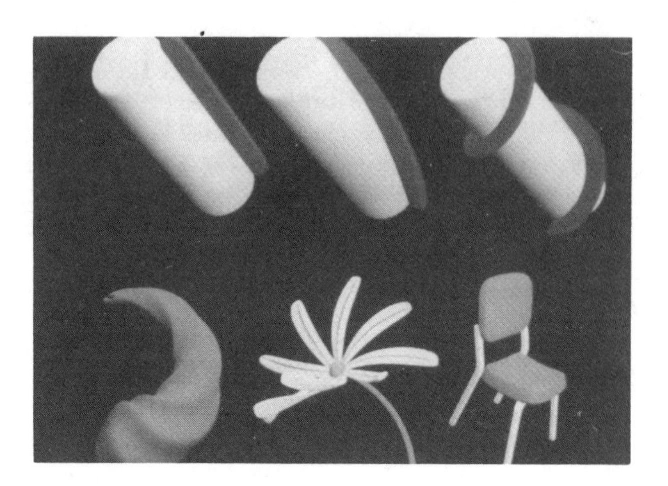

The Laplacian Pyramid as a Compact Image Code

PETER J. BURT, MEMBER, IEEE, AND EDWARD H. ADELSON

Abstract—**We describe a technique for image encoding in which local operators of many scales but identical shape serve as the basis functions. The representation differs from established techniques in that the code elements are localized in spatial frequency as well as in space.**

Pixel-to-pixel correlations are first removed by subtracting a low-pass filtered copy of the image from the image itself. The result is a net data compression since the difference, or error, image has low variance and entropy, and the low-pass filtered image may represented at reduced sample density. Further data compression is achieved by quantizing the difference image. These steps are then repeated to compress the low-pass image. Iteration of the process at appropriately expanded scales generates a pyramid data structure.

The encoding process is equivalent to sampling the image with Laplacian operators of many scales. Thus, the code tends to enhance salient image features. A further advantage of the present code is that it is well suited for many image analysis tasks as well as for image compression. Fast algorithms are described for coding and decoding.

INTRODUCTION

A COMMON characteristic of images is that neighboring pixels are highly correlated. To represent the image directly in terms of the pixel values is therefore inefficient: most of the encoded information is redundant. The first task in designing an efficient, compressed code is to find a representation which, in effect, decorrelates the image pixels. This has been achieved through predictive and through transform techniques (cf. [9], [10] for recent reviews).

In predictive coding, pixels are encoded sequentially in a raster format. However, prior to encoding each pixel, its value is predicted from previously coded pixels in the same and preceding raster lines. The predicted pixel value, which represents redundant information, is subtracted from the actual pixel value, and only the difference, or prediction error, is encoded. Since only previously encoded pixels are used in predicting each pixel's value, this process is said to be causal. Restriction to causal prediction facilitates decoding: to decode a given pixel, its predicted value is recomputed from already decoded neighboring pixels, and added to the stored prediction error.

Noncausal prediction, based on a symmetric neighborhood centered at each pixel, should yield more accurate prediction and, hence, greater data compression. However, this approach does not permit simple sequential coding. Noncausal approaches to image coding typically involve image transforms, or the solution to large sets of simultaneous equations. Rather than encoding pixels sequentially, such techniques encode them all at once, or by blocks.

Both predictive and transform techniques have advantages. The former is relatively simple to implement and is readily adapted to local image characteristics. The latter generally provides greater data compression, but at the expense of considerably greater computation.

Here we shall describe a new technique for removing image correlation which combines features of predictive and transform methods. The technique is noncausal, yet computations are relatively simple and local.

The predicted value for each pixel is computed as a local weighted average, using a unimodal Gaussian-like (or related trimodal) weighting function centered on the pixel itself. The predicted values for all pixels are first obtained by convolving this weighting function with the image. The result is a low-pass filtered image which is then subtracted from the original.

Let $g_0(ij)$ be the original image, and $g_1(ij)$ be the result of applying an appropriate low-pass filter to g_0. The prediction error $L_0(ij)$ is then given by

$$L_0(ij) = g_0(ij) - g_1(ij).$$

Rather than encode g_0, we encode L_0 and g_1. This results in a net data compression because a) L_0 is largely decorrelated, and so may be represented pixel by pixel with many fewer bits than g_0, and b) g_1 is low-pass filtered, and so may be encoded at a reduced sample rate.

Further data compression is achieved by iterating this process. The reduced image g_1 is itself low-pass filtered to yield g_2 and a second error image is obtained: $L_2(ij) = g_1(ij) - g_2(ij)$. By repeating these steps several times we obtain a sequence of two-dimensional arrays $L_0, L_1, L_2, \cdots, L_n$. In our implementation each is smaller than its predecessor by a scale factor of 1/2 due to reduced sample density. If we now imagine these arrays stacked one above another, the result is a tapering pyramid data structure. The value at each node in the pyramid represents the difference between two Gaussian-like or related functions convolved with the original image. The difference between these two functions is similar to the "Laplacian" operators commonly used in image enhancement [13]. Thus, we refer to the proposed compressed image representation as the Laplacian-pyramid code.

The coding scheme outlined above will be practical only if required filtering computations can be performed with an efficient algorithm. A suitable fast algorithm has recently been developed [2] and will be described in the next section.

Paper approved by the Editor for Signal Processing and Communication Electronics of the IEEE Communications Society for publication after presentation in part at the Conference on Pattern Recognition and Image Processing, Dallas, TX, 1981. Manuscript received April 12, 1982; revised July 21, 1982. This work was supported in part by the National Science Foundation under Grant MCS-79-23422 and by the National Institutes of Health under Postdoctoral Training Grant EY07003.

P. J. Burt is with the Department of Electrical, Computer, and Systems Engineering, Rensselaer Polytechnic Institute, Troy, NY 12181.

E. H. Adelson is with the RCA David Sarnoff Research Center, Princeton, NJ 08540.

THE GAUSSIAN PYRAMID

The first step in Laplacian pyramid coding is to low-pass filter the original image g_0 to obtain image g_1. We say that g_1 is a "reduced" version of g_0 in that both resolution and sample density are decreased. In a similar way we form g_2 as a reduced version of g_1, and so on. Filtering is performed by a procedure equivalent to convolution with one of a family of local, symmetric weighting functions. An important member of this family resembles the Gaussian probability distribution, so the sequence of images $g_0, g_1, \cdots, g_n$ is called the Gaussian pyramid.[1]

A fast algorithm for generating the Gaussian pyramid is given in the next subsection. In the following subsection we show how the same algorithm can be used to "expand" an image array by interpolating values between sample points. This device is used here to help visualize the contents of levels in the Gaussian pyramid, and in the next section to define the Laplacian pyramid.

Gaussian Pyramid Generation

Suppose the image is represented initially by the array g_0 which contains C columns and R rows of pixels. Each pixel represents the light intensity at the corresponding image point by an integer I between 0 and $K-1$. This image becomes the bottom or zero level of the Gaussian pyramid. Pyramid level 1 contains image g_1, which is a reduced or low-pass filtered version of g_0. Each value within level 1 is computed as a weighted average of values in level 0 within a 5-by-5 window. Each value within level 2, representing g_2, is then obtained from values within level 1 by applying the same pattern of weights. A graphical representation of this process in one dimension is given in Fig. 1. The size of the weighting function is not critical [2]. We have selected the 5-by-5 pattern because it provides adequate filtering at low computational cost.

The level-to-level averaging process is performed by the function REDUCE.

$$g_k = \text{REDUCE}\,(g_{k-1}) \tag{1}$$

which means, for levels $0 < l < N$ and nodes i, j, $0 \leqslant i < C_l$, $0 \leqslant j < R_l$,

$$g_l(i,j) = \sum_{m=-2}^{2} \sum_{n=-2}^{2} w(m,n) g_{l-1}(2i+m, 2j+n).$$

Here N refers to the number of levels in the pyramid, while C_l and R_l are the dimensions of the lth level. Note in Fig. 1 that the density of nodes is reduced by half in one dimension, or by a fourth in two dimensions from level to level. The dimensions of the original image are appropriate for pyramid construction if integers M_C, M_R, and N exist such that $C = M_C 2^N + 1$ and $R = M_R 2^N + 1$. (For example, if M_C and M_R are both 3 and N is 5, then images measure 97 by 97 pixels.) The dimensions of g_l are $C_l = M_C 2^{N-l} + 1$ and $R_l = M_R 2^{N-l} + 1$.

[1] We will refer to this set of low-pass filtered images as the Gaussian pyramid, even though in some cases it will be generated with a trimodal rather than unimodal weighting function.

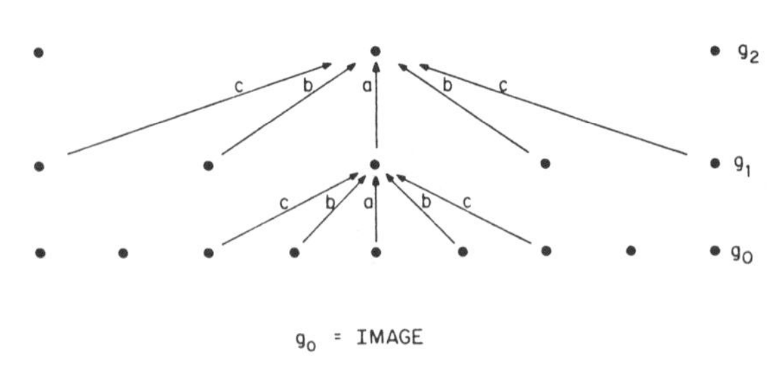

Fig. 1. A one-dimensional graphic representation of the process which generates a Gaussian pyramid. Each row of dots represents nodes within a level of the pyramid. The value of each node in the zero level is just the gray level of a corresponding image pixel. The value of each node in a high level is the weighted average of node values in the next lower level. Note that node spacing doubles from level to level, while the same weighting pattern or "generating kernel" is used to generate all levels.

The Generating Kernel

Note that the same 5-by-5 pattern of weights w is used to generate each pyramid array from its predecessor. This weighting pattern, called the generating kernel, is chosen subject to certain constraints [2]. For simplicity we make w separable:

$$w(m,n) = \hat{w}(m)\hat{w}(n).$$

The one-dimensional, length 5, function $\hat{w}$ is normalized

$$\sum_{m=-2}^{2} \hat{w}(m) = 1$$

and symmetric

$$\hat{w}(i) = \hat{w}(-i) \qquad \text{for } i = 0,\ 1, 2.$$

An additional constraint is called equal contribution. This stipulates that all nodes at a given level must contribute the same total weight (=1/4) to nodes at the next higher level. Let $\hat{w}(0) = a$, $\hat{w}(-1) = \hat{w}(1) = b$, and $\hat{w}(-2) = \hat{w}(2) = c$. In this case equal contribution requires that $a + 2c = 2b$. These three constraints are satisfied when

$$\hat{w}(0) = a$$

$$\hat{w}(-1) = \hat{w}(1) = 1/4$$

$$\hat{w}(-2) = \hat{w}(2) = 1/4 - a/2.$$

Equivalent Weighting Functions

Iterative pyramid generation is equivalent to convolving the image g_0 with a set of "equivalent weighting functions" h_l:

$$g_l = h_l \oplus g_0$$

or

$$g_l(i,j) = \sum_{m=-M_l}^{M_l} \cdot \sum_{n=-M_l}^{M_l} h_l(m,n) g_0(i2^l + m \cdot j2^l + n).$$

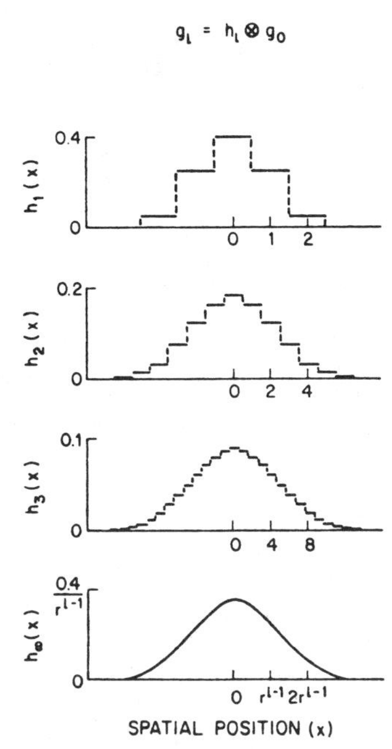

Fig. 2. The equivalent weighting functions $h_l(x)$ for nodes in levels 1, 2, 3, and infinity of the Gaussian pyramid. Note that axis scales have been adjusted by factors of 2 to aid comparison. Here the parameter a of the generating kernel is 0.4, and the resulting equivalent weighting functions closely resemble the Gaussian probability density functions.

The size M_l of the equivalent weighting function doubles from one level to the next, as does the distance between samples.

Equivalent weighting functions for Gaussian-pyramid levels 1, 2, and 3 are shown in Fig. 2. In this case $a = 0.4$. The shape of the equivalent function converges rapidly to a characteristic form with successively higher levels of the pyramid, so that only its scale changes. However, this shape does depend on the choice of a in the generating kernel. Characteristic shapes for four choices of a are shown in Fig. 3. Note that the equivalent weighting functions are particularly Gaussian-like when $a = 0.4$. When $a = 0.5$ the shape is triangular; when $a = 0.3$ it is flatter and broader than a Gaussian. With $a = 0.6$ the central positive mode is sharply peaked, and is flanked by small negative lobes.

Fast Filter

The effect of convolving an image with one of the equivalent weighting functions h_l is to blur, or low-pass filter, the image. The pyramid algorithm reduces the filter band limit by an octave from level to level, and reduces the sample interval by the same factor. This is a very fast algorithm, requiring fewer computational steps to compute a set of filtered images than are required by the fast Fourier transform to compute a single filtered image [2].

Example: Fig. 4 illustrates the contents of a Gaussian pyramid generated with $a = 0.4$. The original image, on the far left, measures 257 by 257. This becomes level 0 on the pyramid. Each higher level array is roughly half as large in each dimension as its predecessor, due to reduced sample density.

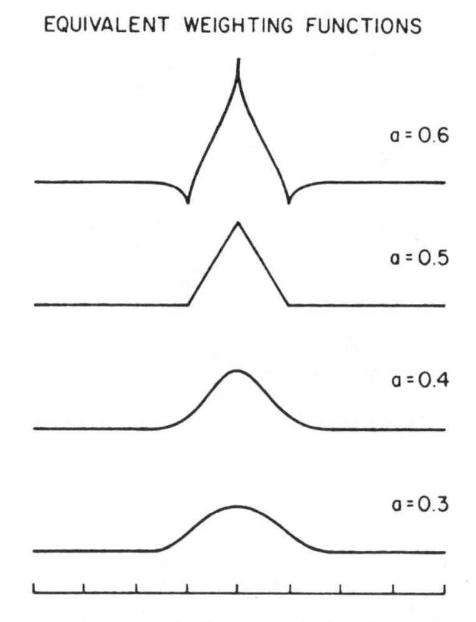

Fig. 3. The shape of the equivalent weighting function depends on the choice of parameter a. For $a = 0.5$, the function is triangular; for $a = 0.4$ it is Gaussian-like, and for $a = 0.3$ it is broader than Gaussian. For $a = 0.6$ the function is trimodal.

Gaussian Pyramid Interpolation

We now define a function EXPAND as the reverse of REDUCE. Its effect is to expand an $(M + 1)$-by-$(N + 1)$ array into a $(2M + 1)$-by-$(2N + 1)$ array by interpolating new node values between the given values. Thus, EXPAND applied to array g_l of the Gaussian pyramid would yield an array $g_{l,1}$ which is the same size as g_{l-1}.

Let $g_{l,n}$ be the result of expanding g_l n times. Then

$$g_{l,0} = g_l$$

and

$$g_{l,n} = \text{EXPAND}\,(g_{l,n-1}).$$

By EXPAND we mean, for levels $0 < l \leqslant N$ and $0 \leqslant n$ and nodes i, j, $0 \leqslant i < C_{l-n}$, $0 \leqslant j < R_{l-n}$,

$$g_{l,n}(ij) = 4 \sum_{m=-2}^{2} \sum_{n=-2}^{2} w(m,n) \cdot g_{l,n-1}\left(\frac{i-m}{2}, \frac{j-n}{2}\right). \tag{2}$$

Only terms for which $(i - m)/2$ and $(j - n)/2$ are integers are included in this sum.

If we apply EXPAND l times to image g_l, we obtain $g_{l,l}$, which is the same size as the original image g_0. Although full expansion will not be used in image coding, we will use it to help visualize the contents of various arrays within pyramid structures. The top row of Fig. 5 shows image $g_{0,0}$, $g_{1,1}$, $g_{2,2}$, ··· obtained by expanding levels of the pyramid in Fig. 4. The low-pass filter effect of the Gaussian pyramid is now shown clearly.

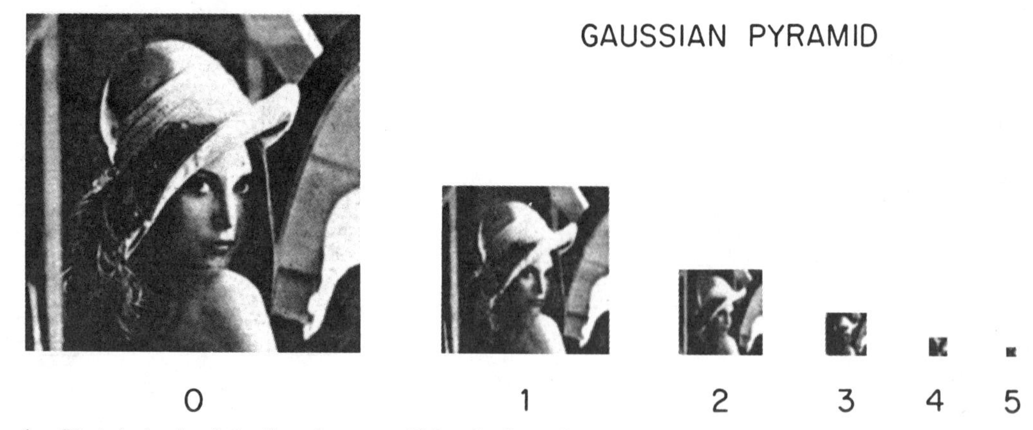

Fig. 4. First six levels of the Gaussian pyramid for the "Lady" image. The original image, level 0, measures 257 by 257 pixels, and each higher level array is roughly half the dimensions of its predecessor. Thus, level 5 measures just 9 by 9 pixels.

THE LAPLACIAN PYRAMID

Recall that our purpose for constructing the reduced image g_1 is that it may serve as a prediction for pixel values in the original image g_0. To obtain a compressed representation, we encode the error image which remains when an expanded g_1 is subtracted from g_0. This image becomes the bottom level of the Laplacian pyramid. The next level is generated by encoding g_1 in the same way. We now give a formal definition for the Laplacian pyramid, and examine its properties.

Laplacian Pyramid Generation

The Laplacian pyramid is a sequence of error images L_0, L_1, ⋯, L_N. Each is the difference between two levels of the Gaussian pyramid. Thus, for $0 \leqslant l < N$,

$$L_l = g_l - \text{EXPAND}\,(g_{l+1})$$

$$= g_l - g_{l+1,1}. \tag{3}$$

Since there is no image g_{N+1} to serve as the prediction image for g_N, we say $L_N = g_N$.

Equivalent Weighting Functions

The value at each node in the Laplacian pyramid is the difference between the convolutions of two equivalent weighting functions h_l, h_{l+1} with the original image. Again, this is similar to convolving an appropriately scaled Laplacian weighting function with the image. The node value could have been obtained directly by applying this operator, although at considerably greater computational cost.

Just as we may view the Gaussian pyramid as a set of low-pass filtered copies of the original image, we may view the Laplacian pyramid as a set of bandpass filtered copies of the image. The scale of the Laplacian operator doubles from level to level of the pyramid, while the center frequency of the pass-band is reduced by an octave.

In order to illustrate the contents of the Laplacian pyramid, it is helpful to interpolate between sample points. This may be done within the pyramid structure by Gaussian interpolation. Let $L_{l,n}$ be the result of expanding L_l n times using (2). Then, $L_{l,l}$ is the size of the original image.

The expanded Laplacian pyramid levels for the "Lady" image of Fig. 4 are shown in the bottom row of Fig. 5. Note that image features such as edges and bars appear enhanced in the Laplacian pyramid. Enhanced features are segregated by size: fine details are prominent in $L_{0,0}$, while progressively coarser features are prominent in the higher level images.

Decoding

It can be shown that the original image can be recovered exactly by expanding, then summing all the levels of the Laplacian pyramid:

$$g_0 = \sum_{l=0}^{N} L_{l,l}.$$

A more efficient procedure is to expand L_N once and add it to L_{N-1}, then expand this image once and add it to L_{N-2}, and so on until level 0 is reached and g_0 is recovered. This procedure simply reverses the steps in Laplacian pyramid generation. From (3) we see that

$$g_N = L_N \tag{4}$$

and for $l = N-1, N-2, \cdots, 0$,

$$g_l = L_l + \text{EXPAND}\,(g_{l+1}).$$

Entropy

If we assume that the pixel values of an image representation are statistically independent, then the minimum number of bits per pixel required to exactly encode the image is given by the entropy of the pixel value distribution. This optimum may be approached in practice through techniques such as variable length coding.

The histogram of pixel values for the "Lady" image is shown in Fig. 6(a). If we let the observed frequency of occurrence $f(i)$ of each gray level i be an estimate of its probability of occurrence in this and other similar images, then the entropy

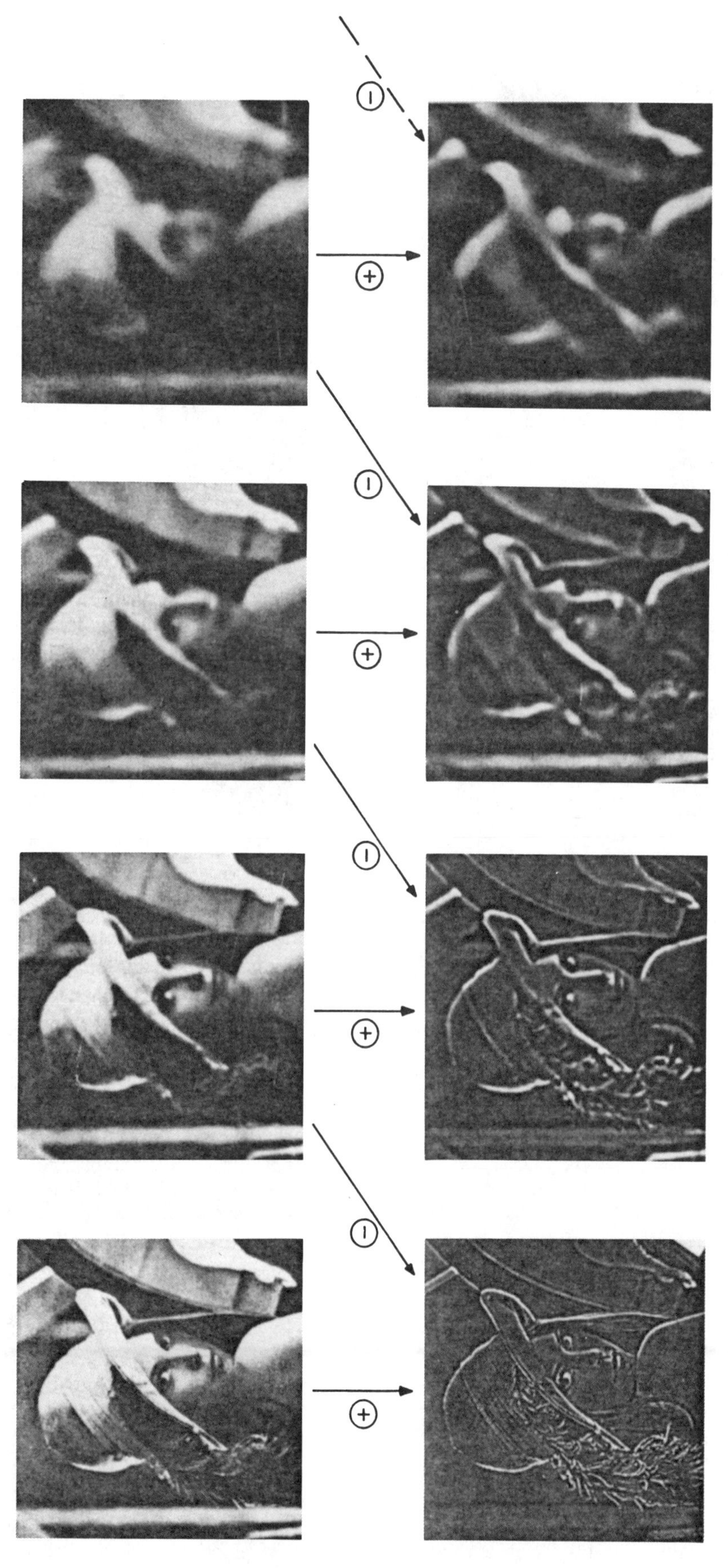

Fig. 5. First four levels of the Gaussian and Laplacian pyramids. Gaussian images, upper row, were obtained by expanding pyramid arrays (Fig. 4) through Gaussian interpolation. Each level of the Laplacian pyramid is the difference between the corresponding and next higher levels of the Gaussian pyramid.

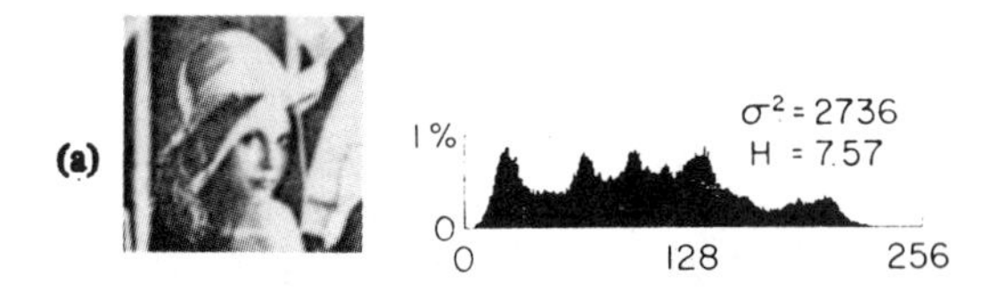

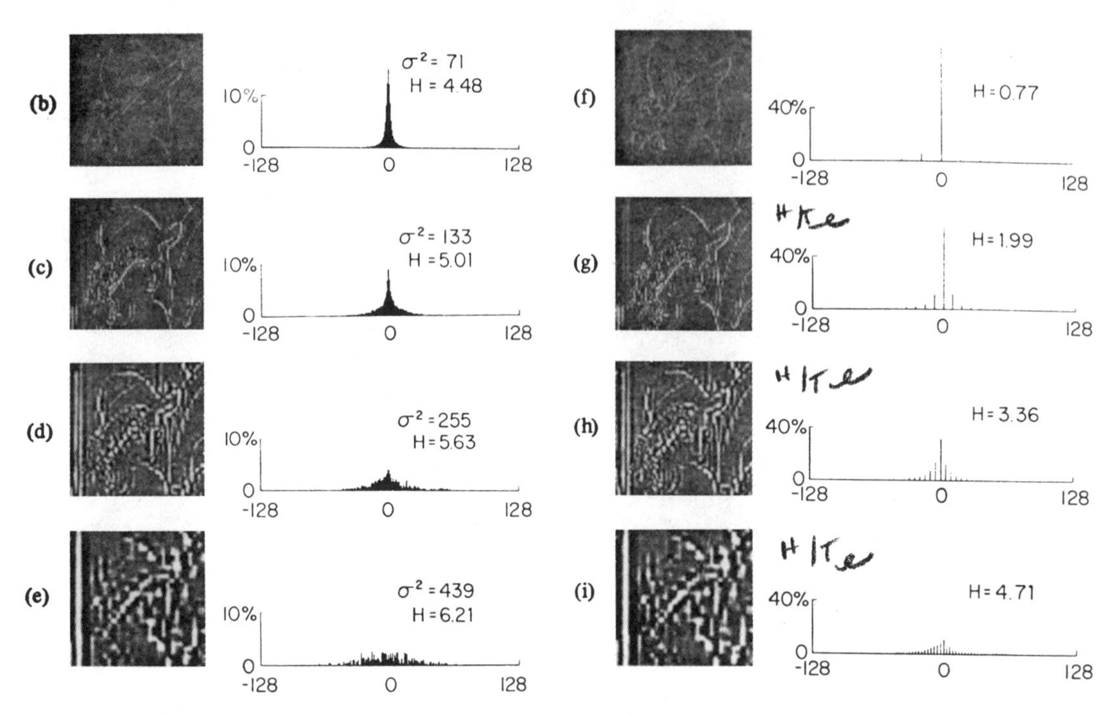

Fig. 6. The distribution of pixel gray level values at various stages of the encoding process. The histogram of the original image is given in (a). (b)–(e) give histograms for levels 0–3 of the Laplacian pyramid with generating parameter $a = 0.6$. Histograms following quantization at each level are shown in (f)–(i). Note that pixel values in the Laplacian pyramid are concentrated near zero, permitting data compression through shortened and variable length code words. Substantial further reduction is realized through quantization (particularly at low pyramid levels) and reduced sample density (particularly at high pyramid levels).

is given by

$$H = -\sum_{i=0}^{255} f(i) \log_2 f(i).$$

The maximum entropy would be 8 in this case since the image is initially represented at 256 gray levels, and would be obtained when all gray levels were equally likely. The actual entropy estimate for "Lady" is slightly less than this, at 7.57.

The technique of subtracting a predicted value from each image pixel, as in the Laplacian pyramid, removes much of the pixel-to-pixel correlation. Decorrelation also results in a concentration of pixel values around zero, and, therefore, in reduced variance and entropy. The degree to which these measures are reduced depends on the value of the parameter "a" used in pyramid generation (see Fig. 7). We found that the greatest reduction was obtained for $a = 0.6$ in our examples. Levels of the Gaussian pyramid appeared "crisper" when generated with this value of a than when generated with a smaller value such as 0.4, which yields more Guassian-like equivalent weighting functions. Thus, the selection $a = 0.6$ had perceptual as well as computational advantages. The first four levels of the corresponding Laplacian pyramid and their histograms are shown in Fig. 6(b)-(e). Variance (σ^2) and entropy (H) are also shown for each level. These quantities generally are found to increase from level to level, as in this example.

QUANTIZATION

Entropy can be substantially reduced by quantizing the pixel values in each level of the Laplacian pyramid. This introduces quantization errors, but through the proper choice of the number and distribution of quantization levels, the degradation may be made almost imperceptible to human observers. We illustrate this procedure with uniform quantization. The range of pixel values is divided into bins of size n, and the quantized value $C_l(i, j)$ for pixel $L_l(i, j)$ is just the middle

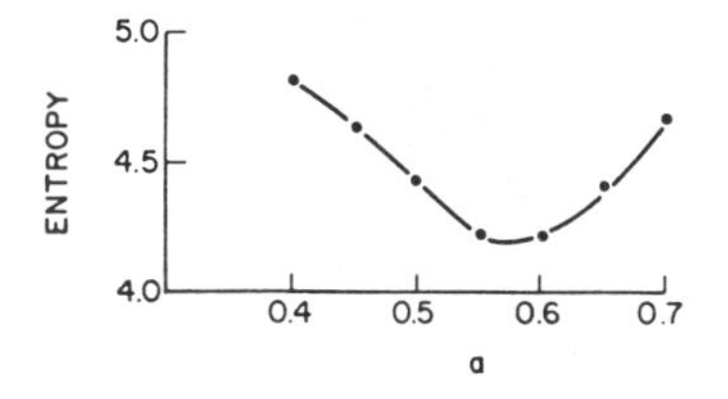

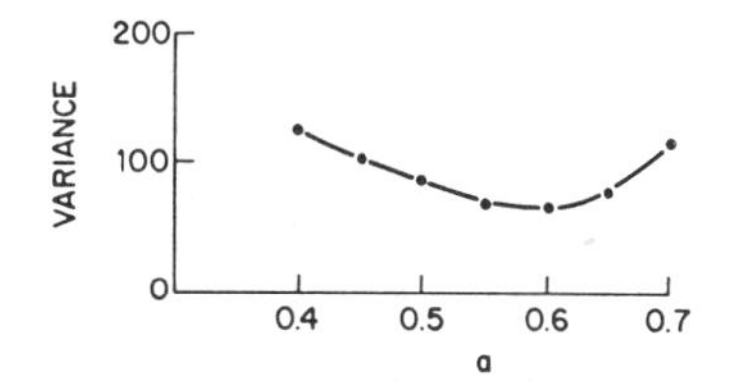

Fig. 7. Entropy and variance of pixel values in Laplacian pyramid level 0 as a function of the parameter "a" for the "Lady" image. Greatest reduction is obtained for $a \cong 0.6$. This estimate of the optimal "a" was also obtained at other pyramid levels and for other images.

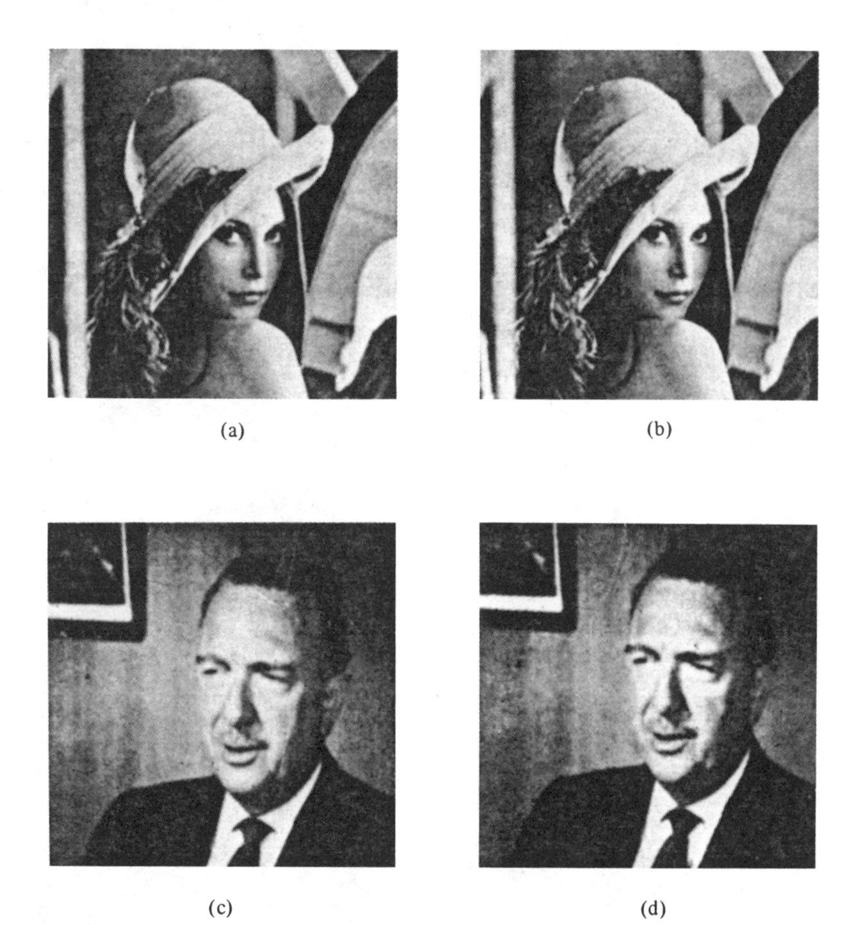

Fig. 8. Examples of image data compression using the Laplacian pyramid code. (a) and (c) give the original "Lady" and "Walter" images, while (b) and (d) give their encoded versions. The data rates are 1.58 and 0.73 bits/pixel for "Lady" and "Walter," respectively. The corresponding mean square errors were 0.88 percent and 0.43 percent, respectively.

value of the bin which contains $L_l(i,j)$:

$$C_l(i,j) = mn \quad \text{if } (m - 1/2)n < L_l(i,j) \leqslant (m + 1/2)n. \qquad (5)$$

The quantized image is reconstructed through the expand and sum procedure (4) using C values in the place of L values.

Results of quantizing the "Lady" image are shown in Fig. 6(f)-(i). The bin size for each level was chosen by increasing n until degradation was just perceptible when viewed from a distance of approximately five times the image width (pixel-pixel separation $\cong$ 3 min arc). Note that bin size becomes smaller at higher levels (lower spatial frequencies). Bin size at a given pyramid level reflects the sensitivity of the human observer to contrast errors within the spatial frequency bands represented at that level. Humans are fairly sensitive to contrast perturbations at low and medium spatial frequencies, but relatively insensitive to such perturbations at high spatial frequencies [3], [4], [7].

This increased observer sensitivity along with the increased data variance noted above means that more quantization levels must be used at high pyramid levels than at low levels. Fortunately, these pixels contribute little to the overall bit rate for the image, due to their low sample density. The low-level (high-frequency) pixels, which are densely sampled, can be coarsely quantized (cf. [6], [11], [12]).

RESULTS

The final result of encoding, quantization, and reconstruction are shown in Fig. 8. The original "Lady" image is shown in Fig. 8(a); the encoded version, at 1.58 bits/pixel, is shown in Fig. 8(b). We assume that variable-length code words are used to take advantage of the nonuniform distribution of node values, so the bit rate for a given pyramid level is its estimated entropy times its sample density, and the bit rate for the image is the sum of that for all levels. The same procedure was performed on the "Walter" image; the original is shown in Fig. 8(c). while the version encoded at 0.73 bits/pixel is shown in Fig. 8(d). In both cases, the encoded images are almost indistinguishable from the originals under viewing conditions as stated above.

PROGRESSIVE TRANSMISSION

It should also be observed that the Laplacian pyramid code is particularly well suited for progressive image transmission. In this type of transmission a coarse rendition of the image is sent first to give the receiver an early impression of image content, then subsequent transmission provides image detail of progressively finer resolution [5]. The observer may terminate transmission of an image as soon as its contents are recognized, or as soon as it becomes evident that the image will not be of interest. To achieve progressive transmission, the topmost level of the pyramid code is sent first, and expanded in the receiving pyramid to form an initial, very coarse image. The next lower level is then transmitted, expanded, and added to the first, and so on. At the receiving end, the initial image appears very blurry, but then comes steadily into "focus." This progression is illustrated in Fig. 9, from left to right. Note that while 1.58 bits are required for each pixel of the full transmission (rightmost image), about half of these, or 0.81 bits, are needed for each pixel for the previous image (second from right, Fig. 9), and 0.31 for the image previous to that (third from right).

SUMMARY AND CONCLUSION

The Laplacian pyramid is a versatile data structure with many attractive features for image processing. It represents an image as a series of quasi-bandpassed images, each sampled at successively sparser densities. The resulting code elements, which form a self-similar structure, are localized in both space and spatial frequency. By appropriately choosing the parameters of the encoding and quantizing scheme, one can substantially reduce the entropy in the representation, and simultaneously stay within the distortion limits imposed by the sensitivity of the human visual system.

Fig. 10 summarizes the steps in Laplacian pyramid coding. The first step, shown on the far left, is bottom-up construction of the Gaussian pyramid images $g_0, g_1, \cdots, g_N$ [see (1)]. The Laplacian pyramid images $L_0, L_1, \cdots, L_N$ are then obtained as the difference between successive Gaussian levels [see (3)]. These are quantized to yield the compressed code represented by the pyramid of values $C_l(ij)$ [see (5)]. Finally, image reconstruction follows an expand-and-sum procedure [see (4)] using C values in the place of L values. Here we designate the reconstructed image by r_0.

It should also be observed that the Laplacian pyramid encoding scheme requires relatively simple computations. The computations are local and may be performed in parallel, and the same computations are iterated to build each pyramid level from its predecessors. We may envision performing Lapla-

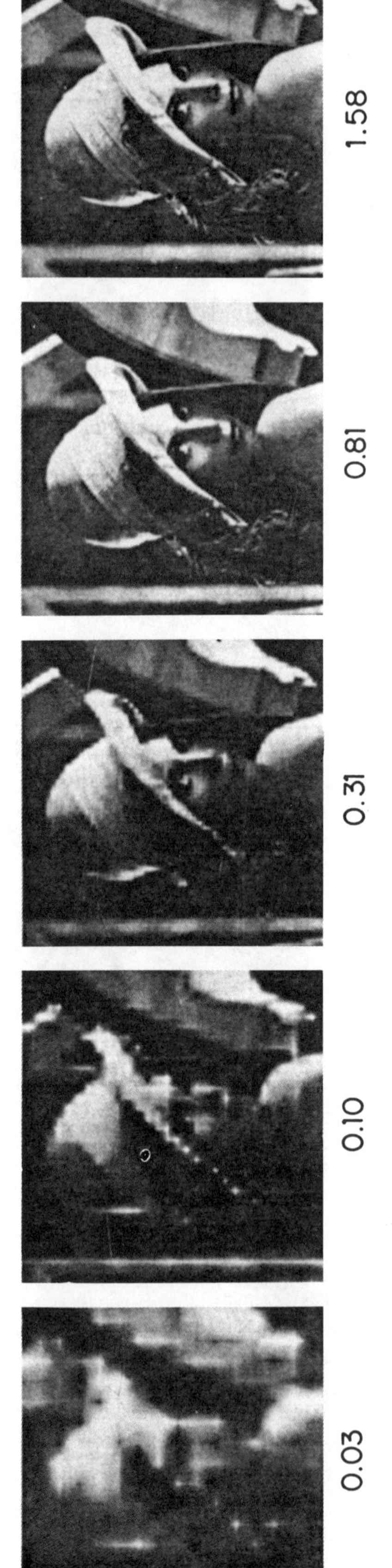

Fig. 9. Laplacian pyramid code applied to progressive image transmission. High levels of the pyramid are transmitted first to give the receiver a quick but very coarse rendition of the image. The receiver's image is then progressively refined by adding successively lower pyramid levels as these are transmitted. In the example shown here, the leftmost figure shows reconstruction using pyramid levels 4–8, or just 0.03 bits/pixel. The following four figures show the reconstruction after pyramid levels 3, 2, 1, and 0 have been added. The cumulative data rates are shown under each figures in bits per pixel.

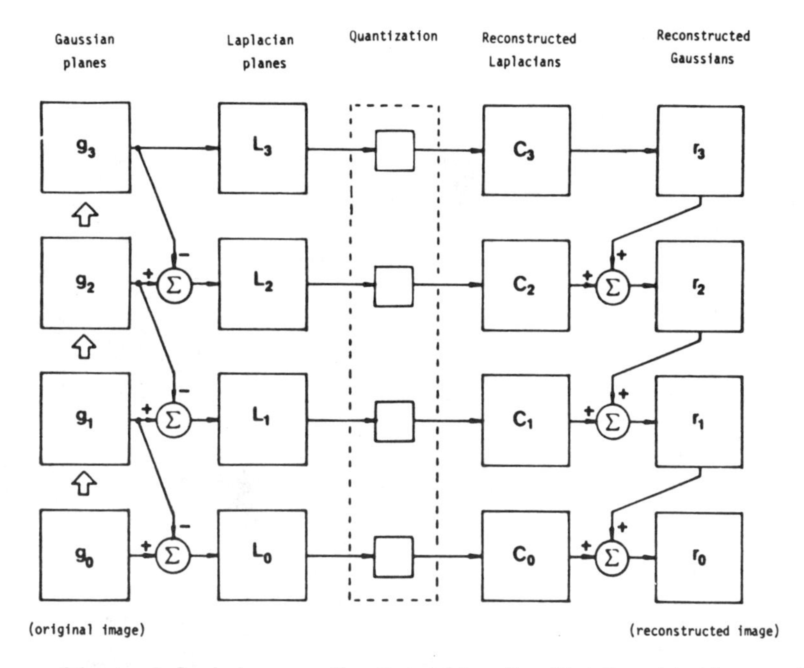

Fig. 10. A summary of the steps in Laplacian pyramid coding and decoding. First, the original image g_0 (lower left) is used to generate Gaussian pyramid levels $g_1, g_2, \cdots$ through repeated local averaging. Levels of the Laplacian pyramid $L_0, L_1, \cdots$ are then computed as the differences between adjacent Gaussian levels. Laplacian pyramid elements are quantized to yield the Laplacian pyramid code $C_0, C_1, C_2, \cdots$. Finally, a reconstructed image r_0 is generated by summing levels of the code pyramid.

cian coding and decoding in real time using array processors and a pipeline architecture.

An additional benefit, previously noted, is that in computing the Laplacian pyramid, one automatically has access to quasi-bandpass copies of the image. In this representation, image features of various sizes are enhanced and are directly available for various image processing (e.g., [1]) and pattern recognition tasks.

REFERENCES

[1] K. D. Baker and G. D. Sullivan, "Multiple bandpass filters in image processing," *Proc. IEE*, vol. 127, pp. 173–184, 1980.

[2] P. J. Burt, "Fast filter transforms for image processing," *Comput. Graphics, Image Processing*, vol. 16, pp. 20–51, 1981.

[3] C. R. Carlson and R. W. Cohen, "Visibility of displayed information," Off. Naval Res., Tech. Rep., Contr. N00014-74-C-0184, 1978.

[4] ——, "A simple psychophysical model for predicting the visibility of displayed information," *Proc. Soc. Inform. Display*, pp. 229–246, 1980.

[5] K. Knowlton, "Progressive transmission of grayscale and binary pictures by simple, efficient, and lossless encoding schemes," *Proc. IEEE*, vol. 68, pp. 885–896, 1980.

[6] E. R. Kretzmer, "Reduced-alphabet representation of television signals," in *IRE Nat. Conv. Rec.*, 1956, pp. 140–147.

[7] J. J. Kulikowski and A. Gorea, "Complete adaptation to patterned stimuli: A necessary and sufficient condition for Weber's law for contrast," *Vision Res.*, vol. 18, pp. 1223–1227, 1978.

[8] A. N. Netravali and B. Prasada, "Adaptive quantization of picture signals using spatial masking," *Proc. IEEE*, vol. 65, pp. 536–548, 1977.

[9] A. N. Netravali and J. O. Limb, "Picture coding: A review," *Proc. IEEE*, vol. 68, pp. 336–406, 1980.

[10] W. K. Pratt, Ed., *Image Transmission Tecniques*. New York: Academic, 1979.

[11] W. F. Schreiber, C. F. Knapp, and N. D. Key, "Synthetic highs, an experimental TV bandwidth reduction system," *J. Soc. Motion Pict. Telev. Eng.*, vol. 68, pp. 525–537, 1959.

[12] W. F. Schreiber and D. E. Troxel, U.S. Patent 4 268 861, 1981.

[13] A. Rosenfeld and A. Kak, *Digital Picture Processing*. New York: Academic, 1976.

Peter J. Burt (M'80) received the B.A. degree in physics from Harvard University, Cambridge, MA, in 1968, and the M.S. and Ph.D. degrees in computer science from the University of Massachusetts, Amherst, in 1974 and 1976, respectively.

From 1968 to 1972 he conducted research in sonar, particularly in acoustic imaging devices, at the USN Underwater Sound Laboratory, New London, CT, and in London, England. As a Postdoctoral Fellow, he has studied both natural vision and computer image understanding at New York University, New York, NY (1976–1978), Bell Laboratories (1978–1979), and the University of Maryland, College Park (1979–1980). He has been a member of the faculty at Rensselaer Polytechnic Institute, Troy, NY, since 1980.

Edward H. Adelson received the B.A. degree in physics and philosophy from Yale University, New Haven, CT, in 1974, and the Ph.D. degree in experimental psychology from the University of Michigan, Ann Arbor, in 1979.

From 1979 to 1981 he was Postdoctoral Fellow at New York University, New York, NY. Since 1981, he has been at RCA David Sarnoff Research Center, Princeton, NJ, as a member of the Technical Staff in the Image Quality and Human Perception Research Group. His research interests center on visual processes in both human and machine visual systems, and include psychophysics, image processing, and artificial intelligence.

Dr. Adelson is a member of the Optical Society of America, the Association for Research in Vision and Ophthalmology, and Phi Beta Kappa.

Perceptual Organization and the Representation of Natural Form

Alex P. Pentland
Artificial Intelligence Center, SRI International,
333 Ravenswood Ave, Menlo Park, CA 94025, U.S.A.
Center for the Study of Language and Information,
Stanford University, Stanford, CA 94038, U.S.A.

Recommended by Daniel G. Bobrow and Pat Hayes

ABSTRACT

To support our reasoning abilities perception must recover environmental regularities—e.g., rigidity, "objectness," axes of symmetry—for later use by cognition. To create a theory of how our perceptual apparatus can produce meaningful cognitive primitives from an array of image intensities we require a representation whose elements may be lawfully related to important physical regularities, and that correctly describes the perceptual organization people impose on the stimulus. Unfortunately, the representations that are currently available were originally developed for other purposes (e.g., physics, engineering) and have so far proven unsuitable for the problems of perception or common-sense reasoning. In answer to this problem we present a representation that has proven competent to accurately describe an extensive variety of natural forms (e.g., people, mountains, clouds, trees), as well as man-made forms, in a succinct and natural manner. The approach taken in this representational system is to describe scene structure at a scale that is similar to our naive perceptual notion of "a part," by use of descriptions that reflect a possible formative history of the object, e.g., how the object might have been constructed from lumps of clay. For this representation to be useful it must be possible to recover such descriptions from image data; we show that the primitive elements of such descriptions may be recovered in an overconstrained and therefore reliable manner. We believe that this descriptive system makes an important contribution towards solving current problems in perceiving and reasoning about natural forms by allowing us to construct accurate descriptions that are extremely compact and that capture people's intuitive notions about the part structure of three-dimensional forms.

1. Introduction

Our world is very highly structured: evolution repeats its solutions whenever possible [1], and inanimate forms are constrained by physical laws to a limited number of basic patterns [2]. The apparent complexity of our environment is produced from this limited vocabulary by compounding these basic forms in myriad different combinations. Indeed, the highly patterned nature of our environment is a necessary precondition for intelligence; for if the apparent complexity of our environment were approximately the same as its intrinsic (Kolmogorov) complexity, then intelligent prediction and planning would be impossible, for there would be no lawful relations. It is this internal structuring of our environment, then, that causes object features to cluster into groups, and allows us to reason successfully using the simplified category descriptions that we typically employ [3].

To support our reasoning abilities, therefore, perception must recover these environmental regularities—e.g., rigidity, "objectness," axes of symmetry—for later use in cognitive processes. This recovery of structure is known as *perceptual organization*, familiar from such research efforts as the Gestalt movement [4], Johansson's [5] study of the organization of motion perception, and more recently Marr and Nishihara's [6, 7] theory of form perception using a description based on generalized cylinders [8]. The problem of perceptual organization is important because the structural regularities that perception recovers are the parts from which we construct our picture of the world; they are the building blocks of all cognitive activities.

FIG. 1. A scene described and generated by the representational system described within: tree leaves and bark, rocks and hair are fractal surfaces, the overall shape is described by Boolean combination of appropriately deformed superquadrics. Only 56 primitives are required (fewer than 500 bytes of information) to specify this scene. The slightly cartoon-like appearance is primarily due to the lack of surface texturing.

In this paper we will approach the problem of visual perceptual organization in a manner similar to Gibson [9] or Marr [10]: we want to construct a theory of how our perceptual concepts—shape, objectness, and the like—are lawfully related to the regularities (structure) in our environment. Like Marr, but unlike Gibson, we desire a *computational* theory: one that details how the physical regularities of our world allow the computation of meaningful descriptions of our surroundings by use of image data. Further, we want these descriptions to match the perceptual organization we impose on the stimulus—the one structuring of the stimulus that we know can support general-purpose cognitive activity. That is, we want a theory that both details how meaningful assertions can be derived from image data and accounts for human perceptual characteristics.

The core of any such theory must be a representation that is isomorphic to our perceptual organization, and whose elements can be computed from the unstructured array of image intensities. Unfortunately, the representations that are currently available were originally developed for other purposes (e.g., the pointwise descriptions of physics or the platonic-solids descriptions of engineering) and are therefore often unsuitable for the problems of perception.

Most current-day vision research, for instance, is based on the pointwise representation used in describing the physics of image formation, and consequently research has focused on analyzing image content on a local, point-by-point basis. Biological visual systems, however, can not recover scene structure from such local information.[1] In fact, biological visual systems are strikingly insensitive to the point-by-point particulars of the image-formation process (e.g., reflectance function or illuminant direction), factors that figure prominently in today's best vision research.

Rather than depending only upon pointwise information, people seem to make heavy use of the larger-scale structure of the scene in order to guide their perceptual interpretation. Similarly, the performance of most current-day vision algorithms depends critically upon assumed larger-scale structural context, e.g., upon assuming smoothness or isotropy. To progress towards general-purpose vision, therefore, we need new representations capable of describing these critical larger-scale structures; the "parts" or "building blocks" that we use to organize the image and provide a framework for perceptual interpretation.

Towards this end Marr and Nishihara [6] proposed a scheme using hierarchies of cylinder-like modeling primitives to describe natural forms. Their proposal is, it seems, the most widely known representation suggested to date; it captures many of our intuitions about axes of symmetry and hierarchical description (see also [11–14]). Further, in recent years representations like theirs have found considerable success in industrial-style machine-vision systems where an exact model of the specific objects that are to be discovered in the image data is available [15, 16]. Unfortunately, such a representation is only capable of an extremely abstracted description of most natural and biological forms, as is illustrated in Fig. 2. It cannot accurately and succinctly[2] describe most natural animate forms or produce a succinct description of complex inanimate forms such as clouds or mountains.

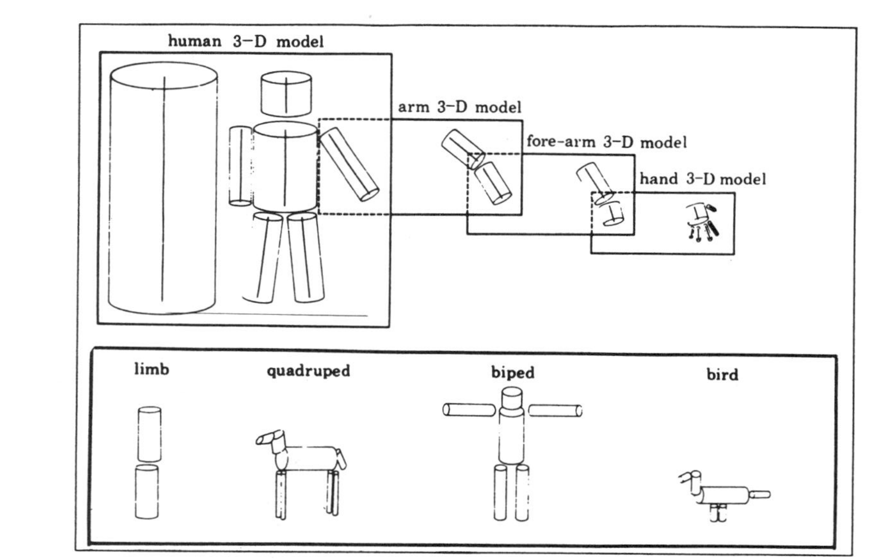

FIG. 2. Marr and Nishihara's scheme for the description of biological form.

In this paper we will present a representational system that has proven competent to accurately describe an extensive variety of natural forms (e.g., people, mountains, clouds, trees), as well as man-made forms, in a succinct and natural manner. Fig. 1 shows an example of a scene described in this representation; only 56 descriptive "parts" (about 500 bytes of information) were employed. We will then present evidence that we can use the special properties of this representational system to recover descriptions of specific objects from image data, and finally we will argue that these recovered descriptions are extremely useful in supporting both common-sense reasoning and man–machine communication.

[1] As you can confirm for yourself by looking through a long, one-inch wide tube such as found in rolls of wrapping paper.

[2] If we retreat from cylinders to generalized cylinders we can, of course, describe such shapes accurately. The cost of such retreat is that we must introduce a 1-D (at least) function describing the sweeping function; which makes the representation neither succinct nor intuitively attractive.

2. Vision, Cognition, and Models of Scene Structure

Perception is the mind's window on the world: its task is to recognize and report objects and relations that are important to the organism. It is this perceptual link between the *objective* environment and our *conception* of the environment that makes our thoughts meaningful; it ensures that they have some correspondence with the surrounding world.

Because the objects and relations recovered by perception are the primitives upon which all cognition is built, the particular way in which our perceptual apparatus organizes sensory data—that is, which regularities are noted and which are ignored—places strong constraints on the ways in which we can think about our environment. When perception organizes the sensory data in a way unsuited to the current task even simple problems can become nearly impossible to solve, as is illustrated by problems where you "see" the solution only when you "look" at them in the right way.

Identifying important environmental regularities and relating them to the primitive elements of cognition is, therefore, crucial to an understanding of cognitive function, and has consequently become the principal goal of research into visual function [9, 10]. The central problem in such research is, of course, that the sensory data underdetermines the scene structure. Image pixels, by themselves, can determine nothing. Some model of image formation and environmental structure is *required* in order to obtain any assertion about the viewed scene.

To construct a theory relating cognitive primitives to environmental structure, therefore, we must view visual perception as the process of recognizing image regularities that are known—on the basis of one's model of the world—to be reliably and lawfully related to cognitive primitives. The need for a model cannot be sidestepped, for it is the model that relates the theory's representations and computations to the state of the real world, and thus explains the semantics—the *meaning*—of the theory. A theory of visual function that has no model of the world also has no meaning.[3]

Understanding the early stages of perception as the interpretation of sensory data by use of models (knowledge) of the world has, of course, become a standard vision research paradigm. To date, however, most models have been of two kinds: high-level, *specific* models, e.g., of people or houses, and low-level models of image formation, e.g., of edges. The reason research has almost exclusively focused on these two types of model is a result more of historical accident than concious decision. The well-developed fields of optics, material science and physics (especially photometry) have provided well worked-out and easily adaptable models of image formation, while engineering, especially recent work in computer-aided design, have provided standard ways of modeling industrial parts, airplanes and so forth.

Both the use of image-formation models and specialized models has been thoroughly investigated. It appears to us that both types of models, although useful for many applications, encounter insuperable difficulties when applied to the problems faced by, for instance, a general-purpose robot. In the next two subsections we will examine both types of models and outline their advantages and disadvantages for recovering important scene information. In the remainder of this section we will then motivate, develop and investigate an alternative category of models.

2.1. Model of image formation

Most recent research in computational vision has focused on using pointwise models of image formation borrowed from optics, material science and physics. This research has been pursued within the general framework originally suggested by Marr [10] and by Barrow and Tenenbaum [17], in which vision proceeds through a succession of levels of representation. The initial level is computed directly from local image features, and higher levels are then computed from the information contained in small regions of the preceding levels. Processing is primarily data-driven (i.e., bottom-up).

In Marr's scheme the initial level is called the "raw primal sketch," and contains a description of significant local image structure, e.g., edges, lines, or flowfield vectors, represented in the form of an array of feature descriptors that preserves the local two-dimensional geometry of the image. The second level is called the "$2\frac{1}{2}$-D sketch," and is intended to describe local surface properties (e.g., color, orientation) and discontinuities in a viewer-centered coordinate frame. Again, the recovered local surface properties are placed in a set of numeric arrays in registration with original image. From this point an object-centered, volumetric representation was to be computed, such as is illustrated by Fig. 2. The rationale for this level of representation is that tasks such as navigation or object recognition seem to require description in a viewpoint-independent coordinate frame.

Despite its prevalence, there are serious problems that seem to be inherent to this research paradigm. Because scene structure is underdetermined by the local image data [18], researchers have been forced to make *unverifiable* assumptions about large-scale structure (e.g., smoothness, isotropy) in order to derive useful information from their local analyses of the image. In the real

[3] Theories of visual function, therefore, are based on models: models of how the world is structured and of how this structure is evidenced by regularities in the image. Much vision research is *not* model-based, of course: research on the mechanisms of vision (e.g., parallel processors, neurons), or on procedures for accomplishing visual tasks (e.g., variational calculus, relaxation methods) need not employ models of the world. But to understand visual *function*—that is, how one can infer information about the world—it is necessary to have a model of the salient world structure and of how that structure evidences itself in the image. Only then can one understand how certain features of the image can allow recognition and recovery of the information of interest.

world, unfortunately, such assumptions are often seriously in error: in natural scenes the image-formation parameters change in fairly arbitrary ways from point to point, making any assumption about local context quite doubtful. As a result, those techniques that rely on strong assumptions such as isotropy or smoothness have proved fragile and error-prone; they are simply not useful for many natural scenes (for a more extended discussion see Witkin and Tenebaum [19]).

That such difficulties have been encountered should not, perhaps, be too surprising. It is easily demonstrated (by looking through a viewing or reduction tube) that people can obtain little information about the world from a local image patch taken out of its context. It is also clear that detailed, analytic models of the image-formation process are not essential to human perception; humans function quite well with range finder images (where brightness is proportional to distance rather than a function of surface orientation), electron-microscope images (which are approximately the reverse of normal images), and distorted and noisy images of all kinds—not to mention paintings and drawings.

Perhaps even more fundamentally, however, even if depth maps and other maps of intrinsic surface properties could be reliably and densely computed, how useful would they be? As Witkin and Tenenbaum point out, industrial vision work [16] using laser range data has demonstrated that the depth maps, reflectance maps and the other maps of the $2\frac{1}{2}$-D sketch are still basically just images. Although useful for obstacle avoidance and other very simple tasks, they still must be segmented, interpreted and so forth before they can be used for any more sophisticated task. The conclusion to be drawn from such work is that image-like measurements of range and other surface properties contribute incrementally, in much the same way as color: they add a dimension that simplifies some decisions, but they do *not* solve the difficult problems encountered in image interpretation.

2.2. Specialized models

The alternative to models of image formation has been engineering-style representations; e.g., CAD-CAM models of specific objects that are to be identified and located. Such detailed, specific models evidence themselves in image data in an extremely complex manner, in part because the models themselves are often complex, but more importantly because it is the object's surface shape, and not the appearance of the object, that is described. As the object's orientation varies, therefore, these models produce a *very* large number of different pixel configurations—to say nothing of what happens when we vary the illumination and imaging conditions. As a consequence, the image regularities that allow reliable recognition across all of the allowable configurations are very subtle and complex.

The large number of possible appearances for such models makes the problem of recognizing them very difficult—unless an extremely simplified representation is employed. The most common type of simplified representation is that of a wireframe model whose components correspond to the imaged edges. Such a simplied representation permits reliable recognition of models with currently available computational resources, given that we are in a restricted environment where the descriptive power of such wireframe models is sufficient, e.g., as in industrial applications. As a result systems based on CAD-like models of specific objects have provided most of the success stories in machine vision.

Despite this success, the use of an impoverished representation generally means that the flexibility, reliability and discriminability of the recognition process is limited. Thus research efforts employing specific object models have floundered whenever the number of objects to be recognized becomes large, when the objects may be largely obscured, or when there are many unknown objects also present in the scene.

An even more substantive limitation of systems that employ *only* high-level, specific models is that there is no way to learn new *types* of objects: new model types must be specially entered, usually by hand, into the database of known models. This is a significant limitation, because the ability to encounter a new type of object, enter it into a catalog of known objects, and thereafter recognize it is an absolute requirement of truly general-purpose vision.

2.3. Part and process models

Some sort of additional constraint is required to oversome the fundamental problem of insufficient information being available from the image. If sufficient constraint is not available from models of image formation, then from where? Human vision seems to function quite well as long as the imaging process preserves the basic spatial structure of the scene. It seems, therefore, that human perception must be exploiting constraints provided by the structure of the scene without reliance on quantitative, pointwise models of the image-formation process. What is required, then, are models of scene structure that capture something about the larger-scale structure of our environment. We cannot, however, appeal to CAD-like models of specific objects because of the impossibility of learning new descriptions.

In response to these seemingly intractable problems some researchers have begun to search for a third type of model, one with a grain size intermediate between the pointwise models of image formation and the complex, specific models of particular objects (see [20]). There is good reason to believe that it may be possible to accurately describe our world by means of such intermediate-grain models; that world can be modeled as a relatively small set of generic processes that occur again and again, with the apparent complexity of our environment being produced from this limited vocabulary by compounding these basic forms in myriad different combinations.

We have known for over a century that evolution repeats its solutions whenever possible [1], resulting in great regularities across all species: there are but a few types of limb, a few types of skin, a few types of leaf, and a few patterns of branching. An amazingly good model of a tree, for instance, is the composition of a simple branching process with three-dimensional texture processes for generating bark and leaves [21]; the same branching models can also serve for rivers, veins, or coral. Similarly, it is now being discovered that inanimate forms may also be constrained by physical laws to a limited number of basic patterns [2,22]. Mandelbrot has shown that such apparently complex forms such as clouds, hills, coastlines or cheese can all be described by simple patterns recursively repeated at all different scales [22], while Stevens presents strong evidence that natural textures occur in but a few basic forms [2].

It is this internal structuring in our environment that allows us to derive lawful relationships [23].[4] It is exactly this internal structuring of our environment that causes object features to cluster into groups, and thus allows us to successfully employ simplified category descriptions for common-sense reasoning [3].

It appears, then, that it may be possible to accurately model the world in terms of *parts*: macroscopic models that, in relatively simple combination, can be used to form rough-and-ready models of the objects in our world and how they behave. If we adopt this view, then the central problem of perception research is *not* Marr's scheme of successively describing images, surfaces, and volumes, with the hope that we will eventually arrive at recognition of high-level models [10]. Rather, the central problem for perception is to find a set of generically applicable part models, discover image regularities that are lawfully associated with the individual parts, and then use these regularities to recognize the content of an image as a combination of these generic primitives. This new proposal, then, is that our theory of perception can dispense entirely with these initial stages of description and begin immediately with recognition of part models: models that are in principle much like models of houses and chairs, but that are more generally applicable and less detailed.

Because such models would be simpler than models of specific objects we would expect that we could more readily characterize how they would appear in an image. On the other hand, because they describe larger-scale structure than pointwise models of image formation, we would expect that they might not suffer from the problems of underdetermination that have forced researchers to make unrealistically strong assumptions such as smoothness or isotropy. Besides offering a good balance between complexity and reliability, such intermediate-grain part models spark considerable interest because they describe the world in the right terms: they speak qualitatively of whole objects and of relations between objects, rather than of local surface patches or of specific objects. Thus, they can potentially provide a vocabulary for describing the world at the grain size that is most often directly useful to us.

The problem with forming such "part" models is that they must be complex enough to be reliably recognizable, and yet simple enough to reasonably serve as building blocks for specific object models. Current 3-D machine-vision systems, for instance, typically use rectangular solids and cylinders to model specific shapes. Using these primitives for the automatic construction of a description for an arbitrary new object has not proven possible, except[5] (as in industrial or urban imagery) when the set of objects that will be encountered is constrained to be simple combinations of rectangular solids or cylinders [24]. To support truly general-purpose vision, therefore, we need to develop new modeling primitives that can be used to build descriptions of arbitrary objects and that are recognizable in standard imagery. Our work towards this goal is the subject of the remainder of this paper.

3. A Representation for Natural Forms

We present here a representational system that has been proven competent to accurately describe an extensive variety of natural forms (e.g., people, mountains, clouds, trees), as well as man-made forms, in a succinct and natural manner (see Fig. 1). The idea behind this representational system is to provide a vocabulary of models and operations that will allow us to model our world as the relatively simple composition of component "parts," parts that are reliably recognizable from image data.

The most primitive notion in this representation may be thought of as a "lump of clay," a modeling primitive that may be deformed and shaped, but which is intended to correspond roughly to our naive perceptual notion of "a part." It is worth noting that this notion of "part" agrees with that used by Konderink and Van Doorn [25, 26] or by Hoffman and Richards [27] in their analysis of how part boundaries impose constraints upon three-dimensional surfaces, although they did not actually propose a model of what constitutes a three-dimensional "part." For this basic modeling element we use a parameterized family of shapes known as a *superquadrics* [28, 29], which are described (adopting the notation $\cos \eta = C_\eta$, $\sin \omega = S_\omega$) by the following equation:

[4] If the apparent complexity of our environment were equal to its intrinsic Kolmogorov complexity, then no lawful relationships would be possible.

[5] A caveat should be noted with respect to laser range finders and the like: in some cases the thousands of range measurements provided by these active sensors can give enough additional constraint to allow recovery of low-level, polygon-like descriptions of novel objects.

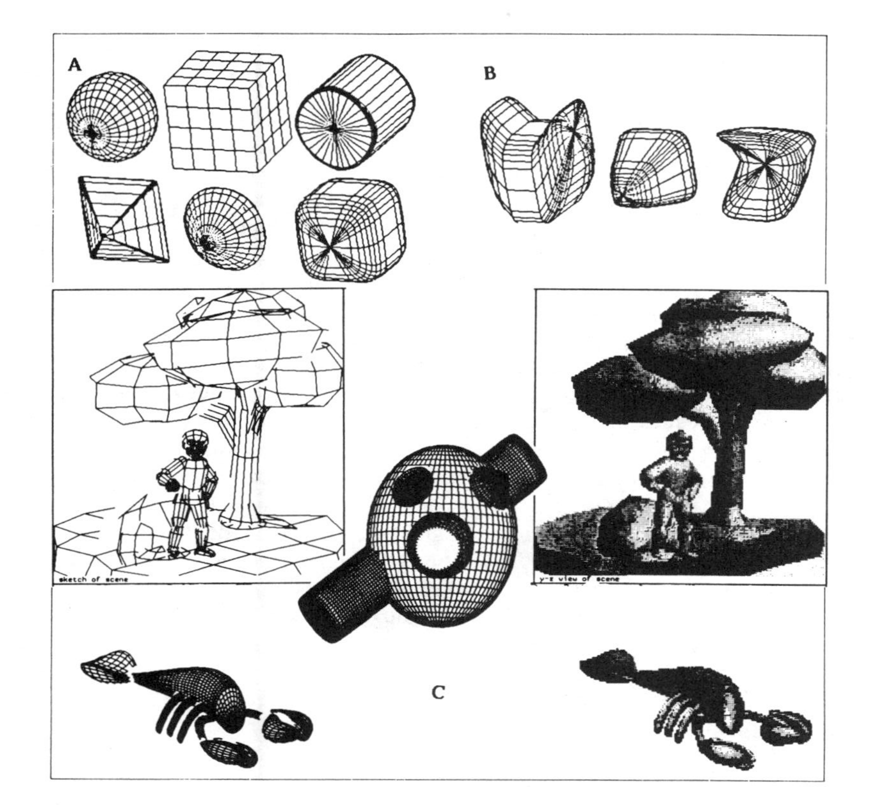

Fig. 3. (a) A sampling of the basic forms allowed. (b) Deformations of these forms. (c) Boolean combination (ors and nots) of the basic forms.

$$X(\eta, \omega) = \begin{pmatrix} C_\eta^{\varepsilon_1} C_\omega^{\varepsilon_2} \\ c_\eta^{\varepsilon_1} S_\omega^{\varepsilon_2} \\ S_\eta^{\varepsilon_1} \end{pmatrix},$$

where $X(\eta, \omega)$ is a three-dimensional vector that sweeps out a surface parameterized in latitude η and longitude ω, with the surface's shape controlled by the parameters ε_1 and ε_2. This family of functions includes cubes, cylinders, spheres, diamonds and pyramidal shapes as well as the round-edged shapes intermediate between these standard shapes. Some of these shapes are illustrated in Fig. 3(a). Superquadrics are, therefore, a superset of the modeling primitives currently in common use.

These basic "lumps of clay" (with various symmetries and profiles) are used as prototypes that are then deformed by stretching, bending, twisting or tapering, and then combined using Boolean operations to form new, complex prototypes that may, recursively, again be subjected to deformation and Boolean combination. As an example, the back of a chair is a rounded-edge cube that has been flattened along one axis, and then bent somewhat to accommodate the rounded human form. The bottom of the chair is a similar object, but rotated 90°, and by "oring" these two parts together with elongated rectangular primitives describing the chair legs we obtain a complete description of the chair, as illustrated in Fig. 4.

This descriptive language is designed to describe shapes in a manner that corresponds to a possible formative history, e.g., how one would create a given shape by combining lumps of clay. Thus the description provides us with an explanation of the image data in terms of the interaction of generic formative processes. This primitive explanation can then be refined by application of specific world knowledge and context, eventually deriving causal connections, affordances, and all of the other information that makes our perceptual experience appear so rich and varied. For instance, if we have parsed the chair in Fig. 4 into its constituent parts we could deduce that the bottom of the chair is a stable platform and thus might be useful as a seat, or we might hypothesize that the back of the chair can rigidly move relative to the supporting rod, given the evidence that they are separate "parts" and thus likely separately formed.

FIG. 4. A chair formed from Boolean combinations of appropriately deformed superquadrics.

The reader is encouraged to consider other examples of how the knowledge of part structure can help in forming hypotheses about function.

We have found that by using such a process-oriented, possible-history representation we force the resulting descriptions to group points that have similar causal histories, thus obtaining "parts" that interact with the world in a relatively simple, holistic manner. This further simplifies many reasoning tasks, because the parameters and components that affect interactions tend to be explicitly represented rather than being some complex or difficult-to-calculate function of the description's variables. For instance, use of this type of representation sufficiently simplfies questions about spatial relationships, intersection, image appearance, and so forth that we have been able to use it to construct a real-time 3-D graphical modeling system, using a Symbolics 3600 computer.[6] This system, called "SuperSketch," was used to make the figures in this paper.

Such descriptions may be written as a predicate calculus formula. We may then use this description, which has a clear model-theoretic semantics, in conjunction with constraint satisfaction or theorem-proving mechanisms, to accomplish whatever reasoning is required. Interestingly, it has been found that when adult human subjects are required to verbally describe imagery with completely novel content, their typical spontaneous strategy is to employ a descriptive system analogous to this one—i.e., form is described by modifying and combining prototypes [30]. The classic work by Rosch [3] supports the view that such a prototype-and-differences descriptive system is common in human reasoning: she showed that even primitive New Guinea tribesmen (who appear to have no concept of regular geometric shapes) form geometric prototypes in much the same manner as people from other cultures and describe novel shapes in terms of differences from these prototypes.

This representational system provides a grammar of form that has surprising descriptive power. Such descriptions have the intuitively satisfying nature of the Marr and Nishihara scheme; they incorporate hierarchies of primitives with axes of symmetry. This new descriptive language, however, is considerably more powerful than other representations that have been suggested. For example, a trivial comparison is that we can describe a wider range of basic shapes, as shown in Fig. 3(a). By allowing deformations of these shapes we greatly expand the range of primitives allowed, as shown in Fig. 3(b) (see also [31, 32, 42] on describing shape using modifications of prototypes). We have, so far, required only stretching, bending, tapering and twisting deformations to construct an extremely wide variety of objects. But the most powerful notion in this language is that of allowing (hierarchical) Boolean combination of these primitives. This intuitively attractive constructive solid-modeling approach—building specific object descriptions by applying the logical set operations "and," "or," and "not" to component *parts*—introduces a language-like generative power that allows the creation of a tremendous variety of form, such as is illustrated by Fig. 3(c) or by Fig. 1.

3.1. Biological forms

Biological forms such as the human body are naturally described by hierarchical Boolean combinations of the basic primitives, allowing the construction of accurate—but quite simple—descriptions of the detailed shape, as illustrated by Fig. 5 (the slightly cartoon-like nature of these illustrations is due primarily to the lack of surface texturing). The entire human body shown in Fig. 5, including face and hands, requires combining only 40 primitives, or approximately 300 bytes of information (these informational requirements are not a function of body position). Similarly, the description for the face requires the combination of only 13 primitives, or fewer than 100 bytes of information. The extreme brevity of these descriptions makes many otherwise difficult reasoning tasks relatively simple, e.g., even NP-complete problems can be easily solved when the size of the problem is small enough.

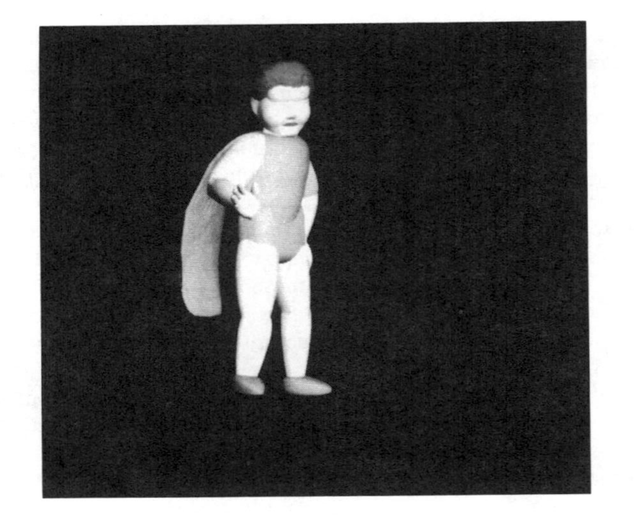

FIG. 5. The human form described (and rendered) by use of this representational system; only 40 primitives are required, approximately 300 bytes of information.

[6] "Real-time" in this case means that a "lump" can be moved, hidden-surface removal accomplished, and drawn as a 100-polygon line-drawing approximation in one eighth of a second, and a complex (full-color) image such as Fig. 1 can be rendered in approximately 20 seconds. The Symbolics speed is roughly comparable to a VAX 11/780, except for being almost an order of magnitude slower on the floating-point operations that are used heavily in this modeling system.

In Fig. 5 (as in all cases examined to date) when we try to model a particular 3-D form we find that we are able to describe—indeed, we are almost *forced* to describe—the shape in a manner that corresponds to the organization our perceptual apparatus imposes upon the image. That is, the components of the description match one-to-one with our naive perceptual notion of the "parts" in the figure, e.g., the face in Fig. 5 is composed of primitives that correspond exactly to the cheeks, chin, nose, forehead, ears, and so forth. Fig. 6 shows how the face is formed from the Boolean sum of several different primitives. The basic form for the head is a slightly tapered ellipsoid. To this basic form is

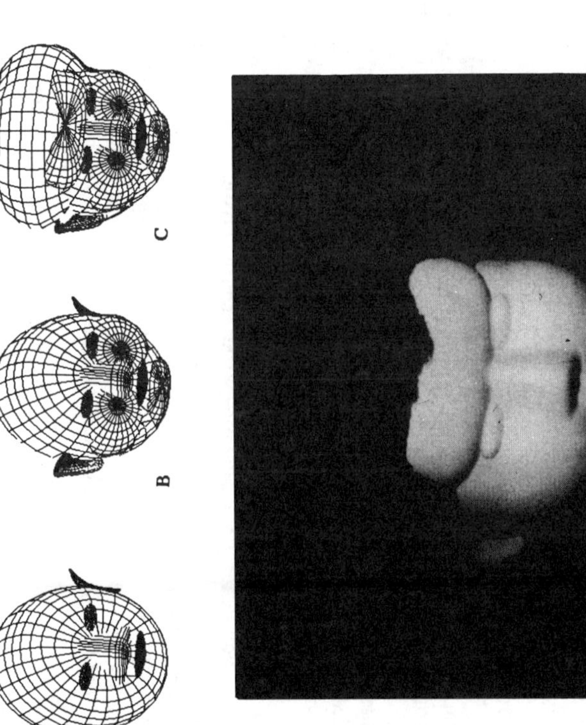

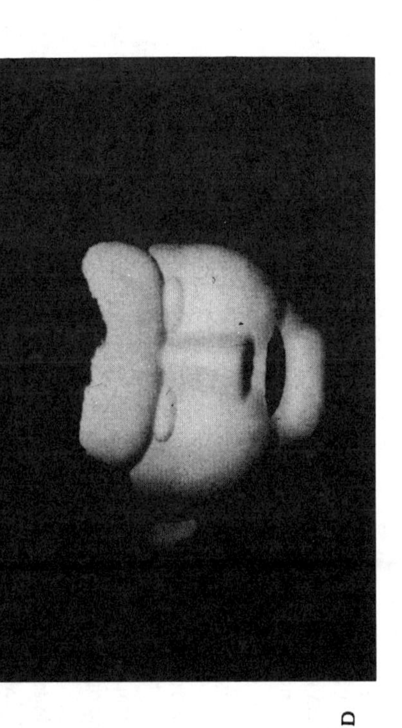

FIG. 6. (a) shows that the basic form for the head is a slightly tapered ellipsoid; to this basic form is added a somewhat cubical nose, bent pancake-like primitives for ears, bent thin ellipsoids for lips, and almond-shaped eyes. (b) shows the addition of rounded cheeks and slightly pointed chin (is this Yoda from Star Wars?), and finally (c) shows the addition of a squarish forehead and slightly fractalized hair. The smoothly shaded result is shown in (d)—it is a reasonably accurate human head, composed of only 13 primitives, specified by slightly less than 100 bytes of information.

added a somewhat cubical nose, bent pancake-like primitives for ears, bent thin ellipsoids for lips, and almond-shaped eyes, as is shown in Fig. 6(a). Fig. 6(b) shows the addition of rounded cheeks and a slightly pointed chin (is this Yoda from Star Wars?), and finally Fig. 6(c) shows the addition of a squarish forehead and slightly fractalized hair. The smoothly shaded result is shown in Fig. 6(d)—it is a reasonably accurate human head, composed of only 13 primitives, specified by slightly less than 100 bytes of information. One should remember that this representation is *not* in any way tailored for describing the human form: it is a general-purpose vocabulary.

The correspondence between the organization of descriptions made in this representation and human perceptual organization is important because it is strong evidence that we are on the right track. The fact that the distinctions made in this representation are very similar to those made by people makes it likely that descriptions couched in this language will be useful in understanding common-sense reasoning tasks, e.g., that the vocabulary of this representation might constitute a good set of primitive predicates for theories of common-sense reasoning such as sought by the Naive Physics [33] research program.[7]

Similarly, the ability to make the right "part" distinctions offers hope that we can form qualitative descriptions of specific objects ("Ted's face") or of classes of objects ("a long, thin face") by specifying constraints on part parameters and on relations between parts, in the manner of Marr and Nishihara [6,7], Winston [46, 47] or Davis [48]. And, of course, such representational correspondence is also important because it provides the basis for useful man–machine interaction.

3.2. Complex inanimate forms

This method for representing the three-dimensional world, although excellent for biological and man-made forms, becomes awkward when applied to complex natural surfaces such as mountains or clouds. The most pronounced difficulty is that, like previously proposed representations, our superquadric lumps-of-clay representation becomes implausibly complex when confronted with the problem of representing, e.g., a mountain, crumpled newspaper, a bush or a field of grass. This makes the technique ill-suited to solving the problem of representing *classes* of such objects, or determining that a particular object is a member of that class.

Why is it that such introspectively simple tasks turn out to be so hard? Intuitively, the main source of difficulty is that there is too much information to

[7] Descriptions that correspond to a possible formative history explicitly group together parts of a form that have a similar causal history, i.e., that came about in the same manner. It appears that such groupings have a strong tendency to *continue* to act as a simple whole. Why this should be true is unclear; perhaps there are only a few basic categories of physical interaction that all may be characterized using the same definition of "part."

deal with. Natural objects are amazingly bumpy and detailed; and classes of such objects seem to include virtually infinite variability. There is simply too much detail, and it is too variable. When we attempt to represent such objects in a detailed, quantitative manner, we are forced to an unwieldy description.

Nor does it suffice to simply introduce error tolerances into the representation, e.g., a mountain is a cone $\pm x$. For not only is such a representation misleading (do we *really* want to say that a cube is a sphere $\pm 0.25r$?), but it does not allow for the ability to distinguish between a mountain (represented as a cone $\pm x$) and a cone with a few dents in it (also represented as a cone $\pm x$).

Experiments in human perception suggest a way out of such problems. When we view a crumpled newspaper (for instance), it seems that the description we store is not accurate enough to recover every detail; rather, it seems that out of the welter of image detail people abstract a few properties such as the general "crumpledness" and a few major features of the shape, e.g., the general outline. The rest of the crumpled newspaper's structure is ignored; it is unimportant, *random*. For the purpose of describing that crumpled newspaper, then, the only important constraints on shape are the crumpledness and general outline.

People escape the trap of overwhelming complexity, it seems, by varying the level of descriptive abstraction—the amount of detail captured—depending on the task. In cases like the crumpled newspaper, or when recognizing classes of objects such as "a mountain" or "a cloud," the level of abstraction is very high. Almost no specific detail is required, only that the crumpledness of the form comply with the general physical properties characteristic of that type of object. In recognizing a specific mountain, however, people will require that all of the major features be identical, although they typically ignore smaller details. Even though these details are "ignored," however, they must still conform to the constraints characteristic of that type of object: we would never mistake a smooth cone for a rough-surfaced mountain even if it had a generally conical shape.

The fractal model of natural surfaces [34, 35] allows us to duplicate this sort of physically meaningful abstraction from the morass of details encountered in natural scenes. It lets us describe a crumpled newspaper by specifying certain structural regularities—its crumpledness, in effect—and leave the rest as variable detail. It lets us specify the qualitative shape—i.e., the surface's roughness—without (necessarily) worrying about the details.

3.2.1. *Fractal-based qualitative description*

Many naturally occurring forms are fractals[8] [22, 34–36]; Mandelbrot, for instance, shows that fractal surfaces are produced by several basic physical processes. One general characterization of naturally occurring fractals is that they are the end result of any physical processes that randomly modifies shape through local action, i.e., they are a generalization of random walks and Brownian motion. After innumerable repetitions, such processes will typically produce a fractal surface shape. Thus clouds, mountains, turbulent water, lightning and even music have all been shown to have a fractal form.

During the last two years we have developed these fractal functions into a statistical model for describing complex, natural surface shapes [34, 35, 37], and have found that it furnishes a good description for many surfaces. Evidence for the descriptive adequacy of this model comes from several sources. Recently conducted surveys of natural imagery [34–36], for instance, have found that this model accurately describes how most homogeneous textured or shaded image regions change over scale (change in resolution). The prevalence of surfaces with fractal statistics is explained by analogy to Brownian motion (the archetypical fractal function): just as when a dust mote randomly bombarded by air molecules produces a fractal Brownian random walk, the complex interaction of processes that locally modify shape produces a fractal Brownian surface.

For our current purposes, perhaps the most important fact is that one of the parameters of this statistical model (specifically, the fractal dimension of the surface) has been found to correspond very closely to people's perceptual notion of *roughness* [38, 39]. We have been able, for instance, to accurately predict a surface's perceptual smoothness or roughness on the basis of knowing it's fractal statistics. The fractal model, therefore, gives us a way of *qualitatively* describing surface shape [34, 35].

The fractal model shows how we may use physically motivated statistical description to abstract away from the overwhelming amount of detail present in many natural forms. To be useful, however, we must combine the fractal model's notion of qualitative description by physically meaningful statistical abstraction together with the quantitative descriptive abilities of the lump-of-clay descriptive language developed in the previous sections.

3.2.2. *Qualitative and quantitative description*

We begin the task of unifying the fractal model's notion of qualitative description with the quantitative lump-of-clay description by considering the basic properties of naturally occurring examples of fractal Brownian surfaces. Such surfaces all have two important properties: (1) each segment is statistically similar to all others; (2) segments at different scales are statistically indistinguishable, i.e., as we examine such a surface at greater or lesser imaging resolution its statistics (curvature, etc.) remain the same. Because of these invariances, the most important *variable* in the description of such a shape is how it varies with scale; in essence, how many large features there are relative to the number of middle-sized and smaller-sized features. For fractal shapes (and thus for many real shapes) the ratio of the number of features of one size to

[8] The defining characteristic of a fractal is that it has a *fractional dimension*, from which we get the word "fractal."

the number of features of the next larger size is a constant—a surprising fact that derives from the property of scale invariance. The fractal model, therefore, leads us to a statistical characterization of a surface in terms of two parameters: the surface's variance (amplitude), and the ratio between the frequency of smaller and larger features (i.e., its fractal dimension).

We may, therefore, construct fractal surfaces by using our superquadric "lumps" to describe the surface's features; specifically, we can use the recursive sum of smaller and smaller superquadric lumps to form a true fractal surface. This construction is illustrated in Figs. 7(a)–(c).

We start by specifying the surface's qualitative appearance—its roughness—by picking a ratio r, $0 \leq r \leq 1$, between the number of features of one size to the number of features that are twice as large. This ratio describes how the surface varies across different scales (resolutions, spatial frequency channels, etc.) and is related to the surface's fractal dimension D by $D = T + r$, where T is the topoligical dimension of the surface.

We then randomly place n^2 large bumps on a plane, giving the bumps a Gaussian distribution of altitude (with variance σ^2), as seen in Fig. 7(a). We then add to that $4n^2$ bumps of half the size, and altitude variance $\sigma^2 r^2$, as shown in Fig. 7(b). We continue with $16n^2$ bumps of one quarter the size, and altitude $\sigma^2 r^4$, then $64n^2$ bumps one-eighth size, and altitude $\sigma^2 r^6$ and so forth, as shown in Fig. 7(c). The final result, shown in Fig. 7(c) is a true Brownian fractal shape. The validity of this construction does not depend on the particular shape of the superquadric primitives employed, the only constraint is that the sum must fill out the Fourier domain. Different shaped lumps will, however, give different appearance or texture to the resulting fractal surface; this is an important and as yet relatively uninvestigated aspect of the fractal model. Figs. 7(d) and 7(e) illustrate the power and generality of this construction; all of the forms and surfaces in these images can be constructed in this manner.

When the placement and size of these superquadric lumps is random, we obtain the classical Brownian fractal surface that has been the subject of our previous research. When the larger components of this sum are matched to a particular object, however, we obtain a description of that object that is exact to the level of detail encompassed by the specified components. This makes it possible to specify a global shape while retaining a qualitative, statistical description at smaller scales: to describe a complex natural form such as a cloud or mountain, we specify the "lumps" down to the desired level of detail by fixing the larger elements of this sum, and then we specify only the fractal statistics of the smaller lumps thus fixing the qualitative appearance of the surface. Fig. 8 illustrates an example of such description. The overall shape is that of a sphere; to this specified large-scale shape, smaller lumps were added randomly. The smaller lumps were added with six different choices of r (i.e., six different choices of fractal statistics) resulting in six qualitatively different surfaces—each with the same basic spherical shape.

The ability to fix particular "lumps" within a given shape provides an elegant

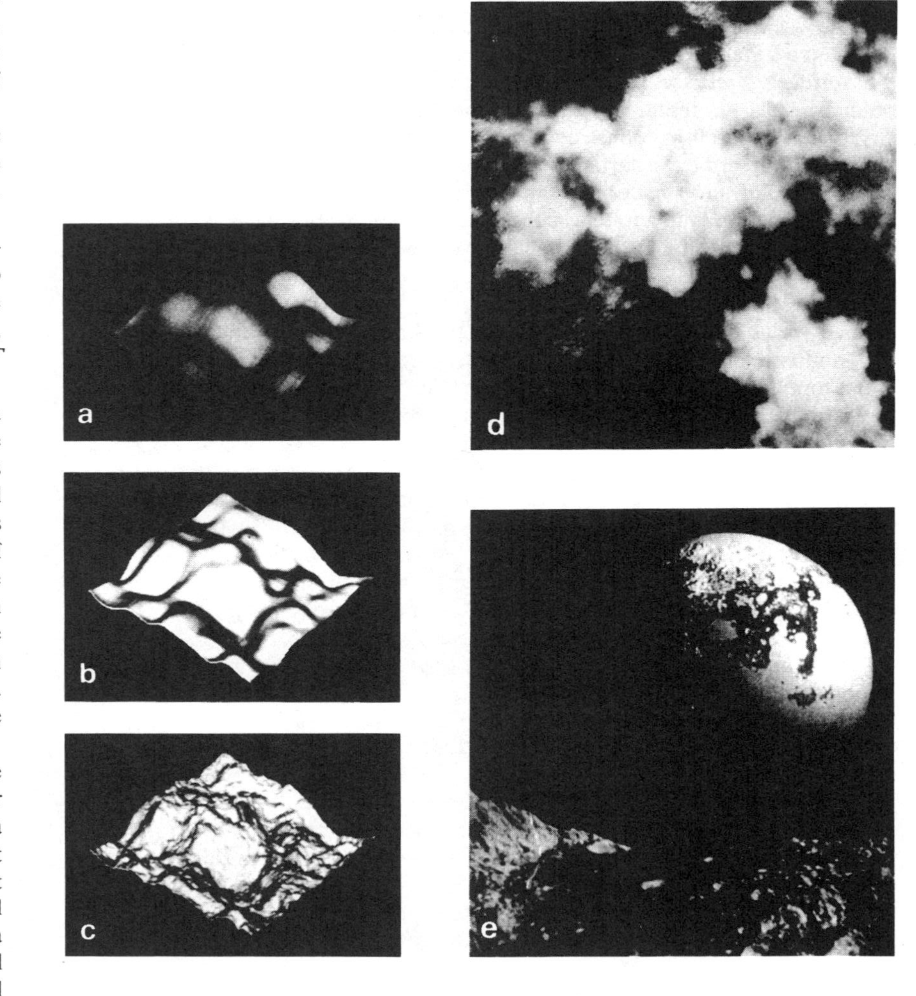

FIG. 7. (a)–(c) show the construction of a fractal shape by successive addition of smaller and smaller features with number of features and amplitudes described by the ratio $1/r$. All of the forms and surfaces shown in (d) and (e) (which are images by Voss and Mandelbrot, see [22]) can be generated in this manner.

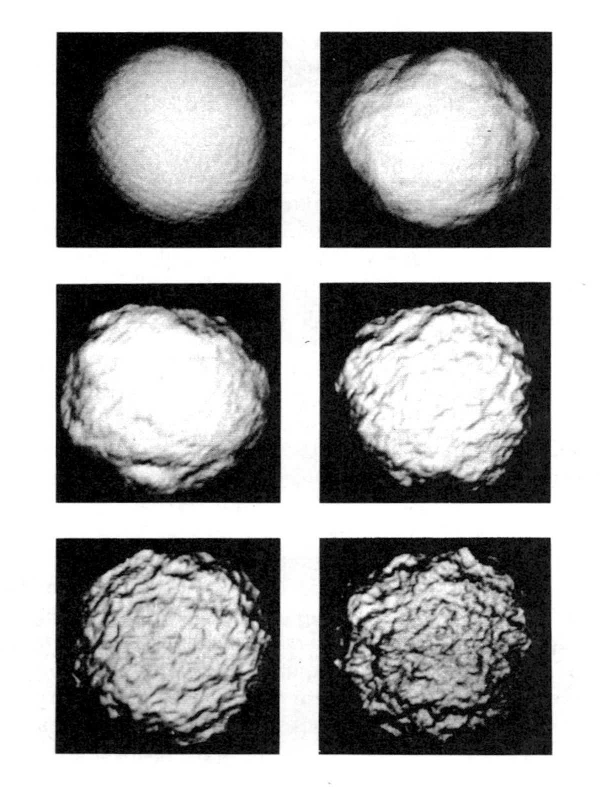

Fig. 8. Spherical shapes with surface crenulations ranging from smooth (fractal dimension = topological dimension, $r \approx 0$) to rough (fractal dimension $\gg$ topological dimension, $r \approx 1$).

way to pass from a qualitative model of a surface to a quantitative one—or *vice versa*. We can refine a general model of the class "a mountain" to produce a model of a *particular* mountain by fixing the position and size of the largest lumps used to build the surface, while still leaving smaller details only statistically specified. Or we can take a very specific model of a shape, discard the smaller constituent lumps after calculating their statistics, and obtain a model that is less detailed than the original but which is still *qualitatively* correct.

4. Primitive Perception: Recognizing Instances of Models

During the last decade, the dominant view of human perception has been that perception proceeds through successive levels of increasingly sophisticated representations until finally, at some point, information is transferred to our general cognitive faculties. And indeed, there *does* seem to be a gradient of sophistication in human perception, ranging from seemingly primitive inferences of shapes, textures, colors, and the like, to the apparently more sophisticated inferences of chairs, trees, affordances[9] and people's emotions. There is significant reason to believe, however, that this is not simply the flow of information through successive levels of representation.

To summarize Fodor's excellent and extended argument for this conclusion [40], we note that the sophisticated end of perception can involve virtually anything we know, and seems to blend smoothly into general cognition—for instance, we speak of perceiving abstract mathematical relationships or people's intentions. There is no principled reason to separate sophisticated perception from general-purpose reasoning. The characteristics of primitive perception, however, are quite different from that of cognition:

- *Informational encapsulation.* Primitive perception proceeds without benefit of intimate access to the full range of our world knowledge. Most visual illusions, for instance, cannot be dispelled merely by recognizing them as illusions [41].
- *Limited extent.* The body of knowledge on which primitive perception draws is of quite limited extent, at least in comparison to our conscious world knowledge. People of all cultures seem to share a common perceptual framework [43]; it is this shared framework that makes possible any communication at all.
- *Functional autonomy.* Primitive perception proceeds with little regard to the particulars of the task at hand, under at most limited voluntary control. We are capable of the same discriminations, regardless of purpose or task, except (perhaps) for a few very practiced tasks, e.g., birdwatchers discriminating between different types of bird. This is not to say that we always *do* make the same discriminations (we can, after all, focus our attention), but rather that whenever we attend to a particular stimulus dimension we are always capable of making the same discriminations.

Primitive perception is at least roughly the realm of perceptual organization, i.e., the pre-attentive organization of sensory data into primitives like texture, color and form. Thus, although we often speak as if perception were a smooth series of progressively more sophisticated inferences (e.g., Marr [10]), it is more likely that there are separate, specialized mechanisms for primitive and sophisticated perception.

This leads to a conception of our perceptual apparatus as containing two distinct parts: the first, a special-purpose, perhaps innate mechanism that supports primitive perception, and the second something that closely resembles general cognition. Most of the time the sensory data are first examined by the mechanisms of primitive perception to discover instances of rigidity,

[9] Affordances are the purpose(s) of an object.

parallelism, part-like groupings and other evidences of causal organization, thus providing an explanation of the image data in terms of generic formative processes. The mechanisms of sophisticated perception then use specific, learned knowledge about the world to refine this primitive, generic explanation into a detailed account of the environment.

It should be noted, however, that for at least the most practiced discriminations things seem to happen somewhat differently. When a percept, even if of a very sophisticated nature, is highly practiced or very important it appears that our minds build up a special-purpose mechanism solely for that purpose. Consider, for instance, our incredible facility at recognizing our own name, or the faces of familiar people. There may be, therefore, a sort of "compiler" for building specialized routines for these oft-repeated, important or time-critical discriminations. How much of our day-to-day perception is handled by such special-purpose routines is very much an open question.

Primitive perception, by our definition, was first seriously addressed by the Gestalt psychologists [4], who noticed that people seem to spontaneously impose a physically meaningful organization upon visual stimuli, through grouping, figure/ground separation, and so forth. They found that the addition of semantic context very rarely affects this spontaneous, pre-attentive organization of the image; somehow the visual system seems able to group an image into the correct, physically meaningful parts *before* contextual knowledge is available.

The Gestalt psychologists described this spontaneous organization as being governed by the principle of *Pragnanz*[10], however their lack of modern notions of computation limited their ability to crisply define *Pragnanz* and thus doomed them to a rather limited success. Nevertheless, their work paved the way for the two-stage model of perception that is enjoying widespread popularity in academic circles today. The first stage, which we are describing here as primitive perception, is spontaneous and pre-attentive. It carves the sensory data into likely-meaningful parts, and presents them to the later stages of perception. The second stage of perception, which we are calling sophisticated perception, is very little (if at all) different from our general cognitive faculty—including the ability to make very efficient, "compiled" routines, presumably by combining the outputs of primitive perception.

4.1. Recognizing our modeling primitives

It is our goal to provide the beginnings of a theory for our faculty of pre-attentive, primitive perception: to present a rigorous mathematical definition for the vague notion of "a part" and to explain how we can, Gestalt-like, carve an image up into meaningful "parts" without need of semantic context or specific a priori knowledge. We have already described a representation that is competent to describe a wide range of natural forms, and whose primitive elements seem to correspond closely to our naive notions of perceptual parts. What remains to be done is to show that these descriptive primitives can be recovered from the image data.

The major difficulty in recovering such descriptions is that image data are mostly a function of surface normals, and not directly a function of the surface shape. This is because image intensity, texture anisotropy, contour shape, and the like—the information we have about surface shape—is largely determined by the direction of the surface normal. To recover the shape of a general volumetric primitive, therefore, we must (typically) first compute a dense depth map from information about the surface normals. The computation of such a depth map has been the major focus of effort in vision research over the last decade and, although the final results are not in, the betting is that such depth maps are impossible to obtain in the general, unconstrained situation. Even given such a depth map, the recovery of a shape description has proven extremely difficult, because the parameterization of the surface given in the depth map is generally unrelated to that of the desired description.

Because image information is largely a function of the surface normal, one of the most important properties of superquadrics is the simple "dual" relation between their surface normal and their surface shape. It appears that this dual relationship can allow us to form an overconstrained estimate of the 3-D parameters of such a shape from noisy or partial image data, as outlined by the following equations.

The surface position vector of a superquadric with length, width and breadth a_1, a_2 and a_3 is (again writing $\cos\eta = C_\eta$, $\sin\omega = S_\omega$)

$$\boldsymbol{X}(\eta, \omega) = \begin{pmatrix} a_1 C_\eta^{\varepsilon_1} C_\omega^{\varepsilon_2} \\ a_2 C_\eta^{\varepsilon_1} S_\omega^{\varepsilon_2} \\ a_3 S_\eta^{\varepsilon_1} \end{pmatrix} \tag{1}$$

and the surface normal at that point is

$$\boldsymbol{N}(\eta, \omega) = \begin{pmatrix} \frac{1}{a_2} C_\eta^{2-\varepsilon_1} C_\omega^{2-\varepsilon_2} \\ \frac{1}{a_2} C_\eta^{2-\varepsilon_1} S_\omega^{2-\varepsilon_2} \\ \frac{1}{a_3} S_\eta^{2-\varepsilon_1} \end{pmatrix} \tag{2}$$

Therefore the surface vector $\boldsymbol{X} = (x, y, z)$ is dual to the surface normal vector $\boldsymbol{N} = (x_n, y_n, z_n)$ in the following sense:

$$\boldsymbol{N}(\eta, \omega) = \begin{pmatrix} \frac{1}{x} C_\eta^2 C_\omega^2 \\ \frac{1}{y} C_\eta^2 S_\omega^2 \\ \frac{1}{z} S_\eta^2 \end{pmatrix} \tag{3}$$

[10]Pragnanz is normally translated as meaning "goodness of form."

From (1) and (3), then we have

$$x_n = \frac{C_\eta^2 C_\omega^2}{x}, \tag{4}$$

$$y_n = \frac{C_\eta^2 S_\omega^2}{y}, \tag{5}$$

so

$$\frac{y_n}{x_n} = \frac{x}{y} \tan^2 \omega . \tag{6}$$

or

$$\left(\frac{y y_n}{x x_n}\right)^{1/2} = \tan \omega . \tag{7}$$

We may also derive alternative expressions for tan ω as follows:

$$x = a_1 C_\eta^{\varepsilon_1} C_\omega^{\varepsilon_2}, \qquad y = a_2 C_\eta^{\varepsilon_1} S_\omega^{\varepsilon_2}, \tag{8}$$

so

$$\frac{x}{y} = \frac{a_1}{a_2} \left(\frac{C_\omega}{S_\omega}\right)^{\varepsilon_2} \tag{9}$$

or

$$\left(\frac{y a_1}{x a_2}\right)^{1/\varepsilon_2} = \tan \omega . \tag{10}$$

Combining these expressions for tan ω we obtain

$$\left(\frac{y y_n}{x x_n}\right)^{1/2} = \left(\frac{y a_1}{x a_2}\right)^{1/\varepsilon_2} \tag{11}$$

or

$$\frac{y_n}{x_n} = \left(\frac{y}{x}\right)^{2/\varepsilon_2 - 1} \left(\frac{a_1}{a_2}\right)^{2/\varepsilon_2} . \tag{12}$$

Letting $\tau = y_n/x_n$, $k = (a_1/a_2)^{2/\varepsilon_2}$ and $\xi = 2/\varepsilon_2 - 1$ we find that

$$\tau = k\left(\frac{y}{x}\right)^{\xi}, \tag{13}$$

$$\frac{d\tau}{dy} = \frac{k\xi}{x} \left(\frac{y}{x}\right)^{\xi-1}, \tag{14}$$

$$\frac{d\tau}{dx} = \frac{-k\xi y}{x^2} \left(\frac{y}{x}\right)^{\xi-1} . \tag{15}$$

This gives us two equations relating the unknown shape parameters to image measurable quantities, i.e.,

$$\frac{\tau}{d\tau/dy} = \frac{y}{\xi} \tag{16}$$

and

$$\frac{\tau}{d\tau/dx} = \frac{-x}{\xi} . \tag{17}$$

Thus (16) and (17) allow us to construct a linear regression to solve for center and orientation of the form, as well as the shape parameter ε_2, given only that we can estimate the surface-tilt direction τ.

4.1.1. *Overconstraint and reliability*

Perhaps the most important aspect of these equations is that we can form an *overconstrained* estimate of the 3-D parameters, we can *check* that the parameters we estimate are correct. This property of overconstraint comes from using models: when we have used some points on a surface to estimate 3-D parameters, we can check if we are correct by examining additional points. The model predicts what these new points should look like; if they match the predictions, then we can be sure that the model applies and that the parameters are correctly estimated. If the predictions do *not* match the new data points, then we know that something is wrong. The ability to check your answer is perhaps the most important property any vision system can have, because only when you can check your answers can you build a reliable vision system. And it is *only* when you have a model that relates many different image points (such as a model of how rigid motion appears in an image sequence, or a CAD-CAM model, or this 3-D shape model) that you can have the overconstraint needed to check your answer.

One other aspect of (16) and (17) that deserves special note is that the only image measurement needed to recover 3-D shape is the surface tilt τ, the component of shape that is unaffected by projection and, thus, is the most reliably estimated parameter of surface shape. It is, for instance, known exactly at smooth occluding contours and both shape-from-shading and shape-from-texture methods produce a more reliable estimate of τ than of slant, the other surface shape parameter. That we need only the (relatively) easily estimated tilt to estimate the 3-D shape parameters makes robust recovery of 3-D shape much more likely.

When we generalize these equations to include unknown orientation and position parameters for the superquadric shape, we obtain a new set of nonlinear equations that can then be solved (in closed form) for the unknown shape parameters ε_1 and ε_2, the center position, and the three angles giving the objects orientation. Once these unknowns are obtained the remaining unknowns (a_1, a_2, and a_3, the three dimensions of the object) are easily obtained.

For the case of rotation and translation in the image plane, the equations become:

$$x^* = C_\theta(x - x_0) + S_\theta(y - y_0), \qquad y^* = -S_\theta(x - x_0) + C_\theta(y - y_0) \tag{18}$$

where θ is the rotation, x_0, y_0 the translation, and (x^*, y^*) the new rotated and translated coordinate system. The tilt τ then becomes

$$\tau = \frac{y^*_n}{x^*_n} = \frac{(-S_\theta x_n + C_\theta y_n)}{(C_\theta x_n + S_\theta y_n)}, \tag{19}$$

and the derivative of (19) is

$$\begin{aligned}\frac{d\tau}{dy^*} &= \left(-S_\theta \frac{dx_n}{dy^*} + C_\theta \frac{dy_n}{dy^*}\right)(C_\theta x_n + S_\theta y_n)^{-1} \\ &\quad -(-S_\theta x_n + C_\theta y_n)(C_\theta x_n + S_\theta y_n)^{-2}\left(C_\theta \frac{dx_n}{dy^*} + S_\theta \frac{dy_n}{dy^*}\right) \\ &= (C_\theta x_n + S_\theta y_n)^{-2}\left(x_n \frac{dy_n}{dy^*} - y_n \frac{dx_n}{dy^*}\right).\end{aligned} \tag{20}$$

Noting that

$$\begin{aligned}\frac{dx_n}{dy^*} &= \frac{dx_n}{dx}\frac{dx}{dy^*} + \frac{dx_n}{dy}\frac{dy}{dy^*} = -\frac{dx_n}{dx} S + \frac{dx_n}{dy} C, \\ \frac{dy_n}{dy^*} &= \frac{dy_n}{dx}\frac{dx}{dy^*} + \frac{dy_n}{dy}\frac{dy}{dy^*} = -\frac{dy_n}{dx} S + \frac{dy_n}{dy} C.\end{aligned} \tag{21}$$

Equation (16) can now be rewritten as

$$\begin{aligned}&(C_\theta x_n + S_\theta y_n)(-S_\theta x_n + C_\theta y_n) \\ &= \frac{1}{\xi}\left[-S_\theta(x - x_0) + C_\theta(y - y_0)\right] \\ &\quad \times \left(x_n\left(-\frac{dy_n}{dx} S + \frac{dy_n}{dy} C\right) - y_n\left(-\frac{dx_n}{dx} S + \frac{dx_n}{dy} C\right)\right).\end{aligned} \tag{22}$$

Our estimates of tilt from local image information typically have considerable noise in them [18, 37, 44]; in order to still obtain good estimate of three-dimensional shape we will formulate the problem of recovering the shape parameters as a linear regression. Collecting the image-measurable terms together (in square brackets), this equation becomes

$$\begin{aligned}0 &= [x_n^2 - y_n^2](\xi C_\theta S_\theta) + [x_n y_n](\xi(S_\theta^2 - C_\theta^2)) \\ &+ \left[x x_n \frac{dy_n}{dy} - x y_n \frac{dx_n}{dy}\right](-S_\theta C_\theta) + \left[x x_n \frac{dy_n}{dx} - x y_n \frac{dx_n}{dx}\right](S_\theta^2) \\ &+ \left[y x_n \frac{dy_n}{dy} - y y_n \frac{dx_n}{dy}\right](C_\theta^2) + \left[y x_n \frac{dy_n}{dx} - y y_n \frac{dx_n}{dx}\right](-S_\theta C_\theta) \\ &+ \left[x_n \frac{dy_n}{dy} - y_n \frac{dx_n}{dy}\right](S_\theta C_\theta x_0 - C_\theta^2 y_0) \\ &+ \left[x_n \frac{dy_n}{dx} - y_n \frac{dx_n}{dx}\right](S_\theta C_\theta y_0 - S_\theta^2 x_0).\end{aligned} \tag{23}$$

This equation, then, can be used for a linear regression to solve for the unknown coefficients (in curved brackets). We have seven unknown coefficients and so we require tilt information at as few as seven points in order to solve for all these unknowns. By combining this equation together with the equivalent version derived from (17) we can obtain closed form solutions for the center of the object (x_0, y_0), the shape parameter ε, and the orientation θ. In fact, things are somewhat better than this, because we have two such equations at each point (one for dx and one for dy) so that fewer points are actually required. The small number of points required opens up the possibility of segmenting images in terms of the parameters of the 3-D surface.

At occluding contours the situation is better yet, because we also know that $y_n^2 + x_n^2 = 1$, and considerable extra constraint is available. This formulation, therefore, reflects the fact that contour information is more powerful than shading or texture information. One of the more interesting aspects of this approach is that contour information and information from shading or texture contribute toward estimating shape in exactly the same manner—by providing information about surface tilt—and therefore we may combine information from all of these sources by use of the same set of equations, those derived from (16) and (17).

Because we have formulated the problem of primitive perception as one of recognizing instances of the "parts" found in a representational vocabulary, we can frame the problem as one in statistical decision theory: we have a range of alternatives that we entertain, and use image data to decide among the alternatives. This gives us a rigorous framework for integrating information from motion, stereo, etc., together with contour, shading and texture information without having to make further assumptions. This is in considerable contrast to approaches that try to apply strong, unverifiable assumptions about the nature of surfaces (e.g., that all surfaces are "smooth") in order to integrate various information sources. Here we are attempting to collect a vocabulary of models that span the space of shape possibilities, so that we can replace unverifiable assumptions with verifiable models. We want perception to proceed by making an overconstrained, statistical determination that a particular model is applicable (rather than simply making an assumption), and then estimate the parameters of that model. If our vocabulary of shape does in fact cover the range of shape that actually occurs, then *we will have made the best shape estimate possible with the available image data.*

Equation (23) does not reflect the full sophistication possible in statistical decision theory; a regression using this equation results in a maximum likelihood estimate of compound parameters such as C_θ and $\xi C_\theta S_\theta$ rather than estimates of the individual parameters ε and θ. Still, the main power of the approach remains. Our modeling primitives provide us with a parameterized range of hypotheses that we can choose among using established statistical tools, thus providing us a rigorous framework for integrating contour with shading and texture information, as well as allowing us to include a priori information that we may have gained from previous views. The power of this framework has been illustrated by the work of Ferrie and Levine [44] who, using a simpler shape vocabulary consisting of ellipsoids and cylinders, have combined our local shape-from-shading technique [18] with motion information to accurately recover 3-D shape.

Although the equations presented here are only for rotations in the image plane, the general equations are similar, although somewhat more complicated. As in the simpler case, information at relatively few points[11] is required in order to solve for the unknowns, and the situation is considerably better along occluding contours.

Fig. 9 illustrates the process of recovering 3-D shape using this technique. Fig. 9(a) shows a half-toned version of an image of a superquadric with shape parameters $\varepsilon_1, \varepsilon_2 = 0.5$. To this image, we applied the local shape-from-shading/texture technique developed by Pentland [18, 37]. The estimation technique employs second-derivative filters with local support to make estimates of surface slant and tilt, with the estimates of tilt being more reliable than the estimates of slant [18, 44]. Fig. 9(b) shows a view of the surface tilt (i.e., y_n/x_n) recovered from the continuous 8-bit image of the shape illustrated by Fig. 9(a); in this figure the image x-axis runs left-right and the y-axis runs up-down. From this estimated tilt surface we can use (16) and (17) to estimate the center of the shape, the shape parameter ε_2, and the width and breadth of the shape. Fig. 9(c) shows two views of the recovered shape; it can be seen that in this simple case a good estimate of the 3-D shape can be made. It appears, then, that (16) and (17) offer a good hope for recovering surface shape; in our future research we hope to extend these preliminary results to natural imagery.

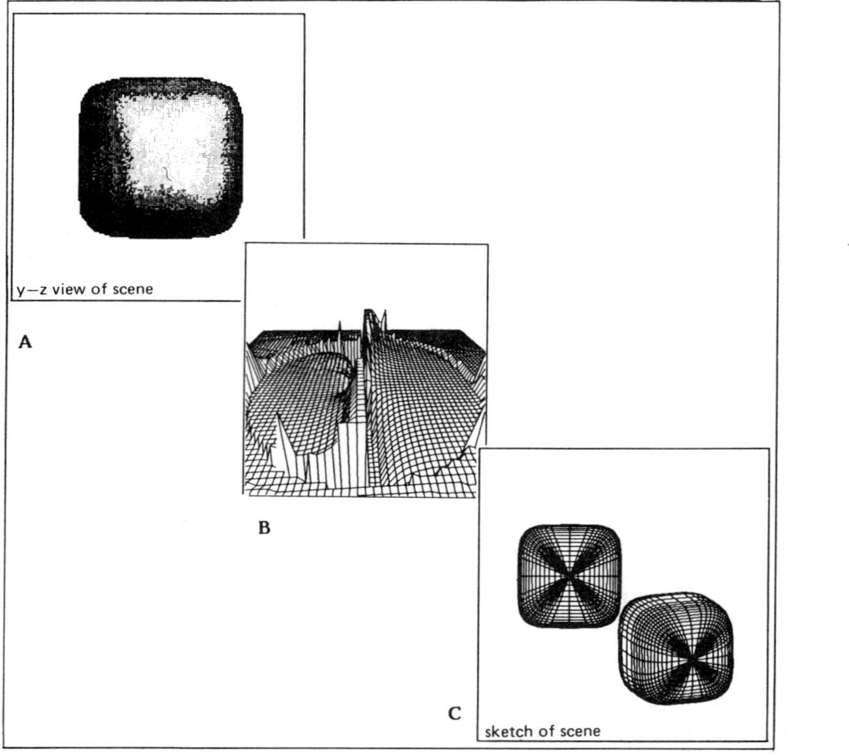

FIG. 9. (a) A half-toned version of an image of a superquadric with shape parameters $\varepsilon_1, \varepsilon_2 = 0.5$. (b) The surface tilts estimated using the local shape-from-shading/texture algorithm described in [18, 37]. (c) Two views of the 3-D shape estimated by use of equations (16) and (17), using the tilt estimates shown in (b).

4.2. Model-based vision, the blocks world, and our effort

The most successful (i.e., working, practical) efforts in machine vision have all been accomplished within two paradigms that are generally lumped together under the rubric of "model-based vision." The first of these paradigms is to

[11] Depending on the exact formulation, 15 points are required.

take a CAD-CAM-type model of a specific object, find configurations of image features that uniquely determine identity and orientation of the object, and then search the image for those configurations. A similar, but fundamentally quite different "model-based vision" paradigm was first employed in the blocks-world research during the 1960s [45], and more recently in such work as the 3-D mosaic program of Hermann and Kanade [24]. In this second paradigm, the models are of the *parts* that make up specific objects, rather than a model of the entire object, and the goal is to identify those component parts. Once the parts have been identified and their spatial layout determined, one can ask if this configuration of parts is an object that has been seen before. This latter approach has the very significant advantage that it can learn new object descriptions by example: it can look at a new object, identify the object's parts, and then use that part-wise description to build up a general model of the specific object in a manner similar to that proposed by Winston [46, 47].

Because this second, find-the-parts approach to model-based vision can learn descriptions of novel objects, it has the potential to support general-purpose vision. The major limitation on the success of this approach is the availability of part models that are individually recognizable and which have the expressive power to describe everything within the domain of interest. What we are attempting to do is develop a vocabulary of just such individually recognizable part models. One may, therefore, think of the research described here as returning to the blocks world, but with models of 3-D structure that are tremendously more sophisticated than simple blocks or polyhedra.

We believe that the modeling language presented here has a good chance of being able to handle most of the forms found in the real world. The images in this paper demonstrate the expressive power of this new vocabulary of models (their cartoon-like nature is primarily due to the lack of surface texturing), and the mathematics in this section of the paper demonstrate the plausibility of recovering such part descriptions from sparse and partial image data. Even if it should turn out that our models aren't yet sophisticated enough to deal with the complexity of real world, we will have *at least* made major progress towards bridging the gap between the present state of the art and that needed to construct a general-purpose, real-world vision system.

5. Using the Representation

The particular models of the world that perception uses to interpret sensory data induce a profound organization on all of our conceptual structures. If we stand in the center of Stonehenge, we can see either a collection of pillars, several irregular walls of pillars, or concentric circular structures with regularly spaced pillars. This is the familiar Gestalt phenomenon of grouping; what is important about it is that *which* grouping you spontaneously see strongly influences what hypotheses you entertain when trying to deduce, for instance, the purpose of Stonehenge. Examples such as this demonstrate that the manner in which perception "carves up" the world—that is, its models of the world—strongly determine the way in which we think about the world.

The issue of perceptual models is, therefore, of more than passing interest to those interested in cognition. It seems reasonable that if we are to develop machines that are able to display common-sense reasoning abilities, for instance, we must have spatial representations that are at least roughly equivalent to those people employ in organizing their picture of the world. Similarly, if we are ever to communicate with machines about our shared environment we must develop spatial representations that are at least isomorphic to the representations that we use. We must have a representation that captures the same sorts of distinctions we make when we carve objects into parts.

Because communication depends upon having a shared representation of the situation, we can use man–machine communication as a fairly sensitive test of whether a particular representation captures the notions of difference and similarity that humans employ. The empirical (and so far informal) finding that the organization of our shape descriptions correspond closely with the human perceptual organization is, as a consequence, quite interesting: the representation seems to offer exciting possibilities for flexible, effective man–machine communication. It was, therefore, of great interest to test how effectively we can use the representation described here as a basis for communication between a computer and its operator concerning image data and 3-D shape.

5.1. Communicating about a digital terrain map

As a first experiment we took the problem of communicating with a computer about a digital terrain map, as might be done in guiding a stereo compilation process or when plotting a path through the terrain. Fig. 7 showed how a mountain-like surface can be built up from the combination of progressively smaller primitives. We can also take a real surface, such as the digital terrain map of Yosemite Valley shown in Fig. 10(a) and decompose it into a canonical lump description by use of a minimum-complexity criterion, that is, we attempt to account for the shape with the fewest number of component parts as is possible (see [49]). One simple mechanism for approximating this decomposition is to form a Laplacian pyramid [50], examine the entries in this pyramid to find those points that most closely correspond to the shape of a single "lump" (by looking at the neighbors of the point in both space and scale), subtract off that lump from the original form. We then repeat this procedure until no entries remain in the pyramid.

If we want to have a "sketch" of the DTM surface (a simplified description that we can use for communication), we can use estimates of the surface's variance and fractal dimension to set an acceptance threshold, so that our decomposition procedure finishes by taking only the 50 or so most prominent surface features. To adequately characterize a DTM we have found that we need to look for only two types of primitive elements: one, a vertically oriented symmetrical peak, and two, a horizontally oriented elongated ridge or valley. The fractal statistics of the surface characterize how features of the surface change across scale and, therefore, gives us the information needed to adjust the acceptance threshold for different scales, so that the prominence of features

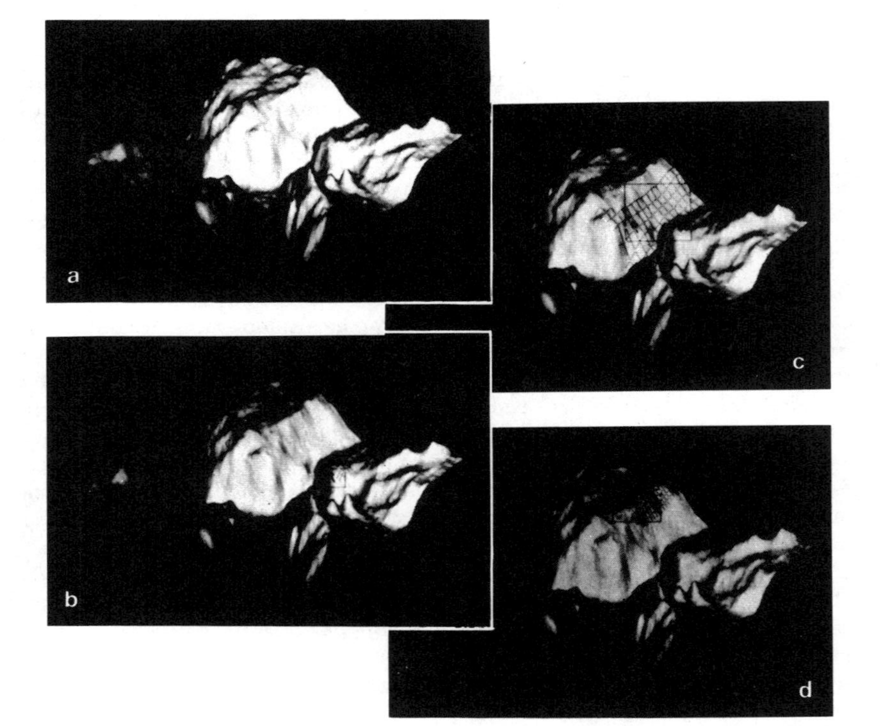

Fig. 10. (a) A digital terrain map of Yosemite Valley, which is automatically decomposed into a "sketch," a description in our representational system that contains terms ("lumps") that correspond roughly to "peaks," "valleys," and "ridges," so that the parts of this description correspond closely with the perceptual organization that we impose on the scene. This is illustrated in (b), (c) and (d), which show a person pointing to a part of the image, and the computer using this sketch to determine what part of the terrain is being gestured at, and highlighting the "part" referred to by covering it with cross-hatching. This decomposition of the scene into perceptually salient "parts" thus fulfills a critical requirement for effective man–machine communication: similar representations of the scene.

accepted at one scale corresponds to the prominence of smaller- or larger-scale features. When we do this, the result is a description that organizes the pixel data into its most prominent components at all scales, in a way that we have found corresponds closely with our naive perceptual organization of the surface—e.g., organizing the surface into peaks ridges, valleys, and the like.

The ability to structure the pixel data in a manner that corresponds to the perceptual organization we impose upon the data allows us to support human–computer communication about the scene. It allows us to point to a part of the scene, say "that thing" and have the machine be able to make a good guess about what part of the surface we want to indicate, as opposed to the current state of the art in which we have to carefully outline the part of the surface that we want manipulate.

This sort of communication is illustrated in Figs. 10(b), (c) and (d), which shows the operation of a program we have constructed that performs this parsing of a digital terrain map (DTM), identifies the 50 or so most prominent perceptual "parts," and then allows the user to interact with the DTM by simply pointing to peaks, valleys, ridges and so forth. These figures show a user pointing, the program interpreting what "feature" the user intended to indicate, and then highlighting that feature by cross-hatching it. The highlighted feature can then be edited to improve the DTM, defined as a primitive object in a path planning calculation, or used in whatever manner the user's purpose demands. As these figures illustrate, we have found a good correspondence between this program's structuring of the image and the structure people impose on the image.

5.2. Building 3-D models

One other example that illustrates using the representation to facilitate man–machine communication is the 3-D modeling system called "SuperSketch" that was used to make most of the images in this paper. In this Symbolics-3600-based modeling system users create "lumps," change their squareness/roundness, stretch, bend, and taper them, and make Boolean combinations of them in real time by moving the mouse through the relevant parameter space, controlling which parameter is being varied by using the mouse buttons. Because these forms have an underlying analytical form, we can use fast, qualitative approximations to accomplish hidden-surface removal, intersection and image-intensity calculations in real time—something that could not be accomplished on a Symbolics 3600 if a polygon-based description were employed. "Real time" in this case means that a "lump" can be moved, hidden-surface removal accomplished, and drawn as a 200-polygon line drawing-approximation in one eighth of a second, and a complex, full color image such as Fig. 1 can be rendered in approximately 20 seconds.

Because the primitives, operations and combining rules used by the computer are very well matched to those of the human operator, we have found that interaction is surprisingly effortless: it took a relatively unskilled operator less than a half hour to assemble the face in Fig. 6, about ten minutes to create the lobster in Fig. 3, and about four hours total to make Fig. 1. This is in rather stark contrast to more traditional 3-D modeling systems that might require days or weeks to build up a scene such as shown in Fig. 1. This performance, perhaps more than any other statistic that could be given, illustrates how the close match between this representational system and the perceptual organization employed by human operators facilitates effective man–machine communication.

6. Summary

To support our reasoning abilities perception must recover environmental regularities—e.g., rigidity, "objectness," axes of symmetry—for later use in cognitive processes. Understanding this recovery of structure is critically important because the structural organization that perception delivers to cognition is the foundation upon which we construct our picture of the world; these regularities are the building blocks of all cognitive activities.

To create a theory of how our perceptual apparatus can produce meaningful cognitive building blocks from an array of image intensities we require a representation whose elements may be lawfully related to important physical regularities, and that correctly describes the perceptual organization people impose on the stimulus. Unfortunately, the representations that are currently available were originally developed for other purposes (e.g., physics, engineering) and have so far proven unsuitable for the problems of perception or common-sense reasoning.

For instance, the complexity of standard descriptions for such common natural forms as clouds, human faces, or trees has been a fundamental block to progress in computational psychology, artificial intelligence and machine vision. It is a fundamental result of mathematics that one cannot recover 3-D shape descriptions from an image when the number of parameters to be recovered is greater than the number of pixels in the image. How, then, can we hope to understand perception when our representational tools force us into the uncomfortable position of knowing a priori that we cannot recover the desired descriptions from image data? Further, even if we *could* recover such descriptions, how can we hope to understand common-sense reasoning if forced to use such overly complex descriptions?

In answer to these problems we have presented a representation that has proven competent to accurately describe an extensive variety of natural forms (e.g., people, mountains, clouds, trees), as well as man-made forms, in a succinct and natural manner. The approach taken in this representational system is to describe scene structure at a scale that is more like our naive perceptual notion of "a part" than the pointwise descriptions typical of current image understanding research, and to use a description that reflects a possible formative history of the object, e.g., how the object might have been constructed from lumps of clay.

Each of the component parts of this representation—superquadric "lumps," deformations, Boolean combination, and the recursive fractal construction—have been previously suggested as elements of various shape descriptions, usually for other purposes. The contribution of this paper is to bring all of these separate descriptive elements together, and employ them as a representation for natural forms and as a theory of perceptual organization. In particular, we believe that the important contributions of this paper are the following.

(1) We have demonstrated that this process-oriented representational system is able to accurately describe a very wide range of natural and man-made forms in an extremely simple, and therefore useful, manner. Further, the representation can be used to support fast, qualitative approximations to determine, e.g., intersection, appearance or relative position. Such qualitative reasoning is employed in SuperSketch allow real-time movement, deformation, Boolean combination, hidden-surface removal, intersection and rendering.

(2) We have found that descriptions couched in this representation are similar to people's (naive) verbal descriptions and appear to match people's (naive) perceptual notion of "a part", this correspondence is strong evidence that the descriptions we form will be good spatial primitives for a theory of common-sense reasoning. Additionally, we hope that this descriptive system will provide the beginnings of a rigorous, mathematical treatment of the still vaguely defined subject of human perceptual organization.

(3) The part-model approach to perception makes the problem of recovering shape descriptions overconstrained and therefore potentially extremely reliable, while still providing the flexibility to learn new object descriptions. Toward this end we have shown that our current descriptive vocabulary is capable of describing a wide range of natural forms, and that the primitive elements of this language can be recovered from partial image data in an overconstrained and apparently noise-insensitive manner.

(4) And finally, we have shown that descriptions framed in the representation have markedly facilitated man–machine communication about both natural and man-made 3-D structures. It appears, therefore, that this representation gives us the right "control knobs" for discussing and manipulating 3-D forms.

The representational framework presented here is *not* complete. It seems clear that additional process-oriented modeling primitives, such as branching structures [21] or particle systems [51], will be required to accurately represent objects such as trees, hair, fire, or river rapids. Further, it seems clear that domain experts form descriptions differently than naive observers, reflecting their deeper understanding of the domain-specific formative processes and their more specific, limited purposes. Thus, accounting for expert descriptions will require additional, more specialized models. Nonetheless, we believe this descriptive system makes an important contribution toward solving current problems in perceiving and reasoning about natural forms, by allowing us to construct accurate models that are still simple enough to be useful, and by providing us with the basis for more effective man–machine communication.

ACKNOWLEDGMENT

I would like to thank Andy Witkin for his collaboration in developing the basic elements of this approach to perception, and for his help in refining my ideas about fractals and superquadrics. I would also like to thank David Heeger, Oscar Firschein, Bob Bolles and Tracy Heibeck for their help and criticism in the writing of this manuscript.

This research was made possible by National Science Foundation, Grant No. DCR-83-12766, by Defense Advanced Research Projects Agency contract no. MDA 903-83-C-0027, and by a grant from the systems Development Foundation.

REFERENCES

1. Thompson, D'Arcy, *On Growth and Form* (University Press, Cambridge, U.K., 2nd ed., 1942).
2. Stevens, S., *Patterns in Nature* (Atlantic-Little, Brown Books, Boston, MA, 1974).
3. Rosch, E., On the internal structure of perceptual and semantic categories, in: T.E. Moore (Ed.), *Cognitive Development and the Acquisition of Language* (Academic Press, New York, 1973).
4. Wertheimer, M., Laws of organization in perceptual forms, in: W.D. Ellis (Ed.), *A Source Book of Gestalt Psychology* (Harcourt Brace, New York, 1923).
5. Johansson, G., *Configurations in Event Perception* (Almqvist and Wiksell, Stockholm, 1950).
6. Marr, D. and Nishihara, K., Representation and recognition of the spatial organization of three-dimensional shapes, *Proc. Roy. Soc. London B* **200** (1978) 269-294.
7. Nishihara, H.K., Intensity, visible-surface and volumetric representations. *Artificial Intelligence* **17** (1981) 265–284.
8. Binford, T.O., Visual perception by computer, in: *Proceedings IEEE Conference on Systems and Control*, Miami, FL, 1971.
9. Gibson, J.J., *The Ecological Approach to Visual Perception* (Houghton Mifflin, Boston, MA, 1979).
10. Marr, D. *Vision* (Freeman, San Francisco, CA, 1982).
11. Agin, G.J. and Binford, T.O., Computer descriptions of curved objects, *IEEE Trans. Comput.* **25** (1976) 439–449.
12. Nevatia, R. and Binford, T.O., Description and recognition of curved objects *Artificial Intelligence* **8** (1977) 77–98.
13. Badler, N. and Bajacsy, R., Three-dimensional representations for computer graphics and computer vision, *Comput. Graphics* **12** (1978) 153–160.
14. Brady, J.M., Describing visible surfaces, in: A. Hanson and E. Riseman (Eds.), *Computer Vision Systems* (Academic Press, New York, 1978).
15. Brooks, R., Model based 3-D interpretation of 2-D images, in: A. Pentland (Ed.), *From Pixels to Predicates* (Ablex, Norwood, NJ, 1985).
16. Bolles, B. and Haroud, R., 3DPO: An inspection system, in A. Pentland (Ed.), *From Pixels to Predicates* (Ablex, Norwood, NJ, 1985).
17. Barrow, H.G. and Tenebaum, J.M., Recovering intrinsic scene characteristics from images, in: A. Hanson and E. Riseman (Eds.), *Computer Vision Systems* (Academic Press, New York, 1978).
18. Pentland, A., Local analysis of the image, *IEEE Trans. Pattern Anal. Machine Intelligence* **6** (1984) 170–187.
19. Witkin, A.P. and Tenebaum, J.M., On perceptual organization, in: A. Pentland (Ed.), *From Pixels to Predicates* (Ablex, Norwood, NJ, 1985).
20. A. Pentland and A. Witkin, On perceptual organization, in: *Proceedings Second Conference on* Perceptual Organization, Pajaro Dunes, CA, 1984.
21. Smith, A.R., Plants, fractals and formal languages, *Comput. Graphics* **18** (3) (1984) 1–11.
22. Mandelbrot, B.B., *The Fractal Geometry of Nature* (Freeman, San Francisco, CA, 1982).
23. Georgeff, M.P. and Wallace, C.S., A general selection criterion for inductive inference, in: *Proceedings Sixth European Conference on Artificial Intelligence*, Pisa, Italy, 1984.
24. Herman, M. and Kanade, T., The 3-D mosaic scene understanding systems, in: A. Pentland (Ed.), *From Pixels to Predicates* (Ablex, Norwood, NJ, 1985).
25. Konderink. J.J. and Van Doorn, A.J., The shape of smooth objects and the way contours end, *Perception* **11** (1982) 129–137.
26. Konderink, J.J. and Van Doorn, A.J., The internal representation of solid shape with respect to vision, *Biol. Cybernet.* **32** (1979) 211–216.
27. Hoffman, D. and Richards, W., Parts of recognition, in: A. Pentland (Ed.), *From Pixels to Predicates* (Ablex, Norwood, NJ, 1985).
28. Barr, A., Superquadrics and angle-preserving transformations, *IEEE Comput. Graphics Appl.* **1** (1981) 1–20.
29. Kauth, R., Pentland, A. and Thomas, G., BLOB: an unsupervised clustering approach to spatial grouping in: *Proceedings Eleventh International Symposium on Remote Sensing of the Environment*, Ann Arbor, MI 1977.
30. Hobbs, J. Final Report on Commonsense Summer, Tech. Note 370, SRI Artificial Intelligence Center, Menlo Park, CA, 1985.
31. Barr, A., Global and local deformations of solid primitives, *Comput. Graphics* **18** (3) (1984) 21–30.
32. Hollerbach, J.M., Hierarchical shape description of objects by selection and modification of prototypes, Ph.D. Theses, AI Tech. Rept. 346, MIT, Cambridge, MA, 1975.
33. Hayes, P., The second naive physics manifesto, in: J. Hobbs and R. Moore (Eds.), *Formal Theories of the Commonsense World* (Ablex, Norwood, NJ, 1985).
34. Pentland, A., Fractal-based description of natural scenes, *IEEE Trans. Pattern Anal. Machine Intelligence* **6** (1984) 661–674.
35. Pentland, A., Fractal-based description in: *Proceedings Eighth International Joint Conference on Artificial Intelligence*, Karlsruhe, F.R.G. (1983) 973–981.
36. Medioni, G. and Yasumoto, Y., A note on using the fractal dimension for segmentation, *IEEE Computer Vision Workshop*, Annapolis, MD, 1984.
37. Pentland, A., Shading into texture in: *Proceedings Fourth National Conference on Artificial Intelligence*, Austin, TX (1984) 269–273.
38. Pentland, A., Fractals: a model for both texture and shading, *Optic News* (October, 1984) 71.
39. Pentland, A., Perception of three-dimensional textures, *Investigative Opthomology and Visual Science* **25** (3) (1984) 201.
40. Fodor, J., *Modularity of Mind: An Essay on Faculty Psychology* (MIT Press, Cambridge MA, 1982).
41. Gregory, R.L., *The Intelligent Eye* (McGraw-Hill, New York, 1970).
42. Leyton, M., Perceptual organization as nested control, *Biol. Cybernet.* **51** (1984) 141–153.
43. Held, R. and Richards, W., (Eds.) *Recent Progress in Perception, Readings from Scientific American* (Freeman, San Francisco, CA, 1975).
44. Ferrie, F.P. and Levine, M.D., Piecing together the 3-D shape of moving objects: an overview, in: *Proceedings IEEE Conference on Vision and Pattern Recognition*, San Francisco, CA, 1985.
45. Roberts, L., Machine perception of three-dimensional solids, in: J.T. Tippet, et al. (Eds.), *Optical and Electrooptical Information Processing* (MIT Press, Cambridge, MA, 1965).

46. Winston, P.H., Learning structural descriptions from examples, in: P.H. Winston (Ed.), *The Psychology of Computer Vision* (McGraw-Hill, New York, 1975).
47. Winston, P., Binford, T., Katz, B. and Lowry, M., Learning physical descriptions from functional definitions, examples, and precedents, in: *Proceedings Third National Conference on Artificial Intelligence*, Washington, DC (1983) 433–439.
48. Davis, E., The MERCATOR representation of spatial knowledge, in: *Proceedings Eighth International Joint Conference on Artificial Intelligence*, Karlsruhe, F.R.G. (1983) 295–301.
49. Pentland, A., On describing complex surfaces, *Image and Vision Computing* **3**(4) (1985) 1–15.
50. Burt, P.J. and Adelson, E.H., The Laplacian pyramid as a compact image code, *IEEE Trans. Communications* **31** (1983) 532–540.
51. Reeves, W.T., Particle systems—a technique for modeling a class of fuzzy objects, *ACM Trans. Graphics* **2** (2) (1983) 91–108.

Codon Constraints on Closed 2D Shapes

WHITMAN RICHARDS AND DONALD D. HOFFMAN*

Natural Computation Group, Massachusetts Institute of Technology, Cambridge, Massachusetts

Received September 27, 1984

Codons are simple primitives for describing plane curves. They thus are primarily image-based descriptors. Yet they have the power to capture important information about the 3D world, such as making part boundaries explicit. The codon description is highly redundant (useful for error-correction). This redundancy can be viewed as a constraint on the number of possible codon strings. For smooth closed strings that represent the bounding contour (silhouette) of many smooth 3D objects, the constraints are so strong that sequences containing 6 elements yield only 33 generic shapes as compared with a possible number of 15,625 combinations.

1. INTRODUCTION

An important task for object recognition is the description of the shape of a bounding contour, such as a silhouette that outlines an object. Although recognition need require only partial segments of such contours, the internal canonical description, against which the image contour is compared, is very likely a closed ring. Our concept of most "objects" should lead us to expect such a closed contour. The description of closed, 2D contours thus is an important ingredient of a system for object recognition. First we present such a scheme, described in more detail elsewhere [3, 4] and then show how the scheme leads to a hierarchical taxonomy of closed, 2D shapes.

2. THE REPRESENTATION

When we view shapes such as those in Fig. 1, we immediately see the ellipse and square as being "simpler" (in some psychological sense) than the lemniscate or epicycloid. Why? If we were to "measure" the simplicity of a shape contour by the degree of its polynomial equation, then the cardioid in the middle would have the simplest form, and the square the most complex, being the highest order polynomial. Clearly a polynomial representation seems quite inappropriate for our visual system, because it does not make explicit the meaningful properties of the shapes.

If we asked a child why the ellipse is "simpler" than the lemniscate, he would probably reply "because the latter has two parts, whereas the ellipse has only one." This simple observation is the basis for our representation for shapes: namely a shape should be described in terms of its natural "parts." Fortunately, the rule for finding "parts" is conceptually simple, for when 3D entities are joined to create complex objects, then concavities almost always are created at the join, as indicated by the small arrows in Fig. 2.

This regularity of natural objects follows a principle of transversality treated more fully elsewhere [5]. In the silhouette, these concavities appear as cusps, or as places

*D. D. Hoffman is now at the University of California, Irvine.

FIG. 1. An ellipse, square, cardioid, lemniscate, and epicycloid. The cardioid has the simplest (lowest order) equation; the square is the highest order polynomial equation.

FIG. 2. Joining parts generally provides concavities in the resulting silhouette.

of maximum negative curvature. Natural parts thus lie between concave cusps. In Fig. 1, the rule specifies that the ellipse and the square have no parts, whereas the lemniscate has two and the epicycloid has three. (The cardioid can not be broken simply into two parts, hence must be "simpler" than the two figures on its right.) Our first rule for representing (2D) shapes is thus as follows:

> *Segment a curve at concave cusps* (*or minima of negative curvature*) *in order to break the shape into its "parts."*

3. PART DESCRIPTORS: CODONS

Having now broken a curve into "parts" our next task is to describe the part. Again, we wish that our description capture some natural property of shapes, rather than an arbitrary mathematical formula, such as a polynomial equation. For example, at some stage in our representation, we would like to know whether it is round or polygonal. But even before such descriptors, is there a still simpler, more abstract, representation? Perhaps first we should represent the "sides" of the part, or its "top." As a step in this direction, we propose a very primitive representation based upon the singular points of curvature, namely the maxima, minima, and zeroes of curvature along the curve. An important property of these descriptors is that their ordinal relations remain invariant under translations, rotations, and dilations. Thus, regardless of the 3D orientation and size of a part to its whole, a

FIG. 3. Minima of curvature are indicated in slashes. Arrows indicate direction of traversal of curve. "Figure" is taken to be to the left of the direction of traversal.

relation between these descriptors is preserved in the 2D image. This property follows because the inflection of a 3D curve is preserved under projection, guaranteeing that at least the ordinal relations between minima, maxima, and zeroes of curvature will be preserved under projection. Our scheme thus provides a very primitive representation for a part, simply in terms of the ordinal relations of the extrema of curvature. This approach yields six different basic primitive shapes, or codons (see Fig. 4).

In order to define the codon types, it is first necessary to define maxima and minima of curvature. These definitions require that a convention be adopted for the sign of curvature. Consider Fig. 3. There are two directions along which the profile of the face may be traversed. In the upward direction (left) the minima of curvature (slashes) correspond to the points where the curve rotates at the greatest rate in the clockwise direction. If the same curve is traversed in the opposite direction, however, then the maxima and minima reverse. Our convention thus places "figure" to the left of the direction of traversal. When the figure is on the left, then the profile indeed looks like a face because the minima of curvature divide the curve into the natural parts—namely forehead, nose, mouth, and chin. (Note that the opposite view yields the "vase" of Rubin's famous figure–ground illusion observed as early as 1819 by Turton [14].) Thus, knowing which side is the figure determines the choice of orientation on a curve, or, conversely, choosing an orientation determines which side is the figure by convention. Minima are then typically associated with the concavities of the figure, whereas maxima are convexities.

To define our basic primitive codons, we first note that all curve segments lying between minima of curvature must have zero, one, or two points of zero curvature. If there are no zeroes (i.e., inflections), then the segment is designated as a type 0 codon (see Fig. 4). Those with two zeroes are called type 2 codons. If a segment has exactly one zero, then the zero may be encountered either before (type 1^-) or after (type 1^+) reaching the maximum point of the segment during traversal in the chosen orientation.

The type 0 codons may be further subdivided into 0^+, 0^- and (∞) to yield six basic codon types. Consider Fig. 3 once again. Note that as the ellipse is traversed in different directions, the minima of curvature change as expected. In the lower ellipse, which corresponds to a "hole" with figure outside, the minima have negative curvature, because the direction of rotation is clockwise. (Thus, the slashes suggest a part boundary by our rule, which will be repaired later when we discuss "holes.") In the upper ellipse, however, the minima have positive curvature (the rotation is always counterclockwise). Thus, the type 0 codon can be subdivided into 0^+ and 0^- with the superscript indicating the sign of curvature. Note that the 0^- codon can constitute a part boundary, whereas the type 0^+ codon must appear only as a shape

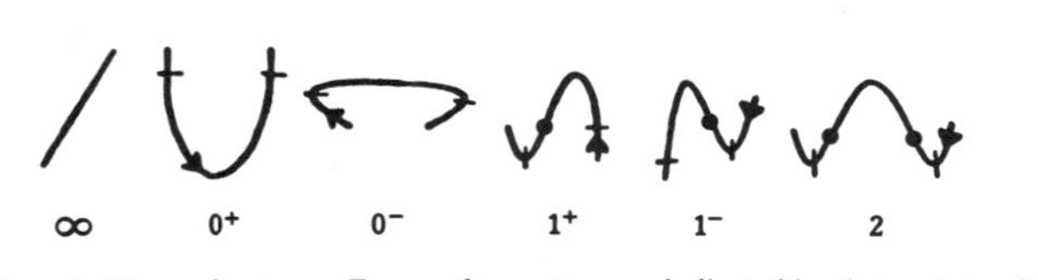

FIG. 4. The primitive codon types. Zeroes of curvature are indicated by dots, minima by slashes. The straight line (∞) is a degenerate case included for completeness, although it is not treated in the text.

descriptor. Finally, the type ∞ codon simply is the degenerate case of a straight line that has an ∞ of zeroes.

4. CONSTRAINTS ON SMOOTH CODON STRINGS

Not all sequences of codons are possible if the curve is smooth. Referring to Fig. 4 once again, note that a 1^- can not follow a 1^- codon unless a cusp is allowed. Similarly, a 1^+ can not follow a 1^+, because if such a join is attempted either a cusp will be created or, if the curve is indeed smooth, the 1^+ codon would have to be transformed into a type 2. To specify all legal smooth codon strings, we will first enumerate all pairs, and then show what pair substitutions are legal for one element in a sequence of pairs, thereby creating all possible triples.

Define the "tail" of a codon as the region about the first minima encountered when traversing the curve. The "head" of the codon is the subsequent minima. A smooth string of two codons is then allowable only if the head of the first codon has the same sign of curvature as the tail of the second codon in the string. Table 1a shows the sign of curvature for each codon type (excluding the degenerate type ∞). Table 1b is constructed simply by multiplying the sign of curvature of the "head" of the first codon (the left-most column) by the "tail" of the second (given in the second row). If the signs agree, then a (+) is entered, indicating a legal smooth join, otherwise the (−) product is an illegal smooth join. Thus, for these five codons, there are 13 legal joins out of a possible 25 combinations.

To enumerate the possible codon triples for a smooth contour, we now require that the curvature of both the head and tail of a middle codon match the tail of its successor or the head of its predecessor in the string. Table 2 provides the signs of the heads (and tails) of the legal pairs (left column) which must match the tail (or head) of the third codon in the string. For each column under the third codon, the legal triplets are indicated by a (+). If two pluses appear in brackets, then the third codon can either precede or follow the pair. Consider first the case where the third codon follows the pair (these are given in the columns headed "tail"). There are 34 legal smooth triplets of this type. Symmetry arguments yield a similar number of triplets when the third codon precedes the pairs (these are given in the columns headed "head"). Thus, there are only 34 legal codon triplets out of a possible $5^3 = 125$.

TABLE 1

Codon Signatures (a) and Legal Smooth Codon Pairs (b)

a — CODON SIGNATURES

CODON	TAIL	HEAD
0^-	−	−
0^+	+	+
1^-	−	+
1^+	+	−
2	−	−

b — LEGAL CODON PAIRS

1st CODON (head)	2nd CODON (tail): $0^-(-)$	$0^+(+)$	$1^-(-)$	$1^+(+)$	$2(-)$
$0^-(-)$	+	−	+	−	+
$0^+(+)$	−	+	−	+	−
$1^-(+)$	−	+	−	+	−
$1^+(-)$	+	−	+	−	+
$2\ (-)$	+	−	+	−	+

TABLE 2

Legal Smooth Codon Triplets

LEGAL CODON PAIRS			THIRD CODON									
			0^-		0^+		1^-		1^{+-}		2	
			−	−	+	+	−	+	+	−	−	−
	TAIL	HEAD	TAIL	HEAD	TAIL	HEAD	TAIL	HEAD	TAIL	HEAD	TAIL	HEAD
0^-0^-	−	−	[+	+]			+			+	[+	+]
0^-1^-	−	+		+	+				[+	+]		+
0^-2	−	−	[+	+]			+			+	[+	+]
0^+0^+	+	+			[+	+]		+	+			
0^+1^+	+	−	+			+	[+	+]			+	
1^-0^+	−	+		+	+				[+	+]		+
1^-1^+	−	−	[+	+]			+			+	[+	+]
1^+0^-	+	−	+			+	[+	+]			+	
1^+1^-	+	+			[+	+]		+	+			
1^+2	+	−	+			+	[+	+]			+	
$2\ 0^-$	−	−	[+	+]			+			+	[+	+]
$2\ 1^-$	−	+		+	+				[+	+]		+
$2\ 2$	−	−	[+	+]			+			+	[+	+]
NUMBER OF LEGAL PAIR SUBSTITUTIONS			5		2		3		3		5	
NUMBER OF PAIR SUBSTITUTIONS			9		4		6		6		9	

Note. The third codon can either follow or precede the pair. A (+) indicates a proper join. Because of symmetry, there are an equal number of total pluses in the head and tail columns.

Of the 34 possible triplets, there are 18 cases where the same codon can be attached to either end of the codon pair, indicated by [+ +]. This subset is particularly useful for establishing legal smooth codon strings of order higher than three. For example, consider a codon sequence C_{j-1}, C_j, C_{j+1}. We now desire to expand the sequence. This can be done simply by replacing C_j by C_iC_k, where C_iC_k is one of the pairs that will accept C_j at either end. To extend the string, we thus have the following rewrite rule:

> *Any individual codon in a smooth string may be replaced by any pair yielding a* [+ +] *for the (third) codon in Table 2.*

Thus, an 0^- codon may be replaced by any one of the following pairs: $0^-0^-, 0^-2, 1^-1^+, 20^-, 22$, etc.

To calculate the number of possible quadruplets, we can determine from Table 2 how many times any given codon type appears in the middle portion of the string.

TABLE 3

A Comparison Showing How the Number of Possible Strings of Codon Elements is Reduced as First Smoothness (Open Strings) and Then Closure Are Imposed as Constraints on a Curve

NUMBER OF CODONS	NUMBER OF:		
IN STRING	COMBINATIONS	OPEN STRINGS	CLOSED STRINGS
1	5	5	(2)
2	25	13	3
3	125	34	5
4	625	89	9
5	3,125	233	17
6	15,625	610	33
7	78,125	1,597	65
8	390,625	4,181	129
9	1,953,125	10,946	257
10	9,765,625	28,657	513

Then we can multiply this number of occurrences by the number of possible pair substitutions. For example, the type 0^- codon appears as the middle codon in rows 1, 8, and 11. In each row, the only legal third codons whose tail curvature matches that of the head of 0^- are 0^-, 1^- and 2 (see Table 1). Thus the O^- codon appears as the middle codon in 9 of all the possible triplets. Similarly we find the following number of occurrences of the other codon types in the middle position of the string:

$$0^- = 9; 0^+ = 4; 1^- = 6; 1^+ = 6; 2 = 9. \Sigma = 34$$

Thus the total number of possible smooth codon quadruples will simply be the sum of each of these numbers times the number of possible pair substitutions for each type (next to last row of Table 2), less any duplicate strings. The total substitutions are 134, but the 0^- and 2's duplicate each other, reducing the total quadruples by 45. The answer is the difference of 89 out of a possible 625. Table 3 shows how the number of possible open strings increases with the number of codon elements. In general there will be less than $5 \cdot 7^{(N-1)/2}$ possible smooth strings of N codons compared with 5^N possible (see Appendix II).

5. CLOSED CODONS

Because most objects have closed bounding contours, matching to closed codon sequences is of greater interest for shape recognition than representing open strings. Clearly this constraint on the codon sequence will further reduce the number of allowable smooth shapes. Indeed, this constraint is so powerful that all closed shapes containing up to four codons will be enumerated shortly.

First, let us examine the generating rules. Closed codon pairs can be noted simply by inspecting Table 2. Here, the signs of the heads and tails of the pairs in the first column must agree. The only cases are 0^-0^-, 0^+0^+, 0^-2, 1^-1^+, and 22. These shapes are depicted in Fig. 5, with figure indicated by cross-hatching. Note that there are only three basic outlines, if figure and ground are ignored. Later, we will address

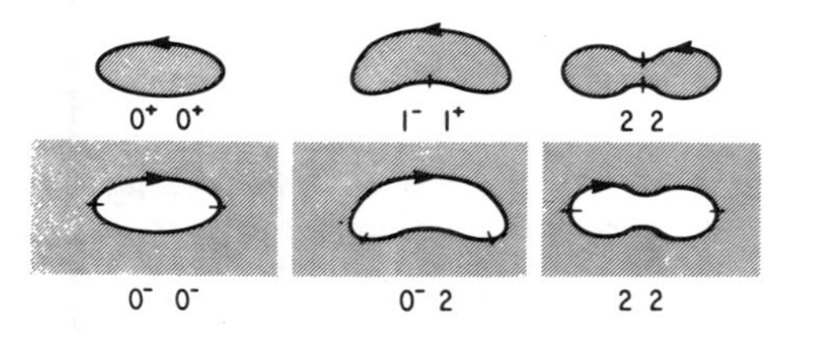

FIG. 5. Legal smooth, closed codon pairs. Figure is indicated by cross hatching. Part boundaries are noted by the slashes.

the problem of indexing identical codon descriptors that have different figure–ground relations (e.g., the 22 pair) and also the observation that the part boundaries (slashes) for the "holes" do not seem appropriate.

From these five legal codon pairs, we can now easily generate the legal closed triples. Simply consider the pair as a string of three elements and then replace each "middle" element by a pair according to Table 2. Thus, the closed pair 1^-1^+ may be rewritten as $1^-1^+1^-$ (or $1^+1^-1^+$) and the 1^+ (or 1^-) can then be replaced by 0^-1^-, 1^-0^+, 21^- (or $0^+1^+, 1^+0^-, 1^+2$). Such substitutions yield a total of 10 different codon triplets, or only 5 different outlines out of a possible 125 if figure and ground are ignored. These shapes are shown in Fig. 6 with their codon labels. Figure 7 shows the result of applying the same rewrite rules to the triples to enumerate all possible codon quadruples. Here there are only 9 outlines out of a total possible combination of 5^4 or 625 sequences. Appendix III shows that the upper bound on the number of closed smooth codons is $2^{N-1} + 1$, where N is the number of codons in the ring. The compression is thus about $2^{N-1}/5^N$, or over 10^4 for a 10-element ring. The reduction comes in part from a propagation of constraints through the closed string, very analogous to the constraint propagation used by Waltz [15] to solve for "blocks-world" shapes using constraints on legal trihedral joins.

6. MIRROR REVERSAL, HOLES, AND FIGURE–GROUND

In Fig. 5, three pairs of codon shapes are possible for a two element ring. For each pair, the outline is the same, but the figure–ground relation is reversed. The situation is similar to a lock–key arrangement, where the shapes in the upper row fit snugly into their "hole" complements in the lower row.

Two problems arise in representing these mirror shapes: First, a lock–key pair may have the same codon description, such as "22" on the right. This is easily rectified by adding an extra index. The second problem is perceptual. For each of

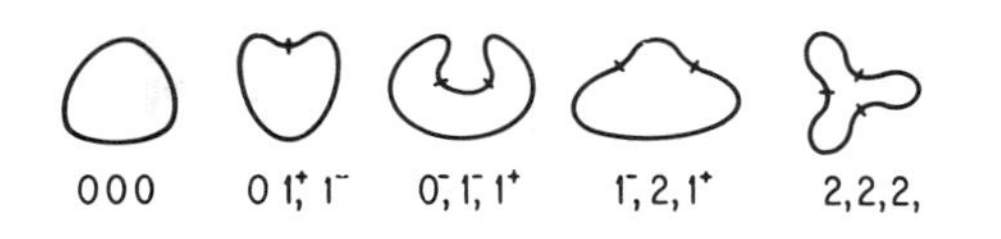

FIG. 6. Legal smooth, closed codon triplets.

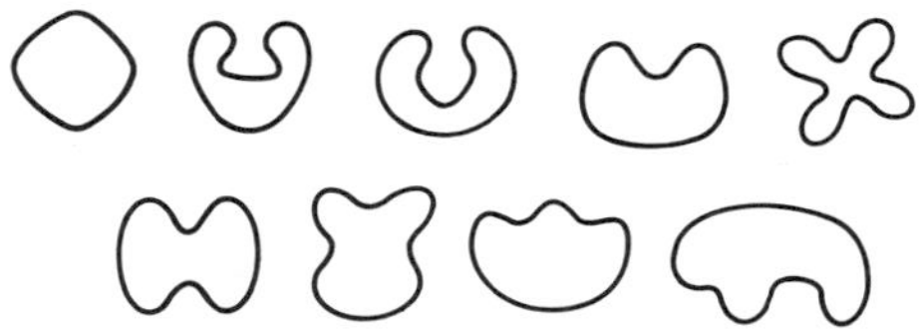

FIG. 7. Legal smooth, closed codon quadruples.

the shapes in the lower row, although "figure" is outside, we strongly prefer the "hole" as the figure. Why?

Some insight into why "holes" are preferred as figure can be obtained by noting that all the shapes in the lower row consist of two parts by our rule. (The slashes indicate the points of maximum negative curvature.) On the other hand, the "ellipse" and "peanut" in the upper row are single entities without parts. Certainly a single entity is "simpler" in some very basic sense than one with parts. Thus, it makes sense to describe the "hole" as the complementary figure if that complement has fewer parts. Note that for the 22 dumbbell shapes, the preference between the hole and its "key" is less strong. In fact, it is not too difficult to regard each "bump" in the hole as a part of the hole, whereas it is almost impossible to view the 0^-0^- elliptical hole as having two parts. Our rule for representing "holes" is thus:

> *Represent a "hole" by its figure–ground complement if that complement has fewer "parts."*

Note that there are many linguistic examples where this rule has been applied. For example, "key-hole," "screw-hole," "oval window," etc., are all descriptions of a hole in terms of the figure–ground complement. (Appendix I shows how a figure–ground complement can be computed easily using a binary representation.)

7. INDEX DEVELOPMENT

The shapes of closed codon rings shown in Figs. 5–7 have all been drawn to preserve symmetry. Furthermore the axes of elongation tend to be straight, and the "parts" are neither too "thin" nor too "thick" for most peoples' taste. What are the rules that underly the canonical representation of these primitive shapes? What regularities of the world are captured here?

Clearly any time a rule is applied to put a codon ring in canonical form, then we have an implicit shape index which is being defaulted. An interesting extension of the abstract codon description is thus to develop indices that are meaningful. The fact that the codon hierarchy is small provides an opportunity for a rational development of such indices—at least for a start. (A more complete and useful set should relate each index to a desired real-world property.)

To give the flavor of this approach, consider the first three primitive codon shapes in Fig. 5: the "ellipse" (= 00), the "peanut" ($= 1^-1^+$), and the "dumbbell" (= 22). What useful properties can now be assigned to an ellipse outline? We have already mentioned the need to specify figure–ground. What about the orientation of the ellipse, or its eccentricity, or even perhaps its size? These four parameters will completely specify the elliptical shape relative to a reference frame. However, if we encounter an ovoidal shape having the same 00 codon description, then still another index may be required. For our single, most primitive closed shape we thus have already the following possible indices:

FIGURE–GROUND	0, 1
ORIENTATION (OF AXIS)	ϕ
ECCENTRICITY (ASPECT RATIO)	ρ
SIZE	Σ
SKEW	ν

As we proceed to the next more complex shape, the "peanut," the axis is now curved, and the left and right portions need not have identical size. Two more parameters thus must be added:

RELATION LEFT AXIS TO RIGHT AXIS	α
RELATIVE SIZE, LEFT TO RIGHT "PART"	σ_{lr}

These indices suffice for the dumbbell.

In a similar vein we may proceed up the closed codon hierarchy, adding additional indices. This procedure automatically provides an ordinal order to the indices (whether this order is perceptually appropriate is a separate issue!). Note that left–right "handedness" does not appear until we encounter four element codon rings. For codon rings greater than four or five, the complexity of the shape undoubtedly prohibits practical use. In sum, the codon hierarchy can thus be used to develop an ordered set of indices to the more metrical properties of shapes.[1]

8. MAPPING 3D ⇔ 2D

Codons are descriptors for 2D plane curves, and hence of necessity are an image-based representation. In Marr's [9, 10] terminology, they are part of the data structure of a primal sketch. An important aspect of the motivation for the codon description comes from the nature of the 3D world, however, namely the rule for locating part boundaries at maxima of negative curvature. This rule for partitioning a curve captures the concavity regularity created when two 3D parts are joined (see Fig. 2), as seen in the 2D image. Thus, the presence of a concavity in silhouette is used to infer a part boundary in the 3D world. (See [5] for a more rigorous treatment of this inference.) Can other inferences about the properties of 3D objects also be made from the codon descriptors?

To explore the kinds of inferences possible about 3D shape from 2D contours, we will consider some of the canonical, primitive shapes generated by codons shown in Figs. 5–7. Our aim is not to exhaust all possible inferences, but rather to indicate a profitable direction for future study.

There are two kinds of inferences to consider: (1) those that lead to the acceptance of a particular 3D shape and (2) those that reject a possible 3D shape [12, 11]. Consider the first three primitive outlines given in Fig 5: the ellipse, the peanut, and the dumbbell. An example of the rejection strategy is that the 2D peanut contour can not arise from a surface of revolution about a straight axis, for if it did, then the concavity in the outline would be eliminated. Similarly, the dumbbell can not be the projection of a 3D surface of revolution about a vertical axis, although it could be a 3D shape created by revolving the outline about the horizontal axis (i.e., a dumbbell in 3D).

An example of the accept strategy is the inference that the elliptical outline represents a 3D ellipsoid. But of course although it is true that a 3D ellipsoid will generate a 2D elliptical contour, so will any planar 2D ellipse or more awkwardly any shape whatsoever that has at least one elliptical cross section.

To infer the 3D shape from the 2D contour thus requires assumptions about (1) what the hidden 3D surface looks like, (2) whether the shape is 2D or 3D, and (3) whether the silhouette arises from a plane curve, etc. [13]. Clearly, then, the accept mode of inference is much more fragile than the rejection strategy. In one case the inference rests on assumptions, whereas in the other, possible assumptions are rejected.

What kinds of 3D properties then can be tested from 2D codons? We have already mentioned the surface of revolution constraint, which is a particularly popular basis for modelling shapes [2, 1, 8]. But still deeper insights into 3D shape can be obtained if the part boundaries are reinforced by spines and cusps which appear in the image [7]. For example, in Fig. 8 the silhouette of all three figures is the same. However, their interpretation is quite different. Assuming general position, outline (A) is seen as planar, (B) as three dimensional, whereas (C) suggests a 2D fin on a 3D ball, otherwise it would be an impossible object [6]. It is clear that as the codon representation is developed, the internal contours must play an important role. Their presence may force (or exclude) a particular 3D interpretation. These differences in interpretation should be reflected in the codon description. For example, the description of (B) should include three separate codon strings rather than just one for the silhouette. Thus the silhouette (A) = ⟨2002⟩ whereas (B) is more correctly depicted as ⟨[2][000][2]⟩, with the two [2] being the "ears" of the head [000]. In this case the transformation from A to B is a simple restructuring of the string. Other cases with similar silhouettes will not be so simple. Yet, just as 3D shapes constrain the 2D codon descriptors and vice versa, so will there be constraints on the transformation of strings of type B into those of type A (and vice versa). Here is an exciting but difficult area for future study.

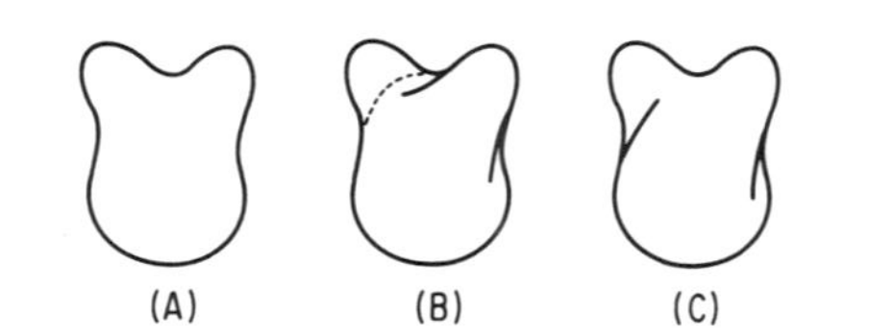

FIG. 8. Internal contours may drastically alter the 3D interpretation as well as the codon description.

[1] It is expected that a mapping between our codon-based and axial-based (or "grassfire") representations for shapes can be made with a suitable list of indexed parameters. We see the advantage of the codon scheme being that crude part descriptions appear at the top level, allowing immediate access.

9. SUMMARY

Codons are simple primitives for describing planar curves. They thus are principally image-based descriptors. Yet they have the power to capture important information about the 3D world, such as making part boundaries explicit. The codon description is highly redundant (useful for error-correction). This redundancy can be viewed as a constraint on the number of possible codon strings. For smooth closed strings that represent the bounding contour (silhouette) of many smooth 3D objects, the constraints are so strong that sequences containing 6 elements yield only 33 generic shapes as compared with a possible number of 15,625 combinations. An intriguing and important question for image understanding is to explore the constraints on the possible 3D configurations that can project into these 33 generic 2D shapes.

APPENDIX I: A BINARY MAPPING FOR CODON STRINGS[2]

1. Mapping Rule

Five basic codon types would normally require at least three bits for a binary encoding. However, there are sufficient constraints on codon joins that the 0^- and 0^+ codons can be distinguished, provided at least one member of the string is not a type 0 codon. By inspecting Tables 1 or 2, we see that if a type 0 codon follows a 1^- codon, then it must be an 0^+, whereas if the type 0 follows a 1^+ or 2 type codon, then it must be an 0^-. Similarly if a type 0 precedes a 1^- or 2, it must be type 0^-, but if it precedes a 1^+ codon, then it is type 0^+. Because adjacent type 0 codons must have the same sign, the designation of the 0 codon type will be completely specified by its neighbors. This redundancy is also reflected in the number of legal codon pairs, which is thirteen and can be mapped into 4 bits or 2 bits per codon.

Our mapping scheme utilizes the constraint that between every minima there is at least one maxima (provided the positive minima of 0^+ are noted). Thus, given the location of a minima in a binary string we only need to encode the position and number of inflections in relation to the maxima. (This was the basis for the codon definitions.) Let "1" represent an inflection and "ϕ" represent no inflection. Then the mapping rule will be as follows:

CODON TYPE		BINARY CODE
0^-	→	$\phi\phi$
1^+	→	$\phi 1$
1^-	→	1ϕ
2	→	11

Note that in this mapping, the position of both the maximum and minimum is implicit. Namely the minima lie at the beginning and end of a binary pair whereas the maximum lies between the pair. This property will be seen to be useful in depicting figure-ground reversals where maxima and minima exchange places. The exchange can be brought about simply by phase shifting the starting point on a string by one element.

2. Mirror Transform

A mirror transform in a codon string occurs most frequently when a shape is symmetric. The rule for effecting a mirror transform is to read the string backwards and change the signs of the 1^+ and 1^- codons (see [4]). The binary rule is simpler: "Read the string backwards."

EXAMPLE.

ORIGINAL STRING	1^+	2	1^-	1^+	0	1^-
BINARY MAPPING	01	11	10	01	00	10
REVERSAL BINARY	01	00	10	01	11	10
MIRROR STRING	1^+	0	1^-	1^+	2	1^-

Note that the mirror codon is simply the original read backwards with the signs changed.

3. Lock–Key or Figure–Ground Transform

The basic idea underlying a figure–ground reversal is that the maxima and minima must be exchanged. This can be accomplished by rotating the binary string by one element. However, because a figure–ground exchange also entails a reversal in the direction of traversing the curve, the order of the binary string must also be reversed. Hence the figure–ground rule is: "Rotate the binary string by one element and read the string backwards."

EXAMPLE.

ORIGINAL STRING (FIGURE)	1 +	2	1^-	1^+	0	1^-
BINARY MAPPING	01	11	10	01	00	10
ROTATE ONE ELEMENT	00	11	11	00	10	01
REVERSED BINARY	10	01	00	11	11	00
COMPLEMENTARY STRING (GROUND)	1^-	1^+	0	2	2	0

APPENDIX II: NUMBERS OF POSSIBLE OPEN, SMOOTH CODON STRINGS

1. Strategy of Proof

An upper bound on the number of legal codon strings with smooth joins, but which are not closed, will be obtained by induction. First we will determine the number of possible strings of length $N + 1$ given the number of strings of length N. This relation will then lead to an obvious sequential pattern for even and odd Ns of low numbers. The sequence will be bounded by $5 \cdot 7^{(N-1)/2}$.

2. Two Types of Strings

Referring to Table 1, we note that a codon string may be extended in only two ways: (a) by adding either a 0^-, 1^-, or 2 or (b) by adding either a 0^+ or 1^+. If the string ends in a 0^-, 1^+, or 2, then there are three choices, namely (a), for the

[2] This mapping and its properties were first noted by Chris Fitch and Steve Schad in WR's class, "Natural Computation" in the Fall of 1983.

addition, but if the string ends in a 0^+ or 1^-, then there are only two choices, namely (b). Let us designate the 0^-, 1^+, and 2 codons as type A and the 0^+, 1^- as type B. (A and B simply specify whether the head of the codon has positive or negative curvature, respectively.) Then if there are A_N strings of length N ending in 0^-, 1^+, or 2, there will be $3A_N$ possible strings of length $N+1$ that are constructed from these A_N codons. Similarly there will be $2B_N$ possible strings of length $N+1$ constructed from these B_N codons. The total number $\Sigma(N+1)$ of strings of length $N+1$ will then be

$$\Sigma(N+1) = 3A_N + 2B_N. \tag{1}$$

Let us now consider a string of length N ending in a type A codon $(0^-, 1^-, 2)$. How did each of these codon types arise from the string of length $N-1$? Again referring to Table 1, we see that the next-to-last codon must have been either an 0^-, 1^+, or 2. Thus we have two instances of type A (namely $0^-, 2$) and one instance of type B (namely 1^+). The number of A_N codon strings is then simply

$$A_N = 2A_{N-1} + B_{N-1}. \tag{2}$$

Similarly, we find that the last type B codon in a string (namely $0^+, 1^-$) must be preceded by either an 0^+, which is type B, or a 1^-, which is type A. Hence we have

$$B_N = A_{N-1} + B_{N-1} \tag{3}$$

We now can solve for A_N and B_N in terms of Σ. But first note that adding (2) and (3) gives us the relation

$$A_N + B_N = 3A_{N-1} + 2B_{N-1} = \Sigma(N) \tag{4a}$$

or

$$B_N = \Sigma(N) - A_N. \tag{4b}$$

Now if the right-hand terms of (4b) are replaced by the appropriate forms of Eqs. (1) and (2), we find that

$$B_N = \Sigma(N-1). \tag{5a}$$

Thus from (4a),

$$A_N = \Sigma(N) - \Sigma(N-1). \tag{5b}$$

Finally, by substitution in (1) and changing the index by one, we obtain

$$\Sigma(N) = 3\Sigma(N-1) - \Sigma(N-2). \tag{6}$$

These totals for N are given in column 3 of Table 3.

3. Upper Bound

An upper bound on these totals can be set by making the negative right-hand term of Eq. (6) smaller and then approximating the sums by the positive term. For example, by algebraic manipulation, the negative term can be reduced to $\Sigma(N-4)$, giving

$$\Sigma(N) = 7 \cdot \Sigma(N-2) - \Sigma(N-4). \tag{7}$$

An upper bound on the possible number of open codon strings is thus $7 \cdot \Sigma(N-2)$. We then observe the following pattern:

		CORRECT
$\Sigma(3) = 7 \cdot 5$	$= 35$	34
$\Sigma(4) = 7 \cdot 13$	$= 91$	89
$\Sigma(5) = 7 \cdot 7 \cdot 5$	$= 245$	233
$\Sigma(6) = 7 \cdot 7 \cdot 13$	$= 637$	610
⋮		

Note that for all odd sums the factor other than 7 is 5, whereas for the even sums, the factor is 13. Thus if N is odd, $\Sigma_{\text{odd}} < 5 \cdot 7^{(N-1)/2}$ whereas if N is even, $\Sigma_{\text{even}} < 13 \cdot 7^{(N-2)/2}$. Now note that 13 may be approximated by $5 \cdot 7^{1/2}$, or more especially $13 < 5 \cdot 7^{1/2}$. Thus the factor 13 may be replaced to yield the single equation for the upper bound on Σ,

$$\Sigma(N) < 5 \cdot 7^{(N-1)/2}. \tag{8}$$

APPENDIX III: Number of Smooth Closed Codon Strings

1. Strategy of Proof

We will use the binary representation presented in Appendix I. Furthermore, we will count only the basic closed shapes without regard to whether figure or ground is specified. The counting proceeds by constructing a binomial tree, subject to three constraining rules:

(i) The sum of the 1s must be even.

(ii) There must be an even number of adjacent binary ϕs in any sequence.

(iii) Only the 1111..., string can end in a 1.

The first rule says that the total number of inflections in a closed string must be even (or zero). If there were an odd number of inflections then the string cannot close on itself because the sign of curvature at the beginning and end of the string would be different.

The second rule follows indirectly from the first. A type 2 codon is represented by "11" because it has two inflections. Thus taking all "2"s from the string will still leave an even number of "1"s. For each remaining "1" there will be an equal number of ϕs. This number will be even. The total number of ϕs in the binary string remains even because all other ϕs in the binary string will come from the type 0 codon, which is represented by a pair of binary ϕs. The requirement that there must be an even number of adjacent ϕs follows from the legal joins shown in Table 1. For example, a $1\phi11$ sequence, corresponding to a 1^-2 string is illegal. The third digit must be a ϕ, changing the string to a 1^-1^+, etc.

TABLE 4
Tree Structure of Binary Codon Strings of Length N

STRING POSTION (or LEVEL, N)	BINARY TREE	LEGAL POSSIBILITIES
0	(1 / 1)	—
1	ϕ ϕ - - - - 1 1	—
2	ϕ ϕ 1 1 ϕ ϕ. 1. 1	3
3	ϕ ϕ 1 1 ϕ ϕ 1 1 ϕ. ϕ 1 1. ϕ ϕ 1 1	5
4	ϕ1 ϕ1 ϕ1 ϕ1 ϕ1 ϕ1 ϕ1 ϕ1 ϕ1 ϕ1 ϕ. 1 ϕ. 1 ϕ1 ϕ1 ϕ1 ϕ1 (w) (y) (z) (w)	9

The third rule simply forces the end position in the string to be a zero to prevent duplication of strings. The exception is a codon string made up only of type "2" codons, which must be included in the count. Otherwise the string is rotated to remove a final binary "1".

2. The Construction

We wish to count all possibilities for the binary ϕ and 1 in any position of the string of codon length N. This will be accomplished by constructing a tree, with two binary positions added at each new level, N, of the tree. The binary positions, of course, represent the five basic codons plus the constraint that allows both the 0^- and 0^+ codons to be represented by the binary pair ϕϕ (see Appendix I).

Table 4 shows the construction of the binary tree. The tree is initiated by the pair of 1s in the first row at level 0. These pairs of 1s then branch to a pair of 0s and 1s at level 1, corresponding to a single codon.

At this level the possible legal binary pairs according to our rules would be ϕϕ and 11, corresponding to the type 0 and 2 codons. However neither of these single codons can close on itself without introducing a maxima. Hence the first legal *closed* binary sequences begin at level 2.

Reading down from the top through level 2 (and ignoring the initializing 1s), we have only three possibilities: ϕϕϕϕ, 11ϕ, ϕ, and 1111 (see Fig. 5). The sequences ϕϕ11 and 1ϕϕ1 are illegal because a sequence cannot end in a "1" unless all members of the sequence are "1" (rule (iii)). Hence there are only three possible closed strings made of two codons.

Moving to the third level, we now need explore only those branchings (and cross-branchings) that end in ϕ. There are five legal strings obtained by directly moving down the branches without crossing over. These are ϕϕϕϕϕϕ, ϕϕ11ϕϕ, 11ϕϕϕϕ, 1111ϕϕ, and 111111. However, note that the second and third strings are simply a rotation of one another and hence are duplicates. Indeed, any right-branching sequence from a ϕϕ to a 11 on the left side of the tree must be duplicated on the right side of the tree, because a simple leftward rotation of the binary string can move the 11 pair into the first position. The initial pair 11 will now correspond to the 11 pair at level one on the right side of the tree, and the preceding pair of binary ϕϕs will have moved into the last position, corresponding to the ϕϕs at the highest level being considered. Hence all direct strings on the left side of the tree can be ignored in the direct string count, with the exception of the sequence ϕϕϕϕ..., consisting solely of binary ϕs. Thus, the number of legal "direct" strings constructed by moving down the tree without "crossovers" (dashed lines) will be

$$\text{Number of direct strings:} = 2 + 2^{N-2}. \tag{9}$$

3. Crossover Strings

The above count does not include "crossings" between branches similar to those indicated by the dashed lines. Given that a "crossing" is made at one level, we are then required by rule (ii) to cross back at the next or later level in such a manner that the number of adjacent ϕs is even, and such that the string ends in ϕ. Thus the first crossing 1ϕϕ1 is illegal because it ends in a "1." However this sequence can be extended at level 3 into a legal string, namely 1ϕϕ1ϕϕ. (Note that the other crossing 1ϕ11ϕϕ is illegal because there is a single "ϕ.") We thus see that there are no legal crossover strings of codon length two, and only one of codon length three, namely 1ϕϕ1ϕϕ.

Moving to level four, we may continue to use the 1ϕϕ1 crossover as a "header." Now a 11 at level three may be used to extend the string, for we have a pair of binary ϕs available at level four. This gives us the two crossover strings ending in columns "y" and "z," as indicated in the last row of Table 4. In addition, we may now also create a new crossover header, namely 1ϕϕϕϕ1, which becomes legal because of the binary ϕs available at level four. This new sequence, 1ϕϕϕϕ1ϕϕ ends in column "x." By considering crossovers on the left side of the tree we have thus added three more strings of codon length four.

Now consider the right-half of the tree. At level two, we may add another crossover at the "1" followed by a period. Crossing over here and back again at level three, we obtain the following sequence ending in column (w): 111ϕϕ1ϕϕ. But this sequence is the mirror transform of a previous string found by an earlier crossover to the left side of the tree, namely 1ϕϕ111ϕϕ. (The mirror transform is the same string read in reverse, and in this case rotated by two to place the two binary ϕs at the end.) It should be obvious that the symmetry of the crossover header, namely 1ϕ...ϕ1, allows one to read it backwards to get the same result. Hence all crossovers on the right half of the tree will be the "mirrors" of strings obtained on the left half.

We are now in a position to count the crossover strings. At each new level, we add one string of the form 1ϕϕ...ϕ1ϕϕ. This sequence doubles at each successive level, giving us 2^{N-3} possible strings from the first possible crossover, 2^{N-4} from the second, etc.:

$$\text{Number of crossover strings} := \sum_{J=3}^{N} 2^{N-J}. \tag{10}$$

4. Final Count

Adding the counts for the direct (9) and crossover (10) strings, we obtain

$$\text{Number of closed strings:} = 2 + 2^{N-2} + \sum_{J=3}^{N} 2^{N-J}. \tag{11}$$

By algebraic manipulation, it can be shown that the above reduces to

$$\text{Number of closed strings:} = 2^{N-1} + 1 \qquad (N \geq 2). \tag{12}$$

ACKNOWLEDGMENTS

This report describes research done at the Department of Psychology and the Artificial Intelligence Laboratory of the Massachusetts Institute of Technology. Support for this work is provided by AFOSR under a grant for Image Understanding, contract F49620-83-C-0135. William Gilson helped with the preparation of the manuscript. The comments of Aaron Bobick and John Rubin were appreciated. Chris Fitch and Steve Schad broke the combination that made Appendix III possible.

REFERENCES

1. G. Agin, *Representation and Description of Curved Objects*, Stanford A.I. Memo No. 173, Stanford Univ., 1974.
2. T. O. Binford, Unusual perception by computer, in *IEEE Conference on Systems and Control*, December, Miami, 1972.
3. D. Hoffman, *Representing Shapes for Visual Recognition*, Ph.D. thesis, MIT, Cambridge, Mass., 1983.
4. D. Hoffman and W. Richards, Representing smooth plane curves for visual recognition: Implications for Figure–Ground Reversal, in *Proceedings of the American Association for Artificial Intelligence*, pp. 5–8, 1982.
5. D. D. Hoffman and W. Richards, Parts of recognition, *Cognition*, in press; MIT A.I. Memo No. 714, 1984.
6. D. A. Huffman, Impossible objects as nonsense sentences, in *Machine Intelligence*, Vol. 6 (B. Meltzer and D. Michie, Eds.), Edinburgh Univ. Press, Edinburgh, 1971.
7. J. Koenderink and A. Van Doorn, The shape of smooth objects and the way contours end, *Perception*, **11** (1982), 129–137.
8. D. Marr, Analysis of occluding contour, *Proc. R. Soc. London Ser. B*, **197**, 1977, 441–475.
9. D. Marr, Early processing of visual information, *Philos. Trans. R. Soc. London*, **275**, 1976, 483–524.
10. D. Marr, *Vision: A Computational Investigation into the Human Representation and Processing of Visual Information*, Freeman, San Francisco, 1982.
11. W. Richards, J. Rubin and D. D. Hoffman, Equation counting and the interpretation of sensory data, *Perception*, **11**, 1983, 557–576.
12. J. M. Rubin and W. A. Richards, Color vision and image intensities: When are changes material? *Biol. Cybern.*, **45**, 1982, 215–226.
13. K. A. Stevens, Visual interpretation of surface contours, *Artif. Intell.*, **17**, 1983, 47–73.
14. W. Turton, *A Conchological Dictionary of the British Islands*, (frontispiece), printed for John Booth, London, (1819). [This early reference was kindly pointed out to us by J. F. W. McOmie.]
15. D. Waltz, Understanding line drawings of scenes with shadows, in *The Psychology of Computer Vision* (P. Winston, Ed.), McGraw–Hill, New York, 1975.

Chapter 7

Matching, Model Fitting, Deduction and Information Integration

An essential attribute of intelligent behavior is the ability to make effective decisions. In computer vision most decision problems are of the form: "Which model best fits or describes some subset of imaged data?" In this chapter we discuss a number of important variants of this problem. First, we look at the classification and image-matching problems—determining the best match of either one of a number of stored models, or some segment of one image, to a segment of a second image. In the image-matching case we assume that both images are different views of the same scene and that corresponding image regions depict the same scene content; either image segment can be considered the model and the other image the data set.

Second, we discuss the model-fitting problem. Here we are given an object model containing free parameters and are required to find instances of this model in a set of imaged data and to instantiate the model parameters for each detected instance. In this problem the free parameters can simply describe the location and orientation of the model (e.g., finding lines in an image). Often additional free parameters of the model will be used to describe shape, size, color, and so on.

The final decision-making problem we consider is that of making the "best" decision, such as choosing the best of a given set of classification alternatives based on evidence obtained from multiple sources—that is, the problem of multisource information integration. For example, frequently a number of distinct operators, such as edge operators, can be run over a given image. Each operator is nominally intended to produce the same interpretation of the image, but the actual outputs are different. How should these outputs be combined in realistic situations where we do not have a quantitative model describing the relationship between the operators?

In summary, the two critical problems we discuss here are:

1. How can we determine when a data set contains instances of one or more stored models?
2. How can we combine information from multiple sources (for example, to make a classification decision) when the relationship between the sources is either qualitative or unknown?

7.1 The Classification and Image-Matching Problems

In this set of problems we wish to find the best "explanation" of our data in terms of a fixed vocabulary of models. These models can range from symbolically encoded prototypes of a number of different classes to a single patch of intensity values extracted from an image.

In general, three attributes characterize a matching or classification technique: the selection of the features and relationships to be matched, the control strategy that specifies how to search for potential matches, and the criteria for evaluating and selecting a best match.

The matching criteria are formal statements of how to measure "similarity" between the data and the model. In the simplest forms of image matching (e.g., matching pixel intensities), the geometric relationships between the features are ignored or dealt with implicitly, and the matching process is one of comparing feature vectors. In more complex matching problems, where the relationships between features are important distinguishing criteria, we must determine similarity between graphs rather than between feature vectors. Comparison of feature vectors is best modeled in terms of statistical decision theory; graph matching is usually treated as a problem in combinatorial optimization.

In correlation matching, one of the most common forms of image matching, the features are fixed-size overlapping patches centered around each pixel in the image. The pixel intensity values in these often square or rectangular-shaped patches are assembled into a linear string (by concatenating successive rows of the image patch) to form the feature vector, and each such vector is compared to all the similarly formed vectors of a second image to find a best match. Thus, a correspondence is set up between the individual pixels of the two images as required for such applications as stereo compilation or motion analysis (see Chapter 1). Fischler [Fischler 71] shows that if the two images differ only due to horizontal and vertical displacement and the addition of independent random noise to the intensity value of each pixel, then "unnormalized" cross-correlation is an optimal decision procedure (in the sense of minimizing error probability). When other differences exist between the two images—for example, amplitude bias or gain differences—then various forms of intensity normalization are required. Image size or rotation differences limit the maximum size of the correlation "patch"—that is, the maximum number of image measurements that can be used in making a local match decision. Some interesting variations of correlation-based image matching can be found in [Anuta 70], [Barnea 72], and [Kuglin 75].

For most applications, where the model is not a second image acquired under almost identical viewing conditions as the given (first) image, intensity patches extracted from the raw image are not reliable features. Changes in viewing conditions allow the introduction of occlusions, perspective distortion, movement of highlights, and changes in reflected energy. If the images are acquired at different times, physical changes may have occurred in the scene. For example, new or altered features such as roads, buildings, floods, or seasonal changes may be present, or objects, such as cars in a parking lot, may have moved. If the images were acquired by different cameras, their distortion, resolution, and spectral characteristics may be different. Thus, we want features invariant under the changes and distortions just described. Such "higher-level" features (e.g., specifically defined shape primitives, or physically defined features such as material or occlusion boundaries) are not densely distributed over the image. A different control strategy is required when searching for matches between sparse (rather than densely-distributed) features.

Kalvin et al. [Kalvin 85] describe a feature-based approach for selecting a relevant set of models from a large collection of models for the purpose of 2-D object classification based on shape matching. The authors note that if one must deal with occlusions and overlapping objects, then the indexing features should be based only on local shape (boundary) information, since such global attributes as area or center of gravity cannot be reliably measured. Given a set of local shape measurements that are invariant under rotation and translation, one extracts such measurements from each model and stores the name of the model in appropriate lists indexed by these measurement sets. (Kalvin suggests a way of obtaining such measurements from the Fourier coefficients, computed over small segments of the given object boundary, that describe turning angle as a function of arc length.) If a particular set of measurements characterizes k different locations on the boundary of a particular prototype, then the name of that prototype is entered k times on the corresponding measurement list. When an unknown object is to be classified, it is described in terms of the same set of characterizing local measurements as the prototypes. Each set of local measurements extracted from the unknown object causes the names of the prototypes on the corresponding measurement list to be accumulated on an identity list; if some set of measurements occurs p times on the unknown object, then the corresponding measurement list is is accumulated p times. Finally, the most frequently occurring model names on the identity list are chosen as the relevant candidates for identifying the unknown object. (There are many similarities between this technique and the generalized Hough transform described in Section 7.2.)

Fischler and Elschlager [Fischler 73] offered perhaps the first computationally feasible approach to the problems of relational image matching and object classification based on the then-novel concept of using a dynamic programming type of optimization technique on a stack of feature arrays extracted from and registered to the input image. The model in this case was a graph structure describing the relationship between the features of the reference object (or objects). The objective function to be optimized was the sum of a term that measured geometric distortion between feature placement in the image as compared to the model, and a term that measured how closely each individual feature detected in the image corresponded to that of the model. An intuitive description of this technique as applied to the problem of matching two images is that one of the images is recorded on a transparent sheet of rubber, which is then placed over the second image and stretched to provide a best fit. The quality of the fit is measured by how much stretching was required and by how well each feature on the rubber sheet overlays the corresponding feature on the image beneath it. When the process of feature detection is separated from that of finding the best configuration of features to match the model, it is more appropriate to consider the model as a set of feature templates connected by springs.

Another form of relational matching, discussed in Chapter 3 (i.e., maximal cliques, [Bolles 82] [Ambler 75]), involves finding in an image those clusters of image features having specified geometric interrelationships that, in a binary sense (i.e., the feature is present or not; the relationship is present or not) best match the ideal graphs of features and relationshps defined for each of a number of objects to be recognized.

If we are dealing with a problem in which the differences that exist between an image and a model are primarily due to projective distortion introduced by the imaging process, then a technique described by Barrow et al. [Barrow 77] can be used simultaneously to adjust an approximately correct camera model, project the object model onto the image plane, and find the required match between the projected object model and the image data.

Recently researchers have proposed a number of relational matching techniques based on a "scale space" representation (see Chapter 2) in which the input data are represented by a continuous hierarchy of approximations. Witkin [**Witkin 83a**] introduced a representation that displays the effects of scale on the inflections of a shape. First, the points of inflection (the "zero crossings") on the curve are found by convolving the curve with a 1-D Gaussian kernel and marking the locations where the second derivative undergoes a sign change. Zero crossings for various scales are obtained by varying the values of σ in the Gaussian kernel, and the zero crossings are plotted in a "scale space" whose coordinates are σ of the Gaussian kernel and arc length along the curve.

A series of papers published in 1986 discuss the theoretical aspects of scale space. Yuille and Poggio [Yuille 86] and Babaud et al. [Babaud 86] prove that the Gaussian is the only smoothing function for which the amplitude of the local maxima of a 1-D signal always increase, and the local minima decrease, as the bandwidth of the filter is increased. This property means that for any well-behaved signal, the peaks and valleys become monotonically more pronounced as the bandwidth increases, and that zero-crossing contours observed at low bandwidth cannot in general vanish with increasing bandwidth. Because the scale space approach is based on tracking the inflection points of a contour, it is not suitable for shapes having long straight lines or for smooth convex curves. [**Hummel 86**] discusses the *completeness* of the scale space representation and shows that if zero-crossings data are supplemented with knowledge of the gradient data at the zero crossings, then reconstruction of at least some of the data is theoretically possible, but not practical, due to numerical instabilities.

Several papers have described the use of scale space for either characterizing or matching shapes. Asada and Brady [Asada 86] use the scale space concept for shape analysis based on describing discontinuities in terms of a limited number of primitives, such as corner, smooth join, end, and so on. The curve is filtered at a variety of scales and the locations of local positive and negative extrema in scale space are used to produce a "tree" that is "parsed" to determine the location and identity of the primitives. This composite description is called the "curvature primal sketch."

Mokhtarian and Mackworth [Mokhtarian 86] use scale space to carry out a matching process that is invariant under rotation, uniform scaling, and translation. They make no assumptions about shape primitives; the scale space plot of the test shape is matched against the corresponding plot of the model. Matching is based on a uniform cost algorithm [Nilsson 71], which is a special case of the A* algorithm [Nilsson 80]. The technique is applied to register a LANDSAT satellite image to a map.

Witkin, Terzopoulus, and Kass [**Witkin 86**] apply the scale space concept to the general signal-matching problem. They suggest that this problem can be viewed as the recovery of the deformation, given a collection of similar signals that have been deformed with respect to each other. The problem is formulated in terms of the minimization of an energy measure that combines a smoothness term and a similarity term (see [**Barnard 86**] in Chapter 1). The computation reduces to the solution of a system of first-order differential equations. The approach is to find an optimal solution at a coarse scale and then track it continuously to a fine scale. The scale parameter is the smoothing filter parameter σ. Witkin et al. describe the general process as follows:

> Imagine the energy landscape at each value of σ as a contoured surface in 3-space. The surfaces are stacked one above the other, so that the topmost surface is very smooth, while the lower ones become increasingly bumpy. Imagine a hole drilled at each local mimimum on each surface. A ball bearing dropped onto the topmost surface will roll down to the bottom of the hill. At this point it falls through to the next level, rolls down again, falls through again, and so on to the bottom.

Their approach addresses issues in constrained optimization, surface reconstruction, and coarse-to-fine matching. Experimental results are presented for 1-D signals, a motion sequence, and a stereo pair of images.

7.2 Model Fitting (Instantiation)

All computer vision is model-based in the sense that some internal description guides and constrains the machine's interpretation of the external world. Perception in the machine consists of selecting models relevant to the context of a sensed scene and then determining the values of the parameters of these models that best describe scene content (instantiating the models). Classes of models frequently dealt with are geometric models, illumination models, sensor or imaging models, and semantic models; usually these models are actually collections of equations with numerical parameters. The basic issues in model fitting are computational efficiency and the ability of the algorithms to deal with deviant (inaccurate or even wildly incorrect) data points.

There are three approaches to assigning values to the numerical parameters of a model based on observed or experimental data. These different approaches depend on whether the data *overconstrain, underconstrain* or *exactly constrain* the parameter values.

The classical approach is to use an optimization technique, such as "least squares," to solve an *overconstrained* set of equations in order to derive an instantiated model that best fits *all* the data—that is, all the data are used simultaneously to solve for the parameters of the model (e.g., [Duda 73] or [Sorenson 85]). Problems arise when the data contain gross errors or intermixed measurements from multiple objects. For example, even a single measurement error, if large enough, can cause least squares to fail [**Fischler 81a**] and there is no general procedure within the least-squares formalism for reliably eliminating such gross errors.

A second approach to model fitting [Duda 72] takes one data point at a time and finds all the parameter sets that instantiate the model so as to be consistent with the data point. That is, we solve an *underconstrained* set of equations to find all parameter sets compatible with the given data point. Each solution of the model for each data point "votes" for a corresponding set of parameter values. After all the data points have been processed, the parameter set receiving the most votes is chosen as the desired solution. Ballard [**Ballard 81a**] describes how this *Hough transform* approach can be used for recognizing objects based on a comparison of their shape to an ideal prototype. The prototype can differ from the imaged shape in size and orientation, its location in the image can be unknown, and parts of the shape of the imaged object can be occluded. While additional uncertainties can easily be handled by the formalism, each additional uninstantiated model parameter that must be evaluated causes an exponential increase in both computation time and storage requirements since we must construct and maintain a histogram containing every possible model solution for every given data point.

The third approach to model fitting, RANSAC, described in Fischler and Bolles [**Fischler 81a**], randomly selects just enough data points to solve the model equations and then attempts to confirm this instantiated model by testing it against the remaining data (a more complete discussion of how this confirmation process should be carried out is contained in Bolles and Fischler [Bolles 81]). If such confirmation fails, the process is repeated with another random selection of data points. RANSAC is as robust as the Hough transform in avoiding the disruptive effects of "wild" points; RANSAC completely dominates the Hough transform in computational efficiency if an error model exists, or if the data errors are systematic rather than uniformly distributed—for example, either very small (i.e., good measurements for the correct model) or very large (wrong model or a grossly inaccurate measurement) but few "intermediate" errors.

It is interesting to note that under the right conditions it is more efficient to solve a problem by a guess-and-check approach than by an exhaustive analysis of the data [Fischler 87].

7.3 Multisource Integration

Sensor data and stored knowledge provide clues about the nature of scene content, but we must still employ some form of reasoning (deduction or induction) to combine the available information into a "best" description of the scene. A number of formalisms exist for combining "facts" (e.g., logical deduction using the propositional or predicate calculus [Nilsson 80]) or "evidence" (e.g., Baysian statistics or the Shafer-Dempster formalism [G. Shafer 76] [Wesley 86]) when certain conditions are satisfied. For example, logical deduction is a powerful tool if we can describe our problem in the formal language it uses and if our observations are not contradictory. In general, low- and intermediate-level problems in scene analysis fail on both these requirements. Pixel-level descriptions, when converted to logical propositions, lead to impractical computational difficulties, and sensor-based observations are often erroneous. Statistical techniques generally require that we can assign prior probabilities to events, that we know relevant statistical distributions, and (for practical reasons) that our information sources are independent of each other. When we have one image of a real scene, rarely are any of these conditions satisfied. Thus, most existing formal procedures for combining information cannot be effectively used in their "pure" form, and modified versions of these methods must be constructed.

Fischler, Tenenbaum, and Wolf [**Fischler 81b**] argue that information provided by a realistic knowledge source, such as an image operator, can generally be viewed as a combination of facts and opinions. Facts are true statements (in practice, statements with a very low probability of error) that can be combined, by formal logical procedures, regardless of their source; opinions (less reliable judgments of the knowledge sources) cannot be combined in any formal way. The technique presented in this paper combines the facts from all the knowledge sources to form a "solution framework." Each knowledge source then produces its own interpretation of the image by merging (only) its opinions into the combined solution framework. A best solution can then be chosen from the set of alternatives offered by the different knowledge sources. This integration approach is combined with a new highly efficient optimization technique (the F* algorithm for finding a best path in a graph) to find linear structures in aerial images. The implemented technique has proven to be very effective and has subsequently been used for such other purposes as path planning.

7.4 Discussion

The problems of image and scene interpretation are often problems in making a best decision. We could have combined Chapters 3 (object recognition) and 4 (relational description) with the current chapter. Instead we chose to make a distinction between papers in which the contribution was in choosing an appropriate set of assumptions or a descriptive formalism that made the interpretation problem tractable versus (in this chapter) contributions to the general problems of matching, classification, model fitting, and information integration. Nevertheless, it is obvious that the problems of decision making and the choice of the representations over which the decisions are made cannot be separated. With the wrong representation the best decision procedure will be rendered computationally infeasible; with a good representation and the appropriate constraints, almost any decision procedure will produce the correct answers. It is important to recognize that most advances in machine perception are the result of a better understanding of the problem domain and the ability to describe this knowledge to the machine; decision-making techniques are effective only to the extent that they can operate within the framework of an effective problem representation.

Papers Included

Ballard, D.H. Generalizing the Hough transform to detect arbitrary shapes. *Pattern Recognition*, 13(2):111–122, 1981.

Fischler, M.A. and R.C. Bolles. Random sample consensus: A paradigm for model fitting with applications to image analysis and automated cartography. *Comm. ACM*, 24(6):381–395, June 1981.

Fischler, M.A., J.M. Tenenbaum, and H.C. Wolf. Detection of roads and linear structures in low resolution aerial imagery using a multisource knowledge integration technique. *CGIP*, 15(3):201–223, March 1981.

Hummel, R.A. Representations based on zero-crossings in scale-space. *Proc. IEEE CVPR '86*, pages 204-209, Miami Beach, Florida, June 22–26, 1986 (IEEE Computer Society Order Number 721).

Witkin, A.D., D. Terzopoulis, and M. Kass. Signal matching through scale space. In *Proc. National Conf. on Artificial Intelligence*, Philadelphia, PA, pp. 714–719, Aug. 1986.

GENERALIZING THE HOUGH TRANSFORM TO DETECT ARBITRARY SHAPES*

D. H. BALLARD

Computer Science Department, University of Rochester, Rochester, NY 14627, U.S.A.

(*Received* 10 *October* 1979; *in revised form* 9 *September* 1980; *received for publication* 23 *September* 1980)

Abstract—The Hough transform is a method for detecting curves by exploiting the duality between points on a curve and parameters of that curve. The initial work showed how to detect both analytic curves[1,2] and non-analytic curves,[3] but these methods were restricted to binary edge images. This work was generalized to the detection of some analytic curves in grey level images, specifically lines,[4] circles[5] and parabolas.[6] The line detection case is the best known of these and has been ingeniously exploited in several applications.[7,8,9]

We show how the boundaries of an *arbitrary* non-analytic shape can be used to construct a mapping between image space and Hough transform space. Such a mapping can be exploited to detect instances of that particular shape in an image. Furthermore, variations in the shape such as rotations, scale changes or figure–ground reversals correspond to straightforward transformations of this mapping. However, the most remarkable property is that such mappings can be composed to build mappings for complex shapes from the mappings of simpler component shapes. This makes the generalized Hough transform a kind of universal transform which can be used to find arbitrarily complex shapes.

Image processing Hough transform Shape recognition Pattern recognition Parallel algorithms

1. INTRODUCTION

In an image, the pertinent information about an object is very often contained in the shape of its boundary. Some appreciation of the importance of these boundary shapes in human vision can be gained from experiments performed on the human visual system, which have shown that crude encodings of the boundaries are often sufficient for object recognition[10] and that the image may be initially encoded as an 'edge image', i.e. an image of local intensity or color gradients. Marr[11] has termed this edge image a 'primal sketch' and suggested that this may be a necessary first step in image processing. We describe a very general algorithm for detecting objects of a specified shape from an image that has been transformed into such an edge representation. In that representation, sample points in the image no longer contain grey level information, but instead each sample point contains a magnitude and direction representing the severity and orientation of the local grey level change.

Operators that transform the image in such a way are known as edge operators, and many such operators are available, all based on different models of the local grey level changes. Two of the most used are the gradient operator (for example, see Prewitt[12]) and the Hueckel operator,[13] which model local grey level changes as a ramp and a step respectively.

Our generalized Hough algorithm uses edge information to define a mapping from the orientation of an edge point to a reference point of the shape. The reference point may be thought of as the origin of a local co-ordinate system for the shape. Then there is an easy way of computing a measure which rates how well points in the image are likely to be origins of the specified shape. Figure 1 shows a few graphic examples of the information used by the generalized Hough transform. Lines indicate gradient directions. A feature of the transform is that it will work even when the boundary is disconnected due to noise or occlusions. This is generally not true for other strategies which track edge segments.

The original algorithm by Hough[2] did not use

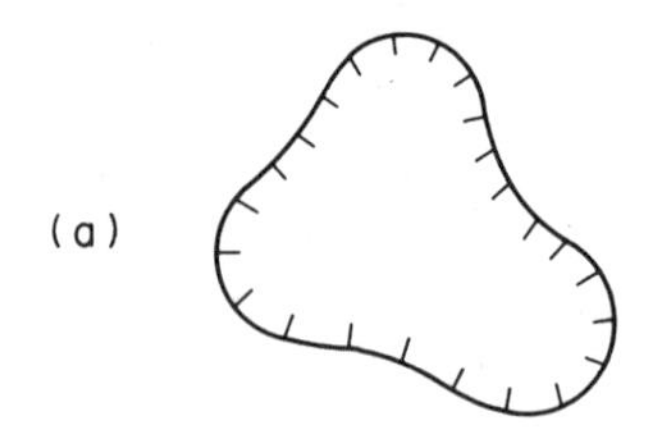

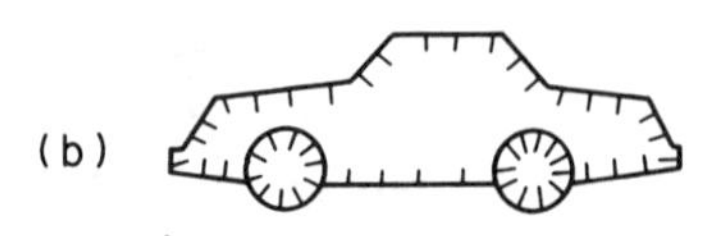

Fig. 1. Kinds of shapes detected with generalized Hough transform. (a) Simple shape; (b) composite shape.

* The research described in this report was supported in part by NIH Grant R23-HL-2153-01 and in part by the Alfred P. Sloan Foundation Grant 78-4-15.

orientation information of the edge, and was considerably inferior to later work using the edge orientation for parametric curves.[5,6,14] Shapiro[15,16,17] has collected a good bibliography of previous work as well as having contributed to the error analysis of the technique.

1.1 *Organization*

Section 2 describes the Hough transform for analytic curves. As an example of the parametric version of the transform, we use the ellipse. This example is very important due to the pervasiveness of circles in images, and the fact that a circle becomes an ellipse when rotated about an axis perpendicular to the viewing angle. Despite the importance of ellipses, not much work has used the Hough transform. The elliptical transform is discussed in detail in Section 3. Section 4 describes the generalized algorithm and its properties. Section 5 describes special strategies for implementing the algorithm and Section 6 summarizes its advantages.

2. THE HOUGH TRANSFORM FOR ANALYTIC CURVES

We consider analytic curves of the form $f(\mathbf{x}, \mathbf{a}) = 0$ where $\mathbf{x}$ is an image point and $\mathbf{a}$ is a parameter vector.

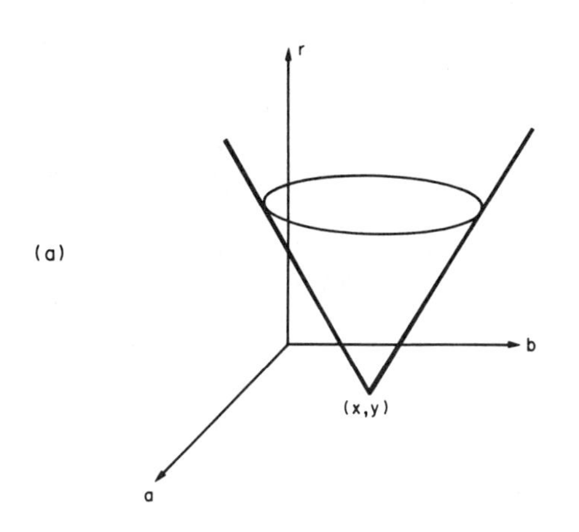

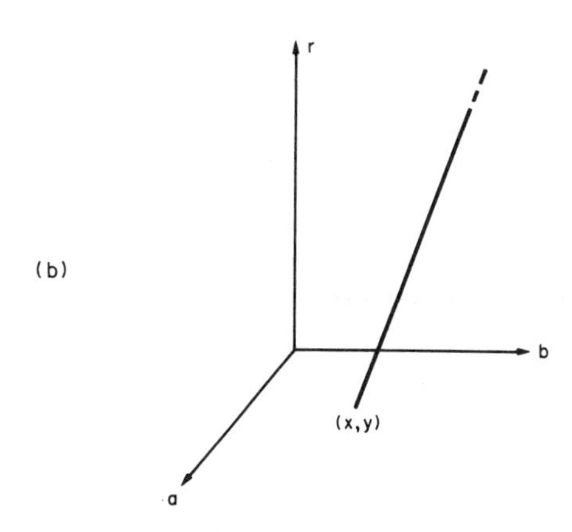

Fig. 2. (a) Locus of parameters with no directional information. (b) Locus of parameters with directional information.

To see how the Hough transform works for such curves, let us suppose we are interested in detecting circular boundaries in an image. In Cartesian coordinates, the equation for a circle is given by

$$(x-a)^2 + (y-b)^2 = r^2. \quad (1)$$

Suppose also that the image has been transformed into an edge representation so that only the magnitude of local intensity changes is known. Pixels whose magnitude exceeds some threshold are termed *edge pixels*. For each edge pixel, we can ask the question: if this pixel is to lie on a circle, what is the locus for the parameters of that circle? The answer is a right circular cone, as shown in Fig. 2(a). This can be seen from equation (1) by treating x and y as fixed and letting, a, b, and r vary.

The interesting result about this locus in parameter space is the following. If a set of edge pixels in an image are arranged on a circle with parameters a_0, b_0, and r_0, the resultant loci of parameters for each such point will pass through the same point (a_0, b_0, r_0) in parameter space. Thus many such right circular cones will intersect at a common point.

2.1 *Directional information*

We see immediately that if we also use the *directional* information associated with the edge, this reduces the parameter locus to a line, as shown in Fig. 2(b). This is because the center of the circle for the point (x, y) must lie r units along the direction of the gradient. Formally, the circle involves 3 parameters. By using the equation for the circle together with its derivative, the number of free parameters is reduced to one. Formally, what happens is the equation

$$\frac{df}{dx}(\mathbf{x}, \mathbf{a}) = 0$$

introduces a term dy/dx which is known since

$$\frac{dy}{dx} = \tan\left[\phi(\mathbf{x}) - \frac{\pi}{2}\right]$$

where $\phi(\mathbf{x})$ is the gradient direction. This suggests the following algorithm.

Hough algorithm for analytic curves in grey level images. For a specific curve $f(\mathbf{x}, \mathbf{a}) = 0$ with parameter vector $\mathbf{a}$, form an array $A(\mathbf{a})$, initially set to zero. This array is termed an accumulator array. Then for each edge pixel $\mathbf{x}$, compute all $\mathbf{a}$ such that $f(\mathbf{x}, a) = 0$ and $df/dx(\mathbf{x}, \mathbf{a}) = 0$ and increment the corresponding accumulator array entries:

$$A(\mathbf{a}) := A(\mathbf{a}) + 1.$$

After each edge pixel $\mathbf{x}$ has been considered, local maxima in the array A correspond to curves of f in the image.

If only the equation $f(\mathbf{x}, \mathbf{a}) = 0$ is used, the cost of the computation is exponential in the number of parameters minus one, that is, where m parameters each have M values, the computation is proportional to

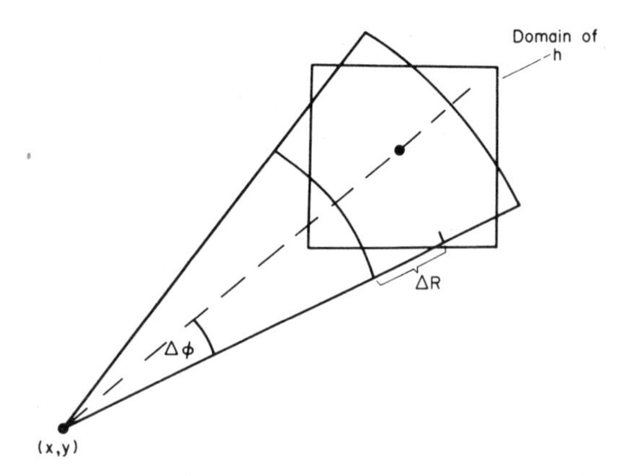

Fig. 3. Using convolution templates to compensate for errors.

M^{m-1}. This is because the equation of the curve can be used to determine the last parameter. The use of gradient directional information saves the cost of another parameter making the total effort proportional to M^{m-2}, for $m \geq 2$.

2.2 *Compensating for errors*

A problem arises in detecting maxima in the array $A(\mathbf{a})$. Many sources of error effect the computation of the parameter vector $\mathbf{a}$ so that in general many array locations in the vicinity of the ideal point $\mathbf{a}$ are incremented instead of the point itself. One way of handling this problem is to use a formal error model on the incrementation step. This model would specify a set of nearby points instead of a single point. Shapiro[15–18] has done extensive work on this subject. Another solution to this problem is to replace uncompensated accumulator values by a function of the values themselves and nearby points after the incrementation step. The effect of this operation is to smooth the accumulator array. We show that, under the assumption of isotropic errors, these methods are equivalent.

Returning to the initial example of detecting circles, the smoothing of the accumulator array is almost equivalent to the change in the incrementing procedure we would use to allow for uncertainties in the gradient direction ϕ and the radius r. If we recognized these uncertainties as:

$$\phi(\mathbf{x}) \pm \Delta\phi$$

$$r \pm \Delta r(r)$$

we would increment all values of $\mathbf{a}$ which fall within the shaded band of Fig. 3. We let Δr increase with r so that uncertainties are counted on a percentage basis. Figure 3 shows the two-dimensional analog of the general three-dimensional case.

Suppose we approximate this procedure by incrementing all values of $\mathbf{a}$ which fall inside the square domain centered about the nominal center shown in Fig. 3, according to some point spread function h. After the first contributing pixel which increments center $\mathbf{a}_0$ has been taken into account, the new accumulator array contents A will be given by

$$A(\mathbf{a}) = h(\mathbf{a} - \mathbf{a}_0) \tag{2}$$

where $\mathbf{a} = (a_1, a_2, r)$ and $\mathbf{a}_0 = (a_{10}, a_{20}, r_0)$. If we include all the contributing pixels for that center, denoted by C, the accumulator is

$$A(\mathbf{a}) = C(\mathbf{a}_0)h(\mathbf{a} - \mathbf{a}_0). \tag{3}$$

Finally for all incremented centers, we sum over $\mathbf{a}_0$:

$$A(\mathbf{a}) = \sum_{\mathbf{a}_0} C(\mathbf{a}_0)h(\mathbf{a} - \mathbf{a}_0). \tag{4}$$

But $C(\mathbf{a}_0) = A(\mathbf{a}_0)$, so that

$$A(\mathbf{a}) = \sum_{\mathbf{a}_0} A(\mathbf{a}_0)h(\mathbf{a} - \mathbf{a}_0)$$
$$= A*h$$
$$\equiv A_s(\mathbf{a}). \tag{5}$$

Thus within the approximation of letting the square represent the shaded band shown in Fig. 3, the smoothing procedure is equivalent to an accommodation for uncertainties in the gradient direction and radius.

3. AN EXAMPLE: ELLIPSES

The description of the algorithm in Section 2.1 is very terse and its implementation often requires considerable algebraic manipulation. We use the example of finding ellipses to show the kinds of calculation which must be done. Ellipses are an important example, as circles, which are a ubiquitous part of many everyday objects, appear as ellipses when viewed from a distant, oblique angle.

We use the center of the ellipse as a reference point and assume that it is centered at x_0, y_0 with major and minor diameters a and b. For the moment, we will assume that the ellipse is oriented with its major axis parallel to the x-axis. Later we will relax this requirement by introducing an additional parameter for arbitrary orientations. For the moment, assume a and b are fixed. Then the equation of the ellipse is:

$$\frac{(x - x_0)^2}{a^2} + \frac{(y - y_0)^2}{b^2} = 1. \tag{6}$$

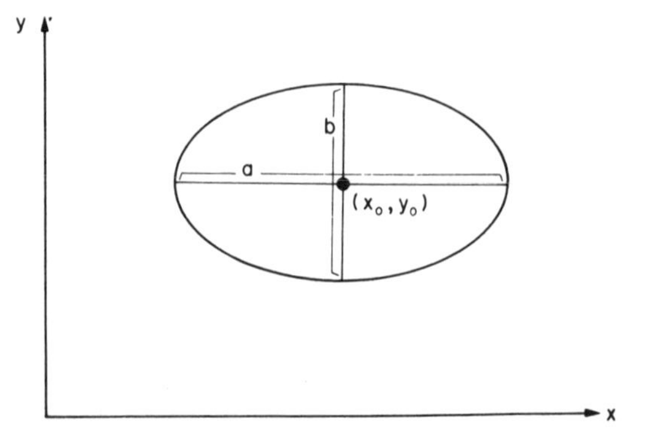

Fig. 4. Parametrization of an ellipse with major axis parallel to x-axis.

Let $X = x - x_0$, $Y = y - y_0$, then

$$\frac{X^2}{a^2} + \frac{y^2}{b^2} = 1 \quad (7)$$

Differentiating with respect to X

$$\frac{2X}{a^2} + \frac{2Y}{b^2}\frac{dY}{dX} = 0. \quad (8)$$

But dY/dX is known from the edge pixel information! Let $dY/dX = \xi$, then from (8)

$$X^2 = \left(\frac{a^2}{b^2}\xi\right)^2 Y^2. \quad (9)$$

Substituting in (7)

$$\frac{Y^2}{b^2}\left(1 + \frac{a^2}{b^2}\xi^2\right) = 1 \quad (10)$$

$$Y = \pm \frac{b^2}{\sqrt{\left(1 + \frac{a^2}{b^2}\xi^2\right)}} \quad (11)$$

so that

$$X = \pm \frac{a^2}{\sqrt{\left(1 + \frac{b^2}{a^2\xi^2}\right)}} \quad (12)$$

and finally, given a, b, x, y and dY/dX, we can determine x_0 and y_0 as:

$$x_0 = x \pm \frac{a^2}{\sqrt{\left(1 + \frac{b^2}{a^2\xi^2}\right)}} \quad (13)$$

$$y_0 = y \pm \frac{b^2}{\sqrt{\left(1 + \frac{a^2\xi^2}{b^2}\right)}}. \quad (14)$$

The four solutions correspond to the four quadrants, as shown in Fig. 5. The appropriate quadrant can be found from the gradient by testing the signed differences dY and dX.

The final step is to handle rotations by introducing a fifth parameter θ. For an arbitrary θ, we calculate (X, Y) using

$$\xi = \tan\left(\phi - \theta - \frac{\pi}{2}\right)$$

and rotate these (X, Y) by θ to obtain the correct (x_0, y_0). In ALGOL we would implement this as:

```
procedure HoughEllipse (integer XminXmax, YminYmax,
θminθmax, aminamax, bminbmax, x, y, x0, y0, dx, dy; real angle, ξ;
integer array A, P);
begin;
for x: = xmin step dx to xmax do
for y: = ymin step dy to ymax do
  begin
    dX:= P(x+delta, y) − P(x, y);
    dY: = P(x, y+delta) − P(x, y);
      for a: = amin step da until amax do
      for b: = bmin step db until bmax do
      for θ: = θmin step dθ until θmax do
        begin;
          angle: = arctan(dY/dX) − θ − π/2;
          ξ: = tan(angle);
          dx: = Sign X(dX, dY) a²/sqrt(1 + b²/(a²ξ²));
          dy: = Sign Y (dX, dY) b²/sqrt(1 + a²/b²)²;
          Rotate-by-Theta(dx, dy);
          x0: = x + dx;
          y0: = y + dy;
          A(x0, y0, θ, a, b): = A(x0, y0, θ, a, b) + 1;
        end;
end.
```

Notice that to determine the appropriate formulae for an arbitrary orientation angle θ, we need only rotate the gradient angle and the offsets dx and dy. Sign X and Sign Y are functions which return ± 1 depending on the quadrant determined by dX and dY.

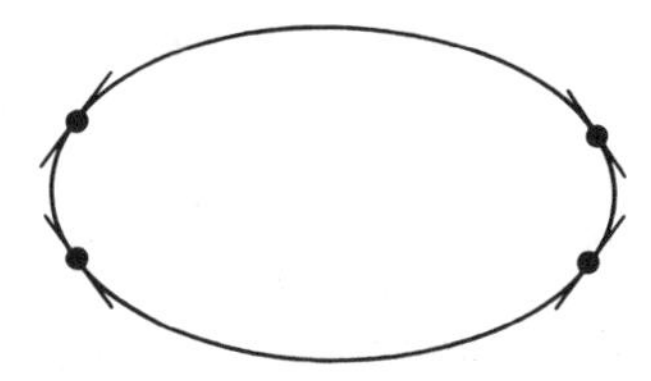

Fig. 5. Four reference point solutions resolvable with gradient quadrant information.

3.1 *Parameter space image space trade-offs*

Tsuji and Matsumoto[19] recognized that a decreased computational effort in parameter space could be traded for an increased effort in edge space. It is our intent to place these ideas on a formal footing. Later we will see that the same kind of trade-off is potentially available for the case of arbitrary shapes, but is impractical to implement.

An ellipse has five parameters. Referring to the basic algorithm in Section 2.1, we use the equation for the ellipse together with its derivative to solve for two of these parameters as a function of the other three. Thus the algorithm examines every edge point and uses a three-dimensional accumulator array so that the computations are of order $0(ed^3)$. Here e is the number of edge pixels and we are assuming d distinct values for each parameters. Suppose we use pairs of edge points in the algorithm. This results in four equations, two involving the equation for an ellipse evaluated at the different points and two for the related derivatives. This leaves one free parameter. Thus the resultant computational effort is now $0(e^2d)$. The detailed derivation of this form of the Hough algorithm is presented in the Appendix.

If parameter space can be highly constrained so that the set of plausible values is small, then the former technique will be more efficient, whereas if there are

Table 1. Analytic curves described in terms of the generalized shape parameters x_r, y_r, S_x, S_y, θ

Analytic form	Parameters	Equation
Line	S, θ	$x \cos\theta + y \sin\theta = S$
Circle	x_r, y_r, S	$(x-x_r)^2 + (y-y_r)^2 = S^2$
Parabola	x_r, y_r, S_x, θ	$(y-y_r)^2 = 4S_x(x-x_r)$*
Ellipse	$x_r, y_r, S_x, S_y, \theta$	$\frac{(y-y_r)^2}{S_y^2} + \frac{(x-x_r)^2}{S_x^2} = 1$*

* Plus rotation by θ.

relatively few edges and large variations in parameters, the latter will be more efficient.

4. GENERALIZING THE HOUGH TRANSFORM

To generalize the Hough algorithm to non-analytic curves we define the following parameters for a generalized shape:

$$\mathbf{a} = \{\mathbf{y}, \mathbf{s}, \theta\},$$

where $\mathbf{y} = (x_r, y_r)$ is a reference origin for the shape, θ is its orientation, and $\mathbf{s} = (s_x, s_y)$ describes two orthogonal scale factors. As before, we will provide an algorithm for computing the best set of parameters $\mathbf{a}$ for a given shape from edge pixel data. These parameters no longer have equal status. The reference origin location, $\mathbf{y}$, is described in terms of a table of possible edge pixel orientations. The computation of the additional parameters $\mathbf{s}$ and θ is then accomplished by straightforward transformations to this table. [To simplify the development slightly, and because of its practical significance, we will work with the four-dimensional sunspace $\mathbf{a} = (y, s, \theta)$, where s is a scalar.]

In a sense this choice of parameters includes the previous analytic forms to which the Hough transform has been applied. Table 1 shows these relationships.

4.1 *Earlier work: arbitrary shapes in binary edge images*

Merlin and Farber[3] showed how to use a Hough algorithm when the desired curves could not be described analytically. Each shape must have a specific reference point. Then we can use the following algorithm for a shape with boundary points B denoted by $\{\mathbf{x}_B\}$ which are relative to some reference origin $\mathbf{y}$.

Merlin–Farber Hough algorithm: non-analytic curves with no gradient direction information $\mathbf{a} = \mathbf{y}$. Form a two-dimensional accumulator array $A(\mathbf{a})$ initialized to zero. For each edge pixel $\mathbf{x}$ and each boundary point $\mathbf{x}_B$, compute $\mathbf{a}$ such that $\mathbf{a} = \mathbf{x} - \mathbf{x}_B$ and increment $A(\mathbf{a})$. Local maxima in $A(\mathbf{a})$ correspond to instances of the shape in the image.

Note that this is merely an efficient implementation of the convolution of the shape template where edge pixels are unity and others are zero with the corresponding image, i.e.,

$$A(\mathbf{x}) = T(\mathbf{x}) * S(\mathbf{x}) \tag{15}$$

where E is the binary edge image defined by

$$E(\mathbf{x}) = \begin{cases} 1 & \text{if } \mathbf{x} \text{ is an edge pixel} \\ 0 & \text{otherwise} \end{cases}$$

and $T(\mathbf{x})$ is the shape template consisting of ones where $\mathbf{x}$ is a boundary point and zeros otherwise, i.e.,

$$T(\mathbf{x}) = \begin{cases} 1 & \text{if } \mathbf{x} \text{ is in } B \\ 0 & \text{otherwise} \end{cases}.$$

This result is due to Sklansky.[20]

The Merlin–Farber algorithm is impractical for real image data. In an image with a multitude of edge pixels, there will be many false instances of the desired shape due to coincidental pixel arrangements. Nevertheless, it is the logical precursor to our generalized algorithm.

4.2 *The generalization to arbitrary shapes*

The key to generalizing the Hough algorithm to arbitrary shapes is the use of directional information. Directional information, besides making the algorithm faster, also greatly improves its accuracy. For example, if the directional information is not used in the circle detector, any significant group of edge points with quite different directions which lie on a circle will be detected. This can be appreciated by comparing Figs 2(a) and 2(b).

Consider for a moment the circular boundary detector with a fixed radius r_0. Now for each gradient point $\mathbf{x}$ with direction ϕ, we need only increment a single point $\mathbf{x}+\mathbf{r}$. For the circle:

$$|\mathbf{r}| = r_0 \tag{16}$$

$$\text{Angle}(\mathbf{r}) = \phi(\mathbf{x}). \tag{17}$$

Now suppose we have an arbitrary shape like the one shown in Fig. 6. Extending the idea of the circle detector with fixed radius to this case, for each point $\mathbf{x}$ on the boundary with gradient direction ϕ, we increment a point $\mathbf{a} = \mathbf{x}+\mathbf{r}$. The difference is that now $\mathbf{r} = \mathbf{a} - \mathbf{x}$ which, in general, will vary in magnitude and direction with different boundary points.

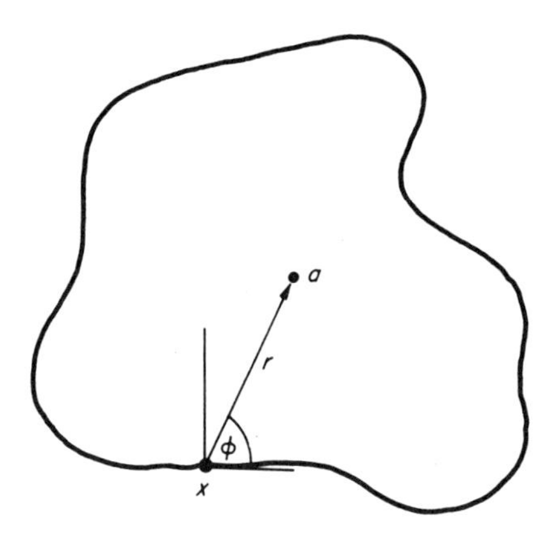

Fig. 6. Geometry for generalized Hough transform.

The fact that **r** varies in an arbitrary way means that the generalized Hough transform for an arbitrary shape is best represented by a table which we call the *R*-table.

4.3 *The* R-*table*

From the above discussion, we can see that the *R*-table is easily constructed by examining the boundary points of the shape. The construction of the table is accomplished as follows.

Algorithm for constructing an R-*table.* Choose a reference point **y** for the shape. For each boundary point x, compute $\phi(\mathbf{x})$ the gradient direction and $\mathbf{r} = \mathbf{y} - \mathbf{x}$. Store **r** as a function of ϕ.

Notice that the mapping the table represents is vector-valued and, in general, an index ϕ may have many values of **r**. Table 2 shows the form of the *R*-table diagrammatically.

The *R*-table is used to detect instances of the shape S in an image in the following manner.

Generalized Hough algorithm for single shapes. For each edge pixel **x** in the image, increment all the corresponding points $\mathbf{x} + \mathbf{r}$ in the accumulator array A where **r** is a table entry indexed by ϕ, i.e., $\mathbf{r}(\phi)$. Maxima in A correspond to possible instances of the shape S.

4.4 *Examples*

Some simple shapes are rotation-invariant, that is, the entries in the incrementation table are invariant functions of the gradient direction ϕ. Figure 7(a) shows an example for washers (or bagels). Here there are exactly two entries for each ϕ, one r units in the gradient direction and one R units in the direction opposite to the gradient direction. In another case the entries may be a simple function of ϕ. Figure 7(b)

Table 2. *R*-table format

i	ϕ_i	R_{ϕ_i}
0	0	$\{\mathbf{r} \mid \mathbf{a} - \mathbf{r} = \mathbf{x},\ \mathbf{x} \text{ in } B,\ \phi(\mathbf{x}) = 0\}$
1	$\Delta\phi$	$\{\mathbf{r} \mid \mathbf{a} - \mathbf{r} = \mathbf{x},\ \mathbf{x} \text{ in } B,\ \phi(\mathbf{x}) = \Delta\phi\}$
2	$2\,\Delta\phi$	$\{\mathbf{r} \mid \mathbf{a} - \mathbf{r} = \mathbf{x},\ \mathbf{x} \text{ in } B,\ \phi(\mathbf{x}) = 2\,\Delta\phi$
...	...	...

shows such an example; hexagons. Irrespective of the orientation of the edge, the reference point locus is on a line of length l parallel to the edge pixel and $(3/2)l$ units away from it.

Another example is shown in Fig. 8. Here the points on the boundary of the shape are shown in Fig. 8(a). A reference point is selected and used to construct the *R*-table. Figure 8(b) shows a synthetic image of four different shapes and Fig. 8(c) shows the portion of the accumulator array for this image which has the correct values of orientation and scale. It is readily seen that edge points on the correct shape have incremented the same point in the accumulator array, whereas edge points on the other shapes have incremented disparate points.

4.5 R-*table properties and the general notion of a shape*

Up to this point we have considered shapes of fixed orientation and scale. Thus the accumulator array was two-dimensional in the reference point co-ordinates. To search for shapes of arbitrary orientation θ and scale s we add these two parameters to the shape description. The accumulator array now consists of four dimensions corresponding to the parameters $(\mathbf{y}, s, \theta)$. The *R*-table can also be used to increment this larger dimensional space since different orientations and scales correspond to easily-computed transformations of the table. Additionally, simple transformations to the *R*-table can also account for figure–ground reversals and changes of reference point.

We denote a particular *R*-table for a shape S by $R(\phi)$. R can be viewed as a multiply-vector-valued function. It is easy to see that simple transformations to this table will allow it to detect scaled or rotated instances of the same shape. For example if the shape is scaled by s and this transformation is denoted by T_s, then

$$T_s[R(\phi)] = sR(\phi) \tag{18}$$

i.e., all the vectors are scaled by s. Also, if the object is rotated by θ and this transformation is denoted by T_θ, then

$$T_\theta[R(\phi)] = \mathrm{Rot}\{R[(\phi - \theta) \bmod 2\pi], \theta\} \tag{19}$$

i.e., all the indices are incremented by $-\theta$ modulo 2π, the appropriate vectors **r** are found, and then they are rotated by θ.

To appreciate that this is true, refer to Fig. 9. In this figure an edge pixel with orientation ϕ may be considered as corresponding to the boundary point $\mathbf{x}_A$, in which case the reference point is $\mathbf{y}_A$. Alternatively, the edge pixel may be considered as $\mathbf{x}_B$ on a rotated instance of the shape, in which case the reference point is at $\mathbf{y}_B$ which can be specified by translating $\mathbf{r}_A$ to $\mathbf{x}_B$ and rotating it through $+\Delta\theta$.

Figure–ground intensity reversals can also be taken into account via a simple *R*-table modification. The indices in the table are changed from ϕ to $(\phi + \pi) \bmod 2\pi$. Of course

$$T_{fg}\{T_{fg}[R(\phi)]\} = R(\phi)$$

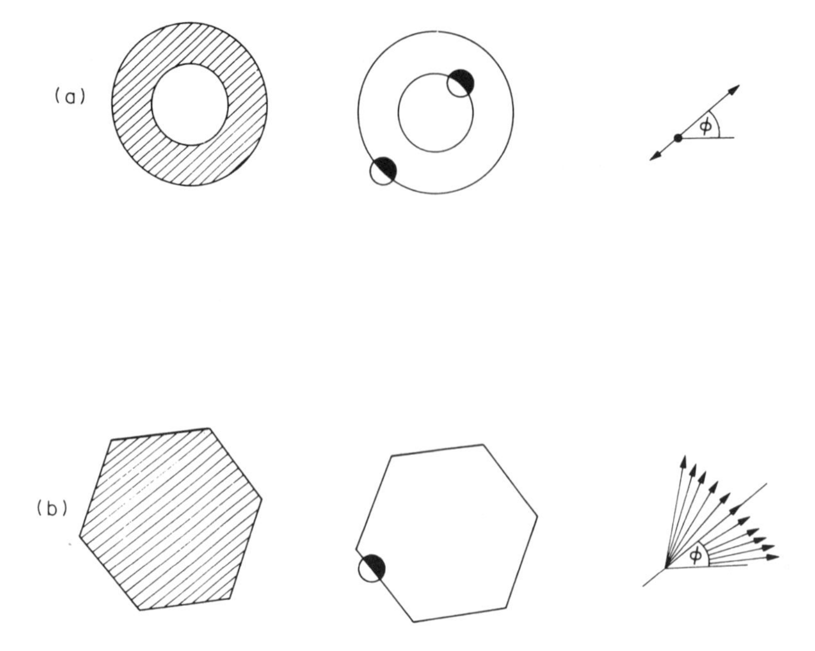

Fig. 7. Simple examples using R-tables; (a) washers; (b) hexagons.

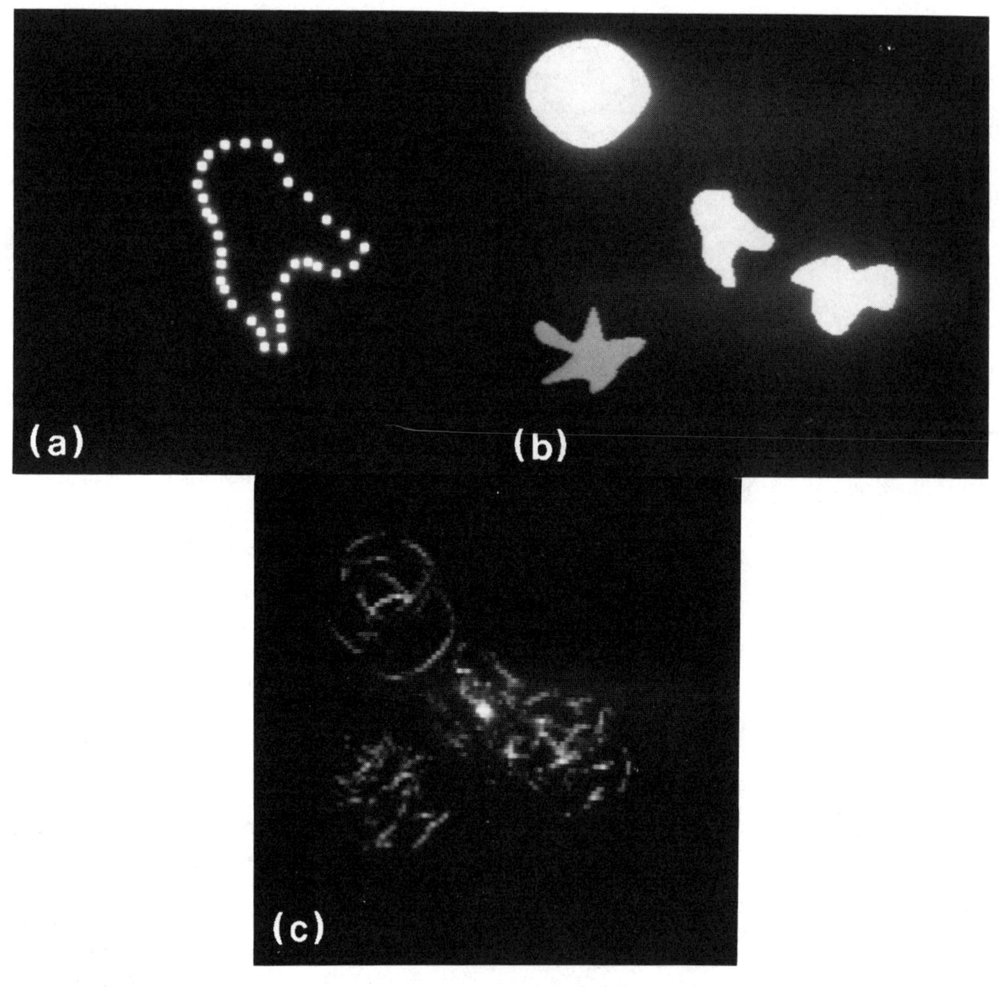

Fig. 8. An example. (a) Points on a shape used to encode R-table. (b) Image containing shape. (c) A plane through the accumulator array $A(x_r, y_r, S_0, \theta_0)$, where S_0 and θ_0 are appropriate for the shape in the image ($S_0 = 64$, $\theta_0 = 0$).

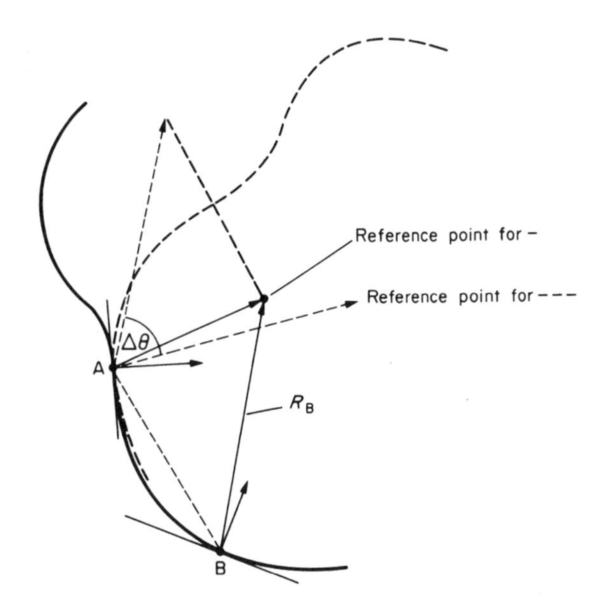

Fig. 9. Construction for visualizing the R-table transformation for a rotation by $\Delta\theta$. Point A can be viewed as: (1) on the shape (——), or (2) as point B on the shape (- - - -), rotated by $\Delta\theta$. If (2) is used then the appropriate $\mathbf{R}$ is obtained by translating $\mathbf{R}_B$ to A and rotating it by $\Delta\theta$ as shown.

where T_{fg} denotes the figure–ground transformations.

Another property which will be useful in describing the composition of generalized Hough transforms is the change of reference point. If we want to choose a new reference point y' such that $\mathbf{y}-\mathbf{y}' = \mathbf{r}$ then the modification to the R-table is given by $R(\phi) + \mathbf{r}$, i.e. $\mathbf{r}$ is added to each vector in the table.

4.6 *Using pairs of edges*

We can also entertain the idea of using pairs of edge pixels to reduce the effort in parameter space. Using the R-table and the properties of the previous section, each edge pixel defines a surface in the four-dimensional accumulator space of $\mathbf{a} = (\mathbf{y}, s, \theta)$. Two edge pixels at different orientations describe the same surface rotated by the same amount with respect to θ. Points where these two surfaces intersect (if any) correspond to possible parameters $\mathbf{a}$ for the shape. Thus in a similar manner to Section 3.1, it is theoretically possible to use the two points in image space to reduce the locus in parameter space to a single point. However, the difficulties of finding the intersection points of the two surfaces in parameter space will make this approach unfeasible for most cases.

4.7 *The Hough transform for composite shapes*

Now suppose we have a composite shape S which has two subparts S_1 and S_2. This shape can be detected by using the R-tables for S_1 and S_2 in a remarkably simple fashion. If $\mathbf{y}$, $\mathbf{y}_1$, $\mathbf{y}_2$ are the reference points for shapes S, S_1 and S_2 respectively, we can compute $\mathbf{r}_1 = \mathbf{y}-\mathbf{y}_1$ and $\mathbf{r}_2 = \mathbf{y}-\mathbf{y}_2$. Then the composite generalized Hough transform $R_S(\phi)$ is given by

$$R_S(\phi) = [R_{S_1}(\phi) + \mathbf{r}_1] \dot{\cup} [R_{S_2}(\phi) + \mathbf{r}_2] \quad (20)$$

which means that for each index value ϕ, $\mathbf{r}_1$ is added to $R_{S_1}(\phi)$, $\mathbf{r}_2$ is added to $R_{S_2}(\phi)$, and the union of these sets is stored in $R_S(\phi)$. Equation 20 is very important as it represents a way of composing transforms.

In a similar manner we can define shapes as the difference between tables with common entries, i.e.,

$$R_S = R_{S_1} \dot{-} R_{S_2} \quad (21)$$

means the shape S defined by S_1 with the common entries with S_2 deleted. The intersection operation is defined similarly. The primary use of the union operation is to detect shapes which are composites of simpler shapes. However, the difference operation also serves a useful function. Using it, R-tables which explicitly differentiate between two similar kinds of shapes can be constructed. An example would be differentiating between the washers and hexagons discussed earlier.

4.8 *Building convolution templates*

While equation (20) is one way of composing Hough transforms, it may not be the best way. This is because the choice of reference point can significantly affect the accuracy of the transform. Shapiro[15,16,17] has shown this, emphasizing analytic forms. This is also graphically shown in Fig. 10. As the reference point becomes distant from the shape, small angular errors in ϕ can produce large errors in the vectors $R(\phi)$.

One solution to this problem is to use the table for each subshape with its own best reference point and to smooth the resultant accumulator array with a composite smoothing template. Recall that for the case of a single shape and isotropic errors (Section 2.2), convolving the accumulator array in this fashion was equivalent to taking account of the errors during the incrementation.

Where $h_i(\mathbf{y}_i)$ denotes the smoothing template for reference point $\mathbf{y}_i$ of shape S_i the composite convolution template is given by

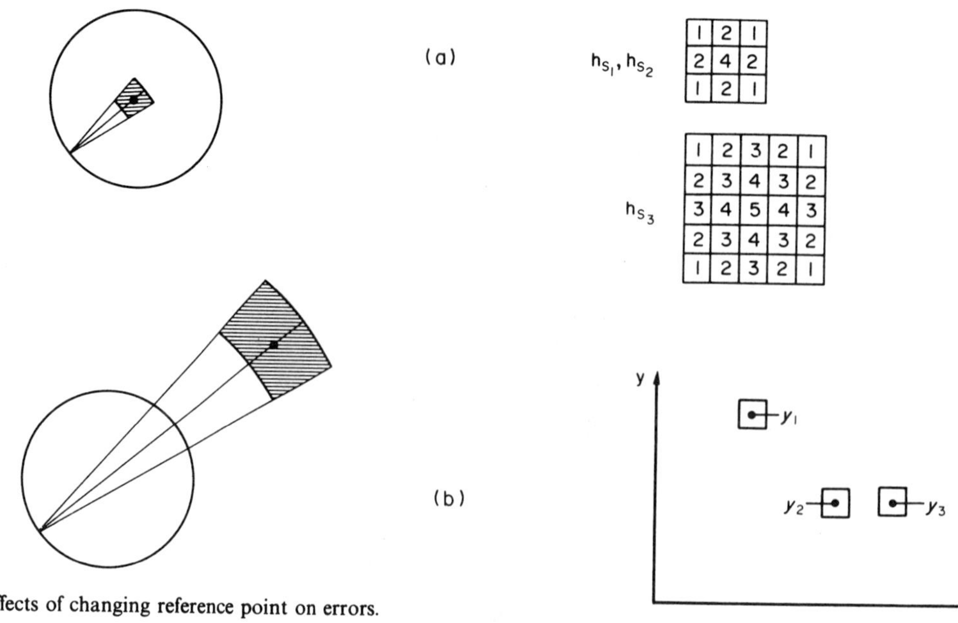

Fig. 10. Effects of changing reference point on errors.

$$H(y) = \sum_{i=1}^{N} h_i(\mathbf{y}-\mathbf{y}_i). \tag{22}$$

So finally, we have the following algorithm for the detection of a shape S which is the composite of subparts $S_1 \ldots S_N$.

Generalized Hough algorithm for composite shapes. 1. For each edge point with direction ϕ and for each value of scale s and orientation θ, increment the corresponding points $\mathbf{x} + \mathbf{r}$ in A where $\mathbf{r}$ is in

$$R_s(\phi) = T_s\left\{T_\theta\left[\bigcup_{k=1}^{N} R_{S_k}(\phi)\right]\right\}.$$

2. Maxima in $A_s = A^*H$ correspond to possible instances of the shape S. Figure 11 shows a simple example of how templates are combined.

If there are n edge pixels and M points in the error point spread function template, then the number of additions in the incrementation procedure is M. Thus this method might at first seem superior to the convolution method, which requires approximately n^2M additions and multiplications where $M < n^2$, the total number of pixels. However, the following heuristic is available for the convolution since A is typicallly very sparse. Compute

$$A_s(\mathbf{a}) \quad \text{only if } A(\mathbf{a}) > 0. \tag{23}$$

This in practice is very effective, although it may introduce errors if the appropriate index has a zero value and is surrounded by high values.

5. INCREMENTATION STRATEGIES

If we use the strategy of incrementing the accumulator array by unity, then the contents of the accumulator array are approximately proportional to the perimeter of the shape that is detectable in the image.

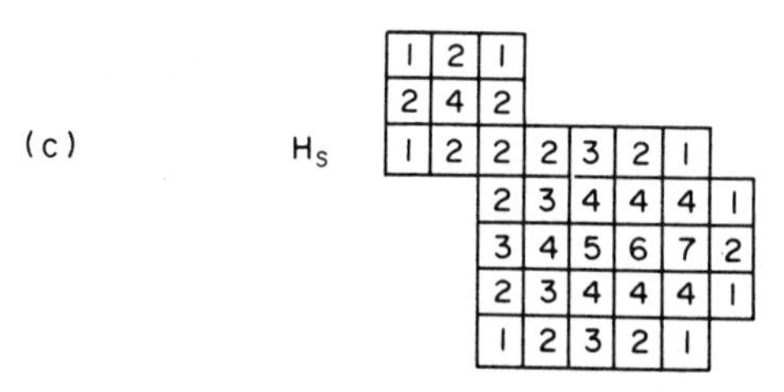

Fig. 11. Example of composite smoothing template construction. (a) Convolution templates for shapes S_1, S_2, S_3. (b) Relationships between reference points y_1, y_2, and y_3 in composite shape S. (c) Combined smoothing template H as a function of h_1, h_2, and h_3 and y_1, y_2, and y_3.

This strategy is biased towards finding shapes where a large portion of the perimeter is detectable. Several different incrementation strategies are available, depending on the different quality of image data. If shorter, very prominent parts of the perimeter are detected, as might be the case in partially occluded objects, then an alternative strategy of incrementing by the gradient modulus value might be more successful, i.e.,

$$A(\mathbf{a}) := A(\mathbf{a}) + g(\mathbf{x}). \tag{24}$$

Of course the two strategies can be combined, e.g.,

$$A(\mathbf{a}) := A(\mathbf{a}) + g(\mathbf{x}) + c \tag{25}$$

where c is a constant.

Another possibility is the use of local curvature information in the incrementation function. Using this strategy, neighboring edge pixels are examined to calculate approximate curvature, K. This requires a more complicated operator than the edge operators we have considered, and complicates the table. Now along with each value of r the corresponding values of curvature must be stored. Then the incrementation

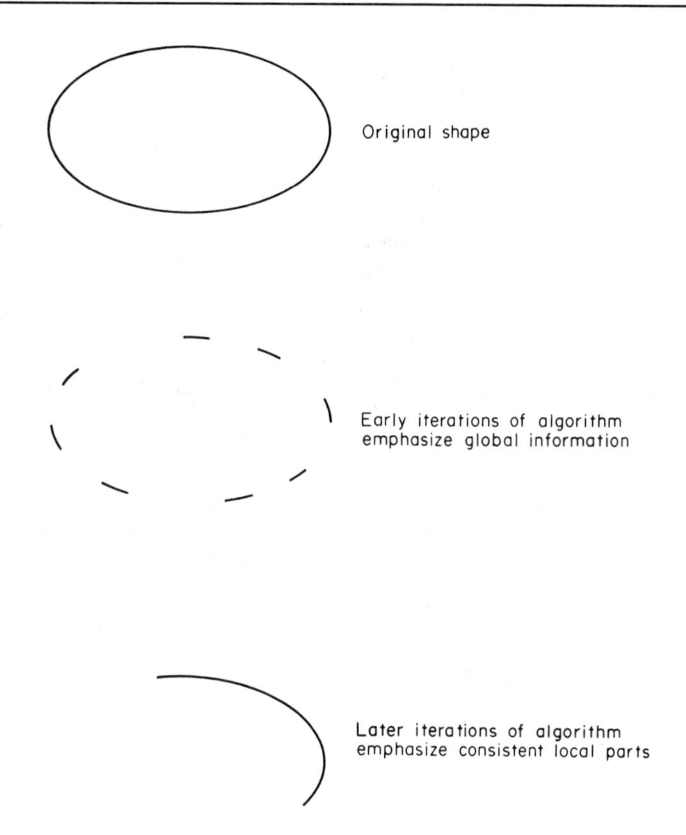

Fig. 12. Dynamic Hough transform.

weights 'informative' high local curvature edge pixels as follows:

$$A(\mathbf{a}) := A(\mathbf{a}) + K. \tag{26}$$

5.1 *Weighting locally consistent information*

Under certain circumstances we may want to weight local information that is consistent. For example, in searching for the boundary of that object, a connected set of edges conforming to the object may be more important than a set of unconnected edges. Figure 12 shows this example. Figure 12(a) might arise in situations with very noisy data. Figure 12(b) is an example where an object is occluded by another object. Wechsler and Sklansky,[6] in the analytic formulation, successfully used the related strategy of increasing the incrementation factor if there were also neighboring edge pixels with the same edge direction. However, we would like to measure local consistency in parameter space.

A simple strategy for handling this case is to explicitly record the reference points for each edge pixel during a first pass. Then on a second pass edge pixels can increment by more than unity if neighboring edge pixels are incrementing the same reference point.

A more complicated strategy is to search for connected curve segments in image space which have compatible parameters. Such an algorithm, based on dynamic programming, is described in Ballard and Sklansky.[14] The appropriate objective function for a curve segment would be

$$h(\mathbf{x}_1, \mathbf{x}_2, \ldots, \mathbf{x}_n) = \sum_{k=1}^{n} g(\mathbf{x}_k) + \sum_{k=1}^{n-1} q(\mathbf{x}_k, \mathbf{x}_{k+1}) \tag{27}$$

where

$$g(\mathbf{x}_k) = \text{the gradient magnitude} \tag{28}$$

and

$$q(\mathbf{x}_k, \mathbf{x}_{k+1}) = 0 \text{ if } |\phi(\mathbf{x}_\mu) - \phi(\mathbf{x}_{k-1})|_{\mathrm{mod}\,\pi} \text{ is small and } -\infty \text{ otherwise} \tag{29}$$

In the dynamic programming algorithm, at each iteration step we can build longer compatible curves from all the edge points. Thus the incrementation function for a point $\mathbf{x}$ would represent the longest compatible curve from that point. (If a longer curve cannot be built at any iteration, we can easily find this out.)

In a parallel implementation of this algorithm the contents of the accumulator array could be made to vary dynamically. Initially the contents would reflect global information, but with successive iterations the contents would be weighted in favor of consistent, local information.

5.2 *More complex strategies*

When searching for a composite object, different parts may have different importance. This is readily accommodated by associating a weight w_i with each table R_{S_i} so that each entry in R_{S_i} increments by a factor w_i instead of unity.

The composite object may be searched for in a sequential manner. Applying the table sequentially could greatly improve the efficiency of the computations by limiting areas for subsequent suitable incrementations. Furthermore, standard methods[21,22] could be used to stop the process once the shape had been located to the desired confidence level.

Even more complex strategies are possible wherein the process is integrated into a larger system. Here contextual information can be used to relegate all the previous operations including (a) building composite templates, (b) choosing weights, (c) choosing application sequences, and (d) adjusting weights in new contexts.

6. CONCLUSIONS

We have described a method for detecting instances of a shape S in an image which is a generalization of the Hough transform. This transform is a mapping from edge space to accumulator space such that instances of S produce local maxima in accumulator space. This mapping is conveniently described as a table of edge-orientation reference-point correspondence termed an R-table. This method has the following properties.

1. Scale changes, rotations, figure–ground reversals, and reference point translation of S can be accounted for by straightforward modifications to the R-table.
2. Given the boundary of the shape, its R-table can be easily constructed and requires a number of operations proportional to the number of boundary points.
3. Shapes are stored as canonical forms; instances of shapes are detected by knowing the transformation from the canonical form to the instance. If this transformation is not known then all plausible transformations must be tried.
4. If a shape S is viewed as a composite of several subparts $S_1 \ldots S_n$ then the generalized Hough transform R-table for S can be simply constructed by combining the R-tables for $S_1 \ldots S_n$.
5. A composite shape S may be efficiently detected in a sequential manner by adding the R-tables for the subparts S_i incrementally to the detection algorithm until a desired confidence level is reached.
6. The accumulator table values can be weighted in terms of locally consistent information.
7. The importance of a subshape S_i may be regulated by associating a weight w_i with the R-table.
8. Last but not least, the generalized Hough transform is a parallel algorithm.

Future work will be directed towards characterizing the computational efficiency of the algorithm and exploring its feasibility as a model of biological perception.

Acknowledgements – Portions of this paper benefitted substantially from discussions with Ken Sloan and Jerry Feldman. Special thanks go to R. Peet and P. Meeker for typing this manuscript. The work herein was supported by National Institutes of Health grant R23-HL21253-02.

REFERENCES

1. R. O. Duda and P. E. Hart, Use of the Hough transform to detect lines and curves in pictures, *Communs Ass. comput. Mach.* **15,** 11–15 (1975).
2. P. V. C. Hough, Method and means for recognizing complex patterns, U.S. Patent 3069654 (1962).
3. P. M. Merlin and D. J. Farber, A parallel mechanism for detecting curves in pictures, *IEEE Trans. Comput.* **C24,** 96–98 (1975).
4. F. O'Gorman and M. B. Clowes, Finding picture edges through collinearity of feature points, Proc. 3rd Int. Joint Conf. Artificial Intelligence, pp. 543–555 (1973).
5. C. Kimme, D. H. Ballard and J. Sklansky, Finding circles by an array of accumulators, *Communs Ass. comput. Mach.* **18,** 120–122 (1975).
6. H. Wechsler and J. Sklansky, Automatic detection of ribs in chest radiographs, *Pattern Recognition* **9,** 21–30 (1977).
7. S. A. Dudani and A. L. Luk, Locating straight-line edge segments on outdoor scenes, Proc. IEEE Computer Society on Pattern Recognition and Image Processing, Rensselaer Polytechnic Institute (1977).
8. C. L. Fennema and W. B. Thompson, Velocity determination in scenes containing several moving objects, Technical Report, Central Research Laboratory, Minnesota Mining and Manufacturing Co. St. Paul (1977).
9. J. R. Kender, Shape from texture: a brief overview and a new aggregation transform, Proc., DARPA Image Understanding Workshop, pp. 79–84. Pittsburgh, November (1978).
10. F. Attneave, Some informational aspects of visual perception, *Psychol. Rev.* **61,** 183–193 (1954).
11. D. Marr, Analyzing natural images: a computational theory of texture vision, MIT-AI-Technical Report 334, June (1975).
12. J. M. S. Prewitt, Object enhancement and extraction, *Picture Processing and Psychopictorics*, B. S. Lipkin and A. Rosenfeld, eds. Academic Press, New York (1970).
13. M. Hueckel, A local visual operator which recognizes edges and lines, *J. Ass. comput. Mach.* **20,** 634–646 (1973).
14. D. H. Ballard and J. Sklansky, A ladder-structured decision tree for recognizing tumors in chest radiographs, *IEEE Trans. Comput.* **C25,** 503–513 (1976).
15. S. D. Shapiro, Properties of transforms for the detection of curves in noisy pictures, *Comput. Graphics Image Process.* **8,** 219–236 (1978).
16. S. D. Shapiro, Feature space transforms for curve detection, *Pattern Recognition* **10,** 129–143 (1978).
17. S. D. Shapiro, Generalization of the Hough transform for curve detection in noisy digital images, *Proc. 4th* Int. Joint Conf. *Pattern Recognition*, pp. 710–714. Kyoto, Japan, November (1978).
18. S. D. Shapiro, Transformation for the computer detection of curves in noisy pictures, *Comput. Graphics Image Process.* **4,** 328–338 (1975).
19. S. Tsuji and F. Matsumoto, Detection of elliptic and linear edges by searching two parameter spaces, Proc. 5th Int. Joint Conf. Artificial Intelligence, Vol. 2, pp. 700–705. Cambridge, MA, August (1977).
20. J. Sklansky, On the Hough technique for curve detection, *IEEE Trans. Comput.*, **C27,** 923–926 (1978).
21. K. S. Fu, *Sequential Methods in Pattern Recognition and Machine Learning*. Academic Press, New York (1968).
22. R. Bolles, Verification vision with a programmable assembly system, Stanford AI Memo, AIM-275, December (1975).

APPENDIX. ANALYTIC HOUGH FOR PAIRS OF EDGE POINTS

To develop an explicit version of the Hough algorithm for ellipses using pairs of edge points, we consider the string-tied-at-two-ends parameterization of an ellipse:

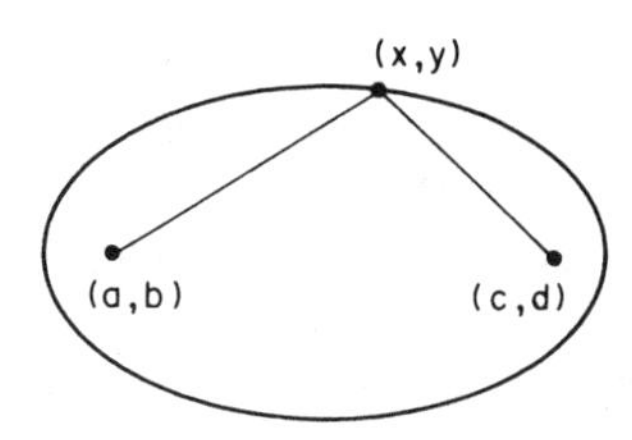

Fig. A1. String-tied-at-both-ends parameterization of an ellipse.

$$(x-a)^2 + (y-b)^2 + (x-c)^2 + (y-d)^2 = l^2$$

where (a, b) and (c, d) are the ends and l is the length of the string, as shown in Fig. A1. Now if we have two edge points (x_1, y_1) and (x_2, y_2) with gradients φ_1 and φ_2, the following equations result:

$$(x_1-a)^2 + (y_1-b)^2 + (x_1-c)^2 + (y_1-d)^2 = l^2 \quad \text{(A1)}$$

$$(x_1-a) + (y_1-b)\varphi_1 + (x_1-c) + (y_1-d)\varphi_1 = 0 \quad \text{(A2)}$$

$$(x_2-a)^2 + (y_2-b)^2 + (x_2-c)^2 + (y_2-d)^2 = l^2 \quad \text{(A3)}$$

$$(x_2-a) + (y_2-b)\varphi_2 + (x_2-c) + (y_2-d)\varphi_2 = 0 \quad \text{(A4)}$$

where in terms of the gradient direction

$$\varphi = \tan\left(\phi - \frac{\pi}{2}\right) = \frac{dy}{dx}.$$

From (A2):

$$a_1 = \varphi_1[(y_1-b) + (y_1-d)] + 2x_1 - c \cdot$$

Substituting in (4):

$$x_2 - \varphi_1[(y_1 - b) + (y_1 - d)] - 2x_1 + (y_2-b)\varphi_2 + (y_2-d)\varphi_2 + x_2 = 0.$$

Rearranging terms:

$$2x_2 - 2x_1 - 2\varphi_1 y_2 + (\varphi_1 - \varphi_2)b + (\varphi_1 - \varphi_2)d = 0.$$

Now where:

$$S \equiv \varphi_1 - \varphi_2$$

and

$$K \equiv 2(x_2 - x_1 - \varphi_1 y_1 + \varphi_2 y_2)$$

and

$$t = \frac{K}{S}$$

we have

$$b = t - d. \quad \text{(A5)}$$

Now we substitute for b in (2)

$$x_1 - a = -(x_1 - c) - \varphi_1(2y_1 - t)$$

so that we have

$$c \equiv \eta - a \quad \text{(A6)}$$

where $n = \varphi_1(2y - t)$

$$(x_1-a)^2 + [y_1-(t-d)]^2 + [x_1 - (\eta-a)]^2 + (y_1-d)^2 = l^2 \quad \text{(A7)}$$

$$(x_2-a)^2 + [y_2 - (t-d)]^2 + [x_2 - (\eta-a)]^2 + (y-d)^2 = l^2 \quad \text{(A8)}$$

Thus our strategy for using two edge points is as follows:

Step 1: choose a.
Step 2: solve equations (5) and (6), a quadratic in d, for d.
Step 3: solve equation (2) for b and equation 3 for c.
Step 4: solve equation (1) for l.

Thus the vector $\mathbf{a} = (a, b, c, d, l)$ has been determined for a pair of edge pixels and can be used to increment the accumulator array.

About the Author—DANA H. BALLARD was born in Holyoke, MA, on October 15, 1946. He received the B.Sc. degree in aeronautics and astronautics from the Massachusetts Institute of Technology, Cambridge, in 1967. He received the M.S.E. degree in information and control engineering from the University of Michigan, Ann Arbor, in 1970 and the Ph.D. degree in information engineering from the University of California, Irvine, in 1974.

From 1968 to 1971 he was a Systems Analyst for Autonetics, Anaheim, Calif. Since 1971 he has been active in computer vision research. During the academic year 1974–1975, he was a Visiting Professor at the Laboratorio Biomediche Technologie, Rome. Italy. He is presently an Assistant Professor of Computer Science and Radiology at the University of Rochester, Rochester, N.Y. His current interests are in artificial intelligence and computer vision, particularly control strategies and geometric models, and their applications to biomedical image processing.

Graphics and Image Processing

J. D. Foley Editor

Random Sample Consensus: A Paradigm for Model Fitting with Applications to Image Analysis and Automated Cartography

Martin A. Fischler and Robert C. Bolles
SRI International

A new paradigm, Random Sample Consensus (RANSAC), for fitting a model to experimental data is introduced. RANSAC is capable of interpreting/smoothing data containing a significant percentage of gross errors, and is thus ideally suited for applications in automated image analysis where interpretation is based on the data provided by error-prone feature detectors. A major portion of this paper describes the application of RANSAC to the Location Determination Problem (LDP): Given an image depicting a set of landmarks with known locations, determine that point in space from which the image was obtained. In response to a RANSAC requirement, new results are derived on the minimum number of landmarks needed to obtain a solution, and algorithms are presented for computing these minimum-landmark solutions in closed form. These results provide the basis for an automatic system that can solve the LDP under difficult viewing and analysis conditions. Implementation details and computational examples are also presented.

Key Words and Phrases: model fitting, scene analysis, camera calibration, image matching, location determination, automated cartography.

CR Categories: 3.60, 3.61, 3.71, 5.0, 8.1, 8.2

The work reported herein was supported by the Defense Advanced Research Projects Agency under Contract Nos. DAAG29-76-C-0057 and MDA903-79-C-0588.

Authors' Present Address: Martin A. Fischler and Robert C. Bolles, Artificial Intelligence Center, SRI International, Menlo Park CA 94025.

I. Introduction

We introduce a new paradigm, Random Sample Consensus (RANSAC), for fitting a model to experimental data; and illustrate its use in scene analysis and automated cartography. The application discussed, the location determination problem (LDP), is treated at a level beyond that of a mere example of the use of the RANSAC paradigm; new basic findings concerning the conditions under which the LDP can be solved are presented and a comprehensive approach to the solution of this problem that we anticipate will have near-term practical applications is described.

To a large extent, scene analysis (and, in fact, science in general) is concerned with the interpretation of sensed data in terms of a set of predefined models. Conceptually, interpretation involves two distinct activities: First, there is the problem of finding the best match between the data and one of the available models (the classification problem); Second, there is the problem of computing the best values for the free parameters of the selected model (the parameter estimation problem). In practice, these two problems are not independent—a solution to the parameter estimation problem is often required to solve the classification problem.

Classical techniques for parameter estimation, such as least squares, optimize (according to a specified objective function) the fit of a functional description (model) to *all* of the presented data. These techniques have no internal mechanisms for detecting and rejecting gross errors. They are averaging techniques that rely on the assumption (the smoothing assumption) that the maximum expected deviation of any datum from the assumed model is a direct function of the size of the data set, and thus regardless of the size of the data set, there will always be enough good values to smooth out any gross deviations.

In many practical parameter estimation problems the smoothing assumption does not hold; i.e., the data contain uncompensated gross errors. To deal with this situation, several heuristics have been proposed. The technique usually employed is some variation of first using all the data to derive the model parameters, then locating the datum that is farthest from agreement with the instantiated model, assuming that it is a gross error, deleting it, and iterating this process until either the maximum deviation is less then some preset threshold or until there is no longer sufficient data to proceed.

It can easily be shown that a single gross error ("poisoned point"), mixed in with a set of good data, can

Fig. 1. Failure of Least Squares (and the "Throwing Out The Worst Residual" Heuristic), to Deal with an Erroneous Data Point.

PROBLEM: Given the set of seven (x,y) pairs shown in the plot, find a best fit line, assuming that no valid datum deviates from this line by more than 0.8 units.

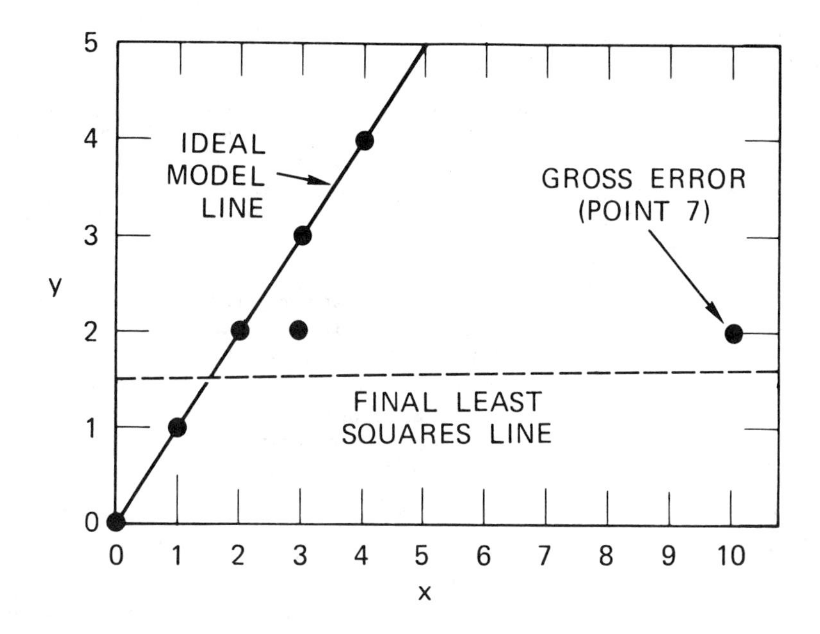

POINT	x	y
1	0	0
2	1	1
3	2	2
4	3	2
5	3	3
6	4	4
7	10	2

COMMENT: Six of the seven points are valid data and can be fit by the solid line. Using Least Squares (and the "throwing out the worst residual" heuristic), we terminate after four iterations with four remaining points, including the gross error at (10,2) fit by the dashed line.

SUCCESSIVE LEAST SQUARES APPROXIMATIONS		
ITERATION	DATA SET	FITTING LINE
1	1, 2, 3, 4, 5, 6, 7	1.48 + .16x
2	1, 2, 3, 4, 5, 7	1.25 + .13x
3	1, 2, 3, 4, 7	0.96 + .14x
4	2, 3, 4, 7	1.51 + .06x

COMPUTATION OF RESIDUALS				
POINT	ITERATION 1 RESIDUALS	ITERATION 2 RESIDUALS	ITERATION 3 RESIDUALS	ITERATION 4 RESIDUALS
1	-1.48	-1.25	-.96*	—
2	-0.64	-0.38	-.10	-.57
3	-0.20	0.49	.76	.37
4	0.05	0.36	.63	.31
5	1.05	1.36*	—	—
6	1.89*	—	—	—
7	-1.06	-0.57	-.33	-.11

cause the above heuristic to fail (for example, see Figure 1). It is our contention that averaging is not an appropriate technique to apply to an unverified data set.

In the following section we introduce the RANSAC paradigm, which is capable of smoothing data that contain a significant percentage of gross errors. This paradigm is particularly applicable to scene analysis because local feature detectors, which often make mistakes, are the source of the data provided to the interpretation algorithms. Local feature detectors make two types of errors—classification errors and measurement errors. Classification errors occur when a feature detector incorrectly identifies a portion of an image as an occurrence of a feature. Measurement errors occur when the feature detector correctly identifies the feature, but slightly miscalculates one of its parameters (e.g., its image location). Measurement errors generally follow a normal distribution, and therefore the smoothing assumption is applicable to them. Classification errors, however, are gross errors, having a significantly larger effect than measurement errors, and do not average out.

In the final sections of this paper the application of RANSAC to the location determination problem is discussed:

Given a set of "landmarks" ("control points"), whose locations are known in some coordinate frame, determine the location (relative to the coordinate frame of the landmarks) of that point in space from which an image of the landmarks was obtained.

In response to a RANSAC requirement, some new results are derived on the minimum number of landmarks needed to obtain a solution, and then algorithms are presented for computing these minimum-landmark solutions in closed form. (Conventional techniques are iterative and require a good initial guess to assure convergence.) These results form the basis for an automatic system that can solve the LDP under severe viewing and analysis conditions. In particular, the system performs properly even if a significant number of landmarks are incorrectly located due to low visibility, terrain changes, or image analysis errors. Implementation details and experimental results are presented to complete our description of the LDP application.

II. Random Sample Consensus

The RANSAC procedure is opposite to that of conventional smoothing techniques: Rather than using as much of the data as possible to obtain an initial solution and then attempting to eliminate the invalid data points, RANSAC uses as small an initial data set as feasible and enlarges this set with consistent data when possible. For example, given the task of fitting an arc of a circle to a set of two-dimensional points, the RANSAC approach would be to select a set of three points (since three points are required to determine a circle), compute the center and radius of the implied circle, and count the number of points that are close enough to that circle to suggest their compatibility with it (i.e., their deviations are small enough to be measurement errors). If there are enough compatible points, RANSAC would employ a smoothing technique such as least squares, to compute an improved estimate for the parameters of the circle now that a set of mutually consistent points has been identified.

The RANSAC paradigm is more formally stated as follows:

Given a model that requires a minimum of n data points to instantiate its free parameters, and a set of data points P such that the number of points in P is greater than n [$\#(P) \geq n$], randomly select a subset $S1$ of n data points from P and instantiate the model. Use the instantiated model $M1$ to determine the subset $S1^*$ of points in P that are within some error tolerance of $M1$. The set $S1^*$ is called the consensus set of $S1$.

If $\#(S1^*)$ is greater than some threshold t, which is a function of the estimate of the number of gross errors in P, use $S1^*$ to compute (possibly using least squares) a new model $M1^*$.

If $\#(S1^*)$ is less than t, randomly select a new subset $S2$ and repeat the above process. If, after some predetermined number of trials, no consensus set with t or more members has been found, either solve the model with the largest consensus set found, or terminate in failure.

There are two obvious improvements to the above algorithm: First, if there is a problem related rationale for selecting points to form the S's, use a deterministic selection process instead of a random one; second, once a suitable consensus set S^* has been found and a model M^* instantiated, add any new points from P that are consistent with M^* to S^* and compute a new model on the basis of this larger set.

The RANSAC paradigm contains three unspecified parameters: (1) the error tolerance used to determine whether or not a point is compatible with a model, (2) the number of subsets to try, and (3) the threshold t, which is the number of compatible points used to imply that the correct model has been found. Methods are discussed for computing reasonable values for these parameters in the following subsections.

A. Error Tolerance For Establishing Datum/Model Compatibility

The deviation of a datum from a model is a function of the error associated with the datum and the error associated with the model (which, in part, is a function of the errors associated with the data used to instantiate the model). If the model is a simple function of the data points, it may be practical to establish reasonable bounds on error tolerance analytically. However, this straightforward approach is often unworkable; for such cases it is generally possible to estimate bounds on error tolerance experimentally. Sample deviations can be produced by perturbing the data, computing the model, and measuring the implied errors. The error tolerance could then be set at one or two standard deviations beyond the measured average error.

The expected deviation of a datum from an assumed model is generally a function of the datum, and therefore the error tolerance should be different for each datum. However, the variation in error tolerances is usually relatively small compared to the size of a gross error. Thus, a single error tolerance for all data is often sufficient.

B. The Maximum Number of Attempts to Find a Consensus Set

The decision to stop selecting new subsets of P can be based upon the expected number of trials k required to select a subset of n good data points. Let w be the probability that any selected data point is within the error tolerance of the model. Then we have:

$$E(k) = b + 2*(1-b)*b + 3*(1-b)^2*b$$
$$\cdots + i*(1-b)^{i-1}*b + \cdots,$$
$$E(k) = b*[1 + 2*a + 3*a^2 \cdots + i*a^{i-1} + \cdots],$$

where $E(k)$ is the expected value of k, $b = w^n$, and $a = (1-b)$.
An identity for the sum of a geometric series is

$$a/(1-a) = a + a^2 + a^3 \cdots + a^i + \cdots.$$

Differentiating the above identity with respect to a, we have:

$$1/(1-a)^2 = 1 + 2*a + 3*a^2 \cdots + i*a^{i-1} + \cdots.$$

Thus,

$$E(k) = 1/b = w^{-n}$$

The following is a tabulation of some values of $E(k)$ for corresponding values of n and w:

w	$n = 1$	2	3	4	5	6
0.9	1.1	1.2	1.4	1.5	1.7	1.9
0.8	1.3	1.6	2.0	2.4	3.0	3.8
0.7	1.4	2.0	2.9	4.2	5.9	8.5
0.6	1.7	2.8	4.6	7.7	13	21
0.5	2.0	4.0	8.0	16	32	64
0.4	2.5	6.3	16	39	98	244
0.3	3.3	11	37	123	412	—
0.2	5.0	25	125	625	—	—

In general, we would probably want to exceed $E(k)$ trials by one or two standard deviations before we give up. Note that the standard deviation of k, $SD(k)$, is given by:

$$SD(k) = sqrt\,[E(k^2) - E(k)^2].$$

Then

$$E(k^2) = \sum_{i=0}^{\infty} (b*i^2*a^{i-1}),$$
$$= \sum_{i=0}^{\infty} [b*i*(i-1)*a^{i-1}] + \sum_{i=0}^{\infty} (b*i*a^{i-1}),$$

but (using the geometric series identity and two differentiations):

$$2a/(1-a)^3 = \sum_{i=0}^{\infty} (i*(i-1)*a^{i-1}).$$

Thus,

$$E(k^2) = (2-b)/(b^2),$$

and

$$SD(k) = [sqrt\,(1-w^n)]*(1/w^n).$$

Note that generally $SD(k)$ will be approximately equal to $E(k)$; thus, for example, if ($w = 0.5$) and ($n = 4$), then $E(k) = 16$ and $SD(k) = 15.5$. This means that one might want to try two or three times the expected number of random selections implied by k (as tabulated above) to obtain a consensus set of more than t members.

From a slightly different point of view, if we want to ensure with probability z that at least one of our random selections is an error-free set of n data points, then we must expect to make at least k selections (n data points per selection), where

$$(1-b)^k = (1-z),$$
$$k = [\log(1-z)]/[\log(1-b)].$$

For example, if ($w = 0.5$) and ($n = 4$), then ($b = 1/16$). To obtain a 90 percent assurance of making at least one error-free selection,

$$k = \log(0.1)/\log(15/16) = 35.7.$$

Note that if $w^n \ll 1$, then $k \approx \log(1-z)E(k)$. Thus if $z = 0.90$ and $w^n \ll 1$, then $k \approx 2.3E(k)$; if $z = 0.95$ and $w^n \ll 1$, then $k \approx 3.0E(k)$.

C. A Lower Bound On the Size of an Acceptable Consensus Set

The threshold t, an unspecified parameter in the formal statement of the RANSAC paradigm, is used as the basis for determining that an n subset of P has been found that implies a sufficiently large consensus set to permit the algorithm to terminate. Thus, t must be chosen large enough to satisfy two purposes: that the correct model has been found for the data, and that a sufficient number of mutually consistent points have been found to satisfy the needs of the final smoothing procedure (which computes improved estimates for the model parameters).

To ensure against the possibility of the final consensus set being compatible with an incorrect model, and assuming that y is the probability that any given data point is within the error tolerance of an incorrect model, we would like y^{t-n} to be very small. While there is no general way of precisely determining y, it is certainly reasonable to assume that it is less than w (w is the *a priori* probability that a given data point is within the error tolerance of the correct model). Assuming $y < 0.5$, a value of $t - n$ equal to 5 will provide a better than 95 percent probability that compatibility with an incorrect model will not occur.

To satisfy the needs of the final smoothing procedure, the particular procedure to be employed must be specified. If least-squares smoothing is to be used, there are many situations where formal methods can be invoked to determine the number of points required to produce a desired precision [10].

D. Example

Let us apply RANSAC to the example described in Figure 1. A value of w (the probability that any selected data point is within the error tolerance of the model) equal to 0.85 is consistent with the data, and a tolerance (to establish datum/model compatibility) of 0.8 units was supplied as part of the problem statement. The RANSAC-supplied model will be accepted without external smoothing of the final consensus set; thus, we would like to obtain a consensus set that contains all seven data points. Since one of these points is a gross error, it is obvious that we will not find a consensus set of the desired size, and so we will terminate with the largest set we are able to find. The theory presented earlier indicates that if we take two data points at a time, compute the line through them and measure the deviations of the remaining points from this line, we should expect to find a suitable consensus set within two or three trials; however, because of the limited amount of data, we might be willing to try all 21 combinations to find the largest consensus set. In either case, we easily find the consensus set containing the six valid data points and the line that they imply.

III. The Location Determination Problem (LDP)

A basic problem in image analysis is establishing a correspondence between the elements of two representations of a given scene. One variation of this problem, especially important in cartography, is determining the location in space from which an image or photograph was obtained by recognizing a set of landmarks (control points) appearing in the image (this is variously called the problem of determining the elements of exterior camera orientation, or the camera calibration problem, or the image-to-database correspondence problem). It is routinely solved using a least-squares technique [11, 8] with a human operator interactively establishing the association between image points and the three-dimensional coordinates of the corresponding control points. However, in a fully automated system, where the correspondences must be based on the decisions of marginally competent feature detectors, least squares is often incapable of dealing with the gross errors that may result; this consideration, discussed at length in Sec. II, is illustrated for the LDP in an example presented in Sec. IV.

In this section a new solution to the LDP is presented based on the RANSAC paradigm, which is unique in its ability to tolerate gross errors in the input data. We will first examine the conditions under which a solution to the LDP is possible and describe new results concerning this question; we then present a complete description of the RANSAC-based algorithm, and finally, describe experimental results obtained through use of the algorithm.

Fig. 2. Geometry of the Location Determination Problem.

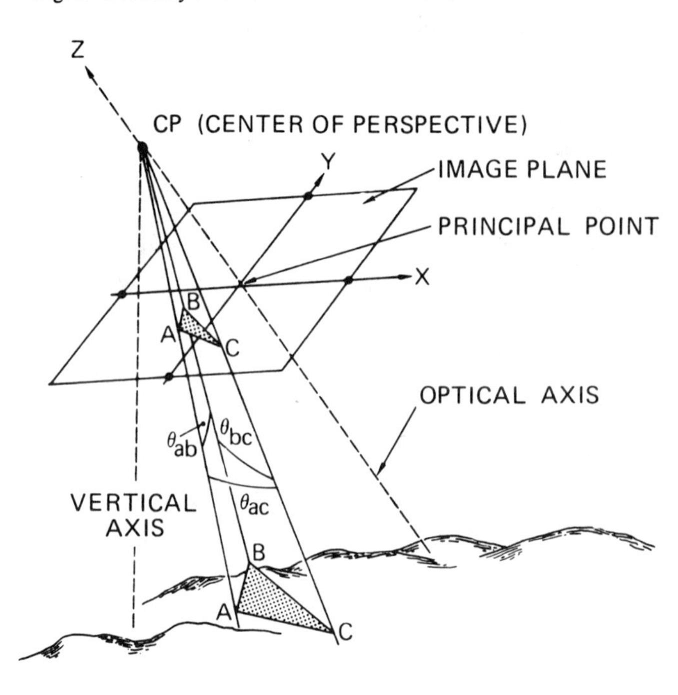

The LDP is formally defined as follows:

Given a set of m control points, whose 3-dimensional coordinates are known in some coordinate frame, and given an image in which some subset of the m control points is visible, determine the location (relative to the coordinate system of the control points) from which the image was obtained.

We will initially assume that we know the correspondences between n image points and control points; later we consider the situation in which some of these correspondences are invalid. We will also assume that both the principal point in the image plane (where the optical axis of the camera pierces the image plane) and the focal length (distance from the center of perspective to the principal point in the image plane) of the imaging system are known; thus (see Figure 2) we can easily compute the angle to any pair of control points from the center of perspective (CP). Finally, we assume that the camera resides outside and above a convex hull enclosing the control points.

We will later demonstrate (Appendix A) that if we can compute the lengths of the rays from the CP to three of the control points then we can directly solve for the location of the CP (and the orientation of the image plane if desired). Thus, an equivalent but mathematically more concise statement of the LDP is

Given the relative spatial locations of n control points, and given the angle to every pair of control points from an additional point called the Center of Perspective (CP), find the lengths of the line segments ("legs") joining the CP to each of the control points. We call this the "perspective-n-point" problem (PnP).

In order to apply the RANSAC paradigm, we wish to determine the smallest value of n for which it is possible to solve the PnP problem.

Fig. 3. Geometry of the P2P Problem.

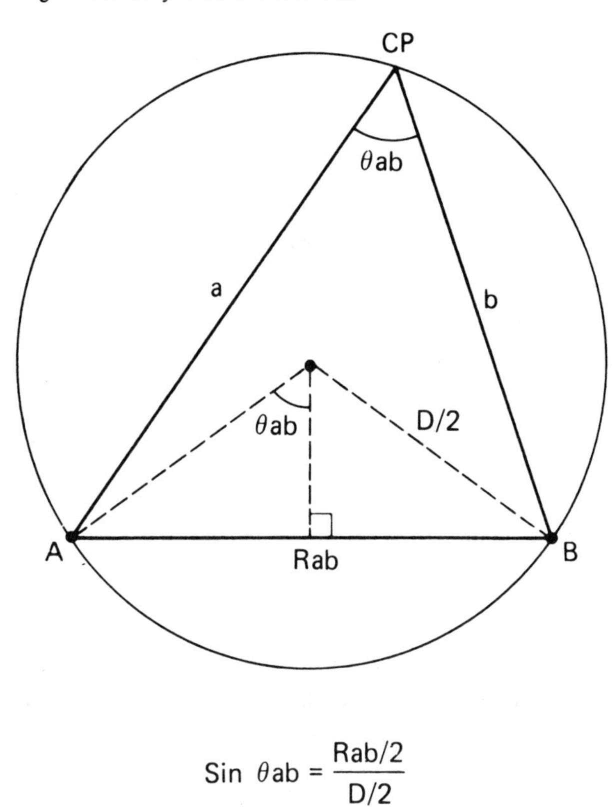

$$\text{Sin } \theta ab = \frac{Rab/2}{D/2}$$

$$D = \frac{Rab}{\text{Sin } \theta ab}$$

A. Solution of the Perspective-*n*-Point Problem

The P1P problem ($n = 1$) provides no constraining information, and thus an infinity of solutions is possible. The P2P problem ($n = 2$), illustrated in Figure 3, also admits an infinity of solutions; the CP can reside anywhere on a circle of diameter Rab/sin(θab), rotated in space about the chord (line) joining the two control points A and B.

The P3P problem ($n = 3$) requires that we determine the lengths of the three legs of a tetrahedron, given the base dimensions and the face angles of the opposing trihedral angle (see Figure 4). The solution to this problem is implied by the three equations [A^*]:

$$(Rab)^2 = a^2 + b^2 - 2{*}a{*}b{*}[\cos(\theta ab)]$$
$$(Rac)^2 = a^2 + c^2 - 2{*}a{*}c{*}\,[\cos(\theta ac)] \qquad [A^*]$$
$$(Rbc)^2 = b^2 + c^2 - 2{*}b{*}c{*}\,[\cos(\theta bc)]$$

It is known that n independent polynomial equations, in n unknowns, can have no more solutions than the product of their respective degrees [2]. Thus, the system A^* can have a maximum of eight solutions. However, because every term in the system A^* is either a constant or of second degree, for every real positive solution there is a geometrically isomorphic negative solution. Thus, there are at most four positive solutions to A^*, and in Figure 5 we show an example demonstrating that the upper bound of four solutions is attainable.

Fig. 4. Geometry of the P3P Problem (L is the Center of Perspective).

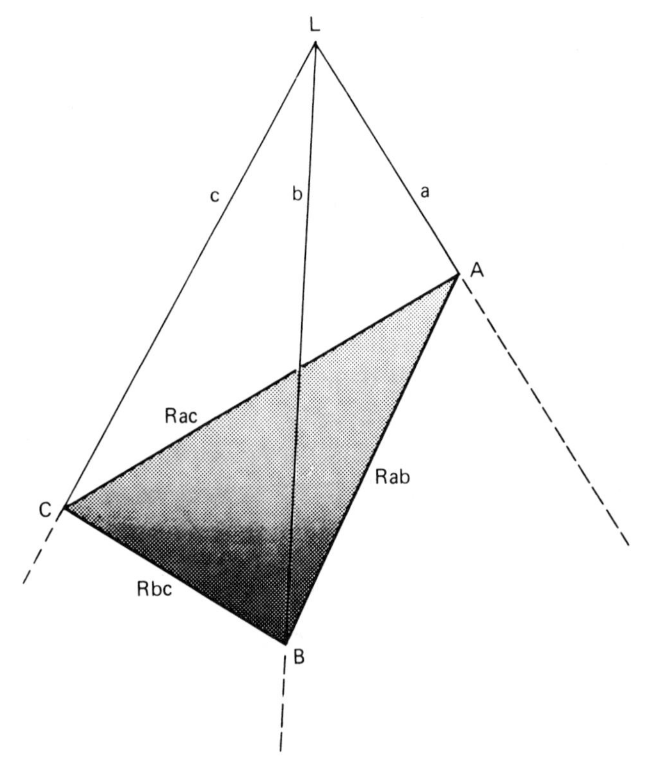

In Appendix A we derive an explicit algebraic solution for the system A^*. This is accomplished by reducing A^* to a biquadratic (quartic) polynomial in one unknown representing the ratio of two legs of the tetrahedron, and then directly solving this equation (we also present a very simple iterative method for obtaining the solutions from the given problem data).

For the case $n = 4$, when all four control points lie in a common plane (not containing the CP, and such that no more than two of the control points lie on any single line), we provide a technique in Appendix B that will always produce a unique solution. Surprisingly, when all four control points do not lie in the same plane, a unique solution cannot always be assured; for example, Figure 6 shows that at least two solutions are possible for the P4P problem with the control points in "general position."

To solve for the location of the CP in the case of four nonplanar control points, we can use the algorithm presented in Appendix A on two distinct subsets of the control points taken three at a time; the solution(s) common to both subsets locate the CP to within the ambiguity inherent in the given information.

The approach used to construct the example shown in Figure 6 can be extended to any number of additional points. It is based on the principle depicted in Figure 3: If the CP and any number of control points lie on the same circle, then the angle between any pair of control points and the CP will be independent of the location on the circle of the CP (and hence the location of the CP cannot be determined). Thus, we are able to construct the example shown in Figure 7, in which five control points in general position imply two solutions to the P5P

Fig. 5. An Example Showing Four Distinct Solutions to a P3P Problem.

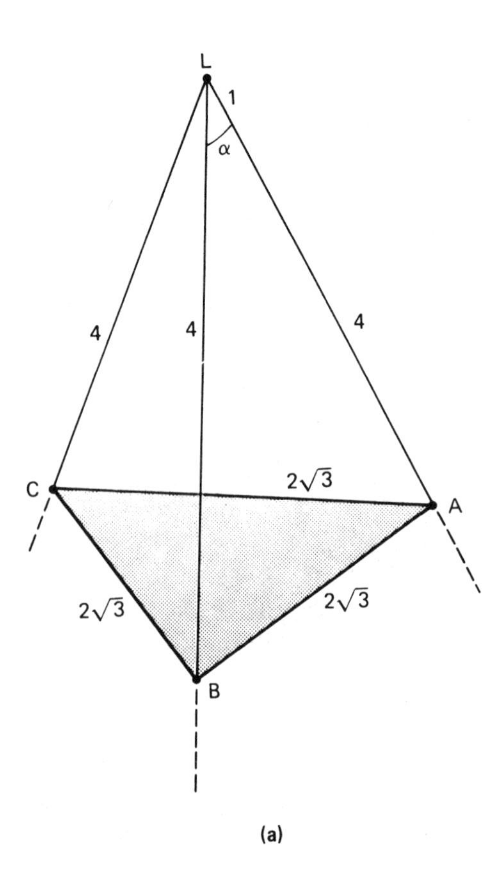

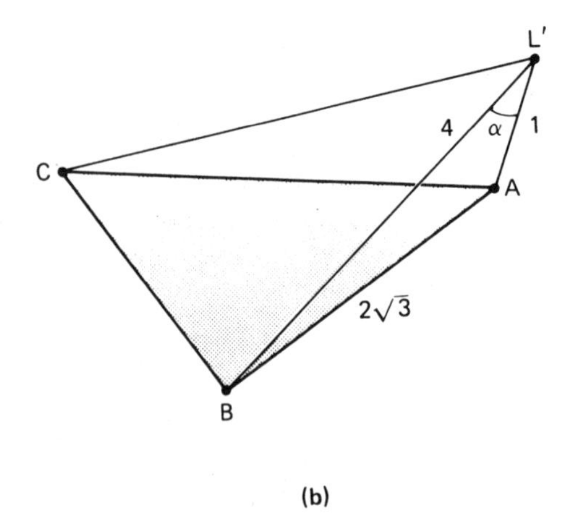

Consider the tetrahedron in Figure 5(a). The base *ABC* is an equilateral triangle and the "legs" (i.e., *LA*, *LB*, and *LC*) are all equal. Therefore, the three face angles at *L* (i.e., <*ALB*, <*ALC*, and <*BLC*) are all equal. By the law of cosines we have:

$\text{Cos}(\alpha) = 5/8.$

This tetrahedron defines one solution to a P3P problem. A second solution is shown in Figure 5(b). It is obtained from the first by rotating *L* about *BC*. It is necessary to verify that the length of *L'A* can be 1, given the rigid triangle *ABC* and the angle alpha. From the law of cosines we have:

$(2 * \sqrt{3})^2 = 4^2 + (L'A)^2 - 2*4*(L'A)*(5/8)$

which reduces to:

$(L'A - 1) * (L'A - 4) = 0.$

Therefore, *L'A* can be either 1 or 4. Figure 5(a) illustrates the *L'A* = 4 case and Figure 5(b) illustrates the *L'A* = 1 case.

Notice that repositioning the base triangle so that its vertices move to different locations on the legs is equivalent to repositioning *L*. Figure 5(c) shows the position of the base triangle that corresponds to the second solution.

Since the tetrahedron in Figure 5(a) is threefold rotationally symmetric, two more solutions can be obtained by rotating the triangle about *AB* and *AC*.

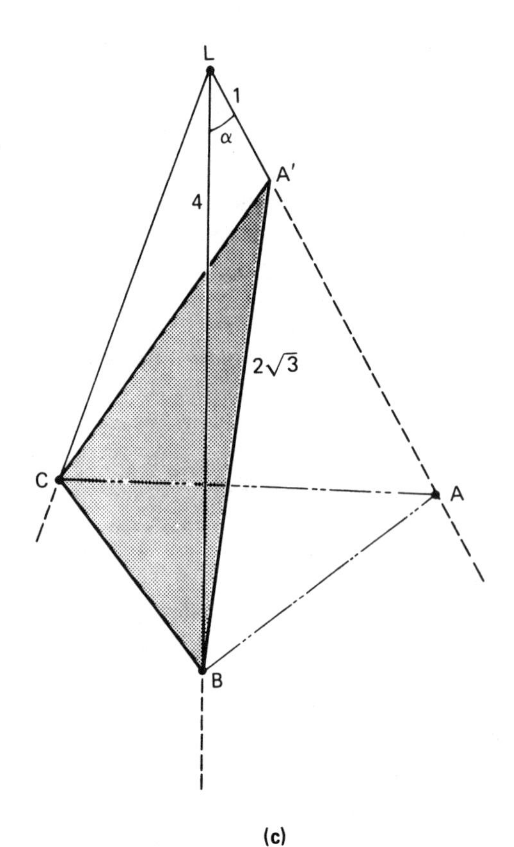

Fig. 6. An Example of a P4P Problem with Two Solutions.

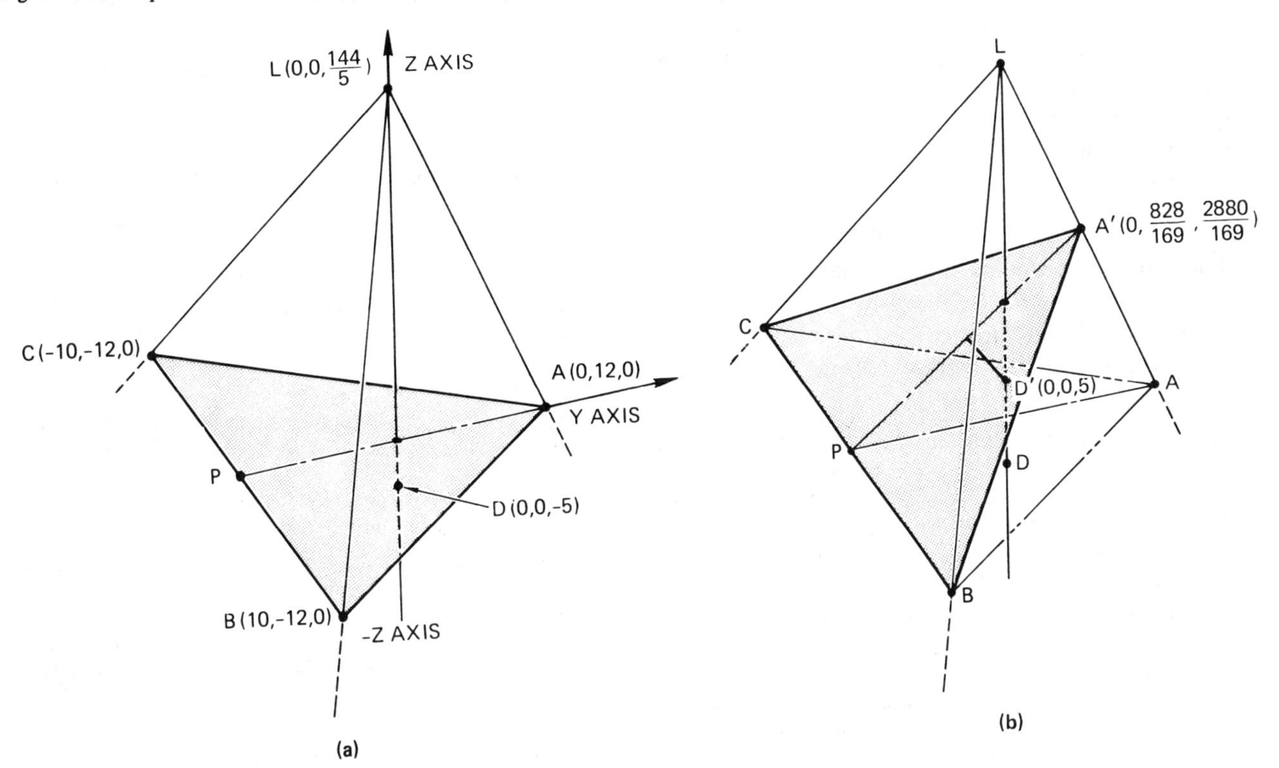

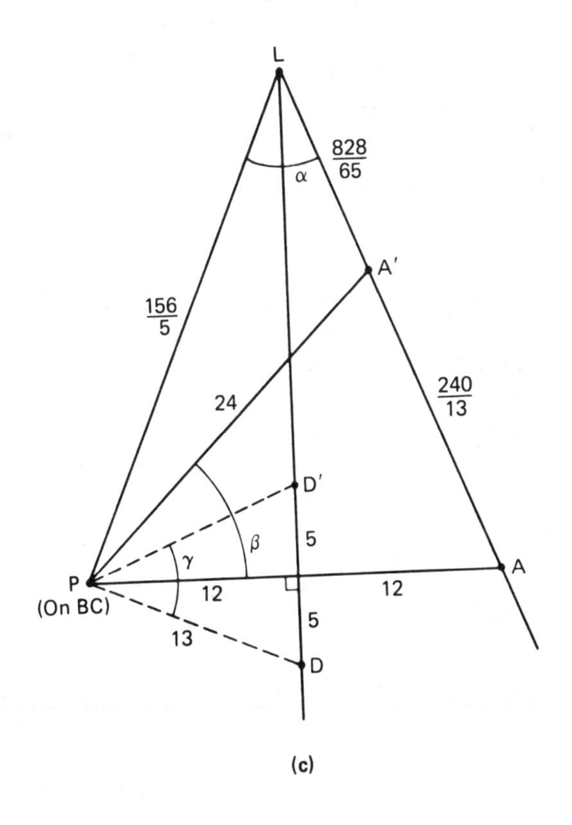

Figure 6(a) specifies a P4P problem and demonstrates one solution. A second solution can be achieved by rotating the base about BC so that A is positioned at a different point on its leg (see Figure 6(b)). To verify that this is a valid solution consider the plane $X = O$, which is normal to BC and contains the points, L, A, and D. Figure 6(c) shows the important features in this plane. The cosine of alpha is 119/169. A rotation of beta about BC repositions A at A'. The law of cosines can be used to verify the position of A'.

To complete this solution it is necessary to verify that the rotated position of D is on LD. Consider the point D' in Figure 6(c). It is at the same distance from P as D is and by the law of cosines we can show that gamma equals beta. Therefore, D', which is on LD, is the rotated position of D. The points A', B, C, and D' form the second solution to the problem.

problem. While the same technique will work for six or more control points, four or more of these points must now lie in the same plane and are thus no longer in general position.

To prove that six (or more) control points in general position will always produce a unique solution to the P6P problem, we note that for this case we can always solve for the 12 coefficients of the 3×4 matrix T that specifies the mapping (in homogeneous coordinates) from 3-space to 2-space; each of the six correspondences provides three new equations and introduces one additional unknown (the homogeneous coordinate scale factor). Thus, for six control points, we have 18 linear equations to solve for the 18 unknowns (actually, it can be shown that, at most, 17 of the unknowns are independent). Given the transformation matrix T, we can construct an additional (synthetic) control point lying in a common plane with three of the given control points and compute its location in the image plane; the technique described in Appendix B can now be used to find a unique solution.

IV. Implementation Details and Experimental Results

A. The RANSAC/LD Algorithm

The RANSAC/LD algorithm accepts as input the following data:

(1) A list L of m 6-tuples—each 6-tuple containing the 3-D spatial coordinates of a control point, its corresponding 2-D image plane coordinates, and an optional number giving the expected error (in pixels) of the given location in the image plane.

(2) The focal length of the imaging system and the image plane coordinates of the principal point.

(3) The probability $(1 - w)$ that a 6-tuple contains a gross mismatch.

(4) A "confidence" number G which is used to set the internal thresholds for acceptance of intermediate results contributing to a solution. A confidence number of one forces very conservative behavior on the algorithm; a confidence number of zero will call almost anything a valid solution.

Fig. 7. An Example of a P5P Problem with Two Solutions.

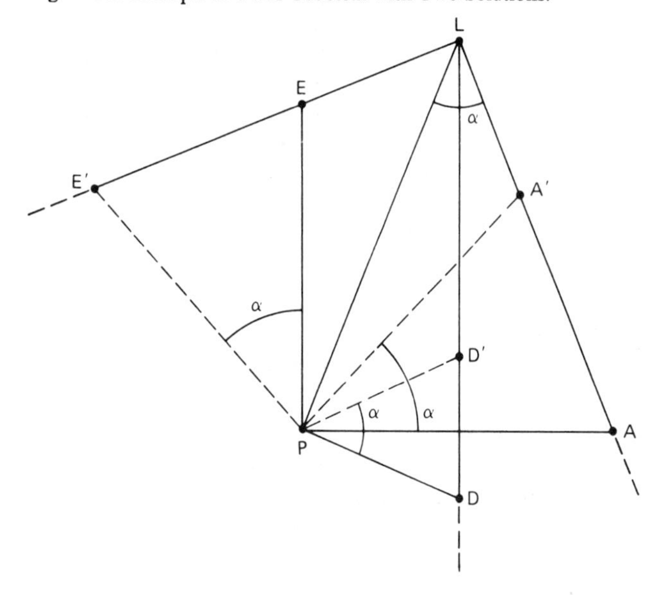

This example is the same as the P4P example described in Figure 6 except that a fifth control point, E, has been added. The initial position for E and its rotated position, E', are shown in Figure 7. The points E and E' were constructed to be the mirror images of A' and A about the line LP; therefore, a rotation of alpha about P repositions E at E'. One solution of the P5P problem is formed by points A, B, C, and D (shown in Figure 6(a)) plus point E. The second solution is formed by points A', B, C, D', and E'. Consequently there are two different positions of L such that all five points lie on their appropriate legs.

The RANSAC/LD algorithm produces as output the following information:

(1) The 3-D spatial coordinates of the lens center (i.e., the Center of Perspective), and an estimate of the corresponding error.

(2) The spatial orientation of the image plane.

The RANSAC/LD algorithm operates as follows:

(1) Three 6-tuples are selected from list L by a quasirandom method that ensures a reasonable spatial distribution for the corresponding control points. This initial selection is called S1.

(2) The CP (called CP1) corresponding to selection S1 is determined using the closed-form solution provided in Appendix A; multiple solutions are treated as if they were obtained from separate selections in the following steps.

(3) The error in the derived location of CP1 is estimated by perturbing the given image plane coordinates of the three selected control points (either by the amount specified in the 6-tuples or by a default value of one pixel), and recomputing the effect this would have on the location of the CP1.

(4) Given the error estimate for the CP1, we use the technique described in [1] to determine error ellipses (dimensions based upon the supplied confidence number) in the image plane for each of the control points specified in list L; if the associated image coordinates reside within the corresponding error ellipse, then the 6-tuple is appended to the consensus set S1/CP1.

(5) If the size of S1/CP1 equals or exceeds some threshold value t (nominally equal to a value between 7 and mw), then the consensus set S1/CP1 is supplied to a least-squares routine (see [1] or [7]) for final determination of the CP location and image plane orientation.[1] Otherwise, the above steps are repeated with a new random selection S2, S3, . . .

(6) If the number of iterations of the above steps exceeds $k = [\log(1 - G)]/[\log(1 - w^3)]$, then the largest consensus set found so far is used to compute the final solution (or we terminate in failure if this largest consensus set contains fewer than six members).

B. Experimental Results

To demonstrate the validity of our theoretical results, we performed three experiments. In the first experiment we found a specific LDP in which the common least-squares pruning heuristic failed, and showed that RANSAC successfully solved this problem. In the second experiment, we applied RANSAC to 50 synthetic problems in order to check the reliability of the approach over a wide range of parameter values. In the third experiment we used standard feature detection techniques to locate landmarks in an aerial image and then used RANSAC to determine the position and orientation of the camera.

C. A Location Determination Problem Example of a Least Squares Pruning Error

The LDP in this experiment was based upon 20 landmarks and their locations in an image. Five of the 20 correspondences were gross errors; that is, their given locations in the image were further than 10 pixels from their actual locations. The image locations for the good

[1] An alternative to least squares would be to average the parameters computed from random triples in the consensus set that fall within (say) the center 50 percent of the associated histogram.

correspondences were normally distributed about their actual locations with a standard deviation of one pixel.

The heuristic to prune gross errors was the following:

* Use all of the correspondences to instantiate a model.
* On the basis of that model, delete the correspondence that has the largest deviation from its predicted image location.
* Instantiate a new model without that correspondence.
* If the new model implies a normalized error for the deleted correspondence that is larger than three standard deviations, assume that it is a gross error, leave it out, and continue deleting correspondences. Otherwise, assume that it is a good correspondence and return the model that included it as the solution to the problem.

This heuristic successfully deleted two of the gross errors; but after deleting a third, it decided that the new model did not imply a significantly large error, so it returned a solution based upon 18 correspondences, three of which were gross errors. When RANSAC was applied to this problem, it located the correct solution on the second triple of selected points. The final consensus set contained all of the good correspondences and none of the gross errors.

D. 50 Synthetic Location Determination Problems

In this experiment RANSAC was applied to 50 synthetic LDPs. Each problem was based upon 30 landmark-to-image correspondences. A range of probabilities were used to determine the number of gross errors in the problems; the image location of a gross error was at least 10 pixels from its actual location. The location of a good correspondence was distributed about its actual location with a normal distribution having a standard deviation of one pixel. Two different camera positions were used—one looking straight down on the landmarks and one looking at them from an oblique angle. The RANSAC algorithm described earlier in this section was applied to these problems; however, the simple iterative technique described in Appendix A was used to locate solutions to the P3P problems in place of the closed form method also described in that appendix, and a second least-squares fit was used to extend the final consensus set (as suggested in Sec. II of this paper). Table I summarizes the results for ten typical problems (RANSAC successfully avoided including a gross error in its final consensus set in all of the problems); in five of these problems the probability of a good correspondence was 0.8, and in the other five problems, it was 0.6. The execution time for the current program is approximately 1 sec for each camera position considered.

E. A "Real" Location Determination Problem

Cross correlation was used to locate 25 landmarks in an aerial image taken from approximately 4,000 ft with a 6 in. lens. The image was digitized on a grid of 2,000 × 2,000 pixels, which implies a ground resolution of approximately 2 ft per pixel. Three gross errors were

Table I. Typical Experimental Results Using RANSAC.

No. of Good Correspondences	No. of Correspondences in Final Consensus Set	No. of Triples Considered	No. of Camera Positions Considered
	$w = 0.8$		
22	19	6	10
23	23	1	3
19	19	2	3
25	25	1	2
24	23	3	8
	$w = 0.6$		
21	20	11	21
17	17	1	1
17	16	6	8
18	16	9	21
21	18	9	15

made by the correlation feature detector. When RANSAC was applied to this problem, it located a consensus set of 17 on the first triple selected and then extended that set to include all 22 good correspondences after the initial least-squares fit. The final standard deviations about the camera parameters were as follows:

X: 0.1 ft	Heading: 0.01°
Y: 6.4 ft	Pitch: 0.10°
Z: 2.1 ft	Roll: 0.12°

V. Concluding Comments

In this paper we have introduced a new paradigm, Random Sample Consensus (RANSAC), for fitting a model to experimental data. RANSAC is capable of interpreting/smoothing data containing a significant percentage of gross errors, and thus is ideally suited for applications in automated image analysis where interpretation is based on the data provided by error-prone feature detectors.

A major portion of this paper describes the application of RANSAC to the Location Determination Problem (LDP): Given an image depicting a set of landmarks with known locations, determine that point in space from which the image was obtained. Most of the results we presented concerning solution techniques and the geometry of the LDP problem are either new or not generally known. The current photogrammetric literature offers no analytic solution other than variants of least squares and the Church method for solving perspective-n-point problems. The Church method, which provides an iterative solution for the P3P problem [3, 11], is presented without any indication that more than one physically real solution is possible; there is certainly no indication that anyone realizes that physically real multiple solutions are possible for more than three control points in general position. (It should be noted that because the multiple solutions can be arbitrarily close together, even when an iterative technique is initialized to a value close to the correct solution there is no assurance that it will converge to the desired value.)

In the section on the LDP problem (and associated appendices) we have completely characterized the P3P problem and provided a closed-form solution. We have shown that multiple physically real solutions can exist for the P4P and P5P problems, but also demonstrated that a unique solution is assured when four of the control points reside on a common plane (solution techniques are provided for each of these cases). The issue of determining the maximum number of solutions possible for the P4P and P5P problems remains open, but we have shown that a unique solution exists for the P6P problem when the control points are in general position.

Appendix A. An Analytic Solution for the Perspective-3-Point Problem

The main body of this paper established that P3P problems can have as many as four solutions. In this appendix a closed form expression for obtaining these solutions is derived. Our approach involves three steps: (1) Find the lengths of the legs of the ("perspective") tetrahedron given the base (defined by the three control points) and the face angles of the opposing trihedral angle (the three angles to the three pairs of control points as viewed from the CP); (2) Locate the CP with respect to the 3-*D* reference frame in which the control points were originally specified; (3) Compute the orientation of the image plane with respect to the reference frame.

1. A Solution for the Perspective Tetrahedron (see Figure 4)

Given the lengths of the three sides of the base of a tetrahedron (*Rab*, *Rac*, *Rbc*), and given the corresponding face angles of the opposing trihedral angle (θab, θac, θbc), find the lengths of the three remaining sides of the tetrahedron (*a*, *b*, *c*).

A solution to the above problem can be obtained by simultaneously solving the system of equations:

$$(Rab)^2 = a^2 + b^2 - 2*a*b*\cos(\theta ab), \quad \text{(A1)}$$
$$(Rac)^2 = a^2 + c^2 - 2*a*c*\cos(\theta ac), \quad \text{(A2)}$$
$$(Rbc)^2 = b^2 + c^2 - 2*b*c*\cos(\theta bc). \quad \text{(A3)}$$

We now proceed as follows:

$$\text{Let } b = x*a \text{ and } c = y*a, \quad \text{(A4)}$$
$$(Rac)^2 = a^2 + (y^2)*(a^2) - 2*(a^2)*y*\cos(\theta ac), \quad \text{(A5)}$$
$$(Rab)^2 = a^2 + (x^2)*(a^2) - 2*(a^2)*x*\cos(\theta ab), \quad \text{(A6)}$$
$$(Rbc)^2 = (x^2)*(a^2) + (y^2)*(a^2) - 2*(a^2)*x*y*\cos(\theta bc). \quad \text{(A7)}$$

From Eqs. (A5) and (A7)

$$[(Rbc)^2]*[1 + (y^2) - 2*y*\cos(\theta ac)] = [(Rac)^2]*[(x^2) + (y^2) - 2*x*y*\cos(\theta bc)]. \quad \text{(A8)}$$

From Eqs. (A6) and (A7)

$$[(Rbc)^2]*[1 + (x^2) - 2*x*\cos(\theta ab)] = [(Rab)^2]*[(x^2) + (y^2) - 2*x*y*\cos(\theta bc)], \quad \text{(A9)}$$

$$\text{Let } \frac{(Rbc)^2}{(Rac)^2} = K1 \quad \text{and} \quad \frac{(Rbc)^2}{(Rab)^2} = K2. \quad \text{(A10)}$$

From Eqs. (A8) and (A9)

$$0 = (y^2)*[1 - K1] + 2*y*[K1*\cos(\theta ac) - x*\cos(\theta bc)] + [(x^2) - K1]. \quad \text{(A11)}$$

From Eqs. (A9) and (A10)

$$0 = (y^2) + 2*y*[-x*\cos(\theta bc)] + [(x^2)*(1 - K2) + 2*x*K2*\cos(\theta ab) - K2]. \quad \text{(A12)}$$

Equations (A11) and (A12) have the form:

$$0 = m*(y^2) + p*y + q, \quad \text{(A13)}$$
$$0 = m'*(y^2) + p'*y + q'. \quad \text{(A14)}$$

Multiplying Eqs. (A13) and (A14) by m' and m, respectively, and subtracting,

$$0 = [p*m' - p'*m]*y + [m'*q - m*q']. \quad \text{(A15)}$$

Multiplying Eqs. (A13) and (A14) by q' and q, respectively, substracting, and dividing by y,

$$0 = [m'*q - m*q']*(y^2) + [p'*q - p*q']*y,$$
$$0 = [m'*q - m*q']*y + [p'*q - p*q']. \quad \text{(A16)}$$

Assuming $m'*q \neq m*q'$,

$$[(x^2) - K1] \neq [(x^2)*(1 - K1)*(1 - K2) + 2*x*K2*(1 - K1)*\cos(\theta ab) - (1 - K1)*K2],$$

then Eqs. (A15) and (A16) are equivalent to Eqs. (A13) and (A14). We now multiply Eqs. (A15) by $(m'*q - m*q')$, and Eq. (A16) by $(p*m' - p'*m)$, and subtract to obtain

$$0 = (m'*q - m*q')^2 - [p*m' - p'*m]*[p'*q - p*q']. \quad \text{(A17)}$$

Expanding Eq. (A17) and grouping terms we obtain a biquadratic (quartic) polynomial in x:

$$0 = G4*(x^4) + G3*(x^3) + G2*(x^2) + G1*(x) + GO, \quad \text{(A18)}$$

where

$$G4 = (K1*K2 - K1 - K2)^2 - 4*K1*K2*[\cos(\theta bc)^2], \quad \text{(A19)}$$

$$G3 = 4*[K1*K2 - K1 - K2]*K2*(1 - K1)*\cos(\theta ab) + 4*K1*\cos(\theta bc)*[(K1*K2 + K2 - K1)*\cos(\theta ac) + 2*K2*\cos(\theta ab)*\cos(\theta bc)], \quad \text{(A20)}$$

$$G2 = [2*K2*(1 - K1)\cos(\theta ab)]^2 + 2*[K1*K2 + K1 - K2]*[K1*K2 - K1 - K2] + 4*K1*[(K1 - K2)*(\cos(\theta bc)^2) + (1 - K2)*K1*(\cos(\theta ac)^2) - 2*K2*(1 + K1)*\cos(\theta ab)*\cos(\theta ac)*\cos(\theta bc)], \quad \text{(A21)}$$

$$G1 = 4*(K1*K2 + K1 - K2)*K2*(1 - K1)*\cos(\theta ab) + 4*K1*[(K1*K2 - K1$$

$$+ K2)^* \cos(\theta ac)^* \cos(\theta bc) + 2^*K1^*K2^* \cos(\theta ab)^* (\cos(\theta ac)^2)], \quad (A22)$$

$$G0 = (K1^*K2 + K1 - K2)^2 - 4^*(K1^2)^*K2^* (\cos(\theta ac)^2). \quad (A23)$$

Roots of Eq. (A18) can be found in closed form [5], or by iterative techniques [4]. For each positive real root of Eq. (A18), we determine a single positive real value for each of the sides a and b. From Eq. (A6) we have

$$a = \frac{Rab}{SQRT[(x^2) - 2^*x^* \cos(\theta ab) + 1]}, \quad (A24)$$

and from Eq. (A4) we obtain

$$b = a^*x. \quad (A25)$$

If $m'^*q \neq m^*q'$, then from Eq. (A16) we have

$$y = \frac{p'^*q - p^*q'}{m^*q' - m'^*q}. \quad (A26)$$

If $m'^*q = m^*q'$, then Eq. (A26) is undefined and we obtain two values of y from Eq. (A5):

$$y = \cos(\theta ac) \pm SQRT\left[(\cos(\theta ac))^2 + \frac{(Rac)^2 - (a^2)}{(a^2)}\right]. \quad (A27)$$

For each real positive value of y, we obtain a value of c from Eq. (A4):

$$c = y^*a \quad (A28)$$

When values of y are obtained from Eq. (A5) rather than Eq. (A26), the resulting solutions can be invalid; they must be shown to satisfy Eq. (A3) before they are accepted.

It should be noted that because each root of Eq. (A18) can conceivably lead to two distinct solutions, the existence of the biquadratic does not by itself imply a maximum of four solutions to the P3P problem; some additional argument, such as the one given in the main body of this paper, is necessary to establish the upper bound of four solutions.

2. Example

For the perspective tetrahedron shown in Figure 5, we have the following parameters:

$$Rab = Rac = Rbc = 2^*SQRT(3),$$
$$\cos(\theta ab) = \cos(\theta ac) = \cos(\theta bc) = \frac{(a^2) + (b^2) - (Rab)^2}{2^*a^*b} = \frac{20}{32}.$$

Substituting these values into Eqs. (A19) through (A23), we obtain the coefficients of the biquadratic defined in Eq. (A18):

[−0.5625, 3.515625, −5.90625, 3.515625, −0.5625]

The roots of the above equation are

[1, 1, 4, 0.25]

For each root

Root	a	b	y	c
1	4	4	1	4
1	4	4	0.25	1
4	1	4	4	4
0.25	4	1	1	4

3. An Iterative Solution for the Perspective Tetrahedron (see Figure 8)

A simple way to locate solutions to P3P problems, which is sometimes an adequate substitute for the more involved procedure described in the preceding subsection, is to slide one vertex of the control-point triangle down its leg of the tetrahedron and look for positions of the triangle in which the other two vertices lie on their respective legs. If vertex A is at a distance a from L (L is the center of perspective), the lengths of the sides Rab and Rac restrict the triangle to four possible positions. Given the angle between legs LA and LB, compute the distance of point A from the line LB and then compute points $B1$ and $B2$ on LB that are the proper distance from A to insert a line segment of length Rab. Similarly, we compute at most two locations for C on its leg. Thus, given a position for A we have found at most four positions for a triangle that has one side of length Rab and one of length Rac. The lengths of the third sides (BC) of the four triangles vary nonlinearly as point A is moved down its leg. Solutions to the problem can be obtained by iteratively repositioning A to imply a third side of the required length.

4. Computing the 3-D Location of the Center of Perspective (see Figure 9)

Given the three-dimensional locations of the three control points of a perspective tetrahedron, and the lengths of the three legs, the 3-D location of the center of perspective can be computed as follows:

(1) Construct a plane $P1$ that is normal to AB and passes through the center of perspective, L. This plane can be constructed without knowing the position of L, which is what we are trying to compute. Consider the face of the tetrahedron that contains vertices A, B, and L. Knowing the lengths of sides LA, LB and AB, we can use the law of cosines to find the angle LAB, and then the projection QA of LA on AB. (Note that angle LQA is a right angle, and the point Q is that point on line AB that is closest to L). Construct a plane normal to AB passing through Q; this plane also passes through L.

(2) Similarly construct a plane $P2$ that is normal to AC and passes through L.

(3) Construct the plane $P3$ defined by the three points A, B, and C.

(4) Intersect planes $P1$, $P2$, and $P3$. By construction, the point of intersection R is the point on $P3$ that is closest to L.

(5) Compute the length of the line AR and use that in conjunction with the length of LA to compute the length of the line RL, which is the distance of L from the plane $P3$.

(6) Compute the cross product of vectors AB and AC to form a vector perpendicular to $P3$. Then scale that vector by the length of RL and add it to R to get the 3-D location of the center of perspective L.

Fig. 8. Geometry for an Iterative Solution to the P3P Problem.

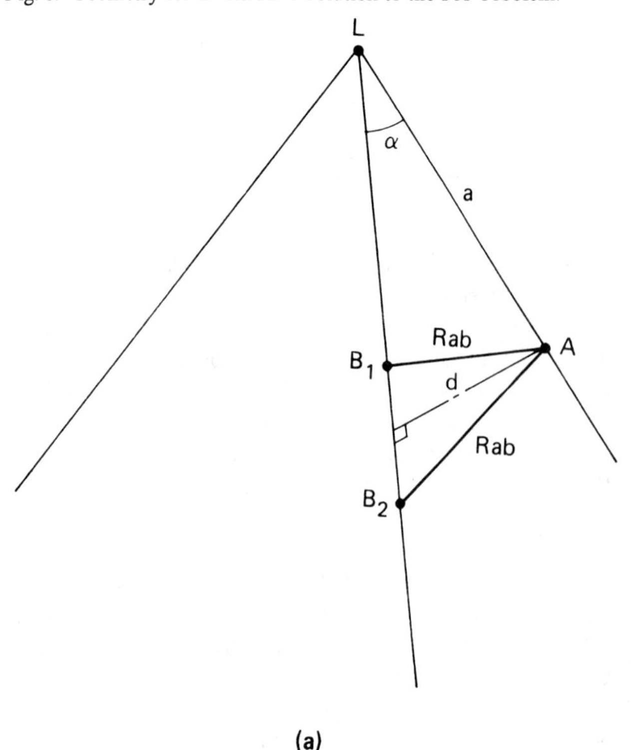

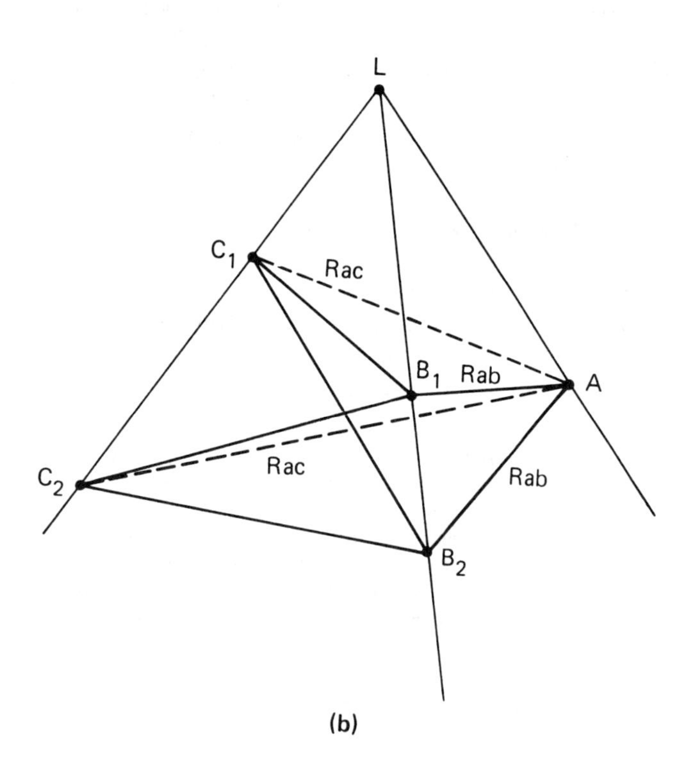

If the focal length of the camera and the principal point in the image plane are known, it is possible to compute the orientation of the image plane with respect to the world coordinate system; that is, the location of the origin and the orientation of the image plane coordinate system with respect to the 3-*D* reference frame. This can be done as follows:

Fig. 9 Computing the 3-*D* Location of the Center of the Perspective (*L*).

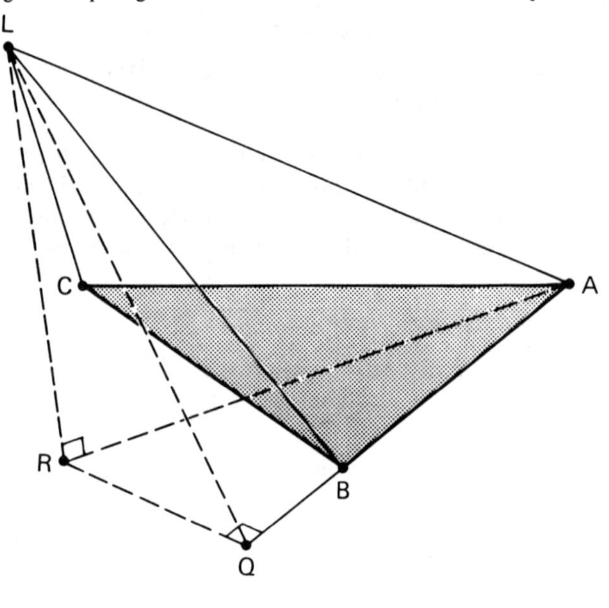

(1) Compute the 3-*D* reference frame coordinates of the center of perspective (as described above).

(2) Compute the 3-*D* coordinates of the image locations of the three control points: since we know the 3-*D* coordinates of the CP and the control points, we can compute the 3-*D* coordinates of the three rays between the CP and the control points. Knowing the focal length of the imaging system, we can compute, and subtract from each ray, the distance from the CP to the image plane along the ray.

(3) Compute the equation of the plane containing the image using the three points found in step (2). The normal to this plane, passing through the CP, gives us the origin of the image plane coordinate system (i.e., the 3-*D* location of the principal point), and the *Z* axis of this system.

(4) The orientation of the image plane about the *Z* axis can be obtained by computing the 3-*D* coordinates of a vector from the principal point to any one of the points found in step (2).

Appendix B. An Analytic Solution for the Perspective-4-Point Problem (with all control points lying in a common plane)

In this appendix an analytic technique is presented for obtaining a unique solution to the P4P problem when the four given control points all lie in a common plane.

1. Problem Statement (see Figure 10)

GIVEN: A correspondence between four points lying in a plane in 3-*D* space (called the object plane), and four points lying in a distinct plane (called the image plane); and given the distance between the center of perspective and the image plane (i.e., the focal length of the imaging system); and also given the principal point in the image plane (i.e., the location, in image plane coordinates, of the point at which the optical axis of the lense pierces the image plane).

FIND: the 3-*D* location of the center of perspective relative to the coordinate system of the object plane.

Fig. 10. Geometry of the P4P Problem (With all Control Points Lying in a Common Plane).

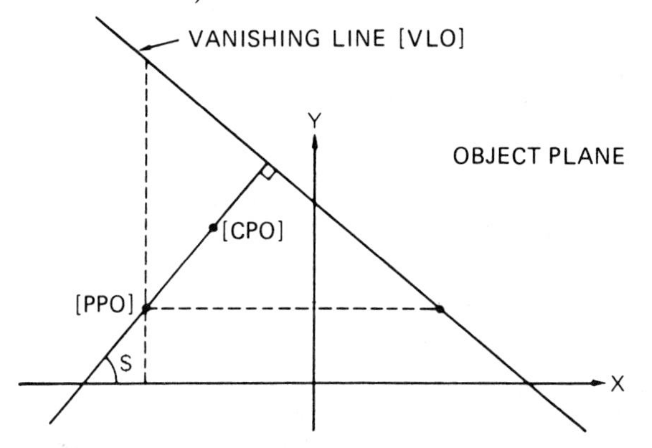

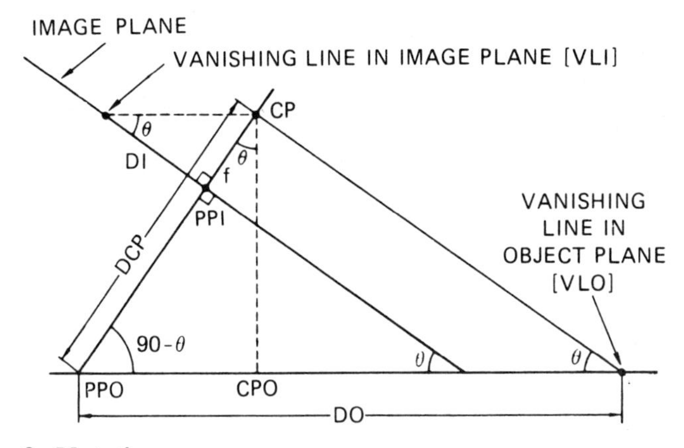

2. Notation

* Let the four given image points be labeled $\{Pi\}$, and the four corresponding object points $\{Qi\}$.
* We will assume that the 2-*D* image plane coordinate system has its origin at the principal point (*PPI*).
* We will assume that the object plane has the equation $Z = 0$ in the reference coordinate system. Standard techniques are available to transform from this coordinate system into a ground reference frame (e.g., see [6] or [9]).
* Homogeneous coordinates will be assumed [12].
* Primed symbols represent transposed structures.

3. Solution Procedure

(a) Compute the 3×3 collineation matrix T which maps points from object plane to image plane (a procedure for computing T is given later):

$$[Pi] = [T]*[Qi],$$

where

$$[Pi] = [ki*xi, ki*yi, ki]', \quad [Qi] = [Xi, Yi, 1]'. \qquad \text{(B1)}$$

(b) The ideal line in the object plane, with coordinates $[0, 0, 1]'$ is mapped into the vanishing line in the image plane $[VLI]$ by the transformation:

$$[VLI] = [\text{inv}[T]]'*[0, 0, 1]'. \qquad \text{(B2)}$$

(c) Determine the distance DI from the origin of the image plane (PPI) to the vanishing line $[VLI] = [a1, a2, a3]'$:

$$DI = \left| \frac{a3}{sqrt[(a1)^2 + (a2)^2]} \right|. \qquad \text{(B3)}$$

(d) Solve for the dihedral (tilt) angle θ between the image and object planes:

$$\theta = \arctan(f/DI), \qquad \text{(B4)}$$

where f = focal length.

(e) The ideal line in the image plane with coordinates $[0, 0, 1]'$ is mapped into the vanishing line in the object plane $[VLO]$ by the transform

$$[VLO] = [T]'*[0, 0, 1]'. \qquad \text{(B5)}$$

(f) Compute the location of point $[PPO]$ in the object plane ($[PPO]$ is the point at which the optical axis of the lense pierces the object plane):

$$[PPO] = [\text{inv}[T]]'*[0, 0, 1]' \qquad \text{(B6)}$$

(g) Compute the distance DO from $[PPO] = [c1, c2, c3]'$ to the vanishing line $[VLO] = [b1, b2, b3]'$ in the object plane:

$$DO = \left| \frac{b1*c1 + b2*c2 + b3*c3}{c3*sqrt[(b1)^2 + (b2)^2]} \right|. \qquad \text{(B7)}$$

(h) Solve for the "pan" angle \$ as the angle between the normal to $[VLO] = [b1, b2, b3]'$ and the X axis in the object plane:

$$\$ = \arctan(-b2/b1). \qquad \text{(B8)}$$

(i) Determine *XSGN* and *YSGN*: If a line (parallel to the X axis in the object plane) through $[PPO]$ intersects $[VLO]$ to the right of $[PPO]$, then XSGN = 1. Otherwise $XSGN = -1$. Thus,

$$\text{if } \frac{b1*c1 + b2*c2 + b3*c3}{b1*c3} < 0,$$

then $XSGN = 1$, otherwise $XSGN = -1$. (B9)

Similarly,

$$\text{if } \frac{b1*c1 + b2*c2 + b3*c3}{b2*c3} < 0$$

then $YSGN = 1$, otherwise $YSGN = -1$. (B10)

(j) Solve for the location of the CP in the object plane coordinate system:

$$DCP = DO*\sin(\theta) \qquad \text{(B11)}$$

$$XCP = XSGN*abs[DCP*\sin(\theta)*\cos(\$)] + c1/c3 \qquad \text{(B12)}$$

$$YCP = YSGN*abs[DCP*\sin(\theta)*\sin(\$)] + c2/c3 \qquad \text{(B13)}$$

$$ZCP = DCP*\cos(\theta) \qquad \text{(B14)}$$

Note: If $[VLI]$, as determined in (b), has the coordinates $[0, 0, k]$, then the image and object planes are parallel ($\theta = 0$). Rather than continuing with the above procedure, we now solve for the desired information using similar triangles and Euclidean geometry.

4. Computing the Collineation Matrix T

Let

$$[Q] = \begin{Vmatrix} X1 & Y1 & 1 \\ X2 & Y2 & 1 \\ X3 & Y3 & 1 \end{Vmatrix} = [[Q1]', [Q2]', [Q3]'],$$

$$[P] = \begin{Vmatrix} x1 & y1 & 1 \\ x2 & y2 & 1 \\ x3 & y3 & 1 \end{Vmatrix} = [[P1]', [P2]', [P3]'],$$

$$[Q4] = [X4,\ Y4,\ 1]',$$
$$[P4] = [x4,\ y4,\ 1]',$$
$$[V] = [\mathrm{inv}[P]]' * [P4] = [v1,\ v2,\ v3]',$$
$$[R] = [\mathrm{inv}[Q]]' * [Q4] = [r1,\ r2,\ r3]',$$
$$w1 = \frac{v1}{r1} * \frac{r3}{v3},$$
$$w2 = \frac{v2}{r2} * \frac{r3}{v3},$$
$$[w] = \begin{Vmatrix} w1 & 0 & 0 \\ 0 & w2 & 0 \\ 0 & 0 & 1 \end{Vmatrix}.$$

Then,

$$[T]' = [\mathrm{inv}[Q]] * [W] * [P]$$

such that

$$[Pi] = ki * [xi,\ yi,\ 1] = [T] * [Qi].$$

5. Example

Given:

$f = 0.3048$ m (12 in.)

$P1 = (-0.071263,\ 0.029665)$	$Q1 = (-30,\ 80)$
$P2 = (-0.053033,\ -0.006379)$	$Q2 = (-100,\ -20)$
$P3 = (-0.014063,\ 0.061579)$	$Q3 = (140,\ 50)$
$P4 = (0.080120,\ -0.030305)$	$Q4 = (-40,\ -240)$

(a)

$$[T]' = \begin{Vmatrix} 0.000212 & 0.000236 & 0.000925 \\ -0.000368 & 0.000137 & 0.000534 \\ -0.025404 & 0.021650 & 0.843879 \end{Vmatrix}$$

$$[\mathrm{inv}[T]]' = \begin{Vmatrix} 1117.14 & -2038.86 & 0.0 \\ 3371.56 & 2302.22 & -5.14991 \\ -51.0636 & -120.442 & 1.31713 \end{Vmatrix}$$

(b) $[VLI] = [0,\ -5.14991,\ 1.31713]'$

(c) $DI = 0.255758$

(d) $\theta = 0.872665$ rad $(50°)$

(e) $[VLO] = [0.000925,\ 0.000534,\ 0.843880]'$

(f) $[PPO] = [-51.0636,\ -120.442,\ 1.31713]'$

(g) $DO = 711.196$

(h) $\$ = -0.523599$ rad $(-30°)$

(i) $XSGN = -1$
$YSGN = -1$

(j) $DCP = 544.8081$
$XCP = -400.202$
$YCP = -300.117$
$ZCP = 350.196$

Received 4/80; revised 1/81; accepted 1/81

References

1. Bolles, R.C., Quam, L.H., Fischler, M.A., and Wolf, H.C. The SRI road expert: Image to database correspondence. In Proc. Image Understanding Workshop, Pittsburgh, Pennsylvania, Nov., 1978.
2. Chrystal, G. *Textbook of Algebra* (Vol 1). Chelsea, New York, New York 1964, p. 415.
3. Church, E. Revised geometry of the aerial photograph. *Bull. Aerial Photogrammetry.* 15, 1945, Syracuse University.
4. Conte, S.D. *Elementary Numerical Analysis.* McGraw Hill, New York, 1965.
5. Dehn, E. *Algebraic Equations.* Dover, New York, 1960.
6. Duda, R.O., and Hart, P.E. *Pattern Classification and Scene Analysis.* Wiley-Interscience, New York, 1973.
7. Gennery, D.B. Least-squares stereo-camera calibration. Stanford Artificial Intelligence Project Internal Memo, Stanford, CA 1975.
8. Keller, M. and Tewinkel, G.C. Space resection in photogrammetry. ESSA Tech. Rept C&GS 32, 1966, U.S. Coast and Geodetic Survey.
9. Rogers, D.P. and Adams, J.A. *Mathematical Elements for Computer Graphics.* McGraw Hill, New York, 1976.
10. Sorensen, H.W. Least-squares estimation: from Gauss to Kalman. *IEEE Spectrum* (July 1970), 63–68.
11. Wolf, P.R. *Elements of Photogrammetry.* McGraw Hill, New York, 1974.
12. Wylie, C.R. Jr. *Introduction to Projective Geometry.* McGraw-Hill, New York, 1970.

Detection of Roads and Linear Structures in Low-Resolution Aerial Imagery Using a Multisource Knowledge Integration Technique*

M. A. FISCHLER, J. M. TENENBAUM, AND H. C. WOLF

Artificial Intelligence Center, Computer Science and Technology Division, SRI International, 333 Ravenswood Avenue, Menlo Park, California 94025

Received October 26, 1979; revised February 4, 1980; accepted March 25, 1980

This paper describes a computer-based approach to the problem of detecting and precisely delineating roads, and similar "line-like" structures, appearing in low-resolution aerial imagery. The approach is based on a new paradigm for combining local information from multiple, and possibly incommensurate, sources, including various line and edge detection operators, map knowledge about the likely path of roads through an image, and generic knowledge about roads (e.g., connectivity, curvature, and width constraints). The final interpretation of the scene is achieved by using either a graph search or dynamic programming technique to optimize a global figure of merit. Implementation details and experimental results are included.

1. INTRODUCTION

Given the problem of producing an overlay showing the clearly visible roads in an aerial image, a person would normally be expected to accomplish this task with little difficulty, even though he may be completely unfamiliar with the terrain depicted in the image. Our purpose in this paper is to clarify the nature of this task and some of its generalizations. In particular, we wish to specify the requirements and mechanisms for a machine to be capable of near-human performance in finding roads and other semantically meaningful linear structures in aerial imagery.

A. Performance Criteria

Our goal is to produce a list of connected points for each segment of road which is tracked in the input image. Each such track is a delineation of the actual road and should have the following properties:

(1) No point on a track should be located outside of the road boundaries when the road is clearly visible.

(2) The track should be smooth where the road is straight or smoothly curving (within the constraints of a digital raster representation).

(3) If parts of the road are occluded, those portions of the continuous track overlaying the occluded segments should be labeled as such.

(4) In areas where the road is partially occluded, the track should follow the actual center of the road (as opposed to the center of the visible portion). If the road is composed of adjacent but separated lanes, then each lane will be considered a separate road for our purposes.

*The work reported herein was supported by the Defense Advanced Research Projects Agency under Contracts DAAG29-76-C-0057 and MDA903-79-C-0588; and by the U.S. Army Engineer Topographic Laboratory under Contract DAAK70-78-C-0114.

B. Contextual Settings for Road Tracking

A "road" is a functionally defined entity whose appearance in an image depends largely on its width and how much internal road detail is visible; i.e., appearance depends largely on image resolution (see Fig. 1). Additional factors having a major effect on visually locating roads in imagery include the visible extent of the road, its contrast with the adjacent terrain, the presence of nearby linear structures, and any prior knowledge about the actual shape of the road and its location in the image.

FIG. 1. Road scenes depicted at a spectrum of resolutions.

We have found that the following contextual settings require significantly different approaches to the road tracking problem:

(1) High vs low resolution (low resolution is defined as roads having an image width of three or fewer pixels).

(2) Clear vs occluded viewing (clear viewing is defined as a situation in which no more than approximately 30% of the road being tracked is occluded by clouds, intervening objects, etc.).

(3) High vs low density of linear detail (nominally, this distinction corresponds to urban vs rural scenes).

In this paper we will mainly be concerned with tracking roads in clear imagery of rural scenes at low resolution. A robust technique for tracking roads in high-resolution imagery was previously reported (Quam [14]). We note that in the case of high-resolution imagery, once the road has been "acquired" and we are able to track features internal to the road boundaries, the surrounding detail is of minor importance (except as it introduces shadows and oeclusions); thus, the distinction between urban and rural scenes is important mainly at low resolution. Where the roads are heavily occluded, road matching rather than road tracking is the appropriate technique; here one needs to have prior knowledge of the geometry of the road networks being searched for. Prior knowledge about the (approximate) location and/or direction of the roads in the imagery is important if a specific road (as opposed to all roads) is to be tracked; some method of indicating which road we are interested in is necessary, and this is typically done by delimiting a search area in the input image. Finally, prior knowledge about terrain type and/or scene elevations can be used to help distinguish low-resolution roads from other linear features by invoking cultural and economic constraints which are known to affect road construction.

In the following section, we argue that there is no currently available single coherent model suitable for reliable detection of local road presence. It is thus essential that some means for integrating information from multiple (incommensurate) image operators and knowledge sources be devised. We present a general paradigm for this multisource integration task, and describe its specific application to the problem of detecting linear structures.

2. LOW-RESOLUTION ROAD TRACKING

At low resolution, roads are often indistinguishable from other linear features appearing in the image (including artifacts, such as scratches). Thus, the low-resolution road tracking problem largely reduces to the general problem of line (as opposed to edge) following. Nevertheless, there are still some weak semantics that can be invoked to specifically tailor a system for road tracking, trading some generality for significant increases in performance.

A. The Basic Paradigm

The basic paradigm we employ is to first evaluate all local evidence for the presence of a road at every location in the search area (a low numeric value indicates a high likelihood that the given image point lies on a road), and then find a single track which, while satisfying imposed constraints (such as continuity), minimizes the sum of the local evaluation scores (costs) associated with every point along the track. While the basic optimization paradigm is not new (e.g., Fischler [5], Montanari [12], Martelli [11], Barrow and Tenenbaum [1], and Rubin [18]), it is incomplete in that it does not provide mechanisms for reconciling incommensurate sources of information. This capability is crucial in problems such as road tracking in which no single coherent model is adequate for reliable detection. In this paper we introduce the following new and relatively simple mechanisms for combining local evidence and constraints in the context of an optimization paradigm for detecting linear structures:

—Partitioning image operators into two classes, based on their error characteristics: Type I operators that will almost never incorrectly classify artifacts as instances of the structure they are searching for, but may often miss correct instances; versus Type II operators that accurately measure relevant parameters of all true instances but may falsely classify and incorrectly parameterize noninstances.

—Differential use of these two classes of operators: operators with good classification ability provide a framework which gets filled in by information supplied by operators providing precise local feature characterization.

—Transformation of individual operator scores to a scale in which differential numerical values are responsive to externally supplied knowledge about the particular scene being processed (e.g., knowledge about road shape or occlusions).

B. Detecting Local Road Presence—Road Operators and Models

At low resolution, roads are line-like structures of essentially constant width, which, in general, are locally constant in intensity in the along-track direction and show significant contrast with the adjacent terrain (generally, they are either uniformly lighter or darker). A specific interpretation of this low-resolution road model is embodied in the Duda Road Operator (DRO)[1] described in the Appendix. In Fig. 2 we show some examples of the scores produced by this operator on a variety of road scenes. It is apparent that the DRO does a good job most of the time but has some significant weaknesses; it is sensitive to (a) road orientation (in directions other than the four principal directions explicitly covered by the masks described in Fig. A1), (b) raster quantization effects (e.g., where a straight line segments "jogs" in crossing a quantization boundary), (c) sharp changes in road direction, and (d) certain contrast problems with the adjacent terrain.

At this point one might wonder if a special road operator is really required; why not simply use a generic edge detector (e.g., Sobel [in Duda and Hart [4], Roberts [15], or Hueckel [9, 10])? Even more to the point, we notice that it is possible to interpret the effect of employing an operator on an image as resulting in the suppression of all detail other than that associated with the entity to be detected; therefore, a high-pass filter might act as a perfectly good road operator. Finally, roads will generally be lighter or darker than the immediately adjacent terrain; why not simply use the actual intensity values (contrast enhanced and possibly inverted, depending on the relative brightness between the road and adjacent terrain)? In

[1]Suggested by R. O. Duda of SRI International; other similar operators specialized for line detection are described in Rosenfeld and Thurston [17] and in VanderBrug [19, 20].

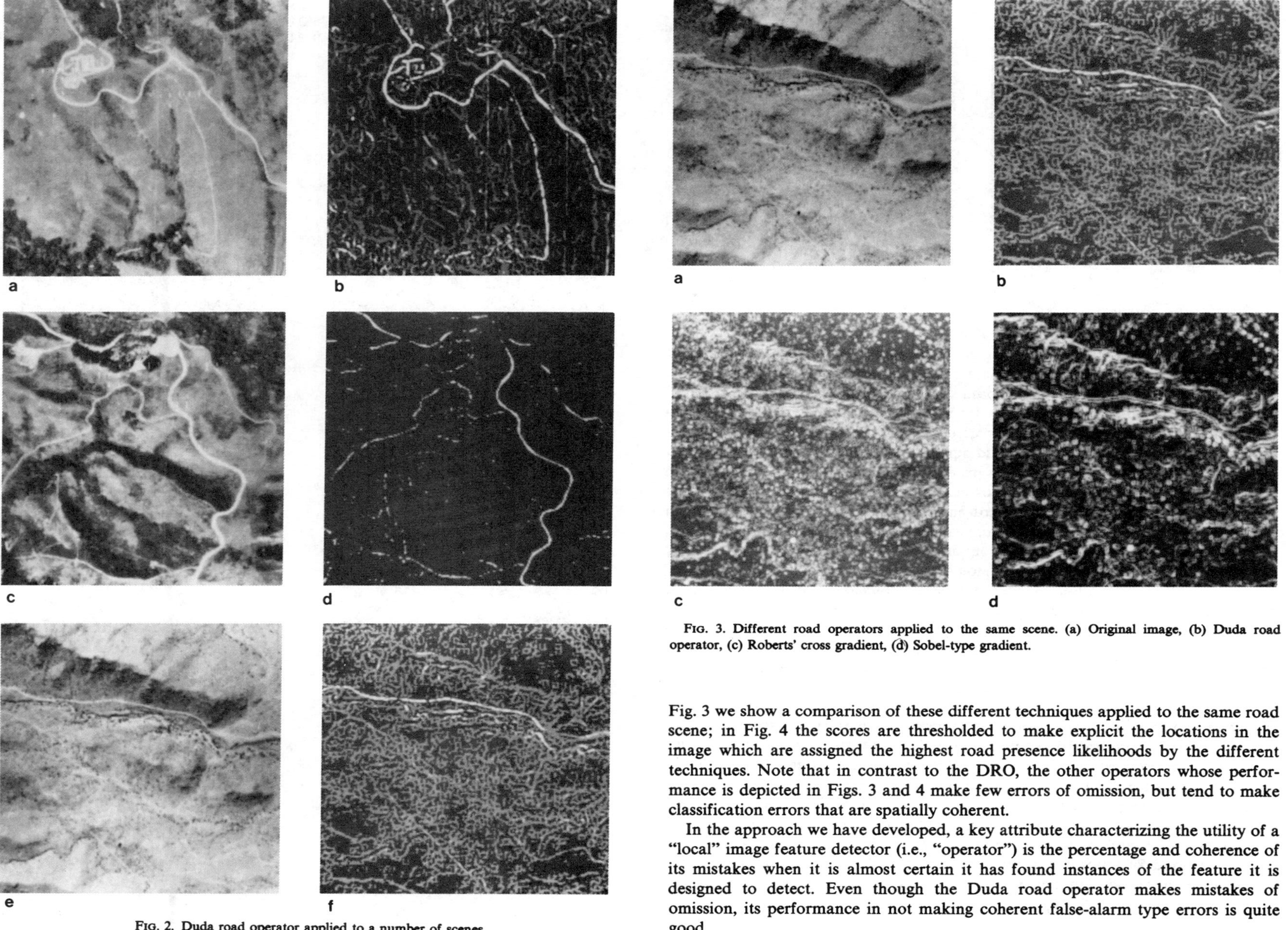

FIG. 2. Duda road operator applied to a number of scenes.

FIG. 3. Different road operators applied to the same scene. (a) Original image, (b) Duda road operator, (c) Roberts' cross gradient, (d) Sobel-type gradient.

Fig. 3 we show a comparison of these different techniques applied to the same road scene; in Fig. 4 the scores are thresholded to make explicit the locations in the image which are assigned the highest road presence likelihoods by the different techniques. Note that in contrast to the DRO, the other operators whose performance is depicted in Figs. 3 and 4 make few errors of omission, but tend to make classification errors that are spatially coherent.

In the approach we have developed, a key attribute characterizing the utility of a "local" image feature detector (i.e., "operator") is the percentage and coherence of its mistakes when it is almost certain it has found instances of the feature it is designed to detect. Even though the Duda road operator makes mistakes of omission, its performance in not making coherent false-alarm type errors is quite good.

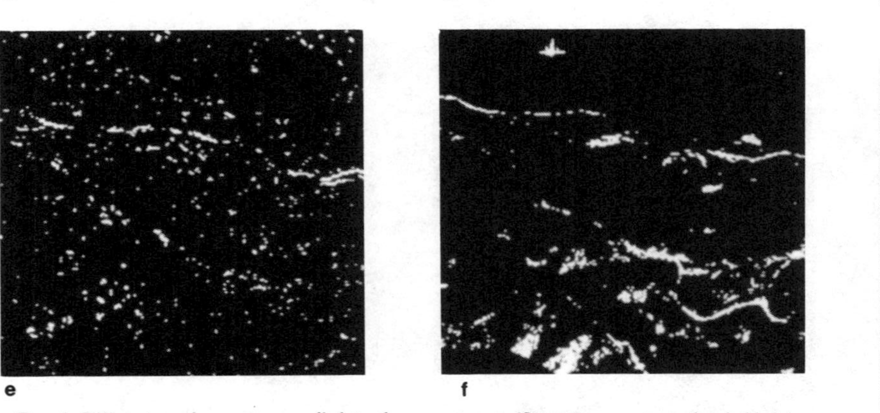

FIG. 4. Different road operators applied to the same scene. (Operator scores are thresholded to highlight the locations assigned the best scores.) (a) Original image, (b) Duda road operator, (c) Roberts' cross gradient, (d) Sobel-type gradient, (e) Hueckel line operator, (f) intensity.

C. Combining Incommensurate Sources of Knowledge—An Elaboration of the Basic Optimization Paradigm

We will now specify a general approach for combining the results deduced by the application of a set of (road) operators, and for introducing prior knowledge and constraints to influence the answer produced by the optimization algorithm.

We partition our inventory of operators into two categories—Type I operators, each of which can be adjusted to make very few coherent errors in detecting instances of the relevant feature when the feature is not present (possibly at the cost of making a large number of omission errors); and Type II operators, each of which can be adjusted to reliably give a quantitative indication of the presences of the feature when it is actually under examination (but these operators might be very unreliable in their assertions when examining something other than the desired feature). Our basic approach is to strongly bias (or even constrain) the desired answer to fit the coherent pattern produced by a superposition of evidence provided by all the Type I operators and to fill in the details locally, using that particular Type II operator which seems to be most certain that it has found the desired feature. (A more comprehensive discussion of methods for combining multisource evidence is given in Fischler and Garvey [6].)

A problem that immediately arises is how to combine the results of several Type I and Type II operators. By considering the output of Type I operators to be valid binary decisions, we have made them commensurate and can logically combine their outputs. In the context of tracking roads (or other linear features), we scan each of our Type I operators over some specified region of interest and create a binary overlay mask containing the logical union of the locations at which one or more of these operators has detected the presence of a road with high likelihood. An example of such a mask, called a "perfect road score" (PRS) mask, is compared in Fig. 5 with the road image from which it was derived.

The problem of combining the results produced by a set of Type II operators has no acceptable solution when the values they return are not probabilities or other commensurate quantities. However, Type I and Type II operator scores can be

FIG. 5. A scene and its perfect road score mask.

combined, since a positive Type I output can always be set to the maximum value (zero cost) on the likelihood scale of any Type II operator.

Our approach is thus to modify the cost array (CA) produced by each Type II operator so that zero cost is incurred at those locations marked in the PRS mask as places at which there is a very high likelihood of local road presence. The optimization algorithm is separately applied to each CA, and a best path through each such array is independently computed. Since there is no way to directly compare the relative quality of these alternative solutions, we employ the following heuristic: the average cost per pixel along the track in each Type II array is calculated. This average cost is normalized by determining its ranking in a histogram of costs obtained by the same operator in a region surrounding the road track. The track with the highest normalized ranking is selected as the primary road track through the given region.

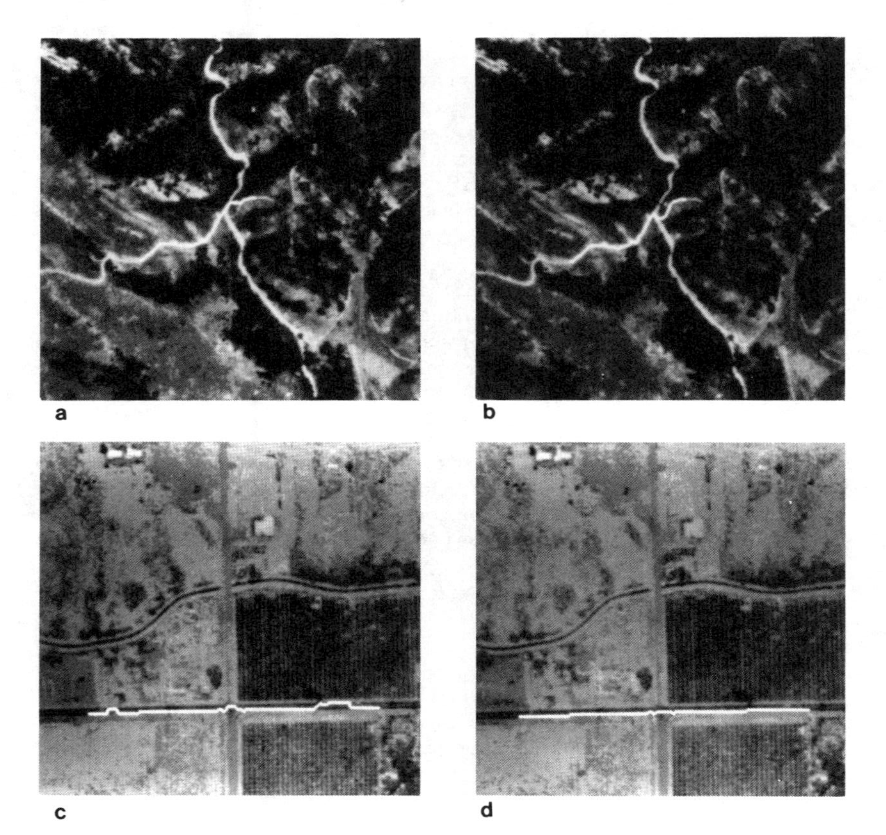

FIG. 6. Examples of how transforming Type II Image operator scores (X) allows us to adjust the trade-off between road smoothness and placing the road track at its locally most probable location. (a) $X' = X\uparrow 1 + 1$, (b) $X' = X\uparrow 2 + 1$, (c) $X' = X\uparrow 2 + 1$, (d) $X' = X\uparrow 2 + 2000$.

1. Introducing A Priori Knowledge

In addition to embedding Type I information into the Type II cost arrays as a framework around which the road track will be constructed, constraints and preferences on shape, and on the balance between local and global information, can also be inserted by appropriate modification of the values in the Type II cost arrays. Adding a constant bias "b" to each cost value tends to smooth and straighten the road track (somewhat like pulling the path taut); this effect occurs because as the bias increases, the length of the track becomes relatively more important in comparison to the local quality as defined by the individual costs returned by an operator. Similarly, raising each cost to a power "a" introduces a very strong inhibition against going through a point having a low likelihood of being on a road (but can result in detouring around shadows, vehicles, etc.).

The above cost transformations provide a convenient means for introducing a priori knowledge into the optimization procedure. For instance, if we are tracking a ragged coastline, we would opt for placing the path through those locations having the best edge scores, as opposed to trying to smooth the result; here we would use a zero bias. However, if the linear feature was known to be fairly straight, a high bias would be appropriate. A bias could also be used where there is evidence that occlusion exists (e.g., due to clouds or to intervening objects), or where there is no significant contrast between the road and the adjacent terrain, to reduce the preference for one path over another. Raising costs to a power would be helpful in tracking a road through a region where other strong linear structures were known to exist since jumps between features would be inhibited. Figure 6 shows some examples of tracking a road with modified costs obtained using the transformation $\text{cost}' = \text{cost}^a + b$.

D. The Optimization Algorithm—Finding a Best Path Through an Array of Local Likelihoods

The result of applying Type I and II operators to an image, and combining the resulting scores via the methods described in the preceding section, is an array of positive costs inversely related to the local likelihood of the presence of a road at the corresponding image location. We now consider the problem of finding a path through this array such that the sum of the costs along this path is minimized. We will initially assume that specific starting and terminating points are given and later show how this constraint can easily be removed.

The connectivity relationships between adjacent pixels in an image array define a graph: the image pixels correspond to vertices, each of which is connected by a directed weighted arc to its eight immediately adjacent array neighbors. If the weight associated with each arc[2] represents the incremental cost of a road track passing through the vertex (pixel) to which it points, then the problem of finding an optimal road track is equivalent to finding a minimum cost path through this graph.

[2]Since the incremental cost incurred by adjoining a specific pixel to a given track is simply the cost attached to that pixel and is independent of any characteristics of the other path members, all arcs entering a given vertex will generally have equal weights. However, arcs from diagonally located neighbors may be scaled by a factor of sqrt(2) to reflect the additional physical distance to these neighbors.

There is a large body of literature describing various algorithmic solutions to the minimum path problem (e.g., see the excellent survey by Dreyfus [3]), and thus our main concern here is algorithmic efficiency under the specific connectivity constraints specified above. Two alternatives were considered: the A^* algorithm presented in Duda and Hart [4] (a generalization of the Moore [13] and Dijkstra [2] algorithms) and used by Martelli [11] for edge tracking; and a variation of an algorithm originally described by Ford [7] (and related to the distance transform described by Rosenfeld [16]), which we will call the F^* algorithm.

1. The A^* Algorithm

The A^* algorithm iteratively constructs a minimal cost path from some "start" node s to a "goal" node g by extending the best partial path available at each step of the iteration (i.e., a best first search). The vertex v selected for "expansion" at each step is chosen according to an evaluation function $f(v)$, which is an estimate of the cost of a minimal cost path from s to g constrained to go through v. The evaluation function $f(v)$ can be expressed as

$$f(v) = f1(v) + f2(v),$$

where $f1(v)$ is the lowest cost path from s to v found so far, and $f2(v)$ is an estimate of the cost of a minimal cost path from v to g ($f1$ is initialized to infinity for all vertices other than s). For optimality, $f2$ must be a lower bound on the true cost. It can be set equal to zero if no better estimate is available; when $f2$ is identically zero for all vertices, A^* and the Dijkstra algorithm are also identical.

If $f2$ is identically zero (the Dijkstra algorithm) or satisfies the "consistency" condition which requires that the difference between the estimated costs to the goal vertex g from any pair of vertices $v1$ and $v2$ must be a lower bound on the true cost of an optimal path from $v1$ to $v2$, then it can be shown that when A^* expands a vertex v, an optimal path from s to v has been found. Thus, since at each iteration (assuming we employ an $f2$ which satisfies the consistency condition) A^* determines the true minimal path length from s to some new vertex v, it will converge in less than N iterations (where N is the number of vertices in the graph being searched).

After each iteration, we examine those vertices still not having final values for $f1$ and select that vertex v' with the smallest current value of f as the next vertex to be expanded: the value of $f1$ at v' cannot be further reduced and is thus labeled as final; each of the eight neighbors of v' is examined to see if a reduced value of $f1$ at these vertices is possible due to a path from s to the neighbor through v'. The algorithm terminates when $v' = g$ and an optimal path can be found by starting at g and iteratively moving to the neighboring vertex with the lowest value[3] of $f1$ until vertex s is reached.

An average of $N/2$ comparisons are needed to find v' after each iteration if we examine all the vertices which still do not have final values. If we are willing to maintain an ordered list of the values of f for those vertices which have already been updated at least once, but still do not have final values, then the average number of comparisons is on the order of $\log(N)$: v' is removed from the top of the list, and each of its eight neighbors must either be inserted or repositioned in the list if they undergo an update of their f value.

Thus, A^* can require on the order of N^2 operations (additions and comparisons) for the simpler version of the algorithm; and it can require on the order of $N \log N$ operations for a worst case situation using the ordered list approach. When the optimum path is clearly distinguished from any of the alternatives, A^* is very efficient and can have a computational requirement proportional to the number of vertices along the optimal path. However, for real imagery and currently available operators, the number of alternatives which must be examined grows very quickly, making A^* undesirable from a computational standpoint.

2. The F^* Algorithm

The F^* algorithm requires the designation of a start vertex s, a goal vertex g, and a cost array C. The first iteration of the algorithm involves a top-to-bottom, row-by-row updating of a path array P (which conceptually overlays C), in which all values were initialized to infinity except the start vertex, which has an initial value equal to its cost. The elements of the ith row of P are subject to two adjustments per iteration. First, all the elements of the ith row are iteratively adjusted from left to right according to the rule

$$\begin{aligned} P(i,j) = \min[\, & P(i-1,j-1) + c(i,j); \\ & P(i-1,j) + c(i,j); \\ & P(i-1,j+1) + c(i,j); \\ & P(i,j-1) + c(i,j); \\ & P(i,j)\,]. \end{aligned} \tag{1}$$

Next, all elements of the ith row of P are iteratively adjusted from right to left according to the rule

$$P(i,j) = \min[\, P(i,j+1) + c(i,j); P(i,j)\,]. \tag{2}$$

All additional iterations (passes), when required, alternate between a bottom-to-top pass (with the row indexing reversed so that the bottom row corresponds to $i = 1$), followed by a top-to-bottom pass (with normal row indexing) using updating rules (1) and (2) given above.

It can easily be shown that if the row indices for the elements along the optimal path increases monotonically, only a single pass through P is needed to assure convergence. In general, finding an optimal path from s to g will require a number of passes equal to the number or row index "reversals" along such a path; if the needed number of passes is not known in advance, then the algorithm can terminate when in a complete pass there is no element of P that had a change to a new path cost less than the current path cost assigned to g. As with A^*, the optimal path can be found by a backtrace, starting at g and iteratively moving to the neighbor with the smallest value of P until vertex s is reached.

For each iteration, the number of operations (comparisons and additions) required for F^* is of order $5N$; and since the number of direction reversals for roads and other linear structures generally tends to be small along the nominal

[3] As long as all the costs are greater than zero, ties can be resolved arbitrarily; if zero costs are allowed, then "looping" can result.

direction of travel, F^* is computationally more attractive than A^* for tracking such structures. (In an interactive setting, the monotonicity condition can be insured by proper partitioning of the image, and thus only a single iteration is required per partition.) For the general case, F^* insures that in each iteration at least one element of P achieves its final value, and thus F^* can require on the order of N^2 operations.

We note that for both A^* and F^* we can find the optimal path from a set S of start vertices, rather than a single vertex, by connecting a pseudo-start vertex s to each member of S by a zero cost path; we can specify a set of goal vertices G, rather than just a single vertex g, by an identical mechanism.

An example showing the operation of the F^* algorithm is presented below:

5	1s	4	8	1	6	1
6	3	1	6	2	7	5
3	1	5	2	8	1	2
5	2	8	6	7	1	2
8	3	8	7	4	2g	3

Cost Array C, start vertex s (1, 2), goal vertex g (5, 6).

6	1	5	13	14	20	21
7	4	2	8	10	17	22
6	3	7	4	12	11	13
8	5	11	10	11	12	13
13	8	13	17	14	13	15

Path Cost Array P after the first iteration (top-to-bottom pass).

6	1	5	10*	7*	12*	13*
7	4	2	8	6*	13*	16*
6	3	7	4	12	11	13
8	5	11	10	11	12	13
13	8	13	17	14	13	15

Path Cost Array P after the second iteration (bottom-to-top pass); elements with updated costs are marked by a following asterisk.

6	1	5	10	7	12	13
7	4	2	8	6	13	16
6	3	7	4	12	7*	9*
8	5	11	10	11	8*	9*
13	8	13	17	12*	10*	11*

Path Cost Array P after the third iteration (top-to-bottom pass).

6	−1	5	10	7	12	13
7	4	−2	8	−6	13	12*
6	3	7	−4	12	−7	9
8	5	11	10	11	−8	9
13	8	13	17	12	−10	11

Path Cost Array P after the fourth iteration (bottom-to-top pass); elements on the minimum cost path are preceded by a minus sign.

The algorithm terminated after the fourth pass because the only element changed (2, 7) had a higher path cost (12) than the current path cost of g (10).

E. A Low-Resolution Road Tracking Algorithm

The ideas presented in preceding sections have been synthesized into an algorithm for precisely tracking the major linear structure visible in a delimited region of an image. This algorithm, known as LRRT, takes as input an image array, a search region defined by a binary mask, and constraints on starting and ending coordinates of the track (defined by two specific regions through which the road is constrained to pass, such as the sides of the search window or a pair of boxes.)

Algorithm LRRT operates as follows:

(1) A selected set of Type I operators are scanned over the region designated by the search mask; and the costs produced by each such operator are histogrammed and thresholded at some preset level (generally operator dependent), so that the number of points below this threshold will not exceed the number of road points estimated to be present in the search window. Selecting 5% of the points in the search window is a typical upper limit for the Duda Road Operator. A PRS mask is generated as the union of those locations at which each Type I operator returns a cost lower than its associated threshold.

(2) A selected set of Type II operators is scanned over the region designated by the search mask. The scores for each operator are self-normalized and stored in individual cost arrays. (Normalization is accomplished by separately histogramming the response of each operator and converting the raw values to percentile ranks on a scale of 1–100.)

(3) The costs in each Type II array are modified using the transform, $\text{cost}' = \text{cost}^a + b$, described earlier, to incorporate a priori knowledge and constraints on road smoothness, visibility, and presence of interfering structures.

(4) Costs in Type II arrays at locations designated in the PRS mask are set to a very small positive value. (True zero values could lead to arbitrary wandering, or even cycling, through regions of zero cost.)

(5) Each Type II cost array is considered to be a graph with each pixel connected to each of its eight neighbors. A minimum cost path is found in each such array between the starting and terminating delimiters, using algorithm F^*. The average cost per pixel along the track in each Type II array is computed and self-normalized by ranking it in a histogram of costs for that operator compiled over the original search region. The track with the highest rank is chosen as the preferred track.

F. Extensions of Algorithm LRRT to Delineation of Complete Road Networks

Algorithm LRRT can be extended to sequentially delineate all road segments contained in a specified search region of an image. This delineation is obtained by making multiple passes through the Type II cost arrays with algorithm F^*. After each pass, a ribbon centered on the detected road track is marked as a forbidden area to allow the next most prominent road segment to be detected. Marking is accomplished by adding a large cost to every pixel located within some number (currently three) of pixels of any point on the original track; a distance propagation algorithm described in Rosenfeld [16] is used for this purpose. This approach will delineate all roads connecting the original start and stop delimiters and can be used, for example, to delineate all paths between opposite sides of a rectangular search region; the process can then be applied to find all paths connecting the alternate opposing sides of the search rectangle.

Since algorithm F^* computes the cost from a given starting delimiter to all locations in the search region, road segments branching out from a primary path can often be found by backtracing from alternate stopping points. In particular, a complete road network can often be found by backtracing from local cost minima along the periphery of the original search region.

3. IMPLEMENTATION DETAILS AND EXPERIMENTAL RESULTS

Algorithm LRRT has been embedded in a complete system for delineating roads in aerial imagery. System operation is partitioned into three phases: initialization, tracking, and smoothing.

The objective of the initialization phase is to select the search region within which LRRT will be applied and to constrain the starting and ending points.

The search region can be automatically specified from a map data base in approximate registration with the image. Alternatively, the search region can be specified interactively on a graphic display.[4] In either case, the desired search region can be specified explicitly or derived from a sketch of the road path. (A sketch is expanded into a search region using the distance transform referenced in Rosenfeld [16].)

In an approach we are currently developing (to be described in detail in a later paper), it should be possible to obtain a fairly good sketch of the major road segments depicted in an arbitrarily large image without any manual intervention or data base knowledge. The approach is first to obtain a PRS mask for the entire image, and then extract clusters of marked points from this mask which lie within some maximum distance of their nearest neighbor. Each such cluster is used to generate a minimum spanning tree, and the major branches of the tree are taken to be the approximate road tracks (i.e., sketches). An example of this process is presented in Fig. 7.

Algorithm LRRT is used in the tracking phase to obtain a precise delineation for each road segment derived in the initialization phase. The resulting path may bridge small occlusions and have gaps in regions of significant occlusion. The smoothing phase explicitly marks those portions of a road track that were inferred from continuity, rather than direct visibility, and links separated segments of the same road network.

While we have addressed the problems associated with each of the above phases for automatically delineating the low-resolution roads and linear structures in an image, most of our current experimental work has been concerned with obtaining a high-performance solution to the problem of precise delineation required in phase two. We have implemented two versions of algorithm LRRT: an INTERLISP/SAIL version for developmental work and a FORTRAN version for more extensive experimentation and evaluation. Both versions run on the SRI PDP-10: the FORTRAN version is also compatible with a CDC 6400 system at the U.S. Army Engineer Topographic Lab (ETL) at Ft. Belvoir. The FORTRAN version has a minimum core requirement of 20,000 60-bit words and will track a road segment 128 pixels long in 15 sec of CPU time; the corresponding numbers for the INTERLISP version are 90,000 36-bit words of core and 60 sec of CPU time.

The FORTRAN version of the LRRT makes some additional assumptions about the roads to be tracked: it assumes that they are generally lighter or darker than the surrounding terrain and that they do not "double back" on themselves in the designated search areas. It uses the F^* algorithm, a single Type II operator (based

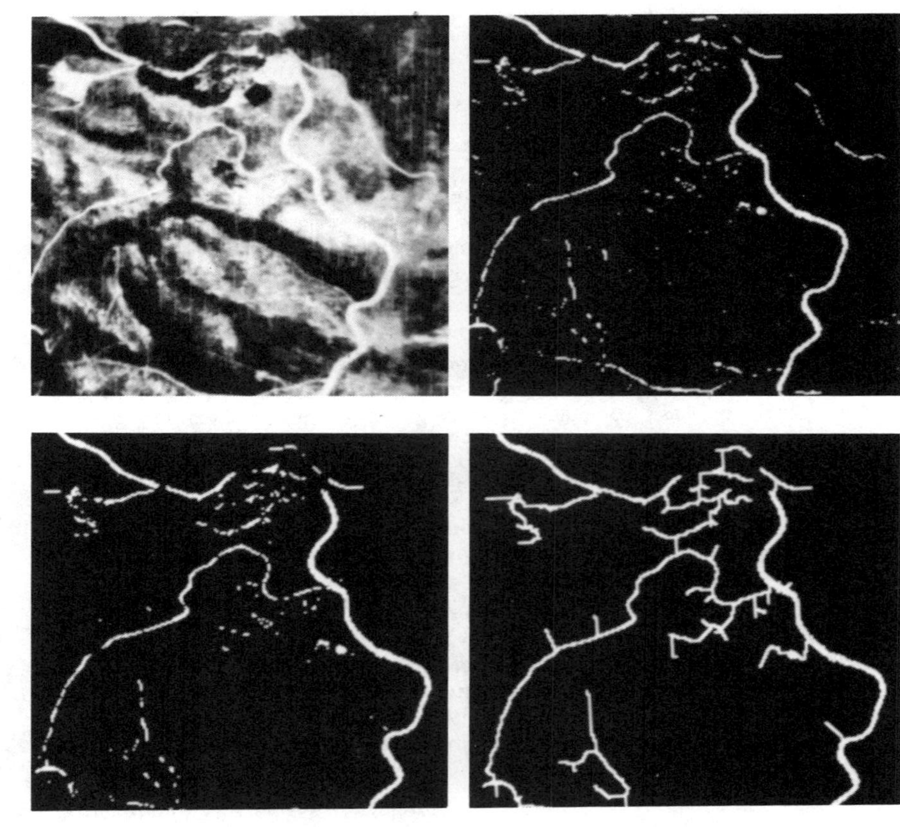

FIG. 7. Automatic road segment extraction. Resulting sketch delimits search areas for precise delineation using algorithm LRRT. (a) Intensity image of road scene. (b) Perfect road score mask (PRS) of image (derived from thresholded intensity, Duda road operator scores, and Rosenfeld–Thurston nonlinear operator scores described in [17–19]). (c) Largest cluster of points (Cluster 1) extracted from PRS (each point is within 10 pixels of some other cluster point. (Cluster is truncated at bottom edge due to program storage limitations.) (d) Minimum spanning tree for Cluster 1. (e) Maximum path through minimum spanning tree for Cluster 1. (f) Minimum spanning trees for all clusters of PRS. (g) Maximum paths through minimum spanning trees for all clusters. (h) Maximum paths with length greater than 60 points for all clusters.

[4] These two methods were used to obtain the experimental results discussed later and illustrated in Fig. 8; widths of typical search regions were 20–30 times road width, and lengths were often equal to the full image dimension.

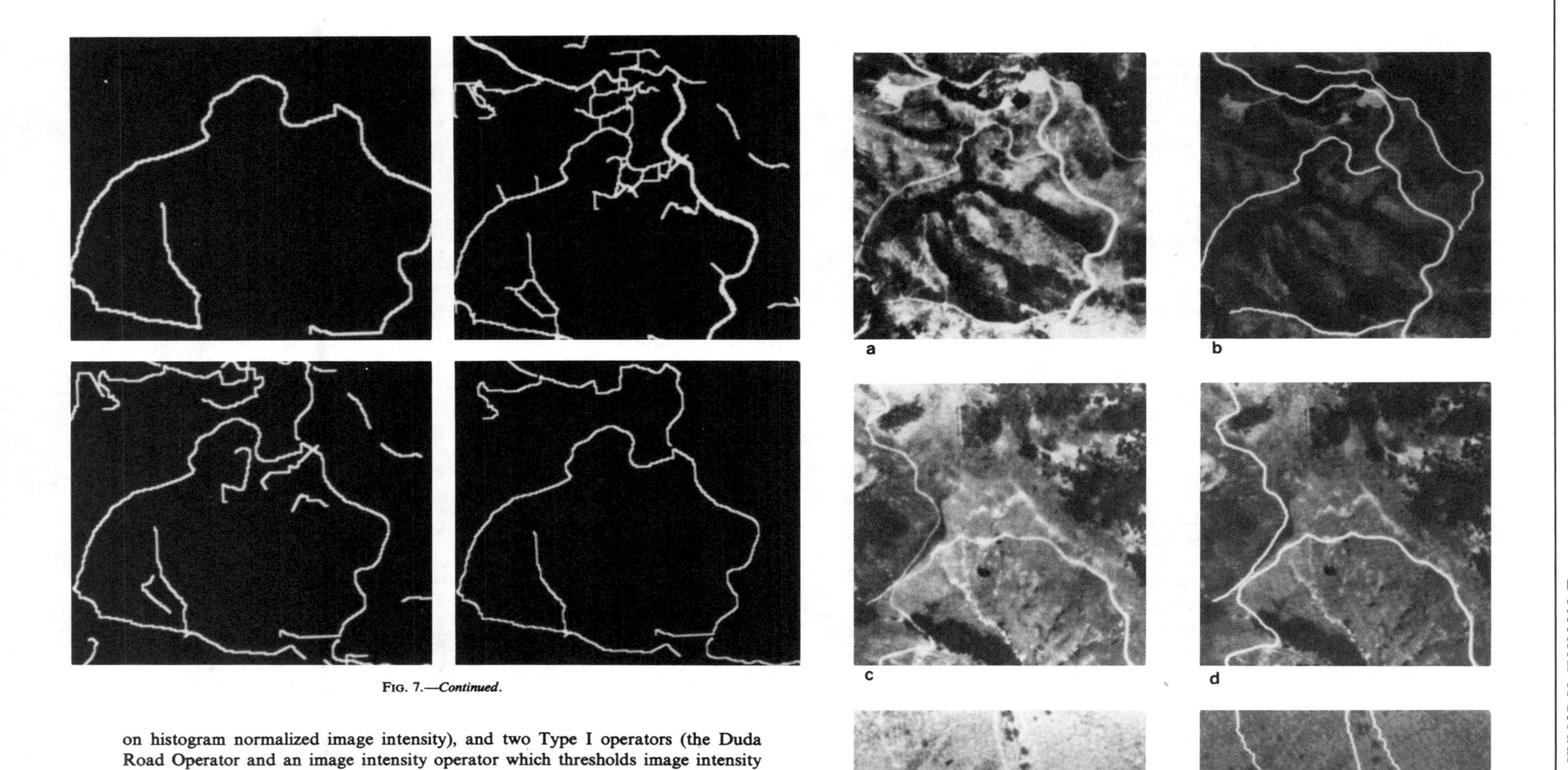

FIG. 7.—*Continued.*

FIG. 8. Examples of road delineation produced by the low-resolution road tracker.

on histogram normalized image intensity), and two Type I operators (the Duda Road Operator and an image intensity operator which thresholds image intensity and also checks whether the width of the above-threshold region at a potential road point is sufficiently narrow).[5] This program has been tested on approximately 50 road segments found in aerial images of seven different geographic locations with no failures, when the assumptions are satisfied and the roads are clearly visible (some examples are shown in Fig. 8).

4. CONCLUDING COMMENTS

In this paper we have addressed the problem of precise delineation of the roads and linear features appearing in aerial photographs, using an approach based on global optimization of locally evaluated evidence. Since there does not appear to

[5]Additional details of the particular variants of the Duda and intensity operators used in the FORTRAN LRRT are described in the Appendix.

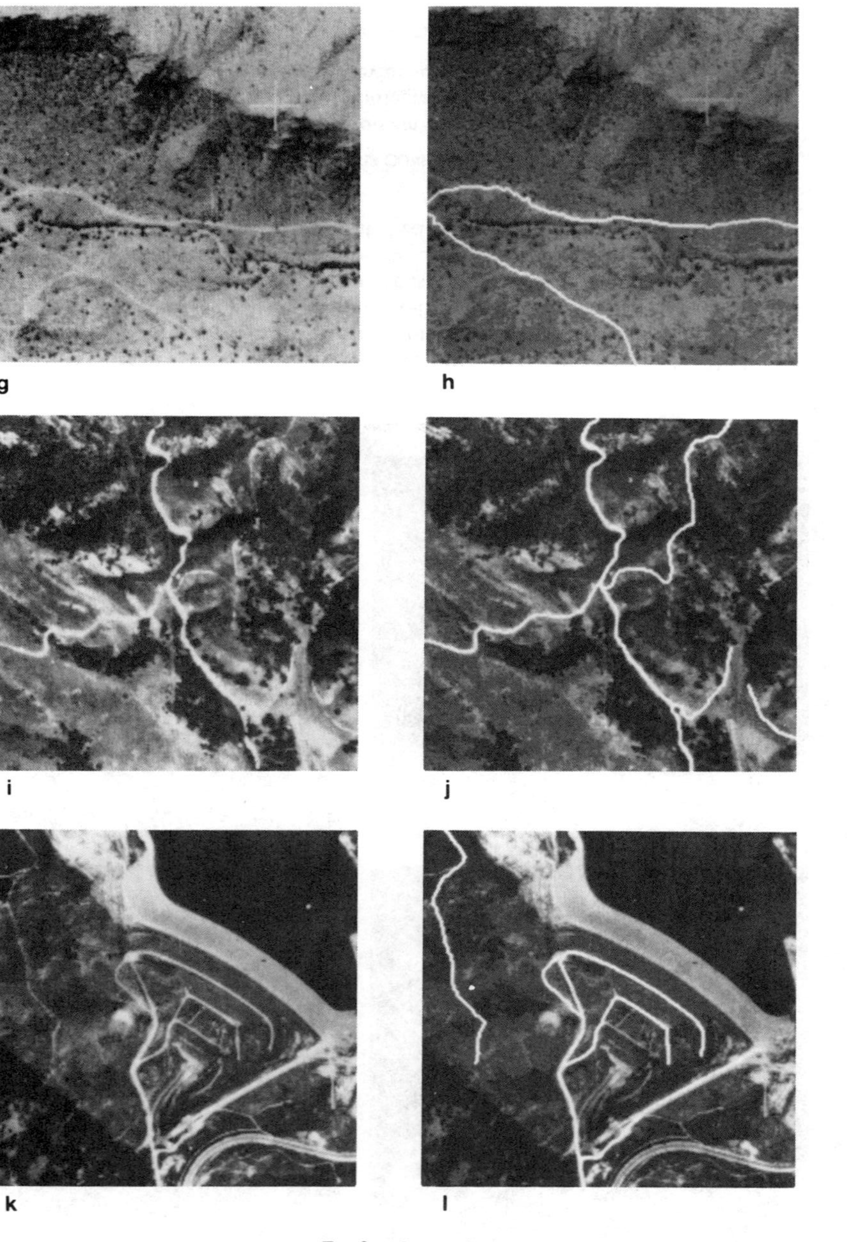

FIG. 8.—*Continued.*

exist a single coherent model suitable for reliable detection of local road presence, it was essential that some means for integrating information from multiple (incommensurate) image operators and knowledge sources be devised—the conventional optimization paradigm does not provide any formal machinery for achieving this task.

Two key points characterize the basis of our approach:

(1) Rather than projecting all image operators on a single linear scale and attempting to use them in the same qualitative manner, we have identified the distinctly different nature and potential use of operators which have strong object detection capabilities, as opposed to those which are useful for object analysis once identification and/or location is known. (Depending on the specific context, a particular operator might switch from one role to the other.) We have provided a simple and uniform mechanism for integrating the information provided by the two classes of operators for the specific task of tracking linear structures, and we believe that the same general approach is applicable in a wider range of problem settings.

(2) We have recognized that the score returned by an image operator usually has little absolute meaning, and yet a monotonic transformation of this score can lead to a significantly different final result in tracking linear structures. We have capitalized on this property by introducing a monotonic transform which provides a simple and uniform mechanism for adjusting the scores to reflect a priori information and semantic constraints.

Our plans for future work include the continuing development of fully automated techniques for road tracking, extension of these techniques to detection and classification of different types of linear structures (e.g., rivers, railroads, runways, etc.), and development of techniques for tracking linear structures in three dimensions using stereo image pairs.

The scientific content of this work lies in discovering effective models for representing and detecting the linear structures of interest and developing paradigms for integrating information from the wide variety of knowledge sources available to the human observer whose performances we are attempting to equal or surpass. Applications of our work include road monitoring for planning and intelligence purposes, delineation of roads and linear features for automated cartography, and detection of roads and linear features as landmarks for autonomous navigation.

APPENDIX

Implementation Details of the Duda and Intensity Operators

1. The Duda Road Operator (DRO)

The essential details of the DRO are shown in Fig. A1. The masks at the top of the figure measure uniformity of intensity along a potential road track (a_1, a_2, a_3) and contrast of this potential track with the adjacent terrain $(b_1, b_2, b_3;\ c_1, c_2, c_3)$. Note that terrain intensity is sampled at a one-pixel offset from the assumed track to allow for slight variations in road width. Scoring function $G(u)$ is designed to ignore slight variations in intensity along the road and to penalize all significant variations equally; similarly, scoring function $F(u)$ rewards all large contrasts equally.

The unspecified parameters in Fig. A1 were assigned the following experimentally optimized values for an intensity range of 1–127:

$$M = 1.5; \quad \theta = 15; \quad \theta 1 = 5; \quad \theta 2 = 20; \quad e = 0.1.$$

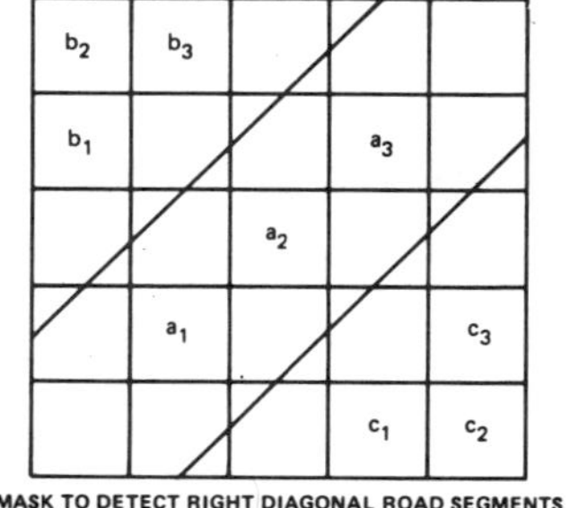

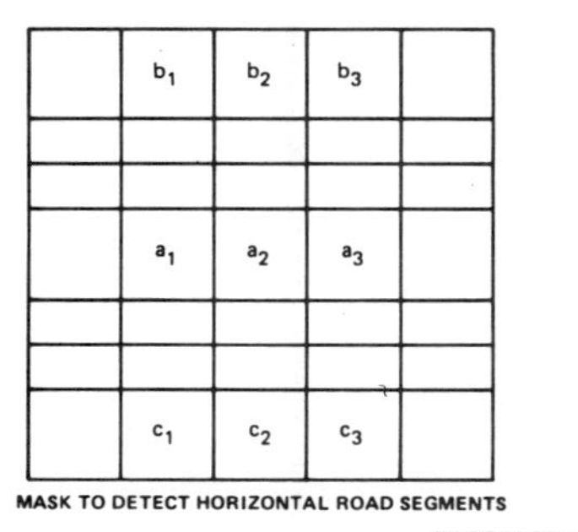

WINDOWS FOR THE ROAD OPERATORS

Two other masks, similar to these, are used to detect vertical and left diagonal road segments.

SCORE: $[G(|a_1 - a_2|) \times G(|a_2 - a_3|) / \sum_{i=1}^{3} F(a_i - b_i) + F(a_i - c_i)]$

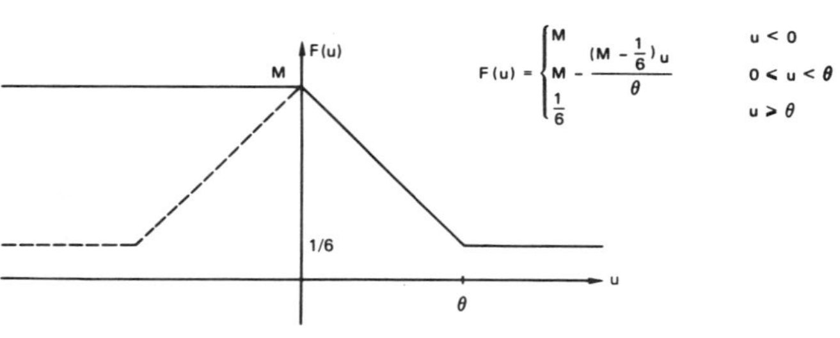

ROAD EDGE SCORING FUNCTION

Function depicted as solid line is used for light roads. F(−u) is used for dark roads. Symmetric form of the function, shown by dashed lines for negative values of u, is used when road to background contrast is unknown.

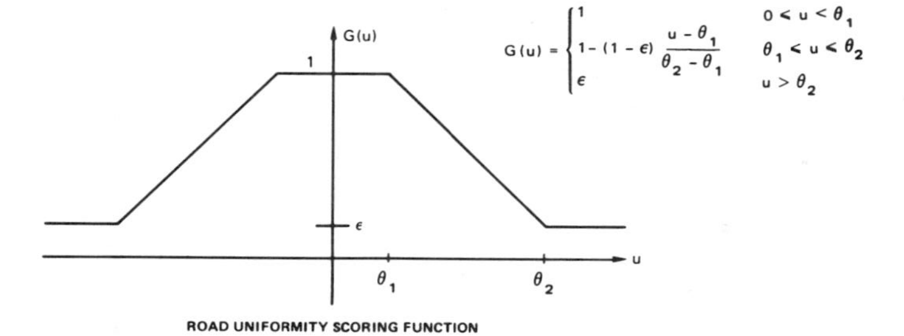

ROAD UNIFORMITY SCORING FUNCTION

FIG. A1. Description of Duda road operator.

When used as a Type I operator, all the scores produced by the DRO over some region of interest in the image are histogrammed; and the best N percent of the scores are interpreted as producing positive Type I responses. N is determined either by the nominal upper bound of 1–2% for images where road content is initially unknown or by an a priori estimate of the number of road pixels in the segment of the image to be analyzed. For example, when tracking a road whose average width is two pixels and assuming that this road is known to lie somewhere in a band 40 pixels wide, we would set N at 3–5%.

2. *The Intensity Operator*

The intensity operator looks for a narrow region of high brightness at a potential road point (or low brightness for dark roads). For dark roads, the score returned by the operator is just the intensity value scaled to a range of 1–127. For light roads, intensities are subtracted from 128 after scaling, so that, in both cases, low values (costs) correspond to likely road points. If at a given point the "width" condition described below is not satisfied, then the score at that point is degraded by adding a constant (25 in our current implementation) to it. Since the width test is applied directionally, the operator itself is directional.

When used as a Type I operator, the scaled intensity values in the entire search region are histogrammed. To produce a positive Type I response, intensity at a potential road point must satisfy the additional condition of being among the lowest K% of histogram values. K is determined in a manner analogous to N (used in the DRO) based on the expected number of road points which can be detected by the operator: typical values range from 5–10%.

The width condition at a point is evaluated by looking at the brightness scores along a line normal to an assumed local road direction at that point. We require that (for a maximum road width of three pixels) in an interval of 15 pixels centered on the point, there should not exist, in a contiguous sequence of five points, more than three points with intensity scores in the lowest (best) K%.

ACKNOWLEDGMENTS

The authors wish to acknowledge the important contributions made by Harry G. Barrow, Richard O. Duda, and Thomas D. Garvey in the formative stages of the work described in this paper.

REFERENCES

1. H. G. Barrow and J. M. Tenenbaum, *The Representation and Use of Knowledge in Vision*, AI Technical Note 108, SRI International, Menlo Park, California, July 1975.
2. E. W. Dijkstra, A note on two problems in connection with graphs, *Numer. Math.* **1**, 1959, 269–271.
3. S. E. Dreyfus, An appraisal of some shortest-path algorithms, *Operations Res.* **17**, No. 3, May–June 1969, 395–412.
4. R. O. Duda and P. E. Hart, *Pattern Classification and Scene Analysis*, Wiley, New York, 1973.
5. M. A. Fischler and R. A. Elschlager, The representation and matching of pictorial structures, *IEEE Trans. Computers* **C-22**, No. 1, 1973, 67–92.
6. M. A. Fischler and T. D. Garvey, A computer-based approach to perceptual reasoning and multisensor integration, in preparation.
7. L. R. Ford, Jr., *Network Flow Theory*, The Rand Corporation, P-923 (August 1956).
8. P. Hart, N. Nilsson, and B. Raphael, A formal basis for the heuristic determination of minimum cost paths, *IEEE Trans. Syst. Sci. Cybern.* **SSC-4**, No. 2, July 1968, 100–107.
9. M. Hueckel, An operator which locates edges in digitized pictures, *J. Assoc. Comput. Mach.* **18**, No. 1, January 1971, 113–125.

10. M. Hueckel, A local visual operator which recognizes edges and lines, *J. Assoc. Comput. Mach.* **20**, No. 4, October 1973, 634–647.
11. A. Martelli, An application of heuristic search methods to edge and curve detection, *Comm. ACM* **19**, No. 2, February 1976.
12. U. Montanari, On optimal detection of lines in noisy pictures, *Comm. ACM* **14**, No. 5, May 1971.
13. E. Moore, The shortest path through a maze, Proc. Intern. Symp. Theory Switching, Part II, The Annals of the Computation Laboratory of Harvard University 30, Harvard University Press, Cambridge, Mass., 1959.
14. L. Quam, Road tracking and anomaly detection, in Proc. Image Understanding Workshop, pp. 51–55 (May 1978).
15. L. G. Roberts, Machine perception of three-dimensional solids, in *Optical and Electro-Optical Information Processing* (Tippett *et al.*, Eds.), pp. 159–197, MIT Press, Cambridge, Mass., 1965.
16. A. Rosenfeld and J. L. Pfaltz, Sequential operations in digital picture processing, *J. Assoc. Comput. Mach.* **13**, No. 4, 1966, 471–494.
17. A. Rosenfeld and M. Thurston, Edge and curve detection for visual scene analysis, *IEEE Trans. Computers* **C-20**, No. 5, 1971, 562–569.
18. S. Rubin, *The ARGOS Image Understanding System*, Ph.D. thesis, Dept. of Computer Science, Carnegie-Mellon University, Pittsburgh, Penn., 1978.
19. G. J. Vanderbrug, Semilinear line detectors, *Computer Graphics and Image Processing* **4**, No. 3, September 1975, 287–293.
20. G. J. Vanderbrug, Line detection in satellite imagery, *IEEE Trans. Geosci. Electron.* **GE-14**, No. 1, 1976, 37–44.

Representations based on Zero-crossings in Scale-Space

Robert A. Hummel

Courant Institute of Mathematical Sciences
New York University

Abstract

Using the Heat Equation to formulate the notion of scale-space filtering, we show that the evolution property of level-crossings in scale-space is equivalent to the maximum principle. We briefly discuss filtering over bounded domains. We then consider the completeness of the representation of data by zero-crossings, and observe that for polynomial data, the issue is solved by standard results in algebraic geometry. For more general data, we argue that gradient information along the zero-crossings is needed, and that although such information more than suffices, the representation is still not stable. We give a simple linear procedure for reconstruction of data from zero-crossings and gradient data along zero-crossings in both continuous and discrete scale-space domains.

1. Scale-space and zero-crossings

The use of multiresolution representations is an important idea for the analysis of signal and image data. Many data structures have been studied, including Gaussian pyramids, difference-of-Gaussian pyramids, Laplacian pyramids, and "scale-space" formulations [3,4,16]. The latter formulation, to be described briefly below, can be used as a continuous model of the other formulations. We will discuss the representation of data by zero-crossings in scale-space, and consider the stability of reconstruction methods.

The natural framework for the analysis of scale-space formulations of multiresolution representation is in terms of the heat equation [9,20]. Specifically, let $f(x)$ be a bounded function defined for $x \in \mathbb{R}^n$. (Arbitrary dimensions can be handled with little additional fuss over the case $n = 1$; we will later comment on the case of a bounded domain $D \subseteq \mathbb{R}^n$). We define $u(x,t)$ to be a bounded solution to the heat equation:

$$\frac{\partial u}{\partial t} = \Delta u, \qquad \textbf{(Heat Equation)}$$

$$u(x,0) = f(x).$$

The solution is given by convolution against the fundamental solution to the Heat Equation, which for the domain $\mathbb{R}^n$ is given by

$$u(x,t) = \int_{\mathbb{R}^n} K(x-y,t) f(y)\, dy,$$

where

$$K(x,t) = (4\pi t)^{-n/2} e^{-|x|^2/4t}.$$

We see that $u(x,t)$ is obtained by blurring $f(x)$ by increasingly diffuse Gaussians, parameterized by $t>0$, with standard deviations σ satisfying $2\sigma^2 = 4t$. In computer vision, scale-space sometimes refers to the (x,σ) variables that can be used to reparameterize the domain of u. We retain the (x,t) parameterization to keep the linear Heat Equation relation for the function u.

Convolution by Gaussians is considered special for many reasons [1,9,19]. We see from the above analysis a relationship between Gaussian convolution, the Heat Equation, and the Laplacian operator. Of course, Gaussian convolution enjoys other properties; for example, the central limit theorem implies that Gaussian convolution is easy to implement by an iterative procedure. However, we also see the extent of the similarity of difference of Gaussians and the Laplacian of the Gaussian; namely, since $K(x,t)$ is itself a solution to the Heat Equation,

$$(\Delta K)\,(x,t) = \frac{\partial K}{\partial t}(x,t) = \lim_{\tau \to 0}(K(x,t+\tau) - K(x,t))/\tau.$$

That is, the difference of Gaussians is a good approximation to ΔK as the separation between the spread of the two Gaussians approaches zero (and the difference is scaled).

Filtering by the Laplacian of a Gaussian can be written in three ways:

$$\Delta K * f = K * \Delta f = \Delta(K * f).$$

If we denote the result by $v(x,t)$, we see that

(1) $v(x,t)$ is the $f(x)$ data filtered by the Laplacian of a Guassian.

(2) $v(x,t)$ is the solution to the Heat Equation with initial data Δf.

(3) $v(x,t)$ is $\Delta u(x,t)$, where u is the solution to the Heat Equation with initial data $f(x)$.

The zero set of $v(x,t)$ is the point set in (x,t) where $v = 0$. The set might be empty (for instance, if f is subharmonic or superharmonic; [5]) everything (if f is harmonic; [14]), or a proper subset of (x,t) space. In the latter case, zeros can be isolated points, lines, and surfaces (but never regions). We distinguish components of the zero set which form manifolds of codimension one:

Definition: The *zero-crossings* of $v(x,t)$ refers to the point set

$$\partial\{(x,t)\,|\,v(x,t) < 0\} \cap \partial\{(x,t)\,|\,v(x,t)>0\}. \blacksquare$$

Zero-crossings have been suggested for segmentation of imagery by edge detection [11], and for stereo matching and motion correspondence between pairs of images (e.g., [10]). It has also been suggested [19] that the zero-crossings are a nearly complete representation of Δf. Finally, Witkin [16] observes that zero-crossings in scale-space evolve as t increases, and are never created at some nonzero t. This property, discussed in [19] and in [1], ensures that zero-crossing surfaces are nested, one within another, enclosing regions containing the face $\{t = 0\}$, or forming a sheet meeting the face $\{t = 0\}$ and extending to $t = \infty$. The property can be given a precise statement:

Evolution property of zero-crossings: Let C be a connected component of the set of zero-crossings in the domain $\{(x,t)\,|\,x \in \mathbb{R}^n, T_1 \leq t \leq T_2\}$, where $0 \leq T_1 < T_2$. Then $C \cap \{(x,t)\,|\,t = T_1\} \neq \emptyset$. ■

In the remainder of this paper, we wish to make two main points. First, we establish the equivalence of the "Evolution property for zero-crossings" and the classical maximum principle for parabolic partial differential equations, thereby allowing us to consider bounded domains and nonstationary convolutions. Second, we consider reconstructibility from zero-crossings. The principle result is that if knowledge of the location of the zero-crossings is supplemented with gradient information of v *at only those points in the zero-crossings*, then there is a simple scheme for reconstructing some of the data, but that even then numerical accuracy of the reconstruction is unstable.

2. The Maximum Principle

The classical maximum principle for the solution to the parabolic equation $\partial u/\partial t = \Delta u$ states (see, e.g., [2,15,8]):

Maximum Principle: Let $D \subseteq \mathbb{R}^n$ be open and bounded. Suppose u is a solution in $T = \{(x,t)\,|\,x \in D, \underline{0}<t<T\}$ of class C^2 which is continuous in the closure $\bar{T}$. Then u assumes its maximum at some point (x,t) for which either $x \in \partial D$ or $t = 0$. ■

Next, suppose that scale-space construction is denoted by the operator $v = Sg$, which is to say that the scaled function $v(x,t)$ is obtained from the initial data $g(x)$. In the previous section, we defined S to be $v(x,t) = K(\cdot,t) * g$, where $g(x) = \Delta f(x)$, but we can imagine more general operators. In any case, it is logical to make certain assumptions about S, although all we will require is that

(1) If $g(x)$ is continuous, then Sg is continuous.

(2) $S(-g) = -Sg$ for all g, and if $v = Sg$ and $\bar{g}(x) = v(x,\tau)$, then $\bar{v} = S\bar{g}$, where $\bar{v}(x,t) = v(x,t+\tau)$.

(3) If $g(x)\to 0$ as $|x|\to\infty$ and $v = Sg$, then for each t, $v(x,t)\to 0$ as $|x|\to\infty$.

We note that if S is defined by convolution with Gaussians as in Section 1, then the maximum principle holds for $v = Sg$ as long as g is continuous; further, conditions (1) - (3) hold.

Our first result is:

Proposition: The following are equivalent:

(i) The maximum principle holds for solutions $v = Sg$ using continuous initial data $g(x)$ satisfying $g(x)\to 0$ as $|x|\to\infty$.

(ii) The evolution property holds for level-crossings of solutions $v = Sg$ using a scale-space operator S satisfying (1) - (3) above and continuous $g(x)$ satisfying $g(x)\to 0$ as $|x|\to\infty$.

Proof: We first show that the maximum principle implies the evolution property. For if the evolution property fails for a level-crossing l for some $v(x,t)$, then by suitably transforming v, g, and l, we can assume that $l \geq 0$, and that there is a solution $v = Sg$, (with $g\to 0$ as $|x|\to\infty$) with a component C of $\{(x,t)\,|\,0\leq t\leq T, v(x,t)>l\}$ disjoint from the plane $t = 0$. Let (x_0,t_0) be a relative maximum in C. Then since C is open, there is a bounded cylinder in C with (x_0,t_0) in the interior, which is in violation of the maximum principle.

Conversely, if the maximum principle fails, then for some cylinder $D\times[0,T]$ and some $v(x,t)$ given by $v = Sg$ with $g(x)\to 0$ as $|x|\to\infty$, the maximum of v occurs either in the interior or the top of the cylinder. Either way, there is a value l less than the maximum but greater than the values on the bottom and sides of the cylinder. Thus there is a component of the level-l crossing within the wedge $0\leq t\leq T$ which lies entirely within the interior of the cylinder, and thus does not meet $\{t = 0\}$. So the evolution property for level-crossings is violated. ■

We illustrate the utility of the proposition with three observations. First, given the equivalence with the maximum principle, any proof of the evolution property for zero-crossings that does not either use the maximum principle or essentially redo the proof of the maximum principle is highly suspect. Since the maximum principle is slightly delicate, especially in the absence of strong regularity assumptions, the former course seems more appropriate.

Second, using the version of the gradient Hopf maximum principle for the Heat Equation [15], it is not hard to show that knowledge of the zero-crossings together with gradient information along with the zero-crossings (or even just one zero-crossing contour) is sufficient to determine $v(x,t)$ uniquely [6]. In section 3, we give a more constructive discussion of this point; but it is interesting that the maximum principle establishes this uniqueness.

Finally, we see that any scaling method obeying the maximum principle will yield the evolution property for zero-crossings. Under fairly severe restrictions, this leads one to Gaussian convolution [1], but more general scaling methods are possible. For example, blurring by a parabolic operator of the form $\partial u/\partial t = Lu$, where L is a uniformly elliptic linear second order differential operator with nonconstant coefficients will certainly still give a maximum principle. In fact, L can be nonlinear [13]. Moreover, suppose we replace $\mathbb{R}^n$ with a bounded domain $D \subseteq \mathbb{R}^n$, and insist on data $f(x)$ with compact support in the interor of D. We may then define $v = Sg$, where $g = \Delta f$, by solving

$$\partial v/\partial t = \Delta v \quad \text{in } D\times(0,\infty),$$

$$v(x,0) = g(x) \quad \text{for } x \in D$$

$$v(x,t) = 0 \quad \text{for } x \in \partial D.$$

This scaling is not given by convolution against a Gaussian, since the domain is bounded, but nonetheless obeys a maximum principle, and gives the same evolution property.

3. Completeness

We return to a consideration of zero-crossings of data $v(x,t)$ obtained by filtering initial data $f(x)$ defined for $x \in \mathbb{R}^n$ by the Laplacian of Gaussians, $\Delta K(\cdot,t)$. The question we wish to address is: to what extent do the zero-crossings represent $f(x)$? Clearly, $f(x)$ can at best be reconstructed to within an arbitrary additive harmonic function and a scalar multiple. However, if we assume that $f(x) \to 0$ as $|x| \to \infty$, then only the multiplicative constant is of concern.

Yuille and Poggio [18] make the observation that if $g(x)$ and hence $f(x)$ is polynomial in x, and if $n=1$, then reconstruction from zero-crossings is theoretically possible. They also refer to the validity of the observation for larger n.

We note, however, that when $g(x)$ is a polynomial in $x \in \mathbb{R}^n$ for any n, then $v(x,t)$ is a polynomial in $(x,t) \in \mathbb{R}^{n+1}$. Accordingly, the zero-crossings are part of the *analytic varieties* of the polynomial v as studied in algebraic geometry. It is well known that the varieties in C^n determine the complex polynomial defined on n complex variables. It is not as commonly used, but nonetheless true, that an n-dimensional subportion of the intersection of the analytic variety with $\mathbb{R}^{n+1}$ also determines the polynomial [12]. Thus the case of polynomial data can be settled with algebraic geometry.

However, since the determination of a polynomial by its varieties is essentially an analytic continuation result, stability of the reconstruction is unlikely. That is, small errors in measurement of the zero-crossings could lead to arbitrarily large errors in the determination of $g(x)$. Put differently, there can be widely different initial data leading to nearly identical zero-crossing data.

Worse, settling the case for polynomial data says little about the case of continuous initial data. Although the Stone-Weierstrass theorem says that a continuous function can be uniformly approximated by a polynomial on a compact set, the zero-crossings depend on the initial data globally, and the dependence can't be localized. Further, the lack of stability means that the approximation is irrelevant. The situation is similar to the fact that a polynomial of a single variable with all real roots is determined by its zeros, but that given all the zeros of a continuous function, one knows nothing more than the zeros.

In fact, there are known examples of pairs of functions $f_1(x)$ and $f_2(x)$ such that the corresponding $v_1(x,t)$ and $v_2(x,t)$ have identical zero-crossings at all levels of resolution. John Daugman supplies the example (for two space dimensions) of $f_1(x_1,x_2) = \sin x_1$, and $f_2(x_1,x_2) = (\sin x_1)(2 + \cos x_2)$.

However, if the zero-crossing data is supplemented with knowledge of the gradient data at the zero-crossings, then reconstruction of at least some of the data $g(x)$ is theoretically possible by a quite easy procedure, given below. Details of these ideas were reported earlier in an unpublished work [6]. The use of gradient data for the representation also appears in [9], but the gradient data there is not limited to the zero-crossings. The use of gradient data along zero-crossings is discussed in [17]. Many researchers have noted from a casual observation of zero-crossings of image data that zero-crossings with large gradient magnitudes are of greater significance than those with low gradient magnitudes.

3.1. Continuous Case

Specifically, let Ω be a *bounded* connected component of $\{(x,t) \mid t \geq 0,\ v(x,t) \neq 0\}$, and denote by D the set $\{x \in \mathbb{R}^n \mid (x,0) \in \Omega\}$, and by Γ the zero-crossing $\partial\Omega \cap \{t>0\}$. Let τ be a value such that $\tau > \sup\{t \mid (x,t) \in \Omega\}$. Next, we set $\tilde{g}(x) = g(x)$ for $x \in D$, and $\tilde{g}(x) = 0$ elsewhere. Finally, let $b(x)$ be the $\tilde{g}(x)$ data blurred to the level τ:

$$b(y) = \int_{\mathbb{R}^n} K(y-x,\tau)\tilde{g}(x)dx.$$

Using Green's theorem, it is easy to show:

Proposition:

$$b(y) = \int_{\Gamma} K(y-x,\tau-t)\nabla v(x,t)\cdot \mathbf{n} d\sigma,$$

where $\mathbf{n}$ is a surface normal to Γ at (x,t), and $d\sigma$ is surface area measure. ∎

Thus given the zero-crossing Γ and $\nabla v(x,t)$ for $(x,t) \in \Gamma$, then the blurred data $b(x)$ can be constructed by a simple linear process. The original data $\tilde{g}(x)$ can be reconstructed by deblurring the $b(x)$ data [7]. Deblurring is, of course, a classic unstable process. The situation is not hopeless, however, since $\tilde{g}(x)$ has known compact support, which might be used to advantage, and also since errors that occur are predominantly in high frequency components, which might not be as essential to visual interpretability.

The lesson of this section, ultimately, is that even for bounded zero-crossings supplemented with gradient data along the zero-crossing, reconstruction is still unstable. We defer a remark on relaxing the constraint that the zero-crossing be bounded until the next subsection, where we consider a discrete version of the result of this section.

3.2. Discrete Data

For simplicity, we treat the case of one unbounded space dimension, although the results extend easily. We are given data f_i, $i = \cdots, -1, 0, 1, \cdots$, and define

$$g_i = \frac{1}{4}f_{i-1} - \frac{1}{2}f_i + \frac{1}{4}f_{i+1}.$$

We define the filtered data $v_{i,k}$ recursively:

$$v_{i,0} = g_i,$$

$$v_{i,k+1} = \frac{1}{4}v_{i-1,k} + \frac{1}{2}v_{i,k} + \frac{1}{4}v_{i+1,k}.$$

We also define the blurring kernel

$$K_{i,k} = \frac{1}{4^k}\binom{2k}{i+k}.$$

Both v and K satisfy a discrete version of the Heat Equation, namely

$$u_{i,k+1} - u_{i,k} = \frac{1}{4}u_{i-1,k} - \frac{1}{2}u_{i,k} + \frac{1}{4}u_{i+1,k}.$$

It is not hard to prove a discrete analogue of the evolution property for zero-crossings. The key, as one might suspect from Section 1, is a discrete version of the maximum principle, which is easy to establish.

Let Ω be a bounded 4-connected collection of pixels (i,k) with a nonempty set $D = \{i \mid (i,0) \in \Omega\}$. Let T be an upper bound $T > \max\{k \mid (i,k) \in \Omega\}$, and define b_i to be the data g_i, $i \in D$, blurred to level T:

$$b_i = \sum_{j \in D} K_{i-j,T} \cdot g_j.$$

Finally, let

$$\partial_{(\pm 1,0)}\Omega = \{(i,k) \in \Omega \mid (i \pm 1, k) \notin \Omega\},$$

$$\partial_{(0,1)}\Omega = \{(i,k) \in \Omega \mid (i, k+1) \notin \Omega\},$$

$$\partial_{(0,-1)}\Omega = \{(i,k) \in \Omega \mid k > 0,\ (i, k-1) \notin \Omega\}.$$

Then simple but messy algebra allows us to show

Proposition:

$$4b_j = \sum_{\epsilon=-1,1} \sum_{(i,k) \in \partial_{(\epsilon,0)}} \Omega$$

$$\left[\frac{v_{i,k} + v_{i+\epsilon,k}}{2}[K_{i+\epsilon-j,T-k-1} - K_{i-j,T-k-1}]\right.$$

$$\left. - \frac{K_{i+\epsilon-j,T-k-1} + K_{i-j,T-k-1}}{2}[v_{i+\epsilon,k} - v_{i,k}]\right]$$

$$+ \sum_{(i,k) \in \partial_{(0,1)}\Omega} 4v_{i,k+1} \cdot K_{i-j,T-k-1}$$

$$- \sum_{(i,k) \in \partial_{(0,-1)}\Omega} 4v_{i,k} \cdot K_{i-j,T-k}.$$

To reconstruct data by the above equation, choose a connected component of $\{(i,k) \mid v(i,k) > 0\}$, (or respectively < 0). If the component extends to infinity in either coordinate, truncate the domain to become a convenient bounded collection of pixels, and denote the result by Ω. We store the sets $\partial_{(\pm 1,0)}\Omega$, $\partial_{(0,\pm 1)}\Omega$, and D as defined earlier. For pixels (i,k) in $\partial_{(1,0)}\Omega$ (respectively $\partial_{(-1,0)}\Omega$), we store the information $v_{i,k}$ and $v_{i+1,k}$, (respectively $v_{i,k}$ and $v_{i-1,k}$). For pixels (i,k) in $\partial_{(0,1)}\Omega$, we store the data $v_{i,k+1}$ and for $\partial_{(0,-1)}\Omega$ pixels we store $v_{i,k}$. Using the above equation, we choose a T and reconstruct the blurred data b_j. To reconstruct the data g_i for $i \in D$, it suffices to deblur the b_i data by solving for g_i in the linear equations defining b_i. In fact, the system is overdetermined, although still poorly conditioned, especially if $|D|$ or T is large.

In order to make the computations feasible, it is necessary to modify the formulas for a bounded spatial domain. We in fact solved a bounded domain problem, with $-N \le i \le N$, setting $v_{i,k} = 0$ for $i = \pm N$. The blurring kernel K is changed by this modification, but the proposition carries over with little change.

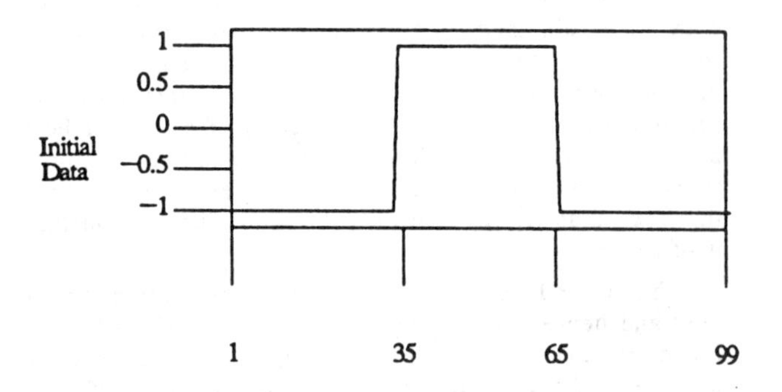

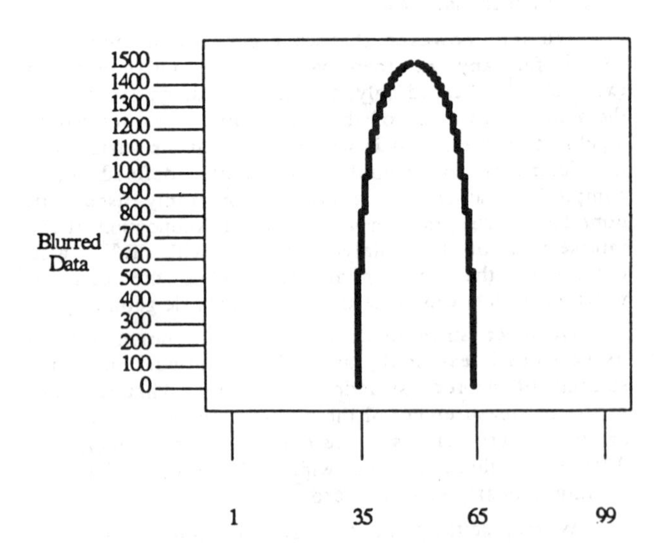

Figure 1. An initial function g_i and the zero crossing pixels in $v_{i,k}$, where $v_{i,k}$ blurs the initial data by scaling in k.

In Figure 1, we show a 1-D signal g_i, and the zero crossing seperating positive and negative regions of the associated $v_{i,k}$. Applying the above procedure to the central positive component Ω, we obtain the reconstructed b_i data shown in Figure 2. The true b_i data is identical to essentially machine precision. Finally, using the method of psuedoinverses to deblur the data shown in Figure 2, we obtain g_i for the i in the middle range, as shown in Figure 3. This is to be compared with the true initial data in Figure 1a. The poor correspondence is due to the fact that the deblurring problem is poorly conditioned; better deblurring results are obtained if the amount of deblurring is very small. However, this requires that the top of zero-crossing contour enclosing the data occurs after not too many blurring steps. The lesson learned here is that although reconstruction is in theory possible, practical reconstruction may be impossible.

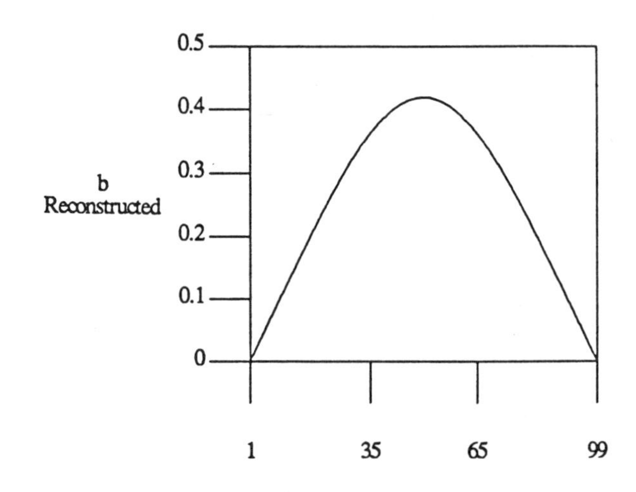

Figure 2. The reconstructed function b_i using the formula from Proposition 2. The data represents the result of blurring the g_i data restricted to the central positive interval, zero extended elsewhere, to a level k above the top of the zero crossing for $v_{i,k}$ shown in Figure 1 above. The reconstruction uses only information about $v_{i,k}$ along the zero-crossing, and is nearly exact to machine precision.

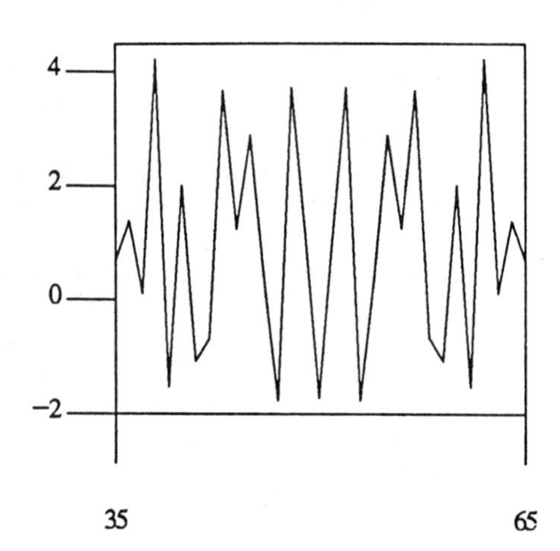

Figure 3. An attempt at reconstructing g_i using the data b_i from Figure 2. The attempt fails because the matrix equation $K\mathbf{g} = \mathbf{b}$ relating the **g** data to the **b** data is poorly conditioned, even though there are many more **b** values than **g** values. Thus the small errors due to round-off in representing the b_i data are magnified when reconstructing g_i. The singular value decomposition software in Creve Moler's "matlab" package was used.

4. Comments

Zero-crossings of scale-space filtered data seems like an unlikely form of representation of data. The results presented here suggest that even when supplemented with gradient data along the zero-crossings, the representation is still unlikely. However, the instability of the representation does not completely deny its utility, since it might happen that the classes of functions mapping to similar representations share properties essential for interpretation. A required step in the validation of the utility of a representation is an analysis of the invariant properties of signals that yield similar representations. A necessary condition is that attempted reconstructions differ from originals in unessential ways, (from the standpoint of interpretation). The methods outlined here should prove useful in verifying or disproving this necessary aspect of establishing a viable representation.

Acknowledgements

Work described in this paper was supported by Office of Naval Research grant N00014-85-K-0077. Many of the results of Section 2 were obtained in joint collaboration with Basilis Gidas in 1983. I thank David Mumford for helpful pointers. Jean Nguyen did the programming for the figures, and Linda Narcowich assisted in document preparation.

References

[1] Babaud, J., A. P. Witkin, M. Baudin, and R. O. Duda, "Uniqueness of the Gaussian kernel for scale-space filtering," *IEEE Transactions on Pattern Analysis and Machine Intelligence* **8**, pp. 26-33 (1986).

[2] Bers, L., F. John, and M. Schechter, *Partial Differential Equations,* American Mathematical Society, Providence, RI (1964).

[3] Burt, Peter J. and Edward H. Adelson, "A multiresolution spline with applications to image mosaics," *ACM Transactions on Graphics* **2**, pp. 217-236 (1983).

[4] Crowley, J., "A representation for visual information," CMU Robotics Institute, Ph.D. Thesis (1982).

[5] Daugman, J. G., "Six formal properties of anisotropic visual filters," *IEEE Transactions on Systems, Man, and Cybernetics* **13**, pp. 882-887 (1983).

[6] Hummel, Robert A. and Basilis C. Gidas, "Zero crossings and the heat equation," NYU Robotics Report 18 (March, 1984).

[7] Hummel, Robert A. and B. Kimia, "Deblurring gaussian blur," *Computer Vision, Graphics, and Image Processing,* (1985). To Appear. Also appears in *IEEE Conference on Computer Vision and Pattern Recognition*, San Francisco, 1985

[8] John, F., *Partial Differential Equations,* Springer-Verlag, New York (1975).

[9] Koenderink, Jan J., "The structure of images," *Biological Cybernetics* **50**, pp. 363-370 (1984).

[10] Marr, D. and T. Poggio, "A computational theory of human stereo vision," *Proceedings Royal Society London (B),* p. 301 (1979).

[11] Marr, D. and E. Hildreth, "Theory of edge detection," *Proceedings Royal Society London (B),* p. 187 (1980).

[12] Mumford,, Personal communication; To be included as an appendix in a larger version of this paper. A copy of Mumford's proof is available from the author of this report.

[13] Nirenberg, Louis, "A Strong Maximum Principle for Parabolic Equations," *Communications on Pure and Applied Mathematics* **6**, pp. 167-177 (1953).

[14] Prazdny, K., "Stereopsis in the absence of zero-crossings in the bandpass filtered images," Memo, Laboratory for AI Research, Fairchild, Schlumberger, 4001 Miranda Ave, Palo Alto, CA 94304 ().

[15] Protter, M. and H. Weinberger, *Maximum Principles in Differential Equations,* Prentice-Hall (1967).

[16] Witkin, A., "Scale space filerting," *Proceedings of the 8th International Joint Conference on Artificial Intelligence,* p. 1019 (1983).

[17] Yuille, A. L. and T. Poggio, "Fingerprints theorems," *Proceedings of the American Association of Artificial Intelligence,* pp. 362-365 (1984).

[18] Yuille, A. L. and T. Poggio, "Fingerprints theorems for zero crossings," *J. Optical Society of America* **2**, pp. 683-692 (1985).

[19] Yuille, A. L. and T. A. Poggio, "Scaling theorems for zero crossings," *IEEE Transactions on Pattern Analysis and Machine Intelligence* **8**, pp. 15-25 (1986).

[20] Zucker, Steven W. and Robert A. Hummel, "Receptive fields and the representation of visual information," *Proceedings of the Seventh International Conference on Pattern Recognition*, (July, 1984).

SIGNAL MATCHING THROUGH SCALE SPACE

Andrew Witkin
Demetri Terzopoulos
Michael Kass

Schlumberger Palo Alto Research
3340 Hillview Ave.
Palo Alto, CA 94304

ABSTRACT

Given a collection of similar signals that have been deformed with respect to each other, the general signal matching problem is to recover the deformation. We formulate the problem as the minimization of an energy measure that combines a smoothness term and a similarity term. The minimization reduces to a dynamic system governed by a set of coupled, first-order differential equations. The dynamic system finds an optimal solution at a coarse scale and then tracks it continuously to a fine scale. Among the major themes in recent work on visual signal matching have been the notions of matching as constrained optimization, of variational surface reconstruction, and of coarse-to-fine matching. Our solution captures these in a precise, succinct, and unified form. Results are presented for one-dimensional signals, a motion sequence, and a stereo pair.

I Introduction

Given a collection of similar signals that have been deformed with respect to each other, the general signal matching problem is to recover the deformation. Important matching problems include stereo vision, motion analysis, and a variety of registration problems such as template matching for speech and vision.

We cast the problem as the minimization of an energy functional $E(\mathbf{V})$ where $\mathbf{V}$ is the deformation. The energy functional is the sum of two terms, one based on the correlation of the deformed signals, and the other based on the smoothness of the deformation.

In general, the energy functional $E(\mathbf{V})$ can be highly nonconvex, so that ordinary optimization methods become trapped in local minima. Optimization by simulated annealing can be attempted, but at severe computational expense. Instead, we rely on *continuation methods* to solve the problem. By introducing a scale parameter σ, the minimization problem is embedded within a larger space. A suitable minimum can be achieved relatively easily for large σ because the signals and hence the energy landscape are very smooth. The solution of the original minimization problem is then obtained by continuously tracking the minimum as σ tends to zero. This is analogous to a coarse-to-fine tracking of extrema through scale-space in the sense of [Witkin, 1983].

The entire procedure consists of solving the first-order dynamic system

$$\dot{\sigma} = -k_1 \exp(-k_2|\nabla E(\mathbf{V},\sigma)|), \quad \dot{\mathbf{V}} = -\nabla E(\mathbf{V},\sigma),$$

where the dot denotes a time derivative, σ is the scale parameter and k_1 and k_2 are constants. Given an initial crude estimate for $\mathbf{V}$ at a coarse scale σ_0, the system minimizes E at σ_0 and follows a trajectory of minima through finer scales, thereby increasing the resolution of V. Any of a number of well known numerical techniques can be used to solve for the trajectory. Through a series of incremental deformations, correlations of deformed signals are optimized and balanced against the smoothness of the deformations while moving from coarse to fine scale. Thus, the first-order system compactly unifies a number of important yet seemingly disparate signal-matching notions.

In the remainder of this section, the relation of our technique to previous work on matching is discussed. Then in section 2, a framework for the minimization problem is introduced. In section 3, the solution of the problem by continuation methods and the resulting single differential equation are also developed. Section 4 describes the specific similarity term employed and, in section 5, the details of the smoothness term are discussed. Finally, section 6 presents several examples of matching results for one- and two-dimensional signals.

A. Background

An enormous amount of work has been done on signal matching, giving precedent for several components of our approach.

Optimization of constained deformations guided by correlation or L_2 metrics can be found in prior work. In speech recognition, the problem of time warping speech segments to match input utterances with stored prototypes has been addressed in this context. Dynamic programming has been used to compute constrained warping functions (see [Rabiner and Schafer, 1978] and [Sankoff and Kruskal, 1983] Part II). This particular optimization technique is readily applicable in matching situations involving sequentially ordered signals, such as speech, and unilateral continuity constraints. However, its stringent requirements on the energy functional appear incompatible with the unordered multi-dimensional signals and isotropic smoothness constraints which are of primary concern to us.

Smoothness constraints have been popular in computational vision. Consider the important problem of stereo matching. In the past, dense disparity maps have been computed through a two step process of local matching followed by smooth [Grimson, 1983], multiresolution [Terzopoulos, 1983], or piecewise continuous [Terzopoulos, 1986] surface reconstruction from the sparse disparities. The approach in the present paper unifies matching and piecewise smooth reconstruction into a single iterative optimization process.

Broit's [1981] work in registering a deformed image to a model image resembles ours, in that matching is explicitly formulated as a minimization problem involving a cost functional that combines both a deformation constraint and a similarity measure. His deformation model, which involves the strain energy of a globally smooth elastic body, is more elaborate than the deformation constraints inherent in the spring loaded subtemplate matching technique of Fischler and Elschlager [1973] or the iterative Gaussian smoothed deformation models proposed by Burr [1981]. Our controlled-continuity deformation model provides us with the additional capability to regulate the order of smoothness and to preserve discontinuities in the deformation.

Coarse-to-fine matching schemes have previously been treated as a multistage process in which a matching operation is performed at each successive level [Mori, et al., 1973; Hannah, 1974; Moravec, 1977; Marr and Poggio, 1979; Gennery, 1980]. We have extended this idea into a matching process which evolves continuously towards finer spatial scale. The idea of progressing continuously through scale space derives from Witkin [1983].

As our matching process computes the deformation iteratively, it is best to perform the similarity measurements by deforming the signals according to the current approximation of the deformation. This concern has also been addressed by the matching algorithms described in [Mori, et al., 1973; Burr, 1981; Broit, 1981; Quam, 1984].

II Framework

Consider a vector of n similar signals $\mathbf{f}^*(\mathbf{x}) = [f_1^*(\mathbf{x}), \ldots, f_n^*(\mathbf{x})]$ defined in d dimensional space $\mathbf{x} = [x_1, \ldots, x_d] \in \Re^d$, and a deformation mapping $\mathbf{V} : \Re^d \mapsto \Re^{nd}$, such that $\mathbf{V}(\mathbf{x}) = [\mathbf{v}_1(\mathbf{x}), \ldots, \mathbf{v}_n(\mathbf{x})]$, where each of the n disparity functions $\mathbf{v}_i : \Re^d \mapsto \Re^d$ is a vector valued function $\mathbf{v}_i(\mathbf{x}) = [v_1(\mathbf{x}), \ldots, v_d(\mathbf{x})]^\top$. Given a set of deformed signals $\mathbf{f}$ such that $\mathbf{f}^*(\mathbf{x}) = \mathbf{f}(\mathbf{V}^*(\mathbf{x}))$, the matching problem is to recover the deformation $\mathbf{V}^*(\mathbf{x})$.

Suppose that the similarity between the signals $\mathbf{f}$ for a given deformation $\mathbf{V}$ is measured by a functional $Q(\mathbf{V}) : \Re^{nd} \mapsto \Re$ bounded from above by a value achieved by the best possible match. A reasonable objective is to find the deformation $\mathbf{U}$ which maximizes the quality of the match; i.e., to minimize $-Q(\mathbf{V})$ over possible deformations $\mathbf{V}$. Thus, $\mathbf{U}$ represents an optimal approximation to $\mathbf{V}^*$.

This minimization problem is clearly ill-posed in the absense of constraints on admissible deformations, since, e.g., degenerate or chaotic deformations can always be contrived that achieve the minimum value. Such constraints may be encoded by a second functional $S(\mathbf{V}) : \mathcal{H}^{nd} \mapsto \Re$, where $\mathcal{H}^{nd} \subset \Re^{nd}$ is the subset of admissible deformations.

Useful instances of similarity and constraint functionals will be formulated shortly. Their combination, however, leads to the following minimization problem: Find the deformation $\mathbf{U} \in \mathcal{H}^{nd}$ such that $E(\mathbf{U}) = \inf_{\mathbf{V} \in \mathcal{H}^{nd}} E(\mathbf{V})$, where the energy functional is given by

$$E = -(1 - \lambda)Q - \lambda S \qquad (1)$$

and where $\lambda \in (0, 1)$ is a weighting parameter.

Stabilization offers a general approach to a numerical solution through the construction of a discrete dynamic system whose fixed points include a discrete solution of the above optimization problem [Bakhvalov, 1977]. A simple dynamic system with this property is characterized by the differential equations

$$\dot{\mathbf{V}} + \nabla E = \mathbf{0}, \qquad (2)$$

where the dot denotes differentiation with respect to time t and ∇E denotes the gradient of E with respect to the free variables of the discrete deformation. Optimization occurs by dissipation of energy; energy cannot increase along the system's trajectory $\mathbf{V}(\mathbf{x}, t)$ in $\mathcal{H}^{nd}$, which follows the direction of the gradient of E. Although the trajectory terminates at a local minimum of E, there is no guarantee that the global minimum $\mathbf{U}$ will be attained by solving this initial value problem starting from an arbitrary initial condition $\mathbf{V}(\mathbf{x}, 0)$.

III Continuation over Scale

The key remaining difficulty is that for obvious choices of Q, such as linear correlation, E is likely to have many local minima, making the minimization problem highly non-convex and therefore extremely difficult to solve. There are two options: solving this hard problem directly (for example by simulated annealing) or simplifying the problem by choosing Q to be convex or nearly so. We pursue the second option because annealing is expensive.

A. Continuation Methods

Q may be smoothed by subjecting $\mathbf{f}$ to a smoothing filter of characteristic width σ. We observe empirically that the best solution for $\mathbf{v}$ as σ increases tends to be an increasingly smoothed version of the correct solution. These means that slightly deblurring $\mathbf{v}$ by reducing σ produces a slightly better solution close to the one just obtained. To the extent this is so, we can solve the problem using equation 2 by means of *continuation methods* [Dahlquist and Björck, 1974].

Continuation methods embed the problem to be solved,

$$\mathbf{g}(\mathbf{v}) = 0,$$

in a family of problems

$$\mathbf{g}(\mathbf{v}, s) = 0,$$

parameterized by s. Let $s_{i+1} = s_i + \Delta s$, $\mathbf{g}(\mathbf{v}, s_n)$ be the problem we wish to solve, presumably difficult, and $\mathbf{g}(\mathbf{v}, s_1)$ a readily solvable member of the family, and let

$$\mathbf{u}(s) = \mathbf{H}(\mathbf{g}, s, \mathbf{v}_0)$$

be the solution for $\mathbf{g}$ at s given $\mathbf{v}_0$ as an initial condition. Then $\mathbf{u}(s_n)$ is obtained from $\mathbf{u}(s_1)$ by the iteration

$$\mathbf{u}(s_{i+1}) = \mathbf{H}(\mathbf{g}, s_{i+1}, \mathbf{u}_i); \qquad i = 1, \ldots, n-1$$

that is, each solution is used as an initial condition to obtain the next one.

For the current problem, the continuation parameter is σ, with $\Delta\sigma < 0$. We continue from an initial coarse scale σ_1 and an initial guess $\mathbf{V}_1$ by

$$\mathbf{V}_{i+1} = \mathbf{H}(\dot{\mathbf{V}} + \nabla E, \sigma_{i+1}, \mathbf{V}_i),$$

to a fine scale σ_n and a final answer $\mathbf{V}_n$. To visualize this method, imagine the energy landscape at each value of σ as a contoured surface in 3-space. The surfaces are stacked one above the other, so that the topmost surface is very smooth, while the lower ones become increasingly bumpy. Imaging a hole drilled at each local minimum on each surface. A ball bearing dropped onto the topmost surface will roll down to the bottom of the hill. At this point, it falls through to the next level, rolls down again, falls through again, and so on to the bottom.

B. A Scale Space Equation

This iteration solves a separate initial value problem at each step. A more attractive alternative is to collapse the continuation over σ into a single differential equation. Ideally, the solution should follow a curve $\mathbf{V}(\sigma)$ satisfying

$$|\nabla E(\mathbf{V}(\sigma))| \equiv 0;$$

i.e., a continuous curve of solutions over scale. A differential equation for this curve is

$$\mathbf{V}_\sigma = -(\nabla E)_\sigma (\nabla\nabla E)^{-1}. \qquad (3)$$

The solution to this equation tracks a given coarse-scale solution continuously to fine scale, in precise analogy with the coarse-to-fine tracking through scale space of [Witkin, 1983]. Unfortunately, it is impractical to solve this equation for arbitrary S and Q, since $\nabla\nabla E$ is high dimensional.

To construct an approximate equation, we introduce the quantity

$$N = -k_1 e^{-k_2|\nabla E|}, \tag{4}$$

so that $N = -k_1$ at a solution to equation 2, diminishing with distance from the solution at a rate determined by the space constant k_2. The equation

$$\dot{\sigma} = N, \quad \dot{\mathbf{V}} = -\nabla E \tag{5}$$

approximates the desired behavior. Far from a solution, where N is small, equation 5 approaches equation 2, changing $\mathbf{V}$ but not σ. Approaching a solution, σ begins to decrease. At a solution, $\dot{\mathbf{V}} = 0$ and $\dot{\sigma} = -k_1$. From an initial $\mathbf{V}(t_0), \sigma(t_0)$, the solution $\mathbf{V}(t), \sigma(t)$ moves through $\mathbf{V}$ at nearly constant scale until a minimum in E is approached, then it begins descending in scale staying close to a solution. [1]

Equation 5 finds a solution at the initial scale, then tracks it continuously to finer scales. To use equation 5, we choose a coarse scale $\sigma(t_0)$, a crude initial guess $\mathbf{V}(t_0)$, and a terminal fine scale σ_T. We then run the equation until $\sigma(t) = \sigma_T$, taking $\mathbf{V}(t)$ as the solution.

C. Ambiguous Solutions

From time to time, we expect to encounter instabilities in the solutions of equation 5, in the sense that a small perturbation of the data induces a large change in the solution curve's trajectory through scale space. These instabilities correspond to bifurcations of the solution curve, analogous to bifurcations that can be observed in Gaussian scale space. We have considered two approaches to dealing with them. First, by adding a suitable noise term to E, equation 5 becomes a hybrid of scale space continution and simulated annealing. We believe that local ambiguities can be favorably resolved using low-amplitude noise, hence with little additional computational cost. A second approach is to regard these instabilities as genuine ambiguities whose resolution falls outside the scope of the method. In that case, a set of alternative solutions can be explored by the addition of externally controlled bias terms to E. These terms can reflect outside constraints of any kind, for example, those imposed by the operation of attentional proceseses.

In the following sections, we turn to specific choices for S and Q.

IV Similarity Functional

In general the similarity measure Q should capture what is known about the specific matching problem. In many cases, the undeformed signals are sufficiently similar that a simple correlation measure suffices. In this section we formulate a generic choice for this class of problems. Note that, by assumption, it is the undeformed signals $\mathbf{f}^*(\mathbf{x})$ which are similar, so the quality of a potential solution $\mathbf{V}$ should be measured by the similarity of the signals $\mathbf{f}(\mathbf{V}(\mathbf{x}))$.

Consider the case of two signals f_i and f_j. A general family of similarity measures is obtained by integrating a local measure of similarity over position. If $K_{i,j}(\mathbf{x})$ is a local measure of the similarity of $f_i(\mathbf{v}_i(\mathbf{x}))$ and $f_j(\mathbf{v}_j(\mathbf{x}))$ around $\mathbf{x}$, then $Q_{i,j} = \int K_{i,j}(\mathbf{x})d\mathbf{x}$ is a global measure of similarity for f_i and f_j. By simply adding up pairwise similarities, a global measure of similarity can be constructed for n signals: $Q = \sum_{i \neq j} Q_{i,j}$.

A number of possibilities exist for the local similarity measure $K_{i,j}(\mathbf{x})$. Normalized cross-correlation produces good results for several matching problems that we have examined. If $W_\gamma(\mathbf{x})$ is a window function where γ denotes the width parameter, $\mu_i(\mathbf{x}) = \int f_i(\mathbf{v}_i(\mathbf{y}-\mathbf{x}))W_\gamma(\mathbf{y})d\mathbf{y}$, and $\nu_i(\mathbf{x}) = \int [f_i(\mathbf{v}_i(\mathbf{x}-\mathbf{y})) - \mu_i(\mathbf{x}-\mathbf{y})]^2 W(\mathbf{y})\,d\mathbf{y}$, then the normalized cross-correlation can be written

$$\begin{aligned} K_{i,j}(\mathbf{x}) &= [\nu_i(\mathbf{x})\nu_j(\mathbf{x})]^{-1/2} \int \Big\{ [f_i(\mathbf{v}_i(\mathbf{x}-\mathbf{y})) - \mu_i(\mathbf{x}-\mathbf{y})] \\ &\quad \times [f_j(\mathbf{v}_j(\mathbf{x}-\mathbf{y})) - \mu_j(\mathbf{x}-\mathbf{y})] W(\mathbf{y}) \Big\}\, d\mathbf{y}. \end{aligned}$$

The resulting functional $Q(\mathbf{V})$ generally has many local minima. In order to apply the continuation method, we compute Q for signals $\mathbf{f}$ which have been smoothed by Gaussians of standard deviation σ. The resulting functional $Q_\sigma(\mathbf{V})$ can then be made as smooth as is desired.

The correlation window size W_γ should be large enough to provide an accurate local estimate of the mean and variance of the signals, but small enough that non-stationarities in the signals do not become a problem. A convenient way to set γ to a reasonable value is to make it a fixed multiple of the average autocorrelation widths of the smoothed signals. Then γ can be regarded as a function of σ.

Note that $Q_\sigma(\mathbf{V})$ must be recomputed at each iteration, with the signals resampled to reflect the current choice of $\mathbf{V}$. If the deformation is very small, the distortion induced by failing to resample can be ignored, but the value of such resampling in stereo matching, for example, is well established [Mori, et al., 1973; Quam, 1984]. The Gaussian signal smoothing should also take place on the resampled functions $f(\mathbf{v}(\mathbf{x}))$.

V Smoothness Functionals

The functional $S(\mathbf{V})$ places certain restricitions on admissible deformations in order to render the minimization problem better behaved. Perhaps the simplest possible restriction, and one that has been used often in the past, is to limit possible disparities between signals to prespecified ranges. A deformation can then be assigned within disparity bounds on a point by point basis according to maximal similarity criteria. Although simple, such limited searches are unfortunately error prone, since they are based on purely local information.

This problem can be resolved by imposing global constraints on the deformation that are more restrictive yet remain generic. Such constraints may be based on *a priori* expectations about deformations; for example, that they are coherent in some sense. In particular, admissible deformations may be characterized according to the *controlled-continuity constraints* defined in [Terzopoulos, 1986]. These constraints, which are based on generalized splines, restrict the admissible space of deformations to a class of piecewise continuous functions. Not only is the deformation's order of continuity controllable, but discontinuities of various orders (e.g., jump, slope, and curvature discontinuities) are permitted to form, subject to an energy penalty.

A general controlled-continuity constraint is imposed on the deformation by the functional

$$S(\mathbf{v}) = -\sum_{m=0}^{p} \int_{\Re^d} w_m(\mathbf{x}) \sum_{j_1+\cdots+j_d=m} \frac{m!}{j_1!\ldots j_d!} \left| \frac{\partial^m \mathbf{v}(\mathbf{x})}{\partial x_1^{j_1} \ldots \partial x_d^{j_d}} \right|^2 d\mathbf{x}.$$

[1] The solution to equation 5 oscillates around the exact solution (equation 3,) with frequency and amplitude controlled by k_1 and k_2. This oscillation can be damped by the addition of second order terms in t, but we have not found it necessary to do so in practice.

Figure 2 shows a more challenging example in which 4 signals are matched simultaneously. The signals are intensity profiles from a complex natural image. On the left the four signals are shown superimposed at several points in the matching process. As before, the original signals appear at the top and the final result is at the bottom. Note that coarse-scale features are aligned first in the matching process while fine scale features are matched later. The four corresponding deformation functions $v_i(x)$ are shown to the right.

C. Motion Sequence

Figure 3 shows two frames from a motion sequence showing M. Kass moving against a stationary background. The frames are separated in time by about 1.5 seconds. Results of the matching process are shown as follows: the original image has been mapped onto a surface that encodes estimated speed as elevation. The raised area shows the region in which the algorithm detects motion.

D. Stereo Matching

Figure 4 contains a stereogram showing a potato partly occluding a pear. The matching results are rendered as two shaded surfaces with depth computed from the disparity. An image coordinate grid is mapped onto the first surface and the left image is mapped onto the second. The reconstructed surfaces are rendered from an oblique viewpoint showing the computed surface discontinuities. Only those portions of the scene visible in the original stereogram are shown.

VII Conclusion

The main contribution of this paper is two-fold. First, we introduced the notion of tracking the solution to the matching problem continuously over scale. Second, we developed a single system of first-order differential equations which characterize this process. The system is governed by an energy functional which balances similarity of the signals against smoothness of the deformation. The effectiveness of this approach has been demonstrated for both one- and two-dimensional signals.

Acknowledgements

We thank Al Barr for introducing us to continuation methods, and for helping us with numerical solution methods for differential equations. Keith Nishihara provided us with stereo correlation data.

References

Bakhvalov, N.S., *"Numerical Methods,"* Mir, Moscow (1977).

Broit, C., "Optimal Registration of Deformed Images," Ph.D. Thesis, Computer and Information Science Dept., University of Pennsylvania, Philadelphia, PA (1981).

Burr, D.J., "A dynamic model for image registration," *Computer Graphics and Image Processing,* **15** (1981) 102–112.

Dahlquist, G., and Björck, A., *"Numerical Methods,"* N. Anderson (trans.), Prentice–Hall, Englewood Cliffs, NJ (1974).

Fischler, M.A., and Elschlager, R.A., "The representation and matching of pictorial structures," *IEEE Trans. Computers,* **C-22** (1973) 67–92.

Gennery, D.B., "Modeling the Environment of an Exploring Vehicle by means of Stereo Vision," Ph.D. thesis, Stanford Artificial Intelligence Laboratory, also *Artificial Intelligence Laboratory Memo* 339 (1980).

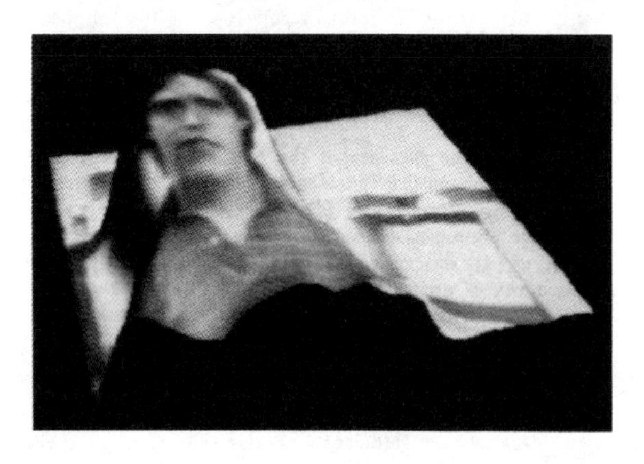

Figure 3: Two frames from a motion sequences about 1.5 seconds apart. The original image has been texture mapped onto a surface that encodes speed as elevation. The raised area is moving while the background remains stationary.

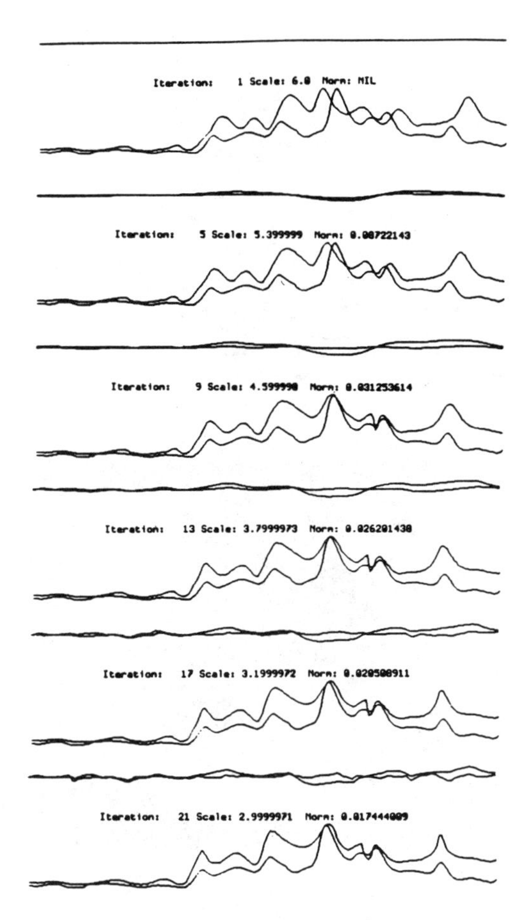

Figure 1: Matching two one-dimensional signals. The signals are measurements of the resistivity of a geological structure as a function of depth at two different locations. From top to bottom, the signals first appear in their original form, then partially deformed at intermediate stages of the matching process, finally showing the end result. Above the signals are shown the deformation function $\mathbf{V}$ and the correlation gradient ∇Q.

The positive integer p indicates the highest order generalized spline that occurs in the functional, and this determines the maximum order of continuity (C^{p-1}) of the admissible deformations. The nonnegative continuity control functions $\mathbf{w}(\mathbf{x}) = [w_0(\mathbf{x}), \ldots, w_p(\mathbf{x})]$ determine the placement of discontinuities. A discontinuity of order $q < p$ is permitted to occur at $\mathbf{x}_0$ by forcing $w_i(\mathbf{x}_0) = 0$ for $i \geq q$ (see [Terzopoulos, 1986] for details).

The $p = 2$ order controlled-continuity constraint is employed in our implementations to date. If, for convenience, a "rigidity" function $\rho(\mathbf{x})$ and a "tension" function $[1 - \tau(\mathbf{x})]$ are introduced such that $w_0(\mathbf{x}) = 0$, $w_1(\mathbf{x}) = \rho(\mathbf{x})[1 - \tau(\mathbf{x})]$, and $w_2(\mathbf{x}) = \rho(\mathbf{x})\tau(\mathbf{x})$, then it is natural to view the functionals as characterizing "generalized piecewise continuous splines under tension." In particular, for the case of n signals in 1 dimension, the functional reduces to

$$S(\mathbf{v}) = -\int_{\Re} \rho(x)\left\{[1 - \tau(x)]|\mathbf{v}_x|^2 + \tau(x)|\mathbf{v}_{xx}|^2\right\} dx,$$

while for the case of 2 signals in 2 dimensions, it becomes

$$\begin{aligned} S(\mathbf{v}) &= -\int\int_{\Re^2} \rho(x,y)\left\{[1 - \tau(x,y)](|\mathbf{v}_x|^2 + |\mathbf{v}|_y^2)\right. \\ &\quad \left. + \tau(x,y)(|\mathbf{v}_{xx}|^2 + 2|\mathbf{v}_{xy}|^2 + |\mathbf{v}_{yy}|^2)\right\} dx\,dy, \end{aligned}$$

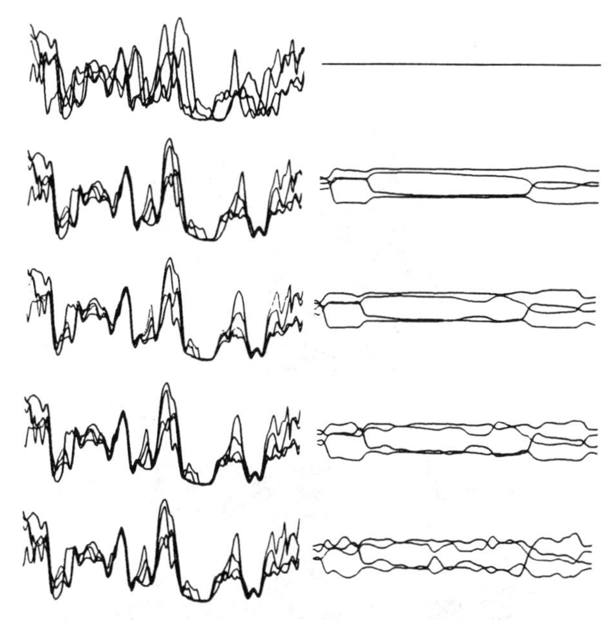

Figure 2: Simultaneous matching of four signals. The signals are intensity profiles from a complex natural image. On the left the four signals are shown superimposed at several points in the matching process. As before, the original signals appear at the top and the final result is at the bottom. Note that coarse-scale features are aligned first in the matching process while fine scale features are matched at the end. The four corresponding deformation functions $v_i(x)$ are shown to the right.

where $x_1 = x$ and $x_2 = y$.

VI Results

A. Implementation Notes

Discretization of the continuous variational form of the matching problem can be carried out using standard methods. Although finite element methods offer the greatest flexibility, for simplicity we employ standard multidimensional finite difference formulas for uniform meshes to approximate the spatial derivatives in $S(\mathbf{V})$. These approximations yield local computations analogous to those in [Terzopoulos, 1983].

Equation 5 is a standard first-order initial value problem, for which solution methods abound. We have employed numerical methods of varying sophistication, each giving satisfactory results. In order of sophistication, these include Euler's method, a fourth-order Runge-Kutta method, and Adams-Moulton predictor-corrector methods. The latter offer the advantage that the step size can be automatically adapted, making them particularly robust [Dahlquist and Björck, 1974].

B. One-dimensional Signals

The method is applicable to matching n signals, each of which is d-dimensional. Figure 1 shows the simplest case, that of matching two one-dimensional signals. The signals are measurements of the resistivity of a geological structure as a function of depth at two different locations. From top to bottom, the signals first appear in their original form, then partially deformed at intermediate stages of the matching process, finally showing the end result. The deformation function $\mathbf{V}$ and the correlation gradient ∇Q are shown above the signals.

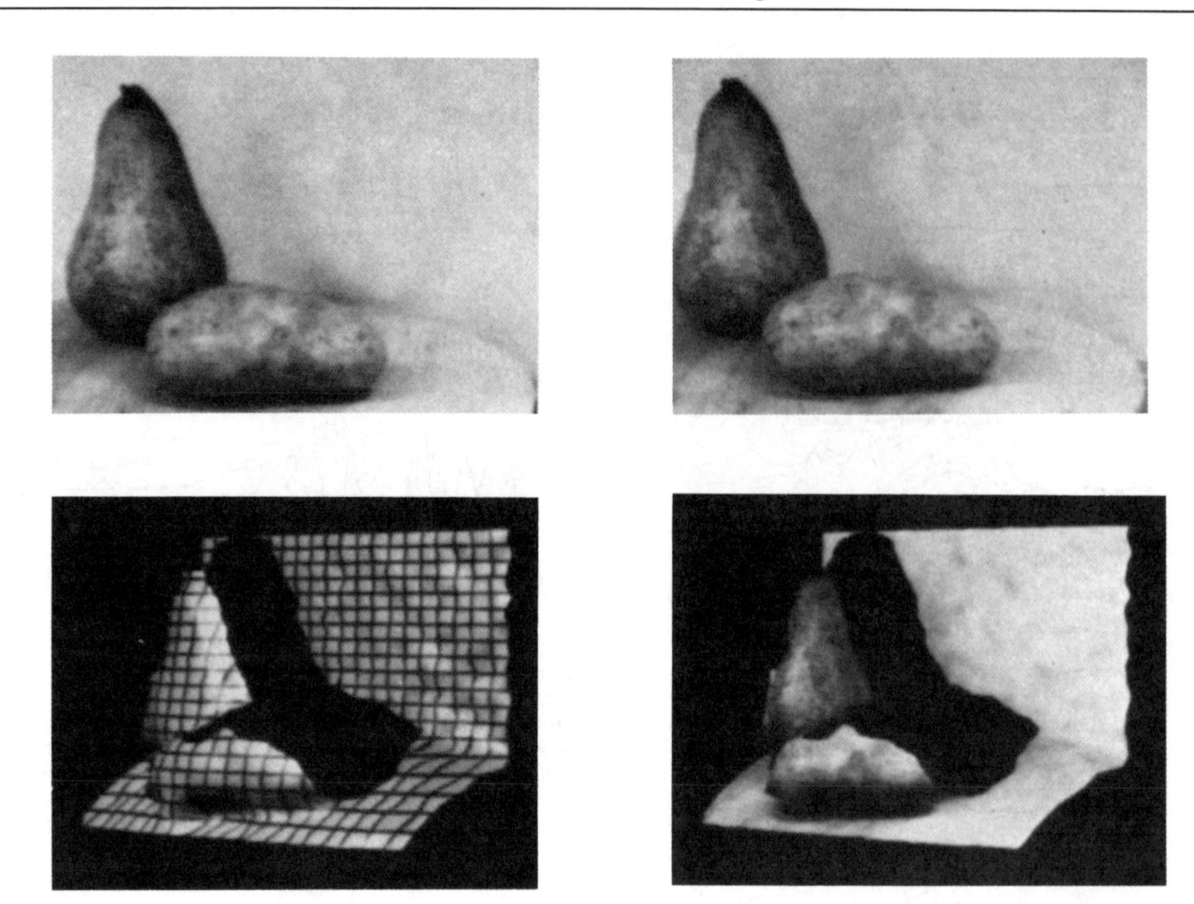

Figure 4: A stereogram showing a potato partly occluding a pear (the images are reversed for free fusing). The matching results are rendered as two shaded surfaces with depth computed from the disparity. An image coordinate grid is mapped onto the first surface and the left image is mapped onto the second. The reconstructed surfaces are rendered from an oblique viewpoint showing the computed surface discontinuities. Only those portions of the scene visible in the original stereogram are shown.

Grimson, W.E.L., "An implementation of a computational theory of visual surface interpolation," *Computer Vision, Graphics, and Image Processing,* **22** (1983) 39–69.

Hannah, M.J., "Computer matching of areas in stereo images," Stanford Artificial Intelligence Laboratory memo, AIM-239, July (1974).

Marr, D., and Poggio, T. "A theory of human stereo vision," *Proc. Roy. Soc. Lond.*B, **204** (1979) 301–328.

Moravec, H.P., "Towards automatic visual obstacle avoidance," *Fifth International Joint Conference on Artificial Intelligence,* Cambridge, Massachusetts (1977) 584.

Mori, K., Kidodi, M., and Asada, H., "An iterative prediction and correction method for automatic stereo comparison," *Computer Graphics and Image Processing,* **2** (1973) 393–401.

Quam, L.H., "Hierarchical warp stereo," Proc. Image Understanding Workshop, New Orleans, LA, October, 1984, 149–155.

Rabiner, L.R., and Schafer, R.W., *Digital Processing of Speech Signals,* Prentice-Hall, NJ (1978).

Sankoff, D., and Kruskal, J.B., (eds.), *Time Warps, String Edits, and Macromolecules: The Theory and Practice of Sequence Comparison,* Addison-Wesley, Reading, MA, (1983).

Terzopoulos, D., "Multilevel computational processes for visual surface reconstruction," *Computer Vision, Graphics, and Image Processing,* **24** (1983) 52–96.

Terzopoulos, D., "Regularization of inverse visual problems involving discontinuities," *IEEE Trans. Pattern Analysis and Machine Intelligence,* **PAMI-8** (1986).

Witkin, A., "Scale Space Filtering," Proceedings of International Joint Conference on Artificial Intelligence, Karlsruhe (1983) 1019-1021.

Appendix A

Key Ideas, Assumptions, and Open Issues in Computer Vision

The comparatively small amount of information contained in an image, or even in many images, cannot encode the full complexity of a natural 3-D scene. Thus, the competence of any vision system must, in large measure, depend on the assumptions that it uses concerning the relationships between the image observables and what is true about the world. In this appendix we list approximately thirty of the key assumptions that are frequently employed in computer vision systems. Few of these assumptions are strictly true, and some even contradict each other, but each one is often a useful approximation to reality.

We also list some of the computational paradigms and open issues that are the focus of current computer vision research.

Assumptions about the Utility of Image Appearance

- It is possible to extract a qualitatively accurate symbolic description of the environment from the information contained in (even) a single image.
- Nature does not conspire to fool us. Appearances are not deceiving; if something looks like a geometric feature in the image, such as a straight line, we can generally assume that it corresponds to that feature in space. (This assumption subsumes the mathematical concept of "general position"—the assumption that a slight change in the location from which we view the scene will not cause us to change our interpretation.)
- Image measurements are (at worse) linear transformations of geometric and photometric world properties, possibly perturbed by independent additive Gaussian noise. First, second, and even third derivatives can be measured with reasonable accuracy.

Assumptions of Coherence and Continuity

- Nature is a mathematician; she allows accurate descriptions of the environment in the vocabulary of algebra, topology, geometry, calculus, and statistics.
- Nature produces coherent, rather than random, structures. Images of natural scenes can be decomposed into a relatively small number of parts corresponding to the coherent structures in the world.
- Coherent objects are relatively uniform in some set of measureable properties; properties measured at any location in an image are good approximations to the same properties at (at least some) nearby points in the image.
- Coherence in the world produces continuity in the image. The surfaces of coherent objects in

the world can be described by continuous analytic functions over portions of the image plane.

Assumptions of Discontinuity and Image Composition

- Intersections or occlusions between 3-D objects map to singularities in the image projections of their bounding contours.
- Intensity discontinuities in an image correspond to object boundaries in the corresponding scene.
- Edges, the boundaries between coherent objects or parts of objects, are continuous closed curves in the image (even though segments of the edge may be occluded and thus not visible in the image).

Assumptions of Stability and Simplicity

- Nature has "islands" (zones) of stability in space, time, and scale of resolution. With the appropriate descriptive language and scale of resolution, certain objects will come into focus. That is, they will have compact descriptions against a random background.
- Objects have measurable properties that remain relatively constant over time.
- Simpler explanations of image composition are more likely to be correct than more complex ones.

Assumptions of Randomness

- Image properties that are not of direct interest, or that are measured at a scale of resolution above or below some ideal level, have a random distribution.
- Random processes are either uniform, Gaussian, or fractal (depending on which best suits our current desires).

Assumptions about Objects and Their Attributes

- The world and the objects in it are describable in terms of a small number of processes (e.g., accretion, deletion, flow, branching), each with a few characterizing parameters.
- The world is composed of discrete objects that can be decomposed into a small number of geometric building blocks—for example, polyhedrons, superquadrics, or generalized cylinders.
- Reflections are Lambertian.
- Scene illumination is relatively constant.
- Surfaces are opaque and piecewise uniform in albedo, color, and texture.

Assumptions about World Dynamics

- Objects are geometrically rigid.
- Objects are static or move with fixed velocity.
- The images of moving objects can be can be put into exact correspondence with each other by translations, rotations, and dilations.
- The imaged contours of moving objects deform in a continuous manner.

Assumptions about the Human Visual System

- The eye is a camera.
- The human visual system analyzes the world as a sequence of static images.
- The human visual system builds an initial description of the world that is independent of context and stored knowledge.
- The higher-level interpretative processes in the human visual system first extract invariant scene properties from the initial (low-level) description, and then invoke reasoning and stored knowledge to obtain a useful model of the environment.
- The human visual system is accurate in its interpretations.

Computational Paradigms

- Signals-to-symbols paradigm.
- Regularization paradigm.
- Relaxation paradigm.
- Production rule paradigm.
- Connectionist paradigm.
- Logic or deduction paradigm.
- Heuristic search (combinatorial optimization) paradigm.
- Statistical decision theory paradigm (including various voting schemes).
- Coarse-fine control paradigm.
- Random selection (guessing) of model parameters and subsequent model evaluation.

Problems and Open Issues

- Multisource integration—including methods for predicting the performance of a given scene analysis algorithm, and criteria for comparing the results of alternative interpretations produced by different algorithms.
- Learning: methods by which vision systems can autonomously improve their performance (or at least avoid repeatedly making the same mistakes).
- Representation and indexing: design of representations for visual description of complex scenes that are also suitable for reasoning and indexing into a large database of stored knowledge.
- Discovery of natural constraints that are left invariant under image transformations.
- Techniques for distinguishing between random and coherent processes without resorting to stored models of explicit image content (criteria for perceptual organization).
- Methods and criteria for determining if the assumptions prerequisite for use of an analysis technique are valid in a given application instance.

Appendix B

Parallel Computer Architectures for Computer Vision

This appendix briefly describes some of the parallel computer architectures and the associated algorithms available for computer vision applications. The significantly reduced computation time provided by such machines over conventional serial architectures (from hours to minutes or fractions of a second in some cases) is important both for increased effectiveness of experimentation as well as for applications involving real-time computation.

The basic idea of a parallel architecture is to trade hardware for speed as effectively as possible. However, as the number of processors increases, the problem of effective communication between processors and strategies for partitioning the problems to take advantage of the parallelism become more difficult. Yalamanchili and Aggarwal [Yalamanchili 85a] model parallel architectures to compare various architectures and communication schemes.

Parallel architectures are typically single-instruction multiple-data (SIMD) in which the same instruction is executed on many data items simultaneously, or multiple-instruction multiple-data (MIMD) in which different instructions are carried out simultaneously on many different data items. The main advantage of the MIMD scheme over the SIMD scheme occurs in high-level image understanding tasks. For example, it is possible to assign a set of processors to analyze a set of segmented objects, where each processor deals with a separate object. Alternatively, several processors could analyze a single object, with each processor executing a different algorithm.

The architectures are often discussed in terms of "grain" of computation—that is, the complexity of the computation assigned to each processor. Thus, an architecture that assigned a single processor to alter the intensity of each pixel of an image based on the intensities of neighboring pixels would be considered a fine-grained design. (Sometimes one sees grain defined in terms of the number of processors in a system; coarse grain is a design with 2 to 128 processors, medium grain 128 to 1000 processors, and fine grain anything above that.) SIMD architectures tend to have small grain size, while MIMDs generally have large grain size. Another important characteristic is whether the processors are grid connected (each processor communicates to a set of neighbor processors) or whether a more general connectivity capability is provided.

Although there are many specialized image-processing configurations, we review those architectures that are, or soon will be, available to the image-understanding community. For more comprehensive descriptions of the field, see A. P. Reeves's survey [A.P. Reeves 84], the IEEE Computer Society tutorial [Wah 85], the Uhr collection [Uhr 84], and special issues of *IEEE Computer* [IEEE 82, 83]. There is also a *Journal of Parallel and Distributed Computing* and an annual IEEE/ACM conference resulting in *Proceedings of the International Conference on Parallel Processing.*

Parallel Architectures

The Butterfly

The Butterfly is a multiple-instruction, multiple-data (MIMD) computer; in each of its processors (up to 256 of them) computations can proceed independently of the instructions or data in other processors. This coarse-grain multiprocessor was developed by Bolt Beranek and Newman (BBN) under DARPA sponsorship and is manufactured by General Electric. It consists of a number of "processor nodes" interconnected by a "butterfly switch" that provides general connectivity. Each processor node consists of a Motorola 68000 microprocessor, memory management hardware, an interface to the switch, and an interface to I/O devices. Each node has 256K bytes of memory expandable to 4 Mbytes. The components of the processor node are controlled by a microcoded coprocessor called the "Processor Node Controller" (PNC), a 16-bit-wide processor that extends the functionality of the MC68000 and coordinates the use of the various internal processor node buses. The Butterfly uses shared memory as the fundamental form of interprocessor communication. While a few Butterfly machines with 64 to 128 processors have been sold, most configurations to date have fewer than 12 processors.

All the software is developed in the C language, running in the object-oriented Chrysalis operating system loosely based on UNIX. Because Chrysalis was developed by BBN for a communication application, there are a number of drawbacks concerned with memory management and resource allocation when using the system for vision applications.

For a tutorial discussion of the use of the Butterfly for vision research, see the *Rochester Computer Science and Computer Engineering Research Review* [Rochester 85].

Cosmic Cube

The Cosmic Cube [Seitz 85], developed at California Institute of Technology (Caltech) under DARPA sponsorship, is a configuration of up to 256 processors that work concurrently on a problem and coordinate their computations by sending messages to each other. Each computer is called a node, and each node is connected through bidirectional, asynchronous, point-to-point communication channels to six other nodes to form a communication network that follows the plan of a six-dimensional hypercube. An operating system kernel in each node schedules and runs processes within that node, provides system calls for processes to send and receive messages, and routes the messages that flow through the node. So far the Cosmic Cube has been used mostly for physics and applied mathematics problems.

The Connection Machine

The Connection Machine (CM)[1] [Hillis 84] consists of thousands of processors in a single-instruction, multiple-data (SIMD) configuration. The CM was designed by MIT and is manufactured and sold by Thinking Machines, Inc. of Cambridge, Mass. The CM is a fine-grained massively parallel machine containing 65K 1-bit serial processors that can communicate with each other by two distinct mechanisms. One of these mechanisms has the topology of a boolean 16-cube and is called the router network. The other mechanism consists of a four-connected x-y grid called the north-east-west-south (NEWS) network. NEWS is used for local communication, as required in convolution or relaxation algorithms. The router is used for global communication, as required for permuting, sorting, merging, summing, histogramming, region-growing, and image sampling. (See the discussion by Hillis [Hillis 84] for the actual connection strategy.) Each processor has 4K bits of memory. The machine is programmed in a SIMD fashion from a host computer such as a lisp machine or a VAX, using extended versions of Common Lisp and C. The computations performed by a particular processor depend on the data contained in its memory.

A convenient way to visualize the CM for image applications is to view each processor as storing a pixel in the picture array. The individual pixels of an 8-bit-per-pixel image might be contained in memory locations 0 to 7 in every processor, and additional images and intermediate results can be placed in subsequent locations, up to 4096 bits in depth.

For a description of the use of the CM to implement a real-time stereo vision system, see [Drumheller 86]. He indicates that a Gaussian convolution using a mask of diameter 49 on a 4K $\times$ 4K image takes about 0.02 sec.

The Massively Parallel Machine

The Massively Parallel Processor (MPP) is a single-instruction-stream, multiple-data (SIMD) processor designed at NASA Goddard and built by Goodyear Aerospace; its intended use is for solving two-dimen-

[1] Connection Machine is a trademark of Thinking Machines Corporation.

sional data processing problems such as those encountered in the processing of satellite imagery. The heart of the MPP is the array unit (ARU) that has 1 staging plane, 1024 memory planes, and 35 processing planes each containing 16384 elements (128 rows by 128 columns). The nominal machine cycle time is 100 nanoseconds. A plane of data can be transferred en masse between the staging plane and a memory plane or en masse between a processing plane and a memory plane. Instructions operate on an entire plane of data in parallel.

As with conventional computers, the ARU can be programmed at different levels. Low-level program modules contain instructions that specify the transfer of information to and from memory planes. These modules are analogous to assembly language routines. High level programs contain instructions that specify how memory planes or arrays of memory planes are to be combined and processed.

For a description of the MPP, see [Batcher 80, 85] or the collection edited by Potter [Potter 85]. For a discussion of image processing on the MPP, see Potter [Potter 83].

The NON-VON Machine

The NON-VON (for non-Von Neumann), which was developed at Columbia University, is a family of massively parallel tree-structured supercomputers. The architectures of all the NON-VON family members include a tree-structured primary processing subsystem (PPS) based on custom VLSI circuits, along with a secondary processing subsystem (SPS) based on a bank of "intelligent" disk drives. The PPS is configured as a binary tree of small processing elements (SPE). Each SPE has a small RAM (64 bytes), a modest amount of processing logic, and an I/O switch. The I/O switch can be set for global bus communication, for communication between parents and children (tree neighbor communication), or for reconfiguration of the binary tree as a linear array of processors (linear neighbor communication). The SPS consists of several rotating storage devices, each of which has logic for dynamically examining the data. There is a parallel transfer of data between the PPS and SPS.

The early members of the NON-VON family function as SIMD machines, but NON-VON 4 will function both as an MIMD and an SIMD machine.

The architecture is described by Shaw [Shaw 84], and algorithms for the NON-VON are given in Ibrahim's thesis [Ibrahim 84].

Array Processors

Array processors are synchronous, parallel, pipelined[2] machines used in combination with a standard host machine. Both specialized hardware and software provide extremely fast vector computations such as addition, subtraction, comparison and matching, and convolution. These machines have processing rates of from one million floating point operation per second (Mflop) to one hundred Mflops. Prices range from a few thousand dollars for a single board device to a quarter of a million dollars for a 100 Mflops machine. For a good review of array processors see Lim and Binford [Lim 84].

Systolic Arrays

In a systolic system, data flows from the computer memory passing through many processing elements before it returns to the memory. As described by Kung [Kung 82], "The system works like an automobile assembly line where different people work on the same car at different times and many cars are assembled simultaneously." A systolic array consists of a pipelinelike structure of gates and processors that can be linear, triangular, or hexagonal to make use of higher degrees of parallelism to compute some special-purpose function as efficiently as possible. The Warp machine developed at Carnegie-Mellon University is a linear array of cells such that each cell is able to read and write from memory, read a word from a previous cell, send a word to the next cell, and complete a floating point add and multiply. All the preceding operations can be carried out simultaneously in the basic cycle time of 200 nanoseconds. The Warp nominally uses ten cells, and each cell is a fairly versatile computer in itself. The capabilities of a cell include: a 4K 32-bit word RAM used for storage of constants and implementation of shift registers, arrays, and so on; a 2K 80-bit word microstore; and special 32-bit floating point add-and-multiply chips that provide results in five pipelined stages (1 microsecond). There is two-way data flow—each cell can read and send data to its predecessor or its successor. Various internal mechanisms are used to coordinate data processed by the various elements of the cell.

In the Warp, a 3×3 convolution applied to 256×256 image requires on the order of 10 milliseconds, and a

[2] A pipeline is a set of processors, strung one after the other, through which data are pumped.

fast Fourier transform for 1024 points requires 600 microseconds. A VAX 11/780 is about a thousand times slower, taking about 10 seconds for the convolution and about 1/2 second for the transform.

Kung [Kung 84] describes several systolic algorithms in the areas of signal and image processing.

Content Addressable Array Parallel Processor

The Image Understanding architecture research group at the University of Massachusetts, Amherst, Mass., has developed the Content Addressable Array Parallel Processor (CAAPP). It embodies the feature of both content addressable processors (such as STARAN) and parallel array processors (such as the ILLIAC IV). The resulting architecture can be used for both associative and array type of operations.

The CAAPP is both a 512×512 SIMD array processor and an associative memory. The design is based on a 64×64 array of custom VLSI chips performing bit-serial arithmetic; it is intended to act as a slave processor for a general-purpose computer system. Each chip contains 64 cells, an instruction decoder, and some miscellaneous logic. There are eight basic instruction types recognized by the chip, each performed in parallel by the constituent cells. Most instructions take one minor cycle time (100 nanoseconds) to execute. Intercell communication is accomplished by a four-way (N,S,E,W) cell interconnect network. Simulated evaluations of the still incomplete machine show that a 3×3 Sobel edge extraction operator can be convolved with a 512×512 image in 50 to 100 microseconds.

Algorithms and Languages for Parallel Machines

We must often rethink the traditional approaches and languages used for sequential computers to take full advantage of the computational power offered by the various parallel computers. Potter [Potter 85] indicates how the conventional programming concepts used for "number crunching" should be revised for a parallel machine. For nonnumerical applications, such as AI and database management, he states that to take advantage of a SIMD machine, the slogan should be "Search, don't sort."

Algorithms

There has been an explosion of activity in algorithms for parallel machines. As yet there is no good single source; papers can be found in the Annual International Conference on Parallel Processing, sponsored by Ohio State University, the IEEE Computer Society, and the ACM (published by IEEE Computer Society Press). The journal *Communications of the ACM* contains papers on parallel algorithms from time to time.

Uhr in Part III of [Uhr 84] discusses parallel algorithms and languages. He examines techniques for restructuring serial code into more parallel form and for revising algorithms, programming languages, and even approaches to problems to make fuller use of parallel architectures.

Languages

A recent *Computer* issue devoted to "domesticating parallelism" is a particularly good source of information [IEEE 86]. This IEEE issue classifies parallel languages into parallel Algol-based languages (Parallel Pascal, Ada), Lisps and logic languages (Multilisp and Concurrent Prolog), and parallel functional languages (Paralfl). Except for Ada and Parallel Pascal, most of the languages are still in a research phase.

Parallel Pascal. A.P. Reeves [A.P. Reeves 85] describes the use of Parallel Pascal, the first high-level language to be implemented on the MPP. Three fundamental classes of operations on array data are discussed: data reduction (sum, product, maximum, minimum of the elements of an array), data permutation (shift, rotate, transpose), and data broadcast.

Ada. Ada,[3] a language developed for the Department of Defense, was designed from the beginning with parallel processing in mind. The parallel mechanism that Ada introduced is the "tasking model." Ada provides high-level mechanisms for task creation and synchronization that are machine and operating system independent. The mechanisms can be efficiently implemented on distributed systems as well as in shared memory, and on multiprocessors as well as uniprocessors. However, a more general synchronization mechanism would have simplified the programming of many applications. See the discussion by Mundie and Fisher [Mundie 86].

Multilisp. Multilisp is a Lisp-like language, developed at M.I.T, that allows the programmer to specify concurrent execution. It is still too early to judge the ultimate success of the programming language ideas in Multilisp. For a discussion of this language, see [Halstead 86].

[3] Ada is a registered trademark of the U.S. Government Ada Joint Program Office.

Concurrent Prolog. Concurrent Prolog is a logic programming language designed for parallel execution. There are various formulations of Concurrent Prolog, and as of 1987 the practicality of the language for programming parallel computers has not yet been demonstrated. See Shapiro's discussion of the problems that must be solved, and the various approaches that are possible [Shapiro 86].

Glossary

The definitions in this glossary have been restricted to terms directly relevant to scene analysis. The definitions have been kept informal to emphasize the underlying concepts.

A* algorithm [Nilsson 80] A "best-first" search technique for finding the least-cost path in a network whose links have associated weights or costs. A* is able to take advantage of information about the approximate cost/distance between a given node and a goal node.

affine transformation [Ahuja 68] A transformation in which every point in the 2-D plane has its x and y coordinates mapped to new values by linear functions. For example, straight lines are mapped into straight lines, parallel lines are mapped into parallel lines, and circles are mapped into circles or ellipses.

albedo The ratio of the total radiant energy returned by a body to the total energy incident on the body (also see **reflectance**).

aliasing [Rosenfeld 82] If a function is to be totally recoverable from its sample values, a sufficient number of samples must be obtained. If too few samples are obtained, or if the samples are improperly chosen, then the difference between the function and the interpolated function is called the aliasing error.

aperture problem [Ullman 81] The tangential component of the image velocity of an object profile cannot be recovered by an analysis technique that examines only a small segment of the profile in the image plane.

back projection [Barnard 83] The recreation of the "bundle of rays" that produced a given image. Back projection requires the availability, or the assumption, of a camera model.

binary image [Rosenfeld 82] An image in which each pixel assumes one of exactly two values; the values are usually designated as 0 and 1.

blocks world [Ballard 82] A simple world consisting of planar-faced solids, such as cubes and pyramids, that is often used as the experimental domain for image analysis problems.

Boltzmann machine [Ackley 85] [Fahlman 87] A network of interconnected "units" with each unit in one of two states—*on* or *off*. A unit adopts one of these states as a probabilistic function of its neighboring units and the weight on the interconnecting links. The weights are determined for a particular application by a "learning" procedure based on simulated annealing.

camera model [Strat 84] A mathematical specification for mapping the 3-D world to a 2-D image, often expressed as a 4x3 matrix when the mapping is described in terms of homogeneous coordinates.

chain code [Freeman 74] A representation of a curve using line segments that must lie on a fixed grid with a fixed set of orientations. A curve can be compactly expressed as the sequence of integers representing the orientation of each line segment.

chamfering [Barrow 77] A transform that weights points in an image in terms of their distance from one or more designated contours. Chamfering is often used to "blur" or spread contour information prior to carrying out a matching operation between two line drawings.

clique [Bolles 82] [Ballard 82] A totally connected subgraph in which each node is connected to every other node of the subgraph by an arc. Used in generalized structure matching where nodes represent features and the links represent relationships between features; a clique indicates a completely consistent grouping of data.

color space [Ballard 82] From a psychological and practical standpoint, color can be represented in a computer by a triple of values in a number of ways. For example: (1) the intensity of the red, green, and blue components; (2) the values of the intensity, saturation, and hue; and (3) a set of differences (red-green), (blue-yellow), and (white-black). Each of these representational systems defines a "space" with different "distance" relationships existing between a given pair of "color vectors."

coarse/fine control strategy [Moravec 77, 79] A general strategy in which computations are first carried out on a "coarse" version of an image. Subsequent computations are carried out on finer and finer versions, with each additional pass more tightly constrained by the preceding ones.

connected component labeling [Firschein 78] [Ballard 82] Given a pixel array, distinct labels are assigned to groupings of pixels that are "adjacent" according to some specified criterion.

connectionism [Fahlman 87] An approach to parallel computation in which algorithmic knowledge is stored as a pattern of connections or connection strengths among the processing elements. Thus, the connections directly determine how the processing elements interact. This is in contrast to the conventional approach to computation in which algorithms are stored in computer memory as symbolic structures and interpreted by a central processing unit when appropriate.

contour map [Merrill 73] [Morse 69] [Ward 78] A two-dimensional representation of the topography of a (terrain) surface using a set of closed curves, each of which represents a constant surface elevation.

contrast [Rosenfeld 82] The intensity difference between an object and its background.

Coons patch [Coons 74] A four-sided surface patch for approximating a piece of a surface. The four sides of the patch are specified by polynomials, and these polynomials are used to interpolate interior points.

cooperative algorithms [Ballard 82] [Marr 82] Iterative techniques, such as relaxation labeling, for finding consistent solutions to a system of constraint equations—for example, a system of inequalities.

correlation [Fischler 71] [Ballard and Brown 82] A numerical measure of similarity between two vector quantities that is computed as the dot (scalar) product of the vectors—that is, the projection of one vector onto the other. The scalar product can be shown to be an approximation to the distance between the vectors represented as points in a Euclidean space. The definition of correlation as used in statistics, sometimes called "normalized correlation," involves adjusting each measurement (vector component) by subtracting its population mean and dividing by its population standard deviation. Under these adjustments, correlation, now a value between +1 and −1, measures the cosine of the angle between two vectors.

corresponding points [Barnard 82] Two points, one in each image of an image pair, that correspond to the same point in space.

cost array [Fischler 81b] A feature array whose elements represent the probability of the presence of the feature at the given pixel location, or the cost of assuming that the feature is present at the given pixel location.

curvature [Hilbert 52] For a curve in a plane, curvature at a point on the curve is a measure of how fast the tangent to the curve at that point is changing direction. If we measure the curvature at the common point on the curves formed by intersecting a surface with all planes containing the normal to the surface at the given point, we find that the maximum and minimum curvatures (called the principal curvatures) correspond to curves at right angles to each other. The product of the principal curvatures at the point is called the Gaussian curvature.

DTM Abbreviation for digital terrain model. A numerical array representing terrain elevation as a function of geographic or spatial location.

digitization See **quantizing**.

disparity [Barnard 82] The relative displacement of corresponding points in the two images of a stereo pair; disparity is inversely related to scene depth.

distance transform [Fischler 80] Given a set of distinguished points in a 2-D plane (image), the distance transform assigns a numerical value to each point in the plane specifying its distance to the nearest distinguished point.

dot product space [Firschein 79] Given a vector A and a span of unit vectors, dot product space is the plot of the dot products of A with the unit vectors versus the unit vector directions. For example, to determine the "diameter" of a region in an image, one can take the dot product of the span of unit vectors and the vector from some arbitrary image location to each pixel in the given region. The diameter is found in dot product space as the largest difference between the maximum dot product and the minimum dot product at a particular unit vector direction.

dynamic programming [Bellman 62] An iterative optimization technique based on the "principle of optimality," which asserts that for any given initial state and feasible initial decision in the iterative optimization process, the resulting (reduced) problem must be a new (simpler) optimization problem, again subject to the principle of optimality. Phrased in terms of the problem of finding the shortest path in a graph between two points, the dynamic programming concept says that when looking for the shortest path from A to B, ignore all paths from A to any intermediate node, I, other than the minimum length path from A to I.

early vision [Marr 82] The first stages of processing in the human visual system. (See **low vision**.)

edge detection [Haralick 84] The location of boundaries across which one or more image attributes —such as intensity, texture, or color—are discontinuous.

epipolar images [Bolles 85] Epipolar images are obtained from a "dense" set of images—images taken close enough together in time to form a solid block of data. An epipolar image is a 2-D plane with one spatial and one time dimension, extracted from the solid block of image data.

epipolar lines [Barnard 82] The corresponding pairs of lines in the two images of a stereo pair that are the intersections of the two image planes and the set of planes passing through the two lens centers. Corresponding points in the two images fall on corresponding epipolar lines. Use of the *epipolar constraint* reduces stereo matching from a 2-D search problem to a search in one dimension.

expert (production rule) systems [McKeown 85] Knowledge about the world is expressed as a collection of rules of the form "If condition A is satisfied, then carry out operation B." A control system examines all the rules to determine those rules whose "if" portion is satisfied and decides which rule is to be activated when more than one is satisfied. Rule-based systems have the attraction of allowing addition and deletion of rules without requiring extensive reprogramming.

feature array [Barrow 78] A pixel array whose elements are the values of an attribute of some image or scene feature—for example, intensity gradient, surface orientation, material type, and so on.

filter [Rosenfeld 82] An operator that replaces the value of a pixel P by a function of P and its neighbors. Filtering is used to smooth, enhance, or detect some property or aspect of an image.

focal length [Strat 84] The distance from the plane in which an image is formed to the lens center for scene objects located at 1"infinite" distance from the imaging system.

focus of expansion [Ballard 82] For straight-line motion of a camera and a forward-looking direction, the projection of each point in the scene flows across the image plane along a straight line emenating from a unique point in the image called the focus of expansion.

Fourier descriptors [Persoon 74] The boundary of a region can be represented as an analytic function, such as tangent angle versus arc length, which can be expanded in a Fourier series. The Fourier or shape descriptors of the boundary are the coefficients of the Fourier series; these descriptors can be used to compare boundary shapes.

frame [Fischler 87] A frame is a data structure that represents some prototypical situation or object. A frame has "slots" that describe important attributes of the prototype. Constraints on possible slot values and their relation to other slot values can be defined. A frame can contain information on how it is to be used, default values for its slots, and what to do if something unexpected happens.

frequency space [Rosenfeld 82] An explicit representation of the spatial variation of some image attribute, such as intensity, in terms of frequency-parameterized basis functions, such as produced by the Fourier transform.

Gaussian sphere [Ikeuchi 81] A representation in which some geometric property of objects or surfaces in a Cartesian space is mapped onto a unit sphere. The finite extent of the Gaussian/unit sphere is an advantage when searching for some property, as opposed to searching the infinite extent of Cartesian space.

generalized cone [Binford 71] A curve in space, called a spine, along which a 2-D shape, called the cross section, is swept. The cross section is kept at a constant angle to the tangent of the spine and is deformed according to a deformation function called a sweeping rule. Configurations of generalized cones are used to form complex objects.

gradient space [Shafer 83] A 2-D space explicitly depicting the orientation of scene surfaces. The axes of the space are the first-order partial derivatives of a surface. Each point in gradient space corresponds to the orientation of a possible surface normal. Gradient space makes explicit many image properties that constrain the geometry of the corresponding 3-D scene objects.

gray level [Rosenfeld 82] The intensity value of an image at a picture element (pixel) location.

heuristic search [Nilsson 80] [Pearl 84] Any strategy for efficiently searching through a very large number of possibilities where the rules guiding the search (called heuristics) will not guarantee an optimal solution, but will generally provide a "good" answer.

high-level vision [Fischler 87] The portion of the signals-to-symbols paradigm concerned with providing a goal-oriented description of a scene in which purpose, context, and space and time relations are all made explicit.

histogram [Ballard 82] A plot of the count of a quantity vs. the quantity. An intensity histogram is the number of pixels in an image lying in specified intensity intervals, plotted on a scale of increasing intensity quantized into "buckets" the size of the specified intervals.

homogeneous coordinates [Ahuja 68] A point in Cartesian n-space is represented as a line in homogeneous $(n+1)$ space. Thus, each unique Cartesian coordinate point corresponds to infinitely many homogeneous coordinate locations. Homogeneous coordinates allow many important geometrical transformations to be represented uniformly and elegantly—for example, as matrix multiplications rather than as nonlinear functional mappings.

Hough space [Duda 72] [Ballard 81] A multidimensional histogram used to estimate model parameters; each histogram bucket corresponds to a complete specification of the model parameters. For example, for each edge point detected in an image, the "Hough transform" will increment the counters for each bucket in Hough space that corresponds to the parameters of a line that could pass through the given point. All the points in the image on the same line contribute a vote to the same histogram bucket, allowing this line to be detected. The Hough transform can be generalized to detect arbitrary shapes.

image velocity See **velocity flow field**.

intermediate-level vision [Fischler 87] The portion of the signals-to-symbols paradigm concerned with producing a description of image and scene attributes in which "global" and goal-oriented information is made explicit. Primary operations involve aggregation (partitioning or linking), matching, labeling (naming), model fitting and instantiation, and shape description.

intrinsic image [Barrow 78] A set of registered "images" describing scene surfaces with respect to depth, surface orientation, reflectance, and incident illumination.

inverse optics [Zucker 86] The recovery of the 3-D scene from the information contained in a 2-D image of the scene.

Lambertian surface [Horn 77] A matte surface that has a reflectance function that is a constant times the cosine of the angle between the incident radiation and the surface normal. The brightness of points in an image of a Lambertian surface is not a function of camera location.

LANDSAT (ERTS) [Ballard 82] A satellite used to obtain images of the earth's surface for the purpose of resource analysis, land-use, and geological studies.

Laplacian [Ballard 82] [Marr 82] A mathematical operator for computing the sum of the second partial derivatives of a multidimensional function. As s sum, the Laplacian is isotropic—it does not provide information about the direction in which the function is changing.

least squares [Duda 73] Selecting the parameters of a model to describe a set of data points so that the sum of the squared differences between the points and the model is minimized.

linking (edge following) [Fischler 81b] Linking takes local evidence for edges and assigns a continuous path so as to satisfy such criteria as maximizing overall contrast or minimizing overall curvature.

logarithmic spiral grid [Weiman 79]. This image is obtained by a transform from rectangular to polar coordinates; the corresponding representation of an object is invariant under object rotation, and a change of scale for the object simply causes a translation of its representation.

low-level vision [Fischler 87] The first phase of the signals-to-symbols paradigm. In low-level vision the processing is local and is independent of purpose or content. The output is a description of intensity variations and anomalies in the image as well as the local intrinsic characteristics of the visible scene surfaces, such as depth, orientation, and velocity.

matte reflectivity [Ballard 82] A surface with a "dull," "flat," and "diffuse" reflectivity. (see **Lambertian**.)

medial axis [Fischler 80] A skeleton or stick figure derived from an image region by "thinning" algorithms that preserve connectivity of the region. (See **skeleton**.)

minimum spanning tree (MST) [Zahn 71] A tree that spans a graph and whose sum of edge lengths is minimum. The MST is often used in image segmentation and edge-linking algorithms.

model fitting (instantiation) [Fischler 81a] Selecting the parameters of a model to describe a set of data points. (See **least squares**.)

Necker cube [Gregory 78] A 2-D line drawing of a 3-D cube with all hidden lines shown. It is possible to see the cube as two different 3-D objects, and both interpretations alternately replace each other as one shares at the figure.

noise [Rosenfeld 82] Random variations in a signal, or an image, caused by the processes and transformations that are employed in sensing the signal or scene and converting it to a computer-usable form.

normalization [Rosenfeld 82] An operation carried out on a set of numerical quantities which "adjusts" them to remove irrelevant influences and allows them to be compared for some specific purpose.

occlusion The hiding from view of part of one object by another.

optical flow [Ullman 81] The "flow line" in an image obtained by tracing the path of a scene surface point of an object moving relative to the camera.

orthographic projection [Ahuja 68] A parallel projection of points in space onto an image plane.

perspective projection [Ahuja 68] A projection in which points in 3-D space are mapped onto an image by lines or "rays" passing through a single point in space called the center of perspective. The projection normally seen by the human eye—for example, a straight road in perspective converges at a point in the distance.

photometric stereo [Woodham 81] Two images obtained from a stationary camera, but using a differently positioned light source for each image, provide sufficient information for depth recovery. The advantage of photometric stereo is that the matching problem is eliminated by the fixed positioning of the camera relative to the scene.

preprocessing [Rosenfeld 82] Preliminary operations carried out on an image to remove degradations such as noise or distortion, or to enhance certain aspects of an image prior to further analysis.

production rule [Fischler 87] A rule of the form "If condition X is satisfied, then carry out operation Y."

projective projection [Ahuja 68] [Rogers 76] A mapping defined by a sequence of perspective projections.

pixel An elemental picture (image) element.

pixel array [Rosenfeld 82] A numerical array in geometric correspondence with the picture (image) elements depicting a scene. The numerical values are typically light intensity, range, or other sensor-derived "point" measurements.

Ponzo illusion [Gregory 78] Given two parallel lines of equal length enclosed by converging lines such as railroad ties enclosed by a pair of rails, the line in the narrower portion of the space enclosed by the converging lines appears to be longer than the other line.

propositional and predicate calculus [Nilsson 84] [Davis 83] Knowledge is represented in a formalism that permits explicit information to be derived from implicit information. The advantage of this formalism is that only consistent results can be deduced—that is, it is not possible to deduce both a fact and its contradiction. Disadvantages are that there is no guarantee that reasonable descriptions of real-world problems can be constructed, and for the more expressive predicate calculus, that the desired explicit information can be deduced, or that practical algorithms can be found to perform the deduction.

pyramid [Uhr 72] [Sammet 80] [Rosenfeld 75, 80] A data structure consisting of the same image information at successively increasing levels of resolution. Quad trees and strip trees are examples of pyramids.

quad tree [Samet 80] Given a $2^n \times 2^n$ array of pixels, a quadtree is constructed by repeatedly subdividing the array into quadrants, subquadrants, and so on until we obtain blocks consisting of a single value. This process is represented as a tree in which the root node corresponds to the entire array, the four sons of the root node correspond to the quadrants, and the terminal nodes correspond to those blocks of the array for which no further subdivision is necessary to describe significant information.

quantizing [Rosenfeld 82] The operation in which a continuous quantity is converted into a discrete quantity—for example, a real number into the closest integer.

random dot stereogram [Julesz 71] Portions of a random dot field depicted in one image are coherently displaced to form a second image of a stereo pair. The random dot stereogram is an important tool in investigating human visual perception because there are no recognizable features in the individual images of the stereogram.

range image [Ballard 82] An image in which the value of each pixel is the distance between the sensor and the object in the scene imaged on that pixel. Also called a depth map.

RANSAC (RANdom SAMple Consensus) [Fischler 81] A technique for obtaining the parameters of a model to describe a data set. A selection containing the exact number of items (e.g., points) required to solve for the model parameters is randomly chosen from the input data. If the entire input data set does not fit the instantiated model, the process is repeated with a new random selection of input points. The process is continued until a suitable model is obtained.

ray tracing [Youssef 86] [Kajiya 83] An approach to creating synthetic images of modeled 3-D scenes in which the intensity at a particular image pixel is determined by tracing all rays from light sources to the 3-D scene surfaces that contribute to the intensity at that pixel.

reflectance [Ballard 82] The ratio of the radiant power per unit area reflected by the surface to the radiant power per unit area incident on the surface. In general, the reflectance function of a surface is a function of the angles between the surface normal and both the light source and the observer. (See **albedo**.)

relational or semantic nets [Ballard 82] Graph structures that represent objects as nodes and relationships between objects as labeled arcs.

relaxation [Hummel 83] A computational approach that uses locally interacting processes in an iterative fashion over an image array such that the current local values are updated to achieve a globally consistent final result.

resolution [Rosenfeld 82] A measurement that indicates how precisely an image can represent an object. The resolution limit is the smallest dimension of an object that can be discriminated or observed.

scale space [Witkin 83] A 1-D signal is convolved with a Gaussian filter, and the zeros of the second derivative (actually, the "zero crossings") are found and followed as the size of the filter increases. Scale space is the plot of the zero-crossing contours on the (x, σ) plane, where σ measures the size of the Gaussian filter and x is the signal variable.

segmentation [Fischler 87] The partitioning of an image into regions or other meaningful shapes.

shape from shading or texture [Strat 86] [Marr 82] [Horn 77] Determination of the 3-D shape of an object shown in a 2-D image by analysis of shading or texture.

signals to symbols [Fischler 87] The currently dominant paradigm for computer-based scene analysis. The raw image is transformed into a meaningful and explicit description of the corresponding scene by a series of inductive steps employing progressively more abstract representations.

simulated annealing [Kirkpatrick 83] A global optimization technique that attempts to avoid being trapped in local minima by (1) testing solutions suggested by a random process that occasionally proposes very large excursions from the current best solution, and (2) using a "temperature" parameter, which is slowly decreased, to control the acceptance of any change in a variable. During the course of this procedure an accepted solution will sometimes be worse than the previous one.

skeleton [Fischler 80] Intuitively, the axis of symmetry of an object, or a reduction of a thin region to a stick figure. More precisely, the set of points interior to an object that are maximally distant from its boundary. (See **medial axis**; see **thinning algorithm**.)

slant [Pentland 84] The angle between a normal to a surface and the normal to the image plane. Slant is a number between 0 and 90 degrees.

smoothing [Rosenfeld 82] Reducing the variation in a 1-D signal or a 2-D image, often by using filtering techniques.

spectral response [Rosenfeld 82] A function describing detector sensitivity versus input signal frequency.

spline [Pavlidis 77] An interpolation technique using piecewise polynomial curves with smoothness constraints between the polynomial segments. The resulting fit "looks good" to people—that is, the interpolation is similar to the freehand fit that people would make.

statistical pattern recognition [Duda 73] A methodology for solving classification (naming) problems as decision problems posed in probabilistic terms; the relevant statistical distributions and prior probabilities must be known.

stereo image pair [Barnard 82] A pair of images that typically view the same scene with 40 to 80 percent overlap to obtain a depth map of the scene.

stereo vision [Barnard 82] The use of two different views of the same scene to recover 3-D scene geometry.

strip tree [Ballard 81] A binary tree used to represent a curve. The datum at each node is an eight-tuple, of which six entries define a rectangle bounding a segment of the curve and two denote the addresses of the sons, if any.

subjective contour [Kanisza 76] [Gregory 78] An illusory edge formed when the visual system is led to believe that an occluding surface is being viewed, even though the supposed occluding surface and background are not physically different in intensity.

symmetry/skew-symmetry [Kanade 81] Symmetric shapes are left unchanged by some specified set of geometric transformations—such as rotation or reflection about a point or axis. A bilaterally symmetric figure in an arbitrarily-oriented plane, viewed under orthography, yields a skew-symmetric figure.

texture [Ballard 82] A spatial distribution of image intensities or a patterning of primitive image shapes.

thinning algorithm [Arcelli 85] An algorithm that reduces an image region to a single-pixel "thickness" while retaining the connectivity of the figure. (See **skeleton**.)

thresholding [Rosenfeld 82] An operation that takes a gray scale image and produces a binary image. A pixel whose gray value is less than the threshold value is set to zero; otherwise, the pixel is set to one.

tilt [Pentland 84] The angle between the projection of the normal to (a 3-D) surface and a fixed coordinate axis in the image plane.

torsion For a mathematical definition of torsion, see pp. 276–77 of Ballard and Brown [Ballard 82]. Torsion can be understood conceptually as follows: Given a pipe that has been formed into a planar circle, keep one end of the pipe fixed while raising the other end. The planar pipe has zero torsion, and as the end is raised above the plane, the torsion increases.

tree [Ballard 82] A representation consisting of nodes and connecting links. Each node in a tree has a unique parent.

triangular irregular networks [Peuker 78] [Fowler 79] [Watson 84] [Boissonnat 84] Triangular patches joined together to approximate a surface.

vanishing point [Barnard 83] The projection onto the image plane of the point at infinity in a given direction.

variational problem [Poggio 85] Many problems in image analysis can be phrased in terms of an optimization problem where some objective function is to be maximized or minimized.

velocity flow field [Ullman 81] A description of the speed and direction, in the image plane, of the projection of each visible point in the scene.

vergence [Marr 82] A movement of one eye in relation to the other.

Waltz filtering [Waltz 72] The use of local constraints rather than global constraints to eliminate possibilities before performing global optimization. Waltz filtering was done in the context of providing 3-D labels, such as "convex" or "concave," to a solid portrayed in a line drawing.

wire frame [Herman 84] A representation of a solid in which an object is shown as if it were constructed of wires for all its edges. This representation is often used to store models of objects to be matched or recognized by a vision system.

zero crossing [Ullman 81] [Marr 82] The locations in an image where the second derivative of the intensity function undergoes a sign change. The "zero crossings" are good candidates for edge pixels. In practice, the zero crossings are obtained by taking the differences of Gaussian-smoothed images.

Bibliography

D.H. Ackley, G.E. Hinton, and T.J. Sejnowski. A learning algorithm for Bolzmann machines. *Cognitive Science*, 9:147–169, 1985.

J.K. Aggarwal, L.S. Davis, and W.N. Martin. Correspondence processes in dynamic scene analysis. *Proc. IEEE*, 69(5):562–572, May 1981.

J.K. Aggarwal, R. Duda, and A. Rosenfeld, editors. *Computer Methods in Image Analysis.* IEEE Press, New York, 1977. Distributed by Wiley.

J.K. Aggarwal and A. Mitichie. Structure and motion from images: Fact and fiction. *Proc. IEEE Third Workshop on Computer Vision: Representation and Control*, pages 127–128, Bellaire, Michigan, October 13–16, 1985.

A.K. Agrawala, editor. *Machine Recognition of Patterns*, IEEE Press, New York, 1976.

D.V. Ahuja and S.A. Coons. Geometry for construction and display. *IBM Systems Journal*, 7(3–4):188–205, 1968

N. Ahuja. Dot pattern processing using Voronoi neighborhoods. *IEEE PAMI*, 4(3):336–343, May 1982.

N. Ahuja and B. Schachter. *Pattern Models.* Wiley, New York, 1983.

A.P. Ambler, H.G. Barrow, C.M. Brown, R.M. Burstall, and R.J. Popplestone. A versatile computer-controlled assembly system. *Artificial Intelligence*, 6(2):129–156, 1975.

H.C. Andrews and B.R. Hunt. *Digital Image Restoration.* Prentice-Hall, Englewood Cliffs, New Jersey, 1977.

P.E. Anuta. Spatial registrations of multispectral and multitemporal digital imagery using fast Fourier transform techniques. *IEEE Trans. on Geoscience Electronics*, 8(4):353–368, October 1970.

C. Arcelli and G.S. di Baja. A width-independent fast thinning algorithm. *IEEE PAMI*, 7(4):463–474, July 1985.

H. Asada and M. Brady. The curvature primal sketch. *IEEE PAMI*, 8(1):2–14, January 1986.

J. Babaud, A.P. Witken, M. Baudin, and R.O. Duda. Uniqueness of the Gaussian kernel for scale-space filtering. *IEEE PAMI*, 8(1):26–33, January 1986.

R. Bajcsy and L. Lieberman. Texture gradient as a depth cue. *CGIP*, 5(1):52–67, March 1976.

D.H. Ballard. Cortical connections and parallel processing: Structure and function. *The Behavioral and Brain Sciences*, 9:67–120, 1986.

D.H. Ballard. Model-directed detection of ribs in chest radiographs. *Proc. 4th IJCPR*, pages 907–910, Kyoto, Japan, 1978.

D.H. Ballard. Parameter nets. *Artificial Intelligence*, 22(3):235–267, April 1984.

D.H. Ballard and D. Sabah. Viewer independent shape recognition. *IEEE PAMI*, 5(6):653–660, November 1983.

D.H. Ballard, U. Shani, and R. B. Schudy. Anatomic models for medical images. *Proc. 3rd Computer Software and Applications Conference*, pages 565–570, Chicago, Ill., November 1979.

D.H. Ballard. Generalizing the Hough transform to detect arbitrary shapes. *Pattern Recognition* 13(2):111–122, 1981a.

D.H. Ballard. Strip trees: A hierarchical representation for curves. *Comm. ACM*, 24(5):310–321, May 1981b.

D.H. Ballard and C.M. Brown. *Computer Vision.* Prentice-Hall, Englewood Cliffs, N.J., 1982.

S. T. Barnard. Hierarchical microcanonical annealing. *Proc. DARPA IU Workshop*, University of Southern California, Los Angeles, Cal., 1987.

S.T. Barnard. Choosing a basis for perceptual space. *CVGIP*, 29:87–99, 1985.

S.T. Barnard. *An Inductive Approach to Figural Perception.* Technical Report TN 325, SRI International, AI Center, September 1984.

S.T. Barnard. Interpreting perspective images. *Artificial Intelligence*, 21(4):435–462, November 1983.

S.T. Barnard. A stochastic approach to stereo vision. *Proc. 5th National Conf. on AI*, pages 676–680, Philadelphia, Pa., August 11–15, 1986.

S.T. Barnard and M.A. Fischler. Computational stereo. *Computing Surveys*, 14(4):553–572, December 1982. Updated in *Encyclopedia of Artificial Intelligence*, Wiley, New York, 1987.

S.T. Barnard and W. Thompson. Disparity analysis of images. *IEEE PAMI*, 2(4):333–340, July 1980.

D.I. Barnea and H.F. Silverman. A class of algorithms for fast digital image registration. *IEEE Trans. on Comm.*, 21(2):179–186, February 1972.

A.H. Barr. Global and local deformations of solid primitives. *ACM Computer Graphics, SIGGRAPH '84*, 18(3):21–30, July 1984.

A.H. Barr. Superquadrics and angle-preserving transformations. *IEEE Comp. Graphics and Applications*, 1:1–20, 1981.

J. Barron. *A Survey of Approaches for Determining Optic Flow, Environmental Layout, and Egomotion.* Technical Report RBCV-TR-84-5, Univ. of Toronto, Dept. of Computer Science, Toronto, Canada, November 1984.

H.G. Barrow and J.M. Tenenbaum. Recovering intrinsic scene characteristics from images. A.R. Hanson and E.M. Riseman, editors, *Computer Vision Systems*, Academic Press, New York, 1978.

H.G. Barrow and J.M. Tenenbaum. Computational vision. *Proc. IEEE*, 69:572–595, 1981a.

H.G. Barrow and J.M. Tenenbaum. Interpreting line drawings as three-dimensional surfaces. *Artificial Intelligence*, 17(1–3):75–116, August 1981.

H.G. Barrow, J.M. Tenenbaum, R.C. Bolles, and H.C. Wolf. Parametric correspondence and chamfer matching: Two new techniques for image matching. *Proc. 5th IJCAI*, pages 659–663, Cambridge, Mass., August 1977.

K.E. Batcher. Design of a massively parallel processor. *IEEE Trans. Computers*, 29:836–840, September 1980.

K.E. Batcher. The massively parallel processor system overview. J.L. Potter, editor, *The Massively Parallel Processor*, MIT Press, Cambridge, Mass., 1985.

J. Beck. Perceptual grouping produced by line figures. *Percept. Psychophys.*, 2:491–495, 1967.

J. Beck, B. Hope, and A. Rosenfeld, editors. *Human and Machine Vision.* Academic Press, New York, 1983.

R. Belknap, E.M. Riseman, and A. Hansen. The information fusion problem and rule-based hypotheses applied to complex aggregations of image events. *IEEE CVPR*, 227–234, 1986.

R.E. Bellman and S.E. Dreyfus. *Applied Dynamic Programming.* Princeton University Press, Princeton, New Jersey, 1962.

G. Beni and S. Hackwood. *Recent Advances in Robotics.* Wiley, New York, 1985.

S.S. Bergstrom. Illumination, color, and three-dimensional form. J. Beck, editor, *Organization and Representation in Perception*, Lawrence Erlbaum Assoc., Hillsdale, New Jersey, 1982.

R. Bernstein, editor. *Digital Image Processing for Remote Sensing.* IEEE Press, New York, 1978.

P.J. Besl and R.C. Jain. Three-dimensional object recognition. *ACM Computing Surveys*, 17(1):75–145, March 1985.

I. Biederman. Human image understanding: Recent research and theory. *CVGIP*, 32:29–73, 1985.

T.O. Binford. Visual perception by computer. *Proc. of the IEEE Conf. on Systems and Control*, Miami, Florida, 1971.

T.O. Binford. Inferring surfaces from images. *Artificial Intelligence*, 17(1–3):205–244, August 1981.

T.O. Binford. Survey of model-based image analysis systems. *Int. J. Robotics Research*, 1:18–64, 1982.

T.O. Binford and G.J. Agin. Computer description of curved objects. *Proc. 3rd Int. Joint Conf. on AI*, pages 629–640, Stanford, California, August 20–23, 1973.

P. Blicher. *Edge Detection and Geometric Methods in Computer Vision.* PhD thesis, Dept. of Mathematics, University of California, Berkeley, California, October 1984.

J. Bloomenthal. Modeling the mighty maple. *Computer Graphics*, 19(3):305–311, July 1985.

H. Blum. Biological shape and visual science. *J. Theoretical Biology*, 38:205–287, 1973.

J. Boissonnat. Geometric structures for three-dimensional shape representation. *ACM Trans. on Graphics*, 3(4):266–286, October 1984.

R.C. Bolles, H.H. Baker, and D.H. Marimont. Epipolar-plane image analysis: An approach to determining structure from motion. *International Journal of Computer Vision*, 1(1), 1987. In press.

R.C. Bolles, P. Horaud, and M.J. Hannah. 3DPO: A three-dimensional part orientation system. *Proc. 8th Int Joint Conf. on AI*, pages 1116–1120, Karlsruhe, West Germany, August 8–12, 1983.

R.C. Bolles and H.H. Baker. Epipolar-plane image analysis: A technique for analyzing motion sequences. *Proc. IEEE 3rd Workshop on Computer vision, Representation, and Control*, pages 168–178, Bellaire, Michigan, October 13-16, 1985.

R.C. Bolles and R.A. Cain. Recognizing and locating partially visible objects: The local feature focus methods. *Int. J. of Robotics Research*, 1(3):57–82, Fall 1982.

R.C. Bolles and M.A. Fischler. A RANSAC-based approach to model fitting and its applications to finding cylinders in range data. *Proc. 7th Int. Joint Conf. on AI*, pages 637–643, Vancouver, B.C., Canada, August 1981.

J.M. Brady, editor. *Artificial Intelligence, 17(1-3): Special Volume on Computer Vision*, August 1981a.

J.M. Brady. Preface – The changing shape of computer vision. *Artificial Intelligence*, 17(1–3), 1981b.

J.M. Brady. Computational approaches to image understanding. *ACM Computing Surveys*, 14(1):3–71, March 1982.

P. Brodatz. *Textures: A Photographic Album for Artists and Designers.* Dover, Toronto, Canada, 1966.

R.A. Brooks. Model-based three-dimensional interpretations of two-dimensional images. *IEEE PAMI*, 5(2):140–149, March 1983.

R.A. Brooks. Symbolic reasoning among 3-D models and 2-D images. *Artificial Intelligence*, 17:285–348, 1981.

R.A. Brooks. Visual map making for a mobile robot. *1985 IEEE International Conference on Robotics and Automation*, pages 824–829, St. Louis, Missouri, March 25–28, 1985. IEEE Cat. 85CH2152-7.

A. Bruss. *Shape from Shading and Bounding Contour.* PhD thesis, Dept. of EE and CS, MIT, Cambridge, Mass., 1981.

J.B. Burns, A.R. Hanson, and E.M. Riseman. Extracting straight lines. *Proc. DARPA IU Workshop*, pages 165–168, New Orleans, La., October 1984.

P.J. Burt and E.H. Adelson. The Laplacian pyramid as a compact image code. *IEEE Trans. on Communications*, 31(4):532–540, April 1983a.

P.J. Burt and E.H. Adelson. A multiresolution spline with applications to image mosaics, *ACM Trans. on Graphics*, 2(4):217–236, October 1983b.

J.F. Canny. A computational approach to edge detection. *IEEE PAMI*, 8(6):679–698, November 1986.

J.F. Canny. A variational approach to edge detection. *Proc. National Conf. on AI*, pages 54–58, Washington, D.C., August 22–26, 1983.

P. Carnevali, L. Coletti, and S. Patarnello. Image processing by simulated annealing. *IBM J of Research and Development*, 29(6):569–579, November 1985.

K.R. Castleman. *Digital Image Processing.* Prentice-Hall, Englewood Cliffs, N.J., 1979.

P. Cavanagh. Reconstructing the third dimension: Interaction between color, texture, motion, binocular disparity, and shape. *CVGIP*, 1987. In press.

R. Chellappa and A.A. Sawchuck, editors. *Digital Image Processing and Analysis: Vol. 2, Digital Image Analysis.* IEEE Computer Society Press, Washington, D.C., 1985. IEEE Order EHO232-9.

M.B. Clowes. On seeing things. *Artificial Intelligence*, 2:79–116, 1971.

S.A. Coons. Surface-patches and B-spline curves. R.E. Barnhill and R.F. Riesenfeld, editors, *Computer-Aided Geometric Design*, Academic Press, New York, 1974.

D.B. Cooper. Maximum likelihood estimation of Markov-process blob boundaries in noisy images. *IEEE PAMI*, 1(4):372–384, October 1979.

E. Davis. The Mercator description of spatial knowledge. *Proc. 8th Int Joint Conf. on AI*, pages 295–301, Karlsruhe, West Germany, August 8–12, 1983.

E. Davis. *Representing and Acquiring Geographic Knowledge.* Pitman, London and Morgan Kaufmann, Los Altos, California, 1986.

L.S. Davis. Shape matching using relaxation techniques. *IEEE PAMI*, 1:60–72, January 1979.

L.S. Davis. A survey of edge detection techniques. *CGIP*, 4:248–270, 1975.

L.S. Davis, S.A. Johns, and J.K. Aggarwal. Texture analysis using generalized cooccurrence matrices. *IEEE PAMI*, 1:251–259, 1979a.

L. Davis and A. Rosenfeld. Cooperating processes for low level vision: A survey. *Artificial Intelligence*, 17(1–3):245–264, August 1981.

H.P. Decell and L. F. Guseman. Linear feature selection with applications. *Pattern Recognition*, 11:55–63, 1979.

S.W. Draper. The use of gradient and dual space in line-drawing Interpretation. *Artificial Intelligence*, 17(1–3):461–496, August 1981.

M. Drumheller. Connection machine stereomatching. *Proc. 5th National Conf. on AI*, pages 748–753, Philadelphia, Pa., August 11–15, 1986.

J. Duchon. Interpolation des fonctions de deux variables suivant le principe de la flexion des plaques minces. *R.A.I.R.O. Analyse Numerique*, 10:5–12, 1976.

R.O. Duda and P.E. Hart. Use of the Hough transform to detect lines and curves in pictures. *Comm. ACM*, 15(1):11–15, 1972.

R.O. Duda and P.E. Hart. *Pattern Classification and Scene Analysis.* Wiley, New York, 1973.

R.O. Duda, D. Nitzan, and P. Barrett. Use of range and reflectance data to find planar surface regions. *IEEE PAMI*, 1:259–271, July 1979.

J.O. Eklundh, H. Yamamoto, and A. Rosenfeld. A relaxation method for multispectral pixel classification. *IEEE PAMI*, 2:72–75, 1980.

T.G. Evans. A heuristic program to solve geometric-analogy problems. *Spring Joint Computer Conference*, American Federation of Information Processing Societies, 1964. Reprinted in *Artificial Intelligence*, O. Firschein (ed.), *Vol. VI, Information Technology Series*, pp. 5–16, American Federation of Information Processing Societies, Inc., Reston, Virginia 22091, 1984.

T.G. Evans. A program for the solution of a class of geometric-analogy intelligence-test questions. M. Minsky, editor, *Semantic Information Processing*, pages 271–353, MIT Press, Cambridge, Mass., 1968.

S.E. Fahlman and G.E. Hinton. Connectionist architectures for artificial intelligence. *Computer*, 20(1):100–109, January 1987.

O.D. Faugeras and M. Hebert. A 3-D recognition and positioning algorithm using geometrical matching between primitive surfaces. *Proc. of the Eighth International Joint Conference on Artificial Intelligence*, pages 996–1002, Karlsruhe, West Germany, August 1983.

I.D. Faux and M.J. Pratt. *Computerized Geometry for Design and Manufacture.* Ellis Horwood Ltd., West Sussex, England, 1979.

J.A. Feldman. Connectionist models and parallelism in high level vision. *Computer Vision, Graphics, and Image Processing*, 31:178–200, 1985.

J.A. Feldman and D.H. Ballard. Connectionist models and their properties. *Cognitive Science*, 6:205–254, 1982.

C.L. Fennema and W.B. Thompson. Velocity determination in scenes containing several moving objects. *Computer Graphics and Image Processing*, 9(4):301–315, 1979.

R.E. Fikes and N.J. Nilsson. STRIPS: A new approach to the application of theorem proving in problem solving. *Artificial Intelligence*, 3(2):189–208, 1971.

O. Firschein, W. Eppler, and M.A. Fischler. A fast defect measurement algorithm and its array processor mechanization. *Proc. IEEE Conf. on Pattern Recognition and Image Processing*, pages 109–113, Chicago, Ill., August 6–8, 1979.

O. Firschein and M.A. Fischler. Association algorithms for digital imagery. *Conference Record of the 12th Annual Asilomar Conference on Circuits, Systems, and Computers*, Pacific Grove, California, November 1978. IEEE Cat. 78CH1369-8C/CAA/CS.

M.A. Fischler. Aspects of the detection of scene congruence. *Proc. 2nd Int. Joint Conf. on AI*, pages 88–100, September 1971.

M.A. Fischler and P. Barrett. An iconic transform for sketch and shape abstraction. *Computer Graphics and Image Processing*, 13(3):334–360, August 1980.

M.A. Fischler and R.C. Bolles. Perceptual organization and curve partitioning. *IEEE PAMI*, 8(1):100–105, January 1986.

M.A. Fischler and R.C. Bolles. Random sample consensus: A paradigm for model fitting with applications to image analysis and automated cartography. *Comm. ACM*, 24(6):381–395, June 1981a.

M.A. Fischler and R.A. Elschlager. The representation and matching of pictorial structures. *IEEE Trans. on Comp.*, 22(1):67–92, January 1973.

M.A. Fischler and O. Firschein. *Intelligence: The Eye, the Brain, and the Computer.* Addison-Wesley, Reading, Mass., 1987.

M.A. Fischler and O. Firschein. Parallel guessing: A strategy for high-speed computation. *Pattern Recognition*, 1987. In press.

M.A. Fischler, J.M. Tenenbaum, and H.C. Wolf. Detection of roads and linear structures in low resolution aerial imagery using a multisource knowledge integration technique. *Computer Graphics and Image Processing*, 15(3):201–223, March 1981b.

M.A. Fischler and H.C. Wolf. Linear delineation. *Proc. IEEE Conf. on Computer Vision and Image Processing*, pages 351–356, Washington, D.C., June 19–23, 1983.

J. Foley and A. van Dam. *Fundamentals of Interactive Computer Graphics.* Addison-Wesley, Reading, Mass., 1982.

R.J. Fowler and J.J. Little. Automatic extraction of irregular network digital terrain models. *Computer Graphics*, 13(2):199–207, August 1979.

H. Freeman. Computer processing of line drawing images. *Computing Surveys*, 6(1):57–98, March 1974.

K.A. Frenkel. Evaluating two massively parallel machines. *Comm. ACM*, 29(8):752–758, August 1986.

S.A. Friedberg. Finding axes of skewed symmetry. *CVGIP*, 34(2):138–155, May 1986.

S.A. Friedberg and C.M. Brown. Symmetry evaluators. *Proc. DARPA Image Understanding Workshop*, pages 90–97, New Orleans, La., October 1984.

J.P. Frisby. *Seeing.* Oxford University Press, New York, 1979.

B.V. Funt. Problem solving with diagrammatic representations. *Artificial Intelligence*, 13:201–230, 1980.

S. Ganapathy. Decomposition of transformation matrices for robot vision. *IEEE Int. Conf. on Robotics*, pages 130–139, Atlanta, Georgia, March 13–15, 1984.

G.Y. Gardner. Simulation of natural scenes using textured quadric surfaces. *Computer Graphics*, 18(3):11–20, July 1984.

T.D. Garvey, *Perceptual Strategies for Purposive Vision.* Technical Report 117, SRI International, Menlo Park, California, September 1976.

D. Gelernter. Domesticating parallelism. *IEEE Computer*, 19(8):12–16, August 1986.

S. Geman and D. Geman. Stochastic relaxation, Gibbs distribution, and the Bayesian restoration of images. *IEEE PAMI*, 6:721–741, 1984.

D.B. Gennery. Stereo-camera calibration. *Proc. DARPA IU Workshop*, pages 101–107. Los Angeles, California, November 1979.

M.P. Georgeff and Christopher S. Wallace. A general selection criterion for inductive inference. *Proc. Sixth European Conference on Artificial Intelligence*, Pisa, Italy, September 1984.

R. Gershon. Aspects of perception and computation in color vision. *CVGIP*, 32:244–277, 1985.

R. Gershon, A.D. Jepson, and J.K. Tsotsos. Ambient illumination and the determination of material changes. *J. Optical Society America A*, 3, October 1986.

J.J. Gibson. *The Perception of the Visual World.* Houghton Mifflin, Boston, Mass., 1950.

A Gilchrist. Perceived lightness depends on perceived spatial arrangement. *Science*, 195:185–187, 1977.

C. Goad. Special purpose automatic programming for 3-D model-based vision. *Proc. Image Understanding Workshop*, pages 94–104, Arlington, Virginia, June 1983. See also, Fast 3D model-based vision. A.P. Pentland, editor, *From Pixels to Predicates*, Ablex, Norwood, New Jersey, 1986.

R.C. Gonzalez and P.A. Wintz. *Digital Image Processing.* Addison-Wesley, Reading, Mass., 1977.

Y. Goto, K. Matsuzaki, I. Kweon, and T. Obatake. CMU sidewalk navigation system: A blackboard-based outdoor navigation system using colored-range images. *Proc. IEEE Fall Joint Conference*, pages 105–113, Dallas, Texas, November 2–6, 1986.

R.L. Gregory. *Eye and Brain: The Psychology of Seeing.* McGraw-Hill, New York, 1978. World Universal Library (3rd edition).

W.E.L. Grimson and T. Lozano-Perez. Model-based recognition and localization from sparse range or tactile data. *Int. J. of Robotics Research*, 3(3):3–35, Fall 1984.

W.E.L. Grimson. Computational experiments with a feature based stereo algorithm. *IEEE PAMI*, 7(1):17–34, January 1985b.

W.E.L. Grimson. *From Images to Surfaces: A Computational Study of the Human Early Visual System.* MIT Press, Cambridge, Mass., 1981.

W.E.L. Grimson. An implementation of a computational theory of visual surface interpolation. *CGIP*, 22:39–69, 1983.

W.E.L. Grimson, E.C. Hildreth, and R. M. Haralick. Comments on 'Digital step edges from zero crossings of second directional derivatives' and author's reply. *IEEE PAMI*, 7(1):121–129, January 1985a.

A. Guzman. Decomposition of a visual scene into three-dimensional bodies. *Fall Joint Computer Conference*, volume 33, 1968. Reprinted in the Information Technology Series, Vol. VI, *Artificial Intelligence*, O. Firschein (editor). AFIPS Press, Reston, Va. 1984.

E.L. Hall. *Computer Image Processing and Recognition.* Academic Press, New York, 1979.

R.H. Halstead. Parallel symbolic computing. *IEEE Computer*, 19(8):35–43, August 1986.

M.J. Hannah. *Evaluation of STEREOSYS Versus Other Stereo Systems.* Technical Report TN 365, SRI International, AI Center, Menlo Park, California, October 10, 1985.

A.R. Hansen and E.M. Riseman. Computer vision systems. A.R. Hansen and E.M. Riseman, editors, *Computer Vision Systems*, Academic Press, New York, 1978.

A.J. Hanson. *Hyperquadrics: Smoothly Deformable Shapes with Convex Polyhedral Bounds.* Technical Report, Artificial Intelligence Center, SRI International, Menlo Park, California, October 1986.

R.M. Haralick. The digital step edge from zero crossings of second directional derivatives. *IEEE PAMI*, 6(1):58–68, January 1984.

R.M. Haralick. Statistical and structural approaches to texture. *Proc. IEEE*, pages 786–804, May 1979.

S.Y. Harmon, G.L. Bianchini, and B.E. Pinz. Sensor fusion through a distributed blackboard. *Proc. of the IEEE Int. Conf. on Robotics and Automation*, pages 2002–2011, San Francisco, California, April 7–10, 1986.

M. Herman and T. Kanade. The 3D MOSAIC scene understanding system: Incremental reconstruction of 3D scenes from complex images. *Proc. DARPA IU Workshop*, pages 137–148, New Orleans, Louisiana, October 3–4, 1984. Also see *Artificial Intelligence* 30(3):289-341, Dec. 1986.

A. Herskovitz and T. Binford. *On Boundary Detection.* Technical Report AI Memo 183, MIT Artificial Intelligence Lab., Cambridge, Mass., 1970.

D.H. Hilbert and S. Cohn-Vossen. *Geometry and the Imagination.* Chelsea Publishing Co., New York, 1952.

E.C. Hildreth. Computations underlying the measurement of visual motion. *Artificial Intelligence*, 23(3):309–354 August 1984.

E.C. Hildreth. Computing the velocity field along contours. *Proc. ACM SIGGRAPH/SIGART Interdisciplinary Workshop on Motion*, pages 26–32, Toronto, Canada, April 1983.

E.C. Hildreth. *Implementation of a Theory of Edge Detection.* AI Report AI-TR-579, MIT Artificial Intelligence Lab., Cambridge, Mass., April 1980.

W.D. Hillis. The connection machine: A computer architecture based on cellular automata. *Physica*, pages 213–228, 1984. Reprinted in: *Computers for Artificial Intelligence Applications*, B. Wah and G.-J. Li (eds.) IEEE Computer Society Press, Washington, D.C., IEEE Computer Society Order No. 706

W. Hoff and N. Ahuja. *Extracting Surfaces from Stereo Images: An Integrated Approach.* Technical Report UILU-ENG-87-2204, Univ. of Illinois, Coordinated Science Lab., Urbana-Champaign, Ill., January 1987.

D.D. Hoffman and W. Richards. Parts of recognition. *Cognition*, 18:65–96, 1985. Reprinted in: *From Pixels to Predicates*, Pentland, A. (ed.), Ablex, Norwood, N.J., 1986.

J.J. Hopfield. Neural networks and physical systems with emergent collective computational abilities. *Proc. National Academy of Sciences USA*, 79:2554–2558, 1982.

R. Horaud. Spatial object perception from an image. *Proc. 10th Int. Joint Conf. on AI*, pages 1116–1119, August 1985.

B.K.P. Horn. *The Binford-Horn Line-Finder.* Technical Report MIT AI Memo 285, MIT Artificial Intelligence Lab., Cambridge, Mass., 1973.

B.K.P. Horn. *Robot Vision.* MIT Press, Cambridge, Mass., 1986.

B.K.P. Horn. Understanding image intensities. *Artificial Intelligence*, 8(2):201–231, 1977.

B.K.P. Horn and B.G. Schunck. Determining optical flow. *Artificial Intelligence*, 17(1–3):185–203, August 1981.

D.H. Hubel and T.N. Wiesel. Functional architecture of macaque monkey visual cortex. *J. Physiol.*, 195:215–242, 1968.

M. Hueckel. An operator which locates edges in digital pictures. *J. ACM*, 18:113–125, 1971.

D.A. Huffman. Impossible objects as nonsense sentences. B. Meltzer and D. Michie, editors, *Machine Intelligence 6*, Edinburgh University Press, Edinburgh, Scotland, 1971.

R.A. Hummel. Representations based on zero-crossings in scale-space. *Proc. IEEE CVPR '86*, pages 204–209, Miami Beach, Florida, June 22–26, 1986. IEEE Computer Society Order Number 721.

R.A. Hummel and S.W. Zucker. On the foundation of relaxation labeling processes. *IEEE PAMI*, 5(3):267–287, May 1983.

H.A.H. Ibrahim. *Tree Machines: Architecture and Algorithms.* PhD thesis, Columbia University, 1984.

IEEE. Domesticating parallelism. *IEEE Computer*, 19(8), August 1986. IEEE Computer Society.

IEEE. Special issue on computer architectures for image processing. *IEEE Computer* 16(1), January 1983. IEEE Computer Society.

IEEE. Special issue on highly parallel computing. *IEEE Computer*, 15(1), January 1982. IEEE Computer Society.

K. Ikeuchi and B.K.P. Horn. Numerical shape from shading and occluding boundaries. *Artificial Intelligence*, 17(1–3):141–184, August 1981.

A.K. Jain. Advances in mathematical models for image processing. *Proc. IEEE*, 69(5):502–528, May 1981.

R.A. Jarvis. A perspective on range finding techniques for computer vision. *IEEE PAMI*, 5(2), March 1983.

M. Jenkins and P.A. Kolers. *Some Problems with Correspondence.* Technical Report RBCV-TR-86-10, University of Toronto, Toronto, Canada, April 1986.

G. Johansson. Visual perception of biological motion and a model for its analysis. *Perception and Psychophysics*, 14(2):201–211, 1973.

G. Johansson and G. Jansson. Perceived rotary motion from changes in a straight line. *Perception and Psychophysics*, 4(3):165–170, 1968.

B. Julesz. Cooperative phenomena in binocular depth perception. *American Scientist*, 62(1):32–43, Jan.–Feb. 1974.

B. Julesz. *Foundations of Cyclopean Perception.* University of Chicago, Chicago, Ill., 1971.

B. Julesz and J.R. Bergen. Textons, the fundamental elements in preattentive vision and the perception of textures. *Bell System Technical Journal*, 62(6):1619–1644, July–August 1983.

J.T. Kajiya. New techniques for ray tracing procedurally defined objects. *ACM Trans. on Graphics*, 2(3):161–181, July 1983.

A. Kalvin, E. Schonberg, J.T. Schwartz, and M. Sharir. *Two Dimensional Model Based Boundary Matching Using Footprints.* Technical Report TR 162, Courant Institute of Mathematics, New York University, May 1985.

T. Kanade. Geometrical aspects of interpreting images as a three-dimensional scene. *Proc. IEEE*, pages 789–802, July 1983a.

T. Kanade. Recovery of the three-dimensional shape of an object from a single view. *Artificial Intelligence*, 17(1–3):409–460, August 1981.

T. Kanade. A theory of origami world. *Artificial Intelligence*, 13(1):279–311, 1980.

T. Kanade and J.P. Kender. Mapping image properties into shape constraints: Skewed symmetry, affine-transformable patterns, and the shape-from-texture paradigm. Beckm, Hope, and Rosenfeld, editors, *Human and Machine Vision*, Academic Press, New York, 1983b.

G. Kanisza. Subjective contours. *Scientific American*, 234:48–52, 1976.

G. Kanisza. *Organization in Vision.* Praeger, New York, 1979.

M. Kass and A.P. Witkin. Analyzing oriented patterns. *Proc. Ninth IJCAI*, pages 944–952, Los Angeles, California, August 18–23, 1985.

M. Kass and A.P. Witkin. On edge detection. *IEEE PAMI*, 8(2):147–163, 1986.

D. Keirsey, J. Mitchell, B. Bullock, T. Nussmeier, and D.Y. Tseng. Autonomous vehicle control using AI technique. *IEEE Trans. on Software engineering*, 11:986–992, 1985.

J.R. Kender. *Shape from Texture.* PhD thesis, Dept. of Computer Science, Carnegie-Mellon Univ., Pittsburgh, Pa., 1980.

J.R. Kender. Surface constraints from linear extents. *Proc. DARPA IU Workshop*, pages 49–53, Arlington, Virginia, June 1983.

E.W. Kent, M.O. Shnerier, and R. Lumia. PIPE (Pipeline Image-Processing Engine). *J. Parallel Distributed Computing*, 2:50–78, 1985.

S. Kirkpatrick, C.D. Gelatt Jr., and M.P. Vecchi. Optimization by simulated annealing. *Science*, 220(4598):671–680, May 13, 1983.

R. Kirsch. Computer determination of the constituent structure of biological images. *Computers and Biomedical Research*, 4(3):315–328, June 1971.

J. Kittler and J. Illingworth. Relaxation labeling algorithms - A review. *Image and Vision Computing*, 3(4):206–216, November 1985.

A. Klinger and C.R. Dyer. Experiments in picture representation using regular decomposition. *CGIP*, 5:68–105, 1976.

J.J. Koenderinck and A. van Dorn. The shape of smooth objects and the way contours end. *Perception*, 11:129–137, 1982.

E. Kruppa. Zur Ermittlung eines Objektes aus zwei Perspektiven mit innerer Orientierung. *Sitzungsberichte der math.-naturw. Kl. der kaiserlichen Akademie der Wissenschaftern*, Vienna, Autria, Abt. IIa(122):1939–1948, 1913.

D.T. Kuan. Terrain map knowledge representation for spatial planning. *First National Conference on Artificial Intelligence Applications*, pages 578–584, IEEE Computer Society, Denver, Colorado, December 1984. IEEE Cat. 84CH2107-1.

C.D. Kuglin and D.C. Hines. The phase correlation image alignment method. *Proc. IEEE 1975 Int. Conf. on Cybernetics and Society*, September 1975.

H.T. Kung. Why systolic architectures? *IEEE Computer*, 15(1):37–46, January 1982.

H.T. Kung. Systolic algorithms for the CMU warp processor. *Proc. 7th Int. Conf. on Pattern Recognition*, pages 570–577, Montreal, 1984.

E.H. Land. The retinex theory of color vision. *Scientific American*, 237:108–128, 1977a.

E.H. Land and J.J. McCann. Lightness and retinex theory. *J. Opt. Soc. America*, 61:1–11, 1977b.

K.I. Laws. Goal-directed textured-image segmentation. *Proc. SPIE Conf. on Applications of AI II*, volume 548, Arlington, Va., August 9–11, 1985.

D.T. Lawton, T.S. Levitt, C. McConnell, and J. Glicksman. Terrain models for an autonomous land vehicle. *Proc. IEEE International Conf. on Robotics and Automation*, pages 2043–2051, San Francisco, California, April 7–10, 1986. IEEE Cat. 86CH2282-2.

D.T. Lawton. Processing translational motion sequences. *CGIP*, 22(1):116–144, April 1983.

Y. Leclerc. Capturing the local structure of image discontinuities in two dimensions. *Proc. IEEE CVPR*, pages 34–38, San Francisco, California, June 19–23, 1985.

H.J. Lee and Z. Chen. Optimal search procedures for 3D human movement determination. *First Conference on Artificial Intelligence Applications*, pages 389–393, IEEE Computer Society, Denver, Colorado, December 5–7, 1984.

M.D. Levine. *Vision in Man and Machine.* McGraw-Hill, New York, 1985.

M. Leyton. *A Process Grammar for Shape.* Technical Report, Dept. of Computer Science and Psychology, State University of N.Y., Buffalo, N.Y., 1986.

H.S. Lim and T.O. Binford. Survey of array processors. *Proc. DARPA Image Understanding Workshop*, pages 334–343, Science Applications International, New Orleans, LA, October 1984.

J.O. Limb and J.A. Murphy. Estimating the velocity of moving objects from television signals. *CGIP*, 4:311–327, 1975.

H.C. Longuet-Higgens. A computer algorithm for reconstructing a scene from two projections. *Nature*, 293:133–135, 1981.

H.C. Longuet-Higgens. The reconstruction of a scene from two projections – configurations that defeat the 8-point algorithm. *Proc. of the First Conf. on Artificial intelligence Applications*, pages 395–397, Denver, Colorado, December 5–7, 1984.

H.C. Longuet-Higgens and K. Prazdny. The interpretation of a moving retinal image. *Proc. Royal Society, London, ser. B*, 208:385–397, 1980.

D.G. Lowe. *Perceptual Organization and Visual Recognition.* Kluwer Academic Publishers, Boston, Mass., 1985.

D.G. Lowe and T.O. Binford. Segregation and aggregation: An approach to figure-ground phenomena. *Proc. DARPA IU Workshop*, pages 168–178, Palo Alto, California, September 1982.

J. Lowrie. Autonomous land vehicle program: 1987 update. *Proc. SPIE: Advances in Intelligent Robotics Systems*, Cambridge, Mass., October 26–31, 1986. SPIE 727–01.

A.K. Mackworth. Interpreting pictures of polyhedral scenes. *Artificial Intelligence*, 4:121–137, 1973.

A.K. Mackworth. *On Reading Sketch Maps.* Technical Report 77-2, Dept. of Computer Science, Univ. of British Columbia, Vancouver, B.C., May 1977.

L.T. Maloney and B.A. Wandell. Color constancy: A method for recovering surface spectral reflectance. *J. Optical Society Am.*, 3:29–33, 1986.

B.B. Mandelbrot. *The Fractal Geometry of Nature.* W.H. Freeman, San Francisco, California, 1982.

B.B. Mandelbrot. *Fractals: Form, Chance, and Dimension.* W.H. Freeman, San Francisco, California, 1977.

M.A. Markowsky and G. Wesley. Fleshing out projections. *IBM J. of R and D*, 25(6):934–954, November 1981.

D. Marr. Early processing of visual information. *Proc. Royal Soc., London, ser. B*, 275:483–534, 1976.

D. Marr. *Vision.* W.H. Freeman, San Francisco, California, 1982.

D. Marr and E. Hildreth. Theory of edge detection. *Proc. Royal Society, London, ser. B*, 270:187–217, 1980.

D. Marr and H.K. Nishihara. Representation and recognition of the spatial organization of three-dimensional shapes. *Proc. Royal Society, London, ser. B*, 200:269–294, 1978a.

D. Marr and H.K. Nishihara. Visual information processing: Artificial intelligence and the sensorium of sight. *Technology Review*, 81(1):2–23, October 1978b.

D. Marr and T. Poggio. *A Theory of Human Stereo Vision.* Technical Report AIL Memo 451, MIT Artificial Intelligence Laboratory, Cambridge, Mass., November 1977.

D. Marr and S. Ullman. Directional selectivity and its use in early visual processing. *Proc. Royal Society, London, ser. B*, 211:151–180, 1981.

D. Marr and T. Poggio. A theory of human stereo vision. *Proc. Royal Society London, ser. B*, 204:301–328, 1979.

A. Martelli. An application of heuristic search methods to edge and contour detection. *Comm. of the ACM*, 19:73–83, February 1976.

A. Martelli. Edge detection using heuristic search methods. *CGIP*, 1:169–182, 1972.

W. Martin and J.K. Aggarwal. Computer analysis of dynamic scenes containing curvilinear figures. *Pattern Recognition*, 11:169–178, 1979.

J.E.W. Mayhew and J.P. Frisby. Psychophysical and computational studies towards a theory of human stereopsis. *Artificial Intelligence*, 17(1–3):349–385, August 1981.

D. McDermott and E. Davis. Planning routes through uncertain territory. *Artificial Intelligence*, 22:107–156, 1984.

D.M. McKeown, Jr., W.A. Harvey, Jr., and J. McDermott. Rule-based interpretation of aerial imagery. *IEEE PAMI*, 7(5):570–585, September 1985.

R.D. Merrill. Representations of contours and regions for efficient computer search. *Comm. ACM*, 16(2):69–82, February 1973.

D. Milgram. Region extraction using convergent evidence. *CGIP*, 11:1–12, September 1979.

W.F. Miller and A.C. Shaw. Linguistic methods in picture processing. *Proc. AFIPS 1968 Fall Joint Computer Conference*, pages 279–290, Washington, D.C., 1968.

M. Minsky and S. Papert. *Perceptrons.* MIT Press, Cambridge, Mass., 1969.

F. Mokhtarian and A. Mackworth. Scale-based description and recognition of planar curves and two-dimensional shapes. *IEEE PAMI*, 8(1):34–43, January 1986.

U. Montanari. On the optimal detection of curves in noisy pictures. *Comm. ACM*, 14:335–345, 1971.

H.P. Moravec. Visual mapping by a robot rover. *Proc. 6th Int. Joint Conf. on AI*, pages 598–600, Tokyo, Japan, 1979.

H.P. Moravec. Towards automatic visual obstacle avoidance. *Proc. 5th Int. Joint Conf. on AI*, page 584, August 1977.

S.P. Morse. Concepts of use in contour map processing. *Comm. ACM*, 12(3):147–152, March 1969.

P.G. Mulgaonkar, L. G. Shapiro, and R. M. Haralick. Matching 'sticks, plates, and blobs' objects using geometric and relational constraints. *Image and Vision Computing*, 2:85–98, May 1984.

D.A. Mundie and D. A. Fisher. Parallel processing in Ada. *IEEE Computer*, 19(8):20–25, August 1986.

M. Nagao and T. Matsuyama. *A Structural Analysis of Complex Aerial Photographs.* Plenum, New York, 1980.

H.H. Nagel. Analytic techniques for image sequences. *Proc. of the 4th Int. Joint Conf. on Pattern Recognition*, pages 186–211, Kyoto, Japan, November 4–10, 1978.

H.H. Nagel and B. Neumann. On 3-D reconstruction from two perspective views. *Proc. IJCAI 81*, August 1981. Also H.H. Nagel, On the derivation of 3-D rigid point configurations from images sequences. *Proc. IEEE PRIP*, pp. 103-108, Dallas, Texas, Aug. 3–5, 1981.

G. Nagy. Optical character recognition – Theory and practice. P.R. Krishnaiah and L.N. Kanal, editors, *Handbook of Statistics, Vol. 2*, pages 621–649, North Holland, 1982.

V.S. Nalwa and T.O. Binford. On detecting edges. *IEEE PAMI*, 8(6):699–714, November 1986.

N. Nandhakumar and J.K. Aggarwal. The artificial intelligence approach to pattern recognition – A perspective and an overview. *Pattern Recognition*, 18(6):383–389, 1985.

A.M. Nazif and M.D. Levine. Low level image segmentation: An expert system. *IEEE PAMI*, 6(5):555–577, September 1984.

U. Neisser. *Cognitive Psychology.* Appleton-Century-Crofts, New York, 1967.

R. Nevatia. *Machine Perception.* Prentice-Hall, Englewood Cliffs, New Jersey, 1982a.

R. Nevatia and R. Babu. Linear feature extraction and description. *CGIP*, 13:257–269, 1980.

R. Nevatia and T.O. Binford. Description and recognition of curved objects. *Artificial Intelligence*, 8:77–98, 1977.

R. Nevatia and K.E. Price. Locating structures in aerial images. *IEEE PAMI*, 4(5):476–484, September 1982b.

H. Niemann, H. Bunke, I. Hofmann, and G. Sagerer. Diagnostic inferences from image sequences – A knowledge based approach. *First Conference on Artificial Intelligence Applications*, pages 610–616, IEEE Computer Society, Denver, Colorado, December 5–7, 1984.

N.J. Nilsson. Artificial intelligence prepares for 2001. *AI Magazine*, 4(4):7–14, Winter 1983.

N.J. Nilsson. *Principles of Artificial Intelligence.* Morgan Kaufmann Publishers, Los Altos, CA, 1980.

N.J. Nilsson. *Problem-Solving Methods in Artificial Intelligence.* McGraw-Hill, New York, 1971.

N.J. Nilsson. *Shakey the Robot.* Technical Report Tech. Note 323, AI Center, SRI International, Menlo Park, California, April 1984.

H.K. Nishihara. Practical real-time imaging stereo matcher. *Optical Engineering*, 23(5):536–545, September–October 1984.

D. Nitzan, A. Brain, and R. Duda. The measurement and use of registered reflectance and range data in scene analysis. *Proc. IEEE*, 65(2):206–220, February 1977.

R. Ohlander. *Analysis of Natural Scenes.* PhD Thesis, Carnegie-Mellon University, Pittsburgh, Pa., 1975.

R. Ohlander, K. Price, and D.R. Reddy. Picture segmentation using a secursive splitting method, *CGIP*, 8:313–333, 1978.

J. O'Rourke and J.K. Badler. Model-based image analysis and human motion using constraints propagation. *IEEE PAMI* 2(6):522–536, November 1980.

M. Oshima and Y. Shirai. Object recognition using three-dimensional information. *IEEE PAMI*, 5:353–361, July 1983.

D. Panton. A flexible approach to digital stereo mapping. *J. Photogrammetric Eng. and Remote Sensing*, 44(12):1499–1512, December 1984.

T. Pavlidis. *Algorithms for Graphics and Image Processing.* Computer Science Press, Rockville, Maryland, 1982.

T. Pavlidis. A minimum storage boundary tracing algorithm and its application to automatic inspection. *IEEE Systems, Man, and Cybernetics*, 8(1):66–69, January 1978.

T. Pavlidis. *Structural Pattern Recognition.* Springer, New York, 1977.

J. Pearl. *Heuristics: Intelligent Strategies for Computer Problem Solving.* Addison-Wesley, Reading, Mass., 1984.

A.P. Pentland. Fractal-based description of natural scenes. *IEEE PAMI*, 6(6):661–674, November 1984a.

A.P. Pentland. Local shading analysis. *IEEE PAMI*, 6(2):170–187, 1984b.

A.P. Pentland. Parts: Structured description of shape. *Proc. 5th National Conference on AI*, pages 695–701, Philadelphia, Pa., August 11–15, 1986.

A.P. Pentland. Perceptual organization and the representation of natural form. *Artificial Intelligence*, 28:293–331, May 1986a.

A.P. Pentland, editor. *Pixels to Predicates.* Ablex, Norwood, New Jersey, 1986b.

A.P. Pentland. Shading into texture. *Artificial Intelligence*, 29(2):147–170, August 1986c.

A.P. Pentland and M.A. Fischler. A more rational view of logic or, up against the wall, logic imperialists! *AI Magazine*, 4(4):15–18, Winter 1983.

W.A. Perkins. Using circular symmetry and intensity profiles for computer vision inspection. *CGIP*, 17:161–172, 1981.

E. Persoon and K.S. Fu. Shape discrimination using Fourier descriptors. *Proc. 2nd IJCPR*, pages 126–130, August 1974.

T.K. Peucker, R.J. Fowler, and J.J. Little. The triangulated irregular network. *Proc. of the ASP-ACSM Symposium on DTMs*, St. Louis, Missouri, October 1978.

T. Poggio, V. Torre, and C. Koch. Computational vision and regularization theory. *Nature*, 317(26):314–319, September 26, 1985.

J.L. Potter. Image processing on the massively parallel processor. *IEEE Computer*, 16(1):62–67, January 1983.

J.L. Potter. Programming the MPP. J. L. Potter, editor, *The Massively Parallel Computer*, MIT Press, 1985.

W.K. Pratt. *Digital Image Processing.* Wiley-Interscience, New York, 1978.

K. Prazdny. Detection of binocular disparities. *Biological Cybernetics*, 52:93–99, 1985a.

K. Prazdny. Egomotion and relative depth map from optical flow. *Biological Cybernetics*, 36:87–102, 1980.

K. Prazdny. On the nature of inducing forms generating perception of illusory contours. *Perception and Psychophysics*, 37(3):237–242, 1985b.

J. Prewitt. Object enhancement and extraction. B. Lipkin and A. Rosenfeld, editors, *Picture Processing and Psychopictorics*, pages 75–149, Academic Press, New York, 1970.

L.H. Quam. Hierarchical warp stereo. *Proc. DARPA Image Understanding Workshop*, pages 149–155, New Orleans, LA, October 1984.

L.H. Quam. Road tracking and anomaly detection. *Proc. DARPA IU Workshop*, pages 51–55, Cambridge, Mass., May 1978.

R. Rashid. Towards a system for the interpretation of moving light displays. *IEEE PAMI*, 2(6):574–581, November 1980.

A.P. Reeves. Parallel computer architectures for image processing. *Comput. Vision Graphics Image Process*, 25:68–88, 1984.

A.P. Reeves. Parallel pascal and the massively parallel processor. J.L.Potter, editor, *The Massively Parallel Computer*, MIT Press, Cambridge, Mass., 1985.

W.T. Reeves. Particle systems – A technique for modeling a class of fuzzy objects. *ACM Trans. on Graphics*, 2(2):91–108, April 1983.

W.T. Reeves and R. Blau. Approximate and probabilistic algorithms for shading and rendering structural particle systems. *Computer Graphics*, 19(3):313–322, July 1985.

W. Richards and D.D. Hoffman. Codon constraints on closed 2D shapes. *CVGIP*, 31(3):265-281, September 1985.

L.G. Roberts. Machine perception of three-dimensional solids. J.T. Tippett, editor, *Optical and Electro-Optical Information Processing*, MIT Press, 1965. Reprinted in: *Computer Methods in Image Analysis*, J.K. Aggarwal, R.O. Duda, and A. Rosenfeld (eds), IEEE Press, N.Y., pp. 285-323, 1977.

University of Rochester. *1984–1985 Computer Science and Computer Engineering Research Review.* Technical Report, Computer Science Department, University of Rochester, Rochester, N.Y., 1985, pages 3–24.

I. Rock. *The Logic of Perception.* MIT Press, Cambridge, Mass., 1983. See, The perception of lightness, pp. 203–212.

D.F. Rogers and J.A. Adams. *Mathematical Elements for Computer Graphics.* McGraw-Hill, New York, 1976.

A. Rosenfeld. Image analysis: Problems, progress, and prospects. *Pattern Recognition*, 17(1):3–12, 1984.

A. Rosenfeld. Quadtrees and pyramids for pattern recognition and image processing. *Proc. 5th Int. Conf. Pattern Recognition*, pages 802–807, 1980.

A. Rosenfeld and L.S. Davis. Image understanding techniques for autonomous vehicle navigation. *DARPA Image Understanding Workshop*, pages 1–3, Miami Beach, Florida, December 1985.

A. Rosenfeld, R.A. Hummel, and S.W. Zucker. Scene labeling by relaxation operations. *IEEE Systems, Man, and Cybernetics*, 6:420–433, 1976a.

A. Rosenfeld and A.C. Kak. *Digital Picture Processing.* Academic Press, New York, 1976b. Revised 1982.

A. Rosenfeld and M. Thurston. Edge and curve detection for visual scene analysis. *IEEE Trans. Computers*, 20:562–569, 1971.

J.M. Rubin and W.A. Richards. Color vision and image intensities: When are changes material? *Biological Cybernetics*, 45:215–226, 1982.

J.M. Rubin and W.A. Richards. *Color Vision: Representing Material Changes.* Technical Report AI Memo 764, MIT AI Laboratory, Cambridge, Mass., 1984.

D.E. Rumelhart, J.L. McClelland, and the PDP Research Group. *Parallel Distributed Processing, Explorations in the Microstructures of Cognition.* Models, MIT Press, Cambridge, Mass., 1986. Vol. 1 *Foundations*, and Vol. 2, *Psychological and Biological Models.*

D. Sabah. *A Connectionist Approach to Visual Recognition.* Technical Report TR107 and Ph.D. thesis, Computer Science Department, University of Rochester, Rochester, N.Y., April 1982.

H. Samet. Region representation: Quadtrees from boundary codes. *Comm. ACM*, 23(3):163–170, March 1980.

C.L. Seitz. The cosmic cube. *Communications of the ACM*, 28(1):22–33, January 1985.

J. Serra. Introduction to mathematical morphology. *CVGIP*, 35(3):283–325, September 1986.

G. Shafer. *A Mathematical Theory of Evidence.* Princeton U. Press, Princeton, N.J., 1976.

S.A. Shafer. *Optical Phenomena in Computer Vision.* Technical Report TR 135, Carnegie-Mellon University, Computer Science Department, Pittsburgh, Pa., March 1984.

S.A. Shafer. *Shadows and Silhouettes in Computer Vision.* Kluwer Academic, Boston, Mass., 1985.

S.A. Shafer, T. Kanade, and J.R. Kender. Gradient apace under orthography and perspective. *Proc. of Workshop on Computer Vision: Representation and Control*, pages 26–34, IEEE Computer Society, Rindge, New Hampshire, August 23–25, 1982. Order 437 (Also CVGIP 24:182–199, 1983).

S.A. Shafer, A. Stentz, and C.E. Thorpe. An architecture for sensor fusion in a mobile robot. *Proc. 1986 IEEE Int. Conf. on Robotics and Automation*, pages 2002–2011. San Francisco, California, April 7–10, 1986. IEEE Cat. CH2282-2/86.

R. Shapira. A note on Sugihara's claim. *IEEE PAMI*, 6(1):122–123, January 1984.

E. Shapiro. Concurrent Prolog: A progress report. *IEEE Computer*, 19(8):44–58, August 1986.

D.E. Shaw. SIMD and MSIMD variants of the NON-VON supercomputer. *Proc. COMPCON*, February 1984.

R.N. Shepard. Ecological constraints on internal representation: Resonant kinematics of perceiving, imaging, thinking, and dreaming. *Psychological Review*, 91:417–447, 1984.

J. Sklansky. On the Hough technique for curve detection. *IEEE Trans. on Computers*, 27(10), October 1978.

K.R. Sloan and R. Bajcy. *World Model-Driven Recognition of Outdoor Scenes.* Technical Report TR-40, University of Rochester, Comp. Sci. Dept., Rochester, N.Y., September, 1979.

A.R. Smith. Plants, fractals, and formal languages. *Computer Graphics*, 18(3):1–10, July 1984.

G.B. Smith and H.C. Wolf. Image-to-image correspondence: Linear-structure matching. *Proc. 2nd Annual NASA Symposium on Mathematical Pattern Recognition and Image Analysis*, pages 467–487, Houston, Texas, June 1984a. Also, SRI International, AI Center, TN 331, July 1984.

G.B. Smith. A fast surface interpolation technique. *Proc. DARPA Image Understanding Workshop*, pages 211–215, New Orleans, Louisiana, Oct. 2–3, 1984b. Also, SRI International, AI Center, TN-333, Menlo Park, California, August 1984.

G.B. Smith. *Shape From Shading: An Assessment.* Technical Report SRI Tech Note 287, SRI International, Menlo Park, California, May 1983.

G.B. Smith. Stereo integral equations. *Proc. 5th National Conf. on AI*, pages 689–694, Philadelphia, PA., August 11–15, 1986.

H.W. Sorenson, editor. *Kalman Filtering: Theory and Application.* IEEE Press, 1985.

S.N. Srihari. Representation of three-dimensional digital images. *ACM Computing Surveys*, 13(4):399–424, December 1981.

K.A. Stevens. *Surface Perception by Local Analysis of Texture and Contour.* Technical Report 512, MIT AI Laboratory, Cambridge, Mass., 1980.

K.A. Stevens. The visual interpretation of surface contours. *Artificial Intelligence*, 17(1–3):47–73, August 1981.

P.S. Stevens. *Patterns in Nature.* Little, Brown, Boston, Mass., 1974.

T.M. Strat. Recovering the camera parameters from a transformation matrix. *Proc. DARPA IU Workshop*, pages 264–271, New Orleans, Louisiana, October 2–3, 1984.

T.M. Strat and M.A. Fischler. One-eyed stereo: A general approach to modeling scene geometry. *IEEE PAMI*, 8(6):730–741, November 1986.

K. Sugihara. An algebraic approach to shape-from-image problems. *Artificial Intelligence*, 23(1):59–95, May 1984a.

K. Sugihara. Mathematical structures of line drawings of polyhedrons – Towards man-machine communication by means of line drawings. *IEEE PAMI*, 4:458–469, September 1982.

K. Sugihara. A necessary and sufficient condition for a picture to represent a polyhedral scene. *IEEE PAMI*, 6(5):578–586, September 1984b.

M.J. Swain and J.L. Mundy. Experiments in using a theorem prover to prove and develop geometrical theorems in computer vision. *1986 IEEE Int. Conf. on Robotics and Automation*, pages 280–285, San Francisco, California, April 7–10, 1986.

S. Tanimoto and T. Pavlidis. A hierarchical data structure for picture processing. *CGIP*, 4(2):104–119, June 1975.

S.L. Tanimoto and A. Klinger. *Structured Computer Vision: Machine Perception Through Hierarchical Computation Structures.* Academic Press, New York, 1980.

J.M. Tenenbaum, H.G. Barrow, R.A. Bolles, M.A. Fischler, and H.C. Wolf. Map-guided interpretation of remotely-sensed imagery. *IEEE National Computer Conference*, pages 391–407, 1980.

D. Terzopoulis. Multilevel computational processes for visual surface reconstruction. *CVGIP*, 24:52–96, 1983.

D. Terzopoulis. Regularization of inverse visual problems involving discontinuities. *IEEE PAMI*, 8(4):413–424, July 1986.

W.B. Thompson. Dynamic occlusion analysis in optical flow fields. *IEEE PAMI*, 7(4):374–383, July 1985.

V. Torre and T.A. Poggio. On edge detection. *IEEE PAMI*, 8(2):147–163, 1986.

J.T. Tou and R.C. Gonzalez. *Pattern Recognition Principles.* Addison-Wesley, Reading, Mass., 1974.

A. Treisman. Preattentive processing in vision. *CVGIP*, 31(2):156–177, August 1985.

R.Y. Tsai and T.S. Huang. Uniqueness and estimation of 3-D motion parameters and surface structures of rigid objects. S. Ullman and W. Richards, editors, *Image Understanding 1984*, pages 135–171, Ablex, Norwood, New Jersey, 1984a.

R.Y. Tsai and T.S. Huang. Uniqueness and estimation of three-dimensional motion parameters of rigid objects with curved surfaces. *IEEE PAMI*, 6(1):13–27, January 1984b.

J.K. Tsotsos. Knowledge of the visual process: Content, form, and use. *Proc. 6th Int. Conf. on Pattern Recognition*, pages 654–669, Munich, Germany, October 1982.

J.K. Tsotsos. Knowledge organization and its role in representation and interpretation for time-varying data: The ALVEN system. *Computational Intelligence*, 1:16–32, 1985.

J.K. Tsotsos. *Representational Axes and Temporal Cooperative Processes.* Technical Report RBCV-TR-2, Dept. of Computer Science, Univ. of Toronto, Toronto, Canada, 1984.

J.K. Tsotsos. Temporal event recognition: An application to left ventricular performance assessment. *Proc. IJCAI*, pages 900–907, Vancouver, Canada, 1981.

J.K. Tsotsos, J. Mylopoulos, H.D. Covey, S.W. Zucke. A framework for visual motion understanding. *IEEE PAMI*, 2(6):563–573, November 1980.

J.D. Tubbs, W.A. Coberly, and D. M. Young. Linear dimension reduction and Bayes classification with unknown parameters. *Pattern Recognition*, 15:167–172, 1982.

L. Uhr. *Algorithm-Structured Computer Arrays and Networks: Architectures and Processes for Images, Percepts, Models, Information.* Academic Press, New York, 1984.

L. Uhr. Layered 'recognition cone' networks that preprocess, classify, and describe. *IEEE Trans. on Computers*, C21(7):758–768, July 1972.

K.A. Ulbricht. Comparative experimental study on the use of original and compressed multispectral LANDSAT data for applied research. *Proc. 14th Int.*

Symp. on Remote Sensing of the Environment, pages 967–977, 1980.

S. Ullman. Analysis of visual motion by biological and computer systems. *IEEE Computer*, 14(8):57–69, August 1981.

S. Ullman. *The Interpretation of Visual Motion.* MIT Press, Cambridge, Mass., 1979.

S. Ullman. Visual routines. *Cognition* 18:97–160, 1984.

B. Wah and G. J. Li, editors. *Computers for Artificial Intelligence Applications.* IEEE Computer Society Press, Washington, D.C., 1985. IEEE Computer Society Order No. 706.

R. Wallace, Matsuzaki, Y. Goto, J. Crisman, J. Webb, and T. Kanade. Progress in road-following. *Proc. IEEE int. Conf. on Robotics and Automation*, pages 1615–1621, San Francisco, California, April 7–10, 1986.

D.A. Waltz. Generating semantic descriptions from drawings of scenes with shadows. P.H. Winston, editor, *The Psychology of Computer Vision*, McGraw-Hill, New York, 1985. See also Ph.D. Thesis, MIT, 1972.

S.A. Ward. Real time plotting of approximate contour maps. *Comm. ACM*, 21(9):788–790, September 1978.

S. Watanabe. *Pattern Recognition: Human and Mechanical.* Wiley, New York, 1985.

D.F. Watson and G.M. Philip. Systematic triangulations. *CVGIP*, 26:217–223, 1984.

A.M. Waxman. An image flow paradigm. *Proc. IEEE Workshop on Computer Vision: Representation and Control*, pages 49–55, Annapolis, Md., April 30–May 2, 1984.

J.A. Webb and J.K. Aggarwal. Structure from motion of rigid and jointed objects. *Artificial Intelligence*, 19:107–130, 1982.

C.F.R. Weiman. Logarithmic spiral grids for image processing and display. *CGIP*, 11:197–226, 1979.

M. Wertheimer. Laws of organization in perceptual forms. W. Ellis, editor, *A Sourcebook of Gestalt Psychology*, pages 301–350, Harcourt, Brace, New York, 1938.

L.P. Wesley. Evidential knowledge-based computer vision. *Optical Engineering*, 25(3):363–379, March 1986.

T.E. Weymouth. *Using Object Descriptions in a Schema Network for Machine Vision.* PhD thesis, Computer and Information Science Department, Univ. of Mass., Amherst, Mass., 1986.

T.E. Weymouth, J.S. Griffith, A.R. Hanson, and E.M. Riseman. Rule based strategies for image interpretation. *Proc. of the 2nd National Conf. on AI*, pages 429–432, Washington, D.C., August 22–26, 1983.

P.H. Winston. *The Psychology of Computer Vision.* McGraw-Hill, New York, 1975.

A.P. Witkin, D. Terzopoulis, and M. Kass. Signal matching through scale space. *Proc. 5th National Conf. on AI*, pages 714–719, Philadelphia, PA, August 11–15, 1986.

A.P. Witkin. Intensity-based edge classification. *Proc. National Conf. on AI*, pages 36–41, Pittsburgh, Pa., August 18–20, 1982.

A.P. Witkin. Recovering surface shape and orientation from texture. *Artificial Intelligence*, 17(1–3):17–45, August 1981.

A.P. Witkin. Scale-space filtering. *Proc. 8th Int. Joint Conf. on AI*, pages 1019–1022, Karlsruhe, West Germany, August 8–12, 1983a.

A.P. Witkin and J.M. Tenenbaum. On the role of structure in vision. Beck, Hope, and Rosenfeld, editors, *Human and Machine Vision*, pages 481–543. Academic Press, 1983b.

R.J. Woodham. Analyzing images of curved surfaces. *Artificial Intelligence*, 17(1–3):117–140, August 1981.

W. Wu. Basic principles of mechanical theorem proving in elementary geometries. *J. Sys. Sci. and Math. Sci.*, 4(3):207–235, 1984.

G. Wyvill, A. Pearce, and B. Wyvill. The representation of water. *Graphics Interface '86/Vision Interface '86*, pages 217–222, Vancouver, Canada, May 1986.

S. Yalamanchili and J.K. Aggarwal. Analysis of a model for parallel image processing. *Pattern Recognition*, 18(1):1–16, 1985a.

S. Yalamanchili and J.K. Aggarwal. A system organization for parallel image processing. *Pattern Recognition*, 18(1):17–29, 1985b.

D.M. Young and P.L. Odell. A formalism and comparison of two linear feature selection techniques applicable to statistical classification. *Pattern Recognition*, 17(3):331–337, 1984.

S. Youssef. A new algorithm for object oriented ray tracing. *CVGIP*, 34(2):125–137, May 1986.

A.L. Yuille and T.A. Poggio. Scaling theorems for zero crossings. *IEEE PAMI*, 8(1):15–25, January 1986.

C.T. Zahn. Graph-theoretical methods for detecting and describing gestalt clusters. *IEEE Trans. on Computers*, C20:68–86, 1971.

S.W. Zucker. Computational and psychophysical experiments in grouping. Beck, Hope, and Rosenfeld, editors, *Human and Machine Vision*, pages 545–567, Academic Press, New York, 1983.

S.W. Zucker. Early orientation selection: Tangent fields and the dimensionality of their support. *CVGIP*, 32(1):74–103, October 1985.

S.W. Zucker. *Early Vision.* Technical Report TR-86-5R, McGill University, Dept. of EE, Computer Vision and Robotics Lab, Montreal, Quebec, Canada, 1986. to be published in the *Encyclopedia of Artificial Intelligence*, Wiley, 1987.

Subject Index

Author Index